Dictionary of Organic Compounds

FIFTH EDITION

EIGHTH SUPPLEMENT

Dictionary

of

Organic

Compounds

FIFTH EDITION

EIGHTH SUPPLEMENT

CHAPMAN AND HALL
Scientific Data Division
LONDON · NEW YORK · TOKYO · MELBOURNE · MADRAS

UK	Chapman and Hall, 2-6 Boundary Row, London SE1 8HN
USA	Chapman and Hall, 29 West 35th Street, New York NY 10001
JAPAN	Chapman and Hall Japan, Thomson Publishing Japan, Hirakawacho Nemoto Building, 7F, 1-7-11 Hirakawa-cho, Chiyoda-ku, Tokyo 102
AUSTRALIA	Chapman and Hall Australia, Thomas Nelson Australia, 480 La Trobe Street, PO Box 4725, Melbourne 3000
INDIA	Chapman and Hall India, R. Seshadri, 32 Second Main Road, CIT East, Madras 600 035

The Fifth Edition of the Dictionary of Organic Compounds
in seven volumes published 1982
The First Supplement published 1983
The Second Supplement published 1984
The Third Supplement published 1985
The Fourth Supplement published 1986
The Fifth Supplement (in two volumes) published 1987
The Sixth Supplement published 1988
The Seventh Supplement published 1989
This Eighth Supplement published 1990

© 1990 Chapman and Hall Ltd

Typeset and printed in Great Britain at the University Press, Cambridge

ISBN 0 412 17080 9

British Library Cataloguing in Publication Data

Dictionary of organic compounds — 5th ed.
Eighth supplement
  1. Organic compounds
  I. Buckingham, J. (John)
  547

  ISBN 0-412-17080-9

Library of Congress Cataloguing in Publication Data

Dictionary of organic compounds
  Eighth supplement.—5th ed.
      p.   cm.
    "Executive editor, J. Buckingham"—P.
    Includes bibliographical references.
    ISBN 0-412-17080-9
    1. Chemistry, Organic—Dictionaries,
    I. Buckingham, J.
QD246.D5   1982   Suppl. 7
547′.003—dc20   89-23854

Note to Readers

Always use the latest Supplements

Supplements are published in the middle of each year and contain new and updated Entries derived from the primary literature of the preceding year. Searching the entire Supplement series is facilitated by consulting first the indexes in the latest Supplement. The Supplement indexes are cumulative to facilitate the rapid location of data.

For full information on Supplements please write to:

The Marketing Manager   *or*   Chapman and Hall
Scientific Data Division   29 West 35th Street
Chapman and Hall   New York, NY 10001
2-6 Boundary Row   U.S.A.
London
SE1 8HN

New Compounds for DOC 5

The Editor is always pleased to receive comments on the selection policy of DOC5, and in particular welcomes specific suggestions for compounds or groups of compounds to be considered for inclusion in the Supplements.

Write to:

The Editor
Dictionary of Organic Compounds
Scientific Data Division
Chapman and Hall
2-6 Boundary Row
London
SE1 8HN

Specialist Dictionaries

Five important specialist publications are now available which greatly extend the coverage of the DOC databank in key specialist areas. These are as follows:

Dictionary of Drugs, 1990, 2 volumes, ISBN 0 412 27300 4

Dictionary of Alkaloids, 1989, 2 volumes, ISBN 0 412 24910 3

Dictionary of Antibiotics and Related Substances, 1987, ISBN 0 412 25450 6

Dictionary of Organophosphorus Compounds, 1987, ISBN 0 412 25790 4

Carbohydrates (a Chapman and Hall Chemistry Sourcebook), 1987,
   ISBN 0 412 26960 0

Caution

Treat all organic compounds as if they have dangerous properties.

The publisher makes no representation, express or implied, with regard to the accuracy of the information contained in this Dictionary, and cannot accept any legal responsibility or liability for any errors or omissions that may be made.
   The specific information in this publication on the hazardous and toxic properties of certain compounds is included to alert the reader to possible dangers associated with the use of those compounds. The absence of such information should not however be taken as an indication of safety in use or misuse.

Eighth Supplement

Introduction

For detailed information about how to use DOC 5, see the Introduction in Volume 1 of the Main Work.

1. Using DOC 5 Supplements

As in the Main Work volumes, every Entry is numbered to assist ready location. The DOC Number consists of a letter of the alphabet followed by a five-digit number. In this eighth supplement the first digit is invariably 8. Cross-references within the text to Entries having numbers beginning with zero refer to Main Work Entries and with 1, 2, 3, 4, 5, 6 or 7 refer to the first seven supplements.

Where a Supplement Entry contains additional or corrected information referring to an Entry in the Main Work or earlier supplements, the whole Entry is reprinted, with the accompanying statement "Updated Entry replacing . . .". In such cases, the new Entry contains all of the information which appeared in the former Entry, except for any which has been deliberately deleted. In such cases there is therefore no necessity for the user to consult the Main Work or previous supplements.

2. Literature Coverage

In compiling this Supplement the primary literature has been surveyed to mid-1989. A considerable number of compounds from the older literature have also been included for the first time.

3. Indexes

The indexes in the Supplement cover the Sixth, Seventh and Eighth supplements. A cumulative index volume to Supplements 1–5 inclusive was issued as part of the Fifth Supplement. In order to find a compound in DOC, look first in the Eighth Supplement, then in the Fifth Supplement cumulative indexes, then in the Main Work indexes.

No CAS Registry Number Index is included in this Supplement, for reasons of pressure on space. A CAS Index was published with the Seventh Supplement, and will next reappear as part of the Tenth Supplement.

4. Symbols

The additional symbol † after a chemical name has been introduced beginning with this supplement, to denote a (trivial) name which appears in the literature more than once for different compounds (usually natural products).

5. Additional registry numbers

Some entries in DOC now carry additional Chemical Abstracts Service (CAS) registry numbers, which have been introduced to assist readers carrying out further searches (especially online searches) on substances included in DOC entries.

A small proportion of these may be duplicate numbers, but the majority are numbers which although clearly referring to the DOC entry under which they are given, cannot be unequivocally assigned to one of the substances covered by that entry, for example unfavoured tautomers and other variants for which no physical properties can be reported.

The additional numbers are given in smaller type after the text of the entry, and immediately before the references.

Contents

A

8(14),13(15)-Abietadien-18-oic acid   A-80001

Neoabietic acid

[471-77-2]

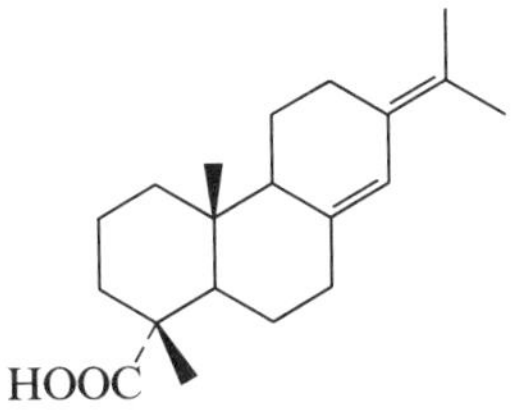

$C_{20}H_{30}O_2$ M 302.456

Isol. from resins of *Pinus* spp., *Abies* spp. and *Agathis microstachys*. Cryst. (EtOH aq.). Mp 173-173.5° (167-169°) (sealed tube under N₂). $[\alpha]_D^{25}$ +161.6° (c, 2.5 in 95% EtOH). Air-sensitive. Dextrosapinic acid was a mixt. of Abietic and Neoabietic acids.

Cyclohexylamine salt: Needles (EtOH aq.). Mp 211-214° dec. $[\alpha]_D^{27}$ +110.2° (c,0.295 in 95% EtOH aq.).

Harris, G.C. *et al, J. Am. Chem. Soc.*, 1948, **70**, 334, 339.
Schuller, W. *et al, J. Am. Chem. Soc.*, 1961, **83**, 2563 (*isol, props, struct*)
Carman, R.M. *et al, Aust. J. Chem.*, 1966, **19**, 2403 (*isol*)

7,13-Abietadien-18-ol   A-80002

Abietinol

$C_{20}H_{32}O$ M 288.472

Isol. from *Pinus sylvestris* and *Abies sibirica*. Cryst. Mp 85.5-87°. $[\alpha]_D$ −132.5° (c, 2 in EtOH).

Erdtman, H. *et al, Acta Chem. Scand.*, 1963, **17**, 1826.
Chirkova, M.A. *et al, Khim. Prir. Soedin.*, 1966, **2**, 99.

6,8,11,13-Abietatriene-2,12-diol   A-80003

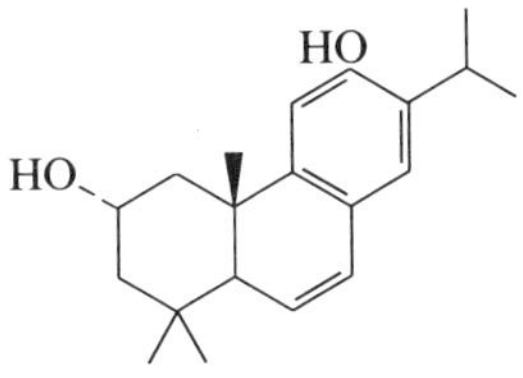

$C_{20}H_{28}O_2$ M 300.440

2α-form

6,7-Dehydrosalviol

Caustic of *Salvia texana*. Yellow oil.

Gonzalez, A.G. *et al, J. Chem. Res. (S)*, 1989, 132.

Abrusin   A-80004

[120727-02-8]

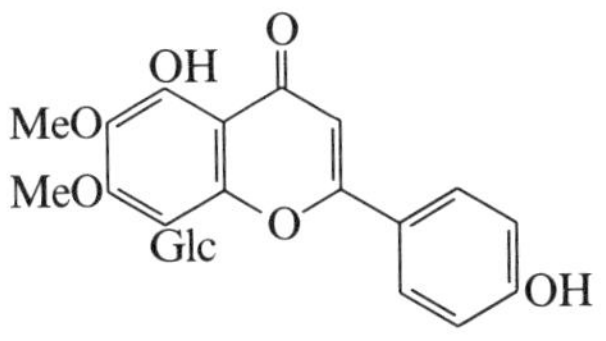

$C_{23}H_{24}O_{11}$ M 476.436

Constit. of *Abrus precatorius*. Cryst. Mp 270-275° dec.

2″-O-β-Apiofuranosyl: [120727-04-0]. **Abrusin 2″-O-apioside**
$C_{28}H_{32}O_{15}$ M 608.552
Constit. of *A. precatorius*. Cryst. Mp 258-263° dec. Substituted in the glucosyl residue.

Markham, K.R. *et al, Phytochemistry*, 1988, **27**, 299.

Abrusogenin   A-80005

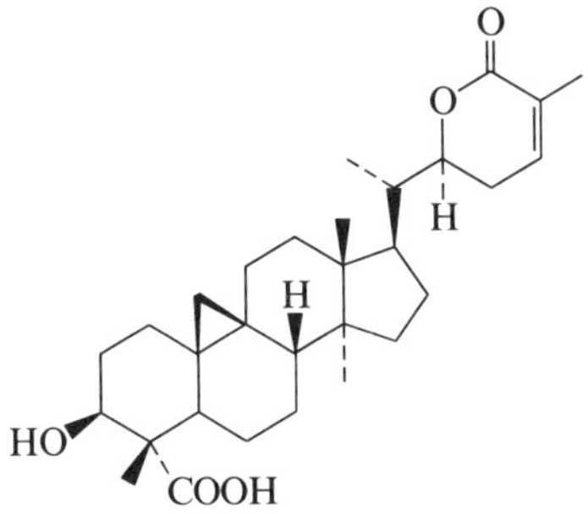

$C_{30}H_{44}O_5$ M 484.675

Constit. of *Abrus precatorius*. Mp 278-280°. $[\alpha]_D$ +37° (c, 0.1 in CHCl₃/MeOH).

3-O-β-D-Glucopyranoside: **Abrusoside A**
$C_{36}H_{54}O_{10}$ M 646.817
Constit. of *A. precatorius*. Cryst. Mp 278-280°. $[\alpha]_D$ +11.2° (c, 0.31 in Py).

3-O-[β-D-Glucopyranosyl-(1→2)-β-D-6-methylglucuronopyranoside]: **Abrusoside B**
$C_{43}H_{64}O_{16}$ M 836.969
Constit. of *A. precatorius*. Cryst. Mp 243-245°. $[\alpha]_D$ +5.8° (c, 0.35 in Py).

3-O-[β-D-Glucopyranosyl-(1→2)-β-D-glucopyranoside]: **Abrusoside C**
$C_{42}H_{64}O_{15}$ M 808.959
Constit. of *A. precatorius*. Cryst. Mp 260-262°. $[\alpha]_D$ +31.4° (c, 0.34 in Py).

3-O-[β-D-Glucopyranosyl-(1→2)-β-D-glucuronopyranoside]: **Abrusoside D**
$C_{42}H_{62}O_{16}$ M 822.942
Constit. of *A. precatorius*. Cryst. Mp 237-239°. $[\alpha]_D$ +9.9° (c, 0.31 in Py).

Choi, Y.-H. *et al, J. Nat. Prod. (Lloydia)*, 1989, **52**, 1118 (*isol, pmr, cmr, cryst struct*)

Abyssinone I   A-80006

7-Hydroxy-2′,2′-dimethyl-[2,6′-bi-2H-1-benzopyran]-4(3H)one

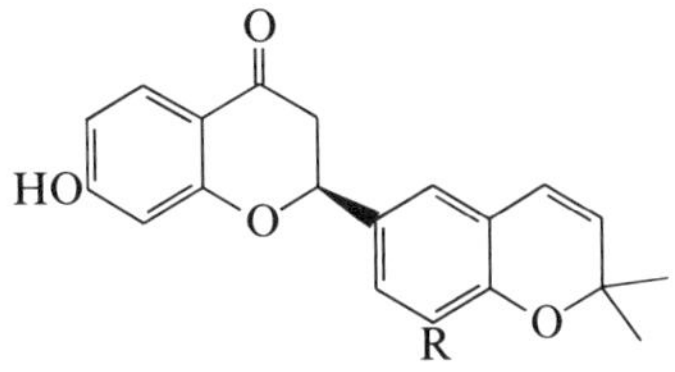

R = H

$C_{20}H_{18}O_4$ M 322.360

(S)-form [77263-07-1]

Isol. from *Erythrina abyssinica*. Shows antimicrobial
activity. CA name incorrect.

Kamat, V.S. *et al, Heterocycles*, 1981, **15**, 1163 (*isol, uv, pmr, cd,
struct, abs config*)

Abyssinone II   A-80007

*2,3-Dihydro-7-hydroxy-2-[4-hydroxy-3-(3-methyl-2-
butenyl)phenyl]-4H-1-benzopyran-4-one,9CI. 4′,7-Dihydroxy-
3′-C-prenylflavanone*

[77263-08-2]

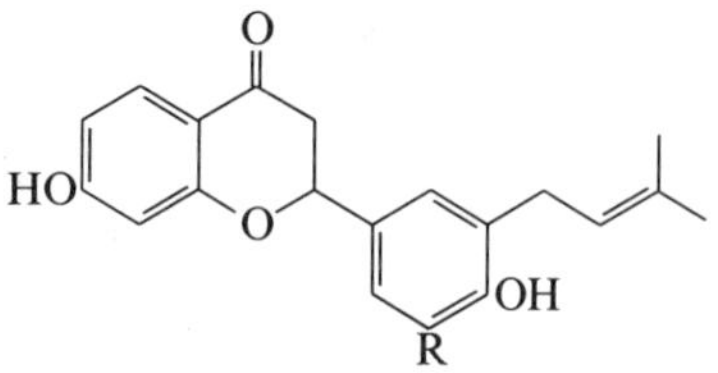

R = H

C$_{20}$H$_{20}$O$_4$ M 324.376

Isol. from *Erythrina abyssinica*. Shows antimicrobial props.
Opt. inactive; CAS no. refers to the (*S*)-form.

Kamat, V.S. *et al, Heterocycles*, 1980, **15**, 1163 (*isol, uv, pmr,
struct*)

Abyssinone III   A-80008

*7-Hydroxy-2,2′-dimethyl-8′-(3-methyl-2-butenyl)-[2,6′-bi-2H-
1-benzopyran]-4(3H)one, 9CI*

As Abyssinone I, A-80006 with

R = —CH$_2$CH=C(CH$_3$)$_2$

C$_{25}$H$_{26}$O$_4$ M 390.478

(*S*)-*form* [77263-09-3]

Isol. from *Erythrina abyssinica*. Shows antimicrobial
activity.

Kamat, V.S. *et al, Heterocycles*, 1980, **15**, 1163 (*isol, uv, pmr, cd,
struct, abs config*)

Abyssinone IV   A-80009

*2,3-Dihydro-7-hydroxy-2-[4-hydroxy-3,5-bis(3-methyl-2-
butenyl)phenyl]-4H-1-benzopyran-4-one, 9CI. 4′,7-
Dihydroxy-3′,5′-di-C-prenylflavanone*

[77263-10-6]

As Abyssinone II, A-80007 with

R = —CH$_2$CH=C(CH$_3$)$_2$

C$_{25}$H$_{28}$O$_4$ M 392.494
Isol. from *Erythrina abyssinica*. Shows antimicrobial props.
CAS no. refers to (*S*)-form but the isolate was opt.
inactive.

5-Hydroxy: [77263-11-7]. **Abyssinone V.** *4′,5,7-Trihydroxy-
3′,5′-di-C-prenylflavanone*
C$_{25}$H$_{28}$O$_5$ M 408.493
Isol. from *E. abyssinica*. Shows antimicrobial props.
CAS no. refers to (S)-form but the isolate was opt.
inactive.

Kamat, V.S. *et al, Heterocycles*, 1980, **15**, 1163 (*isol, uv, pmr,
struct*)

Abyssinone VI   A-80010

*3-[4-Hydroxy-3,5-bis(3-methyl-2-butenyl)phenyl]-1-(4-
hydroxyphenyl)-2-propen-1-one, 9CI. 2′,4,4′-Trihydroxy-3,5-
di-C-prenylchalcone*

[77263-12-8]

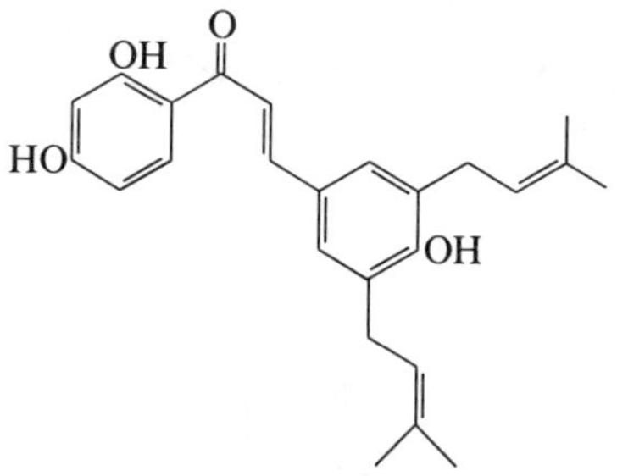

C$_{25}$H$_{28}$O$_4$ M 392.494
Isol. from *Erythrina abyssinica*. Shows antimicrobial props.

Kamat, V.S. *et al, Heterocycles*, 1980, **15**, 1163 (*isol, uv, pmr,
struct*)

Acalycixeniolide B   A-80011

Updated Entry replacing A-60011

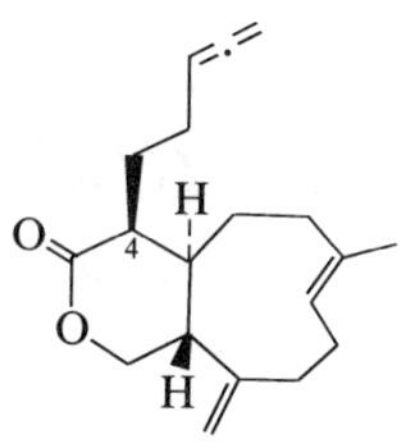

C$_{19}$H$_{26}$O$_2$ M 286.413
Constit. of gorgonian *Acalycigorgia inermis*. Inhibits cell
division in fertilised starfish eggs.

4-epimer: Acalysceniolide B′
C$_{19}$H$_{26}$O$_2$ M 286.413
Constit. of *A.* sp. Oil. [α]$_D$ +15° (c, 0.25 in CHCl$_3$).

Fusetani, Y. *et al, Tetrahedron*, 1989, **45**, 1647.

Acalycixeniolide C   A-80012

[121769-80-0]

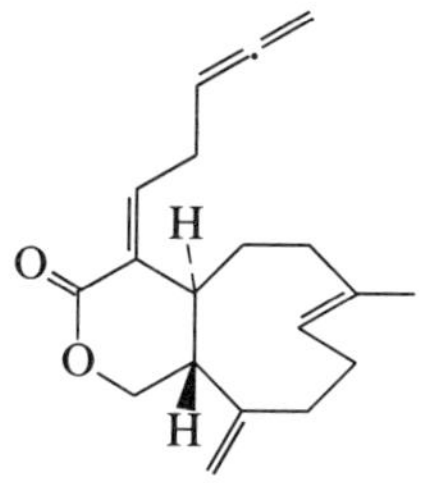

C$_{19}$H$_{24}$O$_2$ M 284.397
Constit. of *Acalycigorgia* sp. Cytotoxic, inhibits cell
division of fertilised starfish eggs. Oil. [α]$_D$ +208° (c,
0.20 in CHCl$_3$).

Fusetani, N. *et al, Tetrahedron*, 1989, **45**, 1647.

N-Acetoxy-4-aminobiphenyl A-80013

N-(*Acetyloxy*)-[*1,1'-biphenyl*]-4-amine, 9CI. O-*Acetyl*-N-(*4-biphenylyl*)hydroxylamine

[119273-47-1]

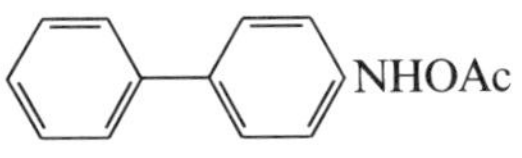

$C_{14}H_{13}NO_2$ M 227.262

Postulated proximal carcinogen of the carcinogenic 4-aminobiphenyl. Bright-yellow cryst. (Et$_2$O/pet. ether at $-30°$). Dec. in 1hr $>10°$.

Famulok, M. *et al*, Angew. Chem., Int. Ed. Engl., 1989, **28**, 337 (*synth, pmr, cmr, struct*)

24-Acetoxy-12-deacetyl-12-epideoxoscalarin A-80014

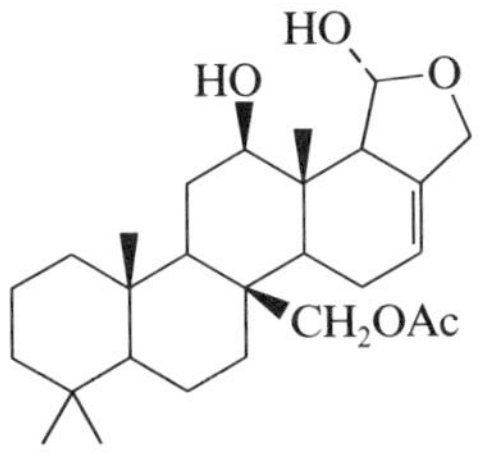

$C_{27}H_{42}O_5$ M 446.626

Constit. of *Hyatella intestinalis*. Waxy oil.

Karuso, P. *et al*, J. Nat. Prod. (*Lloydia*), 1989, **52**, 289 (*isol, pmr, cmr*)

2-Acetylbenzoic acid, 9CI A-80015

Updated Entry replacing A-20016

3-Hydroxy-3-methylphthalide. Acetophenone-2-carboxylic acid

[577-56-0]

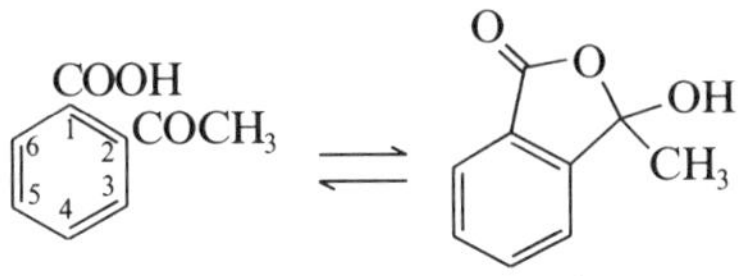

$C_9H_8O_3$ M 164.160

Cryst. (pet. ether). Mp 117-118°. Exists as phthalide in solid state and non-aq. soln., and as 2-acetylbenzoic acid only in aq. alkaline soln.

Me ester: [1077-79-8].
 $C_{10}H_{10}O_3$ M 178.187
 Liq. Bp$_2$ 94-95°.

Nitrile: [91054-33-0]. o-*Cyanoacetophenone*
 C_9H_7NO M 145.160
 Mp 39-42° (48°).

Nitrile, phenylhydrazone: Mp 205-207°.

Riemschneider, R. *et al*, Chem. Ber., 1959, **92**, 1705 (*synth*)
Newman, M.S., J. Chem. Educ., 1977, **54**, 191 (*synth*)
Tyman, J.H.P. *et al*, Spectrochim. Acta, Part A, 1977, **33**, 479 (*struct, pmr, cmr, ms, ir*)
Panetta, C.A. *et al*, Synthesis, 1977, 43 (*use*)
Durrani, A.A. *et al*, J. Chem. Soc., Perkin Trans. 1, 1979, 2069 (*synth, pmr*)
Ebert, G.W. *et al*, J. Org. Chem., 1988, **53**, 4482 (*nitrile*)

2-Acetyl-4*H*-1-benzopyran-4-one, 9CI A-80016

2-Acetylchromone

[65844-02-2]

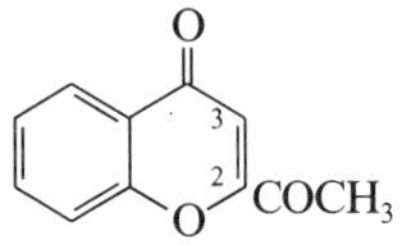

$C_{11}H_8O_3$ M 188.182

Cryst. (EtOH). Mp 137°.

Bevan, P.S. *et al*, J. Chem. Soc., Perkin Trans. 1, 1981, 2552 (*synth, ir, pmr*)

3-Acetyl-4*H*-1-benzopyran-4-one A-80017

3-Acetylchromone

[17422-75-2]

$C_{11}H_8O_3$ M 188.182
Mp 129°.

Eiden, F. *et al*, Arch. Pharm. (*Weinheim, Ger.*), 1972, **305**, 81.
Becket, G.J.P. *et al*, Tetrahedron Lett., 1976, 719 (*synth, ir, pmr*)

1-Acetylbiphenylene A-80018

1-(1-Biphenylenyl)ethanone, 9CI. 1-Biphenylenyl methyl ketone, 8CI

[778-92-7]

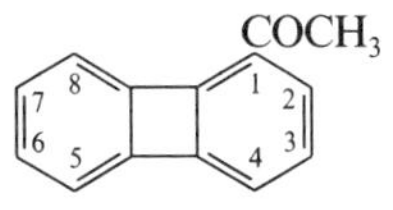

$C_{14}H_{10}O$ M 194.232
Yellow cryst. Mp 58.5°.

Martin, R.H. *et al*, Tetrahedron, 1964, **20**, 2373 (*pmr*)
Boulton, A.J. *et al*, J. Chem. Soc. C, 1968, 328 (*synth*)

2-Acetylbiphenylene A-80019

1-(2-Biphenylenyl)ethanone. 2-Biphenylenyl methyl ketone

[779-26-0]

$C_{14}H_{10}O$ M 194.232
Yellow plates (MeOH). Mp 138-139°.

2,4-Dinitrophenylhydrazone: Red cryst. (AcOH). Mp 257-258°.

Baker, W. *et al*, J. Chem. Soc., 1954, 1476; 1958, 2666 (*synth, uv*)
Droske, J.P. *et al*, Macromolecules, 1984, **17**, 1 (*synth, ir, pmr*)
Sutherlin, D.M. *et al*, Macromolecules, 1986, **19**, 257 (*synth*)

7-Acetyl-6,8-dihydroxy-3-methoxy-4,4-dimethyl-1(4*H*)-naphthalenone A-80020

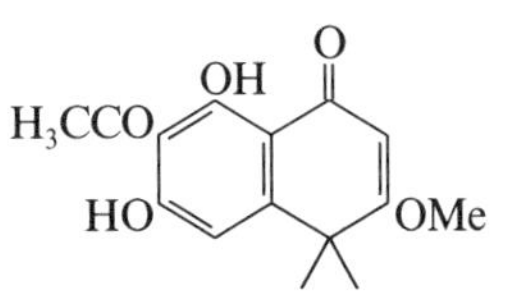

$C_{15}H_{16}O_5$ M 276.288

Pigment from root of *Cassia semicordata*. Cryst. Mp 191-192°.

Delle Monache, G. *et al*, Tetrahedron Lett., 1989, **30**, 6203.

3-[3-Acetyl-5-(3,7-dimethyl-2,6-octadienyl)-2,4,6-trihydroxybenzyl]-4-hydroxy-5,6-dimethyl-2H-pyran-2-one   A-80021

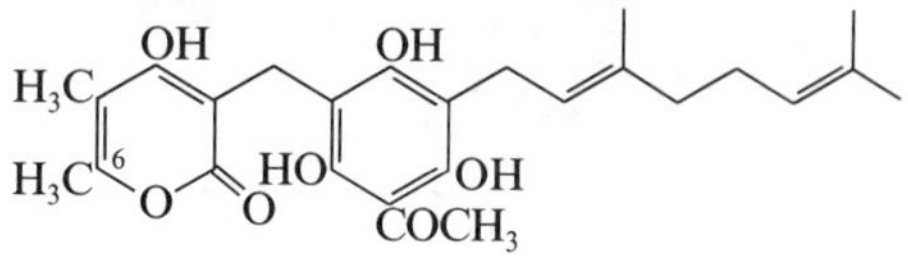

$C_{26}H_{32}O_7$ M 456.535

Constit. of *Helichrysum decumbens*. Has antifungal properties.

6-Ethyl analogue: 3-[3-Acetyl-5-(3,7-dimethyl-2,6-octadienyl)-2,4,6-trihydroxybenzyl]-6-ethyl-4-hydroxy-5-methyl-2H-pyran-2-one
$C_{27}H_{34}O_8$ M 486.561
Constit. of *H. decumbens*. Has antifungal properties.

6-Propyl analogue: 3-[3-Acetyl-5-(3,7-dimethyl-2,6-octadienyl)-2,4,6-trihydroxybenzyl]-4-hydroxy-5-methyl-6-propyl-2H-pyran-2-one
$C_{28}H_{36}O_8$ M 500.588
Constit. of *H. decumbens*. Has antifungal properties.

Tomás-Lorente, F. *et al, Phytochemistry*, 1989, **28**, 1613.

5-Acetyl-2-hydroxybenzaldehyde   A-80022

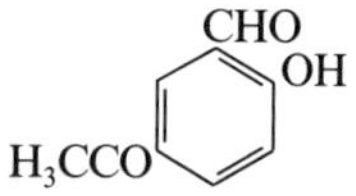

$C_9H_8O_3$ M 164.160

Me ether: 5-Acetyl-2-methoxybenzaldehyde
$C_{10}H_{10}O_3$ M 178.187
Isol. from leaves of *Encelia farinosa*. Needles (Et$_2$O). Mp 143-144°.

Gray, R. *et al, J. Am. Chem. Soc.*, 1948, **70**, 1249 (*isol, struct, synth*)

6-Acetyl-7-hydroxy-8-methoxy-2,2-dimethyl-2H-1-benzopyran   A-80023

Updated Entry replacing A-50030
Ripariochromene A
[20770-16-5]

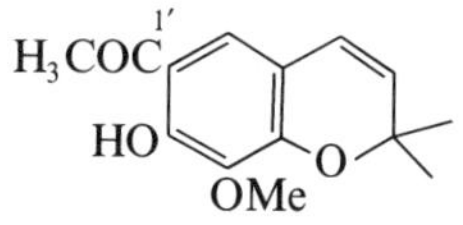

$C_{14}H_{16}O_4$ M 248.278
Constit. of *Eupatorium riparium*. Cryst. Mp 88.5°.

Me ether: [20819-46-9]. *6-Acetyl-7,8-dimethoxy-2,2-dimethyl-2H-1-benzopyran.* **Methylripariochromene A**
$C_{15}H_{18}O_4$ M 262.305
Constit. of *E. riparium* and *Stevia serrata*. Low-melting solid. Mp 35°.

Me ether, 1'-alcohol: 6-(1-Hydroxyethyl)-7,8-dimethoxy-2,2-dimethyl-2H-1-benzopyran. 6-(1-Hydroxyethyl)eupatoriochromene B
$C_{15}H_{20}O_4$ M 264.321
Constit. of *Ageratina riparia*. Gum. [α]$_D^{24}$ −6.29° (c, 0.89 in CHCl$_3$).

Tsukayama, M. *et al, Bull. Chem. Soc. Jpn.*, 1975, **48**, 80 (*isol*)
Kohda, H. *et al, Phytochemistry*, 1976, **15**, 846 (*pmr*)
Banerjee, S. *et al, Phytochemistry*, 1985, **24**, 2681 (*isol*)
Guerin, J.-C. *et al, J. Nat. Prod. (Lloydia)*, 1989, **52**, 171 (*isol, bibl, cryst struct*)

2-Acetyl-3-hydroxy-5-(1-propynyl)thiophene   A-80024

3-Hydroxy-5-(1-propynyl)-2-thienyl methyl ketone, 8CI. 1-[3-Hydroxy-5-(1-propynyl)-2-thiophenyl]ethanone
[5556-30-9]

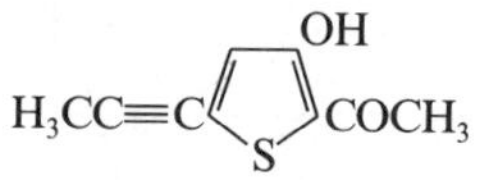

$C_9H_8O_2S$ M 180.227
Isol. from roots of *Artemisia arborescens* and *Liatris pycnostachya*. Cryst. (pet. ether). Mp 100.5° (92-94°).

Me ether: 2-Acetyl-3-methoxy-5-(1-propynyl)thiophene
$C_{10}H_{10}O_2S$ M 194.254
Isol. from roots of *A. arborescens*. Cryst. (pet. ether). Mp 90-91°.

Bohlmann, F. *et al, Chem. Ber.*, 1962, **95**, 2934; 1963, **96**, 584 (*isol, uv, ir, pmr, synth*)
Atkinson, R.E. *et al, Phytochemistry*, 1971, **10**, 454 (*isol*)

1-Acetyl-4-isopropenylcyclopentene   A-80025

1-[4-(1-Methylethenyl)-1-cyclopenten-1-yl]ethanone, 9CI
[2704-76-9]

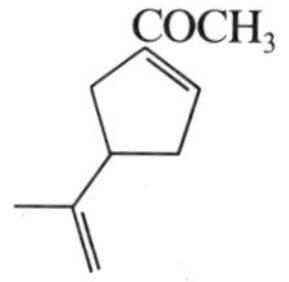

$C_{10}H_{14}O$ M 150.220
Isol. from oil of *Eucalyptus globulus*. Oil. Bp$_{738}$ 225-225.5°, Bp$_{1.5}$ 67-68°. $n_D^{24.5}$ 1.4965.

2,4-Dinitrophenylhydrazone: Red plates (EtOH). Mp 178-180°.

Semicarbazone: Cryst. (MeOH). Mp 195-196°.

Schmidt, H., *Chem. Ber.*, 1947, **80**, 528, 533 (*isol*)
Wolinsky, J. *et al, J. Am. Chem. Soc.*, 1960, **82**, 636 (*struct, synth*)
Conia, J.M. *et al, Bull. Soc. Chim. Fr.*, 1964, 1963 (*synth, uv, ir, pmr*)
Vig, P. *et al, Indian J. Chem.*, 1968, **6**, 564 (*synth*)

1-Acetyl-4-isopropylidenecyclopentene   A-80026

1-[4-(1-Methylethylidene)-1-cyclopenten-1-yl]ethanone, 9CI
[55873-39-7]

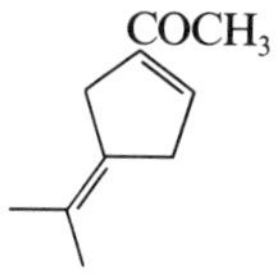

$C_{10}H_{14}O$ M 150.220
Cryst. (pentane). Mp 47-48°. Bp$_8$ 93-94°. Descr. (1947) as a natural prod. from *Eucalyptus globulus*. This was 1-Acetyl-4-isopropenylcyclopentene, A-80025.

Bozzato, G. *et al, J. Chem. Soc., Chem. Commun.*, 1974, 1005 (*synth*)
Thomas, A.F. *et al, Tetrahedron*, 1986, **42**, 3311 (*synth, uv, pmr, ms*)

N-Acetylmethionine, 9CI   A-80027

2-(Acetylamino)-4-(methylthio)butanoic acid. Aminotylon.
Methionamine. Thiomedon-Amp

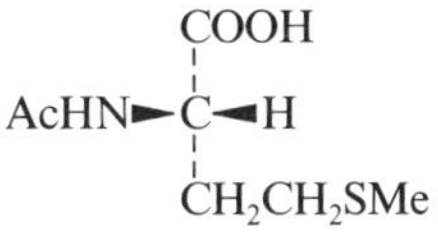

$$AcHN-\underset{\underset{CH_2CH_2SMe}{|}}{\overset{\overset{COOH}{|}}{C}} \blacktriangleleft H \qquad (S)\text{-}form$$

$C_7H_{13}NO_3S$ M 191.251
(*S*)-form illustrated. Methionine therapeutic. Lipotropic
agent.

(*R*)-*form* [1509-92-8]
    Cryst. (EtOAc or H$_2$O). Mp 104-105°. [α]$_D^{25}$ +20.3° (c, 4
    in H$_2$O).

(*S*)-*form* [65-82-7]
    Mp 104°. [α]$_D^{25}$ −20.3°.
▷ PD0480000.

(±)-*form* [1115-47-5]
    Cryst. (H$_2$O). Mp 114-115°.
▷ PD0500000.

Wheeler, G.P. *et al, J. Am. Chem. Soc.*, 1951, **73**, 4604 (*synth*)
U.S. Pat., 3 028 395; *CA*, **57**, 16742 (*resoln*)
Heyns, K. *et al, Justus Liebigs Ann. Chem.*, 1963, **667**, 194 (*ms*)
U.K. Pat., 1 072 876; *CA*, **67**, 114015 (*resoln*)
Jung, G. *et al, Eur. J. Biochem.*, 1973, **35**, 436 (*cd*)
Hawkes, G.E. *et al, Nature* (*London*), 1975, **257**, 767 (*N-15 nmr*)
Rotruck, J.T. *et al, J. Nutr.*, 1975, **105**, 331 (*metab*)
Boggs, R.W., *Adv. Exp. Med. Biol.*, 1978, **105**, 571 (*props*)
Ponnuswamy, M.N. *et al, Acta Crystallogr.*, *Sect. C*, 1985, **41**, 917
    (*cryst struct*)
Martindale, The Extra Pharmacopoeia, 28th/29th Ed.,
    Pharmaceutical Press, London, 1982/1989, 571.

1-Acetyl-2-methylcyclobutene   A-80028

1-(2-Methyl-1-cyclobuten-1-yl)ethanone, 9CI
[67223-99-8]

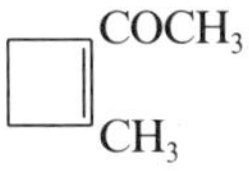

$C_7H_{10}O$ M 110.155
Liq. Bp$_{16}$ 69-72°, Bp$_6$ 47-48°.

Balenkova, E.S. *et al, Zh. Org. Khim.*, 1978, **14**, 1109; *jocet*, 1033
    (*synth, ir, pmr*)
Negishi, E. *et al, J. Am. Chem. Soc.*, 1988, **110**, 5383 (*synth, ir,*
    pmr, cmr)

5-Acetylthiazole   A-80029

1-(5-Thiazolyl)ethanone, 9CI
[91516-28-8]
C_5H_5NOS M 127.167
Cryst. (Et$_2$O/hexane). Mp 68-70° (64.5-65.5°).

Kochetkov, N.K. *et al, J. Gen. Chem. USSR* (*Engl. Transl.*), 1959,
    29, 2294 (*synth*)
Dondoni, A. *et al, J. Org. Chem.*, 1988, **53**, 1748 (*synth, ir, pmr,*
    ms)

15-Acetylthioxyfurodysinin lactone   A-80030

[122771-73-7]

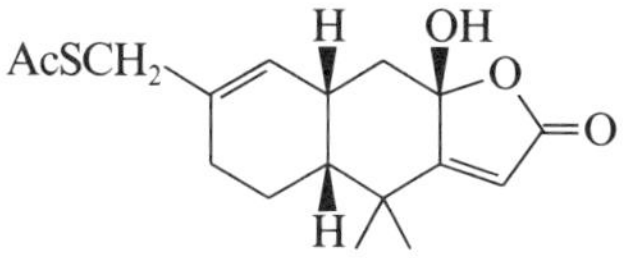

$C_{17}H_{22}O_4S$ M 322.424
Metab. of sponge *Dysidea* sp. Potent LTB$_4$ receptor partial
    agonist. Cryst. Mp 144-145°. [α]$_D$ −178° (CHCl$_3$).

Carte, B. *et al, Tetrahedron Lett.*, 1989, **30**, 2725.

4(5)-Acetyl-1,2,3-triazole   A-80031

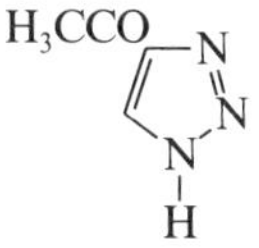

$C_4H_5N_3O$ M 111.103
Cryst. (cyclohexane). Mp 136.5°.

Banert, K., *Chem. Ber.*, 1989, **122**, 1175.

Achilleol A   A-80032

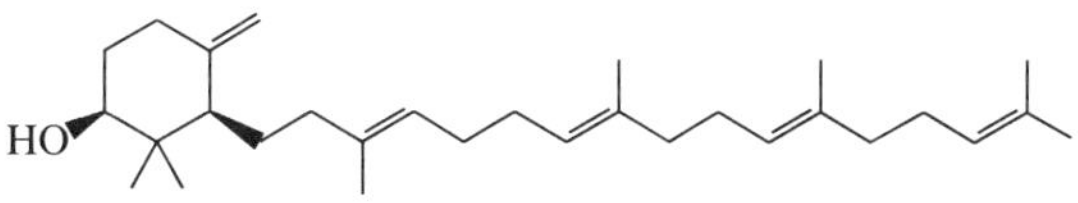

$C_{30}H_{50}O$ M 426.724
Constit. of *Achillea odorata*. [α]$_D$ −10.9° (c, 0.9 in CHCl$_3$).

Barrero, A.F. *et al, Tetrahedron Lett.*, 1989, **30**, 3351.

Achillin   A-80033

Updated Entry replacing A-00432
[5956-04-7]

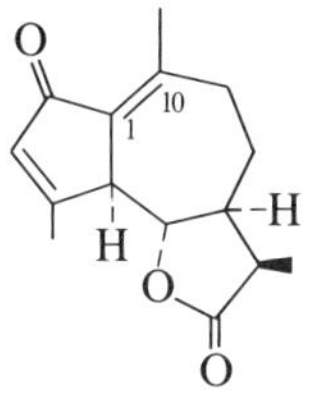

$C_{15}H_{18}O_3$ M 246.305
Constit. of *Achillea* spp. and *Artemisia* spp. Cryst.
    (C$_6$H$_6$/pet. ether). Mp 144-145°. [α]$_D^{29}$ +160° (c, 1.2 in
    EtOH).

1β,10β-Epoxide: 1β,10β-Epoxyachillin
    $C_{15}H_{18}O_4$ M 262.305
    Constit. of *Artemisia lanata* and *A. assoana*. Cryst. Mp
    236-238°. [α]$_D$ +102° (c, 0.18 in CHCl$_3$).

Marx, J.N. *et al, Tetrahedron*, 1969, **25**, 2117 (*isol*)
González, A.G. *et al, An. Quim.*, 1976, **72C**, 695 (*isol*)
Martinéz, V. *et al, J. Nat. Prod.* (*Lloydia*), 1988, **51**, 221 (*pmr,*
    cmr)
Marco, J.A. *et al, J. Nat. Prod.* (*Lloydia*), 1989, **52**, 547 (*cryst*
    struct, pmr, cmr)

Acidissimin   **A-80034**

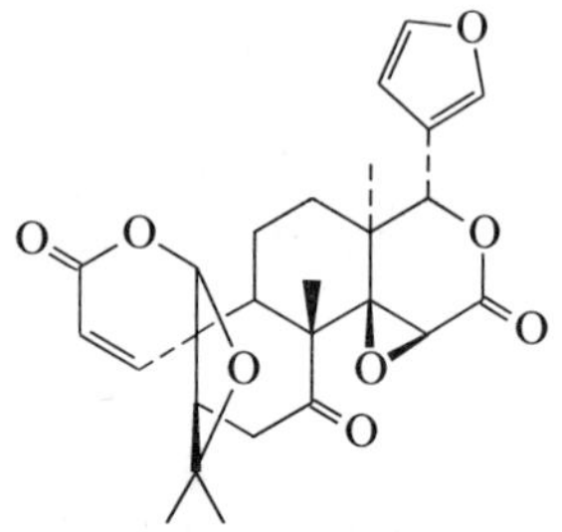

$C_{26}H_{28}O_8$ M 468.502
Constit. of *Limonia acidissima*. Cryst. Mp 284°. [α]$_D$ +22°
(c, 0.5 in Me$_2$CO).

MacLeod, J.K. *et al, J. Nat. Prod.* (*Lloydia*), 1989, **52**, 882 (*isol*,
   pmr, cmr)

Acridine   **A-80035**

Updated Entry replacing A-00457
Dibenzo[b,e]*pyridine. 9-Azaanthracene*
[260-94-6]

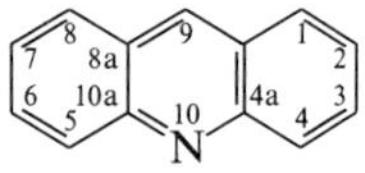

$C_{13}H_9N$ M 179.221
Obt. from coal tar. Parent compd. of many important
   dyes, alkaloids and antibacterials. Needles or prisms.
   Mp 111° (sealed tube)(subl. >100°). Bp ~360°. pK_a
   5.60 (H$_2$O). Solutions show blue fluor. Five cryst.
   modifications known, all melting at ca. 106-111°.
   Triboluminescent.

▷ Toxic, irritant. LD$_{50}$ 400 mg/kg (s.c., mice). Nasal and
   skin irritant. AR7175000.

B,MeI: Red triclinic cryst.

N-*Oxide:* [10399-73-2].
   $C_{13}H_9NO$ M 195.220
   Yellow needles (pet. ether). Mp 169°.

Radulescu, D. *et al, Chem. Ber.*, 1931, **64**, 2233 (*uv*)
Corwin, A.H., *Chem. Heterocycl. Compd.*, (Elderfield, R.C. Ed.),
   Wiley, N.Y., 1950, **1** (*rev*)
Albert, A., *The Acridines*, Arnold, London, 1951.
Kokko, J.P. *et al, Spectrochim. Acta*, 1963, **19**, 1119 (*pmr*)
Nakamura, K. *et al, J. Chem. Soc., Chem. Commun.*, 1970, 1135
   (*cryst struct*)
Acheson, R.M., *Chem. Heterocycl. Compd.*, 2nd Ed. (Weissberger,
   A., Ed.), 1972, **9** (*rev, ir, uv, pmr, ms*)
Noeth, M. *et al, Chem. Ber.*, 1974, **107**, 3070 (*nmr*)
Sax, N.I., *Dangerous Properties of Industrial Materials*, 5th Ed.,
   Van Nostrand-Reinhold, 1979, 342.

Actifolin   **A-80036**

[120931-63-7]

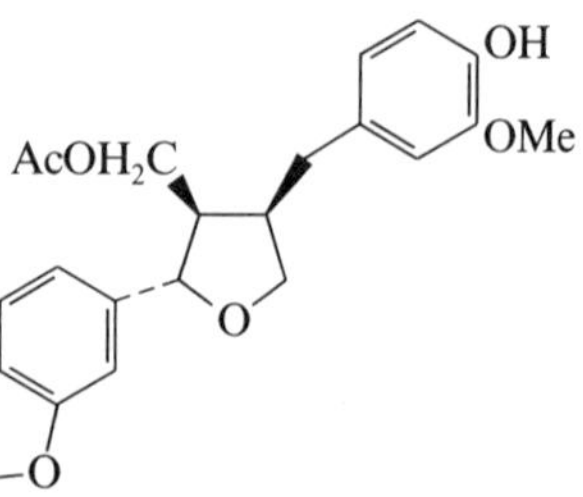

$C_{22}H_{24}O_7$ M 400.427

Constit. of *Actinodaphne longifolia*. Oil. [α]$_D$ +13.3° (c,
0.48 in CHCl$_3$).

Tanaka, H. *et al, Phytochemistry*, 1989, **28**, 952.

1,3,5,7-Adamantanetetracarboxylic acid   **A-80037**
Tricyclo[3.3.1.1^{3,7}]decane-1,3,5,7-tetracarboxylic acid, 9CI
[100884-80-8]

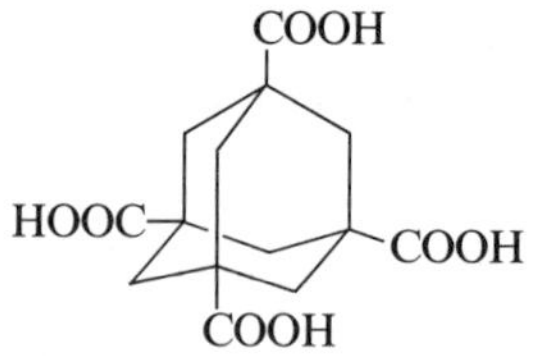

$C_{14}H_{16}O_8$ M 312.276
Cryst. (EtOAc). Mp 390° dec. (browns >350°).

Tetra-Me ester:
   $C_{18}H_{24}O_8$ M 368.383
   Cryst. (MeOH). Mp 168-169°. Bp$_{12}$ 252°.

Tetra-Et ester:
   $C_{22}H_{32}O_8$ M 424.490
   56-57. Mp 56-57°.

Stetter, H. *et al, Chem. Ber.*, 1956, **89**, 1922 (*synth*)
Landa, S. *et al, Collect. Czech. Chem. Commun.*, 1959, **24**, 4004
   (*synth*)
Ermer, O., *J. Am. Chem. Soc.*, 1988, **110**, 3747 (*cryst struct*)

2-Adamantanethiol   **A-80038**
Tricyclo[3.3.1.1^{3,7}]decane-2-thiol. 2-Mercaptoadamantane
[23695-66-1]
$C_{10}H_{16}S$ M 168.302
Cryst. by subl. Mp 160-161°. Bp$_{15}$ 90° subl.

Disulfide: [30979-75-0]. *Di(2-adamantyl) disulfide*
   $C_{20}H_{30}S_2$ M 334.589
   Cryst. (C$_6$H$_6$/MeCN). Mp 198-200°, Mp 295°.

Me thioether: [30979-72-7]. *2-(Methylthio)adamantane*
   $C_{11}H_{18}S$ M 182.329
   Oil. Bp$_4$ 95-97°.

Me thioether, S-oxide: [30979-73-8]. *2-
   (Methylsulfinyl)adamantane*
   $C_{11}H_{18}OS$ M 198.329
   Shiny plates (hexane/C$_6$H$_6$). Mp 122-123°.

Me thioether, S-dioxide: [23695-67-2]. *2-
   (Methylsulfonyl)adamantane*
   $C_{11}H_{18}O_2S$ M 214.328
   Cryst. Mp 114.5-115.5°.

Greidanus, J.W. *et al, Can. J. Chem.*, 1969, **47**, 3715; 1970, **48**,
   3593 (*synth, ir, pmr, deriv*)
Katada, T. *et al, J. Chem. Soc., Perkin Trans. 1*, 1984, 1869 (*deriv,
   synth*)

Aemulin BB   **A-80039**
*1-[3-[[2,6-Dihydroxy-4-methoxy-3-methyl-5-(1-
oxobutyl)phenyl]methyl]-2,4,6-trihydroxy-5-methylphenyl]-1-
butanone, 9CI*
[56226-93-8]

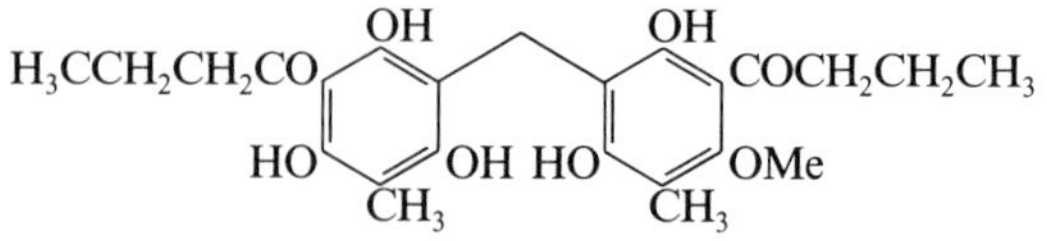

$C_{24}H_{30}O_8$ M 446.496

The BB suffix derives from the two butyryl sidechains. Other aemulins may be nat. prods. Isol. from ferns *Dryopteris aemula* and *D. pseudo-abbreviata*. Mp 90-91°.

Widén, C.-J. *et al*, *Helv. Chim. Acta*, 1975, **58**, 880 (*isol, pmr, ms, struct*)
Widén, C.-J. *et al*, *Helv. Chim. Acta*, 1976, **59**, 1725 (*occur*)

Agelasidine C   A-80040

[92664-78-3]

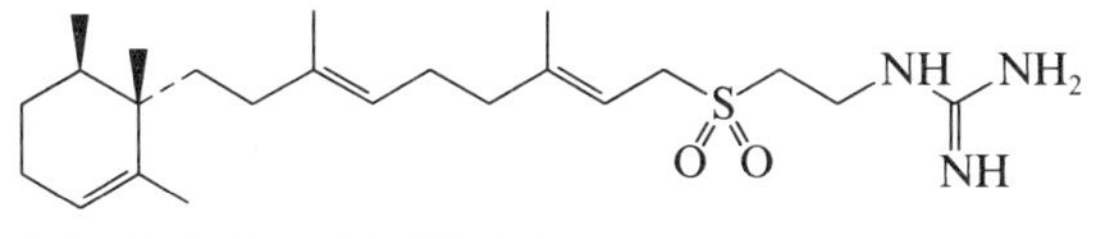

$C_{23}H_{41}N_3O_2S$ M 423.662
Constit. of sponge *Agelas* sp.

Nakamura, H. *et al*, *J. Org. Chem.*, 1985, **50**, 2494 (*isol*)
Asao, K. *et al*, *Chem. Lett.*, 1989, 1813 (*synth*)

Agrostistachin   A-80041

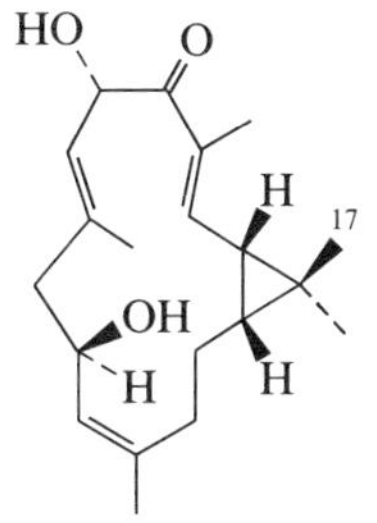

$C_{20}H_{30}O_3$ M 318.455
Constit. of *Agrostistachys hookeri*. Cryst. (EtOAc). Mp 155-157°. $[\alpha]_D$ +248° (c, 0.18 in CHCl$_3$).

17-Hydroxy: **Hydroxyagrostistachin**
$C_{20}H_{30}O_4$ M 334.455
Constit. of *A. hookeri*. Resin. $[\alpha]_D$ +167° (c, 0.09 in CHCl$_3$).

Choi, Y.-H. *et al*, *Tetrahedron Lett.*, 1986, **27**, 5795 (*isol, cryst struct*)
Choi, Y.-H. *et al*, *J. Nat. Prod.* (*Lloydia*), 1988, **51**, 110 (*deriv*)

Alantodiene   A-80042

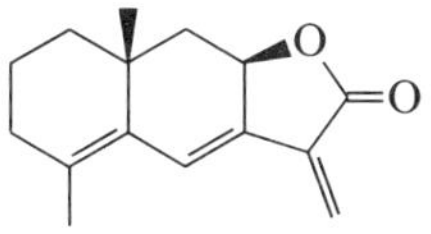

$C_{15}H_{18}O_2$ M 230.306
Constit. of *Inula racemosa*. Oil.

Kalsi, P.S. *et al*, *Phytochemistry*, 1989, **28**, 2093 (*isol, pmr*)

Albaspidins   A-80043

Updated Entry replacing A-00708

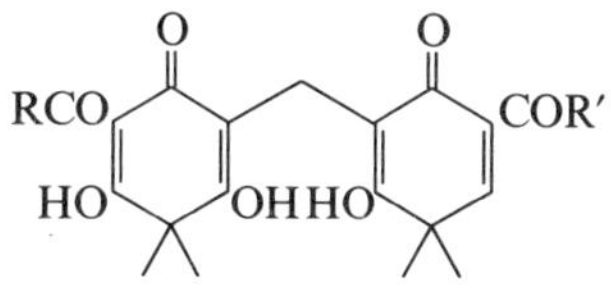

Albaspidins AA, R = R′ = CH$_3$
            AP, R = CH$_3$, R′ = CH$_2$CH$_3$
            AB, R = CH$_3$, R′ = CH$_2$CH$_2$CH$_3$
            PP, R = R′ = CH$_2$CH$_3$
            PB, R = CH$_2$CH$_3$, R′ = CH$_2$CH$_2$CH$_3$
            BB, R = R′ = CH$_2$CH$_2$CH$_3$

The suffix letters denote the side-chains (A = acetyl, P = propionyl, B = butyryl). Constits. of *Dryopteris* and *Aspidium* spp. ferns. Albaspidin 1 was a mixt. of BB, PB and PP. Albaspidin 2 was a mixt. of AB and AP. The various albaspidins disproportionate under mildly alkaline conds.

Albaspidin AA [3570-40-9]
2,2′-*Methylenebis*[6-*acetyl*-3,5-*dihydroxy*-4,4-*dimethyl*-2,5-*cyclohexadien*-1-*one*], 9CI. *Albaspidin 3*
$C_{21}H_{24}O_8$ M 404.416
Cryst. (Me$_2$CO). Mp 170-171°.

Albaspidin AP [59092-91-0]
$C_{22}H_{26}O_8$ M 418.443

Albaspidin AB [3570-35-2]
$C_{23}H_{28}O_8$ M 432.469

Albaspidin PP [3570-39-6]
$C_{23}H_{28}O_8$ M 432.469
Mp 135-137°.

Albaspidin PB [3772-65-4]
$C_{24}H_{30}O_8$ M 446.496

Albaspidin BB [644-61-1]
Polystichalbin. White polystichum acid
$C_{25}H_{32}O_8$ M 460.523
Needles (AcOH). Mp 147-145°.

Penttila, A. *et al*, *Acta Chem. Scand.*, 1964, **18**, 344 (*isol, struct, synth*)
Launasmaa, M. *et al*, *Acta Chem. Scand.*, 1972, **26**, 89 (*synth, ms*)
Widen, C.J. *et al*, *Helv. Chim. Acta*, 1975, **58**, 880 (*isol, struct*)
Puri, H.S. *et al*, *Phytochemistry*, 1976, **15**, 343 (*isol*)
Jayasuriya, H. *et al*, *J. Nat. Prod.* (*Lloydia*), 1989, **52**, 325 (*isol*)

Alloalantolactone   A-80044

[64340-41-6]

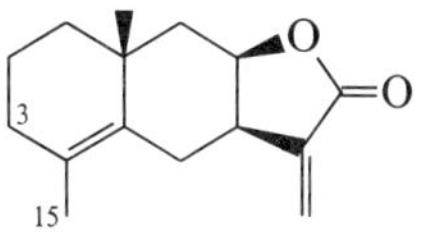

$C_{15}H_{20}O_2$ M 232.322
Constit. of *Inula helenium*. Oil. $[\alpha]_D$ +35.2° (c, 0.5 in CHCl$_3$).

3α,15-Dihydroxy: [119285-38-0]. **3α,15-Dihydroxyalloalantolactone**
$C_{15}H_{20}O_4$ M 264.321
Constit. of *Ambrosia artemisioides*. Oil.

15-Acetoxy, 3α-hydroperoxy: [119285-37-9]. **15-Acetoxy-3α-hydroperoxyalloalantolactone**
$C_{17}H_{22}O_6$ M 322.357
Constit. of *A. artemisioides*. Oil.

15-Hydroxy, 3-oxo: [119285-39-1]. **15-Hydroxy-3-oxoalloalantolactone**
$C_{15}H_{18}O_4$ M 262.305
Constit. of *A. artemisioides*. Oil.

Ananthasubramanian, L. *et al, Indian J. Chem., Sect. B,* 1978, **16**, 191 *(synth)*
Bohlmann, F. *et al, Phytochemistry,* 1978, **17**, 1165 *(isol)*
Jakupovic, J. *et al, Phytochemistry,* 1988, **27**, 3551 *(deriv)*

Allogibberic acid   A-80045

Updated Entry replacing A-00787
Gibberellin B
[427-79-2]

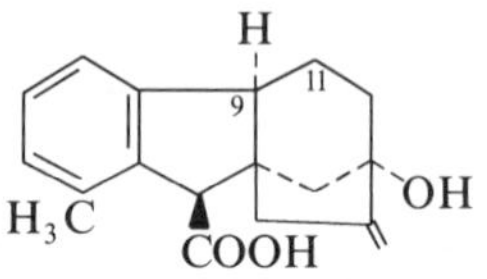

$C_{18}H_{20}O_3$ M 284.354
Isol. from *Gibberella fujikuroi* and *Lemna perpusilla*. Inhibitor of flowering. Cryst. Mp 201-203° (anhyd.). $[\alpha]_D^{20}$ −84° (EtOH).
Me ester: Mp 98-99°. $[\alpha]_D^{22}$ −82° (EtOH).
9,11-Didehydro: **Dehydroallogibberic acid**
$C_{18}H_{18}O_3$ M 282.338
Isol. from cultures of *Gibberella fujikuroi*. Needles (Me$_2$CO/pet. ether). Mp 211-214° dec. $[\alpha]_D^{27}$ −37° (c, 0.2 in EtOH).

Cross, B.E., *J. Chem. Soc.,* 1954, 4670 *(isol)*
Mulholland, T.P.C. *et al, J. Chem. Soc.,* 1958, 2693; 1960, 3007 *(isol)*
Stork, G. *et al, J. Am. Chem. Soc.,* 1959, **81**, 3168 *(stereochem)*
Cross, B.E. *et al, J. Chem. Soc.,* 1963, 2937 *(Dehydroallogibberic acid)*
Pryce, R.J., *Phytochemistry,* 1973, **12**, 1745 *(isol)*
Voigt, D. *et al, Org. Mass Spectrom.,* 1978, **13**, 599 *(ms)*

2-(4-Allyl-2-methoxyphenoxy)-1-(4-hydroxy-3-methoxyphenyl)-1-propanol   A-80046

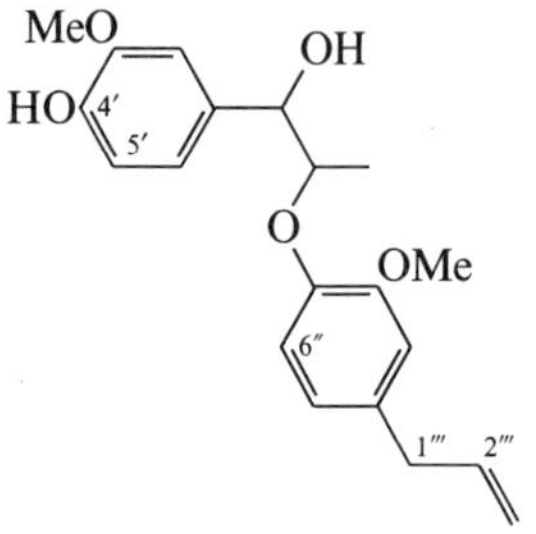

$C_{20}H_{24}O_5$ M 344.407
Constit. of Oil of *Myristica fragrans*. Oil.
5′-Methoxy,Δ$^{1'''}$-isomer(E-): 1-(4-Hydroxy-3,5-dimethoxyphenyl)-2-[2-methoxy-4-(1-propenyl)phenoxy]-1-propanol
$C_{21}H_{26}O_6$ M 374.433
Constit. of *M. fragrans* oil. Oil.
5′,6″-Dimethoxy: 2-(4-Allyl-2.6-dimethoxyphenoxy)-1-(4-hydroxy-3,5-dimethoxyphenyl)-1-propanol
$C_{22}H_{20}O_7$ M 396.396
Constit. of oil of *M. fragrans*. Oil.

O$^{4'}$-Me, 5′-hydroxy,6″-methoxy: 2-(4-Allyl-2,6-dimethoxyphenoxy)-1-(3-hydroxy-4,5-dimethoxyphenyl)-1-propanol
$C_{22}H_{28}O_7$ M 404.459
Constit.of oil of *M. fragrans*. Oil.

Haltori, M. *et al, Chem. Pharm. Bull.,* 1987, **35**, 668.

Almadioxide   A-80047

[119400-88-3]

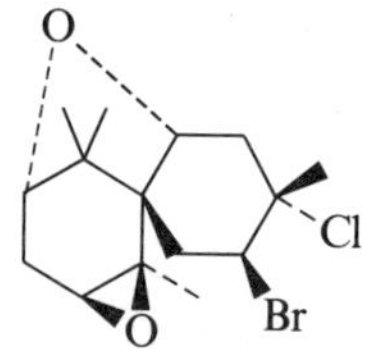

$C_{15}H_{22}BrClO_2$ M 349.694
Constit. of alga *Laurencia intricata*. Cryst. (CH$_2$Cl$_2$/hexane). Mp 206°. $[\alpha]_D$ +4° (c, 0.7 in CHCl$_3$).

Aknin, M. *et al, Tetrahedron Lett.,* 1989, **30**, 559 *(cryst struct)*

Alnusiin   A-80048

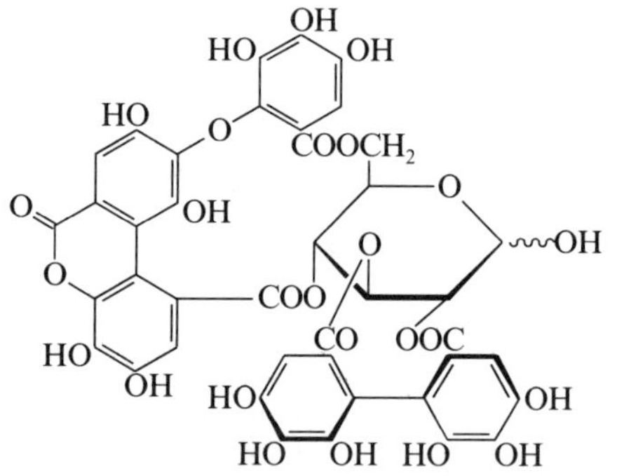

$C_{41}H_{26}O_{26}$ M 934.641
Constit. of *Alnus sieboldiana*. Amorph.

Yoshida, T. *et al, Heterocycles,* 1981, **16**, 1085; 1989, **29**, 861 *(isol, struct, cmr)*

Aloeemodin dianthrone   A-80049

4,4′,5,5′-Tetrahydroxy-2,2′-bis(hydroxymethyl)-[9,9′-bianthracene]-10,10′(9H,9′H)-dione, 9CI. Aloeemodin bianthrone
[4461-75-0]

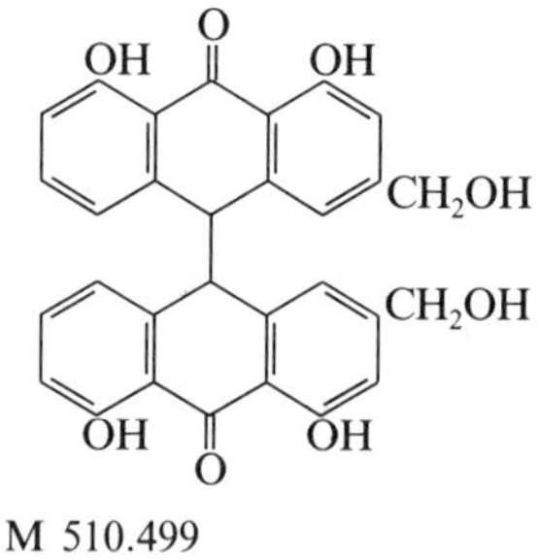

$C_{30}H_{22}O_8$ M 510.499
Isol. from leaves of *Cassia angustifolia* and fruits of *Frangula alnus*. Dark-brown solid. Mp >260° dec. λ_{max} 220, 270 nm.

Auterhoff, H. *et al, Arch. Pharm. (Weinheim, Ger.),* 1960, **293**, 918 *(synth, ir, uv)*
Lemli, J. *et al, Pharm. Weekbl.,* 1964, **99**, 589 *(isol)*
Lemli, J., *J. Nat. Prod. (Lloydia),* 1965, **28**, 63 *(isol)*

Alpinumisoflavone   A-80050

Updated Entry replacing A-00851
5-Hydroxy-7-(4-hydroxyphenyl)-2,2-dimethyl-2H,6H-
benzo[1,2-b:5,4-b']dipyran-6-one, 9CI. 5-Hydroxyerythrinin
A
[34086-50-5]

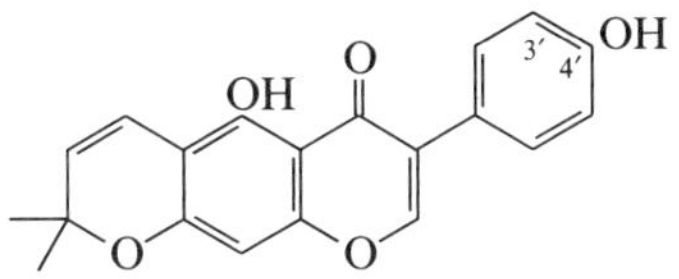

$C_{20}H_{16}O_5$ M 336.343

Constit. of the twigs of *Laburnum alpinum* and of the bark
of *Erythrina variegata*. Also from *Calopogonium
mucunoides* and *Millettia thonningii*. Cryst.
(Me₂CO/hexane or CHCl₃/EtOAc). Mp 213-214°.

Di-Ac: Cryst. (MeOH or EtOAc/pet. ether). Mp 135-137°
(219°).

4'-Me ether: [27762-87-4].
$C_{21}H_{18}O_5$ M 350.370
Isol. from seeds of *C. mucunoides*.

Di-Me ether: [34086-56-1].
$C_{22}H_{20}O_5$ M 364.397
Isol. from seeds of *Derris robusta*. Cryst. (C_6H_6). Mp
119-120°.

4'-Me ether, 3'-hydroxy: **3'-Hydroxyalpinumisoflavone 4'-
methyl ether**
$C_{21}H_{18}O_6$ M 366.370
Isol. from seeds of *M. thonningii*.

Jackson, B. *et al, J. Chem. Soc. C*, 1971, 3389 (*isol, struct*)
Jain, A.C. *et al, J. Org. Chem.*, 1974, **39**, 2215 (*synth*)
Vilain, C. *et al, Bull. Soc. R. Sci. Liege*, 1975, **44**, 306; *CA*, **84**,
    74136e (4'-*O*-Methylalpinumisoflavone)
Deshpande, V.H. *et al, Indian J. Chem., Sect. B*, 1977, **15**, 205
    (*isol*)
Chibber, S.S. *et al, Indian J. Chem., Sect. B*, 1979, **18**, 471 (*isol,
    deriv*)
Ingham, J.L., *Fortschr. Chem. Org. Naturst.*, 1983, **43**, 1 (*rev*)

Alterlosin I   A-80051

5,6,6a,6b-Tetrahydro-3,5,6,6b,10-pentahydroxy-4,9-
perylenedione, 9CI
[120462-02-4]

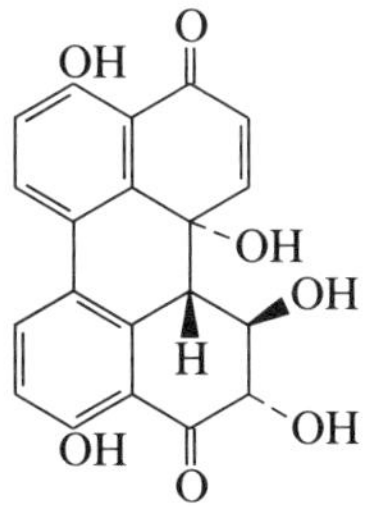

$C_{20}H_{14}O_7$ M 366.326

Metab. of *Alternaria alternata*. Phytotoxin. Cryst. Mp 191-
193°. [α]_D +122° (c, 0.21 in MeOH).

Stierle, A.C. *et al, J. Nat. Prod.* (*Lloydia*), 1989, **52**, 42 (*isol, ir,
    pmr*)

Alterlosin II   A-80052

1,2,6b,7,8,12b-Hexahydro-4,7,8,10,12b-pentahydroxy-3,9-
perylenedione, 9CI
[120462-03-5]

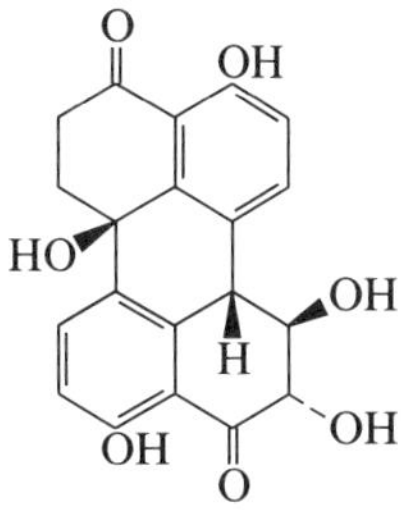

$C_{20}H_{16}O_7$ M 368.342

Metab of *Alternaria alternata*. Phytotoxin. Cryst. Mp 185-
187°. [α]_D +131° (c, 0.2 in MeOH).

Stierle, A.C. *et al, J. Nat. Prod.* (*Lloydia*), 1989, **52**, 42 (*isol, ir,
    pmr*)

Alternanthin   A-80053

[119269-48-6]

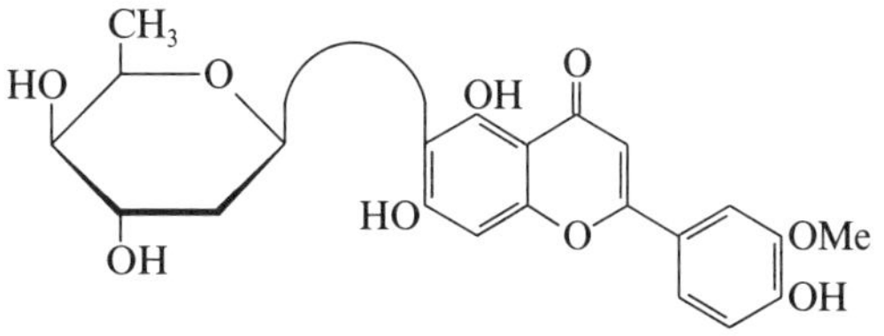

$C_{22}H_{22}O_9$ M 430.410

Constit. of *Alternanthera philoxeroides*. Yellow needles
(MeOH). Mp 225-227°. [α]_D +86.2° (c, 0.23 in MeOH).

Zhou, B.-N. *et al, Phytochemistry*, 1988, **27**, 3633.

Alterporriol C   A-80054

[118074-08-1]

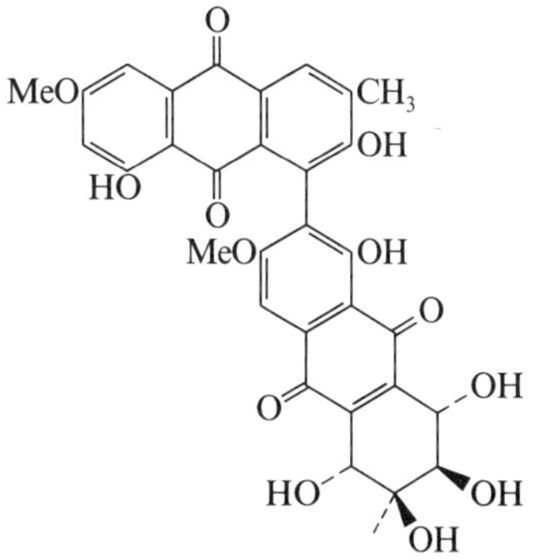

$C_{32}H_{26}O_{13}$ M 618.550

Metab. of *Alternaria porri*. Dark-red amorph. solid. Mp
>300° dec. [α]_D^{25} −268° (c, 0.05 in EtOH).

Suemitsu, R. *et al, Phytochemistry*, 1988, **27**, 3251.

Altertoxin II   A-80055
Stemphyltoxin II
[56257-59-1]

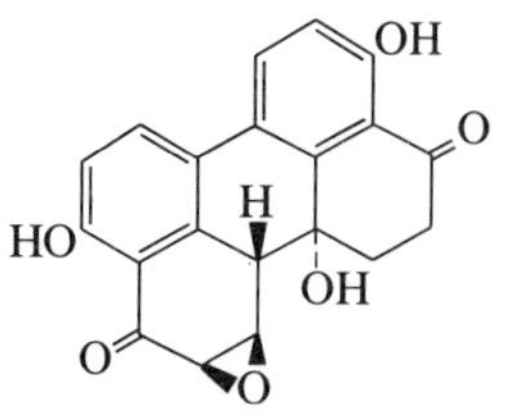

$C_{20}H_{14}O_6$ M 350.327
Isol. from *Alternaria tenuis* and *A. mali* and *Stemphylium botryosum*. Phototoxin. Active against gram-positive bacteria. Orange cryst. solid. Mp 245-250° (300° dec.). $[\alpha]_D$ +636° (c, 0.001 in CHCl$_3$).
▷ BC9625300.

Pero, R.W. *et al*, *EHP Environ. Health Perspect.*, 1973, **4**, 87 (*isol*)
Harvan, D. *et al*, *Adv. Chem. Ser.*, 1976, **149**, 344 (*rev*)
Griffin, G.F. *et al*, *J. Chromatogr.*, 1983, **280**, 363 (*isol, hplc*)
Arnone, A. *et al*, *J. Chem. Soc., Perkin Trans.* 1, 1986, 525 (*Stemphyltoxins*)
Stack, M.E. *et al*, *J. Nat. Prod.* (*Lloydia*), 1986, **49**, 866 (*pmr, cmr, struct*)
Cole, R.J. *et al*, *Handbook of Toxic Fungal Metabolites*, Academic Press , N.Y., 1981, 640.

4-Ambiguen-1-ol   A-80056
1-Hydroxy-4-ambiguene

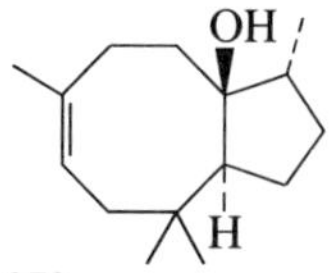

$C_{15}H_{26}O$ M 222.370
Constit. of *Helichrysum bilobum* ssp. *bilobum*. Oil. $[\alpha]_D^{24}$ −45° (c, 0.89 in CHCl$_3$).

Jakupovic, J. *et al*, *Phytochemistry*, 1989, **28**, 543.

Ambofuranol   A-80057
2-(2-Hydroxy-4-methoxyphenyl)-3-methoxy-5-(3-methyl-2-butenyl)-6-benzofuranol, 9CI. 6-Hydroxy-2-(2-hydroxy-4-methoxyphenyl)-3-methoxy-5-C-prenylbenzofuran
[76869-00-6]

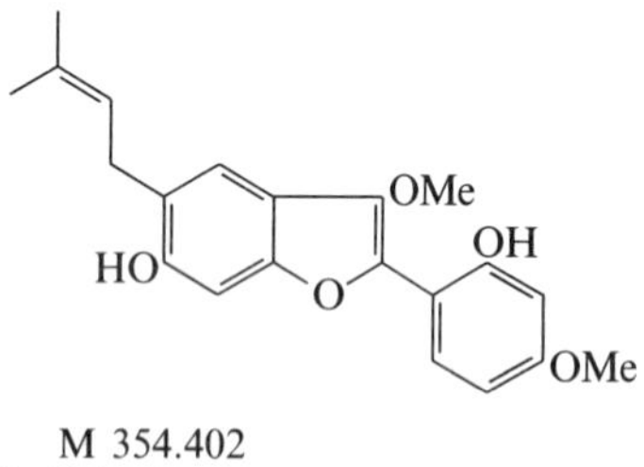

$C_{21}H_{22}O_5$ M 354.402
Isol. from bulb bark of *Neorautanenia amboensis*. Needles (C$_6$H$_6$ or EtOH). Mp 147-148°.

Breytenbach, J.C. *et al*, *Tetrahedron Lett.*, 1980, 4535 (*isol, pmr, ms, struct*)

2-Amino-3-azidopropanoic acid   A-80058
3-Azidoalanine

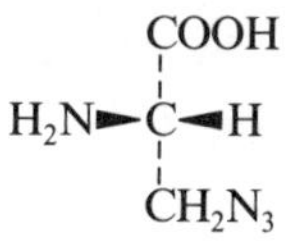

(S)-form

$C_3H_6N_4O_2$ M 130.106
(S)-form [105661-40-3]
L-form
Metabolite isol. from *Salmonella* grown in the presence of azide. Cryst. (H$_2$O/MeOH/Me$_2$CO at −20°). Mp 174-175° dec.
▷ Mutagenic.

Arnold, L.D. *et al*, *J. Am. Chem. Soc.*, 1988, **110**, 2237 (*synth, ir, pmr, cmr, ms*)

4-Aminobenzimidazole   A-80059
Updated Entry replacing A-01018
1H-Benzimidazol-4-amine, 9CI
[4331-29-7]
$C_7H_7N_3$ M 133.152
Prisms (CHCl$_3$). Mp 120-121°.
▷ DD5780000.

B,2HCl: [37724-28-0].
    Cryst. (MeOH). Mp 247-249° dec.
1-Oxide:
    $C_7H_7N_3O$ M 149.152
    Cryst. + 1H$_2$O (H$_2$O). Mp 118-120°.
1-Oxide; B-2HCl: Cryst. (conc. HCl). Mp 224-226°.
3-Oxide:
    $C_7H_7N_3O$ M 149.152
    Cryst. + 1H$_2$O (H$_2$O). Mp 108-110°.
3-Oxide; B,2HCl: Cryst. (conc. HCl). Mp 190-192° dec.

Van der Want, G.M., *Recl. Trav. Chim. Pays-Bas* (*J. R. Neth. Chem. Soc.*), 1948, **67**, 45 (*synth*)
Fisher, E.C. *et al*, *J. Org. Chem.*, 1958, **23**, 1944 (*synth*)
Keiser, S. *et al*, *Pharmazie*, 1972, **27**, 287 (*rev*)
Harvey, I.W. *et al*, *J. Chem. Soc., Perkin Trans.* 1, 1988, 1939 (*oxides*)

1-Aminobicyclo[2.2.1]heptane   A-80060
Bicyclo[2.2.1]heptan-1-amine, 9CI. 1-Aminonorbornane
[21245-51-2]

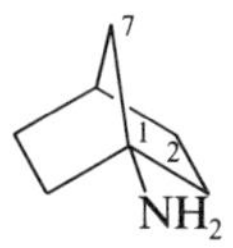

$C_7H_{13}N$ M 111.186
Liq. Bp 142-145°. pK_a 10.20.
B,HCl: [75934-56-4].
    Solid. Mp >350°.
Picrate: Cryst. (EtOAc/pet. ether). Mp 210°.
[107474-88-4, 107474-89-5]

Werner, R.B. *et al*, *J. Med. Chem.*, 1961, **4**, 183 (*synth*)
Wiberg, K.B. *et al*, *J. Am. Chem. Soc.*, 1963, **85**, 3188.
Deutsch, J. *et al*, *Isr. J. Chem.*, 1972, **10**, 51 (*ms*)
Golzke, V. *et al*, *Nouv. J. Chim.*, 1978, **2**, 169; *CA*, **89**, 42069x (*synth*)
Pachter, R. *et al*, *Theochem*, 1986, **30**, 143; *CA*, **105**, 60240z (*cmr*)
Della, E.W. *et al*, *J. Am. Chem. Soc.*, 1987, **109**, 2746 (*synth, N-15 nmr*)

7-Aminobicyclo[2.2.1]heptane   A-80061

Bicyclo[2.2.1]heptan-7-amine, 9CI. 7-Aminonorbornane

[35092-58-1]

$C_7H_{13}N$ M 111.186

Oily solid or liq. Bp 147-152°, Bp_8 66-67.5°.

B,HCl: [35092-59-2].
   Cryst.

N-*Methoxycarbonyl:*
   $C_9H_{15}NO_2$ M 169.223
   Long needles. Mp 68-69°.

N-*Ac:*
   $C_9H_{15}NO$ M 153.224
   Cryst. (hexane). Mp 130-131°.

N-*Chloroacetyl:*
   $C_9H_{14}ClNO$ M 187.668
   Cryst. (pet. ether). Mp 84-85° (80°).

Boehme, W.R. *et al, J. Med. Chem.*, 1961, **4**, 183 (*synth*)
Zalkow, L.H. *et al, J. Org. Chem.*, 1963, **28**, 3303 (*synth, ir, pmr, derivs*)
Stille, J.K. *et al, J. Am. Chem. Soc.*, 1972, **94**, 8489 (*deriv*)
Hermes, M.E. *et al, J. Org. Chem.*, 1972, **37**, 2969 (*synth, pmr*)
Pachter, R. *et al, Theochem*, 1986, **30**, 143; *CA*, **105**, 60240z (*cmr*)
Fukunage, K. *et al, Synthesis*, 1987, 1097 (*synth*)

1-Aminobiphenylene   A-80062

1-Biphenylenamine, 9CI

[16294-10-3]

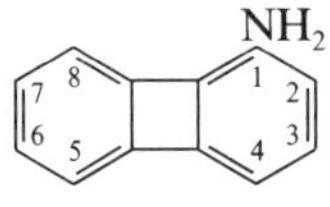

$C_{12}H_9N$ M 167.210

Pale yellow needles (pet. ether). Mp 80-81°.

N-*Ac:* [16294-11-4].
   $C_{14}H_{11}NO$ M 209.247
   Pale yellow needles (C_6H_6). Mp 193-194°.

Barton, J.W. *et al, J. Chem. Soc. C*, 1967, 2097 (*synth, uv, pmr*)

2-Aminobiphenylene   A-80063

2-Biphenylenamine

[55716-75-1]

$C_{12}H_9N$ M 167.210

Polymer intermediate. Yellow needles. Mp 123-124°.

N-*Ac:*
   $C_{14}H_{11}NO$ M 209.247
   Pale yellow plates (EtOH aq.). Mp 146-147°.

Baker, W. *et al, J. Chem. Soc.*, 1954, 1476; 1958, 2666 (*synth, deriv, uv*)
Barton, J.W. *et al, J. Chem. Soc. C*, 1967, 2097 (*pmr*)
Droske, J.P. *et al, Macromolecules*, 1984, **17**, 10.
Takeichi, T. *et al, Macromolecules*, 1986, **19**, 2093, 2103, 2108 (*use*)

4-Aminobutanal, 9CI   A-80064

Updated Entry replacing A-01166
4-Aminobutyraldehyde

[4390-05-0]

$$H_2NCH_2CH_2CH_2CHO$$

C_4H_9NO M 87.121

2,4-Dinitrophenylhydrazone; B,HCl: Solid (EtOH). Mp 198-199°.

Di-Me acetal: [19060-15-2]. *4,4-Dimethoxybutylamine*
   $C_6H_{15}NO_2$ M 133.190

Oil. Bp_{22} 83-85°.

Di-Et acetal: [6346-09-4]. *4,4-Diethoxybutylamine*
   $C_8H_{19}NO_2$ M 161.244
   Bp 196°, $Bp_{0.5}$ 84-85°.

Burckhalter, J.H. *et al, J. Org. Chem.*, 1958, **23**, 1278 (*synth*)
Leonard, N.J. *et al, Tetrahedron Lett.*, 1960, 44 (*synth*)
Gribble, G.W. *et al, J. Org. Chem.*, 1988, **53**, 3164 (*acetal, synth, ir, pmr*)

3-Amino-3-cyclobutene-1,2-dione, 9CI   A-80065

Semisquaric amide

[65842-67-3]

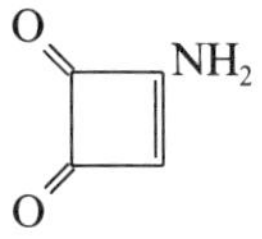

$C_4H_3NO_2$ M 97.073

Cryst. (EtOH). Mp 170° dec. An earlier report of this compd. (mp. 85° dec.) was erroneous.

Schmidt, A.H. *et al, Synthesis*, 1987, 134 (*synth, ir, pmr, cmr*)

4-Amino-4,6-dihydro-3-methyl-1*H*-cyclopenta[*e*]1,2,4-triazine-5,7-dicarboxylic acid   A-80066

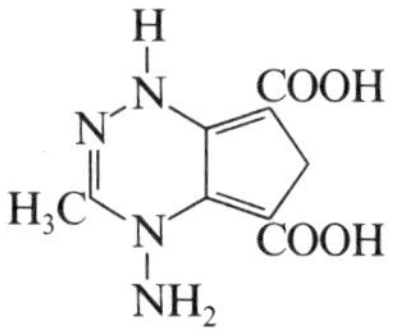

$C_9H_{10}N_4O_4$ M 238.202

Di-Me ester: [95927-46-1].
   $C_{11}H_{14}N_4O_4$ M 266.256
   Reagent for conversion of aldehydes to nitriles in good yield, *via* hydrazone formn. Mp 198-199° dec.

Di-Et ester: [117227-44-8].
   $C_{13}H_{18}N_4O_4$ M 294.310
   Reagent for conversion of aldehydes to nitriles. Mp 185-186° dec.

Neunhoeffer, H. *et al, Justus Liebigs Ann. Chem.*, 1989, 105 (*synth, ir, pmr, use*)

4-Amino-3,4-dihydro-2-phenyl-2*H*-1-benzopyran   A-80067

3,4-Dihydro-2H-1-benzopyran-4-amine, 9CI. 4-Aminoflavan

[6517-15-3]

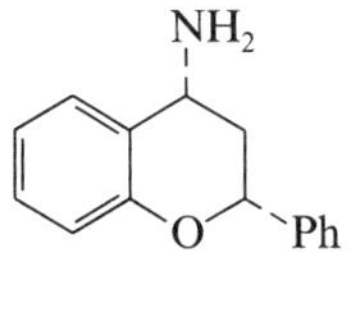

(2R,4R)-form

$C_{15}H_{15}NO$ M 225.290

(2R,4R)-form
   (+)-cis-*form*
   Mp 112° dec. $[\alpha]_D$ +14° (EtOH).

B,HCl: [29834-28-4].
   Mp 281-283°. $[\alpha]_D$ +26.8° (EtOH).

N-*Ac:*
   $C_{17}H_{17}NO_2$ M 267.327

Mp 208-209°. $[\alpha]_D$ +32.1°.

(2S,4S)-form
(−)-cis-*form*
Mp 112° dec. $[\alpha]_D$ −13.8° (EtOH).
B,HCl: [29834-29-5].
Mp 285-286°. $[\alpha]_D$ −26.5° (EtOH).
N-Ac: Mp 224-225°. $[\alpha]_D$ −14.5° (EtOH).

(2R,4S)-form
(−)-trans-*form*
Mp 116° dec. $[\alpha]_D^{20}$ −14° (c, 0.5 in EtOH).
B,HCl: Mp 263-264°. $[\alpha]_D^{20}$ −22° (c, 0.5 in EtOH).

(2S,4R)-form
(+)-trans-*form*
Mp 116°. $[\alpha]_D^{20}$ +14° (c, 0.5 in EtOH).
B,HCl: Mp 260°. $[\alpha]_D^{20}$ +24° (c, 0.5 in EtOH).

(2RS,4RS)-form
(±)-cis-*form*
B,HCl: Mp 267-270°.
N-Ac: Mp 202°.

Tökés, A.L. *et al, Justus Liebigs Ann. Chem.*, 1989, 89 (*resoln, abs config, bibl*)

2-Amino-4,5-diphenyloxazole   A-80068

4,5-Diphenyl-2-oxazolamine, 9CI
[33119-63-0]

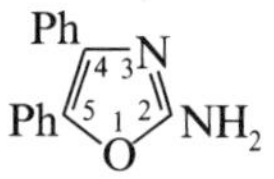

$C_{15}H_{12}N_2O$ M 236.273
Cryst. (pet. ether). Mp 142°.
Picrate: Mp 222°.

Gompper, R. *et al, Chem. Ber.*, 1959, **92**, 1944.

4-Amino-2,5-diphenyloxazole   A-80069

2,5-Diphenyl-4-oxazolamine, 9CI
[121019-48-5]
$C_{15}H_{12}N_2O$ M 236.273
Plates (EtOH). Mp 261-262°. First known simple 4-aminooxazole. Various ring-substd. analogues also prepd.

Lakhan, R. *et al, J. Heterocycl. Chem.*, 1988, **25**, 1413.

5-Amino-2,4-diphenyloxazole   A-80070

2,4-Diphenyl-5-oxazolamine, 9CI
[17960-26-8]
$C_{15}H_{12}N_2O$ M 236.273
Cryst. (CH_2Cl_2). Mp 133-135°.

Poupaert, J. *et al, Synthesis*, 1972, 622.

Aminododecahedrane   A-80071

Tetradecahydro-5,2,1,6,3,4-[2,3]butanediyl[1,4]diylidenedipentaleno[2,1,6-cde:2′,1′,6′-gha]pentalen-1(2H)-amine, 9CI. Dodecahedranamine
[116635-07-5]

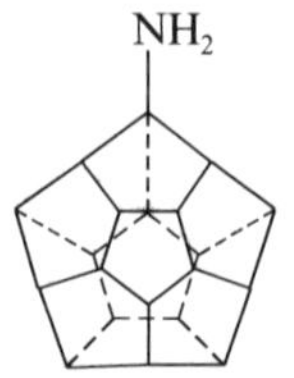

$C_{20}H_{21}N$ M 275.393
B,HCl: [116663-41-3].
Solid. Mp >250°.
N-Ac: [112896-39-6].
$C_{22}H_{23}NO$ M 317.430
Solid. Mp >250°.

Weber, J.C. *et al, J. Org. Chem.*, 1988, **53**, 5315 (*synth, pmr, cmr*)

2-Amino-1,1,2-ethanetricarboxylic acid, 9CI   A-80072

β-Carboxyaspartic acid. Asa
[75898-26-9]

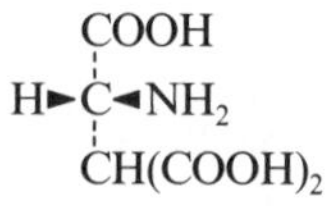

(R)-form

$C_5H_7NO_6$ M 177.113
Obt. only in racemic form from natural systems.

(R)-form [99265-42-6]
D-form
Synthetic.
B,HCl: [111934-05-5].
Solid. $[\alpha]_D^{25}$ −13.25° (c, 0.8 in H_2O).
N-Ac, Tri-Me ester: [99265-41-5].
$C_{10}H_{15}NO_7$ M 261.231
$[\alpha]_D^{20}$ +86.9° (c, 2.0 in MeOH).
*1,1-Di-*tert*-butyl ester, 2-Me ester:* [99166-13-9].
Mp 63-64°. $[\alpha]_D^{25}$ +0.4° (c, 1.0 in MeOH).

(±)-form [78000-84-7]
Isol. from the ribosomal proteins of *E. coli* and human atherosclerotic plaque. pK_a ca 0.8. V. sensitive to decarboxylation so difficult to detect in natural systems.
B,HCl: [78000-85-8].
Unstable.
N-Ac,Tri-Me ester: [78000-86-9].
Cryst. Mp 77-78°.

[90365-27-8]

Christy, M.R. *et al, J. Am. Chem. Soc.*, 1981, **103**, 3935 (*isol, synth, pmr, cmr*)
Henson, E.B. *et al, Tetrahedron*, 1981, **37**, 2561 (*synth, ms*)
Richey, B. *et al, Biochemistry*, 1982, **21**, 4819 (*cryst struct*)
Koch, T.H. *et al, Methods Enzymol.*, 1984, **107**, 563 (*synth, ir, pmr, cmr*)
Van Buskirk, J.J. *et al, Proc. Natl. Acad. Sci. U.S.A.*, 1984, **81**, 722 (*biochem*)
Schöllkopf, U. *et al, Angew. Chem., Int. Ed. Engl.*, 1985, **24**, 1066 (*synth*)
Williams, R.M. *et al, J. Am. Chem. Soc.*, 1988, **110**, 482 (*synth, pmr*)

2-Amino-3-fluorobenzenethiol, 9CI   A-80073

2-Fluoro-6-mercaptoaniline
[73628-29-2]

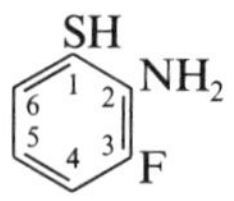

C_6H_6FNS M 143.184
Mp 92°.

Gupta, R.R. *et al, Synth. Commun.*, 1987, **17**, 229.

2-Amino-5-fluorobenzenethiol, 9CI   A-80074

4-Fluoro-2-mercaptoaniline

[33264-82-3]

C_6H_6FNS M 143.184

Fine yellow cryst. (EtOH). Mp 82°.

Mital, R. *et al, Monatsh. Chem.*, 1971, **102**, 760 (*synth*)
Sawhney, S.N. *et al, J. Org. Chem.*, 1979, **44**, 1136 (*cmr*)
Gupta, R.R. *et al, J. Fluorine Chem.*, 1985, **28**, 381 (*synth*)

2-Amino-6-fluorobenzenethiol, 9CI   A-80075

3-Fluoro-2-mercaptoaniline

[100493-32-1]

C_6H_6FNS M 143.184

Eur. Pat., 158 340, (1985); *CA*, **105**, 6526j.

4-Amino-2-fluorobenzenethiol, 8CI   A-80076

3-Fluoro-4-mercaptoaniline

[15178-52-6]

C_6H_6FNS M 143.184

Pohloudek-Fabini, R. *et al, CA*, 1967, **67**, 73287s (*synth*)

4-Amino-3-fluorobenzenethiol, 9CI   A-80077

2-Fluoro-4-mercaptoaniline

[15178-48-0]

C_6H_6FNS M 143.184

[100280-13-5]

Pohlondek-Fabini, R. *et al, CA*, 1967, **67**, 73287s.

8-Aminoguanine   A-80078

2,8-Diamino-1,7-dihydro-6H-purin-6-one, 9CI. 2,8-Diaminohypoxanthine. 2,8-Diamino-6-hydroxypurine

[28128-41-8]

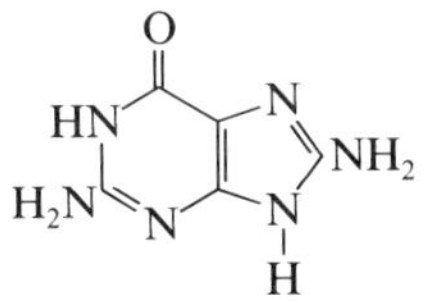

$C_5H_6N_6O$ M 166.142

Solid. Mp >240°.

Holmes, R.E. *et al, J. Am. Chem. Soc.*, 1965, **87**, 1772 (*synth*)
Lin, T.S. *et al, J. Med. Chem.*, 1985, **28**, 1194 (*synth*)

2-Amino-1,3,4-hexadecanetriol   A-80079

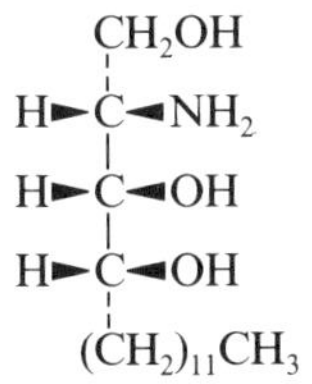

$C_{16}H_{35}NO_3$ M 289.457

(2S,3S,4R)-form

Common phytosphingosine, occurring for example in Acanthocerebrosides.

O,N,N,N-*Tetra-Ac*: [114379-45-2].
$C_{24}H_{43}NO_7$ M 457.606

Viscous liq. $[\alpha]_D^{25}$ +27.9° (c, 1.5 in $CHCl_3$).

Sugiyama, S. *et al, Justus Liebigs Ann. Chem.*, 1988, 619 (*abs config, bibl*)

4-Amino-5-hexenoic acid, 9CI   A-80080

Vigabatrin, BAN, INN. *γ-Vinyl GABA. GVG. MDL 71754. RMI 71754*

[60643-86-9]

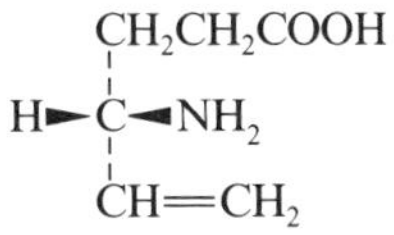

$C_6H_{11}NO_2$ M 129.158

(S)-form [74046-07-4]

Pharmacol active isomer. Anticonvulsant. Inhibitor of GABA transaminase. Cryst. (EtOH/2-propanol). $[\alpha]_D$ +12.4° (c, 0.5 in H_2O).

(±)-form [68506-86-5]

Cryst. (EtOH aq.).

[77162-51-7]

U.S. Pat., 3 960 927; *CA*, **85**, 143512 (*synth*)
Palfreyman, M.G. *et al, Biochem. Pharmacol.*, 1981, **30**, 817 (*rev*)
U.K. Pat., 2 133 002; *CA*, **101**, 231027 (*synth, isomer*)
Chen, T.M. *et al, J. Chromatogr.*, 1984, **314**, 495 (*resoln*)
Hammond, E.J. *et al, Gen. Pharmacol.*, 1985, **16**, 441 (*rev*)
Smithers, J.A. *et al, J. Chromatogr.*, 1985, **341**, 232 (*hplc*)
Martindale, The Extra Pharmacopoeia, 28th/29th Ed., Pharmaceutical Press, London, 1982/1989, 17030.

3-Amino-4-hydroxybutanoic acid   A-80081

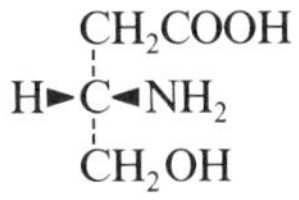

$C_4H_9NO_3$ M 119.120

(S)-form

Lactone: *4-Aminodihydro-2(3H)-furanone. β-Amino-γ-butyrolactone*
$C_4H_7NO_2$ M 101.105
Mp 198-201° (as hydrobromide). $[\alpha]_D^{20}$ −42.6° (c, 1.08 in H_2O).

Lactone, N-*benzoyl:*
$C_{11}H_{11}NO_3$ M 205.213
Mp 123.5-126°. $[\alpha]_D^{20}$ −97.0° (c, 1.41 in $CHCl_3$).

(±)-form

Lactone: Mp 185-186° (as hydrochloride).

Chibnall, A.C. *et al, Biochem. J.*, 1958, **68**, 122 (*synth*)
McGarvey, G. *et al, J. Am. Chem. Soc.*, 1986, **108**, 4943 (*synth, pmr*)
Hvidt, T. *et al, Can. J. Chem.*, 1988, **66**, 779 (*synth, pmr, cmr*)

1-Amino-3-(hydroxymethyl)cyclobutanecarboxylic acid, 9CI   A-80082

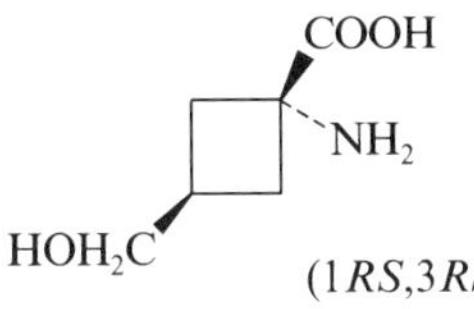

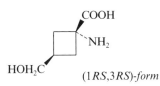

$C_6H_{11}NO_3$ M 145.158

(1RS,3RS)-form [116823-32-6]

trans-*form*

Synthetic. Cryst. Mp 255° dec.

(1RS,3SR)-form
cis-*form*
Minor component of the Costa Rican legume *Atelia herbert smithii* and of other *A.* and *Cyanthiostegia* spp. Cryst. (EtOH aq.). Mp 250-260° dec. (synthetic)(210°).
B,HCl: [109826-20-2].
Cryst.

Austin, G.N. *et al, Tetrahedron*, 1987, **43**, 1857 (*isol, ms, ir, pmr, cryst struct*)
Hughes, P. *et al, J. Org. Chem.*, 1988, **53**, 4793 (*synth, ir, pmr, cmr, ms*)
Fleet, G.W.J. *et al, Tetrahedron*, 1988, **44**, 2077 (*synth, ir, pmr, cmr, ms*)

1-Amino-2-(hydroxymethyl)cyclopropanecarboxylic acid   A-80083

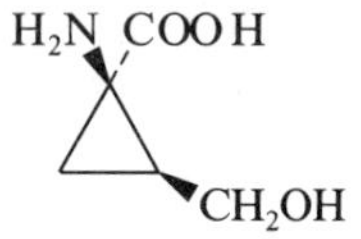

(1*R*,2*S*)-*form*

$C_5H_9NO_3$ M 131.131
Model for ethylene biogeneration in plants.
(1R,2S)-form
$[\alpha]_D^{25}$ +73.8° (c, 0.48 in H_2O).
(1S,2R)-form
Cryst. Mp 240° dec. $[\alpha]_D^{25}$ −74.5° (c, 0.184 in H_2O).
(1RS,2SR)-form
Cryst. (EtOH aq.). Mp 202° dec.
N-tert-*Butyloxycarbonyl:* Rhombs. Mp 151°.

Pirrung, M.C. *et al, Helv. Chim. Acta*, 1989, **72**, 1301 (*synth, ir, pmr, cmr*)

3-Amino-3-(4-hydroxyphenyl)propanoic acid   A-80084

β-Amino-4-hydroxybemenepropanoic acid, 9CI. β-Amino-p-hydroxyhydrocinnamic acid, 8CI. β-Tyrosine. Isotyrosine
[6049-54-3]

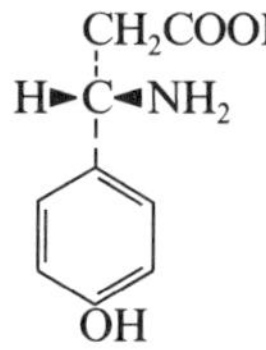

(*S*)-*form*

$C_9H_{11}NO_3$ M 181.191
(S)-form
D-*form*
B,HCl: [54732-46-6].
Needles (EtOH/EtOAc). $[\alpha]_D^{25}$ +7.8° (c, 1.67 in H_2O).
(±)-form
Cryst. (EtOH aq.). Mp 198°.

[52997-29-2, 54732-46-6, 73025-68-0, 73025-69-1]

Posner, T., *Justus Liebigs Ann. Chem.*, 1912, **389**, 1 (*synth*)
Profft, E. *et al, J. Prakt. Chem.*, 1965, **30**, 18 (*synth*)
Roncar, G. *et al, Biochemistry*, 1966, **5**, 2153.
Kurylo-Borowska, Z. *et al, Biochim. Biophys. Acta*, 1972, **264**, 1 (*biosynth*)

Chekhlov, A.N. *et al, Kristallografiya*, 1974, **19**, 981; (*Engl. Transl.*), 608 (*cryst struct*)
Parry, R.J. *et al, J. Am. Chem. Soc.*, 1980, **102**, 836 (*biosynth,abs config*)

4-Amino-4-(3-hydroxypropyl)-1,7-heptanediol   A-80085

Bis-homotris
[116747-79-6]

$$H_2NC(CH_2CH_2CH_2OH)_3$$

$C_{10}H_{23}NO_3$ M 205.297
Solid. Mp 108.5-109.8°. $Bp_{0.5}$ 220-235°.

Newkome, G.R. *et al, J. Org. Chem.*, 1988, **53**, 5552 (*synth, pmr, ir, cmr*)

2-Amino-4(5)-imidazoleacetic acid   A-80086

[73086-08-5]

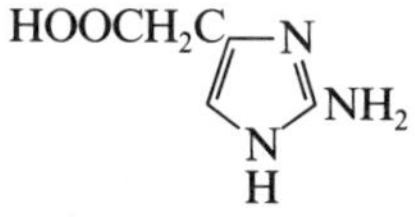

$C_5H_7N_3O_2$ M 141.129
Me ester: [110295-91-5].
$C_6H_9N_3O_2$ M 155.156
Oil (as trifluorocetate).

[110295-92-6]

Bouchet, M.-J. *et al, J. Med. Chem.*, 1987, **30**, 2222 (*synth, pmr, ms*)

7-Amino-3*H*-imidazo[4,5-*b*]pyridine   A-80087

1H-Imidazo[4,5-b]pyridin-7-amine, 9CI. 1-Deazaadenine
[6703-44-2]

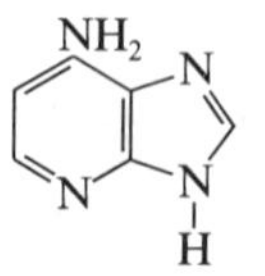

$C_6H_6N_4$ M 134.140
Cryst. (H_2O). Mp 258-260°.
B,HCl: [4261-04-5].
Needles (H_2O). Mp 327-329°.
B,2HCl: Cryst. Mp 325-328°.
Picrate: Mp >260°.
3-β-D-Ribofuranosyl: [14432-09-8]. *1-Deazaadenosine*
$C_{11}H_{14}N_4O_4$ M 266.256
Anti-leukaemic, shows other biological props. Cryst. (H_2O). Mp 262-263° dec. $[\alpha]_D^{25}$ −96.3° (c, 0.85 in DMF).

[57232-66-3]

Kögl, F. *et al, Recl. Trav. Chim. Pays-Bas (J. R. Neth. Chem. Soc.)*, 1948, **67**, 29 (*synth*)
Jain, P.C. *et al, Indian J. Chem.*, 1966, **4**, 403 (*synth, deriv, uv, ir*)
De Roos, K.B. *et al, Recl. Trav. Chim. Pays-Bas (J. R. Neth. Chem. Soc.)*, 1969, **88**, 1263; 1971, **90**, 654 (*deriv, synth, pmr, ir, uv*)
Itoh, T. *et al, J. Heterocycl. Chem.*, 1972, **9**, 465 (*deriv, synth, uv*)
Cristalli, G. *et al, J. Med. Chem.*, 1987, **30**, 1686 (*deriv, synth*)

2-Amino-3-iodopyridine, 8CI   A-80088

3-Iodo-2-pyridinamine, 9CI

[104830-06-0]

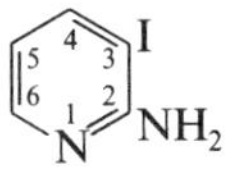

$C_5H_5IN_2$ M 220.012

Needles (hexane). Mp 90-91.5°.

N-Me: [113975-23-8].

$C_6H_7IN_2$ M 234.039

Mp 50°. Bp_{14} 129°.

Sakamoto, T. *et al, Chem. Pharm. Bull.*, 1985, **33**, 4764 (*synth, pmr*)

Estel, L. *et al, J. Org. Chem.*, 1988, **53**, 2740 (*synth, pmr, ir, deriv*)

2-Amino-4-iodopyridine, 8CI   A-80089

4-Iodo-2-pyridinamine, 9CI

$C_5H_5IN_2$ M 220.012

Cryst. (H_2O). Mp 163-164°.

Picrate: Cryst. (H_2O). Mp 253-254°.

Graf, R., *Chem. Ber.*, 1931, **64**, 21, 25.

2-Amino-5-iodopyridine, 8CI   A-80090

5-Iodo-2-pyridinamine, 9CI

[20511-12-0]

$C_5H_5IN_2$ M 220.012

Needles (C_6H_6). Mp 130°.

Picrate: Yellow needles (EtOH or Me_2CO). Mp 240°.

Magidson, O.Yu. *et al, Ber.*, 1928, **58**, 114 (*synth*)

Caldwell, W.T. *et al, J. Am. Chem. Soc.*, 1944, **66**, 1479, 1481 (*synth*)

Shepherd, R.G. *et al, J. Am. Chem. Soc.*, 1948, **70**, 159 (*synth*)

Cumper, C.W.N., *J. Chem. Soc. B*, 1968, 649 (*uv*)

Jovanovic, M.V., *Heterocycles*, 1984, **22**, 1195 (*synth, pmr*)

2-Amino-6-iodopyridine, 8CI   A-80091

6-Iodo-2-pyridinamine, 9CI

$C_5H_5IN_2$ M 220.012

Cryst. (Et_2O/pet.ether). Mp 106-106.5°.

Johnson, F., *J. Org. Chem.*, 1962, **27**, 2473 (*synth*)

3-Amino-5-iodopyridine, 8CI   A-80092

5-Iodo-3-pyridinamine, 9CI

[25391-66-6]

$C_5H_5IN_2$ M 220.012

Cryst. (C_6H_6/pet.ether). Mp 71°.

Picrate: [25391-67-7].

Cryst. Mp 252° (225°).

N-Ac:

$C_7H_7IN_2O$ M 262.050

Cryst. Mp 202°.

N-Benzoyl:

$C_{12}H_9IN_2O$ M 324.120

Cryst. Mp 158°.

Plazek, E. *et al, Pol. J. Chem.* (*Rocz. Chem.*), 1938, **18**, 210, 215 (*synth*)

Batkowski, T., *Pol. J. Chem.* (*Rocz. Chem.*), 1969, **43**, 1623 (*synth, derivs*)

4-Amino-2-iodopyridine, 8CI   A-80093

2-Iodo-4-pyridinamine, 9CI

$C_5H_5IN_2$ M 220.012

Cryst. (EtOH). Mp 98-99°.

N-oxide:

$C_5H_5IN_2O$ M 236.012

Cryst. (H_2O). Mp 110°.

Talik, T., *Pol. J. Chem.* (*Rocz. Chem.*), 1957, **31**, 569, 575 (*synth*)

Talik, Z., *Pol. J. Chem.* (*Rocz. Chem.*), 1961, **35**, 475 (*oxide*)

4-Amino-3-iodopyridine, 8CI   A-80094

3-Iodo-4-pyridinamine, 9CI

$C_5H_5IN_2$ M 220.012

Needles (C_6H_6). Mp 76-77°.

Picrate: Mp 253° dec.

B,MeI: Cryst. (DMF or Et_2O). Mp 254.5-255°.

Profft, E. *et al, J. Prakt. Chem.*, 1959, **8**, 156, 163.

Talik, T., *Pol. J. Chem.* (*Rocz. Chem.*), 1962, **36**, 1049; 1963, **37**, 69.

5-Amino-2-iodopyridine, 8CI   A-80095

6-Iodo-3-pyridinamine, 9CI

[29958-12-1]

$C_5H_5IN_2$ M 220.012

Needles (EtOH). Mp 132°.

Caldwell, W.T. *et al, J. Am. Chem. Soc.*, 1944, **66**, 1479, 1482 (*synth*)

Tomasik, P., *Pol. J. Chem.* (*Rocz. Chem.*), 1970, **44**, 509 (*synth*)

4-Aminoisoxazole   A-80096

4-Isoxazolamine, 9CI

[108511-97-3]

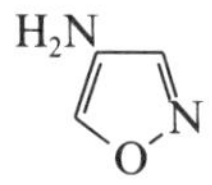

$C_3H_4N_2O$ M 84.077

Dark-brown liq. V. sol. H_2O. Not obt. pure, clearly v. air-sensitive.

B,HCl: [108511-98-4].

Mp 160-163°.

Reiter, L.A., *J. Org. Chem.*, 1987, **52**, 2715 (*synth*)

Pascual, A., *Helv. Chim. Acta*, 1989, **72**, 556 (*synth, ir, pmr*)

3-Amino-2-methylbutanoic acid   A-80097

[32723-74-3]

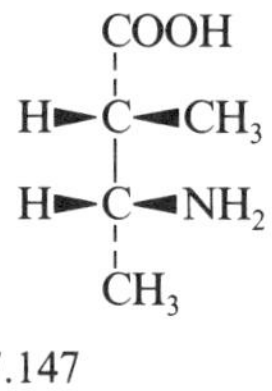

$C_5H_{11}NO_2$ M 117.147

(2RS,3RS)-form

Needles (MeOH/Et_2O). Mp 221-222° dec.

[121054-31-7]

Estermann, H. *et al, Helv. Chim. Acta*, 1988, **71**, 1824 (*synth, ir, pmr, cmr*)

1-Amino-3-methylisoquinoline   A-80098

3-Methyl-1-isoquinolinamine, 9CI

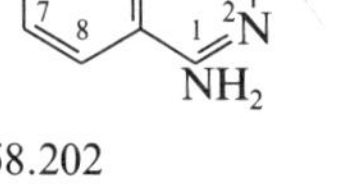

$C_{10}H_{10}N_2$ M 158.202

Cryst. (C_6H_6). Mp 130.5-131°.
B,HCl: Cryst. Mp 272-278°.

Bergstrom, F.W. *et al, J. Org. Chem.*, 1945, **10**, 479 (*synth*)

3-Amino-1-methylisoquinoline   A-80099

1-Methyl-3-isoquinolinamine, 9CI
[15787-16-3]
$C_{10}H_{10}N_2$ M 158.202
Cryst. (EtOH). Mp 122-124°.
Picrate: [15896-92-1].
   Cryst. Mp 232-233°.

Kametani, T. *et al, Chem. Pharm. Bull.*, 1967, **15**, 7 (*synth*)

3-Amino-4-methylisoquinoline   A-80100

4-Methyl-3-isoquinolinamine, 9CI
[7697-66-7]
$C_{10}H_{10}N_2$ M 158.202
Pale yellow cryst. Mp 118-119°.

Win, H. *et al, J. Org. Chem.*, 1967, **32**, 59 (*synth, pmr, uv*)
Neumeyer, J.L., *J. Med. Chem.*, 1970, **13**, 613 (*synth*)

4-Amino-1-methylisoquinoline   A-80101

1-Methyl-4-isoquinolinamine, 9CI
[104704-41-8]
$C_{10}H_{10}N_2$ M 158.202
Yellow prisms (Me_2CO/hexane).

Sawanishi, H. *et al, Chem. Pharm. Bull.*, 1985, **33**, 4564 (*synth, pmr*)

5-Amino-1-methylisoquinoline   A-80102

1-Methyl-5-isoquinolinamine, 9CI
[20335-61-9]
$C_{10}H_{10}N_2$ M 158.202
Cryst. (C_6H_6). Mp 213-214°.
N-Ac:
   $C_{12}H_{12}N_2O$ M 200.240
   Cryst. (dioxan). Mp 226-227°.
N-Me: [53760-81-9].
   $C_{11}H_{12}N_2$ M 172.229
   Pale yellow prisms. (EtOAc/hexane). Mp 161.5-162.5°.
N,N-Di-Me: [84689-38-3].
   $C_{12}H_{14}N_2$ M 186.256
   Liq. $Bp_{0.05}$ 95-98°.

Agrawal, K.C. *et al, J. Med. Chem.*, 1968, **11**, 700.
Mooney, P.D. *et al, J. Med. Chem.*, 1974, **17**, 1145.
Tsuchiya, T. *et al, Chem. Pharm. Bull.*, 1982, **30**, 3757.

5-Amino-3-methylisoquinoline   A-80103

3-Methyl-5-isoquinolinamine, 9CI
[54410-17-2]
Plates (EtOH aq.). Mp 219.5-221°.

Bergstrom, F.W. *et al, J. Org. Chem.*, 1945, **10**, 479 (*synth*)
Elderfield, R.C. *et al, J. Org. Chem.*, 1958, **23**, 435 (*synth*)
Buu-Hoi, N.P. *et al, J. Chem. Soc.*, 1964, 3924 (*synth*)

7-Amino-1-methylisoquinoline   A-80104

1-Methyl-7-isoquinolinamine, 9CI
[31181-24-5]
$C_{10}H_{10}N_2$ M 158.202
Cryst. (C_6H_6). Mp 200-201°.

French, F.A. *et al, J. Med. Chem.*, 1970, **13**, 1117 (*synth*)

8-Amino-1-methylisoquinoline   A-80105

1-Methyl-8-isoquinolinamine, 9CI
[31181-26-7]
$C_{10}H_{10}N_2$ M 158.202
Cryst. (C_6H_6). Mp 141-142°.

French, F.A. *et al, J. Med. Chem.*, 1970, **13**, 1117 (*synth*)

3-Amino-4-methylisoxazole   A-80106

4-Methyl-3-isoxazolamine, 9CI
[1750-43-2]

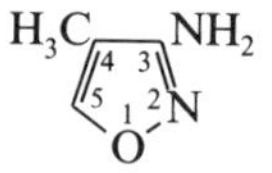

$C_4H_6N_2O$ M 98.104
Cryst. (C_6H_6/Et_2O). Mp 45-50°.

Klotzer, W. *et al, Monatsh. Chem.*, 1970, **101**, 1109 (*synth*)

4-Amino-3-methylisoxazole   A-80107

3-Methyl-4-isoxazolamine, 9CI
$C_4H_6N_2O$ M 98.104
Yellow oil. d_4^{14} 1.172. Bp_{25} 130°. n_D^{14} 1.5069.
B,HCl: [108512-04-5].
   Pink cryst. Mp 188-189°.

Quilico, A. *et al, Gazz. Chim. Ital.*, 1941, **71**, 327; 1942, **72**, 399 (*synth*)
Reiter, L.A., *J. Org. Chem.*, 1987, **52**, 2714 (*synth, ms, pmr*)

4-Amino-5-methylisoxazole   A-80108

5-Methyl-4-isoxazolamine
[87988-94-1]
$C_4H_6N_2O$ M 98.104
Mp 25-31°.
B,HCl: [100499-66-9].
   Cryst. Mp 184-195°.

Reiter, L.A., *J. Org. Chem.*, 1987, **52**, 2714 (*synth, ms, pmr*)
Pascual, A., *Helv. Chim. Acta*, 1989, **72**, 556 (*synth, ir, pmr*)

5-Amino-4-methylisoxazole   A-80109

4-Methyl-5-isoxazolamine, 9CI
[35143-75-0]
$C_4H_6N_2O$ M 98.104

Matsumura, K. *et al, CA*, 1971, **76**, 85737n (*synth*)
Albrecht, H.A. *et al, J. Org. Chem.*, 1979, **44**, 4191 (*synth*)

3-Amino-5-methyl-1*H*-pyrazole-4-carboxylic acid, 9CI   A-80110

5-Amino-3-methyl-4-pyrazolecarboxylic acid
[23286-64-8]

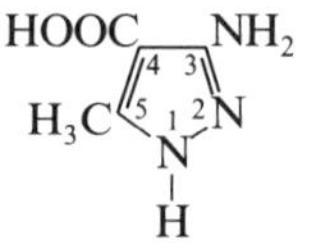

$C_5H_7N_3O_2$ M 141.129
Needles. Mp 124-125°.
Me ester: [23286-71-7].
   $C_6H_9N_3O_2$ M 155.156
   Microcryst. Mp 145-146.5°.
Et ester: [23286-70-6].
   $C_7H_{11}N_3O_2$ M 169.183
   Prisms. Mp 113°.

Et ester, monohydrate: Needles (H₂O). Mp 69°.

Et ester; B,HCl: [113270-87-4].
　　Cryst. Mp 188-189°.

Nitrile: [5453-07-6]. *2-Amino-3-cyano-4-methylpyrazole*
　　$C_5H_6N_4$　　M 122.129
　　Mp 163°.

Beyer, H. *et al, Chem. Ber.,* 1956, **89**, 1652 (*synth, ester*)
Badger, G.M. *et al, Aust. J. Chem.,* 1965, **18**, 1267 (*synth, nitrile*)
Baba, H. *et al, Bull. Chem. Soc. Jpn.,* 1969, **42**, 1653 (*synth*)

4-Amino-5-methyl-1*H*-pyrazole-3-carboxylic acid, 9CI　　　A-80111

[94993-81-4]
$C_5H_7N_3O_2$　　M 141.129
Cryst. (H₂O). Mp 185-190° dec., Mp 208° dec.

B,HCl: Cryst. (H₂O). Mp 213-214°.

Hydrazide: [28668-09-9].
　　$C_5H_9N_5O$　　M 155.159
　　Cryst. (2-Propanol). Mp 187.5-190°.

Et ester: [70015-75-7].
　　$C_7H_{11}N_3O_2$　　M 169.183
　　Cryst. (C₆H₆). Mp 96-96.5°.

N-Ac, Et ester: [70015-77-9].
　　$C_9H_{13}N_3O_3$　　M 211.220
　　Cryst. (EtOAc). Mp 176-177°.

Amide:
　　$C_5H_8N_4O$　　M 140.144
　　Cryst. (H₂O). Mp 209-211°.

Nitrile: [28745-14-4]. *4-Amino-3-cyano-5-methylpyrazole*
　　$C_5H_6N_4$　　M 122.129
　　Cryst. (toluene). Mp 158-160°.

Musante, C. *et al, Gazz. Chim. Ital.,* 1947, **77**, 182 (*synth*)
Robins, R.K. *et al, J. Org. Chem.,* 1956, **21**, 833 (*synth*)
Long, R.A. *et al, J. Heterocycl. Chem.,* 1970, **7**, 863 (*nitrile, hydrazide*)
Takei, H. *et al, Bull. Chem. Soc. Jpn.,* 1979, **52**, 208 (*synth, esters*)

1-Amino-2,6-naphthalenediol, 8CI, 9CI　　　A-80112

2,6-Dihydroxy-1-naphthylamine. 1-Amino-2,6-dihydroxynaphthalene

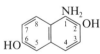

$C_{10}H_9NO_2$　　M 175.187

Di-Me ether: 2,6-Dimethoxy-1-naphthylamine. *1-Amino-2,6-dimethoxynaphthalene*
　　$C_{12}H_{13}NO_2$　　M 203.240
　　Orange needles (C₆H₆). Mp 189°.

Chakravarty, S. .N. *et al, J. Chem. Soc.,* 1937, 1859 (*synth*)

1-Amino-2,7-naphthalenediol, 8CI, 9CI　　　A-80113

Updated Entry replacing D-04799
2,7-Dihydroxy-1-naphthylamine
$C_{10}H_9NO_2$　　M 175.187

O,O,N-Tri Ac:
　　$C_{16}H_{15}NO_5$　　M 301.298
　　Cryst. Mp 183°.

Di-Me ether: [83011-60-3]. *2,7-Dimethoxy-1-naphthylamine*
　　$C_{12}H_{13}NO_2$　　M 203.240
　　Cryst. (C₆H₆/ligroin); Needles (EtOH aq.). Mp 82-83°.

Di-Me ether, N-Ac:
　　$C_{14}H_{15}NO_3$　　M 245.277
　　Cryst. (EtOH). Mp 179-180°.

Clausius, A., *Ber.,* 1890, **23**, 517.
Nietski, R. *et al, Ber.,* 1897, **30**, 1119.
Chakravarti, S.N. *et al, J. Chem. Soc.,* 1937, 1859 (*synth, deriv*)
Soffer, M.D. *et al, J. Am. Chem. Soc.,* 1950, **72**, 3704 (*synth*)

2-Amino-1,4-naphthalenediol, 9CI, 8CI　　　A-80114

Updated Entry replacing D-04794
1,4-Dihydroxy-2-naphthylamine
[75539-34-3]
$C_{10}H_9NO_2$　　M 175.187

O,O,N-Tri-Ac:
　　$C_{16}H_{15}NO_5$　　M 301.298
　　Needles (AcOH). Mp 259-260°.

Di-Me ether: [50885-06-8]. *1,4-Dimethoxy-2-naphthylamine, 8CI*
　　$C_{12}H_{13}NO_2$　　M 203.240
　　Plates (hexane or MeOH). Mp 100-101°.

Kehrmann, F., *Ber.,* 1894, **27**, 3337 (*synth*)
McCaustland, D.J. *et al, J. Med. Chem.,* 1973, **16**, 1311 (*synth*)
Ishii, H. *et al, Chem. Pharm. Bull.,* 1983, **31**, 4391 (*synth, ir*)
Malesani, G. *et al, J. Heterocycl. Chem.,* 1985, **22**, 1141 (*synth, pmr*)

2-Amino-1,6-naphthalenediol, 8CI, 9CI　　　A-80115

Updated Entry replacing D-04796
1,6-Dihydroxy-2-naphthylamine
[4826-64-6]
$C_{10}H_9NO_2$　　M 175.187
Free base is very unstable.

O,O,N-Tri-Ac:
　　$C_{16}H_{15}NO_5$　　M 301.298
　　Cryst. (ligroin or EtOH aq.). Mp 150°.

Fischer, E. *et al, J. Prakt. Chem.,* 1916, **94**, 5.

3-Amino-1,2-naphthalenediol, 8CI, 9CI　　　A-80116

Updated Entry replacing D-04801
3,4-Dihydroxy-2-naphthylamine
[58245-82-2]
$C_{10}H_9NO_2$　　M 175.187
Cryst. (Et₂O or C₆H₆/EtOH). Mp 164°.

N-Ac:
　　$C_{12}H_{11}NO_3$　　M 217.224
　　Mp 170° dec.

O,O-Di-Ac:
　　$C_{14}H_{13}NO_4$　　M 259.261
　　Cryst. (AcOH). Mp >200° dec.

Di-Me ether: 3,4-Dimethoxy-2-naphthylamine
　　$C_{12}H_{13}NO_2$　　M 203.240
　　Prisms (pet. ether). Mp 76°.

Ueno, A. *et al, Chem. Pharm. Bull.,* 1966, **14**, 129 (*synth, deriv*)
Greenland, H. *et al, Aust. J. Chem.,* 1975, **28**, 2655 (*synth, uv, ir, pmr*)

4-Amino-1,2-naphthalenediol, 8CI, 9CI　　　A-80117

Updated Entry replacing D-04800
3,4-Dihydroxy-1-naphthylamine
$C_{10}H_9NO_2$　　M 175.187

N-Ac:
　　$C_{12}H_{11}NO_3$　　M 217.224
　　Cryst. (H₂O). Mp 187° dec.

O,O,N-Tri-Ac:
　　$C_{16}H_{15}NO_5$　　M 301.298
　　Cryst. (EtOH). Mp 206.5-207.5°.

Di-Me ether: [73168-34-0]. *3,4-Dimethoxy-1-naphthylamine*

$C_{12}H_{13}NO_2$ M 203.240
Cryst. (C_6H_6/hexane). Mp 140-142°.
Di-Me ether, N-Ac:
$C_{14}H_{15}NO_3$ M 245.277
Cryst. (C_6H_6). Mp 133-135°.

Kehrmann, F., *Ber.*, 1894, **27**, 3337.
Bogdanov, S.V. *et al, CA*, 1961, **55**, 11372b.
Archer, S. *et al, J. Med. Chem.*, 1980, **23**, 516.

4-Amino-1,3-naphthalenediol, 8CI, 9CI   A-80118

Updated Entry replacing D-04798
2,4-Dihydroxy-1-naphthylamine. 2,4-Dihydroxy-1-naphthalenamine, 9CI
[31209-82-2]
$C_{10}H_9NO_2$ M 175.187
Mp 162°.
B,HCl: [30832-70-3].
Needles (dil. HCl).
O,O,N-Tri-Ac:
$C_{16}H_{15}NO_5$ M 301.298
Plates (C_6H_6). Mp 170-171° (155-156°).

Kehrmann, F. *et al, Chem. Ber.*, 1895, **28**, 345; 1896, **29**, 1415.
Lantz, R., *Bull. Soc. Chim. Fr.*, 1971, 249, 253.

4-Amino-1,5-naphthalenediol, 8CI, 9CI   A-80119

4,8-Dihydroxy-1-naphthylamine. 4-Amino-1,5-dihydroxynaphthalene
$C_{10}H_9NO_2$ M 175.187
Di-Me ether: 4,8-Dimethoxy-1-naphthylamine. 4-Amino-1,5-dimethoxynaphthalene
$C_{12}H_{13}NO_2$ M 203.240
Leaflets (EtOH). Mp 160°.
Di-Me ether, N-Ac:
$C_{14}H_{15}NO_3$ M 245.277
Needles (EtOH). Mp 186°.

Thomson, R.H. *et al, J. Chem. Soc.*, 1947, 350 (*synth, deriv*)

4-Amino-1,7-naphthalenediol, 8CI, 9CI   A-80120

4,6-Dihydroxy-1-naphthylamine. 1-Amino-4,6-dihydroxynaphthalene
[38410-19-4]
$C_{10}H_9NO_2$ M 175.187
Mp 178° dec.
O,O,N-Tribenzoyl:
$C_{31}H_{21}NO_5$ M 487.511
Mp 208°.

Jubault, M. *et al, Bull. Soc. Chim. Fr.*, 1972, 1551 (*synth, deriv, ir*)

5-Amino-1,2-naphthalenediol, 8CI, 9CI   A-80121

5,6-Dihydroxy-1-naphthylamine. 5-Amino-1,2-dihydroxynaphthalene
$C_{10}H_9NO_2$ M 175.187
Di-Me ether: 5,6-Dimethoxy-1-naphthylamine. 5-Amino-1,2-dimethoxynaphthalene
$C_{12}H_{13}NO_2$ M 203.240
Cryst. (EtOH). Mp 97-98°.

Sprenger, W.K. *et al, J. Org. Chem.*, 1966, **31**, 2402.

5-Amino-2,3-naphthalenediol   A-80122

1-Amino-6,7-dihydroxynaphthalene
$C_{10}H_9NO_2$ M 175.187
Di-Me ether: [52401-42-0]. 6,7-Dimethoxy-1-naphthylamine. 1-Amino-6,7-dimethoxynaphthalene

$C_{12}H_{13}NO_2$ M 203.240
Cryst. (EtOH). Mp 130°.
Di-Me ether, N-Ac:
$C_{14}H_{15}NO_3$ M 245.277
Cryst. (C_6H_6). Mp 185°.

Govindachari, T.R. *et al, J. Chem. Soc.*, 1955, 2534.
Kessar, S.V. *et al, Indian J. Chem.*, 1973, **11**, 1105 (*synth*)
Stermitz, F.R. *et al, J. Med. Chem.*, 1975, **18**, 708 (*synth*)

6-Amino-2,3-naphthalenediol, 8CI, 9CI   A-80123

6,7-Dihydroxy-2-naphthylamine. 6-Amino-2,3-dihydroxynaphthalene
$C_{10}H_9NO_2$ M 175.187
Di-Me ether: [56517-29-4]. 6,7-Dimethoxy-2-naphthylamine. 6-Amino-2,3-dimethoxynaphthalene
$C_{12}H_{13}NO_2$ M 203.240
Cryst. Mp 202-203°.

Stermitz, F.R. *et al, J. Med. Chem.*, 1975, **18**, 708 (*synth, deriv*)

6-Amino-1,2-napthalenediol, 8CI, 9CI   A-80124

5,6-Dihydroxy-2-naphthylamine. 6-Amino-1,2-dihydroxynaphthalene
[21489-73-6]
$C_{10}H_9NO_2$ M 175.187
B,HCl: [21538-03-4].
Cryst. (MeOH/Et_2O). Mp 275-277° dec.
Di-Me ether: [23922-96-5]. 5,6-Dimethoxy-2-naphthylamine. 6-Amino-1,2-dimethoxynaphthalene
$C_{12}H_{13}NO_2$ M 203.240
Prisms (EtOH). Mp 148.5-149.5°.

Sprenger, W.K. *et al, J. Med. Chem.*, 1969, **12**, 487 (*synth*)

2-Amino-3-nitrobenzyl alcohol   A-80125

2-Amino-3-nitrobenzenemethanol, 9CI. 2-Hydroxymethyl-6-nitroaniline

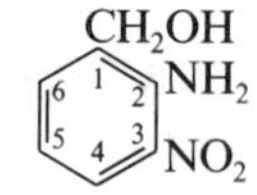

$C_7H_8N_2O_3$ M 168.152
N-Et: [37637-56-2].
$C_9H_{12}N_2O_3$ M 196.205
Liq. $Bp_{0.1}$ 137-138°.
N-Et, Et ether: [37637-57-3].
$C_{11}H_{16}N_2O_3$ M 224.259
Orange liq. $Bp_{0.2}$ 135-140°.

Gawron, O. *et al, J. Am. Chem. Soc.*, 1972, **94**, 5396 (*synth, uv, ms, ir*)

2-Amino-6-nitrobenzyl alcohol   A-80126

2-Amino-6-nitrobenzenemethanol, 9CI. 2-Hydroxymethyl-3-nitroaniline
[98451-51-5]
$C_7H_8N_2O_3$ M 168.152
Yellow needles (CH_2Cl_2/pet. ether). Mp 96.5-97.5°.

Hewgill, F.R. *et al, J. Chem. Soc., Perkin Trans.* 1, 1988, 1305 (*synth, pmr*)

3-Amino-2-nitrobenzyl alcohol   A-80127

3-Amino-2-nitrobenzenemethanol, 9CI. 3-Hydroxymethyl-2-nitroaniline
$C_7H_8N_2O_3$ M 168.152
N-Et: [37637-53-9].

$C_9H_{12}N_2O_3$ M 196.205
Cryst. (H_2O). Mp 105-105.5°.

Gawron, O. et al, J. Am. Chem. Soc., 1972, **94**, 5396 (synth, uv, ms, ir)

3-Amino-4-nitrobenzyl alcohol   A-80128

3-Amino-4-nitrobenzenemethanol, 9CI. 5-Hydroxymethyl-2-nitroaniline

[37637-55-1]

$C_7H_8N_2O_3$ M 168.152
Yellow needles (C_6H_6). Mp 139-140.5°.

N-*Et:* [37637-54-0].
   $C_9H_{12}N_2O_3$ M 196.205
   Orange needles (H_2O). Mp 95-96°.

Gawron, O. et al, J. Am. Chem. Soc., 1972, **94**, 5396 (synth, ir, uv, ms)

3-Amino-2-oxetanone   A-80129

α-Amino-β-propiolactone. Serine β-lactone

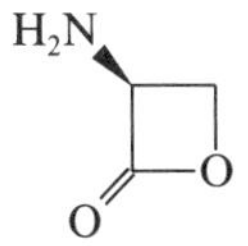

(S)-form

$C_3H_5NO_2$ M 87.078

(*S*)-form

   L-form
   Salts descr. react with nucleophiles to give optically pure amino acids.

Trifluoroacetate salt: [112839-94-8].
   Unstable, used immediately.

4-Methylbenzenesulfonate: [112839-95-9].
   Cryst. (DMF/Et_2O). $[\alpha]_D^{25}$ −15.9° (c, 2.2 in DMF).
   Darkens from 135°, dec. >173°.

N-tert-*Butyloxycarbonyl:* [98541-64-1].
   $C_8H_{13}NO_4$ M 187.195
   Cryst. Mp 119.5-120.5° dec. $[\alpha]_D^{22}$ −26.7° (c, 1 in MeCN).

Arnold, L.D. et al, J. Am. Chem. Soc., 1985, **107**, 7105; 1987, **109**, 4649; 1988, **110**, 2237 (synth, ir, pmr, cmr, ms, use)

8-Amino-7-oxononanoic acid, 9CI   A-80130

[4707-58-8]

$$H_3CCH(NH_2)CO(CH_2)_5COOH$$

$C_9H_{17}NO_3$ M 187.238
Isol. from *Penicillium chrysogeum* as a biotin intermediate.
Cryst. (EtOH/Et_2O). Mp 133-134°.

Suyama, T. et al, CA, 1964, **60**, 4013 (synth)
Iwahara, S. et al, Agric. Biol. Chem., 1966, **30**, 304.
Eisenburg, M.A. et al, Biochemistry, 1970, **9**, 108.

2-Amino-1,5-pentanediol, 9CI   A-80131

Updated Entry replacing A-02457
Glutaminediol
[21926-01-2]

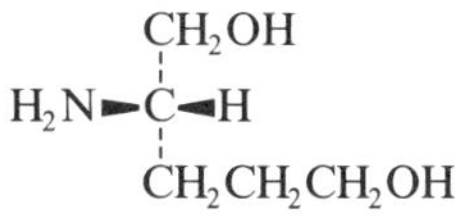

(S)-form

$C_5H_{13}NO_2$ M 119.163
(*S*)-form [21946-71-4]

$Bp_{1.5}$ 150-159°.
Oxalate: Mp 111-113°. $[\alpha]_D^{20}$ +10.0° (c, 2.0 in H_2O).

Jutisz, M. et al, Bull. Soc. Chim. Biol., 1954, **36**, 117; CA, **48**, 10494 (synth)
Herbrandson, H.F. et al, J. Med. Chem., 1969, **12**, 620 (synth)
Prasad, B. et al, Indian J. Chem., 1974, **12**, 290 (synth)
Danklmaier, J. et al, Justus Liebigs Ann. Chem., 1988, 851 (synth, cmr)

5-Amino-1,4-pentanediol   A-80132

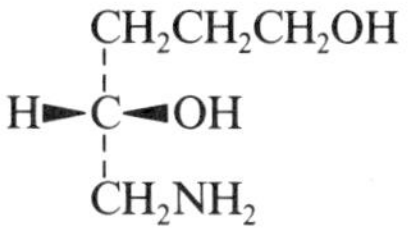

$C_{15}H_{13}NO_2$ M 239.273
(*S*)-form
   $Bp_{0.02}$ 90-100° (bath) . $[\alpha]_D^{20}$ +2.2° (c, 1.5 in AcOH).

Danklmeier, J. et al, Justus Liebigs Ann. Chem., 1988, 1149 (synth, cmr)

3-Amino-2,6-piperidinedione, 9CI   A-80133

2-Aminoglutarimide. Glutamic acid imide

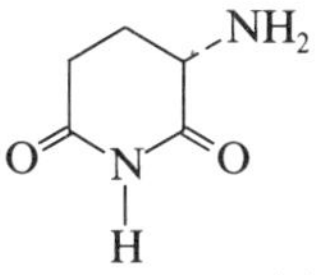

(S)-form

$C_5H_8N_2O_2$ M 128.130
(*S*)-form
   B,HBr: [118061-00-0].
   Mp 279° dec.

Poloński, T., J. Chem. Soc., Perkin Trans. 1, 1988, 639 (synth, cd, uv, pmr)

3-Amino-2-formyl-2-propenoic acid   A-80134

$$H_2NCH{=}\!{=}C(CHO)COOH$$

$C_4H_5NO_3$ M 115.088
Derivs are useful synthetic intermeds.

Me ester: [64944-80-5]. *Methyl 3-amino-2-formyl-2-propenoate*
   $C_5H_7NO_3$ M 129.115
   Cryst. (MeCN). Mp 115°.

N-*Me, Me ester:* [107195-10-8]. *Methyl 2-formyl-3-methylamino-2 propenoate*
   $C_6H_9NO_3$ M 143.142
   Cryst. Mp 69°.

N-tert-*Butyl, Me ester:* [82100-23-0]. *Methyl 3-tert-butylamino-2-formyl-2-propenoate*
   $C_9H_{15}NO_3$ M 185.222
   Cryst. Mp 55°.

N-*Ph, Me ester: Methyl 2-formyl-3-phenylamino-2-propenoate*
   $C_{11}H_{11}NO_3$ M 205.213
   Cryst. Mp 73°.

Tietze, L.F. et al, Chem. Ber., 1989, **122**, 83.

2-Amino-4(3*H*)-pteridinone-6-carboxylic acid   A-80135

2-Amino-1,4-dihydro-4-oxo-6-pteridinecarboxylic acid, 9CI.
Pterin-6-carboxylic acid

[948-60-7]

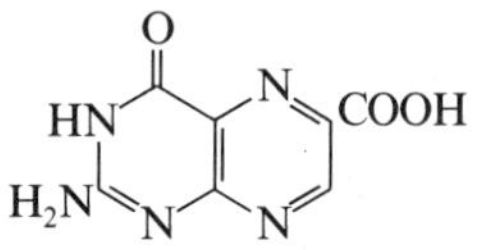

$C_7H_5N_5O_3$ M 207.148
Cream cryst. Mp >300°.

Sato, N. *et al, J. Heterocycl. Chem.*, 1988, **25**, 1737.

2-Amino-3-pyridinethiol, 9CI   A-80136

2-Amino-3-mercaptopyridine

[110402-20-5]

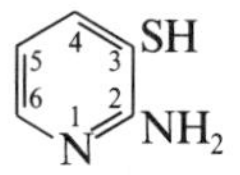

$C_5H_6N_2S$ M 126.182
Has antibacterial props. Intermed. for pharmaceuticals.

Ger. Pat., 3 545 124; 3 545 097; *CA*, **107**, 134219q; **107**, 154346g (*synth, use*)

4-Amino-3-pyridinethiol   A-80137

4-Amino-3-mercaptopyridine

[52334-54-0]

$C_5H_6N_2S$ M 126.182
Semisolid.

Picrate: Yellow needles (EtOH). Mp 228° dec.

Takahashi, T. *et al, Chem. Pharm. Bull.*, 1956, **4**, 216 (*synth*)
Smith, K. *et al, Chem. Ind. (London)*, 1988, 302 (*synth*)

6-Amino-3-pyridinethiol   A-80138

2-Amino-5-mercaptopyridine

[68559-17-1]

$C_5H_6N_2S$ M 126.182

SnCl$_2$/HCl salt: Cryst. (conc. HCl). Mp 265-267°.

S-Et: [71167-00-5]. *5-Ethylthio-2-pyridinamine, 9CI. 2-Amino-5-ethylthiopyridine*
$C_7H_{10}N_2S$ M 154.235
Mp 20°. Bp$_{15}$ 166-169°.

S-Et, picrate: [71167-01-6].
Cryst. (EtOH). Mp 238-240°.

Ger. Pat., 2 812 914, (1978); *CA*, **90**, 6397q (*synth, props*)
Gold'farb, Ya.L. *et al, Chem. Heterocycl. Compd. (Engl. Transl.)*, 1979, **15**, 514 (*synth, deriv*)
U.S. Pat., 4 221 796, (1980); *CA*, **94**, 192328y (*synth, use, deriv*)

2-Amino-4(1*H*)-pyridinethione, 9CI   A-80139

2-Amino-4-pyridinethiol. 2-Amino-4-mercaptopyridine

[38240-25-4]

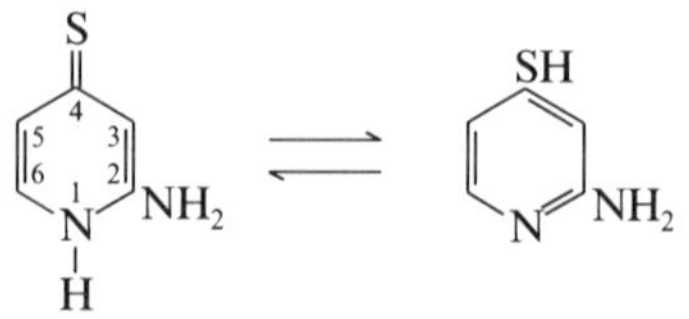

$C_5H_6N_2S$ M 126.182

Intermed. for cephalosporins. Cryst. (Me$_2$CO). Mp 223° dec.

Picrate: Yellow cryst. (H$_2$O). Mp >185° dec.

S-Me: [38240-26-5]. *4-Methylthio-2-pyridinamine, 9CI. 2-Amino-4-methylthiopyridine*
$C_6H_8N_2S$ M 140.209
Cryst. (cyclohexane). Mp 122-123°.

[59243-52-6, 59243-57-1, 59243-61-7, 65872-14-2]

Barlin, G.B., *J. Chem. Soc., Perkin Trans. 2*, 1972, 1459 (*synth, uv*)

3-Amino-4(1*H*)-pyridinethione, 9CI   A-80140

3-Amino-4-pyridinethiol. 3-Amino-4-mercaptopyridine

[89002-13-1]

$C_5H_6N_2S$ M 126.182
Intermed. for cephalosporins. Pale yellow granules (H$_2$O). Sol. aq. alkalies. Mp 213° dec.

S-Me: [38240-24-3]. *4-Methylthio-3-pyridinamine. 3-Amino-4-methylthiopyridine*
$C_6H_8N_2S$ M 140.209
Yellow needles (MeOH)(as picrate). Mp 173-174° (165-166°) (picrate).

S-Et: *4-Ethylthio-3-pyridinamine. 3-Amino-4-ethylthiopyridine*
$C_7H_{10}N_2S$ M 154.235
Needles (EtOH)(as hydrochloride). Mp 213° dec. (hydrochloride).

N-Ac,S-Me:
$C_8H_{10}N_2OS$ M 182.246
Granules $+\frac{1}{2}$H$_2$O (Me$_2$CO). Mp 109° approx.

[38240-23-2, 59243-59-3, 59243-62-8]

Takahashi, T. *et al, Chem. Pharm. Bull.*, 1954, **2**, 196; 1955, **3**, 356 (*synth, derivs*)
Barlin, G.B., *J. Chem. Soc., Perkin Trans. 2*, 1972, 1459 (*synth, uv, props*)

5-Amino-2(1*H*)-pyridinethione, 8CI   A-80141

5-Amino-2-pyridinethiol. 5-Amino-2-mercaptopyridine

[27885-56-9]

$C_5H_6N_2S$ M 126.182
Intermed. for diazo dyes. Golden needles (H$_2$O). Sol. hot H$_2$O. Mp 172-175°. Readily oxidises in air to disulfide.

N-Ac: [27885-57-0].
$C_7H_8N_2OS$ M 168.219
Bright yellow cryst. (H$_2$O). Mp 244-246°.

S-Me: [29958-08-5]. *6-Methylthio-3-pyridinamine, 9CI. 5-Amino-2-methylthiopyridine, 8CI*
$C_6H_8N_2S$ M 140.209
Brown-yellow or light brown needles (C$_6$H$_6$/pet. ether or cyclohexane). Mp 71-72°.

S-Me, picrate: Cryst. (EtOH). Mp 127-129°.

S-Et: [52025-15-7]. *6-Ethylthio-3-pyridinamine. 5-Amino-2-ethylthiopyridine*
$C_7H_{10}N_2S$ M 154.235
Viscous yellow oil. Bp 126°.

S-Et; B,2HCl: [96592-07-3].
Small blunt needles (EtOH). Mp 175-176°.

N-Ac, S-Me: [29958-16-5].
$C_8H_{10}N_2OS$ M 182.246
Cryst. (H$_2$O). Mp 155-155.5°.

[59243-51-5, 59315-59-2, 96592-06-2]

Rath, C., *Justus Liebigs Ann. Chem.*, 1931, **487**, 105 (*synth*)
Caldwell, W.T. *et al, J. Am. Chem. Soc.*, 1942, **64**, 1695 (*synth, props, deriv*)
Takahashi, T. *et al, Yakugaku Zasshi (J. Pharm. Soc. Jpn.)*, 1942, **62**, 140; 1945, **65**, 14 (*synth, derivs*)

Forrest, H.S. *et al*, *J. Chem. Soc.*, 1948, 1939 (*synth, deriv*)
Tomasik, P., *Pol. J. Chem.* (*Rocz. Chem.*), 1970, **44**, 509 (*synth, props, deriv*)
Seydel, J.K., *J. Med. Chem.*, 1971, **14**, 724 (*pmr, deriv*)
Barlin, G.B., *J. Chem. Soc., Perkin Trans. 2*, 1972, 1459 (*synth, uv, derivs*)
Akers, H.A. *et al*, *Can. J. Chem.*, 1987, **65**, 1364 (*synth, ir*)

6-Amino-2(1*H*)-pyridinethione, 9CI   A-80142

6-Amino-2-pyridinethiol, 9CI. 2-Amino-6-mercaptopyridine

[38240-17-4]

$C_5H_6N_2S$ M 126.182

Cryst. (Me$_2$CO). Mp 198°. Prob. yellow, though no colour mentioned.

S-*Me:* [38240-18-5]. *6-Methylthio-2-pyridinamine, 9CI. 2-Amino-6-methylthiopyridine*
$C_6H_8N_2S$ M 140.209
Cryst. (cyclohexane). Mp 67-68°.

[59315-52-5, 59315-60-5]

Barlin, G.B., *J. Chem. Soc., Perkin Trans. 2*, 1972, 1459 (*synth, uv, deriv*)
Kwiatkowski, J.S. *et al*, *Acta Phys. Pol. A*, 1976, **49**, 393; *CA*, **84**, 179475t (*uv*)

3-Aminopyrrole   A-80143

1H-Pyrrole-3-amine

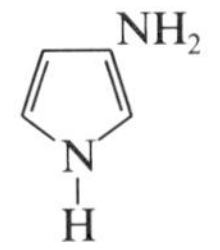

$C_4H_6N_2$ M 82.105

1,N³-Di-tert-*butyl:* [117203-10-8]. *1*-tert-*Butyl-3-(tert-butylamino)-1H-pyrrole*
$C_{12}H_{22}N_2$ M 194.319
Bp$_{0.01}$ 50°.

tom Dieck, H. *et al*, *Chem. Ber.*, 1989, **122**, 129 (*synth, ir, pmr, cmr*)

2-Aminopyrrolo[2,3-*d*]pyrimidine-4-thione   A-80144

7-Deaza-6-thioguanine

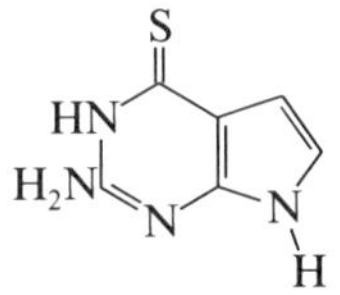

$C_6H_6N_4S$ M 166.206

7-(2-Deoxy-β-ᴅ-ribofuranosyl): [104291-18-1]. *2′-Deoxy-7-deaza-6-thioguanosine*
$C_{11}H_{14}N_4O_3S$ M 282.323
Pale yellow needles (H$_2$O). Mp 196-210° dec. (>200°).

Seela, F. *et al*, *Justus Liebigs Ann. Chem.*, 1987, 15.
Ramasany, K. *et al*, *Tetrahedron Lett.*, 1987, **28**, 5107.

2-Amino-1,7,8,9-tetrahydro-2-thioxo-6*H*-purin-6-one, 9CI   A-80145

8-Mercaptoguanine, 8CI. 2-Amino-6-hydroxy-8-mercaptopurine

[6324-72-7]

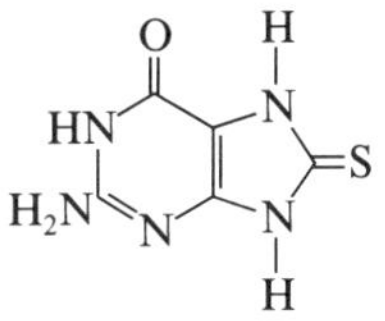

$C_5H_5N_5OS$ M 183.193

[28128-40-7]

Elion, G.B. *et al*, *J. Am. Chem. Soc.*, 1959, **81**, 1898 (*synth*)
Holmes, R.E. *et al*, *J. Am. Chem. Soc.*, 1964, **86**, 1242 (*synth*)

2-Amino-4-thiazoleacetic acid   A-80146

[29676-71-9]

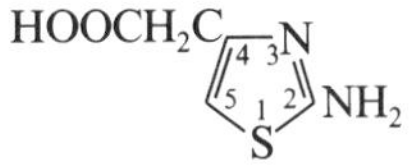

$C_5H_6N_2O_2S$ M 158.181

Mod. active plant growth regulator. Mp 121-122° dec. (130°).

HBF$_4$ salt: [110295-78-8].
Cryst. Mp 180-185°.

Et ester:
$C_7H_{10}N_2O_2S$ M 186.234
Mp 92-94°.

Steude, M., *Justus Liebigs Ann. Chem.*, 1891, **261**, 22 (*synth*)
Garraway, J.L., *Pestic. Sci.*, 1970, **1**, 240 (*props*)
Bouchet, M.-J. *et al*, *J. Med. Chem.*, 1987, **30**, 2222 (*synth, pmr*)

2-Amino-5-thiazoleacetic acid   A-80147

$C_5H_6N_2O_2S$ M 158.181
HBF$_4$ salt: [110295-79-9].
Mp 215° dec.

Et ester:
$C_7H_{10}N_2O_2S$ M 186.234
Plates (C$_6$H$_6$/pet. ether or CHCl$_3$/pet. ether). Mp 100-101°.

Et ester, picrate: Yellow needles (EtOH). Mp 204-205° dec.
Amide:
$C_5H_7N_3OS$ M 157.196
Cryst. Mp 174-175° dec.

Mory, R. *et al*, *Helv. Chim. Acta*, 1950, **33**, 405 (*deriv, synth*)
Bouchet, M.-J. *et al*, *J. Med. Chem.*, 1987, **30**, 2222 (*synth, pmr*)

4-Amino-1,3,5-triazine-2(1*H*)-thione, 9CI   A-80148

4-Amino-1,2-dihydro-1,3,5-triazine-2-thione

[36469-86-0]

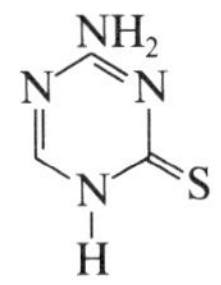

$C_3H_4N_4S$ M 128.157
Cryst. Mp >330°.

Niedballa, V. *et al*, *J. Org. Chem.*, 1974, **39**, 3672 (*synth, uv, pmr*)

4-Amino-1,3,5-triazin-2(1*H*)-one, 9CI   A-80149

4-Amino-s-triazin-2-ol. 2-Amino-4-hydroxy-s-triazine. 5-Azacytosine

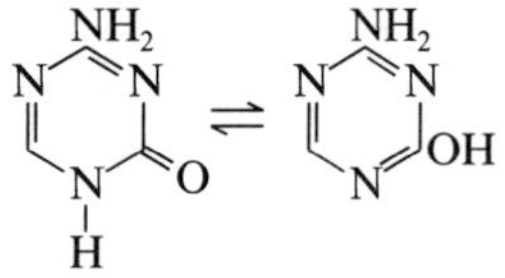

$C_3H_4N_4O$ M 112.091
Microcryst. powder. Insol. org. solvs. Mp >350° dec.

B,HCl: [16352-14-0].
   Solid. Mp >350° (cryst. change at 250°).

1-(2,3-Dideoxy-β-D-ribofuranosyl): [107036-52-2]. *2′,3′-Dideoxy-5-azacytidine*
   $C_8H_{12}N_4O_3$ M 212.208
   Solid. Mp 247° dec.

4-N-Me: [16352-16-2].
   $C_4H_6N_4O$ M 126.118
   Cryst. (H₂O). Mp 310-311°.

1,4-N-Di-Me: [16352-08-2].
   $C_5H_8N_4O$ M 140.144
   Cryst. Mp 270-271°.

4,4-N-Di-Me: [16352-15-1].
   $C_5H_8N_4O$ M 140.144
   Cryst. (MeOH). Mp 268-269°.

1,4,4-N-Tri-Me: [3391-73-9].
   $C_6H_{10}N_4O$ M 154.171
   Cryst. (MeOH). Mp 201-202°.

2-Me ether: [1122-73-2]. *2-Amino-4-methoxy-1,3,5-triazine*
   $C_4H_6N_4O$ M 126.118
   Cryst. (C₆H₆). Mp 188-189°.

[4040-10-2]

Grundmann, C. et al, Chem. Ber., 1954, **87**, 19 (synth)
Piškala, A., Collect. Czech. Chem. Commun., 1967, **32**, 3966, 4271 (synth, derivs)
Hartenstein, R. et al, J. Org. Chem., 1967, **32**, 1653 (synth)
Niedballa, V. et al, J. Org. Chem., 1974, **39**, 3672 (synth)
Kim, C.-H. et al, J. Med. Chem., 1987, **30**, 862 (deriv, synth, pmr, ms)

Amoenumin   A-80150

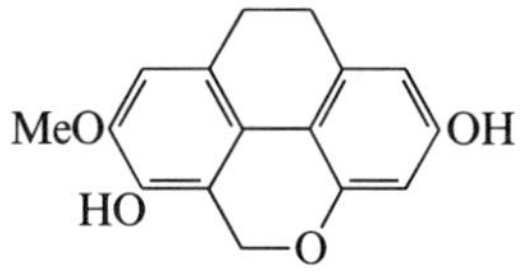

$C_{16}H_{14}O_4$ M 270.284
Constit. of *Dendrobium amoenum*. Cryst. (MeOH). Mp 120°.

Veerraju, P. et al, Phytochemistry, 1989, **28**, 950.

Amorilin   A-80151

Updated Entry replacing A-20156
4′,5,7-Trihydroxy-3′,6,8-triprenylflavanone. Euchrenone a₃
[83474-69-5]

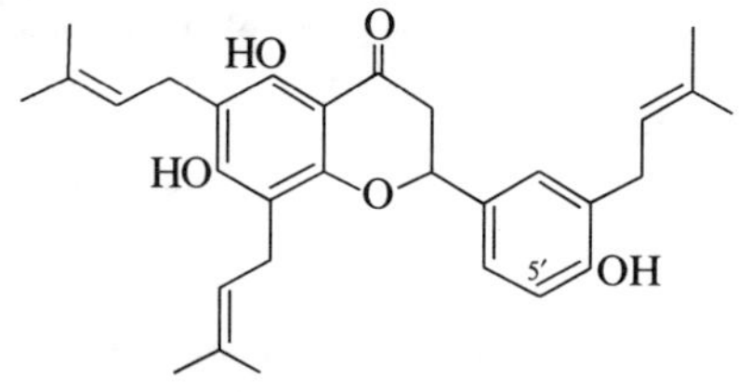

$C_{30}H_{36}O_5$ M 476.611
Constit. of *Amorpha fruticosa* and *Euchresta japonica*. Oil or pale-yellow powder (hexane). Mp 70-72°.

Rózsa, Z. et al, Heterocycles, 1982, **19**, 1793.
Mizuno, M. et al, Phytochemistry, 1988, **27**, 1831 (isol)

Amorphigenin   A-80152

Updated Entry replacing A-02757
8′-Hydroxyrotenone
[4208-09-7]

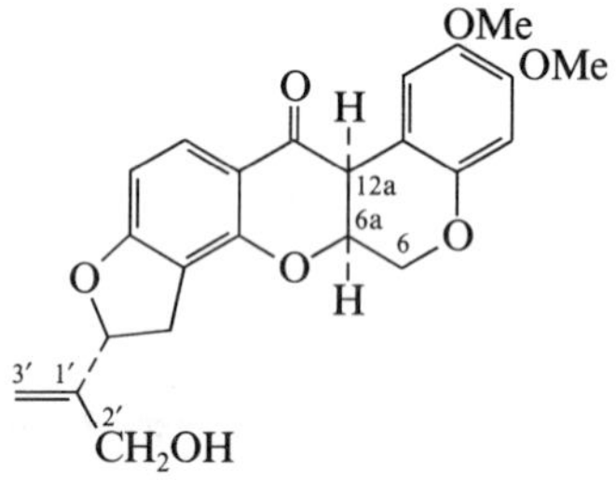

$C_{23}H_{22}O_7$ M 410.423
Various numbering schemes have been used for the side-chain (here numbered 1′, 2′, 3′). Constit. of *Amorpha fruticosa* and of *Dalbergia monetaria*. Cryst. (C₆H₆/CHCl₃, EtOH/CHCl₃ or Me₂CO aq.). Mp 196-197°. $[\alpha]_D^{20}$ −125.6° (c, 2.04 in CHCl₃).

▷ VL1575000.

O-β-D-Glucopyranoside:
   $C_{29}H_{32}O_{12}$ M 572.565
   Isol. from seeds of *A. spp.* and *D. monetaria*. Mp 164°. $[\alpha]_D^{20}$ −122° (c, 0.1 in EtOH).

2′-(6-O-α-L-Arabinopyranosyl-β-D-glucopyranoside): [4207-90-3]. *Amorphin*
   $C_{34}H_{40}O_{16}$ M 704.680
   Glycoside from *A. fruticosa*. Needles. Mp 154-155°. $[\alpha]_D^{18.5}$ −123.6° (c, 1.1 in MeOH), $[\alpha]_D^{24.5}$ −87.9° (c, 2.55 in Py).

1′,3′-Dihydro: [30462-22-7]. *Dihydroamorphigenin. 22,23-Dihydro-24-hydroxyrotenone*
   $C_{23}H_{24}O_7$ M 412.438
   Isol. from seeds of *A. fruticosa*. Mp 189-190°.

1′,3′-Dihydro, 1′-hydroxy: [29360-12-1]. *Amorphigenol*
   $C_{23}H_{24}O_8$ M 428.438
   Constit. of *A. spp.* Needles. Mp 195-196°. $[\alpha]_D^{20}$ −124° (c, 0.47 in EtOH).

1′,3′-Dihydro, 1′-hydroxy, 1′-O-β-D-glucopyranoside: [29360-13-2]. *Amorphigenol glucoside*
   $C_{29}H_{34}O_{13}$ M 590.580
   Isol. from seeds of *A. fruticosa*. Faintly yellow cryst. Mp 189-192° dec. $[\alpha]_D^{20}$ −94.7° (c, 1.3 in MeOH).

1′,3′-Dihydro, 2′-O-[α-L-arabinopyranosyl(1→6)-β-D-glucopyranoside]: [53947-91-4]. *Amorphol*

$C_{34}H_{42}O_{16}$ M 706.696
Constit. of the roots of *A. fruticosa*. Yellowish powder.
Mp 159-162° dec. $[\alpha]_D^{22}$ −96.6° (c, 2.07 in MeOH).
Incorr. descr. as a dioxin deriv. in CA owing to a
drawing error in the paper.

6a,12a-Didehydro: [29444-01-7]. ***Dehydroamorphigenin***
$C_{23}H_{20}O_7$ M 408.407
Isol. from *A.* spp. seeds. Pale-yellow needles. Mp 228-5-
229.5° dec. $[\alpha]_D^{19}$ −50.8° (c, 0.83 in CHCl₃).

Claisse, J. *et al, J. Chem. Soc.*, 1964, 6023 (*isol, struct, bibl*)
Kasymov, A.U. *et al, Khim. Prir. Soedin.*, 1968, **4**, 326; 1970, **6**, 197; 1972, **8**, 115; 1974, **10**, 464; *Chem. Natl. Compd.*, 277, 192, 109, 470 (*Amorphigenol glucoside, Dihydroamorphigenin, Amorphol, Amorphigenol*)
Crombie, L. *et al, J. Chem. Soc., Perkin Trans. 1*, 1973, 1285; 1975, 1497 (*biosynth, pmr*)
Crombie, L. *et al, J. Chem. Soc., Chem. Commun.*, 1979, 1143, 1144; 1986, 1063 (*biosynth*)

Amphidinolide B   A-80153

Updated Entry replacing A-70205
[110786-78-2]

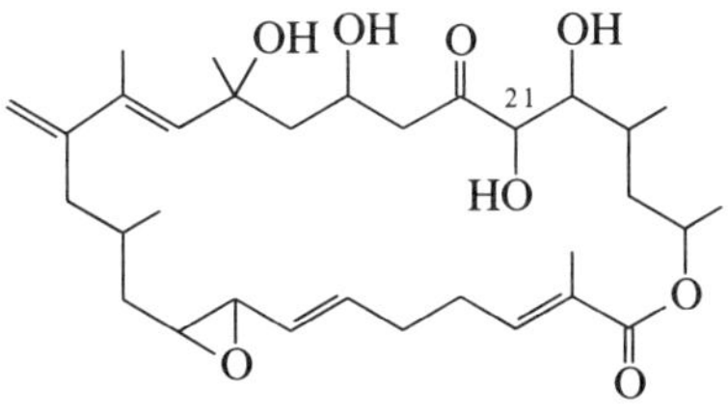

$C_{32}H_{50}O_8$ M 562.742
Macrolide antibiotic. Isol. from *Amphidinium* sp. Shows
antitumour props. Amorph. $[\alpha]_D^{25}$ −45° (c, 1 in CHCl₃).

*21-Epimer: **Amphidinolide D***
$C_{32}H_{50}O_8$ M 562.742
Isol. from an *A.* sp. Shows antitumour props. Amorph.
solid. $[\alpha]_D^{30}$ −30° (c, 0.5 in CHCl₃).

Ishibashi, M. *et al, J. Chem. Soc., Chem. Commun.*, 1987, 1127 (*isol, struct*)
Kobayashi, J. *et al, J. Nat. Prod. (Lloydia)*, 1989, **52**, 1036 (*isol, pmr, cmr*)

Anastomosine   A-80154

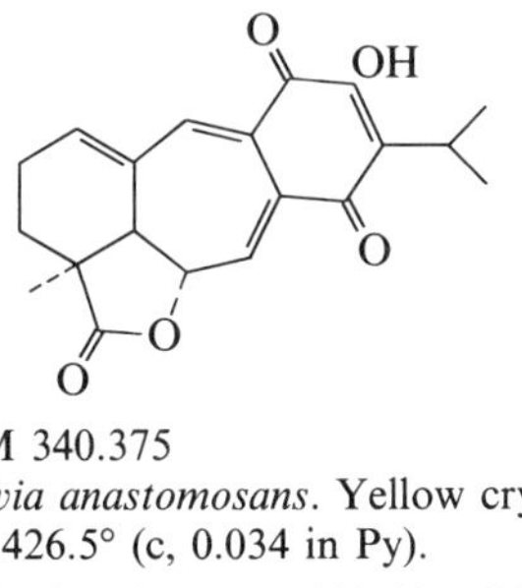

$C_{20}H_{20}O_5$ M 340.375
Constit. of *Salvia anastomosans*. Yellow cryst. Mp 207-
215°. $[\alpha]_D^{20}$ +426.5° (c, 0.034 in Py).

Sánchez, C. *et al, Phytochemistry*, 1989, **28**, 1681.

Angolensin   A-80155

Updated Entry replacing A-02947
*1-(2,4-Dihydroxyphenyl)-2-(4-methoxyphenyl)-1-propanone,
9CI*

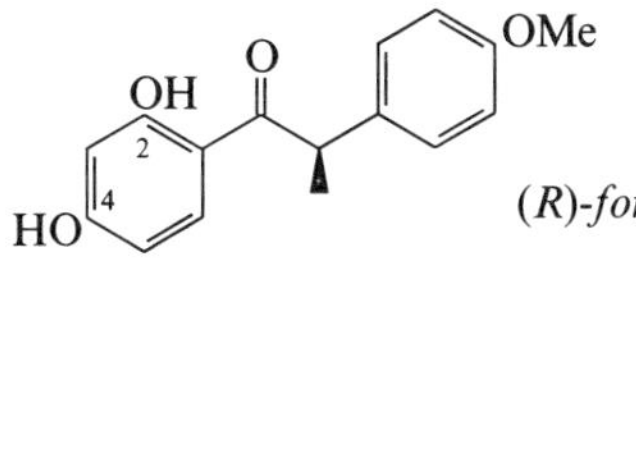

(R)-form

$C_{16}H_{16}O_4$ M 272.300
(R)-form [4842-48-2]
Constit. of *Pericopsis elata, P. mooniana, Pterocarpus
angolensis, P. erinaceus* and *P. indica*. Prisms (pet.
ether). Mp 120-121°. $[\alpha]_D^{24}$ −114° (c, 1 in CHCl₃).

2-Me ether: [58822-06-3]. **2-O-Methylangolensin**
$C_{17}H_{18}O_4$ M 286.327
Constit. of *Pericopsis elata*. Oil. $[\alpha]_D^{24}$ +5.3° (CHCl₃).

4-Me ether: Mp 62°. $[\alpha]_D^{24}$ −142° (c, 2 in CHCl₃).

2,4-Di-Me ether: Mp 49°. $[\alpha]_D^{24}$ +26° (CHCl₃).

(S)-form
4-Me ether: [75946-85-9]. **4-O-Methylangolensin**
$C_{17}H_{18}O_4$ M 286.327
Isol. from heartwood of *P. angolensis*. Needles (MeOH).
Mp 28-30°.

(±)-form
Constit. of *Pericopsis mooniana*. Mp 86-88°.

2,4-Di-Me ether: Mp 70°.

King, F.E. *et al, J. Chem. Soc.*, 1952, 1920 (*isol, struct*)
Foxall, C.D. *et al, J. Chem. Soc.*, 1963, 5573 (*isol, struct*)
Ollis, W.D. *et al, Aust. J. Chem.*, 1965, **18**, 1787 (*abs config*)
Clark-Lewis, J.W. *et al, Aust. J. Chem.*, 1965, **18**, 1791 (*abs config*)
Aggarwal, S.K. *et al, Indian J. Chem.*, 1970, **8**, 478 (*synth*)
Fitzgerald, M.A. *et al, J. Chem. Soc., Perkin Trans. 1*, 1976, 186 (2-O-Methylangolensin)
Bezuidenhondt, B.C.B. *et al, J. Chem. Soc., Perkin Trans. 1*, 1980, 2179 (4-O-Methylangolensin)
Ingham, J.L., *Fortschr. Chem. Org. Naturst.*, 1983, **43**, 1 (*rev, occur*)

Angoletin   A-80156

*1-(2,4-Dihydroxy-6-methoxy-3,5-dimethylphenyl)-3-phenyl-1-
propanone, 9CI. 2′,4′-Dihydroxy-6′-methoxy-3′,5′-
dimethyldihydrochalcone*
[76444-55-8]

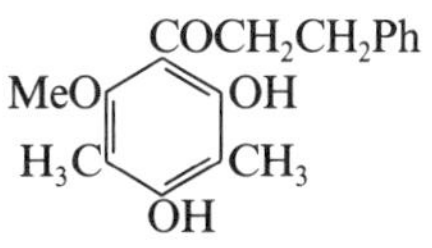

$C_{18}H_{20}O_4$ M 300.354
Isol. from *Uvaria angolensis*. Cryst. (CHCl₃/hexane). Mp
74-76°.

Hufford, C.D. *et al, Phytochemistry*, 1980, **19**, 2036 (*isol, uv, pmr, struct, cmr*)

Anhydroarthrosporone   A-80157

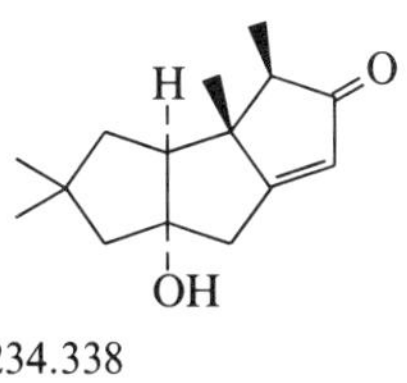

$C_{15}H_{22}O_2$ M 234.338
Metab. of an unidentified fungus (UAMH 4262). Cryst.
(Et₂O/hexane). Mp 118-119°. $[\alpha]_D$ +62° (c, 2 in CHCl₃).

Amouzou, E. *et al, J. Nat. Prod. (Lloydia)*, 1989, **52**, 1042 (*isol, pmr*)

Anisatin A-80158

Updated Entry replacing A-03011

[5230-87-5]

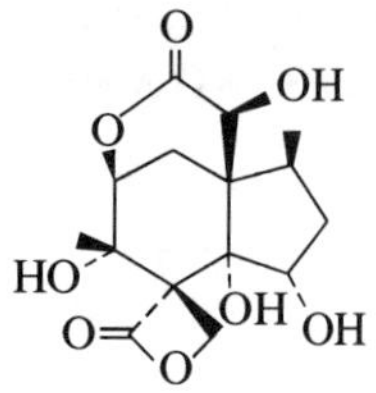

$C_{15}H_{20}O_8$ M 328.318

Isol. from seeds of *Illicium anisatum*. Cryst. (H_2O). Mp 227-228°. $[\alpha]_D^{25}$ −27° (c, 2 in dioxan).

▷ Convulsant poison.

Lane, J.F. *et al, J. Am. Chem. Soc.*, 1952, **74**, 3211 (*isol*)
Yamada, K. *et al, Tetrahedron Lett.*, 1965, 4797 (*struct*)
Wong, M.G. *et al, Aust. J. Chem.*, 1988, **41**, 1071 (*cryst struct, pmr*)

Anisocoumarin A A-80159

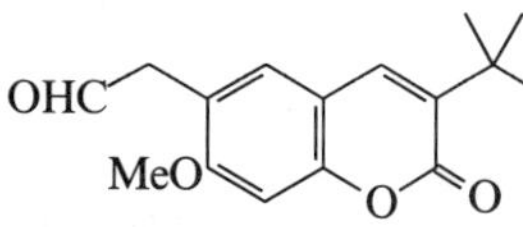

$C_{17}H_{18}O_4$ M 286.327

Constit. of *Clausena anisata*. Yellow oil.

Ngadjui, B.T. *et al, Phytochemistry*, 1989, **28**, 585.

Anisocoumarin C A-80160

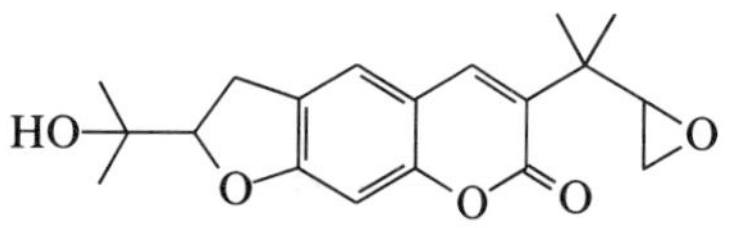

$C_{19}H_{22}O_5$ M 330.380

Constit. of *Clausena anisata*. Oil. $[\alpha]_D^{25}$ +15.5° (c, 1.1 in $CHCl_3$).

Ngadjui, B.T. *et al, Phytochemistry*, 1989, **28**, 585.

Anisocoumarin D A-80161

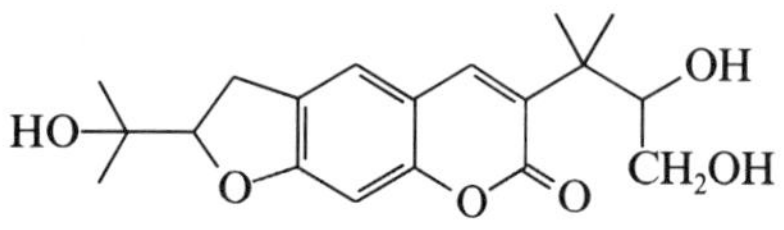

$C_{19}H_{24}O_6$ M 348.395

Constit. of *Clausena anisata*. Cryst. (MeOH). Mp 210-211°. $[\alpha]_D^{25}$ +20.5° (c, 1.2 in MeOH).

Ngadjui, B.T. *et al, Phytochemistry*, 1989, **28**, 585.

Antheliolide A A-80162

Updated Entry replacing A-70213

[114915-36-5]

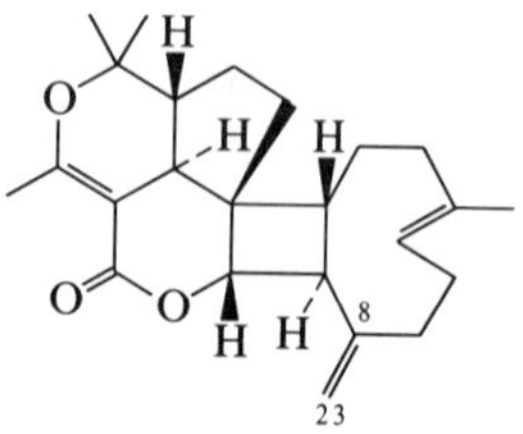

$C_{24}H_{32}O_3$ M 368.515

Novel acetoacetylated diterpenoid skeleton. Constit. of soft coral *Anthelia glauca*. Mp 180°. Exists as mixt. of two conformers.

8ξ,23-Epoxide: [114933-19-6]. **Antheliolide B**
$C_{24}H_{32}O_4$ M 384.514
Constit. of *A. glauca*. Oil.

Green, D. *et al, Tetrahedron Lett.*, 1988, **29**, 1605 (*isol*)
Smith, A.B. *et al, Tetrahedron Lett.*, 1989, **30**, 3363 (*cryst struct*)

9-Anthracenesulfonic acid, 9CI A-80163

Updated Entry replacing A-03081

[22582-76-9]

$C_{14}H_{10}O_3S$ M 258.297

V. sol. H_2O. Dec. in acid soln.

Na salt: [17213-01-3].
Gold plates.

Chloride: [53973-96-9].
$C_{14}H_9ClO_2S$ M 276.743
Used as protecting group in peptide synth. Yellow solid. Mp 133°. Unstable.

Yura, S., *Kogyo Kagaku Zasshi*, 1941, **44**, 731; *CA*, **42**, 6793 (*synth*)
Zorn, H., *Monatsh. Chem.*, 1967, **98**, 2406 (*synth, ir*)
Arzeno, H.B. *et al, Synthesis*, 1988, 32 (*chloride*)

2H-Anthra[1,2-b]pyran-2-one A-80164

[6812-13-1]

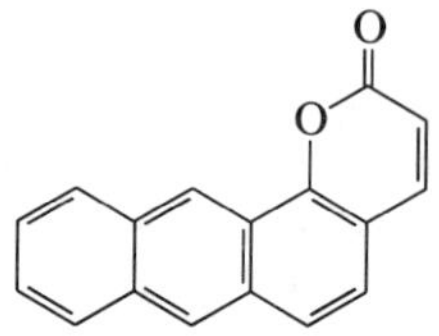

$C_{17}H_{10}O_2$ M 246.265
Cryst. Mp 223-225°.

Harvey, R.G. *et al, J. Org. Chem.*, 1988, **53**, 3936 (*synth, pmr*)

3H-Anthra[2,1-b]pyran-3-one A-80165

[6706-90-7]

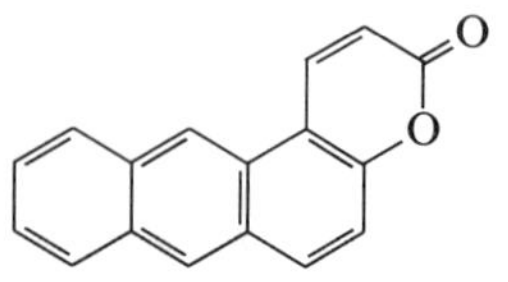

$C_{17}H_{10}O_2$ M 246.265
Cryst. Mp 198-199°.

Harvey, R.G. *et al, J. Org. Chem.*, 1988, **53**, 3936 (*synth, pmr*)

Antiquorin A-80166

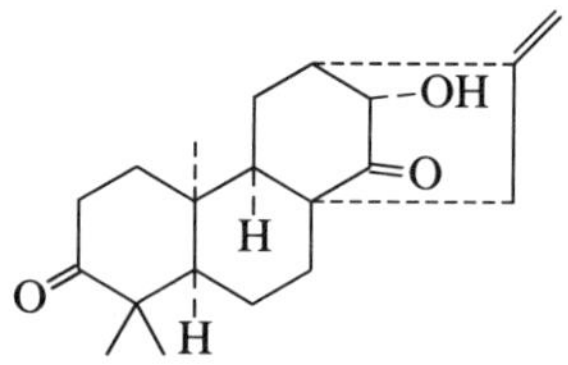

$C_{20}H_{28}O_3$ M 316.439

Constit. of *Euphorbia antiquorum*. Cryst. ($CHCl_3$/MeOH). Mp 163-164°. $[\alpha]_D^{15}$ +70° (c, 0.1 in MeOH).

Zhi-Da, M. *et al, Phytochemistry*, 1989, **28**, 553.

3,24-Aonenadiene A-80167

3,21-Aonenadiene

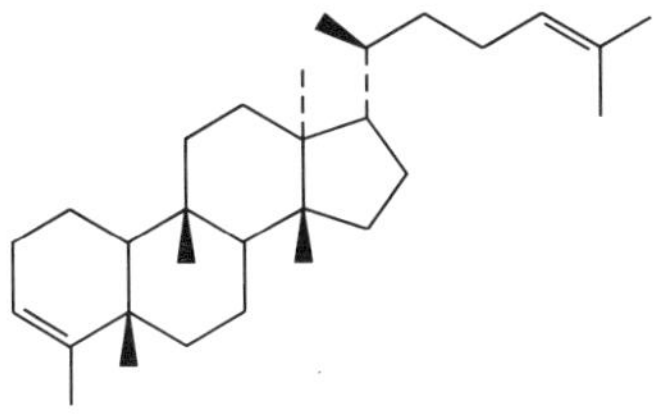

$C_{30}H_{50}$ M 410.725

3,24-Aonenadiene corresponds to the normal numbering system. Constit. of rhizomes of *Polypodioides niponica*. Cryst. Mp 89-91°. $[\alpha]_D^{23}$ +8.9° (c, 0.8 in CHCl₃).

San Feliciano, A. et al, Tetrahedron Lett., 1989, 30, 7209.

Aquatolide A-80168

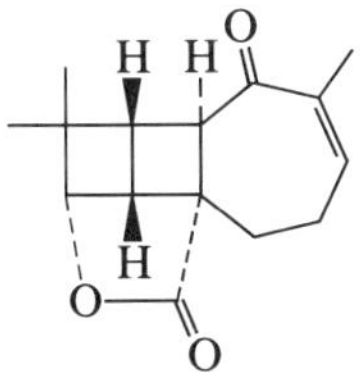

$C_{15}H_{18}O_3$ M 246.305

Constit. of *Asteriscus aquaticus*. Cryst. Mp 142-143°. $[\alpha]_D$ +66.3° (c, 0.45 in CHCl₃).

San Feliciano, A. et al, Tetrahedron Lett., 1989, 30, 2851.

Armillarisin A A-80169

3-Acetyl-5-(hydroxymethyl)-7-hydroxy-2H-1-benzopyran-2-one, 9CI. 3-Acetyl-5-hydroxymethyl-7-hydroxycoumarin
[56696-47-5]

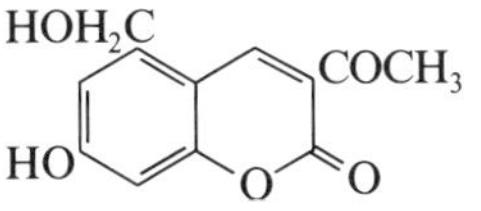

$C_{12}H_{10}O_5$ M 234.208

Isol. from *Armillariella tabescens*. Mp 245-246°.

Anon, CA, 1975, 82, 1919m, 40463c.

Artemorin A-80170

Updated Entry replacing A-20211
[64845-92-7]

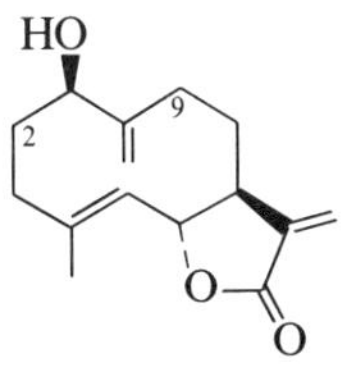

$C_{15}H_{20}O_3$ M 248.321

Constit. of *Artemisia verlotorum*. Cryst. (CH₂Cl₂/Et₂O). Mp 120-121°. $[\alpha]_D$ +89°.

2α-Hydroxy: **2α-Hydroxyartemorin**
$C_{15}H_{20}O_4$ M 264.321
Constit. of *A. hispanica*. Needles (EtOH). Mp 220-221°.

9β-Hydroxy: **9β-Hydroxyartemorin**
$C_{15}H_{20}O_4$ M 264.321
Constit. of *Inula heterolepis*. Gum.

Geissman, T.A., Phytochemistry, 1970, 9, 2377 (isol)

El-Feraly, F.S. et al, Tetrahedron Lett., 1977, 1973 (struct)
Bohlmann, F. et al, Phytochemistry, 1982, 21, 1166 (isol)
Sanz, J.F. et al, Phytochemistry, 1989, 28, 2163 (isol, pmr)

Arthrosporol A-80171

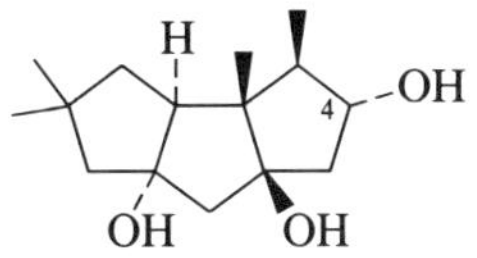

$C_{15}H_{26}O_3$ M 254.369

Metab. of an unidentified fungus (UMAH 4262). Cryst. (Et₂O). Mp 163-164°. $[\alpha]_D$ −29° (c, 2 in CHCl₃).

4-Ketone: **Arthrosporone**
$C_{15}H_{24}O_3$ M 252.353
Metab. of fungus UAMH 4262. Cryst. (Et₂O/hexane). Mp 139-141°. $[\alpha]_D$ −140.8° (c, 0.9 in CHCl₃).

Amouzou, E. et al, J. Nat. Prod. (Lloydia), 1989, 52, 1042 (isol, pmr)

Artobiloxanthone A-80172

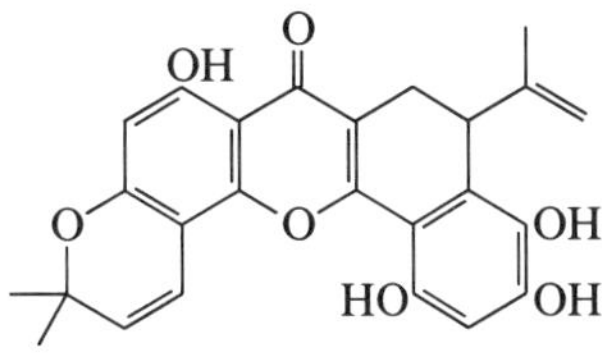

$C_{25}H_{22}O_7$ M 434.445

Constit. of *Artocarpus nobilis*. Yellow solid. Mp 162-164°.

Sultanbawa, M.U.S. et al, Phytochemistry, 1989, 28, 599.

Asacoumarin A A-80173

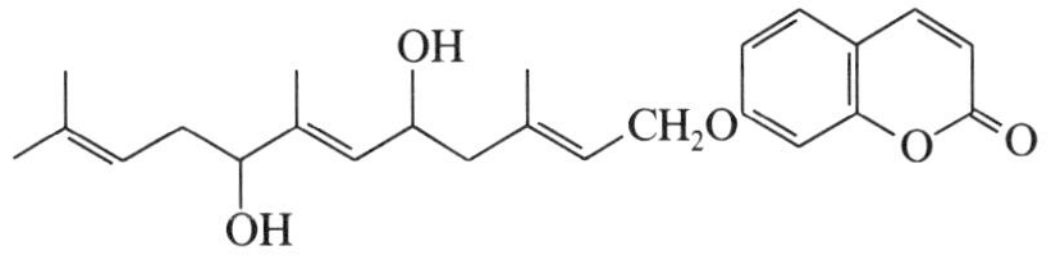

$C_{24}H_{30}O_5$ M 398.498

Constit. of *Ferula assa-foetida*. Oil. $[\alpha]_D$ +7° (c, 0.7 in CHCl₃).

Kazimoto, T. et al, Phytochemistry, 1989, 28, 1761.

Asacoumarin B A-80174

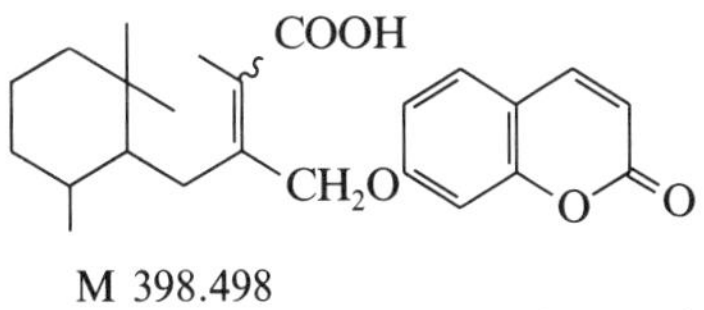

$C_{24}H_{30}O_5$ M 398.498

Constit. of *Ferula assa-foetida*. Amorph. powder. $[\alpha]_D$ −13.3° (c, 0.4 in CHCl₃).

Kajimoto, T. et al, Phytochemistry, 1989, 28, 1761.

Asadisulphide   A-80175

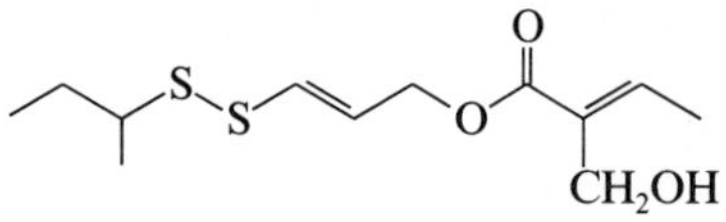

C$_{12}$H$_{20}$O$_3$S$_2$ M 276.420
Constit. of *Ferula assa-foetida*. Oil. [α]$_D$ +24° (c, 0.4 in CHCl$_3$).

Kazimoto, T. *et al*, *Phytochemistry*, 1989, **28**, 1761.

Ascidiatrienolide A   A-80176
[122752-22-1]

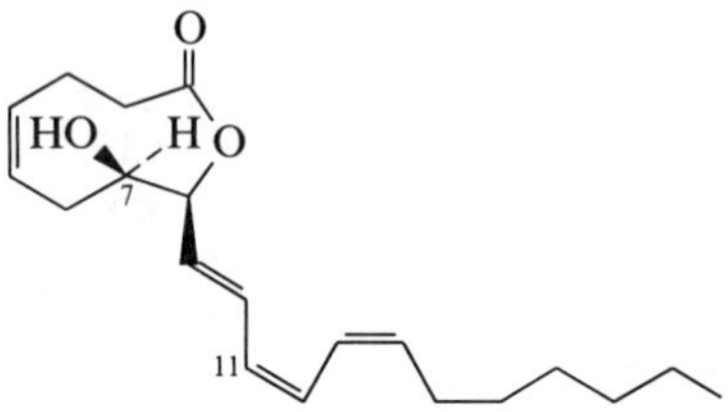

C$_{20}$H$_{30}$O$_3$ M 318.455
Constit. of *Didemnum candidum*. [α]$_D$ −14.8° (c, 4.5 in CHCl$_3$).

7-Epimer: [122799-22-8]. *Ascidiatrienolide B*
C$_{20}$H$_{30}$O$_3$ M 318.455
Constit. *D. candidum*. [α]$_D$ −4.1° (c, 3.4 in CHCl$_3$).

11E-Isomer: [122799-23-9]. *Ascidiatrienolide C*
C$_{20}$H$_{30}$O$_3$ M 318.455
Constit. of *D. candidum*. [α]$_D$ −10.6° (c, 11.3 in CHCl$_3$).

Lindquist, N. *et al*, *Tetrahedron Lett.*, 1989, **30**, 2735.

Asnipyrone A   A-80177

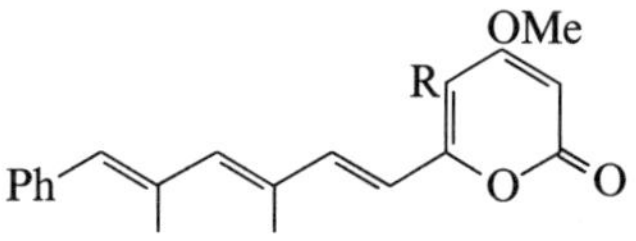

R = CH$_3$

C$_{21}$H$_{22}$O$_3$ M 322.403
Metab. of *Aspergillus niger*. Yellow cryst. (EtOH). Mp 161-163°.

Guang-yi, L. *et al*, *Heterocycles*, 1989, **28**, 899.

Asnipyrone B   A-80178

As Asnipyrone A, A-80177 with

R = H

C$_{20}$H$_{20}$O$_3$ M 308.376
Metab. of *Aspergillus niger*. Yellow cryst. (MeOH/CH$_2$Cl$_2$). Mp 158-160°.

Guang-yi, L. *et al*, *Heterocycles*, 1989, **28**, 899.

Asperpentyne   A-80179
[119483-44-2]

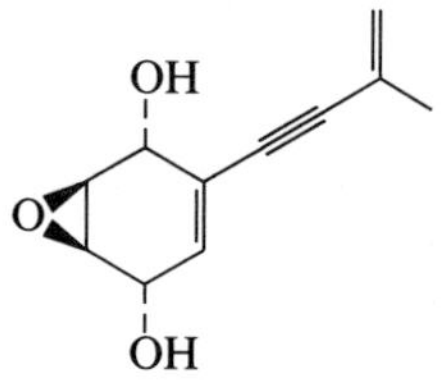

C$_{11}$H$_{12}$O$_3$ M 192.214
Metab. of *Aspergillus duricaulis*. Oil.

Mühlenfeld, A. *et al*, *Phytochemistry*, 1988, **27**, 3853.

Aspidins   A-80180
Updated Entry replacing A-03572

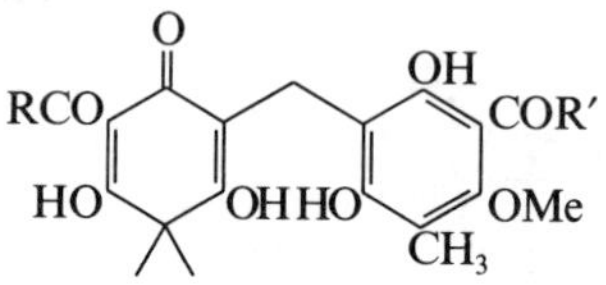

Aspidin AA, R = R' = CH$_3$
    AB, R = CH$_3$, R' = CH$_2$CH$_2$CH$_3$
    PB, R = CH$_2$CH$_3$, R' = CH$_2$CH$_2$CH$_3$
    BB, R = R' = CH$_2$CH$_2$CH$_3$

The suffix letters derive from the acyl side-chains (A = acetyl, P = propionyl, B = butyryl). Constits. of *Dryopteris* spp. ferns, also from *Aspidium* spp.

▷ GU5505000.

Aspidin AA [53526-70-8]
    C$_{21}$H$_{24}$O$_8$ M 404.416
    Constit. of *D. gymnospora*. Yellow needles (hexane). Mp 135-136°.

Aspidin AB [19489-46-4]
    C$_{23}$H$_{28}$O$_8$ M 432.469
    Mp 117-120°.

Aspidin BB [584-28-1]
    Aspidin. Polystichin. *Yellow polystichum acid*
    C$_{25}$H$_{32}$O$_8$ M 460.523
    Pale-yellow cryst. (EtOH). Mp 125°.

Di-Ac: Needles and prisms (MeOH). Mp 108°.

Mitteldorf, R. *et al*, *Chem. Ber.*, 1956, **89**, 2595 (*synth*)
Aebi, A. *et al*, *Helv. Chim. Acta*, 1957, **40**, 569 (*struct*)
Huhtikangas, A. *et al*, *Acta Chem. Scand.*, 1972, **26**, 89 (*ms*)
Hisada, S. *et al*, *Phytochemistry*, 1974, **13**, 655 (*isol, struct*)
Widen, C.-J. *et al*, *Helv. Chim. Acta*, 1976, **59**, 1725 (*isol, pmr*)

Astrogorgiadiol   A-80181

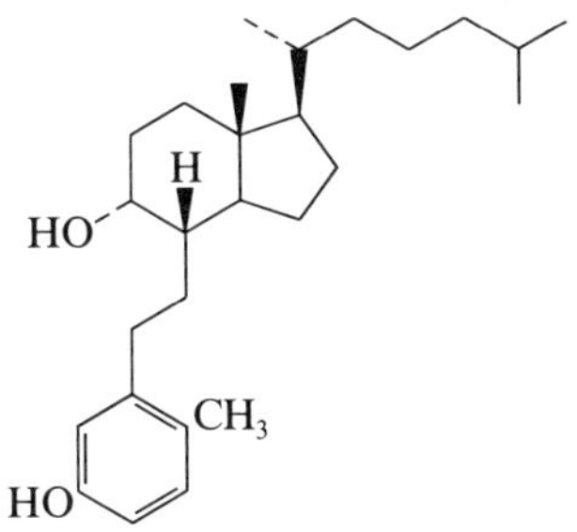

$C_{27}H_{44}O_2$ M 400.643
Metab. of gorgonian *Astrogorgia* sp. Inhibits cell division in fertilised starfish eggs. Solid. $[\alpha]_D$ −16.4° (c, 0.058 in CHCl₃).

Fusetani, N. *et al*, *Tetrahedron Lett.*, 1989, **30**, 7079.

Astrogorgin   A-80182

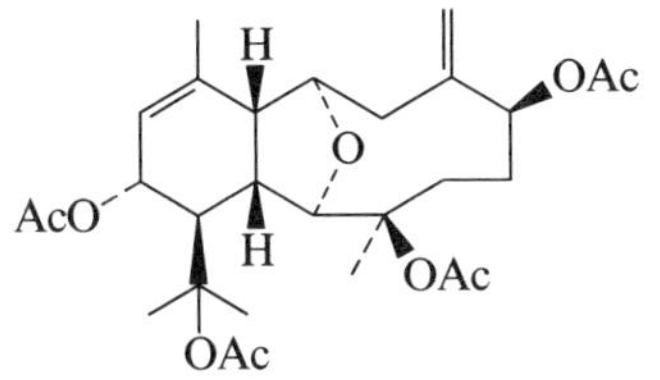

$C_{28}H_{40}O_9$ M 520.619
Metab. of gorgonian *Astrogorgia* sp. Inhibits cell division in fertilised starfish eggs. $[\alpha]_D$ −118° (c, 0.064 in CHCl₃).

Fusetani, N. *et al*, *Tetrahedron Lett.*, 1989, **30**, 7079.

Atalantoflavone   A-80183

[119309-02-3]

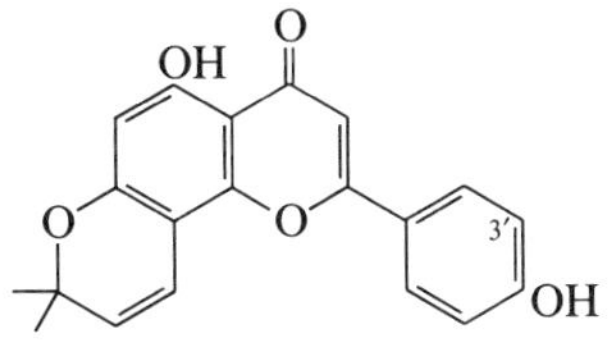

$C_{20}H_{16}O_5$ M 336.343
Constit. of *Atalantia racemosa*. Yellow needles (Me₂CO/hexane). Mp 289-290°.

3′-Methoxy: [106055-12-3]. **Racemoflavone**
$C_{21}H_{18}O_6$ M 366.370
Constit. of *A. racemosa*. Yellow cryst. (Me₂CO/hexane). Mp 236-237°.

Banerji, A. *et al*, *Phytochemistry*, 1988, **27**, 3637 (*isol, struct, synth*)

ent-16α,18-Atisanediol   A-80184

[118243-65-5]

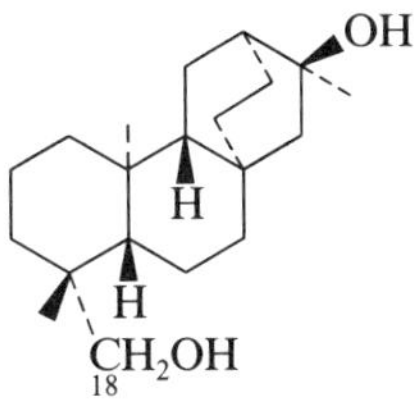

$C_{20}H_{34}O_2$ M 306.487

Constit. of *Xylopia aromatica*. Cryst. (hexane). Mp 160-162°. $[\alpha]_D^{25}$ −45° (c, 0.02 in CHCl₃).

18-Carboxylic acid: [118243-66-6]. **ent-*16α-Hydroxy-18-atisanoic acid***
$C_{20}H_{32}O_3$ M 320.471
Constit. of *X. aromatica*.

18-Carboxylic acid, Me ester: Cryst. (hexane). Mp 140-142°. $[\alpha]_D^{25}$ −24° (c, 0.01 in CHCl₃).

Moraes, M.P.L. *et al*, *Phytochemistry*, 1988, **27**, 3205.

ent-16-Atisene-3,14-dione   A-80185

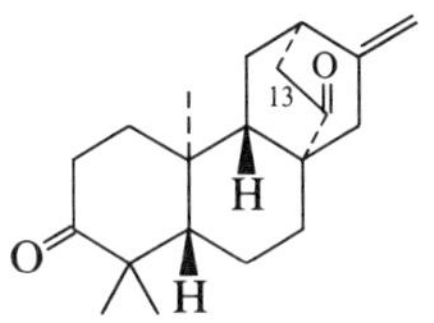

$C_{20}H_{28}O_2$ M 300.440
Constit. of *Euphorbia fidjiana*. Cryst. Mp 160-164°. $[\alpha]_D^{25}$ +5.5° (c, 0.02 in CHCl₃).

13R-Hydroxy: **ent-*13R-hydroxy-16-atisene-3,14-dione***
$C_{20}H_{28}O_3$ M 316.439
Constit. of *E. fidjiana*. Cryst. Mp 175-177°. $[\alpha]_D^{25}$ +44° (c, 0.03 in CHCl₃).

Lal, A.R. *et al*, *Tetrahedron Lett.*, 1989, **30**, 3205 (*cryst struct*)

Aucubigenin   A-80186

Updated Entry replacing A-10277
[64274-28-8]

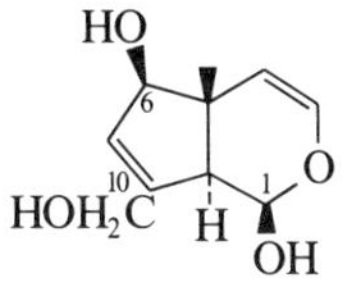

$C_9H_{12}O_4$ M 184.191
Oil.

1-O-β-D-Glucopyranoside: [479-98-1]. *Aucubin. Aucuboside*
$C_{15}H_{22}O_9$ M 346.333
Constit. of *Plantago lanceolata* and *Aucuba japonica*. Cryst. + 1H₂O (EtOH aq.). Mp 181°. $[\alpha]_D^{21}$ −171.4° (H₂O).

1-O-β-Cellobioside:
$C_{21}H_{32}O_{14}$ M 508.475
Constit. of *Odontites verna*. Amorph. powder. $[\alpha]_D^{25}$ −92.0° (c, 0.7 in MeOH).

1-O-β-Gentiobioside:
$C_{21}H_{32}O_{14}$ M 508.475
Constit. of *O. verna*. Cryst. (EtOH). Mp 180-181°. $[\alpha]_D^{25}$ −45° (c, 1.5 in MeOH).

1-O-β-D-(2′-Benzoylglucopyranosyl): 2′-O-Benzoylaucubin
$C_{22}H_{26}O_{10}$ M 450.441
Constit. of *O. verna*. Amorph. powder. $[\alpha]_D^{25}$ −124.8° (c, 0.4 in MeOH).

6-Epimer, 1-O-β-D-glucopyranoside: 6-Epiaucubin
$C_{15}H_{22}O_9$ M 346.333
Constit. of *Tecoma chrysantha*. Amorph. powder. $[\alpha]_D^{25}$ −58.9° (c, 0.7 in MeOH).

1-O-β-D-Glucopyranoside,1′-O-(p-hydroxybenzoyl): **Agnuside. Agnoside**
$C_{22}H_{26}O_{11}$ M 466.441

Isol. from *Vitex agnus-castus*, *V. regundo*, *V. trifolia*, *V. lucens* and other V. spp., also *Rhinanthus* spp. Cryst. (H₂O). Mp 146°. [α]²⁰_D −91.5° (EtOH).

Hänsel, R. *et al*, *Phytochemistry*, 1965, **4**, 19 (*Agnaside*)

Bianco, A. *et al*, *Tetrahedron*, 1977, **33**, 847; 1982, **38**, 362.

Bianco, A. *et al*, *Gazz. Chim. Ital.*, 1981, **111**, 91, 479; 1982, **112**, 227 (*isol*)

Damtoft, S., *Phytochemistry*, 1983, **22**, 1929 (*biosynth*)

Auraptene†   A-80187

7-(3,7-Dimethyl-2,6-octadienyl)oxy-2H-1-benzopyran-2-one, 9CI. O-Geranylumbelliferone. Feronialactone. 7-Geranyloxycoumarin

[495-02-3]

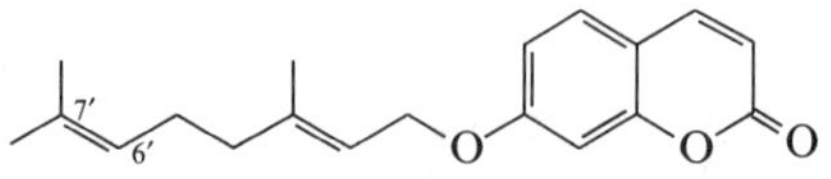

C₁₉H₂₂O₃   M 298.381

Isol. from *Citrus aurantium*, *Feronia elephantum*, *Aegle marmelos*, *Libanotis intermedia* and others. Prisms (EtOH). Mp 68°.

6′R,7-Epoxy: [36414-00-3]. ***6′,7′-Epoxyaurapten.*** 7-[[5-(3,3-Dimethyloxiranyl)-3-methyl-2-pentenyl]oxy]-2H-1-benzopyran-2-one, 9CI

C₁₉H₂₂O₄   M 314.380

Isol. from *Aster yunnanensis*. Oil or cryst. Mp 55-57.5°. [α]²²_D +21° (CHCl₃).

Kariyone, T. *et al*, *Chem. Pharm. Bull.*, 1959, **1**, 119 (*struct, synth, bibl*)

Bohlmann, F. *et al*, *Justus Liebigs Ann. Chem.*, 1968, **717**, 186 (*isol, uv, pmr, struct, deriv*)

Coates, R.M. *et al*, *Tetrahedron*, 1970, **26**, 5699 (*synth*)

Yamada, S. *et al*, *Tetrahedron Lett.*, 1976, **29**, 2557 (*synth, abs config, deriv*)

Aureusidin   A-80188

Updated Entry replacing A-10282

2-[(3,4-Dihydroxyphenyl)methylene]-4,6-dihydroxy-3(2H)-benzofuranone, 9CI. 3′,4,4′,6-Tetrahydroxyaurone. Cernuine

[480-70-6]

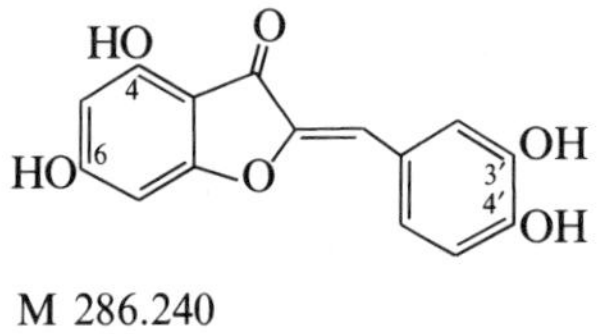

C₁₅H₁₀O₆   M 286.240

Aglucone from *Antirrhinum majus*, also isol. from *Citrus medica*. Present in various members of the Gesneriaceae. Deep-yellow cryst. + 1H₂O (MeOH aq.). Mp 270°, Mp 295° dec. (double Mp).

Tetra-Ac: Yellow needles (pet. ether). Mp 188-189°.

4-O-β-D-Glucopyranoside: [480-69-3]. ***Cernuoside***

C₂₁H₂₀O₁₁   M 448.382

Pigment from *Oxalis cernua*, also present in *Chirita micromusa*, *Limonium bonduellii* and *Petrocosmea kerrii*. Yellow cryst. (EtOH aq.). Mp 250-258°. [α]³⁰_D −13° (Py).

6-O-β-D-Glucopyranoside: [633-15-8]. ***Aureusin***

C₂₁H₂₀O₁₁   M 448.382

Constit. of *A. majus*, also in *Antirrhinum nuttalianum* and *Linaria maroccana*. Needles (EtOAc) (as hepta-Ac). Mp 264.5-265.5° (as hepta-Ac).

6-Me ether: [54826-89-0]. ***Rengasin.*** *3′,4,4′-Trihydroxy-6-methoxyaurone*

C₁₆H₁₂O₆   M 300.267

Constit. of *Melanorrhea* spp. heartwood. Golden-yellow cryst. Mp 314-316° dec.

Tetra-Me ether: Greenish-yellow needles (MeOH aq.). Mp 173° (169-170°).

[38216-54-5]

Geissman, T.A. *et al*, *J. Am. Chem. Soc.*, 1950, **72**, 5725; 1955, **77**, 4622 (*isol, struct*)

Farkas, L. *et al*, *Chem. Ber.*, 1961, **94**, 2221; 1964, **97**, 1044; 1969, **49**, 2221 (*synth*)

Batterham, T.J. *et al*, *Aust. J. Chem.*, 1964, **17**, 428 (*pmr*)

Harborne, J.B., *Phytochemistry*, 1966, **5**, 111, 589 (*occur*)

Hastings, J.S. *et al*, *J. Chem. Soc., Perkin Trans.* 1, 1972, 2128 (*pmr, struct*)

Deshmukh, S.W. *et al*, *Indian J. Chem.*, 1974, **12**, 893 (*synth*)

Avarol   A-80189

Updated Entry replacing A-10285

[55303-98-5]

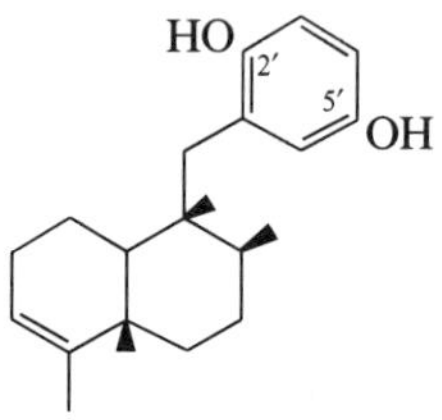

Absolute configuration

C₂₁H₃₀O₂   M 314.467

Constit. of *Dysidea avara*. Cryst. (CHCl₃). Mp 148-150°. [α]_D +6.1°.

5′-Ac:

C₂₃H₃₂O₃   M 356.504

Constit. of *D. avara*. Cryst. (Et₂O/hexane). Mp 151-152°. [α]²⁵_D +11° (c, 3.1 in CHCl₃).

2′,5′-Quinone: [55303-99-6]. ***Avarone***

C₂₁H₂₈O₂   M 312.451

From *D. avara*. Oil.

Minale, L. *et al*, *Tetrahedron Lett.*, 1974, 3401 (*isol*)

de Rosa, S. *et al*, *J. Chem. Soc., Perkin Trans.* 1, 1976, 1408 (*struct*)

Sarma, A.S. *et al*, *J. Org. Chem.*, 1982, **47**, 1727 (*synth*)

Crispino, A. *et al*, *J. Nat. Prod.* (*Lloydia*), 1989, **52**, 646 (*isol, pmr, cmr*)

1-Azabicyclo[3.2.2]nonane   A-80190

[283-20-5]

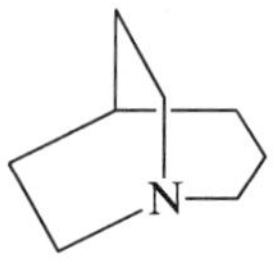

C₈H₁₅N   M 125.213

Cryst. Mp 129°. Bp 170°.

Picrate: Yellow needles (EtOH). Mp 288-289°.

Prelog, V. *et al*, *Justus Liebigs Ann. Chem.*, 1937, **532**, 83 (*synth*)

Ruggles, C.J. *et al*, *J. Am. Chem. Soc.*, 1988, **110**, 5692 (*synth, spectra*)

1-Azabicyclo[3.3.1]nonane, 9CI, 8CI   A-80191

Updated Entry replacing A-10287
Isogranatanine
[280-77-3]

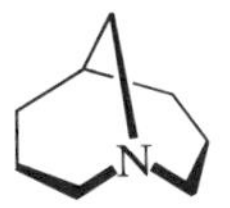

C$_8$H$_{15}$N   M 125.213
Needles. Mp 114°. Bp$_{18}$ 100-105°. Hygroscopic.
B,HCl: Cryst. (EtOH/Et$_2$O). Mp > 350°.
Picrate: Yellow cryst. (EtOH). Mp 282.5-283° dec.

Prelog, V. *et al, Chem. Ber.,* 1939, **72,** 1319 (*synth*)
Miyano, S. *et al, J. Chem. Soc., Perkin Trans.* 1, 1987, 313 (*synth, pmr, cmr*)

3-Azetidinone   A-80192

C$_3$H$_5$NO   M 71.079
Claimed syntheses prior to 1988 were prob. erroneous.
B,HCl: [17557-84-5].
    Solid. Sol. H$_2$O. Mp 110-140°. Cont. ca 20% MeOH.
1-Benzoyl: [25566-02-3].
    C$_{10}$H$_9$NO$_2$ M 175.187
    Cryst.

Baumann, H. *et al, Helv. Chim. Acta,* 1988, **71,** 1035 (*synth, ir, pmr*)

3-Azido-1-butyne   A-80193

[118723-97-0]

HC≡CCH(N$_3$)CH$_3$

C$_4$H$_5$N$_3$ M 95.104
Liq.

Banert, K., *Chem. Ber.,* 1989, **122,** 911 (*synth, ir, pmr, cmr*)

4-Azido-2-butyn-1-ol   A-80194

[118723-98-1]

N$_3$CH$_2$C≡CCH$_2$OH

C$_4$H$_5$N$_3$O   M 111.103
Liq.
▷ Potentially explosive.

Banert, K., *Chem. Ber.,* 1989, **122,** 911 (*synth, ir, pmr, cmr*)

1-Azido-4-chloro-2-butyne   A-80195

[119720-88-6]

N$_3$CH$_2$C≡CCH$_2$Cl

C$_4$H$_4$ClN$_3$ M 129.548
Liq.

Banert, K., *Chem. Ber.,* 1989, **122,** 1175.

2-Azidocyclohexanol   A-80196

[71559-13-2]

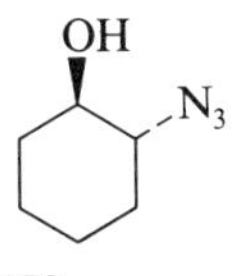

C$_6$H$_{11}$N$_3$O   M 141.172
(*1RS,2RS*)-*form* [10027-78-8]

(±)-trans-*form*
Liq. Bp$_{1.5}$ 70-71°.
Ac: [10027-85-7].
    C$_8$H$_{13}$N$_3$O$_2$ M 183.210
    Oil.

Van der Werf, C.A. *et al, J. Am. Chem. Soc.,* 1954, **76,** 1231 (*synth, ir*)
Cambie, R.C. *et al, J. Chem. Soc., Perkin Trans.* 1, 1976, 840 (*synth, ir, pmr*)
Mereyala, H.B. *et al, Helv. Chim. Acta,* 1986, **69,** 415 (*synth*)

3′-Azido-3′-deoxythymidine, 9CI   A-80197

Updated Entry replacing A-50371
3′-Azidothymidine. **Zidovudine,** BAN, INN, USAN. *BW A509U. Retrovir. AZT*
[30516-87-1]

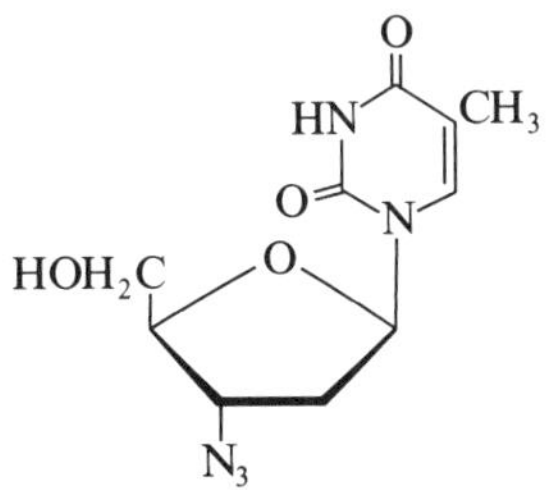

C$_{10}$H$_{13}$N$_5$O$_4$ M 267.244
Antiviral agent active exclusively against retroviruses, esp.
    AIDS virus and leukaemias. Drug of choice (1986) for
    AIDS treatment. Needles (Et$_2$O). Mp 106-112°, Mp 119-
    121° (after drying).

Horowitz, J.P. *et al, J. Org. Chem.,* 1964, **29,** 2076 (*synth, uv*)
Lin, T.-S. *et al, J. Med. Chem.,* 1978, **21,** 109 (*synth*)
Martindale, The Extra Pharmacopoeia, 28th/29th Ed.,
    Pharmaceutical Press, London, 1982/1989, 18797.
Mitsuya, H. *et al, Proc. Natl. Acad. Sci. U.S.A.,* 1985, **82,** 7096 (*pharmacol*)
Dyer, I. *et al, Acta Crystallogr., Sect. C,* 1988, **44,** 767 (*cryst struct*)
Van Roey, P. *et al, J. Am. Chem. Soc.,* 1988, **110,** 2277 (*cryst struct*)
Fleet, G.W.J. *et al, Tetrahedron,* 1988, **44,** 625 (*synth*)

2-Azido-1,3-diiodopropane   A-80198

(ICH$_2$)$_2$CHN$_3$

C$_3$H$_5$I$_2$N$_3$ M 336.902
Liq.

Sehgal, R.K. *et al, J. Med. Chem.,* 1987, **30,** 1626 (*synth, ir, pmr*)

4-Azido-1-fluoro-2-nitrobenzene, 9CI   A-80199

[28166-06-5]

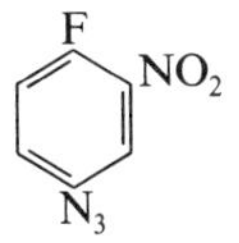

C$_6$H$_3$FN$_4$O$_2$ M 182.114
Used for affinity labelling of antibodies. Orange or straw
    needles (pet. ether). Mp 53-55° (52°).
▷ Explosive when dry.

Fleet, G.W.J. *et al, Nature (London),* 1969, **224,** 511 (*synth, use*)
Fleet, G.W.J. *et al, Biochem. J.,* 1972, **128,** 499 (*synth, uv, haz*)
Gabizon, R. *et al, J. Biol. Chem.,* 1982, **257,** 15145 (*ms*)

2-Azido-3-iodo-1-propanol   A-80200

[109122-91-0]

$$ICH_2CH(N_3)CH_2OH$$

$C_3H_6IN_3O$ M 227.004

(±)-*form*

Yellowish liq.

Sehgal, R.K. *et al*, *J. Med. Chem.*, 1987, **30**, 1626 (*synth, pmr, ir*)

2-Azido-3-methylbenzaldehyde   A-80201

[113302-67-3]

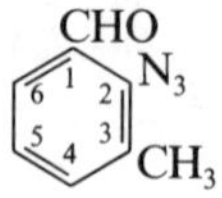

$C_8H_7N_3O$ M 161.163

Pale yellow oil.

Cuevas, J.C. *et al*, *J. Org. Chem.*, 1988, **53**, 2055 (*synth, ir, pmr, ms*)

2-Azido-5-methylbenzaldehyde   A-80202

[113302-66-2]

$C_8H_7N_3O$ M 161.163

Yellow solid. Mp 55.5-56.5°.

Cuevas, J.C. *et al*, *J. Org. Chem.*, 1988, **53**, 2055 (*synth, ir, pmr, ms*)

3-Azido-3-methyl-1-butyne   A-80203

[118723-88-9]

$$N_3C(CH_3)_2C{\equiv}CH$$

$C_5H_7N_3$ M 109.130

Liq.

▷ Potentially explosive.

Banert, K., *Chem. Ber.*, 1989, **122**, 911 (*synth, ir, pmr, cmr, ms*)

1-Azido-1-propanol   A-80204

$$H_3CCH_2CH(OH)N_3$$

$C_3H_7N_3O$ M 101.108

Me ether: [111238-42-7]. *1-Azido-1-methoxypropane*

$C_4H_9N_3O$ M 115.135

Liq. Bp_{625} 100-105°.

Hassner, A. *et al*, *J. Org. Chem.*, 1988, **53**, 22 (*synth, ir, pmr, ms*)

Azocyclopropane   A-80205

Updated Entry replacing A-20255

Dicyclopropyldiazene

[78331-66-5]

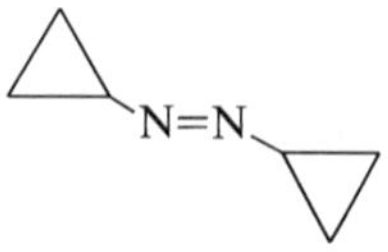

(*E*)-*form*

$C_6H_{10}N_2$ M 110.158

N-Oxide: [33425-51-3]. *Azoxycyclopropane*

$C_6H_{10}N_2O$ M 126.158

Bp_{40} 104-107°, Bp_{10} 80-83°. Geom. isom. not determined.

(*E*)-*form* [80201-75-8]

Bp 139°.

(*Z*)-*form* [80201-76-9]

Mp 38-39.5°.

Iversen, P.E., *Chem. Ber.*, 1971, **104**, 2195 (*oxide, synth, ir*)

Engel, P.S. *et al*, *J. Am. Chem. Soc.*, 1981, **103**, 7689 (*synth, spectra*)

Engel, P.S. *et al*, *J. Org. Chem.*, 1988, **53**, 4748 (*synth, pmr, cmr, uv, ms, props*)

1-Azoniatricyclo[3.3.3.0]undecane   A-80206

Tetrahydro-1H,5H-4,7a-propanopyrrolizinium, 9CI. 1-Azonia[3.3.3]propellane

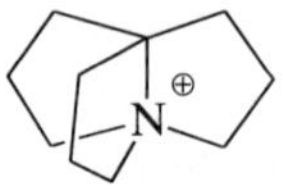

$C_{10}H_{18}N^{\oplus}$ M 152.259

Chloride: [116747-82-1].

$C_{10}H_{18}ClN$ M 187.712

Solid. Mp >350°.

Bromide:

$C_{10}H_{18}BrN$ M 232.163

Prisms. Mp >350°.

Šorm, F. *et al*, *Collect. Czech. Chem. Commun.*, 1954, **19**, 298 (*synth, ir*)

Newkome, G.R. *et al*, *J. Org. Chem.*, 1988, **53**, 5552 (*synth, pmr, ir, pmr*)

Azuleno[1,2-*b*]anthracene-6,13-dione, 9CI   A-80207

Anthra[2,3-a]azulene-6,13-dione

[87121-88-8]

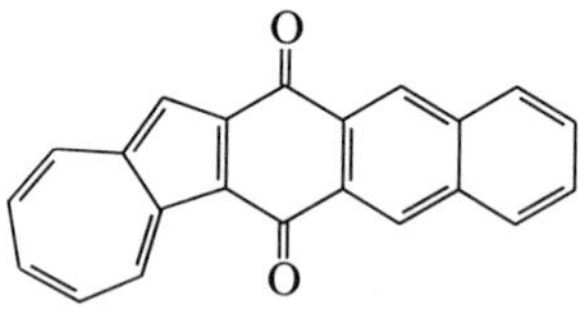

$C_{22}H_{12}O_2$ M 308.336

Green cryst. Mp >310°.

Bindl, J. *et al*, *Chem. Ber.*, 1983, **116**, 2408 (*synth, uv, pmr*)

B

Baccatin   B-80001

Updated Entry replacing B-00002

[66107-60-6]

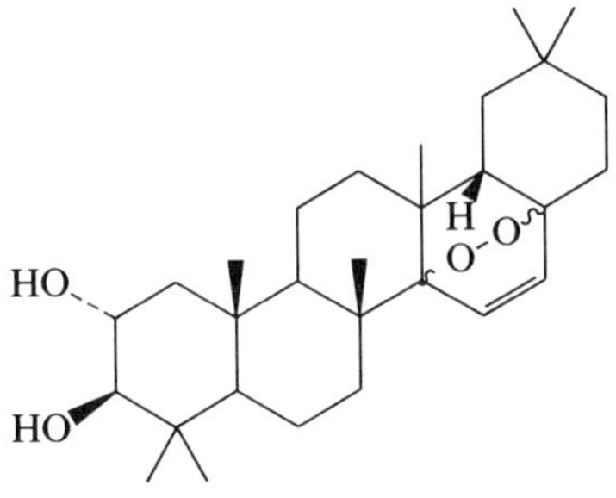

$C_{29}H_{46}O_4$ M 458.680

Constit. of *Sapium baccatum*. Cryst. Mp 228-229° dec. $[\alpha]_D$ −9.09°.

Saha, B. *et al*, *Tetrahedron Lett.*, 1977, 3095.

Baccatin I   B-80002

Updated Entry replacing B-00003

2α,5α,7β,9α,10β,13α-Hexaacetoxy-4β,20-epoxy-11-taxene

[30244-35-0]

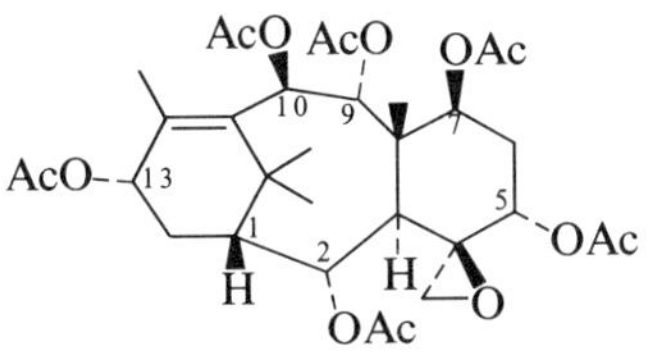

$C_{32}H_{44}O_{13}$ M 636.692

Constit. of *Taxus baccata*. Cryst. Mp 298°. $[\alpha]_D$ +86°.

5-Deacetyl: [30244-36-1].
   $C_{30}H_{42}O_{12}$ M 594.655
   Constit. of *T. baccata*. Cryst. Mp 256-258°.

1β-Hydroxy: [30244-37-2]. ***1β-Hydroxybaccatin I***
   $C_{32}H_{44}O_{14}$ M 652.691
   Constit. of *T. baccata*. Cryst. Mp 273°. $[\alpha]_D$ +102°.

1β-Acetoxy, 5-deacetyl: [119120-27-3]. ***1β-Acetoxy-5-deacetylbaccatin I.*** *1β,2α,7β,9α,10β,13α-Hexaacetoxy-4β,20-epoxy-11-taxen-5α-ol*
   $C_{32}H_{44}O_{14}$ M 652.691
   Constit. of *T. mairei*. Cryst. (EtOH). Mp 240-241°.

de Marcano, D.P.D.C. *et al*, *J. Chem. Soc., Chem. Commun.*, 1970, 1381 (*isol, struct*)
Lian, J.-Y. *et al*, *Phytochemistry*, 1988, **27**, 3674 (*deriv*)

Baccatin IV   B-80003

Updated Entry replacing B-30001

2α,4α,7β,9α,10β,13α-Hexaacetoxy-5β,20-epoxy-11-taxen-1β-ol

[57672-77-2]

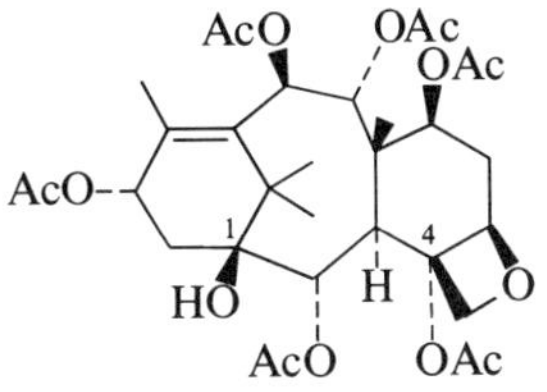

$C_{32}H_{44}O_{14}$ M 652.691

Constit. of *Taxus baccata*. Cryst. Mp 254-255° dec. $[\alpha]_D$ +19°.

1-Deoxy, 4-deacetyl: [119120-26-2]. ***1-Dehydroxy-4-deacetylbaccatin IV.*** *2α,7β,9α,10β,13α-Pentaacetoxy-5β,20-epoxy-11-taxen-4α-ol*
   $C_{30}H_{42}O_{12}$ M 594.655
   Constit. of *T. mairei*. Cryst. (EtOH). Mp 229-230°. $[\alpha]_D^{14}$ +40° (c, 0.5 in Me$_2$CO).

de Marcano, D.P.D.C. *et al*, *J. Chem. Soc., Chem. Commun.*, 1975, 365 (*isol, struct*)
McLaughlin, J.L. *et al*, *J. Nat. Prod. (Lloydia)*, 1981, **44**, 312 (*isol*)
Lian, J.-Y. *et al*, *Phytochemistry*, 1988, **27**, 3674 (*deriv*)

Bacchabolivic acid   B-80004

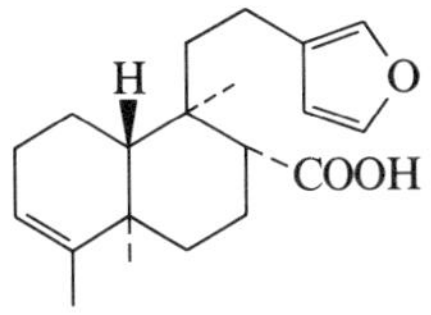

$C_{20}H_{28}O_3$ M 316.439

Constit. of *Baccharis boliviensis*. Cryst. Mp 147°.

Xylopyranoside:
   $C_{25}H_{36}O_7$ M 448.555
   Constit. of *B. boliviensis*.

Zdero, C. *et al*, *Phytochemistry*, 1989, **28**, 531.

Bacchalatifolin   B-80005

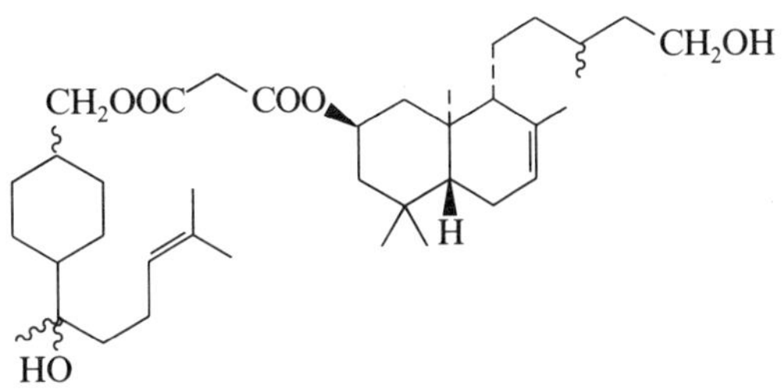

$C_{38}H_{64}O_6$ M 616.920

Constit. of *Baccharis latifolia*.

Zdero, C. *et al*, *Phytochemistry*, 1989, **28**, 531.

Bacchotricuneatin B
B-80006

Updated Entry replacing B-00016

[65596-26-1]

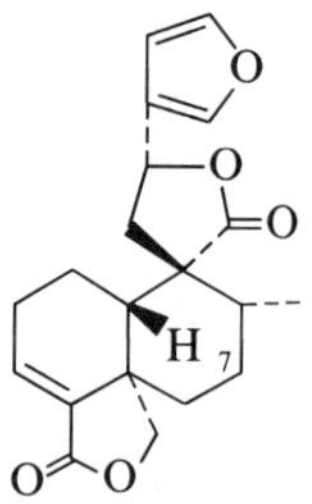

$C_{20}H_{22}O_5$ M 342.391

Constit. of *Baccharis tricuneata*. Cryst. (CH$_2$Cl$_2$/hexane). Mp 192-193°. $[\alpha]_D^{25}$ −102.4° (c, 0.55 in CHCl$_3$).

7α-Acetoxy: [116139-48-1]. *7α-Acetoxybacchotricuneatin B*

$C_{22}H_{24}O_7$ M 400.427

Constit. of *Olearia pimeleoides*. Gum.

Wagner, H. *et al*, *Tetrahedron Lett.*, 1977, 3039 (*isol, struct*)
Warning, U. *et al*, *J. Nat. Prod.* (*Lloydia*), 1988, **51**, 513 (*deriv*)

Badione A
B-80007

[90295-66-2]

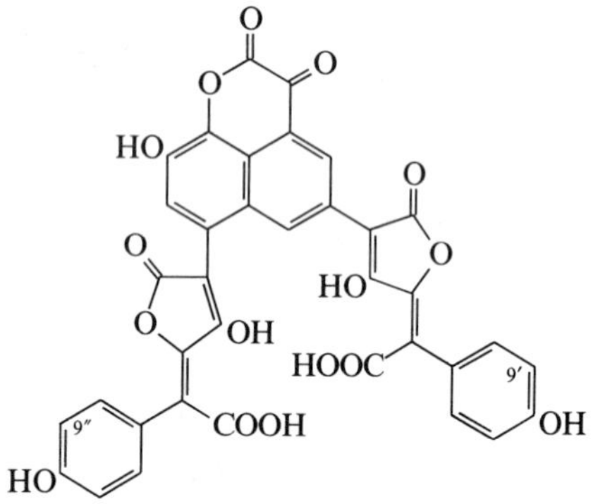

$C_{36}H_{18}O_{16}$ M 706.529

Pigment from the bay boletus *Xerocomus badius* and from *Boletus pinicola*. Shiny blackish-brown microcryst. Dec. >250° without melting. λ_{max} 260, 368, 482sh nm.

9′,9″-Dihydroxy: [97486-15-2]. *Badione B*

$C_{36}H_{18}O_{18}$ M 738.527

Pigment from *B. erythropus*.

Steffan, B. *et al*, *Angew. Chem., Int. Ed. Engl.*, 1984, **23**, 445 (*isol, cmr, struct*)

Balaenol
B-80008

Updated Entry replacing B-70006

[115031-67-9]

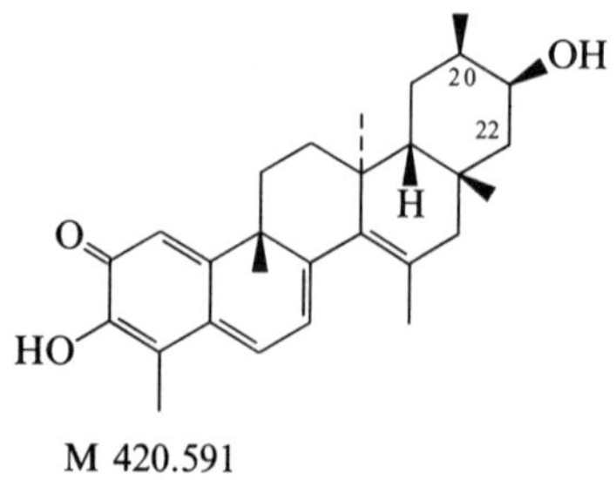

$C_{28}H_{36}O_3$ M 420.591

Constit. of *Cassine balae*. Mp 139-140°.

22-Oxo: [115028-52-9]. *Balaenonol*

$C_{28}H_{34}O_4$ M 434.574

Constit. of *C. balae*. Mp 205-208°.

20-epimer,22β-hyddroxy: *Isobalaendiol*

$C_{28}H_{36}O_4$ M 436.590

Constit. of *C. balae*. Orange-red solid (CH$_2$Cl$_2$/Et$_2$O). Mp 210-213°. $[\alpha]_D$ +65° (c, 4.5 in CHCl$_3$).

Fernando, H.C. *et al*, *Tetrahedron Lett.*, 1988, **29**, 387 (*struct*)
Fernando, H.C. *et al*, *Tetrahedron*, 1989, **45**, 5867 (*Isobalaendiol*)

Barbacenic acid
B-80009

Updated Entry replacing B-40004

[97746-05-9]

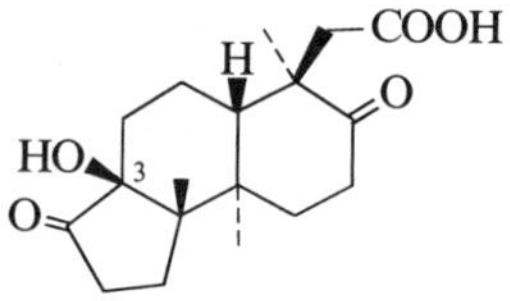

$C_{18}H_{26}O_5$ M 322.400

Constit. of *Barbacenia flava*.

Me ester: Cryst. Mp 137-139°. $[\alpha]_D^{25}$ +40° (c, 0.6 in CHCl$_3$).

3-Deoxy: [119967-64-5]. *3-Deoxybarbacenic acid*

$C_{18}H_{26}O_4$ M 306.401

Constit. of *B. flava*. Cryst. (EtOAc/hexane). Mp 200-202°.

Pinto, A.C. *et al*, *J. Chem. Soc., Chem. Commun.*, 1985, 446 (*isol*)
Pinto, A.C. *et al*, *Phytochemistry*, 1988, **27**, 3917 (*3-Deoxybarbacenic acid*)

Barbaloin
B-80010

Updated Entry replacing B-00074

10-β-D-Glucopyranosyl-1,8-dihydroxy-3-(hydroxymethyl)-9(10H)-anthracenone, 9CI. Aloin. Socaloin. Ugandaloin. Jafaloin. Cafaloin

[1415-73-2]

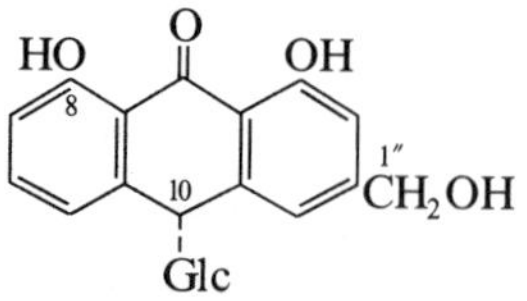

(10R)-form

$C_{21}H_{22}O_9$ M 418.399

Found in Barbados aloes. Yellow needles (EtOH). Sol. H$_2$O. Mp 150°, Mp 146-148°. $[\alpha]_D$ −8.3° (EtOH aq.), $[\alpha]_D$ −10.4° (EtOAc). The 10-config. is as given in CA but does not appear to be clearly stated in the lit.

▷ LZ6520000.

Ac: Cryst. (pet. ether). Mp 95-96°.

Hepta-Me ether: [13954-21-7].
Prisms (EtOH). Mp 177-179°. $[\alpha]_{46}$ −12.05° (CHCl$_3$).

10-Epimer: [28371-16-6]. *Isobarbaloin*

$C_{21}H_{22}O_9$ M 418.399

Present in *A.* spp. Yellow needles (H$_2$O). $[\alpha]_D$ −19.4° (EtOAc).

1″-O-α-L-Rhamnopyranosyl: [11006-91-0]. *Aloinoside B*

$C_{27}H_{32}O_{13}$ M 564.542

Isol. from *Aloe* cf. *perryi*. Cryst. (Me$_2$CO aq.). Mp 233°. $[\alpha]_D^{20}$ −45.3° (c, 1.5 in 50% dioxan aq.).

8-β-D-Glucoside (stereoisomer 1): [53823-08-8]. *Cascaroside A*

$C_{27}H_{32}O_{14}$ M 580.541

Constit. of *Rhamnus purshiana* (*Cascarae sagradae*).
Cryst. + 2H$_2$O. Mp 184-187°. [α]$_D^{30}$ +36.8° (MeOH).
Cascarosides *A* and *B* are a pair of 10-epimers (config.
unknown) which equilibrate *via* the anthranol tautomer
in Py/DMSO soln.

8-β-D-Glucoside (stereoisomer 2): **Cascaroside B**
C$_{27}$H$_{32}$O$_{14}$ M 580.541
From *R. purshianus.* Cryst. + 2H$_2$O. Mp 175-178°. [α]$_D^{30}$
−104.4° (MeOH).

1'-Deoxy, 8-β-D-Glucoside(stereoisomer 1): **Cascaroside C**
C$_{27}$H$_{32}$O$_{13}$ M 564.542
From *R. purshiana.* Cascarosides C and D are a pair of
epimers resembling Cascarosides A and B.

1'-Deoxy, 8-β-D-Glucoside(stereoisomer 2): [53861-35-1].
Cascaroside D
C$_{27}$H$_{32}$O$_{13}$ M 564.542
From *R. purshiana.*

1"-O-α-L-Rhamnopyranosyl: **Aloinoside B**
C$_{27}$H$_{32}$O$_{13}$ M 564.542
Isol. from *Aloe* cf. *perryi.* Cryst. (Me$_2$CO aq.). Mp 233°.
[α]$_D^{20}$ −45.3° (c, 1.5 in 50% dioxan aq.).

Owen, L. *et al, J. Am. Chem. Soc.,* 1942, **64**, 2516 (*struct*)
Mühlemann, H., *Pharm. Acta Helv.,* 1952, **27**, 17 (*struct*)
Böhme, H. *et al, Arch. Pharm.* (*Weinheim, Ger.*), 1955, **286**, 510
    (*synth*)
Birch, A.J. *et al, Aust. J. Chem.,* 1955, **8**, 523 (*struct*)
Hay, J.E. *et al, J. Chem. Soc.,* 1956, 3141 (*struct*)
Hörhammer, L. *et al, Z. Naturforsch.,* B, 1964, **19**, 222 (*Aloinoside B*)
van Oudtshoorn, M.C.B.van.R. *et al, Naturwissenschaften,* 1965,
    52, 186 (*occur*)
Piox, A., *Tetrahedron,* 1968, **24**, 3697 (*ms*)
Wagner, H. *et al, Z. Naturforsch.,* C, 1974, **29**, 444; *Z. Naturforsch.,* B, 1976, **31**, 267 (*Cascarosides*)
Auterhoff, H. *et al, Arch. Pharm.* (*Weinheim, Ger.*), 1980, **313**, 113
    (*resoln*)
Martindale, The Extra Pharmacopoeia, 28th/29th Ed.,
    Pharmaceutical Press, London, 1982/1989, 7502.
Rauwald, H.W. *et al, Arch. Pharm.* (*Weinheim, Ger.*), 1984, **317**, 362 (*pmr*)

Benz[*a*]anthracen-10-o1, 9CI   B-80011

10-Hydroxybenz[a]*anthracene*
[69884-53-3]
C$_{18}$H$_{12}$O   M 244.292
Pale yellow cryst. (C$_6$H$_6$). Mp 219-221°.

O-Benzoyl: [71685-67-1].
C$_{25}$H$_{16}$O$_2$ M 348.400
Pale yellow needles (C$_6$H$_6$/hexane). Mp 174-176°.

Fu, P.P. *et al, J. Org. Chem.,* 1979, **44**, 4265 (*synth, deriv, pmr*)
Pataki, J. *et al, J. Org. Chem.,* 1989, **54**, 840 (*synth*)

Benz[*a*]anthracen-11-o1, 9CI   B-80012

11-Hydroxybenz[a]*anthracene*
[63019-35-2]
C$_{18}$H$_{12}$O   M 244.292
Cryst. (C$_6$H$_6$/hexane). Mp 120-124°.

O-Ac: [71685-69-3].
C$_{20}$H$_{14}$O$_2$ M 286.329
Cryst. (C$_6$H$_6$/hexane). Mp 121-124°.

Fu, P.P. *et al, J. Org. Chem.,* 1979, **44**, 4265 (*synth, deriv, pmr*)

Benzenecarbodithioic acid anhydrosulfide, 9CI   B-80013

Bis(dithiobenzoic) thioanhydride
[22778-02-5]

PhCS-S-CSPh

C$_{14}$H$_{10}$S$_3$ M 274.431
Dark blue oil.

Kato, S. *et al, Justus Liebigs Ann. Chem.,* 1982, 1229 (*synth, ir, uv, nmr*)
Kato, S. *et al, Synthesis,* 1987, 304 (*synth, uv, pmr, cmr, ms*)

Benzenehexamine, 9CI   B-80014

Updated Entry replacing B-60012
Hexaaminobenzene
[4444-26-2]

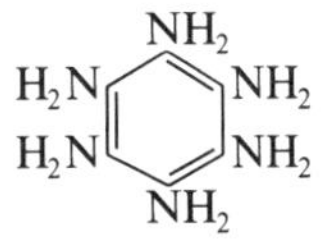

C$_6$H$_{12}$N$_6$ M 168.201
Needles. Mp ca. 255°. Formerly descr. as a dark-brown
cryst. powder. Prepn. under Ar is necessary to obt. a
colourless prod. V. air- and light-sensitive.

▷ Potentially hazardous synth.

N^1,N^1,N^2,N^2,N^3,N^3,N^4,N^4,N^5,N^5-*Deca-Me:* [114422-56-9].
    Pentakis(dimethylamino)aniline
    C$_{16}$H$_{32}$N$_6$ M 308.469
    Cryst. solid (Me$_2$CO). Mp 171-172°.

N-Dodeca-Me: [114396-17-7].
    Hexakis(dimethylamino)benzene
    C$_{18}$H$_{36}$N$_6$ M 336.523
    Cryst. solid (Me$_2$CO). Mp 235-237°.

N-Dodeca-Et: [114396-23-5]. *Hexakis(diethylamino)benzene*
    C$_{30}$H$_{60}$N$_6$ M 504.844
    Cryst. (Me$_2$CO). Mp 211-213°.

Kohne, B. *et al, Justus Liebigs Ann. Chem.,* 1987, 265 (*synth, ir, pmr, cmr, ms*)
Chance, J.M. *et al, J. Org. Chem.,* 1988, **53**, 3226 (*derivs, synth, pmr, cmr, cryst struct*)

Benzo[*b*]benzo[3,4]cyclobuta[1,2-*h*] biphenylene, 9CI   B-80015

Naphtho[2',3':3,4]*cyclobuta*[1,2-b]*biphenylene*
[117965-75-0]

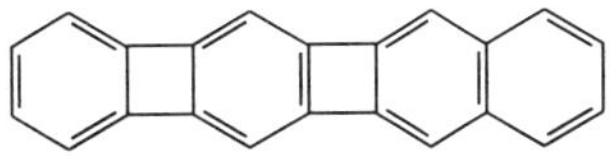

C$_{22}$H$_{12}$ M 276.337
Bright orange microcryst. Highly insol. Mp >305°
dec.(subl. >250°).

Shepherd, M.K., *J. Chem. Soc., Perkin Trans.* 1, 1988, 961 (*synth, ms, pmr*)

Benzo[*h*]benzo[3,4]cyclobuta[1,2-*a*] biphenylene, 9CI   B-80016

Naphtho[2',3':3,4]*cyclobuta*[1,2-a]*biphenylene*
[117965-76-1]

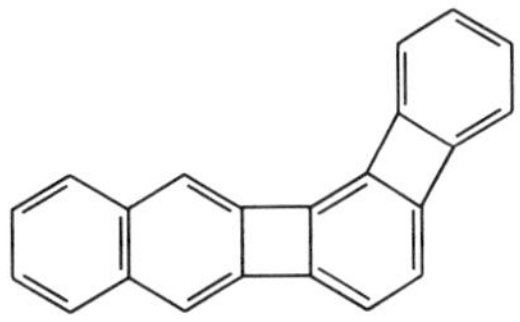

C$_{22}$H$_{12}$ M 276.337

Lemon yellow cryst. Mp 192-194° (subl. > 145°). Blue fluor. in dil. soln.

Shepherd, M.K., *J. Chem. Soc., Perkin Trans.* 1, 1988, 961 (*synth, pmr, cmr, ms*)

11*H*-Benzo[*a*]carbazole-1-carboxylic acid   B-80017

[93328-83-7]

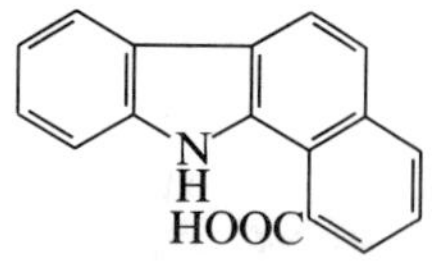

C₁₇H₁₁NO₂   M 261.279
Cryst. (CHCl₃). Mp 230-232°.
Me ester:
 C₁₈H₁₃NO₂   M 275.306
 Light yellow needles (MeOH). Mp 99.5-100.6°.

Dokunikhin, N.S. *et al*, *J. Gen. Chem. USSR (Engl. Transl.)*, 1963, **33**, 962 (*synth*)
Katritzky, A.R. *et al*, *J. Org. Chem.*, 1988, **53**, 794 (*synth, pmr*)

Benzo[*l*]cyclopenta[*cd*]pyrene, 9CI   B-80018

Naphth[1,2,3-mno]acephenanthrylene
[113779-16-1]

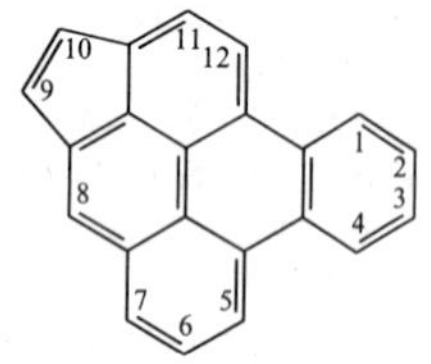

C₂₂H₁₂   M 276.337
Yellow-orange solid. Mp 190-191°.

Sangaiah, R. *et al*, *J. Org. Chem.*, 1988, **53**, 2620 (*synth, pmr, ir, uv, ms*)

1*H*-1,2-Benzodiazepine, 9CI   B-80019

[264-60-8]

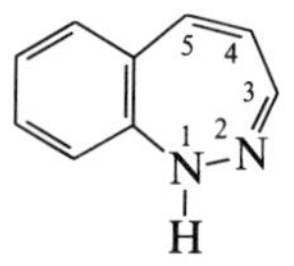

C₉H₈N₂   M 144.176
Red prisms (diisopropyl ether). Mp 63-64°.

Tsuchiya, T. *et al*, *J. Org. Chem.*, 1977, **42**, 1856 (*synth, ir, ms, pmr*)

3*H*-1,2-Benzodiazepine, 9CI   B-80020

[55379-61-8]
Yellow oil. Bp₁ 126-128° (bath).
1-Oxide: [59066-30-7].
 C₉H₈N₂O   M 160.175
 Prisms (C₆H₆/diisopropyl ether). Mp 93-94°.
2-Oxide: [59066-32-9].
 C₉H₈N₂O   M 160.175
 Prisms (C₆H₆/diisopropyl ether). Mp 85-86°.

Tsuchiya, T. *et al*, *Chem. Pharm. Bull.*, 1978, **26**, 1890, 1896 (*synth, pmr, uv, oxides*)

[1,4]Benzodioxino[2,3-*b*]pyridine   B-80021

[72850-33-0]

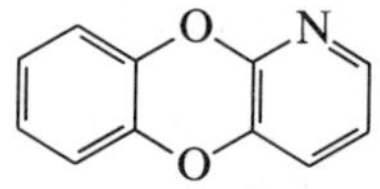

C₁₁H₇NO₂   M 185.182
Greyish solid. Mp 95-97°.

Sutherland, R.G. *et al*, *J. Heterocycl. Chem.*, 1988, **25**, 1911 (*synth, pmr, cmr*)

1,3-Benzodioxole-2-thione, 9CI   B-80022

Thiocarbonic acid cyclic O,O-o-phenylene ester, 8CI. 2-Thioxobenzodioxole
[2231-05-2]

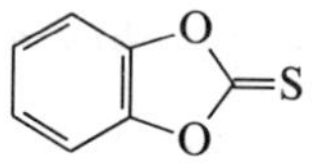

C₇H₄O₂S   M 152.173
Pale-pink needles (EtOH). Mp 152-153°.

Autenreith, A. *et al*, *Ber.*, 1925, **58**, 2151 (*synth*)
Sugimoto, H. *et al*, *J. Org. Chem.*, 1988, **53**, 2263 (*synth*)

Benzofuro[2,3-*b*]pyrazine, 9CI   B-80023

[114954-31-3]

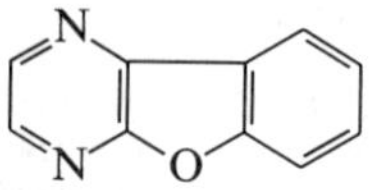

C₁₀H₆N₂O   M 170.170
Needles. Mp 89-90°.

Taylor, E.C. *et al*, *Tetrahedron*, 1987, **43**, 5159 (*synth, ir, pmr*)

4*H*-Benzo[*def*]naphtho[2,3-*b*]carbazole, 9CI   B-80024

4,5-Iminobenzo[k]chrysene. Naphtho[2′,3′:1,2]-4,5-iminophenanthrene (incorr.)
[42130-22-3]

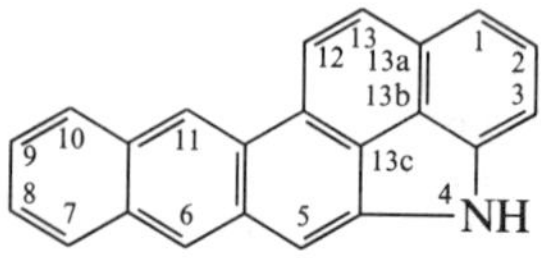

C₂₂H₁₃N   M 291.351
Reg. no. refers to 1*H*-form. Yellow plates (xylene). Mp 379-380°.

Zander, M. *et al*, *Chem. Ber.*, 1966, **99**, 1279 (*synth, uv*)

Benzo[*h*]naphtho[2′,3′:3,4]cyclobuta[1,2-*a*] biphenylene, 9CI   B-80025

Benzo[1‴,2‴:3,4:3″,4″:3′,4′]dicyclobuta[1,2-b:1′,2′-b′]dinaphthalene
[117965-77-2]

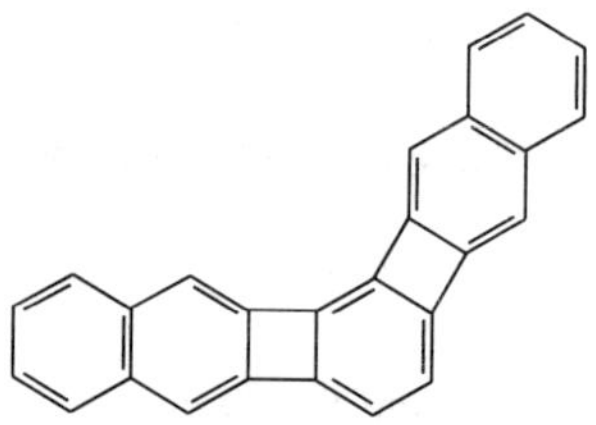

$C_{26}H_{14}$ M 326.397
Pale yellow cryst. by subl. Mp 295-298° (subl. >200°).
  Greenish fluor. in dil. soln.

Shepherd, M.K., *J. Chem. Soc., Perkin Trans. 1*, 1988, 961 (*synth, pmr, ms*)

Benzo[*b*]naphtho[2,3-*e*][1,4]dioxan   B-80026

5,12-Dioxanaphthacene

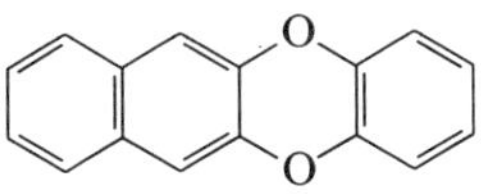

$C_{16}H_{10}O_2$ M 234.254
Silvery microcryst. Mp 202-203°.

Sutherland, R.G. *et al, J. Heterocycl. Chem.*, 1988, **25**, 1911 (*synth, pmr, cmr*)

Benzo[*a*]naphtho[2,1-*d*]fluoren-9-one   B-80027

[114469-09-9]

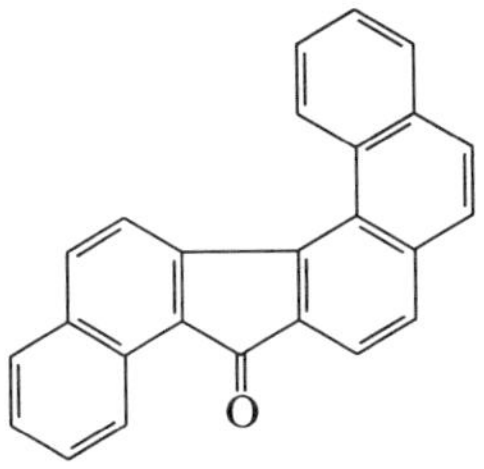

$C_{25}H_{14}O$ M 330.385
Yellow solid. Mp 244-245° (238-239°).

Campbell, A.D., *J. Chem. Soc.*, 1954, 3659 (*synth*)
Smith, J.G. *et al, J. Org. Chem.*, 1988, **53**, 2942 (*synth, pmr*)

Benzo[*b*]naphtho[2,1-*d*]fluoren-9-one   B-80028

[114469-08-8]

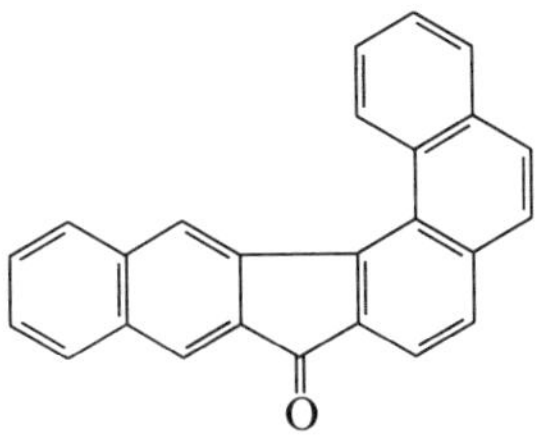

$C_{25}H_{14}O$ M 330.385
Red solid. Mp 203-204° (193-195°).

Campbell, A.D., *J. Chem. Soc.*, 1954, 3659 (*synth*)
Smith, J.G. *et al, J. Org. Chem.*, 1988, **53**, 2942 (*synth, pmr*)

Benzonaphtho[2.2]paracyclophane   B-80029

*6,9:14,17-Diethenobenzo[7,8]cyclododeca[1,2-*b*]naphthalene, 9CI*

[97315-27-0]

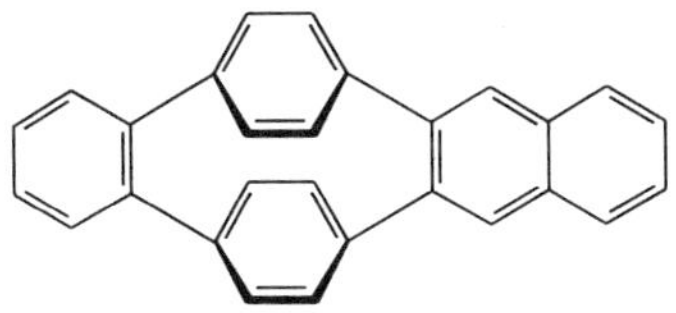

$C_{28}H_{18}$ M 354.450
Cryst. (toluene). Mp 278° dec.

Chan, C.W. *et al, J. Am. Chem. Soc.*, 1988, **110**, 462 (*synth, uv, pmr*)

Benzo[*b*]-[1,5]-naphthyridine   B-80030

1,10-Diazaanthracene

[261-05-2]

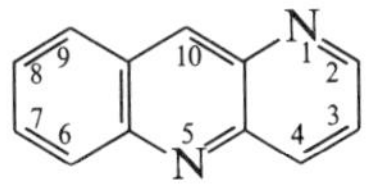

$C_{12}H_8N_2$ M 180.209
Cryst. Mp 106-107°. $Bp_{0.08}$ 65° subl.

Hicks, M.G. *et al, J. Chem. Soc., Perkin Trans. 1*, 1988, 69 (*synth, uv, pmr*)

Benzo[*b*][1,7]naphthyridine   B-80031

2,9-Diazaanthracene

[260-95-7]

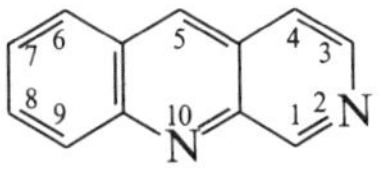

$C_{12}H_8N_2$ M 180.209
Cryst. (cyclohexane). Mp 134-135°. $Bp_{0.02}$ 60°.

Hicks, M.G. *et al, J. Chem. Soc., Perkin Trans. 1*, 1988, 69 (*synth, uv, pmr, ms*)

Benzo[*b*][1,8]naphthyridine   B-80032

1,9-Diazaanthracene

[261-00-7]

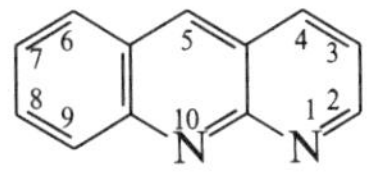

$C_{12}H_8N_2$ M 180.209
Mp 190-192°.

Hamada, Y. *et al, Chem. Pharm. Bull.*, 1974, **22**, 485.
Hicks, M.G. *et al, J. Chem. Soc., Perkin Trans. 1*, 1988, 69 (*synth*)

Benzo[2.2]paracyclophan-9-ene   B-80033

5,8:11,14-Diethenobenzocyclododecene, 9CI

[64586-11-4]

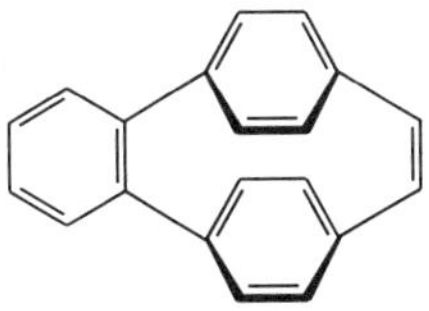

$C_{20}H_{14}$ M 254.331
Platelets (heptane). Mp 227-229°.

Jacobson, N. *et al, Angew. Chem., Int. Ed. Engl.*, 1978, **17**, 46 (*synth, pmr*)
Chan, C.W. *et al, J. Am. Chem. Soc.*, 1988, **110**, 462 (*synth, uv, pmr*)

Benzo[*c*]phenanthrene-3,4-dione   B-80034

Benzo[c]phenanthrene-3,4-quinone

[118658-88-1]

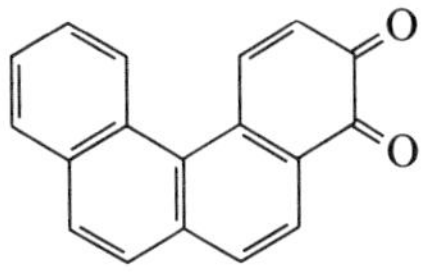

C₁₈H₁₀O₂   M 258.276
Cryst. (C₆H₆). Mp 178-180°.

Pataki, J. *et al, J. Org. Chem.*, 1989, **54**, 840 (*synth, pmr*)

10*H*-[1]Benzopyrano[2,3-*b*]pyrazin-10-one, 9CI   B-80035

1,4-Diazaxanthone
[121246-97-7]

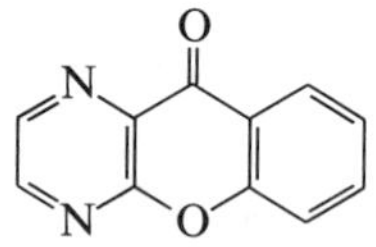

C₁₁H₆N₂O₂   M 198.181
Cryst. Mp 224°.

Turck, A. *et al, Synthesis*, 1988, 881 (*synth, pmr*)

[1]Benzopyrano[3,4-*c*]pyrazol-4(3*H*)-one, 9CI   B-80036

[91166-74-4]

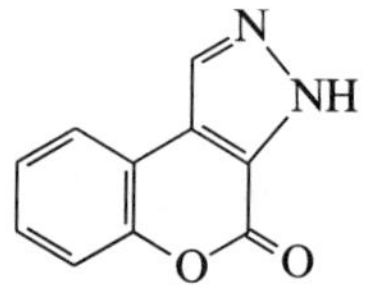

C₁₀H₆N₂O₂   M 186.170
Needles (dioxan). Mp >250°.

Ito, K. *et al, J. Heterocycl. Chem.*, 1988, **25**, 1681 (*synth, pmr*)

10*H*-[1]Benzopyrano[2,3-*d*]pyridazin-10-one, 9CI   B-80037

[115276-46-5]

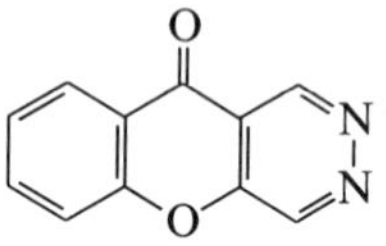

C₁₁H₆N₂O₂   M 198.181
Pale yellow cryst. (MeOH). Mp 189-190°.

Haider, N. *et al, J. Chem. Soc., Perkin Trans.* 1, 1988, 401 (*synth, ir, pmr, ms*)

10*H*-[1]Benzopyrano[3,2-*c*]pyridin-10-one, 9CI   B-80038

2-Azaxanthone
[54629-30-0]

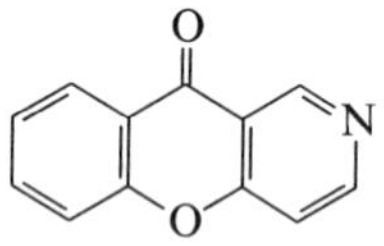

C₁₂H₇NO₂   M 197.193
Cryst. (C₆H₆/pet. ether). Mp 185°.

Ghosh, C.K. *et al, J. Chem. Soc., Perkin Trans.* 1, 1988, 1489 (*synth, ir, uv, pmr, ms*)

Benzo[*h*]quinoline-7,8-dione   B-80039

[113163-25-0]

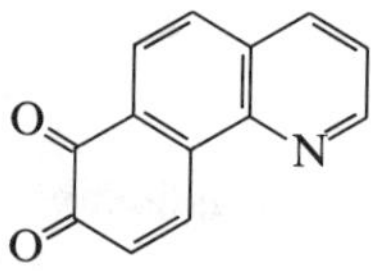

C₁₃H₇NO₂   M 209.204
Dark red solid (C₆H₆). Mp 118°.

Czech, A. *et al, J. Org. Chem.*, 1988, **53**, 1329 (*synth, ms, pmr*)

1,4-Benzothiazepin-5(4*H*)-one   B-80040

[115483-67-5]

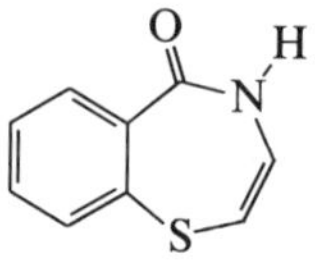

C₉H₇NOS   M 177.226
Cryst. (MeOH). Mp 143.5-145°.
N-*Me*: [115483-68-6].
    C₁₀H₉NOS   M 191.253
    Orange-tinted cryst. (MeOH). Mp 84-84.5°.

Hofmann, H. *et al, Chem. Ber.*, 1988, **121**, 2147 (*synth, pmr*)

2*H*-1,3-Benzothiazine-2,4(3*H*)-dione   B-80041

[10512-65-9]

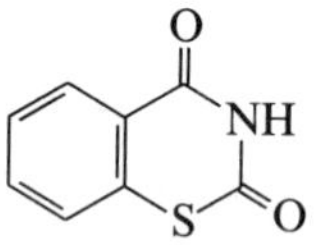

C₈H₅NO₂S   M 179.199
Cryst. (EtOH). Mp 216°.

Palazzo, S. *et al, Gazz. Chim. Ital.*, 1963, **93**, 207 (*synth*)

2*H*-3,1-Benzothiazine-2,4(1*H*)-dione   B-80042

[703-65-1]

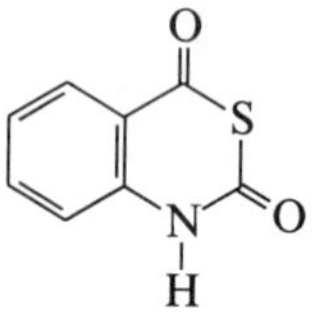

C₈H₅NO₂S   M 179.199
Cryst. (EtOAc). Mp 232-233° (226-235°).
N-*Me*:
    C₉H₇NO₂S   M 193.226
    Fine prisms (EtOH or Me₂CO). Mp 170.5-172°.

Marshall, J.R., *J. Chem. Soc.*, 1965, 938 (*synth*)
Kricheldorf, H.R., *Chem. Ber.*, 1971, **104**, 3156 (*ir*)
Sugimoto, H. *et al, J. Org. Chem.*, 1988, **53**, 2263 (*synth*)

2H-1,4-Benzothiazin-3(4H)-one, 9CI   B-80043

3,4-Dihydro-3-oxo-2H-1,4-benzothiazine

[5325-20-2]

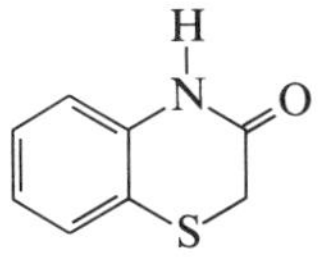

C$_8$H$_7$NOS   M 165.215

Cryst. (DMF/MeCN). Mp 177-179°.

N-Me: [37142-87-3].

C$_9$H$_9$NOS   M 179.242

Cryst. Mp 50-53°.

Krapcho, J. *et al*, *J. Med. Chem.*, 1973, **16**, 776 (*synth*)
Worley, J.W. *et al*, *J. Org. Chem.*, 1975, **40**, 1731 (*deriv, synth, pmr*)

1-Benzothiepin   B-80044

Updated Entry replacing B-00575

[264-82-4]

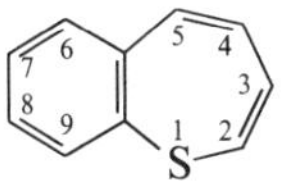

C$_{10}$H$_8$S   M 160.239

Pale-yellow needles. Mp 23.5-24.5°. λ_{max} 217 (log ϵ 4.07), 256(4.07), 295 sh (2.96) and 345 nm (2.55)(cyclohexane).

1-Oxide:

C$_{10}$H$_8$OS   M 176.239

Pale-yellow needles at −40°. Dec. to naphthalene after 1h at 13°.

1,1-Dioxide: [41887-86-9].

C$_{10}$H$_8$O$_2$S   M 192.238

Mp 138-139°.

Traynelis, V.J. *et al*, *J. Org. Chem.*, 1973, **38**, 3978 (*synth*)
Murata, I. *et al*, *Angew. Chem., Int. Ed. Engl.*, 1974, **13**, 142 (*synth*)
Yasuoka, N. *et al*, *Angew. Chem., Int. Ed. Engl.*, 1974, **15**, 297 (*cryst struct, deriv*)
Nishino, K. *et al*, *Angew. Chem., Int. Ed. Engl.*, 1988, **27**, 1717 (*oxide*)

10H-[1]Benzothiopyrano[2,3-d]pyridazin-10-one, 9CI   B-80045

[115276-47-6]

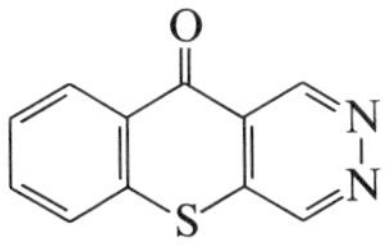

C$_{11}$H$_6$N$_2$OS   M 214.247

Pale yellow needles (Me$_2$CO). Mp 228-229°.

Haider, N. *et al*, *J. Chem. Soc., Perkin Trans. 1*, 1988, 401 (*synth, pmr, ir, ms*)

Benzo[b]triphenylen-1-ol, 9CI   B-80046

1-Hydroxydibenz[a,c]anthracene

[115165-01-0]

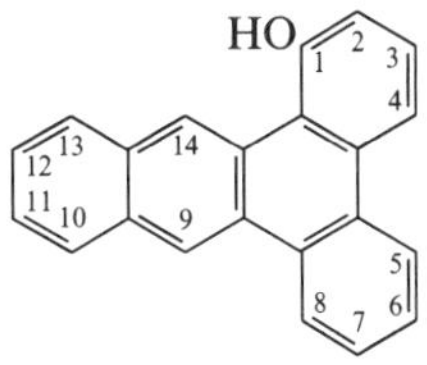

C$_{22}$H$_{14}$O   M 294.352

Cryst. (C$_6$H$_6$). Mp 210-211°.

Kole, P.L. *et al*, *J. Org. Chem.*, 1989, **54**, 845 (*synth, pmr*)

Benzo[b]triphenylen-2-ol, 9CI   B-80047

2-Hydroxydibenz[a,c]anthracene

[113058-38-1]

C$_{22}$H$_{14}$O   M 294.352

Cryst. (Et$_2$O). Mp 287-288°.

Kole, P.L. *et al*, *J. Org. Chem.*, 1989, **54**, 845 (*synth, pmr*)

Benzo[b]triphenylen-3-ol, 9CI   B-80048

3-Hydroxydibenz[a,c]anthracene

[88488-97-5]

C$_{22}$H$_{14}$O   M 294.352

Cryst. (Et$_2$O). Mp 252-253°.

Harvey, R.G. *et al*, *Org. Prep. Proced. Int.*, 1983,, **15**, 335 (*synth*)
Kole, P.L. *et al*, *J. Org. Chem.*, 1989, **54**, 845 (*synth, pmr*)

Benzo[b]triphenylen-4-ol, 9CI   B-80049

4-Hydroxydibenz[a,c]anthracene

[113058-39-2]

C$_{22}$H$_{14}$O   M 294.352

Cryst. (C$_6$H$_6$). Mp 266-267°.

Kole, P.L. *et al*, *J. Org. Chem.*, 1989, **54**, 845 (*synth, pmr*)

Benzo[b]triphenylen-10-ol, 9CI   B-80050

10-Hydroxydibenz[a,c]anthracene

[39081-15-7]

C$_{22}$H$_{14}$O   M 294.352

Cryst. (Et$_2$O/hexane). Mp 239-241° (220-222°).

Sims, P., *Biochem. J.*, 1972, **130**, 27 (*synth, uv*)
Kole, P.L. *et al*, *J. Org. Chem.*, 1989, **54**, 845 (*synth, pmr*)

Benzo[b]triphenylen-11-ol, 9CI   B-80051

11-Hydroxydibenz[a,c]anthracene

[39147-43-8]

C$_{22}$H$_{14}$O   M 294.352

Cryst. (C$_6$H$_6$). Mp 200-201°.

Sims, P., *Biochem. J.*, 1972, **130**, 27 (*synth, uv*)
Kole, P.L. *et al*, *J. Org. Chem.*, 1989, **54**, 845 (*synth, pmr*)

1,2,3-Benzoxathiazin-4(3H)-one, 9CI   B-80052

3,4-Dihydro-4-oxo-1,2,3-benzothiazine

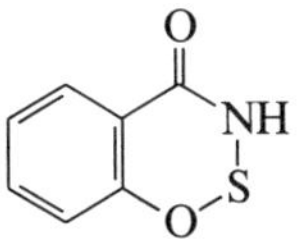

C$_7$H$_5$NO$_2$S   M 167.188

2,2-Dioxide: [51299-06-0].

$C_7H_5NO_4S$ M 199.187
Cryst. (CHCl$_3$/hexane). Mp 218-220° dec.

Kamal, A. *et al*, *J. Org. Chem.*, 1988, **53**, 4112 (*synth, ir, pmr*)

1,5-Benzoxepine-2,4-(3*H*,5*H*)-dione   B-80053
[84755-33-9]

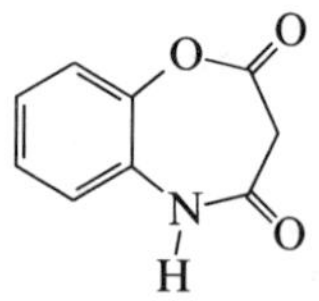

$C_9H_7NO_3$ M 177.159
Cryst. (EtOH). Mp 199-201°. A synthesis claimed in 1982
appears to be erroneous.

Bartsch, H. *et al*, *Justus Liebigs Ann. Chem.*, 1989, 177 (*synth, pmr*)

2-Benzoylbenzoic acid, 9CI   B-80054
Updated Entry replacing B-00632
Benzophenone-2-carboxylic acid. Diphenylmethanone-2-carboxylic acid
[85-52-9]

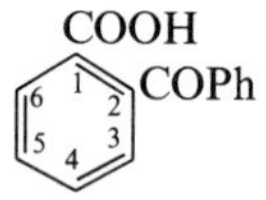

$C_{14}H_{10}O_3$ M 226.231
Needles + 1H$_2$O (H$_2$O), cryst. (C$_6$H$_6$/pet. ether). Mp 95°
(hydrate), Mp 128° (anhyd.). pK_a 3.43.
Me ester: [606-28-0].
   $C_{15}H_{12}O_3$ M 240.258
   Prisms. Mp 52°. Bp 352°.
Et ester: [604-61-5].
   $C_{16}H_{14}O_3$ M 254.285
   Mp 58°.
Chloride: [22103-85-1].
   $C_{14}H_9ClO_2$ M 244.677
   Prisms. Mp 70°.
Anhydride:
   $C_{28}H_{18}O_5$ M 434.447
   Mp 140°.
Amide: [7500-78-9].
   $C_{14}H_{11}NO_2$ M 225.246
   Mp 165°.
Nitrile: [37774-78-0]. o-*Cyanobenzophenone*
   $C_{14}H_9NO$ M 207.231
   Cryst. Mp 83-84°.

Rubidge, C.R. *et al*, *J. Am. Chem. Soc.*, 1914, **36**, 732 (*synth*)
CRC Atlas of Spectral Data and Physical Constants, 1962 (*ir, uv, ms, nmr*)
Delettre, J., *Compt. Rend. Hebd. Seances Acad. Sci.*, 1969, **269**, 113 (*cryst struct*)
Vogel, A.I., *Textbook of Practical Organic Chemistry*, Longman, London, 4th Ed., 1978, 780 (*synth*)
Ebert, G.W. *et al*, *J. Org. Chem.*, 1988, **53**, 4482 (*nitrile*)

Benzoylcyclopentane   B-80055
Cyclopentylphenylmethanone, 9CI. Cyclopentyl phenyl ketone
[5422-88-8]

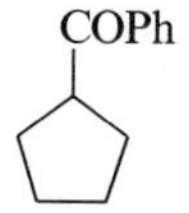

$C_{12}H_{14}O$ M 174.242
Oil. Bp$_{16}$ 138-140°, Bp$_{0.05}$ 89-90°. n$_D^{20}$ 1.5290, 1.5404.
Oxime: [83568-09-6].
   $C_{12}H_{15}NO$ M 189.257
   Needles (pet. ether). Mp 106-108°.
Semicarbazone: Needles (C$_6$H$_6$). Mp 107.5-109.5°.
2,4-Dinitrophenylhydrazone: Yellow cryst. (EtOH). Mp 142-143°.

Hey, D.H. *et al*, *J. Chem. Soc.*, 1949, 3156 (*synth*)
Zhang, P. *et al*, *Synth. Commun.*, 1986, **16**, 957 (*synth*)

6-Benzoyl-5,7-dihydroxy-2-methyl-2-(4-methyl-3-pentenyl)chroman   B-80056
3,4-Dihydro-5,7-dihydroxy-2-methyl-2-(4-methyl-3-pentenyl)-2H-benzopyran-6-ylphenylmethanone

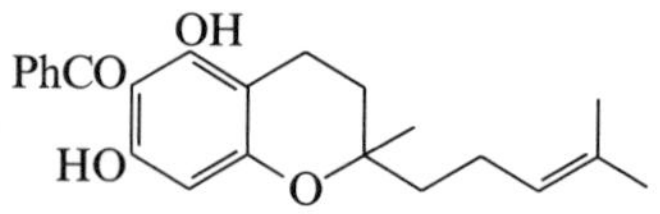

$C_{23}H_{26}O_4$ M 366.456
Isol. from *Helichrysum monticola*. Oil. Not abstracted by
CA. Incorrectly named in the reference.

Bohlmann, F. *et al*, *Phytochemistry*, 1980, **19**, 683.

2-Benzoylpropanal   B-80057
Updated Entry replacing B-00700
α-*Methyl-β-oxobenzenepropanal, 9CI. 2-Benzoylpropionaldehyde, 8CI. 1-Formylethyl phenyl ketone.*
α-*Hydroxymethylenepropiophenone. 2-Formylpropiophenone*
[16837-43-7]

$$PhCOCH(CH_3)CHO \rightleftharpoons PhCOC(CH_3){=}CHOH$$

$C_{10}H_{10}O_2$ M 162.188
Cryst. (EtOH). Mp 118-119°.

Takagi, S. *et al*, *CA*, 1959, **53**, 18003f (*synth*)
Aspart-Pascot, L. *et al*, *Bull. Soc. Chim. Fr.*, 1971, 483 (*synth*)

2-*O*-Benzylglyceraldehyde   B-80058
3-Hydroxy-2-(phenylmethoxy)propanal, 9CI

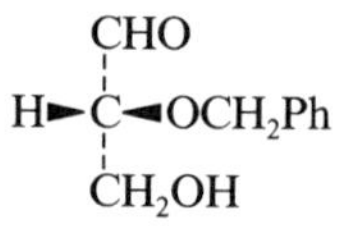

$C_{10}H_{12}O_3$ M 180.203
(*R*)-*form* [76227-09-3]
   Configurationally stable chiral intermed. for synthesis.
   Oil with fruity odour. [α]$_D$ +21.7→+16.1°
   (6h)→ +30.3° (19d) (c, 1.23 in EtOH). Oligomerises on
   standing without loss of configurational homogeneity.
   (*S*)-enantiomer also accessible.

Jäger, V. *et al*, *Angew. Chem., Int. Ed. Engl.*, 1989, **28**, 469 (*synth, use*)

1-Benzylidene-2,3-diphenylindene   B-80059

2,3-Diphenyl-1-(phenylmethylene)-1H-indene, 9CI

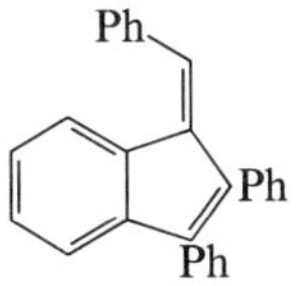

$C_{28}H_{20}$ M 356.466

(*E*)-*form* [121638-18-4]

Unexpected prod. of reacn. of hexaphenylstannole with Br_2. Cryst. Mp 176°.

Muchmore, C.R.A. *et al, Acta Crystallogr., Sect. C*, 1989, **45**, 824 (*cryst struct*)

2-Benzylpyrrole   B-80060

2-(Phenylmethyl)-1H-pyrrole, 9CI

[33234-48-9]

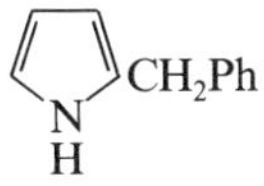

$C_{11}H_{11}N$ M 157.215

$Bp_{2.5}$ 130°, $Bp_{0.12-0.15}$ 85-89°.

Gogan, N.J. *et al, J. Organomet. Chem.*, 1972, **39**, 129 (*synth*)
McGillivray, G. *et al, J. Chem. Soc., Perkin Trans.* 1, 1983, 633 (*synth*)

3-Benzylpyrrole   B-80061

3-(Phenylmethyl)-1H-pyrrole, 9CI

[33234-57-0]

$C_{11}H_{11}N$ M 157.215

No phys. props. reported.

Groves, J.K. *et al, Can. J. Chem.*, 1971, **49**, 2427 (*synth, pmr, ms*)

Benzyl vinyl sulfide, 8CI   B-80062

[(Ethenylthio)methyl]benzene, 9CI

[1822-76-0]

$$PhCH_2SCH=CH_2$$

$C_9H_{10}S$ M 150.244

Liq. d_{25}^{25} 1.038. Bp_{10} 98°, $Bp_{0.4}$ 52-56°. n_D^{25} 1.5773.

S-Oxide: [71841-84-4]. *Benzyl vinyl sulfoxide, 8CI.*
[(Ethylsulfinyl)methyl]benzene, 9CI
$C_9H_{10}OS$ M 166.243
Cryst. Mp 34-35°.

S-Dioxide: [15753-89-6]. *Benzyl vinyl sulfone, 8CI.*
[(Ethenylsulfonyl)methyl]benzene, 9CI
$C_9H_{10}O_2S$ M 182.243
Cryst.

Protiva, M., *Collect. Czech. Chem. Commun.*, 1949, **14**, 354 (*synth*)
Reppe, W. *et al, Justus Liebigs Ann. Chem.*, 1956, **601**, 111 (*synth*)
Schneider, H.J. *et al, J. Org. Chem.*, 1961, **26**, 1980 (*synth, props*)
Annunziata, R. *et al, J. Chem. Soc., Perkin Trans.* 1, 1979, 1684 (*oxide*)
Baum, J.C. *et al, Can. J. Chem.*, 1984, **62**, 1687 (*synth, ir, pmr, ms*)

Besthorn's red   B-80063

14-Hydroxyimidazo[1,2-a: 3,4-a']diquinolin-7-ium hydroxide, inner salt, 9CI

[98779-77-2]

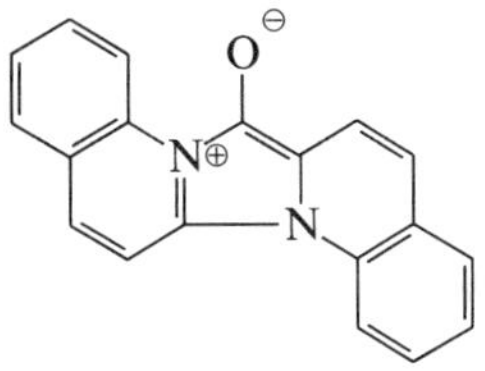

$C_{19}H_{12}N_2O$ M 284.317

Dark red glistening cryst. Mp 280° dec. (sinters from 230°). Forms MeOH solvates.

Besthorn, E. *et al, Ber.*, 1904, **37**, 1236 (*synth*)
Krollpfeiffer, F. *et al, Justus Liebigs Ann. Chem.*, 1937, **530**, 34 (*synth, uv*)
Edward, J.T. *et al, J. Org. Chem.*, 1985, **50**, 4855 (*synth, ir, bibl*)

ent-15-Beyerene-14β,19-diol   B-80064

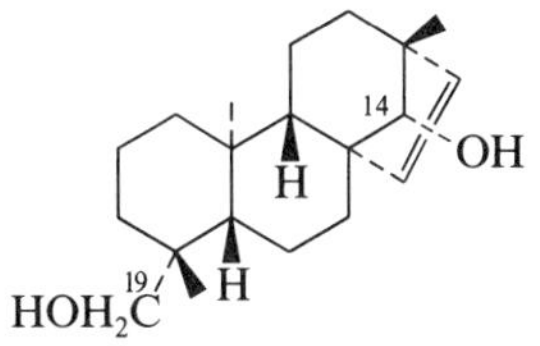

$C_{20}H_{32}O_2$ M 304.472

14-O-Tigloyl: **Elasclepiol**
$C_{25}H_{38}O_3$ M 386.573
Constit. of *Elaeoselinium asclepium*. Needles (hexane).
Mp 142-143°. $[\alpha]_D$ +45.4° (c, 1 in $CHCl_3$).

19-Carboxylic acid, 14-O-tigloyl: **Elasclepic acid**
$C_{25}H_{36}O_4$ M 400.557
Constit. of *E. asclepium*. Oil. $[\alpha]_D$ −8.4° (c, 1.6 in EtOH).

Grande, M. *et al, Phytochemistry*, 1989, **28**, 1955.

ent-15-Beyerene-2,12-dione   B-80065

Yucalexin B-7

[64657-12-1]

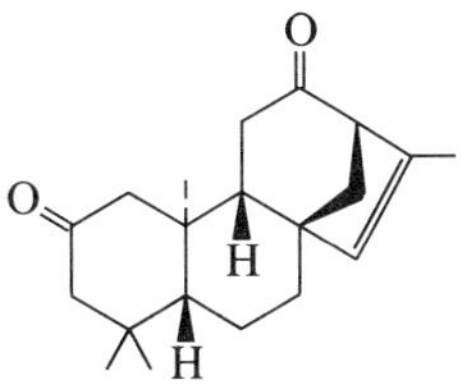

$C_{20}H_{28}O_2$ M 300.440

Constit. of cassava roots (*Manihot esculenta*).

Sakai, T. *et al, Phytochemistry*, 1988, **27**, 3769.

ent-15-Beyeren-18-oic acid   B-80066

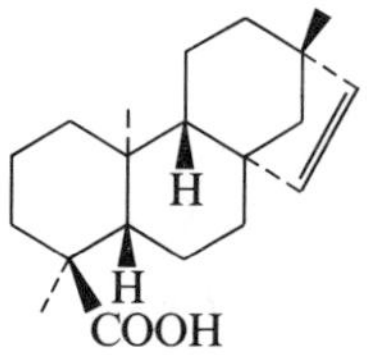

C$_{20}$H$_{30}$O$_2$ M 302.456
Constit. of *Helipterum floribundum*.

Zdero, C. *et al*, *Phytochemistry*, 1989, **28**, 517.

1,1′-Biacenaphthylidene-2,2′-dione   B-80067
2-(2-Oxo-1(2H)-acenaphthalenylidene)-1(2H)-acenaphthylenone, 9CI. Δ$^{1,1'}$-[*Biacenaphthene*]-2,2′-dione, *8CI. Biacenedione*

[13286-14-1]

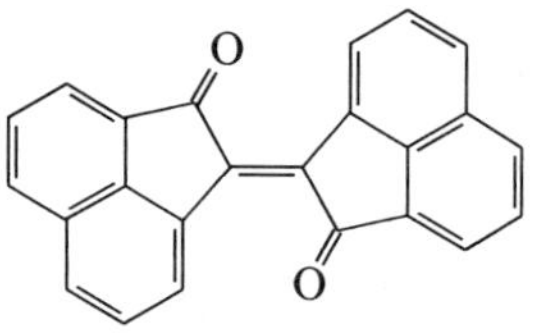

(*E*)-*form*

C$_{24}$H$_{12}$O$_2$ M 332.358
(*E*)-*form* [24141-18-2]
    Red cryst. Mp 300° (289-291°). Most refs. do not specify the *E*-form but that is to be assumed.

Dziewonski, K. *et al*, *Chem. Ber.*, 1925, **58**, 2539 (*synth*)
Rule, H.G. *et al*, *J. Chem. Soc.*, 1937, 1761 (*synth*)
De Jong, D.C. *et al*, *Tetrahedron*, 1972, **28**, 3603 (*synth*)
Huang, N.Z. *et al*, *J. Org. Chem.*, 1987, **52**, 169 (*synth*)

7,7′-Bi-7-azabicyclo[2.2.1]heptane, 9CI   B-80068
[116005-44-8]

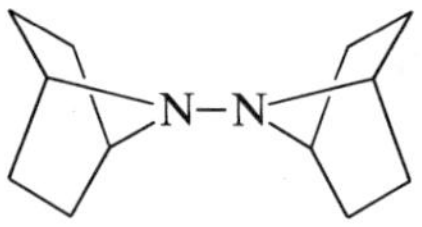

C$_{12}$H$_{20}$N$_2$ M 192.303
Cryst. by subl. Mp 111.5-112°.

[116005-46-0]

Nelsen, S.F. *et al*, *J. Am. Chem. Soc.*, 1988, **110**, 6149 (*synth, pmr, cmr*)

2,2-Bi(1,4-benzoquinone)   B-80069
[*Bi-1,4-cyclohexadien-1-yl*]*-3,3′,6,6′-tetrone, 9CI*
[23783-80-4]

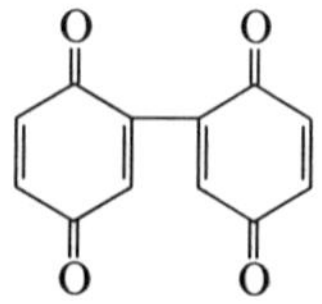

C$_{12}$H$_6$O$_4$ M 214.177
Chestnut cryst. (nitrobenzene). Mp 230°.

Paraskevas, S.M. *et al*, *Synthesis*, 1988, 897 (*synth, ir, pmr*)

4,4′-Bi(1,2-benzoquinone)   B-80070
[*Bi-1,5-cyclohexadien-1-yl*]*-3,3′,4,4′-tetrone, 9CI*
[120811-93-0]

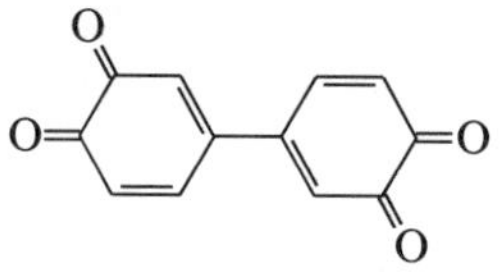

C$_{12}$H$_6$O$_4$ M 214.177
Red cryst. Mp 136° dec.

Paraskevas, S.M. *et al*, *Synthesis*, 1988, 897 (*synth, ir, pmr*)

Bibenzyl-2,2′-dicarboxylic acid   B-80071
Updated Entry replacing B-00989
2,2′-(1,2-Ethanediyl)bisbenzoic acid, 9CI. α,α′-Bi-o-toluic acid, 8CI. 1,2-Bis(2-carboxyphenyl)ethane. Ethylenedibenzoic acid (misleading)
[4281-17-8]

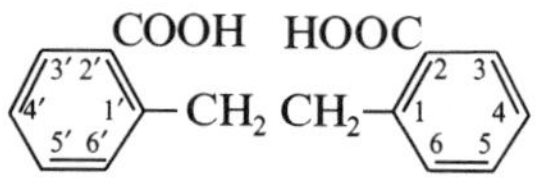

C$_{16}$H$_{14}$O$_4$ M 270.284
Needles (EtOH), cryst. (CHCl$_3$/MeOH). Mp 237-239° (226-228°).
Di-Me ester:
   C$_{18}$H$_{18}$O$_4$ M 298.338
   Mp 100-101°.
Di-Et ester:
   C$_{20}$H$_{22}$O$_4$ M 326.391
   Needles (MeOH). Mp 67-68°.
Dichloride:
   C$_{16}$H$_{12}$Cl$_2$O$_2$ M 307.175
   Liq.
Dinitrile: 2,2′-*Dicyanobibenzyl*
   C$_{16}$H$_{12}$N$_2$O$_2$ M 264.283
   Mp 139°.

Fuson, R.C., *J. Am. Chem. Soc.*, 1926, **48**, 830 (*synth*)
Hannemann, K. *et al*, *Helv. Chim. Acta*, 1988, **71**, 1841 (*synth, ir, pmr*)

1,1′-Bi(bicyclo[1.1.1]pentane)-3-carboxylic acid   B-80072
[*2*]*Staffane-3-carboxylic acid*

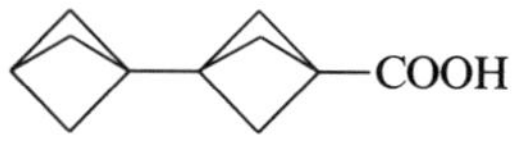

C$_{11}$H$_{14}$O$_2$ M 178.230
Cryst. Mp *ca* 200°.
Me ester:
   C$_{12}$H$_{16}$O$_2$ M 192.257
   Cryst. Mp ∼ 60°.

Kaszynski, P. *et al*, *J. Am. Chem. Soc.*, 1988, **110**, 5225 (*synth, cryst struct, pmr*)

Bicornin   B-80073

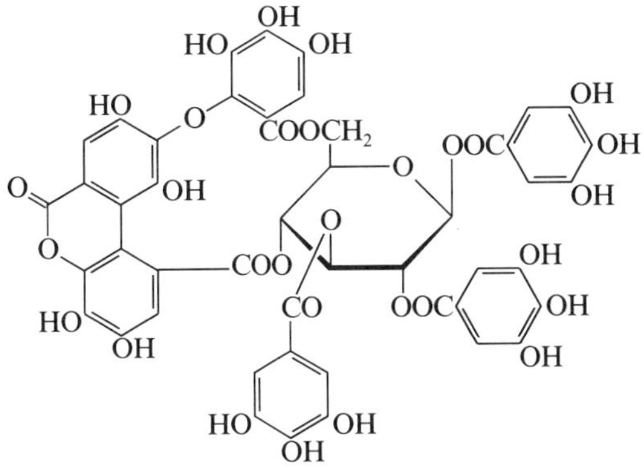

C_{48}H_{32}O_{30}   M 1088.763
Constit. of *Trapa bicornis*. Amorph. powder. [α]_D +4.7°
(MeOH).

Yoshida, T. *et al, Heterocycles*, 1989, **29**, 861.

Bicyclo[6.2.0]dec-9-ene   B-80074

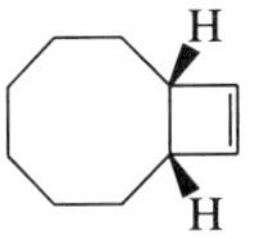

$C_{10}H_{16}$ M 136.236
(*1RS,7SR*)-*form* [17385-24-9]
    Liq.

Dauben, W.G. *et al, J. Org. Chem.*, 1988, **53**, 600 (*synth, ir, pmr, cmr*)

1(10),4-Bicyclogermacradien-13-oic acid   B-80075

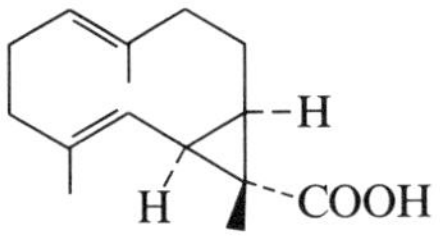

$C_{15}H_{22}O_2$ M 234.338
Constit. of *Helichrysum chionosphaerum*. Oil.

Jakupovic, J. *et al, Phytochemistry*, 1989, **28**, 1119.

13-Bicyclogermacrenal   B-80076

[118855-37-1]

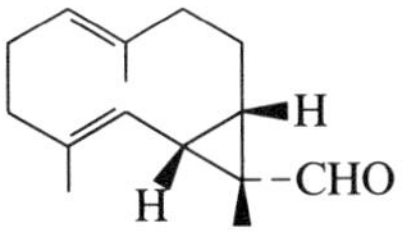

$C_{15}H_{22}O$ M 218.338
Constit. of *Conocephalum conicum*. Oil. $[α]_D^{20}$ −52° (c, 0.4
in CHCl₃).

Toyota, M. *et al, Phytochemistry*, 1988, **27**, 3317.

Bicyclo[14.4.1]heneicosa-1,3,5,7,9,11,13,   B-80077
15,17,19-decaene, 9CI

1,6-Methano[20]annulene
[110977-29-2]

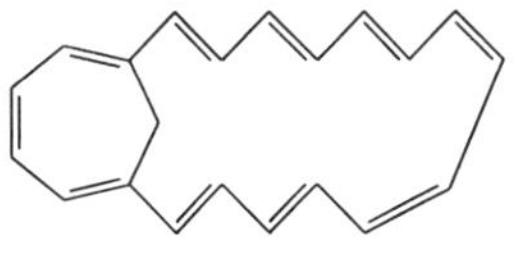

$C_{21}H_{20}$ M 272.389
Black needles (hexane). Mp 108-109°.

Yamamoto, K. *et al, J. Chem. Soc., Perkin Trans. 1*, 1988, 395
    (*synth, uv, ir, pmr*)

Bicyclo[4.1.0]hepta-1(6),3-diene-2,5-dione,   B-80078
9CI

Benzocyclopropene-p-*quinone. Cyclopropa*-p-*benzoquinone*
[111870-65-6]

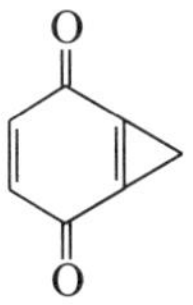

$C_7H_4O_2$ M 120.107
Unstable intermediate trapped as Diels-Alder adduct with
    anthracene.

Watabe, T. *et al, J. Org. Chem.*, 1988, **53**, 216 (*synth, ir*)

Bicyclo[4.1.0]hept-3-ene   B-80079

[16554-83-9]

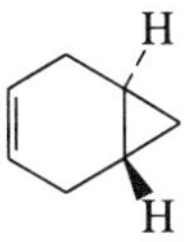

(*1RS,6RS*)-*form*

C_7H_{10} M 94.156
(*1RS,6RS*)-*form* [84194-54-7]
    (±)-trans-*form*
    Liq. Bp₁₄₂ 45-50°.
(*1RS,6SR*)-*form* [88090-52-2]
    cis-*form*
    Liq. Bp 114-115°.

Simmons, H.E. *et al, J. Am. Chem. Soc.*, 1964, **86**, 1347 (*synth*)
Plemenkov, V.V. *et al, J. Org. Chem. USSR (Engl. Transl.)*, 1974,
    10, 1672 (*synth*)
Gassman, P.G. *et al, J. Am. Chem. Soc.*, 1983, **105**, 667 (*synth,
    pmr, cmr, ir*)
Dixon, D.A. *et al, J. Am. Chem. Soc.*, 1988, **110**, 2309 (*struct, bibl*)

Bicyclo[2.2.1]hept-5-en-2-one, 9CI   B-80080

Updated Entry replacing B-10139
    Dehydronorcamphor. 5-Norbornen-2-one
    [694-98-4]

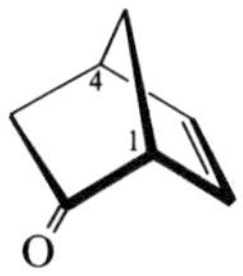

(*1R,4R*)-*form*

C_7H_8O M 108.140

(1R,4R)-form [16346-63-7]
Liq. Bp$_{19}$ 65.5-66.5°. [α]$_D^{22}$ +1012° (c, 1.65 in 2,3,3-trimethylpentane)(90% e.e.).

(±)-*form* [51736-74-4]
Oil. Mp 22-23°. Bp$_{24}$ 65-66.5°.

Oxime: [53838-00-9]
C$_7$H$_9$NO   M 123.154
Solid (C$_6$H$_6$/EtOAc). Mp 79°.

(±)-*form*
Oil. Mp 22-23°. Bp$_{24}$ 65-66.5°.

Oxime:
C$_7$H$_9$NO   M 123.154
Solid (C$_6$H$_6$/EtOAc). Mp 79°.

Mislow, K. *et al, J. Am. Chem. Soc.*, 1962, **84**, 1956 (synth)
Weiss, R.G. *et al, J. Am. Chem. Soc.*, 1978, **100**, 1172 (synth)
Lightner, D.A. *et al, J. Am. Chem. Soc.*, 1980, **102**, 5749 (synth)
Nakazak, N. *et al, J. Org. Chem.*, 1980, **45**, 4437 (synth)
Oppolzer, W. *et al, Helv. Chim. Acta*, 1985, **68**, 2100 (synth)
Eichberger, G. *et al, Tetrahedron Lett.*, 1986, **27**, 2843 (synth)
Mattay, J. *et al, Chem. Ber.*, 1989, **122**, 327 (synth, ir, pmr, cmr)

Bicyclo[12.4.1]nonadeca-1,3,5,7,9,11,13, 15,17-nonaene   B-80081

1,6-Methano[18]annulene

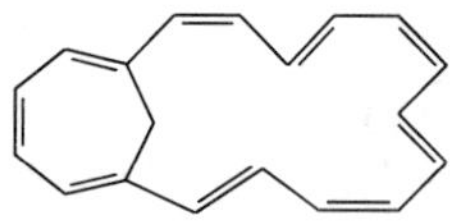

C$_{19}$H$_{18}$ M 246.351
Black needles (hexane). Mp 138-139°.

Yamamoto, K. *et al, J. Chem. Soc., Perkin Trans.* 1, 1988, 395 (synth, uv, pmr)

Bicyclo[3.3.1]nonane-2,4,7-trione   B-80082
[110935-39-2]

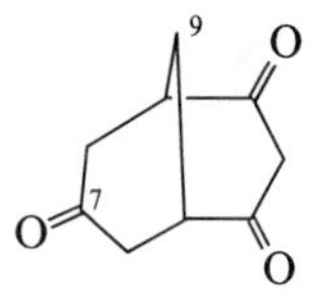

C$_9$H$_{10}$O$_3$ M 166.176
Needles (CHCl$_3$/hexane). Mp 158-160° (154-155°). 100% enolised in CDCl$_3$ soln. at r.t.

Hutmacher, H.-M. *et al, Chem. Ber.*, 1977, **110**, 3118 (synth, nmr, ms)
Yamazaki, T. *et al, Chem. Pharm. Bull.*, 1987, **35**, 3453 (synth, pmr)
Ebel, K. *et al, Chem. Ber.*, 1988, **121**, 323 (deriv)

Bicyclo[3.3.1]nonane-2,4,9-trione, 9CI   B-80083
[94632-45-8]

C$_9$H$_{10}$O$_3$ M 166.176
Mp 158-160°. 100% enolised in soln.

Yogev, A. *et al, J. Org. Chem.*, 1967, **32**, 2162.
Schoenwaelder, K.H. *et al, Chem. Ber.*, 1984, **117**, 3280 (synth, pmr, cryst struct)

Bicyclo[4.2.1]nona-2,4,7-trien-9-one, 9CI   B-80084
[34733-74-9]

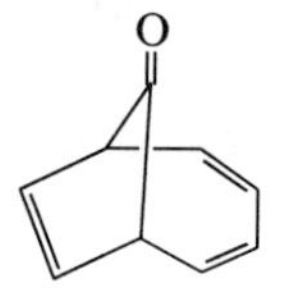

C$_9$H$_8$O   M 132.162
Liq. solidifying on cooling. Bp$_{0.3}$ 46-47°.

Oxime:
C$_9$H$_9$NO   M 147.176
Cryst. (CH$_2$Cl$_2$/hexane). Mp 112-113°.

4-Methylbenzenesulfonylhydrazone: Solid (MeOH). Mp 177-179°.

Antkowiak, T.A. *et al, J. Am. Chem. Soc.*, 1972, **94**, 5366 (synth, derivs)
Kurabayashi, K. *et al, Tetrahedron Lett.*, 1972, 1049 (uv, ms)
Diaz, A. *et al, J. Org. Chem.*, 1975, **40**, 2459 (pmr)

Bicyclo[4.2.0]octa-2,4-diene   B-80085
Updated Entry replacing B-01103
[3725-28-8]

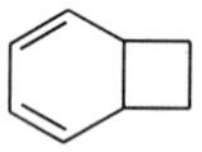

C$_8$H$_{10}$ M 106.167
Bp$_{0.5}$ <30°.

Cope, A.C. *et al, J. Am. Chem. Soc.*, 1952, **74**, 4867 (synth)
Alder, K. *et al, Chem. Ber.*, 1954, **87**, 1492 (synth)
Nefedov, O.M. *et al, CA*, 1972, **76**, 140004y (synth)
Bally, C.T. *et al, Helv. Chim. Acta*, 1989, **72**, 73 (synth, bibl)

Bicyclo[2.2.2]oct-7-ene-2,5-dione   B-80086
Updated Entry replacing B-50196
[17660-74-1]

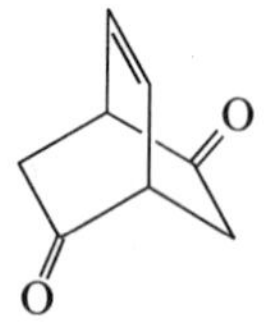

(1S,4S)-*form*

C$_8$H$_8$O$_2$ M 136.150
(1S,4S)-*form*
Cryst. Mp 98-100°. [α]$_D^{29.5}$ −1257° (c, 0.0015 in cyclohexane).

(±)-*form* [94669-40-6]
Cryst. (Et$_2$O/pet. ether or by subl.). Mp 97-99°. Bp$_{11}$ 80° subl. . Hygroscopic.

Grob, C.A. *et al, Helv. Chim. Acta*, 1960, **43**, 1390 (synth, uv)
Tomer, K.B. *et al, Tetrahedron*, 1973, **29**, 3491 (ms)
Jefford, C.W. *et al, J. Org. Chem.*, 1977, **42**, 1654 (synth)
Hill, R.K. *et al, J. Org. Chem.*, 1985, **50**, 5528 (synth, resoln, abs config)
Demuth, M. *et al, J. Am. Chem. Soc.*, 1986, **108**, 4149 (synth, resoln, cd)
Lightner, D.A. *et al, J. Org. Chem.*, 1988, **53**, 1969 (synth, ir, pmr, cmr, resoln, abs config)

Bicyclo[3.2.1]oct-6-en-3-one   B-80087

[3721-60-6]

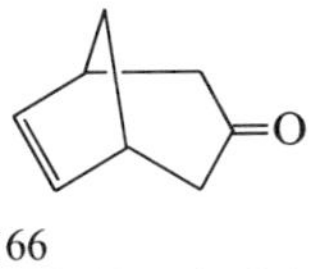

$C_8H_{10}O$ M 122.166
Mp 100°. Bp$_{11}$ 120-124° (Kugelrohr) .

Föhlisch, B. *et al, Chem. Ber.*, 1988, **121**, 1585 (*synth, pmr, cmr*)

Bicyclo[4.2.0]oct-2-en-7-one   B-80088

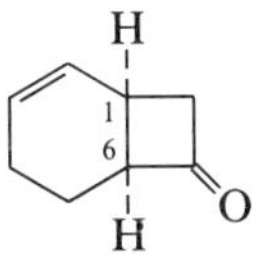

(1R,6S)-form

$C_8H_{10}O$ M 122.166
(1R,6S)-form [52466-04-3]
   (+)-cis-form
   Volatile liq. Bp$_{12}$ 65° (kugelrohr) . [α]$_D^{25}$ +171.4° (0.47 in CHCl$_3$).
(1RS,6SR)-form [52466-03-2]
   (±)-cis-form
   Liq. Bp$_{22}$ 85-90°.

[34896-01-0]

Corey, E.J. *et al, Tetrahedron Lett.*, 1971, 4753 (*synth, ir*)
Kertesz, D.J. *et al, J. Org. Chem.*, 1988, **53**, 4962 (*synth, resoln, ms, pmr, cmr*)

Bicyclo[18.4.1]pentacosa-1,3,5,7,9,11,13,   B-80089
15,17,19,21,23-dodecaene, 9CI

1,6-Methano[24]annulene

[110977-31-6]

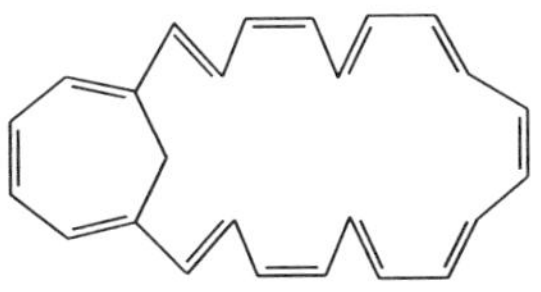

$C_{25}H_{24}$ M 324.465
Black needles (hexane). Mp 152-153°. Paratropic.

Yamamoto, K. *et al, J. Chem. Soc., Perkin Trans.* 1, 1988, 395 (*synth, uv, ir, pmr*)

Bicyclo[1.1.1]pentane-1,3-dicarboxylic acid   B-80090

Updated Entry replacing B-20094

[56842-95-6]

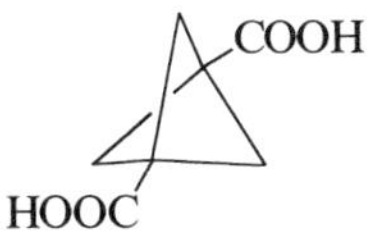

$C_7H_8O_4$ M 156.138
Solid. Subl. at 260-95° without melting.
Mono-Me ester: [83249-10-9].
   $C_8H_{10}O_4$ M 170.165
   Cryst. (heptane/CHCl$_3$). Mp 139.5-140°.
Di-Me ester: [115913-32-1].
   $C_9H_{12}O_4$ M 184.191
   Cryst. Mp 92°. Bp$_{12}$ 125-130°.
Dichloride: [115913-31-0].
   $C_7H_6Cl_2O_2$ M 193.029

Mp 55-57°. Bp$_{12}$ 120°.

Applequist, D.E. *et al, J. Org. Chem.*, 1982, **47**, 4985 (*synth*)
Kaszynski, P. *et al, J. Org. Chem.*, 1988, **53**, 4593 (*synth, derivs, ir, pmr, cmr, ms*)

Bicyclo[6.4.1]trideca-1,3,5,7,9,11-hexaene,   B-80091
9CI

1,6-Methano[12]annulene

[52954-75-3]

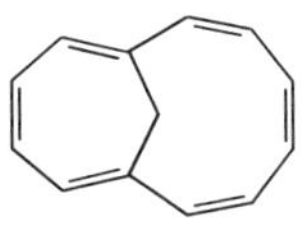

$C_{13}H_{12}$ M 168.238
Pale-yellow liq.

Yamamoto, K. *et al, J. Chem. Soc., Perkin Trans.* 1, 1988, 395 (*synth, uv, pmr*)

2,2'-Bi(4,6-dimethyl-5H-1,3-dithiolo[4,5-c]   B-80092
pyrrolylidene)

2-(4,6-Dimethyl-5H-1,3-dithiolo[4,5-c]pyrrol-2-ylidene)-5H-1,3-dithiolo[4,5-c]pyrrole, 9CI. Bis(2,5-dimethylpyrrolo[3,4-d]tetrathiafulvene)

[117860-14-7]

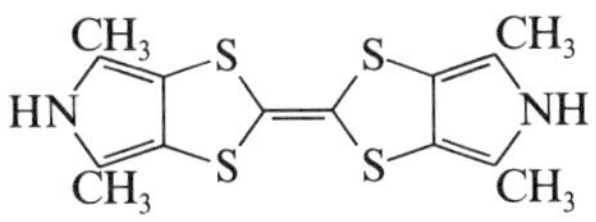

$C_{14}H_{14}N_2S_4$ M 338.542
Electron donor. Mp >300°.

Chen, W. *et al, J. Am. Chem. Soc.*, 1988, **110**, 7903 (*synth, pmr, ms, props*)

Bifurcarenone   B-80093

Updated Entry replacing B-60105

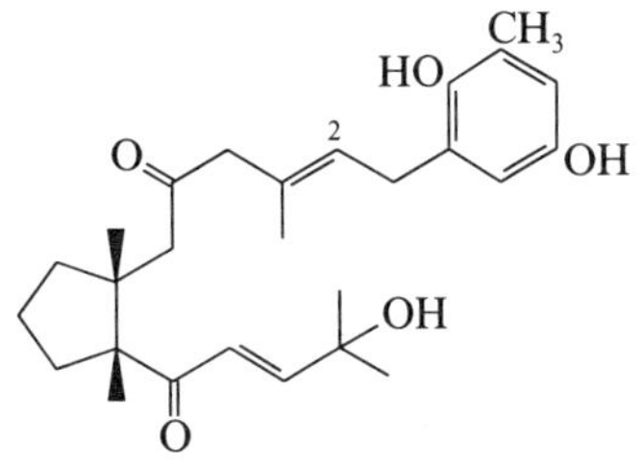

$C_{27}H_{38}O_5$ M 442.594
(2E)-form
   Constit. of *Cystoseira stricta*. Oil. [α]$_D^{20}$ +5.5° (c, 8 in EtOH).
   2'-Me ether:
     $C_{28}H_{40}O_5$ M 456.621
     Constit. of *C. stricta*. Oil. [α]$_D^{20}$ +5.3° (c, 1.15 in EtOH).
(2Z)-form [75872-68-3]
   Constit. of *Bifurcaria galapagensis*. Antibacterial. Oil.
   [α]$_D$ −5.7° (c 0.34 in CHCl$_3$).

Sun, H.H. *et al, Tetrahedron Lett.*, 1980, 3123 (*isol, struct, spectra*)
Amico, V. *et al, Phytochemistry*, 1987, **26**, 1715 (*isol*)
Mori, K. *et al, Tetrahedron*, 1989, **45**, 1945 (*synth*)

2,2′-Bifurylidene-5,5′-dione   B-80094

5-[5-Oxo-2(5H)-furanylidene]-2(5H)-furanone, 9CI

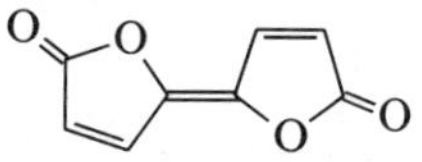

(E)-form

C$_8$H$_4$O$_4$ M 164.117

(E)-form [15061-83-3]
V. pale yellow cryst. (AcOH). Mp 237° dec.

(Z)-form [15061-82-2]
V. pale yellow cryst. Mp 248° dec.

Sauer, J.C. et al, J. Am. Chem. Soc., 1959, 81, 3677 (synth, uv, ir, cryst struct)

1,1′-Biindane-2,2′-dione   B-80095

1,1′,3,3′-Tetrahydro-[1,1′-bi-2H-indene]-2,2′-dione, 9CI

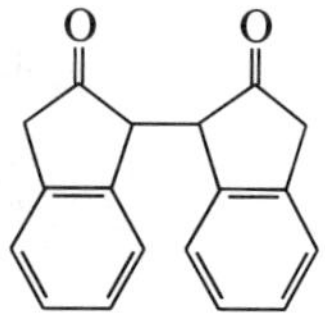

C$_{18}$H$_{14}$O$_2$ M 262.307
Mp 168° dec. Mixt. of meso- and (±)- stereoisomers.
[115679-36-2, 115679-45-3]

Baierweck, P. et al, Chem. Ber., 1988, 121, 2195 (synth, pmr, cmr)

2,2′-Biindane-1,1′-dione   B-80096

2,2′,3,3′-Tetrahydro-[2,2′-bi-1H-indene]-1,1′-dione, 9CI

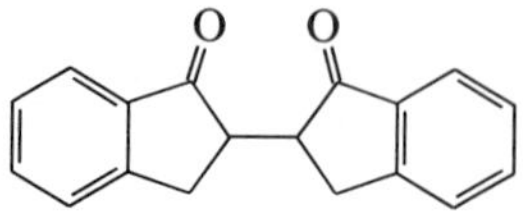

C$_{18}$H$_{14}$O$_2$ M 262.307
Mp 140°. Mixt. of meso- and (±)- isomers.
[115679-35-1, 115679-40-8]

Baierweck, P. et al, Chem. Ber., 1988, 121, 2195 (synth, pmr, cmr)

2,2′-Biindanylidene-1,1′-dione   B-80097

2,3-Dihydro-2-(1,3-dihydro-1-oxo-2H-inden-2-ylidene)-1H-inden-1-one, 9CI

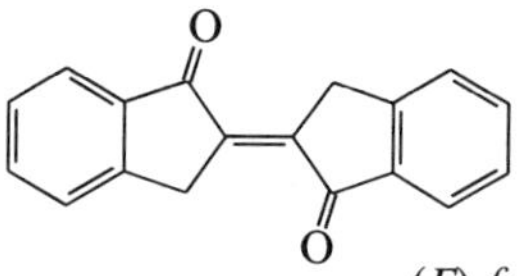

(E)-form

C$_{18}$H$_{12}$O$_2$ M 260.292

(E)-form [115679-41-9]
Mp 225° dec.

(Z)-form [115679-38-4]
Cryst. (MeOH). Mp 200° dec.

Baierweck, P. et al, Chem. Ber., 1988, 121, 2195 (synth, pmr, cmr)

1,1′-Bi-1H-indene, 9CI   B-80098

1,1′-Biindenyl
[2177-49-3]

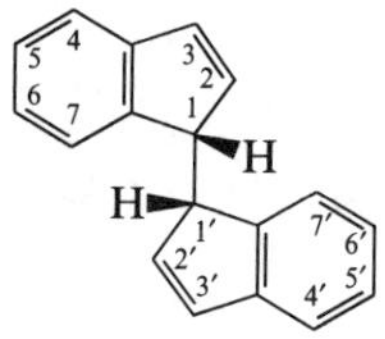

(1RS,1′RS)-form

C$_{18}$H$_{14}$ M 230.309

(1RS,1′RS)-form [81523-14-0]
(±)-form
Yellow needles (EtOH). Mp 99-100°.

(1RS, 1′SR)-form [74339-76-7]
meso-form
Leaflets (Me$_2$CO). Mp 77.5°.

Lustenberger, P. et al, Z. Kristallogr., Kristallogeom., Kristallphys., Kristallchem., 1979, 150, 235 (synth, cryst struct)
Heimer, N.E. et al, J. Org. Chem., 1982, 47, 2593 (synth, ir, pmr, cmr, abs config)
Escher, A. et al, Helv. Chim. Acta, 1986, 69, 1644 (synth, uv, pmr, ms)
Nicolet, P. et al, Synthesis, 1987, 202 (synth, pmr)

2,2′-Bi-1H-indene   B-80099

2,2′-Biindenyl

C$_{18}$H$_{14}$ M 230.309
Leaflets (EtOH). Mp 252°.

Schroth, W. et al, Z. Chem., 1963, 3, 309 (synth)
Krauch, C.H. et al, Chem. Ber., 1965, 98, 2762 (synth)
Baierweck, P. et al, Chem. Ber., 1988, 121, 2195 (synth, pmr, cmr)

2,3′-Bi-1H-indene   B-80100

2,3′-Biindenyl
[17252-90-3]

C$_{18}$H$_{14}$ M 230.309
Prisms (C$_6$H$_6$/pet.ether). Mp 220-224°.

Kemp, W. et al, J. Chem. Soc. C, 1967, 2544 (synth, uv, ir)

3,3′-Bi-1H-indene, 9CI   B-80101

3,3′-Biindenyl
[7530-35-0]

C$_{18}$H$_{14}$ M 230.309
Cryst. (Ac$_2$O). Mp 131°.

Ginsburg, D. et al, J. Chem. Soc., 1961, 1498 (synth, uv)
Majerus, G. et al, Bull. Soc. Chim. Fr., 1967, 4147 (synth, uv, ir)
Höhn, J. et al, Chem. Ber., 1983, 116, 798 (pmr, cmr)
Nicolet, P. et al, Synthesis, 1987, 202 (synth)

2,2′-Bi-1H-indole, 9CI   B-80102

Updated Entry replacing B-01208
2,2′-Biindolyl
[40899-99-8]

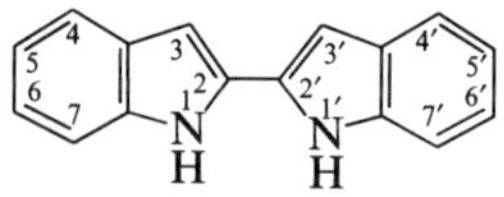

C$_{16}$H$_{12}$N$_2$ M 232.284
Cryst. (EtOAc). Mp 308-310°.

Faseeh, S.A. *et al, J. Chem. Soc.*, 1957, 4141 (*synth*)
Bergman, J. *et al, Tetrahedron*, 1979, **36**, 1439 (*ir, ms*)
Capuano, L. *et al, Chem. Ber.*, 1988, **121**, 2259 (*synth, pmr*)

4,4'-Bi(1,2-naphthaquinone)   B-80103

[1,1'-Binaphthalene]-3,3',4,4'-tetrone, 9CI
[64517-67-5]

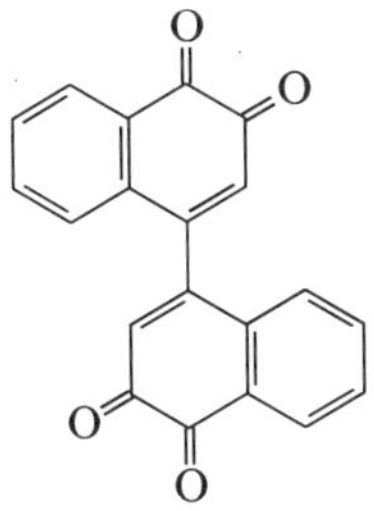

$C_{20}H_{10}O_4$ M 314.297
Yellow to orange-red cryst. Mp 289°.

Rosenhauer, E. *et al, Ber.*, 1937, **70**, 2281 (*synth*)
Paraskevas, S.M. *et al, Synthesis*, 1988, 897 (*synth, ir, pmr*)

2,2'-Bi(1,4-naphthoquinone)   B-80104

[2,2'-Binaphthalene]-1,1',4,4'-tetrone
[3408-13-7]

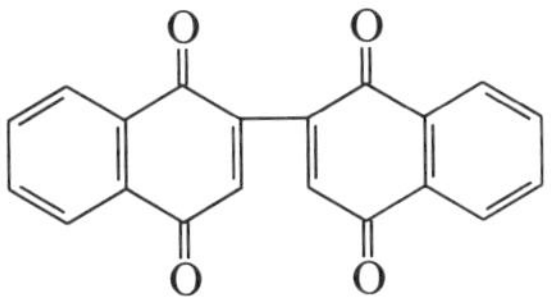

$C_{20}H_{10}O_4$ M 314.297
Yellow needles. Mp 273° dec.(darkens >270°).

Rosenhauer, E. *et al, Ber.*, 1937, **70**, 2281 (*synth*)
Paraskevas, S.M. *et al, Synthesis*, 1988, 897 (*synth, pmr, ir*)

Biperezone   B-80105

[119318-13-7]

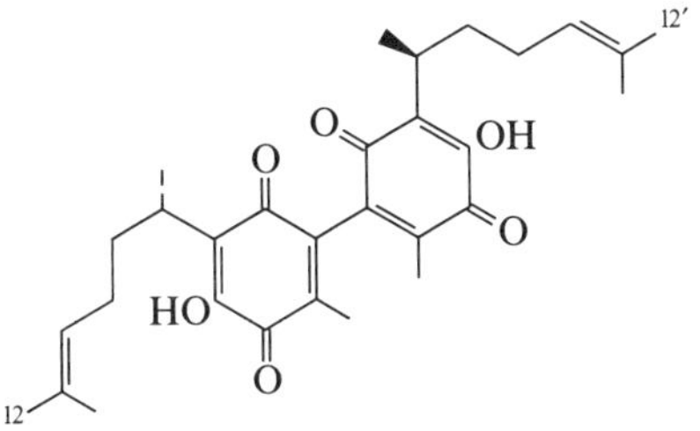

$C_{30}H_{38}O_6$ M 494.627
Constit. of *Coreocarpus arizonicus*.

12,12'-Bis(2-methylbutanoyloxy): [119285-24-4].
    $C_{40}H_{54}O_{10}$ M 694.861
    Constit. of *C. arizonicus*.
12,12'-Bis(3-methylbutanoyloxy): [119318-14-8].
    $C_{40}H_{54}O_{10}$ M 694.861
    Constit. of *C. arizonicus*.

Jolad, S.D. *et al, Phytochemistry*, 1988, **27**, 3545.

3,4'-Biphenyldiol   B-80106

Updated Entry replacing B-01264
3,4'-Dihydroxybiphenyl
[18855-13-5]
$C_{12}H_{10}O_2$ M 186.210

Mp 197-198°.

Di-Me ether: 3,4'-Dimethoxybiphenyl
    $C_{14}H_{14}O_2$ M 214.263
    Mp 60-61°.

Ito, R. *et al, Bull. Chem. Soc. Jpn.*, 1963, **36**, 985 (*synth*)
Itoh, Y. *et al, Helv. Chim. Acta*, 1988, **71**, 1199 (*synth, deriv, ir, pmr*)

1-Biphenylenecarboxaldehyde   B-80107

1-Formylbiphenylene
[1079-55-6]

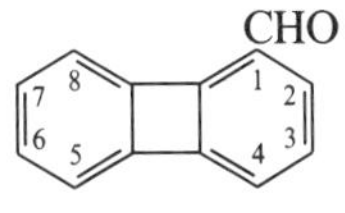

$C_{13}H_8O$ M 180.206
Pale yellow cryst. Mp 44°.

2,4-Dinitrophenylhydrazone: [1103-51-1].
    Orange cryst. Mp 287° dec.

Boulton, A.J. *et al, J. Chem. Soc. C*, 1968, 328 (*synth, ir, pmr*)

2-Biphenylenecarboxaldehyde   B-80108

2-Formylbiphenylene
[65350-41-6]
$C_{13}H_8O$ M 180.206
Yellow cryst. (pet. ether). Mp 80-81°.

2,4-Dinitrophenylhydrazone: Deep red cryst. Mp 312-314° dec.

McOmie, J.F.W. *et al, J. Chem. Soc.*, 1962, 5298 (*synth, uv*)
Buckland, P.R. *et al, Tetrahedron*, 1977, **33**, 1797 (*synth*)
Kurosawa, K. *et al, Bull. Chem. Soc. Jpn.*, 1981, **54**, 3877 (*ir, pmr*)

1-Biphenylenecarboxylic acid   B-80109

[17936-04-8]

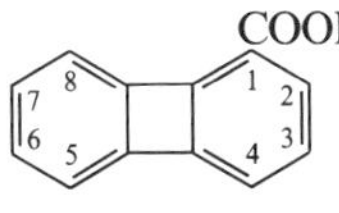

$C_{13}H_8O_2$ M 196.205
Pale yellow plates (H_2O). Mp 224-225°.

Me ester: [1210-95-3].
    $C_{14}H_{10}O_2$ M 210.232
    Yellow needles. Mp 81°.

Nitrile: [35502-53-5]. *1-Cyanobiphenylene*
    $C_{13}H_7N$ M 177.205
    Yellow needles (hexane). Mp 59-59.5°.

Boulton, A.J. *et al, J. Chem. Soc. C*, 1968, 328 (*synth, deriv, uv*)
Barton, J.W. *et al, J. Chem. Soc., Perkin Trans. 1*, 1972, 717 (*synth, deriv, ir*)

2-Biphenylenecarboxylic acid   B-80110

[93103-69-6]
$C_{13}H_8O_2$ M 196.205
Yellow prisms (MeOH aq.). Mp 223-224°.

Me ester: [782-20-7].
    $C_{14}H_{10}O_2$ M 210.232
    Pale yellow plates (MeOH aq.). Mp 114-115°.

Chloride: [75292-39-6].
    $C_{13}H_7ClO$ M 214.650
    Yellow cryst. (C_6H_6). Mp 95.5-96.5°.

Nitrile: 2-Cyanobiphenylene
    $C_{13}H_7N$ M 177.205
    Yellow plates (pet. ether). Mp 99-100°.

Baker, W. *et al, J. Chem. Soc.*, 1954, 1476; 1958, 2666 (*synth, deriv, uv*)
Martin, R.H. *et al, Tetrahedron*, 1964, **20**, 2373 (*deriv, pmr*)
Droske, J.P. *et al, Macromolecules*, 1984, **17**, 10 (*synth, deriv*)

1,5-Biphenylenedicarboxylic acid   B-80111

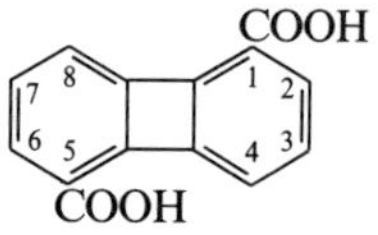

$C_{14}H_8O_4$ M 240.215
Polymer intermediate.
Di-Me ester: [20275-25-6].
   $C_{16}H_{12}O_4$ M 268.268
   Light yellow needles (Me$_2$CO). Mp 187-188°.
Dichloride: [81505-33-1].
   $C_{14}H_6Cl_2O_2$ M 277.106
   Yellow needles (C$_6$H$_6$/hexane). Mp 209-210°.

Newman, M.S. *et al, J. Org. Chem.*, 1971, 1398 (*synth, deriv, uv, ir, pmr*)
Sutter, A. *et al, J. Polym. Sci., Polym. Chem. Ed.*, 1982, **20**, 609 (*synth, deriv, ir, pmr*)

1,8-Biphenylenedicarboxylic acid   B-80112

[119414-44-7]
$C_{14}H_8O_4$ M 240.215
Polymer intermediate. Yellow cryst.
Di-Me ester: [20275-26-7].
   $C_{16}H_{12}O_4$ M 268.268
   Yellow needles (EtOH). Mp 147-148°.
Dichloride: [81505-34-2].
   $C_{14}H_6Cl_2O_2$ M 277.106
   Yellow plates (hexane). Mp 139-140°.

Newman, M.S. *et al, J. Org. Chem.*, 1971, **36**, 1398 (*synth, deriv, uv, ir, pmr*)
Sutter, A. *et al, J. Polym. Sci., Polym. Chem. Ed.*, 1982, **20**, 609 (*synth, deriv, ir, pmr*)
Wilcox, C.F. *et al, J. Org. Chem.*, 1989, **54**, 2190 (*synth*)

2,3-Biphenylenedicarboxylic acid   B-80113

$C_{14}H_8O_4$ M 240.215
Di-Me ester: [84509-69-3].
   $C_{16}H_{12}O_4$ M 268.268
   Yellow cryst. (cyclohexane). Mp 80-81°.

Berris, B.C. *et al, J. Am. Chem. Soc.*, 1985, **107**, 5670 (*synth, uv, ir, pmr, ms*)

2,6-Biphenylenedicarboxylic acid   B-80114

[65330-85-0]
$C_{14}H_8O_4$ M 240.215
Polymer intermediate. Mp 350° dec.
Di-Me ester: [67278-30-2].
   $C_{16}H_{12}O_4$ M 268.268
   Cryst. (C$_6$H$_6$). Mp 223-225°.
Dichloride: [65330-84-9].
   $C_{14}H_6Cl_2O_2$ M 277.106
   Cryst. (C$_6$H$_6$). Mp 242-244°.

Swedo, R.J. *et al, J. Polym. Sci., Polym. Lett. Ed.*, 1977, **15**, 683 (*synth*)
Recca, A. *et al, Macromolecules*, 1978, **11**, 479 (*synth, ir*)
Sutter, A. *et al, J. Polym. Sci., Polym. Chem. Ed.*, 1982, **20**, 609 (*synth*)
Sato, M. *et al, Tetrahedron Lett.*, 1984, **25**, 3603 (*synth*)

2,7-Biphenylenedicarboxylic acid   B-80115

[69417-80-7]
$C_{14}H_8O_4$ M 240.215
Polymer intermediate. Yellow solid. Mp 354-360°.
Di-Me ester: [69417-79-4].
   $C_{16}H_{12}O_4$ M 268.268
   Yellow cryst. (Me$_2$CO). Mp 190-193°.
Dichloride: [69417-81-8].
   $C_{14}H_6Cl_2O_2$ M 277.106
   Yellow needles (C$_6$H$_6$/hexane). Mp 197-198°.

Hanson, M.P. *et al, J. Polym. Sci., Polym. Lett. Ed.*, 1978, **16**, 653 (*synth, uv*)
Sutter, A. *et al, J. Polym. Sci., Polym. Chem. Ed.*, 1982, **20**, 609 (*synth, ir, pmr*)

1,2-Biphenylenediol, 9CI   B-80116

1,2-Dihydroxybiphenylene

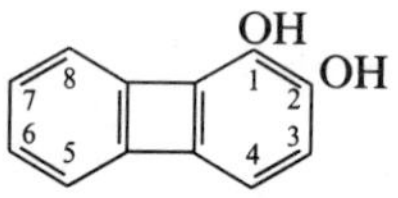

$C_{12}H_8O_2$ M 184.194
O^2-*Me:* [63723-84-2]. *2-Methoxy-1-biphenylenol. 1-Hydroxy-2-methoxybiphenylene*
   $C_{13}H_{10}O_2$ M 198.221
   Yellow needles. Mp 94-95°.
Di-Me ether: [63723-85-3]. *1,2-Dimethoxybiphenylene*
   $C_{14}H_{12}O_2$ M 212.248
   Yellowish leaflets (EtOH). Mp 132.5-133.5°.

Sato, M. *et al, Bull. Chem. Soc. Jpn.*, 1980, **53**, 2334 (*synth, ir, ms, pmr*)

1,5-Biphenylenediol, 9CI   B-80117

1,5-Dihydroxybiphenylene
[98991-02-7]
$C_{12}H_8O_2$ M 184.194
Cryst. (CHCl$_3$). Mp 266-268°.
Di-Me ether: [98991-11-8]. *1,5-Dimethoxybiphenylene*
   $C_{14}H_{12}O_2$ M 212.248
   Pale yellow cryst. (CH$_2$Cl$_2$/hexane). Mp 153-155°.
[98991-11-8]

Hine, J. *et al, J. Org. Chem.*, 1985, **50**, 5092 (*synth, pmr*)

1,8-Biphenylenediol   B-80118

1,8-Dihydroxybiphenylene
[18798-64-6]
$C_{12}H_8O_2$ M 184.194
Straw-coloured prisms (C$_6$H$_6$). Mp 223-224°.
Di-Ac: [18798-65-7].
   $C_{16}H_{12}O_4$ M 268.268
   Needles (hexane). Mp 136-137°.

Blatchly, J.M. *et al, J. Chem. Soc. C*, 1968, 1545 (*synth, uv, pmr*)
Hine, J. *et al, J. Org. Chem.*, 1986, **51**, 577; 1987, **52**, 2083 (*props, uv*)

2,3-Biphenylenediol   B-80119

2,3-Dihydroxybiphenylene
$C_{12}H_8O_2$ M 184.194
Di-Me ether: *2,3-Dimethoxybiphenylene*
   $C_{14}H_{12}O_2$ M 212.248
   Yellow needles by subl. Mp 87-88°. Bp$_{12}$ 85° subl.

Blatchly, J.M. *et al, J. Chem. Soc.*, 1962, 5085, 5090 (*synth, uv*)

2,6-Biphenylenediol   B-80120
2,6-Dihydroxybiphenylene
$C_{12}H_8O_2$ M 184.194
Di-Ac: [13625-15-5].
      $C_{16}H_{12}O_4$ M 268.268
      Faint yellow plates (AcOH), cryst. (Me₂CO). Mp 220°
      dec.
Di-Me ether: [13625-14-4]. *2,6-Dimethoxybiphenylene*
      $C_{14}H_{12}O_2$ M 212.248
      Lemon yellow plates or needles (MeOH). Mp 132-133°.

Blatchly, J.M. *et al, J. Chem. Soc. C*, 1967, 272 (*synth, uv, ir*)

2,7-Biphenylenediol, 9CI   B-80121
2,7-Dihydroxybiphenylene
$C_{12}H_8O_2$ M 184.194
Di-Ac: [13625-17-7].
      $C_{16}H_{12}O_4$ M 268.268
      Needles (hexane), cryst. (EtOH aq.). Mp 141-143°.
Me ether, Ac: [18798-63-5].
      $C_{15}H_{12}O_3$ M 240.258
      Yellow needles (EtOH aq.). Mp 95.5-96°.
Di-Me ether: [18798-68-0]. *2,7-Dimethoxybiphenylene, 9CI*
      $C_{14}H_{12}O_2$ M 212.248
      Bright yellow plates (Et₂O, EtOH, or pentane). Mp
      107.5-109°. Bp₁₂ 130° subl.
Di-Me ether, picrate: Red-black cryst. (EtOH). Mp 125°.

Lothrop, W.C., *J. Am. Chem. Soc.*, 1942, **64**, 1698 (*synth, deriv*)
Baker, W. *et al, J. Chem. Soc.*, 1958, 2688; 1961, 3986 (*synth, uv,
      deriv*)
Farnum, D.G. *et al, J. Org. Chem.*, 1961, **26**, 3204 (*uv, props,
      deriv*)
Blatchly, J.M. *et al, J. Chem. Soc. C*, 1967, 272; 1968, 1545 (*synth,
      uv, pmr, ir, derivs*)

[1,1′-Biphenyl]-2-sulfonic acid   B-80122
2-Phenylbenzenesulfonic acid
[19813-86-6]

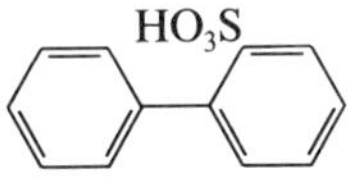

$C_{12}H_{10}O_3S$ M 234.275
Cryst. (tetrachloroethane). Mp 94°.
Me ester:
      $C_{13}H_{12}O_3S$ M 248.302
      Cryst (pet. ether). Mp 83.5-85°.
Chloride:
      $C_{12}H_9ClO_2S$ M 252.721
      Cryst. (hexane/CHCl₃). Mp 103°.

Dannley, R.L. *et al, J. Am. Chem. Soc.*, 1954, **76**, 2997 (*ester*)
Zaraiskii, A.P. *et al, J. Org. Chem. USSR (Engl. Transl.)*, 1973, **9**,
      1017 (*synth*)

[1,1′-Biphenyl]-3-sulfonic acid, 9CI   B-80123
3-Phenylbenzenesulfonic acid
[41949-30-8]
$C_{12}H_{10}O_3S$ M 234.275
Cryst. Mp 38-40°.
Me ester:
      $C_{13}H_{12}O_3S$ M 248.302
      Bp₀.₃ 150-160°.
Chloride:
      $C_{12}H_9ClO_2S$ M 252.721
      Cryst. Mp 44° (37-40°).

Amide:
      $C_{12}H_{11}NO_2S$ M 233.290
      Cryst. Mp 131° (121°).

Finzi, C. *et al, Ann. Chim. (Rome)*, 1950, **40**, 334 (*synth*)
Dannley, R.L. *et al, J. Am. Chem. Soc.*, 1954, **76**, 2997 (*ester*)
Zaraiskii, A.P. *et al, J. Org. Chem. USSR (Engl. Transl.)*, 1973, **9**,
      1017 (*synth*)

[1,1′-Biphenyl]-4-sulfonic acid, 9CI   B-80124
4-Phenylbenzenesulfonic acid
[2113-68-0]
$C_{12}H_{10}O_3S$ M 234.275
Cryst. Mp 138° (117-119°).
Me ester:
      $C_{13}H_{12}O_3S$ M 248.302
      Cryst. (pet.ether). Mp 76-77°.
Chloride: [1623-93-4].
      $C_{12}H_9ClO_2S$ M 252.721
      Mp 115°, Mp 142-143°.
Amide:
      $C_{12}H_{11}NO_2S$ M 233.290
      Cryst. Mp 228°.

Gebauer-Fülnegg, E. *et al, Monatsh. Chem.*, 1928, **49**, 46 (*chloride*)
Pollak, J. *et al, Monatsh. Chem.*, 1930, **35**, 358 (*synth*)
Dannley, R.L. *et al, J. Am. Chem. Soc.*, 1954, **76**, 2997 (*ester*)
Schultz, R.G., *J. Org. Chem.*, 1961, **26**, 5195 (*synth*)
Roemmele, R.C. *et al, J. Org. Chem.*, 1988, **53**, 2367 (*synth, pmr,
      ir*)

2,2′,3,4-Biphenyltetrol   B-80125
2,2′,3,4-Tetrahydroxybiphenyl
$C_{12}H_{10}O_4$ M 218.209
Tetra-Me ether: [119101-10-9]. *2,2′,3,4-
Tetramethoxybiphenyl*
      $C_{16}H_{18}O_4$ M 274.316
      Cryst. (diisopropyl ether). Mp 105°.

Itoh, Y. *et al, Helv. Chim. Acta*, 1988, **7**, 1199 (*synth, pmr*)

2,3,3′,4-Biphenyltetrol   B-80126
Updated Entry replacing B-01303
2,3,3′,4-Tetrahydroxybiphenyl
$C_{12}H_{10}O_4$ M 218.209
Tetra-Me ether: 2,3,3′,4-Tetramethoxybiphenyl
      $C_{16}H_{18}O_4$ M 274.316
      Cryst. (MeOH). Mp 96-100°. Also descr. as oil.

Goto, K. *et al, Bull. Chem. Soc. Jpn.*, 1941, **16**, 170 (*synth*)
Itoh, Y. *et al, Helv. Chim. Acta*, 1988, **71**, 1199 (*synth, deriv, ir,
      pmr*)

2,3,4,4′-Biphenyltetrol   B-80127
2,3,4,4′-Tetrahydroxybiphenyl
$C_{12}H_{10}O_4$ M 218.209
Tetra-Me ether: [6271-59-6]. *2,3,4,4′-Tetramethoxybiphenyl*
      $C_{16}H_{18}O_4$ M 274.316
      Cryst. (diisopropyl ether). Mp 75°.

Itoh, Y. *et al, Helv. Chim. Acta*, 1988, **71**, 1199 (*synth, pmr*)

2,3′,4′,5′-Biphenyltetrol   B-80128
Updated Entry replacing B-20101
2,3′,4′,5′-Tetrahydroxybiphenyl
$C_{12}H_{10}O_4$ M 218.209
Mp 150-151.5°.
*2,3′,5′-Tri-Me ether: 2,6-Dimethoxy-4-(2-
methoxyphenyl)phenol. 4-Hydroxy-2′,3,5-*

$C_{15}H_{16}O_4$ M 260.289
Constit. of the heartwood of *Sorbus aucuparia* and *S. decora*. Mp 120-122°.

Tetra-Me ether: 2′,3,4,5-*Tetramethoxybiphenyl*
$C_{16}H_{18}O_4$ M 274.316
Cryst. (MeOH aq.). Mp 71.5-72°.

Erdtman, H. *et al*, *Acta Chem. Scand.*, 1963, **17**, 1151 (*synth, uv*)
Nilsson, M. *et al*, *Acta Chem. Scand.*, 1963, **17**, 1157 (*Methoxyaucuparin*)

3,4,5-Biphenyltriol, 8CI   B-80129

Updated Entry replacing B-01332
5-*Phenylpyrogallol*. 3,4,5-*Trihydroxybiphenyl*
$C_{12}H_{10}O_3$ M 202.209
Cryst. (MeOH aq.). Mp 196-198°.

3,5-Di-Me ether: 3,5-*Dimethoxy-4-biphenylol. 4-Hydroxy-3,5-dimethoxybiphenyl.* **Aucuparin**
$C_{14}H_{14}O_3$ M 230.263
Isol. from trunkwood of *Sorbus aucuparia*. Cryst. Mp 101-101.5°.

Tri-Me ether: [50816-75-6]. 3,4,5-*Trimethoxybiphenyl*
$C_{15}H_{16}O_3$ M 244.290
Cryst. (MeOH). Mp 84-89°.

Narasimhachari, N. *et al*, *Can. J. Chem.*, 1962, **40**, 1118 (*isol, deriv*)
Erdtman, H. *et al*, *Acta Chem. Scand.*, 1963, **17**, 1151 (*uv, pmr*)
Nilsson, M. *et al*, *Acta Chem. Scand.*, 1963, **17**, 1157 (*synth*)

2,2′-Bi-1H-pyrrole-5-carboxaldehyde   B-80130

5-*Formyl-2,2′-bipyrrole*
[22187-87-7]

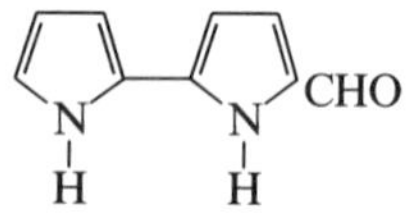

$C_9H_8N_2O$ M 160.175
Cryst. (EtOH aq.). Mp 234-240° dec.

Hearn, W.R. *et al*, *J. Org. Chem.*, 1970, **35**, 142 (*synth, uv, pmr, ms*)
Boger, D.L. *et al*, *J. Org. Chem.*, 1988, **53**, 1405 (*synth, ir, pmr, ms*)

Biquadricyclenylidene   B-80131

3-*Tetracyclo[3.2.0.0^{2,7}.0^{4,6}]hept-3-ylidenetetracyclo[3.2.0.0^{2,7}0^{4,6}]heptane, 9CI*
[53312-09-7]

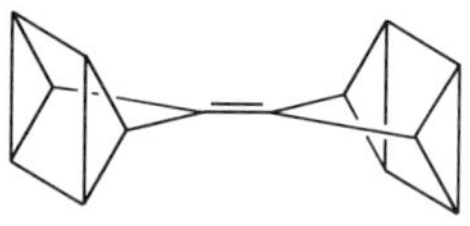

$C_{14}H_{12}$ M 180.249
Prisms. Mp 173° subl.

Sauter, H. *et al*, *Angew. Chem., Int. Ed. Engl.*, 1973, **12**, 991 (*synth, pmr*)
Knothe, L. *et al*, *Justus Liebigs Ann. Chem.*, 1977, 709 (*cmr*)
Prinzbach, H. *et al*, *Tetrahedron Lett.*, 1981, **22**, 2541 (*struct*)
Trah, S. *et al*, *Tetrahedron Lett.*, 1987, **28**, 4399, 4403 (*struct*)

1,10-Bisabolene-3,6-diol   B-80132

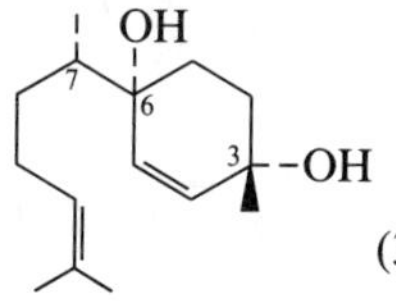

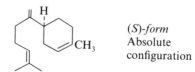

(3R,6R,7S)-*form*

$C_{15}H_{26}O_2$ M 238.369
Incorrect numbering system used in reference.
(**3R,6R,7S**)-*form* [119285-25-5]
Constit. of *Coreocarpus arizonicus*. Oil.
(**3S,4S,7S**)-*form* [119364-23-7]
Constit. of *C. arizonicus*. Oil.

Jolad, S.D. *et al*, *Phytochemistry*, 1988, **27**, 3545.

3,10-Bisaboladiene-7,14-diol   B-80133

[116537-80-5]

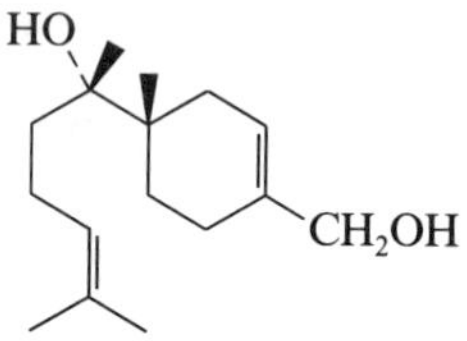

$C_{15}H_{26}O_2$ M 238.369
Constit. of *Vanillosmopsis arborea*. Oil.

Matos, M.E.O. *et al*, *J. Nat. Prod.* (*Lloydia*), 1988, **51**, 780.

β-Bisabolene   B-80134

Updated Entry replacing B-40083
1-*Methyl-4-(5-methyl-1-methylene-4-hexenyl)cyclohexene, 9CI*

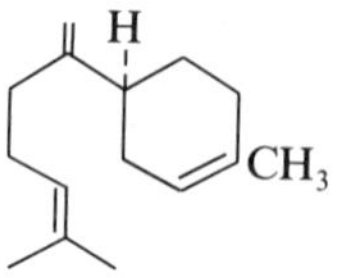

(S)-*form*
Absolute
configuration

$C_{15}H_{24}$ M 204.355
(**R**)-*form*
Isol. from *Chamaecyparis nootkatensis* and *Pinus sibirica*. $[\alpha]_D$ +75°. n_D^{20} 1.4879.
(**S**)-*form* [495-61-4]
Constit. of the essential oils of bergamot, lemon and wild carrot. Oil. $Bp_{10.5}$ 129-130°. $[\alpha]_D^{20}$ −84.4°.

Pentegova, V.A. *et al*, *Collect. Czech. Chem. Commun.*, 1961, **26**, 1362 (*isol*)
Manjarrez, A. *et al*, *J. Org. Chem.*, 1966, **31**, 348 (*synth*)
Andersen, N.H. *et al*, *Phytochemistry*, 1970, **9**, 1325 (*isol*)
Vig, O.P. *et al*, *J. Indian Chem. Soc.*, 1971, **48**, 993 (*synth*)
Crawford, R.J. *et al*, *J. Am. Chem. Soc.*, 1972, **94**, 4298 (*synth*)
Sakane, S. *et al*, *J. Am. Chem. Soc.*, 1983, **105**, 6154 (*synth*)
Sakurai, H. *et al*, *Tetrahedron*, 1983, **39**, 883 (*synth*)
Nabeta, K. *et al*, *Agric. Biol. Chem.*, 1986, **50**, 2915 (*biosynth*)

10-Bisabolene-7,15-diol   B-80135

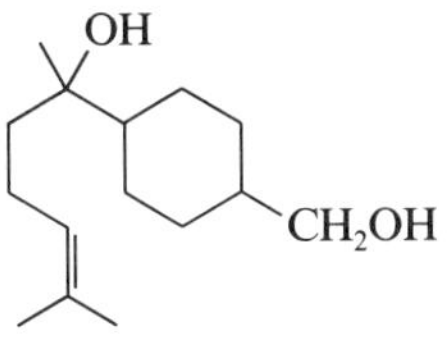

$C_{15}H_{28}O_2$ M 240.385
Constit. of *Baccharis latifolia*. Oil.

Zdero, C. *et al, Phytochemistry*, 1989, **28**, 531.

11-Bisabolene-7,10,15-triol   B-80136

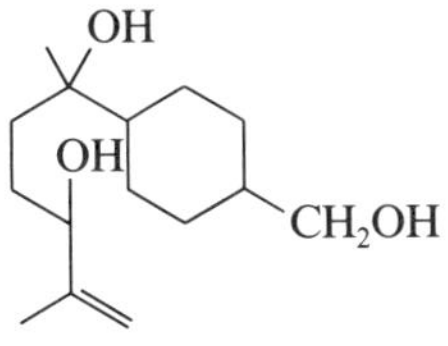

$C_{15}H_{28}O_3$ M 256.384
Constit. of *Baccharis latifolia*.

Zdero, C. *et al, Phytochemistry*, 1989, **28**, 531.

2,3-Bis(bromomethyl)thiopene   B-80137

[13250-86-7]

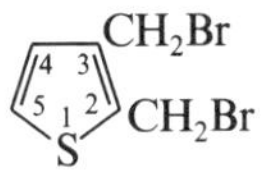

$C_6H_6Br_2S$ M 269.987
Cryst. (hexane). Mp 49-50°.

MacDowell, D.W.H. *et al, J. Org. Chem.*, 1966, **31**, 3592 (*synth, ir, uv, pmr*)
Chadwick, D.J. *et al, Tetrahedron Lett.*, 1987, **28**, 6085 (*synth*)

2,5-Bis(bromomethyl)thiophene   B-80138

[59311-25-0]
$C_6H_6Br_2S$ M 269.987
Cryst. (pet. ether). Mp 78°.

Clarke, J.A. *et al, Tetrahedron Lett.*, 1975, 4183 (*synth*)
Bilger, C. *et al, Synthesis*, 1988, 902 (*synth, ms, ir, pmr*)

3,4-Bis(bromomethyl)thiophene   B-80139

[38447-31-3]
$C_6H_6Br_2S$ M 269.987
Cryst. (cyclohexane/pet. ether). Mp 83°.

Von Reinhard, H., *J. Prakt. Chem.*, 1972, **314**, 334 (*synth, nmr*)

Bis(2-bromophenyl)ethyne   B-80140

2,2'-Dibromotolan. 2,2'-Dibromodiphenylacetylene

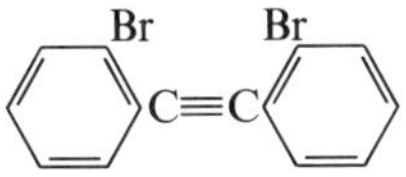

$C_{14}H_8Br_2$ M 336.025
Cryst. (EtOH). Mp 81.0-81.5°.

Letsinger, R.L., *J. Am. Chem. Soc.*, 1959, **81**, 3013 (*synth, ir*)

Bis(4-bromophenyl)ethyne   B-80141

4,4'-Dibromotolan. 4,4'-Dibromodiphenylacetylene
$C_{14}H_8Br_2$ M 336.025
Cryst. (CHCl$_3$). Mp 184.5-185°.

Misumi, S. *et al, Bull. Chem. Soc. Jpn.*, 1962, **35**, 135 (*synth*)

Bis(chlorocarbonyl) disulfide   B-80142

Dithiobis(carbonylchloride). Thioperoxydicarbonic dichloride, 9CI

[51615-88-4]

ClCO-S-S-COCl

$C_2Cl_2O_2S_2$ M 191.058
Liq. Bp$_{20}$ 84°. Stable to storage if kept dry.

Kobayashi, N. *et al, Chem. Lett.*, 1973, 1315 (*synth*)
Haas, A. *et al, J. Fluorine Chem.*, 1973, **3**, 383 (*synth, ir, ms*)
Barany, G. *et al, J. Org. Chem.*, 1983, **48**, 4750 (*synth*)

2,4-Bis(chloromethyl)pyridine   B-80143

[101822-23-5]
$C_7H_7Cl_2N$ M 176.044
Red oil.

[120492-00-4]

Kawashima, T. *et al, Chem. Lett.*, 1988, 1707.
Przybilla, K.J. *et al, Chem. Ber.*, 1989, **122**, 347 (*synth, pmr*)

1,2-Bis(cyclohexylidene)ethene   B-80144

1,1'-(1,2-Ethenediylidene)biscyclohexane, 9CI. 1,4-Bis(pentamethylene)-1,2,3-butatriene

[36204-35-0]

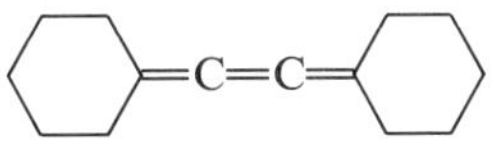

$C_{14}H_{20}$ M 188.312
Cryst. by subl. Mp 81.5-82.0°. Unstable in air, can be stored in CCl$_4$ or hexane soln. under Ar.

Iyoda, M. *et al, J. Am. Chem. Soc.*, 1988, **110**, 8494 (*synth, pmr, cmr, uv, ir*)

Biscyclolobin   B-80145

3,3'-[Oxybis(3-hydroxy-2-methoxy-4,1-phenylene)]-bis[3,4-dihydro-2H-1-benzopyran-7-ol], 9CI

[58219-01-5]

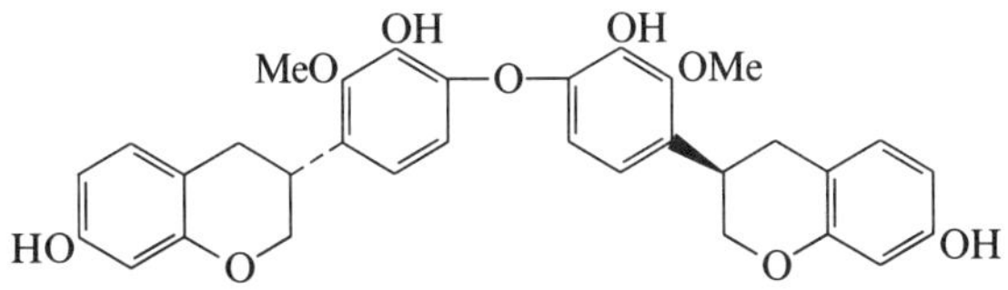

$C_{32}H_{30}O_9$ M 558.584
Isol. from heartwood of *Cyclobium clausseni*. Mp 220-223°.

Gottlieb, O.R. *et al, Phytochemistry*, 1975, **14**, 2495 (*isol, uv, ms, ord, struct*)

2,6-Bis(3,4-Dihydro-2*H*-pyrrol-5-yl) pyridine   **B-80146**

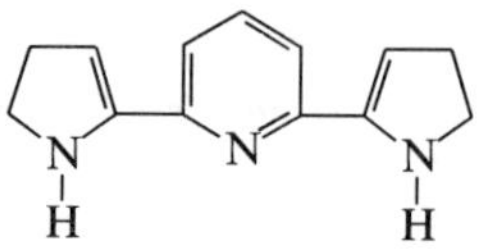

$C_{13}H_{15}N_3$ M 213.282
Cryst. (H$_2$O). Mp 132°.

Bernauer, K. *et al*, *Helv. Chim. Acta*, 1989, **72**, 477 (*synth, pmr*)

1,3-Bis(dimethylamino)pentalene   **B-80147**

N,N,N′,N′-*Tetramethyl-1,3-pentalenediamine, 8CI*
[14749-76-9]

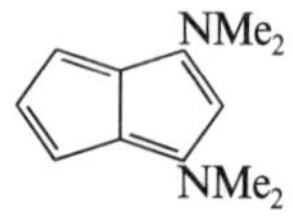

$C_{12}H_{16}N_2$ M 188.272
Simple stabilised pentalene. Dark-blue cryst. Mp 163° dec. Stable in air for several hr. at r.t.

Hafner, K. *et al*, *Angew. Chem., Int. Ed. Engl.*, 1967, **6**, 451 (*synth, uv, pmr*)
Eckert-Maksić, M. *et al*, *Chem. Ber.*, 1988, **121**, 1219 (*pe*)

Bis(2-ethylhexyl)adipate   **B-80148**

Diethylhexyl adipate. DEHA
[103-23-1]

$H_3C(CH_2)_3CHCH_2OOC(CH_2)_4COOCH_2CH(CH_2)_3CH_3$

$C_{22}H_{42}O_4$ M 370.571
Plasticizer; used as component in lube oils. Liq. d$_4^{20}$ 0.927. Bp$_2$ 181-5°. Pour point −70°, glass point −112°. n$_D^{20}$ 1.447.

▷ May be carcinogenic, mutagenic.

Klamann, D. *et al*, *Brennst.-Chem.*, 1964, **45**, 33; *CA*, **60**, 13137e (*synth*)
Cornell, R.R. *et al*, *Arch. Mass Spectral Data*, 1971, **2**, 710 (*ms*)
Takahashi, T. *et al*, *Toxicology*, 1981, **22**, 223; *CA*, **96**, 80969t (*tox*)
Ovchinnikov, E.Yu. *et al*, *Russ. J. Phys. Chem. (Engl. Transl.)*, 1985, **59**, 1693 (*props*)
Castle, L. *et al*, *J. Assoc. Off. Anal. Chem.*, 1988, **71**, 394 (*glc, ms*)

Bis(2-ethylhexyl) decanedioate   **B-80149**

Bis(2-ethylhexyl) sebacate. Di(2-ethylhexyl) sebacate
[122-62-3]

$H_3C(CH_2)_3CH(CH_2CH_3)CH_2OOC(CH_2)_8COOCH_2$
$CH(CH_2CH_3)(CH_2)_3CH_3$

$C_{26}H_{50}O_4$ M 426.679
Used in inks, lubricating oils, hydraulic fluids, plasticizers, coatings. Glc stationary phase. Liq. Sol. EtOH, C$_6$H$_6$, Me$_2$CO. d$_4^{20}$ 0.914. Mp −48°. Bp$_5$ 256°. n$_D^{25}$ 1.4496. The coml. prod. will be a mixt. of diastereoisomers expected to have v. similar props.

Durr, A.M. *et al*, *ACS Div. Petrol. Chem. Preprints*, 1963, **8**, 49 (*synth, props*)
Reith, H. *et al*, *CA*, 1963, **59**, 11159b (*props*)
Wood, G.M. *et al*, *J. Vac. Sci. Technol.*, 1969, **6**, 871; *CA*, **71**, 126859w (*ms*)
Fr. Pat., 2 151 525, (1973); *CA*, **79**, 52813t (*synth, purifn*)
Reglero, G. *et al*, *Chromatographia*, 1986, **22**, 358 (*glc*)

1,2-Bis(4-hydroxy-3-methoxyphenyl)-1,3-propanediol, 8CI   **B-80150**

1,2-Diguaiacyl-1,3-propanediol
[4206-59-1]

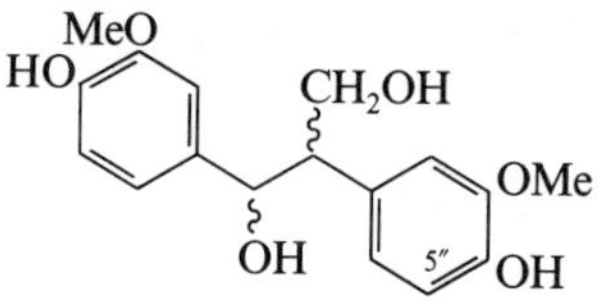

$C_{17}H_{20}O_6$ M 320.341
Obt. by mild hydrol. of wood of *Fagus sylvatica* and *Picea excelsa*. Prisms (EtOH aq.). Mp 143.5°.

5″-Methoxy: 1-(4-Hydroxy-3-methoxyphenyl)-2-(4-hydroxy-3,5-dimethoxyphenyl)-1,3-propanediol
$C_{18}H_{22}O_7$ M 350.368
Obt. by mild hydrol. of lignin of *F. sylvatica*. Needles (MeOH aq.). Mp 192-193°.

Nimz, H., *Chem. Ber.*, 1966, **99**, 469.

7,11-Bis(hydroxymethyl)-3,15-dimethyl-2,6,10,14-hexadecatetraen-1-ol   **B-80151**

18,19-Dihydroxygeranylnerol

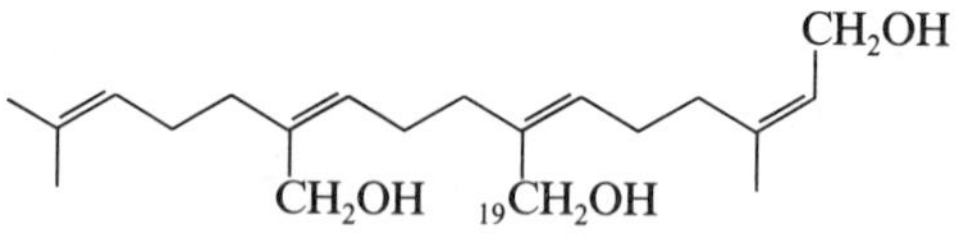

$C_{20}H_{34}O_3$ M 322.487
19-Ac: 19-Acetoxy-18-hydroxygeranylnerol
$C_{22}H_{36}O_4$ M 364.524
Constit. of *Ageratina ligustrina*. Oil.

Tamayo-Catillo, G. *et al*, *Phytochemistry*, 1988, **27**, 2893.

1,3-Bis(4-hydroxyphenyl)-2-propen-1-one   **B-80152**

4,4′-Dihydroxychalcone
[108997-30-4]

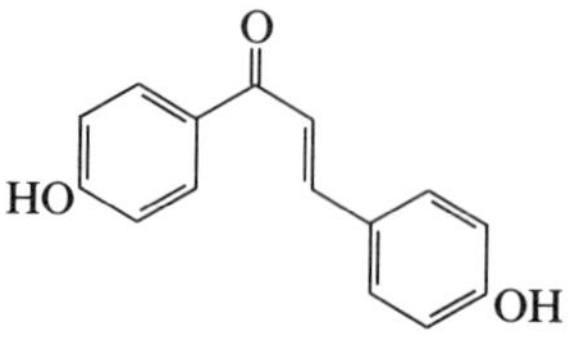

$C_{15}H_{12}O_3$ M 240.258
Constit. of *Chamaecyparis obtusa*. Yellow needles (MeOH aq.). Mp 207-208°.

Klinke, P. *et al*, *Chem. Ber.*, 1961, **94**, 26 (*synth*)
Ohashi, H. *et al*, *Phytochemistry*, 1988, **27**, 3993 (*isol*)

2,5-Bis(iodomethyl)thiophene   **B-80153**

[120651-39-0]

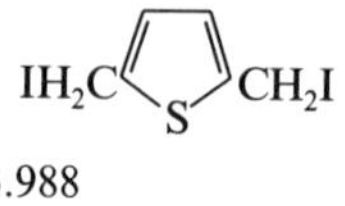

$C_6H_6I_2S$ M 363.988
Solid. Unstable, must be used immediately.

Bilger, C. *et al*, *Synthesis*, 1988, 902 (*synth*)

1,3-Bis[(2-methoxyethoxy)methoxy]-1,3-diphenyl-2-propanone   B-80154

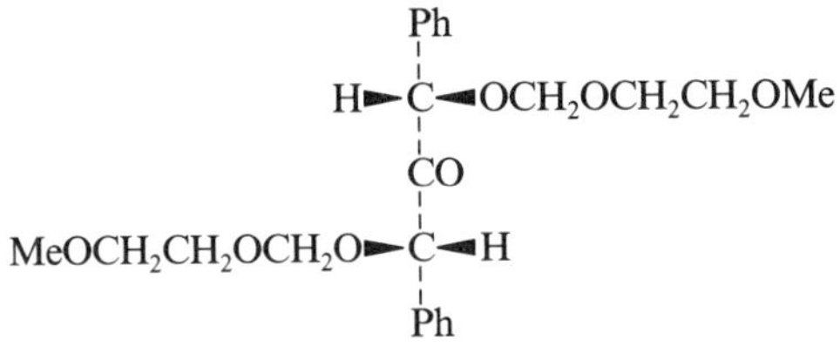

$$Ph$$
$$H\blacktriangleright \overset{|}{C}\blacktriangleleft OCH_2OCH_2CH_2OMe$$
$$\overset{|}{CO}$$
$$MeOCH_2CH_2OCH_2O\blacktriangleright \overset{|}{\underset{|}{C}}\blacktriangleleft H$$
$$Ph$$

$C_{23}H_{30}O_7$ M 418.486

Readily available C_2-symmetric ketone for asymmetric synthesis.

(S,S)-form [119945-33-4]
$[\alpha]_D^{25}$ +212.4° (c, 0.58 in 95% EtOH aq.). *meso*-form also obt.
[119909-46-5]

Braun, M. *et al, Chem. Ber.*, 1989, **122**, 1215 (*synth, ir, pmr*)

2,4-Bis(3-methyl-2-butenyl)phenol   B-80155

2,4-Diprenylphenol
[55824-31-2]

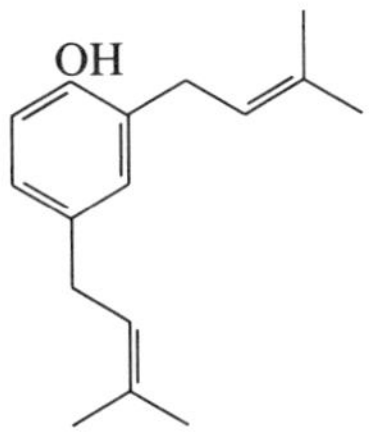

$C_{16}H_{22}O$ M 230.349

Constit. of *Perithalia caudata*. Oil.

Kyogoku, K. *et al, Agric. Biol. Chem.*, 1975, **39**, 133 (*synth*)
Yamada, S. *et al, Synth. Commun.*, 1975, **5**, 181 (*synth*)
Blackmann, A.J. *et al, Phytochemistry*, 1988, **27**, 3686 (*isol*)

3,3-Bis(methylthio)propenal   B-80156

[78263-38-4]

$$(MeS)_2C{=}CHCHO$$

$C_5H_8OS_2$ M 148.250

Dieter, R.K. *et al, J. Org. Chem.*, 1988, **53**, 2031 (*synth, ir, pmr, cmr*)

Bisparthenolidine   B-80157

[112078-76-9]

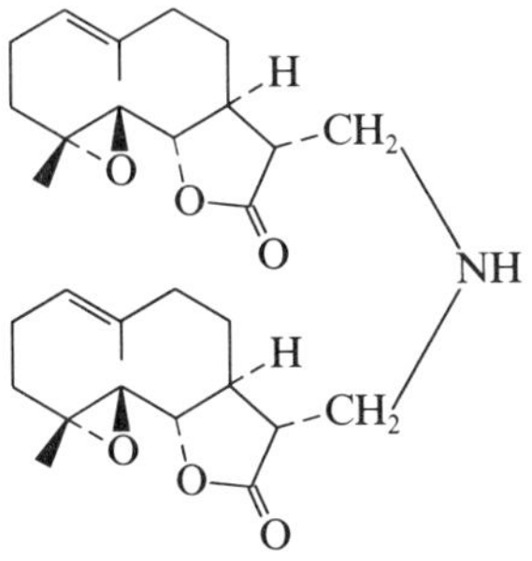

$C_{30}H_{43}NO_6$ M 513.673

Constit. of *Paramichelia baillonii* and *Michelia rajaniana* (Magnoliaceae). Cryst. (CHCl$_3$). Mp 100-103°. $[\alpha]_D^{20}$ –112° (CHCl$_3$).

Ruangrungsi, N. *et al, J. Nat. Prod. (Lloydia)*, 1987, **50**, 891; 1988, **51**, 1230 (*isol, pmr, cmr*)

Bis(pentafluorophenyl)methane   B-80158

[5736-46-9]

$$(C_6F_5)_2CH_2$$

$C_{13}H_2F_{10}$ M 348.143
Cryst. by subl. Mp 63-64°.

Beckert, W.F. *et al, J. Org. Chem.*, 1967, **32**, 582 (*synth, ir, F nmr*)
Filler, R. *et al, J. Chem. Soc., Chem. Commun.*, 1968, 287 (*props*)
Bordwell, F.G. *et al, J. Org. Chem.*, 1988, **53**, 780 (*synth, pmr*)

2,3-Bis(phenylseleno)-2-butenedioic acid   B-80159

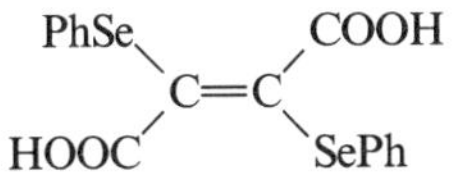

$$PhSe \qquad COOH$$
$$\overset{}{C}{=}\overset{}{C} \qquad (E)\text{-form}$$
$$HOOC \qquad SePh$$

$C_{16}H_{12}O_4Se_2$ M 426.188

(E)-form [114221-67-9]
Pale yellow cryst. (Me$_2$CO/hexane). Mp 240-246°.
Di-Me ester: [114221-61-3].
$C_{18}H_{16}O_4Se_2$ M 454.242
Pale yellow solid (CHCl$_3$/hexane). Mp 109.5-110°.

(Z)-form [114221-68-0]
Bright yellow solid. Mp 169-172°.
Di-Me ester: [114221-62-4].
Bright yellow solid (CHCl$_3$/hexane). Mp 117-118.5°.

Anhydride: [114221-69-1].
$C_{16}H_{10}O_3Se_2$ M 408.173
Bright yellow gum.

Back, T.G. *et al, J. Org. Chem.*, 1988, **53**, 2533 (*synth, ir, pmr, ms*)

2,6-Bis(2-pyrrolidinyl)pyridine   B-80160

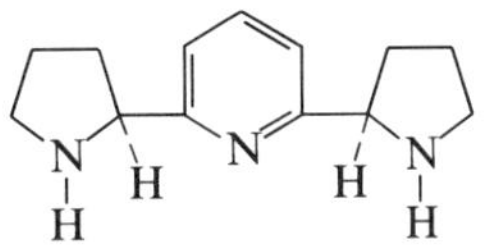

(R*,R*)-form

$C_{13}H_{19}N_3$ M 217.313

Tridentate ligand. Bp$_{0.005}$ 96°. Bp refers to a mixt. of ($\pm$)- and *meso*-isomers, not separately characterised. The ($\pm$)-form was resolved.

(R*,R*)-form
(+)-*form*
$[\alpha]_D$ +126° (c, 0.1 in H$_2$O).

(S*,S*)-form
(–)-*form*
$[\alpha]_D$ –113°.

Bernauer, K. *et al, Helv. Chim. Acta*, 1989, **72**, 477 (*synth, uv, pmr, resoln*)

[Bis(trifluoroacetoxy)iodo]pentafluorobenzene   B-80161

(Pentafluorophenyl)bis(trifluoroacetato-O)-iodine, 9CI
[14353-88-9]

$$C_6F_5I(OOCCF_3)_2$$

$C_{10}F_{11}IO_4$ M 519.994
Reagent for mild oxidative cleavage of alkynes and carbonyl compds. Cryst. by subl. Mp 119-120°.

Schmeisser, M. *et al, Chem. Ber.*, 1967, **100**, 1633 (*synth*)
Moriarty, R.M. *et al, J. Chem. Soc., Chem. Commun.*, 1987, 202 (*use*)
Moriarty, R.M. *et al, J. Org. Chem.*, 1988, **53**, 6124 (*use*)

Bis(trifluoromethyl)sulfene   B-80162

$(F_3C)_2C{=}SO_2$

$C_3F_6O_2S$ M 214.088
Quinuclidine adduct: [118724-94-0].
Cryst. Mp 181° dec.

Hartwig, U. *et al, Angew. Chem., Int. Ed. Engl.*, 1989, **28**, 221 (*synth, cryst struct*)

Bis[(trifluoromethyl)thio]acetylene, 8CI   B-80163

[2069-87-6]

$F_3CSC{\equiv}CSCF_3$

$C_4F_6S_2$ M 226.166
Liq. Bp 83°. n_D^{25} 1.3924. Slowly darkens on standing.

Harris, J.F., *J. Org. Chem.*, 1967, **32**, 2063 (*synth, ir*)
Haas, A. *et al, Chem. Ber.*, 1988, **121**, 1833 (*synth, cmr*)

1,2-Bis[(trifluoromethyl)thio]ethene, 9CI   B-80164

$F_3CSCH{=}CHSCF_3$

$C_4H_2F_6S_2$ M 228.182
Liq. Bp 92-98°. Geom. isomers separable by glc.
[13003-45-7, 13003-46-8]

Harris, J.F., *J. Org. Chem.*, 1967, **32**, 2063 (*synth, ir, uv*)

1,4-Bis(trifluorovinyl)benzene   B-80165

1,4-Bis(trifluoroethenyl)benzene
[113268-53-4]

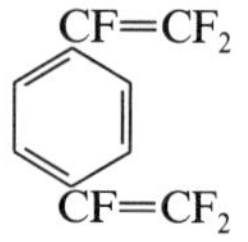

$C_{10}H_4F_6$ M 238.132
Liq. Pure product readily dimerises.

Heinze, P.L. *et al, J. Org. Chem.*, 1988, **53**, 2714 (*synth, F nmr, pmr, ir, ms*)

4-[2,2′-Bithiophen]-5-yl-3-butyne-1,2-diol, 9CI   B-80166

5-(3,4-Dihydroxy-1-butynyl)-2,2′-bithienyl
[1211-45-6]

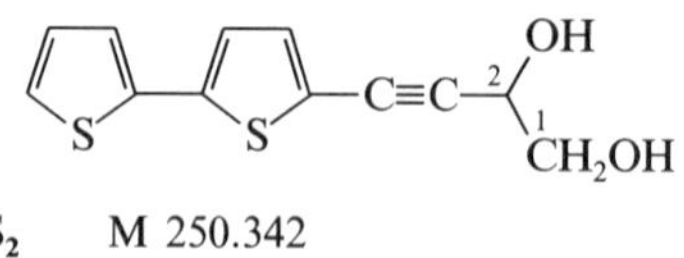

$C_{12}H_{10}O_2S_2$ M 250.342
Isol. from roots of *Echinops sphaerocephalus*. Cryst. (CCl₄/Et₂O). Mp 102°. $[\alpha]_{546}^{25}$ +12° (c, 2.55 in Et₂O).
O^1-Ac: *5-(3-Hydroxy-4-acetoxy-1-butynyl)-2,2′-bithienyl*
$C_{14}H_{12}O_3S_2$ M 292.379
Isol. from *E. sphaerocephalus*. Oil. λ_{max} 325,332 nm.
O^2-Ac: [95910-62-6]. *5-(4-Hydroxy-3-acetoxy-1-butynyl)-*

2,2′-bithienyl. α-*Tertiophene*
$C_{14}H_{12}O_3S_2$ M 292.379
Isol. from *E. sphaerocephalus*.
Di-Ac: [1233-95-0]. *5-(3,4-Diacetoxy-1-butynyl)-2,2′-bithienyl*
$C_{16}H_{14}O_4S_2$ M 334.416
Isol. from *E. sphaerocephalus* and *Tagetes patula*. Oil. $[\alpha]_{546}^{25}$ +183° (c, 0.89 in Et₂O).

Bohlmann, F. *et al, Chem. Ber.*, 1965, **98**, 155, 876 (*isol, uv, ir, pmr, ord, struct, biosynth*)
Pensl, R. *et al, Z. Naturforsch., C*, 1985, **40**, 3 (*isol, biosynth*)

4-[2,2′-Bithiophen-5-yl]-3-butyn-1-ol, 9CI   B-80167

Updated Entry replacing H-01421
4-[5-(2-Thienyl)-2-thienyl]-3-butyn-1-ol, 8CI. 5-(4-Hydroxy-1-butynyl)-2,2′-bithiophene
[1137-87-7]

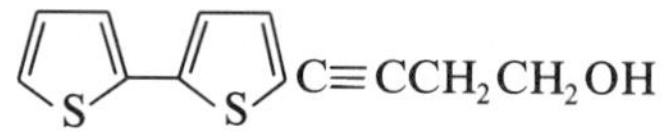

$C_{12}H_{10}OS_2$ M 234.342
Constit. of the roots of *Tagetes minuta, Echinops sphaerocephalus, Cullumia* and *Berkheya* spp. Cryst. (Et₂O/pet. ether). Mp 67° (59°).
Ac:
$C_{14}H_{12}O_2S$ M 244.314
Constit. of *T. patula, T. erecta, T. minuta, E. sphaerocephalus, C.* and *B.* spp. etc. Oil. Bp₀.₀₀₁ 115°.

Bohlmann, F. *et al, Chem. Ber.*, 1962, **95**, 2945; 1964, **97**, 2125; 1965, **98**, 155, 876; 1975, **108**, 515 (*isol, uv, ir, pmr, struct, biosynth*)
Atkinson, R.E. *et al, J. Chem. Soc.*, 1965, 7109 (*isol*)
Bohlmann, F. *et al, Phytochemistry*, 1976, **15**, 1309 (*isol*)

4,4′-Bi-1H-1,2,3-triazole   B-80168

[119720-89-7]

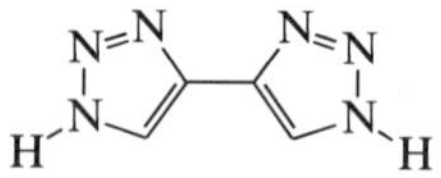

$C_4H_4N_6$ M 136.116
Cryst. (MeOH). Mp 280° dec.

Banert, K., *Chem. Ber.*, 1989, **122**, 1175 (*synth, ir, pmr, cmr*)

9,9′-Bitriptycyl   B-80169

9,9′(10H,10H′)-Bi-9,10[1′,2′]benzenoanthracene, 9CI. 9,9′-Bitryptycene
[7213-66-3]

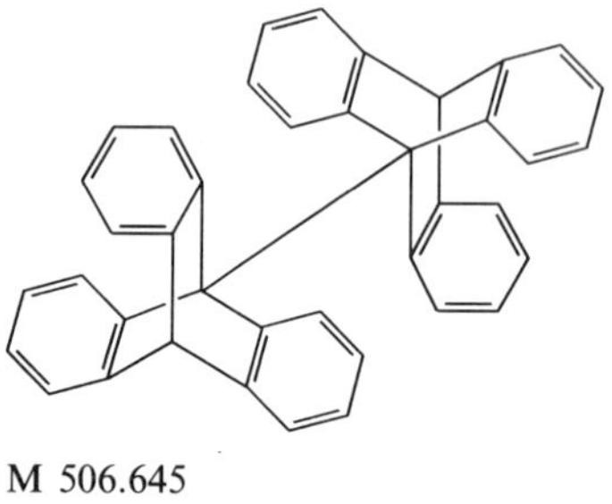

$C_{40}H_{26}$ M 506.645
Cryst. (PhNO₂). Mp 577°.

Bartlett, P.D. *et al, J. Am. Chem. Soc.*, 1954, **76**, 1088 (*synth, ir*)
Koukotas, C. *et al, J. Org. Chem.*, 1966, **31**, 1970 (*synth, ir, uv, ms*)
Ardebili, M.H.P. *et al, J. Am. Chem. Soc.*, 1978, **100**, 7994 (*cryst struct*)

Blastidic acid
B-80170

$$H_2N\text{---}\underset{\underset{Me}{|}}{\overset{\overset{COOH}{|}}{C}}\text{◄}H$$

CH_2CH_2NCH_2

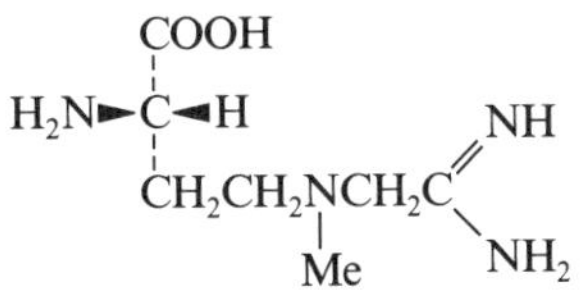

$C_7H_{16}N_4O_2$ M 188.229

(S)-*form*
  L-*form*
  Component of Blasticidin S.
  B,2HCl: Needles (EtOH). Mp 192-192.5° dec. $[\alpha]_D^{15}$ +25.0°
  (c, 1 in H_2O).

Otake, N. *et al, Agric. Biol. Chem.*, 1966, **30**, 126.
Yonehara, H. *et al, Tetrahedron Lett.*, 1966, 3785 (*abs config*)

Bonafousioside
B-80171

[115900-10-2]

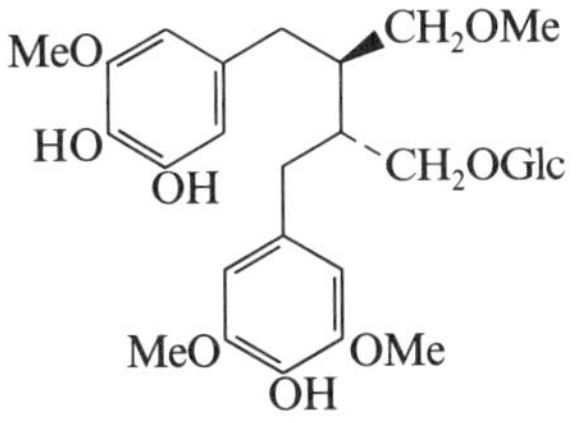

$C_{28}H_{40}O_{13}$ M 584.616
Constit. of *Bonafousia macrocalyx.*

Garnier, J. *et al, J. Nat. Prod. (Lloydia)*, 1988, **51**, 484.

Brianthein V
B-80172

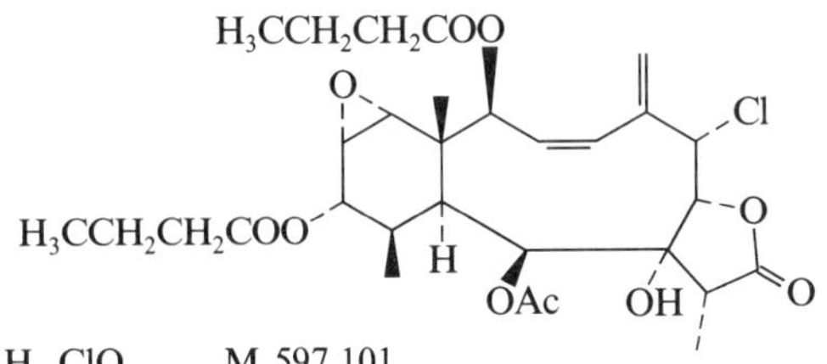

$C_{30}H_{41}ClO_{10}$ M 597.101
Constit. of *Briareum asbestinum.* Pale yellow needles. Mp
  222°. $[\alpha]_D^{20}$ −57° (c, 1.2 in $CHCl_3$).

Coval, S.J. *et al, J. Nat. Prod. (Lloydia)*, 1988, **51**, 981 (*isol, pmr, cmr, cryst struct*)

10-Bromoanthrone, 8CI
B-80173

10-Bromo-9(10H)-anthracenone, 9CI
[1560-32-3]

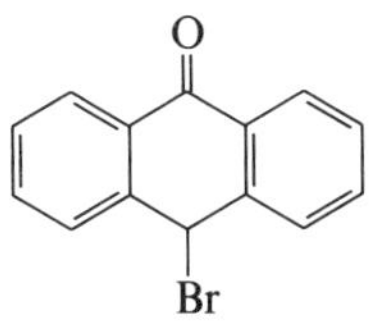

$C_{14}H_9BrO$ M 273.128
Keto-form (illus.) predominates in $CDCl_3$ soln. Yellow
  needles (toluene). Mp 148°.

[51678-94-5]

Meyer, K.H., *Justus Liebigs Ann. Chem.*, 1911, **379**, 37 (*synth*)
Koch, W. *et al, Helv. Chim. Acta*, 1965, **48**, 554 (*pmr*)
Branz, S.E. *et al, Synth. Commun.*, 1986, **16**, 441 (*synth, pmr, tautom*)

1-Bromobenz[a]anthracene, 9CI
B-80174

[78302-32-6]

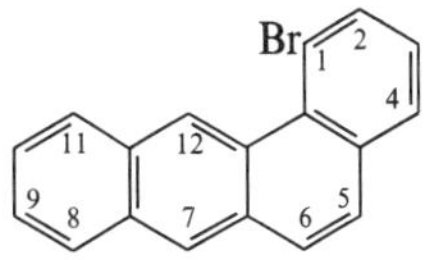

$C_{18}H_{11}Br$ M 307.189
Cryst. (C_6H_6). Mp 167-168°.

Hallmark, R.K., *J. Labelled Compd. Radiopharm.*, 1981, **18**, 331
  (*synth*)

2-Bromobenz[a]anthracene, 9CI
B-80175

[78302-30-4]
$C_{18}H_{11}Br$ M 307.189
Cryst. (C_6H_6). Mp 159.5-161°.

Hallmark, R.K., *J. Labelled Compd. Radiopharm.*, 1981, **18**, 331
  (*synth*)
Carmichael, I. *et al, J. Phys. Chem. Ref. Data*, 1987, **16**, 239 (*uv*)

3-Bromobenz[a]anthracene, 9CI
B-80176

[78302-31-5]
$C_{18}H_{11}Br$ M 307.189
Cryst. (C_6H_6). Mp 189-191°.

Hallmark, R.K., *J. Labelled Compd. Radiopharm.*, 1981, **18**, 331
  (*synth*)

4-Bromobenz[a]anthracene, 9CI
B-80177

[61921-39-9]
$C_{18}H_{11}Br$ M 307.189
Needles (C_6H_6), plates (AcOH). Mp 215-216°.

[61921-52-6]

Badger, G.M. *et al, J. Chem. Soc.*, 1949, 799 (*synth*)
Bodger, E.O. *et al, Acta Crystallogr., Sect. B*, 1977, **33**, 125 (*cryst struct*)
Cho, B.P. *et al, J. Org. Chem.*, 1987, **52**, 5668 (*synth, pmr, uv*)

7-Bromobenz[a]anthracene, 9CI
B-80178

$C_{18}H_{11}Br$ M 307.189
Needles or plates (AcOH), pale orange needles. Mp 150-
  151° (147.5-148.5°).

▷ Weak carcinogen.

Picrate: Blood-red, flat needles (AcOH). Mp 155.5-156.5°.

Badger, G.M. *et al, J. Chem. Soc.*, 1940, 409 (*synth*)
Mikhailov, B.M. *et al, Zh. Obshch. Khim.*, 1951, **21**, 2184; *ca, 46,*
  8080a (*synth*)
Yagi, H. *et al, J. Labelled Compd. Radiopharm.*, 1976, **12**, 127
  (*synth*)
Lee, H.M. *et al, J. Org. Chem.*, 1979, **44**, 4948 (*synth*)

2-Bromobicyclo[2.2.1]hept-2-ene, 9CI
B-80179

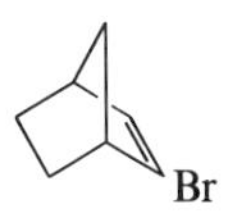

C_7H_9Br M 173.052
(±)-*form* [105231-24-1]
  Oil. Bp_{15} 65-66°.

Paquette, L.A. *et al, J. Am. Chem. Soc.*, 1988, **110**, 879 (*synth, pmr, cmr*)

1-Bromobicyclo[2.2.2]octane   B-80180

[7697-09-8]

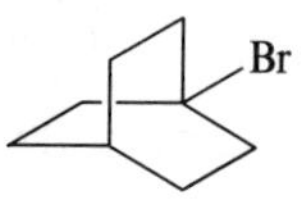

$C_8H_{13}Br$ M 189.095

Cryst. (EtOH aq. or by subl.). Mp 63.5-64.5°.

Grob, C.A. *et al, Helv. Chim. Acta*, 1958, **41**, 1191 (*synth*)
Suzuki, Z. *et al, J. Org. Chem.*, 1967, **32**, 31 (*synth*)

1-Bromobiphenylene   B-80181

[17557-58-3]

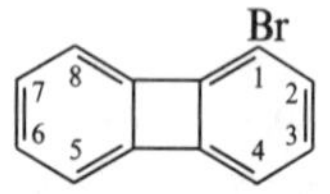

$C_{12}H_7Br$ M 231.091

Pale yellow oil.

2,4,7-Trinitrofluorenone complex: [17557-57-2].
Deep red leaflets (C_6H_6/MeOH). Mp 181-183°.

Barton, J.W. *et al, J. Chem. Soc. C*, 1968, 28 (*synth, uv*)

2-Bromobiphenylene   B-80182

[17573-59-0]

$C_{12}H_7Br$ M 231.091

Pale yellow plates (MeOH). Mp 64-65°.

2,4,7-Trinitrofluorenone complex: Scarlet needles
(EtOH/AcOH). Mp 136-137°.

Baker, W. *et al, J. Chem. Soc.*, 1958, 2658, 2666 (*synth, uv*)
Blatchly, J.M. *et al, J. Chem. Soc. C*, 1969, 2789 (*pmr*)
McKillop, A. *et al, J. Org. Chem.*, 1972, **37**, 88 (*synth*)
Cracknell, M.E. *et al, J. Chem. Soc., Perkin Trans. 1*, 1985, 115
(*synth*)

Bromocarbonimidic difluoride   B-80183

N-*Bromo-1,1-difluoromethanimine, 9CI*

$$F_2C=NBr$$

$CBrF_2N$ M 110.913

Pale yellow gas. Mp −93° to −92°. Bp 14.5°.

O'Brien, B.A. *et al, J. Am. Chem. Soc.*, 1984, **106**, 4266 (*synth, ir, raman, F-19 nmr, uv*)
Bauknight, C.W. *et al, J. Org. Chem.*, 1988, **53**, 4443 (*synth*)

10-Bromo-2,7-chamigradiene   B-80184

10-Bromo-α-chamigrene

[57473-87-7]

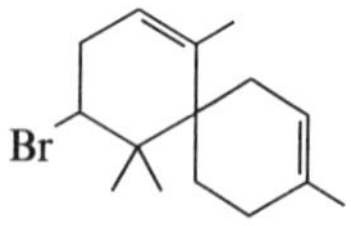

$C_{15}H_{23}Br$ M 283.251

Metab. of *Laurencia pacifica*. Oil. $[\alpha]_D$ −71.1° (c, 0.18 in $CHCl_3$).

Havard, B.M. *et al, Tetrahedron Lett.*, 1976, 2519.

2-Bromo-4-chloro-1-(2-chloroethenyl)-1-methyl-5-methylenecyclohexane, 9CI   B-80185

[119903-45-6]

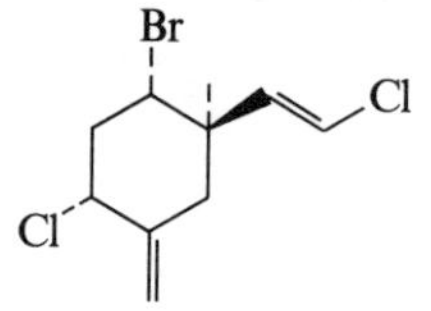

$C_{10}H_{13}BrCl_2$ M 284.022

Constit. of *Plocamium hamatum*. Oil. $[\alpha]_D$ −34° (c, 0.008 in $CHCl_3$).

Coll, J.C. *et al, Aust. J. Chem.*, 1988, **41**, 1743.

1-Bromo-2-chloro-1,1,2-trifluoroethane, 9CI   B-80186

[354-06-3]

$$F_2CBrCHFCl$$

$C_2HBrClF_3$ M 197.382

Inhalation anaesthetic agent. d_4^{25} 1.864. Bp 50-55°, Bp_{615} 46.02°. n_D^{20} 1.3705, n_D^{25} 1.3685.

▷ Irritant.

[74925-63-6]

Haszeldine, R.N. *et al, J. Chem. Soc.*, 1954, 3747 (*synth*)
Lee, J. *et al, Trans. Faraday Soc.*, 1958, **54**, 308 (*F-19 nmr, pmr*)
Kakac, B. *et al, Collect. Czech. Chem. Commun.*, 1965, **30**, 745 (*ir*)
Norris, R.D. *et al, J. Am. Chem. Soc.*, 1973, **95**, 182 (*F-19 nmr*)
Ng, S., *J. Magn. Reson.*, 1975, **17**, 244 (*pmr*)
Kirk-Othmer Encycl. Chem. Technol., 3rd Ed., 1978-1984, *Wiley, NY*, 3rd Ed., Wiley, N.Y., 1978-1984, **4**, 258; **10**, 868 (*props*)
Cascorbi, H.F. *et al, Anaesthesist*, 1980, **29**, 169 (*tox, use*)

1-Bromochrysene   B-80187

[76670-38-7]

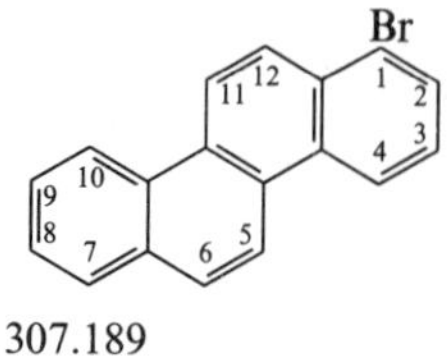

$C_{18}H_{11}Br$ M 307.189

Cryst. (C_6H_6). Mp 240-241°.

Archer, W.J. *et al, J. Chem. Soc., Perkin Trans. 2*, 1980, 1828 (*synth, pmr*)

2-Bromochrysene   B-80188

[55120-48-4]

$C_{18}H_{11}Br$ M 307.189

Photosensitizer. Silvery plates ($CHCl_3$). Mp 261-262°. A 1966 ref. referring to 2-bromochrysene was incorr. numbered and in fact referred to 6-bromochrysene.

Archer, W.J. *et al, J. Chem. Soc., Perkin Trans. 2*, 1980, 1828 (*synth, pmr*)

3-Bromochrysene   B-80189

[56158-60-2]

$C_{18}H_{11}Br$ M 307.189

Cryst. (Me_2CO). Mp 182.5-183° (195-196°).

Masetti, F. *et al, Gazz. Chim. Ital.*, 1975, **105**, 419 (*synth, props*)
Archer, W.J. *et al, J. Chem. Soc., Perkin Trans. 2*, 1980, 1828 (*synth, pmr*)

4-Bromochrysene   B-80190

[76670-41-2]

$C_{18}H_{11}Br$ M 307.189

Oil.

Archer, W.J. *et al, J. Chem. Soc., Perkin Trans.* 2, 1980, 1828 (*synth*)

5-Bromochrysene   B-80191

$C_{18}H_{11}Br$ M 307.189

Fine needles (EtOH). Mp 98°.

Archer, W.J. *et al, J. Chem. Soc., Perkin Trans.* 2, 1980, 1828 (*synth, pmr*)

6-Bromochrysene   B-80192

[7397-93-5]

$C_{18}H_{11}Br$ M 307.189

The 1966 paper incorr. reports this as 2-bromochrysene. Cryst. (C_6H_6). Mp 153.6-155°.

Altschuler, L. *et al, J. Am. Chem. Soc.*, 1966, **88**, 5837 (*synth*)
Archer, W.J. *et al, J. Chem. Soc., Perkin Trans.* 2, 1980, 1828 (*synth, pmr*)
Van Bladeren, P.J. *et al, Tetrahedron Lett.*, 1983, **24**, 4903 (*synth, pmr*)
Cho, B.P. *et al, J. Org. Chem.*, 1987, **52**, 5668 (*synth, pmr, ms, chromatog*)

2-Bromo-1,2-dichloro-1,1-difluoroethane, 9CI   B-80193

[2821-90-1]

$$F_2CClCHBrCl$$

$C_2HBrCl_2F_2$ M 213.836

(±)-*form*

Shows anaesthetic props. Bp 95.5°, Bp 76-78°. n_D^{20} 1.4298.

Di Paola, T. *et al, J. Pharm. Sci.*, 1979, **68**, 39 (*pharmacol*)
Caphiel, P. *et al, Tetrahedron*, 1979, **35**, 2661 (*synth, pmr, F nmr*)

1-Bromo-1,2-difluoroethylene, 8CI   B-80194

1-Bromo-1,2-difluoroethene, 9CI

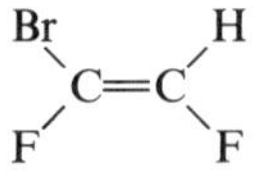

(*E*)-*form*

C_2HBrF_2 M 142.931

(*E*)-*form*

cis-form

Gas. Bp 16°.

(*Z*)-*form* [3685-10-7]

trans-form

Gas. Bp 13°.

Demiel, A., *J. Org. Chem.*, 1965, **30**, 2121 (*synth, pmr, F nmr, ir*)

1-Bromo-3,3-difluoro-2-iodocyclopropene, 9CI   B-80195

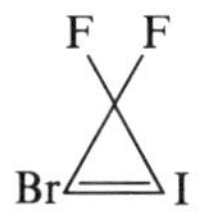

C_3BrF_2I M 280.838

Liq., isol. by glc. n_D^{27} 1.5326.

Sepiol, J. *et al, J. Org. Chem.*, 1975, **40**, 3791 (*synth, ir, F-19 nmr*)

1-Bromo-1,1-difluoro-2-iodoethane, 9CI   B-80196

[420-93-9]

$$F_2CBrCH_2I$$

$C_2H_2BrF_2I$ M 270.843

Bp$_{216}$ 84°. n_D^{20} 1.504.

Haszeldine, R.N., *J. Chem. Soc.*, 1953, 1764; 1954, 923 (*synth, uv*)

1-Bromo-3,4-dihydro-2-naphthalene-carboxaldehyde, 9CI   B-80197

[117582-62-4]

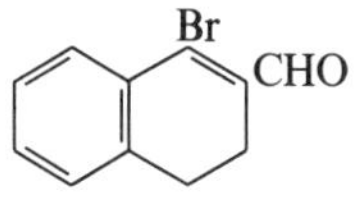

$C_{11}H_9BrO$ M 237.095

Yellow solid (EtOH). Mp 42-44°.

Gilchrist, T.L. *et al, J. Chem. Soc., Perkin Trans.* 1, 1988, 2595 (*synth, ir, pmr, ms*)

2-Bromo-3,4-dihydro-1-naphthalene-carboxaldehyde, 9CI   B-80198

[112631-07-9]

$C_{11}H_9BrO$ M 237.095

Mp 26-27°.

Ethylene acetal:

$C_{13}H_{13}BrO_2$ M 281.148

Cryst. (Et$_2$O/pet. ether). Mp 62-63°.

Gilchrist, T.L. *et al, J. Chem. Soc., Perkin Trans.* 1, 1988, 2595 (*synth, ir, pmr, ms*)

5-Bromo-3,3-dimethylcyclohexene   B-80199

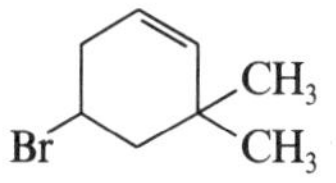

$C_8H_{13}Br$ M 189.095

(±)-*form* [113647-43-1]

Liq. Bp$_{26}$ 90-92°.

McIntosh, J.M. *et al, J. Org. Chem.*, 1988, **53**, 1947 (*synth, pmr*)

3-Bromo-2,2-dimethylpropanol   B-80200

[40894-00-6]

$$BrCH_2C(CH_3)_2CH_2OH$$

$C_5H_{11}BrO$ M 167.045

Bp$_{13}$ 80°.

Searles, S. *et al, J. Org. Chem.*, 1960, **24**, 1839 (*synth*)
Hua, D.Y. *et al, J. Org. Chem.*, 1988, **53**, 507 (*synth, ir, pmr*)

5-Bromo-4,6-dimethyl-1,2,3-triazine   B-80201

[109520-03-8]

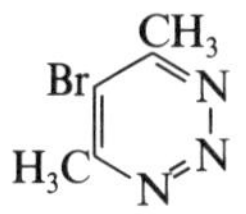

$C_5H_6BrN_3$ M 188.026

Needles (hexane). Mp 104-105°.

Ohsawa, A. *et al, Chem. Pharm. Bull.*, 1988, **36**, 3838 (*synth, pmr*)

Bromododecahedrane   B-80202

1-Bromohexadecahydro-5,2,1,6,3,4-
[2,3]butanediyl[1,4]diylidenepentaleno[2,1,6-cde:2′,1′,6′-
gha]pentalene, 9CI
[112896-27-2]

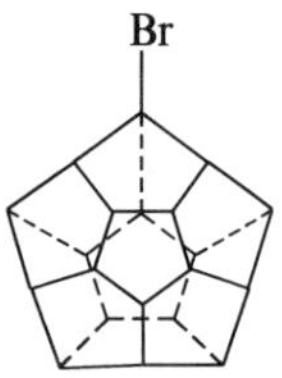

$C_{20}H_{19}Br$ M 339.274
Solid. Mp >240°.

Paquette, L.A. *et al, J. Am. Chem. Soc.,* 1988, **53**, 1303 (*synth, cmr*)

5-Bromo-2-ethynylpyridine   B-80203

[111770-86-6]

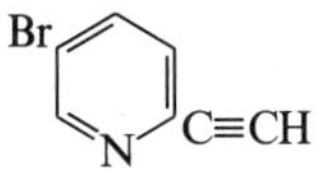

C_7H_4BrN M 182.019
Cryst. by subl. Mp 84-86°.

Tilley, J.W. *et al, J. Org. Chem.,* 1988, **53**, 386 (*synth, ms, pmr*)

5-Bromo-2(5H)-furanone, 9CI   B-80204

[40125-53-9]

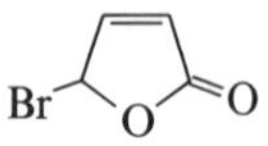

$C_4H_3BrO_2$ M 162.970
(±)-*form*
   $Bp_{0.1}$ 69-70°.

Elming, N. *et al, Acta Chem. Scand.,* 1952, **6**, 565 (*synth*)
Doerr, I.L. *et al, J. Org. Chem.,* 1973, **38**, 3878 (*synth, ir, pmr*)
Wolff, S. *et al, Synthesis,* 1988, 760 (*synth*)

8-Bromoguanine, 8CI   B-80205

2-Amino-8-bromo-1,7-dihydro-6H-purin-6-one, 9CI. 2-
Amino-8-bromo-6-hydroxypurine
[3066-84-0]

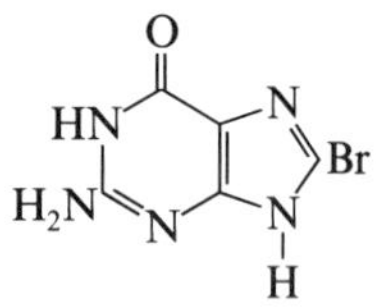

$C_5H_4BrN_5O$ M 230.023
Cryst. (H_2O). Insol. H_2O, EtOH, Et_2O.

Fischer, E. *et al, Justus Liebigs Ann. Chem.,* 1883, **221**, 336 (*synth*)
Holmes, R.E. *et al, J. Am. Chem. Soc.,* 1964, **86**, 1242 (*synth*)

1-Bromo-1,1,2,2,3,3,3-heptafluoropropane, 9CI   B-80206

Heptafluoropropyl bromide. Perfluoropropyl bromide.
Heptafluorobromopropane
[422-85-5]

$$F_3CCF_2CF_2Br$$

C_3BrF_7 M 248.926
d_4^0 1.87. Bp_{742} 15-15.2°, Bp 11-13°. $n_D^{-42.2}$ 1.3131, $n_D^{-29.8}$ 1.3070.

▷ Irritant.

Hauptschein, M. *et al, J. Am. Chem. Soc.,* 1952, **74**, 1347 (*synth, ir*)
Nodiff, E.A. *et al, J. Org. Chem.,* 1953, **18**, 235 (*ir, props*)
Husted, D.R. *et al, J. Am. Chem. Soc.,* 1954, **76**, 5141 (*synth, ir*)
Evans, D.F. *et al, J. Am. Chem. Soc.,* 1963, **85**, 238 (*F-19 nmr*)

2-Bromo-1,1,1,2,3,3,3-heptafluoropropane, 9CI   B-80207

[422-77-5]

$$F_3CCFBrCF_3$$

C_3BrF_7 M 248.926
Mobile liq. Bp 15-16° (11-12°) .

Barlow, G.B. *et al, J. Chem. Soc.,* 1955, 1749 (*synth*)
Naae, D.G. *et al, Org. Mass Spectrom.,* 1974, **9**, 1203 (*ms*)
Eujen, R. *et al, J. Fluorine Chem.,* 1983, **22**, 263 (*F-19 nmr*)

6-Bromo-1-heptene   B-80208

[38334-98-4]

$$H_2C=CH(CH_2)_3CHBrCH_3$$

$C_7H_{13}Br$ M 177.084
(±)-*form*
   Liq. Bp_{17} 59-61°.

Ashby, E.C. *et al, J. Org. Chem.,* 1988, **53**, 6068 (*synth, pmr, ms*)

1-Bromo-3-hexene, 9CI   B-80209

$$H_3CCH_2CH=CHCH_2CH_2Br$$

$C_6H_{11}Br$ M 163.057
(*E*)-*form* [63281-96-9]
   Liq. Bp_{15} 78-80°.
(*Z*)-*form* [5009-31-4]
   Liq. Bp 149-151°, Bp_{15} 65°.

Harper, S.H. *et al, J. Chem. Soc.,* 1955, 1512 (*synth*)
Stetter, H. *et al, Synthesis,* 1975, 379 (*synth*)
Sun, P.E. *et al, Can. J. Chem.,* 1979, **57**, 1475 (*synth, ir, pmr, ms*)
Rao, A.V.R. *et al, Tetrahedron,* 1987, **43**, 4385 (*synth, pmr*)

1-Bromo-5-isopropylcyclopentene   B-80210

1-Bromo-5-(1-methylethyl)cyclopentene, 9CI

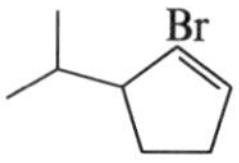

$C_8H_{13}Br$ M 189.095
(±)-*form* [112655-69-3]
   Liq. Bp_7 68-69°.

Paquette, L.A. *et al, J. Am. Chem. Soc.,* 1988, **110**, 890 (*synth, pmr, cmr*)

9-Bromolaurenisol   B-80211

Updated Entry replacing B-02636

[72782-83-3]

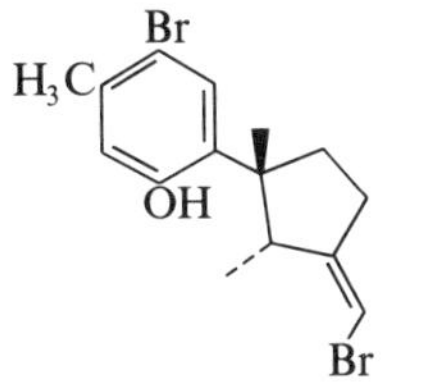

$C_{15}H_{18}Br_2O$ M 374.115

Constit. of *Laurencia glandulifera*. Oil. $[\alpha]_D^{22}$ +74° (c, 0.6 in CHCl$_3$).

Suzuki, M. *et al, Bull. Chem. Soc. Jpn.*, 1979, **52**, 3349 (*isol*)
Capon, R.J. *et al, J. Nat. Prod. (Lloydia)*, 1988, **51**, 1302 (*isol, pmr*)

Bromomethanol   B-80212

[50398-29-3]

$$BrCH_2OH$$

CH_3BrO M 110.938

Ac: [590-97-6].
    $C_3H_5BrO_2$ M 152.975
    Alkylating agent. Oil. Bp 132-138°.

Chloroacetyl:
    $C_3H_4BrClO_2$ M 187.420
    Oil. d_4^{20} 1.783. $Bp_{3.5}$ 57.2°. n_D^{20} 1.4873.

Benzoyl: [40796-19-8].
    $C_8H_7BrO_2$ M 215.046
    Oil. Bp_{19} 135-137°, $Bp_{0.3}$ 62-65°.

4-Methylbenzenesulfonyl: [117184-51-7].
    $C_8H_9BrO_3S$ M 265.127
    Oil or low melting solid, may be recryst. from MeOH at 0-5°. Mp <20°. $Bp_{0.4}$ 112-114°.

Ulich, L.H. *et al, J. Am. Chem. Soc.*, 1921, **43**, 660 (*acetate*)
Euranto, E.K. *et al, Acta Chem. Scand.*, 1966, **20**, 1273 (*chloroacetate*)
Bodor, N. *et al, J. Med. Chem.*, 1980, **23**, 566 (*benzoate, synth, ir, pmr*)
Hahn, R.C. *et al, J. Org. Chem.*, 1988, **53**, 5783 (*tosylate, synth, pmr, cmr*)

3-Bromo-3-methyl-1-butyne   B-80213

[6214-31-9]

$$(H_3C)_2CBrC{\equiv}CH$$

C_5H_7Br M 147.014

d_4^{25} 1.268. Bp_{106} 42.2°. n_D^{25} 1.4118.

Shiner, V.J. *et al, J. Am. Chem. Soc.*, 1962, **84**, 2402.
Jacobs, T.L. *et al, J. Org. Chem.*, 1963, **28**, 1360 (*synth, ir*)

1-Bromomethyl-2,3-diphenylcyclopropane   B-80214

1,1′-[3-(Bromomethyl)-1,2-cyclopropanediyl]bisbenzene, 9CI

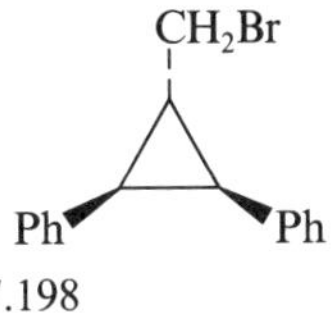

$C_{16}H_{15}Br$ M 287.198

(1α,2β,3β)-form [117021-83-7]

Cryst. (pet. ether). Mp 58.0-58.5°.

Castellino, A.J. *et al, J. Am. Chem. Soc.*, 1988, **110**, 7512 (*synth, pmr*)

1-(Bromomethyl)-2-fluorobenzene, 9CI   B-80215

α-Bromo-o-fluorotoluene, 8CI. o-*Fluorobenzyl bromide*

[446-48-0]

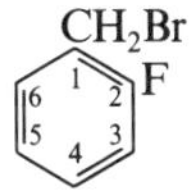

C_7H_6BrF M 189.027

Liq. Bp 195-202°, Bp_{15} 86-87°.

▷ Corrosive, lachrymator.

Olah, G. *et al, J. Org. Chem.*, 1957, **22**, 879 (*synth*)
Bergman, E.D. *et al, Tetrahedron*, 1976, **32**, 2847 (*props*)

1-(Bromomethyl)-3-fluorobenzene, 9CI   B-80216

α-Bromo-m-fluorotoluene, 8CI. m-*Fluorobenzyl bromide*

[456-41-7]

C_7H_6BrF M 189.027

Bp 196-200°, Bp_{15} 84-85°.

▷ Corrosive, lachrymator.

Olah, G. *et al, J. Org. Chem.*, 1957, **22**, 879 (*synth*)
Tait, J.M.S. *et al, J. Am. Chem. Soc.*, 1962, **84**, 4 (*ms*)
Roberts, B.P. *et al, J. Chem. Soc., Chem. Commun.*, 1978, 752 (*synth*)
Happer, D.A.R. *et al, Org. Magn. Reson.*, 1983, **21**, 252 (*F-nmr*)

1-(Bromomethyl)-4-fluorobenzene, 9CI   B-80217

α-Bromo-p-fluorotoluene, 8CI. p-*Fluorobenzyl bromide*

[459-46-1]

C_7H_6BrF M 189.027

Bp_{27} 92-102°. n_D 1.547.

▷ Corrosive, lachrymator.

Olah, G. *et al, J. Org. Chem.*, 1957, **22**, 879 (*synth*)
Tait, J.M.S. *et al, J. Am. Chem. Soc.*, 1962, **84**, 4 (*ms*)
Verdonck, L. *et al, Spectrochim. Acta, Part A*, 1972, **28**, 55 (*ir, raman*)
Roberts, B.P. *et al, J. Chem. Soc., Chem. Commun.*, 1978, 752 (*synth*)
Happer, D.A.R. *et al, Org. Magn. Reson.*, 1983, **21**, 252 (*F nmr*)

3-Bromo-5-methylisoquinoline, 9CI   B-80218

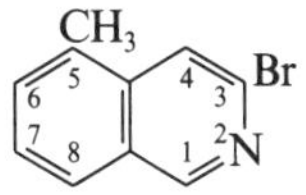

$C_{10}H_8BrN$ M 222.084

Cryst. (C_6H_6/hexane). Mp 90-91°.

Ban, Y. *et al, Chem. Pharm. Bull.*, 1964, **12**, 1296 (*synth*)

4-Bromo-1-methylisoquinoline, 9CI   B-80219

[104704-40-7]

$C_{10}H_8BrN$ M 222.084

Needles (hexane). Mp 48-49°.

Sawanishi, H. *et al, Chem. Pharm. Bull.*, 1985, **33**, 4564 (*synth, pmr*)

5-Bromo-3-methylisoquinoline, 9CI   B-80220

[16552-67-3]

$C_{10}H_8BrN$ M 222.084

Cryst. Mp 40-42°. $Bp_{0.75}$ 35° subl.

Hamilton, H.J., *J. Heterocycl. Chem.*, 1967, **4**, 410 (*synth*)

5-(Bromomethyl)isoxazole   B-80221

[69735-35-9]

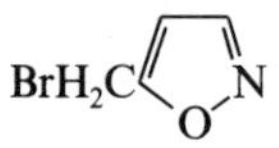

C_4H_4BrNO M 161.986

Liq. $Bp_{0.5}$ 58-62°.

DeShong, P. *et al*, *J. Org. Chem.*, 1988, **53**, 1356 (*synth, ir, pmr, ms*)

(Bromomethyl)pentafluorobenzene, 9CI   B-80222

Pentafluorobenzyl bromide

[1765-40-8]

$(C_6F_5)CH_2Br$

$C_7H_2BrF_5$ M 260.989

Liq. Bp 174-175°.

▷ Extremely lachrymatory.

Barbour, A.K. *et al*, *J. Chem. Soc.*, 1961, 808 (*synth, ir*)

1-Bromo-3-methylpentane   B-80223

[51116-73-5]

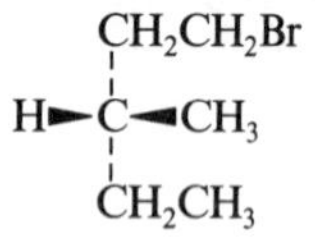

$C_6H_{13}Br$ M 165.073

(**R**)-form

Oil. Bp_{63} 71-72.5°. $[\alpha]_D^{20}$ −19.43° (neat).

[22299-70-3]

Mori, K. *et al*, *Tetrahedron*, 1984, **40**, 299 (*synth, ir, pmr*)
Fuganti, C. *et al*, *J. Chem. Soc., Perkin Trans.* 1, 1988, 3061 (*synth*)

1-(Bromomethyl)phenanthrene, 9CI   B-80224

[42050-05-5]

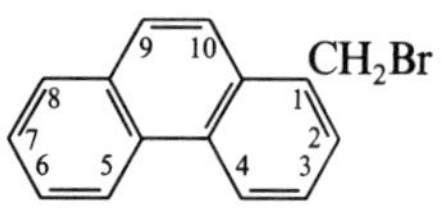

$C_{15}H_{11}Br$ M 271.156

Leaflets (C_6H_6/pet. ether). Mp 97°.

Bachmann, W.E. *et al*, *J. Am. Chem. Soc.*, 1937, **59**, 2207 (*synth*)

2-(Bromomethyl)phenanthrene   B-80225

[2417-66-5]

$C_{15}H_{11}Br$ M 271.156

Needles (C_6H_6). Mp 111-111.5°.

Mosettig, E. *et al*, *J. Am. Chem. Soc.*, 1933, **55**, 2995 (*synth*)

3-(Bromomethyl)phenanthrene   B-80226

[24471-44-1]

$C_{15}H_{11}Br$ M 271.156

Needles (C_6H_6). Mp 117.5° (114.5-115°).

Mosettig, E. *et al*, *J. Am. Chem. Soc.*, 1933, **55**, 2995 (*synth*)
Bachmann, W.E. *et al*, *J. Am. Chem. Soc.*, 1937, **59**, 2207 (*synth*)

9-(Bromomethyl)phenanthrene   B-80227

[24471-57-6]

$C_{15}H_{11}Br$ M 271.156

Cryst. Mp 118.5-119° (103-103.5°).

Mosettig, E. *et al*, *J. Am. Chem. Soc.*, 1933, **55**, 2995 (*synth*)
Bachmann, W.E. *et al*, *J. Am. Chem. Soc.*, 1937, **59**, 2207 (*synth*)
Lambert, J.B. *et al*, *J. Org. Chem.*, 1979, **44**, 1480 (*synth, ir, pmr*)
Sugimoto, A. *et al*, *J. Chem. Soc., Perkin Trans.* 1, 1988, 2579 (*synth, pmr, ms*)

1-Bromo-2-methyl-1-phenylpropene   B-80228

(*1-Bromo-2-methyl-1-propenyl*)*benzene*

$(H_3C)_2C{=}CBrPh$

$C_{10}H_{11}Br$ M 211.101

Liq. $Bp_{0.8}$ 90-100° (Kugelrohr) .

Maloney, M.G. *et al*, *J. Chem. Soc., Perkin Trans.* 1, 1988, 2847 (*synth, uv, pmr*)

2-Bromo-2-methylpropanoic acid, 9CI   B-80229

Updated Entry replacing B-02834

2-Bromoisobutyric acid

[2052-01-9]

$(H_3C)_2CBrCOOH$

$C_4H_7BrO_2$ M 167.002

Cryst. Mp 48-49°. Bp 198-200°, Bp_{20} 115°.

Me ester: [23426-63-3].
    $C_5H_9BrO_2$ M 181.029
    Liq. Bp_9 38°, Bp_{19} 51-52°.

Bromide: [20769-85-1].
    $C_4H_6Br_2O$ M 229.899
    Liq. Bp 162-164°, Bp_{100} 91-98°.

Anhydride: [42069-15-8].
    $C_8H_{12}Br_2O_3$ M 315.989
    Cryst. Mp 63-65°. Bp_{35} 135-140°.

Amide: [7462-74-0].
    C_4H_8BrNO M 166.017
    Cryst. Mp 148°. Bp_{17} 145°.

Anilide:
    $C_{10}H_{12}BrNO$ M 242.115
    Cryst. (EtOH aq.). Mp 83°.

Nitrile: [41658-69-9]. α-*Bromoisobutyronitrile. 2-Bromo-2-cyanopropane*
    C_4H_6BrN M 148.002
    Liq. d_4^{25} 1.38. Bp 137-139°.

Michael, A., *Ber.*, 1901, **34**, 4043 (*synth*)
Stevens, C.L., *J. Am. Chem. Soc.*, 1948, **70**, 165 (*nitrile*)
Ford, M.C. *et al*, *J. Chem. Soc.*, 1951, 1851 (*ester*)
Org. Synth., 1953, **33**, 29 (*bromide*)
Smissman, E., *J. Am. Chem. Soc.*, 1954, **76**, 5805 (*amide*)
Sax, N.I., *Dangerous Properties of Industrial Materials*, 5th Ed., Van Nostrand-Reinhold, 1979, 434.

4-Bromo-6-methylpyrimidine, 9CI　　B-80230
[69543-98-2]

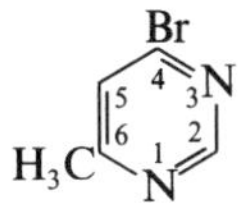

C₅H₅BrN₂ M 173.012
Mp 29-31°. Bp₁₈ 105-110°. Slowly dec. on standing.

Rasmussen, C.A.H. *et al, Recl. Trav. Chim. Pays-Bas* (*J. R. Neth. Chem. Soc.*), 1979, **98**, 5 (*synth*)

5-Bromo-4-methylpyrimidine, 9CI　　B-80231
[1439-09-4]

C₅H₅BrN₂ M 173.012
Liq. Bp₂₃ 90°, Bp₁₄ 72-73°.

van der Plas, H.C., *Recl. Trav. Chim. Pays-Bas* (*J. R. Neth. Chem. Soc.*), 1965, **84**, 1101 (*synth*)
Geerts, J. *et al, Recl. Trav. Chim. Pays-Bas* (*J. R. Neth. Chem. Soc.*), 1974, **93**, 231 (*pmr*)
Yamenaka, H. *et al, Chem. Pharm. Bull.*, 1987, **35**, 3119 (*synth, pmr*)

4-Bromo-5-methyl-1,2,3-triazine　　B-80232

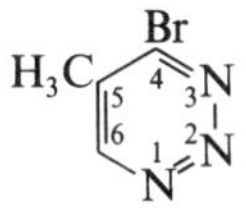

C₄H₄BrN₃ M 174.000
Needles (hexane). Mp 36-38°. Subl. very readily.

Ohsawa, A. *et al, Chem. Pharm. Bull.*, 1988, **36**, 3838 (*synth, pmr*)

5-Bromo-4-methyl-1,2,3-triazine　　B-80233
[114078-90-9]

C₄H₄BrN₃ M 174.000
Needles (hexane). Mp 73-74°.

Ohsawa, A. *et al, Chem. Pharm. Bull.*, 1988, **36**, 3838 (*synth, pmr*)

2-Bromo-2-nitroadamantane　　B-80234
2-Bromo-2-nitrotricyclo[3.3.1.1³,⁷]decane
[40854-02-2]

C₁₀H₁₄BrNO₂ M 260.130
Cryst. (EtOH aq.). Mp 190-191°.

Archibald, T.G. *et al, J. Org. Chem.*, 1988, **53**, 4645 (*synth, ir, pmr*)

4-Bromo-2-nitrobenzaldehyde, 9CI　　B-80235
Updated Entry replacing B-02967
[5551-12-2]
C₇H₄BrNO₃ M 230.017
Yellow cryst. (EtOH). Mp 97-98°.
Oxime:
C₇H₅BrN₂O₃ M 245.032
Mp 164°.
Semicarbazone: Cryst. (AcOH). Mp 276° (270-272°).
Phenylhydrazone: Brown-red cryst. Mp 181-182°.

Sachs, F. *et al, Ber.*, 1904, **37**, 1861 (*synth*)
Barber, H.J. *et al, J. Chem. Soc.*, 1945, 167 (*synth*)

Dandegaonker, S.H., *CA*, 1966, **64**, 17463; *J. Indian Chem. Soc.*, 1969, **46**, 148 (*synth, uv*)
Danieli, R. *et al, CA*, 1968, **69**, 106403 (*synth*)
Voss, G. *et al, Chem. Ber.*, 1989, **122**, 1199 (*synth, uv, pmr, cmr, ms*)

4-Bromo-3-nitrobenzaldehyde, 9CI　　B-80236
Updated Entry replacing B-02968
C₇H₄BrNO₃ M 230.017
Pale-yellow needles (EtOH). Mp 106°. Steam-volatile.
Oxime:
C₇H₅BrN₂O₃ M 245.032
Orange-yellow needles (EtOH). Mp 154°.
p-*Nitrophenylhydrazone:* Yellow needles. Mp 282-283°.
Semicarbazone: Mp 254-255° dec.
Phenylhydrazone: Two forms: red and yellow. Mp 146-147° (both forms).

Hodgson, H.H. *et al, J. Chem. Soc.*, 1927, 20 (*synth*)
Voss, G. *et al, Chem. Ber.*, 1989, **122**, 1199 (*uv, pmr, cmr, ms*)

5-Bromo-2-nitrobenzaldehyde, 9CI　　B-80237
Updated Entry replacing B-02969
[20357-20-4]
C₇H₄BrNO₃ M 230.017
Cryst. (C₆H₆ or EtOH aq.) or needles (C₆H₆/cyclohexane). Mp 77-78°.
▷ CU4875000.
Oxime:
C₇H₅BrN₂O₃ M 245.032
Yellow needles. Mp 113°.
Phenylhydrazone: Red needles (EtOH). Mp 180° dec.

Mettler, C., *Ber.*, 1905, **38**, 2809.
Alford, E.J. *et al, J. Chem. Soc.*, 1952, 2102.
Danieli, R. *et al, CA*, 1968, **69**, 106403 (*synth*)
Voss, G. *et al, Chem. Ber.*, 1989, **122**, 1199 (*synth, uv, pmr, cmr, ms*)

4-Bromo-2-nitrobenzyl alcohol, 8CI　　B-80238
4-Bromo-2-nitrobenzenemethanol, 9CI
[22996-19-6]
C₇H₆BrNO₃ M 232.033
Needles (EtOH). Mp 80-81°.

Cal'bershtam, M. *et al, Zh. Org. Khim.*, 1969, **5**, 953; *jocet*, 1969, 938 (*synth, ir*)

2-Bromo-4-nitrothiophene　　B-80239
[81975-00-0]
C₄H₂BrNO₂S M 208.035
Liq. Bp₅₅ 105-110°, Bp₀.₆ 64-68°.

Elliott, R.L. *et al, Tetrahedron*, 1987, **43**, 3295 (*synth, ir, pmr*)

15-Bromo-9(11)-parguerene-2,7,16,19-tetrol　　B-80240

C₂₀H₃₁BrO₄ M 415.366

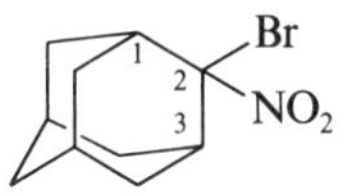

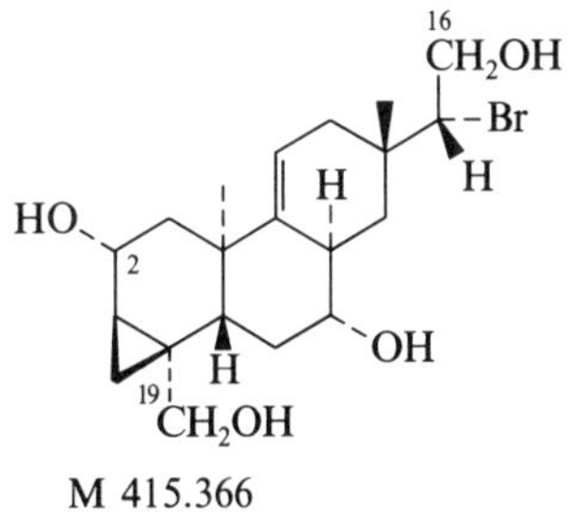

(2α,7α)-form

2-Ac: [83115-36-0]. *2α-Acetoxy-15-bromo-9(11)-parguerene-7α,16,19-triol*. **Parguerene**
$C_{22}H_{33}BrO_5$ M 457.404
Metab. of *Aplysia dactylomela*. Oil. $[\alpha]_D$ −40° (c, 0.03 in $CHCl_3$).

19-Deoxy, 2-Ac: [83115-44-0]. *2α-Acetoxy-15-bromo-9(11)-parguerene-7α,16-diol*. **Deoxyparguerene**
$C_{22}H_{33}BrO_4$ M 441.404
Metab. of *A. dactylomela*. Oil. $[\alpha]_D$ −35.8° (c, 0.62 $CHCl_3$).

19-Deoxy, 2,16-di-Ac: [82668-11-9]. *2α,16-Diacetoxy-15-bromo-9(11)-pargueren-7α-ol*
$C_{24}H_{35}BrO_5$ M 483.441
Metab. of *Laurencia obtusa*. Oil. $[\alpha]_D$ −48° (c, 0.1 in MeOH).

Schmitz, F.J. *et al, J. Am. Chem. Soc.*, 1982, **104**, 6415 *(isol)*
Higgs, M.D. *et al, Phytochemistry*, 1982, **21**, 789 *(isol)*
Suzuki, T. *et al, Chem. Lett.*, 1989, 969 *(deriv)*

15-Bromo-9(11)-parguerene-2,7,16-triol   B-80241

15-Bromo-2,7,16-trihydroxy-9(11)-parguerene

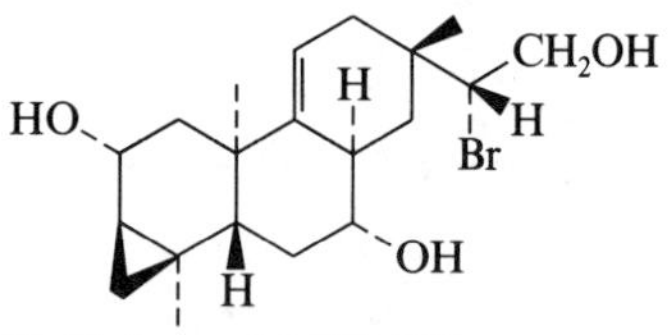

HO — ... — CH₂OH ... H ... Br ... OH ... H   Absolute configuration

$C_{20}H_{31}BrO_3$ M 399.367
Metab. of *Laurencia obtusa*. Cryst. Mp 173-174°. $[\alpha]_D$ −36.4° (c, 1.85 in $CHCl_3$).

Suzuki, T. *et al, Chem. Lett.*, 1989, 969 *(cryst struct)*

2-Bromo-4-phenylbutanoic acid, 9CI   B-80242

Updated Entry replacing B-03154
[16503-46-1]

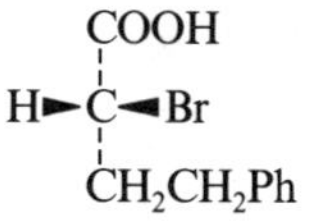

COOH
H — C — Br   *(R)-form*
CH₂CH₂Ph

$C_{10}H_{11}BrO_2$ M 243.100
(R)-form
Oil. $[\alpha]_D^{20}$ +65.5° (c, 1 in $CHCl_3$).
Et ester:
$C_{12}H_{15}BrO_2$ M 271.153
Oil. $[\alpha]_D^{20}$ +55° (c, 1 in $CHCl_3$).
(±)-form
Oil. d_4^{25} 1.44. Bp_1 145-147°.
Et ester: Yellow liq. Bp_{14} 157°.
Amide:
$C_{10}H_{12}BrNO$ M 242.115
Cryst. (EtOAc). Mp 130-131°.

Fischer, E. *et al, Ber.*, 1906, **39**, 2213 *(synth)*
v. Braun, J., *Ber.*, 1923, **56**, 2178 *(ester)*
Stevens, C.A. *et al, J. Org. Chem.*, 1953, **18**, 1112 *(synth)*
Iwasaki, G. *et al, Chem. Pharm. Bull.*, 1989, **37**, 280 *(synth, pmr)*

2-Bromo-1-phenyl-1-propanone, 9CI   B-80243

Updated Entry replacing B-03198
2-Bromopropiophenone, 8CI. 1-Bromoethyl phenyl ketone. 1-Benzoyl-1-bromoethane
[2114-00-3]

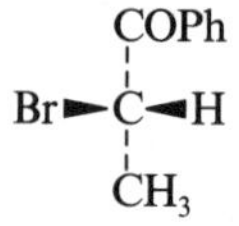

COPh
Br — C — H   *(S)-form*
CH₃

C_9H_9BrO M 213.073
▷ UG8400000.
(S)-form
Oil.
(±)-form
Liq. d_0^0 1.454. Bp 245-250°, Bp_{10} 125-130°. n_D^{25} 1.5686.
2,4-Dinitrophenylhydrazone: Mp 160-161°.

Bauer, D.P. *et al, J. Org. Chem.*, 1975, **40**, 1990 *(synth)*
Castaldi, G. *et al, Synthesis*, 1987, 1039 *(synth, resoln, uv, cd, pmr)*

3-Bromo-3-phenyl-1-propyne   B-80244

(1-Bromo-2-propynyl)benzene, 9CI
[50874-14-1]

$$PhCHBrC{\equiv}CH$$

C_9H_7Br M 195.058
(±)-form
$Bp_{0.003}$ 51-53°.

Kirmse, W. *et al, Chem. Ber.*, 1973, **106**, 3086.
Brooks, J.R. *et al, J. Chem. Soc., Perkin Trans. 1*, 1973, 2588.

2-Bromo-4-phenylpyridine, 9CI, 8CI   B-80245

[54151-74-5]

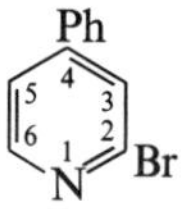

$C_{11}H_8BrN$ M 234.095
Cryst. (pet. ether). Mp 65-66°.

Case, F.H. *et al, J. Am. Chem. Soc.*, 1956, **78**, 5842.

2-Bromo-5-phenylpyridine, 9CI, 8CI   B-80246

[107351-82-6]
$C_{11}H_8BrN$ M 234.095
Cryst. (hexane). Mp 78-79°.

Tilley, J.W. *et al, J. Org. Chem.*, 1988, **53**, 386 *(synth, ms, pmr)*

2-Bromo-6-phenylpyridine, 9CI, 8CI   B-80247

[39774-26-0]
$C_{11}H_8BrN$ M 234.095
Cryst. (pet. ether). Mp 51-52°.

Case, F.H. *et al, J. Am. Chem. Soc.*, 1956, **78**, 5842.

3-Bromo-2-phenylpyridine, 9CI, 8CI   B-80248

$C_{11}H_8BrN$ M 234.095
Pale yellow liq. Bp_{720} 305-307°.

Abramovitch, R.A. *et al, Can. J. Chem.*, 1963, **41**, 1752 *(synth, ir)*

3-Bromo-4-phenylpyridine, 9CI, 8CI   B-80249

[88345-89-5]

$C_{11}H_8BrN$ M 234.095

Cryst. Mp 49-50°.

Comins, D.L. *et al, J. Heterocycl. Chem.*, 1983, **20**, 1239 (*synth, pmr*)

4-Bromo-2-phenylpyridine, 9CI, 8CI   B-80250

[98420-98-5]

$C_{11}H_8BrN$ M 234.095

Oil.

Picrate: Cryst. Mp 174.5-176°.

Comins, D.L. *et al, J. Org. Chem.*, 1985, **50**, 4410 (*synth, ir, pmr*)

5-Bromo-2-phenylpyridine, 9CI, 8CI   B-80251

[27012-25-5]

$C_{11}H_8BrN$ M 234.095

Cryst. (hexane). Mp 74-76°.

Giam, C.S. *et al, J. Chem. Soc., Chem. Commun.*, 1970, 478 (*synth*)

Comins, D.L. *et al, J. Heterocycl. Chem.*, 1983, **20**, 1239 (*synth, pmr*)

Tilley, J.W. *et al, J. Org. Chem.*, 1988, **53**, 386 (*synth*)

5-Bromo-4-phenyl-1,2,3-triazine   B-80252

[114078-92-1]

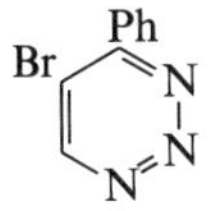

$C_9H_6BrN_3$ M 236.070

Flakes (hexane/Et$_2$O). Mp 125-126° dec.

Ohsawa, A. *et al, Chem. Pharm. Bull.*, 1988, **36**, 3838 (*synth, pmr*)

3(5)-Bromopyrazole, 9CI   B-80253

[14521-80-3]

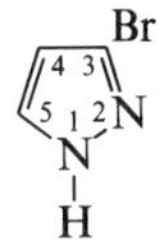

$C_3H_3BrN_2$ M 146.974

Cryst. by subl. Mp 70° (66°). Bp$_{0.1}$ 70-72°.

1H-form

1-Ph: [50877-46-8].

$C_9H_7BrN_2$ M 223.072

No Mp reported.

2H-form

2-Ph: [17635-42-6]. *5-Bromo-1-phenylpyrazole, 9CI*

$C_9H_7BrN_2$ M 223.072

Cryst. (EtOH aq.). Mp 56°.

Reimlinger, H.K. *et al, Chem. Ber.*, 1966, **99**, 3350 (*synth*)

Begtrup, M., *Acta Chem. Scand.*, 1973, **27**, 2051 (*derivs*)

Freche, P. *et al, Tetrahedron*, 1977, **33**, 2069 (*synth, pmr*)

Birkofer, L. *et al, Chem. Ber.*, 1981, **114**, 2293 (*synth, ir, pmr*)

Van Tilborg, M.W.E.M. *et al, Org. Mass Spectrom.*, 1984, **19**, 569 (*ms*)

4-Bromopyrazole, 9CI   B-80254

[2075-45-8]

$C_3H_3BrN_2$ M 146.974

Alcohol dehydrogenase inhibitor. Cryst. by subl. Mp 91-92°.

▷ Possibly mutagenic.

1-Me: [15803-02-8]. *4-Bromo-1-methylpyrazole, 9CI*

$C_4H_5BrN_2$ M 161.001

Oily liq. Bp$_{18}$ 76-78°. n_D^{23} 1.5298.

▷ Possibly mutagenic.

1-Ph: [15115-52-3].

$C_9H_7BrN_2$ M 223.072

Needles (EtOH). Mp 83°.

Hüttel, R. *et al, Justus Liebigs Ann. Chem.*, 1955, **593**, 179 (*synth, deriv*)

Elguero, J. *et al, Bull. Soc. Chim. Fr.*, 1966, 2832, 3727, 3744 (*synth, pmr, uv, deriv*)

Liljefors, S. *et al, Chem. Scr.*, 1980, **15**, 102 (*deriv*)

Claramunt, R.M. *et al, Bull. Soc. Chim. Fr., Part II*, 1983, 5 (*synth, cmr, pmr, deriv*)

Cabildo, P. *et al, Org. Magn. Reson.*, 1984, **22**, 603 (*cmr*)

Van Tilborg, M.W.E.M. *et al, Org. Mass Spectrom.*, 1984, **19**, 569 (*ms*)

5-Bromo-2(1*H*)-pyridinethione, 9CI   B-80255

Updated Entry replacing B-60317

5-Bromo-2-pyridinethiol. 5-Bromo-2-mercaptopyridine

[56673-34-8]

C_5H_4BrNS M 190.063

S-Me: [51933-78-9]. *5-Bromo-2-(methylthio)pyridine*

C_6H_6BrNS M 204.090

Mp 38-39°.

S-Me, S-dioxide: [98626-95-0]. *5-Bromo-2-(methylsulfonyl)pyridine, 9CI*

$C_6H_6BrNO_2S$ M 236.089

Mp 95-96°.

U.K. Pat., 791 190; *CA*, **52**, 15597 (*synth*)

Abramovitch, R.A. *et al, J. Org. Chem.*, 1974, **39**, 2690 (*deriv, synth*)

Davies, J.S. *et al, Tetrahedron Lett.*, 1979, 5035 (*synth*)

Testaferri, L. *et al, Tetrahedron*, 1985, **41**, 1373 (*deriv, synth, pmr*)

6-Bromo-2(1*H*)-pyridinethione   B-80256

2-Bromo-6-mercaptopyridine. 6-Bromo-2-pyridinethiol

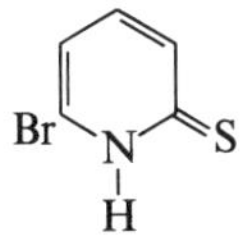

C_5H_4BrNS M 190.063

S-Me: [74134-42-2]. *2-Bromo-6-(methylthio)pyridine*

C_6H_6BrNS M 204.090

Oil.

S-Me, S-dioxide: [98626-92-7]. *2-Bromo-6-(methylsulfonyl)pyridine, 9CI*

$C_6H_6BrNO_2S$ M 236.089

Mp 135-136°.

Furukawa, N. *et al, Tetrahedron Lett.*, 1983, **24**, 3243.

Testaferri, L. *et al, Tetrahedron*, 1985, **41**, 1373 (*deriv, synth, pmr*)

5-Bromo-2,4-(1*H*,3*H*)-pyrimidinedione   B-80257

Updated Entry replacing B-60318
5-Bromouracil, 8CI
[51-20-7]

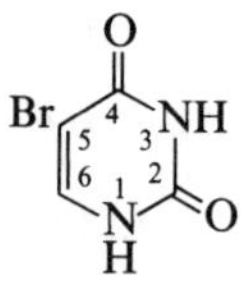

$C_4H_3BrN_2O_2$ M 190.984
Prisms (H_2O). Mp 293°. pK_a 8.06 (H_2O), pK_a 12.7 (DMF).
▷ Mutagen. YQ9060000.

Wheeler, H.L. *et al, Am. Chem. J.*, 1903, **29**, 478 (*synth*)
Levene, P.A. *et al, Ber.*, 1912, **45**, 608 (*synth*)
Ulrich, J. *et al, Org. Mass Spectrom.*, 1969, **2**, 1183 (*ms*)
Ellis, P.D. *et al, J. Am. Chem. Soc.*, 1973, **95**, 4398 (*cmr*)
Sternglanz, H. *et al, Biochim. Biophys. Acta*, 1975, **378**, 1 (*cryst struct*)
Yoshimura, M. *et al, Yakugaku Zasshi (J. Pharm. Soc. Jpn.)*, 1976, **96**, 1094; *CA*, **86**, 155589g (*synth, chromatog*)
Turturo, A. *et al, J. Chromatogr.*, 1982, **252**, 335 (*hplc*)
Merck Index, 10th Ed., 1983, 10th Ed., no. 1414.
Srivastava, S.L. *et al, Spectrochim. Acta, Part A*, 1984, **40**, 675 (*uv, tautom*)
Chandrasekaran, S. *et al, J. Org. Chem.*, 1985, **50**, 829 (*nmr*)
Rabbani, R.S. *et al, J. Magn. Reson.*, 1987, **72**, 422 (*nqr*)

3-Bromo-2-pyrrolidinone   B-80258

[40557-20-8]

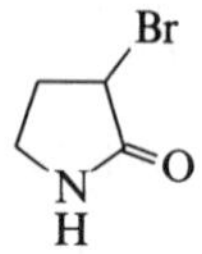

C_4H_6BrNO M 164.001
(±)-*form* [93681-10-8]
    Cryst. (Et$_2$O). Mp 82° (78°).

Yamada, Y. *et al, Agric. Biol. Chem.*, 1973, **37**, 649 (*synth, ir, pmr*)
Ikuta, H. *et al, J. Med. Chem.*, 1987, **30**, 1995 (*synth, pmr*)

1-Bromo-1,1,2,2-tetrafluoroethane, 9CI, 8CI   B-80259

[354-07-4]

$$F_2CBrCHF_2$$

C_2HBrF_4 M 180.927
Shows anaesthetic props. Bp$_{625.6}$ −3.5° to −3.0°, Bp 12.5°.

[30283-90-0]

Park, J.D. *et al, J. Am. Chem. Soc.*, 1949, **71**, 2339 (*synth*)
Klaboe, P. *et al, J. Chem. Phys.*, 1961, **34**, 1819 (*ir, raman*)
Joshi, R.M., *J. Macromol. Sci., Chem.*, 1974, **8**, 861 (*props*)
Di Paolo, T. *et al, J. Pharmacol.*, 1979, **68**, 39 (*pharmacol*)

1-Bromo-1,1,2,2-tetrafluoro-2-iodoethane,   B-80260
9CI

[421-70-5]

$$F_2CICBrF_2$$

C_2BrF_4I M 306.824
Shows anaesthetic activity. Bp 80.5-81°. n_D^{16} 1.433.

[57171-64-9]

Haszeldine, R.N., *J. Chem. Soc.*, 1953, 1764, 2075 (*synth, uv*)
Serboli, G. *et al, Spectrochim. Acta, Part A*, 1971, **27**, 1175 (*ir*)
Cavalli, L., *J. Magn. Reson.*, 1972, **6**, 298 (*F-19 nmr*)

3-Bromo-1,1,2,2-tetramethylcyclopropane,   B-80261
9CI

[3815-06-3]

$C_7H_{13}Br$ M 177.084
Liq. Bp$_{22}$ 51°.

Seyferth, D. *et al, J. Org. Chem.*, 1963, **28**, 703 (*synth, pmr*)
Hackett, M. *et al, J. Am. Chem. Soc.*, 1988, **110**, 1449 (*synth, pmr*)

5-Bromo-1*H*-tetrazole, 9CI, 8CI   B-80262

[42371-37-9]

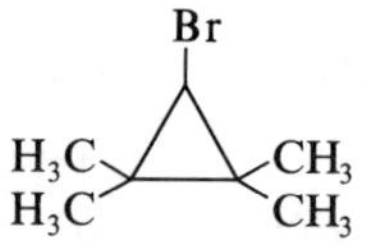

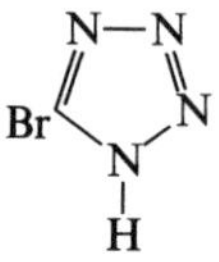

$CHBrN_4$ M 148.950
Needles (C_6H_6). Mp 156° dec. pK_a 4.13.
1-Me: [16681-79-1].
    $C_2H_3BrN_4$ M 162.976
    Yellow cryst. Mp 72-74°.
1-Ph: [18233-34-6].
    $C_7H_5BrN_4$ M 225.047
    Cryst. Mp 151° dec.

Stollé, R. *et al, Ber.*, 1929, **62**, 1123 (*synth*)
Stollé, R. *et al, J. Prakt. Chem.*, 1932, **134**, 282 (*phenyl*)
Lieber, E. *et al, Anal. Chem.*, 1951, **23**, 1594 (*ir*)
Lieber, E. *et al, J. Am. Chem. Soc.*, 1951, **73**, 1792 (*props*)
Hattori, K. *et al, J. Am. Chem. Soc.*, 1956, **78**, 411 (*methyl*)
Ansell, J.B. *et al, J. Chem. Soc., Perkin Trans. 2*, 1973, 2036 (*cryst struct*)

5-Bromo-1,2,3-triazine   B-80263

[114078-88-5]

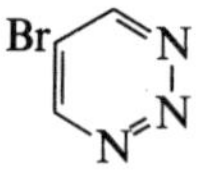

$C_3H_2BrN_3$ M 159.973
Mp 125-126° dec.

Ohsawa, A. *et al, Chem. Pharm. Bull.*, 1988, **36**, 3838 (*synth, pmr*)

1-Bromo-2,3,3-trifluorocyclopropene, 9CI   B-80264

[29777-44-4]

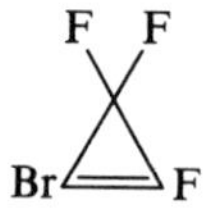

C_3BrF_3 M 172.932
Bp 38°, Bp 44°.

McGlinchey, M.J. *et al, J. Chem. Soc., Dalton Trans.*, 1970, 1264 (*synth, F-19 nmr, ir*)
Law, D.C.F. *et al, J. Org. Chem.*, 1973, **38**, 771 (*synth, ir, F-19 nmr*)

2-Bromo-1,1,1-trifluoroethane, 9CI   B-80265

2,2,2-Trifluoroethyl bromide

[421-06-7]

$$F_3CCH_2Br$$

$C_2H_2BrF_3$ M 162.937

Anaesthetic. d_4^{20} 1.788. Mp $-93.9°$. Bp $26.3°$. n_D^{20} 1.3331.

[30283-91-1]

Robbins, B.H., *J. Pharmacol.*, 1946, **86**, 197 (*use*)
Henne, A.L. *et al, J. Am. Chem. Soc.*, 1948, **70**, 1025 (*synth*)
Nielsen, J.R. *et al, J. Chem. Phys.*, 1957, **27**, 891 (*ir, raman*)
Saur, O. *et al, Spectrochim. Acta, Part A*, 1973, **29**, 243 (*ir*)
Van der Puy, M., *J. Fluorine Chem.*, 1979, **13**, 375 (*synth*)
Kushida, K. *et al, Proton and Fluorine Nuclear Magnetic Resonance Spectral Data*, Japan Halon Co. Ltd, 1988, 028, 029 (*F-19 nmr, pmr*)

5-(2-Bromovinyl)-2′-deoxyuridine   B-80266

Updated Entry replacing B-20211

5-(2-Bromoethenyl)-2′-deoxyuridine, 9CI. 2′-Deoxy-5-(2-bromovinyl)uridine. **Brivudine,** INN. *Bromovinyl deoxyuridine. BVDU*

[82768-44-3]

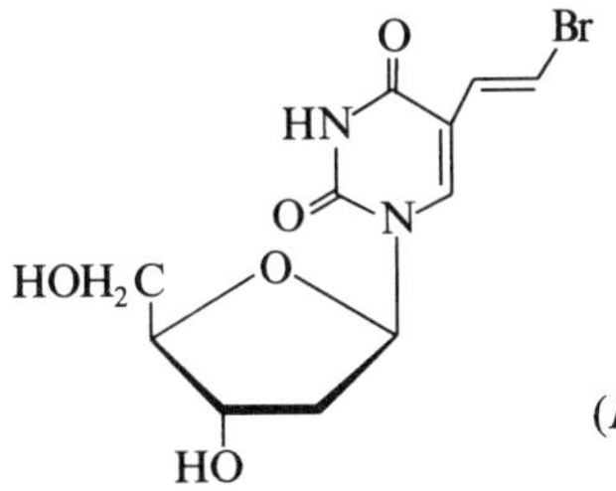

(*E*)-*form*

$C_{11}H_{13}BrN_2O_5$ M 333.138

Antiviral agent highly active against herpes simplex.

(*E*)-*form* [69304-47-8]

Mp 123-125° dec. (141°). λ_{max} 253 (ϵ 13 100) and 295 nm(ϵ 10 300).

▷ YU7355000.

3′,5′-Di-Ac:

$C_{15}H_{17}BrN_2O_7$ M 417.212

Antiviral active against HSV-1. Needles (2-propanol). Mp 156°.

3-Me:

$C_{12}H_{15}BrN_2O_5$ M 347.165

Cryst. (H_2O).

3′-Me ether:

$C_{12}H_{15}BrN_2O_5$ M 347.165

Solid.

(*Z*)-*form* [77530-02-0]

Hemihydrate. λ_{max} 234 (ϵ 11 904) and 295 nm (8 040).

Barr, P.J. *et al, J. Chem. Soc., Perkin Trans. 1*, 1981, 1665 (*synth*)
Jones, A.S. *et al, J. Med. Chem.*, 1981, **24**, 759 (*synth*)
Sim, T.S. *et al, CA*, 1982, **97**, 207422n (*rev*)
Martindale, The Extra Pharmacopoeia, 28th/29th Ed., Pharmaceutical Press, London, 1982/1989, 12455.
Goodchild, J. *et al, J. Med. Chem.*, 1983, **26**, 1252 (*synth, Z and E-forms*)
Robins, M.J. *et al, J. Org. Chem.*, 1983, **48**, 1854 (*synth*)
De Clercq, E. *et al, J. Antimicrob. Chemother.*, 1984, **14**, 85 (*rev, pharmacol*)
De Clercq, E. *et al, Pharmacol. Ther.*, 1984, **26**, 1 (*rev, synth, pharmacol*)
Parkanyi, L. *et al, Nucleic Acids Res.*, 1987, **15**, 4111 (*cryst struct, conformn*)
Reefschlaeger, J. *et al, Pharmazie*, 1987, **42**, 407 (*rev*)
Ashwell, M. *et al, Tetrahedron*, 1987, **43**, 4601 (*synth, props, uv, pmr*)

Brosimone B   B-80267

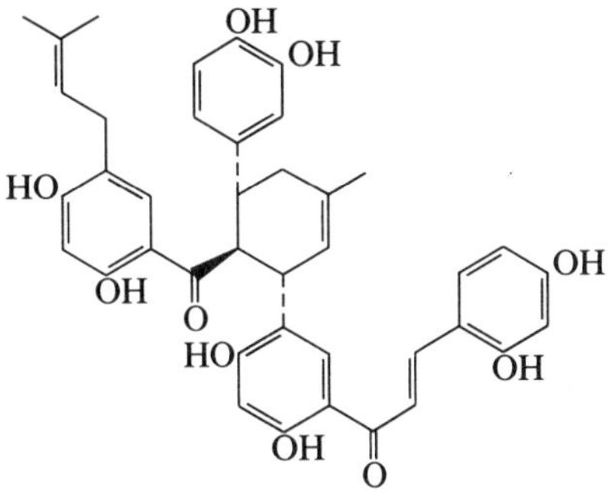

$C_{40}H_{38}O_{10}$ M 678.734

Constit. of *Brosimopsis oblongifolia*. Amorph. powder. $[\alpha]_D$ $-447°$ (c, 0.1 in MeOH).

Messana, I. *et al, Heterocycles*, 1989, **29**, 683.

Brosimone D   B-80268

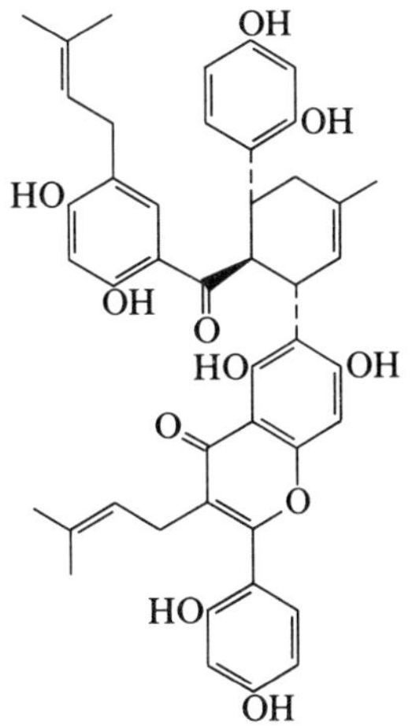

$C_{45}H_{44}O_{11}$ M 760.836

Constit. of *Brosimopsis oblongifolia*. Amorph. powder. $[\alpha]_D$ $-204°$ (c, 0.3 in MeOH).

Messana, I. *et al, Heterocycles*, 1989, **29**, 683.

Bruceantinoside B   B-80269

Updated Entry replacing B-10322

Yadanzioside P

[79439-84-2]

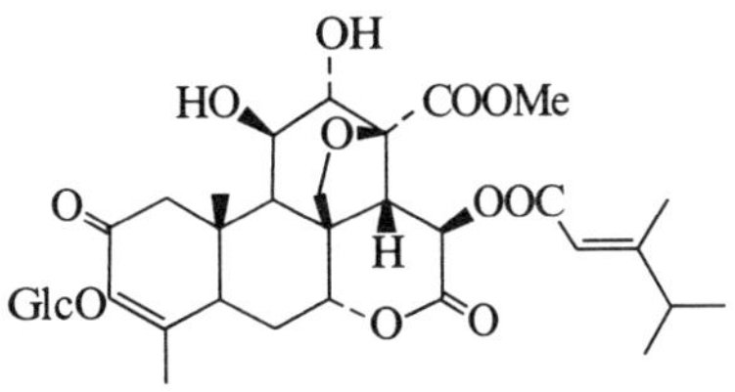

$C_{34}H_{46}O_{16}$ M 710.728

Constit. of *Brucea antidysenterica*. Amorph. powder. Mp 193-198°. $[\alpha]_D^{23}$ $+7.0°$ (c, 0.57 in EtOH).

Okano, M. *et al, J. Nat. Prod. (Lloydia)*, 1981, **44**, 470 (*isol*)
Okano, M. *et al, J. Nat. Prod. (Lloydia)*, 1989, **52**, 398 (*isol, pmr, cmr*)

Bryophollenone B-80270

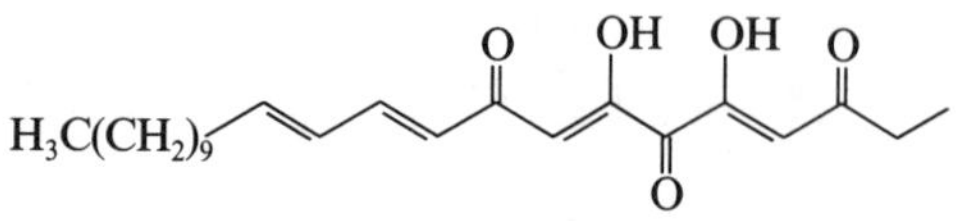

$C_{23}H_{34}O_5$ M 390.519
Constit. of *Bryophyllum pinnatum*.

Siddiqui, S. *et al, Phytochemistry*, 1989, **28**, 2433 (*isol, pmr, ms*)

Bryophollone B-80271

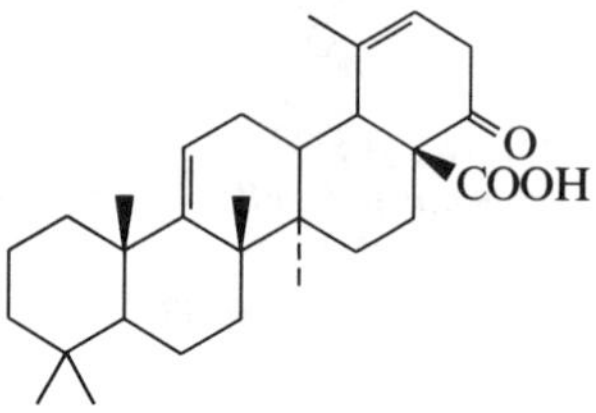

$C_{29}H_{42}O_3$ M 438.649
Constit. of *Bryophyllum pinnatum*. Cryst. (MeOH/C_6H_6).
Mp 210°.

Siddiqui, S. *et al, Phytochemistry*, 1989, **28**, 2433 (*isol, pmr*)

Bryophyllin B B-80272

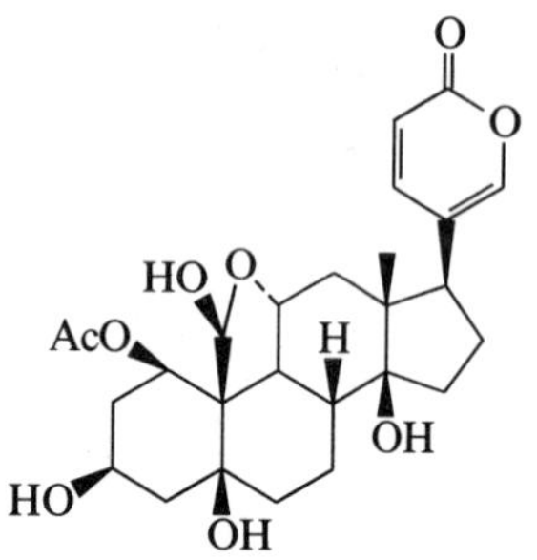

$C_{26}H_{34}O_9$ M 490.549
Constit. of *Bryophyllum pinnatum*. Amorph. powder. Mp
178-180°. $[\alpha]_D^{20}$ +20° (c, 0.1 in $CHCl_3$).

Yamagishi, T. *et al, J. Nat. Prod. (Lloydia)*, 1989, **52**, 1071 (*isol, pmr, cmr*)

Bullatacin B-80273

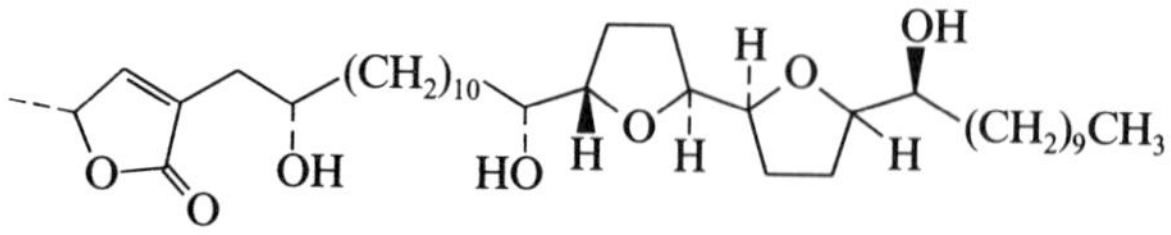

$C_{37}H_{66}O_7$ M 622.924
Constit. of *Annona bullata*. Cytotoxic agent. Cryst.
(EtOAc). Mp 69-70°. $[\alpha]_D^{23}$ +13° (c, 0.004 in $CHCl_3$).

Hui, Y.-H. *et al, J. Nat. Prod. (Lloydia)*, 1989, **52**, 463 (*isol, pmr, cmr*)

Bullatacinone B-80274

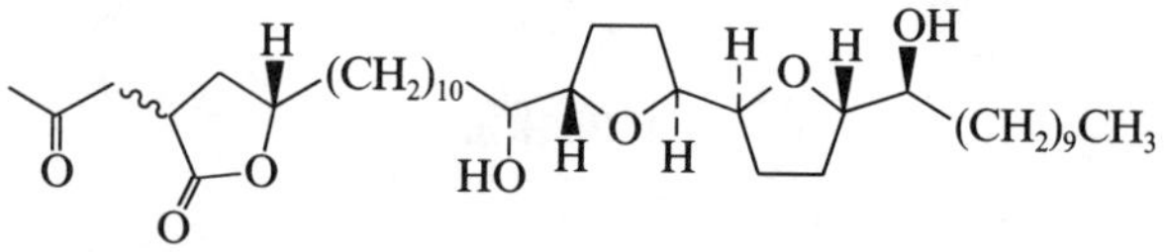

$C_{37}H_{66}O_7$ M 622.924
Constit. of *Annona bullata*. Cytotoxic agent. Cryst.
(MeOH). Mp 90.5-90.7°. $[\alpha]_D^{23}$ +12.0° (c, 0.4 in $CHCl_3$).

Hui, Y.-H. *et al, J. Nat. Prod. (Lloydia)*, 1989, **52**, 463 (*isol, pmr, cmr*)

Bullatalicin B-80275

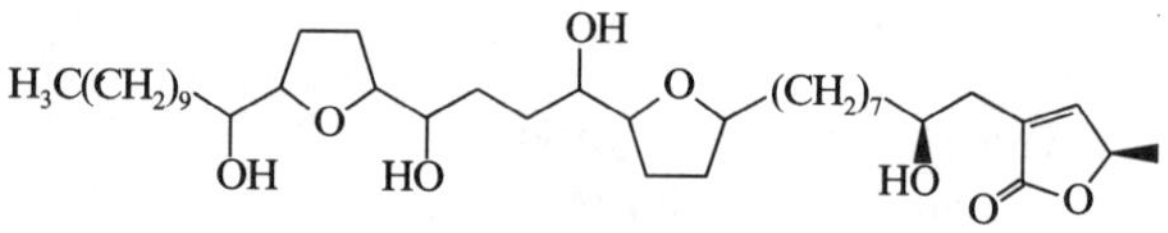

$C_{37}H_{66}O_8$ M 638.924
Constit. of *Annona bullata*. Amorph. solid. Mp 120-121°.
$[\alpha]_D$ +13.25° (c, 4.0 in EtOH).

▷ Cytotoxic.

Hui, Y. *et al, Tetrahedron*, 1989, **45**, 6941.

Buprofezin B-80276

*2-[(1,1-Dimethylethyl)imino]tetrahydro-3-(1-methylethyl)-5-
phenyl-4H-1,3,5-thiadiazin-4-one, 9CI. 2-tert-Butylimino-3-
isopropyl-5-phenyl-1,3,5-thiadiazinan-4-one. Applaud*
[69327-76-0]

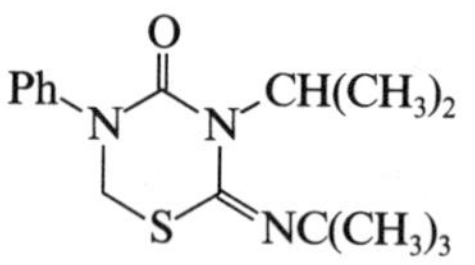

$C_{16}H_{23}N_3OS$ M 305.443
Insecticide, acaricide. Cryst. Mp 104.5-105.5°.

▷ LD_{50} 2,000 mg/Kg (rat).

Ger. Pat., 2 824 126, (1978); *ca*, **90**, 121669s (*synth*)
Kanno, H. *et al, Proc. Brit. Crop Prot. Comb. Pests Dis.*, 1981, **1**,
59 (*use*)
Pesticide Manual, 8th Ed., 1987, 1480.

2-(1,3-Butadienyl)-3-(1,3,5-heptatriynyl) B-80277
oxirane

5,6-Epoxy-1,3-tridecadiene-7,9,11-triyne. **Cotaepoxide**

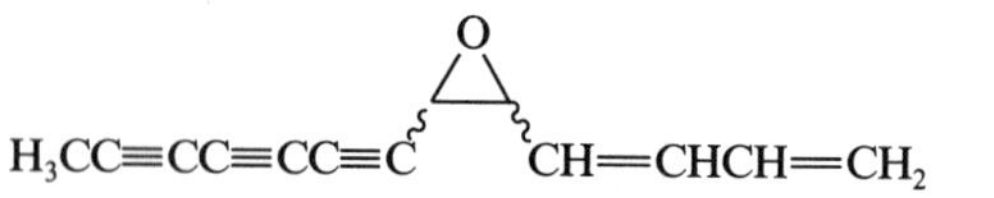

$C_{13}H_{10}O$ M 182.221
Isol. from roots of *Anthemis cota*. Cryst. (pet. ether). Mp
66°. $[\alpha]_{546}^{20}$ −302° (c, 0.4 in Et_2O).

Bohlmann, F. *et al, Chem. Ber.*, 1962, **95**, 1320; 1963, **96**, 1485
(*isol, uv, ir, struct*)

2-Buten-1-amine, 9CI B-80278

Updated Entry replacing A-01176
1-Amino-2-butene. Crotylamine
[21035-54-1]

$$H_3CCH=CHCH_2NH_2$$

C_4H_9N M 71.122
▷ EN0702500.

(E)-form [56930-04-2]
  Bp 83.6-84.2°. n_D^{25} 1.4320.
Picrate: Mp 158.3-159°.
(Z)-form [58887-05-1]
  d^{29} 0.793. Bp 85-87°. n_D^{28} 1.4295.
B,HCl: Flat prisms (EtOH/Et$_2$O). Mp 138°.
Picrate: [55987-11-6].
  Cryst. (EtOH aq.). Mp 131.5-132.5°.

Bookman, S., *Ber.*, 1895, **28**, 3114 (*synth*)
Galand, E. *et al*, *Bull. Soc. Chim. Belg.*, 1930, **39**, 529 (*synth*)
Roberts, J.D. *et al*, *J. Am. Chem. Soc.*, 1951, **73**, 2509 (*synth*)
Kjaer, A. *et al*, *Acta Chem. Scand.*, 1954, **8**, 1335 (*synth*)
Ettlinger, M.G. *et al*, *J. Am. Chem. Soc.*, 1955, **77**, 1831 (*synth*)

5-(3-Buten-1-ynyl)-5′ hydroxymethyl-2,2′-bithienyl   B-80279

5′-(3-Buten-1-ynyl)-[2,2′-bithienyl]-2-methanol
[1211-41-2]

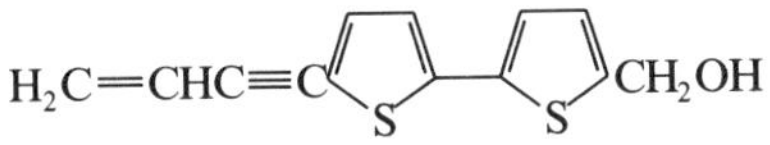

C$_{13}$H$_{10}$OS$_2$ M 246.353
Isol. from *Echinops sphaerocephalus*, *Buphthalmum salicifolium*, *B. grandiflorum* and *Bidens dahlioides*. Yellowish cryst. (CCl$_4$ or Et$_2$O/pet. ether). Mp 98°.

Ac: [1152-21-2]. *5-Acetoxymethyl-5′-(3-buten-1-ynyl)-2,2′-bithienyl*
  C$_{15}$H$_{12}$O$_2$S$_2$ M 288.391
  Isol. from roots of *Flaveria repanda*, *E. sphaerocephalus*, *Buphthalmum salicifolium*, *B. grandiflorum* and *Bidens dahlioides*. Yellowish cryst. (pet. ether). Mp 53.5°.

Deoxy: [1137-83-3]. *5-(3-Buten-1-ynyl)-5′-methyl-2,2′-bithienyl*
  C$_{13}$H$_{10}$S$_2$ M 230.354
  Isol. from *Buphthalmum salicifolium* and *B. grandiflorum*. Light-yellow cryst. (pet. ether). Mp 61° (53-54°).

[1152-21-2]

Bohlmann, F. *et al*, *Chem. Ber.*, 1963, **96**, 1229; 1965, **98**, 155, 883, 1228; 1966, **99**, 984 (*isol, uv, ir, pmr, biosynth*)
Atkinson, R.E. *et al*, *J. Chem. Soc. C*, 1967, 2011 (*synth*)
Curtis, R.F. *et al*, *Tetrahedron*, 1967, **23**, 4419 (*uv, ir*)

2-(3-Buten-1-ynyl)-5-(1,3-pentadiynyl) thiophene   B-80280

[1205-94-3]

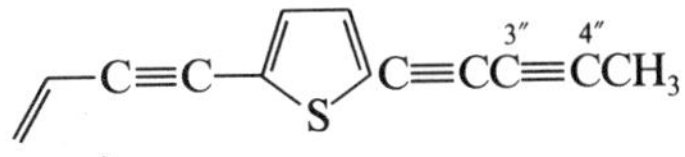

C$_{13}$H$_8$S   M 196.272
Isol. from roots of *Echinops sphaerocephalus*, from *Calocephalus citreus*, *Eriophyllum caespitosum*, *Ambrosia trifida* and *A. trifoliata*. Oil. Mp 5-10°. Bp$_{0.001}$ 110°.

3″,4″-Dihydro (E)-: [16714-43-5]. *2-(3-Buten-1-ynyl)-5-(3-penten-1-ynyl)thiophene*
  C$_{13}$H$_{10}$S   M 198.288
  Isol. from *Lasthenia glaberrima*, *Baeria* spp., *Bidens* spp., and *Centaurea* spp. Liq. Bp$_{0.001}$ 80°. λ_{max} 333.5, 355 nm.

Bohlmann, F. *et al*, *Chem. Ber.*, 1964, **97**, 2125; 1965, **98**, 155, 1228, 3081; 1966, **99**, 984 (*isol, uv, ir, pmr, struct*)
Sörensen, J.S. *et al*, *Acta Chem. Scand.*, 1965, **18**, 2182 (*isol, uv, ir, pmr, struct*)
Atkinson, R.E. *et al*, *Tetrahedron Lett.*, 1965, 297 (*isol, deriv*)

5-[5-(3-Buten-1-ynyl)-2-thienyl]-2-penten-4-yn-1-ol   B-80281

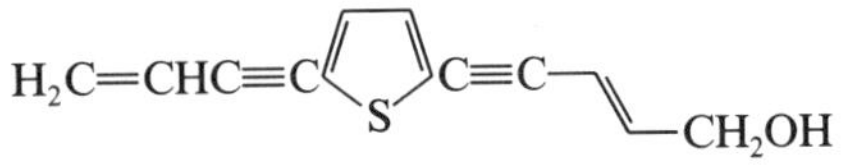

C$_{13}$H$_{10}$OS   M 214.287
(E)-form [14744-66-2]
  Minor constit. of roots of *Baeria* spp. Yellow cryst. (pet. ether). Mp 39°. λ_{max} 335 mm (Et$_2$O).

Ac: [14744-67-3]. *2-(1-Acetoxy-3-penten-1-ynyl)-5-(3-buten-1-ynyl)thiophene. 1-Acetoxy-5-[5-(3-buten-1-ynyl)-2-thienyl]-2-penten-4-yne*
  C$_{15}$H$_{12}$O$_2$S   M 256.325
  Cryst. (pet. ether). Mp 43.5-44°. λ_{max} 335 mm (Et$_2$O).

1-Aldehyde: 5-[5-(3-Buten-1-ynyl)-2-thienyl]-2-penten-4-ynal, 9CI
  C$_{13}$H$_8$OS   M 212.272
  Isol. from roots of *B.* spp. and *Veronia cotoneaster*. Cryst. (pet. ether). Mp 60°.

Bohlmann, F. *et al*, *Chem. Ber.*, 1964, **97**, 2125; 1967, **100**, 1200 (*isol, uv, ir, pmr, struct, synth*)
Bohlmann, F. *et al*, *Phytochemistry*, 1982, **21**, 695 (*isol, deriv*)

1-*tert*-Butoxy-1-hexyne   B-80282

[113279-42-8]

$$H_3C(CH_2)_3C{\equiv}COC(CH_3)_3$$

C$_{10}$H$_{18}$O   M 154.252
Oil.

Pericàs, M.A. *et al*, *Tetrahedron*, 1987, **43**, 2311 (*synth, ir, pmr, cmr, ms*)

3-*tert*-Butyl-2-cyclobuten-1-one   B-80283

3-(1,1-Dimethylethyl)-2-cyclobuten-1-one, 9CI
[95904-85-1]

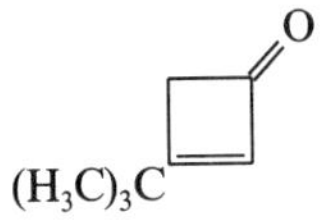

C$_8$H$_{12}$O   M 124.182

Schmit, C. *et al*, *Tetrahedron Lett.*, 1984, **25**, 5043 (*synth*)
Danheiser, R.L. *et al*, *Tetrahedron Lett.*, 1987, **28**, 3299 (*synth*)

6-*tert*-Butylcyclopenta[*c*]pyran   B-80284

6-(1,1-Dimethylethyl)cyclopenta[c]pyran
[115162-91-9]

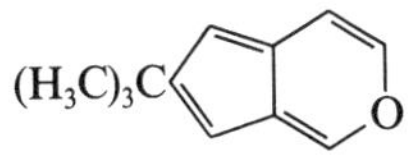

C$_{12}$H$_{14}$O   M 174.242
Stabilised homologue of Cyclopenta[*c*]pyran, C-80192. Lemon-yellow cryst. Mp 62° subl. Stable for several months at 4°.

Kämpchen, T. *et al*, *Justus Liebigs Ann. Chem.*, 1988, 855 (*synth, ir, uv, pmr, cmr*)

tert-Butylcyclopropane   **B-80285**

(1,1-Dimethylethyl)cyclopropane, 9CI

[4741-87-1]

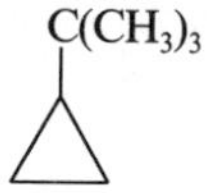

C_7H_{14} M 98.188

Kropp, P.J. *et al, Tetrahedron,* 1981, **37**, 3229 (*synth*)
Engel, P.S. *et al, J. Org. Chem.,* 1988, **53**, 4748 (*synth, ms*)

3-*tert*-Butyl-1,4,2-dioxazol-5-one   **B-80286**

3-(1,1-Dimethylethyl)-1,4,2-dioxazol-5-one, 9CI

[114379-11-2]

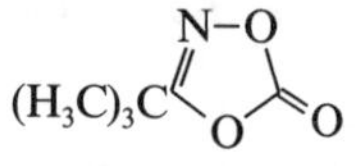

$C_6H_9NO_3$ M 143.142
Liq. $Bp_{3.75}$ 70°.

Mitra, S. *et al, J. Org. Chem.,* 1988, **53**, 2674 (*synth, pmr, ir*)

tert-Butyl hypoiodite   **B-80287**

[917-97-5]

$(H_3C)_3COI$

C_4H_9IO M 200.019
The reagent known by this name is oligomeric and two
different compositions are obt. depending on the
synthetic route. Homolytic and heterolytic source of
iodine. Made and used in soln. V. moisture-sensitive.

Glover, S.A. *et al, Tetrahedron Lett.,* 1980, **21**, 2005 (*synth, use*)
Tanner, D.D. *et al, J. Am. Chem. Soc.,* 1984, **106**, 5261 (*synth, struct, use, bibl*)
Heasley, V.L. *et al, J. Org. Chem.,* 1988, **53**, 198 (*props, bibl*)

4-*tert*-Butyl-2(3*H*)-oxazolethione   **B-80288**

4-(1,1-Dimethylethyl)-2(3H)-oxazolethione, 9CI

[112683-63-3]

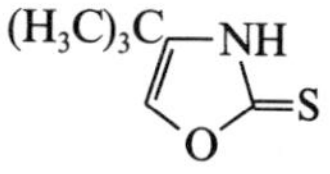

$C_7H_{11}NOS$ M 157.236
Cryst. Mp 145-147°.

Mohanraj, S. *et al, J. Org. Chem.,* 1988, **53**, 1113 (*synth, ir, pmr, cmr, cryst struct*)

2-*tert*-Butylpentanedioic acid   **B-80289**

*2-(1,1-Dimethylethyl)pentanedioic acid. 2-tert-Butylglutaric
acid*

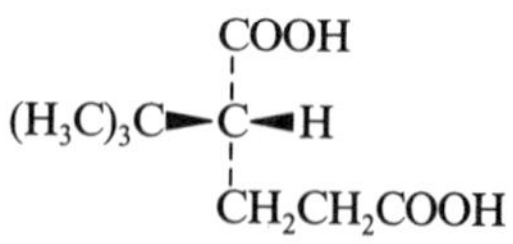

$C_9H_{16}O_4$ M 188.223
(*R*)-*form* [118060-95-0]
    Cryst. (toluene/hexane). Mp 94-95°. $[\alpha]_D^{20}$ +14° (c, 2 in
    Me_2CO).
Anhydride: [118060-96-1].

$C_9H_{14}O_3$ M 170.208
Cryst. (toluene/hexane). Mp 51-52°.
Imide: [118060-97-2]. *3-tert-Butyl-2,6-piperidinedione. 2-
tert-Butylglutarimide*
$C_9H_{15}NO_2$ M 169.223
Cryst. (toluene/hexane). Mp 158-159°. $[\alpha]_D^{20}$ −24° (c, 1
in MeOH).

Poloński, T., *J. Chem. Soc., Perkin Trans. 1,* 1988, 639 (*synth, cd,
pmr, ir*)

2-Butylpiperidine   **B-80290**

[72939-22-1]

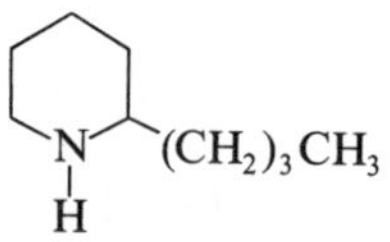

$C_9H_{19}N$ M 141.256
N-Me: **2-Butyl-1-methylpiperidine.** *1-Methyl-2-
butylpiperidine*
$C_{10}H_{21}N$ M 155.283
Trace alkaloid from oil of *Pimpinella acuminata*
(Umbelliferae). Opt. activity not reported.
(+)-*form*
    $[\alpha]_D$ +15.7°.
    B,H_2PtCl_6: Mp 131-132°.
(−)-*form*
    $[\alpha]_D$ −18.7°.
    Tartrate salt: Needles. Mp 41-42°. Forms a dihydrate.
(±)-*form* [68144-45-6]
    Liq. Bp 191-193°, Bp_{14} 75°.
    B,HCl: Mp 191-193° (181-182°).
    B,HBr: Mp 196-197°.
    N-Me: Liq. Bp_{15} 75-80°.
    N-Me, picrate: Mp 88°.

[68474-12-4]

Loffler, K. *et al, Ber.,* 1907, **40**, 1310 (*synth, resoln*)
Diels, O. *et al, Justus Liebigs Ann. Chem.,* 1937, **530**, 68 (*synth*)
Astier, A. *et al, Tetrahedron Lett.,* 1978, 2051 (*synth*)
Ashraf, M. *et al, Pak. J. Sci. Ind. Res.,* 1979, **22**, 79 (*isol, deriv*)
Scully, F.E. *et al, J. Org. Chem.,* 1980, **45**, 1515 (*synth, pmr, ir*)
Pirkle, W.H. *et al, J. Org. Chem.,* 1984, **49**, 2504 (*resoln*)

3(5)-*tert*-Butylpyrazole   **B-80291**

3(5)-(1,1-Dimethylethyl)pyrazole

[15802-80-9]

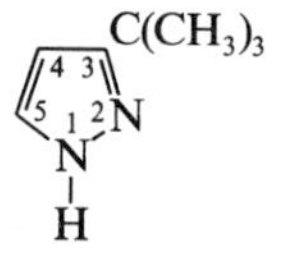

$C_7H_{12}N_2$ M 124.185
Mp 53-54°. $Bp_{0.15}$ 127° ($Bp_{0.3}$ 76-80°) .

Elguero, J. *et al, Bull. Soc. Chim. Fr.,* 1966, 3744; 1968, 707 (*synth,
uv*)
Magee, W.L. *et al, J. Am. Chem. Soc.,* 1977, **99**, 633 (*synth*)
Cabildo, P. *et al, Org. Magn. Reson.,* 1984, **22**, 603 (*cmr*)
Trofimenko, S. *et al, Inorg. Chem.,* 1987, **26**, 1507 (*synth, pmr*)
Sorrell, T.N. *et al, Inorg. Chem.,* 1987, **26**, 1755 (*synth, pmr*)

4-*tert*-Butylpyrazole   **B-80292**

4-(1,1-Dimethylethyl)pyrazole, 9CI

[105285-21-0]

$C_7H_{12}N_2$ M 124.185

Yellowish oil.

Picrate: Mp 100-102°.

Cativiela, C. *et al, Gazz. Chim. Ital.*, 1986, **116**, 119 (*synth, ir, uv, pmr, deriv*)

tert-Butylsulfinic acid   B-80293

2-Methyl-2-propanesulfinic acid

[29099-08-9]

$$(H_3C)_3CSO_2H$$

$C_4H_{10}O_2S$ M 122.188

Acid not isol.

2,4-Dinitrophenyl ester: [112881-95-5].
Cryst. Mp 83-85°.

4-Nitrobenzyl ester: Cryst. (C_6H_6/pet. ether). Mp 178°.

Chloride: [31562-43-3].
C_4H_9ClOS M 140.633
Pale yellowish oil. Bp_{11} 54-57°.

Kader, A.T. *et al, J. Chem. Soc.*, 1962, 3425 (*synth*)
Nooi, J.R. *et al, Tetrahedron Lett.*, 1970, 2531 (*synth, pmr*)
Block, E. *et al, J. Am. Chem. Soc.*, 1974, **96**, 3921 (*deriv, synth*)
Netscher, T. *et al, Synthesis*, 1987, 683 (*deriv, synth, ir, pmr*)

3-Butyn-1-amine, 9CI   B-80294

4-Amino-1-butyne

[14044-63-4]

$$HC{\equiv}CCH_2CH_2NH_2$$

C_4H_7N M 69.106

Liq. Bp 94-95° (99°) .

Dumont, J.L. *et al, Bull. Soc. Chim. Fr.*, 1967, 588 (*synth, ir*)
Taylor, E.C. *et al, Tetrahedron*, 1987, **43**, 5145 (*synth, ir, pmr*)

Butyrylmallotochromene   B-80295

[116979-51-2]

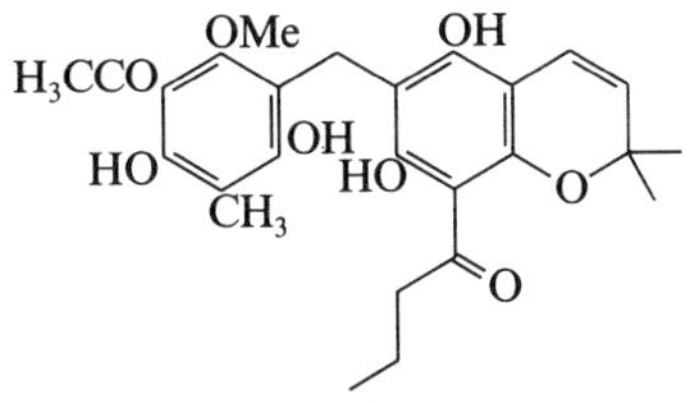

$C_{26}H_{30}O_8$ M 470.518

See also Isobutyrylmallotochromene, I-80050. Constit. of *Mallotus japonicus*. Cytotoxic agent. Yellow needles (MeOH). Mp 175-176°.

Fujita, A. *et al, J. Nat. Prod. (Lloydia)*, 1988, **51**, 708.

Byakangelicin   B-80296

Updated Entry replacing B-03861

9-(2,3-Dihydroxy-2-methylbutoxy)-4-methoxy-7H-furo[3,2-g][1]benzopyran-7-one, 9CI

[19573-01-4]

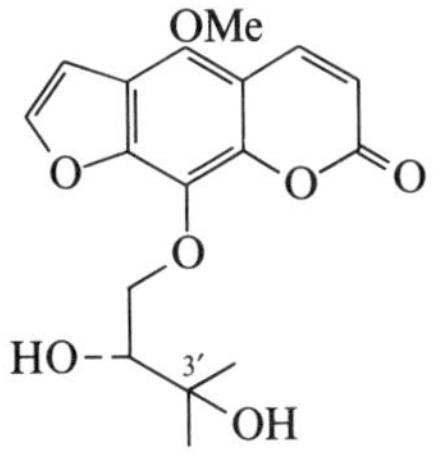

Absolute configuration

$C_{17}H_{18}O_7$ M 334.325

(R)-form [482-25-7]

Constit. of Japanese drug byakusi obt. from *Angelica* spp. Mp 117-118°. Opt. rotn. data inaccessible but is thought to be dextrorotatory.

▷ LV1049000.

O-*Angeloyl:*
$C_{22}H_{24}O_8$ M 416.427
Isol. from *Seseli libanotis*. No opt. rotn. or config.

3′-Deoxy, 3′-chloro:
$C_{17}H_{17}ClO_6$ M 352.770
Isol. from *Heracleum pyrenaicum*. Mp 156-158°. $[\alpha]_D$ −28.3° ($CHCl_3$).

3′-Deoxy, 3′,4′-didehydro: [35214-82-5]. **Neobyakangelicol**
$C_{17}H_{16}O_6$ M 316.310
Isol. from roots of *A. dahurica*. Mp 106-107°. $[\alpha]_D^{20}$ −2.0° (EtOH). Abs. config. of side-chain not known.

3′-Deoxy, 2′-ketone: [35214-81-4]. **Anhydrobyakangelicin.**
Isobyakangelicol
$C_{17}H_{16}O_6$ M 316.310
Isol. from roots of *A. dahurica*. Mp 108-109°.

Perel'son, M.E. *et al, Khim. Prir. Soedin.*, 1971, 576; *CA*, **76**, 45335y (*pmr*)
Saiki, Y. *et al, Yakugaku Zasshi (J. Pharm. Soc. Jpn.)*, 1971, **91**, 1313 (*Neobyakangelicol, Anhydrobyakangelicin*)
Karrer, W. *et al, Konstitution und Vorkommen der Organischen Pflanzenstoffe*, 2nd Ed., Birkhäuser Verlag, Basel, 1972-1985, no. 580 (*occur*)
Fukushima, S. *et al, Chem. Pharm. Bull.*, 1974, **22**, 1227 (*ms*)
Kovalev, I.P. *et al, Farm. Zh. (Kiev)*, 1975, **30**, 44; *CA*, **84**, 104899a (*abs config*)
González, A.G. *et al, An. Quim.*, 1976, **72**, 584 (*deriv*)

C

1(10),4-Cadinadiene-3,9-diol   C-80001

3,9-Dihydroxy-δ-cadinene

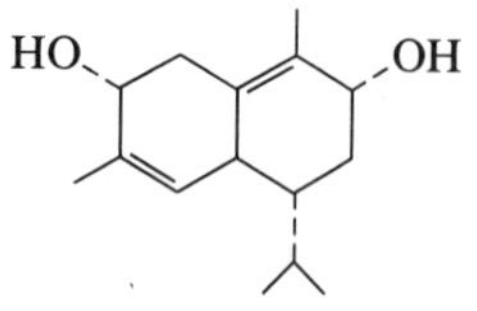

$C_{15}H_{24}O_2$ M 236.353

Constit. of *Helichrysum dasyanthum*. Gum.

Jakupovic, J. *et al*, *Phytochemistry*, 1989, **28**, 1119.

4,11(13)-Cadinadien-12-oic acid   C-80002

Updated Entry replacing C-70004

1,2,3,4,4a,5,6,8a-Octahydro-4,7-dimethyl-α-methylene-1-naphthaleneacetic acid,9CI. Artemisininic acid. Artemisic acid. Qing Hau acid. Arteannuic acid. Artemisinic acid
[80286-58-4]

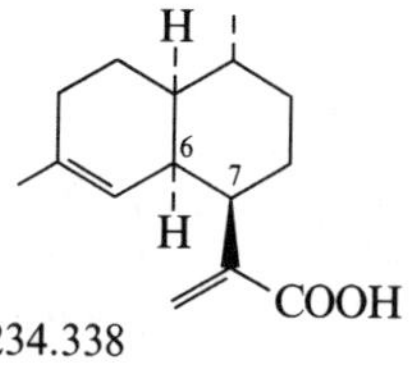

$C_{15}H_{22}O_2$ M 234.338

Isol. from *Artemisia annua*. Shows antibacterial activity. Cubes (pet. ether). Mp 131°. $[\alpha]_D$ +36° (c, 0.01 in $CHCl_3$).

6,7-Didehydro: 4,6,11(13)-Cadinatrien-12-oic acid. 6,7-Dehydroartemisinic acid
$C_{15}H_{20}O_2$ M 232.322
Constit. of *A. annua*. Cryst. (Et_2O/hexane). Mp 129-130°. $[\alpha]_D$ +317° (c,0.05 in MeOH).

Tu, Y.Y. *et al*, *Planta Med.*, 1982, **44**, 143 (*isol*)
Zhou, W. *et al*, *CA*, 1985, **103**, 160709 (*config*)
El-Feraly, F.S. *et al*, *J. Nat. Prod.* (*Lloydia*), 1989, **52**, 196 (*isol, deriv, pmr, cmr*)

1,9-Cadinadien-3-one   C-80003

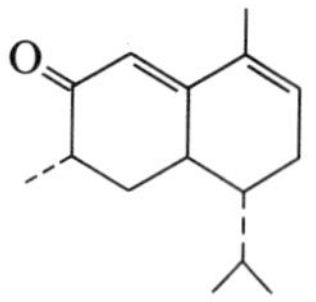

$C_{15}H_{22}O$ M 218.338

Constit. of *Helichrysum petiolare*. Oil.

Jakupovic, J. *et al*, *Phytochemistry*, 1989, **28**, 1119.

4-O-Cadinylangolensin   C-80004

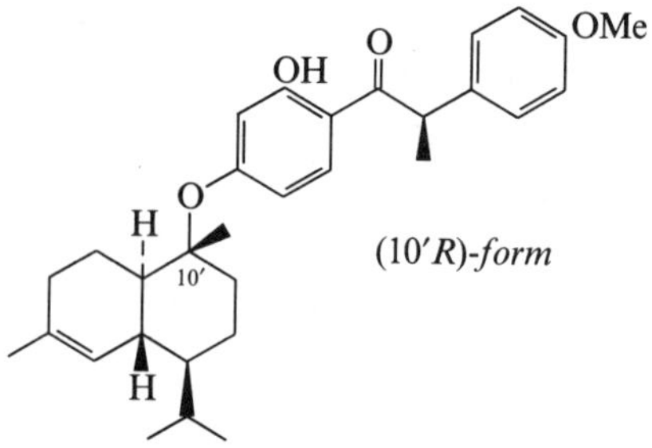

(10'R)-form

$C_{31}H_{40}O_4$ M 476.655

(10'R)-form [75917-91-8]
4-O-α-*Cadinylangolensin*
Isol. from *Pterocarpus angolensis* heartwood. Needles (EtOH). Mp 136°.

(10'S)-form [75872-84-3]
4-O-τ-*Cadinylangolensin*
Isol. from *P. angolensis* heartwood. Noncryst. Mp 44-46°.

Bezuidenhoudt, B.C.B. *et al*, *J. Chem. Soc., Perkin Trans. 1*, 1980, 2179 (*isol, cryst struct, ms, cd, pmr, cmr*)

Caespitol   C-80005

Updated Entry replacing C-00036
[50656-64-9]

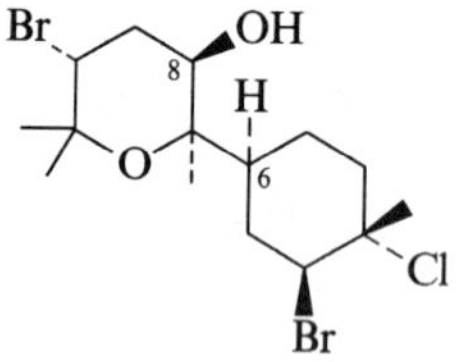

$C_{15}H_{25}Br_2ClO_2$ M 432.622

Constit. of *Laurencia caespitosa*. Cryst. (hexane). Mp 109-111°.

8-Deoxy: Caespitane
$C_{15}H_{25}Br_2ClO$ M 416.623
Constit. of *L. caespitosa*. Cryst. Mp 82-84°. $[\alpha]_D$ +39.8° (c, 0.89 in $CHCl_3$).

6α-Hydroxy: 6-Hydroxycaespitol
$C_{15}H_{25}Br_2ClO_3$ M 448.621
Constit. of *L. caespitosa*. Cryst. Mp 152-153°. $[\alpha]_D$ +11.7° (c, 0.29 in $CHCl_3$).

Gonzales, A.G. *et al*, *Tetrahedron Lett.*, 1974, 1249 (*struct*)
Gonzales, A.G. *et al*, *Tetrahedron Lett.*, 1976, 3051 (*biosynth*)
Gonzales, A.G. *et al*, *Tetrahedron Lett.*, 1979, 2719 (*cmr*)
Chang, M. *et al*, *Phytochemistry*, 1989, **28**, 1417 (*isol, cryst struct, abs config*)

4-O-Caffeoylshikimic acid   C-80006

Isodactylifric acid

$C_{16}H_{16}O_8$ M 336.298

Isol. from unripe dates (*Phoenix dactylifera*). Tentative struct.

Maier, V.P. *et al*, *Biochem. Biophys. Res. Commun.*, 1964, **14**, 124.

5-*O*-Caffeoylshikimic acid   C-80007

Neodactylifric acid

$C_{16}H_{16}O_8$ M 336.298

Isol. from unripe dates (*Phoenix dactylifera*). Tentative struct.

Maier, V.P. *et al, Biochem. Biophys. Res. Commun.*, 1964, **14**, 124.

Cajaisoflavone   C-80008

[72578-99-5]

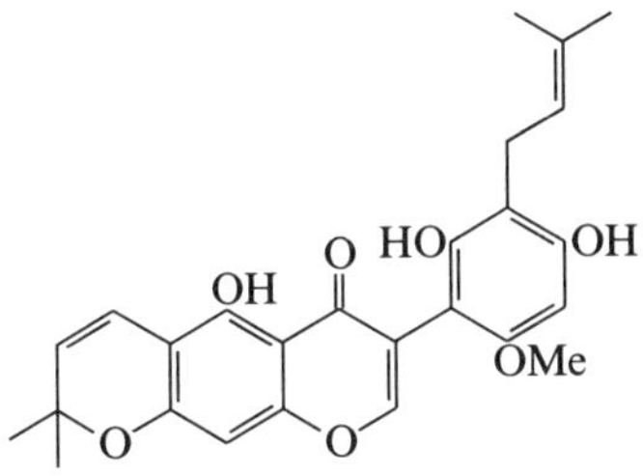

$C_{26}H_{26}O_7$ M 450.487

Isol. from root bark of *Cajanus cajan*. Orange semisolid.

Bhanumati, S. *et al, Phytochemistry*, 1979, **18**, 1254 (*isol, pmr, ms*)

Cajanone   C-80009

Updated Entry replacing C-00048

[63006-48-4]

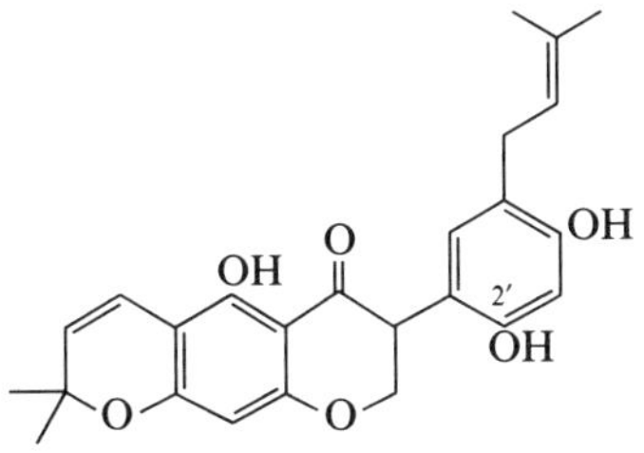

$C_{25}H_{26}O_6$ M 422.477

Constit. of *Cajanus cajan*. Shows antifungal activity.

2′-Me ether: [71765-79-2]. **2′-*O*-Methylcajanone**
    $C_{26}H_{28}O_6$ M 436.504
    Isol. from root bark of *C. cajan*. Yellow cryst. solid (EtOAc/pet. ether). Mp 85°. Opt. inactive.

Preston, N.W., *Phytochemistry*, 1977, **16**, 143 (*isol, uv, pmr, struct*)
Bhanumati, S. *et al, Phytochemistry*, 1979, **18**, 693 (*deriv*)

Caleamyrcenolide   C-80010

Updated Entry replacing C-20013

[84749-85-9]

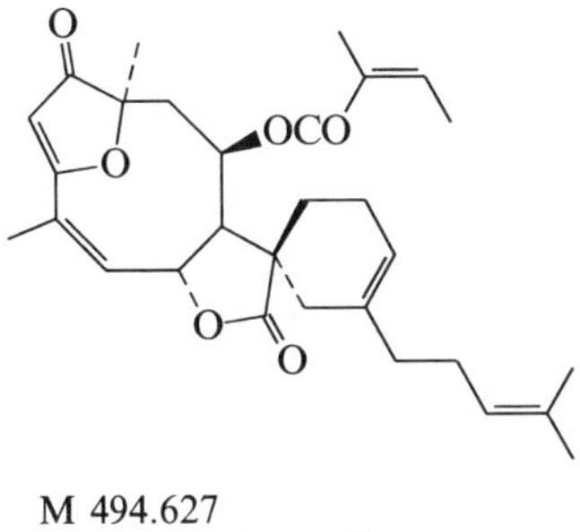

$C_{30}H_{38}O_6$ M 494.627

Constit. of *Calea hymenolepis*. Gum.

Bohlmann, F. *et al, Phytochemistry*, 1982, **21**, 2045.

Caleprunifolin   C-80011

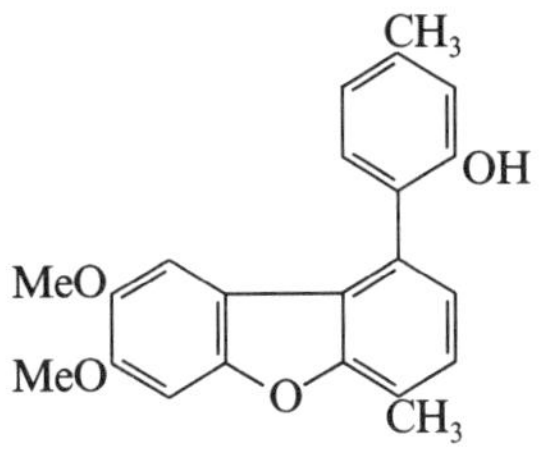

$C_{22}H_{20}O_4$ M 348.398

Constit. of *Calea prunifolia*. Gum.

Castro, V. *et al, Phytochemistry*, 1989, **28**, 2415 (*isol, pmr*)

Caleteucrin   C-80012

Caleurticin

[80453-42-5]

$C_{13}H_{14}O_5$ M 250.251

Constit. of *Calea urticifolia*. Oil.

7-Methoxy: **7-Methoxycaleteucrin**
    $C_{14}H_{16}O_6$ M 280.277
    Constit. of *Calea prunifolia*. Oil.

Bohlmann, F. *et al, Phytochemistry*, 1979, **18**, 119 (*isol, pmr*)
Castro, V. *et al, Phytochemistry*, 1989, **28**, 2415 (*isol, pmr*)

Calopogoniumisoflavone A   C-80013

*3-(4-Methoxyphenyl)-8,8-dimethyl-4*H,8H-*benzo[1,2-b:3,4-b′]dipyran-4-one, 9CI*

[31273-64-0]

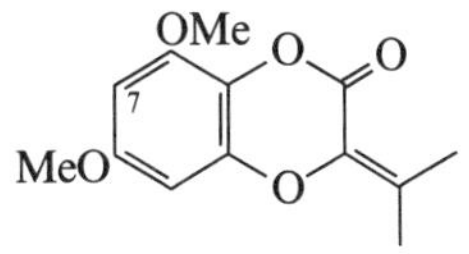

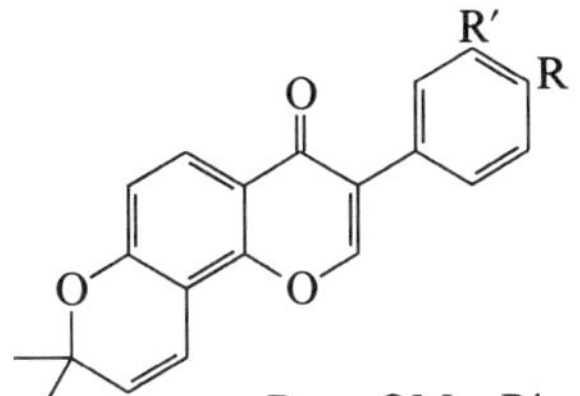

R = OMe, R′ = H

$C_{21}H_{18}O_4$ M 334.371

Isol. from seeds of *Calopogonium mucunoides*.

Vilain, C. *et al, Bull. Soc. R. Sci. Liege*, 1975, **44**, 306; *CA*, **84**, 74136e.

Candesalvone A   C-80014

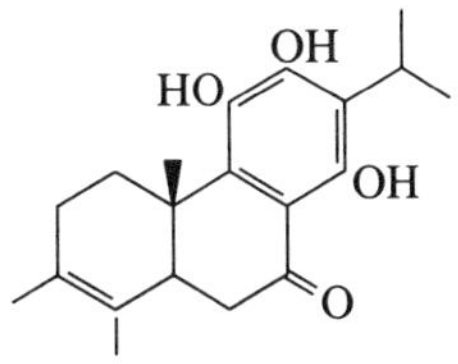

$C_{20}H_{28}O_4$ M 332.439

Constit. of *Salvia candelabrum*.

12-Me ether: Yellow plates (MeOH). Mp 209-212°.

Mendez, E. *et al, Phytochemistry*, 1989, **28**, 1685.

Candesalvone B   C-80015

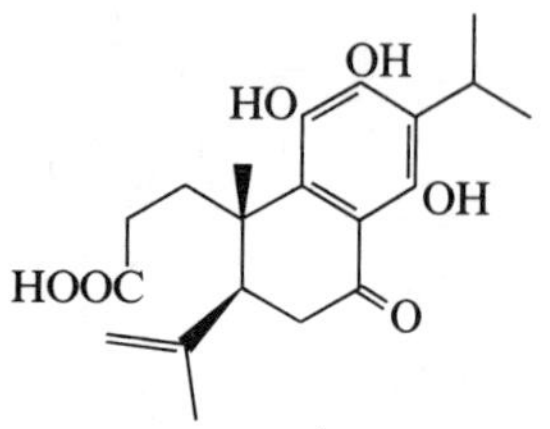

$C_{20}H_{26}O_6$ M 362.422
Constit. of *Salvia candelabrum*. Amorph. powder.
12-Me ether, Me ester: Yellow needles. Mp 85-87°.
Mendes, E. *et al, Phytochemistry*, 1989, **28**, 1685.

Candicanin   C-80016
[36149-85-6]

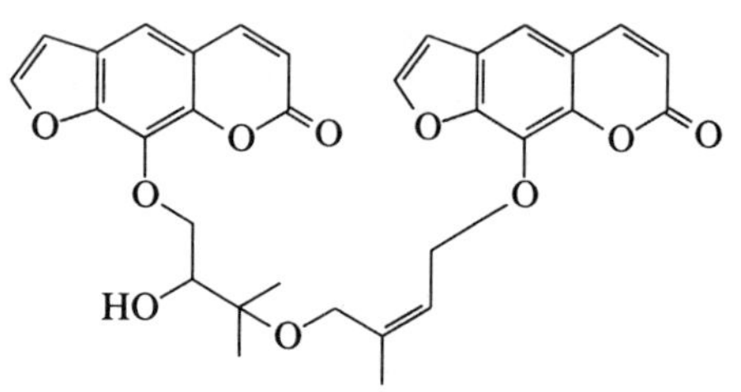

$C_{32}H_{28}O_{10}$ M 572.567
MF incorr. given by the authors as $C_{32}H_{26}O_{10}$. Constit. of the roots of *Heracleum candicans*. Mp 153°.
Bandopadhyay, M. *et al, Tetrahedron Lett.*, 1971, 4221.

Cannogeninic acid   C-80017

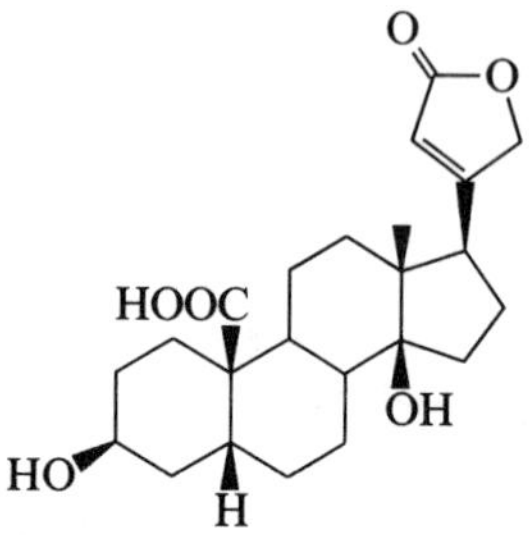

$C_{23}H_{32}O_6$ M 404.502
3-O-(6-Deoxy-3-O-methyl-L-glucopyranoside): **Perusitin**
$C_{30}H_{44}O_{10}$ M 564.672
Isol. from seeds of *Thevetia peruviana*. Cryst. Mp 168-170°. $[\alpha]_D^{28.5}$ −45.5° (c, 1.37 in MeOH).
Lang, H.Y. *et al, CA*, 1965, **62**, 9465.

Capsianside A   C-80018
[116107-40-5]

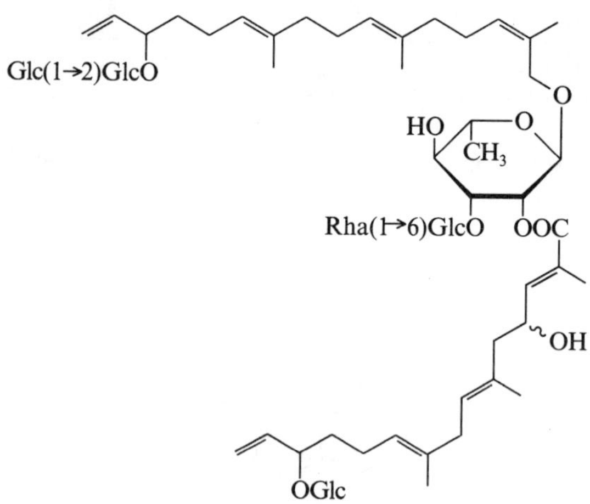

$C_{76}H_{124}O_{33}$ M 1565.796
Constit. of fruits of *Capsicum annuum*. Amorph. powder.
$[\alpha]_D$ −25.1° (MeOH).
Yahara, S. *et al, Tetrahedron Lett.*, 1988, **29**, 1943 (*struct*)

Carbonotrithioic acid, 9CI   C-80019
Updated Entry replacing C-00253
Trithiocarbonic acid
[594-08-1]

$$S=C(SH)_2$$

CH_2S_3 M 110.225
Liq. Mp −26.9°. Salts give red aq. solns.
Di-Na salt: Yellow needles (EtOH). Mp 80°.
Di-Me ester: [2314-48-9].
   $C_3H_6S_3$ M 138.278
   Yellow oil. Bp_{10} 90°.
Di-Et ester:
   $C_5H_{10}S_3$ M 166.332
   Yellow oil. Bp_7 130°.
Yeoman, E.W., *J. Chem. Soc.*, 1921, 38 (*salts*)
Gattow, G. *et al, Z. Anorg. Allg. Chem.*, 1963, **321**, 143 (*synth, ir, uv*)
Philippot, E., *Rev. Chim. Miner.*, 1967, **4**, 643 (*salts, rev*)
Tomita, K. *et al, Chem. Pharm. Bull.*, 1972, **20**, 2302 (*deriv*)
Barbero, M. *et al, Synthesis*, 1988, 22 (*Di-Me ester, synth, pmr, use*)

N,N'-Carbonylbisglycine, 9CI   C-80020
N,N'-*Bis(carboxymethyl)urea. Urea N,N'-diacetic acid*

$$HOOCCH_2NHCONHCH_2COOH$$

$C_5H_8N_2O_5$ M 176.129
Cryst. (H_2O). Mp 204-206° (188-189°).
Di-Me ester: [72129-70-5].
   $C_7H_{12}N_2O_5$ M 204.182
   Cryst. ($CHCl_3$/hexane). Mp 155-156°.
Di-Et ester: [7150-63-2].
   $C_9H_{16}N_2O_5$ M 232.236
   Cryst. ($CHCl_3$). Mp 147-148° (143-145°).
Wessely, F. *et al, Hoppe Seyler's Z. Physiol. Chem.*, 1928, **174**, 306 (*synth*)
Kondo, K. *et al, Synthesis*, 1979, 735 (*synth, ir*)
Buntain, I.G. *et al, J. Chem. Soc., Perkin Trans. 1*, 1988, 3175 (*synth, ir, pmr*)

3,3′-Carbonylbis[5-phenyl-1,3,4-oxadiazole-2(3H)-thione], 9CI   C-80021

[122350-19-0]

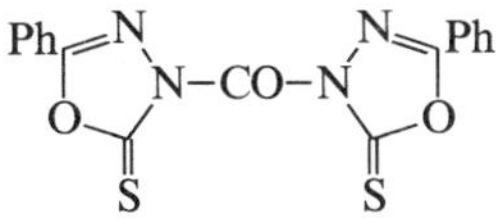

$C_{17}H_{10}N_4O_3S_2$ M 382.423

Stable one-pot condensing agent. Pale-yellow granular cryst. (C_6H_6). Mp 202°.

Saesuga, Y. *et al, Bull. Chem. Soc. Jpn.*, 1989, **62**, 539 (synth, use)

Carbosulfan   C-80022

2,3-Dihydro-2,2-dimethyl-7-benzofuranyl [(di-butylamino)thio]methylcarbamate, 9CI. Marshal. Advantage

[55285-14-8]

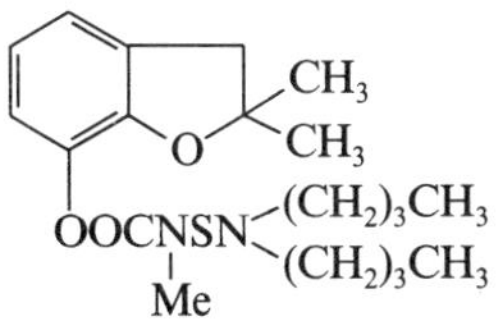

$C_{20}H_{32}N_2O_3S$ M 380.550

Systemic insecticide for control of soil pests. Brown viscous liq.

▷ LD$_{50}$ 200 mg/kg (Rats).

Ger. Pat., 2 433 680, (1975); *CA*, **82**, 156050 (synth)
Maitlen, E.C. *et al, Proc. Brit. Crop Prot. Comb. Pests Dis.*, 1979, **2**, 557 (use)
Pesticide Manual, 8th Ed., 1987, 2060.

2-Carboxycyclohexaneacetic acid, 9CI   C-80023

Hexahydrohomophthalic acid

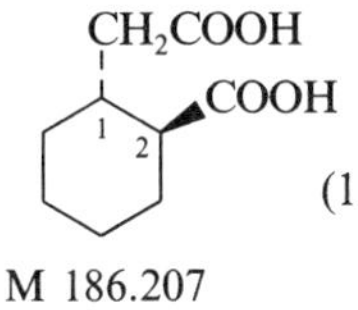

(1R,2S)-form

$C_9H_{14}O_4$ M 186.207

(1R,2S)-form
  (+)-trans-*form*
  Mp 116°. $[\alpha]_D^{20}$ +40.1° (in dioxan).
  Anhydride: *Hexahydro-1H-2-benzopyran-1,3(4H)-dione, 9CI*
    $C_9H_{12}O_3$ M 168.192
    Cryst. (toluene/hexane). Mp 111-112°. $[\alpha]_D^{20}$ −81° ($CHCl_3$).
(1R*,2R*)-form
  (+)-cis-form
  Cryst. Mp 82°. $[\alpha]_D^{20}$ +12.0°.
  Anhydride: Cryst. Mp 88°. $[\alpha]_D^{20}$ −30.3°.
(1RS,2RS)-form
  (±)-cis-*form*
  Mp 147°.
(1RS,2SR)-form
  (±)-trans-*form*
  Mp 161°.

[14715-37-8]

Benner, E. *et al, Acta Chem. Scand.*, 1954, **8**, 64 (synth, resoln, bibl)
Poloński, T., *J. Chem. Soc., Perkin Trans.* 1, 1988, 639 (synth, cd)

(2-Carboxyethyl)dimethylsulfonium(1+)   C-80024

Dimethyl-β-propriothetin

[6708-36-7]

$$Me_2S^{\oplus}CH_2CH_2COOH$$

$C_5H_{11}O_2S^{\oplus}$ M 135.207 (ion)

Isol. from green and red algae, e.g. *Enteromorpha intestinalis*, *Ulva lactuca*. Also from *Spartina anglica*. Biol. precursor of dimethyl sulfide.

Chloride: [4337-33-1].
  $C_5H_{11}ClO_2S$ M 170.659
  Needles (EtOH). Mp 134° dec.

Challenger, F. *et al, J. Chem. Soc.*, 1948, 1591 (isol)
Greene, R.C. *et al, J. Biol. Chem.*, 1962, **237**, 2251 (biosynth)
Larher, F. *et al, Phytochemistry*, 1977, **16**, 2019 (isol)
Blunden, G. *et al, Magn. Reson. Chem.*, 1986, **24**, 965 (nmr)

4-(2-Carboxyethyl)-4-nitroheptanedioic acid, 9CI   C-80025

[59085-15-3]

$$O_2NC(CH_2CH_2COOH)_3$$

$C_{10}H_{15}NO_8$ M 277.230

Solid. Mp 186°.

Tri-Et ester:
  $C_{16}H_{27}NO_8$ M 361.391
  Mp 186°.
Trinitrile: [1466-48-4]. *Tris(2-cyanoethyl)nitromethane*
  $C_{10}H_{12}N_4O_2$ M 220.230
  Mp 116.0-118.0°.

▷ Synth v. exothermic.

Bruson, H.A. *et al, J. Am. Chem. Soc.*, 1943, **65**, 23 (nitrile)
Šorm, F. *et al, Collect. Czech. Chem. Commun.*, 1954, **19**, 298 (synth)
Newkome, G.R. *et al, J. Org. Chem.*, 1988, **53**, 5552 (synth, nitrile, pmr, cmr)

Careyagenolide   C-80026

Updated Entry replacing C-20049
2α,3β-Dihydroxy-28,20β-taraxastanolide

[84749-88-2]

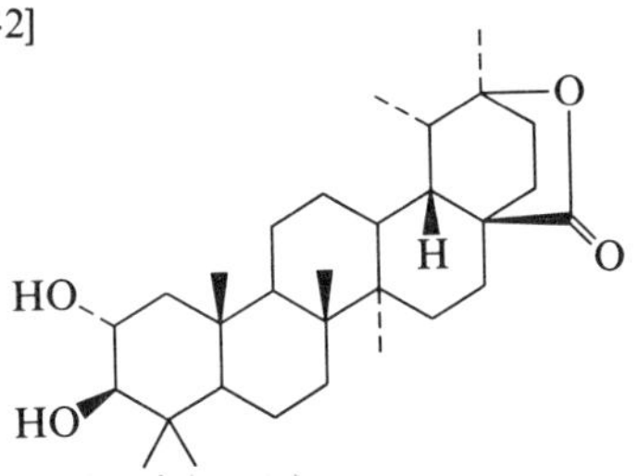

$C_{30}H_{48}O_4$ M 472.707

Constit. of *Careya arborea*. Cryst. (MeOH). Mp 299° dec.

Das, M.C. *et al, Phytochemistry*, 1982, **21**, 2069.

Carpesialactone   C-80027

[82460-83-1]

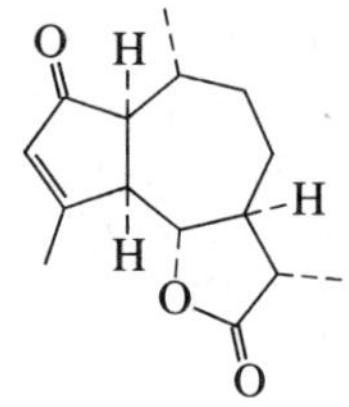

$C_{15}H_{20}O_3$ M 248.321

The identity of Bohlmann's sample having stereochem. as illus. with an earlier sample of undetd. stereochem. is not certain. Isol. from seeds of *Carpesium abrotanoides* and roots of *Hypochoeris oligocephala*. Oil or gum.

Kariyone, T. *et al*, *Yakugaku Zasshi (J. Pharm. Soc. Jpn.)*, 1955, **75**, 39; *ca*, **50**, 890 (*struct*)
Naito, S., *Yakugaku Zasshi (J. Pharm. Soc. Jpn.)*, 1955, **75**, 93, 325; *ca*, **50**, 891, 1680 (*struct, synth*)
Bohlmann, F. *et al*, *Phytochemistry*, 1982, **21**, 460 (*isol, struct*)

β-Caryophyllene alcohol   C-80028

[472-97-9]

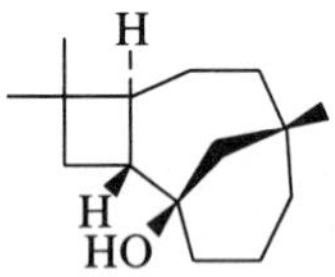

$C_{15}H_{26}O$ M 222.370

Constit. of Palmarosa and other oils. Oil.
Barrow, C.J. *et al*, *Aust. J. Chem.*, 1988, **41**, 1755 (*pmr, cmr*)

Caucalol   C-80029

Updated Entry replacing C-00445

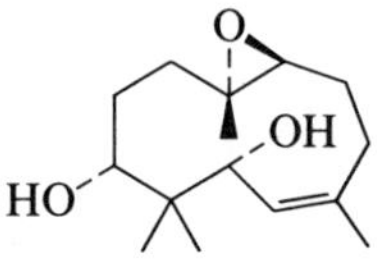

$C_{15}H_{26}O_3$ M 254.369
Di-Ac:
  $C_{19}H_{30}O_5$ M 338.443
  Constit. of *Caucalis scabra*. Cryst. Mp 121-122°. $[\alpha]_D^{15.5}$ +33.4° (CHCl₃).

Sasaki, S. *et al*, *Tetrahedron Lett.*, 1966, 623.

1(14),4,8,12-Cembratetraene-2,3-diol   C-80030

Updated Entry replacing S-50015
  Sarcophytol B
[72629-68-6]

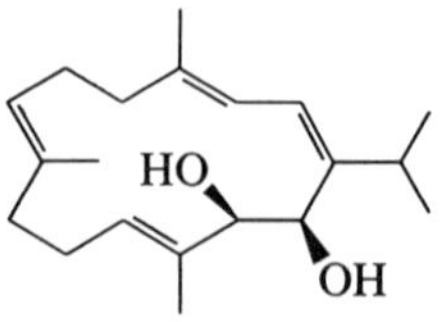

$C_{20}H_{32}O_2$ M 304.472
Constit. of *Sarcophyton glaucum* and the coral *Alcyonium flaccidum*. Cryst. (Me₂CO). Mp 125-126.5°. $[\alpha]_D$ +164° (c, 1 in CHCl₃).

Kobayashi, M. *et al*, *Chem. Pharm. Bull.*, 1979, **27**, 2382 (*isol*)

Kashman, Y. *et al*, *J. Org. Chem.*, 1981, **46**, 3592 (*isol*)
Czarkie, D. *et al*, *Tetrahedron*, 1985, **41**, 1049 (*isol*)
McMurry, J.E. *et al*, *Tetrahedron Lett.*, 1989, **30**, 1173 (*synth, cryst struct*)

2,7,11-Cembratriene-4,6-diol   C-80031

Updated Entry replacing C-70030
  4,8,13-Duvatriene-1,3-diol
[57605-80-8]

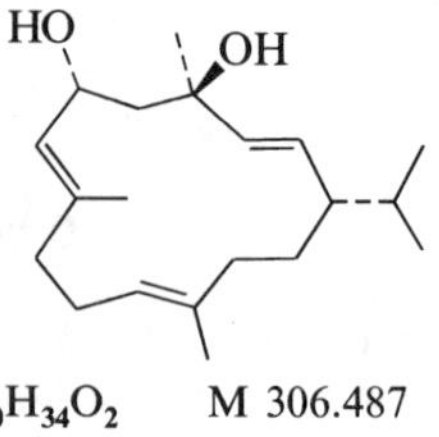

(1S,2E,4R,6R,7E,11E)-*form*

$C_{20}H_{34}O_2$ M 306.487
Cryst.

(*1S,2E,4R,6R,7E,11E*)-*form* [58190-98-0] *β-Cembrenediol*
  Constit. of tobacco. Plant growth inhibitor. Cryst. Mp 150-152° (123°). $[\alpha]_D$ +40° (CHCl₃).
(*1S,2E,4S,6R,7E,11E*)-*form* [75282-01-8] *α-Cembrenediol*
  Constit. of tobacco. Mp 118-120°. $[\alpha]_D$ +100° (CHCl₃).

Springer, J.P. *et al*, *Tetrahedron Lett.*, 1975, 2737 (*isol*)
Chang, S.Y. *et al*, *Phytochemistry*, 1976, **15**, 961 (*isol*)
Wahlberg, I. *et al*, *Acta Chem. Scand.*, *Ser. B*, 1982, **36**, 443 (*isol*)
Crombie, L. *et al*, *Phytochemistry*, 1988, **27**, 1685 (*biosynth*)
Begley, M.J. *et al*, *Phytochemistry*, 1988, **27**, 1695 (*cryst struct*)
Marshall, J.A. *et al*, *Tetrahedron Lett.*, 1989, **30**, 1055 (*synth, abs config*)

Centrolobin   C-80032

4-[2-[Tetrahydro-6-(4-methoxyphenyl)-2H-pyran-2-yl]ethyl]phenol, 9CI

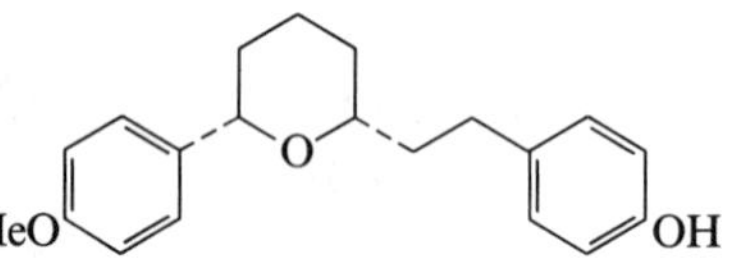

$C_{20}H_{24}O_3$ M 312.408
(+)-*form* [30358-99-7]
  Isol. from wood of *Centrolobium robustum*. Cryst. Mp 87°. $[\alpha]_D^{20}$ +97° (c, 0.4 in CHCl₃).
(±)-*form*
  Oil. Bp₀.₀₃ 190°.

de Albuquerque, I.L. *et al*, *Gazz. Chim. Ital.*, 1964, **94**, 287 (*isol, ir, pmr, struct*)
Galeffi, C. *et al*, *Gazz. Chim. Ital.*, 1965, **95**, 95 (*synth*)
Craveiro, A.A. *et al*, *Phytochemistry*, 1970, **9**, 1975 (*isol*)
Nagal, M. *et al*, *Chem. Pharm. Bull.*, 1986, **34**, 1056 (*abs config*)

Cerberalignan D   C-80033

[119403-22-4]

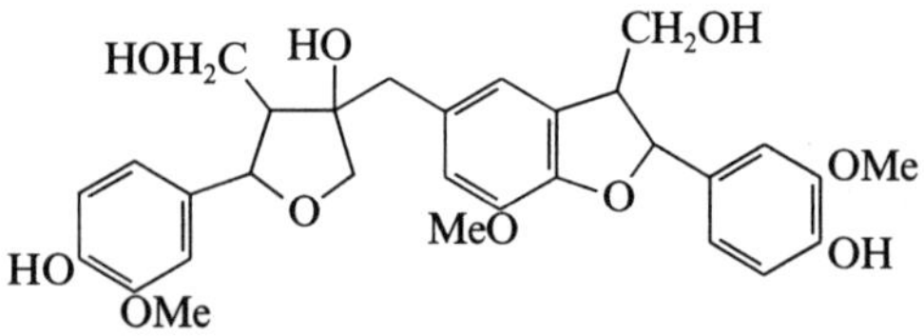

$C_{30}H_{34}O_{10}$ M 554.593
Constit. of *Cerbera manghas* and *C. odollam*. Solid. $[\alpha]_D^{25}$ -50.6° (c, 0.98 in MeOH).

Abe, F. *et al*, *Phytochemistry*, 1988, **27**, 3627.

Cerberalignan E   C-80034

[119403-23-5]

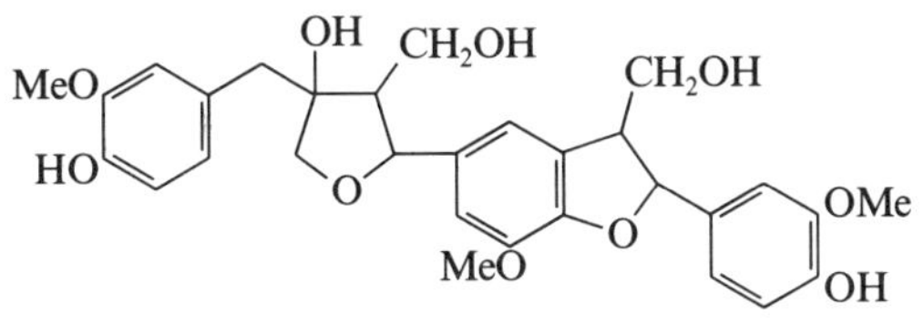

$C_{30}H_{34}O_{10}$ M 554.593

Constit. of *Cerbera manghas* and *C. odollam*. Solid. $[\alpha]_D^{24}$ −35.9° (c, 0.4 in MeOH).

Abe, F. *et al*, *Phytochemistry*, 1988, **27**, 3627.

Cerberalignan F   C-80035

[119420-37-0]

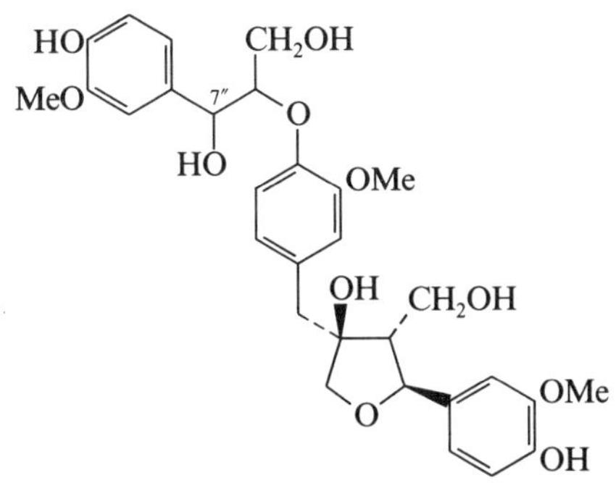

$C_{30}H_{36}O_{11}$ M 572.608

Constit. of *Cerbera manghas* and *C. odollam*. Solid. $[\alpha]_D^{28}$ −49° (c, 0.4 in MeOH).

7″-Epimer: [119478-88-5]. **Cerberalignan G**
  $C_{30}H_{36}O_{11}$ M 572.608
  Constit. of *C. manghas* and *C. odollam*. Solid. $[\alpha]_D^{28}$ −68.4° (c, 0.5 in MeOH).

Abe, F. *et al*, *Phytochemistry*, 1988, **27**, 3627.

Cerberalignan H   C-80036

[119420-38-1]

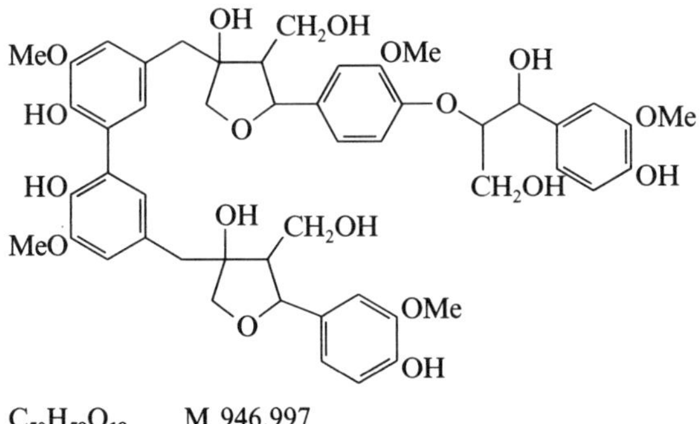

$C_{50}H_{58}O_{18}$ M 946.997

Constit. of *Cerbera manghas* and *C. odollam*. Solid. $[\alpha]_D^{25}$ −62° (c, 0.39 in MeOH).

Abe, F. *et al*, *Phytochemistry*, 1988, **27**, 3627.

Cerberalignan I   C-80037

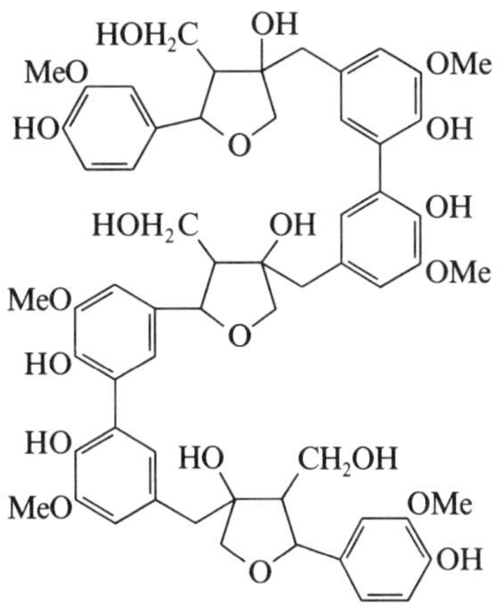

$C_{60}H_{68}O_{21}$ M 1125.185

Constit. of *Cerbera manghas* and *C. odollam*. Solid. $[\alpha]_D^{25}$ −76.7° (c, 0.72 in MeOH).

Abe, F. *et al*, *Phytochemistry*, 1988, **27**, 3627.

Ceroptene   C-80038

3-Hydroxy-5-methoxy-4,4-dimethyl-2-(1-oxo-3-phenyl-2-propenyl)-2,5-cyclohexadien-1-one. 2-Cinnamoyl-3-hydroxy-5-methoxy-4,4-dimethyl-2,5-cyclohexadien-1-one

[56015-03-3]

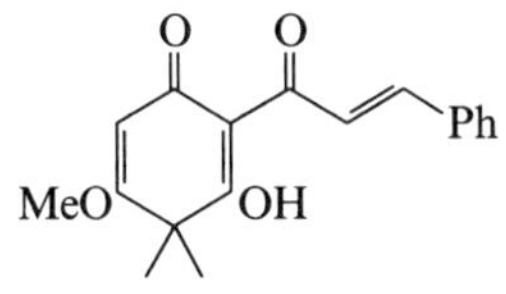

$C_{18}H_{18}O_4$ M 298.338

Mixt. of tautomers in soln. Probable struct. Isol. from the yellow leaf deposit of ferns *Pitryogramma triangularis*. Yellow prisms or plates (MeOH). Mp 137-140°.

Forsen, S. *et al*, *Acta Chem. Scand.*, 1959, **13**, 750, 1383 (*isol, ir, ms, pmr, struct*)
Dreyer, D.L. *et al*, *Tetrahedron*, 1975, **31**, 287 (*cmr*)

1-Chlorobenzo[*a*]carbazole   C-80039

[111960-32-8]

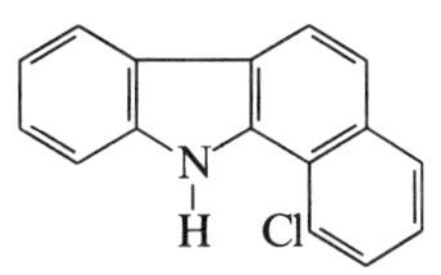

$C_{16}H_{10}ClN$ M 251.714
Cryst. (EtOH). Mp 114-115°.

Katritzky, A.R. *et al*, *J. Org. Chem.*, 1988, **53**, 794 (*synth, pmr*)

2-Chloro-1,4-butanediol, 9CI   C-80040

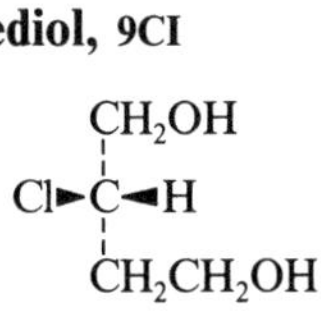

$C_4H_9ClO_2$ M 124.567
(*S*)-*form* [101021-55-0]
  Pale yellow oil. $Bp_{0.5}$ 95-104°. $[\alpha]_D^{19}$ −45.0° (c, 1.2 in MeOH).

[101021-57-2]
Kunec, E.K. *et al*, *J. Chem. Soc., Perkin Trans. 1*, 1987, 1089 (*synth, ir, pmr*)

4-Chloro-2-buten-1-amine, 9CI   C-80041

Updated Entry replacing C-50094
1-Amino-4-chloro-2-butene

$$ClCH_2CH=CHCH_2NH_2$$

C_4H_8ClN M 105.567
(*E*)-*form*
  B,HCl: [100350-86-5].
    Cryst. (Me$_2$CO). Mp 152-153° dec.
(*Z*)-*form*
  B,HCl: [7153-66-4].
    Sticky cryst. prod.
Gajda, T. *et al*, *Justus Liebigs Ann. Chem.*, 1986, 992 (*synth, ir, pmr*)
Brandänge, S. *et al*, *Synthesis*, 1988, 347 (*synth, pmr, cmr*)

3-Chloro-1-butene   C-80042

Updated Entry replacing C-00843
[563-52-0]

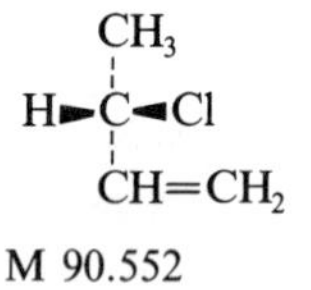

(*S*)-*form*
Absolute configuration

C_4H_7Cl M 90.552
(*S*)-*form* [35729-37-4]
  $[\alpha]_D^{25}$ +15.1° (neat) (25% opt. pure).
(±)-*form*
  Bp$_{766}$ 64°.
Kharasch, M.S. *et al*, *J. Org. Chem.*, 1938, **2**, 489 (*synth*)
Young, W.G. *et al*, *J. Org. Chem.*, 1961, **26**, 245 (*abs config*)
Normant, J.F. *et al*, *Bull. Soc. Chim. Fr.*, 1972, **7**, 2854 (*synth*)
Snyder, E.I. *et al*, *J. Org. Chem.*, 1972, **37**, 1466 (*synth*)
Magid, R.M. *et al*, *Tetrahedron Lett.*, 1977, 2999 (*synth*)

2-Chloro-2-butenedioic acid, 9CI   C-80043

Updated Entry replacing C-00845

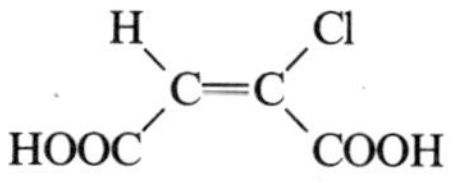

(*E*)-*form*

$C_4H_3ClO_4$ M 150.518
(*E*)-*form* [617-43-6]
  Chloromaleic acid
  Cryst. (Et$_2$O/CHCl$_3$). Insol. pet. ether. Mp 114° (sinters at 96°). Heat at 180° → anhydride.
  Di-Me ester: [19393-45-4].
    $C_6H_7ClO_4$ M 178.572
    Bp$_{18}$ 106.5°.
  ▷ ON0525000.
  Anhydride: 3-Chloro-2,5-furandione
    C_4HClO_3 M 132.503
    Mp 33°. Bp 196°, Bp$_8$ 78°.
(*Z*)-*form* [617-42-5]
  Chlorofumaric acid
  Plates (AcOH). Sol. H$_2$O, Et$_2$O. Mp 191-192°. pK_{a1} 7.76, pK_{a2} 5.81. Sublimes.
  ▷ LT1050000.
  Di-Me ester: [5331-33-9].
    Bp 224°, Bp$_{15}$ 108°.

Di-Et ester: [10302-94-0]. *Diethyl chlorofumarate*
  $C_8H_{11}ClO_4$ M 206.625
  Fungicide, algicide, lubricant for nylon fibre finishing and in textile pastes. Bp$_{0.75}$ 100-105°.
Dichloride: [17096-37-6].
  $C_4Cl_2O_2$ M 150.948
  Bp 185-187° part. dec.
Dinitrile: 1-Chloro-1,2-dicyanoethylene
  C_4HClN_2 M 112.518
  Bp 172°.
[6910-76-5]
van der Riet, B., *Justus Liebigs Ann. Chem.*, 1894, **280**, 229 (*synth*)
v. Auwers, K. *et al*, *Ber.*, 1929, **62**, 1685 (*synth*)
Roedig, A. *et al*, *Justus Liebigs Ann. Chem.*, 1965, **683**, 30 (*synth*)
Maas, G. *et al*, *Justus Liebigs Ann. Chem.*, 1965, **686**, 55 (*synth*)
Mai, K. *et al*, *Chem. Ind.* (*London*), 1986, 670 (*synth, ester, use*)
Akhtar, M. *et al*, *Tetrahedron*, 1987, **43**, 5899 (*synth, ir, pmr, cmr, ms*)
Sax, N.I., *Dangerous Properties of Industrial Materials*, 5th Ed., Van Nostrand-Reinhold, 1979, 839.

4-Chloro-2-butyn-1-ol   C-80044

[13280-07-4]

$$ClCH_2C\equiv CCH_2OH$$

C_4H_5ClO M 104.536
Dupont, G. *et al*, *Bull. Soc. Chim. Fr.*, 1954, 816.

1′-(Chlorocarbonyl)-1′,2′-dihydro-1,2′-bipyridinium (1+), 9CI   C-80045

1-[2-(Chloroformyl)-2-azacyclohexa-3,5-dienyl]pyridinium

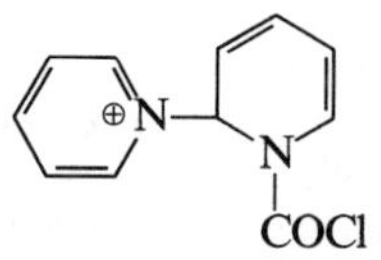

$C_{11}H_{10}ClN_2O^\oplus$ M 221.665 (ion)
Chloride: [117371-69-4]. *Phosgene-in-a-can*
  $C_{11}H_{10}Cl_2N_2O$ M 257.118
  Prod. of reaction of excess pyridine with phosgene. Useful source of phosgene. Light yellow solid. Mp 84-87° dec. Thermally stable, liberates phosgene in soln.
King, J.A. *et al*, *J. Org. Chem.*, 1988, **53**, 6145 (*synth, ir, pmr, cmr*)

2-Chloro-1,3-cyclohexanedione   C-80046

[932-23-0]

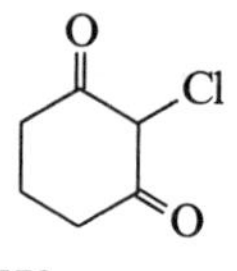

$C_6H_7ClO_2$ M 146.573
Me ether: [18369-67-0]. *2-Chloro-3-methoxy-2-cyclohexen-1-one, 9CI*
  $C_7H_9ClO_2$ M 160.600
  Cryst. (EtOAc). Mp 90-92°.
Shepherd, R.G. *et al*, *J. Chem. Soc., Perkin Trans. 1*, 1987, 2153 (*synth, deriv, pmr, ir*)

6-Chlorodibenzo[*c,g*]carbazole   C-80047

[111960-36-2]

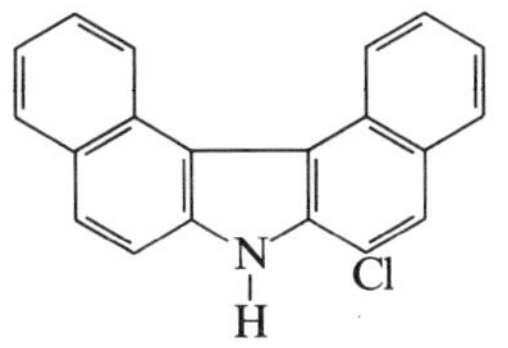

C$_{20}$H$_{12}$ClN   M 301.774
Cryst. (EtOH). Mp 193-195°.

Katritzky, A.R. *et al*, *J. Org. Chem.*, 1988, **53**, 794 (*synth, pmr*)

1-Chloro-1,2-difluoroethene, 9CI   C-80048

[359-04-6]

$$FClC{=}CHF$$

C$_2$HClF$_2$ M 98.479
−15° to −12° (−6° to −4°) . Prob. mixt. of geom.
  isomers.

[2837-86-7, 27156-04-3, 30860-28-7]

Birchall, J.M. *et al*, *J. Chem. Soc.*, 1961, 2204 (*synth*)
Fuller, G. *et al*, *Tetrahedron*, 1962, **18**, 123 (*synth*)
Barlow, M.G., *J. Chem. Soc., Chem. Commun.*, 1966, 703 (*F nmr*)
Craig, N.C. *et al*, *J. Phys. Chem.*, 1971, **75**, 1453 (*isom*)
Syrvatka, B.G. *et al*, *Zh. Org. Khim.*, 1972, **8**, 1553 (*ms*)
Ruehl, E., *Can. J. Chem.*, 1985, **63**, 1949 (*uv*)

2-Chloro-1,1-difluoroethene, 9CI   C-80049

2,2-Difluorovinyl chloride
[359-10-4]

$$F_2C{=}CHCl$$

C$_2$HClF$_2$ M 98.479
Gas. Mp −138.5°. Bp −19.0°.

▷ Flammable.

Henne, A.L. *et al*, *J. Am. Chem. Soc.*, 1948, **70**, 1025 (*synth*)
Chandra, S., *J. Phys. Chem.*, 1967, **71**, 1927 (*microwave*)
Torkelson, T.R. *et al*, *Toxicol. Appl. Pharmacol.*, 1971, **19**, 1 (*tox*)
Scott, J.D. *et al*, *J. Am. Chem. Soc.*, 1972, **94**, 2634 (*uv*)
Mackenzie, M.W., *Spectrochim. Acta, Part A*, 1984, **40**, 279 (*ir*)
Osten, H.J. *et al*, *J. Chem. Phys.*, 1985, **83**, 5434 (*F-19 nmr*)

2-Chloro-4,5-dihydro-1*H*-imidazole, 9CI   C-80050

2-Chloro-2-imidazoline
[54255-11-7]

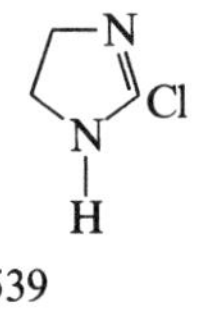

C$_3$H$_5$ClN$_2$ M 104.539
Free base decomposes at r.t.
B,HCl: [54255-14-0].
  Cryst. (EtOH). Mp 189-190°.
B,H$_2$SO$_4$: [54255-12-8].
  Cryst. (MeOH). Mp 161-162° dec.

Trani, A. *et al*, *J. Heterocycl. Chem.*, 1974, **11**, 257 (*synth, ir, pmr*)

5-Chloro-5,6-dihydro-2-(2,4,6-octatriynylidene)-2*H*-pyran, 9CI   C-80051

Updated Entry replacing C-00961
[2271-20-7]

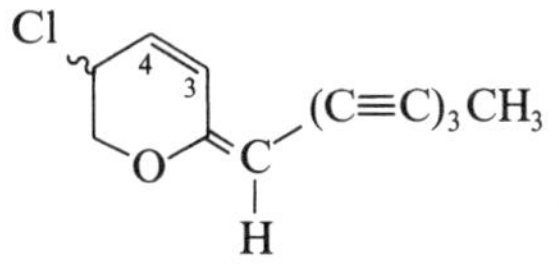

(*E*)-*form*

C$_{13}$H$_9$ClO   M 216.666
(*E*)-*form* [37940-06-0]
  Constit. of roots of *Anaphalis triplinervis*. Yellow cryst.
  (pet. ether). [α]$_D^{20}$ +526° (c, 0.118 in EtOH). Dec. >80°.
  λ$_{max}$ 269.5 (ε 25 300), 341.5 (23 900).
  3,4-Epoxide: [6558-99-2]. *5-Chloro-2-(2,4,6-
  octatriynylidene)-3,7-dioxabicyclo[4.1.0]heptane, 9CI. 5-
  Chloro-5,6-dihydro-3,4-epoxy-2-(2,4,6-octatriynylidene)-
  2H-pyran*
  C$_{13}$H$_9$ClO$_2$ M 232.666
  Isol. from roots of *A. triplinervis*. Yellowish cryst. (pet.
  ether). Mp 87°.
(*Z*)-*form*
  Constit. of *A. margaritacea* and *A. triplinervis*. Light-
  yellow cryst. or oil. Mp 73°. [α]$_D^{20}$ −219° (c, 0.49 in
  Et$_2$O).

Bohlmann, F. *et al*, *Chem. Ber.*, 1965, **98**, 1416; 1966, **99**, 1648;
  1972, **105**, 3036; 1973, **106**, 1337 (*isol, uv, ir, pmr, struct, synth*)
Bohlmann, F. *et al*, *Phytochemistry*, 1978, **17**, 1917 (*isol, deriv*)

5-Chloro-4,6-dimethyl-1,2,3-triazine   C-80052

[109520-02-7]

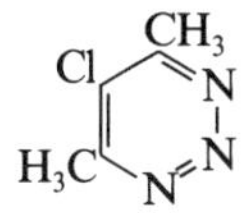

C$_5$H$_6$ClN$_3$ M 143.575
Needles (hexane). Mp 89-90°.

Ohsawa, A. *et al*, *Chem. Pharm. Bull.*, 1988, **36**, 3838 (*synth, pmr*)

2-Chloro-1,3-dinitro-5-(trifluoromethyl)benzene, 9CI   C-80053

*4-Chloro-α,α,α-trifluoro-3,5-dinitrotoluene. 3,5-Dinitro-4-
chlorobenzotrifluoride*
[393-75-9]

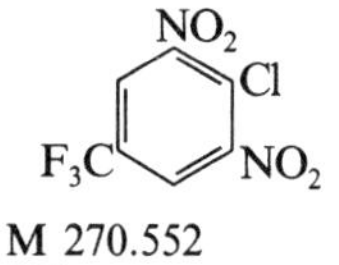

C$_7$H$_2$ClF$_3$N$_2$O$_4$ M 270.552
Cryst. Mp 58° (53-57°).

▷ Acute oral toxin, mutagenic.

Jurgens, H.R. *et al*, *J. Org. Chem.*, 1960, **25**, 1710 (*synth*)
Hill, J.R. *et al*, *Org. Magn. Reson.*, 1977, **9**, 589 (*C-13 nmr, F-19
  nmr*)
Bucchi, A.R. *et al*, *Ann. Ist. Super. Sanita*, 1983, **19**, 351 (*tox*)
Schaefer, T. *et al*, *Can. J. Chem.*, 1983, **61**, 2779 (*F-19 nmr*)
Stemmler, E.A. *et al*, *Biomed. Environ. Mass Spectrom.*, 1987, **14**,
  417 (*ms*)

5-Chloro-4,6-diphenyl-1,2,3-triazine   C-80054

[114078-87-4]

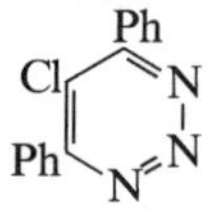

$C_{15}H_{10}ClN_3$ M 267.717
Needles (hexane/Et_2O). Mp 159-160°.

Ohsawa, A. *et al, Chem. Pharm. Bull.*, 1988, **36**, 3838 (*synth, pmr*)

Chlorododecahedrane   C-80055

*1-Chlorohexadecahydro-5,2,1,6,3,4-
[2,3]butanediyl[1,4]diylidenepentaleno[2,1,6-cde:2',1',6'-
gha]pentalene, 9CI*
[112896-28-3]

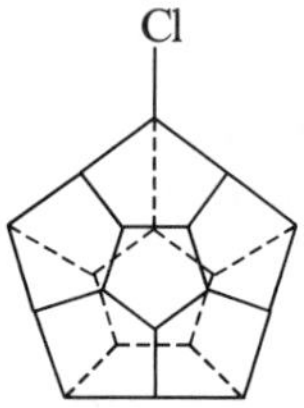

$C_{20}H_{19}Cl$ M 294.823
Solid. Mp >280°.

Paquette, L.A. *et al, J. Am. Chem. Soc.*, 1988, **110**, 1303 (*synth, cmr*)

Chlorofluoroiodomethane   C-80056

[1512-28-3]

CHClFI

CHClFI   M 194.375
Light sensitive liq. Bp 76°, Bp_{150} 35°.

Hazeldine, R.N., *J. Chem. Soc.*, 1952, 4259 (*synth*)
Samajajulu, G.R. *et al, J. Magn. Reson.*, 1979, **33**, 559 (*C-13 nmr*)
Kudchadker, A.P. *et al, J. Phys. Chem. Ref. Data*, 1979, **8**, 499 (*props*)

8-Chloroguanine, 8CI   C-80057

*2-Amino-8-chloro-1,7-dihydro-6H-purin-6-one, 9CI. 2-Amino-
8-chloro-6-hydroxypurine*
[22052-03-5]

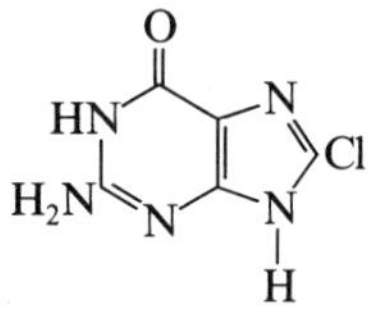

$C_5H_4ClN_5O$ M 185.572
B,HCl: [34618-09-2].
   Cryst. (2M HCl).

Wölcke, U. *et al, Tetrahedron Lett.*, 1969, 785 (*synth*)
Birdsall, N.J.M. *et al, Tetrahedron*, 1971, **27**, 5969 (*synth*)
Ryu, E.K. *et al, J. Org. Chem.*, 1981, **46**, 2819 (*synth*)

6-Chloro-1-hexene   C-80058

[928-89-2]

$H_2C{=}CH(CH_2)_3CH_2Cl$

$C_6H_{11}Cl$ M 118.606

Liq. Bp 128-130°, Bp_{32} 55°.

Black, H.K. *et al, J. Chem. Soc.*, 1953, 1785 (*synth*)
Jenkins, C.L. *et al, J. Org. Chem.*, 1971, **36**, 3103 (*synth*)
Ashby, E.C. *et al, J. Org. Chem.*, 1988, **53**, 6068 (*synth, pmr, ms*)

2-Chloro-1-hexen-3-ol, 9CI, 8CI   C-80059

[29279-85-4]

$H_3CCH_2CH_2CH(OH)CCl{=}CH_2$

$C_6H_{11}ClO$ M 134.605
(±)-*form*
   Liq. Bp_3 47.5-49°. n_D^{20} 1.4588.

Bianchini, J.P. *et al, Tetrahedron*, 1970, **26**, 3401 (*synth*)

2-Chloro-4-hexen-3-ol   C-80060

$H_3CCH{=}CHCH(OH)CHClCH_3$

$C_6H_{11}ClO$ M 134.605
Cryst. (pet. ether), Liq. Mp 109-110°. Bp_{17} 72-79°. n_D^{22}
1.4669 Mixt. of stereoisomers.

[70040-78-7, 70041-11-1, 70041-12-2, 70041-13-3]
Sauleau, J., *Bull. Soc. Chim. Fr.*, Part II, 1978, 474 (*synth, ir*)

4-Chloro-4-hexen-3-ol, 9CI, 8CI   C-80061

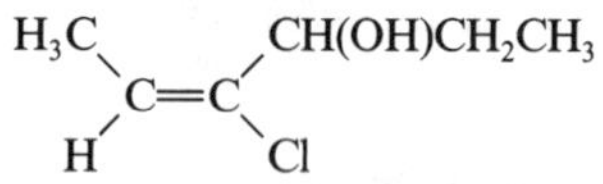

$C_6H_{11}ClO$ M 134.605
(±)-(*E*)-*form* [96943-38-3]
   Liq. $Bp_{0.1}$ 45-48°.

Barluenga, J. *et al, J. Chem. Soc., Perkin Trans. 1*, 1985, 447 (*synth, pmr*)

6-Chloro-2-hexen-1-ol, 9CI, 8CI   C-80062

$ClCH_2CH_2CH_2CH{=}CHCH_2OH$

$C_6H_{11}ClO$ M 134.605
(*Z*)-*form* [76047-81-9]
   Liq. $Bp_{0.3}$ 70-72°, Bp_1 54-58°.

Holton, R.A. *et al, J. Am. Chem. Soc.*, 1985, **107**, 2124 (*synth, pmr, ir*)
Brennan, J.P. *et al, Tetrahedron*, 1986, **42**, 6719 (*synth, ir, pmr, ms*)

6-Chloro-3-hexen-1-ol, 9CI, 8CI   C-80063

$ClCH_2CH_2CH{=}CHCH_2CH_2OH$

$C_6H_{11}ClO$ M 134.605
(*Z*)-*form*
   Liq. $Bp_{0.5}$ 63°. n_D^{26} 1.4742.

Meinwald, J. *et al, J. Am. Chem. Soc.*, 1960, **82**, 4087 (*synth*)

6-Chloro-4-hexen-1-ol, 9CI, 8CI   C-80064

$ClCH_2CH{=}CHCH_2CH_2CH_2OH$

$C_6H_{11}ClO$ M 134.605
(*Z*)-*form* [104410-97-1]
   Liq.

Nguyen, V.B. *et al, Tetrahedron Lett.*, 1986, **27**, 841 (*synth*)

6-Chloro-5-hexen-3-ol, 9CI   C-80065

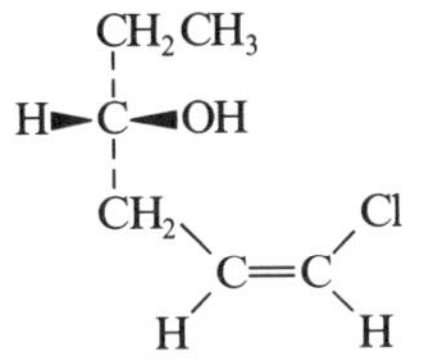

$$CH_2CH_3$$
$$H\!-\!C\!-\!OH \qquad (S,Z)\text{-}form$$
$$CH_2$$
$$C\!=\!C$$
$$H \qquad H$$

$C_6H_{11}ClO$ M 134.605

(S,Z)-form [90760-63-7]
Liq. $[\alpha]_D^{20}$ +9.2° (c, 9.15 in EtOH).

$(\pm)$-(Z)-form [87921-54-8]
Liq.

Hoffmann, R.W. *et al*, *Chem. Ber.*, 1986, **119**, 1039, 2013 (synth, pmr, cmr)

2-Chloro-6-hydroxy-1,4-naphthoquinone   C-80066

2-Chloro-6-hydroxy-1,4-naphthalenedione

[76665-65-1]

$C_{10}H_5ClO_3$ M 208.600
Cryst. (1,2-dichloroethane). Mp 229-230°.

Brisson, C. *et al*, *J. Org. Chem.*, 1981, **46**, 1810 (synth, uv, ir, pmr)
Boisvert, L. *et al*, *J. Org. Chem.*, 1988, **53**, 4052 (synth, ir, uv, pmr)

2-Chloro-7-hydroxy-1,4-naphthoquinone   C-80067

2-Chloro-7-hydroxy-1,4-naphthalenedione

[69119-29-5]

$C_{10}H_5ClO_3$ M 208.600
Cryst. (dichloroethane). Mp 216-217°.

Brisson, C. *et al*, *J. Org. Chem.*, 1981, **46**, 1810 (synth, uv, ir, pmr)
Boisvert, L. *et al*, *J. Org. Chem.*, 1988, **53**, 4052 (synth, uv, ir, pmr)

Chloroiodomethane   C-80068

[593-71-5]

$$ClCH_2I$$

CH_2ClI M 176.384
Versatile organic synthon. Liq. Bp 108-109° (104-105°) .

Miyano, S. *et al*, *Bull. Chem. Soc. Jpn.*, 1971, **44**, 2864 (synth, pmr)
Hahn, R.C., *J. Org. Chem.*, 1988, **53**, 1331 (synth, bibl, use)

1-Chloro-2-isocyanatoethane, 9CI   C-80069

2-Chloroethyl isocyanate, 8CI

[1943-83-5]

$$ClCH_2CH_2NCO$$

C_3H_4ClNO M 105.523
Cryst. d_{20} 1.237. Fp 56°. Bp 135°, Bp_{16} 42°. n_D^{20} 1.4493.

▷ Corrosive, lachrymator.

Sadtler Standard Infrared Spectra, 23442, 42442 (ir)
Sadtler Standard NMR Spectra, 13504 (pmr)
Wenker, H., *J. Am. Chem. Soc.*, 1936, **58**, 2608 (synth)
Johnson, C.K., *J. Org. Chem.*, 1967, **32**, 1508 (synth)

Chloromethanol   C-80070

[15454-33-8]

$$ClCH_2OH$$

CH_3ClO M 66.487

Ac: [625-56-9].
$C_3H_5ClO_2$ M 108.524
Alkylating agent. Oil. Bp_{748} 113-115°, Bp_5 28°. n_D^{20} 1.4102.

Chloroacetyl: [6135-23-5].
$C_3H_4Cl_2O_2$ M 142.969
Oil. Bp_{745} 130-132°, Bp_{21} 77.5°. n_D^{20} 1.4603.

Trichloroacetyl:
$C_3H_2Cl_4O_2$ M 211.858
Oil. d_4^{20} 1.60. Bp 170°, Bp_{10} 68°. n_D^{20} 1.5988.

Benzoyl: [5335-05-7].
$C_8H_7ClO_2$ M 170.595
Oil. d_4^{25} 1.233. Bp_8 114-115°, Bp_6 98-99°. n_D^{25} 1.5328.

4-Methylbenzenesulfonyl: [117184-50-6].
$C_8H_9ClO_3S$ M 220.676
Oil. $Bp_{0.4}$ 99-101°.

Ulich, L.H. *et al*, *J. Am. Chem. Soc.*, 1921, **43**, 660 (acetate, chloroacetate)
Bavin, P.M.G., *Can. J. Chem.*, 1964, **42**, 704 (acetate)
Euranto, E.K. *et al*, *Acta Chem. Scand.*, 1966, **20**, 1273 (trichloroacetate)
Bodor, N. *et al*, *J. Med. Chem.*, 1980, **23**, 566 (benzoate)
Korhonen, I.O.O., *Org. Mass Spectrom.*, 1984, **19**, 34 (ms)
Hahn, R.C. *et al*, *J. Org. Chem.*, 1988, **53**, 5783 (synth, pmr, cmr)

N-Chloro-4-methylbenzenesulfonamide   C-80071

N-*Chloro-p-toluenesulfonamide*

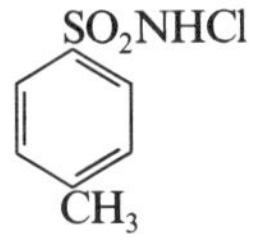

$$SO_2NHCl$$
(benzene ring with CH_3 para)

$C_7H_8ClNO_2S$ M 205.664

Na Salt: [127-65-1]. *Chloramine T*
Oxidising agent. Cryst. + $3H_2O$ (H_2O or EtOH). Mp 176-180° (explodes). Explodes on melting. Respiratory irritant, may cause emphysema.

Heintzelman, R.W., *Synthesis*, 1976, 731 (synth)
Hutchins, M.G. *et al*, *J. Org. Chem.*, 1982, **47**, 4847 (synth)
Bremner, D.H., *Synth. Reagents*, 1985, **6**, 9 (rev)
Olmstead, M.P. *et al*, *Inorg. Chem.*, 1986, **25**, 4057 (cryst struct)

(Chloromethyl)pentafluorobenzene, 9CI   C-80072

Pentafluorobenzyl chloride

[653-35-0]

$$(C_6F_5)CH_2Cl$$

$C_7H_2ClF_5$ M 216.537
Liq. Bp 160-163°, Bp_{47} 80°.

▷ Slightly lachrymatory.

Barbour, A.K. *et al*, *J. Chem. Soc.*, 1961, 808 (synth, ir)

1-(Chloromethyl)phenanthrene   C-80073

[20485-61-4]

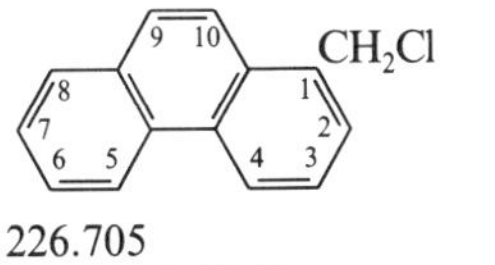

$C_{15}H_{11}Cl$ M 226.705
Cryst. (pet. ether). Mp 89-90.5°.

Fierens, P.J.C. *et al*, *Helv. Chim. Acta*, 1955, **38**, 2005 (synth)

2-(Chloromethyl)phenanthrene   C-80074

[885-21-2]
$C_{15}H_{11}Cl$ M 226.705
Cryst. (pet. ether). Mp 98°.

Fierens, P.J.C. *et al*, *Helv. Chim. Acta*, 1955, **38**, 2005 (synth)

3-(Chloromethyl)phenanthrene   C-80075
[20485-60-3]
$C_{15}H_{11}Cl$ M 226.705
Cryst. (pet. ether). Mp 81-82°.

Fierens, P.J.C. *et al*, *Helv. Chim. Acta*, 1955, **38**, 2005 (*synth*)

4-(Chloromethyl)phenanthrene   C-80076
[38021-49-7]
$C_{15}H_{11}Cl$ M 226.705
Cryst. (pet. ether). Mp 72-73°.

Fierens, P.J.C. *et al*, *Helv. Chim. Acta*, 1955, **38**, 2005 (*synth*)

9-(Chloromethyl)phenanthrene, 9CI   C-80077
[951-05-3]
$C_{15}H_{11}Cl$ M 226.705
Cryst. (hexane). Mp 101-102°. Bp₁ 186-188°.

Cook, J.W. *et al*, *J. Chem. Soc.*, 1935, 1319 (*synth*)
Tarbell, D.S. *et al*, *J. Am. Chem. Soc.*, 1943, **65**, 2149 (*synth*)
Fierens, P.J.C. *et al*, *Helv. Chim. Acta*, 1955, **38**, 2005 (*synth*)
Fernández, F. *et al*, *Synthesis*, 1988, 802 (*synth*)

4-Chloro-3-methylpyridazine, 9CI   C-80078
[101346-01-4]

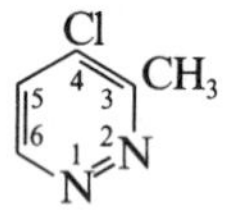

$C_5H_5ClN_2$ M 128.561
1-Oxide:
  $C_5H_5ClN_2O$ M 144.560
  Needles (EtOH). Mp 132.5-133°.

Ogata, M., *Chem. Pharm. Bull.*, 1963, **11**, 1511.

4-Chloro-5-methylpyridazine, 9CI   C-80079
$C_5H_5ClN_2$ M 128.561
1-Oxide:
  $C_5H_5ClN_2O$ M 144.560
  Needles (C_6H_6/cyclohexane). Mp 61-62°.

Ogata, M., *Chem. Pharm. Bull.*, 1963, **11**, 1511.

2-Chloro-1-naphthylamine   C-80080
2-Chloro-1-naphthalenamine, 9CI. 1-Amino-2-chloronaphthalene
[13711-39-2]
$C_{10}H_8ClN$ M 177.633
Needles. Mp 56°, Mp 60°. Steam-volatile.
N-*Ac:* [13711-40-5].
  $C_{12}H_{10}ClNO$ M 219.670
  Needles (EtOH). Mp 195°.
N,N-*Di-Ac:*
  $C_{14}H_{12}ClNO_2$ M 261.707
  Prisms (EtOH aq.). Mp 88°.

Cleve, P.T., *Ber.*, 1887, **20**, 450.
Bowers, G.W., *J. Am. Chem. Soc.*, 1936, **58**, 1573.
Hodgson, H. *et al*, *J. Chem. Soc.*, 1942, 744; 1944, 538.
Stepanov, B.I., *Zh. Obshch. Khim.*, 1960, **30**, 2008.
Godfrey, K.E. *et al*, *J. Chem. Soc. C*, 1967, 400.

3-Chloro-3-(1-naphthylmethyl)diazirine   C-80081
[112399-65-2]

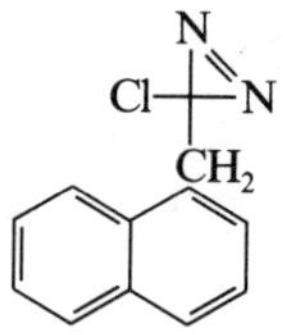

$C_{12}H_9ClN_2$ M 216.669
Cryst. (pentane). Mp 41°.

Linden, A. *et al*, *J. Org. Chem.*, 1988, **53**, 1085 (*synth, ir, uv, pmr, ms, cryst struct*)

2-Chloro-2-nitroadamantane   C-80082
2-Chloro-2-nitrotricyclo[3.3.1.1³,⁷]decane

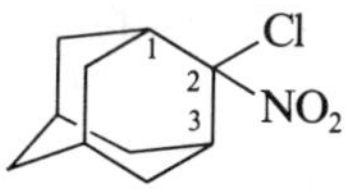

$C_{10}H_{14}ClNO_2$ M 215.679
Cryst. (EtOH aq.). Mp 200-201°.

Archibald, T.G. *et al*, *J. Org. Chem.*, 1988, **53**, 4645 (*synth, ir, pmr*)

1-Chloro-2-nitrosobenzene, 9CI   C-80083
[932-33-2]

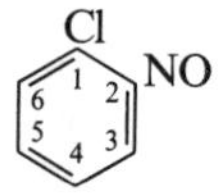

C_6H_4ClNO M 141.556
Needles (EtOH). Mp 56-57°.

Haworth, R.D. *et al*, *J. Chem. Soc.*, 1921, **119**, 768 (*synth*)

1-Chloro-3-nitrosobenzene, 9CI   C-80084
[932-78-5]
C_6H_4ClNO M 141.556
Needles (C_6H_6). Mp 72°.

Haworth, R.D. *et al*, *J. Chem. Soc.*, 1921, **119**, 768 (*synth*)

1-Chloro-4-nitrosobenzene, 9CI   C-80085
[932-98-9]
C_6H_4ClNO M 141.556
Cream cryst. (EtOH). Mp 130-140° dec. (92-93°). Cont. *ca.* 5% corresponding azoxy compd.

Ingold, C.K., *J. Chem. Soc.*, 1924, 87 (*synth*)
Defoin, A. *et al*, *Helv. Chim. Acta*, 1989, **72**, 1199 (*synth, ir, pmr*)

4-Chloro-1-nitro-2-(trifluoromethyl)benzene   C-80086
5-Chloro-α,α,α-trifluoro-2-nitrotoluene
[118-83-2]

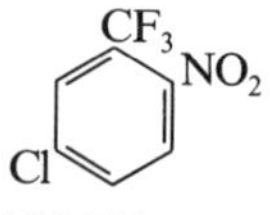

$C_7H_3ClF_3NO_2$ M 225.554
Pale yellow oil. Mp 21-22°. Bp 224°, Bp₄.₅ 86-87°.

Whalley, W.B., *J. Chem. Soc.*, 1949, 3016 (*synth*)
Newmark, R.A. *et al*, *Org. Magn. Reson.*, 1977, **9**, 589 (*C-13 nmr*)

1-Chloro-2,3,3,4,4-pentafluorocyclobutene, C-80087
9CI

[377-94-6]

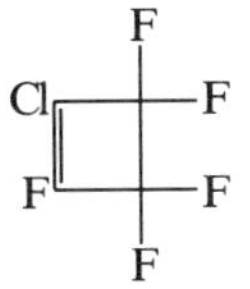

C$_4$ClF$_5$ M 178.489
Bp 33-33.4°. n_D^{25} 1.3208.

Newmark, R.A. *et al, J. Magn. Reson.*, 1969, **1**, 418 (*F-19 nmr*)
Chia, L.S. *et al, Can. J. Chem.*, 1974, **52**, 3484 (*Cl-nqr*)
Bauer, G. *et al, Z. Naturforsch., B*, 1979, **34**, 1249 (*synth*)

1-Chloro-1,1,3,3,3-pentafluoropropane, 9CI   C-80088

[460-92-4]

$$F_3CCH_2CClF_2$$

C$_3$H$_2$ClF$_5$ M 168.493
Shows anaesthetic props. d_4^{20} 1.4372. Fp $-107°$. Bp 28.4°.
n_D^{20} 1.2875.

[108662-83-5]

Henne, A.L. *et al, J. Am. Chem. Soc.*, 1946, **68**, 496 (*synth*)
Vorob'ev, V.N. *et al, Russ. J. Phys. Chem.*, 1974, **48**, 147 (*props*)
Di Paolo, T. *et al, J. Pharm. Sci.*, 1979, **68**, 39 (*pharmacol*)
Gachegov, Yu.N. *et al, Zh. Prikl. Spektrosk.*, 1979, **30**, 497 (*cryst struct*)

1-Chloro-2,2,3,3,3-pentafluoropropane   C-80089

2,2,3,3,3-Pentafluoropropyl chloride
[422-02-6]

$$F_3CCF_2CH_2Cl$$

C$_3$H$_2$ClF$_5$ M 168.493
Shows anaesthetic props. d_4^{25} 1.395. Bp 25-27°. n_D^{20} 1.292.

McBee, E.T. *et al, J. Am. Chem. Soc.*, 1955, **77**, 3149 (*synth*)
Haszeldine, R.N. *et al, J. Chem. Soc.*, 1957, 2193 (*synth*)
Paleta, O. *et al, Collect. Czech. Chem. Commun.*, 1971, **36**, 1867 (*synth*)

1-Chloro-1,2,3,3,3-pentafluoro-1-propene, C-80090
9CI

[2804-49-1]

$$F_3CCF{=}CFCl$$

C$_3$ClF$_5$ M 166.478
Fp $-156.8°$ to $-159.6°$. Bp 7.7-7.9°, Bp 8.4°. Mixt. of
stereoisomers.

[14003-57-7, 14003-62-4, 89331-22-6]

Henne, A.L. *et al, J. Am. Chem. Soc.*, 1948, **70**, 130 (*synth*)
Park, J.D. *et al, J. Org. Chem.*, 1958, **23**, 1169 (*synth*)
Fried, J.H. *et al, J. Am. Chem. Soc.*, 1959, **81**, 2078 (*synth*)
Paleta, O. *et al, Bull. Soc. Chim. Fr.*, 1986, 920 (*F nmr*)

4-Chloro-4-penten-1-ol   C-80091

[1647-18-3]

$$H_2C{=}CClCH_2CH_2CH_2OH$$

C$_5$H$_9$ClO   M 120.578
Oil. Bp$_{3.1}$ 61.5-62.5°.

Wender, P.A. *et al, J. Am. Chem. Soc.*, 1988, **110**, 2218 (*synth, ir, pmr, ms*)

2-Chloro-1*H*-perimidine   C-80092

[30837-50-4]

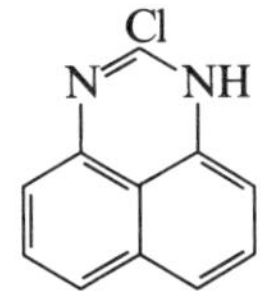

C$_{11}$H$_7$ClN$_2$ M 202.642
Pale green plates or yellow needles (Me$_2$CO). Mp 202-
203.5° (194°).

Sachs, F., *Justus Liebigs Ann. Chem.*, 1909, **365**, 135 (*synth*)
Pozharskii, A.F. *et al, Chem. Heterocycl. Compd. (Engl. Transl.)*,
1970, **6**, 1055 (*synth*)
Herbert, J.M. *et al, J. Med. Chem.*, 1987, **30**, 2081 (*synth, pmr*)

2-Chloro-1-phenazinecarboxylic acid, 9CI,   C-80093
8CI

[106976-13-0]

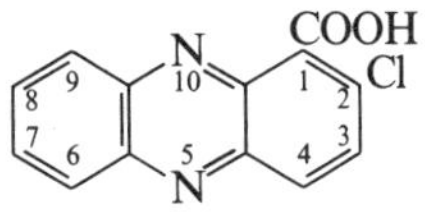

C$_{13}$H$_7$ClN$_2$O$_2$ M 258.663
Cryst. (MeOH). Mp 247-249°.

Rewcastle, G.W. *et al, J. Med. Chem.*, 1987, **30**, 843 (*synth*)

3-Chloro-1-phenazinecarboxylic acid, 9CI,   C-80094
8CI

[103942-85-4]
C$_{13}$H$_7$ClN$_2$O$_2$ M 258.663
Cryst. (MeOH). Mp 255-257°.

Rewcastle, G.W. *et al, J. Med. Chem.*, 1987, **30**, 843 (*synth*)

6-Chloro-1-phenazinecarboxylic acid, 9CI,   C-80095
8CI

[103942-87-6]
C$_{13}$H$_7$ClN$_2$O$_2$ M 258.663
Cryst. (EtOH). Mp 310° dec.

Rewcastle, G.W. *et al, J. Med. Chem.*, 1987, **30**, 843 (*synth*)

7-Chloro-1-phenazinecarboxylic acid, 9CI,   C-80096
8CI

[103942-92-3]
C$_{13}$H$_7$ClN$_2$O$_2$ M 258.663
Cryst. (MeOH). Mp 274° dec.

Rewcastle, G.W. *et al, J. Med. Chem.*, 1987, **30**, 843 (*synth*)

8-Chloro-1-phenazinecarboxylic acid, 9CI,   C-80097
8CI

[103942-82-1]
C$_{13}$H$_7$ClN$_2$O$_2$ M 258.663
Cryst. (MeOH). Mp 286-289°.

N(5)-Oxide: [29137-34-6].
Cryst. Mp 116-117°.

Ger. Pat., 1 935 705, (1970); *CA*, **72**, 134173 (*oxide*)
Rewcastle, G.W. *et al, J. Med. Chem.*, 1987, **30**, 843 (*synth*)

9-Chloro-1-phenazinecarboxylic acid, 9CI, C-80098
8CI

[103942-93-4]

$C_{13}H_7ClN_2O_2$ M 258.663
Cryst. (EtOH). Mp 272-273°.

Rewcastle, G.W. *et al, J. Med. Chem.*, 1987, **30**, 843 (*synth*)

2-Chloro-1-phenylethanol   C-80099
Updated Entry replacing C-02052
α-(*Chloromethyl)benzenemethanol, 9CI. α-
Chloromethylbenzyl alcohol, 8CI. Styrene chlorohydrin.
Phenylethylene chlorohydrin. β-Chloro-α-
hydroxyethylbenzene. Chloromethylphenylcarbinol*

[1674-30-2]

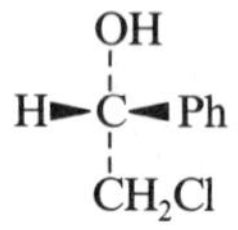

(R)-*form*

C_8H_9ClO M 156.611
(R)-*form* [56751-12-3]
Bp_{11} 119-120°. $[\alpha]_D^{25}$ −47.8° (c, 2.8 in hexane).
(S)-*form*
Liq. Bp_6 112-4°. $[\alpha]_D^{24}$ +49.6° (c, 2.81 in
cyclohexane)(96.5% e.e.).
(±)-*form*
$d_4^{20.5}$ 1.16. Bp_6 110-111°. n_D^{20} 1.5400.
Ac: [79465-05-7].
$C_{10}H_{11}ClO_2$ M 198.648
Oil. Bp_3 101-104°.

4-Nitrobenzoyl: Cryst. (EtOH). Mp 81°.

[829-23-2, 33942-01-7]

Emerson, W., *J. Am. Chem. Soc.*, 1945, **67**, 516 (*synth*)
Hanby, W.E. *et al, J. Chem. Soc.*, 1946, 114 (*synth*)
Sumrell, G. *et al, Can. J. Chem.*, 1964, **42**, 2896.
Gounelle, Y. *et al, Bull. Soc. Chim. Fr.*, 1968, 2815 (*ir*)
Hartgerink, J.W. *et al, Tetrahedron*, 1971, **27**, 4323.
Corey, E.J. *et al, J. Org. Chem.*, 1988, **53**, 2861 (*synth*)

2-Chloro-4-phenylpyrimidine, 9CI   C-80100
[13036-50-5]

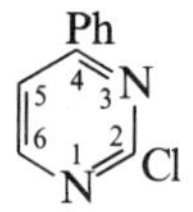

$C_{10}H_7ClN_2$ M 190.631
Cryst. (pet. ether). Mp 88.5-89.5°.

[56734-25-9]

Matsukawa, T. *et al, Yakugaku Zasshi (J. Pharm. Soc. Jpn.*), 1950,
70, 134 (*synth*)
Lythgoe, B. *et al, J. Chem. Soc.*, 1951, 2323 (*synth*)
Kato, T. *et al, Org. Mass Spectrom.*, 1974, **9**, 981 (*ms*)
Kroon, A.P. *et al, Recl. Trav. Chim. Pays-Bas (J. R. Neth. Chem.
Soc.*), 1974, **93**, 325 (*use*)
Sedova, V.F. *et al, Khim. Geterotsikl. Soedin.*, 1977, **5**, 678 (*synth*)
Baram, S.G. *et al, Izv. Sib. Otd. Akad. Nauk SSSR, Ser. Khim.
Nauk*, 1982, 135; *ca*, **98**, 4208v (*cmr*)

2-Chloro-5-phenylpyrimidine, 9CI   C-80101
[22536-62-5]
Cryst. (C_6H_6/pet. ether). Mp 124-128° (123°).

Brown, D.J. *et al, J. Chem. Soc. C*, 1970, 214 (*synth, pmr, uv*)
Allen, D.W. *et al, J. Chem. Soc., Perkin Trans. 1*, 1977, 621
(*synth*)

Ivanovskaya, L.Y. *et al, Izv. Sib. Otd. Akad. Nauk SSSR, Ser.
Khim. Nauk*, 1979, 71; *CA*, **90**, 203080g (*ms*)
Coppola, G.M. *et al, J. Heterocycl. Chem.*, 1980, **17**, 1479 (*synth,
ir, pmr*)

4-Chloro-2-phenylpyrimidine, 9CI   C-80102
[14790-42-2]

$C_{10}H_7ClN_2$ M 190.631
Cryst. (EtOH). Mp 74°.

[23149-84-0]

Van Meeteren, H.W. *et al, Recl. Trav. Chim. Pays-Bas (J. R. Neth.
Chem. Soc.*), 1967, **86**, 15 (*synth*)
Geerts, J.P. *et al, Recl. Trav. Chim. Pays-Bas (J. R. Neth. Chem.
Soc.*), 1973, **92**, 1232 (*pmr*)
Geerts, J.P. *et al, Org. Magn. Reson.*, 1975, **7**, 86 (*cmr*)
Ivanovskaya, L.Y. *et al, Izv. Sib. Otd. Akad. Nauk SSSR, Ser.
Khim. Nauk*, 1979, 71; *CA*, **90**, 203080g (*ms*)

4-Chloro-5-phenylpyrimidine, 9CI   C-80103
[60122-80-7]

$C_{10}H_7ClN_2$ M 190.631
Cryst. (pet. ether). Mp 71-72°. Bp_{19} 156-158°.

Davies, W.H. *et al, J. Chem. Soc.*, 1945, 347 (*synth*)
Boarland, M.P.V. *et al, J. Chem. Soc.*, 1952, 3722 (*uv*)

4-Chloro-6-phenylpyrimidine, 9CI   C-80104
[3435-26-5]

$C_{10}H_7ClN_2$ M 190.631
Solid. Mp 96-97.5°.

[39189-96-3, 40889-35-8]

van der Plas, H.C. *et al, Tetrahedron Lett.*, 1964, 2093 (*synth*)
De Valk, J. *et al, Recl. Trav. Chim. Pays-Bas (J. R. Neth. Chem.
Soc.*), 1972, **91**, 1414 (*synth*)
Kato, T. *et al, Org. Mass Spectrom.*, 1974, **9**, 981 (*ms*)
U.S. Pat., 3 908 012, (1975); *CA*, **84**, 59528r (*synth*)
Baram, S.G. *et al, Izv. Sib. Otd. Akad. Nauk SSSR, Ser. Khim.
Nauk*, 1982, 135; *CA*, **98**, 4208v (*cmr*)

5-Chloro-2-phenylpyrimidine, 9CI   C-80105
[34771-50-1]

$C_{10}H_7ClN_2$ M 190.631
Small needles (EtOH). Mp 96°.

Kunckell, F. *et al, Chem. Ber.*, 1902, **35**, 3164 (*synth*)
Wagner, R.M. *et al, Chem. Ber.*, 1971, **104**, 2975 (*synth, pmr*)
Baram, S.G. *et al, Izv. Akad. Sci. USSR, Ser. Sci. Khim.*, 1983,
299 (*cmr*)

5-Chloro-4-phenylpyrimidine, 9CI   C-80106
[72261-64-4]

$C_{10}H_7ClN_2$ M 190.631
Prisms (hexane). Mp 87.5-88.5°.

Yamanaka, H. *et al, Chem. Pharm. Bull.*, 1987, **35**, 3119 (*synth,
pmr*)

2-Chloro-4-phenylthiazole, 9CI   C-80107
[1826-23-9]

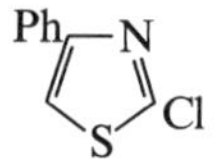

C_9H_6ClNS M 195.672
Cryst. (EtOH). Mp 53-54°.
B,MeFSO_3: Mp 138-139°.
B,EtBF_4: Mp 120-121°.

Bariana, D.S. *et al*, *J. Indian Chem. Soc.*, 1955, **32**, 427 (*synth*)
Vernin, G. *et al*, *Bull. Soc. Chim. Fr.*, 1963, 2498 (*synth*)
Aune, J.P. *et al*, *Bull. Soc. Chim. Fr.*, 1972, 2679 (*pmr*)
Sugimoto, H. *et al*, *J. Org. Chem.*, 1988, **53**, 2263 (*synth*)

2-Chloropyrazine   C-80108

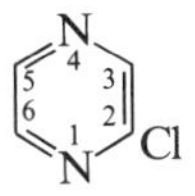

$C_4H_3ClN_2$ M 114.534
Liq. Bp_{26} 60.5°.

1-Oxide: [16025-16-4].
   $C_4H_3ClN_2O$ M 130.533
   Mp 140-146° (131-132°).

4-Oxide:
   $C_4H_3ClN_2O$ M 130.533
   Cryst. (EtOH aq.). Mp 95-96°.

Klein, B. *et al*, *J. Am. Chem. Soc.*, 1951, **73**, 2949 (*synth*)
Bernadi, L. *et al*, *Gazz. Chim. Ital.*, 1961, **91**, 1431 (*oxides*)
Klein, B. *et al*, *J. Org. Chem.*, 1963, **28**, 1682 (*synth, oxide, uv*)
Elisa, A.S. *et al*, *J. Heterocycl. Chem., USSR (Engl. Transl.)*, 1967, **3**, 127 (*synth*)
Hashimoto, M. *et al*, *J. Heterocycl. Chem.*, 1988, **25**, 1705 (*synth, pmr, uv*)

1-Chloro-2,3,3,3-tetrafluoro-1-propene, 9CI   C-80109

$$F_3CCF{=}CHCl$$

C_3HClF_4 M 148.487
Fumigant. Fp −115.8°. Bp 15.0°. Isomeric composition unknown.

Henne, A.L. *et al*, *J. Am. Chem. Soc.*, 1946, **68**, 496 (*synth*)

2-Chlorotetrahydro-2*H*-pyran, 9CI   C-80110

[3136-02-5]

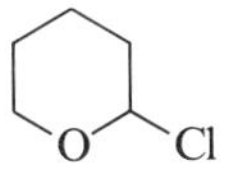

C_5H_9ClO M 120.578
(±)-*form*
   Liq. Bp_{15} 44-45°, Bp_{12} 35-36°.

Booth, H. *et al*, *Tetrahedron*, 1987, **43**, 4699 (*synth, pmr, cmr, conformn*)

3-Chlorotetrahydro-2*H*-pyran   C-80111

[6581-54-0]
C_5H_9ClO M 120.578
(±)-*form*
   Liq. Bp 140-143°.

Crombie, L. *et al*, *J. Chem. Soc.*, 1956, 136 (*synth*)

4-Chlorotetrahydro-2*H*-pyran   C-80112

[1768-64-5]
C_5H_9ClO M 120.578
Liq. d_4^{20} 1.11. Bp_{12} 42°.

Hanschke, E., *Chem. Ber.*, 1955, **88**, 1053 (*synth*)

5-Chlorotetrazole   C-80113

[55011-47-7]

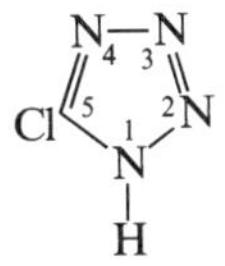

$CHClN_4$ M 104.498
Needles (C_6H_6). Mp 73°. pK_a 4.07.

Na salt: [96107-81-2].
   Needles + $2H_2O$ (H_2O).
   ▷ Explodes when heated to 335°.

1H-form
   1-tert-Butyl: [59772-99-5].
      $C_5H_9ClN_4$ M 160.606
      Cryst. (cyclohexane). Mp 114-115°.
   1-Benzyl:
      $C_8H_7ClN_4$ M 194.623
      Mp 38°.
   1-Ph:
      $C_7H_5ClN_4$ M 180.596
      Cryst. Mp 124°.

2H-form
   2-tert-Butyl:
      $C_5H_9ClN_4$ M 160.606
      Oil. $Bp_{0.001}$ 25°.

Stollé, R. *et al*, *Chem. Ber.*, 1929, **62**, 1113 (*synth*)
Stollé, R. *et al*, *J. Prakt. Chem.*, 1932, **134**, 282 (*Phenyl*)
Lieber, E. *et al*, *J. Am. Chem. Soc.*, 1951, **73**, 1792 (*props*)
Henry, R.A. *et al*, *J. Am. Chem. Soc.*, 1954, **76**, 290 (*synth*)
Henry, R.A. *et al*, *J. Heterocycl. Chem.*, 1976, **13**, 391 (*tert-Butyl*)
Spear, R.J., *Aust. J. Chem.*, 1984, **37**, 2453 (*synth*)
Klich, M. *et al*, *Tetrahedron*, 1986, **42**, 2677 (*Benzyl, ir, pmr*)

6-Chloro-1,2,4-triazine-3,5(1*H*,3*H*)-dione, 9CI   C-80114

6-Chloro-3,5-dihydroxy-1,2,4-triazine. 5-Chloro-6-azauracil

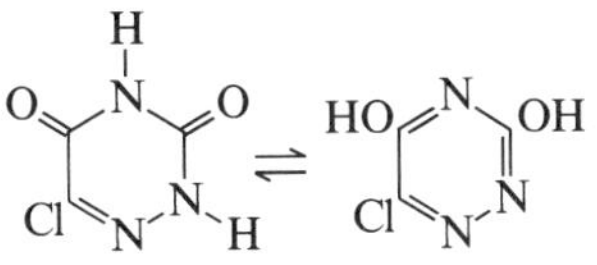

$C_3H_2ClN_3O_2$ M 147.520
Cryst. (H_2O). Mp 232-233° (225-227°). pK_a 5.80. Softens >200°.

Benzylammonium salt: Cryst. (H_2O). Mp 232-233° (212-213°).

Chang, P.K., *J. Org. Chem.*, 1961, **26**, 1118 (*synth*)
Pískala, A. *et al*, *Collect. Czech. Chem. Commun.*, 1975, **40**, 2680 (*deriv*)
Farkaš, J., *Collect. Czech. Chem. Commun.*, 1983, **48**, 2676 (*deriv*)

6-Chloro-1,3,5-triazine-2,4(1*H*,3*H*)-dione, 9CI   C-80115

2-Chloro-4,6-dihydroxy-1,3,5-triazine, 8CI
[69125-10-6]

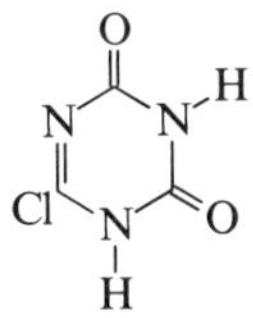

$C_3H_2ClN_3O_2$ M 147.520

Known as salts.

Di-Me ether: [3140-73-6]. *2-Chloro-4,6-dimethoxy-1,3,5-triazine, 9CI*

$C_5H_6ClN_3O_2$ M 175.574

Coupling reagent for peptide synth. Cryst. (pet. ether). Mp 74.2-76.2°.

[32998-00-8, 54268-35-8, 67410-55-3]

Bovee, W.M.M.J. *et al*, *Recl. Trav. Chim. Pays-Bas (J. R. Neth. Chem. Soc.)*, 1978, **97**, 107 (*synth, ir, pmr*)

1-Chloro-3,11-tridecadiene-5,7,9-triyn-2-ol   C-80116

$H_3CCH{=}CHC{\equiv}CC{\equiv}CC{\equiv}CCH{=}CHCH(OH)CH_2Cl$

$C_{13}H_{11}ClO$ M 218.682

(E,E)-form

Isol. from *Carthamus tinctorius*. Yellowish cryst (Et$_2$O/pet. ether). Mp 83°. $[\alpha]_{546}$ −3° (c, 0.33 in Et$_2$O).

Bohlmann, F. *et al*, *Chem. Ber.*, 1966, **99**, 3433 (*isol, uv, ir, pmr, ord, struct*)

2-Chloro-3,11-tridecadiene-5,7,9-triyn-1-ol   C-80117

[2060-58-4]

$H_3CCH{=}CHC{\equiv}CC{\equiv}CC{\equiv}CCH{=}CHCHClCH_2OH$

$C_{13}H_{11}ClO$ M 218.682

(E,E)-form [16863-64-2]

Isol. from plants of the Asteraceae incl. *Centaurea*, *Gnaphalium*, and *Dicoma* spp. Has some pesticide/herbicide properties. Mp 68°. $[\alpha]_D^{20}$ −45.5° (CHCl$_3$), $[\alpha]_D^{20}$ −54.5° (CHCl$_3$).

Ac: [16982-47-1].

$C_{15}H_{13}ClO_2$ M 260.719

In *Centaurea, Dicoma, Coreopsis* spp. Cryst. (pet. ether). Mp 49°. $[\alpha]_D^{20}$ −10.3° (MeOH).

Bohlmann, F. *et al*, *Chem. Ber.*, 1958, **91**, 1642; 1961, **94**, 3179; 1967, **100**, 3201 (*isol, ir, uv*)

Anderson, A.B. *et al*, *Phytochemistry*, 1977, **16**, 1829 (*isol*)

Bohlmann, F. *et al*, *Phytochemistry*, 1978, **17**, 570; 1980, **19**, 71; 1983, **22**, 2858 (*isol*)

Jente, R. *et al*, *Phytochemistry*, 1979, **18**, 829 (*pmr*)

Picman, A.K. *et al*, *J. Chromatogr.*, 1980, **189**, 187 (*chromatog*)

Arnason, T. *et al*, *Can. J. Bot.*, 1981, **59**, 54; *CA*, **94**, 169071a (*tox*)

McLacklan, D. *et al*, *Biochem. Syst. Ecol.*, 1986, **14**, 17; *CA*, **105**, 1789n.

1-Chloro-2,3,3-trifluorocyclopropene, 9CI   C-80118

[24921-89-9]

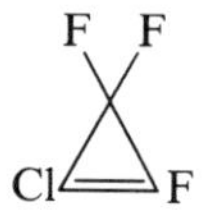

C_3ClF_3 M 128.481

Bp 20-21°, Bp 28°.

Camaggi, G. *et al*, *J. Chem. Soc. C*, 1970, 178 (*synth, ir, F-19 nmr, ms*)

Law, D.C.F. *et al*, *J. Org. Chem.*, 1973, **38**, 768 (*synth, ir, F-19 nmr, ms*)

1-Chloro-1,1,2-trifluoro-2-iodoethane, 9CI   C-80119

[354-26-7]

$F_2CClCHFI$

C_2HClF_3I M 244.382

Shows anaesthetic props. d_4^{20} 2.181. Bp 82-83°. n_D^{20} 1.433.

Griffin, C.E. *et al*, *J. Chem. Soc.*, 1960, 1398 (*synth, uv*)

Norris, R.D. *et al*, *J. Am. Chem. Soc.*, 1973, **95**, 182 (*F-19 nmr*)

Buchet, R. *et al*, *Biophys. Chem.*, 1985, **22**, 249 (*use*)

6-Chloro-4′,5,7-trihydroxyisoflavone   C-80120

6-Chloro-5,7-dihydroxy-3-(4-hydroxyphenyl)-4H-1-benzopyran-4-one, 9CI. 6-Chlorogenistein

[64545-64-8]

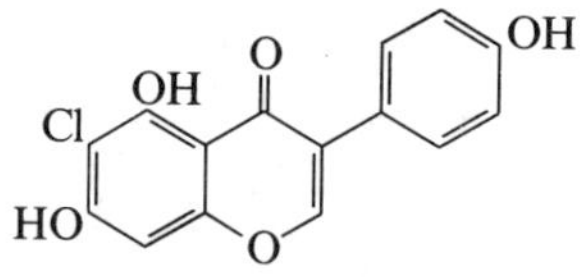

$C_{15}H_9ClO_5$ M 304.686

Metab. of *Streptomyces griseus*. All isolations of isoflavonoids from microorganisms are considered dubious. Prob. an artifact derived from the soybean medium.

König, W.A. *et al*, *Helv. Chim. Acta*, 1977, **60**, 2071 (*isol, ir, pmr, ms, struct*)

2-Chloro-3,5,6-trimethylpyridine   C-80121

[121767-77-9]

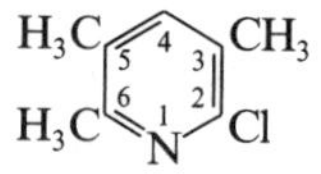

$C_8H_{10}ClN$ M 155.626

Mp 60°.

Mittelbach, M. *et al*, *Acta Chem. Scand., Ser. B*, 1988, **42**, 524 (*synth, pmr*)

3-Chloro-2,4,6-trimethylpyridine   C-80122

$C_8H_{10}ClN$ M 155.626

Mp 150-151°. Bp 190-192°.

Batkowski, T. *et al*, *Pol. J. Chem. (Rocz. Chem.)*, 1962, **36**, 51 (*synth*)

4-Chloro-2,3,5-trimethylpyridine, 9CI   C-80123

[109371-18-8]

$C_8H_{10}ClN$ M 155.626

Oil.

N-Oxide:

$C_8H_{10}ClNO$ M 171.626

Mp 148°.

Japan. Pat., 6 272 666, (1987); *CA*, **107**, 96598a (*synth*)

Joiner, K.A. *et al*, *Tetrahedron Lett.*, 1987, **28**, 3733 (*use*)

Mittelbach, M. *et al*, *Acta Chem. Scand., Ser. B*, 1988, **42**, 524 (*synth, pmr*)

Cholesta-7,22-diene-3,6-diol   C-80124

$C_{27}H_{44}O_2$ M 400.643

(3β,5α,6α,22E)-form

Constit. of *Spongionella gracillis*. Cryst. (MeOH/pet. ether). Mp 172-174°.

Madaio, A. *et al*, *J. Nat. Prod. (Lloydia)*, 1989, **52**, 952 (*isol, pmr*)

Cholestane-3,6,7,8,15,16,26-heptol   C-80125

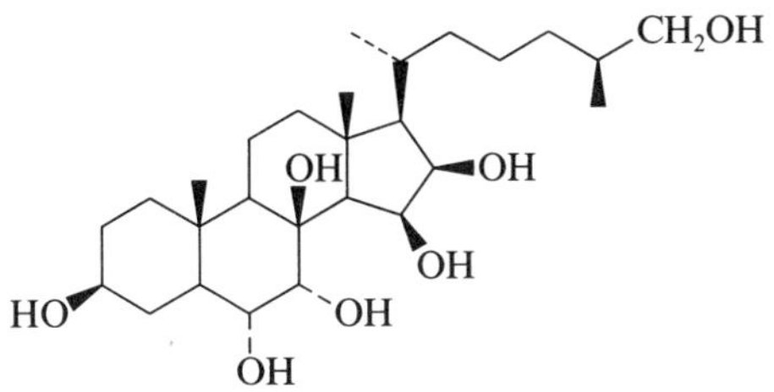

$C_{27}H_{48}O_7$ M 484.672

(3β,5α,6α,7α,8β,15β,16β,25S)-form

Constit. of *Pyncopodia helianthoides*. $[α]_D$ +16°.

Bruno, I. *et al, J. Nat. Prod.* (*Lloydia*), 1989, **52**, 1022.

Cholesta-7,9(11),22-triene-3,6-diol   C-80126

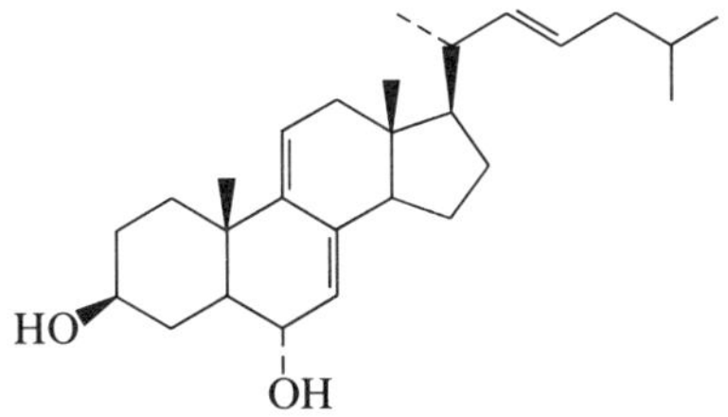

$C_{27}H_{42}O_2$ M 398.628

(3β,5α,6α,22E)-form

Constit. of *Spongionella gracilis*. Cryst. (MeOH/pet. ether). Mp 184-186°.

Madaio, A. *et al, J. Nat. Prod.* (*Lloydia*), 1989, **52**, 952 (*isol, pmr*)

Chrysartemin A   C-80127

Updated Entry replacing C-02435

Canin

[24959-84-0]

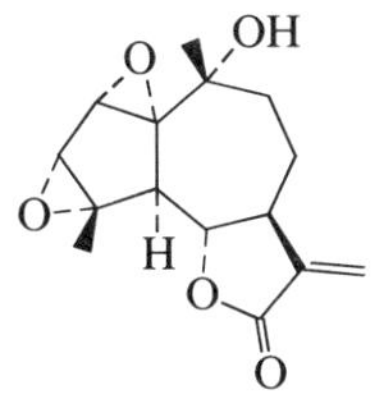

$C_{15}H_{18}O_5$ M 278.304

There is considerable confusion in the literature concerning the structure of this compound. Constit. of *Chrysanthemum parthenium, C. morifolium, Artemisia mexicana, A. klotzchiana* and *Tanacetum parthenium*. Cryst. (Me_2CO). Mp 250°. $[α]_D^{20}$ −13.4° (c, 0.35 in CHCl_3). The structs. of closely related compds. Chrysartemin B and Artecanin require clarification at present.

[29431-83-2, 29431-84-3]

Begley, M.J. *et al, Phytochemistry*, 1989, **28**, 940 (*isol, cryst struct, bibl*)

1-Chrysenecarboxylic acid   C-80128

[87717-16-6]

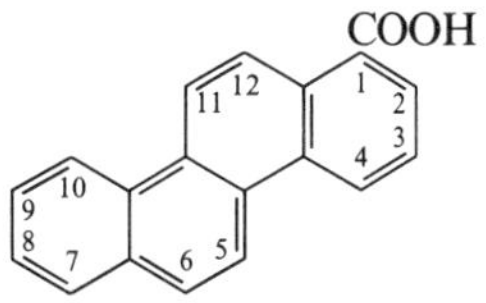

$C_{19}H_{12}O_2$ M 272.303
Mp >350°.

Me ester: [90340-68-4].
$C_{20}H_{14}O_2$ M 286.329
Cryst. (Me_2CO). Mp 175-176°.

Nitrile: [68723-52-4]. *1-Cyanochrysene*
$C_{19}H_{11}N$ M 253.303
Cryst. (C_6H_6/EtOH). Mp 228-229°.

Gore, P.H. *et al, Synthesis*, 1978, 773 (*synth, deriv, ir, pmr*)
Gore, P.H. *et al, Synth. Commun.*, 1980, **10**, 319 (*synth, deriv*)
Jones, A.J. *et al, Aust. J. Chem.*, 1984, **37**, 561 (*synth, deriv, pmr, ms*)

2-Chrysenecarboxylic acid   C-80129

[96403-26-8]

$C_{19}H_{12}O_2$ M 272.303
Pale yellow needles (dioxan). Mp 325-327° dec.

Me ester:
$C_{20}H_{14}O_2$ M 286.329
Pale yellow plates (C_6H_6). Mp 219°.

Carruthers, W. *et al, J. Chem. Soc.*, 1953, 3486; 1954, 2047.

3-Chrysenecarboxylic acid   C-80130

[96403-20-2]

$C_{19}H_{12}O_2$ M 272.303
Pale yellow needles (dioxan aq.). Mp 295° dec.

Me ester:
$C_{20}H_{14}O_2$ M 286.329
Plates. Mp 146-147°.

Nitrile: [36288-23-0]. *3-Cyanochrysene*
$C_{19}H_{11}N$ M 253.303
Cryst. (EtOH). Mp 200-201°.

Carruthers, W., *J. Chem. Soc.*, 1953, 3486 (*synth, deriv*)
Leznoff, C.C. *et al, Can. J. Chem.*, 1972, **50**, 528 (*synth, deriv, uv, pmr*)
Gore, P.H. *et al, Synthesis*, 1978, 773 (*synth, deriv, ir, pmr*)

5-Chrysenecarboxylic acid   C-80131

Updated Entry replacing C-40154

[68723-48-8]

$C_{19}H_{12}O_2$ M 272.303
Buff needles (C_6H_6/Me_2CO). Mp 225-226° (212-224°).

Me ester: [71431-97-5].
$C_{20}H_{14}O_2$ M 286.329
Prisms (C_6H_6). Mp 159-160°.

Nitrile: [68723-47-7]. *5-Cyanochrysene*
$C_{19}H_{11}N$ M 253.303
Cryst. (MeOH). Mp 165.5°.

Fieser, L.F. *et al, J. Am. Chem. Soc.*, 1940, **60**, 1211 (*synth*)
Gore, P.H. *et al, Synthesis*, 1978, 773 (*synth, deriv, ir, pmr*)
Gore, P.H. *et al, Tetrahedron*, 1979, **35**, 2927 (*nitrile*)
Lee-Ruff, E. *et al, J. Org. Chem.*, 1984, **49**, 553 (*synth*)

6-Chrysenecarboxylic acid   C-80132

[95646-96-1]

$C_{19}H_{12}O_2$ M 272.303

Needles (dioxan). Mp 314°.

Me ester: [99762-96-6].

$C_{20}H_{14}O_2$ M 286.329

Needles (EtOH). Mp 148°.

Nitrile: [68723-50-2]. *6-Cyanochrysene*

$C_{19}H_{11}N$ M 253.303

Cryst. (C_6H_6/EtOH).

Dewar, M.J.S. *et al*, *J. Chem. Soc.*, 1957, 2946 (*synth, deriv*)
Gore, P.H. *et al*, *Synthesis*, 1978, 773 (*synth, deriv, ir, pmr*)
Gore, P.H. *et al*, *Tetrahedron*, 1979, **35**, 2927 (*synth, deriv*)

Cinnamolide   C-80133

Updated Entry replacing C-02480

[23599-47-5]

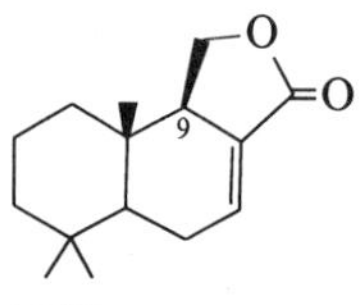

$C_{15}H_{22}O_2$ M 234.338

Constit. of *Cinnamosa fragrans*. Cryst. Mp 125-126°. $[\alpha]_D^{20}$ −29.4° (c, 1 in $CHCl_3$).

9α-Hydroxy: 9α-Hydroxycinnamolide

$C_{15}H_{22}O_3$ M 250.337

Constit. of *Canella winterana*. Cryst. Mp 150-155°. $[\alpha]_D$ −10° (c, 0.44 in $CHCl_3$).

Suzuki, T. *et al*, *Bull. Chem. Soc. Jpn.*, 1970, **43**, 1268 (*synth*)
Yamagawa, H. *et al*, *Synthesis*, 1970, 257 (*synth*)
Nakanishi, K. *et al*, *Isr. J. Chem.*, 1977, **16**, 28 (*isol*)
Kioy, D. *et al*, *J. Nat. Prod.* (*Lloydia*), 1989, **52**, 174 (*isol, deriv, pmr, cmr*)

[7]Circulene   C-80134

Updated Entry replacing C-30165

Dinaphtho[2,1,8,7-ghij:2′,2′,8′,7′-nopq]pleiadene, 9CI

[76276-09-0]

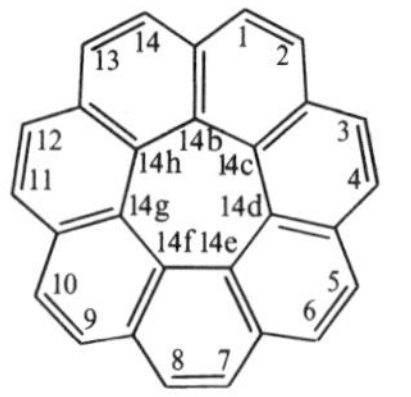

$C_{28}H_{14}$ M 350.419

Yellow plates. Mp 295-296°.

Yamamoto, K. *et al*, *J. Am. Chem. Soc.*, 1988, **110**, 3578 (*synth, uv, ir, pmr, cmr, ms, cryst struct*)

Citlalitrione   C-80135

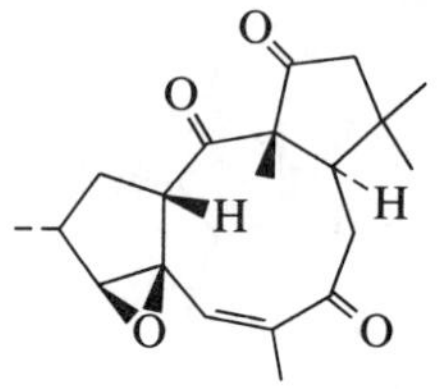

$C_{20}H_{26}O_4$ M 330.423

Constit. of *Jatropha dioica* var. *sessiliflora*. Cryst. (MeOH/hexane). Mp 194-196°. $[\alpha]_D$ −146° (c, 7.5 in $CHCl_3$).

Villarreal, A.M. *et al*, *J. Nat. Prod.* (*Lloydia*), 1988, **51**, 749 (*isol, cryst struct*)

Citreofuran   C-80136

[122400-13-9]

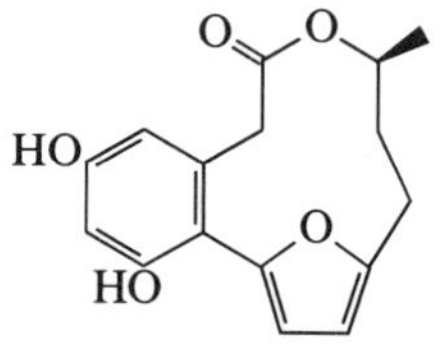

$C_{16}H_{16}O_5$ M 288.299

Metab. of hybrid strain of *Penicillium citreo-viride B*. Cryst. Mp 203-205°. $[\alpha]_D^{27}$ +112° (c, 0.18 in EtOH).

Lai, S. *et al*, *Tetrahedron Lett.*, 1989, **30**, 2241.

Cladiellin   C-80137

Updated Entry replacing C-02522

[66873-35-6]

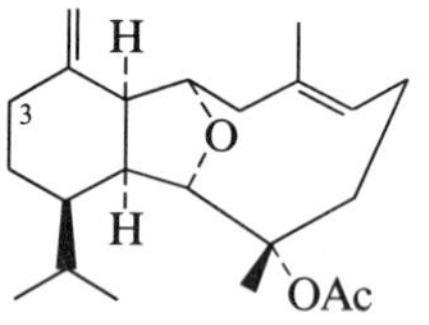

$C_{22}H_{34}O_3$ M 346.509

Constit. of *Cladiella* spp.

Δ^3-*Isomer, O-de-Ac:*

$C_{20}H_{32}O_2$ M 304.472

Constit. of *C*. spp. Cryst. Mp 48-52°. $[\alpha]_D$ −22.7° (c, 0.3 in $CHCl_3$).

Kazlauskas, R. *et al*, *Tetrahedron Lett.*, 1977, 4643 (*isol*)
Hochlowski, J.E. *et al*, *Tetrahedron Lett.*, 1980, 4055 (*isol, deriv*)

Clandestacarpin   C-80138

[79002-16-7]

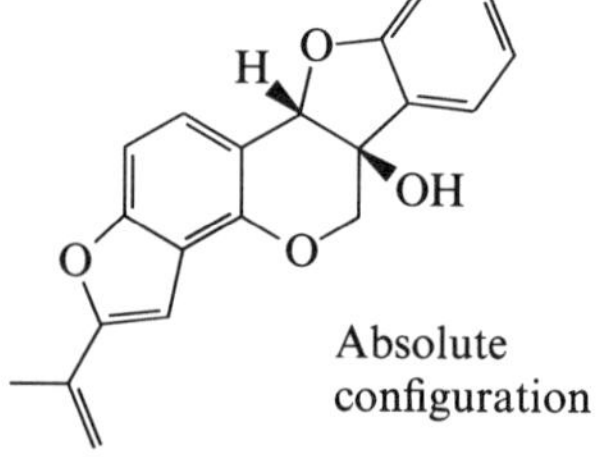

Absolute
configuration

$C_{20}H_{16}O_5$ M 336.343

Phytoalexin from fungus-infected leaves of *Glycine clandestina*, *G. tabacina* and *G. tomentella*. Phytoalexin.

Lyne, R.L. *et al*, *Tetrahedron Lett.*, 1981, 2483 (*isol, struct, abs config*)

Ingham, J.L., *Fortschr. Chem. Org. Naturst.*, 1983, **43**, 1 (*isol*)

Clausenarin   C-80139

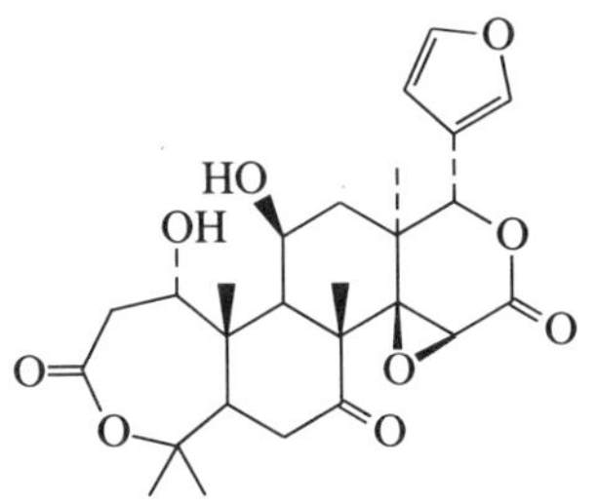

$C_{26}H_{32}O_9$ M 488.533

Constit. of *Clausena anisata*. Cryst. (DMSO aq.). Mp 293-294°. $[\alpha]_D^{25}$ −87.5° (c, 1.04 in Me_2CO).

Ngadjui, B.T. *et al*, *J. Nat. Prod. (Lloydia)*, 1989, **52**, 832 (*isol, pmr, cmr*)

Clausenolide   C-80140

Updated Entry replacing C-02534

[71899-58-6]

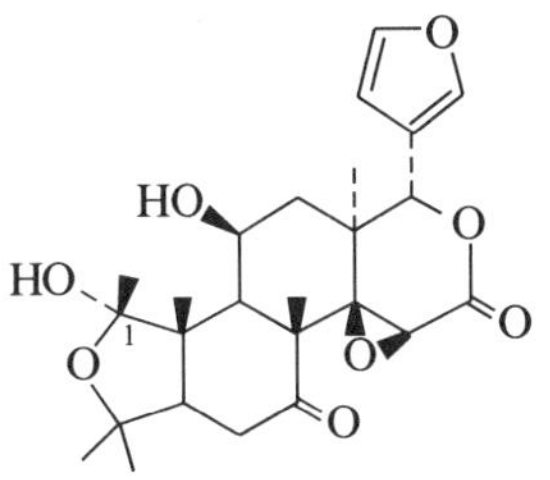

$C_{25}H_{32}O_8$ M 460.523

Constit. of *Clausena heptaphylla*. Cryst. Mp 150°. $[\alpha]_D^{27}$ −70.7° (c, 1 in $CHCl_3$).

1-Et ether:
   $C_{27}H_{36}O_8$ M 488.577
   Constit. of *C. anisata*. Cryst. (Et_2O). Mp 134-135°. $[\alpha]_D$ −63.5° (c, 1.02 in $CHCl_3$).

Chakraborty, D.P. *et al*, *J. Chem. Soc., Chem. Commun.*, 1979, 246 (*isol*)

Ngadjui, B.T. *et al*, *J. Nat. Prod. (Lloydia)*, 1989, **52**, 832 (*isol, pmr, cmr*)

ent-3,13-Clerodadiene-15,16,17-triol   C-80141

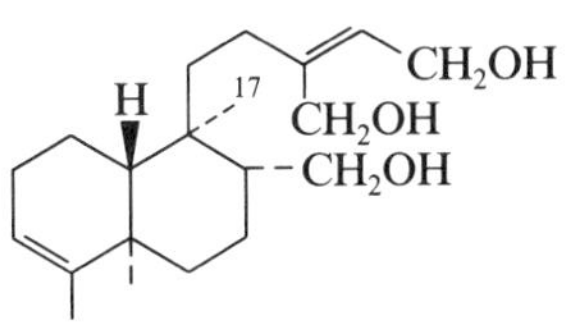

$C_{20}H_{34}O_3$ M 322.487

Constit. of *Baccharis boliviensis*.

17-O-β-Xylopyranoside:
   $C_{25}H_{40}O_8$ M 468.586
   Constit. of *B. boliviensis*.

Zdero, C. *et al*, *Phytochemistry*, 1989, **28**, 531.

ent-4(18),-13-Clerodadien-15-oic acid   C-80142

[120552-37-6]

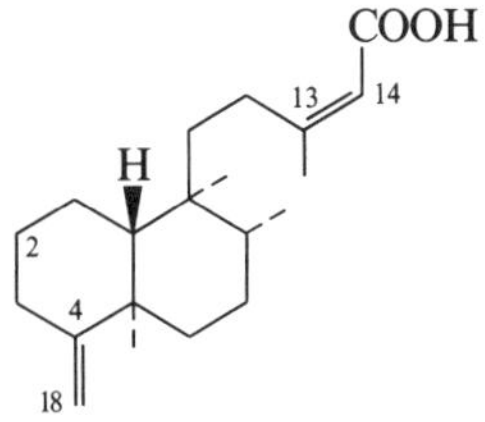

$C_{20}H_{32}O_2$ M 304.472

Constit. of *Ageratina ixiocladon*.

4α,18-Epoxide: [120552-39-8]. **ent-4β,18-Epoxy-13-cleroden-15-oic acid**
   $C_{20}H_{32}O_3$ M 320.471
   Constit. of *A. ixiocladon*.

4β,18-Epoxide: [120662-27-3]. **ent-4α,18-Epoxy-13-cleroden-15-oic acid**
   $C_{20}H_{32}O_3$ M 320.471
   Constit. of *A. ixiocladon*.

13,14-Dihydro: [120552-38-7]. **ent-4(18)-Cleroden-15-oic acid**
   $C_{20}H_{34}O_2$ M 306.487
   Constit. of *A. ixiocladon*.

13,14-Dihydro, 4α,18-epoxide: [120552-40-1]. **ent-4β,18-Epoxy-15-clerodanoic acid**
   $C_{20}H_{34}O_3$ M 322.487
   Constit. of *A. ixiocladon*.

13,14-Dihydro, 4β,18-epoxide: [120662-28-4]. **ent-4α,18-Epoxy-15-clerodanoic acid**
   $C_{20}H_{34}O_3$ M 322.487
   Constit. of *A. ixiocladon*.

4,18-Dihydro,4α-hydroxy: [120662-29-5]. **ent-4β-Hydroxy-13-cleroden-15-oic acid**
   $C_{20}H_{34}O_3$ M 322.487
   Constit. of *A. ixiocladon*.

$\Delta^{3,4}$-*Isomer, 2-oxo:* [120662-30-8]. **ent-2-oxo-3,13-clerodadien-15-oic acid**
   $C_{20}H_{30}O_3$ M 318.455
   Constit. of *A. ixiocladon*.

Tanayo-Castillo, G. *et al*, *Phytochemistry*, 1989, **28**, 139.

ent-2,4(18),13-Clerodatriene-15,16-diol   C-80143

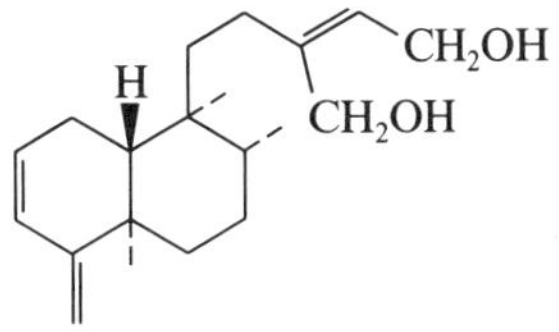

$C_{20}H_{32}O_2$ M 304.472

Constit. of *Baccharis boliviensis*.

Zdero, C. *et al*, *Phytochemistry*, 1989, **28**, 531.

ent-13-Clerodene-2β,3β,4α,15,16-pentol   C-80144

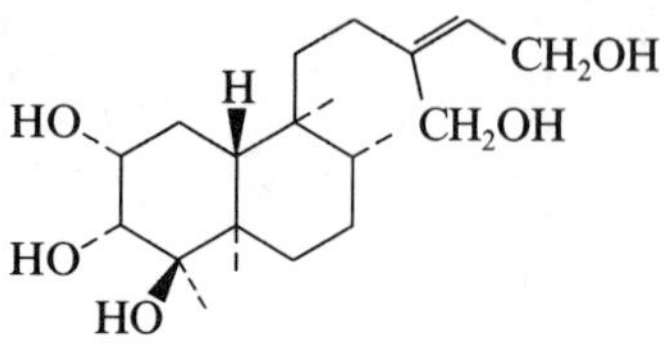

$C_{20}H_{36}O_5$ M 356.501

Constit. of *Baccharis boliviensis*.

Zdero, C. *et al*, *Phytochemistry*, 1989, **28**, 531.

Coenzyme I, 8CI   C-80145

Updated Entry replacing C-02663

Adenosine 5'-(trihydrogen diphosphate) (5'→5') ester with 3-(aminocarbonyl)-1-β-D-ribofuranosylpyridinium hydroxide inner salt, 9CI. Nicotinamide adenine dinucleotide. **Nadide, BAN, USAN, INN.** *Codehydrase I. Codehydrogenase I. Cozymase. Diphosphopyridine nucleotide. Enzopride. DPN. NAD. NSC 20272. Co I. NAD⁺*

[53-84-9]

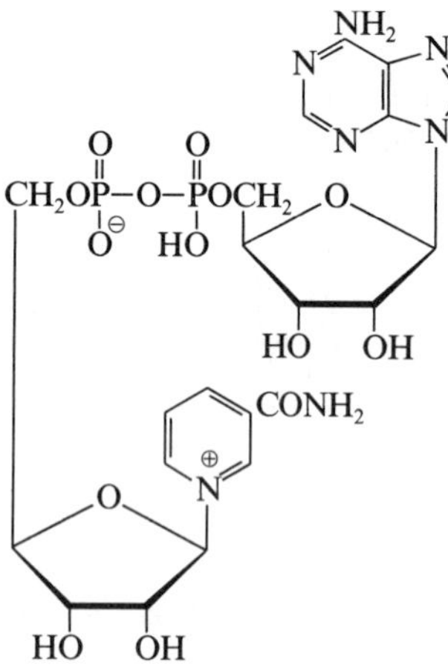

$C_{21}H_{27}N_7O_{14}P_2$ M 663.430

Isol. from bakers' yeast. Antagonist to alcohol and narcotic analgesics. Biological hydrogen acceptor. Hygroscopic powder. Sol. H_2O. λ_{max} 260 (ϵ 17.6 × 10⁶) (H_2O), 340 nm (6.2 × 10⁶)(reduced form). Forms a complex with alkaline cyanide which may be used for its estimation. Stable for weeks in cold neutral soln. Less stable in acid soln., rapidly dec. by alkali.

▷ UU3450000.

Dihydro: [58-68-4]. *Adenosine 5'-(trihydrogen diphosphate), 5'→5'-ester with 1,4-dihydro-1-β-D-ribofuranosyl-3-pyridinecarboxamide, 9CI. NADH*
$C_{21}H_{29}N_7O_{14}P_2$ M 665.446
Reduced form of Coenzyme I in biological systems.

Biochem. Prep., 1949, **1**, 28 (*isol*)
Biochem. Prep., 1953, **3**, 20 (*isol, synth*)
Hughes, N.A. *et al*, *J. Chem. Soc.*, 1957, 3733 (*synth*)
Lemieux, R.U. *et al*, *Can. J. Chem.*, 1963, **41**, 889 (*stereochem*)
Biochem. Prep., 1966, **11**, 84 (*purifn*)
Sarma, R.H. *et al*, *Biochemistry*, 1970, **9**, 557 (*conformn, pmr*)
Blumenstein, M. *et al*, *Biochemistry*, 1973, **12**, 3585 (*cmr*)
Nishizuka, Y. *et al*, *Method. Chim.*, 1977, **11**, 84 (*rev*)
Walsh, C., *Annu. Rev. Biochem.*, 1978, **47**, 881 (*rev*)
Martindale, The Extra Pharmacopoeia, 28th/29th Ed., Pharmaceutical Press, London, 1982/1989, 12991.
Kemal, C., *Comprehensive Heterocyclic Chemistry*, Katritzky A.R. and Rees C.W., Eds., Pergamon, 1984, **1**, 247 (*rev, bibl*)
Walt, D.R. *et al*, *J. Am. Chem. Soc.*, 1984, **106**, 234 (*synth*)

Coenzyme II, 9CI, 8CI   C-80146

Updated Entry replacing C-02664

Adenosine 5'-(trihydrogen diphosphate) 2'-(dihydrogen phosphate) (5'→5') ester with 3-(aminocarbonyl)-1-β-D-ribofuranosylpyridinium hydroxide inner salt, 9CI. Nicotinamide adenine dinucleotide phosphate. Codehydrase II. Codehydrogenase II. Nadifosfate. Phosphocozymase. Triphosphopyridine nucleotide. NADP. TPN. NADP⁺

[53-59-8]

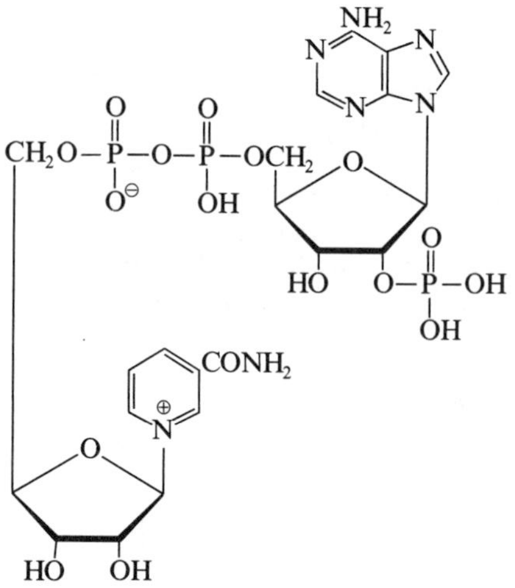

$C_{21}H_{28}N_7O_{17}P_3$ M 743.410

Wide occurrence in living matter, particularly in the liver and in red blood corpuscles, mainly in the reduced form. A component of vitamin B_2 complex. Hydrogen carrier in biochemical redox systems. In the hexose monophosphoric acid system it is reduced to Dihydrocoenzyme II and reoxid. in the presence of flavoproteins. Greyish-white powder. Sol. H_2O. pK_{a1} 3.9, pK_{a2} 6.1. λ_{max} 260 (ϵ 18 × 10⁶)(H_2O), 340 nm (ϵ 6.2 × 10⁶)(reduced form).

Mono-Na salt: Cryst. + H_2O. Mp 175-178° dec.

Dihydro: [53-57-6]. *NADPH*
$C_{21}H_{30}N_7O_{17}P_3$ M 745.426
Reduced form of Coenzyme II in biological systems.

Todd, A.R., *J. Chem. Soc.*, 1941, 427 (*rev*)
Lepage, G.A., *J. Biol. Chem.*, 1949, **180**, 775 (*isol*)
Biochem. Prep., 1953, **3**, 24 (*isol, synth*)
Sund, R., *The Pyridine Nucleotide Coenzymes*, in *Biological Oxidations*, (Singer, T.,Ed.), Interscience, N.Y., 1968, 603 (*rev*)
Blumenstein, M. *et al*, *Biochemistry*, 1973, **12**, 3585 (*cmr*)
Sarma, R.H. *et al*, *Can. J. Chem.*, 1973, **51**, 1843 (*nmr*)
Nishizuka, Y., *Method. Chim.*, 1977, **11**, 84 (*rev*)
Wood, H.C.S., *Compr. Org. Chem.*, Eds., Barton D.H.R. and Ollis, W.D., Pergamon, 1979, **5**, 490 (*rev, bibl*)
Walt, D.R. *et al*, *J. Am. Chem. Soc.*, 1984, **106**, 234 (*synth*)

Collybolidol   C-80147

Updated Entry replacing I-00948

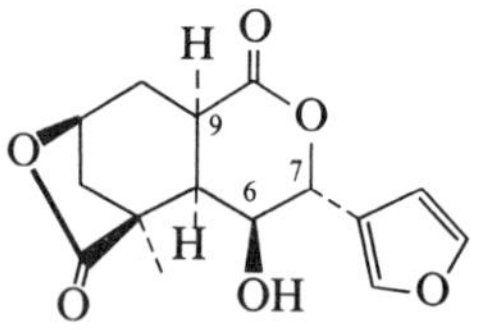

Absolute configuration

$C_{15}H_{16}O_6$ M 292.288

Parent compd. unknown. Hydrol. of Collybolide causes opening of the lactone ring.

O-Benzoyl: [33340-30-6]. **Collybolide**

$C_{22}H_{20}O_7$ M 396.396
Constit. of the basidiomycete *Collybia maculata*. Cryst. (EtOH). Mp 210°. $[\alpha]_D$ +17° (CHCl₃).

9-Epimer, O-benzoyl: [31199-75-4]. **Isocollybolide**
$C_{22}H_{20}O_7$ M 396.396
Constit. of *C. maculata*. Mp 193°. $[\alpha]_D$ −17° (c, 1.7 in CHCl₃).

7,9-Diepimer, 6-deoxy: [106794-13-2]. **Deoxycollybolidol**
$C_{15}H_{16}O_5$ M 276.288
Constit. of *C. peronata*. Cryst. (EtOAc/hexane). Mp 189-190°. $[\alpha]_D$ +21° (c, 0.1 in CHCl₃).

Pascard-Billy, C., *Acta Crystallogr., Sect. C*, 1972, **28**, 331 (*cryst struct, Isocollybolide*)
Bui, A.M. *et al, Tetrahedron*, 1974, **30**, 1327 (*isol, pmr, cmr, uv, struct*)
Fogedal, M. *et al, Phytochemistry*, 1986, **25**, 2661 (*Deoxycollybolidol*)

Coniferaldehyde   C-80148

3-(4-Hydroxy-3-methoxyphenyl)-2-propenal, 9CI. 4-Hydroxy-3-methoxycinnamaldehyde. Ferulaldehyde. Hadromal

[458-36-6]

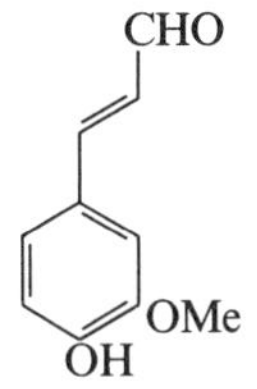

$C_{10}H_{10}O_3$ M 178.187
(E)-form [20649-42-7]
  Occurs in wood, free and as component of lignin. Found extensively in brandies etc. as leachingprod. from wood. Light-yellow needles (pet. ether). Mp 84°.

2,4-Dinitrophenylhydrazene: Mp 265-266°.

Tiemann, F., *Ber.*, 1885, **18**, 3481 (*synth*)
Adler, E. *et al, Acta Chem. Scand.*, 1948, **2**, 839 (*isol*)
Black, R.A. *et al, J. Am. Chem. Soc.*, 1953, **75**, 5344 (*isol, uv*)
Nakanura, Y. *et al, CA*, 1977, **86**, 139537z (*synth*)
Liptaj, T. *et al, Collect. Czech. Chem. Commun.*, 1980, **45**, 33 (*pmr*)
Kutsuki, H. *et al, CA*, 1981, **95**, 132414j (*synth*)

2-Copaen-8-one   C-80149

α-Copaen-8-one

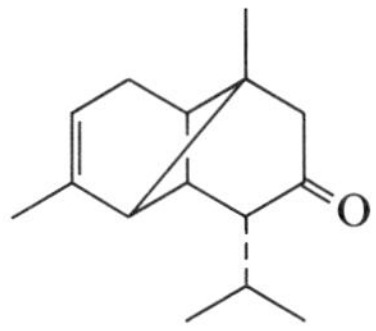

$C_{15}H_{22}O$ M 218.338
Constit. of *Neomirandea guevarii*. Oil. $[\alpha]_D^{24}$ +14° (c, 0.53 in CHCl₃).

Tamayo-Castillo, G. *et al, Phytochemistry*, 1989, **28**, 938.

Corallistin A   C-80150

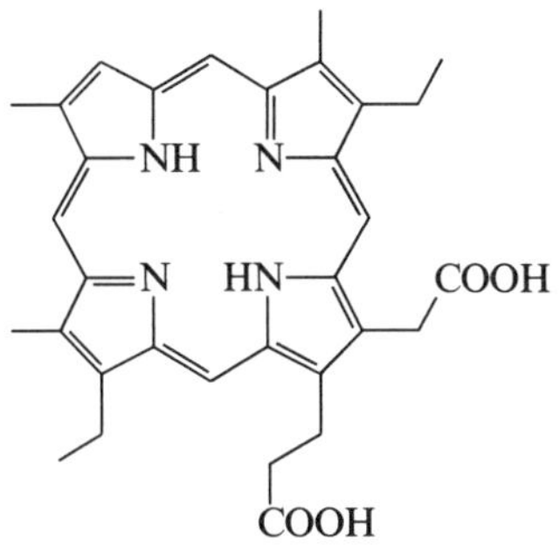

$C_{32}H_{34}N_4O_4$ M 538.645
Isol. from a Coral Sea desmosponge *Corallistes* sp. Rare example of a free porphyrin from a natural organism.

D'Ambrosio, M. *et al, Helv. Chim. Acta*, 1989, **72**, 1451 (*isol, uv, pmr, cmr, struct*)

Cornus-tannin 2   C-80151

1,2,3-Tri-O-galloyl-4,6-O-hexahydroxydiphenoylglucose
[58970-75-5]

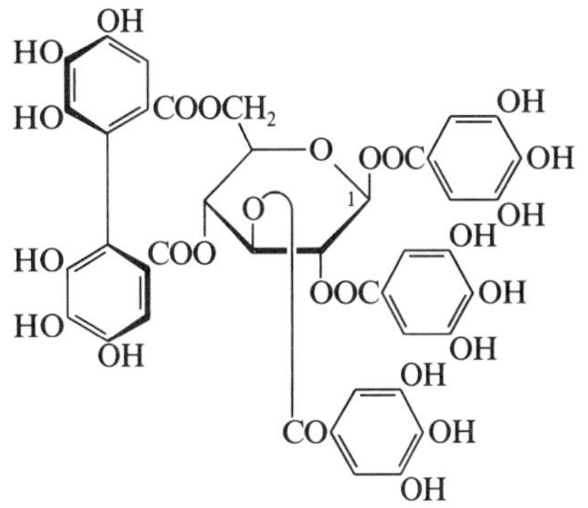

$C_{41}H_{30}O_{25}$ M 922.673
Isol. from fruits of *Cornus officinalis* and from *Tellima grandiflora*.

1-O-Degalloyl: [30737-92-9]. **Cornus-tannin 3**
$C_{34}H_{26}O_{22}$ M 786.566
Isol. from fruits of *C. officinalis* and from *T. grandiflora*.

Wilkins, C.K. *et al, Phytochemistry*, 1976, **15**, 211.
Okuda, T. *et al, Heterocycles*, 1981, **16**, 1321.

Corylidin   C-80152

Updated Entry replacing C-20231
[63109-31-9]

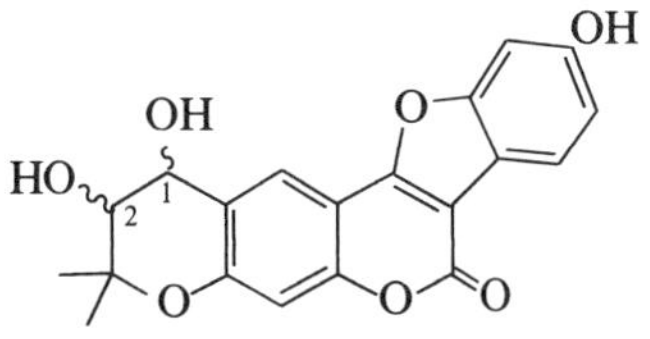

$C_{20}H_{16}O_7$ M 368.342
CA numbering shown. Constit. of the fruits of *Psoralea corylifolia*. Needles (EtOH/Me₂CO). Mp 349-351°.

1,2-Dideoxy: [3564-61-2]. **Isopsoralidin**
$C_{20}H_{16}O_5$ M 336.343
Isol. from seeds of *P. corylifolia*. Cryst. (MeOH). Mp 284-287°. Violet fluor in EtOH soln.

Jain, A.C. *et al, Indian J. Chem.*, 1974, **12**, 659 (*Isopsoralidin*)
Gupta, G.K. *et al, Phytochemistry*, 1977, **16**, 403 (*isol*)

Corylinal   C-80153

2-Hydroxy-5-(7-hydroxy-4-oxo-4H-1-benzopyran-3-yl)benzaldehyde, 9CI. 3′-Formyl-4′,7-dihydroxyisoflavone

[65615-46-5]

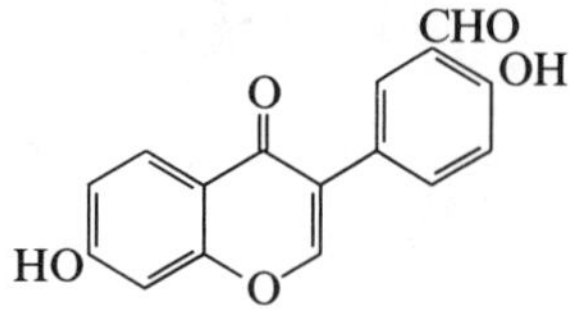

$C_{16}H_{10}O_5$ M 282.252

Isol. from seeds of *Psoralea corylifolia*. Isol. as Me ether after methylation.

7-Me ether: Needles (MeOH). Mp 223-224°.

Gupta, G.K. *et al, Phytochemistry*, 1978, **17**, 164.

Costatolide   C-80154

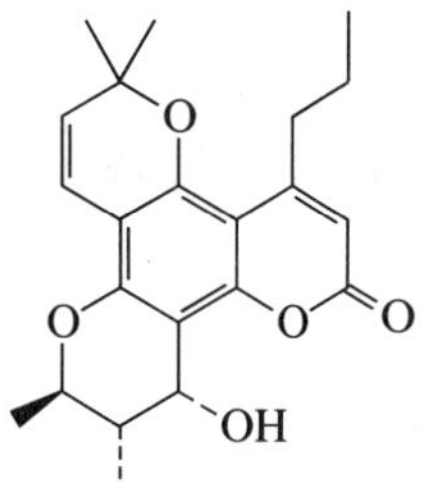

$C_{22}H_{26}O_5$ M 370.444

Isol. from resin of *Calophyllum costatum*. Cryst. (C_6H_6). Mp 181-182°. $[\alpha]_D^{25}$ −19.9° (c, 0.42 in $CHCl_3$), $[\alpha]_D^{25}$ −50.4° (c, 1.55 in Me_2CO).

Stout, G.H. *et al, J. Org. Chem.*, 1964, **29**, 3604 (*isol, uv, ir, pmr, ms, struct*)

Crassifolioside   C-80155

[120693-47-2]

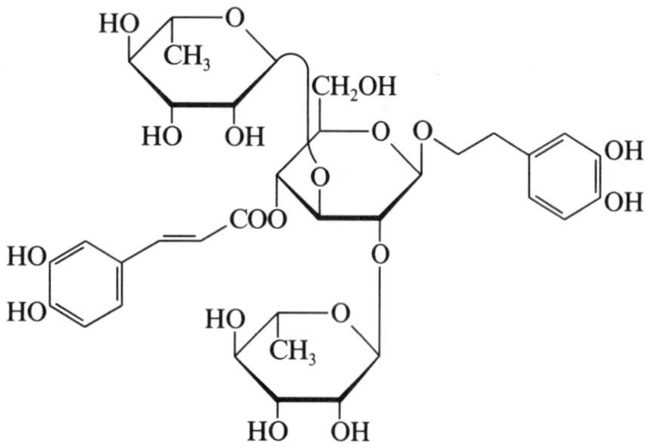

$C_{35}H_{46}O_{19}$ M 770.737

Constit. of *Plantago crassifolia*. Pale yellow amorph. powder. $[\alpha]_D^{20}$ −110.9° (c, 0.32 in MeOH).

Andary, C. *et al, Phytochemistry*, 1989, **28**, 288.

Cristacarpin   C-80156

9-Methoxy-10-(3-methyl-2-butenyl)-6H-benzofuro[3,2-c][1]benzopyran-3,6-(11aH)-diol, 9CI. 3,6a-Dihydroxy-9-methoxy-10-prenylpterocarpan. Erythrabyssin I

[74515-47-2]

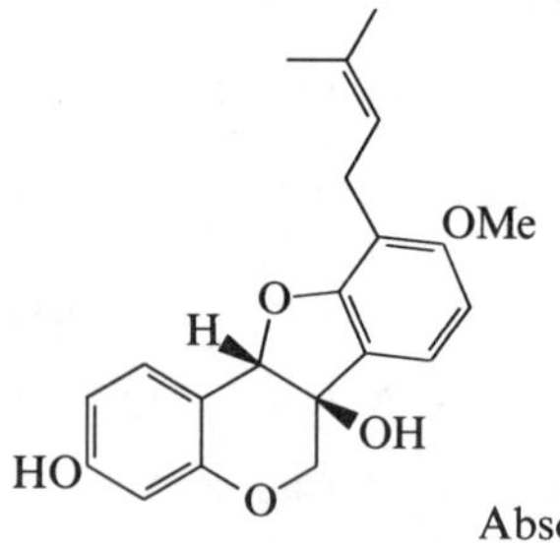

Absolute configuration

$C_{21}H_{22}O_5$ M 354.402

Isol. from *Erythrina abyssinica, E. crista-galli, E. sandwicensis* and *Psophocarpus tetragonolobus*. Phytoalexin. $[\alpha]_D$ −220° (c, 0.02 in MeOH).

O-De-Me: [74515-45-0]. **Sandwicarpin**
    $C_{20}H_{20}O_5$ M 340.375
    Isol. from leaves of *E. sandwicensis*. Phytoalexin. $[\alpha]_D$ −278° (MeOH).

Ingham, J.L. *et al, Phytochemistry*, 1980, **19**, 1203 (*isol, uv, ms, struct, abs config*)
Ingham, J.L., *Z. Naturforsch., C*, 1980, **35**, 384 (*Sandwicarpin*)
Kamat, V.S. *et al, Heterocycles*, 1981, **15**, 1163 (*isol, uv, cd, abs config*)

Crotmadine   C-80157

Updated Entry replacing C-30199

[92662-86-7]

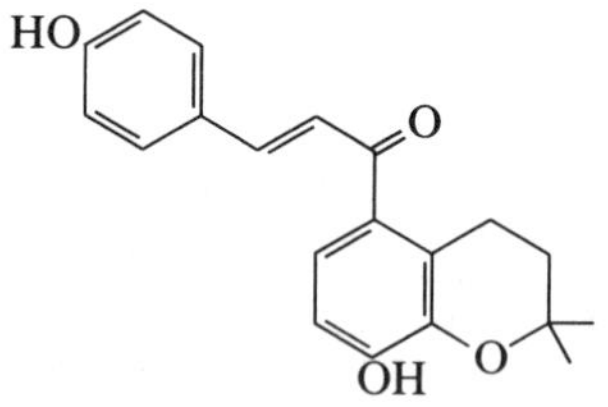

$C_{20}H_{20}O_4$ M 324.376

Constit. of *Crotalaria madurensis*. Cryst. ($CHCl_3$/MeOH). Mp 191°.

Bhakuni, D.S. *et al, J. Nat. Prod. (Lloydia)*, 1984, **47**, 585 (*isol*)
Malhotra, S. *et al, J. Nat. Prod. (Lloydia)*, 1988, **51**, 578 (*struct*)

Crotonin   C-80158

Updated Entry replacing C-30201

[17633-81-7]

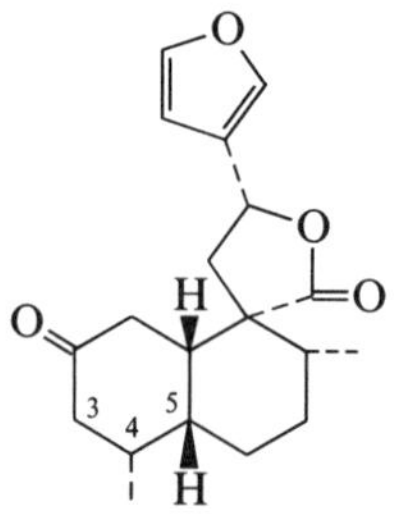

$C_{19}H_{24}O_4$ M 316.396

Constit. of *Croton lucidus*. Cryst. (EtOAc/hexane). Mp 148-149°. $[\alpha]_D$ −1.4° ($CHCl_3$).

5-Epimer: **t-Crotonin**

$C_{19}H_{24}O_4$ M 316.396
Constit. of *C. cajucara*. Needles. Mp 131-132°. $[\alpha]_D$ +2.3° (c, 3.29 in $CHCl_3$).

5-Epimer, 3,4-didehydro: **t-Dehydrocrotonin**
$C_{19}H_{22}O_4$ M 314.380
Constit. of *C. cajucara*. Needles. Mp 138.5-140.5°. $[\alpha]_D$ +11.9° (c, 0.3 in $CHCl_3$).

Chan, W.R. *et al, J. Chem. Soc. C*, 1968, 2781 (*isol*)
Blount, J.F. *et al, J. Chem. Res. (S)*, 1984, 114 (*cryst struct*)
Itokawa, H. *et al, Phytochemistry*, 1989, **28**, 1667 (*isol*)

Crustaxanthin   C-80159

β,β-Carotene-3,3′,4,4′-tetrol
[6094-35-5]

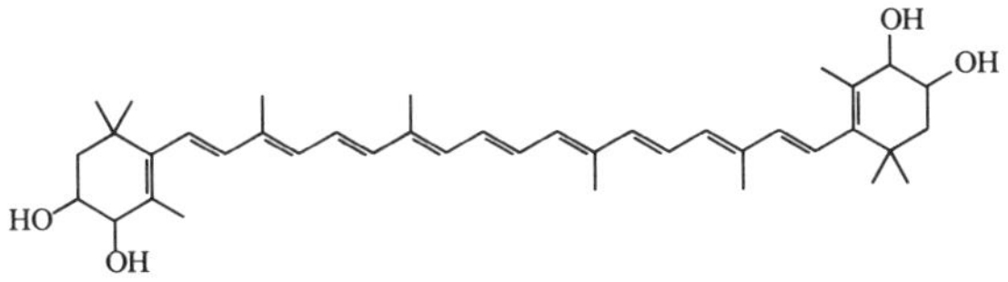

$C_{40}H_{56}O_4$ M 600.880
Isol. from the crustacean *Arctodiaptomus salinus*, also from the copepod *Euchaeta russelli* and in aplanospores of *Haematococcus pluvialis* and fish skins.

Bartlett, L. *et al, J. Chem. Soc. C*, 1969, 2527 (*ord*)
Andrewes, A.G. *et al, Acta Chem. Scand., Ser. B*, 1974, **28**, 730 (*synth, uv*)
Englert, G., *Helv. Chim. Acta*, 1975, **58**, 2367 (*cmr*)
Bandaranayake, W.M. *et al, Comp. Biochem. Physiol. B: Comp. Biochem.*, 1982, **72**, 409; *CA*, **98**, 14534n (*isol*)

Cubene   C-80160

Pentacyclo[4.2.0.0²,⁵.0³,⁸.0⁴,⁷]oct-1-ene, 9CI. 1,2-Dehydrocubane
[77478-10-5]

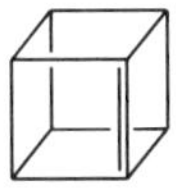

C_8H_6 M 102.135
Transient intermediate trapped by Diels Alder addn.

Eaton, P.E. *et al, J. Am. Chem. Soc.*, 1988, **110**, 7230 (*synth*)

Cucurbic acid   C-80161

Updated Entry replacing C-50302
3-Hydroxy-2-(2-pentenyl)cyclopentaneacetic acid

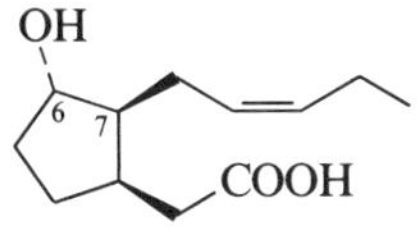

$C_{12}H_{20}O_3$ M 212.288
Constit. of *Cucurbita pepo*. Gum.

6,7-Diepimer: **6-Epi-7-isocucurbic acid**
$C_{12}H_{20}O_3$ M 212.288
Constit. of *Vicia faba*. Oil. $[\alpha]_D^{22}$ +6.1° (c, 0.1 in EtOH).

Fukui, H. *et al, Agric. Biol. Chem.*, 1977, **41**, 175 (*isol*)
Kitahara, T. *et al, Agric. Biol. Chem.*, 1987, **51**, 1129 (*synth*)
Miersch, O. *et al, Phytochemistry*, 1989, **28**, 339 (*isol*)

Cucurbitacin T   C-80162

[118172-78-4]

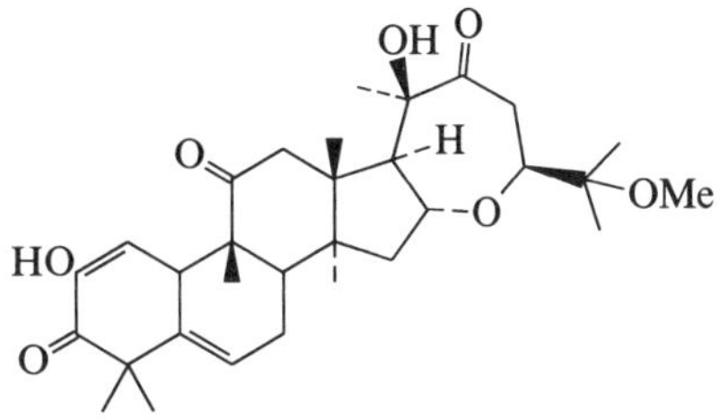

$C_{31}H_{44}O_7$ M 528.684
Constit. of *Colocynthis vulgaris*. Pale yellow semisolid. $[\alpha]_D^{28}$ −53.5° (c, 1.44 in $CHCl_3$).

Gamlath, C.B. *et al, Phytochemistry*, 1988, **27**, 3225.

Cudraniaxanthone   C-80163

2-(1,1-Dimethyl-2-propenyl)-1,5,6-trihydroxy-3-methoxy-9H-xanthen-9-one, 9CI
[38710-42-8]

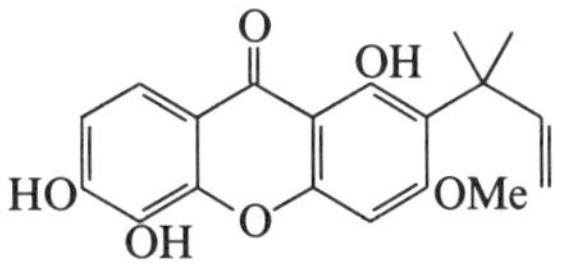

$C_{19}H_{18}O_6$ M 342.348
Constit. of the roots of *Cudrania javanensis*. Mp 308°.
Tri-Me ether: Mp 122-124°.

Murti, V.V.S. *et al, Phytochemistry*, 1972, **11**, 2089.

Cumambranolide   C-80164

Updated Entry replacing C-40184
10α-Hydroxy-3,11(13)-guaiadien-12,6α-olide. 8-Deoxycumambrin B
[21956-65-0]

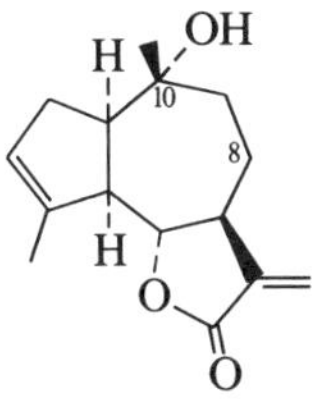

$C_{15}H_{20}O_3$ M 248.321
Constit. of *Artemisia nova* and *A. tripartita* ssp. *rupicola*. Prisms (Et_2O/pet. ether). Mp 117-118°.

8α-Hydroxy: [21982-83-2]. **Cumambrin B**
$C_{15}H_{20}O_4$ M 264.321
Constit. of *A. nova*, *A. tripartita* ssp. *rupicola*, *Ambrosia cumanensis* and *A. acanthicarpa*. Cryst. (EtOAc/Et_2O). Mp 178-180°. $[\alpha]_D$ +92.5° (c, 5.58 in $CHCl_3$).

8α-Acetoxy: [20482-33-1]. **Cumambrin A**
$C_{17}H_{22}O_5$ M 306.358
Constit. of *Artemesia nova*, *A. tripartita* ssp. *rupicola*, *Ambrosia cumanensis* and *A. acanthicarpa*. Cryst. Mp 188-190°. $[\alpha]_D$ +103° (c, 2.79 in $CHCl_3$).

8α-(2-Methylpropanoyloxy):
$C_{19}H_{26}O_5$ M 334.411
Constit. of *Eremanthus incanus*. Cryst. (Et_2O/pet. ether). Mp 130°. $[\alpha]_D^{24}$ +84.1° (c, 0.22 in $CHCl_3$).

8α-(2-Methylpropenoyloxy):
$C_{19}H_{24}O_5$ M 332.396
Constit. of *Lychnophora blanchetti*. Gum.

8α-Tigloyloxy:
$C_{20}H_{26}O_5$ M 346.422
Constit. of *L. blanchetti.* Gum.

8α-Angeloyloxy:
$C_{20}H_{26}O_5$ M 346.422
Constit. of *Chrysanthemum ornatum.* Antimicrobial. Oil.
$[\alpha]_D^{20}$ +100° (c, 0.2 in MeOH).

8β-Hydroxy: Cryst. (MeOH). Mp 91-92°.

8β-Angeloyloxy:
$C_{20}H_{26}O_5$ M 346.422
Constit. of *Helianthus maximiliani.* Cryst. (EtOAc). Mp 160-161°.

8β-(Epoxyangeloyloxy):
$C_{20}H_{26}O_6$ M 362.422
From *H. maximiliani.* Oil.

8β-(3-Hydroxy-2-methylenebutanoyloxy):
$C_{20}H_{26}O_6$ M 362.422
From *H. maximiliani.* Oil.

8β-Sarracinoyloxy:
$C_{20}H_{26}O_6$ M 362.422
From *H. maximiliani.* Cryst. (diisopropyl ether/MeOH). Mp 129-130°.

8α-Acetoxy, 11β,13-Dihydro: **Dihydrocumambrin A**
$C_{17}H_{24}O_5$ M 308.374
From *C. coronarium.* Cryst. Mp 174°. $[\alpha]_D^{24}$ +44° (c, 0.08 in CHCl₃).

10-O-β-D-Glucopyranoside: [100187-65-3]. **Macrocliniside G**
$C_{21}H_{30}O_8$ M 410.463
Constit. of *Macroclinidium trilobum.* Needles (MeOH aq.). Mp 128-129°. $[\alpha]_D^{25}$ +39.8° (c, 1.86 in MeOH).

10-Epimer: **10-Epicumambranolide.** *10-Epi-8-deoxycumambrin B*
$C_{15}H_{20}O_3$ M 248.321
Constit. of *Stevia yaconensis.* Cryst. (Et₂O). Mp 123-124°.

Romo, J. et al, Tetrahedron, 1968, **24**, 5625 (isol)
Irwin, M.A. et al, Phytochemistry, 1969, **8**, 305 (isol)
Bohlmann, F.B. et al, Phytochemistry, 1980, **19**, 2663, 2669 (isol)
Haruna, M. et al, Phytochemistry, 1981, **20**, 2583 (isol)
Watson, W.H. et al, Acta Crystallogr., Sect. B, 1982, **38**, 1608 (cryst struct)
Gershenzon, J. et al, Phytochemistry, 1984, **23**, 1959 (isol)
El-Masry, S. et al, Phytochemistry, 1984, **23**, 2953 (isol)
Miyase, T. et al, Chem. Pharm. Bull., 1985, **33**, 4445 (isol)
Sosa, V.E. et al, Phytochemistry, 1989, **28**, 1925 (isol, cryst struct)

Curcumin   C-80165

Updated Entry replacing C-20253
1,7-Bis(4-hydroxy-3-methoxyphenyl)-1,6-heptadiene-3,5-dione, 9CI. Curcuma yellow. Diferuloylmethane
[458-37-7]

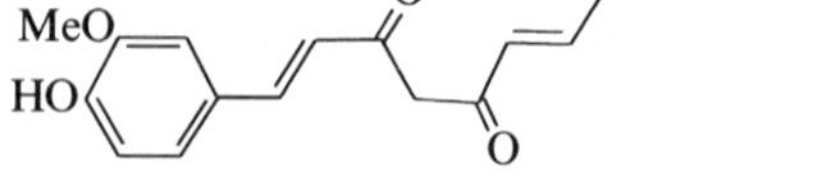

$C_{21}H_{20}O_6$ M 368.385
Isol. from *Curcuma* spp. (turmeric). Natural colouring matter used extensively in Indian curries etc. Orange prisms. Mp 183°.

Di-Ac: Cryst. Mp 170-171°.

Dibenzoyl: Cryst. Mp 210°.

Demethoxy: [22608-11-3]. *1-(4-Hydroxy-3-methoxyphenyl)-7-(4-hydroxyphenyl)-1,6-heptadiene-3,5-dione, 9CI.* p-

Hydroxycinnamoylferuloylmethane. **Demethoxycurcumin**
$C_{20}H_{18}O_5$ M 338.359
Isol. from *C. longa, C. domestica, C. xanthorrhiza.* Orange-yellow powder. Mp 168°.

Bis-demethoxy: [22608-12-4]. *1,7-Bis(4-hydroxyphenyl)-1,6-heptadiene-3,5-dione, 9CI. p,p'-Dihydroxydicinammoylmethane.* **Bisdemethoxycurcumin**
$C_{19}H_{16}O_4$ M 308.333
Isol. from *C. longa, C. domestica* and *C. aromatica.* Absent in some *C.* spp. Yellow plates + H₂O (EtOH). Mp 224°.

[24939-16-0, 33171-05-0]

Lampe, V., Ber., 1918, **51**, 1347 (synth)
Srinivasan, K.R., J. Pharm. Pharmacol., 1953, **5**, 448 (derivs)
Jentzsch, K. et al, Sci. Pharm., 1968, **36**, 251; CA, **70**, 90793h (isol)
Sastry, B.S., Res. Ind., 1970, **15**, 258; CA, **75**, 75063e (isol)
Kuroyagi, M. et al, Yakugaku Zasshi (J. Pharm. Soc. Jpn.), 1970, **90**, 1467; CA, **74**, 61612a (isol)
Roughley, P.J. et al, J. Chem. Soc., Perkin Trans. 1, 1973, 2379 (biosynth)
Karig, F., Dtsch. Apoth.-Ztg., 1975, **115**, 325; ca, **83**, 65372f (derivs)
Kashina, C. et al, Heterocycles, 1977, **7**, 241 (synth)
Wahlstrom, B. et al, Acta Pharmacol. Toxicol., 1978, **43**, 86 (metab)
Holder, G.M. et al, Xenobiotica, 1978, **8**, 761 (metab)
Toønnesen, H.H. et al, Acta Chem. Scand., Ser. B, 1982, **36**, 475 (cryst struct, bibl)

Curvularin   C-80166

Updated Entry replacing C-03021

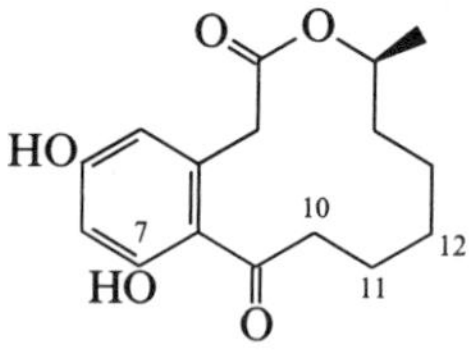

$C_{16}H_{20}O_5$ M 292.331

(S)-form [10140-70-2]
Isol. from *Curvularia* spp., *Penicillium gilmanii, Alternaria cinerariae* and *A. macrospora.* Phytotoxin. Cryst. (MeOH/C₆H₆). Mp 206°. $[\alpha]_D^{18}$ −36.3°.

Di-Me ether: Cryst. (MeOH aq.). Mp 72°, Mp 105-108°. $[\alpha]_D^{18}$ −2.9° (c, 2.7 in CHCl₃).

Dibenzoyl: Cryst. (C₆H₆/pet. ether). Mp 133-134°. $[\alpha]_D^{18}$ −10.8° (c, 1.9 in CHCl₃).

10,11-Didehydro(E-): [21178-57-4]. **trans-Dehydrocurvularin**
$C_{16}H_{16}O_5$ M 288.299
From *C.* spp., *A. cinerariae* and *A. macrospora.* Phytotoxin. Mp 224°. $[\alpha]_D^{22}$ −79.8° (c, 3.0 in EtOH).

10,11-Didehydro(E-), 7-Ac:
$C_{18}H_{18}O_6$ M 330.337
From *C. scirpicola.* Mp 105-110°. $[\alpha]_D^{20}$ −31.7° (c, 0.4 in MeOH).

10,11-Didehydro(Z-): **cis-Dehydrocurvularin**
$C_{16}H_{18}O_5$ M 290.315
Metabolite of hybrid strain of *Penicillium citreo-viride B.* Amorph. powder. $[\alpha]_D^{22}$ +7.3° (c, 0.78 in EtOH).

12-Oxo: **12-Oxocurvularin**
$C_{16}H_{18}O_6$ M 306.315
Metabolite of *P. citreo-viride B.* Amorph. powder. $[\alpha]_D^{29}$ −43.5° (c, 0.47 in EtOH).

11α-Hydroxy: **11α-Hydroxycurvularin**
$C_{16}H_{20}O_6$ M 308.330
Metabolite of *P. citreo-viride B.* Cryst. Mp 150-152°. $[\alpha]_D^{26}$ −29.4° (c, 0.33 in EtOH).

11β-Hydroxy: 11β-Hydroxycurvularin
$C_{16}H_{20}O_6$ M 308.330
Metabolite of *P. citreo-viride B*. Cryst. Mp 138-140°.
$[\alpha]_D^{24}$ −10.9° (c, 0.19 in EtOH).
(±)-*form* [65690-37-1]
Mp 182-184°.

Birch, A.J. *et al, J. Chem. Soc.*, 1959, 3146 *(isol)*
Baker, P.M. *et al, J. Chem. Soc. C*, 1967, 1913 *(synth)*
Munro, H.D. *et al, J. Chem. Soc. C*, 1967, 947 *(deriv)*
Coombe, R.G. *et al, Aust. J. Chem.*, 1968, **21**, 783 *(isol)*
Raistrick, H. *et al, J. Chem. Soc. C*, 1971, 3069 *(isol)*
Gerlach, H., *Helv. Chim. Acta*, 1977, **60**, 3039 *(synth)*
Assante, G. *et al, Phytochemistry*, 1977, **16**, 243 *(deriv)*
Takahashi, T. *et al, Tetrahedron Lett.*, 1980, **21**, 3885 *(synth)*
Robeson, D.J. *et al, Z. Naturforsch., C*, 1981, **36**, 1081; *J. Nat. Prod. (Lloydia)*, 1985, **48**, 139 *(isol)*
Lai, S., *Tetrahedron Lett.*, 1989, **30**, 2241 *(derivs)*

Cyanic acid   C-80167

Updated Entry replacing C-03036
Hydrogen cyanate
[420-05-3]

$$HOC{\equiv}N$$

CHNO   M 43.025
Is not in equilib. with Isocyanic acid, I-80057. Refs. are freq. confused in CA. Obt. in Ar matrix. Kinetically unstable.

Jacox, M.E. *et al, J. Chem. Phys.*, 1964, **40**, 2457.
Groving, N. *et al, Acta Chem. Scand.*, 1965, **19**, 1768 *(synth)*
Fieser and Fieser's Reagents for Organic Synthesis, Wiley, 1967, **1**, 170.
Bondybey, V.E. *et al, J. Mol. Spectrosc.*, 1982, **92**, 431 *(ir, uv)*
Teles, J.H. *et al, Chem. Ber.*, 1989, **122**, 753 *(struct, bibl)*
Sax, N.I., *Dangerous Properties of Industrial Materials*, 5th Ed., Van Nostrand-Reinhold, 1979, 526.

N-Cyanobenzenecarboximidoyl chloride   C-80168

[107129-18-0]

$$PhCCl{=}NCN$$

$C_8H_5ClN_2$ M 164.594
Solid. Mp 81°. $Bp_{0.2}$ 110°.

Hartke, K. *et al, Chem. Ber.*, 1989, **122**, 657 *(synth, pmr, cmr)*

Cyanogen fluoride, 9CI, 8CI   C-80169

Updated Entry replacing C-20257
Fluoroformonitrile
[1495-50-7]

$$FCN$$

CFN   M 45.016
Mp −72° subl.
▷ Can polymerise explosively.

Fawcett, F.S. *et al, J. Am. Chem. Soc.*, 1960, **82**, 1509 *(synth, props)*
Bieri, G., *Chem. Phys. Lett.*, 1977, **46**, 107 *(uv)*
Shimanoachi, T., *J. Phys. Chem. Ref. Data*, 1977, **6**, 993 *(ir)*
Bauknight, C.W. *et al, J. Org. Chem.*, 1988, **53**, 4443 *(synth, haz)*
Bretherick, L., *Handbook of Reactive Chemical Hazards*, 2nd Ed., Butterworths, London and Boston, 1979, 283.
Sax, N.I., *Dangerous Properties of Industrial Materials*, 5th Ed., Van Nostrand-Reinhold, 1979, 528.

Cycloaraneosene   C-80170

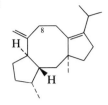

$C_{20}H_{32}$ M 272.473
Constit. of *Sordaria araneosa*. Oil. $[\alpha]_D^{27}$ −37.5° (c, 0.21 in $CHCl_3$).

8β-Hydroxy: 8β-Hydroxycycloaraneosene
$C_{20}H_{32}O$ M 288.472
Constit. of *S. araneosa*. Oil. $[\alpha]_D$ +15°.

Borschberg, H.J., *Ph.D. Thesis*, ETH, Zurich, 1975 *(isol)*
Kato, N. *et al, Bull. Chem. Soc. Jpn.*, 1988, **61**, 3231 *(synth)*
Kato, N. *et al, Chem. Lett.*, 1989, 91 *(synth)*

23-Cycloartene-3,25-diol   C-80171

9,19-Cyclolanost-23-ene-3,25-diol, 9CI

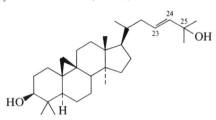

$C_{30}H_{50}O_2$ M 442.724
3β-form [14599-48-5]
   Isol. from *Tillandsia usneoides* and *Tricholepsis glaberrima*. Prisms (EtOAc). Mp 200-204°. $[\alpha]_D^{27}$ +38° (c, 0.85 in $CHCl_3$).
3-Ac: [26531-71-5].
   $C_{32}H_{52}O_3$ M 484.761
   Isol. from roots of *Sapium insigne*. Prisms (hexane). Mp 148-150°. $[\alpha]_D^{22}$ +43° (c, 0.75 in $CHCl_3$).
Di-Ac: Cryst. (Et_2O). Mp 105-108°.
O^{25}-*Me: 25-Methoxy-23-cycloarten-3β-ol*
   $C_{31}H_{52}O_2$ M 456.751
   Isol. from *Tillandsia usneoides*. Isol. as acetate.
O^{25}-*Me,O^3-Ac:* Needles (MeOH/Et_2O). Mp 152-154°. $[\alpha]_D^{28}$ +48° (c, 0.75 in $CHCl_3$).

Djerassi, C. *et al, J. Chem. Soc.*, 1962, 4034 *(isol, ir, pmr, struct)*
Fourrey, J.L. *et al, Tetrahedron*, 1970, **26**, 3839 *(synth)*
Chawla, A.S. *et al, Planta Med.*, 1978, **34**, 109 *(isol)*
Srivastava, S.K. *et al, J. Nat. Prod. (Lloydia)*, 1985, **48**, 496 *(acetate, isol, pmr)*

24-Cycloartene-1,3-diol   C-80172

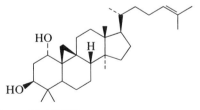

$C_{30}H_{50}O_2$ M 442.724
(1α,3β)-*form*
   1-Ac: [119765-93-4]. *1α-Acetoxy-24-cycloarten-3β-ol*
   $C_{32}H_{52}O_3$ M 484.761

Constit. of *Commiphora incisa*. Amorph. solid.
[119765-93-4]

Provan, G.J. *et al*, *Phytochemistry*, 1988, **27**, 3841.

Cycloartobiloxanthone   C-80173

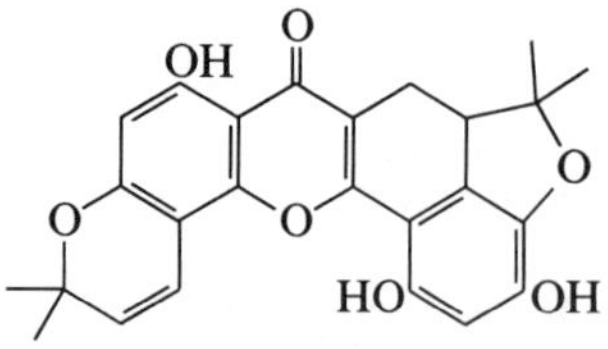

C25H22O7   M 434.445
Constit. of *Artocarpus nobilis*. Dark yellow solid (MeOH).
Mp 285-287°.

Sultanbawa, M.U.S. *et al*, *Phytochemistry*, 1989, **28**, 599.

Cyclobuta[*b*]anthracen-1(2*H*)one   C-80174

1,2-Dihydrocyclobuta[b]anthracen-1-one

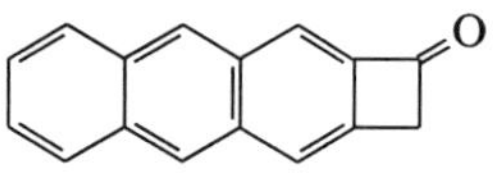

C16H10O   M 218.254
Yellow solid. Mp 230-251° dec.

Müller, P. *et al*, *Helv. Chim. Acta*, 1989, **72**, 1618 (*synth, uv, ir, pmr, ms*)

Cyclobuta[*b*]naphthalen-1(2*H*)-one   C-80175

Naphtho[b]cyclobutenone
[89185-31-9]

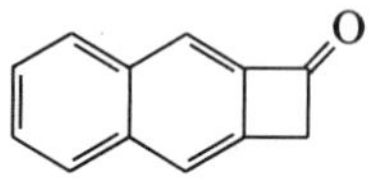

C12H8O   M 168.195
Cryst. (CCl4). Mp 163-164°.

2,4-Dinitrophenylhydrazone: Orange cryst. (EtOAc). Mp
266-269°.

Abou-Teim, O. *et al*, *J. Chem. Soc., Perkin Trans. 1*, 1983, 2659 (*synth, uv*)

Cyclocaducinol   C-80176

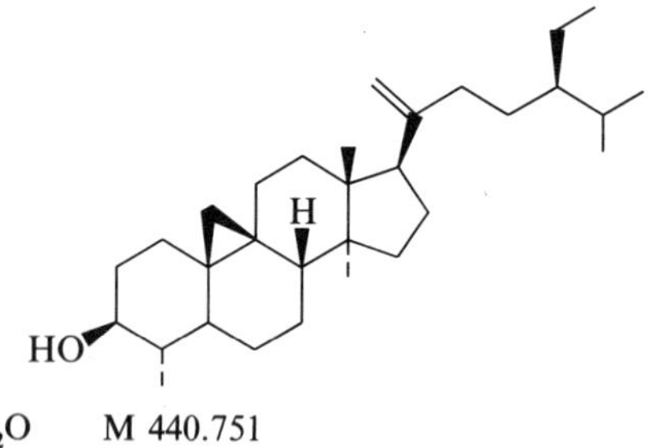

C31H52O   M 440.751
Constit. of *Euphorbia caducifolia*. Needles (Me2CO). Mp
96-98°. [α]D +48.9° (c, 0.16 in CHCl3).

Afza, N. *et al*, *Phytochemistry*, 1989, **28**, 1982.

3-Cyclodecene-1,5-diyne, 9CI   C-80177

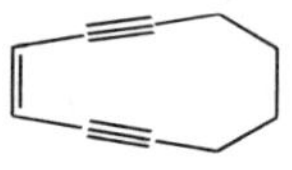

C10H10   M 130.189
(**Z**)-*form* [115227-60-6]
Slowly dec. on standing.

Nicolaou, K.C. *et al*, *J. Am. Chem. Soc.*, 1988, **110**, 4866 (*synth, pmr, cmr, uv*)

1-Cycloheptenecarboxylic acid   C-80178

[4321-25-9]

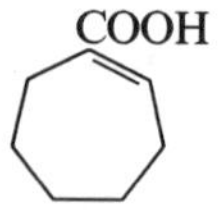

C8H12O2   M 140.182
Mp 54°. Bp2.5 124-125°.

Me ester:
C9H14O2   M 154.208
Liq. Bp0.15 42-46°.

Et ester:
C10H14O2   M 166.219
Bp22 116-119°.

Amide:
C8H13NO   M 139.197
Needles (MeOH/EtOH). Mp 125-127°.

Nitrile: [20343-19-5]. *1-Cyanocycloheptene*
C8H11N   M 121.182
Liq. Bp12 95-96°, Bp0.25 47-51°.

Ayerst, G.G. *et al*, *J. Chem. Soc.*, 1960, 3445 (*nitrile, amide, Et ester, synth*)
Braidy, R., *Bull. Soc. Chim. Fr.*, 1967, 3489 (*synth, ir, cryst struct*)
Mathur, K.C. *et al*, *Indian J. Chem.*, 1968, **6**, 248 (*synth, nitrile, Et ester*)
Pawlak, J.L. *et al*, *J. Org. Chem.*, 1988, **53**, 4063 (*nitrile, Me ester, synth, pmr*)

2-Cycloheptene-1,4-diol   C-80179

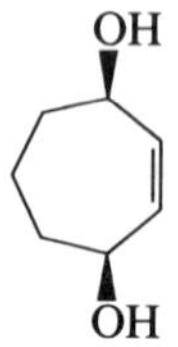

C7H12O2   M 128.171
(*1RS,4SR*)-*form*
cis-form
Di-Me ether: [116910-22-6]. *2,7-Dimethoxycycloheptene*
C9H16O2   M 156.224
Oil. Bp2 130° (kugelrohr) .
Di-Et ether: [90786-03-1]. *2,7-Diethoxycycloheptene*
C11H20O2   M 184.278
Oil. Bp2 140° (kugelrohr) .

Bäckvall, J.E. *et al*, *J. Org. Chem.*, 1988, **53**, 5695 (*synth, pmr, cmr, ir*)

4-Cyclohexene-1,2-diamine, 9CI   C-80180

4,5-Diaminocyclohexene

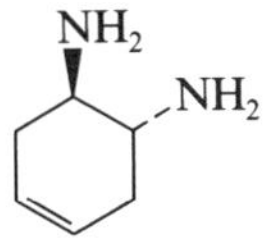

(1*RS*,2*RS*)-*form*

$C_6H_{12}N_2$ M 112.174

(1*RS*,2*RS*)-*form*

(±)trans-form

B,2HCl: [108796-57-2].
Powder. Mp >280°.

(1*RS*,2*SR*)-*form*

cis-form

B,2HCl: [105249-35-2].
Powder. Mp 255-256°.

Witiak, D.T. *et al, J. Med. Chem.*, 1987, **30**, 1327 (*synth, pmr, ir*)

2-Cyclohexene-1,4-diol, 9CI   C-80181

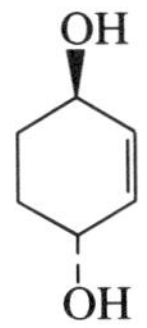

(1*RS*,4*RS*)-*form*

$C_6H_{10}O_2$ M 114.144

(1*RS*,4*RS*)-*form* [41513-32-0]

(±)-*trans form*
Cryst. Mp 83-84°.

Di-Ac: [78776-44-0].
Cryst. (hexane). Mp 49-50°.

(1*RS*,4*SR*)-*form* [53762-85-9]

cis-form
Cryst. Mp 57-59°.

Di-Ac: [78776-45-1].
$C_{10}H_{14}O_4$ M 198.218
Liq.

Di-Me ether: [59415-77-9]. *3,6-Dimethoxycyclohexene*
$C_8H_{14}O_2$ M 142.197
Oil. Bp$_2$ 120°.

Di-Et ether: [90786-01-9]. *3,6-Diethoxycyclohexene*
$C_{10}H_{18}O_2$ M 170.251
Oil. Bp$_1$ 75° (kugelrohr) .

Dibenzyl ether: [116910-21-5].
$C_{20}H_{22}O_2$ M 294.393
Oil. Bp$_{0.3}$ 230°.

Kaneko, C. *et al, Synthesis*, 1974, 876 (*synth, pmr*)
Bäckvall, J.-E. *et al, J. Org. Chem.*, 1984, **49**, 4619; 1988, **53**, 5695 (*synth, pmr, cmr, ir*)

4-Cyclohexene-1,2-diol   C-80182

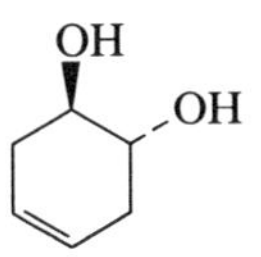

(1*R*,2*R*)-*form*

$C_6H_{10}O_2$ M 114.144

(1*R*,2*R*)-*form*

Mono-Ac:
$C_8H_{12}O_3$ M 156.181

Mp 71°. [α]$_D^{22}$ −121.6° (c, 0.86 in CHCl$_3$).

(1*RS*,2*RS*)-*form*

(±)-trans-*form*
Needles (EtOAc/hexane). Mp 95°.

Di-Ac:
$C_{10}H_{14}O_4$ M 198.218
Oil.

Suemune, H. *et al, Chem. Pharm. Bull.*, 1989, **37**, 1379.

2-(1-Cyclohexenyl)propanal   C-80183

α-Methyl-1-cyclohexene-1-acetaldehyde, 9CI

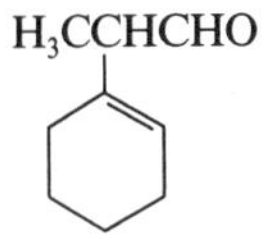

H₃CCHCHO

$C_9H_{14}O$ M 138.209

(±)-*form* [114221-81-7]
Oil.

Tsukamoto, M. *et al, J. Chem. Soc., Chem. Commun.*, 1986, 880 (*synth*)
Maruoka, K. *et al, J. Am. Chem. Soc.*, 1988, **110**, 3588 (*synth, pmr, ir*)

2-Cyclohexylpropanal   C-80184

α-Methylcyclohexaneacetaldehyde, 9CI

[2109-22-0]

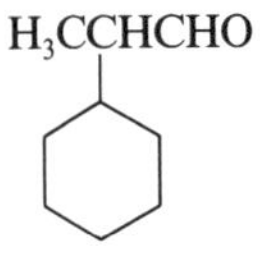

H₃CCHCHO

$C_9H_{16}O$ M 140.225

(±)-*form* [97859-57-9]
Oil.

Meyers, A.I. *et al, Tetrahedron*, 1985, **41**, 5089 (*synth, pmr, ir*)
Maruoka, K. *et al, J. Am. Chem. Soc.*, 1988, **110**, 3588 (*synth, pmr*)

Cyclo(leucylprolyl)   C-80185

Hexahydro-3-(2-methylpropyl)pyrrolo[1,2a]pyrazine-1,4-dione. Gancidin W

[5654-86-4]

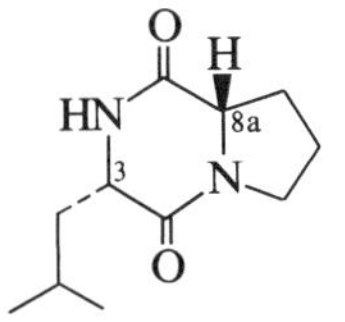

$C_{11}H_{18}N_2O_2$ M 210.275
Diketopiperazine antibiotic.

(3*S*,8a*S*)-*form* [2873-36-1]

L,L-*form*
Prod. by *Streptomyces gancidicus, Nocardia restricta, Candida albicans, Guignardia laricina* and several *Ceratocystis* spp. Also isol. from the sponge *Tedania ignis*. Bitter component of sake and contributes to the flavour of beer. Shows antitumour and antifungal props. Phytotoxin associated with fungal diseases of trees. Cryst. Mp 168-172° (158-159°). [α]$_D^{25-28}$ −133° (c, 1 in EtOH), [α]$_D^{21}$ −144° (c, 0.5 in H$_2$O).
▷ UY8708800.

(3*R*,8a*S*)-*form* [36238-67-2]

Cryst. Mp 148-149° (119-122°). [α]$_D^{31}$ −105.5° (c, 0.5 in EtOH).

[19943-30-7, 43041-29-8]

Kodaira, Y. *et al, Agric. Biol. Chem.,* 1961, **25**, 261.
Nitecki, D.E. *et al, J. Org. Chem.,* 1968, **33**, 864 (*synth*)
Siemion, I.Z. *et al, Org. Magn. Reson.,* 1971, **3**, 545 (*synth, nmr*)
Vicar, J. *et al, Collect. Czech. Chem. Commun.,* 1972, **37**, 4060; 1973, **38**, 1940 (*synth, ir, pmr*)
Karle, I.L., *J. Am. Chem. Soc.,* 1972, **94**, 81 (*cryst struct*)
Takahashi, K. *et al, Agric. Biol. Chem.,* 1974, **38**, 927.
Bycroft, B.W. *et al, J. Chem. Soc., Chem. Commun.,* 1975, 988 (*synth*)
Davies, D.B. *et al, J. Chem. Soc., Perkin Trans. 2,* 1976, 187 (*conformn*)
Pancoska, P. *et al, Collect. Czech. Chem. Commun.,* 1977, **44**, 1296 (*cd*)
Jain, T.C. *et al, Heterocycles,* 1977, **7**, 341 (*isol, struct, uv, ms*)
Bjoerkman, S. *et al, J. Med. Chem.,* 1979, **22**, 931 (*synth*)
Suzuki, K. *et al, Chem. Pharm. Bull.,* 1981, **29**, 233 (*synth*)
Schmidtz, F.J. *et al, J. Org. Chem.,* 1983, **48**, 3941 (*isol, ms, pmr, synth, bibl*)
Ayer, W.A. *et al, Can. J. Chem.,* 1986, **64**, 904 (*isol*)
Kricheldorf, H.R. *et al, Magn. Reson. Chem.,* 1986, **24**, 21 (*cmr*)

Cycloocta[*def*]biphenylene-1,4-dione   C-80186

[119392-56-2]

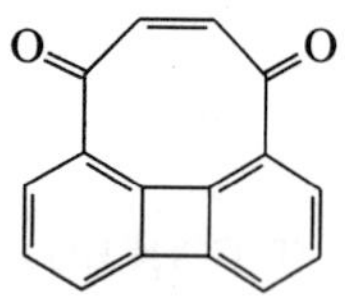

C$_{16}$H$_8$O$_2$ M 232.238

Yellow solid. Mp 162-165°. Non-planar structure.

Wilcox, C.F. *et al, J. Org. Chem.,* 1989, **54**, 2190 (*synth, pmr, cmr, ms, cryst struct*)

Cycloocta[*def*]phenanthrene   C-80187

[111409-91-7]

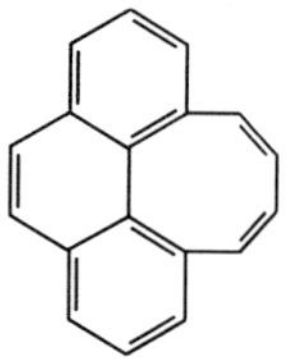

C$_{18}$H$_{12}$ M 228.293

Heinz, W. *et al, Angew. Chem., Int. Ed. Engl.,* 1987, **26**, 1291 (*synth, pmr, cmr, uv*)

1,3,5-Cyclooctatriene, 9CI   C-80188

Updated Entry replacing C-03516

[1871-52-9]

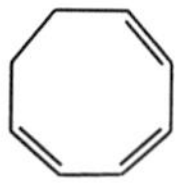

C$_8$H$_{10}$ M 106.167

Liq. d$_4^{25}$ 0.897. Bp 145-146°, Bp$_{90}$ 76°.

AgNO$_3$ complex: Cryst. (EtOH). Mp 125-126°.

Cope, A.C. *et al, J. Am. Chem. Soc.,* 1952, **74**, 4867 (*synth*)
Tarakanova, A.V. *et al, Zh. Org. Khim.,* 1972, **8**, 1619 (*synth*)
Anet, F.A.L. *et al, Tetrahedron Lett.,* 1975, 4221 (*conformn*)
Bally, T. *et al, Helv. Chim. Acta,* 1989, **72**, 73 (*synth, bibl*)

1,2-Cyclopentanediol, 9CI   C-80189

Updated Entry replacing C-03608

[4065-92-3]

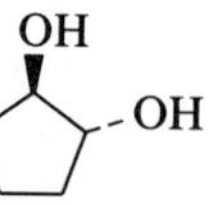

C$_5$H$_{10}$O$_2$ M 102.133

(1R,2R)-form [63261-45-0]

(−)-trans-*form*

Mp 48-48.5°. [α]$_D^{20}$ −24.88° (c, 6.0 in EtOH).

Phenylurethane: Mp 213-215°.

(1S,2S)-form [107492-82-0]

Fragrant solid. Mp 47.0-48.0°. [α]$_D^{20}$ +24.54° (c, 5.4 in EtOH). Highly hygroscopic.

(1RS,2RS)-form [5057-99-8]

(±)-trans-*form*

Mp 55°. Bp$_{22}$ 136°.

Di-Ac: [65783-98-4].

C$_9$H$_{14}$O$_4$ M 186.207

Mp −4°. Bp$_{3.5}$ 86°.

Dibenzoyl:

C$_{19}$H$_{18}$O$_4$ M 310.349

Mp 63°.

Bis(4-methylbenzenesulfonyl): Mp 109°.

(1RS,2SR)-form [5057-98-7]

cis-*form*

Mp 30°. Bp$_{29}$ 123.5°.

Di-Ac: [37854-32-3].

Mp −5°. Bp$_2$ 80°.

Dibenzoyl: [37854-33-4].

Mp 47°.

[86703-52-8]

Verkade, P.E., *Justus Liebigs Ann. Chem.,* 1930, **477**, 279, 289 (*synth*)
Owen, L.N. *et al, J. Chem. Soc.,* 1952, 4026 (*synth*)
Singy, G.A. *et al, Helv. Chim. Acta,* 1971, **54**, 537 (*ms*)
Kono, H. *et al, Org. Prep. Proced. Int.,* 1974, **6**, 19 (*synth*)
Mangoni, L. *et al, Gazz. Chim. Ital.,* 1975, **105**, 377 (*synth*)
Cunningham, A.F. *et al, J. Org. Chem.,* 1988, **53**, 1823 (*synth, ir, pmr*)

1*H*-Cyclopenta[*l*]phenanthrene, 9CI   C-80190

[235-92-7]

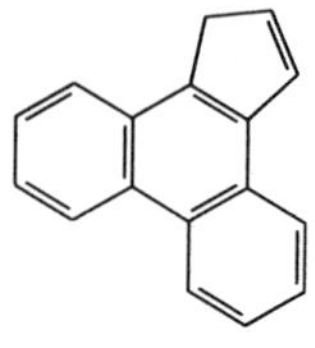

C$_{17}$H$_{12}$ M 216.282

Cryst. (2-propanol). Mp 150-152°.

Cope, A.C. *et al, J. Am. Chem. Soc.,* 1956, **78**, 2547 (*synth*)

Cyclopenta[*cd*]pleiadene, 9CI   C-80191

Benz[h]*acepleiadylene. Acepleiadylene*

[193-60-2]

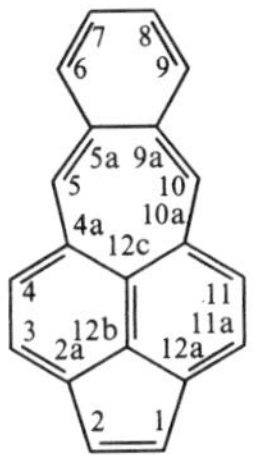

$C_{20}H_{12}$ M 252.315

Diels-Alder diene. Obt. as a deep blue soln., could not be isol. but dec. to polymeric material and a dimer.

Cava, M.P. *et al, Tetrahedron*, 1965, **21**, 3051, 3059, 3073 (*synth, uv, props*)

Cyclopenta[*c*]pyran   C-80192

[270-61-1]

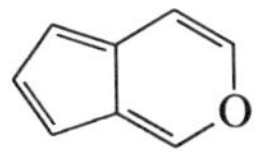

C_8H_6O M 118.135

Unstable oil.

Kämpchen, T. *et al, Justus Liebigs Ann. Chem.*, 1988, 855 (*synth, uv, pmr, cmr*)

4-Cyclopentene-1,3-diol, 9CI   C-80193

Updated Entry replacing C-50374

3,5-Dihydroxycyclopentene

[4157-01-1]

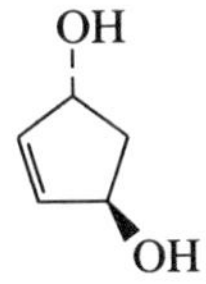

(1*R*,3*R*)-*form*

$C_5H_8O_2$ M 100.117

(1*R*,3*R*)-*form*

Di-Ac: [60410-14-2].

  $C_9H_{12}O_4$ M 184.191

  $[\alpha]_D^{20}$ +208° (c, 0.025 in MeOH).

(1*RS*,3*RS*)-*form* [694-47-3]

  (±)-*trans-form*

  $Bp_{0.5}$ 100-105°. n_D^{25} 1.5024. V. hygroscopic.

Dibenzoyl:

  $C_{19}H_{16}O_4$ M 308.333

  Mp 124°.

Bis(3,5-dinitrobenzoyl): Mp 218-219°.

(1*RS*,3*SR*)-*form* [29783-26-4]

  cis-*form*

  Cryst. (EtOAc/pet. ether). Mp 59-61°. An achiral diol having synthetically important chiral monoacyl etc. blocked derivs.

Mono-Ac:

  $C_9H_{12}O_4$ M 184.191

  Cryst (Et₂O/pentane). Mp 47.5-48°. $[\alpha]_D^{22}$ +75.0° (c, 1.16 in CHCl₃).

Dibenzoyl: Mp 58-60°.

Bis(3,5-dinitrobenzoyl): Mp 208-209°.

[54664-61-8, 59415-74-6, 60389-71-1, 61826-75-3]

Sable, H.Z. *et al, Helv. Chim. Acta*, 1962, **41**, 371; 1963, **46**, 1157 (*synth, bibl*)

Cocu, F.G. *et al, Helv. Chim. Acta*, 1970, **53**, 739 (*pmr*)

Singy, G.A. *et al, Helv. Chim. Acta*, 1973, **56**, 449 (*ms*)

Org. Synth., Coll. Vol., 5, 1973, 416 (*synth*)

Kaneko, C. *et al, Synthesis*, 1974, 876 (*synth*)

Muira, S. *et al, Tetrahedron*, 1976, **32**, 1893 (*synth*)

Laumen, K. *et al, Tetrahedron Lett.*, 1984, **25**, 5875; 1985, **26**, 407 (*synth*)

Duhamel, L. *et al, Tetrahedron Lett.*, 1985, **26**, 3099 (*chiral deriv*)

Laumen, K. *et al, J. Chem. Soc., Chem. Commun.*, 1986, 1298 (*derivs*)

Sugai, T. *et al, Synthesis*, 1988, 19 (*synth, pmr*)

Theil, F. *et al, Synthesis*, 1988, 540 (*synth, pmr*)

Cyclo(prolylalanyl)   C-80194

Hexahydro-3-methylpyrrolo[1,2-a]pyrazine-1,4-dione, 9CI

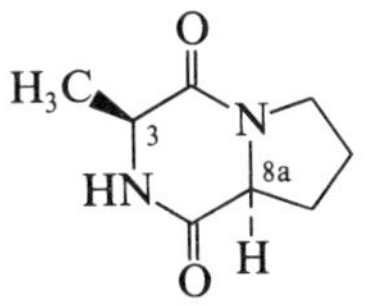

(3*S*,8a*S*)-*form*

$C_8H_{12}N_2O_2$ M 168.195

(3*S*,8a*S*)-*form* [36357-32-1]

  L,L-*form*. (3S-trans)-*form*

  Isol. from cocoa; metab. product of the Caribbean sponge *Tedania ignis*. Mp 153-156°. $[\alpha]_{546}^{20}$ −160.0° (c, 1 in EtOH).

(3*S*,8a*R*)-*form* [19943-29-4]

  D,L-*form*. (3S-cis-*form*)

  From *Beauveria bassiana*. Mp 162-166°. $[\alpha]_D^{22}$ −85°.

(3*R*,8a*S*)-*form* [36238-64-9]

  L,D-*form*. (3*R*-cis)-*form*

  Mp 139-142°.

[19943-28-3, 65556-33-4]

Westley, J.W. *et al, Anal. Chem.*, 1968, **40**, 1888 (*glc, chromatog, pmr*)

Siemion, I.Z., *Org. Magn. Reson.*, 1971, **3**, 545 (*synth, pmr*)

Young, P.E. *et al, J. Am. Chem. Soc.*, 1976, **98**, 5365 (*synth, pmr, cmr*)

Izumiya, N. *et al, Pept. Chem.*, 1977, **15**, 49 (*pharmacol*)

Cotrait, M. *et al, Cryst. Struct. Commun.*, 1979, **8**, 819 (*cryst struct*)

Grove, J.F. *et al, Phytochemistry*, 1981, **20**, 815 (*isol*)

Schmitz, F.J. *et al, J. Org. Chem.*, 1983, **48**, 3941 (*isol, pmr, ms*)

Langhammer, M. *et al, Fresenius' Z. Anal. Chem.*, 1986, **324**, 5 (*ms*)

Cyclopropadodecahedrane   C-80195

Tetradecahydro-4a,2,1,5,3,3c-[2,3]butanediyl[1,4]diylidene-4H-cyclopropa[4,5]pentaleno[2,1,6-cde]pentaleno[2,1,6-gha]pentalene, 9CI

[112896-42-1]

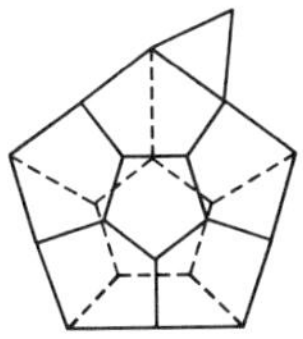

$C_{21}H_{20}$ M 272.389

Paquette, L.A. *et al, J. Am. Chem. Soc.*, 1988, **110**, 1305 (*synth*)

1,1-Cyclopropanedicarboxylic acid, 9CI   C-80196

Updated Entry replacing C-03707

Ethylenemalonic acid. Vinaconic acid

[598-10-7]

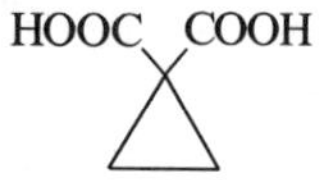

HOOC COOH

$C_5H_6O_4$ M 130.100

Prisms or needles (CHCl₃), prisms + 1H₂O (H₂O). Sol. H₂O, org. solvs. Mp 140°. pK_{a1} 7.70 (7.67), pK_{a2} 2.92.

Di-Me ester: [6914-71-2].
  $C_7H_{10}O_4$ M 158.154
  Bp₇₆₄ 198°.

Di-Et ester: [1559-02-0].
  $C_9H_{14}O_4$ M 186.207
  Liq. Bp₂₂ 115°.

Mononitrile: [6914-79-0]. *1-Cyanocyclopropanecarboxylic acid*
  $C_5H_5NO_2$ M 111.100
  Mp 149° (140°).

Nitrile, Me ester: [6914-73-4]. *Methyl 1-cyanocyclopropanecarboxylate*
  $C_6H_7NO_2$ M 125.127
  Liq. Bp₁₈ 92°.

Nitrile, Et ester: [1558-81-2]. *Ethyl 1-cyanocyclopropanecarboxylate*
  $C_7H_9NO_2$ M 139.154
  Bp 210-211°, Bp₈₀ 137°.

Nitrile, amide:
  $C_5H_6N_2O$ M 110.115
  Mp 160°.

Dinitrile: [1559-03-1]. *1,1-Dicyanocyclopropane*
  $C_5H_4N_2$ M 92.100
  Bp₂₀ 103°. n_D^{25} 1.4463.

Perkin, W.H., *J. Chem. Soc.*, 1885, **47**, 807, 817.
Stewart, J.M. *et al*, *J. Org. Chem.*, 1965, **30**, 1951 (*dinitrile*)
Meester, M.A.M. *et al*, *Acta Crystallogr.*, *Sect. B*, 1971, **27**, 630 (*cryst struct*)
Danishefsky, S. *et al*, *J. Am. Chem. Soc.*, 1975, **97**, 3239 (*synth*)
Jones, P.G. *et al*, *Acta Crystallogr.*, *Sect. C*, 1987, **43**, 1576 (*cryst struct, mononitrile*)
Heiszman, J. *et al*, *Synthesis*, 1987, 738 (*deriv, synth, bibl*)

1H-Cyclopropa[l]phenanthrene   C-80197

[278-91-1]

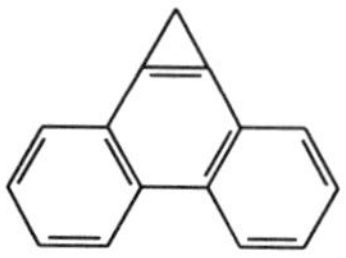

$C_{15}H_{10}$ M 190.244

Generated and trapped as adducts.

Dent, B.R. *et al*, *Aust. J. Chem.*, 1987, **40**, 925.

4-Cyclotetradecen-1-one, 9CI   C-80198

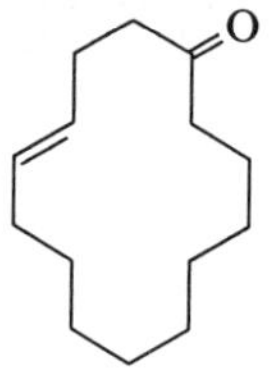

$C_{14}H_{24}O$ M 208.343

[101999-70-6]

Porter, N.A. *et al*, *J. Am. Chem. Soc.*, 1988, **110**, 3554 (*synth, ir, pmr, cmr*)

5-Cyclotetradecen-1-one   C-80199

$C_{14}H_{24}O$ M 208.343

[75426-30-1, 114157-30-1]

Porter, N.A. *et al*, *J. Am. Chem. Soc.*, 1988, **110**, 3554 (*synth, ir, pmr, cmr*)

Cycloundecene   C-80200

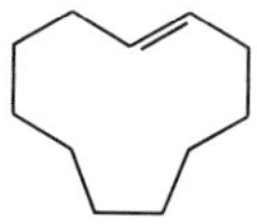

(E)-form

$C_{11}H_{20}$ M 152.279

(E)-form [13151-60-5]
  Oil. Bp₁₂ 88-90°, Bp₀.₀₁ 25-30°.

(Z)-form [13151-61-6]
  Bp₁₂ 100°, Bp₀.₀₁ 25-30°.

Fawcett, R.W. *et al*, *J. Chem. Soc.*, 1954, 2673 (*synth, uv*)
Prelog, V. *et al*, *Helv. Chim. Acta*, 1955, **38**, 1776 (*synth*)
Vilsmaier, E. *et al*, *J. Org. Chem.*, 1988, **53**, 1806 (*synth, ir, pmr*)

3-Cycloundecene-1,5-diyne, 9CI   C-80201

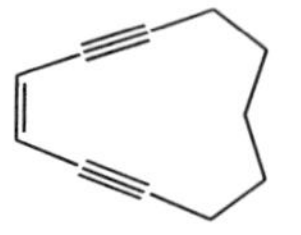

$C_{11}H_{12}$ M 144.216

(Z)-form [115227-61-7]
  Large plates (pentane). Mp 36-36.5°.

Nicolaou, K.C. *et al*, *J. Am. Chem. Soc.*, 1988, **110**, 4866 (*synth, cryst struct*)

Cymobarbatol   C-80202

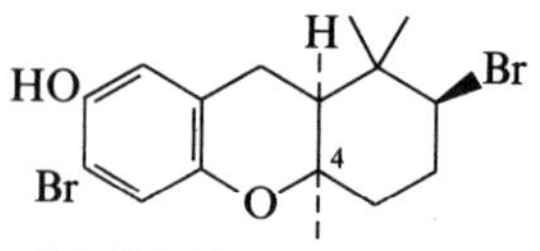

$C_{16}H_{20}Br_2O_2$ M 404.141

Metab. of *Cymopolia barbata*. Cryst. Mp 166°. $[\alpha]_D^{23}$ −15.4°.

4-Epimer: 4-Isocymobarbatol
  $C_{16}H_{20}Br_2O_2$ M 404.141
  Metab. of *C. barbata*. Cryst. Mp 147°. $[\alpha]_D^{23}$ −51.4°.

Wall, M.E. *et al*, *J. Nat. Prod.* (*Lloydia*), 1989, **52**, 1092 (*pmr, cmr, cryst struct*)

Cyperanic acid   C-80203

[117869-89-3]

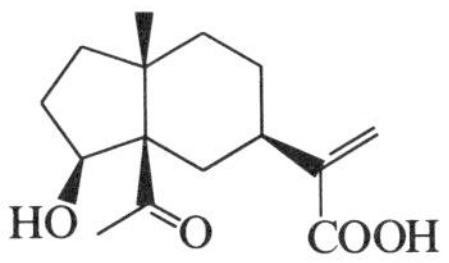

$C_{15}H_{22}O_4$ M 266.336

Constit. of *Dittrichia viscosa*. Oil. $[\alpha]_D$ +26.3° (c, 0.004 in $CHCl_3$).

Me ester: Cryst. (Et_2O). Mp 62-64°.

Ceccherelli, P. *et al, J. Nat. Prod.* (*Lloydia*), 1988, **51**, 1006 (*isol, pmr, cmr*)

Cytidine-5′-diphosphate   C-80204

Cytidine 5′-(trihydrogen diphosphate), 9CI. Cytidine pyrophosphate

[63-38-7]

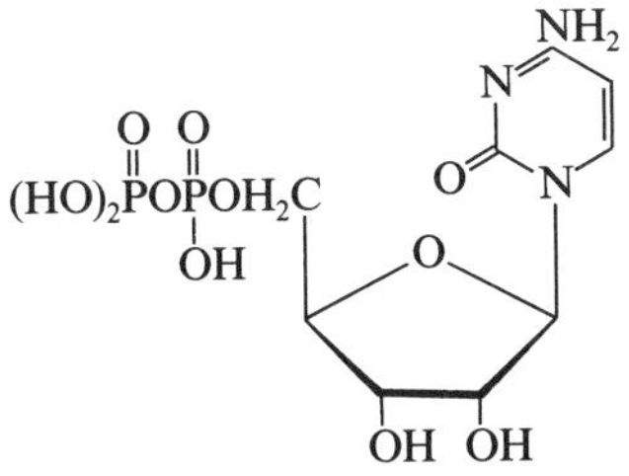

$C_9H_{15}N_3O_{11}P_2$ M 403.178

Exists as zwitterion. Prod. by *Plasmodium berghei* and *Bacillus brevis*. Found in heart, liver, leukaemia cells. Affects bladder contraction. Powder.

Kimura, A. *et al, Agric. Biol. Chem.*, 1975, **39**, 1469 (*manuf*)
Viswamitra, M.A. *et al, Nature* (*London*), 1975, **258**, 497 (*cryst struct*)

Labotka, R.J. *et al, J. Am. Chem. Soc.*, 1976, **98**, 3699 (*nmr*)
Iio, M. *et al, Agric. Biol. Chem.*, 1977, **41**, 155 (*synth*)
Kim, C.H. *et al, J. Am. Chem. Soc.*, 1978, **100**, 1571 (*conformn*)
Scheller, K.H. *et al, J. Am. Chem. Soc.*, 1983, **105**, 5891 (*pmr*)
Veda, T. *et al, J. Chromatogr.*, 1987, **386**, 273 (*hplc*)
Davisson, V.J. *et al, J. Org. Chem.*, 1987, **52**, 1794 (*synth, cmr, nmr, pmr*)

Cytidine 5′-triphosphate   C-80205

Cytidine 5′-(tetrahydrogen triphosphate), 9CI

[65-47-4]

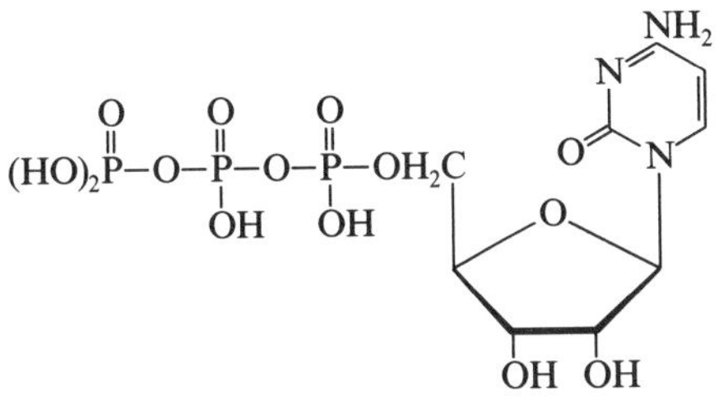

$C_9H_{16}N_3O_{14}P_3$ M 483.158

Prod. by *Leishmania* and *Streptomyces griseus*; found in cheese, brain, liver, erythrocytes and lymphocytes. Enzyme inhibitor. Implicated in purine and pyrimidine metabolism disorders. Contracts arteries.

Na salt: Cryst. powder.

Tanaka, K. *et al, Chem. Pharm. Bull.*, 1962, **10**, 220 (*synth*)
Lee, G.C.Y. *et al, J. Am. Chem. Soc.*, 1972, **94**, 951 (*tautom, pmr*)
Labotka, R.J. *et al, J. Am. Chem. Soc.*, 1976, **98**, 3699 (*nmr*)
Iio, M. *et al, Agric. Biol. Chem.*, 1977, **41**, 155 (*synth*)
Jost, W. *et al, Anal. Biochem.*, 1983, **135**, 120 (*chromatog*)
Kehr, J. *et al, Fresenius' Z. Anal. Chem.*, 1986, **325**, 466 (*hplc*)
Veda, T. *et al, J. Chromatogr.*, 1987, **386**, 273 (*hplc*)
Hammer, D.F. *et al, Anal. Biochem.*, 1988, **169**, 300 (*hplc*)

D

Dactylomelol D-80001

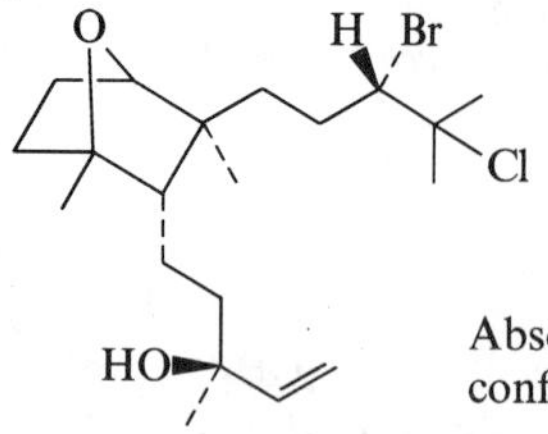

Absolute configuration

$C_{20}H_{34}BrClO_2$ M 421.844

Metab. of *Aplysia dactylomela*. Cryst. (CH_2Cl_2/hexane). Mp 85-86°. $[\alpha]_D^{20}$ −31.3° (c, 0.7 in $CHCl_3$).

Estrada, D.M. *et al, Tetrahedron Lett.*, 1989, **30**, 6219 (*cryst struct*)

Dactylospongenone A D-80002

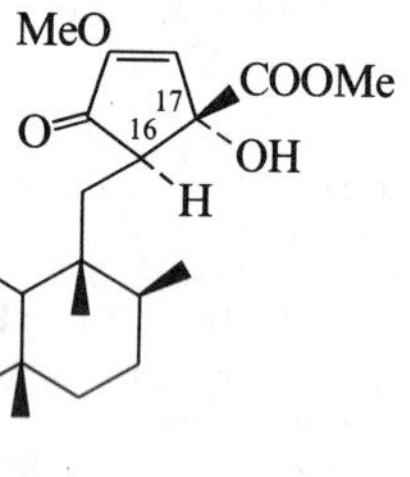

$C_{23}H_{34}O_5$ M 390.519

Metab. of *Dactylospongia* sp. Cryst. $[\alpha]_D$ −167.7° (c, 0.062 in MeOH).

16-Epimer: Dactylospongenone D
$C_{23}H_{34}O_5$ M 390.519
Metab. of *D.* sp. $[\alpha]_D$ −121.7° (c, 0.14 in MeOH).

17-Epimer: Dactylospongenone C
$C_{23}H_{34}O_5$ M 390.519
Metab. of *D.* sp. $[\alpha]_D$ +25.5° (c, 0.20 in MeOH).

16,17-Diepimer: Dactylospongenone B
$C_{23}H_{34}O_5$ M 390.519
Metab. of *D.* sp. $[\alpha]_D$ +96.4° (c, 0.22 in MeOH).

Kushlan, D.M. *et al, Tetrahedron*, 1989, **45**, 3307 (*cryst struct*)

Dalpanol D-80003

Updated Entry replacing D-00015
6′,7′-Dihydro-6′-hydroxyrotenone

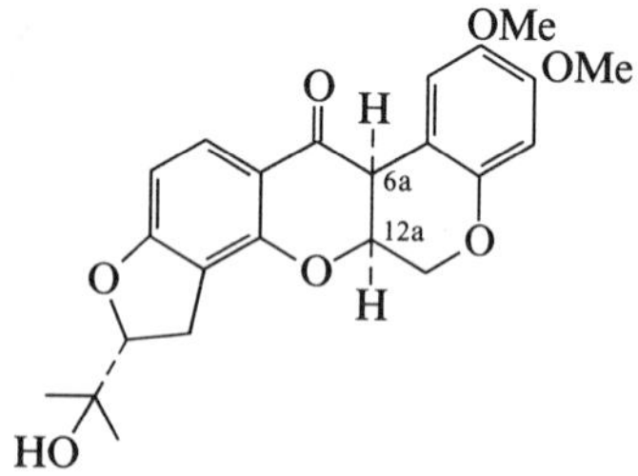

$C_{23}H_{24}O_7$ M 412.438

CA numbering shown. Isol. from ripe seeds of *Dalbergia paniculata*. Also from *Amorpha fruticosa*. Cryst. (C_6H_6). Mp 196°. $[\alpha]_D^{22}$ −136.3° ($CHCl_3$).

O-Glucoside:

$C_{29}H_{34}O_{12}$ M 574.580

Isol. from *D. paniculata* seeds. Cryst. solid. $[\alpha]_D^{34}$ −215.4° (c, 0.26 in 80% MeOH aq.).

6a,12a-Didehydro: Dehydrodalpanol
$C_{23}H_{22}O_7$ M 410.423
Isol. from seeds of *D. paniculata*. Minute yellow needles ($CHCl_3$/MeOH). Mp 238-240°.

Adinarayana, D. *et al, J. Chem. Soc. C*, 1971, 29 (*isol*)
Crombie, L. *et al, J. Chem. Soc., Perkin Trans. 1*, 1973, 1277 (*synth*)
Radhakrishniah, M. *et al, Phytochemistry*, 1973, **12**, 3003 (*glucoside*)
Adinarayana, D. *et al, Indian J. Chem.*, 1975, **13**, 425 (*Dehydrodalpanol*)
Crombie, L. *et al, J. Chem. Soc., Chem. Commun.*, 1979, 1143 (*biosynth*)

8-Daucene-5,7-diol D-80004
Carotdiol

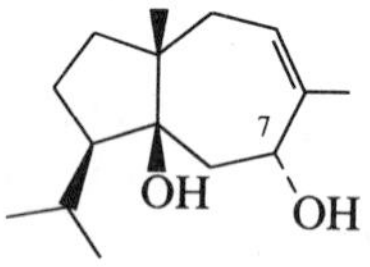

$C_{15}H_{26}O_2$ M 238.369
Cryst. Mp 76-78°.

7-Ac: Carotdiol acetate
$C_{17}H_{28}O_3$ M 280.406
Constit. of *Ferula linkii*. Oil.

7-(3,4-Dimethoxybenzoyl): Carotdiol veratrate
$C_{24}H_{34}O_5$ M 402.530
Constit. of *F. linkii*. Oil.

Fraga, B.M. *et al, Phytochemistry*, 1989, **28**, 1649.

Daviciin M₁ D-80005

[119139-62-7]

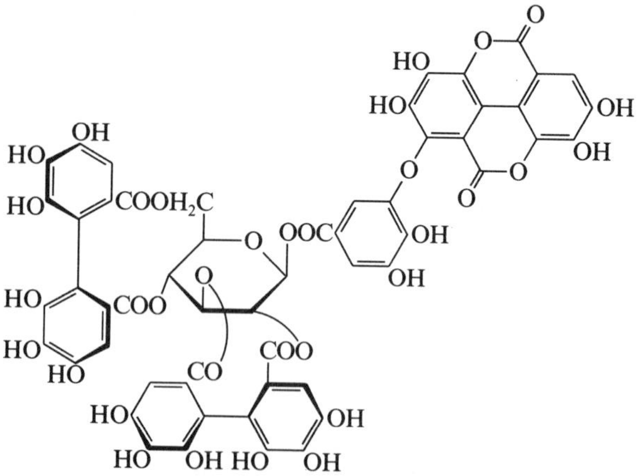

$C_{55}H_{32}O_{34}$ M 1236.837

Constit. of *Rosa davurica*. Amorph. powder. $[\alpha]_D$ −58° (c, 1 in MeOH).

Yoshida, T. *et al, Phytochemistry*, 1989, **28**, 2177 (*isol, pmr, cmr*)

8-Deacyl-15-deoxypunctatin   D-80006

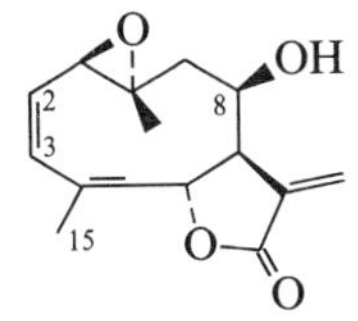

$C_{15}H_{18}O_4$ M 262.305

8-O-(2-Methyl-2,3-epoxybutanoyl):
$C_{20}H_{24}O_6$ M 360.406
Constit. of *Trichogonia santosii*. Cryst. (Et₂O). Mp 173°.
[α]$_D^{24}$ −422° (c, 0.09 in CHCl₃).

2α,2α-Epoxy, 8-O-angeloyl: 15-Deoxyliscundin
$C_{20}H_{24}O_6$ M 360.406
Constit. of *T. santosii*. Oil.

2α,3α-Epoxy, 8-O-(2-methyl-2,3-epoxybutanoyl):
$C_{20}H_{24}O_7$ M 376.405
Constit. of *T. santosii*. Cryst. (Et₂O). Mp 198°. [α]$_D$
−104° (C, 0.65 in CHCl₃).

2α,3α-Epoxy, 8-O-(2-hydroxymethyl-2Z-butenoyl):
$C_{20}H_{24}O_7$ M 376.405
Constit. of *T. santosii*. Cryst. (Et₂O). Mp 172°. [α]$_D$
−135° (c, 0.2 in CHCl₃).

15-Hydroxy, 8-O-(2-hydroxymethyl-2Z-butenoyl): [49776-54-7]. **Punctaliatrin.** *Punctatin*
$C_{20}H_{24}O_7$ M 376.405
Constit. of *Liatris punctata*. Cryst. Mp 163-165°.

15-Hydroxy, 2α,3α-epoxy, 8-O-angeloyl: [57498-86-9].
Liscundin
$C_{20}H_{24}O_7$ M 376.405
Constit. of *L. secunda*. Oil. [α]$_D^{22}$ −75° (c, 2 in CHCl₃).

15-Hydroxy, 2α,3α-epoxy, 8-O-(2-acetoxymethyl-2Z-butenoyl): [57498-88-1]. **Liscunditrin**
$C_{22}H_{26}O_9$ M 434.442
Constit. of *L. secunda*. Amorph. solid. Mp 161°. [α]$_D^{22}$
−5° (CHCl₃).

Herz, W. *et al*, *Phytochemistry*, 1973, **12**, 1421 (*Punctaliatrin*)
Herz, W. *et al*, *Phytochemistry*, 1975, **14**, 1561 (*Liscundin, Liscunditrin*)
Bohlmann, F. *et al*, *Justus Liebigs Ann. Chem.*, 1984, 162 (*isol*)

8-Deacyltrichogoniolide   D-80007

Updated Entry replacing T-02890

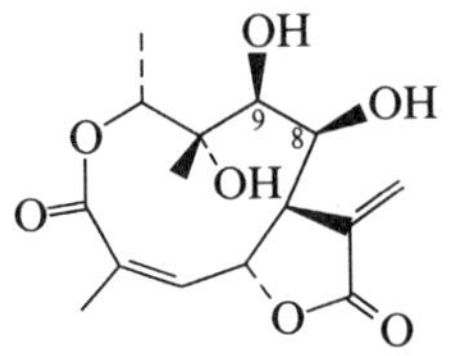

$C_{15}H_{20}O_7$ M 312.319

8-O-Angeloyl:
$C_{20}H_{26}O_8$ M 394.421
Constit. of *Trichogonia villosa*. Oil.

8-O-Angeloyl, 9-Ac:
$C_{22}H_{28}O_9$ M 436.458
Constit. of *T. histifolia*. Oil.

8-O-(2-Methylpropenoyl): [79694-65-8]. **Trichogoniolide**
$C_{19}H_{24}O_8$ M 380.394
Constit. of *T. prancii*. Gum.

8-O-(2-Methylpropenoyl), 9-Ac:
$C_{21}H_{26}O_9$ M 422.431

Constit. of *T. prancii*. Cryst. (Et₂O). Mp 215°. [α]$_D^{24}$
−246° (c, 0.7 in CHCl₃).

Bohlmann, F. *et al*, *Phytochemistry*, 1981, **20**, 1323 (*isol*)
Bohlmann, F. *et al*, *Justus Liebigs Ann. Chem.*, 1984, 162 (*isol*)

2,8-Decadiene-4,6-diyn-1-al, 9CI   D-80008
Matricarianal
[30377-44-7]

$$H_3CCH=CHC≡CC≡CCH=CHCHO$$

$C_{10}H_8O$ M 144.173
Isol. from underground parts of *Pimpinella magna*.

(E,E)-form [59950-66-2]
Isol. from above-ground parts of *Aethusa cynapium*.
Yellowish cryst. (pet. ether). Mp 68.5°.

(2E,8Z)-form [13894-66-1]
Bp$_{0.1}$ 60-70°.

(2Z,8E)-form [29576-73-6]
Isol. from *Grindelia* sp.

Bohlmann, F. *et al*, *Chem. Ber.*, 1955, **88**, 1330; 1964, **97**, 2598; 1966, **99**, 3194; 1970, **103**, 2245 (*isol, uv, ir, pmr, synth*)
Schulte, K.E. *et al*, *Arch. Pharm. (Weinheim, Ger.)*, 1970, **303**, 912 (*isol, uv, ir*)

2a,3,4,5,5a,6,7,8,8a,8b-Decahydroacenaphthene   D-80009
Tricyclo[7.2.1.0⁵,¹²]dodec-10-ene

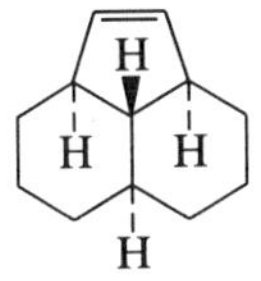

$C_{12}H_{18}$ M 162.274
(2aα,5aα,9aα,9bβ)-form [115650-98-1]
Ufolene
Bp$_{20}$ 102°.

Boldt, P. *et al*, *Chem. Ber.*, 1980, **121**, 2029 (*synth, ir, pmr, cmr, ms*)

Decahydro-1*H*-dicyclopenta[*b,d*]pyrrole, 9CI   D-80010

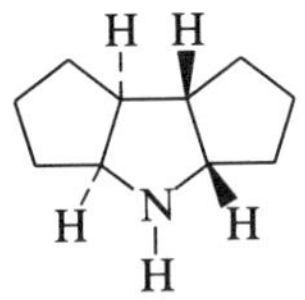

$C_{10}H_{17}N$ M 151.251
(3aS,4aS,7aS,7bS)-form [117064-12-7]
Reagent for asymmetric induction. [α]$_D$ −339.3° (c, 1.04 in EtOH). Enantiomer also prepd.

[116971-46-1, 117064-10-5, 117064-11-6]

Whitesell, J.K. *et al*, *J. Org. Chem.*, 1988, **53**, 5383 (*synth, resoln, use, abs config*)

2,3,4,5,6,7,8,9,10,11-Decahydro-1,3a,5a,7a,9a,12-hexaazatricyclopenta[a,ef,j]heptalene, 9CI   D-80011

[111161-01-4]

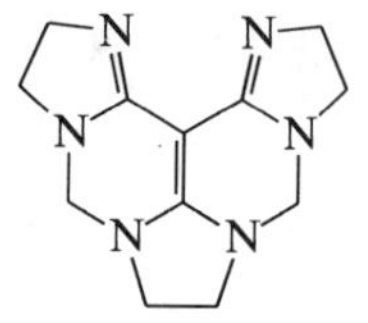

$C_{14}H_{20}N_6$ M 272.352

Mod. sol. aprotic solvs. Mp 228-231°. pK_a 29.22 (MeCN). Mod. stable. Analogues also prepd.

Schwesinger, R. *et al*, *Angew. Chem., Int. Ed. Engl.*, 1987, **26**, 1165 (*synth, pmr, cryst struct*)

Decamethylbiphenyl   D-80012

Permethylbiphenyl

[18356-20-2]

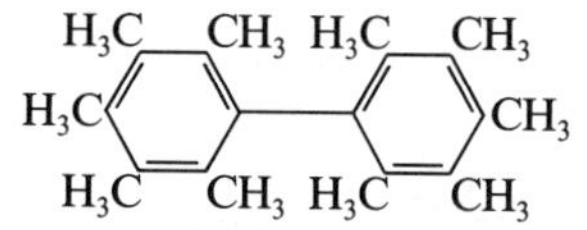

$C_{22}H_{30}$ M 294.479

Mp 239-241°.

Smith, L.I. *et al*, *J. Am. Chem. Soc.*, 1934, **56**, 2169 (*synth*)
Adcock, J.L. *et al*, *J. Organomet. Chem.*, 1974, **72**, 323 (*synth, pmr*)
Biali, S.E. *et al*, *J. Am. Chem. Soc.*, 1988, **110**, 1917 (*synth*)

2-Decaprenyl-6-methoxy-1,4-benzoquinone   D-80013

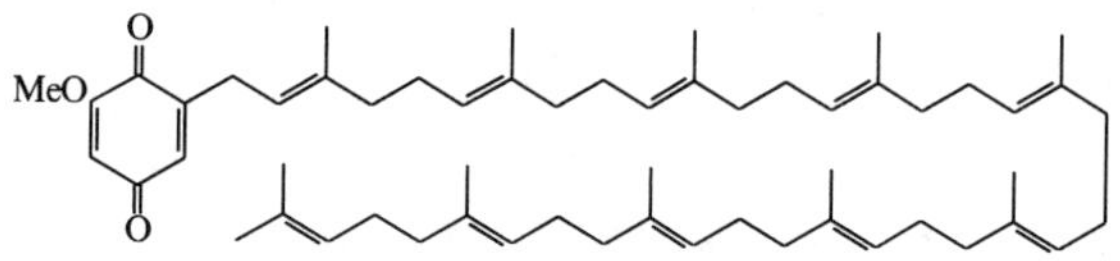

$C_{57}H_{86}O_3$ M 819.305

Isol. from *Rhodospirillum rubrum*. Intermed. in biosynth. of ubiquinones. Not obt. pure.

Friis, P. *et al*, *J. Am. Chem. Soc.*, 1966, **88**, 4754.

Decarboxythamnolic acid   D-80014

[4356-38-1]

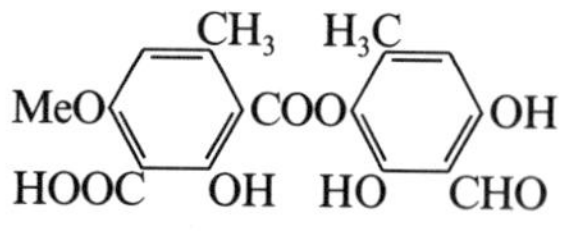

$C_{18}H_{16}O_9$ M 376.319

Detected in *Siphula decumbens* and in *Alectoria* sp. Prisms (Me₂CO). Mp 225°.

Asahina, Y. *et al*, *Ber.*, 1936, **69**, 330; 1939, **72**, 1402 (*synth*)
Bendz, G. *et al*, *Acta Chem. Scand.*, 1965, **19**, 1250.
Solberg, Y., *Acta Chem. Scand., Ser. B*, 1975, **29**, 145 (*isol*)

2,5,8-Decatriene-1,10-diol   D-80015

HOCH₂CH=CHCH₂CH=CHCH₂CH=CHCH₂OH

$C_{10}H_{16}O_2$ M 168.235

(**Z,Z,Z**)-*form* [115482-90-1]
Oil. Bp₀.₀₁ 142°.

Bis(4-methylbenzenesulfonyl): [115482-91-2].

Yellowish oil.

Brudermüller, M. *et al*, *Chem. Ber.*, 1988, **121**, 2255 (*synth, ir, pmr*)

2,5,8-Decatriyne-1,10-diol   D-80016

[13002-25-0]

HOCH₂C≡CCH₂C≡CCH₂C≡CCH₂OH

$C_{10}H_{10}O_2$ M 162.188

Pale brownish powder (CHCl₃/pentane). Mp 109°.

Brudermüller, M. *et al*, *Chem. Ber.*, 1988, **121**, 2255 (*synth*)

4-Decenoic acid   D-80017

Updated Entry replacing D-00185

[26303-90-2]

H₃C(CH₂)₄CH=CHCH₂CH₂COOH

$C_{10}H_{18}O_2$ M 170.251

(**E**)-*form*
Bp₁₃ 145-146°.

(**Z**)-*form* [505-90-8]
Obtusilic acid
Constit. of seed oil of *Lindera obtusiloba* and *L. umbellata*. Bp₁₃ 148-150°.

4-Bromophenacyl ester: Mp 43.3°.

Komori, S. *et al*, *Bull. Chem. Soc. Jpn.*, 1937, **12**, 226, 433 (*isol*)
Iwakiri, M., *Yakugaku Zasshi (J. Pharm. Soc. Jpn.)*, 1951, **78**, 1460; *ca*, **53**, 21633 (*synth, config*)
Hopkins, C.Y. *et al*, *Lipids*, 1966, **1**, 118 (*isol*)
Bus, J. *et al*, *Chem. Phys. Lipids*, 1976, **17**, 501 (*cmr*)
Furukawa, K. *et al*, *Yukagaku*, 1976, **25**, 190, 358 (*isol*; 296 (*glc, ms, ir, nmr*)

6-Decen-1-ol   D-80018

H₃CCH₂CH₂CH=CH(CH₂)₄CH₂OH

$C_{10}H_{20}O$ M 156.267

(**E**)-*form* [38421-92-0]
Liq. Bp₀.₈ 78-79°.

(**Z**)-*form* [68760-59-8]
Liq. Bp₀.₃ 74-6°.

Svirskaya, P.I. *et al*, *J. Chem. Eng. Data*, 1979, **24**, 152 (*synth, ir, pmr*)
Brown, H.C. *et al*, *J. Org. Chem.*, 1988, **53**, 246 (*synth, ir, pmr, cmr*)

2-(9-Decenyl)phenanthrene   D-80019

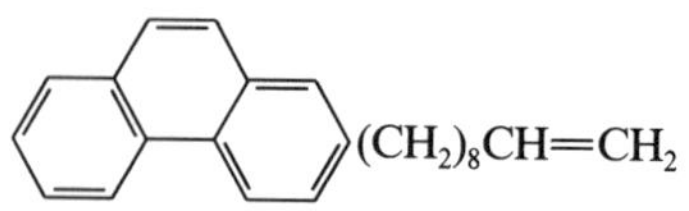

$C_{24}H_{28}$ M 316.485

Constit. of *Bryophyllum pinnatum*.

Siddiqui, S. *et al*, *Phytochemistry*, 1989, **28**, 2433 (*isol, pmr, ms*)

Deguelin   D-80020

Updated Entry replacing D-20021
13,13a-Dihydro-9,10-dimethoxy-3,3-dimethyl-3H-bis[1]benzopyrano[3,4-b:6′,5′-e]pyran-7(7aH)one, 9CI.
Isodeguelin
[522-17-8]

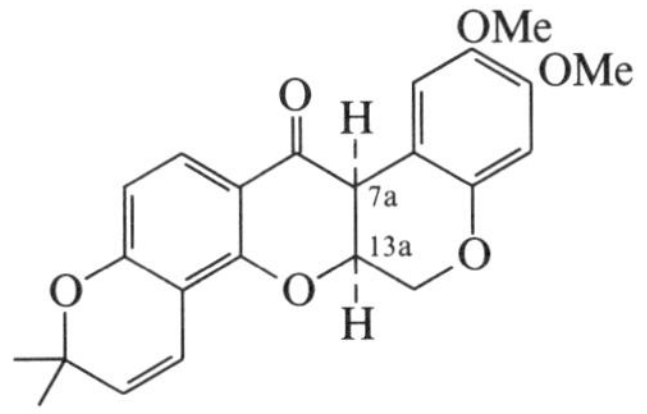

$C_{23}H_{22}O_6$ M 394.423
CA numbering shown. Isol. from *Crotalaria burhia, Derris elliptica, D. malaccensis, Lonchocarpus* spp., *Millettia ferruginea, M. pachycarpa, Mundulea sericea, Piscidia mollis* and *Tephrosia*spp. (all Leguminosae, Papilionoideae). Cryst. (EtOH). Mp 171°.

▷ Strong irritant, toxic by inhalation.

7a,13a-Didehydro: **Dehydrodeguelin**
$C_{23}H_{20}O_6$ M 392.407
Isol. from *Amorpha fruticosa, D. elliptica, Millettia dura, T. candida* and *T. vogelii.* Yellow cryst. or thick yellow prisms (CHCl₃/MeOH). Mp 232-233° (217°).

Clark, E.P. *et al, J. Am. Chem. Soc.,* 1933, **55**, 422 (*Dehydrodeguelin*)
Merz, K.W. *et al, Arch. Pharm.* (*Weinheim, Ger.*), 1935, **273**, 1 (*Dehydrodeguelin*)
Fukami, H. *et al, Agric. Biol. Chem.,* 1961, **25**, 252 (*synth*)
Carlson, D.G. *et al, Tetrahedron,* 1973, **29**, 2731 (*pmr*)
Omokawa, H. *et al, Agric. Biol. Chem.,* 1974, **38**, 1731 (*synth*)
Ingham, J.L., *Fortschr. Chem. Org. Naturst.,* 1983, **43**, 1 (*rev, occur*)
Crombie, L. *et al, J. Chem. Soc., Chem. Commun.,* 1986, 352 (*biosynth*)
Begley, M.J. *et al, J. Chem. Soc., Chem. Commun.,* 1986, 353 (*biosynth*)
Sax, N.I., *Dangerous Properties of Industrial Materials,* 5th Ed., Van Nostrand-Reinhold, 1979, 537.

1″,2″-Dehydrocyclokievitone   D-80021

3-(2,4-Dihydroxyphenyl)-2,3-dihydro-5-hydroxy-8,8-dimethyl-4H,8H-benzo[1,2-b:3,4-b′]dipyran-4-one, 9CI
[74175-82-9]

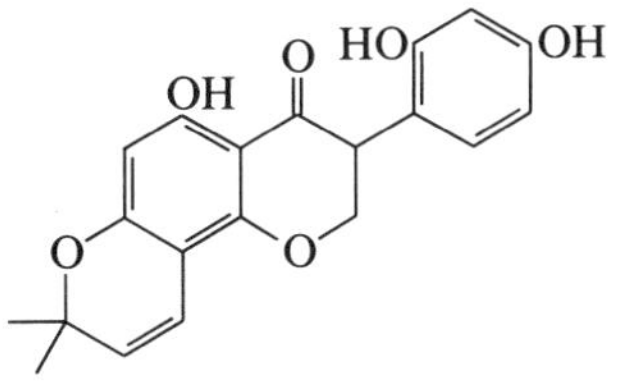

$C_{20}H_{18}O_6$ M 354.359
Tentative struct. Isol. from pods of *Phaseolus vulgaris.*

Woodward, M.D., *Phytochemistry,* 1979, **18**, 2007 (*isol, uv, pmr, ms*)

Dehydrodendrolasin   D-80022

Updated Entry replacing D-00234
3-(4,8-Dimethyl-2,4,6-nonatrienyl)furan
[41060-02-0]

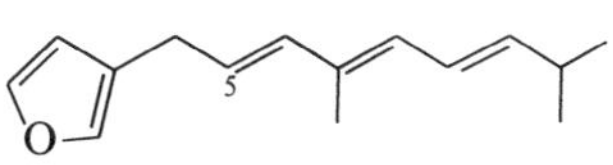

$C_{15}H_{20}O$ M 216.322
Constit. of *Pleraphysilla spinifera.* Oil.

5Z-Isomer: [117569-45-6].
$C_{15}H_{20}O$ M 216.322
Constit. of *Ceratosoma brevicaudata.* Oil. Reg. no. refers to (*Z,Z,E*) isomer, apparently incorrectly.

Cimino, G. *et al, Tetrahedron,* 1972, **28**, 4761 (*isol, struct*)
Ksebati, M.B. *et al, J. Nat. Prod.* (*Lloydia*), 1988, **51**, 857 (*isol, struct, pmr, cmr*)

Dehydrodieugenol   D-80023

3,3′-Dimethoxy-5,5′-di-2-propenyl[1,1′-biphenyl]-2,2′-diol, 9CI. 5,5′-Diallyl-2,2′-dihydroxy-3,3′-dimethoxybiphenyl. Δ⁸,⁸′-4,4′-Dihydroxy-5,5′-dimethoxy-3,3′-neolignan.
Bieugenol
[4433-08-3]

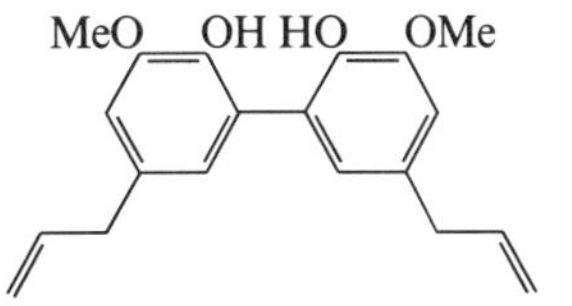

$C_{20}H_{22}O_4$ M 326.391
Isol. from Java citronella oil, from *Litsea turfosa, Ocotea cymbanum, Virola carinata* and *Nectandra polita.* Tablets (EtOH). Mp 106°.

Mono-Me ether: [75225-84-2]. **O-Methyldehydrodieugenol**
$C_{21}H_{24}O_4$ M 340.418
Isol. from *O. cymbanum, V. carinata* and *N. polita.* Oil.

Di-Me ether: [13417-56-6]. **Dimethyldehydrodieugenol**
$C_{22}H_{26}O_4$ M 354.445
Isol. from *N. polita.* Oil.

Jones, H.A. *et al, J. Am. Chem. Soc.,* 1940, **62**, 2558 (*isol*)
Holloway, D.M. *et al, Phytochemistry,* 1973, **12**, 1503 (*isol*)
de Diaz, A.M.P. *et al, Phytochemistry,* 1980, **19**, 681 (*isol*)
Kawanishi, K. *et al, Phytochemistry,* 1981, **20**, 1166 (*isol, pmr, deriv*)
Suarez, M. *et al, Phytochemistry,* 1983, **22**, 609 (*isol, cmr, derivs*)

Dehydrodieugenol B   D-80024

2-Methoxy-6-[2-methoxy-4-(2-propenyl)phenoxy]-4-(2-propenyl)phenol, 9CI. 4′,5-Diallyl-2-hydroxy-2′,3-dimethoxybiphenyl ether
[75225-33-1]

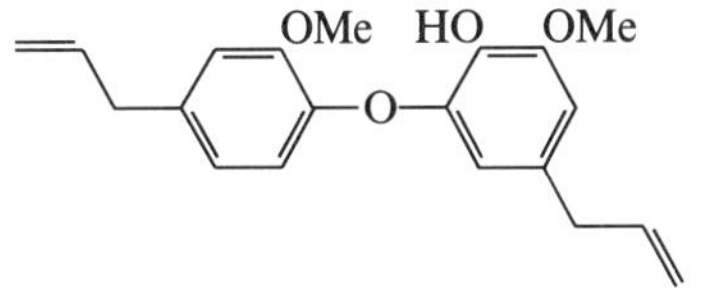

$C_{20}H_{22}O_4$ M 326.391
Isol. from trunkwood of *Octotea cymbarum.* Oil.

de Diaz, A.M.P. *et al, Phytochemistry,* 1980, **19**, 681 (*isol, uv, ir, pmr, ms, struct*)

Dehydroguaiaretic acid
D-80025

8-(4-Hydroxy-3-methoxyphenyl)-3-methoxy-6,7-dimethyl-2-naphthol

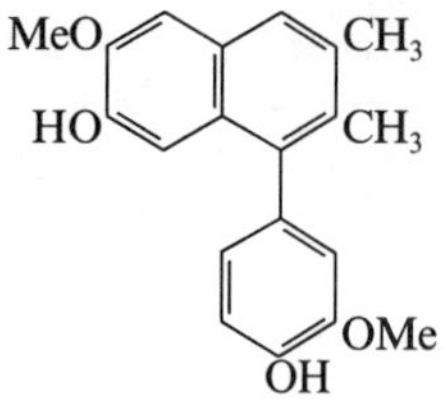

$C_{20}H_{20}O_4$ M 324.376

Minor constit. of wood of *Guaiacum officinale*.

Di-Et ester: Prisms (EtOAc). Mp 162-163°.

King, F.E. *et al, J. Chem. Soc.*, 1964, 4011 (*isol, uv, synth*)

Dehydrorotenone
D-80026

1,2-Dihydro-8,9-dimethoxy-2-(1-methylethenyl)[1]benzopyrano[3,4-b]furo[2,3-h][1]benzopyran-6(12H)one, 9CI. Didehydrorotenone

[30990-44-4]

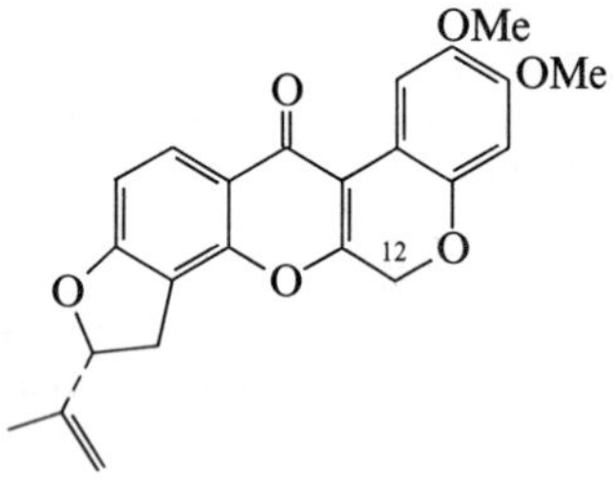

$C_{23}H_{20}O_6$ M 392.407

(*R*)-*form* [3466-09-9]

Isol. from *Derris elliptica, D. ferruginea, D. negrensis, D. uliginosa Lonchocarpus longifolius, L. urucu, Millettia ferruginea, M. pachycarpa, Mundulea sericea, Neorautanenia amboensis, Tephrosia falciformis* and *T. virginiana*. Pale yellow cryst. (C_6H_6/pet. ether). Mp 212-213°. $[\alpha]_D$ −27.5° (c, 0.8 in $CHCl_3$).

12-Hydroxy: [57103-58-9]. **Amorpholone**
$C_{23}H_{20}O_7$ M 408.407
Constit. of the roots of *Amorpha canescens*. Yellow needles ($CHCl_3$). Mp 255-260°. No opt. rotn. or chirality reported.

12-Oxo: [4439-62-7]. **Rotenonone**
$C_{23}H_{18}O_7$ M 406.391
Isol. from *A. canescens* and *N. amboensis*. Bright-yellow needles (dichloroethane). Mp 295-308° dec.

Piatak, D.M. *et al, Phytochemistry*, 1975, **14**, 1391 (*Amorpholone*)
Bose, P.C. *et al, Indian J. Chem., Sect. B*, 1976, **14**, 1012 (*isol, uv, ms, struct*)
Oberholzer, M.E. *et al, Phytochemistry*, 1976, **15**, 1283 (*Rotenonone*)
Ingham, J.L., *Fortschr. Chem. Org. Naturst.*, 1983, **43**, 1 (*occur*)

Deltamethrin
D-80027

3-(2,2-Dibromoethenyl)-2,2-dimethylcyclopropanecarboxylic acid cyano(3-phenoxyphenyl)methyl ester, 9CI. Decamethrin. Decis. K-Othrine. Butox. Butoflin

[52918-63-5]

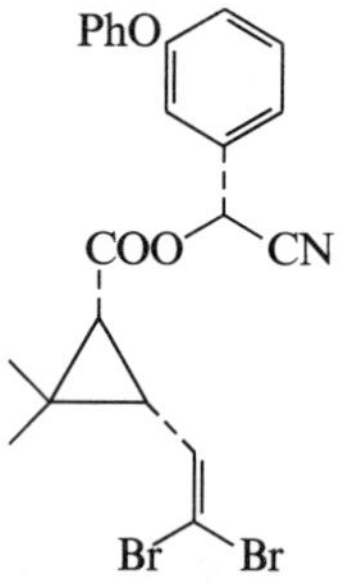

Absolute configuration

$C_{22}H_{19}Br_2NO_3$ M 505.205

Potent, photostable pyrethroid used as contact and stomach poison against a wide range of insects. Powder. Mp 100°. $[\alpha]_D$ +61° (c, 4 in C_6H_6).

Elliott, M. *et al, Nature (London)*, 1974, **248**, 710 (*synth, pmr*)
Lidgurd, R.O. *et al, Biomed. Mass Spectrom.*, 1986, **13**, 677; *CA*, **106**, 133627p (*ms*)
Pesticide Manual, 8th Ed., 1987, 3890.

Dendroserin
D-80028

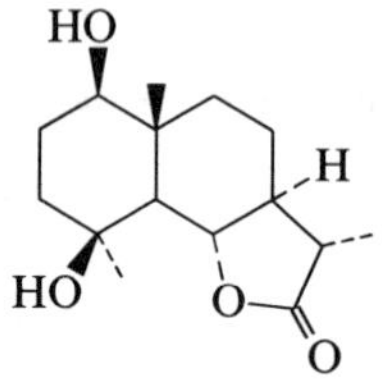

$C_{15}H_{24}O_4$ M 268.352

Constit. of *Dendroseris neriifolia*. Needles (MeOH). Mp 178-182°.

Campos, V. *et al, Heterocycles*, 1989, **28**, 779.

2'-Deoxyinosine
D-80029

9-(2-Deoxyribofuranosyl)hypoxanthine. Hypoxanthine 2-deoxyriboside

[29868-32-4]

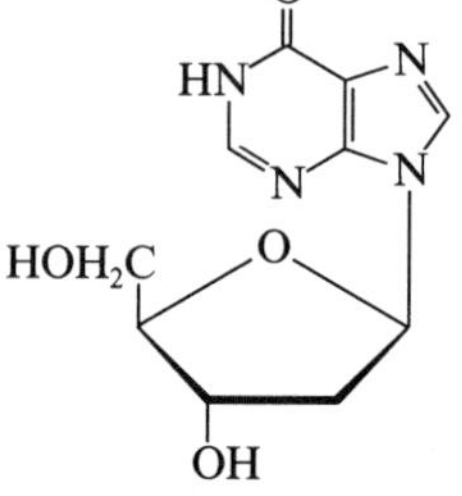

$C_{10}H_{12}N_4O_4$ M 252.229

Isol. from herring sperm DNA, from *Phaseolus vulgaris, Laminaria saccharina, Furcellaria fastigiata, Lactobacillus* spp. etc. Needles (MeOH). Mp 218° dec. $[\alpha]_D^{27}$ +7.92° (c, 0.53 in 0.1M NaOH).

Manson, L.A. *et al, J. Biol. Chem.*, 1951, **191**, 87 (*isol*)
Banhidi, Z.G. *et al, Acta Chem. Scand.*, 1953, **7**, 713 (*isol*)
Venner, H., *Chem. Ber.*, 1960, **93**, 140 (*synth*)
Robins, M.J. *et al, J. Am. Chem. Soc.*, 1965, **87**, 4934 (*pmr*)
Rousseau, R.J. *et al, J. Heterocycl. Chem.*, 1970, **7**, 367 (*synth, uv*)

Deoxyokamurallene   D-80030

Updated Entry replacing D-10036

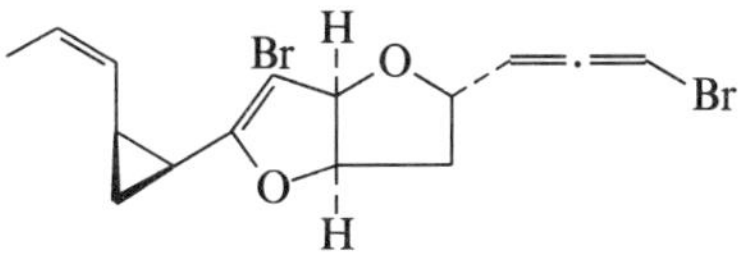

$C_{15}H_{16}Br_2O_2$ M 388.098
Constit. of *Laurencia okamurai*. Oil. $[\alpha]_D^{23}$ +220° (c, 0.84 in $CHCl_3$).

Suzuki, M. *et al, Chem. Lett.*, 1982, 289 (*isol*)
Suzuki, M. *et al, Phytochemistry*, 1989, **28**, 2145 (*struct*)

Derrone   D-80031

5-Hydroxy-3-(4-hydroxyphenyl)-8,8-dimethyl-4H,8H-
benzo[1,2-b:3,4-b′]dipyran-4-one, 9CI
[76166-59-1]

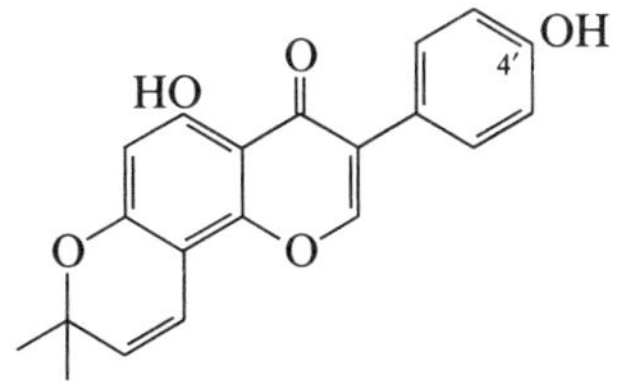

$C_{20}H_{16}O_5$ M 336.343
Isol. from seeds of *Derris robusta*. Cryst. Mp 216-218°.

Di-Ac: Needles. Mp 225-227°.

4′-O-Me: [34086-52-7]. ***4′-O-Methylderrone***
   $C_{21}H_{18}O_5$ M 350.370
   Isol. from *D. robusta* and from *Calopogonium*
   mucunoides. Mp 170-172°.

Jackson, B. *et al, J. Chem. Soc. C*, 1971, 3389 (*synth, deriv*)
Vilain, C. *et al, Bull. Soc. R. Sci. Liege*, 1975, **44**, 306; *CA*, **84**,
   74136e (*isol, deriv*)
Chibber, S.S., *Phytochemistry*, 1980, **19**, 1857 (*isol, uv, pmr*)

Desaspidins   D-80032

Updated Entry replacing D-00465

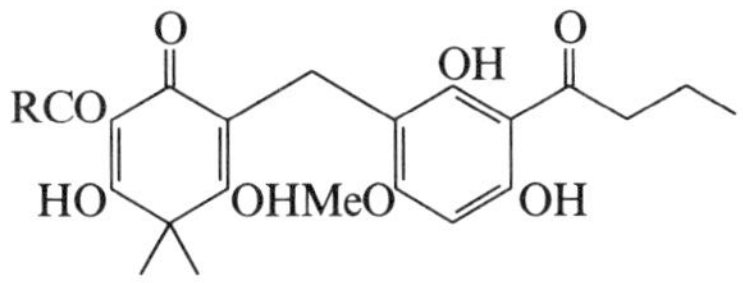

Desaspidin AB, R = CH_3
         PB, R = CH_2CH_3
         BB, R = $CH_2CH_2CH_3$

The alphabetical suffixes refer to the acyl chains (A =
acetyl, P = propionyl, B = butyryl). Isol. from male
ferns *Dryopteris* spp.
▷ EU1225000.

Desaspidin AB [3572-07-4]
   $C_{22}H_{26}O_8$ M 418.443
   Cryst. (cyclohexane). Mp 145-147° (synthetic).

Desaspidin PB
   $C_{23}H_{28}O_8$ M 432.469
   Mp 141-142° (synthetic).

Desaspidin BB [114-43-2]
   Desapidin. Resapin
   $C_{24}H_{30}O_8$ M 446.496
   Anthelmintic. Cryst. (Et_2O/pet. ether). Mp 150-150.5°.

Aebi, A. *et al, Helv. Chim. Acta*, 1957, **40**, 266 (*isol*)
Penttilä, A. *et al, Acta Chem. Scand.*, 1964, **18**, 344 (*isol, synth*)
Lounasmaa, M. *et al, Acta Chem. Scand.*, 1972, **26**, 89 (*ms*)
Widen, C.J. *et al, Helv. Chim. Acta*, 1975, **58**, 880; 1976, **59**, 1725
   (*isol, bibl*)
Ayras, P. *et al, Org. Magn. Reson.*, 1981, **16**, 209 (*cmr*)
Martindale, The Extra Pharmacopoeia, 28th/29th Eds.,
   Pharmaceutical Press, London, 1982/1989, 770.

Des-A-26,27-dinoroleana-5,7,9,11,13-   D-80033
   pentaene

$C_{22}H_{28}$ M 292.463
Constit. of surface sediments.

Trendel, J.M. *et al, Tetrahedron*, 1989, **45**, 4457.

Des-A-26,27-dinor-5,7,9,11,13-   D-80034
   ursapentaene

$C_{22}H_{28}$ M 292.463
Constit. of surface sediments.

Trendel, J.M. *et al, Tetrahedron*, 1989, **45**, 4457.

Des-A-lup-9-ene   D-80035

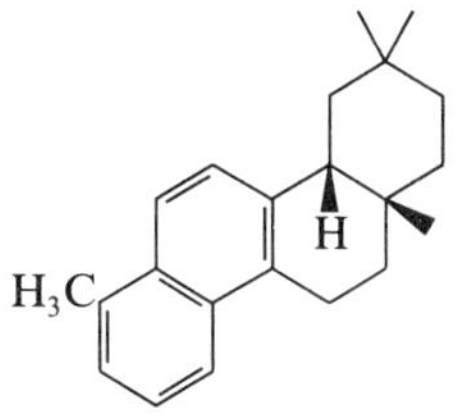

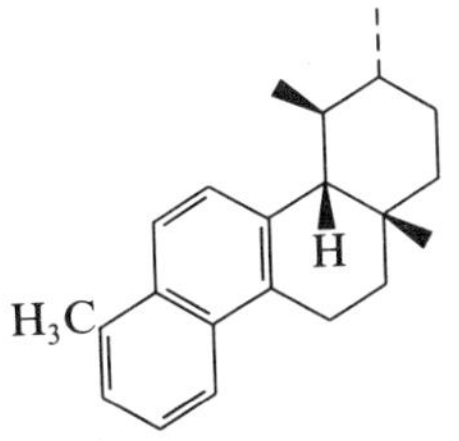

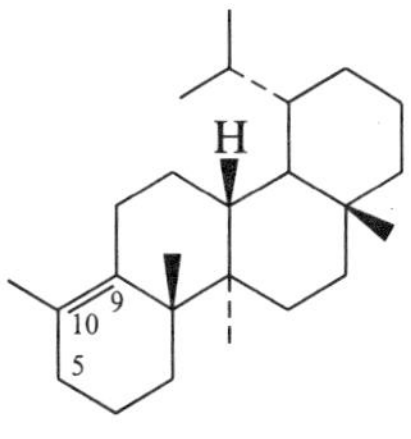

$C_{24}H_{40}$ M 328.580
Constit. of surface sediments. Cryst. ($MeOH/CH_2Cl_2$). Mp
117.5°. $[\alpha]_D$ −107° (c, 0.586 in $CHCl_3$).

9α,10β-Dihydro: ***10βH-Des-A-lupane***
   $C_{24}H_{42}$ M 330.596
   Constit. of surface sediments. Cryst. ($MeOH/CH_2Cl_2$).
   Mp 94-95°.

$\Delta^{5(10)}$-*Isomer:* ***Des-A-lup-5(10)-ene***
   $C_{24}H_{40}$ M 328.580
   Constit. of surface sediments. Cryst. ($MeOH/CH_2Cl_2$).
   Mp 174°.

Corbet, B. *et al, J. Am. Chem. Soc.*, 1980, **102**, 1171 (*synth*)
Trendel, J.M. *et al, Tetrahedron*, 1989, **45**, 4457 (*isol, synth, cryst
   struct*)

Des-A-26-nor-5,7,9-lupatriene   D-80036

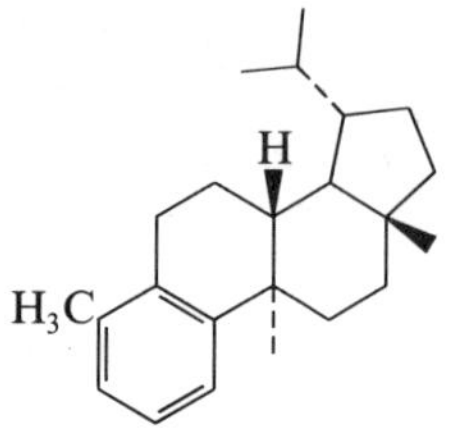

$C_{23}H_{34}$ M 310.522
Constit. of sediments.

Wolff, G.A. *et al, Tetrahedron*, 1989, **45**, 6721 (*synth*)

Des-A-5(10),12-oleanadiene   D-80037

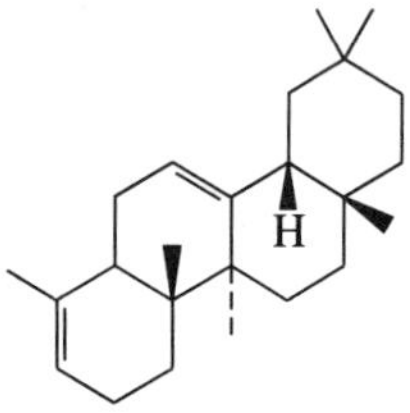

$C_{24}H_{38}$ M 326.564
Constit. of surface sediments.

Trendel, J.M. *et al, Tetrahedron*, 1989, **45**, 4457.

Des-A-26,27,28-trisnorursa-5,7,9,11,13,15,17-heptaene   D-80038

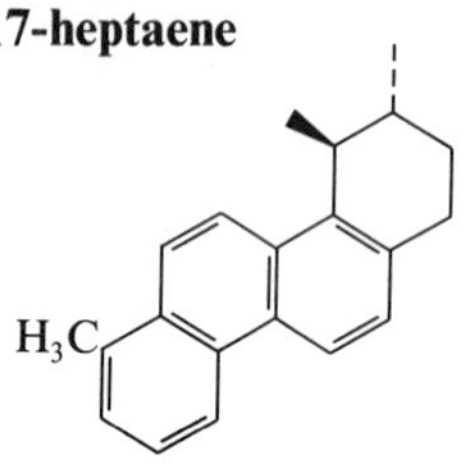

$C_{21}H_{22}$ M 274.405
Constit. of surface sediments. Cryst. (MeOH/CH$_2$Cl$_2$).

Trendel, J.M. *et al, Tetrahedron*, 1989, **45**, 4457 (*cryst struct*)

Des-A-5(10),12-ursadiene   D-80039

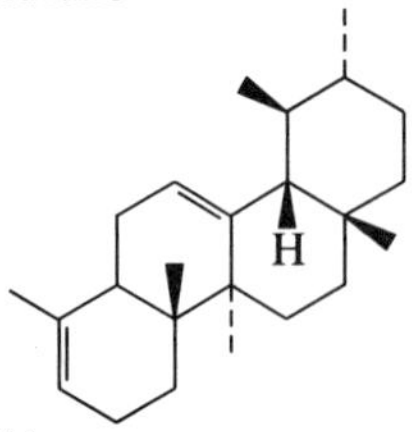

$C_{24}H_{38}$ M 326.564
Constit. of surface sediments.

Trendel, J.M. *et al, Tetrahedron*, 1989, **45**, 4457.

Detoxinine   D-80040

Updated Entry replacing D-20037

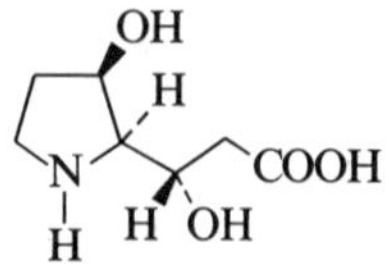

$C_7H_{13}NO_4$ M 175.184

(−)-*form*
Structural component of Detoxin D$_1$, D-00484. $[\alpha]_D^{23}$ −4.1° (c, 0.5 in H$_2$O).

(±)-*form*
Prisms. Mp >200° dec.

Häusler, J., *Justus Liebigs Ann. Chem.*, 1983, 982 (*synth*)
Ohfune, Y. *et al, Tetrahedron Lett.*, 1984, **25**, 4133 (*synth*)
Ewing, W.R. *et al, Tetrahedron*, 1986, **42**, 2421 (*synth, ir, pmr*)

1,3-Diacetylbicyclo[1.1.1]pentane   D-80041

[115913-30-9]

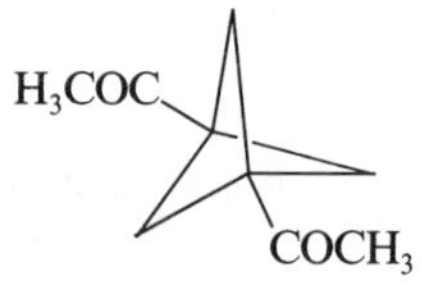

$C_9H_{12}O_2$ M 152.193
Cryst. (heptane). Mp 67-69°.

Kaszynski, P. *et al, J. Org. Chem.*, 1988, **53**, 4593 (*synth, pmr, cmr, ir, ms*)

2,3-Diacetyl-2-butenedioic acid, 9CI   D-80042

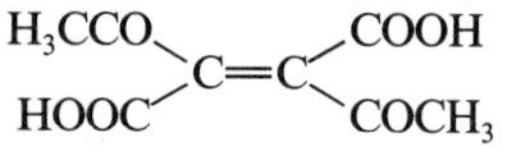

(E)-*form*

$C_8H_8O_6$ M 200.148
(E)-*form*
Di-Et ester: [69622-58-8]. *Diethyl diacetylfumarate*
$C_{12}H_{16}O_6$ M 256.255
Needles (Et$_2$O). Mp 93-95°. Readily cyclises.

[69622-59-9, 92236-43-6]

Adembri, G. *et al, Gazz. Chim. Ital.*, 1983, **113**, 489.

Di-2-(7-acetyl-1,4-dihydro-3,6,8-trihydroxy-4,4-dimethyl-1-oxonaphthyl)methane   D-80043

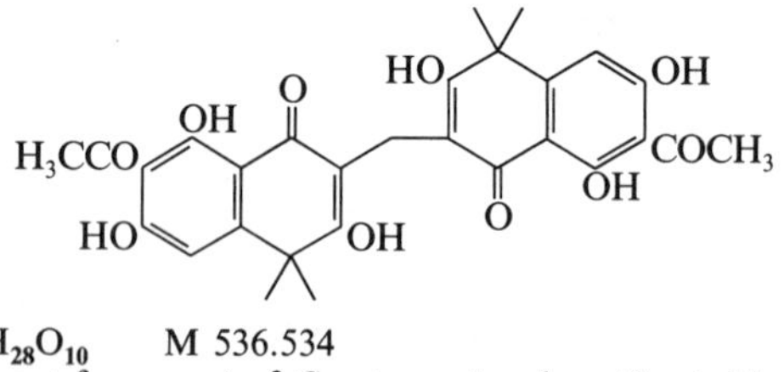

$C_{29}H_{28}O_{10}$ M 536.534
Pigment from root of *Cassia semicordata*. Cryst. Mp 266-267°.

Delle Monache, G. *et al, Tetrahedron Lett.*, 1989, **30**, 6203.

3,6-Diacetyl-1,4,2,5-dioxadiazine   D-80044

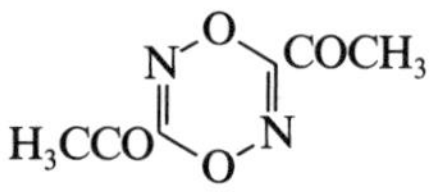

$C_6H_6N_2O_4$ M 170.124
Dioxime: [116303-45-8]. *Schmitz's compound*
$C_6H_8N_4O_4$ M 200.154
Cryst. (EtOH). Mp 191°.

Fruttero, R. *et al, Justus Liebigs Ann. Chem.*, 1988, 1017 (*struct, cmr, N-15 nmr, bibl*)

3,4-Diacetylfurazan   D-80045

1,1′-(3,4-Furazandiyl)bisethanone, 9CI

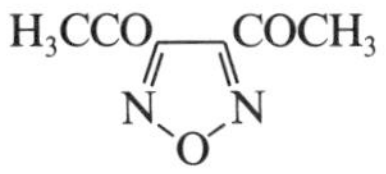

H$_3$CCO—COCH$_3$

$C_6H_6N_2O_3$ M 154.125
Oil.
N-*Oxide:* [6103-08-8]. *3,4-Diacetylfuroxan*
   $C_6H_6N_2O_4$ M 170.124
   Yellow oil. Bp$_1$ 100°.
▷ Explodes on dist. unless pure.
N-*Oxide, dioxime*(E,E-)*:* [116303-46-9]. *3,4-Diacetylfuroxan*
dioxime. Tryller's compound
   $C_6H_8N_4O_4$ M 200.154
   Cryst. (CHCl$_3$). Mp 145°.

Peterson, L.I., *Tetrahedron Lett.*, 1966, 1727 (*synth, oxide, pmr*)
Fruttero, R. *et al*, *Justus Liebigs Ann. Chem.*, 1988, 1017 (*struct, cmr, N-15 nmr, bibl*)

2,2′-Diamino-4,4′-biphenyldicarboxylic acid   D-80046

2,2′-Diamino-4,4′-diphenic acid
[41738-56-1]

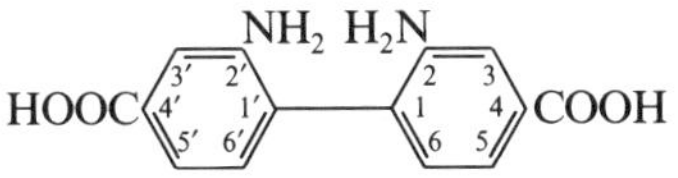

HOOC〈 〉—〈 〉COOH
(with NH$_2$ H$_2$N substituents)

$C_{14}H_{12}N_2O_4$ M 272.260
Yellow platelets (EtOH or H$_2$O). Mp 307-309°.
Di-Me ester: [23933-57-5].
   $C_{16}H_{16}N_2O_4$ M 300.313
   Mp 174-176°.
Di-Et ester:
   $C_{18}H_{20}N_2O_2$ M 296.368
   Needles or pale-yellow scales. Mp 99-100°, Mp 112-115° (dimorph.).
N,N′-Di-Ac:
   $C_{18}H_{16}N_2O_6$ M 356.334
   Mp 250°.

Jakubowski, Z. *et al*, *Ber.*, 1909, **42**, 634.
Adams, R. *et al*, *J. Am. Chem. Soc.*, 1933, **55**, 1649.
Cheung King Ling, C., *J. Chem. Soc.*, 1964, 1825.

4,4′-Diamino-2,2′-biphenyldicarboxylic acid   D-80047

4,4′-Diaminodiphenic acid, 8CI. Benzidine-2,2′-dicarboxylic acid
[17557-76-5]
$C_{14}H_{12}N_2O_4$ M 272.260
Polymer intermediate. Needles. Mp 265°.
Anhydride, N,N′-Di-Ac:
   $C_{18}H_{14}N_2O_5$ M 338.319
   Needles (Ac$_2$O). Mp >300°.

Schmidt, J. *et al*, *Ber.*, 1905, **28**, 3771 (*synth*)
Adkins, H. *et al*, *J. Am. Chem. Soc.*, 1917, **46**, 1924 (*synth*)
Patel, H.R. *et al*, *J. Am. Pharm. Assoc.*, 1957, **46**, 51 (*synth*)

4,4′-Diamino-2,3′-biphenyldicarboxylic acid   D-80048

Benzidine-2,3′-dicarboxylic acid
$C_{14}H_{12}N_2O_2$ M 240.261

Paal, C. *et al*, *Ber.*, 1892, **25**, 3590.

4,4′-Diamino-3,3′-biphenyldicarboxylic acid   D-80049

Benzidine-3,3′-dicarboxylic acid
[2130-56-5]
$C_{14}H_{12}N_2O_4$ M 272.260
Polymer intermed. Needles. Mp 300° (275°) dec.
▷ Weak carcinogen.
N,N′-Di-Ac:
   $C_{18}H_{16}N_2O_6$ M 356.334
   Mp 300° (unsharp).
N,N′-Dibenzoyl:
   $C_{28}H_{20}N_2O_6$ M 480.476
   Mp 302-304°.
Diamide: [15420-64-1].
   $C_{14}H_{14}N_4O_4$ M 302.289
   Mp 340°.

Bülow, C. *et al*, *Ber.*, 1898, **31**, 2574 (*derivs*)
Schultz, G. *et al*, *Justus Liebigs Ann. Chem.*, 1907, **352**, 111 (*synth*)
Pliss, G.B., *CA*, 1969, **71**, 48163h (*tox*)
Rabilloud, G. *et al*, *J. Heterocycl. Chem.*, 1980, **17**, 1065 (*synth*)
Niume, K. *et al*, *J. Polym. Sci., Part A*, 1981, **19**, 1745 (*synth*)

4,6′-Diamino-2,2′-biphenyldicarboxylic acid   D-80050

4,6′-Diaminodiphenic acid, 8CI
[67003-04-7]
$C_{14}H_{12}N_2O_4$ M 272.260
Needles. V. insol. org. solvs. Mp 327-330° dec. (312-313°).

Schad, P., *Ber.*, 1893, **26**, 216 (*synth*)
Adkins, H. *et al*, *J. Am. Chem. Soc.*, 1917, **46**, 1924 (*synth*)
Patel, H.R. *et al*, *J. Am. Pharm. Assoc.*, 1957, **46**, 51 (*synth*)

5,5′-Diamino-2,2′-biphenyldicarboxylic acid   D-80051

5,5′-Diaminodiphenic acid
$C_{14}H_{12}N_2O_2$ M 240.261
Mp 265° dec.
Di-Me ester:
   $C_{16}H_{16}N_2O_2$ M 268.315
   Mp 220-222°.
N,N′-Di-Ac:
   $C_{18}H_{16}N_2O_4$ M 324.335
   Mp >300° (darkens).

Pufahl, F., *Ber.*, 1929, **62**, 2817.

6,6′-Diamino-2,2′-biphenyldicarboxylic acid   D-80052

6,6′-Diamino-2,2′-diphenic acid. 2,7-Diaminodiphenic acid (obsol.)
$C_{14}H_{12}N_2O_4$ M 272.260
(±)-*form*
   Mp 265° (250-251°).

Adkins, H. *et al*, *J. Am. Chem. Soc.*, 1924, **46**, 1917 (*synth*)

1,5-Diaminobiphenylene   D-80053

1,5-Biphenylenediamine, 9CI

[17532-90-0]

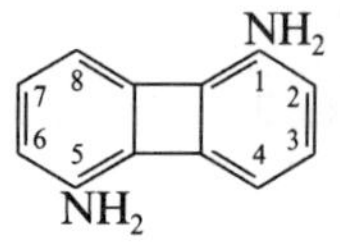

$C_{12}H_{10}N_2$ M 182.224

Pale yellow needles (MeOH). Mp 233-235°.

Iqbal, K. *et al*, *J. Chem. Soc. C*, 1967, 1690 (*synth, pmr*)

2,3-Diaminobiphenylene   D-80054

2,3-Biphenylenediamine

[32281-46-2]

$C_{12}H_{10}N_2$ M 182.224

Yellow-brown plates (C_6H_6/hexane). Mp 198-200° subl.

N,N'-Di-Ac: [32281-47-3].

   $C_{16}H_{14}N_2O_2$ M 266.299

   Pale yellow needles (MeOH). Mp 285-287°.

Barton, J.W. *et al*, *J. Chem. Soc. C*, 1971, 1384 (*synth, deriv*)

2,6-Diaminobiphenylene   D-80055

2,6-Biphenylenediamine

[70468-22-3]

$C_{12}H_{10}N_2$ M 182.224

Polymer intermediate. Deep yellow plates (pet. ether). Mp 218-220° dec.

N,N'-Di-Ac:

   $C_{16}H_{14}N_2O_2$ M 266.299

   Orange needles (EtOH). Mp 328-330° dec.

Baker, W. *et al*, *J. Chem. Soc.*, 1960, 414 (*synth, deriv, uv*)

Swedo, R.J. *et al*, *J. Polym. Sci., Polym. Chem. Ed.*, 1978, **16**, 2711 (*synth*)

1,8-Diaminocarbazole   D-80056

9H-Carbazole-1,8-diamine

[117110-85-7]

$C_{12}H_{11}N_3$ M 197.239

Silver plates (C_6H_6). Mp 230° dec.

1,8-N-Di-Ac:

   $C_{16}H_{15}N_3O_2$ M 281.313

   Needles (EtOH). Mp 320-322° dec.

Takahashi, K. *et al*, *J. Chem. Soc., Perkin Trans. 1*, 1988, 1869 (*synth, ir, pmr*)

2,4-Diamino-5-chloropyrimidine, 8CI   D-80057

5-Chloro-2,4-pyrimidinediamine, 9CI

[18620-64-9]

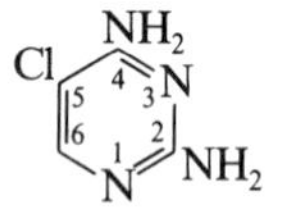

$C_4H_5ClN_4$ M 144.563

Cryst. (H_2O). Mp 200-202°. pK_a 5.58.

3-Oxide: [91481-97-9].

   $C_4H_5ClN_4O$ M 160.562

   Cryst. (EtOH). Mp 222-224°. pK_a 1.99.

Roth, B. *et al*, *J. Org. Chem.*, 1969, **34**, 821 (*uv*)

Cowden, W.B. *et al*, *Aust. J. Chem.*, 1979, **32**, 2049; 1984, **37**, 1195 (*synth, props, uv, oxide*)

2,4-Diamino-6-chloropyrimidine, 8CI   D-80058

6-Chloro-2,4-pyrimidinediamine, 9CI

[156-83-2]

$C_4H_5ClN_4$ M 144.563

Cryst. (Me_2CO). Mp 200-201°. pK_a 3.57.

▷ Irritant.

3-Oxide: [35139-67-4].

   $C_4H_5ClN_4O$ M 160.562

   Cryst. (EtOH). Mp 190° dec. pK_a 2.25.

Aldrich Library of IR Spectra, No.1363A.

Aldrich Library of NMR Spectra, 2nd Ed., **2**, 703D.

Hull, R. *et al*, *J. Chem. Soc.*, 1947, 41 (*synth*)

Roth, B. *et al*, *J. Org. Chem.*, 1969, **34**, 821 (*uv*)

Delia, T.J. *et al*, *J. Heterocycl. Chem.*, 1972, **9**, 73 (*oxide, synth, pmr, uv*)

Staedeli, W. *et al*, *Helv. Chim. Acta*, 1980, **63**, 504 (*N-15 nmr*)

Cowden, W.B. *et al*, *Aust. J. Chem.*, 1984, **37**, 1195 (*props, oxide, synth, pmr, uv*)

3,4-Diaminohexanoic acid   D-80059

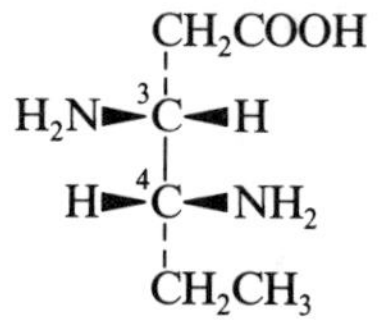

$$CH_2COOH$$
$$H_2N \blacktriangleright \overset{3}{C} \blacktriangleleft H$$
$$H \blacktriangleright \overset{4}{C} \blacktriangleleft NH_2$$
$$CH_2CH_3 \qquad (3R,4R)\text{-}form$$

$C_6H_{14}N_2O_2$ M 146.189

(3R,4R)-form [120452-75-7]

   N^3-tert-*Butyloxycarbonyl:* Cryst. (MeOH/Et$_2$O). $[\alpha]_D^{20}$ +7.0° (c, 0.9 in MeOH).

(3R,4S)-form [120452-77-9]

   N^3-tert-*Butyloxycarbonyl:* Cryst. (MeOH/Et$_2$O). Mp 151-153°. $[\alpha]_D^{20}$ + ±0.0° (c, 1.0 in MeOH). Incorr. descr. as (3R,4R).

Kano, S. *et al*, *Chem. Pharm. Bull.*, 1988, **36**, 3341 (*synth, pmr*)

5,6-Diaminohexanoic acid   D-80060

δ-Lysine

$$H_2NCH_2CH(NH_2)CH_2CH_2CH_2COOH$$

$C_6H_{14}N_2O_2$ M 146.189

(±)-form

   B,2HCl: [118894-93-2].

   Mp 163° dec.

Altman, J. *et al*, *Justus Liebigs Ann. Chem.*, 1989, 493 (*synth, pmr*)

2,5-Diamino-2-methylpentanoic acid   D-80061

2-Methylornithine, 9CI

[48047-94-5]

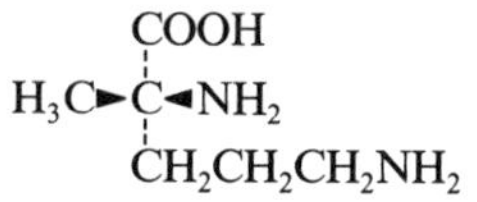

$$COOH$$
$$H_3C \blacktriangleright C \blacktriangleleft NH_2$$
$$CH_2CH_2CH_2NH_2$$

$$(R)\text{-}form$$

$C_6H_{14}N_2O_2$ M 146.189

(R)-form [59574-27-5]

   D- *form*

   Mp 197-199° dec. $[\alpha]_D$ −10.21° (c, 0.7 in 4M HCl).

   B,HCl: Mp 214°. $[\alpha]_D^{20}$ −10° (c, 1 in H_2O).

(S)-form [48047-95-6]

   L- *form*

   B,HCl: [59545-84-5].

Cryst. (EtOH aq.). Mp 218°. [α]$_D^{20}$ +8.4° (c, 0.5 in H$_2$O).
(±)-*form*
Potent reversible inhibitor of ornithine decarboxylase.
B,HCl: Cryst. + 1H$_2$O (Me$_2$CO aq.). Dehydrates at 250°.

Abdel-Monem, M.M. *et al, J. Med. Chem.*, 1974, **17**, 447.
Bey, P. *et al, J. Med. Chem.*, 1978, **21**, 50 (*synth, abs config*)
Gander-Coquoz, M. *et al, Helv. Chim. Acta*, 1988, **71**, 224 (*synth, ir, pmr, ms*)

Diaporthin   D-80062

Updated Entry replacing D-70050
*8-Hydroxy-3-(2-hydroxypropyl)-6-methoxy-1H-2-
benzopyran-1-one, 9CI. 8-Hydroxy-3-(2-hydroxypropyl)-6-
methoxyisocoumarin*

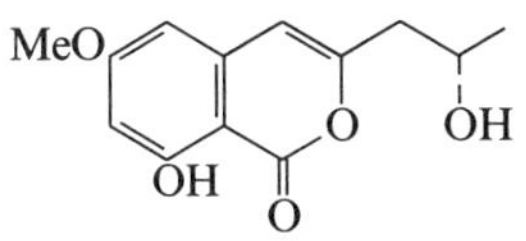

C$_{13}$H$_{14}$O$_5$ M 250.251
(*S*)-*form* [10532-39-5]
Isol. from cultures of *Endothia parasitica*. Phytotoxin.
Needles or plates. Mp 90-92° (83-85°). [α]$_D$ +54° (c,
10.87 in CHCl$_3$).

O-De-Me: Orthosporin. De-O-methyldiaporthin
C$_{12}$H$_{12}$O$_5$ M 236.224
Phytotoxin from *Drechslera siccans* and *Rhynchosporium
orthosporum*. Needles. Mp 183-184°. [α]$_D^{22}$ +61.8° (c, 1 in
MeOH).

Hardegger, E. *et al, Helv. Chim. Acta*, 1966, **49**, 1283 (*isol, props*)
Hallock, Y.F. *et al, Phytochemistry*, 1988, **27**, 3123 (*isol*)
Ichihara, A. *et al, Chem. Lett.*, 1989, 1495 (*struct, synth, abs config*)

3,4-Diazido-3,5-dihydro-thiophene 1,1-   D-80063
dioxide

3,4-Diazido-3-sulfolene
[114397-29-4]

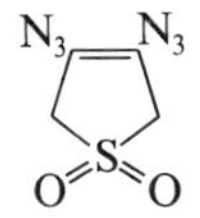

C$_4$H$_4$N$_6$O$_2$S   M 200.181
Yellow solid. Mp 82-83.5°. Gradually dec. on standing.

Chou, T. *et al, J. Org. Chem.*, 1988, **53**, 3027 (*synth, pmr, ir, ms*)

7-Diazo-7H-benzocycloheptene, 9CI   D-80064

4,5-Benzodiazocycloheptatriene
[111743-26-1]

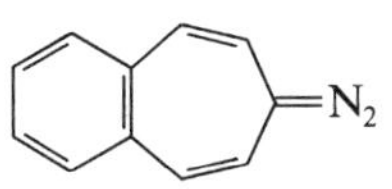

C$_{11}$H$_8$N$_2$ M 168.198
Precursor to the triplet 4,5-benzocycloheptatrienylidene
carbene and 4,5-benzocycloheptatrienylradical. Brown
solid. Thermally unstable. Could not be cryst.

Chateauneuf, J.E. *et al, J. Am. Chem. Soc.*, 1988, **110**, 539 (*synth, ir, uv, pmr*)

Diazomethanesulfonic acid   D-80065

N$_2$=CHSO$_3$H

CH$_2$N$_2$O$_3$S   M 122.104
Esters prepd. and characterised spectroscopically.
[120263-45-8, 120263-47-0]

Berkessel, A. *et al, Chem. Ber.*, 1989, **122**, 1147.

2-Diazopropanal, 9CI   D-80066
[19547-06-9]

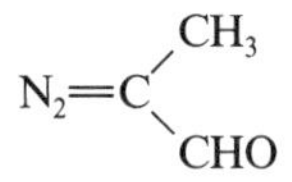

C$_3$H$_4$N$_2$O   M 84.077
Oil. Bp$_{12}$ 35°.

Regitz, M. *et al, Chem. Ber.*, 1968, **101**, 2622 (*synth*)
Kučera, J. *et al, Collect. Czech. Chem. Commun.*, 1970, **35**, 3618 (*synth, uv*)
Menicagli, R. *et al, Tetrahedron*, 1987, **43**, 171 (*synth, ir, pmr*)

11H-Dibenzo[a,def]carbazole   D-80067

1,12-Iminobenz[a]anthracene
[40831-21-8]

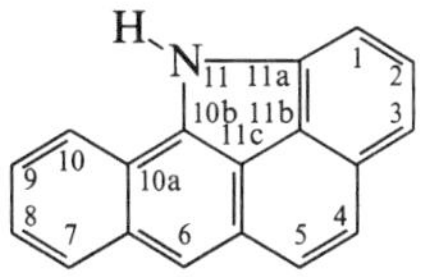

C$_{18}$H$_{11}$N   M 241.292
Reg. no. refers to 1*H*-form. Pale-yellow cryst. (C$_6$H$_6$). Mp
217-219.5°. Intense blue fluor. in soln.
Picrate: Dark brown needles. Mp 191-192°.

Badger, G.M. *et al, J. Chem. Soc.*, 1949, 799.

Dibenzo[c,g]carbazole-6-carboxylic acid   D-80068
[111960-35-1]

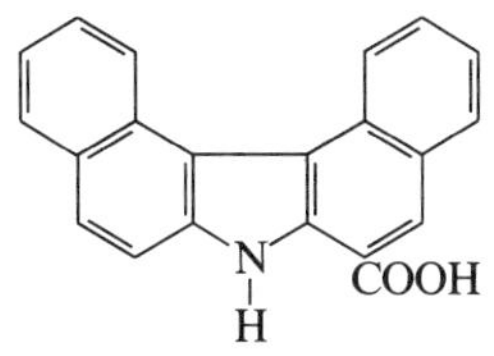

C$_{21}$H$_{13}$NO$_2$ M 311.339
Cryst. (CHCl$_3$). Mp 286-288°.

Katritzky, A.R. *et al, J. Org. Chem.*, 1988, **53**, 794 (*synth, pmr*)

1,2-Di-(1-Benzofuranyl)ethylene   D-80069

2,2′-(1,2-Ethenediyl)bisbenzofuran, 9CI

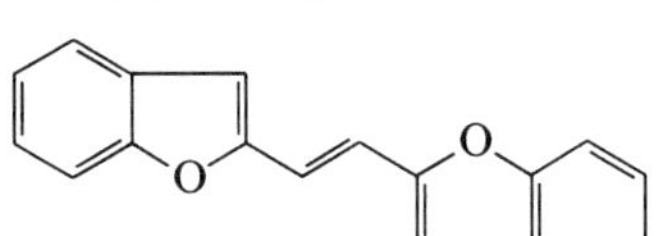

C$_{18}$H$_{12}$O$_2$ M 260.292
(*E*)-*form* [115983-55-6]
Cryst. (EtOH). Mp 185°.

Capuano, L. *et al, Chem. Ber.*, 1988, **121**, 2259 (*synth, pmr, uv*)

Dibenzo[f,h]furo[2,3-b]quinoline, 9CI   D-80070

Phenanthro[9,10-e]furo[2,3-b]pyridine
[115170-34-8]

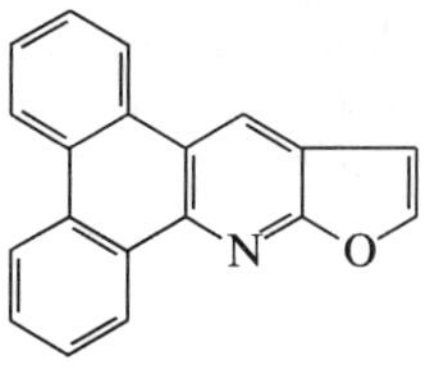

$C_{19}H_{11}NO$ M 269.302
Powder. Mp 161.5-163°.

Taylor, E.C. *et al, Tetrahedron*, 1987, **43**, 5145 (*synth, ir, pmr, cmr*)

Dibenzo[b,g][1,8]naphthyridine-11,12-diamine, 9CI   D-80071

11,12-Diaminodibenzo[b,g][1,8]naphthyridine
[117096-00-1]

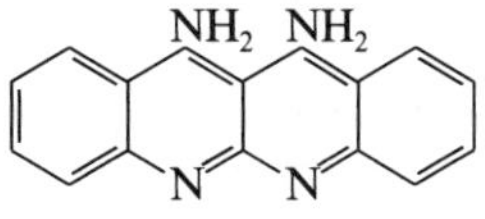

$C_{16}H_{12}N_4$ M 260.298
Yellow solid + $0.4H_2O$ by subl. Mp 314-316°. pK_a 3.3.

Shi, S. *et al, J. Org. Chem.*, 1988, **53**, 5379 (*synth, ir, pmr, cmr, ms*)

Dibenzo[fg,mn]octalene   D-80072

[111409-90-6]

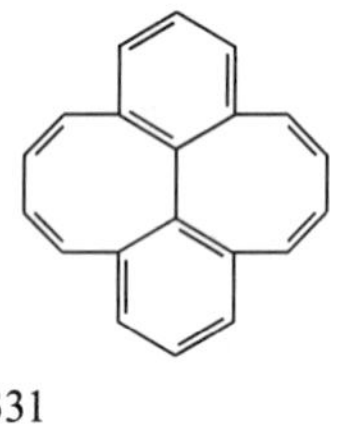

$C_{20}H_{14}$ M 254.331
Needles (hexane). Mp 188°.

Heinz, W. *et al, Angew. Chem., Int. Ed. Engl.*, 1987, **26**, 1291 (*synth, pmr, cmr, uv*)

Dibenzo[fg,ij]pentaphene-15,16-dione, 9CI   D-80073

2,3:10,11-Dibenzperylene-1,12-quinone (*obsol.*)
[116897-91-7]

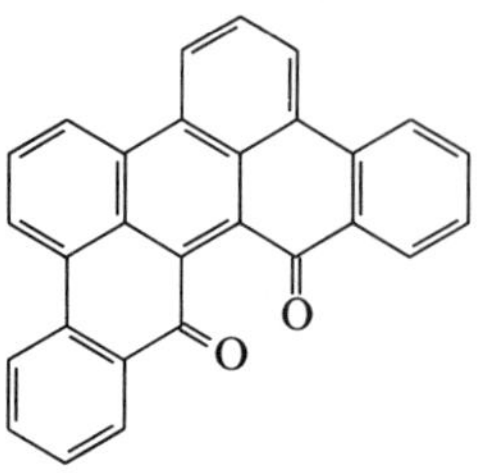

$C_{28}H_{14}O_2$ M 382.417
Mp 336° (330°).

Zinke, A. *et al, Ber.*, 1941, **74**, 115 (*synth*)
Hewgill, F.R. *et al, J. Chem. Soc., Perkin Trans. 1*, 1988, 1305 (*synth, ir, pmr, cmr*)

Di-2-benzothiazolylamine   D-80074

N-2-Benzothiazolyl-2-benzothiazolamine, 9CI. 2,2'-Iminobis[benzo-1,3-thiazole]
[34997-17-6]

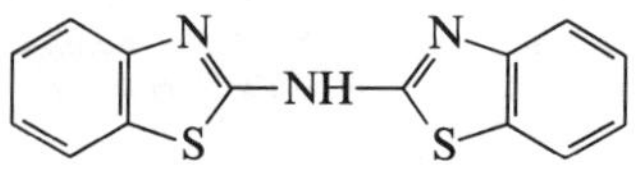

$C_{14}H_9N_3S_2$ M 283.377
Cryst. (1-butanol). Mp 256-257°.

Neidlein, R. *et al, Synthesis*, 1971, 540 (*synth*)
Garín, J. *et al, Synthesis*, 1987, 368 (*synth, pmr, ms*)

Dibenzo[b,e]thiepin-11(6H)-one, 9CI   D-80075

6,11-Dihydrodibenzo[b,e]thiepin-11-one
[1531-77-7]

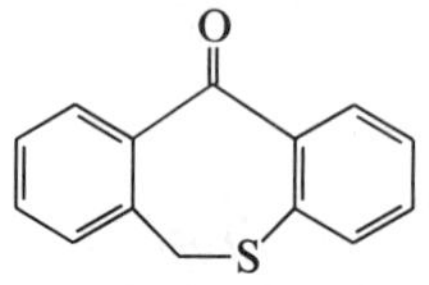

$C_{14}H_{10}OS$ M 226.298
Intermediate in synth. of tranquilizers, antidepressants, antiemetics and antihistamines. Cryst. (EtOH or Et_2O/pet. ether). Mp 86-88°. Bp_1 175-180°.

Stach, K. *et al, Angew. Chem.*, 1962, **74**, 31, 752.
Stach, K. *et al, Monatsh. Chem.*, 1962, **93**, 889 (*synth, uv, ir, pmr*)
Volanschi, E. *et al, Rev. Roum. Chim.*, 1974, **19**, 755; *CA*, **81**, 62865s (*esr*)
Sindelar, K. *et al, Collect. Czech. Chem. Commun.*, 1983, **48**, 1898 (*pharmacol*)
Dolivka, Z. *et al, Collect. Czech. Chem. Commun.*, 1985, **50**, 1078; 1987, **52**, 1566 (*use*)
Wyatt, D.K. *et al, Appl. Spectrosc.*, 1986, **40**, 369 (*cmr*)

Dibenzo[b,f]thiepin-10(11H)-one, 9CI   D-80076

10,11-Dihydrodibenzo[b,f]thiepin-10-one
[1898-85-7]

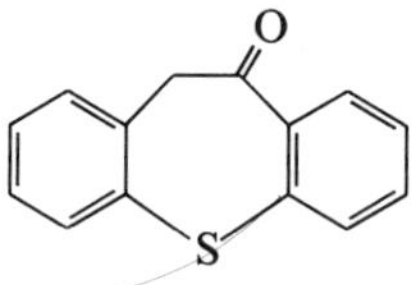

Used in synth. of antidepressants and analgesics. Solid. Mp 72-73°.

Org. Synth., Coll.Vol.1, 552 (*synth*)
Jelek, J.O. *et al, Monatsh. Chem.*, 1965, **96**, 182 (*synth, ir*)
Ueda, I., *Bull. Chem. Soc. Jpn.*, 1975, **48**, 2306 (*synth*)
Sindelar, K. *et al, Collect. Czech. Chem. Commun.*, 1983, **48**, 1187 (*synth*)

1,3-Dibenzyl-1,3-dihydroisoindole   D-80077

2,3-Dihydro-1,3-bis(phenylmethyl)-1H-isoindole, 9CI. 1,3-Dibenzylisoindoline
[117185-28-1]

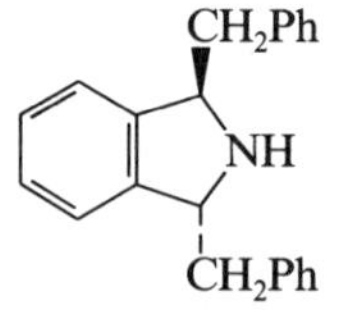

$C_{22}H_{21}N$ M 299.415
(*R,R*)-*form*

Reagent for asymmetric synthesis. $[\alpha]_D$ −5.4° (c, 2.15 in EtOH).

Gawley, R.E. *et al, J. Org. Chem.*, 1988, **53**, 5381 (*synth, pmr*)

1,4-Dibenzyl-6,7-dihydroxy-1,4-diazocane-5,8-dione   D-80078

1,2,3,4,6,7-Hexahydro-6,7-dihydroxy-1,4-bis(phenylmethyl)-1,4-diazocine-5,8-dione, 9CI. N,N′-Dibenzyl-N,N′-ethylenetartramide

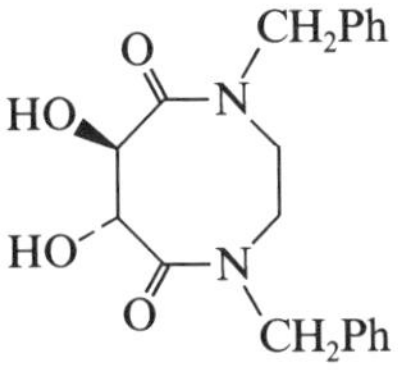

(*R,R*)-*form*

$C_{20}H_{22}N_2O_4$ M 354.405

(*R,R*)-*form* [112897-01-5]

Chiral auxiliary for allylboronation. Cryst. (EtOH aq.). Mp 202-203°. $[\alpha]_D^{25}$ −73.9° (c, 1 in dioxan).

Roush, W.R. *et al, J. Am. Chem. Soc.*, 1988, **110**, 3979 (*synth, pmr, ir, use*)

1,1-Dibromo-2,2-bis(chloromethyl) cyclopropane   D-80079

[98577-44-7]

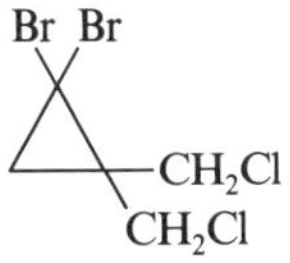

$C_5H_6Br_2Cl_2$ M 296.816

Cryst. (Et$_2$O). Mp 45-46°.

Belzner, J. *et al, Chem. Ber.*, 1989, **122**, 397 (*synth, pmr, cmr, ms*)

2,4-Dibromobutanoic acid   D-80080

[63164-16-9]

$$BrCH_2CH_2CHBrCOOH$$

$C_4H_6Br_2O_2$ M 245.898

(±)-*form*

Bp$_{15}$ 130-131°.

Chloride: [82820-87-9].
$C_4H_5Br_2ClO$ M 264.344
Bp$_1$ 60-75°.

Amide: [54882-37-0].
$C_4H_7Br_2NO$ M 244.913
Solid.

Nicolet, B.H. *et al, J. Am. Chem. Soc.*, 1927, **49**, 2066 (*synth*)
U.S. Pat., 4 247 468, (1981); *CA*, **94**, 174376h.
Ikuta, H. *et al, J. Med. Chem.*, 1987, **30**, 1995 (*chloride, amide, pmr*)

2,10-Dibromo-3-chloro-7-chamigren-9-ol   D-80081

[73494-22-1]

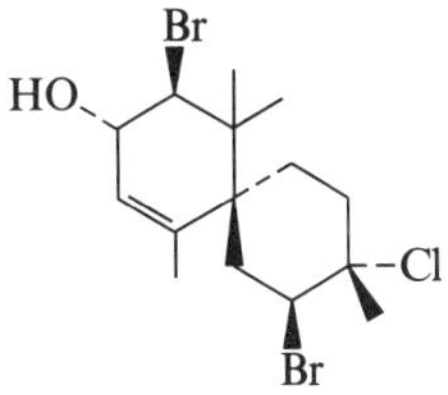

$C_{15}H_{23}Br_2ClO$ M 414.607

Metab. of *Laurencia nipponica*. Cryst. (diisopropyl ether). Mp 119-120°. $[\alpha]_D^{19}$ +25.3° (c, 1.39 in CHCl$_3$).

Suzuki, M. *et al, Bull. Chem. Soc. Jpn.*, 1988, **61**, 3371.

1,1-Dibromo-2-chlorocyclopropane   D-80082

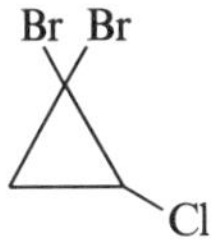

$C_3H_3Br_2Cl$ M 234.317

(±)-*form* [77628-95-6]

Bp$_{0.001}$ 80°.

Weber, A. *et al, Helv. Chim. Acta*, 1988, **71**, 2026 (*synth, ir, pmr, ms*)

4,10-Dibromo-3-chloro-7,8-epoxy-9-hydroxy-α-chamigrane   D-80083

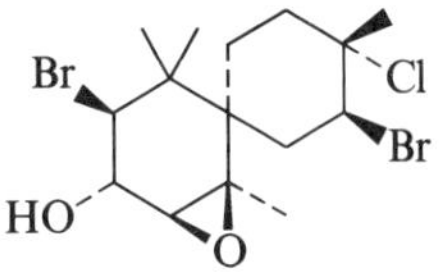

$C_{15}H_{23}Br_2ClO_2$ M 430.606

Constit. of *Laurencia glomerata*. Gum. $[\alpha]_D^{25}$ +32.7° (c, 0.91 in CHCl$_3$).

Elsworth, J.F. *et al, J. Nat. Prod. (Lloydia)*, 1989, **52**, 893 (*isol, pmr, cmr*)

1,1-Dibromo-1-chloro-2,2,2-trifluoroethane, 9CI   D-80084

[754-17-6]

$$F_3CCBr_2Cl$$

$C_2Br_2ClF_3$ M 276.278

Shows fire-extinguishing props. Mp 42-45°. Bp 92-95°.

▷ Irritant.

[29256-79-9]

Paleta, O. *et al, Collect. Czech. Chem. Commun.*, 1970, **35**, 1302 (*synth*)
Naae, D.G. *et al, Org. Mass Spectrom.*, 1974, **9**, 1203 (*ms*)
Okuhara, K., *J. Org. Chem.*, 1978, **43**, 2745 (*synth, F-19 nmr*)

1,2-Dibromo-1-chloro-1,2,2-trifluoroethane, 9CI   D-80085

2-Chloro-1,2-dibromo-1,1,2-trifluoroethane. Freon 113B2

[354-51-8]

$$F_2CBrCBrClF$$

$C_2Br_2ClF_3$ M 276.278

Refrigerant (CFC). Aerosol propellant. d_4^{20} 2.248. Mp $-72.8°$. Bp 93-94°. n_D^{20} 1.427.

Glockler, G. *et al, J. Chem. Phys.*, 1941, **9**, 387 (*raman*)
Drysdale, J.J. *et al, J. Am. Chem. Soc.*, 1957, **79**, 319 (*synth, F-19 nmr*)
Dean, R.R. *et al, Trans. Faraday Soc.*, 1968, **64**, 1409 (*pmr, F-19 nmr*)
Norris, R.D. *et al, J. Am. Chem. Soc.*, 1973, **95**, 182 (*F-19 nmr*)
Naae, D.G. *et al, Org. Mass Spectrom.*, 1974, **9**, 1203 (*ms*)
Majer, V. *et al, Collect. Czech. Chem. Commun.*, 1981, **46**, 817 (*props*)
Kushida, K. *et al, Proton and Fluorine Nuclear Magnetic Resonance Spectral Data*, Japan Halon Co. Ltd, 1988, 084 (*F-19 nmr*)

4,8-Dibromo-3,7-dichloro-3,7-dimethyl-1,5-octadiene   D-80086

[119903-44-5]

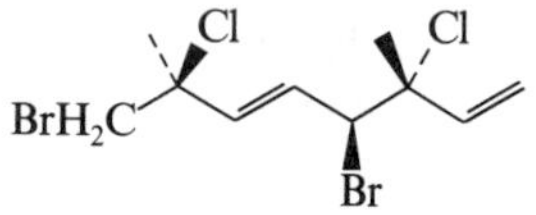

$C_{10}H_{14}Br_2Cl_2$ M 364.934

Constit. of *Plocamium hamatum*. Oil. $[\alpha]_D$ $-7°$ (c, 0.006 in $CHCl_3$).

Coll, J.C. *et al, Aust. J. Chem.*, 1988, **41**, 1743.

1,2-Dibromo-4,5-difluorobenzene, 9CI   D-80087

[64695-78-9]

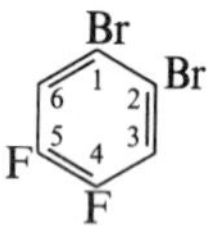

$C_6H_2Br_2F_2$ M 271.887
Cryst. (Et$_2$O). Mp 32.5-33.5°. Readily subl.

[101051-60-9]

Cervena, I. *et al, Collect. Czech. Chem. Commun.*, 1977, **42**, 2001 (*synth*)

1,4-Dibromo-2,5-difluorobenzene, 9CI   D-80088

[327-51-5]
$C_6H_2Br_2F_2$ M 271.887
Cryst. (EtOH). Mp 64.5-65.5°. Bp_{20} 96°. Subl. slowly on standing.

Finger, G.C. *et al, J. Am. Chem. Soc.*, 1951, **73**, 145 (*synth*)
Gatowsky, H.S. *et al, J. Am. Chem. Soc.*, 1952, **74**, 4809 (*F-19 nmr*)
Wasylishen, R. *et al, Can. J. Chem.*, 1970, **48**, 2885 (*pmr*)
Anderson, J.M. *et al, J. Magn. Reson.*, 1981, **44**, 1 (*pmr*)

2,2-Dibromo-2,3-dihydro-1,1,3,3-tetramethyl-1H-indene, 9CI   D-80089

2,2-Dibromo-1,1,3,3-tetramethylindane
[118617-97-3]

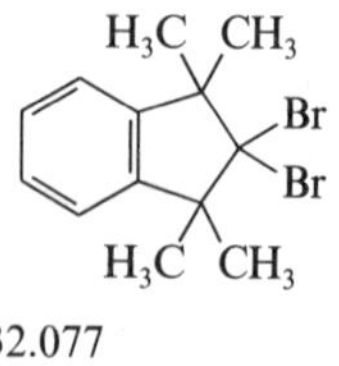

$C_{13}H_{16}Br_2$ M 332.077
Cryst. Mp 43-44°.

Guziec, F.S. *et al, Synthesis*, 1988, 547 (*synth, ir, pmr, cmr*)

4,5-Dibromo-3,6-diiodo-1,2-benzenediol, 9CI   D-80090

4,5-Dibromo-3,6-diiodocatechol

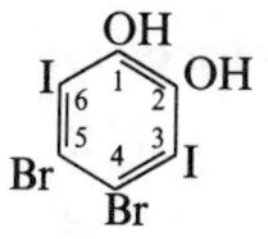

$C_6H_2Br_2I_2O_2$ M 519.698
Di-Me ether: [120231-43-8]. *1,2-Dibromo-3,6-diiodo-4,5-dimethoxybenzene. 4,5-Dibromo-3,6-diiodoveratrole*
$C_8H_6Br_2I_2O_2$ M 547.751
Needles (CHCl$_3$/MeOH). Mp 144-145°.

Meador, M.A. *et al, J. Org. Chem.*, 1989, **54**, 2336 (*synth, ms*)

2,2′-Dibromo-3,5′-dinitrobiphenyl   D-80091

[87682-47-1]

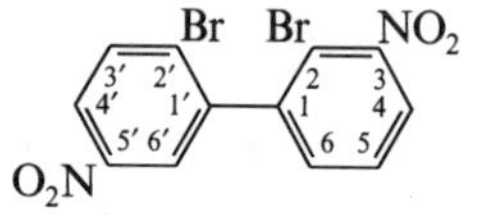

$C_{12}H_6Br_2N_2O_4$ M 401.998
Yellow prisms (MeOH). Mp 124-125°.

Baker, W. *et al, J. Chem. Soc.*, 1960, 414 (*synth*)
Dougherty, T.K. *et al, J. Org. Chem.*, 1983, **48**, 5273 (*synth*)

2,2′-Dibromo-5,5′-dinitrobiphenyl   D-80092

[52026-22-9]
$C_{12}H_6Br_2N_2O_4$ M 401.998
Pale-yellow prisms (EtOH/Me$_2$CO). Mp 220-221°.

Baker, W. *et al, J. Chem. Soc.*, 1960, 414 (*synth*)
Dougherty, T.K. *et al, J. Org. Chem.*, 1983, **48**, 5273 (*synth, ir, pmr*)

5,5′-Dibromo-2,2′-dinitrobiphenyl   D-80093

$C_{12}H_6Br_2N_2O_4$ M 401.998
Yellow prisms (EtOAc). Mp 204-205°.

Murakami, M. *et al, Nippon Kagaku Kaishi (J. Chem. Soc. Jpn.)*, 1949, **70**, 236 (*synth*)
Forrest, J., *J. Chem. Soc.*, 1960, 594.

1,3-Dibromo-1,1,2,2,3,3-hexafluoro-propane, 9CI   D-80094

[4259-29-4]

$$F_2CBrCF_2CF_2Br$$

$C_3Br_2F_6$ M 309.831
Liq. d_4^{20} 2.172. Bp 74.2°. n_D^{20} 1.3590, $n_D^{30.2}$ 1.3526.

Hauptschein, M. *et al, J. Am. Chem. Soc.*, 1952, **74**, 848 (*synth*)
Nodiff, E.A. *et al, J. Org. Chem.*, 1953, **18**, 235 (*props*)

1,6-Dibromo-3,4-hexanediol, 9CI   D-80095

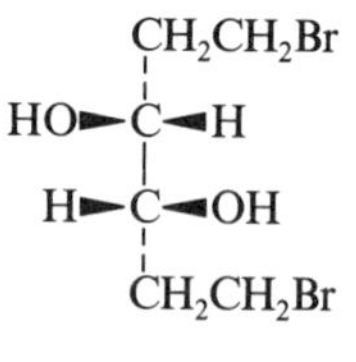

(3R,4R)-form

$C_6H_{12}Br_2O_2$ M 275.968
(3R,4R)-form [82188-39-4]

Mp 90-91°. $[\alpha]_D^{27}$ +58.4° (c, 2.16 in $CHCl_3$).

Bis-4-methylbenzenesulfonyl: Plates (EtOH aq.). Mp 139.5-141°. $[\alpha]_D^{25}$ +74.3° (c, 3.2 in $CHCl_3$).

[53399-76-1]

Cope, A.C. *et al, J. Am. Chem. Soc.,* 1956, **78**, 5916 (*synth*)
Cerè, V. *et al, J. Org. Chem.,* 1988, **53**, 5689 (*synth, cmr*)

4,5-Dibromo-6-methyl-1,2,3-triazine   D-80096

[114078-91-0]

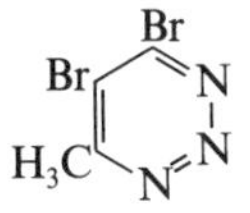

$C_4H_3Br_2N_3$ M 252.896
Granules (hexane/Et_2O). Mp 130-132° dec.

Ohsawa, A. *et al, Chem. Pharm. Bull.,* 1988, **36**, 3838 (*synth, pmr*)

4,5-Dibromo-6-phenyl-1,2,3-triazine   D-80097

[114078-93-2]

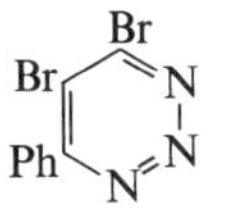

$C_9H_5Br_2N_3$ M 314.967
Needles (hexane/Et_2O). Mp 110-112° dec.

Ohsawa, A. *et al, Chem. Pharm. Bull.,* 1988, **36**, 3838 (*synth, pmr*)

4,9-Dibromopyrene   D-80098

$C_{16}H_8Br_2$ M 360.047
Cryst. ($CHCl_3$). Mp 247°.

Blatter, K. *et al, Synthesis,* 1989, 356 (*synth, ir, pmr, cmr, ms*)

2,6-Dibromo-4(1*H*)-pyridone   D-80099

2,6-Dibromo-4-hydroxypyridine
$C_5H_5Br_2NO$ M 254.909

Me ether: [117873-72-0]. *2,6-Dibromo-4-methoxypyridine*
  $C_6H_5Br_2NO$ M 266.920
  Needles. Mp 136°.

Den Hertog, H.J., *Recl. Trav. Chim. Pays-Bas* (*J. R. Neth. Chem. Soc.*), 1948, **67**, 381.
Neumann, U. *et al, Chem. Ber.,* 1989, **122**, 589.

4,6-Dibromo-2(1*H*)pyridone   D-80100

2,4-Dibromo-6-hydroxypyridine
$C_5H_3Br_2NO$ M 252.893

Me ether: [117873-73-1]. *2,4-Dibromo-6-methoxypyridine*
  $C_6H_5Br_2NO$ M 266.920
  Cryst. Mp 92°.

Neumann, U. *et al, Chem. Ber.,* 1989, **122**, 589.

2,2-Dibromo-1,1,3,3-tetramethylcyclo-hexane, 9CI   D-80101

[118617-98-4]

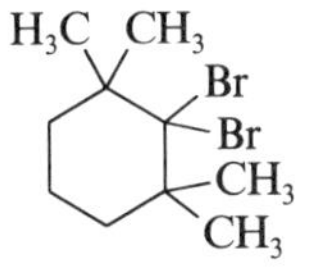

$C_{10}H_{18}Br_2$ M 298.060
Cryst. Mp 185°.

Guziec, F.S. *et al, Synthesis,* 1988, 547 (*synth, ir, pmr, cmr*)

2,2-Dibromo-1,1,3,3-tetramethylcyclo-pentane, 9CI   D-80102

[118617-99-5]

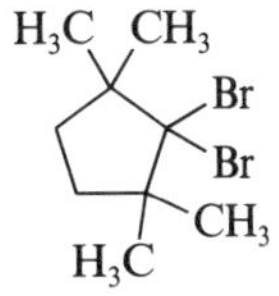

$C_9H_{16}Br_2$ M 284.033
Cryst. Mp 159-161°. Unstable, must be kept in dark and cold.

Guziec, F.S. *et al, Synthesis,* 1988, 547 (*synth, ir, pmr, cmr*)

1,1-Dibromo-2,2,3,3-tetramethylcyclo-propane, 9CI   D-80103

[22715-57-7]

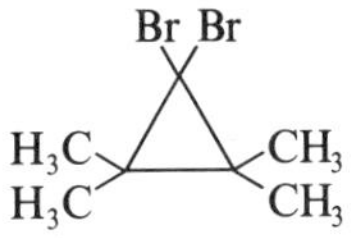

$C_7H_{12}Br_2$ M 255.980
Plates (EtOH aq.). Mp 78-80°.

Hackett, M. *et al, J. Am. Chem. Soc.,* 1988, **110**, 1449 (*synth, pmr*)

3,3-Dibromo-2,2,4,4-tetramethylpentane, 9CI   D-80104

Dibromodi-tert-butylmethane
[84679-81-2]

$$[(H_3C)_3C]_2CBr_2$$

$C_9H_{18}Br_2$ M 286.049
Cryst. by subl. Mp 183-184° (175°).

Kalinowski, H. *et al, Org. Magn. Reson.,* 1983, **21**, 64 (*synth, cmr, pmr*)
Guziec, F.S. *et al, Synthesis,* 1988, 547 (*synth*)

4,5-Dibromo-1,2,3-triazine   D-80105

[114078-89-6]

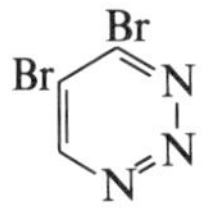

$C_3HBr_2N_3$ M 238.869
Needles (hexane/Et_2O). Mp 138-139° dec.

Ohsawa, A. *et al, Chem. Pharm. Bull.,* 1988, **36**, 3838 (*synth, pmr*)

2,2-Dibromo-1,1,1-trifluoroethane, 9CI   D-80106

[354-30-3]

$$F_3CCHBr_2$$

$C_2HBr_2F_3$ M 241.833
Shows anaesthetic activity, component of fire extinguishers. d_4^{24} 2.22. Bp 73°. n_D^{26} 1.4029.

[108662-84-6]

Robbins, B.H., *J. Pharmacol.,* 1946, **86**, 197 (*use*)

McBee, E.T. *et al*, *Ind. Eng. Chem.*, 1947, **39**, 420 (*synth*)
Kushida, K. *et al*, *Proton and Fluorine Nuclear Magnetic Resonance Spectral Data*, Japan Halon Co Ltd, Japan, 1980, 030, 031 (*F-19 nmr, pmr*)

1,11-Dibromo-6-undecanone, 9CI　　　D-80107

[116288-13-2]

$$BrCH_2(CH_2)_4CO(CH_2)_4CH_2Br$$

$C_{11}H_{20}Br_2O$　　M 328.086
Oil. Bp_2 153°.

Momenteau, M. *et al*, *J. Chem. Soc., Perkin Trans. 1*, 1988, 283 (*synth, pmr*)

2,3-Di-*tert*-butyl-1,3-butadiene　　　D-80108

Updated Entry replacing D-02189
2,2,5,5-Tetramethyl-3,4-bis(methylene)hexane, 9CI
[3378-20-9]

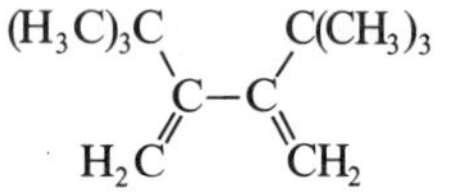

$C_{12}H_{22}$　　M 166.306
Liq. Bp_{12} 58°.

Lenoir, D., *Chem. Ber.*, 1978, **111**, 411 (*synth, ir, pmr*)
Roth, W.R. *et al*, *J. Am. Chem. Soc.*, 1988, **110**, 1883 (*props*)

2,6-Di-*tert*-butyl-1-nitronaphthalene　　　D-80109

2,6-Bis(1,1-dimethylethyl)-1-nitronaphthalene
[20870-30-8]

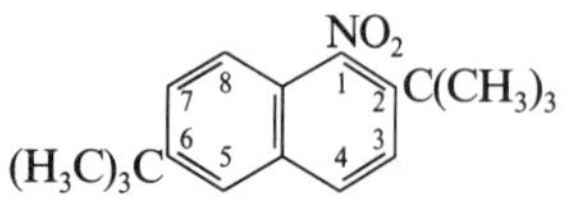

$C_{18}H_{23}NO_3$　　M 301.385
Light yellow needles (EtOH). Mp 80-81°. Light-sensitive.

Döpp, D. *et al*, *Chem. Ber.*, 1988, **121**, 2045 (*synth, pmr, cmr, uv, ms*)

3,7-Di-*tert*-butyl-1-nitronaphthalene　　　D-80110

3,7-Bis(1,1-dimethylethyl)-1-nitronaphthalene
[20820-31-9]
$C_{18}H_{23}NO_2$　　M 285.385
Yellow needles (EtOH). Mp 64-65°. Light-sensitive.

Döpp, D. *et al*, *Chem. Ber.*, 1988, **121**, 2045 (*synth, pmr, cmr, uv, ms*)

2,6-Di-*tert*-butylpiperidine　　　D-80111

2,6-Bis(1,1-dimethylethyl)piperidine, 9CI

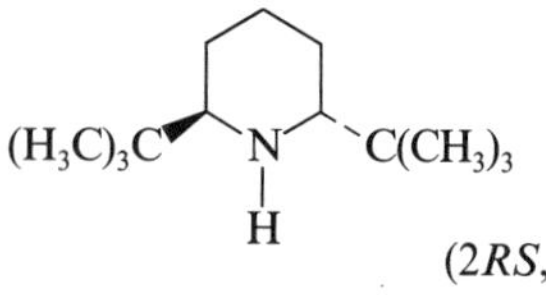

(*2RS,6RS*)-*form*

$C_{13}H_{27}N$　　M 197.363
(**2RS,6RS**)-**form** [101166-52-3]
　(±)-trans-form
Picrate: Mp 165.5-166°.
(**2RS,6SR**)-**form** [66922-18-7]
　cis-form
　Bp_{10} 95-97°.
Picrate: Mp 224-225°.

Day, J.C., *J. Org. Chem.*, 1978, **43**, 3646 (*synth, pmr, cmr, N-15 nmr*)
Francis, R.F. *et al*, *J. Org. Chem.*, 1986, **51**, 1889 (*synth, pmr, cmr*)

2,4-Di-*tert*-butylthiophene　　　D-80112

2,4-Bis(1,1-dimethylethyl)thiophene, 9CI
[33369-81-2]

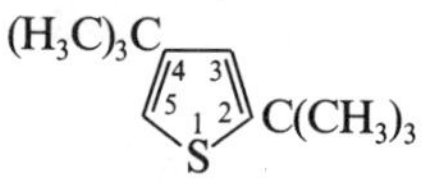

$C_{12}H_{20}S$　　M 196.356
Mp −1°. Bp_{752} 220.4°.

Wynberg, H. *et al*, *J. Org. Chem.*, 1965, **30**, 1058 (*synth, uv, ir, pmr*)
Nakayama, J. *et al*, *Tetrahedron Lett.*, 1988, **29**, 1161 (*synth*)

3,4-Di-*tert*-butylthiophene　　　D-80113

Updated Entry replacing D-02263
3,4-Bis(1,1-dimethylethyl)thiophene, 9CI
[22808-03-3]
$C_{12}H_{20}S$　　M 196.356
Mp 43-43.5°. Bp_4 75° (bulb) . Highly strained.
1,1-Dioxide: [116375-45-2].
　　$C_{12}H_{20}O_2S$　　M 228.355
　　Cryst. Mp 132.5-133°.

Brandsma, L. *et al*, *J. Chem. Soc., Chem. Commun.*, 1980, 922 (*synth, cryst struct, pmr*)
Nakayama, J. *et al*, *J. Am. Chem. Soc.*, 1988, **110**, 6598 (*use*)
Nakayama, J. *et al*, *Tetrahedron Lett.*, 1988, **29**, 1161 (*dioxide*)

Dicarbon dioxide　　　D-80114

Ethenedione, 9CI
[4363-38-6]

$$OC{=}CO$$

C_2O_2　　M 56.021
Not known. Calcns. suggest extreme instability.

Hoffman, R.W. *et al*, *Tetrahedron*, 1965, **21**, 891.
Herberhold, M. *et al*, *Z. Naturforsch., B*, 1976, **31**, 35.

6-(1,2-Dicarboxyethylamino)purine　　　D-80115

N-1H-Purin-6-ylaspartic acid, 9CI
[26511-42-2]

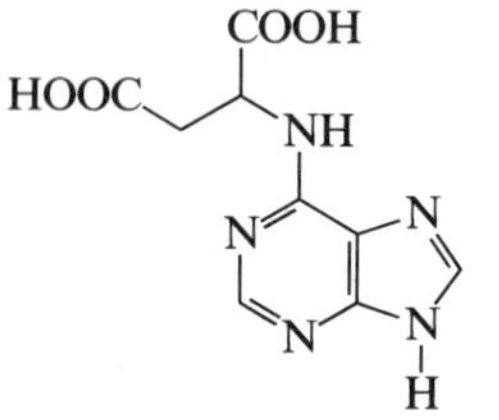

$C_9H_9N_5O_4$　　M 251.201
Isol. from *Fusarium* sp.

[26287-66-1]

Ballio, A. *et al*, *Gazz. Chim. Ital.*, 1964, **94**, 156 (*isol*)
McCloskey, J.A. *et al*, *Biomed. Mass Spectrom.*, 1975, **2**, 90 (*ms*)

1,1-Dichloroethylene, 8CI   D-80116

1,1-Dichloroethene, 9CI. Vinylidene chloride

[75-35-4]

$$H_2C=CCl_2$$

$C_2H_2Cl_2$ M 96.943

Intermediate in production of plastics as comonomer with vinyl chloride, acrylonitrile, acrylates etc. Reacts with alcohols and halides to give carboxylic acids. Volatile liq. d^{20} 1.218. Fp $-122.5°$. Bp 31.7°, Bp °(37°) . n_D^{20} 1.4249. Readily polymerises.

▷ Mod. toxic by inhalation, TLV 40. Exp. carcinogen. Extremely flammable. Rapidly absorbs O_2 forming violently explosive peroxide. KV9275000.

Kunichika, S. *et al*, *Nippon Kagaku Kaishi* (*J. Chem. Soc. Jpn.*), 1950, **53**, 407; *CA*, **47**, 5871 (*synth*)
Miyajima, G. *et al*, *J. Phys. Chem.*, 1971, **75**, 331 (*cmr*)
Fieser and Fieser's Reagents for Organic Synthesis, Wiley, 1969, **2**, 118.
Bretherick, L., *Handbook of Reactive Chemical Hazards*, 2nd Ed., Butterworths, London and Boston, 1979, 354.
Sax, N.I., *Dangerous Properties of Industrial Materials*, 5th Ed., Van Nostrand-Reinhold, 1979, 1088.
Hazards in the Chemical Laboratory, (Bretherick, L., Ed.), 3rd Ed., Royal Society of Chemistry, London, 1981, 279.

1,1-Dichloro-1-fluoroethane, 9CI   D-80117

1-Fluoro-1,1-dichloroethane. HCFC 141B

[1717-00-6]

$$H_3CCCl_2F$$

$C_2H_3Cl_2F$ M 116.950

Alternative refrigerant. Shows anaesthetic props. d_4^{10} 1.250. Fp $-103.5°$. Bp 32°. n_D^{10} 1.360.

Brown, J.H. *et al*, *J. Soc. Chem. Ind.*, *London*, 1948, **67**, 331 (*synth*)
Smith, D.C. *et al*, *J. Chem. Phys.*, 1952, **20**, 473 (*ir, raman*)
Durig, J.R. *et al*, *Spectrochim. Acta, Part A*, 1976, **32**, 175 (*ir, raman*)

1,1-Dichloro-2-fluoroethene, 9CI   D-80118

[359-02-4]

$$Cl_2C=CHF$$

C_2HCl_2F M 114.934

d_4^{20} 1.383. Mp $-108.8°$. Bp 37-39°. n_D^{20} 1.4036.

Henne, A.L. *et al*, *J. Am. Chem. Soc.*, 1936, **58**, 402 (*synth*)
Yanwood, J., *J. Chem. Soc.*, 1965, 7481 (*ir, raman*)
Craig, N.C., *J. Mol. Spectrosc.*, 1967, **23**, 307 (*ir, raman*)
Moehlmann, J.G. *et al*, *J. Chem. Phys.*, 1975, **62**, 3052 (*synth, ir*)

2,2-Dichloro-6-heptenoic acid, 9CI   D-80119

[105764-09-8]

$$H_2C=CH(CH_2)_3CCl_2COOH$$

$C_7H_{10}Cl_2O_2$ M 197.060

$Bp_{<0.1}$ 60-65° (bulb) .

Et ester: [105764-08-7].
   $C_9H_{14}Cl_2O_2$ M 225.114
   Liq. $Bp_{0.5}$ 80-85°.

Hayes, T.K. *et al*, *J. Am. Chem. Soc.*, 1988, **110**, 5533 (*synth, pmr, ir, ms, cmr*)

1,2-Dichloro-1,1,2,3,3,3-hexafluoro-propane, 9CI   D-80120

[661-97-2]

$$F_3CCClFCClF_2$$

$C_3Cl_2F_6$ M 220.929

Refrigerant. Solv. in the manuf. of polyethers and for disperse dyes. d_4^{20} 1.590, d_4^{20} 1.598. Fp $-136°$ ($-131.1°$). Bp 34.5°. n_D^{20} 1.3029, n_D^0 1.3110.

Park, J.D. *et al*, *J. Org. Chem.*, 1961, **26**, 4017 (*synth*)
Fainberg, A.H. *et al*, *J. Org. Chem.*, 1965, **30**, 864 (*synth, props*)
Schmeisser, M. *et al*, *Angew. Chem., Int. Ed. Engl.*, 1967, **6**, 627 . (*synth*)
De Marco, A. *et al*, *J. Magn. Reson.*, 1972, **6**, 208 (*F nmr*)
Capriel, P. *et al*, *Tetrahedron*, 1979, **35**, 2661 (*F nmr*)

1,3-Dichloro-1,1,2,2,3,3-hexafluoro-propane, 9CI   D-80121

Freon 216

[662-01-1]

$$F_2CClCF_2CClF_2$$

$C_3Cl_2F_6$ M 220.929

Refrigerant, working fluid for rankine cycle electric power systems, shows anaesthetic props. d_4^{20} 1.573. Fp $-125.3°$ ($-193.7°$). Bp 33.8° (36.1°) . $n_D^{22.2}$ 1.3022, n_D^{20} 1.3030.

Hauptschein, M. *et al*, *J. Am. Chem. Soc.*, 1952, **74**, 848 (*synth*)
Nodiff, E.A. *et al*, *J. Org. Chem.*, 1953, **18**, 235 (*props*)
Angell, C.L., *J. Chem. Eng. Data*, 1964, **9**, 341 (*ir, raman*)
Fainberg, A.H. *et al*, *J. Org. Chem.*, 1965, **30**, 864 (*props*)
White, H.F., *Anal. Chem.*, 1966, **38**, 625 (*F nmr*)
Schmeisser, M. *et al*, *Angew. Chem., Int. Ed. Engl.*, 1967, **6**, 627 (*synth*)
Fernholt, L. *et al*, *Acta Chem. Scand., Ser. A*, 1978, **32**, 225 (*ed*)

2,2-Dichloro-1,1,1,3,3,3-hexafluoro-propane, 9CI   D-80122

[1652-80-8]

$$F_3CCCl_2CF_3$$

$C_3Cl_2F_6$ M 220.929

Solv. for disperse dyes. Shows anaesthetic props. Mp 3° (-10 to $-13°$). Bp 32.5-33°. n_D^{20} 1.3050 (1.3032).

Maynard, J.T., *J. Org. Chem.*, 1963, **28**, 112 (*synth, ir*)
Farah, B.S. *et al*, *J. Org. Chem.*, 1965, **30**, 1241 (*synth*)
Weigert, F.J. *et al*, *J. Am. Chem. Soc.*, 1972, **94**, 5314 (*F nmr*)
Naae, D.G. *et al*, *Org. Mass Spectrom.*, 1974, **9**, 1203 (*ms*)
Buerger, H. *et al*, *Spectrochim. Acta, Part A*, 1979, **35**, 525 (*ir, raman*)
Eujen, R. *et al*, *J. Fluorine Chem.*, 1983, **22**, 263 (*F nmr*)

2,2-Dichloro-7-octenoic acid   D-80123

[105764-15-6]

$$H_2C=CH(CH_2)_4CCl_2COOH$$

$C_8H_{12}Cl_2O_2$ M 211.087

$Bp_{<0.1}$ 90-95° (bulb) .

Et ester: [105764-14-5].
   $C_{10}H_{16}Cl_2O_2$ M 239.141
   $Bp_{<0.1}$ 84-88°.

Amide: [115119-89-6].
   $C_8H_{13}Cl_2NO$ M 210.102
   Pale yellow solid.

Nitrile: [115119-91-0].
   $C_8H_{11}Cl_2N$ M 192.087

Oil.

Hayes, T.K. *et al, J. Am. Chem. Soc.*, 1988, **110**, 5533 (*synth, derivs, pmr, ir, ms, cmr*)

1,5-Dichloro-3-pentanol   D-80124

[72548-31-3]

$$ClCH_2CH_2CH(OH)CH_2CH_2Cl$$

$C_5H_{10}Cl_2O$ M 157.039
Oil. $Bp_{0.1}$ 64-70°.

Reese, C.B. *et al, J. Chem. Soc., Perkin Trans. 1*, 1988, 2881 (*synth, pmr, cmr*)

1,5-Dichloro-3-pentanone, 9CI   D-80125

Di(2-chloroethyl) ketone. 2,2'-Dichlorodiethyl ketone
[3592-25-4]

$$ClCH_2CH_2COCH_2CH_2Cl$$

$C_5H_8Cl_2O$ M 155.023
Liq. $Bp_{0.8}$ 78°. Stable at 0°, polym. at r.t.

Cardwell, H.M.E. *et al, J. Chem. Soc.*, 1949, 708 (*synth*)
Bowden, K. *et al, J. Chem. Soc.*, 1952, 1164 (*synth*)
Arbuzov, Y.A. *et al, Zh. Obshch. Khim.*, 1962, **32**, 3681; (*synth*)
Owen, G.R. *et al, J. Chem. Soc. C*, 1970, 2401 (*synth, ir, pmr, ms*)

2,5-Dichloro-1-pentene   D-80126

[58568-26-6]

$$H_2C=CClCH_2CH_2CH_2Cl$$

$C_5H_8Cl_2$ M 139.024
Oil. Bp_{20} 58-60°.

Wender, P.A. *et al, J. Am. Chem. Soc.*, 1988, **110**, 2218 (*synth, pmr, ir, ms*)

2,3-Dichloro-2-propenal   D-80127

α,β-Dichloroacrolein
[26910-68-9]

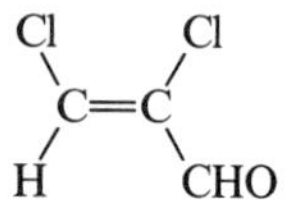

$C_3H_2Cl_2O$ M 124.954
(Z)-form [6695-22-3]
d_4^{20} 1.390. Bp_9 44°. n_D^{25} 1.5110. The *E*-form could not be obt.

2,4-Dinitrophenylhydrazone: Deep red cryst. (EtOH). Mp 244° dec. (235°).

[99414-75-2, 99569-10-5]

Tobey, W. *et al, J. Am. Chem. Soc.*, 1966, **88**, 2478 (*synth, ir, pmr*)
Annenkova, V. *et al, J. Org. Chem. USSR (Engl. Transl.)*, 1970, 223 (*synth*)
Wille, F. *et al, Z. Naturforsch., B*, 1977, 733 (*synth, pmr*)
Roedig, A. *et al, Justus Liebigs Ann. Chem.*, 1979, 194 (*synth, ir, pmr, config*)
Rosen, J.D. *et al, Mutat. Res.*, 1980, **78**, 113 (*synth, pmr, ms*)
Ruzo, L.O. *et al, J. Agric. Food Chem.*, 1985, **33**, 272 (*pmr*)
Segall, Y. *et al, Mutat. Res.*, 1985, **158**, 61 (*tox*)
Bostroem, G.O. *et al, J. Mol. Struct.*, 1988, **172**, 227 (*struct*)

3,3-Dichloro-2-propenal, 9CI   D-80128

β,β-Dichloroacrolein

$$Cl_2C=CHCHO$$

$C_3H_2Cl_2O$ M 124.954

Pale yellow liq. Sol. H_2O. Bp 124°, Bp_{21} 38-39°. n_D^{24} 1.5062.
▷ Lachrymator.

2,4-Dinitrophenylhydrazone: Mp 164-165°.

Kharasch, M.S. *et al, J. Am. Chem. Soc.*, 1947, **69**, 1105 (*synth, deriv*)
Soulen, R.L. *et al, J. Org. Chem.*, 1971, **36**, 3386 (*synth*)
Pochat, F. *et al, Bull. Soc. Chim. Fr.*, 1972, 3145 (*ir*)
Ellern, J.B. *et al, J. Org. Chem.*, 1972, **37**, 4485 (*synth, deriv, ir*)
Coleman, W.E. *et al, Environ. Sci. Technol.*, 1984, **18**, 674 (*glc, ms*)
Hanack, M. *et al, Synthesis*, 1987, 944 (*synth*)

5,6-Dichloro-2,3-pyrazinedicarboxylic acid   D-80129

[59715-45-6]

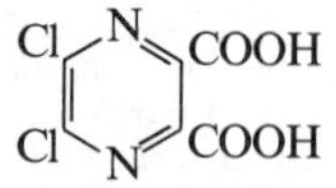

$C_6H_2Cl_2N_2O_4$ M 236.998
Dinitrile: [56413-95-7]. *2,3-Dichloro-5,6-dicyanopyrazine*
    $C_6Cl_2N_4$ M 198.998
Readily accessible intermed. for polycyclic heteroaromatics. Needles ($CHCl_3$). Mp 179-180°.

Suzuki, T. *et al, J. Heterocycl. Chem.*, 1986, **23**, 1419.
Ried, W. *et al, Justus Liebigs Ann. Chem.*, 1988, 1197 (*use*)

2,6-Dichloro-4(3H)-pyrimidinone   D-80130

[120977-94-8]

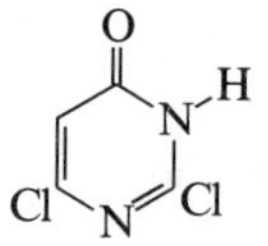

$C_4H_2Cl_2N_2O$ M 164.978
Cryst. (H_2O). Mp 170-171°. Reg. no. refers to the (prob. unfavoured) 1*H*-form.

Hübsch, W. *et al, Helv. Chim. Acta*, 1989, **72**, 738.

4,6-Dichloro-2(1H)-pyrimidinone   D-80131

4,6-Dichloro-2-hydroxypyrimidine
[6297-80-9]

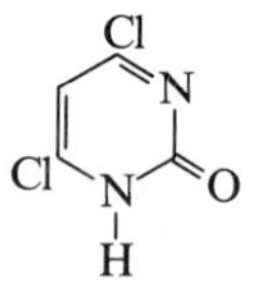

$C_4H_2Cl_2N_2O$ M 164.978
Cryst. powder. Mp 157°. Higher Mps previously reported are said to indicate a high content of the Na salt.

Hübsch, W. *et al, Helv. Chim. Acta*, 1989, **72**, 738.

1,2-Dichloro-3,3,4,4-tetrafluorocyclo-   D-80132
butene, 9CI

[377-93-5]

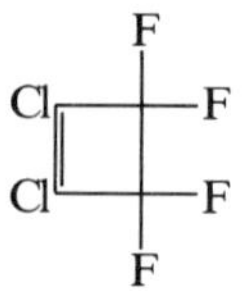

$C_4Cl_2F_4$ M 194.943
d_4^{25} 1.534. Mp −43.4° (−48°). Bp 67.1°. n_D^{25} 1.3699.

Henne, A.L. *et al, J. Am. Chem. Soc.*, 1947, **69**, 281 (*synth, props*)

Harris, R.K. *et al*, *J. Magn. Reson.*, 1969, **1**, 362 (*F-19 nmr*)
King, B.R. *et al*, *J. Fluorine Chem.*, 1972, **1**, 283 (*synth*)
Chia, L.S. *et al*, *Can. J. Chem.*, 1974, **52**, 3484 (*Cl-nqr*)
Bauer, G. *et al*, *Z. Naturforsch.*, *B*, 1979, **34**, 1249 (*synth*)

3,4-Dichloro-1,2,3,4-tetrafluorocyclo- butene, 9CI   D-80133

[425-62-7]

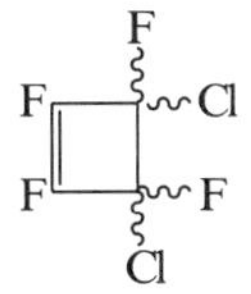

$C_4Cl_2F_4$ M 194.943
Bp 65.5°. n_D^{20} 1.375. No stereoisomers reported.

Haszeldine, R.N. *et al*, *J. Chem. Soc.*, 1955, 3880 (*synth*)
Fainberg, A.H. *et al*, *J. Org. Chem.*, 1965, **30**, 864.

1,1-Dichloro-1,2,2,2-tetrafluoroethane   D-80134

[374-07-2]

$$F_3CCFCl_2$$

$C_2Cl_2F_4$ M 170.921
Refrigerant component. d_4^{25} 1.455. Fp −56.6°, Mp −94°.
Bp 3-4°. n_D^0 1.3092.

▷ Irritant.

[1320-37-2]

Nielsen, J.R. *et al*, *J. Chem. Phys.*, 1953, **21**, 383 (*ir, raman*)
Goldwhite, H. *et al*, *J. Chem. Soc.*, 1961, 3825 (*synth, ir*)
Vecchio, M. *et al*, *J. Fluorine Chem.*, 1974, **4**, 117 (*synth*)
Naae, D.G. *et al*, *Org. Mass Spectrom.*, 1974, **9**, 1203 (*ms*)
Eujen, R. *et al*, *J. Fluorine Chem.*, 1983, **22**, 263 (*F-19 nmr*)

1,1-Dichloro-2,3,3,3-tetrafluoro-1-propene, 9CI   D-80135

[2804-55-9]

$$Cl_2C{=}CFCF_3$$

$C_3Cl_2F_4$ M 182.932
Fumigant. d_4^{20} 1.539. Fp −139.6°. Bp 45-46.5° (43.5°) . n_D^{20} 1.3504.

[111548-56-2]

Fainberg, A.H. *et al*, *J. Org. Chem.*, 1965, **30**, 864 (*props*)
Paleta, O. *et al*, *Collect. Czech. Chem. Commun.*, 1971, **36**, 2257 (*synth*)
Paleta, O. *et al*, *Bull. Soc. Chim. Fr.*, 1986, 920 (*synth, F nmr*)

1,2-Dichloro-1,3,3,3-tetrafluoro-1-propene, 9CI   D-80136

[431-53-8]

$$F_3CCCl{=}CFCl$$

$C_3Cl_2F_4$ M 182.932
Fumigant. Fp −137°. Bp 47.3°. Bp refers to mixt. of stereoisomers.

[73562-83-1, 73562-84-2, 111548-56-2]

Henne, A.L. *et al*, *J. Am. Chem. Soc.*, 1946, **68**, 496 (*synth*)
McConnell, H.M. *et al*, *J. Chem. Phys.*, 1956, **25**, 184 (*F nmr*)
Swalden, J.D. *et al*, *J. Chem. Phys.*, 1961, **34**, 2122 (*F nmr*)

3,3-Dichloro-2,2,4,4-tetramethylpentane, 9CI   D-80137

Di-tert-*butyldichloromethane*

[79991-69-8]

$$[(H_3C)_3C]_2CCl_2$$

$C_9H_{18}Cl_2$ M 161.694
Cryst. by subl. Mp 103-104°.

Kalinowski, H. *et al*, *Org. Magn. Reson.*, 1983, **21**, 64 (*synth, cmr, pmr*)

1,4-Dichloro-3,3,4-trifluorocyclobutene, 9CI   D-80138

[2927-72-2]

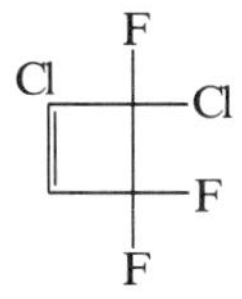

$C_4HCl_2F_3$ M 176.953
Bp 91-92°. n_D^{25} 1.3942.

Raasch, M.S. *et al*, *J. Am. Chem. Soc.*, 1959, **81**, 2678 (*synth*)
Park, J.D. *et al*, *J. Org. Chem.*, 1965, **30**, 400 (*synth, ir, F-19 nmr*)
Newmark, R.A. *et al*, *J. Magn. Reson.*, 1969, **1**, 418 (*F-19 nmr*)

1,1-Dichloro-1,2,2-trifluoroethane, 9CI   D-80139

[812-04-4]

$$Cl_2CFCF_2H$$

$C_2HCl_2F_3$ M 152.931
Bp 59-61°, Bp 30.2°.

Haszeldine, R.N. *et al*, *J. Chem. Soc.*, 1960, 4503 (*synth*)
Fuller, G. *et al*, *Tetrahedron*, 1962, **18**, 123 (*synth*)

1,2-Dichloro-1,1,2-trifluoroethane, 9CI   D-80140

[354-23-4]

$$F_2CClCHClF$$

$C_2HCl_2F_3$ M 152.931
Refrigerant. Shows anaesthetic properties. Gas. $d_4^{27.4}$ 1.498. Bp 28-30°. n_D^{15} 1.3371.

▷ Irritant, convulsant.

[34077-87-7, 90454-18-5]

Glockler, G. *et al*, *J. Chem. Phys.*, 1941, **9**, 387 (*raman*)
Park, J.D. *et al*, *J. Am. Chem. Soc.*, 1951, **73**, 711 (*synth, ir*)
Fuller, G. *et al*, *Tetrahedron*, 1962, **18**, 123 (*tox*)
Doucet, J. *et al*, *J. Chem. Phys.*, 1975, **62**, 355 (*uv, pe*)
Kirk-Othmer Encycl. Chem. Technol., 3rd Ed., 1978-1984, *Wiley, NY*, 3rd Ed., Wiley, N.Y., 1978-1984, **10**, 861 (*props*)
Capnel, P. *et al*, *Tetrahedron*, 1979, **35**, 2661 (*F-19 nmr*)
Service, C.F. *et al*, *J. Fluorine Chem.*, 1982, **20**, 135 (*synth*)

2,2-Dichloro-1,1,1-trifluoroethane, 9CI   D-80141

HCFC 123. Freon 123

[306-83-2]

$$F_3CCHCl_2$$

$C_2HCl_2F_3$ M 152.931
Used in aerosol formulations, refrigerant, degreasing solvent. Shows anaesthetic props. Gas. d_4^{15} 1.475. Mp −107°. Bp 28.7°, Bp_{747} 27.1°. n_D^{15} 1.333.

McBee, E.T. *et al*, *Ind. Eng. Chem.*, 1947, **39**, 409 (*synth*)
Nielsen, J.R. *et al*, *J. Chem. Phys.*, 1953, **21**, 1060 (*ir, raman*)

Ng, S., *J. Magn. Reson.*, 1972, **7**, 370 (*pmr*)
Burns, T.H.S. *et al*, *Anaesthesia*, 1982, **37**, 278 (*use*)
Kushida, K. *et al*, *Proton and Fluorine Nuclear Magnetic Resonance Spectral Data*, Japan Halon Co., 1988, 016, 017 (*F-19 nmr, pmr*)

1,1-Dichloro-1,2,2-trifluoro-2-iodoethane, 9CI   D-80142

[661-66-5]

$$ICF_2CCl_2F$$

$C_2Cl_2F_3I$ M 278.827
Light-sensitive liq. d_4^{25} 2.189. Bp 103.1°. n_D^{25} 1.444.

▷ Lachrymator.

Chambers, R.D. *et al*, *J. Chem. Soc.*, 1961, 3779 (*ir*)
Bissell, E.R. *et al*, *J. Org. Chem.*, 1962, **27**, 1482 (*synth*)
Dean, R.R. *et al*, *Trans. Faraday Soc.*, 1968, **64**, 1409 (*F-nmr*)

1,2-Dichloro-1,1,2-trifluoro-2-iodoethane, 9CI   D-80143

[354-61-0]

$$F_2CClCClFI$$

$C_2Cl_2F_3I$ M 278.827
Liq. d_4^{25} 2.199. Bp 103.9°, Bp_{100} 43-44°. $n_D^{25.2}$ 1.448. Turns pink in light.

[57171-62-7]

Hazeldine, R.N., *J. Chem. Soc.*, 1953, 1764 (*uv*)
Hauptschein, M. *et al*, *J. Am. Chem. Soc.*, 1961, **83**, 2495 (*synth, ir*)
Bissell, E.R. *et al*, *J. Org. Chem.*, 1962, **27**, 1482 (*synth*)
Capnel, P. *et al*, *Tetrahedron*, 1979, **35**, 2661 (*F-19 nmr*)

2,2-Dichloro-1,1,1-trifluoropropane, 9CI   D-80144

[7126-01-4]

$$F_3CCCl_2CH_3$$

$C_3H_3Cl_2F_3$ M 166.957
Shows anaesthetic props. d_4^{20} 1.3842. Mp 13.8°. Bp 48.8°. n_D^{20} 1.348.

Henne, A.L. *et al*, *J. Am. Chem. Soc.*, 1942, **64**, 1157 (*synth*)
McBee, E.T. *et al*, *J. Am. Chem. Soc.*, 1947, **69**, 944 (*synth*)

2,3-Dichloro-1,1,1-trifluoropropane, 9CI   D-80145

[338-75-0]

$$F_3CCHClCH_2Cl$$

$C_3H_3Cl_2F_3$ M 166.957
Shows anaesthetic props. Bp 76-77°. n_D^{20} 1.3671.

Haszeldine, R.N. *et al*, *J. Chem. Soc.*, 1953, 1199, 3371 (*synth, ir*)
Di Paolo, T. *et al*, *J. Pharm. Sci.*, 1979, **68**, 39 (*pharmacol*)

3,3-Dichloro-1,1,1-trifluoropropane, 9CI   D-80146

[460-69-5]

$$F_3CCH_2CHCl_2$$

$C_3H_3Cl_2F_3$ M 166.957
Shows anaesthetic props. d_4^{20} 1.4408. Bp 72-74°. n_D^{20} 1.3531.

McBee, E.T. *et al*, *J. Am. Chem. Soc.*, 1947, **69**, 944 (*synth*)
Haszeldine, R.N. *et al*, *J. Chem. Soc.*, 1953, 1199 (*synth, ir*)
Varushchenko, R.M. *et al*, *Russ. J. Phys. Chem.*, 1972, **46**, 755 (*props*)
Di Paolo, T. *et al*, *J. Pharm. Sci.*, 1979, **68**, 39 (*pharmacol*)

3',6-Dichloro-4',5,7-trihydroxyisoflavone   D-80147

6-Chloro-3-(3-chloro-4-hydroxyphenyl)-5,7-dihydroxy-4H-1-benzopyran-4-one. 3',6-Dichlorogenistein

[64545-65-9]

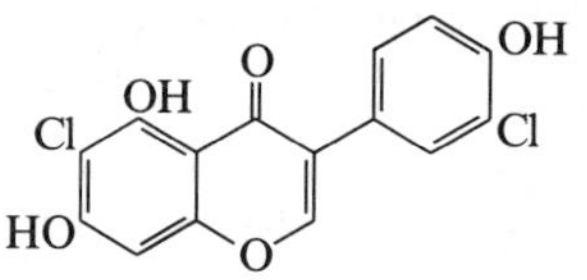

$C_{15}H_8Cl_2O_5$ M 339.131
Metab. of *Streptomyces griseus*. All isolations of isoflavonoids from microorganisms are considered dubious. Prob. artifact derived from the soybean medium.

König, W.A. *et al*, *Helv. Chim. Acta*, 1977, **60**, 2071 (*isol, ir, pmr, ms, struct*)

Dictyoceratin   D-80148

[123062-43-1]

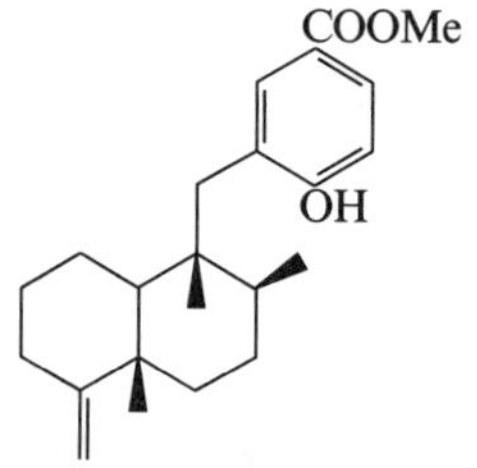

$C_{23}H_{32}O_3$ M 356.504
Metab. of *Dactylospongia* sp. Solid.

Kushlan, D.M. *et al*, *Tetrahedron*, 1989, **45**, 3307.

Dictyodendrillolide   D-80149

5-(Acetyloxy)-3-(4,8,12-trimethyl-3,7,11-tridecatrienyl)-2(5H)-furanone, 9CI

[117585-47-4]

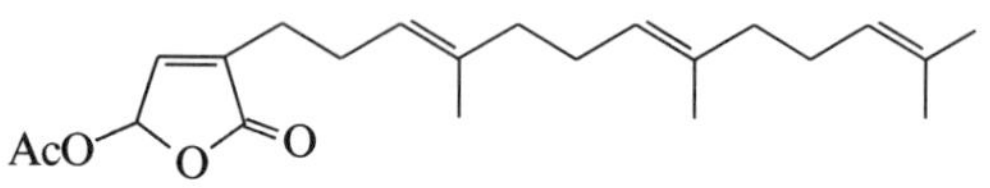

$C_{22}H_{32}O_4$ M 360.492
Constit. of a *Dictyodendrilla* sp. Oil. $[\alpha]_D^{23}$ +21.9° (c, 0.43 in $CHCl_3$).

Cambie, R.C. *et al*, *J. Nat. Prod. (Lloydia)*, 1988, **51**, 1014.

Dicyclopropylmethanethione, 9CI   D-80150

Dicyclopropyl thioketone

[38381-24-7]

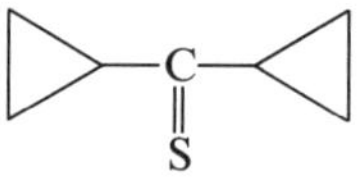

$C_7H_{10}S$ M 126.222
Orange liq.

Fournier, C. *et al*, *Bull. Soc. Chim. Fr.*, 1975, 2753; 1980, 463 (*synth, uv, pmr*)
Andrieu, C.G. *et al*, *Org. Magn. Reson.*, 1978, **11**, 528 (*cmr*)
Mieloszynski, J.L. *et al*, *Recl. Trav. Chim. Pays-Bas (J. R. Neth. Chem. Soc.)*, 1985, **104**, 9 (*cmr*)

3′,4′-Didehydro-1′,2′-dihydro-1′-methoxy-β,ψ-caroten-3-ol   D-80151

[119188-86-2]

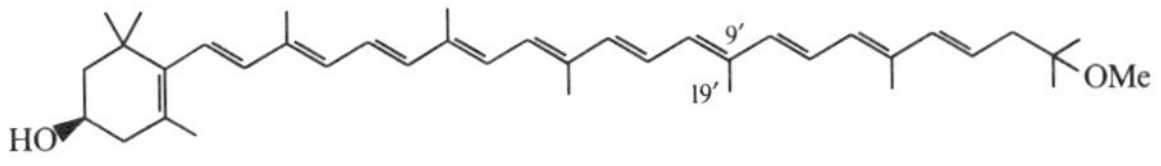

$C_{41}H_{58}O_2$ M 582.908

Constit. of *Erythrobacter longus*.

19′-Aldehyde, 9′-E-isomer: [119188-87-3]. *(9′E)-3′,4′-Didehydro-1′,2′-dihydro-3R-hydroxy-1′-methoxy-β,ψ-caroten-19′-al*

$C_{41}H_{56}O_3$ M 596.892

Constit. of *E. longus*.

Takaichi, S. *et al, Phytochemistry*, 1988, **27**, 3605.

2′,3′-Dideoxycytidine, 9CI   D-80152

DDC

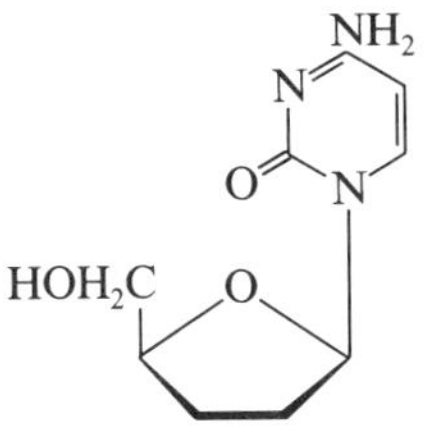

$C_9H_{13}N_3O_3$ M 211.220

Antiviral agent. Potentially of use against HIV infection. Mp 231-232° (214-217°). $[\alpha]_D^{20}$ +90° (c, 0.5 in H_2O), $[\alpha]_D$ +105.9° (c, 0.53 in MeOH).

Horwitz, J.P. *et al, J. Org. Chem.*, 1967, **32**, 817 (*synth*)
Mitsuya, H. *et al, Proc. Natl. Acad. Sci. U.S.A.*, (*London*), 1986, **83**, 1911 (*pharmacol*)
Mitsuya, H. *et al, Nature* (*London*), 1987, **325**, 773.
Okabe, M. *et al, J. Org. Chem.*, 1988, **53**, 4780 (*synth, ir, pmr, ms*)

ent-3β,4β;15,16-Diepoxy-13(16),14-clerodadien-2β-ol   D-80153

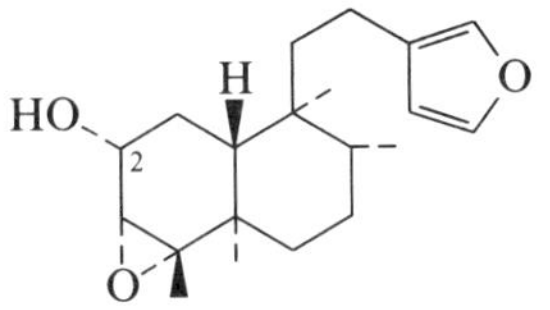

$C_{20}H_{30}O_3$ M 318.455

Constit. of *Baccharis boliviensis*. Gum. $[\alpha]_D^{24}$ +8° (c, 5.65 in $CHCl_3$).

2-Me ether: ent-3β,4β;15,16-*Diepoxy-2β-methoxy*-13(16),14-*clerodadiene*

$C_{21}H_{32}O_3$ M 332.482

Constit. of *B. boliviensis*. Gum.

Zdero, C. *et al, Phytochemistry*, 1989, **28**, 531.

4,7:12,15-Diethenobenzo[7,8]cyclododeca[1,2-c]furan, 9CI   D-80154

Benzofurano[c][2.2]paracyclophane

[111615-65-7]

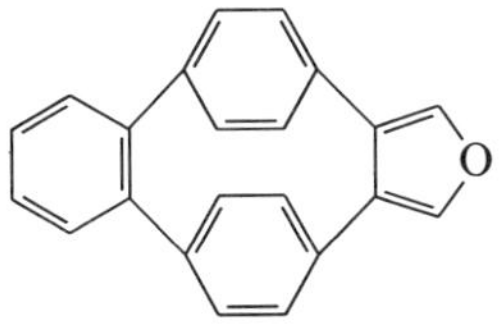

$C_{22}H_{14}O$ M 294.352

Cryst. (heptane). Mp 275° dec.

Chan, C.W. *et al, J. Am. Chem. Soc.*, 1988, **110**, 462 (*synth, uv, pmr*)

1,8-Diethynylanthracene   D-80155

[78053-58-4]

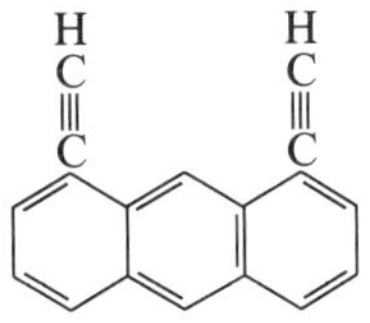

$C_{18}H_{10}$ M 226.277

Yellow plates (toluene). Mp 150° dec.

Katz, H.E., *J. Org. Chem.*, 1989, **54**, 2179 (*synth, pms, cmr*)

2,4-Diethynylthiazole   D-80156

[113705-26-3]

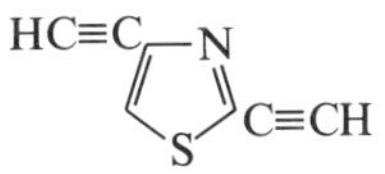

C_7H_3NS M 133.173

Oil. Unstable, darkened on standing.

▷ Potentially explosive.

Neenan, T.X. *et al, J. Org. Chem.*, 1988, **53**, 2489 (*synth, pmr, ms, haz*)

2,5-Diethynylthiophene   D-80157

[79109-72-1]

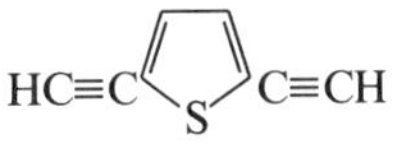

C_8H_4S M 132.186

Liq. $Bp_{0.1}$ 40°.

▷ Potentially explosive.

Okuhara, K., *Bull. Chem. Soc. Jpn.*, 1981, **54**, 2045 (*synth, ms, ir*)
Neenan, T.X. *et al, J. Org. Chem.*, 1988, **53**, 2489 (*synth, ir, pmr, cmr, haz*)

2,2-Difluoro-3,3-bis(trifluoromethyl)oxirane, 9CI   D-80158

1,2-Epoxy-1,1,3,3,3-pentafluoro-2-(trifluoromethyl)propane, 8CI. 1,2-Epoxyperfluoroisobutene

[707-13-1]

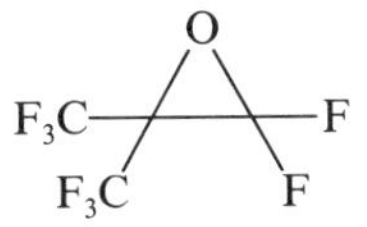

C_4F_8O M 216.031

Mp −122°. Bp 3°, Bp 6-7°.

Knunyants, I.L. *et al, Khim. Geterotsikl. Soedin.*, 1966, 873 (*synth, F-19 nmr*)
Croft, T.S., *J. Fluorine Chem.*, 1976, **7**, 438 (*synth, F-19 nmr*)

1,2-Difluoro-2-butenedioic acid   D-80159

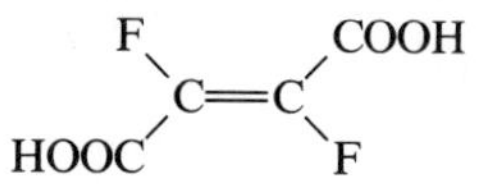

(E)-form

$C_4H_2F_2O_4$ M 152.054

(E)-form [2714-32-1]
Difluorofumaric acid
Cryst. (H_2O). Mp 268-270°.

(Z)-form [685-64-3]
Difluoromaleic acid
Cryst. (Me_2CO/C_6H_6). Mp 219-220°.
Anilinium salt: Mp 177-178.5°.
S-Benzylthiouronium salt: [15514-25-7].
Mp 231-232°.
Di-Me ester: [39774-02-2].
$C_6H_6F_2O_4$ M 180.108
Bp 153°.
Di-Et ester: [815-97-4].
$C_8H_{10}F_2O_4$ M 208.161
Bp_3 67°.
Anhydride: [669-78-3]. *3,4-Difluoro-2,5-furandione, 9CI.*
Difluoromaleic anhydride
$C_4F_2O_3$ M 134.039
Mp 20°. Bp 128°. n_D^{25} 1.4197.

Raasch, M.S. *et al, J. Am. Chem. Soc.*, 1959, **81**, 2678 (*anhydride*)
Kobrina, L.S. *et al, J. Org. Chem. USSR (Engl. Transl.)*, 1972, **8**, 2209 (*synth, deriv, uv, F nmr*)

3,3-Difluoro-1,2-diiodocyclopropene, 9CI   D-80160

1,2-Diiodo-3,3-difluorocyclopropene
[56830-73-0]

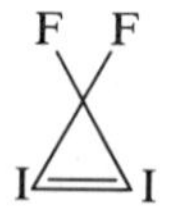

$C_3F_2I_2$ M 327.839
Bp_{35} 82-85°. n_D^{25} 1.5920.

Sepiol, J. *et al, J. Org. Chem.*, 1975, **40**, 3791 (*synth, F-19 nmr, ir*)

3,3′-Difluoro-4,4′-dinitrobiphenyl   D-80161

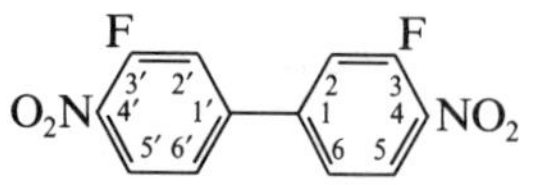

$C_{12}H_6F_2N_2O_4$ M 280.187
Cryst. (MeOH). Mp 197.5-198.5°.

Fletcher, T.L. *et al, J. Org. Chem.*, 1960, **25**, 1342.

4,4′-Difluoro-2,2′-dinitrobiphenyl   D-80162

[316-00-7]
$C_{12}H_6F_2N_2O_4$ M 280.187
Cryst. (C_6H_6). Mp 168-169°.

Shaw, F.R. *et al, J. Chem. Soc.*, 1932, 509 (*synth*)
Protiva, M. *et al, Collect. Czech. Chem. Commun.*, 1986, **51**, 698 (*uv, ir, pmr, ms*)

5,5′-Difluoro-2,2′-dinitrobiphenyl   D-80163

[41860-49-5]
$C_{12}H_6F_2N_2O_4$ M 280.187
Cryst. (EtOH). Mp 121°.

Dickerson, D.R. *et al, J. Fluorine Chem.*, 1973, **3**, 113 (*synth*)

2,3-Difluoro-3-phenyl-2-propenal, 9CI   D-80164

2,3-Difluorocinnamaldehyde

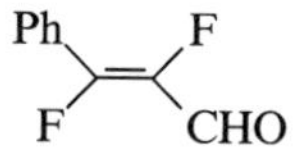

(E)-form

$C_9H_6F_2O$ M 168.143
(E)-form [113894-39-6]
Yellow cryst. (hexane). Mp 32.0-33.0°.

Matsuo, N. *et al, J. Org. Chem.*, 1988, **53**, 2304 (*synth, pmr, F-19 nmr, ms*)

2,2-Difluoro-3-(trifluoromethyl)-2H-azirine, 9CI, 8CI   D-80165

Perfluoro(3-methyl-2H-azirine)
[3291-42-7]

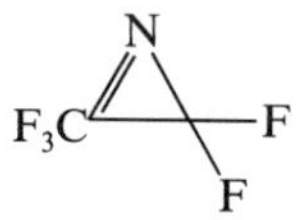

C_3F_5N M 145.032
Pale yellow oil. Bp −19.8°.

Cleaver, C.S. *et al, J. Am. Chem. Soc.*, 1965, **87**, 3716 (*synth, ir, F-19 nmr, ms*)
Banks, R.E. *et al, J. Chem. Soc. C*, 1966, 2304 (*synth, ir, F-19 nmr*)

2,3-Difluoro-2-(trifluoromethyl)-2H-azirine, 8CI   D-80166

Perfluoro(2-methyl-2H-azirine). Pentafluoro-2-azetine
[3291-41-6]

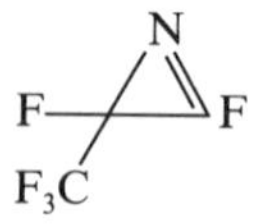

C_3F_5N M 145.032
Bp −15.5°, Bp −17°. Isomerises in glass or with moisture. Polymerises in contact with metal containers.

Cleaver, C.S. *et al, J. Am. Chem. Soc.*, 1965, **87**, 3716 (*synth, ir, F-19 nmr, ms*)
Banks, R.E. *et al, J. Chem. Soc. C*, 1966, 2304 (*synth, F-19 nmr*)

5-(2,2-Difluorovinyl)uracil   D-80167

[108711-79-1]

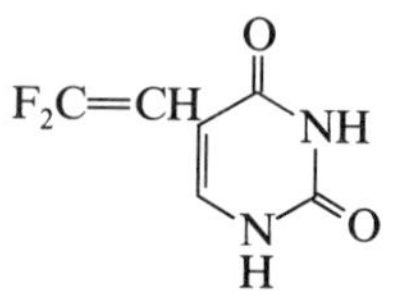

$C_6H_4F_2N_2O_2$ M 174.107
Cryst. (MeOH/EtOH). Mp 288-292°.

1-(2-Deoxy-α-D-ribofuranosyl): [108742-87-6].
$C_{11}H_{12}F_2N_2O_5$ M 290.223
Mp 173-174°.

1-(2-Deoxy-β-D-ribofuranosyl): [108711-80-4]. *5-(2,2-Difluorovinyl)-2′-deoxyuridine*
$C_{11}H_{12}F_2N_2O_5$ M 290.223

Active against Herpes simplex virus type 1 (HSV-1). Mp 152°.

Babek, M. *et al, J. Med. Chem.*, 1987, **30**, 1494 (*synth, pmr*)

6,8-Di-*C*-glucosyl-3′,4′,5,7-tetrahydroxy-isoflavone   D-80168

*6,8-Di-*C*-glucopyranosylorobol*

[76436-69-6]

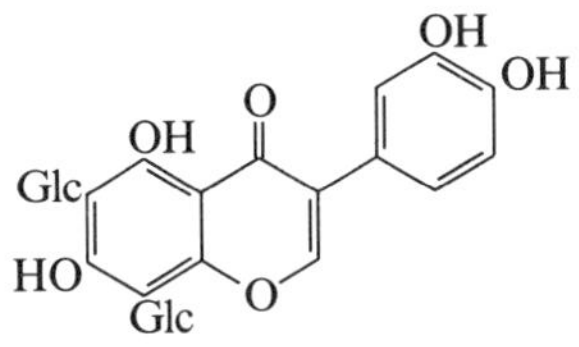

$C_{27}H_{30}O_{18}$ M 642.523

Isol. from bark of *Dalbergia nitidula*. Yellow amorph. solid. Mp 183-186°.

Van Heerden, F.R. *et al, J. Chem. Soc., Perkin Trans.* 1, 1980, 2463 (*isol, cmr, struct*)

2,6-Dihydroaceanthrylene   D-80169

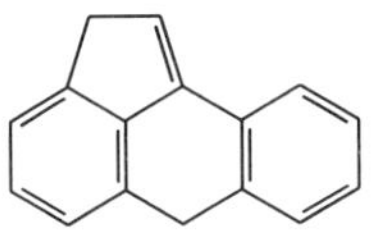

$C_{16}H_{12}$ M 204.271

Pale yellow cryst. $(CH_2Cl_2/hexane)$. Mp 111°.

Rabideau, P.W. *et al, J. Org. Chem.*, 1988, **53**, 589 (*synth, pmr*)

1,2-Dihydrobenz[f]isoquinoline   D-80170

[23950-51-8]

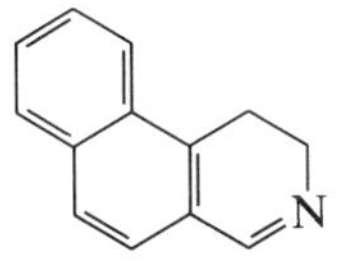

$C_{13}H_{11}N$ M 181.237

B,HCl: [112576-40-6].
Cryst. Mp 190-191°.

Young, S.D. *et al, J. Org. Chem.*, 1988, **53**, 1114 (*synth, pmr, cmr, ms*)

3,4-Dihydrobenz[*g*]isoquinoline   D-80171

[112576-38-2]

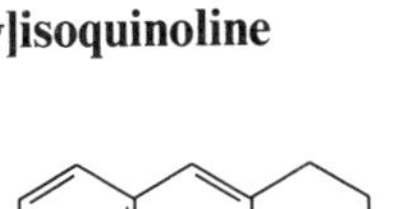

$C_{13}H_{11}N$ M 181.237

Cryst.

B,HCl: [112576-39-3].
Cryst. Mp 225-229°.

Young, S.D. *et al, J. Org. Chem.*, 1988, **53**, 1114 (*synth, pmr, ms*)

5,6-Dihydrobenzo[*c*]cinnoline, 9CI   D-80172

[52125-81-2]

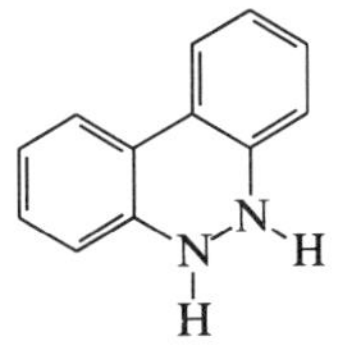

$C_{12}H_{10}N_2$ M 182.224

Cryst. Mp 122°.

Mono-Ac:
$C_{14}H_{12}N_2O$ M 224.262
Needles (cyclohexane). Mp 114°.

Di-Ac:
$C_{16}H_{14}N_2O_2$ M 266.299
Leaflets (EtOH aq.). Mp 170°.

Wittig, G. *et al, Chem. Ber.*, 1955, **88**, 234 (*derivs, synth*)
Etienne, A. *et al, Bull. Soc. Chim. Fr.*, 1962, 292 (*synth, derivs*)
Mugnier, Y. *et al, J. Org. Chem.*, 1988, **53**, 5781 (*synth*)

2,3-Dihydro-1*H*-1,4-benzodiazepine, 9CI   D-80173

[5945-91-5]

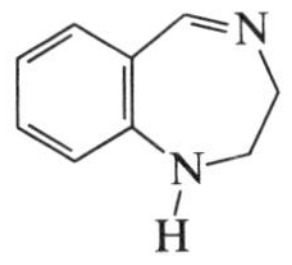

$C_9H_{10}N_2$ M 146.191

Cryst. (EtOH). Mp 253° (244-246°).

Uskokovic, M. *et al, J. Org. Chem.*, 1962, **27**, 3606 (*synth, ir, uv*)
Betakis, E. *et al, Synthesis*, 1988, 820 (*synth, ir, uv, pmr*)

2,3-Dihydro-1,4-benzodioxin-5,8-dione, 9CI   D-80174

2,3-Ethylenedioxy-1,4-benzoquinone

[42965-39-9]

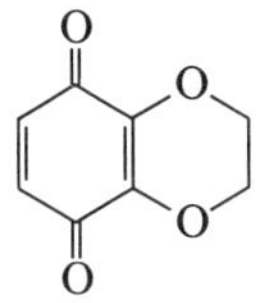

$C_8H_6O_4$ M 166.133

Red needles (EtOAc/hexane). Mp 158-159.5° (153-154°).

Bowman, C. *et al, J. Med. Chem.*, 1973, **16**, 988 (*synth*)
Hayakawa, K. *et al, J. Chem. Soc., Perkin Trans.* 1, 1988, 511 (*synth, pmr, ms*)

1,4- Dihydro-2,3-benzodithiin, 9CI   D-80175

[3886-39-3]

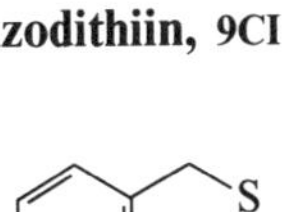

$C_8H_8S_2$ M 168.283

Needles (MeOH). Mp 77-78° (80°).

Lüttringhaus, A. *et al, Angew. Chem.*, 1955, **67**, 304 (*synth*)
Milligan, B. *et al, J. Chem. Soc.*, 1965, 2901 (*synth, pmr*)

1,5-Dihydro-2,3,4-benzotrithiepin, 9CI   D-80176

2,3,4-Benzotrithiepin

[3354-86-7]

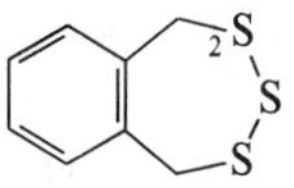

$C_8H_8S_3$ M 200.349

Plates (pet. ether). Mp 101-102°.

2-Oxide:

$C_8H_8OS_3$ M 216.349

Yellow prisms (EtOAc). Mp 133°.

Milligan, B. *et al, J. Chem. Soc.*, 1965, 2901 (*synth, ir, pmr*)

2,3-Dihydro-1*H*-cyclopenta[*l*] phenanthrene, 9CI   D-80177

9,10-Cyclopentenophenanthrene

[723-98-8]

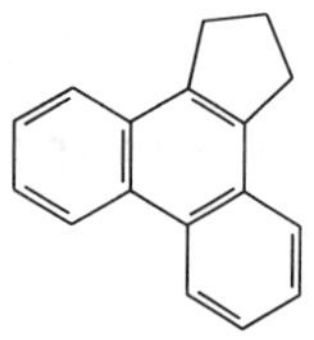

$C_{17}H_{14}$ M 218.298

Cryst. (EtOH). Mp 152.5-153.5°.

Cope, A.C. *et al, J. Am. Chem. Soc.*, 1956, **78**, 2547 (*synth*)
Wood, C.S. *et al, J. Org. Chem.*, 1964, **29**, 3373 (*synth*)
Penn, J.H. *et al, J. Org. Chem.*, 1989, **54**, 601 (*synth*)

15,16-Dihydro-17*H*-cyclopenta[*a*] phenanthren-17-one, 8CI   D-80178

[786-66-3]

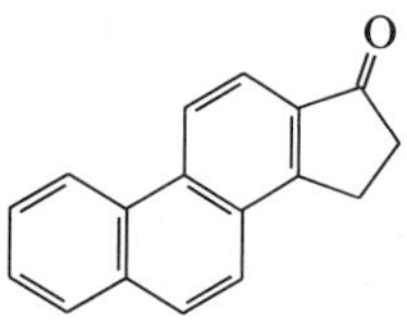

$C_{17}H_{20}O$ M 240.344

Solid (EtOAc). Mp 201-202°.

Coombs, M.M., *J. Chem. Soc. C*, 1966, 955 (*synth*)
Lee, H. *et al, J. Org. Chem.*, 1988, **53**, 4253 (*synth, pmr*)

10,11-Dihydro-5*H*-dibenzo[*a,d*] cyclohepten-5,10-imine, 9CI   D-80179

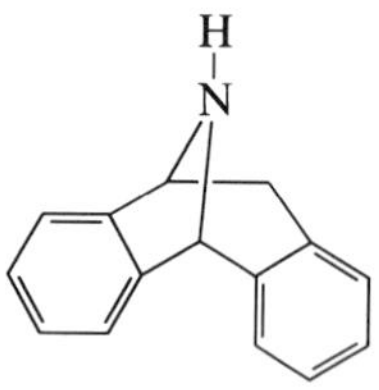

$C_{15}H_{13}N$ M 207.274

(±)-*form* [50537-17-2]

Cryst. (CH₂Cl₂/pet. ether). Mp 111.5-114° (120°).

Lamanec, T.R. *et al, J. Org. Chem.*, 1988, **53**, 1768 (*synth*)

1,4-Dihydrodibenzothiophene   D-80180

[109644-16-8]

$C_{12}H_{10}S$ M 186.277

Cryst. Mp 76°. Bp₆ 160-165°.

Gilman, H. *et al, J. Org. Chem.*, 1938, **3**, 108 (*synth*)
Francisco, M.A. *et al, J. Org. Chem.*, 1988, **53**, 596 (*synth, pmr, ms*)

3,4-Dihydro-4,8-dihydroxy-1(2*H*)- naphthalenone   D-80181

Regiolone

[54712-38-8]

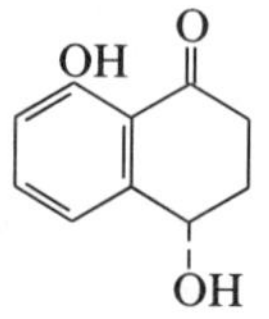

$C_{10}H_{10}O_3$ M 178.187

Constit. of *Juglans regia*. Cryst. (pet. ether). Mp 73°. [α]_D −3.3° (c, 0.077 in EtOH).

Talapatra, S.K. *et al, Phytochemistry*, 1988, **27**, 3929.

10,11-Dihydro-5*H*-diindeno[1,2-*b*;2′,1′-*d*] pyrrole, 9CI   D-80182

[7099-31-2]

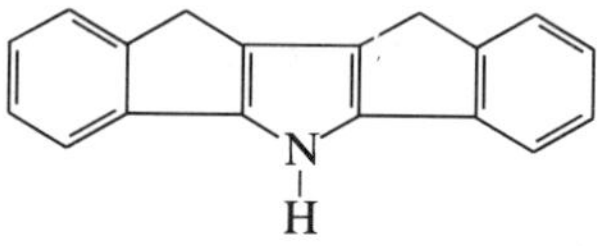

$C_{18}H_{13}N$ M 243.307

Cryst. (toluene under Ar). Mp 275° dec.

N-*Ph:* [115679-37-3].

$C_{24}H_{17}N$ M 319.405

Mp 169°.

Baierweck, P. *et al, Chem. Ber.*, 1988, **121**, 2195 (*synth, pmr, cmr*)

10,11-Dihydrodiindeno[1,2-*b*:2′,1′-*d*] thiophene, 9CI   D-80183

[7099-33-4]

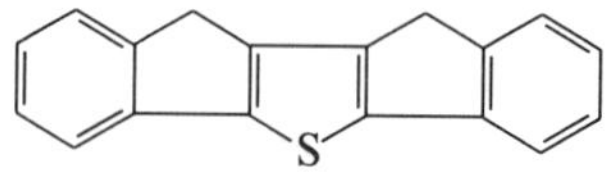

$C_{18}H_{12}S$ M 260.359

Cryst. (CHCl₃). Mp 288°.

Baierweck, P. *et al, Chem. Ber.*, 1988, **121**, 2195 (*synth, cmr*)
Czogalla, C.D. *et al, Phosphorus Sulfur*, 1988, **35**, 127 (*synth*)

3,4-Dihydro-2,2-dimethyl-2*H*-1-benzo- pyran, 9CI   D-80184

2,2-Dimethylchroman, 8CI

[1198-96-5]

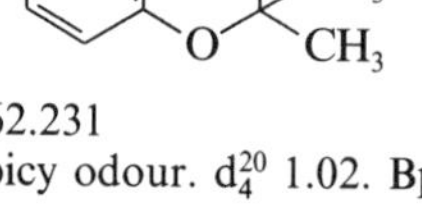

$C_{11}H_{14}O$ M 162.231

Yellow oil with spicy odour. d_4^{20} 1.02. Bp₁₁.₅ 98-98.5°.

Claisen, L., *Ber.*, 1921, **54**, 200 (*synth*)
Shriner, R.L. *et al, J. Org. Chem.*, 1939, **4**, 575 (*synth*)

Frater, G. *et al, Helv. Chim. Acta*, 1967, **50**, 255 (*synth*)
Miller, J.A. *et al, J. Chem. Soc. C*, 1968, 1837 (*synth*)
Bolzoni, L. *et al, Angew. Chem., Int. Ed. Engl.*, 1978, **17**, 684 (*synth*)
Camps, F. *et al, Synthesis*, 1979, 126 (*synth*)
Fatope, M.O. *et al, J. Med. Chem.*, 1987, **30**, 1973 (*synth, ir, pmr*)

3,4-Dihydro-2,2-dimethyl-2*H*-1-benzo-pyran-6-carboxaldehyde, 9CI   D-80185

2,2-Dimethyl-6-chromancarboxaldehyde. 6-Formyl-2,2-dimethylchroman

[61370-75-0]

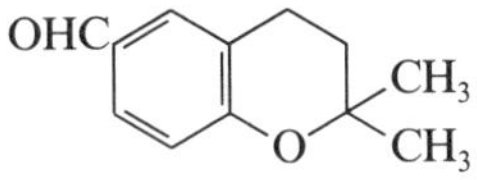

$C_{12}H_{14}O_2$ M 190.241
Liq. Bp_2 110-114°.

Knight, D.W. *et al, J. Chem. Soc., Perkin Trans. 1*, 1979, 30 (*synth, ir, pmr*)
Fatope, M.O. *et al, J. Med. Chem.*, 1987, **30**, 1973 (*synth, pmr*)

3,4-Dihydro-2,2-dimethyl-2*H*-1-benzo-pyran-6-carboxylic acid, 9CI   D-80186

2,2-Dimethyl-6-chromancarboxylic acid

[2039-47-6]

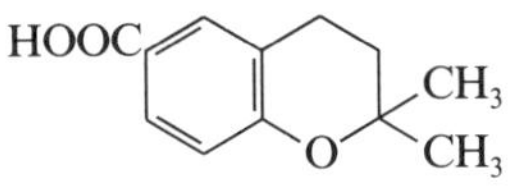

$C_{12}H_{14}O_3$ M 206.241
Cryst. (MeOH aq.). Mp 184° (177-178°).
p-*Bromophenacyl ester:* Mp 147-148°.

Lauer, W.M. *et al, J. Am. Chem. Soc.*, 1943, **65**, 289 (*synth*)
Fatope, M.O. *et al, J. Med. Chem.*, 1987, **30**, 1973 (*synth, ir, pmr, ms*)

1,2-Dihydro-1,2-dimethylenenaphthalene   D-80187

1,2-Dihydro-1,2-bis(methylene)naphthalene, 9CI. 1,2-Naphthalenequinodimethide

[41182-10-9]

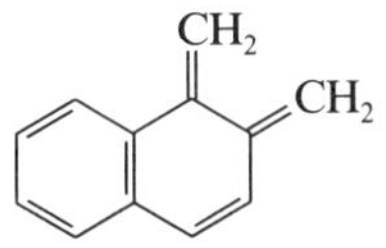

$C_{12}H_{10}$ M 154.211
Transient species detected by flow pmr.

Cava, M.P. *et al, J. Org. Chem.*, 1962, **27**, 755.
Trahanovsky, W.S. *et al, J. Am. Chem. Soc.*, 1988, **110**, 6579.

1,2-Dihydro-1,4-dimethylenenaphthalene   D-80188

1,4-Dihydro-1,4-bis(methylene)naphthalene, 9CI. 1,4-Naphthalenequinodimethide

[7545-39-3]

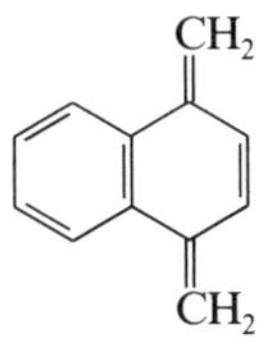

$C_{12}H_{10}$ M 154.211

Unstable intermediate trapped by Diels-Alder reaction or obs. by low temp. spectroscopy.

Cram, D.J. *et al, J. Am. Chem. Soc.*, 1963, **85**, 1088 (*synth*)
Williams, D.J. *et al, J. Am. Chem. Soc.*, 1970, **92**, 1436 (*synth, pmr, bibl*)
Pearson, J.M. *et al, J. Am. Chem. Soc.*, 1971, **93**, 5034 (*synth, uv, ir*)

2,3-Dihydro-2,3-dimethylenenaphthalene, 8CI   D-80189

2,3-Dihydro-2,3-bis(methylene)naphthalene, 9CI

[39638-07-8]

$C_{12}H_{10}$ M 154.211
Reactive intermediate isolable at low temperature. Diene for cycloaddition reactions.

Cava, M. .P. *et al, J. Org. Chem.*, 1962, **27**, 755.
Gisin, M. *et al, Helv. Chim. Acta*, 1976, **59**, 2273 (*synth, uv*)
Bell, T.W. *et al, Tetrahedron Lett.*, 1980, **21**, 3299 (*dimers*)
Chew, S. *et al, J. Chem. Soc., Chem. Commun.*, 1984,, 911 (*synth, use*)

2,3-Dihydro-4,5-dimethylfuran   D-80190

[1487-16-7]

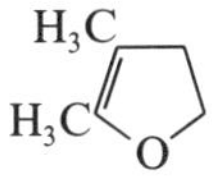

$C_6H_{10}O$ M 98.144
Liq. Bp 106°.

McGreer, D.E. *et al, Can. J. Chem.*, 1965, **43**, 1398 (*synth*)
Adam, W. *et al, J. Org. Chem.*, 1988, **53**, 1492 (*synth*)

Dihydro-4,5-dimethyl-2(3*H*)-furanone, 9CI   D-80191

Updated Entry replacing D-03708
β-Methyl-γ-valerolactone. β,γ-Dimethylbutyrolactone. Tetrahydro-4,5-dimethyl-2-furanone

[6971-63-7]

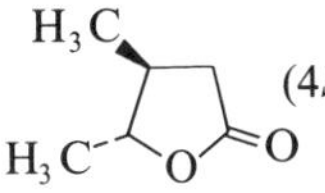

$C_6H_{10}O_2$ M 114.144
Tobacco constit.

(4*S*,5*R*)-*form* [90026-46-3]
  (+)-trans-*form*
  $[\alpha]_D^{20}$ +56.9° (c, 1.8 in $CHCl_3$).

(4*RS*,5*RS*)-*form* [10150-95-5]
  (±)-cis-*form*
  Bp_{12} 95-97°.

(4*RS*,5*SR*)-*form* [10150-96-6]
  (±)-trans-*form*
  Bp_{12} 101-103°.

[110171-22-7]

Tokuda, M. *et al, J. Org. Chem.*, 1972, **37**, 1859 (*synth*)
Kimland, B. *et al, Phytochemistry*, 1973, **12**, 835 (*isol*)
Sucrow, W. *et al, Chem. Ber.*, 1975, **108**, 48 (*synth*)
Pyysalo, H. *et al, Finn. Chem. Lett.*, 1975, 129 (*cmr*)
Tsuzuki, K. *et al, Tetrahedron Lett.*, 1976, 4745 (*synth*)
Mattes, H. *et al, J. Med. Chem.*, 1987, **30**, 1948 (*synth, ir, pmr*)

**Dihydro-4,5-dimethyl-3-methylene-2(3*H*)-   D-80192
 furanone**

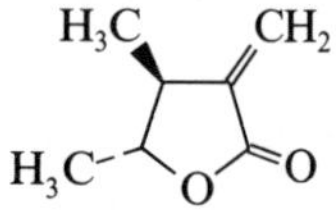

C$_7$H$_{10}$O$_2$ M 126.155
(4*S*,5*R*)-*form* [110171-20-5]
   (+)-trans-*form*
   [α]$_D^{20}$ +16.4° (c, 1.10 in CHCl$_3$).
   [110171-21-6]
   Mattes, H. *et al*, *J. Med. Chem.*, 1987, **30**, 1948 (*synth, ir, pmr*)

**1,4-Dihydro-2,6-dimethyl-3,5-pyridine-   D-80193
 dicarboxylic acid, 9CI**
[51595-68-7]

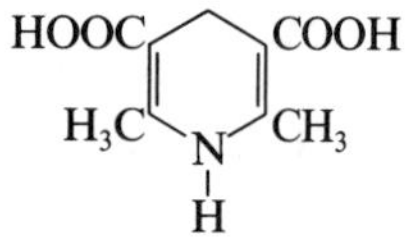

C$_9$H$_{11}$NO$_4$ M 197.190
Derivs. are vasodilators, hypotensives and oxidation
 inhibitors, heat stabilisers, etc.
Di-Me ester: [17438-14-1].
   C$_{11}$H$_{15}$NO$_4$ M 225.244
   Yellow needles (MeOH). Mp 225-227° (219-220°).
Di-Et ester: [1149-23-1].
   C$_{13}$H$_{19}$NO$_4$ M 253.297
   Yellow needles with green fluorescence (EtOH). Mp 183-
   185° (176-183°).
*Di-*tert-*butyl ester:* [55536-71-5].
   C$_{17}$H$_{27}$NO$_4$ M 309.405
   Cryst. (EtOH). Mp 142-144°.
Dibenzyl ester: [36138-79-1].
   C$_{23}$H$_{23}$NO$_4$ M 377.439
   Cryst. (EtOH). Mp 119-121°.
Dinitrile: [3274-36-0].
   C$_9$H$_9$N$_3$ M 159.190
   Yellowish plates (MeOH). Mp 232-233° (222°).
1-Me, di-Et ester: [14258-07-2].
   C$_{14}$H$_{21}$NO$_4$ M 267.324
   Cryst. (pet. ether). Mp 84-85°.
1-Me, dinitrile: [14258-16-3].
   C$_{10}$H$_{11}$N$_3$ M 173.217
   Cryst. (C$_6$H$_6$/pet. ether). Mp 132.5-133°.

Meyer, E., *J. Prakt. Chem.*, [2], 1908, **78**, 497 (*nitrile*)
Org. Synth., Coll. Vol., 2, 1943, 214 (*Et ester, synth*)
Kuthan, J. *et al*, *Collect. Czech. Chem. Commun.*, 1964, **29**, 1654
   (*dinitrile*)
Brignell, P.J. *et al*, *J. Chem. Soc. B*, 1966, 1083 (*derivs, uv*)
Roomi, M.W., *J. Med. Chem.*, 1975, **18**, 457 (*esters, pharmacol*)
Lenstra, A.T.H. *et al*, *Bull. Soc. Chim. Belg.*, 1979, **88**, 133 (*Et
   ester, cryst struct*)
Kurfürst, A. *et al*, *Collect. Czech. Chem. Commun.*, 1983, **48**, 1422;
   1984, **49**, 2893 (*derivs, uv, pmr*)

2,3-Dihydro-2,2-dimethylthiazole   D-80194
2,2-Dimethylthiazolidine

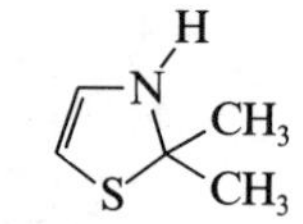

C$_5$H$_9$NS   M 115.199
Potent inhibitor of platelet aggregation.
B,HCl: Mp 174-176°.
Mimura, T. *et al*, *Chem. Pharm. Bull.*, 1988, **36**, 1110 (*synth, pmr,
   pharmacol*)

2,3-Dihydro-4,5-dimethylthiophene   D-80195
[113379-98-9]

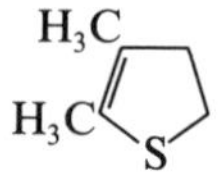

C$_6$H$_{10}$S   M 114.211
Liq. Bp$_{12}$ 42°. Stored at 0° under N$_2$.
1-Oxide: [113380-05-5].
   C$_6$H$_{10}$OS   M 130.210
   Liq.
   Adam, W. *et al*, *J. Org. Chem.*, 1988, **53**, 1492 (*synth, pmr, ir, uv*)

**4,5-Dihydro-5,5-dimethyl-3*H*-1,2,4-triazol-   D-80196
 3-one, 9CI**
5,5-Dimethyl-Δ^1-1,2,4-triazolin-3-one
[112700-85-3]

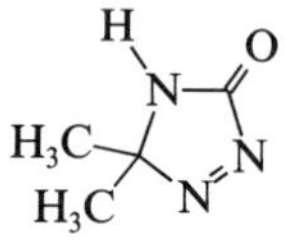

C$_4$H$_7$N$_3$O   M 113.119
Cryst. (Et$_2$O). Mp 84-87°.
N-*Benzoyl:*
   C$_{11}$H$_{11}$N$_3$O$_2$ M 217.227
   Cryst. (EtOH). Mp 107-107.5°.
Schantl, J.G. *et al*, *Heterocycles*, 1987, **26**, 1439 (*synth*)
Schantl, J.G. *et al*, *J. Heterocycl. Chem.*, 1987, **24**, 1401 (*cryst
   struct, ir, uv, pmr, cmr, ms*)
Schantl, J.G. *et al*, *Synthesis*, 1987, 986 (*synth*)

Dihydro-2,2-diphenyl-2(3*H*)-furanone, 9CI   D-80197
[101110-06-9]

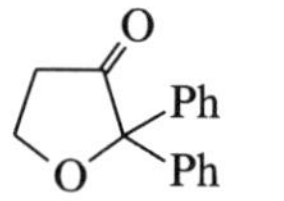

C$_{16}$H$_{14}$O$_2$ M 238.285
Needles (EtOH). Mp 74.5-75.5°.
Flavin, M.T. *et al*, *J. Med. Chem.*, 1987, **30**, 278 (*synth, pmr, ir,
   ms*)

Dihydro-3,3-diphenyl-2(3*H*)-furanone, 9CI   D-80198

α,α-Diphenyl-γ-butyrolactone

[956-89-8]

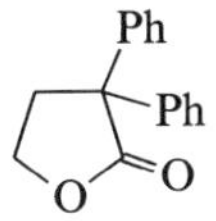

$C_{16}H_{14}O_2$ M 238.285

Prisms (MeOH/Me₂CO). Mp 80-81° (77-79°).

Morand, P. *et al, J. Chem. Soc., Chem. Commun.*, 1982, **8**, 458 (*synth*)
Bocelli, G. *et al, J. Mol. Struct.*, 1982, **81**, 113 (*cryst struct*)
Kayser, M.M. *et al, Can. J. Chem.*, 1983, **61**, 439 (*synth, uv, pmr*)
Shibata, I. *et al, Bull. Chem. Soc. Jpn.*, 1986, **59**, 4000 (*synth, ir, ms, pmr*)

Dihydro-3,4-diphenyl-2(3*H*)-furanone, 9CI   D-80199

α,β-Diphenyl-γ-butyrolactone

[25109-89-1]

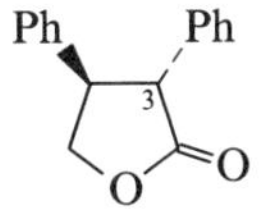

(3R,4R)-form

$C_{16}H_{14}O_2$ M 238.285

(3R,4R)-form [24435-71-0]

(−)-trans-*form*

Cryst. (EtOH). Mp 96.5-98°. $[\alpha]_D^{23}$ − 196.7° (c, 0.231 in CHCl₃).

(3RS,4RS)-form

(±)-trans-*form*

Cryst. Mp 96-96.5°.

Berova, N. *et al, Tetrahedron*, 1969, **25**, 2301 (*synth, ir, uv, abs config*)
Fukuoka, S. *et al, J. Org. Chem.*, 1970, **35**, 3184 (*synth, ir*)

Dihydro-3,5-diphenyl-2(3*H*)-furanone, 9CI   D-80200

α,γ-Diphenyl-γ-butyrolactone

[21034-29-7]

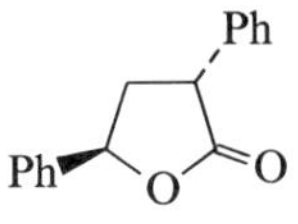

(3RS,5RS)-form

$C_{16}H_{14}O_2$ M 238.285

(3RS,5RS)-form [20272-26-8]

(±)-trans-*form*

Cryst. (Et₂O/hexane). Mp 74° (68-69°).

(3RS,5SR)-form [20272-24-6]

(±)-cis-*form*

Cryst. (Me₂CO/hexane). Mp 107-107.5°.

[94848-88-1]

Givens, R.S. *et al, J. Org. Chem.*, 1972, **37**, 4325 (*synth*)
Fujita, T. *et al, Aust. J. Chem.*, 1974, **27**, 2205 (*synth*)
Hussain, T.S.A.M. *et al, J. Chem. Soc., Perkin Trans. 1*, 1975, 1480 (*synth, pmr*)

Dihydro-4,4-diphenyl-2(3*H*)-furanone, 9CI   D-80201

β,β-Diphenyl-γ-butyrolactone

[67390-35-6]

$C_{16}H_{14}O_2$ M 238.285

Mp 109-110°.

Legagneur, F.S., *Compt. Rend. Hebd. Seances Acad. Sci.*, 1948, **227**, 437 (*synth*)
Morand, P. *et al, J. Chem. Soc., Chem. Commun.*, 1982, 458 (*synth*)
Kayser, M.M. *et al, Can. J. Chem.*, 1983, **61**, 439 (*synth, uv, pmr*)

Dihydro-4,5-diphenyl-2(3*H*)-furanone, 9CI   D-80202

2,3-Diphenyl-4-butanolide

[20453-83-2]

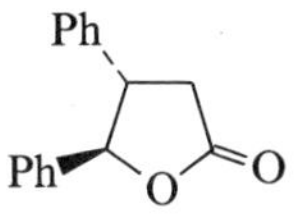

(3RS,4RS)-form

$C_{16}H_{14}O_2$ M 238.285

(3RS,4RS)-form [53861-52-2]

(±)-trans-*form*

No phys. props. reported.

(3RS,4SR)-form [5635-41-6]

(±)-cis-*form*

Needles(EtOAc). Mp 127-128°.

[69573-46-2]

Burns, W.D.P. *et al, J. Chem. Soc. C*, 1967, 1554 (*synth, uv, pmr*)
Heiba, E.I. *et al, J. Am. Chem. Soc.*, 1968, **90**, 5905; 1974, **96**, 7977 (*synth*)
Schine, M. *et al, Bull. Chem. Soc. Jpn.*, 1982, **55**, 224 (*synth, pmr*)

Dihydro-5,5-diphenyl-2(3*H*)-furanone, 9CI   D-80203

4,4-Diphenylbutyrolactone. 4,4-Diphenyl-4-butanolide

[7746-94-3]

$C_{16}H_{14}O_2$ M 238.285

Microcryst. (EtOH). Mp 91-92°.

Staab, H.A. *et al, Justus Liebigs Ann. Chem.*, 1966, **694**, 78 (*synth*)
King, G.S. *et al, J. Chem. Soc., Perkin Trans. 1*, 1974, 1499 (*synth*)
Loozen, H.J.J. *et al, J. Org. Chem.*, 1975, **40**, 892 (*synth*)
Sturtz, G. *et al, Tetrahedron Lett.*, 1976, 47 (*synth*)
Miller, R.D. *et al, J. Am. Chem. Soc.*, 1984, **106**, 1508 (*synth*)
Fujimoto, N. *et al, Bull. Chem. Soc. Jpn.*, 1986, **59**, 3161 (*synth*)
Fukuzawa, S. *et al, J. Chem. Soc., Chem. Commun.*, 1986, 475, 624 (*synth*)
Lehmann, J. *et al, Synthesis*, 1987, 1064 (*synth, ir, pmr*)

2,5-Dihydro-3,6-diphenylpyrrolo[3,4-c] pyrrole-1,4-dione, 9CI   D-80204

Updated Entry replacing D-70204

3,6-Diphenyl-2,5-diaza-1,4(2H,5H)-pentalenedione

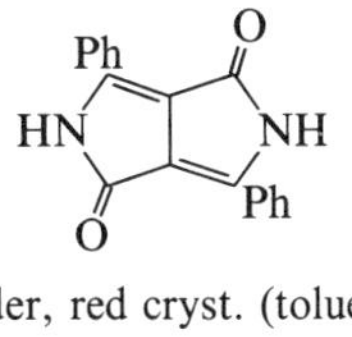

Pigment. Red powder, red cryst. (toluene or DMF). Mp >350°.

2,5-Di-Me: [96159-17-0].

$C_{20}H_{16}N_2O_2$ M 316.359

Orange cryst. (PhNO₂). Mp 236-238°.

2,5-Di-Et: [96159-13-6].

$C_{22}H_{20}N_2O_2$ M 344.412

Orange yellow needles (butanol). Mp 229-230°.

2-Butyl: [96159-00-1].

$C_{22}H_{20}N_2O_2$ M 344.412

Yellow cryst. (MeOH). Mp 250-252°.

2,5-Dibenzyl: [96159-02-3].

$C_{32}H_{24}N_2O_2$ M 468.554
Orange yellow cryst. (DMF). Mp 290-292°.

Farnum, D.G. *et al*, *Tetrahedron Lett.*, 1974, 2549 (*synth*)
Eur. Pat., 133 156, (1984); *CA*, **102**, 186667 (*derivs*)
Closs, F. *et al*, *Angew. Chem., Int. Ed. Engl.*, 1987, **26**, 552 (*synth, uv*)
Potrawa, T. *et al*, *Chem. Ber.*, 1987, **120**, 1075 (*synth, derivs, uv, pmr*)

5,6-Dihydro-1,3-dithiolo[4,5-*b*][1,4]dithiin-2-one, 9CI   D-80205

2,5,7,9-Tetrathiabicyclo[4.3.0]non-1(6)-en-8-one
[74962-29-1]

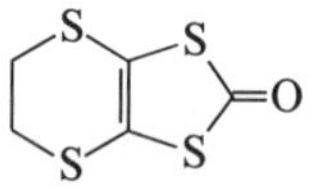

$C_5H_4OS_4$ M 208.350
Cryst. Mp 127-128°.

Hartke, K. *et al*, *Chem. Ber.*, 1980, **113**, 1898 (*synth, pmr, cmr, ir, uv*)
Varma, S.K. *et al*, *Synthesis*, 1987, 837 (*synth, pmr, cmr, ir, uv, ms*)

5,6-Dihydro-1,3-dithiolo[4,5-*b*][1,4]dithiin-2-thione, 9CI   D-80206

2,5,7,9-Tetrathiabicyclo[4.3.0]non-1(6)-ene-8-thione
[59089-89-3]

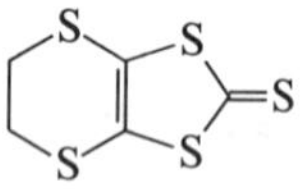

$C_5H_4S_5$ M 224.417
Cryst. (CHCl$_3$/EtOH). Mp 119-120°.

Maruno, M. *et al*, *J. Org. Chem.*, 1976, **41**, 1484 (*synth*)
Hartke, K. *et al*, *Chem. Ber.*, 1980, **113**, 1898 (*synth, pmr, cmr, ir*)
Varma, S.K. *et al*, *Synthesis*, 1987, 837 (*synth, pmr, cmr, ir, uv, ms*)

1,4-Dihydro-1,4-ethanonaphtho[1,8-*de*][1,2]diazepine, 9CI   D-80207

[52720-26-0]

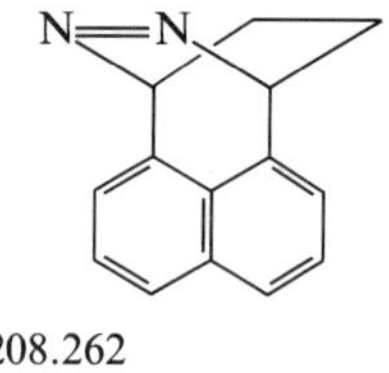

$C_{14}H_{12}N_2$ M 208.262
Pale yellow cryst. (CHCl$_3$/hexane) or glassy solid. Mp 123-125°.

Watson, C.R. *et al*, *J. Am. Chem. Soc.*, 1976, **98**, 2551 (*synth, uv, pmr*)
Burnett, M.N. *et al*, *J. Am. Chem. Soc.*, 1988, **110**, 2527 (*synth, uv, use*)

Dihydro-3-fluoro-2(3*H*)-furanone   D-80208

2-Fluoro-γ-butyrolactone

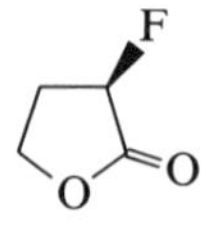

(R)-form

$C_4H_5FO_2$ M 104.081

(*R*)-form [86677-74-9]
Volatile oil. [α]$_D^{22}$ +50.3° (c, 0.95 in CHCl$_3$).

[86677-82-9]

Shiuey, S.-J. *et al*, *J. Org. Chem.*, 1988, **53**, 1040 (*synth, ir, pms, ms*)

2,3-Dihydro-3′,4′,4″,5,5″,7,7″-hepta-hydroxy-3,8′-biflavone   D-80209

3′,4′,4″,5,5″,7,7″-Heptahydroxy[3,8]flavanonylflavone
[119459-71-1]

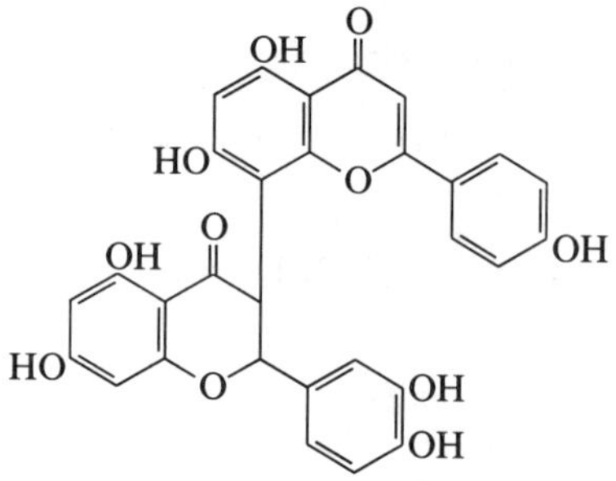

$C_{30}H_{20}O_{11}$ M 556.481
Constit. of *Garcinea nervosa*. Yellow cryst. (MeOH). Mp 232-234°.

Babu, V. *et al*, *Phytochemistry*, 1988, **27**, 3332.

Dihydro-3,3,4,4,5,5-hexamethyl-2*H*-pyran-2,6(3*H*)-dione, 9CI   D-80210

3,3,4,4,5,5-Hexamethyltetrahydropyran-2,6-dione

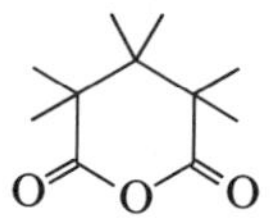

$C_{11}H_{18}O_3$ M 198.261
Cryst. by subl. Mp 174-175°.

Baran, J. *et al*, *J. Org. Chem.*, 1988, **53**, 4626 (*synth, ir, pmr, cmr, ms*)

Dihydro-3-hydroxy-2,5-furandione, 9CI   D-80211

Hydroxysuccinic anhydride. Malic anhydride
[1121-34-2]

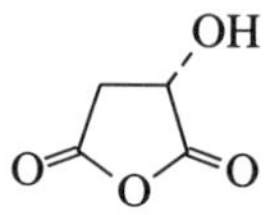

(S)-form

$C_4H_4O_4$ M 116.073
Freq. confused in CA with maleic anhydride.

(*S*)-form
Me ether: [117237-88-4]. *Dihydro-3-methoxy-2,5-furandione, 9CI. 2-Methoxysuccinic anhydride*
$C_5H_6O_4$ M 130.100
Bp$_{15}$ 125°. [α]$_D^{20}$ −104° (c, 2 in C$_6$H$_6$).
tert-Butyl ether: [117237-89-5]. *3-(1,1-Dimethylethoxy)dihydro-2,5-furandione, 9CI. 2-tert-Butoxysuccinic anhydride*
$C_8H_{12}O_4$ M 172.180
Cryst. (toluene/hexane). Mp 64°. [α]$_D^{20}$ −51.5°.

(±)-form
Cryst. (Et$_2$O). Sol. H$_2$O. Mp 75-76°.

Purdie, T. *et al*, *J. Chem. Soc.*, 1910, **97**, 1524 (*derivs*)
Denhorn, W.S., *J. Chem. Soc.*, 1913, **103**, 1861 (*synth*)
Polonski, T., *J. Chem. Soc., Perkin Trans. 1*, 1988, 629 (*derivs*)

Dihydro-3-hydroxy-2(3*H*)-furanone   D-80212

2-Hydroxy-γ-butyrolactone

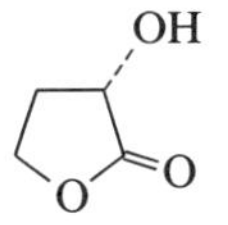

(S)-form

$C_4H_6O_3$ M 102.090
(*S*)-form [52079-23-9]
  Liq. $Bp_{0.1}$ 95°. $[\alpha]_D^{25}$ −65.2°.

Abdallah, M.A. *et al*, *J. Chem. Soc., Perkin Trans. 1*, 1975, 888 (*synth*)
Shiuey, S.-J. *et al*, *J. Org. Chem.*, 1988, **53**, 1040 (*synth, ir, pmr, ms*)

2,3-Dihydro-5-hydroxy-8-methoxy-2,4-dimethylnaphtho[1,2-b]furan-6,9-dione   D-80213

[115401-70-2]

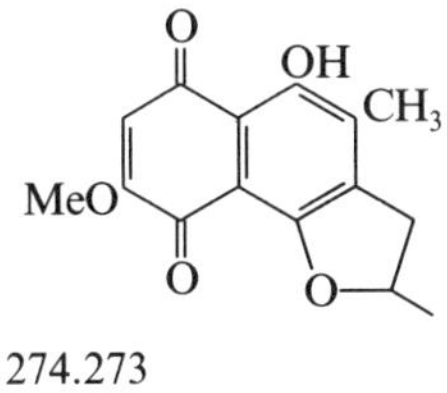

$C_{15}H_{14}O_5$ M 274.273
Metab. of *Fusarium solani*. Red cryst. (EtOAc). Mp 220-224°.

Tatum, J.H. *et al*, *Phytochemistry*, 1989, **28**, 283.

Dihydro-4-hydroxy-3-methyl-2(3*H*)-furanone   D-80214

2-Methyl-3,4-dihydroxybutanoic acid 1,4-lactone

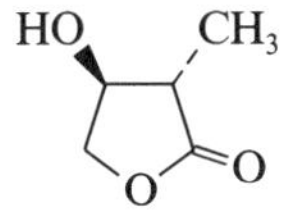

$C_5H_8O_3$ M 116.116
(3*S*,4*S*)-form [112138-02-0]
  (−)-*trans*-form
  Liq. Bp_{15} 163-165°. $[\alpha]_D^{20}$ −58° (c, 2.87 in MeOH).

Larcheveque, M. *et al*, *Tetrahedron*, 1987, **43**, 2303 (*synth, pmr, cmr, ms*)

1,3-Dihydro-2*H*-imidazole-2-thione, 9CI   D-80215

Imidazole-2-thione. 2-Mercaptoimidazole. Imidazole-2-thiol, 8CI

[872-35-5]

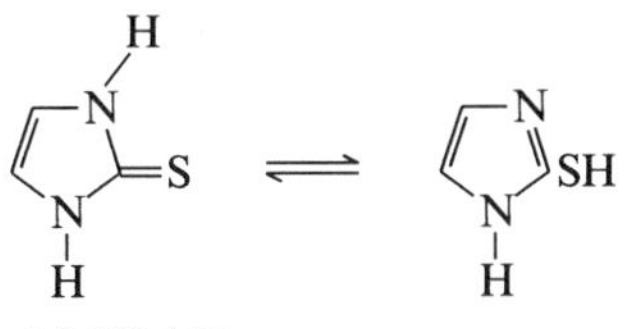

$C_3H_4N_2S$ M 100.144
Cryst. (H_2O). Mp 226-228°.

1-Me: see 1,3-*Dihydro-1-methyl*-2H-*imidazole-2-thione*, D-80221

S-*Me:* [7666-04-8]. *2-(Methylthio)-1H-imidazole, 9CI*
$C_4H_6N_2S$ M 114.171
Mp 142°.

1,3-Di-Me: see 1,3-*Dihydro-1-methyl*-2H-*imidazole-2-thione*, D-80221

Marckwald, W., *Ber.*, 1892, **25**, 2354 (*synth*)
Akobori, S., *Ber.*, 1933, **66**, 151 (*synth*)
Simon, I.B. *et al*, *J. Gen. Chem. USSR (Engl. Transl.)*, 1955, **25**, 1173 (*synth, deriv*)
Freeman, F. *et al*, *J. Am. Chem. Soc.*, 1988, **110**, 2586 (*ir, pmr, cmr, uv, ms, cryst struct*)

5,10-Dihydroindeno[1,2-*b*]indole, 9CI   D-80216

*Indeno[1,2-*b*]indole*

[3254-91-9]

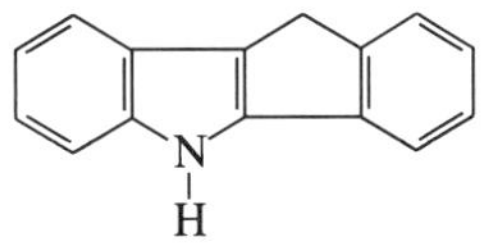

$C_{15}H_{11}N$ M 205.259
Needles (MeOH). Mp 227-228°.
N-Me:
  $C_{16}H_{13}N$ M 219.285
  Cryst. (EtOH). Mp 153-154°.

Armit, J.W. *et al*, *J. Chem. Soc.*, 1922, 827.
Kahaoka, Y. *et al*, *Bull. Chem. Soc. Jpn.*, 1966, **14**, 934 (*synth*)

2,3-Dihydro-1,4-isoquinolinedione   D-80217

Tetrahydroisoquinolinedione

[31053-30-2]

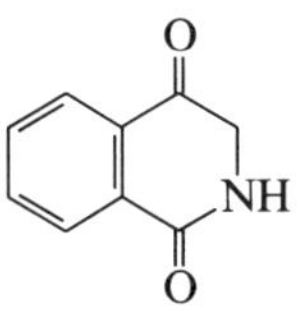

$C_9H_7NO_2$ M 161.160
Mp >300°.

Caswell, R.L. *et al*, *J. Org. Chem.*, 1961, **26**, 4175 (*synth*)
Caswell, R.L. *et al*, *J. Heterocycl. Chem.*, 1970, **7**, 1205 (*uv*)

2,3-Dihydro-1,2,3-metheno-1*H*-phenalene, 9CI   D-80218

*Naphtho[1,8]tricyclo[4.1.0.0^{2,7}]heptene.
Naphthobicyclobutane*

[40480-63-5]

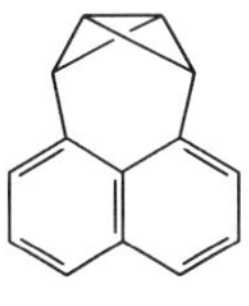

$C_{14}H_{10}$ M 178.233
Needles (pet. ether). Mp 76.5-77°.

Murata, I. *et al*, *Tetrahedron Lett.*, 1973, 47 (*synth, ms, uv, ir, pmr*)
Pagni, R.M. *et al*, *Tetrahedron Lett.*, 1973, 59 (*synth*)
Watson, C.R. *et al*, *J. Am. Chem. Soc.*, 1976, **98**, 2557 (*synth*)
Gleiter, R. *et al*, *Helv. Chim. Acta*, 1981, **64**, 1312 (*pe, struct*)

10,11-Dihydro-5-methylene-5*H*-dibenzo [*a*,*d*]cycloheptene, 9CI　　　D-80219

3-Methylenedibenzsuberane

[2732-90-3]

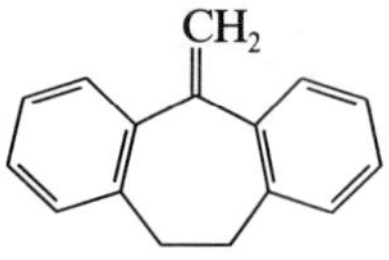

C$_{16}$H$_{14}$　　M 206.287
Cryst. (pet. ether). Mp 65-67°.

Triebs, W. *et al*, *Ber*., 1950, **83**, 367 (*synth*)
Lamanec, T.R. *et al*, *J. Org. Chem*., 1988, **53**, 1768 (*synth, pmr*)

Dihydro-4-methyl-2(3*H*)-furanone　　　D-80220

3-Methyl-4-butanolide. 3-Methyl-γ-butyrolactone

[1679-49-8]

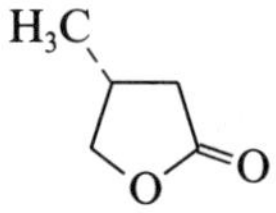

(*R*)-*form*

C$_5$H$_8$O$_2$　　M 100.117
(*R*)-*form* [65284-00-6]
　Bp$_{45}$ 120°. [α]$_D^{20}$ +24.0° (c, 3.32 in Et$_2$O).
(*S*)-*form* [64190-48-3]
　Bp$_{15}$ 87-89°. [α]$_D$ −24.6° (neat).
(±)-*form* [70470-05-2]
　d$_4^{20}$ 1.058. Bp$_{11}$ 76-76.5°. n$_D^{20}$ 1.4339.

Seidel, C.F. *et al*, *Helv. Chim. Acta*, 1959, **42**, 1830 (*synth, ir*)
Leuenberger, H.G.W. *et al*, *Helv. Chim. Acta*, 1979, **62**, 455 (*synth, pmr*)
Mukaiyama, T. *et al*, *Chem. Lett*., 1980, 635 (*synth*)
Mori, K. *et al*, *Tetrahedron*, 1983, **39**, 3107 (*synth, ir, pmr*)
Mattes, H. *et al*, *J. Med. Chem*., 1987, **30**, 1948 (*synth, pmr, ir*)
Fuganti, C. *et al*, *J. Chem. Soc., Perkin Trans. 1*, 1988, 3061 (*synth*)

1,3-Dihydro-1-methyl-2*H*-imidazole-2-thione, 9CI　　　D-80221

Updated Entry replacing D-03892

1-Methylimidazole-2-thiol, 8CI. 2-Mercapto-1-methylimidazole. **Methimazole, BAN.** *Thiamazole, INN. Mercazolyl. Metimazol. Tapazole. Other proprietary names*

[60-56-0]

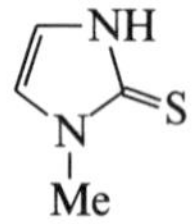

C$_4$H$_6$N$_2$S　　M 114.171
Used in treatment of hyperthyroidism. Leaflets (EtOH).
　Mp 146-148°. Bp 280° dec.
▷ NI8615000.

Wohl, A. *et al*, *Ber*., 1889, **22**, 1354 (*synth*)
Jones, R.G. *et al*, *J. Am. Chem. Soc*., 1949, **71**, 4000 (*synth*)
Bowie, J.H. *et al*, *Aust. J. Chem*., 1967, **20**, 1613 (*ms*)
Langer, P., *Endocrinology*, 1968, **83**, 1268 (*pharmacol*)
Aboul-Enein, H.Y. *et al*, *Anal. Profiles Drug Subst*., 1979, **8**, 351 (*rev*)
Skellern, G.G. *et al*, *Xenobiotica*, 1981, **11**, 627 (*metab*)
Martindale, The Extra Pharmacopoeia, 28th/29th Ed., Pharmaceutical Press, London, 1982/1989, 835.

Raper, E.S. *et al*, *Acta Crystallogr., Sect. B*, 1983, **39**, 355 (*cryst struct*)
Anjaneyulu, B. *et al*, *J. Labelled Compd. Radiopharm*., 1983, **20**, 951 (*synth*)
Merck Index, 10th Ed. Nos. 1983, 5844, 1983.
Cooper, D.S. *et al*, *Endocrinology*, 1984, **114**, 786 (*pharmacol*)
Balestrero, R.S. *et al*, *Magn. Reson. Chem*., 1986, **24**, 651 (*N-15 nmr, cmr*)
Negwer, M., *Organic-Chemical Drugs and their Synonyms*, 6th Ed., Akademie-Verlag, Berlin, 1987, 180 (*synonyms*)

Dihydro-4-methyl-3-methylene-2(3*H*)-furanone　　　D-80222

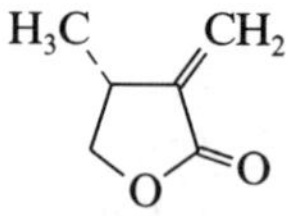

(*R*)-*form*

C$_6$H$_8$O$_2$　　M 112.128
(*R*)-*form* [62322-50-3]
　[α]$_D^{20}$ +73.3° (6.17 in CHCl$_3$).
[110098-10-7]

Mattes, H. *et al*, *J. Med. Chem*., 1987, **30**, 1948 (*synth, ir, pmr*)

7-[[4-(2,5-Dihydro-4-methyl-5-oxo-2-furanyl)-3-methyl-2-butenyl]oxy]-2*H*-1-benzopyran-2-one, 9CI　　　D-80223

[53771-63-4]

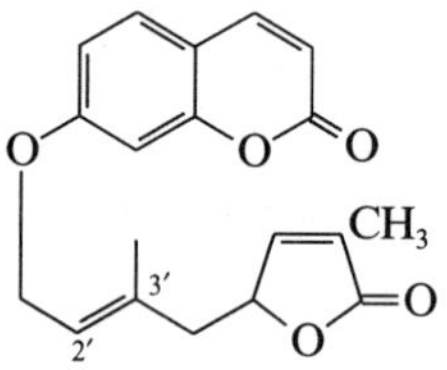

C$_{19}$H$_{18}$O$_5$　　M 326.348
Isol. from *Capnophyllum peregrinum*. Cryst. (MeOH). Mp 126°. [α]$_D^{25}$ −26°.
2′,3′-Epoxide: [51559-34-3]. *7-[[3-[(2,5-Dihydro-4-methyl-5-oxo-2-furanyl)methyl]-3-methyloxiranyl]methoxy]-2H-1-benzopyran-2-one, 9CI*
C$_{19}$H$_{18}$O$_6$　　M 342.348
Mp 125°.
[51559-33-2]

Bohlmann, F. *et al*, *Chem. Ber*., 1970, **103**, 3619; 1974, **107**, 1780 (*isol, uv, ir, pmr, struct*)

1,4-Dihydro-6-methyl-4-oxo-3-pyridine-carboxylic acid, 9CI　　　D-80224

6-Methyl-4(1H)-pyridone-3-carboxylic acid. 4-Hydroxy-6-methyl-3-pyridinecarboxylic acid

[33821-58-8]

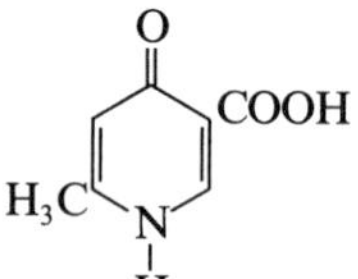

C$_7$H$_7$NO$_3$　　M 153.137
Cryst. (H$_2$O). Mp 267°.
1-Me: [33821-59-9].
C$_8$H$_9$NO$_3$　　M 167.164

Cryst. (EtOH aq.). Mp 231-232°.

Kilbourn, E. *et al, J. Org. Chem.,* 1972, **37**, 1145 (*synth, deriv*)
Mittelbach, M. *et al, Synthesis,* 1988, 479 (*synth*)

1,6-Dihydro-3-methyl-6-oxo-2-pyridine-carboxylic acid, 9CI   D-80225

3-Methyl-6-carboxy-2-pyridone

[115185-81-4]

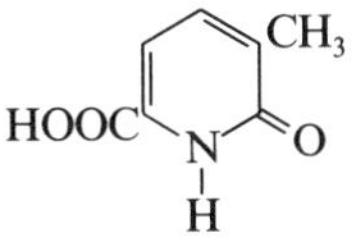

$C_7H_7NO_3$ M 153.137
Cryst. (MeOH). Mp 209-210°.

Konno, K. *et al, J. Am. Chem. Soc.,* 1988, **110**, 4807 (*synth, ir, uv, pmr*)

1,6-Dihydro-5-methyl-6-oxo-2-pyridine-carboxylic acid, 9CI   D-80226

5-Methyl-6-carboxy-2-pyridone

[115185-79-0]

$C_7H_7NO_3$ M 153.137
Cryst. Mp 295° dec.

Konno, K. *et al, J. Am. Chem. Soc.,* 1988, **110**, 4807 (*synth, ir, uv, pmr*)

Dihydro-4-methyl-5-pentyl-2(3H)-furanone   D-80227

Tetrahydro-4-methyl-5-pentyl-2-furanone. 3-Methyl-4-nonanolide. Cognac lactone

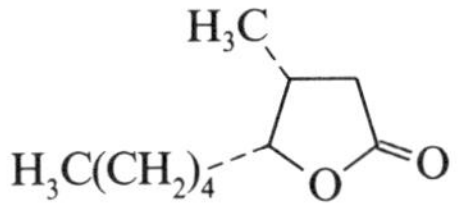

$C_{10}H_{18}O_2$ M 170.251
(3RS,4RS)-form
(±)-cis-*form*
Found in cognac, presumably in an enantiomeric form. Oil.

Jefford, C.W. *et al, Helv. Chim. Acta,* 1989, **72**, 1362 (*synth, ir, pmr, cmr, bibl*)

5,6-Dihydro-3-methyl-2H-pyran-2-one   D-80228

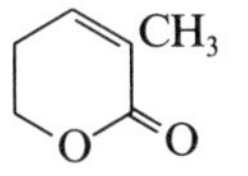

$C_6H_8O_2$ M 112.128
Liq. Bp$_2$ 42-44° (bulb) .

Hofstraat, R.G. *et al, J. Chem. Soc., Perkin Trans.* 1, 1988, 2315 (*synth, ms, ir, pmr*)

3,4-Dihydro-6-methyl-2(1H)-pyridinone, 9CI   D-80229

1,2,3,4-Tetrahydro-6-methyl-2-oxopyridine

[10333-14-9]

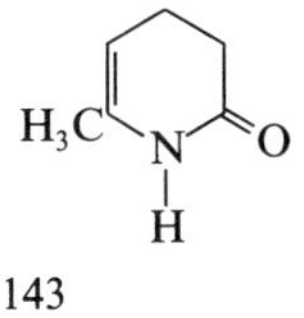

C_6H_9NO M 111.143

Cryst. (Et$_2$O). Mp 108°.

Schumann, D. *et al, Justus Liebigs Ann. Chem.,* 1983, 220 (*synth, ir, uv, pmr, cmr, ms*)
Knapp, S. *et al, J. Org. Chem.,* 1988, **53**, 4006 (*synth, ir*)

2H-3,4-Dihydronaphtho[2,3-b]pyran-5,10-dione   D-80230

[106261-85-2]

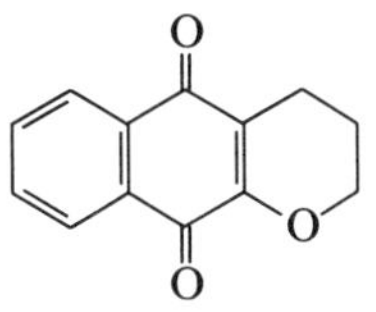

$C_{13}H_{10}O_3$ M 214.220
Yellow needles (MeOH). Mp 219-220°.

Hayashi, T. *et al, J. Med. Chem.,* 1987, **30**, 2005 (*synth, ir, uv, pmr*)

1,2-Dihydro-3H-naphtho[2,1-b]pyran-3-one, 9CI   D-80231

3,4-Dihydro-5,6-benzocoumarin

[5690-03-9]

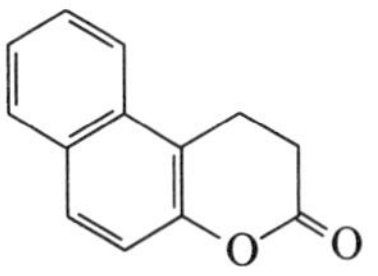

$C_{13}H_{10}O_2$ M 198.221
Solid (hexane). Mp 54-55°.

Das Gupta, A.K. *et al, J. Chem. Soc. C,* 1969, 2618 (*synth, ir, uv*)

2,3-Dihydro-1H-naphtho[2,1-b]pyran-1-one, 9CI   D-80232

7,8-Benzochromanone

[4707-36-2]

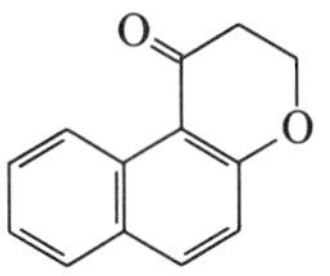

$C_{13}H_{10}O_2$ M 198.221
Light yellow needles (EtOH) or rhombic tablets. Mp 108° (104.5°, 96-98°). Bp$_{0.5}$ 156-160°.

Oxime:
    $C_{13}H_{11}NO_2$ M 213.235
    Plates (EtOH). Mp 144-144.5° (134°).

Semicarbazone: Needles. Mp 276° (259-260°)dec.

2,4-Dinitrophenylhydrazone: Red cryst. Mp 302°.

Pfeiffer, P. *et al, J. Chem. Soc.,* 1917, **112**, 661 (*synth*)
Colonge, J. *et al, Bull. Soc. Chim. Fr.,* 1958, 325 (*synth*)
Cagniant, P. *et al, Bull. Soc. Chim. Fr.,* 1966, 3249 (*synth*)
Kasturi, T.R. *et al, Indian J. Chem.,* 1970, **8**, 203 (*synth, ir*)
Sharma, S.D. *et al, Indian J. Chem., Sect. B,* 1984, **23**, 518 (*synth, ir, pmr*)
Subramanian, L.M. *et al, Synthesis,* 1984, 1063 (*synth*)
Ravikumar, K. *et al, Acta Crystallogr., Sect. C,* 1986, **42**, 1423 (*cryst struct*)

2,3-Dihydro-4*H*-naphtho[1,2-*b*]pyran-4-one, 9CI   D-80233

6,7-Benzochromanone

[16563-51-2]

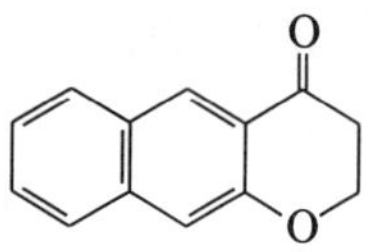

$C_{13}H_{10}O_2$ M 198.221

Pale yellow needles (MeOH). Mp 131-132°.

2,4-Dinitrophenylhydrazone: Red cryst. (CHCl$_3$/MeOH). Mp 252-254° dec.

Bell, K.H. *et al*, *Aust. J. Chem.*, 1963, **16**, 690 (*synth*)

2,3-Dihydro-4*H*-naphtho[2,3-*b*]pyran-4-one, 9CI   D-80234

5,6-Benzochromanone

[4707-39-5]

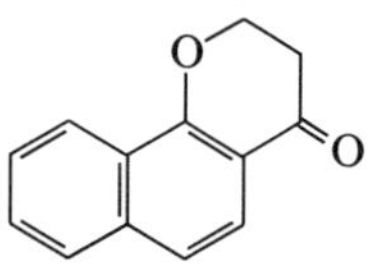

$C_{13}H_{10}O_2$ M 198.221

Yellow needles (EtOH), cryst. (C$_6$H$_6$/pet. ether). Mp 50-51° (44°). Bp$_{2.5}$ 160-161°. $n_D^{23.2}$ 1.6556.

Oxime:

$C_{13}H_{11}NO_2$ M 213.235

Cryst. (EtOH). Mp 112°.

2,4-Dinitrophenylhydrazone: Brilliant scarlet solid. Mp 312°.

Semicarbazone: Light yellow plates (EtOH). Mp 227° dec.

Bachman, G.B. *et al*, *J. Am. Chem. Soc.*, 1947, **69**, 2341 (*synth*)
·Colonge, J. *et al*, *Bull. Soc. Chim. Fr.*, 1958, 325 (*synth*)
Bell, K.H. *et al*, *Aust. J. Chem.*, 1963, **16**, 101 (*synth*)
Cagniant, P. *et al*, *Bull. Soc. Chim. Fr.*, 1966, 3249 (*synth*)
Shenker, G.C. *et al*, *Synthesis*, 1983, 310 (*synth, ir, uv*)

3,4-Dihydro-2*H*-naphtho[1,2-*b*]pyran-2-one, 8CI   D-80235

3,4-Dihydro-7,8-benzocoumarin

[5690-05-1]

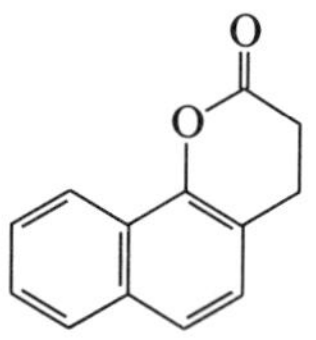

$C_{13}H_{10}O_2$ M 198.221

Mp 75-76°.

Das Gupta, A.K. *et al*, *J. Chem. Soc. C*, 1969, **18**, 2618 (*synth, uv, ir*)
Sarma, A.S., *Indian J. Chem.*, 1971, **9**, 93 (*synth, uv*)

3,4-Dihydro-1*H*-naphtho[1,2-*c*]pyran-3-one   D-80236

3,4-Dihydro-7,8-benzoisocoumarin

[54959-66-9]

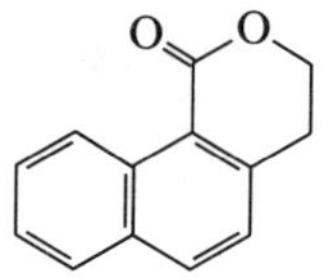

$C_{13}H_{10}O_2$ M 198.221

Solid. Mp 106°.

Bhide, B.H. *et al*, *Chem. Ind.* (*London*), 1974, 773 (*synth, uv, pmr*)

3,4-Dihydro-1*H*-naphtho[2,3-*c*]pyran-1-one, 9CI   D-80237

3,4-Dihydro-6,7-benzoisocoumarin

[54959-67-0]

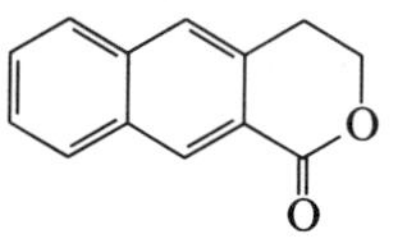

$C_{13}H_{10}O_2$ M 198.221

Solid. Mp 101°.

Bhide, B.H. *et al*, *Chem. Ind.* (*London*), 1974, 773 (*synth, uv, pmr*)

2,5-Dihydro-2-(4,6,8-nonatrien-2-ynylidene)furan, 9CI   D-80238

2-(4,6,8-Nonatrien-2-ynylidene)-2,5-dihydrofuran

[13894-85-4]

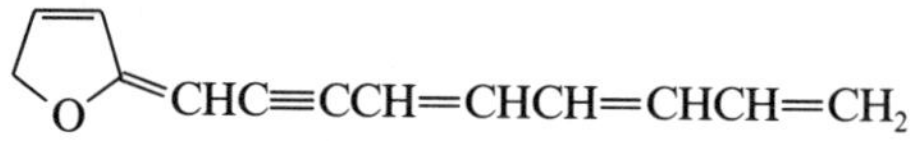

CHC≡CCH=CHCH=CHCH=CH$_2$

$C_{13}H_{12}O$ M 184.237

Isol. from roots of *Centaurea pullata* and *C. ferax*. Yellowish oil. λ_{max} 359, 379 nm (Et$_2$O).

Bohlmann, F. *et al*, *Chem. Ber.*, 1966, **99**, 3544 (*isol, uv, ir, pmr, ms, struct*)

Dihydro-5-phenyl-2(3*H*)-furanone, 9CI   D-80239

4-Phenyl-4-butanolide. γ-Phenylbutyrolactone

[1008-76-0]

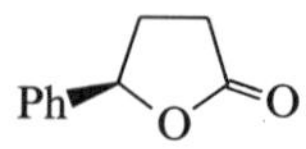

Ph   (*R*)-form

$C_{10}H_{10}O_2$ M 162.188

(*R*)-*form* [111138-03-5]

[α]$_D$ +16.3° (c, 0.89 in EtOH).

(±)-*form* [69814-97-7]

Solid. Mp 38°. Bp$_{11}$ 171-172°, Bp$_{0.6}$ 115°.

Cromwell, N.H. *et al*, *J. Am. Chem. Soc.*, 1956, **78**, 4412 (*synth*)
Francotte, E. *et al*, *Helv. Chim. Acta*, 1987, **70**, 1569 (*resoln, abs config*)
Ballini, R. *et al*, *Synthesis*, 1987, 711 (*synth, ir, pmr*)

5,6-Dihydro-6-phenyl-2,4(1*H*, 3*H*)-pyrimidinedione, 9CI   **D-80240**

6-Phenyldihydrouracil

[6300-95-4]

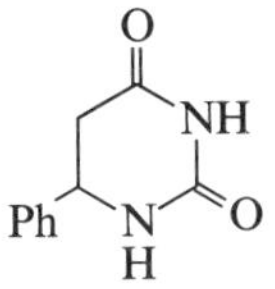

C$_{10}$H$_{10}$N$_2$O$_2$ M 190.201

(±)-*form*

Cryst. Mp 216-218°.

Sandakhchiev, L.S. *et al, Izv. Sib. Otd. Akad. Nauk SSSR, Ser. Khim. Nauk*, 1961, 72; *CA*, **56**, 2448.
Strekowski, L. *et al, Synthesis*, 1987, 579 (*synth, pmr*)

1,2-Dihydro-5-phenyl-3*H*-1,2,4-triazol-3-one, 9CI   **D-80241**

5-Phenyl-Δ^{1,5}-1,2,4-triazolin-3-one

[939-07-1]

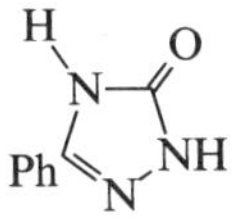

C$_8$H$_7$N$_3$O   M 161.163

Cryst. (butanol). Mp 323-324° (314-317°) dec.

Young, G. *et al, J. Chem. Soc.*, 1900, 224 (*synth*)
George, B. *et al, J. Org. Chem.*, 1976, **41**, 3233 (*synth, ir*)
Schantl, J.G. *et al, Synthesis*, 1987, 986 (*synth, ir, pmr*)

3,6-Dihydro-2*H*-pyran-2-carboxylic acid, 9CI   **D-80242**

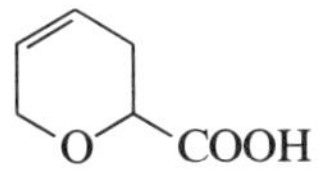

C$_6$H$_8$O$_3$ M 128.127

(±)-*form* [116233-54-6]

Oil. Bp$_{10}$ 100°.

Et ester:
   C$_8$H$_{12}$O$_3$ M 156.181
   Liq. Bp$_2$ 57-58°. d$_4^{20}$ 1.07.

Amide:
   C$_6$H$_9$NO$_2$ M 127.143
   Cryst. (diisopropyl ether). Mp 110-111°.

Arbuzov, A. *et al, Dokl. Akad. Nauk SSSR, Ser. Khim.*, 1962, **142**, 341 (*ester*)
Mochalins, V.B. *et al, J. Gen. Chem. USSR (Engl. Transl.)*, 1974, **44**, 2288 (*amide*)
Knapp, S. *et al, J. Org. Chem.*, 1988, **53**, 4773 (*synth, pmr, ir*)

3,6-Dihydro-2*H*-pyran-2,2-dicarboxylic acid, 9CI   **D-80243**

[57668-92-5]

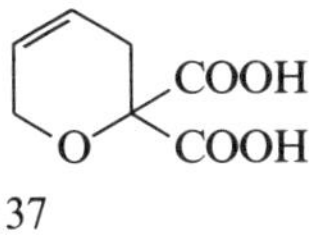

C$_7$H$_8$O$_5$ M 172.137

Golden oil.

Di-Et ester:
   C$_{11}$H$_{16}$O$_5$ M 228.244

Oil. Bp$_{0.8}$ 100°.

Ruden, R.A. *et al, J. Am. Chem. Soc.*, 1975, **97**, 6892 (*synth*)
Bonjouklian, R. *et al, J. Org. Chem.*, 1977, **42**, 4095 (*synth*)
Knapp, S. *et al, J. Org. Chem.*, 1988, **53**, 4773 (*synth, pmr, ir*)

3,4-Dihydro-2*H*-pyrano[2,3-*b*]pyridine, 9CI   **D-80244**

8-Azachroman

[26267-89-0]

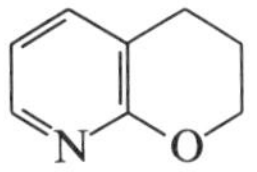

C$_8$H$_9$NO   M 135.165

Pale yellow liq. Mp 30°. Bp$_1$ 56°.

Sliwa, H., *Bull. Soc. Chim. Fr.*, 1970, 631 (*synth, ir, pmr*)
Taylor, E.C. *et al, Tetrahedron*, 1987, **43**, 5145 (*synth, ir, pmr, cmr*)

3,5-Dihydro-4*H*-pyrazolo[3,4-*c*]quinolin-4-one, 9CI   **D-80245**

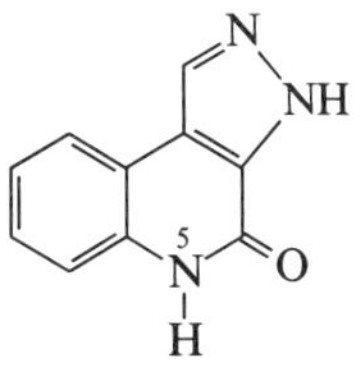

C$_{10}$H$_7$N$_3$O   M 185.185

N^5-Me: [120757-42-8].
   C$_{11}$H$_9$N$_3$O   M 199.212
   Pale yellow prisms (2-propanol). Mp >250°.

Ito, K. *et al, J. Heterocycl. Chem.*, 1988, **25**, 1681 (*synth, pmr*)

7,12-Dihydropyrido[3,2-*b*:5,4-*b'*]diindole   **D-80246**

*Pyrido[3,2-*b:5,4-*b'*]diindole*

[98263-45-7]

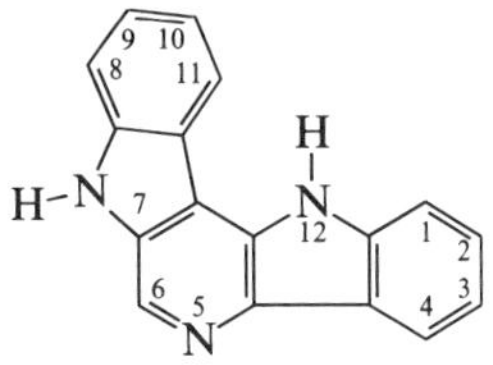

C$_{17}$H$_{11}$N$_3$ M 257.294

7-Me: [116130-39-3].
   C$_{18}$H$_{13}$N$_3$ M 271.321
   Yellow solid. Mp >300°.

7-Me; B,HCl: [116130-40-6].
   Solid. Mp >300°.

12-Me; B,HCl: [116130-42-8].
   C$_{18}$H$_{13}$N$_3$ M 271.321
   Mp >300°.

[116130-41-7, 116130-46-2]

Fukada, N. *et al, Tetrahedron Lett.*, 1985, **26**, 2139 (*synth*)
Trudell, M.L. *et al, J. Med. Chem.*, 1987, **30**, 456 (*synth*)
Trudell, M.L. *et al, J. Org. Chem.*, 1988, **53**, 4873 (*N*-Me, ms, pmr, ir)

3,4-Dihydro-2*H*-pyrrole, 9CI   D-80247

Updated Entry replacing D-04055
1-Pyrroline
[5724-81-2]

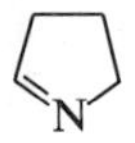

C$_4$H$_7$N   M 69.106
Stable in dil. soln., trimerises readily.

Poisel, H., *Monatsh. Chem.*, 1978, **109**, 925 (*synth, pmr*)
Bock, H. *et al, Chem. Ber.*, 1987, **120**, 1961 (*synth, pe, bibl*)
Guillemin, J.-C. *et al, Tetrahedron*, 1988, **44**, 4447 (*synth, pmr*)

3,4-Dihydro-2*H*-pyrrole-2-carboxylic acid, 9CI   D-80248

1-Pyrroline-5-carboxylic acid, 8CI. 1,5-Didehydroproline
[2906-39-0]

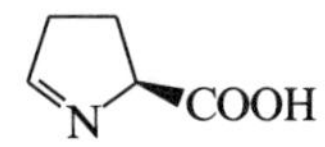

(*S*)-*form*

C$_5$H$_7$NO$_2$ M 113.116
Dimerises in solution.

(*S*)-*form* [64199-88-8]
  L-form
  Intermed. in biosynth. and biodegradn. of L-proline.
(±)-*form* [23141-14-2]
  Cryst. (H$_2$O). Mp 140-142° dec.
Et ester:
  C$_7$H$_{11}$NO$_2$ M 141.169
  Liq. Bp$_{17}$ 110°.

Vogel, H.J. *et al, J. Am. Chem. Soc.*, 1952, **74**, 109 (*synth, bibl*)
Osugi, K., *Yakugaku Zasshi (J. Pharm. Soc. Jpn.)*, 1958, **78**, 1332; *CA*, **53**, 8109 (*synth, ester*)
Strecker, H.J. *et al, Methods Enzymol., B*, 1971, **17**, 254 (*synth*)
Mezl, V.A. *et al, Anal. Biochem.*, 1976, **74**, 430 (*synth*)

3,4-Dihydro-2*H*-pyrrole-5-carboxylic acid, 9CI   D-80249

1-Pyrroline-2-carboxylic acid, 8CI
[2139-03-9]

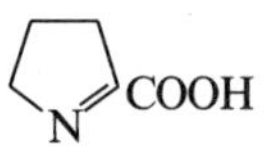

C$_5$H$_7$NO$_2$ M 113.116
Na salt: [72978-16-6].
  Cryst. Mp 265° dec.
B,HBr: Cryst. (Et$_2$O). Mp 136° dec.
Me ester: [57224-14-3].
  C$_6$H$_9$NO$_2$ M 127.143
  Liq. Bp$_{0.13}$ 43°.
tert-*Butyl ester:* [74409-25-9].
  C$_9$H$_{15}$NO$_2$ M 169.223
  Liq. Bp$_{0.27}$ 43°.

Meister, A., *J. Biol. Chem.*, 1954, **206**, 577 (*synth*)
Macholan, L. *et al, Chem. Ber.*, 1963, **96**, 237 (*synth*)
Poisl, H. *et al, Chem. Ber.*, 1975, **108**, 2547 (*synth*)
Hänglen, J. *et al, Justus Liebigs Ann. Chem.*, 1981, 1073 (*esters, synth, pmr*)

2,3-Dihydro-1*H*-pyrrolizin-6(5*H*)-one, 9CI   D-80250

2,3,5,6-Tetrahydro-1H-pyrrolizin-6-one
[97181-96-9]

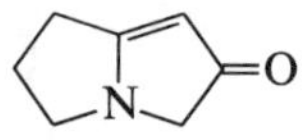

C$_7$H$_9$NO   M 123.154
Oil. Bp$_{0.0006}$ 104°.

Pommelet, J.C. *et al, J. Org. Chem.*, 1988, **53**, 5680 (*synth, ir, pmr, cmr, ms*)

2,3-Dihydro-1*H*-pyrrolo[3,2-*c*] pyridine, 9CI   D-80251

5-Azaindoline
[23596-28-3]

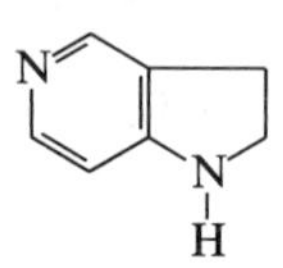

C$_7$H$_8$N$_2$ M 120.154
Cryst (cyclohexane). Mp 102-103°.
B,HCl: [5912-19-6].
  Cryst. (EtOH/EtOAc). Mp 188-189°.

Yakhontov, L.N. *et al, J. Heterocycl. Chem. USSR (Engl. Transl.)*, 1969, **5**, 410 (*synth, uv, ir, pmr*)

1,6-Dihydro-7*H*-pyrrolo[2,3-*c*]pyridin-7-one, 9CI   D-80252

1H-Pyrrolo[2,3-c]pyridin-7(6H)-one

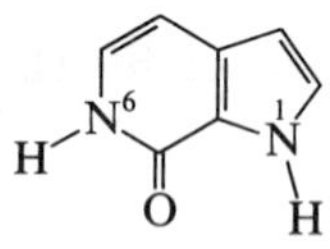

C$_7$H$_6$N$_2$O   M 134.137
6-Me: [116212-46-5].
  C$_8$H$_8$N$_2$O   M 148.164
  Clusters (toluene). Mp 190-192°.

Dumas, D.J., *J. Org. Chem.*, 1988, **53**, 4650 (*synth, ir, pmr*)

1,4-Dihydropyrrolo[3,2-*b*]pyrrole, 9CI   D-80253

Updated Entry replacing D-30286
1,4-Dihydro-1,4-diazapentalene. Pyrrolo[3,2-b]pyrrole. 1,4-Diazapentalene
[63156-05-8]

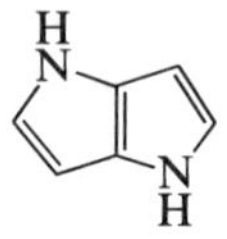

C$_6$H$_6$N$_2$ M 106.127
Needles (Me$_2$CO/hexane). Mp 123-124° dec. Unstable to air and acids.
1,4-Bis(methoxycarbonyl):
  C$_{10}$H$_{10}$N$_2$O$_4$ M 222.200
  Plates. Mp 163-164°.
1,4-Di-Me: [54724-95-7].
  C$_8$H$_{10}$N$_2$ M 134.180
  Mp 110°.
1,4-Diisopropyl: [117203-13-1].
  C$_{12}$H$_{18}$N$_2$ M 190.288
  Bp$_3$ 80°.

1,4-Di-tert-*butyl:* [117203-12-0].
$C_{14}H_{22}N_2$ M 218.341
Cryst.

Prinzbach, H. *et al*, *Angew. Chem., Int. Ed. Engl.*, 1975, **14**, 347 (*deriv*)
Kumagai, T. *et al*, *Tetrahedron Lett.*, 1984, **25**, 5669 (*synth, ir, pmr, cmr, uv*)
tom Dieck, H. *et al*, *Chem. Ber.*, 1989, **122**, 129 (*synth, cryst struct, ir, raman, pmr, derivs*)

3,4-Dihydro-3,3,6,8-tetramethyl-1(2*H*)-   D-80254
napthalenone, 8CI, 9CI

3,3,6,8-Tetramethyl-1-tetralone
[5409-55-2]

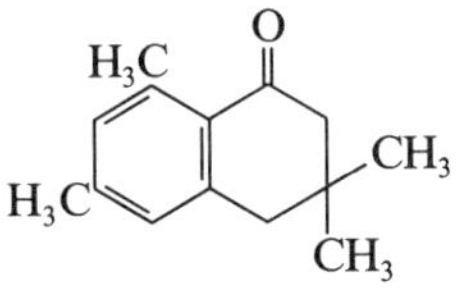

$C_{14}H_{18}O$ M 202.296
Condensation prod. of acetone. Cryst. (MeOH). Mp 56.5-57°. Bp 292-294°, Bp$_{12}$ 146°.

2,4-Dinitrophenylhydrazone: [16695-83-3].
Red cryst. (EtOH/EtOAc). Mp 234°.

[38157-22-1]

Roger, A. *et al*, *Bull. Soc. Chim. Fr.*, 1967, 3030 (*synth, uv, pmr, ir*)
Frank, H.-G. *et al*, *Justus Liebigs Ann. Chem.*, 1969, **724**, 94 (*synth*)
Riand, J., *Bull. Soc. Chim. Fr.*, 1975, 1339 (*ms*)
Haag, R. *et al*, *Helv. Chim. Acta*, 1977, **60**, 2595 (*pmr, props*)
Bertrand, J.A. *et al*, *J. Org. Chem.*, 1977, **42**, 1600 (*uv, pmr, ms*)

3,4-Dihydro-3,3,4,4-tetramethyl-5-thioxo-   D-80255
2(1*H*)pyrrolidinone

2,2,3,3-Tetramethylmonothiosuccinimide

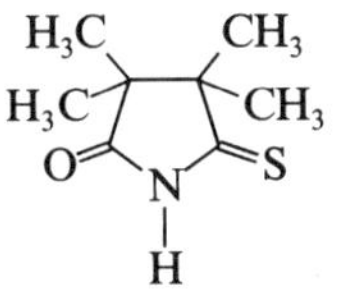

$C_8H_{13}NOS$ M 171.263
Pale yellow needles. Mp 102-105° subl.

Battersby, A.R. *et al*, *J. Chem. Soc., Perkin Trans. 1*, 1988, 1577 (*synth, ir, pmr, ms*)

7,8-Dihydrotetrazolo[1,5-*b*][1,2,4]triazine   D-80256

1,7-Dihydrotetrazolo[1,5-b][1,2,4]triazine
[117039-67-5]

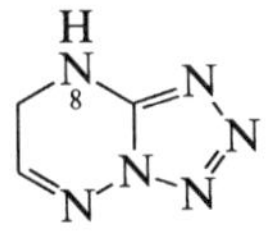

$C_3H_4N_6$ M 124.105
CAS no. refers to 1,7-dihydro. Cryst. (MeOH). Mp 161°.

8-Nitro: [117039-76-6].
$C_3H_3N_7O_2$ M 169.102
Cryst. Mp 124-126°. Dec. on recryst.

Willer, R. *et al*, *J. Org. Chem.*, 1988, **53**, 5371 (*synth, pmr, cmr*)

6,7-Dihydro-5*H*-thieno[3,2-*b*]thiopyran   D-80257

[7677-36-3]

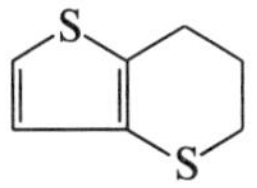

$C_7H_8S_2$ M 156.272
Oil. d$_4^{20}$ 1.26. Bp$_{17.6}$ 136.5-137°.

Cagniant, P. *et al*, *Bull. Soc. Chim. Fr.*, 1966, 2172 (*synth*)

6,7-Dihydro-4*H*-thieno[3,2-*c*]thiopyran   D-80258

[16401-43-7]

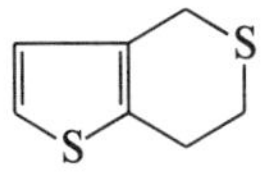

$C_7H_8S_2$ M 156.272
Strong smelling liq. d$_4^{20}$ 1.25. Bp$_{17}$ 141-142°.

Cagniant, P. *et al*, *Bull. Soc. Chim. Fr.*, 1967, 2597 (*synth, uv, ir, pmr*)

2,3-Dihydro-4*H*-thieno[2,3-*b*]thiopyran-4-   D-80259
one

[7675-04-9]

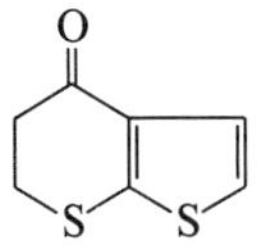

$C_7H_6OS_2$ M 170.256
Oil, slowly cryst. giving needles. Mp 61°. Bp$_{15}$ 170°.

Oxime:
$C_7H_7NOS_2$ M 185.270
Cryst. (C_6H_6/pet. ether). Mp 128°.

Semicarbazone: Cryst. (EtOH). Mp 249.5°.

2,4-Dinitrophenylhydrazone: Red cryst. (C_6H_6/EtOH). Mp 260°.

Cagniant, P. *et al*, *Bull. Soc. Chim. Fr.*, 1966, 2172 (*synth*)
Ponticello, G.S. *et al*, *J. Org. Chem.*, 1988, **53**, 9 (*synth, pmr*)

5,6-Dihydro-7*H*-thieno[3,2-*b*]thiopyran-7-   D-80260
one

[7677-33-0]

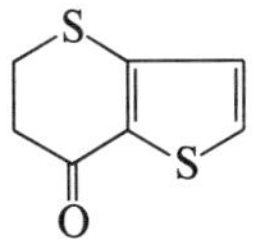

$C_7H_6OS_2$ M 170.256
Plates (EtOH aq.). Mp 71°. Bp$_{14}$ 175°.

Oxime:
$C_7H_7NOS_2$ M 185.270
Cryst. (C_6H_6/pet. ether). Mp 178°.

Semicarbazone: Cryst. (EtOH). Mp 272°.

2,4-Dinitrophenylhydrazone: Red plates (EtOH/C_6H_6). Mp 210° and 245° (double Mp).

Cagniant, P. *et al*, *Bull. Soc. Chim. Fr.*, 1966, 2172 (*synth*)

1,2-Dihydro-2-thioxo-4*H*-3,1-benzoxazin-4-one, 9CI   D-80261

2-Thioisatoic anhydride

[703-64-0]

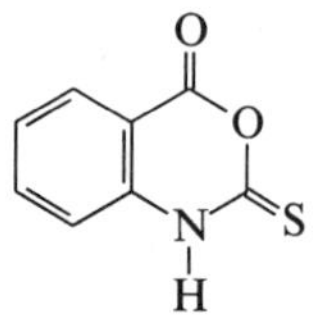

C$_8$H$_5$NO$_2$S   M 179.199

Cryst. (THF/CCl$_4$). Mp 239-241° (230-235°).

Marshall, J.R., *J. Chem. Soc.*, 1965, 938 (*synth, ir*)
Kricheldorf, H.R., *Chem. Ber.*, 1971, **104**, 3156 (*synth*)
Beam, C.F. *et al, J. Heterocycl. Chem.*, 1976, **13**, 421 (*synth*)

1,4-Dihydrotriphenylene   D-80262

[39935-60-9]

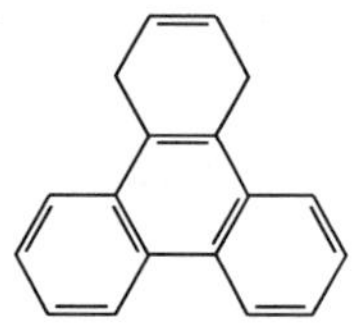

C$_{18}$H$_{14}$ M 230.309

Cryst. (EtOH). Mp 203-204°.

Marcinow, Z. *et al, J. Org. Chem.*, 1988, **53**, 3603 (*synth, pmr*)

4,5-Dihydro-2,4,5-triphenyl-1*H*-imidazole, 9CI   D-80263

2,4,5-Triphenyl-2-imidazoline, 8CI. 2,4,5-Triphenyldihydroglyoxaline

[37134-88-6]

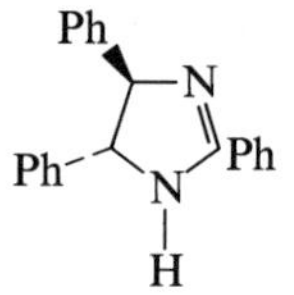

(4*R**,5*R**)-*form*

C$_{21}$H$_{18}$N$_2$ M 298.387

(4*R,5*R**)-*form***

(+)-trans-*form*

Mp 180°. [α]$_D$ +46.3°.

(+)-*Tartrate salt:* Prisms (EtOH). Mp 175-176°. [α]$_D^{13}$ +55.0° (c, 1.5 in EtOH).

(4*S,5*S**)-*form***

(−)-trans-*form*

Mp 180°. [α]$_D$ −46.9°.

(+)-*Tartrate salt:* Prisms (EtOAc). Mp 174-175°. [α]$_D^{13}$ −67.1° (c, 1.5 in EtOH).

(4*RS*,5*RS*)-*form* [24427-33-6]

(±) trans-*form. Isoamarin*

Light tan cryst. (EtOH). Mp 198-201°.

1-Ac:

C$_{23}$H$_{20}$N$_2$O   M 340.424

Cryst. (pet. ether). Mp 114-116°.

1-Benzoyl:

C$_{28}$H$_{22}$N$_2$O   M 402.495

Cryst. (pet. ether). Mp 177-178°.

(4*RS*,5*SR*)-*form* [573-33-1]

(±)-cis-*form. Amarin*

Cryst. (C$_6$H$_6$/pet. ether). Mp 128-131°.

B,HCl: [42078-44-4].

Cryst. Mp 260-265°.

1-Ac: Cryst. (pet. ether). Mp 140-142°.

1-Benzoyl: Cryst. (pet. ether). Mp 179°.

1-Benzyl:

C$_{28}$H$_{24}$N$_2$ M 388.511

Needles (EtOH). Mp 123-124°.

Claus, R. *et al, Chem. Ber.*, 1885, **18**, 3079 (*derivs*)
Snape, H.L., *J. Chem. Soc.*, 1900, **77**, 781 (*resoln*)
Williams, O.F. *et al, J. Am. Chem. Soc.*, 1959, **81**, 4464 (*synth*)
White, D.M. *et al, J. Org. Chem.*, 1964, **29**, 1926 (*ir*)
Hunter, D.H. *et al, J. Am. Chem. Soc.*, 1969, **91**, 6202 (*synth*)
Hunter, D.H. *et al, Can. J. Chem.*, 1972, **50**, 669 (*synth, pmr, deriv*)

Dihydroxanthommatin   D-80264

α-Amino-3-carboxydihydro-1-hydroxy-γ,5-dioxo-5H-pyrido[3,2-a]phenoxazine-11-butanoic acid, 9CI

[25705-16-2]

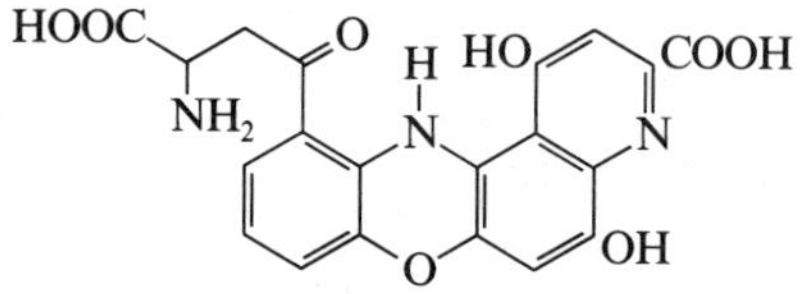

C$_{20}$H$_{15}$N$_3$O$_8$ M 425.354

Found in eyes and skin of *Loligo vulgaris, Sepia officinalis* and other molluscs, also in heads of flies *Musca domestica* and bees *Apis mellifera*. The *O*-sulfate and *O*-glucoside prob. occur also.

[121727-27-3]

Bolognese, A. *et al, Experientia*, 1974, **30**, 225.
Bolognese, A. *et al, J. Heterocycl. Chem.*, 1988, **25**, 1247.

7,12-Dihydroxy-8,12-abietadiene-11,14-dione   D-80265

Updated Entry replacing D-60304

7-Hydroxyroyleanone

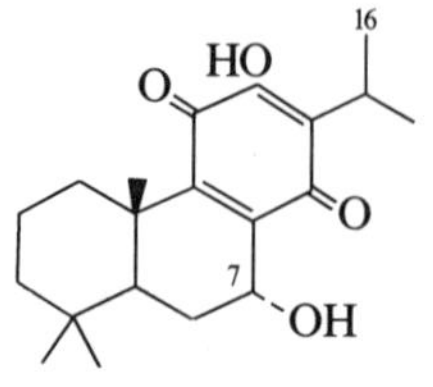

C$_{20}$H$_{28}$O$_4$ M 332.439

7α-form [21887-01-4] **Horminone**

Constit. of *Plectranthus* spp. Golden cryst. (MeOH aq.). Mp 178°. [α]$_D$ −132° (c, 1.24 in CHCl$_3$).

7-Ac: 7α-Acetoxyroyleanone. 7-Acetylhorminone

C$_{22}$H$_{30}$O$_5$ M 374.476

Isol. from whole plants of *Horminum pyrenaicum*, roots of *Salvia* spp. and leaf glands of *Coleus carnosus*.

7-Formyl ester: 7-O-Formylhorminone

C$_{21}$H$_{28}$O$_5$ M 360.449

Constit. of *Plectranthus sanguineus*. Yellow-orange cryst. (hexane). Mp 152.5-154°.

7-Me ether: 7-O-Methylhorminone. 7α-Methoxyroyleanone

C$_{21}$H$_{30}$O$_4$ M 346.466

Constit. of *Lepechinia bullata*. Yellow cryst. (CHCl$_3$). Mp 126-128°.

12-Ac, 16-acetoxy: 16-Acetoxy-12-O-acetylhorminone

C$_{24}$H$_{32}$O$_7$ M 432.513

Constit. of *R. lophanthoides*. Yellow needles. Mp 158°.

7β-form [21764-41-0] *Taxoquinone*
  Isol. from *Taxodium distichum*. Mp 214°. $[\alpha]_D^{25}$ +240° (c, 1.4 in CHCl₃).

Kupchan, S.M. *et al*, *J. Org. Chem.*, 1969, **34**, 3912 (*isol*)
Hensch, M. *et al*, *Helv. Chim. Acta*, 1975, **58**, 1921 (*isol*)
Bhat, S.V. *et al*, *Tetrahedron*, 1975, **31**, 1001 (*isol, deriv*)
Matsumoto, T. *et al*, *Bull. Chem. Soc. Jpn.*, 1979, **52**, 1459 (*synth*)
Hueso-Rodriguez, J.A. *et al*, *Phytochemistry*, 1983, **22**, 2005 (*isol, derivs*)
Matloubi-Moghadam, F. *et al*, *Helv. Chim. Acta*, 1987, **70**, 975 (*isol, deriv*)
Jonathan, L.T. *et al*, *J. Nat. Prod.* (*Lloydia*), 1989, **52**, 571 (*isol, deriv, pmr, cmr*)

11,12-Dihydroxy-5,8,11,13-abietatetraen-7-one   D-80266

Salvinolone

[120278-22-0]

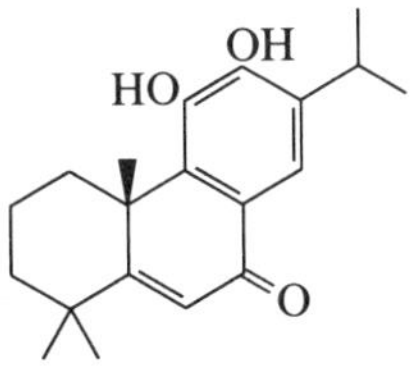

C₂₀H₂₆O₃   M 314.424
Constit. of *Salvia prionitis*. Cryst. Mp 253-254°. $[\alpha]_D$ +35.8° (c, 0.12 in MeOH).

Lin, L.-Z. *et al*, *Phytochemistry*, 1989, **28**, 177.

5,12-Dihydroxy-6,8,12-abietatriene-11,14-dione   D-80267

Hypargenin F

[119767-16-7]

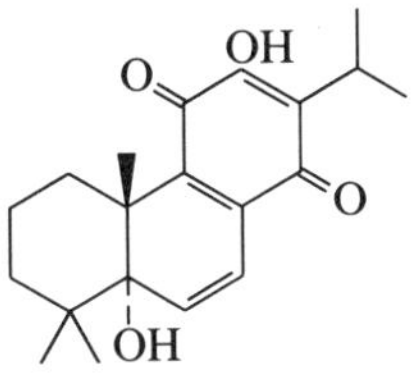

C₂₀H₂₆O₄   M 330.423
Constit. of *Salvia hypargeia*. Antibacterial agent. Dark red amorph. solid.

Ulubelen, A. *et al*, *J. Nat. Prod.* (*Lloydia*), 1988, **51**, 1178 (*isol, pmr, cmr*)

6,12-Dihydroxy-8,11,13-abietatriene-1,7-dione   D-80268

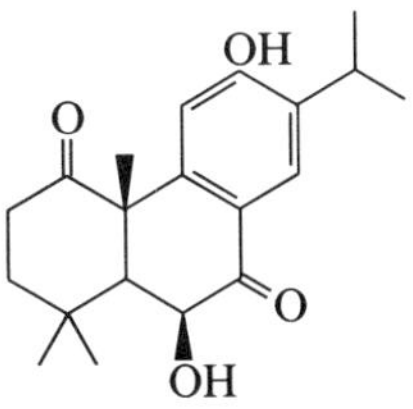

[119767-16-7]

C₂₀H₂₆O₄   M 330.423
6β-form [119767-12-3] *Hypargenin A*
  Constit. of *Salvia hypargeia*. Antibacterial agent. Dark yellow amorph. solid.

Ulubelen, A. *et al*, *J. Nat. Prod.* (*Lloydia*), 1988, **51**, 1178 (*isol, pmr,*)

11,12-Dihydroxy-8,11,13-abietatrien-20-oic acid   D-80269

Carnosic acid. *Deoxypicrosalvinic acid*

[3650-09-7]

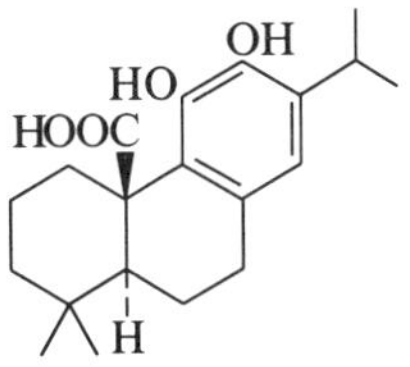

C₂₀H₂₈O₄   M 332.439
Isol. from *Salvia officinalis* and *Rosmarinus officinalis*. Cryst. (hexane). Mp 185-190° dec. $[\alpha]_D^{23}$ +191° (c, 1.07 in MeOH).
Di-Ac: Needles (hexane). Mp 212-217°. $[\alpha]_D^{22}$ +139° (c, 1.15 in CHCl₃).

Linde, H. *et al*, *Helv. Chim. Acta*, 1964, **47**, 1234 (*isol, uv, ir, pmr*)
Wenkert, E. *et al*, *J. Org. Chem.*, 1965, **30**, 2931 (*isol, ir, pmr, config*)
Meyer, W.L. *et al*, *Tetrahedron Lett.*, 1966, 4261 (*synth*)

12,15-Dihydroxy-8,11,13-abietatrien-7-one   D-80270

Hypargenin B

[119767-13-4]

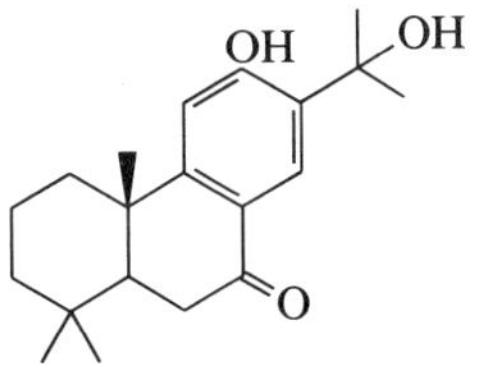

C₂₀H₂₈O₃   M 316.439
Constit. of *Salvia hypargeia*. Antibacterial agent. Orange amorph. solid.

Ulubelen, A. *et al*, *J. Nat. Prod.* (*Lloydia*), 1988, **51**, 1178 (*isol, pmr*)

6,14-Dihydroxy-8,11,13-abietatrien-1-one   D-80271

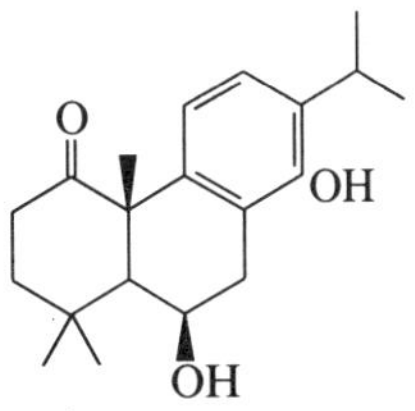

C₂₀H₂₈O₃   M 316.439
6β-form [119767-15-6] *Hypargenin E*
  Constit. of *Salvia hypargeia*. Amorph.

Ulubelen, A. *et al*, *J. Nat. Prod.* (*Lloydia*), 1988, **51**, 1178 (*isol, pmr*)

ent-3β,16α-Dihydroxy-13-atisen-2-one   D-80272

Yucalexin A-19

[119642-83-0]

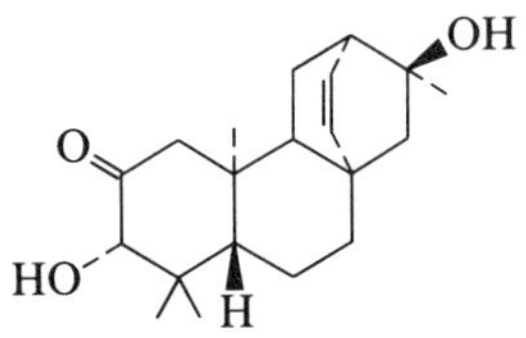

$C_{20}H_{30}O_3$ M 318.455
Constit. of cassava roots (*Manihot esculenta*). $[\alpha]_D^{23}$ +8.28°
(c, 0.35 in $CHCl_3$).

Sakai, T. *et al*, *Phytochemistry*, 1988, **27**, 3769.

4',6-Dihydroxyaurone   D-80273

*6-Hydroxy-2-[(4-hydroxyphenyl)methylene]-3(2H)-
benzofuranone, 9CI. 6-Hydroxy-2-(p-hydroxybenzylidene)-
3(2H)benzofuranone, 8CI.* **Hispidol**

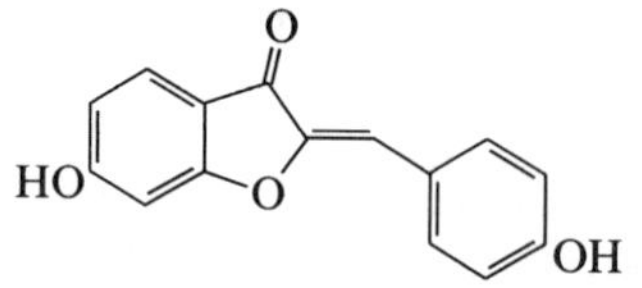

$C_{15}H_{10}O_4$ M 254.242
(Z)-form [5786-54-9]
     Isol. from seedlings of *Glycine max* (*Soja hispida*) and
     from *Lygos raetam*. Yellow needles (EtOH aq.). Mp
     288° dec.
Di-Ac: Pale-yellow needles (MeOH/AcOH). Mp 162-163°.
6-O-β-D-Glucopyranosyl: [20550-08-7].
     $C_{21}H_{20}O_9$ M 416.384
     Isol. from seedlings of *G. max* and from *L. raetam*.
     Yellow cryst. (H_2O). Mp 211-212° (191-192°).

Geissman, T.A. *et al*, *J. Am. Chem. Soc.*, 1956, **78**, 832 (*synth, uv*)
Wong, E., *Phytochemistry*, 1966, **5**, 463 (*isol*)
Farkas, L. *et al*, *Tetrahedron*, 1968, **24**, 4213 (*synth*)
Huke, M. *et al*, *Arch. Pharm.* (*Weinheim, Ger.*), 1969, **302**, 401,
     423 (*uv, pmr*)
El Sherbeiny, A.E.A. *et al*, *Planta Med.*, 1978, **34**, 335 (*isol*)

3,4-Dihydroxybenzocyclobutene-1,2-dione   D-80274

2,3-Dihydroxybicyclo[4.2.0]octa-1,3,5-triene-7,8-dione

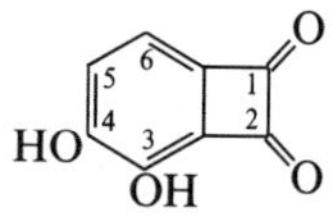

$C_8H_4O_4$ M 164.117
Di-Me ether: [118112-21-3]. *3,4-Dimethoxbenzocyclobutene-
1,2-dione*
     $C_{10}H_8O_4$ M 192.171
     Cryst. (H_2O). Mp 146°.
Methylene ether: [118112-23-5]. *3,4-
Methylenedioxybenzocyclobutene-1,2-dione*
     $C_9H_4O_4$ M 176.128
     Cryst. (Me_2CO aq.). Mp 164-166°.

Liebeskind, L.S. *et al*, *J. Org. Chem.*, 1989, **54**, 1435 (*synth, ir,
     pmr, cmr*)

3,5-Dihydroxybenzocyclobutene-1,2-dione   D-80275

2,4-Dihydroxybicyclo[4.2.0]octa-1,3,5-triene-7,8-dione
$C_8H_4O_4$ M 164.117
Di-Me ether: [118112-22-4]. *3,5-
Dimethoxybenzocyclobutene-1,2-dione*
     $C_{10}H_8O_4$ M 192.171
     Cryst. (hexane). Mp 147-148°.

Liebeskind, L.S. *et al*, *J. Org. Chem.*, 1989, **54**, 1435 (*synth, ir,
     pmr*)

3,6-Dihydroxybenzocyclobutene-1,2-dione   D-80276

2,5-Dihydroxybicyclo[4.2.0]octa-1,3,5-triene-7,8-dione
[75833-49-7]
$C_8H_4O_4$ M 164.117
Cryst. (H_2O). Mp 225-228°. pK_{a1} 5.3 (25°), pK_{a2} 9.3.
Di-Me ether: [75833-47-5]. *3,6-Dimethoxybenzocyclobutene-
1,2-dione*
     $C_{10}H_8O_4$ M 192.171
     Yellow plates (cyclohexane). Mp 190-192°.

Abou-Teim, I. *et al*, *J. Chem. Soc., Perkin Trans. 2*, 1980, 1841
     (*synth, deriv, uv, ir, pmr*)
Liebeskind, L.S. *et al*, *J. Org. Chem.*, 1989, **54**, 1435 (*synth, deriv,
     ir, pmr, cmr*)

4,5-Dihydroxybenzocyclobutene-1,2-dione   D-80277

3,4-Dihydroxybicyclo[4.2.0]octa-1,3,5-triene-7,8-dione
[41737-30-8]
$C_8H_4O_4$ M 164.117
Benzologue of squaric acid. Yellow prisms (H_2O). Mp 243-
245° dec. pK_{a1} 4.48 (20°), pK_{a2} 8.05.
Di-Me ether: [41634-27-9]. *4,5-Dimethoxybenzocyclobutene-
1,2-dione*
     $C_{10}H_8O_4$ M 192.171
     Needles (cyclohexane). Mp 222-223°.

Abou-Teim, O. *et al*, *J. Chem. Soc., Perkin Trans. 1*, 1980, 1841
     (*synth, uv, ir, pmr*)

3,6-Dihydroxybenzocyclobuten-1-one   D-80278

Updated Entry replacing D-20333
     2,5-Dihydroxybicyclo[4.2.0]octa-1,3,5-trien-7-one

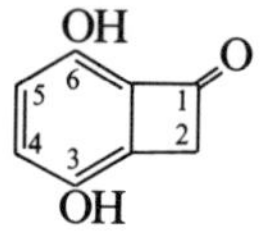

$C_8H_6O_3$ M 150.134
Di-Me ether: [75833-45-3]. *3,6-Dimethoxybenzocyclobuten-
1-one. 2,5-Dimethoxybicyclo[4.2.0]octa-1,3,5-trien-7-one*
     $C_{10}H_{10}O_3$ M 178.187
     Cryst. (pet. ether). Mp 107-108°.

McOmie, J.F.W. *et al*, *J. Chem. Soc., Perkin Trans. 1*, 1980, 1841
     (*synth, ir, pmr*)
Liebeskind, L.S. *et al*, *J. Org. Chem.*, 1989, **54**, 1435 (*synth, ir,
     pmr, cmr*)

4,5-Dihydroxybenzocyclobuten-1-one   D-80279

Updated Entry replacing D-20334
     3,4-Dihydroxybicyclo[4.2.0]octa-1,3,5-trien-7-one
     $C_8H_6O_3$ M 150.134
Di-Me ether: [55171-76-1]. *4,5-Dimethoxybenzocyclobuten-
1-one. 3,4-Dimethoxybicyclo[4.2.0]octa-1,3,5-trien-7-one,
9CI*
     $C_{10}H_{10}O_3$ M 178.187
     Plates. Mp 142-143°. $Bp_{0.05}$ 140° subl.

Kametani, T. *et al*, *J. Chem. Soc., Perkin Trans. 1*, 1974, 1712
     (*synth, ir, uv, pmr*)

4,6-Dihydroxybenzocyclobuten-1-one   D-80280

3,5-Dihydroxybicyclo[4.2.0]octa-1,3,5-trien-7-one
$C_8H_6O_3$ M 150.134
Di-Me ether: [118112-18-8]. *4,6-Dimethoxybenzocyclobuten-
1-one*
     $C_{10}H_{10}O_3$ M 178.187

Cryst. (H$_2$O). Mp 51-52°.

Liebeskind, L.S. *et al*, *J. Org. Chem.*, 1989, **54**, 1435 (*synth, ir, pmr, cmr*)

5,6-Dihydroxybenzocyclobuten-1-one   D-80281

Updated Entry replacing D-20335
4,5-Dihydroxybicyclo[4.2.0]octa-1,3,5-trien-7-one
C$_8$H$_6$O$_3$ M 150.134
Di-Me ether: [81447-58-7]. *5,6-Dimethoxybenzocyclobuten-1-one. 4,5-Dimethoxybicyclo[4.2.0]octa-1,3,5-trien-7-one, 9CI*
C$_{10}$H$_{10}$O$_3$ M 178.187
Cryst. (pet. ether). Mp 86-87°.

Stevens, R.V. *et al*, *J. Org. Chem.*, 1982, **47**, 2393 (*synth, ir, pmr*)
Liebeskind, L.S. *et al*, *J. Org. Chem.*, 1989, **54**, 1435 (*synth, ir, pmr, cmr*)

5,7-Dihydroxy-2*H*-1-benzopyran-2-one,   D-80282
9CI

Updated Entry replacing D-04285
5,7-Dihydroxycoumarin
[2732-18-5]
C$_9$H$_6$O$_4$ M 178.144
Found in *Rumex conglomeratus*. Prisms (AcOH aq.), yellow needles (H$_2$O). Mp 285-286°.

▷ Exp. carcinogen.

Di-Ac: [21524-17-4].
  C$_{13}$H$_{10}$O$_6$ M 262.218
  Prisms (EtOH). Mp 140°.
Di-Me ether: [487-06-9]. *5,7-Dimethoxy-2H-1-benzopyran-2-one. 5,7-Dimethoxycoumarin*. **Citropten**. *Limettin*
  C$_{11}$H$_{10}$O$_4$ M 206.198
  Found in lime and bergamot oils, *Eriostema obovalis*, *Xanthoxylum inerme* and *Glycosmis cyanocarpa*. Prisms (MeOH). Mp 147°. Fluorescent in EtOH.
O^5-*(2-Propenyl):* [23053-60-3]. *7-Hydroxy-5-(2-propenyloxy)-4H-1-benzopyran-2-one, 9CI. 5-Allyloxy-7-hydroxycoumarin*. **Lacoumarin**
  C$_{12}$H$_{10}$O$_4$ M 218.209
  Constit. of *Lawsonia inermis*. Mp 162-164°.
O^5-*(3-Methyl-2-butenyl),* O^7-*Me:*
  C$_{15}$H$_{16}$O$_4$ M 260.289
  Isol. from *Citrus limon*. Mp 90-92°.
O^5-*(3,7-Dimethyl-2,6-octadienyl),* O^7-*Me:*
  C$_{20}$H$_{24}$O$_4$ M 328.407
  Isol. from *C. aurantifolia*. Mp 86-87°.
O^5-*(6-Hydroxy-3,7-dimethyl-2,6-octadienyl),* O^7-*Me:*
  C$_{20}$H$_{24}$O$_5$ M 344.407
  Isol. from *C. aurantifolia*.
O^7-*(3-Methyl-2-butenyl):* **Anisocoumarin B**
  C$_{14}$H$_{14}$O$_4$ M 246.262
  Constit. of *Clausena anisata*. Granules. Mp 94-95°.

Heyes, R.G. *et al*, *J. Chem. Soc.*, 1936, 1831 (*synth*)
Gerphagnon, M.C. *et al*, *Compt. Rend. Hebd. Seances Acad. Sci.*, 1958, **246**, 1201 (*uv*)
Crombie, L. *et al*, *J. Chem. Soc. C*, 1971, 788 (*synth, ir, pmr, ms*)
Talapatra, B. *et al*, *Indian J. Chem.*, 1975, **13**, 835 (*isol, deriv*)
Günther, D.H. *et al*, *Org. Magn. Reson.*, 1975, **7**, 339 (*cmr, deriv*)
Bhardwaj, D.K. *et al*, *Phytochemistry*, 1976, **15**, 1789 (*Lacoumarin*)
Gray, A.I. *et al*, *J. Chem. Soc., Perkin Trans. 2*, 1978, 391 (*pmr*)
Joseph-Nathan, P. *et al*, *J. Heterocycl. Chem.*, 1984, **21**, 1141 (*pmr, deriv*)
Ngadjai, B.J. *et al*, *Phytochemistry*, 1989, **28**, 585 (*Anisocoumarin B*)
Sax, N.I., *Dangerous Properties of Industrial Materials*, 5th Ed., Van Nostrand-Reinhold, 1979, 594.

2,3-Dihydroxybenzyl alcohol   D-80283

3-(Hydroxymethyl)-1,2-benzenediol, 9CI
[14235-77-9]

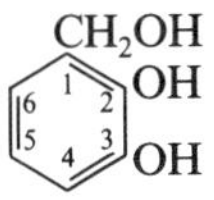

C$_7$H$_8$O$_3$ M 140.138
O^2-*β-D-Glucoside:* [20331-85-5]. **Idesin**
  C$_{13}$H$_{18}$O$_8$ M 302.280
  Isol. from leaves of *Idesia polycarpa*. Mp 98-100°.

Kubota, T. *et al*, *Bot. Mag.*, 1966, **79**, 770; *CA*, **68**, 49982 (*Idesin*)

3,4-Dihydroxybenzyl alcohol, 8CI   D-80284

Updated Entry replacing D-04308
3,4-Dihydroxybenzenemethanol, 9CI. α,3,4-Trihydroxytoluene. Protocatechuyl alcohol. 4-Hydroxymethyl-1,2-benzenediol
[3897-89-0]
C$_7$H$_8$O$_3$ M 140.138
Prisms (H$_2$O). Mp 137-138°. Highly air-sensitive.
Tri-Ac:
  C$_{13}$H$_{14}$O$_6$ M 266.250
  Bp$_{0.05}$ 143-147°.
4-O-β-D-Glucopyranoside: [20300-53-2]. **Calleryanin**
  C$_{13}$H$_{18}$O$_8$ M 302.280
  Isol. from leaves of *Pyrus calleryana*. Occurs partly as various aromatic esters, not fully characterised.
O^4-*(6-O-Benzoyl-β-D-glucopyranoside):* [41942-94-3].
  Lacticolorin
  C$_{10}$H$_{22}$O$_9$ M 286.278
  Isol. from leaves of *Protea lacticolor*. Cryst. (H$_2$O). Mp 192-195°. [α]$_D$ − 58° (c, 0.59 in MeOH).
3-Me ether: [498-00-0]. *4-Hydroxy-3-methoxybenzenemethanol, 9CI. α,4-Dihydroxy-3-methoxytoluene. Vanillyl alcohol*
  C$_8$H$_{10}$O$_3$ M 154.165
  Cryst. (H$_2$O). Mp 114-115°.
4-Me ether: [4383-06-6]. *3-Hydroxy-4-methoxybenzenemethanol, 9CI. 3-Hydroxy-4-methoxybenzyl alcohol, 8CI. Isovanillyl alcohol*
  Cryst. (toluene). Mp 131-132°.

Rosenmund, K.W. *et al*, *Arch. Pharm. (Weinheim, Ger.)*, 1926, **264**, 458.
Pearl, I.A., *J. Org. Chem.*, 1947, **12**, 79 (*deriv*)
Cook, P.L. *et al*, *J. Org. Chem.*, 1962, **27**, 3873 (*deriv*)
Kaemmerer, H. *et al*, *Makromol. Chem.*, 1963, **67**, 167 (*synth*)
Challice, J.S. *et al*, *Phytochemistry*, 1968, **7**, 119 (*Calleryanin*)
Perold, G.W. *et al*, *J. Chem. Soc., Perkin Trans. 1*, 1973, 638 (*Lacticolorin*)
Danishevsky, S. *et al*, *J. Org. Chem.*, 1976, **41**, 1081 (*deriv*)
Krohn, K. *et al*, *J. Chem. Soc., Perkin Trans. 1*, 1977, 1186 (*Lacticolorin*)
Becher, H.D. *et al*, *J. Org. Chem.*, 1980, **45**, 1596 (*deriv*)

ent-2α,3β-Dihydroxy-15-beyeren-12-one   **D-80285**
Yucalexin B-22

[119626-55-0]

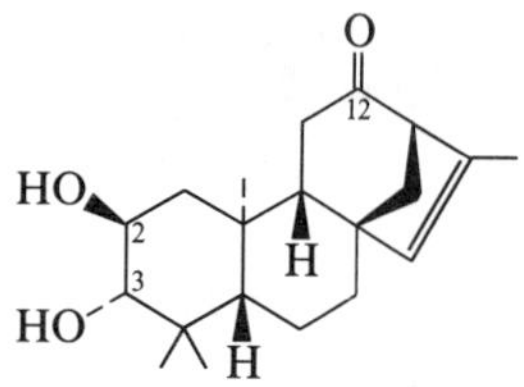

$C_{20}H_{30}O_3$ M 318.455

Constit. of cassava roots (*Manihot esculenta*).

2-Ketone: [35470-61-2]. ent-3β-*Hydroxy-15-beyerene-2,12-dione.* **Yucalexin B-9**
$C_{20}H_{28}O_3$ M 316.439
Constit. of *M. esculenta*. $[\alpha]_D^{23} -73.9°$ (c, 0.83 in $CHCl_3$).

2,3-Diketone: [50719-31-8]. ent-15-*Beyerene-2,3,12-trione.* ent-2-*Hydroxy-1,15-beyeradiene-3,12-dione.* **Yucalexin B-5**
$C_{20}H_{26}O_3$ M 314.424
Constit. of *M. esculenta*. Exists as an enol form.

2-Deoxy: [119626-51-6]. ent-3β-*Hydroxy-15-beyeren-12-one.* **Yucalexin B-14**
$C_{20}H_{30}O_2$ M 302.456
Constit. of *M. esculenta*.

2-Deoxy, 3-ketone: [88048-00-4]. ent-15-*Beyerene-3,12-dione.* **Yucalexin B-6**
$C_{20}H_{28}O_2$ M 300.440
Constit. of *M. esculenta*.

Sakai, T. *et al, Phytochemistry*, 1988, **27**, 3769.

ent-3β,12α-Dihydroxy-15-beyeren-2-one   **D-80286**
Yucalexin B-20

[119679-04-8]

$C_{20}H_{30}O_3$ M 318.455
Constit. of cassava roots (*Manihot esculenta*).

12-Epimer: [119626-53-8]. ent-3β,12β-*Dihydroxy-15-beyeren-2-one.* **Yucalexin B-18**
$C_{20}H_{30}O_3$ M 318.455
Constit. of *M. esculenta*. $[\alpha]_D^{26} -9.4°$ (c, 0.17 in $CHCl_3$).

Sakai, T. *et al, Phytochemistry*, 1988, **27**, 3769.

2,2′-Dihydroxy-3,3′-bipyridine   **D-80287**
[*3,3′-Bipyridine*]-2,2′(1H,1′H)-*dione, 9CI*. [*3,3′-Bipyridine*]-2,2′-diol. 3,3′-*Bi-2-pyridone*

[97033-24-4]

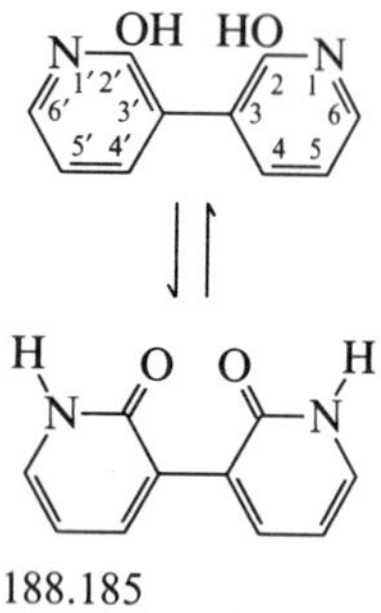

$C_{10}H_8N_2O_2$ M 188.185
Pyridone tautomers prob. predominate where feasible in this series. Entries are given under the dihydroxydipyridine names for ease of presentation.
Cryst. Mp >300°.

Di-Me ether: [95881-80-4]. 2,2′-*Dimethoxy-3,3′-bipyridine*, *9CI*

$C_{12}H_{12}N_2O_2$ M 216.239
Mp 117-118°.

Becalski, A. *et al, Bull. Pol. Acad. Sci., Chem.*, 1984, **32**, 105 (*synth, ir, ms, pmr*)
Tiecco, M. *et al, Synthesis*, 1984, 736 (*synth, deriv, pmr*)
Kaczmarek, L., *Pol. J. Chem.* (*Rocz. Chem.*), 1985, **59**, 1141 (*synth*)

2,2′-Dihydroxy-4,4′-bipyridine   **D-80288**
[*4,4′-Bipyridine*]-2,2′-diol. [*4,4′-Bipyridine*]-2,2′(1H,1′H)-*dione*. 4,4′-*Bi-2-pyridone*

[58426-03-2]

$C_{10}H_8N_2O_2$ M 188.185
Cryst. (H_2O). Mp >300°.

Abramovitch, R.A. *et al, J. Org. Chem.*, 1976, **41**, 1717 (*synth, ir, ms, pmr*)

2,4′-Dihydroxy-1,1′-bipyridine   **D-80289**
[*1,1′(2H,4′H)-Bipyridine*]-2,4′-*dione, 9CI*

[68304-41-6]

$C_{10}H_8N_2O_2$ M 188.185
Prisms (EtOH). Mp 233-236°.

B, HCl: [68304-44-9].
Prisms (EtOH). Mp 210-212°.

Katritzky, A.R. *et al, Pak. J. Sci. Ind. Res.*, 1978, **21**, 1 (*synth, ir, pmr, uv*)
Afridi, A.S. *et al, J. Chem. Soc. Pak.*, 1982, **4**, 55 (*ms*)

3,3′-Dihydroxy-2,2′-bipyridine   **D-80290**
[*2,2′-Bipyridine*]-3,3′-diol, *9CI*

[36145-03-6]

$C_{10}H_8N_2O_2$ M 188.185
Cryst. by subl. Mp 196° (188-190°).

Mono-Me ether: [87385-13-5]. 3-*Hydroxy-3′-methoxy-2,2′-bipyridine*
$C_{11}H_{10}N_2O_2$ M 202.212
Yellow cryst. ($CHCl_3$/pet. ether). Mp 85°.

Di-Me ether: [93560-59-9]. 3,3′-*Dimethoxy-2,2′-bipyridine*
$C_{12}H_{12}N_2O_2$ M 216.239
Cryst. (CCl_4). Mp 138°.

Siemanowski, W. *et al, Justus Liebigs Ann. Chem.*, 1984, 1731 (*synth, deriv, pmr*)
Tiecco, M. *et al, Synthesis*, 1984, 736 (*deriv, synth, pmr*)
Tiecco, M. *et al, Tetrahedron*, 1986, **42**, 1475 (*synth, pmr, cmr*)

3,3′-Dihydroxy-4,4′-bipyridine   **D-80291**
4,4′-*Bipyridine-3,3′-diol*

$C_{10}H_8N_2O_2$ M 188.185
Cryst. (MeOH). Mp 190-191°.

Ziyaev, A.A. *et al, J. Gen. Chem. USSR* (*Engl. Transl.*), 1964, **34**, 349 (*synth*)

3,4′-Dihydroxy-2,3′-bipyridine   **D-80292**
[*2,3′-Bipyridine*]-3,4′-diol, *9CI*

[89945-98-2]

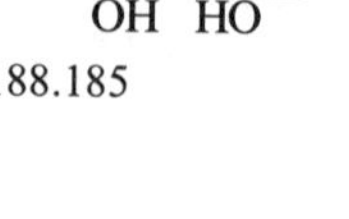

$C_{10}H_8N_2O_2$ M 188.185

Alkaloid from the timber of *Broussonetia zeylanica* (Moraceae). Deep-yellow cryst. (CHCl₃). Mp 171-173°, Mp 223-224°. Struct. of the nat. prod. in doubt; authentic synthetic material (Mp 171-173°) is not identical with the alkaloid assigned this struct.

Di-Ac: [89945-99-3].
Cryst. (C₆H₆/pet. ether). Mp 160-161°.

Gunatilaka, A.A.L. *et al, Phytochemistry*, 1983, **22**, 2847 (*isol, uv, ir, pmr, cmr, ms, struct*)
Dehmlow, E.V. *et al, Justus Liebigs Ann. Chem.*, 1987, 1123 (*synth, uv, ir, pmr, cmr*)

4,4′-Dihydroxy-1,1′-bipyridine   D-80293

[1,1′(4H,4′H)-Bipyridine]-4,4′-dione, 9CI. 1,1′-Bipyridine-4,4′-diol. 1,1′-Bi-4-pyridone

[68304-42-7]

$C_{10}H_8N_2O_2$ M 188.185
Prisms (EtOH). Mp 205-210° dec.

B,HCl: [68304-43-8].
Prisms (EtOH). Mp 210-230° dec.

Katritzky, A.R. *et al, Pak. J. Sci. Ind. Res.*, 1978, **21**, 1 (*synth, ir, pmr, uv*)
Afridi, A.S. *et al, J. Chem. Soc. Pak.*, 1982, **4**, 55 (*ms*)

4,4′-Dihydroxy-2,2′-bipyridine   D-80294

[2,2′-Bipyridine]-4,4′-diol. [2,2′-Bipyridine]-4,4′(1H,1′H)-dione. 2,2′-Bi-4-pyridone

$C_{10}H_8N_2O_2$ M 188.185
Cryst. (H₂O). Mp 342-343°. Forms a hemihydrate.

1,1′-Dioxide:
$C_{10}H_8N_2O_4$ M 220.184
Mp 308° dec.

Di-Me ether: [17217-57-1]. *4,4′-Dimethoxy-2,2′-bipyridine, 9CI*
$C_{12}H_{12}N_2O_2$ M 216.239
Cryst. (EtOH). Mp 170-172°.

Di-Me ether, 1,1′-dioxide: [84175-10-0].
$C_{12}H_{12}N_2O_4$ M 248.238
Cryst. Mp 229° dec.

Di-Me ether, 1,1′-dioxide, picrate: Cryst. Mp 190.5-192°.

Haginiwa, J., *Yakugaku Zasshi (J. Pharm. Soc. Jpn.)*, 1955, **75**, 733 (*dioxide*)
Case, F.H., *J. Org. Chem.*, 1962, **27**, 640 (*synth*)
Stoesser, R. *et al, Z. Chem.*, 1978, **18**, 187 (*esr*)
Wenkert, D. *et al, J. Org. Chem.*, 1983, **48**, 283 (*synth derivs, pmr, ir, uv*)
Connor, J.A. *et al, J. Organomet. Chem.*, 1985, **282**, 349 (*pmr, deriv*)

4,4′-Dihydroxy-3,3′-bipyridine   D-80295

[3,3′-Bipyridine]-4,4′-diol, 9CI. [3,3′-Bipyridine]-4,4′(1H,1′H)-dione, 9CI. 3,3′-Bi-4-pyridone

[27353-34-0]

$C_{10}H_8N_2O_2$ M 188.185
Needles.

[109324-77-8]

Kosuge, T. *et al, Chem. Pharm. Bull.*, 1970, **18**, 1068 (*synth, ir, uv*)
Kaczmarek, L., *Pol. J. Chem. (Rocz. Chem.)*, 1985, **59**, 1141 (*synth*)

4,6-Dihydroxy-2,2′-bipyridine   D-80296

[2,2′-Bipyridine]-4,6-diol
$C_{10}H_8N_2O$ M 172.186
O⁴-Me: [14759-78-5]. *4-Methoxy-2,2′-bipyridin-6-ol, 8CI. 6-Hydroxy-4-methoxy-2,2′-bipyridine*

$C_{11}H_{10}N_2O_2$ M 202.212
Needles (C₆H₆). Mp 172.5°.

Divekar, P.V. *et al, Can. J. Chem.*, 1967, **45**, 1215 (*synth, ir, uv*)

5,5′-Dihydroxy-2,2′-bipyridine   D-80297

2,2′-Bipyridine-5,5′-diol
$C_{10}H_8N_2O_2$ M 188.185
Cryst. Mp 250°.

[39858-87-2]

Otroshchenko, O.S. *et al, CA*, 1964, **63**, 4248e (*synth*)

5,5′-Dihydroxy-3,3′-bipyridine   D-80298

3,3′-Bipyridine-5,5′-diol
$C_{10}H_8N_2O_2$ M 188.185
Di-Me ether: [95881-81-5]. *5,5′-Dimethoxy-3,3′-bipyridine, 9CI*
$C_{12}H_{12}N_2O_2$ M 216.239
Mp 161-162°.

Tiecco, M. *et al, Synthesis*, 1984, 736 (*synth, pmr, deriv*)

6,6′-Dihydroxy-2,2′-bipyridine   D-80299

[2,2′-Bipyridine]-6,6′-(1H,1′H)-dione, 9CI. 2,2′-Bipyridine-6,6′-diol. 6,6′-Bi-2-pyridone
$C_{10}H_8N_2O_2$ M 188.185
Sesquihydrate. Mp 325-326° dec.

Di-Ac:
$C_{14}H_{12}N_2O_4$ M 272.260
Mp 175-176° dec.

Di-Me ether: [39858-88-3]. *6,6′-Dimethoxy-2,2′-bipyridine, 9CI*
$C_{12}H_{12}N_2O_2$ M 216.239
Mp 118°.

Dihydrazone: [53301-38-5].
$C_{10}H_{12}N_6$ M 216.245
Yellow needles (H₂O). Mp 208°.

Murase, I., *Nippon Kagaku Zasshi (Jpn. J. Chem.)*, 1956, **77**, 682; *CA*, **52**, 91006 (*synth*)
Lewis, J. *et al, J. Chem. Soc., Dalton Trans.*, 1977, 734 (*hydrazone*)
Tiecco, M. *et al, Synthesis*, 1984, 736 (*synth, deriv, pmr*)
Vanderesse, R. *et al, Tetrahedron Lett.*, 1986, **27**, 5483 (*synth, deriv*)

6,6′-Dihydroxy-3,3′-bipyridine   D-80300

3,3′-Bipyridine-6,6′-diol. [3,3′-Bipyridine]-6,6′(1H,1′H)-dione. 5,5′-Bi-2-pyridone
$C_{10}H_8N_2O_2$ M 188.185
Di-Me ether: [95881-82-6]. *6,6′-Dimethoxy-3,3′-bipyridine, 9CI*
$C_{12}H_{12}N_2O_2$ M 216.239
Mp 104-105°.

Tiecco, M. *et al, Synthesis*, 1984, 736 (*synth, pmr, deriv*)

4,5-Dihydroxy-1,3-bis(hydroxymethyl)-2-imidazolidinone, 9CI   D-80301

Dihydroxydimethylethyleneurea.
Dimethyloldihydroxyethyleneurea. Dihydroxy DMEU.
DMDHEU

[1854-26-8]

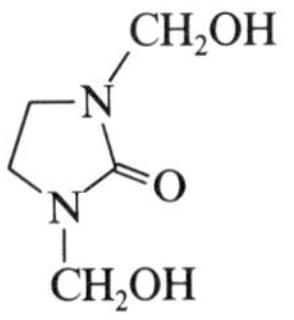

$C_5H_{10}N_2O_5$ M 178.144

Cross-linking agent. Textile-finishing agent, especially for cotton garments. Hygroscopic cryst.

U.K. Pat., 720 386, (1954); *CA*, **49**, 9326i (*manuf*)

Soignet, D.M. *et al*, *Appl. Spectrosc.*, 1970, **24**, 272 (*pmr*)

Beck, K.R. *et al*, *J. Chromatogr.*, 1980, **190**, 226 (*chromatog*)

Andrews, B.A.K., *J. Chromatogr.*, 1984, **288**, 101 (*cmr, hplc*)

Hermanns, K. *et al*, *Ind. Eng. Chem. Prod. Res. Dev.*, 1986, **25**, 469 (*cmr, bibl*)

Eur. Pat., 221 537, (1987); *CA*, **107**, 238841c (*manuf*)

3,6-Dihydroxy-7,25-cholestadien-24-one   D-80302

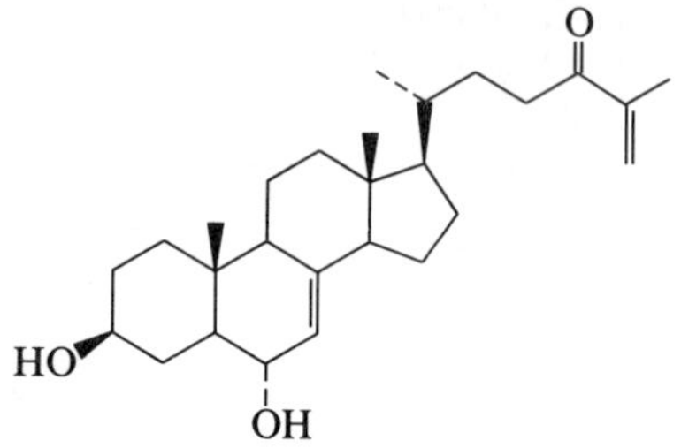

$C_{27}H_{42}O_3$ M 414.627

(3β,5α,6α)-form

Constit. of *Spongionella gracilis*.

Madaio, A. *et al*, *J. Nat. Prod. (Lloydia)*, 1989, **52**, 952 (*isol, pmr*)

ent-3β,4α-Dihydroxy-13-cleroden-15,16-olide   D-80303

[118173-00-5]

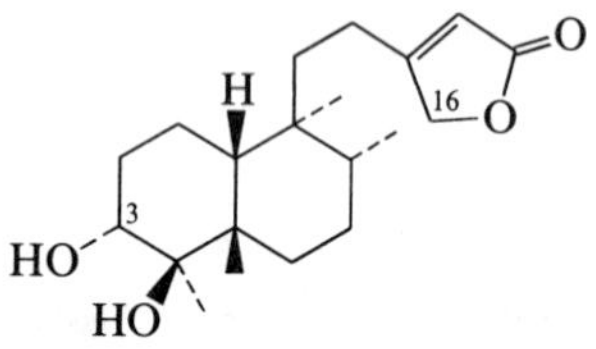

$C_{20}H_{32}O_4$ M 336.470

Constit. of *Ageratina saltillensis*.

3-Me ether: [118173-01-6]. **ent-4α-Hydroxy-3β-methoxy-13-cleroden-15,16-olide**

$C_{21}H_{34}O_4$ M 350.497

Constit. of *A. saltillensis*.

3-Ketone: [118173-02-7]. **ent-4α-Hydroxy-3-oxo-13-cleroden-15,16-olide**

$C_{20}H_{30}O_4$ M 334.455

Constit. of *A. saltillensis*.

16ξ-Hydroxy: [118172-99-9]. **ent-3β,4α,16ξ-Trihydroxy-13-cleroden-15,16-olide**

$C_{20}H_{32}O_5$ M 352.470

Constit. of *A. saltillensis*.

Fang, N. *et al*, *Phytochemistry*, 1988, **27**, 3187.

3,22-Dihydroxy-24-cycloarten-26-oic acid   D-80304

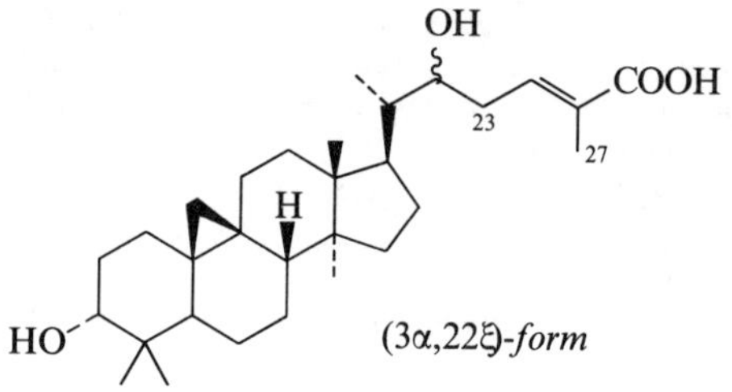

(3α,22ξ)-*form*

$C_{30}H_{48}O_4$ M 472.707

(3α,22ξ)-form

Constit. of *Mangifera indica*. Cryst. (CHCl₃/MeOH). Mp 218-220°. [α]$_D^{30}$ +27.5° (c, 0.8 in CHCl₃).

(3β,22ξ)-form

Constit. of *M. indica*.

Anjaneyulu, V. *et al*, *Phytochemistry*, 1989, **28**, 1471.

3,23-Dihydroxy-24-cycloarten-26-oic acid   D-80305

$C_{30}H_{48}O_4$ M 472.707

(3β,23ξ)-form

Constit. of *Mangifera indica*.

Anjaneyulu, V. *et al*, *Phytochemistry*, 1989, **28**, 1471.

3,27-Dihydroxy-24-cycloarten-26-oic acid   D-80306

$C_{30}H_{48}O_4$ M 472.707

3α-form

Constit. of *Mangifera indica*. Cryst. (C₆H₆/EtOAc). Mp 205-207°. [α]$_D^{30}$ +21.5° (c, 0.8 in CHCl₃).

Anjaneyulu, V. *et al*, *Phytochemistry*, 1989, **28**, 1471.

2,3-Dihydroxy-2,4,6-cycloheptatrien-1-one   D-80307

2,3-Dihydroxytropone. 3-Hydroxytropolone. 2,7-Dihydroxytropone. 7-Hydroxytropolone

[34777-04-3]

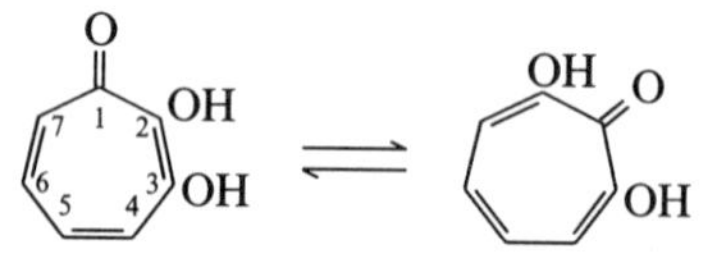

$C_7H_6O_3$ M 138.123

Prod. by a *Pseudomonas* strain. Shows antibacterial activity. Mp 143-144° (139°).

Korth, H. *et al*, *Z. Naturforsch., C*, 1981, **36**, 728 (*isol, ms, uv, pmr*)

Takeshita, H. *et al*, *Synthesis*, 1986, 578 (*synth, cmr, bibl*)

2,4-Dihydroxy-2,4,6-cycloheptatrien-1-one   D-80308

2,4-Dihydroxytropone. 4-Hydroxytropolone. 2,6-Dihydroxytropone. 6-Hydroxytropolone

[7009-19-0]

$C_7H_6O_3$ M 138.123

Needles. Mp 229-230° (227°).

Toda, T., *Bull. Chem. Soc. Jpn.*, 1967, **40**, 588 (*synth, ir, uv*)

2,5-Dihydroxy-2,4,6-cycloheptatrien-1-one, 9CI   D-80309

2,5-Dihydroxytropone. 5-Hydroxytropolone

[15852-34-3]

$C_7H_6O_3$ M 138.123

Cryst. by subl. Mp 246-247°.

Russell, G.A. *et al*, *J. Org. Chem.*, 1978, **43**, 3278 (*synth, pmr*)

ent-1α,10β-Dihydroxy-9α,15α-cyclo-20-nor-16-gibberellene-7,19-dioic acid 10,19-lactone   D-80310

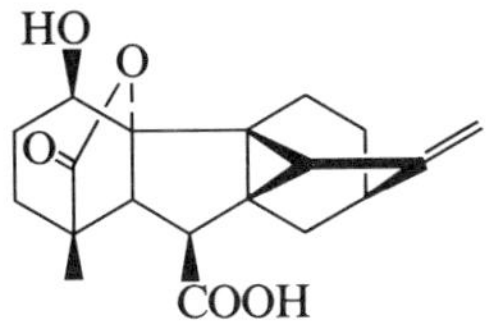

$C_{19}H_{22}O_5$ M 330.380

Antheridiogenin from the fern *Anemia mexicana*. Needles. Mp 213-214°. $[\alpha]_D^{21}$ −72° (c, 0.013 in MeOH).

Furber, M. *et al, J. Am. Chem. Soc.*, 1988, **110**, 4084 (*synth*)
Furber, M. *et al, Phytochemistry*, 1989, **28**, 63 (*isol*)

1,18-Dihydroxy-14-decipien-19-oic acid   D-80311

[76337-44-5]

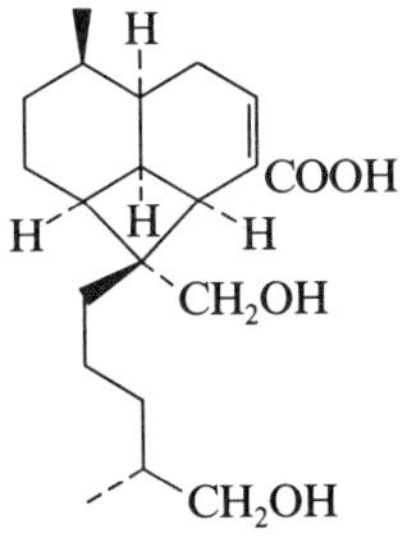

$C_{20}H_{32}O_4$ M 336.470

Isol. from an *Eremophila* sp. Needles (Me₂CO). Mp 133-134°. $[\alpha]_D$ +7° (c, 1.02 in CHCl₃).

Croft, K.D. *et al, Aust. J. Chem.*, 1980, **33**, 1529 (*isol, pmr, cmr, ms, cryst struct*)

2,3-Dihydroxy-2,3-diphenylbutanedioic acid, 9CI   D-80312

Diphenyltartaric acid

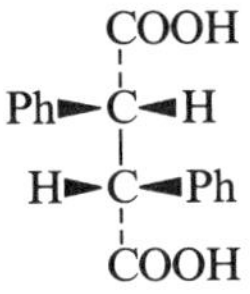

$C_{16}H_{14}O_6$ M 302.283

(2RS,3RS)-form

(±)-threo-*form*

Di-Et ester: [120915-13-1].
$C_{20}H_{22}O_6$ M 358.390
Cryst. (C₆H₆/hexane). Mp 122.5-123°.

Ohgo, Y. *et al, J. Heterocycl. Chem.*, 1988, **25**, 1583 (*synth, cryst struct*)

5,6-Dihydroxy-7,9,11,14-eicosatetraenoic acid   D-80313

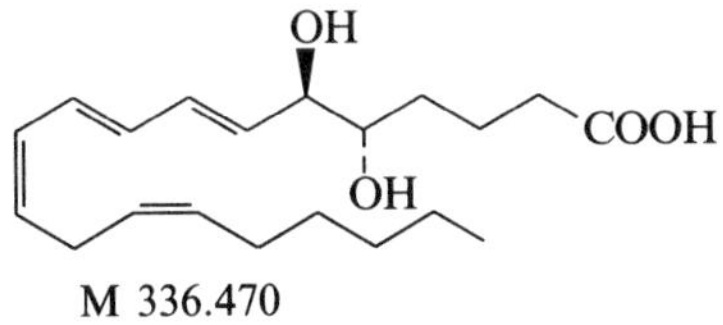

$C_{20}H_{32}O_4$ M 336.470

(5S,6R,7E,9E,11Z,14Z)-form

Metab. of arachidonic acid derived from Leucotriene A₄.

Me ester:
$C_{21}H_{34}O_4$ M 350.497
Oil. $[\alpha]_D^{20}$ +0.91° (c, 0.44 in CHCl₃). Other diastereoisomers also prepd.

Nicolaou, K.C. *et al, Angew. Chem., Int. Ed. Engl.*, 1989, **28**, 587 (*synth, ir, uv, pmr*)

1,10-Dihydroxy-11(13)-eremophilen-12,8-olide   D-80314

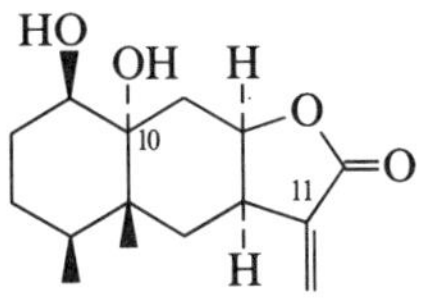

$C_{15}H_{22}O_4$ M 266.336

(1β,8β,10α)-form

Constit. of *Ondetia linearis*. Oil. $[\alpha]_D^{24}$ +43° (c, 1.14 in CHCl₃).

11β,13-Dihydro: 1β,10α-Dihydroxy-11βH-*eremophilan*-12,8β-*olide*
$C_{15}H_{24}O_4$ M 268.352
Constit. of *O. linearis*. Gum.

10-Deoxy, 10α-chloro: 10α-Chloro-1β-hydroxy-11(13)-*eremophilen*-12,8β-*olide*
$C_{15}H_{21}ClO_3$ M 284.782
Constit. of *O. linearis*. Cryst. Mp 203°. $[\alpha]_D^{24}$ +41° (c, 0.46 in CHCl₃).

Zdero, C. *et al, Phytochemistry*, 1989, **28**, 1653.

1,8-Dihydroxy-4(15),11(13)-eudesmadien-12,6-olide   D-80315

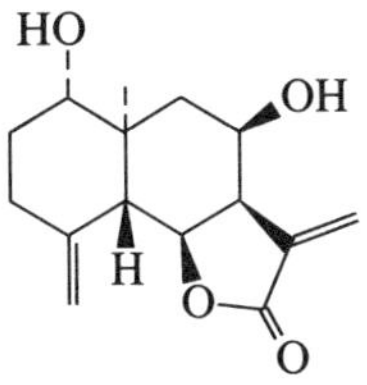

$C_{15}H_{20}O_4$ M 264.321

(1α,5β,6β,8β,10α)-form

Constit. of *Pegolettia oxydonta*. Gum.

Zdero, C. *et al, Phytochemistry*, 1989, **28**, 1949.

1,8-Dihydroxy-4(15)-eudesmen-12,6-olide   D-80316

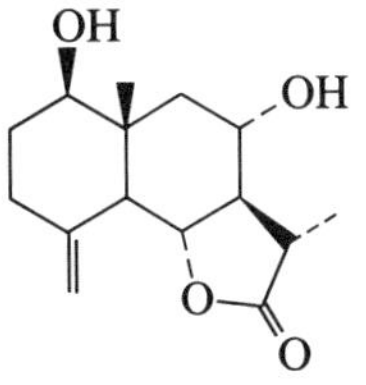

$C_{15}H_{22}O_4$ M 266.336

(1β,6α,8α,11S)-form

Constit. of *Artemisia gypsacea*.

Fernandez, I. *et al, Tetrahedron*, 1987, **43**, 805 (*synth*)
Rustaiyan, A. *et al, Phytochemistry*, 1989, **28**, 1535 (*isol*)

1,3-Dihydroxy-4,10(14)-germacradien-12,6-olide   D-80317

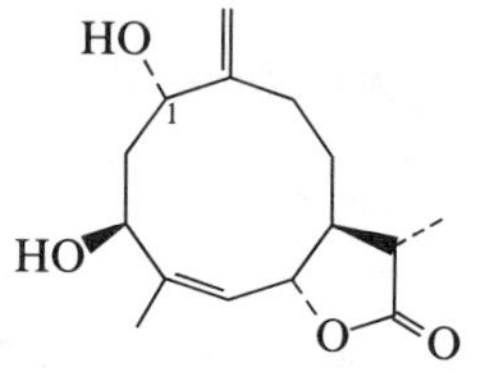

(1α,3β,4Z,6α,11βH)-form

$C_{15}H_{22}O_4$ M 266.336

(1α,3β,4Z,6α,11βH)-form [105304-94-7]
Constit. of *Pyrethrum santalinoides*.

(1β,3β,4Z,6α,11βH)-form [120163-08-8]
Constit. of *P. santalinoides*. Gum. $[\alpha]_D$ +46° (c, 0.18 in CHCl$_3$).

El-Sebakhy, Y. *et al*, *Pharmazie*, 1986, **41**, 525 (*isol*)
Abdel-Mogib, M. *et al*, *Phytochemistry*, 1989, **28**, 268 (*isol*)

1,3-Dihydroxy-4,9-germacradien-12,6-olide   D-80318

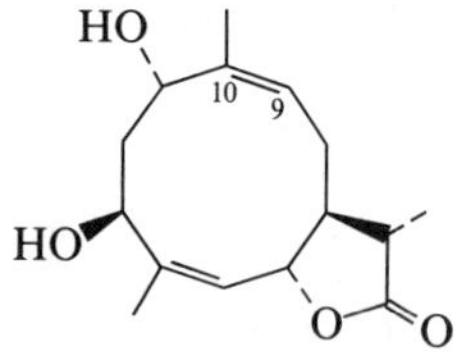

$C_{15}H_{22}O_4$ M 266.336

(1α,3β,4Z,6α,9Z,11βH)-form [120172-53-4]
Constit. of *Pyrethrum santalinoides*. Cryst. Mp 163°.
$[\alpha]_D^{24}$ +154° (c, 0.13 in CHCl$_3$).

9β,10β-Epoxide: [120181-04-6]. 9β,10β-Epoxy-1α,3β-dihydroxy-4Z-germacren-12,6α-olide
$C_{15}H_{22}O_5$ M 282.336
Constit. of *P. santalinoides*. Gum. $[\alpha]_D^{24}$ +34° (c, 0.21 in CHCl$_3$).

Abdel-Mogib, M. *et al*, *Phytochemistry*, 1989, **28**, 268.

2,8-Dihydroxy-4,10(14)-germacradien-12,6-olide   D-80319

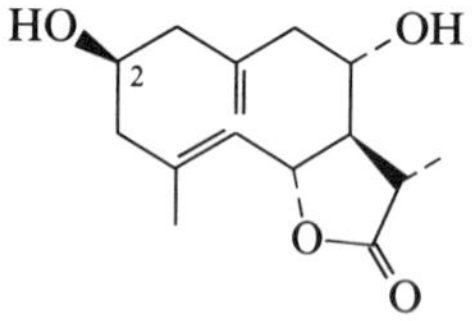

$C_{15}H_{20}O_4$ M 264.321

(2β,6α,8α,11S)-form
11β,13-Dihydroanhydroverlotorin
Constit. of *Artemisia gypsacea*. Gum. $[\alpha]_D^{24}$ +83° (c, 0.45 in CHCl$_3$).

2-Ketone: 8α-Hydroxy-2-oxo-4,10(14)-germacradien-12,6α-olide. **11β,13-Dihydroartemorin**
$C_{15}H_{18}O_4$ M 262.305
Constit. of *A. gypsacea*. Gum. $[\alpha]_D^{24}$ +48° (c, 0.3 in CHCl$_3$).

Rustaiyan, A. *et al*, *Phytochemistry*, 1989, **28**, 1535.

8,14-Dihydroxy-1(10),4-germacradien-12,6-olide   D-80320

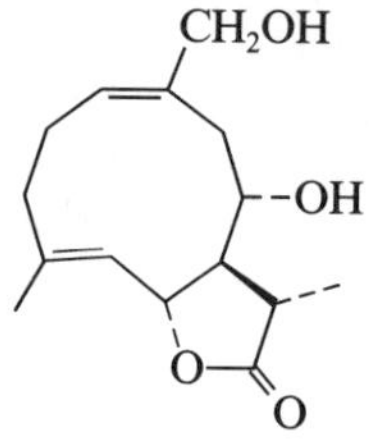

$C_{15}H_{22}O_4$ M 266.336

(1(10)E,4E,6α,8α,11S)-form
8α,14-Dihydroxy-11,13-dihydromelampolide
Constit. of *Artemisia gypsacea*. Gum. $[\alpha]_D^{24}$ +16° (c, 0.2 in CHCl$_3$).

Rustaiyan, A. *et al*, *Phytochemistry*, 1989, **28**, 1535.

8,9-Dihydroxy-4,6-heptadecadiyn-3-one   D-80321
Panaxacol

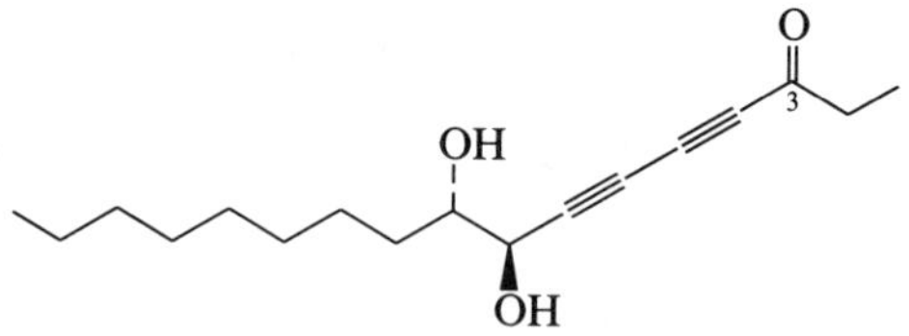

$C_{17}H_{26}O_3$ M 278.391
Isol. from callus of *Panax ginseng*.

3-Alcohol: 4,6-Heptadecadiyne-3,8,9-triol. **Dihydropanaxacol**
$C_{17}H_{28}O_3$ M 280.406
Isol. from *P. ginseng*.

Fujimoto, Y. *et al*, *Phytochemistry*, 1987, **26**, 2850.

4′,7-Dihydroxyisoflavanone   D-80322
Updated Entry replacing D-04539
2,3-Dihydro-7-hydroxy-3-(4-hydroxyphenyl)-4H-1-benzopyran-4-one, 9CI. Dihydrodaidzein
[17238-05-0]
$C_{15}H_{12}O_4$ M 256.257

(R)-form [58865-02-4]
Constit. of *Pericopsis* spp. Cryst. Mp 201°.
Di-Ac:
$C_{19}H_{16}O_6$ M 340.332
Cryst. (MeOH). Mp 153°. $[\alpha]_D^{24}$ −20.7° (CHCl$_3$).

4′-Me ether: [4626-22-6]. 7-Hydroxy-4′-methoxyisoflavanone. *Dihydroformononetin*
$C_{16}H_{14}O_4$ M 270.284
Constit. of *Myroxylon balsamum* trunkwood. Cryst. (EtOH). Mp 185-188°. No opt. rotn. given.

4′-Me ether, 7-O-glucoside: 2,3-Dihydroononin
$C_{22}H_{24}O_9$ M 432.426
Constit. of roots of *Ononis spinosa*. Mp 269-271°.

Di-Me ether: Cryst. (EtOH). Mp 129-130°.
[71815-35-5]

Donnelly, D.M.X. *et al*, *J. Chem. Soc., Perkin Trans. 1*, 1976, 186 (*isol, struct, synth*)
Haźnagy, A. *et al*, *Arch. Pharm. (Weinheim, Ger.)*, 1978, **311**, 318 (*Dihydroononin*)
de Oliveira, A.B. *et al*, *Phytochemistry*, 1978, **17**, 593 (*Dihydroformononetin*)
Jain, A.C. *et al*, *Indian J. Chem., Sect. B*, 1987, **26**, 136 (*synth*)

4′,7-Dihydroxyisoflavone   D-80323

Updated Entry replacing D-60340
Daidzein. Dimethylbiochanin B. *Daizeol*
[486-66-3]
$C_{15}H_{10}O_4$ M 254.242
Widespread isoflavone in the Leguminosae
 (Papilionoideae) eg. in *Chamaecytisus* spp., *Cytisus* spp.,
 Phaseolus spp. Also from *Streptomyces xanthophaeus*.
 Pale-yellow cryst. (EtOH aq.). Mp 330° (323°). Tatoin
 was impure Daidzein.

Di-Ac: [3682-01-7].
 $C_{19}H_{14}O_6$ M 338.316
 Cryst. (EtOH). Mp 188-189°.

7-O-β-D-Glucopyranosyl: [552-66-9]. **Daidzin**
 $C_{21}H_{20}O_9$ M 416.384
 Isol. from soya bean meal, roots of *P. lobata* and shoots
 of *Piptanthus nepalensis*. Also from *Baptisia* spp.,
 Colycine max, Medicago sativa, Thermopsis spp.
 andothers in Leguminosae. Cryst. + $1H_2O$ (H_2O)
 becoming anhyd. at 120°. Mp 233-235°. $[\alpha]_D^{20}$ −36.4°
 (0.02N KOH).

4′,7-Di-O-β-D-glucopyranosyl: [53681-67-7].
 $C_{27}H_{30}O_{14}$ M 578.526
 Stress metab. of cell cultures of *Vigna angularis*. Isol.
 also from *P. nepalensis*. Needles. Mp 241°.

7-O-(Apiosyl-(1→6)-β-D-glucopyranoside): [108044-04-8].
 Ambonin
 $C_{26}H_{28}O_{13}$ M 548.499
 Constit. of *Neorautanenia amboensis*. Rosettes (Me₂CO
 aq.). Mp 225-227°. $[\alpha]_D$ −71° (c, 0.014 in H_2O).

4′-Apioside, 7-β-D-glucopyranoside: [108069-01-8]. **Neobanin**
 $C_{26}H_{28}O_{13}$ M 548.499
 From *N. amboensis*. Rosettes (CHCl₃/MeOH). Mp 211-
 212°. $[\alpha]_D$ −72.6° (c, 0.018 in H_2O).

4′-Me ether: see 7-Hydroxy-4′-methoxyisoflavone, H-80190

Di-Me ether: [1157-39-7]. *4′,7-Dimethoxyisoflavone. 4′,7-Di-
 O-methyldaidzein*
 $C_{19}H_{14}O_4$ M 306.317
 Isol. from *Dalbergia violacea* and *Pterodon apparicioi*
 heartwoods. Cryst. (EtOH). Mp 162-164° (154-156°).

7-O-(6-O-Acetyl-β-D-glucopyranoside): **6″-O-Acetyldaidzin**
 $C_{23}H_{22}O_{10}$ M 458.421
 Isol. from seeds of *G. max*. Needles. Mp 186-189°.

4′-O-β-D-Glucopyranoside:
 $C_{21}H_{20}O_9$ M 416.384
 Isol. from twigs of *P. nepalensis*.

7-O-Rhamnoside:
 $C_{21}H_{20}O_8$ M 400.384
 Metab. of *Streptomyces xanthophaeus*. Isolation
 dubious.

7-O-(Rhamnosylglucoside):
 $C_{27}H_{30}O_{13}$ M 562.526
 Isol. from leaves and stems of *Baptisia* spp.

4′,7-Di-O-rhamnoside:
 $C_{27}H_{30}O_{12}$ M 546.527
 Metab. of *S. xanthophaeus*.

2,3-Dihydro: see 4′,7-Dihydroxyisoflavanone, D-80322

Farkas, L. *et al, Chem. Ber.,* 1959, **92**, 819 (*synth*)
Markham, K.R. *et al, Phytochemistry,* 1968, **7**, 791 (*isol*)
Gupta, S.R. *et al, Phytochemistry,* 1971, **10**, 877 (*synth*)
Dement, W.A., *Phytochemistry,* 1972, **11**, 1089 (*isol*)
Inone, T. *et al, Chem. Pharm. Bull.,* 1974, **22**, 1422 (*biosynth*)
Deshpande, V.H. *et al, Indian J. Chem., Sect. B,* 1977, **15**, 201
 (*isol*)
Nakayama, M. *et al, Bull. Chem. Soc. Jpn.,* 1978, **51**, 2398 (*synth*)
Ohta, N. *et al, Agric. Biol. Chem.,* 1979, **43**, 1415 (6″-*O*-
 Acetyldaidzin)
Jha, H.C. *et al, Can. J. Chem.,* 1980, **58**, 1211 (*cmr*)
Ayabe, S. *et al, J. Chem. Soc., Perkin Trans.* 1, 1982, 2725
 (*biosynth, ms*)
Ingham, J.L., *Fortschr. Chem. Org. Naturst.,* 1983, **43**, 1 (*rev,
 occur*)
Kobayashi, M. *et al, Phytochemistry,* 1983, **22**, 1257 (*isol*)
Jain, A.C. *et al, J. Chem. Soc., Perkin Trans.* 1, 1986, 215 (*synth*)
Breytenbach, J.C., *J. Nat. Prod.* (*Lloydia*), 1986, **49**, 1003 (*isol,
 deriv*)

ent-7β,14α-Dihydroxy-1,16-kauradiene-   D-80324
3,15-dione
 Liangshanin A

$C_{20}H_{26}O_4$ M 330.423
Constit. of *Rabdosia liangshanica*. Needles. Mp 238-240°.
 $[\alpha]_D$ −192.5° (c, 0.52 in CHCl₃).

7-Ac: ent-7β-*Acetoxy*-14α-*hydroxy*-1,16-*kauradiene*-3,15-
 dione. **Liangshanin C**
 $C_{22}H_{28}O_5$ M 372.460
 Constit. of *R. liangshanica*. Cryst. Mp 204-210°.

14-Ac: ent-14α-*Acetoxy*-7β-*hydroxy*-1,16-*kauradiene*-3,15-
 dione. **Liangshanin B**
 $C_{22}H_{28}O_5$ M 372.460
 Constit. of *R. liangshanica*. Cryst. Mp 204-210°.

Fenglei, Z. *et al, Phytochemistry,* 1989, **28**, 1671.

ent-3α,11α-Dihydroxy-16-kaurene-6,15-   D-80325
dione

$C_{20}H_{28}O_4$ M 332.439

3-Ac: ent-3α-*Acetoxy*-11α-*hydroxy*-16-*kaurene*-6,15-*dione*.
 Inflexarabdonin C
 $C_{22}H_{30}O_5$ M 374.476
 Constit. of *Rabdosia inflexa*. Cryst. (MeOH). Mp 179-
 181°. $[\alpha]_D^{24}$ −55.2° (c, 0.87 in MeOH).

Takeda, Y. *et al, Phytochemistry,* 1989, **28**, 2423 (*isol, pmr, cmr*)

ent-13,16-Dihydroxy-8(17),14-labdadien-   D-80326
18-oic acid

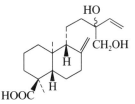

$C_{20}H_{32}O_4$ M 336.470
Constit. of fruits of *Xylopia aromatica*.

Moraes, M.P.L. *et al, Phytochemistry,* 1988, **27**, 3205.

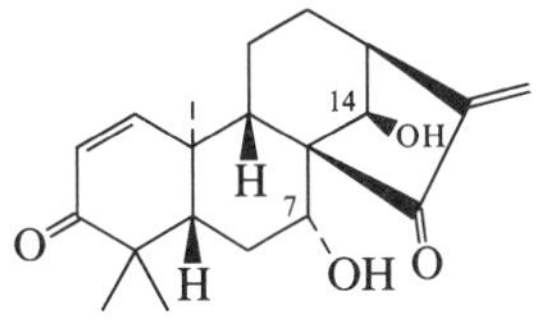

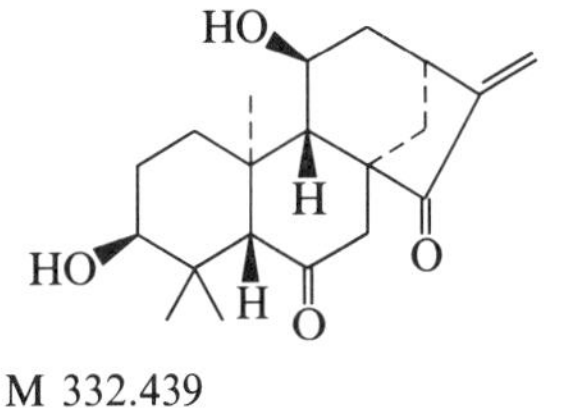

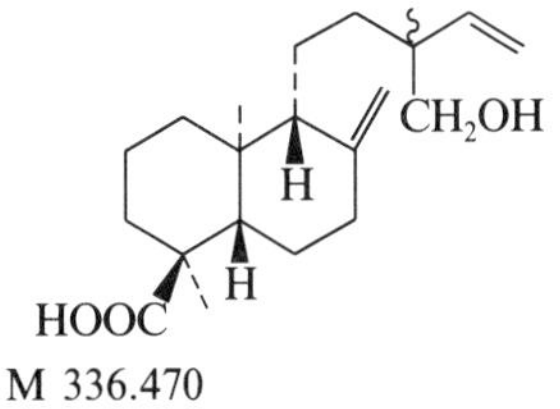

2,13-Dihydroxy-8,14-labdadien-7-one   D-80327

$C_{20}H_{32}O_3$ M 320.471

(2α,13R)-form

Constit. of *Waitzia acuminata.*

2-Ketone: **13R-*Hydroxy-8,14-labdadiene-2,7-dione***
$C_{20}H_{30}O_3$ M 318.455
Constit. of *W. acuminata.*

Jakupovic, J. *et al, Phytochemistry,* 1989, **28**, 1943.

ent-3β,15-Dihydroxy-7,13-labdadien-2-one   D-80328

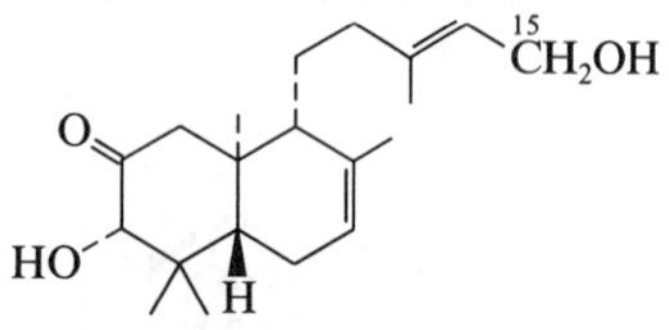

$C_{20}H_{32}O_3$ M 320.471

Constit. of *Baccharis boliviensis.* Gum.

15-Aldehyde: **ent-3β-*Hydroxy-2-oxo-7,13-labdadien-15-al***
$C_{20}H_{30}O_3$ M 318.455
Constit. of *B. boliviensis.* Oil. Both *E*- and *Z*-isomers occur.

15-Aldehyde, 3-deoxy: **ent-*2-Oxo-7,13-labdadien-15-al***
$C_{20}H_{30}O_2$ M 302.456
Constit. of *B. boliviensis.* Oil.

Zdero, C. *et al, Phytochemistry,* 1989, **28**, 531.

6,15-Dihydroxy-8-labden-17-oic acid   D-80329

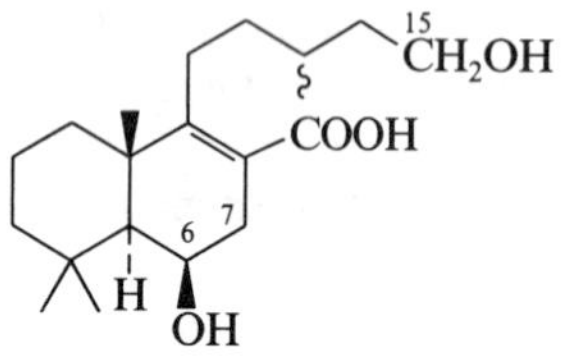

$C_{20}H_{34}O_4$ M 338.486

6β-form

Constit. of *Halimium verticillatum.*

Urones, J.G. *et al, Phytochemistry,* 1989, **28**, 557.

7,15-Dihydroxy-8-labden-17-oic acid   D-80330

$C_{20}H_{34}O_4$ M 338.486

7α-form

7-Me ether: **15-*Hydroxy-7α-methoxy-8-labden*-17-oic acid**
$C_{21}H_{36}O_4$ M 352.513
Constit. of *Halimium verticillatum.*

7-Me ether, 15-Ac, Me ester: Oil. $[\alpha]_D^{22}$ +21.36° (c, 1.1 in CHCl₃).

7β-form

Di-Ac: **7β,15-*Diacetoxy-8-labden*-17-oic acid**
$C_{24}H_{38}O_6$ M 422.561
Constit. of *H. verticillatum.*

Di-Ac, Me ester: Oil. $[\alpha]_D^{22}$ +24.0° (c, 1.32 in CHCl₃).

Urones, J.G. *et al, Phytochemistry,* 1989, **28**, 557.

3,15-Dihydroxy-8,24-lanostadien-26-oic acid   D-80331

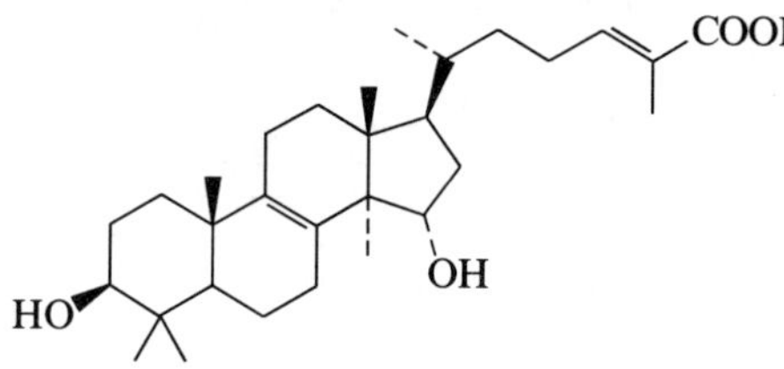

$C_{30}H_{48}O_4$ M 472.707

(3β,15α,24E)-form

Di-Ac: [117842-20-3]. *3β,15α-Diacetoxy-8,24-lanostadien-26-oic acid*
$C_{34}H_{52}O_6$ M 556.781
Constit. of *Ganoderma lucidum.*

Lin, L.-J. *et al, J. Nat. Prod. (Lloydia),* 1988, **51**, 919.

7,9-Dihydroxy-2-longipinen-1-one   D-80332

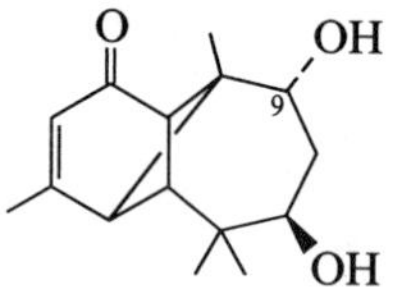

(7β,9α)-*form*

$C_{15}H_{22}O_3$ M 250.337

(7β,9α)-form [97335-21-2]

Constit. of *Lavandula stoechas* and *Stevia salicifolia.*
Cryst. (CHCl₃). Mp 183-184°. $[\alpha]_D$ +48° (c, 2 in EtOH).

9-Ac: *9α-Acetoxy-7β-hydroxy-2-longipinen-1-one*
$C_{17}H_{24}O_4$ M 292.374
Constit. of *L. stoechas.*

Roman, L.U. *et al, J. Org. Chem.,* 1985, **50**, 3965 (*isol, struct*)
Ulubelen, A. *et al, Phytochemistry,* 1988, **27**, 3966 (*isol, deriv*)

2,4-Dihydroxy-7-methoxy-2*H*-1,4-benzoxazin-3(4*H*)-one   D-80333

Updated Entry replacing D-10388
DIMBOA
[15893-52-4]

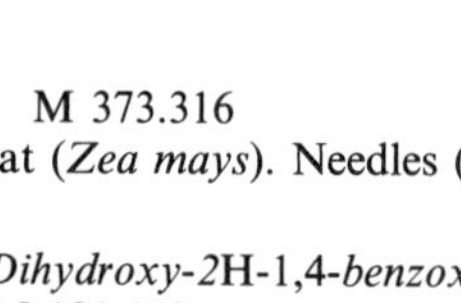

$C_9H_9NO_5$ M 211.174

Isol. from wheat, in which it is present mainly as glucoside. Appears to be a natural aphicide and fungicide. Involved in the *in vivo* detoxification of herbicides, e.g. Simazine. Pink needles. Mp 156-157° dec.

O-β-D-Glucoside:
$C_{15}H_{19}NO_{10}$ M 373.316
Isol. from wheat (*Zea mays*). Needles (EtOH). Mp 262-263° dec.

Demethoxy: *2,4-Dihydroxy-2H-1,4-benzoxazin-3(4H)-one*
$C_8H_7NO_4$ M 181.148
Isol. from rye (*Secale cereale*) and maize (*Z. mays*). Cryst. (Et₂O/cyclohexane). Mp 152°.

Demethoxy, 2-O-β-D-glucoside:
$C_{14}H_{17}NO_9$ M 343.290
Isol. from *S. cereale* and *Z. mays.* Cryst. Mp 186.5-187°.

Wahlroos, O. *et al, Acta Chem. Scand.,* 1959, **13**, 1906 (*isol, struct*)

Virtanen, A.L. *et al, Acta Chem. Scand.*, 1960, **14**, 499, 502, 504 (*isol, synth, derivs*)
Reimann, J.E. *et al, Biochemistry*, 1964, **3**, 847 (*biosynth*)
Hofman, J. *et al, Tetrahedron Lett.*, 1969, 5001 (*isol, uv, glucoside*)
Argandoã, V.H. *et al, Phytochemistry*, 1980, **19**, 1665; 1981, **20**, 673 (*isol*)

2,2′-Dihydroxy-3-methoxy-5,5′-di-2-propenylbiphenyl   D-80334

5,5′-Diallyl-2,2′-hydroxy-3-methoxybiphenyl

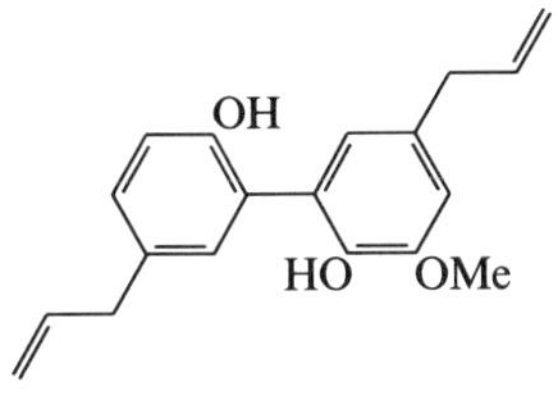

$C_{19}H_{20}O_3$ M 296.365
Constit. of *Magnolia henryi*. Gum.

Kijjoa, A. *et al, Phytochemistry*, 1989, **28**, 1284.

4′,5-Dihydroxy-7-methoxyisoflavone   D-80335

Updated Entry replacing D-10392
Prunetin. *Padmakastein. Prunusetin*
[26015-63-4]

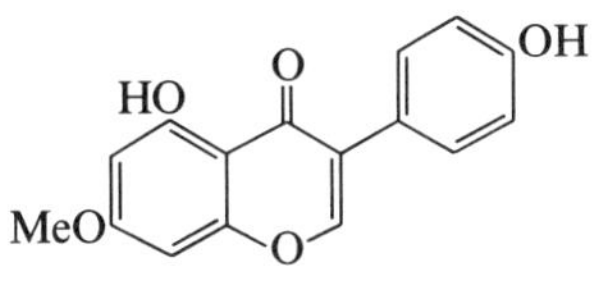

$C_{16}H_{12}O_5$ M 284.268
Occurs in bark of *Prunus puddum* and in several other *P.* spp. (Rosaceae) and in *Genista violacea, G. carinalis, Glycyrrhiza glabra* and *Pterocarpus angolensis* (Leguminosae). Also prod. by cultures of *Mycobacterium phlei*. Cryst. (EtOH). Mp 236-238°. All isolations of isoflavones from microorganism cultures are considered doubtful.

4′-O-β-D-Glucopyranoside: [154-36-9]. **Prunitrin.** *Prunitroside. Trifoside. Prunetrin*
$C_{22}H_{22}O_{10}$ M 446.410
Isol. from *P. mahaleb, P. puddum, Trifolium medium, T. pratense* and *Genista carinalis*. Reported to have antispasmodic, expectorant and hypolipidaemic props. Mp 181° (sinters from 160°). $[\alpha]_D$ −82.5° (Py). Padmakastin was prob. impure prunitrin.

Hasegawa, M., *J. Am. Chem. Soc.*, 1957, **79**, 1738 (*isol*)
Plouvier, V. *et al, Compt. Rend. Hebd. Seances Acad. Sci.*, 1960, **250**, 594 (*isol, deriv*)
Hudson, A.T. *et al, J. Chem. Soc., Chem. Commun.*, 1969, 830 (*isol*)
Farkas, L. *et al, Tetrahedron*, 1969, **25**, 1013 (*struct, synth*)
Ingham, J.L., *Fortschr. Chem. Org. Naturst.*, 1983, **43**, 1 (*rev, occur*)

4′,7-Dihydroxy-5-methoxyisoflavone   D-80336

7-Hydroxy-3-(4-hydroxyphenyl)-5-methoxy-4H-1-benzopyran-4-one, 9CI. **Isoprunetin.** *5-O-Methylgenistein*
[4569-98-6]
$C_{16}H_{12}O_5$ M 284.268

Isol. from a wide variety of spp. in Leguminosae subf. Papilionoideae, eg. *Adenocarpus, Chamaecytisus, Cytisus, Genista* and *Ulex* spp. Needles. Mp 302° (290°) dec.

Chopin, J. *et al, Bull. Soc. Chim. Fr.*, 1964, 1038 (*isol, ir, uv, struct, synth*)
Harborne, J.B., *Phytochemistry*, 1969, **8**, 1449 (*occur*)
Ingham, J.L., *Fortschr. Chem. Org. Naturst.*, 1983, **43**, 1 (*occur*)

5,7-Dihydroxy-4′-methoxyisoflavone   D-80337

Updated Entry replacing D-04603
5,7-Dihydroxy-3-(4-methoxyphenyl)-4H-1-benzopyran-4-one, 9CI. **Biochanin** A. *Genistein 4′-methyl ether. Pratensol*
[491-80-5]
$C_{16}H_{12}O_5$ M 284.268
Constit. of clover (*Trifolium repens*), *Dalbergia* and *Cotoneaster* spp. Widely distributed in the Leguminosae (Papilionoideae), also in *Cotoneaster pannosa* and *C. serotina* (Rosaceae) and *Virola caducifolia* (Myricaceae). Sl. oestrogenic. Needles (MeOH or EtOH aq.). Mp 215-216°.

7-Ac: Rectangular plates (MeOH). Mp 155-156°.
5,7-Di-Ac: Needles (MeOH). Mp 190°.
7-β-D-Glucoside: [5928-26-7]. **Sissotrin.** *Astroside*
$C_{22}H_{22}O_{10}$ M 446.410
Constit. of *Thermopsis* and *C.* spp., also from *Astragalus austriacus, Baptisia leucantha, Cicer arietinum, Pueraria thumbergiana, Sophora japonica* and *Trifolium* spp. Mp 220°. $[\alpha]_D^{30}$ −35.3° (c, 1.644 in DMF).
7-(β-D-Apiosyl(1→2)-β-D-glucopyranoside): [15914-68-8]. **Lanceolarin**
$C_{27}H_{30}O_{14}$ M 578.526
Constit. of the root bark of *D. lanceolaria*. Cryst. (MeOH). Mp 165-167° (unsharp, previous sintering). $[\alpha]_D^{22}$ −96.9° (c, 1.032 in 80% MeOH aq.).
7-O-Rhamnosylglucoside:
$C_{28}H_{32}O_{14}$ M 592.552
Isol. from *B.* spp.
7-O-Rutinoside:
$C_{28}H_{32}O_{14}$ M 592.552
Isol. from *D. paniculata* bark. Thin rods + $3H_2O$ (MeOH). Mp 153-154°. $[\alpha]_D^{20}$ −51.9° (c, 1 in MeOH).
7-O-Gentiobioside:
$C_{28}H_{32}O_{15}$ M 608.552
Isol. from wood of *S. japonica*. Prisms (MeOH aq.). Mp 224-226°. $[\alpha]_D^{24}$ −38.7° (c, 1.79 in 80% MeOH aq.).
7-O-(Xylosylglucoside): Isol. from *S. japonica* wood. Needles (MeOH aq.). Mp 228-230°. $[\alpha]_D^{24}$ −79.5° (c, 0.52 in 80% MeOH aq.).
7-O-(6-O-Malonyl-D-glucoside):
$C_{25}H_{24}O_{13}$ M 532.457
Isol. from leaves of *Trifolium pratense* and *T. subterranium*. Needles (MeOH). Mp 217°. Revised struct.
7-O-(6-O-Malonyl-D-glucoside), Me ester:
$C_{26}H_{26}O_{13}$ M 546.484
Isol. from leaves of *T. pratense* and *T. subterraneum*. Needles (MeOH). Mp 195°.

Baker, W. *et al, J. Chem. Soc.*, 1953, 1852 (*synth*)
Geissman, T.A. *et al, Phytochemistry*, 1965, **4**, 89 (*Sissotrin*)
Malhotra, A. *et al, Tetrahedron*, 1967, **23**, 405 (*Lanceolarin*)
Beck, A.B. *et al, Aust. J. Chem.*, 1971, **24**, 1509 (*malonates*)
Parthasarathy, M.R. *et al, Phytochemistry*, 1976, **15**, 1025 (*rutinoside*)
Foot, S. *et al, Synthesis*, 1976, 326 (*synth*)
Diedrich, D.F. *et al, J. Chem. Eng. Data*, 1977, **22**, 448 (*pmr*)
Takeda, T. *et al, Phytochemistry*, 1977, **16**, 619 (*xylosylglucoside, gentiobioside*)

Bass, R.J. *et al*, *J. Chem. Soc., Perkin Trans.* 1, 1978, 666 (*cmr*)
Ashdown, D.H.J. *et al*, *Synthesis*, 1978, 843 (*synth*)
Ingham, J.L., *Fortschr. Chem. Org. Naturst.*, 1983, **43**, 1 (*rev, occur*)

2′,6′-Dihydroxy-4′-methoxy-3′-(1-p-menthen-3-yl)chalcone   D-80338

1-[2,6-Dihydroxy-4-methoxy-3-[3-methyl-6-(1-methylethyl)-2-cyclohexen-1-yl]phenyl]-3-phenyl-2-propen-1-one, 9CI. 2′,6′-Dihydroxy-4′-methoxy-3′-(3-methyl-6-methylethyl-2-cyclohexenyl)chalcone

[119628-53-4]

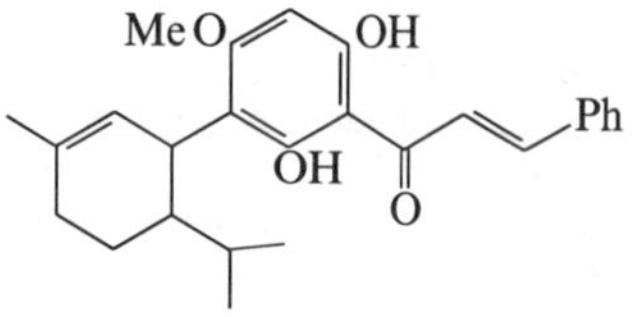

$C_{26}H_{30}O_4$ M 406.521
Constit. of *Lindera umbellata*. Yellow powder. $[\alpha]_D^{20}$ +119.7° (c, 0.76 in $CHCl_3$).

Shimomura, H. *et al*, *Phytochemistry*, 1988, **27**, 3937.

4,5-Dihydroxy-2-methylbenzoic acid   D-80339

$C_8H_8O_4$ M 168.149
Di-Me ether: [20736-28-1]. *4,5-Dimethoxy-2-methylbenzoic acid*
$C_{10}H_{12}O_4$ M 196.202
Cryst. (EtOH aq.). Mp 139-141°.
Di-Me ether, N,N-diethylamide: [113975-92-1].
$C_{14}H_{21}NO_3$ M 251.325
Oil.

Clark, R.D. *et al*, *J. Org. Chem.*, 1988, **53**, 2378 (*derivs, synth, pmr*)

5,7-Dihydroxy-7-(3-methyl-2-butenyl)-8-(2-methyl-2-oxopropyl)-3-propyl-2H-1-benzopyran-2-one   D-80340

5,7-Dihydroxy-8-(2-methyl-1-oxopropyl)-7-prenyl-3-propylcoumarin

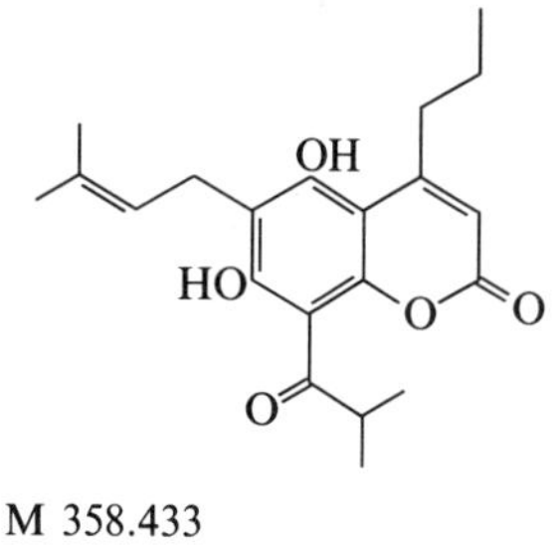

$C_{21}H_{26}O_5$ M 358.433
Isol. from *Mammea americana*. Related to the Mammeins.

Games, D.E. *et al*, *Tetrahedron Lett.*, 1972, 3187.

4′,5-Dihydroxy-6,7-methylenedioxyisoflavone   D-80341

Updated Entry replacing D-04712
9-Hydroxy-7-(4-hydroxyphenyl)-8H-1,3-dioxolo[4,5-g][1]benzopyran-8-one, 9CI. **Irilone**

[41653-81-0]

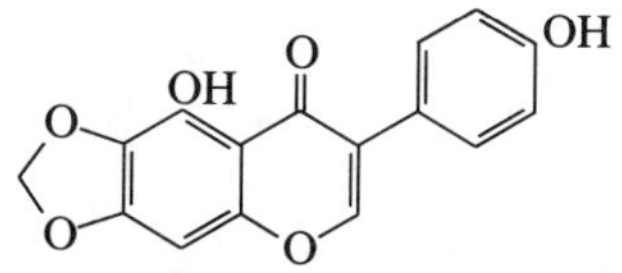

$C_{16}H_{10}O_6$ M 298.251
Constit. of *Iris germanica* and of *Trifolium pratense*. Yellow cryst. Mp 231°.

4′-O-Glucoside: [50868-47-8]. *Irilone-4′-glucoside*
$C_{22}H_{20}O_{11}$ M 460.393
Isol. from rhizomes of *Iris florentina* and roots of *Trifolium pratense*. Needles ($CHCl_3$/MeOH)(as pentaacetate). Mp 233-235° (pentaacetate).

5-Me ether: [3301-68-6]. *4′-Hydroxy-5-methoxy-6,7-methylenedioxyisoflavone.* **Irisolone**
$C_{17}H_{12}O_6$ M 312.278
Isol. from *I. florentina, I. germanica* and *I. nepalensis*. Plates (EtOAc/2-methoxyethanol). Mp 269-270°.

5-Me ether, 4′-O-glucosylglucoside: [50938-05-1]. *Irisolone-4′-bioside*
$C_{29}H_{32}O_{16}$ M 636.562
Isol. from rhizomes of *I. florentina*. Cryst (MeOH)(as heptaacetate). Mp 173-175° (heptaacetate).

Di-Me ether: [3405-76-3]. *4′,5-Dimethoxy-6,7-methylenedioxyisoflavone. Irisolone methyl ether*
$C_{18}H_{14}O_6$ M 326.305
Isol. from bulbs of *I. tingitana*. Mp 180-183°.

Gopinath, K.W. *et al*, *Tetrahedron*, 1961, **16**, 201 (*isol, struct*)
Arisawa, M. *et al*, *Chem. Pharm. Bull.*, 1973, **21**, 2323 (*isol, deriv*)
Pailer, M. *et al*, *Monatsh. Chem.*, 1973, **104**, 1394 (*isol, deriv*)
Tsukida, K. *et al*, *Phytochemistry*, 1973, **12**, 2318 (*glycosides*)
Dhar, K.L. *et al*, *Phytochemistry*, 1973, **12**, 734 (*isol, uv*)
El-Emary, N.A. *et al*, *Phytochemistry*, 1980, **19**, 1878 (*isol, deriv*)

5,7-Dihydroxy-8-(2-methyl-1-oxobutyl)-4-pentyl-2H-1-benzopyran-4-one, 9CI   D-80342

5,7-Dihydroxy-8-(2-methylbutyryl)-4-pentylcoumarin

[36478-56-5]

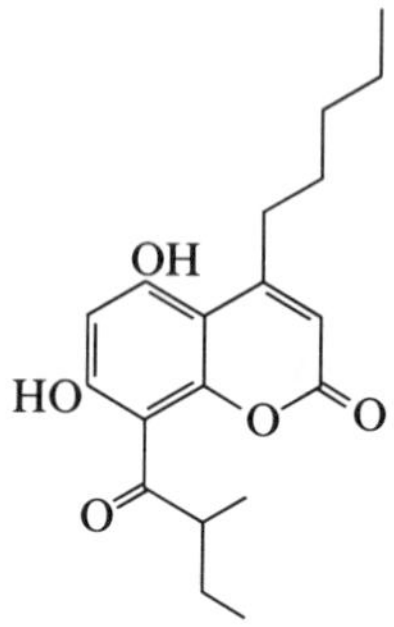

$C_{19}H_{24}O_5$ M 332.396
Isol. from *Mammea africana*. Mp 218-220°. Related to the Mammeins.

Carpenter, I. *et al*, *J. Chem. Soc. C*, 1971, 3783.

2-(1,4-Dihydroxy-4-methylpentyl)-5,8-dihydroxy-1,4-naphthoquinone   D-80343

Updated Entry replacing A-00173

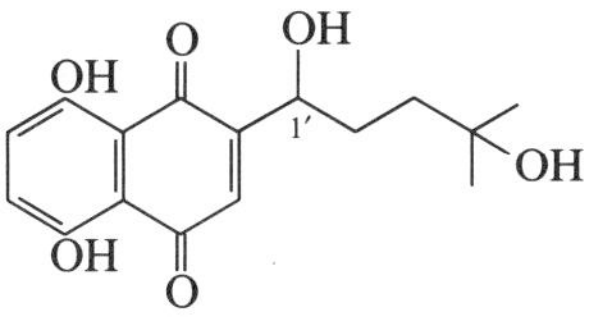

$C_{16}H_{18}O_6$ M 306.315

1'-Ac: [34232-25-2]. **Arnebin VI.** *2-(1-Acetoxy-4-hydroxy-4-methylpentyl)-5,8-dihydroxy-1,4-naphthoquinone*
$C_{18}H_{20}O_7$ M 348.352
Constit. of *Arnebia nobilis.* Mp 88-90°.

1'-(4-Methyl-2-butenoyl): [36883-11-1]. **Arnebin II**
$C_{21}H_{24}O_7$ M 388.416
Major quinone from roots of *A. nobilis.* Dark red needles. Mp 92-94°.

1'-Deoxy: **Arnebin V.** *5,8-Dihydroxy-2-(4-hydroxy-4-methylpentyl)naphthoquinone*
$C_{16}H_{18}O_5$ M 290.315
Isol. from *A. nobilis.* Mp 111-112°.

Shukla, Y.N. *et al, Phytochemistry,* 1971, **10**, 1909 (*isol, struct*)

3,24-Dihydroxy-30-nor-12,20(29)-oleanadien-28-oic acid   D-80344

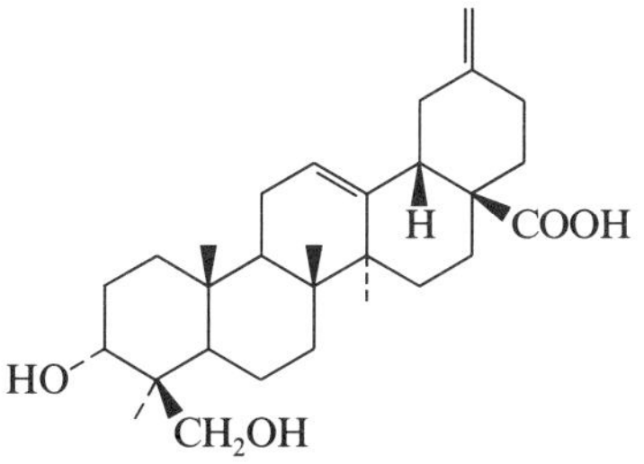

$C_{29}H_{44}O_4$ M 456.664

3α-form [119863-89-7] **Quinatic acid**
Constit. of *Akebia quinata* callus tissue. Needles. Mp 269-272°. $[\alpha]_D^{18}$ +66.6° (c, 0.375 in Py).

Ikuta, A. *et al, Phytochemistry,* 1988, **27**, 3809.

4,14-Dihydroxyoctadecanoic acid   D-80345

$H_3C(CH_2)_3CH(OH)(CH_2)_9CH(OH)CH_2CH_2COOH$

$C_{18}H_{34}O_4$ M 314.464

14-O-Diglucoside, Et ester: **Muricatin A**
$C_{32}H_{21}O_{10}$ M 565.512
Isol. from *Ipomoeia muricata.* Mp 117-118°. $[\alpha]_D^{28}$ −29.8° (EtOH). Complete struct. not known. Muricatin A is not homogeneous.

Khanna, S.N. *et al, Phytochemistry,* 1967, **6**, 735.

2,3-Dihydroxy-12-oleanene-23,28-dioic acid   D-80346

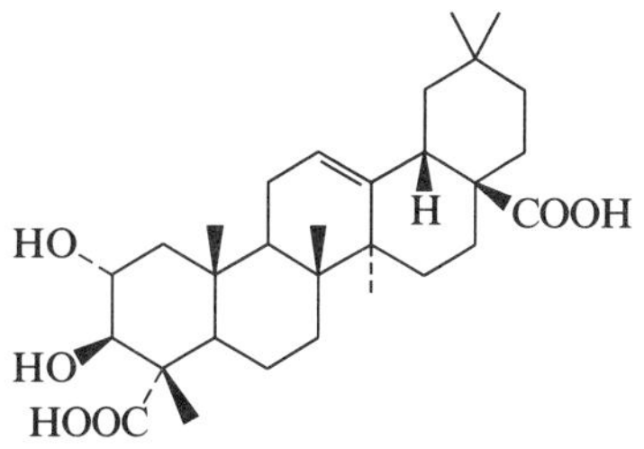

$C_{30}H_{46}O_6$ M 502.690

(2α,3β)-form [471-58-9] **Barringtogenic acid**
Isol. from *Barringtonia* spp. Prisms (MeOH). Mp 332-334° dec. $[\alpha]_D$ +72° (MeOH).

Di-Me ester: Mp 253-254°. $[\alpha]_D^{30}$ +61°.

Anantaraman, R. *et al, J. Chem. Soc.,* 1956, 4369.
Sastry, C.S.P. *et al, Tetrahedron,* 1967, **23**, 3837.

3,29-Dihydroxy-12-oleanen-28-oic acid   D-80347

Updated Entry replacing D-04896

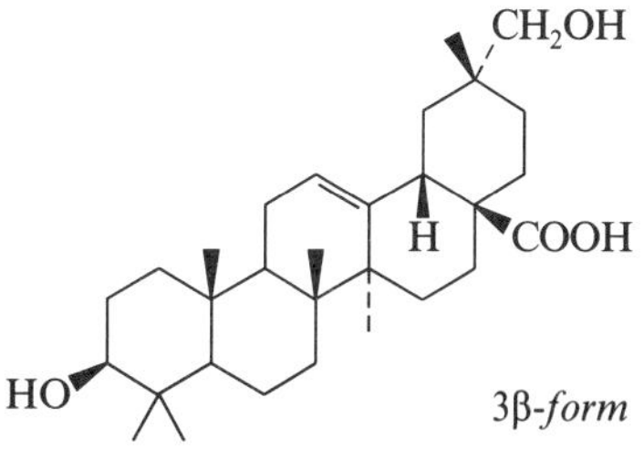

$C_{30}H_{48}O_4$ M 472.707
3α-form
    3-Epimesembryanthemoidigenic acid
    Constit. of *S. hexaphylla.*
Ac:
    $C_{32}H_{50}O_5$ M 514.744
    Constit. of *S. hexaphylla.* Amorph. $[\alpha]_D^{18}$ +106.7° (c, 0.03 in CHCl₃).
3β-form [4871-87-8] **Mesembryanthemoidigenic acid**
    Constit. of *Rhipsalis mesembryanthemoides.* Cryst. (MeOH). Mp 305-309°. $[\alpha]_D$ +70.7°.

3-Ac: Constit. of *Stauntonia hexaphylla.*

Tursch, B. *et al, Tetrahedron Lett.,* 1965, 4161 (*isol, struct*)
Ikuta, A. *et al, J. Nat. Prod. (Lloydia),* 1989, **52**, 623 (*isol, pmr, cmr*)

10,15-Dihydroxy-4-oplopanone   D-80348

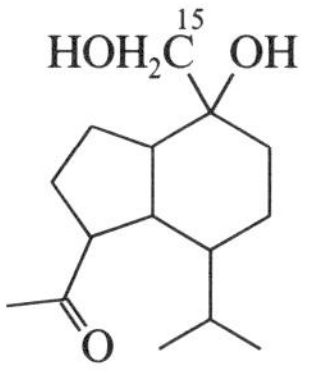

$C_{15}H_{36}O_3$ M 264.448
15-Angeloyl: *15-Angeloyloxy-10-hydroxy-4-oplopanone*
    $C_{20}H_{32}O_4$ M 336.470
    Constit. of *Senecio mexicanus.* Oil. $[\alpha]_D$ −22° (c, 0.17 in CHCl₃).

Joseph-Nathan, P. *et al, Phytochemistry,* 1989, **28**, 2397 (*isol, pmr, ms*)

ent-13*R*-14*R*-Dihydroxy-3-oxo-15-atisen-17-al   D-80349

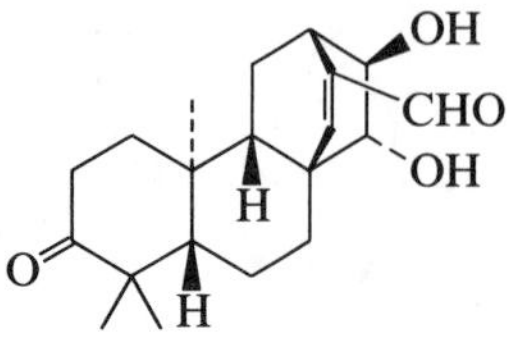

$C_{20}H_{28}O_4$ M 332.439
Constit. of *Euphorbia fidjiana*. Oil.

Lal, A.R. *et al*, *Tetrahedron Lett.*, 1989, **30**, 3205.

3,5-Dihydroxy-19-oxo-14,20(22)-cardadienolide, 9CI   D-80350

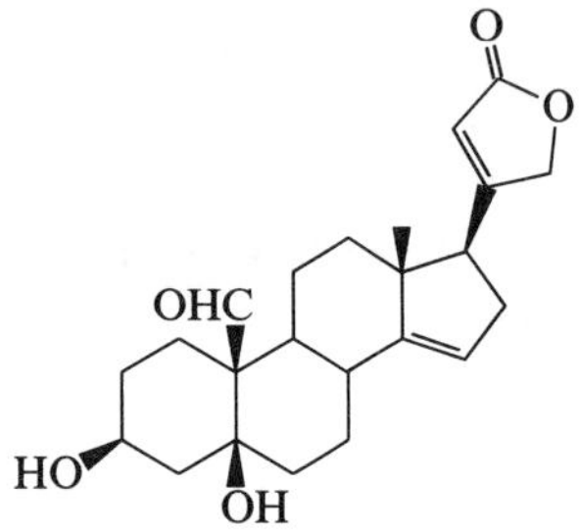

$C_{23}H_{30}O_5$ M 386.487

(*3β,5β*)-*form* [806-02-0] **Diffugenin**. 14-*Anhydrostrophanthidin*
Aglycone isol. from seeds of *Erysimum diffusum* after
hydrol. Needles (EtOH/Et$_2$O). Mp 186-188°. $[\alpha]_D^{20}$ +5.4°
(c, 1.24 in MeOH).

Makarichev, G.K. *et al*, *Zh. Obshch. Khim.*, 1962, **32**, 2372 (*isol, struct*)
Rashkes, Ya.V. *et al*, *Khim. Prir. Soedin.*, 1971, **7**, 747; *Chem. Nat. Compd.*, 724 (*ms*)

2,3-Dihydroxy-6-oxo-23,24-dinor-1,3,5(10),7-friedelatetraen-29-oic acid   D-80351
23-Nor-6-oxodemethylpristimerol
[118172-80-8]

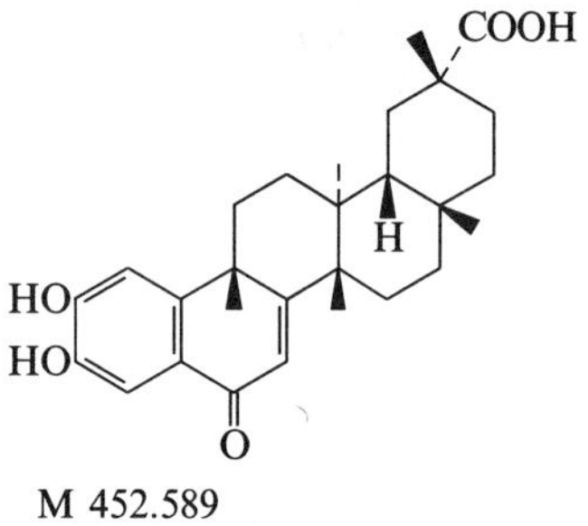

$C_{28}H_{36}O_5$ M 452.589
Constit. of *Kokoona zeylanica*. Yellow solid. Mp 228-230°.
$[\alpha]_D^{27}$ −102° (CHCl$_3$).
Me ester: [118172-79-5]. **23-Nor-6-oxopristimerol**
     $C_{29}H_{38}O_5$ M 466.616
     Constit. of *K. zeylanica*. Pale yellow solid. Mp 135-138°.
     $[\alpha]_D^{27}$ −74.9° (CHCl$_3$).

Gamlath, C.B. *et al*, *Phytochemistry*, 1988, **27**, 3221.

2,19-Dihydroxy-3-oxo-1,12-ursadien-28-oic acid   D-80352

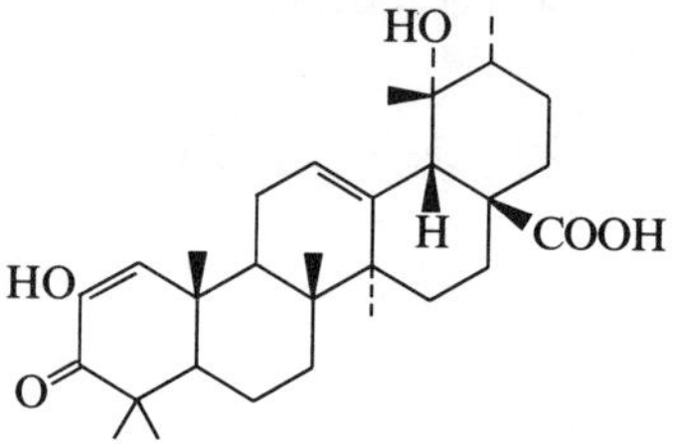

$C_{30}H_{44}O_5$ M 484.675
19α-form
     Fupenzic acid
     Constit. of *Rubus chingii*. Needles. Mp 189-190°. $[\alpha]_D^{25}$
     +45.6° (c, 0.16 in MeOH).

Hattori, M. *et al*, *Phytochemistry*, 1988, **27**, 3975.

3,4-Dihydroxypentanoic acid   D-80353

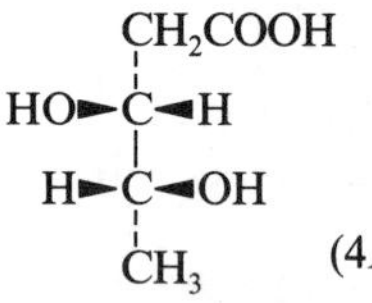

(4*R*,5*R*)-*form*

$C_5H_{10}O_4$ M 134.132
(*4R,5R*)-*form*
     D-threo-form
     1,4-Lactone: [38996-24-6]. *Dihydro-4-hydroxy-5-methyl-2(3H)furanone. 2,5-Dideoxy-γ-pentonolactone*
     $C_5H_8O_3$ M 116.116
     Oil. $[\alpha]_D$ +75° (c, 0.25 in EtOH).
(*4R,5S*)-*form*
     L-erythro-form
     1,4-Lactone: [98512-76-6].
     Liq. Bp$_{12}$ 125°. $[\alpha]_D^{20}$ −10.81° (c, 1.85 in CHCl$_3$).
(*4S,5S*)-*form*
     L-threo-form
     1,4-Lactone: [105881-47-8].
     Oil. $[\alpha]_D^{20}$ −73.7° (c, 1.6 in EtOH).
(*4S,5R*)-*form*
     D-erythro-form
     1,4-Lactone: [88400-20-8].
     Liq. Bp$_{0.025}$ 108-110°. $[\alpha]_D$ +10.87° (c, 2.42 in CHCl$_3$).

Chen, S. *et al*, *J. Org. Chem.*, 1984, **49**, 2168 (*synth, pmr, cmr, ir, ms*)
Moore, R.E. *et al*, *J. Org. Chem.*, 1984, **49**, 2484 (*synth, pmr, cd*)
Ortuño, R.M. *et al*, *Tetrahedron*, 1987, **43**, 2191 (*synth, pmr, ir, ms*)

6,7-Dihydroxy-1,12-perylenedione, 9CI   D-80354
[117156-88-4]

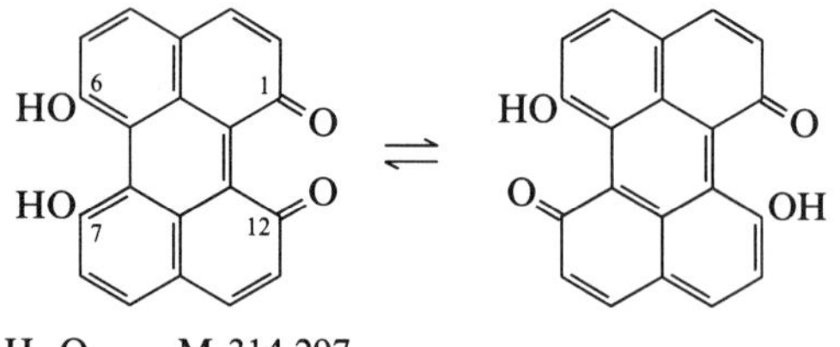

$C_{20}H_{10}O_4$ M 314.297
Tautomeric.
6,7-Dihydroxy-form
     Di-Ac: [117156-93-1].

146

$C_{24}H_{14}O_6$ M 398.371
Plates (CH_2Cl_2/MeOH). Mp 218-219° dec.

Di-Me ether: [117156-90-8]. *6,7-Dimethoxy-1,12-perylenedione*
$C_{22}H_{14}O_4$ M 342.350
Dark red needles (CH_2Cl_2/MeOH). Mp 277-279°.

6,12-Dihydroxy-form
6,12-Dihydroxy-1,7-perylenedione

Di-Ac: [117156-92-0].
Red plates (CH_2Cl_2/MeOH). Mp >280° dec.

Di-Me ether: [117156-91-9]. *6,12-Dimethoxy-1,7-perylenedione*
$C_{22}H_{14}O_4$ M 342.350
Dark red microcryst. (CH_2Cl_2/pet. ether). Mp 285-290° dec.

[117156-89-5]

Calderon, J.S. *et al, J. Chem. Soc., Perkin Trans.* 1, 1988, 583 (*synth, ir, uv, pmr, ms*)

2-(2,4-Dihydroxyphenyl)-5,6-dihydroxy-benzofuran   D-80355

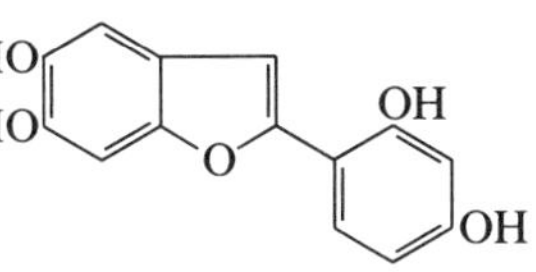

$C_{14}H_{10}O_5$ M 258.230

5,6-Di-Me ether: [67492-33-5]. *2-(2,4-Dihydroxyphenyl)-5,6-dimethoxybenzofuran*
$C_{16}H_{14}O_5$ M 286.284
Isol. from trunkwood of *Myroxylon balsamum*. Mp 178-180°.

5,6-Methylene ether: [67121-26-0]. *2-(2,4-Dihydroxyphenyl)-5,6-methylenedioxybenzofuran*
$C_{15}H_{10}O_5$ M 270.241
Isol. from aerial parts of *Sophora tomentosa*. Plates. Mp 235-237°.

4′-Me, 5,6-methylene ether: 2-(2-Hydroxy-4-methoxyphenyl)-5,6-methylenedioxybenzofuran
$C_{16}H_{12}O_5$ M 284.268
Isol. from aerial parts of *S. tomentosa*. Needles (MeOH). Mp 179-181°.

[67236-31-1]

Komatsu, M. *et al, Chem. Pharm. Bull.*, 1978, **26**, 1274 (*isol, uv, pmr, struct*)
Braga de Oliviera, A. *et al, Phytochemistry*, 1978, **17**, 593 (*isol, ir, uv, pmr, ms, struct*)

4-(3,4-Dihydroxyphenyl)-5,7-dihydroxy-2H-1-benzopyran-2-one   D-80356

4-(3,4-Dihydroxyphenyl)-5,7-dihydroxycoumarin. 3′,4′,5,7-Tetrahydroxyneoflavone

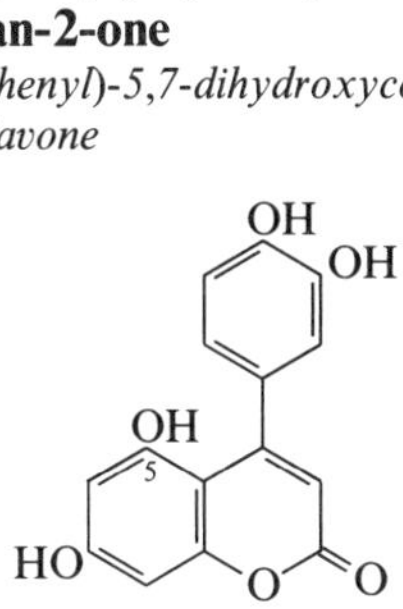

$C_{15}H_{10}O_6$ M 286.240

5-Me ether: 4-(3,4-Dihydroxyphenyl)-7-hydroxy-5-methoxycoumarin. 3′,4′,7-Trihydroxy-5-methoxyneoflavone
$C_{16}H_{12}O_6$ M 300.267
Constit. of *Coutarea hexandra*. Cryst. (MeOH). Mp 269-270°.

4′,7-Di-Me ether: 5-Hydroxy-4-(3-hydroxy-4-methoxyphenyl)-7-methoxycoumarin. 3′,5-Dihydroxy-4′,7-dimethoxyneoflavone
$C_{17}H_{14}O_6$ M 314.294
Constit. of *Exostema caribaeum*. Cryst. Mp 225-226°.

7-Me ether, 5-O-β-D-glucopyranosyl: [116310-58-8]. *5-O-β-D-Glucopyranosyl-3′,4′-dihydroxy-7-methoxyneoflavone*
$C_{22}H_{22}O_{11}$ M 462.409
Constit. of *E. caribaeum*. Cryst. (MeOH). Mp 237-238°.

7-Me ether, 5-O-(6-acetyl-β-D-galactopyranosyl): 5-O-(6-Acetyl-β-D-galactopyranosyl)-3′,4′-dihydroxy-7-methoxyneoflavone
$C_{24}H_{24}O_{11}$ M 488.447
Constit. of *E. caribaeum*. Cryst. Mp 215-220°.

Mata, R. *et al, J. Nat. Prod. (Lloydia)*, 1988, **51**, 851 (*isol*)
Delle Monache, G. *et al, Heterocycles*, 1989, **29**, 355 (*isol*)

1-(2,4-Dihydroxyphenyl)-3-[3,4-dihydroxy-2-(3,7-dimethyl-2,6-octadienyl)phenyl]-1-propanone   D-80357

AC-5-1

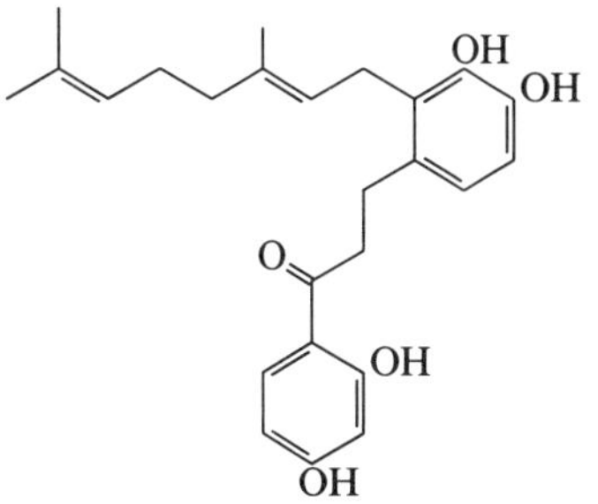

$C_{25}H_{30}O_5$ M 410.509
Constit. of *Artocarpus communis*. 5-Lipoxygenase inhibitor. Powder. Mp 132-136°.

Nakano, J. *et al, Heterocycles*, 1989, **29**, 427.

4-(3,4-Dihydroxyphenyl)-6,7-dihydroxy-2-naphthalenecarboxylic acid, 9CI   D-80358

[121242-02-2]

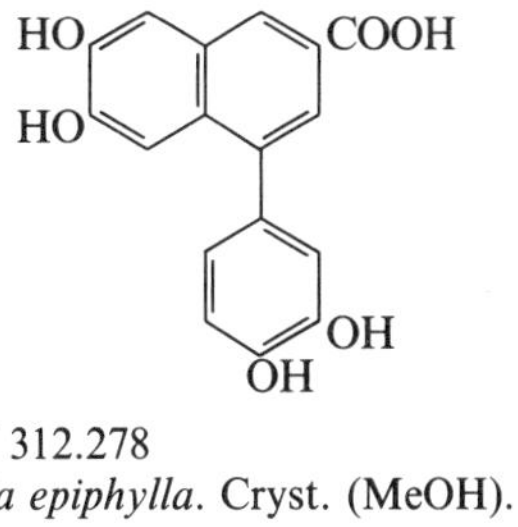

$C_{17}H_{12}O_6$ M 312.278
Constit. of *Pellia epiphylla*. Cryst. (MeOH). Mp 325-327° dec.

Rischmann, M. *et al, Phytochemistry*, 1989, **28**, 867.

1-(3,4-Dihydroxyphenyl)-2-(3,5-dihydroxyphenyl)ethane   D-80359

4-[2-(3,5-Dihydroxyphenyl)ethyl]-1,2-benzenediol

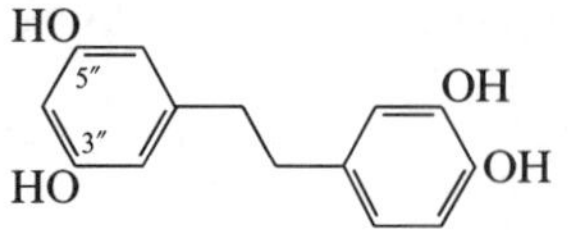

$C_{14}H_{14}O_4$ M 246.262

3″,5″-Di-Me ether: [116518-75-3]. *1-(3,4-Dihydroxyphenyl)-2-(3,5-dimethoxyphenyl)ethane.* **Combretastatin B4**
$C_{16}H_{18}O_4$ M 274.316
Constit. of *Combretum caffrum.* Viscous oil.

Pettit, G.R. *et al, J. Nat. Prod. (Lloydia),* 1988, **51**, 517.

1-(3,4-Dihydroxyphenyl)-2-(3,4,5-dihydroxyphenyl)ethylene   D-80360

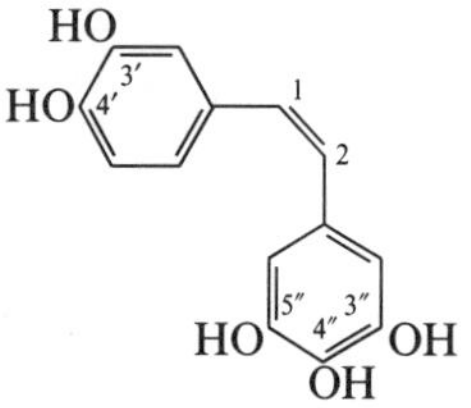

$C_{14}H_{12}O_5$ M 260.246
(Z)-form

3′,4′-Di-Me, 4″,5″-methylene ether: [111394-44-6].
Combretastatin A2
$C_{17}H_{16}O_5$ M 300.310
Constit. of *Combretum caffrum.* Oil.

3″,4′,4″-Tri-Me ether: [111394-45-7]. **Combretastatin A3**
$C_{17}H_{18}O_5$ M 302.326
Constit. of *C. caffrum.* Oil.

3′,4′,4″-Tri-Me ether, 1,2-dihydro: [111394-46-8]. *3,5-Dihydroxy-3′,4,4′-trimethoxybibenzyl.* **Combretastatin B2**
$C_{17}H_{20}O_5$ M 304.342
Constit. of *C. caffrum.* Oil.

3′,3″,5″-Tri-Me ether, 1,2-dihydro: [108853-14-1]. *4,4′-Dihydroxy-3,3′,5-trimethoxybibenzyl*
$C_{17}H_{20}O_5$ M 304.342
Constit. of *Dendrobium chrysanthum.* Cryst. (Et₂O). Mp 87-89°.

Pettit, G.R. *et al, Can. J. Chem.,* 1987, **65**, 2390.
Min, Z.-D. *et al, J. Nat. Prod. (Lloydia),* 1987, **50**, 1189.

1-(3,4-Dihydroxyphenyl)-2-(3,5-dihydroxyphenyl)ethylene   D-80361

Updated Entry replacing D-10420
4-[2-(3,5-Dihydroxyphenyl)ethenyl]-1,2-benzenediol, 9CI.
3,3′,4,5′-Stilbenetetrol, 8CI. 3,3′,4,5′-Tetrahydroxystilbene.
Piceatannol. Astringenin

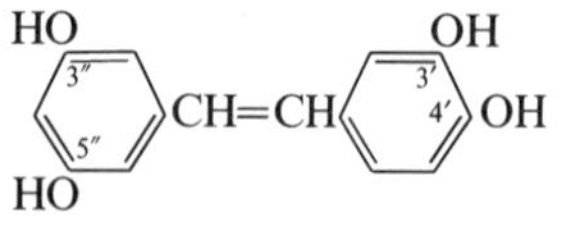

$C_{14}H_{12}O_4$ M 244.246
(E)-form [10083-24-6]

Found in Norway spruce, *Vouacapoua macropetala, Pericopsis angolensis* and other spp. Fungal inhibitor. Plant growth inhibitor and ichthyotoxin. Needles (EtOAc/hexane). Mp 231-232° (216°).

Tetra-Ac: Needles (MeOH). Mp 115°, Mp 124° (double Mp).

Tetrabenzoyl: Mp 153-154°.

Tetra-Me ether: Needles (MeOH aq.). Mp 67-69°.

3′-Glucoside: [29884-49-9]. **Astringin**
$C_{20}H_{22}O_9$ M 406.388
Constit. of *Picea* spp. Also from *Eucalyptus dundasii.*

3′-Me ether: [32507-66-7]. *5-[2-(3-Hydroxy-4-methoxyphenyl)ethenyl]-1,3-benzenediol, 9CI. 3′-Methoxy-3,4′5-stilbenetriol, 8CI. Isorhapontigenin*
$C_{15}H_{14}O_4$ M 258.273
Cryst. (EtOAc/pet. ether). Mp 182-183°.

3′-Me ether, 3″-glucoside: [32727-29-0]. **Isorhapontin**
$C_{21}H_{24}O_9$ M 420.415
Constit. of *Picea glauca, P. koraensis, P. obovata* and *P. glehnii.* White-yellow amorph. powder (MeOH/CHCl₃). Mp 193-194°. $[\alpha]_D^{21}$ − 54.1° (c, 2.24 in Me₂CO).

4′-Me ether: [500-65-2]. *1-(3,5-Dihydroxyphenyl)-2-(3-hydroxy-4-methoxyphenyl)ethylene. 3,3′,5-Trihydroxy-4′-methoxystilbene.* **Rhapontigenin.** Pontigenin
$C_{15}H_{14}O_4$ M 258.273
Isol. from rhizomes of *Rheum undulatum.* Mp 186-187°.

4′-Me ether, 3′-glucoside: [155-58-8]. **Rhapontin.** Rhaponticin. Ponticin
$C_{21}H_{24}O_9$ M 420.415
Isol. from seeds of *R. rhaponticum, R. emodi* and *Euclalyptus sideroxylon.* Mp 236-237° dec. $[\alpha]_D^{32}$ − 59.5° (Me₂CO).

1,2-Dihydro, 3′,3″-di-Me ether: [67884-30-4]. *1-(3-Hydroxy-5-methoxyphenyl)-2-(4-hydroxy-3-methoxyphenyl)ethane.* **Gigantol**
$C_{16}H_{18}O_4$ M 274.316
Constit. of *Cymbidium giganteum.* Cryst. (CHCl₃/pet. ether). Mp 94-95°.

Grassman, W. *et al, Chem. Ber.,* 1956, **89**, 2523.
Cunningham, J. *et al, J. Chem. Soc.,* 1963, 2875 (*struct, synth, pmr*)
Grassman, W. *et al, J. Chem. Soc.,* 1965, 4579 (*isol, struct*)
Andrews, D.H. *et al, Can. J. Chem.,* 1968, **46**, 2525 (*isol, uv*)
Harper, S.H. *et al, J. Chem. Soc. C,* 1969, 1109 (*isol*)
Banks, H.J. *et al, Aust. J. Chem.,* 1971, **24**, 2427 (*deriv*)
Reimann, G., *Justus Liebigs Ann. Chem.,* 1971, **750**, 109 (*synth*)
Manners, G.D. *et al, Phytochemistry,* 1971, **10**, 607 (*occur, isol, uv*)
Nakajima, K. *et al, Chem. Pharm. Bull.,* 1978, **26**, 3050 (*isol*)
Hata, K. *et al, Chem. Pharm. Bull.,* 1979, **27**, 984 (*isol*)
Inamori, Y. *et al, Chem. Pharm. Bull.,* 1984, **32**, 213 (*props, pharmacol*)
Juneja, R.K. *et al, Phytochemistry,* 1985, **24**, 321 (*deriv*)

1-(2,4-Dihydroxyphenyl)-3-(3,4-dihydroxyphenyl)-2-propen-1-one, 9CI   D-80362

Updated Entry replacing D-20298
2′,3,4,4′-Tetrahydroxychalcone. **Butein**
[21849-70-7]

$C_{15}H_{12}O_5$ M 272.257

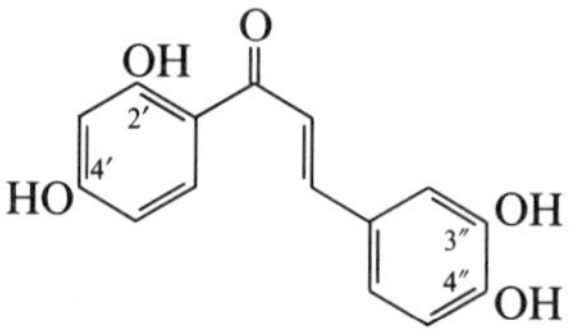

(*E*)-*form* [487-52-5]
Constit. of *Dalbergia stevensonii* and *Dahlia variabilis*, *Butea frondosa*, *Coreopsis* spp., *Cosmos sulphureus* and others. Orange-yellow cryst. + 1H$_2$O (EtOH aq.). Mp 198°, Mp 213-215°.

Tetra-Ac: Mp 129-131°.

3-O-β-D-Glucopyranoside: [30382-19-5]. **Monospermoside**
C$_{21}$H$_{22}$O$_{10}$ M 434.399
Constit. of the flowers of *Butea monosperma*. Yellow cryst. (MeOH aq.). Mp 194-195°.

4″-O-β-D-Glucopyranoside: [499-29-6]. **Coreopsin**. *Choreopsin*
C$_{21}$H$_{22}$O$_{10}$ M 434.399
Obt. from flowers of *Cosmos sulphureus, C. gigantea, Bidens frondosa* and other spp. Yellow needles (EtOH aq.). Mp 190-195° dec.

3′,4″-Di-O-β-D-glucopyranoside: **Isobutrin**. *Isobutyroin*
C$_{27}$H$_{32}$O$_{15}$ M 596.541
Isol. from flowers of *B. frondosa*. Bright-yellow cryst. (MeOH). Mp 190-191° dec. (sinters at 155°).

3″-Me ether: [34000-39-0]. *1-(2,4-Dihydroxyphenyl)-3-(4-hydroxy-3-methoxyphenyl)-2-propen-1-one, 9CI. 2′,4,4′-Trihydroxy-3-methoxychalcone.* **Homobutein**
C$_{16}$H$_{14}$O$_5$ M 286.284
Isol. from Subterranean clover *Trifolium subterraneum*. Mp 210-211°.

Price, J.R. *et al, J. Chem. Soc.,* 1939, 1017 (*isol, struct*)
Geissman, T.A., *J. Am. Chem. Soc.,* 1941, **63**, 2689; 1942, **64**, 1704 (*isol, deriv*)
Burkhart, W. *et al, Justus Liebigs Ann. Chem.,* 1942, **550**, 146 (*synth*)
Shimokoriyama, M., *J. Am. Chem. Soc.,* 1953, **75**, 1900 (*isol, deriv*)
Puri, B. *et al, J. Chem. Soc.,* 1955, 1589 (*Isobutrin*)
Wong, E. *et al, Phytochemistry,* 1968, **7**, 2123 (*Homobutein*)
Gupta, S.R. *et al, Phytochemistry,* 1970, **9**, 2231 (*Monospermoside*)
Saito, N. *et al, Bull. Chem. Soc. Jpn.,* 1972, **45**, 2274 (*cryst struct*)
Karrer, W. *et al, Konstitution und Vorkommen der Organischen Pflanzenstoffe,* 2nd Ed., Birkhäuser Verlag, Basel, 1972-1985, nos. 494, 495 (*occur, bibl*)

2-(2,4-Dihydroxyphenyl)-6-hydroxy-benzofuran   D-80363

Updated Entry replacing V-20015
4-(6-Hydroxy-2-benzofuranyl)-1,3-benzenediol, 9CI
[67736-22-5]

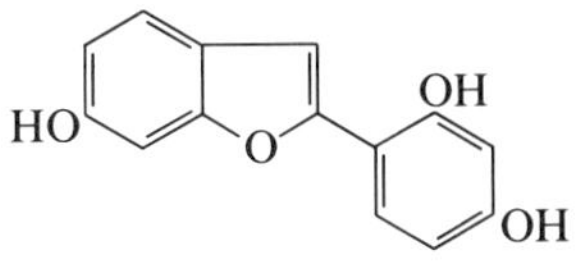

C$_{14}$H$_{10}$O$_4$ M 242.231
Isol. from heartwood of *Lespedeza cyrtobotrya*. Brownish needles. Mp 207-208°.

Tri-Ac: Mp 154-156°.

2′-Me ether: [74048-90-1]. *6-Hydroxy-2-(4-hydroxy-2-methoxyphenyl)benzofuran. 6-Demethylvignafuran*
C$_{15}$H$_{12}$O$_4$ M 256.257
Constit. of *Anthyllis vulneraria, Coronilla emerus* and *Tetragonolobus maritimus*leaves.

2′,6-Di-Me ether: [57800-41-6]. *2-(4-Hydroxy-2-methoxyphenyl)-6-methoxybenzofuran.* **Vignafuran**
C$_{16}$H$_{14}$O$_4$ M 270.284
Constit. of leaves of *Vignia unguiculata* infected with *Colletotrichum lindenuthianium* and of *Lablab niger*. Phytoalexin. Glass.

2′,6-Di-Me ether, Ac: [57800-42-7].

Needles (MeOH). Mp 94-94.5°.

Preston, N.W. *et al, Phytochemistry,* 1975, **14**, 1843 (*isol, struct, uv, ms, pmr, synth, Vignafuran*)
Duffley, R.P. *et al, J. Chem. Soc., Perkin Trans.* 1, 1977, 802 (*synth, pmr, uv*)
Ingham, J.L. *et al, Phytochemistry,* 1978, **17**, 535; 1980, **19**, 289 (*Demethylvignafuran*)
Miyase, T. *et al, Chem. Pharm. Bull.,* 1981, **29**, 2205 (*isol, uv, ms*)

3-(2,4-Dihydroxyphenyl)-7-hydroxy-2*H*-1-benzopyran-2-one   D-80364

3-(2,4-Dihydroxyphenyl)-7-hydroxycoumarin

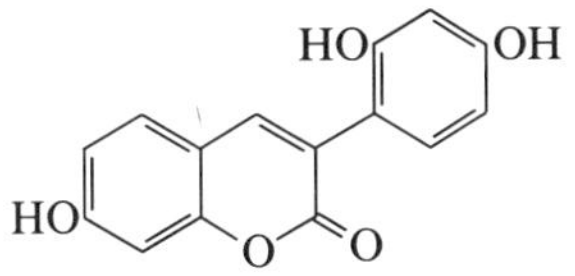

C$_{15}$H$_{10}$O$_5$ M 270.241
4′-Me ether: [54300-95-7]. *7-Hydroxy-3-(2-hydroxy-4-methoxyphenyl)coumarin. 2′,7-Dihydroxy-4′-methoxy-3-phenylcoumarin*
C$_{16}$H$_{12}$O$_5$ M 284.268
Isol. from heartwood of *Dalbergia olivieri*. Yellow rhombs (EtOH aq.). Mp 258-260°.

Donnelly, D.M.X. *et al, Phytochemistry,* 1974, **13**, 2587 (*isol, uv, pmr, struct*)

1-(3,4-Dihydroxyphenyl)-5-(4-hydroxy-phenyl)-1-penten-4-yne   D-80365

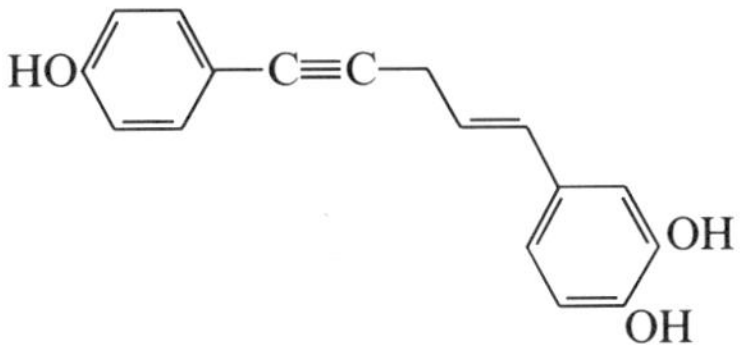

C$_{20}$H$_{20}$O$_3$ M 308.376
Constit. of *Hypoxis rooperi*.

Drewes, S.E. *et al, Phytochemistry,* 1989, **28**, 153.

3-(2,4-Dihydroxyphenyl)-1-(4-hydroxy-phenyl)-1-propanone   D-80366

2,4,4′-Trihydroxydihydrochalcone

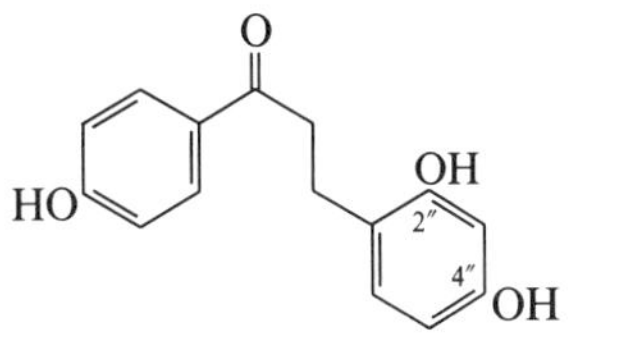

C$_{15}$H$_{14}$O$_4$ M 258.273
2″-Me ether: [116384-24-8]. *3-(4-Hydroxy-2-methoxyphenyl)-1-(4-hydroxyphenyl)-1-propanone.* **Loureirin C**
C$_{16}$H$_{16}$O$_4$ M 272.300
Constit. of *Dracaena loureiri*. Yellow foam. Mp 157-158°.

2″,-4″-Di-Me ether: [119425-89-7]. *3-(2,4-Dimethoxyphenyl)-1-(4-hydroxyphenyl)-1-propanone.* **Loureirin A**
C$_{17}$H$_{18}$O$_4$ M 286.327
Constit. of *D. loureiri*. Cryst. Mp 124°.

Meksuriyen, D. *et al, J. Nat. Prod. (Lloydia),* 1988, **51**, 1129 (*isol, pmr, cmr*)

1-(2,4-Dihydroxyphenyl)-3-(4-hydroxyphenyl)-2-propen-1-one   D-80367

2′,4,4′-Trihydroxychalcone. **Isoliquiretigenin**

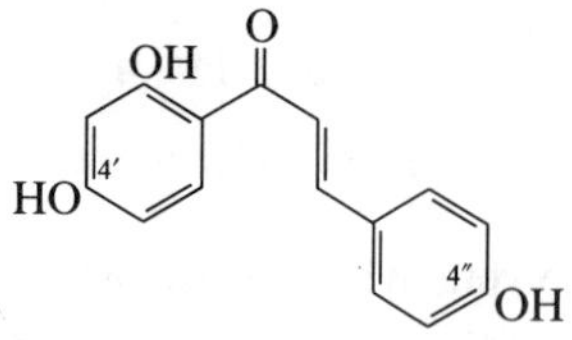

$C_{15}H_{12}O_4$ M 256.257

Isol. from *Dahlia variabilis*, also from *Pterocarpus* spp., *Robinia pseudoacacia*, *Cicer arietinum*, *Glycyrrhiza uralensis*, *Pterocarbus indicus* and other leguminosae. Yellow cryst. (EtOH aq.). Mp 200-204°.

4′-O-β-D-Glucopyranoside: [59122-93-9]. *Neoisoliquiritin*
$C_{21}H_{22}O_9$ M 418.399
Isol. from *G. glabra*, *G. uralensis*, *Cicer arietinum* and *Trifolium subterraneum*. Yellow needles (EtOH). Mp 230-232° (217-218°). $[\alpha]_D^{18}$ −61.5° (c, 0.5 in EtOH).

4″-O-β-D-Glucopyranoside: **Isoliquiritin.** *Isoliquiritoside*
$C_{21}H_{22}O_9$ M 418.399
Isol. from roots of *G. glabra* and from *D.* spp. Mp 185-186° dec.

4′-O-Glucosylglucoside:
$C_{27}H_{32}O_{14}$ M 580.541
Isol. from *Dahlia variabilis* and *Ulex europaeus*. λ_{max} 242, 372 nm (95% EtOH).

4″-O-[β-D-Glucopyranosyl(1→2)-β-D-apioside]: **Liquiraside**
$C_{26}H_{30}O_{13}$ M 550.515
Isol. from *G. glabra* and *G. uralensis*. Gold cryst. (EtOH aq.). Mp 150-151°. $[\alpha]_D^{18}$ −125° (c, 0.8 in EtOH).

4′-O-Glucopyranosylglucopyranoside, 4″-O-glucopyranoside:
$C_{33}H_{42}O_{19}$ M 742.683
Isol. from *U. europaeus*. λ_{max} 237, 361 nm (95% EtOH).

4′-O-(Rhamnopyranosylglucopyranoside): **Rhamnoisoliquiritin**
$C_{27}H_{32}O_{13}$ M 564.542
Isol. from liquorice root. Mp 127°.

$O^{4′}$-*Me:* **2′,4-Dihydroxy-4′-methoxychalcone**
$C_{16}H_{14}O_4$ M 270.284
Isol. from *Xanthorrhoea australis*. Yellow needles (EtOH aq.). Mp 172-176°.

4′-O-α-L-Rhamnopyranoside: [78795-34-3].
$C_{27}H_{32}O_{13}$ M 564.542
Isol. from *Viburnum cortinifolium*. Yellow needles (MeOH aq.). Mp 60-63°.

[5041-81-6, 23141-04-0, 27233-96-1]

Geissman, T.A. *et al, J. Am. Chem. Soc.*, 1946, **68**, 697 (*bibl*)
Bate-Smith, E.C. *et al, J. Chem. Soc.*, 1953, 2183 (*isol*)
Duewell, H., *J. Chem. Soc.*, 1954, 2562 (*isol, deriv*)
Harborne, J.B., *Phytochemistry*, 1962, **1**, 203 (*glycosides*)
Litvinenko, V.I. *et al, Dokl. Akad. Nauk SSSR, Ser. Khim.*, 1964, **155**, 600; 1966, **169**, 347 (*Neoisoliquiritin, Liquiraside*)
Kubota, T. *et al, Nippon Kagaku Zasshi (Jpn. J. Chem.)*, 1966, **87**, 1201; *CA*, **66**, 94951y (*synth*)
Wong, E. *et al, Phytochemistry*, 1968, **7**, 2123 (*isol*)
Winger, G. *et al, Biochemistry*, 1969, **8**, 2067 (*synth*)
Van Hulle, C. *et al, Planta Med.*, 1971, **20**, 278 (*Rhamnoisoliquiritin*)
Markham, K.R. *et al, Tetrahedron*, 1976, **32**, 2607 (*cmr*)
Srivastava, S. *et al, Indian J. Chem., Sect. B*, 1981, **20**, 347 (*isol, deriv, uv*)
Maurya, R. *et al, J. Nat. Prod. (Lloydia)*, 1984, **47**, 179 (*isol*)

1-(2,6-Dihydroxyphenyl)-4-methyl-4-tridecen-1-one   D-80368

[119736-94-6]

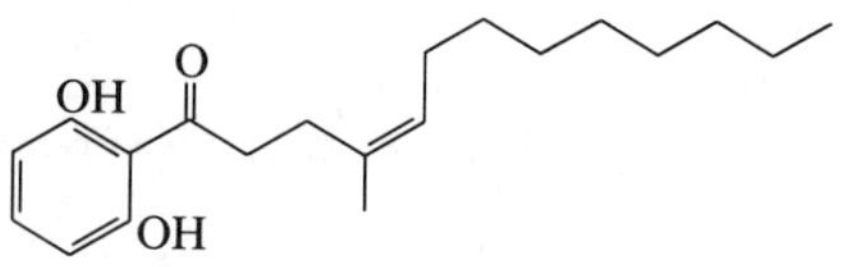

$C_{20}H_{30}O_3$ M 318.455

(*Z*)-*form*
Constit. of *Horsfieldia glabra*. Cryst. (MeOH). Mp 42-44°.

Pinto, M.M.M. *et al, Phytochemistry*, 1988, **27**, 3988.

3,5-Dihydroxy-4-phenylpyrazole   D-80369

*1,2-Dihydro-5-hydroxy-4-phenyl-3*H-*pyrazol-3-one, 9CI. 4-Phenyl-3,5-pyrazolediol*

[106367-54-8]

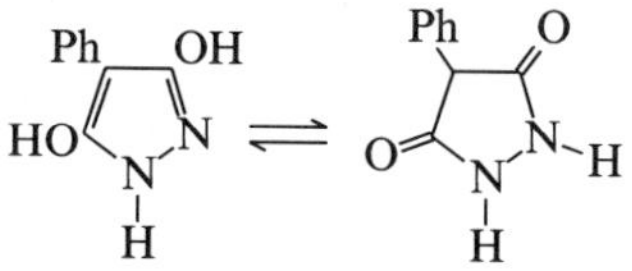

$C_9H_8N_2O_2$ M 176.174

Several other tautomers possible including zwitterions. In neutral solvs. the diol form appearsto predominate with extensive ionisation to the anion in polar solvs. Cryst. (THF). Mp 229°. pK_{a1} 3.70, pK_{a2} 12.71 (H_2O).

Zvilichovsky, G. *et al, J. Org. Chem.*, 1982, **47**, 295 (*synth, uv*)
Zvilichovsky, G. *et al, J. Heterocycl. Chem.*, 1988, **25**, 1307 (*tautom*)

4-(3,4-Dihydroxyphenyl)-5,7,8-trihydroxycoumarin   D-80370

3′,4′,5,7,8-Pentahydroxyneoflavone

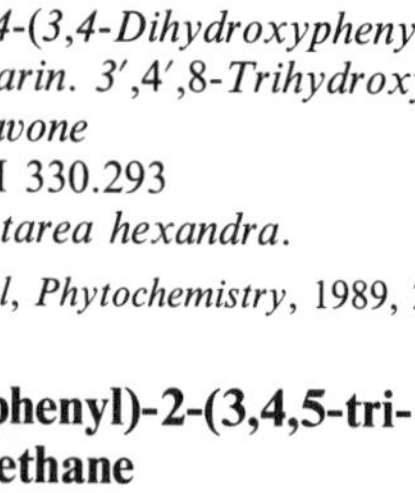

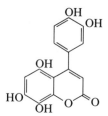

$C_{15}H_{10}O_7$ M 302.240

5,7-Di-Me ether: 4-(3,4-Dihydroxyphenyl)-8-hydroxy-5,7-dimethoxycoumarin. *3′,4′,8-Trihydroxy-5,7-dimethoxyneoflavone*
$C_{17}H_{14}O_7$ M 330.293
Constit. of *Coutarea hexandra*.

D'Agostino, M. *et al, Phytochemistry*, 1989, **28**, 1773.

1-(3,4-Dihydroxyphenyl)-2-(3,4,5-trihydroxyphenyl)ethane   D-80371

5-[2-(3,4-Dihydroxyphenyl)ethyl]-1,2,3-benzenetriol

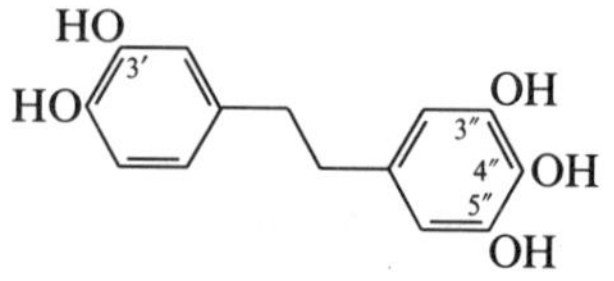

$C_{14}H_{14}O_5$ M 262.262

3″,4″,5″-Tri-Me ether: [116518-76-4]. *1-(3,4-Dihydroxyphenyl)-2-(3,4,5-trimethoxyphenyl)ethane.*
Combretastatin B3
$C_{17}H_{20}O_5$ M 304.342
Constit. of *Combretum caffrum*. Powder (EtOH/Et$_2$O). Mp 113-115°.

3′,3″,4″,5″-Tetra-Me ether: 1-(4-Hydroxy-3-methoxyphenyl)-2-(3,4,5-trimethoxyphenyl)-ethane. **Crepidatin**
$C_{18}H_{22}O_5$ M 318.369
Constit. of *Dendrobium crepidatum*. Cryst. (EtOAc/pet. ether). Mp 99°.

Pettit, G.R. *et al*, *J. Nat. Prod. (Lloydia)*, 1988, **51**, 517.
Majumder, P.L. *et al*, *Phytochemistry*, 1989, **28**, 1986.

3-(3,4-Dihydroxyphenyl)-1-(2,3,4-tri-hydroxyphenyl)-2-propen-1-one   D-80372
2′,3,3′,4,4′-Pentahydroxychalcone

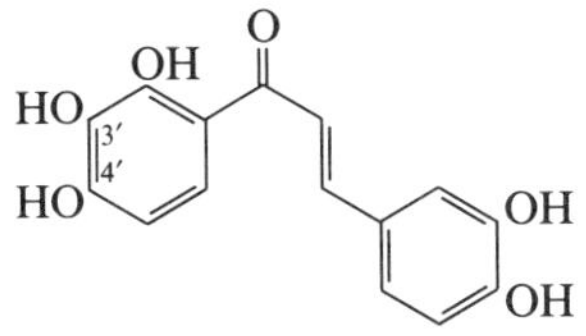

$C_{15}H_{12}O_6$ M 288.256
3′-Me ether: [6542-59-2]. *1-(2,4-Dihydroxy-3-methoxyphenyl)-3-(3,4-dihydroxyphenyl)-2-propen-1-one. 2′,3,4,4′-Tetrahydroxy-3′-methoxychalcone.* **Lanceoletin**
$C_{16}H_{14}O_6$ M 302.283
Rare chalcone present in floral tissues of some Compositae.

3′-Me ether, tetra-Ac: Pale-yellow prisms (MeOH). Mp 162-166°.

3′-Me ether, 4′-O-β-D-glucopyranoside: [64181-95-9].
Lanceolin†
$C_{22}H_{24}O_{11}$ M 464.425
Isol. from *Coreopsis lanceolata*.

Shinokoriyama, M. *et al*, *J. Am. Chem. Soc.*, 1953, **75**, 1900 (*isol, struct*)
Nicholls, K.W. *et al*, *Phytochemistry*, 1979, **18**, 1076 (*isol*)
Crawford, D.J. *et al*, *Am. J. Bot.*, 1983, **70**, 355; *CA*, **98**, 212856g (*occur*)

ent-3β,14α-Dihydroxy-7,9(11),15-pimaratriene-2,12-dione   D-80373
Yucalexin P-15
[119626-52-7]
$C_{20}H_{26}O_4$ M 330.423
Constit. of cassava roots (*Manihot esculenta*).

Sakai, T. *et al*, *Phytochemistry*, 1988, **27**, 3769.

12,13-Dihydroxy-8,11,13-podocarpatriene-3,7-dione   D-80374

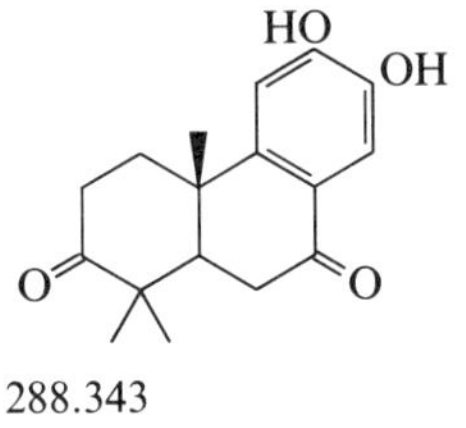

$C_{17}H_{20}O_4$ M 288.343
Di-Me ether: [119269-83-9]. *12,13-Dimethoxy-8,11,13-podocarpatriene-3,7-dione.* **Methyl nimbionone**
$C_{19}H_{24}O_4$ M 316.396

Constit. of *Azadirachta indica*. Needles (hexane). Mp 116-119°.

Ara, I. *et al*, *J. Nat. Prod. (Lloydia)*, 1988, **51**, 1054.

3,9-Dihydroxypterocarpan   D-80375
Updated Entry replacing D-60373
Demethylmedicarpin
[61135-91-9]

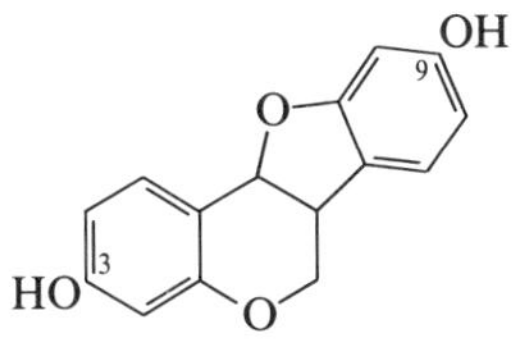

$C_{15}H_{12}O_4$ M 256.257
Isol. from *Albizia prozera, Erythrina crista-galli, E. sandwicensis, Melilotus alba, Pachyrrhizus erosus, Phaseolus vulgaris, Psophocarpus tetragonolobus* and *Trifolium repens*.Fungal metab. of Medicarpin. Pale yellow amorph. powder.

Di-Ac: Needles. Mp 122°. [α]$_D$ +126° (c, 0.25 in CHCl$_3$).

3-Me ether: [38822-00-3]. *9-Hydroxy-3-methoxypterocarpan.* **Isomedicarpin**
$C_{16}H_{14}O_4$ M 270.284
Constit. of *Psophocarpus tetragonolobus*. Cryst. Mp 106°. [α]$_D^{18}$ −201° (CHCl$_3$).

McMurry, T.B.H. *et al*, *Phytochemistry*, 1972, **11**, 3283 (*synth*)
Deshpande, V.H. *et al*, *Indian J. Chem., Sect. B*, 1977, **15**, 201 (*isol*)
Preston, N.W. *et al*, *Phytochemistry*, 1977, **16**, 2044 (*Isomedicarpin*)
Ingham, J.L. *et al*, *Phytochemistry*, 1980, **19**, 1203 (*Isomedicarpin*)
Ingham, J.L., *Fortschr. Chem. Org. Naturst.*, 1983, **43**, 1 (*rev, occur*)
Prasad, A.V.K. *et al*, *J. Chem. Soc., Perkin Trans. 1*, 1986, 1561 (*synth*)

3,9-Dihydroxypterocarpene   D-80376
6H-Benzofuro[3,2-c][1]benzopyran-3,9-diol, 9CI.
Anhydroglycinol
[67685-22-7]

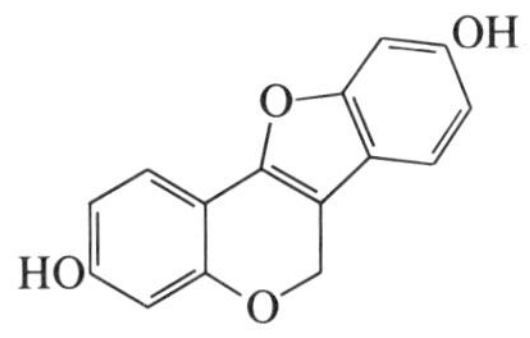

$C_{15}H_{10}O_4$ M 254.242
Isol. from heartwood of *Lespedeza cyrtobotrya*; phytoalexin from *Tetragonolobus maritimus* R1. Cryst. (MeOH/CHCl$_3$). Mp 207-209° dec.

Di-Me ether: [1433-08-5]. *3,9-Dimethoxypterocarpene. 3,9-Dimethoxy-6H-benzofuro[3,2-c][1]benzopyran, 9CI.* **Anhydrovariabilin**
$C_{17}H_{14}O_4$ M 282.295
Isol. from heartwoods of *Dalbergia decipularis* and *Swartzia madagascariensis*. Prisms (EtOH). Mp 116° (110-112°).

Bowyer, W.J. *et al*, *J. Chem. Soc.*, 1964, 4212 (*synth, deriv*)
Harper, S.H. *et al*, *J. Chem. Soc. C*, 1969, 1109 (*isol, deriv*)
De Alencar, R. *et al*, *Phytochemistry*, 1972, **11**, 1517 (*isol, deriv*)
Miyase, T. *et al*, *Chem. Pharm. Bull.*, 1980, **28**, 1172 (*isol, uv, ir, pmr*)

3,4-Dihydroxy-2,5-pyrrolidinedione, 9CI D-80377

Tartrimide, 8CI. 2,3-Dihydroxysuccinimide

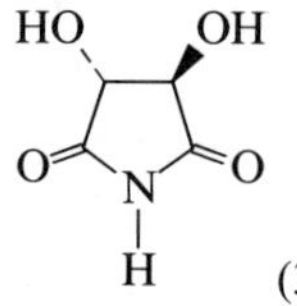

(3R,4R)-form

$C_4H_5NO_4$ M 131.088

(3R,4R)-form [18366-18-2]

Mp 200° dec. (205-206°). $[\alpha]_D^{20}$ +202° (c, 0.5 in H_2O).

Di-tert-butyl ether: [117237-94-2].

$C_{12}H_{21}NO_4$ M 243.302

Cryst. (toluene/hexane). Mp 154°. $[\alpha]_D^{20}$ +205° (c, 1 in $CHCl_3$).

N-Benzyl: [19728-93-9].

Cryst. (H_2O). Mp 196-198°. $[\alpha]_D^{25}$ +126° (MeOH).

Dave, H.R. *et al, J. Chem. Soc., Chem. Commun.,* 1967, 743 (*synth, cd*)
Wong, C.M. *et al, Can. J. Chem.,* 1968, **46**, 3091 (*deriv*)
Poloński, T., *J. Chem. Soc., Perkin Trans.* 1, 1988, 629 (*synth, ir, uv, pmr, cd*)

2,2'-Dihydroxystilbene D-80378

Updated Entry replacing B-40108

2,2'(1,2-Ethenediyl)bisphenol, 9CI. 2,2'-Stilbenediol, 8CI. 1,2-Bis(2-hydroxyphenyl)ethylene

[4752-75-4]

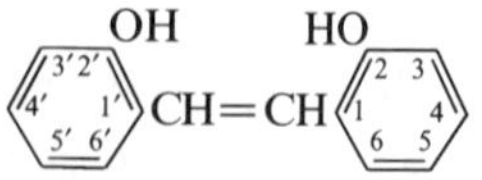

$C_{14}H_{12}O_2$ M 212.248

(E)-form [18221-50-6]

Needles (EtOH). Mp 197°.

Di-Ac: [51042-12-7].

$C_{18}H_{16}O_4$ M 296.322

Mp 141-142°.

Di-Me ether: [20516-15-8].

$C_{16}H_{16}O_2$ M 240.301

Mp 136°.

(Z)-form

Mp 95°.

Dibenzoyl:

$C_{28}H_{20}O_4$ M 420.464

Mp 107-108°.

Buu-Hoï, Ng.Ph. *et al, Bull. Soc. Chim. Fr.,* 1967, 955 (*synth*)
Lapkin, I.I. *et al, J. Gen. Chem. USSR (Engl. Transl.),* 1973, **43**, 1968 (*synth*)
Tirado-Rives, J. *et al, J. Org. Chem.,* 1984, **49**, 1627 (*synth, pmr, cmr, ir, ms, uv, cryst struct*)

2,4'-Dihydroxystilbene D-80379

2-[2-(4-Hydroxyphenyl)ethenyl]phenol, 9CI. 2,4'-Stilbenediol. 1-(2-Hydroxyphenyl)-2-(4-hydroxyphenyl)ethylene

$C_{14}H_{12}O_2$ M 212.248

(E)-form [110983-43-2]

Mp 193-194°.

Mylona, A. *et al, J. Chem. Res. (S),* 1986, 433 (*synth, ir, uv*)

2,5-Dihydroxystilbene D-80380

2-(2-Phenylethenyl)-1,4-benzenediol, 9CI. 2,5-Stilbenediol. 1-(2,5-Dihydroxyphenyl)-2-phenylethylene

$C_{14}H_{12}O_2$ M 212.248

(E)-form [34701-58-1]

Mp 165-166°. $Bp_{0.001}$ 150-155° subl.

(Z)-form [34701-63-8]

Cryst. (Et_2O). Mp 90-92°.

Bruce, J.M. *et al, J. Chem. Soc. C,* 1971, **22**, 3749 (*synth, ir, uv, pmr*)

3,3'-Dihydroxystilbene D-80381

3,3'-(1,2-Ethenediyl)bisphenol, 9CI. 1,2-Bis(3-hydroxyphenyl)ethylene. 3,3'-Stilbenediol

[70709-65-8]

$C_{14}H_{12}O_2$ M 212.248

Needles (hexane). Mp 111-112°.

Hata, K. *et al, Chem. Pharm. Bull.,* 1979, **27**, 984 (*synth, ir, pmr*)

3,4-Dihydroxystilbene D-80382

4-(2-Phenylethenyl)-1-2-benzenediol, 9CI. 3,4-Stilbenediol. 1-(3,4-Dihydroxyphenyl)-2-phenylethylene

$C_{14}H_{12}O_2$ M 212.248

(E)-form [19826-29-0]

Platelets (Me_2CO/C_6H_6). Mp 168-169°.

Gorham, J. *et al, Phytochemistry,* 1980, **19**, 2059.
Donnelly, D.M.X. *et al, J. Chem. Soc., Perkin Trans.* 1, 1987, 2719 (*synth, ir, uv, pmr*)

3,4'-Dihydroxystilbene D-80383

3-[2-(4-Hydroxyphenyl)ethenyl]phenol, 9CI. 3,4'-Stilbenediol. 1-(3-Hydroxyphenyl)-2-(4-hydroxyphenyl)ethylene

[62574-04-3]

$C_{14}H_{12}O_2$ M 212.248

Isolated from roots of *Hydrangea macrophylla.*

[63877-76-9]

Gorham, J., *Phytochemistry,* 1977, **16**, 249 (*isol*)

3,5-Dihydroxystilbene D-80384

Updated Entry replacing P-01174

5-(2-Phenylethenyl)-1,3-benzenediol, 9CI. 1-(3,5-Dihydroxyphenyl)-2-phenylethylene. 5-Styrylresorcinol.

Pinosylvin

[102-61-4]

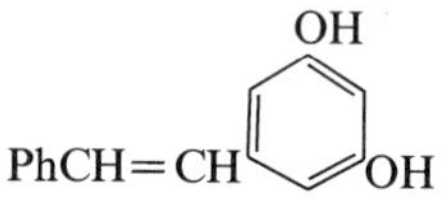

$C_{14}H_{12}O_2$ M 212.248

(E)-form [22139-77-1]

Constit. of *Pinus* spp. and *Dalbergia sissoo.* Shows fungistatic props. Needles (AcOH). Mp 155.5-156°.

▷ WJ5580000.

Di-Ac:

$C_{18}H_{16}O_4$ M 296.322

Mp 100-101°.

Mono-Me ether: [5150-38-9]. *3-Methoxy-5-(2-phenylethenyl)phenol, 9CI. 3-Hydroxy-5-methoxystilbene. 5-Methoxy-3-stilbenol*

$C_{15}H_{14}O_2$ M 226.274

Isol. from wood of many *P.* spp. Cryst. Mp 122-123°.

Di-Me ether: [21956-56-9].

$C_{16}H_{16}O_2$ M 240.301

Isol. from wood of *P. palustris, P. griffithii, P. formosana* and other *P.* spp. Prisms (MeOH). Mp 56-57°.

(Z)-form [106325-78-4]

Di-Me ether: [21956-55-8].
Isol. from Jack pine (*P. banksiana*). Bp$_{0.02}$ 123°.

Erdtmann, H., *Justus Liebigs Ann. Chem.*, 1939, **539**, 116 (*isol, struct*)
Cox, R.F.B., *J. Am. Chem. Soc.*, 1940, **62**, 3512 (*isol, struct*)
Liebherr, E. *et al, Ber.*, 1941, **74**, 869 (*synth*)
Erdtmann, H. *et al, Phytochemistry*, 1966, **5**, 927 (*isol*)
Rowe, J.W. *et al, Phytochemistry*, 1969, **8**, 235 (*isol, derivs*)
Bachelor, F.W. *et al, Can. J. Chem.*, 1970, **48**, 1554 (*synth*)
Cardona, L. *et al, Tetrahedron*, 1986, **42**, 2725 (*synth, pmr, cmr*)

4,4′-Dihydroxystilbene   D-80385

Updated Entry replacing B-01588
4,4′-(1,2-Ethenediyl)bisphenol, 9CI. 4,4′-Stilbenediol, 8CI.
1,2-Bis(4-hydroxyphenyl)ethylene
[659-22-3]
$C_{14}H_{12}O_2$ M 212.248
Needles (AcOH). Mp 293-294° (284°).

Di-Ac:
$C_{18}H_{16}O_4$ M 296.322
Mp 213°.

Di-Me ether: [4705-34-4].
$C_{16}H_{16}O_2$ M 240.301
Mp 214-215°. Sublimes.

Laarhoven, W.H., *Recl. Trav. Chim. Pays-Bas (J. R. Neth. Chem. Soc.)*, 1961, **80**, 775 (*synth*)
Sieber, R.W., *Justus Liebigs Ann. Chem.*, 1969, **730**, 31 (*synth*)

3,9-Dihydroxytetradecanoic acid   D-80386

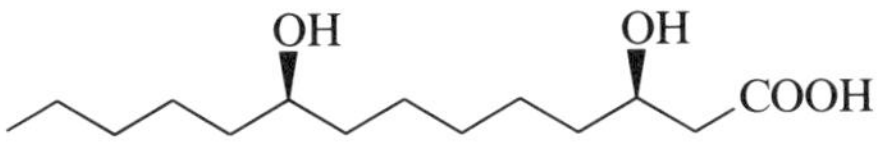

$C_{14}H_{28}O_4$ M 260.373
(3R,9R)-form

Me ester: [115482-43-4].
$C_{15}H_{30}O_4$ M 274.400
Solid. Mp 57-59°. $[\alpha]_D^{27}$ −10.55° (c, 1.45 in CHCl$_3$).

Schreiber, S.L. *et al, J. Am. Chem. Soc.*, 1988, **110**, 6210 (*synth, pmr, cmr, ms, ir*)

3,8-Dihydroxy-13,14,15,16-tetranor-12-   D-80387
labdanoic acid

$C_{16}H_{28}O_4$ M 284.395
(3α,8α)-form

8-Ac, Me ester: Methyl 8α-acetoxy-3α-hydroxy-13,14,15,16-
tetranorlabdan-12-oate
$C_{19}H_{32}O_5$ M 340.459
Constit. of *Salvia aethiopis*. Gum. $[\alpha]_D$ −0.79° (c, 1.05 in
CHCl$_3$).

Gonzalez, M.S. *et al, Tetrahedron*, 1989, **45**, 3575.

ent-3β,18-Dihydroxy-19-trachylobanoic   D-80388
acid

$C_{20}H_{30}O_4$ M 334.455
Constit. of *Jungermannia exsertifolia* subsp. *cordifolia*.
Me ester: Cryst. (hexane). Mp 134-135°. $[\alpha]_D$ −60.8° (c,
0.75 in CHCl$_3$).

Harrison, L.J. *et al, Phytochemistry*, 1989, **28**, 1533.

5,6-Dihydroxy-2-(2,3,4-trihydroxy-   D-80389
phenyl)benzofuran

$C_{14}H_{10}O_6$ M 274.229
3′-Me, 5,6-methylene ether: [79295-81-1]. *2-(2,4-Dihydroxy-
3-methoxyphenyl)-5,6-methylenedioxybenzofuran. 5-
Furo[2,3-f]-1,3-benzodioxol-6-yl-2-methoxy-1,3-
benzenediol, 9CI.* **Sophorafuran A**
$C_{16}H_{12}O_6$ M 300.267
Isol. from roots of *Sophora franchetiana*. Needles
(C$_6$H$_6$). Mp 145-146°.

Komatsu, M. *et al, Chem. Pharm. Bull.*, 1981, **29**, 2069 (*isol, uv, ms, pmr, cmr, struct*)

3,15-Dihydroxy-7,11,23-trioxo-8,20-   D-80390
lanostadien-26-oic acid

$C_{30}H_{42}O_7$ M 514.658
(3β,15α,20E)-form [120462-49-9] *Ganoderenic acid I*
Constit. of *Ganoderma applantum*.
Me ester: $[\alpha]_D^{23}$ +96° (c, 0.2 in EtOH).
3-Ketone: [120481-73-4]. *15α-Hydroxy-3,7,11,23-tetraoxo-
8,20E-lanostadien-26-oic acid.* **Ganoderenic acid G**
$C_{30}H_{40}O_7$ M 512.642
Constit. of *G. applantum*. $[\alpha]_D^{23}$ +189° (c, 0.2 in EtOH).
15-Ketone: [120462-48-8]. *3β-Hydroxy-7,11,15,23-tetraoxo-
8,20E-lanostadien-26-oic acid.* **Ganoderenic acid H**
$C_{30}H_{40}O_7$ M 512.642
Constit. of *G. applantum*.
15-Ketone, Me ester: $[\alpha]_D^{23}$ +61° (c, 0.2 in EtOH).
3,15-Diketone: [120462-47-7]. *3,7,11,15,23-Pentaoxo-8,20E-
lanostadien-26-oic acid.* **Ganoderenic acid F**
$C_{30}H_{38}O_7$ M 510.626
Constit. of *G. applantum*. $[\alpha]_D^{23}$ +93° (c, 0.2 in EtOH).

Nishitoba, T. *et al, Phytochemistry*, 1989, **28**, 193.

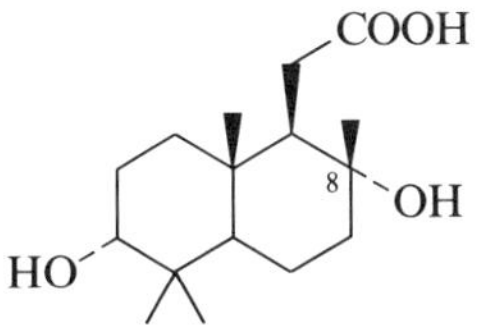

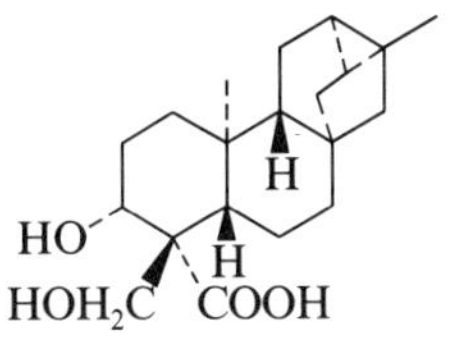

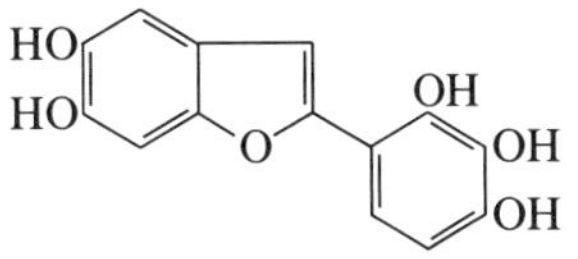

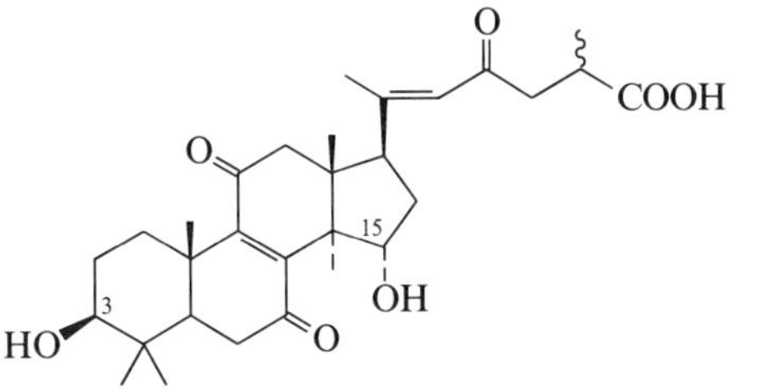

2,10-Dihydroxy-11,12,13-trisnor-6-eremophilen-8-one   D-80391

2,10-Dihydroxyondetianone

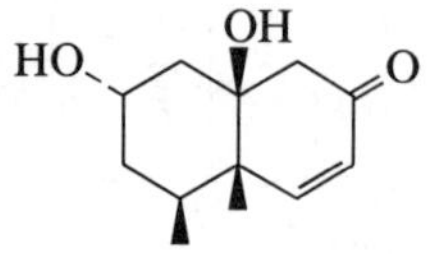

$C_{12}H_{18}O_3$ M 210.272

(2α,10β)-form

Constit. of *Ondetia linearis*. Cryst. Mp 148°. $[\alpha]_D^{24}$ +27° (c, 0.23 in CHCl₃).

Zdero, C. *et al*, *Phytochemistry*, 1989, **28**, 1653.

1,11-Dihydroxy-6-undecanone, 9CI   D-80392

[116288-12-1]

$$HOCH_2(CH_2)_4CO(CH_2)_4CH_2OH$$

$C_{11}H_{22}O_3$ M 202.293
Solid. Mp 58.5°.

Momenteau, M. *et al*, *J. Chem. Soc., Perkin Trans.* 1, 1988, 283 (*synth, pmr*)

3,19-Dihydroxy-12-ursene-23,28-dioic acid   D-80393

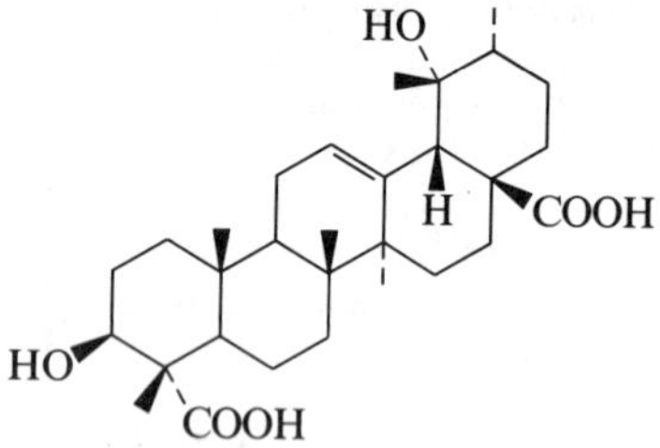

$C_{30}H_{46}O_6$ M 502.690

(3β,19α)-form

Rotundioic acid

Constit. of *Ilex rotunda*. Cryst. (MeOH). Mp 295-298° dec. $[\alpha]_D$ +50° (MeOH).

Nakatani, M. *et al*, *Phytochemistry*, 1989, **28**, 1479.

Di-2-indenyl sulfide   D-80394

2,2′-Thiobis-1H-indene

[115679-49-7]

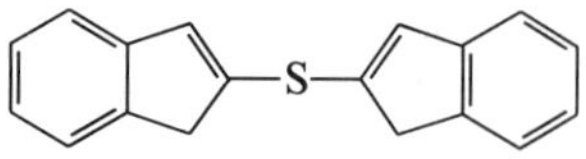

$C_{18}H_{14}S$ M 262.375
Mp 96°.

Baierweck, P. *et al*, *Chem. Ber.*, 1988, **121**, 2195 (*synth, pmr, cmr*)

1,2-Di-2-indolylethylene   D-80395

2,2′-(1,2-Ethenediyl)bis-1H-indole, 9CI

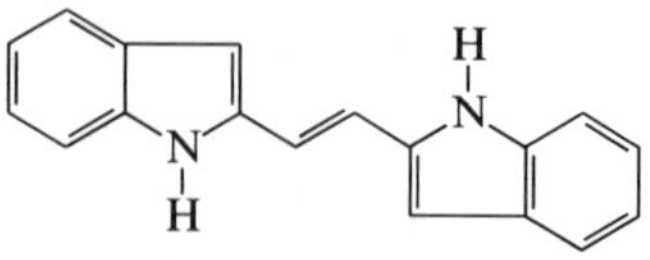

$C_{18}H_{14}N_2$ M 258.322

(E)-form [115983-53-4]

Cream cryst. (EtOAc). Mp >300° (resinifies above 280°).

N,N′-*Di-Me:* [115983-54-5].
   $C_{20}H_{18}N_2$ M 286.376
   Yellow needles (EtOAc). Mp 251°.

Capuano, L. *et al*, *Chem. Ber.*, 1988, **121**, 2259 (*synth, pmr, uv*)

2,3-Diiodobenzoic acid, 9CI, 8CI   D-80396

Updated Entry replacing D-05161

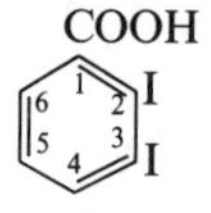

$C_7H_4I_2O_2$ M 373.916
Cryst. (EtOH). Mp 183-185° (178-181°).

Twiss, D. *et al*, *J. Org. Chem.*, 1950, **15**, 496 (*synth*)
Newman, M.S. *et al*, *J. Org. Chem.*, 1971, **36**, 1398 (*synth, pmr*)

1,4-Diiodobicyclo[2.2.2]octane   D-80397

[10364-05-3]

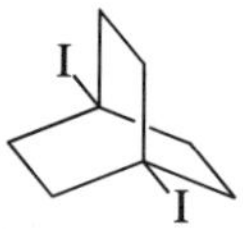

$C_8H_{12}I_2$ M 361.992
Cryst. (hexane/EtOH). Mp 239-240°.

McKinley, J.W. *et al*, *J. Am. Chem. Soc.*, 1973, **95**, 2030 (*synth, pmr*)
Zimmerman, H.E. *et al*, *J. Org. Chem.*, 1980, **45**, 3933 (*synth*)
Adcock, W. *et al*, *J. Org. Chem.*, 1988, **53**, 5259 (*synth*)

1,2-Diiodo-3,5-dinitrobenzene   D-80398

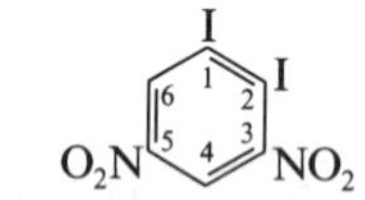

$C_6H_2I_2N_2O_4$ M 419.902
Yellow cryst. (CCl₄/pet. ether). Mp 110°.

Deorha, D.S. *et al*, *J. Indian Chem. Soc.*, 1965, **42**, 101 (*synth*)

1,2-Diiodo-4,5-dinitrobenzene   D-80399

[29270-47-1]
$C_6H_2I_2N_2O_4$ M 419.902
Pale yellow needles (AcOH). Mp 183-184°.

Deorha, D.S. *et al*, *J. Indian Chem. Soc.*, 1965, **42**, 101 (*synth*)
Arotsky, J. *et al*, *J. Chem. Soc. C*, 1970, 1480 (*synth, pmr*)

1,3-Diiodo-4,6-dinitrobenzene   D-80400

[37923-51-6]
$C_6H_2I_2N_2O_4$ M 419.902
Mp 161-162°.

Siddalingaiah, K.S. *et al*, *Curr. Sci.*, 1972, **41**, 451 (*synth*)
Shamanna, D. *et al*, *Indian J. Chem.*, 1974, **12**, 510.

2,2′-Diiodo-4,5′-dinitrobiphenyl   D-80401

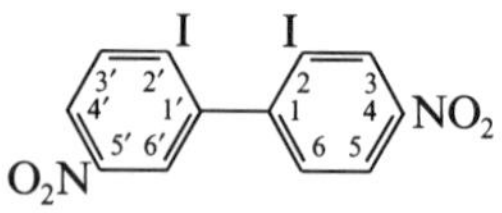

$C_{12}H_6I_2N_2O_4$ M 495.999
Pale yellow prisms (EtOH). Mp 167-168°.

Baker, W. *et al*, *J. Chem. Soc.*, 1958, 2658 (*synth*)

2,2′-Diiodo-5,5′-dinitrobiphenyl   D-80402

[61837-21-6]

$C_{12}H_6I_2N_2O_4$ M 495.999

Pale yellow cryst. (EtOH/Me₂CO). Mp 236-238°.

Dougherty, T.K. *et al*, *J. Org. Chem.*, 1983, **48**, 5273 (*synth, ir, pmr*)

2,2′-Diiodo-6,6′-dinitrobiphenyl   D-80403

[17532-87-5]

$C_{12}H_6I_2N_2O_4$ M 495.999

Pale yellow prisms (C₆H₆). Mp 152-153°.

Iqbal, K. *et al*, *J. Chem. Soc. C*, 1967, 1690 (*synth, pmr*)

4,4′-Diiodo-3,3′-dinitrobiphenyl   D-80404

$C_{12}H_6I_2N_2O_4$ M 495.999

Yellow needles (EtOH). Mp 151-152°.

Wirth, H.O. *et al*, *Makromol. Chem.*, 1964, **77**, 90 (*synth*)

1,12-Diiodododecane   D-80405

Dodecamethylene diiodide

[24772-65-4]

$$ICH_2(CH_2)_{10}CH_2I$$

$C_{12}H_{24}I_2$ M 422.131

Platelets (EtOAc). Mp 41°. Bp₀.₅ 192-198°.

Nineham, A.W., *J. Chem. Soc.*, 1953, 2601 (*synth*)
Ainscow, T.A. *et al*, *Tetrahedron*, 1987, **43**, 115 (*synth, pmr*)

1,4-Diisocyanatocubane   D-80406

1,4-Diisocyanatopentacyclo[4.2.0.0²,⁵.0³,⁸.0⁴,⁷]octane, 9CI

[116635-16-6]

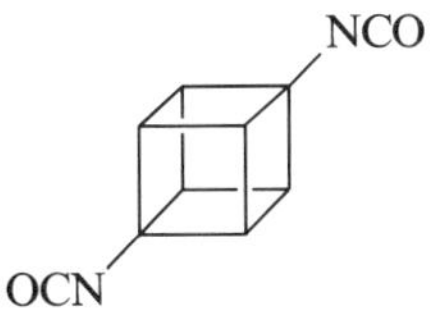

$C_{10}H_6N_2O_2$ M 186.170

Solid by subl. Mp 113-115°. Moisture sensitive, stable under N₂ in cold.

▷ Hazardous prepn., can explode.

Eaton, P.E. *et al*, *J. Org. Chem.*, 1988, **53**, 5353 (*synth, pmr, cmr, ir*)

Dilophus enone   D-80407

[121961-78-2]

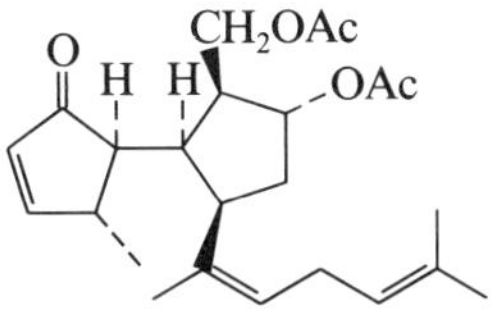

$C_{24}H_{34}O_5$ M 402.530

Constit. of alga *Dilophus okamurai*. Antifeedant. [α]_D −46.0° (c, 1.30 in CHCl₃).

Kurata, K. *et al*, *Tetrahedron Lett.*, 1989, **30**, 1567.

Dilophus ether   D-80408

[121940-50-9]

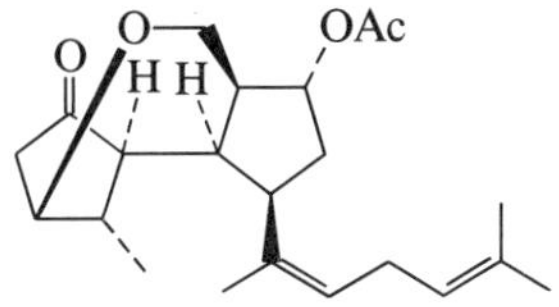

$C_{22}H_{32}O_4$ M 360.492

Constit. of alga *Dilophus okamurai*. Antifeedant. [α]_D −69.0° (c, 0.860 in CHCl₃).

Kurata, K. *et al*, *Tetrahedron Lett.*, 1989, **30**, 1567.

4,5-Dimercapto-1,3-dithiole-2-thione   D-80409

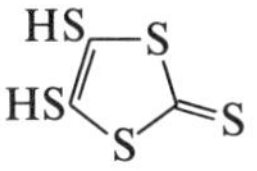

$C_3H_2S_5$ M 198.379

Di-Na salt: [54995-24-3].
     Crimson red solid. Fairly stable under Ar in dark.

Di-Me thioether: [49638-64-4].
     $C_5H_6S_5$ M 226.432
     Mp 99-100°.

Di-Et thioether: [59065-21-3].
     $C_7H_{10}S_5$ M 254.486
     Mp 36.5-37°.

Dibenzyl thioether: [54995-25-4].
     $C_{17}H_{14}S_5$ M 378.628
     Mp 86°.

Wawzonek, S. *et al*, *J. Org. Chem.*, 1974, **39**, 511 (*synth, ir, raman, pmr, uv, ms*)
Steinecke, G. *et al*, *Z. Chem.*, 1975, **15**, 28 (*synth*)
Reuter, U. *et al*, *Z. Anorg. Allg. Chem.*, 1976, **421**, 143 (*synth, deriv, ms, uv, pmr*)
Hartke, K. *et al*, *Chem. Ber.*, 1980, **113**, 1898 (*deriv, synth*)
Varma, S.K. *et al*, *Synthesis*, 1987, 837 (*synth*)

2,3-Dimercapto-2-propene-1-thione   D-80410

Trithiodeltic acid

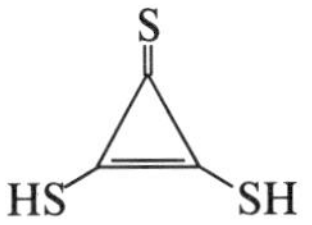

$C_3H_2S_3$ M 134.247

Di-Na salt: Beige solid. Mp >300°. Mesomeric.

Baum, G. *et al*, *Angew. Chem., Int. Ed. Engl.*, 1987, **26**, 1163 (*synth, cryst struct*)

3,4-Dimethoxybenzyl isothiocyanate   D-80411

4-(Isothiocyanatomethyl)-1,2-dimethoxybenzene. Veratryl isothiocyanate

[14596-50-0]

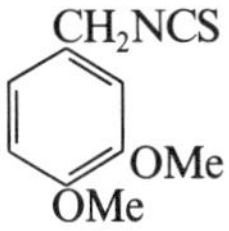

$C_{10}H_{11}NO_2S$ M 209.268

Isol. from seeds of *Heliophila longifolia*. Viscous pale-yellow oil or cryst. Mp 22-24°. Bp 135-137°. n_D^{25} 1.5960.

Ettlinger, M.G. *et al*, *Acta Chem. Scand.*, 1966, **20**, 1778 (*isol, struct, synth*)

3,3-Dimethoxy-3*H*-diazirine, 9CI   D-80412

[114980-39-1]

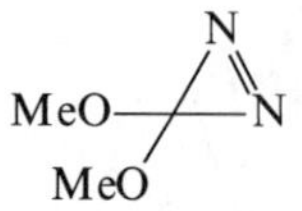

$C_3H_6N_2O_2$ M 102.093

Precursor for Dimethoxycarbene. Obt. as *ca* 0.7 M
pentane soln.

Moss, R.A. *et al, J. Am. Chem. Soc.*, 1988, **110**, 4443 (*synth, uv, use*)

9,10-Dimethoxy-2-methyl-1,4-anthra-quinone   D-80413

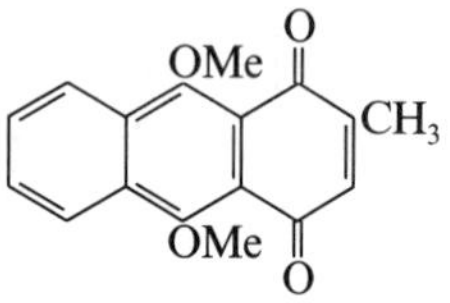

$C_{17}H_{14}O_4$ M 282.295

Constit. of *Tectonia grandis*. Yellow cryst.

Singh, P. *et al, Phytochemistry*, 1989, **28**, 1258.

1-(3,4-Dimethoxyphenyl)-2,3-bis(hydroxy-methyl)-6,7-methylenedioxynaphthalene   D-80414

[42923-60-4]

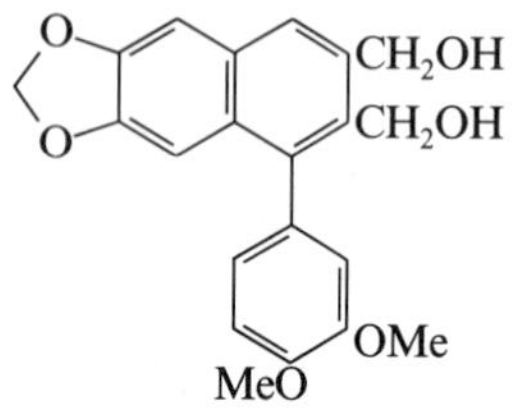

$C_{21}H_{20}O_6$ M 368.385

Constit. of *Jatropha gossypifolia*. Cryst. (C_6H_6/pet. ether).
Mp 184°.

Das, B. *et al, Phytochemistry*, 1988, **27**, 3684.

7-(2,4-Dimethoxyphenyl)-3,4-dihydro-5-hydroxy-10-(3-hydroxy-3-methylbutyl)-2,2-dimethyl-2*H*,6*H*-benzo[1,2-*b*:5,4-*b'*]dipyran-6-one, 9CI   D-80415

[78876-32-1]

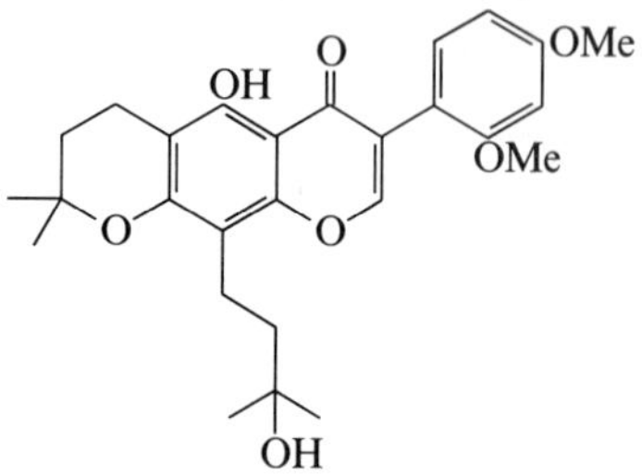

$C_{27}H_{32}O_7$ M 468.546

Isol. from leaves of *Millettia pachycarpa*. Cryst. (pet.
ether). Mp 150°.

Singhal, A.K. *et al, Phytochemistry*, 1981, **20**, 803 (*isol, uv, pmr, ms, struct*)

7-(3,4-Dimethoxyphenyl)-3,4-dihydro-5-hydroxy-10-(3-hydroxy-3-methylbutyl)-2,2-dimethyl-2*H*,6*H*-benzo[1,2-*b*:5,6-*b'*]dipyran-6-one, 9CI   D-80416

[78876-31-0]

$C_{27}H_{32}O_7$ M 468.546

Isol. from leaves of *Millettia pachycarpa*. Cryst.
(EtOAc/pet. ether). Mp 140°.

Singhal, A.K. *et al, Phytochemistry*, 1981, **20**, 803 (*isol, uv, pmr, ms, struct*)

3,9-Dimethyladenine   D-80417

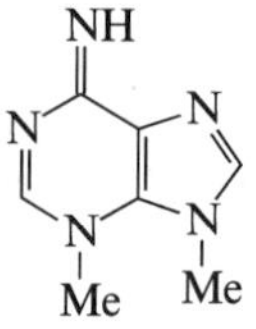

$C_7H_9N_5$ M 163.182

B,HCl: Prisms. Mp 281-282° dec.

B,HClO_4: Prisms (H_2O). Mp > 300°. Other 3,9-dialkyladenines also prepd.

Fujii, T. *et al, Chem. Pharm. Bull.*, 1989, **37**, 1504 (*synth, uv, pmr*)

3,3-Dimethyl-2(3*H*)-benzofuranone   D-80418

[13524-76-0]

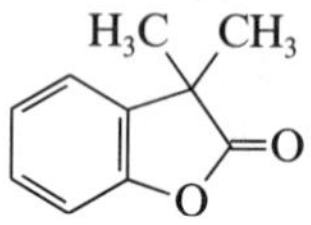

$C_{10}H_{10}O_2$ M 162.188

Liq. Bp_2 101-102°, $Bp_{0.4}$ 63°.

Gripenberg, J. *et al, Acta Chem. Scand.*, 1966, **20**, 1561 (*synth, uv, ir, pmr*)

Amyes, T.L. *et al, J. Am. Chem. Soc.*, 1988, **110**, 6505 (*synth, ir, pmr, ms*)

7,8-Dimethylbenzo[*no*]naphtho[2,1,8,7-*ghij*]pleiadene, 9CI   D-80419

1,16-Dehydro-2,15-dimethylhexahelicene

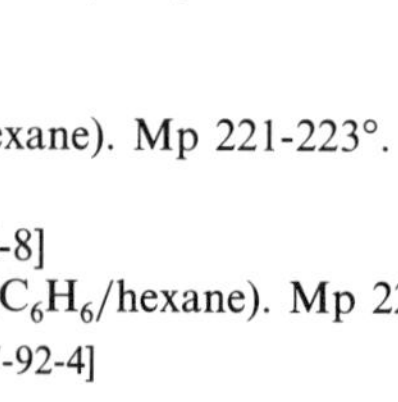

$C_{28}H_{18}$ M 354.450

(+)-***form***

Cryst. (C_6H_6/hexane). Mp 220-222°. $[\alpha]_D^{25}$ +1879°
(CHCl_3).

(−)-***form***

Cryst. (C_6H_6/hexane). Mp 221-223°. $[\alpha]_D^{25}$ −1882°
(CHCl_3).

(±)-***form*** [114417-96-8]

Yellow prisms (C_6H_6/hexane). Mp 226-228°.

[114324-63-9, 114417-92-4]

Yamamoto, K. *et al, J. Am. Chem. Soc.*, 1988, **110**, 3578 (*synth, ir, pmr, uv, resoln, abs config*)

3-(2,2-Dimethyl-2*H*-1-benzopyran-6-yl)-2-propenoic acid   D-80420

Updated Entry replacing D-50601
2,2-Dimethylchromene-6-propenoic acid
[104387-05-5]

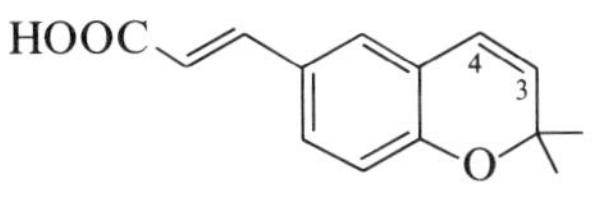

C$_{14}$H$_{14}$O$_3$ M 230.263
Constit. of *Baccharis* spp. Cryst. Mp 191-192°.

3,4-Dihydro: [12772-83-7]. *3-(3,4-Dihydro-2,2-dimethyl-2H-1-benzopyran-6-yl)-2-propenoic acid, 9CI.* **Drupacin**
C$_{14}$H$_{16}$O$_3$ M 232.279
Isol. from fruits of *Psoralea drupacea*. Acicular cryst. (MeOH). Mp 205-206°.

Golovina, L.A. *et al, Khim. Prir. Soedin.*, 1973, **9**, 700; *Chem. Nat. Compd.*, 672 (*Drupacin*)
Labbe, C. *et al, J. Nat. Prod. (Lloydia)*, 1986, **49**, 517.

3,3′-Dimethyl-2,2′-biindazole   D-80421

[113302-76-4]

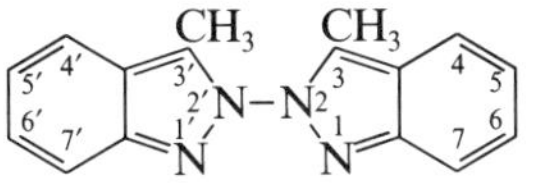

$$\text{structure}$$

C$_{16}$H$_{14}$N$_4$ M 262.313
Cryst. Mp 143-144°.

Cuevas, J.C. *et al, J. Org. Chem.*, 1988, **53**, 2055 (*synth, ir, ms, pmr, cmr*)

5,5′-Dimethyl-2,2′-biindazole   D-80422

[113302-77-5]
C$_{16}$H$_{14}$N$_4$ M 262.313
Cryst. Mp 218.5-219.5°.

Cuevas, J.C. *et al, J. Org. Chem.*, 1988, **53**, 2055 (*synth, ir, ms, pmr, cmr*)

7,7′-Dimethyl-2,2′-biindazole   D-80423

[113302-78-6]
C$_{16}$H$_{14}$N$_4$ M 262.313
Cryst. Mp 165-166°.

Cuevas, J.C. *et al, J. Org. Chem.*, 1988, **53**, 2055 (*synth, ir, pmr, cmr, ms*)

1,4-Dimethylbiphenylene   D-80424

[64472-47-5]

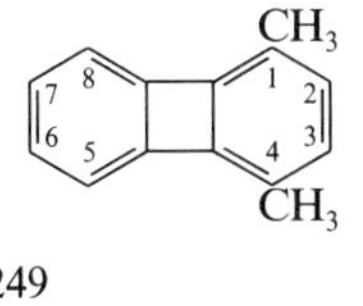

$$\text{structure}$$

C$_{14}$H$_{12}$ M 180.249
Cryst. (EtOH). Mp 28°.

Hellwinkel, D. *et al, Justus Liebigs Ann. Chem.*, 1977, 1013 (*synth, pmr*)

1,5-Dimethylbiphenylene   D-80425

[55277-56-0]
C$_{14}$H$_{12}$ M 180.249
Cryst. Mp 51-53°.

Campbell, C.D. *et al, J. Chem. Soc. C*, 1969, 742 (*synth*)
Wilcox, C.F. *et al, J. Am. Chem. Soc.*, 1975, **97**, 1914 (*synth*)

1,8-Dimethylbiphenylene   D-80426

[36230-17-8]
C$_{14}$H$_{12}$ M 180.249
Pale yellow plates (MeOH). Mp 79-80°.

Picrate: Crimson needles (EtOH). Mp 126°.

Lothrop, W.C. *et al, J. Am. Chem. Soc.*, 1942, **64**, 1698 (*synth*)
Campbell, C.D. *et al, J. Chem. Soc. C*, 1969, 742 (*synth*)
Wilcox, C.F. *et al, J. Org. Chem.*, 1988, **53**, 4333; 1989, **54**, 2190 (*synth, pmr, cmr, bibl*)

2,3-Dimethylbiphenylene   D-80427

[65350-34-7]
C$_{14}$H$_{12}$ M 180.249
Pale yellow cryst. (MeOH). Mp 111-112°.

Buckland, P.R. *et al, Tetrahedron*, 1977, **33**, 1797 (*synth, pmr*)

2,6-Dimethylbiphenylene   D-80428

[2918-97-0]
C$_{14}$H$_{12}$ M 180.249
Pale yellow cryst. (EtOH). Mp 139-141°.

Hart, F.A. *et al, J. Chem. Soc.*, 1957, 3939 (*synth, uv*)
Campbell, C.D. *et al, J. Chem. Soc. C*, 1969, 742 (*synth*)

2,7-Dimethylbiphenylene   D-80429

[4874-81-1]
C$_{14}$H$_{12}$ M 180.249
Pale yellow plates (EtOH). Mp 115°.

Picrate: Crimson needles (EtOH). Mp 110°.

Lothrop, W.C., *J. Am. Chem. Soc.*, 1941, **63**, 1187 (*synth, deriv*)
Constantine, P.R. *et al, J. Chem. Soc. C*, 1966, 1767 (*synth, uv, pmr*)
Campbell, C.D. *et al, J. Chem. Soc. C*, 1969, 742 (*synth*)

2,2′-Dimethyl-3,3′-bithiophene, 9CI   D-80430

[15940-57-5]

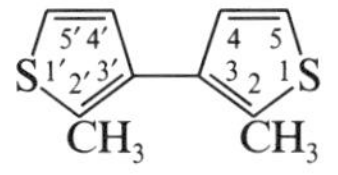

C$_{10}$H$_{10}$S$_2$ M 194.321
Bp$_{17}$ 141-144°.

Gronowitz, S. *et al, Ark. Kemi*, 1967, **27**, 153 (*synth, pmr*)
Wiklund, E. *et al, Chem. Scr.*, 1973, **3**, 226 (*pmr*)
Gronowitz, S. *et al, Tetrahedron*, 1978, **34**, 587 (*synth*)

3,3′-Dimethyl-2,2′-bithiophene, 9CI   D-80431

C$_{10}$H$_{10}$S$_2$ M 194.321
Liq. Bp$_{25}$ 160-162°, Bp$_{11}$ 131°.

Uhlenbroek, J.H. *et al, Recl. Trav. Chim. Pays-Bas (J. R. Neth. Chem. Soc.)*, 1960, **79**, 1181 (*synth, uv*)
Gronowitz, S. *et al, Tetrahedron*, 1978, **34**, 587 (*synth, pmr*)
Cunningham, D.D. *et al, J. Chem. Soc., Chem. Commun.*, 1987, 1021 (*uv*)
Krische, B. *et al, J. Chem. Soc., Chem. Commun.*, 1987, 1476 (*synth, pmr*)

3,4′-Dimethyl-2,2′-bithiophene, 9CI   D-80432

[113386-74-6]

$C_{10}H_{10}S_2$ M 194.321

Cunningham, D.D. *et al, J. Chem. Soc., Chem. Commun.*, 1987, 1021 (*uv*)

3,5-Dimethyl-2,3′-bithiophene, 9CI   D-80433

[90655-30-4]

$C_{10}H_{10}S_2$ M 194.321
Bp_1 115-117°.

Gronowitz, S. *et al, Chem. Scr.*, 1983, **22**, 265 (*synth, pmr*)

4,4′-Dimethyl-2,2′-bithiophene, 9CI   D-80434

[111372-97-5]

$C_{10}H_{10}S_2$ M 194.321
Mp 67°.

Cunningham, D.D. *et al, J. Chem. Soc., Chem. Commun.*, 1987, 1021 (*uv*)
Krische, B. *et al, J. Chem. Soc., Chem. Commun.*, 1987, 1476 (*synth, pmr*)

4,4′-Dimethyl-3,3′-bithiophene, 9CI   D-80435

[15940-59-7]

$C_{10}H_{10}S_2$ M 194.321
Viscous oil. Bp_{10} 133-134°.

Gronowitz, S. *et al, Ark. Kemi*, 1967, **27**, 153 (*synth, pmr*)
Wiklund, E. *et al, Chem. Scr.*, 1973, **3**, 226 (*pmr*)

5,5′-Dimethyl-2,2′-bithiophene, 9CI   D-80436

$C_{10}H_{10}S_2$ M 194.321
Shows nematocidal props. Cryst. (MeOH), plates (pentane). Mp 68° (64-65°).

Lescot, E. *et al, J. Chem. Soc.*, 1959, 3234 (*synth*)
Uhlenbroek, J.H. *et al, Recl. Trav. Chim. Pays-Bas (J. R. Neth. Chem. Soc.)*, 1960, **79**, 1181 (*uv, pharmacol*)
Curtis, R.F. *et al, J. Chromatogr.*, 1962, **9**, 366 (*chromatog*)
Atkinson, R.E. *et al, J. Chem. Soc. C*, 1967, 2011 (*synth, ir, chromatog*)
Abu-Eittah, R.H. *et al, Bull. Chem. Soc. Jpn.*, 1985, **58**, 2126 (*synth, uv*)
Krische, B. *et al, J. Chem. Soc., Chem. Commun.*, 1987, 1476 (*synth, pmr*)

5,5′-Dimethyl-3,3′-bithiophene, 9CI   D-80437

[38418-27-8]

$C_{10}H_{10}S_2$ M 194.321
Flakes (ligroin). Mp 135-136°.

Gronowitz, S. *et al, Acta Chem. Scand.*, 1962, **16**, 1127 (*synth, pmr*)
Norden, B. *et al, Acta Chem. Scand.*, 1972, **26**, 429 (*uv, conformn*)
Wiklund, E. *et al, Chem. Scr.*, 1973, **3**, 226 (*pmr*)

2,3-Dimethyl-3-butenoic acid   D-80438

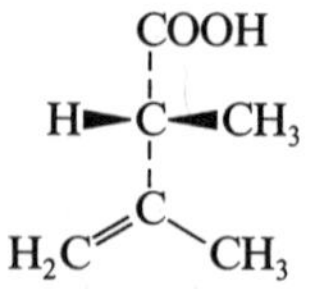

(R)-form

$C_6H_{10}O_2$ M 114.144

(R)-form

Liq. Bp_{42} 120°. $[\alpha]_D^{25}$ −27° (0.73 in $CHCl_3$) (71% e.e.).

Walba, D.M. *et al, J. Org. Chem.*, 1988, **55**, 1046 (*synth, ir, pmr, cmr, ms*)

2,3-Dimethyl-3-buten-1-ol   D-80439

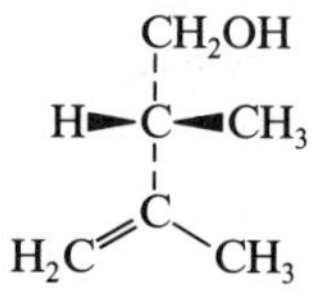

$C_6H_{12}O$ M 100.160

(R)-form

Liq. Bp_{78} 87°.

Walba, D.M. *et al, J. Org. Chem.*, 1988, **53**, 1046 (*synth, ir, pmr, cmr, ms*)

3,3-Dimethyl-1,4-cyclohexadiene   D-80440

[35934-83-9]

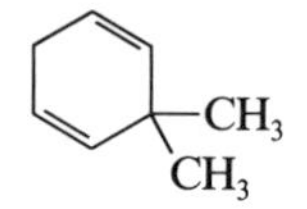

C_8H_{12} M 108.183
Liq. Bp 110-113° (91% pure) .

Stoos, F. *et al, J. Am. Chem. Soc.*, 1972, **94**, 2719 (*synth*)
Jacobson, B.M. *et al, J. Org. Chem.*, 1988, **53**, 3247 (*synth*)

4,4-Dimethyl-2,5-cyclohexadien-1-one   D-80441

[1073-14-9]

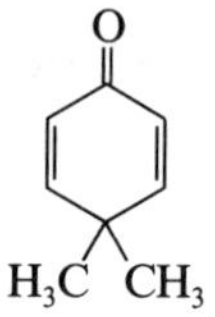

$C_8H_{10}O$ M 122.166
$Bp_{0.15}$ 41°.

Zimmerman, H.E. *et al, J. Am. Chem. Soc.*, 1967, **89**, 6589; 1971, **93**, 3653 (*synth*)

6,6-Dimethyl-2,4-cyclohexadien-1-one   D-80442

[21428-63-7]

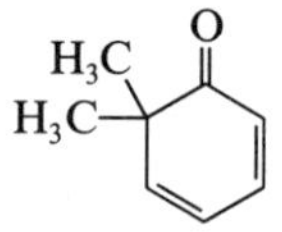

$C_8H_{10}O$ M 122.166
Liq. d_4^{20} 0.964. n_D^{20} 1.4992. Rearr. on htg. to phenols.

Alder, K. *et al, Chem. Ber.*, 1957, **90**, 1709.

2,2-Dimethyl-1,3-cyclohexanedione   D-80443

[562-13-0]

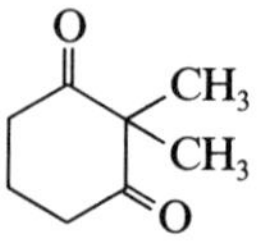

$C_8H_{12}O_2$ M 140.182
Cryst. Mp 39-40°. Bp_{11} 103-105°.

Dioxime:

$C_8H_{14}N_2O_4$ M 202.210
Cryst. (EtOH). Mp 195-195.5°.

Swaminathan, S. *et al, Tetrahedron*, 1964, **20**, 1119 (*synth, ir*)
Jacobson, B.M. *et al, J. Org. Chem.*, 1988, **53**, 3247 (*synth*)

2,2-Dimethylcyclopropanecarboxylic acid   D-80444

[931-26-0]

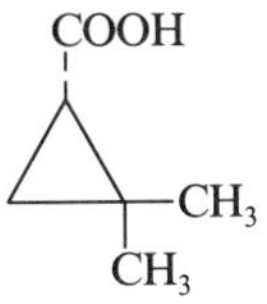

COOH

(*S*)-*form*

CH₃
CH₃

C₆H₁₀O₂   M 114.144

(*S*)-*form* [14590-53-5]

Liq. [α]₂₀ᴅ +132° (c, 1.01 in MeOH), [α]₂₀ᴅ +142° (c, 1.01 in CHCl₃).

Amide:

C₆H₁₁NO   M 113.159

Mp 136-137.5°. [α]₂₀ᴅ +101.4° (c, 1.0 in CHCl₃).

(±)-*form* [75885-59-5]

Liq. Bp 198-201°, Bp₅ 83-84°.

Amide: Needles (EtOAc). Mp 177-177.5°.

Nitrile: 2-*Cyano-1,1-dimethylcyclopropane*

C₆H₉N   M 95.144

Liq. d²⁴·⁶₂₀ 0.85. Bp 154.5-155.5°.

Nelson, E.R. *et al*, *J. Am. Chem. Soc.*, 1957, **79**, 3467 (*synth, deriv*)
Graham, D.W. *et al*, *J. Med. Chem.*, 1987, **30**, 1074 (*resoln, abs config, cryst struct, pmr*)

1,3-Dimethyl-2,9-dioxabicyclo-[3.3.1]nonane   D-80445

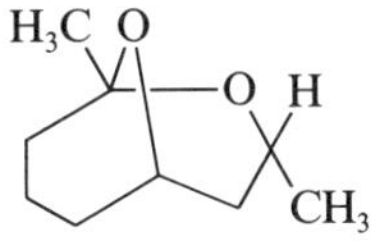

H₃C O

O H

CH₃

C₉H₁₆O₂   M 156.224

(1*R*,3*R*)-*form* [76740-34-6]

Isol. from Norway spruce infected with the ambrosia beetle *Trypodendron lineatum*. Host-specific pheromone. Liq. [α]²⁴ᴅ −37.9° (c, 0.5 in pentane).

[63013-24-1, 63013-25-2, 76334-10-6, 76740-35-7, 76740-36-8]

Mori, Y. *et al*, *Chem. Pharm. Bull.*, 1989, **37**, 1078 (*synth, pmr, bibl*)

3,3-Dimethyl-1,2-dioxolane   D-80446

[67393-70-8]

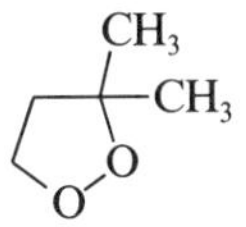

CH₃
CH₃
O

C₅H₁₀O₂   M 102.133

Bp₁₆ 52-54°.

Nixon, J.R. *et al*, *J. Org. Chem.*, 1978, **43**, 4048 (*synth, pmr*)
Bloodworth, A.J. *et al*, *J. Org. Chem.*, 1986, **51**, 2110 (*synth, pmr, cmr*)

3,5-Dimethyl-2,6-diphenylpyridine, 9CI   D-80447

[14435-89-3]

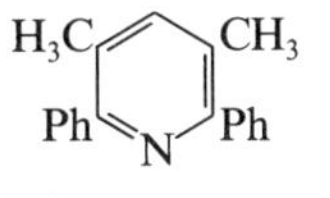

H₃C   CH₃

Ph   N   Ph

C₁₉H₁₇N   M 259.350

Cryst. solid (hexane/CHCl₃). Mp 136-137°.

Newkome, G.R. *et al*, *J. Org. Chem.*, 1972, **37**, 1329 (*synth, pmr, uv*)
Barluenga, J. *et al*, *J. Org. Chem.*, 1988, **53**, 5960 (*synth, pmr, cmr*)

2,6-Dimethylenebicyclo[3.3.1]nonane   D-80448

2,6-Bis(methylene)bicyclo[3.3.1]nonane

[26069-13-6]

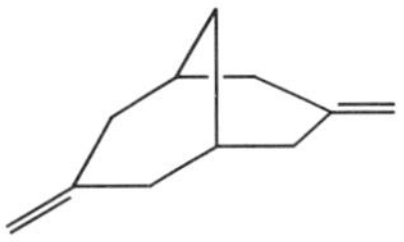

C₁₁H₁₆   M 148.247

Liq. Bp 185-190°, Bp₁₅ 73-75°.

Landa, S. *et al*, *Collect. Czech. Chem. Commun.*, 1970, **35**, 1005 (*synth*)
Bishop, R. *et al*, *Aust. J. Chem.*, 1979, **32**, 2675 (*synth, ir, pmr, cmr*)

4,5-Dimethyl-2,3-furandicarboxylic acid   D-80449

[111301-74-7]

C₈H₈O₅   M 184.148

Powder. Mp 243-244°.

Mann, J. *et al*, *Tetrahedron*, 1987, **43**, 2533 (*synth, ir*)

4,4-Dimethyl-6-heptenoic acid, 9CI   D-80450

[92611-60-4]

H₂C=CHCH₂C(CH₃)₂CH₂CH₂COOH

C₉H₁₆O₂   M 156.224

Oil. Bp₁ 80-90°.

S-Benzylisothiouronium salt: [116914-27-3].

Cryst. (H₂O). Mp 140-141°.

Me ester: [113086-97-8].

C₁₀H₁₈O₂   M 170.251

Oil. Bp₁₂ 78-82°.

Pattenden, G. *et al*, *Tetrahedron*, 1987, **43**, 5637 (*synth, ir, pmr, cmr*)

1,3-Dimethyl-1*H*-imidazolium(1+), 9CI   D-80451

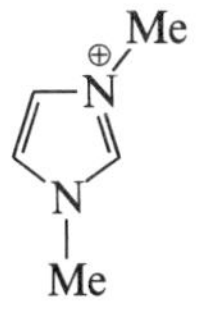

C₅H₉N₂⊕   M 97.139

Iodide: [4333-62-4].

C₅H₉IN₂   M 224.044

Prisms (EtOAc). Mp 86.5-88°.

Picrate: Yellow cryst. (EtOH). Mp 100-103°.

Overberger, C.G. *et al*, *J. Org. Chem.*, 1965, **30**, 3580 (*synth*)
Karkhanis, D.W. *et al*, *Phosphorus Sulfur*, 1985, **22**, 49 (*synth*)
Org. Synth., 1986, **64**, 92 (*synth, pmr, cmr*)

1,4-Di-*O*-methyl-*myo*-inositol   **D-80452**
Liriodendritol

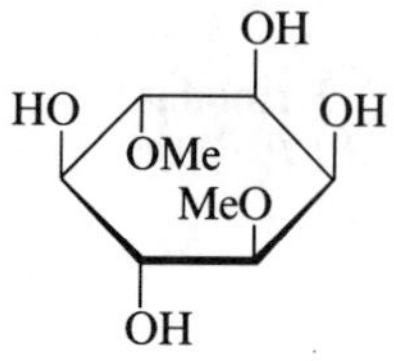

(−)-*form*

$C_8H_{16}O_6$ M 208.211

(−)-*form* [22006-88-8]
   Isol. from *Liriodendron tulipifera*, *L. chinense* and other plants. Prisms (EtOH). Mp 224° (anhyd.). $[\alpha]_D$ −25° (H_2O).
   Tetra-Ac:
      $C_{16}H_{24}O_{10}$ M 376.360
      Mp 139°. $[\alpha]_D$ −24° ($CHCl_3$).
(±)-*form*
   Mp 203°.
   Tetra-Ac: Cryst. (EtOH aq.). Mp 155-156°.

Angyal, S.J. *et al*, *J. Chem. Soc.*, 1961, 4718 (*struct, synth*)
Kindl, H. *et al*, *Fortschr. Chem. Org. Naturst.*, 1966, **24**, 149 (*rev*)

1,5-Dimethyl-8-(5-methyl-1-methylene-4-hexenyl)-1,5-cyclodecadiene, 9CI   **D-80453**
13-Prenyl-1(10),4,11-germacratriene
[122616-70-0]

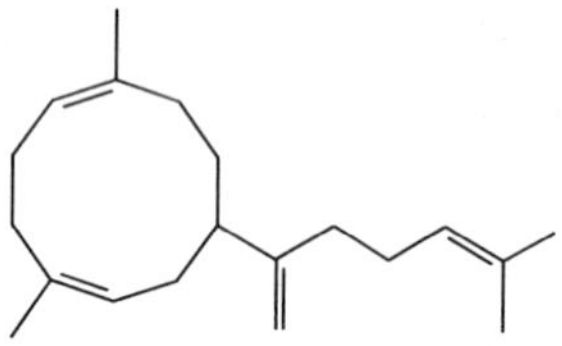

$C_{20}H_{32}$ M 272.473
Constit. of *Helichrysum argyrophyllum*. Oil.

Jakupovic, J. *et al*, *Phytochemistry*, 1989, **28**, 1119.

2,6-Dimethyl-3-morpholinone   **D-80454**

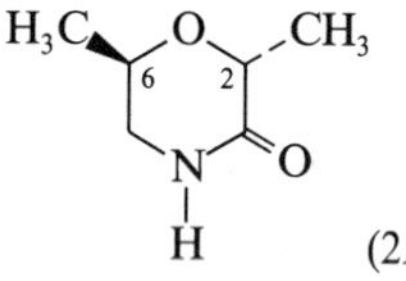

$C_6H_{11}NO_2$ M 129.158
(2R,6R)-*form* [117465-49-3]
   Cryst. (CH_2Cl_2/C_6H_6). Mp 83-85°. $[\alpha]_D^{20}$ −20.8° (c, 1.0 in $CHCl_3$). (2*R*,6*S*) and (2*S*,6*S*)-isomers also prepd., as a diastereoisomericmixt.

[117465-50-6, 117465-51-7, 117465-54-0]

Danklmaier, J. *et al*, *Justus Liebigs Ann. Chem.*, 1988, 1149 (*synth, pmr, cmr*)

4,8-Dimethyl-1,3,7-nonatriene   **D-80455**
Updated Entry replacing D-50644

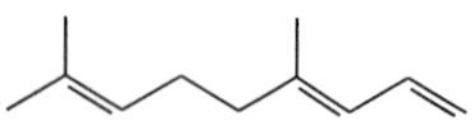

$C_{11}H_{18}$ M 150.263
(*E*)-*form*
   Constit. of essential oil of *Elettaria cardamomum*, *Rosa* spp., *Magnolia liliflora*. Oil.

[21214-62-0]
Pattenden, G. *et al*, *J. Chem. Soc. C*, 1968, 1984 (*synth*)
Maurer, B. *et al*, *Tetrahedron Lett.*, 1986, **27**, 2111 (*struct, synth*)
Boland, W. *et al*, *Helv. Chim. Acta*, 1989, **72**, 247 (*biosynth, bibl*)

2,6-Dimethyl-2,7-octadiene-1,6-diol   **D-80456**

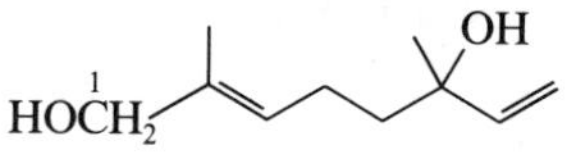

(*E*)-*form*

$C_{10}H_{18}O_2$ M 170.251
(*E*)-*form*
   9-Hydroxylinalool
   1-O-β-D-Glucoside: [64776-96-1]. **Betulalbuside A**
      $C_{16}H_{28}O_7$ M 332.393
      Constit. of *Betula alba* and *Chaenomeles japonica*. Oil. 6-config not detd.
(*Z*)-*form*
   1-Hydroxylinalool
   1-O-β-D-Glucoside: [64813-08-7]. **Betulalbuside B**
      $C_{16}H_{28}O_7$ M 332.393
      From *B. alba* and *C. japonica*. Oil. 6-config. not detd.

Tschesche, R. *et al*, *Chem. Ber.*, 1977, **110**, 3111.

3,7-Dimethyl-2,5-octadiene-1,7-diol   **D-80457**

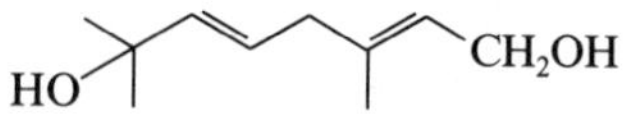

$C_{10}H_{18}O_2$ M 170.251
1-Ac: [33766-42-6]. *8-Acetoxy-2,6-dimethyl-3,6-octadien-2-ol*
   $C_{12}H_{20}O_3$ M 212.288
   Constit. of *Jasonia montana*. Oil.

Ahmed, A.A. *et al*, *Phytochemistry*, 1988, **27**, 3875.

2-(3,7-Dimethyl-2,6-octadienyl)-4-hydroxy-6-methoxyacetophenone   **D-80458**
1-[2-(3,7-Dimethyl-2,6-octadienyl)-4-hydroxy-6-methoxyphenyl]ethanone, 9CI
[121379-44-0]

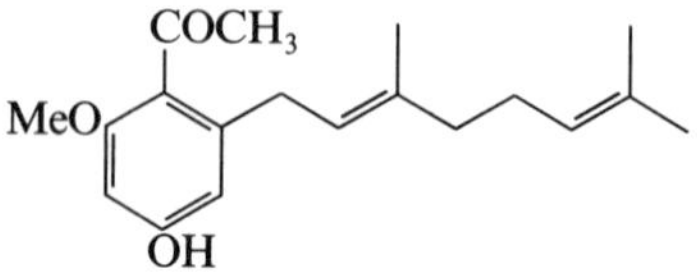

$C_{19}H_{26}O_3$ M 302.413
Constit. of *Dioscorea bulbifera*. Cryst. Mp 225°.

Gupta, D. *et al*, *Phytochemistry*, 1989, **28**, 947.

1-[3-(3,7-Dimethyl-2,6-octadienyl)-2,4,5,6-tetrahydroxyphenyl]-3-(4-hydroxy-phenyl)-1-propanone,9CI   D-80459

2′,4,4′,5′,6-Pentahydroxy-3′-neryldihydrochalcone.
2′,4,4′,5′,6′-Pentahydroxy-3′-geranyldihydrochalcone

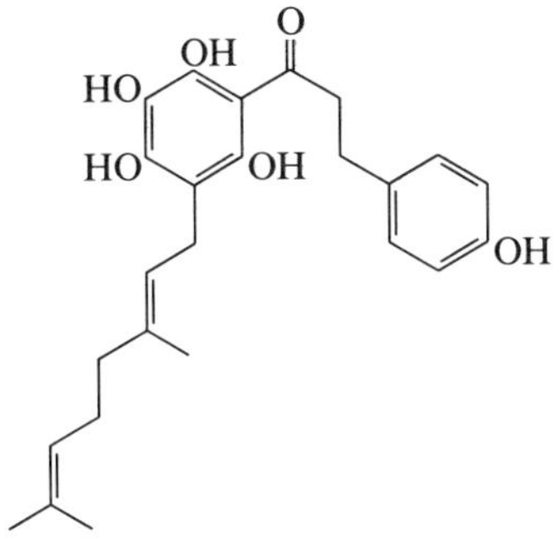

$C_{25}H_{30}O_6$ M 426.508
Isol. from *Helichrysum monticola*. Oil. Unseparated mixt. of (*E*)- and (*Z*)- forms.

[76015-50-4, 76015-51-5]

Bohlmann, F. *et al, Phytochemistry*, 1980, **19**, 683.

3-(3,7-Dimethyl-2,6-octadienyl)-2,4,6-trihydroxybenzophenone   D-80460

[3-(3,7-Dimethyl-2,6-octadienyl)-2,4,6-trihydroxyphenyl]phenylmethanone, 9CI. 3-Geranyl-2,4,6-trihydroxybenzophenone

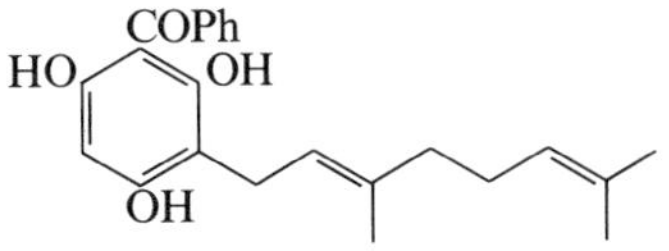

(*E*)-*form*

$C_{23}H_{26}O_4$ M 366.456
(*E*)-*form* [70219-87-3]
    2-Benzoyl-6-geranylphloroglucinol
    Isol. from *Leontonyx* spp. and *Helichrysum* spp. Oil.
(*Z*)-*form* [76015-48-0]
    2-Benzoyl-6-nerylphloroglucinol
    $C_{23}H_{26}O_4$ M 366.456
    Isol. from *H. monticola*. Oil.

Bohlmann, F. *et al, Phytochemistry*, 1978, **17**, 1929; 1979, **18**, 2046; 1980, **19**, 683.

2,5-Dimethyl-4-oxazolecarboxylic acid, 9CI   D-80461

[23000-14-8]

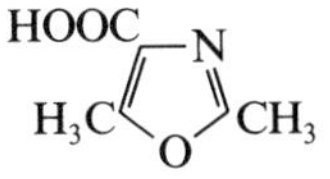

$C_6H_7NO_3$ M 141.126
Cryst. (MeOH). Mp 236-237° (220° dec.).
Et ester:
    $C_8H_{11}NO_3$ M 169.180
    Bp$_{22}$ 125°, Bp$_{11}$ 117°.
Amide:
    $C_6H_8N_2O_2$ M 140.141
    Cryst. (H$_2$O). Mp 234°.
Nitrile: 4-Cyano-2,5-dimethyloxazole
    $C_6H_6N_2O$ M 122.126

Oil. Bp$_{18}$ 84-85°.

Treibs, A. *et al, Chem. Ber.*, 1951, **84**, 96 (*synth, ester, amide*)
Brown, D.J. *et al, J. Chem. Soc. B*, 1969, 270 (*pmr*)
Doleschall, G. *et al, J. Chem. Soc., Perkin Trans.* 1, 1988, 1875 (*synth, ir, pmr*)

2,3-Dimethyloxiranemethanol   D-80462

2,3-Epoxy-2-methyl-1-butanol

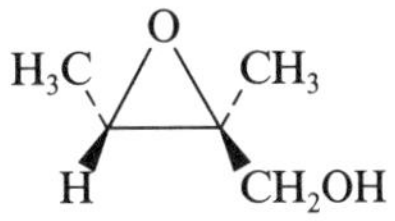

$C_5H_{10}O_2$ M 102.133
(*2S,3S*)-*form* [113531-34-3]
    Oil. Bp$_5$ 120° (oven) . [α]$_D$ −22.2° (c, 3.0 in CH$_2$Cl$_2$). 94% ee.

Evans, D.A. *et al, J. Am. Chem. Soc.*, 1988, **110**, 2506 (*synth, ir, pmr, cmr*)

7-(3,3-Dimethyloxiranyl)methoxy-5,6-methylenedioxycoumarin   D-80463

4-(3,3-Dimethyloxiranyl)methoxy-7H-1,3-dioxolo[4,5-f][1]benzopyran-7-one, 9CI

[57419-59-7]

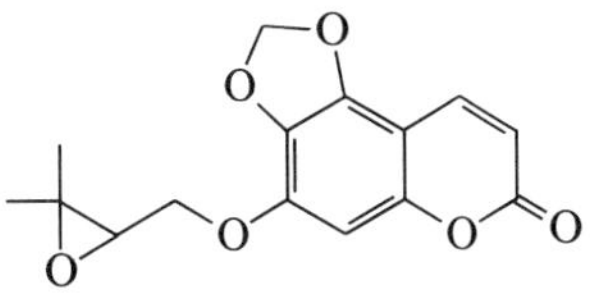

$C_{15}H_{14}O_6$ M 290.272
Isol. from *Pteronia glabrata*. Mp 81°. [α]$_D$ +22° (CHCl$_3$).

Bohlmann, F. *et al, Chem. Ber.*, 1975, **108**, 2955.

2,2-Dimethyl-4-oxo-1-pentanal   D-80464

[61031-76-3]

$$H_3CCOCH_2C(CH_3)_2CHO$$

$C_7H_{12}O_2$ M 128.171
Bp$_{20}$ 84-86°.

Magnus, P.D. *et al, Synth. Commun.*, 1980, **10**, 273 (*synth, ir, pmr*)
Hudlicky, T. *et al, J. Am. Chem. Soc.*, 1988, **110**, 4735 (*synth, ir, pmr*)

6,12-Dimethyl-2-pentadecanone   D-80465

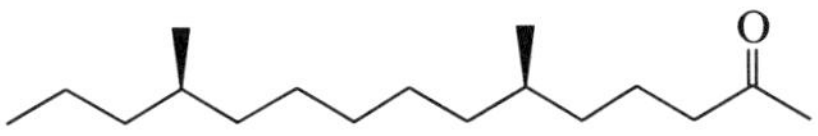

(*6R,12R*)-*form*

$C_{17}H_{34}O$ M 254.455
Sex pheromone of the banded cucumber beetle *Diabrotica balteata* (abs. config. of nat. pheromone not yet known).
(*6R,12R*)-*form*
    Bp$_{0.29}$ 108-111°. [α]$_D^{22}$ −0.5° (c, 1.13 in CHCl$_3$). n$_D^{22}$ 1.4388.
(*6R,12S*)-*form*
    Bp$_{0.29}$ 104-107°. [α]$_D^{20}$ +1.9° (c, 1.16 in CHCl$_3$). n$_D^{20}$ 1.4392.
(*6S,12R*)-*form*
    Bp$_{0.28}$ 114-117°. [α]$_D^{20}$ −1.9° (c, 1.14 in CHCl$_3$). n$_D^{19.5}$ 1.4392.
(*6S,12S*)-*form*

Bp$_{0.2}$ 108-113°. [α]$_D^{21}$ +0.5° (c, 1.06 in CHCl$_3$). n$_D^{21}$ 1.4393.

Mori, K. *et al*, *Justus Liebigs Ann. Chem.*, 1988, 717 (*synth, pmr, cmr, abs config, bibl*)

2,4-Dimethyl-1,3-pentanediol   D-80466

[60712-38-1]

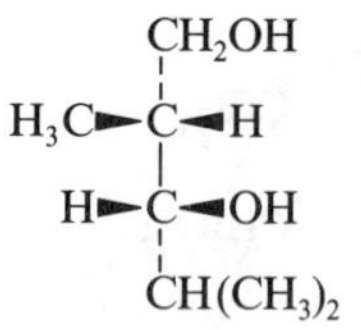

H$_3$C—C◀H with CH$_2$OH above, H▶C◀OH, CH(CH$_3$)$_2$ *(2R,3R)-form*

C$_7$H$_{16}$O$_2$ M 132.202

(2R,3R)-form [82335-99-7]
Mp 84-85°. [α]$_D$ −16.4° (c, 0.04 in CHCl$_3$).

(2S,3R)-form [97906-32-6]
Cryst. Mp 43-45°. [α]$_D^{20}$ +16.5° (c, 1.1 in CH$_2$Cl$_2$). Not optically pure.

3-O-Benzoyl: [113778-42-0].
C$_{14}$H$_{20}$O$_3$ M 236.310
Oil. Bp$_{0.5}$ 115°.

(2S,3S)-form [98391-91-4]
[α]$_D^{25}$ +11.3° (c, 0.6 in CHCl$_3$).

(2RS,3RS)-form [103729-84-6]
(±)-erythro-*form*
Prisms. Mp 53.0-54.5°. The *erythro* label is ambiguous.

[3876-47-9, 98391-91-4, 103729-88-0]

Wood, R.D. *et al*, *Tetrahedron Lett.*, 1982, **23**, 707 (*synth, pmr, ms*)
Helmchen, G. *et al*, *Angew. Chem., Int. Ed. Engl.*, 1985, **24**, 874 (*synth, abs config*)
Masamune, S. *et al*, *J. Am. Chem. Soc.*, 1986, **108**, 8279 (*synth*)
Evans, D.A. *et al*, *J. Am. Chem. Soc.*, 1988, **110**, 3560 (*synth, ir, pmr, cmr*)
Baker, R. *et al*, *J. Chem. Soc., Perkin Trans. 1*, 1988, 85 (*synth, uv, pmr, ms*)

3,3-Dimethyl-4-pentynal, 9CI   D-80467

[35849-64-0]

HC≡CC(CH$_3$)$_2$CH$_2$CHO

C$_7$H$_{10}$O   M 110.155
Liq. Bp 132-134°, Bp$_{140}$ 75-80°.

Oxime:
C$_7$H$_{11}$NO   M 125.170
Bp$_{42}$ 120-122°.

2,4-Dinitrophenylhydrazone: Cryst. (EtOH). Mp 100-101°.

Di-Me acetal: [82850-96-2]. *5,5-Dimethoxy-3,3-dimethyl-1-pentyne, 9CI*
C$_9$H$_{16}$O$_2$ M 156.224
Liq. Bp$_{15}$ 75-80°.

[34718-63-3, 35849-90-2]

Stevens, R.V. *et al*, *J. Am. Chem. Soc.*, 1971, **93**, 6629 (*synth, ir, pmr*)
McMurry, J.E. *et al*, *Tetrahedron*, 1987, **43**, 5489 (*synth, ir, pmr*)

4,4-Dimethyl-1-pentyn-3-ol, 9CI   D-80468

tert-*Butylethynylcarbinol*

[19115-28-7]

C≡CH, H▶C◀OH, C(CH$_3$)$_3$

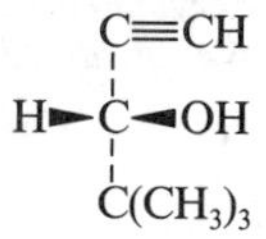

R-form

C$_7$H$_{12}$O   M 112.171

(R)-form [61317-72-4]
Liq. [α]$_D^{20}$ +17.95° (neat).

(±)-form [61348-37-6]
Liq. Bp$_{45}$ 75°.

Barrelle, M. *et al*, *Ann. Chim. (Paris)*, 1967, **2**, 243 (*synth*)
Vigneson, J.-P. *et al*, *Tetrahedron Lett.*, 1979, 2683 (*synth*)
Henderson, M.A. *et al*, *J. Org. Chem.*, 1988, **53**, 4736 (*synth, ir, pmr, cmr, resoln*)

2,4-Dimethyl-5-phenyloxazole, 9CI   D-80469

[23012-31-9]

H$_3$C—, Ph, CH$_3$ oxazole ring (positions 1,2,3,4,5)

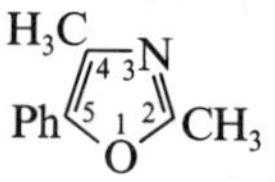

C$_{11}$H$_{11}$NO   M 173.214
Mp 51-52°. Bp$_{12}$ 129-131°, Bp$_{0.15}$ 75°.

Picrate: [56965-53-8].
Mp 170-172°.

Bachstez, M., *Chem. Ber.*, 1914, **47**, 3163 (*synth*)
Bredereck, H. *et al*, *Chem. Ber.*, 1960, **93**, 1389 (*synth*)
Brown, D.J. *et al*, *J. Chem. Soc. B*, 1969, 270 (*pmr, props*)

3-(1,1-Dimethyl-2-propenyl)-7,8-dimethoxy-2H-1-benzopyran-2-one, 9CI   D-80470

Updated Entry replacing D-10609
3-(1,1-Dimethylallyl)-7,8-dimethoxycoumarin
[30310-54-4]

MeO, MeO, 8, benzopyranone structure

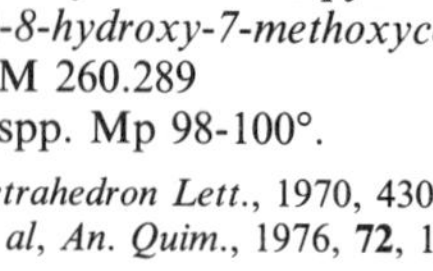

C$_{16}$H$_{18}$O$_4$ M 274.316
Isol. from roots of *Ruta graveolens*. Cryst. (hexane). Mp 85-86°.

8-O-De-Me: [61899-42-1]. *3-(1,1-Dimethyl-2-propenyl)-8-hydroxy-7-methoxy-2H-1-benzopyran-2-one. 3-(1,1-Dimethylallyl)-8-hydroxy-7-methoxycoumarin*
C$_{15}$H$_{16}$O$_4$ M 260.289
Isol. from *R. spp.* Mp 98-100°.

Reisch, J. *et al*, *Tetrahedron Lett.*, 1970, 4305 (*isol*)
Gonzalez, A.G. *et al*, *An. Quim.*, 1976, **72**, 191 (*isol, synth, deriv*)

4-(1,1-Dimethyl-2-propenyl)-2-(3-methyl-2-butenyl)phenol   D-80471

[73215-04-0]

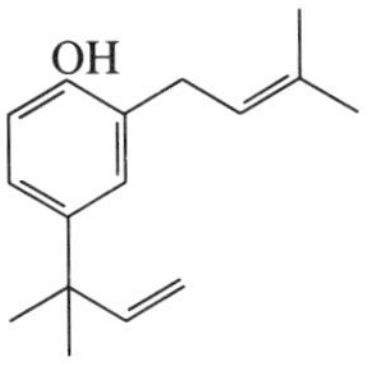

$C_{16}H_{22}O$ M 230.349

Constit. of *Perithalia caudata*. Oil.

Blackman, A.J. *et al, Aust. J. Chem.*, 1979, **32**, 2783.

6-(1,1-Dimethyl-2-propenyl)-4′,5,7-trihydroxyisoflavone   D-80472

6-(1,1-Dimethyl-2-propenyl)-5,7-dihydroxy-3-(4-hydroxyphenyl)-4H-1-benzopyran-4-one, 9CI

[77390-42-2]

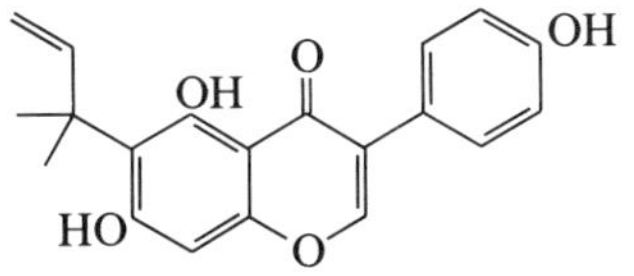

$C_{20}H_{18}O_5$ M 338.359

Incorr. reported as $C_{21}H_{18}O_5$. May be the 8-alkyl isomer. Isol. from wood of *Moghania macrophylla*. Prisms (CHCl$_3$/MeOH). Mp 123-124°.

Krishnamurty, H.G. *et al, Phytochemistry*, 1980, **19**, 2797 (*isol, ir, pmr, ms, struct*)

2-(3,5-Dimethyl-1H-pyrazol-1-yl)pyridine   D-80473

N-(*2-Pyridinyl*)-*3,5-dimethylpyrazole*

[21018-71-3]

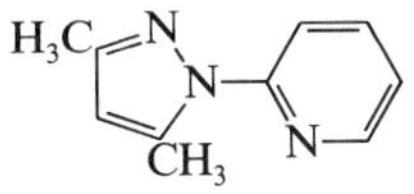

$C_{10}H_{11}N_3$ M 173.217

Bidentate ligand; extraction agent for Cd(II), Hg(II) and Pb(II). Light yellow oil. Bp$_{0.4}$ 74°.

B,HCl: Mp 140-144°.

Saha, N. *et al, J. Inorg. Nucl. Chem.*, 1977, **39**, 1236 (*synth*)
Baker, A.T. *et al, Aust. J. Chem.*, 1989, **42**, 623 (*bibl*)

3,4-Dimethyl-1H-pyrrole-2,5-dione, 9CI   D-80474

3,4-Dimethylmaleimide. 2,5-Dioxo-3,4-dimethyl-Δ^3-pyrroline

[17825-86-4]

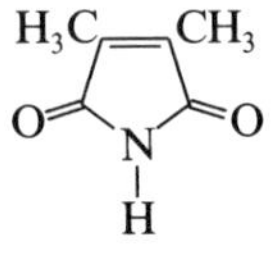

$C_6H_7NO_2$ M 125.127

Cryst. (EtOH). Mp 118-119°.

Dioxime: [14445-80-8].
    $C_6H_9N_3O_2$ M 155.156
    Cryst. (EtOH). Mp >300° dec.

Otto, R. *et al, Ber.*, 1885, **18**, 825 (*synth*)
Bischoff, C.A., *Ber.*, 1900, **33**, 1416 (*synth*)

Plieninger, H. *et al, Justus Liebigs Ann. Chem.*, 1968, **711**, 130 (*synth*)
Alonso Garrido, D.O. *et al, J. Org. Chem.*, 1988, **53**, 403 (*dioxime, synth, pmr*)

Dimethylsulfonium 9-fluorenylide   D-80475

[72393-25-0]

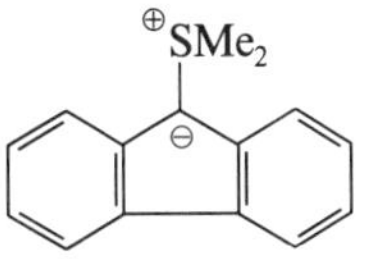

$C_{15}H_{14}S$ M 226.342

Pale yellow solid. Mp 126-127° (70-75°). Unstable in air.

B,HBr: Hexagonal plates (MeNO$_2$). Mp 133°.

Picrate: Needles (H$_2$O). Mp 149-150°.

Ingold, C.K. *et al, J. Chem. Soc.*, 1930, 713 (*synth, props*)
Johnson, A.W. *et al, J. Am. Chem. Soc.*, 1961, **83**, 417 (*synth*)
Zhang, J.-J. *et al, J. Org. Chem.*, 1988, **53**, 716 (*synth, pmr, cmr*)

3,5-Dimethyl-2,4,6-tribromobenzaldehyde   D-80476

[114634-34-3]

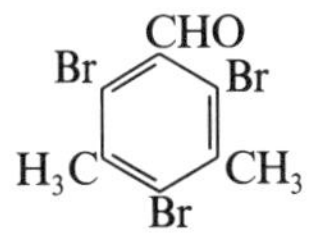

$C_9H_7Br_3O$ M 370.866

Cryst. Mp 224-225°.

Banfi, S. *et al, J. Org. Chem.*, 1988, **53**, 2863 (*synth, pmr*)

3,5-Dimethyl-2,4,6-trichlorobenzaldehyde   D-80477

[114634-35-4]

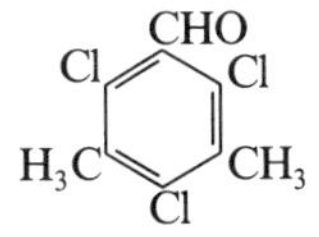

$C_9H_7Cl_3O$ M 237.512

Cryst. Mp 192-194°.

Banfi, S. *et al, J. Org. Chem.*, 1988, **53**, 2863 (*synth, pmr*)

3,5-Dimethyl-2,4,6-triphenylpyridine   D-80478

[78018-66-3]

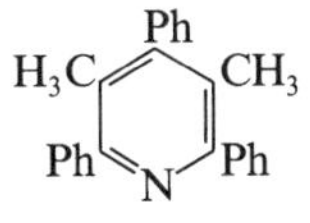

$C_{25}H_{21}N$ M 335.448

Cryst. solid (hexane/CHCl$_3$). Mp 154-156°.

Galiah, V. *et al, J. Indian Chem. Soc.*, 1955, **32**, 274 (*synth*)
Barluenga, J. *et al, J. Org. Chem.*, 1988, **53**, 5960 (*synth, ir, pmr, cmr*)

1,2-Di-2-naphthalenyl-1,2-ethanediol, 9CI   D-80479

2-Hydronaphthoin

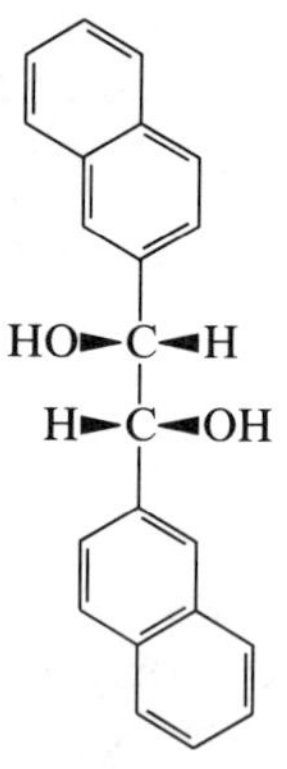

$(1R^*,2R^*)$-form

$C_{22}H_{18}O_2$ M 314.383

(1R*,2R*)-form [113469-20-8]
Cryst. + H_2O. Mp 237-238°. $[\alpha]_D^{23}$ +212° (c, 1.04 in THF). (−)-form also descr.

(1RS,2RS)-form [3408-28-4]
(±)-*form*
Cryst. + H_2O (C_6H_6). Mp 218-219°.

Di-Ac:
$C_{26}H_{22}O_4$ M 398.457
Cryst. (EtOAc). Mp 187-188°.

(1RS,2SR)-form
meso-*form*
Plates (xylene), microcryst. (dioxan). Mp 261-262°.

Di-Ac: Cryst. (Py or xylene). Mp 205-207°.

Badger, G.M., *Nature* (*London*), 1950, **165**, 647 (*synth*)
Mondodoev, G.T. *et al*, *J. Org. Chem. USSR* (*Engl. Transl.*), 1965, **1**, 1257 (*synth*)
Chênevert, R. *et al*, *Synthesis*, 1987, 739 (*synth, resoln, ir, pmr*)

Dinaphtho[1,2-*b*:1′,2′-*d*]thiophene   D-80480

[207-94-3]

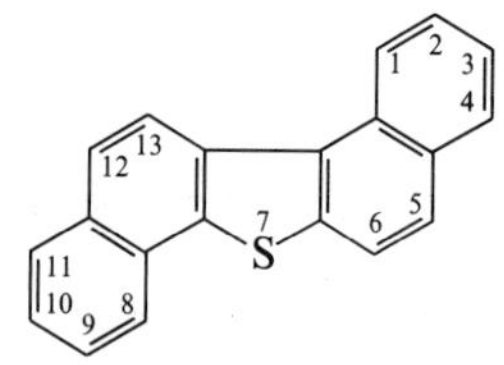

$C_{20}H_{12}S$ M 284.381
Leaflets. Mp 162°.

Tedjamulia, M.L. *et al*, *J. Heterocycl. Chem.*, 1983, **20**, 1143 (*synth, pmr*)

Dinaphtho[1,2-*b*:2′,1′-*d*]thiophene   D-80481

[239-72-5]

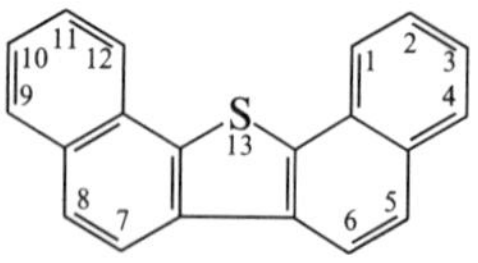

$C_{20}H_{12}S$ M 284.381
Cryst. ($CHCl_3$). Mp 255°.

Tedjamulia, M.L. *et al*, *J. Heterocycl. Chem.*, 1983, **20**, 1143 (*synth, pmr*)
Klemm, L.H. *et al*, *J. Heterocycl. Chem.*, 1988, **25**, 1111 (*synth, uv, pmr*)

Dinaphtho[1,2-*b*:2′,3′-*d*]thiophene   D-80482

[239-94-1]

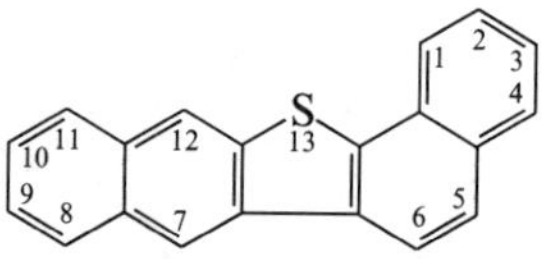

$C_{20}H_{12}S$ M 284.381
Beige solid ($CHCl_3$), pale yellow flakes. Mp 321°.

Tedjamulia, M.L. *et al*, *J. Heterocycl. Chem.*, 1983, **20**, 1143 (*synth, pmr*)
Klemm, L.H. *et al*, *J. Heterocycl. Chem.*, 1988, **25**, 1111 (*synth, uv, pmr*)
Johnston, M.D. *et al*, *J. Heterocycl. Chem.*, 1988, **25**, 1593 (*pmr*)

Dinaphtho[2,1-*b*:1′,2′-*d*]thiophene   D-80483

[194-65-0]

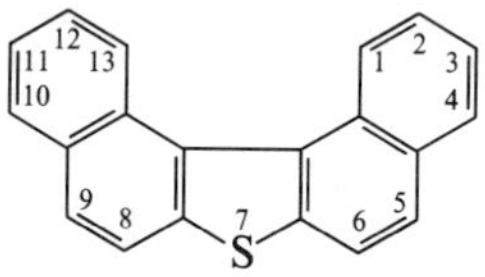

$C_{20}H_{12}S$ M 284.381
Cryst. Mp 206-207°.

Tedjamulia, M.L. *et al*, *J. Heterocycl. Chem.*, 1983, **20**, 1143 (*synth, pmr*)

Dinaphtho[2,1-*b*:2′,3′-*d*]thiophene   D-80484

[204-93-3]

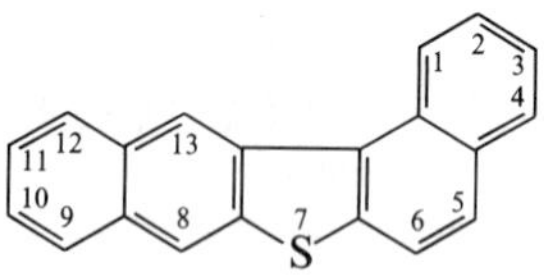

$C_{20}H_{12}S$ M 284.381
Pale yellow flakes. Mp 196° (189°).

Tedjamulia, M.L. *et al*, *J. Heterocycl. Chem.*, 1983, **20**, 1143 (*synth, pmr*)

Dinaphtho[2,3-*b*:2′,3′-*d*]thiophene   D-80485

[242-53-5]

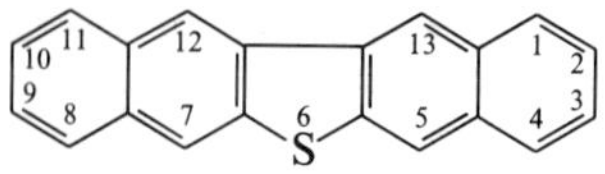

$C_{20}H_{12}S$ M 284.381
Fluorescent yellow leaves. Mp 265-266°.

Tedjamulia, M.L. *et al*, *J. Heterocycl. Chem.*, 1983, **20**, 1143 (*synth, pmr*)

2,2-Di-(1-naphthyl)ethanal   D-80486

α-1-Naphthalenyl-1-naphthaleneacetaldehyde, 9CI. 2,2-Di-(α-naphthyl)acetaldehyde

[98585-92-3]

CH–CHO

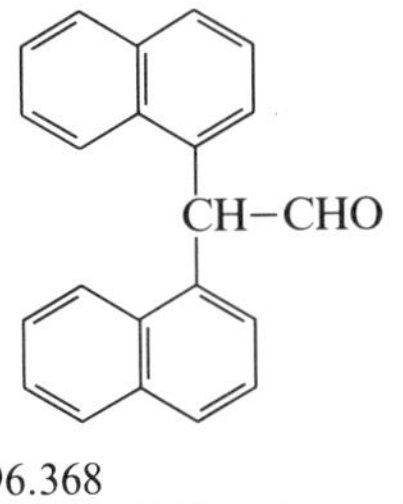

$C_{22}H_{16}O$ M 296.368
Cream-coloured cryst. (CHCl₃). Mp 188-189°.

Zimmerman, H.E. *et al, J. Am. Chem. Soc.*, 1985, **107**, 7732 (*synth, ir, pmr, cmr*)

2,2-Di-(2-naphthyl)ethanal   D-80487

α-2-Naphthalenyl-2-naphthaleneacetaldehyde. 2,2-Di-(β-naphthyl)acetaldehyde

[98586-00-6]

$C_{22}H_{16}$ M 280.368
Light yellow cryst. (CH₂Cl₂/hexane). Mp 163-164°.

Zimmerman, H.E. *et al, J. Am. Chem. Soc.*, 1985, **107**, 7732 (*synth, ir, pmr*)

2,2-Dinitroadamantane   D-80488

Updated Entry replacing D-50699
2,2-Dinitrotricyclo[3.3.1.1³,⁷]decane, 9CI

[88381-75-3]

$C_{10}H_{14}N_2O_4$ M 226.232
Cryst. (hexane). Mp 212-213°.

George, C. *et al, Acta Crystallogr., Sect. C*, 1983, **39**, 1674 (*cryst struct*)
Archibald, T.G. *et al, J. Org. Chem.*, 1988, **53**, 4645 (*synth, ir, pmr*)

2,4-Dinitroadamantane   D-80489

2,4-Dinitrotricyclo[3.3.1.1³,⁷]decane

$C_{10}H_{14}N_2O_4$ M 226.232
Cryst. (EtOH). Mp 175-180°.

Archibald, T.G. *et al, J. Org. Chem.*, 1988, **53**, 4645 (*synth, ir, pmr*)

2,6-Dinitroadamantane   D-80490

2,6-Dinitrotricyclo[3.3.1.1³,⁷]decane

$C_{10}H_{14}N_2O_4$ M 226.232
Cryst. (EtOH). Mp 165-170°.

Archibald, T.G. *et al, J. Org. Chem.*, 1988, **53**, 4645 (*synth, ir, pmr*)

4,6-Dinitro-1,3-benzenediol, 9CI   D-80491

Updated Entry replacing D-70450
4,6-Dinitroresorcinol, 8CI

[616-74-0]

$C_6H_4N_2O_6$ M 200.107
Pale-yellow cryst. Mp 215°. pK_a 3.98.

▷ Synthesis produces explosive byprods.

Di-Ac:
$C_{10}H_8N_2O_8$ M 284.182
Mp 139°.

Me ether: [51652-35-8]. *5-Methoxy-2,4-dinitrophenol*

$C_7H_6N_2O_6$ M 214.134
Mp 113°.

Et ether: 5-Ethoxy-2,4-dinitrophenol
$C_8H_8N_2O_6$ M 228.161
Mp 77°.

Di-Me ether: [1210-96-4]. *1,5-Dimethoxy-2,4-dinitrobenzene, 9CI, 8CI*
$C_8H_8N_2O_6$ M 228.161
Mp 157° (154°).

Di-Et ether: 1,5-Diethoxy-2,4-dinitrobenzene
$C_{10}H_{12}N_2O_6$ M 256.215
Mp 133°.

Pantlischenko, M. *et al, Monatsh. Chem.*, 1950, **81**, 293 (*synth*)
Rauner, W. *et al, Z. Chem.*, 1968, **8**, 338 (*synth*)
Granzhan, V.A. *et al, Zh. Strukt. Khim.*, 1971, **12**, 809; *CA*, **76**, 24579 (*ir*)
Crampton, M.R. *et al, J. Chem. Soc., Perkin Trans. 2*, 1972, 1178 (*deriv*)
Schmitt, R.J. *et al, J. Org. Chem.*, 1988, **53**, 5568 (*synth*)

1,5-Dinitrobiphenylene   D-80492

[17532-86-4]

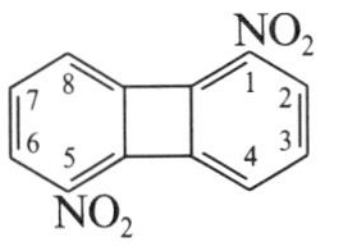

$C_{12}H_6N_2O_4$ M 242.190
Yellow needles (Me₂CO). Mp 272-274°.

Iqbal, K. *et al, J. Chem. Soc. C*, 1967, 1690 (*synth*)

2,6-Dinitrobiphenylene   D-80493

$C_{12}H_6N_2O_4$ M 242.190
Deep yellow needles (EtOH). Subl. > 260°.

Baker, W. *et al, J. Chem. Soc.*, 1958, 2658, 2666 (*synth, uv*)

1,2-Dinitrocyclopropane   D-80494

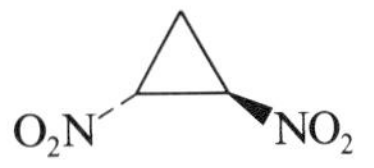

$C_3H_4N_2O_4$ M 132.076
(1RS,2RS)-form [109744-96-9]
(±)-trans-*form*
Cryst. (CCl₄). Mp 60-61°.

Wade, P.A. *et al, J. Am. Chem. Soc.*, 1987, **109**, 5452 (*synth, ir, pmr, cryst struct*)

1,12-Dinitrododecane   D-80495

[110843-41-9]

$$O_2NCH_2(CH_2)_{10}CH_2NO_2$$

$C_{12}H_{24}N_2O_4$ M 260.333
Cryst. (MeOH). Mp 59-60°.

Ainscow, T.A. *et al, Tetrahedron*, 1987, **43**, 115 (*synth, ir, pmr*)

24,25-Dinor-1,3,5(10)-lupatriene   D-80496

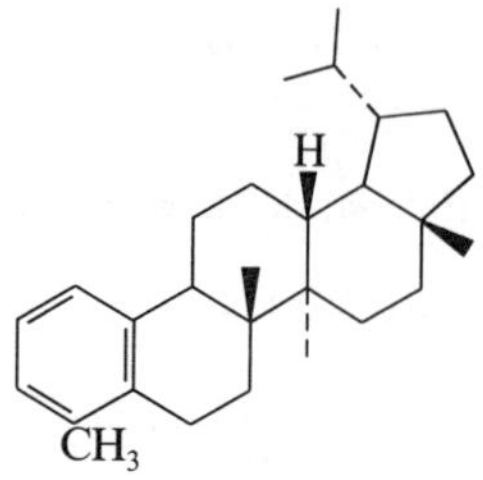

$C_{28}H_{42}$ M 378.640
Constit. of sediments.

Wolff, G.A. *et al, Tetrahedron*, 1989, **45**, 6721 (*synth*)

24,25-Dinor-1,3,5(10),12-oleanatetraene   D-80497

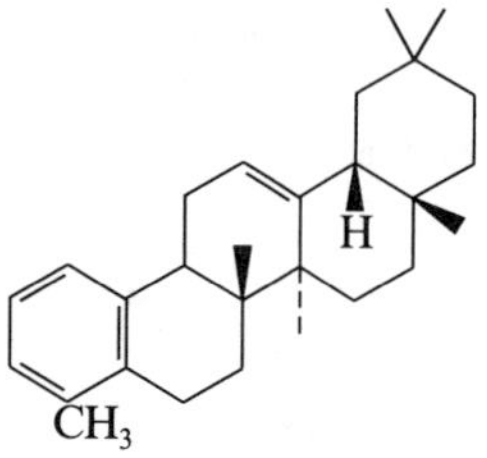

$C_{28}H_{40}$ M 376.624
Constit. of sediments.

Wolff, G.A. *et al, Tetrahedron*, 1989, **45**, 6721 (*synth*)

24,25-Dinor-1,3,5(10),12-ursatetraene   D-80498

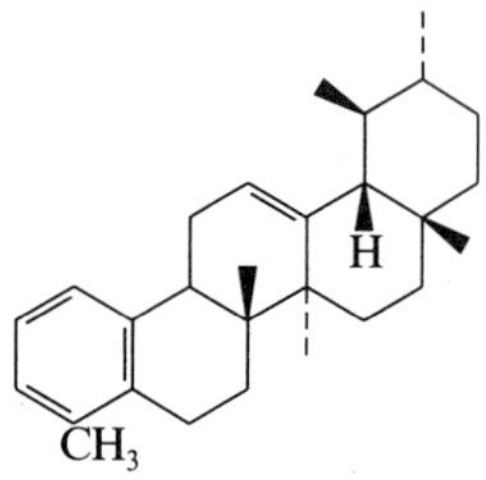

$C_{28}H_{40}$ M 376.624
Constit. of sediments.

Wolff, G.A. *et al, Tetrahedron*, 1989, **45**, 6721 (*synth*)

4,9-Dioxa-1,6,11-dodecatriyne   D-80499

1,4-Bis(2-propynyloxy)-2-butyne, 9CI
[83469-21-0]

$$HC{\equiv}CCH_2OCH_2C{\equiv}CCH_2OCH_2C{\equiv}CH$$

$C_{10}H_{10}O_2$ M 162.188
Oil. $Bp_{1.0}$ 78-81°.

Grigg, R. *et al, J. Chem. Soc., Perkin Trans. 1*, 1988, 1357 (*synth, pmr, ir, ms*)

1,4-Dioxane-2,6-dione, 9CI   D-80500

Diglycollic anhydride
[4480-83-5]

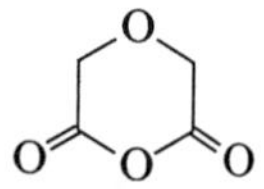

$C_4H_4O_4$ M 116.073
Cryst. (C_6H_6). Mp 91-93°. Bp_{12} 120°.

Anschutz, R., *Justus Liebigs Ann. Chem.*, 1890, **259**, 190 (*synth*)
Hurd, C.D. *et al, J. Am. Chem. Soc.*, 1939, **61**, 3490 (*synth*)
Brisse, F. *et al, Acta Crystallogr., Sect. B*, 1975, **31**, 2829 (*cryst struct*)

1,6-Dioxaspiro[4,5]decane   D-80501

[177-23-1]

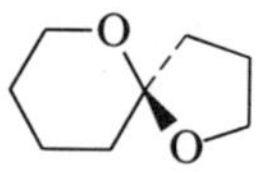

(*R*)-*form*

$C_8H_{14}O_2$ M 142.197
(*R*)-*form* [112996-24-4]
    Oil. Bp_{30} 110-120° (bath) . $[\alpha]_D^{25}$ −44.4° (c, 0.635 in pentane).
(*S*)-*form* [112996-25-5]
    Bp_{30} 110-120°. $[\alpha]_D$ +43.9° (c, 0.76 in pentane).
(±)-*form*
    Oil.

Ley, S.V. *et al, Tetrahedron*, 1985, **41**, 3825 (*synth*)
Iwata, C. *et al, Chem. Pharm. Bull.*, 1988, **36**, 4785 (*synth*)

1,3,2-Dioxathiane, 9CI   D-80502

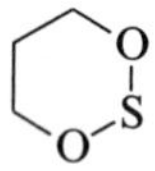

$C_3H_6O_2S$ M 106.145
2-Oxide: [4176-55-0]. *Trimethylene sulfite*
    $C_3H_6O_3S$ M 122.145
    d_4^{25} 1.33. Bp_{14} 66-67°.
2,2-Dioxide: [1073-05-8]. *Trimethylene sulfate*
    $C_3H_6O_4S$ M 138.144
    Mp 60°.

de la Mare, P.B.D. *et al, J. Chem. Soc.*, 1956, 1813 (*synth, ir*)
Kaiser, E.T. *et al, J. Am. Chem. Soc.*, 1963, **85**, 602 (*oxides, synth, ir*)
Van Woerden, H.F. *et al, Recl. Trav. Chim. Pays-Bas (J. R. Neth. Chem. Soc.)*, 1967, **86**, 353 (*oxide, synth*)
Lowe, G. *et al, J. Chem. Soc., Chem. Commun.*, 1983, 1392 (*synth, pmr*)
Lowe, G. *et al, J. Am. Chem. Soc.*, 1988, **110**, 8512 (*dioxide, cryst struct*)

2,4-Dioxatricyclo[3.3.1.1³,⁷]decane, 9CI   D-80503

2,4-Dioxaadamantane
[13288-29-4]

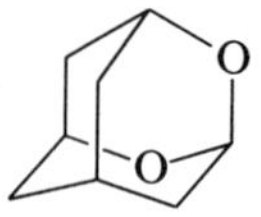

$C_8H_{12}O_2$ M 140.182
Cryst. (pentane or by subl.). Mp 210-212° (softens at 204°).

Briggs, A.J. *et al, Synthesis*, 1988, 66 (*synth, ms, ir, pmr*)

2,2-Diphenylaziridine D-80504

[25564-63-0]

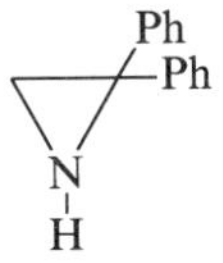

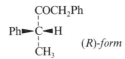

$C_{14}H_{13}N$ M 195.263
Oil. $Bp_{0.17}$ 111-115°.

Hahn, C.S. *et al*, *CA*, 1970, **72**, 78779s (*synth*)
Hassner, A. *et al*, *J. Am. Chem. Soc.*, 1970, **92**, 3733 (*synth, pmr*)
Perlman, M.E. *et al*, *J. Org. Chem.*, 1988, **53**, 1761 (*synth, ir, pmr*)

2,3-Diphenyl-2,3-butanediol, 9CI D-80505

Acetophenonepinacol

[1636-34-6]

$$HO–\underset{CH_3}{\overset{CH_3}{C}}–Ph$$
$$Ph–\underset{CH_3}{\overset{|}{C}}–OH$$

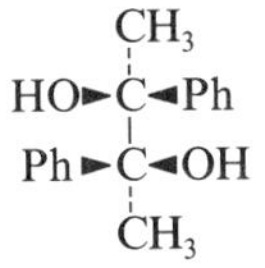

(2R,3R)-form

$C_{16}H_{18}O_2$ M 242.317
(2R,3R)-form [33603-65-5]
$[\alpha]_D^{22}$ +31.4° (c, 0.5 in Et_2O).
(2S,3S)-form [63902-56-7]
Cryst. (EtOH). Mp 104.5-105°. $[\alpha]_D^{25}$ −34.4° (c, 2.7 in EtOH).
Di-Me ether: 2,3-Dimethoxy-2,3-diphenylbutane
$C_{18}H_{22}O_2$ M 270.371
Fine needles. Mp 93-94°. $[\alpha]_D^{24}$ +71.6° (c, 3.1 in $CHCl_3$).
(2RS,3RS)-form [22985-90-6]
(±)-*form*
Mp 125-126° (122-124°).
Di-Me ether: Mp 111-111.5°.
(2RS,3SR)-form [4217-65-6]
meso-*form*
Cryst. (EtOH). Mp 120°.
Di-Me ether: Cubes. Mp 171-173° (164-166°).

Cram, D.J. *et al*, *J. Am. Chem. Soc.*, 1959, 2748 (*synth, config, ir, deriv*)
Stocker, J.H. *et al*, *J. Am. Chem. Soc.*, 1960, 3913 (*synth*)
Stocker, J.H. *et al*, *J. Org. Chem.*, 1968, **33**, 294 (*synth*)
Seebach, D. *et al*, *Chem. Ber.*, 1977, **110**, 2316 (*synth*)
Reichel, L.W. *et al*, *Can. J. Chem.*, 1984, **62**, 424 (*synth*)
Nakayama, J. *et al*, *Tetrahedron Lett.*, 1987, **28**, 1799 (*synth, pmr, cmr*)

2,3-Diphenylbutanoic acid D-80506

Updated Entry replacing D-07796
β-Methyl-α-phenylbenzenepropanoic acid, 9CI. β-Methyl-α-phenylhydrocinnamic acid

$$\underset{CH_3}{\overset{COOH}{\underset{|}{H–\overset{2}{C}–Ph}}} \\ H–\overset{3}{C}–Ph$$

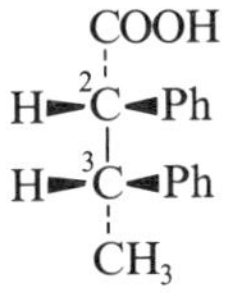

(2S,3R)-form
Absolute
configuration

$C_{16}H_{16}O_2$ M 240.301
(2S,3R)-form [52195-02-5]
(+)-erythro-*form*
Cryst. (Et_2O/pet. ether). Mp 185-186°. $[\alpha]_D^{26}$ +25.5° (c, 4 in EtOH).

(2RS,3SR)-form [33398-55-9]
(±)-erythro-*form*
Needles (EtOH). Mp 187-189°.
Amide: [16213-71-1].
$C_{16}H_{17}NO$ M 239.316
Cryst. Mp 193°.
Nitrile:
$C_{16}H_{15}N$ M 221.301
Bp_{13} 188-192°.
(2RS,3RS)-form [5350-86-7]
(±)-threo-*form*
Cryst. Mp 125-130° (133-134°).
Amide: [16213-72-2].
Cryst. Mp 173-174°.
Nitrile: [15645-55-3].
Needles. Mp 35-36°. Bp_{16} 210-212°.

[5558-38-3, 52195-01-4, 56648-88-5]

Plentl, A.A. *et al*, *J. Am. Chem. Soc.*, 1941, **63**, 989 (*synth, derivs*)
Chambers, W.J. *et al*, *J. Am. Chem. Soc.*, 1957, **79**, 875 (*abs config*)
Auerbach, R.A. *et al*, *Tetrahedron*, 1971, **27**, 2069 (*synth, pmr*)
Ochiai, Y. *et al*, *Bull. Chem. Soc. Jpn.*, 1976, **49**, 2525 (*synth*)

1,3-Diphenyl-2-butanone D-80507

Updated Entry replacing D-10653

[13363-25-2]

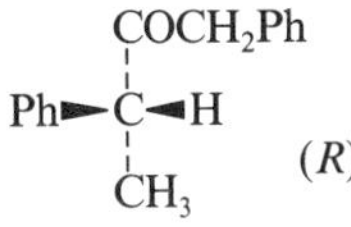

$$Ph–\underset{CH_3}{\overset{COCH_2Ph}{\underset{|}{C}–H}} \quad (R)\text{-}form$$

$C_{16}H_{16}O$ M 224.302
(R)-form [67504-36-3]
$[\alpha]_D$ −12.0° (c, 4.3 in C_6H_6, 41% ee).
(S)-form [59983-38-9]
$[\alpha]_D^{20}$ +88° (c, 1.0 in C_6H_6).
(±)-*form*
Bp_{40} 205-206°, Bp_2 138-139°.
Oxime:
$C_{16}H_{17}NO$ M 239.316
Mp 82-83°.
Semicarbazone: Mp 143-145°.

Levy, J. *et al*, *Bull. Soc. Chim. Fr.*, 1929, **45**, 94 (*synth*)
Perez-Ossorio, R. *et al*, *CA*, 1966, **65**, 19985 (*synth*)
Posner, G. *et al*, *J. Am. Chem. Soc.*, 1973, **95**, 3076 (*synth*)
Thompson, H.W. *et al*, *J. Chem. Soc., Chem. Commun.*, 1973, 636 (*synth*)
Enders, D. *et al*, *Angew. Chem., Int. Ed. Engl.*, 1976, **15**, 549 (*synth*)
Meyers, A.I. *et al*, *J. Org. Chem.*, 1978, **43**, 3245 (*abs config*)

2,4-Diphenyl-3-butyn-1-ol D-80508

[86100-15-4]

$$PhC{\equiv}CCHPhCH_2OH$$

$C_{16}H_{14}O$ M 222.286
(±)-*form*
Oil, cryst. on standing. Mp 51°.

Krause, N. *et al*, *Chem. Ber.*, 1988, **121**, 1315 (*synth, ir, pmr, ms*)

1,2-Diphenyl-1,2-cyclopropanedicarboxylic acid, 9CI   D-80509

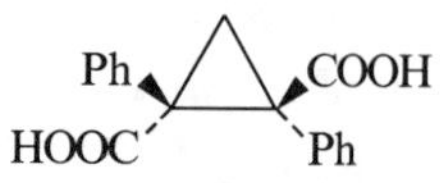

(1*RS*,2*RS*)-form

$C_{17}H_{14}O_4$ M 282.295

(**1RS,2RS**)-form [896-59-3]
($\pm$)-trans-*form*
Cryst. (EtOH aq.). Mp 273°.
Di-Me ester: [111977-16-3].
$C_{19}H_{18}O_4$ M 310.349
Cryst. (EtOH). Mp 141-143°.
(**1RS,2SR**)-form [736-88-9]
cis-*form*
Cryst. (H₂O). Mp 175°.
Di-Me ester: [17160-34-8].
Cryst. (EtOH). Mp 138°.
Di-Et ester:
$C_{21}H_{22}O_4$ M 338.402
Cryst. (EtOH). Mp 76°.
Anhydride:
$C_{17}H_{12}O_3$ M 264.280
Cryst. Mp 186°.

Bonavent, G. *et al, Bull. Soc. Chim. Fr.,* 1964, 2462 (*synth, anhydride*)
Broser, W. *et al, Chem. Ber.,* 1967, **100**, 3472 (*synth*)
Hill, R.K. *et al, J. Am. Chem. Soc.,* 1988, **110**, 497 (*synth, pmr*)

2,5-Diphenylhexanedioic acid   D-80510

2,5-Diphenyladipic acid
[6622-43-1]

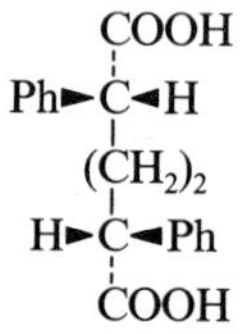

(2*RS*,5*RS*)-form

$C_{18}H_{18}O_4$ M 298.338
(**2RS,5RS**)-form
Mp 208°.
Dinitrile:
$C_{18}H_{16}N_2$ M 260.338
Cryst. (Et₂O). Mp 103-105°. Isomerises to *meso*-form on boiling in EtOH.
(**2RS,5SR**)-form [4070-63-7]
meso-*form*
Cryst. (AcOH). Mp 255-260° (266-267°).
Dinitrile: Cryst. (EtOH). Mp 173-175°.

Frank, C.E. *et al, Tetrahedron,* 1961, **26**, 303, 307 (*synth*)
U.S. Pat., 3 351 646, (1967); *CA,* **68**, 49723f (*synth*)
Casini, G. *et al, Tetrahedron,* 1972, **28**, 1497 (*synth, deriv*)

3,4-Diphenylhexanedioic acid, 9CI   D-80511

3,4-Diphenyladipic acid

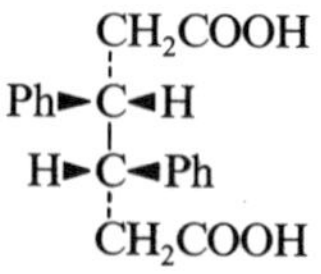

(3*R*,4*R*)-form

$C_{18}H_{18}O_4$ M 298.338
(**3R,4R**)-form [108391-57-7]

Cryst. (EtOH/AcOH). Mp 200-201°. $[\alpha]_D^{22}$ +13.1° (EtOH).
Dinitrile:
$C_{18}H_{16}N_2$ M 260.338
Mp 125-126°. $[\alpha]_D^{22}$ +20.3° (CHCl₃).
(**3S,4S**)-form [108391-56-6]
Mp 198-199°. $[\alpha]_D^{22}$ −11.0° (EtOH).
(**3RS,4RS**)-form [65610-00-6]
Mp 187°.
(**3RS,4SR**)-form [65610-01-7]
Needles (EtOH). Mp 270-271° (264-266°).
Dichloride:
$C_{18}H_{16}Cl_2O_2$ M 335.229
Cryst. (C₆H₆). Mp 187°.

[25347-44-8]

Oomen, M.P. *et al, J. Chem. Soc.,* 1930, 2148 (*synth*)
Shitov, G.G. *et al, Zh. Org. Khim.,* 1977, **13**, 2146; *CA,* **88**, 89229u (*synth*)
Toki, S. *et al, Nippon Kagaku Kaishi (J. Chem. Soc. Jpn.),* 1984, **1**, 152; *CA,* **100**, 191505 (*synth*)
Naemura, K. *et al, J. Chem. Soc., Chem. Commun.,* 1986, 1675 (*synth, abs config, deriv synth*)

Diphenylketazine   D-80512

Diphenylmethanone (diphenylmethylene)hydrazone, 9CI.
Benzophenone azine. Bisdiphenylazimethylene
[983-79-9]

$$Ph_2C=N-N=CPh_2$$

$C_{26}H_{20}N_2$ M 360.457
Bright yellow cryst. solid (EtOH). Mp 162°.

Curtius, T. *et al, J. Prakt. Chem.,* 1891, **44**, 192 (*synth*)
Morton, A.A. *et al, J. Am. Chem. Soc.,* 1931, **53**, 2769 (*synth*)

5,5-Diphenyl-4-pentenoic acid, 9CI, 8CI   D-80513

Updated Entry replacing D-08032
[5747-00-2]

$$Ph_2C=CHCH_2CH_2COOH$$

$C_{17}H_{16}O_2$ M 252.312
Cryst. Mp 82-83° (77-79°).
Me ester:
$C_{18}H_{18}O_2$ M 266.339
Mp 120°.

Fecht, H., *Ber.,* 1908, **41**, 2983 (*synth*)
Osman, A.M., *J. Prakt. Chem.,* 1969, **311**, 266 (*synth, pmr*)
Lehmann, J. *et al, Justus Liebigs Ann. Chem.,* 1988, 827 (*synth, pmr*)

4,6-Diphenylthieno[2,3-c]furan   D-80514

[105480-49-7]

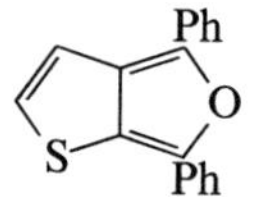

$C_{18}H_{12}O_2S$ M 292.358
Fine yellow needles. Mp 139°.

Schöning, A. *et al, Chem. Ber.,* 1989, **122**, 1119 (*synth, uv, cmr, ms, cryst struct*)

Di-2-propenyl tetrasulfide, 9CI   D-80515

Allyl tetrasulfide, 8CI. Diallyl tetrasulfide
[2444-49-7]

$$H_2C=CHCH_2-S-S-S-S-CH_2CH=CH_2$$

$C_6H_{10}S_4$ M 210.409
Isol. from garlic oil (*Allium sativum*) and from flowers and leaves of *Adenocalymna alliacea*. Oil with penetrating odour. Bp_{16} 122°.

Milligan, B. *et al, J. Chem. Soc.*, 1963, 3608 (*synth*)
Martin, D.J. *et al, Anal. Chem.*, 1966, **38**, 1604 (*pmr*)
Apparao, M. *et al, Phytochemistry*, 1978, **17**, 1660 (*isol*)
Block, E. *et al, J. Am. Chem. Soc.*, 1988, **110**, 7813 (*ms, cmr*)

Dipyrido[1,2-*a*: 1′,2′-*c*]imidazolium-11-thiolate   D-80516

*11-Mercaptodipyrido[1,2-*a*: 1′,2′-c]imidazol-5-ium hydroxide, inner salt, 9CI*
[98779-37-4]

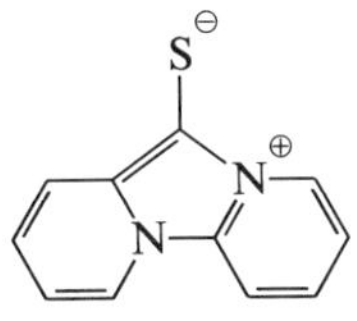

$C_{11}H_8N_2S$ M 200.264
Lustrous, dark maroon needles (EtOH). Mp 265-267° (under N_2), Mp 260-263° dec. (in air). Unstable to light.
B, HPF₆: Orange-yellow cryst. Mp 221-225° (under N_2).
B, HClO₄: [98779-74-9].
  Cryst. Mp 237-239° (under N_2).
B,HBF₄: [98779-75-0].
  Mp 216-217.5° (under N_2).
B,HI: [98779-76-1].
  Mp 215.5-218° (under N_2).

Edward, J.T. *et al, J. Org. Chem.*, 1985, **50**, 4855 (*synth, ms*)

Disain   D-80517

3,3′-Dibenzo[b,e][1,4]dioxin-2,8-diylbis[5-hydroxy-8,8-dimethyl-6-(3-methyl-2-butenyl)-4H,8H-benzo[1,2-b:3,4-b′]dipyran-4-one], 9CI
[53755-80-9]

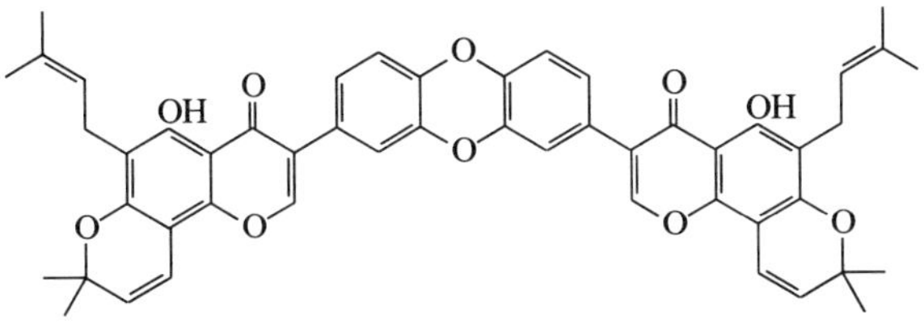

$C_{50}H_{44}O_{10}$ M 804.892
Isol. from *Maclura aurantiaca*. Oxidn. prod. of Osajin.

Nikonov, G.K. *et al, CA*, 1975, **82**, 13975q (*struct*)

2,18-Dithia[3.1.3.1]paracyclophane   D-80518

8,20-Dithiapentacyclo[20.2.2.2³,⁶.2¹⁰,¹³.2¹⁵,¹⁸]dotriaconta-3,5,10,12,15,17,22,24,25,27,29,31-dodecaene, 9CI
[114067-21-9]

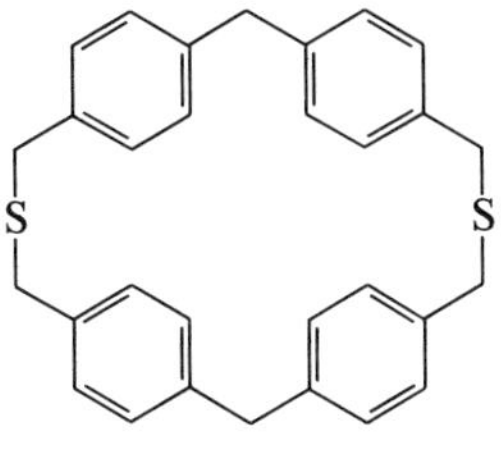

$C_{30}H_{28}S_2$ M 452.683

Needles (CH_2Cl_2).

Grützmacher, H.-F. *et al, Tetrahedron*, 1987, **43**, 3205 (*synth, pmr, ir, ms*)

1,2-Dithiocane   D-80519

Updated Entry replacing D-50815
Hexamethylene disulfide. 1,2-Dithiacyclooctane
[6008-69-1]

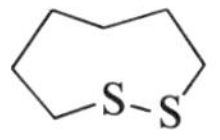

$C_6H_{12}S_2$ M 148.293
Bp_2 65.5°. Readily polym.

Schoberl, A. *et al, Justus Liebigs Ann. Chem.*, 1958, **614**, 66 (*synth*)
Cragg, R.H. *et al, Tetrahedron Lett.*, 1973, 655 (*synth*)
Harpp, D.N. *et al, Tetrahedron Lett.*, 1986, **27**, 441 (*synth*)

2,3-Dodecadien-1-ol   D-80520

[34656-66-1]

$H_3C(CH_2)_7CH=C=CHCH_2OH$

$C_{12}H_{22}O$ M 182.305

Landor, P.D. *et al*, 1971, 1638 (*synth, ir*)
Djahanbini, D. *et al, Tetrahedron*, 1987, **43**, 3441 (*synth, ir, pmr*)

Dodecahedranecarboxaldehyde   D-80521

Tetradecahydro-5,2,1,6,3,4-[2,3]butanediyl[1,4]diylidenedipentaleno[2,1,6-cde:2′,1′,6′-gha]pentalene-1(2H)-carboxaldehyde, 9CI
Formyldodecahedrane
[112896-33-0]

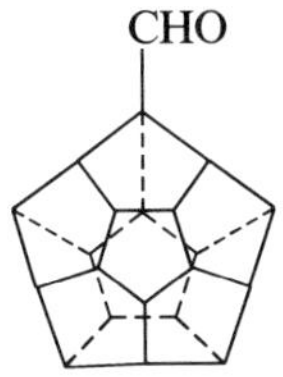

$C_{21}H_{20}O$ M 288.388
Solid. Mp >250°.

Paquette, L.A. *et al, J. Am. Chem. Soc.*, 1988, **110**, 1303 (*synth, cmr*)

Dodecahedranecarboxylic acid   D-80522

[112896-30-7]
$C_{21}H_{20}O_2$ M 304.388
Cryst. Mp >250°.
Me ester: [112896-31-8].
  $C_{22}H_{22}O_2$ M 318.415
  Solid. Mp 192-193°.
Amide: [112896-34-1].
  $C_{21}H_{21}NO$ M 303.403
  Solid. Mp >250°.
Nitrile: [116635-01-9]. *Cyanododecahedrane*
  $C_{21}H_{19}N$ M 285.388
  Solid. Mp >250°.

Paquette, L.A. *et al, J. Am. Chem. Soc.*, 1988, **110**, 1303 (*synth, cmr, ms*)
Weber, J.C. *et al, J. Org. Chem.*, 1988, **53**, 5315 (*deriv, synth*)
Gallucci, J.C. *et al, Acta Crystallogr., Sect. C*, 1989, **45**, 893 (*cryst struct, ester*)

Dodecahedranol D-80523

Tetradecahydro-5,2,1,6,3,4-
[2,3]butanediyl[1,4]diylidenepentaleno[2,1,6-cde:2′,1′,6′-
gha]pentalen-1(2H)-ol, 9CI. Hydroxydodecahedrane
[112896-36-3]

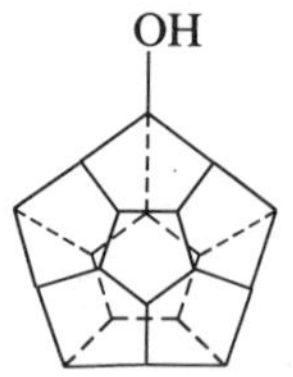

C₂₀H₂₀O   M 276.377

Solid. Mp >250°.

Trifluoroacetyl: [112925-47-0].
$C_{22}H_{19}F_3O_2$ M 372.386
Solid. Mp 183-185°.

O-Nitrate: [112896-38-5].
$C_{20}H_{19}NO_3$ M 321.375
Solid. Mp 220°.

Paquette, L.A. *et al, J. Am. Chem. Soc.*, 1988, **110**, 1303 (*synth, cmr*)

2,3,4,5,6,7,8,10,11,12,13,?-Dodecahydro-1,4:5,8:10,13-triethano-1*H*-tribenzo[*a,c,e*]cycloheptenylium(1+), 9CI D-80524

1,2:3,4:5,6-Tris(bicyclo[2.2.2]octeno)tropylium

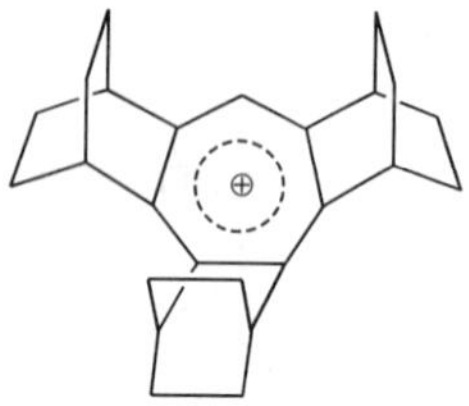

$C_{25}H_{31}^{\oplus}$ M 331.520

Thermodynamically very stable carbocation.

Hexafluoroantimonate: [112440-02-5].
$C_{25}H_{31}Cl_6Sb$ M 665.986
Mp 290-292° dec.

Komatsu, K. *et al, J. Am. Chem. Soc.*, 1988, **110**, 633 (*synth, uv, pmr, cmr*)

1,2,3,4,5,6,7,8,9,10,11,12-Dodecahydro-1,4:5,8:9,12-triethanotriphenylene, 9CI D-80525

[112439-96-0]

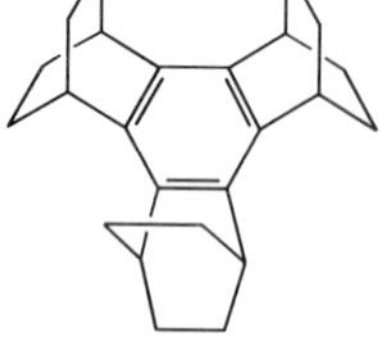

C₂₄H₃₀   M 318.501

Cryst. Mp 277-279°.

Komatsu, K. *et al, J. Am. Chem. Soc.*, 1988, **110**, 633 (*synth, uv, pmr, cmr*)

1-*O*-(4,6,8-Dodecatrienyl)glycerol D-80526

3-(4,6,8-Dodecatrienyloxy)-1,2-propanediol

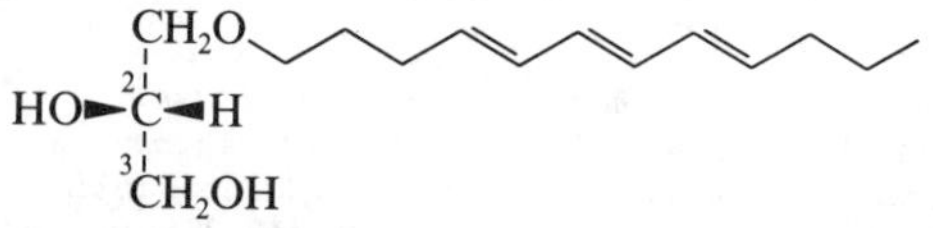

$C_{15}H_{26}O_3$ M 254.369

Closely related to the fecapentaenes.

(*S, all-E*)-form

2-O-(4-Hydroxybenzoyl): **Bretonin A**
$C_{22}H_{30}O_5$ M 374.476
Isol. from an unidentified marine demosponge. Oil. Opt. rotn. too small to measure.

2-O-(4-Hydroxybenzoyl), di-Ac: $[\alpha]_D^{25}$ −7.0° (c, 0.13 in CHCl₃).

3-O-(4-Hydroxybenzoyl): **Isobretonin A**
$C_{22}H_{30}O_5$ M 374.476
Isol. from an unidentified marine demosponge. Oil. $[\alpha]_D^{25}$ +7.3° (CHCl₃). Has (*R*-) config.

Guella, G. *et al, Helv. Chim. Acta*, 1989, **72**, 1121 (*isol, uv, pmr, cmr, struct*)

2-Dodecyl-3-methylbutanedioic acid, 9CI D-80527

Updated Entry replacing C-00305
2-Dodecyl-3-methylsuccinic acid

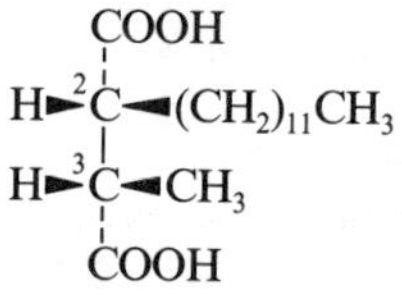

$C_{17}H_{32}O_4$ M 300.437

(*2R,3S*)-form [22139-54-4] *Roccellic acid*

Occurs in *Lecanora* spp., *Roccella tinctoria, Lepraria latebrarum, Lecidea aglaeotara, Pannaria languinosa* and other lichens. Cryst. (pet. ether). Mp 131-132°. $[\alpha]_D$ +17.4° (EtOH).

Di-Me ester: Cryst. (Et₂O). Mp 28-29°.

Phenylimide: Cryst. (MeOH). Mp 57-58°.

Åkermark, B., *Acta Chem. Scand.*, 1962, **16**, 599 (*abs config*)
Huneck, S. *et al, Phytochemistry*, 1969, **8**, 1301 (*isol*)
Huneck, S. *et al, Z. Naturforsch., B*, 1969, **24**, 750 (*isol*)
Åkermark, B., *Acta Chem. Scand.*, 1970, **24**, 1456 (*isol*)
Ollis, W.D. *et al, J. Chem. Soc. C*, 1971, 1318 (*isol*)

Dolineone D-80528

Updated Entry replacing D-08628
Dolichone
[10065-28-8]

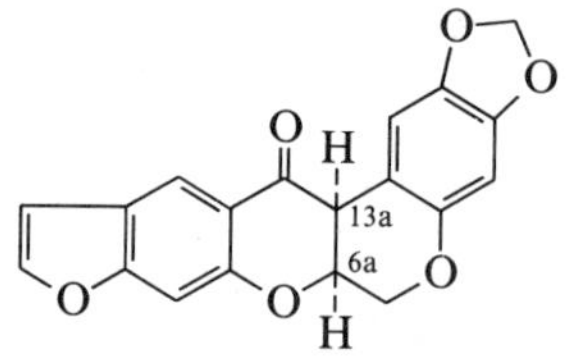

$C_{19}H_{12}O_6$ M 336.300

CA numbering shown. Isol. from roots of *Neorautanenia pseudopachyrrhiza*, also from *Pachyrrhizus erosus*. Needles (MeOH). Mp 233-235°. $[\alpha]_D^{20}$ +135° (c, 1.3 in CHCl₃).

6a,13a-Didehydro: [28570-72-1]. **Dehydrodolineone**
$C_{19}H_{10}O_6$ M 334.284

Isol. from *Neorautanenia amboensis* roots. Yellow needles. Mp 290-292°. Poss. artifact.

Crombie, L. *et al*, *J. Chem. Soc.*, 1963, 1569 (*isol, struct*)
Oberholzer, M.E. *et al*, *Tetrahedron Lett.*, 1974, 2211 (*Dehydrodolineone*)

7,8,9,10,24,25,27,28,30,31,33,34,36,37,39, 40,54,55,56,57,71,72,74,75,77,78,80,1, 83,84,86,87-Dotriacontahydro-1,93:4, 6:11,13:16,18:19,22:42,45:46,48:51, 53:58,60:63,65:66,69:89,92-dodecatheno-tetrabenzo[z,j₁,q₂,a₃][1,4,7,10,13,16, 19,44,47,50,53,56,59,62,25,28,35,38, 68,71,78,81]tetradecaoxaotaaza-cyclohexaoctacontine   D-80529

[118798-04-2]

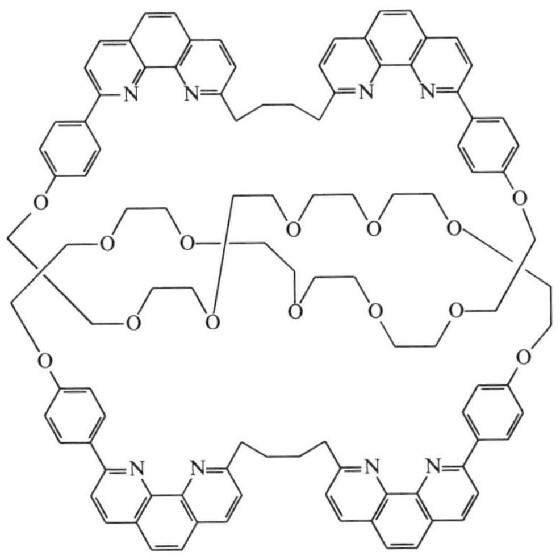

$C_{104}H_{104}N_8O_{14}$ M 1690.011
First molecule with trefoil knot topology. Glass.

Dietrich-Buchecker, C.O. *et al*, *Angew. Chem., Int. Ed. Engl.*, 1989, **28**, 189.

Drummondin A   D-80530

[119171-76-5]

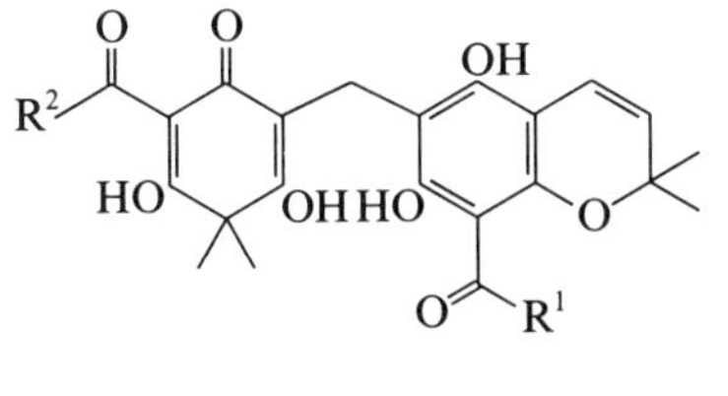

$R^1 = R^2 = CH_2CH_3$

$C_{26}H_{30}O_8$ M 470.518
Constit. of *Hypericum drummondii*. Antimicrobial and cytotoxic agent. Yellow cryst. (hexane). Mp 130-132°.

Jayasuriya, H. *et al*, *J. Nat. Prod.* (*Lloydia*), 1989, **52**, 325 (*isol, pmr, cmr*)

Drummondin B   D-80531

[119171-77-6]

As Drummondin A, D-80530 with

$R^1 = CH_3, R^2 = CH_2CH_3$

$C_{25}H_{28}O_8$ M 456.491

Constit. of *Hypericum drummondii*. Antimicrobial and cytotoxic agent. Yellow cryst. (hexane). Mp 136-138°.

Jayasuriya, H. *et al*, *J. Nat. Prod.* (*Lloydia*), 1989, **52**, 325 (*isol, pmr, cmr*)

Drummondin C   D-80532

[119171-78-7]

As Drummondin A, D-80530 with

$R^1 = R^2 = CH_3$

$C_{24}H_{26}O_8$ M 442.465
Constit. of *Hypericum drummondii*. Antimicrobial and cytotoxic agent. Yellow cryst. (EtOAc). Mp 184-187°.

Jayasuriya, H. *et al*, *J. Nat. Prod.* (*Lloydia*), 1989, **52**, 325 (*isol, pmr, cmr*)

Drummondin F   D-80533

[122127-73-5]

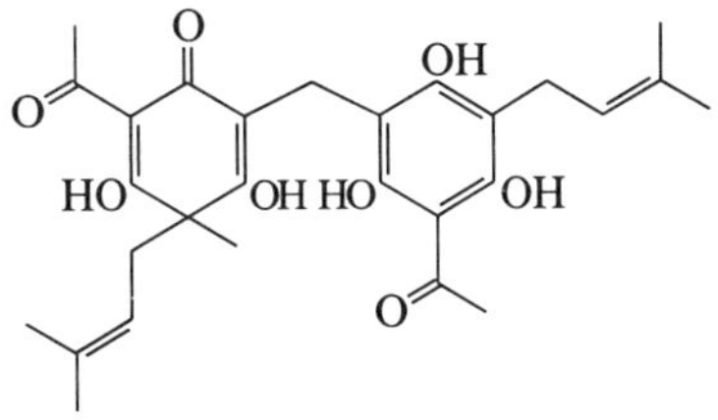

$C_{28}H_{34}O_8$ M 498.572
Constit. of *Hypericum drummondii*. Antimicrobial and cytotoxic agent. Yellow cryst. (hexane). Mp 106-107°. $[\alpha]_D$ +56.4° (c, 2.5 in MeOH).

Jayasuriya, H. *et al*, *J. Nat. Prod.* (*Lloydia*), 1989, **52**, 325 (*isol, pmr, cmr*)

Drupanin   D-80534

3-[4-Hydroxy-3-(3-methyl-2-butenyl)phenyl]-2-propenoic acid, 9CI. 4-Hydroxy-2-prenylcinnamic acid
[53755-58-1]

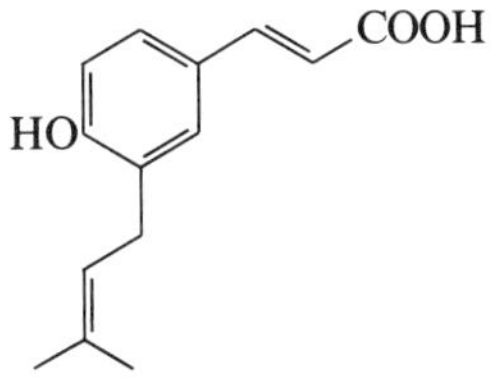

$C_{14}H_{16}O_3$ M 232.279
Constit. of the fruits of *Psoralea drupaecea*. Acicular cryst. Mp 147-148°.

Golovina, L.A. *et al*, *Khim. Prir. Soedin.*, 1973, **9**, 700; *cnc*, 672 (*isol, uv, pmr, struct*)

Dryocrassin   D-80535

[12777-70-7]

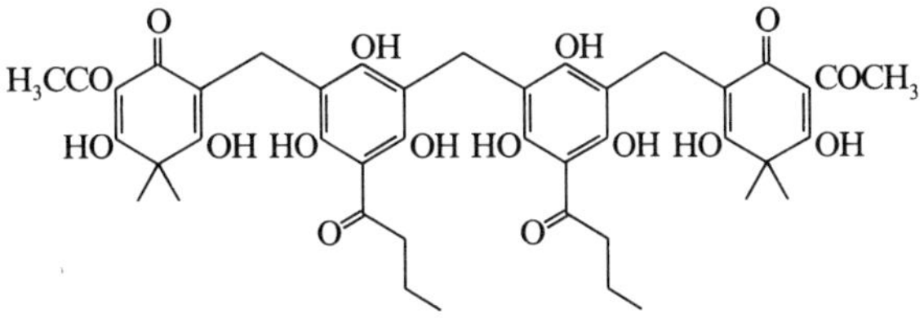

$C_{43}H_{48}O_{16}$ M 820.843

Isol. from *Dryopteris crassirhizoma* and *D. polylepis*. Yellow cryst. (Me$_2$CO). Mp 209-214°.

Noro, Y. *et al*, *Phytochemistry*, 1973, **12**, 1491 (*isol, uv, pmr, struct, synth*)

Dumortierigenin D-80536

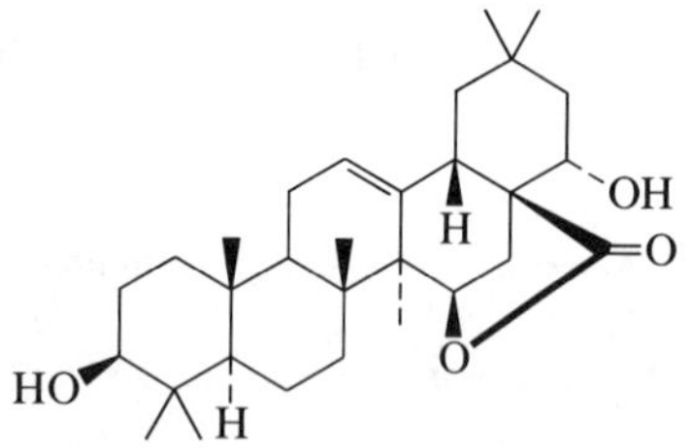

C$_{30}$H$_{46}$O$_4$ M 470.691

Isol. from the cactus *Lemaireocereus dumortieri*. Cryst. (EtOH). Mp 292-295°. [α]$_D^{30}$ −18.6° (CHCl$_3$).

Djerassi, C. *et al*, *J. Am. Chem. Soc.*, 1954, **76**, 2969; 1956, **78**, 5685 (*isol, struct*)
Shamma, M. *et al*, *J. Org. Chem.*, 1962, **27**, 4512 (*pmr*)

Durmillone D-80537

Updated Entry replacing D-08685

[24211-35-6]

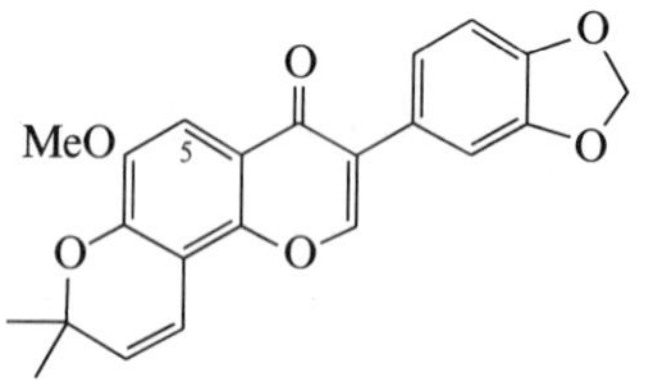

C$_{22}$H$_{18}$O$_6$ M 378.381

Obt. from seeds of *Millettia dura*. Also from *M. ferruginea* and *m. rubiginosa*. Fine needles (MeOH). Mp 182-185°.

5-Methoxy: 5-Methoxydurmillone
C$_{23}$H$_{20}$O$_7$ M 408.407
Constit. of *Millettia ferruginea*. Needles (MeOH). Mp 142-143°.

Ollis, W.D. *et al*, *Tetrahedron*, 1967, **23**, 4741 (*isol*)
Highet, R.J. *et al*, *J. Org. Chem.*, 1967, **32**, 1055 (*isol, struct*)
Krishnamurthi, M. *et al*, *Indian J. Chem.*, 1972, **10**, 914 (*synth*)
Desai, H.K. *et al*, *Indian J. Chem., Sect. B*, 1977, **15**, 291 (*isol*)
Dagne, E. *et al*, *Phytochemistry*, 1989, **28**, 1897 (*derivs*)

Dysidazirine D-80538

[113507-74-7]

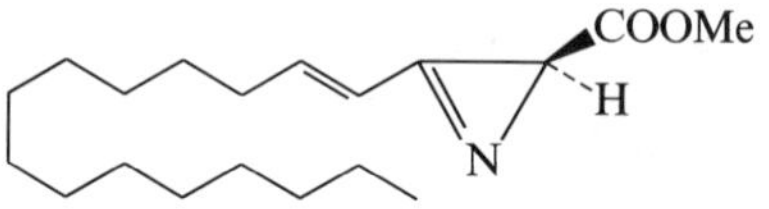

C$_{19}$H$_{33}$NO$_2$ M 307.475

Isol. from Fijian marine sponge, *Dysidea fragilis*. Low melting solid. [α]$_D$ −165° (c, 0.5 in MeOH).

▷ Cytotoxic.

Molinski, T.F. *et al*, *J. Org. Chem.*, 1988, **53**, 2103 (*isol, uv, pmr, ms, abs config*)

Dysoxysulfone D-80539

$$MeSO_2CH_2\text{—}S\text{—}S\text{—}CH_2\text{—}S\text{—}CH_2SO_2Me$$

C$_5$H$_{12}$O$_4$S$_5$ M 296.477

Constit. of leaves of *Dysoxylum richii*. Shows antibiotic props. Prisms (CHCl$_3$). Mp 97-99°.

Jogia, M.K. *et al*, *Tetrahedron Lett.*, 1989, **30**, 4919 (*cryst struct*)

E

Eburicoic acid　　　　　　　　　　　　E-80001

Updated Entry replacing E-00002

3β-Hydroxy-24-methylene-8-lanosten-21-oic acid. Eburicolic acid

[560-66-7]

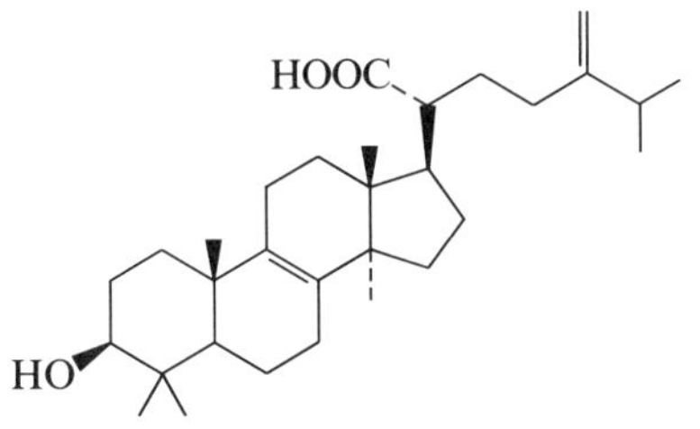

$C_{31}H_{50}O_3$　　M 470.734

Constit. of *Polyporus* spp., *Lenzites* spp., *Fomes officinalis* and other fungi. Cryst. (CHCl$_3$/EtOH). Mp 292-293°. $[\alpha]_D^{17}$ +34° (c, 1.2 in Py).

Ac:

$C_{33}H_{52}O_4$　　M 512.771

Constit. of *P. anthracophilus*. Cryst. Mp 259°. $[\alpha]_D^{19}$ +42° (c, 0.42 in CHCl$_3$).

Me ester: [1259-85-4].

$C_{32}H_{52}O_3$　　M 484.761

Constit. of *Basidiomycetes* spp. Cryst. (EtOAc). Mp 146-147°. $[\alpha]_D^{17}$ +45° (c, 1.8 in CHCl$_3$).

Holker, J.S.E. *et al, J. Chem. Soc.,* 1953, 2422 (*struct*)
Cohen, A.I. *et al, Tetrahedron,* 1965, **21**, 3171 (*pmr*)
Lawrie, W. *et al, J. Chem. Soc. C,* 1967, 2002 (*biosynth*)
Batey, I.L. *et al, Aust. J. Chem.,* 1972, **25**, 2511 (*isol*)
Yokoyama, A. *et al, Chem. Pharm. Bull.,* 1974, **22**, 877 (*isol*)

Eccremocarpol A　　　　　　　　　　E-80002

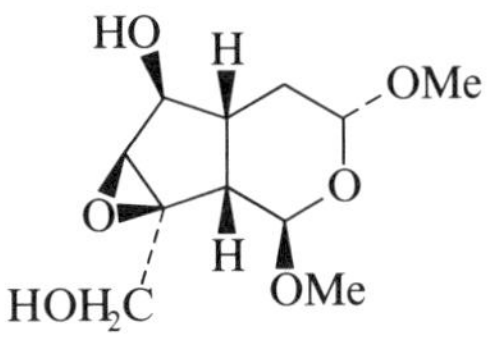

$C_{11}H_{18}O_6$　　M 246.260

Constit. of *Eccremocarpus scaber.* Oil. $[\alpha]_D^{25}$ −29° (MeOH).

Garbarino, J.A. *et al, Heterocycles,* 1989, **28**, 697.

Eccremocarpol B　　　　　　　　　　E-80003

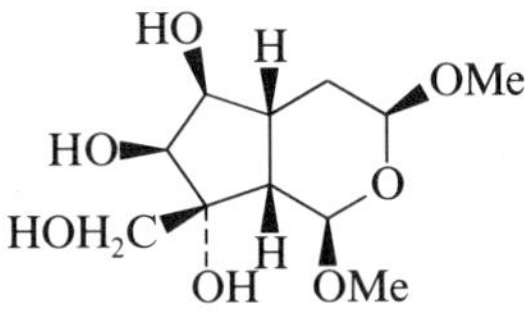

$C_{11}H_{20}O_7$　　M 264.275

Constit. of *Eccremocarpus scaber.* Oil. $[\alpha]_D^{25}$ −18.5° (MeOH).

Garbarino, J.A. *et al, Phytochemistry,* 1989, **28**, 697.

Echinacoside　　　　　　　　　　　　E-80004

[82854-37-3]

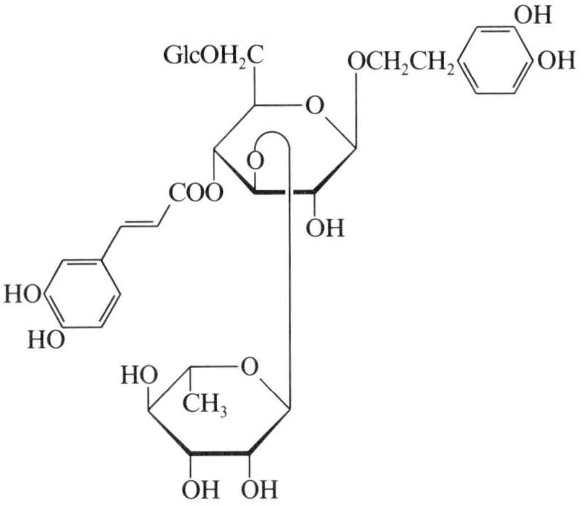

$C_{35}H_{46}O_{20}$　　M 786.736

Isol. from roots of *Echinacea angustifolia* and from *Cistanche salsa.* Needles (EtOH). $[\alpha]_D^{20}$ −56.5° (H$_2$O).

Stoll, A. *et al, Helv. Chim. Acta,* 1950, **33**, 1877 (*isol*)
Becker, H. *et al, Z. Naturforsch., C,* 1982, **37**, 351 (*pmr, cmr, struct*)
Kobayashi, H. *et al, Chem. Pharm. Bull.,* 1984, **32**, 3009 (*isol*)

Ecklonialactone A　　　　　　　　　E-80005

[121923-95-3]

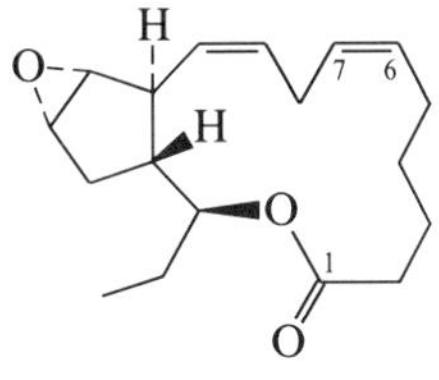

$C_{18}H_{26}O_3$　　M 290.402

Metab. of brown alga *Ecklonia stolonifera.* Cryst. (EtOH). Mp 96-98°. $[\alpha]_D$ −87.7° (c, 1.02 in CHCl$_3$).

6,7-Dihydro: [121923-96-4]. ***Ecklonialactone B***

$C_{18}H_{28}O_3$　　M 292.417

Metab. of *E. stolonifera.* Cryst. (EtOH). Mp 64-66°. $[\alpha]_D$ −49.3° (c, 1.08 in CHCl$_3$).

Kurata, K. *et al, Chem. Lett.,* 1989, 267 (*cryst struct*)

Edgeworin　　　　　　　　　　　　　E-80006

3-(7-Coumarinyloxy)-7-hydroxycoumarin

[120028-43-5]

$C_{18}H_{10}O_6$　　M 322.273

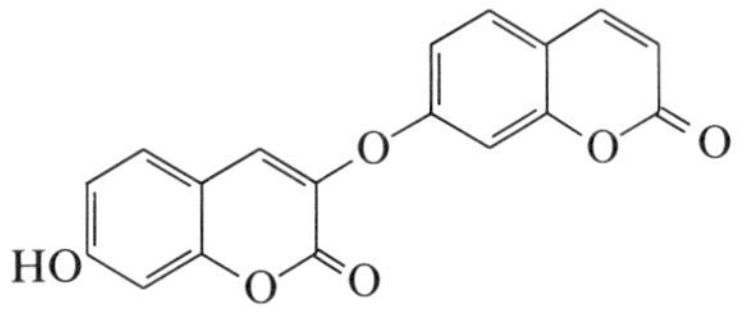

Constit. of *Edgeworthia chrysantha*. Powder. Mp 284-296°
dec.

Ac: Cryst. (EtOAc/hexane). Mp 236-238°.

Baba, K. *et al, Phytochemistry*, 1989, **28**, 221.

Edgeworoside *A*   E-80007

*3-(7-Coumarinyloxy)-8-(7-hydroxycoumarin-8-yl)-7-α-L-
rhamnopyranosyloxycoumarin*

[120040-21-3]

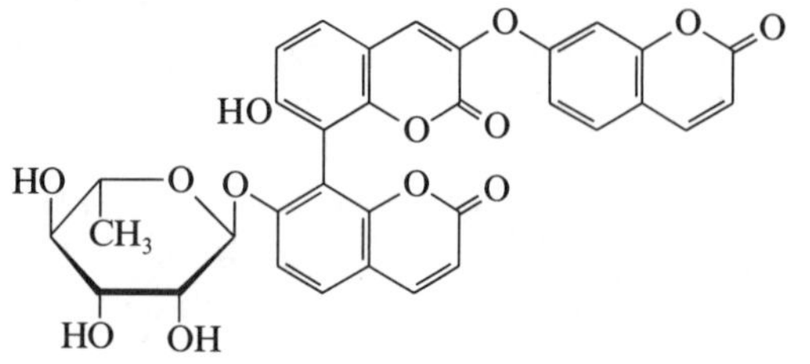

$C_{33}H_{24}O_{13}$ M 628.545

Constit. of *Edgeworthia chrysantha*. Powder. Mp 208-209°.

Baba, K. *et al, Phytochemistry*, 1989, **28**, 221.

Edulin   E-80008

Neodulin

[13401-64-4]

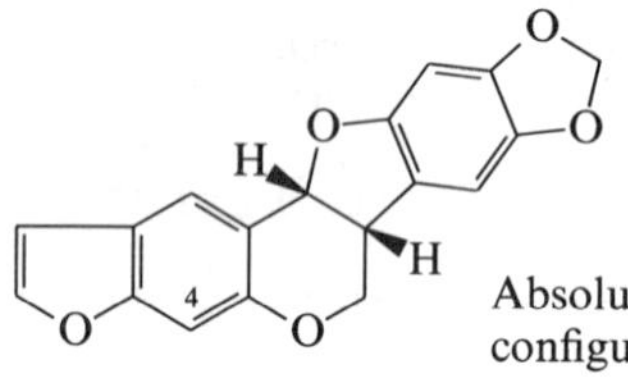

Absolute
configuration

$C_{18}H_{12}O_5$ M 308.290

Pterocarpan numbering shown. Isol. from *Neorautanenia
edulis*. Needles (CHCl₃/Et₂O). Mp 225°. $[\alpha]_D^{25}$ −265.3°
(CHCl₃).

4-Methoxy: [10338-03-1]. ***Ficinin***

   $C_{19}H_{14}O_6$ M 338.316

   Isol. from *N. ficifolia*. Cryst. Mp 180-180.5°. $[\alpha]_D^{20}$
   −199.7°. No abs. config. assigned; given here by prob.
   analogy with Edulin.

Van Duuren, B.L., *J. Org. Chem.*, 1961, **26**, 5013 (*isol, uv, ir, pmr,
   struct*)
Itô, S. *et al, J. Chem. Soc., Chem. Commun.*, 1965, 595 (*abs config*)
Suginome, H., *Bull. Chem. Soc. Jpn.*, 1966, **39**, 1544 (*abs config*)
Brink, C. *et al, J. S. Afr. Chem. Inst.*, 1966, **19**, 24; *ca*, **65**, 13677
   (*Ficinin*)

5,8,11,14,16-Eicosapentaenoic acid, 9CI   E-80009

*5,8,11,14,16-Icosapentaenoic acid. 16,17-Dehydroarachidonic
acid*

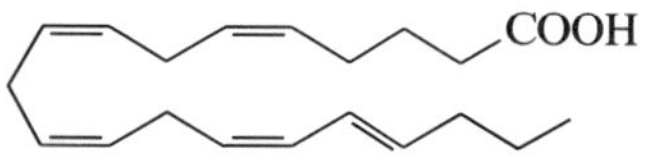

$C_{20}H_{30}O_2$ M 302.456

(5Z,8Z,11Z,14Z,16E)-form

   Substrate for soybean lipoxygenase.

Corey, E.J. *et al, Tetrahedron Lett.*, 1987, **28**, 5391 (*synth, uv, ms*)

1,3,11-Elematriene-9,14-diol   E-80010

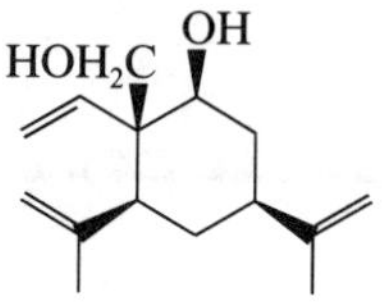

$C_{15}H_{24}O_2$ M 236.353

9β-form

   Periplocadiol

   Constit. of *Periploea laevigata*. Oil. $[\alpha]_D$ −19° (c, 12.4 in
   CHCl₃).

Askri, M. *et al, J. Nat. Prod. (Lloydia)*, 1989, **52**, 792 (*isol, pmr,
   cmr*)

Eloxanthin   E-80011

*5,6-Epoxy-5,6-dihydro-β,ε-carotene-3,3′-diol, 9CI.
Xanthophyll epoxide. Lutein epoxide*

[28368-08-3]

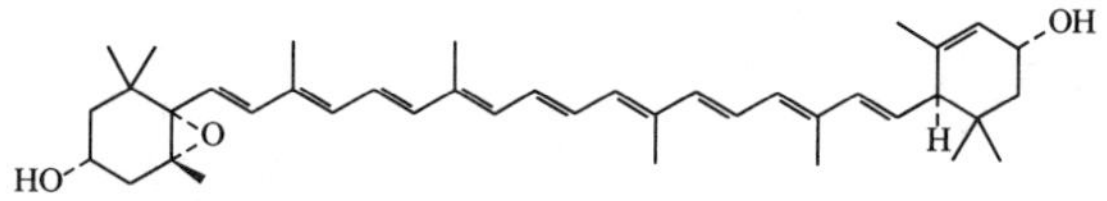

$C_{40}H_{56}O_3$ M 584.881

Isol. from a variety of higher plants and from algae.
Orange cryst. (Et₂O/MeOH). Mp 192°. $[\alpha]_{644}^{18}$ +225°
(C₆H₆).

Karrer, P. *et al, Helv. Chim. Acta*, 1945, **28**, 1146, 1526, 1528;
   1946, **29**, 1539; 1947, **30**, 537, 1158, 1774; 1948, **31**, 113, 802;
   1950, **33**, 300 (*isol, struct*)

Emehetin   E-80012

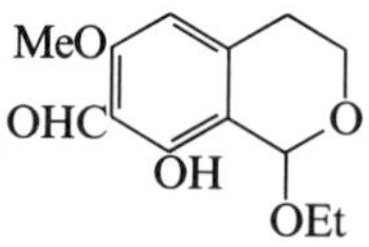

$C_{13}H_{16}O_5$ M 252.266

Metab. of *Emericella heterothallica*. Pale yellow cryst.
(cyclohexane). Mp 109-110°. $[\alpha]_D^{25}$ −33.5° (c, 1.18 in
CHCl₃).

Kawahara, N. *et al, Phytochemistry*, 1989, **28**, 1546.

Enukokurin   E-80013

[122148-80-5]

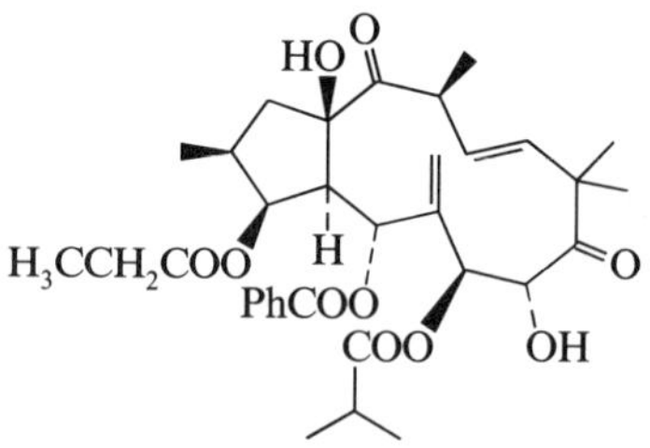

$C_{34}H_{44}O_{10}$ M 612.716

Constit. of *Euphorbia lateriflora*. Needles (Et₂O). Mp 193-
195°. $[\alpha]_D^{25}$ +30.8° (c, 12.4 in CHCl₃).

Fakunle, C.O. *et al, J. Nat. Prod. (Lloydia)*, 1989, **52**, 279 (*isol,
   pmr, cmr*)

10,11-Epoxy-2,7(14)-bisaboladien-4-one   E-80014
[119765-82-1]

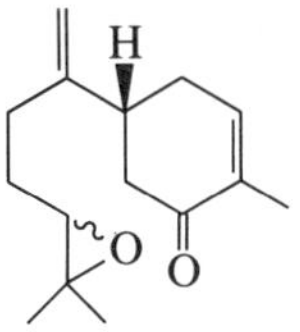

C$_{15}$H$_{22}$O$_2$ M 234.338
Constit. of *Mikania shushunensis*. Oil. [α]$_D^{23}$ −18.9° (c, 2.1
in CHCl$_3$).

Gutierrez, A.B. *et al, Phytochemistry*, 1988, **27**, 3871.

2-(3,4-Epoxy-1-butynyl)-5-(1,3-penta-diynyl)thiophene, 8CI   E-80015
[1209-16-1]

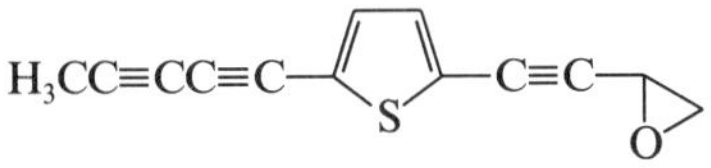

H$_3$CC≡CC≡C——C≡C

C$_{13}$H$_8$OS   M 212.272
Isol. from roots of *Echinops sphaerocephalus*. Cryst. (pet.
ether). Mp 54°. [α]$_{546}^{25}$ +36° (c, 0.885 in Et$_2$O).

Bohlmann, F. *et al, Chem. Ber.*, 1965, **98**, 155, 876; 1968, **101**,
4163 (*isol, uv, ir, pmr, ord, struct, biosynth, synth*)

ent-15,16-Epoxy-13(16),14-clerodadiene-2β,3β,4α-triol   E-80016

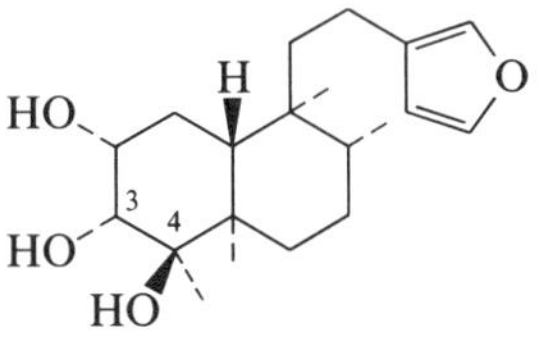

C$_{20}$H$_{32}$O$_4$ M 336.470
Constit. of *Baccharis boliviensis*. Cryst. Mp 155°. [α]$_D^{24}$
+15° (c, 0.43 in CHCl$_3$).

4-Deoxy, 3-ketone: **ent-15,16-Epoxy-2β-hydroxy-13(16),14-clerodadien-3-one**
C$_{20}$H$_{30}$O$_3$ M 318.455
Constit. of *B. boliviensis*. Gum.

Zdero, C. *et al, Phytochemistry*, 1989, **28**, 531.

ent-3α,4α-Epoxy-13,15-clerodanediol   E-80017
[118173-06-1]

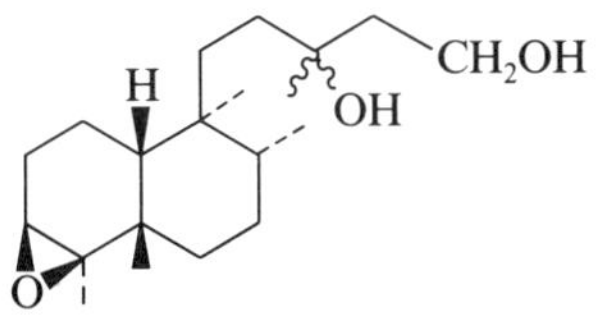

C$_{20}$H$_{36}$O$_3$ M 324.503
Constit. of *Ageratina saltillensis*.

Fang, N. *et al, Phytochemistry*, 1988, **27**, 3187.

ent-3α,4α-Epoxy-15-clerodanoic acid   E-80018
[118173-05-0]

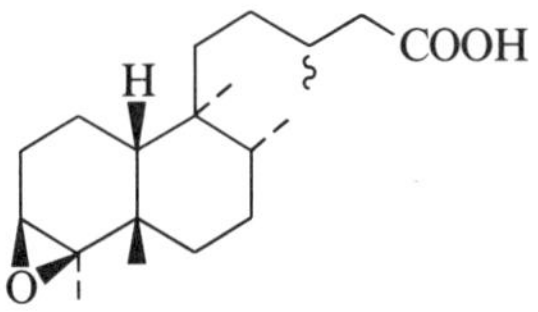

C$_{20}$H$_{34}$O$_3$ M 322.487
Constit. of *Ageratina saltillensis*.

Fang, N. *et al, Phytochemistry*, 1988, **27**, 3187.

ent-3α,4α-Epoxy-13-cleroden-15,16-olide   E-80019
[118243-67-7]

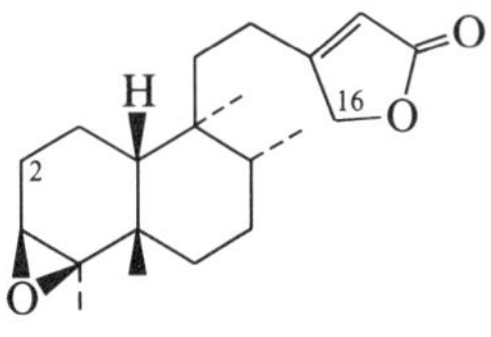

C$_{20}$H$_{30}$O$_3$ M 318.455
Constit. of *Solidago shortii*, *Ageratina cronquisti* and *A.
saltillensis*. Cryst. Mp 91°.

2β-Hydroxy: [118243-68-8]. ent-3α,4α-*Epoxy-2α-hydroxy-*
13-*cleroden*-15,16-*olide*
C$_{20}$H$_{30}$O$_4$ M 334.455
Constit. of *A. saltillensis*.

16ξ-Hydroxy: [118172-98-8]. ent-3α,4α-*Epoxy-16ξ-hydroxy-*
13-*cleroden*-15,16-*olide*
C$_{20}$H$_{30}$O$_4$ M 334.455
Constit. of *A. saltillensis*.

Anthonsin, T. *et al, Acta Chem. Scand.*, 1975, **25**, 1924 (*isol*)
Fang, N. *et al, Phytochemistry*, 1988, **27**, 3187 (*isol, cryst struct*)

3′,6′-Epoxycycloauraptene   E-80020
[21499-19-4]

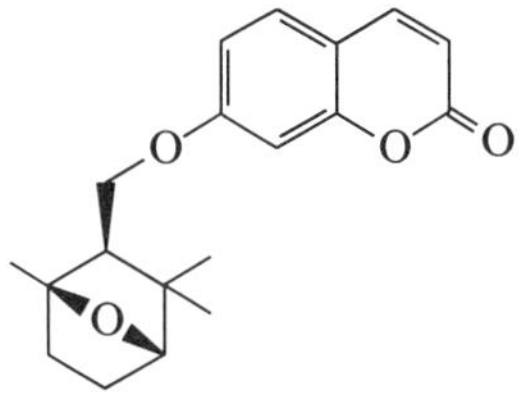

C$_{19}$H$_{22}$O$_4$ M 314.380
Isol. from *Aster yunnanensis*. Mp 187°. [α]$_D$ +18.9°
(CHCl$_3$).

(±)-*form* [31947-00-9]
Mp 155-156°.

Bohlmann, F. *et al, Justus Liebigs Ann. Chem.*, 1968, **717**, 186
(*isol*)
Coates, R.M. *et al, Tetrahedron*, 1970, **26**, 5699 (*synth*)

5,8-Epoxydaucane   E-80021

C$_{15}$H$_{26}$O   M 222.370
(*5β,8β*)-*form*

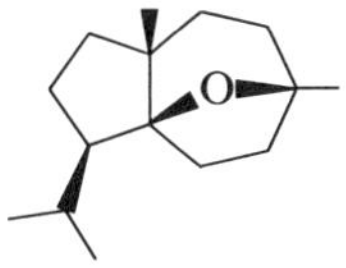

Carota-1,4β-oxide
Constit. of *Daucus carota*. Oil.

Dhillon, R.S. *et al*, *Phytochemistry*, 1989, **28**, 639.

7,8-Epoxy-4,6,9-daucanetriol   E-80022

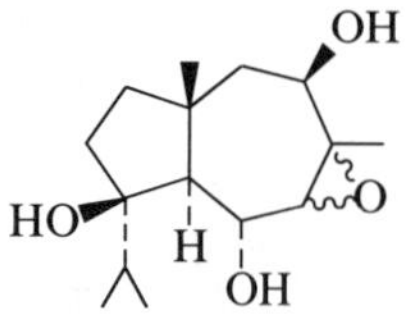

$C_{15}H_{26}O_4$ M 270.368
(4β,6α,7ξ,8ξ,9β)-form
    2β-Hydroxy-3,4-epoxyjaeschkeanadiol
    Constit. of *Ferula jaeschkeana*. Viscous mass. [α]$_D$ +47°
    (c, 1.5 in CHCl$_3$).

Garg, S.N. *et al*, *J. Nat. Prod. (Lloydia)*, 1988, **51**, 771.

1,10-Epoxy-11(13)-eremophilen-12-oic acid   E-80023

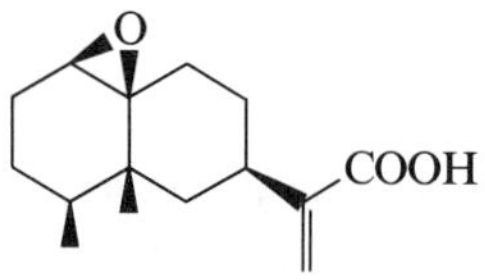

$C_{15}H_{22}O_3$ M 250.337
(1β,10β)-form
    Oil.

Zdero, C. *et al*, *Phytochemistry*, 1989, **28**, 1653.

1,10-Epoxy-11(13)-eremophilen-12,8-olide   E-80024

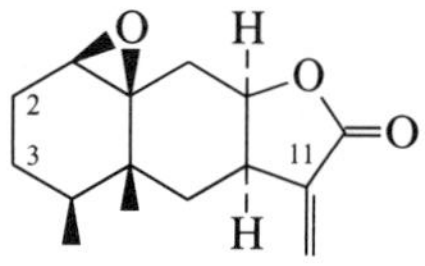

$C_{15}H_{20}O_3$ M 248.321
(1β,8β,10β)-form
    Constit. of *Ondetia linearis*. [α]$_D^{24}$ −17° (c, 3.68 in
    CHCl$_3$).
    11α,13-Dihydro: *1β,10β-Epoxy-11αH-eremophilen-12,8β-*
    olide
    $C_{15}H_{22}O_3$ M 250.337
    Constit. of *O. linearis*. Cryst. Mp 98°.
    2α-Hydroxy: *1β,10β-Epoxy-2α-hydroxy-11(13)-eremophilen-*
    12,8β-olide
    $C_{15}H_{20}O_4$ M 264.321
    Constit. of *O. linearis*. Gum.
    2α-Hydroxy, 11α,13-dihydro: *1β,10β-Epoxy-2α-hydroxy-*
    11αH-eremophilen-12,8β-olide
    $C_{15}H_{22}O_4$ M 266.336
    Constit. of *O. linearis*. Cryst. Mp 145°. [α]$_D^{24}$ −74° (c, 0.4
    in CHCl$_3$).
    2α-Hydroxy, 11β,13-dihydro: *1β,10β-Epoxy-2α-hydroxy-*
    11βH-eremophilen-12,8β-olide
    $C_{15}H_{22}O_4$ M 266.336
    Constit. of *O. linearis*. Gum.
    3α-Hydroxy, 11β,13-dihydro: *1β,10β-Epoxy-3α-hydroxy-*
    11βH-eremophilen-12,8β-olide

$C_{15}H_{22}O_4$ M 266.336
Constit. of *O. linearis*. Gum.

Zdero, C. *et al*, *Phytochemistry*, 1989, **28**, 1653.

6,15-Epoxy-1,4-eudesmanediol   E-80025

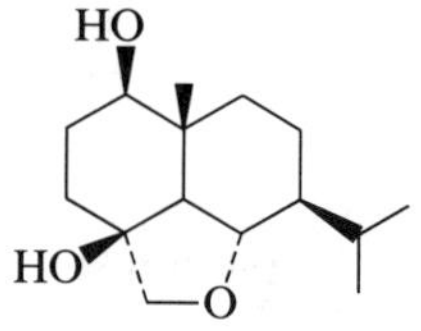

$C_{15}H_{26}O_3$ M 254.369
(1β,4β)-form [118243-71-3]
    Constit. of *Ageratina saltillensis*.

Fang, N. *et al*, *Phytochemistry*, 1988, **27**, 3187.

1,4-Epoxy-6-eudesmanol   E-80026

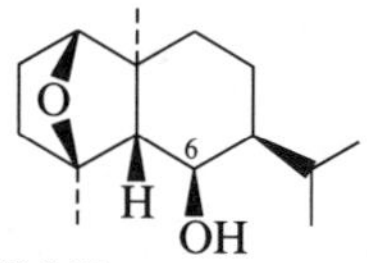

$C_{15}H_{26}O_2$ M 238.369
(1β,4β,5β,6β,10α)-form
    6-Cinnamoyl: [119364-26-0]. *6α-Cinnamoyloxy-1β,4β-epoxy-*
    1β,10α-eudesmane
    $C_{24}H_{32}O_3$ M 368.515
    Constit. of *Ambrosia artemisioides*. Oil.
    6-(4-Methoxybenzoyl): [119285-47-1]. *6α-Anisoyloxy-1β,4β-*
    epoxy-1β,10α-eudesmane
    $C_{23}H_{32}O_4$
    Constit. of *A. artemisioides*. Oil.

Jakupovic, J. *et al*, *Phytochemistry*, 1988, **27**, 3551.

6,13-Epoxy-3-eunicellene-8,9,12-triol   E-80027
[123012-55-5]

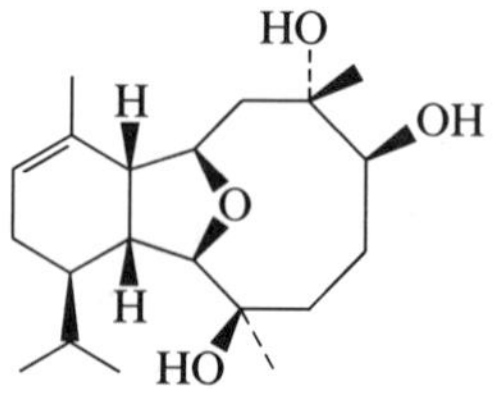

$C_{20}H_{34}O_4$ M 338.486
Metab. of *Cladiella* sp. Prisms (CH$_2$Cl$_2$/hexane). Mp
205.5-206°. [α]$_D$ −16.1° (c, 0.75 in CHCl$_3$). Cembrane
numbering.

Uchio, Y. *et al*, *Tetrahedron Lett.*, 1989, **30**, 3331 (*cryst struct*)

11,12-Epoxy-10(14)-guaien-4-ol   E-80028

$C_{15}H_{22}O$ M 218.338
(1α,4α,5α)-form

$C_{15}H_{22}O_4$ M 266.336
Constit. of *O. linearis*. Gum.

Zdero, C. *et al*, *Phytochemistry*, 1989, **28**, 1653.

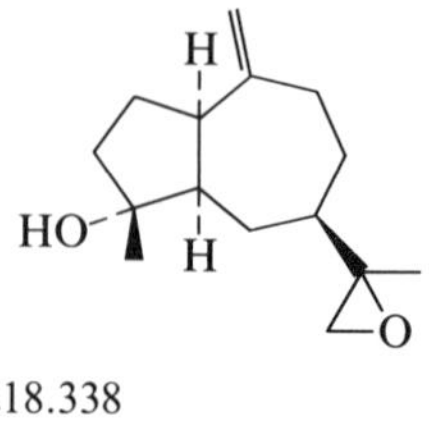

Constit. of *Pleocarphus revolutus*. Oil. $[\alpha]_D^{24}$ +5° (c, 0.89 in CHCl$_3$).

Zdero, C. *et al*, *J. Nat. Prod.* (*Lloydia*), 1988, **51**, 509.

ent-15,16-Epoxy-4α-hydroxy-13(16),14-clerodadiene-3,12-dione   E-80029

[118607-72-0]

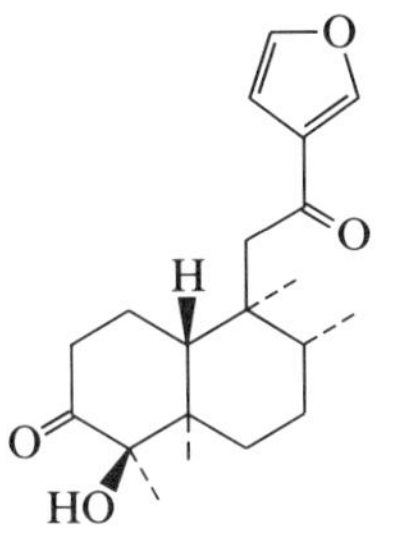

C$_{20}$H$_{28}$O$_4$ M 332.439
Constit. of *Croton argyrophylloides*. Cryst.
(MeOH/hexane). Mp 122-124°. $[\alpha]_D^{25}$ +41° (c, 1.07 in CHCl$_3$).

Monte, F.J.Q. *et al*, *Phytochemistry*, 1988, **27**, 3209.

3,4-Epoxy-5-hydroxy-1-cyclohexene-carboxylic acid   E-80030

5-Hydroxy-7-oxabicyclo[4.1.0]hept-2-ene-3-carboxylic acid.
Anhydroshikimic acid

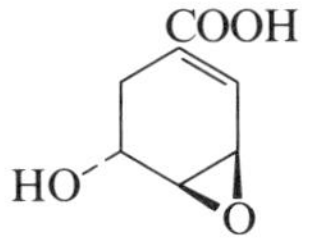

C$_7$H$_8$O$_4$ M 156.138
Me ester: [78961-89-4].
   C$_8$H$_{10}$O$_4$ M 170.165
   Metabolic of fungus *Chalara microspora*. Unstable oil.
   $[\alpha]_D$ +248° (c, 0.5 in EtOH).

[78844-86-7]

Fex, T. *et al*, *Acta Chem. Scand., Ser. B*, 1981, **35**, 91 (*isol*)
Wood, H.B. *et al*, *Tetrahedron Lett.*, 1989, **30**, 6257 (*synth*)

1,10-Epoxy-13-hydroxy-7(11)-eremo-philen-12,8-olide   E-80031

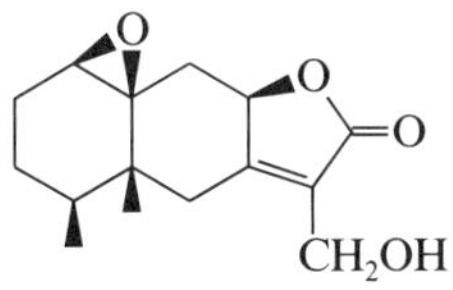

C$_{15}$H$_{20}$O$_4$ M 264.321
(1β,8β,10β)-form
   Constit. of *Ondetia linearis*. Cryst. Mp 129°. $[\alpha]_D^{24}$ −75° (c, 0.2 in CHCl$_3$).

Zdero, C. *et al*, *Phytochemistry*, 1989, **28**, 1653.

ent-8α,14α-Epoxy-3β-hydroxy-9(11),15-pimaradiene-2,12-dione   E-80032
Yucalexin P-8
[119626-47-0]

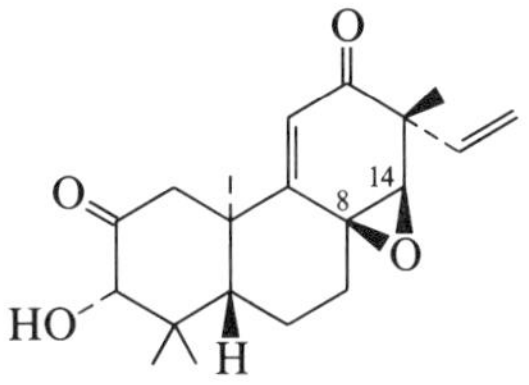

C$_{20}$H$_{26}$O$_4$ M 330.423
Constit. of cassava roots (*Manihot esculenta*).

8,14-Diepimer: *ent*-8β,14β-*Epoxy*-3β-*hydroxy*-9(11),15-*pimaradiene*-2,12-*dione*. **Yucalexin P-10**
   C$_{20}$H$_{26}$O$_4$ M 330.423
   Constit. of *M. esculenta*.

Sakai, T. *et al*, *Phytochemistry*, 1988, **27**, 3769.

15,16-Epoxy-8-isopimaren-7-one   E-80033
Nanuzone
[120019-36-5]

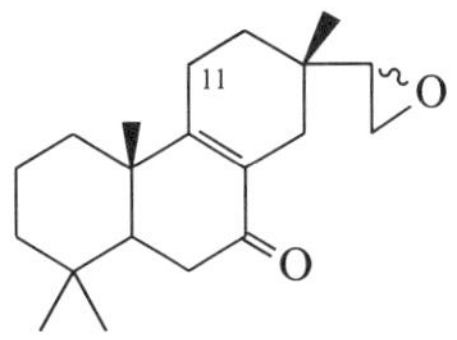

C$_{20}$H$_{30}$O$_2$ M 302.456
Constit. of *Vellozia nanuzae*. Needles (hexane). Mp 132-134°. $[\alpha]_D$ +50° (c, 0.78 in CHCl$_3$).

11β-Hydroxy: [119944-13-7]. 15,16-*Epoxy*-11β-*hydroxy*-8-*isopimaren*-7-*one*. **11β-Hydroxynanuzone**
   C$_{20}$H$_{30}$O$_3$ M 318.455
   Constit. of *V. nanuzae*. Needles (EtOAc/hexane). Mp 149-151°.

Pinto, A.C. *et al*, *Phytochemistry*, 1988, **27**, 3909.

ent-11α,16α-Epoxy-17-kauranol   E-80034
Liangshanin G

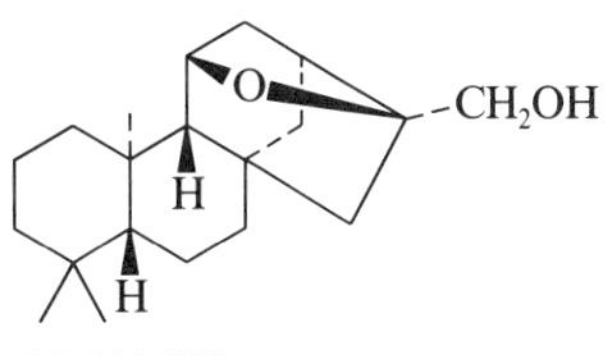

C$_{20}$H$_{32}$O$_2$ M 304.472
Constit. of *Rabdosia liangshanica*. Cryst. Mp 146-151°.

Fenglei, Z. *et al*, *Phytochemistry*, 1989, **28**, 1671.

9,13-Epoxy-7,14-labdadiene-2,3,20-triol   E-80035

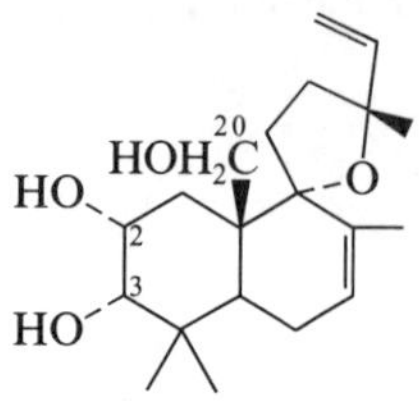

$C_{20}H_{32}O_4$ M 336.470

(2α,3α,9α,13R)-form

2-Angeloyl, 3,20-di-Ac: 3α,20-Diacetoxy-2α-angeloyloxy-9α,13R-epoxy-7,14-labdadiene
$C_{29}H_{42}O_7$ M 502.647
Constit. of *Waitzia acuminata*.

Jakupovic, J. *et al, Phytochemistry*, 1989, **28**, 1943.

8,20-Epoxy-14-labdene-2,3,7,13-tetrol   E-80036

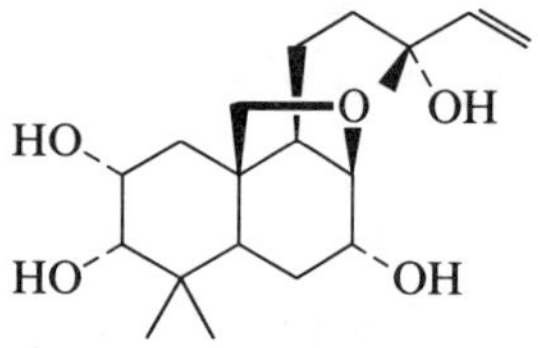

$C_{20}H_{34}O_5$ M 354.486

(2α,3α,7α,8β,13R)-form

2-Angeloyl: 2α-Angeloyloxy-8β,20-epoxy-14-labdene-3α,7α,13R-triol
$C_{25}H_{40}O_6$ M 436.587
Constit. of *Waitzia acuminata*.

2-Angeloyl, 3-Ac: 3α-Acetoxy-2α-angeloyloxy-8β,20-epoxy-14-labdene-7α,13R-diol
$C_{27}H_{42}O_7$ M 478.625
Constit. of *W. acuminata*.

Jakupovic, J. *et al, Phytochemistry*, 1989, **28**, 1943.

8,13-Epoxy-14-labden-11-one   E-80037

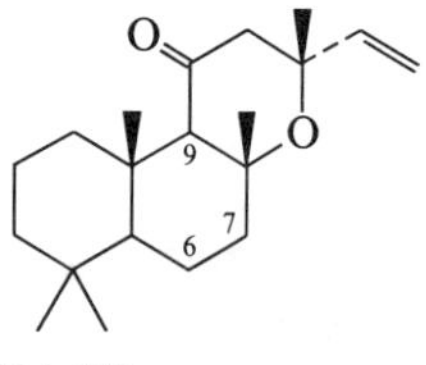

$C_{20}H_{32}O_2$ M 304.472

(8α,13R)-form [61242-47-5]
Constit. of *Coleus forskohlii*. Cryst. (2-propanol). Mp 96-97°. $[\alpha]_D^{20}$ − 103.2° (c, 0.2 in CHCl).

6β-Hydroxy: [121817-29-6]. **8α,13R-Epoxy-6β-hydroxy-14-labden-11-one**
$C_{20}H_{32}O_3$ M 320.471
Constit. of *C. forskohlii*. Cryst. (EtOAc/cyclohexane). Mp 120-121°. $[\alpha]_D^{20}$ − 133° (c, 0.1 in CHCl₃).

6β,7β,9α-Trihydroxy: [121606-18-6]. **8α,13R-Epoxy-6β,7β,9-trihydroxy-14-labden-11-one**
$C_{20}H_{32}O_5$ M 352.470
Constit. of *C. forskohlii*. Cryst. (2-propanol). Mp 148-149°. $[\alpha]_D^{20}$ − 25.9° (c, 0.1 in CHCl₃).

6β,9α-Dihydroxy,7β-acetoxy: [72963-77-0]. **7β-Acetoxy-8α,13R-epoxy-6β,9α-dihydroxy-14-labden-11-one**
$C_{22}H_{34}O_6$ M 394.507

Constit. of *C. forskohlii*. Cryst. (hexane). Mp 135-136°. $[\alpha]_D^{20}$ − 23.1° (c, 0.1 in CHCl₃).

Gabetta, B. *et al, Phytochemistry*, 1989, **28**, 859.

2,9-Epoxylactarotropone   E-80038

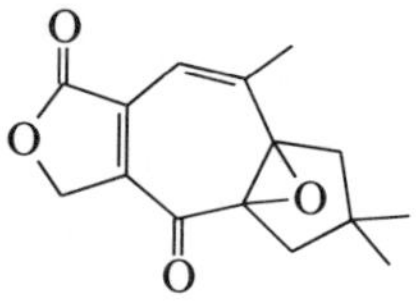

$C_{15}H_{16}O_4$ M 260.289
Constit. of *Lactarius scrobiculatus*. Oil.

Bosetti, A. *et al, Phytochemistry*, 1989, **28**, 1427.

22S,25-Epoxy-24-methyl-3,11,20-furo-stanetriol   E-80039

Updated Entry replacing E-10055
Hippuristanol
[80442-78-0]

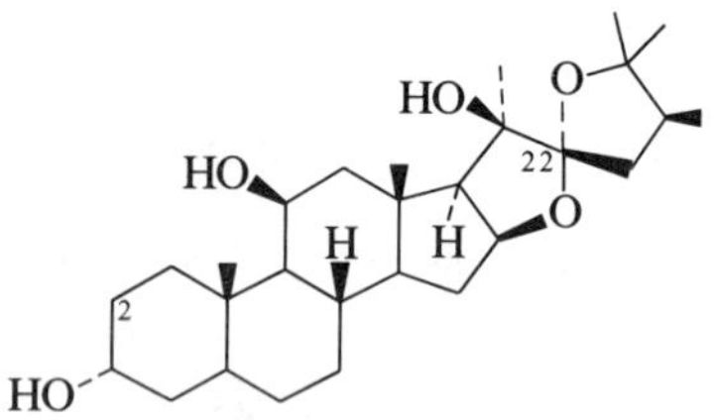

$C_{28}H_{46}O_5$ M 462.668
Constit. of *Isis hippuris*. Shows *in vivo* anticancer activity. Cryst. (MeOH). Mp 188-190°.

22-Epimer: [80442-79-1]. **22-Epihippuristanol**
Constit. of *Isis hippuris*. Cryst. (MeOH). Mp 248-249°.

22-Epimer,di-Ac: [80442-84-8]. **Hippurin-2**
$C_{32}H_{50}O_7$ M 546.743
Constit. of *I. hippuris*. Cryst. (MeOH). Mp 252-253°.

2α-Hydroxy: **2α-Hydroxyhippuristanol**
$C_{28}H_{46}O_6$ M 478.668
Constit. of *I. hippuris*. Shows *in vivo* anticancer activity. Amorph. solid.

2α-Acetoxy: [66536-82-1]. **Hippurin 1**
$C_{30}H_{48}O_7$ M 520.705
Constit. of *I. hippuris*. Cryst. (H₂O/pet. ether). Mp 183-185°. $[\alpha]_D$ +36.2° (c, 1 in CHCl₃).

2α-Acetoxy, 22-epimer: **22-Epihippurin 1**
Constit. of *I. hippuris*. Cryst. (MeOH). Mp 243-245°.

2α,3,11-Di-Ac: **3,11-Diacetylhippurin-1**
$C_{34}H_{52}O_9$ M 604.779
Constit. of *I. hippuris*. Oil.

2α-Hydroxy,22-epimer: **2-Desacetyl-22-epihippurin 1**
$C_{28}H_{46}O_6$ M 478.668
Constit. of *I. hippuris*. Cryst. Mp 260-261°.

22-Epimer,2α-Hydroxy-3-Ac: **3-Acetyl-2-desacetyl-22-epihippurin 1**
$C_{30}H_{48}O_7$ M 520.705
Constit. of *I. hippuris*. Cryst. Mp 248-250°.

22-Epimer,2α-acetoxy,3-Ac: **3-Acetyl-22-epihippurin-1**
$C_{32}H_{50}O_8$ M 562.742
Constit. of *I. hippuris*. Cryst. Mp 240-243°.

22-Epimer,2α-Acetoxy,3,11-di-Ac: **3,11-Diacetyl-22-epihippurin 1**
$C_{34}H_{52}O_9$ M 604.779
Constit. of *I. hippuris*. Needles. Mp 178-180°.

Kazlauskas, A. *et al, Tetrahedron Lett.*, 1977, 4439.
Higa, T. *et al, Chem. Lett.*, 1981, 1647.
Rao, C.B. *et al, J. Nat. Prod. (Lloydia)*, 1988, **51**, 954.

8,19-Epoxy-17-methyl-1(15)-trinervitene-2,3,7,9,14,17-hexol   E-80040

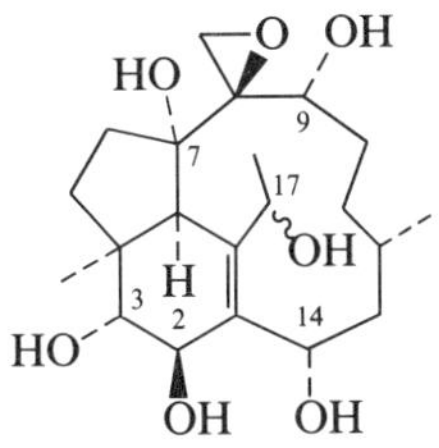

$C_{21}H_{34}O_6$ M 382.496
Unusual 17-methyltrinervitane skeleton.

(2β,3α,7α,8β,9α,14α)-form

2,3,9,14-Tetrapropanoyl:
$C_{35}H_{52}O_{12}$ M 664.789
Constit. of *Hospitalitermes umbrinus*. Gluey liquid.

Goh, S.H. *et al, Tetrahedron Lett.*, 1988, **29**, 113 (*struct*)

1,3-Epoxy-2-nor-1,10-seco-1(5),3-aroma-dendradien-10-one   E-80041

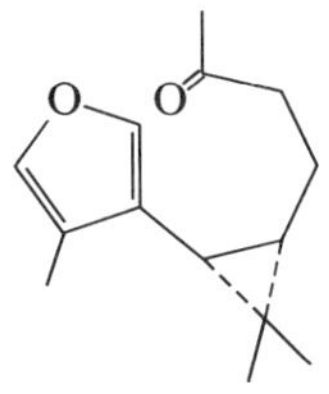

$C_{14}H_{20}O_2$ M 220.311
Constit. of *Mylia taylorii*. Oil. [α]$_D$ −25.8° (c, 2.98 in CHCl$_3$).

Harrison, L.J. *et al, Phytochemistry*, 1989, **28**, 1261.

2,3-Epoxy-5,7-octadien-4-ol   E-80042

6,7-Epoxy-5-hydroxyocimene
[119308-11-1]

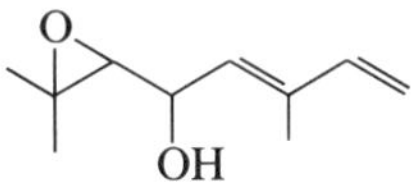

$C_{10}H_{16}O_2$ M 168.235
Constit. of *Cineraria britteniae*. Oil.

Lehmann, L. *et al, Phytochemistry*, 1988, **27**, 3307.

11,15-Epoxy-3(20)-phytene-1,2-diol   E-80043

11,15-Epoxy-1,2-dihydroxy-3(20)-phytene

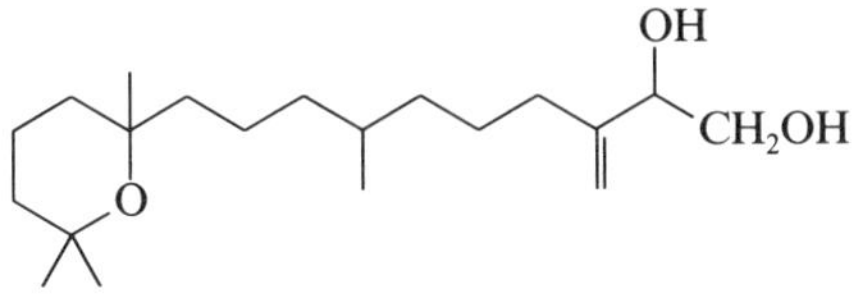

$C_{20}H_{38}O_3$ M 326.518
Constit. of *Anisopappus pinnatifidus*.

Zdero, C. *et al, Phytochemistry*, 1989, **28**, 1155.

11,12-Epoxy-14-taraxeren-3-ol   E-80044

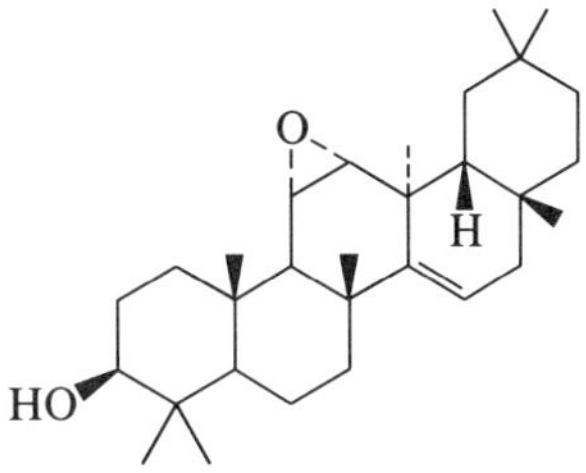

$C_{30}H_{48}O_2$ M 440.708
(3β,11α,12α)-form [3866-77-1] *11α,12α-Oxidotaraxerol*
Constit. of *Euphorbia supina*. Needles (MeOH/CHCl$_3$).
Mp 286-288°. [α]$_D^{23}$ −38.9° (c, 0.98 in CHCl$_3$).

Tanaka, R. *et al, Phytochemistry*, 1988, **27**, 3579.

14,15-Epoxy-3,7,11,15-tetramethyl-2,6,10-hexadecatrienal   E-80045

13-(3,3-Dimethyloxiranyl)-3,7,11-trimethyl-2,6,10-tridecatrienal, 9CI

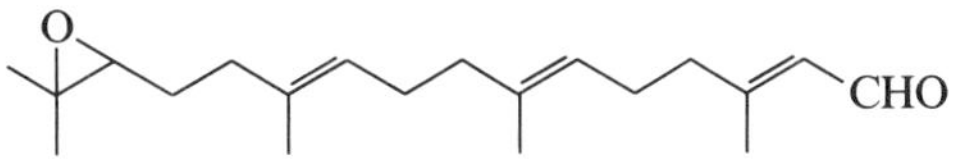

(2E,6E,10E)-form

$C_{20}H_{32}O_2$ M 304.472
(2E,6E,10E)-form [108613-62-3]
Constit. of *Cystophora moniliformis*. Oil. [α]$_D$ −17° (c, 0.98 in CHCl$_3$).
(2Z,6E,10E)-form [115075-54-2]
Constit. of *C. moniliformis*. Oil. [α]$_D$ −22° (c, 0.4 in CHCl$_3$).

van Altena, I.A. *et al, Aust. J. Chem.*, 1988, **41**, 49.

ent-7β,20-Epoxy-1β,14α,20-trihydroxy-16-kauren-15-one   E-80046

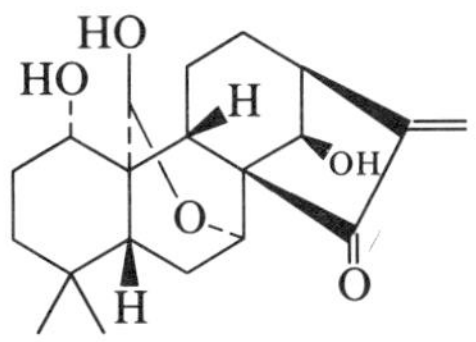

$C_{20}H_{28}O_5$ M 348.438
Constit. of *Rabdosia umbrosa*. Amorph. powder. [α]$_D^{22}$ −55.8° (c, 1.54 in MeOH).

Takeda, Y. *et al, Phytochemistry*, 1989, **28**, 1691.

5,6-Epoxy-6,10,14-trimethyl-9,13-pentadecadien-2-one   E-80047

4-[3-(4,8-Dimethyl-3,7-nonadienyl)-3-methyloxiranyl]-2-butanone, 9CI
[115028-53-0]

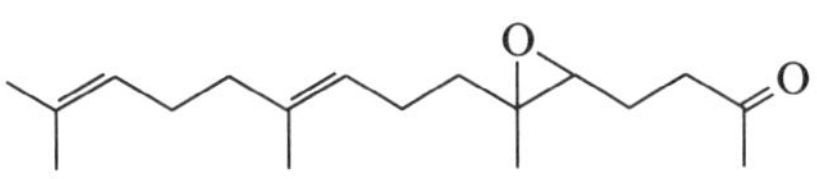

$C_{18}H_{30}O_2$ M 278.434
Constit. of *Cystophora moniliformis*. Unstable oil.

van Altena, I.A., *Aust. J. Chem.*, 1988, **41**, 49.

8,19-Epoxy-1(15)-trinervitene-2,3,7,9,14, 17-hexol   E-80048

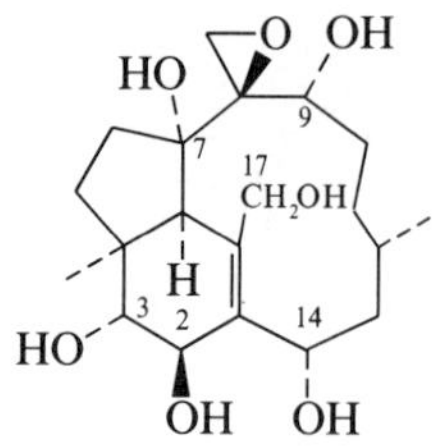

$C_{20}H_{32}O_7$ M 384.469

(2β,3α,7α,8β,9α,14α)-form

  2,3,9,14,17-Pentapropanoyl:
    $C_{35}H_{52}O_{12}$ M 664.789
    Constit. of *Hospitalitermes umbrinus*. Gluey liquid.

  Goh, S.H. *et al*, *Tetrahedron Lett.*, 1988, **29**, 113 (*struct*)

Ergocerebrin   E-80049

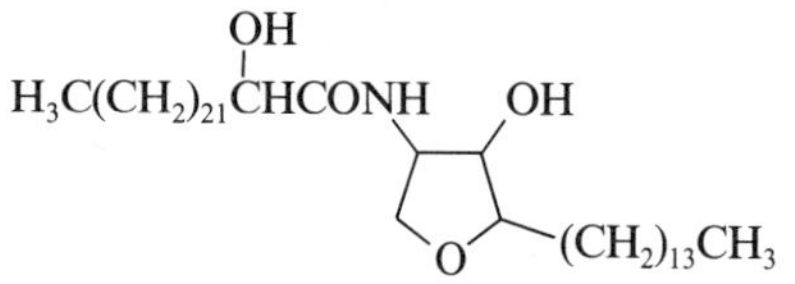

$C_{42}H_{83}NO_4$ M 666.122

Isol. from cultures of *Claviceps purpurea*. Needles. Mp 106-107°. $[\alpha]_D^{20}$ +16.3° (c, 0.17 in tetrachloroethane). Accompanied by its lower homologues with dodecyl and tridecyl sidechains.

Simon, J.W.A. *et al*, *J. Chem. Soc.*, 1965, 4164.

10(20),16-Ericacadien-19-oic acid   E-80050

[122585-76-6]

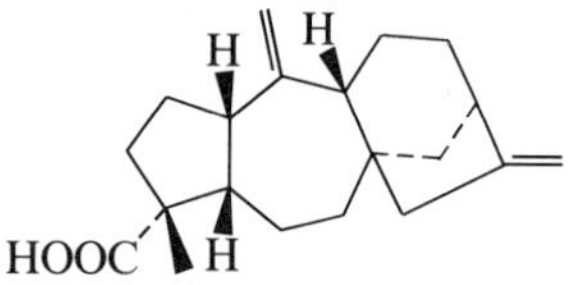

$C_{20}H_{28}O_2$ M 300.440

Constit. of *Helichrysum chionosphaerum*. Gum.

Jakupovic, J. *et al*, *Phytochemistry*, 1989, **28**, 1119.

Erybraedin D   E-80051

[119269-72-6]

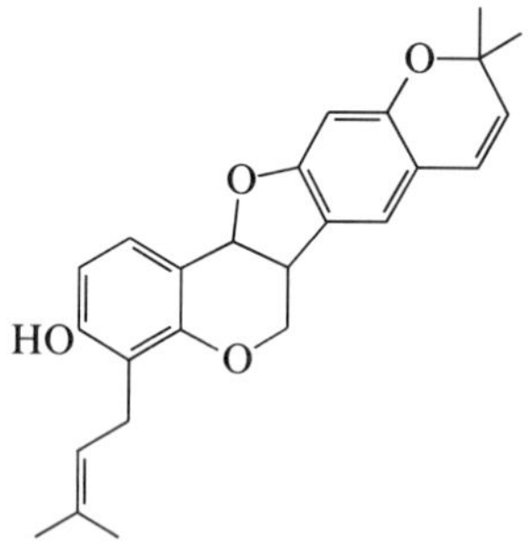

$C_{25}H_{26}O_4$ M 390.478

Constit. of *Erythrina mildbraedii*. $[\alpha]_D^{25}$ −67° (c, 0.23 in MeOH).

Mitscher, L.A. *et al*, *Heterocycles*, 1988, **27**, 2517.

Erybraedin E   E-80052

[119269-73-7]

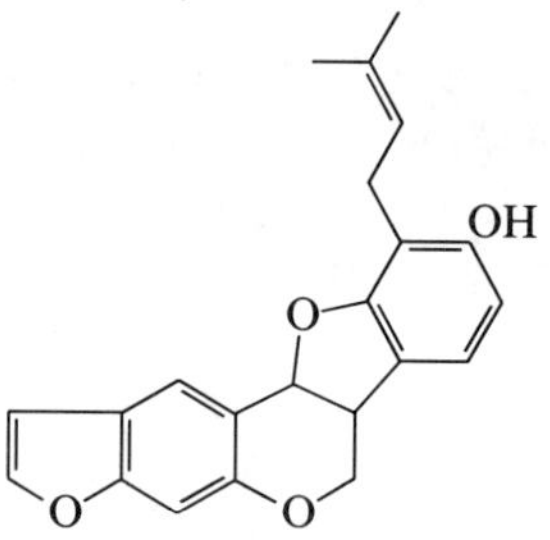

$C_{22}H_{20}O_4$ M 348.398

Constit. of *Erythrina mildbraedii*. $[\alpha]_D^{25}$ −104° (c, 0.29 in MeOH).

Mitscher, L.A. *et al*, *Heterocycles*, 1988, **27**, 2517.

Erythrabyssin II   E-80053

3,9-Dihydroxy-2,10-diprenylcoumestan

[77263-06-0]

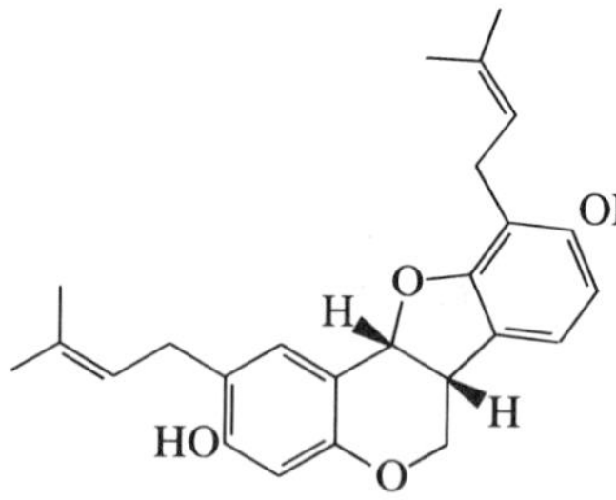

Absolute configuration

$C_{25}H_{28}O_4$ M 392.494

Isol. from roots of *Erythrina abyssinica*. Shows antimicrobial props.

Kamat, V.S. *et al*, *Heterocycles*, 1981, **15**, 1163 (*isol, uv, pmr, struct, abs config*)

Erythrinin C   E-80054

2,3-Dihydro-4-hydroxy-2-(1-hydroxy-1-methylethyl)-6-(4-hydroxyphenyl)-5H-furo[3,2-g][1]benzopyran-5-one, 9CI

[63807-85-2]

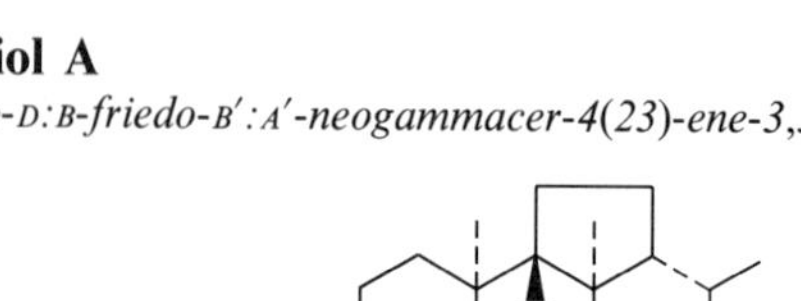

$C_{20}H_{18}O_6$ M 354.359

Isol. from bark of *Erythrina variegata*. Yellow needles (Me_2CO/C_6H_6). Mp 198°.

Deshpande, V.H. *et al*, *Indian J. Chem., Sect. B*, 1977, **15**, 205.

Espinendiol A   E-80055

3,4-Seco-D:B-friedo-B':A'-neogammacer-4(23)-ene-3,5α-diol

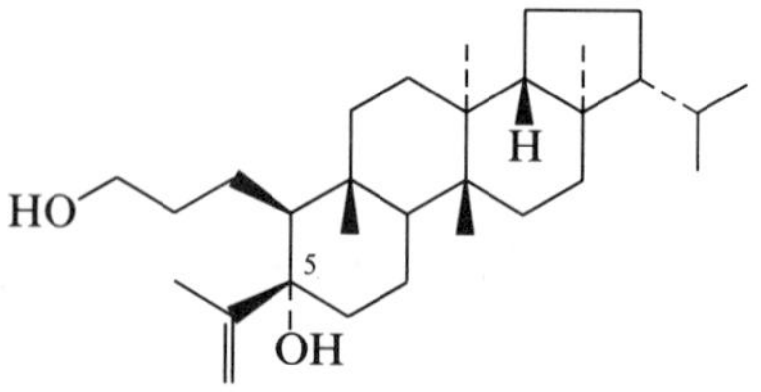

$C_{30}H_{52}O_2$ M 444.740

Constit. of *Euphorbia supina.* Cryst. Mp 194-196°. $[\alpha]_D^{23}$ +90.7° (c, 0.76 in CHCl₃).

5-Epimer: **Espinendiol B.** *3,4-Seco-ᴅ:ʙ-friedo-ʙ′:ᴀ′-neogammacer-4(23)-ene-3,5β-diol*
Constit. of *E. supina.* Cryst. Mp 192-193.5°. $[\alpha]_D^{23}$ −17.1° (c, 0.54 in CHCl₃).

Tanaka, R. *et al, Tetrahedron Lett.,* 1989, **30**, 1661 (*cryst struct*)

Espinenoxide   E-80056

3,4-Seco-ᴅ:ʙ-friedo-ʙ′:ᴀ′-neogammacer-4(23)-ene 3,5β-oxide

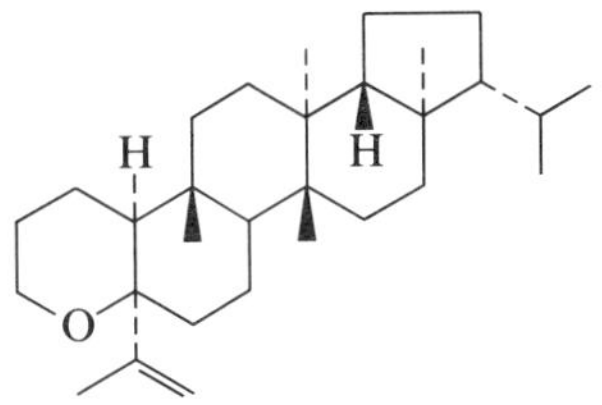

$C_{30}H_{50}O$ M 426.724
Constit. of *Euphorbia supina.* Cryst. Mp 215-218°. $[\alpha]_D^{23}$ +7.8° (c, 0.32 in CHCl₃).

Tanaka, R. *et al, Tetrahedron Lett.,* 1989, **30**, 1661 (*cryst struct*)

Estafiatin   E-80057

Updated Entry replacing E-10077
[10180-89-9]

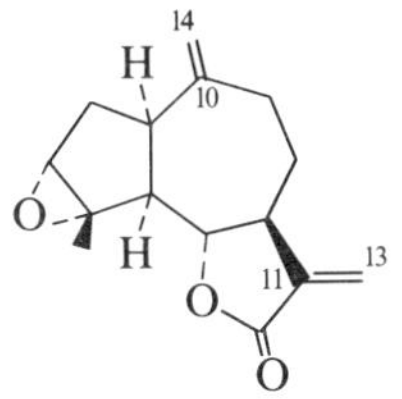

$C_{15}H_{18}O_3$ M 246.305
Constit. of *Artemisia mexicana* and from *Cotula coronopifolia.* Cryst. (Et₂O/hexane). Mp 104-106°. $[\alpha]_D^{20}$ −9.9° (CHCl₃).

11α,13-Dihydro: **11α,13-Dihydroestafialin.** *3α,4α-Epoxy-10(14)-guaien-6α,12-olide*
$C_{15}H_{20}O_3$ M 248.321
Constit. of *A. sieversiana.* Oil. $[\alpha]_D^{24}$ +94° (c, 0.05 in CHCl₃).

10α,14-Epoxide: **10α,14-Epoxyestafiatin**
$C_{15}H_{18}O_4$ M 262.305
Constit. of *Stevia alpina.* Cryst. (EtOAc/CHCl₃). Mp 113-119°.

10β,14-Epoxide: **10β,14-Epoxyestafiatin**
$C_{15}H_{18}O_4$ M 262.305
Constit. of *S. alpina.* Cryst. (EtOAc/CHCl₃). Mp 159-161°.

Sánchez-Viesca, F. *et al, Tetrahedron,* 1963, **19**, 1285 (*isol, uv, ir, pmr*)
Bohlmann, F. *et al, Chem. Ber.,* 1966, **99**, 2828 (*isol*)
Edgar, M.T. *et al, J. Org. Chem.,* 1979, **44**, 159 (*synth*)
Macaira, L.A. *et al, Tetrahedron Lett.,* 1980, 773 (*synth*)
Demuznek, M. *et al, Tetrahedron Lett.,* 1982, **23**, 2501 (*synth*)
Bohlmann, F. *et al, Phytochemistry,* 1985, **24**, 1009 (*isol*)
de Heluani, C.S. *et al, Phytochemistry,* 1989, **28**, 1931 (*isol, cryst struct*)

Ethenimine, 9CI   E-80058

Vinylideneamine. Keteneimine
[17619-22-6]

$$H_2C{=}C{=}NH$$

C_2H_3N M 41.052
Transient species, rapidly tautomerizes to acetonitrile. Obs. at 14K and 77K.

Jacox, M.E., *Chem. Phys.,* 1979, **43**, 157 (*synth, ir, uv*)
Guillemin, J.-C. *et al, J. Chem. Soc., Chem. Commun.,* 1983, 238 (*synth, ir*)
Rodler, M. *et al, Chem. Phys. Lett.,* 1984, **110**, 447 (*synth, bibl*)
Wentrup, C. *et al, J. Am. Chem. Soc.,* 1988, **110**, 1337 (*props, bibl*)

3-Ethoxy-2-propenenitrile, 9CI   E-80059

β-Ethoxyacrylonitrile
[61310-53-0]

$$EtOCH{=}CHCN$$

C_5H_7NO M 97.116
Inexpensive, comly. available synthetic reagent. Synthon for cytosines.

(E)-form [58243-08-6]
Liq. Bp₁₅ 76-78°. n_D^{25} 1.4510.

(Z)-form [60036-64-8]
Liq. Bp₁₅ 85-87°. n_D^{25} 1.4530.

McElvain, S.M. *et al, J. Am. Chem. Soc.,* 1947, **69**, 2657 (*synth, props*)
Scotti, F. *et al, J. Org. Chem.,* 1964, **29**, 1800 (*synth, ir, pmr, uv*)
Ford, G.P. *et al, J. Chem. Soc., Perkin Trans.* 2, 1975, 1371 (*synth, ir, props*)
Movsum-Zade, E.M. *et al, J. Org. Chem. USSR (Engl. Transl.),* 1977, **13**, 1694 (*synth, ir*)
Smirnow, D. *et al, Synth. Commun.,* 1986, **16**, 1187 (*use*)

24-Ethyl-7,22-cholestadiene-3,5,6-triol   E-80060

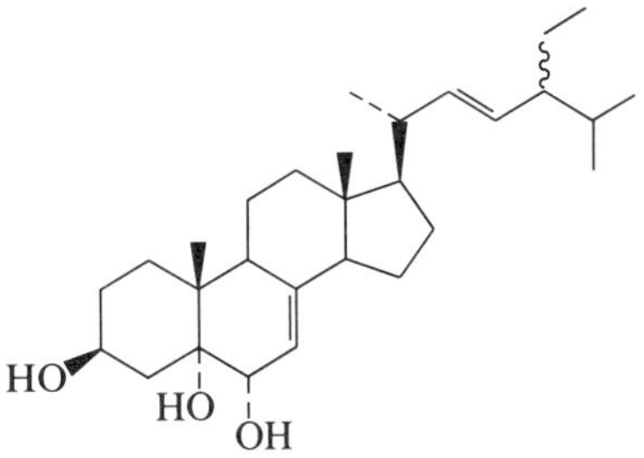

$C_{29}H_{48}O_3$ M 444.696
(3β,5α,6α,22E,24ξ)-form
Constit. of *Hippospongia communis, Spongia officinalis* and *Ircinia variabilis.*

Madaio, A. *et al, J. Nat. Prod. (Lloydia),* 1989, **52**, 952 (*isol, pmr*)

24-Ethyl-7-cholestene-3,6-diol E-80061

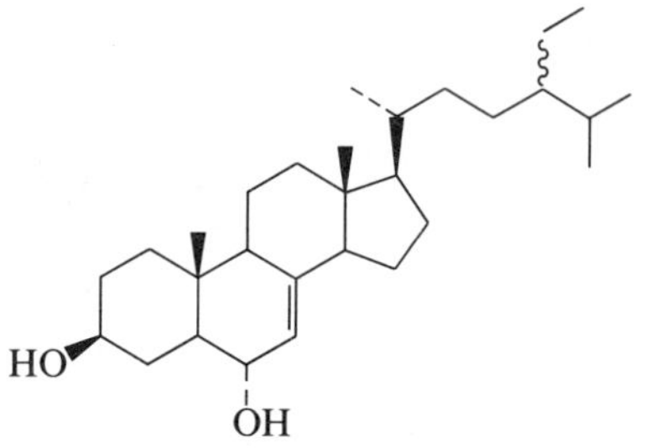

$C_{29}H_{50}O_2$ M 430.713

($3\beta,5\alpha,6\alpha,24\xi$)-*form*

Constit. of *Spongionella gracilis*.

Madaio, A. *et al*, *J. Nat. Prod. (Lloydia)*, 1989, **52**, 952 (*isol, pmr*)

6-Ethyl-2,3-dihydro-2-methyl-4H-pyran-4-one E-80062

Updated Entry replacing E-50071

[96998-54-8]

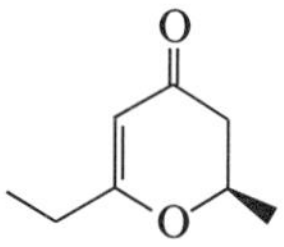

$C_8H_{12}O_2$ M 140.182

(*R*)-*form*

Component of pheromone of male swift moth *Hepialus hecta*. Bp$_{20}$ 105°. [α]$_D^{22}$ +189° (c, 2.1 in pentane).

Sinnwell, V. *et al*, *Tetrahedron Lett.*, 1985, **26**, 1707 (*isol*)
Descheneaux, P.F. *et al*, *Helv. Chim. Acta*, 1989, **72**, 1259 (*synth, pmr, bibl*)

Ethyl ethoxyiminoacetate, 9CI E-80063

Ethyl 1-carboethoxyformimidate

[816-27-3]

$$HN{=}C(OEt)COOEt$$

$C_6H_{11}NO_3$ M 145.158

Reagent for heterocyclic synthesis. Liq. Bp 175° dec. , Bp$_{18}$ 73°.

Lander, G.D., *J. Chem. Soc.*, 1895, **79**, 202; 1899, **83**, 411 (*synth*)
Cornforth, J.W. *et al*, *J. Chem. Soc.*, 1947, 96 (*synth*)
Braz, G.I. *et al*, *Zh. Obshch. Khim.*, 1964, **34**, 2980 (*synth*)
Sugiyama, Y. *et al*, *J. Org. Chem.*, 1978, **43**, 4485 (*use*)
Gomez, E. *et al*, *Tetrahedron*, 1986, **42**, 2625 (*use*)

1-Ethyl-2,3,4,6,7,12-hexahydroindolo[2,3-a]quinolizine, 9CI E-80064

Wenkert's enamine

[40163-47-1]

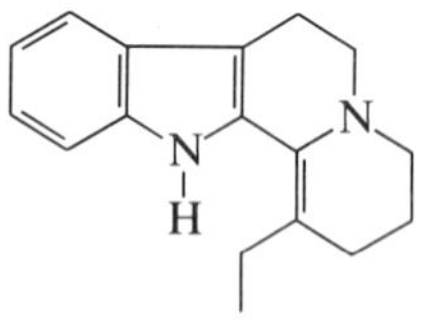

$C_{17}H_{20}N_2$ M 252.358

Key intermed. for eburnane alkaloid synth.
$B,HClO_4$: Mp 175-177°.

Wenkert, E. *et al*, *J. Am. Chem. Soc.*, 1965, **87**, 1580 (*synth*)

Husson, H.P. *et al*, *J. Chem. Soc., Chem. Commun.*, 1972, 930 (*synth*)
Chevolot, L. *et al*, *Bull. Soc. Chim. Fr.*, 1976, 1222 (*synth*)
Szantay, C. *et al*, *Tetrahedron*, 1977, **33**, 1803 (*synth, use*)
Danieli, B. *et al*, *J. Chem. Soc., Chem. Commun.*, 1980, 109 (*synth, uv, ir*)
Lounasmaa, M. *et al*, *Tetrahedron*, 1987, **43**, 2135 (*synth*)

6-Ethyl-4-hydroxy-3,5-dimethyl-2H-pyran-2-one E-80065

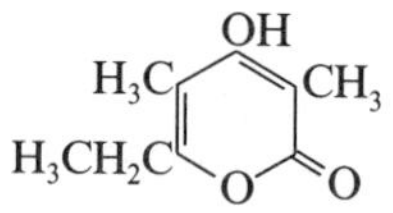

$C_9H_{12}O_3$ M 168.192

Metab. of *Emericella heterothallica*. Prisms (C_6H_6). Mp 152-154°.

Kawahara, N. *et al*, *Phytochemistry*, 1989, **28**, 1546.

24-Ethylidene-7-cholestene-3,5,6-triol E-80066

24-Ethyl-7,24(28)-cholestadiene-3,5,6-triol

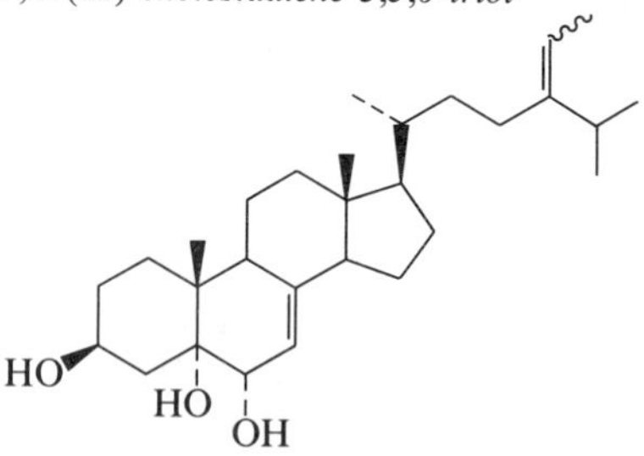

$C_{29}H_{48}O_3$ M 444.696

($3\beta,5\alpha,6\alpha$)-*form*

Constit. of *Hippospongia communis*, *Spongia officinalis*, and *Ircinia variabilis*.

Madaio, A. *et al*, *J. Nat. Prod. (Lloydia)*, 1989, **52**, 952 (*isol, pmr*)

3-Ethyl-1H-pyrrole E-80067

[1551-16-2]

C_6H_9N M 95.144

Liq. Bp$_{34}$ 88°. The 3-ethylpyrrole descr. in the old lit. is said to be the 2-isomer.

Fischer, H. *et al*, *Justus Liebigs Ann. Chem.*, 1935, **519**, 1 (*synth*)
Soth, S. *et al*, *Bull. Soc. Chim. Fr.*, 1975, 2511 (*synth, pmr*)
Alonso Garrido, D.O. *et al*, *J. Org. Chem.*, 1988, **53**, 403 (*synth, cmr, ms*)

2-Ethyl-4,6,6-trimethyl-2-cyclohexen-1-one E-80068

Angustifolenone

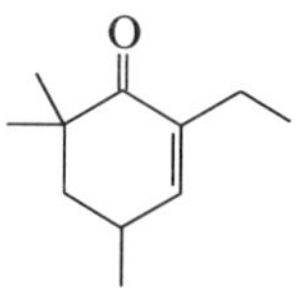

$C_{11}H_{18}O$ M 166.263

Constit. of *Backhousia angustifolia*. Oil.

Brophy, J.J. *et al*, *Phytochemistry*, 1989, **28**, 1259.

6,6′-(1,2-Ethynediyl)bis-2(1*H*)-pyridinone, 9CI   E-80069

[117068-70-9]

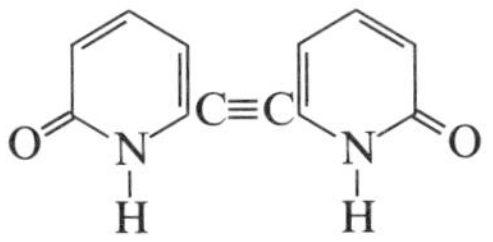

$C_{12}H_8N_2O_2$ M 212.207

Ducharme, Y. *et al*, *J. Org. Chem.*, 1988, **53**, 5387 (*synth*)

Ethynol, 9CI   E-80070

Updated Entry replacing E-50084
Hydroxyacetylene
[32038-79-2]

$$HC{\equiv}COH$$

C_2H_2O M 42.037
Obt. as stable sp. in gas phase.
tert-*Butyl ether:* [89489-28-1]. tert-*Butoxyethyne. 2-
(Ethynyloxy)-2-methylpropane*
$C_6H_{10}O$ M 98.144
Bp_{22} ~20°.
Adamantyl ether: [113279-41-7]. *(1-Adamantyloxy)ethyne.
1-(Ethynyloxy)adamantane*
$C_{12}H_{16}O$ M 176.258
Cryst. (Et$_2$O/pentane). Mp 49-51°.

van Baar, B. *et al*, *Angew. Chem., Int. Ed. Engl.*, 1986, **25**, 282
(*synth, ms*)
Pericàs, M.A. *et al*, *Tetrahedron*, 1987, **43**, 2311 (*deriv, synth, ir,
pmr, cmr, ms*)

2-Ethynyl-2*H*-azirine   E-80071

[119720-93-3]

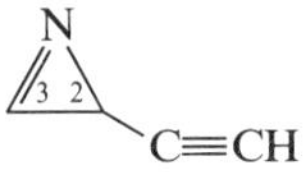

C_4H_3N M 65.074

Kurtz, P. *et al*, *Justus Liebigs Ann. Chem.*, 1959, **624**, 1 (*synth*)
Banert, K., *Chem. Ber.*, 1989, **122**, 1175 (*synth, pmr, cmr*)

3-Ethynyl-2*H*-azirine   E-80072

[119720-91-1]
C_4H_3N M 65.074

Banert, K., *Chem. Ber.*, 1989, **122**, 1175 (*synth, pmr, cmr*)

2-Ethynyl-1,3-butadiene   E-80073

[6929-96-0]

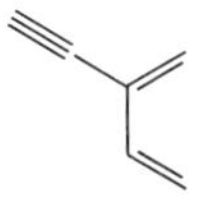

C_6H_6 M 78.113
Extremely mobile liq. Bp 72°. n_D^{20} 1.4669 Dec. slowly at r.t.

Hopf, H., *Chem. Ber.*, 1971, **104**, 1499 (*synth, uv, pmr*)
Bader, H. *et al*, *Chem. Ber.*, 1989, **122**, 1193 (*synth, bibl*)

4-Ethynylindole   E-80074

[102301-81-5]
$C_{10}H_7N$ M 141.172
Cryst. (EtOAc/hexane). Mp 66-66.5°.

Kozikowski, A.P. *et al*, *J. Org. Chem.*, 1988, **53**, 863 (*synth, ir,
pmr, ms*)

Ethynylpentafluorobenzene   E-80075

Updated Entry replacing E-70053
(Pentafluorophenyl)acetylene
[5122-07-6]

$$(C_6F_5)C{\equiv}CH$$

C_7HF_5 M 180.077
Liq. Bp 130-131°, Bp_{37} 50-52°.
▷ Potentially explosive.

Coe, P.L. *et al*, *J. Chem. Soc. C*, 1966, 597 (*synth, uv, ir, pmr*)
Waugh, F. *et al*, *J. Organomet. Chem.*, 1972, **39**, 275 (*synth, uv, ir,
pmr*)
Neenan, T.X. *et al*, *J. Org. Chem.*, 1988, **53**, 2489 (*synth, ir, pmr,
haz*)

2-Ethynylpyridine, 9CI   E-80076

(2-Pyridyl)acetylene
[1945-84-2]

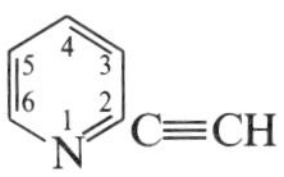

C_7H_5N M 103.123
Liq. Bp_{14} 77-78°. n_D^{21} 1.553.

Haug, V. *et al*, *Chem. Ber.*, 1960, **93**, 593 (*synth*)
Leaver, D. *et al*, *J. Chem. Soc.*, 1963, 6053 (*synth*)
Brandsma, L., *Prep. Acetylenic Chem.*, Elsevier, Amsterdam, 1971,
117 (*synth*)
Ames, D.E. *et al*, *Synthesis*, 1981, 364 (*synth*)
Sakamoto, T. *et al*, *Synthesis*, 1983, 312 (*synth, ir, pmr*)

3-Ethynylpyridine, 9CI   E-80077

[2510-23-8]
C_7H_5N M 103.123
Cryst. (ligroin). Mp 38.5°. Bp_{30} 83-84°.
N-*oxide:* [49836-11-5].
C_7H_5NO M 119.123
Mp 104.5-105°.

Alberts, A. *et al*, *J. Am. Chem. Soc.*, 1935, **57**, 1286 (*synth*)
Kozhevnikova, A.N., *Izv. Akad. Sci. USSR, Ser. Sci. Khim.*, 1973,
1168 (*oxide*)

4-Ethynylpyridine, 9CI   E-80078

(4-Pyridyl)acetylene
[2510-22-7]
C_7H_5N M 103.123
Cryst. (pentane). Mp 96-96.5°. Bp_{15} 70° subl.
N-*Oxide:* [49836-10-4].
C_7H_5NO M 119.123
No Mp reported.

Haug, U. *et al*, *Chem. Ber.*, 1960, **93**, 593 (*synth*)
Gray, A.P. *et al*, *J. Org. Chem.*, 1968, **33**, 3007 (*synth*)
Kozhevnikova, A.N. *et al*, *Izv. Akad. Sci. USSR, Ser. Sci. Khim.*,
1973, 1168 (*oxide*)
Della Ciana, L. *et al*, *J. Heterocycl. Chem.*, 1984, **21**, 607 (*synth,
pmr*)

4(5)-Ethynyl-1,2,3-triazole　　　　　　E-80079
[119720-87-5]

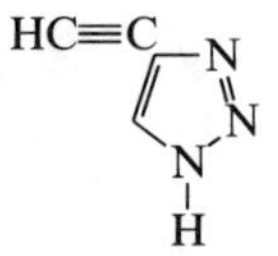

$C_4H_3N_3$　　M 93.088
Solid (CH_2Cl_2 or by subl.). Mp 127.5°.

Banert, K., *Chem. Ber.*, 1989, **122**, 1175 (*synth, pmr, cmr*)

Euchretin A　　　　　　　　　　　　E-80080
[119459-83-5]

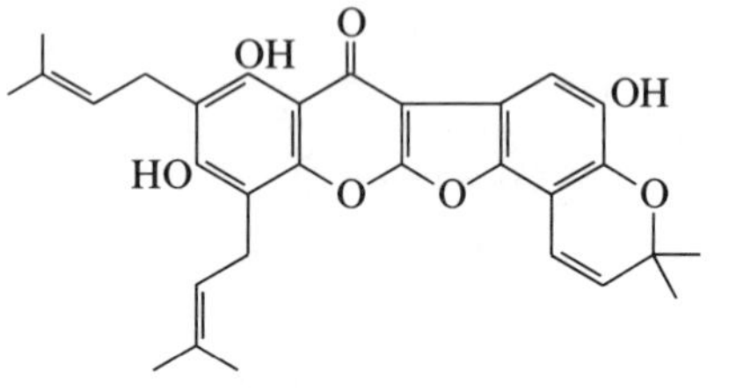

$C_{30}H_{30}O_7$　　M 502.563
Constit. of *Euchresta japonica*. Pale yellow needles. Mp 208-209°.

Mizuno, M. *et al*, *Heterocycles*, 1988, **27**, 2047.

4,11-Eudesmadien-3-one　　　　　　E-80081
Updated Entry replacing C-03796
　　α-Cyperone
[473-08-5]

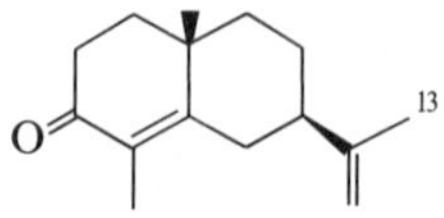

Absolute configuration

$C_{15}H_{22}O$　　M 218.338
Constit. of *Cyperus rotundus*, *C. scariosus* and *Juniperus horizontalis*. Bp 177°. $[\alpha]_D$ +138°.

Semicarbazone: Cryst. Mp 216°.

13-Hydroxy: [56423-80-4]. *12-Hydroxy-4,11(13)-eudesmadien-3-one*. **12-Hydroxy-α-cyperone**
$C_{15}H_{22}O_2$　　M 234.338
Isol. from *Artemisia afra*. Oil.

Adamson, P.S. *et al*, *J. Chem. Soc.*, 1937, 1576 (*struct*)
Naves, Y.-R., *Bull. Soc. Chim. Fr.*, 1954, 332 (*isol*)
Howe, R. *et al*, *J. Chem. Soc.*, 1955, 2423 (*synth*)
Caine, D. *et al*, *J. Org. Chem.*, 1974, **39**, 2154 (*synth*)
Jakupovic, J. *et al*, *Phytochemistry*, 1988, **27**, 1129 (*deriv*)

5,11-Eudesmanediol　　　　　　　　E-80082

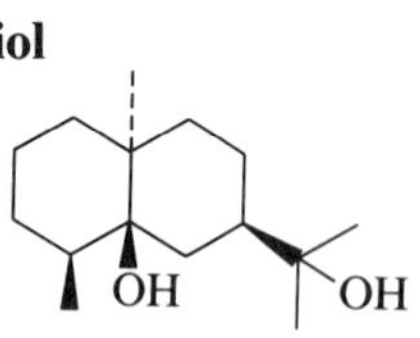

$C_{15}H_{28}O_2$　　M 240.385
(4β,5β,7β,10α)-form [121817-19-4]
　　Constit. of *Cymbopogon distans*. Cryst. (Me_2CO). Mp 120-121°. $[\alpha]_D^{25}$ −9.0° (c, 0.8 in $CHCl_3$).

Mathela, C.S. *et al*, *Phytochemistry*, 1989, **28**, 936 (*isol, cryst struct*)

3-Eudesmene-1,6-diol　　　　　　　E-80083
Updated Entry replacing E-01299
　　3-Selinene-1,6-diol

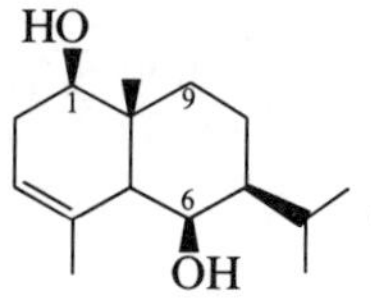

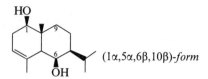

(1α,5α,6β,10β)-*form*

$C_{15}H_{26}O_2$　　M 238.369
(1β,5α,6β,10β)-form
　　6-Cinnamoyl: [77355-61-4].
　　　Constit. of *Solidago nemoralis*.
(1α,5β,6β,10α)-form
　　1-Cinnamoyl:
　　　$C_{24}H_{32}O_3$　　M 368.515
　　　Constit. of *Ambrosia artemisioides*. Oil.
　　6-Cinnamoyl:
　　　$C_{24}H_{32}O_3$　　M 368.515
　　　Constit. of *A. artemisioides*. Oil.
　　6-Anisoyl:
　　　$C_{23}H_{32}O_4$　　M 372.503
　　　Constit. of *A. artemisioides*. Oil.
　　6-O-β-D-Glucopyranoside:
　　　$C_{21}H_{36}O_7$　　M 400.511
　　　Constit. of *A. artemisioides*. Gum.
[77355-61-4]

Bohlmann, F. *et al*, *Phytochemistry*, 1980, **19**, 2655 (*isol*)
Jakupovic, J. *et al*, *Phytochemistry*, 1988, **27**, 3551 (*deriv*)

4(15)-Eudesmene-1,6-diol　　　　　　E-80084
Updated Entry replacing E-30080
　　1,6-Dihydroxy-4(15)-eudesmene

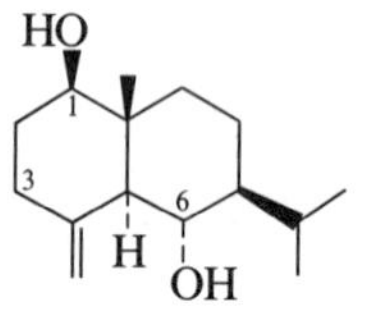

Absolute configuration

$C_{15}H_{26}O_2$　　M 238.369
(1β,5α,6α,10β)-form
　　Constit. of *Senecio microglossus* and *Helianthus microcephalus*. Gum. $[\alpha]_D^{24}$ +7° (c, 0.1 in $CHCl_3$).
　　6-Me ether: *6α-Methoxy-4(15)-eudesmen-1β-ol*. *1β-Hydroxy-6α-methoxy-4(15)-eudesmene*
　　　$C_{16}H_{28}O_2$　　M 252.396
　　　Constit. of *Torilis japonica*. Amorph. $[\alpha]_D$ +7.8°.
(1α,5β,6β,10α)-form
　　6-Angeloyl: *6β-Angeloyloxy-1α-hydroxy-5,10-bisepi-4(15)-eudesmene*
　　　$C_{20}H_{32}O_3$　　M 320.471
　　　Constit. of *Ambrosia artemisioides*. Oil.
　　6-Cinnamoyl, 3β-hydroperoxy: *6β-Cinnamoyloxy-3β-hydroperoxy-1α-hydroxy-5,10-bisepi-4(15)-eudesmene*
　　　$C_{24}H_{32}O_5$　　M 400.514
　　　Constit. of *A. artemisioides*. Oil.

Itokawa, H. *et al*, *Chem. Lett.*, 1983, 1253 (*isol*)
Bohlmann, F. *et al*, *Phytochemistry*, 1983, **22**, 1675 (*isol*)
Gutierrez, A.B. *et al*, *Phytochemistry*, 1988, **27**, 2225 (*isol*)
Jakupovic, J. *et al*, *Phytochemistry*, 1988, **27**, 3551 (*deriv*)

4(15)-Eudesmene-2,11-diol   E-80085

Updated Entry replacing E-01301
4(15)-Selinene-2,11-diol

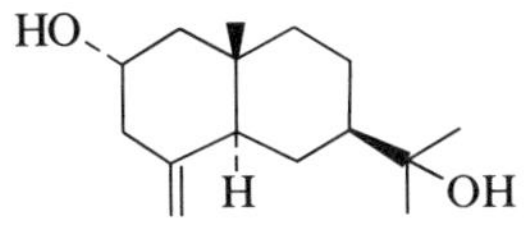

$C_{15}H_{26}O_2$ M 238.369

2α-form [21677-80-5] **Pterocarpol**
   Constit. of the heartwood of *Pterocarpus macrocarpus*
   and other *P.* spp. Cryst. (C_6H_6). Mp 104-105°. $[\alpha]_D$
   +39° (c, 1.05 in $CHCl_3$).

Bahl, C.P. *et al*, *Tetrahedron*, 1968, **24**, 6231 (*isol, struct*)
Kukla, A.S. *et al*, *Indian J. Chem., Sect. B*, 1976, **14**, 905 (*abs config*)

3-Eudesmen-11-ol   E-80086

Updated Entry replacing E-01309
   α-Eudesmol. Selinelol. Atractylol. 3-Selinen-11-ol

[473-16-5]

$C_{15}H_{26}O$ M 222.370
Occurs with *β*-Eudesmol in various eucalyptus oils. Cryst.
   by subl. Mp 75°. Bp_{10} 156°. $[\alpha]_D$ +28.6° (c, 1.86 in
   $CHCl_3$).

7-Epimer: 7-Epi-α-eudesmol
   $C_{15}H_{26}O$ M 222.370
   Constit. of *Amyris balsamifera* oil. Oil.

McQuillan, F.J. *et al*, *J. Chem. Soc.*, 1956, 2973 (*isol, struct*)
Humber, D.C. *et al*, *Tetrahedron Lett.*, 1966, 4985 (*synth*)
Kutney, J.P. *et al*, *Can. J. Chem.*, 1984, **62**, 1407 (*synth*)
Schwartz, M.A. *et al*, *J. Org. Chem.*, 1985, **50**, 1359 (*synth*)
Van Beek, T.A. *et al*, *Phytochemistry*, 1989, **28**, 1909 (*isol*)

6-Eudesmen-4-ol   E-80087

$C_{15}H_{26}O$ M 222.370
4α-form [118173-08-3]
   Constit. of *Ageratina saltillensis*.

Fang, N. *et al*, *Phytochemistry*, 1988, **27**, 3187.

Eudesobovatol A   E-80088

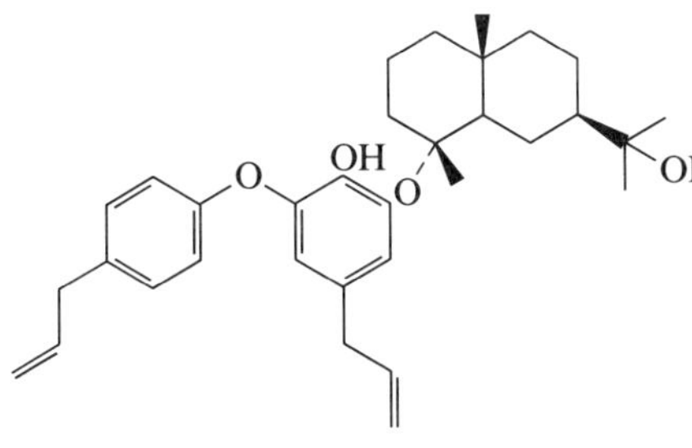

$C_{33}H_{44}O_4$ M 504.708
Constit. of *Magnolia obovata*. Neurotrophic agent. $[\alpha]_D$
   −40.3° (c, 2.5 in $CHCl_3$).

Fukuyama, Y. *et al*, *Tetrahedron Lett.*, 1989, **30**, 5907.

Eudesobovatol B   E-80089

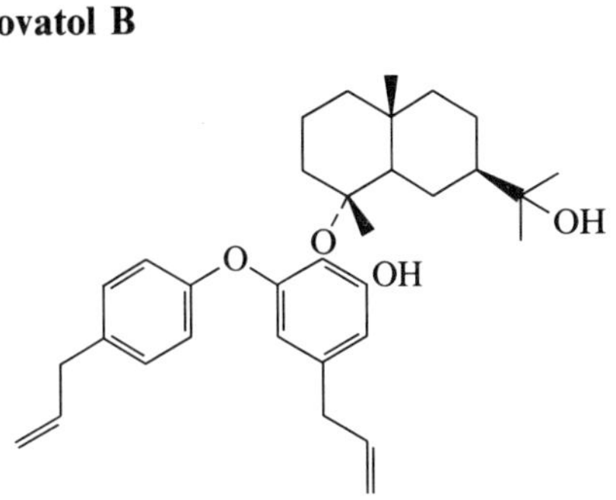

$C_{33}H_{44}O_4$ M 504.708
Constit. of *Magnolia obovata*. Neurotropic agent. $[\alpha]_D$
   −26.1° (c, 1.15 in $CHCl_3$).

Fukuyama, Y. *et al*, *Tetrahedron Lett.*, 1989, **30**, 5907.

Euparotin   E-80090

Updated Entry replacing E-01338
[10191-01-2]

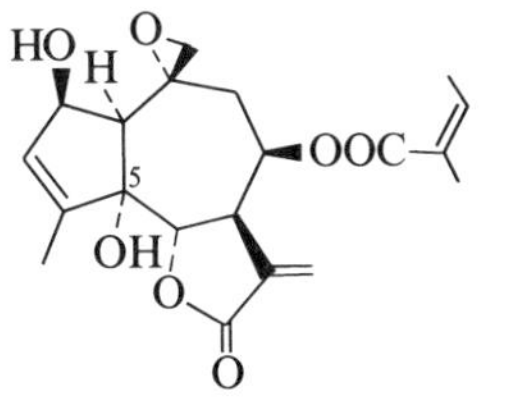

$C_{20}H_{24}O_7$ M 376.405
Constit. of *Eupatorium rotundifolium*. Tumour growth
   inhibitor. Cryst. (EtOAc/pet. ether). Mp 199-200° (vac.).
   $[\alpha]_D^{32}$ −124° (c, 1.25 in EtOH).

Ac: [10215-89-1]. **Euparotin acetate**
   $C_{22}H_{26}O_8$ M 418.443
   Constit. of *E. rotundifolium*. Tumour inhibitor. Cryst.
   ($CHCl_3/C_6H_6$/pet. ether). Mp 156-157° (vac.). $[\alpha]_D^{30}$
   −191° (c, 0.54 in EtOH).

5-Deoxy: [89396-25-8]. **5-Desoxyeuparotin**
   $C_{20}H_{24}O_6$ M 360.406
   Constit. of *Trichogonia santosii*. Oil.

Kupchan, S.M. *et al*, *J. Org. Chem.*, 1969, **34**, 3876 (*struct, isol*)
Kupchan, S.M. *et al*, *J. Med. Chem.*, 1971, **14**, 1147 (*pharmacol*)
McPhail, A.T. *et al*, *Tetrahedron*, 1973, **29**, 1751 (*abs config*)
Bohlmann, F. *et al*, *Justus Liebigs Ann. Chem.*, 1984, 162 (*deriv*)

Evodione   E-80091

*1-(5,7,8-Trimethoxy-2,2-dimethyl-2H-1-benzopyran-6-
yl)ethanone, 9CI. 6-Acetyl-5,7,8-trimethoxy-2,2-dimethyl-
2H-1-benzopyran*

[482-07-5]

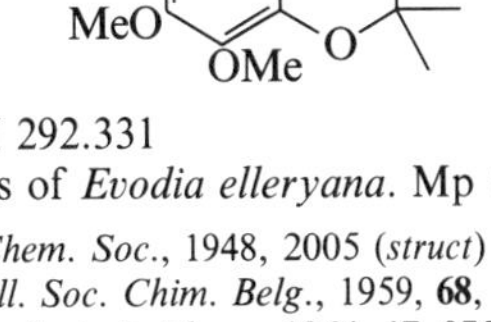

$C_{16}H_{20}O_5$ M 292.331
Isol. from leaves of *Evodia elleryana*. Mp 57°.

Wright, S.E., *J. Chem. Soc.*, 1948, 2005 (*struct*)
Huls, R. *et al*, *Bull. Soc. Chim. Belg.*, 1959, **68**, 325 (*synth*)
Barnes, C.S. *et al*, *Aust. J. Chem.*, 1964, **17**, 975 (*ms*)

Excelsaoctaphenol

E-80092

[121747-80-6]

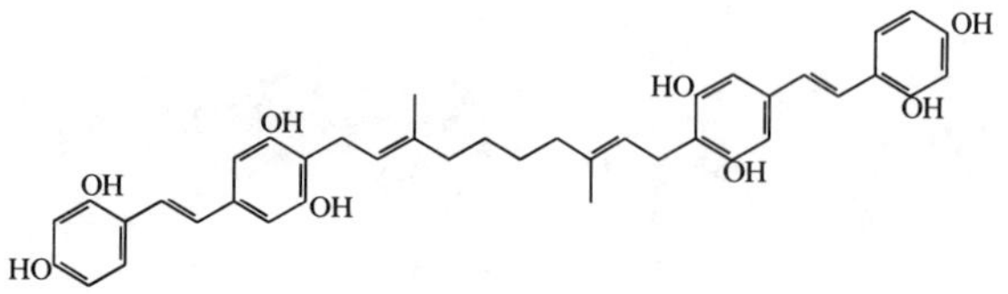

$C_{40}H_{42}O_8$ M 650.767

Constit. of *Chlorophora excelsa*. Light brown amorph. solid. Mp 210-216°.

Christensen, L.P. *et al*, *Phytochemistry*, 1989, **28**, 917.

Exidonin

E-80093

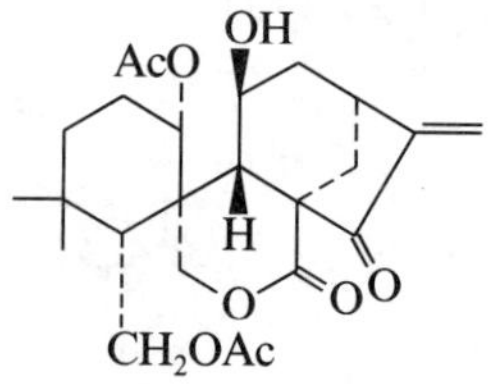

$C_{24}H_{32}O_8$ M 448.512

Constit. of *Rabdosia henryi*. Cryst. Mp 206-207°. $[\alpha]_D^{25}$ +101.6° (c, 0.37 in Py). A cleaved kaurane.

Meng, X. *et al*, *J. Nat. Prod.* (*Lloydia*), 1988, **51**, 812.

F

2,6,10-Farnesatriene-1,15;13,9-diolide   F-80001

2-Methyl-5-[2-methyl-5-(5-oxo-2H-furan-3-yl)-2-pentenyl]-2(5H)-furanone

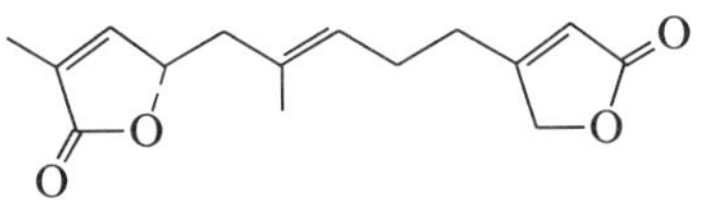

$C_{15}H_{18}O_4$ M 262.305

Compd. not named by authors. Constit. of *Eremophila miniata*. Amorph. solid. $[\alpha]_D$ +21.5° (c, 3 in $CHCl_3$).

Dastlik, K.A. *et al*, *Phytochemistry*, 1989, **28**, 1425.

7,9(11)-Fernadien-3-ol   F-80002

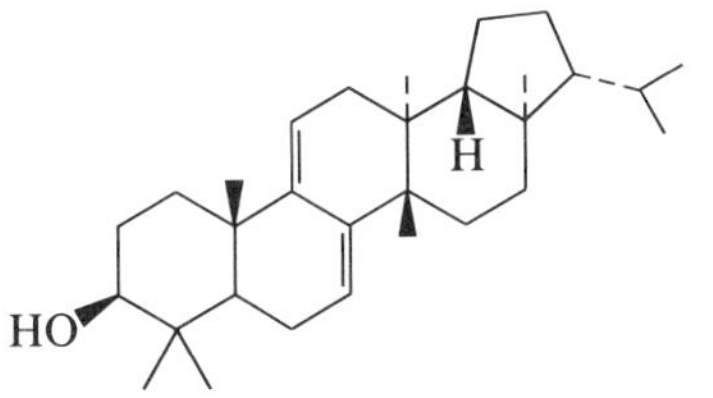

$C_{30}H_{48}O$ M 424.709

3β-form [53527-38-1]

Constit. of *Euphorbia supina*. Cryst. (MeOH/CHCl₃). Mp 193-196.5°.

Tanaka, R. *et al*, *Phytochemistry*, 1988, **27**, 3579.

Fernolin   F-80003

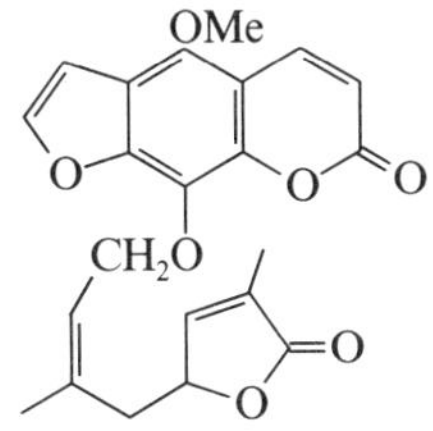

$C_{22}H_{20}O_7$ M 396.396

Constit. of *Fernonia limonia*. Cryst. Mp 262°.

Agrawal, A. *et al*, *Phytochemistry*, 1989, **28**, 1229.

Feronolide   F-80004

4,5-Dihydro-5-(1-oxotetradecyl)-2(3H)furanone. Dihydro-5-myristoyl-2(3H)-furanone, 8CI. 5-Oxo-4-octadecanolide
[21566-75-6]

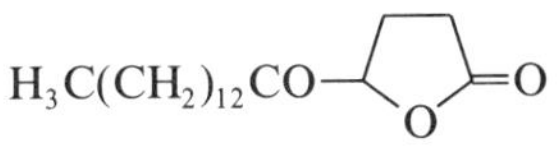

$$H_3C(CH_2)_{12}CO$$

$C_{18}H_{32}O_3$ M 296.449

The proposed struct. (illus.) has been synthesised and was not identical with Feronolide. Currently the struct. of Feronolide is unknown. Isol. from cortex of *Feronia elephantum*. Cryst. (EtOH). Mp 115°. Mp. given refers to natural Feronolide. Synthetic compd. had Mp 66.5-68.5° (2,4-dinitrophenylhydrazone Mp. 61-63°).

Tiwari, R.D. *et al*, *Arch. Pharm. (Weinheim, Ger.)*, 1964, **297**, 236.
Crundwell, E. *et al*, *J. Med. Chem.*, 1969, **12**, 547.

Ferprenin   F-80005

Updated Entry replacing F-70003
[114727-96-7]

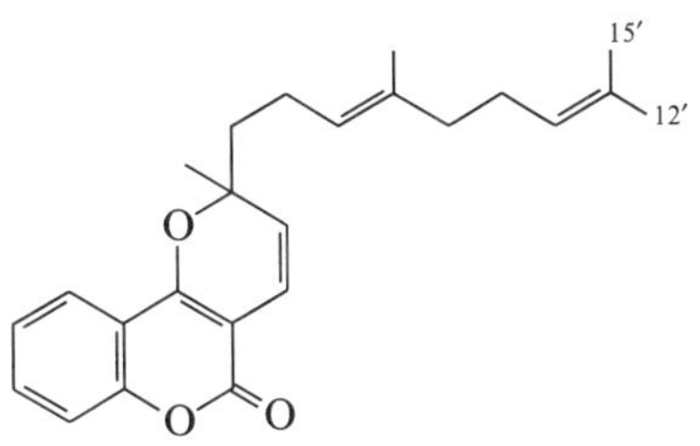

$C_{24}H_{28}O_3$ M 364.483

Constit. of *Ferula communis*. Oil. $[\alpha]_D^{25}$ +10° (c, 0.9 in $CHCl_3$).

12'-Hydroxy: (**E**)-ω-**Hydroxyferprenin**
$C_{24}H_{28}O_4$ M 380.483
Constit. of *F. communis*. Yellow oil.

15'-Hydroxy: (**Z**)-ω-**Hydroxyferprenin**
$C_{24}H_{28}O_4$ M 380.483
Constit. of *F. communis*. Yellow oil.

12'-Acetoxy: (**E**)-ω-**Acetoxyferprenin**
$C_{26}H_{30}O_5$ M 422.520
Constit. of *F. communis*. Yellow oil.

15'-Acetoxy: (**Z**)-ω-**Acetoxyferprenin**
$C_{26}H_{30}O_5$ M 422.520
Constit. of *F. communis*. Yellow oil.

12'-Oxo: (**E**)-ω-**Oxoferprenin**
$C_{24}H_{26}O_4$ M 378.467
Constit. of *F. communis*. Yellow oil.

Appendino, G. *et al*, *Phytochemistry*, 1988, **27**, 3619 (*deriv*)

Ferrugone   F-80006

3-(4,7-Dimethoxy-1,3-benzodioxol-5-yl)-8,8-dimethyl-4H,8H-benzo[1,2-b:3,4-b']dipyran-4-one, 9CI
[7731-08-0]

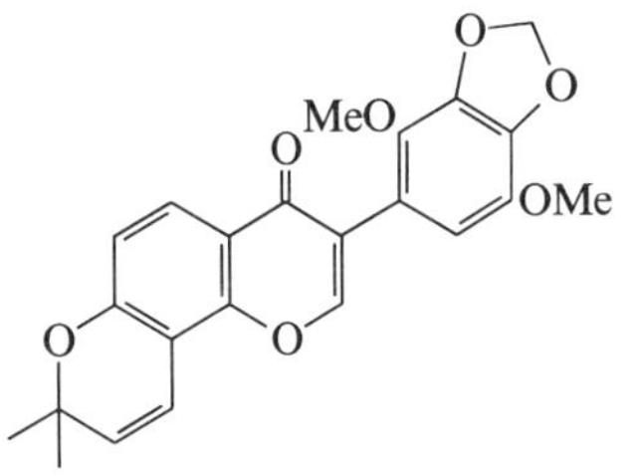

$C_{23}H_{20}O_7$ M 408.407

Isol. from *Millettia ferruginea* seeds. Needles (MeOH). Mp 167-169°.

Highet, R.J. *et al*, *J. Org. Chem.*, 1967, **32**, 1055 (*isol, uv, ir, pmr, ms, struct*)

Ferulenol
F-80007

Updated Entry replacing F-60005

4-Hydroxy-3-[(3,7,11-trimethyl-2,6,10-dodecatrienyl)]-2H-1-benzopyran-2-one, 9CI. 3-(1-Farnesyl)-4-hydroxycoumarin

[6805-34-1]

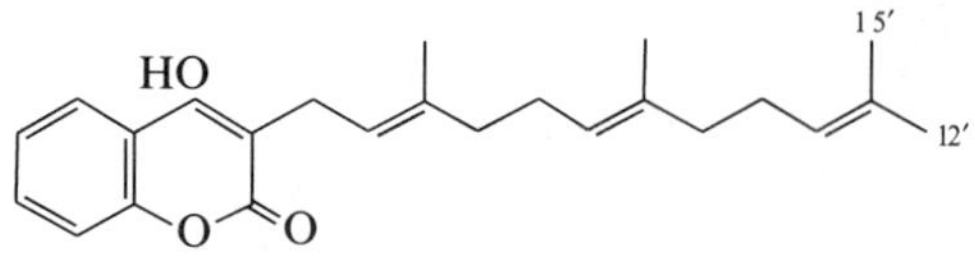

$C_{24}H_{30}O_3$ M 366.499

Constit. of the latex of *Ferula communis*. Shows haemorrhagic action. Mp 64-65°.

12′-Hydroxy: [106909-08-4]. (*E*)-ω-*Hydroxyferulenol*
$C_{24}H_{30}O_4$ M 382.499
Constit. of *F. communis*. Shows haemorrhagic action. Yellow oil.

15′-Hydroxy: [110107-13-6]. (*Z*)-ω-*Hydroxyferulenol*
$C_{24}H_{30}O_4$ M 382.499
Constit. of *F. communis* and *F. communis* var. *genuina*. Yellow oil.

12′-Acetoxy: [119234-02-5]. (*E*)-ω-*Acetoxyferulenol*
$C_{26}H_{32}O_5$ M 424.536
Constit. of *F. communis*. Oil.

15′-Acetoxy: [119259-89-1]. (*Z*)-ω-*Acetoxyferulenol*
$C_{26}H_{32}O_5$ M 424.536
Constit. of *F. communis*. Oil.

12′-Oxo: [106894-38-6]. (*E*)-ω-*Oxoferulenol*
$C_{24}H_{28}O_4$ M 380.483
Constit. of *F. communis*. Oil.

Carboni, S. *et al, Tetrahedron Lett.*, 1964, 2783 (*struct, pmr*)
Lamnaouer, D. *et al, Phytochemistry*, 1987, **26**, 1613 (*isol, pmr, cmr*)
Valle, M.G. *et al, Phytochemistry*, 1987, **26**, 253 (*isol*)
Appendino, G. *et al, Phytochemistry*, 1988, **27**, 3619 (*derivs*)

Filixic acids
F-80008

Updated Entry replacing F-20013

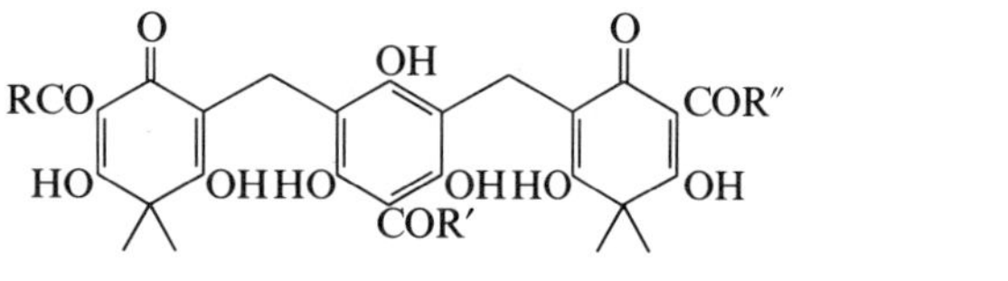

Filixic acid
ABA, R = R″ = CH₃, R′ = CH₂CH₂CH₃
ABP, R = CH₃, R′ = CH₂CH₂CH₃, R″ = CH₂CH₃
ABB, R = CH₃, R′ = R″ = CH₂CH₂CH₃
PBP, R = R″ = CH₂CH₃, R′ = CH₂CH₂CH₃
PBB, R = CH₂CH₃, R′ = R″ = CH₂CH₂CH₃
BBB, R = R′ = R″ = CH₂CH₂CH₃

The alphabetical suffixes refer to the acyl sustituents (A = acetyl, P = propionyl, B = butyryl). Isol. from ferns *Dryopteris* and *Aspidium* spp.

▷ LK4250000.

Filixic acid ABA [38226-84-5]
$C_{32}H_{36}O_{12}$ M 612.629
From *D. dickinsii*. Yellow needles (Me₂CO). Mp 163-166°.

Filixic acid ABP [57765-54-5]
$C_{33}H_{38}O_{12}$ M 626.656

Filixic acid ABB [37318-24-4]
$C_{34}H_{40}O_{12}$ M 640.683

Filixic acid PBP [51005-85-7]
$C_{34}H_{40}O_{12}$ M 640.683
Light-yellow needles (Me₂CO). Mp 192-194°. MF incorr. given as C_{30} by Hisada *et al*.

Filixic acid PBB [49582-09-4]
$C_{35}H_{42}O_{12}$ M 654.710
Mp 184-186°.

Filixic acid BBB [4482-83-1]
Filixic acid. Filicin
$C_{36}H_{44}O_{12}$ M 668.736
Cryst. (Me₂CO), light yellow tablets (EtOAc). Mp 184-185° (168-170°).

Penttila, A. *et al, Acta Chem. Scand.*, 1963, **17**, 191 (*isol, struct*)
Hisada, S. *et al, Phytochemistry*, 1972, **11**, 1850, 2881; 1973, **12**, 1493 (*isol*)
Lounasmaa, M. *et al, Helv. Chim. Acta*, 1973, **56**, 1133 (*ms*)
Widen, C.J. *et al, Planta Med.*, 1975, **28**, 144 (*isol*)
Lounasmaa, M., *Planta Med.*, 1978, **33**, 173 (*cmr*)

Flaccidinin
F-80009

2,6-Dihydroxy-7-methoxy-5H-phenanthro[4,5-bcd]pyran-5-one, 9CI

[121817-23-0]

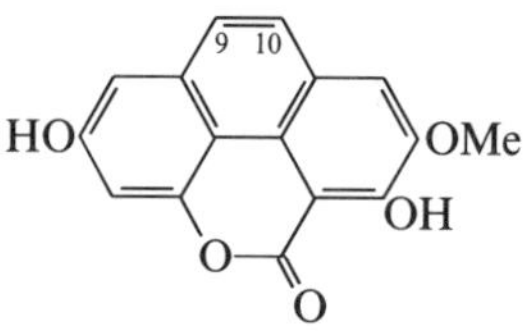

$C_{16}H_{10}O_5$ M 282.252

Constit. of *Coelogyne flaccida*. Cryst. (EtOAc/pet. ether). Mp 360° dec.

9,10-Dihydro: [121817-24-1]. *Oxoflaccidin*
$C_{16}H_{12}O_5$ M 284.268
Constit. of *C. flaccida*.

9,10-Dihydro, deoxo: [115531-76-5]. *Flaccidin*
$C_{16}H_{14}O_4$ M 270.284
Constit. of *C. flaccida*. Cryst. (CHCl₃). Mp 200°.

Majumder, P.L. *et al, Phytochemistry*, 1988, **27**, 899; 1989, **28**, 887.

Flavanthrone, 8CI
F-80010

Benzo[h]benz[5,6]acridino[2,1,9,8-klmna]acridine-8,16-dione, 9CI. C.I. Pigment Yellow 24. C.I. Vat Yellow 1

[475-71-8]

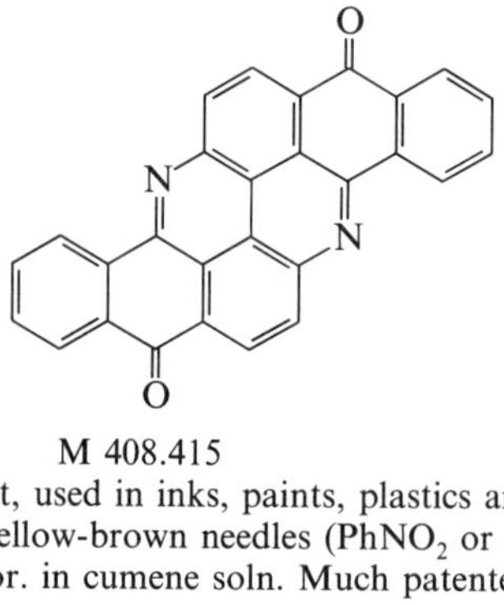

$C_{28}H_{12}N_2O_2$ M 408.415

Dye, pigment, used in inks, paints, plastics and coloured textiles. Yellow-brown needles (PhNO₂ or by subl.). Green fluor. in cumene soln. Much patented.

Colour Index, 3, 3274.
Krepelka, V. *et al, Collect. Czech. Chem. Commun.*, 1937, **9**, 29 (*synth*)
Stadler, H.P., *Acta Crystallogr.*, 1953, **6**, 540 (*cryst struct*)
Durie, R.A. *et al, Aust. J. Chem.*, 1957, **10**, 429 (*ir*)
Moran, J.J. *et al, J. Chem. Soc.*, 1957, 765 (*purifn, uv*)
Sramek, J., *J. Soc. Dyers Colour.*, 1962, **78**, 326 (*chromatog*)

Kratochvil, V. *et al*, *Collect. Czech. Chem. Commun.*, 1974, **39**, 2814 (*synth, uv*)
Kirk-Othmer Encycl. Chem. Technol., 3rd Ed., 1978-1984, *Wiley, NY*, 3rd Ed., Wiley, N.Y., 1984, **8**, 242.

Flavaspidic acids   F-80011

Updated Entry replacing F-00164

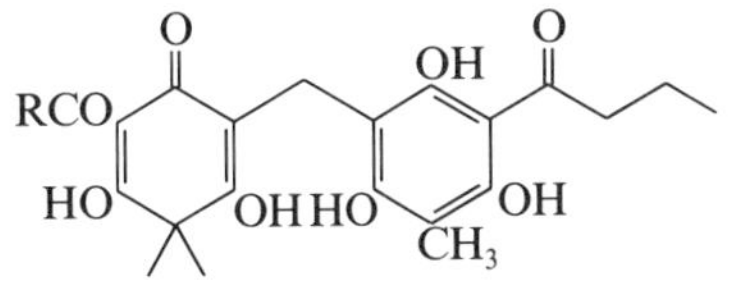

Flavaspidic acid   AB, R = CH₃
                   PB, R = CH₂CH₃
                   BB, R = CH₂CH₂CH₃

The suffixes refer to the acyl substituents (A = acetyl, P = propionyl, B = butyryl). Isol. from ferns, esp. *Dryopteris* spp.

▷ EU1400000.

Flavaspidic acid AB [3761-64-6]
$C_{22}H_{26}O_8$ M 418.443
Mp 210-212°.

Flavaspidic acid PB [3773-25-9]
$C_{23}H_{28}O_8$ M 432.469
Mp 170-171°.

Di-Ac:
$C_{28}H_{34}O_{10}$ M 530.571
Mp 142-143°.

Flavaspidic acid BB [114-42-1]
Flavaspidic acid. Polystichocitrin. Toxifren
$C_{24}H_{30}O_8$ M 446.496
Yellow or colourless cryst. Mp 92°, Mp 156° (double Mp). The two forms are considered to be tautomers which interconvert on heating.

Di-Ac: Mp 142-143°.

McGookin, A. *et al*, *J. Chem. Soc.*, 1953, 1828 (*synth*)
Penttilä, A. *et al*, *Acta Chem. Scand.*, 1964, **18**, 344 (*isol, synth*)
Erämetsä, O. *et al*, *Acta Chem. Scand.*, 1970, **24**, 3335 (*cryst struct*)
Lounasmaa, M. *et al*, *Helv. Chim. Acta*, 1973, **56**, 1133 (*ms*)
Widén, C.-J. *et al*, *Helv. Chim. Acta*, 1976, **59**, 1725 (*isol, bibl*)
Lounasmaa, M., *Planta Med.*, 1978, **33**, 173 (*cmr*)

Flindercarpin 2   F-80012

[61903-01-3]

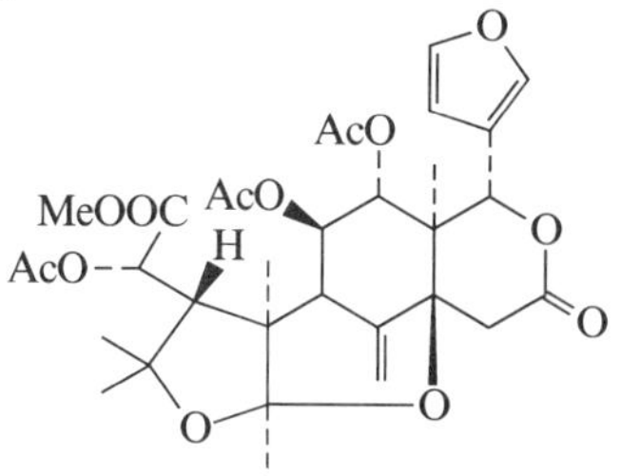

$C_{32}H_{40}O_{13}$ M 632.660
Isol. from the bark of *Flindersia laevicarpa*. Shows insect antifeedant activity. Mp 232°. The structs of Flindercarpin 1, $C_{25}H_{28}O_{11}$ and Flindercarpin 3, $C_{26}H_{32}O_9$ have not yet been detd.

Picker, K. *et al*, *Aust. J. Chem.*, 1976, **29**, 2023 (*isol*)
Breen, G.W. *et al*, *Aust. J. Chem.*, 1984, **37**, 1461 (*cryst struct*)

Floribundone 2   F-80013

[118555-83-2]

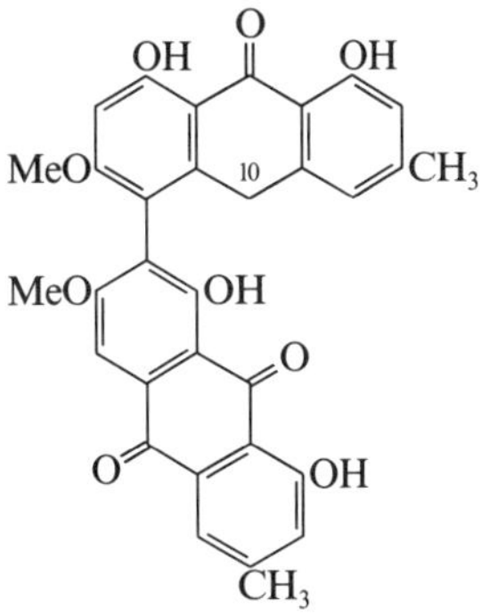

$C_{32}H_{24}O_9$ M 552.536
Constit. of leaves of *Cassia floribunda*.

10-Oxo: [118555-84-3]. *Floribundone 1*
$C_{32}H_{22}O_{10}$ M 566.520
Orange cryst. (MeOH). Mp >260°. $[\alpha]_D^{18}$ +130° (c, 0.05 in CHCl₃).

Alemayehu, G. *et al*, *Phytochemistry*, 1988, **27**, 3255.

Floridic acid   F-80014

3β,6β,19α-Trihydroxy-23-nor-4(24),12-ursadien-28-oic acid

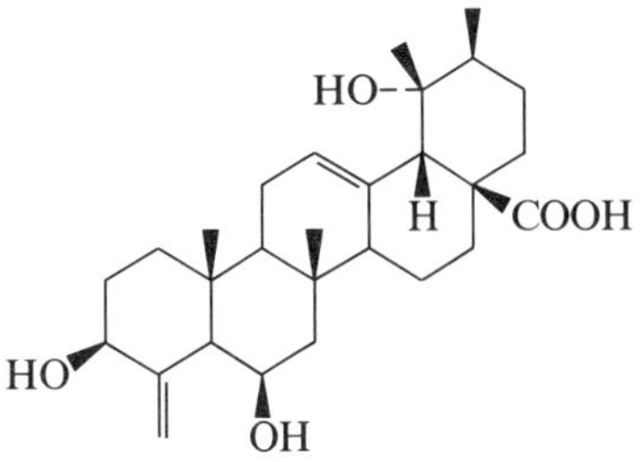

$C_{29}H_{44}O_5$ M 472.664
Constit. of *Uncaria florida*.

Me ester: Prisms (Me₂CO aq.). Mp 146-148°.

Aimi, N. *et al*, *Tetrahedron*, 1989, **45**, 4125.

9*H*-Fluorene-9-thiol   F-80015

9-Fluorenylmercaptan. 9-Mercaptofluorene
[19552-08-0]

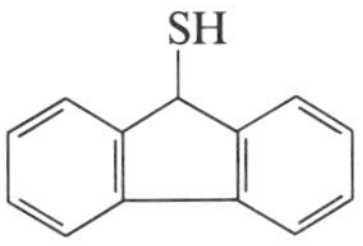

$C_{13}H_{10}S$ M 198.288
Plates (MeOH/AcOH). Mp 105-106°.

S-Me: [59431-17-3]. *9-(Methylthio)-9H-fluorene*
$C_{14}H_{12}S$ M 212.315
Cryst. (MeOH). Mp 46-47°.

S-Me, S-oxide: 9-(Methylsulfinyl)-9H-fluorene
$C_{14}H_{12}OS$ M 228.314
Mp 83-84°.

S-Me, S-dioxide: [31859-90-2]. *9-(Methylsulfonyl)-9H-fluorene*
$C_{14}H_{12}O_2S$ M 244.314
Long needles (MeOH or AcOH). Mp 188-189°.

S-Ph: [28114-92-3]. *9-(Phenylthio)-9H-fluorene*
$C_{19}H_{14}S$ M 274.386
Mp 48-49°.

S-*Ph*, S-*oxide:* 9-(*Phenylsulfinyl*)-9H-*fluorene*
$C_{19}H_{14}OS$ M 290.385
Mp 104-105°.

S-*Ph*, S-*dioxide:* [22010-78-2]. 9-(*Phenylsulfonyl*)-9H-*fluorene*
$C_{19}H_{14}O_2S$ M 306.384
Prisms (Me_2CO/heptane). Mp 182-183°.

Klerk, M.M. *et al, J. Am. Chem. Soc.*, 1948, **70**, 3846 (*synth*)
Bavin, P.M.G., *Can. J. Chem.*, 1960, **38**, 917; 1962, **40**, 220 (*deriv*)
Pan, H.-L. *et al, Chem. Ind.* (*London*), 1968, 546 (*synth*)
Kice, J.L. *et al, J. Org. Chem.*, 1988, **53**, 3593 (*synth, pmr, deriv*)

9*H*-Fluoren-1-ol, 9CI   F-80016

Updated Entry replacing F-00259
1-Hydroxyfluorene
[6344-61-2]

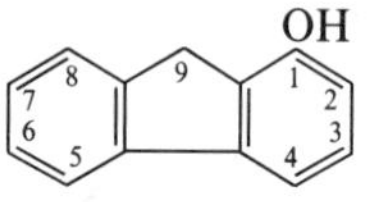

$C_{13}H_{10}O$ M 182.221
A quoted synthesis by Kumar *et al* (1983) is prob. erroneous. Mp 119.9-120.5°.

Me ether: [26060-14-0]. *1-Methoxyfluorene*
$C_{14}H_{12}O$ M 196.248
Mp 85-86°.

Weisburger, E.K. *et al, J. Org. Chem.*, 1953, **18**, 864 (*synth, derivs*)
Shapiro, M.J., *J. Org. Chem.*, 1978, **43**, 3769 (*nmr*)
Kumar, B. *et al, Synthesis*, 1983, 115 (*synth*)
Blokker, V. *et al, J. Org. Chem.*, 1988, **53**, 5567 (*synth*)

Fluorescamine   F-80017

4-Phenylspiro[furan-2(3H),1′(3′H)-isobenzofuran]-3,3′-dione, 9CI. 4-Phenylspiro[furan-2(3H),1′phthalan]-3,3′-dione. Fluram
[38183-12-9]

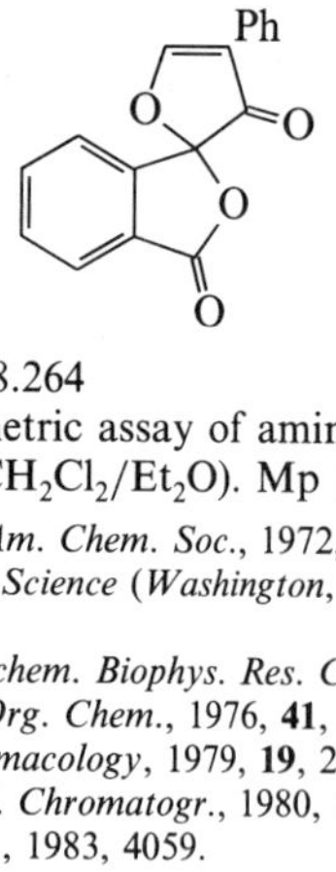

$C_{17}H_{10}O_4$ M 278.264
Reagent for fluorimetric assay of amino acids and primary amines. Cryst. (CH_2Cl_2/Et_2O). Mp 154-155°.

Weigele, M. *et al, J. Am. Chem. Soc.*, 1972, **94**, 5927 (*synth*)
Undenfriend, S. *et al, Science* (*Washington, D.C.*), 1972, **178**, 871 (*use*)
Weigele, M. *et al, Biochem. Biophys. Res. Commun.*, 1973, **50**, 352.
Weigele, M. *et al, J. Org. Chem.*, 1976, **41**, 388 (*synth, ir, uv*)
Undenfriend, S., *Pharmacology*, 1979, **19**, 223 (*rev, use*)
Doetsch, P.W. *et al, J. Chromatogr.*, 1980, **189**, 79 (*use, bibl*)
Merck Index, 10th Ed., 1983, 4059.

2-Fluorobenzenethiol, 9CI   F-80018

1-Fluoro-2-mercaptobenzene. 2-Fluorothiophenol
[2557-78-0]

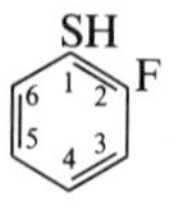

C_6H_5FS M 128.170
Bp_{15} 65°. n_D^{26} 1.5395.

Sharghi, N., *J. Chem. Eng. Data*, 1963, **8**, 276 (*synth*)
Cervena, I. *et al, Collect. Czech. Chem. Commun.*, 1979, **44**, 2139 (*synth*)

3-Fluorobenzenethiol, 9CI   F-80019

1-Fluoro-3-mercaptobenzene. 3-Fluorothiophenol
[2557-77-9]

C_6H_5FS M 128.170
Bp 156°, Bp_{11} 53°. n_D^{27} 1.5481.

Taft, R.W., *J. Am. Chem. Soc.*, 1963, **85**, 709 (*pmr*)
Sharghi, N. *et al, J. Chem. Eng. Data*, 1963, **8**, 276 (*synth*)
Rajšner, M. *et al, Collect. Czech. Chem. Commun.*, 1967, **32**, 2021 (*synth*)
Cutress, N.C. *et al, J. Chem. Soc., Perkin Trans.* 2, 1974, 263 (*ir*)
Protiva, M. *et al, Collect. Czech. Chem. Commun.*, 1986, **51**, 2598 (*synth*)

4-Fluorobenzenethiol, 9CI   F-80020

1-Fluoro-4-mercaptobenzene. 4-Fluorothiophenol
[371-42-6]

C_6H_5FS M 128.170
Bp 162°, $Bp_{2.3}$ 29-30°. n_D^{25} 1.5485.

Oae, S. *et al, Tetrahedron Lett.*, 1972, 1283 (*synth*)
Cutress, N.C. *et al, J. Chem. Soc., Perkin Trans.* 2, 1974, 263 (*ir*)
Rajsner, M. *et al, Collect. Czech. Chem. Commun.*, 1975, **40**, 719 (*synth*)
Senoff, C.V. *et al, Inorg. Chem.*, 1975, **14**, 278 (*cmr*)
Pradip, H. *et al, Proc. Natl. Acad. Sci. U.S.A.*, 1982, **79**, 7056 (*nmr*)

3-Fluorobenzocyclobutene-1,2-dione   F-80021

2-Fluorobicyclo[4.2.0]octa-1,3,5-triene-7,8-dione
[118112-24-6]

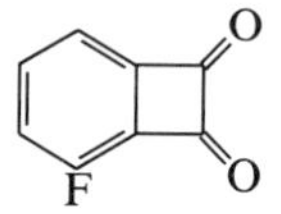

$C_8H_3FO_2$ M 150.109
Cryst. (hexane). Mp 95°.

Liebeskind, L.S. *et al, J. Org. Chem.*, 1989, **54**, 1435 (*synth, ir, pmr, cmr*)

6-Fluorobenzocyclobuten-1-one   F-80022

5-Fluorobicyclo[4.2.0]octa-1,3,5-trien-7-one
[118112-20-2]

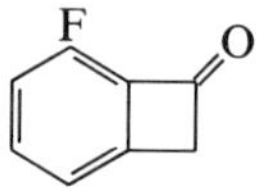

C_8H_5FO M 136.125
Cryst. by subl. Mp 39-42°.

Liebeskind, L.S. *et al, J. Org. Chem.*, 1989, **54**, 1435 (*synth, ir, pmr, cmr*)

6-Fluorobenzo[*a*]pyrene   F-80023

[59417-86-6]

$C_{20}H_{11}F$ M 270.305

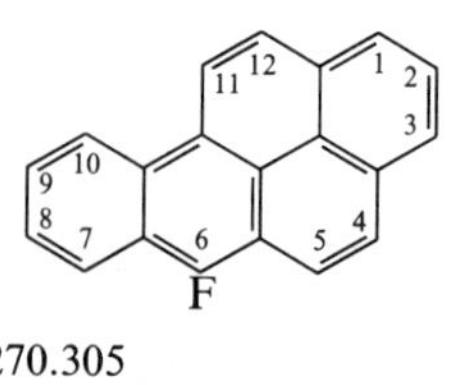

Cryst. (Me$_2$CO/MeOH). Mp 167-169°.

Agranat, I. *et al, Experientia,* 1976, **32**, 417 (*synth*)
Cremonesi, P. *et al, J. Org. Chem.,* 1989, **54**, 3561 (*synth, pmr*)

7-Fluorobenzo[*a*]pyrene   F-80024

[71511-38-1]
C$_{20}$H$_{11}$F   M 270.305
Yellow-green fluorescent cryst. (Et$_2$O/pet.ether). Mp 174-175°.

Paulsen, H. *et al, Chem. Ber.,* 1979, **112**, 2907 (*synth, uv, ir*)
Newman, M.S. *et al, J. Am. Chem. Soc.,* 1979, **101**, 6788 (*synth, pmr, F nmr*)
Sardella, D.J. *et al, Magn. Reson. Chem.,* 1986, **24**, 287 (*cmr*)

8-Fluorobenzo[*a*]pyrene   F-80025

[71171-92-1]
C$_{20}$H$_{11}$F   M 270.305
Cryst. by subl. Mp 167.5-168°.

Newman, M.S. *et al, J. Org. Chem.,* 1979, **44**, 3388 (*synth*)
Sardella, D.J. *et al, Magn. Reson. Chem.,* 1986, **24**, 287 (*cmr*)

9-Fluorobenzo[*a*]pyrene   F-80026

[71171-93-2]
C$_{20}$H$_{11}$F   M 270.305
Yellow-green fluorescent cryst. (Et$_2$O/pet.ether). Mp 157.5-158.5°.

Paulsen, H. *et al, Chem. Ber.,* 1979, **112**, 2907 (*synth, uv, ir*)
Newman, M.S. *et al, J. Org. Chem.,* 1979, **44**, 3388 (*synth*)
Sardella, D.J. *et al, Magn. Reson. Chem.,* 1986, **24**, 287 (*cmr*)

10-Fluorobenzo[*a*]pyrene   F-80027

[74018-58-9]
C$_{20}$H$_{11}$F   M 270.305
Cryst. (C$_6$H$_6$/EtOH). Mp 152-153°.

Newman, M.S. *et al, Bull. Soc. Chim. Belg.,* 1979, **88**, 871 (*synth*)
Sardella, D.J. *et al, Magn. Reson. Chem.,* 1986, **24**, 287 (*cmr*)

1-Fluorobicyclo[2.2.2]octane   F-80028

[20277-22-9]

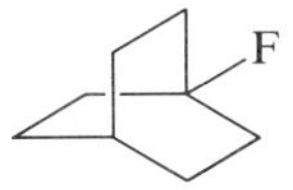

C$_8$H$_{13}$F   M 128.189
Cryst. Mp 174° (182-184°). Mostly sublimes at 152°.

Suzuki, Z. *et al, Bull. Chem. Soc. Jpn.,* 1968, **41**, 1724 (*synth*)
Kopecky, J. *et al, Chem. Ind. (London),* 1969, 271 (*synth*)
Maciel, G.E. *et al, J. Am. Chem. Soc.,* 1971, **93**, 1268 (*cmr*)
Rozen, S. *et al, J. Org. Chem.,* 1988, **53**, 2803 (*synth*)

2-Fluorobicyclo[3.2.1]octane   F-80029

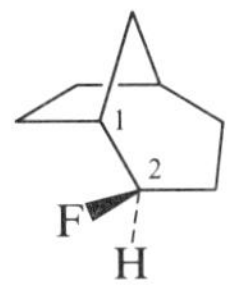

C$_8$H$_{13}$F   M 128.189
(*1RS,2RS*)-*form* [114423-47-1]
  (±)-exo-*form*
  Cryst. (pentane/MeOH). Mp 131°.

Rozen, S. *et al, J. Org. Chem.,* 1988, **53**, 2803 (*synth, pmr, F nmr, cmr*)

2-Fluoro-2-butenedioic acid, 9CI   F-80030

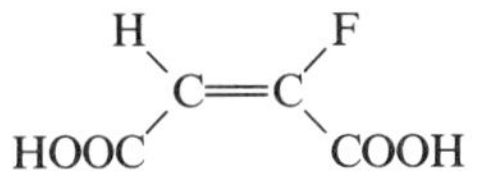

H$_3$C──C══C──F   (*E*)-*form*
HOOC──────COOH

C$_4$H$_3$FO$_4$ M 134.064
(*E*)-*form* [760-82-7]
  Fluoromaleic acid
  Cryst. (Me$_2$CO/C$_6$H$_6$). Mp 132-133°.
  Anhydride: [2714-23-0]. *3-Fluoro-2,5-furandione, 9CI.*
  Fluoromaleic anhydride
  C$_4$HFO$_3$ M 116.048
  Bp 162°. n$_D^{25}$ 1.4452.
(*Z*)-*form* [672-18-4]
  Fluorofumaric acid
  Cryst. (Et$_2$O/pet. ether). Mp 206-208° dec., Mp 236-237°.

Raasch, M. *et al, J. Am. Chem. Soc.,* 1959, **81**, 2678 (*anhydride*)
Akhtar, M. *et al, Tetrahedron,* 1987, **43**, 5899 (*synth, ir, pmr, cmr, ms*)

1-Fluorocyclohexanecarboxylic acid   F-80031

[117169-31-0]

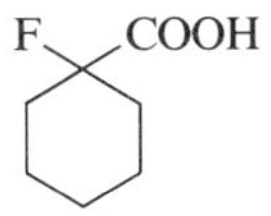

F──COOH

C$_7$H$_{11}$FO$_2$ M 146.161
Oily solid.
Me ester: [117169-32-1].
  C$_8$H$_{13}$FO$_2$ M 160.188
  Oil.

Mongelli, N. *et al, Synthesis,* 1988, 310 (*synth, ir, pmr, cmr*)

N-Fluoro-2,3-dihydro-3,3-dimethyl-1,2-benzothiadiazole 1,1-dioxide   F-80032

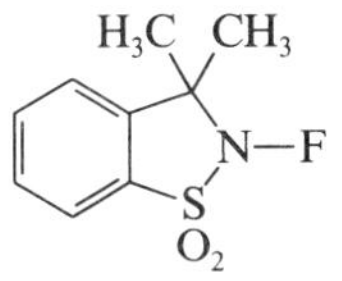

H$_3$C   CH$_3$

N──F

S

O$_2$

C$_9$H$_{10}$FNO$_2$S   M 215.248
Superior fluorinating agent for synth. of α-fluorocarbonyl compds. Cryst. Mp 114-116°.

Differding, E. *et al, Helv. Chim. Acta,* 1989, **72**, 1248 (*synth, pmr, cmr, ms, use*)

Fluorododecahedrane   F-80033

1-Fluorohexadecahydro-5,2,1,6,3,4-[2,3]butanediyl[1,4]diylidenepentaleno[2,1,6-cde:2′,1′,6′-gha]pentalene, 9CI
[112896-29-4]

C$_{20}$H$_{19}$F   M 278.368

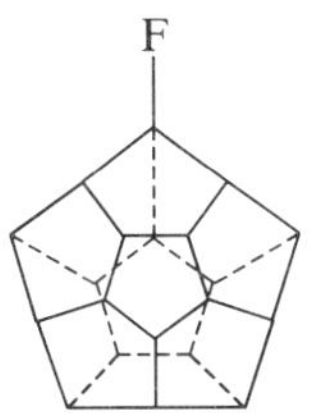

Solid. Mp >260°.

Paquette, L.A. *et al*, *J. Am. Chem. Soc.*, 1988, **101**, 1303 (*synth*, *cmr*)

1-Fluoro-9H-fluorene   F-80034

[343-25-9]

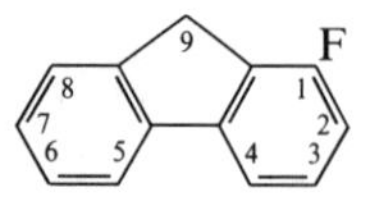

$C_{13}H_9F$ M 184.212
Cryst. (EtOH aq.). Mp 83-84°.

Suzuki, K. *et al*, *J. Org. Chem.*, 1959, **24**, 1511 (*synth*, *ir*)
Sardella, D.J. *et al*, *J. Fluorine Chem.*, 1984, **24**, 249.

2-Fluoro-9H-fluorene   F-80035

[343-43-1]

$C_{13}H_9F$ M 184.212
Cryst. (C_6H_6). Mp 101-102°. pK_a 21.0 (DMSO,25°).

Bergmann, E.D. *et al*, *J. Am. Chem. Soc.*, 1956, **78**, 6037 (*synth*)
Fletcher, T.L. *et al*, *Chem. Ind.* (*London*), 1961, 179 (*synth*)
Bordwell, F.G. *et al*, *J. Org. Chem.*, 1976, **41**, 2391; 1988, **53**, 780 (*props*)
Kitching, W. *et al*, *Org. Magn. Reson.*, 1980, **14**, 502 (*cmr*)

3-Fluoro-9H-fluorene   F-80036

[343-40-8]

$C_{13}H_9F$ M 184.212
Cryst. (cyclohexane). Mp 80°. pK_a 22.07 (DMSO, 25°C).

Suzuki, K. *et al*, *J. Org. Chem.*, 1959, **24**, 1511; 1961, **26**, 2236 (*synth*)
Bordwell, F.G. *et al*, *J. Org. Chem.*, 1988, **53**, 780 (*props*)

4-Fluoro-9H-fluorene   F-80037

[317-71-5]

$C_{13}H_9F$ M 184.212
Cryst. (cyclopentane). Mp 38-39°.

Fletcher, T.L. *et al*, *J. Org. Chem.*, 1960, **25**, 996 (*synth*)
Buckle, D.R. *et al*, *J. Chem. Soc.*, *Perkin Trans.* 1, 1979, 3004 (*synth*)

9-Fluoro-9H-fluorene   F-80038

[20825-90-5]

$C_{13}H_9F$ M 184.212
Cryst. (MeOH at −60°). Mp 60-60.5°. Polymerises with loss of HF at r.t.

Streitwieser, A. *et al*, *J. Am. Chem. Soc.*, 1968, **90**, 2444 (*synth*, *uv*, *ir*, *pmr*)
Johnson, A.L., *J. Org. Chem.*, 1982, **47**, 5220 (*F nmr*)

1-Fluoro-9H-fluoren-9-one, 9CI   F-80039

[1514-16-5]

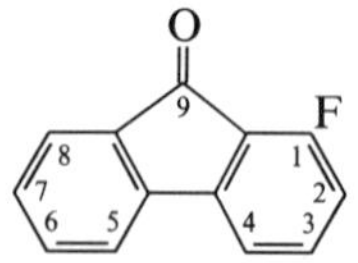

$C_{13}H_7FO$ M 198.196
Cryst. (pet. ether or EtOH). Mp 110-111.5°.

Fletcher, T.L. *et al*, *J. Org. Chem.*, 1960, **25**, 996 (*synth*)
Fletcher, T.L. *et al*, *Chem. Ind.* (*London*), 1961, 179 (*synth*)
Agranat, I. *et al*, *J. Am. Chem. Soc.*, 1977, **99**, 7068 (*synth*, *F nmr*)

2-Fluoro-9H-fluoren-9-one   F-80040

[343-01-1]

$C_{13}H_7FO$ M 198.196
Cryst. (MeOH). Mp 117-118°. Bp_{10} 185°.

Bergmann, E. *et al*, *Chem. Ber.*, 1933, **66**, 46 (*synth*, *deriv*)
Fletcher, T.L. *et al*, *Chem. Ind.* (*London*), 1961, 179 (*synth*)
Schaefer, T. *et al*, *Can. J. Chem.*, 1986, **64**, 2162 (*cmr*, *F nmr*)

3-Fluoro-9H-fluoren-9-one   F-80041

[1514-15-4]

$C_{13}H_7FO$ M 198.196
Yellow needles (C_6H_6/pet. ether). Mp 129-130°.

Suzuki, K. *et al*, *J. Org. Chem.*, 1961, **26**, 2239 (*synth*, *uv*)
Dewar, M.J.S. *et al*, *J. Org. Chem.*, 1963, **28**, 1759 (*synth*)
Agranat, I. *et al*, *J. Am. Chem. Soc.*, 1977, **99**, 7068 (*F nmr*)

4-Fluoro-9H-fluoren-9-one   F-80042

[1514-18-7]

$C_{13}H_7FO$ M 198.196
Yellow needles (EtOH). Mp 161-162.5°.

Fletcher, T.L. *et al*, *J. Org. Chem.*, 1960, **25**, 996 (*synth*)
Fletcher, T.L. *et al*, *Chem. Ind.* (*London*), 1961, 179 (*synth*)

3-Fluoro-1H-indole, 9CI   F-80043

[66946-81-4]

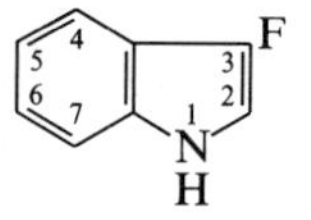

C_8H_6FN M 135.140
Cryst. or liq. Mp ∼20°.

Barton, D.H.R. *et al*, *J. Chem. Soc.*, *Perkin Trans.* 1, 1977, 2604 (*synth*, *uv*, *ir*, *pmr*)
Ermolenko, M.S. *et al*, *Khim. Geterotsikl. Soedin.*, 1978, 933; *CA*, **89**, 146707v (*synth*)

4-Fluoro-1H-indole, 9CI   F-80044

[387-43-9]

C_8H_6FN M 135.140
Prisms (MeOH aq.). Mp 29-30°. pK_a 16.30.

Picrate: Orange-red cryst. (C_6H_6). Mp 164-166°.

1,3,5-Trinitrobenzene complex: Orange-yellow cryst. (EtOH). Mp 172°.

Allen, F.L. *et al*, *J. Chem. Soc.*, 1955, 1283 (*synth*, *uv*, *derivs*)
Bentov, N. *et al*, *Isr. J. Chem.*, 1964, **2**, 25 (*synth*)
Yagil, G., *Tetrahedron*, 1967, **23**, 2855 (*props*)
Asher, K. *et al*, *Isr. J. Chem.*, 1968, **6**, 927.
Sugasawa, T. *et al*, *J. Org. Chem.*, 1979, **44**, 578 (*synth*, *ir*)
Somei, M. *et al*, *Chem. Pharm. Bull.*, 1985, **33**, 3696 (*synth*, *ir*, *pmr*)

5-Fluoro-1H-indole, 9CI   F-80045

[399-52-0]

C_8H_6FN M 135.140
Cryst. (C_6H_6). Mp 46-47°. pK_a 16.30.

Picrate: Orange cryst. (C_6H_6). Mp 155-156°.

Allen, F.L. *et al*, *J. Chem. Soc.*, 1955, 1283 (*synth*, *uv*, *derivs*)
Pelchowicz, Z. *et al*, *J. Chem. Soc.*, 1961, 5418 (*synth*)
Hoffmann, E. *et al*, *J. Heterocycl. Chem.*, 1965, **2**, 298 (*synth*)
Yagil, G., *Tetrahedron*, 1967, **23**, 2855 (*props*)
Lallemand, J.Y. *et al*, *Bull. Soc. Chim. Fr.*, 1970, **11**, 4091 (*pmr*)
Sirowe, H. *et al*, *Synthesis*, 1972, 84 (*synth*)

Rosenberg, E. *et al*, *Org. Magn. Reson.*, 1976, **8**, 117 (*N-15 nmr*, *cmr*)
Ayachit, N.H. *et al*, *Spectrochim. Acta, Part A*, 1986, **42**, 781 (*ir*, *uv*)

6-Fluoro-1*H*-indole, 9CI   F-80046

[399-51-9]

C_8H_6FN M 135.140
Cryst. (Et$_2$O/pet. ether). Mp 75-76°.
Picrate: Orange-red cryst. (C$_6$H$_6$). Mp 147-148°.

Allen, F.L. *et al*, *J. Chem. Soc.*, 1955, 1283, 1286 (*synth, uv, derivs*)
Bergmann, E.D. *et al*, *J. Chem. Soc.*, 1959, 1913 (*synth*)
Bentov, M. *et al*, *J. Chem. Soc.*, 1962, 2825 (*synth*)
Sugasawa, T. *et al*, *J. Org. Chem.*, 1979, **44**, 478 (*synth*)
Guo, W. *et al*, *Magn. Reson. Chem.*, 1986, **24**, 75 (*F-19 nmr, cmr*)
Ayachit, N.H. *et al*, *Spectrochim. Acta, Part A*, 1986, **42**, 781 (*uv*, *ir*)

7-Fluoro-1*H*-indole, 9CI   F-80047

C_8H_6FN M 135.140
Cryst. (EtOH aq.). Mp 61-62°.
Picrate: Orange-red cryst. (C$_6$H$_6$). Mp 154-155°.

Allen, F.L. *et al*, *J. Chem. Soc.*, 1955, 1283 (*synth, uv*)

2-Fluoro-3-iodopyridine   F-80048

[113975-22-7]

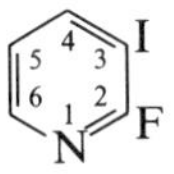

C_5H_3FIN M 222.988
Cryst. Mp 42°. Bp$_{17}$ 97-99°.

Estel, L. *et al*, *J. Org. Chem.*, 1988, **53**, 2740 (*synth, ir, pmr*)

1-Fluoro-4-isothiocyanatobenzene, 9CI   F-80049

p-*Fluorophenyl isothiocyanate, 8CI. 4-Fluorophenyl thiocarbimide*

[1544-68-9]

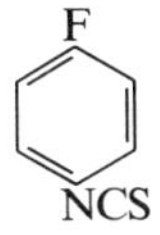

C_7H_4FNS M 153.180
Has fungicidal props. Oil. Mp 23.5-25.5° (12°). Bp 228°, Bp$_{1.3}$ 60°. n_D^{25} 1.6178.

Craig, C. *et al*, *Aust. J. Chem.*, 1960, **13**, 341 (*synth*)
Ham, N.S. *et al*, *Spectrochim. Acta, Part A*, 1960, **16**, 279 (*ir*)
Kinoshita, Y. *et al*, *Agric. Biol. Chem.*, 1966, **30**, 447 (*synth*)
Weigert, F.J. *et al*, *J. Org. Chem.*, 1976, **41**, 4006 (*F-nmr*)

2-Fluoro-4-methylquinoline, 9CI   F-80050

2-Fluorolepidine, 8CI

$C_{10}H_8FN$ M 161.178
Liq. Bp$_{16}$ 140°.

U.K. Pat., 845 062, (1960); CA, **55**, 5544 (*synth*)

3-Fluoro-4-methylquinoline, 9CI   F-80051

3-Fluorolepidine, 8CI

[72473-78-0]

$C_{10}H_8FN$ M 161.178
Oil.

Dehmlow, E.V. *et al*, *Justus Liebigs Ann. Chem.*, 1979, 1456 (*synth, pmr, F nmr*)

4-Fluoro-2-methylquinoline, 9CI   F-80052

4-Fluoroquinaldine, 8CI

$C_{10}H_8FN$ M 161.178
Cryst. Mp 40° (hydrate). Bp$_{16}$ 114°.

U.K. Pat., 845 062, (1960); CA, **55**, 5544 (*synth*)
Bellas, M. *et al*, *J. Chem. Soc.*, 1965, 2096 (*synth*)

6-Fluoro-2-methylquinoline, 9CI   F-80053

6-Fluoroquinaldine, 8CI

[1128-61-6]

$C_{10}H_8FN$ M 161.178
Cryst. (Et$_2$O/hexane). Mp 57-59°.

Leir, C.M., *J. Org. Chem.*, 1977, **42**, 911 (*synth*)

6-Fluoro-5-methylquinoline, 9CI   F-80054

[107224-22-6]

$C_{10}H_8FN$ M 161.178
Oil.

Gerster, J.F. *et al*, *J. Med. Chem.*, 1987, **30**, 839 (*synth*)

7-Fluoro-2-methylquinoline, 9CI   F-80055

7-Fluoroquinaldine, 8CI

[1128-74-1]

$C_{10}H_8FN$ M 161.178
Bp$_{15}$ 120°.
Hydrate: Mp 38°.
B,HNO$_3$: Mp 137° dec.

Haigh, C.W., *J. Chem. Soc.*, 1965, 6004 (*synth, pmr*)

2-Fluorooctadecanoic acid   F-80056

2-Fluorostearic acid

[1578-61-6]

$$H_3C(CH_2)_{15}CHFCOOH$$

$C_{18}H_{35}FO_2$ M 302.472
(±)-*form*
Cryst. Mp 88-90°.
Me ester:
$C_{19}H_{37}FO_2$ M 316.498
Cryst. (MeOH). Mp 42-44°.

Pattison, F.L. *et al*, *Can. J. Chem.*, 1965, **43**, 1700 (*synth*)
Pogány, S.A. *et al*, *Synthesis*, 1987, 718 (*synth, ir, pmr, cmr, F nmr*)

18-Fluoro-9-octadecenoic acid   F-80057

ω-Fluorooleic acid

$$FCH_2(CH_2)_7CH{=}CH(CH_2)_7COOH$$

$C_{18}H_{33}FO_2$ M 300.456
(Z)-*form*
Isol. from seeds of *Dichapetalum toxicarium*. Oil or cryst. Mp 13.5°.

Peters, R.A. *et al*, *Biochem. J.*, 1960, **71**, 17 (*isol*)
Dean, R.E.A. *et al*, *J. Am. Chem. Soc.*, 1963, **85**, 612 (*synth*)

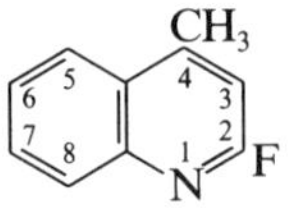

1-Fluorooctane, 9CI   F-80058

[463-11-6]

$$H_3C(CH_2)_6CH_2F$$

$C_8H_{17}F$ M 132.221
Liq. d 0.81. Bp 142-144°. n_D^{20} 1.3955.

Macey, W.A.T., *J. Phys. Chem.*, 1960, **64**, 254 (*synth*)
Landini, D. *et al*, *Synthesis*, 1974, 428 (*synth*)
Badone, D. *et al*, *Synthesis*, 1987, 920 (*synth*)

1-Fluoro-2-phenylethane   F-80059

(*2-Fluoroethyl)benzene, 9CI*

[458-87-7]

$$PhCH_2CH_2F$$

C_8H_9F M 124.157
Liq. Bp_{10} 53-54°.

DePuy, C.H. *et al*, *J. Am. Chem. Soc.*, 1960, **82**, 2535 (*synth*)
Koch, M.F. *et al*, *J. Am. Chem. Soc.*, 1981, **103**, 5423 (*synth, pmr*)
Badone, D. *et al*, *Synthesis*, 1987, 920 (*synth, pmr*)

5-Fluoro-3-pyridinecarboxaldehyde   F-80060

[39891-04-8]

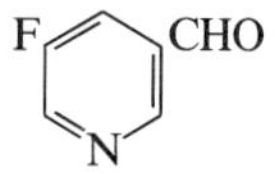

C_6H_4FNO M 125.102
Liq. Bp_{22} 90°, Bp_{10} 71-76°.

Kyba, E.P. *et al*, *J. Org. Chem.*, 1988, **53**, 3513 (*synth, pmr, F nmr, ir*)

5-Fluoropyrimidine, 9CI   F-80061

[675-21-8]

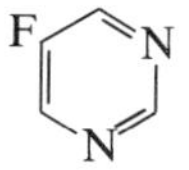

$C_4H_3FN_2$ M 98.079
Bp_{733} 101-102°. n_D^{19} 1.4693.

Reichardt, C. *et al*, *Justus Liebigs Ann. Chem.*, 1975, 470 (*synth, F nmr, pmr*)

2-Fluoroquinoline, 9CI   F-80062

[580-21-2]

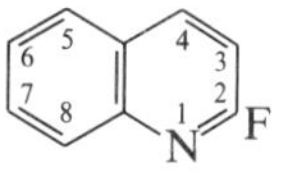

C_9H_6FN M 147.151
Bp_{30} 130-133°, Bp_2 75°. n_D^{25} 1.5827.
B,HNO_3: Cryst. (EtOH). Mp 137-141°.
Picrate: Mp 227-229°.
N-*oxide:*
   C_9H_6FNO M 163.151
   Mp 67°.

Roe, A. *et al*, *J. Am. Chem. Soc.*, 1949, **71**, 1785 (*synth*)
Miller, W.K. *et al*, *J. Am. Chem. Soc.*, 1950, **72**, 1629 (*uv*)
Hamer, J. *et al*, *Recl. Trav. Chim. Pays-Bas* (*J. R. Neth. Chem. Soc.*), 1962, **81**, 1058 (*synth*)
Dewar, M.J.S. *et al*, *J. Chem. Phys.*, 1968, **49**, 499 (*F nmr*)

3-Fluoroquinoline, 9CI   F-80063

[396-31-6]

C_9H_6FN M 147.151
d_4^{25} 1.19. Bp_{15} 102°. n_D^{25} 1.5902. Nonmutagenic. Forms solid hydrate.
Picrate: [448-47-5].
   Mp 185-186° (183-185°).
N-*oxide:*
   C_9H_6FNO M 163.151
   Mp 118°.
N-*oxide, picrate:* Cryst. (EtOH). Mp 130°.

Roe, A. *et al*, *J. Am. Chem. Soc.*, 1949, **71**, 1785 (*synth*)
Miller, W.K. *et al*, *J. Am. Chem. Soc.*, 1950, **72**, 1629 (*uv*)
Bellas, M. *et al*, *J. Chem. Soc.*, 1963, 4007 (*oxide*)
Dewar, M.J.S. *et al*, *J. Chem. Phys.*, 1968, **49**, 499 (*F nmr*)
Doddrell, D. *et al*, *J. Chem. Soc., Perkin Trans. 2*, 1976, 402 (*cmr*)
Takahashi, K. *et al*, *Chem. Pharm. Bull.*, 1988, **36**, 4630 (*tox*)

4-Fluoroquinoline, 9CI   F-80064

[394-70-7]

C_9H_6FN M 147.151
Synth. not certain. Bp_{30} 119° (?) .

Roe, A. *et al*, *J. Am. Chem. Soc.*, 1949, **71**, 1785 (*synth*)

5-Fluoroquinoline, 9CI   F-80065

[394-69-4]

C_9H_6FN M 147.151
d_4^{25} 1.19. Bp_{30} 123°, Bp_{15} 107-108°. n_D^{25} 1.5916.
▷ Mutagenic.
Picrate: Mp 199-200.5°.
N-*oxide:*
   C_9H_6FNO M 163.151
   Mp 176°.
N-*oxide, picrate:* Mp 136°.

Roe, A. *et al*, *J. Am. Chem. Soc.*, 1949, **71**, 1785 (*synth*)
Miller, W.K. *et al*, *J. Am. Chem. Soc.*, 1950, **72**, 1629 (*uv*)
Palmer, M.H., *J. Chem. Soc.*, 1962, 3645 (*synth*)
Bellas, M. *et al*, *J. Chem. Soc.*, 1963, 4007 (*oxide*)
Haigh, C.W. *et al*, *J. Chem. Soc.*, 1965, 6004 (*pmr*)
Dewar, M.J.S. *et al*, *J. Chem. Phys.*, 1968, **49**, 499 (*F nmr*)
Doddrell, D. *et al*, *J. Chem. Soc., Perkin Trans. 2*, 1976, 402 (*cmr*)
Takahashi, K. *et al*, *Chem. Pharm. Bull.*, 1988, **36**, 4630 (*tox*)

6-Fluoroquinoline, 9CI   F-80066

[396-30-5]

C_9H_6FN M 147.151
d_4^{25} 1.20. Bp_{30} 124-126°. n_D^{25} 1.5908. Forms solid hydrate.
▷ Weakly mutagenic.
Picrate: Mp 218-219°.
N-*oxide:* [2338-74-1].
   C_9H_6FNO M 163.151
   Mp 100°.
N-*oxide, picrate:* Mp 130°.

Roe, A. *et al*, *J. Am. Chem. Soc.*, 1949, **71**, 1785 (*synth*)
Miller, W.K. *et al*, *J. Am. Chem. Soc.*, 1950, **72**, 1629 (*uv*)
Sveinbjornsson, A. *et al*, *J. Org. Chem.*, 1951, **16**, 1450 (*synth*)
Bellas, M. *et al*, *J. Chem. Soc.*, 1963, 4007 (*oxide*)
Dewar, M.J.S. *et al*, *J. Chem. Phys.*, 1968, **49**, 499 (*F nmr*)
Doddrell, D. *et al*, *J. Chem. Soc., Perkin Trans. 2*, 1976, 402 (*C-13 nmr*)
Takahashi, K. *et al*, *Chem. Pharm. Bull.*, 1988, **36**, 4630 (*tox*)

7-Fluoroquinoline, 9CI
F-80067

[396-32-7]

C_9H_6FN M 147.151

Shows herbicidal props. Bp_{15} 107-108° (112°) . n_D^{25} 1.5845. n_D^{20} 1.5915.

Roe, A. *et al*, *J. Am. Chem. Soc.*, 1949, **71**, 1785 (*synth*)
Miller, W.K. *et al*, *J. Am. Chem. Soc.*, 1950, **72**, 1629 (*uv*)
Palmer, M.H., *J. Chem. Soc.*, 1962, 3645 (*synth*)
Bellas, M. *et al*, *J. Chem. Soc.*, 1963, 4007 (*oxide*)
Haigh, C.W. *et al*, *J. Chem. Soc.*, 1965, 6004 (*pmr*)
Dewar, M.J.S. *et al*, *J. Chem. Phys.*, 1968, **49**, 499 (*F nmr*)

8-Fluoroquinoline, 9CI
F-80068

[394-68-3]

C_9H_6FN M 147.151

d_4^{25} 1.21. Bp_{30} 148-150°. n_D^{25} 1.6028.

▷ Mutagenic.

Picrate: [394-62-7].
Mp 170-172°.

N-oxide: [2795-43-9].
C_9H_6FNO M 163.151
Hygroscopic solid. Mp 64°.

N-oxide, picrate: Cryst. (EtOH). Mp 128°.

Roe, A. *et al*, *J. Am. Chem. Soc.*, 1949, **71**, 1785 (*synth*)
Miller, W.K. *et al*, *J. Am. Chem. Soc.*, 1950, **72**, 1629 (*uv*)
Bellas, M. *et al*, *J. Chem. Soc.*, 1963, 4007 (*oxide*)
Dewar, M.J.S. *et al*, *J. Chem. Phys.*, 1968, **49**, 499 (*F nmr*)
Doddrell, D. *et al*, *J. Chem. Soc., Perkin Trans. 2*, 1976, 402 (*cmr*)
Sibi, M.P. *et al*, *Org. Magn. Reson.*, 1980, **14**, 494 (*N-15 nmr*)
Takahashi, K. *et al*, *Chem. Pharm. Bull.*, 1988, **36**, 4630 (*tox*)

5-Fluoro-1H-tetrazole, 9CI, 8CI
F-80069

[60471-49-0]

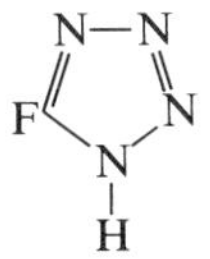

$CHFN_4$ M 88.044

1-Benzyl:
$C_8H_7FN_4$ M 178.168
Yellow oil.

U.S. Pat., 530 260, (1974); *CA*, **85**, 102223 (*synth*)
Klich, M. *et al*, *Tetrahedron*, 1986, **42**, 2677 (*deriv, synth, ir, pmr*)

2-Fluoro-3-(trifluoromethyl)oxirane
F-80070

2,3-Epoxy-1,1,1,3-tetrafluoropropane, 8CI. 1,3,3,3-Tetrafluropropylene oxide

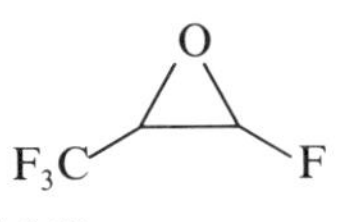

$C_3H_2F_4O$ M 130.042

Bp 37°. Prob. mixt. of stereoisomers.

McBee, E.T. *et al*, *J. Am. Chem. Soc.*, 1953, **75**, 4091 (*synth*)

Folicur
F-80071

α-[2-(4-Chlorophenyl)ethyl]-α-(1,1-dimethylethyl)-1H-1,2,4-triazole-1-ethanol, 9CI. Raxil. HWG-1608

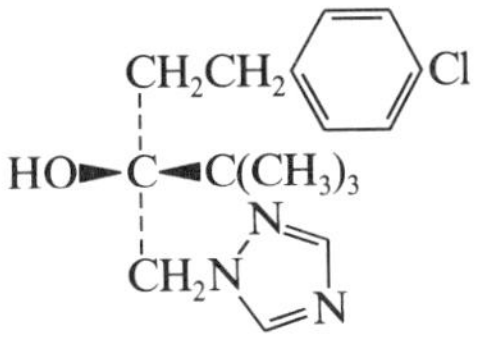

$C_{16}H_{22}ClN_3O$ M 307.822

(S)-form [119364-85-1]
Fungicide functioning as ergosterol-biosynthesis inhibitor. $[\alpha]_D^{20}$ −34.0° (c, 1 in $CHCl_3$). This is the more active enantiomer.

[107534-96-3]

Kaulen, J., *Angew. Chem., Int. Ed. Engl.*, 1989, **28**, 462 (*synth, abs config, bibl*)

Foliosate
F-80072

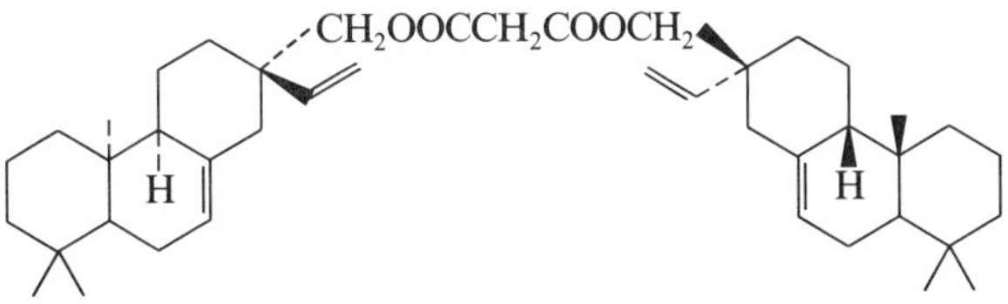

$C_{43}H_{64}O_4$ M 644.976

Constit. of *Calceolaria foliosa*. Gum. $[\alpha]_D^{25}$ −133.7° (c, 1.07 in $CHCl_3$).

Chamy, M.C. *et al*, *Phytochemistry*, 1989, **28**, 571.

Fonsecin
F-80073

Updated Entry replacing F-00654

2,3-Dihydro-2,5,8-trihydroxy-6-methoxy-2-methyl-4H-naphtho[2,3-b]pyran-4-one, 9CI

[3748-39-8]

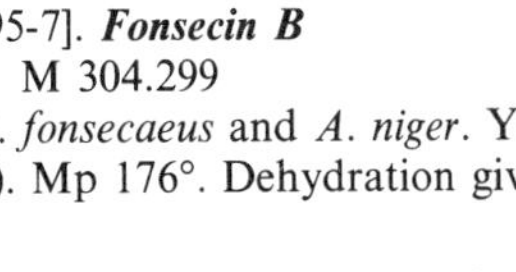

$C_{15}H_{14}O_6$ M 290.272

Pigment from a mutant of *Aspergillus fonsecaeus* and *A. carbonarius*. Mycotoxin. Bright-yellow irregular prisms. Mp 198° dec. λ_{max} 233 (log ε 4.42), 275(4.50), 321 (3.86), 332 (3.90) and 400 nm (3.94).

O^8-*Me:* [1856-95-7]. ***Fonsecin B***
$C_{16}H_{16}O_6$ M 304.299
Metab. of *A. fonsecaeus* and *A. niger*. Yellow needles (Me_2CO aq.). Mp 176°. Dehydration gives Rubrofusarin B.

Raper, K.B., *J. Elisha Mitchell Sci. Soc.*, 1953, **69**, 1 (*isol*)
Galmanini, O.L. *et al*, *Nature (London)*, 1962, **195**, 502 (*isol*)
Galmanini, O.L. *et al*, *Experientia*, 1974, **30**, 586 (*deriv*)
Ehrlich, K.C. *et al*, *Appl. Environ. Microbiol.*, 1984, **48**, 1 (*deriv*)
Bloomer, J.L. *et al*, *J. Org. Chem.*, 1984, **49**, 5027 (*isol, cmr, biosynth*)
Priestap, H.A., *Tetrahedron*, 1984, **40**, 3617 (*pmr*)

3-Formyl-2,2-dimethyl-1-cyclopropane-carboxylic acid   F-80074

Caronaldehydic acid

[25312-79-2]

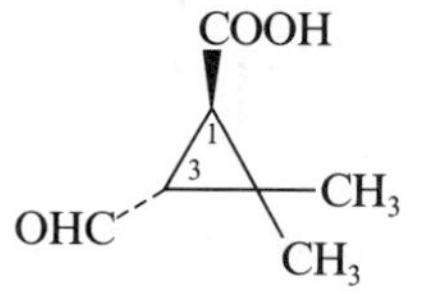

$C_7H_{10}O_3$ M 142.154

Intermed. for synth. of pyrethroids.

(*1R,3R*)-form [31062-21-2]

Me ester: [27335-33-7].

  $C_8H_{12}O_3$ M 156.181

Gum. $[\alpha]_D^{19}$ +19.3° (c, 0.38 in Me$_2$CO), $[\alpha]_D^{22}$ +15.3° (c, 12 in CHCl$_3$).

(*1S,3S*)-form [26883-41-0]

Me ester: [26770-99-0].

  $[\alpha]_D$ −0.45° (c, 2.93 in CHCl$_3$).

(*1R,3S*)-form [28075-68-5]

Me ester: [55701-02-5].

  $[\alpha]_D^{22}$ −72°, $[\alpha]_D^{25}$ −23.1° (c, 3.2 in CHCl$_3$).

(*1S,3R*)-form [40556-99-8]

Me ester: [84710-48-5].

  $[\alpha]_D^{20}$ +79.8° (c, 0.18 in Me$_2$CO).

(*1RS,3RS*)-form [54984-59-7]

(±)-trans-*form*

Cryst. (EtOAc/hexane). Mp 80.5°.

2,4-Dinitrophenylhydrazone: Cryst. (EtOH-EtOAc). Mp 206°, Mp 208°.

Me ester: [41301-44-4].

Oil. Bp$_{0.75}$ 47-49°.

(*1RS,3SR*)-form [56711-70-7]

(±)-cis-*form*

Prisms (hexane), plates (EtOAc/hexane). Mp 84-85° (79°). In equilib. with hemiacetal.

2,4-Dinitrophenylhydrazone: Orange-yellow prisms (EtOH/EtOAc). Mp 207-209° dec.

Me ester: [51424-62-5].

Pale yellow oil.

[27335-34-8, 54984-59-7, 67528-52-3, 67968-42-7]

Matsui, M. *et al*, *Agric. Biol. Chem.*, 1963, **27**, 554; 1964, **28**, 32 (*synth, deriv*)

Sugiyama, T. *et al*, *Agric. Biol. Chem.*, 1975, **39**, 1483 (*synth*)

Ortiz de Montellano, P.R. *et al*, *J. Org. Chem.*, 1978, **43**, 4323 (*synth*)

Devos, M.J. *et al*, *Tetrahedron Lett.*, 1978, 1847 (*synth, deriv*)

Crombie, L. *et al*, *J. Chem. Soc., Perkin Trans. 1*, 1980, 1711 (*synth, ir, pmr, ester*)

Martel, J.J. *et al*, *Pestic. Sci.*, 1980, **11**, 188 (*synth, resoln*)

Jakovuc, I.J. *et al*, *J. Am. Chem. Soc.*, 1982, **104**, 4659 (*synth, deriv*)

Gupta, D. *et al*, *Tetrahedron*, 1982, **38**, 3013 (*synth, pmr*)

Franck-Neumann, M. *et al*, *Tetrahedron Lett.*, 1982, **23**, 3493 (*synth, deriv*)

Tessier, J., *Chem. Ind.* (London), 1984, 199 (*manuf, resoln*)

Matsuo, A. *et al*, *J. Chem. Soc., Perkin Trans. 1*, 1984, 203 (*synth, ir, pmr, deriv*)

Takano, S. *et al*, *Heterocycles*, 1985, **23**, 2859 (*synth, deriv*)

Frauenrath, H. *et al*, *Justus Liebigs Ann. Chem.*, 1985, 1303 (*synth, ir, pmr, deriv*)

Mujiani, Z. *et al*, *Synth. Commun.*, 1988, **18**, 135 (*synth, ir, pmr*)

2-Formyl-1*H*-indole-3-acetic acid, 9CI   F-80075

[118361-82-3]

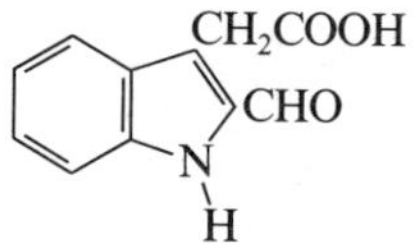

$C_{11}H_9NO_3$ M 203.197

Pale yellow solid. Mp 197-200°.

Et ester: [118361-81-2].

  $C_{13}H_{13}NO_3$ M 231.251

Mp 79-80°.

Moody, C.J. *et al*, *J. Chem. Soc., Perkin Trans. 1*, 1988, 1407 (*synth, ir, pmr, ms*)

Formylmethanesulfonic acid   F-80076

$$OHCCH_2SO_3H$$

$C_2H_4O_4S$ M 124.117

Esters prepd. and used as Li enolates and characterised as dinitrophenylhydrazones.

[120263-38-9, 120263-39-0, 120263-42-5]

Berkessel, A. *et al*, *Chem. Ber.*, 1989, **122**, 1147.

2-Formyl-1*H*-pyrrole-3-carboxylic acid   F-80077

[51361-92-3]

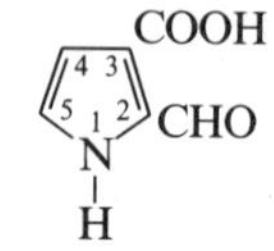

$C_6H_5NO_3$ M 139.110

Cryst. (H$_2$O). Mp 245°.

Me ester: [19075-68-4].

  $C_7H_7NO_3$ M 153.137

Needles (H$_2$O). Mp 129°.

Et ester: [19076-57-4].

  $C_8H_9NO_3$ M 167.164

Cryst. (C$_6$H$_6$/hexane). Mp 120° (117.5°).

Bisagni, E. *et al*, *Bull. Soc. Chim. Fr.*, 1968, 637 (*synth*)

Farnier, M. *et al*, *Bull. Soc. Chim. Fr.*, 1975, 2335 (*synth, pmr*)

4-Formyl-1*H*-pyrrole-2-carboxylic acid   F-80078

$C_6H_5NO_3$ M 139.110

Cryst. (H$_2$O). Mp 220°.

Et ester: [7126-57-0].

  $C_8H_9NO_3$ M 167.164

Cryst. (pet. ether). Mp 104-106°. Bp$_{0.1}$ 115-136°.

Khan, M.K.A. *et al*, *Tetrahedron*, 1966, **22**, 2095 (*synth, ir, pmr*)

5-Formyl-1*H*-pyrrole-2-carboxylic acid   F-80079

[7126-51-4]

$C_6H_5NO_3$ M 139.110

Prod. in cultures of *Erwinia oxoidiae*. Cryst. (H$_2$O). Mp 202-203° (212-215° dec.).

Me ester: [1197-13-3].

  $C_7H_7NO_3$ M 153.137

Cryst. (pet. ether). Mp 92-93°.

Et ester: [7126-50-3].

  $C_8H_9NO_3$ M 167.164

Cryst. (pet. ether). Mp 75°. Bp$_{0.1}$ 96-100°.

1-Me, Me ester:

$C_8H_9NO_3$ M 167.164
Cryst. (pet. ether). Mp 100-102°.

Reichstein, T., *Helv. Chim. Acta*, 1930, **13**, 349 (*synth*)
Khan, M.K.A. *et al*, *Tetrahedron*, 1966, **22**, 2095 (*synth, ir, pmr*)

Fremontin   F-80080

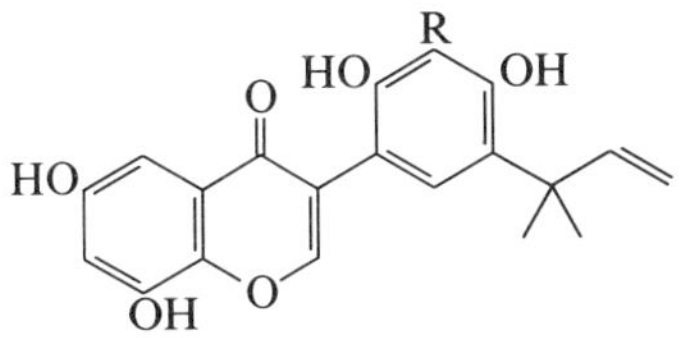

R = H

$C_{20}H_{18}O_6$ M 354.359
Constit. of *Psorothamnus fremontii*. Cryst. Mp 244-247°.

Manikumar, G. *et al*, *J. Nat. Prod.* (*Lloydia*), 1989, **52**, 769 (*isol, pmr, cmr*)

Fremontone   F-80081

As Fremontin, F-80080 with

R = —CH₂CH=C(CH₃)₂

$C_{25}H_{26}O_6$ M 422.477
Constit. of *Psorothamnus fremontii*. Cryst. Mp 124-126°.

Manikumar, G. *et al*, *J. Nat. Prod.* (*Lloydia*), 1989, **52**, 769 (*isol, pmr, cmr*)

3,29-Friedelanediol   F-80082

D:A-Friedooleanane-3,29-diol, 9CI

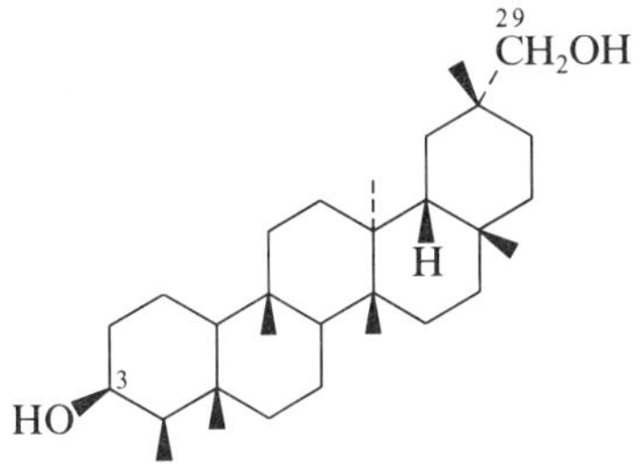

$C_{30}H_{52}O_2$ M 444.740
3β-form
   Constit. of *Mortonia palmerii*. Cryst. Mp 270-272°.
   3-Ketone, 29-aldehyde: [39903-23-6]. *3-Oxo-29-friedelanal*
      $C_{30}H_{48}O_2$ M 440.708
      Constit. of *M. diffusa*. Cryst. Mp 273-275°.

Domínguez, X.A. *et al*, *Phytochemistry*, 1974, **13**, 1292 (*isol*)
Martínez, V.M. *et al*, *J. Nat. Prod.* (*Lloydia*), 1988, **51**, 793 (*struct*)

3,30-Friedelanediol   F-80083

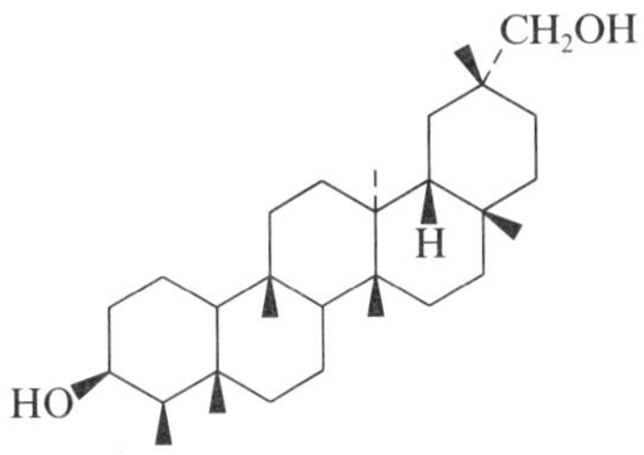

$C_{30}H_{52}O_2$ M 444.740
3β-form

Needles (CHCl₃/MeOH). Mp 294-296°. $[\alpha]_D^{30}$ +13.9° (c, 1 in CHCl₃).
3-Ac: 3β-Acetoxy-30-friedelanol
   $C_{32}H_{54}O_3$ M 486.777
   Constit. of *Euphorbia antiquorum*. Plates
   (CHCl₃/MeOH). Mp 311-312°. $[\alpha]_D^{30}$ +62.8° (c, 0.8 in CHCl₃).
30-Ac: 30-Acetoxy-3β-friedelanol
   $C_{32}H_{54}O_3$ M 486.777
   Constit. of *E. antiquorum*. Plates (CHCl₃/MeOH). Mp 238-240°. $[\alpha]_D^{30}$ +27.5° (c, 0.8 in CHCl₃).
Di-Ac: 3β,30-Diacetoxyfriedelane
   $C_{34}H_{56}O_4$ M 528.814
   Constit. of *E. antiquorum*. Plates (CHCl₃/MeOH). Mp 268-269°. $[\alpha]_D^{30}$ +23.5° (c, 0.85 in CHCl₃).

Anjaneyulu, V. *et al*, *Phytochemistry*, 1989, **28**, 1695.

Frutescin†   F-80084

Methyl 2-(2,4-hexadiynyl)-6-methoxybenzoate, 9CI
[4368-08-5]

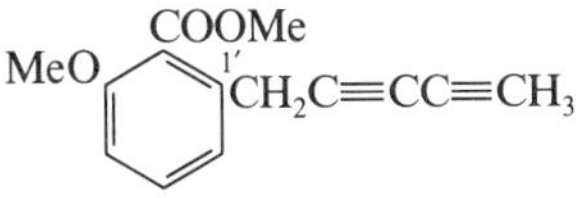

$C_{15}H_{14}O_3$ M 242.274
Isol. from roots of *Chrysanthmum frutescens* and *C. foeniculatum*. Cryst. (pet. ether). Mp 69°. (pet. ether). Mp 69°.
*1′-Acetoxy: **Fruscinol acetate***
   $C_{17}H_{16}O_5$ M 300.310
   Isol. from *C. frutescens* and *C. foeniculatum*. Cryst. (Et₂O/pet. ether). $[\alpha]_D^{20}$ −10° (c, 1.9 in Et₂O).
*1′-Oxo: **Frutescinone***
   $C_{15}H_{12}O_4$ M 256.257
   Isol. from roots of *C. frutescens*. Yellowish cryst. (Et₂O). Mp 121°.

Bohlmann, F. *et al*, *Chem. Ber.*, 1962, **95**, 602; 1964, **97**, 1176, 1179; 1966, **99**, 995, 2822 (*isol, uv, ir, struct, synth, biosynth*)
Bohlmann, F. *et al*, *Chem. Ber.*, 1969, **102**, 3298 (*synth*)

Fruticulin B   F-80085

Updated Entry replacing F-60077
[106664-41-9]

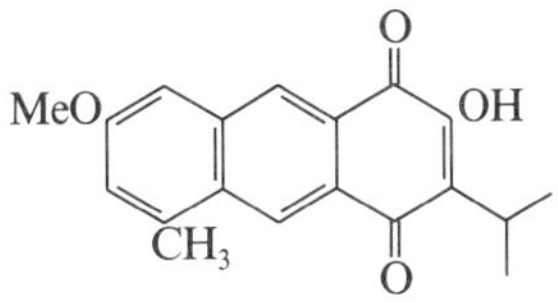

$C_{19}H_{18}O_6$ M 342.348
Constit. of *Salvia fruticulosa*. Cryst. (Me₂CO). Mp 173-174°.

Rodriguez-Hahn, L. *et al*, *Phytochemistry*, 1989, **28**, 567 (*isol, cryst struct*)

Fulminic acid, 9CI, 8CI   F-80086

Updated Entry replacing F-00777
Carbyloxime
[51060-05-0]

$$HC\equiv N^{\oplus}—O^{\ominus}$$

CHNO   M 43.025

Shown to have the struct. shown rather than $C^{\ominus}\equiv N^{\oplus}OH$ which was prev. postulated. Metal salts are widely used in detonators. The free acid is stable for a time only in Et_2O sol. at low temps.; it polymerises rapidly.

▷ Salts highly explosive, shock-sensitive, Hg salt is highly toxic. LS9320000.

Trimolecular Me ester:
C_2H_3NO M 57.052
Needles (H_2O). Mp 149°.

[506-85-4]

Wieland, H., *Ber.*, 1910, **43**, 3362 (*synth*)
Wöhler, L. *et al*, *Ber.*, 1929, **62**, 2742, 2748 (*salts*)
Beck, W. *et al*, *Chem. Ber.*, 1971, **104**, 533 (*ir, struct*)
Grundmann, C. *et al*, *Justus Liebigs Ann. Chem.*, 1973, 898 (*synth*)
Rodd's Chem. Carbon Compd. (2nd Ed.), 2nd Ed., 1973, 143 (*rev*)
De Sarlo, F. *et al*, *Tetrahedron*, 1985, **41**, 5181 (*synth, props*)
Teles, J.H. *et al*, *Chem. Ber.*, 1989, **122**, 753 (*struct, bibl*)
Bretherick, L., *Handbook of Reactive Chemical Hazards*, 2nd Ed., Butterworths, London and Boston, 1979, 276, 405.
Sax, N.I., *Dangerous Properties of Industrial Materials*, 5th Ed., Van Nostrand-Reinhold, 1979, 696, 972.
Hazards in the Chemical Laboratory, (Bretherick, L., Ed.), 3rd Ed., Royal Society ofChemistry, London, 1981, 383, 471.

2,4-(3*H*,5*H*)-Furandione, 9CI   F-80087

Updated Entry replacing F-00825
3,4-Dihydroxycrotonolactone. 3-Oxobutyrolactone. 4-Hydroxyacetoacetic lactone. Tetronic acid
[4971-56-6]

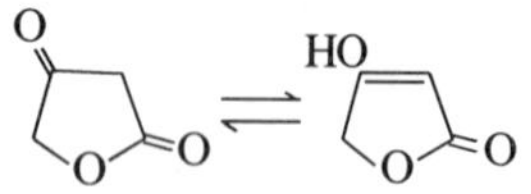

$C_4H_4O_3$ M 100.074
Plates (EtOH/ligroin). V. sol. H_2O. Mp 141°. Sinters at 135°. Strong monobasic acid.

Oxime:
$C_4H_5NO_3$ M 115.088
Plates (EtOH). Dec. at 146°.

Phenylhydrazone: Prisms (EtOH aq.). Mp 128°.

Me ether: [69556-70-3]. *4-Methoxy-2(5H)-furanone*
$C_5H_6O_3$ M 114.101
Cryst. Mp 66°.

Et ether: [69556-72-5]. *4-Ethoxy-2(5H)-furanone*
$C_6H_8O_3$ M 128.127
Faint yellow liq.

Mulholland, T.P.C. *et al*, *J. Chem. Soc., Perkin Trans. 1*, 1972, 1225 (*synth*)
Greenhill, J.V. *et al*, *Tetrahedron Lett.*, 1974, 2683 (*synth*)
Tanaka, K. *et al*, *Chem. Pharm. Bull.*, 1979, **27**, 1901 (*synth*)
Schmidt, D.G. *et al*, *Synth. Commun.*, 1981, **11**, 385 (*synth*)
Grob Schmidt, D. *et al*, *J. Org. Chem.*, 1983, **48**, 1914 (*use, bibl*)
Zimmer, H. *et al*, *J. Org. Chem.*, 1988, **53**, 3368 (*ethers, synth, pmr, ir*)

Furanoganoderic acid   F-80088

[120481-74-5]

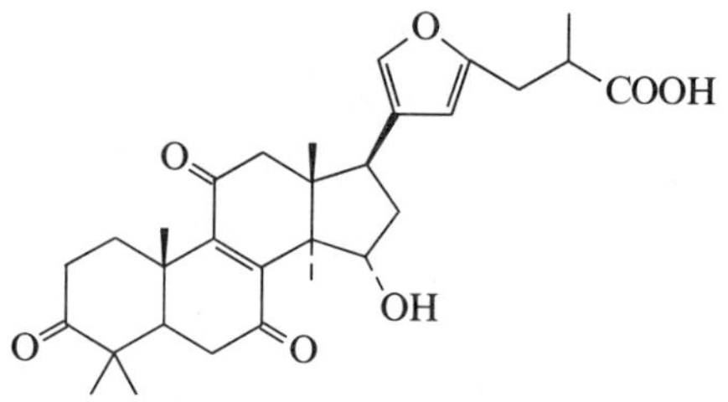

$C_{30}H_{38}O_7$ M 510.626
Constit. of *Ganoderma applantum*. $[\alpha]_D^{23}$ +70° (c, 0.2 in EtOH).

Nishitoba, T. *et al*, *Phytochemistry*, 1989, **28**, 193.

2-Furanpropanal, 9CI   F-80089

2-Furanpropionaldehyde, 8CI. 3-(2-Furyl)propanal. Furfurpropionaldehyde
[4543-51-5]

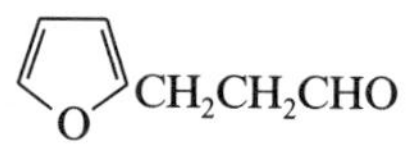

$C_7H_8O_2$ M 124.139
Liq. d_{19}^{19} 1.06. Bp_{17} 81°, Bp_4 59-61°.
Semicarbazone: Mp 82-83°.
Di-Me acetal: 2-(3,3-*Dimethoxypropyl)furan*
$C_9H_{14}O_3$ M 170.208
Liq. d_{20} 1.03. Bp_4 77°.

Weinhaus, H. *et al*, *CA*, 1930, **24**, 2127 (*synth*)
Sherlin, S.M. *et al*, *J. Gen. Chem. USSR (Engl. Transl.)*, 1938, **8**, 7 (*synth*)

3-Furanpropanal, 9CI   F-80090

3-(3-Furyl)propionaldehyde
[56859-93-9]
$C_7H_8O_2$ M 124.139
Liq. Bp_{10} 90-95°.
2,4-Dinitrophenylhydrazone: [56859-95-1].
Mp 118°.

Vig, O.P. *et al*, *J. Indian Chem. Soc.*, 1975, **52**, 199 (*synth, ir*)

2-Furanpropanol, 9CI   F-80091

3-(2-Furyl)-1-propanol
[26908-23-6]

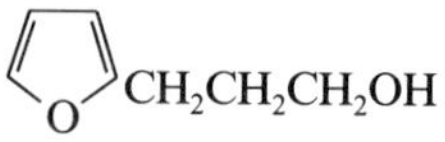

$C_7H_{10}O_2$ M 126.155
Oil. d_{20} 1.03. Bp_6 92-94°, $Bp_{0.2}$ 43-44°.

Bel'skii, I.F. *et al*, *Izv. Akad. Sci. USSR, Ser. Sci. Khim.*, 1964, 326 (*synth*)
DeShong, P. *et al*, *J. Am. Chem. Soc.*, 1988, **110**, 1901 (*synth, ir, pmr*)

3-Furanpropanol, 9CI   F-80092

3-(3-Furyl)-1-propanol
[56859-92-8]
$C_7H_{10}O_2$ M 126.155
Liq. Bp_8 115-120°.

Vig, O.P. *et al*, *J. Indian Chem. Soc.*, 1975, **52**, 199 (*synth, ir*)

5-[4-(2-Furanyl)-3-buten-1-ynyl]-2-thio-phenemethanol   F-80093

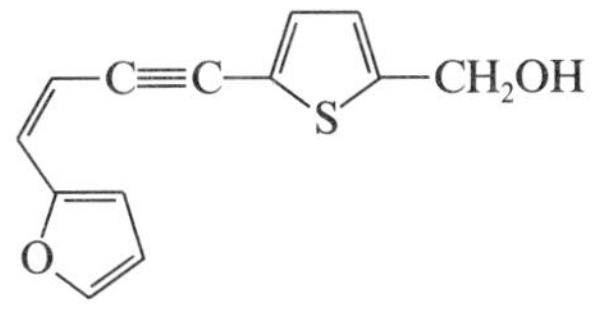

$C_{13}H_{10}O_2S$ M 230.287

(*Z*)-*form*

  Ac: 1-(2-Furyl)-4-(5-acetoxymethyl-2-thienyl)-1-buten-3-yne
    $C_{15}H_{12}O_3S$ M 272.324
    Isol. from *Santolina* sp. Cryst. (Et$_2$O/pet. ether). Mp
    52.5-53°.

  O-(3-Methylbutanoyl): [41628-63-1]. *1-(2-Furyl)-4[5-*
  (isovaleryloxymethyl)-2-thienyl]-1-buten-3-yne
    $C_{18}H_{18}O_3S$ M 314.404
    Constit. of *S. rosmarinifolia.* Oil.

Bohlmann, F. *et al, Chem. Ber.*, 1966, **99**, 135; 1973, **106**, 845 (*isol, uv, ir, pmr, struct*)

1-(2-Furanyl)ethanol   F-80094

Updated Entry replacing H-01785
α-*Methyl-2-furanmethanol, 9CI. Methyl-2-furylcarbinol.* α-*Methylfurfuryl alcohol. 2-(1-Hydroxyethyl)furan*
[4208-64-4]

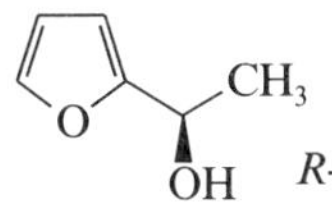

R-form

$C_6H_8O_2$ M 112.128

(*R*)-*form*

  Bp$_{15}$ 70°. [α]$_D^{25}$ +19.09° (c, 1.10 in CHCl$_3$)(94% e.e.). n_D^{15}
  1.4828.

  Ac:
    $C_8H_{10}O_3$ M 154.165
    Bp$_{11}$ 77°. n_D^{16} 1.4616.

(*S*)-*form* [27948-61-4]

  Bp$_{15}$ 70°. [α]$_D^{25}$ −18.4° (c, 6 in EtOH)(89% e.e.). n_D^{15}
  1.4827.

(±)-*form*

  d$_{25}^{25}$ 1.077. Bp 162-163°, Bp$_{23}$ 76-77°. n_D^{15} 1.4827.

  Ac: Bp$_{760}$ 177°, Bp$_{18}$ 85°. n_D^{25} 1.4618.

  Acid phthalate: Mp 80°.

Peters, F.N. *et al, J. Am. Chem. Soc.*, 1930, **52**, 2081 (*synth*)
Duveen, D.I. *et al, J. Chem. Soc.*, 1936, 621 (*synth*)
Crombie, L. *et al, J. Solution Chem.*, 1950, 1707 (*synth*)
Pascal, Y. *et al, Bull. Soc. Chim. Fr.*, 1965, 2211 (*pmr*)
Drueckhammer, D.G. *et al, J. Org. Chem.*, 1988, **53**, 1607 (*synth, resoln, pmr*)

2-(2-Furanyl)-1*H*-indole, 9CI   F-80095

2-(2-Indolyl)furan
[54864-36-7]

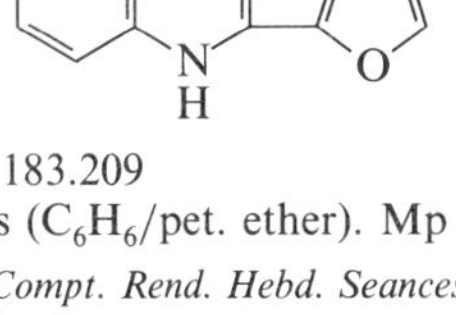

$C_{12}H_9NO$ M 183.209
Pale yellow flakes (C$_6$H$_6$/pet. ether). Mp 123-124°.

Calvaire, A. *et al, Compt. Rend. Hebd. Seances Acad. Sci.*, 1960, **250**, 3194 (*synth*)

Shivarama, H.B. *et al, Indian J. Chem., Sect. B*, 1976, **14**, 579
  (*synth*)
Shivarama, H.B. *et al, Indian J. Chem., Sect. B*, 1979, **17**, 187 (*ms*)

3-(2-Furanyl)-1*H*-indole   F-80096

2-(3-Indolyl)furan
[112616-94-1]

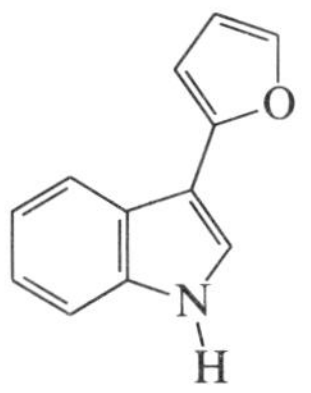

$C_{12}H_9NO$ M 183.209
Pale brown solid. Mp 90-91°.

Campbell, M.M. *et al, Tetrahedron*, 1987, **43**, 1117 (*synth, uv, ir, pmr, ms*)

Furospongin I   F-80097

Updated Entry replacing F-00923
[35075-74-2]

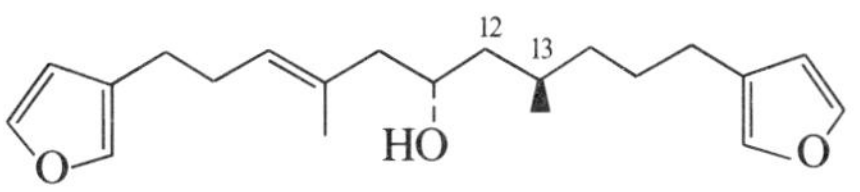

$C_{21}H_{30}O_3$ M 330.466
Constit. of *Spongia officinalis* and *Hippospongia communis.*
  Cryst. (pet. ether). Mp 35°. [α]$_D$ +8.8° (c, 1 in CHCl$_3$).

12,13-Didehydro: 12,13-Didehydrofurospongin I
  $C_{21}H_{28}O_3$ M 328.450
  Constit. of *Carteriospongia flabellifera.*

Cimino, G. *et al, Tetrahedron*, 1971, **27**, 4673 (*isol, struct*)
Schmitz, F.J. *et al, J. Nat. Prod. (Lloydia)*, 1988, **51**, 745 (*deriv*)

2-Furyl 2-thiazolyl ketone   F-80098

2-Furanyl 2-thiazolylmethanone. 2-(2-Furoyl)thiazole
[112969-91-2]

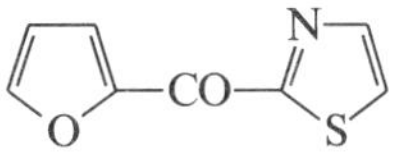

$C_8H_5NO_2S$ M 179.199
Oil.

Dondoni, A. *et al, J. Org. Chem.*, 1988, **53**, 1748 (*synth, ir, pmr*)

Fusalanipyrone   F-80099

3-Methyl-6-(1-methyl-1-propenyl)-2H-pyran-2-one, 9CI
[118169-29-2]

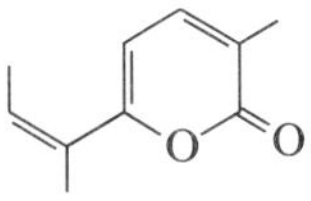

$C_{10}H_{12}O_2$ M 164.204
Metab. of *Fusarium solani.* Oil.

Abraham, W.-R. *et al, Phytochemistry*, 1988, **27**, 3310.

G

6-Galloylglucose G-80001

β-D-Glucopyranose 6-(3,4,5-trihydroxybenzoate), 9CI

[34781-46-9]

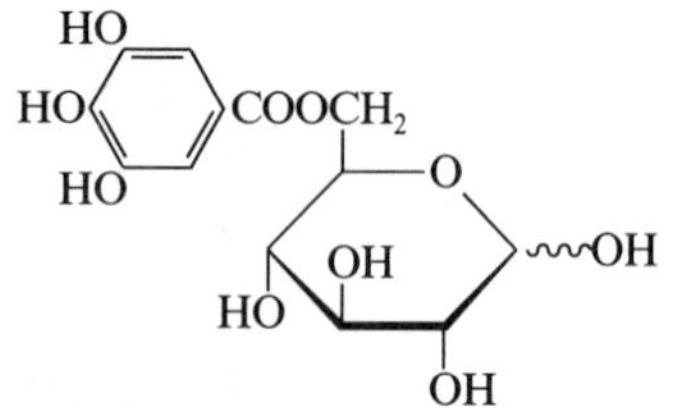

$C_{13}H_{16}O_{10}$ M 332.263

Isol. from rhizomes of *Geranium pratense* and *Polygonum bistorta*. Cryst. + 2H$_2$O (H$_2$O). Mp 166° (sinters from 110°). [α]$_D$ +54.5° → +36.8° (12h) (c, 1.9 in H$_2$O).

[13186-19-1, 33040-89-0]

Schmidt, O.T. *et al, Justus Liebigs Ann. Chem.*, 1951, **571**, 29 (*synth*)
Gstirner, F. *et al, Arch. Pharm. (Weinheim, Ger.)*, 1962, **295**, 23; 1966, **299**, 640 (*isol*)

Gambogic acid G-80002

Guttic acid. Guttatic acid

[2752-65-0]

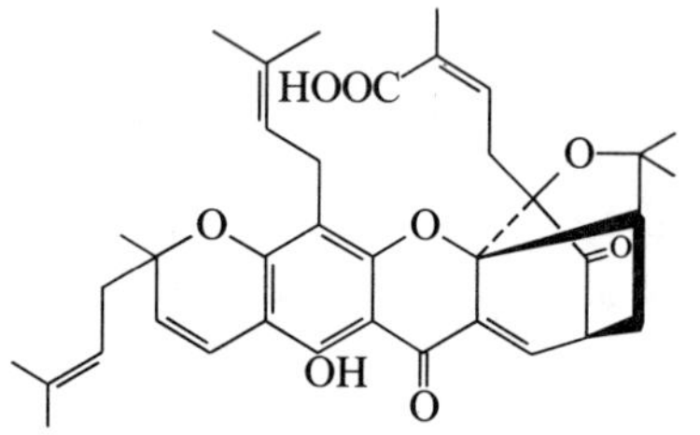

$C_{37}H_{42}O_8$ M 614.734

The name Guttic (Guttatic) acid is a transliteration of the French/Italian names for the substance. The English equiv. is stated to be Gambogic acid. Gamboge is also known as Gummi Gutta. Isol. from Gamboge resin.

Amorosa, M. *et al, Ann. Chim. (Rome)*, 1966, **56**, 232.
Arnone, A. *et al, Tetrahedron Lett.*, 1967, 4201 (*pmr, struct*)

16-Gammaceren-3-ol G-80003

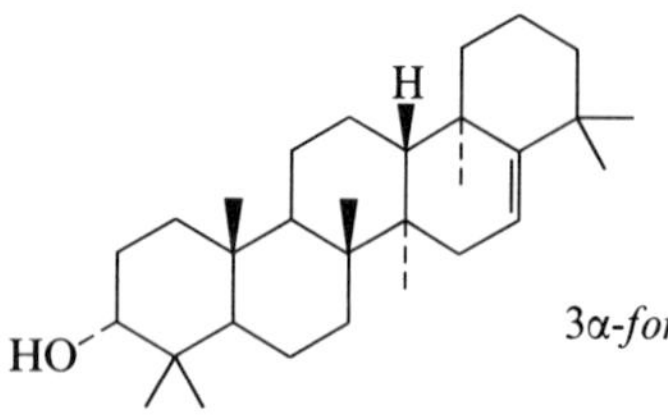

3α-form

$C_{30}H_{50}O$ M 426.724

3α-form

Constit. of *Picris hieracioides*. Cryst. Mp 269-270°. [α]$_D^{23}$ +99° (c, 0.6 in CHCl$_3$).

3β-form

Constit. of roots of *P. hieracioides*. Cryst. Mp 254-256°. [α]$_D^{23}$ +36.6° (c, 0.3 in CHCl$_3$).

Ac:
$C_{32}H_{52}O_2$ M 468.762
Constit. of *P. hieracioides*. Mp 287-288°. [α]$_D^{23}$ +36.0° (c, 0.5 in CHCl$_3$).

Shiojima, K. *et al, Tetrahedron Lett.*, 1989, **30**, 4977 (*cryst struct*)

Ganervosin A G-80004

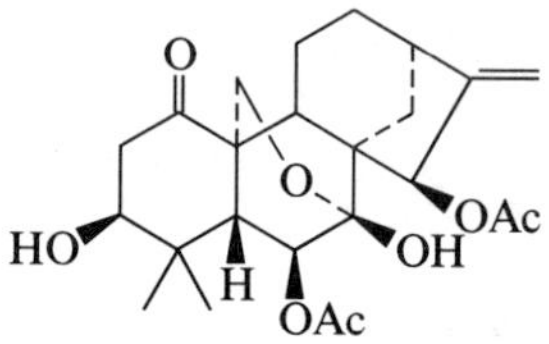

$C_{24}H_{32}O_8$ M 448.512

Constit. of *Rabdosia nervosa*. Cryst. (Me$_2$CO). Mp 224-226°.

Wang, Q.-G. *et al, J. Nat. Prod. (Lloydia)*, 1988, **51**, 775 (*isol, cryst struct*)

Ganschisandrine G-80005

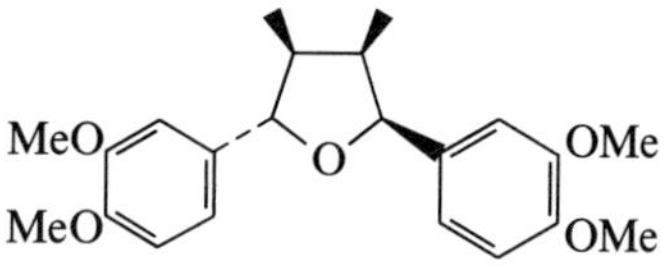

$C_{22}H_{28}O_5$ M 372.460

Constit. of *Schisandra sphenanthera*. Cryst. (Et$_2$O/hexane). Mp 114-115°.

Yue, J.-M. *et al, Phytochemistry*, 1989, **28**, 1774.

Gelidene G-80006

1,2,4-Trichloro-5-(2-chloroethenyl)-1,5-dimethylcyclohexane, 9CI

[123805-38-9]

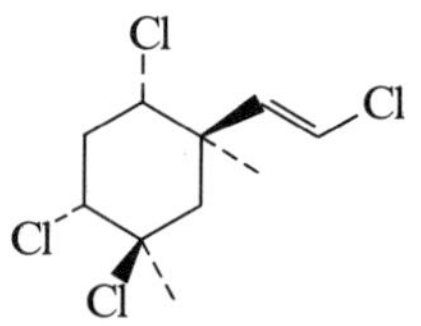

$C_{10}H_{14}Cl_4$ M 276.031

Metab. of *Gelidium sesquipedale*. Cryst. Mp 82°. [α]$_D^{20}$ −55°.

[106621-87-8, 106621-88-9, 106621-89-0, 106621-90-3]

Aazizi, M.A. *et al, J. Nat. Prod. (Lloydia)*, 1989, **52**, 829 (*isol, pmr, cmr*)

Gelomulide B   G-80007

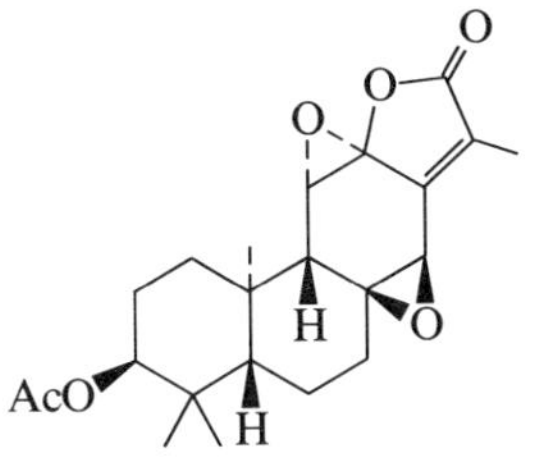

$C_{22}H_{28}O_6$ M 388.460

Constit. of *Gelonium multiflorum*. Needles. Mp 252°. $[\alpha]_D$ +126.7° (c, 0.015 in CHCl$_3$).

Talapatra, S.K. *et al, Phytochemistry*, 1989, **28**, 1181.

Gelomulide C   G-80008

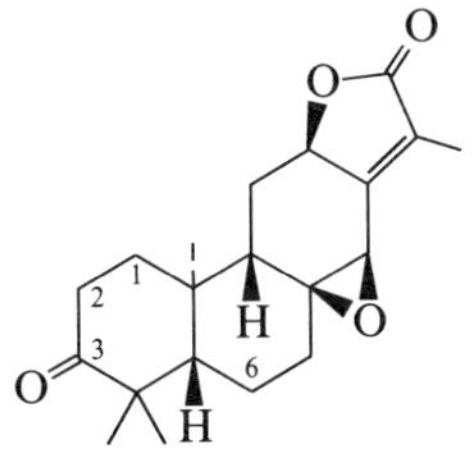

$C_{20}H_{26}O_4$ M 330.423

Constit. of *Gelonium multiflorum*. Needles. Mp 173°. $[\alpha]_D$ −27° (c, 0.077 in CHCl$_3$).

1β-Acetoxy: **Gelomulide F**
$C_{22}H_{28}O_6$ M 388.460
Constit. of *G. multiflorum*. Needles. Mp 220°. $[\alpha]_D$ −11° (c, 0.082 in CHCl$_3$).

3-Alcohol, 3-O-Ac: **Gelomulide A**
$C_{22}H_{30}O_5$ M 374.476
Constit. of *G. multiflorum*. Needles. Mp 240°. $[\alpha]_D$ +95° (c, 0.08 in CHCl$_3$). 3-Config. undetd.

1,2-Didehydro: **Gelomulide D**
$C_{20}H_{24}O_4$ M 328.407
From *G. multiflorum*. Needles. Mp 220°. $[\alpha]_D$ +7.6° (c, 0.08 in CHCl$_3$).

1,2-Didehydro, 6β-acetoxy: **Gelomulide E**
$C_{22}H_{26}O_6$ M 386.444
Constit. of *G. multiflorum*. Needles. Mp 241°. $[\alpha]_D$ −12.5° (c, 0.08 in CHCl$_3$).

Talapatra, S.K. *et al, Phytochemistry*, 1989, **28**, 1181.

Gentiopicroside   G-80009

Updated Entry replacing G-30003
Gentiopicrin
[20831-76-9]

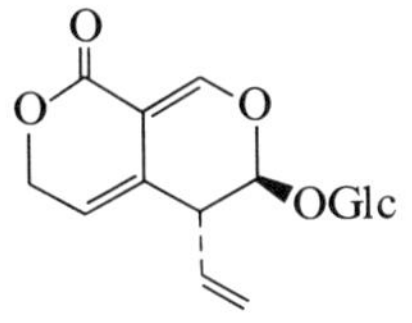

$C_{16}H_{20}O_9$ M 356.329

Widespread in the Gentianaceae, eg. *Swertia* and *Gentiana* spp. Cryst. Mp 191° (from EtOAc), Mp 122° (from H$_2$O). $[\alpha]_D$ −196.3° (H$_2$O). The nat. prods. Erytaurin, Gentiamarin and Sabbatin were said to be identical with Gentiopicroside (Korte, 1954-55). However this work is unreliable (Tomita 1960).

Aglucone: Mesogentiogenin
$C_{10}H_{10}O_4$ M 194.187
Oil.

Tomita, Y., *Nippon Kagaku Kaishi (J. Chem. Soc. Jpn.)*, 1960, **81**, 1726 (*pmr, struct*)
Inouye, H. *et al, Tetrahedron Lett.*, 1968, 4429 (*pmr, struct*)
Karrer, W. *et al, Konstitution und Vorkommen der Organischen Pflanzenstoffe*, 2nd Ed., Birkhäuser Verlag, Basel, 1972-1985 (*occur*)
Das, S. *et al, Phytochemistry*, 1984, **23**, 908 (*isol*)

1(10),5-Germacradiene-3,4-diol   G-80010

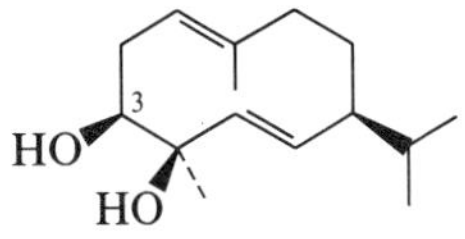

$C_{15}H_{26}O_2$ M 238.369
Leaflets (hexane). Mp 128°. $[\alpha]_D^{25}$ −201° (c, 1.3 in CH$_2$Cl$_2$).

3-Ac: 3-Acetoxy-1(10),5-germacradien-4-ol
$C_{17}H_{28}O_3$ M 280.406
Constit. of *Pallenis spinosa*. Needles (hexane). Mp 113°. $[\alpha]_D^{25}$ −145° (c, 0.8 in CH$_2$Cl$_2$).

Appendino, G. *et al, Phytochemistry*, 1989, **28**, 849.

1(10),4(15)-Germacradien-6-one   G-80011

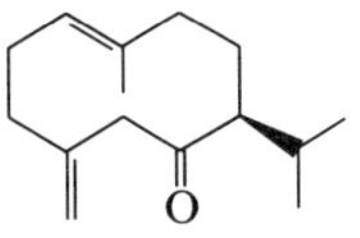

$C_{15}H_{24}O$ M 220.354
Constit. of *Baccharis latifolia*. Oil.

Zdero, C. *et al, Phytochemistry*, 1989, **28**, 531.

Gerontoxanthone A   G-80012

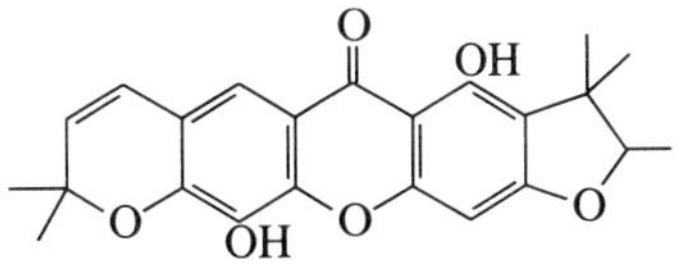

$C_{23}H_{22}O_6$ M 394.423
Constit. of *Cudrania cochinchinensis*. Yellow needles (EtOAc). Mp 236-238°. $[\alpha]_D$ −22.12° (c, 0.6 in CHCl$_3$).

Chang, C.-H. *et al, Phytochemistry*, 1989, **28**, 595.

Gerontoxanthone B   G-80013

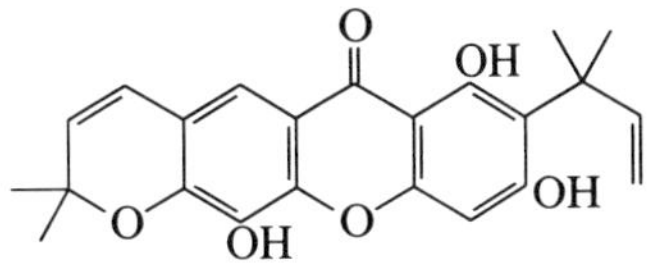

$C_{23}H_{22}O_6$ M 394.423
Constit. of *Cudrania cochinchinensis*. Yellow needles (EtOAc). Mp 216-218°.

Chang, C.-H. *et al, Phytochemistry*, 1989, **28**, 595.

Gerontoxanthone C   G-80014

C₂₃H₂₄O₆   M 396.439

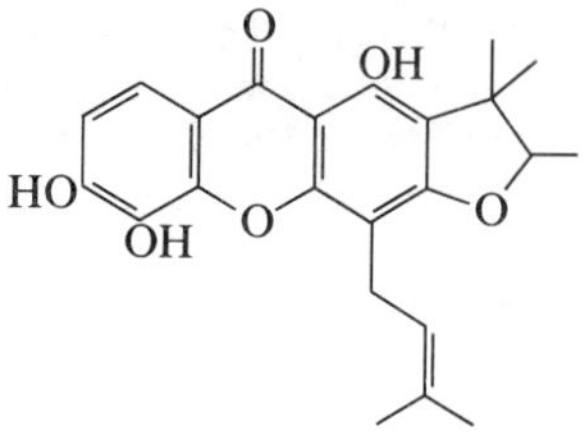

$C_{23}H_{24}O_6$ M 396.439

Constit. of *Cudrania cochinchinensis*. Pale yellow needles (MeOH). Mp 204-206°.

Chang, C.-H. *et al, Phytochemistry*, 1989, **28**, 595.

Gerontoxanthone D   G-80015

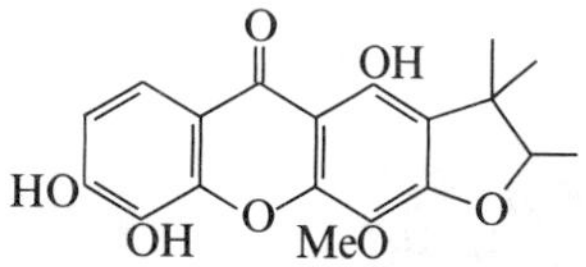

$C_{19}H_{18}O_7$ M 358.347

Constit. of *Cudrania cochinchinensis*. Yellow needles (MeOH). Mp 300°.

Chang, C.-H. *et al, Phytochemistry*, 1989, **28**, 595.

Ghalakinoside   G-80016

[119459-76-6]

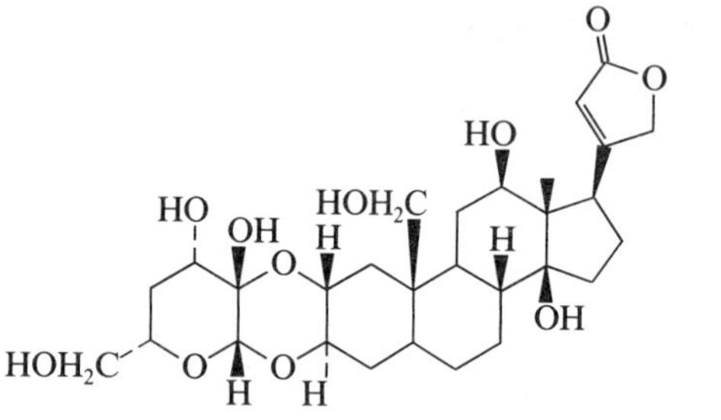

$C_{29}H_{42}O_{11}$ M 566.644

Constit. of *Pergularia tomentosa*. A cytotoxic cardiac glycoside. Cryst. (MeOH). Mp 215-220°. $[\alpha]_D^{25}$ +60°.

Al-Said, M.-S. *et al, Phytochemistry*, 1988, **27**, 3245 (*isol, cryst struct*)

Ginamallene   G-80017

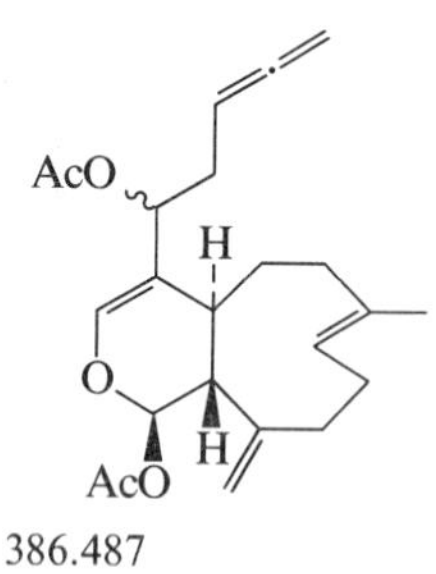

$C_{23}H_{30}O_5$ M 386.487

Constit. of *Acalycigorgia* spp. Oil. $[\alpha]_D$ +53° (c, 1.05 in CHCl₃).

Fusetani, N. *et al, Tetrahedron*, 1989, **45**, 1647.

Glaciolide   G-80018

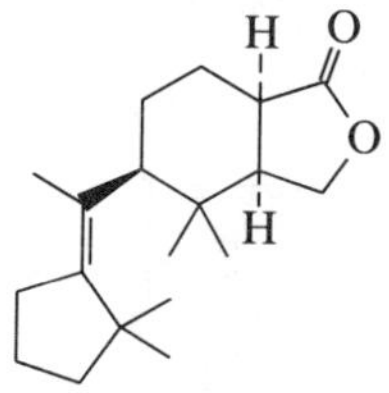

$C_{19}H_{30}O_2$ M 290.445

Constit. of nudibranch *Cadlina luteomarginata* and the sponge *Aplysilla glacialis*. Needles (hexane). Mp 102-103°. $[\alpha]_D$ +18.9° (c, 0.14 in EtOH).

Tischler, M. *et al, Tetrahedron Lett.*, 1989, **30**, 5717.

Glaucocalactone   G-80019

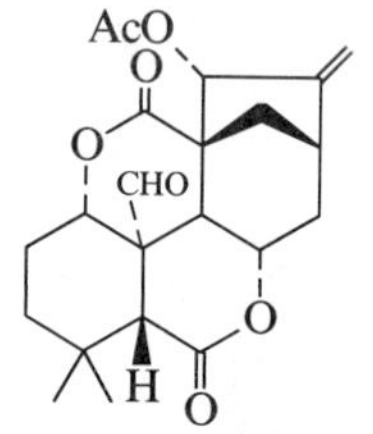

$C_{22}H_{26}O_7$ M 402.443

Constit. of *Rabdosia japonica*. Prisms (CH₂Cl₂). Mp 318-320° dec.

Meng, X. *et al, Phytochemistry*, 1989, **28**, 1163 (*isol, cryst struct*)

Globuxanthone   G-80020

4-(1,1-Dimethyl-2-propenyl)-1,2,5-trihydroxy-9H-xanthen-9-one, 9CI

[13586-28-2]

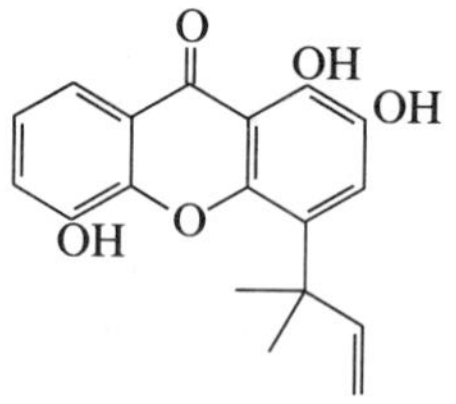

$C_{18}H_{16}O_5$ M 312.321

Isol. from heartwood of *Symphonia globulifera*. Orange needles (C₆H₆). Mp 162-163°.

Tri-Me ether: [13797-87-0].
    Mp 125.5°.

Locksley, H.D. *et al, J. Chem. Soc. C*, 1966, 430, 2186 (*isol, uv, ir, pmr, ms*)

Glucocleomin   G-80021

[36286-64-3]

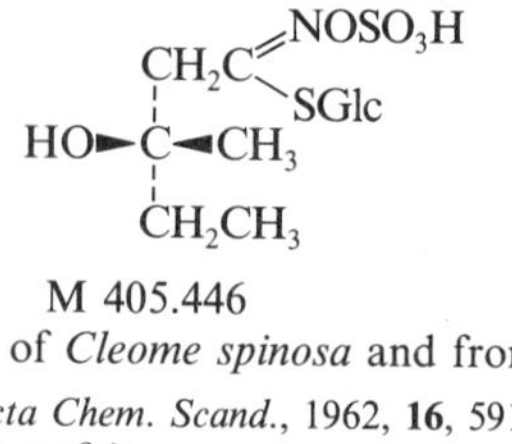

$C_{12}H_{23}NO_{10}S_2$ M 405.446

Isol. from seeds of *Cleome spinosa* and from *Capparis* sp.

Kjaer, A. *et al, Acta Chem. Scand.*, 1962, **16**, 591; 1963, **17**, 279 (*isol, struct, abs config*)

Ahmed, Z.F. *et al, Phytochemistry*, 1972, **11**, 251 (*isol*)

Glucoiberin

G-80022

1-Thio-β-D-glucopyranoside 1-[4-(methylsulfinyl)]N-(sulfooxy)butanimidate, 9CI

[554-88-1]

$$CH_2CH_2CH_2N=C\begin{array}{c}SGlc\\OSO_2OH\end{array}$$

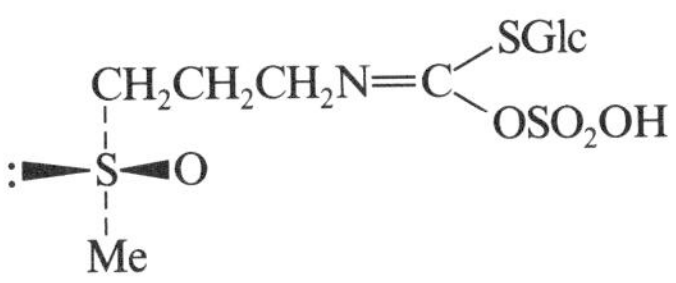

$C_{11}H_{21}NO_{10}S_3$ M 423.486

Isol. from seeds of *Iberis amata, Brassica oleracea* and other crucifers. Needles (EtOH aq.). Mp 142-144°. $[\alpha]_D^{20}$ −55.3° (H₂O).

Schutz, O.E. *et al, Arch. Pharm. (Weinheim, Ger.),* 1954, 404 (*isol, struct*)
Benkert, K., *Naturwissenschaften,* 1966, **53**, 200 (*isol*)

1-*O*-β-D-Glucopyranosyl-*N*-(2-hydroxy-hexadecanoyl)-4,8-sphingadiene

G-80023

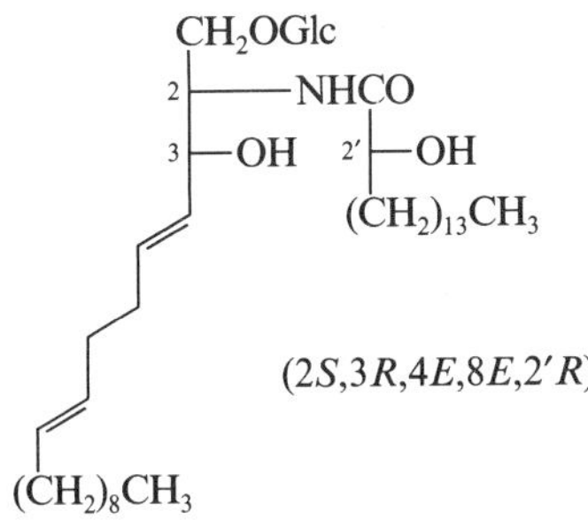

$$\begin{array}{c}CH_2OGlc\\2\!-\!NHCO\\3\!-\!OH \quad 2'\!-\!OH\\(CH_2)_{13}CH_3\end{array}$$

(2S,3R,4E,8E,2′R)-form

$$(CH_2)_8CH_3$$

$C_{40}H_{75}NO_9$ M 714.034

(2S,3R,4E,8E,2′R)-form [114297-20-0]
Cerebroside isol. from *Tetragonia tetragonoides* (Tetragoniae Herba). Antiulcerogenic principle. Hygroscopic granules. Mp 184-186°. $[\alpha]_D^{24}$ +10.5° (c, 0.30 in MeOH/CHCl₃, 3:2).

(2S,3R,4E,8Z,2′R)-form [115074-93-6]
Isol. from *T. tetragonoides.* Antiulcerogenic principle. Hygroscopic granules. Mp 192-194° (183°). $[\alpha]_D^{24}$ +13.4° (c, 0.43 in MeOH/CHCl₃ 3:2)(+4.6°).

Mori, K. *et al, Justus Liebigs Ann. Chem.,* 1988, 807 (*abs config, ir, pmr, cmr, synth, bibl*)

8-*C*-Glucosyl-3′,4′,5,7-tetrahydroxy-isoflavone

G-80024

Orobol 8-C-glucoside. 8-C-β-D-Glucopyranosylorobol

[66026-81-1]

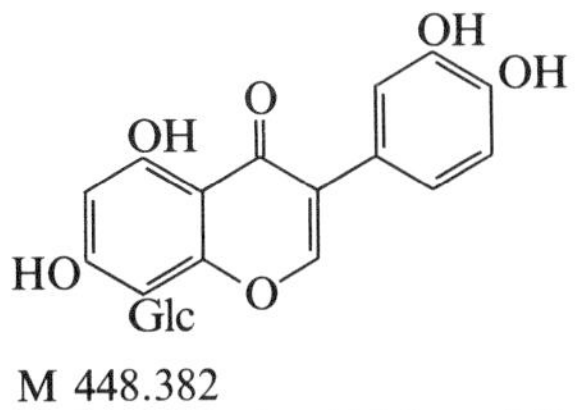

$C_{21}H_{20}O_{11}$ M 448.382

Isol. from bark of *Dalbergia nitidula* and flowers of *Lupinus luteus.* Light-yellow amorph. solid. Mp 190-193°, Mp 265°. $[\alpha]_D$ + ±0° (MeOH/Py 1:1).

Octa-Ac: Amorph. solid. Mp 120°, Mp 132-134°.

3′-Me ether: [40522-83-6]. 8-C-*Glucosyl-4′,5,7-trihydroxy-3′-methoxyisoflavone.* **Dalpanitin**
$C_{22}H_{22}O_{11}$ M 462.409

Isol. from seeds of *D. paniculata.* Rectangular plates + 1½H₂O. Mp 213-214° dec. $[\alpha]_D$ +35° (c, 0.73 in EtOH).

Adinarayana, D. *et al, Tetrahedron,* 1972, **28**, 5377 (*Dalpanitin*)
Zapesochnaya, G.G. *et al, Khim. Prir. Soedin.,* 1977, **13**, 862; *cnc,* 729 (*isol, pmr, struct*)
Van Heerden, F.R. *et al, J. Chem. Soc., Perkin Trans. 1,* 1980, 2463 (*isol, cmr*)

8-*C*-Glucosyl-4′,5,7-trihydroxyisoflavone

G-80025

Genistein 8-C-glucoside. 8-C-Glucopyranosylgenistein

[66026-80-0]

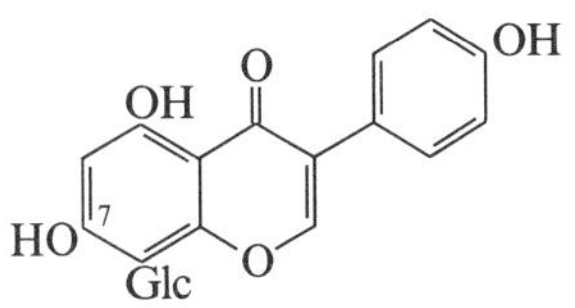

$C_{21}H_{20}O_{10}$ M 432.383

Isol. from bark of *Dalbergia nitidula* and from most organs of *Lupinus luteus.* Light-yellow amorph. solid. Monohydrate (MeOH aq.). Mp 185-189°, Mp >350°. $[\alpha]_D^{20}$ +24° (c, 1 in MeOH). Wide discrepancy in Mp's, no comparison of samples.

Hepta-O-Ac: Light-yellow amorph. solid. Mp 108-110°.

7-Me ether: [52448-12-1]. 8-C-*Glucosyl-4′,5-dihydroxy-7-methoxyisoflavone. Prunetin 8-C-glucoside. 8-C-Glucosylprunetin*
$C_{22}H_{22}O_{10}$ M 446.410

Isol. from bark of *D. paniculata.* Plates + 1H₂O (MeOH). Mp 286-287°.

Parthasarathy, M.R. *et al, Phytochemistry,* 1976, **15**, 1025 (*isol, uv, ir, ms, deriv*)
Zapesochnaya, G.G. *et al, Khim. Prir. Soedin.,* 1977, 862; *Chem. Nat. Compd. (Engl. Transl.),* 729 (*isol, pmr*)
Van Heerden, F.R. *et al, J. Chem. Soc., Perkin Trans. 1,* 1980, 2463 (*isol, cmr*)

5-Gluten-3-ol

G-80026

Updated Entry replacing G-70020
D:B-Friedoolean-5-en-3-ol, 9CI

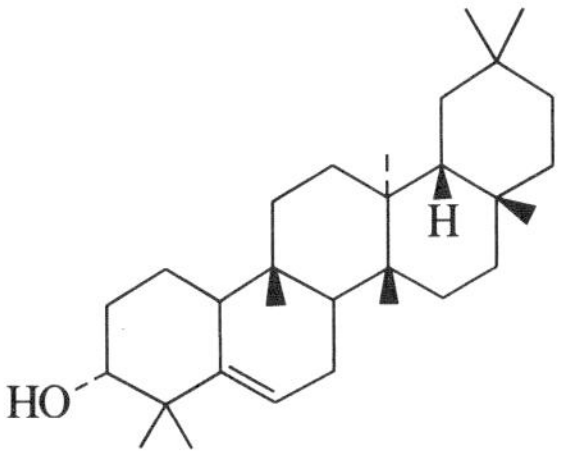

$C_{30}H_{50}O$ M 426.724

3α-form [14554-13-3] **Alnusenol.** *Glutinol*
Constit. of *Euphorbia cyparissias* and *E. royleana.* Cryst. (MeOH). Mp 210.5-211.5° (203-205°). $[\alpha]_D$ +61° (c, 1 in CHCl₃).

3β-form [545-24-4]
Constit. of *E. spp.* Cryst. (MeOH/CHCl₃). Mp 210-213°. $[\alpha]_D^{23}$ +63.3° (c, 0.71 in CHCl₃).

Ac:
$C_{32}H_{52}O_2$ M 468.762
Constit. of *E. maculata.* Needles (MeOH/CHCl₃). Mp 190-191.5°. $[\alpha]_D^{23}$ +76.8° (c, 0.5 in CHCl₃).

Sengupta, P. *et al, J. Indian Chem. Soc.,* 1965, **42**, 543 (*isol*)
Starratt, A.N., *Phytochemistry,* 1966, **5**, 1341 (*isol*)
Matsunaga, S. *et al, Phytochemistry,* 1988, **27**, 535 (*isol*)

Glyceollidin I   G-80027

3,6a,9-Trihydroxy-4-(3-methyl-2-butenyl)pterocarpan. 4-(Dimethylallyl)trihydroxypterocarpan. 4-Isopentenyl-3,6a,9-trihydroxypterocarpan

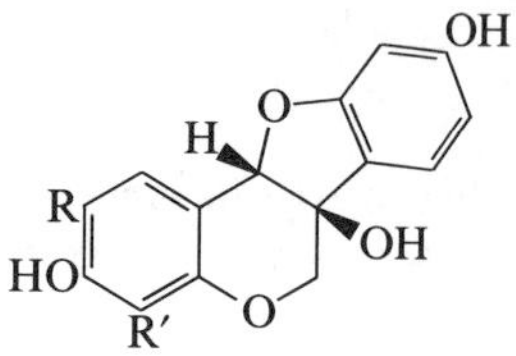

R = H, R′ = — CH$_2$CH = C(CH$_3$)$_2$

Absolute configuration

C$_{20}$H$_{20}$O$_5$ M 340.375

No opt. rotn. reported. Phytoalexin from *Glycine max.*
Zähringer, U. *et al, Z. Naturforsch., C*, 1981, **36**, 234.

Glyceollidin II   G-80028

2-(3-Methyl-2-butenyl)-6H-benzofuro[3,2-c][1]benzopyran-3,6a,9(11aH)triol, 9CI. 2-Isopentenyl-3,6a,9-trihydroxypterocarpan. 2-(Dimethylallyl)trihydroxypterocarpan. Glyceocarpin
[77979-22-7]

As Glyceollidin I, G-80027 with

R = — CH$_2$CH=C(CH$_3$)$_2$, R′ = H

C$_{20}$H$_{20}$O$_5$ M 340.375

Phytoalexin from *Glycine max.*

Ingham, J.L. *et al, Phytochemistry*, 1981, **20**, 795 (*isol, struct, abs config*)
Zähringer, U. *et al, Z. Naturforsch., C*, 1981, **36**, 234 (*isol*)

Glyceollin III   G-80029

Updated Entry replacing G-30013
[67314-02-7]

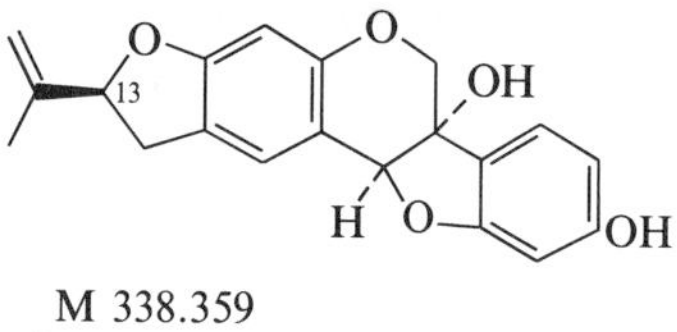

C$_{20}$H$_{18}$O$_5$ M 338.359

Phytoalexin from *Glycine max.* Also from *G. gracilis, G. latrobeana* and *G. soja.* Cryst. (EtOH aq.). Mp 149-153°.

13-Epimer: [79082-46-5]. **Canescacarpin**
C$_{20}$H$_{18}$O$_5$ M 338.359
Isol. from leaves of *Glycine canescens.* Cryst. (EtOH). Mp 164-167°.

Lyne, R.L. *et al, J. Chem. Soc., Chem. Commun.*, 1976, 497 (*isol*)
Lyne, R.L. *et al, Tetrahedron Lett.*, 1981, 2483 (*Canescacarpin*)
Ingham, J.L., *Fortschr. Chem. Org. Naturst.*, 1983, **43**, 1 (*occur*)
Banks, S.W. *et al, Phytochemistry*, 1983, **22**, 2729 (*biosynth*)

Glycolloglycollic acid   G-80030

Carboxymethyl hydroxyacetate, 9CI. Glycolic acid bimol. ester, 8CI
[30450-85-2]

HOCH$_2$COOCH$_2$COOH

C$_4$H$_6$O$_5$ M 134.088

Prisms (EtOAc). Sol. H$_2$O, AcOH. Mp 99-100°.

Wolf, L. *et al, Justus Liebigs Ann. Chem.*, 1900, **312**, 146 (*synth*)
Fischer, H.O.L. *et al, Helv. Chim. Acta*, 1933, **16**, 1130 (*synth*)
Micheau, J.C. *et al, Bull. Soc. Chim. Fr.*, 1970, 4018 (*synth, pmr, ir*)

Glycyrin   G-80031

3-(2,4-Dihydroxyphenyl)-5,7-dimethoxy-6-(3-methyl-2-butenyl)-2H-1-benzopyran-2-one, 9CI. 3-(2,4-Dihydroxyphenyl)-5,7-dimethoxy-6-prenylcoumarin
[66056-18-6]

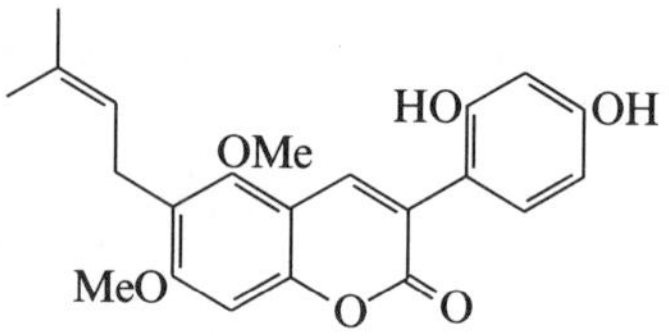

C$_{22}$H$_{22}$O$_6$ M 382.412

Isol. from roots of *Glycyrrhiza* sp. (licorice). Yellow needles. Mp 209-211°.

Kinoshita, T. *et al, Chem. Pharm. Bull.*, 1978, **26**, 135 (*isol, uv, pmr, struct*)

Gold's reagent   G-80032

N-[[[(Dimethylamino)methylene]amino]methylene]-N-methylmethanaminium(1+), 9CI. [[[(Dimethylamino)methylene]amino]methylene]dimethylammonium(1+), 8CI

Me$_2$N$^{\oplus}$=CHN=CHNMe$_2$

C$_6$H$_{14}$N$_3$$^{\oplus}$ M 128.197
Intermed. for triazoles, triazines.

Chloride: [20353-93-9].
  C$_6$H$_{14}$ClN$_3$ M 163.649
  Tan solid (Me$_2$CO). Mp 103°.

Hexachloroantimonate: [104543-47-7].
  C$_6$H$_{14}$Cl$_6$N$_3$Sb   M 462.663
  Yellow prisms (MeOH). Mp 133-135°.

Perchlorate: [87112-29-6].
  C$_6$H$_{14}$ClN$_3$O$_4$ M 227.647
  Cryst. (EtOH). Mp 110-111°.

[50781-48-1]

Krchnak, V. *et al, Collect. Czech. Chem. Commun.*, 1974, **39**, 3327; 1975, **40**, 1384 (*synth, ir, uv*)
Kantlehner, W. *et al, Justus Liebigs Ann. Chem.*, 1979, 528 (*synth, pmr*)
Gupton, J.T., *J. Org. Chem.*, 1980, **45**, 4522 (*synth, pmr*)
Gould, R.O. *et al, Acta Crystallogr., Sect. C*, 1983, **39**, 1097 (*cryst struct*)
Mueller, E. *et al, Tetrahedron*, 1985, **41**, 5901 (*synth, ir, pmr, uv*)
Gupton, J.T., *Aldrichimica Acta*, 1986, **19**, 43 (*rev*)

1,3,5,6,11-Gorgostanepentol G-80033

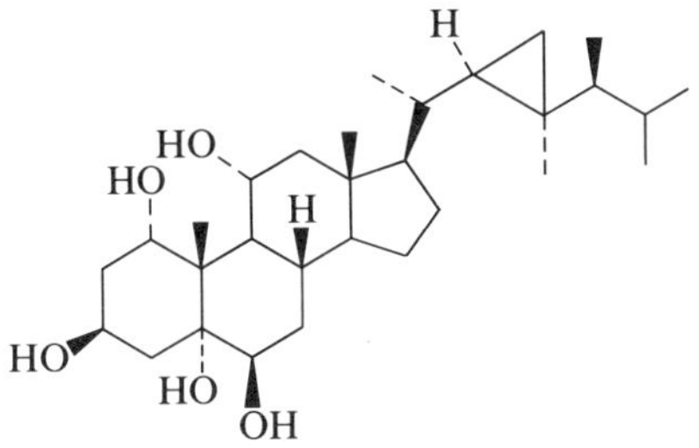

$C_{30}H_{52}O_5$ M 492.738

(1α,3β,5α,6β,11α)-form [117569-43-4]
Constit. of *Iris hippuris*. Cryst. Mp 295-297°.

Rao, C.B. *et al, J. Nat. Prod.* (*Lloydia*), 1988, **51**, 954.

Graciosallene G-80034

[119285-22-2]

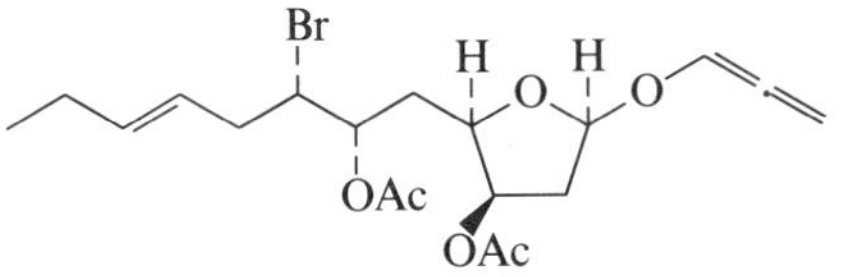

$C_{19}H_{27}BrO_6$ M 431.323
Metab. of *Laurencia obtusa*. Oil. $[\alpha]_D^{25}$ −83.3° (c, 0.018 in CHCl$_3$).

Norté, M. *et al, Phytochemistry*, 1988, **27**, 3537.

Graciosin G-80035

[119285-19-7]

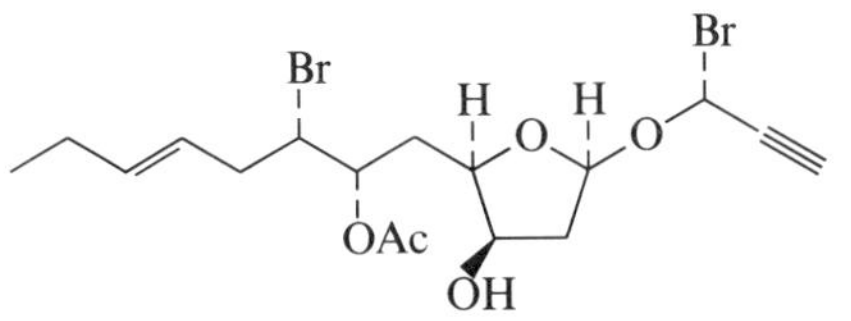

$C_{17}H_{24}Br_2O_5$ M 468.182
Metab. of *Laurencia obtusa*. Oil. $[\alpha]_D^{25}$ −14.5° (c, 0.62 in CHCl$_3$).

González, A.G. *et al, Tetrahedron*, 1984, **40**, 3443 (*isol*)
Norte, M. *et al, Phytochemistry*, 1988, **27**, 3537 (*isol, cryst struct*)

Grangolide G-80036

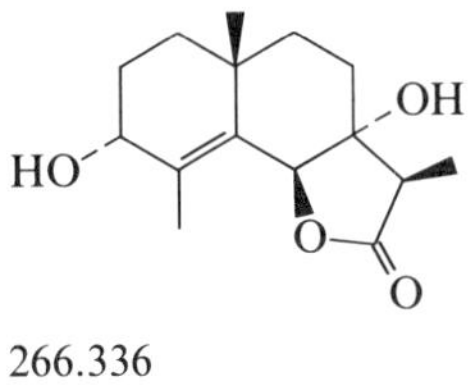

$C_{15}H_{22}O_4$ M 266.336
Constit. of *Grangea maderspatana*. Cryst. Mp 135-139°.
$[\alpha]_D^{24}$ +12° (c, 2.76 in CHCl$_3$).

Ruangrungsi, N. *et al, J. Nat. Prod.* (*Lloydia*), 1989, **52**, 130.

Gravolenic acid G-80037

[13781-46-9]

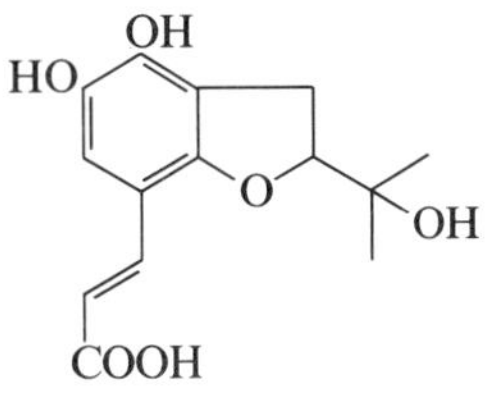

$C_{14}H_{16}O_6$ M 280.277
Isol. from above-ground parts of *Ruta graveolens*. Light-yellow cryst. (EtOAc or Me$_2$CO aq.). Mp 194°.

Reisch, J. *et al, Acta Pharm. Suec.*, 1966, **3**, 423 (*pmr, ord*)
Novák, I. *et al, Planta Med.*, 1966, **14**, 151 (*isol, uv, ir, struct*)

Grayanotoxin II G-80038

Updated Entry replacing G-00665
10(20)-Grayanotoxene-3β,5β,6β,14R,16α-pentol
[4678-44-8]

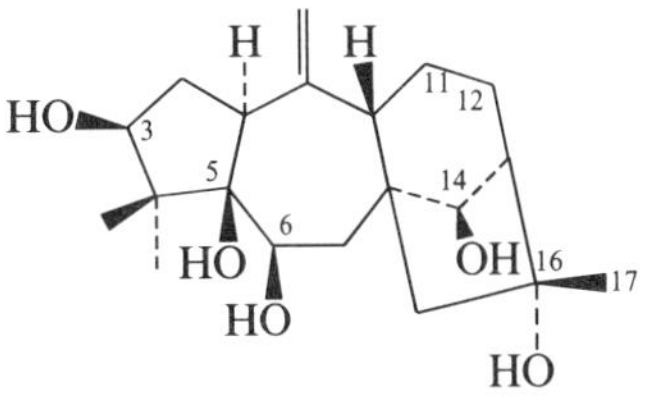

$C_{20}H_{32}O_5$ M 352.470
Toxic substance from the leaves of *Leucothoe grayana*.
Cryst. (EtOAc). Mp 197-198°. $[\alpha]_D^{28}$ −41.9°. Substances known as Asebotoxin, Rhodotoxin and Andromedotoxin as well as Grayanotoxin were isolated from various plants before 1960 and mostly consisted of Grayanotoxins or mixtures.

▷ PB9220000.

6-Ac: [59236-87-2]. ***Grayanotoxin XVI***
$C_{22}H_{34}O_6$ M 394.507
Constit. of *L. grayana*. Cryst. (Et$_2$O). Mp 116-118°. $[\alpha]_D$ +28° (c, 1 in MeOH).

14-Ac: [30272-17-4]. ***Grayanotoxin IV***
$C_{22}H_{34}O_6$ M 394.507
Constit. of *L. grayana*. Cryst. (EtOAc). Mp 174-175°.
$[\alpha]_D$ −18.6° (c, 1 in MeOH).
▷ PB9221000.

O^6-Ac, O^3-β-D-Glucopyranoside: [69842-18-8]. ***Grayanoside A***
$C_{28}H_{44}O_{11}$ M 556.649
Constit. of *L. grayana*. Syrup. $[\alpha]_D^{25}$ −19.1° (c, 2.35 in MeOH).

11α-Hydroxy: [35928-08-6]. ***Grayanotoxin XII***. *10(20)-Grayanotoxene-3β,5β,6β,11α,14R,16α-hexol*
$C_{20}H_{32}O_6$ M 368.469
Constit. of *L. grayana*. Cryst. Mp 208-209°.

12β-Hydroxy: [33880-98-7]. ***Grayanotoxin XI***. *10(20)-Grayanotoxene-3β,5β,6β,12β,14R,16α-hexol*
$C_{20}H_{32}O_6$ M 368.469
Constit. of *L. grayana*. Cryst. Mp 176-177°.

12β-Hydroxy,14-Ac: [35928-07-5]. ***Grayanotoxin XIII***
$C_{22}H_{34}O_7$ M 410.506
Constit. of *L. grayana*. Cryst. Mp 150-151°.

Karrer, W. *et al, Konstitution und Vorkommen der Organischen Pflanzenstoffe*, 2nd Ed., Birkhäuser Verlag, Basel, 1972-1985, no. 3678 (*bibl*)
Kakisawa, H. *et al, Tetrahedron Lett.*, 1962, 215 (*struct*)
Yasue, M. *et al, Chem. Pharm. Bull.*, 1970, **18**, 2586 (*stereochem*)

Okuno, T. *et al*, *Tetrahedron*, 1970, **26**, 4765 (*struct*)
Hikino, H. *et al*, *Chem. Pharm. Bull.*, 1971, **19**, 1289; 1972, **20**, 422 (*Grayanotoxin XI, Grayanotoxin XII, Grayanotoxin XIII*)
Gasa, S. *et al*, *Bull. Chem. Soc. Jpn.*, 1976, **49**, 835 (*isol*)
Gasa, S. *et al*, *Tetrahedron Lett.*, 1976, 553 (*synth*)
Sakakibara, J. *et al*, *Phytochemistry*, 1978, **17**, 1672 (*Grayanoside A*)
Furusaki, A. *et al*, *Bull. Chem. Soc. Jpn.*, 1980, **53**, 1956 (*cryst struct*)

Grayanotoxin III

G-80039

Updated Entry replacing G-00666
3β,5β,6β,10α,14R,16α-Grayanotoxanehexol. Andromedol
[4678-45-9]

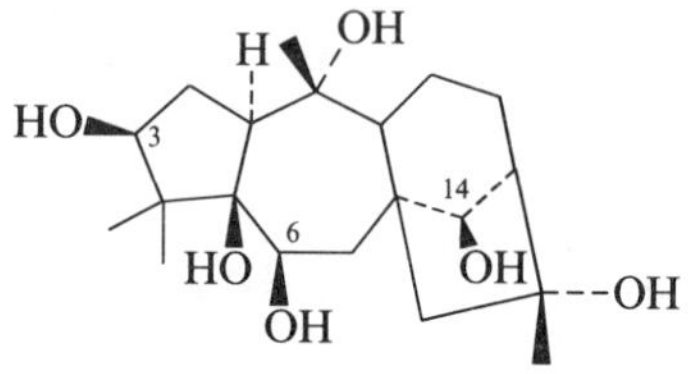

$C_{20}H_{34}O_6$ M 370.485
Toxic constit. of *Leucothoe grayana*. Cryst. $+ \frac{1}{2}$ EtOAc. Mp 218° dec. $[\alpha]_D^{15} -12°$.
▷ Toxic. PB9190000.

14-Ac: [4720-09-6]. ***Grayanotoxin I***. *Acetylandromedol*
$C_{22}H_{36}O_7$ M 412.522
Toxic agent from *Rhododendron*, *Kalmin*, *Leucothoe* and *Lyonia* spp. Cryst. (EtOAc/pentane). Mp 267-270°. $[\alpha]_D^{25} -8.8°$ (c, 2.3 in EtOH).
▷ PB9195000.

6,14-Di-Ac: [30460-34-5]. ***Rhodojaponin IV***
$C_{24}H_{38}O_8$ M 454.559
Constit. of *Rhododendron* spp. Cryst. Mp 245-247°.

3-Ketone: [30272-18-5]. ***Grayanotoxin V***.
5β,6β,10α,14R,16α-Pentahydroxygrayanotoxan-3-one
$C_{20}H_{32}O_6$ M 368.469
Constit. of *L. grayana* leaves. Cryst. (EtOAc). Mp 230-232° dec. $[\alpha]_D -61.5°$ (c, 1 in MeOH).

3-Ketone,14-Ac: [39012-12-9]. ***Grayanotoxin XIV***
$C_{32}H_{34}O_7$ M 530.616
Constit. of *L. grayana*. Cryst. Mp 242-243°. $[\alpha]_D -51°$ (c, 1 in MeOH).

3,6-Diketone: [59740-27-1]. ***Grayanotoxin XVII***.
5β,10α,14R,16α-Tetrahydroxy-3,6-grayanotoxanedione
$C_{20}H_{30}O_6$ M 366.453
Constit. of *L. grayana*. Cryst. (MeOH). Mp 268-270° dec. $[\alpha]_D -157.3°$ (c, 1 in MeOH).

Tallent, W.H. *et al*, *J. Am. Chem. Soc.*, 1957, **79**, 4548 (*struct*)
Kakisawa, H. *et al*, *Tetrahedron Lett.*, 1962, 215 (*struct*)
Hikino, H. *et al*, *Chem. Pharm. Bull.*, 1970, **18**, 1071, 2357 (*stereochem, isol*)
Okuno, T. *et al*, *Tetrahedron*, 1970, **26**, 4765 (*isol*)
Narayanan, P. *et al*, *Tetrahedron Lett.*, 1970, 3943 (*cryst struct*)
Hamanaka, N. *et al*, *Chem. Lett.*, 1972, 779 (*isol*)
Gasa, S. *et al*, *Bull. Chem. Soc. Jpn.*, 1976, **49**, 835 (*Grayanotoxin XVII*)
Jawad, F.H. *et al*, *Biomed. Mass Spectrom.*, 1977, **4**, 331 (*ms*)
Masutani, T. *et al*, *Agric. Biol. Chem.*, 1979, **43**, 631 (*cmr*)

Grayanotoxin VII

G-80040

Updated Entry replacing G-00669
10(20),15-Grayanotoxadiene-3β,5β,6β,14R-tetrol
[30460-59-4]

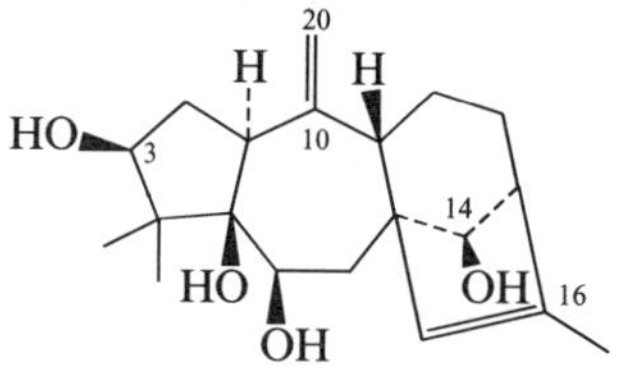

$C_{20}H_{30}O_4$ M 334.455
Constit. of *Leucothoe grayana*. Cryst. Mp 187-189°.
▷ PB9240000.

14-Ac: [30460-58-3]. ***Grayanotoxin IX***
$C_{22}H_{32}O_5$ M 376.492
Constit. of *L. grayana*. Cryst. Mp 151-152°.

Δ¹⁶-Isomer: [30460-60-7]. ***Grayanotoxin VIII***. *10(20),16-Grayanotoxadiene-3β,5β,6β,14R-tetrol*
$C_{20}H_{30}O_4$ M 334.455
Constit. of *L. grayana*. Cryst. Mp 190-193°.

Δ¹⁶-Isomer,14-Ac: [36660-75-0]. ***Grayanotoxin X***
$C_{22}H_{32}O_5$ M 376.492
Constit. *L. grayana*. Mp 165.5-166.5°.

10,20-Dihydro,10α-hydroxy: [30460-36-7]. ***Grayanotoxin VI***. *15-Grayanotoxene-3β,5β,6β,10α,14R-pentol*
$C_{20}H_{32}O_5$ M 352.470
Constit. of *L. grayana*. Cryst. Mp 219-221°.

14-Deoxy: [75829-08-2]. ***Grayanotoxin XIX***. *10(20),15-Grayanotoxadiene-3β,5β,6β-triol*
$C_{20}H_{30}O_3$ M 318.455
Constit. of *L. grayana*. Cryst. (Et₂O). Mp 132-133°. $[\alpha]_D -12°$ (c, 1 in MeOH).

3-Ketone: [78534-58-4]. ***Grayanotoxin XX***. *5β,6β,14R-Trihydroxy-10(20),15-grayanotoxadien-3-one*
$C_{20}H_{28}O_4$ M 332.439
Data inaccessible.

Hikino, H. *et al*, *Chem. Pharm. Bull.*, 1970, **18**, 2357; 1971, **19**, 1289 (*isol*)
Furusaki, A. *et al*, *Bull. Chem. Soc. Jpn.*, 1981, **54**, 1622 (*cryst struct, Grayanotoxin XIX*)

Grayanotoxin XVIII

G-80041

Updated Entry replacing G-00675
10(20)Grayanotoxene-3β,5,6β,16-tetrol
[70474-76-9]

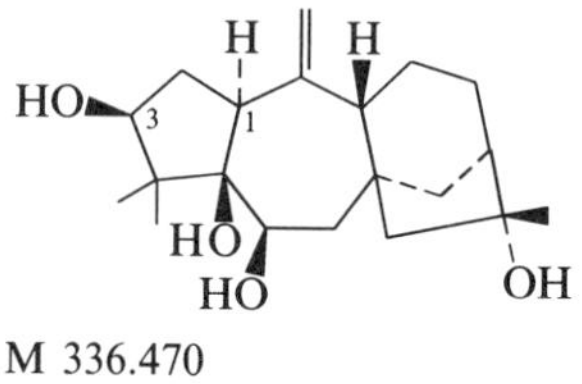

$C_{20}H_{32}O_4$ M 336.470
Constit. of *Leucothoe grayana*. Cryst. (diisopropyl ether). Mp 162-164°. $[\alpha]_D^{25} -6.8°$ (c, 2.2 in MeOH).

3-O-β-D-Glucopyranoside: [70474-75-8]. ***Grayanoside B***
$C_{26}H_{42}O_9$ M 498.612
Constit. of *L. grayana*. Viscous syrup. $[\alpha]_D^{16} -14.1°$ (c, 2.6 in MeOH).

1-Epimer, 3-O-β-D-glucopyranoside: [74311-31-2]. ***Grayanoside C***
$C_{26}H_{42}O_9$ M 498.612
Isol. from *L. grayana*. Syrup. Mp 220-222° (as pentaacetate). $[\alpha]_D^{24} +6.0°$ (c, 2 in MeOH).

Sakakibara, J. *et al*, *Phytochemistry*, 1979, **18**, 135; 1980, **19**, 1495 (*isol, struct, pmr*)
Furusaki, A. *et al*, *Bull. Chem. Soc. Jpn.*, 1981, **54**, 657 (*cryst struct*)

Grosheimin   G-80042

Updated Entry replacing G-00696

[22489-66-3]

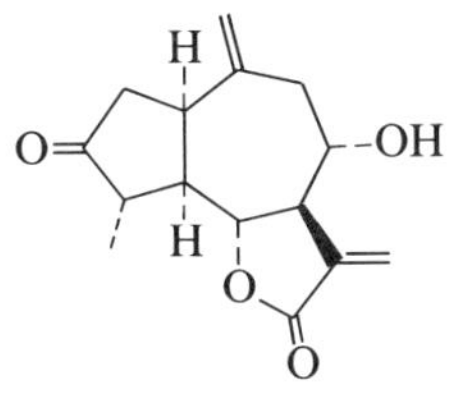

$C_{15}H_{18}O_4$ M 262.305

Constit. of *Chartolepsis intermedia* and *Grossheimia macrocephala*. Cryst. (MeOH). Mp 205°. $[\alpha]_D^{20}$ +137.7° (c, 0.2 in MeOH).

3α-Alcohol: **Grosheiminol**

$C_{15}H_{20}O_4$ M 264.321

Constit. of *G. macrocephala*. Oil. $[\alpha]_D^{25}$ +56.4° (c, 0.1 in Py).

Samek, Z. *et al*, *Collect. Czech. Chem. Commun.*, 1972, **37**, 2611.
Daniewski, W. *et al*, *Collect. Czech. Chem. Commun.*, 1982, **47**, 3160 (*Grosheiminol*)

4,11(13)-Guaiadien-3-one   G-80043

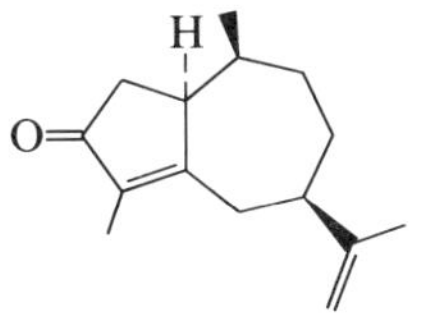

$C_{15}H_{22}O$ M 218.338

Constit. of *Baccharis boliviensis*. Oil. $[\alpha]_D^{24}$ +63° (c, 0.23 in CHCl$_3$).

Zdero, C. *et al*, *Phytochemistry*, 1989, **28**, 531.

4,11-Guaiadien-3-one   G-80044

[7764-53-6]

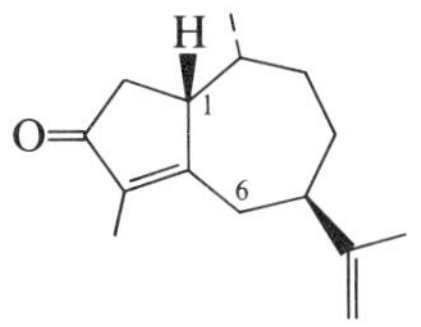

$C_{15}H_{22}O$ M 218.338

Constit. of *Pleocarphus revolutus*. Oil. $[\alpha]_D^{24}$ −63° (c, 0.75 in CHCl$_3$).

1β-Hydroxy, 6β-acetoxy: [116085-10-0]. *6β-Acetoxy-1β-hydroxy*-4,11-guaiadien-3-one

$C_{17}H_{24}O_4$ M 292.374

Constit. of *P. revolutus*.

Büchi, G. *et al*, *Proc. Chem. Soc., London*, 1962, 280 (*synth*)
Zdero, C. *et al*, *J. Nat. Prod.* (*Lloydia*), 1988, **51**, 509.

Gutiesolbriolide   G-80045

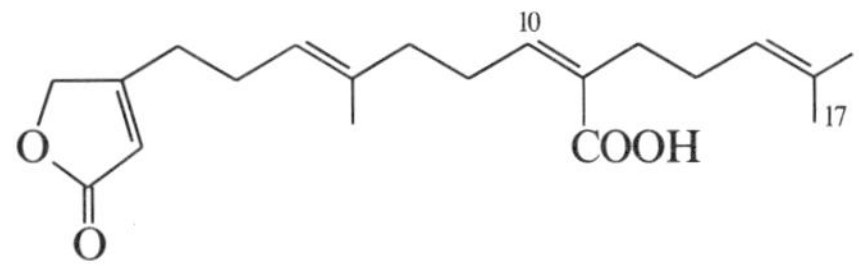

$C_{20}H_{28}O_4$ M 332.439

Parent compd. unknown.

17-Hydroxy: **17-Hydroxygutiesolbriolide**

$C_{20}H_{28}O_5$ M 348.438

Constit. of *Gutierrezia solbrigii*.

17-Hydroxy, 10E-isomer:

$C_{20}H_{28}O_5$ M 348.438

Condit. of *G. solbrigii*.

17-Acetoxy: **17-Acetoxygutiesolbriolide**

$C_{22}H_{30}O_6$ M 390.475

Constit. of *G. solbrigii*.

Jakupovic, J. *et al*, *Tetrahedron*, 1985, **41**, 4537.

α-Guttiferin   G-80046

Updated Entry replacing G-60040

[11048-92-3]

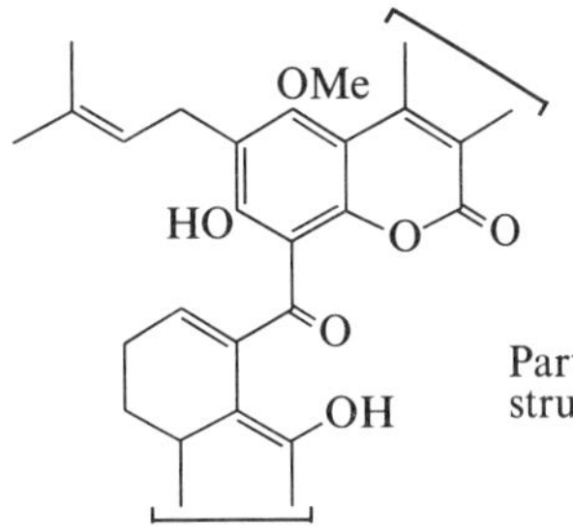

Partial structure

$C_{33}H_{38}O_8$ M 562.658

In *CA*, α-Guttiferin is confused with α_2- Guttiferin which is identical with Morellic acid. Constit. of Gamboge the resinous exudation of *Garcinia morella*. Mp 113-115°. $[\alpha]_D^{26}$ −475° (c, 1.5 in CHCl$_3$).

▷ MG1578700.

Py complex: Mp 115-117°. $[\alpha]_D^{26}$ −561.2° (c, 1.496 in CHCl$_3$).

Nageswara Rao, K.V. *et al*, *Experientia*, 1961, **17**, 213 (*struct*)

Gypothamniol   G-80047

Updated Entry replacing G-70037

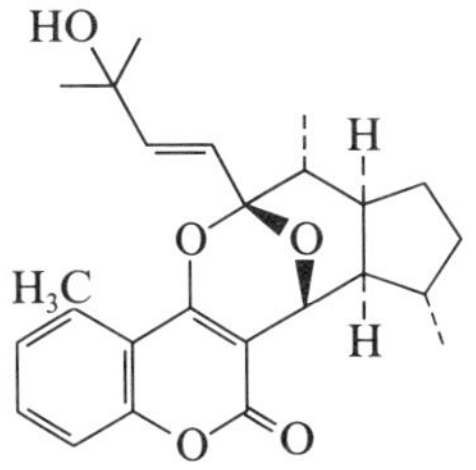

$C_{25}H_{30}O_5$ M 410.509

Constit. of *Gypothamnium pinifolium*. Oil. $[\alpha]_D^{24}$ +58° (c, 0.44 in CHCl$_3$).

Zdero, C. *et al*, *Phytochemistry*, 1988, **27**, 2953.

H

Haenkeanoside

H-80001

[119944-56-8]

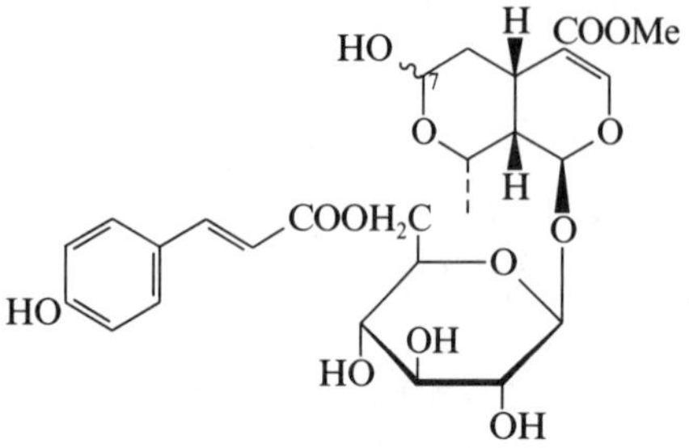

$C_{26}H_{32}O_{13}$ M 552.531
Constit. of *Isertia haenkeana*.

7-Me ether: [119944-57-9]. **O-Methylhaenkeanoside**
$C_{27}H_{34}O_{13}$ M 566.558
Constit. of *I. haenkeana*.

Z-isomer: [119943-44-1]. **Isohaenkeanoside**
$C_{26}H_{32}O_{13}$ M 552.531
Constit. of *I. haenkeana*.

Z-isomer, 7-Me ether: [119864-05-0]. **O-Methylisohaenkeanoside**
$C_{27}H_{34}O_{13}$ M 566.558
Constit. of *I. haenkeana*.

[119864-04-9, 119943-43-0, 119943-47-4, 119944-58-0]

Rumbero-Sánchez, A. *et al, Heterocycles*, 1988, **27**, 2863.

Halicholactone

H-80002

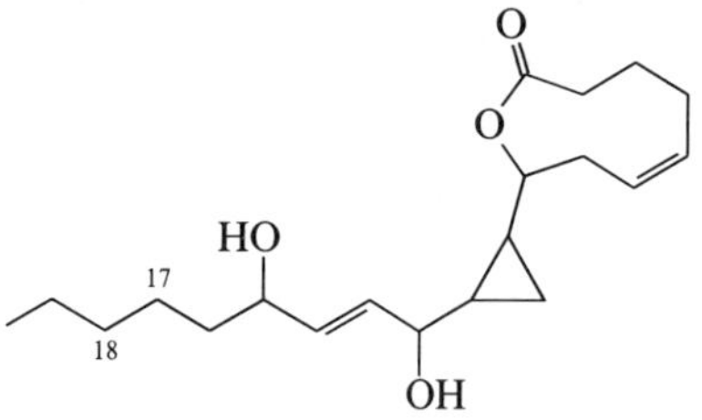

$C_{20}H_{32}O_4$ M 336.470
Metab. of sponge *Halichondria okadai*. Oil. $[\alpha]_D^{23}$ −85.4° (c, 1.16 in $CHCl_3$).

17Z,18-Didehydro: **Neohalicholactone**
$C_{20}H_{30}O_4$ M 334.455
Metab. of *H. okadai*. Cryst. (Et_2O/pentane). Mp 69-70°. $[\alpha]_D^{16}$ −54.2° (c, 0.73 in $CHCl_3$).

Niwa, H. *et al, Tetrahedron Lett.*, 1989, **30**, 4543.

Heaxahydroxydiphenic acid α-L-arabinosediyl ester

H-80003

Arabinopyranose cyclic 3,4-(4,4′,5,5′,6,6′-hexahydroxydiphenate), 8CI
[19833-16-0]

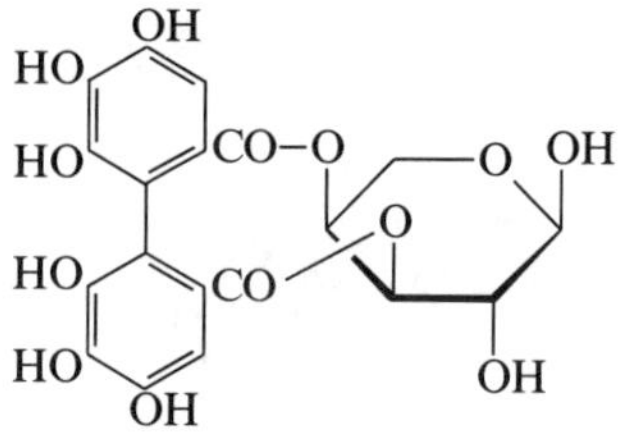

$C_{19}H_{16}O_{13}$ M 452.328
Isol. from fruits of *Psidium guava*. Plates (EtOH aq.). Mp 230-235° dec. $[\alpha]_D^{25}$ −24.2° (c, 0.4 in Py).

Misra, K. *et al, Phytochemistry*, 1968, **7**, 641.

Helicascolide A

H-80004

[121350-98-9]

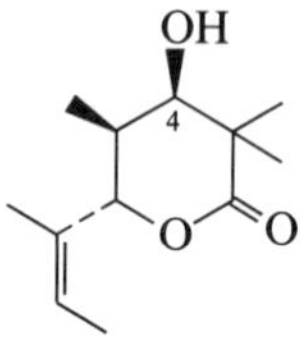

$C_{12}H_{20}O_3$ M 212.288
Metab. of *Helicascus kanaloanus*. Cryst. Mp 97-98°. $[\alpha]_D^{31}$ −25.0° (c, 1.4 in $CHCl_3$).

4-Epimer: [121325-38-0]. **Helicascolide B**
$C_{12}H_{20}O_3$ M 212.288
Metab. of *H. kanoloanus*. Cryst. Mp 61-62°. $[\alpha]_D^{31}$ −27.6° (c, 0.4 in $CHCl_3$).

Poch, G.K. *et al, J. Nat. Prod.* (*Lloydia*), 1989, **52**, 257 (*isol, pmr, cmr*)

Helichromanochalcone

H-80005

[72247-81-5]

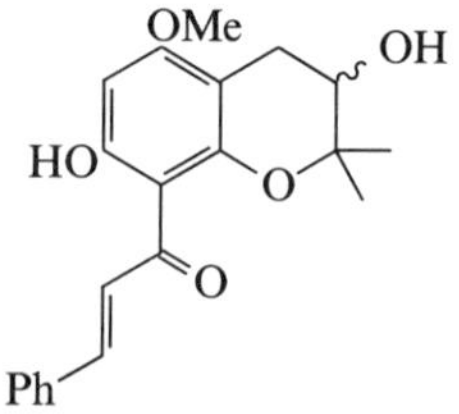

$C_{21}H_{22}O_5$ M 354.402
Isol. from roots of *Helichrysum cymosum*. Yellowish oil.

Bohlmann, F. *et al, Phytochemistry*, 1979, **18**, 1033.

Helihumulone   H-80006

[72247-83-7]

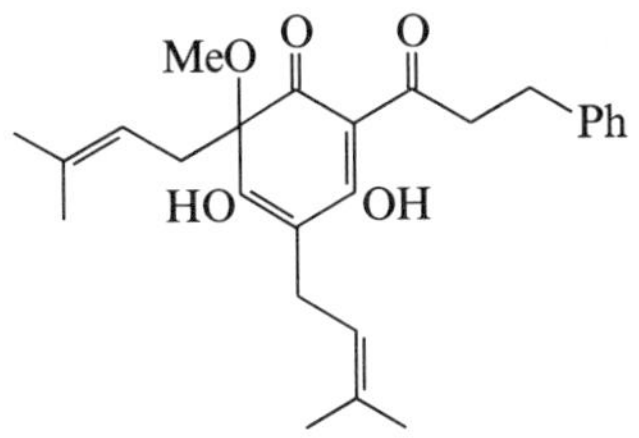

$C_{26}H_{32}O_5$ M 424.536

Isol. from roots of *Helichrysum cymosum*. Oil. $[\alpha]_D^{24}$ +57.8° (c, 1.4 in $CHCl_3$).

Bohlmann, F. *et al*, *Phytochemistry*, 1979, **18**, 1033.

Helikrausichalcone   H-80007

[122585-71-1]

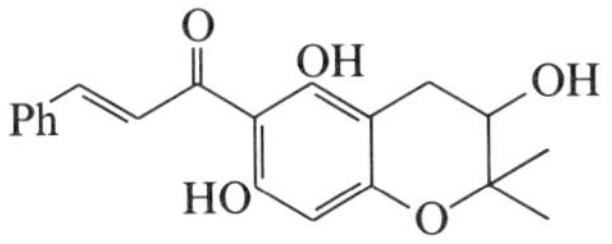

$C_{20}H_{20}O_5$ M 340.375

Constit. of *Helichrysum krausii*. Gum.

Jakupovic, J. *et al*, *Phytochemistry*, 1989, **28**, 1119.

Helilupulone   H-80008

[72247-85-9]

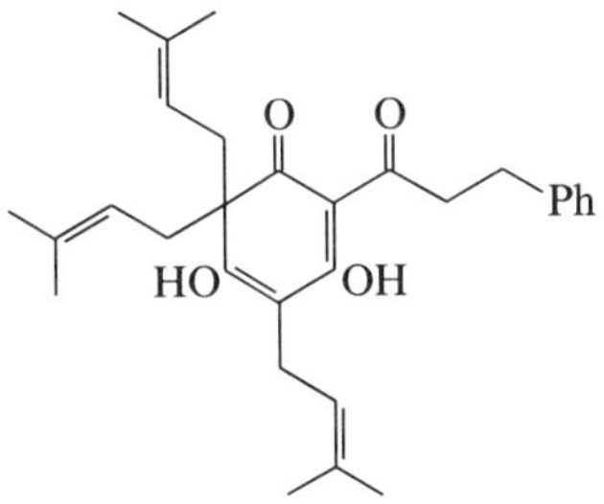

$C_{30}H_{38}O_4$ M 462.628

Isol. from roots of *Helichrysum tenuiculum*. Oil.

Bohlmann, F. *et al*, *Phytochemistry*, 1979, **18**, 1033.

Helipterol   H-80009

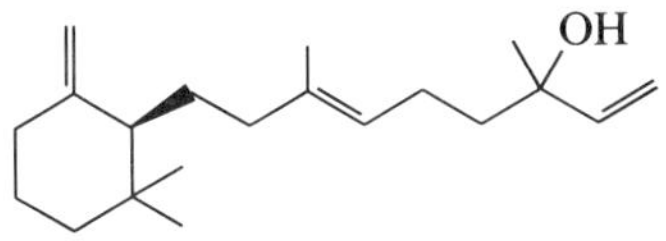

$C_{20}H_{34}O$ M 290.488

Constit. of *Helipterum venustum*. Oil. $[\alpha]_D^{24}$ +13° (c, 0.31 in $CHCl_3$).

Zdero, C. *et al*, *Phytochemistry*, 1989, **28**, 517.

Helisplendidilactone   H-80010

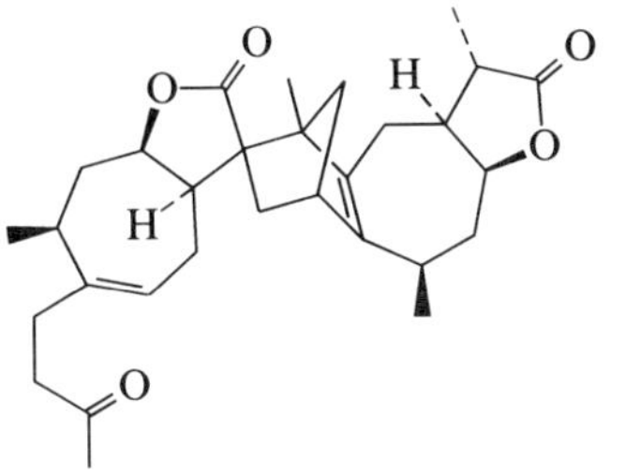

$C_{30}H_{40}O_5$ M 480.643

Constit. of *Helichrysum splendidum*. Cryst. Mp 172°. $[\alpha]_D^{24}$ −64° (c, 0.21 in $CHCl_3$).

Jakupovic, J. *et al*, *Phytochemistry*, 1989, **28**, 1119.

Helvolic acid   H-80011

Updated Entry replacing H-00097

6,16-Bis(acetyloxy)-3,7-dioxo-29-nordammara-1,17(20),24-trien-21-oic acid, 9CI. Fumigacin

[29400-42-8]

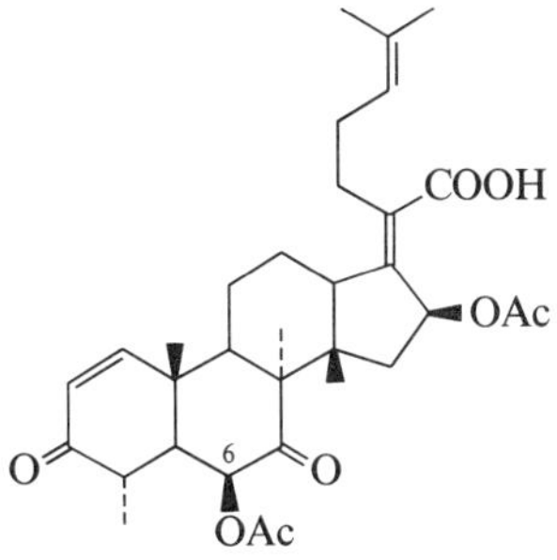

$C_{33}H_{44}O_8$ M 568.706

Produced by *Aspergillus fumigatus* mut.*helvola*. Shows antibiotic activity against a wide range of organisms, incl. *Staphyloccocus, Streptococcus, Micrococcus, Clostridium, Pseudomonas, Myobacterium, Shigella, Salmonella* and*Corynebacterium*. Needles (MeOH). Mp 215°. $[\alpha]_D^{25}$ −121° (c, 1 in $CHCl_3$).

▷ RC1370000.

Me ester: Cryst. (MeOH). Mp 262°. $[\alpha]_D^{23}$ −140° ($CHCl_3$).

Semicarbazone: Mp 240-250° dec.

Dioxime: Needles. Mp 189°.

O⁶-De-Ac: [10072-61-4]. **Helvolinic acid**
$C_{31}H_{42}O_7$ M 526.669
Isol. from *Cephalosporium caerulens*, also obt. by mild hydrol. of Helvolic acid. Needles. Mp 202.5-203°. $[\alpha]_D^{24}$ −88.9° (c, 1 in $CHCl_3$).

O⁶-Deacetoxy: **6-Deacetoxyhelvolic acid**
$C_{31}H_{42}O_6$ M 510.669
Isol. from culture broth of *C. caerulens*. Needles. Mp 214.5-215°. $[\alpha]_D$ −51.7° (c, 0.5 in $CHCl_3$). Originally descr. as 7-Deacetoxyhelvolic acid prior to the revision of struct. for Helvolic acid.

Allinger, N.L. *et al*, *J. Org. Chem.*, 1961, **26**, 4522 (*uv, ir, pmr*)
Okuda, S. *et al*, *Chem. Pharm. Bull.*, 1966, **14**, 436 (6-Deacetoxyhelvolic acid, Helvolinic acid)
Oxley, P., *J. Chem. Soc., Chem. Commun.*, 1966, 729 (*Helvolic acid*)
Okuda, S. *et al*, *Tetrahedron Lett.*, 1967, 2295 (*Helvolic acid, struct*)
v. Daehne, W. *et al*, *Tetrahedron Lett.*, 1968, 4843 (*isol, nmr*)
Iwasaki, S. *et al*, *J. Chem. Soc., Chem. Commun.*, 1970, 1119 (*struct, pmr*)
Cole, R.J. *et al*, *Handbook of Toxic Fungal Metabolites*, Academic Press , N.Y., 1981, 806.

Heneicosanedioic acid   H-80012

Updated Entry replacing H-00108

[505-55-5]

$$HOOC(CH_2)_{19}COOH$$

$C_{21}H_{40}O_4$ M 356.545

Little known. Was descr. (1900) as nat. prod. under the name Japanic acid but this was erroneous.

Lamberton, J.A., *Aust. J. Chem.*, 1961, **14**, 323.

Heneicosanoic acid, 9CI   H-80013

[2363-71-5]

$$H_3C(CH_2)_{19}COOH$$

$C_{21}H_{42}O_2$ M 326.562

Isol. from wood of *Ulex europaeus*, bark of *Pinus contorta*, from *Shorea maranti*, olive oil (*Olea europaea*) etc. Needles (Me$_2$CO). Mp 73-74°.

[28898-67-1, 59252-40-3]

Le Sueur, H.R. *et al*, *J. Chem. Soc.*, 1915, **107**, 736 (*synth*)
Cocker, W. *et al*, *Perfum. Essent. Oil Rec.*, 1963, **54**, 235 442 (*isol*)
Org. Synth., 1981, **60**, 11 (*synth*)

12,14-Hentriacontanedione, 9CI   H-80014

[71149-64-9]

$$H_3C(CH_2)_{16}COCH_2CO(CH_2)_{10}CH_3$$

$C_{31}H_{60}O_2$ M 464.814

Isol. from wax of *Dianthus caryophyllus* and other plants. Incorr. descr. in one ref. as the 14,16-dione.

Horn, D.H.S. *et al*, *Chem. Ind. (London)*, 1962, 1036 (*isol*)
Barbieri, G. *et al*, *J. Nat. Prod. (Lloydia)*, 1987, **50**, 646 (*synth, ir, pmr, cmr, ms*)

14,16-Hentriacontanedione, 9CI   H-80015

[24724-84-3]

$$H_3C(CH_2)_{14}COCH_2CO(CH_2)_{12}CH_3$$

$C_{31}H_{60}O_2$ M 464.814

Isol. from wax of *Eucalyptus risdoni*, *E. coccifera* and other plants.

Horn, D.H.S. *et al*, *Aust. J. Chem.*, 1964, **17**, 464 (*isol*)
Trka, A. *et al*, *Collect. Czech. Chem. Commun.*, 1974, **39**, 468 (*ms*)
Barbieri, G. *et al*, *J. Nat. Prod. (Lloydia)*, 1987, **50**, 646 (*synth, ir, pmr, cmr, ms*)

Hentriacontanoic acid, 9CI   H-80016

[38232-01-8]

$$H_3C(CH_2)_{29}COOH$$

$C_{31}H_{62}O_2$ M 466.830

Isol. from leaf wax of *Agave sisalana* and other plants. Cryst. (Me$_2$CO). Mp 93-93.2°.

[77630-51-4]

Piper, S.H. *et al*, *Biochem. J.*, 1934, **28**, 2175 (*synth*)
Razafindrazaka, J. *et al*, *Bull. Soc. Chim. Fr.*, 1963, 1633 (*isol*)
Rao, S.J. *et al*, *Indian J. Chem., Sect. B*, 1987, **26**, 208 (*synth*)

Heptacyclo[19.3.0.0^{1,5}.0^{5,9}.0^{9,13}.0^{13,17}.0^{17,21}] tetracosane   H-80017

[6.5]*Coronane.* [3.3.3.3.3.3]*Hexannulane*

[106115-42-8]

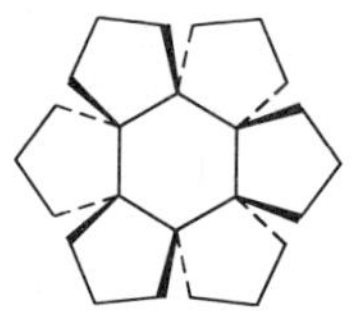

$C_{24}H_{36}$ M 324.548

Mp 222-224° (subl. from 200°).

Wehle, D. *et al*, *Chem. Ber.*, 1988, **121**, 2171 (*synth, pmr, cmr*)

8,10,12,14-Heptadecatetraen-1-al   H-80018

[13894-87-6]

$$H_3CCH_2(CH=CH)_4(CH_2)_6CHO$$

$C_{17}H_{26}O$ M 246.392

Isol. from leaves of *Centaurea* spp. Oil. λ_{max} 279, 291, 304, 319 nm.

(all-E)-form [32507-73-6]

Present in *Zoegea baldshuanica*.

Bohlmann, F. *et al*, *Chem. Ber.*, 1966, **99**, 3544; 1971, **104**, 961 (*isol, uv, ir, pmr, struct*)

1,7,9,15-Heptadecatetraene-11,13-diyne, 9CI   H-80019

Centaur X$_4$

[62787-38-6]

$$H_3CCH=CHC\equiv CC\equiv CCH=CHCH=CH(CH_2)_4$$
$$CH=CH_2$$

$C_{17}H_{20}$ M 224.345

(E,E,E)-form [34219-14-2]

Isol. from *Dahlia, Caiea, Gynoxys, Heterotheca, Bidens, Centaurea, Senecio, Pleiotaxis, Coreopsis, Calotis* and *Greenmaniella* spp. (all-E config. not always specifically indicated). Phytotoxic. Mp 29-30°.

(7E,9E,15Z)-form [22626-80-8]

Isol. from *Aster bellidiastrum*. Oil.

[71608-97-4]

Bohlmann, F. *et al*, *Chem. Ber.*, 1958, **91**, 1624; 1961, **94**, 3179; 1964, **97**, 2135; 1969, **102**, 1034 (*isol, ir, ms, pmr, uv*)
Lam, J. *et al*, *Phytochemistry*, 1971, **10**, 2227 (*occur*)
Bedford, C.T. *et al*, *J. Chem. Soc., Perkin Trans. 1*, 1976, 735 (*occur*)
Andersen, A.B. *et al*, *Phytochemistry*, 1977, **16**, 1829, 1185 (*isol*)
Bohlmann, F. *et al*, *Phytochemistry*, 1977, **16**, 285, 774, 965, 1065; 1979, **18**, 336, 1185; 1982, **21**, 1434 (*isol*)
Metwally, M.A. *et al*, *Phytochemistry*, 1985, **24**, 182 (*occur*)
Zdero, C. *et al*, *Phytochemistry*, 1987, **26**, 1999; 1988, **27**, 1105 (*occur*)

1,9,15-Heptadecatriene-11,13-diyn-8-ol   H-80020

$$H_3CCH=CHC\equiv CC\equiv CCH=CHCH(OH)(CH_2)_5$$
$$CH=CH_2$$

$C_{17}H_{22}O$ M 242.360

(Z,Z)-form

Isol. from roots of *Cousinia hystrix*. λ_{max} 227, 294, 313 nm (Et$_2$O).

(9E,15Z)-form [23369-07-5]

Ac: [73631-06-8].

$C_{19}H_{24}O_2$ M 284.397
Found in *Athanasia tridens*. Yellow oil.
Et ether: Yellow oil.

Bohlmann, F. *et al*, *Chem. Ber.*, 1966, **99**, 590; 1967, **100**, 1915 (*isol, struct, synth*)
Bohlmann, F. *et al*, *Phytochemistry*, 1979, **18**, 1736.

1,9,16-Heptadecatriene-4,6-diyn-3-ol   H-80021

[88153-63-3]

$$H_2C=CH(CH_2)_5CH=CHCH_2C≡CC≡$$
$$CCH(OH)CH=CH_2$$

$C_{17}H_{22}O$ M 242.360

(-)-(**Z**)-*form* [36150-08-0] ***Dehydrofalcarinol***
Isol. from roots of *Artemisia atrata*. Pale-yellow oil. $[\alpha]_D^{23}$ − 5.0° (c, 13.9 in Et$_2$O).

Ac: [78516-98-0].
  $C_{19}H_{24}O_2$ M 284.397
  Isol. from *Viguiera* sp. Oil.

[13894-97-8]

Bohlmann, F. *et al*, *Chem. Ber.*, 1966, **99**, 3552 (*isol, ir, pmr, struct*)
Bohlmann, F. *et al*, *Phytochemistry*, 1981, **20**, 113 (*isol*)
Harada, R. *et al*, *Phytochemistry*, 1982, **21**, 2009 (*isol*)

2,9,16-Heptadecatriene-4,6-diyn-1-ol   H-80022

$$H_2C=CH(CH_2)_5CH=CHCH_2C≡CC≡CCH=CHCH_2OH$$

$C_{17}H_{22}O$ M 242.360

(**Z,Z**)-*form*
Isol. from roots of *Anthemis tinctoria*. Oil. λ_{max} 251.5, 267, 282.5 nm.

Bohlmann, F. *et al*, *Chem. Ber.*, 1966, **99**, 2096.

2,9,16-Heptadecatriene-4,6-diyn-8-ol   H-80023

$$H_2C=CH(CH_2)_5CH=CHCH(OH)C≡CC≡$$
$$CCH=CHCH_3$$

$C_{17}H_{22}O$ M 242.360

(**Z,Z**)-*form* [13894-72-9]
Isol. from fresh roots of *Silybum marianum*. Oil. λ_{max} 246.5 (ε 4000), 254 (7200), 268.5 (10,400), 284 nm (8000) (Et$_2$O).

Bohlmann, F. *et al*, *Chem. Ber.*, 1966, **99**, 3201 (*isol, uv, ir, pmr, struct*)
Schulz, K.E. *et al*, *Arch. Pharm.* (*Weinheim, Ger.*), 1970, **301**, 7 (*isol*)

1-Heptadecene-4,6-diyne-3,9,10-triol, 9CI   H-80024
Panaxytriol

[87005-03-6]

$$H_2C=CHCH(OH)C≡CC≡CCH_2CH(OH)CH(OH)(CH_2)_6$$
$$CH_3$$

$C_{17}H_{26}O_3$ M 278.391
Isol. from *Panax ginseng*. Shows antitumour activity.
Needles (H$_2$O).

Matsunaga, H. *et al*, *Chem. Pharm. Bull.*, 1989, **37**, 1279 (*isol, uv, pmr, cmr, synth*)

1,5-Heptadiyne, 8CI   H-80025

Updated Entry replacing H-00254
[764-56-7]

$$HC≡CCH_2CH_2C≡CCH_3$$

C_7H_8 M 92.140
Bp$_{140}$ 85-90°.

Walba, D.M. *et al*, *J. Org. Chem.*, 1980, **45**, 2259 (*synth, pmr*)

1,1,2,2,3,3,3-Heptafluoro-1-iodopropane,   H-80026
9CI
Perfluoro-1-iodopropane
[27636-85-7]

$$F_3CCF_2CF_2I$$

C_3F_7I M 295.926
Liq. d$_4^{20}$ 2.06. Bp 41.2° (39.5°) . n_D^{20} 1.3272. Turns pink in air.

Hauptschein, M. *et al*, *J. Am. Chem. Soc.*, 1951, **73**, 2461 (*synth*)
Haszeldine, R.N., *J. Chem. Soc.*, 1953, 2622 (*uv*)
Nodiff, E.A. *et al*, *J. Org. Chem.*, 1953, **18**, 235 (*props*)
Evans, D.F.E. *et al*, *J. Am. Chem. Soc.*, 1963, **85**, 238 (*F nmr*)

1,2,4,6,8,9,14-Heptahydroxydihydro-β-   H-80027
agarofuran

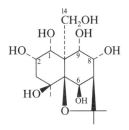

$C_{15}H_{26}O_8$ M 334.366

(*1α,2α,6β,8α,9α*)-*form*
9-*O-Benzoyl, 1,2,6,8,14-penta-Ac:* ***Celangulin***
  $C_{32}H_{40}O_{14}$ M 648.660
  Constit. of *Celastrus angulatus*. Insect antifeedant. Amorph. powder.

1,9-*Dibenzoyl, 2,6,8,14-tetra-Ac:* ***2α,6β,8α,14-Tetraacetoxy-1α,9α-dibenzoyloxy-4β-hydroxydihydro-β-agarofuran***
  $C_{37}H_{42}O_{14}$ M 710.730
  Constit. of *Maytenus canariensis*. Amorph. solid. Mp 50-52°.

Wakabayashi, N. *et al*, *J. Nat. Prod.* (*Lloydia*), 1988, **51**, 537.
González, A.G. *et al*, *Phytochemistry*, 1989, **28**, 173.

3,3′,4′,5,6,7,8-Heptahydroxyflavone   H-80028

Updated Entry replacing H-60024
2-(3,4-Dihydroxyphenyl)-3,5,6,7,8-pentahydroxy-4H-1-benzopyran-4-one, 9CI

$C_{15}H_{10}O_9$ M 334.239

3′,6,7,8-*Tetra-Me ether: 3,4′,5-Trihydroxy-3′,6,7,8-tetramethoxyflavone*
  $C_{19}H_{18}O_9$ M 390.346
  Constit. of *Gymnosperma glutinosum*.

3′,4,5,6,7,8-*Hexa-Me ether:* [35154-55-3]. *3-Hydroxy-3′,4′,5,6,7,8-hexamethoxyflavone*. ***Natsudaidan***
  $C_{21}H_{22}O_9$ M 418.399
  Constit. of peel oil of *C. natsudaidai*. Yellow cryst. (EtOH). Mp 154-156° (141-143°).

3′,4,5,6,7,8-*Hexa-Me ether, Ac:* [35154-56-4].
  Mp 125°.

3,3′,4′,6,7,8-*Hexa-Me ether:* [1176-88-1]. *5-Hydroxy-3,3′,4′,6,7,8-hexamethoxyflavone*
  $C_{21}H_{22}O_9$ M 418.399
  Isol. from peel of *Citrus sinensis*. Yellow cryst. (MeOH). Mp 110-111°.

3,3',4,5,8-Penta-Me, 6,7-methylene ether: 3,3',4,5,8-
Pentamethoxy-6,7-methylenedioxyflavone. **Melicophyllin**
$C_{21}H_{20}O_9$ M 416.384
Constit. of *Melicope triphylla.* Cryst. (Me$_2$CO). Mp 175-
177°.

Sastry, G.P. *et al, Tetrahedron,* 1961, **15**, 111 (*synth*)
Row, L.R. *et al, Indian J. Chem.,* 1963, **1**, 207 (*synth*)
Gentili, B. *et al, Tetrahedron,* 1964, **20**, 2313 (*synth*)
Kinoshita, K. *et al, Yakugaku Zasshi (J. Pharm. Soc. Jpn.),* 1971,
   91, 1105 (*isol*)
Tatum, J.H. *et al, Phytochemistry,* 1972, **11**, 2283 (*isol, uv*)
Bittner, M. *et al, Phytochemistry,* 1983, **22**, 1523 (*isol*)
Yu, S. *et al, Phytochemistry,* 1988, **27**, 171 (*isol*)
Jong, T.-T. *et al, Phytochemistry,* 1989, **28**, 245 (*isol*)

3',4',5,5',6,7,8-Heptahydroxyflavone   H-80029

Updated Entry replacing H-40022
*5,6,7,8-Tetrahydroxy-2-(3,4,5-trihydroxyphenyl)-4H-1-
benzopyran-4-one, 9CI*
$C_{15}H_{10}O_9$ M 334.239

4',5',6,8-Tetra-Me ether: [18398-74-8]. *3',5,7-Trihydroxy-
4',5',6,8-tetramethoxyflavone.* **Scaposin**
$C_{19}H_{18}O_9$ M 390.346
Isol. from *Hymenoxys scaposa.* Yellow needles
(CHCl$_3$/C$_6$H$_6$). Mp 210-212°.

4',5',6,8-Tetra-Me ether, Tri-Ac: [18398-75-9].
Prisms (MeOH). Mp 179-180°.

4',6,7,8-Tetra-Me ether: [29550-07-0]. *3',5,5'-Trihydroxy-
4',6,7,8-tetramethoxyflavone.* **Gardenin E**
$C_{19}H_{18}O_9$ M 390.346
Isol. from *Gardenia lucida, G. turgida* and Dikamali
gum. Yellow needles (EtOH). Mp 234°.

4',6,7,8-Tetra-Me ether, Tri-Ac: [51863-86-6].
Platelets (EtOAc/pet. ether). Mp 163-164°.

3',4',5',6,8-Penta-Me ether: [53950-56-4]. *5,7-Dihydroxy-
3',4',5',6,8-pentamethoxyflavone.* **Luiselizondin**
$C_{20}H_{20}O_9$ M 404.373
Isol. from aerial parts of *Gymnosperma glutinosum.*
Yellow needles (MeOH/CCl$_4$). Mp 185-186° (178-179°).
Struct. uncertain.

4',5',6,7,8-Penta-Me ether: [29550-05-8]. *3',5-Dihydroxy-
4',5',6,7,8-pentamethoxyflavone.* **Gardenin C**
$C_{20}H_{20}O_9$ M 404.373
Isol. from roots of *Gardenia lucida.* Yellow flakes
(EtOAc/pet. ether). Mp 179-180°.

3',4',5',6,7,8-Hexa-Me ether: [21187-73-5]. *5-Hydroxy-
3',4',5',6,7,8-hexamethoxyflavone.* **Gardenin A**
$C_{21}H_{22}O_9$ M 418.399
Isol. from roots of *Gardenia turgida* and lead buds of *G.
lucida.* Yellow needles (MeOH). Mp 164-165°.

3',5,5',6,7,8-Hexa-Me ether: [85644-03-7]. *4'-Hydroxy-
3',5,5',6,7,8-hexamethoxyflavone*
$C_{21}H_{22}O_9$ M 418.399
Constit. of *Eupatorium leucolepis.* Cryst.
(CHCl$_3$/MeOH). Mp 191-193°.

4',5,5',6,7,8-Hexa-Me ether: [42557-19-7]. *3'-Hydroxy-
4',5,5',6,7,8-hexamethoxyflavone*
$C_{21}H_{22}O_9$ M 418.399
Constit. of *E. leucolepis.* Gum.

5,5',6,8-Tetra-Me, 3',4'-methylene ether: 7-Hydroxy-
5,5',6,8-tetramethoxy-3',4'-methylenedioxyflavone
$C_{20}H_{18}O_9$ M 402.357
Constit. of *Ageratum tomentosum* var. *bracteatum.*
Yellow cryst. Mp 188-190°.

3,4-Dihydro, hepta-Me ether: 3',4',5,5',6,7,8-
Heptamethoxyflavanone
$C_{22}H_{26}O_9$ M 434.442

Constit. of *A. tomentosum* var. *bracteatum.* Yellow gum.

Rao, A.V.R. *et al, Indian J. Chem.,* 1968, **6**, 677; 1970, **8**, 398 (*isol,
   struct, pmr, ms*)
Thomas, M.B. *et al, Tetrahedron,* 1968, **24**, 3675 (*isol, synth, uv, ir,
   pmr, Scaposin*)
Kalra, A.J. *et al, Indian J. Chem.,* 1970, **8**, 398; 1973, **11**, 1092
   (*isol, synth, uv, pmr*)
Kamalam, M. *et al, Indian J. Chem.,* 1970, **8**, 573 (*synth*)
Domínguez, X.A. *et al, Phytochemistry,* 1974, **13**, 1624 (*isol, uv,
   pmr, ms*)
Gupta, S. *et al, Indian J. Chem.,* 1975, **13**, 785 (*isol, uv*)
Chhabra, S.C. *et al, Indian J. Chem., Sect. B,* 1977, **15**, 421 (*synth,
   uv, ir, pmr*)
Joshi, K.C. *et al, J. Indian Chem. Soc.,* 1979, **56**, 327 (*isol*)
Herz, W. *et al, Phytochemistry,* 1982, **21**, 2683 (*isol*)
Voirin, B., *Phytochemistry,* 1983, **22**, 2107 (*uv*)
Vázquez, M.M. *et al, Phytochemistry,* 1988, **27**, 3706 (*isol*)

1,5-Heptanediol, 9CI   H-80030

Updated Entry replacing H-00291
[60096-09-5]

$$H_3CCH_2CH(OH)CH_2CH_2CH_2CH_2OH$$

$C_7H_{16}O_2$ M 132.202
(±)-form
Liq. Insol. H$_2$O. d_4^{20} 0.971. Bp$_{11}$ 135-136°. n_D^{22} 1.4571.

Pierce, J.S. *et al, J. Am. Chem. Soc.,* 1925, **47**, 1098.
Paul, R., *Bull. Soc. Chim. Fr.,* 1935, **2**, 311.
Bihovsky, R. *et al, J. Org. Chem.,* 1988, **53**, 4026 (*synth*)

3-Heptynoic acid   H-80031

[59862-93-0]

$$H_3CCH_2CH_2C\equiv CCH_2COOH$$

$C_7H_{10}O_2$ M 126.155
Liq. Bp$_{22}$ 140-141°.

Bigley, D.B. *et al, J. Chem. Soc., Perkin Trans. 2,* 1972, 592
   (*synth*)

4-Heptynoic acid   H-80032

[42441-83-8]

$$H_3CCH_2C\equiv CCH_2CH_2COOH$$

$C_7H_{10}O_2$ M 126.155
Needles (pet. ether). Mp 57-59°.

4-Bromophenacyl ester: Plates (EtOH). Mp 99-100°.

Nitrile: [18719-33-0]. *4-Heptynenitrile, 9CI. 1-Cyano-3-
hexyne*
C_7H_9N M 107.155
Bp$_{16}$ 82-83°.

Ansell, M.F. *et al, J. Chem. Soc. C,* 1968, 217 (*synth*)

6-Heptynoic acid   H-80033

[30964-00-2]

$$HC\equiv C(CH_2)_4COOH$$

$C_7H_{10}O_2$ M 126.155
Liq. Bp$_1$ 93-94°.

Me ester:
$C_8H_{12}O_2$ M 140.182
Bp$_{11}$ 80°.

Nitrile: [15295-69-9]. *6-Heptynenitrile, 9CI. 6-Cyano-1-
hexyne*
C_7H_9N M 107.155
Bp$_{14}$ 87-93°.

Taylor, W.R. *et al, J. Am. Chem. Soc.,* 1950, **72**, 4263 (*synth*)

Ferrier, R.J. *et al*, *J. Chem. Soc.*, 1957, 1435 (*synth, nitrile*)
Gautier, J.A. *et al*, *Bull. Soc. Chim. Fr.*, 1967, 1551 (*nitrile*)
Moody, C.J. *et al*, *J. Chem. Soc., Perkin Trans.* 1, 1988, 3249 (*synth, anhydride, ir, pmr, ms*)

Heritonin   H-80034

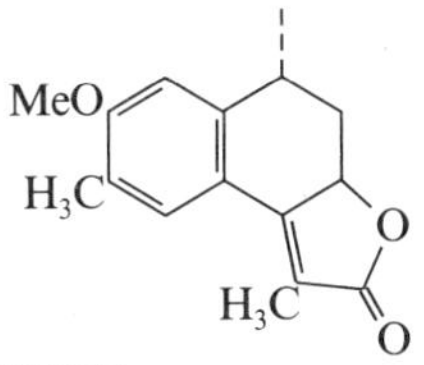

$C_{16}H_{18}O_3$ M 258.316
Isol. from *Heritiera littoralis*. Piscicide. Cryst. (C_6H_6/Et_2O). Mp 115-116°.

Miles, P.H. *et al*, *J. Nat. Prod.* (*Lloydia*), 1989, **52**, 896 (*isol, pmr*)

2a,4a,6a,8a,10a,12a-Hexaazacoronene-1,3,5,7,9,11(2H,4H,6H,8H,10H,12H)-hexone, 9CI   H-80035

[92187-33-2]

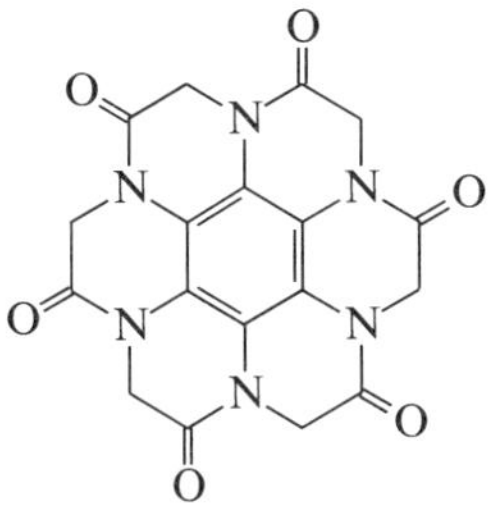

$C_{18}H_{12}N_6O_6$ M 408.329
Fine powder. Mp >350°.

Thomaides, J. *et al*, *J. Am. Chem. Soc.*, 1988, **110**, 3970 (*synth, pmr, ms, ir*)

3,9,17,23,29,30-Hexaaza-6,20-dioxa-tricyclo[23.3.1.1^{11,15}]triaconta-1(28),11(29),12,14,25,27-hexaene   H-80036

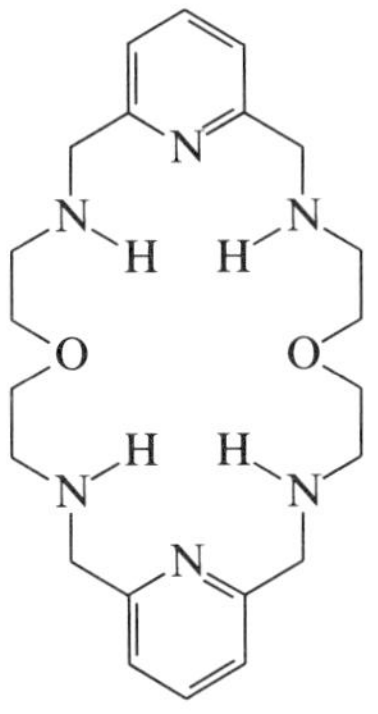

$C_{20}H_{34}N_6O_2$ M 390.528
B,HCl: V. hygroscopic solid + 2EtOH. Mp 160°.
N-*Tetrakis(4-methylbenzenesufonyl):* Cryst. (CH_2Cl_2/hexane). Mp 185°.

Hosseini, M.W. *et al*, *Helv. Chim. Acta*, 1989, **72**, 1066 (*synth*)

Hexa[7]circulene   H-80037

Updated Entry replacing H-00485
1,14-Ethenonaphtho[1′,8′:5,6,7]cyclohepta[1,2,3,4-def]phenanthrene, 9CI. 1,16-Didehydrohexahelicene
[56594-61-7]

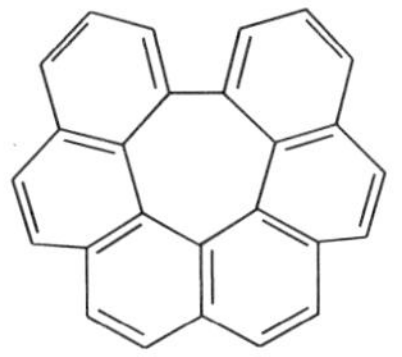

$C_{26}H_{14}$ M 326.397
Pale-yellow plates (EtOH aq.). Mp 162-163°. May be resolved chromatographically at low temp. V. optically labile.

Jessup, P.J. *et al*, *Tetrahedron Lett.*, 1975, 1453; *Aust. J. Chem.*, 1976, **29**, 173 (*synth, ms, pmr, uv*)
Yamamoto, K. *et al*, *J. Am. Chem. Soc.*, 1988, **110**, 3578 (*resoln*)

Hexacosanal   H-80038

[26627-85-0]

$$H_3C(CH_2)_{24}CHO$$

$C_{26}H_{52}O$ M 380.696
Isol. from *Vitis vinifera* var. *sultana*. Cryst. (MeOH). Mp 73-73.5°.

Crabbé, P. *et al*, *Bull. Soc. Chim. Belg.*, 1961, **70**, 168 (*synth, ir*)
Radler, F. *et al*, *Aust. J. Chem.*, 1965, **18**, 1059 (*isol*)

Hexacyclo[6.5.1.0^{2,7}.0^{3,11}.0^{4,9}.0^{10,14}]tetradeca-5,12-diene   H-80039

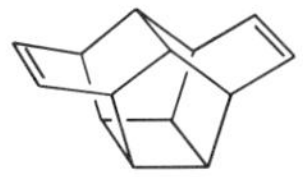

$C_{14}H_{14}$ M 182.265
Liq.

Chou, T.-C. *et al*, *J. Org. Chem.*, 1988, **53**, 5168 (*synth, ir, pmr, cmr, ms*)

7,9,11,13-Hexadecatetraen-1-al   H-80040

[65398-27-8]

$$H_3CCH_2(CH=CH)_4(CH_2)_5CHO$$

$C_{16}H_{24}O$ M 232.365
Present in leaves of *Centaurea* spp. Not isol. in pure form.
(*all-E*)-*form* [32507-72-5]
     Present in *Zoegea baldshuanica*. Not isol. in pure form.

Bohlmann, F. *et al*, *Chem. Ber.*, 1966, **99**, 3544; 1971, **104**, 961 (*isol, struct*)
Andersen, A.B. *et al*, *Phytochemistry*, 1977, **16**, 1829 (*isol*)

6,8,12,14-Hexadecatetraen-10-yn-1-ol, 9CI   H-80041

[77319-35-8]

$$H_3CCH=CHCH=CHC\equiv CCH=CHCH=CH(CH_2)_4CH_2OH$$

$C_{16}H_{22}O$ M 230.349
(*all-E*)-*form* [1540-99-4]

Isol. from above ground parts of *Dahlia merckii*, from roots of *Bupleurum gibraltaricum*, from *Zoegea baldshuanica* and above ground parts of *Serrulata gmelini*. Shows pesticidal and herbicidal props. Needles or leaflets (pet. ether). Mp 76-77° (65-71°).

Ac: 1-Acetoxy-6,8,12,14-hexadecatetraen-10-yne
$C_{18}H_{24}O_2$ M 272.386
Isol. from *D. merckii*. Leaflets (pet. ether). Mp 52-53°.

Bohlmann, F. *et al*, *Chem. Ber.*, 1965, **98**, 872; 1967, **100**, 1915, 1936; 1971, **104**, 961; 1975, **108**, 2822 (*isol, uv, ir, pmr, struct, ms, synth*)
Canadian Pat., 1 172 460, (1984); *CA*, **102**, 57830z (*use*)

6,8,14-Hexadecatriene-10,12-diyn-1-al   H-80042

[65398-29-0]

$$H_3CCH=CHC\equiv CC\equiv CCH=CHCH=CH(CH_2)_4CHO$$

$C_{16}H_{18}O$ M 226.318
(**all-E**)-**form** [13894-88-7]
Isol. from leaves of *Centaurea involucrata*. Oil. λ_{max} 249, 266, 296, 314, 337 nm (Et$_2$O).

Bohlmann, F. *et al*, *Chem. Ber.*, 1966, **99**, 3544 (*isol, struct*)
Andersen, A.B. *et al*, *Phytochemistry*, 1977, **16**, 1829 (*isol*)

1,6,8-Hexadecatriene-10,12,14-triyne   H-80043

[6112-04-5]

$$H_3CC\equiv CC\equiv CC\equiv CCH=CHCH=CH(CH_2)_3CH=CH_2$$

$C_{16}H_{16}$ M 208.302
(**E,E**)-**form** [3513-88-0]
Isol. from above-ground parts of *Chrysanthemum ircutianum* and *C. maximum*. Cryst. Mp ~20°. λ_{max} 258, 269, 288, 305, 325, 348 nm (Et$_2$O).

Bohlmann, F. *et al*, *Chem. Ber.*, 1965, **98**, 2596; 1966, **99**, 586 (*isol, uv, ir, struct*)
Wrang, P.A. *et al*, *Phytochemistry*, 1975, **14**, 1027 (*isol*)

7,12,14-Hexadecatrien-10-yn-1-ol   H-80044

$$H_3CCH=CHCH=CHC\equiv CCH_2CH=CH(CH_2)_5CH_2OH$$

$C_{16}H_{24}O$ M 232.365
(**7Z,12E,14E**)-**form** [16697-19-1]
Isol. from roots of *Dahlia merckii*.
p-*Phenylazobenzoyl*: Cryst. (pet. ether). Mp 51-51.5°.

Bohlmann, F. *et al*, *Chem. Ber.*, 1965, **98**, 872; 1967, **100**, 1936 (*isol, struct, synth*)
Bedford, C.T. *et al*, *J. Chem. Soc., Perkin Trans.* 1, 1976, 735 (*isol*)

5-Hexadecenoic acid, 9CI   H-80045

[20057-14-1]

$$H_3C(CH_2)_9CH=CH(CH_2)_3COOH$$

$C_{16}H_{30}O_2$ M 254.412
(**E**)-**form** [7056-84-0]
Isol. from seed oil of *Thalictrum venulosum*.
(**Z**)-**form** [7056-90-8]
Isol. from *T. venulosum* and *Dioscoreophyllum cumminsii* seed oils and from a strain of *Bacillus megaterium* (biosynthetically from hexadecanoic acid).

Fulco, A.J. *et al*, *J. Biol. Chem.*, 1964, **239**, 998 (*isol*)
Bhatty, M.K. *et al*, *Can. J. Biochem.*, 1966, **44**, 311 (*isol*)
Spencer, G.F. *et al*, *Lipids*, 1972, **7**, 435 (*isol*)
Starratt, A.N., *Chem. Phys. Lipids*, 1976, **16**, 215 (*synth*)

3-[5-(2,4-Hexadiynyl)-2-furanyl]-2-propenal   H-80046

3-[2-(2,4-Hexadiynyl)-5-furyl]acrolein

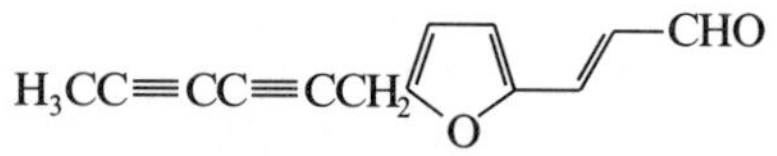

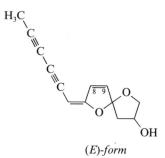

$$H_3CC\equiv CC\equiv CCH_2$$

$C_{13}H_{10}O_2$ M 198.221
(**E**)-**form**
Isol. from *Chrysanthemum silvaticum*. Cryst. (Et$_2$O). Mp 106.5°.

Bohlmann, F. *et al*, *Chem. Ber.*, 1965, **98**, 2596 (*isol, uv, ir, pmr, struct, synth*)

7-(2,4-Hexadiynylidene)-1,6-dioxaspiro[4.4]non-8-en-3-ol, 9CI   H-80047

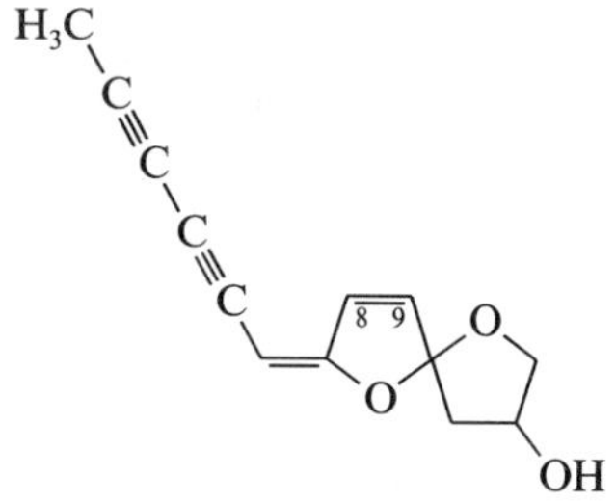

(*E*)-*form*

$C_{13}H_{12}O_3$ M 216.236
(**E**)-**form**
Ac: [85799-14-0]. *8-Acetoxy-2-(2,4-hexadiynylidene)-1,6-dioxaspiro[4.4]non-3-ene*
$C_{15}H_{14}O_4$ M 258.273
Isol. from roots of *Chrysanthemum parthenium*. Cryst. (Et$_2$O/pet. ether). Mp 98.5°.
(**Z**)-**form**
Ac: [85799-16-2].
Isol. from roots of *C. mawii*. Amorph.
Deoxy, 8,9-epoxy: *3,4-Epoxy-2-(2,4-hexadiynylidene)-1,6-dioxaspiro[4.4]nonane*
$C_{13}H_{12}O_2$ M 200.237
Isol. from roots of *C. parthenifolium*. Cryst. (pet. ether). Mp 126°.

Bohlmann, F. *et al*, *Chem. Ber.*, 1965, **98**, 1411, 2596 (*isol, uv, ir, pmr, synth*)

2-(2,4-Hexadiynylidene)-5-(propionylmethylidene)-2,5-dihydrofuran   H-80048

5,8-Epoxy-4,6,8-tetradecatriene-10,12-diyn-3-one

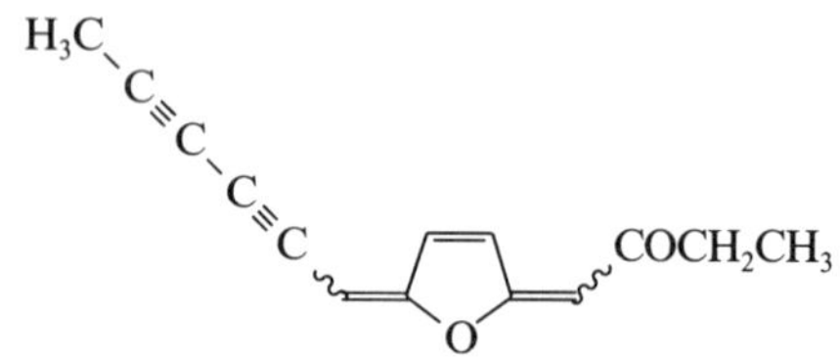

$C_{14}H_{12}O_2$ M 212.248
Isol. from roots of *Anacyclus radiatus* and *Chrysanthemum pyrethrum*. Cryst. (pet. ether). Mp 78.5°.

Bohlmann, F. *et al*, *Chem. Ber.*, 1963, **96**, 588; 1964, **97**, 1179 (*isol, uv, pmr, struct*)

1,1,1,2,2,3-Hexafluoropropane, 9CI   H-80049
[677-56-5]

$$F_3CCF_2CH_2F$$

C$_3$H$_2$F$_6$ M 152.039
Bp 1.2°.

[27070-61-7]

Edgell, W.F. *et al, J. Am. Chem. Soc.*, 1955, **77**, 4899 (*synth*)
Haszeldine, R.N. *et al, J. Chem. Soc.*, 1957, 2800 (*synth*)
Lelleman, D.D. *et al, J. Mol. Spectrosc.*, 1961, **7**, 322 (*F nmr*)

1,1,1,3,3,3-Hexafluoropropane, 9CI, 8CI   H-80050
2,2-Dihydroperfluoropropane
[690-39-1]

$$F_3CCH_2CF_3$$

C$_3$H$_2$F$_6$ M 152.039
Shows anaesthetic props. Fp −93.62°. Bp −0.5° (0.5-1.0°)

[27070-61-7]

Tarrant, P. *et al, J. Am. Chem. Soc.*, 1955, **77**, 2783 (*synth*)
Lelleman, D.D. *et al, J. Mol. Spectrosc.*, 1961, **7**, 322 (*F nmr*)
Cheburkov, Yu.A. *et al, Izv. Akad. Sci. USSR, Ser. Sci. Khim.*, 1963, 1573 (*synth*)
Naae, D.G. *et al, Org. Mass Spectrom.*, 1974, **9**, 1203 (*ms*)
Di Paolo, T. *et al, J. Pharm. Sci.*, 1979, **68**, 39 (*pharmacol*)
Buerger, H. *et al, Spectrochim. Acta, Part A*, 1979, **35**, 517 (*ir, raman, F nmr*)
Eujen, R. *et al, J. Fluorine Chem.*, 1983, **22**, 263 (*F nmr*)
Bloshchitsa, F.A. *et al, Zh. Org. Khim.*, 1985, **21**, 1414 (*synth, F nmr, pmr*)

2,3,4,6,7,8-Hexahydro-1,5-anthracene-dione, 9CI   H-80051
[82817-91-2]

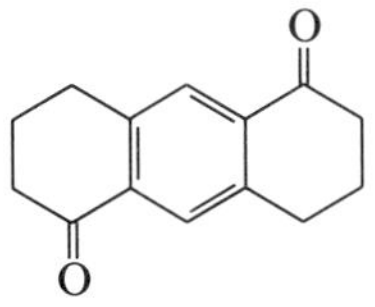

C$_{14}$H$_{14}$O$_2$ M 214.263
Cryst. (EtOAc). Mp 160°.

Caluwe, P. *et al, J. Org. Chem.*, 1988, **53**, 1786 (*synth, pmr, cmr*)

1,2,3,4,7,8-Hexahydrobenzo[*c*]phenanthrene, 9CI   H-80052
[56922-96-4]

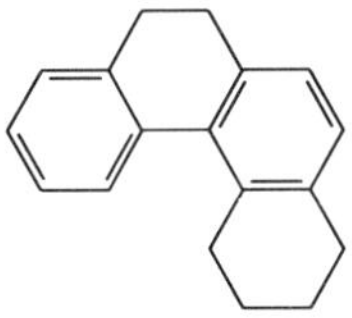

C$_{18}$H$_{18}$ M 234.340
Solid. Mp 77-78°. Bp$_{0.1}$ 155°.

Muller, D. *et al, Tetrahedron*, 1975, **31**, 1449 (*synth, ir, pmr, uv*)
Gilchrist, T.L. *et al, J. Chem. Soc., Perkin Trans. 1*, 1988, 2595 (*synth, pmr, ms*)

1,3,4,6,7,11*b*-Hexahydro-2*H*-benzo[*a*]quinolizine   H-80053
Benzo[a]*quinolizidine*
[55302-24-4]

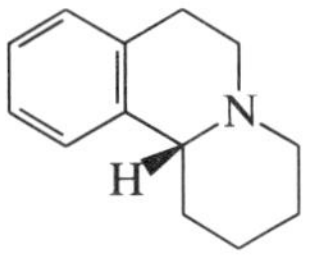

(*R*)-*form*

C$_{13}$H$_{17}$N   M 187.284
(*R*)-*form* [2737-54-4]
Bp$_7$ 127-128°. [α]$_D$ +222° (c, 0.022 in Py).
B,HCl: [58613-60-8].
Cryst. (2-propanol). Mp 160-161°.
B,HClO$_4$· [2737-55-5].
Cryst. (2-propanol). Mp 160-161°.
B,MeI: Cryst. (2-propanol). Mp 173-174° dec.
(*S*)-*form* [17279-34-4]
Oil. Bp$_{0.03}$ 75°. [α]$_D^{25}$ −206° (c, 0.017 in pyridine).
Optically pure.
B,HClO$_4$: Cryst. (2-propanol). Mp 160-162°. [α]$_D$ −75.3°.
B,HCl: [17279-35-5].
[α]$_D$ −140° (c, 0.425 in EtOH).
(±)-*form*
B,HCl: [20072-13-3].
Mp 242-243°.

Horii, Z. *et al, Tetrahedron*, 1963, **19**, 2101 (*synth, abs config, ord*)
Craig, J.C. *et al, Tetrahedron*, 1967, **23**, 3573 (*ord, abs config*)
Parello, J., *Bull. Soc. Chim. Fr.*, 1968, 1117 (*synth, uv*)
Meyers, A.I. *et al, Tetrahedron*, 1987, **43**, 5095 (*synth, ir, pmr*)

3,4,7,8,11,12-Hexahydro-2*H*,6*H*,10*H*-benzo[1,2-*b*:3,4-*b*':5,6-*b*″]tris[1,4]oxazine, 9CI   H-80054
[114491-24-6]

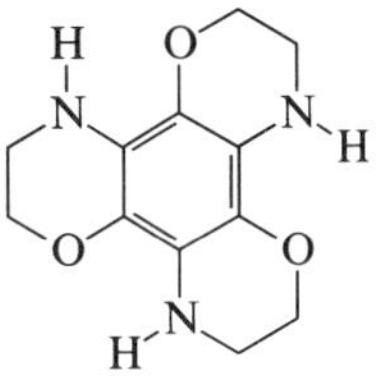

C$_{12}$H$_{15}$N$_3$O$_3$ M 249.269
Solid. Mp 216-219°. Mildly air sensitive.
Tri-N-Me: [114491-20-2].
C$_{15}$H$_{21}$N$_3$O$_3$ M 291.349
Solid. Mp 148-150°. Forms a cation but salts unstable.

Thomaides, J. *et al, J. Am. Chem. Soc.*, 1988, **110**, 3970 (*synth, pmr, ir, esr*)

3,4,7,8,11,12-Hexahydro-2*H*,6*H*,10*H*-benzo[1,2-*b*:3,4-*b*':5,6-*b*″]tris[1,4]thiazine, 9CI   H-80055
[114491-30-4]

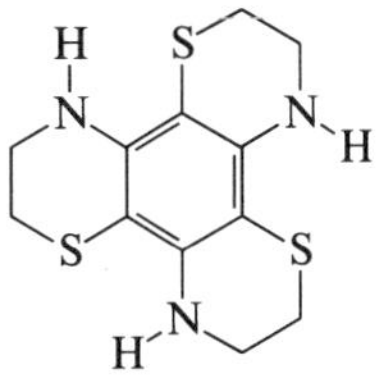

$C_{12}H_{15}N_3S_3$　　M 297.469
Solid. Mp 202-204°.
Tri-N-Me: [114491-28-0].
　$C_{15}H_{21}N_3S_3$　　M 339.549
　Solid. Mp 252-253°.

Thomaides, J. *et al, J. Am. Chem. Soc.*, 1988, **110**, 3970 (*synth, pmr, ir, ms, esr*)

1,2,3,4,4a,10a-Hexahydro-dibenzo[b,e][1,4]dioxin　　H-80056

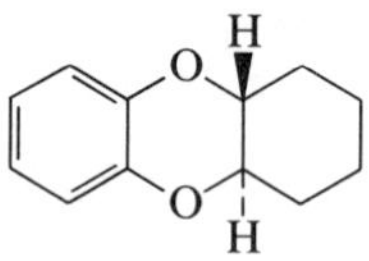

(4aRS,10aRS)-*form*

$C_{12}H_{14}O_2$　　M 190.241
(**4aRS,10aRS**)-*form* [75768-17-1]
　(±)-trans-*form*
　Cryst. (hexane). Mp 115-117°.
(**4aRS,10aSR**)-*form* [75459-45-9]
　cis-*form*
　Cryst. (MeOH). Mp 52-54° (43-44°).

Antus, S. *et al, Chem. Ber.*, 1989, **122**, 1017 (*synth, ir, pmr, cmr*)

2,3,4,6,7,8-Hexahydro-4,8-dihydroxy-2-(1-hydroxyheptyl)-5H-1-benzopyran-5-one, 9CI　　H-80057

4,8-Dihydroxy-2-(1-hydroxyheptyl)-3,4,5,6,7,8-hexahydro-2H-1-benzopyran-5-one
[119903-56-9]

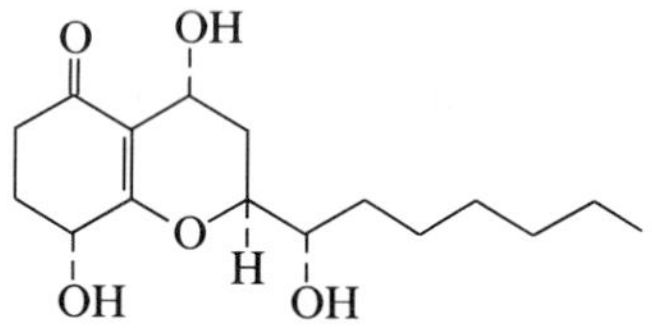

$C_{16}H_{26}O_5$　　M 298.378
Metab. of *Trichoderma koningii.* Antibiotic. Needles (CHCl$_3$/pentane). Mp 122-123°. [α]$_D$ +166.9° (c, 0.3 in CHCl$_3$).

Dunlop, R.W. *et al, J. Nat. Prod.* (*Lloydia*), 1989, **52**, 67 (*isol, pmr, cmr*)

2,3,5,6,8,9-Hexahydro-1H-diimidazo[1,2-d:2′,1′-g][1,4]diazepine　　H-80058

[111062-19-2]

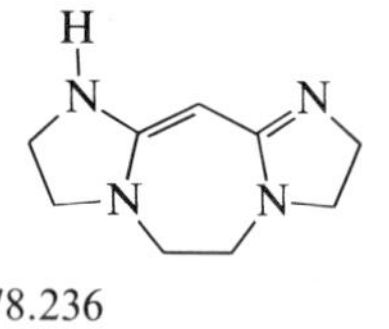

$C_9H_{14}N_4$　　M 178.236
Readily obt. highly basic proton sponge. Mp 131-132°.
pK_a 26.22 (MeCN).
B,HCl: Mp ~290° dec. Delocalised.
B,HClO$_4$: Mp 260-265°.
N-Me: [111062-21-6].
　$C_{10}H_{16}N_4$　　M 192.263
　pK_a 26.95 (MeCN).

Schwesinger, R., *Angew. Chem., Int. Ed. Engl.*, 1987, **26**, 1164 (*synth, pmr*)

Hexahydro-2H-furo[3,2-b]pyrrol-2-one, 9CI　　H-80059

2-Oxa-6-azabicyclo[3.3.0]octan-3-one. 3-Hydroxypyrrolidine-2-acetic acid lactone. Geissman-Waiss lactone

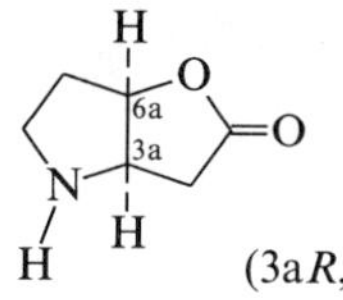

(3aR,6aR)-*form*

$C_6H_9NO_2$　　M 127.143
Intermed. in the synth. of retronecine and related pyrrolizidine alkaloids.
(**3aR,6aR**)-*form* [81445-21-8]
　(+)-cis-*form*
　B,HCl: Needles (EtOH). Mp 185-186.5°. [α]$_D^{25}$ +48.8° (c, 0.20 in MeOH).
(**3aS,6aS**)-*form*
　(−)-cis-*form*
　B,HCl: [118710-52-4].
　Mp 182-184°. [α]$_D$ −42.9°.

Geissman, T.A. *et al, J. Org. Chem.*, 1962, **27**, 139 (*synth*)
Rueger, H. *et al, Heterocycles*, 1982, **19**, 23 (*synth*)
Buchanan, J.G. *et al, J. Chem. Soc., Perkin Trans. 1*, 1987, 2377 (*synth, ir, pmr*)
Shishido, K. *et al, J. Chem. Soc., Perkin Trans. 1*, 1987, 993 (*synth*)
Cooper, J. *et al, J. Chem. Soc., Chem. Commun.*, 1988, 509 (*synth*)

Hexahydro-1,3(2H,4H)-isoquinolinedione, 9CI　　H-80060

2-Carboxycyclohexaneacetic acid imide. 3-Azabicyclo[4.4.0]decane-2,4-dione

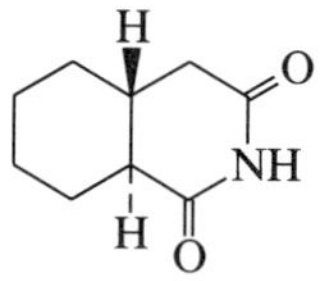

$C_9H_{13}NO_2$　　M 167.207
(**4aR,8aS**)-*form* [116673-41-7]
　Cryst. (toluene/hexane). Mp 186-187°. [α]$_D^{20}$ −73° (c, 1.2 in CHCl$_3$).

Poloński, T., *J. Chem. Soc., Perkin Trans. 1*, 1988, 639 (*synth, uv, ir, pmr*)

3,4,4a,5,8,8a-Hexahydro-1(2H)-naphthalenone　　H-80061

Δ^6-1-Octalone

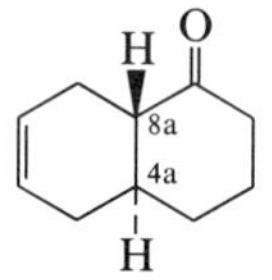

(4aS,8aS)-*form*

$C_{10}H_{14}O$　　M 150.220
(**4aS,8aS**)-*form*
　(+)-trans-*form*
　Cryst. by subl. Mp 57-58°. [α]$_D$ +88° (c, 0.693 in pet. ether).
(**4aRS,8aRS**)-*form* [70749-11-0]
　(±)trans-*form*
　Mp 46-47° (32-34°). Bp$_{14}$ 111-112°.
Oxime:

$C_{10}H_{15}NO_2$ M 181.234
Cryst. (EtOH). Mp 154-156°.

2,4-Dinitrophenylhydrazone: [83586-08-7].
Mp 195-196°.

(4aRS,8aSR)-form [70749-10-9]
($\pm$)-cis-*form*
Liq.

[78341-47-6]

Ireland, R.E. *et al, J. Org. Chem.*, 1962, **27**, 1620 (*synth*)
Acklin, W. *et al, Helv. Chim. Acta*, 1965, **48**, 1725 (*synth, abs config*)
Fringuelli, F. *et al, J. Org. Chem.*, 1982, **47**, 5056 (*synth, cmr*)
Angell, E.C. *et al, J. Org. Chem.*, 1988, **53**, 1424 (*synth, ir, pmr, cmr*)

4a,5,6,7,8,8a-Hexahydro-2(1*H*)-naphtha-lenone   H-80062

Δ^3-*2-Octalone*
[2384-50-1]

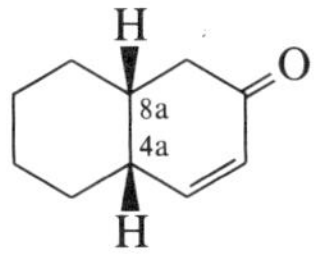

(4aRS,8aRS)-*form*

$C_{10}H_{14}O$ M 150.220

(4aRS,8aRS)-form [55999-54-7]
($\pm$)-cis-form
Oil. $Bp_{0.1}$ 75-77°.

(4aRS,8aSR)-form [71956-17-7]
($\pm$)-trans-form
Liq. Bp_{12} 127-130°, $Bp_{0.1}$ 85°.

Oxime:
$C_{10}H_{15}NO$ M 165.235
Needles (Et$_2$O). Mp 145°.

Semicarbazone: Cryst. (EtOH aq.). Mp 188-190°.

2,4-Dinitrophenylhydrazone: Mp 194-195°.

[18317-63-0]

Lehmann, E. *et al, Ber.*, 1934, **67**, 1867 (*synth*)
Mühle, H. *et al, Helv. Chim. Acta*, 1962, **45**, 1475 (*synth, uv, ir*)
Corey, E.J. *et al, Tetrahedron Lett.*, 1978, 4597 (*synth*)
Angell, E.C. *et al, J. Org. Chem.*, 1988, **53**, 1424 (*synth, ir, pmr, cmr*)

1,2,3,9,10,10a-Hexahydrophenanthrene   H-80063

[62690-96-4]
$C_{14}H_{16}$ M 184.280
Oil.

Chini, M. *et al, J. Org. Chem.*, 1989, **54**, 3930 (*synth, pmr*)

1,2,3,6,7,8-Hexahydro-4,5-phenanthrene-dione   H-80064

[110028-98-3]

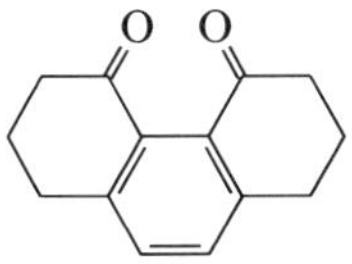

$C_{14}H_{14}O_2$ M 214.263
Cryst. (hexane). Mp 130°.

Caluwe, P. *et al, J. Org. Chem.*, 1988, **53**, 1786 (*synth, pmr, cmr*)

2,3,4,5,6,7-Hexahydro-1,8-phenanthrene-dione, 9CI   H-80065

[82817-89-8]

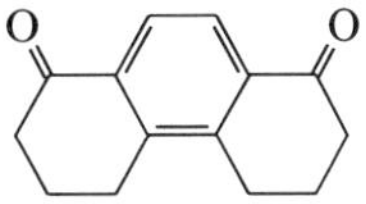

$C_{14}H_{14}O_2$ M 214.263
Cryst. (Et$_2$O). Mp 167-168°.

Caluwe, P. *et al, J. Org. Chem.*, 1988, **53**, 1786 (*synth, ir, pmr*)

5,5a,6,11,11a,12-Hexahydroquinoxalino [2,3-b]quinoxaline   H-80066

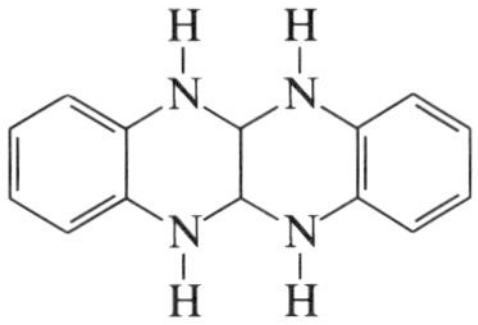

$C_{14}H_{14}N_4$ M 238.291
N-Tetra-Me: [13784-23-1].
$C_{18}H_{22}N_4$ M 294.399
Cryst. Mp 167-169°.

Tauer, E. *et al, Angew. Chem., Int. Ed. Engl.*, 1989, **28**, 338 (*synth, pmr, uv*)

3,3′,4′,5,6,7-Hexahydroxyflavone   H-80067

Updated Entry replacing H-20066
Quercetagetin
[90-18-6]
$C_{15}H_{10}O_8$ M 318.239
Isol. from *Tagetes patula* (African marigold), *Gmelina arborea*, *G. asiatica*, *Artemesia taurica* and *Eriocaulon* spp. Pale-yellow cryst. + 2H$_2$O (EtOH aq.). Mp 318-320° (232-234°).

7-Glucoside: [548-75-4]. *Quercetagitrin*
$C_{21}H_{20}O_{13}$ M 480.381
Isol. from petals of *T. erecta* (African marigold) and *Lepidophorum*, *Tetragonotheca* and *Anacyclus* spp. Cryst. (Py aq.). Mp 236-238° dec.

3,3′-Di-Me ether: [36034-36-3]. *4′,5,6,7-Tetrahydroxy-3,3′-dimethoxyflavone*
$C_{17}H_{14}O_8$ M 346.293
Isol. from *Parthenium tomentosum*. Cryst. (C$_6$H$_6$/Me$_2$CO). Mp 214-215° (242-244°).

3,3′,6-Tri-Me ether:
Found in *Prunus avium*, *Fluorensia ilicitolia* and *Centaurea hyssopifolia*. Pale yellow needles. Mp 127-133°.

3,3′,6-Tri-Me ether, 7-O-β-D-glucopyranoside: [35305-11-4].
Jacein
$C_{24}H_{26}O_{13}$ M 522.462
Constit. of *Centaurea jacea* and *C. hyssopifolia*. Cryst. Mp 205-207°. $[\alpha]_D^{20}$ −73.1° (c, 1.48 in MeOH).

3,3′,5,6,7-Penta Me ether: 4′-Hydroxy-3,3′,5,6,7-pentamethoxyflavone
$C_{20}H_{20}O_8$ M 388.373
Constit. of *Jasonia montana*.

3,3′,6,7-Tetra-Me ether: [603-56-5]. *4′,5-Dihydroxy-3,3′,6,7-tetramethoxyflavone*. *Chrysosplenetin*

$C_{19}H_{18}O_8$ M 374.346
Constit. of leaves of *Plectranthus marrubioides*. Cryst. (MeOH or EtOH). Mp 182-183°.

3,3′,4′,5,6-Penta-Me ether: [57393-68-7]. *7-Hydroxy-3,3′,4′,5,6-pentamethoxyflavone*
$C_{20}H_{20}O_8$ M 388.373
Constit. of *Ambrosia grayi*. Mp 132-135°.

Hexa-Me ether: [1251-84-9]. *3,3′,4′,5,6,7-Hexamethoxyflavone. Hexamethylquercetagenin*
$C_{21}H_{22}O_8$ M 402.400
Cryst. (Me₂CO). Mp 142-143°.

3,3′,6,7-Tetra Me ether: 4′,5-Dihydroxy-3,3′,6,7-tetramethoxyflavone
$C_{19}H_{18}O_8$ M 374.346
Constit. of *B. malcomii*. Cryst. Mp 189-191°. Originally described as 5′,6-Dihydroxy-2′,3,5,7-trimethoxyflavone.

3,3′,4′,6,7-Penta-Me ether: 5-Hydroxy-3,3′,4′,6,7-pentamethoxyflavone
$C_{20}H_{20}O_8$ M 388.373
Constit. of *B. malcomii*. Cryst. Mp 166-168°. Originally described as 6-Hydroxy-2′,3,5,5′,7-pentamethoxyflavone.

3,6,7-Tri-Me ether: 3′,4′,5-Trihydroxy-3,6,7-trimethoxyflavone
$C_{18}H_{16}O_6$ M 328.321
Constit. of *Blumea malconnii*. Cryst. (Me₂CO). Mp 252°. Originally described as 2′,5′,6-Trihydroxy-3,5,7-trimethoxyflavone.

Baker, W. et al, J. Chem. Soc., 1929, 74 (isol)
Farkas, L. et al, Chem. Ber., 1964, 97, 610 (isol)
Bate-Smith, E.C. et al, Phytochemistry, 1969, 8, 1035 (isol)
González, A.G. et al, An. Quim., 1971, 67, 795 (isol)
Tarpo, E., CA, 1971, 19, 25 (isol)
Hensch, M. et al, Helv. Chim. Acta, 1972, 55, 1610 (deriv)
Rodriguez, E. et al, Phytochemistry, 1972, 11, 1507 (isol)
Ickes, G.R. et al, J. Pharm. Sci., 1973, 62, 1009 (isol)
Nair, A.G.R. et al, Phytochemistry, 1975, 14, 1135 (isol)
Herz, W. et al, Tetrahedron, 1975, 31, 1577 (deriv)
Oganesyan, E.S. et al, Khim. Prir. Soedin., 1976, 599 (isol)
Wagner, H. et al, Tetrahedron Lett., 1976, 67 (synth)
Bacon, J.D. et al, Phytochemistry, 1978, 17, 1939 (isol)
Kiso, Y. et al, Heterocycles, 1982, 19, 1615 (isol)
Kulkarni, M.M. et al, Phytochemistry, 1987, 26, 2079 (isol)
Shilin, Y. et al, Phytochemistry, 1989, 28, 1509 (isol)
Markham, K.R., Phytochemistry, 1989, 28, 243 (struct)
Ahmed, A.A. et al, Phytochemistry, 1989, 28, 665 (isol)

2′,3′,4′,6,7,8-Hexahydroxyisoflavan   H-80068

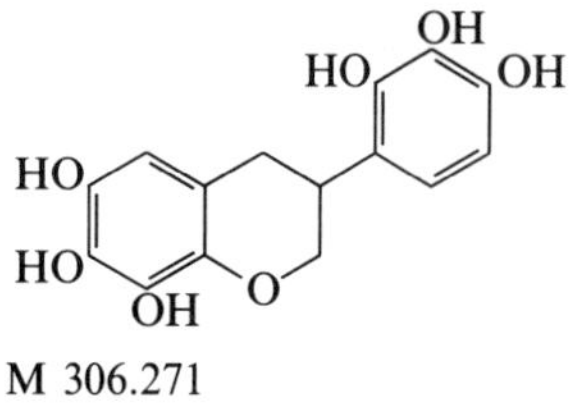

$C_{15}H_{14}O_7$ M 306.271

3′,4′,7-Tri-Me ether: [51798-41-5]. *2′,6,8-Trihydroxy-3′,4′,7-trimethoxyisoflavan. Machaerol C*
$C_{18}H_{20}O_7$ M 348.352
Isol. from heartwood of *Machaerium pedicellatum*. Tentative struct.

3′,4′,7,8-Tetra-Me ether: [51798-42-6]. *2′,6-Dihydroxy-3′,4′,7,8-tetramethoxyisoflavan. Machaerol B*
$C_{19}H_{22}O_7$ M 362.379
Isol. from *M. pedicellatum* heartwood. Tentative struct.

Hexa-Me ether: 2′,3′,4′,6,7,8-Hexamethoxyisoflavan

Mp 77-79°.

Ogiyama, K. et al, Phytochemistry, 1973, 12, 2544 (isol, pmr)
Ingham, J.L., Fortschr. Chem. Org. Naturst., 1983, 43, 1 (struct)

2′,4′,5,5′,6,7-Hexahydroxyisoflavone   H-80069

Updated Entry replacing D-20312
$C_{15}H_{10}O_8$ M 318.239

2′,4′,5′,6-Tetra-Me ether: [4935-92-6]. *5,7-Dihydroxy-2′,4′,5′,6-tetramethoxyisoflavone. 5,7-Dihydroxy-6-methoxy-3-(2,4,5-trimethoxyphenyl)-4H-1-benzopyran-4-one, 9CI. Caviunin*
$C_{19}H_{18}O_8$ M 374.346
Isol. from *Dalbergia nigra* and several other *D.* spp. Needles (EtOH). Mp 191-193°.

2′,4′,5′,6-Tetra-Me ether, 7-O-glucoside: [50299-68-8].
$C_{25}H_{28}O_{13}$ M 536.488
Isol. from *D. paniculata* roots and *D. sissoo* pods. Fibrous cryst. (EtOH). Mp 236-237°.

2′,4′,5′,6-Tetra-Me ether, 7-O-rhamnosylglucoside: [75883-12-4].
$C_{31}H_{38}O_{17}$ M 682.631
Isol. from roots of *D. paniculata*. Dihydrate. Mp 138-139°.

2′,4′,5′,6-Tetra-Me ether, 7-O-gentiobioside: [72578-98-4].
$C_{31}H_{38}O_{18}$ M 698.630
Isol. from pods of *D. sissoo*. Mp 210-212°.

Farkas, L. et al, Chem. Ber., 1961, 94, 2501 (synth)
Gottlieb, O.R. et al, J. Org. Chem., 1961, 26, 2449 (isol)
Dyke, S.F. et al, J. Org. Chem., 1961, 26, 2453 (synth)
Várady, J., Tetrahedron Lett., 1965, 4273 (synth)
Gottlieb, H.E., Phytochemistry, 1977, 16, 1811 (cmr)
Radhakrishniah, M., J. Indian Chem. Soc., 1979, 46, 81 (glucoside)
Sharma, A. et al, Phytochemistry, 1979, 18, 1253 (gentiobioside)
Rajulu, K.G. et al, Phytochemistry, 1980, 19, 1563 (rhamnosylglucoside)

2′,4′,5,5′,7,8-Hexahydroxyisoflavone   H-80070

5,7,8-Trihydroxy-3-(2,4,5-trihydroxyphenyl)-4H-1-benzopyran-4-one
$C_{15}H_{10}O_8$ M 318.239

2′,4′,5′,8-Tetra-Me ether: [4968-78-9]. *5,7-Dihydroxy-2′,4′,5′,8-tetramethoxyisoflavone. Isocaviunin*
$C_{19}H_{18}O_8$ M 374.346
Isol. from pods of *Dalbergia sissoo*. Mp 193-194°.

2′,4′,5′,8-Tetra-Me ether, 7-O-glucoside: [74148-49-5]. *Isocaviudin*
$C_{25}H_{28}O_{13}$ M 536.488
Isol from *D. sissoo* and *D. paniculata*. Cryst. (EtOH). Mp 220-222° (189-190°). [α]$_D^{25}$ −30.7° (DMSO).

2′,4′,5′,8-Tetra-Me ether, 7-O-gentiobioside: [74517-73-0].
$C_{31}H_{38}O_{18}$ M 698.630
Isol. from pods of *D. sissoo*. Yellow. Mp 170-172°. [α]$_D^{25}$ −65.6° (DMSO).

Sharma, A. et al, Indian J. Chem., Sect. B, 1979, 18, 472; 1980, 19, 237 (isol, struct, uv, pmr, ms)
Sharma, A. et al, Phytochemistry, 1980, 19, 715 (gentiobioside)

3′,4′,5,5′,6,7-Hexahydroxyisoflavone   H-80071

Updated Entry replacing T-03532
5,6,7-Trihydroxy-3-(3,4,5-trihydroxyphenyl)-4H-1-benzopyran-4-one. Irigenol
$C_{15}H_{10}O_8$ M 318.239

3′,4′,6-Tri-Me ether: [548-76-5]. *3′,5,7-Trihydroxy-4′,5′,6-trimethoxyisoflavone. Irigenin*
$C_{18}H_{16}O_8$ M 360.320

Constit. of several *Iris* spp. and of *Juniperus macropoda*. Shows antifungal props. Pale-yellow needles or plates (EtOH aq.). Mp 185°.

3′,4′,6-Tri-Me ether, 7-O-β-D-Glucopyranoside: [491-74-7].
Iridin
$C_{24}H_{26}O_{13}$ M 522.462
From *I.* spp., *Belamcanda chinensis* and *J. macropoda*. Needles (MeOH aq.). Mp 208°.

3′,4′,5′,6-Tetra-Me ether: [78134-86-8]. *5,7-Dihydroxy-3′,4′,5′,6-tetramethoxyisoflavone.* **Junipegenin C**
$C_{19}H_{18}O_8$ M 374.346
Isol. from leaves of *J. macropoda*. Fine needles (EtOAc). Mp 233-235°.

3′,4′,5,5′-Tetra-Me, 6,7-methylenedioxy ether: [41743-73-1]. *3′,4′,5,5′-Tetramethoxy-6,7-methylenedioxyisoflavone.*
Irisflorentin
$C_{20}H_{18}O_8$ M 386.357
Isol. from *I. florentina, I. germanica* and *I. tingitana*. Needles (MeOH). Mp 175°.

Baker, W., *J. Chem. Soc.*, 1928, 1022 (*bibl*)
Baker, W. *et al, J. Chem. Soc. C,* 1970, 1219 (*synth*)
Dhar, K.L. *et al, Phytochemistry,* 1972, 3097 (*isol*)
Arisawa, M. *et al, Chem. Pharm. Bull.,* 1973, **21**, 2323 (*deriv*)
Morita, N. *et al, Chem. Pharm. Bull.,* 1973, **21**, 600 (*Irisflorentin*)
Johnson, G. *et al, Physiol. Plant Pathol.,* 1976, **8**, 225 (*pharmacol*)
Sethi, M.L. *et al, Phytochemistry,* 1981, **20**, 341 (*Junipegenin C*)
Ingham, J.L., *Fortschr. Chem. Org. Naturst.,* 1983, **43**, 1 (*rev, occur*)

2,3,4,8,9,10-Hexahydroxypterocarpan   H-80072

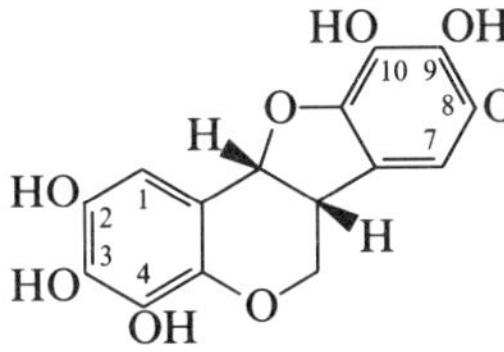

$C_{15}H_{12}O_8$ M 320.255
3,4,9,10-Tetra-Me ether: [89675-61-6]. *2,8-Dihydroxy-3,4,9,10-tetramethoxypterocarpan*
$C_{19}H_{20}O_8$ M 376.362
Isol. from *Swartzia laevicarpa* trunkwood. Cryst. (hexane). Mp 188-190°. Cas no. refers to racemate.

Braz Filho, R. *et al, Phytochemistry,* 1980, **19**, 2003 (*isol, uv, pmr, ord, ms, struct, abs config*)

Hexakis(bromomethylene)cyclohexane, 9CI   H-80073

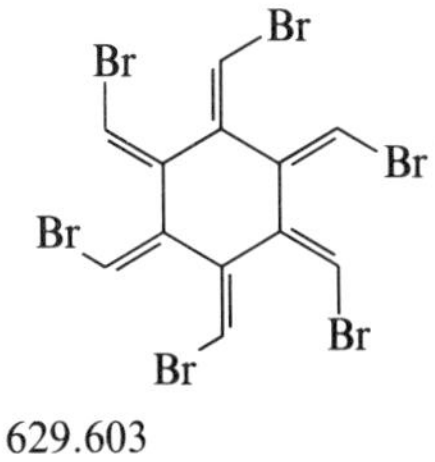

$C_{12}H_6Br_6$ M 629.603
(all-E)-form
Cryst. Mp >110° slow dec.
[117965-68-1, 118099-26-6]

Shepherd, M.K., *J. Chem. Soc., Perkin Trans.* 1, 1988, 961 (*synth, pmr, cmr, ms*)

3,3,4,4,5,5-Hexamethyl-1,2-cyclo-pentanedione   H-80074
[16980-19-1]

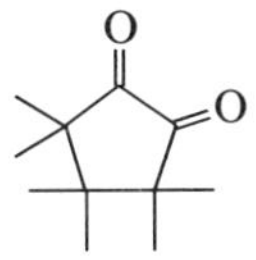

$C_{11}H_{18}O_2$ M 182.262
Orange cryst. solid. Mp 105-110°.

Crandall, J.K. *et al, J. Org. Chem.,* 1968, **33**, 3291 (*synth, ir, uv, pmr*)
Baran, J. *et al, J. Org. Chem.,* 1988, **53**, 4626 (*synth*)

1,1,2,2,3,3-Hexamethylindane   H-80075
2,3-Dihydro-1,1,2,2,3,3-hexamethyl-1H-indene
[91324-94-6]

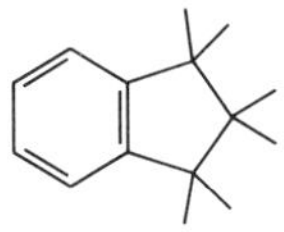

$C_{15}H_{22}$ M 202.339
Liq. $Bp_{0.38}$ 52-55°.

Baran, J. *et al, J. Org. Chem.,* 1988, **53**, 4626 (*synth, ir, pmr, cmr, ms*)

Hexamethylpentanedioic acid, 9CI   H-80076
Hexamethylglutaric acid
[115942-55-7]

$$HOOCC(CH_3)_2C(CH_3)_2C(CH_3)_2COOH$$

$C_{11}H_{20}O_4$ M 216.277
Cryst. (MeOH). Mp 188-189° dec.
Di-Me ester: [115942-56-8].
$C_{13}H_{24}O_4$ M 244.330
Viscous liq. $Bp_{0.42}$ 90° (bath) .
Anhydride: [115942-54-6]. *Dihydro-3,3,4,4,5,5-hexamethyl-2H-pyran-2,6(3H)dione, 9CI*
$C_{11}H_{18}O_3$ M 198.261
Cryst. by subl. Mp 174-175°.

Baran, J. *et al, J. Org. Chem.,* 1988, **53**, 4626 (*synth, ir, pmr, cmr*)

3,7,11,15,19,23-Hexamethyl-2,6,10,14,18, 22-tetracosahexaen-1-ol   H-80077

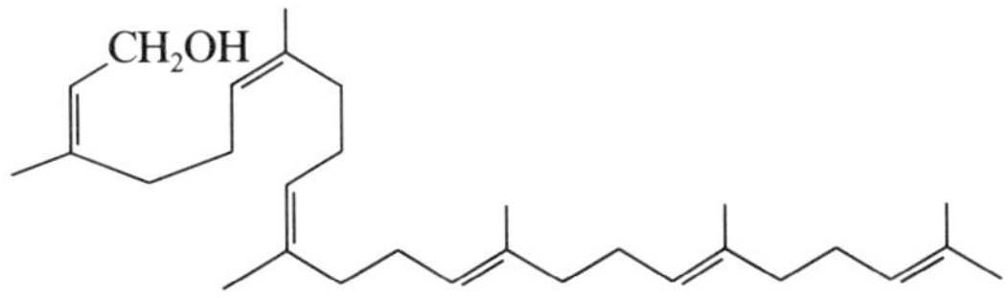

$C_{30}H_{50}O$ M 426.724
(2Z,6Z,10Z,14E,18E,22E)-form
Constit. of *Phyllanthus niruri*. Viscous mass.

Moiseenkov, A.M. *et al, Tetrahedron Lett.,* 1981, **22**, 3309 (*synth*)
Singh, B. *et al, Phytochemistry,* 1989, **28**, 1980 (*isol*)

5-Hexen-1-amine, 9CI   H-80078
6-Amino-1-hexene
[34825-70-2]

$$H_2C=CHCH_2CH_2CH_2CH_2NH_2$$

$C_6H_{13}N$ M 99.175
Liq. Bp 130-132°, Bp$_{55}$ 58°.
Phenylthiourea deriv.: Cryst. (EtOAc/pet. ether or EtOH).
Mp 57°.
N-*Me:*
$C_7H_{15}N$ M 113.202
Bp$_{750}$ 129-130°. n_D^{25} 1.4229.
Cogdell, T.J., *J. Org. Chem.*, 1972, **37**, 2541 (*synth, pmr*)
Abd El Samii, Z.K.M. *et al, J. Chem. Soc., Perkin Trans.* 1, 1988,
2517 (*synth*)

2-Hexenedial   H-80079
[4216-41-5]

$$OHCCH_2CH_2CH{=}CHCHO$$

$C_6H_8O_2$ M 112.128
(*Z*)-*form*
Bp$_{0.35}$ 60° (Kugelrohr) .
Hudlicky, T. *et al, J. Am. Chem. Soc.*, 1988, **110**, 4735 (*synth, pmr,
ir, ms*)

3-Hexenedial   H-80080
[6820-21-9]

$$OHCCH_2CH{=}CHCH_2CHO$$

$C_6H_8O_2$ M 112.128
[72530-32-6]
Uchida, M. *et al, Agric. Biol. Chem.*, 1979, **43**, 1919 (*synth, ir*)

2-Hexene-1,6-diol   H-80081

$$HOCH_2CH{=}CHCH_2CH_2CH_2OH$$

$C_6H_{12}O_2$ M 116.160
(*E*)-*form* [63974-05-0]
Bp$_{0.01}$ 100-105°.
(*Z*)-*form* [38796-35-9]
Bp$_{0.3}$ 96°.
Paquette, L.A. *et al, J. Am. Chem. Soc.*, 1972, **94**, 6751 (*synth,
pmr, ir*)
Egmond, G.J.N. *et al, Recl. Trav. Chim. Pays-Bas (J. R. Neth.
Chem. Soc.)*, 1977, **96**, 172 (*synth, pmr, ir*)

3-Hexene-1,6-diol   H-80082
[67077-43-4]

$$HOH_2CCH_2CH{=}CHCH_2CH_2OH$$

$C_6H_{12}O_2$ M 116.160
(*Z*)-*form* [72530-33-7]
Bp$_{0.6}$ 105-110°.
[71655-17-9]
Uchida, M. *et al, Agric. Biol. Chem.*, 1979, **43**, 1919 (*synth, ir,
pmr, ms*)

3-Hexene-2,5-diol, 9CI   H-80083
Updated Entry replacing H-00862
[7319-23-5]

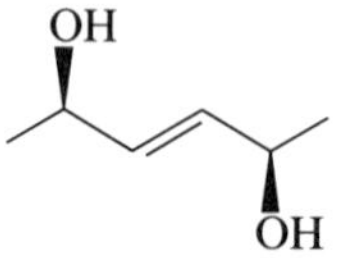

(*2RS,5RS*)-(*E*)-*form*

$C_6H_{12}O_2$ M 116.160

(*2RS,5RS*)-(*E*)-*form*
(±)-trans-*form*
Cryst. (Et$_2$O). Mp 97°.
Di-Me ether: [98721-08-5]. *2,5-Dimethoxy-3-hexene*
$C_8H_{16}O_2$ M 144.213
Slightly yellow oil. Bp$_{2-3}$ 110°.
Di-Et ether: [116910-25-9]. *2,5-Diethoxy-3-hexene*
$C_{10}H_{20}O_2$ M 172.267
Slightly yellow oil. Bp$_{2-3}$ 125°.
(*2RS,5SR*)-(*E*)-*form*
meso-trans-*form*
Di-Me ether: [98721-09-6].
Oil. Bp$_{2-3}$ 105° (kugelrohr) .
Di-Et ether: [116910-26-0].
Oil. Bp$_{2-3}$ 110° (kugelrohr) . Cont. 15% (±)-form.
(*Z*)-*form*
Di-Ac:
$C_{10}H_{16}O_4$ M 200.234
Bp$_{12}$ 102°. meso- and (±)- stereoisomers not
differentiated.
Alder, K. *et al, Justus Liebigs Ann. Chem.*, 1957, **608**, 195 (*synth*)
Bäckvall, J.-E. *et al, J. Org. Chem.*, 1988, **53**, 5695 (*ether, synth,
pmr, cmr, ir*)

3-Hexene-2,5-dione, 9CI   H-80084
Updated Entry replacing H-00863
1,2-Diacetylethylene
[4436-75-3]

$$H_3CCOCH{=}CHCOCH_3$$

$C_6H_8O_2$ M 112.128
(*E*)-*form*
Pale-yellow or colourless cryst. (pet. ether). Mp 77°.
Bp$_{15}$ 90°.
Bis-2,4-dinitrophenylhydrazone: Red needles (Py). Mp 291-
292°.
(*Z*)-*form* [17559-81-8]
Pale-yellow liq. Isom. to (*E*)-form after several weeks at
0°.
Goldberg, M.W. *et al, Helv. Chim. Acta*, 1938, **21**, 1699 (*synth*)
Bestman, H.-J. *et al, Chem. Ber.*, 1969, **102**, 2259 (*synth*)
Severin, T. *et al, Chem. Ber.*, 1975, **108** (*synth*)
Williams, P.D. *et al, J. Org. Chem.*, 1981, **46**, 4143 (*Z*-form)
Lepage, L. *et al, Synthesis*, 1983, 1018 (*synth*)

2-Hexyne-1,6-diol   H-80085
[32114-34-4]

$$HOCH_2CH_2CH_2C{\equiv}CCH_2OH$$

$C_6H_{10}O_2$ M 114.144
Bp$_4$ 145-147°, Bp$_{0.3}$ 103°.
Paquette, L.A. *et al, J. Am. Chem. Soc.*, 1972, **94**, 6751 (*synth,
pmr, ir*)

5-Hexynoic acid   H-80086
Updated Entry replacing H-40084
[53293-00-8]

$$HC{\equiv}C(CH_2)_3COOH$$

$C_6H_8O_2$ M 112.128
Liq. Bp$_{20}$ 120-125°, Bp$_6$ 95-96°. Claimed in one paper to be
cryst., Mp 41-43.5°.
Me ester: [77758-51-1].
$C_7H_{10}O_2$ M 126.155
Bp$_{10}$ 56°.

Chloride: [55183-45-4].
C_6H_7ClO M 130.573
Liq. $Bp_{0.5}$ 50-55°.
Nitrile: [14918-21-9]. *5-Cyano-1-pentyne*
C_6H_7N M 93.128
Bp_7 65.5°.
Anhydride:
$C_{12}H_{14}O_3$ M 206.241
Liq. $Bp_{0.6}$ 122-124°.

Eglington, G. *et al, J. Chem. Soc.,* 1953, 3052 (*synth*)
Just, G. *et al, Synth. Commun.,* 1979, **9**, 613 (*synth*)
Gerlach, H. *et al, Helv. Chim. Acta,* 1980, **63**, 2312 (*synth, pmr, ir, ms*)
Krafft, G.A. *et al, J. Am. Chem. Soc.,* 1981, **103**, 5459 (*synth, pmr*)
Millar, J.G. *et al, J. Org. Chem.,* 1984, **49**, 2332 (*synth, pmr, ms*)
Earl, R.A. *et al, J. Org. Chem.,* 1984, **49**, 4786 (*synth*)
Atkinson, R.S. *et al, J. Chem. Soc., Perkin Trans. 1,* 1986, 1215 (*synth*)
Moody, C.J. *et al, J. Chem. Soc., Perkin Trans. 1,* 1988, 3249 (*anhydride*)

Hieracin I   H-80087

[117259-18-4]

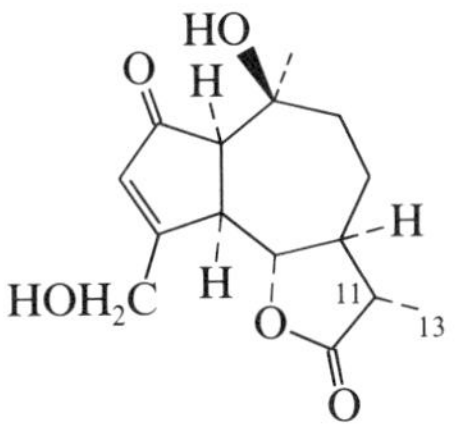

$C_{15}H_{20}O_5$ M 280.320
Constit. of *Picris hieracioides.*

11,13-Didehydro: [117259-19-5]. **Hieracin II**
$C_{15}H_{18}O_5$ M 278.304
Constit. of *P. hieracioides.*

Kanayama, T. *et al, Bull. Chem. Soc. Jpn.,* 1988, **61**, 2971.

Hinesene   H-80088

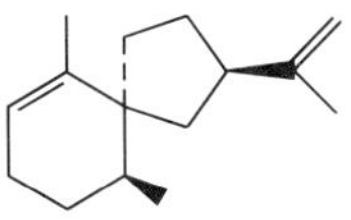

$C_{15}H_{24}$ M 204.355
Constit. of *Rolandra fruticosa.* Oil. $[\alpha]_D^{24}$ −44° (c, 0.1 in $CHCl_3$).

Jakupovic, J. *et al, Phytochemistry,* 1989, **28**, 1937.

Hirsutinolide   H-80089

Updated Entry replacing D-00512
[68024-72-6]

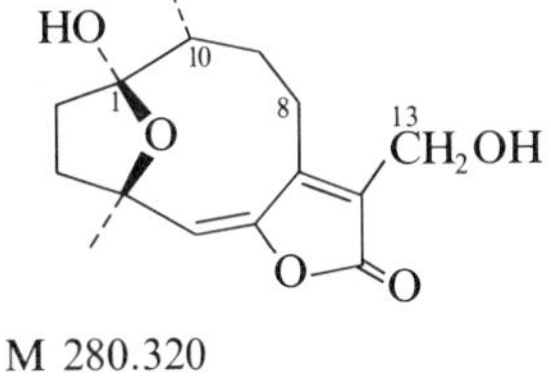

$C_{15}H_{20}O_5$ M 280.320
Parent compd. not known.

10β-Hydroxy-8β-acetoxy, 13-Ac: [72712-41-5]. *8β-Acetoxy-10β-hydroxyhirsutinolide 13-O-acetate. Hirsutolide*
$C_{19}H_{24}O_9$ M 396.393

Constit. of *Vernonia saltensis.* Oil. $[\alpha]_D^{24}$ +36.3° (c, 4.1 in $CHCl_3$). Identity of the two samples not certain.

10β-Hydroxy, 8β-acetoxy, 1,13-di-Ac: [72712-42-6]. *8β-Acetoxy-10β-hydroxyhirsutinolide 1,13-O-diacetate*
$C_{21}H_{26}O_{10}$ M 438.430
Constit. of *V. saltensis.* Oil. $[\alpha]_D$ −3° (c, 0.6 in $CHCl_3$).

10β-Hydroxy, 8β-propanoyloxy, 13-Ac: [71305-86-7]. *8β-Propionyloxy-10β-hydroxyhirsutinolide-13-O-acetate*
$C_{20}H_{26}O_9$ M 410.420
Constit. of *V. scorpioides.* Oil. $[\alpha]_D^{24}$ +63.5° (c, 0.5 in $CHCl_3$).

10β-Hydroxy, 8β-propanoyloxy, 1,13-di-Ac: [79081-71-3]. *8β-Propionyloxy-10β-hydroxyhirsutinolide 1,13-di-O-acetate*
$C_{22}H_{28}O_{10}$ M 452.457
Constit. of *V. scorpioides.* Oil.

10β-Hydroxy, 8-propanoyloxy, O^1-Me, 13-Ac: [71305-76-5]. *8-Propionoyloxy-10β-hydroxy-1-O-methylhirsutinolide-13-O-acetate*
$C_{21}H_{28}O_9$ M 424.447
Constit. of *V. scorpioides.* Oil. C8-config. uncertain.

8β,10β-Diacetoxy, 13-Ac: [72748-29-9]. *8β,10β-Diacetoxyhirsutinolide 13-O-acetate*
$C_{21}H_{26}O_{10}$ M 438.430
Constit. of *Stokesia laevis.* Oil. Occurs with the 1-epimer.

8β,10β-Diacetoxy, O^1-Me, 13-Ac: [72778-10-0]. *8β,10β-Diacetoxy-1-O-methylhirsutinolide 13-O-acetate*
$C_{22}H_{28}O_{10}$ M 452.457
Constit. of *S. laevis.* Oil. Occurs with the 1-epimer.

8β-(2-Methylpropenoyloxy): *8β-(2-Methylacryloxyloxy)hirsutinolide*
$C_{19}H_{24}O_7$ M 364.394
Isol. from *V.* spp. Cryst. (Et_2O/pet. ether). Mp 184°. $[\alpha]_D^{24}$ −6.9° (c, 1.1 in $CHCl_3$).

8β-(2,3-Epoxy-2-methylpropanoyloxy): *8β-(2-Methyl-2,3-epoxypropionyloxy)hirsutinolide*
$C_{19}H_{24}O_8$ M 380.394
Isol. from *V.* spp. Cryst. (Et_2O/pet. ether). Mp 170°. $[\alpha]_D^{24}$ +0.7° (c, 1.1 in $CHCl_3$).

8β-(2-Hydroxymethylpropenoyloxy), 13-Ac: *8β-(2-Hydroxymethylacryloyloxy)hisutinolide 13-O-acetate*
$C_{21}H_{26}O_9$ M 422.431
Constit. of *V.* spp. Cryst. (Et_2O/pet. ether). Mp 72°. $[\alpha]_D^{24}$ +111.2° (c, 1.3 in $CHCl_3$).

8β-(2-Hydroxymethylpropenoyloxy), 15-hydroxy, 13-Ac: *8β-(2-Methylacryloyloxy)-15-hydroxyhisutinolide 13-acetate*
$C_{21}H_{26}O_9$ M 422.431
Isol. from *V.* spp. Oil. Misnamed in the paper.

8β-(2-Methylpropenoyloxy), 13-Ac: *8β-(2-Methylacryloyloxy)hirsutinolide 13-O-acetate*
$C_{21}H_{26}O_8$ M 406.432
Isol. from *V.* spp. Oil. $[\alpha]_D^{24}$ +19.5° (c, 1.0 in $CHCl_3$).

8β-(2,3-Epoxy-2-methylpropanoyloxy), 13-Ac: *8β-(2-Methyl-2,3-epoxypropionyloxy)hirsutinolide 13-O-acetate*
$C_{21}H_{26}O_9$ M 422.431
Isol. from *V.* spp. Oil. $[\alpha]_D^{24}$ +18.3° (c, 2.05 in $CHCl_3$).

Bohlmann, F. *et al, Phytochemistry,* 1978, **17**, 475; 1979, **18**, 289, 987 (*isol, pmr*)
Rustaiyan, A.N. *et al, Fitoterapia,* 1979, **50**, 243 (*Hirsutolide*)

Hiyodorilactone D   H-80090

Updated Entry replacing H-00999
[72493-40-4]

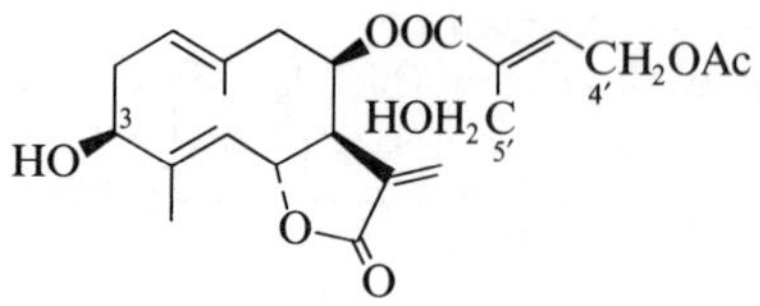

$C_{22}H_{28}O_8$ M 420.458

Constit. of *Eupatorium sachalinense*. Yellow oil. $[\alpha]_D^{28}$ +81°
(c, 0.36 in EtOH).

4'-Deacetyl, 5'-Ac: [72493-41-5]. **Hiyodorilactone E.**
Santhemoidin C
$C_{22}H_{28}O_8$ M 420.458

Constit. of *E. sachalinense* and of *Schkuhria
anthemoidea*. Cryst. Mp 149° dec., Mp 169-172°. $[\alpha]_D^{18}$
+119° (c, 0.22 in EtOH)(+75°). Hiyodorilactone E and
Santhemoidin C have not been compared and their
recorded props. are different.

3,5'-Di-Ac: [72493-42-6]. **Hiyodorilactone F**
$C_{26}H_{32}O_{10}$ M 504.533

Constit. of *E. sachalinense*. Yellow oil. $[\alpha]_D^{30}$ −141° (c,
0.21 in EtOH).

Takahashi, T. *et al, Chem. Pharm. Bull.*, 1979, **27**, 2539 (*isol*)
Pérez, A.L. *et al, Phytochemistry*, 1984, **23**, 2399 (*isol*)

HM-1   H-80091

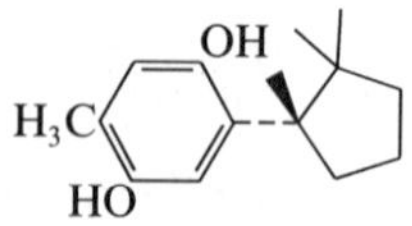

$C_{15}H_{22}O_2$ M 234.338

Metab. of fungus *Helicobasidium mompa*. Needles. Mp
139.5-140.0°. $[\alpha]_D$ −55.0° (CHCl₃).

Kajimoto, T. *et al, Chem. Lett.*, 1989, 527.

HM-2   H-80092

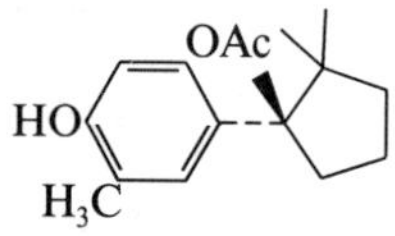

$C_{17}H_{24}O_3$ M 276.375

Metab. of *Helicobasidium mompa*. Oil. $[\alpha]_D$ −37.0°
(CHCl₃).

Kajimoto, T. *et al, Chem. Lett.*, 1989, 527.

HM-3   H-80093

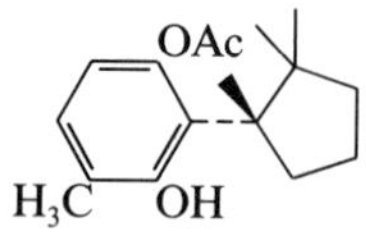

$C_{17}H_{24}O_3$ M 276.375

Metab. of *Helicobasidium mompa*. Needles. Mp 158.5-
160.0°. $[\alpha]_D$ −8.1° (CHCl₃).

Kajimoto, T. *et al, Chem. Lett.*, 1989, 527.

HM-4   H-80094

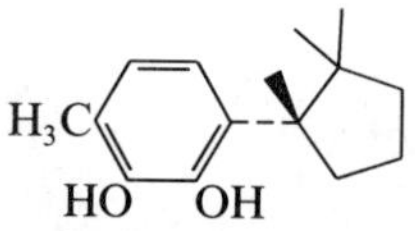

$C_{15}H_{22}O_2$ M 234.338

Metab. of *Helicobasidium mompa*. Needles. Mp 72.0-73.0°.
$[\alpha]_D$ −57.4° (CHCl₃).

Kajimoto, T. *et al, Chem. Lett.*, 1989, 527.

5-Homoverrucosanol   H-80095

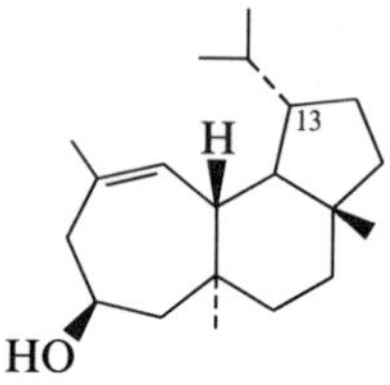

$C_{20}H_{34}O$ M 290.488

5β-form

Constit. of *Schistochila rigidula*.

13-Epimer: [119242-42-1]. **13-Epihomoverrucosan-5β-ol**
Constit. of *S. nobilis*. Cryst. Mp 123-124°. $[\alpha]_D$ +47.1°
(c, 0.55 in CHCl₃).

Asakawa, Y. *et al, Phytochemistry*, 1988, **27**, 3509.

16,22-Hopanediol   H-80096

$C_{30}H_{52}O_2$ M 444.740

16β-form

6-Desoxyleucotylin

Constit. of *Parmelia entotheiochroa*. Cryst. Mp 268°.
$[\alpha]_D$ +68° (CHCl₃).

16-Ac:
$C_{32}H_{54}O_3$ M 486.777

Constit. of *P. entotheiochroa*. Cryst. Mp 228°. $[\alpha]_D$ +52°
(CHCl₃).

Yosioka, I. *et al, Chem. Pharm. Bull.*, 1966, **14**, 804.

21-Hopene-3,20-dione   H-80097

Lageflorin

[119403-28-0]

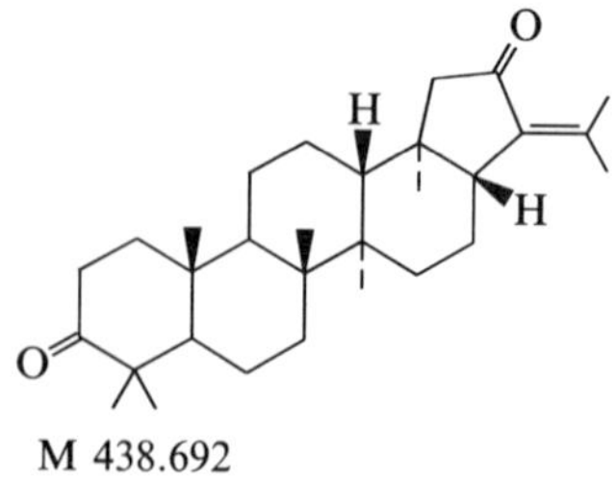

$C_{30}H_{46}O_2$ M 438.692

Constit. of *Lagerstroemia parviflora*. Needles (C_6H_6/pet.
ether). Mp 346-348°.

Barik, B.R. *et al, Phytochemistry*, 1988, **27**, 3679.

Hormothamnione   H-80098

Updated Entry replacing H-50127

[103654-49-5]

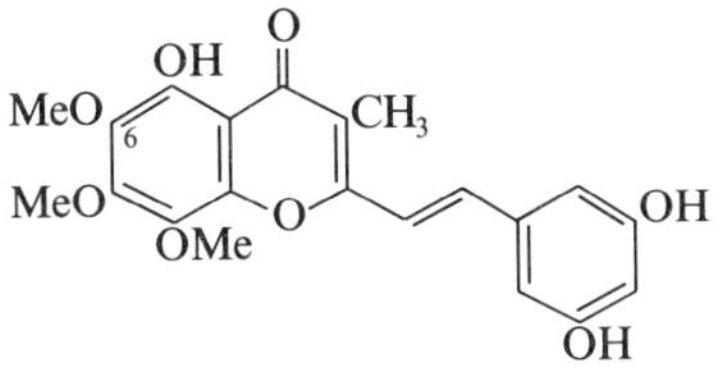

$C_{21}H_{20}O_8$ M 400.384

Constit. of the alga *Chrysophaeum taylori*. Potent cytotoxin. Yellow solid. Mp 270° dec.

6-Demethoxy: **6-Desmethoxyhormothamnione**

$C_{20}H_{18}O_7$ M 370.358

Constit. of *C. taylori*. Cytotoxic agent.

6-Demethoxy, tri-Ac: Cryst. Mp 184-186°.

Gerwick, W.H. *et al*, *Tetrahedron Lett.*, 1986, **27**, 1979 (*cryst struct*)

Gerwick, W.H., *J. Nat. Prod. (Lloydia)*, 1989, **52**, 252 (*isol, deriv*)

Hydrallmanol A   H-80099

[121079-02-5]

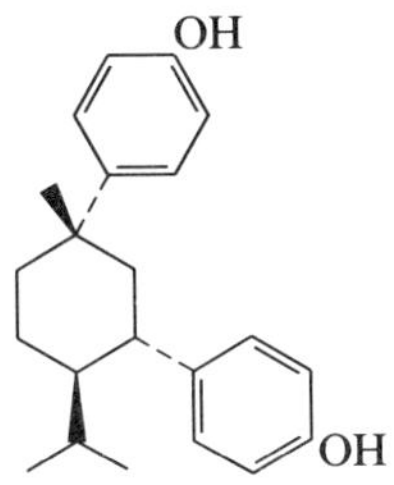

$C_{22}H_{28}O_2$ M 324.462

Metab. of *Hydrallmania falcata*. Cytotoxic agent. Pale yellow oil.

Pathirana, C. *et al*, *Tetrahedron Lett.*, 1989, **30**, 1487.

Hydrobenzamide   H-80100

1-Phenyl-N,N'-bis(phenylmethylene)methanediamine, 9CI.
N,N'-Dibenzylidenetoluene-α,α-diamine, 8CI

[92-29-5]

$$(PhCH{=}N)_2CHPh$$

$C_{21}H_{18}N_2$ M 298.387

Prisms (EtOH). Mp 107-108° (101-102°).

Sadtler Standard Infrared Spectra, No. 20378 (*ir*)
Pirrone, F., *Gazz. Chim. Ital.*, 1937, **67**, 529 (*synth*)
Ogata, Y. *et al*, *J. Org. Chem.*, 1964, **29**, 1985 (*synth*)
Nicholson, I. *et al*, *J. Chem. Soc.*, 1965, 3067 (*uv*)
Hunter, D.H. *et al*, *Can. J. Chem.*, 1972, **50**, 669 (*tautom*)
Nishiyama, K. *et al*, *Bull. Chem. Soc. Jpn.*, 1988, **61**, 609 (*synth, pmr*)

3-Hydroperoxy-4-eudesmene-1,6-diol   H-80101

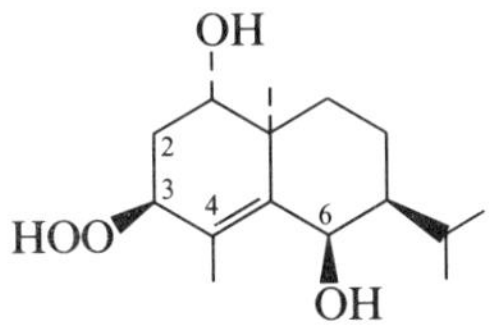

$C_{15}H_{26}O_4$ M 270.368

(1α,3β,5β,6β,10α)-form

6-Cinnamoyl: [119364-25-9]. *6β-Cinnamoyloxy-3β-hydroperoxy-5,10-bisepi-4-eudesmen-1α-ol*

$C_{24}H_{32}O_5$ M 400.514

Constit. of *Ambrosia artemisioides*. Oil.

3-Ketone, 6-cinnamoyl: [119285-46-0]. *6β-Cinnamoyloxy-1α-hydroxy-5,10-bisepi-4-eudesmen-3-one*

$C_{24}H_{30}O_4$ M 382.499

Constit. of *A. artemisioides*. Oil.

Jakupovic, J. *et al*, *Phytochemistry*, 1988, **27**, 3551.

4-Hydroperoxy-2-eudesmene-1,6-diol   H-80102

$C_{15}H_{26}O_4$ M 270.368

(1α,4β,5β,6β,10α)-form

6-Cinnamoyl: [119285-45-9]. *6β-Cinnamoyloxy-4β-hydroperoxy-5,10-bisepi-2-eudesmen-1α-ol*

$C_{24}H_{32}O_5$ M 400.514

Constit. of *Ambrosia artemisioides*. Oil.

Jakupovic, J. *et al*, *Phytochemistry*, 1988, **27**, 3551.

12-Hydroxy-8,11,13-abietatriene-3,7-dione   H-80103
Nimosone

[61494-73-3]

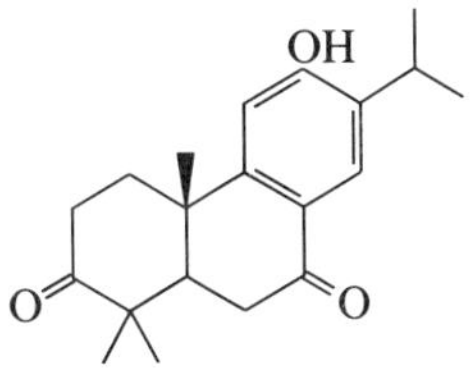

$C_{20}H_{26}O_3$ M 314.424

Constit. of *Azadirachta indica*. Needles ($CHCl_3$). Mp 72-73°.

Ara, I. *et al*, *J. Nat. Prod. (Lloydia)*, 1988, **51**, 1054 (*isol, pmr, cmr*)

12-Hydroxy-8,11,13-abietatriene-6,7-dione   H-80104
Hypargenin C

[119817-27-5]

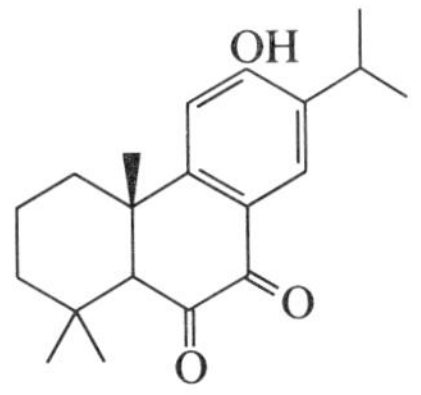

$C_{20}H_{26}O_3$ M 314.424

Constit. of *Salvia hypargeia*. Antibacterial agent. Orange amorph. solid.

Ulubelen, A. *et al*, *J. Nat. Prod. (Lloydia)*, 1988, **51**, 1178 (*isol, pmr, cmr*)

12-Hydroxy-6,8,11,13-abietatrien-3-one   H-80105
Hypargenin D

[119767-14-5]

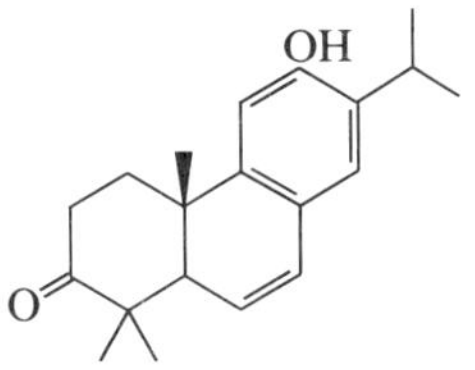

$C_{20}H_{26}O_2$ M 298.424

Constit. of *Salvia hypargeia*. Antibacterial agent. Orange amorph. solid.

Ulubelen, A. *et al, J. Nat. Prod. (Lloydia)*, 1988, **51**, 1178 (*isol, pmr, cmr*)

1-Hydroxyacenaphthene   H-80106

Updated Entry replacing H-01162

1,2-Dihydro-1-acenaphthylenol, 9CI. 1-Acenaphthenol, 8CI

[6306-07-6]

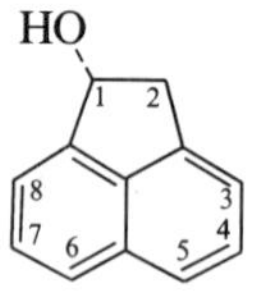

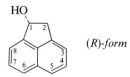

(*R*)-*form*

$C_{12}H_{10}O$ M 170.210

(*R*)-*form*

Mp 127-127.5°. $[\alpha]_D$ – 1.46° (c, 1.23 in $CHCl_3$).

(±)-*form*

Needles (EtOH or C_6H_6). Mp 148° (144.5-145.5°).

Cason, J. *et al, J. Am. Chem. Soc.*, 1940, **62**, 432 (*synth*)
Org. Synth., 1941, **21**, 1 (*synth*)
Pratt, E.F. *et al, J. Org. Chem.*, 1961, **26**, 2973 (*synth*)
Gupta, M.P. *et al, Acta Crystallogr., Sect. B*, 1975, **31**, 7 (*cryst struct*)
Hu, Y. *et al, Can. J. Chem.*, 1989, **67**, 60 (*resoln, cryst struct, abs config*)

2-Hydroxy-3(2*H*)-benzofuranone, 9CI, 8CI   H-80107

2-Hydroxy-3-coumaranone. 2,3-Dihydro-2-hydroxybenzo[b]furan-3-one

[17392-15-3]

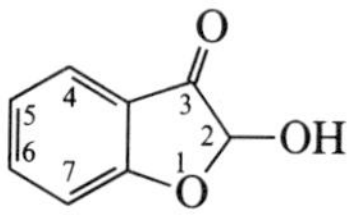

$C_8H_6O_3$ M 150.134

Prisms (Et_2O). Mp 108°.

Me ether: [75335-07-8]. *2-Methoxy-3(2H)-benzofuranone, 9CI*

$C_9H_8O_3$ M 164.160

Oil.

Howe, R. *et al, J. Chem. Soc. C*, 1967, 2510 (*synth, pmr*)
Donnelly, J.A. *et al, Tetrahedron*, 1973, **29**, 3979 (*uv*)
Malaitong, M. *et al, Chem. Lett.*, 1980, 305 (*deriv*)
Antus, S. *et al, Justus Liebigs Ann. Chem.*, 1980, 1271 (*synth, pmr, deriv*)

4-Hydroxy-3(2*H*)-benzofuranone, 9CI, 8CI   H-80108

4-Hydroxy-3-coumaranone

[19278-81-0]

$C_8H_6O_3$ M 150.134

Cryst. (C_6H_6). Mp 120° (subl. at 85°).

Me ether: *4-Methoxy-3(2H)-benzofuranone*

$C_9H_8O_3$ M 164.160

Cryst. (C_6H_6). Mp 151°.

Et ether: *4-Ethoxy-3(2H)-benzofuranone*

$C_{10}H_{10}O_3$ M 178.187

Cryst. (EtOH). Mp 96°.

Shriner, R.L. *et al, J. Am. Chem. Soc.*, 1939, **61**, 2328 (*synth*)
Farmer, V.C. *et al, J. Chem. Soc.*, 1956, 3600 (*deriv*)
Dallacker, F. *et al, Justus Liebigs Ann. Chem.*, 1966, **694**, 98 (*derivs*)

Schenck, G. *et al, Tetrahedron Lett.*, 1968, 2375 (*uv*)
Huke, M. *et al, Arch. Pharm. (Weinheim, Ger.)*, 1969, **302**, 401; **302**, 423 (*uv, pmr*)

5-Hydroxy-3(2*H*)-benzofuranone, 9CI, 8CI   H-80109

5-Hydroxy-3-coumaranone

[19278-82-1]

$C_8H_6O_3$ M 150.134

Yellow prisms (diisopropyl ether). Mp 152-153°.

Ac:

$C_{10}H_8O_4$ M 192.171

Plates (Et_2O/pet. ether). Mp 95-96°.

Me ether: [39581-55-0]. *5-Methoxy-3(2H)-benzofuranone*

$C_9H_8O_3$ M 164.160

Cryst. (MeOH). Mp 88.5-90°.

Kloetzel, M.K. *et al, J. Org. Chem.*, 1955, **20**, 38 (*synth, derivs*)
Schenck, G. *et al, Tetrahedron Lett.*, 1968, 2375 (*uv*)
Huke, M. *et al, Arch. Pharm. (Weinheim, Ger.)*, 1969, **302**, 423 (*pmr, uv*)

6-Hydroxy-3(2*H*)-benzofuranone, 9CI, 8CI   H-80110

6-Hydroxy-3-coumaranone

[6272-26-0]

$C_8H_6O_3$ M 150.134

Yellow plates (EtOH). Mp 245° dec.

Ac:

$C_{10}H_8O_4$ M 192.171

Pale yellow cryst. (EtOH). Mp 125-126°. $Bp_{0.5}$ 110° subl.

Me ether: [15832-09-4]. *6-Methoxy-3(2H)-benzofuranone*

$C_9H_8O_3$ M 164.160

Yellow needles (EtOH) or rose red cryst. (MeOH). Mp 124-125° (yellow form), Mp 170-171° (red form). The 2 variants are reported in different refs. and may be polymorphs, or one may be spurious.

Barltrop, J.A., *J. Chem. Soc.*, 1946, 958 (*deriv*)
Horning, E.C. *et al, J. Am. Chem. Soc.*, 1948, **70**, 3619 (*synth*)
Davies, J.S.H. *et al, J. Chem. Soc.*, 1950, 3206 (*synth*)
Dawkins, A.W. *et al, J. Chem. Soc.*, 1959, 2203 (*deriv*)
Logemann, W., *Chem. Ber.*, 1963, **96**, 1680 (*synth, ir*)
Doifode, K.B. *et al, J. Org. Chem.*, 1964, **29**, 2025 (*deriv*)
Schenck, G. *et al, Tetrahedron Lett.*, 1967, 2059 (*uv*)
Huke, M. *et al, Arch. Pharm. (Weinheim, Ger.)*, 1969, **302**, 423 (*pmr, uv*)

7-Hydroxy-3(2*H*)-benzofuranone, 9CI, 8CI   H-80111

7-Hydroxy-3-coumaranone

[19397-70-7]

$C_8H_6O_3$ M 150.134

Needles (subl. at 125°). Mp 186-188° dec.

Me ether: [7169-37-1]. *7-Methoxy-3(2H)-benzofuranone*

$C_9H_8O_3$ M 164.160

Cryst. (Et_2O). Mp 83-84°.

Mosimann, W. *et al, Chem. Ber.*, 1916, **49**, 1261 (*synth*)
Richtzenhain, H. *et al, Ber.*, 1956, **89**, 378 (*synth, deriv*)
Schenck, G. *et al, Tetrahedron Lett.*, 1968, 2375 (*synth, uv*)
Jung, M.E. *et al, J. Org. Chem.*, 1988, **53**, 423 (*synth, ir, ms, pmr*)

1-(4-Hydroxy-5-benzofuranyl)-3-phenyl-2-propen-1-one   H-80112

Updated Entry replacing H-70107

5-Cinnamoyl-4-hydroxybenzofuran

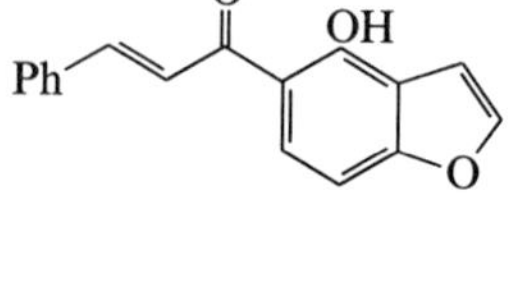

$C_{17}H_{12}O_3$ M 264.280
Constit. of *Milletia ovalifolia*. Yellow needles (EtOH). Mp 124°.

Saxena, D.B. *et al, Indian J. Chem., Sect. B*, 1987, **26**, 704.

3-(6-Hydroxy-5-benzofuranyl)-2-propenoic acid   H-80113

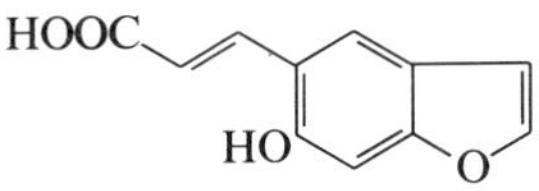

HOOC

HO

$C_{11}H_8O_4$ M 204.182
(E)-form
    Furocoumaric acid. Psoralic acid
    Mp 221°.
    O-β-D-Glucopyranoside: Obt. by isom. of the natural (*Z*)-isomer in hot H_2O. Mp 244°. $[\alpha]_D^{20}$ −80.0° (H_2O).
    Me ether: Mp 222°.
(Z)-form
    Furocoumarinic acid
    Free acid cyclises to psoralen.
    O-β-D-Glucopyranoside:
    $C_{17}H_{18}O_9$ M 366.324
    Isol. from seeds of *Coronilla* spp. Biogenetic precursor to psoralen. Mp 125°. $[\alpha]_D^{20}$ −61.3° (H_2O).
    Me ether: Mp 166°.

Stoll, A. *et al, Helv. Chim. Acta*, 1950, **33**, 1637.

4-Hydroxybenzyl alcohol   H-80114

Updated Entry replacing H-01352
4-Hydroxybenzenemethanol, 9CI. α,4-Dihydroxytoluene. p-Hydroxymethylphenol
[623-05-2]
$C_7H_8O_2$ M 124.139
Constit. of muskmelon (*Cucurbita moschata*) seedlings and from infected bulbs of *Orchis militaris* and *Loroglossum hircinum*. Cofactor for indoleacetic acid oxidase. Prisms or needles (H_2O). Mp 124.5-125.5°.
α-Ac:
    $C_9H_{10}O_3$ M 166.176
    Needles (H_2O). Mp 84°.
Di-Ac: [2937-64-6].
    $C_{11}H_{12}O_4$ M 208.213
    Needles. Mp 75°. Bp$_{11}$ 155-157°.
4-Me ether: [105-13-5]. *4-Methoxybenzyl alcohol. Anisyl alcohol*
    $C_8H_{10}O_2$ M 138.166
    Simple esters are used in flavouring. d_4^{20} 1.115. Mp 24-25°, Mp 45°. Bp$_{12}$ 135-136°.
    ▷ Mod. toxic. DO8925000.
4-Me ether, Ac: [104-21-2]. *Anisyl acetate*
    $C_{10}H_{12}O_3$ M 180.203
    Used in flavour industry. Mp 84°. Bp$_{12}$ 135-136°.
1′-Me ether: 4-*(Methoxymethyl)phenol. p-Hydroxybenzyl methyl ether*
    $C_8H_{10}O_2$ M 138.166
    Isol. from unripe fruits of *Citrullus colocynthis*. Plates. Mp 85-86°.
Di-Me ether: [1515-81-7].
    $C_9H_{12}O_2$ M 152.193
    Bp 225-226°.

Späth, E., *Monatsh. Chem.*, 1914, **35**, 319 (*synth*)
Carothers, W.H. *et al, J. Am. Chem. Soc.*, 1924, **46**, 1675 (*synth*)
Ofner, A., *Helv. Chim. Acta*, 1935, **18**, 951 (*synth*)

Watanabe, H. *et al, CA*, 1961, **55**, 16697 (*isol, deriv*)
Shannon, J.S., *Aust. J. Chem.*, 1962, **15**, 265 (*ms*)
Hardegger, E. *et al, Helv. Chim. Acta*, 1963, **46**, 1171 (*isol*)
Mumford, F.E. *et al, Phytochemistry*, 1963, **2**, 215 (*isol*)
Torii, S. *et al, J. Org. Chem.*, 1979, **44**, 3305 (*synth*)
Sax, N.I., *Dangerous Properties of Industrial Materials*, 5th Ed., Van Nostrand-Reinhold, 1979, 381.

ent-2-Hydroxy-1,15-beyeradien-3-one   H-80115

2-Hydroxy-1,15-stachadien-3-one

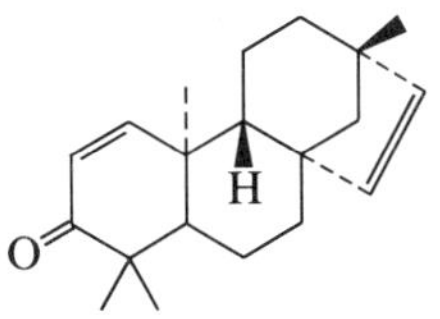

O

H

$C_{20}H_{28}O_2$ M 300.440
Isol. from wood of *Spirostachys africana*. Prisms (hexane). Mp 132°. $[\alpha]_D$ +49° (c, 3.2 in $CHCl_3$).

Baarschers, W.H. *et al, J. Chem. Soc.*, 1962, 4046 (*isol, uv, ir, pmr*)

ent-3-Hydroxy-15-beyeren-2-one   H-80116

3-Hydroxy-15-stachen-2-one
$C_{20}H_{30}O_2$ M 302.456
Isol. from wood of *Spirostachys africana*. Needles (hexane). Mp 129°. $[\alpha]_D$ +30° (c, 3.3 in $CHCl_3$).

Baarschers, W.H. *et al, J. Chem. Soc.*, 1962, 4046 (*isol, uv, ir, pmr, ord*)

2′-Hydroxy-2,3′-bipyridine   H-80117

[2,3′-Bipyridine]-2′(1′H)-one, 9CI. 3-(2-Pyridyl)-2-pyridone
[101002-01-1]
$C_{10}H_8N_2O$ M 172.186
Yellow cryst. (CH_2Cl_2/hexane). Mp 149-150°.

Moran, D.B. *et al, J. Heterocycl. Chem.*, 1986, **23**, 1071 (*synth*)

2′-Hydroxy-2,4′-bipyridine   H-80118

[2,4′-Bipyridin]-2′(1′H)-one, 9CI. 4-(2-Pyridyl)-2-pyridone
$C_{10}H_8N_2O$ M 172.186
[101001-99-4]

MacBride, J.A.H. *et al, Tetrahedron Lett.*, 1982, **23**, 1109.
U.S. Pat., 4 450 166, (1985); *CA*, **104**, 129911t.

2-Hydroxy-3,3′-bipyridine   H-80119

[3,3′-Bipyridine]-2(1H)-one, 9CI. 3-(3-Pyridyl)-2-pyridone
[82407-65-6]

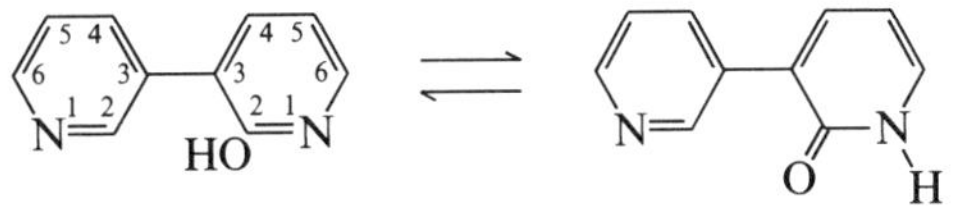

$C_{10}H_8N_2O$ M 172.186
Pyridone tautomers prob. predominate in this series where appropriate. Off-white cryst. (2-propanol). Mp 189-192°.

MacBride, J.A.H. *et al, J. Chem. Res. (S)*, 1984, **10**, 328 (*synth*)
Moran, D.B. *et al, J. Heterocycl. Chem.*, 1986, **23**, 1071 (*synth*)

2-Hydroxy-4,4′-bipyridine   H-80120

[4,4′-Bipyridin]-2(1H)-one, 9CI. 4-(4-Pyridyl)-2-pyridone
$C_{10}H_8N_2O$ M 172.186
Hydrazone: [101002-00-0].
    $C_{10}H_{10}N_4$ M 186.216

Yellow cryst. (CH$_2$Cl$_2$). Mp 109-110°.

U.S. Pat., 4 450 166, (1985); CA, **104**, 129911t (synth)
Moran, D.B. et al, J. Heterocycl. Chem., 1986, **23**, 1071 (synth)

4-Hydroxy-2,2'-bipyridine   H-80121

[2,2'-Bipyridin]-4-ol, 8CI. 2-(2-Pyridyl)-4-pyridone
[14712-32-4]
C$_{10}$H$_8$N$_2$O   M 172.186
Prisms (Me$_2$CO). Mp 145°.
Me ether: [14162-97-1]. 4-Methoxy-2,2'-bipyridine
   C$_{11}$H$_{10}$N$_2$O   M 186.213
   Prisms (CHCl$_3$/hexane). Mp 67.5-68.5°.
Me ether, N-oxide: [14163-05-4].
   C$_{11}$H$_{10}$N$_2$O$_2$ M 202.212
   Cryst. (CH$_2$Cl$_2$/hexane). Mp 117.5-118°.
Me ether, N,N'-dioxide: [84175-06-4].
   C$_{11}$H$_{10}$N$_2$O$_3$ M 218.212
   Mp 242-246° dec. Bp$_{0.03}$ 180° subl.

Divekar, P.V. et al, Can. J. Chem., 1967, **45**, 1215 (synth, ir, uv)
Wenkert, D. et al, J. Org. Chem., 1983, **48**, 283 (synth, ir, pmr, uv)

4-Hydroxy-3,3'-bipyridine   H-80122

[3,3'-Bipyridin]-4(1H)-one, 9CI. 3-(3-Pyridyl)-4-pyridone
[82407-66-7]
C$_{10}$H$_8$N$_2$O   M 172.186
Brownish plates (MeCN). Mp 222-226°. Bp$_{0.0001}$ 200° subl.

MacBride, J.A.H. et al, Tetrahedron Lett., 1982, **23**, 1109 (synth, ir, pmr, uv)
Hull, R. et al, J. Chem. Res. (S), 1984, 328; J. Chem. Res. (M), 1984, 3001 (synth, pmr, uv)

5-Hydroxy-2,2'-bipyridine   H-80123

[2,2'-Bipyridin]-5-ol, 9CI
[1802-32-0]
C$_{10}$H$_8$N$_2$O   M 172.186
Needles (CHCl$_3$/pet. ether). Mp 157° (228-229°).
Ac: [58792-48-6].
   C$_{12}$H$_{10}$N$_2$O$_2$ M 214.223
   Cryst. (ligroin). Mp 102°.
Me ether: [58792-50-0]. 5-Methoxy-2,2'-bipyridine, 9CI
   C$_{11}$H$_{10}$N$_2$O   M 186.213
   Yellow oil which darkens in air. Mp 25°.

Otroshchenko, O.S., CA, 1964, **63**, 4248e (synth)
Kurbatov, Yu.V. et al, CA, 1966, **67**, 82059w (synth)
Keats, N.G. et al, J. Heterocycl. Chem., 1976, **13**, 513 (ms)
Pirzada, N.H. et al, Z. Naturforsch., B, 1976, **31**, 115 (synth, uv, ir, pmr,)

5-Hydroxy-2,3'-bipyridine   H-80124

[2,3'-Bipyridin]-5-ol, 8CI
[1748-06-7]
C$_{10}$H$_8$N$_2$O   M 172.186
Cryst. (MeOH). Mp 184-185° (177-179°).
Picrate: Cryst. (H$_2$O). Mp 217-219°.

Otroshchenko, O.S. et al, J. Gen. Chem. USSR (Engl. Transl.), 1954, **24**, 1661 (synth)
Otroshchenko, O.S. et al, CA, 1964, **63**, 3008e (synth)
Leont'ev, V.B. et al, J. Gen. Chem. USSR (Engl. Transl.), 1965, **35**, 298 (uv)
Leont'ev, V.B. et al, Chem. Heterocycl. Compd. (Engl. Transl.), 1966, **2**, 433 (ir, struct)

5-Hydroxy-3,3'-bipyridine   H-80125

[3,3'-Bipyridin]-5-ol, 8CI
[15862-23-4]
C$_{10}$H$_8$N$_2$O   M 172.186
Mp 146-147°.

Kurbatov, Yu.V. et al, CA, 1966, **67**, 82059w (synth)
Otroshchenko, O.S. et al, CA, 1966, **67**, 99969r (synth)

6-Hydroxy-2,2'-bipyridine   H-80126

[2,2'-Bipyridin]-6(1H)one, 9CI. 6-(2-Pyridyl)-2-pyridone
[101001-90-5]
C$_{10}$H$_8$N$_2$O   M 172.186
Off-white cryst. (CH$_2$Cl$_2$/hexane). Mp 124-127°.

Moran, D.B. et al, J. Heterocycl. Chem., 1986, **23**, 1071 (synth, pmr)

6-Hydroxy-2,3'-bipyridine   H-80127

[2,3'-Bipyridin]-6(1H)-one, 9CI. 6-(3-Pyridyl)-2-pyridone
[39883-44-8]
C$_{10}$H$_8$N$_2$O   M 172.186
[39883-46-0]

U.K. Pat., 1 322 318, (1973); CA, **79**, 105231k.

6-Hydroxy-2,4'-bipyridine   H-80128

[2,4'-Bipyridin]-6(1H)-one, 9CI. 6-(4-Pyridyl)-2-pyridone
[39883-33-5]
C$_{10}$H$_8$N$_2$O   M 172.186
[39883-35-7]

U.K. Pat., 1 322 318, (1973); CA, **79**, 105231k.

6-Hydroxy-3,3'-bipyridine   H-80129

[3,3'-Bipyridin]-6(1H)-one, 9CI. 5-(3-Pyridyl)-2-pyridone
C$_{10}$H$_8$N$_2$O   M 172.186
Hydrazone synthesised.
[101001-98-3]

U.S. Pat., 4 450 166, (1985); CA, **104**, 129911t (synth)

6-Hydroxy-3,4'-bipyridine   H-80130

[3,4'-Bipyridin]-6(1H)-one. 5-(4-Pyridyl)-2-pyridone
[62749-34-2]
C$_{10}$H$_8$N$_2$O   M 172.186
Mp 259-260°. Several patented syntheses in addition to the ref. quoted.
Me ether: [106154-27-2]. 6-Methoxy-3,4'-bipyridine, 9CI
   C$_{11}$H$_{10}$N$_2$O   M 186.213
   Cryst. (CH$_2$Cl$_2$/hexane). Mp 60-61°.
[82718-36-3]

Fang, W.-P. et al, Heterocycles, 1986, **24**, 1585 (synth, ir, pmr)

11-Hydroxy-2,7(14)-bisaboladien-4-one   H-80131

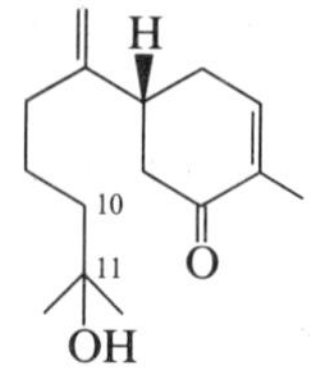

C$_{15}$H$_{24}$O$_2$ M 236.353
Constit. of Mikania shushunensis. Oil. [α]$_D^{24}$ −13.2° (c, 1.8 in CHCl$_3$).

Gutierrez, A.B. et al, Phytochemistry, 1988, **27**, 3871.

10-Hydroxy-2,7(14),11-bisabolatrien-4-one   H-80132

[119765-83-2]

$C_{15}H_{22}O_3$ M 250.337

Constit. of *Mikania shushunensis*. Oil. $[\alpha]_D^{25}$ − 16.2° (c, 0.5 in CHCl$_3$).

Gutierrez, A.B. *et al, Phytochemistry*, 1988, **27**, 3871.

2-Hydroxy-14,15-bisnor-7-labden-13-one   H-80133

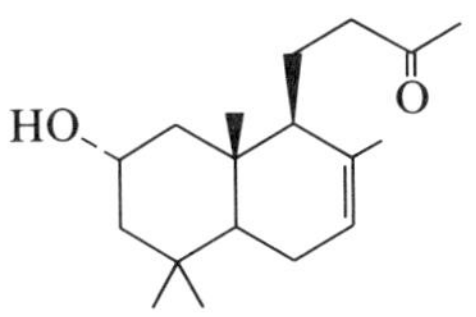

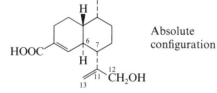

$C_{18}H_{30}O_2$ M 278.434

2α-form

Constit. of *Waitzia acuminata*.

Jakupovic, J. *et al, Phytochemistry*, 1989, **28**, 1943.

4-Hydroxy-2-butenoic acid, 9CI   H-80134

Updated Entry replacing H-01416

4-Hydroxycrotonic acid, 8CI

$$HOCH_2CH{=}CHCOOH$$

$C_4H_6O_3$ M 102.090

(E)-form [24587-49-3]

Cryst. (EtOAc). Mp 107°. Readily undergoes polycondensation.

Et ester: [10080-68-9].

$C_6H_{10}O_3$ M 130.143

d_4^{17} 1.077. Bp$_{13}$ 119-120°. n_D^{23} 1.461.

Laporte, J.F. *et al, Bull. Soc. Chim. Fr.*, 1969, 1340 (*synth, ir, pmr*)
Takano, S. *et al, Synthesis*, 1974, 42 (*lactone*)

12-Hydroxy-4,11(13)-cadinadien-15-oic acid   H-80135

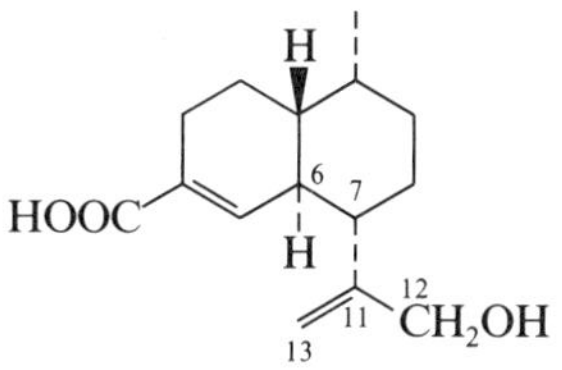

Absolute configuration

$C_{15}H_{22}O_3$ M 250.337

Constit. of *Eremophila virgata*. New stereochemical class in the cadinane series.

11,13-Epoxy, 12-deoxy: 11,13-Epoxy-4-cadinen-15-oic acid

$C_{15}H_{22}O_3$ M 250.337

Constit. of *E. virgata*. Needles (Et$_2$O). Mp 180-181°. $[\alpha]_D$ +117° (c, 0.9 in CHCl$_3$).

11R,13-Dihydro: 12-Hydroxy-4-cadinen-15-oic acid

$C_{15}H_{24}O_3$ M 252.353

Constit. of *E. virgata*.

6,7-Diepimer:

$C_{15}H_{22}O_3$ M 250.337

Constit. of *E. virgata*.

Ghisalberti, E.L. *et al, Tetrahedron*, 1989, **45**, 6297 (*cryst struct*)

4-Hydroxy-1,9-cadinadien-3-one   H-80136

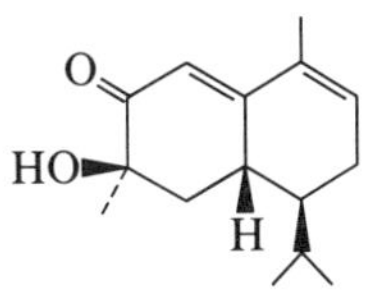

$C_{15}H_{22}O_2$ M 234.338

Me ether: 4-Methoxy-1,9-cadinadien-3-one

$C_{16}H_{24}O_2$ M 248.364

Constit. of *Senecio tomentosus*. Oil. $[\alpha]_D^{24}$ +28° (c, 0.36 in CHCl$_3$).

Mericli, A.H. *et al, Phytochemistry*, 1989, **28**, 1149.

1-Hydroxycarbazole   H-80137

9H-Carbazol-1-ol, 9CI

[61601-54-5]

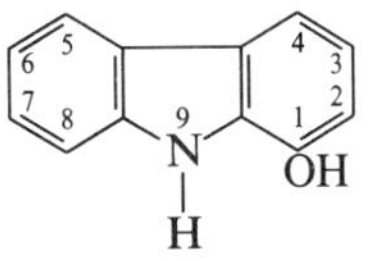

$C_{12}H_9NO$ M 183.209

Long needles (H$_2$O). Mp 160°.

Picrate: Orange needles (H$_2$O). Mp 191-192°.

Ac:

$C_{14}H_{11}NO_2$ M 225.246

Small prisms (H$_2$O). Mp 132-133°.

Me ether: [4544-87-0]. *1-Methoxy-9H-carbazole*

$C_{13}H_{11}NO$ M 197.236

Long needles (EtOH). Mp 69-70°.

Et ether: [115663-18-8]. *1-Ethoxy-9H-carbazole*

$C_{14}H_{13}NO$ M 211.263

Needles (pet. ether). Mp 93.5-94.5°. Bp$_{2.4}$ 185-200°.

Barnes, C.S. *et al, J. Chem. Soc.*, 1949, 1381 (*synth, deriv*)
Martin, T. *et al, J. Chem. Soc., Perkin Trans. 1*, 1988, 235 (*ethers, uv, ir, ms*)

11-Hydroxy-3,6-cholestanedione   H-80138

$C_{27}H_{44}O_3$ M 416.643

(5α,11α)-form

Constit. of *Acantophora spicifera*. Cryst. (MeOH). Mp 145°. $[\alpha]_D^{25}$ −11.1°.

Prakash, O. *et al, J. Nat. Prod. (Lloydia)*, 1989, **52**, 686 (*isol, cmr, pmr*)

2-Hydroxy-1,4-cineole   H-80139

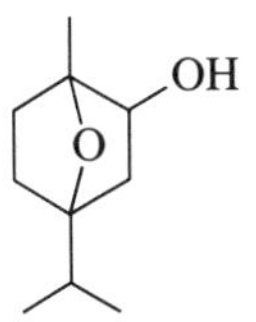

$C_{10}H_{18}O_2$ M 170.251

Constit. of *Ferula jaeschkaena*. Cryst. Mp 75°.

Garg, S.N. *et al, Phytochemistry*, 1989, **28**, 634.

14-Hydroxy-8,12-cleistanthadiene-7,11-dione   H-80140

[119725-17-6]

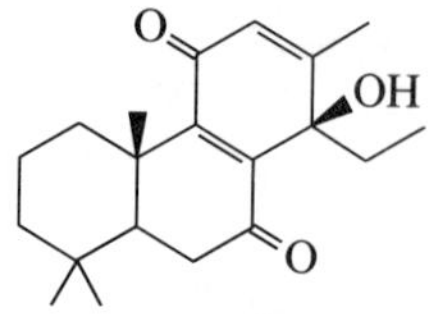

C$_{20}$H$_{28}$O$_3$ M 316.439
Constit. of *Vellozia nivea*. Yellow cryst. Mp 113-115°. [α]$_D^{25}$ −4.7° (c, 0.85 in MeOH).

Pinto, A.C. *et al*, *Phytochemistry*, 1988, **27**, 3973.

11-Hydroxy-8,11,13-cleistanthatrien-7-one   H-80141

[119725-18-7]

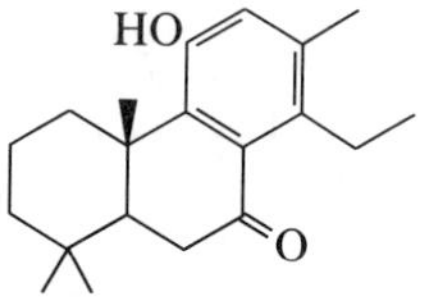

C$_{20}$H$_{28}$O$_2$ M 300.440
Constit. of *Vellozia nivea*. Cryst. Mp 236-238°. [α]$_D^{25}$ +70° (c, 0.5 in CHCl$_3$).

Pinto, A.C. *et al*, *Phytochemistry*, 1988, **27**, 3973.

ent-2α-Hydroxy-3,13-clerodadiene-16,15:18,19-diolide   H-80142

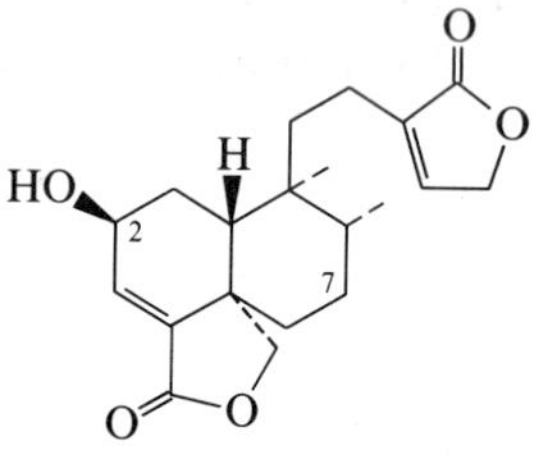

C$_{20}$H$_{26}$O$_5$ M 346.422
Ac: ent-2α-*Acetoxy*-3,13-*clerodadiene*-16,15:18,19-*diolide*
C$_{22}$H$_{28}$O$_6$ M 388.460
Constit. of *Salvia melissodora*. Cryst. (CH$_2$Cl$_2$/MeOH). Mp 202-204°. [α]$_D^{20}$ −209.5° (c, 0.16 in CHCl$_3$).
7α-Hydroxy: ent-2α,7β-*Dihydroxy*-3,13-*clerodadiene*-16,15:18,19-*diolide*
C$_{20}$H$_{26}$O$_6$ M 362.422
Constit. of *S. melissodora*. Cryst. (Me$_2$CO/hexane). Mp 190-192°. [α]$_D^{20}$ −177.6° (c, 0.2 in MeOH).
7α-Acetoxy: ent-7β-*Acetoxy*-2α-hydroxy-3,13-*clerodadiene*-16,15:18,19-*diolide*
C$_{22}$H$_{28}$O$_7$ M 404.459
Constit. of *S. melissodora*. Cryst. (Me$_2$CO/hexane). Mp 236-239°. [α]$_D^{25}$ −123.9° (c, 0.184 in CHCl$_3$).
7α-Hydroxy, 2-Ac: ent-2α-*Acetoxy*-7β-hydroxy-3,13-*clerodadiene*-16,15:18,19-*diolide*
C$_{22}$H$_{28}$O$_7$ M 404.459
Constit. of *S. melissodora*. Cryst. (CH$_2$Cl$_2$/diisopropyl ether). Mp 197-199°. [α]$_D^{20}$ −207° (c, 0.167 in CHCl$_3$).
7-Oxo: ent-2α-*Acetoxy*-7-oxo-3,13-*clerodadiene*-16,15:18,19-*diolide*
C$_{20}$H$_{24}$O$_6$ M 360.406

Constit. of *S. melissodora*. Cryst. (CH$_2$Cl$_2$/hexane). Mp 238-240°. [α]$_D^{20}$ −239.8° (c, 0.171 in CHCl$_3$).

Esquivel, B. *et al*, *Phytochemistry*, 1989, **28**, 561.

ent-7β-Hydroxy-3,13-clerodadiene-16,15:18,19-diolide   H-80143

C$_{20}$H$_{26}$O$_5$ M 346.422
Constit. of *Salvia melissodora*. Cryst. (Me$_2$CO/diisopropyl ether). Mp 178-179°. [α]$_D^{20}$ −154.5° (c, 0.165 in CHCl$_3$).
Ac: ent-7β-*Acetoxy*-3,13-*clerodadiene*-16,15:18,19-*diolide*
C$_{22}$H$_{28}$O$_6$ M 388.460
Constit. of *S. melissodora*. Cryst. (CH$_2$Cl$_2$/diisopropyl ether). Mp 178-179°. [α]$_D^{20}$ −110.7° (c, 0.17 in CHCl$_3$).
7-Epimer: ent-7α-*Hydroxy*-3,13-*clerodadiene*-16,15:18,19-*diolide*
C$_{20}$H$_{26}$O$_5$ M 346.422
Constit of *S. melissodora*. Cryst. (EtOAc/MeOH). Mp 196-197°. [α]$_D^{20}$ −93.3° (c, 0.15 in MeOH).
7-Ketone: ent-7-*Oxo*-3,13-*clerodadiene*-16,15:18,19-*olide*
C$_{20}$H$_{24}$O$_5$ M 344.407
Constit. of *S. melissodora*. Cryst. (EtOAc). Mp 89-93°. [α]$_D^{20}$ −168.4° (c, 0.152 in CHCl$_3$).

Esquivel, B. *et al*, *Phytochemistry*, 1989, **28**, 561.

ent-2α-Hydroxy-3,13-clerodadien-15-oic acid   H-80144

[118173-04-9]

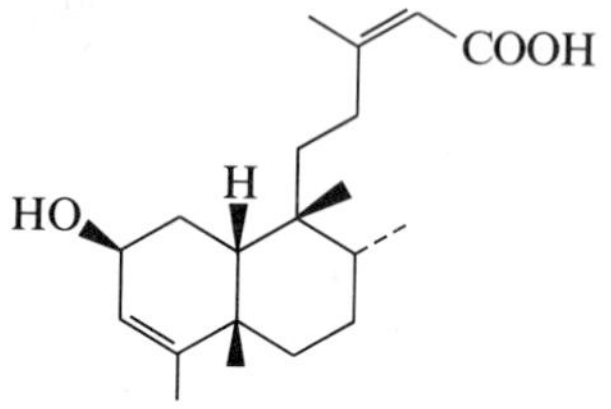

C$_{20}$H$_{32}$O$_3$ M 320.471
Constit. of *Ageratina saltillensis*.
2-Ketone: [118243-69-9]. **ent-2-Oxo-3,13-clerodadien-15-oic acid**
C$_{20}$H$_{30}$O$_3$ M 318.455
Constit. of *A. saltillensis*.

Fang, N. *et al*, *Phytochemistry*, 1988, **27**, 3187.

ent-2α-Hydroxy-3,13-clerodadien-15,16-olide   H-80145

[118173-03-8]

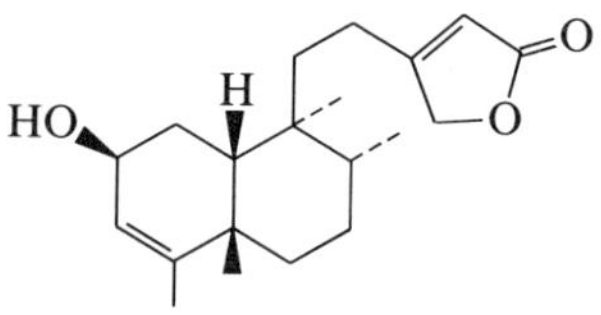

C$_{20}$H$_{30}$O$_3$ M 318.455
Constit. of *Ageratina saltillensis*.

Fang, N. *et al*, *Phytochemistry*, 1988, **27**, 3187.

1-Hydroxy-3-cyclohexene-1-carbox-aldehyde   H-80146

1-Formyl-3-cyclohexenol

HO CHO

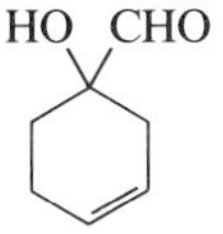

$C_7H_{10}O_2$ M 126.155

Ac: [55638-24-9].
    $C_9H_{12}O_3$ M 168.192
    Oil. $Bp_{0.1}$ 90°.

Me ether: [117370-82-8]. *1-Methoxy-3-cyclohexene-1-carboxaldehyde. 3-Formyl-3-methoxycyclohexene*
    $C_8H_{12}O_2$ M 140.182
    Liq.

Me ether, oxime: [117370-83-9].
    $C_8H_{13}NO_2$ M 155.196
    Solid. Mp 77-78°.

Hassner, A. *et al, J. Org. Chem.*, 1975, **40**, 3427 (*synth, pmr*)
Rubottom, G.M. *et al, Tetrahedron*, 1983, **39**, 861 (*synth*)
Wu, P.-L. *et al, J. Org. Chem.*, 1988, **53**, 5998 (*synth, pmr, ir, cmr*)

6-Hydroxydecanoic acid, 9CI   H-80147

[16899-10-8]

$$H_3C(CH_2)_3CH(OH)(CH_2)_4COOH$$

$C_{10}H_{20}O_3$ M 188.266
Present in lipid fraction of royal jelly.

Lactone:
    $C_{10}H_{18}O_2$ M 170.251
    $Bp_{1.3}$ 96°. n_D^{20} 1.4601.

Netherlands Pat., 100 368, (1962); *CA*, **60**, 2783c (*synth, lactone*)
Japan. Pat., 74 85 276, (1974); *CA*, **82**, 71817a (*use*)
Lercker, G. *et al, Lipids*, 1981, **16**, 912 (*anal*)

7-Hydroxydecanoic acid, 9CI   H-80148

[2034-55-1]

$$H_3CCH_2CH_2CH(OH)(CH_2)_5COOH$$

$C_{10}H_{20}O_3$ M 188.266
Present in hydrolysate of leaf resin of *Ipomoea fistulosa* and saponified cells of *Mucor* spp. Liq.

[73276-73-0]

Legler, G., *Phytochemistry*, 1965, **4**, 29 (*isol*)
Satoshi, T. *et al, Agric. Biol. Chem.*, 1980, **44**, 193 (*synth, ms, isol, ir, glc*)

9-Hydroxydecanoic acid, 9CI   H-80149

[1422-27-1]

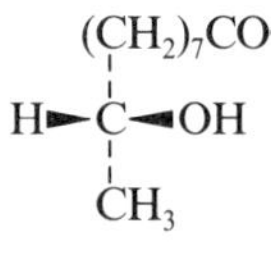

(CH₂)₇COOH
H►C◄OH   (*R*)-form
CH₃

$C_{10}H_{20}O_3$ M 188.266
Present in *Agaricus campestris, Mucor hiemalis* and royal jelly.

(*R*)-*form* [35433-73-9]
    Oil. Mp 32°. $[\alpha]_D$ −7.5° (MeOH), $[\alpha]_D^{23}$ −8.1° (c, 0.11 in CHCl₃).

Me ester:
    $C_{11}H_{22}O_3$ M 202.293
    $Bp_{0.005}$ 78-80°. $[\alpha]_D$ −9.5° (c, 1.45 in cyclohexane).

tert-Butyl ester:
    $C_{14}H_{28}O_3$ M 244.373
    $[\alpha]_D$ −7.0° (c, 1.73 in cyclohexane).

Lactone: [74183-94-1]. *10-Methyl-2-oxecanone, 9CI.*
    Phoracantholide I
    $C_{10}H_{18}O_2$ M 170.251
    Present in metasternal gland of *Phoracantha synonyma*.
    $[\alpha]_D^{22}$ −37.4° (c, 0.02 in CHCl₃).

(*S*)-*form* [35433-72-8]
    $[\alpha]_D^{22}$ +8.14° (c, 0.14 in CHCl₃).

Me ester: Oil. $Bp_{0.002}$ 80°. $[\alpha]_D$ +9.3° (c, 2.0 in cyclohexane).

Lactone: $[\alpha]_D$ +38.38° (c, 0.16 in CHCl₃).

(±)-*form* [40151-97-1]
    Me ester: $Bp_{0.01}$ 85°.

    tert-Butyl ester: $Bp_{0.01}$ 72-74°.

Moore, B.P. *et al, Aust. J. Chem.*, 1976, **29**, 1365 (*isol*)
Gerlach, H. *et al, Helv. Chim. Acta*, 1978, **61**, 1226 (*synth, ir, pmr, ms*)
Fink, M. *et al, Helv. Chim. Acta*, 1982, **65**, 2563 (*derivs*)
Tressl, R. *et al, J. Agric. Food Chem.*, 1982, **30**, 89 (*glc, ms*)
Voss, G. *et al, Helv. Chim. Acta*, 1983, **66**, 2294 (*synth*)
Clarke, M.J. *et al, J. Liq. Chromatogr.*, 1986, **9**, 1711 (*hplc*)
Naoshima, Y. *et al, Chem. Lett.*, 1987, 2379 (*synth, lactone*)

8-Hydroxy-2-decene-4,6-diynoic acid, 9CI   H-80150

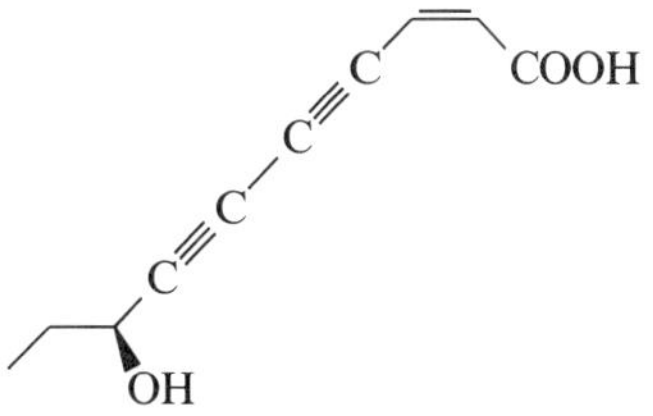

$C_{10}H_{10}O_3$ M 178.187

(*S,E*)-*form* [56085-68-8]
    Isol. from *Aster novi belgii*. Needles (Et₂O/pet. ether). Mp 97-100°.

O-Angeloyl, Me ester: [24486-36-0]. *8-Angeloyloxylachnophyllum ester*
    $C_{16}H_{18}O_4$ M 274.316
    Liq. $Bp_{0.001}$ 120-130° part dec. . $[\alpha]_D^{20}$ −164° (c, 0.85 in Et₂O).

Bohlmann, F. *et al, Chem. Ber.*, 1964, **97**, 3469; 1965, **98**, 2608 (*isol, uv, ir, pmr, ms, abs config*)
Ord, M.R. *et al, J. Chem. Soc., Perkin Trans. 1*, 1975, 687 (*synth, uv, ir, pmr*)

4-Hydroxy-25-desoxyneorollinicin   H-80151

H₃C(CH₂)₉ ... (CH₂)₁₀ ...

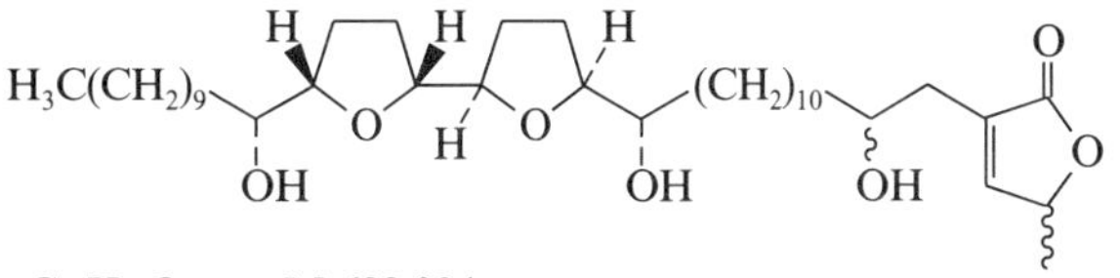

$C_{37}H_{66}O_7$ M 622.924
Constit. of *Rollinia papilionella*. Wax. Mp 25°.

Abreo, M.J. *et al, J. Nat. Prod. (Lloydia)*, 1989, **52**, 822 (*isol, pmr, cmr*)

6-Hydroxy-9,9-dimethyl-5-(3-methyl-1-oxobutyl)-1-propyl-3H,9H-[1,2]-dioxolo-[3′,4′:4,5]furo[2,3-f][1]benzopyran-3-one, 9CI   H-80152

[30390-08-0]

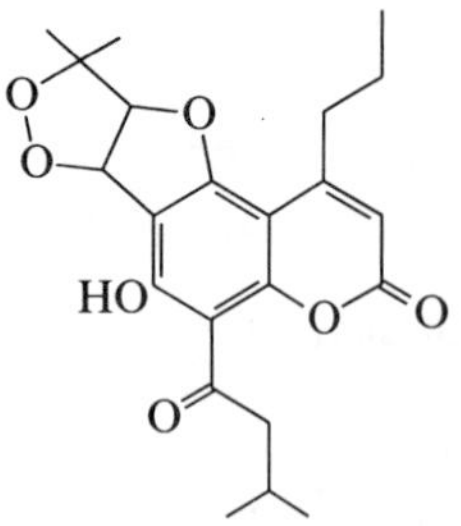

$C_{22}H_{26}O_7$ M 402.443

Isol. from *Mammea americana*. An oxidn. prod. of Mammein. Mp 181-182°. The isomer having a 2-methylbutanoyl side-chain cooccurs.

[30390-09-1]

Crombie, L. *et al*, *J. Chem. Soc., Perkin Trans.* 1, 1972, 2241.
Finnegan, R.A. *et al*, *J. Pharm. Sci.*, 1972, **61**, 1603 (*isol, ir, uv, pmr*)

2-(3-Hydroxy-3,7-dimethyl-6-octenyl)-1,4-benzenediol   H-80153

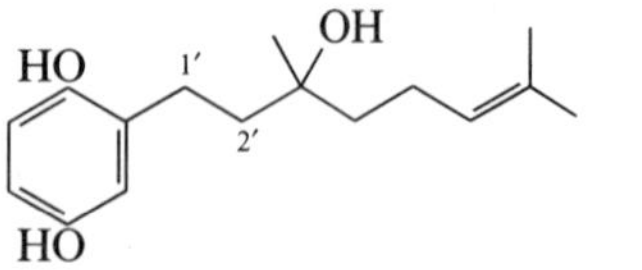

$C_{16}H_{24}O_3$ M 264.364

Metab. of *Amaroucium multiplicatum*. Oil.

1′,2′-Didehydro (E-): 2-(3-Hydroxy-3,7-dimethyl-2,6-octadienyl)-1,4-benzenediol
$C_{16}H_{22}O_3$ M 262.348
Metab. of *A. multiplicatum*. Oil.

Sato, A. *et al*, *J. Nat. Prod.* (*Lloydia*), 1989, **52**, 975 (*isol, pmr, cmr*)

3-Hydroxy-4,4-dimethyl-2-pentanone, 9CI   H-80154
*Acetyl-*tert-*butylcarbinol*

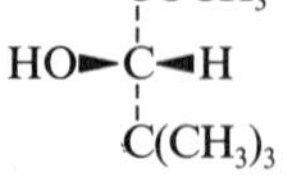

(S)-form

$C_7H_{14}O_2$ M 130.186

(S)-form [97869-13-1]
Oil. Bp_1 45-50°. $[\alpha]_D^{25}$ +133.2° (c, 1 in $CHCl_3$).

(±)-form
d_4^{25} 0.92. $Bp_{0.35}$ 26°.

Semicarbazone: Cryst. (EtOH aq.). Mp 192-193°.

Stacy, G.W. *et al*, *J. Org. Chem.*, 1966, **31**, 1753.
Zhang, W.-Y. *et al*, *J. Am. Chem. Soc.*, 1988, **110**, 4652 (*synth, ir, pmr, cmr*)

5-Hydroxy-5,5-diphenylpentanoic acid   H-80155
[68719-02-8]

$$Ph_2C(OH)CH_2CH_2CH_2COOH$$

$C_{17}H_{18}O_3$ M 270.327
Cryst. Mp 132-134°.

Lehmann, J. *et al*, *Justus Liebigs Ann. Chem.*, 1988, 827 (*synth, pmr*)

7-Hydroxydodecanoic acid, 9CI   H-80156
[70393-62-3]

$$H_3C(CH_2)_4CH(OH)(CH_2)_5COOH$$

$C_{12}H_{24}O_3$ M 216.320

(±)-form [73276-74-1]
Mp 55.5-56.5°. (R)-form also known.

[58262-37-6, 78737-61-8]

Tahera, S. *et al*, *Agric. Biol. Chem.*, 1980, **44**, 193 (*synth, ir, ms, glc*)
Midland, M.M. *et al*, *J. Org. Chem.*, 1981, **46**, 3933 (*synth, pmr*)

11-Hydroxydodecanoic acid, 9CI   H-80157
[32459-66-8]

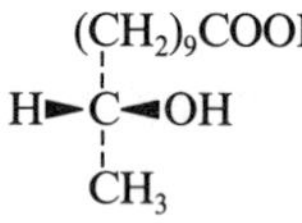

(R)-form

$C_{12}H_{24}O_3$ M 216.320
Constit. of royal jelly. Metab. of *Fusarium oxysporum*.

(R)-form
Me ester: [88785-25-5].
Oil. $Bp_{0.002}$ ~100°. $[\alpha]_D$ −7.7° (c, 2.15 in cyclohexane).
(±)-form [83515-89-3]
Cryst. (Et_2O/pet. ether). Mp 54°.
Me ester: [109718-22-1].
$C_{13}H_{26}O_3$ M 230.347
Liq. or cryst. Bp_7 160°.

Anilide: Small prisms (pet. ether). Mp 87°.

Lactone: 11-Decanolide. Dihydrorecifeiolide
$C_{12}H_{22}O_2$ M 198.305
Oil. $Bp_{0.001}$ 65-70°.

[66605-95-6]

Ellin, A. *et al*, *Arch. Biochem. Biophys.*, 1973, **158**, 597 (*synth, derivs*)
Voss, G. *et al*, *Helv. Chim. Acta*, 1983, **66**, 2294 (*synth, ir, pmr, cmr*)
Hesse, B. *et al*, *Helv. Chim. Acta*, 1984, **67**, 1713; 1986, **69**, 1323 (*lactone*)
Makita, A. *et al*, *J. Antibiot.*, 1986, **39**, 1257 (*synth, derivs*)

7-O-(3-Hydroxy-7-drimen-11-yl) isofraxidin   H-80158
[52418-68-5]

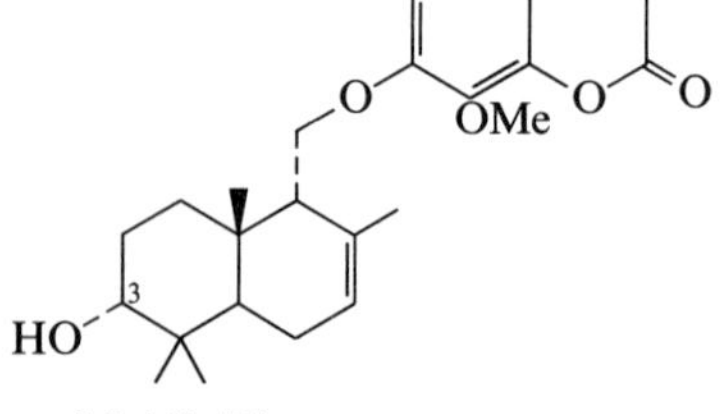

$C_{26}H_{34}O_6$ M 442.551

Isol. from *Artemisia pontica*. Mp 129.5°. $[\alpha]_D^{23}$ +201°
(Me$_2$CO).

3-Ketone: [57194-40-8].
   $C_{26}H_{32}O_6$ M 440.535
   Isol. from *Anthemis aciphylla*. Mp 130°. $[\alpha]_D^{22}$ +156.3°
   (CHCl$_3$).

[52329-14-3, 52329-15-4]

Bohlmann, F. *et al, Chem. Ber.*, 1974, **107**, 644; 1975, **108**, 1902.

10-Hydroxy-1,11(13)-eremophiladien-12,8-olide   H-80159

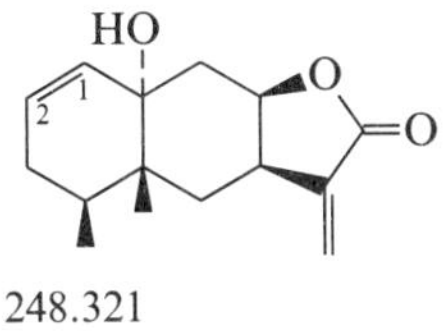

$C_{15}H_{20}O_3$ M 248.321
(8β,10α)-form
   Constit. of *Ondetia linearis*. Oil.

   1α,2α-Epoxide: 1α,2α-Epoxy-10α-hydroxy-11(13)-
      eremophilen-12,8β-olide
      $C_{15}H_{20}O_4$ M 264.321
      Constit. of *O. linearis*. Oil.

Zdero, C. *et al, Phytochemistry*, 1989, **28**, 1653.

2-Hydroxy-4,11(13)-eudesmadien-15-al   H-80160

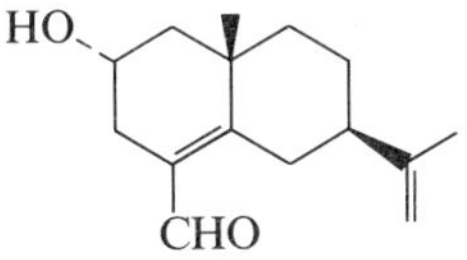

$C_{15}H_{22}O_2$ M 234.338
2α-form [119863-97-7]
   $C_{15}H_{22}O_2$ M 234.338
   Constit. of *Tetragonotheca ludoviciana*. Oil. $[\alpha]_D^{24}$ +51°
   (c, 0.7 in CHCl$_3$).

Jakupovic, J. *et al, Phytochemistry*, 1988, **27**, 3881.

2-Hydroxy-3,11(13)-eudesmadien-12,8-olide   H-80161

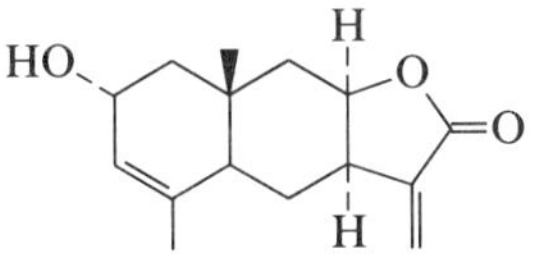

$C_{15}H_{20}O_3$ M 248.321
(2α,8β)-form
   Constit. of *Ondetia linearis*.
   Ac: Cryst. Mp 153°. $[\alpha]_D^{24}$ +71° (c, 0.27 in CHCl$_3$).

Zdero, C. *et al, Phytochemistry*, 1989, **28**, 1653.

7-Hydroxy-4,11(13)-eudesmadien-12,6-olide   H-80162

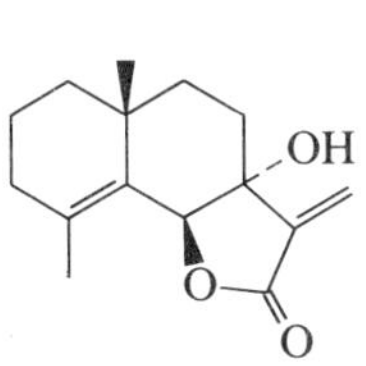

$C_{15}H_{20}O_3$ M 248.321

(6β,7α)-form
   7-Hydroxyfrullanolide
   Constit. of *Sphaeranthus indicus*. Shows antimicrobial
   props. Cryst. (pet. ether). Mp 59-60°. $[\alpha]_D^{24}$ −39° (c, 4.0
   in MeOH).

Atta-ur-Rahman, *J. Chem. Res. (S)*, 1989, 68.

2-Hydroxy-1,4,6-eudesmatriene-3,8-dione   H-80163

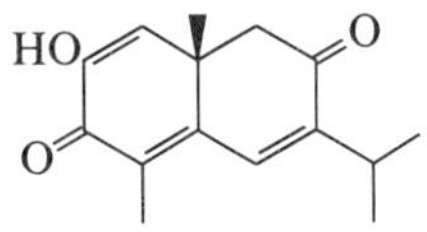

$C_{15}H_{18}O_3$ M 246.305
Constit. of *Baccharis boliviensis*. Cryst. Mp 96°. $[\alpha]_D^{24}$
   +123° (c, 0.49 in CHCl$_3$).

Zdero, C. *et al, Phytochemistry*, 1989, **28**, 531.

3-Hydroxy-1,4(15),11(13)-eudesmatrien-12-oic acid   H-80164

3-Hydroxy-1,2-dehydrocostic acid

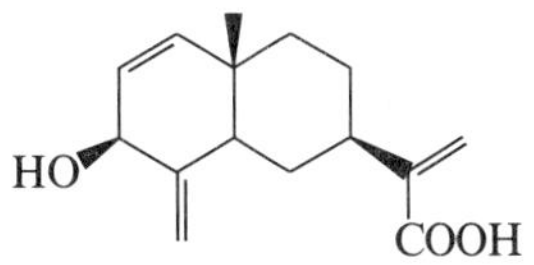

$C_{15}H_{20}O_3$ M 248.321
3β-form
   Me ester:
      $C_{16}H_{22}O_3$ M 262.348
      Constit. of *Centaurea arguta*. Oil.
   3-Ketone, Me ester: Methyl 3-oxo-1,4(15),11(13)-
      eudesmatrien-12-oate
      $C_{16}H_{20}O_3$ M 260.332
      Constit. of *C. arguta*. Oil.

Gadeschi, E. *et al, Phytochemistry*, 1989, **28**, 2204 (*isol, pmr*)

2-Hydroxy-5-eudesmen-12,8-olide   H-80165

$C_{15}H_{22}O_3$ M 250.337
(2α,8β,11αH)-form
   2α-Hydroxy-11α,13-dihydroalantolactone
   Constit. of *Ondetia linearis*. Cryst. Mp 147°.

   11α-Hydroxy: 2α,11α-Dihydroxy-5-eudesmen-12,8β-olide.
      2α,11α-Dihydroxy-11,13-dihydroalantolactone
      $C_{15}H_{22}O_4$ M 266.336
      Constit. of *O. linearis*. Cryst. Mp 186°. $[\alpha]_D^{24}$ −26° (c,
      0.16 in CHCl$_3$).
(2α,8β,11βH)-form
   2α-Hydroxy-11β,13-dihydroalantolactone
   Constit. of *O. linearis*. Cryst. Mp 134°. $[\alpha]_D^{24}$ +9° (c, 0.18
   in CHCl$_3$).

Zdero, C. *et al, Phytochemistry*, 1989, **28**, 1653.

3-Hydroxy-4-eudesmen-12,8-olide   H-80166

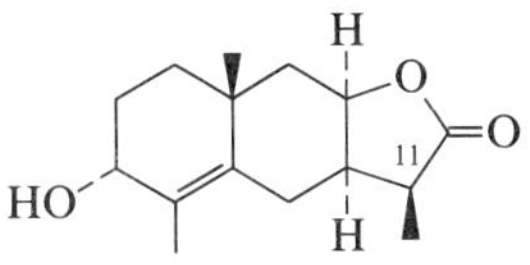

(3α,8β,11αH)-form

$C_{15}H_{22}O_3$ M 250.337
(3α,8β,11αH)-form

Constit. of *Ondetia linearis*. Cryst. Mp 123°. $[\alpha]_D^{24}$ +99° (c, 0.54 in CHCl₃).

(3β,8β,11βH)-form
Constit. of *O. linearis*. Gum. $[\alpha]_D^{24}$ +145° (c, 0.12 in CHCl₃).

Zdero, C. *et al*, *Phytochemistry*, 1989, **28**, 1653.

4-Hydroxy-1(10),11(13)-guaiadien-12,8-olide   H-80167

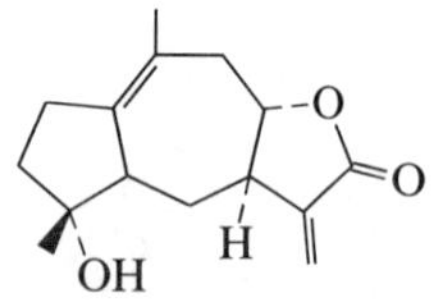

C₁₅H₂₀O₃   M 248.321
Constit. of *Helichrysum dasyanthum*. Gum.

Jakupovic, J. *et al*, *Phytochemistry*, 1989, **28**, 1119.

4-Hydroxy-10(14)-guaien-12,8-olide   H-80168

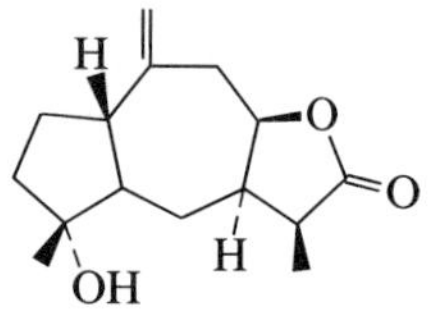

C₁₅H₂₂O₃   M 250.337
Constit. of *Helichrysum splendidum*. Gum.

Jakupovic, J. *et al*, *Phytochemistry*, 1989, **28**, 1119.

4-Hydroxy-9-guaien-12,8-olide   H-80169

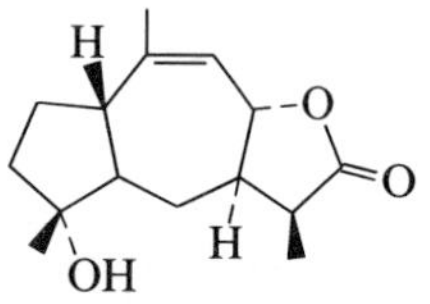

C₁₅H₂₂O₃   M 250.337
Constit. of *Helichrysum splendidum*. Cryst. Mp 113°. $[\alpha]_D^{24}$ −99° (c, 0.19 in CHCl₃).

Jakupovic, J. *et al*, *Phytochemistry*, 1989, **28**, 1119.

10-Hydroxy-16-hentriacontanone   H-80170

10-Hydroxypalmitone
[87264-34-4]

$$H_3C(CH_2)_8CH(OH)(CH_2)_5CO(CH_2)_{14}CH_3$$

C₃₁H₆₂O₂   M 466.830
Isol. from leaf wax of *Santalum album* and from *Machilus glaucescens*. Plates (Me₂CO). Mp 96-97°.

Ac: Mp 39.5-40.0°.

Oxime: Mp 63.4-63.6°.

Chibnall, A.C. *et al*, *Biochem. J.*, 1937, **31**, 1981 (*isol, struct, synth*)
Talapatra, B. *et al*, *J. Indian Chem. Soc.*, 1982, **59**, 1364 (*isol*)

8-Hydroxy-5-hentriacontanone   H-80171

$$H_3C(CH_2)_{22}CH(OH)CH_2CH_2CO(CH_2)_3CH_3$$

C₃₁H₆₂O₂   M 466.830
Constit. of roots of *Costus speciosus*. Cryst. (MeOH). Mp 84-85°.

Gupta, M.M. *et al*, *Phytochemistry*, 1981, **20**, 2553.

11-Hydroxyhexadecanoic acid   H-80172

Updated Entry replacing H-01877
11-Hydroxypalmitic acid. Jalapinolic acid. Scammonolic acid

$$H_3C(CH_2)_4CH(OH)(CH_2)_9COOH$$

C₁₆H₃₂O₃   M 272.427
(+)-form [502-75-0]
Occurs in jalap and Scammony resins (*Ipomoea orizabensis* and *Convolvulus scammonia*) and other *I.* spp., in glycosidic form. Cryst. (EtOAc). Mp 68-69°. $[\alpha]_D$ +0.79° (CHCl₃).

Me ester: [60368-18-5].
C₁₇H₃₄O₃   M 286.454
Cryst. (pet. ether or EtOAc). Mp 40.5-41.5°, Mp 47-49°. Bp₃ 183-186°.

Et ester:
C₁₈H₃₆O₃   M 300.481
Cryst. Mp 47-48°.

Ac:
C₁₈H₃₄O₄   M 314.464
Oil. Bp₅₀ 224-225°.

O-[α-L-*Rhamnopyranosyl*(*1→4*)-α-L-*rhamnopyranoside*]: [68124-11-8]. **Muricatin B**
C₂₈H₅₂O₁₁   M 564.712
Isol. from *I. muricata*. Syrup. $[\alpha]_D^{26}$ −56.6° (c, 0.495 in EtOH)(synthetic).

O-Glycoside: Jalapinic acid. Scammoninic acid. Scammonic acid
Isol. from *I. orizabensis* and *C. scammonia*. Hygroscopic solid. Mp 66°. Gives Jalapinolic acid, glucose, rhamnose and rhodeose on hydrol. May not be homogeneous.

(±)-form
Cryst. (EtOAc). Mp 68-69°.

Me ester: Cryst. (pet. ether). Mp 40.5-41.5°. Bp₃ 183-186°.

Power, F.B. *et al*, *J. Chem. Soc.*, 1912, **101**, 398 (*isol*)
Davies, C.A. *et al*, *J. Am. Chem. Soc.*, 1928, **50**, 1754 (*synth*)
Bauer, K.H. *et al*, *Arch. Pharm. (Weinheim, Ger.)*, 1934, **272**, 841 (*Jalapinic acid*)
Smith, C.R. *et al*, *Phytochemistry*, 1964, **3**, 289 (*isol*)
Khanna, S.N. *et al*, *Phytochemistry*, 1967, **6**, 735.
Tulloch, A.P., *Phytochemistry*, 1976, **15**, 1153 (*isol*)
Wagner, H. *et al*, *Phytochemistry*, 1977, **16**, 715 (*glc, ms*)
Liptak, A. *et al*, *Phytochemistry*, 1978, **17**, 997 (*Muricatin B*)

15-Hydroxyhexadecanoic acid   H-80173

Updated Entry replacing H-00541

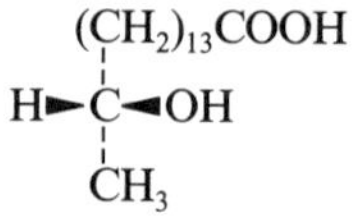

(R)-form

C₁₆H₃₂O₃   M 272.427
(R)-form

Lactone: [69297-56-9]. *15-Hexadecanolide. 16-Methyloxacyclohexadecan-2-one, 9CI*
C₁₆H₃₀O₂   M 254.412

Constit. of Galbanum absolute (from *Ferula galbaniflua* and *F. rubicaulis*). $[\alpha]_D^{20} -16.5°$ (c, 1.03 in CHCl$_3$).

(*S*)-*form*

Lactone: Oil. Bp$_{0.05}$ 110-115°. $[\alpha]_D^{20} +2.41°$ (c, 1.66 in CHCl$_3$).

(±)-*form*

Lactone: Bp$_{0.06}$ 130°.

Becker, J. *et al*, *Helv. Chim. Acta*, 1971, **54**, 2889 (*synth*)
Kaiser, R. *et al*, *Helv. Chim. Acta*, 1978, **61**, 2671 (*isol, synth*)
Kostova, K. *et al*, *Helv. Chim. Acta*, 1984, **67**, 1713 (*synth, pmr, cmr*)
Stanchev, S. *et al*, *Helv. Chim. Acta*, 1989, **72**, 1052 (*synth, abs config, bibl*)

5-Hydroxy-4-(4-hydroxyphenyl)-2(5*H*)-furanone   H-80174

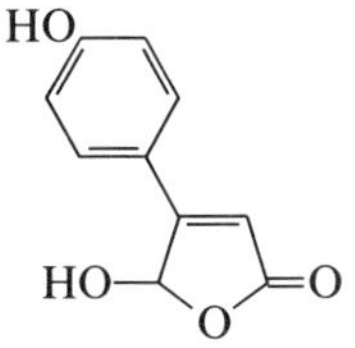

$C_{10}H_8O_4$ M 192.171

Constit. of *Sphagnum* spp. Needles (H$_2$O). Mp 228°. $[\alpha]_D^{20} -3.5°$ (c, 0.55 in MeOH).

Wilschke, J. *et al*, *Phytochemistry*, 1989, **28**, 1725.

7-Hydroxy-6-[2-hydroxy-2-(tetrahydro-2-methyl-5-oxo-2-furyl)ethyl]-5-methoxy-4-methylphthalide,8CI   H-80175

7-Hydroxy-6-[2-hydroxy-2-(2-methyl-5-oxotetrahydro-2-furyl)ethyl]-5-methoxy-4-methyl-1-phthalanone

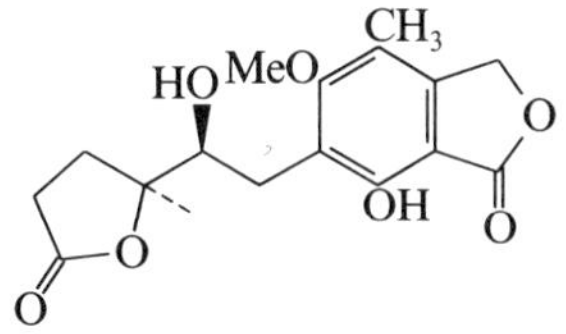

Relative configuration

$C_{17}H_{20}O_7$ M 336.341

Prob. rel. config. shown (descr. as 'threo'). Isol. from *Penicillium brevi-compactum* cultures. Cryst. (Me$_2$CO/pet. ether). Mp 218-220°.

[26644-06-4]

Campbell, I.M. *et al*, *Tetrahedron Lett.*, 1966, 5107 (*isol, uv, ir, pmr, ms, struct*)
Jones, D.F. *et al*, *J. Chem. Soc. C*, 1970, 1725 (*isol, pmr*)

19-Hydroxyicosanoic acid   H-80176

19-Hydroxyeicosanoic acid, 9CI. 19-Hydroxyarachidic acid

[34383-23-8]

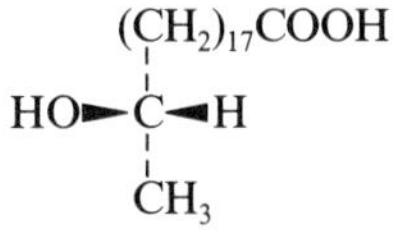

$C_{20}H_{40}O_3$ M 328.534

(*S*)-*form* [34370-60-0]

Isol. from oil of *Torulopsis magnoliae*. Cryst. (Me$_2$CO). Mp 86.5-87.5°. $[\alpha]_D^{25} +5.2°$ (c, 1.4 in AcOH).

Me ester:

$C_{21}H_{42}O_3$ M 342.561

Cryst. (pet. ether). Mp 58-60°. $[\alpha]_D^{25} +3.6°$ (c, 2.4 in MeOH).

Tulloch, A.P. *et al*, *Can. J. Chem.*, 1962, **40**, 1362.

2′-Hydroxyisolupalbigenin   H-80177

2′,4′,5,7-Tetrahydroxy-3′,8-diprenylisoflavone

[121747-94-2]

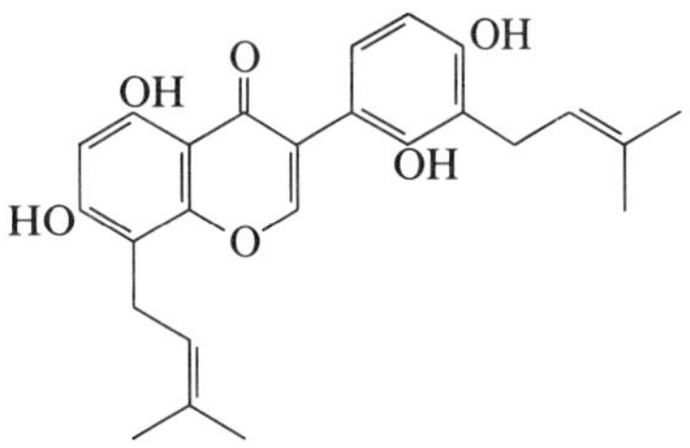

$C_{25}H_{26}O_6$ M 422.477

Constit. of *Lupinus albus*. Gum.

Tahara, S. *et al*, *Phytochemistry*, 1989, **28**, 901.

6-Hydroxyisoquinoline   H-80178

6-Isquinolinol

[7651-82-3]

C_9H_7NO M 145.160

Cryst. (2-propanol). Mp 220°.

Me ether: [52986-70-6]. *6-Methoxyisoquinoline*

$C_{10}H_9NO$ M 159.187

Noncryst. or low-melting cryst. Mp 36°.

Me ether, picrate: Mp 225°.

Me ether, N-oxide:

$C_{10}H_9NO_2$ M 175.187

Cryst. (as hydrochloride). Mp 197° (hydrochloride).

Robinson, R.A., *J. Am. Chem. Soc.*, 1947, **69**, 1939, 1944 (*synth, deriv*)
Mason, S.F., *J. Chem. Soc.*, 1957, 4874, 5010 (*ir, uv, props, tautom*)
Hendrickson, J.B. *et al*, *J. Org. Chem.*, 1983, **48**, 3344 (*synth, deriv*)
Peet, N.P. *et al*, *J. Heterocycl. Chem.*, 1987, **24**, 715 (*ms, deriv*)

ent-1α-Hydroxy-16-kauren-19-oic acid   H-80179

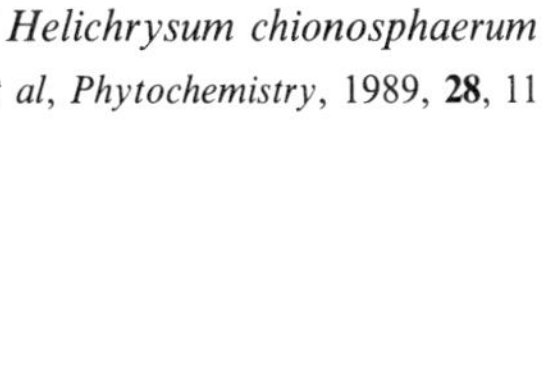

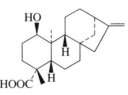

$C_{20}H_{30}O_3$ M 318.455

Ac: [122585-74-4]. *ent*-1α-*Acetoxy*-16-kauren-19-oic acid

$C_{22}H_{32}O_4$ M 360.492

Constit. of *Helichrysum chionosphaerum*. Gum.

Jakupovic, J. *et al*, *Phytochemistry*, 1989, **28**, 1119.

ent-12β-Hydroxy-15-kauren-19-oic acid   H-80180

ent-*12β-Hydroxyisokauren-19-oic acid*

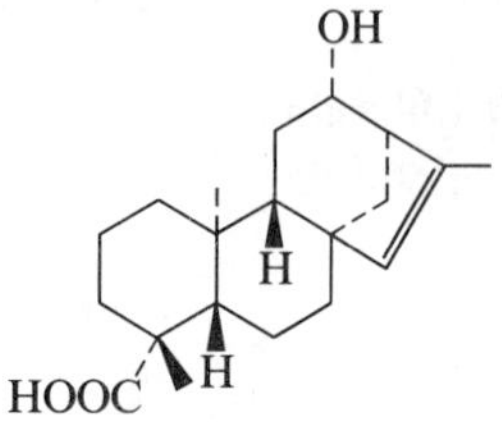

C$_{20}$H$_{30}$O$_3$ M 318.455
Constit. of *Helichrysum davenportii*. Gum. [α]$_D^{24}$ −62° (c, 1.03 in CHCl$_3$).

Jacupovic, J. *et al*, *Phytochemistry*, 1989, **28**, 543.

ent-3β-Hydroxy-16-kauren-18-oic acid   H-80181

C$_{20}$H$_{30}$O$_3$ M 318.455

Ac: ent-3β-*Acetoxy-16-kauren-18-oic acid*. 3-*Acetoxykaurenic acid*
C$_{22}$H$_{32}$O$_4$ M 360.492
Constit. of *Trachylobium verrucosum*.

Ac, Me ester: Mp 130-135°. [α]$_D$ −70° (c, 1.5 in CHCl$_3$).

Hugel, G. *et al*, *Bull. Soc. Chim. Fr.*, 1963, 1974; 1965, 2882, 2888 (*isol, ir, pmr, ms, struct*)

16-Hydroxy-8(17),13-labdadien-15,16-olid-19-oic acid   H-80182

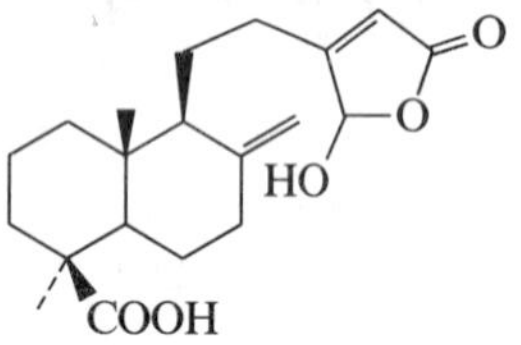

C$_{20}$H$_{28}$O$_5$ M 348.438
Constit. of *Calocedrus formosana*. Cryst. Mp 175-176°.

Fang, J.-M. *et al*, *Phytochemistry*, 1989, **28**, 1173.

15-Hydroxy-7-labden-17-oic acid   H-80183

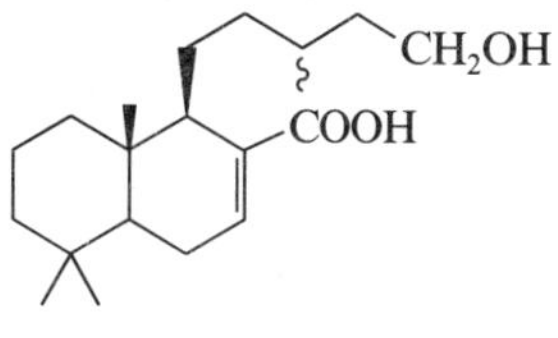

C$_{20}$H$_{34}$O$_3$ M 322.487
Constit. of *Halimium verticillatum*.
Me ether, Me ester: Oil. [α]$_D^{22}$ −21.8° (c, 0.67 in CHCl$_3$).
Ac:
C$_{22}$H$_{36}$O$_4$ M 364.524
Constit of *H. verticillatum*.
O-Z-Cinnamoyl: 15Z-Cinnamoyloxy-7-labden-17-oic acid
C$_{29}$H$_{42}$O$_4$ M 454.648
Constit. of *H. verticillatum*.
O-Z-Cinnamoyl, Me ester: Oil. [α]$_D^{22}$ −25.5° (c, 2.1 in CHCl$_3$).

Urones, J.G. *et al*, *Phytochemistry*, 1989, **28**, 557.

2-Hydroxyligustrin   H-80184

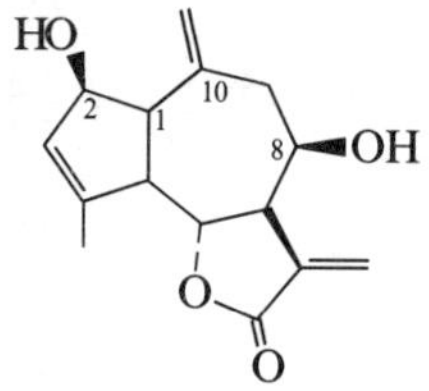

C$_{15}$H$_{18}$O$_4$ M 262.305
2β-form
8-O-(*2-methyl-2,3-epoxybutanoyl*):
C$_{20}$H$_{24}$O$_6$ M 360.406
Constit. of *Trichogonia santosii*. Oil. [α]$_D^{24}$ −31° (c, 0.52 in CHCl$_3$). Diastereoisomeric epoxide also isolated.
2-Ketone, 8-O-(*2-methyl-2,3-epoxybutanoyl*):
C$_{20}$H$_{22}$O$_6$ M 358.390
Constit. of *T. santosii*. Oil. [α]$_D^{24}$ +41° (c, 0.65 in CHCl$_3$).
Δ$^{1(10)}$-*Isomer, 2-ketone, 8-O-(2-methyl-2,3-epoxybutanoyl)*:
8β-(2-methyl-2,3-epoxybutyroyloxy)-dehydroleucodin
C$_{20}$H$_{22}$O$_6$ M 358.390
Constit. of *T.santosii*. Oil. [α]$_D^{24}$ −96° (c, 0.69 in CHCl$_3$).

Bohlmann, F. *et al*, *Justus Liebigs Ann. Chem.*, 1984, 162.

28-Hydroxy-20(29)-lupene-3,7-dione   H-80185
Kanerodione

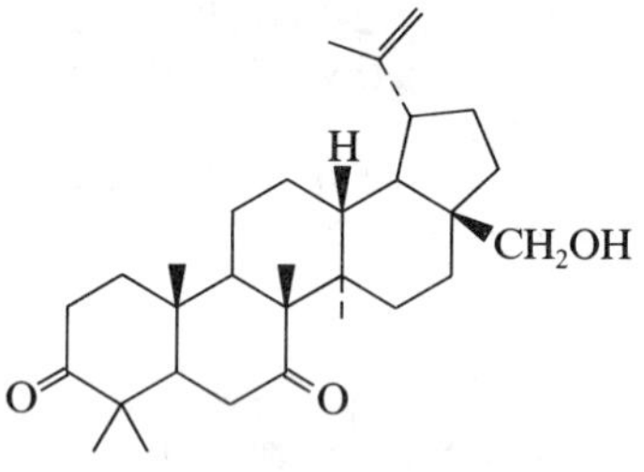

C$_{30}$H$_{46}$O$_3$ M 454.692
Constit. of *Nerium oleander*. Cryst. (EtOAc/C$_6$H$_6$). Mp 178-180°. [α]$_D^{24}$ −36.36° (c, 0.11 in CHCl$_3$).

Siddiqui, S. *et al*, *Phytochemistry*, 1989, **28**, 1187.

3-Hydroxy-20(29)-lupen-11-one   H-80186

C$_{30}$H$_{48}$O$_2$ M 440.708
3α-form
11-*Oxoepilupeol*
Isol. from leaves and twigs of *Fluorensia resinosa*. Cryst. Mp 208°. [α]$_D^{25}$ +34.7°.

Estrada, H. *et al*, *CA*, 1966, **64**, 3617.

7-Hydroxy-20(29)-lupen-3-one   H-80187

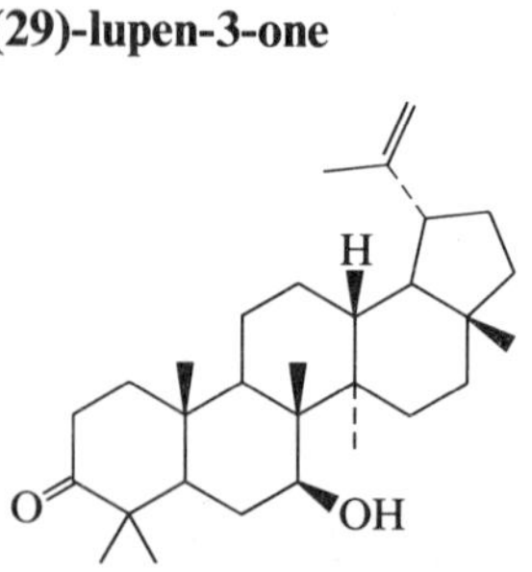

C$_{30}$H$_{48}$O$_2$ M 440.708
7β-form

Constit. of *Salvia pratensis.* Cryst. (MeOH). Mp 217-219°. $[\alpha]_D$ +10.9° (c, 0.8 in MeOH).

Anaya, J. *et al, Phytochemistry*, 1989, **28**, 2206 (*isol, pmr, cmr*)

6-Hydroxy-7-methoxy-2*H*-1-benzopyran-2-one   H-80188

6-Hydroxy-7-methoxycoumarin. **Isoscopoletin**

[776-86-3]

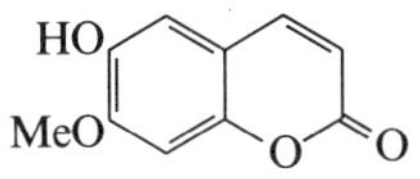

$C_{10}H_8O_4$ M 192.171

Isol. from *Bupleurum fruticosa* and *Artemisia messerschmidiana.* Yellow cryst. (H$_2$O). Mp 185°.

O-β-D-Glucoside: [20186-29-2]. **Magnolioside**
$C_{16}H_{18}O_9$ M 354.313
Isol. from *Magnolia macrophylla.* Mp 227°. $[\alpha]_D$ −28° (Py).

Head, F.S.H. *et al, J. Chem. Soc.*, 1930, 2434 (*isol*)
Plouvier, V., *Compt. Rend. Hebd. Seances Acad. Sci. Sect. D*, 1968, **266**, 1526 (*Magnolioside*)
González, A.G. *et al, An. Quim.*, 1975, **71**, 109 (*isol*)

7-Hydroxy-6-methoxy-2*H*-1-benzopyran-2-one, 9CI   H-80189

Updated Entry replacing H-70153

7-Hydroxy-6-methoxycoumarin. **Scopoletin.** *Aesculetin 6-methyl ether. Chrysatropic acid. Gelseminic acid. β-Methylaesculetin*

[92-61-5]

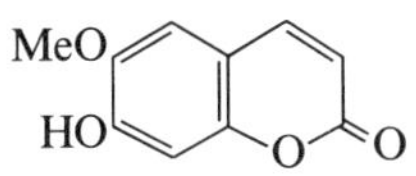

$C_{10}H_8O_4$ M 192.171

Occurs widely in the plant world, for example, the root of *Gelsemium sempervirens, Atropa belladonna, Convolvulus scammonia, Ipomaea orizabensis, Prunus serotina, Fabiana imbricata* and also *Diospyros* spp., *Peucedanum* spp., *Heracleum* spp., *Skimmea* spp. Also occurs in the Chinese crude drug Toki(from *Angelica acutiloba*). Needles or prisms (EtOH). Mp 204°. Reduces Fehling's and Tollen's reagents.

▷ GN6930000.

Ac: Mp 177°.

O-β-D-Glucopyranosyl: [531-44-2]. **Scopolin.** *Murrayin*
$C_{16}H_{18}O_9$ M 354.313
Isol. from *Scopolia japonica, Hedera helix* and others. Mp 217-219°.

O-(6O-β-D-Xylopyranosyl-β-D-glucopyranoside): [18309-73-4]. **Fabiatrin**
$C_{16}H_{18}O_9$ M 354.313
Isol. from leaves of *Fabiana imbricata.* Needles + 2H$_2$O (H$_2$O). Mp 226-228°.

O-(3-Methyl-2-butenyl): [13544-37-1]. *6-Methoxy-7-[(3-methyl-2-butenyl)oxy]coumorin, 8CI*
$C_{15}H_{16}O_4$ M 260.289
Isol. from *Ptaeroxylon obliquum.* Mp 81-82°.

O-(2,3-Dihydroxy-3-methylbutoxy): [75082-72-3]. **Obtusinin.** *7-(2′,3′-Dihydroxy-3′-methylbutyloxy)-6-methoxy-2H-1-benzopyran-2-one*
$C_{15}H_{18}O_6$ M 294.304
Isol. from *Haplophyllum obtusifolium.* Mp 136-138°.

O-(3,7-Dimethyl-2,6-octadienyl): [28587-43-1]. *7-Geranyloxy-6-methoxycoumarin*
$C_{20}H_{24}O_4$ M 328.407
Isol. from *Thapsia gargonica.* Mp 84-84.5°.

Seka, R. *et al, Ber.*, 1931, **64**, 909.
Head, F.G.H. *et al, J. Chem. Soc.*, 1931, 1241 (*synth*)
McCabe, P.H. *et al, J. Chem. Soc. C*, 1967, 145 (3-*Methyl-2-butenyl*)
Larsen, P.K. *et al, Acta Chem. Scand.*, 1970, **24**, 1113 (2,7-*Dimethyl-2,6-octadienyl*)
Karrer, W. *et al, Konstitution und Vorkommen der Organischen Pflanzenstoffe*, 2nd Ed., Birkhäuser Verlag, Basel, 1972-1985, nos. 1328, 1329 (*occur*)
Gonzalez, A.G. *et al, An. Quim.*, 1973, **69**, 1013 (*pmr*)
Tanaka, S. *et al, Arzneim.-Forsch.*, 1977, **27**, 2039 (*isol*)
Herath, W.H.M. *et al, Phytochemistry*, 1978, **17**, 1007 (*isol*)
Ishibura, N. *et al, Z. Naturforsch., C*, 1979, **34**, 628.
Ahluwalia, V.K. *et al, Monatsh. Chem.*, 1982, **113**, 197 (*Obtusinin*)
Abu-Eittah, R.H. *et al, Can. J. Chem.*, 1985, **63**, 1173 (*uv*)
Koul, S.K. *et al, Indian J. Chem., Sect. B*, 1987, **26**, 574 (*synth*)

7-Hydroxy-4′-methoxyisoflavone   H-80190

Updated Entry replacing H-02182

7-Hydroxy-3-(4-methoxyphenyl)-4H-1-benzopyran-4-one, 9CI. **Formononetin.** *Formoononetin. Biochanin B. Neochanin. Pratol*

[485-72-3]

$C_{16}H_{12}O_4$ M 268.268

Constit. of red and subterranean clovers (*Trifolium pratense* and *T. subterraneum*) and of Chana (*Cicer arietinum*). Constit. of the heartwood of *Pterocarpus indicanus.* Found also in *Baptisia australis.* Widely distributed in the Leguminosae (Papilionoideae). Also in *Virola caducifolia* and *V. multinerva* (Myricaceae). Cryst. Mp 265-266° (257°).

▷ Exp. carcinogen.

Ac: Mp 171-172°.

7-O-β-D-Glucoside: [486-62-4]. **Ononin**
$C_{22}H_{22}O_9$ M 430.410
Widely distributed in the Leguminosae subfamily Papilionoideae, eg. in *Amorpha fruticosa, Baptisia* spp., *Cicer arietinum, Cladrastis* spp., *Dalbergia paniculata, Genista patula, Medicago sativa, Ononis* spp., *Piptanthus* spp. *Pueraria thunbergiana, Spartium junceum, Thermopsis* spp., *Trifolium* spp. and *Wisteria floribunda.* Needles (dioxan aq.). Mp 218-219°.

7-O-Laminarabioside: [56222-47-0].
$C_{28}H_{32}O_{14}$ M 592.552
Isol. from *Cladrastis platycarpa* and *C. shikokiana.* Needles. Mp 175-176°.

7-O-Rhammosylglucoside:
$C_{28}H_{32}O_{13}$ M 576.553
Isol. from *B.* spp.

7-O-Rutinoside:
$C_{28}H_{32}O_{13}$ M 576.553
Isol. from bark of *D. paniculata.* Cryst. (C$_6$H$_6$/pet. ether) (as acetate). Mp 120-122° (acetate).

7-O-(6-O-Malonylglucoside): [34232-16-1].
$C_{25}H_{24}O_{12}$ M 516.457
Isol. from leaves of *Trifolium pratense* and *T. subterraneum.* Needles +$\frac{1}{2}$ H$_2$O (MeOH).

7-O-(6-O-Malonylglucoside), Me ester: [34232-18-3].
$C_{26}H_{26}O_{12}$ M 530.484
Isol. from leaves of *T. pratense* and *T. subterraneum.* Needles (MeOH). Mp 198°.

Bradbury, R.B., *J. Chem. Soc.*, 1951, 3447 (*isol*)

Bose, J.L., *J. Sci. Ind. Res., Sect. B*, 1951, **10**, 291 (*isol, struct*)
Baker, W. *et al, J. Chem. Soc.*, 1953, 1852 (*synth*)
Cooke, R.G., *Aust. J. Chem.*, 1964, **17**, 379 (*isol, struct*)
Lebreton, P. *et al, Phytochemistry*, 1967, **6**, 1675 (*isol*)
Benn, M.H., *Can. J. Chem.*, 1970, **48**, 1624 (*isol*)
Markham, K.R. *et al, Phytochemistry*, 1970, **9**, 2359
    (*Rhamnosylglucoside*)
Beck, A.B. *et al, Aust. J. Chem.*, 1971, **24**, 1509 (*malonyl derivs*)
Imamura, H. *et al, Phytochemistry*, 1974, **13**, 757
    (*Laminarabioside*)
Parthusarathy, M.R. *et al, Phytochemistry*, 1976, **15**, 1025
    (*Rutinoside*)
Braz Filho, R. *et al, J. Nat. Prod. (Lloydia)*, 1977, **40**, 236 (*isol*)
Ingham, J.L., *Fortschr. Chem. Org. Naturst.*, 1983, **43**, 1 (*rev,
    occur*)
Al-Ani, H.A.M. *et al, J. Chem. Soc., Perkin Trans. 1*, 1984, 2831
    (*biosynth*)
Sax, N.I., *Dangerous Properties of Industrial Materials*, 6th Ed.,
    Van Nostrand-Reinhold, 1979, 736.

6-Hydroxy-7-methoxy-2-methyl-2-(4-methyl-3-pentenyl)-2*H*-1-benzopyran   H-80191

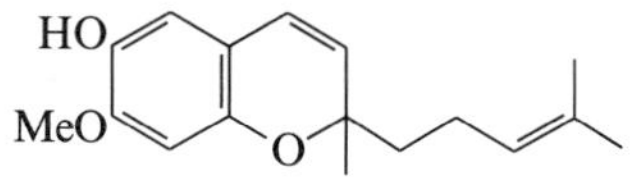

$C_{17}H_{22}O_3$ M 274.359
Metabolite of *Amaroucium multiplicatum*. Oil.

Sato, A. *et al, J. Nat. Prod. (Lloydia)*, 1989, **52**, 975 (*isol, pmr,
    cmr, synth*)

5-Hydroxy-7-methoxy-2-pentyl-4*H*-1-benzopyran-4-one   H-80192

5-Hydroxy-7-methoxy-2-pentylchromone

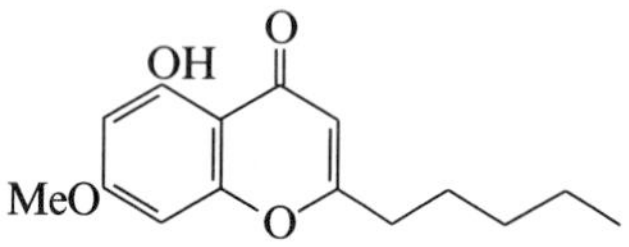

$C_{15}H_{18}O_4$ M 262.305
Constit. of *Zanthoxylum microcarpum* and *Z. valens*. Cryst.
    (EtOH). Mp 55-57°.

Jiménez, C. *et al, Phytochemistry*, 1989, **28**, 1992.

1-(4-Hydroxy-3-methoxyphenyl)-2-[4-(3-hydroxy-1-propenyl)-2-methoxyphenoxy]-1,3-propanediol, 9CI   H-80193

Guaiacylglycerol β-coniferyl ether
[1103-58-8]

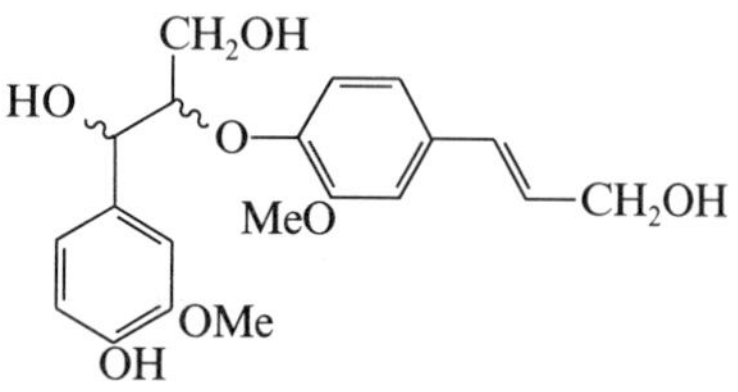

$C_{20}H_{24}O_7$ M 376.405
Prod. of lignin degradation. Isol. from wood of *Picea
    excelsa* and *Larix leptolepsis*. Syrup. $[\alpha]_D^{22}$ −6.7° (c, 0.53
    in MeOH).

Freudenberg, K. *et al, Chem. Ber.*, 1955, **88**, 617, 626 (*synth, uv,
    ir*)
Nimz, H., *Chem. Ber.*, 1965, **98**, 533 (*isol*)
Miki, K. *et al, Phytochemistry*, 1980, **19**, 449 (*isol, pmr*)

2-(Hydroxymethyl)aziridine   H-80194

2-Aziridinemethanol, 9CI
[88419-36-7]

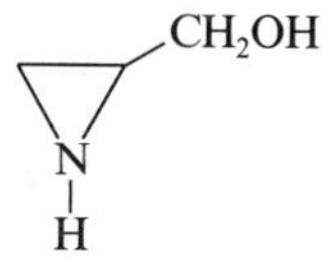

C_3H_7NO M 73.094
(±)-*form*
    Liq. Bp$_{0.005}$ 45° (bath) .

Trapensier, P.T. *et al, Chem. Heterocycl. Compd., Engl. Transl.*,
    1983, **19**, 982 (*synth, pmr*)
Sehgal, R.K. *et al, J. Med. Chem.*, 1987, **30**, 1626 (*synth, pmr*)

3-Hydroxy-5-methylbenzaldehyde   H-80195

Updated Entry replacing H-02277
    5-Hydroxy-m-tolualdehyde. 3,5-Cresotaldehyde. m-*Cresol-5-
    aldehyde. 3-Formyl-5-methylphenol*
[60549-26-0]
$C_8H_8O_2$ M 136.150
Cryst. (CHCl$_3$). Mp 107-108°.

Me ether: [90674-26-3]. *3-Methoxy-5-methylbenzaldehyde*
    $C_9H_{10}O_2$ M 150.177
    Liq. Bp$_{10}$ 95°. n_D^{20} 1.5475.

Brown, P.M. *et al, J. Chem. Soc., Perkin Trans. 1*, 1976, 997
    (*synth*)
Paisdor, B. *et al, Chem. Ber.*, 1988, **121**, 1307 (*deriv*)

5-(Hydroxymethyl)-1,2,3-benzenetriol   H-80196

Updated Entry replacing H-02308
    α,3,4,5-*Tetrahydroxytoluene. 3,4,5-Trihydroxybenzyl
    alcohol. Gallyl alcohol*
[68325-64-4]
$C_7H_8O_4$ M 156.138
Unstable, readily polymerises.

*2-O-β-*D-*Glucopyranoside:* **MP-10**
    $C_{13}H_{18}O_9$ M 318.280
    Isol. from chestnut galls. Hygroscopic powder.

*1′-O-(3,4,5-Trihydroxybenzoyl), 2-O-β-*D-*glucopyranoside:*
    MP-2
    $C_{20}H_{22}O_{13}$ M 470.386
    Isol. from chestnut galls. Needles. Mp 146-148°.

1,2,3-Tri-Me ether: *3,4,5-Trimethoxybenzyl alcohol*
    $C_{10}H_{14}O_4$ M 198.218
    Viscous oil. Bp$_{2-3}$ 160-165°.

1,2,3-Tri-Me ether, 1′-(3,5-dinitrobenzoyl): Mp 143-144°.

Rosemund, K.W. *et al, Ber.*, 1922, **55**, 2369 (*synth*)
Rosemund, K.W. *et al, Arch. Pharm. (Weinheim, Ger.)*, 1927, **264**,
    448; *CA*, **21**, 2886 (*synth*)
Drake, N.L. *et al, J. Am. Chem. Soc.*, 1955, **77**, 1204 (*synth*)
Ozawa, T. *et al, Agric. Biol. Chem.*, 1977, **41**, 1257 (*glycosides*)

2-(Hydroxymethyl)-1,4-butanediol, 9CI   H-80197

Tris(hydroxymethyl)ethane
[6482-32-2]

$$HOCH_2CH_2CH(CH_2OH)_2$$

$C_5H_{12}O_3$ M 120.148
Liq.

Harnden, M.R. *et al, J. Med. Chem.*, 1987, **30**, 1636 (*synth, pmr*)

2-Hydroxy-3-methyl-2-butenenitrile, 9CI   H-80198

1-Cyano-2-methyl-1-propen-1-ol

[116130-35-9]

$$(H_3C)_2C{=}C(OH)CN$$

C_5H_7NO M 97.116

Enol form of 3-Methyl-2-oxobutanenitrile, (see 3-Methyl-2-oxobutanoic acid, M-80129). Isolable enol. Cryst. (pentane). Mp 30-32°.

Trifluoroacetyl: [116130-34-8].
   $C_7H_6F_3NO_2$ M 193.125
   Mp −2°. $Bp_{\sim 3}$ 40°.

Jaroszewski, J.W. *et al, J. Org. Chem.*, 1988, **53**, 4635 (*synth, uv, pmr, cmr, ir*)

5-(Hydroxymethyl)-2-cyclopenten-1-one, 9CI   H-80199

[76909-98-3]

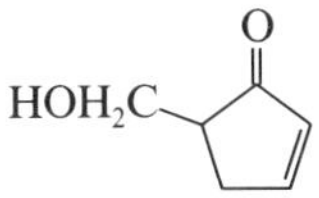

$C_6H_8O_2$ M 112.128

(±)-*form* [110864-96-5]
   Yellow oil.

Paulsen, H. *et al, Chem. Ber.*, 1981, **114**, 346.
Biggadike, K. *et al, J. Chem. Soc., Perkin Trans. 1*, 1988, 549 (*synth, ir, pmr*)

16-Hydroxy-24-methyl-12,24-dioxo-25-scalaranal   H-80200

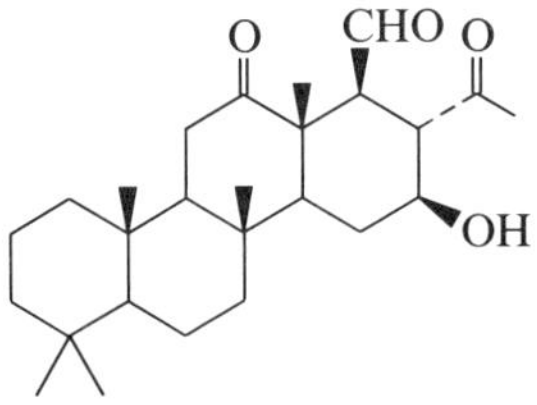

$C_{26}H_{40}O_4$ M 416.600

16β-form

Ac: [116331-46-5]. *16β-Acetoxy-24-methyl-12,24-dioxo-25-scalaranal*
   $C_{28}H_{42}O_5$ M 458.637
   Constit. of *Carteriospongia flabellifera*. Cryst. Mp 227-229°. $[\alpha]_D$ +135° (c, 0.74 in $CHCl_3$).

Schmitz, F.J. *et al, J. Nat. Prod. (Lloydia)*, 1988, **51**, 745.

(Hydroxymethyl)dodecahedrane   H-80201

Tetradecahydro-5,2,1,6,3,4-[2,3]butanediyl[1,4]diylidenepentaleno[2,1,6-cde:2′,1′,6′-gha]pentalene-1(2H)-methanol, 9CI. Dodecahedranemethanol

[112896-32-9]

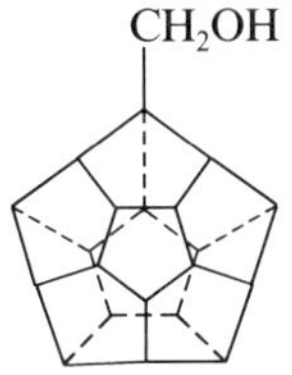

$C_{21}H_{22}O$ M 290.404
Solid. Mp >250°.

Paquette, L.A. *et al, J. Am. Chem. Soc.*, 1988, **110**, 1303 (*synth, cmr*)
Weber, J.C. *et al, J. Org. Chem.*, 1988, **53**, 5315 (*synth, pmr, cmr*)

2-Hydroxy-4,5-methylenedioxybenzoic acid   H-80202

Updated Entry replacing H-60181
   6-Hydroxy-1,3-benzodioxole-5-carboxylic acid, 9CI. 6-Hydroxypiperonylic acid, 8CI

$C_8H_6O_5$ M 182.132
Cryst. ($CHCl_3$). Mp 238° (216-217°).

Me ether: [7168-93-6]. *6-Methoxy-1,3-benzodioxole-5-carboxylic acid, 9CI. 6-Methoxy-3,4-methylenedioxybenzoic acid*
   $C_9H_8O_5$ M 196.159
   Needles ($CHCl_3$). Mp 153-153.2°.

Me ether, chloride:
   $C_9H_7ClO_4$ M 214.605
   Mp 118-121°.

Amide: [120781-07-9].
   $C_8H_7NO_4$ M 181.148
   Cryst. (C_6H_6). Mp 243-245°.

Dallacker, F. *et al, Justus Liebigs Ann. Chem.*, 1966, **694**, 98 (*deriv*)
Moron, J. *et al, Bull. Soc. Chim. Fr.*, 1967, 130 (*synth*)
Stout, G.H. *et al, Tetrahedron*, 1969, **25**, 5295 (*deriv*)
Rall, G.J.H. *et al, Tetrahedron*, 1970, **26**, 5007 (*deriv*)
Dallacker, F. *et al, Z. Naturforsch., C*, 1978, **33**, 465 (*deriv*)
Sartori, G. *et al, Synthesis*, 1988, 783 (*synth, uv, ir, ms, pmr*)

7-Hydroxy-3′,4′-methylenedioxyisoflavone   H-80203

Updated Entry replacing B-60006
   3-(1,3-Benzodioxol-5-yl)-7-hydroxy-4H-1-benzopyran-4-one, 9CI. 7-Hydroxy-3-(3,4-methylenedioxyphenyl)chromone. Pseudobaptigenin. ψ-Baptigenin

[90-29-9]

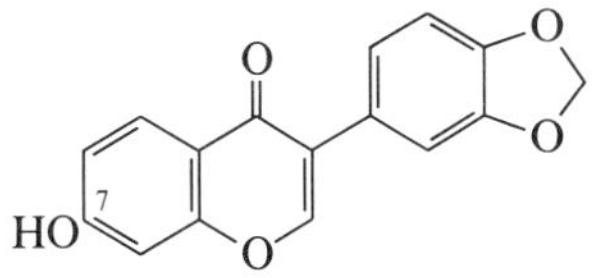

$C_{16}H_{10}O_5$ M 282.252
Isol. from *Baptisia* spp., *Cladrastis platycarpa, C. shikokiana*, several *Dalbergia* spp., *Maackia amurensis, Pisum sativum, Pterocarpus erinaceus, Trifolium hybridum* and *T. pratense*. Mp 296-298°. λ_{max} 226 (log ε 4.4), 248 sh and 294 nm (4.4) (MeOH).

Ac: Mp 173°.

O-β-D-Glucopyranoside: [63347-43-3]. *Rothindin*
   $C_{22}H_{20}O_{10}$ M 444.394
   Isol. from *Rothia indica, Cladrastis platycarpa, C. shikokiana, Ononis spinosa* and *Trifolium pratense*. Needles (MeOH). Mp 236-237°. $[\alpha]_D^{32}$ −37.5° (Py).

O-D-Glucosylrhamnoside: ψ-*Baptisin. Pseudobaptisin*
   $C_{22}H_{20}O_{10}$ M 444.394
   Isol. from *B. tinctoria* and other *B.* sp. Cryst. + $3H_2O$. Mp 148-150° (resolidifies at 180-210° and remelts at 249-251°).

O-Laminarabioside:
   $C_{28}H_{30}O_{15}$ M 606.536
   Isol. from *Cladrastis platycarpa* and *C. shikokiana*. Tentative identification.

Me ether: [4253-04-7]. *7-Methoxy-3′,4′-methylenedioxyisoflavone. Pseudobaptigenin methyl ether*
   $C_{17}H_{12}O_5$ M 296.279

Isol. from seeds of *Calopogonium mucunoides*. Mp 180-182°.

O-(*3-Methyl-2-butenyl*): **Maximaisoflavone B**
$C_{21}H_{18}O_5$ M 350.370
Constit. of *Tephrosia maxima*. Cryst. (C_6H_6/pet. ether). Mp 132-133° (126-128°).

Schmidt, O. *et al*, *Monatsh. Chem.*, 1929, **53**, 454 (*isol, struct*)
Baker, W. *et al*, *J. Chem. Soc.*, 1953, 1582 (*synth*)
Kukla, A.S. *et al*, *Tetrahedron*, 1962, **18**, 1443 (*deriv*)
Bevan, C.W.L. *et al*, *J. Chem. Soc. C*, 1966, 509 (*pmr*)
Dhoubhadel, S.P. *et al*, *Indian J. Chem., Sect. B*, 1975, **52**, 440 (*synth*)
Nair, A.G.R. *et al*, *Indian J. Chem., Sect. B*, 1976, **14**, 801 (*Rothindin*)
Ingham, J.L., *Fortschr. Chem. Org. Naturst.*, 1983, **43**, 1 (*rev, occur*)
Schuda, P.F. *et al*, *J. Org. Chem.*, 1987, **52**, 1972 (*synth*)

Hydroxymethylenepropanedioic acid, 9CI   H-80204

Updated Entry replacing H-02414
Hydroxymethylenemalonic acid, 8CI. Formylmalonic acid, 8CI

$$OHCCH(COOH)_2 \rightleftharpoons HOCH{=}C(COOH)_2$$

$C_4H_4O_5$ M 132.073
Enol-form [50850-50-5]
The free acid has not been isol.

Ba salt: Cryst. Mp 119°.

Mono-Me ester: Methyl hydrogen 2-formylmalonate
$C_5H_6O_5$ M 146.099
Prisms. Mp 38-40°.

Monobenzyl ester: Benzyl hydrogen 2-formylmalonate
$C_{11}H_{10}O_5$ M 222.197
Needles (hexane). Mp 54-56°.

Di-Et ester: [20734-18-3].
$C_8H_{12}O_5$ M 188.180
Liq. Bp_{12} 107-109°.
Aldo-form
The free acid has not been isol.

Di-Et ester, Di-Et acetal: [7251-32-3]. *Diethyl diethoxymethylmalonate*
$C_{12}H_{22}O_6$ M 262.302
Liq. Bp_{50} 70°.

v. Auers, K., *Justus Liebigs Ann. Chem.*, 1918, **415**, 169 (*ester, synth*)
Fuson, R.C. *et al*, *J. Org. Chem.*, 1946, **11**, 194 (*synth*)
Vinokurov, V.G. *et al*, *Zh. Fiz. Khim.*, 1968, **42**, 1052 (*spectra*)
Kabusz, S. *et al*, *Synthesis*, 1971, 312 (*acetal*)
Matrosov, E.I. *et al*, *Spectrochim. Acta, Part A*, 1972, **28**, 191 (*spectra*)
Alaimo, R.J., *J. Heterocycl. Chem.*, 1973, **10**, 769 (*synth*)
Sato, M. *et al*, *Chem. Pharm. Bull.*, 1989, **37**, 665 (*monoesters*)

7-Hydroxy-2-methylisoflavone   H-80205

7-Hydroxy-2-methyl-3-phenyl-4H-1-benzopyran-4-one, 9CI
[2859-88-3]

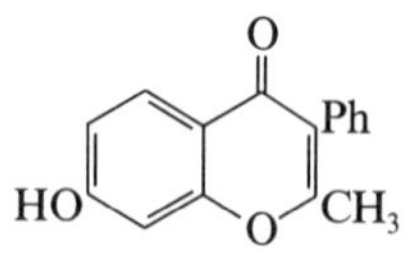

$C_{16}H_{12}O_3$ M 252.269

Isol. from roots of *Glycyrrhiza glabra*. Mp 240°.
Ac: [3211-63-0].
$C_{18}H_{14}O_4$ M 294.306
Isol. from *G. glabra*. Mp 161-162°.

Me ether: [19725-44-1]. *7-Methoxy-2-methylisoflavone*
$C_{17}H_{14}O_3$ M 266.296
Isol. from *G. glabra*. Mp 142-143°.

Woods, L.L. *et al*, *CA*, 1965, **62**, 6454f (*synth*)
Bhardwaj, D.K. *et al*, *Phytochemistry*, 1976, **15**, 352 (*isol*)

3-Hydroxymethyl-2-methylfuran   H-80207

2-Methyl-3-furanmethanol
[5554-99-4]
$C_6H_8O_2$ M 112.128
Liq. Bp_7 70°. n_D^{16} 1.4876.
Ac:
$C_7H_{10}O_3$ M 142.154
Bp_{13} 77-79°.

Janda, M. *et al*, *Collect. Czech. Chem. Commun.*, 1964, **29**, 1731 (*deriv, synth*)
Valenta, M. *et al*, *Collect. Czech. Chem. Commun.*, 1966, **31**, 2410 (*synth*)
Scarpa, J.S. *et al*, *Helv. Chim. Acta*, 1966, **49**, 858 (*synth*)
Mann, J. *et al*, *Tetrahedron*, 1987, **43**, 2533 (*synth*)

3-(5-Hydroxymethyl-5-methyl-2-oxo-5H-furan-3-yl)-2-methylpropanoic acid   H-80208

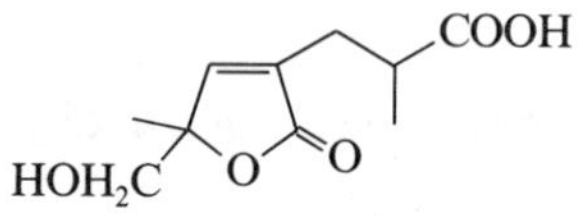

$C_{10}H_{14}O_5$ M 214.218
Constit. of *Crotalaria verrucosa*. Cryst. (C_6H_6). Mp 93°.

Suri, O.P. *et al*, *J. Nat. Prod.* (*Lloydia*), 1989, **52**, 178 (*isol, pmr, cmr*)

3-Hydroxy-2-methylpentanoic acid, 9CI   H-80209

Updated Entry replacing H-60192
3-Hydroxy-2-methylvaleric acid, 8CI
[28892-73-1]

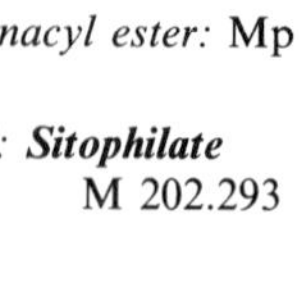

$C_6H_{12}O_3$ M 132.159
(2R,3S)-form [77405-43-7]
Degradn. prod. of Mycobactin. Oil. $Bp_{0.1}$ 90-100°. $[\alpha]_D^{25}$ −14.8° (c, 4.3 in MeOH).

4-Bromophenacyl ester: Mp 89.5-90°. $[\alpha]_D^{18}$ −15° (c, 2.5 in MeOH).

3-Pentanoyl: **Sitophilate**
$C_{11}H_{22}O_3$ M 202.293

Aggregation pheromone of granary weevil *Sitophilus granarius* (stereochem. of pheromone not detd.). Liq. $[\alpha]_D^{24}$ +3.0° (c, 1.5 in CHCl$_3$)(synthetic).

(2*S*,3*S*)-form

Me ester: Bp$_5$ 69-70°. $[\alpha]_D^{23}$ +12.2° (c, 1.12 in CHCl$_3$).

(2*RS*,3*SR*)-form

Me ester: [67498-21-9].
C$_7$H$_{14}$O$_3$ M 146.186
Oil. Bp$_4$ 68-70°.

[60665-94-3, 67498-06-0, 78655-79-5, 78655-82-0, 84277-55-4, 100992-75-4]

Snow, G.A., *J. Chem. Soc.*, 1954, 4080 (*isol, synth*)
Snow, G.A., *Biochem. J.*, 1965, **94**, 160 (*abs config*)
Kirmse, W. *et al, Justus Liebigs Ann. Chem.*, 1976, 1333 (*synth*)
Aten, R.W. *et al, Synthesis*, 1978, 400 (*synth*)
Heathcock, C.H. *et al, J. Org. Chem.*, 1979, **44**, 4294 (*cmr*)
Mori, K. *et al, Tetrahedron*, 1986, **42**, 4685 (*synth, ir, pmr*)
Phillips, J.K. *et al, Tetrahedron Lett.*, 1987, **28**, 6145 (*Sitophilate*)
Chong, J.M., *Tetrahedron*, 1989, **45**, 623 (*Sitophilate*)

3-Hydroxy-3-methylpentanoic acid, 9CI   H-80210

3-Hydroxy-3-methylvaleric acid. β-Methyl-β-ethylhydracrylic acid

[150-96-9]

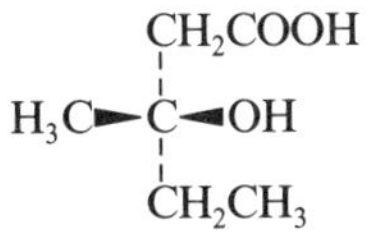

$$CH_2COOH$$
$$H_3C-C-OH$$
$$CH_2CH_3$$ (*R*)-form

C$_6$H$_{12}$O$_3$ M 132.159

(*R*)-form [36567-73-4]
Oil. $[\alpha]_D$ +1.07° (c, 2.25).

(±)-form
Isol. from Turkish tobacco leaves. Liq. Bp$_{19}$ 142-144°.

Et ester: [31033-23-5].
C$_8$H$_{16}$O$_3$ M 160.213
Liq. Bp$_{19}$ 88-100°.

[114926-67-9]

Pocrowsky, A., *J. Prakt. Chem.*, 1900, **62**, 301 (*synth*)
Bohnsack, H. *et al, Ber.*, 1941, **74**, 1575 (*ester, synth*)
Papa, D. *et al, J. Am. Chem. Soc.*, 1954, **76**, 4441 (*synth*)
Kenyon, J. *et al, J. Chem. Soc.*, 1954, 2129 (*synth*)
Fukuzumi, T. *et al, Agric. Biol. Chem.*, 1966, **30**, 513 (*isol, ir*)
Seebach, P. *et al, J. Am. Chem. Soc.*, 1988, **110**, 4763 (*synth, ir, pmr, cmr, ms*)

5-Hydroxy-3-methyl-2-pentenoic acid, 9CI   H-80211

Δ^2-*Anhydromevalonic acid*

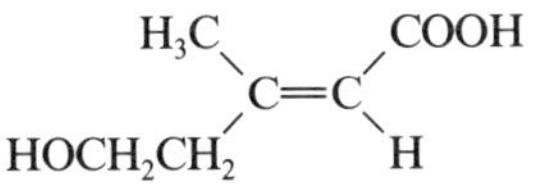

$$H_3C \quad COOH$$
$$C=C$$
$$HOCH_2CH_2 \quad H$$ (*E*)-form

C$_6$H$_{10}$O$_3$ M 130.143
Metabolite of *Fusarium* and other fungi.

(*E*)-form [19710-84-0]
Component of Ferrirubin. Cryst. (C$_6$H$_6$). Mp 64-65°. pK_a 7.42.

Me ester: [35066-36-5].
C$_7$H$_{12}$O$_3$ M 144.170
Oil. Bp$_9$ 115°.

(*Z*)-form [17880-06-7]
Component of Ferrirhodin. Prisms (C$_6$H$_6$). Mp 72-73°.

Me ester: [32775-50-1].
Oil. Bp$_{15}$ 114°.

Lactone: [2381-87-5]. *5,6-Dihydro-4-methyl-2H-pyran-2-one, 9CI. Anhydromevalonolactone*

C$_6$H$_8$O$_2$ M 112.128
Key intermediate for synthesis of pheromones. Oil. Bp$_{14}$ 118-120°, Bp$_{0.4}$ 65°. n_D^{21} 1.4840.

Cornforth, J.W. *et al, Biochem. J.*, 1958, **69**, 146 (*synth*)
Keller-Schierlein, W., *Helv. Chim. Acta*, 1963, **46**, 1920 (*isol*)
Dieckmann, H., *Arch. Microbiol.*, 1968, **62**, 322 (*isol*)
Keller-Schierlein, W. *et al, Helv. Chim. Acta*, 1972, **55**, 198 (*synth*)
Widmer, J. *et al, Helv. Chim. Acta*, 1974, **57**, 1904 (*synth*)
Bonadies, F. *et al, J. Org. Chem.*, 1984, **49**, 1647 (*synth, lactone*)

2-Hydroxymethyl-1,3-propanediol, 9CI   H-80212

Tris(hydroxymethyl)methane

[4704-94-3]

$$CH(CH_2OH)_3$$

C$_4$H$_{10}$O$_3$ M 106.121
Chunky prisms (Me$_2$CO). Mp 67-68° (60-63°). Bp$_{2.5}$ 130-135° subl.

Tribenzoyl: [113967-50-3].
C$_{25}$H$_{22}$O$_6$ M 418.445
Needles (hexane). Mp 75-76°.

Tris(methanesulfonyl): Cryst. (EtOH aq.). Mp 83-84°.

Dekmizian, A.H. *et al, Synth. Commun.*, 1979, **9**, 431 (*synth, ir, pmr*)
Nielsen, A.T. *et al, Pol. J. Chem. (Rocz. Chem.)*, 1981, **55**, 1393 (*synth, pmr, cmr*)
Latour, S. *et al, Synthesis*, 1987, 742 (*synth, ms, ir, pmr*)

6-(Hydroxymethyl)-2,4(1*H*,3*H*)pteridinedione   H-80213

6-Hydroxymethyllumazine, 8CI

[10129-99-4]

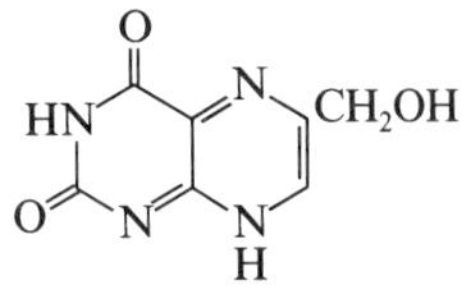

C$_7$H$_6$N$_4$O$_3$ M 194.149
Isol. from leaves of *Spinacia oleracea*. Mp 260-262° dec.

Sugiura, K. *et al, CA*, 1966, **65**, 15375, 20508 (*isol, uv, synth, struct*)

3-Hydroxy-2-methyl-4(1*H*)-pyridinone, 9CI   H-80214

3,4-Dihydroxy-2-methylpyridine

[17184-19-9]

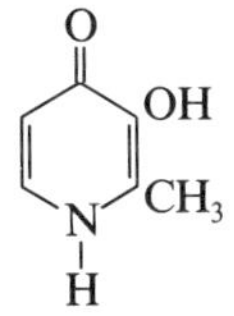

C$_6$H$_7$NO$_2$ M 125.127
Cryst. (H$_2$O). Mp 285-288°.

B,HCl: [17184-18-8].
Prisms (EtOAc/MeOH). Mp 185°.

O-Ac: [17184-20-2]. *3-Acetoxy-2-methyl-4(1H)-pyridinone*
C$_8$H$_9$NO$_3$ M 167.164
Cryst. (EtOAc). Mp 205-207°.

Me ether: [76015-11-7]. *3-Methoxy-2-methyl-4(1H)-pyridinone, 9CI*
C$_7$H$_9$NO$_2$ M 139.154
Cryst. (Me$_2$CO/MeOH). Mp 155-156°.

Benzyl ether: [61160-18-7].

$C_{13}H_{13}NO_2$ M 215.251
Prisms (EtOH). Mp 162-163°.

O^3-β-D-Glucopyranosyl: [15266-34-9]. *Innovanamine*
$C_{12}H_{17}NO_7$ M 287.269
Isol. from leaves of *Evodiopanax innovans*. Prisms +
$2H_2O(H_2O)$. Mp 116-118°. $[\alpha]_D^{25}$ −61.6° (c, 10 in H_2O).

Yasue, M. *et al, Chem. Pharm. Bull.*, 1966, **14**, 1443 (*isol*)
Harris, R.L.N., *Aust. J. Chem.*, 1976, **29**, 1329 (*synth*)
Hwang, D.R. *et al, J. Pharm. Sci.*, 1980, **69**, 1074 (*synth*)

3-Hydroxymethyl-7,11,15-trimethyl-2,6,10-hexadecatriene-1,14,15-triol   H-80215

[119877-40-6]

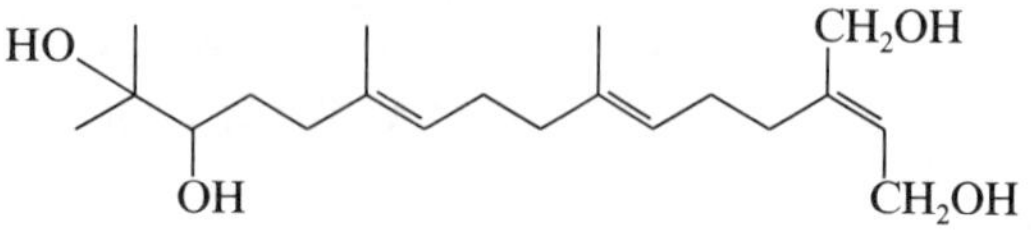

$C_{20}H_{36}O_4$ M 340.502
Constit. of *Tithonia pedunculata*. Oil. $[\alpha]_{241}$ +15.67° (c, 0.185 in MeOH).

Pérez, A.-L. *et al, Phytochemistry*, 1988, **27**, 3897.

ent-1β-Hydroxy-3-nor-15-beyerene-2,12-dione   H-80216

Yucalexin B'-11

[119642-80-7]

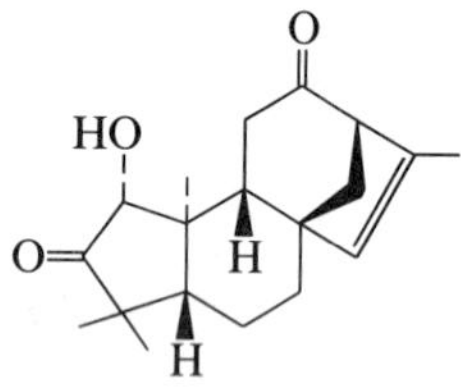

$C_{19}H_{26}O_3$ M 302.413
Constit. of cassava roots (*Manihot esculenta*).

Sakai, T. *et al, Phytochemistry*, 1988, **27**, 3769.

11-Hydroxy-20-nor-8,15-isopimaradien-7-one   H-80217

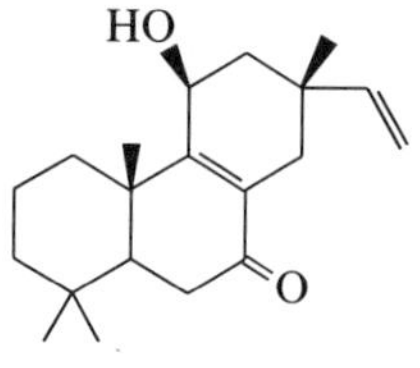

$C_{19}H_{28}O_2$ M 288.429
11β-form [83037-63-2]
Constit. of *Vellozia pusilla*. Cryst. (EtOAc/hexane). Mp 158-160°. $[\alpha]_D^{24}$ +116.1° (c, 0.58 in $CHCl_3$).
11-Ketone: [83037-65-4]. *20-Nor-8,15-isopimaradiene-7,11-dione*
$C_{19}H_{26}O_2$ M 286.413
Constit. of *V. pusilla*. Cryst. (hexane). Mp 80-82°.

Pinto, A.C. *et al, Phytochemistry*, 1988, **27**, 3913.

3-Hydroxy-30-nor-12,18-oleanadien-29-oic acid   H-80218

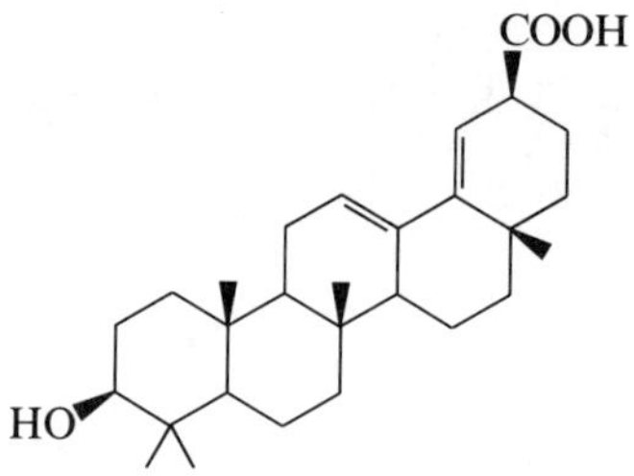

$C_{29}H_{44}O_3$ M 440.665
3β-form
Constit. of *Boussingaultia gracilis*.
Et ester: Cryst. (MeOH). Mp 107-108°.

Lin, H.-Y. *et al, J. Nat. Prod. (Lloydia)*, 1988, **51**, 797.

3-Hydroxy-16-nor-7-pimaren-15-oic acid   H-80219

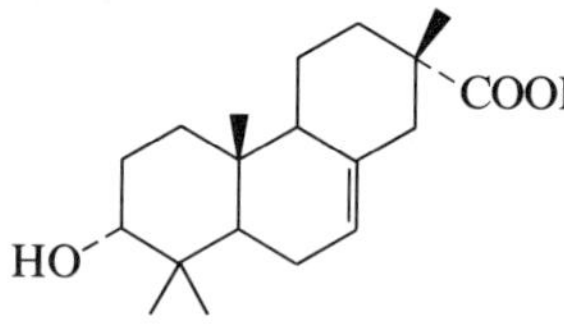

$C_{19}H_{30}O_3$ M 306.444
3α-form
Constit. of *Agathis vitiensis*. Cryst. (C_6H_6). Mp 208-210°.
$[\alpha]_D^{18}$ −10° (c, 1.1 in $CHCl_3$).

Cambie, R.C. *et al, Phytochemistry*, 1989, **28**, 1675.

17-Hydroxyoctadecanoic acid   H-80220

Updated Entry replacing H-02942
17-Hydroxystearic acid
[4552-19-6]

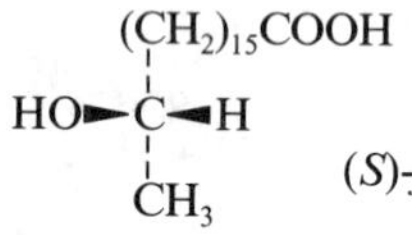

$$(CH_2)_{15}COOH$$
$$HO{-}C{-}H$$
$$CH_3 \qquad (S)\text{-}form$$

$C_{18}H_{36}O_3$ M 300.481
Occurs in beeswax (probably in an enantiomeric form).
(S)-form
Found as sophorose glycoside in *Torulopsis magnoliae*.
Cryst. (Me_2CO). Mp 78-80°. $[\alpha]_D$ +4.4° (c, 7.9 in AcOH).
(±)-form
Mp 76.4-76.6°, Mp 81-81.4° (dimorph.).
Me ester: [2380-14-5].
$C_{19}H_{38}O_3$ M 314.507
Mp 52.0-52.5°, Mp 64.1-64.3° (dimorph.).

Bergström, S. *et al, Acta Chem. Scand.*, 1952, **6**, 1157 (*synth*)
Gorin, P.A.J. *et al, Can. J. Chem.*, 1961, **39**, 846 (*isol*)
Tulloch, A.P., *Chem. Phys. Lipids*, 1971, **6**, 235 (*isol*)
Hammerstrom, S. *et al, Anal. Biochem.*, 1973, **52**, 169 (*resoln*)
Tulloch, A.P. *et al, Org. Magn. Reson.*, 1978, **11**, 109 (*cmr*)

3-Hydroxy-28,13-oleananolide   H-80221

Updated Entry replacing H-60200
$C_{30}H_{48}O_3$ M 456.707
(3β,13β)-form
Oleanolic lactone

Constit. of *Salvia lanigera*.

Ac: 3β-Acetoxy-28,13β-*oleanolide*
$C_{32}H_{50}O_4$ M 498.745
Constit. of *Hyptis mutabilis*. Needles (CHCl₃). Mp 282°.

Bellavista, V. *et al, Phytochemistry*, 1974, **13**, 289 (*isol*)
Al-Hazimi, H.M.G. *et al, Phytochemistry*, 1987, **26**, 1091 (*isol*)
Pereda-Miranda, R. *et al, J. Nat. Prod. (Lloydia)*, 1988, **51**, 996 (*isol*)

3-Hydroxy-12-oleanen-30-oic acid   H-80222

$C_{30}H_{48}O_3$ M 456.707
3β-form [564-16-9]
   11-*Deoxoglycyrrhetinic acid*
   Cryst. (AOH). Isol. from roots of *Glycyrrhiza glabra*.
   Cryst. (AcOH). Mp 330°. $[\alpha]_D$ +148° (c, 0.473 in CHCl₃).

Canonica, L. *et al, Gazz. Chim. Ital.*, 1966, **96**, 833 (*isol, struct*)
Chakrabarti, P. *et al, J. Indian Chem. Soc.*, 1969, **46**, 626 (*synth*)

16-Hydroxy-18-oleanen-3-one   H-80223

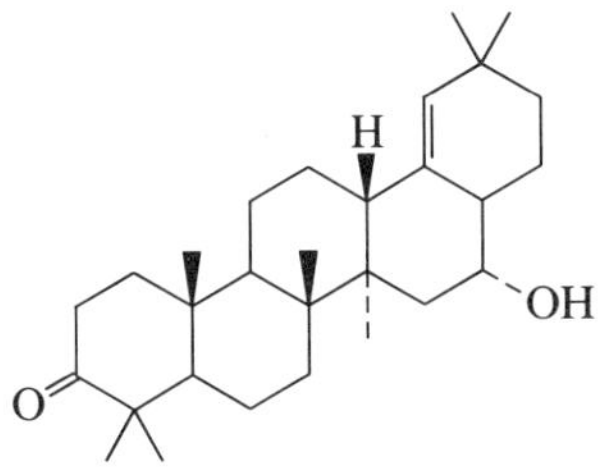

16α-*form*

$C_{30}H_{48}O_2$ M 440.708
16α-form
   Constit. of *Schaefferia cuneifolia*.
   Ac: Cryst. (CHCl₃). Mp 208-210°. $[\alpha]_D^{20}$ +31.5° (c, 0.14 in CHCl₃).
16β-form
   Constit. of *S. cuneifolia*.
   Ac: Cryst. (CHCl₃). Mp 210-212°. $[\alpha]_D^{20}$ +37° (c, 0.13 in CHCl₃).

Gonzalez, A.G. *et al, J. Nat. Prod. (Lloydia)*, 1989, **52**, 567 (*isol, pmr, cmr*)

3-Hydroxy-12-oleanen-1-one   H-80224

[120374-29-0]
$C_{30}H_{48}O_2$ M 440.708
3α-form
   Constit. of *Randia dumetorum*. Needles (MeOH). Mp 268-270°. $[\alpha]_D$ +68°.

Murty, Y.L.N. *et al, Phytochemistry*, 1989, **28**, 276.

4-Hydroxy-10(15)-oplopen-3-one   H-80225

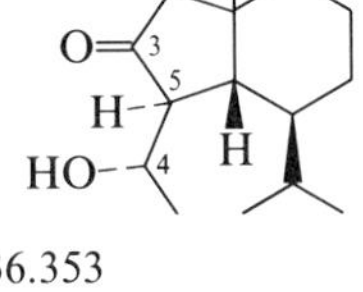

$C_{15}H_{24}O_2$ M 236.353
Cadinane numbering used. Constit. of *Senecio mexicanus*.
Cryst. Mp 100-102°. $[\alpha]_D$ −108° (c, 0.14 in CHCl₃).

Joseph-Nathan, P. *et al, Phytochemistry*, 1989, **28**, 2347 (*isol, pmr, cmr, cryst struct*)

3-Hydroxy-2-oxobutanoic acid   H-80226

[1944-42-9]

$$H_3CCH(OH)COCOOH$$

$C_4H_6O_4$ M 118.089
Prod. by microorganisms e.g. *Aspergillus niger*. Isol. from berries of *Vaccinium vitis-idaea*. Component of cheese aroma.

Et ester, phenylhydrazone: Prisms. Mp 93-94°.

Virtanen, A.I. *et al, Acta Chem. Scand.*, 1955, **9**, 188 (*isol*)
Bowman, R.E. *et al, J. Chem. Soc.*, 1957, 1583 (*synth, deriv*)

4-Hydroxy-2-oxobutanoic acid   H-80227

γ-Hydroxy-α-ketobutyric acid
[22136-38-5]

$$HOCH_2CH_2COCOOH$$

$C_4H_6O_4$ M 118.089
Isol. from fruits of *Vaccinium vitis-idaea* and from *V. oxycoccus*. Unstable, stabilised in frozen soln. at neutral pH.

2,4-Dinitrophenylhydrazone: [22136-40-9].
   Long yellow needles (EtOH). Mp 238-239°.

Lactone: Dihydro-2,3(2H,3H)-furandione
   $C_4H_4O_3$ M 100.074
   Needles. Mp 106-107°.

Virtanen, A.I. *et al, Acta Chem. Scand.*, 1955, **9**, 188 (*isol*)
Lane, A.S. *et al, Biochemistry*, 1969, **8**, 2958 (*synth, ir, uv*)

2-Hydroxy-12-oxo-4,11(13)-cadinadien-15-oic acid   H-80228

Panal
[115374-28-2]

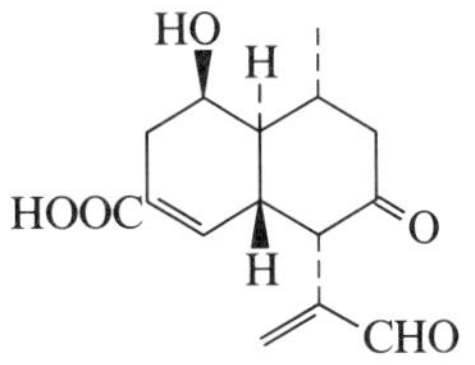

$C_{15}H_{18}O_5$ M 278.304
Constit. of luminous mushroom *Panellus stipticus*.
   Amorph. solid. $[\alpha]_D$ −17° (c, 0.9 in MeOH).

Nakamura, H. *et al, Tetrahedron*, 1988, **44**, 1597 (*isol*)

3-Hydroxy-8-oxo-2(9),6-lactaradien-5,13-olide   H-80229

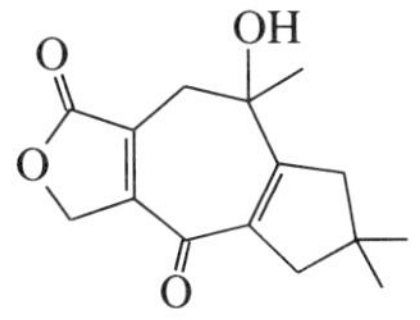

$C_{15}H_{18}O_4$ M 262.305
Constit. of *Lactarius serobiculatus*.

Bosetti, A. *et al, Phytochemistry*, 1989, **28**, 1427.

3-Hydroxy-23-oxo-12-oleanen-28-oic acid   H-80230

Updated Entry replacing H-03015
$C_{30}H_{46}O_4$ M 470.691
3β-form [639-14-5]
   Gypsogenin

Cryst. (MeOH). Mp 274-276°. $[\alpha]_D$ +91° (EtOH).

▷ RK0178000.

28-O-[[D-Xylopyranosyl(1→3)-O-6-deoxy-D-galactopyranosyl(1→2)]-6-deoxy-L-mannopyranoside], 3-O-[L-arabinopyranosyl(1→3)-O-(O-D-galactopyranosyl(1→4)-D-glucopyranosyl(1→4)-D-glucuronate)]: [15588-68-8]. **Gypsoside**
$C_{80}H_{126}O_{44}$ M 1791.849
Constit. of *Gypsophila pacifica* and *G. paniculata.* $[\alpha]_D^{20}$ +47.5° (c, 3.4 in CHCl$_3$) (as Me ether).

3-O-α-D-Glucuronate: **Vaccaroside**
$C_{36}H_{54}O_{10}$ M 646.817
Isol. from seeds of *Vaccaria segetalis.* Cryst. + 2H$_2$O(EtOH). Mp 204-206°. $[\alpha]_D$ +16.5° (c, 1.75 in EtOH).

3-O-[D-Xylopyranosyl(1→4)-β-D-glucopyranosyl(1→3)-β-D-glucopyranosyl(1→4)-β-D-glucuronate]: **Vacsegoside**
$C_{75}H_{118}O_{40}$ M 1659.733
Isol. from seeds of *V. segetalis.* Cryst. (MeOH/butanol). Mp 217-219°. $[\alpha]_D$ +7.5° (c, 3.16 in 30% MeOH aq.).

Ruzicka, L. *et al, Helv. Chim. Acta,* 1937, **20**, 299 (*isol*)
Vogel, A. *et al, Helv. Chim. Acta,* 1951, **34**, 2321 (*isol, struct*)
Kochetkov, N.K. *et al, Tetrahedron Lett.,* 1963, 477 (*Gypsoside*)
Abubakirov, N.K. *et al, Zh. Obshch. Khim.,* 1964, **34**, 1661 (*Vaccaroside*)
Amanmuradov, K. *et al, Khim. Prir. Soedin.,* 1965, **1**, 372; *cnc,* 292 (*Vacsegoside*)

2-Hydroxy-6-(2,4-pentadiynyl)benzoic   H-80231
acid, 9CI

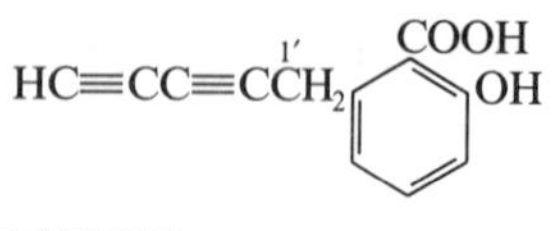

$C_{12}H_8O_3$ M 200.193
Me ether, Me ester: [4345-77-1]. **Demethylfrutescin.**
Desmethylfrutescin
$C_{14}H_{12}O_3$ M 228.247
Isol. from roots of *Chrysanthemum frutescens, C. segetum* and *C. foeniculaceum.* Mp 68.5°.

Me ether, Me ester, 1'-acetoxy: **Demethylfrutescin 1'-ylacetate.** *Desmethylfrutescinol acetate*
$C_{16}H_{14}O_5$ M 286.284
Isol. from *C. frutescens* and *C. foeniculaceum.* Cryst. (Et$_2$O/pet. ether). $[\alpha]_D^{20}$ −1.6° (c, 1.9 in Et$_2$O). λ_{max} 284 nm (ϵ 3300)(Et$_2$O).

Bohlmann, F. *et al, Chem. Ber.,* 1962, **95**, 602; 1963, **26**, 226; 1964, **97**, 1179; 1966, **99**, 995 (*isol, uv, ir, struct, biosynth*)

5-Hydroxy-4-phenanthrenecarboxylic acid   H-80232

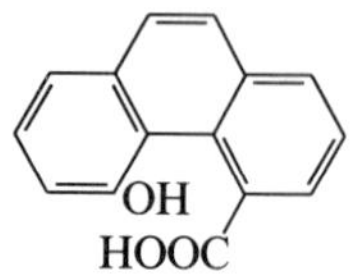

$C_{15}H_{10}O_3$ M 238.242
Lactone: 5H-*Phenanthro*[4,5-bcd]*pyran-5-one*
$C_{15}H_8O_2$ M 220.227
Pale-yellow needles (dioxan aq.). Mp 190°. Lactone ring opened with difficulty.

Me ether: 5-*Methoxy-4-phenanthrenecarboxylic acid*
$C_{16}H_{12}O_3$ M 252.269
Ppt. Mp 220°. Obt. only in low yield.

Gillis, R.G. *et al, Aust. J. Chem.,* 1989, **42**, 1007 (*synth, uv, pmr, ms*)

4-Hydroxy-1-phenyl-2-butyn-1-one   H-80233
3-Benzoyl-2-propyn-1-ol. 4-Hydroxy-2-butynophenone

$$PhCOC{\equiv}CCH_2OH$$

$C_{10}H_8O_2$ M 160.172
Pale-yellow oil.

Obrecht, D., *Helv. Chim. Acta,* 1989, **72**, 447 (*synth, ir, pmr*)

7-(3-Hydroxyphenyl)-2-heptene-4,6-diyn-1-   H-80234
ol
3-(7-Hydroxy-5-heptene-1,3-diynyl)phenol

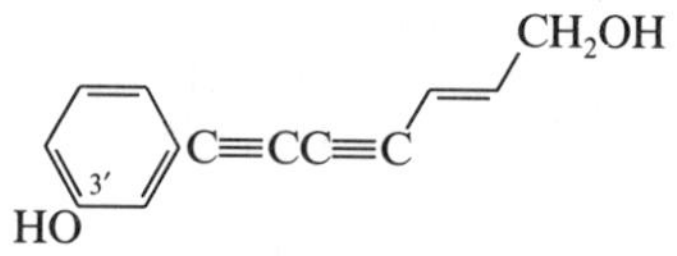

$C_{13}H_{10}O_2$ M 198.221
(E)-form

3'-Ac: 7-*(3-Acetoxyphenyl)-2-heptene-4,6-diyn-1-ol*
$C_{15}H_{12}O_3$ M 240.258
Isol. from roots of *Coreopsis tinctoria.* Not obt. fully pure.

Di-Ac: 1-*Acetoxy-7-(3-acetoxyphenyl)-2-heptene-4,6-diyne*
$C_{17}H_{14}O_4$ M 282.295
Isol. from roots of *C. tinctoria.* Needles (Et$_2$O/pet. ether). Mp 90.5-91°.

Bohlmann, F. *et al, Chem. Ber.,* 1966, **99**, 1223, 2822 (*isol, uv, ir, pmr, struct, synth*)

6-Hydroxy-1-phenyl-2,4-hexadiyn-1-one,   H-80235
9CI
5-Benzoyl-2,4-pentadiyn-1-ol
[7358-89-6]

$$PhCOC{\equiv}CC{\equiv}CCH_2OH$$

$C_{12}H_8O_2$ M 184.194
Isol. from *Lonas annua.* Cryst. (Et$_2$O/pet. ether). Mp 93°.
O-(3-Methyl-2-butenoyl): 1-*(β-Methylcrotonoyloxy)-5-benzoyl-2,4-pentadiyne*
$C_{17}H_{14}O_3$ M 266.296
Isol. from *L. annua.* Yellowish cryst. (pet. ether). Mp 36.5°.
[56558-91-9]

Bohlmann, F. *et al, Chem. Ber.,* 1966, **99**, 2413 (*isol, uv, ir, pmr, synth*)
Fukumaru, T. *et al, Agric. Biol. Chem.,* 1975, **39**, 519 (*synth*)

2-(4-Hydroxyphenyl)-5-(2-Hydroxy-   H-80236
propyl)-3-methylbenzofuran

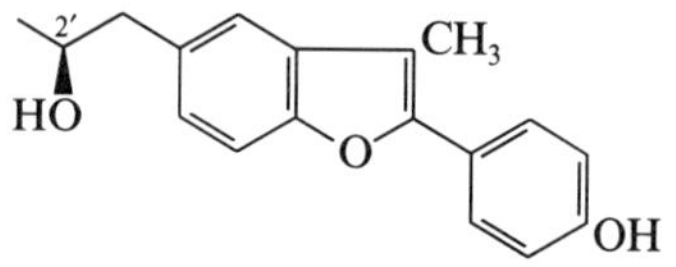

$C_{18}H_{18}O_3$ M 282.338
Constit. of *Krameria lanceolata.* Oil. $[\alpha]_D^{21}$ +9° (c, 0.3 in MeOH).
2'-Ketone: 2-*(4-Hydroxyphenyl)-3-methyl-5-(2-oxopropyl)benzofuran*
$C_{18}H_{16}O_3$ M 280.323
Constit. of *K. lanceolata.* Amorph.

Achenbach, H. *et al, Phytochemistry,* 1989, **28**, 1959.

1-(4-Hydroxyphenyl)-2-(2-methoxy-5-propenylphenyl)-2-propanol   H-80237

4-[2-Hydroxy-2-(2-methoxy-5-propenylphenyl)propyl]phenol

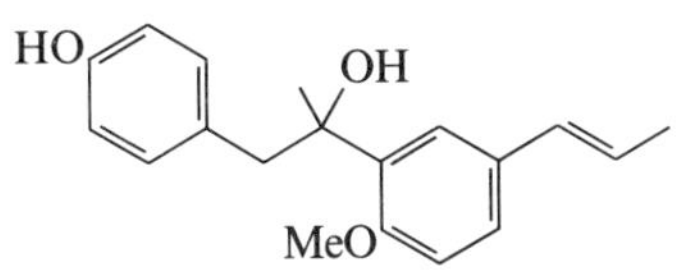

$C_{19}H_{22}O_3$ M 298.381

Constit. of *Krameria lanceolata*. Amorph. $[\alpha]_D^{21}$ +7° (c, 0.2 in MeOH).

Achenbach, H. *et al, Phytochemistry*, 1989, **28**, 1959.

3-(4-Hydroxyphenyl)-2-oxopropanoic acid   H-80238

4-Hydroxy-α-oxobenzenepropanoic acid. p-*Hydroxyphenylpyruvic acid*

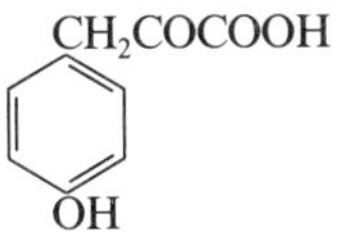

$C_9H_8O_4$ M 180.160

Isol. from cell cultures of *Aerobacter aerogenes* and *E. coli.* Also obt. from *Phlox* spp. and *Thuja plicata*. Cryst. Mp 218°.

Phenylhydrazone: Mp 161-162°.

Doy, C.H. *et al, Biochim. Biophys. Acta*, 1961, **50**, 495 (*isol, synth*)
Brandner, G. *et al, Acta Chem. Scand.*, 1964, **18**, 574 (*isol*)

1-(2-Hydroxyphenyl)-3-phenyl-1,3-propanedione   H-80239

2′,β-Dihydroxychalcone

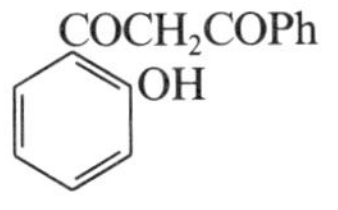

$C_{15}H_{12}O_3$ M 240.258

Enolised β-diketone. Constit. of *Primula pulverulenta*. Yellow prisms (MeOH). Mp 114-117°.

Wollenweber, E. *et al, Phytochemistry*, 1989, **28**, 295.

2-Hydroxy-2-phenylpropanal   H-80240

α-Hydroxy-α-methylbenzeneacetaldehyde, 9CI. α-Methylmandelaldehyde, 8CI

[4361-50-6]

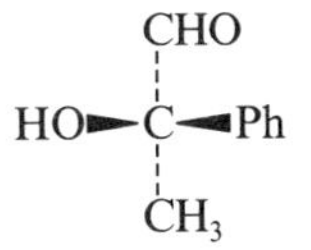

R-form

$C_9H_{10}O_2$ M 150.177

(R)-form

Bp$_{0.3}$ 108-110°, Bp$_{0.1}$ 90-110° (kugelrohr) . $[\alpha]_D^{20}$ +80.2° (c, 1.73 in CHCl$_3$, 31% opt. pure).

(±)-form

Bp$_{0.7}$ 116-120°.

Dimer: [21504-26-7]. *2,5-Dihydroxy-3,6-dimethyl-3,6-diphenyl-1,4-dioxan*
$C_{18}H_{20}O_4$ M 300.354
Mp <27°.

Russell, G.A. *et al, J. Org. Chem.*, 1969, **34**, 3618 (*synth*)

Oldenziel, O.H. *et al, Tetrahedron Lett.*, 1974, 167 (*synth*)
Hundscheid, F.J.A. *et al, Tetrahedron*, 1987, **43**, 5073 (*synth, ir, pmr*)

1-(4-Hydroxyphenyl)-2-propanone   H-80241

p-*Hydroxyphenylacetone*

[770-39-8]

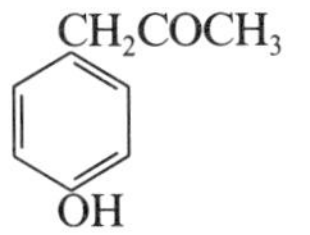

$C_9H_{10}O_2$ M 150.177

Yellowish oil or cryst. Mp 40°. Bp$_{11}$ 168-170°.

Semicarbazone: Mp 212°.

Me ether: [122-84-9]. *1-(4-Methoxyphenyl)-2-propanone. Anisylacetone. Anisyl ketone*
$C_{10}H_{12}O_2$ M 164.204
Isol. from anise oil, fennel oil and oil of *Illicium verum*. Oil. Bp 261-265°.

Me ether, semicarbazone: [7306-38-9].
Mp 183-184° (174°).

Tardy, E., *Bull. Soc. Chim. Fr.*, 1897, 580, 660; 1902, 990 (*isol, deriv*)
Hoover, E.W. *et al, J. Org. Chem.*, 1947, **12**, 501 (*synth, deriv*)
Jones, D.D. *et al, J. Org. Chem.*, 1967, **32**, 1402 (*synth*)
Bricout, J., *Bull. Soc. Chim. Fr.*, 1974, 1901 (*isol, deriv*)
Franke, A. *et al, Helv. Chim. Acta*, 1975, **58**, 278 (*synth*)

3-(4-Hydroxyphenyl)-2-propenoic acid, 9CI   H-80242

Updated Entry replacing H-50300

p-*Hydroxycinnamic acid, 8CI.* p-*Cumaric acid.* p-*Coumaric acid. Naringenic acid. Naringeninic acid*

[7400-08-0]

$C_9H_8O_3$ M 164.160

▷ GD9094000.

(E)-form [501-98-4]

Widespread in plants, e.g. peel of *Prunus serotina*, and from *Trifolium pratense* and *Daviesia latifolia..* Cryst. + 1H$_2$O (cold H$_2$O), anhyd. cryst. (hot H$_2$O). Sol. Et$_2$O, hot EtOH, spar. sol. C$_6$H$_6$, insol. ligroin.

Me ester: [19367-38-5].
$C_{10}H_{10}O_3$ M 178.187
Needles (H$_2$O), tablets (EtOH). Mp 137°.

Et ester: [7362-39-2].
$C_{11}H_{12}O_3$ M 192.214
Mp 83°. Bp$_{0.08}$ 175-184°.

Glycoside: **Pajaneelin**
Glycoside from peel of *Pajanelia rheedii*. Mp 237-239°. $[\alpha]_D^{30}$ −173°. Gives p-Coumaric acid and fructose on hydrol.

Me ester, glucoside: **Linocinnamarin**
$C_{16}H_{20}O_8$ M 340.329
Glycoside from seeds of *Linum usitatissimum*. Fine needles (MeOH aq.). Mp 167° (anhyd.). $[\alpha]_D^{27}$ −73° (c, 1 in MeOH).

Triacontyl ester: **Defuscin**
$C_{39}H_{68}O_3$ M 584.964
Constit. of *Dendrobium fuscescens*. Needles (CHCl$_3$/pet. ether). Mp 101°.

Me ether: [943-89-5]. *3-(4-Methoxyphenyl)-2-propenoic acid*
$C_{10}H_{10}O_3$ M 178.187

Isol. from oil of *Andropogon odoratus, Curcuma aromatica* and other plants. Needles (EtOH). Mp 170-172° (sinters).

Me ether, Et ester: [24393-56-4].
$C_{12}H_{14}O_3$ M 206.241
Major constit. of oil of *Kaempferia galanga*. Large prisms. Mp 49-50°.

O-(3,7-Dimethyl-2,6-octadienyl), Me ester: Methyl p-geranyloxycinnamate
$C_{20}H_{26}O_3$ M 314.424
Isol. from *Acronychia baueri*. Platelets (MeOH). Mp 65-67°.

(Z)-form [4501-31-9]
Found in *Larix sibirica* bark and *Miscanthus floridulus*. Cryst. (C_6H_6 or H_2O). Mp 134-134.5° dec. (subl. from 124°).

[830-09-1, 1929-30-2, 2979-06-8, 3943-97-3, 51507-22-3]

Zincke, T., *Justus Liebigs Ann. Chem.*, 1902, **322**, 224 (*isol*)
Ogawa, S., *Bull. Chem. Soc. Jpn.*, 1927, **2**, 25 (*isol*)
Klosterman, H.J. *et al, J. Am. Chem. Soc.*, 1954, **76**, 1229; 1955, **77**, 420 (*Linocinnamarin*)
U.S. Pat., 3 094 471, (1963); *CA*, **60**, 1655b (*isom*)
Prager, R.H. *et al, Aust. J. Chem.*, 1966, **19**, 451 (*Methyl p-Geranyloxycinnamate*)
Utsumi, H. *et al, Bull. Chem. Soc. Jpn.*, 1967, **40**, 426 (*cryst struct*)
Aulin-Erdtman, G. *et al, Acta Chem. Scand.*, 1968, **22**, 1187 (*props, uv*)
Hartley, R.D., *J. Chromatogr.*, 1975, **107**, 213 (*isom, gc*)
Talapatra, B. *et al, Phytochemistry*, 1989, **28**, 290 (*Defuscin*)

1-(4-Hydroxyphenyl)-3-(2,4,6-trihydroxy-phenyl)-1-propanone   H-80243

2,4,4',6-Tetrahydroxydihydrochalcone

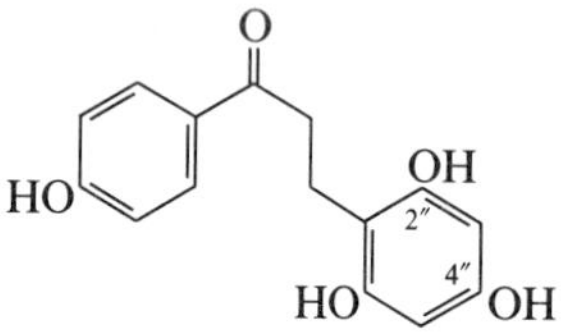

$C_{15}H_{14}O_5$ M 274.273

2"-Me ether: [119425-91-1]. *3-(2,4-Dihydroxy-6-methoxyphenyl)-1-(4-hydroxyphenyl)-1-propanone.*
Loureirin D
$C_{16}H_{16}O_5$ M 288.299
Constit. of *Dracaena loueiri*. Powder (MeOH). Mp 176-178°.

2",4"-Di-Me ether: 4-(2-Hydroxy-4,6-dimethoxyphenyl)-1-(4-hydroxyphenyl)-1-propanone. **Loureirin B**
$C_{17}H_{18}O_5$ M 302.326
Constit. of *D. loueiri*. Needles (MeOH). Mp 132-133°.

Meksuriyen, D. *et al, J. Nat. Prod. (Lloydia)*, 1988, **51**, 1129 (*isol, pmr, cmr*)

3-(4-Hydroxyphenyl)-1-(2,4,6-trihydroxy-phenyl)-1-propanone   H-80244

Updated Entry replacing H-60219
2,4,4',6-Tetrahydroxydihydrochalcone. Dihydronarigenin. Asebogenol. Phloretin. Phloretol
[60-82-2]

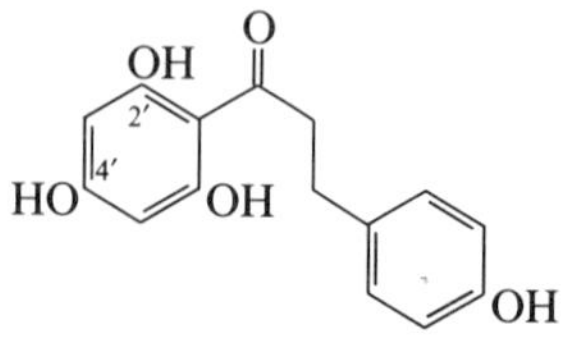

$C_{15}H_{14}O_5$ M 274.273
Occurs in apple (*Pyrus malus*) as glycosides and in leaves of lemon (*Citrus limon*). Needles (EtOH aq.). Insol. Et_2O. Mp ca. 262-264° dec.

2'-Glucoside: [60-81-1]. **Phloridzin.** *Phlorrhizin. Phlorhizin*
$C_{21}H_{24}O_6$ M 372.417
Occurs in *Micromelum teprocarpum*, in apple(*Malus*), *R.* spp., *Kalmia latifolia* and *Piperis japonica*. Produces glucosuria in man. Herbivore antifeedant. Needles. $[\alpha]_D^{18}$ −52.1° (EtOH aq.). Various Mp's, some double, recorded between 108° and 170°.
▷ UC2080000.

2'-Rhamnoside: [19253-17-9]. **Glycyphyllin**
$C_{21}H_{24}O_9$ M 420.415
Isol. from leaves of *Smilax glycyphylla*. Large prisms (H_2O). Mp 175-180° dec.

4'-Me ether: [520-42-3]. *1-(2,6-Dihydroxy-4-methoxyphenyl)-3-(4-hydroxyphenyl)-1-propanone.* **Asebogenin.** *Asebotol*
$C_{16}H_{16}O_5$ M 288.299
Found in *Piperis japonica, Rhododendron* spp. and *Pitryogramma calomelanos*. Cryst. (EtOH). Mp 168°.

4'-Me ether, 2'-O-β-D-glucopyranoside: [11075-15-3]. **Asebotin.** *Asebotoside*
$C_{22}H_{26}O_6$ M 386.444
Constit. of *Andromeda japonica, R.* spp.,*K. latifolia* and *P. japonica*. Mp 148° (135-136°). $[\alpha]_D^{25}$ −46.2° (EtOH).

Eykman, J.F., *Ber.*, 1883, **16**, 2769 (*isol*)
Tamura, K., *Nippon Kagaku Kaishi (J. Chem. Soc. Jpn.)*, 1936, **57**, 1141 (*synth, Asebogenin*)
Zemplén, R. *et al, Ber.*, 1942, **75**, 645, 1298 (*synth*)
Murakami, S., *Yakugaku Zasshi (J. Pharm. Soc. Jpn.)*, 1955, **75**, 573, 603 (*isol*)
Batterham, T.J. *et al, Aust. J. Chem.*, 1964, **17**, 428 (*pmr*)
Williams, A.H. *et al, Nature (London)*, 1964, **202**, 824 (*isol, Glycyphyllin*)
Rice, G.L., *Allelopathy*, Academic Press, N.Y., 1974 (*deriv*)
King, B. *et al, Phytochemistry*, 1975, **14**, 1448 (*deriv*)
Mancini, S.D. *et al, J. Nat. Prod. (Lloydia)*, 1979, **42**, 483 (*isol*)
Hutz, C. *et al, Z. Naturforsch., C*, 1982, **37**, 337 (*isol*)

3-(4-Hydroxyphenyl)-1-(2,3,4-trihydroxy-phenyl)-2-propen-1-one, 9CI   H-80245

Updated Entry replacing H-03269
2',3',4,4'-Tetrahydroxychalcone, 8CI. 4-Hydroxystyryl 2,3,4-trihydroxyphenyl ketone

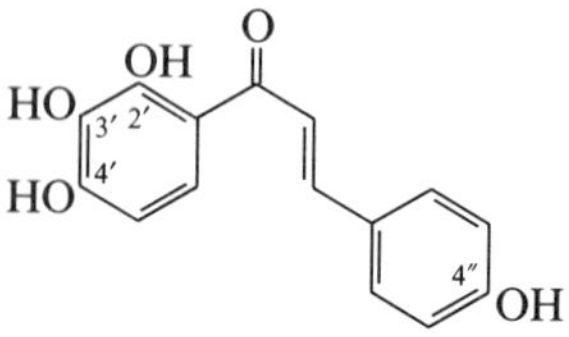

$C_{15}H_{12}O_5$ M 272.257
Isol. from trunkwood of *Acacia auriculiformis*. Red-brown needles (EtOH aq.). Mp 225-225.5°.

4"-Me ether: 3-(4-Methoxyphenyl)-1-(2,3,4-trihydroxyphenyl)-2-propen-1-one. 2',4,4'-Trihydroxy-3'-methoxychalcone. **Kukulkanin B**
$C_{16}H_{14}O_5$ M 286.284
Constit. of *Mimosa tenuefolia*. Cryst. Mp 215°.

3',4"-Di-Me ether: 1-(2,4-Dihydroxy-3-methoxyphenyl)-3-(4-methoxyphenyl)-2-propen-1-one. 2',4'-Dihydroxy-3',4-dimethoxychalcone. **Kukulkanin A**
$C_{17}H_{16}O_5$ M 300.310
Constit. of *N. tenuefolia*. Cryst. Mp 172°.

Tetra-Me ether: Pale-yellow needles (EtOH). Mp 94°.

Russell, A. *et al*, *J. Chem. Soc.*, 1934, 220; 1937, 421 (*synth, uv*)
Geissman, T.A. *et al*, *J. Am. Chem. Soc.*, 1946, **68**, 697 (*synth*)
Drewes, S.E. *et al*, *Biochem. J.*, 1966, **98**, 493 (*isol*)

3-(4-Hydroxyphenyl)-1-(2,4,6-trihydroxy-phenyl)-2-propen-1-one, 9CI   H-80246

Updated Entry replacing H-60220
2′,4,4′,6′-Tetrahydroxychalcone, 8CI. 4-Hydroxystyryl 2,4,6-trihydroxyphenyl ketone. **Chalconaringenin**
[73692-50-9]

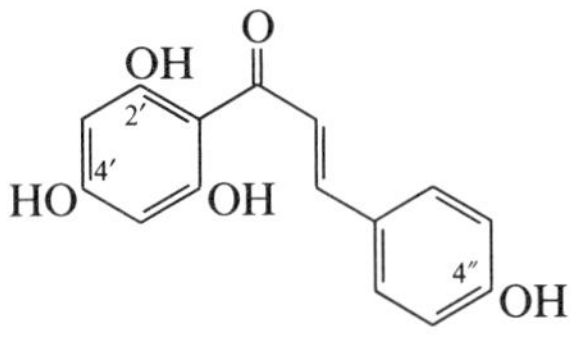

$C_{15}H_{12}O_5$ M 272.257

2′-Glucoside: [4547-85-7]. **Isosalipurposide**
$C_{21}H_{22}O_{10}$ M 434.399
Isol. from *Salix purpurea, Paeonia trollioides, Dianthus caryophyllus, Aeschynanthus pariflorus* and *Asystasia gangetica*. Mp 172-174° (anhyd.). [α]$_D$ −20° (EtOH).

2′-[O-Rhamnosyl(1→4)xyloside]: [82344-84-1].
$C_{26}H_{30}O_{13}$ M 550.515
Pigment from *Acacia dealbata*.

2′-Me ether, 4′-glucoside: [61826-89-9]. **Helichrysin.**
Dehydro-p-asebotin
$C_{22}H_{24}O_{10}$ M 448.426
Constit. of the flowers of *Helichrysum* spp. and of *Gnaphalium affine*. Yellow cryst. (MeOH). Mp 246° dec. (199-200°). [α]$_D^{30}$ −85.7° (MeOH).

4′-O-(3-Methyl-2-butenyl):
$C_{20}H_{20}O_5$ M 340.375
Constit. of *H. athrisciifolium*. Gum.

4′,4″-Di-Me ether: 1-(2,6-Dihydroxy-4-methoxyphenyl)-3-(4-methoxyphenyl)-2-propen-1-one. 2′,6′-Dihydroxy-4,4′-dimethoxychalcone. **Gymnogrammene**
$C_{17}H_{16}O_5$ M 300.310
Isol. from *Pitryogramma chrysophylla*. Red prisms (C_6H_6). Mp 156-157°.

2′,4,4″-Tri-Me ether: [3420-72-2]. *3-(4-Hydroxyphenyl)-1-(2-hydroxy-4,6-dimethoxyphenyl)-2-propen-1-one. 2′-Hydroxy-4,4′,6′-trimethoxychalcone.* **Flavokawin A**
$C_{18}H_{18}O_5$ M 314.337
Constit. of *Piper methysticum* and *Dahlia tenuicaulis*. Yellow needles (EtOH). Mp 113°.

Tetra-Me ether: [25163-67-1].
$C_{19}H_{20}O_5$ M 328.364
Pale-yellow cryst. (EtOH aq.). Mp 119-121°.

Dihydro, 6′-Me ether: 3-(4-Hydroxyphenyl)-1-(2,4-dihydroxy-6-methoxyphenyl)-1-propanone, 9CI. 2′,4,4′-Trihydroxy-6′-methoxydihydrochalcone
$C_{16}H_{16}O_5$ M 288.299
Constit. of *Coptis japonica* var. *dissecta*. Needles (EtOH). Mp 195°.

Kostanecki, S. *et al*, *Ber.*, 1904, **37**, 792 (*synth*)
Mosimann, W. *et al*, *Ber.*, 1916, **49**, 1701 (*synth*)
Nilsson, M. *et al*, *Acta Chem. Scand.*, 1961, **15**, 211 (*Gymnogrammene*)
Guise, G.B. *et al*, *Aust. J. Chem.*, 1962, **15**, 314 (*isol, deriv*)
Mahanthy, P., *Indian J. Chem.*, 1965, **3**, 121 (*ir*)
Harborne, J.B. *et al*, *Phytochemistry*, 1966, **5**, 111 (*Isosalipurposide*)
Ramakrishnan, V.T. *et al*, *J. Org. Chem.*, 1970, **35**, 2901 (*synth*)
Aritomi, M. *et al*, *Chem. Pharm. Bull.*, 1974, **22**, 1800 (*Dehydroasebotin*)

Lam, J. *et al*, *Phytochemistry*, 1975, **14**, 1621 (*isol, deriv*)
Wright, W.G., *J. Chem. Soc., Perkin Trans.* 1, 1976, 1819 (*Helichrysin*)
Duddeck, H. *et al*, *Phytochemistry*, 1978, **17**, 1369 (*pmr*)
Imperato, F., *Phytochemistry*, 1982, **21**, 480 (*rhamnosylxyloside*)
Bohlmann, F. *et al*, *Phytochemistry*, 1984, **23**, 1338 (*isol*)
Babber, S. *et al*, *Indian J. Chem., Sect. B*, 1987, **26**, 797 (*synth*)
Mizuno, M. *et al*, *Phytochemistry*, 1987, **26**, 2071 (*deriv*)

2-Hydroxy-1(10)-pinen-5-one   H-80247

2-Hydroxy-5-oxo-β-pinene

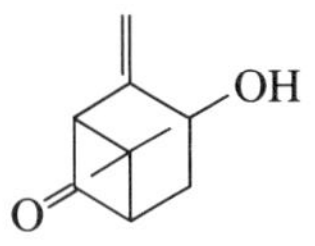

$C_{10}H_{14}O_2$ M 166.219
Constit. of *Artemisia gypsacea*. Oil. [α]$_D^{24}$ −24° (c, 1.2 in $CHCl_3$).

Rustaiyan, A. *et al*, *Phytochemistry*, 1989, **28**, 1535.

3-Hydroxy-1*H*-pyrrole-2-carboxylic acid   H-80248

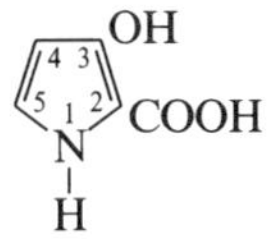

$C_5H_5NO_3$ M 127.099

Me ether, Me ester: [112373-17-8].
$C_7H_9NO_3$ M 155.153
Needles (EtOAc/hexane). Mp 112-114°.

Me ether, Et ester:
$C_8H_{11}NO_3$ M 169.180
Cryst. by subl. Mp 94°.

Rapoport, H. *et al*, *J. Am. Chem. Soc.*, 1962, **84**, 635 (*synth, uv*)
Boger, D.L. *et al*, *J. Org. Chem.*, 1988, **53**, 1405 (*synth, pmr, cmr, ir, uv, ms*)

4-Hydroxy-1*H*-pyrrole-2-carboxylic acid   H-80249

Updated Entry replacing H-10288
Hydroxyminaline

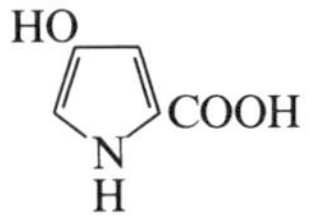

$C_5H_5NO_3$ M 127.099
Component of the molecule pectase found in *Penicillium* and *Aspergillus*. Grey-brown amorph. powder which cannot be recryst. or subl. Dec. readily in soln.

Me ether: 4-Methoxy-2-pyrrolecarboxylic acid
$C_6H_7NO_3$ M 141.126
Cryst. by subl. Mp 179-180°.

Me ether, Me ester:
$C_7H_9NO_3$ M 155.153
Cryst. (MeOH). Mp 85-86°.

Me ether, Et ester:
$C_8H_{11}NO_3$ M 169.180
Cryst. (C_6H_6). Mp 55-58°.

Minagawa, T., *Proc. Imp. Acad.* (*Tokyo*), 1945, **21**, 33, 37; *CA*, **47**, 151i, 152b (*isol*)
Kuhn, R. *et al*, *Chem. Ber.*, 1956, **89**, 1423 (*synth, uv*)
Rapoport, H. *et al*, *J. Am. Chem. Soc.*, 1962, **84**, 630 (*derivs, synth, uv, ir*)

4-Hydroxy-1*H*-pyrrole-3-carboxylic acid   H-80250

$C_5H_5NO_3$ M 127.099

Me ether: 4-Methoxy-1H-pyrrole-3-carboxylic acid
$C_6H_7NO_3$ M 141.126
Cryst. (MeOH/C_6H_6). Mp 203-204°.

Me ether, Me ester:
$C_7H_9NO_3$ M 155.153
Cryst. by subl. Mp 115-117°.

Me ether, Et ester:
$C_8H_{11}NO_3$ M 169.180
Cryst. (C_6H_6 or by subl.). Mp 107-109°.

Rapoport, H. *et al*, *J. Am. Chem. Soc.*, 1962, **84**, 630 (*synth, uv, ir*)

3-Hydroxy-1*H*-pyrrole-2,4-dicarboxylic acid   H-80251

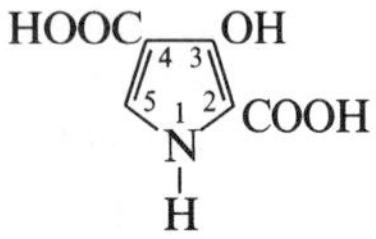

$C_6H_5NO_5$ M 171.109

Di-Et ester:
$C_{10}H_{13}NO_5$ M 227.216
Cryst. (C_6H_6/hexane, MeOH aq. or by subl.). Mp 121°.

Me ether, Di-Et ester:
$C_{11}H_{15}NO_5$ M 241.243
Cryst. (C_6H_6/hexane or by subl.). Mp 83°.

N-*Me, Me ether, Di-Et ester:*
$C_{12}H_{17}NO_5$ M 255.270
Cryst. (hexane or by subl.). Mp 50°.

Rapoport, H. *et al*, *J. Am. Chem. Soc.*, 1962, **84**, 635 (*synth, ir, uv*)

3-Hydroxy-1*H*-pyrrole-2,5-dicarboxylic acid   H-80252

$C_6H_5NO_5$ M 171.109

Me ether, 2-Me ester: [112373-15-6].
$C_8H_9NO_5$ M 199.163
Platelets (MeOH). Mp 182-184°.

Me ether, Di-Me ester: [92144-13-3].
$C_9H_{11}NO_5$ M 213.190
Needles (MeOH). Mp 149.5-150.5°.

Boger, D.L. *et al*, *J. Org. Chem.*, 1988, **53**, 1405 (*synth, pmr, ir, ms*)

N-Hydroxy-4-quinolinamine, 9CI   H-80253

4-Hydroxylaminoquinoline
[13442-05-2]

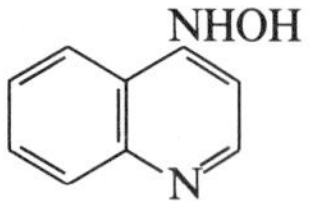

$C_9H_8N_2O$ M 160.175

B,HCl: [21626-47-1].
Mp 161-162°.

1-Oxide: [4637-56-3].
$C_9H_8N_2O_2$ M 176.174
Intermed. in carcinogenic activity of 4-aminoquinoline 1-oxide and related compds. Pale-brown needles or powder (MeOH). Mp 190° dec. Exists mainly as its hydroxyimino tautomer.

[1010-61-3]

Itai, T. *et al*, *Chem. Pharm. Bull.*, 1961, **9**, 87 (*oxide*)

Hamana, M. *et al*, *Yakugaku Zasshi (J. Pharm. Soc. Jpn.)*, 1962, **84**, 42 (*synth*)
Kawazoe, Y. *et al*, *Tetrahedron*, 1980, **36**, 2933 (*uv, pmr, struct, oxide*)

4-Hydroxysapriparaquinone   H-80254

3-Hydroxy-5-(4-hydroxy-4-methylpentyl)-6-methyl-2-(1-methylethyl)-1,4-naphthalenedione, 9CI. 3-Hydroxy-5-(4-hydroxy-4-methylpentyl)-2-isopropyl-6-methyl-1,4-naphthoquinone

[120278-25-3]

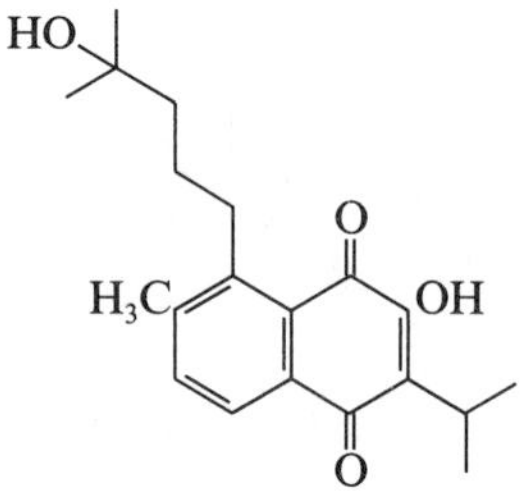

$C_{20}H_{26}O_4$ M 330.423
Constit. of *Salvia prionitis*. Cryst. Mp 61-62°. A 4,5-seco-5,10-friedoabietane.

Lin, L.-Z. *et al*, *Phytochemistry*, 1989, **28**, 177.

6-Hydroxy-5-[(4-sulfophenyl)azo]-2-naphthalenesulfonic acid, 9CI   H-80255

1-(p-Sulfophenylazo)-2-naphthol-6-sulfonic acid. C.I. Food Yellow 3, 8CI. C.I. 15985

[5859-11-0]

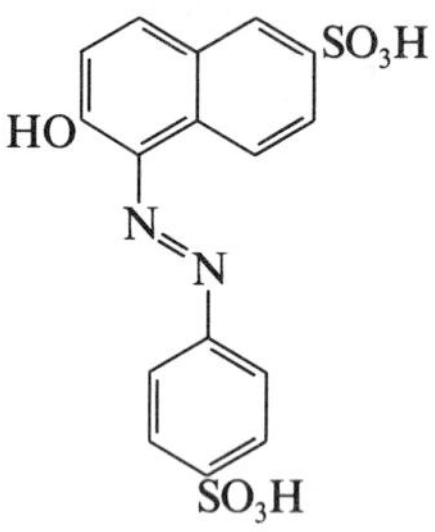

$C_{16}H_{12}N_2O_7S_2$ M 408.412
Acid-base indicator. Orange cryst. Sol. H_2O, EtOH.

Di-Na salt: [2783-94-0]. *Sunset Yellow FCF*
Food dye. Sol. H_2O, v. sl. sol. EtOH.
▷ Mutagenic, carcinogenic, allergenic.

Ba salt: Deep orange-red, microscopic needles. V. sl. sol. H_2O.

Stein, C., *J. Assoc. Off. Agric. Chem.*, 1949, **32**, 672 (*synth, uv, anal*)
Berkman, Yu.P. *et al*, *Dokl. Akad. Nauk SSSR, Ser. Khim.*, 1956, **106**, 693 (*synth, use*)
Reich, G. *et al*, *Zh. Anal. Khim.*, 1960, **177**, 274 (*anal*)
Marmino, D.M., *J. Assoc. Off. Anal. Chem.*, 1974, **57**, 495 (*pmr*)
Saeva, F.D. *et al*, *J. Am. Chem. Soc.*, 1975, **97**, 5631 (*synth*)
Chudy, J. *et al*, *J. Chromatogr.*, 1978, **154**, 306 (*chromatog*)
Gelbcke, M. *et al*, *Anal. Lett. Sect. A, Sect. A*, 1980, **13**, 975 (*cmr*)
Broadbent, P. *et al*, *J. Chem. Res. (S)*, 1980, 160 (*raman*)
Sugiura, T. *et al*, *J. Chem. Res. (S)*, 1980, 164 (*ms, uv*)
Saint-Martin, P. *et al*, *Analusis*, 1987, **15**, LV (*hplc*)
Young, M. .L., *J. Assoc. Off. Anal. Chem.*, 1988, **71**, 458 (*tlc*)

3-Hydroxy-14-taraxeren-28-oic acid   H-80256

Updated Entry replacing H-20250

3-Hydroxy-D-friedoolean-14-en-28-oic acid, 9CI

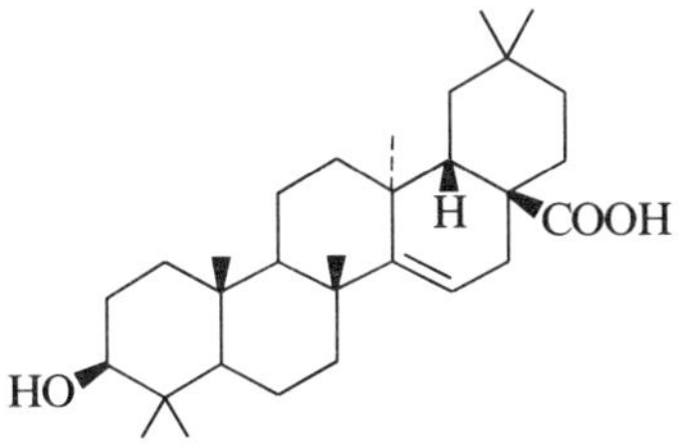

3β-*form*

$C_{30}H_{48}O_3$ M 456.707

3α-form

Epialeuritolic acid

Ac: **Epiacetylaleuritolic acid**

$C_{32}H_{50}O_4$ M 498.745

Constit. of *Phytolacca acinosa*. Cryst. Mp 291-294°.

3β-form [26549-17-7] **Aleuritolic acid**. Maprounic acid

Constit. of *Aleuritus montana* and *Maprounea africana*. Cryst. (CHCl₃/MeOH). $[\alpha]_D^{20}$ +12.8° (c, 0.4 in Py).

Ac: [28937-85-1]. **Acetylaleuritolic acid**

$C_{32}H_{50}O_4$ M 498.745

Constit. of *P. americana*. Cryst. Mp 300-302° dec., Mp 301-302°. $[\alpha]_D^{25}$ +23.1° (c, 0.6 in CHCl₃).

3-(4-Hydroxybenzoyl):

$C_{37}H_{52}O_5$ M 576.815

Constit. of *M. africana*. Cryst. Mp 308-311°. $[\alpha]_D^{20}$ +32.5° (c, 0.79 in Py).

Misra, D.R. *et al*, Tetrahedron, 1970, **26**, 3017 (isol)
Woo, W.S. *et al*, Phytochemistry, 1977, **16**, 1845 (isol)
Razdan, T.K. *et al*, Phytochemistry, 1982, **21**, 2339 (isol)
Wani, M.C. *et al*, J. Nat. Prod. (Lloydia), 1983, **46**, 537 (isol)
McPhail, A.T. *et al*, J. Nat. Prod. (Lloydia), 1989, **52**, 212 (cryst struct, bibl)

3-Hydroxy-11-triacontanone   H-80257

[52656-91-4]

$$H_3C(CH_2)_{18}CO(CH_2)_7CH(OH)CH_2CH_3$$

$C_{30}H_{60}O_2$ M 452.803

Isol. from leaves of *Marsilea minuta*. Plates. Mp 102°. $[\alpha]_D^{25}$ −9° (CHCl₃).

Chakravarti, D. *et al*, J. Indian Chem. Soc., 1974, **51**, 260.

7-Hydroxy-5-triacontanone   H-80258

$$H_3C(CH_2)_{22}CH(OH)CH_2CO(CH_2)_3CH_3$$

$C_{30}H_{60}O_2$ M 452.803

Constit. of roots of *Costus speciosus*. Cryst. (Me₂CO/MeOH). Mp 81°.

Gupta, M.M. *et al*, Phytochemistry, 1981, **20**, 2553.

2-Hydroxytricosanoic acid, 9CI   H-80259

[2718-37-8]

$$H_3C(CH_2)_{20}CH(OH)COOH$$

$C_{23}H_{46}O_3$ M 370.615

Present in cerebrosides.

[26632-12-2]

Downing, D.T., Aust. J. Chem., 1961, **14**, 150 (isol)

6-Hydroxy-2-(2,3,4-trihydroxy-phenyl)benzofuran   H-80260

Updated Entry replacing P-02640

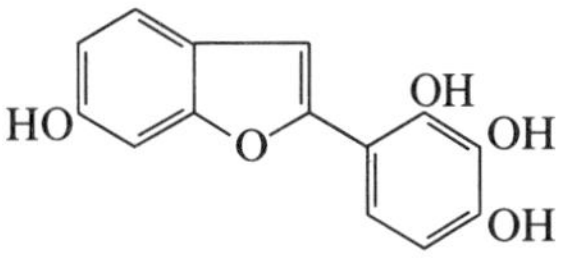

$C_{14}H_{10}O_5$ M 258.230

2′,3′-Di-Me ether: [74048-95-6]. *6-Hydroxy-2-(4-hydroxy-2,3-dimethoxyphenyl)benzofuran. 2-(4-Hydroxy-2,3-dimethoxyphenyl)-6-benzofuranol, 9CI.* **Isopterofuran**

$C_{16}H_{14}O_5$ M 286.284

Isol. from leaves of *Coronilla emerus*.

2′,4′-Di-Me ether: [3784-75-6]. *2-(3-Hydroxy-2,4-dimethoxyphenyl)-6-benzofuranol, 9CI. 6-Hydroxy-2-(3-hydroxy-2,4-dimethoxyphenyl)benzofuran.* **Pterofuran**

$C_{16}H_{14}O_5$ M 286.284

Constit. of the heartwood of *Pterocarpus indicus*. Needles (C₆H₆/EtOH). Mp 208-208.5°.

2′,4′-Di-Me ether, di-Ac: Needles (MeOH). Mp 135-136°.

Tetra-Me ether: Prisms (pet. ether). Mp 86-87°.

Cooke, R.G. *et al*, Aust. J. Chem., 1964, **17**, 379 (isol, struct)
Cooke, R.G. *et al*, Aust. J. Chem., 1969, **22**, 2395 (synth)
Duffley, R.P. *et al*, J. Chem. Soc., Perkin Trans. 1, 1977, 802 (synth, pmr)
Dewick, P.M. *et al*, Phytochemistry, 1980, **19**, 289 (Isopterofuran)

7-Hydroxy-3-(2,4,5-trihydroxyphenyl)-2H-1-benzopyran-2-one   H-80261

7-Hydroxy-3-(2,4,5-trihydroxyphenyl)coumarin

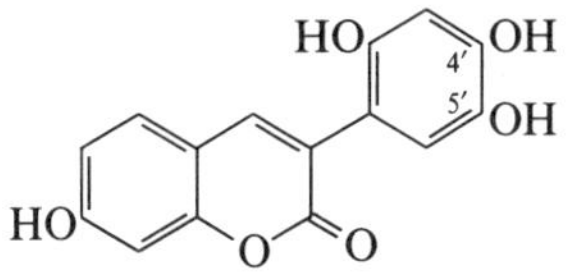

$C_{15}H_{10}O_6$ M 286.240

4′,5′-Methylene ether: [54300-96-8]. *7-Hydroxy-3-(2-hydroxy-4,5-methylenedioxyphenyl)coumarin. 2′,7-Dihydroxy-4′,5′-methylenedioxy-3-phenylcoumarin*

$C_{16}H_{10}O_6$ M 298.251

Isol. from heartwood of *Dalbergia oliveri*.

4′,5′-Methylene ether, di-Ac: Needles (EtOH aq.). Mp 195-196°.

Donnelly, D.M.X. *et al*, Phytochemistry, 1974, **13**, 2587 (isol, uv, pmr, struct, synth)

2-Hydroxy-2,3,3-trimethylbutanoic acid   H-80262

2-tert-Butyllactic acid

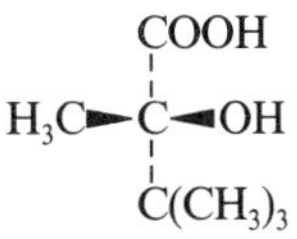

$C_7H_{14}O_3$ M 146.186

(R)-form

Cryst. + ½H₂O. $[\alpha]_D^{20}$ +1.4° (c, 8.36 in EtOH). Abs. config. has been uncertain – now prob. settled.

(S)-form

Needles + ½H₂O (pet. ether). Mp 67-69° (clears at 140°). $[\alpha]_D^{20}$ −1.5° (c, 9.83 in EtOH).

(±)-form

Mp 99-100° (clears at 140°).

Evans, R.J.D. *et al*, *J. Chem. Soc.*, 1963, 1506; 1965, 2553.
Eliel, E.L. *et al*, *Tetrahedron Lett.*, 1987, **28**, 4813 (*abs config, bibl*)

1-Hydroxy-11,12,13-trisnor-9-eremo-philen-8-one   H-80263

1-Hydroxyisoondetianone

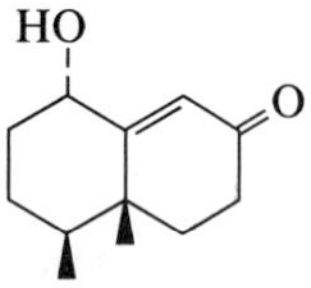

$C_{12}H_{18}O_2$ M 194.273

(1α)-form

Constit. of *Ondetia linearis*. Oil. $[\alpha]_D^{24}$ +140° (c, 0.27 in CHCl₃).

Zdero, C. *et al*, *Phytochemistry*, 1989, **28**, 1653.

2-Hydroxy-6-undecylbenzoic acid, 9CI   H-80264

Anagigantic acid

[14155-32-9]

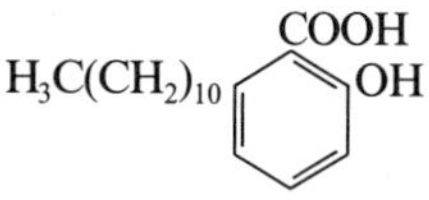

H₃C(CH₂)₁₀ —[benzene ring]— COOH, OH

$C_{18}H_{28}O_3$ M 292.417

Isol. from the pericarp of *Anacardium giganteum* and from *Schistochila appendiculata*. Needles (hexane). Mp 81-82°.

K salt: [108049-75-8].

Mp 201-204° dec.

Sharma, P.V.K. *et al*, *Indian J. Chem.*, 1966, **4**, 99 (*isol, struct*)
Durrani, A.A. *et al*, *J. Chem. Soc., Perkin Trans. 1*, 1979, 2069 (*synth, pmr, ir*)
Asakawa, Y. *et al*, *Phytochemistry*, 1987, **26**, 735 (*isol*)

3-Hydroxy-28-ursanoic acid   H-80265

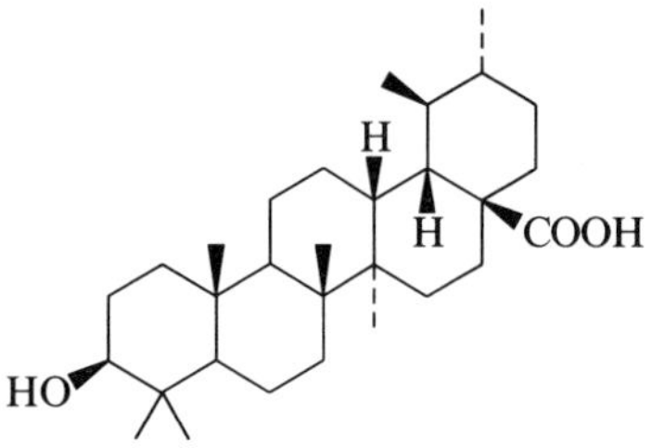

$C_{30}H_{50}O_3$ M 458.723

3β-form [120521-95-1]

Constit. of leaves of *Nerium oleander*. Cryst. (CHCl₃). Mp 150-152°. $[\alpha]_D$ +6.0° (c, 1 in CHCl₃).

Siddiqui, S. *et al*, *J. Nat. Prod.* (*Lloydia*), 1989, **52**, 57 (*isol, pmr, cmr*)

8-Hydroxyxanthone-1-carboxylic acid   H-80266

8-Hydroxy-9-oxo-9H-xanthene-1-carboxylic acid, 9CI

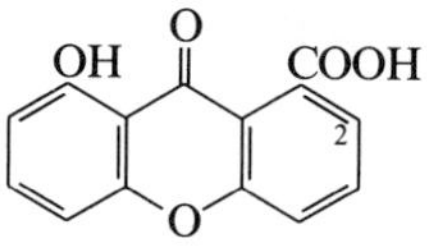

$C_{14}H_8O_5$ M 256.214

Me ester: [120461-93-0]. **Vertixanthone**
$C_{15}H_{10}O_5$ M 270.241
Metab. of *Leptographium wageneri*. Yellow cryst. (EtOAc/pet. ether). Mp 152-154°.

2-Hydroxy, Me ester: [120461-95-2]. **Hydroxyvertixanthone**
$C_{15}H_{10}O_6$ M 286.240
Metab. of *L. wageneri*. Yellow cryst. (Me₂CO). Mp 244-245°.

Ayer, W.A. *et al*, *J. Nat. Prod.* (*Lloydia*), 1989, **52**, 119 (*isol, ir, pmr, cmr*)

Hyperolactone   H-80267

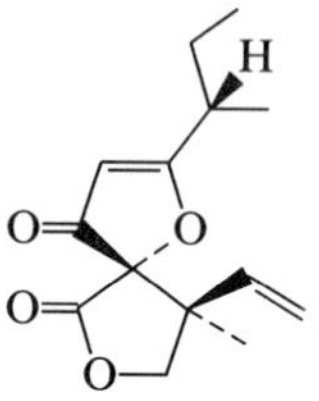

Relative configuration

$C_{14}H_{18}O_4$ M 250.294

Constit. of *Hypericum chinense*. Cryst. Mp 57°. $[\alpha]_D$ −228.93° (c, 0.13 in MeOH).

Tada, M. *et al*, *Chem. Lett.*, 1989, 683 (*cryst struct*)

I

Ichthynone
I-80001

Updated Entry replacing I-00016

6-Methoxy-3-(6-methoxy-1,3-benzodioxol-5-yl)-8,8-dimethyl-4H,8H-benzo[1,2-b:3,4b′]dipyran-4-one, 9CI

[24340-62-3]

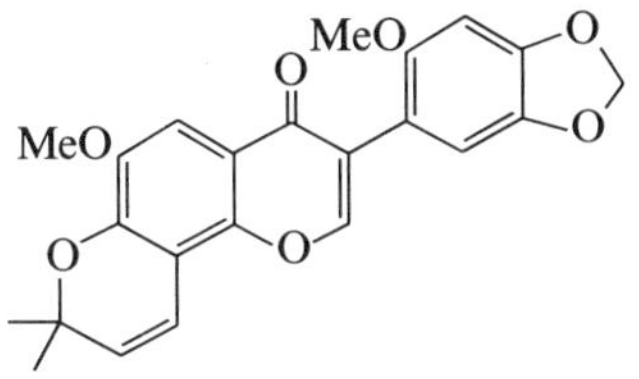

$C_{23}H_{20}O_7$ M 408.407

Present in root bark of Jamaican Dogwood (*Piscidia erythrina* and from *Millettia rubininosa*). Fish poison. Rods (EtOH). Mp 203-204°.

Schwarz, J.S.P. *et al*, *Tetrahedron*, 1964, **20**, 1317 (*isol, uv, pmr, struct*)
Campbell, R.V.M. *et al*, *J. Chem. Soc. C*, 1969, 1787 (*isol, ms*)
Kirshnamurti, M. *et al*, *Indian J. Chem.*, 1972, **10**, 914 (*synth*)
Desai, H.K. *et al*, *Indian J. Chem., Sect. B*, 1977, **15**, 291 (*isol*)

2-Icosanol
I-80002

2-Eicosanol, 9CI

[4340-76-5]

$$H_3C(CH_2)_{17}CH(OH)CH_3$$

$C_{20}H_{42}O$ M 298.551

(+)-*form*

Isol. from bacterial lipids *Mycobacterium* spp. Needles (MeOH). $[\alpha]_D$ +4.2° (CHCl$_3$), $[\alpha]_D$ +6.79° (Et$_2$O).

[34019-45-9]

Anderson, R.J. *et al*, *J. Am. Chem. Soc.*, 1936, **58**, 10; 1937, **59**, 858 (*isol*)
Serck-Hanssen, K. *et al*, *CA*, 1954, **48**, 5071 (*synth*)
Elémadi, A.H. *et al*, *Bull. Soc. Chim. Biol.*, 1965, **47**, 2095 (*biosynth*)
Pearce, P.J. *et al*, *J. Chem. Soc., Perkin Trans. 1*, 1972, 1655 (*synth*)

3-Icosene
I-80003

3-Eicosene, 9CI

[42448-86-2]

$$H_3C(CH_2)_{15}CH=CHCH_2CH_3$$

$C_{20}H_{40}$ M 280.536

Isol. in small amt. from rose oil (*Rosa* spp.).

[74685-33-9]

Wollrab, V. *et al*, *Collect. Czech. Chem. Commun.*, 1965, **30**, 1654.

Ilimaquinone
I-80004

Updated Entry replacing I-70002

[71678-03-0]

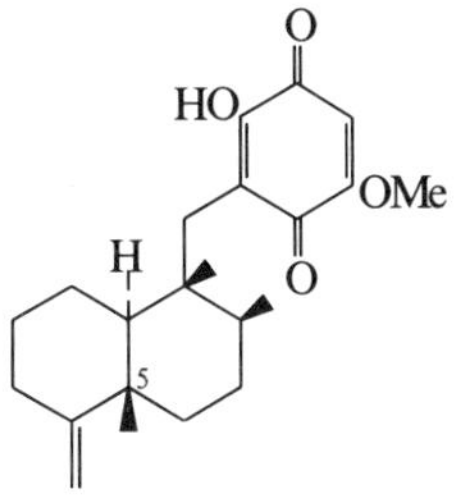

Absolute configuration

$C_{22}H_{30}O_4$ M 358.477

Constit. of *Hippiospongia metachromia*. Cryst. (hexane). Mp 113-114°. $[\alpha]_D^{23}$ −23.2° (c, 1.12 in CHCl$_3$).

5-Epimer: [96806-31-4]. **5-Epiilimaquinone**
$C_{22}H_{30}O_4$ M 358.477
Constit. of a *Fenestraspongia* sp.

O-De-Me: **Smenoquinone**
$C_{21}H_{28}O_4$ M 344.450
Constit. of *Smenospongia* sp. Cryst. Mp >350°.

Luibrand, R.T. *et al*, *Tetrahedron*, 1979, **35**, 609 (*isol*)
Carté, B. *et al*, *J. Org. Chem.*, 1985, **50**, 2785 (*Epiilimaquinone*)
Capon, R.J. *et al*, *J. Org. Chem.*, 1987, **52**, 5059 (*abs config*)
Kondracki, M.-L. *et al*, *Tetrahedron*, 1989, **45**, 1995 (*Smenoquinone*)

1H-Imidazole-4,5-dicarboxylic acid, 9CI
I-80005

Updated Entry replacing I-00115

[570-22-9]

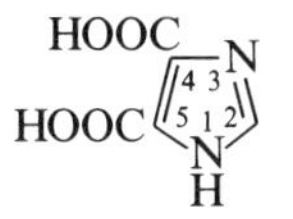

$C_5H_4N_2O_4$ M 156.098
Mp 288° dec.
▷ NI4760000.

Di-Me ester: [3304-70-9].
$C_7H_8N_2O_4$ M 184.151
Mp 200-203°.

Di-Et ester: [1080-79-1].
$C_9H_{12}N_2O_4$ M 212.205
Mp 151-152°.

Diamide: [83-39-6].
$C_5H_6N_4O_2$ M 154.128
Mp >360°.

Dihydrazide: [5423-20-1].
$C_5H_8N_6O_2$ M 184.157
Mp >375°.

Dinitrile: [1122-28-7]. **4,5-Dicyanoimidazole**
$C_5H_2N_4$ M 118.098
Cryst. (H$_2$O). Mp 172-173°.

1-Me: [19485-38-2].
$C_6H_6N_2O_4$ M 170.124
Cryst. Mp 259-260°.

1-Me, di-Me ester: [42545-22-2].

$C_8H_{10}N_2O_4$ M 198.178
Mp 46.5-47°.
1-Me, diamide:
   $C_6H_8N_4O_2$ M 168.155
   Mp 268° (263-266°).
1-Me, dinitrile: [19485-35-9].
   $C_6H_4N_4$ M 132.124
   Cryst. Mp 87-89°.

Baxter, R.A. *et al, J. Chem. Soc.*, 1945, 232; 1947, 378 (*derivs*)
Castle, R.N. *et al, J. Org. Chem.*, 1958, **23**, 1534 (*deriv*)
Bauer, L. *et al, J. Heterocycl. Chem.*, 1964, **1**, 275 (*derivs*)
Yamada, Y. *et al, Bull. Chem. Soc. Jpn.*, 1968, **41**, 1237 (*synth*)
Yasuda, N., *J. Heterocycl. Chem.*, 1985, **22**, 413 (*derivs*)
O'Connell, J.F. *et al, Synthesis*, 1988, 767 (*synth, pmr*)

4-Imidazolidinone   I-80006
[1704-79-6]

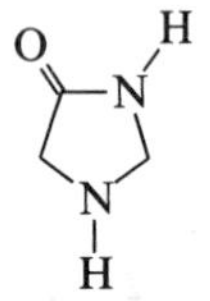

$C_3H_6N_2O$ M 86.093
Amorph. hygroscopic solid. Mp 64-67°.
3-Benzyl: [114981-12-3].
   $C_{10}H_{12}N_2O$ M 176.218
   Mp 164-169° (as hydrochloride).

Pfeiffer, U. *et al, Justus Liebigs Ann. Chem.*, 1988, 993 (*synth, pmr, cmr, bibl*)

Imidazo[2,1-*c*][1,2,4]triazin-4(1*H*)-one   I-80007
[59214-43-6]

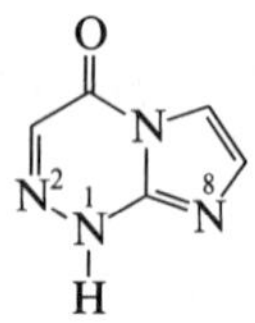

$C_5H_4N_4O$ M 136.113
Mp 265° dec.
1-Me: [112298-51-8].
   $C_6H_6N_4O$ M 150.140
   Mp 159-160°.
8-Me: [112298-48-3].
   $C_6H_6N_4O$ M 150.140
   Mp 280-283°.
2-Me: [112298-52-9].
   $C_6H_6N_4O$ M 150.140
   Mp 247-250° dec. Zwitterionic.

Farràs, J. *et al, J. Org. Chem.*, 1988, **53**, 887 (*synth, pmr, cmr, ir*)

1,1'-Iminobis[3,3-dimethyl-2-butanone],   I-80008
9CI
5-Aza-2,2,8,8-tetramethyl-3,7-nonanedione
[88686-46-8]

$(H_3C)_3CCOCH_2NHCH_2COC(CH_3)_3$

$C_{12}H_{23}NO_2$ M 213.319
Ligand. Cryst. by subl. Mp 61-62°.

Arduengo, A.J. *et al, J. Am. Chem. Soc.*, 1987, **109**, 627 (*synth, pmr, ir, use*)

3-Imino-1(3*H*)-isobenzofuranone, 9CI   I-80009
3-Iminophthalide
[484-90-2]

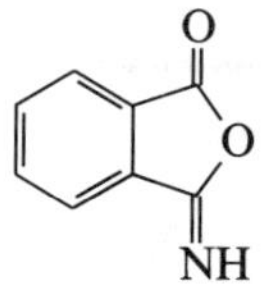

$C_8H_5NO_2$ M 147.133
Solid. Mp 112-116°. Exists largely as open chain tautomer.

Takeuchi, H. *et al, J. Chem. Soc., Perkin Trans.* 1, 1988, 2149 (*synth, pmr, ir*)

Indeno[1,2,3-*cd*]fluoranthene, 9CI, 8CI   I-80010
Dibenz[a,g]*pyracyclene. Bis*-peri-*phenylenenaphthalene*
[193-43-1]

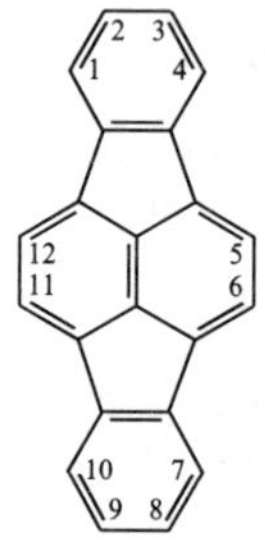

$C_{22}H_{12}$ M 276.337
Orange cryst. (C_6H_6). Mp 261-262°.
Dipicrate: Cryst. Mp 200°.

Clar, E. *et al, Nature* (*London*), 1950, **166**, 1075 (*uv*)
Stubbs, H.W.D. *et al, J. Chem. Soc.*, 1951, 2936 (*synth, uv*)
Beaton, J.M. *et al, J. Chem. Soc.*, 1952, 3870.
Davies, A. *et al, J. Chem. Soc. B*, 1968, 1337 (*props*)
Elschenbroich, C. *et al, Helv. Chim. Acta*, 1970, **53**, 838 (*esr*)

Indeno[2,1,7,6-*ghij*]naphtho[2,1,8,7-*nopq*]   I-80011
pleiadene, 9CI
[114324-69-5]

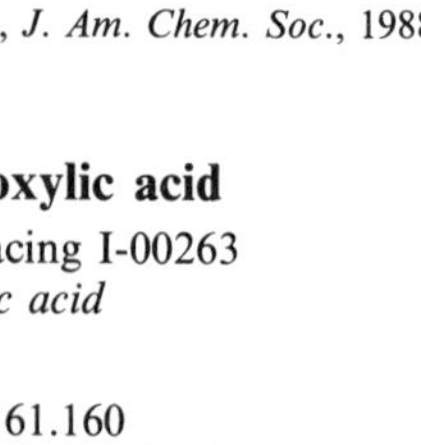

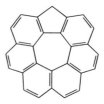

$C_{27}H_{14}$ M 338.408
Yellow needles (C_6H_6). Mp 260° dec.

Yamamoto, K. *et al, J. Am. Chem. Soc.*, 1988, **110**, 3578 (*synth, ir, pmr, uv*)

1*H*-Indole-3-carboxylic acid   I-80012
Updated Entry replacing I-00263
Indole-β-carboxylic acid
[771-50-6]
$C_9H_7NO_2$ M 161.160
Metab. of *Lasiodiplodia theobromae* and the marine algae *Undaria pinnatifida* and *Botryocladia leptopoda*. Mp 210-218° (198-200°).
Me ester: [942-24-5].
   $C_{10}H_9NO_2$ M 175.187

Isol. from the red alga *B. leptopoda*. Mp 147-148°
(140°). Probably an artifact.

Et ester: [776-41-0].
$C_{11}H_{11}NO_2$ M 189.213
Mp 122-124°.

Nitrile: [5457-28-3]. *3-Cyanoindole*
$C_9H_6N_2$ M 142.160
Mp 178°.

Nitrile, N-Ac:
$C_{11}H_8N_2O$ M 184.197
Mp 202°.

N-COOEt, Et ester:
$C_{14}H_{15}NO_4$ M 261.277
Cryst. (EtOH). Mp 102-104°.

N-Me:
$C_{10}H_9NO_2$ M 175.187
Cryst. (2-propanol). Mp 212° (200-201° dec.).

N-Benzyl:
$C_{16}H_{13}NO_2$ M 251.284
Cryst. (EtOAc/hexane). Mp 198-204° (194-196°).

Kasparek, S. *et al*, *Can. J. Chem.*, 1966, **40**, 2805 (synth)
Katner, A.S., *Org. Prep. Proced. Int.*, 1970, **2**, 297 (synth)
Aldridge, D.C. *et al*, *J. Chem. Soc. C*, 1971, 1623 (isol)
Abe, H. *et al*, *Agric. Biol. Chem.*, 1972, **36**, 2259 (isol, uv, ms)
Kakehi, A. *et al*, *J. Org. Chem.*, 1976, **41**, 1570 (esters)
Bano, S. *et al*, *Planta Med.*, 1987, **53**, 117 (isol, uv, pmr, cmr, ms, esters)
Baiocchi, L. *et al*, *J. Heterocycl. Chem.*, 1988, **25**, 1905 (synth)

1*H*-Indole-2,3-dicarboxylic acid   I-80013

Updated Entry replacing I-00268

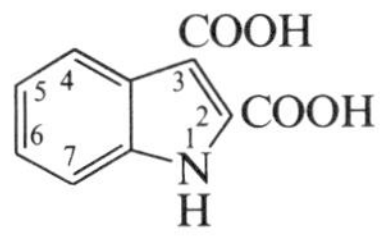

$C_{10}H_7NO_4$ M 205.170

3-Me ester: [34998-78-2].
$C_{11}H_9NO_4$ M 219.196
Cryst. (dioxan). Mp 254-255°.

Di-Me ester: [54781-93-0].
$C_{12}H_{11}NO_4$ M 233.223
Mp 112-112.5°.

N-Me:
$C_{11}H_9NO_4$ M 219.196
Cryst. (2-propanol). Mp 218° (208-209° dec.).

N-Benzyl:
$C_{17}H_{13}NO_4$ M 295.294
Cryst. + $1H_2O$ (EtOH). Mp 198° (194-196°).

N-Ph:
$C_{16}H_{11}NO_4$ M 281.267
Cryst. (MeOH). Mp 207-210° dec.

Huntress, E.H. *et al*, *J. Am. Chem. Soc.*, 1956, **78**, 419 (synth)
Baiocchi, L. *et al*, *J. Heterocycl. Chem.*, 1988, **25**, 1905 (synth, pmr)

5*H*-Indolo[1,7-*ab*][1]benzazepine, 9CI   I-80014

[202-01-7]

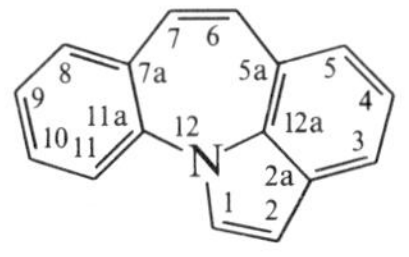

$C_{16}H_{11}N$ M 217.270

Yellow cryst. Mp 113-114°.

Hallberg, A. *et al*, *J. Heterocycl. Chem.*, 1981, **18**, 1255; 1983, **20**, 37; 1984, **21**, 1893 (synth, uv, ms, pmr, cmr)

myo-Inositol, 8CI   I-80015

Updated Entry replacing I-10038
(*1α,2α,3α,4β,5α,6β*)-*Cyclohexanehexol.* meso-*Inositol.*
Dambose. Nucitol. 1,2,3,5/4,6-Inositol. i-Inositol. Inositol.
Mesoinositol. Phaseomannitol. Myoinositol. Other
proprietary names
[87-89-8]

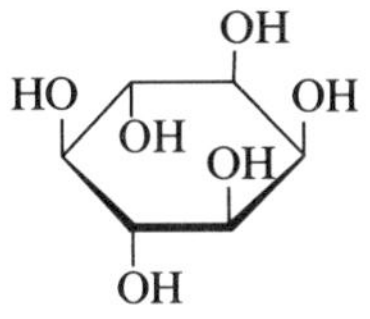

OH
HO   OH
OH OH
OH

$C_6H_{12}O_6$ M 180.157
The most widely distributed member of the group of
stereoisomers. Of common occurrence in plants and
animals. Obt. comly. from phytic acid in corn steep
liquor. Growth factor for animals and microorganisms.
Toxic and lipotropic. Mp 225°. Opt. inactive (*meso*-).

1,2-O-Cyclohexylidene: [6763-47-9]. *1,2-O-Cyclohexylidene-*
myo-*inositol*
$C_{12}H_{20}O_6$ M 260.286
Mp 179°.

1,2:4,5-Di-O-isopropylidene: [34379-30-1]. *1,2:4,5-Di-O-*
*isopropylidene-*myo-*inositol*
$C_{12}H_{20}O_6$ M 260.286
Mp 174°.

1,2:5,6-Di-O-cyclohexylidene: [40773-57-7]. *1,2:5,6-Di-O-*
*cyclohexylidene-*myo-*inositol*
$C_{18}H_{30}O_6$ M 342.431
Mp 133°, Mp 153-156°. $[\alpha]_D^{20}$ −7.4° (c, 0.5 in C_6H_6).

1,4,5,6-Tetra-Ac: *1,2,5,6-Tetra-O-acetyl-*myo-*inositol*
$C_{14}H_{20}O_{10}$ M 348.306
Mp 100°.

1,2,3,4,6-Penta-Ac:
$C_{16}H_{22}O_{11}$ M 390.343
Mp 178°.

Hexa-Ac: [1254-38-2].
$C_{18}H_{24}O_{12}$ M 432.380
Mp 216-217°.

Hexabenzoyl: *1,2,3,4,5,6-Hexa-O-benzoyl-*myo-*inositol*
$C_{48}H_{36}O_{12}$ M 804.805
Mp 258°.

Hexa(3-pyridinecarboxyl): [6556-11-2]. **Inositol nicotinate,**
BAN, INN. *Inositol niacinate, USAN.*
Hexanicotinoylinositol. Hexopal. Mesonex. Lipoflavonoid.
Numerous proprietary names
$C_{42}H_{30}N_6O_{12}$ M 810.732
Peripheral vasodilator, antilipidaemic agent. Mp 254-
255°.
▷ NM7535400.

1-Phosphate: [573-35-3].
$C_6H_{11}O_9P$ M 258.121
Wheat bran phytase hydrol. prod. of sodium phytate.
Mp 195-197° dec.

1-Phosphate, biscyclohexylammonium salt: Mp 202-204°.
$[\alpha]_D^{20}$ +3.4° (c, 1.0 in H_2O).

1,4-Diphosphate:
$C_6H_{10}O_{12}P_2$ M 336.085
Mp 206-207° (as bis(cyclohexylammonium) salt).

Hexaphosphate: [83-86-3]. **Phytic acid.** *Fytic acid, INN.*
Alkalovert
$C_6H_{18}O_{24}P_6$ M 660.036
Complexing agent for the removal of traces of heavy metals; therapeutic hypocalcaemic agent, freq. employed as nona-Na salt (Sodium phytate, USAN). Straw-coloured syrup.
▷ NM7525000.

[7205-52-9, 17211-15-3]

Badgett, C.O. *et al, J. Am. Chem. Soc.*, 1947, **69**, 2907 (*nicotinate*)
Biochem. Prep., 1952, **2**, 65.
Angyal, S.J. *et al, Adv. Carbohydr. Chem.*, 1959, **14**, 135 (*rev*)
Angyal, S.J. *et al, J. Chem. Soc.*, 1961, 4116, 4122, 4718.
Karrer, W. *et al, Konstitution und Vorkommen der Organischen Pflanzenstoffe*, 2nd Ed., Birkhäuser Verlag, Basel, 1972-1985, no. 280 (*bibl*)
Shvets, V.I. *et al, Tetrahedron*, 1973, **29**, 331.
Isbrandt, L.R. *et al, J. Am. Chem. Soc.*, 1980, **102**, 3144.
Martindale, The Extra Pharmacopoeia, 28th/29th Ed., Pharmaceutical Press, London, 1982/1989, 1056, 7877, 9243.
Wells, W.W., *Kirk-Othmer Encycl. Chem. Technol., 3rd Ed.*, 1978-1984, *Wiley, NY*, 3rd Ed., Wiley, N.Y., 1983, **24**, 50 (*rev*)
Bleasdale, J.E. *et al, Eds., Inositol and Phosphoinositides:Metabolism and Regulation*, Humane Press,Clifton, N.J., 1985 (*book*)
Holub, B.J., *Annu. Rev. Nutr.*, 1986, **6**, 563 (*rev*)
Negwer, M., *Organic-Chemical Drugs and their Synonyms*, 6th Ed., Akademie-Verlag, Berlin, 1987, 540, 588, 8197 (*synonyms*)

Intricata bromoallene   I-80016
[123297-18-7]

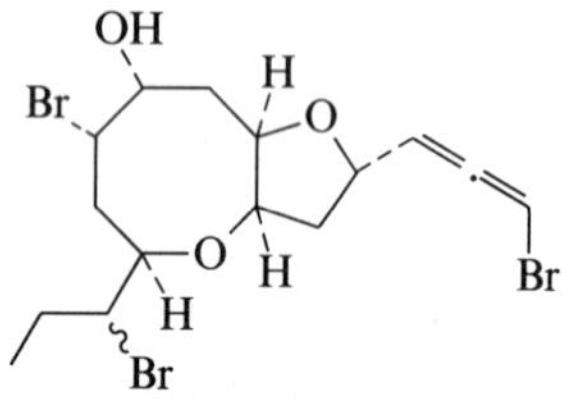

$C_{15}H_{21}Br_3O_3$ M 489.041
Constit. of *Laurencia intricata*. Oil. $[\alpha]_D^{21}$ +95.8° (c, 0.6 in $CHCl_3$).

Suzuki, M. *et al, Phytochemistry*, 1989, **28**, 2145 (*isol, pmr, cmr*)

Intricatinol   I-80017

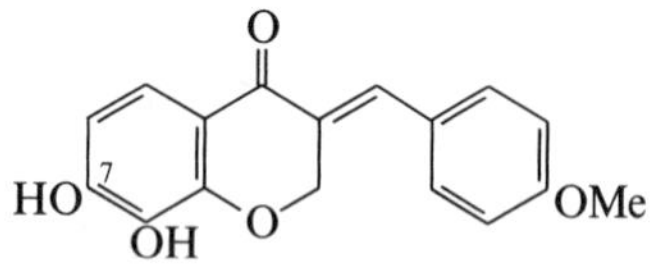

$C_{17}H_{14}O_5$ M 298.295
Constit. of *Hoffmanosseggia intricata*. Yellow needles (MeOH aq.). Mp 196-198°.
7-Me ether: **Intricatin**
$C_{18}H_{16}O_5$ M 312.321
Constit. of *H. intricata*. Yellow cryst. (MeOH aq.). Mp 157-159°.

Wall, M.E. *et al, J. Nat. Prod. (Lloydia)*, 1989, **28**, 774 (*isol, pmr*)

6-Iodobenzo[a]pyrene   I-80018
[39000-82-3]

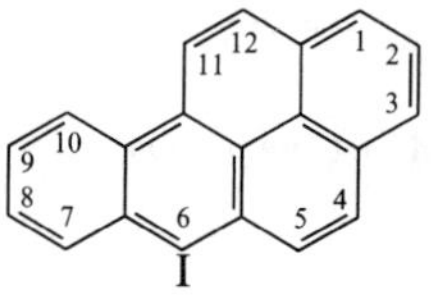

$C_{20}H_{11}I$ M 378.211
Cryst. Mp 214-215°.

Tye, R. *et al, Anal. Chem.*, 1955, **27**, 248 (*synth, uv*)
Johnson, M.D. *et al, Nature (London)*, 1973, **241**, 271 (*synth, pmr*)

1-Iodobicyclo[2.2.2]octane   I-80019
[931-98-6]

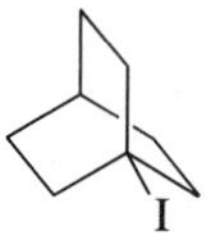

$C_8H_{13}I$ M 236.095
Cryst. (MeOH). Mp 27.5-28.5°.

Suzuki, Z. *et al, J. Org. Chem.*, 1967, **32**, 31 (*synth*)
Adcock, W. *et al, Organometallics*, 1987, **6**, 155 (*synth*)

1-Iodobiphenylene   I-80020
[17557-56-1]

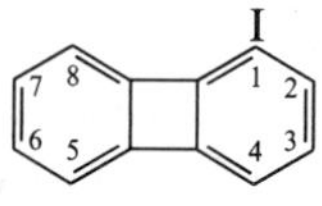

$C_{12}H_7I$ M 278.092
Yellow prisms (hexane). Mp 44-45°.
2,4,7-Trinitrofluorenone complex: [17557-55-0].
Deep red leaflets (C_6H_6/MeOH). Mp 177-179°.

Barton, J.W. *et al, J. Chem. Soc. C*, 1968, 28 (*synth, uv*)
Boulton, A.J. *et al, J. Chem. Soc. C*, 1968, 328 (*synth*)

2-Iodobiphenylene   I-80021
[96694-91-6]
$C_{12}H_7I$ M 278.092
Pale yellow plates (MeOH). Mp 64.5-65.5°.
2,4,7-Trinitrofluorenone complex: Scarlet cryst. (EtOH/AcOH). Mp 132.5-134°.

Baker, W. *et al, J. Chem. Soc.*, 1958, 2666 (*synth, uv*)
Cracknell, M.E. *et al, J. Chem. Soc., Perkin Trans. 1*, 1985, 115.

1-Iodocyclobutanecarboxylic acid   I-80022
[115420-18-3]

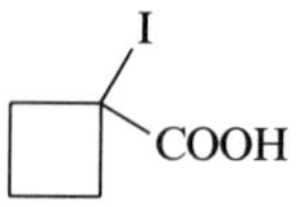

$C_5H_7IO_2$ M 226.014
Cryst. (Et_2O/pentane). Mp 48-50°.

Renaud, P. *et al, J. Org. Chem.*, 1988, **53**, 3745 (*synth, pmr, cmr*)

5-Iodo-4,4-dimethyl-1-pentene, 9CI   **I-80023**

[95106-80-2]

$$H_2C{=}CHCH_2C(CH_3)_2CH_2I$$

$C_7H_{13}I$ M 224.084
Light yellow oil. Bp_{12} 56-57°.

Pattenden, G. *et al*, *Tetrahedron*, 1987, **43**, 5637 (*synth, ir, pmr*)

1-Iodo-9*H*-fluorene   **I-80024**

[54366-33-5]

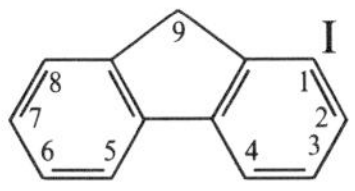

$C_{13}H_9I$ M 292.119
Mp 40-42°.

Minabe, M. *et al*, *J. Org. Chem.*, 1975, **40**, 1298 (*synth*)

2-Iodo-9*H*-fluorene   **I-80025**

[2523-42-4]

$C_{13}H_9I$ M 292.119
Cryst. (EtOH). Mp 129°.

Barnett, M.D. *et al*, *J. Am. Chem. Soc.*, 1959, **81**, 4583 (*synth*)
Chardonnens, L. *et al*, *Helv. Chim. Acta*, 1969, **52**, 1091 (*synth*)
Shapiro, M.J., *J. Org. Chem.*, 1978, **43**, 3769 (*cmr*)

9-Iodo-9*H*-fluorene   **I-80026**

[64421-01-8]

$C_{13}H_9I$ M 292.119
Light yellow cryst. (pet. ether). Mp 148-149° dec.
  Unstable, readily loses iodine.

Dickinson, J.D. *et al*, *J. Chem. Soc.*, 1959, 2337 (*synth*)
Shapiro, M.J. *et al*, *J. Org. Chem.*, 1978, **43**, 3769 (*synth, pmr, cmr*)
Hauser, A. *et al*, *J. Prakt. Chem.*, 1985, **327**, 433 (*pmr, cmr*)

1-Iodo-9*H*-fluoren-9-one   **I-80027**

[52086-21-2]

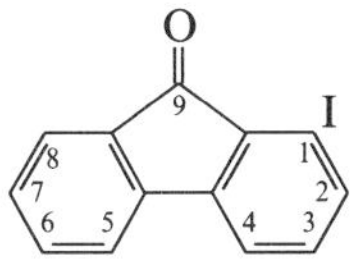

$C_{13}H_7IO$ M 306.102
Yellow needles (EtOH). Mp 146.5-147°.

Huntress, E.H. *et al*, *J. Am. Chem. Soc.*, 1942, **64**, 2845 (*synth*)
Hojo, M. *et al*, *J. Org. Chem.*, 1985, **50**, 1478.

2-Iodo-9*H*-fluoren-9-one   **I-80028**

[3096-46-6]

$C_{13}H_7IO$ M 306.102
Bright yellow needles (EtOH). Mp 152-153°.

Barnett, M.D. *et al*, *J. Am. Chem. Soc.*, 1959, **81**, 4583 (*synth*)
Chardonnens, L. *et al*, *Helv. Chim. Acta*, 1967, **52**, 1091 (*synth*)
Fukuyama, N. *et al*, *Bull. Chem. Soc. Jpn.*, 1987, **60**, 4363 (*synth*)

3-Iodo-9*H*-fluoren-9-one   **I-80029**

[19063-38-8]

$C_{13}H_7IO$ M 306.102
Yellow needles (EtOH). Mp 160.5-161°.

Harget, A.J. *et al*, *J. Chem. Soc. B*, 1968, 214 (*synth*)

4-Iodo-9*H*-fluoren-9-one   **I-80030**

[883-33-0]

$C_{13}H_7IO$ M 306.102
Yellow prisms (EtOH). Mp 121-122°.

Newman, M.S. *et al*, *J. Am. Chem. Soc.*, 1964, **86**, 5601 (*synth*)

8-Iodoguanine, 8CI   **I-80031**

2-Amino-1,7-dihydro-8-iodo-6H-purin-6-one. 2-Amino-6-hydroxy-8-iodopurine

[19690-21-2]

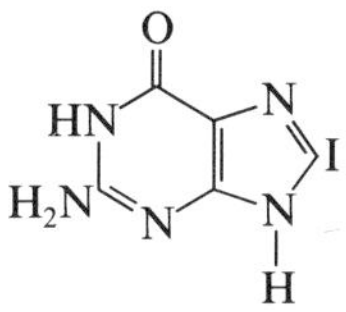

$C_5H_4IN_5O$ M 277.024
B,HCl: [21323-46-6].
  Cryst.

Lipkin, D. *et al*, *J. Biol. Chem.*, 1963, **238**, 6 (*synth*)
Holmes, R.E. *et al*, *J. Am. Chem. Soc.*, 1964, **86**, 1242 (*synth*)
Koda, R.T. *et al*, *J. Polym. Sci.*, 1968, **57**, 2056 (*synth*)

7-Iodo-1,3-heptadiene   **I-80032**

$$H_2C{=}CHCH{=}CHCH_2CH_2CH_2I$$

$C_7H_{11}I$ M 222.068
(*E*)-*form*
  Liq. $Bp_{3.5}$ 55-60°.

Vedejs, E. *et al*, *J. Org. Chem.*, 1988, **53**, 2220 (*synth, ir, pmr*)

Iodomethanol   **I-80033**

[50398-30-6]

$$ICH_2OH$$

CH_3IO M 157.939
Ac: [13398-11-3].
  $C_3H_5IO_2$ M 199.976
  Alkylating agent. Oil. d_4^{20} 1.953. Bp_{14} 49-50°. n_D^{20} 1.5130.
  ▷ Lachrymator.
Benzoyl: [13943-33-4].
  $C_8H_7IO_2$ M 262.047
  Cryst. (pet. ether). Mp 32-33°.
4-Methylbenzenesulfonyl: [117184-52-8].
  $C_8H_9IO_3S$ M 312.128
  Oil. $Bp_{0.2}$ 115-120°.

Renshaw, R.R. *et al*, *J. Am. Chem. Soc.*, 1925, **47**, 2989 (*acetate*)
Euranto, E.K. *et al*, *Acta Chem. Scand.*, 1966, **20**, 1966 (*acetate*)
Wittig, G. *et al*, *Justus Liebigs Ann. Chem.*, 1967, **702**, 24 (*benzoate*)
Hahn, R.C. *et al*, *J. Org. Chem.*, 1988, **53**, 5783 (*tosylate, synth, pmr, cmr*)

2-Iodomethyl-3-methyloxirane   **I-80034**

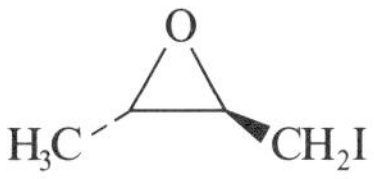

C_4H_7IO M 198.003
(2*RS*,3*SR*)-*form*

(±)-trans-form
Bp$_{4.2}$ 39-40°.

Evans, R.D. *et al*, *Synthesis*, 1988, 862 (*synth*, *pmr*)

2-(Iodomethyl)naphthalene   I-80035

[24515-49-9]
C$_{11}$H$_9$I   M 268.097
Cryst. (EtOH). Mp 79° (72-73.5°). Unstable.

Daub, G.H. *et al*, *J. Org. Chem.*, 1954, **19**, 1571 (*synth*)
Bilger, C. *et al*, *Synthesis*, 1988, 902 (*synth*)

6-(Iodomethyl)-2-piperidinone, 9CI   I-80036

[100556-63-6]

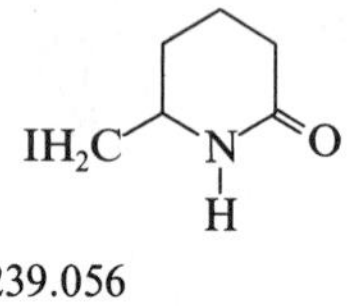

C$_6$H$_{10}$INO   M 239.056
(±)-*form*
Cryst. (Et$_2$O/hexane). Mp 157-158°.

Knapp, S. *et al*, *J. Org. Chem.*, 1988, **53**, 4006 (*synth*, *ir*, *pmr*)

5-(Iodomethyl)-2-pyrrolidinone, 9CI   I-80037

[5831-75-4]

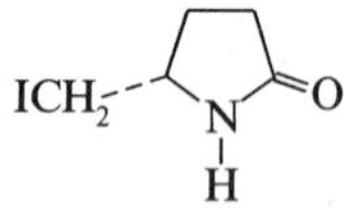

(S)-form

C$_5$H$_8$INO   M 225.029
(S)-*form* [29266-73-7]
  L-form
  Cryst. (EtOAc). Mp 86-87°. [α]$_D$ −55° (c, 1.24 in EtOH).
N-*Ac*:
  C$_7$H$_{10}$INO$_2$ M 267.066
  Yellowish oil. [α]$_D$ −82° (c, 3.6 in EtOH).
(±)-*form*
  Cryst. (Et$_2$O/hexane). Mp 68-70°.

Hardegger, E. *et al*, *Helv. Chim. Acta*, 1955, **38**, 312.
Molin-Case, J.A. *et al*, *J. Am. Chem. Soc.*, 1970, **92**, 4728 (*cryst struct*, *abs config*)
Knapp, S. *et al*, *J. Org. Chem.*, 1988, **53**, 4006 (*synth*, *ir*, *pmr*)

2-Iodo-6-nitrobenzaldehyde   I-80038

[117847-41-3]
C$_7$H$_4$INO$_3$ M 277.018
Cryst. (EtOH). Mp 120-122°.

Hewgill, F.R. *et al*, *J. Chem. Soc.*, *Perkin Trans.* 1, 1988, 1305 (*synth*, *ir*, *pmr*)

2-Iodo-3-phenyl-2-propenal, 9CI   I-80039

α-*Iodocinnamaldehyde*

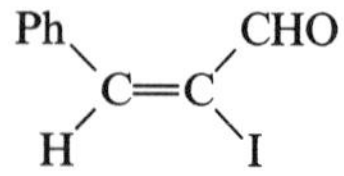

C$_9$H$_7$IO   M 258.058
(E)-*form* [52741-39-6]
  Yellow prisms (EtOH). Mp 89° (85-86°).

[116544-98-0]
Coulomb, F. *et al*, *Bull. Soc. Chim. Fr.*, 1973, 3352 (*synth*, *ir*, *pmr*)
Suzuki, H. *et al*, *Synthesis*, 1988, 236 (*synth*, *ir*, *pmr*, *ms*)

3(5)-Iodopyrazole, 8CI   I-80040

[4522-35-4]

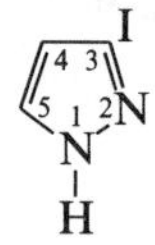

C$_3$H$_3$IN$_2$ M 193.975
Needles (H$_2$O). Mp 72-73°.
1-Me: [92525-10-5]. *3-Iodo-1-methylpyrazole, 9CI*
  C$_4$H$_5$IN$_2$ M 208.001
  Bp$_{11}$ 100-102°.
2-Me: [34091-51-5]. *5-Iodo-1-methylpyrazole, 9CI*
  C$_4$H$_5$IN$_2$ M 208.001
  Cryst. (ligroin). Mp 76-76.5°. Sublimes at 50°/2mm.

Reimlinger, H.K. *et al*, *Chem. Ber.*, 1962, **94**, 1036 (*synth*)
Vasilevskii, S.F. *et al*, *Bull. Acad. Sci. USSR, Div. Chem. Sci.* (*Engl. Transl.*), 1971, 1655 (*synth*, *deriv*)
Effenberger, F. *et al*, *J. Org. Chem.*, 1984, **49**, 4687 (*synth*, *pmr*, *derivs*)

4-Iodopyrazole, 9CI   I-80041

[3469-69-0]
C$_3$H$_3$IN$_2$ M 193.975
Needles (cyclohexane). Mp 109-110°.
1-Me: [39806-90-1]. *4-Iodo-1-methylpyrazole, 9CI*
  C$_4$H$_5$IN$_2$ M 208.001
  Crysts. (C$_6$H$_6$) or prisms (hexane), with odour like that of collidine. Mp 64-65°. Bp$_{15}$ 101-105°.

Hüttel, R. *et al*, *Justus Liebigs Ann. Chem.*, 1955, **593**, 200 (*synth*, *deriv*)
Elguero, J. *et al*, *Bull. Soc. Chim. Fr.*, 1966, 3727, 3744 (*pmr*, *uv*)
Vasilevskii, S.F. *et al*, *Bull. Acad. Sci. USSR, Div. Chem. Sci.* (*Engl. Transl.*), 1972, 2453 (*deriv*)
Rydberg, V. *et al*, *J. Chromatogr.*, 1972, **64**, 170 (*glc*, *ms*)
Liljefors, S. *et al*, *Chem. Scr.*, 1980, **15**, 102 (*synth*, *deriv*)
Cabildo, P. *et al*, *Org. Magn. Reson.*, 1984, **22**, 603 (*cmr*)

5-Iodo-1*H*-tetrazole, 9CI, 8CI   I-80042

[66924-15-0]

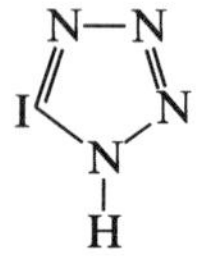

CHIN$_4$ M 195.950
Cryst. (EtOH). Sol. hot H$_2$O, spar. sol. Et$_2$O. Mp 190°. pK_a 4.66.
1-Ph:
  C$_7$H$_5$IN$_4$ M 272.048
  Cryst. (EtOH). Mp 140° dec.

Stollé, R. *et al*, *Ber.*, 1929, **62**, 1118 (*synth*)
Stollé, R. *et al*, *J. Prakt. Chem.*, 1932, **134**, 282 (*deriv*)
Leiber, E. *et al*, *J. Am. Chem. Soc.*, 1951, **73**, 1792 (*props*)

Iresin   I-80043

Updated Entry replacing I-00832

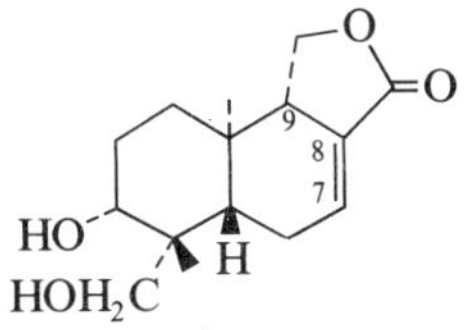

C$_{15}$H$_{22}$O$_4$ M 266.336
Constit. of *Iresine celosioides*. Cryst. (Me$_2$CO/hexane). Mp
140-142°. [α]$_D^{28}$ +21° (CHCl$_3$).

7,8β-Dihydro: **Dihydroiresin**
C$_{15}$H$_{24}$O$_4$ M 268.352
Minor constit. of *I. celosioides*. Mp 212-213° (as
diacetate).

Δ^{8,9}-*Isomer:* [561-89-7]. **Isoiresin**
C$_{15}$H$_{22}$O$_4$ M 266.336
Constit. of *I. celosioides*. Oil. Mp 166-168° (as
diacetate). [α]$_D$ −73.0° (CHCl$_3$) (diacetate).

7,8β-Dihydro, 3-ketone: **Dihydroiresone**
C$_{15}$H$_{22}$O$_4$ M 266.336
Isol. from *I. celosioides*. Needles (C$_6$H$_6$/MeOH). Mp
215-219°.

Crabbé, P. *et al, Bull. Soc. Chim. Belg.,* 1958, **67**, 632 (*derivs*)
Djerassi, C. *et al, J. Am. Chem. Soc.,* 1958, **80**, 2593 (*struct*)
Rossmann, H.G. *et al, Tetrahedron,* 1958, **4**, 275 (*cryst struct*)
Pettetier, S.W. *et al, J. Am. Chem. Soc.,* 1968, **90**, 5318 (*synth*)

Irispurinol   I-80044

[119459-73-3]

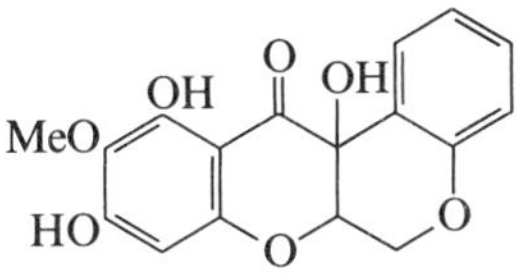

C$_{17}$H$_{14}$O$_7$ M 330.293
Constit. of *Iris spuria*. Pale yellow needles (EtOAc). Mp
255-256°. [α]$_D^{22}$ −75.2° (MeOH).

Shawl, A.S. *et al, Phytochemistry,* 1988, **27**, 3331.

Isoalantodiene   I-80045

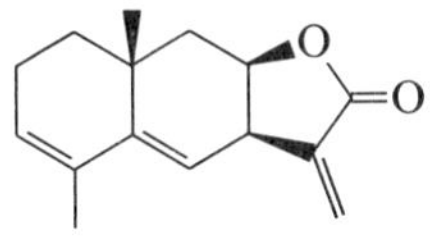

C$_{15}$H$_{18}$O$_2$ M 230.306
Constit. of *Inula racemosa*. Cryst. Mp 82°.

Kalsi, P.S. *et al, Phytochemistry,* 1989, **28**, 2093 (*isol, pmr*)

Isoalantolactone   I-80046

Updated Entry replacing I-30064
Isohelenin
[470-17-7]

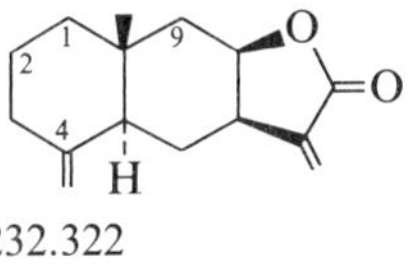

C$_{15}$H$_{20}$O$_2$ M 232.322

Constit. of the essential oil of *Inula helenium, I. racemosa*
and *Telekia speciosa*. Cryst. (EtOH aq.). Mp 109-110°
(115°). [α]$_D$ +172° (CHCl$_3$).

1α,2α-Diacetoxy: **1α,2α-Diacetoxyalantolactone**
C$_{19}$H$_{24}$O$_6$ M 348.395
Constit. of *Inezia integrifolia*. Gum. [α]$_D^{24}$ +68° (c, 0.11
in CHCl$_3$).

1-Oxo: **1-Oxoisoalantolactone**
C$_{15}$H$_{18}$O$_3$ M 246.305
Constit. of *Flourensia macrophylla*. Cryst. Mp 164°.
Incorrectly called 1-Oxoalantolactone in the lit.

!9α-Hydroxy: **9α-Hydroxyisoalantolactone**
C$_{15}$H$_{20}$O$_3$ M 248.321
Constit. of *F. macrophylla*. Oil. Incorrectly called 9-
Hydroxyalantolactone.

9β-Hydroxy: **9β-Hydroxyisoalantolactone**
C$_{15}$H$_{20}$O$_3$ M 248.321
Isol. from *F. macrophylla*. Cryst. Mp 171°. [α]$_D^{24}$ +78° (c,
1.9 in CHCl$_3$).

9-Oxo: **9-Oxoisoalantolactone**
C$_{15}$H$_{18}$O$_3$ M 246.305
Isol. from *F. macrophylla*. Cryst. Mp 118°. Incorrectly
called 9-Oxoalantolactone.

4α,15-Epoxy: **4α,15-Epoxyisoalantolactone**
C$_{15}$H$_{20}$O$_3$ M 248.321
Constit. of *Ambrosia artemisioides*. Oil.

4α,15-Epoxy, 3α-hydroxy: **4α,15-Epoxy-3α-
hydroxyisoalantolactone**
C$_{15}$H$_{20}$O$_4$ M 264.321
Constit. of *A. artemisioides*. Oil.

11,12-Dihydro: **Dihydroisoalantolactone**
C$_{15}$H$_{22}$O$_2$ M 234.338
Isol. from roots of *I. helenium*. Needles (EtOH). Mp
168-174°.

Asselineau, C. *et al, Compt. Rend. Hebd. Seances Acad. Sci.,* 1958,
  246, 1874 (*struct*)
Marshall, J.A. *et al, J. Org. Chem.,* 1964, **29**, 3727 (*pmr, struct*)
Miller, R.B. *et al, Tetrahedron,* 1974, **30**, 2961 (*synth*)
Tada, M., *Chem. Lett.,* 1982, 441 (*synth*)
Bohlmann, F. *et al, Phytochemistry,* 1982, **21**, 2743 (*derivs*)
Bohlmann, F. *et al, Phytochemistry,* 1984, **23**, 1445 (*derivs*)
Jakupovic, J. *et al, Phytochemistry,* 1988, **27**, 3551 (*derivs*)

Isoalloalantolactone   I-80047

Updated Entry replacing I-40044
[64395-76-2]

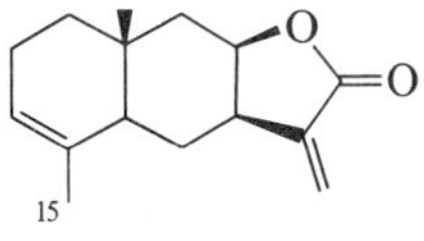

C$_{15}$H$_{20}$O$_2$ M 232.322
Constit. of *Inula racemosa*. Oil.

15-Hydroxy: [119285-33-5]. **15-Hydroxyisoalloalantolactone**
C$_{15}$H$_{20}$O$_3$ M 248.321
Constit. of *Ambrosia artemisioides*. Oil.

15-Acetoxy: [119285-34-6]. **15-Acetoxyisoalloalantolactone**
C$_{17}$H$_{22}$O$_4$ M 290.358
Constit. of *A. artemisioides*. Oil.

15-Cinnamoyloxy (E-): [119285-35-7]. **15-
Cinnamoyloxyisoalloalantolactone**
C$_{24}$H$_{26}$O$_4$ M 378.467
Constit. of *A. artemisioides*. Oil.

15-Cinnamoyloxy (Z-): [119285-36-8].
C$_{24}$H$_{26}$O$_4$ M 378.467

Constit. of *A. artemisioides*. Oil.

Kaur, B. *et al*, *Phytochemistry*, 1985, **24**, 2007 (*isol*)
Jakupovic, J. *et al*, *Phytochemistry*, 1988, **27**, 3551 (*derivs*)

Isobharangin   I-80048

3β,11-Dihydroxy-5,7,9,13-abietatetraene-2,12-dione
[121688-14-0]

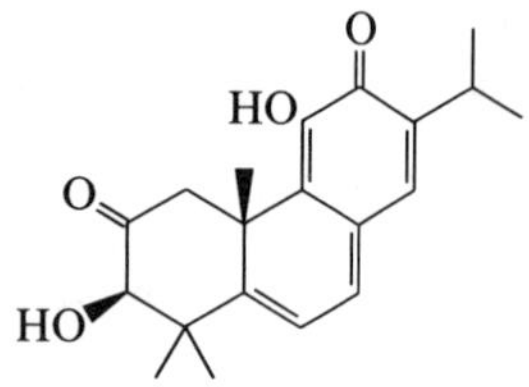

$C_{20}H_{24}O_4$ M 328.407
Constit. of *Pygmacopremna herbacea*. Brown solid. Mp 74-76°. $[\alpha]_D^{26}$ −465° (c, 0.115 in $CHCl_3$).

Sankaram, A.V.B. *et al*, *Tetrahedron Lett.*, 1989, **30**, 867.

Isobuteine   I-80049

Updated Entry replacing I-30069
S-(2-Carboxypropyl)cysteine, 9CI. 3-[(2-Carboxypropyl)thio]alanine
[4746-40-1]

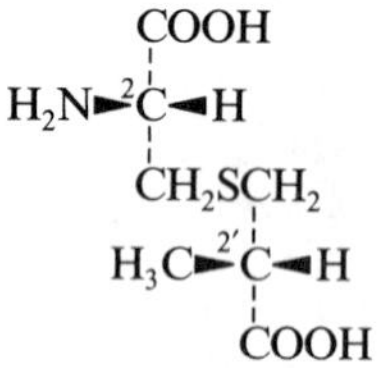

(2R,2′S)-form

$C_7H_{13}NO_4S$ M 207.250
(2R,2′S)-form [66512-75-2]
Occurs naturally as a component of the tripeptide *S*-(2-carboxypropyl)glutathione in onion and in garlic. Also isol. from urine of cystathionuric patients. Intermed. in biosynth. of cycloalliin. Cryst. (Me_2CO aq.). Mp 196-198° dec. $[\alpha]_D^{25}$ −66.1° (c, 2.5 in H_2O), −44.3° (c, 2.4 in 2.5M HCl). Diastereoisomers also known.
S,S-Dioxide: [66403-99-4]. *3-[(2-Carboxypropyl)sulfonyl]alanine, 9CI*
$C_7H_{13}NO_6S$ M 239.249
Blades (H_2O). Mp 174° dec. (gas evolution). $[\alpha]_D^{25}$ −24° (c, 1.7 in H_2O).

[6852-42-2, 66429-71-8, 66512-76-3]

Virtanen, A.I. *et al*, *Hoppe Seyler's Z. Physiol. Chem.*, 1960, **322**, 8 (*isol*)
Granroth, B. *et al*, *Acta Chem. Scand.*, 1967, **21**, 1654 (*bisynth*)
Carson, J.F., *J. Chem. Soc., Perkin Trans. 1*, 1977, 1964 (*synth, pmr*)
Kasai, T. *et al*, *Agric. Biol. Chem.*, 1981, **45**, 433 (*occur*)
Parry, R.J. *et al*, *J. Am. Chem. Soc.*, 1985, **107**, 2512 (*synth, abs config*)

Isobutyrylmallotochromene   I-80050

[116964-16-0]

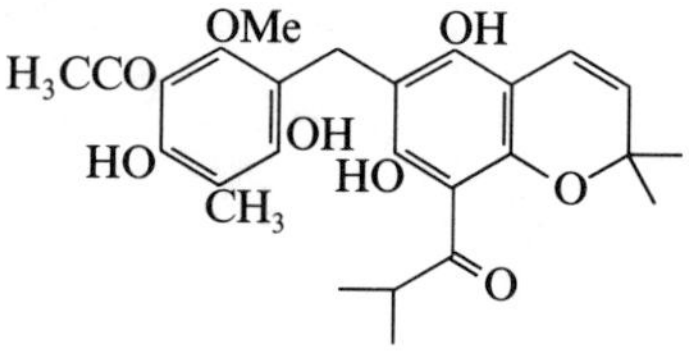

$C_{26}H_{30}O_8$ M 470.518
Constit. of *Mallotus japonicus*. Cytotoxic agent. Yellow needles (MeOH). Mp 180-181°.

Fujita, A. *et al*, *J. Nat. Prod. (Lloydia)*, 1988, **51**, 708.

Isobyakangelicolic acid   I-80051

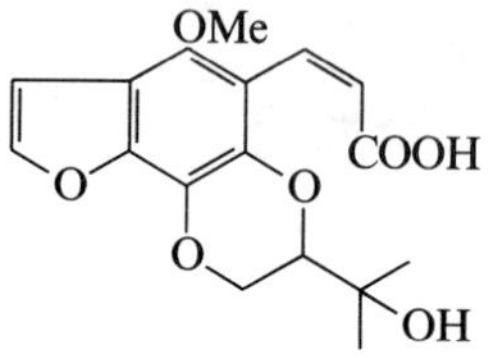

$C_{17}H_{18}O_7$ M 334.325
Doubtful struct. Isol. from *Angelica japonica* and *A. glabra*. Tablets (EtOH). Mp 220°. $[\alpha]_D^{16}$ +5.3° (Py).

Noguchi, T. *et al*, *Yakugaku Zasshi (J. Pharm. Soc. Jpn.)*, 1941, **61**, 77 (*isol*)
Hata, K. *et al*, *Yakugaku Zasshi (J. Pharm. Soc. Jpn.)*, 1960, **54**, 25581 (*isol*)

Isochandalone   I-80052

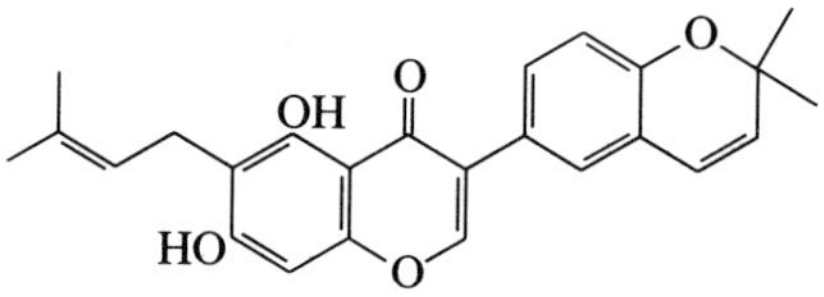

$C_{25}H_{24}O_5$ M 404.462
Constit. of *Lupinus albus*. Amorph. powder.

Tahara, S. *et al*, *Phytochemistry*, 1989, **28**, 901.

1-Isocyanatoadamantane   I-80053

1-Isocyanatotricyclo[3.3.1.1^{3,7}]decane, 9CI. 1-Adamantyl isocyanate
[4411-25-0]

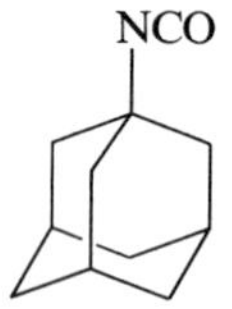

$C_{11}H_{15}NO$ M 177.246
Cryst. Mp 147°.

Stetter, H. *et al*, *Chem. Ber.*, 1962, **95**, 2302 (*synth*)
Boyer, J.H. *et al*, *Synthesis*, 1987, 907 (*synth*)

Isocyanatodiphenylmethane   I-80054

1,1′-(Isocyanatomethylene)bisbenzene, 9CI. Diphenylmethyl isocyanate
[3066-44-2]

Ph_2CHNCO

$C_{14}H_{11}NO$ M 209.247
Liq. Bp_4 148°.

Donleavy, J.J. *et al*, *J. Am. Chem. Soc.*, 1940, **62**, 218 (*synth*)
Boyer, J.H. *et al*, *Synthesis*, 1987, 907 (*synth*)

(Isocyanatomethyl)benzene, 9CI   I-80055

Benzyl isocyanate
[3173-56-6]

$$PhCH_2NCO$$

C_8H_7NO M 133.149
Liq. Bp_{30} 137°, Bp_4 65°.

Ferstandig, L.L. *et al*, *J. Am. Chem. Soc.*, 1959, **81**, 4838 (*synth*)
Boyer, J.H. *et al*, *Synthesis*, 1987, 907 (*synth*)

Isocyanatotriphenylmethane   I-80056

1,1′,1″-(Isocyanatomethylidyne)trisbenzene, 9CI.
Triphenylmethyl isocyanate. Trityl isocyanate
[4737-21-7]

$$Ph_3CNCO$$

$C_{20}H_{15}NO$ M 285.345
Cryst. (pet. ether). V. sol. org. solvs. Mp 91° (85-87°).

Jones, L.W. *et al*, *J. Am. Chem. Soc.*, 1921, **43**, 2422 (*synth*)
Boyer, J.H. *et al*, *Synthesis*, 1987, 907 (*synth*)

Isocyanic acid, 8CI   I-80057

Updated Entry replacing I-20052
[75-13-8]

$$HN{=}C{=}O$$

CHNO   M 43.025
Contrary to popular belief, is not in equilib. with Cyanic
acid, C-80167. Is freq. confused with cyanic acid in CA.
Reagent for conversion of alcohols to allophanates.
Also adds to carbonyl compds. and alkenes. Liq. or gas
with strongly acrid odour. d^0 1.140. Mp −86°. Bp 23.5°.
pK_a 3.66. For esters (isocyanates) see the individual
entries.

▷ Lachrymator, vesicant.

v. Dohlen, W.C. *et al*, *Acta Crystallogr.*, 1955, **8**, 646 (*struct,
synth, ir*)
Steyemark, P.R., *J. Org. Chem.*, 1963, **28**, 586 (*synth*)
Groving, N. *et al*, *Acta Chem. Scand.*, 1965, **19**, 1768 (*synth*)
Belson, D.J. *et al*, *Chem. Soc. Rev.*, 1982, **11**, 41 (*rev*)
Teles, J.H. *et al*, *Chem. Ber.*, 1989, **122**, 753 (*struct, bibl*)

1-Isocyanobutane, 9CI   I-80058

Butyl isocyanide
[2769-64-4]

$$H_3CCH_2CH_2CH_2NC$$

C_5H_9N M 83.133
Extremely disagreeable and pervasive smelling liq. d 0.80.
Bp 124-125°.

▷ Causes headaches and nausea.

Davis, T.L. *et al*, *J. Am. Chem. Soc.*, 1937, **59**, 1998 (*synth*)
Malatesta, L., *Gazz. Chim. Ital.*, 1947, **77**, 238 (*synth*)
Ugi, I. *et al*, *Chem. Ber.*, 1960, **93**, 239 (*synth, ir*)
Kamer, P.C.J. *et al*, *J. Am. Chem. Soc.*, 1988, **110**, 6818 (*synth, ir,
pmr*)

3-Isocyanopentane, 9CI   I-80059

3-Pentyl isocyanide
[115591-41-8]

$$(H_3CCH_2)_2CHNC$$

$C_6H_{11}N$ M 97.160
Kamer, P.C.J. *et al*, *J. Am. Chem. Soc.*, 1988, **110**, 6818 (*synth, ir,
pmr*)

Isoderrone   I-80060

[121747-89-5]

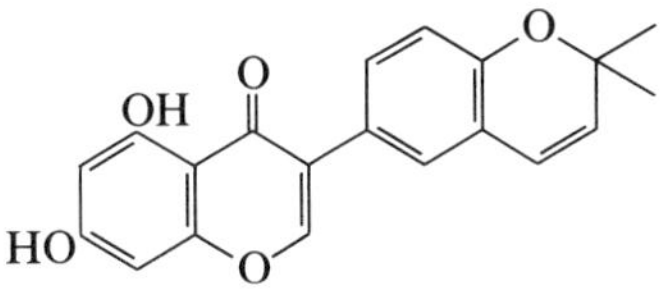

$C_{20}H_{16}O_5$ M 336.343
Constit. of *Lupinus albus*. Amorph. solid.

Tahara, S. *et al*, *Phytochemistry*, 1989, **28**, 901.

Isoglycyrol   I-80061

*3,4-Dihydro-11-hydroxy-5-methoxy-2,2-dimethyl-2H,8H-
benzofuro[3,2-c]pyrano[2,3-f][1]benzopyran-8-one, 9CI*
[23013-86-7]

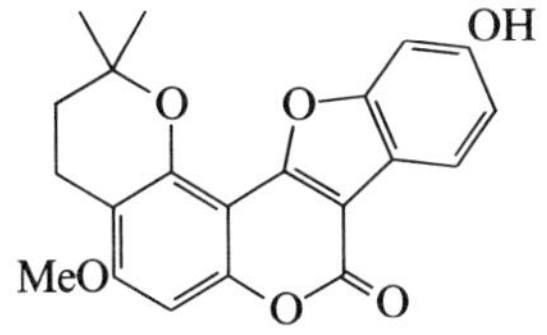

$C_{21}H_{18}O_6$ M 366.370
Isol. from roots of *Glycyrrhiza* sp. Needles (Me_2CO). Mp
298-300° dec.

Ac: Needles (MeOH). Mp 175-176°, Mp 197-197.5°
(double Mp).

Me ether: Pale-yellow needles (Me_2CO). Mp 225.5-224°.

Saitoh, T. *et al*, *Chem. Pharm. Bull.*, 1969, **17**, 729.

Isogutiesolbriolide   I-80062

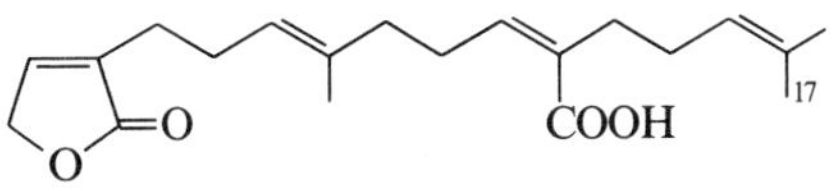

$C_{20}H_{28}O_4$ M 332.439
Parent compd. unknown. L540.

17-Hydroxy: **17-Hydroxyisogutiesolbriolide**
$C_{20}H_{28}O_5$ M 348.438
Constit. of *Gutierrezia solbrigii*.

17-Acetoxy: **17-Acetoxyisogutiesolbriolide**
$C_{22}H_{30}O_6$ M 390.475
Constit. of *G. solbrigii*.

Jakupovic, J. *et al*, *Tetrahedron*, 1985, **41**, 4537.

Isoindole   I-80063

Updated Entry replacing I-20061
Benzo[c]pyrrole
[270-68-8]

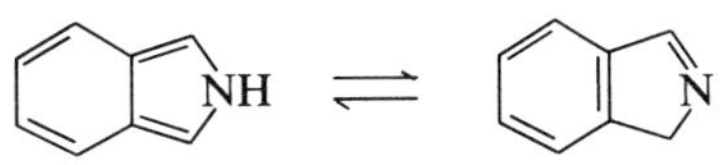

C_8H_7N M 117.150

Solid obt. by pyrolysis and freezing at $-196°$. λ_{max} 263.5, 275, 286.5, 294 (infl.), 300, 306.5, 312.5, 320, 326.5 and 335 nm(hexane).Solid dec. rapidly at r.t., mod. stable in soln. under N_2.

N-tert-*Butyl*: [55023-87-5].
$C_{12}H_{15}N$ M 173.257
Bp_{11} 97-98°.

N-tert-*Butyl*; *B,HClO₄*: [73286-77-8].
Cryst. Mp 241-243° dec.

Kreher, R. *et al*, *Z. Naturforsch., B*, 1965, **20**, 75 (*synth*)
White, J.D. *et al*, *Adv. Heterocycl. Chem.*, 1969, **10**, 113 (*rev*)
Dewar, M.J.S. *et al*, *Tetrahedron*, 1970, **26**, 4505 (*props*)
Bonnett, R. *et al*, *J. Chem. Soc., Perkin Trans. 1*, 1973, 1432 (*synth*)
Chacko, E. *et al*, *Tetrahedron Lett.*, 1977, 1095 (*props*)
Bonnett, R. *et al*, *Adv. Heterocycl. Chem.*, 1981, **29**, 341 (*rev*)
Kreher, R.P. *et al*, *Chem. Ber.*, 1989, **122**, 337 (*derivs*)

Isoindolo[1,2,3-*de*]quinolizinium(1 +), 9CI   I-80064

10c-Azoniafluoranthene

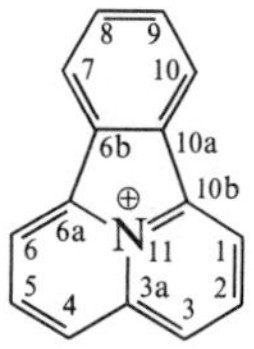

$C_{15}H_{10}N^{\oplus}$ M 204.251 (ion)

Hexafluorophosphate: [120229-35-8].
$C_{15}H_{10}F_6NP$ M 349.215
Transparent needles (MeCN).

Fourmigué, M. *et al*, *Angew. Chem., Int. Ed. Engl.*, 1989, **28**, 588 (*synth, pmr, struct, cmr, uv*)

Isojamaicin   I-80065

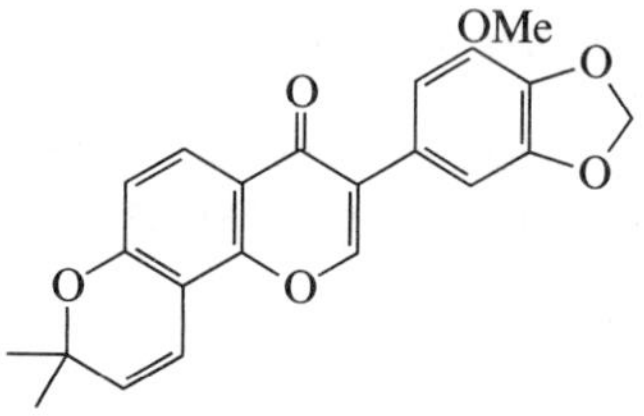

$C_{22}H_{18}O_6$ M 378.381
Constit. of *Millettia ferruginea*. Amorph.

Dagne, E. *et al*, *Phytochemistry*, 1989, **28**, 1897.

Isolonchocarpin   I-80066

Updated Entry replacing I-01090
2,3-Dihydro-8,8-dimethyl-2-phenyl-4H,8H-benzo[1,2-b:3,4-b']dipyran-4-one

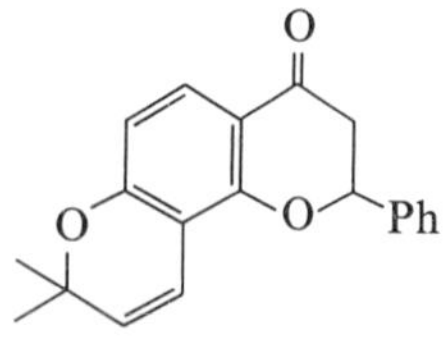

$C_{20}H_{18}O_3$ M 306.360
$(-)$-*form* [34198-88-4]

Constit. of the seed oil of *Pongamia glabra* and of the roots of *Tephrosia purpurea* and *Derris sericea*. Needles (hexane). Mp 115°. $[\alpha]_D^{24}$ $-125°$ (CHCl₃).

4'-Hydroxy: **4'-Hydroxyisolonchocarpin**
$C_{20}H_{18}O_4$ M 322.360
Constit. of *Millettia ferruginea*. Amorph.

Satam, P.G.N. *et al*, *Indian J. Chem.*, 1973, **11**, 209 (*isol, struct*)
Subrahmanyam, K. *et al*, *Indian J. Chem., Sect. B*, 1977, **15**, 105 (*synth*)
Jain, A.C. *et al*, *Proc.-Indian Acad. Sci., Sect. A*, 1978, **87**, 247 (*synth*)
Rao, E.V. *et al*, *Phytochemistry*, 1979, **18**, 1581 (*isol*)
Dagne, E. *et al*, *Phytochemistry*, 1989, **28**, 1897 (*deriv*)

Isookamurallene   I-80067

Updated Entry replacing I-10105

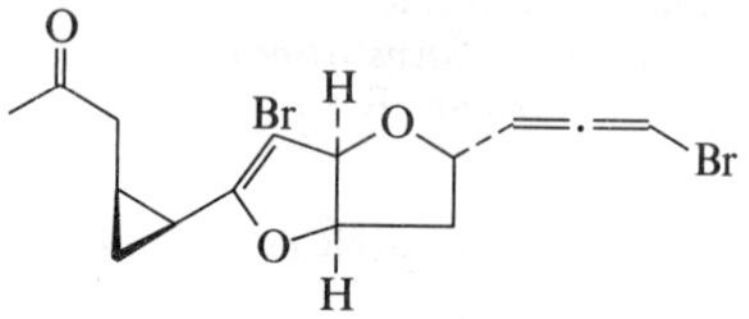

$C_{15}H_{16}Br_2O_3$ M 404.098
Oil. $[\alpha]_D^{27}$ $+130°$ (c, 1.00 in CHCl₃).

Suzuki, M. *et al*, *Chem. Lett.*, 1982, 289 (*isol*)
Suzuki, M. *et al*, *Phytochemistry*, 1989, **28**, 2145 (*struct*)

Isoorientin   I-80068

Updated Entry replacing I-10106
2-(3,4-Dihydroxyphenyl)-6-β-D-glucopyranosyl-5,7-dihydroxy-4H-1-benzopyran-4-one, 9CI. 6-C-Glucopyranosyl-3',4',5,7-tetrahydroxyflavone. 6-C-β-D-Glucopyranosylluteolin. Homoorientin. Lespecapitoside. Lutonaretin
[4261-42-1]

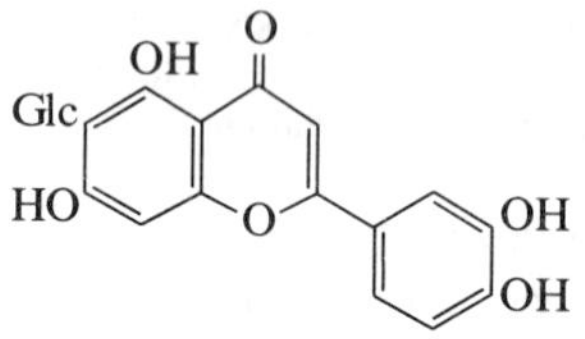

$C_{21}H_{20}O_{11}$ M 448.382
Widespread occurrence in *Adonis, Gentiana, Helenium* and *Passiflora* spp., etc. Light-yellow needles. Mp 235°. $[\alpha]_D^{20}$ $+30.8°$ (c, 1.2 in Py).

4'-Glucoside:
$C_{27}H_{30}O_{16}$ M 610.524
Constit. of the leaves of *G. verna*. Cryst. (H₂O). Mp 216° dec.

7-Glucoside: [35450-86-3]. **Lutonarin**
$C_{27}H_{30}O_{16}$ M 610.524
Found in barley (*Hordeum vulgare*) and other plants in Graminae, Leguminosae and Lemnaceae. Bright yellow cryst. Mp 235° dec.

2''-(4-Hydroxybenzoyl), 4'-β-D-glucopyranoside:
$C_{34}H_{34}O_{18}$ M 730.632
Constit. of *Gentiana asclepiadea*. Cryst. (MeOH). Mp 204-205°.

2''-O-Glucosyl: *3',4',5,7-Tetrahydroxy-6-C-sophorosylflavone*
$C_{27}H_{30}O_{16}$ M 610.524
Constit. of the leaves of *G. verna*. Cryst. (MeOH). Mp 212°.

2'',4'-Diglucosyl:
$C_{33}H_{40}O_{21}$ M 772.666

Constit. of the leaves of *G. asclepiadea*. Mp 211-213° dec.

3-Me ether, 7-O-β-D-glucopyranoside: 3′-O-Methyllutonarin
$C_{28}H_{32}O_{16}$ M 624.551
Isol. from leaves of *Hordeum vulgare*. Amorph. light-yellow powder. Mp 190-195° dec. Struct. revision resulting from revised struct. of Isoorientin.

2″-(4-Hydroxybenzoyl):
$C_{28}H_{24}O_{13}$ M 568.490
Constit. of *Gentiana asclepiadea*.

Seikel, M.K. *et al, Arch. Biochem. Biophys.*, 1962, **99**, 451 (3′-O-Methyllutonarin)
Koeppen, B.H. *et al, Biochem. J.*, 1965, **97**, 444 (*struct*)
Chopin, J. *et al, Tetrahedron Lett.*, 1966, 3657 (*synth*)
Hostettmann, K. *et al, Helv. Chim. Acta*, 1975, **58**, 130 (2″-Glucosyl)
Goetz, M. *et al, Helv. Chim. Acta*, 1977, **60**, 1322; 1978, **61**, 1373 (*deriv*)
Hostettmann, M. *et al, Phytochemistry*, 1978, **17**, 2083 (*rev*)
Congora, C. *et al, Helv. Chim. Acta*, 1986, **69**, 251 (*isol*)

8(14),15-Isopimaradiene-3,18-diol   I-80069
8,(14),15-Sandaracopimaradiene-3,18-diol
$C_{20}H_{32}O_2$ M 304.472
3β-form
Isol. from wood of *Xylia dolabriformis*. Platelets (pet. ether). Mp 152-153°. $[\alpha]_D$ −18.5° (c, 4 in $CHCl_3$).

Laidlaw, R.A. *et al, J. Chem. Soc.*, 1963, 644.

8(14),15-Isopimaradiene-2,3,18,19-tetrol   I-80070
8(14),15-Sandaracopimaradiene-2,3,18,19-tetrol
$C_{20}H_{32}O_4$ M 336.470
(2α,3β)-form
Dacrydol D$_1$
Isol. from *Dacrydium colensoi*. Mp 210-240°.

Carman, R.H. *et al, Tetrahedron Lett.*, 1966, 3173.

8(14),15-Isopimaradiene-2,18,19-triol   I-80071
8(14),15-Sandaracopimaradiene-2,18,19-triol
$C_{20}H_{32}O_3$ M 320.471
2α-form
Dacrydol D$_2$
Isol. from *Dacrydium colensoi*. Mp 224-227°.

Carman, R.N. *et al, Tetrahedron Lett.*, 1966, 3173.

8(14),15-Isopimaradiene-3,18,19-triol   I-80072
8(14),15-Sandaracopimaradiene-3,18,19-triol
$C_{20}H_{32}O_3$ M 320.471
3β-form
Dacrydol D$_3$
Isol. from *Dacrydium colensoi*. Mp 163-166°.

Carman, R.M. *et al, Tetrahedron Lett.*, 1966, 3173.

8,15-Isopimaradien-18-oic acid   I-80073
$\Delta^{8(9)}$-*Isopimaric acid*
$C_{20}H_{30}O_2$ M 302.456
Found in resin of *Pinus* spp. esp. *P. edulis*; also in *Tetraclinis* spp., *Callitris* spp. and *Dacrydium biforme*. Cryst. Mp 105-106°. $[\alpha]_D$ +94° (c, 1.14 in $CHCl_3$).
Me ester: Cryst. (MeOH). Mp 68-70°. $[\alpha]_D$ +118° (c, 1.0 in EtOH).

Edwards, O.E. *et al, Can. J. Chem.*, 1959, **37**, 760 (*synth*)

ApSimon, J.W. *et al, J. Chem. Soc., Chem. Commun.*, 1966, 361 (*pmr*)
Joye, N.M. *et al, J. Org. Chem.*, 1966, **31**, 320 (*occur*)

8(14),15-Isopimaradien-18-ol   I-80074
8(14),15-Sandaracopimaradien-18-ol. **Sandaracopimarinol**
$C_{20}H_{32}O$ M 288.472
Isol. from *Cryptomeria japonica* and *Agathis australis*. Cryst. Mp 63-65°. $[\alpha]_D$ −11° (c, 2.34 in $CHCl_3$).

Nagahama, S. *et al, Bull. Chem. Soc. Jpn.*, 1964, **37**, 886.
Thomas, B.R., *Acta Chem. Scand.*, 1966, **20**, 1074.

ent-9βH-Isopimara-7,15-dien-17-ol   I-80075

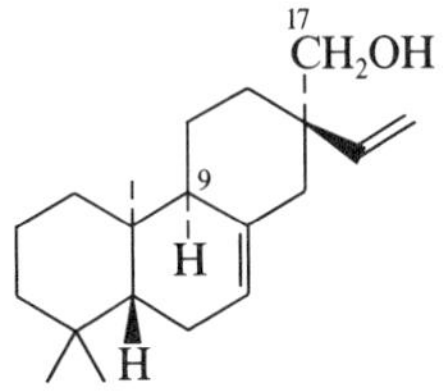

$C_{20}H_{32}O$ M 288.472
Constit. of *Calceolaria foliosa*. Cryst. Mp 75-76°. $[\alpha]_D^{25}$ −160.6° (c, 1 in $CHCl_3$).
Ac: ent-17-Acetoxy-9βH-isopimara-7,15-diene
$C_{22}H_{34}O_2$ M 330.509
Constit. of *C. foliosa*. Oil. $[\alpha]_D^{25}$ −120.6° (c, 1.02 in $CHCl_3$).
Malonyl: ent-17-Malonyloxy-9βH-isopimara-7,15-diene
$C_{23}H_{34}O_4$ M 374.519
Constit. of *C. foliosa*. Oil. $[\alpha]_D^{25}$ −127.5° (c, 1 in $CHCl_3$).
17-Carboxylic acid: **ent-9βH-Isopimara-7,15-dien-17-oic acid**
$C_{20}H_{30}O_2$ M 302.456
Constit. of *C. foliosa*. Cryst. Mp 160-161°. $[\alpha]_D^{25}$ −174.2° (c, 1 in $CHCl_3$).

Chamy, M.C. *et al, Phytochemistry*, 1989, **28**, 571.

8(14),15-Isopimaradien-3-one   I-80076
8(14),15-Sandaracopimaradien-3-one
$C_{20}H_{30}O$ M 286.456
Isol. from heartwood of *Xylia dolabriformis*. Platelets (MeOH). Mp 59-60°. $[\alpha]_D$ −56° (c, 2 in $CHCl_3$).

Laidlaw, R.A. *et al, J. Chem. Soc.*, 1963, 644 (*isol, ir, struct*)
Grant, P.K. *et al, Tetrahedron Lett.*, 1965, 3729 (*synth*)

Isopregomisin   I-80077
[122222-00-8]

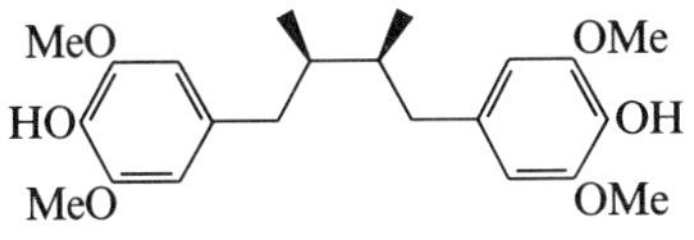

$C_{22}H_{30}O_6$ M 390.475
Constit. of *Porlieria chilensis*. Cryst. Mp 110-112°. *meso*.

Torres, R. *et al, J. Nat. Prod. (Lloydia)*, 1989, **52**, 402 (*isol, pmr, cmr*)

4-Isopropylbenzyl alcohol   I-80078
Updated Entry replacing I-01224
4-(1-Methylethyl)benzenemethanol, 9CI. p-*Cymen-7-ol, 8CI.* *Cuminyl alcohol. Cumic alcohol*
[536-60-7]
$C_{10}H_{14}O$ M 150.220
Bp 246°, Bp_{20} 140°.

▷ GZ7260000.

Ac: [59230-57-8].
$C_{12}H_{16}O_2$ M 192.257
$Bp_{16.5}$ 136°.

O-[β-D-*Glucopyranosyl*-(1→2)-β-D-*galactopyranoside*]:
Coleoside
$C_{22}H_{34}O_{11}$ M 474.504
Constit. of *Coleus forskohlii*. Yellow solid.

Bert, L., *Bull. Soc. Chim. Fr.*, 1925, **37**, 1397, 1577 (*synth*)
Palfray, L. *et al*, *Compt. Rend. Hebd. Seances Acad. Sci.*, 1936, **203**, 1523 (*synth*)
v. Tamelen, E.E. *et al*, *J. Am. Chem. Soc.*, 1974, **96**, 5290 (*synth*)
Ahmed, B. *et al*, *Phytochemistry*, 1988, **27**, 3309 (*Coleoside*)

7-Isopropyl-2,10-dimethylspiro[4.5]dec-1-en-6-ol   I-80079

2,10-Dimethyl-7-(1-methylethyl)spiro[4.5]dec-1-en-6-ol, 9CI
[119432-99-4]

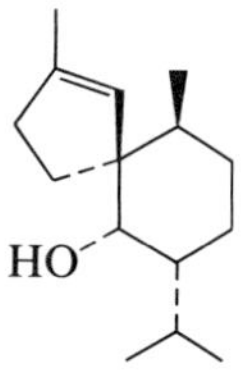

$C_{15}H_{26}O$ M 222.370
Constit. of a *Eurypon* sp. Oil. $[\alpha]_D$ +1.3° (c, 0.6 in $CHCl_3$).
Related to Axisisonitrile 3

Barrow, C.J. *et al*, *Aust. J. Chem.*, 1988, **41**, 1755.

2-Isopropylidene-4,6-dimethoxy-3(2H)-benzofuranone   I-80080

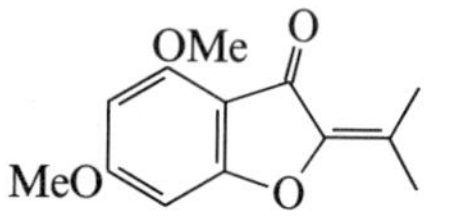

$C_{13}H_{14}O_4$ M 234.251
Constit. of *Calea peckii*. Oil.

Castro, V. *et al*, *Phytochemistry*, 1989, **28**, 2415 (*isol, pmr*)

3-Isopropylidene-6-methylcyclohexene   I-80081

Updated Entry replacing I-01304
2,4(8)-p-Menthadiene. **Menogene**
[586-63-0]

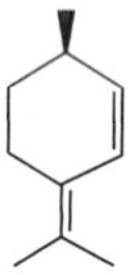

$C_{10}H_{16}$ M 136.236
(R)-form
Isol. from leaf oil of *Juniperus horizontalis*. Oil. $Bp_{764.5}$ 184-186°. $[\alpha]_D^{17}$ +49.1°. n_D^{20} 1.5026.

Ohloff, G. *et al*, *Justus Liebigs Ann. Chem.*, 1959, **625**, 206 (*synth, uv, abs config*)
Couchman, F.M. *et al*, *Can. J. Chem.*, 1965, **43**, 1017 (*isol*)
Verghese, J., *Perfum. Essent. Oil Rec.*, 1967, **58**, 868 (*rev*)
Bark, S. *et al*, *J. Org. Chem.*, 1968, **33**, 221 (*synth*)
Bohlmann, F. *et al*, *Org. Magn. Reson.*, 1975, **7**, 426 (*cmr*)

1-Isopropyl-4-methyl-1,4-cyclohexanediol   I-80082

p-*Menthane-1,4-diol*

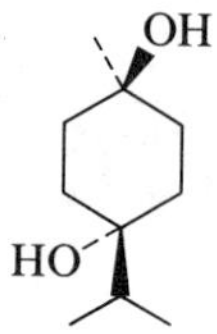

$C_{10}H_{20}O_2$ M 172.267
(1RS,4RS)-form
(1β,4α)-*form*
Constit. of *Ferula jaeshkeana*. Cryst. Mp 153°.

Garg, S.N. *et al*, *Phytochemistry*, 1989, **28**, 634.

2-Isopropyl-5-methylfuran   I-80083

2-Methyl-5-(1-methylethyl)furan
[10504-05-9]

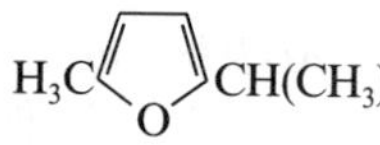

$C_8H_{12}O$ M 124.182
Important natural flavour. Oil with strong herbaceous odour. Bp 136°, Bp_{60} 58°.

Heyns, K. *et al*, *Tetrahedron*, 1966, **22**, 2223 (*ms*)
Eichenberger, H. *et al*, *Helv. Chim. Acta*, 1976, **59**, 1253 (*cmr*)
Weyerstahl, P. *et al*, *Justus Liebigs Ann. Chem.*, 1988, 1015 (*synth, pmr, bibl*)

4-Isopropyl-7-methyl-1-naphthalene-carboxaldehyde   I-80084

7-Methyl-4-(1-methylethyl)-1-naphthalenecarboxaldehyde.
14-Cadalenal

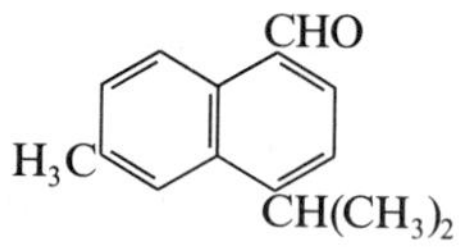

$C_{15}H_{16}O$ M 212.291
Constit. of *Helichrysum ambiguum*. Oil.

Jakupovic, J. *et al*, *Phytochemistry*, 1989, **28**, 543.

Isopropyl phenyl ether   I-80085

(1-Methylethoxy)benzene, 9CI. 2-Phenoxypropane.
Isopropoxybenzene
[2741-16-4]

$PhOCH(CH_3)_2$

$C_9H_{12}O$ M 136.193
d^{20} 0.978. Fp −33.05°. Bp 178°. n_D^{20} 1.4975.

Sadtler Standard Infrared Spectra, No. 11619 (*ir*)
Smith, R.A., *J. Am. Chem. Soc.*, 1934, **56**, 717 (*synth*)
Olson, W.T. *et al*, *J. Am. Chem. Soc.*, 1947, **69**, 2451 (*synth*)
Vogel, A.I., *J. Chem. Soc.*, 1948, 616 (*props*)
Baddeley, G. *et al*, *J. Chem. Soc.*, 1956, 2455 (*uv*)

Isoquinacridone   I-80086

Quino-[3,2-b]acridine-12,14(5H,7H)-dione, 9CI. 5,7-
Dihydroquino[3,2-b]acridine-12,14-dione. lin-cis-
Quinacridone

[3164-44-1]

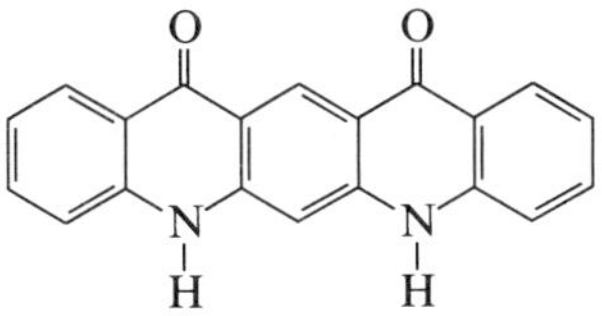

$C_{20}H_{12}N_2O_2$ M 312.327
Brownish-yellow needles (PhNO$_2$). Prac. insol. most solvs.

Eckert, A. *et al, J. Prakt. Chem.*, 1921, **102**, 338 (*synth*)
Leandri, G. *et al, Boll. Sci. Fac. Chim. Ind. Bologna*, 1951, **9**, 32;
  CA, **45**, 10245e (*synth, props*)
Labana, S.S. *et al, Chem. Rev.*, 1967, **67**, 1 (*rev*)
Altiparmakian, R.H. *et al, Helv. Chim. Acta*, 1972, **55**, 85 (*pmr*)

Isorolandrolide   I-80087

Updated Entry replacing I-01549

[76010-19-0]

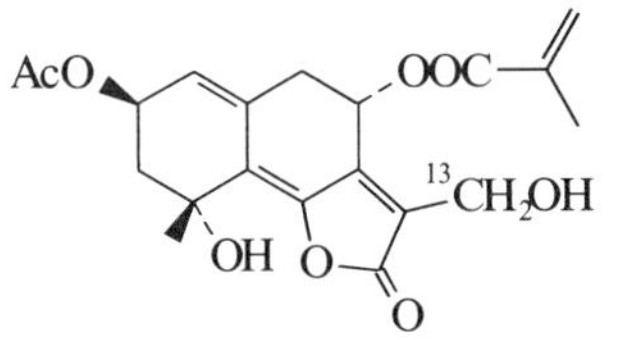

$C_{21}H_{26}O_8$ M 406.432
Constit. of *Rolandra fruticosa*. Gum.

13-Et ether: [76010-18-9].
  $C_{23}H_{30}O_8$ M 434.485
  Constit. of *R. fruticosa*. Noncryst.

Herz, W. *et al, J. Org. Chem.*, 1981, **46**, 761 (*isol*)
Jakupovic, J. *et al, Phytochemistry*, 1989, **28**, 1937 (*struct*)

Isorugosin B   I-80088

[119139-61-6]

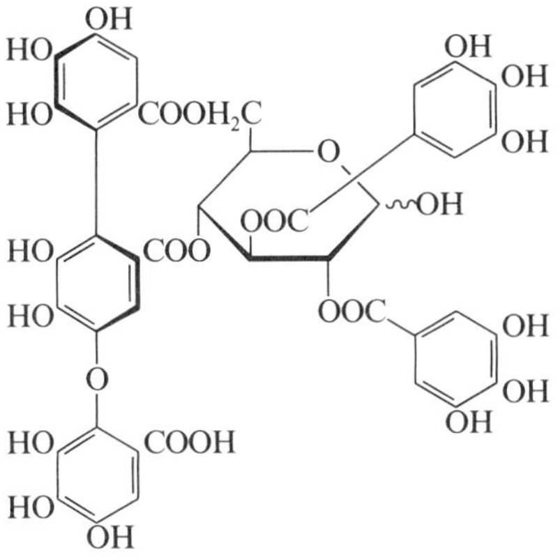

$C_{41}H_{30}O_{27}$ M 954.672
Constit. of *Liquidamber formosa*. Light brown amorph.
  powder. [α]$_D$ +28° (c, 1 in MeOH).

Hatano, T. *et al, Heterocycles*, 1988, **27**, 2081.

Isosafrole   I-80089

Updated Entry replacing M-30101
5-(1-Propenyl)-1,3-benzodioxole, 9CI. 1,2-(Methylenedioxy)-
4-(1-propenyl)benzene, 8CI

[120-58-1]

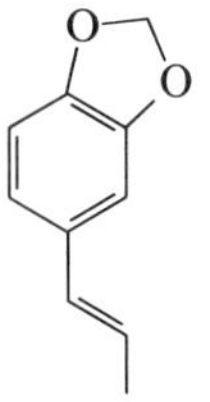

(*E*)-*form*

$C_{10}H_{10}O_2$ M 162.188
Isol. from star anise oil (*Illicium religiosum*), seed oil of
  Ligusticum acutilobum and leaves of *Murraya koenigi*.

(*E*)-*form* [4043-71-4]
  β-Isosafrole
  Perfumery intermed. Fp 6.7-6.8°. Bp$_{768}$ 252.4-252.7°,
  Bp$_{11.5}$ 123°.
▷ DA5951000.

(*Z*)-*form* [17627-76-8]
  α-Isosafrole
  Present in essential oils. Liq.

Crymble, C.R. *et al, J. Chem. Soc.*, 1911, 451 (*uv*)
Robinson, R. *et al, J. Chem. Soc.*, 1927, 2489 (*synth*)
Dutt, S., *CA*, 1958, **52**, 20904 (*isol*)
Willhalm, B. *et al, Tetrahedron*, 1964, **20**, 1185 (*ms*)
Shulgin, A.T., *J. Chromatogr.*, 1967, **30**, 54 (*glc*)
Torii, S. *et al, J. Org. Chem.*, 1984, **49**, 1830 (*synth*)

Isosalicin   I-80090

[7724-09-6]

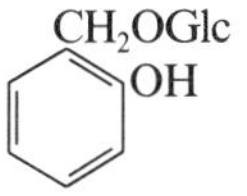

$C_{13}H_{18}O_7$ M 286.281
Isol. from flowers of *Filipendula ulmaria*. Needles + 3H$_2$O
  (Me$_2$CO/pet. ether). Mp 66-68°. [α]$_D^{20}$ −45.2° (anhyd.).

Bourquelot, E. *et al, Compt. Rend. Hebd. Seances Acad. Sci.*, 1913,
  156, 1790 (*synth*)
Thieme, H., *Pharmazie*, 1966, **21**, 123 (*isol*)

Isosphaerodiene 1   I-80091

[80981-70-0]

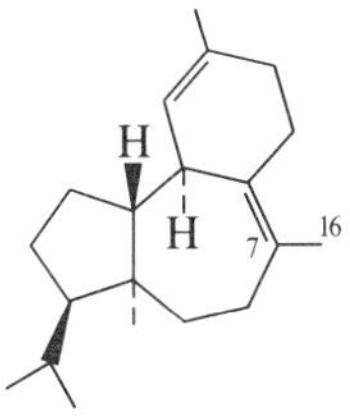

$C_{20}H_{32}$ M 272.473
Constit. of *Sphaerococcus coronopifolius*. Oil. [α]$_D$ +28°
  (CHCl$_3$).

$\Delta^{7(16)}$-*Isomer:* [108907-24-0]. *Isosphaerodiene 2*
  $C_{20}H_{32}$ M 272.473
  Constit. of *S. coronopifolius*. Oil. [α]$_D$ +18° (CHCl$_3$).

Cafieri, F. *et al, Phytochemistry*, 1987, **26**, 471.

Isoterchebin   I-80092

1,2,3-Tri-O-galloyl-4,6-O-dehydrohexahydroxydiphenoyl-β-
D-glucopyranose. Cornus-tannin 1
[58690-20-3]

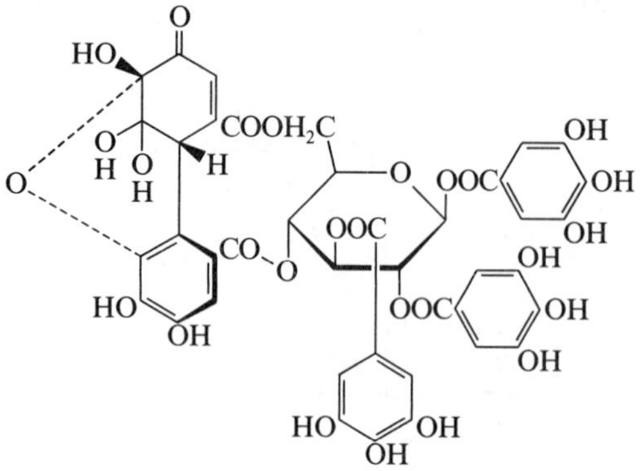

$C_{41}H_{30}O_{26}$ M 938.672
Isol. from *Cytinus hypocistis* and *Cornus officinalis.*
Hygroscopic solid + 6H$_2$O.

Fürstenwerth, H. *et al, Justus Liebigs Ann. Chem.*, 1976, 112 (*isol, pmr*)
Okuda, T. *et al, Heterocycles*, 1981, **16**, 1321 (*struct*)

Isoteucrin H$_4$ I-80093

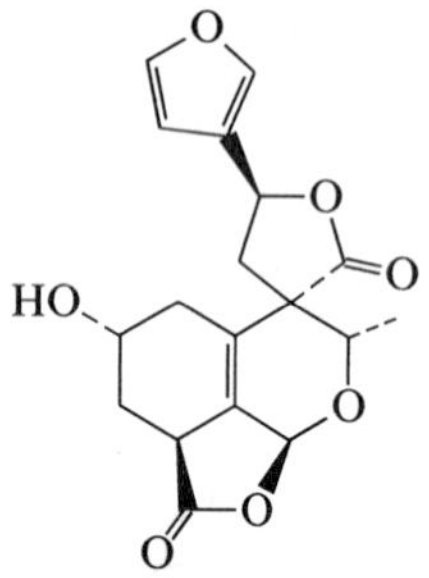

$C_{19}H_{20}O_6$ M 344.363
Constit. of *Teucrium kotschyanum.*
Ac: Cryst. (EtOAc/hexane). Mp 191-194°. [α]$_D^{22}$ +128.6° (c, 0.168 in CHCl$_3$).

Simões, F. *et al, Heterocycles*, 1989, **28**, 111.

1-Isothiocyanato-4-cadinene   I-80094

4-Cadinene-1-isothiocyanate
[117605-65-9]

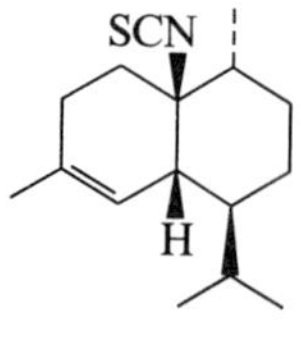

$C_{16}H_{25}NS$ M 263.446
Metab. of the sponge *Acanthella pulcherrima.*
Capon, R.J. *et al, Aust. J. Chem.*, 1988, **41**, 979.

2-Isothiocyanatoquinoxaline   I-80095

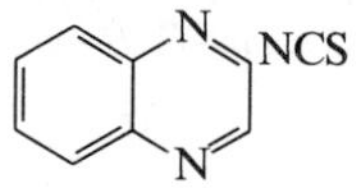

$C_9H_5N_3S$ M 187.225
Pale yellow needles (pet. ether). Mp 48-48.5°.

Ijima, C. *et al, Yakugaku Zasshi (J. Pharm. Soc. Jpn.)*, 1988, **108**, 437.

Isotriptiliocoumarin   I-80096

[118627-71-7]

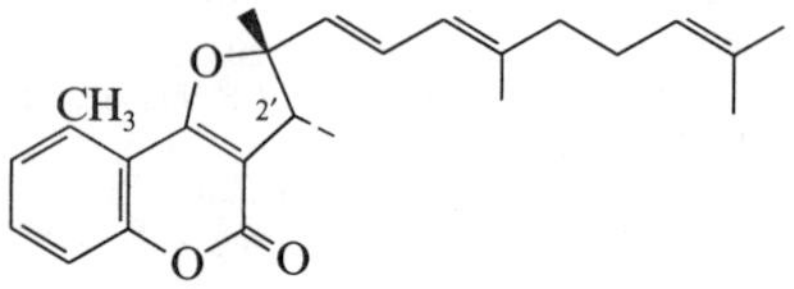

$C_{25}H_{30}O_3$ M 378.510
Constit. of *Triptilion benaventei.* Gum.
2′-Epimer: [118555-94-5]. *2′-Epiisotriptiliocoumarin*
  $C_{25}H_{30}O_3$ M 378.510
  Constit. of *T. benaventei.* Gum.

Bittner, M. *et al, Phytochemistry*, 1988, **27**, 3263.

Isotriptospinocoumarin   I-80097

[119875-30-8]

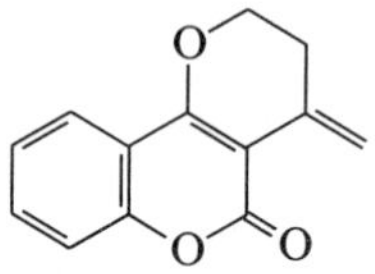

$C_{14}H_{12}O_3$ M 228.247
Constit. of *Nassauvia magellanica.* Oil.

Bittner, M. *et al, Phytochemistry*, 1988, **27**, 3845.

Isovolubilin   I-80098

6-(6-Deoxy-α-L-mannopyranosyl)-5-hydroxy-7-methoxy-3-(4-
methoxyphenyl)-4H-1-benzopyran-4-one, 9CI. 7-O-
Methylbiochanin A 6-C-rhamnoside. 5-Hydroxy-4′,7-
dihydroxy-6-rhamnosylisoflavone
[56317-19-2]

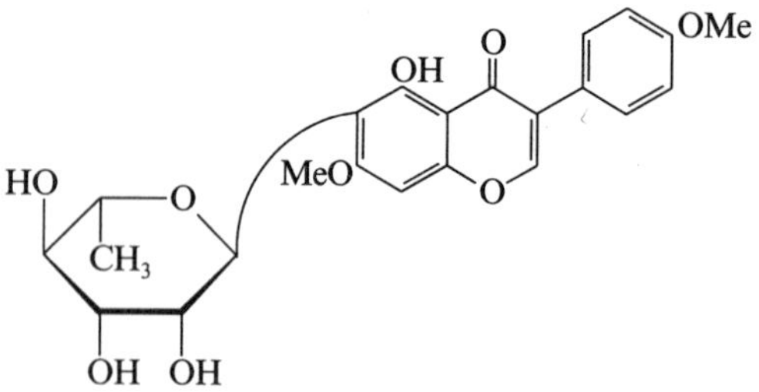

$C_{23}H_{24}O_9$ M 444.437
Isol. from flowers of *Dalbergia volubilis.* Hemihydrate. Mp 298-301°. [α]$_D^{30}$ +17° (c, 1.15 in MeOH).

Chawla, H.M. *et al, Indian J. Chem.*, 1975, **13**, 444.

Isoxanthohumol†   I-80099

Updated Entry replacing I-01631
4′,7-Dihydroxy-5-methoxy-8-C-prenylflavanone. Humulol⁺

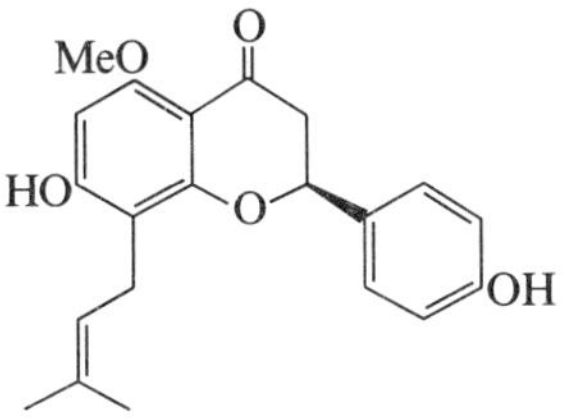

$C_{21}H_{22}O_5$ M 354.402

The name Humulol has historical precedence but Isoxanthohumol has been preferred to avoid confusion with I-01631. However another nat. prod. has since been named Isoxanthohumol; not i see I-80100.

(S)-form [70872-29-6]
Isol. from hop (*Humulus lupulus*). Light yellow needles (AcOH). Mp 198°.

Di-Me ether: Mp 174-175°.

Power, F.B. *et al, J. Chem. Soc.*, 1913, **103**, 1267 (*isol*)
Verzele, M. *et al, Bull. Soc. Chim. Belg.*, 1957, **66**, 452 (*isol*)
Jain, A.C. *et al, Tetrahedron*, 1978, **34**, 3563 (*synth*)

Isoxanthohumol†   I-80100

1-[2,6-Dihydroxy-4-methoxy-3-(3-methyl-2-butenyl)phenyl]-3-phenyl-2-propen-1-one. 2′,6′-Dihydroxy-4′-methoxy-3′-prenylchalcone
[72247-79-1]

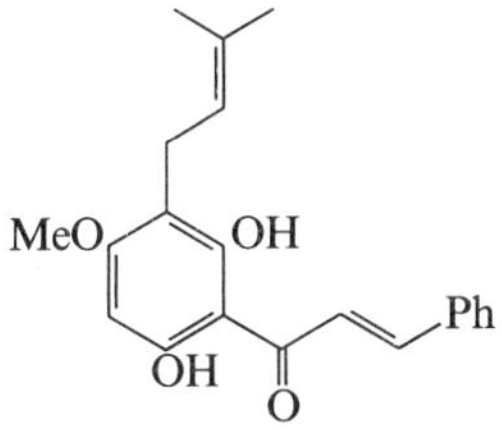

$C_{21}H_{22}O_4$ M 338.402

Not the same as Isoxanthohumol†, I-80099. The name Isoxanthohumol is misleading; not isomeric with X-00027 which has an addnl. OH group. Isol. from *Helichrysum cymosum*. Yellowish oil.

O-De-Me: [72247-78-0]. **Desmethylxanthohumol**
$C_{20}H_{20}O_4$ M 324.376
Isol. from roots of *Helichrysum tenuiculum*. Yellow cryst. (Et₂O). Mp 168-169°. Incorrectly named. See above.

Bohlmann, F. *et al, Phytochemistry*, 1979, **18**, 1033.

4-Isoxazolecarboxylic acid   I-80101

Updated Entry replacing I-01635
$C_4H_3NO_3$ M 113.073
Plates (toluene), cryst. (CHCl₃ or by subl.). Mp 123-124° (121-123°). Dec. at 140° or on keeping.

Et ester:
$C_6H_7NO_3$ M 141.126
Liq. Bp₁₃₋₁₄ 93°.

Nitrile: [68776-58-9]. *4-Cyanoisoxazole*
$C_4H_2N_2O$ M 94.073
Sticky pale yellow solid.

Panizzi, L., *Gazz. Chim. Ital.*, 1947, **77**, 206 (*synth, ester*)
Maggioni, P. *et al, Gazz. Chim. Ital.*, 1966, **96**, 443 (*synth, ir*)
Angus, R.O. *et al, Synthesis*, 1988, 746 (*nitrile*)

Ixerin A   I-80102
[91486-96-3]

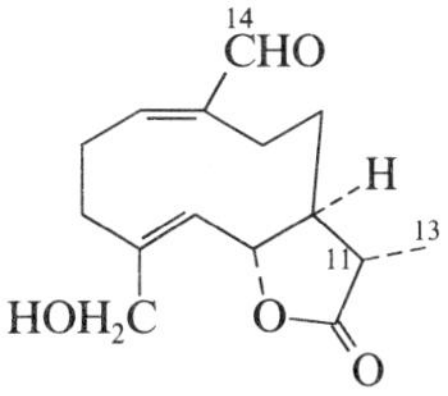

$C_{15}H_{20}O_4$ M 264.321
Isol. from *Ixera tamagawensis*. Prisms (MeOH). Mp 134-138°. $[\alpha]_D^{23}$ −54.6° (c, 0.18 in CHCl₃).

O-β-D-Glucopyranoside: [94356-09-9]. **Ixerin J**
$C_{21}H_{30}O_9$ M 426.463
Constit. of *I. tamagawensis*. Prisms. Mp 113.5-115.5°. $[\alpha]_D^{25}$ −54.0° (c, 0.68 in MeOH).

14-Alcohol: [91486-99-6]. **Ixerin K**
$C_{15}H_{22}O_4$ M 266.336
Constit. of *I. tamagawensis*. Amorph. powder. $[\alpha]_D^{25}$ −40.6° (c, 0.62 in MeOH).

14-Alcohol, 11-epimer: [94356-08-8]. **Ixerin L**
$C_{15}H_{22}O_4$ M 266.336
Constit. of *I. tamagawensis*. Amorph. powder. $[\alpha]_D^{25}$ +23.5° (c, 0.37 in MeOH).

11,13-Didehydro: **8-Desoxyurospermal A**
$C_{15}H_{18}O_4$ M 262.305
Isol. from *I. tamagawensis*. Amorph. powder. $[\alpha]_D^{23}$ −40.6° (c, 0.32 in CHCl₃).

11,13-Didehydro, 15-O-β-D-glucopyranoside: [91486-97-4]. **Ixerin B**
$C_{21}H_{28}O_9$ M 424.447
Constit. of *I. tamagawensis*. Amorph. $[\alpha]_D^{23}$ −11.9° (c, 2.0 in MeOH).

11,13-Didehydro, 15-O-β-D-(6-O-p-hydroxyphenylacetyl)glucopyranoside: [91486-98-5]. **Ixerin C**
$C_{29}H_{34}O_{11}$ M 558.581
Constit. of *I. tamagawensis*. Amorph. $[\alpha]_D^{23}$ +3.8° (c, 0.40 in MeOH).

Asada, H. *et al, Chem. Pharm. Bull.*, 1984, **32**, 3403.

J

Jaborosalactol *M*   J-80001

[120160-86-3]

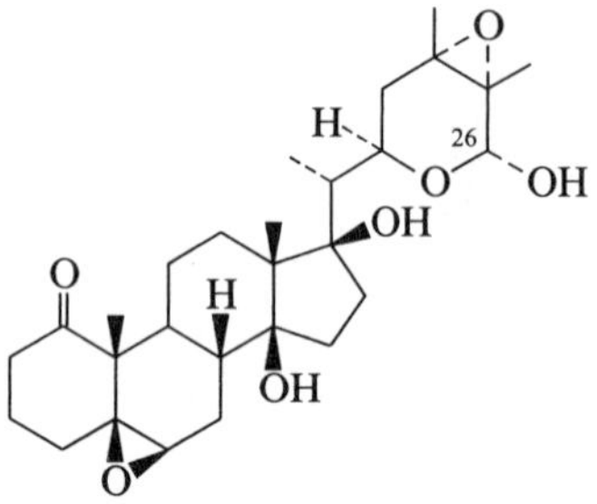

C$_{28}$H$_{42}$O$_7$ M 490.636
Constit. of *Jaborosa bergii*. Cryst. (EtOAc). Mp 234-236°.

26-Ketone: [120160-85-2]. **Jaborosalactone M**
C$_{28}$H$_{40}$O$_7$ M 488.620
Constit. of *J. bergii*.

Monteagudo, E.S. *et al*, *Phytochemistry*, 1988, **27**, 3925.

Jaborosalactol N   J-80002

[120160-87-4]

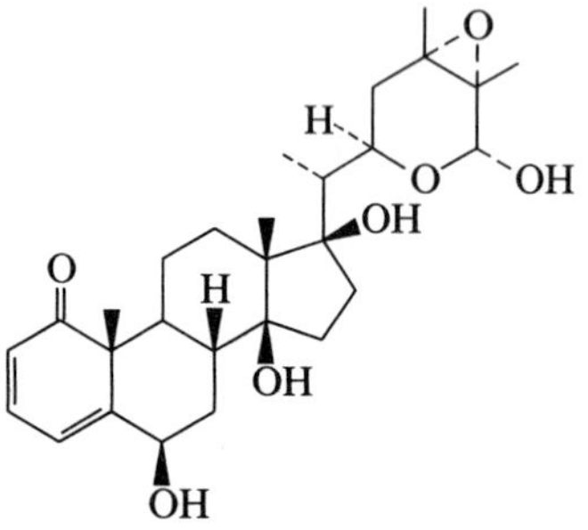

C$_{28}$H$_{40}$O$_7$ M 488.620
Constit. of *Jaborosa bergii*. Cryst. (EtOAc). Mp 183-185°
dec.

Monteagudo, E.S. *et al*, *Phytochemistry*, 1988, **27**, 3925.

Jasionone   J-80003

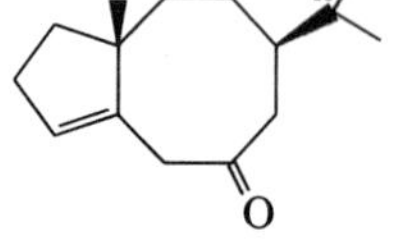

C$_{15}$H$_{24}$O   M 220.354
Parent compd. apparently unknown.

11-Hydroxy: [119765-85-4]. *11-Hydroxyjasionone*
C$_{15}$H$_{24}$O$_2$ M 236.353
Constit. of *Jasonia montana*. Oil.

Ahmed, A.A. *et al*, *Phytochemistry*, 1988, **27**, 3875.

Jasmesosidic acid   J-80004

Updated Entry replacing J-50009

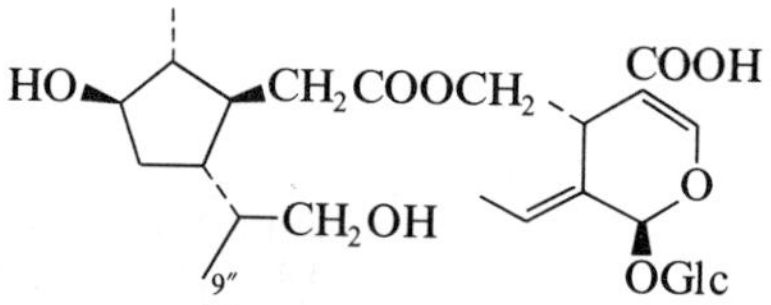

C$_{26}$H$_{40}$O$_{13}$ M 560.594
Me ester: [97777-70-3]. **Jasmesoside**
C$_{27}$H$_{42}$O$_{13}$ M 574.621
Constit. of *Jasminum mesnyi*. Powder. [α]$_D^{24}$ −156° (c,
0.73 in MeOH).

9″-Hydroxy: **9″-Hydroxyjasmesosidic acid**
C$_{26}$H$_{40}$O$_{14}$ M 576.594
Constit. of *J. mesnyi*. Powder. [α]$_D$ −161° (c, 0.65 in
MeOH).

9″-Hydroxy, Me ester: **9″-Hydroxyjasmesoside**
C$_{27}$H$_{42}$O$_{14}$ M 590.620
Constit. of *J. mesnyi*. Powder. [α]$_D$ −164.9° (c, 0.49 in
MeOH).

Inoue, K. *et al*, *Phytochemistry*, 1985, **24**, 1299 (*isol, struct*)
Tanahashi, T. *et al*, *Phytochemistry*, 1989, **28**, 1413 (*isol, struct*)

Jasmine ketolactone   J-80005

Updated Entry replacing J-40003
*1,4,5,8,8a,10,11,11*a-*Octahydrocyclopent*[d]*oxecin-2,9-dione,
9CI. Jasmine oil ketolactone*
[70981-24-7]

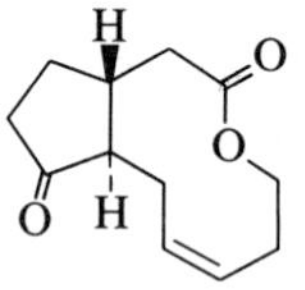

C$_{12}$H$_{16}$O$_3$ M 208.257
Constit. of Italian jasmine oil (*Jasminium grandiflorum*).
Small needles (pet. ether). Mp 104°. [α]$_D^{20}$ −260° (c, 3.05
in MeOH).

Demole, E. *et al*, *Helv. Chim. Acta*, 1964, **47**, 1152 (*isol, ir, pmr,
ms, struct*)
Kitahara, T. *et al*, *Agric. Biol. Chem.*, 1984, **48**, 1731 (*synth*)

Jasmolactone A   J-80006

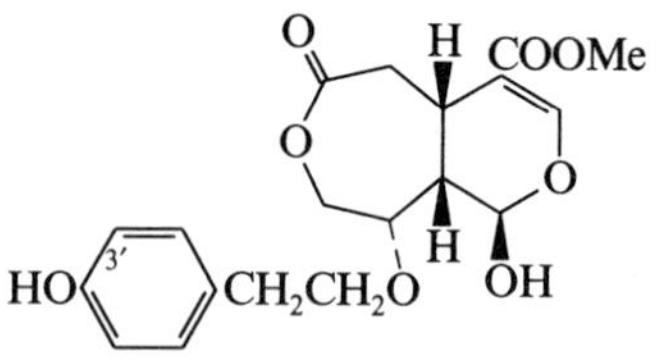

C$_{19}$H$_{22}$O$_8$ M 378.378
Constit. of *Jasminum multiflorum*. Pale yellow solid. [α]$_D^{28}$
+122.4° (c, 1 in CHCl$_3$).

3′-Hydroxy: **Jasmolactone B**
C$_{19}$H$_{22}$O$_9$ M 394.377

Constit. of *J. multiflorum*. Pale yellow gum. $[\alpha]_D^{28}$ +100.1° (c, 0.1 in MeOH).

Shen, Y.-C. *et al, J. Nat. Prod.* (*Lloydia*), 1989, **52**, 1060 (*isol, pmr, cmr*)

Jasmolactone C   J-80007

O

COOCH$_2$CH$_2$

OH

OH

O

HO

CH$_2$CH$_2$O

OH

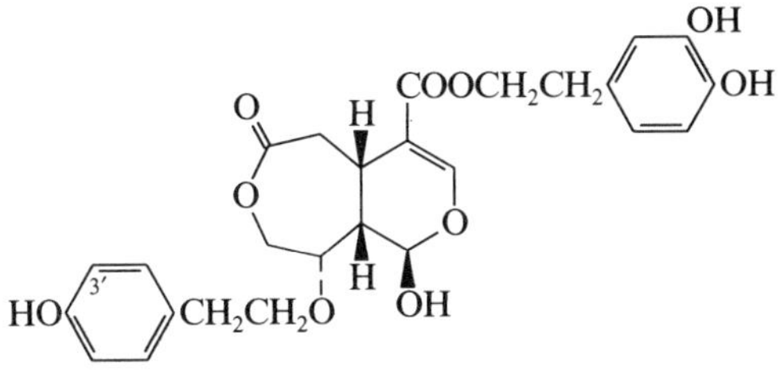

$C_{26}H_{28}O_{10}$ M 500.501

Constit. of *Jasmium multiflorum*. Gum. $[\alpha]_D^{28}$ +48.6° (c, 1 in MeOH).

3'-Hydroxy: Jasmolactone D

Constit. of *J. multiflorum*. Gum. $[\alpha]_D^{28}$ +28.5° (c, 1 in MeOH).

Shen, Y.-C. *et al, J. Nat. Prod.* (*Lloydia*), 1989, **52**, 1060 (*isol, pmr, cmr*)

Jayantinin   J-80008

7,7'-Dimethoxy-8,8'-bi[2H-1-benzopyran-2-one]. 7,7'-Dimethoxy-8,8'-bicoumarin

MeO

O

O

MeO

O

O

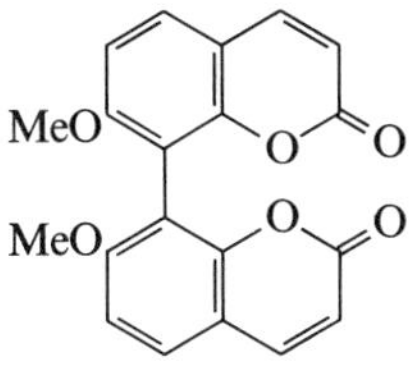

$C_{20}H_{14}O_6$ M 350.327

Constit. of *Boenninghausenia albiflora*. Green powder. Mp 255-256°.

Joshi, P.C. *et al, Phytochemistry*, 1989, **28**, 1281.

Jiofuran   J-80009

HO

HO

H

CH$_2$OH

O

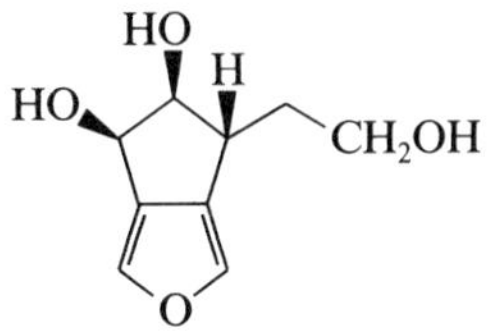

$C_9H_{12}O_4$ M 184.191

Constit. of *Rehmannia glutinosa*. Amorph. powder. $[\alpha]_D^{24}$ −30.4° (c, 0.19 in MeOH).

Morota, T. *et al, Phytochemistry*, 1989, **28**, 2385 (*isol, pmr, cmr*)

Jioglutin A   J-80010

HO

H

OMe

R—

3

O

HO

H

O

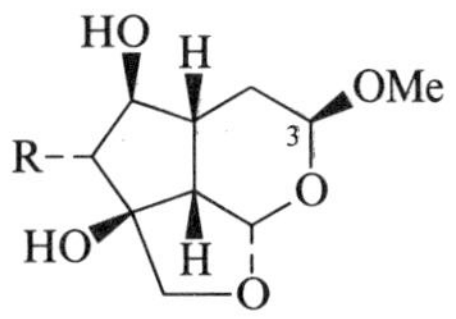

R = Cl

$C_{10}H_{15}ClO_5$ M 250.678

Constit. of *Rehmannia glutinosa*. Amorph. powder. $[\alpha]_D$ +63.3° (c, 1 in MeOH).

3-Epimer: Jioglutin B

$C_{10}H_{15}ClO_5$ M 250.678

Constit. of *Rehmannia glutinosa*. Amorph. powder. $[\alpha]_D$ −63.2° (c, 0.94 in MeOH).

Morota, T. *et al, Phytochemistry*, 1989, **28**, 2385 (*isol, pmr, cmr*)

Jioglutin C   J-80011

As Jioglutin A, J-80010 with

R = OH

$C_{10}H_{16}O_6$ M 232.233

Constit. of *Rehmannia glutinosa*. Amorph. powder. $[\alpha]_D$ +58.1° (c, 0.89 in MeOH).

Morota, T. *et al, Phytochemistry*, 1989, **28**, 2385 (*isol, pmr, cmr*)

Jioglutolide   J-80012

OH

H

O

O

HO

H

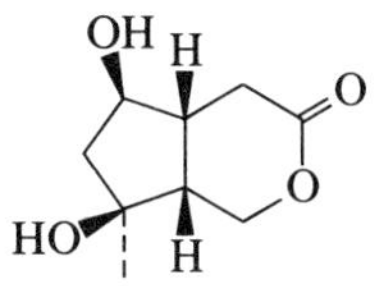

$C_9H_{14}O_4$ M 186.207

Constit. of *Rehmannia glutinosa*. Needles (Me$_2$CO). Mp 141-142°. $[\alpha]_D$ −8.4° (c, 1.19 in MeOH).

Morota, T. *et al, Phytochemistry*, 1989, **28**, 2385 (*isol, pmr, cmr, cryst struct*)

Jodrellin A   J-80013

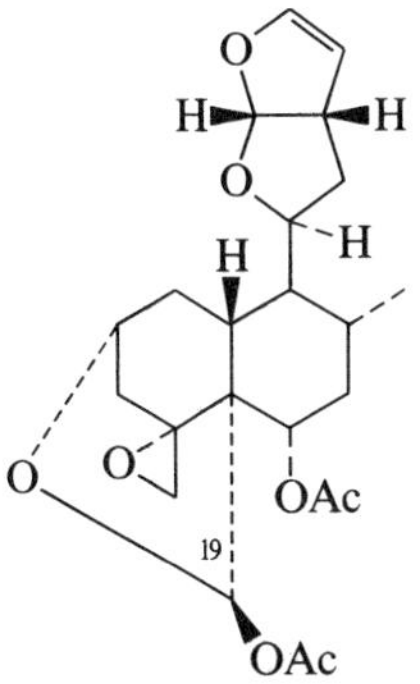

$C_{24}H_{32}O_8$ M 448.512

Constit. of *Scutellaria woronowii*. Insect antifeedant. Oil.

19-De-Ac, 19-O-(2-methylpropanoyl): Jodrellin B

$C_{26}H_{36}O_8$ M 476.566

Constit. of *S. woronowii*. Insect antifeedant. Oil. $[\alpha]_D^{20}$ −11.4° (c, 0.07 in CHCl$_3$).

Anderson, J.C. *et al, Tetrahedron Lett.*, 1989, **30**, 4737.

Junceellin   J-80014

[92508-09-3]

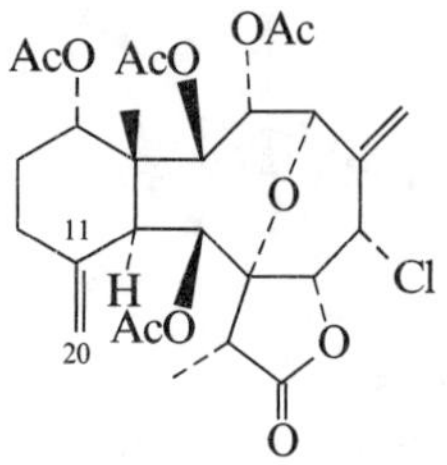

$C_{28}H_{35}ClO_{11}$ M 583.031
Constit. of gorgonians *Junceella squamata* and *J. fragilis*.

11α,20-Epoxide: [86425-67-4]. **Praelolide**
   $C_{28}H_{35}ClO_{12}$ M 599.030
   Constit. of *Plexaureides praelonga* and *J. fragilis*.

Dai, J. *et al, Sci. Sin.* (*Engl. Ed.*), 1985, **28**, 1131.
Shin, J. *et al, Tetrahedron*, 1989, **45**, 1633.

Junceellolide A   J-80015

[121769-81-1]

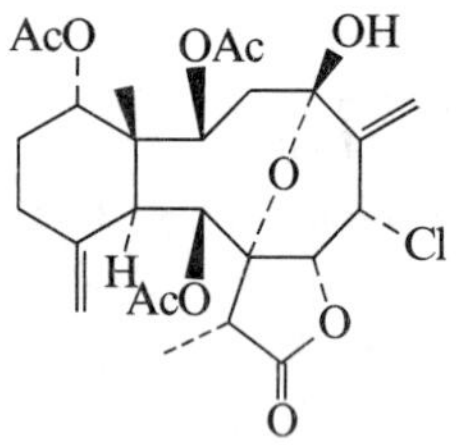

$C_{26}H_{33}ClO_{10}$ M 540.993
Constit. of gorgonian *Junceella fragilis*. Cryst. Mp 115-116°. $[\alpha]_D$ −7.9° (c, 0.6 in $CHCl_3$).

Shin, J. *et al, Tetrahedron*, 1989, **45**, 1633.

Junceellolide B   J-80016

[121798-80-9]

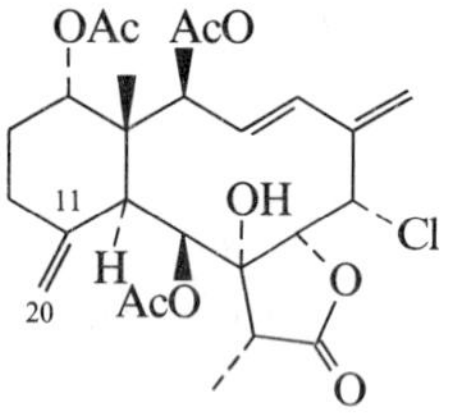

$C_{26}H_{33}ClO_9$ M 524.994
Constit. of *Junceella fragilis*. Cryst. Mp 95-96°. $[\alpha]_D$ +9.4°
(c, 1.3 in $CHCl_3$).

11α,20-Epoxide: [121769-82-2]. **Junceellolide C**
   $C_{26}H_{33}ClO_{10}$ M 540.993
   Constit. of *J. fragilis*. Cryst. Mp 120-125°. $[\alpha]_D$ +36.1°
   (c, 1.2 in $CHCl_3$).

Shin, J. *et al, Tetrahedron*, 1989, **45**, 1633.

Junceellolide D   J-80017

[121798-81-0]

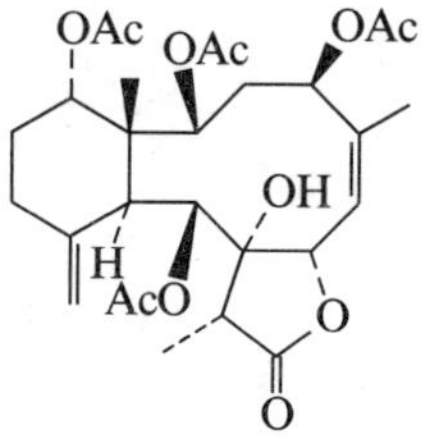

$C_{28}H_{38}O_{11}$ M 550.602
Constit. of *Junceella fragilis*. Cryst. Mp 208-209°. $[\alpha]_D$
   −7.7° (c, 2.5 in $CHCl_3$).

Shin, J. *et al, Tetrahedron*, 1989, **45**, 1633.

Juvenile hormone O   J-80018

Methyl 10,11-epoxy-3,7-diethyl-11-methyl-2,6-tridecadienoate

[117019-48-4]

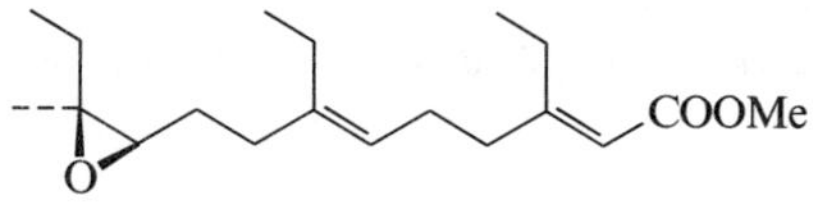

$C_{19}H_{32}O_3$ M 308.460
Isol. from developing embryos of the tobacco hornworm
   moth *Manduca sexta*. Part of the juvenile hormone
   complex essential to the devolopment of insects. Oil.
   $[\alpha]_D^{18}$ +13.8° (c, 0.92 in MeOH). n_D^{18} 1.4752. The nat.
   hormone is known to be (*E,E*) but the abs. config. of
   the epoxide function is currently unknown. The
   synthetic prod. was (10*R*, 11*S*) as shown.

Mori, K. *et al, Justus Liebigs Ann. Chem.*, 1989, 41.

K

Kanerin   K-80001

3β,5α-Dihydroxy-24-nor-4(23),18-ursadien-28-oic acid, 9CI
[120396-45-4]

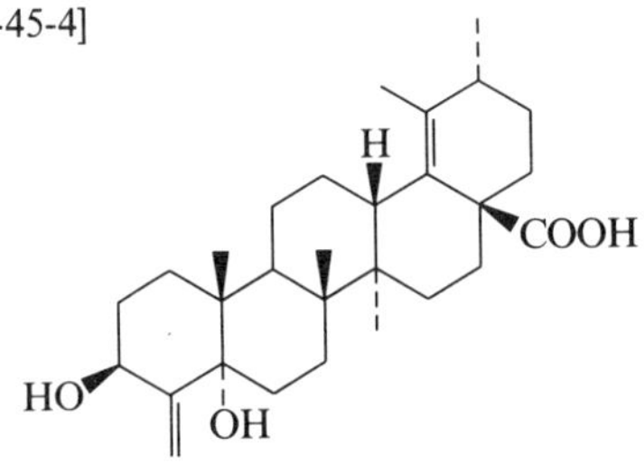

$C_{29}H_{44}O_4$ M 456.664

Constit. of leaves of *Nerium oleander*. Cryst. (MeOH). Mp 280-281°. $[\alpha]_D$ +14.28° (c, 0.14 in $CHCl_3$).

Siddiqui, S. *et al, J. Nat. Prod.* (*Lloydia*), 1989, **52**, 57 (*isol, pmr, cmr*)

Kanjone   K-80002

6-Methoxy-2-phenyl-4H-furo[2,3-h]-1-benzopyran-4-one, 9CI.
6-Methoxyfuro[2,3-h]flavone
[1094-12-8]

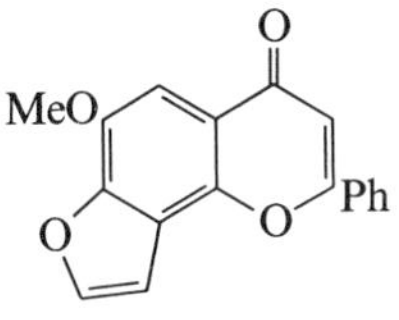

$C_{18}H_{12}O_4$ M 292.290

Isol. from seeds of *Pongamia glabra*. Cryst. (MeOH). Mp 190-191°.

Aneja, R. *et al, J. Chem. Soc.*, 1963, 163 (*isol, uv, ir, synth*)
Talapatra, K.S. *et al, J. Indian Chem. Soc.*, 1982, **59**, 534 (*cmr*)

Kanshone D   K-80003

[119403-26-8]

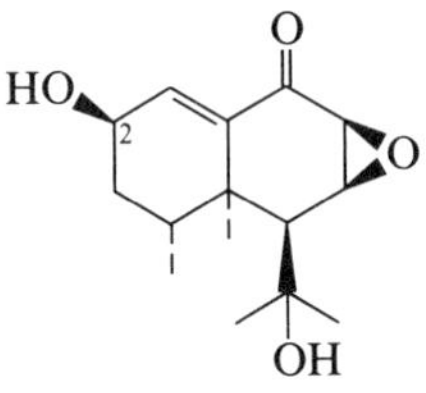

$C_{15}H_{22}O_4$ M 266.336

Constit. of *Nardostachys chinensis*. Needles. Mp 152-153°. $[\alpha]_D$ +57.9° (c, 0.52 in $CHCl_3$).

2-Ketone: [119403-27-9]. **Kanshone E**
$C_{15}H_{20}O_4$ M 264.321

Constit. of *N. chinensis*. Gum. $[\alpha]_D$ −42.2° (c, 1.29 in $CHCl_3$).

Bagchi, A. *et al, Phytochemistry*, 1988, **27**, 3667.

Karatavicinol   K-80004

Updated Entry replacing K-00021
7-[(10,11-Dihydroxy-3,7,11-trimethyl-2,6-dodecadienyl)oxy]-2H-1-benzopyran-2-one, 9CI. 7-(10,11-Dihydroxy-3,7,11-trimethyl-2,6-dodecadienyl)coumarin. Karatavikinol
[25374-66-7]

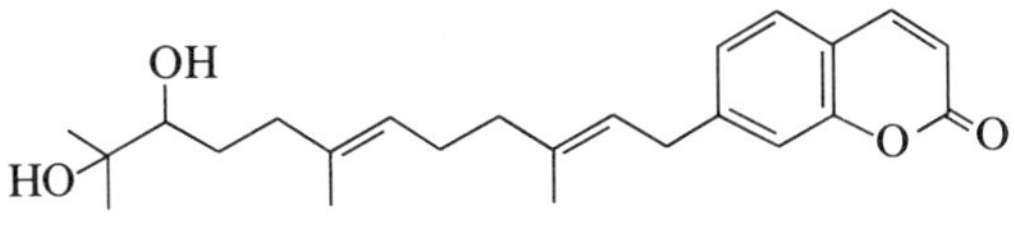

$C_{24}H_{32}O_4$ M 384.514

Constit. of the root of *Ferula karatavica*. Mp 52-53°. $[\alpha]_D^{20}$ −12° (c, 7.5 in EtOH).

10-O-β-Sophoroside: [53011-70-4]. **Reoselin A.** *Reoselin*
$C_{36}H_{52}O_{15}$ M 724.798

Isol. from *F. pseudooreoselinum* and *F. korshinskyi*. Mp 160-161° (155-156°). $[\alpha]_D$ −73.5° (MeOH), $[\alpha]_D$ −24.4° (EtOH).

Kir'yalov, N.P. *et al, Khim. Prir. Soedin.*, 1969, **5**, 225; 1974, **10**, 84; 1975, **11**, 87; *Chem. Nat. Compd.*, pp. 191, 82, 88.

3,13,16-Kauranetriol   K-80005

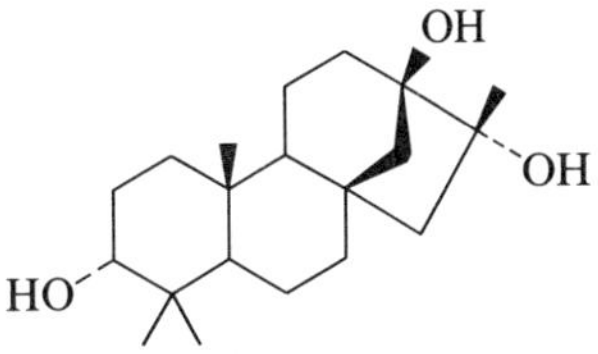

$C_{20}H_{34}O_3$ M 322.487

(3α,16α)-form

Constit. of *Agathis vitiensis*. Needles (EtOAc). Mp 124-126°.

Cambie, R.C. *et al, Phytochemistry*, 1989, **28**, 1675.

ent-7α,16β,18-Kauranetriol   K-80006

Distanol

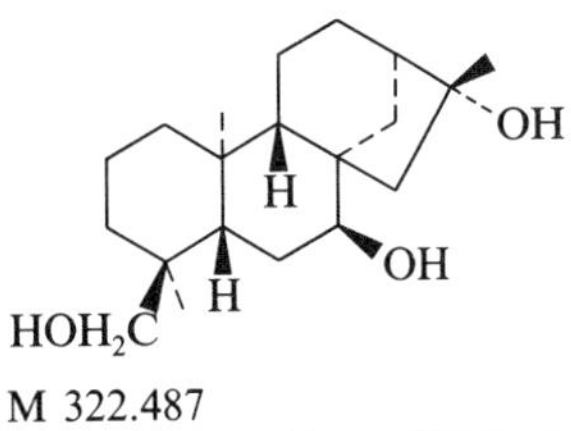

$C_{20}H_{34}O_3$ M 322.487

Constit. of *Sideritis distans*. Cryst. (EtOAc). Mp 260-265°.

Venturella, P. *et al, Phytochemistry*, 1989, **28**, 1976.

16-Kaurene-3,13-diol K-80007

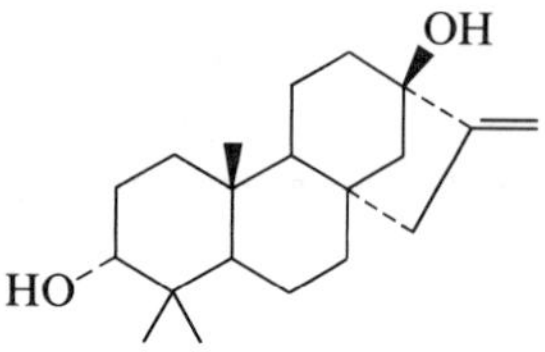

$C_{20}H_{32}O_2$ M 304.472

3α-form

Constit. of *Agathis vitiensis*. Needles (EtOAc). Mp 205-208°. $[\alpha]_D^{17} +28°$ (c, 1 in MeOH).

Cambie, R.C. *et al*, *Phytochemistry*, 1989, **28**, 1675.

ent-16-Kaurene-1β,3α,6β,11α,15α-pentol K-80008

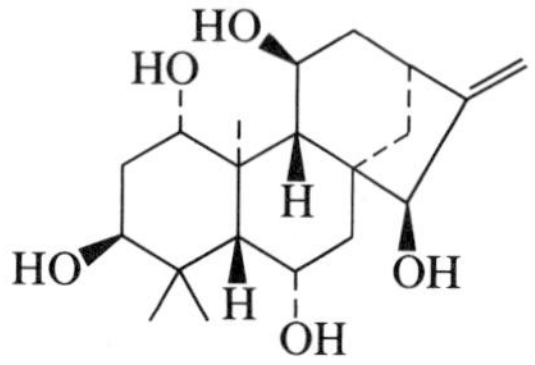

$C_{20}H_{32}O_5$ M 352.470

1,3-Di-Ac: ent-1β,3α-Diacetoxy-16-kaurene-6β,11α,15α-triol. **Inflexarabdonin A**
$C_{24}H_{36}O_7$ M 436.544
Constit. of *Rabdosia inflexa*. Cryst. (CHCl₃/hexane). Mp 118-120°. $[\alpha]_D^{27.5} -5.5°$ (c, 1.95 in MeOH).

6-Ketone, 1,3-Di-Ac: ent-1β,3α-Diacetoxy-11α,15α-dihydroxy-16-kauren-6-one. **Inflexarabdonin B**
$C_{24}H_{34}O_7$ M 434.528
Constit. of *R. inflexa*. Amorph. powder. $[\alpha]_D^{24} -16.9°$ (c, 0.83 in MeOH).

Takeda, Y. *et al*, *Phytochemistry*, 1989, **28**, 851.

ent-15-Kauren-18-oic acid K-80009

Isokaurenic acid

$C_{20}H_{30}O_2$ M 302.456
Isol. from resin of *Trachylobium verrucosum*.

Me ester: Cryst. (Et₂O/MeOH). Mp 123°. $[\alpha]_D \pm 0°$ (c, 1.2 in CHCl₃).

Hugel, G. *et al*, *Bull. Soc. Chim. Fr.*, 1963, 1974; 1965, 2882, 2888 (*isol, ir, pmr, ms, struct*)

Kievitone K-80010

Updated Entry replacing K-00113
3-(2,4-Dihydroxyphenyl)-2,3-dihydro-5,7-dihydroxy-8-(3-methyl-2-butenyl)-4H-1-benzopyran-4-one, 9CI. 8-(3,3-Dimethylallyl)-2′,4′,5,7-tetrahydroxyisoflavanone. 2′,4′,5,7-Tetrahydroxy-8-C-prenylisoflavone. Vignatin. Phaseolus substance II
[40105-60-0]

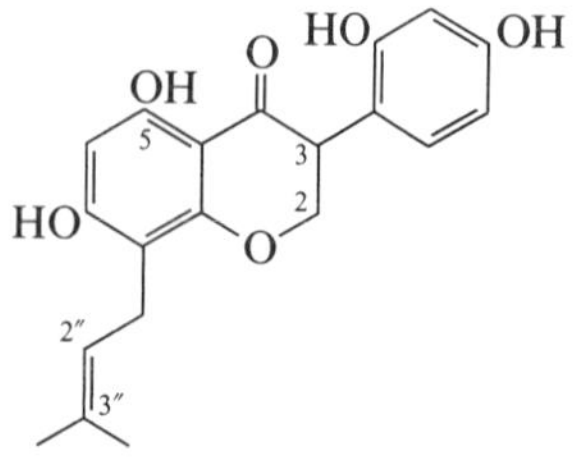

$C_{20}H_{20}O_6$ M 356.374

Isol. from *Dolichos biflorus, Lablab niger, Macroptilium atropurpureum, Macrotyloma axillare, Mucuna utilis, Phaseolus aureus, P. calcaratus, P. lunatus, P. vulgaris, Stizolobium deeringianum* and *Vigna unguiculata* (all Leguminosae, Papilionoideae).

2,3-Didehydro: [74161-25-4]. **2,3-Dehydrokievitone**
$C_{20}H_{18}O_6$ M 354.359
Isol. from pods of *P. vulgaris*.

5-Deoxy: [74161-24-3]. **5-Deoxykievitone**
$C_{20}H_{20}O_5$ M 340.375
Isol. from pods of *P. vulgaris*.

Burden, R.S. *et al*, *Tetrahedron Lett.*, 1972, 4175 (*isol*)
Smith, D.A. *et al*, *Physiol. Plant Pathol.*, 1973, **3**, 293 (*ms, ir, uv*)
Granamanickam, S.S., *Experientia*, 1979, **35**, 323 (*isol*)
Woodward, M.D., *Phytochemistry*, 1979, **18**, 2007 (2,3-Dehydrokievitone, 5-Deoxykievetone)
Woodward, M.D., *Phytochemistry*, 1979, **18**, 363 (*occur*)
Ingham, J.L., *Fortschr. Chem. Org. Naturst.*, 1983, **43**, 1 (*rev, occur*)

Kobophenol A K-80011

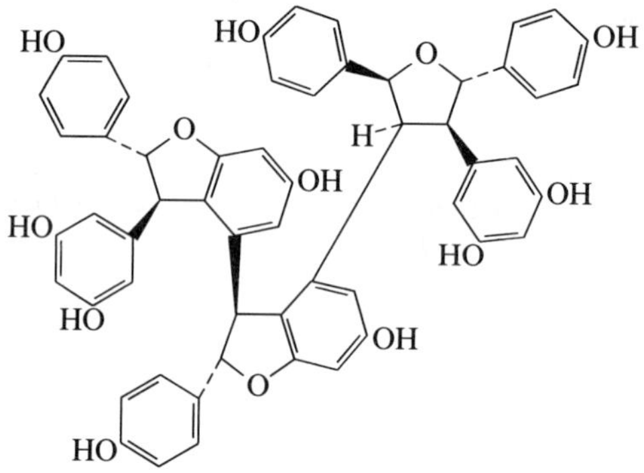

$C_{56}H_{44}O_{13}$ M 924.956
Constit. of *Carex kobomugi*. Pale yellow oil. $[\alpha]_D^{20} +227°$ (c, 0.17 in MeOH).

Kawabata, J. *et al*, *Tetrahedron Lett.*, 1989, **30**, 3785.

Kopeolin K-80012

Updated Entry replacing K-00148
7-[[5-(3,6-Dihydroxy-2,2,6-trimethylcyclohexyl)-3-methyl-2-pentyl]oxy]-2H-1-benzopyran-2-one, 9CI
[51005-86-8]

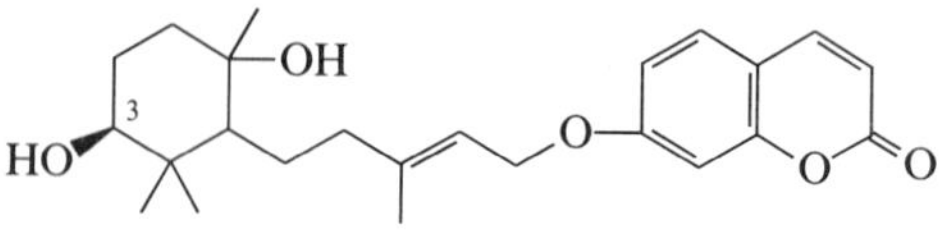

$C_{24}H_{32}O_5$ M 400.514
Constit. of the roots of *Ferula kopetdaghensis*. Cryst. (C₆H₆). Mp 146-147°. $[\alpha]_D^{25} -15.9°$ (c, 0.98 in EtOH).

3-O-β-D-Glucopyranoside: [50982-41-7]. **Kopeoside**
$C_{30}H_{42}O_{10}$ M 562.656
Constit. of *F. kopetdaghensis*. Cryst. (MeOH aq.). Mp 177-178°. $[\alpha]_D^{25} -22.1°$ (c, 1.47 in EtOH).

3-Ketone: [81758-04-5]. **Kopeolone**
$C_{24}H_{30}O_5$ M 398.498
Constit. of *F. kopetdaghensis*. Cryst. Mp 125-126°. $[\alpha]_D +170°$ (c, 1 in EtOH).

Kamilov, Kh.M. *et al*, *Khim. Prir. Soedin.*, 1973, 308; *CA*, **79**, 134321 (*isol*)
Nabiev, A.A. *et al*, *Khim. Prir. Soedin.*, 1982, **18**, 48 (*Kopeolone*)

Kumepaloxane K-80013

[122666-10-8]

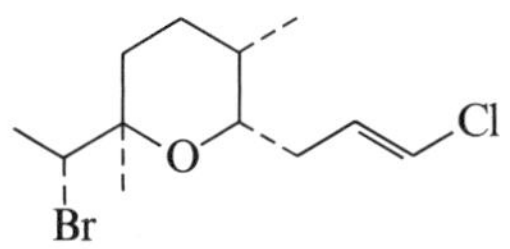

$C_{12}H_{20}BrClO$ M 295.646

Constit. of bubble shell *Haminoea cymbalum*. Fish antifeedant. Oil. $[\alpha]_D$ +22.6° (c, 0.32 in $CHCl_3$).

Poiner, A. *et al*, *Tetrahedron*, 1989, **45**, 617.

Kuwanol C K-80014

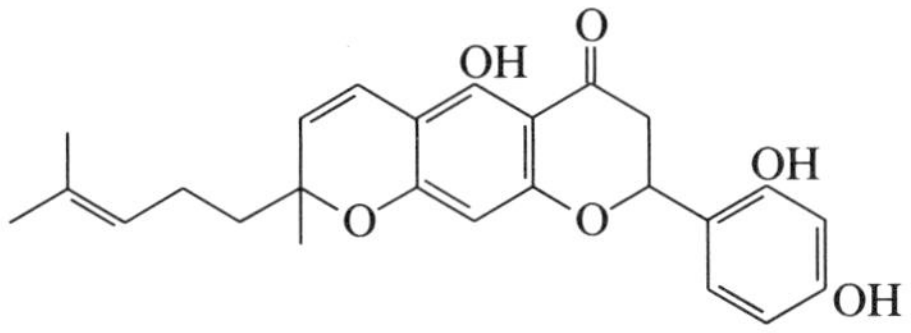

$C_{25}H_{26}O_6$ M 422.477

Constit. of *Morus alba*. Amorph. powder. $[\alpha]_D^{22}$ −10° (c, 0.31 in EtOH).

Hano, Y. *et al*, *Heterocycles*, 1989, **29**, 807.

Kuwanol D K-80015

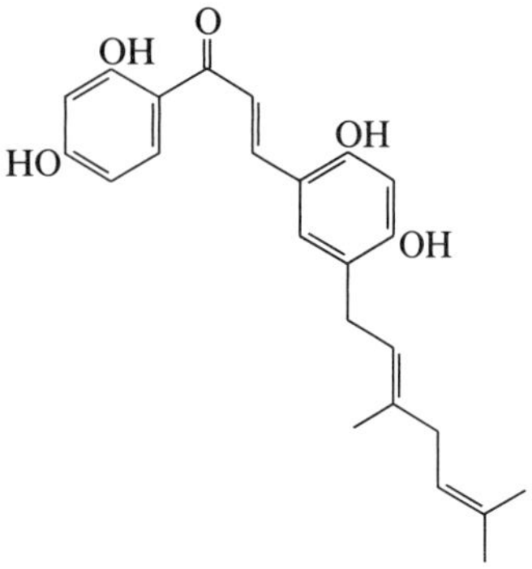

$C_{25}H_{28}O_5$ M 408.493

Constit. of *Morus alba*. Yellow amorph. powder.

Hano, Y. *et al*, *Heterocycles*, 1989, **29**, 807.

L

7,14-Labdadiene-3,13-diol L-80001

$C_{20}H_{34}O_2$ M 306.487

(3α,13R)-form

Constit. of *Waitzia acuminata*.

Jakupovic, J. *et al*, *Phytochemistry*, 1989, **28**, 1943.

8(17),12-Labdadiene-15,16-dial L-80002

[104263-88-7]

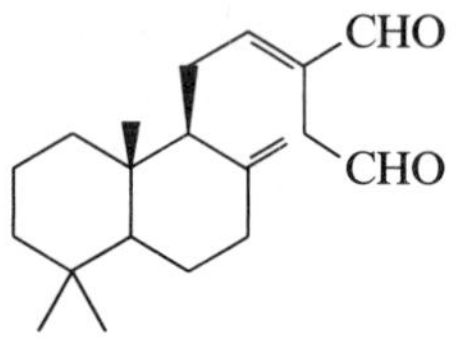

$C_{20}H_{30}O_2$ M 302.456

Constit. of *Curcuma heyneana*. Oil. $[\alpha]_D^{20}$ +16° (c, 0.23 in MeOH).

Firman, K. *et al*, *Phytochemistry*, 1988, **27**, 3887.

ent-7,13-Labdadiene-2α,15-diol L-80003

$C_{20}H_{34}O_2$ M 306.487

Constit. of *Ophryosporus heptanthus*. Gum.

15-Ac: *ent*-15-*Acetoxy-7,13-labdadien-2α-ol*

$C_{22}H_{36}O_3$ M 348.525

Constit. of *O. heptanthus*. Gum.

Ferracini, V.L. *et al*, *Phytochemistry*, 1989, **28**, 1463.

ent-7,13-Labdadiene-2β,15-diol L-80004

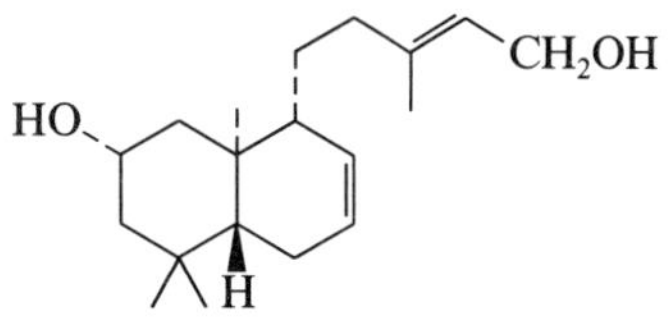

$C_{20}H_{34}O_2$ M 306.487

Constit. of *Baccharis sternbergiana*. Oil. $[\alpha]_D^{24}$ −5° (c, 0.8 in $CHCl_3$).

15-Ac: *ent*-15-*Acetoxy-7,13-labdadien-2β-ol*

$C_{22}H_{36}O_3$ M 348.525

Constit. of *B. sternbergiana*. Oil. $[\alpha]_D^{24}$ −9.5° (c, 2.13 in $CHCl_3$).

2-Ketone: *ent*-15-*Hydroxy-7,13-labdadien-2-one*

$C_{20}H_{32}O_2$ M 304.472

Constit. of *B. sternbergiana*.

Zdero, C. *et al*, *Phytochemistry*, 1989, **28**, 531.

8,14-Labdadiene-2,3,7,13,20-pentol L-80005

$C_{20}H_{34}O_5$ M 354.486

(2α,3α,7α,13R)-form

2-Angeloyl, 3,20-di-Ac: *3α,20-Diacetoxy-2α-angeloyloxy-8,14-labdadiene-7α,13R-diol*

$C_{29}H_{44}O_8$ M 520.662

Constit. of *Waitzia acuminata*.

Jakupovic, J. *et al*, *Phytochemistry*, 1989, **28**, 1943.

8(17),14-Labdadiene-2,3,7,13,20-pentol L-80006

$C_{20}H_{34}O_5$ M 354.486

(2α,3α,7α,13R)-form

2-Angeloyl, 20-Ac: *20-Acetoxy-2α-angeloyloxy-8(17),14-labdadiene-3α,7α,13R-triol*

$C_{27}H_{42}O_7$ M 478.625

Constit. of *Waitzia acuminata*.

2-Angeloyl, 20-(2-methylbutanoyl): *2α-Angeloyloxy-20-(2-methylbutanoyloxy)-8(17),14-labdadiene-3α,7α,13R-triol*

$C_{30}H_{48}O_7$ M 520.705

Constit. of *W. acuminata*.

2-Angeloyl, 3,20-di-Ac: *3α,20-Diacetoxy-2α-angeloyloxy-8(17),14-labdadiene-7α,13R-diol*

$C_{29}H_{44}O_8$ M 520.662

Constit. of *W. acuminata*.

Jakupovic, J. *et al*, *Phytochemistry*, 1989, **28**, 1943.

7,14-Labdadiene-2,3,13,20-tetrol L-80007

$C_{20}H_{34}O_4$ M 338.486

(2α,3α,13R)-form

2-Angeloyl, 20-Ac: *20-Acetoxy-2α-angeloyloxy-7,14-labdadiene-3α,13R-diol*

$C_{27}H_{42}O_6$ M 462.625

Constit. of *Waitzia acuminata*.

2-Angeloyl, 20-(2-methylbutanoyl): *2α-Angeloyloxy-20-(2-methylbutanoyloxy)-7,14-labdadiene-3α,13S-diol*

$C_{30}H_{48}O_6$ M 504.706

Constit. of *W. acuminata*.

Jakupovic, J. *et al*, *Phytochemistry*, 1989, **28**, 1943.

8(17),14-Labdadiene-2,7,13,20-tetrol L-80008

$C_{20}H_{34}O_4$ M 338.486

(2α,7α,13R)-form

2-Angeloyl, 20-Ac: *20-Acetoxy-2α-angeloyloxy-8(17),14-labdadiene-7α,13R-diol*

$C_{27}H_{42}O_6$ M 462.625

Constit. of *Waitzia acuminata*.

Jakupovic, J. *et al*, *Phytochemistry*, 1989, **28**, 1943.

ent-7,13-Labdadiene-2β,3β,15-triol L-80009

$C_{20}H_{34}O_3$ M 322.487

Constit. of *Baccharis sternbergiana*.

Zdero, C. *et al*, *Phytochemistry*, 1989, **28**, 531.

8(17),13(16),14-Labdatrien-19-oic acid L-80010

Updated Entry replacing L-00040

Myrceocommunic acid. $\Delta^{13(16)}$-*Communic acid*

[31323-66-7]

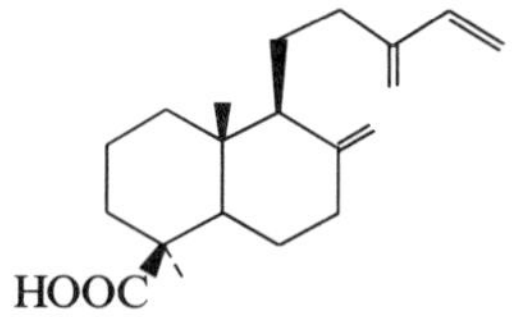

$C_{20}H_{30}O_2$ M 302.456

Constit. of *Callitris columellaris* and *Juniperus oxycedrus*.
Cryst. (MeOH). Mp 129-130° (115-117°). $[\alpha]_D$ +53.6°
(CHCl₃).
19-Aldehyde: [56709-03-6]. *8(17),13(16),14-Labdatrien-19-al.*
Myrceocommunal
$C_{20}H_{30}O$ M 286.456
Isol. from *J. oxycedrus*.

Atkinson, P.W. *et al*, *Tetrahedron*, 1970, **26**, 1935 (*isol, ms*)
Carman, R.M. *et al*, *Aust. J. Chem.*, 1971, **24**, 353 (*isol, pmr*)
Teresa, J.P. *et al*, *An. Quim.*, 1972, **68**, 1061; 1978, **74**, 966 (*isol*)

7-Labdene-15,17-dioic acid   L-80011
$C_{20}H_{32}O_4$ M 336.470
Constit. of *Halimium verticillatum*.
Di-Me ester: Oil. $[\alpha]_D^{22}$ −21.8° (c, 0.77 in CHCl₃).

Urones, J.G. *et al*, *Phytochemistry*, 1989, **28**, 557.

7-Labdene-15,17-diol   L-80012
$C_{20}H_{36}O_2$ M 308.503
Di-Ac: 15,17-Diacetoxy-7-labdene
$C_{24}H_{40}O_4$ M 392.578
Constit. of *Halimium verticillatum*. Oil. $[\alpha]_D^{22}$ −9.7° (c, 0.83 in CHCl₃).

Urones, J.G. *et al*, *Phytochemistry*, 1989, **28**, 557.

8(20)-Labdene-3,15-diol   L-80013
$C_{20}H_{36}O_2$ M 308.503
3β-form
Isol. from *Araucaria imbricata*. Platelets (pet. ether). Mp 114°. $[\alpha]_D$ +29° (c, 0.6 in CHCl₃).

Chandra, G. *et al*, *J. Chem. Soc.*, 1964, 3648 (*isol, uv, ir, ms*)

Laciniatoside V   L-80014

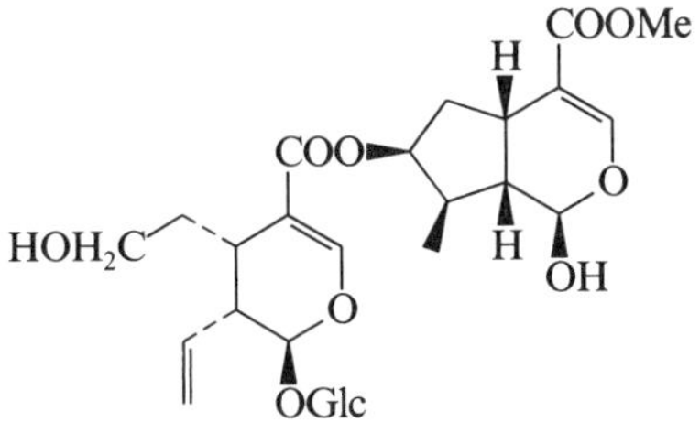

$C_{27}H_{38}O_{14}$ M 586.589
Constit. of *Dipsacus laciniatus*. Amorph. $[\alpha]_D^{27}$ −98° (c, 0.67 in MeOH).

Podányi, B. *et al*, *J. Nat. Prod. (Lloydia)*, 1989, **52**, 135 (*isol, pmr, cmr*)

Lacinolide A   L-80015

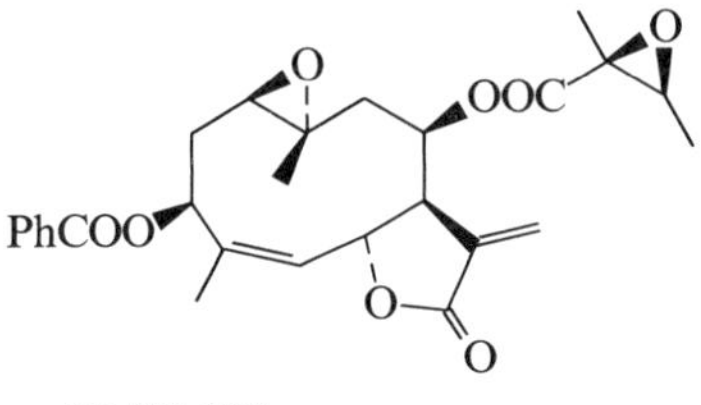

$C_{27}H_{30}O_8$ M 482.529
Constit. of *Viguiera laciniata*. Cryst.

Gao, F. *et al*, *Phytochemistry*, 1989, **28**, 2409 (*isol, pmr, cmr, cryst struct*)

Lacinolide B   L-80016

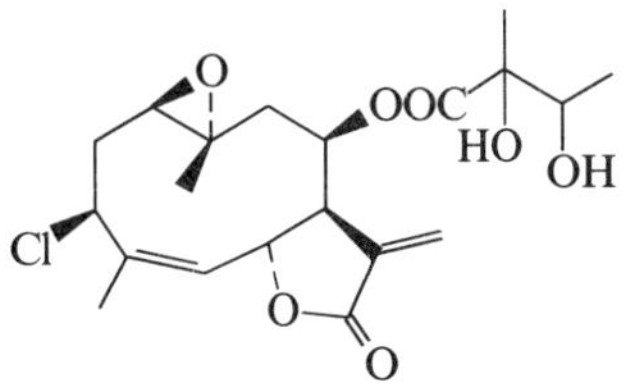

$C_{20}H_{27}ClO_7$ M 414.882
Constit. of *Viguiera laciniata*.

Gao, F. *et al*, *Phytochemistry*, 1989, **28**, 2409 (*isol, pmr, cmr*)

Lancifolide   L-80017

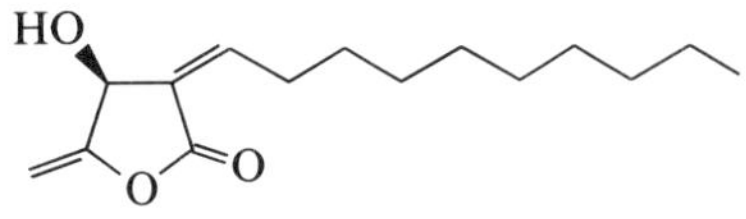

$C_{15}H_{24}O_3$ M 252.353
Constit. of *Actinodaphne lancifolia*. Oil. $[\alpha]_D$ −49.0° (c, 0.58 in CHCl₃).
E-Isomer: Isolancifolide
$C_{15}H_{24}O_3$ M 252.353
Constit. of *A. lancifolia*. Oil. $[\alpha]_D$ −59.0° (c, 0.5 in CHCl₃).

Tanaka, H. *et al*, *Phytochemistry*, 1989, **28**, 626.

Languiduline   L-80018
[114391-81-0]

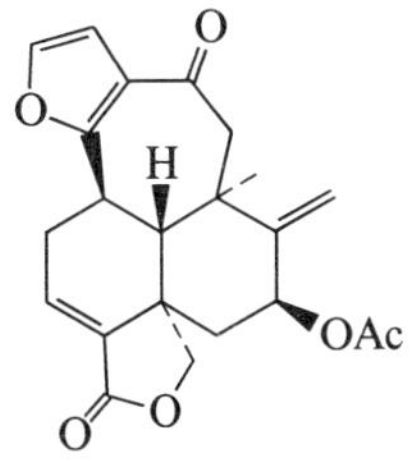

$C_{22}H_{22}O_6$ M 382.412
Constit. of *Salvia languidula*. Cryst. Mp 243-245°. $[\alpha]_D^{20}$ −193.4° (c, 0.2 in CHCl₃).

Córdenas, J. *et al*, *Heterocycles*, 1988, **27**, 1809 (*isol, cryst struct, cmr*)

8,23-Lanostadiene-3,22,25-triol   L-80019

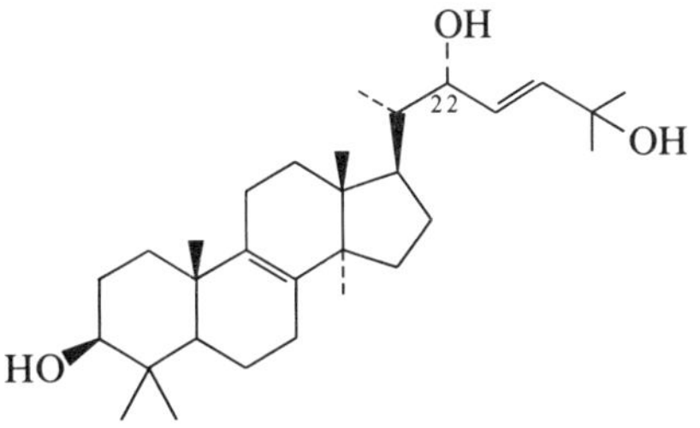

$C_{30}H_{50}O_3$ M 458.723
(3α,22S,23E)-form
22-Ac: [119513-66-5]. *22-Acetoxy-8,23-lanostadiene-3α,25-diol*
$C_{32}H_{52}O_4$ M 500.760
Metab. of *Pisolithus tinctorius*. Cryst. (MeOH). Mp 187-190°.

3,22-Di-Ac: Cryst. (MeOH). Mp 190-192°. [α]$_D$ +6° (c, 1.5 in CHCl$_3$).

Lobo, A.M. *et al*, *Phytochemistry*, 1988, **27**, 3569 (*isol, cryst struct*)

Laucopyranoid A   L-80020

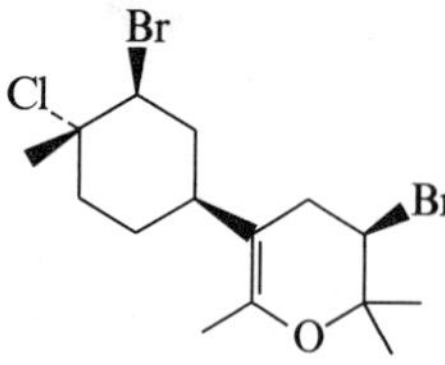

C$_{15}$H$_{23}$Br$_2$ClO   M 414.607

Constit. of *Laurencia caespitosa*. Cryst. Mp 118-120°. [α]$_D$ +0.5° (c, 0.87 in CHCl$_3$).

Chang, M. *et al*, *Phytochemistry*, 1989, **28**, 1417.

Laucopyranoid B   L-80021

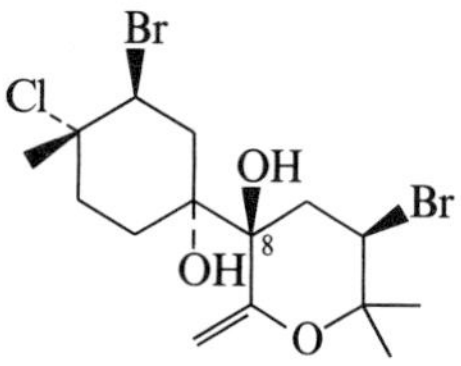

C$_{15}$H$_{23}$Br$_2$ClO$_2$ M 430.606

Constit. of *Laurencia caespitosa*. Cryst. Mp 97-98°. [α]$_D$ +11.0° (c, 1.12 in CHCl$_3$).

8-Epimer: Laucopyranoid C
C$_{15}$H$_{23}$Br$_2$ClO$_2$ M 430.606
Constit. of *L. caespitosa*. Unstable.

Chang, M. *et al*, *Phytochemistry*, 1989, **28**, 1417.

Laureoxolane   L-80022

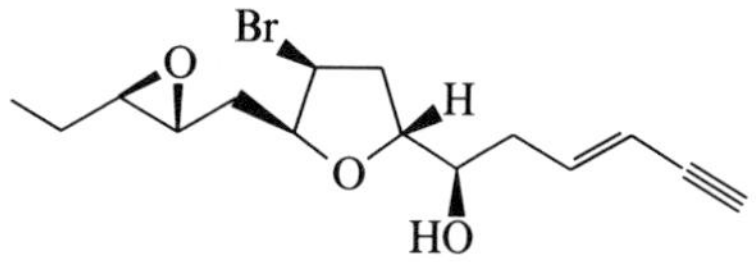

C$_{15}$H$_{21}$BrO$_3$ M 329.233

Metab. of *Laurencia nipponica*. Unstable oil. [α]$_D^{24}$ +21.9° (c, 0.32 in CHCl$_3$).

Fukuzawa, A. *et al*, *Tetrahedron Lett.*, 1989, **30**, 3665.

Lepidissipyrone   L-80023

[122585-68-6]

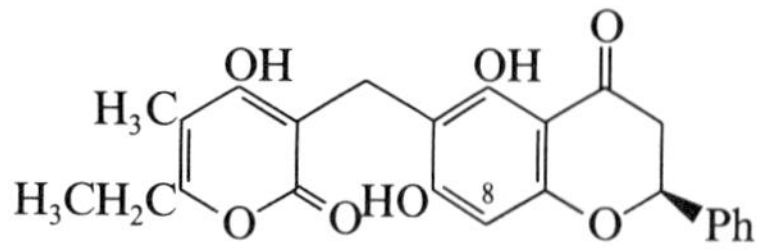

C$_{24}$H$_{22}$O$_7$ M 422.434

Constit. of *Helichrysum lepidissimum*. Gum.

8-(3-Methyl-2-butenyl): [122616-68-6]. *8-Prenyllepidissipyrone*
C$_{29}$H$_{30}$O$_7$ M 490.552
Constit. of *H. lepidissimum*. Gum.

Jakupovic, J. *et al*, *Phytochemistry*, 1989, **28**, 1119.

Lespedezic acid   L-80024

[123955-02-2]

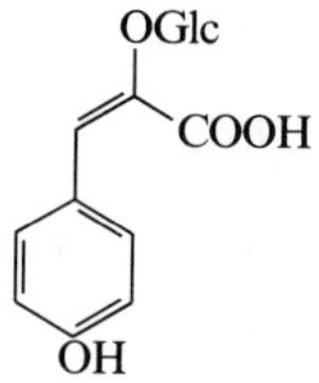

C$_{15}$H$_{18}$O$_9$ M 342.302

Constit. of *Lesperzeza cuneata*. Leaf-opening factor. Powder (as K salt). [α]$_D^{25}$ −57.4° (c, 0.65 in H$_2$O).

Z-Isomer: [123955-03-3]. *Isolespedezic acid*
C$_{15}$H$_{18}$O$_9$ M 342.302
Constit. of *L. cuneata*. Powder (as K salt). [α]$_D^{25}$ +53.1° (c, 1.0 in H$_2$O).

Shigemori, H. *et al*, *Tetrahedron Lett.*, 1989, **30**, 3991.

Libanotic acid   L-80025

4-(1-Methyl-3-oxobutyl)-1-cyclohexene-1-carboxylic acid, 9CI. 11,12,13-Trinor-9-oxo-2-bisabolen-15-oic acid

[119765-97-8]

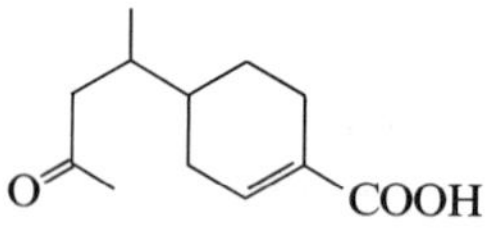

C$_{12}$H$_{18}$O$_3$ M 210.272

Constit. of *Cedrus libani*. Oil.

Avcibasi, H. *et al*, *Phytochemistry*, 1988, **27**, 3967.

Licarin E   L-80026

[51020-87-2]

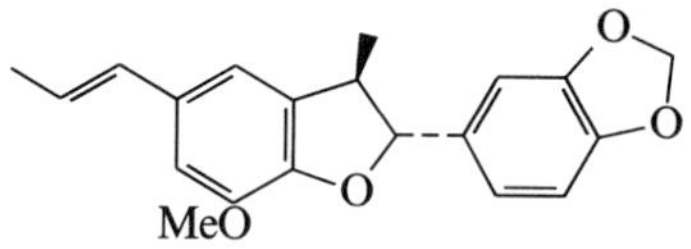

C$_{20}$H$_{20}$O$_4$ M 324.376

Constit. of *Nectandra glabrescens*. Cryst. (MeOH). Mp 83-85°.

Barbosa-Filho, J.M. *et al*, *Phytochemistry*, 1989, **28**, 1991.

Licobenzofuran   L-80027

2-[4-Hydroxy-3-(3-methyl-2-butenyl)phenyl]-5,6-dimethoxy-3-benzofuranol, 9CI. 3-Hydroxy-2-(4-hydroxy-3-C-prenylphenyl)-5,6-dimethoxybenzofuran

[82209-75-4]

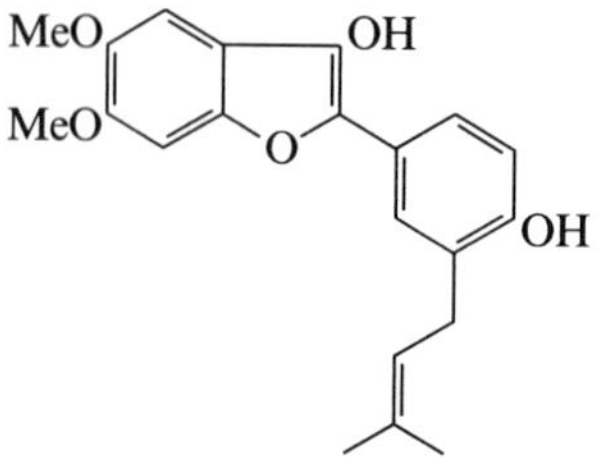

C$_{21}$H$_{22}$O$_5$ M 354.402

Isol. from *Glycyrrhiza* sp. Shows antibacterial props.

Chang, X. *et al*, *Zhongcaoyao*, 1981, **12**, 530; *CA*, **97**, 20701k.

Licoisoflavone A   L-80028

Updated Entry replacing L-40024

[2,4-Dihydroxy-3-(3-methyl-2-butenyl)phenyl]-5,7-dihydroxy-4H-1-benzopyran-4-one, 9CI. 3'-Isopentenyl-2',4',5,7-tetrahydroxyisoflavone. 2',4',5,7-Tetrahydroxy-3-C-prenylisoflavone. Phaseoluteone

[66056-19-7]

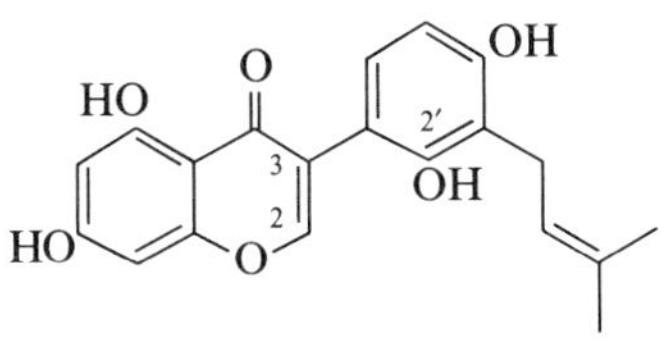

$C_{20}H_{18}O_6$ M 354.359

Constit. of the root of *Glycyrrhiza* spp., also from *Hardenbergia violacea* and *Phaseolus vulgaris*. Pale-yellow prisms (MeOH aq.).

Tetra-Ac: Mp 136-138°.

2,3-Dihydro, 2'-Me ether: [69573-59-7]. **Sophoraisoflavanone A**

$C_{21}H_{22}O_6$ M 370.401

Isol. from aerial parts of *Sophora tomentosa*. Shows antifungal activity. Needles. Mp 178-180°. [α]$_D^{22}$ −17.3° (EtOH).

Kinoshita, T. *et al, Chem. Pharm. Bull.*, 1978, **26**, 141 (*isol, struct*)
Komatsu, M. *et al, Chem. Pharm. Bull.*, 1978, **26**, 3863 (*Sophoraisoflavanone A*)
Woodward, M.D., *Phytochemistry*, 1979, **18**, 363 (*isol, uv, pmr, ir*)
Tsukayama, M. *et al, Bull. Chem. Soc. Jpn.*, 1985, **58**, 136 (*synth*)

Licoricidin   L-80029

Updated Entry replacing L-60025

2',4',7-Trihydroxy-5-methoxy-3',6-diprenylisoflavan

[30508-27-1]

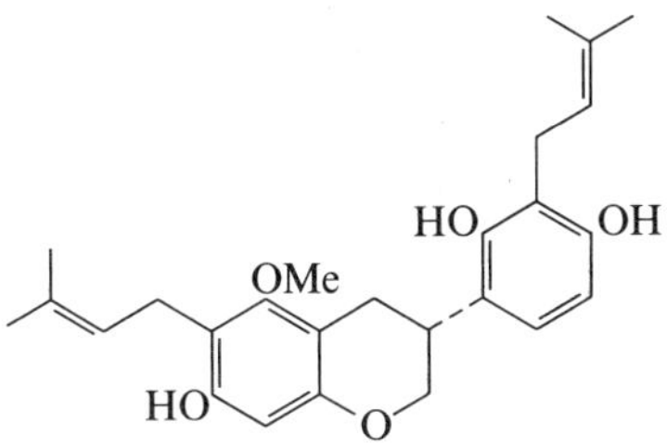

$C_{26}H_{32}O_5$ M 424.536

Struct. revised in 1988 (Me group previously considered to be at 7-posn.).

(R)-form

Constit. of *Glycyrrhiza glabra* and *G. uralensis*. Cryst. (CHCl₃/Et₂O). Mp 154-156°. [α]$_D^{22.5}$ +20° (c, 1 in MeOH).

7-Me ether:

$C_{27}H_{34}O_5$ M 438.563

Constit. of *G. uralensis*. Called '5-O-Methyllicoricidin'.

Shibata, S. *et al, Chem. Pharm. Bull.*, 1968, **16**, 1932 (*isol*)
Kinoshita, T. *et al, Chem. Pharm. Bull.*, 1978, **26**, 141 (*isol*)
Shih, T.L. *et al, J. Org. Chem.*, 1987, **52**, 2029 (*synth*)
Fukai, T. *et al, Heterocycles*, 1988, **27**, 2309 (*isol, struct, cmr*)

Ligustrin   L-80030

Updated Entry replacing L-50046

[21677-87-2]

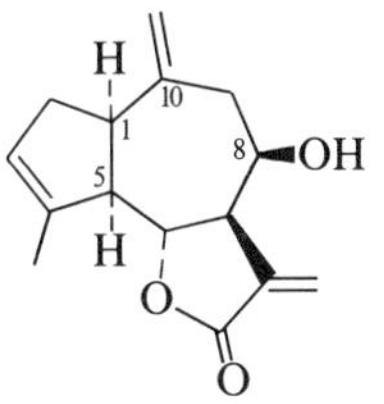

$C_{15}H_{18}O_3$ M 246.305

Constit. of *Eupatorium ligustrinum* and of *Eriophyllum staechadifolium*. Cryst. (Me₂CO/Et₂O). Mp 135-137°. [α]$_D$ +56°.

8-(2-Hydroxymethyl-4-hydroxy-2E-butenoyl): Constit. of *Stevia setifera*. Oil. [α]$_D^{24}$ +4.9° (c, 0.7 in CHCl₃).

8-O-(4-Hydroxytigloyl): [100045-33-8].

$C_{20}H_{24}O_5$ M 344.407

Constit. of *Stevia mercedensis*. Oil.

8-O-(2,5-Dihydro-5-hydroxy-3-furancarboxylate): [100045-34-9].

$C_{20}H_{22}O_6$ M 358.390

Constit. of *S. mercedensis*. Oil. Mixt. of 5'-Epimers.

Δ¹⁽¹⁰⁾-Isomer, 8-O-(2,5-dihydro-5-hydroxy-3-furancarboxylate): [100045-37-2].

$C_{20}H_{22}O_6$ M 358.390

Constit. of *S. mercedensis*. Oil. Mixt. of 5'-epimers.

8-Epimer: **Epiligustrin**

$C_{15}H_{18}O_3$ M 246.305

Constit. of *Leucanthemopsis pulverulenta*. Cryst. (Et₂O). Mp 132°. [α]$_D$ +77.4° (c, 0.9 in CHCl₃). Not interrelated with Ligustrin.

[100045-35-0, 100045-38-3]

Romo, J. *et al, Tetrahedron*, 1968, **24**, 6087 (*isol*)
Bohlmann, F. *et al, Phytochemistry*, 1979, **18**, 673 (*isol*)
Bohlmann, F. *et al, Phytochemistry*, 1981, **20**, 2239 (*isol, stereochem*)
Ito, K. *et al, Phytochemistry*, 1982, **21**, 715 (*synth, pmr, epimer*)
De Pascual Teresa, J. *et al, Phytochemistry*, 1984, **23**, 1178 (*epimer*)
Bohlmann, F. *et al, Justus Liebigs Ann. Chem.*, 1986, 799 (*derivs*)

Limonexic acid   L-80031

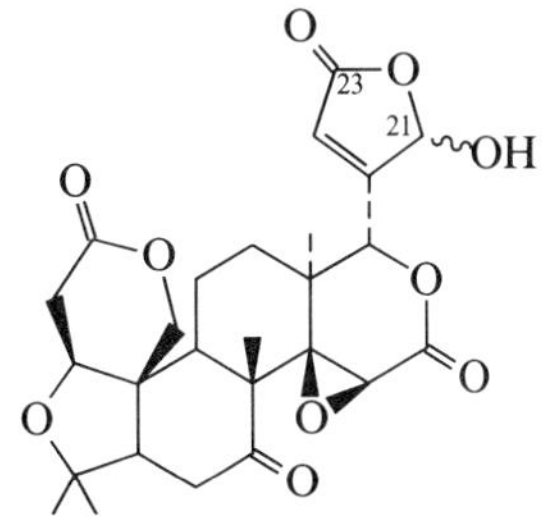

$C_{26}H_{30}O_{10}$ M 502.517

Constit. of *Phellodendron amurense*. Mp 285-286° dec. [α]$_D^{25}$ −127.0°.

21-Oxo, 23ξ-alcohol: **Isolimonexic acid**

$C_{26}H_{30}O_{10}$ M 502.517

Constit. of *Tetradium glabrifolium*. Needles (EtOAc). Mp 295°. $[\alpha]_D$ −65° (c, 0.13 in MeOH). There is an error in the structure diagrams of the reference.

Kondo, Y. *et al*, *Yakugaku Zasshi* (*J. Pharm. Soc. Jpn.*), 1985, **105**, 742 (*Limonexic acid*)

Ng, K.M. *et al*, *J. Nat. Prod.* (*Lloydia*), 1987, **50**, 1160 (*Isolimonexic acid*)

Linderachalcone L-80032

[120948-04-1]

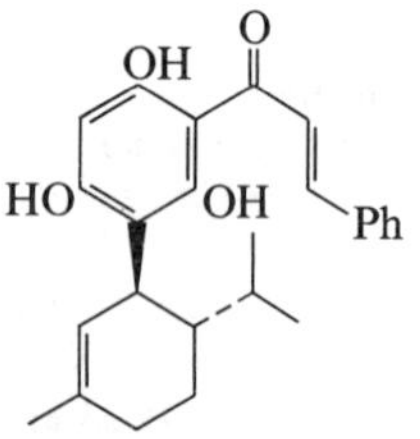

$C_{25}H_{28}O_4$ M 392.494

Constit. of *Lindera umbellata*. Yellow oil. $[\alpha]_D$ +17.8° (c, 0.4 in CHCl₃).

Ichino, K., *Phytochemistry*, 1989, **28**, 955.

Linderatin L-80033

Updated Entry replacing L-50048

[94530-80-0]

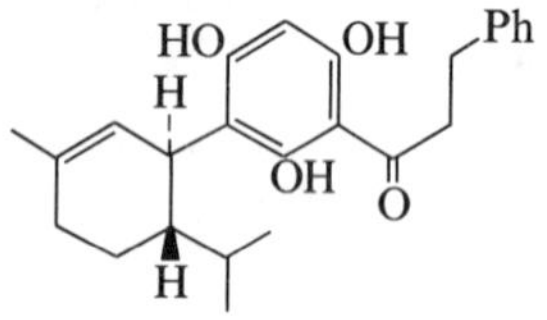

$C_{25}H_{30}O_4$ M 394.510

Constit. of leaves of *Lindera umbellata* var. *lancea*. Oil. $[\alpha]_D$ +19.1° (c, 0.45 in CHCl₃).

4′-Me ether: Methyllinderatin
$C_{26}H_{32}O_4$ M 408.536

Constit. of *L. umbellata*. Oil. $[\alpha]_D$ +41.0° (c, 0.4 in CHCl₃).

Tanaka, H. *et al*, *Chem. Pharm. Bull.*, 1984, **32**, 3747 (*isol, struct*)

Ichino, K., *Phytochemistry*, 1989, **28**, 955 (*deriv*)

Lindleyin L-80034

[59282-56-3]

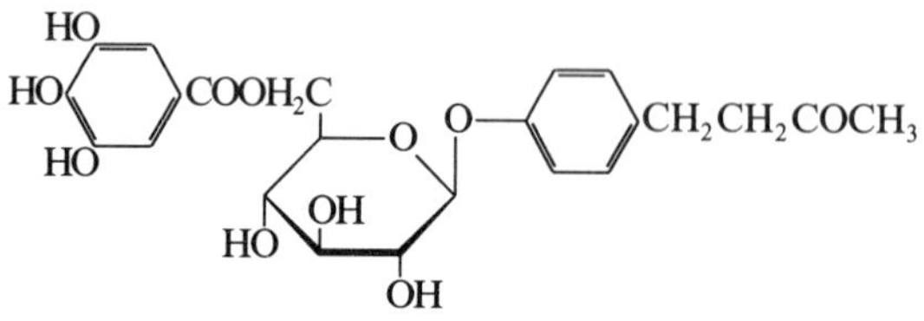

$C_{23}H_{26}O_{11}$ M 478.452

Constit. of *Aeonium lindleyi*. Mp 210-211°. $[\alpha]_D$ +14° (c, 0.38 in Py).

González, A.G. *et al*, *Phytochemistry*, 1976, **15**, 344 (*isol, ir, uv, ms, struct*)

Liriolignal L-80035

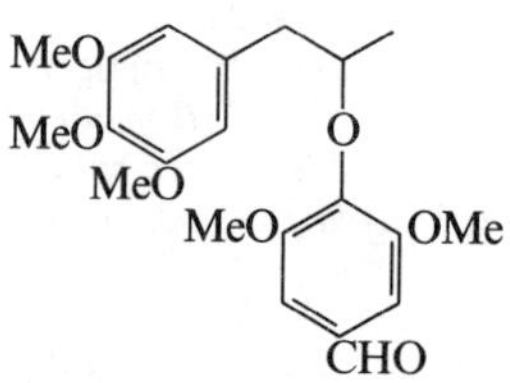

$C_{21}H_{26}O_7$ M 390.432

Constit. of *Liriodendron tulipifera*. Oil. $[\alpha]_D^{25}$ +10.6° (c, 0.01 in CHCl₃).

Muhammad, I. *et al*, *J. Nat. Prod.* (*Lloydia*), 1989, **52**, 1177 (*isol, pmr, cmr*)

Lophirochalcone L-80036

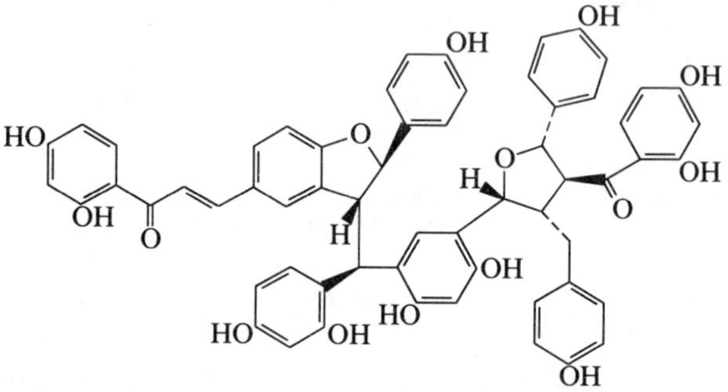

$C_{60}H_{48}O_{15}$ M 1009.030

Constit. of *Lophira lanceolata*. Amorph. brown solid. $[\alpha]_D^{25}$ +181° (c, 0.6 in Me₂CO).

Tih, R.G. *et al*, *Tetrahedron Lett.*, 1989, **30**, 1807.

Lophirone B L-80037

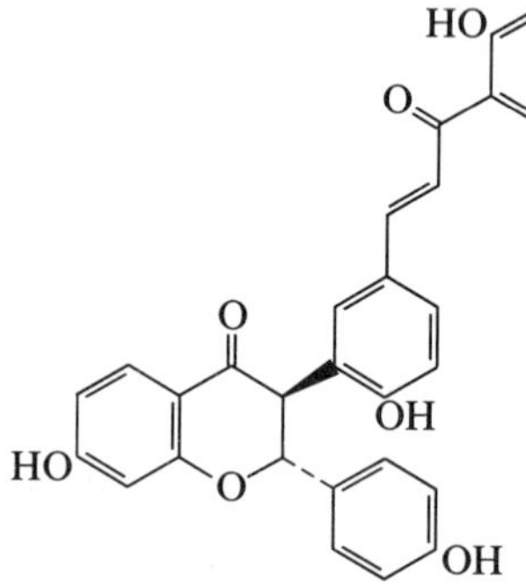

$C_{30}H_{22}O_8$ M 510.499

Constit. of *Lophira lanceolata*. Yellow cryst. (Me₂CO). Mp 251-253° dec. $[\alpha]_D^{25}$ +7° (c, 0.4 in Me₂CO).

Ghogoma, R. *et al*, *Phytochemistry*, 1989, **28**, 1557.

Lophirone C L-80038

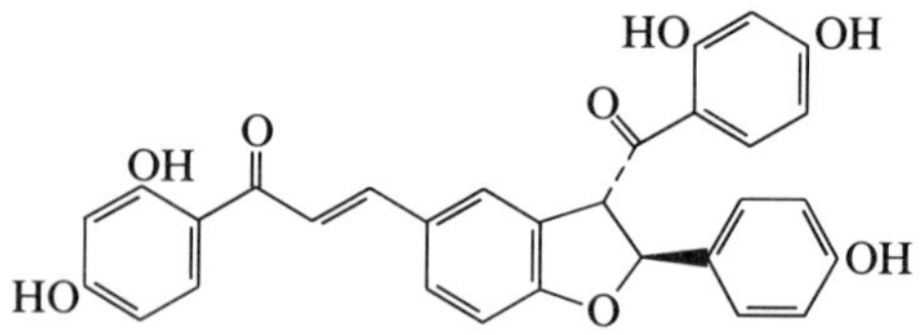

$C_{30}H_{22}O_8$ M 510.499

Constit. of *Lophira lanceolata*. Yellow cryst. (Me₂CO). Mp 191-193° dec. $[\alpha]_D^{25}$ −16.3° (c, 0.5 in Me₂CO).

Ghogoma, R. *et al*, *Phytochemistry*, 1989, **28**, 1557.

Lophirone D   L-80039

[122127-70-2]

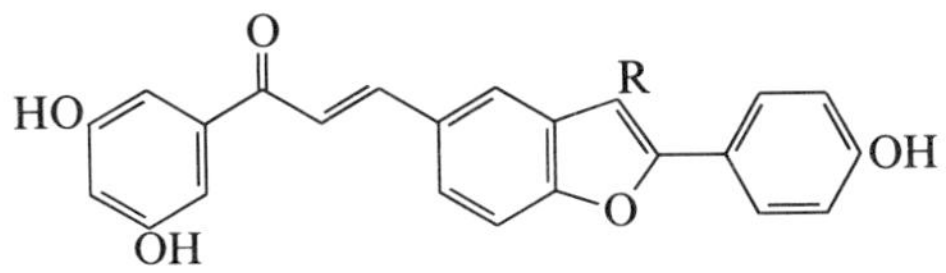

R = CHO

$C_{24}H_{16}O_6$ M 400.387

Constit. of *Lophira lanceolata*. Yellow cryst. (Me$_2$CO). Mp 270-276° dec.

Ghogomu, R. *et al*, *J. Nat. Prod.* (*Lloydia*), 1989, **52**, 284 (*isol, pmr, cmr*)

Lophirone E   L-80040

[122127-72-4]

As Lophirone D, L-80039 with

R = H

$C_{23}H_{16}O_5$ M 372.376

Constit. of *Lophira lanceolata*. Yellow cryst. (Me$_2$CO). Mp 266-268°.

Ghogomu, R. *et al*, *J. Nat. Prod.* (*Lloydia*), 1989, **52**, 284 (*isol, pmr, cmr*)

20(29)-Lupene-3,16-diol   L-80041

$C_{30}H_{50}O_2$ M 442.724

(3β,16β)-form

Isol. from leaves of *Beyera leschenaultii*. Needles (Me$_2$CO). Mp 218-219°. [α]$_D$ +23° (c, 1.18 in CHCl$_3$).

Baddeley, G.V. *et al*, *Aust. J. Chem.*, 1964, **17**, 908.

20(29)-Lupene-3,24-diol   L-80042

$C_{30}H_{50}O_2$ M 442.724

3β-form [119285-48-2]

Constit. of *Phyllanthus flexuosus*. Needles (MeOH/CHCl$_3$). Mp 249-250°. [α]$_D^{23}$ +40.5° (c, 0.58 in CHCl$_3$).

Tanaka, R. *et al*, *Phytochemistry*, 1988, **27**, 3563.

20(29)-Lupen-7-ol   L-80043

$C_{30}H_{50}O$ M 426.724

7β-form [123617-37-8]

Constit. of *Psoralea plicata*. Cryst. (C$_6$H$_6$/MeOH). Mp 192°. [α]$_D$ +19° (c, 0.3 in CHCl$_3$).

Rasool, N. *et al*, *J. Nat. Prod.* (*Lloydia*), 1989, **52**, 749 (*isol, pmr, cmr*)

Lupinisoflavone G   L-80044

[121747-91-9]

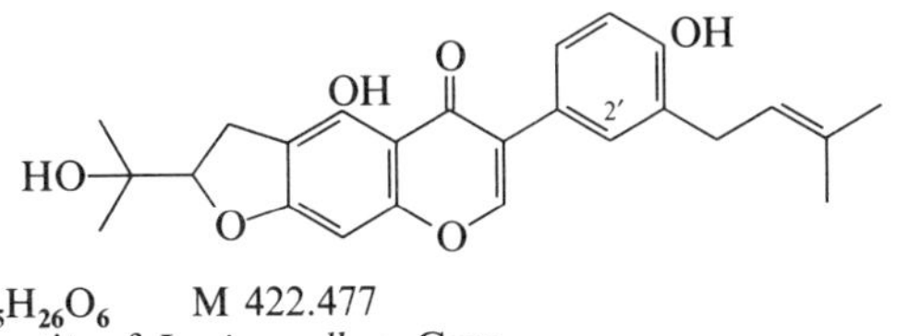

$C_{25}H_{26}O_6$ M 422.477

Constit. of *Lupinus albus*. Gum.

2′-Hydroxy: [121747-92-0]. **Lupinisoflavone H**

$C_{25}H_{26}O_7$ M 438.476

Constit. of *L. albus*. Amorph. powder.

Tahara, S. *et al*, *Phytochemistry*, 1989, **28**, 901.

Lupinisoflavone I   L-80045

[121768-55-6]

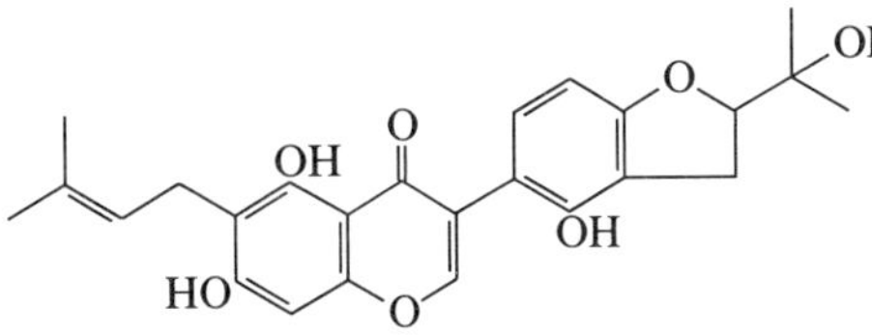

$C_{25}H_{26}O_7$ M 438.476

Constit. of *Lupinus albus*. Amorph. powder.

Tahara, S. *et al*, *Phytochemistry*, 1989, **28**, 901.

Lupinisoflavone J   L-80046

[121747-95-3]

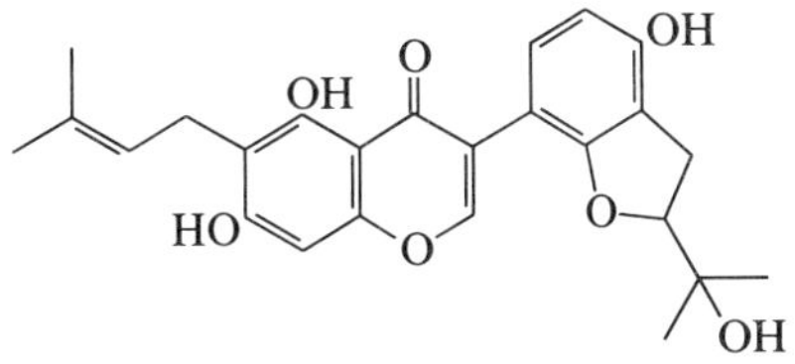

$C_{25}H_{26}O_7$ M 438.476

Constit. of *Lupinus albus*. Amorph. powder.

Tahara, S. *et al*, *Phytochemistry*, 1989, **28**, 901.

Lupinisol A   L-80047

[121747-99-7]

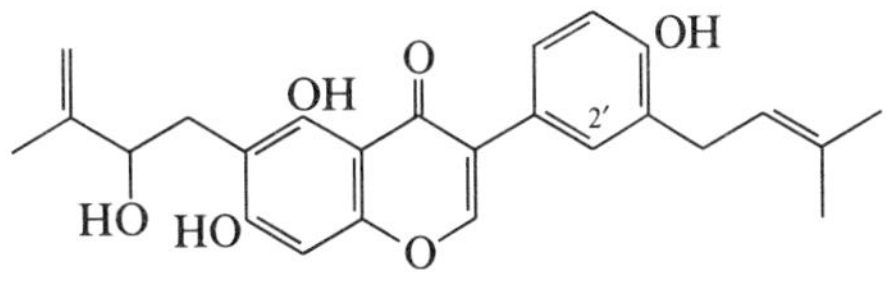

$C_{25}H_{26}O_6$ M 422.477

Constit. of *Lupinus albus*. Gum.

2′-Hydroxy: [121748-00-3]. **Lupinisol B**

$C_{25}H_{26}O_7$ M 438.476

Constit. of *L. albus*. Gum.

Tahara, S. *et al*, *Phytochemistry*, 1989, **28**, 901.

Lupinisol C   L-80048

[121748-01-4]

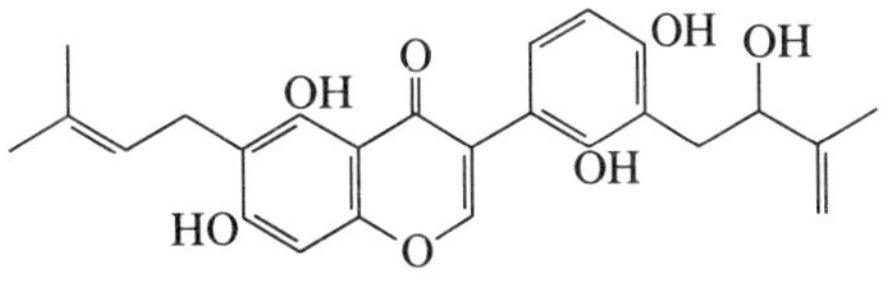

$C_{25}H_{26}O_7$ M 438.476

Constit. of *Lupinus albus*. Gum.

Tahara, S. *et al*, *Phytochemistry*, 1989, **28**, 901.

Lupinisolone A

L-80049

[121747-96-4]

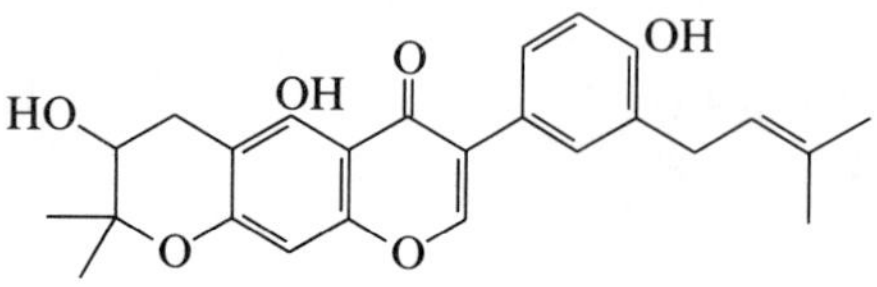

$C_{25}H_{26}O_6$ M 422.477

Constit. of *Lupinus albus*. Gum.

Tahara, S. *et al*, *Phytochemistry*, 1989, **28**, 901.

Lupinisolone B

L-80050

[121747-97-5]

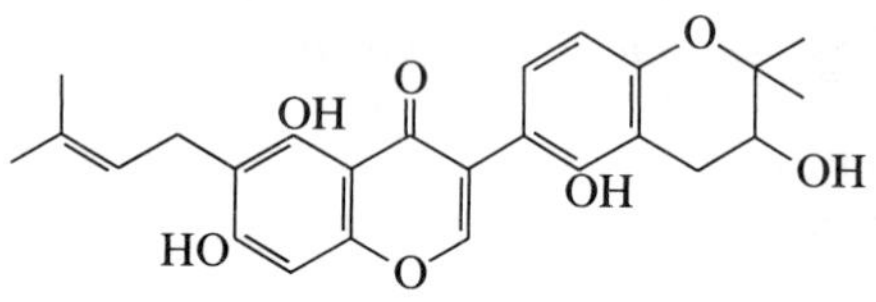

$C_{25}H_{26}O_7$ M 438.476

Constit. of *Lupinus albus*. Gum.

Tahara, S. *et al*, *Phytochemistry*, 1989, **28**, 901.

Lupinisolone C

L-80051

[121747-98-6]

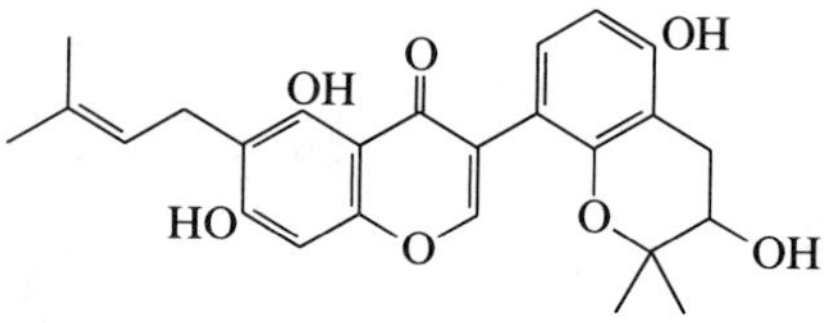

$C_{25}H_{26}O_7$ M 438.476

Constit. of *Lupinus albus*. Glassy solid.

Tahara, S. *et al*, *Phytochemistry*, 1989, **28**, 901.

Luteolic acid

L-80052

3,4,8,9,10-Pentahydroxy-6-oxo-6H-dibenzo[b,d]*pyran-1-carboxylic acid, 9CI*

[476-67-5]

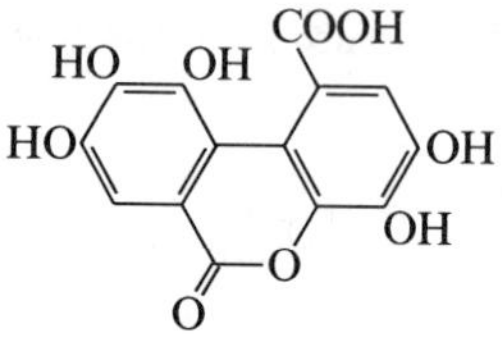

$C_{14}H_8O_9$ M 320.212

Isol. from fruits of *Terminalia chebula*. Reddish needles (AcOH). Mp 338-342° dec.

Diglucoside:

$C_{26}H_{28}O_{19}$ M 644.496

Isol. from leaf tannin of *Lagerstroemia subcostata*. $[\alpha]_D^{20}$ +40.8° (c, 1 in H_2O). Exact struct. unknown.

Nierenstein, M., *Ber.*, 1908, **41**, 3015; 1909, **42**, 353.
Ishii, M., *CA*, 1951, **45**, 2698 (*glucoside*)

M

MAB 6   M-80001

Updated Entry replacing M-00009

5-Hydroxy-8,8-dimethyl-6-(2-methyl-1-oxobutyl)-4-propyl-2H,8H-benzo[1,2-b:3,4-b′]dipyran-2-one

[30390-19-3]

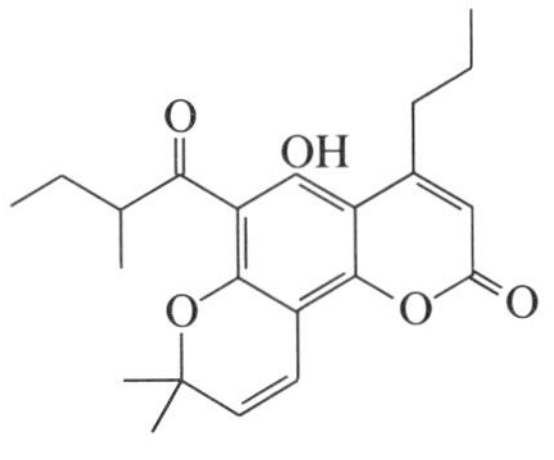

C$_{22}$H$_{26}$O$_5$ M 370.444

Constit. of *Mammea africana*. Gum. λ_{max} 227 (log ϵ 4.25), 285(4.20) and 330 nm (3.72) (0.1N HCl).

Analogue (1): [34107-36-3]. **5-Hydroxy-8,8-dimethyl-6-(2-methyl-1-oxopropyl)-4-propyl-2H,8H-benzo[1,2-b, 3,4-b′]dipyran-2-one, 9CI**
C$_{21}$H$_{24}$O$_5$ M 356.418
Isol. from *M. americana*. Has 2-Methyl-1-oxopropyl in place of the 2-methyl-1-oxobutyl residue.

Analogue (2): [34107-35-2]. **5-Hydroxy-8,8-dimethyl-6-(1-oxobutyl)-4-propyl-2H,8H-benzo[1,2-b:3,4-b′]dipyran-2-one, 9CI**
C$_{21}$H$_{24}$O$_5$ M 356.418
Isol. from *M. americana*. Has 1-oxobutyl in place of 2-methyl-1-oxobutyl.

Analogue (3): [39036-48-1]. **5-Hydroxy-8,8-dimethyl-6-(2-methyl-1-oxobutyl)-4-pentyl-2H,8H-benzo[1,2-b:3,4-b′]dipyran-2-one, 9CI**
C$_{24}$H$_{30}$O$_5$ M 398.498
Isol. from *M. americana*. Has *n*-pentyl in place of the *n*-propyl residue.

Analogue (4): [34124-91-9]. **5-Hydroxy-8,8-dimethyl-6-(3-methyl-1-oxobutyl)-4-propyl-2H,8H-benzo[1,2-b:3,4-b′]dipyran-2-one, 9CI**
C$_{22}$H$_{26}$O$_5$ M 370.444
Isol. from *M. americana* and *M. africana*. Has 3-methyl-1-oxobutyl in place of 2-methyl-1-oxobutyl.

Analogue (5): [39036-47-0]. **5-Hydroxy-8,8-dimethyl-6-(3-methyl-1-oxobutyl)-4-pentyl-2H,8H-benzo[1,2-b:3,4-b′]dipyran-2-one, 9CI**
C$_{24}$H$_{30}$O$_5$ M 398.498
Isol. from *M. americana*. Has *n*-butyl in place of propyl and 3-methyl-1-oxobutyl in place of 2-methyl-1-oxobutyl.

Carpenter, I. *et al*, *J. Chem. Soc. C*, 1971, 3783 (*isol, struct*)
Games, D.E., *Tetrahedron Lett.*, 1972, 3187 (*analogues*)

Maesol   M-80002

3,3′-(1,2-Dodecanediyl)bis[5-methoxy-6-methylphenol], 9CI.
1,12-Bis(3-hydroxy-5-methoxy-4-methylphenyl)dodecane

[119766-98-2]

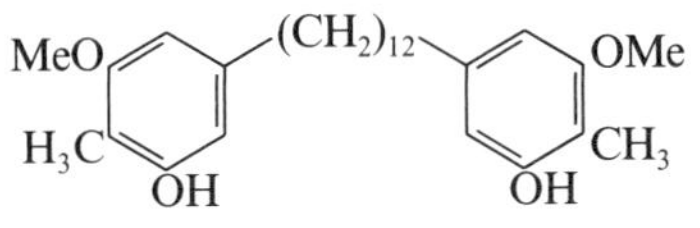

C$_{28}$H$_{42}$O$_4$ M 442.637

Constit. of *Maesa montana* and *M. indica*. Cryst. Mp 76-78°.

Wall, M.E. *et al*, *J. Nat. Prod. (Lloydia)*, 1988, **51**, 1227 (*isol, pmr, cmr*)

Mahuannin A   M-80003

[82765-89-7]

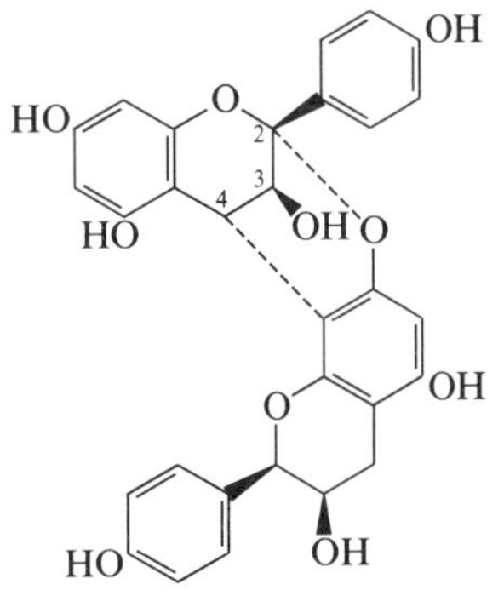

Absolute configuration

C$_{30}$H$_{24}$O$_{10}$ M 544.514

Constit. of *Ephedra* spp. Amorph. powder.

2,3,4-Triepimer: [82796-37-0]. **Mahuannin B**
C$_{30}$H$_{24}$O$_{10}$ M 544.514
Constit. of *E.* spp. Amorph. powder.

Hikino, H. *et al*, *Heterocycles*, 1982, **19**, 1381.

Mandassidione   M-80004

3-Methyl-2-[2-(1-methylethenyl)-5-oxohexyl]-2-cyclopenten-1-one, 9CI

[118855-38-2]

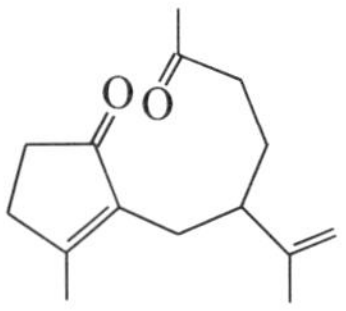

C$_{15}$H$_{22}$O$_2$ M 234.338

Constit. of *Cyperus articulatus*. Oil. $[\alpha]_D^{25}$ −9.10° (c, 0.74 in CHCl$_3$). A cleaved guaiane.

Nyasse, B. *et al*, *Phytochemistry*, 1988, **27**, 3319.

Mangicrocin   M-80005

[122575-51-3]

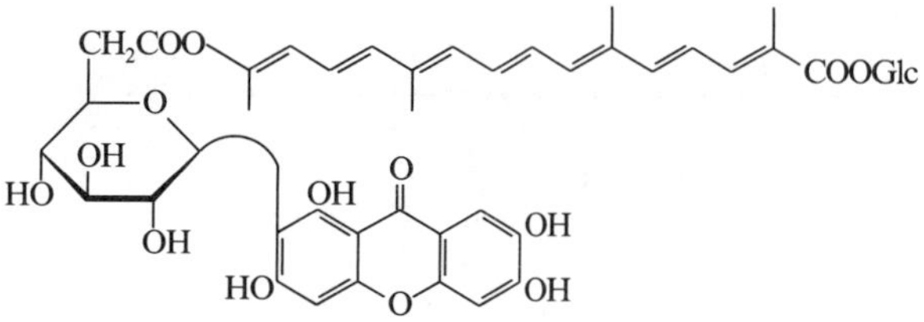

$C_{45}H_{50}O_{19}$ M 894.879

Constit. of saffron. $[\alpha]_D^{28}$ +17.5° (c, 0.53 in MeOH).

Ghosal, S. *et al, J. Chem. Res. (S)*, 1989, 70.

Mansonone A   M-80006

Updated Entry replacing M-00135

5,6,7,8-Tetrahydro-3,8-dimethyl-5-(1-methylethyl)-1,2-naphthalenedione, 9CI

[7715-94-8]

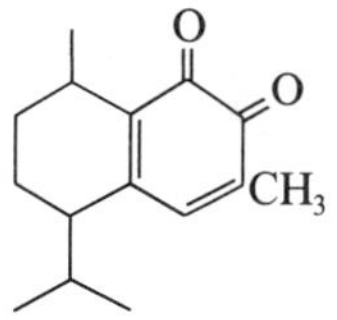

$C_{15}H_{20}O_2$ M 232.322

Constit. of *Mansonia altissima* heartwood. Red cryst. (cyclohexane). Mp 117-118°. $[\alpha]_D$ +680° (c, 0.2 in $CHCl_3$). λ_{max} 209 (log ϵ 4.32) and 430 nm (2.93) (EtOH).

Marini Bettolo, G.B. *et al, Tetrahedron Lett.*, 1965, 4857 (*isol, uv, pmr*)
Tanaka, N. *et al, Tetrahedron Lett.*, 1966, 2767 (*isol, uv*)
Dumas, M.T. *et al, Experientia*, 1983, **39**, 1089 (*isol*)
Murali, D. *et al, Indian J. Chem., Sect. B*, 1987, **26**, 668 (*synth*)

Maoecrystal I   M-80007

ent-*19-Acetoxy-7β,20-epoxy-1α,3α,6α,7α-tetrahydroxy-16-kauren-15-one*

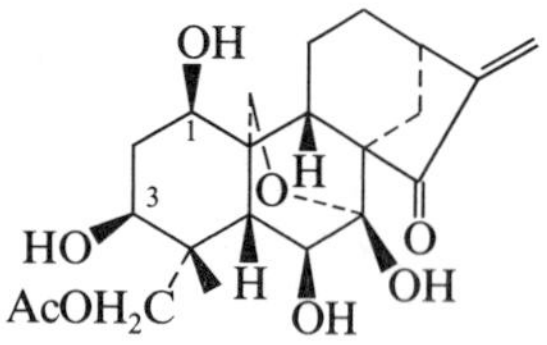

$C_{22}H_{30}O_8$ M 422.474

Constit. of *Rabdosia eriocalyx*. Cryst. (Me_2CO/hexane). Mp 205-206°.

1-Deoxy, 3-Ac: Maeocrystal J
$C_{24}H_{32}O_8$ M 448.512
Constit. of *R. eriocalyx*. Cryst. (Me_2CO/hexane). Mp 249-250°. $[\alpha]_D$ -49.7° (c, 1 in MeOH).

Shen, X. *et al, Phytochemistry*, 1989, **28**, 855.

Maoecrystal K   M-80008

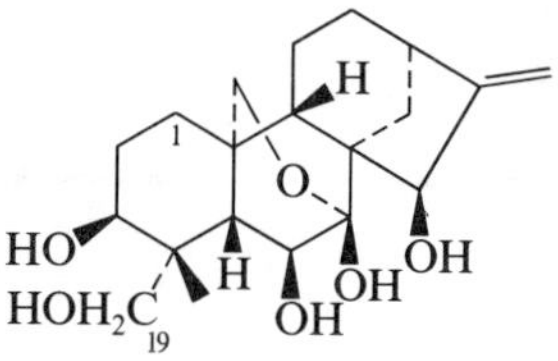

$C_{20}H_{30}O_6$ M 366.453

Constit. of *Rabdosia eriocalyx*. Cryst. Mp 191.5-193°. $[\alpha]_D^{26.5}$ -1.3° (c, 1 in MeOH).

19-O-β-D-Glucopyranoside: Rabdoside 1
$C_{26}H_{40}O_{11}$ M 528.595
Constit. of *R. eriocalyx*. Cryst. Mp 179-180°. $[\alpha]_D^{26.5}$ -4.6° (c, 1 in MeOH).

1β-Hydroxy-19-O-β-D-glucopyranoside: Rabdoside 2
$C_{26}H_{40}O_{12}$ M 544.595
Constit. of *R. eriocalyx*. Cryst. Mp 170-171°. $[\alpha]_D^{26.5}$ -4.5° (c, 1 in MeOH).

Isogai, A. *et al, Phytochemistry*, 1989, **28**, 2427 (*isol, pmr, cmr*)

Maritimetin   M-80009

Updated Entry replacing D-04380

6,7-Dihydroxy-2-(3′,4′-dihydroxyphenylmethylene)-3(2H)-benzofuranone, 9CI. 3′,4′,6,7-Tetrahydroxyaurone

[576-02-3]

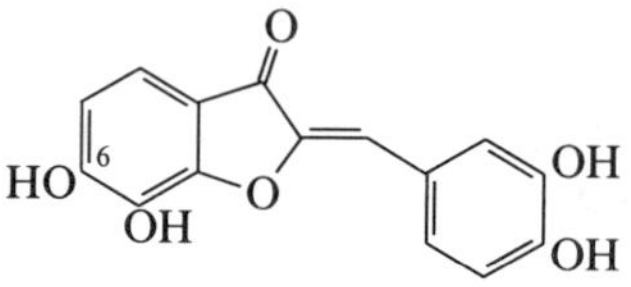

$C_{15}H_{10}O_6$ M 286.240

Isol. from flowers of *Coreopsis maritima, C. gigantea, C. tinctoria* and *Baeria chrysostoma*. Used in flavouring. Orange needles (H_2O). Mp 292° dec.

Tetra-Ac: Cryst. (EtOH). Mp 190-192°.

Tetra-Me ether: Yellow needles (EtOH aq.). Mp 156-157°.

6-O-β-D-Glucopyranoside: [490-54-0]. *Maritimein*
$C_{21}H_{20}O_{11}$ M 448.382
Pigment present in the flowers of *Coreopis maritima, C. tinctoria* and *Baeria chrysotoma*. Yellow cryst (EtOAc). Mp 208-214°.

Harborne, J.B. *et al, J. Am. Chem. Soc.*, 1956, **78**, 825, 829 (*isol, struct*)
Shimokoriyama, M. *et al, J. Org. Chem.*, 1960, **25**, 1956 (*isol*)
Farkas, L. *et al, Chem. Ber.*, 1965, **98**, 2103 (*synth*)
Huke, M. *et al, Arch. Pharm. (Weinheim, Ger.)*, 1969, **302**, 401 (*synth*)

Matricarin   M-80010

Updated Entry replacing M-00188

Artilesin A

[5989-43-5]

$C_{17}H_{20}O_5$ M 304.342

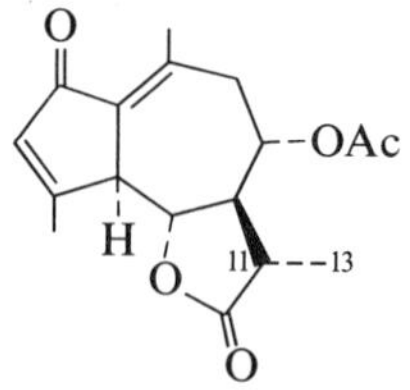

Constit. of *Artemisia* spp. Also from *Matricaria chamomilla* and *Achillea lanulosa*. Cryst. (C_6H_6/pet. ether). Mp 193-195°. $[\alpha]_D^{23}$ +23.5° (c, 0.65 in $CHCl_3$).

O-De-Ac: **Austrisin**. *Desacetylmatricarin*
$C_{15}H_{18}O_4$ M 262.305
Isol. from *Artemisia tilesii*, *A. austriaca*, *A. leucodes*, *A. juncea* and *Achillea lanulosa*. Cryst. + $1H_2O$ (C_6H_6/Me_2CO). Mp 123-125°, Mp 143-146° (double Mp, hydrate).

Deacetoxy: **Leucomisin**. *Leucodin*. *Desacetoxymatricarin*
$C_{15}H_{18}O_3$ M 246.305
Isol. from *Artemisia leucodes*. Prisms (EtOH). Mp 202-203°. $[\alpha]_D^{20}$ +55.9° (c, 2.86 in $CHCl_3$).

11,13-Didehydro: [69904-98-9]. **11,13-Dehydromatricarin**
$C_{17}H_{18}O_5$ M 302.326
Constit. of *Anthanasia* and *Pentzia* spp. Cryst. (Et_2O/pet. ether). Mp 146°. $[\alpha]_D^{24}$ +120.8° (c, 1 in $CHCl_3$).

Herz, W. *et al*, *J. Am. Chem. Soc.*, 1961, **83**, 1139 (*isol*)
Rybalko, K.S., *Zh. Obshch. Khim.*, 1963, **33**, 2734 (*Leucomisin*)
White, E.H. *et al*, *Tetrahedron*, 1969, **25**, 2099 (*struct*)
Bohlmann, F. *et al*, *Phytochemistry*, 1978, **17**, 1595 (*isol*)
Martinez, V. *et al*, *J. Nat. Prod.* (*Lloydia*), 1988, **51**, 211 (*pmr*, *cmr*)

Mbamichalcone   M-80011

Updated Entry replacing M-70014
[119264-65-2]

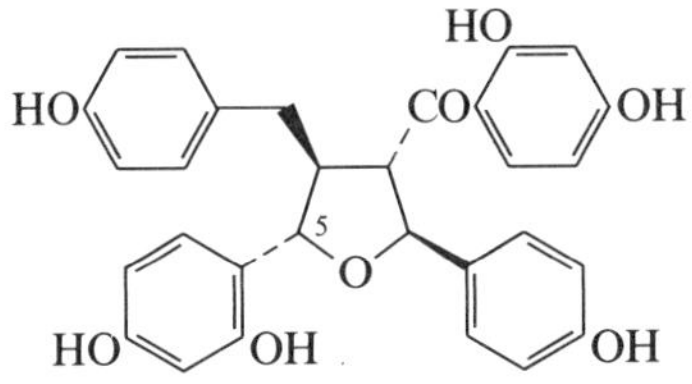

Relative configuration

$C_{30}H_{26}O_8$ M 514.531
Constit. of *Lophira alata*. Amorph. solid. $[\alpha]_D^{25}$ −47.8° (c, 0.65 in MeOH).

5-Epimer: [122621-92-5]. **Isombamichalcone**
$C_{30}H_{26}O_8$ M 514.531
Constit. of bark of *L. lanceolata*. Amorph. powder. $[\alpha]_D^{20}$ −129° (c, 0.5 in Me_2CO). Abs. config. not detd.

Tih, A.E. *et al*, *Tetrahedron Lett.*, 1988, **29**, 5797 (*struct*)
Tih, R.G. *et al*, *Tetrahedron Lett.*, 1989, **30**, 1807 (*Isombamichalcone*)

Melicophyllone A   M-80012

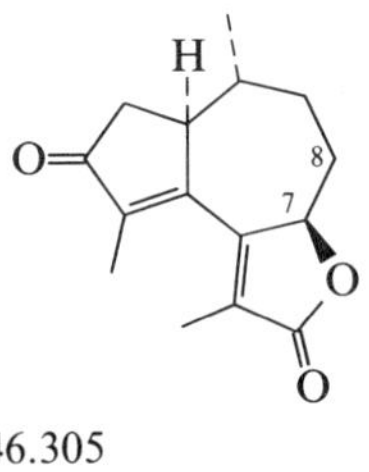

$C_{15}H_{18}O_3$ M 246.305
Constit. of *Melicope triphylla*. Plates. Mp 106-107°. $[\alpha]_D$ −371° ($CHCl_3$).

7-Hydroxy: **Melicophyllone B**
$C_{15}H_{18}O_4$ M 262.305
Constit. of *M. triphylla*. Needles. Mp 171-172°. $[\alpha]_D$ −38° ($CHCl_3$).

7α,8β-Dihydroxy: **Melicophyllone C**
$C_{15}H_{18}O_5$ M 278.304

Constit. of *M. triphylla*. Needles. Mp 254°. $[\alpha]_D$ −56.23° (Me_2CO).

Wu, T.-S. *et al*, *J. Chem. Soc., Chem. Commun.*, 1988, 956 (*cryst struct*)
Jong, T.-T. *et al*, *J. Chem. Res. (S)*, 1989, 273; *J. Chem. Res. (M)*, 1701 (*isol*)

Melitensin   M-80013

Updated Entry replacing M-00281
8α,15-Dihydroxy-1,3-elemadien-12,6α-olide
[38049-38-6]

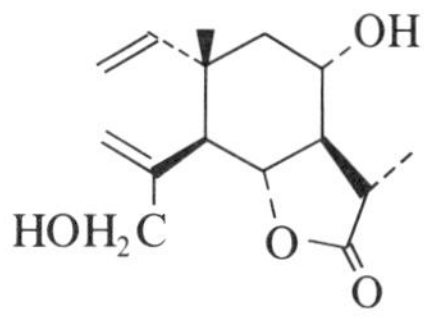

$C_{15}H_{22}O_4$ M 266.336
Constit. of *Centaurea melitensis* and *C. aspera* var. *stenophylla*. Cryst. Mp 167-168°. $[\alpha]_D$ +85°.

11,13-Didehydro: **Dehydromelitensin**
$C_{15}H_{20}O_4$ M 264.321
Constit. of *Onopordon corymbosum*. Oil.

11,13-Didehydro, 8-(2-hydroxymethyl-2-propenoyl): **Dehydromelitensin 4-hydroxymethacrylate**
$C_{19}H_{24}O_6$ M 348.395
Constit. of *O. corymbosum*. Oil.

González, A.G. *et al*, *An. Quim.*, 1971, **67**, 1243; 1974, **70**, 158 (*isol*, *struct*, *synth*)
Tortajada, A. *et al*, *Phytochemistry*, 1988, **27**, 3549 (*cryst struct*)
Cardona, Ma.L. *et al*, *Phytochemistry*, 1989, **28**, 1264 (*deriv*)

p-Menth-4-ene-1,2-diol   M-80014

1-Methyl-4-(1-methylethyl)-4-cyclohexene-1,2-diol, 9CI. 4-Isopropyl-1-methyl-4-cyclohexene-1,2-diol
[4031-37-2]

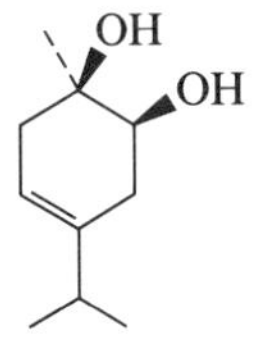

$C_{10}H_{18}O_2$ M 170.251
Constit. of *Ferula jaeschkeana*. Oil.
[21473-37-0, 94492-01-0]

Kapahi, B.K. *et al*, *Pafai*, 1985, **7**, 23.

p-Menth-3-ene-1,2-diol   M-80015

4-Isopropyl-1-methyl-3-cyclohexene-1,2-diol

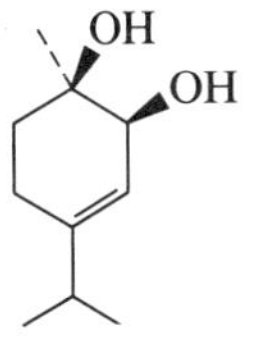

$C_{10}H_{18}O_2$ M 170.251
Constit. of *Ferula jaeschkeana*. Oil.

Kapahi, B.K. *et al*, *Pafai*, 1985, **7**, 23.

p-Menth-7-ene-1,2-diol   M-80016

4-Isopropenyl-1-methyl-1,2-cyclohexanediol

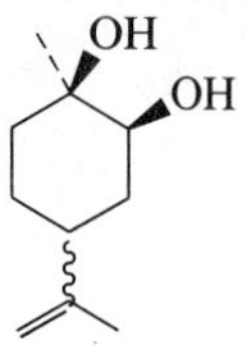

$C_{10}H_{18}O_2$ M 170.251
Constit. of *Ferula jaeschkeana*. Oil.

Kapahi, B.K. *et al, Pafai*, 1985, **7**, 23; *CA*, **103**, 92620y.

Mercaptoacetaldehyde   M-80017

Updated Entry replacing M-50049
[4124-63-4]

$$HSCH_2CHO$$

C_2H_4OS M 76.119
Known only in soln. Readily dimerises

Di-Et acetal: [53608-94-9]. *2,2-Diethoxyethanethiol, 9CI.*
Mercaptoacetal
$C_6H_{14}O_2S$ M 150.241
Liq. with acetal odour. $Bp_{0.3}$ 81°, $Bp_{0.12}$ 59-60°. n_D^{25}
1.4391. Dec. slowly on standing.

S-Me: *(Methylthio)acetaldehyde*
C_3H_6OS M 90.146
Bp 129-134°, Bp_{10} 35°.

S-Me, 2,4-dinitrophenylhydrazone: Cryst. (EtOH aq.). Mp
131.0-131.5°.

S-Me, semicarbazone: Mp 133.5-134°.

S-Me, di-Me acetal: [40015-15-4]. *1,1-Dimethoxy-
2(methylthio)ethane, 9CI*
$C_5H_{12}O_2S$ M 136.215
Bp_{10} 55°.

S-Et: [33672-79-6]. *(Ethylthio)acetaldehyde*
C_4H_8OS M 104.173
Bp_{10} 46.5°.

S-Et, 2,4-dinitrophenylhydrazone: Mp 125.5-126°.

S-Et, semicarbazone: Mp 93-93.5°.

S-Me, di-Et acetal: [51517-03-4]. *2-(Ethylthio)-1,1-
dimethoxyethane, 9CI*
$C_6H_{14}O_2S$ M 150.241
Bp_{11} 63-64°.

Hesse, G. *et al, Chem. Ber.*, 1952, **85**, 924, 933.
Wick, E.H. *et al, J. Agric. Food Chem.*, 1961, **9**, 289 *(deriv, synth)*
Org. Synth., Coll. Vol., 4, 1963, 295.
Gassman, P.G. *et al, J. Am. Chem. Soc.*, 1974, **96**, 5495 *(deriv, synth)*
Mikolajczyk, M. *et al, Synthesis*, 1987, 659 *(synth, pmr, ms, bibl)*

2-Mercaptobenzaldehyde   M-80018

2-Formylbenzenethiol
[29199-11-9]

CHO
SH

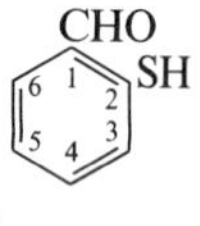

C_7H_6OS M 138.190
Yellow oil.

Me ether: [7022-45-9]. *2-(Methylthio)benzaldehyde*
C_8H_8OS M 152.217
Pale yellow oil. Bp_{19} 149-150°.

Me ether, 2,4-dinitrophenylhydrazone: [52369-78-5].

Mp 243°.

Eistert, B. *et al, Chem. Ber.*, 1964, **97**, 1470 *(synth, deriv)*
Ohno, S. *et al, Chem. Pharm. Bull.*, 1986, **34**, 1589 *(synth, deriv, pmr)*
Kasmai, H.S. *et al, Synthesis*, 1989, 763 *(synth, deriv, pmr, bibl)*

3-Mercaptobenzaldehyde   M-80019

3-Formylbenzenethiol
C_7H_6OS M 138.190

Me ether: [73771-35-4]. *3-(Methylthio)benzaldehyde*
Oil. $Bp_{0.05}$ 82-83°.

Eurby, M. *et al, Synth. Commun.*, 1981, **11**, 849 *(synth, pmr, ms)*

4-Mercaptobenzaldehyde   M-80020

4-Formylbenzenethiol
[91358-96-2]
C_7H_6OS M 138.190
Oil.

Me ether: [3446-89-7]. *4-(Methylthio)benzaldehyde*
C_8H_8OS M 152.217
Pale yellow oil. Bp 273°, Bp_1 101-103°.

Friedländer, P. *et al, Chem. Ber.*, 1912, **45**, 2083 *(synth)*
Rivett, D.E., *Aust. J. Chem.*, 1979, **32**, 1601 *(synth, deriv)*
Young, R.N. *et al, Tetrahedron Lett.*, 1984, **25**, 1753 *(synth)*
Creary, X. *et al, J. Org. Chem.*, 1986, **57**, 1110 *(synth, deriv, pmr)*

2-Mercaptobenzenemethanol, 9CI   M-80021

2-Mercaptobenzyl alcohol. 2-(Hydroxymethyl)thiophenol
[4521-31-7]

CH₂OH
SH

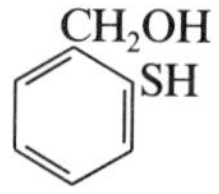

C_7H_8OS M 140.206
Needles (pentane). Mp 31-32°.

S-Me: [33384-77-9].
$C_8H_{10}OS$ M 154.232
$Bp_{0.001}$ 88°.

S-Benzyl:
$C_{14}H_{14}OS$ M 230.330
Needles (pet. ether). Mp 48.5-49.5°.

Disulfide:
$C_{14}H_{14}O_2S_2$ M 278.395
Cryst. Mp 141-142°.

Grice, R. *et al, J. Chem. Soc.*, 1963, 1947 *(deriv)*
Stacy, G.W. *et al, J. Org. Chem.*, 1965, **30**, 4074 *(synth, ir)*
Arnoldi, A. *et al, Synthesis*, 1988, 155 *(synth)*

3-Mercaptobenzenemethanol   M-80022

3-Mercaptobenzyl alcohol. 3-(Hydroxymethyl)thiophenol
[83794-86-9]
C_7H_8OS M 140.206
$Bp_{0.001}$ 88°.

S-Me: [59083-33-9].
$C_8H_{10}OS$ M 154.232
$Bp_{0.001}$ 98°.

Disulfide:
$C_{14}H_{14}O_2S_2$ M 278.395
Prisms (Et₂O/pet. ether).

Grice, R. *et al, J. Chem. Soc.*, 1963, 1947 *(synth)*
Thurber, T.C. *et al, J. Heterocycl. Chem.*, 1982, **19**, 961 *(synth, pmr)*

4-Mercaptobenzenemethanol   M-80023

4-Mercaptobenzyl alcohol. 4-(Hydroxymethyl)thiophenol

[53339-53-0]

C_7H_8OS M 140.206

Plates (CCl_4/pet. ether). Mp 52-54°.

S-*Me:* [3446-90-0].
    $C_8H_{10}OS$ M 154.232
    Cryst. Mp 37-39°. $Bp_{0.01}$ 100°.

Disulfide:
    $C_{14}H_{14}O_2S_2$ M 278.395
    Needles (Me_2CO aq.). Mp 130-132°.

Grice, R. *et al, J. Chem. Soc.,* 1963, 1947 (*synth*)

2-Mercaptodiphenylmethane   M-80024

2-(Phenylmethyl)benzenethiol, 9CI. 2-Benzylbenzenethiol. 1-Benzyl-2-mercaptobenzene

[120454-34-4]

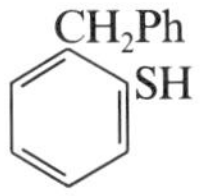

$C_{13}H_{12}S$ M 200.304

Pale brown solid (pet. ether). Mp 44-47°.

Cadogan, J.I.G. *et al, J. Chem. Soc., Perkin Trans.* 1, 1988, 2875 (*synth, pmr, cmr, ms*)

5-Mercaptohistidine, 9CI   M-80025

5-Thiolhistidine

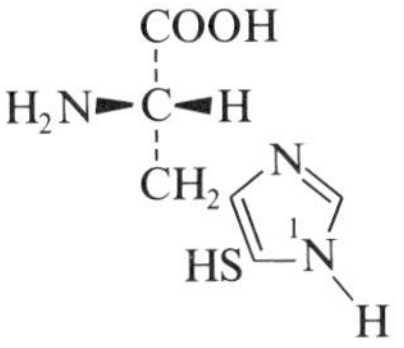

$C_6H_9N_3O_2S$ M 187.222

(*S*)-*form*
    L-*form*

1-*Me:* [62982-24-5]. *5-Mercapto-1-methylhistidine*
    $C_7H_{11}N_3O_2S$ M 201.249
    Constit. of the unfertilised eggs of the sea urchin *Paracentrotus lividus*. Also a component of the struct. of adrenochrome, an Fe(III) contg. peptide from *Octopus vulgaris*.

1-*Me, disulfide:* [83471-81-2]. *5,5′-Dithiobis[1-methylhistidine], 9CI*
    $C_{14}H_{20}N_6O_4S_2$ M 400.482
    Isol. from the unfertilised eggs of *P. lividus*. Prisms (EtOH aq.). Mp 202-205° (darkens). $[\alpha]_D^{20}$ +76° (c, 1.2 in 0.1*M* HCl).

Palumbo, A. *et al, Tetrahedron Lett.,* 1982, **23**, 3207 (*isol, uv, pmr, cmr, struct, deriv*)

2-Mercapto-3-methylpentanoic acid   M-80026

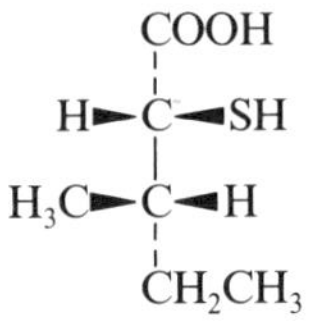

$C_6H_{12}O_2S$ M 148.226

(*2R,3S*)-*form*
    $Bp_{0.02}$ 60°. $[\alpha]_D^{20}$ +10.3° (c, 2.4 in $CHCl_3$). Optically pure.

S-*Benzoyl:*
    $C_{13}H_{16}O_3S$ M 252.334
    Thick oil, solidifies on standing. Mp *ca* 35°. $[\alpha]_D^{20}$ +24.2° (c, 2 in C_6H_6).

Strijtreen, B. *et al, Tetrahedron,* 1987, **43**, 5039 (*synth, pmr, cmr*)

2-Mercapto-4-methylpentanoic acid   M-80027

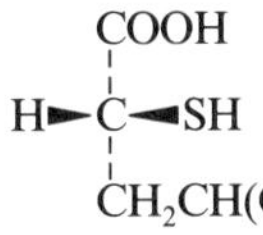

$C_6H_{12}O_2S$ M 148.226

(*R*)-*form*
    $Bp_{0.01}$ 70°. $[\alpha]_D^{22}$ +25.1° (c, 11 in $CHCl_3$).

(*S*)-*form* [66386-08-1]
    $[\alpha]_D^{20}$ −23.8° (c, 1.8 in Et_2O).

Yankeelov, J. *et al, J. Org. Chem.,* 1978, **43**, 1623 (*synth, abs config*)

Strijtveen, B. *et al, Tetrahedron,* 1987, **43**, 5039 (*synth, pmr, cmr*)

3-Mercapto-6(1*H*)-pyridazinethione   M-80028

3,6-Dimercaptopyridazine. 1,2-Dihydro-3,6-pyridazinedithione, 9CI

[1445-58-5]

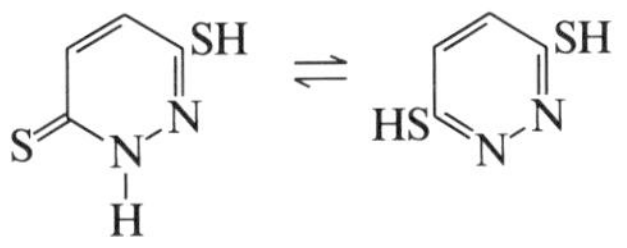

$C_4H_4N_2S_2$ M 144.221

Exists mainly in thione form. Mp >230° dec. Also reported with double Mp 190° and 246° dec.

S-*Ac:*
    $C_6H_6N_2OS_2$ M 186.258
    Mp 152-153°.

S,S-*Di-Ac:*
    $C_8H_8N_2O_2S_2$ M 228.295
    Mp 123-124°.

1,S-*Di-Me:* [54320-90-0]. *2-Methyl-6-(methylthio)-3(2H)-pyridazinethione, 9CI*
    $C_6H_8N_2S_2$ M 172.275
    Mp 86-87° (73-74°).

Di-S-*Me:* [37813-54-0]. *3,6-Bis(methylthio)pyridazine*
    $C_6H_8N_2S_2$ M 172.275
    Needles (hexane). Mp 125-126.5°.

Di-S-*Me; B,MeI:* Cryst. (MeOH/2-methyl-2-propanol). Mp 161-163°.

Druey, J. *et al, Helv. Chim. Acta,* 1954, **37**, 121 (*synth, deriv*)
Stanovnik, B. *et al, Croat. Chim. Acta,* 1964, **36**, 81 (*synth, tautom*)
Pollak, A. *et al, Can. J. Chem.,* 1966, **44**, 829 (*synth*)
Barlin, G.B., *J. Chem. Soc., Perkin Trans.* 2, 1974, 1199 (*synth, deriv, tautom*)
Boger, D.L. *et al, J. Org. Chem.,* 1988, **53**, 1415 (*deriv, synth, pmr, ir, ms*)

4-Mercapto-2(1*H*)-pyridinethione, 9CI   M-80029

2,4-Dimercaptopyridine. 2,4-Pyridinedithiol

[28508-52-3]

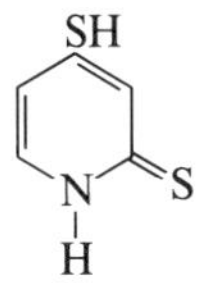

$C_5H_5NS_2$ M 143.233
Yellow solid. Mp 135-137°.

4-S-Me: [71506-84-8]. *4-(Methylthio)-2(1H)-pyridinethione*
$C_6H_7NS_2$ M 157.260
Yellow solid. Mp 220-223° (rapid htg.).

N,S-Di-Me: [71506-86-0]. *1-Methyl-4-(methylthio)-2(1H)-pyridinethione*
$C_7H_9NS_2$ M 171.287
Pale yellow solid. Mp 155-156°.

Di-S-Me: [71506-85-9]. *2,4-Bis(methylthio)pyridine*
$C_7H_9NS_2$ M 171.287
Bp_2 130°, $Bp_{0.05}$ 99°.

Di-S-Me, S-tetroxide: [62916-42-1]. *2,4-Bis(methylsulfonyl)pyridine*
$C_7H_9NO_4S_2$ M 235.284
Mp 167-168°.

Krowicki, K., *Pol. J. Chem.* (*Rocz. Chem.*), 1979, **53**, 701 (*synth, pmr, derivs*)
Woods, S.G. *et al*, *J. Heterocycl. Chem.*, 1984, **21**, 97 (*deriv*)

6-Mercapto-2,4(1*H*,3*H*)-pyrimidine-dithione, 9CI   M-80030

2,4,6-Trithiobarbituric acid, 8CI. 2,4,6-Trimercaptopyrimidine
[6308-40-3]

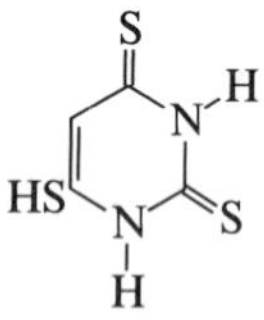

$C_4H_4N_2S_3$ M 176.287
Several tautomers possible. Orange-yellow cryst. Mp >360°.

2-S-Me:
$C_5H_6N_2S_3$ M 190.314
Pale yellow cryst. Mp >360°.

[5658-04-8]

Koppel, H.C. *et al*, *J. Org. Chem.*, 1961, **26**, 792 (*synth*)
Netzel, D.A. *et al*, *Appl. Spectrosc.*, 1968, **22**, 170 (*ir*)
Hübsch, W. *et al*, *Helv. Chim. Acta*, 1989, **72**, 744 (*synth*)

3-Mercapto-2-thiophenecarboxylic acid   M-80031

[90033-62-8]

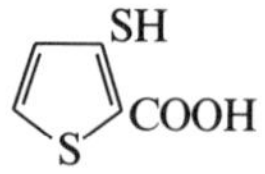

$C_5H_4O_2S_2$ M 160.217
Solid (H_2O). Mp 152-153°.

Corral, C. *et al*, *Synthesis*, 1984, 172 (*synth*)
Corral, C. *et al*, *Org. Prep. Proced. Int.*, 1985, **17**, 163 (*synth, pmr*)

4-Mercapto-3-thiophenecarboxylic acid   M-80032

[98077-07-7]
$C_5H_4O_2S_2$ M 160.217
Cryst. (AcOH). Mp 215-216°.

Corral, C. *et al*, *Org. Prep. Proced. Int.*, 1985, **17**, 163 (*synth, pmr*)

Meridinol   M-80033

[120051-54-9]

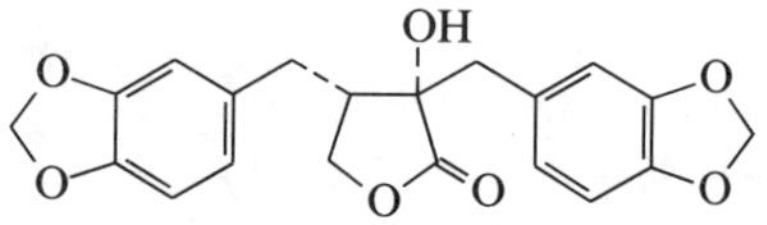

$C_{20}H_{18}O_7$ M 370.358
Constit. of *Zanthoxylum fagara*. Needles(EtOAc/pet. ether). Mp 122-123°. $[\alpha]_D^{25}$ −30° (c, 0.1 in $CHCl_3$).

Amaro-Luis, J.M. *et al*, *Phytochemistry*, 1988, **27**, 3933 (*isol, cryst struct*)

Merrillin   M-80034

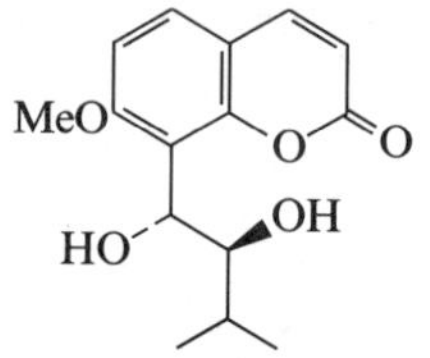

$C_{15}H_{18}O_5$ M 278.304
Constit. of *Merrillia caloxylon*. Plates. Mp 68°. $[\alpha]_D^{20}$ −15.6° (c, 0.135 in $CHCl_3$).

Zakaria, M.B. *et al*, *Phytochemistry*, 1989, **28**, 657.

Mertensene   M-80035

Updated Entry replacing B-30236
2-Bromo-4,5-dichloro-1-(2-chlorovinyl)-1,5-dimethylcyclohexane
[66389-40-0]

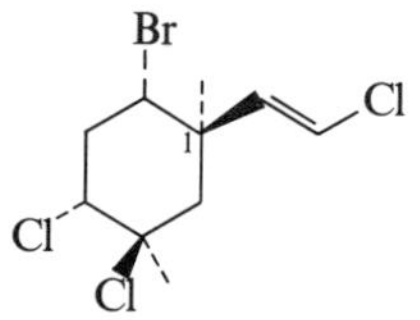

$C_{10}H_{14}BrCl_3$ M 320.483
Constit. of *Plocamium mertensii* and *P. hamatum*. Needles (hexane). Mp 105.5-106°. $[\alpha]_D$ +20° (c, 0.3 in $CHCl_3$).

1-Epimer: [95044-70-5]. ***Coccinene***†
$C_{10}H_{14}BrCl_3$ M 320.483
Constit. of *P. coccineum*. Cryst. (hexane). Mp 65°. $[\alpha]_D^{25}$ −24° (c, 0.82 in $CHCl_3$).

Norton, R.S. *et al*, *Tetrahedron Lett.*, 1977, 3905 (*isol*)
Capon, R.J. *et al*, *Aust. J. Chem.*, 1984, **37**, 537 (*isol*)
Castedo, L. *et al*, *J. Nat. Prod.* (*Lloydia*), 1984, **47**, 724 (*isol, Coccinene*)
Crews, P. *et al*, *J. Org. Chem.*, 1984, **49**, 1371 (*cmr, struct*)
Sardina, F.J. *et al*, *Chem. Lett.*, 1985, 697 (*struct*)
Coll, J.C. *et al*, *Aust. J. Chem.*, 1988, **41**, 1743 (*struct, cmr*)

Metachromin C   M-80036

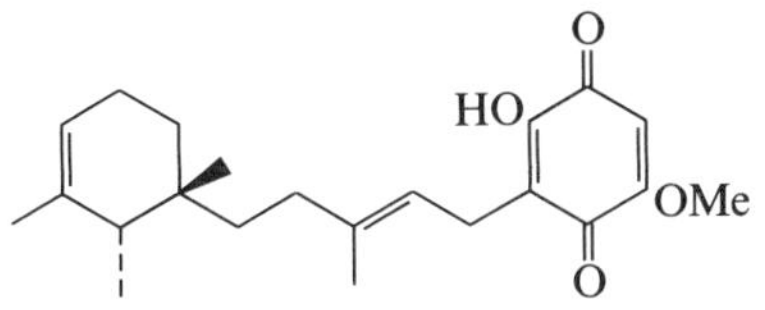

C$_{22}$H$_{30}$O$_4$ M 358.477
Constit. of *Hippospongia metachromia*. Yellow solid
(hexane). Mp 90-91°. [α]$_D^{26}$ −29.7° (c, 0.2 in CHCl$_3$).

Kabayashi, J. *et al*, *J. Nat. Prod.* (*Lloydia*), 1989, **52**, 1173 (*isol, pmr, cmr*)

Methanetetrol   M-80037

Tetra-O-*benzyl:* Tetrabenzyloxymethane.
*Tetrakis(phenylmethoxy)methane. Tetrabenzyl
orthocarbonate. Benzyl orthocarbonate*
C$_{29}$H$_{28}$O$_4$ M 440.538
Cryst. Mp 68-70°.

Latimer, D.R. *et al*, *Can. J. Chem.*, 1989, **67**, 143 (*synth, cryst struct, tetrabenzyl ester*)

Methanetricarboxaldehyde, 9CI   M-80038

Triformylmethane
[18655-47-5]

HC(CHO)$_3$⇌HOCH=C(CHO)$_2$

C$_4$H$_4$O$_3$ M 100.074
Enol. from probable in solid state and Me$_2$CO but not
CHCl$_3$ soln. Solid by subl. Mp 101-103°. pK_a 2.0.

Arnold, Z. *et al*, *Collect. Czech. Chem. Commun.*, 1960, **25**, 1318; 1961, **26**, 3051; 1965, **30**, 2125 (*synth, props*)
Arnold, Z. *et al*, *J. Org. Chem.*, 1988, **53**, 5352 (*synth*)
Keshavarz, K.M. *et al*, *Synthesis*, 1988, 641 (*synth, ir, uv, pmr, cmr, use, tautom, bibl*)

6,12-Methanobenzocyclododecene, 9CI   M-80039

Updated Entry replacing B-60102
Benzo[c]-*1,7-methano*[12]*annulene. Benz*[b]*homoheptalene*
[84537-62-2]

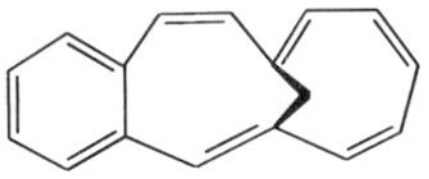

C$_{17}$H$_{14}$ M 218.298
Stable orange cryst. Mp 53-54°.

Scott, L.T. *et al*, *J. Am. Chem. Soc.*, 1983, **105**, 1372 (*synth, pmr, cmr, ir, uv*)
Günther, M.-G. *et al*, *Chem. Ber.*, 1986, **119**, 2942 (*pmr*)

5,11-Methanodibenzo[a,e]cyclooctene-6,12(5H,11H)dione, 9CI   M-80040

2,3;6,7-Dibenzobicyclo[3.3.1]*nona-2,6-diene-4,8-dione*

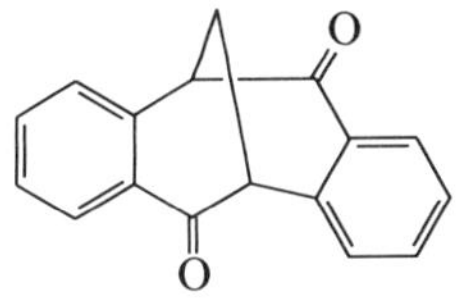

(+)-form

C$_{17}$H$_{12}$O$_2$ M 248.281
(+)-*form* [59147-66-9]
[α]$_D$ +139.5° (C, 0.021 in EtOH). 45% opt. pure.

(−)-*form*
Mp 195-197°. [α]$_D^{22}$ −340° (c, 0.022 in EtOH).
(±)-*form* [59147-65-8]
Needles (MeOH). Mp 147.7-147.8°.

Tatemitsu, H. *et al*, *Bull. Chem. Soc. Jpn.*, 1975, **48**, 2473 (*synth, abs config*)
Naemura, K. *et al*, *Bull. Chem. Soc. Jpn.*, 1989, **62**, 83.

3-Methoxy-2-cyclohexen-1-one   M-80041
[16807-60-6]

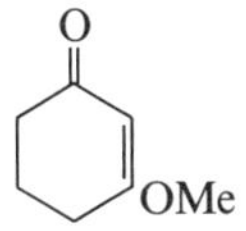

C$_7$H$_{10}$O$_2$ M 126.155
Liq. or cryst. (hexane). Mp 30-31°. Bp$_{14}$ 114° ; Bp$_{0.7}$ 52-53° . n$_D^{20}$ 1.5225.
(Z)-*oxime:* [29900-17-2].
Cryst. (H$_2$O). Mp 86.5-87°.
(E)-*4-Methylbenzenesulfonyloxime:* [32643-77-9].
Cryst. (ligroin). Mp 93-95°.
(Z)-*4-Methylbenzenesulfonyloxime:* [32643-76-8].
Powder (ligroin). Mp 78-82°.

Tamura, Y. *et al*, *Chem. Pharm. Bull.*, 1971, **19**, 523, 529 (*synth, ir, pmr*)
Cant, E. *et al*, *Bull. Soc. Chim. Belg.*, 1974, **83**, 93 (*ms*)
Lessard, J. *et al*, *Can. J. Chem.*, 1977, **55**, 1015 (*synth, pmr*)
Taskinen, E. *et al*, *Tetrahedron*, 1982, **38**, 613.
Pearson, A.J. *et al*, *J. Org. Chem.*, 1984, **49**, 3887 (*synth, ir, ms, pmr*)
Laurence, C. *et al*, *Spectrochim. Acta, Part A*, 1985, **41**, 883 (*ir, props*)
Imashiro, F. *et al*, *J. Am. Chem. Soc.*, 1987, **109**, 5213 (*cmr*)

5-Methoxy-2,2-dimethyl-6-(2-methyl-1-oxo-2-butenyl)-10-propyl-2H,8H-benzo-[1,2-b:3,4-b′]dipyran-8-one, 9CI   M-80042
[39811-39-7]

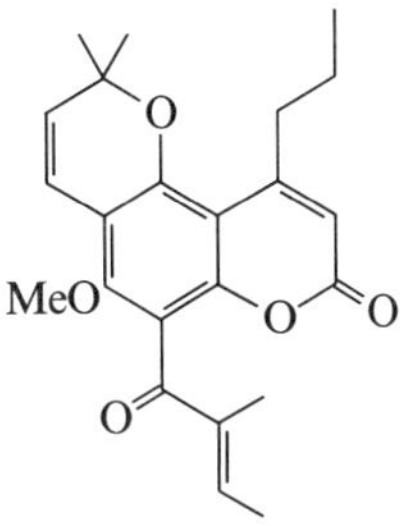

C$_{23}$H$_{26}$O$_5$ M 382.455
Isol. from *Calophyllum inophyllum*. Mp 121-123°.

Cavé, A. *et al*, *Compt. Rend. Hebd. Seances Acad. Sci. Sect. C*, 1972, **275**, 1105.

3-Methoxy-2(5H)-furanone   M-80043
2-Methoxy-2-butenolide
[35214-62-1]

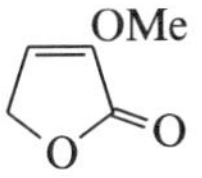

C$_5$H$_6$O$_3$ M 114.101

Isol. from *Narthecium ossifragum* flowering parts. Cryst. (CHCl₃). Mp 57°.

Stabursvik, A., *Acta Chem. Scand.*, 1954, **8**, 525 (*isol, struct, synth*)
Bonini, C.C. *et al, Org. Mass Spectrom.*, 1981, **16**, 68 (*ms*)

4(5)-Methoxyimidazole   M-80044

MeO⏤N
        N
        H

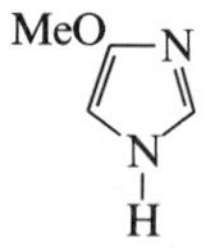

$C_4H_6N_2O$ M 98.104
Cryst. Mp 115-116°.

Hosmane, R.S. *et al, J. Org. Chem.*, 1984, **49**, 1212 (*synth, pmr*)

Methoxy isocyanate   M-80045

[117775-56-1]

MeONCO

$C_2H_3NO_2$ M 73.051
Stable substance at −183°.

Teles, J.H., *Chem. Ber.*, 1989, **122**, 745 (*synth, ir*)

5-Methoxy-2-methyl-3-polyprenyl-1,4-benzoquinone   M-80046

2-Polyprenyl-6-methoxy-3-methyl-1,4-benzoquinone

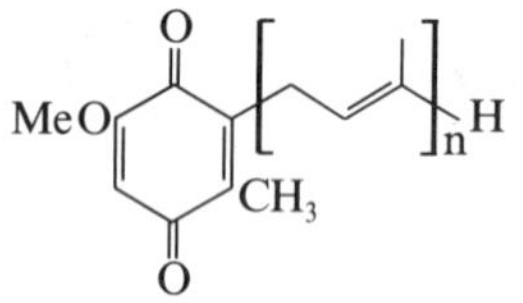

Isol. from *Rhodospirillum rubrum*. Intermed. in ubiquinone biosynth.

5-Methoxy-2-methyl-3-octaprenyl-1,4-benzoquinone [7200-29-5]
     $C_{48}H_{72}O_3$ M 697.095
     No phys. props. reported. n = 8.

5-Methoxy-2-methyl-3-nonaprenyl-1,4-benzoquinone [18735-22-3]
     $C_{53}H_{80}O_3$ M 765.213
     No phys. props. reported. n = 9.

3- Decaprenyl-5-methoxy-2-methyl-1,4-benzoquinone [7200-27-3]
     $C_{58}H_{88}O_3$ M 833.331
     No phys. props. reported. n = 10. λ_{max} 266, 274sh nm.

Friis, P. *et al, J. Am. Chem. Soc.*, 1966, **88**, 4754 (*isol, ms, struct*)

2-(5-Methoxy-5-methyltetrahydro-2-furanyl)-6,10-dimethyl-5,9-undecadien-2-ol   M-80047

α-(4,8-Dimethyl-3,7-nonadienyl)tetrahydro-5-methoxy-α,5-dimethyl-2-furanmethanol, 9CI
[115028-54-1]

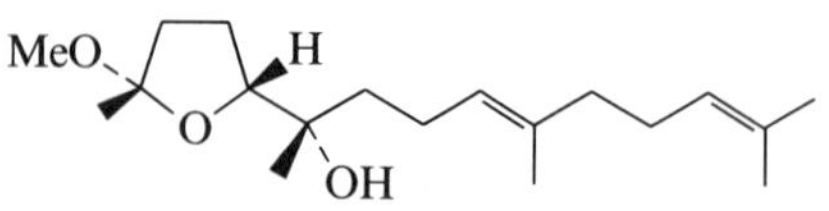

$C_{19}H_{34}O_3$ M 310.476
Constit. of *Cystophora moniliformis*. Oil. [α]$_D$ +11° (c, 1.3 in CHCl₃).

van Altena, I.A., *Aust. J. Chem.*, 1988, **41**, 49.

3-Methyl-1,2,4,5-benzenetetrol   M-80048

2,3,5,6-Tetrahydroxytoluene
[700-19-6]

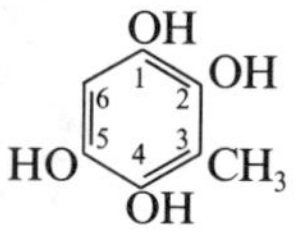

$C_7H_8O_4$ M 156.138
1-Me ether: 5-*Methoxy-3-methyl-1,2,4-benzenetriol. 2,3,6-Trihydroxy-5-methoxytoluene*
     $C_8H_{10}O_4$ M 170.165
     Prisms by subl. Mp 102-103°. Darkens in air.

1-Me ether, Tri-Ac:
     $C_{14}H_{16}O_7$ M 296.276
     Hexagonal plates (MeOH). Mp 155°.

Anslow, W.K. *et al, J. Chem. Soc.*, 1939, 1446.

4-Methyl-1,2,3,5-benzenetetrol   M-80049

2,3,4,6-Tetrahydroxytoluene
$C_7H_8O_4$ M 156.138
1-Me ether: 6-*Methoxy-3-methyl-1,2,4-benzenetriol. 2,3,6-Trihydroxy-4-methoxytoluene*
     $C_8H_{10}O_4$ M 170.165
     Cryst. by subl. Mp 150°.

3-Me ether: 6-*Methoxy-5-methyl-1,2,4-benzenetriol. 3,4,6-Trihydroxy-2-methoxytoluene*
     $C_8H_{10}O_4$ M 170.165
     Cryst. by subl. Mp 146-147°.

3-Me ether, tri-Ac:
     $C_{14}H_{16}O_7$ M 296.276
     Plates (MeOH). Mp 98-99°.

Anslow, W.K. *et al, J. Chem. Soc.*, 1938, 469.

5-Methyl-1,2,3,4-benzenetetrol, 9CI   M-80050

Updated Entry replacing M-00966
2,3,4,5-Tetrahydroxytoluene. 5-Methylapionol
[700-20-9]
$C_7H_8O_4$ M 156.138
Isol. from submerged cultures of *Aspergillus fumigatus*. Cryst. (toluene). Mp 170-171° (140°). Darkens in air.

Tetra-Ac:
     $C_{15}H_{16}O_8$ M 324.287
     Cryst. (MeOH). Mp 132-133°.

2-Me ether: 3-*Methoxy-6-methyl-1,2,4-benzenetriol.*
     Dihydrofumigatin
     $C_8H_{10}O_4$ M 170.165
     Metab. of *A. fumigatus*. Needles by subl.

2-Me, 1,3,4-tri-Ac: Mp 91-92°.

Tetra-Me ether: [35896-58-3]. *1,2,3,4-Tetramethoxy-5-methylbenzene*
     $C_{11}H_{16}O_4$ M 212.245
     Mp 51-52°.

Thiele, J. *et al, Justus Liebigs Ann. Chem.*, 1900, **311**, 341 (*synth*)
Erdtman, H.G.H., *Proc. R. Soc. London, A*, 1934, **143**, 177; *CA*, **38**, 1337 (*synth*)
Anslow, W.K. *et al, Biochem. J.*, 1938, **32**, 687 (*Dihydrofumigatin*)
Packter, N.M., *Biochem. J.*, 1966, **98**, 353 (*isol, synth, deriv*)
Syper, L. *et al, Tetrahedron*, 1980, **36**, 123 (*deriv*)
Keinan, E. *et al, J. Org. Chem.*, 1987, **52**, 3872 (*synth, deriv*)

3-Methyl-2*H*-1-benzothiopyran, 9CI   M-80051

3-Methyl-2H-1-benzothiin. 3-Methyl-2H-thiochromene

[6157-11-5]

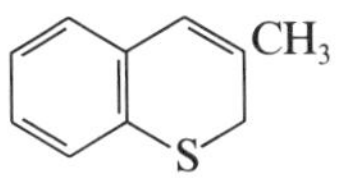

$C_{10}H_{10}S$ M 162.255

Oil. Bp_5 110-115°.

Tilak, B.D. *et al*, *Tetrahedron*, 1966, **22**, 7 (*synth*)
Arnoldi, A. *et al*, *Synthesis*, 1988, 155 (*synth, pmr*)

4-Methyl-1,2,3-benzotriazine, 9CI   M-80052

Updated Entry replacing M-30058

[33334-15-5]

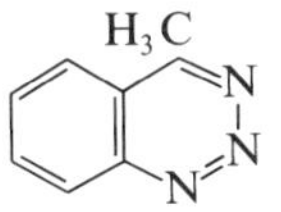

$C_8H_7N_3$ M 145.163

Prisms (EtOH). Mp 120-121°. λ_{max} 207 (log ϵ 3.58), 2.27(4.0) and 275 nm (2.83) (EtOH).

2-Oxide:
   $C_8H_7N_3O$ M 161.163
   Cryst. (CH_2Cl_2/pet. ether or by subl.). Mp 176-178°.

3-Oxide: [41608-06-4].
   $C_8H_7N_3O$ M 161.163
   Orange-yellow plates (EtOH) or cryst. (C_6H_6). Mp 185-188° dec. (175-177°). Dec. in daylight.

Meisenheimer, J. *et al*, *Ber.*, 1927, **60**, 1736 (*oxide*)
Adger, B.M. *et al*, *J. Chem. Soc., Perkin Trans.* 1, 1975, 31 (*synth, ir, uv, pmr*)
Boulton, A.J. *et al*, *J. Chem. Soc., Perkin Trans.* 1, 1988, 1509 (*oxides*)

2-Methyl-1,3-benzoxathiazolium(1+)   M-80053

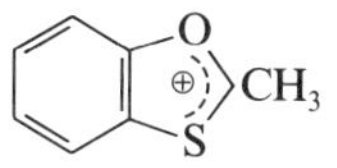

$C_8H_7OS^{\oplus}$ M 151.209 (ion)

Tetrafluoroborate: [112816-51-0].
   $C_8H_7BF_4OS$ M 238.013
   Cryst. Mp 87-88° dec. Moisture sensitive.

Barbero, M. *et al*, *J. Org. Chem.*, 1988, **53**, 2245 (*synth, pmr, cmr*)

2-Methyl-1,3-butanediol   M-80054

[684-84-4]

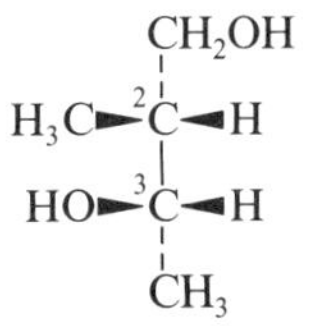

(2R,3S)-form

$C_5H_{12}O_2$ M 104.149

(2R,3S)-form [87678-97-5]
   (+)-erythro-*form*
   $[\alpha]_D^{20}$ +3.5° (c, 1.7 in EtOH).

(2S,3R)-form [90026-43-0]
   (−)-erythro-*form*
   Bp_{15} 110°. $[\alpha]_D$ −6.3°.

(2RS,3RS)(?)-form [16897-85-1]
   (±)-threo-*form*
   n_D^{20} 1.4465. May be the (±)-erythro-form. The 2 racemates have been distinguished and characterised spectroscopically but no phys. props. were given.

Bis-p-nitrobenzoyl: Mp 128°.

(2RS,3SR)-(?)-form [16897-83-9]
   (±)-erythro-*form*
   n_D^{20} 1.4500.

Bis-p-nitrobenzoyl: Mp 113°.

[90026-54-3, 98048-91-0]

Fremaux, B. *et al*, *Bull. Soc. Chim. Fr.*, 1967, 4243 (*synth*)
Dolby, L.J. *et al*, *J. Org. Chem.*, 1968, **33**, 3060 (*synth, pmr*)
Nájera, C. *et al*, *Helv. Chim. Acta*, 1984, **67**, 289 (*synth, pmr*)
White, J.D. *et al*, *J. Org. Chem.*, 1988, **53**, 5909 (*synth, ir, pmr, cmr, ms*)

3-Methyl-1,2-butanediol   M-80055

Isopropylethyleneglycol

[50468-22-9]

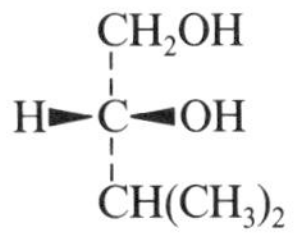

(R)-form

$C_5H_{12}O_2$ M 104.149

(R)-form [31612-62-1]
   Liq. Bp_7 88°. $[\alpha]_D^{20}$ −6.4° (c, 1.46 in cyclohexane), $[\alpha]_D$ −10.95° (c, 1 in $CHCl_3$) (100% opt. purity).

(S)-form [24347-56-6]
   Liq. Bp_3 75.5°. $[\alpha]_D^{25}$ +6.32° (c, 2 in cyclohexane).

[63163-20-2]

Bartlett, P.D. *et al*, *J. Biol. Chem.*, 1937, **118**, 503 (*synth*)
Fuganti, C. *et al*, *Gazz. Chim. Ital.*, 1969, **99**, 316 (*abs config*)
Guetté, J.-P. *et al*, *Bull. Soc. Chim. Fr.*, 1972, 4217 (*synth, abs config*)
Tsuji, K. *et al*, *Makromol. Chem. Suppl.*, 1975, **1**, 55 (*synth*)
White, J.D. *et al*, *J. Org. Chem.*, 1988, **53**, 5909 (*synth, ir, pmr, cmr, ms*)

2-Methyl-2-butenal, 9CI   M-80056

Updated Entry replacing M-01205

2,3-Dimethylacrolein. 2-Methylcrotonaldehyde

[1115-11-3]

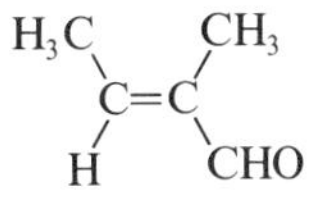

(E)-form

C_5H_8O M 84.118

(E)-form [497-03-0]
   Tiglic aldehyde. Tiglaldehyde
   Isol. from *Allium schoenoprasum* (chives). Spar. sol. H_2O. Bp 116.5-117.5°, Bp_{119} 63.2-65.0°.
   2,4-Dinitrophenylhydrazone: Red cryst. (EtOAc/Me_2CO). Mp 215.5-217.5° (222°).
   Oxime:
     C_5H_9NO M 99.132
     Cryst. (Et_2O). Mp 43°.
   Semicarbazone: Needles (EtOH). Mp 219° (234°).

(Z)-form [6038-09-1]
   Angelic aldehyde
   Liq. Isom. to *E*-form after 2 weeks at r.t.

[6038-09-1]

Wahlroos, O. *et al*, *Acta Chem. Scand.*, 1965, **19**, 1327 (*isol*)
Satsumabayashi, S. *et al*, *Bull. Chem. Soc. Jpn.*, 1970, **43**, 1586 (*synth, ir*)
Botteghi, C. *et al*, *Tetrahedron Lett.*, 1974, 4285 (*synth*)
Vögeli, U. *et al*, *Org. Magn. Reson.*, 1975, **7**, 617 (*synth, pmr, cmr, isom*)
Loewenthal, H.J.E., *Synth. Commun.*, 1975, **5**, 201 (*synth*)
Eletti-Bianchi, G. *et al*, *J. Org. Chem.*, 1976, **41**, 1648 (*synth*)
Kieczykowski, G.R. *et al*, *Tetrahedron Lett.*, 1976, 597 (*synth*)

2-Methyl-2-buten-1-ol, 9CI   M-80057

[4675-87-0]

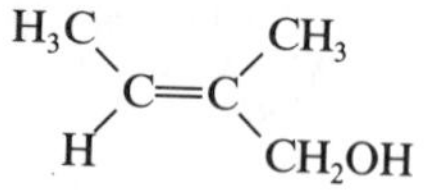

H₃C, CH₃, C=C, H, CH₂OH

(E)-form

C₅H₁₀O   M 86.133

Widespread nat. occurrence, e.g. in *Ochromonas danica*, in *Anthemis nobilis* (as acetate), in fruit juices and animal sources (unspecified stereochem.). Bp_{30} 68-70°.

(E)-form [497-02-9]
  Tiglic alcohol. Tiglyl alcohol
  Formed in *Anthemis nobilis* (together with esters). Bp 122-127° (135°) , Bp_{33} 66°.

(Z)-form [19319-26-7]
  Angelic alcohol. Angelyl alcohol
  Bp 136-137°, Bp_{25} 61-63°. n_D^{20} 1.4410.

Katzenellenbogen, J.A. *et al*, *J. Am. Chem. Soc.*, 1976, **98**, 4925 (*synth, ir, pmr*)
Jarolim, V. *et al*, *Collect. Czech. Chem. Commun.*, 1977, **42**, 3490 (*synth*)
Zaidlewicz, M. *et al*, *Synthesis*, 1979, 62 (*synth, pmr*)
Depezay, J.-C. *et al*, *Bull. Soc. Chim. Fr.*, 1981, 306 (*synth, ms, pmr*)
Smith, P.A.S. *et al*, *J. Org. Chem.*, 1981, **46**, 3970 (*synth, ir, pmr*)
Bury, A. *et al*, *J. Chem. Soc., Perkin Trans. 1*, 1982, 645 (*synth, pmr*)
Masami, Y. *et al*, *Tetrahedron*, 1984, **40**, 3481 (*synth, ir, pmr*)
Ono, N. *et al*, *Tetrahedron Lett.*, 1984, **25**, 5319 (*synth*)
Trost, B.M. *et al*, *J. Am. Chem. Soc.*, 1985, **107**, 1293 (*synth, pmr*)
Kurth, M.J. *et al*, *J. Org. Chem.*, 1985, **50**, 5769 (*synth*)
Ashcroft, M.R. *et al*, *J. Organomet. Chem.*, 1985, **289**, 403 (*synth*)
Simon, H. *et al*, *Angew. Chem., Int. Ed. Engl.*, 1987, **26**, 785 (*synth*)
Bicchi, C. *et al*, *J. Chromatogr.*, 1987, **411**, 237 (*glc, ms*)

24-Methyl-7,22-cholestadiene-3,6-diol   M-80058

7,22-Ergostadiene-3,6-diol

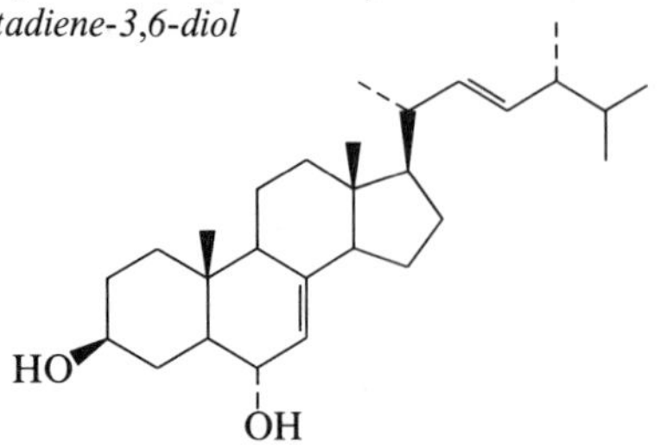

HO, OH

C₂₈H₄₆O₂   M 414.670

(3β,5α,6α,22E,24R)-form
  Constit. of *Spongionella gracilis*. Cryst. (MeOH/pet. ether). Mp 206-208°.

(3β,5α,6α,22E,24S)-form
  Constit. of *S. gracilis*. Cryst. (MeOH/pet. ether). Mp 196-198°.

Madaio, A. *et al*, *J. Nat. Prod. (Lloydia)*, 1989, **52**, 952 (*isol, pmr*)

24-Methyl-7,22-cholestadien-3-ol   M-80059

Updated Entry replacing M-01279
  7,22-Ergostadien-3-ol

C₂₈H₄₆O   M 398.671

(3β,5α,22E,24R)-form [2465-11-4]
  Stellasterol
  Constit. of wood-rotting fungi. Cryst. (CH₂Cl₂/MeOH). Mp 159.5-161°. [α]_D −17.8° (CHCl₃).

(3β,5α,22E,24S)-form [50364-22-2] **5,6-Dihydroergosterol**
  Constit. of *Tylopitus neofelleus* and *Polyporus umbellatus*. Constit. of yeast and grapefruit oil. Cryst. (MeOH). Mp 173.5-174°. [α]_D −23.4°.

  3-Ketone: 24-Methyl-7,22-cholestadien-3-one. 7,22-Ergostadien-3-one
  C₂₈H₄₄O   M 396.655
  Constit. of *Fomes formentarius*. Cryst. (Me₂CO). Mp 184-186°. [α]_D +6° (CHCl₃).

  3-O-β-D-Glucopyranoside:
  C₃₄H₅₆O₆   M 560.813
  Constit. of *T. neofelleus*. Needles (MeOH). Mp 272-274°. $[α]_D^{22}$ +31.1° (c, 0.9 in Py).

Callow, R.K., *Biochem. J.*, 1931, **25**, 87 (*isol*)
Arthur, H.R. *et al*, *J. Chem. Soc.*, 1958, 2603 (*isol*)
Kobayashi, M. *et al*, *Tetrahedron*, 1973, **29**, 1193 (*isol*)
Bélanger, P. *et al*, *Can. J. Chem.*, 1973, **51**, 3294 (*biosynth*)
Rubinstein, I. *et al*, *Phytochemistry*, 1976, **15**, 195 (*pmr*)
Lu, W. *et al*, *Chem. Pharm. Bull.*, 1985, **33**, 5083 (*isol*)
Takaishi, Y. *et al*, *Phytochemistry*, 1989, **28**, 945 (*isol, deriv*)

4-Methylcholesta-8,14-dien-23-ol   M-80060

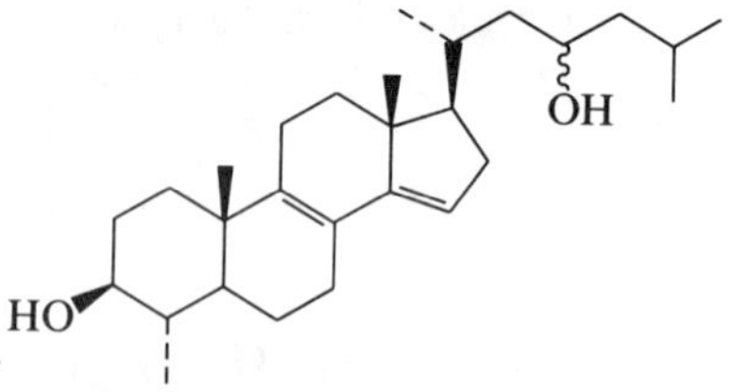

OH, HO

C₂₈H₄₆O₂   M 414.670

(4α,23ξ)-form
  3-O-[β-D-Galactopyranosyl(1→2)-β-D-galactopyranoside]: [119760-82-6]. **Eryloside A**
  C₄₀H₆₆O₁₂   M 738.954
  Constit. of *Erylus lendenfeldi*. Shows antitumour and antifungal props. Amorph. powder. Mp 214-219°. [α]_D +11° (c, 1.5 in CHCl₃).

Carmely, S. *et al*, *J. Nat. Prod. (Lloydia)*, 1989, **52**, 167 (*isol, pmr, cmr*)

24-Methyl-7-cholestene-3,5,6-triol   M-80061

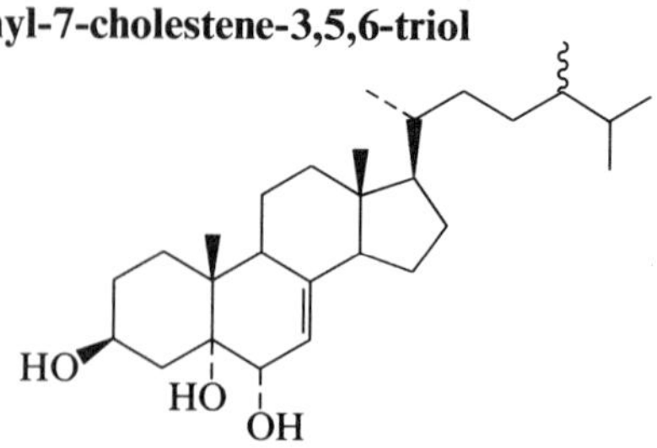

HO, HO, OH

C₂₈H₄₈O₃   M 432.685

(3β,5α,6α,24ξ)-form
  Constit. of *Hippospongia communis*, *Spongia officinalis* and *Ircinia variabilis*.

Madaio, A. *et al*, *J. Nat. Prod. (Lloydia)*, 1989, **52**, 952 (*isol, pmr*)

3-Methyl-3-cyclobutene-1,2-dione   M-80062

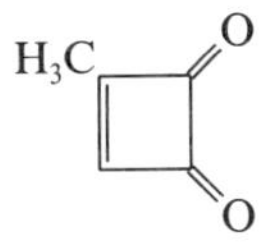

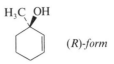

C$_5$H$_4$O$_2$ M 96.085
Golden liq. Bp$_{3.6}$ 68-70°.

Liebeskind, L.S. *et al, J. Org. Chem.*, 1988, **53**, 2482 (*synth, ir, pmr*)

3-Methyl-3,5-cyclohexadiene-1,2-diol, 9CI   M-80063

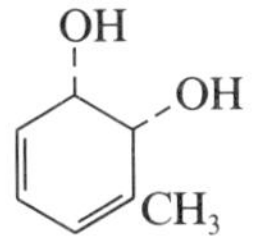

(1*S*,2*R*)-*form*

C$_7$H$_{10}$O$_2$ M 126.155
(**1S,2R**)-*form* [41977-20-2]
 (+)-cis-*form*
 Product of oxidation of toluene by a mutant of *Pseudomonas putida*, a soil bacterium. Prostaglandin synthon. Cryst. (pet. ether). Mp 59°. [α]$_D^{25}$ +26.4° (c, 0.38 in MeOH).

Gibson, D.T. *et al, Biochemistry*, 1970, **9**, 1626 (*isol, uv, ir, pmr*)
Ziffer, H. *et al, Tetrahedron*, 1977, **33**, 2491 (*abs config*)
Hudlicky, T. *et al, J. Am. Chem. Soc.*, 1988, **110**, 4735 (*synth, ir, pmr*)

3-Methyl-1,2-cyclohexanediol, 9CI   M-80064
[23477-91-0]

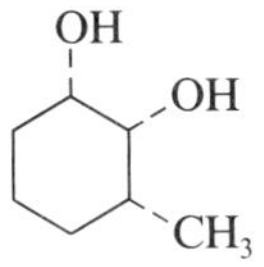

(1*S*,2*R*,3*R*)-*form*

C$_7$H$_{14}$O$_2$ M 130.186
(**1S,2R,3R**)-*form* [41977-21-3]
 (−)-cis,cis-form
 Oil. [α]$_D^{20}$ −37° (c, 0.97 in MeOH).
 1-Benzoyl: Cryst. Mp 84-85°. [α]$_D^{20}$ −25° (c, 1.05 in CHCl$_3$).
 1,2-Dibenzoyl: Oil.
(**1RS,2RS,3SR**)-*form* [15806-70-9]
 (±)-trans,trans-form
 Mp 39-40°.
(**1RS,2RS,3RS**)-*form* [19700-12-0]
 (±)-trans,cis-form
 Cryst. (2,2-dimethylpentane). Mp 94-95°.
(**1RS,2SR,3RS**)-*form* [19700-14-2]
 (±)-cis,trans-form
 Cryst. (hexane/EtOAc). Mp 81-82°.
(**1RS,2SR,3SR**)-*form* [52730-58-2]
 (±)-cis,cis-form
 Cryst. (hexane/EtOAc). Mp 64-65°.

[41977-22-4]

Klein, J. *et al, Tetrahedron*, 1968, **24**, 5701 (*synth, pmr, ir*)
Ziffer, H. *et al, J. Org. Chem.*, 1974, **39**, 3698 (*synth, cmr*)
Ziffer, H. *et al, Tetrahedron*, 1977, **33**, 2491 (*synth, pmr, abs config*)
Hudlicky, T. *et al, J. Am. Chem. Soc.*, 1988, **110**, 4735 (*synth, ir, pmr*)

1-Methyl-2-cyclohexen-1-ol, 9CI   M-80065
Updated Entry replacing M-01411
[23758-27-2]

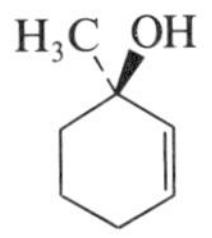

(*R*)-*form*

C$_7$H$_{12}$O   M 112.171
Component of the aggregation pheromone of the douglas fir beetle *Dendroctonus psuedotsugae*.
(**R**)-*form* [88494-89-7]
 Bp$_{16}$ 60°. [α]$_D^{23}$ +75.8° (c, 1.0 in Et$_2$O). n$_D^{20}$ 1.4730.
(**S**)-*form* [112837-29-3]
 Bp$_{16}$ 60°. [α]$_D^{23}$ −76.5° (c, 8.50 in Et$_2$O). n^{20} 1.4705.
(±)-*form*
 Bp$_{23}$ 110°, Bp$_{15}$ 60-61°.
Ac:
 C$_9$H$_{14}$O$_2$ M 154.208
 Bp$_{0.4}$ 85°.
4-Nitrobenzoyl: [38313-13-2].
 Cryst. (hexane). Mp 101-102°.

Mandrou, A.-M. *et al, Bull. Soc. Chim. Fr.*, 1962, 1546 (*synth*)
Magnusson, G. *et al, J. Org. Chem.*, 1973, **38**, 1380 (*synth, cmr*)
Beckwith, A.L.J. *et al, Aust. J. Chem.*, 1976, **29**, 1277 (*synth*)
Mori, K. *et al, Justus Liebigs Ann. Chem.*, 1988, 903 (*synth, bibl*)

4-Methyl-2-cyclohexen-1-one, 9CI   M-80066
Updated Entry replacing M-20108
[5515-76-4]

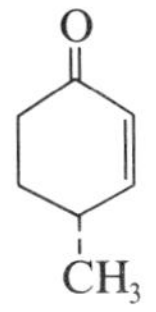

(*R*)-*form*

C$_7$H$_{10}$O   M 110.155
(**R**)-*form* [75337-05-2]
 Oil. Bp 198°. [α]$_D^{22}$ +105° (c, 9.2 in CHCl$_3$).
(**S**)-*form*
 Chiral synthon. [α]$_D^{22}$ −199° (c, 0.37 in EtOH).
(±)-*form*
 Liq. Bp 175-176°, Bp$_{13}$ 81-85°. Steam-volatile.
2,4-Dinitrophenylhydrazone: [3280-41-9].
 Dark-red cryst. Mp 173-174°.

Kötz, A. *et al, Justus Liebigs Ann. Chem.*, 1913, **400**, 86 (*synth*)
Birch, A.J., *J. Chem. Soc.*, 1946, 593 (*synth*)
Torri, J. *et al, Tetrahedron Lett.*, 1973, 3251 (*uv*)
Barieux, J.J. *et al, Bull. Soc. Chim. Fr.*, 1974, 1020 (*cmr*)
Torri, J. *et al, Bull. Soc. Chim. Fr.*, 1974, 1633 (*cmr*)
Silvestri, M.G., *J. Org. Chem.*, 1983, **48**, 2419 (*synth, pmr*)
Hiroi, K. *et al, Synthesis*, 1985, 635 (*synth*)
Hua, D.H. *et al, J. Org. Chem.*, 1988, **53**, 1095 (*synth, ir, pmr, cmr*)

3-Methyl-1-cyclopentenecarboxaldehyde   M-80067
1-Formyl-3-methylcyclopentene
[90049-64-2]

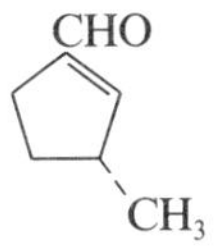

(*R*)-*form*

C$_7$H$_{10}$O   M 110.155

Terpene synthon.

(**R**)-**form** [114818-64-3]

Bp$_2$ 30° (Kugelrohr) . [α]$_D^{25}$ +193.3° (c, 1.058 in CHCl$_3$).

(**S**)-**form** [114818-65-4]

[α]$_D^{25}$ −214.9° (c, 0.1 in CHCl$_3$).

(±)-**form**

Liq. Bp$_{16}$ 56-57°.

2,4-Dinitrophenylhydrazone: Mp 175-183°.

Semicarbazone: Mp 178-180°.

Garanti, L. *et al, Ann. Chim.* (Rome), 1963, **53**, 1619 (*synth*)
Hudlicky, T. *et al, J. Am. Chem. Soc.*, 1988, **110**, 4735 (*synth, ir, pmr*)

Methyl 12,17;15,16-diepoxy-6-hydroxy-19-nor-17-oxo-4,13(16),14-clerodatrien-18-oate   M-80068

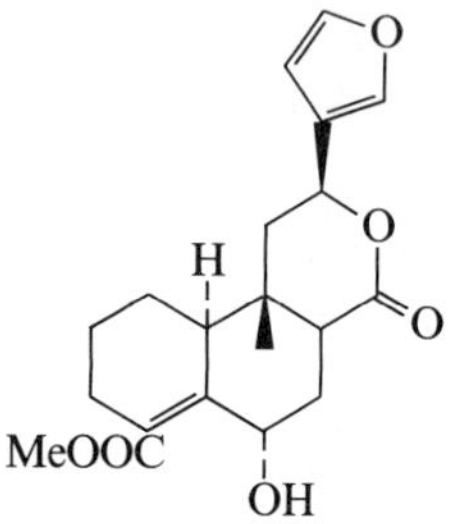

C$_{20}$H$_{24}$O$_6$ M 360.406

Not named in the paper.

O-β-D-Glucopyranoside:

C$_{26}$H$_{34}$O$_{11}$ M 522.548

Constit. of *Tinospora cordifolia*. Cryst. Mp 199-201°.
[α]$_D^{70}$ +23.8° (c, 3.14 in MeOH).

Bhatt, R.K. *et al, Phytochemistry*, 1989, **28**, 2419.

7-Methyl-1,6-dioxaspiro[4.5]decane   M-80069

Updated Entry replacing M-20115

[68108-91-8]

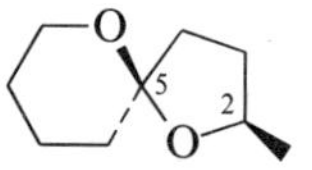

(2R,5R)-**form**

C$_9$H$_{16}$O$_2$ M 156.224

Main component of the odour of the common wasp, *Paravespula vulgaris* which appears to act as aggression inhibitor. Ident. by glc/ms.

(**2R,5R**)-**form**

Bp$_{20}$ 100-110° (bath) . [α]$_D^{20}$ −83.4° (c, 0.718 in pentane).

(**2S,5R**)-**form**

Bp$_{20}$ 100-110° (bath) . [α]$_D^{20}$ −79.1° (c, 0.392 in pentane).
The enantiomeric (2S,5S) and (2R,5S)-forms also prepd.

Erdmann, H., *Justus Liebigs Ann. Chem.*, 1885, **228**, 176 (*synth*)
Francke, W. *et al, Angew. Chem., Int. Ed. Engl.*, 1978, **17**, 862 (*isol, synth*)
Jacobson, R. *et al, J. Org. Chem.*, 1982, **47**, 3140 (*synth*)
Ley, S.V. *et al, Tetrahedron Lett.*, 1984, **25**, 113 (*synth*)
Iwata, C. *et al, Chem. Pharm. Bull.*, 1988, **36**, 4785 (*synth, pmr*)

2-Methyl-1,3-diselenole   M-80070

[115118-34-8]

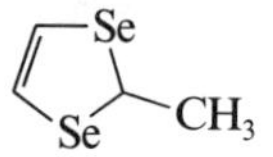

C$_4$H$_6$Se$_2$ M 212.011

Low melting solid. Mp 30°.

Lakshmikantham, M.V. *et al, J. Org. Chem.*, 1988, **53**, 3529 (*synth, pmr, ms*)

2-Methyldotriacontane, 9CI   M-80071

[1720-11-2]

H$_3$C(CH$_2$)$_{29}$CH(CH$_3$)$_2$

C$_{33}$H$_{68}$ M 464.900

Isol. from leaf wax of *Aeonium lindleyi* and *Nicotiana* sp.. Present in surface lipids of pupal tobacco budworms (*Heliothis virescens*), in sex pheromones of insect spp. and in camomile wax.

[69067-11-4]

Eglington, D. *et al, Phytochemistry*, 1966, **5**, 1349 (*isol, ir, ms*)
Guelz, P.G. *et al, Phytochemistry*, 1968, **7**, 1009 (*isol, ms*)
Gamou, K. *et al, Agric. Biol. Chem.*, 1980, **44**, 2119 (*isol, glc*)
Heeman, V. *et al, Phytochemistry*, 1983, **22**, 133 (*isol, ms*)

3-Methyldotriacontane, 9CI   M-80072

[20129-49-1]

H$_3$C(CH$_2$)$_{28}$CH(CH$_3$)CH$_2$CH$_3$

C$_{33}$H$_{48}$ M 444.742

Present in leaves and chloroplasts of *Antirrhinum majus*, in *Nicotiana* sp. and in cuticular membranes of insect spp.

Guelz, P.G. *et al, Phytochemistry*, 1968, **7**, 1009 (*isol, glc, ms*)
Heeman, V. *et al, Phytochemistry*, 1983, **22**, 133 (*isol, ms*)

6-Methyldotriacontene, 9CI   M-80073

[92122-78-6]

H$_3$C(CH$_2$)$_{25}$CH(CH$_3$)CH$_2$CH$_2$CH$_2$CH$_3$

C$_{33}$H$_{68}$ M 464.900

Isol. from leaves and stems of *Duboisia myoporoides*.

Shukla, Y.N. *et al, Phytochemistry*, 1984, **23**, 799 (*isol, struct*)

2-Methylene-1-azabicyclo[2.2.2]octan-3-one   M-80074

2-Methylene-3-quinuclidinone

[5291-26-9]

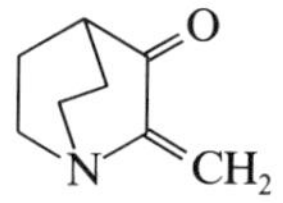

C$_8$H$_{11}$NO   M 137.181

Slightly yellow oil. Bp$_7$ 91-92°, Bp$_{0.5}$ 80°.

B,HCl: Cryst. (Et$_2$O). Mp 243-244° dec.

Hansen, A.R. *et al, J. Heterocycl. Chem.*, 1966, **3**, 109 (*synth, ir*)
Morgan, T.K. *et al, J. Med. Chem.*, 1987, **30**, 2259 (*synth, pmr*)

2-Methylenebicyclo[2.1.0]pentane, 9CI   M-80075

[116377-11-8]

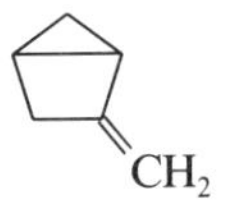

C_6H_8 M 80.129

Andrews, G.D. *et al*, *J. Org. Chem.*, 1988, **53**, 4624 (*synth, pmr, props*)

2,2'-Methylenebis[6-*tert*-butyl-4-methyl-phenol]   M-80076

2,2'-Methylenebis[6-(1,1-dimethylethyl)-4-methylphenol], 9CI.
2,2'-Methylenebis[6-tert-butyl-p-cresol], 8CI

[119-47-1]

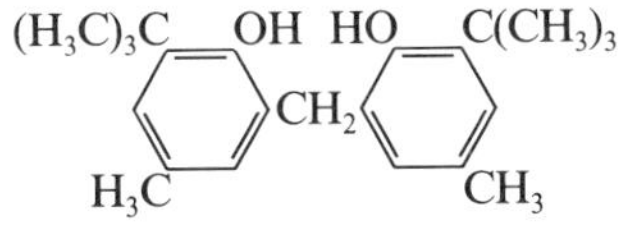

$C_{23}H_{32}O_2$ M 340.505

Polymer antioxidant. Needles (pet. ether). Mp 131°.

Hultzsch, K., *J. Prakt. Chem.*, 1941, **159**, 155, 178 (*synth*)
Coggeshall, N.D. *et al*, *J. Am. Chem. Soc.*, 1948, **70**, 3283; 1950, **72**, 2836 (*ir, uv*)
Wextler, A.S., *Anal. Chem.*, 1963, **35**, 1936 (*uv*)
Kirk-Othmer Encycl. Chem. Technol., 3rd Ed., 1978-1984, *Wiley, NY*, 3rd Ed., Wiley, N.Y., 1968-1984, **3**, 130 (*use*)
Lesko, J. *et al*, *Collect. Czech. Chem. Commun.*, 1969, **34**, 2836 (*ms*)

Methylenebisdesaspidinol BB   M-80077

1,1'-[Methylenebis(2,6-dihydroxy-4-methoxy-3,1-phenylene)]bis-1-butanone, 9CI

[32190-32-2]

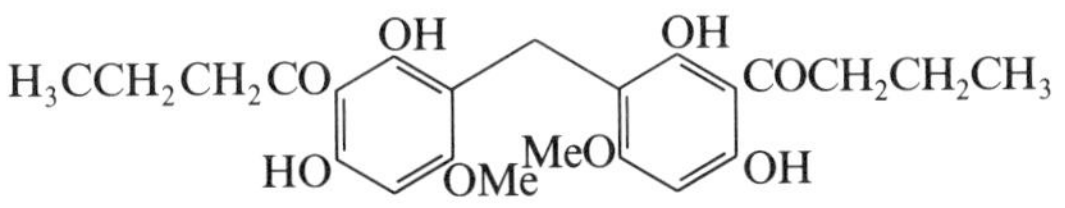

$C_{23}H_{28}O_8$ M 432.469

The BB suffix refers to the two butanoyl side-chains. In theory other methylenebisaspidinols could occur naturally. Isol. from *Dryopteris* spp. ferns. Mp 176-179°.

Penttilä, A. *et al*, *Acta Chem. Scand.*, 1963, **17**, 1886.
Widén, C.-J. *et al*, *Helv. Chim. Acta*, 1971, **54**, 2824.

3,3'-Methylenebisindole, 9CI   M-80078

3,3'-Methylenediindole. Di-3-indolylmethane

[1968-05-4]

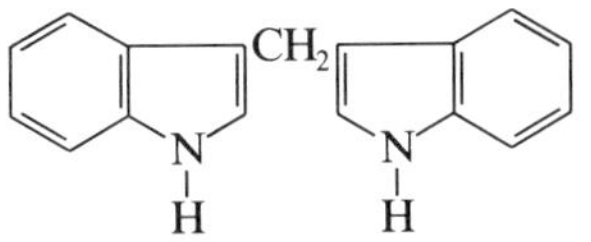

$C_{17}H_{14}N_2$ M 246.311

Needles (EtOH aq.). Mp 165-166°.

1,1'-Di-Me: [31896-75-0].
$C_{19}H_{18}N_2$ M 274.365
Plates (EtOH). Mp 110-112°.

Thesing, J., *Chem. Ber.*, 1954, **87**, 692 (*synth*)
Jackson, A.H. *et al*, *J. Chem. Soc., Perkin Trans. 1*, 1987, 2543 (*synth, pmr*)

Methylenecycloheptane   M-80079

[2505-03-5]

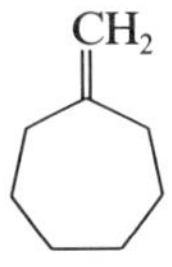

C_8H_{14} M 110.199
d_{20} 0.83. Bp 133° (138-140°) .

Wallach, O., *Justus Liebigs Ann. Chem.*, 1906, **345**, 139 (*synth*)
Šorm, F. *et al*, *Chem. Listy*, 1953, **47**, 708 (*synth*)
Donaldson, W.A., *Tetrahedron*, 1987, **43**, 2901 (*synth*)

2-Methylenecyclohexanone, 9CI, 8CI   M-80080

[3045-98-5]

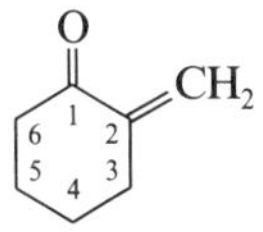

$C_7H_{10}O$ M 110.155

Liq. Dimerises rapidly at r.t.

Semicarbazone: Needles (EtOH). Mp 210° dec.

Di-Et ketal: [23153-78-8]. *1,1-Diethoxy-2-methylenecylohexane*
$C_{11}H_{20}O_2$ M 184.278
Bp_{12} 77-78°.

Mannich, C. *et al*, *Chem. Ber.*, 1941, **74**, 555 (*synth*)
Erskine, R.L. *et al*, *J. Chem. Soc.*, 1960, 3425 (*ir*)
Mühlstädt, M. *et al*, *J. Prakt. Chem.*, 1965, **29**, 158; 1970, **312**, 292 (*synth, pmr*)
Kirmse, W. *et al*, *Chem. Ber.*, 1969, **102**, 2440 (*ketal, synth, pmr*)
Bravo, P. *et al*, *Gazz. Chim. Ital.*, 1969, **99**, 549 (*oxime, synth, ir, pmr, ms*)
Ksander, G.M. *et al*, *J. Org. Chem.*, 1977, **42**, 1180 (*synth*)
Fleming, I. *et al*, *J. Chem. Soc., Perkin Trans. 1*, 1980, 1493 (*synth*)

3-Methylenecyclohexanone, 9CI, 8CI   M-80081

$C_7H_{10}O$ M 110.155

Di-Me ketal: [104598-80-3]. *1,1-Dimethoxy-3-methylenecyclohexane*
$C_9H_{16}O_2$ M 156.224
Liq.

Ethylene ketal: [104598-81-4]. *7-Methylene-1,4-dioxaspiro[4.5]decane*
$C_9H_{14}O_2$ M 154.208
Liq.

Lambert, J.B. *et al*, *J. Am. Chem. Soc.*, 1986, **108**, 7575 (*synth, pmr, uv*)

4-Methylenecyclohexanone, 9CI, 8CI   M-80082

[29648-66-6]
$C_7H_{10}O$ M 110.155
Bp_{15} 70-80°.

2,4-Dinitrophenylhydrazone: Cryst. Mp 154-155°.

Ooba, S. *et al*, *Bull. Chem. Soc. Jpn.*, 1970, **43**, 1782 (*ir, ms, pmr*)
Sonoda, A. *et al*, *Bull. Chem. Soc. Jpn.*, 1972, **45**, 1777 (*synth*)
Coughlin, D.J. *et al*, *J. Org. Chem.*, 1979, **44**, 3784 (*synth, pmr*)

2-Methylene-1,3-cyclopentanedione   M-80083

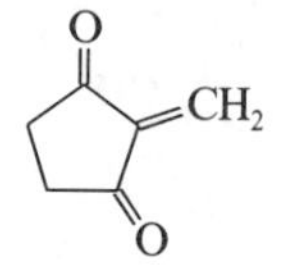

$C_6H_6O_2$ M 110.112

Intermediate trapped as Diels-Alder adduct.

Eaton, P.E. et al, Can. J. Chem., 1984, **62**, 2612 (synth)
Bunnelle, W.H. et al, Tetrahedron, 1987, **43**, 2005 (synth)

5-Methylene-5H-dibenzo[a,d]cycloheptene,   M-80084
9CI

1-Methylene-2,3:6,7-dibenz-4-suberene

[2975-79-3]

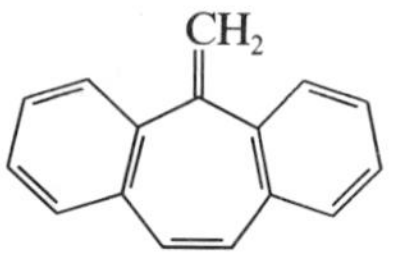

$C_{16}H_{12}$ M 204.271
Cryst. (EtOH). Mp 119-121°.

Triebs, W. et al, Ber., 1951, **84**, 671 (synth)
Granoth, I. et al, J. Org. Chem., 1976, **41**, 3682 (synth)
Lamanec, T.R. et al, J. Org. Chem., 1988, **53**, 1768 (synth, pmr)

5-(3,4-Methylenedioxyphenyl)pentanoic   M-80085
acid

1,3-Benzodioxole-5-pentanoic acid, 9CI. Piperhydronic acid.
Tetrahydropiperinic acid

[41917-45-7]

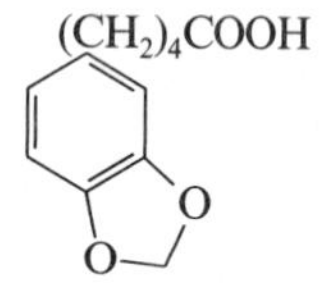

$C_{12}H_{14}O_4$ M 222.240
Isol. from fruits of Piper longum. Tablets or platelets
(EtOH). Mp 100-101° (90-91°).

Et ester: Bp₁ 155°.

Gokhale, V.G. et al, CA, 1949, **43**, 1085 (isol)
Viswanathan, N. et al, Helv. Chim. Acta, 1975, **58**, 2026 (synth)

6-(3,4-Methylenedioxystyryl)-α-pyrone   M-80086

6-[2-(1,3-Benzodioxol-5-yl)ethenyl]-2H-pyran-2-one, 9CI
[1219-50-7]

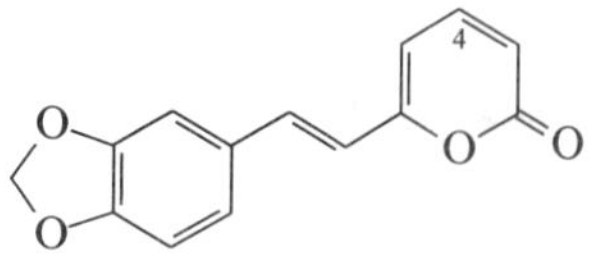

$C_{14}H_{10}O_4$ M 242.231
Isol. from wood of Aniba parviflora. Yellow cryst. (C_6H_6).
Mp 173-174°.

4-Methoxy: 4-Methoxy-6-(3,4-methylenedioxystyryl)-α-
pyrone. 5,6-Dehydromethysticin
$C_{15}H_{12}O_5$ M 272.257
Isol. from wood of A. heringeri. Light yellow-green
cryst. Mp 233-234°.

[3983-44-5]

Mors, W.B. et al, An. Assoc. Quim. Bras., 1962, **21**, 7 (deriv)

Gottlieb, O.R. et al, Ann. Acad. Bras. Cienc., 1964, **36**, 29 (isol, ir,
struct)
Mahajan, J.R. et al, J. Org. Chem., 1971, **36**, 1832 (synth)

24-Methylene-8-lanostene-3,22-diol   M-80087

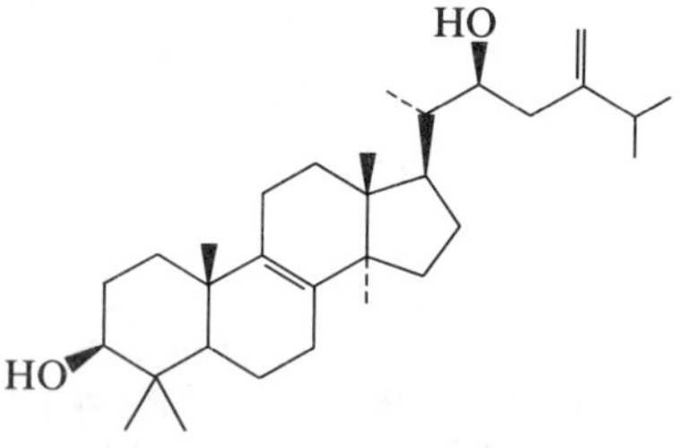

$C_{31}H_{52}O_2$ M 456.751
(3β,22S)-form [119513-64-3]
Metab. of Pisolithus tinctorius. Cryst. (MeOH). Mp 161-
165°. [α]_D +29° (c, 0.16 in CHCl₃).

Lobo, A.M. et al, Phytochemistry, 1988, **27**, 3569.

2-Methylene-3-oxocyclobutanecarboxylic   M-80088
acid

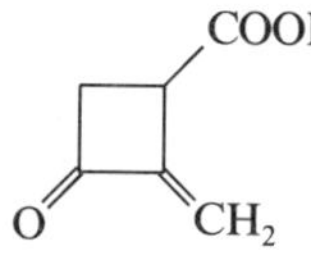

$C_6H_6O_3$ M 126.112
(±)-form
Me ester: [112139-41-0].
$C_7H_8O_3$ M 140.138
Oil.

Vidal, J. et al, J. Org. Chem., 1988, **53**, 611 (synth, pmr, cmr, ir,
uv, ms)

2-Methylenepentanedinitrile, 9CI   M-80089

2-Methyleneglutaronitrile, 8CI. 2,4-Dicyano-1-butene
[1572-52-7]

$$H_2C=C(CN)CH_2CH_2CN$$

$C_6H_6N_2$ M 106.127
Dimer of acrylonitrile. Liq. d²⁰ 0.976. Mp −9.6°. Bp 103°.
n_D^{20} 1.4559.

Aldrich Library of FT-IR Spectra, 1st Ed., **1**, 844D (ir)
Aldrich Library of NMR Spectra, 2nd Ed., **2**, 702D (pmr)
Sadtler Standard Infrared Spectra, No. 34514 (ir)
Sadtler Standard NMR Spectra, No. 5428 (pmr)
Baizer, M.M. et al, J. Org. Chem., 1965, **30**, 1357 (synth, props)
Price, P. et al, Anal. Chem., 1976, **48**, 494 (ms)
Basaraiah, D. et al, Tetrahedron Lett., 1987, **28**, 4591 (synth)

1-Methylene-2-(phenylthio)cyclopropane   M-80090

[(Methylenecyclopropyl)thio]benzene, 9CI. 1-Methylidene-2-
(phenylthio)cyclopropane
[78656-80-1]

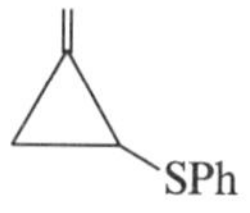

$C_{10}H_{10}S$ M 162.255
(±)-form
Light-yellow oil. Bp₀.₀₁ 65°.

S-*Oxide:* [94923-10-1]. *1-Methylene-2-
(phenylsulfinyl)cyclopropane. 1-Methylidene-2-
(phenylsulfinyl)cyclopropane.
[(Methylenecyclopropyl)sulfinyl]benzene, 9CI*
$C_{10}H_{10}OS$ M 178.254
Oil. Separable by careful chromatography into
diastereoisomers.

S-*Dioxide:* [78656-83-4]. *1-Methylene-2-
(phenylsulfonyl)cyclopropane. 1-Methylidene-2-
(phenylsulfonyl)cyclopropane.
[(Methylenecyclopropyl)sulfonyl]benzene, 9CI*
$C_{10}H_{10}O_2S$ M 194.254
Cryst. Mp 47.5-48.5°.

Weber, A. *et al, Helv. Chim. Acta*, 1988, **71**, 2026 (*synth, ir, pmr, ms*)

2-Methylheneicosane, 9CI   M-80091

$$H_3C(CH_2)_{18}CH(CH_3)_2$$

$C_{22}H_{46}$ M 310.605
Found in orange oil (*Citrus sinensis*) and hops (*Humulus lupulus*).

Steibl, M. *et al, Collect. Czech. Chem. Commun.*, 1964, **29**, 2522 (*synth*)
Wollrab, V. *et al, Collect. Czech. Chem. Commun.*, 1965, **30**, 1670 (*isol*)
Hunter, G.L.K. *et al, Phytochemistry*, 1966, **5**, 807 (*isol*)

3-Methylheptacosane, 9CI   M-80092

[14167-66-9]

$$H_3C(CH_2)_{23}CH(CH_3)CH_2CH_3$$

$C_{28}H_{58}$ M 394.766
Isol. from tobacco (*Nicotiana tabacum*) hop oil (*Humulus lupulus*) orange oil (*Citrus sinensis*), and insect cuticle.
n_D^{50} 1.4407.

[55194-24-6]

Mold, J.D. *et al, Biochemistry*, 1963, **2**, 605 (*isol*)
Wollrab, V. *et al, Collect. Czech. Chem. Commun.*, 1965, **30**, 1670 (*isol*)
Hunter, G.L.K. *et al, Phytochemistry*, 1966, **5**, 807 (*isol*)

3-Methyl-4-hepten-2-one   M-80093

[75343-95-2]

$$H_3CCH_2CH=CHCH(CH_3)COCH_3$$

$C_8H_{14}O$ M 126.198

Dana, G. *et al, Can. J. Chem.*, 1980, **58**, 1451 (*synth, ir, pmr*)
Sato, T. *et al, J. Org. Chem.*, 1988, **53**, 1894 (*synth, ms, pmr*)

6-Methyl-6-hepten-2-one, 9CI   M-80094

[10408-15-8]

$$H_2C=C(CH_3)CH_2CH_2CH_2COCH_3$$

$C_8H_{14}O$ M 126.198
Isol. from fruit oil of *Litsea citrata*. Liq. d_4^{19} 0.846. Bp_{10} 45°. n_D^{19} 1.4344.

2,4-Dinitrophenylhydrazone: [17123-67-0].
Mp 76-78°.

Dimethylhydrazone: [111285-97-3].
Bp_{8-10} 60°.

Kappeler, H. *et al, Helv. Chim. Acta*, 1954, **37**, 957 (*synth, ir*)
Sood, V.K., *Perfum. Essent. Oil Rec.*, 1966, **57**, 285 (*isol*)
Viola, A. *et al, J. Am. Chem. Soc.*, 1967, **89**, 3462 (*synth*)

Bartlett, P.A. *et al, J. Chem. Educ.*, 1984, **61**, 816 (*synth*)
Trehan, I.R. *et al, Indian J. Chem., Sect. B*, 1986, **25**, 1243 (*synth, pmr*)

2-Methylhexacosane, 9CI   M-80095

[1561-02-0]

$$H_3C(CH_2)_{23}CH(CH_3)_2$$

$C_{27}H_{56}$ M 380.739
Isol. from hop oil (*Humulus lupulus*), orange oil (*Citrus sinensis*) and insect cuticle.

Wollrab, V. *et al, Collect. Czech. Chem. Commun.*, 1965, **30**, 1670 (*isol*)
Hunter, G.L.K. *et al, Phytochemistry*, 1966, **5**, 807 (*isol*)
Streibl, M. *et al, Fette, Seifen, Anstrichm.*, 1968, **70**, 543 (*synth*)

2-Methylhexanedial   M-80096

2-Methyladipaldehyde

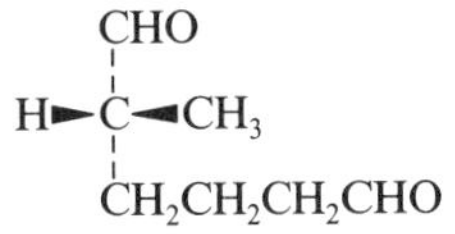

CHO
H►C◄CH₃   (R)-*form*
CH₂CH₂CH₂CHO

$C_7H_{12}O_2$ M 128.171
(R)-*form* [114763-39-2]
Oil.
[114763-40-5]

Hudlicky, T. *et al, J. Am. Chem. Soc.*, 1988, **110**, 4735 (*synth, ir, pmr*)

4-Methyl-3-hexanone, 9CI   M-80097

Updated Entry replacing M-02018
sec-*Butyl ethyl ketone*
[17042-16-9]

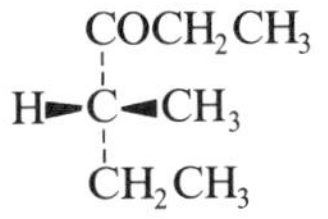

COCH₂CH₃
H►C◄CH₃   (R)-*form*
CH₂CH₃

$C_7H_{14}O$ M 114.187
(R)-*form* [77858-08-3]
Liq. Bp_{100} 74-77°. $[\alpha]_D^{23}$ −30.8° (c, 4 in Et₂O)(96% e.e.).
(S)-*form* [20086-34-4]
$[\alpha]_D^{25}$ +8.5°.
(±)-*form*
d^{19} 0.825. Bp 134-135°.

Semicarbazone: Mp 137°.

Bartlett, P.D. *et al, J. Am. Chem. Soc.*, 1935, **57**, 2580 (*synth, abs config*)
Hudson, B.E. *et al, J. Am. Chem. Soc.*, 1941, **63**, 3163 (*synth*)
Lardicci, L. *et al, J. Chem. Soc., Chem. Commun.*, 1968, 381 (*ord, cd*)
Brown, H.C. *et al, J. Am. Chem. Soc.*, 1988, **110**, 1529 (*synth, pmr, cmr, ir*)

2-Methyl-4-hexenal, 9CI

M-80098

CHO
H►C◄CH₃
CH₂
C=C
H   CH₃
H

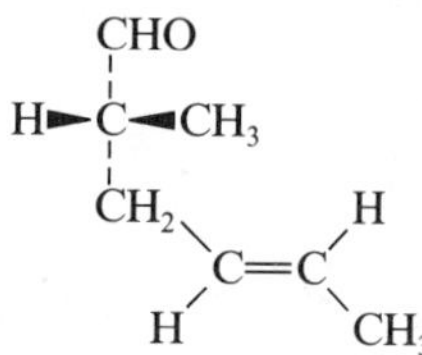

$C_7H_{12}O$ M 112.171

(R,E)-form [104372-54-5]

Liq. Bp_{46} 60° (kugelrohr) . Racemises on heating $>70°$.

Evans, D.A. et al, J. Am. Chem. Soc., 1986, **108**, 6757 (synth, pmr)
Deyo, D.T. et al, Synthesis, 1988, 608 (synth, pmr)

5-Methyl-4-hexen-1-amine

M-80099

6-Amino-2-methyl-2-hexene

[115610-15-6]

$(H_3C)_2C=CHCH_2CH_2CH_2NH_2$

$C_7H_{15}N$ M 113.202
Bp_{20} 51-53°.

Tietze, L.F. et al, Chem. Ber., 1989, **122**, 997 (synth, pmr, cmr)

2-Methyl-4-hexenoic acid

M-80100

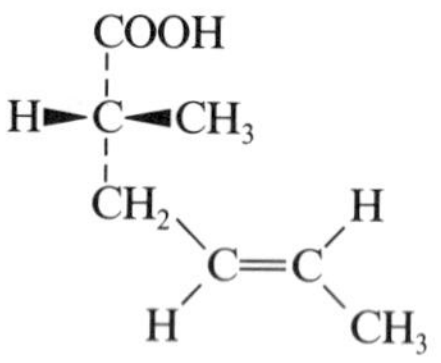

COOH
H►C◄CH₃
CH₂   (R,E)-form
C=C
H   CH₃
H

$C_7H_{12}O_2$ M 128.171

(R,E)-form [93553-73-2]
$[\alpha]_D^{20}$ $-10.2°$ (c, 1.2 in C_6H_6).

(±-E)-form [118492-11-8]
Yellow oil.

Et ester: [118942-48-6].
$C_9H_{16}O_2$ M 156.224
Oil. Bp_{18} 70-75°.

Deyo, D.T. et al, Synthesis, 1988, 608 (synth, pmr, ir)

4-Methyl-2-hexenoic acid, 9CI

M-80101

Updated Entry replacing M-02032
4-Ethyl-4-methylcrotonic acid
[37549-83-0]

CH=CHCOOH
H₃C►C◄H
CH₂CH₃

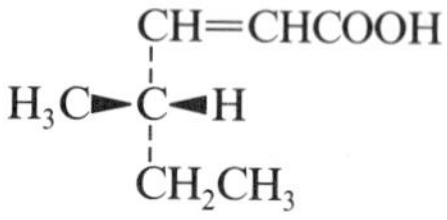

S-form

$C_7H_{12}O_2$ M 128.171

(S,E)-form [82166-24-3]
Prod. of hydrolysis of Leucinostatin A. $[\alpha]_D^{20}$ $+49.7°$ (c, 0.25 in $CHCl_3$).

Me ester:
$C_8H_{14}O_2$ M 142.197
Liq. Bp_{14} 70°. $[\alpha]_D^{20}$ $+37°$ (c, 0.4 in CH_2Cl_2).

(±)-form
Bp_{13} 125°. $n_D^{20.3}$ 1.4526.

Chloride:
$C_7H_{11}ClO$ M 146.616

Bp_{11} 65-66°.

Linstead, R.P. et al, J. Chem. Soc., 1930, 2072 (synth)
Mori, Y. et al, J. Chem. Soc., Chem. Commun., 1982, 94 (isol, ir, pmr)
Baker, R. et al, J. Chem. Soc., Perkin Trans. 1, 1988, 85 (ester, ir, pmr, ms)

2-Methyl-4-hexen-1-ol

M-80102

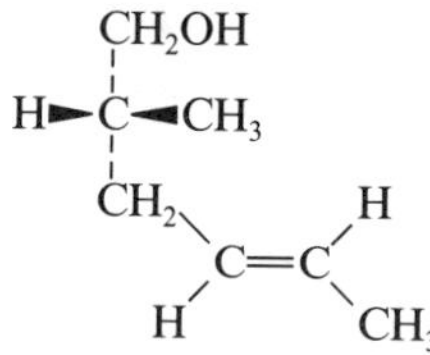

CH₂OH
H►C◄CH₃
CH₂   (R,E)-form
C=C
H   CH₃
H

$C_7H_{14}O$ M 114.187

(R,E)-form
Oil. Bp_{20} 85° (Kugelrohr) . $[\alpha]_D$ $+2.5°$ (c, 1.08 in CH_2Cl_2).

(±)-(E)-form [58927-95-0]
Characterised spectroscopically.

(±)-(Z)-form [58927-94-9]
Characterised spectroscopically.

Chantegnel, B. et al, Bull. Soc. Chim. Fr., 1975, 2639 (synth, ir)
Evans, D.A. et al, J. Am. Chem. Soc., 1986, **108**, 6757 (synth, ir, pmr, cmr)
Deyo, D.T. et al, Synthesis, 1988, 608 (synth, pmr, cmr)

2-Methyl-4-hexen-2-ol, 9CI

M-80103

[70910-33-7]

$(H_3C)_2C(OH)CH_2CH=CHCH_3$

$C_7H_{14}O$ M 114.187
Bp_{747} 139-140°.

(E)-form [19639-97-5]
n_D^{20} 1.4321.

(Z)-form [19639-96-4]
n_D^{20} 1.4370.

Zurquiyah, A. et al, J. Org. Chem., 1969, **34**, 1504 (synth)
Hosomi, A. et al, Tetrahedron Lett., 1976, 1295 (synth)
Lupton, M.F. et al, J. Org. Chem., 1978, **43**, 1409 (synth)
Fujita, K. et al, Chem. Lett., 1982, 1819 (synth)
Satoh, S. et al, Bull. Chem. Soc. Jpn., 1983, **56**, 1791 (synth, ir, pmr, ms)

2-Methyl-5-hexen-1-ol, 9CI

M-80104

[25913-88-6]

$H_2C=CHCH_2CH_2CH(CH_3)CH_2OH$

$C_7H_{14}O$ M 114.187

(±)-form
Odorous liq. d_4^{20} 0.843. Bp_{736} 166-168°, Bp_{25} 88°. n_D^{22} = 1.438.

Allophanate: Cryst. (EtOH). Mp 136°.

Ac:
$C_9H_{16}O_2$ M 156.224
Liq. Bp_{747} 182-184°. n_D^{22} 1.425.

Me ether: 6-Methoxy-5-methyl-1-hexene
$C_8H_{16}O$ M 128.214
Liq. d_4^{25} 0.875. Bp_{20} 76°. n_D^{25} 1.425.

Colonge, J. et al, Bull. Soc. Chim. Fr., 1962, 177 (synth)
Brettle, R. et al, J. Chem. Soc., 1962, 4836 (synth, uv)
Brown, H.C. et al, J. Org. Chem., 1983, **48**, 644 (synth, pmr, cmr)

3-Methyl-5-hexen-1-ol, 9CI   M-80105

[25913-87-5]

$$H_2C=CHCH_2CH(CH_3)CH_2CH_2OH$$

$C_7H_{14}O$ M 114.187

(±)-*form*

Liq. Bp 160°, Bp_{18} 77°. n_D^{20} = 1.4407.

α-*Naphthylurethane:* Cryst. (pet. ether). Mp 45-46°.

Closson, W.D. *et al, J. Org. Chem.,* 1970, **35**, 3737 (*synth, pmr*)
Courtois, G. *et al, J. Organomet. Chem.,* 1973, **52**, 241 (*synth, ir, raman*)
Beckwith, A.L.J. *et al, Aust. J. Chem.,* 1983, **36**, 545 (*synth, pmr*)
Carr, S.A. *et al, J. Org. Chem.,* 1985, **50**, 2782 (*synth*)

4-Methyl-2-hexen-1-ol, 9CI   M-80106

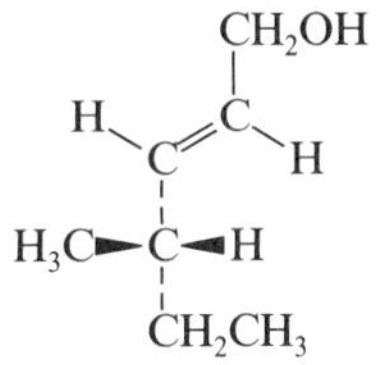

(S,E)-form

$C_7H_{14}O$ M 114.187

(S,E)-*form* [103064-73-9]

Bp_{14} 90°. $[\alpha]_D^{20}$ +35.2°.

(±)-(E)-*form* [87219-85-0]

Liq.

Miller, B.R. *et al, Tetrahedron Lett.,* 1983, **24**, 2055 (*synth*)
Hanessian, S. *et al, J. Am. Chem. Soc.,* 1986, **108**, 2776 (*synth*)
Baker, R. *et al, J. Chem. Soc., Perkin Trans. 1,* 1988, 85 (*synth, uv, pmr, ms*)

4-Methyl-3-hexen-2-ol, 9CI   M-80107

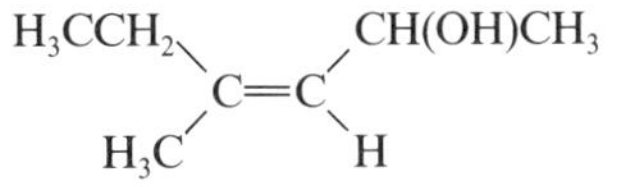

$C_7H_{14}O$ M 114.187

(±)-(Z)-*form* [42998-32-3]

Liq. Bp_{35} 76-78°. n_D^{20} = 1.4437.

Normant, J.F. *et al, J. Organomet. Chem.,* 1974, **77**, 269 (*synth, ir, raman*)

4-Methyl-4-hexen-1-ol, 9CI   M-80108

[59518-07-9]

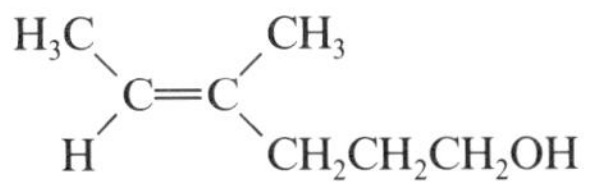

(E)-form

$C_7H_{14}O$ M 114.187

(E)-*form* [57253-27-7]

Bp_{12} 79°. n_D^{20} 1.4500 (1.4520).

α-*Naphthylurethane:* Plates (pet. ether). Mp 75.5-76.5°.

(Z)-*form* [57253-26-6]

Bp_{15} 82-83°. n_D^{20} = 1.4499.

α-*Naphthylurethane:* Cryst. (pet. ether). Mp 65-67° (61-65°).

Diphenylurethane: Laths (pet. ether). Mp 99.5-101° (93-95°).

Ansell, M.F. *et al, J. Chem. Soc.,* 1958, 3388 (*synth, ir*)
Julia, M. *et al, Tetrahedron,* 1975, **31**, 1737 (*synth, ir, raman*)

Hartmann, J. *et al, Helv. Chim. Acta,* 1976, **59**, 453 (*synth, ms, ir, pmr*)
Depezay, J.C. *et al, Bull. Soc. Chim. Fr.,* (*Part II*), 1981, 435 (*synth, ir, raman, ms*)

4-Methyl-4-hexen-2-ol, 9CI   M-80109

$$H_3CCH=C(CH_3)CH_2CH(OH)CH_3$$

$C_7H_{14}O$ M 114.187

Liq. $Bp_{0.1}$ 28-30°. Mixt. of (E)- and (Z)-forms.

[108686-09-5, 108686-15-3]

Barluenga, J. *et al, Synthesis,* 1987, 318 (*synth, ir, pmr, ms, cmr*)

4-Methyl-5-hexen-1-ol, 9CI   M-80110

[25906-56-3]

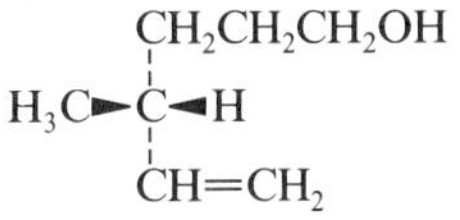

(R)-form

$C_7H_{14}O$ M 114.187

(R)-*form* [69274-97-1]

Bp_{25} 79-81°. $[\alpha]_D^{19}$ −15.5° (neat). n_D^{19} 1.4371.

(±)-*form*

Liq. Bp_{12} 68°. n_D^{22} 1.4375.

α-*Naphthylurethane:* Cryst. (pet. ether). Mp 55.5-56.0°.

[69274-92-6]

Closson, W.D. *et al, J. Org. Chem.,* 1970, **35**, 3737 (*synth, pmr*)
Mori, K. *et al, Tetrahedron Lett.,* 1978, **37**, 3447 (*synth*)
Naef, F. *et al, Helv. Chim. Acta,* 1981, **64**, 1387 (*synth, pmr, ms*)
Utimoto, K. *et al, Heterocycles,* (*Spec. issue*), 1982, **18**, 149 (*synth, uv, pmr, ms*)
Beckwith, A.L.J. *et al, Aust. J. Chem.,* 1983, **36**, 545 (*synth, pmr*)
Carr, S.A. *et al, J. Org. Chem.,* 1985, **50**, 2782 (*synth, cmr*)

5-Methyl-4-hexen-2-ol, 9CI   M-80111

[3695-33-8]

$$(H_3C)_2C=CHCH_2CH(OH)CH_3$$

$C_7H_{14}O$ M 114.187

Baker, R. *et al, J. Chem. Soc., Chem. Commun.,* 1975, 727 (*synth*)

5-Methyl-4-hexen-3-ol, 9CI   M-80112

[53555-59-2]

$$(H_3C)_2C=CHCH(OH)CH_2CH_3$$

$C_7H_{14}O$ M 114.187

Intermed. for Chrysanthemic acid.

(±)-*form*

Pleasant smelling liq. d_4^{25} 0.837. Bp_{22} 63-65°. n_D^{25} 1.4380.

Colonge, J. *et al, Bull. Soc. Chim. Fr.,* 1956, 188 (*synth*)
Deno, N.C. *et al, J. Org. Chem.,* 1975, **40**, 514 (*synth, pmr*)
Japan. Pat., 76 86 410, (1976); *CA,* **86**, 120799q (*synth, use*)
Kondo, K. *et al, Tetrahedron Lett.,* 1976, **48**, 4359 (*use*)

5-Methyl-5-hexen-1-ol, 9CI   M-80113

[5212-80-6]

$$H_2C=C(CH_3)CH_2CH_2CH_2CH_2OH$$

$C_7H_{14}O$ M 114.187

Liq. d_4^{20} 0.845. Bp_{12} 80-82°. n_D^{20} 1.4470 (1.4415).

Phenylurethane: Mp 52-53°.

Eschenmoser, A. *et al, Helv. Chim. Acta,* 1952, **35**, 1660 (*synth, ir*)
Zweifel, G. *et al, J. Am. Chem. Soc.,* 1962, **84**, 190 (*synth, ir*)

Sato, F. *et al, J. Organomet. Chem.*, 1977, **142**, 71 (*synth, pmr*)
Sato, F. *et al, Chem. Lett.*, 1978, 999 (*synth*)
Carr, S.A. *et al, J. Org. Chem.*, 1985, **50**, 2782 (*synth, cmr*)
Gardette, M. *et al, Tetrahedron*, 1985, **41**, 5887 (*synth, pmr, cmr*)

5-Methyl-5-hexen-3-ol, 9CI   M-80114

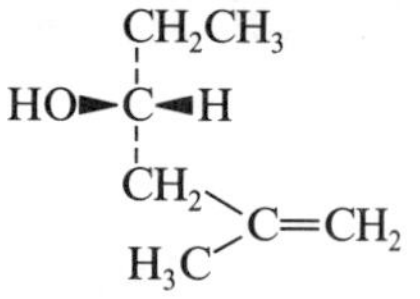

(*R*)-*form*

$C_7H_{14}O$ M 114.187
(*R*)-*form* [77118-89-9]
  $[\alpha]_D^{20}$ +5.13° (c, 3.51 in Et_2O).
(*S*)-*form* [67760-89-8]
  Bp_{60} 78-80°. $[\alpha]_D^{23}$ −3.07° (neat).
(±)-*form*
  Sharp smelling liq. d_4^{22} 0.836. Bp_{30} 63°. n_D^{20} 1.4384.

Herold, T. *et al, Angew. Chem., Int. Ed. Engl.*, 1978, **17**, 768 (*synth*)
Hoffmann, R.W. *et al, Chem. Ber.*, 1981, **114**, 375 (*synth, pmr*)
Brown, H.C. *et al, Tetrahedron Lett.*, 1984, **25**, 5111 (*synth*)
Jadhev, P.K. *et al, J. Org. Chem.*, 1986, **51**, 432 (*synth*)

3-Methyl-4-hexen-2-one   M-80115

[72189-24-3]

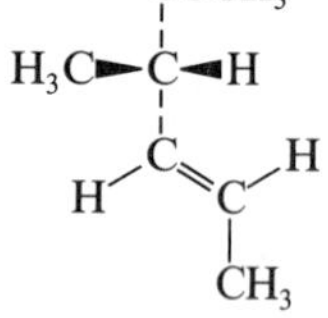

$C_7H_{12}O$ M 112.171
(*S,E*)-*form* [101223-88-5]
  $[\alpha]_D^{20}$ +282°.

Bartik, T. *et al, J. Organomet. Chem.*, 1985, **291**, 253 (*synth*)
Sato, T. *et al, J. Org. Chem.*, 1988, **53**, 1894 (*synth, pmr, ir, ms*)

1-Methyl-1*H*-imidazole-5-carboxylic acid, 9CI   M-80116

Updated Entry replacing M-02077
3-Methyl-4-imidazolecarboxylic acid
[41806-40-0]

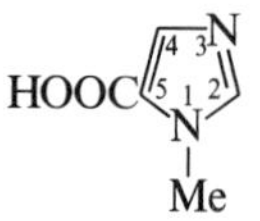

$C_5H_6N_2O_2$ M 126.115
Mp 256-257° (245-248°) dec.
Picrate: Leaflets (H_2O). Spar. sol. H_2O, EtOH. Mp 198-199°.
Me ester:
  $C_6H_8N_2O_2$ M 140.141
  Prisms (MeOH). Sol. H_2O, EtOH, $CHCl_3$, spar. sol. cold Et_2O. Mp 68-70°.

Hubball, W. *et al, J. Chem. Soc.*, 1928, 28 (*synth*)
O'Connell, J.F. *et al, Synthesis*, 1988, 767 (*synth, ir, pmr, cmr*)

5-Methyl-4-isoxazolecarboxylic acid   M-80117

[42831-50-5]

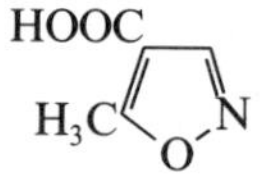

$C_5H_5NO_3$ M 127.099
Prepn. well descr. but no phys. props. reported.
Et ester: [51135-73-0].
  $C_7H_9NO_3$ M 155.153
  No phys. props. reported.
Chloride:
  $C_5H_4ClNO_2$ M 145.545
  Pungent liq. Bp_{14} 78-79°.
▷ Explosion haz. on dist.

Tamura, Y. *et al, J. Chem. Soc., Perkin Trans. 1*, 1973, 2580 (*synth*)
Doleschall, G. *et al, J. Chem. Soc., Perkin Trans. 1*, 1988, 1875 (*synth, chloride, ir*)

7-Methyl-3-methylene-1-octanol, 9CI   M-80118

Bupleurol
[57197-03-2]

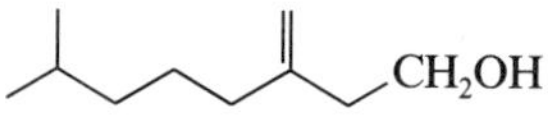

$C_{10}H_{20}O$ M 156.267
Isol. from ethereal oil of *Bupleurum fruticosum* and from *B. linearifolium* seed oils. Bp 209-210°.

[1407-75-6]

Wolinsky, J. *et al, J. Org. Chem.*, 1976, **41**, 278 (*synth, ir, pmr*)
Ashraf, M. *et al, Pak. J. Sci. Ind. Res.*, 1979, **22**, 325; *ca*, **93**, 245254j (*isol*)

4-Methyl-3-methylene-2-oxetanone   M-80119

[117203-16-4]

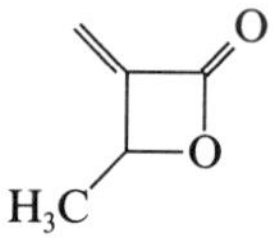

$C_5H_6O_2$ M 98.101
First α-methylene-β-propionolactone.
(±)-*form*
  Pale-yellow liq. with pungent odour. Et and isopropyl analogues also prepd.
▷ Potential carcinogen.

[117203-17-5, 117203-18-6]

Adam, W. *et al, Angew. Chem., Int. Ed. Engl.*, 1988, **27**, 1536 (*synth*)

3-Methyl-1-naphthaldehyde, 8CI   M-80120

3-Methyl-1-naphthalenecarboxaldehyde, 9CI. 1-Formyl-3-methylnaphthalene
[63409-02-9]
$C_{12}H_{10}O$ M 170.210

Sydnes, L.K. *et al, Tetrahedron*, 1985, **41**, 5205 (*synth, ir, pmr, ms*)
Gupta, R.B. *et al, J. Org. Chem.*, 1989, **54**, 1097 (*synth, ir, pmr, cmr*)

5-Methyl-1-naphthaldehyde, 8CI   M-80121

5-Methyl-1-naphthalenecarboxaldehyde, 9CI. 1-Formyl-5-methylnaphthalene
[104306-72-1]

$C_{12}H_{10}O$ M 170.210

2,4-Dinitrophenylhydrazone: Red cryst. Mp 245-246° dec.

Sydnes, L.K. *et al, Tetrahedron*, 1985, **41**, 5703 (*synth, ir, pmr, ms*)

3-Methyl-1-nitrocyclopentene   M-80122

[112683-42-8]

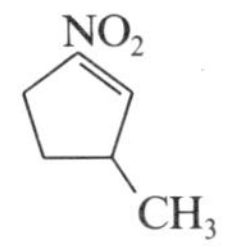

$C_6H_9NO_2$ M 127.143

(±)*-form*
    Liq. Bp$_{0.5}$ 80°.

Denmark, S.E. *et al, J. Org. Chem.*, 1988, **53**, 1251 (*synth, pmr, cmr, ms*)

24-Methyl-27-nor-7,22-cholestadiene-3,6-diol   M-80123

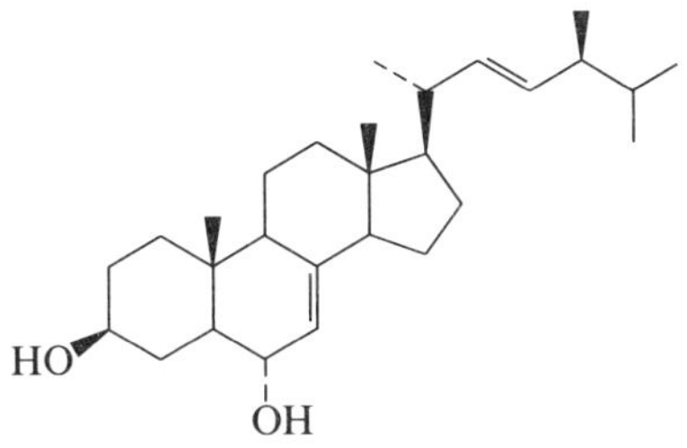

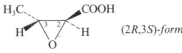

$C_{27}H_{44}O_2$ M 400.643

(*3β,5α,6α,22E*)*-form*
    Constit. of *Spongionella gracilis*. Cryst. (MeOH/pet. ether). Mp 185-187°.

Madaio, A. *et al, J. Nat. Prod. (Lloydia)*, 1989, **52**, 952 (*isol, pmr*)

2-Methyloctacosane   M-80124

[1560-98-1]

$$H_3C(CH_2)_{25}CH(CH_3)_2$$

$C_{29}H_{60}$ M 408.793

Present in hop, tobacco and orange oil. Found in many insects. Liq.

Mold, J.D. *et al, Biochemistry*, 1963, **2**, 605.
Wollrab, V. *et al, Collect. Czech. Chem. Commun.*, 1965, **30**, 1670.
Hunter, G.L.K. *et al, Phytochemistry*, 1966, **5**, 807.

1-Methyl-7-oxabicyclo[4.1.0]heptane, 9CI   M-80125

1-Methylcyclohexene oxide. 1,2-Epoxy-1-methylcyclohexane

[1713-33-3]

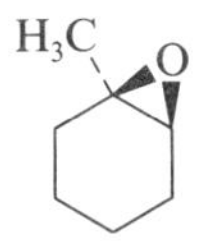

(*1R,2S*)*-form*

$C_7H_{12}O$ M 112.171

(*1R,2S*)*-form* [56246-60-7]
    $[\alpha]_D^{25}$ − 18-65° (neat).

(*1RS,2SR*)*-form* [60363-27-1]
    Bp 130-131° (138-142°) , Bp$_{35}$ 52°.

[56246-59-4]

Tani, K. *et al, Tetrahedron Lett.*, 1979, 3017 (*synth*)
Curci, R. *et al, J. Chem. Soc., Chem. Commun.*, 1984, 155 (*synth*)

Itoi, Y. *et al, Chem. Pharm. Bull.*, 1985, **33**, 1583 (*synth*)
Robinson, P.L. *et al, J. Am. Chem. Soc.*, 1985, **107**, 5210 (*synth, cmr*)
De Poorter, B. *et al, J. Chem. Soc., Perkin Trans. 2*, 1985, 1735 (*synth*)
Yamomoto, H. *et al, J. Chem. Soc., Perkin Trans. 1*, 1986, 173 (*synth, nmr*)
Robinson, P.L. *et al, Phosphorus Sulfur*, 1986, **26**, 15 (*synth*)
Nwaukwa, S.O. *et al, Synth. Commun.*, 1986, **16**, 309 (*synth, ir, pmr*)
Davis, F.A. *et al, Tetrahedron Lett.*, 1986, **27**, 5079 (*synth*)

1-Methyl-6-oxabicyclo[3.1.0]hexane   M-80126

1-Methylcyclopentene oxide. 1,2-Epoxy-1-methylcyclopentane

[16240-42-9]

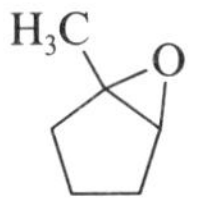

$C_6H_{10}O$ M 98.144
Bp 110-111°.

Rebek, J. *et al, J. Org. Chem.*, 1979, **44**, 1485 (*synth*)
Johnson, W.S. *et al, J. Am. Chem. Soc.*, 1980, **102**, 7800 (*synth*)
Blau, K. *et al, J. Prakt. Chem.*, 1980, **322**, 915 (*synth, pmr*)
Nwaukura, S.O. *et al, Synth. Commun.*, 1986, **16**, 309 (*synth, ir, pmr*)

3-Methyloxiranecarboxylic acid, 9CI   M-80127

Updated Entry replacing M-02836
    2-Methylglycidic acid. 2,3-Epoxybutyric acid

[2443-40-5]

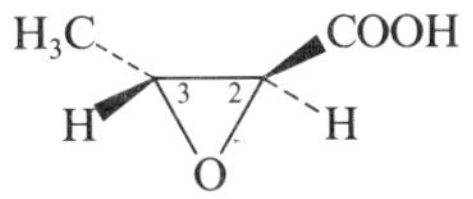

(*2R,3S*)*-form*

$C_4H_6O_3$ M 102.090

(*2R,3R*)*-form*
    Me ester: Bp$_{1.4}$ 64°. $[\alpha]_D^{22}$ + 7.3° (c, 1.26 in CHCl$_3$), $[\alpha]_D^{22}$ − 19.0° (c, 1.6 in MeOH).
    Et ester: Bp$_{1.5}$ 77°. $[\alpha]_D^{25}$ − 1.6° (c, 5.0 in MeOH).
(*2S,3S*)*-form*
    Me ester: $[\alpha]_D^{22}$ + 18.9° (c, 1.8 in MeOH).
    Et ester: $[\alpha]_D^{25}$ + 1.8° (c, 4.8 in MeOH).
(*2R,3S*)*-form* [50468-19-4]
    (−)-trans*-form*
    Cryst. (C$_6$H$_6$). Mp 61°. $[\alpha]_D^{25}$ − 82.5° (c, 0.59 in C$_6$H$_6$).
    Me ester: [50468-20-7].
    $C_5H_8O_3$ M 116.116
    Liq. Bp 150°.
    Et ester:
    $C_6H_{10}O_3$ M 130.143
    Liq. Bp$_{73}$ 106.5°.
    ▷ Toxic. ET0875000.
(*2S,3R*)*-form*
    (+)-trans*-form*
    Cryst. Mp 62°. $[\alpha]_D^{20}$ + 79° (c, 0.6 in C$_6$H$_6$).
(*2RS,3SR*)*-form* [13737-02-5]
    (±)-trans*-form*
    Cryst. (C$_6$H$_6$). Mp 84.5-85°.
    Me ester: [2980-48-5].
    Liq.
    Et ester: [19780-35-9].
    $C_6H_{10}O_3$ M 130.143
    Liq. Bp$_{73}$ 106.5°.

▷ Toxic. ET0875000.

Harada, K. *et al*, *Bull. Chem. Soc. Jpn.*, 1966, **39**, 2311 (*synth, abs config*)
Baldwin, J.E. *et al*, *J. Am. Chem. Soc.*, 1973, **95**, 5261 (*synth*)
Aberhart, D.J. *et al*, *J. Am. Chem. Soc.*, 1973, **95**, 7859 (*synth*)
Aberhart, D.J. *et al*, *J. Chem. Soc., Perkin Trans. 1*, 1974, 2320 (*synth*)
Séquin, U., *Tetrahedron Lett.*, 1979, 1833 (*synth, cmr*)
Petit, Y. *et al*, *Synthesis*, 1988, 538 (*synth, ir, pmr*)
Sax, N.I., *Dangerous Properties of Industrial Materials*, 5th Ed., Van Nostrand-Reinhold, 1979, 663.

3-Methyloxiranemethanol, 9CI   M-80128

Updated Entry replacing M-20168
2,3-Epoxy-1-butanol, 8CI. 2-Hydroxymethyl-3-methyloxirane
[872-38-8]

(*2R,3R*)-*form*

$C_4H_8O_2$ M 88.106

(*2R,3R*)-*form* [58845-50-4]
  (+)-trans-*form*
  $[\alpha]_D^{25}$ +47° (C_6H_6).
(*2S,3S*)-*form* [50468-21-8]
  (−)-trans-*form*
  $[\alpha]_D$ −49° (c, 5 in C_6H_6).
(*2R,3S*)-*form*
  (−)-cis-form
  Oil. Bp_{15} 64°. $[\alpha]_D^{20}$ −33.6° (c, 0.1 in CH_2Cl_2). 90% enantiomeric excess.
(*2S,3R*)-*form*
  (+)-cis-form
  $[\alpha]_D^{20}$ +29.1° (c, 0.06 in CH_2Cl_2).
(*2RS,3RS*)-*form*
  (±)-trans-*form*
  Bp_{10} 58-59°. n_D^{25} 1.4250.
(*2RS,3SR*)-*form*
  (±)-cis-*form*
  Bp_{10} 69-70°. n_D^{25} 1.4308.

Payne, G.B., *J. Org. Chem.*, 1962, **27**, 3819 (*synth*)
Pierre, J.L. *et al*, *Bull. Soc. Chim. Fr.*, 1970, 4459 (*pmr*)
Aberhart, D.J. *et al*, *J. Am. Chem. Soc.*, 1973, **95**, 7859.
Corey, E.J. *et al*, *J. Am. Chem. Soc.*, 1978, **100**, 4618 (*synth*)
Reynolds, K.A. *et al*, *J. Chem. Soc., Perkin Trans. 1*, 1988, 3195 (*synth, pmr, ms*)

3-Methyl-2-oxobutanoic acid   M-80129

Updated Entry replacing M-02849
2-Oxoisopentanoic acid. α-Ketoisovaleric acid. Dimethylpyruvic acid
[759-05-7]

$$(H_3C)_2CHCOCOOH$$

$C_5H_8O_3$ M 116.116
Metab. of *Aspergillus niger* and other microorganisms.
  Cryst. Mp 31°. Bp 170.5°.
Oxime: [13010-50-9].
  $C_5H_9NO_3$ M 131.131
  Cryst. Mp 163-165° dec.
Phenylhydrazone: Long yellow needles (MeOH aq.). Mp 156-157°.
Me ester: [3952-67-8].
  $C_6H_{10}O_3$ M 130.143
  Bp_{12} 70-80°.
Me ester, 2,4-dinitrophenylhydrazone: Cryst. Mp 176-178°.

Et ester: [20201-24-5].
  $C_7H_{12}O_3$ M 144.170
  Bp_{10} 58-60°.
Et ester, oxime:
  $C_7H_{13}NO_3$ M 159.185
  Needles (Et_2O/pet. ether). Mp 57°.
Et ester, semicarbazone: Needles (Et_2O/pet. ether). Mp 95-96°.
Amide: [34906-95-1].
  $C_5H_9NO_2$ M 115.132
  Mp 106-107°.
Nitrile: [42867-39-0]. *3-Methyl-2-oxobutanenitrile, 9CI. Isobutyryl cyanide*
  C_5H_7NO M 97.116
  Bp 117-118°.

Moritz, E., *J. Chem. Soc.*, 1881, 13 (*amide*)
Bouveault, L. *et al*, *Bull. Soc. Chim. Fr.*, 1901, 1031 (*synth, derivs*)
Tschelinzeff, W. *et al*, *Chem. Ber.*, 1929, **62**, 2210 (*nitrile*)
Fischer, R. *et al*, *Chem. Ber.*, 1960, **93**, 1387 (*Et ester*)
Zbiral, E. *et al*, *Monatsh. Chem.*, 1966, **97**, 1797 (*Me ester*)
Arfin, S.M., *J. Biol. Chem.*, 1969, **244**, 2250 (*biosynth*)
Clement, B.A. *et al*, *J. Org. Chem.*, 1974, **39**, 97 (*nitrile*)
Cooper, A.J.C. *et al*, *J. Biol. Chem.*, 1975, **250**, 527 (*pmr, bibl*)

6-Methyl-5-oxoheptanoic acid   M-80130

[40564-61-2]

$$(H_3C)_2CHCO(CH_2)_3COOH$$

$C_8H_{14}O_3$ M 158.197
$Bp_{0.06}$ 100°.
Et ester: [24071-98-5].
  $C_{10}H_{18}O_3$ M 186.250
  Bp_{17} 130°.

Iijima, A. *et al*, *Chem. Pharm. Bull.*, 1973, **21**, 215 (*synth*)
Berthelot, P. *et al*, *J. Heterocycl. Chem.*, 1988, **25**, 1525 (*synth, pmr*)

3-Methylpentacosane, 9CI   M-80131

[6902-54-1]

$$H_3C(CH_2)_{21}CH(CH_3)CH_2CH_3$$

$C_{26}H_{54}$ M 366.713
Isol. from hop oil (*Humulus lupulus*), orange oil (*Citrus sinensis*), and insect cuticle. Liq. Bp_{10} 248°. n_D^{50} 1.4391.
[55194-22-4]

Wollrab, V. *et al*, *Collect. Czech. Chem. Commun.*, 1965, **30**, 1670 (*isol*)
Hunter, G.L.K. *et al*, *Phytochemistry*, 1966, **5**, 807 (*isol*)
Streibl, M. *et al*, *Fette, Seifen, Anstrichm.*, 1968, **70**, 543; *CA*, **70**, 30318x (*synth*)

5-Methyl-5-pentacosanol, 9CI   M-80132

Grandiflorol

$$H_3C(CH_2)_{19}\underset{\underset{OH}{|}}{\overset{\overset{CH_3}{|}}{C}}(CH_2)_3CH_3$$

$C_{26}H_{54}O$ M 382.712
Isol. from leaves of *Sesbania grandiflora*. Mp 83-84°. $[\alpha]_D$ −20° ($CHCl_3$).

Tiwari, R.D. *et al*, *Arch. Pharm. (Weinheim, Ger.)*, 1964, **297**, 310.

2-Methyl-1,3-pentadiene   M-80133

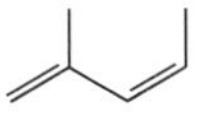

(Z)-form

Coml. samples apparently consist mainly of the Z-form.

Werstiuk, N.H. et al, Can. J. Chem., 1988, **66**, 2954 (pe, config)

4-Methyl-4-pentenal   M-80134

[3973-43-1]

$$H_2C{=}C(CH_3)CH_2CH_2CHO$$

$C_6H_{10}O$ M 98.144
Liq. Bp 99-102°.

2,4-Dinitrophenylhydrazone: Orange cryst. (EtOH). Mp 135-136°.

Vig, O.P. et al, Indian J. Chem., 1969, **7**, 1111 (synth, ir)
Baker, R. et al, J. Chem. Soc., Perkin Trans. 1, 1988, 125 (synth)

2-Methyl-4-penten-1-ol   M-80135

[5673-98-3]

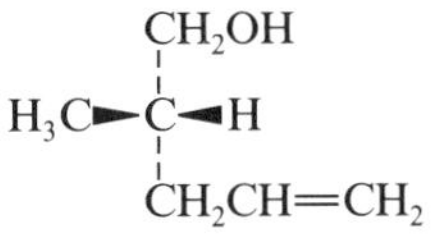

(S)-form

$C_6H_{12}O$ M 100.160
(S)-form [63501-26-8]
    d_4^{18} 0.85. Bp 145-146°. $[\alpha]_D$ −2.3° (c, 1.0 in CHCl₃).
(±)-form
    Bp₁₂ 57-58°.
    Ac:
        $C_8H_{14}O_2$ M 142.197
        Bp₁₇ 60-65°.
    Tetrahydropyranyl ether:
        $C_{11}H_{20}O_2$ M 184.278
        Bp₃ 66-68°.

[17142-57-3, 34288-66-9]

Fray, G.I. et al, J. Chem. Soc., 1956, 2036 (synth)
Fanta, W.I. et al, J. Org. Chem., 1972, **37**, 1624 (synth, ir, pmr, deriv)
Evans, D.A. et al, J. Am. Chem. Soc., 1988, **110**, 2506 (synth, ir, pmr, cmr)
Granatica, P. et al, Tetrahedron, 1988, **44**, 1299 (synth)

3-Methyl-1-phenazinecarboxylic acid   M-80136

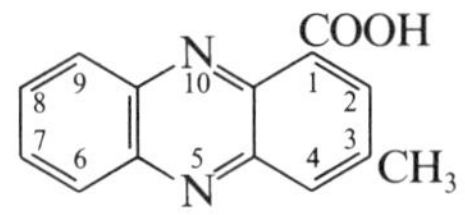

$C_{14}H_{10}N_2O_2$ M 238.245
Mp 212-213°.

Rewcastle, G.W. et al, J. Med. Chem., 1987, **30**, 843 (synth)

4-Methyl-1-phenazinecarboxylic acid, 9CI, 8CI   M-80137

[106976-14-1]

$C_{14}H_{10}N_2O_2$ M 238.245
Cryst. (MeOH). Mp 239-241°.
Me ester: [106976-29-8].

$C_{15}H_{12}N_2O_2$ M 252.272
Cryst. (MeOH). Mp 151-152°.

Rewcastle, G.W. et al, J. Med. Chem., 1987, **30**, 843.

6-Methyl-1-phenazinecarboxylic acid, 9CI, 8CI   M-80138

[103942-86-5]

$C_{14}H_{10}N_2O_2$ M 238.245
Cryst. (MeOH). Mp 239-241°.
Me ester: [73113-61-8].
    $C_{15}H_{12}N_2O_2$ M 252.272
    Cryst. (MeOH). Mp 150-151°.

Budzikiewicz, H. et al, J. Heterocycl. Chem., 1979, **16**, 1307 (ester, synth, uv, pmr, ms)
Roemer, A., Org. Magn. Reson., 1982, **19**, 66 (ester, pmr)
Rewcastle, G.W. et al, J. Med. Chem., 1987, **30**, 843 (synth)

7-Methyl-1-phenazinecarboxylic acid, 9CI, 8CI   M-80139

[103942-88-7]

$C_{14}H_{10}N_2O_2$ M 238.245
Cryst. (MeOH). Mp 260° dec.

Rewcastle, G.W. et al, J. Med. Chem., 1987, **30**, 843 (synth)

8-Methyl-1-phenazinecarboxylic acid, 9CI, 8CI   M-80140

[106976-15-2]

$C_{14}H_{10}N_2O_2$ M 238.245
Cryst. (MeOH). Mp 238° dec.
Me ester: [83297-74-9].
    $C_{15}H_{12}N_2O_2$ M 252.272
    Mp 111°.

Budzikiewicz, H. et al, J. Heterocycl. Chem., 1979, **16**, 1307 (ester, synth, pmr, ms)
Roemer, A., Org. Magn. Reson., 1982, **19**, 66 (ester, pmr)
Rewcastle, G.W. et al, J. Med. Chem., 1987, **30**, 843 (synth)
Rewcastle, G.W. et al, Synth. Commun., 1987, **17**, 1171 (synth)

9-Methyl-1-phenazinecarboxylic acid, 9CI, 8CI   M-80141

[58718-46-0]

$C_{14}H_{10}N_2O_2$ M 238.245
Cryst. (MeOH). Mp 237-238°.
Me ester: [58718-45-9].
    $C_{15}H_{12}N_2O_2$ M 252.272
    Cryst. (2-propanol). Mp 110-112°.

Breitmaier, E. et al, J. Org. Chem., 1976, **41**, 2104 (synth, uv, pmr, cmr)
Roemer, A., Org. Magn. Reson., 1982, **19**, 66 (ester, pmr)
Rewcastle, G.W. et al, J. Med. Chem., 1987, **30**, 843 (synth)

1-Methyl-4-phenyl-1λ⁴-1-benzothiopyran-2-carbonitrile, 9CI   M-80142

2-Cyano-1-methyl-4-phenyl-1-thianaphthalene
[120400-36-4]

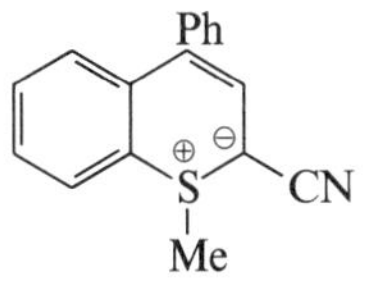

$C_{17}H_{13}NS$ M 263.362

Stable crystalline 1-thianaphthalene. Yellow prisms
(CH$_2$Cl$_2$/Et$_2$O). Mp 137-139° dec.

Hori, M. *et al, Chem. Pharm. Bull.*, 1988, **36**, 3816 (*synth, pmr*)

3-Methyl-3-phenylcyclopentanone   M-80143

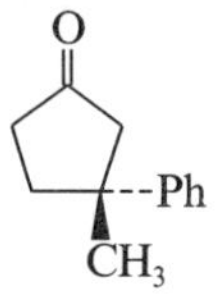

(R)-form

C$_{12}$H$_{14}$O   M 174.242

(R)-form [110508-59-3]
Liq. [α]$_D^{26}$ +4.8° (c, 1.02 in CHCl$_3$) (30% e.e.).

Paquette, L.A. *et al, J. Org. Chem.*, 1988, **53**, 4978 (*synth, ir, pmr*)

3-Methyl-8-phenyl-3,5,7-octatrien-2-one   M-80144

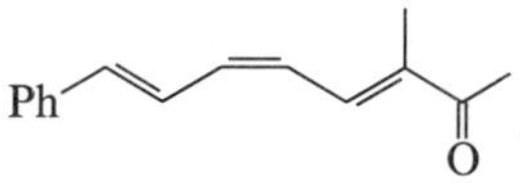

(3E,5Z,7E)-form

C$_{15}$H$_{16}$O   M 212.291

(3E,5Z,7E)-form
Lignarenone A
Metab. of *Scaphander lignarius.*

(3E,5E,7E)-form
Lignarenone B
Isol. from *S. lignarius.*

Cimino, G. *et al, Tetrahedron Lett.*, 1989, **30**, 5003.

2-Methyl-5-phenyloxazole, 9CI   M-80145

[3969-09-3]

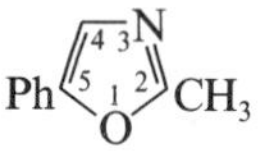

C$_{10}$H$_9$NO   M 159.187
Mp 58-59°. Bp$_{13}$ 128-130°, Bp$_3$ 101-103°.
Picrate: [14895-06-8].
Mp 155-156°.

Gabriel, S., *Chem. Ber.*, 1910, **43**, 1283 (*synth*)
Houwing, H.A. *et al, J. Heterocycl. Chem.*, 1981, **18**, 1133 (*synth, cmr*)

2-Methyl-1-phenyl-2-propen-1-one, 9CI   M-80146

*α-Methylacrylophenone. 2-Benzoylpropene. Phenyl
isopropenyl ketone*
[769-60-8]

$$H_2C=C(CH_3)COPh$$

C$_{10}$H$_{10}$O   M 146.188
Liq. d$_{20}^{20}$ 1.02. Bp$_3$ 60°.

Burckhalter, J.H. *et al, J. Am. Chem. Soc.*, 1948, **70**, 4184 (*synth*)
Sugita, K. *et al, J. Polym. Sci., Polym. Chem. Ed.*, 1976, **14**, 1901 (*synth, ir, pmr*)
Jackson, L.B. *et al, J. Chem. Soc., Perkin Trans. 1*, 1988, 1791 (*synth, ir, pmr*)

2-Methyl-5-phenylpyrrolidine   M-80147

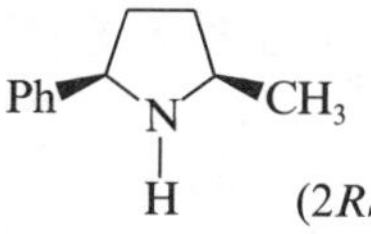

(2RS,5RS)-form

C$_{11}$H$_{15}$N   M 161.246
(2RS,5RS)-form
(±)-cis-*form*
1-Me: [113579-87-6]. *1,2-Dimethyl-5-phenylpyrrolidine*
C$_{12}$H$_{17}$N   M 175.273
Oil.
(2RS,5SR)-form
(±)-trans-*form*
1-Me:B,MeI: Needles (Me$_2$CO/Et$_2$O). Mp 151-152°.

Tokuda, M. *et al, Tetrahedron*, 1987, **43**, 281 (*synth, ir, pmr, cryst struct*)

Methyl phenyl sulfoximine   M-80148
Updated Entry replacing M-40125

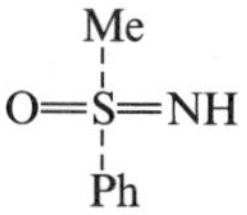

(R)-form

(R)-form [60933-65-5]
Low melting solid. Mp 32-34°. [α]$_D^{25}$ −17.9° (c, 2.62 in
MeOH).
N-Me: [80482-67-3]. *N,S-Dimethyl-S-phenylsulfoximine*
C$_8$H$_{11}$NOS   M 169.247
Reagent for asymmetric synthesis, resoln. and
mechanistic studies. Oil. Bp$_1$ 120-125° (bulb) . [α]$_D^{25}$
−135.9° (c, 4.60 in MeOH).
(±)-form [81162-81-4]
Mp 34-35°.
N-Me: Bp$_{0.4}$ 115-118°.
N-Me, picrate: Mp 189-191°.

[33903-50-3]

Johnson, C.R. *et al, J. Am. Chem. Soc.*, 1970, **92**, 6594; 1973, **95**,
7418 (*synth, N-Me, pmr, ir, resoln, use*)
Oae, S. *et al, Int. J. Sulfur Chem., Part A*, 1972, **2**, 491 (*synth*)
Furukawa, N. *et al, Synthesis*, 1982, 77 (*synth*)
Shiner, C.S. *et al, J. Org. Chem.*, 1988, **53**, 5542 (*synth, resoln, bibl*)

Methylpropanedioic acid, 9CI   M-80149
Updated Entry replacing M-03354
Methylmalonic acid. Isosuccinic acid
[516-05-2]

$$H_3CCH(COOH)_2$$

C$_4$H$_6$O$_4$ M 118.089
Needles (EtOAc/pet. ether), prisms (Et$_2$O/C$_6$H$_6$). Sol.
EtOH, Et$_2$O, EtOAc, spar. sol. H$_2$O, C$_6$H$_6$. Mp 135°
(120°). pK_a 3.06 (25°).
▷ OO1400000.
Mono-Me ester: [3097-74-3].
C$_5$H$_8$O$_4$ M 132.116
Bp$_{16}$ 131°.
Di-Me ester: [609-02-9].
C$_6$H$_{10}$O$_4$ M 146.143
d$_4^{20}$ 1.095. Bp 178°.
Dichloride: [39619-07-3].
C$_4$H$_4$Cl$_2$O$_2$ M 154.980
Bp$_{50}$ 75°.

Diamide: [1113-63-9].
$C_4H_8N_2O_2$ M 116.119
Cryst. (H_2O). Insol. Et_2O. Mp 206°.

Dianilide: [6833-01-8].
$C_{16}H_{16}N_2O_2$ M 268.315
Leaflets (EtOH). Mp 182° (214°).

Dinitrile: Methylpropanedinitrile, 9CI. Methylmalononitrile.
1,1-Dicyanoethane
$C_4H_4N_2$ M 80.089
Needles or oil. Mp 32-34°. Bp_{21} 88-89°.

Meyer, R. *et al, Justus Liebigs Ann. Chem.,* 1906, **347**, 94 (*synth*)
Org. Synth., Coll. Vol., 2, 1943, 279 (*synth*)
Aloskalova, N.I. *et al, Zh. Org. Khim.,* 1966, **2**, 2132 (*synth*)
Fieser and Fieser's Reagents for Organic Synthesis, Wiley, 1975, **5**, 216.
Hosmane, R.S. *et al, J. Am. Chem. Soc.,* 1982, **104**, 235 (*dinitrile*)
Yokoyama, M. *et al, Synthesis,* 1988, 813 (*synth, dinitrile*)

(1-Methylpropyl)butanedioic acid, 9CI   M-80150

Updated Entry replacing M-03392
sec-*Butylsuccinic acid. 3-Methylpentane-1,2-dicarboxylic acid. 3-Carboxy-4-methylhexanoic acid*
[5653-98-5]

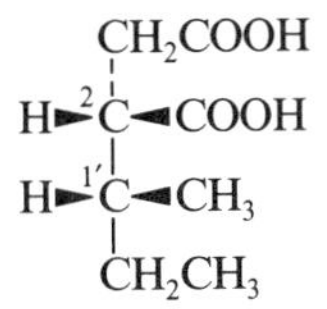

$C_8H_{14}O_4$ M 174.196
Di-Et ester:
$C_{12}H_{22}O_4$ M 230.303
Liq. $Bp_{0.2}$ 75-76°.
(2S,1'R)-form [54976-96-4]
(3S,4R)-*form*
Mp 101-102°. $[\alpha]_D$ +27.2° (c, 0.934 in $CHCl_3$).
(2S,1'S)-form [39542-12-6]
(3S,4S)-*form*
Mp 81-82°. $[\alpha]_D$ +30.6° (c, 0.66 in $CHCl_3$).
(2RS,1'RS)-form [55029-18-0]
(3RS,4RS)-*form*
Cryst. (H_2O). Mp 83-84°.
(2RS,1'SR)-form [55029-19-1]
(3RS,4SR)-*form*
Cryst. (H_2O). Mp 135-136°.

[39497-75-1]

Pini, D. *et al, Gazz. Chim. Ital.,* 1974, **104**, 1295.

6-Methyl-4(1H)-pyridazinone, 9CI   M-80151

6-Methyl-4-pyridazinol. 4-Hydroxy-6-methylpyridazine
[17417-58-2]

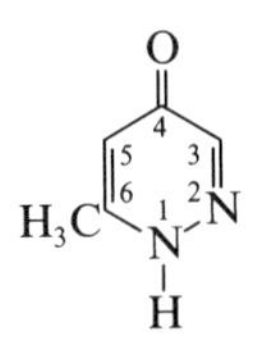

$C_5H_6N_2O$ M 110.115
Needles (EtOH), flakes (H_2O). Mp 238° (231-233°).

[22390-44-9]

Itsuo, I. *et al, Agric. Biol. Chem.,* 1967, **31**, 979 (*synth, uv, pmr*)
Becker, H.G.O. *et al, J. Prakt. Chem.,* 1969, **311**, 286 (*synth, uv, ir*)

5-Methyl-3-pyridinecarboxaldehyde   M-80152

5-Methylnicotinaldehyde
[100910-66-5]

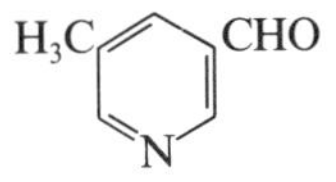

C_7H_7NO M 121.138
Low melting solid. Mp 37°. Bp_{20} 130°, $Bp_{0.5}$ 68°.
Semicarbazone: Cryst. (H_2O). Mp 244°.
Phenylhydrazone: Yellow cryst. (MeOH). Mp 145°.
Oxime:
$C_7H_8N_2O$ M 136.153
Cryst. (H_2O). Mp 153°.

Mathes, W. *et al, Chem. Ber.,* 1960, **93**, 286 (*synth, deriv*)
Kyba, E.P. *et al, J. Org. Chem.,* 1988, **53**, 3513 (*synth, pmr, ir*)

5-Methyl-2-pyridinecarboxylic acid, 9CI   M-80153

Updated Entry replacing M-03479
5-Methylpicolinic acid
[4434-13-3]
$C_7H_7NO_2$ M 137.138
Cryst. (C_6H_6). Mp 167-168°.
▷ TJ7556000.
1-Oxide: [31283-69-9].
$C_7H_7NO_3$ M 153.137
Cryst. (MeOH). Mp 160-161°.
Me ester: [29681-38-7].
$C_8H_9NO_2$ M 151.165
Cryst. (pet. ether). Mp 54-55°.
Amide: [20970-77-8].
$C_7H_8N_2O$ M 136.153
Mp 178°.
Nitrile: 2-Cyano-5-methylpyridine
$C_7H_6N_2$ M 118.138
Cryst. by subl. Mp 70°.

Ferles, M. *et al, Collect. Czech. Chem. Commun.,* 1968, **33**, 3848 (*synth, amide, nitrile*)
Brzezinski, B. *et al, Bull. Acad. Pol. Sci., Ser. Sci. Chim.,* 1970, **18**, 247 (*ir*)
Deady, L.W. *et al, Aust. J. Chem.,* 1971, **24**, 385 (*synth*)
Brzezinski, B. *et al, Pol. J. Chem. (Rocz. Chem.),* 1973, **47**, 445 (*pmr*)
Deady, L.W. *et al, Org. Magn. Reson.,* 1975, **7**, 41 (*nmr*)
Dormagen, W. *et al, Synthesis,* 1988, 636 (*amide, synth, pmr, cnr*)

5-Methyl-3-pyridinecarboxylic acid, 9CI   M-80154

Updated Entry replacing M-03480
5-Methylnicotinic acid. β-Picoline-5-carboxylic acid
[3222-49-9]
$C_7H_7NO_2$ M 137.138
Cryst. (EtOH). Mp 215-216° (208-212°). pK_a 5.2.
Me ester: [29681-45-6].
$C_8H_9NO_2$ M 151.165
Cryst. (hexane). Mp 45-46°.
Amide: [70-57-5].
$C_7H_8N_2O$ M 136.153
Cryst. (EtOAc). Mp 163-165°.
Nitrile: [42885-14-3]. *3-Cyano-5-methylpyridine*
$C_7H_6N_2$ M 118.138
Cryst. (EtOH/pet. ether). Mp 83-84°.

McElvain, S.M. *et al, J. Am. Chem. Soc.,* 1943, **65**, 2233 (*synth*)
Deady, L.W. *et al, Aust. J. Chem.,* 1971, **24**, 385 (*synth*)

Deady, L.W. *et al, Org. Magn. Reson.*, 1975, **7**, 41 (*nmr*)
Kyba, E.P. *et al, J. Org. Chem.*, 1988, **53**, 3513 (*ester, amide, pmr, ir*)

3-Methyl-2,5-pyrrolidinedione, 9CI   M-80155
2-Methylsuccinimide

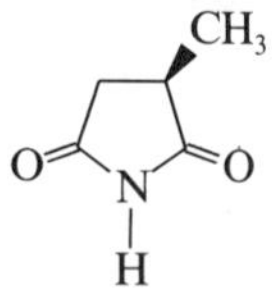

(R)-form

$C_5H_7NO_2$ M 113.116
(R)-form [117307-07-0]
   Cryst. (toluene). Mp 78°. $[\alpha]_D^{20}$ +29.4° (c, 2 in $CHCl_3$).
(±)-form
   Mp 62°.

Turner, D.W., *J. Chem. Soc.*, 1957, 4555.
Poloński, T., *J. Chem. Soc., Perkin Trans.* 1, 1988, 629 (*synth, ir, pmr, cd*)

2-Methyl-1H-pyrrolo[2,3-b]pyridine   M-80156
2-Methyl-7-azaindole
[23612-48-8]

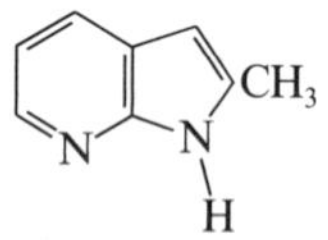

$C_8H_8N_2$ M 132.165
Cryst. (Et_2O). Mp 142° (134-135°).
Picrate: Pale-yellow needles (Me_2CO). Mp 229°.
N-*Benzoyl:*
   $C_{15}H_{12}N_2O$ M 236.273
   Mp 94-95°.

Clemo, G.R. *et al, J. Chem. Soc.*, 1945, 603 (*synth*)
Herbert, R. *et al, J. Chem. Soc. C*, 1969, 1505 (*synth, uv, pmr*)
Estel, L. *et al, J. Org. Chem.*, 1988, **53**, 2740 (*synth, ir, pmr*)

2-Methyl-1H-pyrrolo[3,2-b]pyridine   M-80157
2-Methyl-4-azaindole

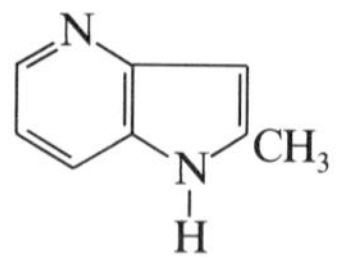

$C_8H_8N_2$ M 132.165
Prisms(C_6H_6). Mp 193-194°.
Picrate: Prisms (Me_2CO). Mp 262° dec.

Clemo, G.R. *et al, J. Chem. Soc.*, 1948, 198 (*synth*)

2-Methyl-1H-pyrrolo[2,3-c]pyridine   M-80158
2-Methyl-6-azaindole
[65645-56-9]

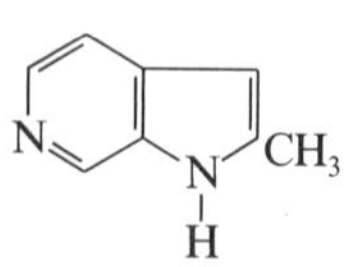

$C_8H_8N_2$ M 132.165
Cryst. by subl. Mp 185°.

B,HCl: Mp 225°.
Picrate: Orange needles. Mp 193°.

Koenigs, E. *et al, Ber.*, 1927, **60**, 2106 (*synth*)
Estel, L. *et al, J. Org. Chem.*, 1988, **53**, 2740 (*synth, pmr, ir*)

2-Methyl-1H-pyrrolo[3,2-c]pyridine   M-80159
2-Methyl-5-azaindole
[113975-37-4]

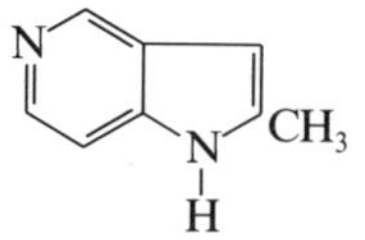

$C_8H_8N_2$ M 132.165
Cryst. by subl. Mp 210°.
Picrate: Bright yellow needles. Mp 213-214° (sinters at 210°).

Clemo, G.R. *et al, J. Chem. Soc.*, 1948, 198 (*synth*)
Estel, L. *et al, J. Org. Chem.*, 1988, **53**, 2740 (*synth, pmr, ir*)

2-Methyl-4,5,8-quinazolinetrione   M-80160
[117498-11-0]

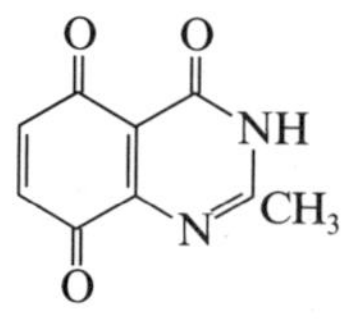

$C_9H_6N_2O_3$ M 190.158
Cryst. Mp 165-168° dec.

Lemus, R.H. *et al, J. Org. Chem.*, 1988, **53**, 6099 (*synth, ir, pmr*)

5-[(5-Methyl-2-thienyl)methylene]-2(5H)-furanone, 9CI   M-80161
5-(5-Methyl-2-thienyl)-2,4-pentadien-4-olide

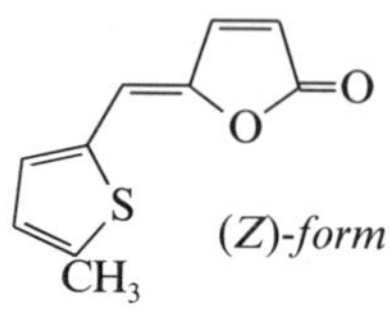

(Z)-form

$C_{10}H_8O_2S$ M 192.238
(E)-form [35304-91-7]
   Needles. Mp 105° approx. Appears to isomerise on heating.
(Z)-form [5705-62-4]
   Minor constit. of the roots of *Chamaemelum nobile*. Yellow cryst. (pet. ether). Mp 117° (113-115°).

Bohlmann, F. *et al, Chem. Ber.*, 1966, **99**, 1226 (*isol, uv, ir, pmr, struct, synth*)
Yamada, K. *et al, Tetrahedron*, 1971, **27**, 5445 (*synth, uv, pmr*)

5-(5-Methyl-2-thienyl)-2-penten-4-ynoic acid   M-80162

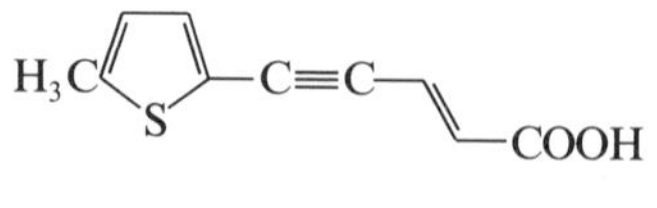

(E)-form

$C_{10}H_8O_2S$ M 192.238
(E)-form
   Me ester: Yellowish cryst. (pet. ether). Mp 49°.

(*Z*)-*form* [23050-78-4]

> *Me ester:*
> $C_{11}H_{10}O_2S$ M 206.265
> Isol. from roots of *Artemisia vulgaris, Anthemis nobilis* and *Anacyclus radiatus*. Light-yellow oil. λ_{max} 338 mm (Et₂O).

> Bohlmann, F. *et al, Chem. Ber.*, 1962, **95**, 1733; 1963, **96**, 584, 588 (*isol, uv, ir, struct, synth*)
> Schulte, K.E., *Phytochemistry*, 1966, **5**, 949 (*biosynth*)

4-Methylthio-2,4-decadiene-6,8-diynoic acid   M-80163

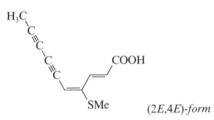

(2*E*,4*E*)-*form*

$C_{11}H_{10}O_2S$ M 206.265

(2*E*,4*E*)-*form*
trans,cis-*form*

> *Me ester:* [13030-72-3].
> $C_{12}H_{12}O_2S$ M 220.292
> Constit. of roots of *Anthemis tinctoria* and *A. cretica*. Yellow cryst. (Et₂O/pet. ether). Mp 112°.

(2*Z*,4*Z*)-*form*
cis,trans-*form*

> *Me ester:* [13030-70-1].
> Isol. from roots of *A. tinctoria*. Yellow needles (pet. ether). Mp 70.5°.

(2*Z*,4*E*)-*form*
cis,cis-*form*

> *Me ester:* [13030-69-8].
> Isol. from roots of *A. tinctoria*. Amorph.

(2*E*,4*Z*)-*form*
trans,trans-*form*

> *Me ester:* [13030-68-7].
> Not obt. naturally. Mp 75°.

[57110-47-1, 57110-48-2, 77915-50-5]

> Bohlmann, F. *et al, Chem. Ber.*, 1966, **99**, 2096; 1968, **101**, 2506; 1975, **108**, 1902 (*isol, uv, ir, struct, synth*)

5-Methylthio-2,4-decadiene-6,8-diynoic acid   M-80164

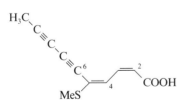

(2*Z*,4*E*)-*form*

$C_{11}H_{10}O_2S$ M 206.265

(2*Z*,4*E*)-*form*
cis,cis-*form*

> *Me ester:*
> $C_{12}H_{12}O_2S$ M 220.292
> Isol. from roots of *Anthemis carpatica, A. ruthenica* and other *A*. sp. Light yellow needles (pet. ether). Mp 40°.

> *Me ester, S-dioxide:* Needles (Et₂O/pet. ether). Mp 96°.

> 6,7-(*Z*)-*dihydro, Me ester:* 5-*Methylthio-2Z,4E,6Z-decatetrien-8-ynoic acid methyl ester*
> $C_{12}H_{14}O_2S$ M 222.307
> Isol. from roots of *Anthemis tinctoria* and other *A*. sp. Liq. Bp₀.₀₀₁ 120°.

(2*Z*,4*Z*)-*form*
cis,trans-*form*

> *Me ester:* Isol. from roots of *A. ruthenica*.

> *Me ester, S-dioxide:* Cryst. (CHCl₃). Mp 86°.

> Bohlmann, F. *et al, Chem. Ber.*, 1963, **96**, 1485; 1965, **98**, 1616; 1966, **99**, 2096 (*isol, uv, ir, pmr, struct*)

7-Methylthio-2,6-decadiene-4,8-diynoic acid   M-80165

[17159-29-4]

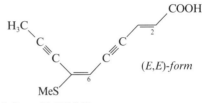

(*E,E*)-*form*

$C_{11}H_{10}O_2S$ M 206.265

(*E,E*)-*form*
trans,cis-*form*

> *Me ester:* [2908-10-3].
> $C_{12}H_{12}O_2S$ M 220.292
> Isol. from roots of *Anthemis* spp. Yellow cryst. (pet. ether). Mp 46°.

(2*E*,6*Z*)-*form*
cis,cis-*form*

> *Me ester:* Isol. from roots of *A. fuscata*.

> *Me ester, S-dioxide:* Mp 99°.

(2*Z*,6*E*)-*form*
cis,cis-*form*

> *Me ester:* [3102-40-7].
> Isol. from roots of *A. brachycentros, A. tinctoria* and *A. ruthenica*.

> *Me ester, S-dioxide:* Cryst. (Et₂O/pet. ether). Mp 58°.

(2*Z*,6*Z*)-*form*
cis,trans-*form*

> *Me ester:* Isol. from various *A.* spp.

> *Me ester, S-dioxide:* Cryst. (Et₂O/pet. ether). Mp 65°.

> Bohlmann, F. *et al, Chem. Ber.*, 1962, **95**, 1320; 1963, **96**, 1485; 1965, **98**, 1616; 1966, **99**, 2096 (*isol, uv, ir, pmr, struct*)

9-Methylthio-2,8-decadiene-4,6-diynoic   M-80167
acid

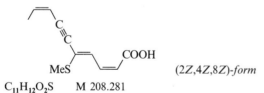

(2E,8E)-form

C₁₁H₁₀O₂S   M 206.265

(2E,8E)-form
  trans,cis-*form*

Me ester:
  C₁₂H₁₂O₂S   M 220.292
  Isol. from *Anthemis tinctoria.* Cryst. (Et₂O/pet. ether).
  Mp 75°.

Me ester, S-dioxide: Cryst. (Et₂O/pet. ether). Mp 119-120°.
(2Z,8E)-form
  cis,cis-*form*

Me ester: Isol. from roots of *A. tinctoria, A. nigrescens*
  and *A. helwegii.* Cryst. (Et₂O/pet. ether). Mp 40.5°.
Me ester, S-dioxide: Cryst. (Et₂O/pet. ether). Mp 89.5°.
(2E,8Z)-form
  trans,trans-*form*

Me ester: Isol. from *A. tinctoria.* Cryst. (Et₂O/pet. ether).
  Mp 43°.
Me ester, S-dioxide: Cryst. (Et₂O/pet. ether). Mp 101°.
(2Z,8Z)-form
  cis,trans-*form*

Me ester: Isol. from *A. tinctoria* and roots of *A.*
  triumfettii. Cryst. (Et₂O/pet. ether). Mp 41°.

Me ether, S-dioxide: Cryst. (Et₂O/pet. ether). Mp 129.5°.

Bohlmann, E. *et al, Chem. Ber.,* 1963, **96,** 1485; 1965, **98,** 1616,
  1736; 1966, **99,** 2096 (*isol, uv, ir, pmr, synth, biosynth*)

5-Methylthio-2,4,8-decatrien-6-ynoic acid   M-80168

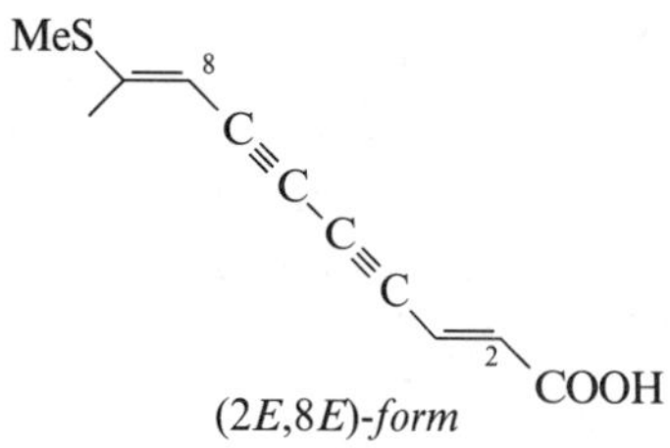

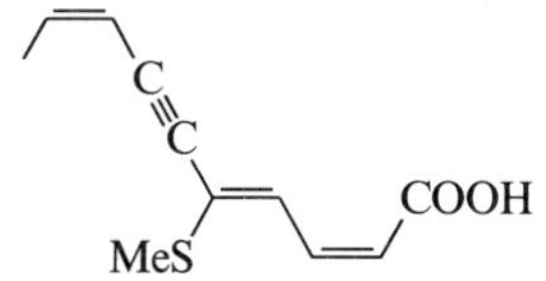

C₁₁H₁₂O₂S   M 208.281

(2Z,4Z,8Z)-*form*

(2Z,4Z,8Z)-form
  cis, trans, cis-*form*

Me ester: [13103-21-4].
  C₁₂H₁₄O₂S   M 222.307
  Isol. from roots of *Anthemis austriaca.*

Me ester, S,S-dioxide: Cryst. (Et₂O/pet. ether). Mp 63-64°.
(2Z,4E,8Z)-form
  cis,cis,cis-*form*

Me ester: [13103-22-5].
  Isol. from roots of *A. austriaca.*

Me ester, S,S-dioxide: Oil.

Bohlmann, F. *et al, Justus Liebigs Ann. Chem.,* 1964, **694,** 149
  (*isol, struct*)
Bohlmann, F. *et al, Chem. Ber.,* 1967, **100,** 1200 (*synth*)

[1-(Methylthio)ethenyl]benzene, 9CI   M-80169
  α-*(Methylthio)styrene. 1-(Methylthio)-1-phenylethylene*
  [18624-64-1]

H₂C=CPhSMe

C₉H₁₀S   M 150.244
Oil.

Lecadet, D. *et al, Compt. Rend. Hebd. Seances Acad. Sci. Sect. C,*
  1973, 875 (*synth*)
Leger, L. *et al, Bull. Soc. Chim. Fr.,* 1975, 657 (*pmr*)
Schaumann, E. *et al, Chem. Ber.,* 1988, **121,** 1159.

[2-(Methylthio)ethenyl]benzene, 9CI   M-80170
  β-*(Methylthio)styrene. Methyl styryl sulfide. 1-(Methylthio)-*
  2-phenylethylene
  [7715-02-8]

PhCH=CHSMe

C₉H₁₀S   M 150.244
(E)-form [15436-06-3]
  Oil. Bp₃₅ 140-142°, Bp₁₂ 119°. n_D^{20} 1.6400.

S-Dioxide: [15436-11-0]. [2-(Methylsulfonyl)ethenyl]benzene,
  9CI. β-(Methylsulfonyl)styrene. Methyl styryl sulfone. 1-
  (Methylsulfonyl)-2-phenylethylene
  C₉H₁₀O₂S   M 182.243
  Solid. Mp 76-78°.
(Z)-form [35822-50-5]
  Oil. Bp₁₇ 130-140°. n_D^{20} 1.6251.

S-Dioxide: [37630-43-6].
  Solid. Mp 64.5-66°.

Truce, W.E. *et al, J. Am. Chem. Soc.,* 1956, **78,** 695 (*synth*)
Wittig, G. *et al, Chem. Ber.,* 1961, **94,** 1373 (*synth*)
Caserio, M.C., *J. Am. Chem. Soc.,* 1966, **88,** 5747 (*synth, uv, pmr*)
Bock, H. *et al, Angew. Chem., Int. Ed. Engl.,* 1972, **11,** 150 (*pe*)
Petrov, M.L. *et al, J. Org. Chem. USSR (Engl. Transl.),* 1978, 197
  (*synth, ir, pmr*)
Ogura, K. *et al, Bull. Chem. Soc. Jpn.,* 1982, **55,** 3669 (*synth, ir,
  pmr*)
Oida, T. *et al, Bull. Chem. Soc. Jpn.,* 1983, **56,** 959 (*synth, ms*)
Tiecco, M. *et al, J. Org. Chem.,* 1983, **48,** 4795 (*synth, deriv*)
Oida, T. *et al, J. Chem. Soc., Perkin Trans. 1,* 1986, 1715 (*synth,
  pmr*)
Fugami, K. *et al, Tetrahedron,* 1988, **44,** 4277 (*synth*)

5-(3-Methylthio-2-hexen-4-ynylidene)-2(5H)-furanone   M-80171

7-Methylthio-2,4,6-decatrien-8-yn-4-olide

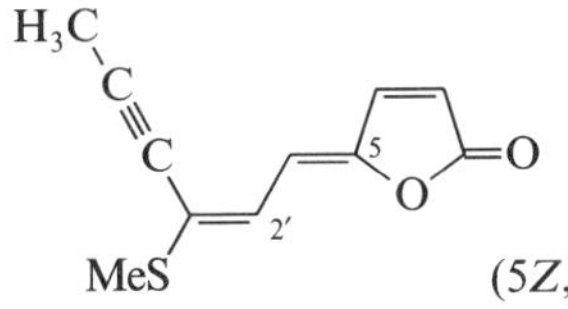

$C_{11}H_{10}O_2S$ M 206.265

(5Z,2'E)-form

trans,cis-*form*
Isol. from roots of *Anthemis kelwayi* and *A. rigescens*.
Yellow needles (pet. ether). Mp 67°. Not obt. fully pure.

(5Z, 2'Z)-form

trans,trans-*form*
Isol. from *A. rigescens* and *A. kelwayi*. Yellow platelets
(pet. ether). Mp 102°.

Bohlmann, F. *et al, Chem. Ber.*, 1965, **98**, 1616 (*isol, uv, ir, pmr, struct*)

5-(5-Methylthio-4-hexen-2-ynylidene)-2(5H)-furanone   M-80172

9-Methylthio-2,4,8-decatriene-6-yn-4-olide. 4-Hydroxy-9-(methylthio)-2,4,8-decatrien-6-ynoic acid γ-lactone

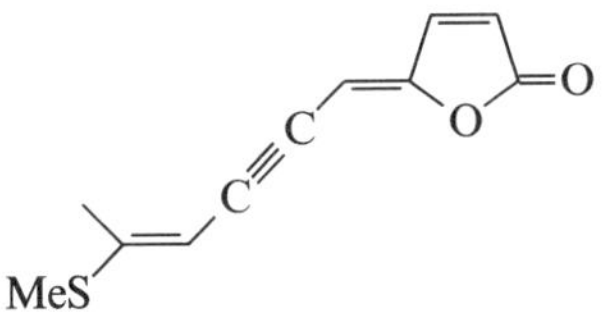

$C_{11}H_{10}O_2S$ M 206.265
Isol. from roots of *Anthemis rigescens*.

S-Dioxide: Yellowish cryst. (Et₂O). Mp 118.5°.

Bohlmann, F. *et al, Chem. Ber.*, 1965, **98**, 1616; 1966, **99**, 3437
(*isol, uv, ir, pmr, struct, synth*)

(Methylthio)oxirane   M-80173

(Mercaptomethyl)oxirane. 1,2-Epoxy-3-mercaptopropane.
2,3-Epoxy-1-propanethiol
[67757-42-0]

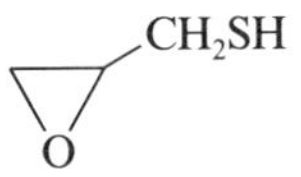

C_3H_6OS M 90.146

(±)-form

S-Ph: [5296-21-9]. [(*Phenylthio*)*methyl*]*oxirane, 9CI.* 1-
Phenylthio-2,3-epoxypropane
$C_9H_{10}OS$ M 166.243
Oil. Bp₀.₂ 82-85°.

Apparao, S. *et al, Synthesis*, 1987, 896 (*synth, pmr*)

6-[4-(Methylthio)-1,2,3-pentatrienyl]-2H-pyran-2-one, 9CI   M-80174

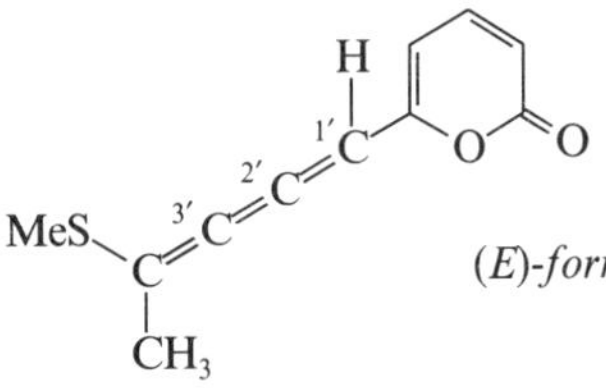

$C_{11}H_{10}O_2S$ M 206.265

(E)-form [52061-38-8]
Isol. from fresh roots of *Anthemis austriaca*. Orange-red
needles (Et₂O/pet. ether). Mp 108.5-109.5°.

1',2'-Didehydro, 2',3'-dihydro (E-): [20903-34-8]. **6-(4-**
Methylthio-3-penten-1-ynyl)-2H-pyran-2-one
Yellow cryst. (CCl₄). Mp 85°.

(Z)-form [20903-32-6]
Isol. from *A. austriaca*. λ_{max} 410 nm.

1',2'-Didehydro, 2',3'-dihydro (Z-): [20903-31-5].
$C_{11}H_{10}O_2S$ M 206.265
Isol. from roots of *A. austriaca*. Yellow cryst. (Et₂O/pet.
ether). Mp 68-69°.

Bohlmann, F. *et al, Justus Liebigs Ann. Chem.*, 1966, **694**, 149
(*isol, uv, ir, pmr, ms, struct*)
Bohlmann, F. *et al, Chem. Ber.*, 1968, **101**, 3562; 1973, **106**, 3772
(*synth, biosynth*)

2-(5-Methylthio-4-penten-2-ynylidene)-1,6-dioxaspiro[4.4]non-3-ene, 9CI   M-80175

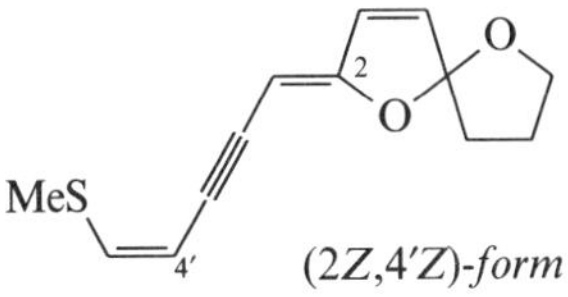

$C_{13}H_{14}O_2S$ M 234.318

(2Z,4'Z)-form
Isol. from above-ground parts of *Chrysanthemum coronarium*. Oil.

S,S-Dioxide: Cryst. (Et₂O). Mp 98.5°. $[\alpha]_{546}^{20}$ −6.1° (c, 3.3 in CHCl₃).

(2Z,4'E)-form
Isol. from *C. coronarium*. Yellowish cryst. (pet. ether).
Mp 71°. $[\alpha]_{546}^{20}$ +1.2° (c, 7.3 in CHCl₃).

S-Oxide: **2-[(5-Methylsulfinyl)-4-penten-2-ynylidene]-1,6-**
dioxaspiro[4.4]non-3-ene
$C_{13}H_{14}O_3S$ M 250.318
Isol. from *C. coronarium*. λ_{max} 338 nm (Et₂O).

Bohlmann, F. *et al, Chem. Ber.*, 1964, **97**, 1179; 1965, **98**, 2605
(*isol, uv, ir, pmr, struct, synth*)

3-Methyl-2-thiophenemethanol, 9CI   M-80176

2-Hydroxymethyl-3-methylthiophene
[63826-56-2]

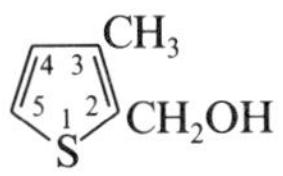

C_6H_8OS M 128.195
Liq. Bp₁.₅ 70-74°. n_D^{25} 1.5524.

α-Naphthylurethane: Glistening plates. Mp 144°.

Campaigne, E. *et al, J. Org. Chem.*, 1963, **28**, 914.

4-Methyl-2-thiophenemethanol, 9CI   M-80177

2-Hydroxymethyl-4-methylthiophene

[74395-18-9]

C_6H_8OS M 128.195

Liq. Bp_1 78-79°. n_D^{20} 1.5512.

Lozanova, A.V. *et al, Izv. Akad. Sci. USSR, Ser. Sci. Khim.*, 1980, 958; 1981, 838; *ca*, **93**, 70917y; **95**, 80025m (*synth, ir, uv, pmr*)

5-Methyl-2-thiophenemethanol, 9CI   M-80178

2-Hydroxymethyl-5-methylthiophene

[63826-59-5]

C_6H_8OS M 128.195

Liq. Bp_3 97-100°. n_D^{20} 1.5090.

Kharchenko, V.G. *et al, J. Org. Chem. USSR (Engl. Transl.)*, 1982, 343 (*synth*)

Misharina, T.A. *et al, Zh. Anal. Khim.*, 1986, **41**, 1876; *ca*, **106**, 131010h (*ms, glc*)

3-[(4-Methylthio)phenyl]-2-butenoic acid   M-80179

p-*Methylthio-β-methylcinnamic acid*

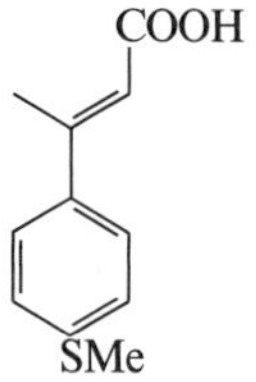

$C_{11}H_{12}O_2S$ M 208.281

(E)-form

Me ester:

$C_{12}H_{14}O_2S$ M 222.307

Isol. from roots of *Anthemis tinctoria*.

Me ester, S-Dioxide: Cryst. (Et_2O/pet. ether).

Bohlmann, F. *et al, Chem. Ber.*, 1966, **99**, 1834 (*isol, biosynth*)

1-Methylthio-5-phenyl-1-penten-3-yne   M-80180

[5-(*Methylthio)-4-penten-2-ynyl*]benzene, 9CI

$PhCH_2C≡CCH=CHSMe$

$C_{12}H_{12}S$ M 188.293

(Z)-form

Isol. from roots of *Chrysanthemum segetum* and *C. coronarium*. Light-yellow oil. $Bp_{0.1}$ 110°. λ_{max} 274 (ε18400), 281 nm (18400)(Et_2O).

Bohlmann, F. *et al, Chem. Ber.*, 1963, **96**, 226; 1964, **97**, 1179 (*isol, uv, ir, pmr, struct*)

5-(Methylthio)-1-phenyl-4-penten-2-yn-1-one, 9CI   M-80181

1-Methylthio-5-oxo-5-phenyl-1-penten-3-yne

$PhCOC≡CCH=CHSMe$

$C_{12}H_{10}OS$ M 202.276

(E)-form [54668-03-0]

Isol. from roots of *Chrysanthemum segetum*. Oil. λ_{max} 258(ε 14400), 336.5 nm (20,000) (Et_2O).

(Z)-form [54668-02-9]

Isol. from roots of *C. segetum* and *C. coronarium*. Cryst. Mp 42.5°. λ_{max} 259(ε 15200), 350(13600) nm (Et_2O).

Bohlmann, F. *et al, Chem. Ber.*, 1963, **96**, 226, 584; 1964, **97**, 809, 1179 (*isol, uv, ir, pmr, struct, synth, biosynth*)

Mikhelashvili, I.L. *et al, Zh. Org. Khim.*, 1974, **10**, 2524 (*synth*)

2-[2-(Methylthio)propenyl]benzoic acid   M-80182

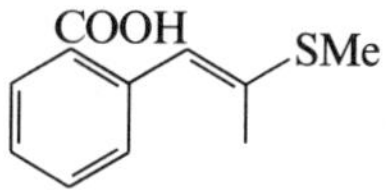

$C_{11}H_{12}O_2S$ M 208.281

(E)-form

cis-*form*

Me ester:

$C_{12}H_{14}O_2S$ M 222.307

Isol. from roots of *Anthemis tinctoria*. $Bp_{0.01}$ 64-74°.

Me ester, S-dioxide: Cryst. (Et_2O/pet. ether). Mp 62°.

(Z)-form

trans-*form*

Me ester: Isol. from roots of *A. tinctoria*. $Bp_{0.1}$ 81-84°.

Me ester, S-dioxide: Cryst. (Et_2O/pet. ether). Mp 77.5-78°.

Bohlmann, F. *et al, Chem. Ber.*, 1963, **96**, 1485; 1965, **98**, 3087 (*isol, uv, ir, pmr, struct, derivs*)

4-[2-(Methylthio)propenyl]benzoic acid   M-80183

$C_{11}H_{12}O_2S$ M 208.281

Me ester: Isol. from roots of *Anthemis tinctoria*. Cryst. (pet. ether). Mp 54°.

Me ester, S-dioxide: Cryst. (Et_2O/pet. ether). Mp 101°.

Bohlmann, F. *et al, Chem. Ber.*, 1963, **96**, 1485; 1965, **98**, 3087 (*isol, uv, ir, pmr, synth*)

8-Methylthio-1,7-tridecadiene-3,5,9,11-tetrayne   M-80184

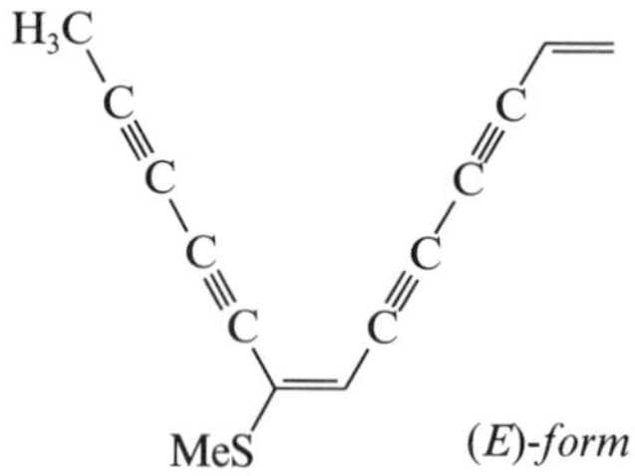

$C_{14}H_{10}S$ M 210.299

(E)-form

cis-*form*

Isol. from roots of *Flaveria repanda* and *Guzotia abyssinica*. Cryst. (pet. ether). Mp 80° dec.

(Z)-form

trans-*form*

Isol. from roots of *F. repanda* and *G. abyssinica*. Cryst. (Et_2O/pet. ether). Mp 75° dec.

Bohlmann, F. *et al, Chem. Ber.*, 1963, **96**, 1229; 1965, **98**, 876, 3015 (*isol, uv, ir, pmr, biosynth, synth*)

3-Methyltricosane, 9CI   M-80185

[13410-45-2]

$H_3C(CH_2)_{19}CH(CH_3)CH_2CH_3$

$C_{24}H_{50}$ M 338.659

Isol. from hop oil (*Humulus lupulus*) and orange oil (*Citrus sinensis*) and from insect cuticles. Bp_{10} 233°. n_D^{50} 1.4375.

[55194-20-2]

Wollrab, V. *et al*, *Collect. Czech. Chem. Commun.*, 1965, **30**, 1670 (*isol*)
Hunter, G.L.K. *et al*, *Phytochemistry*, 1966, **5**, 807 (*isol*)
Streibl,, M. *et al*, *Fette, Seifen, Anstrichm.*, 1968, **70**, 543 (*synth*)

22-Methyl-22-tritetracontanol   M-80186

$$[H_3C(CH_2)_{20}]_2C(OH)CH_3$$

$C_{44}H_{90}O$ M 635.194
Isol. from skin of *Erodium cicutarium*. Mp 65.3-65.6°.

Becker, H.J. *et al*, *Recl. Trav. Chim. Pays-Bas* (*J. R. Neth. Chem. Soc.*), 1940, **59**, 933 (*synth*)
van Eijk, J.L., *CA*, 1952, **46**, 9262 (*isol*)

β-Methyltryptophan, 9CI   M-80187

Updated Entry replacing M-03847
2-Amino-3-(3-indolyl)butanoic acid

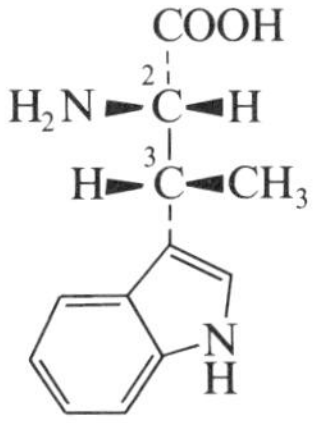

(*2S,3R*)-*form*

$C_{12}H_{14}N_2O_2$ M 218.255
(2S,3R)-form
The rel. config. was detd. by Turchin *et al* and the abs. config. follows from the known2S-centre present in Telomycin.
(2RS,3RS)-form
Isomer B
Cryst. (H$_2$O). Mp 247-251° dec.
Me ester:
$C_{13}H_{16}N_2O_2$ M 232.282
Mp 115-118°.
(2RS,3SR)-form
Isomer A
Cryst. (H$_2$O). Mp 248-250° dec. Cryst. very slowly (several months).
N-*Ac:*
$C_{14}H_{16}N_2O_3$ M 260.292
Cryst. (DMF aq. or EtOH/EtOAc). Mp 216-218°.
Me ester: Mp 121-122°.

Snyder, H.R. *et al*, *J. Am. Chem. Soc.*, 1957, **79**, 2217 (*synth*)
Sheehan, J.C. *et al*, *J. Am. Chem. Soc.*, 1963, **85**, 2867 (*isol*)
Turchin, K.F. *et al*, *J. Org. Chem.*, USSR (*Engl. transl.*), 1971, **7**, 1329 (*config*)
Behforouz, M. *et al*, *J. Heterocycl. Chem.*, 1989, **25**, 1627 (*synth*)

10-Methylundecanoic acid   M-80188

Isododecanoic acid
[2724-56-3]

$$(H_3C)_2CH(CH_2)_8COOH$$

$C_{12}H_{24}O_2$ M 200.320
Isol. from wool fat, from lipids of *Sarcina lutea* and resin of *Shorea maranti*. Tablets (Me$_2$CO or C$_6$H$_6$). Mp 41.2°. Bp_3 140-145°.

[5129-56-6]

Weitkamp, A.W., *J. Am. Chem. Soc.*, 1945, **67**, 447 (*isol*)
Fieser, L.F. *et al*, *J. Am. Chem. Soc.*, 1948, **70**, 3174 (*synth*)
Cocker, W. *et al*, *Perfum. Essent. Oil Rec.*, 1964, **55**, 442 (*isol*)

3-Methyl-1-vinylcyclopentene   M-80189

1-Ethenyl-3-methylcyclopentene, 9CI
[114614-93-6]

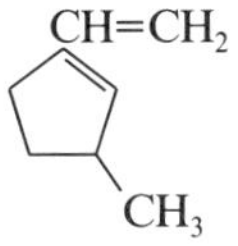

C_8H_{12} M 108.183
(±)-*form*
Oil.

Fisher, M.J. *et al*, *J. Am. Chem. Soc.*, 1988, **110**, 4625 (*synth, pmr, ir, ms*)

Micropubescin   M-80190

Updated Entry replacing M-40154
[23560-52-3]

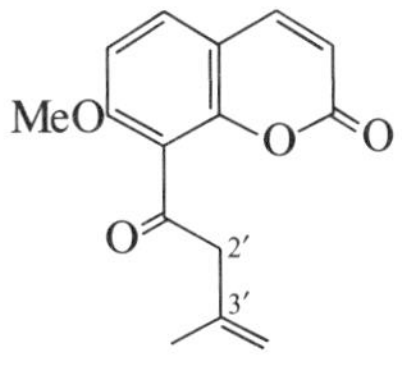

$C_{15}H_{14}O_4$ M 258.273
Constit. of *Micromelum pubescens*. Cryst. Mp 130-131°.
$\Delta^{2',3'}$-*Isomer: 8-(3-Methyl-2-butenoyl)coumarin*
$C_{15}H_{14}O_4$ M 258.273
Constit. of *Ligusticum hultenii*. Mp 131-132°.

Chatterjee, A. *et al*, *Sci. Cult.*, 1967, **33**, 371; 1968, **34**, 366; *CA*, **69**, 2887; **70**, 114946 (*isol*)
Murray, R.D.H. *et al*, *Tetrahedron Lett.*, 1977, 3077 (*synth*)

Millettone   M-80191

Updated Entry replacing M-03960
[50376-38-0]

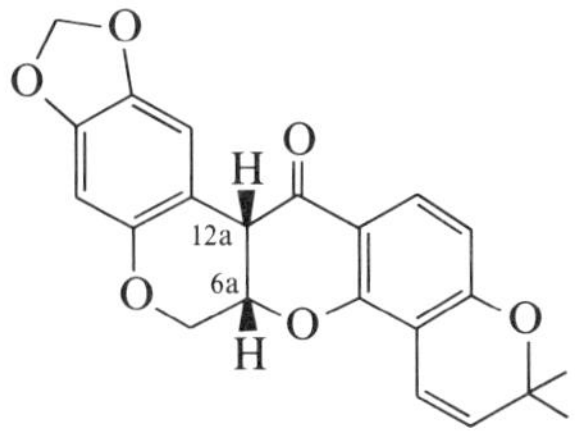

Absolute configuration

$C_{22}H_{18}O_6$ M 378.381
natural-form
Constit. of *Millettia dura*, *Piscida erythrina* and *Lonchocarpus rugosus*. Cream-coloured needles (MeOH). Mp 180-181°. λ_{max} (EtOH) 243 (log ϵ 4.27), 271 (4.35) and 305 nm (4.10).
6a,12a-Didehydro: Dehydromillettone
$C_{22}H_{16}O_6$ M 376.365
Isol. from root of *P. erythrina*. Yellow. Mp 358° dec.
(±)-*form*
Cryst. (C$_6$H$_6$/hexane). Mp 245-246°.

Falshaw, C.P. *et al*, *Tetrahedron, Suppl.* 7, 1966, 333 (*isol*)
Ollis, W.D. *et al*, *Tetrahedron*, 1967, **23**, 4741 (*struct, uv, pmr, ord, cd*)
Fujita, F. *et al*, *Agric. Biol. Chem.*, 1973, **37**, 1737 (*synth*)

Mitorubrinic acid   M-80192

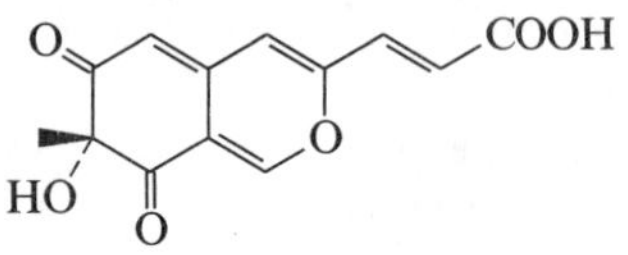

(R)-form

$C_{21}H_{16}O_9$ M 412.352

(R)-form

Isol. from *Penicillium funiculosum*. Morphogenic substance inducing chlamydospore-like cells. Yellow cryst. Mp 222-225°. $[\alpha]_D^{22}$ −450° (c, 0.025 in EtOH).

(S)-form [58958-07-9]

Prod. by *Hypoxylon fragiforme*. Yellow microcryst. Mp 225-227° dec. $[\alpha]_D^{25}$ +500° (c, 0.4 in dioxan).

Locci, R. *et al, G. Microbiol.*, 1967, **15**, 93 (*isol*)
Chong, R. *et al, J. Chem. Soc. C*, 1971, 3571 (*synth*)
Steglich, W. *et al, Phytochemistry*, 1974, **13**, 2874 (*isol*)
Natsume, M. *et al, Agric. Biol. Chem.*, 1985, **49**, 2517 (*isol, abs config*)

2,3-Morpholinedione   M-80193

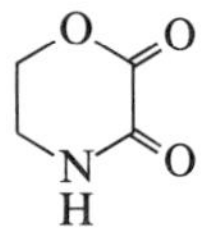

$C_4H_5NO_3$ M 115.088

N-Et: [73978-21-9].
$C_6H_9NO_3$ M 143.142
Cryst. Mp 70.0-70.5°.

Murahashi, S.I. *et al, J. Chem. Soc., Chem. Commun.*, 1987, 125.

Moskachan B   M-80194

Updated Entry replacing M-50261
6-(1,3-Benzodioxol-5-yl)-2-hexanone, 9CI. 6-(3,4-Methylenedioxyphenyl)-2-hexanone
[75787-96-1]

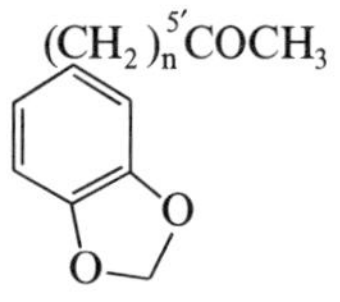

n = 4

$C_{13}H_{16}O_3$ M 220.268
Constit. of *Ruta angustifolia*. Oil.

5′-Alcohol: [105317-65-5]. **Moskachan C.** 6-(3,4-Methylenedioxyphenyl)-2-hexanol
$C_{13}H_{18}O_3$ M 222.283
From *R. angustifolia*. Oil.

5′-Deoxo, 5′,6′-didehydro: 6-(3,4-Methylenedioxylphenyl)-1-hexene. **Dehydromoskachan C**
$C_{13}H_{16}O_2$ M 204.268
Constit. of *R. chalepensis* var. *latifolia*. Oil. The name Dehydromoskachan C is misleading.

Borges Del Castillo, J. *et al, Phytochemistry*, 1986, **25**, 2209 (*isol*)
Ulubelen, A. *et al, J. Nat. Prod.* (*Lloydia*), 1988, **51**, 1012 (*isol*)

Mulberrofuran S   M-80195

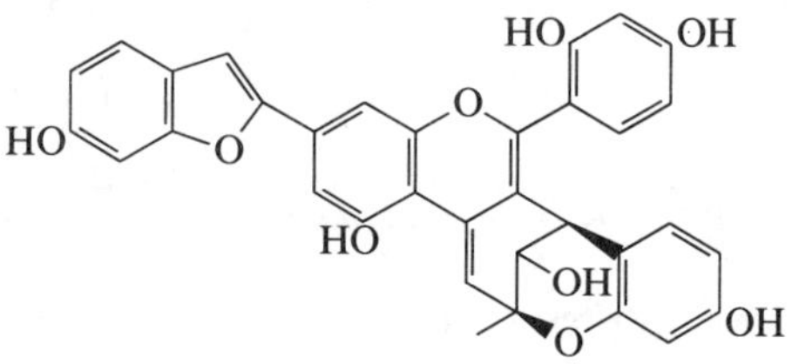

$C_{34}H_{24}O_9$ M 576.558
Isolated from the powder on the surface of *Morus alba* bark. Red amorph. powder. $[\alpha]_D^{23}$ +166° (c, 0.13 in EtOH).

Hano, Y. *et al, Heterocycles*, 1989, **28**, 697.

Munduserone   M-80196

Updated Entry replacing M-04093
6a,12a-Dihydro-2,3,9-trimethoxy[1]benzopyrano[3,4-b][1]benzopyran-12(6H)-one, 9CI

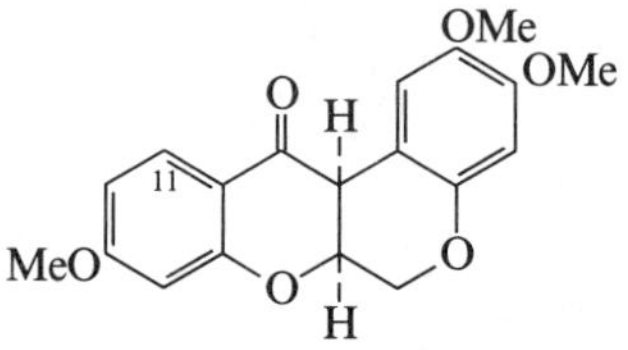

$C_{19}H_{18}O_6$ M 342.348

(+)-form [3564-85-0]

Constit. of the bark of *Mundulea sericea*. Mp 162°. $[\alpha]_D$ +103° (CHCl₃).

11-Hydroxy: [41743-42-4]. **Sermundone**
$C_{19}H_{18}O_7$ M 358.347
Isol. from root bark of *M. sericea*. Details apparently not publ.

(±)-form [19737-92-9]
Cryst. (C_6H_6/hexane). Mp 171.5-172°.

Finch, N. *et al, Proc. Chem. Soc., London*, 1960, 176 (*struct*)
Dyke, S.F. *et al, J. Chem. Soc. C*, 1966, 749 (*Sermundone*)
Chandrashekar, V. *et al, Tetrahedron*, 1967, **23**, 2505 (*synth*)
Nakatani, N. *et al, Agric. Biol. Chem.*, 1968, **32**, 769 (*synth*)
Omokawa, H. *et al, Agric. Biol. Chem.*, 1973, **37**, 1717 (*synth*)
Crombie, L. *et al, J. Chem. Soc., Perkin Trans.* 1, 1973, 1277 (*synth*)
Amos, P.C. *et al, J. Chem. Soc., Chem. Commun.*, 1987, 510 (*synth*)

Murracarpin   M-80197

[120786-75-6]

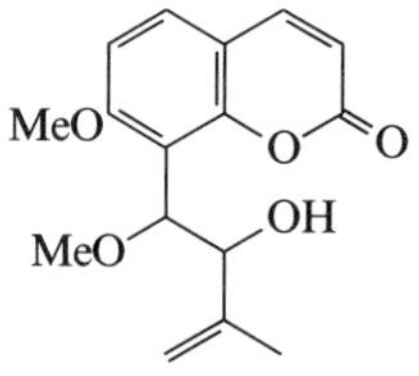

$C_{16}H_{18}O_5$ M 290.315
Constit. of *Murraya paniculata*. Granules (hexane). Mp 164-165°. $[\alpha]_D$ −15.6° (c, 0.068 in CHCl₃).

Wu, T.-S. *et al, Phytochemistry*, 1989, **28**, 293.

Murrayacarpin A — M-80198

8-(Hydroxymethyl)-7-methoxy-2H-1-benzopyran-2-one. 8-Hydroxymethyl-7-methoxycoumarin

[120693-43-8]

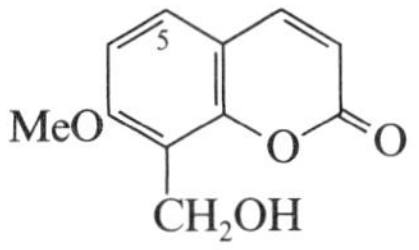

MeO — (structure) — O — O
CH₂OH

$C_{11}H_{10}O_4$ M 206.198

Constit. of *Murraya paniculata*. Needles (CHCl₃). Mp 163-166°.

5-Methoxy: [120693-44-9]. *8-(Hydroxymethyl)-5,7-dimethoxy-2H-1-benzopyran-2-one. 8-Hydroxymethyl-5,7-dimethoxycoumarin.* **Murrayacarpin B**
$C_{12}H_{12}O_5$ M 236.224
Constit. of *M. paniculata*. Granules (CHCl₃). Mp 199-201°.

Wu, T.-S. *et al*, *Phytochemistry*, 1989, **28**, 293.

Musangic acid — M-80199

16α,19α-Dihydroxy-2,3-seco-12-ursene-2,3,28-trioic acid, 9CI

[120396-55-6]

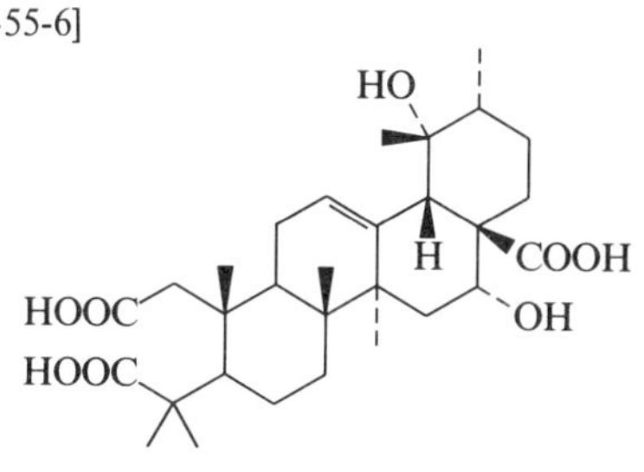

(structure)

$C_{30}H_{46}O_8$ M 534.689

Constit. of *Musanga cecropioides*.

Lontsi, D. *et al*, *J. Nat. Prod. (Lloydia)*, 1989, **52**, 52 (*isol, pmr, cmr*)

4,10(14)-Muuroladiene-1,3,9-triol — M-80200

1,3,9-Trihydroxymuurolene

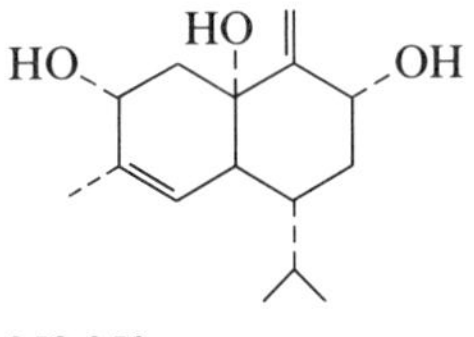

(structure)

$C_{15}H_{24}O_3$ M 252.353

Constit. of *Helichrysum dasyanthum*. Gum.

Jakupovic, J. *et al*, *Phytochemistry*, 1989, **28**, 1119.

Muzigadial — M-80201

Updated Entry replacing M-50280
Canellal
[66550-09-2]

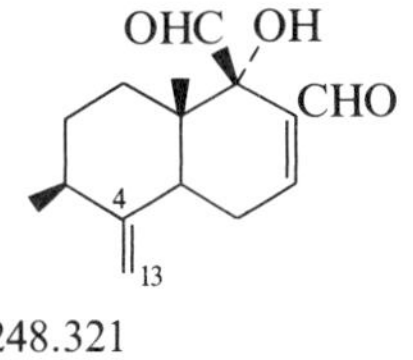

(structure)

$C_{15}H_{20}O_3$ M 248.321

Isol. from *Warburgia ugandensis* and *Canella winteriana*. Antifeedant. Cryst. Mp 127-128°. [α]$_D^{25}$ −193° (c, 0.2 in CHCl₃).

4α,13-Epoxide: **4α,13-Epoxymuzigadial**
$C_{15}H_{20}O_4$ M 264.321
Constit. of *C. winteriana*. Cryst. (Me₂CO/Et₂O). Mp 140-142°. [α]$_D$ −240° (c, 0.5 in CHCl₃).

Kubo, I. *et al*, *Tetrahedron Lett.*, 1977, 4553 (*isol*)
El-Feraly, F.S. *et al*, *J. Chem. Soc., Chem. Commun.*, 1978, 75 (*isol, cryst struct*)
Bosch, M.P. *et al*, *J. Org. Chem.*, 1986, **51**, 773 (*synth*)
Al-Said, M.S. *et al*, *Phytochemistry*, 1989, **28**, 297 (*deriv*)

Mycosinol — M-80202

7-(2,4-Hexadiynylidene)-1,6-dioxaspiro[4.4]nona-2,8-dien-4-ol, 9CI

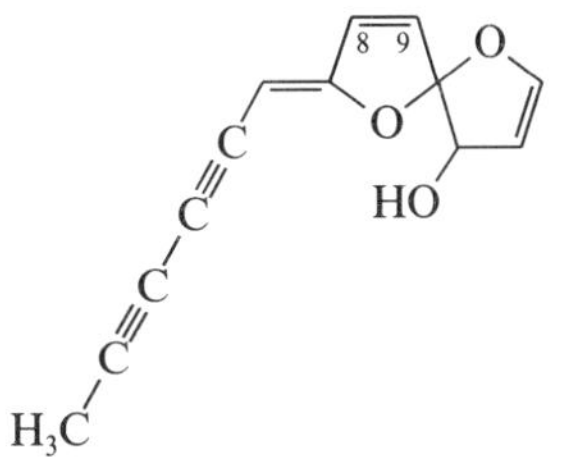

$C_{13}H_{10}O_3$ M 214.220

Variable amts. of information on stereochem. of this compd. and its derivs. Isol. from roots of *Chrysanthemum foeniculaceum* and as phytoalexin of *Coleostephus myconis*. Cryst. (Et₂O/pet. ether). Mp 102°. [α]$_D^{20}$ +44° (c, 0.84 in Et₂O).

Ac (E-):
$C_{15}H_{12}O_4$ M 256.257
From *Chrysanthemum lencanthemum, C. segetum, C. coronarium* and *C. ircutianum*. Cryst. (Et₂O/pet. ether). Mp 123°.

Ac (Z-):
$C_{15}H_{12}O_4$ M 256.257
Isol. from *C. leucanthemum, C. segetum, C. coronarium, C. ircutianum* and *C. silvaticum*. Mp 91°.

8,9-Epoxide: **8,9-Epoxy-7-(2,4-hexadiynylidene)-1,6-dioxaspiro[4.4]non-2-en-4-ol**
$C_{13}H_{10}O_4$ M 230.220
Isol. from above-ground parts of *C. ircutianum*. Cryst. (Et₂O/pet. ether). Mp 139°. [α]$_{546}^{20}$ −145.5° (c, 0.3 in Et₂O).

8,9-Epoxide, Ac: *4-Acetoxy-8,9-epoxy-7-(2,4-hexadiynylidene)-1,6-dioxaspiro[4.4]non-2-ene*
$C_{15}H_{12}O_5$ M 272.257
Isol. from *C. leucanthemum* and *C. maximum*. Cryst. (Et₂O/pet. ether). Mp 153.5°. [α]$_{546}^{20}$ −78° (c, 0.88 in Et₂O). A tentative complete stereochem. was assigned to this compd.

6,7-Epoxide, O-angeloyl: *4-Angeloyloxy-8,9-epoxy-7-(2,4-hexadiynylidene)-1,6-dioxaspiro[4.4]non-2-ene*
$C_{18}H_{16}O_5$ M 312.321
Isol. from roots of *C. maximum* and above-ground parts of *C. ircutianum*. Cryst. (Et₂O/pet. ether). Mp 133°. [α]$_{546}^{20}$ +47.6° (c, 1.2 in Et₂O).

[111768-19-5]

Bohlmann, F. *et al*, *Chem. Ber.*, 1961, **94**, 3193; 1963, **96**, 226; 1964, **97**, 1179; 1965, **98**, 2596 (*isol, uv, ir, pmr, ord*)
Bohlmann, F. *et al*, *Justus Liebigs Ann. Chem.*, 1963, **668**, 51 (*config*)
Bohlmann, F. *et al*, *Chem. Ber.*, 1971, **104**, 11 (*ms*)
Zeisberg, R. *et al*, *Chem. Ber.*, 1974, **107**, 3800 (*cmr*)
Marshall, P.S. *et al*, *Phytochemistry*, 1987, **26**, 2493 (*isol*)

3-Myodeserten-1-ol   M-80203

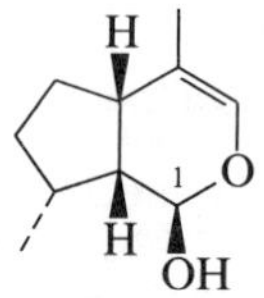

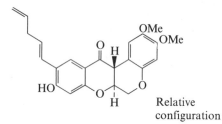

(1*R*)-*form*

$C_{10}H_{16}O_2$ M 168.235

(1*R*)-*form*

    Ac: [95585-89-0]. ***1S-Acetoxy-3-myodesertene***
    $C_{12}H_{18}O_3$ M 210.272
    Constit. of *Myoporum deserti*. Oil. $[\alpha]_D$ −183° (c, 7.7 in
    CHCl₃).

    Me ether: [19479-28-8]. ***1-Methoxy-3-myodesertene***.
    Myodesertin (*obsol.*)
    $C_{11}H_{18}O_2$ M 182.262
    Isol. from *M. deserti*. Bp₂ 67°. $[\alpha]_D$ −165° (neat). n_D^{25}
    1.4752.

(1*S*)-*form*

    Ac: [95585-88-9]. ***1R-Acetoxy-3-myodesertene***
    $C_{12}H_{18}O_3$ M 210.272
    From *M. deserti*. Cryst. (pentane). Mp 17°. $[\alpha]_D$ +273°
    (c, 2.7 in CHCl₃).

    Grant, H.G. *et al, Aust. J. Chem.*, 1980, **33**, 853; 1985, **38**, 325.

Myriconol   M-80204

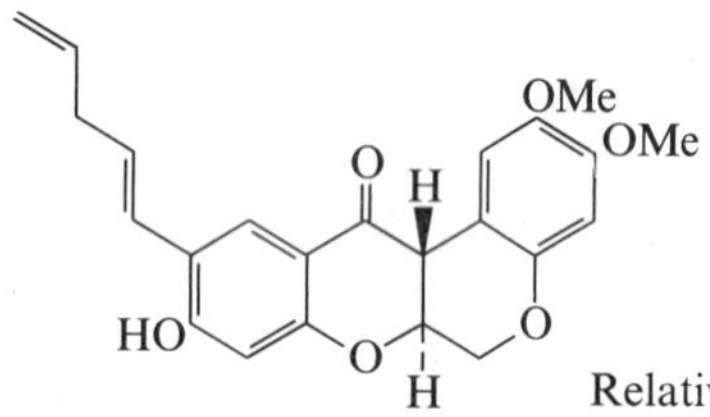

Relative
configuration

$C_{23}H_{22}O_6$ M 394.423
Struct. dubious, esp. the side-chain. Isol. from stem bark
of *Myrica nagi*. Mp 114°. $[\alpha]_D^{34}$ −60.8° (EtOH).

Krishnamoorthy, V. *et al, Curr. Sci.*, 1963, **32**, 16.

Myrtenol   M-80205

Updated Entry replacing M-04173
    2-Pinen-10-ol. Darwinol
    [515-00-4]

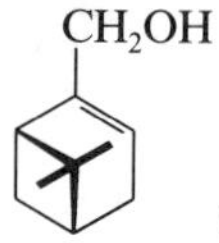

CH₂OH

(+)-*form*

$C_{10}H_{16}O$ M 152.236

(+)-*form* [6712-78-3]
    Isol. from oil of *Myrtus communis, Darwinia grandiflora,
    Leptospermum lanigerum, Chamaecyparis formosensis*
    and others. Oil. Bp₁₁ 103-104°. $[\alpha]_D$ +45.5°.

    10-Aldehyde: [23727-16-4]. ***Myrtenal***
    $C_{10}H_{14}O$ M 150.220
    Occurs in eucalyptus and other plant oils. Unstable oil
    with cinnamon odour. Bp₁₅ 99-100°. $[\alpha]_D^{20}$ +14.75°.

    10-Carboxylic acid: [601-74-1]. *2-Pinen-10-oic acid.
    Myrtenic acid*
    $C_{10}H_{14}O_2$ M 166.219
    Cryst. Mp 53.5-55°. $[\alpha]_D^{20}$ +55° (c, 0.8 in EtOH).

(−)-*form*

    Bp₁₁ 101-102°. $[\alpha]_D^{20}$ −47.1° (EtOH).

    Ac:
    $C_{12}H_{18}O_2$ M 194.273
    Isol. from *Valeriana officinalis*.

    3-Methylbutanoyl: (−)-*Myrtenyl isovalerate*
    Isol. from oil of *V. officinalis*.

    10-Aldehyde: [18486-69-6].
    Unstable oil. Bp 220-221°. $[\alpha]_D$ −15.3°.

(±)-*form*
    Oil. n_D^{20} 1.4963.

    Hydrogen phthalate: Mp 120.5°.

    [19250-17-0]

Stoll, A. *et al, Helv. Chim. Acta*, 1957, **40**, 1205 (*acetate,
    isovalerate*)
Couchman, F.M. *et al, Tetrahedron*, 1964, **20**, 2037 (*isol*)
Banthorpe, D.V. *et al, Chem. Rev.*, 1966, **66**, 643 (*rev*)
Bates, R.B. *et al, J. Org. Chem.*, 1968, **33**, 1730 (*pmr*)
Borowiecki, L. *et al, Pol. J. Chem. (Rocz. Chem.)*, 1971, **45**, 573
    (*Myrtenic acid*)
Kergomard, A. *et al, Bull. Soc. Chim. Fr.*, 1974, 2572 (*synth*)
Rauchschwalbe, G. *et al, Helv. Chim. Acta*, 1975, **58**, 1094 (*synth*)
Bohlmann, F. *et al, Org. Magn. Reson.*, 1975, **7**, 426 (*cmr*)

N

2,3-Naphthalenedicarboxylic acid, 9CI   **N-80001**

Updated Entry replacing N-60003

[2169-87-1]

$C_{12}H_8O_4$ M 216.193

Prisms (AcOH or by subl.). Sol. hot EtOH, spar. sol. Et₂O. Mp 246°, Mp 239-241°.

Di-Me ester: [13728-34-2].
  $C_{14}H_{12}O_2$ M 212.248
  Large plates (Et₂O/pet. ether). Mp 47°. Bp₁ 141-145°.

Di-Et ester: [50919-54-5].
  $C_{16}H_{16}O_2$ M 240.301
  Cryst. (Me₂CO). Mp 88.5-89.5°.

Mononitrile: 3-Cyano-2-naphthoic acid
  $C_{12}H_7NO_2$ M 197.193
  Yellow cryst. Mp 273-274°.

Dinitrile: [22856-30-0]. 2,3-Dicyanonaphthalene
  $C_{12}H_6N_2$ M 178.193
  Needles (EtOH). Mp 251°.

Imide: 2H-Naphtho[2,3-c]pyrrole-1,3-dione
  $C_{12}H_7NO_2$ M 197.193
  Microneedles (CHCl₃/EtOH). Mp 275°. Softens at 250°.

Anhydride: [716-39-2]. Naphtho[2,3-c]furan-1,3-dione, 9CI.
  Naphthalic anhydride
  $C_{12}H_6O_3$ M 198.178
  Light yellow cryst. (Me₂CO/pet. ether). Mp 250-252°.

Freund, M. *et al*, *Justus Liebigs Ann. Chem.*, 1913, **402**, 68.
Waldmann, H. *et al*, *Ber.*, 1931, **64**, 1713.
Bradbrook, E.F. *et al*, *J. Chem. Soc.*, 1936, 1739.
Patton, J.W. *et al*, *J. Org. Chem.*, 1965, **30**, 3869.
Cava, M.P. *et al*, *J. Org. Chem.*, 1969, **34**, 538 (anhydride)
Davies, C., *Fuel*, 1973, **52**, 270; 1974, **53**, 105 (ir, uv)
Org. Synth., Coll. Vol., 5, 1973, 810.
Carlson, R.G. *et al*, *J. Org. Chem.*, 1986, **51**, 3978 (synth)
Fier, S. *et al*, *J. Org. Chem.*, 1988, **53**, 2353 (anhydride)
Patney, H.K., *J. Org. Chem.*, 1988, **53**, 6106 (anhydride)

1,8-Naphthalenediseleninic acid   **N-80002**

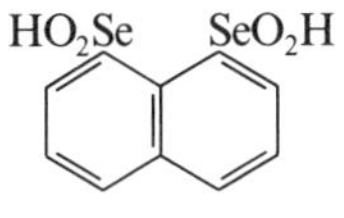

$C_{10}H_8O_4Se_2$ M 350.091

Na salt: Prepd. in soln.

Anhydride: [114031-51-5]. Naphth[1,8-cd][1,2,6]oxadiselenin 1,3-dioxide
  $C_{10}H_6O_3Se_2$ M 332.076
  Solid. Mp 153° dec. Insol. in all common solvents, could not be recryst. nor nmr obtained.

Kice, J.L. *et al*, *J. Org. Chem.*, 1988, **53**, 2435 (synth, ir, uv)

1,4,5,7-Naphthalenetetrol, 9CI   **N-80003**

1,4,5,7-Tetrahydroxynaphthalene

$C_{10}H_8O_4$ M 192.171

5-Me ether, 1,7-Di-Ac: [119098-66-7].
  $C_{15}H_{14}O_6$ M 290.272
  Cubes (EtOH). Mp 137.5-138.5°.

4,5-Di-Me ether, di-Ac: [119098-67-8].
  $C_{16}H_{16}O_6$ M 304.299

Needles (EtOH). Mp 135.5-136°.

1,4,5-Tri-Me ether, Ac: [119098-76-9].
  $C_{15}H_{16}O_5$ M 276.288
  Light brown needles (CH₂Cl₂/pet. ether). Mp 75°.

Giles, R.G.F. *et al*, *J. Chem. Soc., Perkin Trans.* 1, 1988, 2459 (synth, ir, pmr, ms)

4H-Naphtho[1,2,3,4-def]carbazole, 9CI   **N-80004**

1,12-Iminotriphenylene

[109606-75-9]

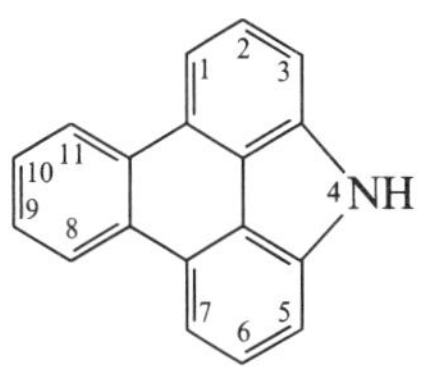

$C_{18}H_{11}N$ M 241.292

Fluffy needles (C₆H₆ or by subl.). Mp 250°. Subl. readily.

Klemm, L.H. *et al*, *J. Heterocycl. Chem.*, 1988, **25**, 1427 (synth, uv, pmr, ms)

Naphtho[1,8-cd]-1,2-diselenole   **N-80005**

[36579-71-2]

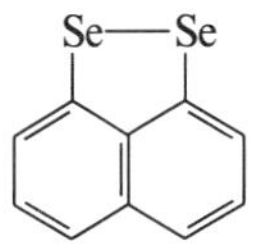

$C_{10}H_6Se_2$ M 284.077

Lustrous purple needles (CH₂Cl₂/hexane). Mp 127-129°.

Meinwald, J. *et al*, *J. Am. Chem. Soc.*, 1977, **99**, 255 (synth, ir, uv, pmr, cmr, ms)
Kice, J.L. *et al*, *J. Org. Chem.*, 1988, **53**, 2435 (synth)

Naphtho[1,8-cd]1,2-ditellurole, 9CI   **N-80006**

[52875-49-7]

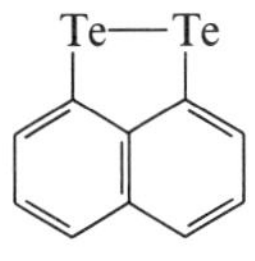

$C_{10}H_6Te_2$ M 381.357

Metallic greenish needles (hexane/CS₂). Mp 212-214°.

Meinwald, J. *et al*, *J. Am. Chem. Soc.*, 1977, **99**, 255 (synth, uv, pmr, cmr, ir, ms)

2H-Naphtho-[2,3-b]pyran-5,10-dione   **N-80007**

[110271-46-0]

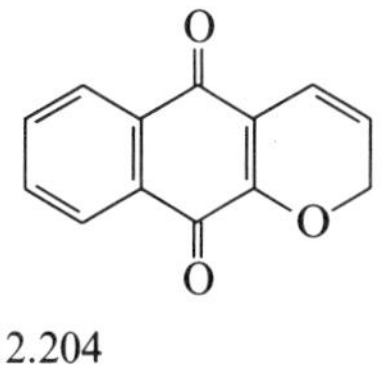

$C_{13}H_8O_3$ M 212.204

Orange prisms. Mp 160-163°.

Hayashi, T. *et al, J. Med. Chem.*, 1987, **30**, 2005 (*synth, ir, uv, pmr*)

1*H*-Naphtho[2,3-*c*]pyran-5,10-dione   N-80008

[106261-83-0]

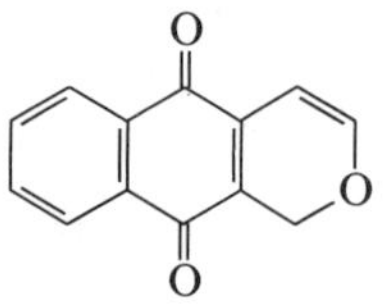

$C_{13}H_8O_3$ M 212.204
Orange cryst.

Hayashi, T. *et al, J. Med. Chem.*, 1987, **30**, 2005 (*synth, ir, uv, pmr*)

Naphtho[1,8-*de*]-1,3-thiazine   N-80009

[204-07-9]

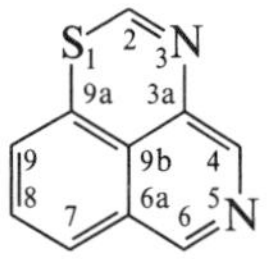

$C_{11}H_7NS$ M 185.249
Green needles melting and resolidifying to prisms. Mp
104-110°, Mp >320° (double Mp).

Herbert, J.M. *et al, Heterocycles*, 1987, **26**, 1037 (*synth, pmr*)

Naphtho[2′,1′:5,6][1,2,4]triazino[4,3-*b*]indazole, 9CI   N-80010

[221-14-7]

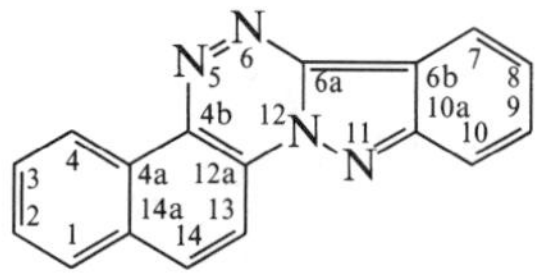

$C_{17}H_{10}N_4$ M 270.293
Orange cryst. (EtOH). Mp 250-251°.

Neidlein, R. *et al, Chem. Ber.*, 1988, **121**, 1359 (*synth, ir, uv, pmr, cmr, ms*)

2-(1-Naphthyl)-2-(2-naphthyl)ethanal   N-80011

α-2-*Naphthalenyl-1-naphthaleneacetaldehyde, 9CI. 2-(α-Naphthyl)-2-(β-naphthyl)acetaldehyde*
[91879-84-4]

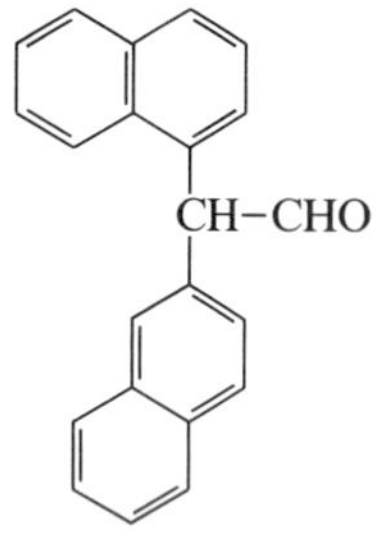

$C_{22}H_{16}O$ M 296.368
Unstable yellow oil.

Zimmerman, H.E. *et al, J. Org. Chem.*, 1989, **54**, 2125 (*synth, ir, pmr*)

1-(1-Naphthyl)-2-propanone   N-80012

1-(1-Naphthalenyl)-2-propanone. 1-Acetonylnaphthalene. 1-Naphthylacetone
[33744-50-2]

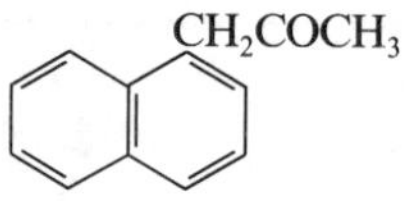

$C_{13}H_{12}O$ M 184.237
Liq. Bp$_{0.75}$ 121-126°, Bp$_{0.4}$ 111-114°.
Oxime (E-)*:* [19534-15-7].
  Mp 96-97°.
Semicarbazone: Mp 189.5-190.5°.

[20557-52-2, 85629-15-8]

Kotera, K. *et al, Tetrahedron*, 1968, **24**, 5677 (*deriv, synth*)
Zimmerman, H.E. *et al, J. Am. Chem. Soc.*, 1971, **93**, 3638 (*synth ir, pmr*)
Barcus, R.L. *et al, J. Am. Chem. Soc.*, 1986, **108**, 3928 (*synth, ir, pmr*)

1-(2-Naphthyl)-2-propanone   N-80013

1-(2-Naphthalenyl)-2-propanone, 9CI. 2-Acetonylnaphthalene. 2-Naphthylacetone
[21567-68-0]
$C_{13}H_{12}O$ M 184.237
Mp 36-37°.

Newman, M.S. *et al, J. Am. Chem. Soc.*, 1949, **71**, 3342 (*synth*)
Inaba, S. *et al, J. Org. Chem.*, 1985, **50**, 1373 (*synth, ir, pmr*)

Nassauvirevolutin A   N-80014

[119889-33-7]

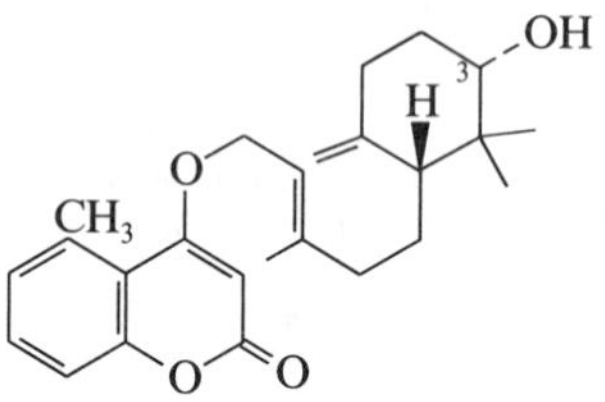

$C_{25}H_{32}O_4$ M 396.525
Constit. of *Nassauvia revoluta*. Gum.

3-Ketone: [119875-28-4]. **Nassauvirevolutin B**
  $C_{25}H_{30}O_4$ M 394.510
  Constit. of *N. revoluta*. Gum.

Bittner, M. *et al, Phytochemistry*, 1988, **27**, 3845.

Nassauvirevolutin C   N-80015

[119875-29-5]

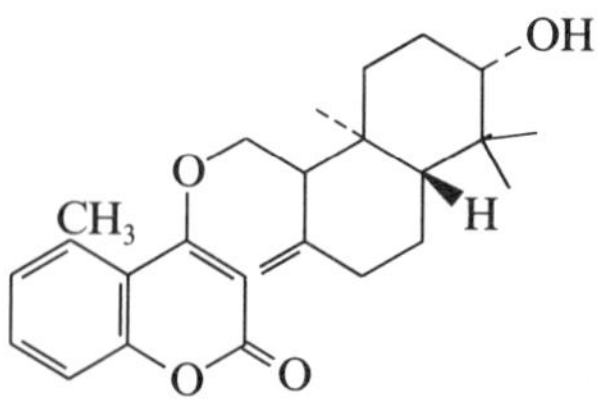

$C_{25}H_{32}O_4$ M 396.525
Constit. of *Nassauvia revoluta*. Gum.

Bittner, M. *et al, Phytochemistry*, 1988, **27**, 3845.

Naviculide   N-80016

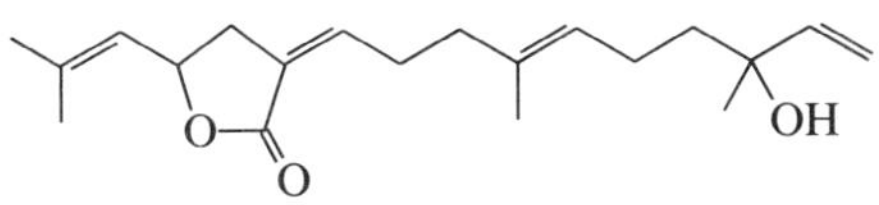

$C_{20}H_{30}O_3$ M 318.455
Constit. of *Porella navicularis*. Oil.

Toyota, M. *et al*, *Phytochemistry*, 1989, **28**, 1661.

Naviculol   N-80017

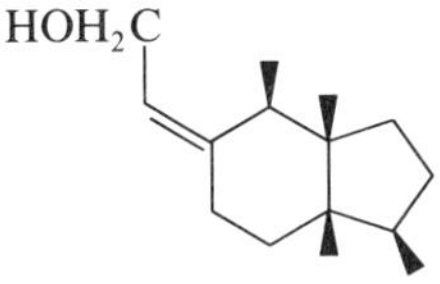

$C_{15}H_{26}O$ M 222.370
Constit. of *Porella navicularis*. Oil. $[\alpha]_D$ +48.5° (c, 1.14 in CHCl₃).

Toyota, M. *et al*, *Phytochemistry*, 1989, **28**, 1661.

Neobalearone   N-80018

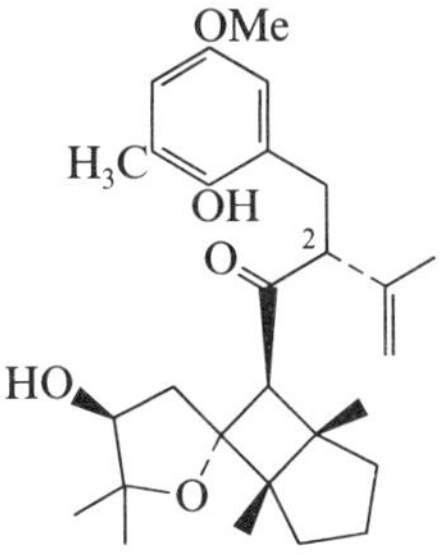

$C_{28}H_{40}O_5$ M 456.621
Metab. of *Cystoseira stricta*. Oil. $[\alpha]_D^{25}$ +120.0° (c, 1.6 in EtOH).

2-Epimer: **Epineobalearone**
$C_{28}H_{40}O_5$ M 456.621
Metab. of *C. stricta*. Oil. $[\alpha]_D^{25}$ −77.1° (c, 1.86 in EtOH).

Amico, V. *et al*, *J. Nat. Prod.* (*Lloydia*), 1989, **52**, 962 (*isol, pmr, cmr*)

Neocynaponogenin A   N-80019

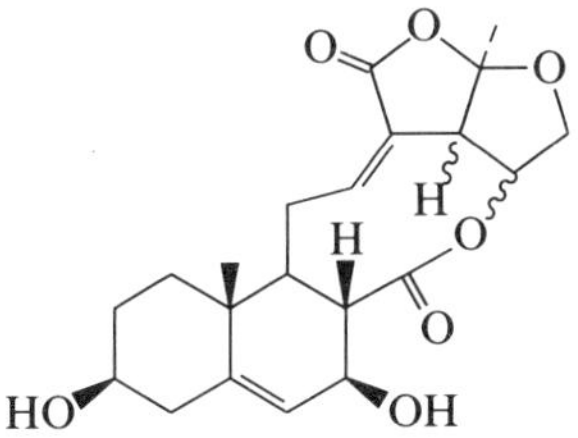

$C_{21}H_{26}O_7$ M 390.432
A discopregnane derivative.

3-O-[α-L-Cymaropyranosyl-(1→4)-β-D-digitoxopyranosyl-(1→4)-β-D-oleandropyranoside]: **Neocynaponoside A**
$C_{41}H_{60}O_{16}$ M 808.915
Constit. of *Cynanchum paniculatum*. Amorph. powder. Mp 105-108°. $[\alpha]_D$ −57.3° (c, 1.54 in CHCl₃).

Sugama, K. *et al*, *Phytochemistry*, 1988, **27**, 3984.

Neodunol   N-80020

Updated Entry replacing N-50092
6a,11a-Dihydro-6H-benzofuro[3,2-c]furo[3,2-g][1]benzopyran-9-ol, 9CI
[53766-53-3]

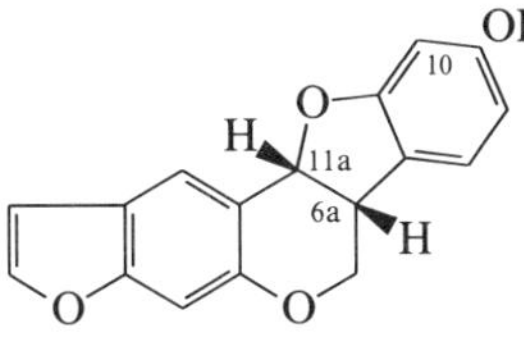

$C_{17}H_{12}O_4$ M 280.279
Pterocarpan numbering shown. Isol. from *Neorautanenia edulis*, also from *N. amboensis* and *Pachyrrhizus erosus*. Cryst. (C₆H₆/hexane). Mp 170.5-171.5°. $[\alpha]_D^{22}$ −284.9° (c, 0.8 in CHCl₃).

Me ether: [53833-80-0].
$C_{18}H_{14}O_4$ M 294.306
Isol. from *Echinopora lamellosa*. Mp 175°. $[\alpha]_D^{25}$ −217° (c, 0.415 in CHCl₃).

Me ether, 10-methoxy: [76165-15-6]. **Ambonane**
$C_{19}H_{16}O_5$ M 324.332
Isol. from *N. amboenis* root bark. Light-yellow needles (Et₂O/hexane). Mp 125-127°. $[\alpha]_D$ −214° (c, 0.01 in CHCl₃).

6a,11a-Didehydro: [53766-54-4]. **Neorauteen**. *6H-Benzofuro[3,2-c]furo[3,2-g][1]benzopyran-9-ol, 9CI*
$C_{17}H_{10}O_4$ M 278.264
Isol. from *N. edulis* root bark. Plates (MeOH). Mp 202.5-204°.

6a,11a-Didehydro, Ac: Needles. Mp 197.5-199°.

Brink, A.J. *et al*, *Phytochemistry*, 1972, **16**, 273 (*isol*)
Brink, A.J. *et al*, *Phytochemistry*, 1974, **13**, 1581 (*Neorauteen*)
Ingham, J.L., *Z. Naturforsch., C*, 1979, **34**, 683 (*isol*)
Breytenbach, J.C. *et al*, *J. Chem. Soc., Perkin Trans. 1*, 1980, 1804 (*Ambonane*)
Sanduja, R. *et al*, *J. Heterocycl. Chem.*, 1984, **21**, 845 (*isol, deriv, cryst struct, abs config*)

Neolinderatin   N-80021

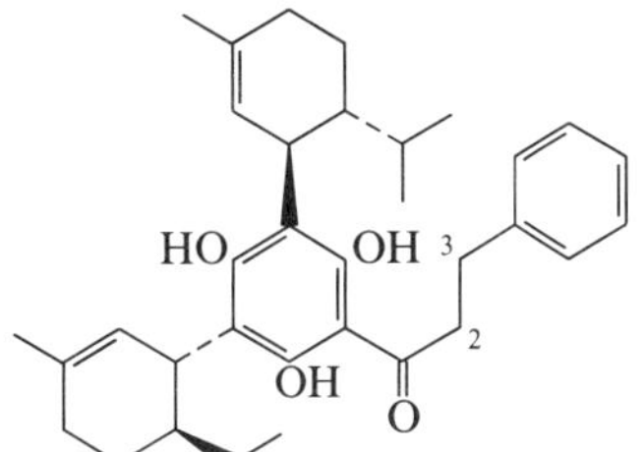

$C_{35}H_{46}O_4$ M 530.746
Constit. of *Lindera umbellata*. Viscous oil. $[\alpha]_D$ +20.3° (c, 0.9 in CHCl₃).

2,3-Didehydro: [120217-40-5]. **Neolinderachalcone**
$C_{35}H_{44}O_4$ M 528.730
Constit. of *L. umbellata*. Viscous oil. $[\alpha]_D$ +26.5° (c, 0.15 in CHCl₃).

Ichino, K. *et al*, *Chem. Lett.*, 1989, 363 (*struct, synth*)

Neorautane   N-80022

Updated Entry replacing N-10030

[36284-97-6]

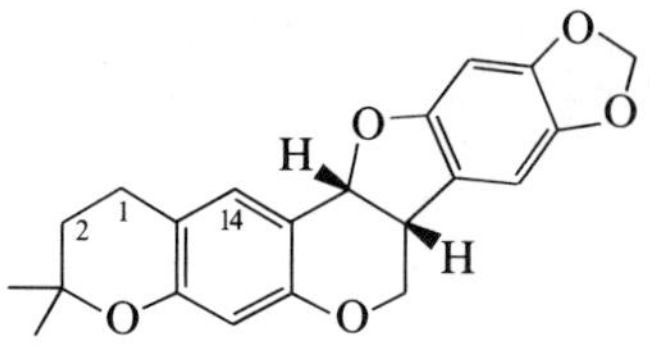

Absolute
configuration

$C_{21}H_{20}O_5$ M 352.386

CA numbering shown. Constit. of root bark of
Neorautanenia edulis. Cryst. (MeOH). Mp 204°. $[\alpha]_D^{23}$
$-255°$ (c, 1.4 in $CHCl_3$).

14-Methoxy: [63343-93-1]. ***Neorautanin***
$C_{22}H_{22}O_6$ M 382.412
From *N. edulis*. Needles (C_6H_6/hexane). Mp 202-204°.
$[\alpha]_D^{21}$ $-303°$ (c, 0.2 in $CHCl_3$).

1,2-Didehydro: [65418-33-9]. ***Neorautenane***. *Neorautenaan*
$C_{21}H_{18}O_5$ M 350.370
Isol. from root bark of *N. amboensis*. Needles (EtOH).
Mp 165-166°. $[\alpha]_D$ $-200°$ (c, 0.1 in $CHCl_3$).

1,2-Didehydro, 14-hydroxy: [76175-39-8]. ***Neorautenanol***
$C_{21}H_{18}O_6$ M 366.370
Isol. from *N. amboensis*. Glass.

2ξ-Hydroxy: [76165-13-4]. ***Neorautanol***
$C_{21}H_{20}O_6$ M 368.385
Isol. from *N. amboensis* root bark. Glass. Mp 93-95°.

Rall, G.J. *et al, J. S. Afr. Chem. Inst.*, 1972, **25**, 25 (*isol*)
Brink, A.J. *et al, Phytochemistry*, 1977, **16**, 273 (*Neorautanin*)
Breytenbach, J.C. *et al, J. Chem. Soc., Perkin Trans.* 1, 1980, 1804
    (*Neorautenane, Neorautanol, Neorautenanol*)

Neorautenol   N-80023

Updated Entry replacing N-10031

[53766-52-2]

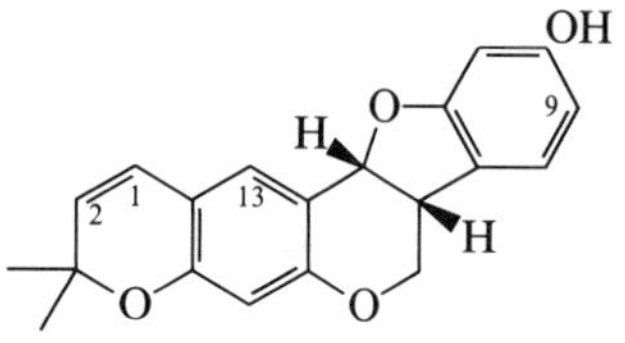

$C_{20}H_{18}O_4$ M 322.360

Isol. from *Neorautanenia edulis*. Cryst. (MeOH). Mp
168.5-170°. $[\alpha]_D^{22}$ $-188.2°$ (c, 1.5 in $CHCl_3$).

13-Methoxy: [76165-14-5]. ***Edulenanol***
$C_{21}H_{20}O_5$ M 352.386
Constit. of *N. amboensis* root bark. Yellow glass. $[\alpha]_D$
$-224°$ (c, 0.8 in $CHCl_3$).

13-Methoxy, Me ether: [53734-77-3]. ***Edulenane***. *Edulaan*
$C_{22}H_{22}O_5$ M 366.413
Constit. of *N. amboensis* root bark. Cubes (EtOH or
Me_2CO), needles (EtOH). Mp 185-186°.

9,13-Dimethoxy: [56257-27-3]. ***Desmodin***
$C_{22}H_{22}O_6$ M 382.412
Isol. from *Desmodium gangeticum* roots. Cryst. Mp 236-
238°.

13-Methoxy, Me ether, 1,2-dihydro: [37706-59-5]. ***Edulane***
$C_{22}H_{24}O_5$ M 368.429
Constit. of *N. edulis* root bark. Needles (C_6H_6). Mp 174-
175°. $[\alpha]_D^{20}$ $-258°$ (c, 0.2 in $CHCl_3$).

Brink, A.J. *et al, Phytochemistry*, 1974, **13**, 1581 (*isol, struct*)
Purushothaman, K.K. *et al, Phytochemistry*, 1975, **14**, 1129
    (*Desmodin*)

Brink, A.J. *et al, Phytochemistry*, 1977, **16**, 273 (*Edulane*)
Chalmers, A.A. *et al, Tetrahedron*, 1977, **33**, 1735 (*cmr*)
Breytenbach, J.C. *et al, J. Chem. Soc., Perkin Trans.* 1, 1980, 1804
    (*Edulenane, Edulenanol*)

5-Neoverrucosanol   N-80024

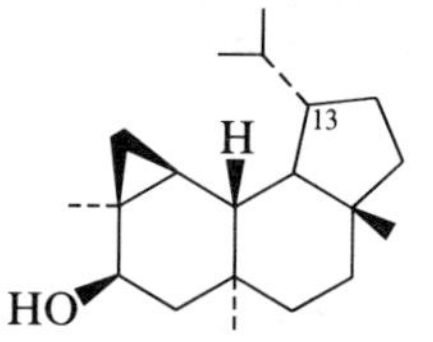

$C_{20}H_{34}O$ M 290.488

5β-form

Constit. of *Schistochila rigidula*.

13-Epimer: [116502-06-8]. ***13-Epi-5β-neoverrucosanol***
$C_{20}H_{34}O$ M 290.488
Constit. of *S. nobilis* and *Plagiochila stephensoniana*.
Cryst. (hexane). Mp 151-153.5°. $[\alpha]_D$ $+45.5°$ (c, 0.8 in
$CHCl_3$).

Wu, C.-L. *et al, J. Hattori Bot. Lab.*, 1988, **64**, 151 (*isol*)
Fukuyama, Y. *et al, Phytochemistry*, 1988, **27**, 1797 (*isol, cryst
    struct*)
Asakawa, Y. *et al, Phytochemistry*, 1988, **27**, 3509 (*isol*)

Nepseudin   N-80025

Updated Entry replacing N-00537

6,7-Dihydro-6-(2,3,4-trimethoxyphenyl)-5H-furo[3,2-
g][1]benzopyran-5-one, 9CI

[20848-57-1]

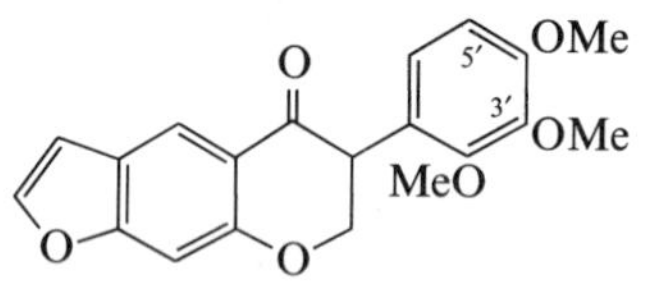

$C_{20}H_{18}O_6$ M 354.359

Isol. from roots of *Neorautanenia pseudopachyrrhiza*.
    Cryst. ($CHCl_3$/MeOH). Mp 115-116°.

3'-Demethoxy: [76165-16-7]. ***Neoraunone***
$C_{19}H_{16}O_5$ M 324.332
Constit. of root bark of *N. amboensis*. Light-yellow
needles (EtOH). Mp 154-155°. Opt. inactive.

3'-Demethoxy, 5'-methoxy: [76165-17-8]. ***Ambonone***
$C_{20}H_{18}O_6$ M 354.359
Constit. of root bark of *N. amboensis*. Light-yellow
rosettes (C_6H_6). Mp 153-155°. Opt. inactive.

Fukui, K. *et al, Experientia*, 1963, **19**, 621 (*isol*)
Crombie, L. *et al, J. Chem. Soc.*, 1963, 1569 (*isol*)
Fukui, K. *et al, Bull. Chem. Soc. Jpn.*, 1968, **41**, 1385 (*synth*)
Breytenbach, J.C. *et al, J. Chem. Soc., Perkin Trans.* 1, 1980, 1804
    (*Neoraunone, Ambonone*)

Nimbanal N-80026

[120462-51-3]

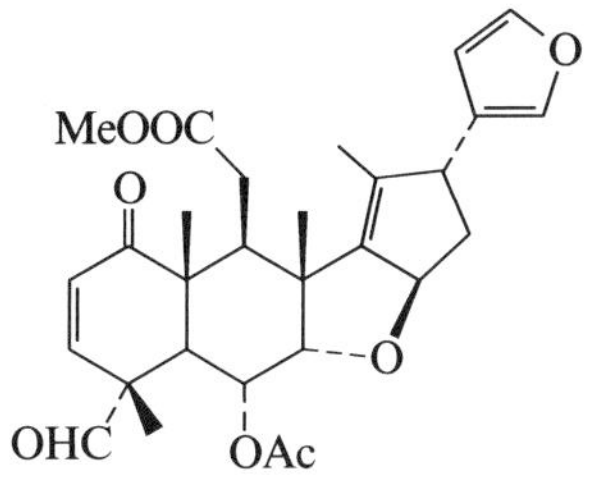

C$_{29}$H$_{34}$O$_8$ M 510.583

Constit. of *Azadirachta indica*. Cryst. (Me$_2$CO/pet. ether). Mp 195-197°. [α]$_D$ +46° (CHCl$_3$).

Rajatkar, S.R. *et al*, *Phytochemistry*, 1989, **28**, 203.

Nimbiol N-80027

Updated Entry replacing N-00629

12-Hydroxy-13-methyl-8,11,13-podocarpatrien-7-one

[561-95-5]

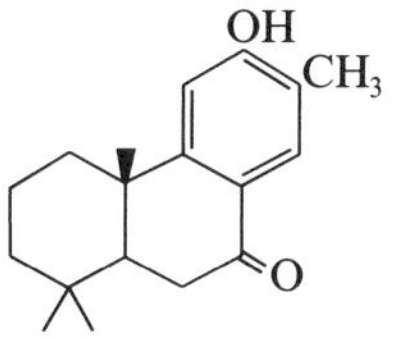

C$_{18}$H$_{24}$O$_2$ M 272.386

Constit. of *Melia azadirachta*. Cryst. Mp 244°. [α]$_D$ +32.3° (CHCl$_3$).

Me ether: [123123-15-9]. **Methylnimbiol**

    C$_{19}$H$_{26}$O$_2$ M 286.413

Constit. of *Azadirachta indica*. Plates (hexane). Mp 142-143°.

Meyer, W.L. *et al*, *J. Org. Chem.*, 1975, **40**, 3686 (*isol, pmr*)
Ara, I. *et al*, *J. Nat. Prod.* (*Lloydia*), 1988, **51**, 1054 (*Methylnimbiol*)

Nimbionol N-80028

3β,12-Dihydroxy-13-methoxy-8,11,13-podocarpatrien-7-one

C$_{18}$H$_{24}$O$_4$ M 304.385

Constit. of *Azadirachta indica*. Needles (CHCl$_3$). Mp 127-129°. [α]$_D$ +10° (c, 0.001 in CHCl$_3$).

3-Ketone: *12-Hydroxy-13-methoxy-8,11,13-podocarpatriene-3,7-dione*. **Nimbionone**

    C$_{18}$H$_{22}$O$_4$ M 302.369

Constit. of *A. indica*. Plates (CHCl$_3$). Mp 78-79°. [α]$_D$ +0.03° (c, 0.036 in CHCl$_3$).

Siddiqui, S. *et al*, *Phytochemistry*, 1988, **27**, 3903.

Nimbolicin N-80029

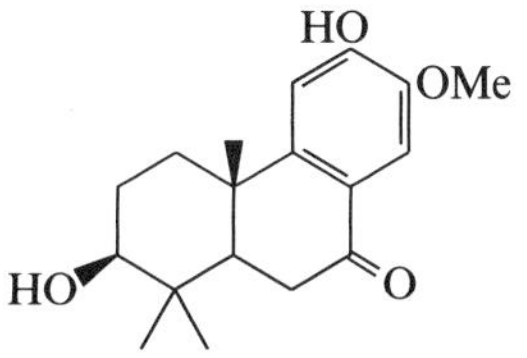

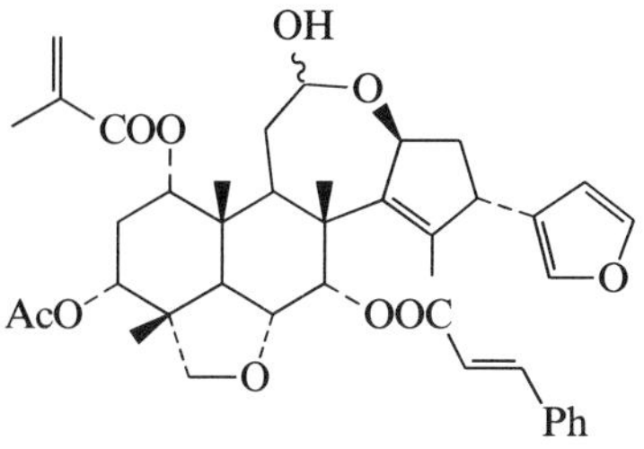

C$_{41}$H$_{48}$O$_{10}$ M 700.824

Constit. of *Azadirachta indica*. Plates. Mp 121-122°. [α]$_D^{24}$ −33.3° (CHCl$_3$).

Ara, I. *et al*, *Heterocycles*, 1989, **29**, 729.

Nimbonolone N-80030

12-Ethyl-13-methoxy-8,11,13-podocarpatrien-3-one

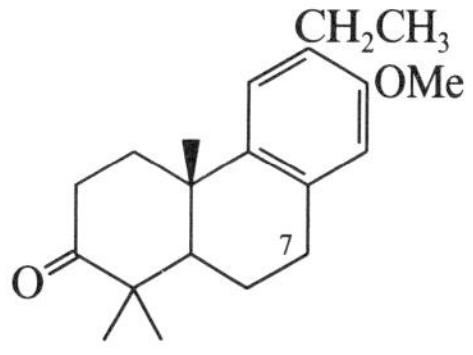

C$_{20}$H$_{28}$O$_2$ M 300.440

Constit. of *Azadirachta indica*. Cryst. (hexane). Mp 73-74°.

3-Deoxo, 7-oxo: *12-Ethyl-13-methoxy-8,11,13-podocarpatrien-7-one*. **Nimbonone**

    C$_{20}$H$_{28}$O$_2$ M 300.440

Constit. of *A. indica*. Cryst. (hexane). Mp 68-69°. [α]$_D$ +15° (c, 0.06 in CHCl$_3$).

Ara, I. *et al*, *Phytochemistry*, 1989, **28**, 1177.

Nimbosone N-80031

12-Methoxy-17-nor-8,11,13-abietatrien-15-one. 13-Acetyl-12-methoxy-8,11,13-podocarpatriene

[19889-21-5]

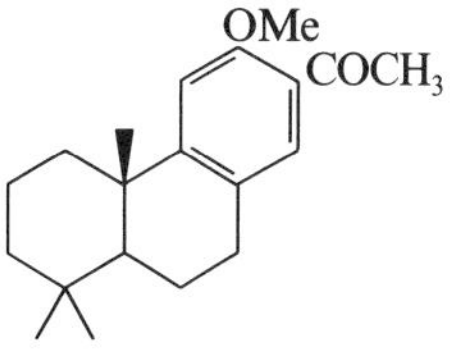

C$_{20}$H$_{28}$O$_2$ M 300.440

Constit. of *Azadirachta indica*. Needles (hexane). Mp 137-139°.

Ohashi, M. *et al*, *Tetrahedron Lett.*, 1968, 719 (*synth*)
Ara, I. *et al*, *J. Nat. Prod.* (*Lloydia*), 1988, **51**, 1054 (*isol, pmr*)

Nirphyllin
N-80032

[120396-54-5]

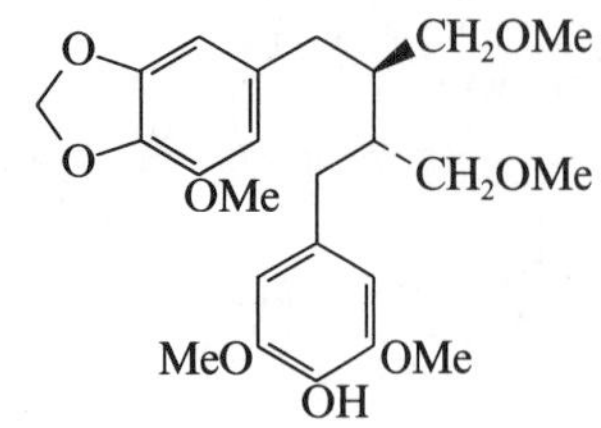

C$_{24}$H$_{32}$O$_8$ M 448.512
Constit. of *Phyllanthus niruri*.

Singh, B. *et al, J. Nat. Prod. (Lloydia)*, 1989, **52**, 48 (*isol, ir, pmr*)

1-Nitroadamantane
N-80033

1-Nitrotricyclo[3.3.1.1^{3,7}]decane, 9CI

[7575-82-8]

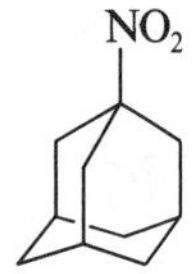

C$_{10}$H$_{15}$NO$_2$ M 181.234
Waxy solid by subl. Mp 173° (158.5-159.0°).

Stetter, H. *et al, Chem. Ber.*, 1960, **93**, 226 (*synth*)
Smith, G.W. *et al, J. Org. Chem.*, 1961, **26**, 2207 (*synth, ir*)

2-Nitroadamantane
N-80034

2-Nitrotricyclo[3.3.1.1^{3,7}]decane, 9CI

[54564-31-7]

C$_{10}$H$_{15}$NO$_2$ M 181.234
Cryst. (EtOH aq.). Mp 165-166°.

Bowan, W.R. *et al, J. Chem. Soc., Perkin Trans. 2*, 1980, 731
    (*cmr*)
Archibald, T.G. *et al, J. Org. Chem.*, 1988, **53**, 4645 (*synth, ir, pmr*)

2-Nitro-1,3-benzenedicarboxylic acid, 9CI
N-80035

Updated Entry replacing N-00712
2-Nitroisophthalic acid, 8CI

[21161-11-5]

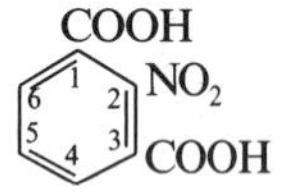

C$_8$H$_5$NO$_6$ M 211.131
Prisms (MeOH). Mp 315°.

Mono-Me ester:
    C$_9$H$_7$NO$_6$ M 225.157
    Needles (H$_2$O). Mp 197°.

Di-Me ester: [57052-99-0].
    C$_{10}$H$_9$NO$_6$ M 239.184
    Needles (H$_2$O). Mp 135°.

Monoamide:
    C$_8$H$_6$N$_2$O$_5$ M 210.146
    Plates (MeOH). Mp 252°.

Diamide:
    C$_8$H$_7$N$_3$O$_4$ M 209.161
    Cryst. (H$_2$O). Mp 278-280°.

Dichloride:
    C$_8$H$_3$Cl$_2$NO$_4$ M 248.021
    Cryst. Mp 128-130°.

Nölting, E. *et al, Ber.*, 1906, **39**, 73 (*synth*)
Wohl, A., *Ber.*, 1910, **43**, 3474 (*synth, derivs*)
Isensee, R.W. *et al, J. Am. Chem. Soc.*, 1948, **70**, 4061.
Andrews, B.D. *et al, Aust. J. Chem.*, 1971, **24**, 413.
Harvey, I.W. *et al, J. Chem. Soc., Perkin Trans.* 1, 1988, 1939
    (*synth, derivs*)

6-Nitro-4H-1-benzopyran-4-one
N-80036

6-Nitrochromone

[51484-05-0]

C$_9$H$_5$NO$_4$ M 191.143
Yellow solid (C$_6$H$_6$). Mp 173-175° (170-171°).

Ellis, G.P. *et al, J. Chem. Soc., Perkin Trans.* 1, 1973, 2781 (*synth, pmr*)

7-Nitrobicyclo[2.2.1]heptane
N-80037

7-Nitronorbornane

[100859-83-4]

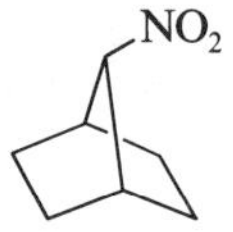

C$_7$H$_{11}$NO$_2$ M 141.169
Low-melting solid. Bp$_7$ 170° (oven) .

Fukunaga, K. *et al, Synthesis*, 1987, 1097 (*synth, ir, pmr, ms*)

2-Nitrobicyclo[4.4.1]undeca-1,3,5,7,9-pentaene
N-80038

2-Nitro-1,6-methano[10]annulene

C$_{11}$H$_9$NO$_2$ M 187.198
(±)-*form* [118372-00-2]
    Yellow oil.

Neidlein, R. *et al, Chem. Ber.*, 1989, **122**, 493 (*synth, pmr*)

3-Nitrobicyclo[4.4.1]undeca-1,3,5,7,9-pentaene
N-80039

3-Nitro-1,6-methano-10-annulene

[118372-01-3]

C$_{11}$H$_9$NO$_2$ M 187.198

Neidlein, R. *et al, Chem. Ber.*, 1989, **122**, 493 (*synth, pmr*)

1-Nitrobiphenylene
N-80040

[16294-08-9]

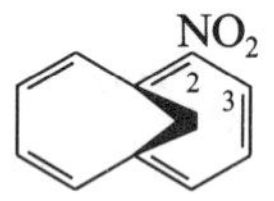

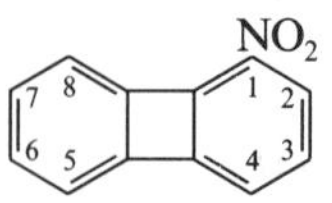

C$_{12}$H$_7$NO$_2$ M 197.193
Yellow needles (pet. ether). Mp 87-88°.

2,4,7-Trinitrofluorenone complex: [16294-07-8].
    Red-brown plates (AcOH/EtOH). Mp 173-174°.

Barton, J.W. *et al, J. Chem. Soc. C*, 1967, 2097 (*synth, pmr*)

2-Nitrobiphenylene   N-80041

[18931-53-8]

$C_{12}H_7NO_2$ M 197.193
Yellow needles (pet. ether). Mp 107-108°.

Baker, W. *et al*, *J. Chem. Soc.*, 1958, 2658, 2666 (*synth, uv*)

2-Nitrocycloheptanone   N-80042

[13154-27-3]

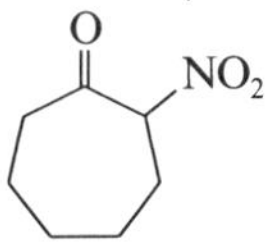

$C_7H_{11}NO_3$ M 157.169

(±)-*form*

Cryst. (2-propanol). Mp 37.5-38.0°.
2,4-Dinitrophenylhydrazone: Mp 173-174°.

Feuer, H. *et al*, *J. Org. Chem.*, 1966, **31**, 3152 (*synth, ir*)

2-Nitrocyclohexanone   N-80043

[4883-67-4]

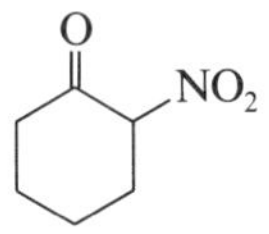

$C_6H_9NO_3$ M 143.142

(±)-*form*

Cryst. solid. Mp 38°.

Wieland, W. *et al*, *Justus Liebigs Ann. Chem.*, 1928, **461**, 295 (*synth*)
Feuer, H. *et al*, *J. Org. Chem.*, 1966, **31**, 3152 (*synth, ir, pmr*)
Griswold, A.A. *et al*, *J. Org. Chem.*, 1966, **31**, 357 (*synth, ir*)
Dampawan, P. *et al*, *Synthesis*, 1983, 545 (*synth, ir*)

2-Nitrocyclopentanone   N-80044

[22498-31-3]

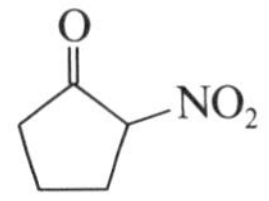

$C_5H_7NO_3$ M 129.115

(±)-*form*

Yellow solid. Darkens on standing at r.t.

Elfehail, F. *et al*, *Synth. Commun.*, 1980, **10**, 929 (*synth, ir, pmr, uv*)

N-Nitrodimethylamine, 8CI   N-80045

Updated Entry replacing N-00898
N-*Methyl*-N-*nitromethanamine, 9CI. Dimethylnitramine*
[4164-28-7]

$$Me_2NNO_2$$

$C_2H_6N_2O_2$ M 90.082
Needles (ligroin). Sol. H_2O. Mp 58°. Bp 187°. Steam-volatile.

▷ Explosive >150°. Synth. produces carcinogenic byproducts. Causes exp. neoplasms. IQ0450000.

Robson, J.H. *et al*, *J. Am. Chem. Soc.*, 1955, **77**, 107 (*synth*)
Norris, W.P., *J. Am. Chem. Soc.*, 1959, **81**, 3346 (*synth*)
Boehme, H. *et al*, *Justus Liebigs Ann. Chem.*, 1965, **688**, 78 (*synth*)

Krebs, B. *et al*, *Acta Crystallogr., Sect. B*, 1979, **35**, 402 (*cryst struct*)
Bottaro, J.C. *et al*, *J. Org. Chem.*, 1988, **53**, 4140 (*synth, haz*)
Sax, N.I., *Dangerous Properties of Industrial Materials*, 5th Ed., Van Nostrand-Reinhold, 1979, 859.

3-Nitro-1,2-diphenylcyclopropene   N-80046

1,1'-(3-Nitro-1-cyclopropene-1,2-diyl)bisbenzene, 9CI
[4949-78-4]

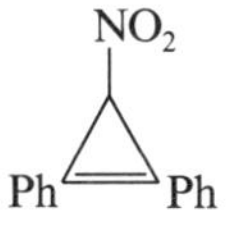

$C_{15}H_{11}NO_2$ M 237.257
Cryst. ($CHCl_3$/pet. ether). Mp 119° dec.

▷ Hazardous synth.

Jones, W.M. *et al*, *J. Org. Chem.*, 1965, **30**, 4389 (*synth, ir*)
Cheer, C.J. *et al*, *J. Am. Chem. Soc.*, 1988, **110**, 226 (*cryst struct, haz*)

1-Nitro-2,3-naphthalenediol, 9CI   N-80047

Updated Entry replacing N-01050
2,3-Dihydroxy-1-nitronaphthalene

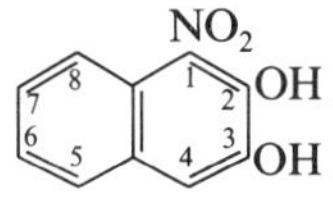

$C_{10}H_7NO_4$ M 205.170
Cryst. (EtOH aq.). Mp 210°.

Di-Me ether: [7311-21-9]. *2,3-Dimethoxy-1-nitronaphthalene*
$C_{12}H_{11}NO_4$ M 233.223
Pale yellow plates (EtOH). Mp 88.5-89°.

Bell, F., *J. Chem. Soc.*, 1959, 519 (*synth*)
Chang, C.W.J. *et al*, *J. Chem. Soc. C*, 1967, 840 (*deriv, pmr*)

1-Nitro-2,6-naphthalenediol, 9CI   N-80048

2,6-Dihydroxy-1-nitronaphthalene
[104750-82-5]
$C_{10}H_7NO_4$ M 205.170
Mp 173°.

6-Me ether: [110387-94-5]. *6-Methoxy-1-nitro-2-naphthol*
$C_{11}H_9NO_4$ M 219.196
Mp >250°.

Di-Me ether: [39077-18-4]. *2,6-Dimethoxy-1-nitronaphthalene*
$C_{12}H_{11}NO_4$ M 233.223
Orange needles (C_6H_6). Mp 189°.

Chakravarti, S.N. *et al*, *J. Chem. Soc.*, 1937, 1859 (*synth, deriv*)
Mechin, B. *et al*, *Org. Magn. Reson.*, 1980, **14**, 79 (*cmr*)
Roussel, J. *et al*, *Tetrahedron Lett.*, 1986, **27**, 27 (*synth*)
Lemaire, M. *et al*, *Tetrahedron*, 1987, **43**, 835 (*synth, ir, pmr, deriv*)

1-Nitro-2,7-naphthalenediol, 9CI   N-80049

Updated Entry replacing N-01051
2,7-Dihydroxy-1-nitronaphthalene
$C_{10}H_7NO_4$ M 205.170
Mp 198°.

Di-Me ether: [4614-14-6]. *2,7-Dimethoxy-1-nitronaphthalene*
$C_{12}H_{11}NO_4$ M 233.223
Yellow prisms (EtOH). Mp 141°.

Janczewski, M. *et al*, *Pol. J. Chem. (Rocz. Chem.)*, 1961, **35**, 953 (*synth*)

Richer, J.C. *et al*, *Can. J. Chem.*, 1965, **43**, 3443 (*deriv, pmr*)
Mechin, B. *et al*, *Org. Magn. Reson.*, 1980, **14**, 79 (*cmr*)
Roussel, J. *et al*, *Tetrahedron Lett.*, 1986, **27**, 27 (*synth*)

2-Nitro-1,4-naphthalenediol, 9CI   N-80050

1,4-Dihydroxy-2-nitronaphthalene

$C_{10}H_7NO_4$ M 205.170

Di-Me ether: [50885-07-9]. *1,4-Dimethoxy-2-nitronaphthalene*
$C_{12}H_{11}NO_4$ M 233.223
Cryst. Mp 97-98°.

Inoue, A. *et al*, *CA*, 1961, **55**, 13853e (*synth*)
Malesani, G. *et al*, *J. Heterocycl. Chem.*, 1985, **22**, 1141 (*pmr*)

2-Nitro-1,5-naphthalenediol   N-80051

1,5-Dihydroxy-2-nitronaphthalene

[104750-78-9]

$C_{10}H_7NO_4$ M 205.170
Mp 226°.

Lemaire, M. *et al*, *Tetrahedron*, 1987, **43**, 835 (*synth, ir, pmr*)

2-Nitro-1,7-naphthalenediol   N-80052

1,7-Dihydroxy-2-nitronaphthalene

[104750-80-3]

$C_{10}H_7NO_4$ M 205.170
Mp 220°.

Lemaire, M. *et al*, *Tetrahedron*, 1987, **43**, 835 (*synth, pmr, ir*)

4-Nitro-1,2-naphthalenediol, 9CI   N-80053

1,2-Dihydroxy-4-nitronaphthalene

$C_{10}H_7NO_4$ M 205.170

Di-Me ether: [42589-91-3]. *1,2-Dimethoxy-4-nitronaphthalene*
$C_{12}H_{11}NO_4$ M 233.223
Cryst. (EtOH). Mp 93-95.5°.

Archer, S. *et al*, *J. Med. Chem.*, 1980, **23**, 516 (*synth*)

4-Nitro-1,5-naphthalenediol, 9CI   N-80054

1,5-Dihydroxy-4-nitronaphthalene

$C_{10}H_7NO_4$ M 205.170
Mp 187°.

5-Me ether: [110387-91-2]. *5-Methoxy-4-nitro-1-naphthol*
$C_{11}H_9NO_4$ M 219.196
Mp 165°.

Di-Me ether: *1,5-Dimethoxy-4-nitronaphthalene*
$C_{12}H_{11}NO_4$ M 233.223
Yellow plates (AcOH). Mp 167°.

Thomson, R.H. *et al*, *J. Chem. Soc.*, 1947, 350 (*synth, deriv*)
Roussel, J. *et al*, *Tetrahedron Lett.*, 1986, **27**, 27 (*synth*)
Lemaire, M. *et al*, *Tetrahedron*, 1987, **43**, 835 (*synth, ir, pmr, deriv*)

4-Nitro-1,7-naphthalenediol, 9CI   N-80055

4,6-Dihydroxy-1-nitronaphthalene

$C_{10}H_7NO_4$ M 205.170
Mp 240°.

Di-Me ether: [1143-67-5]. *4,6-Dimethoxy-1-nitronaphthalene*
$C_{12}H_{11}NO_4$ M 233.223
Cryst. (MeOH). Mp 127-128°.

Roussel, J. *et al*, *Tetrahedron Lett.*, 1986, **27**, 27 (*synth*)
Richer, J.C. *et al*, *Can. J. Chem.*, 1965, **43**, 715 (*synth, deriv, pmr*)
Mechin, B. *et al*, *Org. Magn. Reson.*, 1980, **14**, 79 (*cmr*)
Ferlin, M.G. *et al*, *J. Heterocycl. Chem.*, 1989, **26**, 245 (*synth, deriv*)

6-Nitro-2,3-naphthalenediol, 9CI   N-80056

2,3-Dihydroxy-6-nitronaphthalene

$C_{10}H_7NO_4$ M 205.170

Di-Me ether: [14597-04-7]. *2,3-Dimethoxy-6-nitronaphthalene*
$C_{12}H_{11}NO_4$ M 233.223
Bright yellow needles (CCl$_4$). Mp 162-163°.

Chang, C.W.J. *et al*, *J. Chem. Soc. C*, 1967, 840 (*synth, deriv, pmr*)

1-Nitro-1-phenylpropene   N-80057

(1-Nitro-1-propenyl)benzene, 9CI

[25236-39-9]

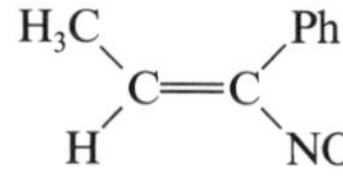

(*E*)-form

$C_9H_9NO_2$ M 163.176
Bp$_{0.5}$ 115°.

(*E*)-*form* [114390-79-3]
Yellow oil. Cont. 10% *Z*-isomer.

[114390-80-6]

Severin, T. *et al*, *Chem. Ber.*, 1969, **102**, 2966 (*synth*)
Asaro, F. *et al*, *Tetrahedron*, 1987, **43**, 3279 (*synth, ir, pmr, cmr*)

2-Nitro-1-pyrenol, 9CI   N-80058

1-Hydroxy-2-nitropyrene

[113093-71-3]

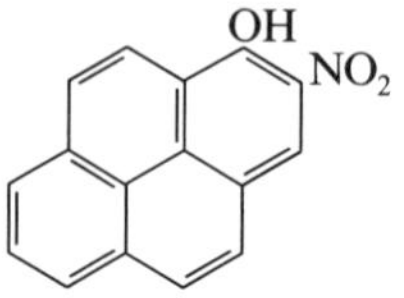

$C_{16}H_9NO_3$ M 263.252
1983 and 1984 refs. descr. this compound as 1-nitro-2-pyrenol. Deep red needles (C$_6$H$_6$). Mp 227-229°.

▷ Potent mutagen.

Yashura, A. *et al*, *Chem. Lett.*, 1983, 347 (*synth, ir, pmr, ms*)
Löfroth, G. *et al*, *Z. Naturforsch., C*, 1984, **39**, 193.
Van den Braken-van Leersum, A.M. *et al*, *J. Chem. Soc., Chem. Commun.*, 1987, 1156 (*synth, pmr, ms, struct*)

2-Nitrosoimidazole   N-80059

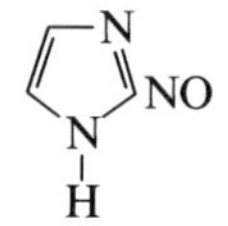

$C_3H_3N_3O$ M 97.076

N-Me: [116169-87-0]. *1-Methyl-2-nitroso-1H-imidazole*
$C_4H_5N_3O$ M 111.103
Green solid.

▷ Highly cytotoxic.

Noss, M.B. *et al*, *Biochem. Pharmacol.*, 1988, **37**, 2585 (*synth*)
Bolton, J.L. *et al*, *Can. J. Chem.*, 1988, **66**, 3044 (*struct, pmr*)

6-Nitro-1-[[(2,3,5,6-tetramethylphenyl) sulfonyl]oxy]-1*H*-benzotriazole, 9CI   N-80060

1-(3-Durenesulfonyloxy)-6-nitrobenzotriazole
[117678-83-8]

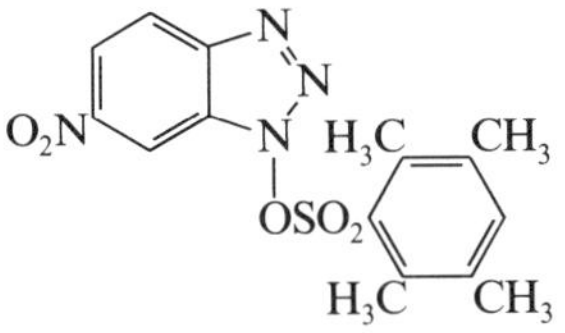

$C_{16}H_{16}N_4O_5S$ M 376.392
Improved reagent for ribonucleotide synth. by the triester method. Yellow cryst. (cyclohexane/toluene). Mp 158°.

Losse, G. *et al, Justus Liebigs Ann. Chem.*, 1989, 19 (*synth, use*)

4-Nitro-2-thiophenecarboxylic acid, 9CI   N-80061

[13138-70-0]

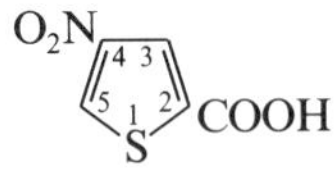

$C_5H_3NO_4S$ M 173.149
Glistening platelets (C_6H_6). Mp 154°.

Me ester: [24647-78-7].
   $C_6H_5NO_4S$ M 187.176
   Cryst. (Et_2O). Mp 98-100°.

Rinkes, I.J., *Recl. Trav. Chim. Pays-Bas (J. R. Neth. Chem. Soc.)*, 1932, **51**, 1134; 1933, **52**, 538 (*synth*)
Gever, G., *J. Am. Chem. Soc.*, 1953, **75**, 4585 (*synth, deriv*)
Noto, R. *et al, J. Chem. Soc., Perkin Trans. 2*, 1987, 689 (*cmr*)
Elliott, R. *et al, Tetrahedron*, 1987, **43**, 3295 (*synth, ir, pmr*)

5-Nitro-2-thiophenecarboxylic acid   N-80062

[6317-37-9]
$C_5H_3NO_4S$ M 173.149
Needles (H_2O). Mp 158°.

Me ester: [5832-01-9].
   $C_6H_5NO_4S$ M 187.176
   Needles (pet. ether). Mp 76°.

Rinkes, I.J., *Recl. Trav. Chim. Pays-Bas (J. R. Neth. Chem. Soc.)*, 1932, **51**, 1134; 1933, **52**, 538 (*synth*)
Noto, R. *et al, J. Chem. Soc., Perkin Trans. 2*, 1987, 689 (*cmr*)

Nivegin   N-80063

4,5-Dihydroxy-7-(4-hydroxyphenyl)-2H-1-benzopyran-2-one
[114020-35-8]

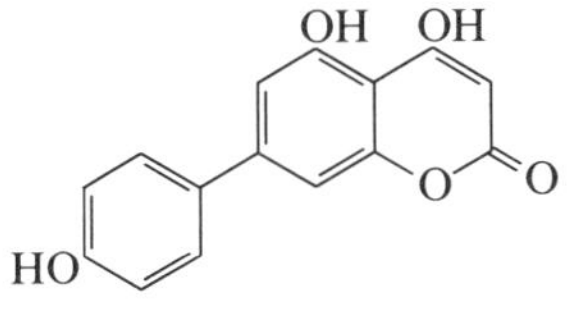

$C_{15}H_{10}O_5$ M 270.241
Constit. of *Echinops niveus*. Cryst. Mp 262-264°.

Parmar, V.S. *et al, Tetrahedron*, 1989, **45**, 1839.

15-Nonacosanol, 9CI   N-80064

[2764-81-0]

$$H_3C(CH_2)_{13}CH(OH)(CH_2)_{13}CH_3$$

$C_{29}H_{60}O$ M 424.792

Isol. from wax of *Arbutus unedo* and Brussels sprouts. Mp 83.1-83.8°.

Sahai, P.N. *et al, Biochem. J.*, 1932, **26**, 403 (*isol*)
Soza, A., *Compt. Rend. Hebd. Seances Acad. Sci.*, 1950, **230**, 995 (*isol*)

1-Nonacosanol   N-80065

[6624-76-6]

$$H_3C(CH_2)_{27}CH_2OH$$

$C_{29}H_{60}O$ M 424.792
Isol. from leaf wax of *Agave sisalana*. Cryst. (C_6H_6). Mp 85°.

Piper, S.H. *et al, Biochem. J.*, 1934, **28**, 2175 (*synth*)
Razafindrazaka, J. *et al, Bull. Soc. Chim. Fr.*, 1963, 1633 (*isol*)
Rao, S.J. *et al, Indian J. Chem., Sect. B*, 1987, **26**, 208 (*synth*)

15-Nonacosanone, 9CI   N-80066

Ditetradecyl ketone. Dimyristyl ketone
[2764-73-0]

$$H_3C(CH_2)_{13}CO(CH_2)_{13}CH_3$$

$C_{29}H_{58}O$ M 422.777
Isol. from cabbage leaves (*Brassica oleracea*). Platelets (C_6H_6/EtOH). Mp 80.5-81°.

Collison, D.L. *et al, Biochem. J.*, 1931, **25**, 606 (*isol*)
Kolattukudy, P.E., *Biochemistry*, 1965, **4**, 1844 (*biosynth*)
Matsota, N.P. *et al, Zh. Prikl. Khim. (Leningrad)*, 1971, **44**, 1823; *ca*, **76**, 3358c (*synth*)

Nonacyclo[11.7.1.1^{2,18}.0^{3,16}.0^{4,13}.0^{5,10}. 0^{6,14}.0^{7,11}.0^{15,20}]docosane   N-80067

Hexadecahydro-10a,1,4-ethanylylidene-7,5,9,11-[1,2]propanediyl[3]ylidene-10aH-benzo[b]fluorene, 9CI.
Bastardane
[20497-81-8]

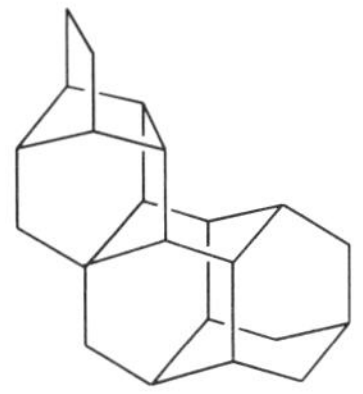

$C_{22}H_{28}$ M 292.463
Cryst. (Me_2CO/hexane or by subl.). Mp 144.5-146.5°.

von Schleyer, P.R. *et al, J. Am. Chem. Soc.*, 1968, **90**, 5034 (*synth, pmr, cryst struct*)
Osawa, E. *et al, J. Org. Chem.*, 1980, **45**, 2985 (*synth*)

6,8-Nonadien-1-ol   N-80068

[74206-16-9]

$$H_2C{=}CHCH{=}CH(CH_2)_4CH_2OH$$

$C_9H_{16}O$ M 140.225
(E)-form
   Oil.

Yasuda, H. *et al, Bull. Chem. Soc. Jpn.*, 1980, **53**, 1089 (*synth, ir, pmr*)
Vedejs, E. *et al, J. Org. Chem.*, 1988, **53**, 2220 (*synth, ir, pmr*)

1,3-Nonadiyne

N-80069

[77657-83-1]

$$H_3C(CH_2)_4C{\equiv}CC{\equiv}CH$$

C₉H₁₂   M 120.194

Kende, A.S. *et al, J. Org. Chem.*, 1988, **53**, 2655 (*synth, ir, pmr, ms*)

2-(3,7,11,15,19,23,27,31,35-Nonamethyl-2,6,10,14,18,22,26,30,34-hexatria-contanonenyl)-1,4-benzoquinone, 9CI

N-80070

2-Solanesyl-1,4-naphthoquinone. Desmethylmenaquinone-9

[2197-56-0]

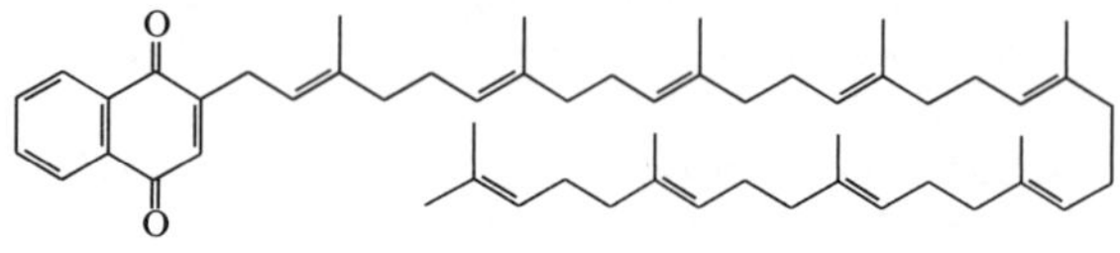

C₅₅H₇₈O₂   M 771.220

Prod. by *Streptococcus faecalis*. Yellow.

[20741-09-7]

Baum, R.H. *et al, J. Biol. Chem.*, 1963, **238**, PC4109; 1965, **240**, 3425 (*isol, uv, pmr, struct*)

2,3,5-Nonatriene

N-80071

[115942-23-9]

$$H_3CCH_2CH_2CH{=}CHCH{=}C{=}CHCH_3$$

C₉H₁₄   M 122.210

Reich, H.J. *et al, J. Am. Chem. Soc.*, 1988, **110**, 6432 (*synth, pmr, ir, cmr*)

4-Nonene-6,8-diyn-1-ol, 9CI

N-80072

$$HC{\equiv}CC{\equiv}CCH{=}CHCH_2CH_2CH_2OH$$

C₉H₁₀O   M 134.177

(E)-form

Isol. from *Aleurodiscus roseus*. Amorph. λ_{max} 229, 238, 251, 265, 281 nm. Not obt. completely pure.

Cambie, R.C. *et al, J. Chem. Soc.*, 1963, 4120 (*isol, uv, ir*)

3-Nonen-1-yne

N-80073

$$H_3C(CH_2)_4CH{=}CHC{\equiv}CH$$

C₉H₁₄   M 122.210

(E)-form

Liq. Bp₉₀ 91.5-92°.

Andreini, B.P. *et al, Tetrahedron*, 1987, **43**, 4591 (*synth, pmr*)

2-Nonyn-1-ol, 9CI

N-80074

Updated Entry replacing N-01595

[5921-73-3]

$$H_3C(CH_2)_5C{\equiv}CCH_2OH$$

C₉H₁₆O   M 140.225

Liq. with weak odour. Fp −21° (−16°). Bp₁₇ 114-116°, Bp₁₁ 109°.

3,5-Dinitrobenzoyl: Cryst. (ligroin). Mp 50°.

Me ether: [34498-02-7]. *1-Methoxy-2-nonyne, 9CI*
C₁₀H₁₈O   M 154.252
Oil.

Conia, J.M., *Bull. Soc. Chim. Fr.*, 1955, 1449 (*synth*)

Jones, V.K. *et al, J. Org. Chem.*, 1965, **30**, 3978 (*synth, spectra*)
Ermilova, E.V. *et al, Zh. Org. Khim.*, 1975, **11**, 520 (*synth*)
Fortunato, J.M. *et al, J. Org. Chem.*, 1976, **41**, 2194 (*synth*)
Zhai, D. *et al, J. Am. Chem. Soc.*, 1988, **110**, 2501 (*ether, synth, pmr, ir, ms*)

Norambreinolide

N-80075

Updated Entry replacing N-40065

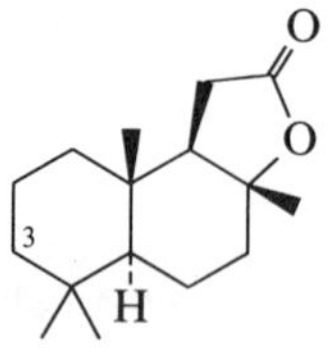

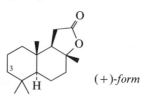

(+)-*form*

C₁₆H₂₆O₂   M 250.380

(+)-form [564-20-5]

Degradn. prod. of ambrein, etc. Found in tobacco. Cryst. (pet. ether or MeOH aq.). Mp 122-125°. [α]_D +42° (c, 0.25 in CHCl₃).

3α-Hydroxy: [123844-63-3]. *3α-Hydroxynorambreinolide*
C₁₆H₂₆O₃   M 266.380
Constit. of *Salvia aethiopis*. Cryst. (Et₂O/hexane). Mp 190-191°. [α]_D +22.2° (c, 1.08 in CHCl₃).

(−)-form [95406-08-9]

ent-*Norambreinolide*

Constit. of *Sideritis nutans*. Cryst. (EtOAc/pet. ether). Mp 120-122°.

Carman, R.M., *Aust. J. Chem.*, 1966, **19**, 1535 (*synth, bibl*)
Fernández, C. *et al, Phytochemistry*, 1985, **24**, 188 (*isol*)
Gonzalez, M.S. *et al, Tetrahedron*, 1989, **45**, 3575 (*3α-Hydroxynorambreinolide*)

Norasperenal A

N-80076

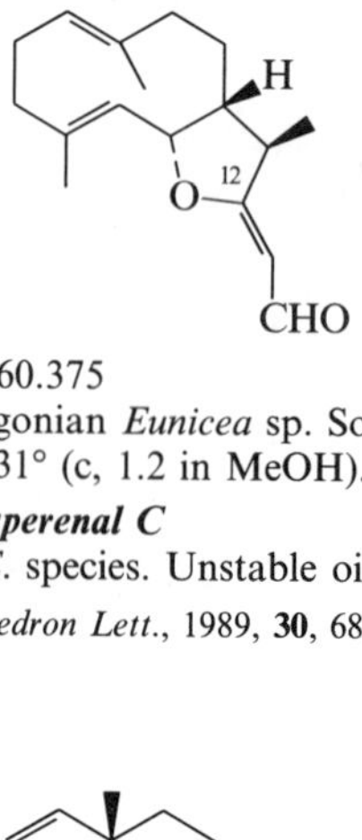

C₁₇H₂₄O₂   M 260.375

Metabolite of gorgonian *Eunicea* sp. Solid. Mp 101.5-102.5°. [α]_D +131° (c, 1.2 in MeOH).

12E-isomer: **Norasperenal C**

Metabolite of *E.* species. Unstable oil.

Shin, J. *et al, Tetrahedron Lett.*, 1989, **30**, 6821.

Norasperenal B

N-80077

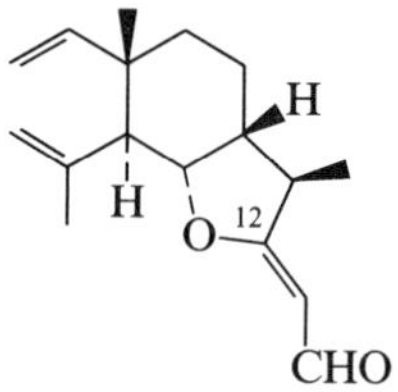

C₁₇H₂₄O₂   M 260.375

Metabolite of gorgonian *Eunicea* sp. Solid. Mp 123-124.5°. [α]_D −23° (c, 0.5 in MeOH).

12E-Isomer: **Norasperenal D**
C₁₇H₂₄O₂   M 260.375
Metab. of *E.* sp. Unstable oil.

Shin, J. *et al, Tetrahedron Lett.*, 1989, **30**, 6821.

24-Nor-7,22-cholestadiene-3,6-diol   N-80078

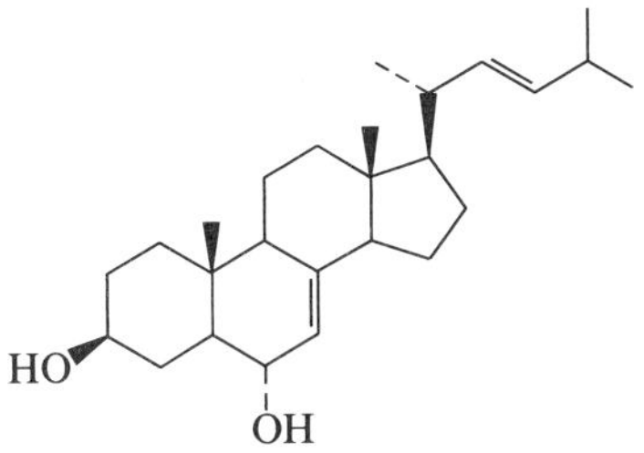

$C_{26}H_{42}O_2$ M 386.617

(3β,5α,6α,22E)-form

Constit. of *Spongionella gracilis*. Cryst. (MeOH/pet. ether). Mp 188-190°.

Madaio, A. *et al, J. Nat. Prod.* (*Lloydia*), 1989, **52**, 952 (*isol, pmr*)

Nordentatin   N-80079

Updated Entry replacing P-30169

10-(1,1-Dimethyl-2-propenyl)-5-hydroxy-8,8-dimethyl-2H,8H-benzo[1,2-b:5,4-b']dipyran-2-one, 9CI

[17820-07-4]

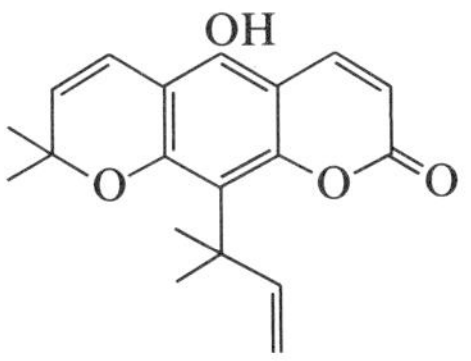

$C_{19}H_{20}O_4$ M 312.365

Constit. of roots of *Clausena dentata*. Mp 182°. Struct. revised in 1973.

Me ether: [22980-57-0]. **Ponicitrin.** *Dentatin*

From *C. dentata* and from roots of *Poncirus trifoliata*. Pillars (EtOH). Mp 93-94°.

O(1,1-Dimethyl-2-propenyl): **Ponfolin**

Isol. from roots of *P. trifoliata*. Oil.

Tomimatsu, T. *et al, Tetrahedron*, 1972, **28**, 2003 (*isol, uv, pmr*)
Mahey, S. *et al, Indian J. Chem.*, 1973, **11**, 983 (*synth*)
Mowat, D. *et al, Tetrahedron*, 1973, **29**, 2943 (*synth*)
Tomimatsu, T. *et al, Tetrahedron*, 1974, **30**, 939 (*struct*)
Murray, R.D.H. *et al, Tetrahedron Lett.*, 1983, **24**, 3773 (*synth*)
Furukawa, H. *et al, Chem. Pharm. Bull.*, 1986, **34**, 3922 (*Ponfolin*)

8-Norlactaranelactone   N-80080

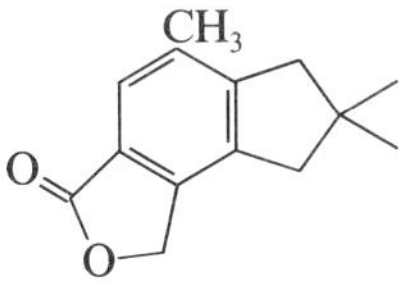

$C_{14}H_{16}O_2$ M 216.279

Constit. of *Lactarius serobiculatus*. Cryst. Mp 93-96°.

Bosetti, A. *et al, Phytochemistry*, 1989, **28**, 1427.

29-Nor-8,24-lanostadiene-1,2,3-triol   N-80081

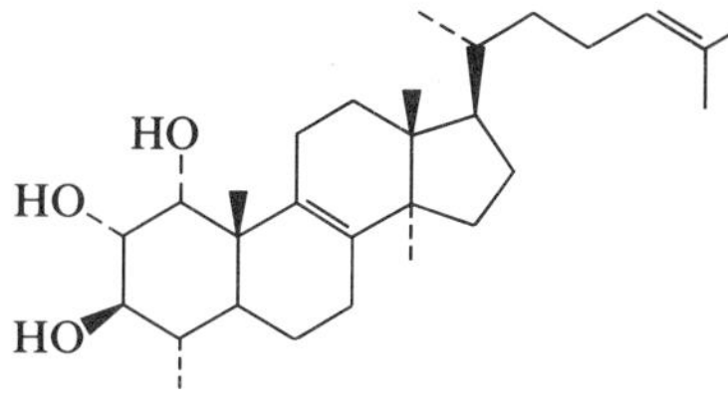

$C_{29}H_{48}O_3$ M 444.696

(1α,2α,3β)-form [119765-92-3]

Constit. of *Commiphora incisa*. Amorph. $[\alpha]_D$ +38.2° (c, 0.1 in $CHCl_3$).

Provan, G.J. *et al, Phytochemistry*, 1988, **27**, 3841.

Norpinguisanolide   N-80082

[119285-62-0]

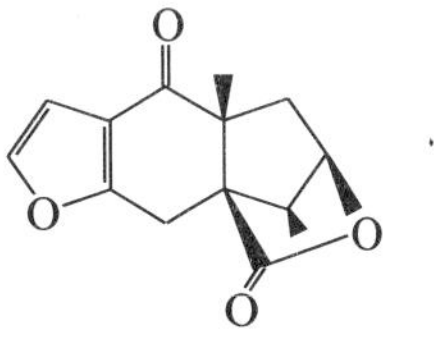

$C_{14}H_{14}O_4$ M 246.262

Constit. of *Porella elegantula*. Plates (hexane). Mp 132-134°. $[\alpha]_D^{22}$ −125° (c, 0.47 in $CHCl_3$).

Fukuyama, Y. *et al, Phytochemistry*, 1988, **27**, 3557 (*isol, pmr*)

Norpinguisone   N-80083

[62121-27-1]

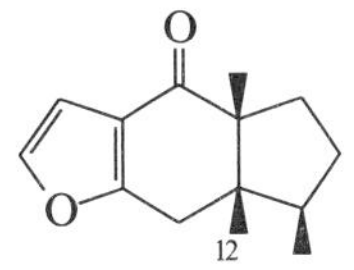

$C_{14}H_{18}O_2$ M 218.295

Constit. of *Porella vernicosa* and *P. elegantula*. Cryst. Mp 126-127°. $[\alpha]_D$ +2° ($CHCl_3$).

12-Carboxylic acid, Me ester: [119285-56-2]. *Methyl 4-oxonorpinguisan-12-oate. Norpinguisone methyl ester (incorr.)*

$C_{15}H_{18}O_4$ M 262.305

Constit. of *P. vernicosa* and *P. elegantula*. Oil. $[\alpha]_D^{16}$ −60.2° (c, 0.91 in $CHCl_3$).

[67594-77-8]

Asakawa, Y. *et al, Bull. Soc. Chim. Fr.*, 1976, 1469 (*isol*)
Fukuyama, Y. *et al, Phytochemistry*, 1988, **27**, 3557 (*isol, struct*)

Numersterol A   N-80084

24-Methylene-1α,3β,5α,6β-cholestanetetrol

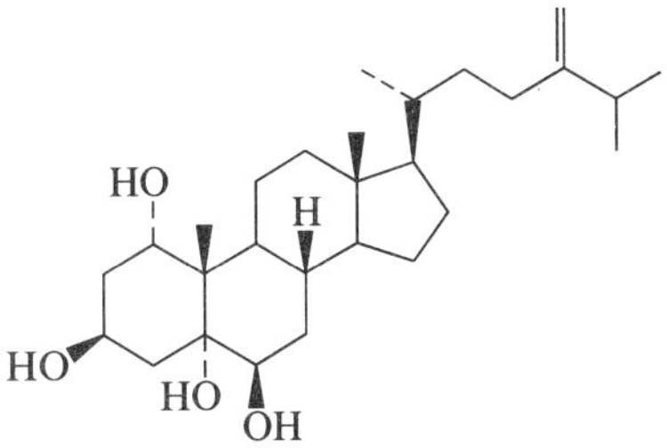

$C_{28}H_{48}O_4$ M 448.685

Constit. of *Sinularia numerosa*. Cryst. (MeOH). Mp 297-
299°. $[\alpha]_D^{25}$ +4.5° (c, 0.332 in MeOH).

Su, J. *et al*, *J. Nat. Prod.* (*Lloydia*), 1989, **52**, 934 (*isol, pmr, cmr, cryst struct*)

Numersterol B   N-80085

*25-Methylene-22-homocholestane-1β,3β,5α-triol. 27,27-
Dimethyl-25-cholestene-1β,3β,5α-triol*

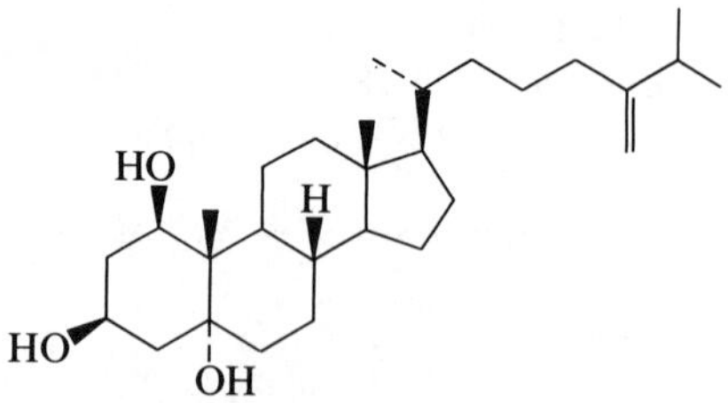

$C_{29}H_{50}O_3$ M 446.712

Constit. of *Sinularia numerosa*. Cryst. (Me₂CO/pet. ether).
Mp 121-122°.

Su, J. *et al*, *J. Nat. Prod.* (*Lloydia*), 1989, **52**, 934 (*isol, pmr, cmr*)

O

Obionin A   O-80001

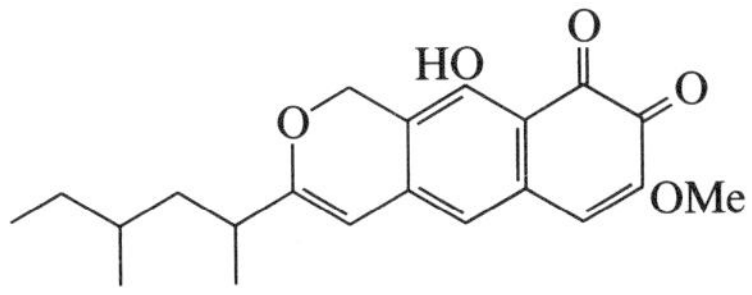

$C_{21}H_{24}O_5$ M 356.418

Metab. of marine fungus *Leptosphaeria obiones*. Brown-red solid. Mp 168-169°. $[\alpha]_D$ +28.5° (c, 0.01 in $CHCl_3$).

Poch, G.K. *et al*, *Tetrahedron Lett.*, 1989, **30**, 3483.

3,11,19,27,33,34,35,36-Octaazapentacyclo [27.3.1.1^{5,9}.1^{13,17}.1^{21,25}]hexatriaconta-1(33),5(34),6,8,13,(35),14,16,21(36),22, 24,29,31-dodecaene   O-80002

Cyclotetra(2,6-pyridinemethyleneimine)

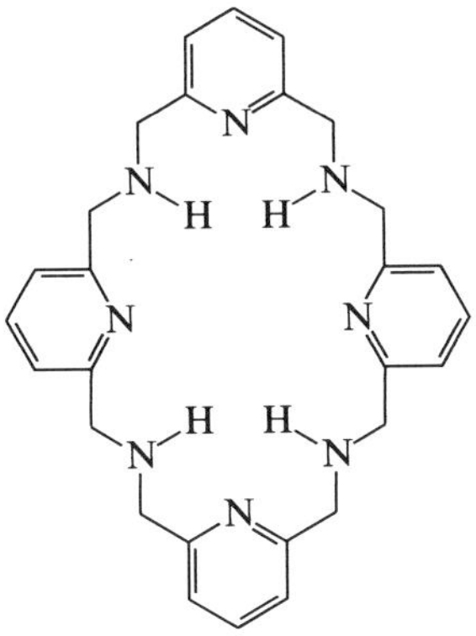

$C_{28}H_{32}N_8$ M 480.614

B,5HCl: Mp >260°.

N-*Tetrakis(4-methylbezenesulfonyl):* Mp >260°.

Hosseini, M.W. *et al*, *Helv. Chim. Acta*, 1989, **72**, 1066 (*synth*)

3,6,9,17,20,23,29,30-Octaazatricyclo [23.3.1.1^{11,15}]triaconta-1(29),11(30),12, 14,25,27-hexaene   O-80003

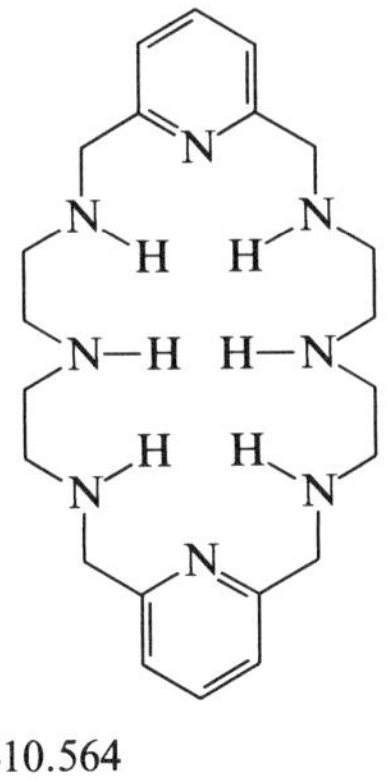

$C_{22}H_{34}N_8$ M 410.564

B,8HCl: Tetrahydrate. Mp 228°.

N-*Hexakis-(4-methylbenzenesulfonyl):* Cryst. (CH_2Cl_2/hexane/DMF). Mp 142°.

Hosseini, M.W. *et al*, *Helv. Chim. Acta*, 1989, **72**, 1066 (*synth*)

Octachloropropane   O-80004

Perchloropropane

[594-90-1]

$$Cl_3CCCl_2CCl_3$$

C_3Cl_8 M 319.655

Oil. Bp 280°, Bp_{734} 268-269°.

Kraft, F. *et al*, *Ber.*, 1875, **8**, 1296 (*synth*)
Farah, B.S. *et al*, *J. Org. Chem.*, 1965, **30**, 1241 (*synth*)

Octacosanal   O-80005

[22725-64-0]

$$H_3C(CH_2)_{26}CHO$$

$C_{28}H_{56}O$ M 408.750

Isol. from wax of grapes (*Vitis vinifera*) and from cabbage leaves (*Brassica oleracea*).

Radler, F. *et al*, *Aust. J. Chem.*, 1965, **18**, 1059 (*isol*)
Schmid, H.H.O. *et al*, *Hoppe Seyler's Z. Physiol. Chem.*, 1969, **350**, 462 (*isol*)

11,17-Octadecadien-9-ynoic acid, 9CI   O-80006

$$H_2C=CH(CH_2)_4CH=CHC\equiv C(CH_2)_7COOH$$

$C_{18}H_{28}O_2$ M 276.418

(*E*)-*form* [5910-97-4]

Isol. from seed oil of *Acanthosyris spinescens* and *Pyrucaria pubera*. Cryst. (hexane). Mp 30.0-34.5°.

Me ester: [20714-98-1].
$C_{19}H_{30}O_2$ M 290.445
Liq.

4-*Phenylphenacyl ester:* [20714-99-2].
Cryst. (EtOH). Mp 53-54°.

Powell, R.G. *et al*, *Biochemistry*, 1966, **5**, 625 (*isol, uv, ir, pmr, struct*)
Hopkins, C.Y. *et al*, *J. Chem. Soc. C*, 1968, 2462 (*isol, uv, ir, pmr*)

2-Octadecanol, 9CI   O-80007

[593-32-8]

$$H_3C(CH_2)_{15}CH(OH)CH_3$$

$C_{18}H_{38}O$ M 270.498

(+)-*form* [23559-98-0]

Isol. from avian tubercle bacillus wax. Needles (MeOH). Mp 56°. $[\alpha]_D^{25}$ +5.7° ($CHCl_3$).

Phenylurethane: Long fine needles. Mp 76-77°. $[\alpha]_D^{22}$ +7.9° ($CHCl_3$).

3,5-*Dinitrobenzoyl:* Cryst. (EtOH). Mp 71-72°. $[\alpha]_D$ +25.3° ($CHCl_3$).

Pangborn, M.C. *et al*, *J. Am. Chem. Soc.*, 1936, **58**, 10 (*isol*)
Reeves, R.E. *et al*, *J. Am. Chem. Soc.*, 1937, **59**, 858 (*isol*)
Tulloch, A.P., *Can. J. Chem.*, 1968, **46**, 3727 (*synth, pmr*)

4,8,12,15-Octadecatetraenoic acid, 9CI   O-80008

[67329-10-6]

$$H_3CCH_2CH=CHCH_2CH=CHCH_2CH_2CH=$$
$$CHCH_2CH_2CH=CHCH_2CH_2COOH$$

$C_{18}H_{28}O_2$ M 276.418

Isol. from sardine oil, from seeds of *Lithospermum officinale* and from *Salpa thomsoni*.

[106440-10-2]

Toyama, Y. *et al, Bull. Chem. Soc. Jpn.*, 1935, **10**, 232 (*isol*)
Hörhammer, L. *et al, Arzneim.-Forsch.*, 1964, **14**, 34 (*isol*)
Mimura, T. *et al, Chem. Pharm. Bull.*, 1986, **34**, 4562 (*isol*)

15-Octadecene-9,11,13-triynoic acid   O-80009

$$H_3CCH_2CH=CH(C≡C)_3(CH_2)_7COOH$$

$C_{18}H_{22}O_2$ M 270.371

Major component of the lipids of *Santalum acuminatum*.
λ_{max} 212, 223, 232, 244, 258, 274, 291, 310, 332 nm.

Bu'Lock, J.D. *et al, Phytochemistry*, 1963, **2**, 289.

17-Octadecen-9-ynoic acid   O-80010

$$H_2C=CH(CH_2)_6C≡C(CH_2)_7COOH$$

$C_{18}H_{30}O_2$ M 278.434

Isol. from seed oil of *Acanthosyris spinescens*.

Powell, R.G. *et al, Biochemistry*, 1966, **5**, 625.

2,6-Octadienoic acid   O-80011

$$H_3CCH=CHCH_2CH_2CH=CHCOOH$$

$C_8H_{12}O_2$ M 140.182

(E,E)-form [116565-36-7]
Waxy solid. Mp 44-46°.
Me ester: [25172-05-8].
$C_9H_{14}O_2$ M 154.208
Liq.

Crandall, J.K. *et al, J. Org. Chem.*, 1970, **35**, 3049 (*synth, ir, pmr*)
Snider, B.B., *J. Org. Chem.*, 1988, **53**, 5320 (*synth, pmr, cmr, ir*)

Octahydro-1H-2-benzopyran-1-one   O-80012

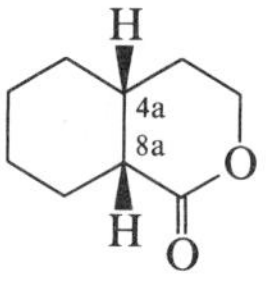

(4aRS,8aRS)-form

$C_9H_{14}O_2$ M 154.208

(4aRS,8aRS)-form
(±)-cis-*form*
Liq. Bp$_{4.5}$ 118°.
Hydrazide: Needles (EtOH/Et$_2$O). Mp 109-110°.
(4aRS,8aSR)-form
(±)-trans-*form*
Prisms. Mp 48°. Bp$_3$ 103°.
Hydrazide: Prisms. Mp 123-124°.

Fujiwara, Y. *et al, Chem. Pharm. Bull.*, 1989, **37**, 1458 (*synth*)

Octahydro-2H-1-benzothiopyran, 9CI   O-80013

Updated Entry replacing O-20017
1-Thiadecalin
[29100-30-9]

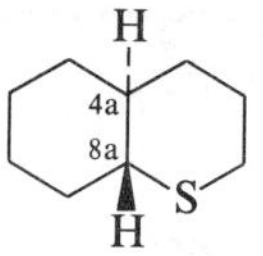

(8aR,4aS)-form

$C_9H_{16}S$ M 156.291

(8aR,4aS)-form [116782-11-7]
(+)-trans-form
Solid. Mp 34.5-35.0°. Bp$_1$ 110°. [α]$_D^{26}$ +24.4° (c, 6.15 in MeOH).
1,1-Dioxide: [116782-13-9].
$C_9H_{16}O_2S$ M 188.290
Cryst. (hexane). Mp 103-103.2°. [α]$_D^{29}$ +3.5° (c, 6.39 in MeOH).
(4aRS,8aRS)-form [57259-80-0]
(±)-cis-*form*
Mp −1° to 1°.
HgCl$_2$ complex: [63714-82-9].
Mp 176-177°.
(4aRS,8aSR)-form [54340-73-7]
(±)-trans-*form*. 4aα,8aβ-*form*
Mp 17-18°.
HgCl$_2$ complex: [63743-83-9].
Mp 170-171°.
1α-Oxide: [67530-10-3].
$C_9H_{16}OS$ M 172.291
Mp 87-88°.
1β-Oxide: [67530-09-0].
$C_9H_{16}OS$ M 172.291
Mp 71-72°.

Claus, P.K. *et al, J. Org. Chem.*, 1977, **42**, 4016 (*synth*)
Oae, S. *et al, Bull. Chem. Soc. Jpn.*, 1983, **56**, 270 (*oxides*)
Cerè, V. *et al, J. Org. Chem.*, 1988, **53**, 5689 (*synth, cmr, pmr*)

5,6,15,16,19,20,29,30-Octahydro-2,17:3,18:7,11:10,14:21,25:24,28-hexamethenobenzocyclooctacosene, 9CI   O-80014

[88903-18-8]

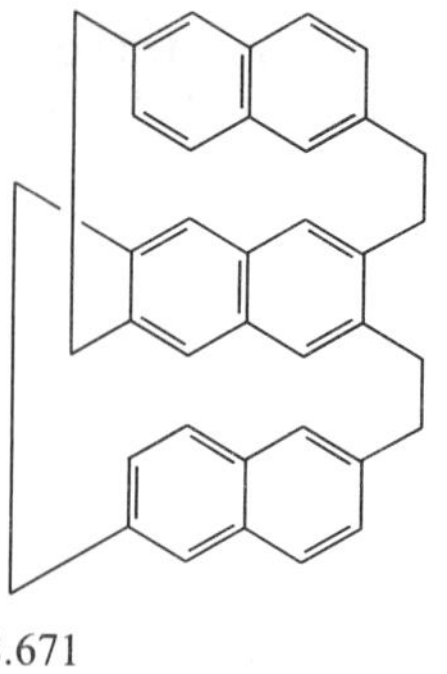

$C_{38}H_{32}$ M 488.671
Prisms. Mp 230° dec.

Otsubo, T. *et al, Bull. Chem. Soc. Jpn.*, 1989, **62**, 164 (*synth, cryst struct*)

1,2,3,4,5,6,7,8-Octahydro-8-oxo-2-naphth-alenecarboxylic acid, 9CI   O-80015

$\Delta^{9,10}$-Octal-1-one-7-carboxylic acid. 3,4,5,6,7,8-Hexahydronaphthalen-1(2H)-one-2-carboxylic acid

[32178-64-6]

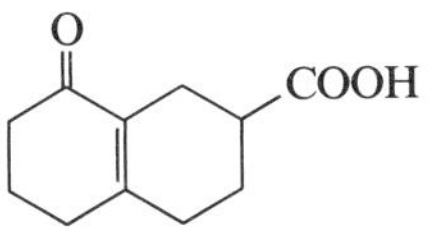

$C_{11}H_{14}O_3$ M 194.230

(±)-form [112400-28-9]
 Needles (MeCN aq.). Mp 145-147°.

Me ester:
 $C_{12}H_{16}O_3$ M 208.257
 Intermed. for terpene synth. Bp$_{0.11}$ 119-122°.

Me ester, 2,4-dinitrophenylhydrazone: Red cryst. (MeOH/EtOAc). Mp 194.5-196.5°.

Et ester: [112400-24-5].
 $C_{13}H_{18}O_3$ M 222.283
 Oil.

[34407-92-6]

Huffman, J.W. et al, J. Org. Chem., 1972, **37**, 13 (synth, ir, uv, pmr)
Carlson, R.G. et al, J. Org. Chem., 1972, **37**, 2468 (synth, ir, uv, pmr)
Strekowski, L. et al, J. Org. Chem., 1988, **53**, 901 (synth)

Octahydropyridazino[1,2-a]pyridazine, 9CI   O-80016

1,6-Diazabicyclo[3.3.0]decane. 4a,8a-Diazadecalin. 10-Azaquinolizidine

[3661-15-2]

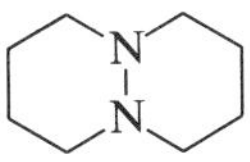

$C_8H_{16}N_2$ M 140.228
Oil. Bp$_5$ 40°. Forms a cryst. monohydrate.
B,HClO$_4$: Mp 231°.
Picrate: Mp 155°.

Hedaya, E. et al, J. Am. Chem. Soc., 1963, **85**, 3052 (synth)
Koopmann, H.P. et al, Spectrochim. Acta, Part A, 1976, **32**, 157 (ir, raman)
Nelsen, S.F. et al, J. Am. Chem. Soc., 1978, **100**, 4004 (cmr)
Nelsen, S.F. et al, J. Org. Chem., 1980, **45**, 3609 (nmr)
White, J.M. et al, Acta Crystallogr., Sect. C, 1988, **44**, 1777 (cryst struct)

2,2′,4,4′,7,7′,8,8′-Octahydroxy-1,1′-biphenanthrene   O-80017

[1,1′-Biphenanthrene]-2,2′,4,4′,7,7′,8,8′-octol

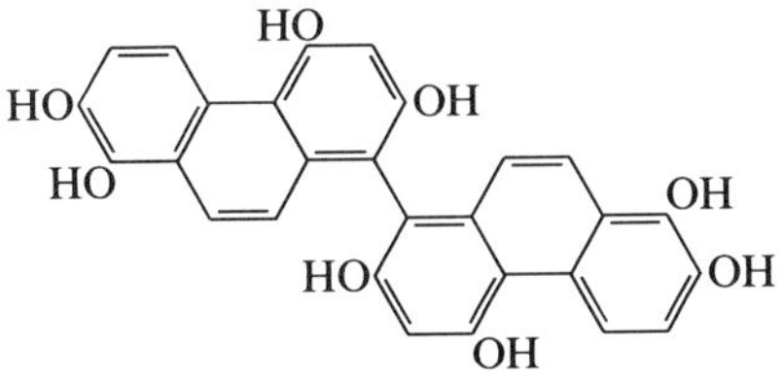

$C_{28}H_{18}O_8$ M 482.445
4,4′,8,8′-Tetra-Me ether: [118201-23-3]. 2,2′,7,7′-Tetrahydroxy-4,4′8,8′-tetramethoxy-1,1′-biphenanthrene. 4,4′,8,8′-Tetramethoxy-[1,1′-biphenanthrene]-2,2′,7,7′-tetrol
$C_{32}H_{26}O_8$ M 538.553

Constit. of Eulophia nuda. Pink needles (Et$_2$O/CHCl$_3$/Me$_2$CO). Mp 310° dec.

Tuchinda, P. et al, Phytochemistry, 1988, **27**, 3267.

1,2,3,4,5,6,7,8-Octamethylanthracene   O-80018

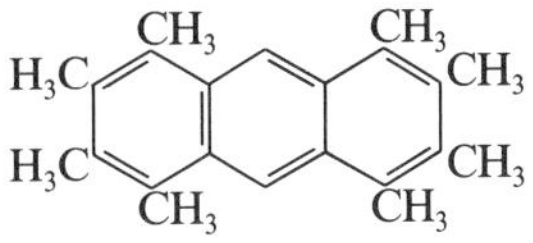

$C_{22}H_{26}$ M 290.447
Yellow cryst. powder (chlorobenzene). Mp 299-300°. Fluor. in soln.

Backer, H.J. et al, Recl. Trav. Chim. Pays-Bas (J. R. Neth. Chem. Soc.), 1939, **58**, 761 (synth)
Krysin, A.P. et al, J. Org. Chem. USSR (Engl. Transl.), 1977, **13**, 1183 (synth, uv, pmr)
Hart, H. et al, J. Org. Chem., 1981, **46**, 1251 (synth)
Meador, M.A. et al, J. Org. Chem., 1989, **54**, 2336 (synth, pmr)

1,2,4-Octatriene   O-80019

$$H_3CCH_2CH_2CH{=}CHCH{=}C{=}CH_2$$

C_8H_{12} M 108.183
(E)-form [115942-28-4]
 Liq.

Reich, H.J. et al, J. Am. Chem. Soc., 1988, **110**, 6432 (synth, pmr, ir, cmr)

2-(3,5,7-Octatrien-1-ynyl)thiophene   O-80020

8-(2-Thienyl)-1,3,5-octatriene-7-yne

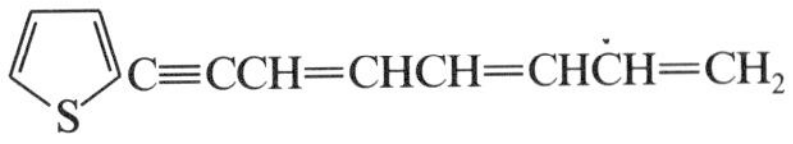

$C_{12}H_{10}S$ M 186.277
Isol. from roots of Matricaria invdora, M. oreades and Xeranthemum cylindraceum. Cryst. or oil. Mp 48°. Prob. a mixt. of geom. isomers as isolated. Mp. refers to synthetic material (geom. isom. not reported).

Sörensen, N.A., Proc. Chem. Soc., London, 1961, 98 (struct, synth)
Bohlmann, F. et al, Chem. Ber., 1964, **97**, 1193, 2125 (isol, struct, uv, ir, pmr, synth)

2-Octenedial   O-80021

[105582-16-9]

$$OHC(CH_2)_4CH{=}CHCHO$$

$C_8H_{12}O_2$ M 140.182
[41980-05-6]

Carlson, R.G. et al, J. Chem. Soc., Chem. Commun., 1973, 223 (synth)
Hudlicky, T. et al, J. Am. Chem. Soc., 1988, **110**, 4735 (synth, ir, pmr)

5-Octenoic acid   O-80022

[63892-00-2]

$$H_3CCH_2CH{=}CHCH_2CH_2CH_2COOH$$

$C_8H_{14}O_2$ M 142.197
(E)-form [16424-53-6]
 Bp$_{1.5}$ 120°.
[41653-97-8]

Scholz, D., Justus Liebigs Ann. Chem., 1984, 264.
Kawashima, M. et al, Bull. Chem. Soc. Jpn., 1988, **61**, 3255.

Octopinic acid — O-80023

Updated Entry replacing O-00426
N^2-(1-Carboxyethyl)ornithine, 9CI
[20197-09-5]

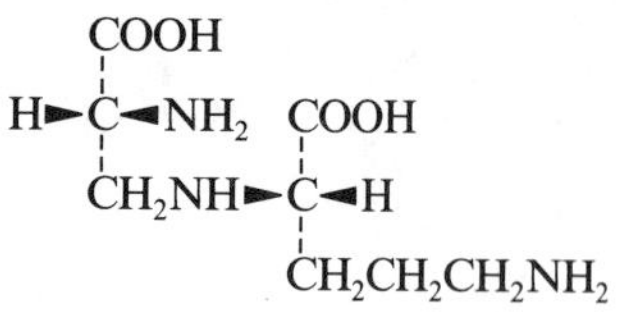

$C_8H_{16}N_2O_4$ M 204.225

Amino acid found in crown gall tissue of *Scorzonera* root infected with *Agrobacterium tumefaciens*. Mp 270-271° dec. $[\alpha]_D^{20}$ +18.5° (c, 0.5 in H_2O).

Lejeune, B., *Compt. Rend. Hebd. Seances Acad. Sci. Sect. D*, 1967, **265**, 1753 (synth)
Goto, K. *et al*, *Bull. Chem. Soc. Jpn.*, 1982, **55**, 261 (synth)

7-Octynoic acid — O-80024

[10297-09-3]

$HC\equiv C(CH_2)_5COOH$

$C_8H_{12}O_2$ M 140.182
Liq. $Bp_{0.4}$ 96-97°.
Anhydride: [119837-88-6].
$C_{16}H_{22}O_3$ M 262.348
Liq. Bp_1 160-166°.

Moody, C.J. *et al*, *J. Chem. Soc., Perkin Trans. 1*, 1988, 3249 (synth, anhydride, ir, pmr, ms)

1-Octyn-3-ol, 9CI — O-80025

Updated Entry replacing O-00469
[818-72-4]

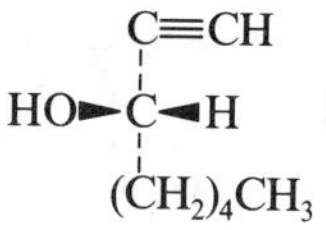

(S)-form

(*S*)-form [32556-71-1]
Oil. Bp_{20} 88-89°. $[\alpha]_D^{20}$ −6.79° ($CHCl_3$).
(±)-form
Bp_{19} 83°.

Midland, M.M., *J. Org. Chem.*, 1975, **40**, 2250 (synth)
Amos, R.A. *et al*, *J. Org. Chem.*, 1978, **43**, 555 (synth)
Larock, R.C. *et al*, *Tetrahedron*, 1987, **43**, 2013 (resoln, pmr)

Odonticin — O-80026

[122398-17-8]

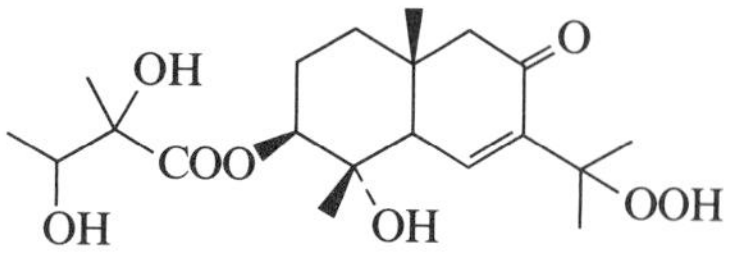

$C_{20}H_{32}O_8$ M 400.468
Constit. of *Pluchea arguta*. Syrup. $[\alpha]_D^{20}$ +178° (c, 0.01 in MeOH).

Ahmad, V.U. *et al*, *J. Nat. Prod. (Lloydia)*, 1989, **52**, 861 (isol, pmr)

Odontin — O-80027

7α-Hydroperoxy-3β,4α-dihydroxy-11-eudesmen-8-one
[122398-16-7]

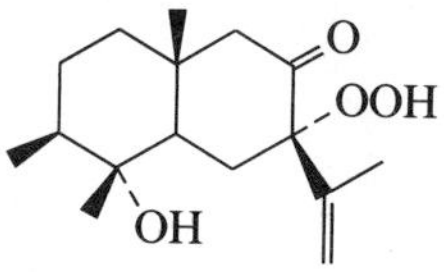

$C_{15}H_{24}O_5$ M 284.352
Constit. of *Pluchea arguta*. Gum. $[\alpha]_D^{20}$ −180° (c, 0.04 in MeOH).

Ahmad, V.U. *et al*, *J. Nat. Prod. (Lloydia)*, 1989, **52**, 561 (isol, pmr)

Okamurallene — O-80028

Updated Entry replacing O-10035
[80539-33-9]

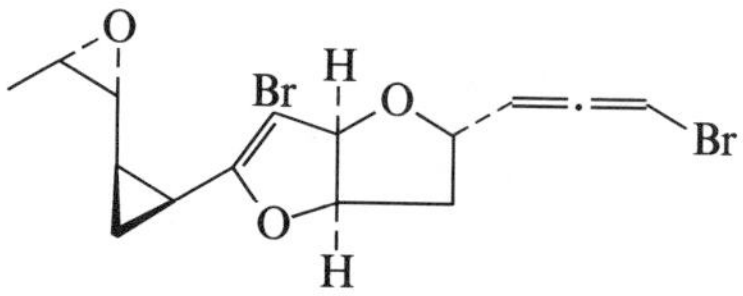

$C_{15}H_{16}Br_2O_3$ M 404.098
Constit. of *Laurencia okamurai*. Oil. $[\alpha]_D^{26}$ +160° (c, 1.74 in $CHCl_3$).

Suzuki, M. *et al*, *Tetrahedron Lett.*, 1981, **22**, 3853 (isol)
Suzuki, M. *et al*, *Phytochemistry*, 1989, **28**, 2145 (cmr, struct)

Okamurallene chlorohydrin — O-80029

[123297-19-8]

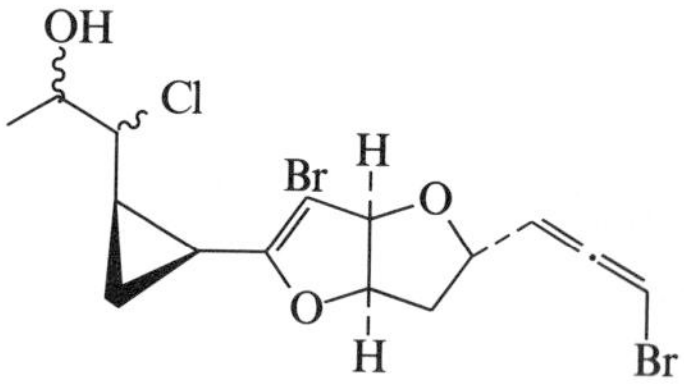

$C_{15}H_{17}Br_2ClO_3$ M 440.558
Metab. of *Laurencia intricata*. Cryst. (hexane). Mp 71-72°. $[\alpha]_D^{18}$ +205° (c, 0.325 in CCl_4).

Suzuki, M. *et al*, *Phytochemistry*, 1989, **28**, 2145 (isol, pmr, cmr)

11,13(18)-Oleanadiene-3,24-diol — O-80030

$C_{30}H_{48}O_2$ M 440.708
3β-form [119285-49-3]
Constit. of *Phyllanthus flexuosus*. Needles (MeOH/CHCl_3). Mp 302-305°.

Tanaka, R. *et al*, *Phytochemistry*, 1988, **27**, 3563.

12-Oleanene-3,15-diol — O-80031

$C_{30}H_{50}O_2$ M 442.724
(3β,15α)-form [61497-68-5]
Constit. of *Phyllanthus flexuosus*. Needles (MeOH/CHCl_3). Mp 245-246.5°. $[\alpha]_D^{23}$ +82.5° (c, 1.21 in $CHCl_3$).

Hui, W.H. *et al*, *Phytochemistry*, 1977, **15**, 1313 (synth)
Tanaka, R. *et al*, *Phytochemistry*, 1988, **27**, 3563 (isol)

12-Oleanene-3,24-diol   O-80032

$C_{30}H_{50}O_2$ M 442.724

3β-form [119318-15-9]

Constit. of *Phyllanthus flexuosus*. Needles
(MeOH/CHCl$_3$). Mp 251-253°. [α]$_D^{23}$ +84.7° (c, 0.46 in
CHCl$_3$).

Ogihara, K. *et al*, *Phytochemistry*, 1987, **26**, 783 (*synth*)
Tanaka, R. *et al*, *Phytochemistry*, 1988, **27**, 3563 (*isol*)

18-Oleanene-3,16-diol   O-80033

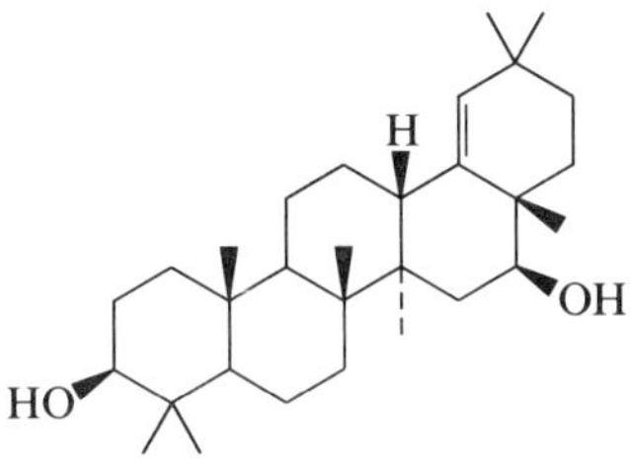

$C_{30}H_{50}O_2$ M 442.724

(3β,16β)-form [122577-94-0]

Constit. of *Schaefferia cuneifolia*. Cryst. Mp 237-238°.
[α]$_D^{20}$ +8.5° (c, 0.07 in CHCl$_3$).

Di-Ac: [122577-95-1].
$C_{34}H_{54}O_4$ M 526.798
Constit. of *S. cuneifolia*. Cryst. (EtOAc/hexane). Mp
230-232°. [α]$_D^{20}$ +6.35° (c, 0.236 in CHCl$_3$).

3-Ketone: [122577-96-2]. *16β-Hydroxy-18-oleanen-3-one*
$C_{30}H_{48}O_2$ M 440.708
Occurs in *S. cuneifolia* as inseparable mixture with 16α-
hydroxy-18-oleanen-3-one.

Gonzalez, A.G. *et al*, *J. Nat. Prod. (Lloydia)*, 1989, **52**, 567 (*isol,
pmr, cmr*)

12-Oleanene-3,16,21,22,23,28-hexol   O-80034

Updated Entry replacing O-00508

$C_{30}H_{50}O_6$ M 506.721

(3β,16β,21β,22α)-form [22467-07-8] *Gymnemagenin*

Aglycone from *Gymnema sylvestre*. Cryst.
(CHCl$_3$/MeOH). Mp 328-335°. [α]$_D^{25}$ +53.1° (c, 1 in
MeOH).

3-O-β-D-Glucuronopyranosyl: Deacylgymnemic acid
Cryst. Mp 230-235°.

3-O-β-D-Glucuronopyranosyl, 21-O-tigloyl: Gymnemic acid IV
$C_{41}H_{64}O_{13}$ M 764.949
Constit. of *G. sylvestre*. Antisweet compound. Cryst. Mp
220-221° (213-215°). [α]$_D$ +8.8° (c, 5.4 in MeOH).

*3-O-β-Glucuronopyranosyl, 21-O-tigloyl, 28-Ac: Gymnemic
acid I*
$C_{43}H_{66}O_{14}$ M 806.986
Constit. of *G. sylvestre*. Antisweet compound. Cryst. Mp
211-212°. [α]$_D$ +36.7° (c, 2.4 in MeOH).

*3-O-β-D-Glucuropyranosyl, 21-O-(2-methylbutanoyl):
Gymnemic acid III*
$C_{41}H_{66}O_{13}$ M 766.965
Constit. of *G. sylvestre*. Antisweet compound. Mp 218-
219° (209.5-212°). [α]$_D$ +7.6° (c, 2.9 in MeOH).

*3-O-β-D-Glucuronopyranosyl, 21-O-(S-2-methylbutanoyl),
28-Ac: Gymnemic acid II*
$C_{43}H_{68}O_{14}$ M 809.002
Constit. of *G. sylvestre*. Antisweet compound. Cryst. Mp
212-213°. [α]$_D$ +36.3° (c, 1.5 in MeOH).

Subba Rao, G. *et al*, *J. Chem. Soc., Chem. Commun.*, 1968, 1681
(*struct*)

Hoge, R. *et al*, *Acta Crystallogr., Sect. B*, 1974, **30**, 1435 (*cryst
struct*)
Yoshikawa, K. *et al*, *Tetrahedron Lett.*, 1989, **30**, 1103 (*isol*)
Maeda, M. *et al*, *Tetrahedron Lett.*, 1989, **30**, 1547 (*isol*)

12-Oleanene-3,16,21,23,28-pentol   O-80035

Updated Entry replacing O-50037

3,16,21,23,28-Pentahydroxy-12-oleanene

$C_{30}H_{50}O_5$ M 490.722

(3β,16β,21β)-form [19942-02-0] *Gymnestrogenin*

Constit. of *Gymnema sylvestre*. Cryst. Mp 288-289°. [α]$_D^{24}$
+53.5° (c, 0.71 in MeOH).

16-Benzoyl: Cryst. (CHCl$_3$/MeOH). Mp 282-284°. [α]$_D$
+59° (c, 1 in EtOH).

*3-O-β-D-Glucopyranosyl-(1→3)-α-L-arabinopyranoside, 16-
benzoyl:* [107110-06-5]. *Glochidioside*
$C_{48}H_{72}O_{15}$ M 889.088
Constit. of *Glochidion heyneanum*. Amorph. powder. [α]$_D$
+42° (c, 1 in EtOH).

3-O-β-D-Glucopyranoside, 16-benzoyl: Glochidioside N
$C_{43}H_{64}O_{11}$ M 756.972
Constit. of *G. heyneanum*. Amorph. powder.

*3-O-(β-D-Glycopyranosyl-(1→2)-β-D-glucopyranoside), 16-
benzoyl: Glochidioside Q*
$C_{49}H_{74}O_{16}$ M 919.114
Constit. of *G. heyneanum*. Amorph. powder.

Stocklin, W., *Helv. Chim. Acta*, 1968, **51**, 1235 (*isol*)
Srivastava, R. *et al*, *Phytochemistry*, 1986, **25**, 2672; 1988, **27**, 3575
(*derivs*)

12-Oleanene-3,21,22,28-tetrol   O-80036

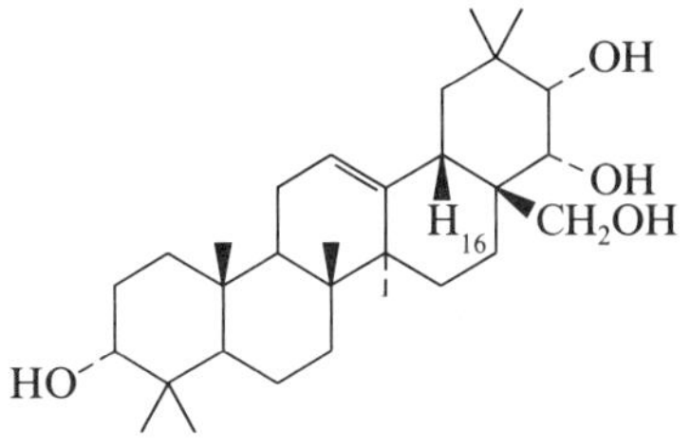

$C_{30}H_{50}O_4$ M 474.723

(3α,21α,22α)-form

Cryst. Mp 304°.

28-β-D-Xylopyranoside:
$C_{35}H_{58}O_8$ M 606.838
Constit. of *Centipeda minima*. Cryst. Mp 126°.

*16α-Hydroxy: 12-Oleanene-3α,16α,21α,22α,28-pentol.
3α,16α,21α,22α,28-Pentahydroxy-12-oleanene*
$C_{30}H_{50}O_5$ M 490.722
Cryst. Mp 315°.

16α-Hydroxy, 28-β-D-xylopyranoside:
$C_{35}H_{58}O_9$ M 622.838
Constit. of *C. minima*. Cryst. Mp 145°.

Gupta, D. *et al*, *Phytochemistry*, 1989, **28**, 1197.

12-Oleanene-2,3,23-triol   O-80037

$C_{30}H_{50}O_3$ M 458.723

(2α,3β)-form

Constit. of *Commiphora merkeri*. Shows antiflammatory
and analgesic properties. Needles. Mp 256°. [α]$_D^{22}$ +48°
(c, 1.8 in dioxan).

Fourie, T.G. *et al*, *J. Nat. Prod. (Lloydia)*, 1989, **52**, 1128 (*isol,
pmr, cmr*)

12-Oleanene-3,15,24-triol   O-80038

$C_{30}H_{50}O_3$ M 458.723

(3β,15α)-form [119285-50-6]

Constit. of *Phyllanthus flexuosus*. Needles. Mp 251-253°.
$[\alpha]_D^{23}$ +59.9° (c, 0.58 in Py).

Tanaka, R. *et al, Phytochemistry*, 1988, **27**, 3563.

12-Oleanene-3,9,11-triol   O-80039

$C_{30}H_{50}O_3$ M 458.723

(3β,9α,11α)-form

Constit. of *Euphorbia supina*. Cryst. (MeOH/CHCl₃).
Mp 242-244°. $[\alpha]_D$ −11° (c, 0.33 in CHCl₃).

Tanaka, R. *et al, Phytochemistry*, 1989, **28**, 1699.

13(18)-Oleanene-3,22,24-triol   O-80040

Updated Entry replacing S-00535

$C_{30}H_{50}O_3$ M 458.723

(3β,22α)-form

O^{22}-Me: [65892-76-4]. *22α-Methoxy-13(18)-oleanen-3β,24-
diol.* **Soyasapogenol D.** *Soyasapogenol M₃*
$C_{31}H_{52}O_3$ M 472.750
Soya-bean saponin (from *Glycine max*). Probably an
artifact. Cryst. (Me₂CO/MeOH). Mp 298-299°. $[\alpha]_D^{31}$
−60.8° (CHCl₃). Shown to be 21- or 22-methoxy by
Cainelli *et al*. CA uses the 21-methoxy struct. but
themost recent assignment seems to be that of Růžička.

Meyer, A. *et al, Helv. Chim. Acta*, 1950, **33**, 687 (*isol*)
Cainelli, G. *et al, Helv. Chim. Acta*, 1958, **41**, 2053 (*struct*)
Ružička, L., *Pure Appl. Chem.*, 1963, **6**, 505 (*struct, biosynth*)
Jurzysta, M. *et al, J. Chromatogr.*, 1978, **148**, 517 (*isol*)

18-Oleanen-3-ol   O-80041

$C_{30}H_{50}O$ M 426.724

3β-form

Me ether: [5945-45-9]. *3β-Methoxy-18-oleanene.* **Miliacin.**
Prosol. Panicol
$C_{31}H_{52}O$ M 440.751
Isol. from millet (*Panicum miliaceum*), various
Chionochloa spp. and other spp. in the Gramineae.
Cryst. (CHCl₃). Mp 282-283°. $[\alpha]_D^{20}$ +22° (c, 0.88 in
CHCl₃) (+8°).

Sugiyama, N. *et al, Nippon Kagaku Zasshi (Jpn. J. Chem.)*, 1961,
     82, 1051, 1054, 1057; *CA*, **57**, 13810n (*isol, struct*)
Bryce, T.A. *et al, Tetrahedron*, 1967, **23**, 1283 (*ms*)
Russell, G.B. *et al, Phytochemistry*, 1976, **15**, 1933 (*isol*)

Oleaxillaric acid   O-80042

[116139-49-2]

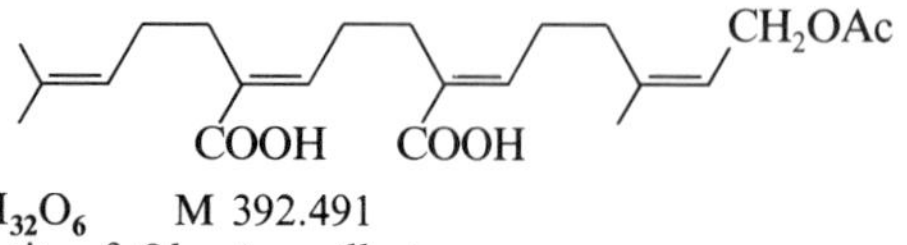

$C_{22}H_{32}O_6$ M 392.491
Constit. of *Olearia axillaris*.

Warning, U. *et al, J. Nat. Prod. (Lloydia)*, 1988, **51**, 513.

Olivin   O-80043

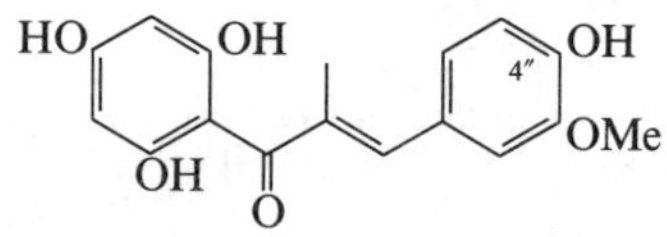

$C_{17}H_{16}O_6$ M 316.310
Isol. from leaves of *Olea europaea*. Cryst. (EtOH/EtOAc).
Mp >300°.

4″-O-(Glucosylglucoside): Olivin 4-diglucoside
$C_{29}H_{36}O_{16}$ M 640.594
Isol. from *O. europaea*. Cryst. (MeOH aq.). Mp 220-
225°. $[\alpha]_D^{20}$ −61.7° (Py).

Bočková, H. *et al, Collect. Czech. Chem. Commun.*, 1964, **29**, 1484
     (*isol, uv, ir, struct*)

Omphalocarpin   O-80044

[120693-45-0]

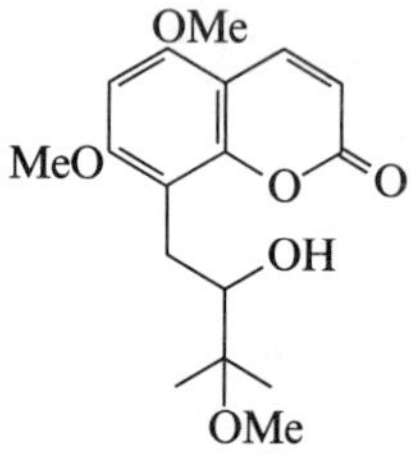

$C_{17}H_{22}O_6$ M 322.357
Constit. of *Murraya paniculata*. Needles (CHCl₃/Me₂CO).
Mp 159-160°. $[\alpha]_D$ −44.1° (c, 0.3 in CHCl₃).

Wu, T.-S. *et al, Phytochemistry*, 1989, **28**, 293.

α-Onocerol   O-80045

Updated Entry replacing O-10050
8(26),14(27)-Onceradiene-3β,21α-diol. α-Onocerin
[511-01-3]

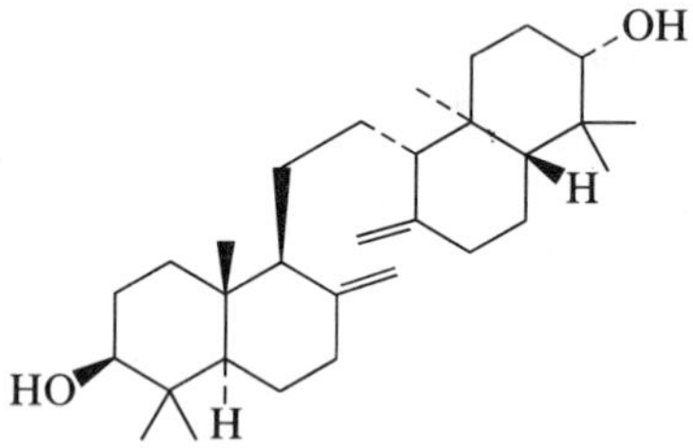

$C_{30}H_{50}O_2$ M 442.724
Constit. of *Ononis spinosa, O. arvensis, O. hircina* and
Lycopodium clavatum. Cryst. (EtOH). Mp 232°. $[\alpha]_D$
+1° (Py), $[\alpha]_D^{20}$ +18° (CHCl₃).

Barton, D.H.R. *et al, J. Chem. Soc.*, 1955, 2639 (*isol, struct*)
Ageta, H. *et al, Chem. Pharm. Bull.*, 1962, **10**, 637 (*isol*)
Stork, G. *et al, J. Am. Chem. Soc.*, 1963, **85**, 3419 (*synth*)
Danieli, N. *et al, Tetrahedron*, 1967, **23**, 509 (*synth*)
Rowan, M.G. *et al, FEBS Lett.*, 1971, **12**, 229 (*biosynth*)
Tsuda, Y. *et al, Chem. Pharm. Bull.*, 1981, **29**, 3424 (*synth*)

4,10(14)-Oplopadien-3-ol   O-80046

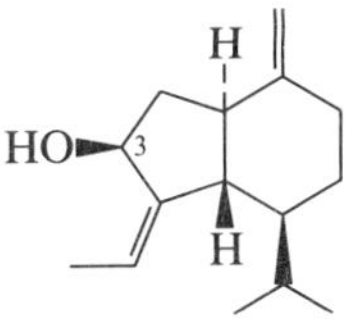

C$_{15}$H$_{24}$O   M 220.354
Biogenetic numbering (cf. cadinanes). Constit. of *Senecio mexicanus*. Cryst. Mp 54-56°. [α]$_D$ +29° (c, 0.1 in CHCl$_3$).

3-Ketone: 4(10),14-Oplopadien-3-one
   Constit. of *S. mexicanus*. Cryst. Mp 51-53°. [α]$_D$ −194° (c, 2 in CHCl$_3$).

Joseph-Nathan, P. *et al*, *Phytochemistry*, 1989, **28**, 1207, 2397 (*isol, pmr, cmr, cryst struct*)

Oplopenone   O-80047

Updated Entry replacing O-50045
   Anhydrooplopanone
   [28305-60-4]

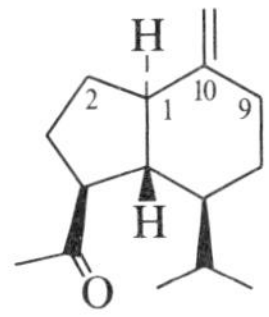

C$_{15}$H$_{24}$O   M 220.354
Different numbering system used in references. Cadinane system used here. Constit. of *Euryops pedunculatus*. Cryst. (pet. ether). Mp 68°. [α]$_D^{24}$ −15.2° (c, 0.4 in CHCl$_3$).

2α-Angeloyloxy: [119285-27-7]. *2α-Angeloxyanhydrooplopanone. 1α-Angeloxyanhydrooplopanone* (*incorr.*)
   C$_{20}$H$_{30}$O$_3$ M 318.455
   Constit. of *Ambrosia artemisioides*. Oil.

Δ^{1,10}-*Isomer, 2-oxo:* [119285-28-8]. *2-Oxoisoanhydrooplopanone. 1-Oxoisoanhydrooplopanone* (*incorr.*)
   C$_{15}$H$_{22}$O$_2$ M 234.338
   Constit. of *A. artemisioides*. Oil.

Δ^{1,10}-*Isomer, 2α-angeloyloxy, 9-oxo:* [119285-29-9]. *2α-Angeloyloxy-9-oxoisoanhydrooplopanone. 1α-Angeloyloxy-7-oxoisoanhydrooplopanone* (*incorr.*)
   C$_{20}$H$_{28}$O$_4$ M 332.439
   Constit. of *A. artemisioides*. Oil.

Δ^{1,10}-*Isomer, 2α-cinnamoyloxy, 9-oxo:* [119285-30-2]. *2α-Cinnamoyloxy-9-oxoisoanhydrooplopanone. 1α-Cinnamoyloxy-7-oxoisoanhydrooplopanone* (*incorr.*)
   C$_{24}$H$_{28}$O$_4$ M 380.483
   Constit. of *A. artemisioides*. Oil.

Δ^{1,10}-*Isomer, 2α-anisoyloxy, 9-oxo:* [119285-31-3]. *2α-Anisoyloxy-9-oxoisoanhydrooplopanone. 1α-Anisoyloxy-7-oxoisoanhydrooplopanone* (*incorr.*)
   C$_{23}$H$_{28}$O$_5$ M 384.471
   Constit. of *A. artemisioides*. Oil.

Bohlmann, F. *et al*, *Phytochemistry*, 1978, **17**, 1135 (*isol*)
Jakupovic, J. *et al*, *Phytochemistry*, 1988, **27**, 3551 (*derivs*)

[2.2]Orthometacyclophane   O-80048

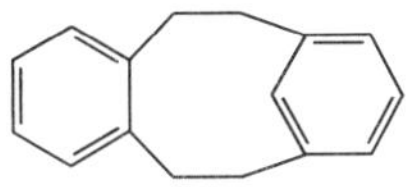

C$_{12}$H$_{12}$ M 156.227
Obt. as *syn-anti*-mixt.

[119392-52-8, 119477-47-3]

Bodwell, G. *et al*, *Angew. Chem., Int. Ed. Engl.*, 1989, **28**, 455 (*synth, pmr, cmr*)

Osthenone   O-80049

[112789-90-9]

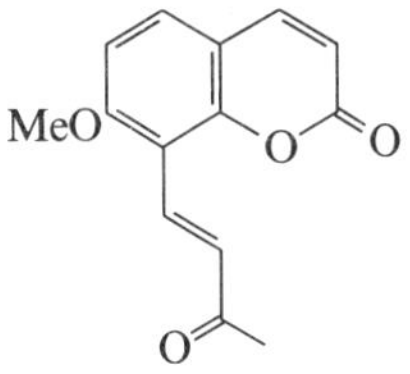

C$_{14}$H$_{12}$O$_4$ M 244.246
Constit. of *Citrus* spp. Yellow prisms. Mp 141-142°.

Ju-ichi, M. *et al*, *Heterocycles*, 1988, **27**, 2197.

Ougenin   O-80050

3-(2,4-Dihydroxyphenyl)-2,3-dihydro-5-hydroxy-7-methoxy-6-methyl-4H-1-benzopyran-4-one. 2′,4′,5-Trihydroxy-7-methoxy-6-methylisoflavanone

[1236-43-7]

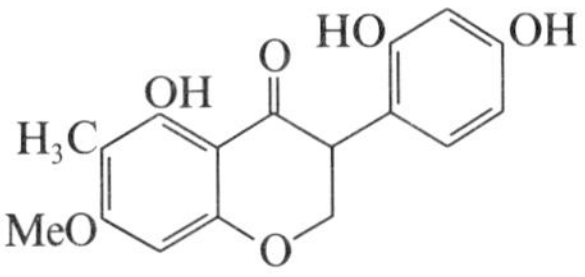

C$_{17}$H$_{16}$O$_6$ M 316.310
Isol. from *Ougeinia dalbergioides*. Pale-yellow needles (EtOAc/pet. ether). Mp 239-241°.

Balakrishna, S.J.D. *et al*, *Proc. R. Soc. London, A*, 1962, **268**, 1.
Ahluwalia, V.K. *et al*, *Indian J. Chem.*, 1966, **4**, 250 (*isol*)

8-Oxabicyclo[3.2.1]oct-6-en-3-one   O-80051

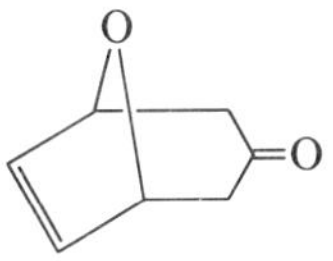

C$_7$H$_8$O$_2$ M 124.139
Oil, cryst. on standing. Mp 38°. Bp$_{11}$ 110-130° (kugelrohr)

Föhlisch, B. *et al*, *Chem. Ber.*, 1988, **121**, 1585 (*synth, pmr, cmr*)

1-Oxa[2.2](2,7)naphthalenophane   O-80052

2-Oxapentacyclo[11.5.3.3^{4,10}.0^{7,23}0^{16,20}]tetracosa-1(19),4,6,8,10(22),13,15,17,20,23-decaene, 9CI

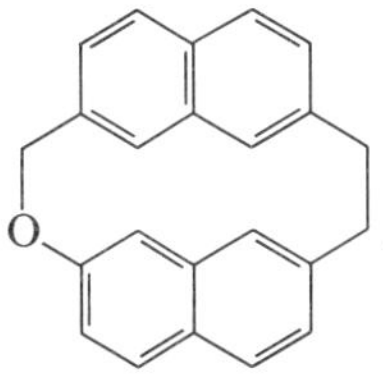

C$_{23}$H$_{18}$O   M 310.395

(+)-*form* [118797-89-0]

[α]$_{365}$ +72° (c, 0.79 in 1-butanol).

(−)-*form* [118797-90-3]

[α]$_{365}$ −102.4° (c, 0.41 in CHCl$_3$).

(±)-*form* [118797-88-9]

Needles (hexane). Mp 206-208°.

Billen, S. *et al, Chem. Ber.*, 1989, **122**, 1113 (*synth, pmr, cmr, uv, cd*)

2-Oxetanecarboxaldehyde   O-80053

2-Formyloxetane

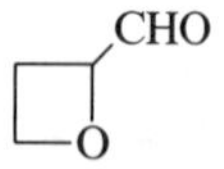

C$_4$H$_6$O$_2$ M 86.090

(±)-*form*

2,4-Dinitrophenylhydrazone: Orange needles (Me$_2$CO). Mp 234-235°.

Di-Et acetal: 2-(*Diethoxymethyl*)*oxetane*
C$_8$H$_{16}$O$_3$ M 160.213
Liq. Bp$_{10}$ 88-90°.

Fitton, A.O. *et al, Synthesis*, 1987, 1140 (*synth, ir, pmr*)

2-Oxetanemethanol, 9CI   O-80054

2-Hydroxymethyloxetane

[61266-70-4]

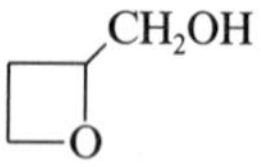

C$_4$H$_8$O$_2$ M 88.106

(±)-*form* [84921-88-0]
Liq. Bp$_{15}$ 94-98°.

4-Methylbenzenesulfonyl: Solid. Mp 58-59°.

Fitton, A.O. *et al, Synthesis*, 1987, 1140 (*synth, ir, pmr*)

3-Oxetanemethanol, 9CI   O-80055

3-Hydroxymethyloxetane

[6246-06-6]
C$_4$H$_8$O$_2$ M 88.106

U.S. Pat., 3 301 923, (1967); *CA*, **66**, 67673c.

3-Oxiranyl-7-oxabicyclo[4.1.0]heptane, 9CI   O-80056

3-(Epoxyethyl)-7-oxabicyclo[4.1.0]heptane, 8CI. 1,2-Epoxy-4-(epoxyethyl)cyclohexane. 1-Epoxyethyl-3,4-epoxycyclohexane. 4-Vinylcyclohexene dioxide

[106-87-6]

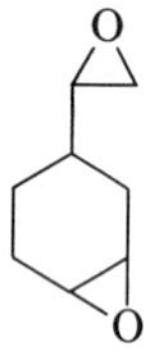

C$_8$H$_{12}$O$_2$ M 140.182

Coml. samples are presumably mixts. of diastereoisomers. Used to make water-resistant adhesives, corrosion inhibitors and polymer stabilizers; cross-linking agent; ingredient in rocket fuels. Liq. V. sol. H$_2$O. d$_{20}^{20}$ 1.099. Mp < −55°. Bp 227°, Bp$_5$ 88-92°. n$_D^{27}$ 1.476.

▷ Mutagenic. Carcinogenic, toxic to ovaries.

Everett, J.L. *et al, J. Chem. Soc.*, 1950, 3131 (*synth*)

Kodama, J.K. *et al, Arch. Environ. Health*, 1961, **2**, 50; *ca*, **55**, 13687i (*tox*)

Watabe, T. *et al, Biochem. Pharmacol.*, 1976, **25**, 601 (*metab*)

Fr. Pat., 2 393 799, (1979); *ca*, **91**, 158325y (*manuf*)

Turchi, G. *et al, Mutat. Res.*, 1981, **83**, 419; *ca*, **96**, 1577n (*tox*)

Bloch, R. *et al, J. Org. Chem.*, 1985, **50**, 1544 (*synth*)

Maronpot, R.R., *EHP Environ. Health Perspect.*, 1987, **13**, 125; *ca*, **107**, 213022y (*rev, bibl, tox*)

3-Oxo-2-azabicyclo[2.1.1]hexane-1-carboxylic acid, 9CI   O-80057

2,4-Methanopyroglutamic acid

[116128-98-4]

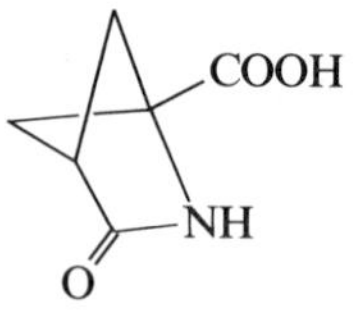

C$_6$H$_7$NO$_3$ M 141.126
Unstable.

Me ester: [116129-05-6].
C$_7$H$_9$NO$_3$ M 155.153
Cryst. Mp 76.5-77.5°.

Hughes, P. *et al, J. Org. Chem.*, 1988, **53**, 4793 (*synth, pmr, cmr, ms, ir*)

7-Oxo-7*H*-benz[*de*]anthracene-1-carboxylic acid, 9CI, 8CI   O-80058

Benzanthrone-1-carboxylic acid

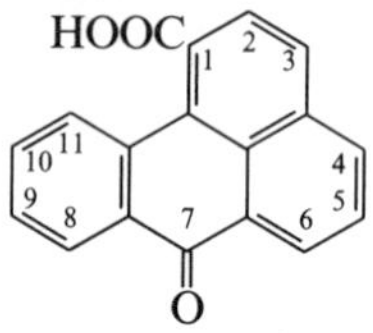

C$_{18}$H$_{10}$O$_3$ M 274.275
Cryst. (PhNO$_2$). Mp 285°.

Boyes, W.H.D. *et al, J. Chem. Soc.*, 1938, 1833 (*synth*)

7-Oxo-7*H*-benz[*de*]anthracene-2-carboxylic acid, 9CI, 8CI   O-80059

Benzanthrone-2-carboxylic acid
C$_{18}$H$_{10}$O$_3$ M 274.275
Yellow needles (PhNO$_2$). Mp 348-349°.

Et ester:
C$_{20}$H$_{14}$O$_3$ M 302.329
Yellow cryst. (EtOH). Mp 172-173°.

Chloride:
C$_{18}$H$_9$ClO$_2$ M 292.721
Yellow prisms (chlorobenzene). Mp 252-253°.

Amide:
C$_{18}$H$_{11}$NO$_2$ M 273.290
Cryst. (PhNO$_2$). Mp 306°.

Nitrile: 2-Cyanobenzanthrone
C$_{18}$H$_9$NO   M 255.275
Yellow cryst. (PhNO$_2$). Mp 285-286°.

Heilbron, I.M. *et al, J. Chem. Soc.*, 1936, 781 (*nitrile, amide*)

Baddar, F.G. *et al, J. Chem. Soc.*, 1941, 310 (*synth*)

Vollmann, H., *Justus Liebigs Ann. Chem.*, 1963, **669**, 22 (*chloride*)

7-Oxo-7*H*-benz[*de*]anthracene-3-carboxylic acid, 9CI, 8CI   O-80060

Benzanthrone-3-carboxylic acid

[67971-99-7]

$C_{18}H_{10}O_3$ M 274.275
Yellow needles (PhNO$_2$). Mp 343-344°.

Chloride:
$C_{18}H_9ClO_2$ M 292.721
Golden yellow needles. Mp 260-261°.

Nitrile: 3-Cyanobenzanthrone
$C_{18}H_9NO$ M 255.275
Yellow cryst. (PhNO$_2$). Mp 242-243°.

Heilbron, I.M. *et al, J. Chem. Soc.*, 1936, 781 (*nitrile*)
Vollmann, H., *Justus Liebigs Ann. Chem.*, 1963, **669**, 22 (*synth, chloride*)
Gorelik, M.V. *et al, J. Org. Chem. USSR (Engl. Transl.)*, 1978, **14**, 1433; 1979, **15**, 2322 (*synth*)

7-Oxo-7*H*-benz[*de*]anthracene-4-carboxylic acid, 9CI, 8CI   O-80061

Benzanthrone-4-carboxylic acid

$C_{18}H_{10}O_3$ M 274.275
Yellow cryst. (PhNO$_2$). Mp 314-315°.

Me ester:
$C_{19}H_{12}O_3$ M 288.302
Yellow cryst. (MeOH). Mp 215-216°.

Et ester:
$C_{20}H_{14}O_3$ M 302.329
Yellow cryst. (EtOH). Mp 134-135°.

Nitrile: 4-Cyanobenzanthrone
$C_{18}H_9NO$ M 255.275
Yellow cryst. (PhNO$_2$). Mp 234°.

U.S. Pat., 1 874 547, (1928) (*nitrile*)
Copp, F.C. *et al, J. Chem. Soc.*, 1942, 209 (*synth*)

7-Oxo-7*H*-benz[*de*]anthracene-8-carboxylic acid, 9CI, 8CI   O-80062

Benzanthrone-8-carboxylic acid

[73975-18-5]

$C_{18}H_{10}O_3$ M 274.275
Yellow cryst. (chlorobenzene). Mp 262-263°. Tautomerises to cyclic hydroxylactone.

Me ester:
$C_{19}H_{12}O_3$ M 288.302
Yellow cryst. (MeOH/MeOAc). Mp 173.5-174.5°.

Copp, F.C. *et al, J. Chem. Soc.*, 1942, 209 (*synth*)
Baddar, F.G. *et al, J. Chem. Soc.*, 1944, 450 (*synth*)
Gomès, L.M. *et al, Compt. Rend. Hebd. Seances Acad. Sci. Sect. C*, 1980, **290**, 29 (*synth, ir*)

7-Oxo-7*H*-benz[*de*]anthracene-9-carboxylic acid, 9CI, 8CI   O-80063

Benzanthrone-9-carboxylic acid

$C_{18}H_{10}O_3$ M 274.275
Yellow cryst. (PhNO$_2$). Mp 352-354°.

Me ester:
$C_{19}H_{12}O_3$ M 288.302
Yellow cryst. (MeOH). Mp 188-189°.

Et ester:
$C_{20}H_{14}O_3$ M 302.329
Yellow cryst. (EtOH). Mp 172.5-173.5°.

Copp, F.C. *et al, J. Chem. Soc.*, 1942, 209 (*synth*)

7-Oxo-7*H*-benz[*de*]anthracene-10-carboxylic acid, 9CI, 8CI   O-80064

Benzanthrone-10-carboxylic acid

$C_{18}H_{10}O_3$ M 274.275
Yellow cryst. (PhNO$_2$). Mp 326-327°.

Me ester:
$C_{19}H_{12}O_3$ M 288.302
Yellow cryst. (MeOH). Mp 168-169°.

Et ester:
$C_{20}H_{14}O_3$ M 302.329
Yellow-gold cryst. (EtOH). Mp 136-138°.

Copp, F.C. *et al, J. Chem. Soc.*, 1942, 209.
Bachmann, W.E. *et al, J. Am. Chem. Soc.*, 1949, **71**, 3062.

7-Oxo-7*H*-benz[*de*]anthracene-11-carboxylic acid, 9CI, 8CI   O-80065

Benzanthrone-11-carboxylic acid

$C_{18}H_{10}O_3$ M 274.275
Yellow cryst. (EtOH). Mp 273°.

Me ester:
$C_{19}H_{12}O_3$ M 288.302
Yellow cryst. (MeOH). Mp 160-161°.

Amide:
$C_{18}H_{11}NO_2$ M 273.290
Yellow cryst. (PhNO$_2$). Mp 325-327°.

Rull, H.G. *et al, J. Chem. Soc.*, 1935, 571; 1937, 1096 (*synth*)
Bigelow, L.A. *et al, J. Chem. Soc.*, 1935, 573 (*synth*)
Boyes, W.H.D. *et al, J. Chem. Soc.*, 1938, 1833 (*amide*)
Copp, F.C. *et al, J. Chem. Soc.*, 1942, 209 (*synth, ester, amide*)
Craig, J.T. *et al, Aust. J. Chem.*, 1966, **19**, 1927 (*synth*)

9-Oxo-2,7-bisaboladien-15-oic acid   O-80066

4-(1,5-Dimethyl-3-oxo-1-hexenyl)-1-cyclohexene-1-carboxylic acid, 9CI. **Atlantonic acid**

[120019-19-4]

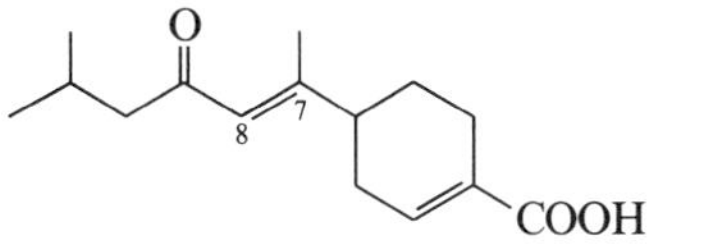

$C_{15}H_{22}O_3$ M 250.337
Constit. of *Cedrus libani*. Oil.

7,8-Dihydro: [120019-18-3]. *9-Oxo-2-bisabolen-15-oic acid.* **Dihydroatlantonic acid**
$C_{15}H_{24}O_3$ M 252.353
Constit. of *C. libani*. Oil.

Avcibasi, H. *et al, Phytochemistry*, 1988, **27**, 3967.

2-Oxobutanal, 9CI   O-80067

2-Oxobutyraldehyde, 8CI. Ethylglyoxal

[4417-81-6]

$$H_3CCH_2COCHO$$

$C_4H_6O_2$ M 86.090
Liq. Sol. H$_2$O. Bp 90°. n_D^{25} 1.4006.

1-Oxime:
$C_4H_7NO_2$ M 101.105
Hygroscopic needles (C$_6$H$_6$). Mp 55°.

Bis(2,4-dinitrophenylhydrazone): [1179-34-6].
Red needles (C$_6$H$_6$). Mp 247°.

Di-Me acetal: [6342-57-0]. *1,1-Dimethoxy-2-butanone*
$C_6H_{12}O_3$ M 132.159
Liq. Bp 125-128°, Bp$_{13}$ 49°.

Di-Et acetal: 1,1-Diethoxy-2-butanone

$C_8H_{16}O_3$ M 160.213
Liq. Bp_{55} 98-100°. n_D^{20} 1.408.

Riley, R. *et al, J. Chem. Soc.*, 1932, 1877 (*synth*)
Ingold, C.K. *et al, J. Chem. Soc.*, 1934, 85 (*synth*)
Sharp, W. *et al, J. Chem. Soc.*, 1948, 1863 (*oxime*)
Royals, E.E. *et al, J. Am. Chem. Soc.*, 1956, **78**, 4161 (*synth*)
Dulou, R. *et al, Compt. Rend. Hebd. Seances Acad. Sci.*, 1966, **262**, 564 (*acetal*)

4-Oxobutanoic acid, 9CI   O-80068

Updated Entry replacing O-00777
Succinaldehydic acid, 8CI. Succinic semialdehyde. 3-Formylpropionic acid. γ-Oxobutanoic acid
[692-29-5]

$$OHCCH_2CH_2COOH$$

$C_4H_6O_3$ M 102.090
Found in *Proteus vulgaris*, and in mouse brains. Has anaesthetic props. Oil with rancid odour. Sol. H_2O, EtOH, Et_2O, C_6H_6. Bp_{14} 134-136°. Steam-volatile.
▷ Mod. toxic. Emits toxic fumes on contact with acids or heating to dec.. WM3380000.

Trimer: [26817-25-4]. *s-Trioxane-2,4,6-tripropionic acid. 1,3,5-Tris-(β-carboxyethyl)-s-trioxane*
$C_{12}H_{18}O_9$ M 306.269
Prisms. Sl. sol. Et_2O, EtOAc, aq. Na_2CO_3. Mp 147°. Bp_{14} 134-136°. On dist. → monomer.

Me ester: [13865-19-5].
$C_5H_8O_3$ M 116.116
Oil. Bp_{15} 69-71°.

Me ester, semicarbazone: Cryst. Mp 130-131°.

Me ester, 2,4-dinitrophenylhydrazone: Cryst. Mp 131-132°.

Nitrile: [3515-93-3]. *4-Oxobutanenitrile, 9CI. Succinaldehydonitrile, 8CI. 3-Cyanopropanal*
C_4H_5NO M 83.090
Bp 77°.

Oxime:
$C_4H_7NO_3$ M 117.104
Cryst. (EtOH). Mp 102-103° (155°).

Semicarbazone: Prisms or needles (H_2O). Mp 190-191° dec.

2,4-Dinitrophenylhydrazone: [4093-65-6].
Yellow cryst. Mp 201°.

Nitrile, 2,4-dinitrophenylhydrazone: Mp 198-199°.

Nitrile, di-Me acetal: 4,4-Dimethoxybutanenitrile
$C_6H_{11}NO_2$ M 129.158
Bp_{20} 110°.

Nitrile, di-Et acetal: [18381-45-8]. *4,4-Diethoxybutanenitrile*
$C_8H_{15}NO_2$ M 157.212
Bp_{10} 97°.

Wislicenus, W. *et al, Justus Liebigs Ann. Chem.*, 1908, **363**, 340 (*synth, derivs*)
Kato, J. *et al, CA*, 1962, **57**, 2064 (*nitrile*)
Nawar, W.W. *et al, J. Agric. Food Chem.*, 1971, **19**, 1039 (*ester*)
Watanabe, Y. *et al, Bull. Chem. Soc. Jpn.*, 1975, **48**, 2490 (*synth*)
Doleschall, G., *Tetrahedron*, 1976, **32**, 2549 (*ester*)
Pirkle, W.H. *et al, J. Org. Chem.*, 1978, **43**, 2091 (*nitrile*)
Manzocchi, A. *et al, Synthesis*, 1983, 324 (*synth*)
Gannett, P.M. *et al, J. Org. Chem.*, 1988, **53**, 1064 (*Me ester, synth, ir, pmr, cmr, ms*)
Gribble, G.W. *et al, J. Org. Chem.*, 1988, **53**, 3164 (*nitrile di-Me acetal*)
Sax, N.I., *Dangerous Properties of Industrial Materials*, 5th Ed., Van Nostrand-Reinhold, 1979, 664.

2-Oxo-3-butynoic acid, 9CI   O-80069

[56842-75-2]

$$HC\equiv CCOCOOH$$

$C_4H_2O_3$ M 98.058
Unstable. Reactive alkylating agent *in vivo*.
▷ Possible mutagen.

Kaczorowski, G. *et al, Biochemistry*, 1975, **14**, 3903.
Pompon, D. *et al, Eur. J. Biol.*, 1982, **129**, 143 (*synth, purifn*)

4-Oxo-2-butynoic acid   O-80070

$$OHCC\equiv CCOOH$$

$C_4H_2O_3$ M 98.058
Et ester: [78076-22-9].
$C_6H_6O_3$ M 126.112
Pale yellow liq. Bp_3 57-58°.

Dimethylamide: [104387-58-8].
$C_6H_7NO_2$ M 125.127
Liq.

Nitrile: [90108-94-4]. *4-Oxo-2-butynenitrile*
C_4HNO M 79.058
Pale yellow liq. Not obt. pure.

Barbot, F. *et al, Bull. Soc. Chim. Fr.*, 1968, 313 (*synth*)
Epsztein, R. *et al, Bull. Soc. Chim. Fr., Part II*, 1983, 41 (*synth*)
Gorgues, A. *et al, Tetrahedron*, 1986, **42**, 351 (*synth, pmr, ir*)

3-Oxo-17-carboxy-3,18-secobarbacenic acid   O-80071

5-(Carboxymethyl)decahydro-1,5,8a-trimethyl-2,6-dioxo-1-naphthalenepropanoic acid, 9CI
[119967-65-6]

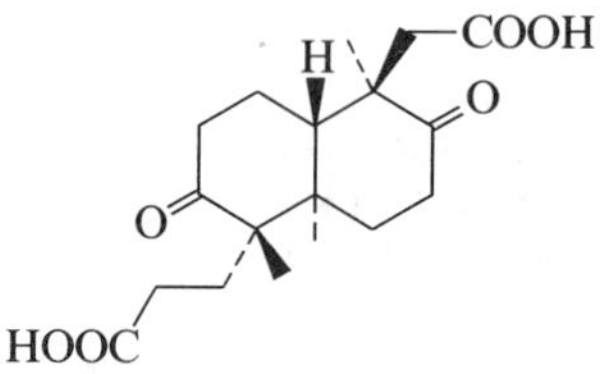

$C_{18}H_{26}O_6$ M 338.400
Constit. of *Barbacenia flava*. Cryst. (EtOAc/hexane). Mp 214-216°.

Pinto, A.C. *et al, Phytochemistry*, 1988, **27**, 3917.

3-Oxo-24-cycloarten-21-oic acid   O-80072

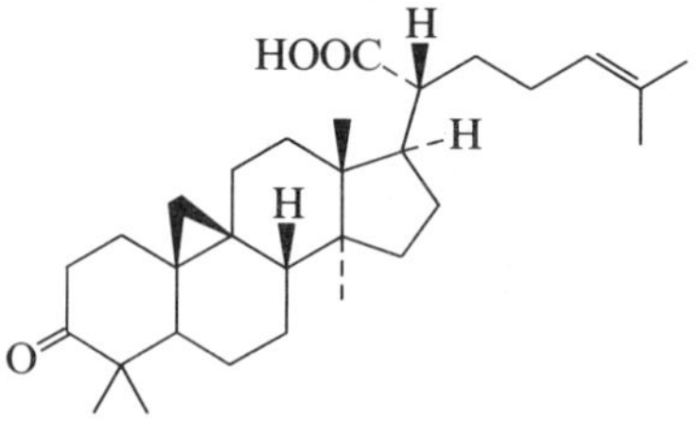

$C_{30}H_{46}O_3$ M 454.692
Constit. of leaves of *Lansium domesticum*. Skin tumour inhibitor. Cryst. Mp 185-186°. $[\alpha]_D^{12}$ +18.7° (c, 1.16 in $CHCl_3$).

Nishizawa, M. *et al, Tetrahedron Lett.*, 1989, **30**, 5615 (*cryst struct*)

3-Oxocyclobutanecarboxylic acid   O-80073

[23761-23-1]

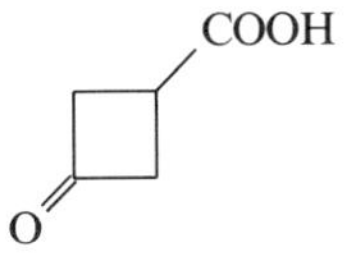

$C_5H_6O_3$ M 114.101
Needles (C_6H_6/hexane). Mp 69-70°.

N,N-*Dimethylamide:*
 $C_7H_{11}NO_2$ M 141.169
 Bp_1 108-111°.

N,N-*Dimethylamide, ethylene ketal:* Bp_1 114°.

N,N-*Dimethylamide, Di-Et ketal:*
 $C_{11}H_{21}NO_3$ M 215.292
 Bp_1 110°.

Avram, M. et al, Chem. Ber., 1957, **90**, 1424 (deriv, synth)
Caserio, F.F. et al, J. Am. Chem. Soc., 1958, **80**, 5837 (synth, deriv)
Pigou, P.E. et al, J. Org. Chem., 1988, **53**, 3841 (synth, pmr, cmr)

2-Oxocyclohexanecarboxaldehyde, 9CI,   O-80074
8CI

2-(*Hydroxymethylene*)*cyclohexanone, 9CI, 8CI.* 2-
Formylcyclohexanone

[823-45-0]

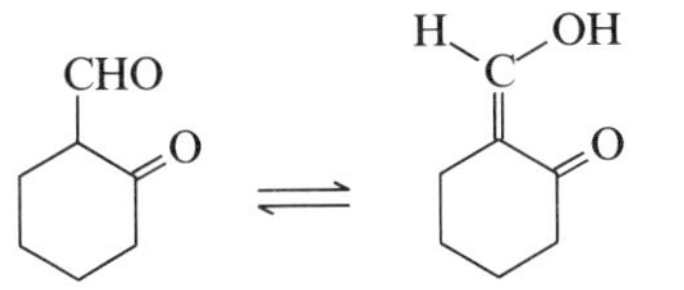

oxo-form   (*Z-enol*)*-form*

$C_7H_{10}O_2$ M 126.155
(Z)-Enol form predominates. Liq. Bp_{14} 86-87°, Bp_5 70-72°.
pK_a 6.35 (20°). n_D^{25} 1.5110. Polymerises slowly at r.t.

Me *ether:* [15839-18-6]. 2-
 (*Methoxymethylene*)*cyclohexanone, 9CI*
 $C_8H_{12}O_2$ M 140.182
 Liq. Bp_{13} 75-80°, Bp_2 84-86°. n_D^{18} 1.4854.

[1193-63-1, 39695-64-2, 72036-60-3]

Garbisch, E.W., J. Am. Chem. Soc., 1963, **85**, 1696; 1965, **87**, 505 (tautom, pmr, cmr)
Org. Synth., Coll. Vol., 4, 1963, 536 (synth)
Barna, J.C.J. et al, Tetrahedron Lett., 1979, 1455 (cmr, deriv)
Brady, W.T. et al, J. Org. Chem., 1983, **48**, 5337 (synth, ir, pmr, deriv)
Jones, R.A. et al, Tetrahedron, 1984, **40**, 1051 (synth, tautom, ir, cmr)

3-Oxocyclohexanecarboxaldehyde, 9CI,   O-80075
8CI

3-*Formylcyclohexanone*

[69814-26-2]

$C_7H_{10}O_2$ M 126.155
Liq. Polymerises on heating.

Baisted, D.J. et al, J. Chem. Soc., 1961, 4089 (synth)
Cohen, T. et al, Tetrahedron Lett., 1978, 3533 (synth)
Lucchetti, A. et al, Synth. Commun., 1983, **13**, 1153 (synth)
Ricci, A. et al, Angew. Chem., Int. Ed. Engl., 1985, **24**, 1068 (synth, ir)
Kerdesky, F.A.J. et al, J. Med. Chem., 1987, **30**, 1177 (synth, pmr)

4-Oxocyclohexanecarboxaldehyde, 9CI,   O-80076
8CI

4-*Formylcyclohexanone*

[55882-00-3]

$C_7H_{10}O_2$ M 126.155
Liq. d_4^{20} 1.091. Bp_{10} 113-113.5°. n_D^{20} 1.4760.

Bissemicarbazone: Cryst. (MeOH). Mp 199°.

Petrov, A.A., J. Gen. Chem. USSR (Engl. Transl.), 1941, **11**, 661; ca, **36**, 1593 (synth)
Fetizon, M. et al, Tetrahedron, 1975, **31**, 171 (synth)

5-Oxo-1-cyclohexene-1-carboxylic acid   O-80077

[37051-60-8]

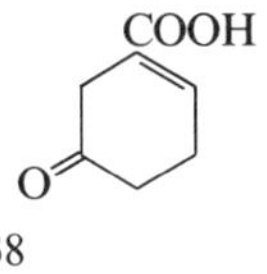

$C_7H_8O_3$ M 140.138
Cryst. (toluene). Mp 99-100°.

Me *ester:* [66838-90-2].
 $C_8H_{10}O_3$ M 154.165
 Liq. $Bp_{0.1}$ 80-83°.

Biffin, M.E.C. et al, Aust. J. Chem., 1972, **25**, 1329 (synth, ir, pmr)
Webster, F.X. et al, Synthesis, 1987, 922 (synth, pmr)

5-Oxo-2-cyclohexene-1-carboxylic acid   O-80078

[37051-56-2]

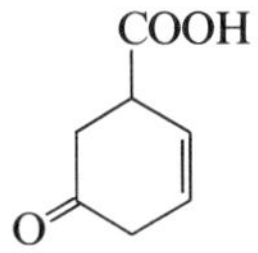

$C_7H_8O_3$ M 140.138

(±)*-form*
 Cryst. (toluene). Mp 104-105°.

Me *ester:*
 $C_8H_{10}O_3$ M 154.165
 Bp_2 134-137°.

Me *ester, 2,4-dinitrophenylhydrazone:* Cryst. Mp 144-145°.

Biffin, M.E.C. et al, Aust. J. Chem., 1972, **25**, 1329 (synth, ir, pmr, deriv)
Webster, F.X. et al, Synthesis, 1987, 922 (synth, pmr)

6-Oxo-1-cyclohexene-1-carboxylic acid   O-80079

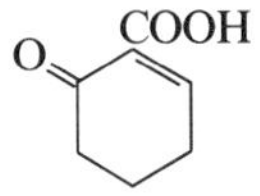

$C_7H_8O_3$ M 140.138
Me *ester:* [57039-20-0]. 2-*Carbomethoxy-2-cyclohexen-1-one*
 $C_8H_{10}O_3$ M 154.165
 Undergoes Diels-Alder reacns. Liq.

Liotta, D. et al, J. Org. Chem., 1981, **46**, 2920 (synth)
Liu, H.J. et al, Can. J. Chem., 1988, **66**, 3143 (use, ester)

5-Oxo-6,8,11,14-eicosatetraenoic acid   O-80080
5-KETE

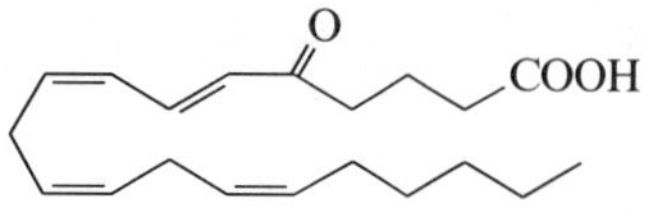

$C_{20}H_{30}O_3$ M 318.455

Toda, M. *et al, J. Med. Chem.*, 1983, **26**, 72 (*synth*)
Kerdesky, F.A.J. *et al, J. Med. Chem.*, 1987, **30**, 1177 (*synth, pmr*)

2-Oxo-1(10),11(13)-eremophiladien-12,8-olide   O-80081

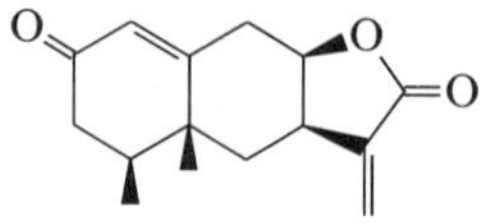

$C_{15}H_{18}O_3$ M 246.305
8β-form
Constit. of *Ondetia linearis*. Oil.
Zdero, C. *et al, Phytochemistry*, 1989, **28**, 1653.

4-Oxo-2-hexenoic acid, 8CI   O-80082
[5636-58-8]

$$H_3CCH_2COCH{=}CHCOOH$$

$C_6H_8O_3$ M 128.127
(*E*)-*form*
Cryst. (pet. ether). Mp 109-110°.
Me ester: [90670-10-3].
 $C_7H_{10}O_3$ M 142.154
 Mp 30-32°. $Bp_{16.5}$ 80-81.5°.
Et ester: [61454-94-2].
 $C_8H_{12}O_3$ M 156.181
 $Bp_{11.5}$ 95-100°.

[70038-66-3, 83349-25-1]

Noltes, A.W. *et al, Recl. Trav. Chim. Pays-Bas (J. R. Neth. Chem. Soc.)*, 1961, **80**, 1334 (*deriv*)
Reinheckel, H. *et al, Angew. Chem., Int. Ed. Engl.*, 1966, **5**, 511 (*synth*)
Nakayama, M. *et al, Bull. Chem. Soc. Jpn.*, 1979, **52**, 184 (*deriv*)
Shnol, R.T. *et al, Zh. Obshch. Khim.*, 1982, **52**, 1351 (*Z-form*)
Miyashita, M. *et al, J. Org. Chem.*, 1984, **49**, 2857 (*deriv*)
Schuda, P.F. *et al, Synthesis*, 1987, 1055 (*deriv, synth, ir, pmr*)

5-Oxo-3-hexenoic acid, 8CI   O-80083
[32794-12-0]

$$H_3CCOCH{=}CHCH_2COOH$$

$C_6H_8O_3$ M 128.127
(*E*)-*form* [28845-68-3]
Pale yellow oil.
2,4-Dinitrophenylhydrazone: Orange needles (EtOAc). Mp 177.5-178°.

Harris, T.M. *et al, J. Org. Chem.*, 1971, **36**, 2181 (*synth, nmr*)
Dieter, R.K. *et al, J. Org. Chem.*, 1988, **53**, 2031 (*synth, nmr*)

6-Oxo-4-hexenoic acid   O-80084
[82934-89-2]

$$OHCCH{=}CHCH_2CH_2COOH$$

$C_6H_8O_3$ M 128.127
Maurer, B. *et al, Helv. Chim. Acta*, 1982, **65**, 462 (*synth, ir, nmr*)

2-Oxo-3-indolineglyoxylic acid   O-80085
(*1,2-Dihydro-2-oxo-*3H-*indol-3-ylidene*)*hydroxyacetic acid, 9CI. 3-Oxindolylglyoxylic acid*
[14370-71-9]

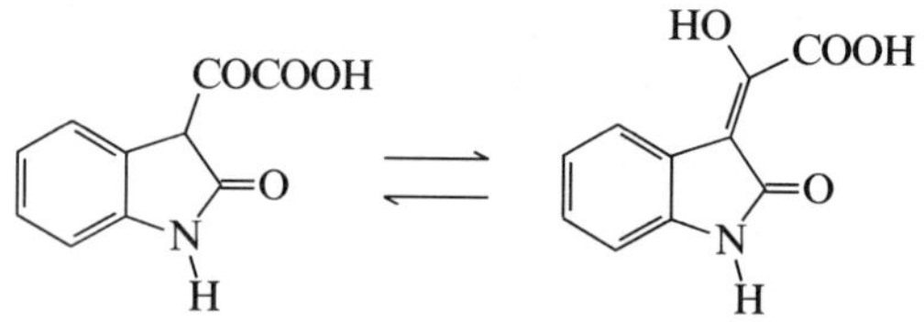

$C_{10}H_7NO_4$ M 205.170
Completely enolised. Cryst. (Me_2CO). Mp 265° (256-257°).
Et ester: [14370-70-8].
 $C_{12}H_{11}NO_4$ M 233.223
 Cryst. (Me_2CO). Mp 186-187°.
Benzyl ester: [110655-07-7].
 $C_{17}H_{13}NO_4$ M 295.294
 Yellow needles (Me_2CO). Mp 205-206°.

Wislicenus, W. *et al, Justus Liebigs Ann. Chem.*, 1924, **436**, 113 (*synth*)
Horner, L. *et al, Justus Liebigs Ann. Chem.*, 1941, **548**, 117 (*synth*)
Julian, P.L. *et al, J. Am. Chem. Soc.*, 1953, **75**, 5305 (*derivs*)
Harley-Mason, J. *et al, J. Chem. Soc.*, 1958, 3639 (*synth*)
Lewer, P., *J. Chem. Soc., Perkin Trans. 1*, 1987, 753 (*synth, pmr, uv, derivs*)

ent-15-Oxo-16-kauren-18-oic acid   O-80086
[71493-37-3]

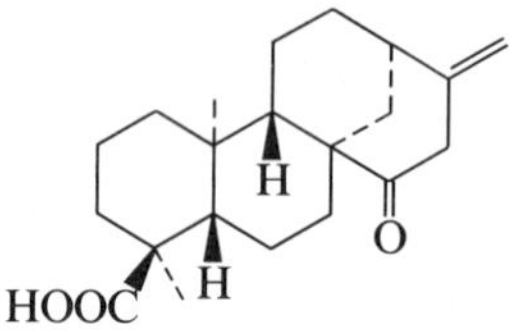

$C_{20}H_{28}O_3$ M 316.439
Constit. of *Croton argyrophylloides*. Cryst. (hexane/MeOH). Mp 202-204°. $[\alpha]_D^{25}$ −73° (c, 2.21 in CHCl_3).

Monte, F.J.Q. *et al, Phytochemistry*, 1988, **27**, 3209.

3-Oxo-8,24-lanostadien-21-oic acid   O-80087
Updated Entry replacing P-01917
Pinicolic acid A. α-Pinicolic acid
[466-05-7]

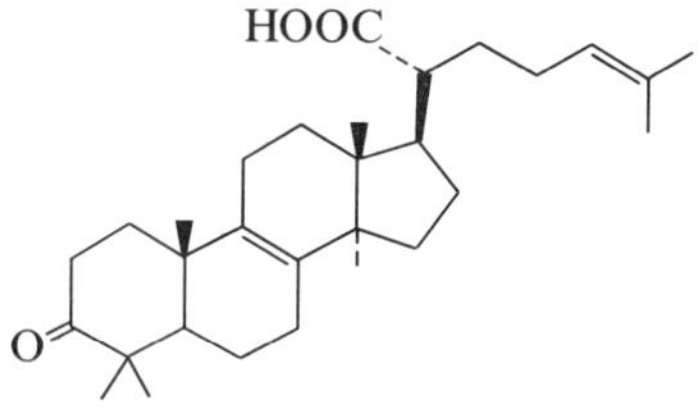

$C_{30}H_{46}O_3$ M 454.692
Produced by *Polyporus pinicola, Heterobasidion tasmanica* and *Fomes senex*. Cryst. Mp 205-207°. $[\alpha]_D$ +68° (c, 0.83 in CHCl_3).

Guider, J.M. *et al, J. Chem. Soc.*, 1954, 4471 (*isol*)
Batta, A.K. *et al, J. Chem. Soc., Perkin Trans. 1*, 1975, 451 (*isol, pmr*)

4-Oxo-4*H*-naphtho[2,3-*b*]pyran-3-carbox-aldehyde — O-80088

[57051-28-2]

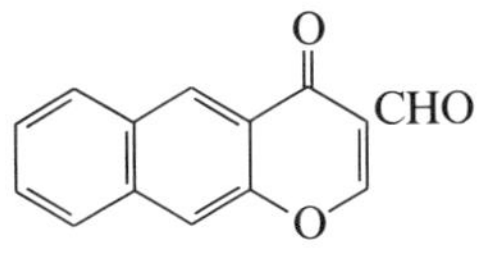

$C_{14}H_8O_3$ M 224.215
Plates (EtOAc). Mp 222-224°.

Cremins, P.J. *et al, Tetrahedron*, 1987, **43**, 3075 (*synth, ir, pmr*)

5-Oxopentanoic acid, 9CI — O-80089

Updated Entry replacing O-20065
4-Formylbutanoic acid. Glutaric semialdehyde
[5746-02-1]

$$OHCCH_2CH_2CH_2COOH$$

$C_5H_8O_3$ M 116.116
Me ester: [6026-86-4].
   $C_6H_{10}O_3$ M 130.143
   Synthon. Liq. Bp₁₀ 90-98°.
Nitrile: [3350-74-1]. 4-Cyanobutanal
   C_5H_7NO M 97.116
   Liq. Bp₂ 83-84°.
Nitrile, 2,4-dinitrophenylhydrazone: Mp 117-118°.

Brown, H.C. *et al, J. Am. Chem. Soc.*, 1969, **91**, 4606 (*nitrile, synth*)
Huckstep, M. *et al, Synthesis*, 1982, 881.
Gannett, P.M. *et al, J. Org. Chem.*, 1988, **53**, 1064 (*ester, synth, ir, pmr, cmr, ms*)
Bergeron, R.J. *et al, J. Org. Chem.*, 1988, **53**, 3131 (*nitrile, pmr, ir*)

3-Oxo-2-(2-pentenyl)-1-cyclopenteneacetic acid — O-80090

3,7-Didehydrojasmonic acid

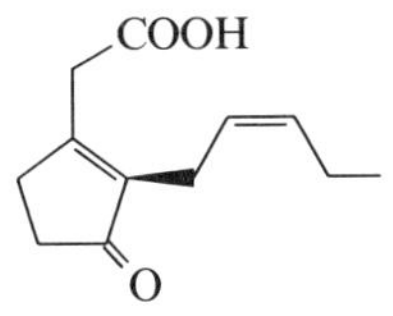

(*Z*)-form [120282-76-0]
   $C_{12}H_{16}O_3$ M 208.257
   Constit. of *Vicia faba*. Oil.

Miersch, O. *et al, Phytochemistry*, 1989, **28**, 339.

3-Oxo-2-pentylcyclopentaneacetic acid — O-80091

9,10-Dihydrojasmonic acid
[98674-52-3]

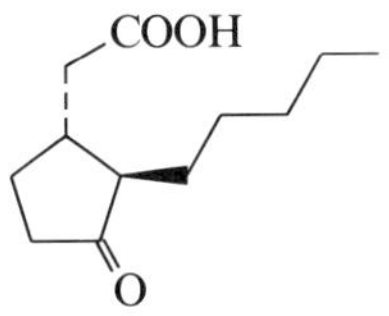

$C_{12}H_{20}O_3$ M 212.288
Constit. of *Vicia faba*. Oil. [α]²²_D −28.3° (c, 0.1 in MeOH).
[76968-33-7]

Miersch, O. *et al, Phytochemistry*; 1989, **28**, 339.

3-Oxo-5-phenylpentanal — O-80092

β-Oxobenzenepentanal, 9CI. 3-Oxo-5-phenylvaleraldehyde, 8CI

$$PhCH_2CH_2COCH_2CHO$$

$C_{11}H_{12}O_2$ M 176.215
Readily oligomerises, isol. as Cu chelate.
Cu chelate: Blue cryst. (EtOH/CHCl₃). Mp 176-178°.

Harris, T.M. *et al, J. Am. Chem. Soc.*, 1963, **85**, 3273 (*synth*)

4-Oxo-5-phenylpentanal — O-80093

γ-Oxobenzenepentanal, 9CI. 4-Oxo-5-phenylvaleraldehyde, 8CI
[53707-97-4]

$$PhCH_2COCH_2CH_2CHO$$

$C_{11}H_{12}O_2$ M 176.215
Liq. Bp₀.₀₂ 60-65°.

Brown, E. *et al, Bull. Soc. Chim. Fr.*, 1974, 1001 (*synth, ir, pmr*)
Baldwin, S.W. *et al, Tetrahedron Lett.*, 1982, **23**, 3883 (*synth*)

5-Oxo-5-phenylpentanal — O-80094

δ-Oxobenzenepentanal, 9CI. 5-Oxo-5-phenylvaleraldehyde, 8CI. 4-Benzoylbutyraldehyde
[75424-63-4]

$$PhCOCH_2CH_2CH_2CHO$$

$C_{11}H_{12}O_2$ M 176.215
Liq.
Di-Me acetal:
   $C_{13}H_{16}O_2$ M 204.268
   Oil.

Masahino, M. *et al, J. Chem. Soc., Perkin Trans. 1*, 1980, 1950, 2909 (*synth, ir, pmr*)
Ounsworth, J. *et al, J. Chem. Soc., Chem. Commun.*, 1986, 232 (*props*)

5-Oxo-3-thiomorpholinecarboxylic acid, 9CI — O-80095

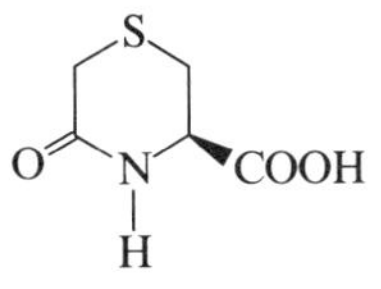

$C_5H_7NO_3S$ M 161.181
(*R*)-form
   Oil.
Me ester: [118903-82-5].
   $C_6H_9NO_3S$ M 175.208
   Cryst. (Et₂O). Mp 70-71°. [α]_D −4.0° (c, 1.0 in CHCl₃).

Zanotti, G. *et al, J. Chem. Soc., Perkin Trans. 1*, 1988, 2647 (*synth, ir, pmr*)

3-Oxotricyclo[2.1.0.0²,⁵]pentane-1,5-di-carboxylic acid — O-80096

[114534-19-9]

$C_7H_4O_5$ M 168.106

333

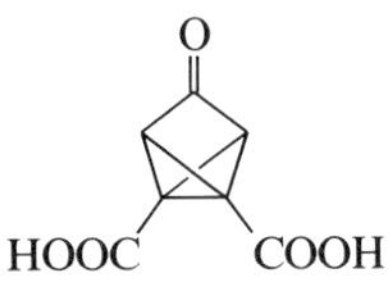

Solid. Mp *ca.* 170° dec. (discolours from 120°).
Di-Me ester: [81710-09-0].
  C₉H₈O₅   M 196.159
  Cryst. (CH₂Cl₂/hexane). Mp 130-131°.
Di-Me ester, ethylene ketal: [114534-20-2].
  C₁₁H₁₂O₆   M 240.212
  Cryst. (Et₂O/pentane). Mp 83-86°.

Irngartinger, H. *et al, J. Org. Chem.,* 1988, **53**, 3046 (*synth, deriv, pmr, ir, ms*)

1,8-Oxybis(ethyleneoxyethyleneoxy)-9,10- O-80097
anthracenedione

6,7,9,10,12,13,15,16-Octahydro-1,21-methano-22H-dibenzo[n,q][1,4,7,10,13]pentaoxacyclooctadecin-22,23-dione, 9CI

[111959-53-6]

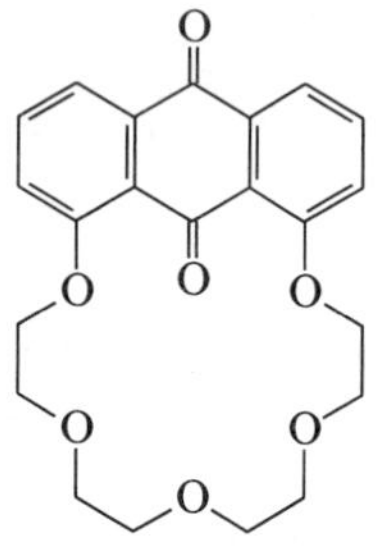

C₂₂H₂₂O₇   M 398.412
Strong cation binder. Yellow cryst. solid (toluene). Mp 155-156°.

Delgado, M. *et al, J. Am. Chem. Soc.,* 1988, **110**, 119 (*synth, esr, pmr, ir*)

3,3′-Oxybispropanoic acid O-80098

Bis(2-carboxyethyl)ether. Diethyl ether 2,2′-dicarboxylic acid

[5961-83-1]

$$HOOCCH_2CH_2OCH_2CH_2COOH$$

C₆H₁₀O₅   M 162.142
Cryst. (Me₂CO/pet. ether). Mp 60-61° (48-49°).
Di-Me ester:
  C₈H₁₄O₅   M 190.196
  Bp₃ 90-91°.
Dichloride: [44995-78-0].
  C₆H₈Cl₂O₃   M 199.033
  Oil. Bp₁ 85-88°.

Diamide:
  C₆H₁₂N₂O₃   M 160.172
  Mp 146°.
Dinitrile: Bis(2-cyanoethyl)ether
  C₆H₈N₂O   M 124.142
  Oil. d₂₅²⁵ 1.05. Bp₅ 161-162°.

Bruson, H.E., *J. Am. Chem. Soc.,* 1943, **65**, 23 (*nitrile, amide*)
Christian, R.V. *et al, J. Am. Chem. Soc.,* 1948, **70**, 1333 (*amide*)
Laing, D.K. *et al, J. Chem. Soc., Dalton Trans.,* 1975, 2297 (*nitrile*)
Sargsian, M.S. *et al, Arm. Khim. Zh.,* 1987, **40**, 548 (*synth, pmr, ester*)
Pratt, J.A.E. *et al, J. Chem. Soc., Perkin Trans.* 1, 1988, 13 (*synth, chloride, ir, pmr*)

2,4′-Oxybispyridine O-80099

2-(4-Pyridinyloxy)pyridine, 9CI. 2,4′-Dipyridyl oxide. 2,4′-Dipyridyl ether

[53258-97-2]

C₁₀H₈N₂O   M 172.186
Needles (pet. ether). Mp 40°.

Rockley, J.E. *et al, Z. Naturforsch., B,* 1982, **37**, 933 (*synth, nmr, uv, ms*)
Summers, L.A., *J. Heterocycl. Chem.,* 1987, **24**, 533 (*rev*)

3,4′-Oxybispyridine O-80100

3-(4-Pyridinyloxy)pyridine, 9CI. 3-Pyridyl 4-pyridyl ether

[53258-98-3]

C₁₀H₈N₂O   M 172.186
Cryst. (hexane). Mp 48-50°.
Picrate: Cryst. Mp 145°.

Bodusek, B. *et al, Pol. J. Chem. (Rocz. Chem.),* 1976, **50**, 2167 (*synth, ir, pmr*)
Summers, L.A., *J. Heterocycl. Chem.,* 1987, **24**, 533 (*rev*)

Oxycurcumenol O-80101

[119205-46-8]

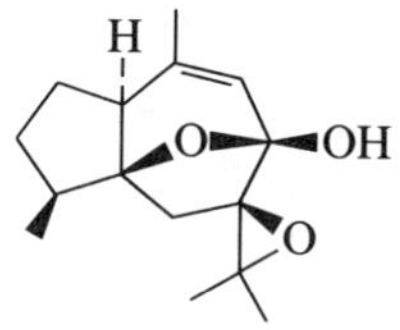

C₁₅H₂₂O₃   M 250.337
Constit. of *Curcuma heyneana.* Prisms (MeOH). Mp 114-116°. [α]ᴅ²⁰ −133° (c, 0.1 in CHCl₃).

Firman, K. *et al, Phytochemistry,* 1988, **27**, 3887.

P

Pachypophyllin
P-80001

[114542-60-8]

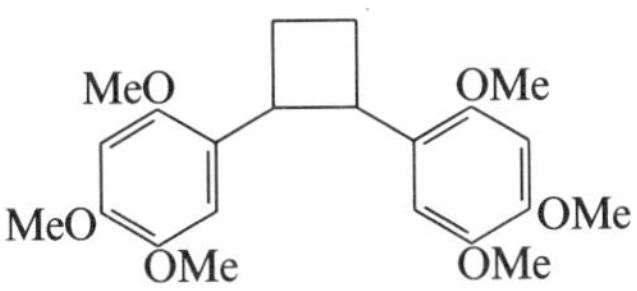

C$_{22}$H$_{28}$O$_6$ M 388.460
Constit. of *Pachypodanthium staudtii*. Cryst. (Et$_2$O/pet. ether). Mp 158°.

Ngadjui, B.T. *et al, Phytochemistry*, 1989, **28**, 231.

Pachypostaudin B
P-80002

[114542-59-5]

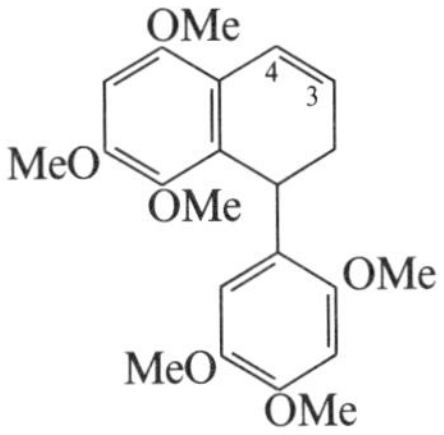

C$_{22}$H$_{26}$O$_6$ M 386.444
Constit. of *Pachypodanthium staudtii*. Needles (EtOAc/hexane). Mp 120°.

3,4-Dihydro: [114542-58-4]. *Pachypostaudin A*
C$_{22}$H$_{28}$O$_6$ M 388.460
Constit. of *P. staudtii*. Granules (EtOAc/hexane). Mp 130°.

Ngadjui, B.T. *et al, Phytochemistry*, 1989, **28**, 231.

Pachyrrhizin
P-80003

Updated Entry replacing P-00005
6-(6-Methoxy-1,3-benzodioxol-5-yl)-7H-furo[3,2-g][1]benzopyran-7-one, 9CI. 6-(2-Methoxy-4,5-methylenedioxyphenyl)furocoumarin. Neorautone
[10091-01-7]

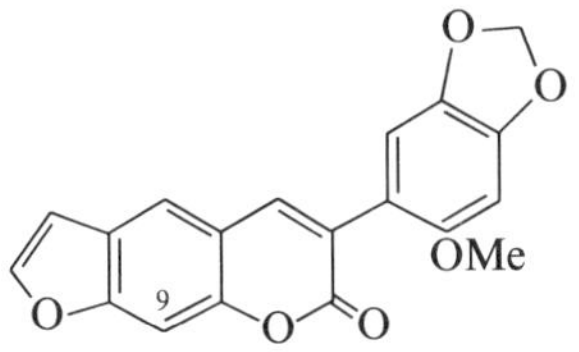

Constit. of *Pachyrrhizus erosus, Neoratanenia pseudopacchyrrhiza* and *N. edulis*. Yellow-green needles. Mp 207-209°.

9-Methoxy: Neofolin
C$_{20}$H$_{14}$O$_7$ M 366.326
Isol. from roots of *N. ficifolia*. Greenish-yellow needles (Me$_2$CO, EtOAc or C$_6$H$_6$); white amorph. solid (MeOH); brownish cubes (1,2-dichloroethylene) (trimorph.). Mp 155.5-157.5° (amorph.), Mp 167-168.5° (cubes), Mp 189.5-190.5° (needles).

Norton, L.B. *et al, J. Am. Chem. Soc.*, 1945, **67**, 1609 (*isol*)

Simanitsch, E. *et al, Monatsh. Chem.*, 1957, **88**, 541 (*struct*)
Rajogopalan, P. *et al, Tetrahedron Lett.*, 1959, 5 (*synth*)
Abrams, C. *et al, CA*, 1963, **59**, 2917 (*isol, struct*)
Crombie, L. *et al, J. Chem. Soc.*, 1963, 1569 (*pmr*)
Brink, C.M. *et al, J. S. Afr. Chem. Inst.*, 1966, **19**, 24 (*Neofolin*)

Pachyrrhizone
P-80004

Updated Entry replacing P-00006
6a,13a-Dihydro-8-methoxy-1,3-dioxolo[6,7][1]benzopyrano[3,4-b]furo[3,2-g]benzopyran-13(16H)-one, 9CI

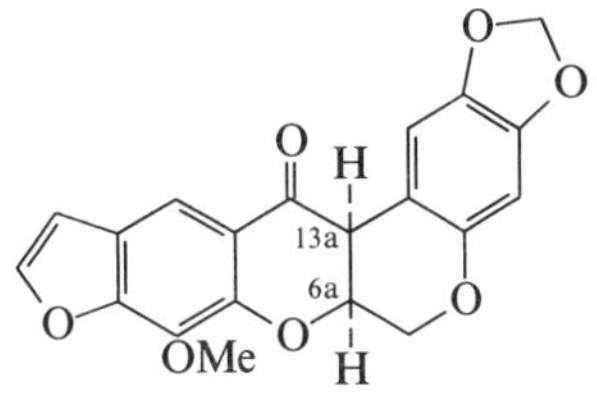

C$_{20}$H$_{14}$O$_7$ M 366.326

(+)-*form* [42485-00-7]
    Constit. of the yam bean *Pachyrrhizus erosus*. Needles. Mp 232-240° dec., Mp 250-251° (sealed tube), Mp 272° (block). [α]$_D^{20}$ +100° (c, 0.39 in CHCl$_3$).

Ac: Mp 147°.

Oxime: Cryst. Mp 267-268° dec. (279° block).

6a,13a-Didehydro: [28607-94-5]. *Dehydropachyrrhizone*
    C$_{20}$H$_{12}$O$_7$ M 364.311
    Isol. from seeds of *P. erosus*. Pale-yellow needles (CHCl$_3$/MeOH). Mp 260-262° dec.

(±)-*form*
    Mp 250° dec. (265° block).

Bickel, H. *et al, Helv. Chim. Acta*, 1953, **36**, 664 (*uv, ir*)
Krishnamurti, M. *et al, Tetrahedron*, 1970, **26**, 3023 (*Dehydropachyrrhizone*)
Uchiyama, M. *et al, Agric. Biol. Chem.*, 1973, **37**, 1227 (*synth*)

Palmidin B
P-80005

4,4',5,5'-Tetrahydroxy-2-(hydroxymethyl)-2'-methyl-[9,9'-bianthracene]-10,10'(9H,9'H)-dione, 9CI. Aloeemodin-chrysophanol bianthrone
[17062-56-5]

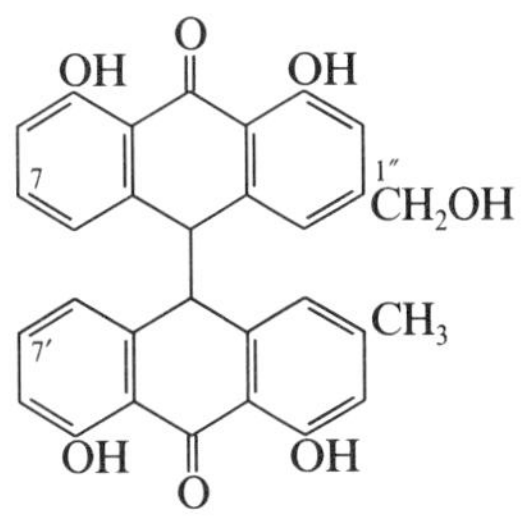

C$_{30}$H$_{22}$O$_7$ M 494.500
Isol. from roots of *Rheum palmatum* and fruits of *Frangula alnus*, also cascara (*Rhamnus purshiana*).

7'-Hydroxy: [17062-55-4]. *Palmidin A. Aloeemodin-emodin bianthrone*
    C$_{30}$H$_{22}$O$_8$ M 510.499
    From roots of *Rheum palmatum* and from cascara.

7-Hydroxy, 1″-deoxy: [17177-86-5]. **Palmidin C.** *Emodin-chrysophanol bianthrone*
$C_{30}H_{22}O_7$ M 494.500
From roots of *R. palmatum* and fruits of *F. alnus*, also cascara. λ_{max} 220, 272, 362 nm.

Lemli, J. *et al, Planta Med.*, 1964, **12**, 107.
Lemli, J., *J. Nat. Prod. (Lloydia)*, 1965, **28**, 63 (*isol, uv*)
Kinget, R., *Planta Med.*, 6th Ed., 1967, **15**, 233 (*isol, uv*)

Panaxydol   P-80006

9,10-Epoxy-1-heptadecene-4,6-diyn-3-ol. 8-(3-Heptyloxiranyl)-1-octene-4,6-diyn-3-ol, 9CI
[72800-72-7]

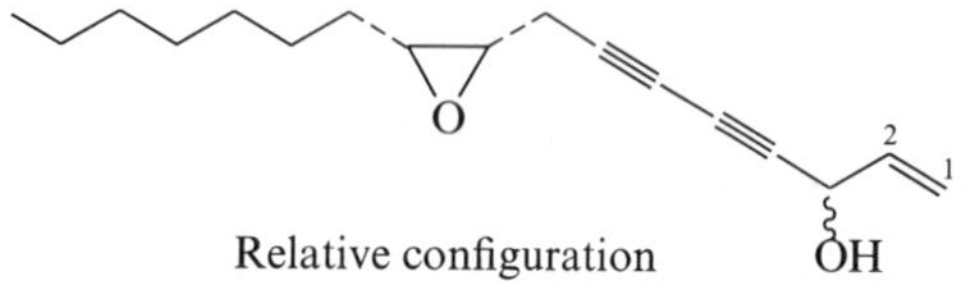

Relative configuration

$C_{17}H_{24}O_2$ M 260.375
Isol. from *Panax ginseng*. Cytotoxic agent. $[\alpha]_D$ −19.5° (c, 0.6 in MeOH).

1,2-Dihydro, 1-chloro, 2-hydroxy: [114687-51-3]. *1-Chloro-9,10-epoxy-4,6-heptadecadiyne-2,3-diol*. **Chloropanaxydiol**
$C_{17}H_{25}ClO_3$ M 312.835
$[\alpha]_D$ −37.2° (c, 0.2 in MeOH).

Fujimoto, Y. *et al, Chem. Pharm. Bull.*, 1988, **36**, 4206 (*isol, pmr, cmr, struct*)

Paniculadiol   P-80007

[122279-89-4]

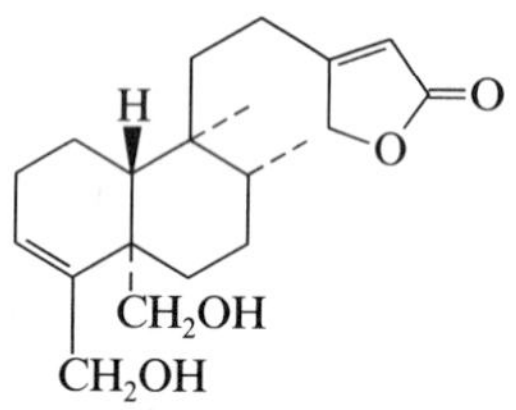

$C_{20}H_{30}O_4$ M 334.455
Constit. of *Baccharis paniculata*. Oil. $[\alpha]_D^{25}$ −1.72° (c, 0.29 in $CHCl_3$).

Di-Ac: [122297-37-4].
$C_{24}H_{34}O_6$ M 418.529
Constit. of *B. paniculata*. Oil.

Rivera, A.P. *et al, J. Nat. Prod. (Lloydia)*, 1989, **52**, 433 (*isol, pmr, cmr*)

Paniculatin   P-80008

6,8-Di-C-glucosyl-4′,5,7-trihydroxyisoflavone. Genistein 6,8-di-C-glucoside
[32361-88-9]

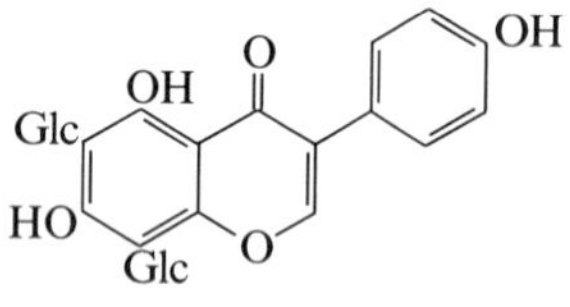

$C_{27}H_{30}O_{15}$ M 594.525
Isol. from *Dalbergia nitidula* and *D. paniculata*. Amorph. powder + $2H_2O$. Mp 225-227°.

Narayanan, V. *et al, Indian J. Chem.*, 1971, **9**, 14.

Pantethine   P-80009

N,N′-[Dithiobis[2,1-ethanediylimino(3-oxo-3,1-propanediyl)]]bis[2,4-dihydroxy-3,3-dimethylbutanamide], 9CI. N,N′-[Dithiobis(ethyleneiminocarbonylethylene)]bis[2,4-dihydroxy-3,3-dimethylbutyramide], 8CI
[303-05-9]

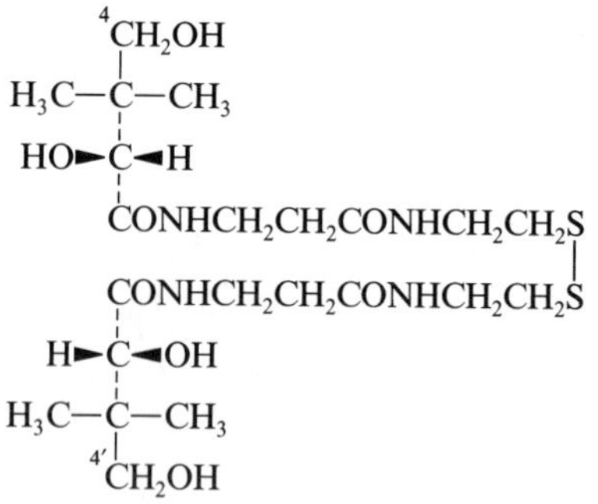

$C_{22}H_{42}N_4O_8S_2$ M 554.728
(R,R)-form [16816-67-4]
D-form
Found in foods. Bacterial growth factor. Synth. intermed. for coenzyme *A*. Cryst. ($MeOH/Me_2CO$), glass (H_2O). $[\alpha]_D^{22}$ +17.9° (c,3.2 in H_2O). Hydrol. by acids.

4,4′-Diphosphate: [17451-69-3].
$C_{22}H_{40}N_4O_{14}P_2S_2$ M 710.656
Intermediate for coenzyme *A* synthesis. Powder (MeOH)(as Ba salt). $[\alpha]_D^{20}$ +12.2° (c, 1 in H_2O).

Brown, G.M. *et al, J. Am. Chem. Soc.*, 1953, **75**, 1691 (*occur, purifn*)
Viscontini, M. *et al, Helv. Chim. Acta*, 1954, **37**, 375 (*synth, tlc, deriv*)
Bowman, R.E. *et al, J. Chem. Soc.*, 1954, 1171 (*anal, rev, bibl*)
Moffatt, J.G. *et al, J. Am. Chem. Soc.*, 1961, **83**, 663 (*chromatog*)
Shimizu, M. *et al, Chem. Pharm. Bull.*, 1965, **13**, 180 (*synth, chromatog, ir, uv*)
Nagase, O., *Chem. Pharm. Bull.*, 1967, **15**, 648 (*phosphate, synth*)
Zhdanovich, E.S. *et al, J. Gen. Chem. USSR (Engl. Transl.)*, 1967, **37**, 337 (*synth, ir*)
Hashimoto, M. *et al, Chem. Lett.*, 1972, 595 (*synth*)

[2.2.2.2.2]Paracyclophane   P-80010

Hexacyclo[26.2.2.2^{4,7}.2^{10,13}.2^{16,19}.2^{22,25}]tetraconta-4,6,10,12,16,18,22,24,28,30,31,33,35,37,39-pentadecaene, 9CI
[43082-13-9]

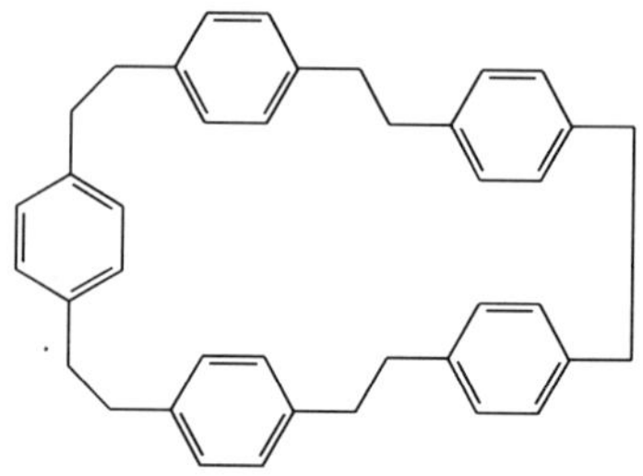

$C_{40}H_{40}$ M 520.756
Mp 172°.

Schmidbaur, H. *et al, Chem. Ber.*, 1988, **121**, 1341 (*synth, ir, ms, cryst struct*)

Parthenin   P-80011

Updated Entry replacing P-30015
[508-59-8]

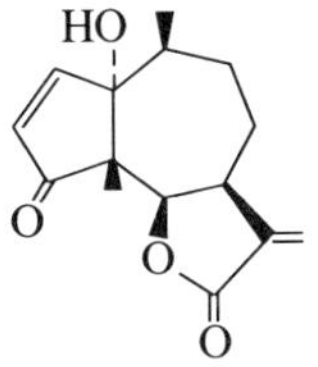

C₁₅H₁₈O₄   M 262.305

$C_{15}H_{18}O_4$ M 262.305

Bitter principle of *Parthenium hysterophorum* and *Iva nevadensis*. Cryst. (H₂O). Mp 163-166°. [α]²⁵_D +7.02° (c, 2.71 in CHCl₃).

▷ BD9635000.

11,13-Dihydro: Dihydroparthenin
$C_{15}H_{20}O_4$ M 264.321
Constit. of *P. hysterophorus*. Cryst. Mp 145°.

Herz, W. *et al, J. Am. Chem. Soc.*, 1962, **84**, 2601 (*isol, struct*)
Kok, P. *et al, Bull. Soc. Chim. Belg.*, 1978, **87**, 615 (*synth*)
Heathcock, C.H. *et al, J. Am. Chem. Soc.*, 1982, **104**, 6081 (*synth*)
Talwar, K.K. *et al, Phytochemistry*, 1989, **28**, 1091 (*deriv*)

Parthenolidine   P-80012

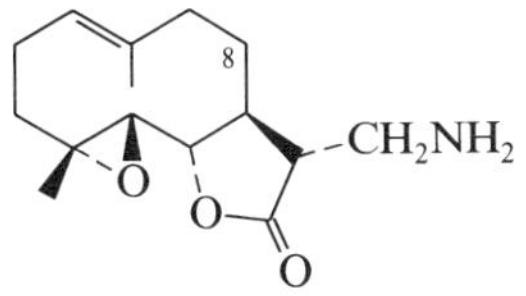

$C_{15}H_{23}NO_3$ M 265.352

N-Ac: [119766-96-0]. **N-Acetylparthenolidine**
$C_{17}H_{25}NO_4$ M 307.389
Constit. of *Michelia rajaniana* (Magnoliaceae). Oil. [α]²⁴_D +23° (c, 0.67 in CHCl₃).

Ruangrungsi, N. *et al, J. Nat. Prod. (Lloydia)*, 1988, **51**, 1220 (*isol, pmr, cmr*)

Parvifoline   P-80013

Updated Entry replacing P-10011
[62706-41-6]

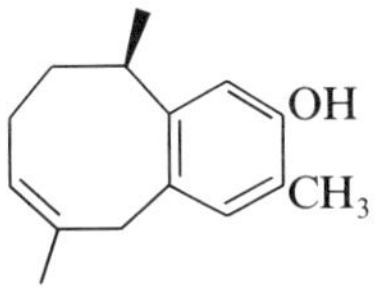

$C_{15}H_{20}O$ M 216.322

Constit. of *Coreopsis parvifolia* and *Perezia carpholepis*. Needles (Me₂CO/hexane). Mp 85-86°. [α]_D −173° (c, 1.73 in CHCl₃).

Ac:
$C_{17}H_{22}O_2$ M 258.360
Cryst. (Me₂CO/hexane). Mp 59-61°. [α]_D −98° (CHCl₃).

Bohlmann, F. *et al, Chem. Ber.*, 1977, **110**, 468 (*struct*)
Joseph-Naphan, P. *et al, Phytochemistry*, 1982, **21**, 669 (*struct*)
Joseph-Naphan, P. *et al, J. Nat. Prod. (Lloydia)*, 1988, **51**, 675 (*abs config*)

Parvisoflavone A   P-80014

3-(2,4-Dihydroxyphenyl)-5-hydroxy-8,8-dimethyl-4H, 8H-benzo[1,2-b:3,4-b']dipyran-4-one, 9CI
[50277-01-5]

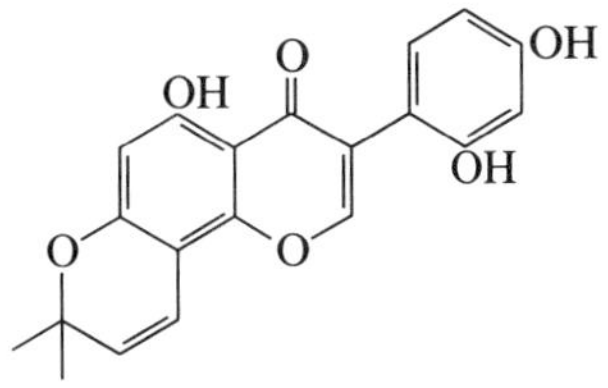

$C_{20}H_{16}O_6$ M 352.343

Obt. only as a mixt. with Parvisoflavone B, P-80015. Isol. from trunkwood of *Poecilanthe parviflora*. Yellow cryst. (Me₂CO/CHCl₃) (mixt.). Mp 235-239°.

Tri-Me ether: Mp 166-168°. Pure.

Assumpção, R.M.V. *et al, Phytochemistry*, 1973, **12**, 1188 (*isol, uv, ir, pmr, ms, struct*)

Parvisoflavone B   P-80015

Updated Entry replacing I-00887
7-(2,4-Dihydroxyphenyl)-5-hydroxy-2,2-dimethyl-2H,6H-benzo[1,2-b:5,4-b']dipyran-6-one, 9CI
[50277-02-6]

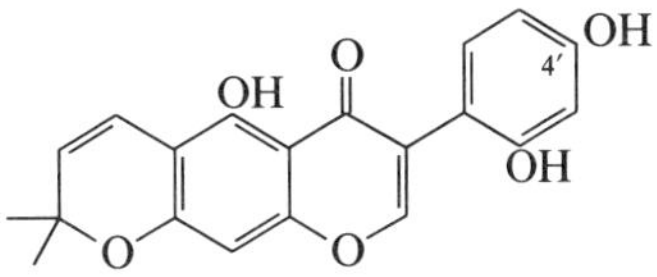

$C_{20}H_{16}O_6$ M 352.343

Obt. only as a mixt. with Parvisoflavone A, P-80014. Isol. from trunkwood of *Poecilanthe parviflora*.

Tri-Me ether: Cryst. Mp 149-151°. This deriv. obt. in pure state.

4'-O-(3-Methyl-2-butenyl): [30431-67-5]. **Isoauriculatin**
$C_{25}H_{24}O_6$ M 420.461
Constit. of *Milletia auriculata*. Pale-yellow plates. Mp 132-134°.

Assumpção, R.M.V. *et al, Phytochemistry*, 1973, **12**, 1188 (*isol, uv, ir, pmr, ms, struct*)
Minhaj, T.V. *et al, Tetrahedron*, 1976, **32**, 749 (*Isoauriculatin*)

Pedunculagin   P-80016

[7045-42-3]

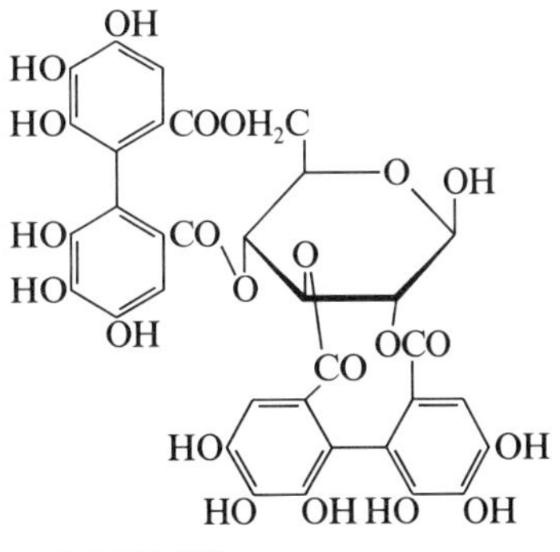

$C_{34}H_{24}O_{22}$ M 784.550

Isol. from *Quercus robor* and *Q. sessiliflora*. Amorph. sandy solid. [α]²⁵_D +106° (c, 2 in MeOH).

Schmidt, O.T. *et al, Justus Liebigs Ann. Chem.*, 1965, **690**, 150.

Pedunculol   P-80017

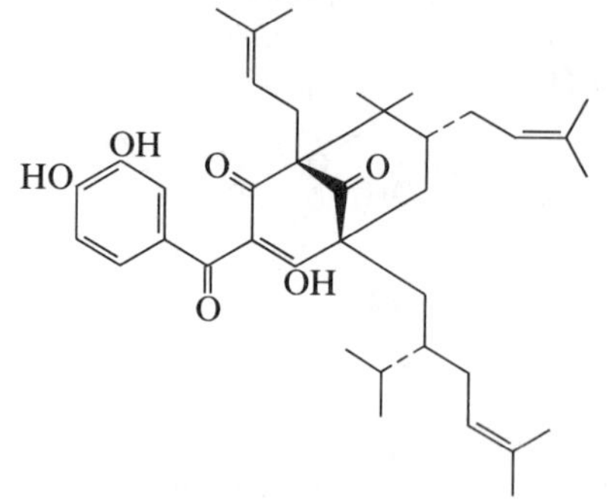

$C_{38}H_{52}O_6$ M 604.825

Constit. of *Garcinia pedunculata*. Yellow cryst. (C_6H_6). Mp 125°. $[\alpha]_D^{25}$ −159° (EtOH).

Sahu, A. *et al*, *Phytochemistry*, 1989, **28**, 1239.

Pendulone   P-80018

5-(3,4-Dihydro-7-hydroxy-2H-1-benzopyran-3-yl)-2,3-dimethoxy-2,5-cyclohexadiene-1,4-dione, 9CI. 7-Hydroxy-3′,4′-dimethoxyisoflavanquinone

[69359-09-7]

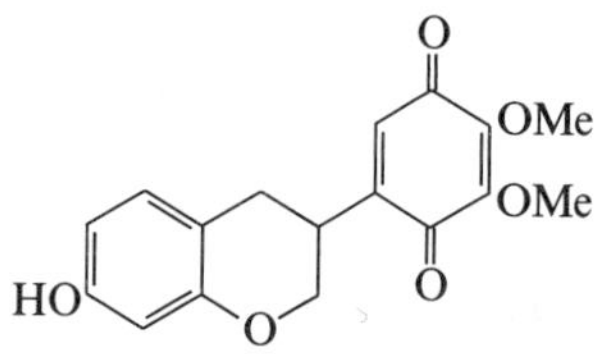

$C_{17}H_{16}O_6$ M 316.310

Tentative struct. Isol. from heartwood of *Millettia pendula*.

Hayashi, Y. *et al*, *Mokuzai Gakkaishi*, 1978, **24**, 898; *CA*, **90**, 100111p.

Penicillide   P-80019

Updated Entry replacing P-00207

11-Hydroxy-3-(1-hydroxy-3-methylbutyl)-4-methoxy-9-methyl-5H,7H-dibenzo[b,g][1,5]dioxocin-5-one, 9CI

[55303-92-9]

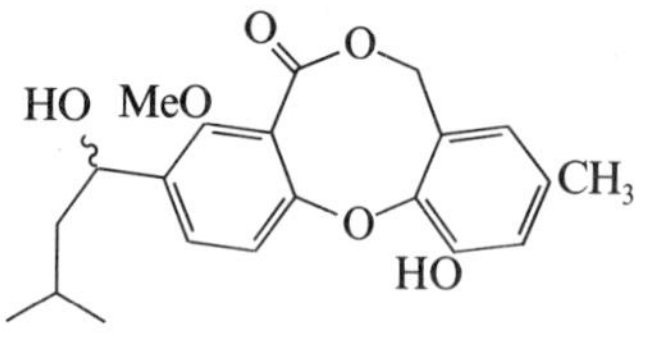

$C_{21}H_{24}O_6$ M 372.417

Metab. of *Penicillium* spp. Plant growth inhibitor.

Me ether: Mp 128-129.5°. $[\alpha]_D^{27}$ +14.4° (MeOH).

Sassa, T. *et al*, *Tetrahedron Lett.*, 1974, 3941 (*isol, struct, uv, ms*)

Pentacyclo[5.1.0.0²,⁴.0³,⁵.0⁶,⁸]octane   P-80020

Octabisvalene

[35434-67-4]

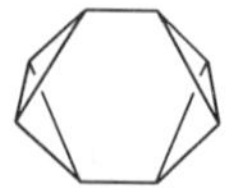

C_8H_8 M 104.151

Solid. Mp 27-30°.

Rücker, C. *et al*, *Chem. Ber.*, 1987, **120**, 1629.
Rücker, C. *et al*, *J. Am. Chem. Soc.*, 1988, **110**, 4828 (*synth, ms, ir, pmr, cmr*)

6,8,10,12-Pentadecatetraenal, 9CI   P-80021

[13894-86-5]

$$H_3CCH_2CH{=}CHCH{=}CHCH{=}CHCH{=}CH(CH_2)_4CHO$$

$C_{15}H_{22}O$ M 218.338

Isol. from leaves of *Centaurea* spp. Cryst. Mp <20°. An (*E,E,E,Z*)-form with exact config. unknown.

Bohlmann, F. *et al*, *Chem. Ber.*, 1966, **99**, 3544 (*isol, uv, pmr*)
Andersen, A.B. *et al*, *Phytochemistry*, 1977, **16**, 1829 (*isol*)

1,8,10,14-Pentadecatetraene-4,6-diyn-3-ol   P-80022

$$H_2C{=}CHCH_2CH_2CH{=}CHCH{=}CHC{\equiv}CC{\equiv}CCH(OH)CH{=}CH_2$$

$C_{15}H_{16}O$ M 212.291

(*E,E*)-*form* [17486-46-3]

Isol. from *Cotula coronopifolia*. Oil. λ_{max} 226, 236, 294, 309 nm. Not obt. pure.

p-*Phenylazobenzoyl:* Cryst. (pet. ether). Mp 44.5-46°.

Bohlmann, F. *et al*, *Chem. Ber.*, 1966, **99**, 2828; 1967, **100**, 3450 (*isol, struct, synth*)

8,10,14-Pentadecatriene-4,6-diyn-3-ol   P-80023

$$H_2C{=}CHCH_2CH_2CH{=}CHCH{=}CHC{\equiv}CC{\equiv}CCH(OH)CH_2CH_3$$

$C_{15}H_{18}O$ M 214.307

(*E,E*)-*form* [13028-53-0]

Isol. from above-ground parts of *Cotula coronopifolia*. Oil. λ_{max} 226, 236, 294, 309 nm (Et₂O). Not obt. pure.

Bohlmann, F. *et al*, *Chem. Ber.*, 1966, **99**, 2828; 1967, **100**, 3450 (*isol, struct, synth*)

10-Pentadecenal   P-80024

[60671-80-9]

$$H_3C(CH_2)_3CH{=}CH(CH_2)_8CHO$$

$C_{15}H_{28}O$ M 224.386

Kubo, I. *et al*, *Tetrahedron*, 1987, **43**, 2653 (*synth, ir, pmr*)

4-[5-(1,3-Pentadiynyl)-7-thienyl]-3-butyn-1-ol, 9CI   P-80025

[1209-21-8]

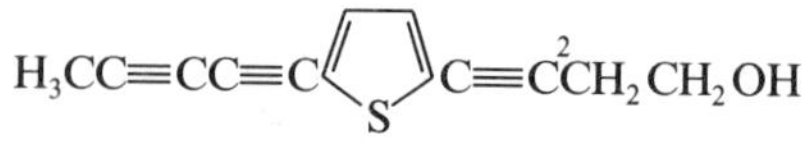

$C_{13}H_{10}OS$ M 214.287

Isol. from oil of *Echinops sphaerocephalus*. Oil. λ_{max} 319, 339 nm.

p-*Phenylazobenzoyl:* [26968-51-4].
Cryst. (Et₂O/pet. ether). Mp 126°.

Ac: 1-Acetoxy-4-[5-(1,3-pentadiynyl)-2-thienyl]-3-butyne
$C_{15}H_{12}O_2S$ M 256.325
Isol. from *E. sphaerocephalus*. Oil. λ_{max}232, 245, 319, 339 nm.

2-*Hydroxy:* [1212-60-8]. *4-[5-(1,3-Pentadiynyl)-2-thienyl]-3-butyne-1,2-diol, 9CI*
$C_{13}H_{10}O_2S$ M 230.287

Isol. from *E. sphaerocephalus* and other plants. Cryst. (CCl₄). Mp 112°.

2-Hydroxy, 1-Ac: [26905-72-6].
$C_{15}H_{12}O_3S$ M 272.324
Isol. from *Eclipta erecta*. Yellowish oil. λ_{max} 340, 320 nm.

2-Hydroxy, 2-Ac:
$C_{15}H_{12}O_3S$ M 272.324
Isol. from *Pluchea suaveolens*.

2-Hydroxy, di-Ac: [1161-52-0]. *2-(1,3-Pentadiynyl)-5-(3,4-diacetoxy-1-butynyl)thiophene*
$C_{17}H_{14}O_4S$ M 314.361
Isol. from *E. sphaerocephalus*. Oil.

2-Chloro: [1142-35-4]. **2-Chloro-4-[5-(1,3-pentadiynyl)-2-thienyl]-3-butyn-1-ol, 9CI**
$C_{13}H_9ClOS$ M 248.732
Isol. from *E. sphaerocephalus* and *Pluchea* sp. Oil.

2-Chloro, Ac: [1152-72-3]. *2-(3-Chloro-4-acetoxy-1-butynyl)-5-(1,3-pentadiynyl)thiophene*
$C_{15}H_{11}ClO_2S$ M 290.769
Isol. from *E. sphaerocephalus* and from roots of *Centaurea cristata*. Oil. λ_{max} 207.5, 233, 245, 320, 340.5 nm.

Bohlmann, F. *et al, Chem. Ber.*, 1965, **98**, 155, 876 (*isol, uv, ir, pmr, struct, biosynth*)
Bohlmann, F. *et al, Chem. Ber.*, 1966, **99**, 3544; 1968, **101**, 4163; 1970, **103**, 834 (*isol, synth, derivs*)
Bohlmann, F. *et al, Phytochemistry*, 1980, **19**, 969 (*deriv*)
Dawidar, A. *et al, Chem. Pharm. Bull.*, 1985, **33**, 5068 (*deriv*)

Pentafluoronitrobenzene, 9CI   P-80026

[880-78-4]

$(C_6F_5)NO_2$

$C_6F_5NO_2$ M 213.063
Golden yellow liq. Bp 158-161°, Bp_{52} 91-94°. Lachrymator.

Brooke, G.M. *et al, J. Chem. Soc.*, 1961, 802 (*synth*)
Bruce, M.I., *J. Chem. Soc. A*, 1968, 1459 (*F nmr*)
Briggs, J.M. *et al, J. Chem. Soc., Perkin Trans. 2*, 1973, 1789 (*cmr*)
Trudell, B.G. *et al, Can. J. Chem.*, 1978, **56**, 538 (*pe*)
Funn, G.G. *et al, J. Fluorine Chem.*, 1985, **28**, 241 (*cryst struct*)
Korobeinicheva, I.K. *et al, Izv. Akad. Sci. USSR, Ser. Sci. Khim.*, 1987, 1766 (*raman*)

Pentafluorophenylacetic acid   P-80027

2,3,4,5,6-Pentafluorobenzeneacetic acid, 9CI
[653-21-4]

$(C_6F_5)CH_2COOH$

$C_8H_3F_5O_2$ M 226.102
Cryst. (pet. ether). Mp 109°.

Et ester: [784-35-0].
$C_{10}H_7F_5O_2$ M 254.156
Bp_{14} 94-96°.

Chloride: [832-72-4].
$C_8H_2ClF_5O$ M 244.548
Bp_{29} 90-95°.

Amide:
$C_8H_4F_5NO$ M 225.118
Cryst. (EtOH/C₆H₆). Mp 187-188°.

Nitrile: [653-30-5]. *Pentafluorophenylacetonitrile*.
(Cyanomethyl)pentafluorobenzene
$C_8H_2F_5N$ M 207.102
Mp 38-38.5°. Bp_8 105°.

Barbour, A.K. *et al, J. Chem. Soc.*, 1961, 808 (*synth, ir*)
Org. Synth., 1977, **57**, 80 (*nitrile*)
Bordwell, F.G. *et al, J. Org. Chem.*, 1988, **53**, 780 (*nitrile*)

1,1,3,3,3-Pentafluoro-1-propene, 9CI   P-80028

2H-Pentafluoropropylene
[690-27-7]

$$F_3CCH{=}CF_2$$

C_3HF_5 M 132.033
Mp −160.9°. Bp −21° (−17°, −29°).

Banks, R.E. *et al, Proc. Chem. Soc., London*, 1964, 121 (*synth*)
Sianesi, D. *et al, Ann. Chim. (Rome)*, 1965, **55**, 872 (*synth, ir*)

1,2,3,8,9-Pentahydroxycoumestan   P-80029

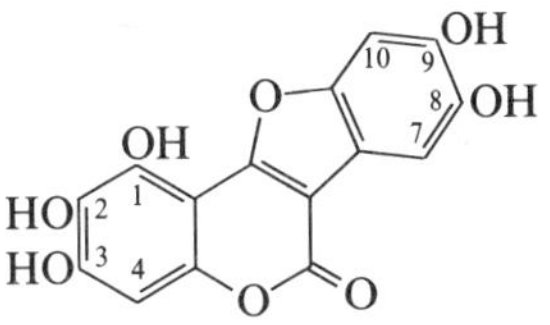

$C_{15}H_8O_8$ M 316.223

1,3-Di-Me, 8,9-methylene ether: [35930-41-7]. **2-Hydroxy-1,3-dimethoxy-8,9-methylenedioxycoumestan**
$C_{18}H_{12}O_8$ M 356.288
Isol. from heartwood of *Swartzia leiocalycina*.

Donnelly, D.M.X. *et al, Phytochemistry*, 1971, **10**, 3147.

1,2,4,6,9-Pentahydroxydihydro-β-agaro-furan   P-80030

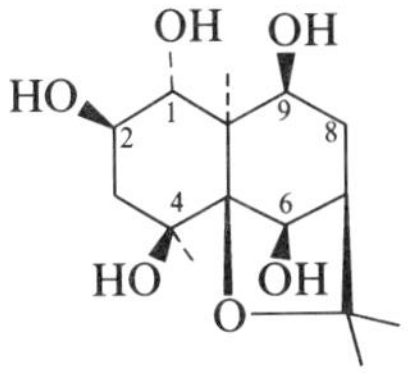

$C_{15}H_{26}O_6$ M 302.367
(1α,2β,6β,9β)-form

1-Cinnamoyl, 6-(3-pyridinecarbonyl), 9-Ac: 9β-Acetoxy-1α-cinnamoyloxy-2β,4β-dihydroxy-6β-nicotinoyloxydihydro-β-agarofuran. 1α-Cinnamoyloxy-6β-nicotinoyloxy-9β-acetoxy-2β,4β-dihydroxydihydro-β-agarofuran
$C_{30}H_{33}NO_7$ M 519.593
Alkaloid from the aerial parts of *Orthosphaenia mexicana* (Celastraceae). Amorph. solid.

6,9-Di-Ac, 1-cinnamoyl: 6β,9β-Diacetoxy-1α-cinnamoyloxy-2β,4β-hydroxydihydro-β-agarofuran
$C_{28}H_{36}O_9$ M 516.587
Constit. of *Orthosphenia mexicana*.

González, A.G. *et al, Phytochemistry*, 1987, **26**, 2133; 1988, **27**, 473 (*isol, ir, pmr, ms, struct*)

1,4,6,8,9-Pentahydroxydihydro-β-agaro-furan   P-80031

$C_{15}H_{26}O_6$ M 302.367
(1α,6β,8β,9α)-form

1-Benzoyl, 6-(3-pyridinecarbonyl), 9-Ac: 9α-Acetoxy-1α-benzoyloxy-4β,8β-dihydroxy-6β-nicotinoyloxydihydro-β-agarofuran. 1α-Benzoyloxy-6β-nicotinoyloxy-9α-acetoxy-4β,8β-dihydroxydihydro-β-agarofuran
$C_{30}H_{35}NO_9$ M 553.608
Alkaloid from the aerial parts of *Orthosphaenia mexicana* (Celastraceae). Mp 184-186°.

1-Benzoyl, 6-(3-pyridinecarbonyl), 8,9-Di-Ac: 8β,9α-Diacetoxy-1α-benzoyloxy-4β-hydroxy-6β-nicotinoyl-β-agarofuran.

nicotinoyloxy-8β,9α-diacetoxy-4β-hydroxydihydro-β-agarofuran
$C_{32}H_{37}NO_{10}$ M 595.645
Alkaloid from the root bark of *Rzedowskia tolantonguensis* (Celastraceae). Mp 114-116°. *Rzedowskia* is not a generally recognised genus.

6,9-Di-Ac, 1-benzoyl: 6β,9α-Diacetoxy-1α-benzoyloxy-4β,8β-dihydroxydihydro-β-agarofuran
$C_{26}H_{34}O_9$ M 490.549
Constit. of *Orthosphenia mexicana*. Cryst. Mp 194-198°.

6,8,9-Tri-Ac, 1-benzoyl: 6β,8β,9α-Triacetoxy-1α-benzoyloxy-4β-hydroxydihydro-β-agarofuran
$C_{28}H_{36}O_{10}$ M 532.586
Constit. of *O. mexicana*. Cryst. Mp 76-80°.

(1α,6β,8β,9β)-form

*1-Benzoyl, 6β-(3-pyridinecarbonyl), 9-Ac: **9β-Acetoxy-1α-benzoyloxy-4β-hydroxy-6β-nicotinoyloxydihydro-β-agarofuran**. 1α-Benzoyloxy-6β-nicotinoyloxy-9β-acetoxy-4β-hydroxydihydro-β-agarofuran*
$C_{30}H_{35}NO_8$ M 537.608
Alkaloid from the aerial parts of *O. mexicana* and the root bark of *R. tolantonguensis* (Celastraceae). Mp 139-148°.

González, A.G. *et al, Phytochemistry*, 1987, **26**, 2133; 1988, **27**, 473 (*isol, ir, uv, pmr, ms, struct*)

1,4,6,9,15-Pentahydroxydihydro-β-agaro-furan   P-80032

$C_{15}H_{26}O_6$ M 302.367

(1α,4β,6β,9β)-form

6,9,15-Tri-Ac, 1-benzoyl: 6β,9β,15-Triacetoxy-1α-benzoyloxy-4β-hydroxydihydro-β-agarofuran
$C_{28}H_{36}O_{10}$ M 532.586
Constit. of *Orthosphenia mexicana*. Cryst. Mp 188-190°.

González, A.G. *et al, Phytochemistry*, 1988, **27**, 473.

2′,4′,5,5′,7-Pentahydroxyflavone   P-80033

$C_{15}H_{10}O_7$ M 302.240

4′,5′-Di-Me ether: 2′,5,7-Trihydroxy-4′,5′-dimethoxyflavone. 2′-Hydoxy-5′-methoxybiochanin A
$C_{17}H_{14}O_7$ M 330.293
Constit. of *Erythrina eriotriocha*. Needles (MeOH/CH$_2$Cl$_2$). Mp 250°.

Nkengfack, A.E. *et al, J. Nat. Prod.* (*Lloydia*), 1989, **52**, 320 (*isol, pmr, cmr*)

2′,4′,5,7,8-Pentahydroxyflavone   P-80034

$C_{15}H_{10}O_7$ M 302.240

2′,4′,7,8-Tetra-Me ether: 5-Hydroxy-2′,4′,7,8-tetramethoxyflavone
$C_{19}H_{18}O_7$ M 358.347
Constit. of *Limnophila rugosa*. Cryst. Mp 188-189°.

Mukherjee, K.S. *et al, Phytochemistry*, 1989, **28**, 1778.

3,4′,5,6,7-Pentahydroxyflavone   P-80035

Updated Entry replacing P-60047
[4324-55-4]
$C_{15}H_{10}O_7$ M 302.240
Yellow cryst. (EtOH/Me$_2$CO). Mp 328-330°.

5,6-Di-Me ether: 3,4′,7-Trihydroxy-5,6-dimethoxyflavone
$C_{17}H_{14}O_7$ M 330.293
Constit. of *Adenostoma sparsifolium*.

*6,7-Di-Me ether: [29536-41-2]. 3,4′,5-Trihydroxy-6,7-dimethoxyflavone. **Eupalitin***

$C_{17}H_{14}O_7$ M 330.293
Constit. of *Eupatorium* spp. Cryst. Mp 291-292°.

*6,7-Di-Me ether, O³-rhamnoside: [29617-75-2]. **Eupalin***
$C_{23}H_{24}O_{11}$ M 476.436
Constit. of the aerial parts of *Eupatorium ligustrinum* and of *Rudbeckia bicolor*. Cryst. (MeOH). Mp 207-210° (189°). [α]$_D$ −129.6° (Py).

*3,6,7-Tri-Me ether: [569-80-2]. 4′,5-Dihydroxy-3,6,7-trimethoxyflavone. **Penduletin***
$C_{18}H_{16}O_7$ M 344.320
Constit. of *Tephrosia candida*. Yellow cryst. (MeOH). Mp 222°.

*4′,6,7-Tri-Me ether: [4324-53-2]. 3,5-Dihydroxy-4′,6,7-trimethoxyflavone. **Mikanin***
$C_{18}H_{16}O_7$ M 344.320
Constit. of *Mikania cordata*. Bright-yellow cryst. (C$_6$H$_6$ or CHCl$_3$/MeOH). Mp 222-224°.

*5,6,7-Tri-Me ether: [35286-55-6]. 3,4′-Dihydroxy-5,6,7-trimethoxyflavone. **Candidol***
$C_{18}H_{16}O_7$ M 344.320
Constit. of *Tephrosia candida*. Cryst. Mp 253-254°.

*3,4′,6-Tri-Me ether: [27782-63-4]. 5,7-Dihydroxy-3,4′-6-trimethoxyflavone. **Santin***
$C_{18}H_{16}O_7$ M 344.320
Isol. from *Chrysanthemum parthenium*. Cryst. (C$_6$H$_6$/EtOAc). Mp 164-165°.

3,4′,6,7-Tetra-Me ether: [14787-34-9]. 5-Hydroxy-3,4′,6,7-tetramethoxyflavone
$C_{19}H_{18}O_7$ M 358.347
Constit. of *Blumea malcomii, Dodonaea lobulata* and *D. viscosa*. Cryst. Mp 178-180° (151.5-153°).

Penta-Me ether: [4472-73-5]. 3,4′,5,6,7-Pentamethoxyflavone
$C_{20}H_{20}O_7$ M 372.374
Needles. Mp 157-158° (151-153°).

Kiang, A.K. *et al, J. Chem. Soc.*, 1965, 6371 (*synth*)
Wagner, H. *et al, Tetrahedron Lett.*, 1965, 3849 (*synth*)
Dawson, R.M. *et al, Aust. J. Chem.*, 1966, **19**, 2133 (*isol*)
Wagner, H. *et al, Chem. Ber.*, 1967, **100**, 1768 (*synth*)
Sim, K.Y., *J. Chem. Soc. C*, 1967, 976 (*synth*)
Quijano, L. *et al, Tetrahedron*, 1970, **26**, 2851 (*Eupalin*)
Southwick, L. *et al, Phytochemistry*, 1972, **11**, 2351 (*isol, struct*)
Rodriguez, E. *et al, Phytochemistry*, 1972, **11**, 3509 (*isol, struct*)
Wagner, H. *et al, Chem. Ber.*, 1974, **107**, 1049 (*synth*)
Jauhari, P.K. *et al, Phytochemistry*, 1979, **18**, 359 (*isol*)
Proksch, M. *et al, Phytochemistry*, 1982, **21**, 2893 (*isol*)
Sachdev, K. *et al, Phytochemistry*, 1983, **22**, 1253 (*isol*)
Dutt, S.K. *et al, Phytochemistry*, 1983, **22**, 325 (*isol*)
Kulkarni, M.M. *et al, Phytochemistry*, 1987, **26**, 2079 (*isol*)
Parmar, V.S. *et al, Tetrahedron*, 1987, **43**, 4241 (*cryst struct*)
Horie, T. *et al, Phytochemistry*, 1988, **27**, 1491 (*synth, struct*)
Markham, K.R., *Phytochemistry*, 1989, **28**, 243 (*struct*)

3′,4′,5,5′,7-Pentahydroxyflavylium   P-80036
Tricetinidin

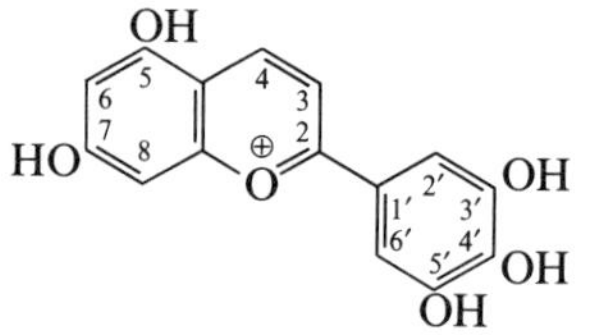

$C_{15}H_{11}O_6^{\oplus}$ M 287.248
Isol. from black tea (*Thea sinensis*) and prob. as glycoside from *Dryopteris erythrosora*.

Chloride:
$C_{15}H_{11}ClO_6$ M 322.701

λ_{max} 281, 513 nm (MOH + HCl); λ_{max} 564 nm (MeOH + NaOH).

Roberts, E.A.H. *et al*, *J. Sci. Food Agric.*, 1957, **8**, 72; 1958, **9**, 217.
Harborne, J.B., *Phytochemistry*, 1966, **5**, 589.

3′,4′,5,7,8-Pentahydroxyflavylium   P-80037
Columnidin

$C_{15}H_{11}O_6^{\oplus}$ M 287.248

Chloride:
$C_{15}H_{11}ClO_6$ M 322.701
λ_{max} 275, 511 nm (MeOH + HCl); λ_{max} 560 nm (MeOH + NaOH).

O-Glucoside: **Columnin**
$C_{21}H_{21}O_{11}^{\oplus}$ M 449.390
Isol. from *Columnea banksii*. Dark-red cryst. (EtOH) (as chloride).

Harborne, J.B., *Phytochemistry*, 1966, **5**, 589.

2′,3′,4′,7,8-Pentahydroxyisoflavan   P-80038
Updated Entry replacing D-05095

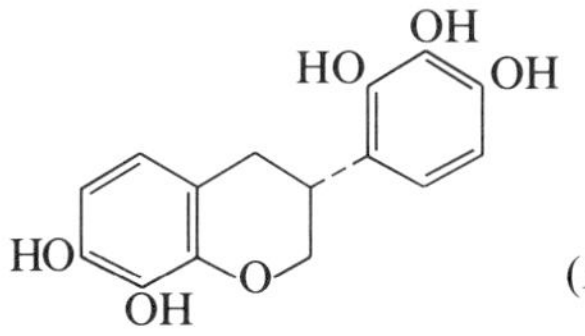

(R)-form

$C_{15}H_{14}O_6$ M 290.272
(R)-form

2′,4′-Di-Me ether: [50439-57-1]. 3′,7,8-Trihydroxy-2′,4′-dimethoxyisoflavan. *8-Demethylduartin*
$C_{17}H_{18}O_6$ M 318.326
Isol. from wood of *Dalbergia ecastophyllum*.

2′,4′-Di-Me ether, tri-Ac: Needles (MeOH). Mp 223.5-225°. $[\alpha]_D^{22}$ +22.3° (CHCl₃).
(S)-form

2′,4′,8-Tri-Me ether: [52305-04-1]. 3′,7-Dihydroxy-2′,4′,8-trimethoxyisoflavan. **Duartin**
$C_{18}H_{20}O_6$ M 332.352
Constit. of the wood of *Machaerium acutifolium*, *M. mucronulatum*, *M. opacum* and *M. villosium*. Needles (MeOH). Mp 145°. $[\alpha]_D$ −25° (Me₂CO), $[\alpha]_D^{20}$ −18.5° (c, 1.68 in CHCl₃).

2′,4′,8-Tri-Me ether, di-Ac: Cryst. (EtOH). Mp 101°. $[\alpha]_D^{20}$ −22° (c, 2.08 in CHCl₃).
(±)-form

2′,4′-Di-Me ether, tri-Ac: Mp 236-237°.

2′,4′,8-Tri-Me ether: Prisms (EtOH). Mp 199-201° (195°).

Kurosawa, K. *et al*, *J. Chem. Soc., Chem. Commun.*, 1968, 1263, 1265 (*Duartin*)
Donnelly, D.M.X. *et al*, *Phytochemistry*, 1973, **12**, 1157 (*8-Demethylduartin*)
Farkas, L. *et al*, *J. Chem. Soc., Perkin Trans. 1*, 1974, 305 (*synth*)
Wenkert, E. *et al*, *Phytochemistry*, 1977, **16**, 1811 (*cmr*)
Ollis, D.W. *et al*, *Phytochemistry*, 1978, **17**, 1401 (*isol, synth*)
Kurosawa, K. *et al*, *Phytochemistry*, 1978, **17**, 1423 (*isol, abs config*)

2′,3′,4′,5,7-Pentahydroxyisoflavanone   P-80039
2,3-Dihydro-5,7-dihydroxy-3-(2,3,4-trihydroxyphenyl)-4H-1-benzopyran-4-one, 9CI

$C_{15}H_{12}O_7$ M 304.256
2′,3′-Di-Me ether: [49776-79-6]. 4′,5,7-Trihydroxy-2′,3′-dimethoxyisoflavanone. **Parvisoflavanone**

$C_{17}H_{16}O_7$ M 332.309
Isol. from trunkwood of *Poecilanthe parviflora*. Prisms (MeOH). Mp 203-205°. Incorr. referred to also as Parvisoflavone.

Penta-Me ether: [49776-80-9]. 2′,3′,4′,5,7-Pentamethoxyisoflavanone
$C_{20}H_{22}O_7$ M 374.390
Needles (MeOH). Mp 159-162°. Incorr. referred to as Di-O-methylparvisoflavone.

Assumpcão, R.M.V. *et al*, *Phytochemistry*, 1973, **12**, 1188.

2′,3′,4′,6,7-Pentahydroxyisoflavene   P-80040

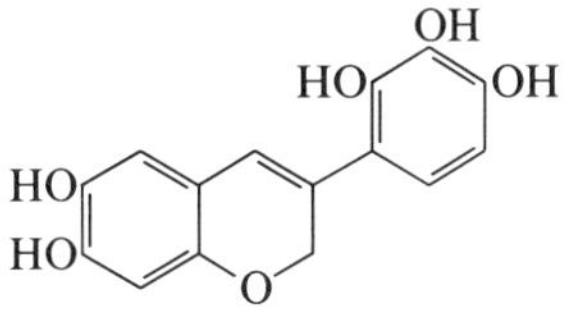

$C_{15}H_{12}O_6$ M 288.256
2′,4′ or 3′,4′-Di-Me ether: [78900-89-7].
$C_{17}H_{16}O_6$ M 316.310
Isol. from *Baphia nitida* heartwood.

Di-Me ether, tri-Ac: Mp 52-55°.

Arnone, A. *et al*, *Phytochemistry*, 1981, **20**, 799 (*isol, pmr*)

2′,4′,5,5′,7-Pentahydroxyisoflavone   P-80041
5,7-Dihydroxy-3-(2,4,5-trihydroxyphenyl)-4H-1-benzopyran-4-one, 9CI

$C_{15}H_{10}O_7$ M 302.240
2′,5′,7-Tri-Me ether: [73428-16-7]. 4′,5-Dihydroxy-2′,5′,7-trimethoxyisoflavone. **Derrugenin**
$C_{18}H_{16}O_7$ M 344.320
Isol. from seed hulls of *Derris robusta*. Mp 218-219°. Revised struct.

2′,4′,5′,7-Tetra-Me ether: [72545-39-2]. 5-Hydroxy-2′,4′,5′,7-tetramethoxyisoflavone. **Robustigenin**
$C_{19}H_{18}O_7$ M 358.347
Isol. from *D. robusta* seed hulls. Cryst.

2′,4′,5′,7-Tetra-Me ether, Ac: Needles. Mp 168°.

Penta-Me ether: [72545-41-6]. 2′,4′,5,5′,7-Pentamethoxyisoflavone. Robustigenin methyl ether
$C_{20}H_{20}O_7$ M 372.374
Isol. from *D. robusta* seed hulls. Needles. Mp 192-193°.

Chibber, S.S. *et al*, *Indian J. Chem., Sect. B*, 1979, **17**, 649 (*Robustigenin methyl ether*)
Chibber, S.S. *et al*, *Phytochemistry*, 1979, **18**, 1082, 1583 (*Robustigenin, Derrugenin*)

2′,4′,5′,6,7-Pentahydroxyisoflavone   P-80042
6,7-Dihydroxy-3-(2,4,5-trihydroxyphenyl)-4H-1-benzopyran-4-one, 9CI

$C_{15}H_{10}O_7$ M 302.240
2′,6-Di-Me, 4′,5′-methylenedioxy ether: [40009-88-9]. 7-Hydroxy-2′,6-dimethoxy-3′,4′-methylenedioxyisoflavone. **Dalpatein**
$C_{18}H_{14}O_7$ M 342.304
Isol. from seeds of *Dalbergia paniculata*. Needles (CHCl₃/MeOH). Mp 253-254°.

2′,6-Di-Me, 4′,5′-methylenedioxy ether, 7-O-β-D-glucopyranoside: see 7-O-β-D-Glucopyranoside [40522-40-5]. **Dalpatin**
$C_{24}H_{24}O_{12}$ M 504.446
Isol. from *D. paniculata*. Needles (EtOH). Mp 261-263° dec.

2',7-Di-Me, 4',5'-methylenedioxy ether: [51986-37-9]. *6-Hydroxy-2',7-dimethoxy-4',5'-methylenedioxyisoflavone*
$C_{18}H_{14}O_7$ M 342.304
Isol. from *Cordyla africana, Dalbergia assamica* and *Mildbraedeodendron excelsa.* Needles (MeOH). Mp 252-253°.

2',6,7-Tri-Me, 4',5'-methylenedioxy ether: [24195-15-1]. *2',6,7-Trimethoxy-4',5'-methylenedioxyisoflavone.* **Milldurone**
$C_{19}H_{16}O_7$ M 356.331
Isol. from *C. africana, D. paniculata, M. excelsa, Millettia dura, Pterodon apparicioi* and *P. pubescens.* Fine cryst. (MeOH). Mp 233-234°.

2',4',5',6-Tetra-Me ether: [22773-72-4]. *7-Hydroxy-2',4',5',6-tetramethoxyisoflavone*
$C_{19}H_{18}O_7$ M 358.347
Isol. from *P. apparicioi* trunkwood. Needles. Mp 205-207°.

Penta-Me ether: [24203-68-7]. *2',4',5',6,7-Pentamethoxyisoflavone*
$C_{20}H_{20}O_7$ M 372.374
Isol. from *C. africana, Mildbraeodendron excelsa, P. apparicioi* and *P. pubescens.* Needles (MeOH). Mp 171-172°.

Ollis, W.D. *et al, Tetrahedron,* 1967, **23**, 4741 (*Milldurone*)
Campbell, R.V.M. *et al, J. Chem. Soc. C,* 1969, 1787 (*isol*)
Nógrádi, M. *et al, Chem. Ber.,* 1970, **103**, 999 (*synth*)
Adinarayana, D. *et al, Tetrahedron,* 1972, **28**, 5377 (*Dalpatin*)
Campbell, R.V.M. *et al, J. Chem. Soc., Perkin Trans. 1,* 1973, 2222 (*6-hydroxy-2',7-dimethoxy-4',5'-methylenedioxyisoflavone*)
Galina, E. *et al, Phytochemistry,* 1974, **13**, 2593 (*isol*)
Meegan, M.J. *et al, Phytochemistry,* 1975, **14**, 2283 (*isol*)
Ingham, J.L., *Fortschr. Chem. Org. Naturst.,* 1983, **43**, 1 (*rev, occur*)

3',4',5,6,7-Pentahydroxyisoflavone   P-80043

Updated Entry replacing T-03329
3-(3,4-Dihydroxyphenyl)-5,6,7-trihydroxy-4H-1-benzopyran-4-one, 9CI

$C_{15}H_{10}O_6$ M 286.240

3',6-Di-Me ether: [39012-01-6]. *4',5,7-Trihydroxy-3',6-dimethoxyisoflavone.* **Iristectorigenin B**
$C_{17}H_{14}O_7$ M 330.293
Isol. from rhizomes of *I. florentina* and *I. germanica.* Pale-yellow needles (MeOH). Mp 153°.

3',6-Di-Me ether, glucoside: **Iristectorin B**
$C_{23}H_{24}O_{12}$ M 492.435
Isol. from rhizomes of *I. tectorum.* Posn. of glycosyl residue not sure.

4',6-Di-Me ether: [37744-62-0]. *3',5,7-Trihydroxy-4',6-dimethoxyisoflavone.* **Iristectorigenin A**
$C_{17}H_{14}O_7$ M 330.293
Isol. from *Iris germanica, I. unguicularis, Monopteryx inpae* and from cultures of *Streptomyces* sp. Shows antihypertensive props. Mp 231°. All isolations of isoflavones from microorganisms are considered suspect.

4',6-Di-Me ether, 7-O-β-D-glucopyranoside: [37744-61-9]. **Iristectorin A**
$C_{23}H_{24}O_{12}$ M 492.435
Constit. of *I. tectorum.* Mp 212-214°.

3',4',6-Tri-Me ether: [78134-85-7]. *5,7-Dihydroxy-3',4',7-trimethoxyisoflavone.* **Junipegenin B. Dalspinosin**
$C_{18}H_{16}O_7$ M 344.320
Isol. from leaves of *Juniperis macropoda* and roots of *Dalbergia spinosa.* Golden-yellow plates (MeOH). Mp 188-189°.

4',5-Di-Me, 6,7-methylenedioxy ether: [69618-04-8]. *3'-Hydroxy-4',5-dimethoxy-6,7-methylenedioxyisoflavone.* **Iriskumaonin**
$C_{18}H_{14}O_7$ M 342.304
Isol. from *I. kumaonensis, I. germanica* and *I. tingitana.* Minute prisms (MeOH), silky solid (EtOAc/pet. ether). Mp 207-208° (203-205°).

6-Me, 3',4'-methylene ether: [83162-85-0]. *5,7-Dihydroxy-6-methoxy-3',4'-methylenedioxyisoflavone.* **Dalspinin**
$C_{17}H_{12}O_7$ M 328.278
Obt. from roots of *Dalbergia spinosa.* Mp 200-202°.

5,6-Di-Me, 3',4'-methylenedioxy ether: [68862-19-1]. *7-Hydroxy-5,6-dimethoxy-3',4'-methylenedioxyisoflavone.* **Isoplatycarpanetin.** *Dipteryxin†*
$C_{18}H_{14}O_7$ M 342.304
Isol. from *Cladrastis platycarpa, C. shikokiana* and *Dipteryx odorata.* Cryst. (CHCl$_3$/Et$_2$O). Mp 235-237°. The name Dipteryxin has also been given to 7,8-Dihydroxy-4',6-dimethoxyisoflavone from *D. odorata* therefore the name Isoplatycarpanetin is preferred here.

5,6-Di-Me, 3',4'-methylenedioxy ether, 7-O-β-D-glucopyranoside:
$C_{24}H_{24}O_{12}$ M 504.446
Isol. from *Cladrastis platycarpa* and *C. shikokiana.*

5,6,7-Tri-Me, 3',4'-methylenedioxy ether: [51986-39-1]. *5,6,7-Trimethoxy-3',4'-methylenedioxyisoflavone.* **Odoratine†**
$C_{19}H_{16}O_7$ M 356.331
Isol. from *Cordyla africana* and *Dipteryx odorata.* Cryst. (CHCl$_3$/Et$_2$O). Mp 172-174°. The name Odoratin(e) has also been used for 3',7-Dihydroxy-4',6-dimethoxy-isoflavone from *D.* spp., and additionally for at least 3 other nat. prods.

[86849-77-6]

Morita, N. *et al, Chem. Pharm. Bull.,* 1972, **20**, 730 (*Iristectorigenin A*)
Arisawa, M. *et al, Chem. Pharm. Bull.,* 1973, **21**, 600, 2323 (*Iristectorigenin B*)
Campbell, R.V.M. *et al, J. Chem. Soc., Perkin Trans. 1,* 1973, 2222 (*Dalspinin*)
Antus, S. *et al, Chem. Ber.,* 1975, **108**, 3883 (*synth, pmr*)
Kalla, A.K. *et al, Phytochemistry,* 1978, **17**, 1441 (*Iriskumaonin*)
Nakano, T. *et al, J. Chem. Soc., Perkin Trans. 1,* 1979, 2107 (*Odoratine, Dipteryxine*)
Sethi, M.L. *et al, Phytochemistry,* 1981, **20**, 341 (*Junipegenin B*)
Dasan, R.G. *et al, Indian J. Chem., Sect. B,* 1982, **21**, 385 (*Junipegenin B*)
Ingham, J.L., *Fortschr. Chem. Org. Naturst.,* 1983, **43**, 1 (*rev, occur*)
Sethi, M.L. *et al, Phytochemistry,* 1983, **22**, 289 (*synth*)
Shawl, A.S. *et al, Phytochemistry,* 1984, **23**, 2405 (*isol*)

3',4',5',6,7-Pentahydroxyisoflavone   P-80044

Updated Entry replacing T-03614
6,7-Dihydroxy-3-(3,4,5-trihydroxyphenyl)-4H-1-benzopyran-4-one

$C_{15}H_{10}O_7$ M 302.240

3',6,7-Tri-Me, 4',5'-methylenedioxy ether: [24203-70-1]. *3',6,7-Trimethoxy-4',5'-methylenedioxyisoflavone*
$C_{19}H_{16}O_7$ M 356.331
Constit. of *Cordyla africana* and *Mildbraeodendron excelsa* heartwoods. Needles (MeOH). Mp 211-212°.

Penta-Me ether: [58523-19-6]. *3',4',5',6,7-Pentamethoxyisoflavone*
$C_{20}H_{20}O_7$ M 372.374
Isol. from trunkwood of *Pterodon apparicioi.* Mp 210-212°.

Campbell, R.V.M. *et al, J. Chem. Soc. C*, 1969, 1787 (*isol, ir, uv, pmr, ms*)
Meegan, M.J. *et al, Phytochemistry*, 1975, **14**, 2283 (*isol*)
Leite de Almeida, M.E. *et al, Phytochemistry*, 1975, **14**, 2716 (*isol*)

3′,4′,5,7,8-Pentahydroxyisoflavone   P-80045

Updated Entry replacing T-01267
3-(3,4-Dihydroxyphenyl)-5,7,8-dihydroxy-4H-1-benzopyran-4-one, 9CI
$C_{15}H_{10}O_7$ M 302.240
Prod. by *Aspergillus niger* and *Streptomyces neyagawaensis* var. *orobolare*. DOPA decarboxylase inhibitor. Brownish cryst. (MeOH/C_6H_6). Mp 252° dec.

8-Me ether: [58262-89-8]. *3′,4′,5,7-Tetrahydroxy-8-methoxyisoflavone. 8-Methoxyorobol*
$C_{16}H_{12}O_7$ M 316.267
Isol. from cultures of *Aspergillus niger*. DOPA decarboxylase inhibitor. Brownish cryst. (MeOH/C_6H_6). Mp 252° dec. All isolations of isoflavones from microorganisms are considered dubious.

3′,8-Di-Me ether: [41744-53-0]. *4′,5,7-Trihydroxy-3′,8-dimethoxyisoflavone. Homotectorigenin*
$C_{17}H_{14}O_7$ M 330.293
Not yet isol. from nature.

3′,8-Di-Me ether, 7-O-glucoside: [41744-52-9].
Homotectoridin
$C_{23}H_{24}O_{12}$ M 492.435
Isol. from rhizomes of *Iris germanica*. Needles (EtOH aq.). Mp 186°.

4′,8-Di-Me ether: [56419-18-2]. *3′,5,7-Trihydroxy-4′,8-dimethoxyisoflavone*
$C_{17}H_{14}O_7$ M 330.293
Isol. from *Monopteryx inpae* and from cultures of *Streptomyces* sp. All isolations of isoflavones from microorganisms are considered dubious.

3′,4′,8-Tri-Me ether: [78182-91-9]. *5,7-Dihydroxy-3′,4′,8-trimethoxyisoflavone*
$C_{18}H_{16}O_7$ M 344.320
Isol. from trunkwood of *M. inpae*. Yellow cryst. Mp 168-169°.

5,8-Di-Me, 3′,4′-methylene ether: [53505-60-5]. *7-Hydroxy-5,8-dimethoxy-3′,4′-methylenedioxyisoflavone.*
Platycarpanetin
$C_{18}H_{14}O_7$ M 342.304
Isol. from *Cladrastis platycarpa* and *C. shikokiana*.

5,8-Di-Me, 3′,4′-methylene ether, 7-O-glucoside: [52783-55-8]. *Platycarpanetin 7-O-glucoside*
$C_{24}H_{24}O_{12}$ M 504.446
Isol. from *C. platycarpa* and *C. shikokiana*. Needles. Mp 142-144°.

5,8-Di-Me, 3′,4′-methylene ether, 7-O-laminaribioside:
Platycarpanetin 7-O-laminaribioside
$C_{30}H_{34}O_{17}$ M 666.588
Isol. from *C. platycarpa* and *C. shikokiana*. Struct. of glycoside residue not certain.

Penta-Me ether: [57800-12-1]. *3′,4′,5,7,8-Pentamethoxyisoflavone*
$C_{20}H_{20}O_7$ M 372.374
Mp 167-169° (163°).

Kawase, A. *et al, Agric. Biol. Chem.*, 1973, **37**, 145.
Imamura, H. *et al, Phytochemistry*, 1974, **13**, 757 (*Platycarpanetin*)
Chimura, H. *et al, J. Antibiot.*, 1975, **28**, 619 (*isol*)
Umezawa, H. *et al, J. Antibiot.*, 1975, **28**, 947 (*isol*)
Tobe, H. *et al, J. Antibiot.*, 1976, **29**, 623 (*isol*)
Albuquerque, F.B. *et al, Phytochemistry*, 1981, **20**, 235 (*5,7-Dihydroxy-3′,4′,8-trimethoxyisoflavone*)

3′,4′,6,7,8-Pentahydroxyisoflavone   P-80046

Updated Entry replacing T-20284
3-(3,4-Dihydroxyphenyl)-6,7,8-trihydroxy-4H-1-benzopyran-4-one
$C_{15}H_{10}O_7$ M 302.240

4′,7,8-Tri-Me ether: *3′,8-Dihydroxy-4′,7,8-trimethoxyisoflavone*
$C_{18}H_{16}O_7$ M 344.320
Isol. from cultures of *Streptomyces* sp. All isolations of isoflavonoids from microorganisms are considered dubious.

6,7,8-Tri-Me, 3′,4′-methylenedioxy ether: [71339-42-9]. *6,7,8-Trimethoxy-3′,4′-methylenedioxyisoflavone.*
Petalostetin
$C_{19}H_{16}O_7$ M 356.331
Isol. from *Petalostemon candidum*. Needles (EtOAc/pet. ether). Mp 169-170°.

3′,4′,8-Tri-Me, 6,7-methylenedioxy ether: *3′,4′,8-Trimethoxy-6,7-methylenedioxyisoflavone*
$C_{19}H_{16}O_7$ M 356.331
Isol. from *Xanthocercis zambesiaca*. Probable struct. Not obt. pure.

Chimura, H. *et al, J. Antibiot.*, 1975, **28**, 619 (*3′,8-Dihydroxy-4′,7,8-trimethoxyisoflavone*)
Harper, S.H. *et al, Phytochemistry*, 1976, **15**, 1019 (*3′,4′,8-Trimethoxy-6,7-methylenedioxyisoflavone*)
Torrance, S.J. *et al, Phytochemistry*, 1979, **18**, 366 (*Petalostetin*)
Bhardwaj, D.K. *et al, Indian J. Chem., Sect. B*, 1982, **21**, 493 (*synth, Petalostetin*)

ent-2α,3α,6β,7β,11β-Pentahydroxy-16-kauren-15-one   P-80047

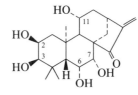

$C_{20}H_{30}O_6$ M 366.453

2,3,6,11-Tetra-Ac: [93078-70-7]. *ent*-2α,3α,6β,11β-Tetraacetoxy-7β-hydroxy-16-kauren-15-one.
Lushanrubescensin
$C_{28}H_{38}O_{10}$ M 534.602
Isol. from leaves of *Rabdosia rubescens*.

7-Epimer, 2,3,6-tri-Ac: [110325-77-4]. *ent*-2α,3α,6β-Triacetoxy-7α,11β-dihydroxy-16-kauren-15-one.
Lushanrubescensin B
$C_{26}H_{36}O_8$ M 476.566
Isol. from *R. rubescens*.

7-Deoxy, 3-Ac: [110325-79-6]. *ent*-3α-Acetoxy-2α,6β,11β-trihydroxy-16-kauren-15-one. **Lushanrubescensin D**
$C_{22}H_{32}O_6$ M 392.491
Isol. from *R. rubescens*.

7-Deoxy, 3,6-Di-Ac: *ent*-3α,6β-Diacetoxy-2α,11β-dihydroxy-16-kauren-15-one. **Lushanrubescensin E**
$C_{24}H_{34}O_7$ M 434.528
Isol. from *Rabdosia rubescens*, f. *lushanensis*. Shows antineoplastic props.

7-Deoxy, tetra-Ac: [110325-78-5]. *ent*-2α,3α,6β,11β-Tetraacetoxy-16-kauren-15-one. **Lushanrubescensin C**
$C_{28}H_{38}O_9$ M 518.603
Isol. from *R. rubescens*.

Qin, C. *et al, CA*, 1984, **101**, 226852v; 1987, **107**, 130873a, 130874b.
Li, J. *et al, CA*, 1988, **108**, 164740r.

ent-1β,3α,6β,7α,11α-Pentahydroxy-16-kauren-15-one   P-80048

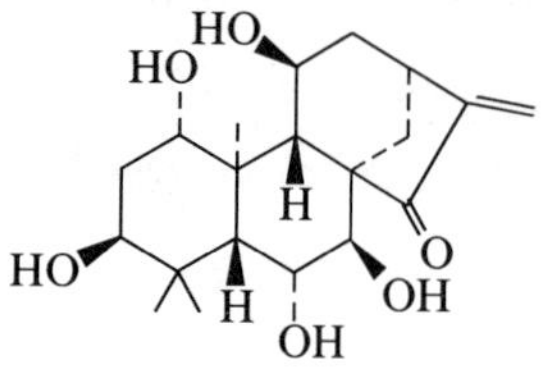

$C_{20}H_{30}O_6$ M 366.453

1,7,11-Tri-Ac: **Weisiensin A**
$C_{26}H_{36}O_9$ M 492.565
Constit. of *Rabdosia weisiensis*. Needles. Mp 298-300°.

Yunlong, X. *et al*, *Phytochemistry*, 1989, **28**, 1978.

1,2,5,6,7-Pentahydroxyphenanthrene   P-80049

Updated Entry replacing P-60050
1,2,5,6,7-Phenanthrenepentol

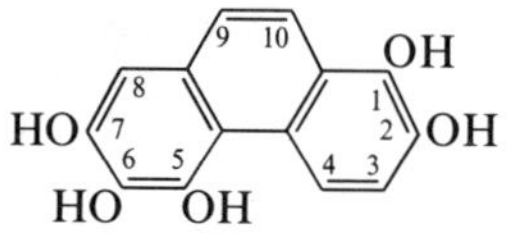

$C_{14}H_{10}O_5$ M 258.230

1,5,6-Tri-Me ether: [108909-02-0]. *1,5,6-Trimethoxy-2,7-phenanthrenediol, 9CI. 2,7-Dihydroxy-1,5,6-trimethoxyphenanthrene*. **Confusarin**
$C_{17}H_{16}O_5$ M 300.310
Constit. of *Eria confusa*. Cryst. (EtOAc/pet. ether). Mp 185°.

1,5,7-Tri-Me ether: 1,5,7-Trimethoxy-2,6-phenanthrenediol. 2,6-Dihydroxy-1,5,7-trimethoxyphenanthrene
$C_{17}H_{16}O_5$ M 300.310
Constit. of *Eulophia nuda*. Plates (CH_2Cl_2). Mp 152-153°.

Majumder, P.L. *et al*, *Phytochemistry*, 1987, **26**, 1127 (*isol*)
Tuchinda, P. *et al*, *Phytochemistry*, 1988, **27**, 3267 (*isol*)

1-(2,3,4,5,6-Pentahydroxyphenyl)-3-phenyl-2-propen-1-one   P-80050

2′,3′,4′,5′,6′-Pentahydroxychalcone

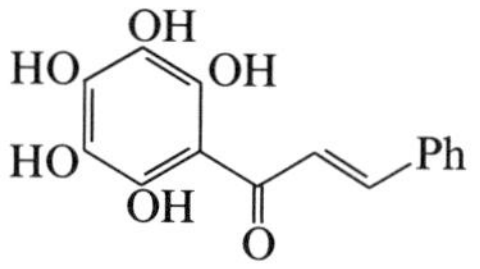

$C_{15}H_{12}O_6$ M 288.256

2′,4′,5′-Tri-Me ether: [521-51-7]. *1-(2,5-Dihydroxy-3,4,6-trimethoxyphenyl)-3-phenyl-2-propen-1-one. 2′,5′-Dihydroxy-3′,4′,6′-trimethoxychalcone*. **Pedicin**
$C_{18}H_{18}O_6$ M 330.337
Isol. from leaves of *Didymocarpus pedicellata* and in various Gesneraceae spp. Orange-red cryst. (Et_2O). Mp 143-145°.

Penta-Me ether: [518-58-1]. *1-(2,3,4,5,6-Pentamethoxyphenyl)-3-phenyl-2-propen-1-one. 2′,3′,4′,5′,6′-Pentamethoxychalcone*. **Pedicellin†**
$C_{20}H_{22}O_6$ M 358.390
Isol. from leaves of *D. pedicellata* and in various Gesneraceae spp. Tablets or needles (Et_2O or pet. ether). Mp 93°.

[27619-60-9]

Siddiqui, S., *Compt. Rend. Hebd. Seances Acad. Sci.*, 1939, 673 (*isol*)
Baker, W., *J. Chem. Soc.*, 1941, 662 (*synth*)
Seshadri, J.R., *J. Indian Chem. Soc.*, 1965, **42**, 343 (*struct, synth, Pedicin*)
Harborne, J.B., *Phytochemistry*, 1966, **5**, 589 (*occur*)

1,2,3,8,9-Pentahydroxypterocarpan   P-80051

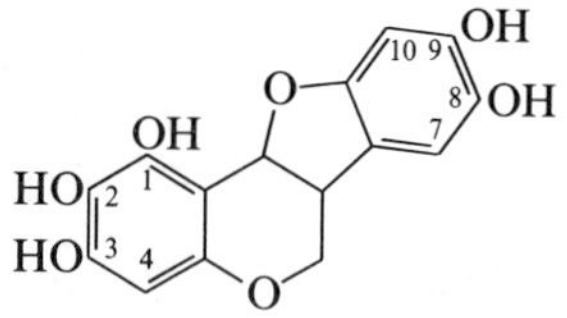

$C_{15}H_{12}O_7$ M 304.256

2-Me, 8,9-methylene ether: [73520-83-9]. *1,3-Dihydroxy-2-methoxy-8,9-methylenedioxypterocarpan*. **Trifolian**
$C_{17}H_{14}O_7$ M 330.293
Isol. from roots of *Trifolium pratense*. Cryst. (MeOH). Mp 132-134°. $[\alpha]_D$ −240° (c, 0.5 in MeOH).

Fraishtat, P.D. *et al*, *Bioorg. Khim.*, 1979, **5**, 1879; 1981, **7**, 927.

2,3,6*a*,8,9-Pentahydroxypterocarpan   P-80052

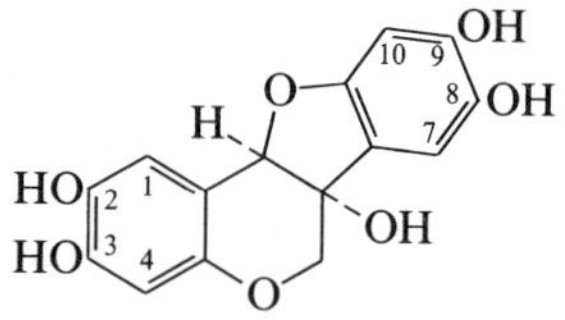

$C_{15}H_{12}O_7$ M 304.256

2,3-Di-Me, 8,9-methylene ether: [83159-18-6]. *6a-Hydroxy-2,3-dimethoxy-8,9-methylenedioxypterocarpan*. **Lathycarpin**
$C_{18}H_{16}O_7$ M 344.320
Phytoalexin from leaves of *Lathyrus sativus*. $[\alpha]_D^{21}$ +232° (c, 0.012 in MeOH).

Ingham, J.L. *et al*, *Z. Naturforsch., C*, 1982, **37**, 724 (*isol, uv, pmr, struct, abs config*)

3,4,6*a*,8,9-Pentahydroxypterocarpan   P-80053

Updated Entry replacing A-20005
$C_{15}H_{12}O_7$ M 304.256

3,4:8,9-Bismethylene ether: [70285-12-0]. *6a-Hydroxy-3,4:8,9-bis(methylenedioxy)pterocarpan*. **Acanthocarpan**
$C_{17}H_{12}O_7$ M 328.278
Stress metab. from leaves of *Caragana acanthophylla* and *Tephrosia bidwilli*. $[\alpha]_D$ −259° (c, 0.23 in MeOH). Revised struct.

4-Me, 8,9-Methylene ether: [87402-98-0]. *3,6a-Dihydroxy-4-methoxy-8,9-methylenedioxypterocarpan*. **Tephrocarpin**
Stress metab. from leaves of *T. bidwilli*. $[\alpha]_D$ −267° (c, 0.04 in MeOH).

Ingham, J.L. *et al*, *Phytochemistry*, 1982, **21**, 2969.

2,3,8,9,10-Pentahydroxypterocarpan   P-80054

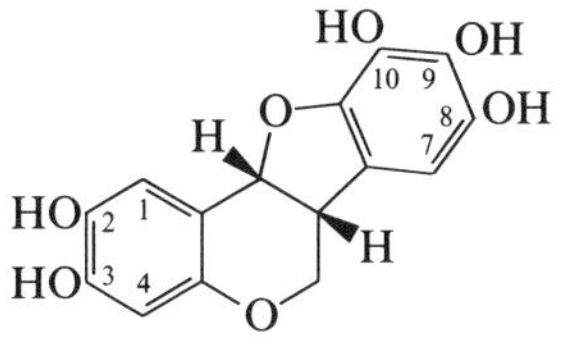

Absolute configuration

$C_{15}H_{12}O_7$ M 304.256

3,9,10-Tri-Me ether: [76474-66-3]. ***2,8-Dihydroxy-3,9,10-trimethoxypterocarpan***
$C_{18}H_{18}O_7$ M 346.336
Isol. from *Swartzia laevicarpa* trunkwood. Cryst. (C_6H_6/Me_2CO). Mp 126-127°.

Braz Filho, R. *et al, Phytochemistry*, 1980, **19**, 2003 (*isol, ir, uv, pmr, cmr, struct*)

3,4,8,9,10-Pentahydroxypterocarpan   P-80055

$C_{15}H_{12}O_7$ M 304.256

3,4,9,10-Tetra-Me ether: [89675-59-2]. *8-Hydroxy-3,4,9,10-tetramethoxypterocarpan*
$C_{19}H_{20}O_7$ M 360.363
Isol. from trunkwood of *Swartzia laevicarpa*. Cryst. ($C_6H_6/pet.$ ether). Mp 196-198°.

3,4,9,10-Tetra-Me ether, Ac: [89675-60-5].
Cryst. (pet. ether). Mp 192-194°.

Braz Filho, R. *et al, Phytochemistry*, 1980, **19**, 2003 (*isol, pmr, ord, ms, struct*)

3,6a,7,8,9-Pentahydroxypterocarpan   P-80056

$C_{15}H_{12}O_7$ M 304.256

8,9-Methylene ether: 3,6a,7-Trihydroxy-8,9-methylenedioxypterocarpan. 6a,7-Dihydroxymaackiain
$C_{16}H_{12}O_7$ M 316.267
Isol. from fungus-infected leaves of *Trifolium pratense*.

Ingham, J.L., *Phytochemistry*, 1976, **15**, 1489.

Pentamethylbenzaldehyde   P-80057

[17432-38-1]

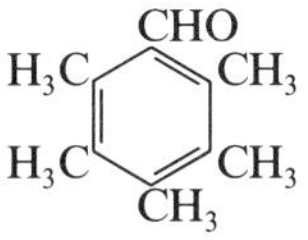

$C_{12}H_{16}O$ M 176.258
Cryst. (hexane). Mp 148-150°.

Wasserman, H.H. *et al, J. Org. Chem.*, 1971, **36**, 1765 (*synth, ir, pmr, ms*)
Bjørgo, J. *et al, J. Chem. Soc., Perkin Trans. 1*, 1977, 254 (*synth*)

Pentamethylbenzyl alcohol   P-80058

2,3,4,5,6-Pentamethylbenzenemethanol, 9CI
[484-66-2]

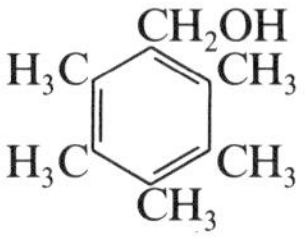

$C_{12}H_{18}O$ M 178.274
Cryst. ($C_6H_6/hexane$). Mp 162-163°.
Ac: [19936-85-7].

$C_{14}H_{20}O_2$. Cryst. Mp 85°.

Wasserman, H.H. *et al, J. Org. Chem.*, 1971, **36**, 1765 (*synth, ir, pmr, ms*)
Baciocchi, E. *et al, J. Org. Chem.*, 1977, **42**, 3682 (*deriv, pmr*)
Rudenko, A.P. *et al, J. Org. Chem. USSR (Engl. Transl.)*, 1985, **21**, 1609 (*synth, deriv, pmr*)

Pentamethyl(nitromethyl)benzene, 9CI   P-80059

Pentamethylphenylnitromethane
[29328-77-6]

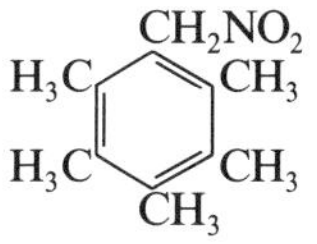

$C_{12}H_{17}NO_2$ M 207.272
Prisms (pet.ether). Mp 86-88°.

Suzuki, H., *Bull. Chem. Soc. Jpn.*, 1970, **43**, 879 (*synth, ir, pmr*)
Chiba, K. *et al, Bull. Chem. Soc. Jpn.*, 1976, **49**, 2614 (*synth*)
Kim, E.K. *et al, J. Org. Chem.*, 1989, **54**, 1692 (*synth, ir, pmr*)

2,3-Pentanediol, 9CI   P-80060

Updated Entry replacing P-00478
[42027-23-6]

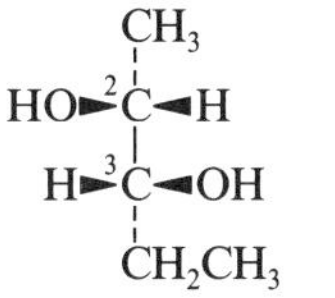

(2RS,3RS)-form

$C_5H_{12}O_2$ M 104.149
(2RS,3RS)-form [61828-36-2]
   (±)-*threo-form*
   Liq. d_4^{25} 0.965. Bp_{10} 83°. n_D^{20} 1.4320.
   Bis(4-methylbenzenesulfonyl): Cryst. ($Et_2O/pentane$). Mp 88-89°.
   Di-Ac:
     $C_9H_{16}O_4$ M 188.223
     Bp_{10} 89°. n_D^{20} 1.4195.
(2RS,3SR)-form [61828-35-1]
   (±)-*erythro-form*
   Liq. d_4^{25} 0.978. Bp_{10} 89°. n_D^{20} 1.4431.
   Bis(4-methylbenzenesulfonyl): Cryst. (pentane). Mp 81.5-83°.
   Di-Ac: Bp_{10} 85°. n_D^{20} 1.4167.
[97847-54-6, 97847-55-7]

Lucas, H.J. *et al, J. Am. Chem. Soc.*, 1941, **63**, 22 (*synth*)
Emmons, W.D. *et al, J. Am. Chem. Soc.*, 1954, **76**, 3472 (*synth*)
Rebrovic, L. *et al, J. Org. Chem.*, 1984, **49**, 2462 (*synth*)

4-Pentyn-1-amine, 9CI   P-80061

4-Pentynylamine, 8CI. 5-Amino-1-pentyne
[15252-44-5]

$$HC{\equiv}CCH_2CH_2CH_2NH_2$$

C_5H_9N M 83.133
Bp 124°.

Dumont, J.-L. *et al, Bull. Soc. Chim. Fr.*, 1967, 588 (*synth, ir*)

4-Pentynoic acid, 9CI   P-80062

Updated Entry replacing P-00592
[6089-09-4]

$$HC{\equiv}CCH_2CH_2COOH$$

$C_5H_6O_2$ M 98.101
Cryst. Mp 57°.
▷ SC4751000.
Me ester: [21565-82-2].
  $C_6H_8O_2$ M 112.128
  Bp 143-144°, Bp$_{175}$ 101-102°.
p-*Bromophenacyl ester:* Long needles. Mp 103°.

Colonge, J. *et al*, *Bull. Soc. Chim. Fr.*, 1954, 797 (*synth*)
Schulte, K.E. *et al*, *Chem. Ber.*, 1954, **87**, 964 (*synth*)
Holland, B.C. *et al*, *Synth. Commun.*, 1974, **4**, 203 (*synth, ir, pmr*)
Martel, J. *et al*, *Bull. Soc. Chim. Fr.*, 1978, 131 (*synth*)
Wulff, W.D. *et al*, *J. Am. Chem. Soc.*, 1988, **110**, 7419 (*synth, pmr*)

Peperinic acid   P-80063

Updated Entry replacing H-00678
2,4,5,6,7,7a-Hexahydro-7a-hydroxy-3,6-dimethyl-2-benzofuranone
[514-93-2]

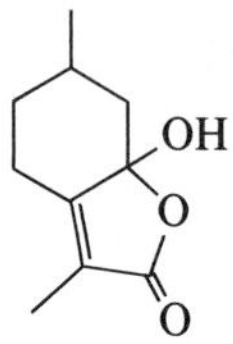

$C_{10}H_{14}O_3$ M 182.219
Constit. of *Bursera graveolens* and *Bystropogon mollis*.
Also isol. from aged peppermint oil. Cryst. (MeOH or C_6H_6). Mp 191° (188-189°). $[\alpha]_D^{21}$ +60.7°. Prob. artifact of oxidn.

Fester, G.A. *et al*, *An. Asoc. Quim. Argent.*, 1949, **37**, 197; 1952, **40**, 246; *CA*, **40**, 4636; **48**, 1974 (*isol*)
Woodward, R.B. *et al*, *J. Am. Chem. Soc.*, 1950, **72**, 399 (*isol*)
Crowley, K.J. *et al*, *J. Chem. Soc.*, 1964, 4254 (*isol*)

Peperomin A   P-80064

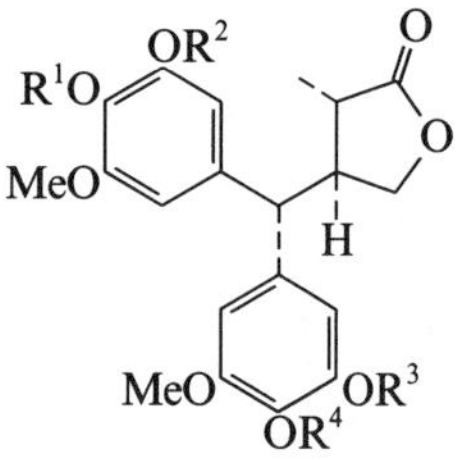

$$R^1R^2 = R^3R^4 = -CH_2-$$

$C_{22}H_{22}O_8$ M 414.411
Constit. of *Peperoma japonica*. Cryst. (MeOH aq.). Mp 143-145°. $[\alpha]_D^{27}$ +20.6° (c, 0.136 in CHCl$_3$).

Chen, C.-M. *et al*, *Heterocycles*, 1989, **29**, 411.

Peperomin B   P-80065

As Peperomin A, P-80064 with

$$R^1 = R^2 = Me, R^3R^4 = -CH_2-$$

$C_{23}H_{26}O_8$ M 430.454
Constit. of *Peperoma japonica*. Cryst. (MeOH aq.). Mp 143-145°. $[\alpha]_D^{27}$ +28.9° (c, 0.444 in CHCl$_3$).

Chen, C.-M. *et al*, *Heterocycles*, 1989, **29**, 411 (*isol, cryst struct*)

Peperomin C   P-80066

As Peperomin A, P-80064 with

$$R^1 = R^2 = R^3 = R^4 = Me$$

$C_{24}H_{30}O_8$ M 446.496
Constit. of *Peperoma japonica*. Cryst. (MeOH). Mp 158-160°. $[\alpha]_D^{27}$ +42.7° (c, 0.059 in CHCl$_3$).

Chen, C.-M. *et al*, *Heterocycles*, 1989, **29**, 411.

1*H*-Perimidine-2-sulfonic acid   P-80067

[110448-71-0]

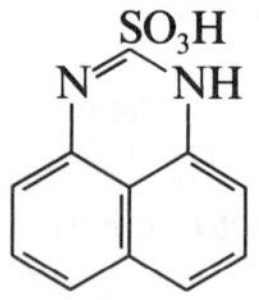

$C_{11}H_8N_2O_3S$ M 248.262
Pale yellow needles. Mp >300°.

Herbert, J.M. *et al*, *Heterocycles*, 1987, **26**, 1043 (*synth, ir, pmr*)

1*H*-Perimidine-2(3*H*)-thione, 9CI   P-80068

Updated Entry replacing P-10051
3H-Benzo[de]quinazoline-2-thione. 2-Mercapto-1H-perimidine
[30837-62-8]

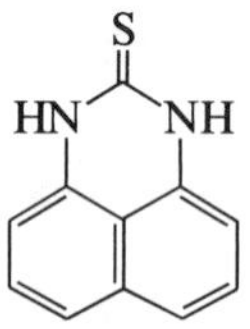

$C_{11}H_8N_2S$ M 200.264
Thione-form (shown) predominates. Readily obt. from 1,8-naphthalenediamine and CS$_2$ in 95% yield. Cryst. (CHCl$_3$). Mp 265° dec.
S-Me: [92972-05-9]. *2-(Methylthio)perimidine*
  $C_{12}H_{10}N_2S$ M 214.290
  Yellow prisms (Me$_2$CO). Mp 199-200.5°.
S-Me; B,HI: [89473-00-7].
  Yellow cryst. Mp >300°.
S-Me, S-oxide: [110191-88-3]. *2-(Methylsulfinyl)perimidine*
  $C_{12}H_{10}N_2OS$ M 230.290
  Yellow needles by subl. Mp 181-182°.

Liu, K.-C. *et al*, *Arch. Pharm. (Weinheim, Ger.)*, 1976, **309**, 928.
Liu, J. *et al*, *J. Heterocycl. Chem.*, 1984, **21**, 911 (*synth, uv, ir, ms*)
Liu, K.-C. *et al*, *J. Heterocycl. Chem.*, 1984, **21**, 911 (*synth, deriv, bibl*)
Herbert, J.M. *et al*, *J. Med. Chem.*, 1987, **30**, 2081 (*synth, deriv, pmr, ms*)

Periplanone D$_1$ P-80069

[123163-72-4]

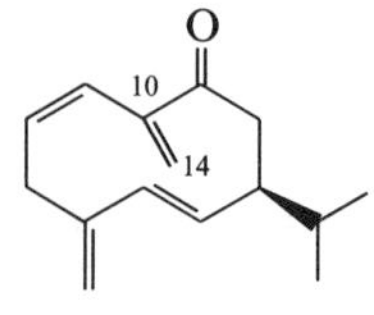

$C_{15}H_{20}O$ M 216.322
Constit. of faeces of female cockroaches *Periplaneta americana*.
10S,14-Dihydro: [123062-72-6]. **Periplanone D$_2$**

$C_{15}H_{22}O$ M 218.338
Metab. of *P. americana*.

[111917-97-6]

Biendl, M. *et al*, *Tetrahedron Lett.*, 1989, **30**, 2367.

Perlatolic acid   P-80070

Updated Entry replacing P-00637

2-Hydroxy-4-[(2-hydroxy-4-methoxy-6-pentylbenzoyl)oxy]-6-pentylbenzoic acid, 9CI. Perlatolinic acid

[529-47-5]

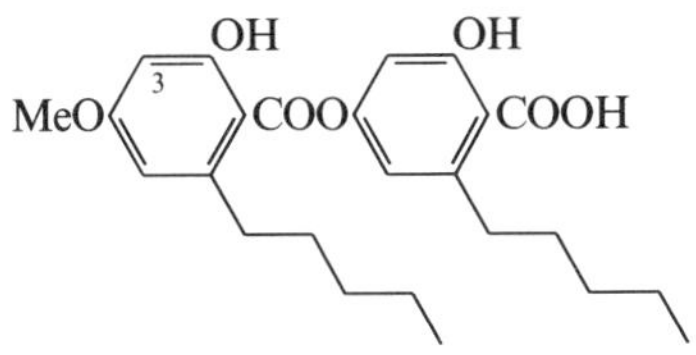

$C_{25}H_{32}O_7$ M 444.524
Constit. of many lichens including *Cladonia evansii* and *Ramalina stenospora*. Needles (C_6H_6). Mp 108°.

Di-Me ether, Me ester: Needles (MeOH). Mp 57°.

Benzyl ester: [53530-28-2].
  Needles (pentane). Mp 56.5-57°.

3-Chloro: [120091-97-6]. **3-Chloroperlatolic acid**
  $C_{25}H_{31}ClO_7$ M 478.968
  Constit. of *Dimelaena calcifornica*. Cryst. (EtOAc/pet. ether). Mp 128°.

Elix, J.A. *et al*, *Aust. J. Chem.*, 1974, **27**, 1767 (*synth, pmr, ms, bibl*)
Elix, J.A. *et al*, *Aust. J. Chem.*, 1988, **41**, 1789 (*isol, struct, synth*)

Petroformyne 1   P-80071

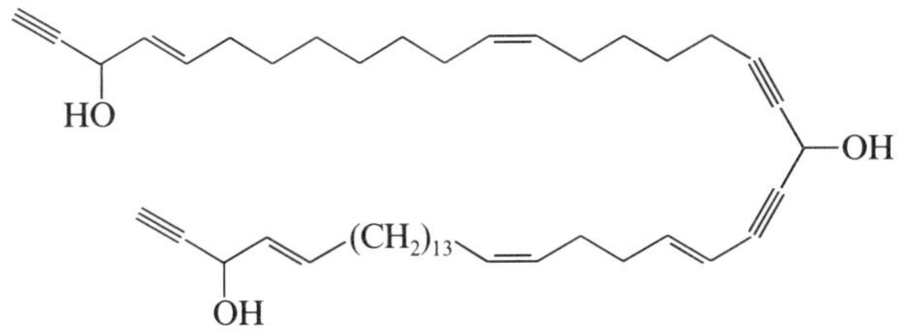

$C_{46}H_{68}O_3$ M 669.041
Constit. of *Petrosia ficiformis*. Cytotoxic, inhibits sea urchin egg development. $[\alpha]_D^{25}$ +12.5° (c, 10 in $CHCl_3$).

Cimino, G. *et al*, *Tetrahedron Lett.*, 1989, **30**, 3563.

Petroformyne 2   P-80072

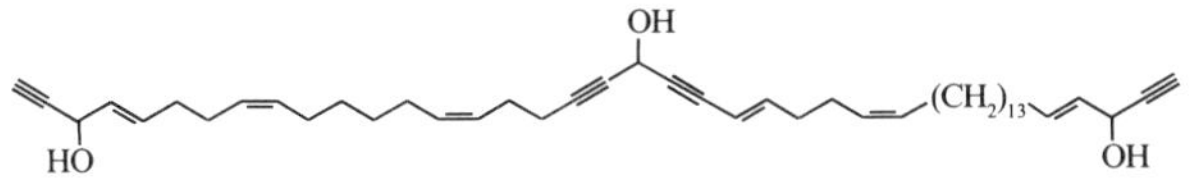

$C_{46}H_{66}O_3$ M 667.026
Constit. of *Petrosia ficiformis*. Cytotoxic, inhibits sea urchin egg development. $[\alpha]_D^{25}$ +15° ($CHCl_3$).

Cimino, G. *et al*, *Tetrahedron Lett.*, 1989, **30**, 3563.

Petroformyne 3   P-80073

$C_{46}H_{70}O_2$ M 655.058
Constit. of *Petrosia ficiformis*. Cytotoxic, inhibits sea urchin egg development. $[\alpha]_D$ +10.0° ($CHCl_3$).

43Z,44-Didehydro: **Petroformyne 4**
  $C_{46}H_{68}O_2$ M 653.042
  Constit. of *P. ficiformis*. Cytotoxic, inhibits sea urchin egg development. $[\alpha]_D$ +6.0° ($CHCl_3$).

Cimino, G. *et al*, *Tetrahedron Lett.*, 1989, **30**, 3563.

Petuniasterone N   P-80074

[123458-54-8]

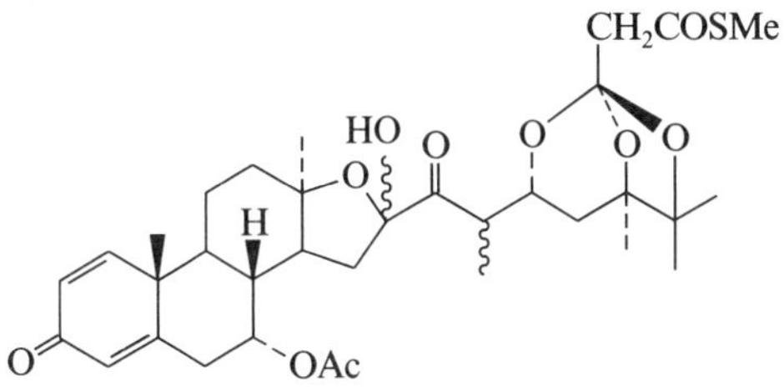

$C_{34}H_{46}O_{10}S$ M 646.797
Constit. of *Petunia* spp.

Elliger, C.A. *et al*, *J. Nat. Prod. (Lloydia)*, 1989, **52**, 576 (*isol, pmr, cmr*)

Peucelinendiol   P-80075

Updated Entry replacing P-00678

7-Hydroxymethyl-2,6,10,14-tetramethyl-2,9E,13E-pentadecatrien-6-ol

[72776-48-8]

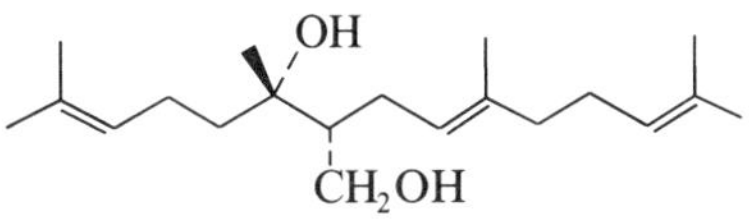

$C_{20}H_{36}O_2$ M 308.503
Constit. of *Peucedanum oreoselinum*. Oil. $[\alpha]_D^{24}$ +5.5° (c, 1.2 in $CHCl_4$).

Lemmich, E., *Phytochemistry*, 1979, **18**, 1195 (*isol*)
Moran, J.R. *et al*, *Bull. Chem. Soc. Jpn.*, 1988, **61**, 4435 (*synth, struct*)

Phaseollidin   P-80076

Updated Entry replacing P-00709

6a,11a-Dihydro-10-(3-methyl-2-butenyl)-6H-benzofuro[3,2-c][1]benzopyran-3,9-diol, 9CI. 3,9-Dihydroxy-10-prenylpterocarpan

[37831-70-2]

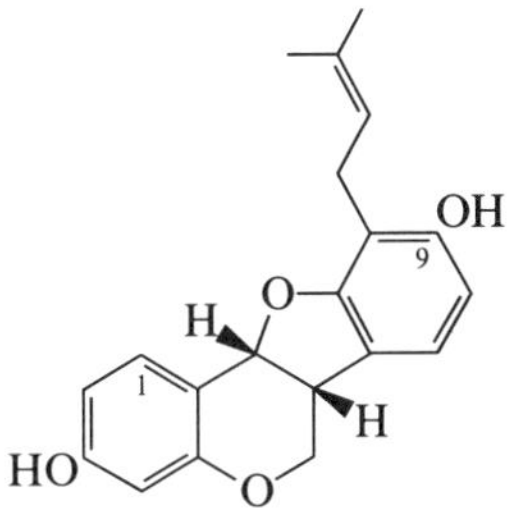

Absolute configuration

$C_{20}H_{20}O_4$ M 324.376

Isol. from *Phaseolus vulgaris*, *P. aureus*, *P. calcaratus*, *Dolichos biflorus*, *Erythrina abyssinica*, *E. crista-galli*, *E. sandwicensis*, *Lablab niger*, *Macroptilium atropurpureum*, *Psophocarpus tetragonolobus* and *Vigna unguiculata* (all Leguminosae, Papilionoideae). Antifungal antibiotic. Green bean phytoalexin. Mp 67-69°.

9-Me ether: [74515-46-1]. **Sandwicensin**
$C_{21}H_{22}O_4$ M 338.402
Isol. from leaves of *Erythrina sandwicensis*. $[\alpha]_D$ −190° (c, ca. 0.02 in MeOH).

1-Methoxy: [65428-13-9]. **1-Methoxyphaseollidin.** *3,9-Dihydroxy-1-methoxy-10-prenylpterocarpan*
$C_{21}H_{22}O_5$ M 354.402
Isol. from seeds of *Psophocarpus tetragonolobus*. $[\alpha]_D^{18}$ −225° (CHCl$_3$). Probable struct.

Scheffer, T.C., *Annu. Rev. Phytopathol.*, 1966, **4**, 147 (*rev*)
Perrin, D.R. *et al, Tetrahedron Lett.*, 1972, 1673 (*struct, nmr, uv, ms*)
Burden, R.S. *et al, Tetrahedron Lett.*, 1972, 4175 (*isol, struct*)
Perrin, D.R. *et al, Aust. J. Chem.*, 1974, **27**, 1607 (*struct*)
Preston, N.W., *Phytochemistry*, 1977, **16**, 2044 (1-*Methoxyphaseollidin*)
Ingham, J.L., *Z. Naturforsch., C*, 1980, **35**, 384 (*Sandwicensin*)
Ingham, J.L., *Fortschr. Chem. Org. Naturst.*, 1983, **43**, 1 (*rev, occur*)

Phaseollidinisoflavan   P-80077

2',4',7-Trihydroxy-3'-prenylisoflavan

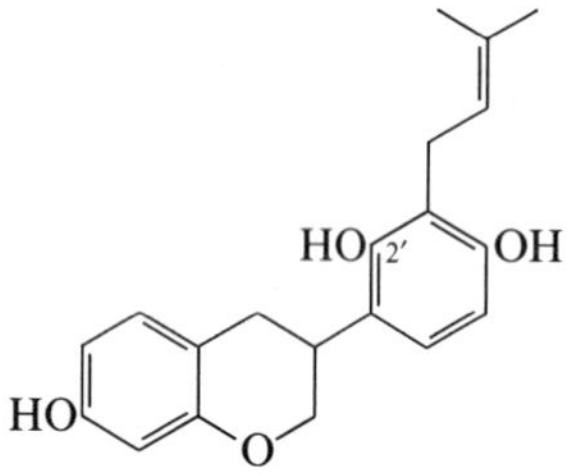

$C_{20}H_{22}O_4$ M 326.391

2'-Me ether: [56257-28-4]. **2'-O-Methylphaseollidinisoflavan**
$C_{21}H_{24}O_4$ M 340.418
Isol. from stems of *Vigna unguiculata*.

Preston, N.W., *Phytochemistry*, 1975, **14**, 1131 (*isol, pmr, struct*)

1,8-Phenanthrenedicarboxylic acid   P-80078

[59795-49-2]
$C_{16}H_{10}O_4$ M 266.253
Cryst. (AcOH). Mp 365° dec.

Di-Me ester: [73049-18-0].
$C_{18}H_{14}O_2$ M 262.307
Cryst. (MeOH). Mp 146-147°.
[38378-77-7]

Rubin, M.B. *et al, J. Org. Chem.*, 1980, **45**, 1847 (*synth, uv, ir, pmr*)

9-Phenanthrenemethanol, 9CI   P-80079

Updated Entry replacing P-00778
9-Hydroxymethylphenanthrene. 9-Phenanthrylcarbinol
[4707-72-6]
$C_{15}H_{12}O$ M 208.259
Needles (C_6H_6/pet. ether). Mp 149-149.5°.

Ac: [53440-11-2].
$C_{17}H_{14}O_2$ M 250.296
Needles (hexane). Mp 79-80°.

Bachmann, W.E., *J. Am. Chem. Soc.*, 1934, **56**, 1366 (*synth*)
Fernández, F. *et al, Synthesis*, 1988, 802 (*synth, ir, pmr*)

2,4,7-Phenanthrenetriol   P-80080

Updated Entry replacing P-20086
2,4,7-Trihydroxyphenanthrene
$C_{14}H_{10}O_3$ M 226.231

Tri-Me ether: [53077-33-1]. *2,4,7-Trimethoxyphenanthrene*
$C_{17}H_{16}O_3$ M 268.312
Mp 113-114°.

2-Me ether, 9,10-dihydro: [87530-30-1]. *9,10-Dihydro-7-methoxy-2,5-phenanthrenediol. 4,7-Dihydroxy-2-methoxy-9,10-dihydrophenanthrene*
$C_{15}H_{14}O_3$ M 242.274
Obt. from *Bletilla striata*. Shows antimicrobial props. Needles (Me$_2$CO). Mp 72°.

4-Me ether, 9,10-dihydro: *9,10-Dihydro-4-methoxy-2,7-phenanthrenediol. 2,7-Dihydroxy-4-methoxy-9,10-dihydrophenanthrene.* **Coelonin**
$C_{14}H_{12}O_3$ M 228.247
Constit. of *Coelogyne ochracea*, *C. elata* and *Eulophia nuda*. Amorph. powder or needles (CHCl$_3$/hexane). Mp 95-96°.

Hardegger, E. *et al, Helv. Chim. Acta*, 1963, **46**, 1171; 1974, **57**, 790, 796 (*synth*)
Majumder, P. *et al, Phytochemistry*, 1982, **21**, 478 (*deriv*)
Takagi, S. *et al, Phytochemistry*, 1983, **22**, 1011 (*deriv*)
Tuchinda, P. *et al, Phytochemistry*, 1988, **27**, 3267 (*isol*)

1-Phenanthridenemethanamine   P-80081

1-(Aminomethyl)phenanthridine
[120616-63-9]

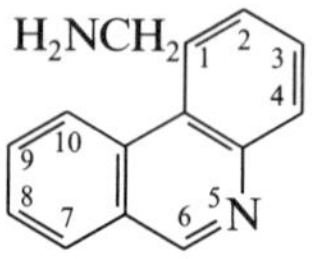

$C_{14}H_{12}N_2$ M 208.262
Mp 94-96°.

Freter, K.R. *et al, J. Heterocycl. Chem.*, 1988, **25**, 1701 (*synth, pmr*)

4-Phenanthridinemethanamine   P-80082

4-(Aminomethyl)phenanthridine
[120616-65-1]
$C_{14}H_{12}N_2$ M 208.262
B,HCl: [120616-66-2].
Cryst. (MeOH). Mp 259-260°.

Freter, K.R. *et al, J. Heterocycl. Chem.*, 1988, **25**, 1701 (*synth, pmr*)

10-Phenanthridinemethanamine   P-80083

10-(Aminomethyl)phenanthridine
[120616-64-0]
$C_{14}H_{12}N_2$ M 208.262
Cryst. (CH$_2$Cl$_2$/pet. ether). Mp 92-93°.

Freter, K.R. *et al, J. Heterocycl. Chem.*, 1988, **25**, 1701 (*synth, pmr*)

1*H*-Phenanthro[1,10,9,8-*cdefg*]carbazole, 9CI　　　P-80084

11,12-Iminoperylene

[35337-22-5]

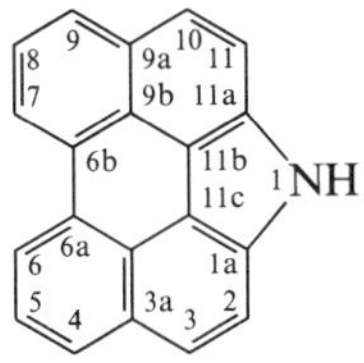

$C_{20}H_{11}N$　　M 265.314
Yellow-brown cryst. Mp 360°.

Looker, J.J., *J. Org. Chem.*, 1972, **37**, 3379.

1,10-Phenanthroline-2,9-dicarboxaldehyde　　　P-80085

2,9-Diformyl-1,10-phenanthroline

[57709-62-3]

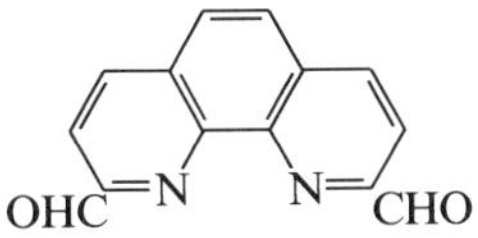

$C_{14}H_8N_2O_2$　　M 236.229
Yellow cryst. (THF). Mp 231-232°.

Chandler, C.J. *et al*, *J. Heterocycl. Chem.*, 1981, **18**, 599.
Lüning, U. *et al*, *Justus Liebigs Ann. Chem.*, 1989, 367 (*synth*)

Phenothiatellurin　　　P-80086

Thiophenoxtellurine

[262-29-3]

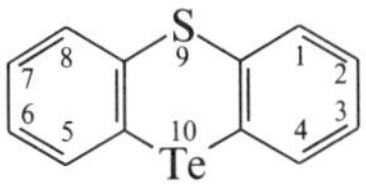

$C_{12}H_8STe$　　M 311.861
Light yellow needles (EtOH). Mp 123-124°.

Petragnani, N., *Tetrahedron*, 1960, **11**, 15 (*synth*)
Gioba, A. *et al*, *Rev. Roum. Chim.*, 1970, **15**, 1967 (*synth*)

Phenoxaselenin, 9CI　　　P-80087

[262-22-6]

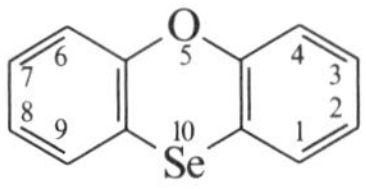

$C_{12}H_8OSe$　　M 247.155
Long rectangular prisms with faint floral odour (EtOH or AcOH). Mp 87-88°.

Se-Oxide:
　$C_{12}H_8O_2Se$　　M 263.154
　Cryst. powder. Mp 171-172°. Reforms phenoxaselenin at Mp.

Drew, H.D.K., *J. Chem. Soc.*, 1928, 511 (*synth, derivs*)

4-Phenoxybenzoic acid, 9CI　　　P-80088

Updated Entry replacing P-00890
Diphenyl ether 4-carboxylic acid
[2215-77-2]
$C_{13}H_{10}O_3$　　M 214.220
Mp 160°. Bp 260°.
Me ester: [21218-94-0].

$C_{14}H_{12}O_3$　　M 228.247
Cryst. (EtOH). Mp 60°.

Nitrile: [3096-81-9]. *4-Phenoxybenzonitrile*
　$C_{13}H_9NO$　　M 195.220
　Cryst. solid. Mp 22-24°.

West, R. *et al*, *J. Am. Chem. Soc.*, 1952, **74**, 3960 (*ester*)
Manly, D.G. *et al*, *J. Org. Chem.*, 1957, **22**, 323 (*ester*)
Sammes, P.G. *et al*, *J. Chem. Soc., Perkin Trans.* 1, 1988, 3229 (*nitrile*)

10-Phenylanthrone, 8CI　　　P-80089

10-Phenyl-9-anthracenol. 9-Hydroxy-10-phenylanthracene

[14596-70-4]

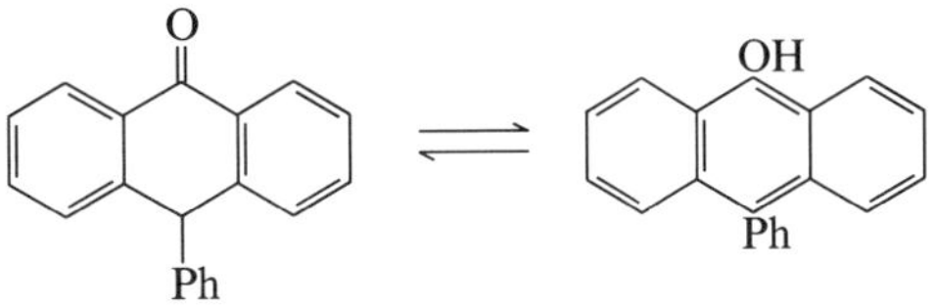

$C_{20}H_{14}O$　　M 270.330
Approx. equal amounts of tautomers in soln. Light yellow powder. Mp 139-140°. Readily oxid. by air.

[60079-98-3]

Branz, S.E. *et al*, *Synth. Commun.*, 1986, **16**, 441 (*synth, pmr, tautom*)

2-Phenylbenzoselenazole, 9CI　　　P-80090

[32586-68-8]

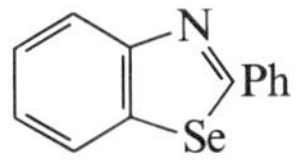

$C_{13}H_9NSe$　　M 258.181
Needles (EtOH). Mp 116-117° subl.

Bauer, H., *Ber.*, 1913, **46**, 92 (*synth*)
Croisy, A. *et al*, *Org. Mass Spectrom.*, 1972, **6**, 1321 (*ms*)

3-Phenyl-2*H*-1-benzothiopyran, 9CI　　　P-80091

3-Phenyl-2H-1-benzothiin. 3-Phenyl-2H-1-thiochromene

[53844-17-0]

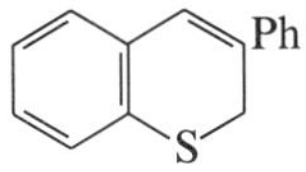

$C_{15}H_{12}S$　　M 224.326
Cryst. Mp 64-65°.

Hortmann, A.G. *et al*, *J. Am. Chem. Soc.*, 1974, **96**, 6119 (*synth, pmr, ir*)
Arnoldi, A. *et al*, *Synthesis*, 1988, 155 (*synth, pmr*)

4-Phenyl-2*H*-1-benzothiopyran, 9CI　　　P-80092

4-Phenyl-2H-thiochromene

[35813-98-0]

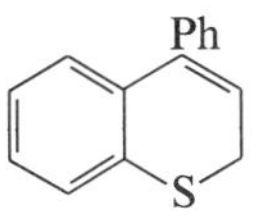

$C_{15}H_{12}S$　　M 224.326
Cryst. (MeOH). Mp 83°.

S-Me, perchlorate: [120400-18-2].
　$C_{16}H_{15}ClO_4S$　　M 338.811

Pale-yellow leaflets. Mp 162-164°.

Lüttringhaus, A. *et al, Justus Liebigs Ann. Chem.*, 1962, **654**, 189.
Hori, M. *et al, Chem. Pharm. Bull.*, 1988, **36**, 3816.

4-Phenyl-1,2,3-benzotriazine, 9CI   P-80093

Updated Entry replacing P-30088
[33334-16-6]

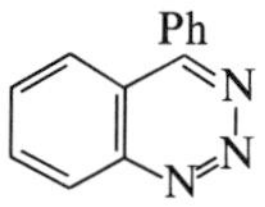

$C_{13}H_9N_3$ M 207.234
Needles (EtOH). Mp 159-160°. λ_{max} 206, 232 (log ϵ 4.1)
and 293 nm (2.95) in EtOH.
2-Oxide:
  $C_{13}H_9N_3O$ M 223.234
  Yellow needles (MeOH). Mp 156.5-157.5°.
3-Oxide: [41572-12-7].
  $C_{13}H_9N_3O$ M 223.234
  Yellow plates (EtOH). Mp 154° dec. Dec. in light.

Meisenheimer, J. *et al, Ber.*, 1927, **60**, 1736 (3-*oxide*)
Adger, B.M. *et al, J. Chem. Soc., Perkin Trans.* 1, 1975, 31 (*synth, uv, ir*)
Boulton, A.J. *et al, J. Chem. Soc., Perkin Trans.* 1, 1988, 1509 (2-*oxide*)

4-Phenyl-2H-1,3-benzoxazin-2-one, 9CI   P-80094

[115438-13-6]

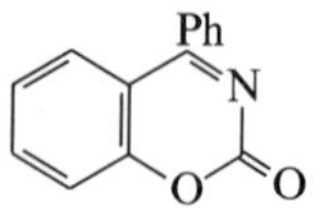

$C_{14}H_9NO_2$ M 223.231
Cryst. Mp 252-255°.

Kamal, A. *et al, J. Org. Chem.*, 1988, **53**, 4112 (*synth, ir*)

1-Phenyl-1,3-butadiyne   P-80095

1,3-Butadiynylbenzene, 9CI
[5701-81-5]

$PhC\equiv CC\equiv CH$

$C_{10}H_6$ M 126.157
Yellow oil. n_D^{20} 1.6230. Highly unstable to polymerisation.

Negishi, E. *et al, J. Org. Chem.*, 1984, **49**, 2629 (*synth*)
Hänninen, E. *et al, Acta Chem. Scand., Ser. B*, 1988, **42**, 614 (*synth*)

2-Phenyl-3-butenoic acid, 8CI   P-80096

Updated Entry replacing P-50120
α-Ethenylbenzeneacetic acid, 9CI. Phenylvinylacetic acid
[30953-22-1]

$H_2C=CHCHPhCOOH$

$C_{10}H_{10}O_2$ M 162.188
(+)-*form*
  Oil. $[\alpha]_D^{25}$ +85.65° (c, 0.704 in EtOH).
(±)-*form*
  Cryst. (pet. ether). Mp 32-33°.

Friedrich, L.E. *et al, J. Org. Chem.*, 1971, **36**, 3011 (*synth, uv, ir, pmr*)
Aggarwal, S.K. *et al, Tetrahedron*, 1987, **43**, 451 (*resoln*)

4-Phenyl-3-buten-1-ol, 9CI   P-80097

Updated Entry replacing P-01064
[937-58-6]

$PhCH=CHCH_2CH_2OH$

$C_{10}H_{12}O$ M 148.204
Mp 36°. Bp_{12} 137-138°.
(*E*)(*?*)-*form*
  Bp_{12} 137-138°.
  Ac: [20473-78-3].
    $C_{12}H_{14}O_2$ M 190.241
    Bp_{12} 149°.
  Phenylurethane: Mp 70°.
Z-form
  Oil. $Bp_{0.15}$ 65° (bulb) .
  Methylsulfonyl: Pale orange oil. Bp_1 80° (bulb) .
  [7515-42-6]

Hands, A.R. *et al, J. Chem. Soc. C*, 1968, 2448 (*synth, pmr*)
Marvell, E.N. *et al, Synthesis*, 1973, 457 (*synth*)
Bachi, M.D. *et al, J. Chem. Soc., Perkin Trans.* 1, 1988, 1517 (*synth, ir, pmr, ms*)

3-Phenyl-1,1-cyclopentanedicarboxylic acid   P-80098

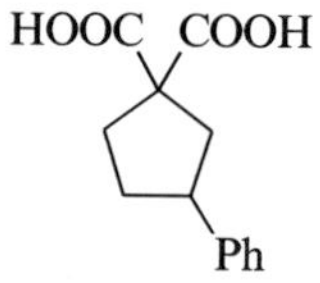

$C_{13}H_{14}O_4$ M 234.251
(±)-*form* [115340-09-5]
  Needles. Mp 167-169°.
Di-Me ester: [115340-08-4].
  $C_{15}H_{18}O_4$ M 262.305
  Oil. $Bp_{0.5}$ 145-155° (bulb) .

Grunewald, G.L. *et al, J. Org. Chem.*, 1988, **53**, 4021 (*synth, ir, pmr, cmr, ms*)

3-Phenylcyclopentanone, 9CI   P-80099

Updated Entry replacing P-01126
[64145-51-3]

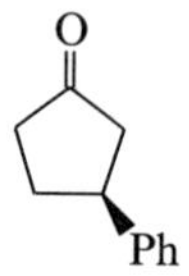

(*S*)-*form*
$C_{11}H_{12}O$ M 160.215
(*S*)-*form* [86505-50-2]
  $[\alpha]_D^{25}$ −84.9° (c, 0.72 in CHCl$_3$).
(±)-*form*
  Liq. Bp_{10} 154-155°, Bp_1 120-122°.
Semicarbazone: Leaflets (EtOH). Mp 181° dec.

Winternitz, F. *et al, Bull. Soc. Chim. Fr.*, 1953, 190 (*synth*)
Kolobielski, M. *et al, J. Am. Chem. Soc.*, 1957, **79**, 5820 (*synth*)
Paquette, L.A. *et al, J. Org. Chem.*, 1988, **53**, 4978 (*synth, ir, pmr*)

5-(2-Phenylethyl)-1,3-benzenediol   P-80100

5-Phenethylresorcinol, 8CI. 1-(3,5-Dihydroxyphenyl)-2-phenylethane. 3,5-Dihydroxybibenzyl. **Dihydropinosylvin**

[14531-52-3]

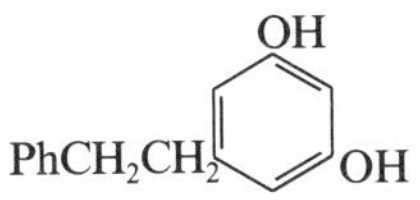

OH

PhCH₂CH₂ —⬡— OH

C₁₄H₁₄O₂   M 214.263

Prob. trace constit. of *Pinus kremphii* wood. Also in fungus-infected *Dioscorea batatus* as phytoalexin. Oil.

Mono-Me ether: [17635-59-5]. *3-Methoxy-5-(2-phenylethyl)phenol. 3-Hydroxy-5-methoxybibenzyl.* **Dihydropinosylvin methyl ether**

C₁₅H₁₆O₂   M 228.290

Isol. from wood of *P. albicaulis* and other *P.* spp. Mp 50-52°.

Me ether, phenylurethane: Mp 124-125°.

Lindstedt, G., *Acta Chem. Scand.*, 1950, **4**, 1246 (*isol, struct, synth, deriv*)
Erdtman, H. *et al, Phytochemistry*, 1966, **5**, 927 (*isol*)
Mitscher, L.A. *et al, Phytochemistry*, 1981, **20**, 781 (*synth*)
Takasugi, M. *et al, Phytochemistry*, 1987, **76**, 371 (*isol, props*)

7-Phenyl-2,4,6-heptatriyn-1-ol, 9CI   P-80101

[60214-15-5]

PhC≡CC≡CC≡CCH₂OH

C₁₃H₈O   M 180.206

Isol. from *Bidens pilosa* and *B. leucantha*. Shows antimicrobial activity. Cryst. (pet. ether or CCl₄). Mp 66.5° (64°).

Ac: [23414-58-6]. *1-Acetoxy-7-phenyl-2,4,6-heptatriyne*

C₁₅H₁₀O₂   M 222.243

Isol. from *B. pilosa, B. leucantha* and *B. dahlioides*, also in trace amts. from *Coreopsis tinctoria*. Long pale yellow needles (pet. ether). Mp 50°.

Prévost, S. *et al, Bull. Soc. Chim. Fr.*, 1961, 2171 (*synth*)
Bohlmann, F. *et al, Chem. Ber.*, 1964, **97**, 2135; 1965, **98**, 1228; 1966, **99**, 1223 (*isol, uv, ir, pmr*)
Smirnov, V.V. *et al, CA*, 1984, **101**, 122607c.

2-Phenyl-2-heptene   P-80102

1-Methyl-1-hexenylbenzene, 9CI

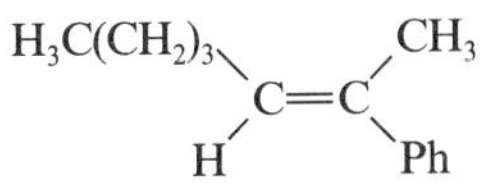

H₃C(CH₂)₃   CH₃
         \\   /
          C=C
         /   \\
        H   Ph

C₁₃H₁₈   M 174.285

(*E*)-*form* [83021-58-3]

Liq. Bp₀.₀₁ 50-52°.

Brown, H.C. *et al, J. Org. Chem.*, 1988, **53**, 6009 (*synth, ir, pmr*)

7-Phenyl-2-heptene-4,6-diynal, 9CI   P-80103

[20252-42-0]

PhC≡CC≡CCH=CHCHO

C₁₃H₈O   M 180.206

(*E*)-*form* [13894-73-0]

Isol. from roots of *Onopordon acanthium*. Yellowish cryst. (pet. ether). Mp 63.5°.

Bohlmann, F. *et al, Chem. Ber.*, 1966, **99**, 3201 (*isol, uv, ir, pmr, synth*)

1-Phenyl-2,4-hexadiyn-1-ol   P-80104

α-1,3-Pentadiynylbenzenemethanol, 9CI. Capillol

[1574-95-4]

PhCH(OH)C≡CC≡CCH₃

C₁₂H₁₀O   M 170.210

Ac: O-Acetylcapillol. Capillol acetate

C₁₄H₁₂O₂   M 212.248

Isol. from roots of *Chrysanthemum frutescens, C. foeniculatum* and *Lonas annua*. Oil. λ_{max} 243, 257 nm (ϵ 1230, 830)(Et₂O).

Bohlmann, F. *et al, Chem. Ber.*, 1962, **95**, 39, 602; 1964, **97**, 1179; 1966, **99**, 995, 2413 (*isol, uv, ir, ord, biosynth*)

1-Phenyl-1,3,5-hexatriyne   P-80105

1,3,5-Hexatriynylbenzene, 9CI

[17814-74-3]

PhC≡CC≡CC≡CH

C₁₂H₆   M 150.179

Oil. Unstable, prepd. in soln.

▷ Explosive.

Bohlmann, F. *et al, Chem. Ber.*, 1964, **97**, 2586 (*synth, ir, uv*)
Eastmond, R. *et al, Tetrahedron*, 1972, **28**, 4591 (*synth*)
Vereshchapin, L.I. *et al, J. Org. Chem. USSR (Engl. Transl.)*, 1976, **12**, 1174 (*synth, ir, haz*)
Shim, S.C. *et al, J. Org. Chem.*, 1988, **53**, 2410 (*synth*)

1-Phenyl-1-octene, 8CI   P-80106

1-Octenylbenzene, 9CI

PhCH=CH(CH₂)₅CH₃

C₁₄H₂₀   M 188.312

(*E*)-*form* [28665-60-3]

Oil. Bp₄ 91-94°.

(*Z*)-*form* [42036-72-6]

Oil.

Yamane, T. *et al, Tetrahedron*, 1973, **29**, 955 (*synth, ir, pmr*)
Kikukawa, K. *et al, J. Org. Chem.*, 1981, **46**, 4885; 1985, **50**, 299 (*synth*)
Lebedev, S.A. *et al, J. Org. Chem. USSR (Engl. Transl.)*, 1985, **21**, 652 (*synth, pmr*)

1-Phenyl-2-octene, 8CI   P-80107

2-Octenylbenzene, 9CI

PhCH₂CH=CH(CH₂)₄CH₃

C₁₄H₂₀   M 188.312

(*E*)-*form* [42079-83-4]

Oil.

Yamane, T. *et al, Tetrahedron*, 1973, **29**, 955 (*synth, ir, pmr*)
Kikukawa, K. *et al, J. Org. Chem.*, 1981, **46**, 4885; 1985, **50**, 299 (*synth*)

2-Phenyl-1-octene, 8CI   P-80108

(1-Methyleneheptyl)benzene, 9CI. 1-(Hexylethenyl)benzene

[5698-49-7]

H₂C=CPh(CH₂)₅CH₃

C₁₄H₂₀   M 188.312

Oil. Bp₁₅ 118°.

Hayashi, T. *et al, Tetrahedron Lett.*, 1980, **21**, 3915 (*synth*)
Avasthi, K. *et al, Tetrahedron Lett.*, 1980, **21**, 945 (*synth*)
Blatter, K. *et al, Synthesis*, 1989, 356 (*synth, ir, pmr, cmr*)

2-Phenyl-4-oxazolecarboxylic acid, 9CI   P-80109

[23012-16-0]

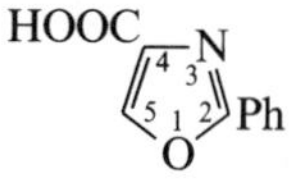

$C_{10}H_7NO_3$ M 189.170
pK_a 3.41 (H_2O, 20°).

Cornforth, J.W., *The Chemistry of Penicillin*, Clarke H.T. *et al*,
   Eds, Princeton University Press, 1949, 688.
Brown, D.J. *et al*, *J. Chem. Soc. B*, 1969, **3**, 270 (*pmr*)
Ger. Pat., 2 459 380, (1975); *CA*, **83**, 131577e (*synth*)

2-Phenyl-5-oxazolecarboxylic acid, 9CI   P-80110

[106833-79-8]
$C_{10}H_7NO_3$ M 189.170
Mp 217-218°.

Belen'kii, L.I. *et al*, *Khim. Geterotsikl. Soedin.*, 1986, 826; *Chem.
   Heterocycl. Compd.* (*Engl. Transl.*), 654.

5-Phenyl-2-oxazolecarboxylic acid, 9CI   P-80111

[1014-14-8]
$C_{10}H_7NO_3$ M 189.170
Mp 96-98° dec. pK_a -1.87.

Tanaka, C., *Yakugaku Zasshi* (*J. Pharm. Soc. Jpn.*), 1965, **85**, 186
   (*synth*)
Brown, D.J. *et al*, *J. Chem. Soc. B*, 1969, **3**, 270 (*pmr*)
Ger. Pat., 3 530 213, (1986); *CA*, **105**, 52229n (*synth*)

5-Phenyl-4-oxazolecarboxylic acid, 9CI   P-80112

[99924-18-2]
$C_{10}H_7NO_3$ M 189.170
Mp 163-164°.
Me ester: [38061-18-6].
   $C_{11}H_9NO_3$ M 203.197
   Cryst. (hexane). Mp 91-93°.
Et ester: [32998-97-3].
   $C_{12}H_{11}NO_3$ M 217.224
   $Bp_{0.1}$ 132°.

Schollkopf, U. *et al*, *Angew. Chem., Int. Ed. Engl.*, 1971, **10**, 333
   (*ester*)
Yasumasa, H. *et al*, *Tetrahedron Lett.*, 1982, **23**, 235 (*ester*)
Moriya, T. *et al*, *J. Med. Chem.*, 1986, **29**, 333 (*synth, ir*)

5-Phenyl-1,3-pentadiyne   P-80113

2,4-Pentadiynylbenzene, 9CI
[41268-41-1]

$$PhCH_2C{\equiv}CC{\equiv}CH$$

$C_{11}H_8$ M 140.184
Isol. from roots of *Artemisia dracunculus* and
   Chrysanthemum segetum. Oil. $Bp_{0.001}$ 45-50°. n_D^{22} 1.5726.

Bohlmann, F. *et al*, *Chem. Ber.*, 1962, **95**, 39; 1963, **96**, 226 (*isol,
   uv, ir, struct*)
Klein, J. *et al*, *J. Chem. Soc., Perkin Trans. 2*, 1973, 599 (*nmr*)
Harada, R. *et al*, *Phytochemistry*, 1982, **21**, 2009 (*isol, ir, uv, nmr*)

1-Phenyl-2,4-pentadiyn-1-one, 9CI   P-80114

Benzoylbutadiyne. Demethylcapillone. Desmethylcapillone
[29743-36-0]

$$PhCOC{\equiv}CC{\equiv}CH$$

$C_{11}H_6O$ M 154.168

Isol. from roots of *Chrysanthemum segetum, C. coronarium*
   and *C. foeniculaceum*. Cryst. (pet. ether). Mp 106°.

Bohlmann, F. *et al*, *Chem. Ber.*, 1963, **96**, 226; 1964, **97**, 1179 (*isol,
   uv, ir, synth*)
Walton, D.R.M. *et al*, *J. Organomet. Chem.*, 1972, **37**, 45 (*synth*)

1-Phenyl-1,2-pentanedione, 9CI, 8CI   P-80115

[20895-66-3]

$$PhCOCOCH_2CH_2CH_3$$

$C_{11}H_{12}O_2$ M 176.215
Liq. Bp_3 82-84°, $Bp_{0.05}$ 60-61°. n_D^{25} 1.5206.

Emmons, W.D. *et al*, *J. Am. Chem. Soc.*, 1955, **77**, 4415 (*synth*)
Wagner, P.J. *et al*, *J. Am. Chem. Soc.*, 1976, **98**, 8125 (*ir, uv, ms*)
De Kimpe, N. *et al*, *J. Org. Chem.*, 1978, **43**, 2933 (*synth*)
Lee, D.G. *et al*, *J. Org. Chem.*, 1979, **44**, 2726 (*synth*)
Ballistrevi, F.P. *et al*, *Tetrahedron Lett.*, 1986, **27**, 5139 (*synth*)

1-Phenyl-1,3-pentanedione, 9CI, 8CI   P-80116

1-Benzoyl-2-butanone
[5331-64-6]

$$PhCOCH_2COCH_2CH_3$$

$C_{11}H_{12}O_2$ M 176.215
Liq. Bp_1 92-94°. n_D^{25} 1.5730.

Hauser, C.R. *et al*, *J. Am. Chem. Soc.*, 1944, **66**, 1220 (*synth*)
Schamp, N. *et al*, *Bull. Soc. Chim. Belg.*, 1966, **75**, 539 (*ms*)
Muir, W.M. *et al*, *J. Org. Chem.*, 1966, **31**, 3790 (*synth*)
Bogovac, M. *et al*, *Bull. Soc. Chim. Fr.*, 1969, 4437 (*synth*)
Koshimura, H. *et al*, *Bull. Chem. Soc. Jpn.*, 1973, **46**, 632 (*pmr,
   tautom*)

1-Phenyl-1,4-pentanedione, 9CI, 8CI   P-80117

4-Benzoyl-2-butanone
[583-05-1]

$$PhCOCH_2CH_2COCH_3$$

$C_{11}H_{12}O_2$ M 176.215
Liq. Bp_{20} 168-172°, Bp_{12} 158°.
4-Mono-(2,4-dinitrophenylhydrazone): [71094-27-4].
   Cryst. (CHCl₃). Mp 197-198°.

Nimgirawath, S. *et al*, *Aust. J. Chem.*, 1976, **29**, 339 (*rev, synth,
   pmr*)
Faragher, R. *et al*, *J. Chem. Soc., Perkin Trans. 1*, 1979, 249
   (*synth, ir, pmr, ms*)
Clark, J.H. *et al*, *J. Chem. Soc., Perkin Trans. 1*, 1983, 2253
   (*synth, cmr, pmr*)
Degl'Innocenti, A. *et al*, *Gazz. Chim. Ital.*, 1987, **117**, 645 (*synth*)

1-Phenyl-2,4-pentanedione, 9CI, 8CI   P-80118

[3318-61-4]

$$PhCH_2COCH_2COCH_3$$

$C_{11}H_{12}O_2$ M 176.215
Liq. Bp_{10} 133-136°.
Cu chelate: Blue-green cryst. (MeOH). Mp 227-229°.

Hauser, C.R. *et al*, *J. Am. Chem. Soc.*, 1958, **80**, 6360.
Hampton, K.G. *et al*, *J. Org. Chem.*, 1964, **29**, 3511 (*synth, bibl*)
Org. Synth., 1971, **51**, 128.

2-Phenyl-2-pentene   P-80119

Updated Entry replacing P-01445
1-Methyl-1-butenylbenzene, 9CI
[53172-84-2]

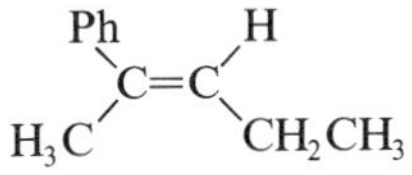

$$(E)\text{-form}$$

$C_{11}H_{14}$ M 146.232
Bp_{16} 90°.

(E)-form [70303-28-5]
Liq. $Bp_{9.5}$ 66-68°.

(Z)-form [53172-85-3]
Isol. by glc.

Ando, W. *et al, J. Chem. Soc., Chem. Commun.*, 1975, 145 (*synth*)
Zioudrou, C. *et al, Tetrahedron*, 1978, **34**, 3181 (*synth, glc, pmr*)
Brown, H.C. *et al, J. Org. Chem.*, 1988, **53**, 6009 (*synth, ir, pmr*)

1-Phenyl-2-(phenylthio)ethanone, 9CI   P-80120

*2-(Phenylthio)acetophenone, 8CI. ω-
Phenylmercaptoacetophenone. Phenacyl phenyl sulfide.
Benzoyl(phenylthio)methane*
[16222-10-9]

$$PhCOCH_2SPh$$

$C_{14}H_{12}OS$ M 228.314
Needles (EtOH). Mp 53-54°.

Phenylhydrazone: Orange needles (MeOH). Mp 116°.

2,4-Dinitrophenylhydrazone: Bright-red needles
(EtOH/EtOAc). Mp 150-150.5°.

S-Oxide: [6099-23-6]. *1-Phenyl-2-(phenylsulfinyl)ethanone,
9CI. 2-(Phenylsulfinyl)acetophenone, 8CI. Phenacyl phenyl
sulfoxide*
$C_{14}H_{12}O_2S$ M 244.314
Cryst. (C_6H_6/heptane). Mp 79-80°.

S-Dioxide: [3406-03-9]. *1-Phenyl-2-(phenylsulfonyl)ethanone,
9CI. 2-(Phenylsulfonyl)acetophenone, 8CI. Phenacyl
phenyl sulfone*
$C_{14}H_{12}O_3S$ M 260.313
Cryst. Mp 96°.

Tröger, J. *et al, J. Prakt. Chem.*, 1913, **287**, 289 (*dioxide*)
Kroehnke, F. *et al, J. Prakt. Chem.*, 1960, **11**, 256 (*synth*)
Kenny, W.J. *et al, J. Am. Chem. Soc.*, 1961, **83**, 4019 (*synth, oxide*)
Griesbaum, K. *et al, J. Am. Chem. Soc.*, 1963, **85**, 1969 (*ir, pmr,
oxide, dioxide*)
Krawiec, M. *et al, Acta Crystallogr., Sect. C*, 1989, **45**, 354
(*dioxide, cryst struct*)

1-Phenyl-1,2-propanediol, 9CI   P-80121

Updated Entry replacing P-01494
1-Phenylpropylene glycol
[1855-09-0]

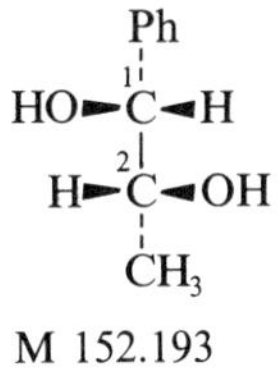

(1R,2R)-form
Absolute
configuration

$C_9H_{12}O_2$ M 152.193

(1R,2R)-form [40421-51-0]
D-*threo-form*
Cryst. (pet. ether). Mp 62° (51-53°). $[\alpha]_D^{22}$ − 61.5° (c, 4.3
in CHCl$_3$).

Dibenzoyl:

$C_{23}H_{20}O_4$ M 360.409
Needles. Mp 101° (89.5-91°). $[\alpha]_D$ + 0°.

(1R,2S)-form
Oil. $[\alpha]_D^{21}$ − 38.6° (c, 3.2 in CHCl$_3$).

(1S,2R)-form [40421-52-1]
D-erythro-*form*

Dibenzoyl: [66841-45-0].
Mp 95-97°. $[\alpha]_D$ − 60.9° (c, 0.46 in CHCl$_3$).

(1RS,2RS)-form [1075-05-4]
(±)-threo-*form*
Cryst. (pet. ether). Mp 52-54°.

Di-Ac: [21145-70-0].
$C_{13}H_{16}O_4$ M 236.267
Bp_1 106-112°.

Dibenzoyl: [21759-66-0].
Mp 96-97°.

(1RS,2SR)-form [1075-04-3]
(±)-erythro-*form*
Powder (C_6H_6/pet. ether). Mp 91-92.5° (89-91°).

Di-Ac: [21145-69-7].
$Bp_{0.4}$ 109°.

Dibenzoyl: [21759-65-9].
Cryst. (MeOH). Mp 96-97°.

[1075-05-4, 40560-98-3, 54826-39-0]

Witkop, B. *et al, J. Am. Chem. Soc.*, 1957, **79**, 197 (*abs config*)
Bowlus, S.B. *et al, J. Org. Chem.*, 1974, **39**, 3309 (*synth*)
Cowles, C.R. *et al, J. Org. Chem.*, 1975, **40**, 1302 (*synth, pmr*)
Fuganti, C. *et al, Chem. Ind. (London)*, 1977, 983 (*synth*)
Chrysochou, P. *et al, Tetrahedron*, 1977, **33**, 2103 (*synth*)
Zioudrou, C. *et al, Tetrahedron*, 1977, **33**, 2103 (*synth*)
Imuta, M. *et al, J. Am. Chem. Soc.*, 1979, **101**, 3990 (*synth*)
Takeshita, M. *et al, Chem. Pharm. Bull.*, 1989, **37**, 1085 (*synth,
bibl*)

1-Phenyl-1,3-propanediol, 9CI   P-80122

Updated Entry replacing P-01495
[4850-49-1]

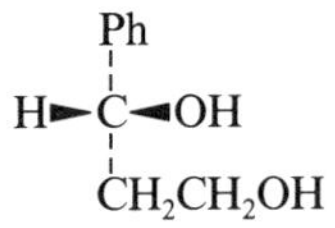

$$(R)\text{-form}$$

$C_9H_{12}O_2$ M 152.193

R-form [103548-16-9]
Oil.

3-O-Methanesulfonyl: [115290-78-3].
Oil.

(±)-form
Oil. Bp_2 126°.

Dibenzoyl:
$C_{23}H_{20}O_4$ M 360.409
Needles (ligroin). Mp 51°.

Bis-4-nitrobenzoyl: Mp 110-110.5°.

Di-Me ether: [26278-67-1]. *1,3-Dimethoxy-1-phenylpropane*
$C_{11}H_{16}O_2$ M 180.246
Bp 215-217° part. dec. , Bp_{15} 94-95°.

[96854-34-1, 115290-77-2]

Searles, S. *et al, J. Org. Chem.*, 1959, **24**, 1770 (*synth*)
Schaal, C., *Bull. Soc. Chim. Fr.*, 1973, 3083 (*synth*)
Gao, Y. *et al, J. Org. Chem.*, 1988, **53**, 4081 (*synth, pmr*)

2-Phenyl-5-(1-propynyl)thiophene, 8CI   P-80123
[1204-82-6]

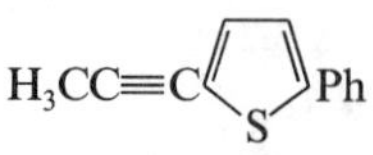

$C_{13}H_{10}S$ M 198.288

Isol. from leaf oil of *Coreopsis grandiflora* and from *C. lanceolata*. Shows nematocidal and insecticidal props. Prismatic needles (C_6H_6/pet. ether). Mp 44°.

Sörensen, J.S. *et al, Acta Chem. Scand.*, 1958, **12**, 771 (*isol, struct*)
Cymerman-Craig, J. *et al, J. Chem. Soc.*, 1963, 3907 (*synth*)
Atkinson, R. *et al, J. Chem. Soc. C*, 1967, 578 (*synth, pmr*)
Nakajima, S. *et al, Agric. Biol. Chem.*, 1980, **44**, 1529.
Carpita, A. *et al, Tetrahedron*, 1985, **41**, 621 (*synth*)

3-Phenyl-4(1*H*)-pyridazinone, 9CI   P-80124
[71109-29-0]

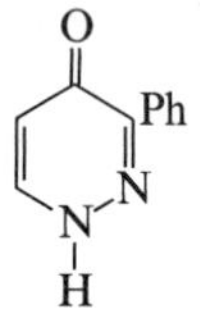

$C_{10}H_8N_2O$ M 172.186
Mp 198-200°.

Gompper, R. *et al, Synthesis*, 1979, 385 (*synth, ir, pmr*)

2-Phenyl-3-pyridinecarboxylic acid, 9CI   P-80125
2-Phenylnicotinic acid, 8CI
[33421-39-5]

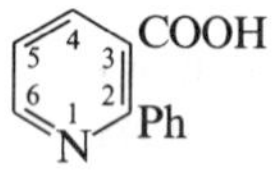

$C_{12}H_9NO_2$ M 199.209
Cryst. (Et_2O). Mp 166-168°.

Abramovitch, R.A. *et al, Can. J. Chem.*, 1960, **38**, 761 (*synth*)
DuPriest, M.T. *et al, J. Org. Chem.*, 1986, **51**, 2021 (*synth, ir, pmr*)

2-Phenyl-4-pyridinecarboxylic acid, 9CI   P-80126
2-Phenylisonicotinic acid
[55240-51-2]
$C_{12}H_9NO_2$ M 199.209
Mp 268-269°.

Usui, Y. *et al, Heterocycles*, 1975, **3**, 155 (*synth, pmr*)

4-Phenyl-2-pyridinecarboxylic acid, 9CI   P-80127
4-Phenylpicolinic acid
[52565-56-7]
$C_{12}H_9NO_2$ M 199.209
Mp 154-155°.

Takahashi, K. *et al, J. Heterocycl. Chem.*, 1978, **15**, 893 (*synth, ir*)

5-Phenyl-2-pyridinecarboxylic acid, 9CI   P-80128
5-Phenylpicolinic acid
[75754-04-0]
$C_{12}H_9NO_2$ M 199.209
Cryst. (pet. ether). Mp 156-157°.

Farley, C.P. *et al, J. Am. Chem. Soc.*, 1956, **78**, 3477 (*synth*)

5-Phenyl-3-pyridinecarboxylic acid, 9CI   P-80129
5-Phenylnicotinic acid
[10177-12-5]
$C_{12}H_9NO_2$ M 199.209
Cryst. (EtOH). Mp 267-269° (260-263°).

Me ester: [10177-13-6].
   $C_{13}H_{11}NO_2$ M 213.235
   Mp 48°. $Bp_{0.01}$ 108-115°.
Amide:
   $C_{12}H_{10}N_2O$ M 198.224
   Mp 177-178°.
Amide;B,HCl: Mp 220-224°.
Dimethylamide:
   $C_{14}H_{14}N_2O$ M 226.277
   Mp 84°. $Bp_{0.01}$ 160-170°.
Dimethylamide;B,HCl: Mp 187-190°.
Diethylamide: [10211-39-9].
   $C_{16}H_{18}N_2O$ M 254.331
   CNS depressant; lysergic acid analogue. Mp 82°. $Bp_{0.04}$ 146-151°.
Diethylamide;B,HCl: Mp 152-155°.

Farley, C.P. *et al, J. Am. Chem. Soc.*, 1956, **78**, 3477 (*synth*)
Julia, M. *et al, Bull. Soc. Chim. Fr.*, 1966, 2387 (*synth, derivs, uv*)
Johnson, F.A. *et al, J. Pharm. Sci.*, 1973, **62**, 1881 (*synth, derivs*)

4-Phenyl-2,6-pyridinedicarboxylic acid, 9CI   P-80130
[83463-12-1]
$C_{13}H_9NO_4$ M 243.218
Cryst. + $2.5H_2O$. Mp 218-220° dec.
Di-Me ester: [117095-81-5].
   $C_{15}H_{13}NO_4$ M 271.272
   Mp 149-150°.
Di-Et ester: [117095-82-6].
   $C_{17}H_{17}NO_4$ M 299.326
   Cryst. Mp 118-120°.

van Staveren, C.J. *et al, J. Am. Chem. Soc.*, 1988, **110**, 8134 (*synth, pmr, cmr*)

2-Phenylpyrimido[2,1,6-*de*]quinolizine, 9CI   P-80131
2-Phenyl-1,9b-diazaphenalene. 2-Phenyl-1-azacycl[3.3.3]azine
[121902-49-6]

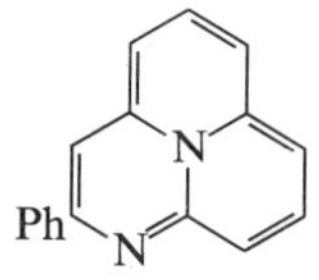

$C_{17}H_{12}N_2$ M 244.295
Unstable greenish-brown ppt.
B,HBr: Mp >300°.

Matsuda, Y. *et al, Chem. Pharm. Bull.*, 1988, **36**, 4307 (*synth, uv, pmr*)

3-Phenyl-2,5-pyrrolidinedione   P-80132

3-Phenylsuccinimide

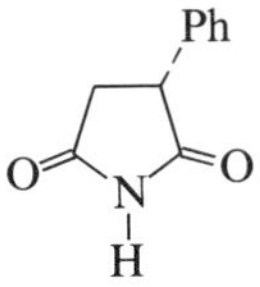

(R)-form

▷ WN3325000.

(R)-form

Mp 83-94°. $[\alpha]_D^{20}$ +151° (c, 2 in EtOH). Not opt. pure.

(±)-form

Mp 90°.

N-*Me*: [86-34-0]. *1-Methyl-3-phenyl-2,5-pyrrolidinedione, 9CI. N-Methyl-2-phenylsuccinimide.* **Phensuximide, BAN, INN.** *Epimid. Lepsol. Lifene. Milontin. Milonton. Mirontin. Petimid. Succitimal. PM 334. Fensuximide* $C_{11}H_{11}NO_2$ M 189.213
Anticonvulsant. Fine cryst. (EtOH). Mp 71-73°.

Wegschieder, R. *et al, Monatsh. Chem.,* 1903, **24**, 422 (*synth*)
Miller, C.A. *et al, J. Am. Chem. Soc.,* 1951, **73**, 4895 (*synth, pharmacol, deriv*)
Glazko, A.J. *et al, J. Pharmacol. Exp. Ther.,* 1954, **111**, 413 (*pharmacol, deriv*)
Turczan, J.W. *et al, J. Pharm. Sci.,* 1973, **62**, 1705 (*pmr*)
Kracmar, J. *et al, Pharmazie,* 1976, **31**, 363 (*uv*)
Ferrendelli, J.A. *et al, Adv. Neurol.,* 1980, **27**, 587 (*rev, deriv*)
Martindale, The Extra Pharmacopoeia, 28th/29th Ed., Pharmaceutical Press London, 1982, 6618 (*deriv*)
Poloński, T., *J. Chem. Soc., Perkin Trans.* 1, 1988, 629 (*synth, ir, pmr, cd*)

3-Phenyl-2(1*H*)-quinolinethione   P-80133

[85274-02-8]

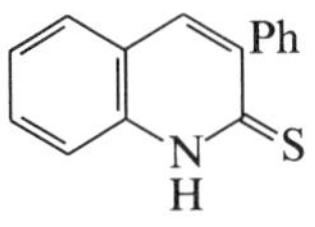

$C_{15}H_{11}NS$ M 237.325
Solid. Mp 242-244°.

Blackburn, T.P. *et al, J. Med. Chem.,* 1987, **30**, 2252 (*synth, pmr*)

2-Phenyl-1-(selenocyanato)acetylene   P-80134

Selenocyanic acid phenylethynyl ester, 9CI

[114908-26-8]

$$PhC\equiv CSeCN$$

C_9H_5NSe M 206.105
Pale yellow oil.

Meinke, P.T. *et al, J. Org. Chem.,* 1988, **53**, 3632 (*synth, pmr, cmr, ir, ms*)

4-Phenyl-2(3*H*)-thiazolethione, 9CI   P-80135

[2103-88-0]

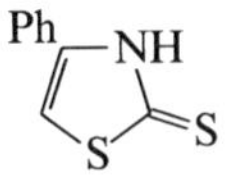

$C_9H_7NS_2$ M 193.293
Mp 141°, Mp 173°.

Vernin, G. *et al, Bull. Soc. Chim. Fr.,* 1963, 2498 (*synth*)
Nalini, V. *et al, J. Chem. Soc., Chem. Commun.,* 1987, 1046 (*cryst struct*)

(Phenylthio)acetaldehyde   P-80136

[66303-55-7]

$$PhSCH_2CHO$$

C_8H_8OS M 152.217
Oxime:
  C_8H_9NOS M 167.231
  Pale-yellow viscous oil.

Di-Et acetal:
  $C_{12}H_{18}OS$ M 210.340
  Liq. $Bp_{0.8}$ 103-104°.

Kanemasa, S. *et al, Chem. Pharm. Bull.,* 1988, **61**, 3973.

2-(Phenylthio)propene   P-80137

[(1-Methylethenyl)thio]benzene, 9CI. Isopropenyl phenyl sulfide

[7594-43-6]

$$H_2C=C(SPh)CH_3$$

Liq. Bp_9 81-86°, $Bp_{0.8}$ 47°.

S-*Dioxide: Isopropenyl phenyl sulfone*
  $C_9H_{10}O_2S$ M 182.243
  Light-sensitive oil. d^{20} 1.19. $Bp_{4.5}$ 142°.

Groen, S.H. *et al, J. Org. Chem.,* 1968, **33**, 2218 (*synth, pmr*)
Weringa, W.D., *Tetrahedron Lett.,* 1969, 273 (*ms*)
Parham, W.E. *et al, Bull. Chem. Soc. Jpn.,* 1972, **45**, 509 (*synth, ir, uv, pmr*)
Groebel, B.T. *et al, Chem. Ber.,* 1977, **110**, 867 (*synth, pmr, ir*)
Reich, H.J. *et al, J. Org. Chem.,* 1981, **46**, 2775 (*synth, pmr*)
Cory, R.M. *et al, J. Org. Chem.,* 1984, **49**, 3898 (*dioxide*)
Fiandanese, V. *et al, Synthesis,* 1987, 1034 (*synth, pmr*)

9-Phenyl-1-(2,4,6-trihydroxyphenyl)-1-nonanone   P-80138

[119736-95-7]

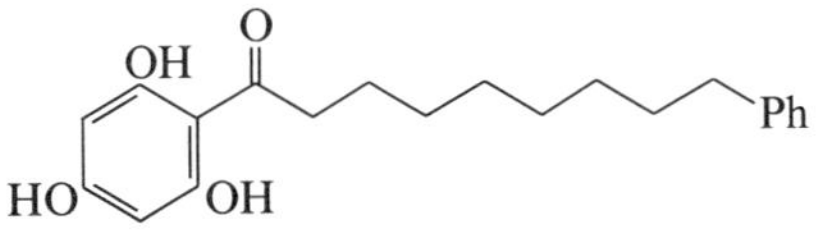

$C_{21}H_{26}O_4$ M 342.434
Constit. of *Horsfieldia glabra.* Cryst. (MeOH). Mp 71-72°.

Pinto, M.M.M. *et al, Phytochemistry,* 1988, **27**, 3988.

3-Phenyl-1-(2,4,6-trihydroxyphenyl)-2-propen-1-one   P-80139

Updated Entry replacing P-20120
2′,4′,6′-Trihydroxychalcone

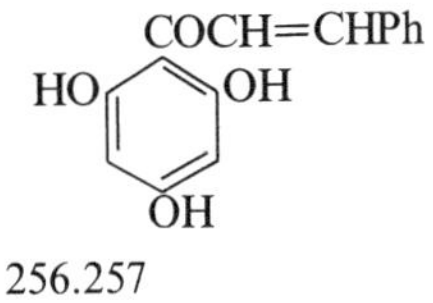

$C_{15}H_{12}O_4$ M 256.257
2-*Me ether:* [18956-16-6]. *1-(2,4-Dihydroxy-6-methoxyphenyl)-3-phenyl-2-propen-1-one, 9CI. 2′,4′-Dihydroxy-6′-methoxychalcone.* **Cardamonin**
$C_{16}H_{14}O_4$ M 270.284
Isol. from *Boesenbergia pandurata.* Yellow needles (CHCl₃/hexane). Mp 199-200°.

4-*Me ether: 1-(2,6-Dihydroxy-4-methoxyphenyl)-3-phenyl-2-propen-1-one. 2′,6′-Dihydroxy-4′-methoxychalcone*
$C_{16}H_{14}O_4$ M 270.284

Isol. from *B. pandurata*. Orange needles (CHCl₃/hexane). Mp 163.5-164.5°.

4'-Me ether: [18956-15-5]. *1-(2,6-Dihydroxy-4-methoxyphenyl)-3-phenyl-2-propen-1-one. 2',6'-Dihydroxy-4'-methoxychalcone*
$C_{16}H_{14}O_4$ M 270.284
Constit. of the leaves of *Lindera umbellata*. Red-brown needles. Mp 161-162°.

2',4'-Di-Me ether: [1775-97-9]. *1-(2-Hydroxy-4,6-dimethoxyphenyl)-3-phenyl-2-propen-1-one. 2'-Hydroxy-4',6'-dimethoxychalcone*
$C_{17}H_{16}O_4$ M 284.311
Constit. of *Piper methysticum*. Cryst. (MeOH aq.). Mp 91°.

Tri-Me ether: 2',4',6'-*Trimethoxychalcone*
$C_{18}H_{18}O_4$ M 298.338
Yellow needles. Mp 82-84°.

2-Dihydro: see 1-(2,4,6-*Trihydroxyphenyl*)-3-*phenyl-1-propanone*, T-80332

Haensel, R. *et al, Z. Naturforsch., B*, 1963, **18**, 370 (*Flavokawin B*)
Kimura, Y. *et al, Yakugaku Zasshi (J. Pharm. Soc. Jpn.)*, 1968, **88**, 239 (*isol*)
Hayashi, N. *et al, Chem. Ind. (London)*, 1969, 1779 (*isol*)
Jaipetch, T. *et al, Aust. J. Chem.*, 1982, **35**, 351 (*isol*)

Phrymarolin II   P-80140

Updated Entry replacing P-60142
1-Acetoxy-6-(2-methoxy-4,5-methylenedioxyphenyl)-2-(3,4-methylenedioxyphenoxy)-3,7-dioxabicyclo[3.3.0]octane
[23720-86-7]

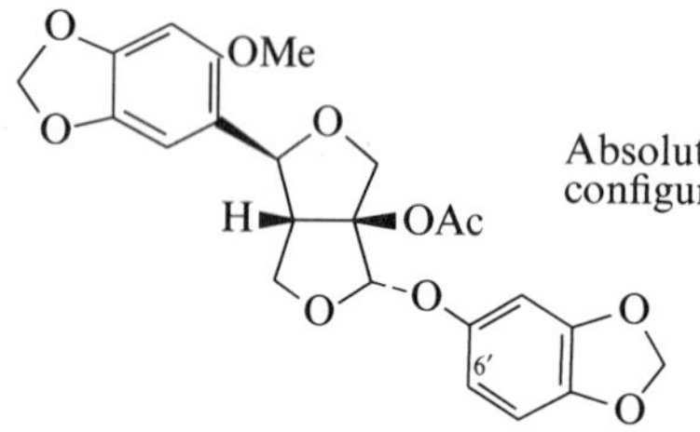

Absolute configuration

$C_{23}H_{22}O_{10}$ M 458.421
Isol. from root of *Phyrma leptostachya*. Cryst. (Et₂O and EtOAc). Mp 161-162°. [α]_D +117.6° (c, 2.72 in dioxan).

6'-Methoxy: [38303-95-6]. ***Phrymarolin I***
$C_{24}H_{24}O_{11}$ M 488.447
From *P. leptostachya*. Cryst. Mp 155-157°. [α]_D +131.3° (dioxan).

Taniguchi, E. *et al, Agric. Biol. Chem.*, 1969, **33**, 466; 1972, **36**, 1489 (*isol*)
Ishibachi, F. *et al, Bull. Chem. Soc. Jpn.*, 1988, **61**, 4361 (*synth, abs config*)

Phyllanthol   P-80141

13,27-Cycloursan-3β-ol, 9CI
[546-51-0]

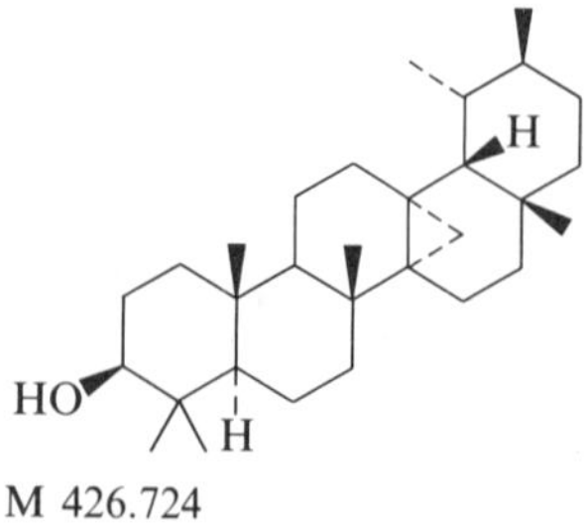

$C_{30}H_{50}O$ M 426.724

Isol. from *Phyllanthus engleri*. Needles (CHCl₃/pet. ether). Mp 233-234°. [α]_D^{15} +43° (CHCl₃).

Cole, A.R.H., *J. Chem. Soc.*, 1954, 3810 (*struct*)
Sengupta, P. *et al, Phytochemistry*, 1966, **5**, 531 (*isol*)

Phyllnirurin   P-80142

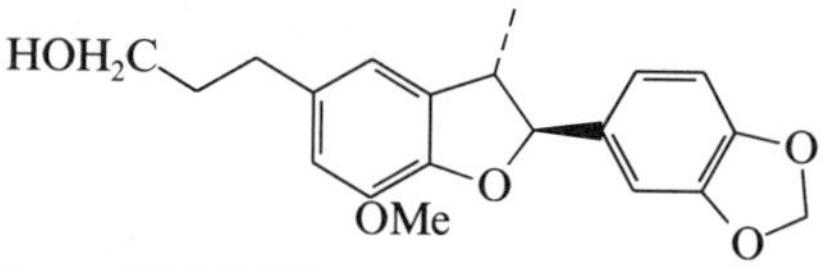

$C_{20}H_{22}O_5$ M 342.391
Constit. of *Phyllanthus niruri*.

Singh, B. *et al, J. Nat. Prod. (Lloydia)*, 1989, **52**, 48 (*isol, ir, pmr*)

Physalin L   P-80143

Updated Entry replacing P-60147
[113146-74-0]

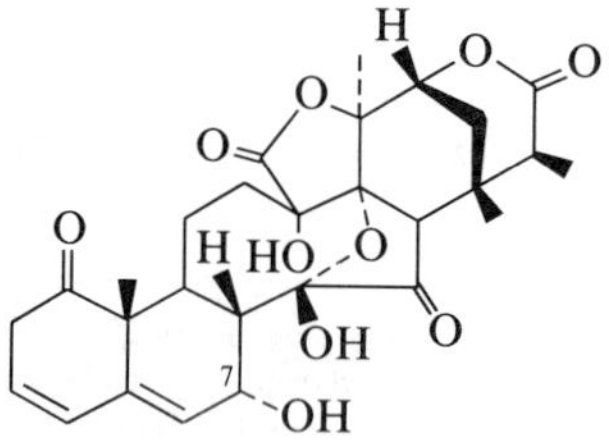

$C_{28}H_{32}O_{10}$ M 528.555
Constit. of *Physalis alkekengi*. Cryst. (2-propanol). Mp 248-249°. [α]_D^{24} −118° (c, 0.3 in Me₂CO).

7-Deoxy: [117591-92-1]. ***Physalin M***
$C_{28}H_{32}O_9$ M 512.555
Constit. of *P. alkekengi*. Cryst. (2-propanol). Mp 224-227°. [α]_D^{24} −106° (c, 0.34 in Me₂CO).

Kawai, M. *et al, Phytochemistry*, 1987, **26**, 3313 (*Physalin L*)
Kawai, M. *et al, Bull. Chem. Soc. Jpn.*, 1988, **61**, 2696 (*Physalin M*)

Phytylplastoquinone   P-80144

[1177-24-8]

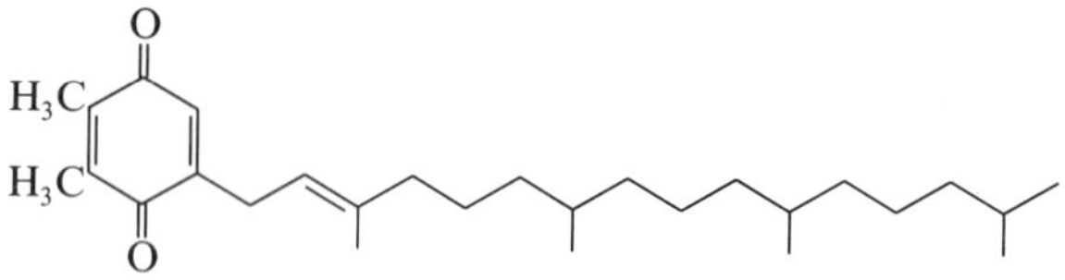

$C_{28}H_{46}O_2$ M 414.670
Isol. from two strains of *Euglena gracilis*. Oil. λ_{max} 254, 261 nm (hexane).

Whistance, G.R. *et al, Phytochemistry*, 1970, **9**, 213.

Pichierenol   P-80145

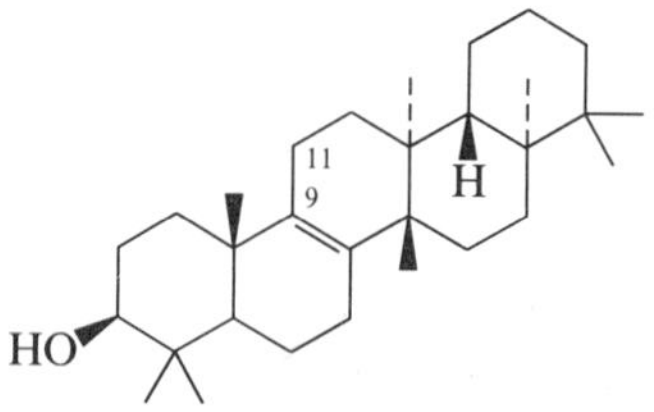

$C_{30}H_{50}O$ M 426.724
Ac: ***Pichierenyl acetate***

$C_{32}H_{52}O_2$ M 468.762
Constit. of roots of *Picris hieracioides*. Cryst. Mp 272.5-273.5°. $[\alpha]_D^{23}$ −31.2° (c, 0.3 in CHCl₃).

$\Delta^{9(11)}$-*Isomer, Ac: Isopichierenyl acetate*
$C_{32}H_{52}O_2$ M 468.762
Constit. of *P. hieracioides*. Cryst. Mp 248.5-249.5°. $[\alpha]_D^{23}$ −2.6° (c, 0.1 in CHCl₃).

Shiojima, K. *et al*, *Tetrahedron Lett.*, 1989, **30**, 6873.

ent-8(14),15-Pimaradiene-3β,12β-diol   P-80146
Yucalexin P-21

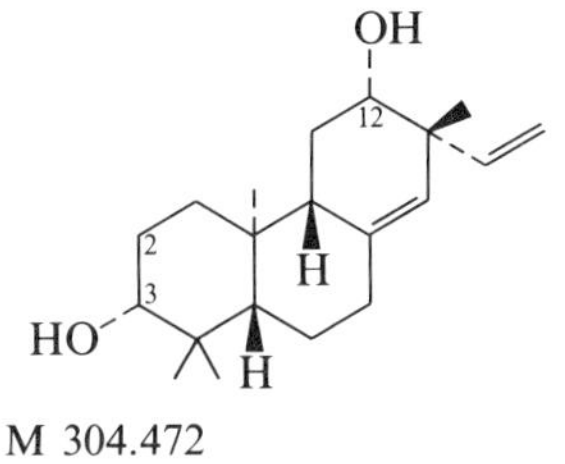

$C_{20}H_{32}O_2$ M 304.472
Constit. of cassava roots (*Manihot esculenta*).

3-Ketone: [119626-50-5]. *ent*-12β-*Hydroxy*-8(14),15-*pimaradien-3-one.* **Yucalexin P-13**
$C_{20}H_{30}O_2$ M 302.456
Constit. of *M. esculenta*.

12-Ketone: [119626-49-2]. *ent*-3β-*Hydroxy*-8(14),15-*pimaradien-12-one.* Yucalexin *P*-12
$C_{20}H_{30}O_2$ M 302.456
Constit. of *M. esculenta*.

3,12-Diketone: [95523-20-9]. *ent*-8(14),15-*Pimaradiene*-3,12-*dione.* **Yucalexin P-4**
$C_{20}H_{28}O_2$ M 300.440
Constit. of *M. esculenta*.

2-Oxo: [119642-82-9]. *ent*-3β,12β-*Dihydroxy*-8(14),15-*pimaradien-2-one.* **Yucalexin P-17**
$C_{20}H_{30}O_3$ M 318.455
Constit. of *M. esculenta*.

Sakai, T. *et al*, *Phytochemistry*, 1988, **27**, 3769.

ent-9(11),15-Pimaradien-19-oic acid   P-80147
[119290-87-8]

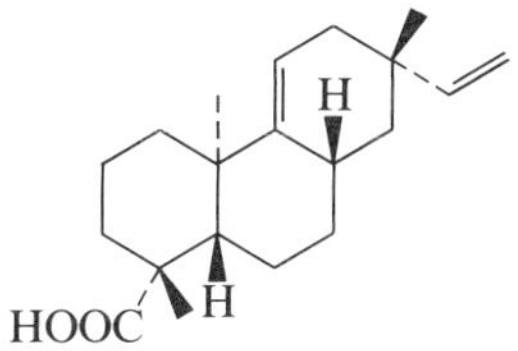

$C_{20}H_{30}O_2$ M 302.456
Constit. of *Acanthopanax koreanum*. Amorph. powder. Mp 135-136°.

Kim, Y.H. *et al*, *J. Nat. Prod.* (*Lloydia*), 1988, **51**, 1080.

ent-9,(11),15-Pimaradien-19-ol   P-80148
$C_{20}H_{32}O$ M 288.472
Constit. of *Acanthopanax koreanum*. Needles (hexane). Mp 73-74°. $[\alpha]_D^{25}$ −14.94° (c, 0.19 in CHCl₃).

Ac: 19-Acetoxy-9,(11),15-pimaradiene
$C_{22}H_{34}O_2$ M 330.509
Constit. of *A. koreanum*. Oil. $[\alpha]_D^{25}$ −8.82° (c, 0.34 in CHCl₃).

Kim, Y.H. *et al*, *J. Nat. Prod.* (*Lloydia*), 1988, **51**, 1080 (*isol, pmr, cmr*)

7-Pimarene-3,15,16-triol   P-80149

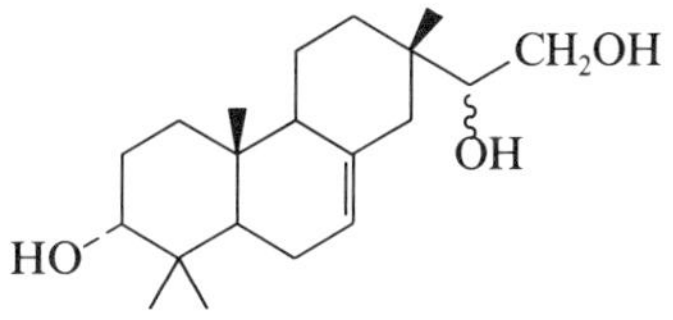

$C_{20}H_{36}O_3$ M 324.503
3α-form
Constit. of *Agathis vitiensis*. Needles (EtOAc). Mp 170-175°. $[\alpha]_D^{17}$ −22° (c, 1 in MeOH).

Cambie, R.C. *et al*, *Phytochemistry*, 1989, **28**, 1675.

Pinnatazone   P-80150
[119765-94-5]

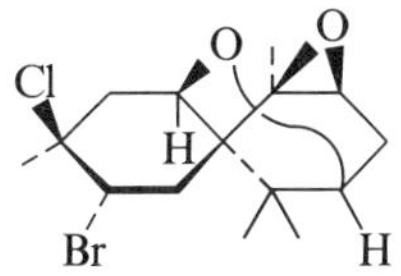

$C_{15}H_{22}BrClO_2$ M 349.694
Constit. of *Laurencia pinnatifida*. Cryst. (MeOH). Mp 190-192°. $[\alpha]_D$ +4.34° (c, 0.23 in CHCl₃).

Atta-ur-Rahman, *et al*, *Phytochemistry*, 1988, **27**, 3879 (*isol, cryst struct*)

Pinthuamide   P-80151
[122279-83-8]

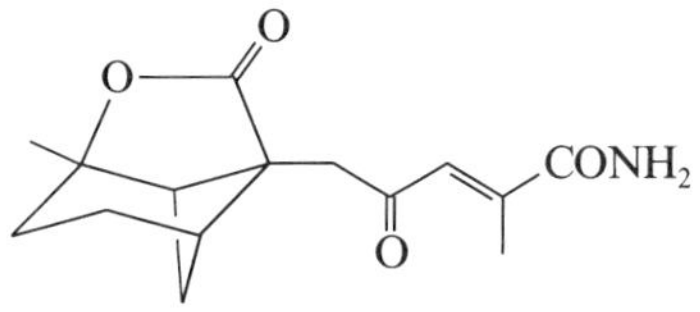

$C_{15}H_{19}NO_4$ M 277.319
Metab. of fungus *Ampulliferina* sp. Needles (EtOAc). Mp 190-193°. $[\alpha]_D^{20}$ +73.4° (c, 1.0 in EtOH).

Kimura, Y. *et al*, *Tetrahedron Lett.*, 1989, **30**, 1267 (*cryst struct*)

Piperdial   P-80152
[100288-36-6]

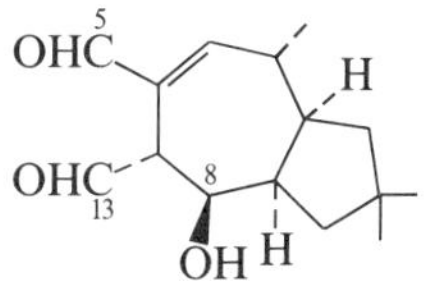

$C_{15}H_{22}O_3$ M 250.337
Constit. of *Lactarius necator*. $[\alpha]_D^{23}$ +77° (c, 0.8 in Et₂O).

13-Alcohol: [100288-35-5]. *Piperalol*
$C_{15}H_{24}O_3$ M 252.353
Constit. of *L. necator*. $[\alpha]_D^{23}$ +57° (c, 4.3 in Et₂O).

5,13-Bisalcohol, 8-epimer: [118918-26-6]. *8-Epipertriol*
$C_{15}H_{26}O_3$ M 254.369
Constit. of *L. necator*. Cryst. Mp 108-114°. $[\alpha]_D^{20}$ −36.6° (c, 1 in CHCl₃).

Sterner, O. *et al*, *Tetrahedron Lett.*, 1985, **26**, 3163 (*isol, struct*)
Daniewski, W.M. *et al*, *Phytochemistry*, 1988, **27**, 3315 (*isol, deriv*)

Piperitone oxide   P-80153

6-Methyl-3-(1-methylethyl)-7-oxabicyclo[4.1.0]heptan-2-one,
9CI. 1,2-Epoxy-p-menthan-3-one, 8CI. 2,3-Epoxy-6-
isopropyl-3-methylcyclohexanone

[5286-38-4]

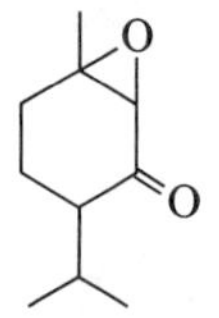

$C_{10}H_{16}O_2$ M 168.235

Stereochem. not clear. Isol. from *Mentha sylvestris* oil and
other essential oils. Cryst. (hexane). d_{25}^{25} 1.01. Mp 14.5-
15.5°. $[\alpha]_D^{22}$ −177.0° (c, 0.96 in EtOH). n_D^{20} 1.4624. Poss.
identical with Dihydrolippione.

Semicarbazone: Cryst. Mp 199-200°. $[\alpha]_{25}$ +213° (c, 0.29 in
95% EtOH aq.).

2,4-Dinitrophenylhydrazone: Cryst. Mp 115-117°.

[4713-37-5, 4713-38-6, 20303-83-7, 57130-28-6, 103476-50-2]

Reitsema, R.H. *et al*, *J. Am. Chem. Soc.*, 1956, **78**, 3792 (*isol*)
Nigam, I.C. *et al*, *Can. J. Chem.*, 1963, **41**, 1535 (*occur*)
Klein, E. *et al*, *Tetrahedron*, 1963, **19**, 1091 (*synth*)
Gacs-Baitz, E. *et al*, *Org. Magn. Reson.*, 1984, **22**, 738 (*pmr, cmr*)

Piscidinol B   P-80154

Updated Entry replacing P-50204

7-Tirucallene-3β,23ξ,24ξ,25-tetrol

[100295-68-9]

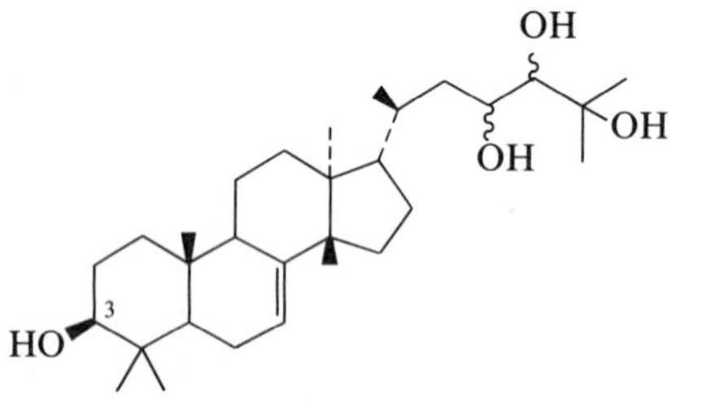

$C_{30}H_{52}O_4$ M 476.738

Constit. of *Walsura piscidia*. Cryst. (MeOH). Mp 240°.

3-Ketone: [100198-09-2]. *23ξ,24ξ,25-Trihydroxy-7-*
tirucallen-3-one. **Piscidinol A**
$C_{30}H_{50}O_4$ M 474.723
From *W. piscidia*, also from *Phellodendron chinense* and
Turrea nilotica. Cryst. (MeOH). Mp 195°. $[\alpha]_D^{25}$ −90° (c,
1 in CHCl₃). The isolate from *T. nilotica* (amorph., no
further details) could be a stereoisomer.

[115404-56-3]

Purushothaman, K.K. *et al*, *Phytochemistry*, 1985, **24**, 2349 (*isol*)
Mulholland, D.A. *et al*, *Phytochemistry*, 1988, **27**, 1220, 1809 (*isol,*
deriv)
Kaur, K.J. *et al*, *Phytochemistry*, 1988, **27**, 1809 (*isol, deriv*)

Pisolactone   P-80155

Updated Entry replacing P-20140

[87164-33-8]

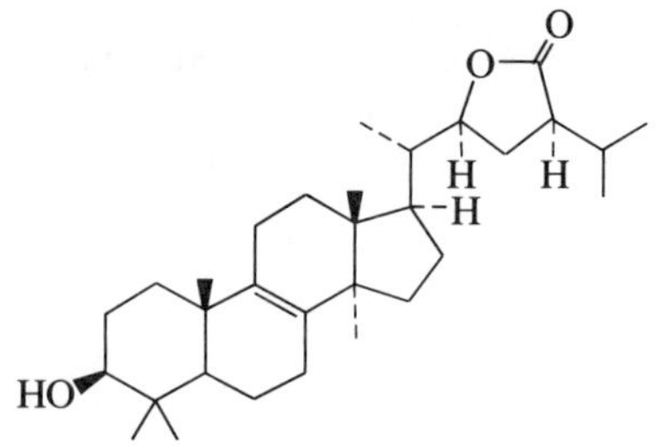

$C_{31}H_{50}O_3$ M 470.734

Constit. of fungus *Pisolithus tinctorius*. Prisms (MeOH).
Mp 279-280°. $[\alpha]_D^{28}$ +60° (c, 1 in CHCl₃).

3-Ketone: [119539-75-2]. **3-Oxopisolactone**
$C_{31}H_{48}O_3$ M 468.718
Metab. of *P. tinctorius*. Cryst. (MeOH). Mp 248-250°.
$[\alpha]_D$ +79° (c, 1.4 in CHCl₃).

Lobo, A.M. *et al*, *Tetrahedron Lett.*, 1983, **24**, 2205 (*isol, cryst*
struct)
Lobo, A.M. *et al*, *Phytochemistry*, 1988, **27**, 3569 (*deriv*)

Plagiospirolide A   P-80156

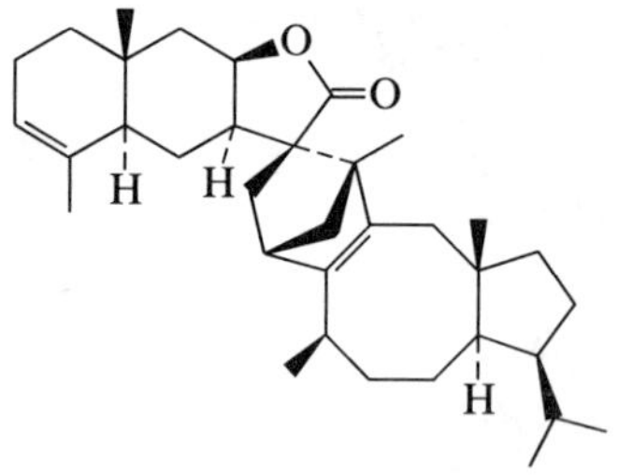

$C_{35}H_{52}O_2$ M 504.795

Constit. of liverwort *Plagiochila moritziana*. Needles
(hexane). Mp 196-198°. $[\alpha]_D^{20}$ +411.9° (c, 0.36 in CHCl₃).

Spörle, J. *et al*, *Tetrahedron*, 1989, **45**, 5003 (*cryst struct*)

Plagiospirolide B   P-80157

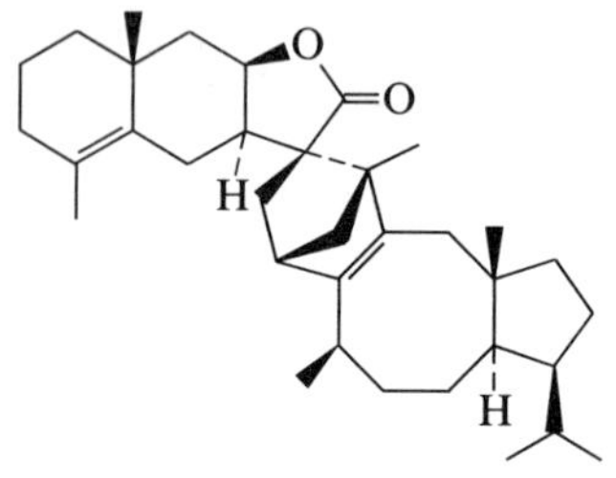

$C_{35}H_{52}O_2$ M 504.795

Constit. of liverwort *Plagiochila moritziana*. Oil. $[\alpha]_D^{20}$
+59.2° (c, 1.176 in CHCl₃).

Spörle, H. *et al*, *Tetrahedron*, 1989, **45**, 5003.

Planaic acid   P-80158

4-Carboxy-3-methoxy-5-pentylphenyl 2,4-dimethoxy-6-pentylbenzoate, 9CI. 4-Hydroxy-6-pentyl-o-anisic acid 2,4-dimethoxy-6-pentylbenzoate, 8CI

[5366-07-4]

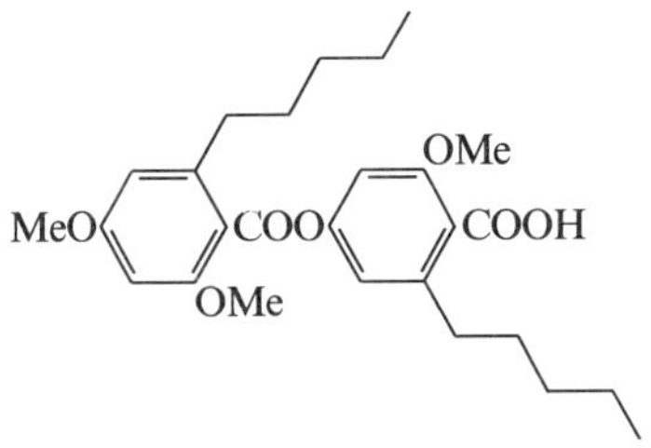

$C_{27}H_{36}O_7$ M 472.577

Isol. from *Lecidea plana* and *L. lithophila*. Needles (MeOH). Mp 110-111°.

Huneck, S. *et al, Z. Naturforsch., B*, 1965, **20**, 1119, 1137 (*isol, ir, pmr, struct*)
Huneck, S. *et al, Tetrahedron*, 1968, **24**, 2707 (*ms*)
Elix, J.A., *Aust. J. Chem.*, 1974, **27**, 1767 (*synth, pmr, ms*)

Plastoquinone 4   P-80159

Plastoquinone 20

$C_{28}H_{40}O_2$ M 408.623

Isol. from leaves of *Aesculus hippocastanum*. Unstable orange oil.

Eck, H. *et al, Z. Naturforsch., B*, 1963, **18**, 446 (*isol, uv, ir, pmr*)
Das, B.C. *et al, Biochem. Biophys. Res. Commun.*, 1965, **21**, 318 (*uv*)
Misiti, D. *et al, J. Am. Chem. Soc.*, 1965, **87**, 1407 (*struct*)

Plastoquinones   P-80160

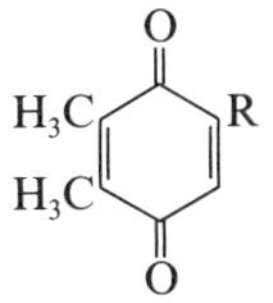

A family of 2,3-dimethyl-5-polyprenyl-1,4-benzoquinones found in plant tissues. R represents a prenyl side-chain which may be unsubstituted (Plastoquinones), monohydroxylated (Plastoquinone C), monoacylated (Plastoquinone B), or monohydroxylated and monoacylated (Plastoquinone Z).The latter classes appear to be confined to higher plant tissues.

Whistance, G.R. *et al, Phytochemistry*, 1970, **9**, 213 (*isol*)
Pennock, J.F., *Biochem. Soc. Trans.*, 1983, **11**, 504 (*rev, biosynth*)

Platypodantherone   P-80161

[118855-44-0]

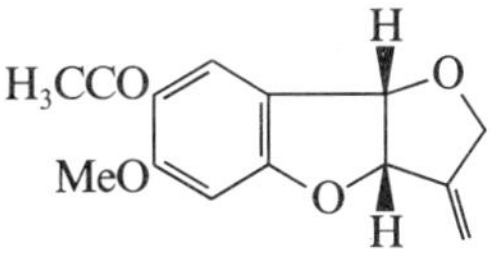

$C_{14}H_{14}O_4$ M 246.262

Constit. of *Platypodanthera melissaefolia*. Oil.

Paredes, L. *et al, Phytochemistry*, 1988, **27**, 3329.

Pleurotin   P-80162

Geogenine

[1404-23-5]

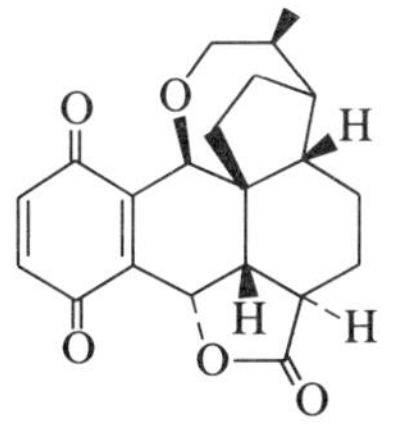

$C_{21}H_{22}O_5$ M 354.402

Benzoquinone antibiotic. Identity of Pleurotin and Geogenine not fully established. Isol. from *Pleurotus griseus*, *Geopetalum geogerium* and *Hohenbuehalia geogenius*. Active against gram-positive bacteria, mycobacteria and possesses antitumour activity. Amber cryst. Mp 200-215° dec. $[\alpha]_D^{23}$ −20° (CHCl$_3$).

[77550-91-5]

Robbins, W.J. *et al, Proc. Natl. Acad. Sci. U.S.A.*, 1947, **33**, 171 (*isol*)
Arigoni, D., *Pure Appl. Chem.*, 1968, **17**, 331 (*struct*)
Grandjean, J. *et al, Tetrahedron Lett.*, 1974, 1893 (*pmr, cmr, struct*)
Fr. Pat., 2 407 718, (1979); *CA*, **91**, 173367 (*isol*)
Riondel, J. *et al, Acta Crystallogr., Sect. C*, 1981, **37**, 1309 (*cryst struct*)
Riondel, J. *et al, Arzneim.-Forsch.*, 1981, **31**, 293 (*isol, uv, ir, pmr, cmr, ms*)

Plexaurolone   P-80163

[75248-46-3]

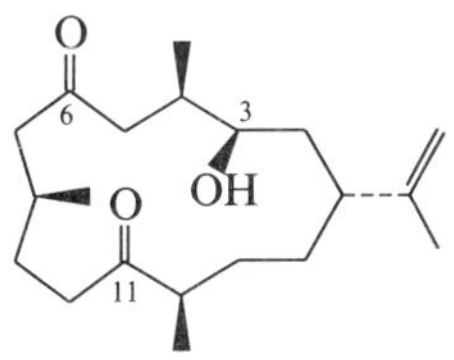

$C_{20}H_{34}O_3$ M 322.487

Metab. of *Plexaura* sp. Cryst. (Me$_2$CO/pet. ether). Mp 110-111°.

11R-Alcohol: [119979-75-8]. **Dihydroplexaurolone**
   $C_{20}H_{36}O_3$ M 324.503
   Metab. of *P.* sp. Prisms (Me$_2$CO/hexane). Mp 125-126°. $[\alpha]_D$ +42° (c, 0.12 in CHCl$_3$).

3-Ketone: [119979-76-9]. **Dehydroplexaurolone**
   $C_{20}H_{32}O_3$ M 320.471
   Metab. of *P.* sp. Needles (Me$_2$CO/pet. ether). Mp 93-95°. $[\alpha]_D$ +48° (c, 0.27 in CHCl$_3$).

3-Ketone, stereoisomer: [120053-65-8].
   Isodehydroplexaurolone
   $C_{20}H_{32}O_3$ M 320.471

Metab. of *P.* sp. Cubes (Me₂CO/pet. ether). Mp 196-199°. [α]_D −11° (c, 0.09 in CHCl₃). Stereoisomeric at C-4 and/or C-12.

Ealick, S.E. *et al, Acta Crystallogr., Sect. B,* 1980, **36**, 1901 (*cryst struct*)
Chan, W.R. *et al, Tetrahedron,* 1989, **45**, 103 (*cryst struct*)

7,17,21-Podiodatriene   P-80164

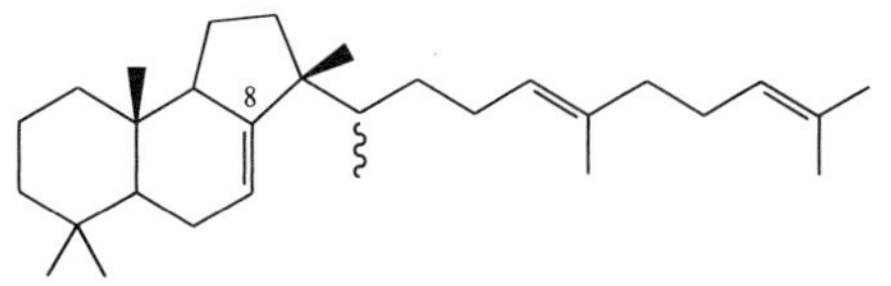

C₃₀H₅₀   M 410.725
Constit. of rhizomes of *Polypodioides niponica.* Oil. [α]_D^23 −11.5° (c, 0.2 in CHCl₃).

Δ⁸-*Isomer:* **8,17,21-Podiodatriene**
Constit. of *P. niponica.* Oil. [α]_D^23 +11.7° (c, 0.2 in CHCl₃).

Arai, Y. *et al, Tetrahedron Lett.,* 1989, **30**, 7209.

Podophyllotoxin   P-80165

Updated Entry replacing P-50227
[518-28-5]

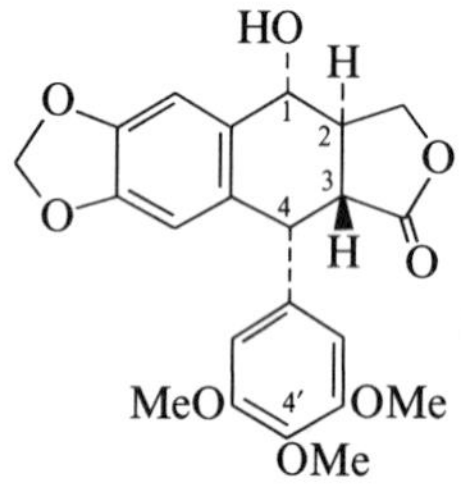

C₂₂H₂₂O₈   M 414.411
Lignan from *Podophyllum emodi* rhizomes. Also isol. from *P. peltatum,* other *P.* spp., *Diphylleia grayi, Callitris drummondii* and *Juniperus* spp. Potent cytotoxin. Cell division inhibitor. Cryst. (Me₂CO aq. or MeOH). Mp 117-118°. [α]_D^14 −101.3° (EtOH). 4 Crystal modifications reported. Mp's up to 188-189° have been reported for carefully dried samples.

▷ Highly toxic orally, exp. carcinogen.

Ac: [1180-34-3].
Cryst. (MeOH). Mp 209.5-210.5°. [α]_D^20 −143° (c, 1.0 in CHCl₃).

1-β-D-Glucoside:
C₂₈H₃₂O₁₃   M 576.553
Isol. from *P. emodi.* Amorph. solid. Mp 149-152°. [α]_D^20 −65° (c, 0.5 in H₂O), [α]_D −75° (c, 0.6 in MeOH), [α]_D −117° (c, 0.67 in Py).

1-Epimer: [4375-07-9]. *Epipodophyllotoxin*
Needles (EtOH aq.). Mp 159.4-161.2°. [α]_D^20 −75° (c, 1 in CHCl₃).
▷ LV2520000.

3-Epimer: [477-47-4]. **Picropodophyllin**
C₂₂H₂₂O₈   M 414.411
Isol. from *P. emodi, Diphylleia grayi* and others. Needles (MeOH or C₆H₆). Mp 226-227°. [α]_D +5.3° (CHCl₃).

3-Epimer, O-β-D-Glucopyranoside: Picropodophyllin glucoside
C₂₈H₃₂O₁₃   M 576.553
Isol. from *P. emodi.* Needles (H₂O). Mp 237-238°. [α]_D^20 −11.5° (Py).

4'-O-De-Me: **4'-Demethylpodophyllotoxin.** *4'-Desmethylpodophyllotoxin.*
C₂₁H₂₀O₈   M 400.384
Isol. from *P. emodi.* Prisms or plates (EtOH aq.). Mp 250-251.6°. [α]_D^20 −130° (CHCl₃).

4'-O-De-Me, 1-O-β-D-glucoside:
C₂₇H₃₀O₁₃   M 562.526
Isol. from *P. emodi.* Powder. Mp 165-170°. [α]_D^20 −75° (H₂O).

1,2,3,4-Tetradehydro: **Dehydropodophyllotoxin**
C₂₂H₁₈O₈   M 410.379
Minor constit. of *P. peltatum.* Needles (EtOH). Mp 272-274°. Achiral.

Hartwell, J.L. *et al, J. Am. Chem. Soc.,* 1951, **73**, 2909 (*isol*)
Nadkarni, M.V. *et al, J. Am. Chem. Soc.,* 1953, **75**, 1308 (*isol, derivs*)
Kofod, H. *et al, Acta Chem. Scand.,* 1954, **8**, 1296 (*Dehydropodophyllotoxin*)
Stoll, A. *et al, J. Am. Chem. Soc.,* 1954, **76**, 3103 (*Glucosides*)
Schrecker, A.W. *et al, J. Org. Chem.,* 1956, **21**, 288, 381 (*props, abs config*)
Gensler, W.J. *et al, J. Am. Chem. Soc.,* 1962, **84**, 1748 (*synth*)
Gensler, W.J. *et al, J. Org. Chem.,* 1966, **31**, 3224 (*synth*)
Duffield, A.M., *J. Heterocycl. Chem.,* 1967, **4**, 16 (*ms*)
Ayres, D.C. *et al, J. Chem. Soc., Perkin Trans. 1,* 1972, 1343 (*pmr*)
Petcher, T.J. *et al, J. Chem. Soc., Perkin Trans. 2,* 1973, 288 (*cryst struct*)
Cannetta, R. *et al, Cancer Chemother. Pharmacol.,* 1982, **7**, 103 (*pharmacol*)
Jackson, D.E. *et al, Phytochemistry,* 1984, **23**, 1029, 1037 (*biosynth*)
Inamori, Y. *et al, Chem. Pharm. Bull.,* 1986, **34**, 3928 (*props*)
Macdonald, D.I. *et al, J. Org. Chem.,* 1986, **51**, 4749 (*synth*)
Kamil, W.M. *et al, Phytochemistry,* 1986, **25**, 2093 (*biosynth*)
Van der Eycken, J. *et al, Tetrahedron,* 1986, **42**, 4297 (*synth*)
Vyas, D.M. *et al, Tetrahedron Lett.,* 1986, **27**, 3099 (*synth*)
Jones, D.W. *et al, J. Chem. Soc., Chem. Commun.,* 1987, 1797 (*synth*)
Kaneko, T. *et al, Tetrahedron Lett.,* 1987, **28**, 517 (*synth*)
Sax, N.I., *Dangerous Properties of Industrial Materials,* 5th Ed., Van Nostrand-Reinhold, 1979, 919.

Polyanthin   P-80166

Updated Entry replacing P-10228
[54631-86-6]

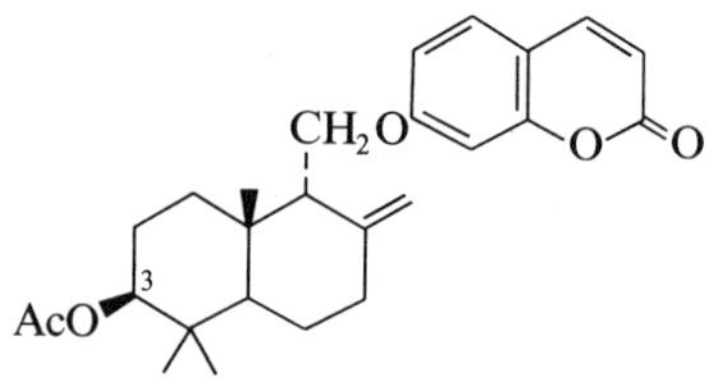

C₂₆H₃₂O₅   M 424.536
Isol. from *Ferula polyantha.* Mp 148-149°.

3-Epimer: [54631-85-5]. **Polyanthinin**
C₂₆H₃₂O₅   M 424.536
From *F. polyantha.* Mp 127-129°.

3-Epimer, O-de-Ac: [51819-92-2]. **Gummosin**
C₂₄H₃₀O₄   M 382.499
From *F. gummosa.* Mp 176-177°. [α]_D −54° (CHCl₃).

Perel'son, M.E., *Khim. Prir. Soedin.,* 1973, **9**, 490; 1974, **10**, 15; 1975, **11**, 244.
Khasanov, T.Kh. *et al, Khim. Prir. Soedin.,* 1974, **10**, 517.

Polypodoside A   P-80167

[119784-25-7]

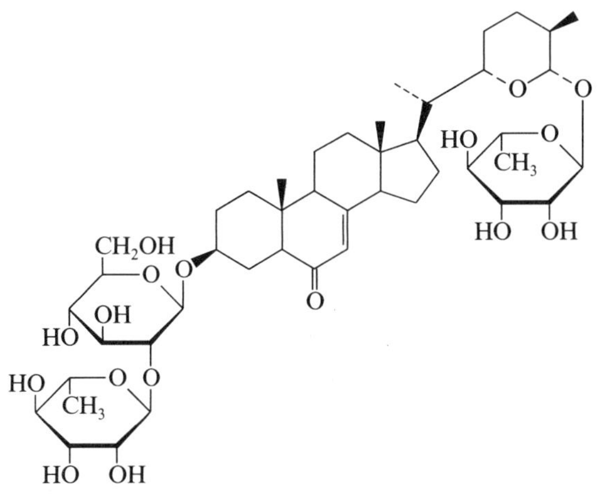

$C_{45}H_{72}O_{17}$ M 885.054

Constit. of rhizomes of *Polypodium glycyrrhiza*. Intensely sweet substance. Needles (EtOH). Mp 198-200°. $[\alpha]_D$ −37° (c, 0.3 in MeOH).

Kim, J. *et al*, *J. Nat. Prod. (Lloydia)*, 1988, **51**, 1166 (*isol, pmr, cmr*)

Pongachalcone II   P-80168

1-(5-Hydroxy-2,2-dimethyl-2H-1-benzopyran-6-yl)-3-(4-hydroxy-3-methoxyphenyl)-2-propen-1-one, 9CI. 5-Hydroxy-6-(4-hydroxy-3-methoxycinnamoyl)-2,2-dimethyl-2H-1-benzopyran

[64173-09-7]

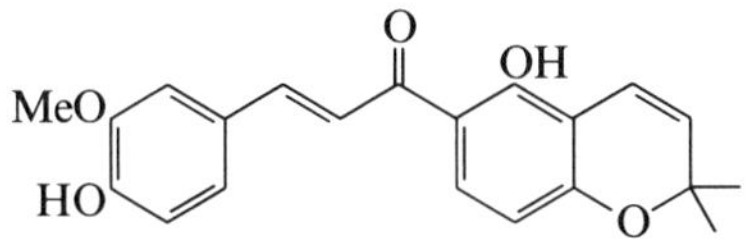

$C_{21}H_{20}O_5$ M 352.386

Isol. from heartwood of *Pongamia glabra*. Orange-red solid. Mp 175-176° (synthetic).

Subrahmanyam, K. *et al*, *Indian J. Chem.*, *Sect. B*, 1977, **15**, 12 (*isol, struct, synth*)

Pongamol   P-80169

Updated Entry replacing P-02168

1-(4-Methoxy-5-benzofuranyl)-3-phenyl-1,3-propanedione, 9CI. 5-Benzoylmethoxy-6-methoxybenzofuran. 5-Benzoylacetyl-4-methoxybenzofuran, 8CI

[484-33-3]

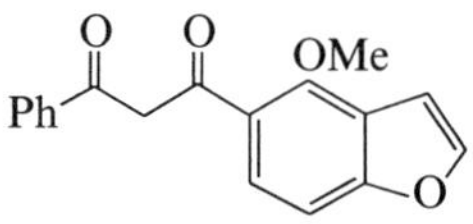

$C_{18}H_{14}O_4$ M 294.306

Isol. from *Pongamia glabra* and *Tephrosia* spp. Used in insecticides and pesticides. Yellow rhombic prisms (EtOH/AcOH). Mp 135-136°. Exists in the enolised form.

Cu complex: Green rhombohedral tablets (CHCl₃/Et₂O). Mp 220-222°.

Rangaswami, S. *et al*, *CA*, 1941, **35**, 5733 (*isol*)
Mukerjee, S.K. *et al*, *J. Chem. Soc.*, 1955, 2048 (*synth*)
Roy, D. *et al*, *Curr. Sci.*, 1977, **46**, 743; *CA*, **88**, 19047 (*isol*)

Sinka, B. *et al*, *Phytochemistry*, 1982, **21**, 1468 (*isol*)
Sighamony, S. *et al*, *Int. Pest Control*, 1983, **25**, 120 (*use*)
Rajain, P. *et al*, *Phytochemistry*, 1988, **27**, 648 (*isol*)
Parmar, V.S. *et al*, *Phytochemistry*, 1989, **28**, 591 (*cryst struct*)

Ponticaepoxide   P-80170

2-Ethenyl-3-(1-nonen-3,5,7-triynyl)oxirane, 9CI. 2-(1-Nonen-3,5,7-triynyl)-3-vinyloxirane. 3,4-Epoxy-1,5-tridecadiene-7,9,11-triyne

[3562-36-5]

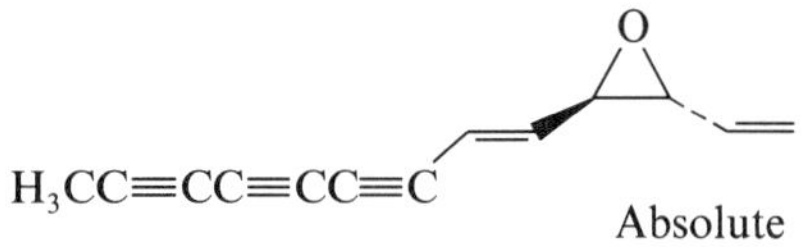

Absolute configuration

$C_{13}H_{10}O$ M 182.221

(+)-**form**

Isol. from various Compositae spp., e.g. *Artemisa, Tanacetum, Chrysanthemum, Cladanthus, Achillea* spp. Cryst. (pet. ether). Mp 66°. $[\alpha]_D^{23}$ +201° (c, 1.0 in Me₂CO).

(±)-**form**

Mp 52°.

Bohlmann, F. *et al*, *Chem. Ber.*, 1960, **93**, 1937; 1962, **95**, 1742; 1964, **97**, 809, 1179; 1966, **99**, 1830 (*isol, uv, struct, abs config*)
Hemmer, E. *et al*, *Acta Chem. Scand.*, 1961, **15**, 691 (*isol*)
Bohlmann, F. *et al*, *Tetrahedron Lett.*, 1965, 1385 (*synth*)

Potentillanin   P-80171

[118555-81-0]

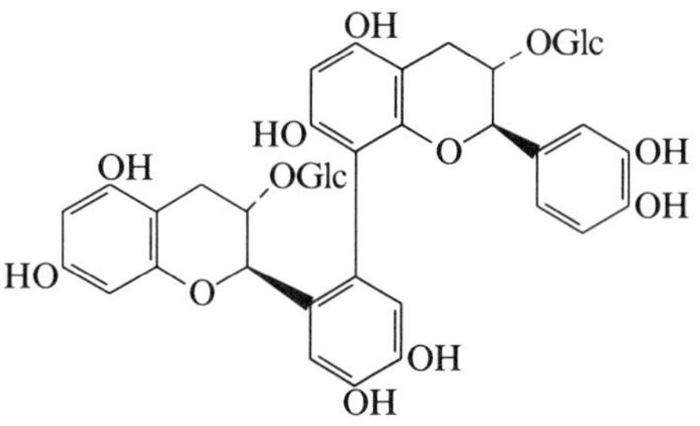

$C_{42}H_{46}O_{22}$ M 902.812

Constit. of *Potentilla viscosa*. Amorph. powder. $[\alpha]_D^{18}$ −80° (c, 1 in Me₂CO aq.).

Zhang, B. *et al*, *Phytochemistry*, 1988, **27**, 3277.

Praealtin A   P-80172

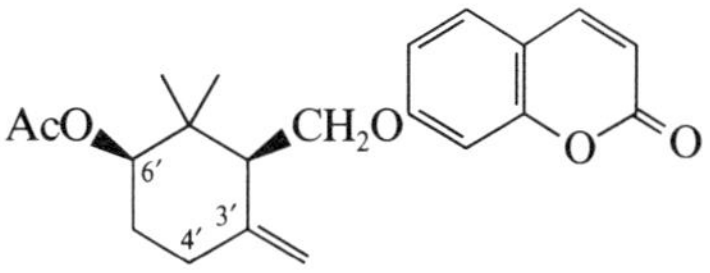

$C_{21}H_{24}O_5$ M 356.418

Constit. of *Aster praealtus*. Cryst. (C₆H₆). Mp 163.3°.

$\Delta^{3'4'}$-*Isomer, 6'-de-Ac, 6'-(2-methylbutanoyl)*: **Praealtin B**
$C_{24}H_{30}O_5$ M 398.498
Constit. of *A. praealtus*.

$\Delta^{3'4'}$*Isomer, 6'-de-Ac, 6'-(2-methylbutanoyl)*: **Praealtin C**
$C_{24}H_{30}O_5$ M 398.498
Constit. of *A. praealtus*.

Wilzer, K.A. *et al*, *Phytochemistry*, 1989, **28**, 1729 (*cryst struct, isol*)

Praealtin D   P-80173

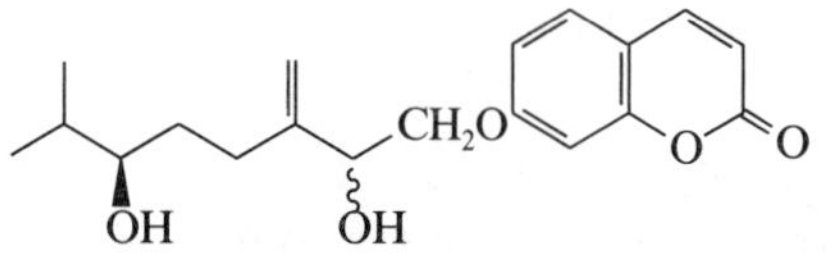

$C_{19}H_{24}O_6$ M 348.395
Constit. of *Aster praealtus*. Gum.

Wilzer, K.A. *et al, Phytochemistry*, 1989, **28**, 1729.

Primnatrienone   P-80174

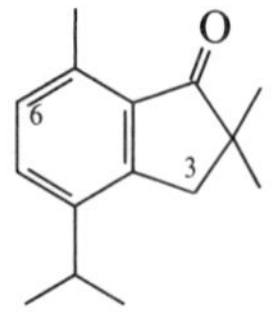

$C_{15}H_{20}O$ M 216.322

6-Hydroxy: [116173-40-1]. **6-Hydroxyprimnatrienone**
$C_{15}H_{20}O_2$ M 232.322
Constit. of a *Primnoeides* sp. Cryst. (CH_2Cl_2/hexane).
Mp 137-138°.

6-Methoxy: [116173-41-2]. **6-Methoxyprimnatrienone**
$C_{16}H_{22}O_2$ M 246.349
Constit. of a *P.* sp. Oil.

6-Methoxy, 3-hydroxy: [116173-38-7]. **3-Hydroxy-6-methoxyprimnatrienone**
$C_{16}H_{22}O_3$ M 262.348
Constit. of *P.* sp. Cryst. (EtOAc/hexane). Mp 122-124°.

6-Methoxy, 3-acetoxy: [116173-39-8]. **3-Acetoxy-6-methoxyprimnatrienone**
$C_{18}H_{24}O_4$ M 304.385
Constit. of a *P.* sp. Oil. $[\alpha]_D^{20}$ +1.5° (c, 0.3 in $CHCl_3$).

6-Methoxy, 3-oxo: [116173-37-6]. **6-Methoxy-1,3-primnatrienedione**
$C_{16}H_{20}O_3$ M 260.332
Pale yellow microcryst. Mp 92-93°.

Cambie, R.C. *et al, Aust. J. Chem.*, 1988, **41**, 365.

Procerin†   P-80175

Updated Entry replacing P-02289
2-Hydroxy-5-(3-methyl-2-butenyl)-4-(1-methylethenyl)-2,4,6-cycloheptatrien-1-one, 9CI. 4-Isopropenyl-5-prenyltropolone.
6-Isopropenyl-5-prenyltropolone
[552-96-5]

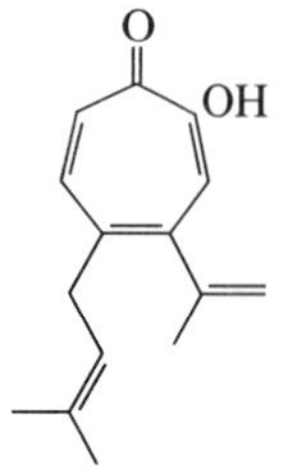

$C_{15}H_{18}O_2$ M 230.306
Present in *Juniperus procera*. Mp 71-72°.

Runeberg, J., *Acta Chem. Scand.*, 1961, **15**, 645 (*ir, synth*)
Kitahara, Y. *et al, Bull. Chem. Soc. Jpn.*, 1964, **37**, 895 (*synth*)

2-(2-Propenyl)cycloheptanone, 9CI   P-80176

2-Allylcycloheptanone
[58105-24-1]

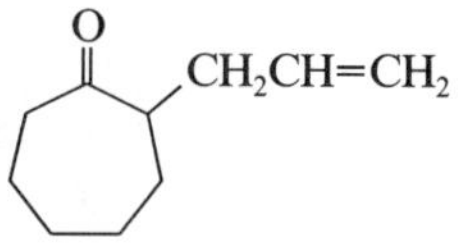

$C_{10}H_{16}O$ M 152.236

(±)-*form* [115094-62-7]
Liq. with odour of menthol. d_4^{22} 0.93. Bp_{18} 103°.

Semicarbazone: Cryst. Mp 145°.

Colonge, J. *et al, Bull. Soc. Chim. Fr.*, 1953, 1074 (*synth*)
Dzhemilev, U.M. *et al, Izv. Akad. Sci. USSR, Ser. Sci. Khim.*, 1988, 378 (*synth*)

2-(2-Propenyl)cyclohexanone, 9CI   P-80177

2-Allylcyclohexanone
[94-66-6]

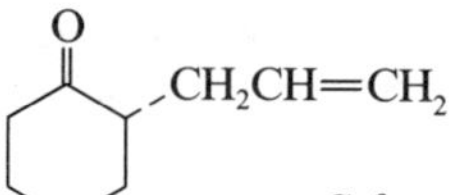

S-form

C_9H_4O M 128.130

(*S*)-*form* [36302-35-9]
Bp_{36} 100°. $[\alpha]_D^{23}$ −2.7° (c, 4.11 in $CHCl_3$). Optical purity 20%.

(±)-*form* [115182-22-4]
Liq. Bp_{12} 79-80°.

Oxime: [59239-07-5].
$C_9H_{15}NO$ M 153.224
Needles. Mp 71°.

2,4-Dinitrophenylhydrazone: Orange needles (EtOH). Mp 149-150°.

[1044-31-1, 59239-07-5]

Cope, A.C. *et al, J. Am. Chem. Soc.*, 1941, **63**, 1842 (*synth*)
Optitz, G. *et al, Justus Liebigs Ann. Chem.*, 1961, **649**, 47 (*synth, ir*)
Hiroi, K. *et al, Chem. Pharm. Bull.*, 1972, **20**, 246 (*synth, cd, ir, ord, abs config*)
Ganem, B., *J. Org. Chem.*, 1975, **40**, 146 (*synth*)
Meyers, A.I. *et al, J. Am. Chem. Soc.*, 1976, **98**, 3032 (*synth*)
Araujo, H.C. *et al, Synthesis*, 1978, **3**, 228 (*synth*)
Murahashi, S.-I. *et al, J. Org. Chem.*, 1988, **53**, 4489 (*synth, ir, pmr*)

2-(2-Propenyl)cyclopentanone, 9CI   P-80178

2-Allylcyclopentanone
[30079-93-7]

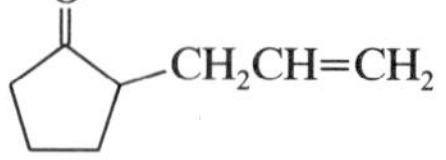

$C_8H_{12}O$ M 124.182

(±)-*form* [55038-60-3]
Liq. with menthyl odour. 188-190°, Bp_{19} 80°.

Semicarbazone: Cryst. Mp 190° (180°).

Colonge, J. *et al, Bull. Soc. Chim. Fr.*, 1953, 1074 (*synth*)
Lorette, N.B. *et al, J. Org. Chem.*, 1961, **26**, 3112 (*synth*)
Oritani, T. *et al, Agric. Biol. Chem.*, 1975, **39**, 89 (*synth, ir*)
Dzhemilev, U.M. *et al, Izv. Akad. Sci. USSR, Ser. Sci. Khim.*, 1988, 378 (*synth*)
Toru, T. *et al, J. Am. Chem. Soc.*, 1988, **110**, 4815 (*synth*)

4-(1-Propenyl)phenol, 9CI   P-80179

Updated Entry replacing P-30187

1-(4-Hydroxyphenyl)-2-propene. p-*Hydroxy-β-methylstyrene.*
Anol

[539-12-8]

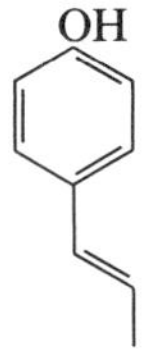

(*E*)-form

$C_9H_{10}O$ M 134.177

(*E*)-form

Leaflets (H_2O). Mp 93°. Bp_{14} 140°. Dec. on standing in
air.

Me ether: [4180-23-8]. *1-Methoxy-4-(1-propenyl)benzene.*
Anethole. *Anistearoptene*
$C_{10}H_{12}O$ M 148.204
Present in anise, fennel and other plant oils. Extensively
used in flavour industry. Leaflets (EtOH). Mp 22.5°. Bp
235°.

▷ Mod. toxic. BZ9275000.

3-Methylbutanoyl: Anol isovalerate
$C_{14}H_{18}O_2$ M 218.295
Constit. of *Coreopsis gigantea*. Oil. $Bp_{0.01}$ 60-65° (bath) .
O-(*3-Methyl-2-butenyl*): [78259-41-3]. **Foeniculin.** *Feniculin*
$C_{14}H_{18}O$ M 202.296
Isol. from fennel and Chinese star anise oils. Mp 23.5-
24.5°.

(*Z*)-form

Me ether: [25679-28-1].
Liq. d_4^{20} 0.989. $Bp_{2.5}$ 63-65°. n_D^{20} 1.5550 (93% pure). Low
level only permitted in flavours.

[104-46-1]

Späth, E. *et al, Ber.*, 1938, **71**, 2708 (*Foeniculin*)
Naves, Y.R., *Helv. Chim. Acta*, 1960, **53**, 230 (*cis*-Anethole)
Briner, E. *et al, Helv. Chim. Acta*, 1963, **46**, 2249 (*ir*)
Klamann, D. *et al, Chem. Ber.*, 1964, **97**, 2534 (*synth*)
Soørensen, J.S. *et al, Acta Chem. Scand.*, 1966, **20**, 992 (*ir*)
Benassi, R. *et al, Org. Magn. Reson.*, 1973, **5**, 391 (*pmr*)
Hamacher, H., *Arch. Pharm. (Weinheim, Ger.)*, 1974, **307**, 309
(*synth*)
Manitto, P. *et al, Tetrahedron Lett.*, 1974, 1567 (*biosynth*)
Okely, H.M. *et al, J. Chem. Soc., Perkin Trans. 1*, 1981, 897
(*Foeniculin*)
Wenkert, E. *et al, Synthesis*, 1983, 701 (*synth*)
Sax, N.I., *Dangerous Properties of Industrial Materials*, 5th Ed.,
Van Nostrand-Reinhold, 1979, 379.

2-(2-Propenyl)-1,3-propanediol, 9CI   P-80180

2-Allyl-1,3-propanediol. 2-Hydroxymethyl-4-penten-1-ol

[42201-43-4]

$$H_2C{=}CHCH_2CH(CH_2OH)_2$$

$C_6H_{12}O_2$ M 116.160
Liq. Bp_{15} 136-139°, $Bp_{1.0}$ 95-98°.

Kharasch, M.S. *et al, J. Org. Chem.*, 1949, **14**, 84 (*synth*)
Wasson, B.K. *et al, Can. J. Chem.*, 1961, **39**, 923 (*synth*)
Courtois, G. *et al, J. Organomet. Chem.*, 1973, **52**, 241 (*synth, ir,
pmr*)
Vite, G.D. *et al, J. Org. Chem.*, 1988, **53**, 2555 (*synth*)

3-(1-Propenyl)pyridine, 9CI   P-80181

1-(3-Pyridyl)-1-propene

C_8H_9N M 119.166

(*E*)-form [52248-76-7]
Bp_{20} 88-90°. n_D^{20} 1.5250.

Brown, H.C. *et al, J. Org. Chem.*, 1988, **53**, 239 (*synth, ir, pmr,
cmr*)

5-(1-Propenyl)-5′-vinyl-2,2′-bithienyl   P-80182

5-Ethenyl-5′-(1-propenyl)-2,2′-bithienyl

[17257-06-6]

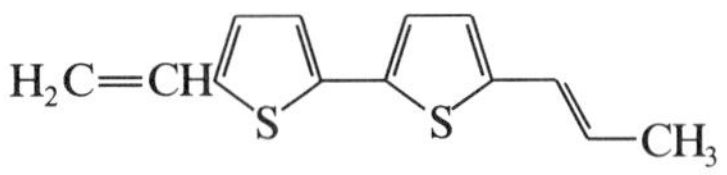

$C_{13}H_{12}S_2$ M 232.370
Isol. from above-ground parts of *Bidens connata*. Light-
yellow cryst. (pet. ether). Mp 46.5-47°.

Bohlmann, F. *et al, Chem. Ber.*, 1965, **98**, 1228; 1966, **99**, 984 (*isol,
uv, ir, pmr, struct, biosynth*)
D'Auria, M. *et al, Gazz. Chim. Ital.*, 1986, **116**, 747 (*synth*)

Prosopidione   P-80183

2,4,4-Trimethyl-6-(3-oxo-1-butenyl)cyclohexanone, 9CI

[120166-32-7]

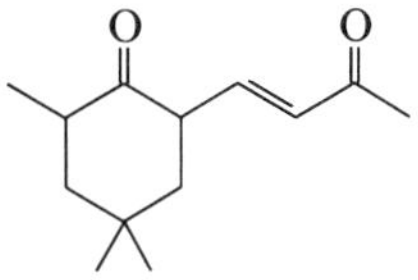

$C_{13}H_{20}O_2$ M 208.300
Constit. of *Prosopis juliflora*. Amorph. powder. Mp 202°
dec. $[\alpha]_D$ −19.2° (MeOH).

Ahmad, V.U. *et al, Phytochemistry*, 1989, **27**, 278.

Pruyanoside A   P-80184

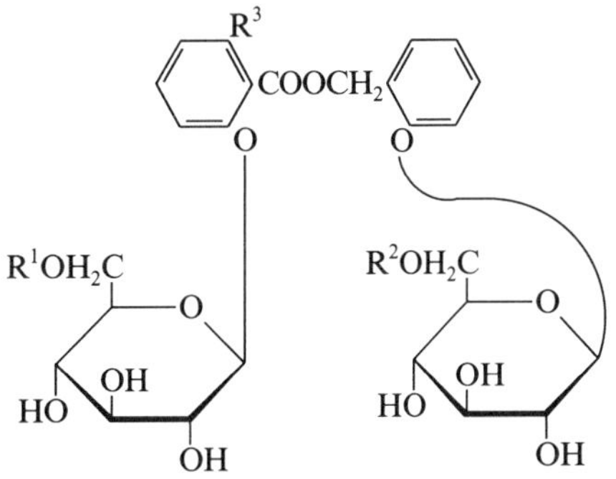

R^1 = PhCO, R^2 = R^3 = H

$C_{33}H_{36}O_{15}$ M 672.638
Constit. of *Prunus grayana*. Needles. Mp 195°. $[\alpha]_D^{28}$ −39.6°
(c, 0.39 in MeOH).

Shimomura, H. *et al, Phytochemistry*, 1989, **28**, 1499.

Pruyanoside B   P-80185

As Pruyanoside A, P-80184 with

R^1 = H, R^2 = PhCO, R^3 = OH

$C_{33}H_{36}O_{16}$ M 688.638

Constit. of *Prunus grayana*. Amorph. powder. $[\alpha]_D^{24}$ −19.3° (c, 1.05 in MeOH).

Shimomura, H. *et al*, *Phytochemistry*, 1989, **28**, 1499.

Psoralidin
P-80186

Updated Entry replacing P-02620
3,9-Dihydroxy-2-prenylcoumestan
[18642-23-4]

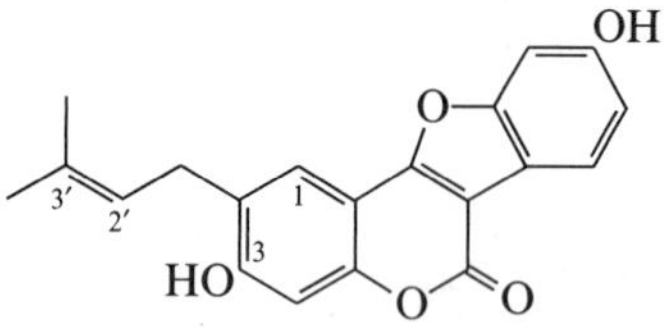

$C_{20}H_{16}O_5$ M 336.343

Constit. of *Psoralea corylifolia*, *Dolichos biflorus* and *Phaseolus lunatus*. Cryst. (Me$_2$CO). Mp 290-292° dec.

Di-Ac: Mp 221-223°.

Di-Me ether: Mp 190-191°.

2′,3′-Epoxide: **Psoralidin oxide**
$C_{20}H_{16}O_6$ M 352.343
Isol. from seeds of *Psoralea corylifolia*. Needles (EtOH). Mp 232-234°. MF incorr. given as $C_{24}H_{20}O_8$.

1-Hydroxy, 3-Me ether: [23013-84-5]. **Glycyrol**
$C_{21}H_{18}O_6$ M 366.370
Isol. from *Glycyrrhiza* sp. root (licorice). Needles (EtOH). Mp 243.5-245°.

1-Methoxy, 3-Me ether: [23013-85-6]. **1-O-Methylglycyrol.**
5-O-Methylglycyrol
$C_{22}H_{20}O_6$ M 380.396
Isol. from *G*. spp. roots. Needles (Me$_2$CO). Mp 259-260.5°.

1-Hydroxy, 3-Me ether, di-Ac: Needles (MeOH). Mp 195-196°.

Chakravarti, K.K. *et al*, *CA*, 1948, **42**, 7492 (*isol*)
Khastgir, H.N. *et al*, *Tetrahedron*, 1961, **14**, 275 (*struct, uv, ir*)
Saitoh, T. *et al*, *Chem. Pharm. Bull.*, 1969, **17**, 729 (*Glycyrol, 1-O-Methylglycyrol*)
Gupta, B.K. *et al*, *Phytochemistry*, 1980, **19**, 2232 (*Psoralidin oxide*)
Keen, N.T. *et al*, *Z. Naturforsch., C*, 1980, **35**, 923 (*isol*)

Puertitol A
P-80187

[119736-67-3]

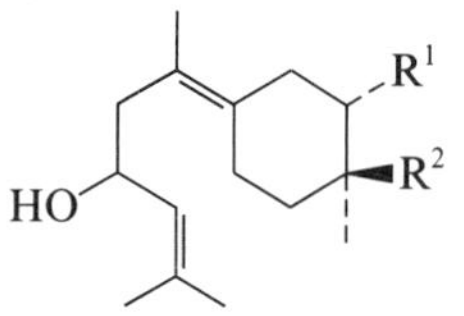

R^1 = Br, R^2 = Cl

$C_{15}H_{24}BrClO$ M 335.711

Constit. of *Laurencia obtusa*. Oil. $[\alpha]_D$ −37° (c, 1.62 in CHCl$_3$).

Vazquez, J.T. *et al*, *J. Nat. Prod. (Lloydia)*, 1988, **51**, 1257 (*isol, pmr, cmr, cd*)

Puertitol B
P-80188

[119736-69-5]

As Puertitol A, P-80187 with

R^1 = Cl, R^2 = Br

$C_{15}H_{24}BrClO$ M 335.711

Constit. of *Laurencia obtusa*. Oil. $[\alpha]_D$ +66° (c, 0.25 in CHCl$_3$).

Vazquez, J.T. *et al*, *J. Nat. Prod. (Lloydia)*, 1988, **51**, 1257 (*isol, pmr, cmr, cd*)

Pulchellin A
P-80189

Updated Entry replacing P-70138
Pulchellin
[6754-35-4]

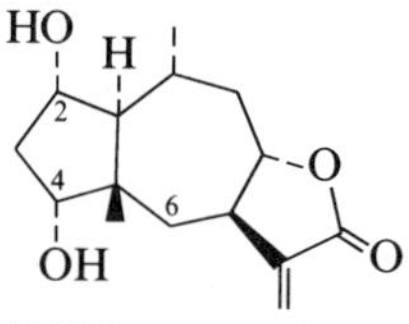

$C_{15}H_{22}O_4$ M 266.336

Constit. of *Gaillardia pulchella*. Cryst. (EtOAc). Mp 166-167°. $[\alpha]_D^{26}$ −36.2° (c, 2.43 in CHCl$_3$).

2,4-Di-Ac: [23754-36-1]. **Pulchellin diacetate**
$C_{19}H_{26}O_6$ M 350.411
From *Loxochysanus sinuatus*. Oil.

2-O-(3-Methylbutanoyl): **Pulchellin 2-O-isovalerate**
$C_{20}H_{32}O_5$ M 352.470
From *L. sinuatus*. Oil. $[\alpha]_D^{26}$ −13° (c, 0.43 in CHCl$_3$).

2-Tigloyl: [97605-30-6]. **Pulchellin 2-O-tiglate**
$C_{20}H_{28}O_5$ M 348.438
From *L. sinuatus*. Oil.

6β-Hydroxy, 2-(3-Methylbutanoyl): [97605-31-7]. **6β-Hydroxypulchellin 2-O-isovalerate**
$C_{20}H_{30}O_6$ M 366.453
From *L. sinuatus*. Cryst. Mp 127°.

6β-Hydroxy, 2-Ac: [97643-92-0]. **6β-Hydroxypulchellin 2-O-acetate**
$C_{17}H_{24}O_6$ M 324.373
From *L. sinuatus*. Cryst. Mp 87°. $[\alpha]_D^{24}$ +14° (c, 0.97 in MeOH).

4-Epimer: **4-Epipulchellin**
$C_{15}H_{24}O_4$ M 268.352
Constit. of *G. pulchella*. Oil.

4-Epimer, di-Ac: Cryst. Mp 133-134°.

6α-Hydroxy: **6α-Hydroxypulchellin**
$C_{15}H_{22}O_5$ M 282.336
Constit. of *G. pulchella*.

6α-Hydroxy, O^4-angeloyl: **6α-Hydroxypulchellin 4-O-angelate**
$C_{20}H_{28}O_6$ M 364.438
Constit. of *G. pulchella*.

6β-Acetoxy, O^4-angeloyl: **6β-Acetoxypulchellin 4-O-angelate**
$C_{22}H_{30}O_7$ M 406.475
Constit. of *G. pulchella*.

2-Deoxy, 4-epimer: **2-Deoxy-4-epipulchellin**
$C_{15}H_{22}O_3$ M 250.337
Constit. of *Ondetia linearis*. Cryst. Mp 135°. $[\alpha]_D^{24}$ +43° (c, 0.59 in CHCl$_3$).

Herz, W. *et al*, *Tetrahedron*, 1963, **19**, 483 (*isol, uv, ir, pmr*)
Aota, K. *et al*, *J. Org. Chem.*, 1970, **35**, 1448 (*isol*)
Inayama, S. *et al*, *Heterocycles*, 1985, **23**, 377 (*synth*)
Bohlmann, F. *et al*, *Phytochemistry*, 1985, **24**, 1021 (*isol*)
Harimaya, K. *et al*, *Heterocycles*, 1988, **27**, 83 (*epimer*)
Yu, S. *et al*, *Phytochemistry*, 1988, **27**, 2887 (*derivs*)
Zdero, C. *et al*, *Phytochemistry*, 1989, **28**, 1653 (*derivs*)

Pulicaral
P-80190

*5,6-Dihydro-1,1,4-trimethyl-9-oxo-1*H,4H-3a,6a-
propanopentalene-3-carboxaldehyde, 9CI

[119767-05-4]

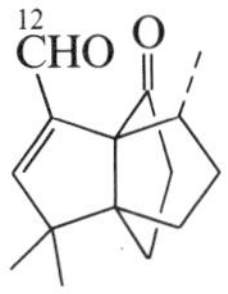

$C_{15}H_{20}O_2$ M 232.322

Constit. of *Pulicaria paludosa*. Cryst. (CH_2Cl_2). Mp 70°.
$[\alpha]_D^{24}$ −58.9° (c, 0.3 in $CHCl_3$).

12-Carboxylic acid: [119767-06-5]. **Pulicaric acid**
$C_{15}H_{20}O_3$ M 248.321
Constit. of *P. paludosa*. Oil. $[\alpha]_D^{24}$ −57.6° (c, 0.7 in
$CHCl_3$).

San Feliciano, A. *et al, J. Nat. Prod.* (*Lloydia*), 1988, **51**, 1153
(*isol, pmr, cmr*)

Pulverulide
P-80191

Updated Entry replacing P-30207

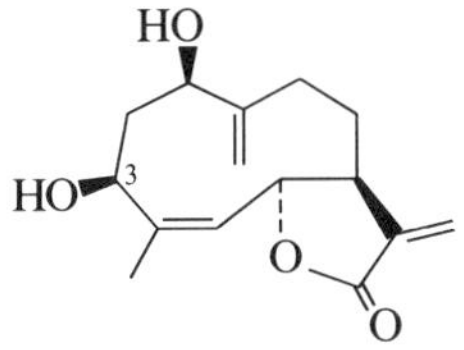

$C_{15}H_{20}O_4$ M 264.321

Constit. of *Leucanthemopsis pulverulenta*. Oil. $[\alpha]_D$ +8.4°
(c, 0.83 in $CHCl_3$).

3-Ac: 3-Acetylpulverulide
$C_{17}H_{22}O_5$ M 306.358
Constit. of *Stevia alpina*. Gum.

De Pascual Teresa, J. *et al, Phytochemistry*, 1984, **23**, 1178 (*isol,
struct*)
de Heluani, C.S. *et al, Phytochemistry*, 1989, **28**, 1931 (*deriv*)

Pulviquinone A
P-80192

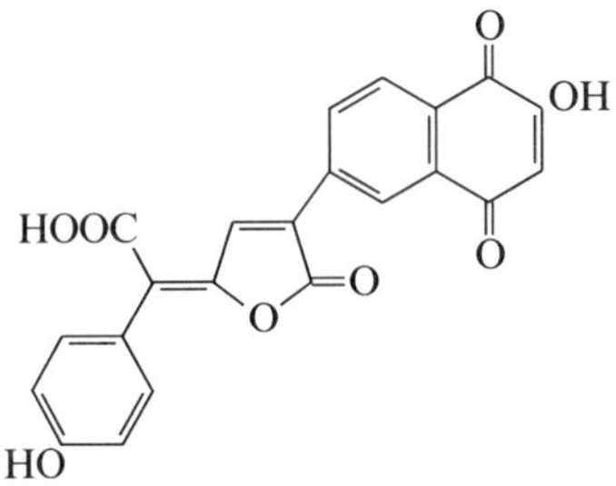

$C_{22}H_{12}O_9$ M 420.331

Pigment from the bay boletus *Xerocomus badius*. λ_{max}
265sh, 278, 312, 354, 422 nm. Isol. as the Me ester, an
artifact of methylation.

[90295-71-9]

Steffan, B. *et al, Angew. Chem., Int. Ed. Engl.*, 1984, **23**, 445 (*isol,
struct*)

Pumilaisoflavone C
P-80193

4′,6,8-Trihydroxy-3′,5′-dimethoxy-2′,7-diprenylisoflavone

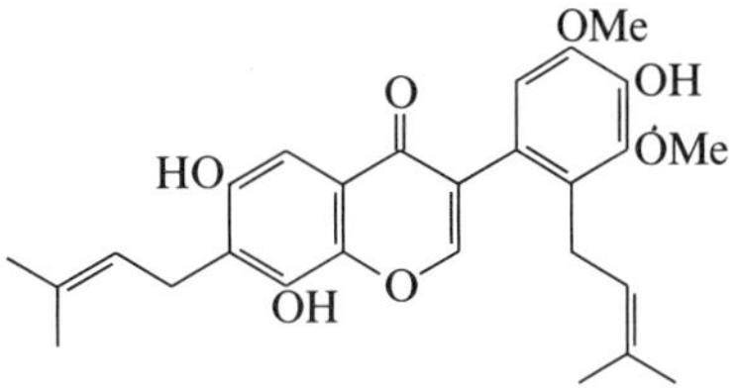

$C_{27}H_{30}O_7$ M 466.530
Constit. of *Tephrosia pumila*. Needles (MeOH). Mp 180-
182°.

Yenesew, A. *et al, Phytochemistry*, 1989, **28**, 1291.

Pumilaisoflavone D
P-80194

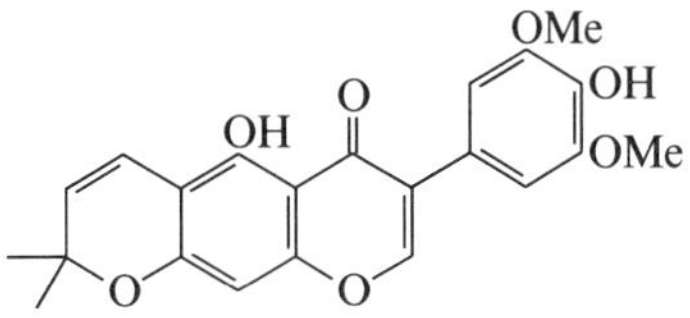

$C_{22}H_{20}O_7$ M 396.396
Constit. of *Tephrosia pumila*. Needles (MeOH). Mp 174-
176°.

Yenesew, A. *et al, Phytochemistry*, 1989, **28**, 1291.

Pygmaeocin A
P-80195

[122590-05-0]

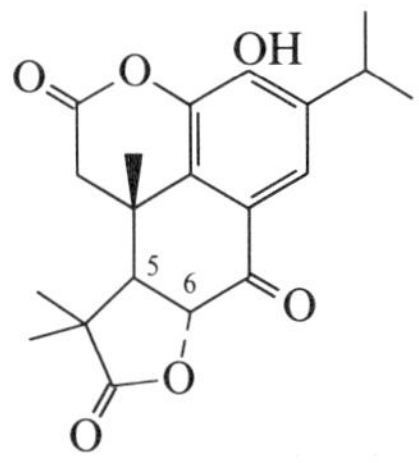

$C_{20}H_{22}O_6$ M 358.390
Constit. of *Pygmaeopremna herbacea*. Cryst. (MeOH). Mp
281-283°.

5,6-Didehydro: [122590-06-1]. *5,6-Didehydropygmaeocin A*
$C_{20}H_{20}O_6$ M 356.374
Constit. of *P. herbacea*. Yellow cryst. (MeOH). Mp 211-
213°.

Chen, W. *et al, J. Nat. Prod.* (*Lloydia*), 1989, **52**, 581 (*isol, pmr,
cmr*)

Pyrano[3,4-*b*]indol-3(9*H*)-one, 9CI
P-80196

[4375-21-7]

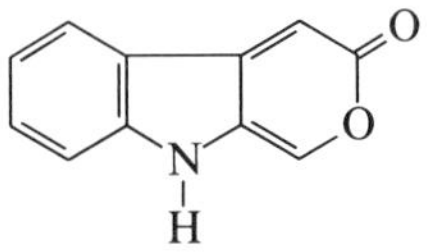

$C_{11}H_7NO_2$ M 185.182
Mp *ca.* 165° (darkens). Numerous 1-subst. derivs.
prepared.

Moody, C.J. *et al, J. Chem. Soc., Perkin Trans. 1*, 1988, 1407
(*synth, ir, pmr, ms*)

1*H*-Pyrazole-3,4-dicarboxylic acid, 9CI   P-80197

Updated Entry replacing P-02758

[35344-99-1]

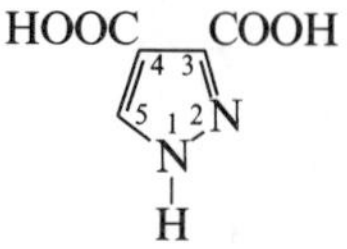

$C_5H_4N_2O_4$ M 156.098

Needles + $1H_2O$ (dil. HNO_3). Mp >300° (anhyd.), Mp 260° dec. (monohydrate). Forms gels on pptn. from H_2O.

Di-Me ester: [33090-46-9].
$C_7H_8N_2O_4$ M 184.151
Needles (C_6H_6/pet. ether). Mp 141°.

Diamide: [21272-56-0].
$C_5H_6N_4O_2$ M 154.128
Cryst. (AcOH). Mp 260° dec.

Dinitrile: [33146-98-4]. *3,4-Dicyanopyrazole*
$C_5H_2N_4$ M 118.098
Cryst. (EtOAc/pet. ether). Mp 194°.

N-Ac:
$C_7H_6N_2O_5$ M 198.135
Mp 169°.

3-Et ester:
$C_7H_8N_2O_4$ M 184.151
Cryst. (EtOH). Mp 263° dec.

1-Benzyl, di-Me ester: [7189-02-8].
$C_{14}H_{14}N_2O_4$ M 274.276
Light yellow oil. $Bp_{.001}$ 185°.

1-Me, di-Et ester: [10514-60-0].
$C_{10}H_{14}N_2O_4$ M 226.232
Yellow oil. $Bp_{0.5}$ 135-145°.

1-Me: [10505-21-2].
$C_6H_6N_2O_4$ M 170.124
Cryst. (H_2O). Mp 239-241° (233°).

1-Me, di-Me ester: [22050-80-2].
$C_8H_{10}N_2O_4$ M 198.178
Needles (pet. ether). Mp 68-69° (67°).

4-Me ester, 3-nitrile: [33090-69-6]. *Methyl 3-cyanopyrazole-4-carboxylate*
$C_6H_5N_3O_2$ M 151.124
Cryst. (EtOAc/pet. ether). Mp 186°.

1-Benzyl: [19311-81-0].
$C_{12}H_{10}N_2O_4$ M 246.222
Mp 195.5-197° dec.

Di-Et ester: [37687-26-6].
$C_9H_{12}N_2O_4$ M 212.205
Cryst. (C_6H_6/pet. ether). Mp 69-70°.

[34982-39-3]

Jones, R.G. et al, J. Org. Chem., 1955, **20**, 1342 (synth, Et esters)
Acheson, R.M. et al, J. Chem. Soc., 1963, 1008 (Me ester)
Huisgen, R. et al, Chem. Ber., 1968, **101**, 536 (N-Ph, N-Benzyl)
Bastide, J. et al, Bull. Soc. Chim. Fr., 1971, 1336 (nitriles, amide, N-Me)
Greco, C.V. et al, J. Chem. Soc., Perkin Trans. 1, 1972, 720 (esters)
Winter, W. et al, Chem. Ber., 1974, **107**, 2127 (N-Me)

1*H*-Pyrazole-3,5-dicarboxylic acid, 9CI   P-80198

Updated Entry replacing P-02759

Mp >360°.

Monohydrate: Needles (HCl aq.). Mp 287-290° dec.

Di-Me ester: [4077-76-3].

$C_7H_9N_2O_4$ M 185.159
Plates (Et_2O). Mp 151-155°.

N-Me: [75092-39-6].
$C_6H_6N_2O_4$ M 170.124
Prisms (H_2O). Mp 266-268° (249.5-250° dec.).

N-Me, 3-Me ester: [117860-55-6].
$C_7H_8N_2O_4$ M 184.151
Needles (MeOH aq.). Mp 129-130.5°.

N-Me, 5-Me ester: [117860-56-7].
$C_7H_8N_2O_4$ M 184.151
Cryst. (EtOAc). Mp 172-172.5°.

N-Me, Di-Me ester: [33146-99-5].
$C_8H_{10}N_2O_4$ M 198.178
Needles (pet. ether). Mp 72-73.5°.

N-Et:
$C_7H_8N_2O_4$ M 184.151
Mp 243°.

N-Et, Di-Me ester:
$C_8H_{10}N_2O_4$ M 198.178
Mp 66°.

N-Ac, Di-Me ester:
$C_9H_{10}N_2O_5$ M 226.188
Mp 84.5-85°.

Di-Et ester: [37687-24-4].
$C_9H_{12}N_2O_4$ M 212.205
Cryst. (C_6H_6/EtOAc). Mp 49-51°.

Buchner, E. et al, Justus Liebigs Ann. Chem., 1893, **273**, 232, 246 (synth, Di-Me ester)
Knorr, L. et al, Justus Liebigs Ann. Chem., 1894, **279**, 218.
Gray, T., Ber., 1900, **33**, 1220.
v. Auwers, K. et al, J. Prakt. Chem., 1930, **126**, 196; 177 (N-Ac)
Eidebenz, E. et al, Arch. Pharm. (Weinheim, Ger.), 1943, **281**, 171 (N-Et derivs, amides)
Daeniker, H.U. et al, Helv. Chim. Acta, 1963, **46**, 805 (amide-ester)
Makabe, O. et al, Bull. Chem. Soc. Jpn., 1975, **48**, 3210 (Di-Et ester)
Denny, W.A., J. Org. Chem., 1989, **54**, 428 (synth, N-Me and esters)

1*H*-Pyrazole-4,5-dicarboxylic acid   P-80199

[31962-35-3]

$C_5H_4N_2O_4$ M 156.098

1-Unsubstituted derivs. tautomeric with the 3,4-dicarboxylic acid,

1-Me: [10505-19-8].
$C_6H_6N_2O_4$ M 170.124
Cryst. (H_2O). Mp 182-184° (176-177°). Early work refers to a dihydrate, mp. 233-235°; this is apparently incorrect.

1-Me, Di-Me ester: [33090-52-7].
$C_8H_{10}N_2O_4$ M 198.178
Cryst. (pet. ether). Mp 32-33°.

1-Me, Di-Et ester: [10514-61-1].
$C_{10}H_{14}N_2O_4$ M 226.232
Oil. $Bp_{0.5}$ 100-105°.

Bauer, L. et al, J. Org. Chem., 1966, **31**, 2491 (N-Me, Et ester)
Winter, W. et al, Chem. Ber., 1974, **107**, 2127 (N-Me)

3,5-Pyrazolidinedione, 9CI, 8CI   P-80200

N,N′-*Malonylhydrazine*

[4744-71-2]

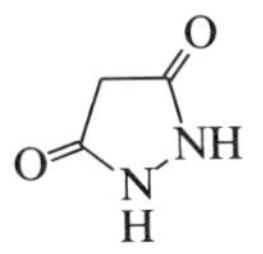

$C_3H_4N_2O_2$ M 100.077
Cryst. (MeOH). Mp 290° dec.

1-Ph: [19933-22-3].
  $C_9H_8N_2O_2$ M 176.174
  Yellow plates (EtOH). Mp 192°.

1,2-Dipropyl: [93847-23-5].
  $C_9H_{16}N_2O_2$ M 184.238
  Yellow, oily solid.

1,2-Dibenzyl: [93847-22-4].
  $C_{17}H_{16}N_2O_2$ M 280.326
  Yellow cryst. (MeCN). Mp 128.5°.

1,2-Di-Ph: [2652-77-9].
  $C_{15}H_{12}N_2O_2$ M 252.272
  Plates (EtOH). Mp 178-179°.

Conrad, M. *et al, Chem. Ber.*, 1906, **39**, 2283 (*phenyl*)
McGee, M.A. *et al, J. Chem. Soc.*, 1960, 1989 (*diphenyl*)
Selva, A. *et al, Org. Mass Spectrom.*, 1975, **10**, 606 (*ms, deriv*)
Dubau, F.-P., *Chem. Ber.*, 1983, **116**, 2714 (*synth, pmr*)
Fritsch, G. *et al, Arch. Pharm.* (*Weinheim, Ger.*), 1986, **319**, 70 (*synth, cryst struct*)
Lawton, G. *et al, J. Chem. Soc., Perkin Trans.* 1, 1987, 877 (*derivs, synth, ir, pmr*)

Pyrazolo[5,1-*c*][1,2,4]triazin-4(1*H*)-one   P-80201

[112298-50-7]

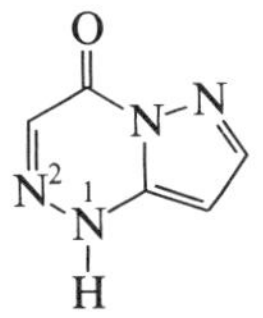

$C_5H_4N_4O$ M 136.113
Mp >300°.

1-Me: [112298-53-0].
  $C_6H_6N_4O$ M 150.140
  Mp 185-187°.

2-Me: [112298-54-1].
  $C_6H_6N_4O$ M 150.140
  Mp 180-183°. Zwitterionic.

Farràs, J. *et al, J. Org. Chem.*, 1988, **53**, 887 (*synth, pmr, cmr*)

1,3-Pyrenedicarboxaldehyde   P-80202

1,3-Diformylpyrene

[120915-14-2]

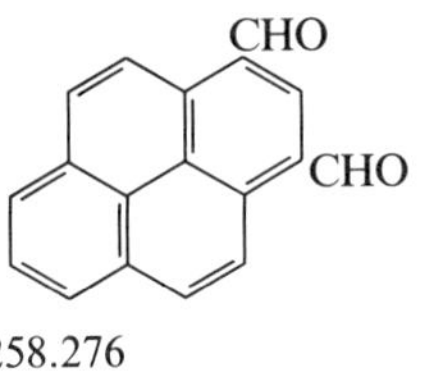

$C_{18}H_{10}O_2$ M 258.276
Yellow needles (CH_2Cl_2). Mp 245-246° (sealed tube).

Goltz, M. *et al, Bull. Chem. Soc. Jpn.*, 1988, **61**, 3767 (*synth, uv, ms, pmr*)

Pyrenocine A   P-80203

4-*Methoxy-6-methyl-5-(1-oxo-2-butenyl)*2-H-*pyran-2-one*, *9CI. Citreopyrone*

[76868-97-8]

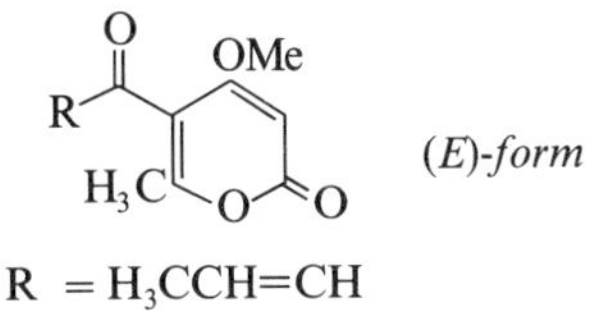

R = $H_3CCH=CH$ (E)-*form*

$C_{11}H_{12}O_4$ M 208.213
Formerly assigned furanoid struct. Isol. from *Pyrenochaeta terrestris* and *Penicillium citreoviride* and *Alternaria* sp. nov. Plant growth inhibitor. Cryst. (MeOH). Mp 110.9-111.6°.

1′-Alcohol: [94474-69-8]. ***Pyrenocine C***
  $C_{11}H_{14}O_4$ M 210.229
  From *P. terrestris*. Weak phytotoxin. Needles (Et_2O). Mp 81.5-82.0°. $[\alpha]_D^{20}$ +0.3° (c, 0.37 in CHCl$_3$). Virtually racemic.

2′,3′-Dihydro, 3′-hydroxy: [72674-29-4]. ***Pyrenocine B***
  $C_{11}H_{14}O_5$ M 226.229
  Metab. of *P. terrestris*. Phytotoxin. Cryst. (Et_2O/hexane). Mp 103-103.5°.

[72674-30-7]

Sato, H. *et al, Agric. Biol. Chem.*, 1979, **43**, 2409 (*isol*)
Niwa, M. *et al, Tetrahedron Lett.*, 1980, 4483 (*isol*)
Sato, H. *et al, Agric. Biol. Chem.*, 1981, **45**, 795 (*cryst struct*)
Sparace, S.A. *et al, Phytochemistry*, 1984, **23**, 2693 (*deriv*)
Shizuri, Y. *et al, Tetrahedron Lett.*, 1984, **25**, 1583 (*biosynth*)
Tal, B. *et al, Z. Naturforsch., C*, 1986, **41**, 1032 (*isol*)
Sporace, S.A. *et al, Can. J. Microbiol.*, 1987, **33**, 327 (*props*)
Ichihara, A. *et al, Tetrahedron*, 1987, **43**, 5245 (*synth*)

8*H*-Pyreno[2,1-*b*]pyran-8-one   P-80204

[115560-48-0]

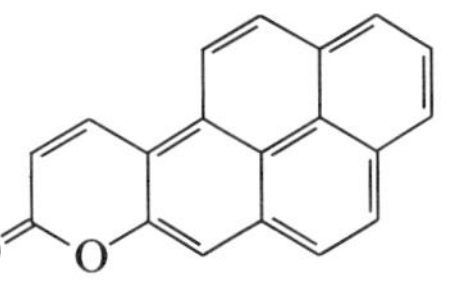

$C_{19}H_{10}O_2$ M 270.287
Cryst. Mp 245-246°.

Harvey, R.G. *et al, J. Org. Chem.*, 1988, **53**, 3936 (*synth, pmr*)

9*H*-Pyreno[1,2-*b*]pyran-9-one   P-80205

[115560-47-9]

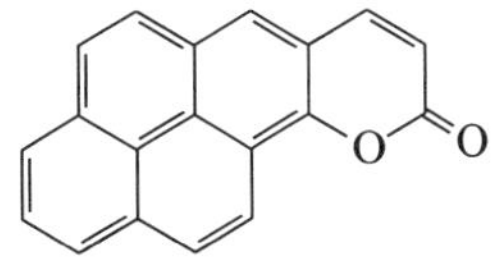

$C_{19}H_{10}O_2$ M 270.287
Cryst. (C_6H_6/hexane). Mp 208-209°.

Harvey, R.G. *et al, J. Org. Chem.*, 1988, **53**, 3936 (*synth, pmr*)

Pyrenoside
P-80206

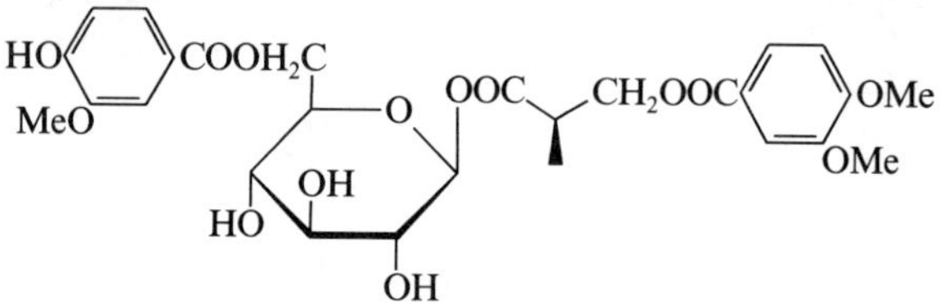

$C_{27}H_{32}O_{14}$ M 580.541
Constit. of *Gentiana pyrenaica*. Amorph. powder.

Garcia, J. *et al, J. Nat. Prod.* (*Lloydia*), 1989, **52**, 996 (*isol, pmr, cmr*)

Pyridazino[4,5-*b*]quinolin-10(5*H*)-one, 9CI
P-80207
[115276-33-0]

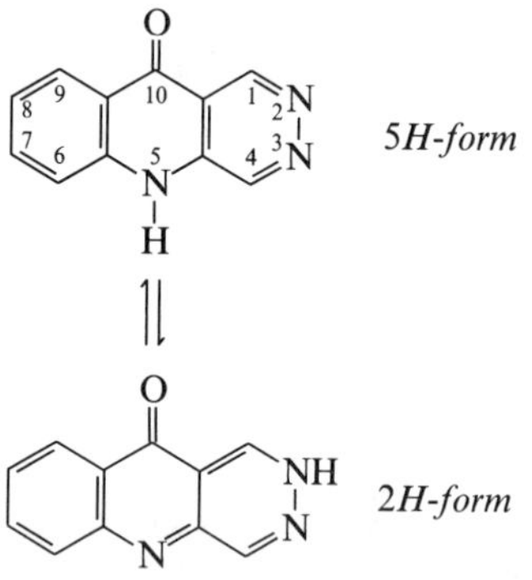

5*H*-form

2*H*-form

$C_{11}H_7N_3O$ M 197.196
5*H*-form
Solid (DMF/MeOH). Mp >300° dec.
5-*Me*: [115276-39-6].
$C_{12}H_9N_3O$ M 211.223
Pale yellow cryst. (butanone). Mp 279-280°.
5-*Et*: [115276-40-9].
$C_{13}H_{11}N_3O$ M 225.249
Pale yellow needles (Me$_2$CO/diisopropyl ether). Mp 214-215°.
2*H*-form
2-*Me*: [115276-34-1].
$C_{12}H_9N_3O$ M 211.223
Yellow needles (propanol). Mp 275-278° dec.
2-*Benzyl*: [115276-35-2].
$C_{18}H_{13}N_3O$ M 287.320
Yellow needles (EtOH). Mp 240-243°.

Haider, N. *et al, J. Chem. Soc., Perkin Trans.* 1, 1988, 401 (*synth, deriv, pmr, ir, uv*)

2-Pyridinesulfinic acid
P-80208
[24367-66-6]

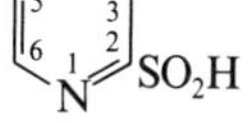

$C_5H_5NO_2S$ M 143.166
Obt. as cryst. Na salt.

Kamiyama, T. *et al, Chem. Pharm. Bull.*, 1988, **36**, 2652.

4-Pyridinesulfinic acid
P-80209
[116008-37-8]
$C_5H_5NO_2S$ M 143.166
Obt. as cryst. Na salt.

Kamiyama, T. *et al, Chem. Pharm. Bull.*, 1988, **36**, 2652.

Pyridinium fluorochromate
P-80210

$$C_5H_5NH^{\oplus} \ CrO_3F^{\ominus}$$

$C_5H_6CrFNO_3$ M 199.102
Selective oxidising agent. Bright orange cryst. (106-108°).
▷ Hazardous synth.

Bhattacharjee, M.N. *et al, Synthesis*, 1982, 588 (*synth.*)
Bhattacharjee, M.N. *et al, Tetrahedron*, 1987, **43**, 5389 (*use*)

[6](2,6)Pyridinophane
P-80211
12-Azabicyclo[6.3.1]dodeca-1(12),8,10-triene, 9CI.
[6](2,6)Pyridophane
[56912-79-9]

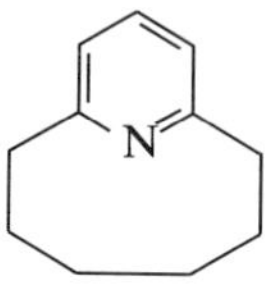

$C_{11}H_{15}N$ M 161.246
Oil. Bp$_{1.9}$ 65°.
B,HClO$_4$: Mp 90-93° (77-78°).
N-*Oxide:* [118537-97-6].
$C_{11}H_{15}NO$ M 177.246
Cryst. (pentane). Mp 83°.

Weber, H. *et al, Chem. Ber.*, 1985, **118**, 4259; 1989, **122**, 945.

[7](2,6)Pyridinophane
P-80212
13-Azabicyclo[7.3.1]trideca-1(13),9,11-triene, 9CI.
[7](2,6)Pyridophane
[22776-72-3]
$C_{12}H_{17}N$ M 175.273
Oil. Bp$_{1.5}$ 90-100°.
B,HClO$_4$: Mp 132-133°.
N-*Oxide:* [118537-98-7].
$C_{12}H_{17}NO$ M 191.272
Cryst. (pentane). Mp 63-65°.

Weber, H. *et al, Chem. Ber.*, 1985, **118**, 4259; 1989, **122**, 945.

[8](2,6)Pyridinophane
P-80213
14-Azabicyclo[8.3.1]tetradeca-1(14),10,12-triene, 9CI.
[8](2,6)Pyridophane
[56912-80-2]
$C_{13}H_{19}N$ M 189.300
Oil. Bp$_{1.0}$ 100-110°.
B,HClO$_4$: Mp 181-182°.
N-*Oxide:* [118537-99-8].
$C_{13}H_{19}NO$ M 205.299
Cryst. (pentane). Mp 76°.

Weber, H. *et al, Chem. Ber.*, 1985, **118**, 4259; 1989, **122**, 945.

[9](2,6)Pyridinophane
P-80214
15-Azabicyclo[9.3.1]pentadeca-1(15),11,13-triene, 9CI.
[9](2,6)Pyridophane
[56929-81-8]
$C_{14}H_{21}N$ M 203.327

$Bp_{1.0}$ 120-130°.

B,HClO$_4$: Mp 174°.

N-*Oxide:*
  $C_{14}H_{21}NO$ M 219.326
  Cryst. (pentane). Mp 70°.

Weber, H. *et al, Chem. Ber.,* 1985, **118**, 4259; 1989, **122**, 945.

[10](2,6)Pyridinophane   P-80215

16-Azabicyclo[10.3.1]hexadeca-1(16),12,14-triene, 9CI.
[10](2,6)*Pyridophane*
[4432-68-2]
$C_{15}H_{23}N$ M 217.353
Oil. $Bp_{1.5}$ 140°.

B,HClO$_4$: Mp 174-175°.

N-*Oxide:* [118538-01-5].
  $C_{15}H_{23}NO$ M 233.353
  Cryst. (pentane). Mp 78-80° (73-74°).

Weber, H. *et al, Chem. Ber.,* 1985, **118**, 4259; 1989, **122**, 945.

[12](2,6)Pyridinophane   P-80216

18-Azabicyclo[12.3.1]octadeca-1(18),14,16-triene, 9CI.
[12](2,6)*Pyridophane*
[56912-81-3]
$C_{17}H_{27}N$ M 245.407
$Bp_{1.2}$ 170-190°.

B,HClO$_4$: Cryst. (Me$_2$CO/Et$_2$O). Mp 146-148°.

Weber, H. *et al, Chem. Ber.,* 1985, **118**, 4259.

1*H*-Pyrido[3,4,5-*de*]quinazoline   P-80217

[110748-36-2]

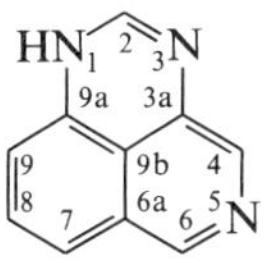

$C_{10}H_7N_3$ M 169.185
Yellow cryst. (MeOH aq.). Mp 246-249°.

Woodgate, P.D. *et al, Heterocycles,* 1987, **26**, 1029 (*synth, pmr, ms*)

1*H*-Pyrido[4,3,2-*de*]quinazoline   P-80218

6-Azaperimidine
[111493-32-4]

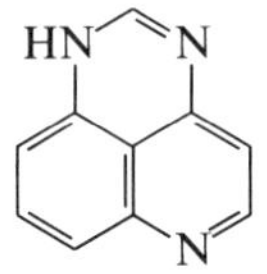

$C_{10}H_7N_3$ M 169.185
Bright yellow needles by subl. Mp 241-244° dec.

Woodgate, P.D. *et al, Heterocycles,* 1987, **26**, 1029 (*synth, ir, pmr, ms*)

1-(2-Pyridyl)-2-pyridone   P-80219

[1(2H),2′-Bipyridin]-2-one, 8CI
[3480-65-7]

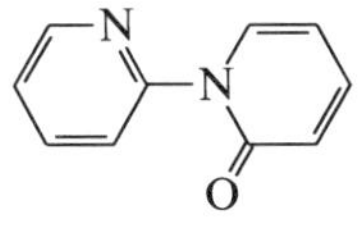

$C_{10}H_8N_2O$ M 172.186
Needles (pet. ether). Mp 55°. $Bp_{0.7}$ 120-130°.

Picrate: Mp 116-118°.

Ramirez, F. *et al, Chem. Ind. (London),* 1957, 46 (*synth*)
de Villiers, P.A. *et al, Recl. Trav. Chim. Pays-Bas (J. R. Neth. Chem. Soc.),* 1957, **76**, 647 (*synth, uv*)
Wibaut, J.P. *et al, Bull. Soc. Chim. Fr.,* 1958, 424 (*synth*)
Ramirez, F. *et al, J. Am. Chem. Soc.,* 1959, **81**, 156 (*synth, ir, uv*)
von Ostwalden, P.W. *et al, J. Org. Chem.,* 1971, **36**, 3792 (*synth, pmr*)
Del Giudice, M.R. *et al, Tetrahedron,* 1984, **40**, 4067 (*ir, pmr*)

1-(2-Pyridyl)-4-pyridone   P-80220

[1(4H),2′-Bipyridin]-4-one, 9CI
[76520-27-9]
$C_{10}H_8N_2O$ M 172.186
Needles (C$_6$H$_6$). Mp 164-165°.

de Villiers, P.A. *et al, Recl. Trav. Chim. Pays-Bas (J. R. Neth. Chem. Soc.),* 1957, **76**, 647 (*synth, uv*)
Matsumura, E. *et al, Bull. Chem. Soc. Jpn.,* 1980, **53**, 2891; 1984, **57**, 1961 (*synth*)

1-(3-Pyridyl)-4-pyridone   P-80221

[1(4H),3′-Bipyridin]-4-one, 9CI
[34847-11-5]
$C_{10}H_8N_2O$ M 172.186
Cryst. (CHCl$_3$). Mp 189-191°.

B, 2HCl: [34847-10-4].
  Needles (H$_2$O). Mp 192-195°.

Fleet, G.W.J. *et al, J. Chem. Soc. C,* 1971, 3948 (*synth, ms, pmr*)

1-(4-Pyridyl)-2-pyridone   P-80222

[1(2H),4′-Bipyridin]-2-one
[62150-42-9]
$C_{10}H_8N_2O$ M 172.186
Blunt needles (C$_6$H$_6$). Mp 155-158°. $Bp_{0.1}$ 125-135°.

de Villiers, P.A. *et al, Recl. Trav. Chim. Pays-Bas (J. R. Neth. Chem. Soc.),* 1957, **76**, 647 (*synth, deriv*)
Witek, W., *Pol. J. Chem. (Rocz. Chem.),* 1976, **50**, 1625 (*synth*)

1-(4-Pyridyl)-4-pyridone   P-80223

[1(4H),4′-Bipyridin]-4-one
[3881-38-7]
$C_{10}H_8N_2O$ M 172.186
Cryst. (C$_6$H$_6$, EtOH, or Me$_2$CO/pet. ether). Mp 177-178°.
Forms a monohydrate.

B, 2HCl: Mp 185-187°.

Thyagarajan, B. *et al, Chem. Ber.,* 1966, **99**, 368 (*synth, ir, uv*)
Oae, S. *et al, Bull. Chem. Soc. Jpn.,* 1967, **40**, 1420 (*synth*)
Kobayashi, Y. *et al, Chem. Pharm. Bull.,* 1970, **18**, 2489 (*synth, pmr, uv*)
Fleet, G.W.J. *et al, J. Chem. Soc. C,* 1971, 3948 (*synth, pmr*)
Kappe, T. *et al, Monatsh. Chem.,* 1972, **103**, 426 (*synth, ir*)
Del Giudice, M.R. *et al, Tetrahedron,* 1984, **40**, 4067 (*cmr, ir, pmr*)

3,4-Pyridyne   P-80224

3,4-Didehydropyridine, 9CI
[7129-66-0]

C_5H_3N M 77.085

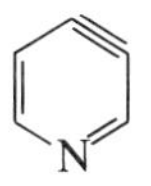

Transient intermediate. Obs. by ir at 13K in N_2 or Ar matrices.

Reinecke, M.G., *Tetrahedron*, 1982, **38**, 427 (*rev*)
Nam, H.-H. *et al*, *J. Am. Chem. Soc.*, 1988, **110**, 4096 (*synth, ir*)

11H-Pyrimidino[4,5-a]carbazole   P-80225

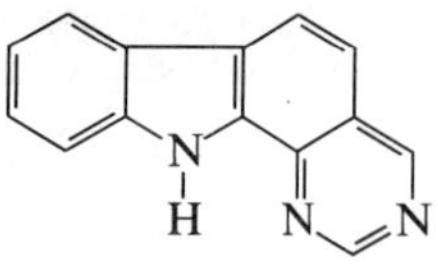

$C_{14}H_9N_3$ M 219.245
Yellow cryst. Mp 218°.

Ganzengel, J.M. *et al*, *Chem. Pharm. Bull.*, 1989, **37**, 1500 (*synth*)

2-Pyrrolethiolcarboxylic acid   P-80226

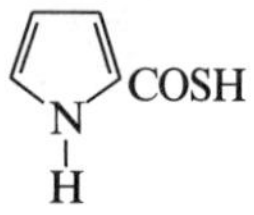

C_5H_5NOS M 127.167
Me ester: [118792-46-4].
  C_6H_7NOS M 141.193
  Cryst. (pet. ether). Mp 79-80°.
1-Me, Me ester: [118792-47-5].
  C_7H_9NOS M 155.220
  Liq. Bp_{24} 131-132°.

Barbero, M. *et al*, *Synthesis*, 1988, 300 (*synth, pmr, cmr, ir*)

3H-Pyrrolo[1,2-a]indol-3-one   P-80227
[24009-76-5]

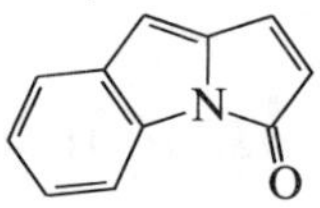

$C_{11}H_7NO$ M 169.182
Cryst. by subl. Mp 94-95°.

Auerbach, J. *et al*, *J. Org. Chem.*, 1971, **36**, 31 (*synth*)
Flitsch, W. *et al*, *Justus Liebigs Ann. Chem.*, 1989, 239 (*synth*)

2H-Pyrrol-2-one   P-80228
2-Aza-2,4-cyclopentadienone

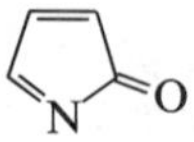

C_4H_3NO M 81.074
Transient species, which acts as diene or dienophile in Diels-Alder reactions.

Gaviña, F. *et al*, *J. Am. Chem. Soc.*, 1988, **110**, 4017.

Q

Quinacridone   Q-80001

5,12-Dihydroquino[2,3-b]acridine-7,14-dione. lin-trans-*Quinacridone. CI Pigment Violet 19*

[1047-16-1]

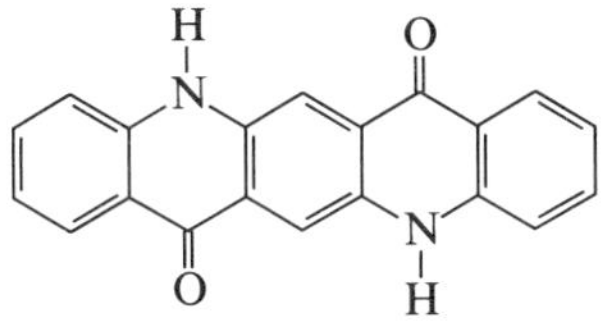

$C_{20}H_{12}N_2O_2$ M 312.327

Pigment, and parent of a family of pigments. Red-violet cryst., yellow in vapour phase. Insol. most solvents. Mp 390°.

Labana, S.S. *et al, Chem. Rev.,* 1967, **67**, 1 (*rev, uv*)
Kimura, S., *Makromol. Chem.,* 1968, **117**, 203 (*synth, ir*)
Chung, F.H. *et al, J. Appl. Crystallogr.,* 1971, **4**, 506 (*cryst struct*)
Altiparmakian, R.H. *et al, Helv. Chim. Acta,* 1972, **55**, 85 (*pmr*)
Yokoyama, Y., *Bull. Chem. Soc. Jpn.,* 1983, **56**, 1775 (*pmr, uv*)

2,3-Quinolinedicarboxylic acid, 9CI   Q-80002

Updated Entry replacing Q-40004

Acridinic acid

[643-38-9]

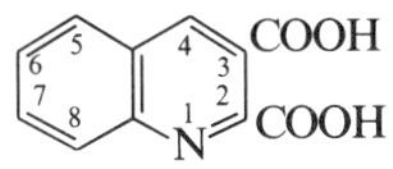

$C_{11}H_7NO_4$ M 217.181

Needles + 2 or 3H_2O (H_2O or EtOH). Spar. sol. cold H_2O, Et_2O, sol. EtOH. Loses CO_2 at 105°.

Di-Me ester: [17507-03-8].
  $C_{13}H_{11}NO_4$ M 245.234
  Prisms (MeOH or C_6H_6). Mp 107-108°.

Di-Et ester: [32413-08-4].
  $C_{15}H_{15}NO_4$ M 273.288
  Cryst. (hexane). Mp 54-55°.

Di-Et ester, N-*oxide:* [92525-68-3].
  $C_{15}H_{15}NO_5$ M 289.287
  Cryst. ($CHCl_3$/diisopropyl ether). Mp 111-112°.

Anhydride: [4945-42-0].
  $C_{11}H_5NO_3$ M 199.165
  Mp 225°.

Imide:
  $C_{11}H_6N_2O_2$ M 198.181
  Mp 316°.

Moriconi, E.J. *et al, J. Am. Chem. Soc.,* 1964, **86**, 38 (*synth*)
Godard, A. *et al, Bull. Soc. Chim. Fr.,* 1971, 906 (*synth*)
Kadin, S.B. *et al, J. Org. Chem.,* 1984, **49**, 4999 (*derivs*)
Ladner, D.W., *Synth. Commun.,* 1986, **16**, 157 (*synth*)
Maulding, D.R., *J. Heterocycl. Chem.,* 1988, **25**, 1777 (*synth, bibl*)

Quinolino[1″,2″:1′,2′]imidazo [4′,5′:4,5]imidazo[1,2-a]quinoline   Q-80003

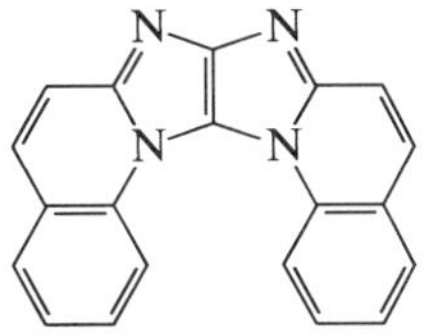

$C_{20}H_{12}N_4$ M 308.342

Solid + ½H_2O. Mp >250°.

Pereira, D.E. *et al, Tetrahedron,* 1987, **43**, 4931 (*synth, pmr, cmr, uv, ms*)

2-(1H)-Quinoxalinethione   Q-80004

2-Mercaptoquinoxaline. 2-Quinoxalinethiol

[6962-54-5]

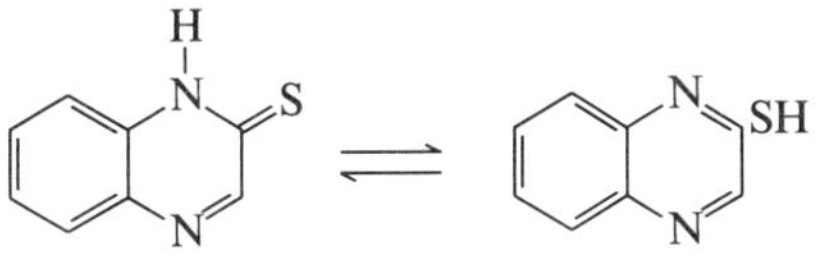

$C_8H_6N_2S$ M 162.215

Orange needles (MeOH). Mp 204-205°.

Wolf, F.J. *et al, J. Am. Chem. Soc.,* 1954, **76**, 2266 (*synth*)
Ijima, C. *et al, Chem. Pharm. Bull.,* 1989, **37**, 618 (*synth, pmr*)

Quinoxalino[1″,2″:1′,2′]imidazo [4′,5′:4,5]imidazo[1,2-a]quinoxaline   Q-80005

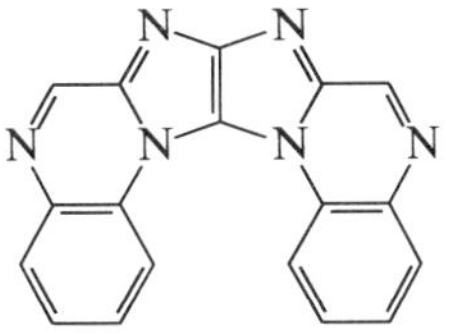

$C_{18}H_{10}N_6$ M 310.317

Solid. Mp >250°.

Pereira, D.E. *et al, Tetrahedron,* 1987, **43**, 4931 (*synth, pmr, cmr, ir, uv, ms*)

R

Radiatoside B
[111514-49-9]

R-80001

$C_{31}H_{42}O_{19}$ M 718.661
Constit. of *Argylia radiata*. Amorph. powder. $[\alpha]_D^{20}$ −138.6° (c, 1 in MeOH).

10-Deoxy, $\Delta^{8,10}$-isomer: [111514-50-2]. **Radiatoside C**
   $C_{31}H_{42}O_{18}$ M 702.662
   Constit. of *A. radiata*. Amorph. powder. $[\alpha]_D^{20}$ −125.3° (c, 1 in MeOH).

Bianco, A. *et al*, *Planta Med.*, 1987, **53**, 385.

Ramosin
[120163-26-0]

R-80002

$C_{36}H_{32}O_{13}$ M 672.641
Constit. of *Asphodelus ramosus*. Yellow amorph. solid. $[\alpha]_D$ −74° (c, 1.1 in MeOH).

Adinolfi, M. *et al*, *Phytochemistry*, 1989, **28**, 284.

Ratanhiaphenol III

R-80003

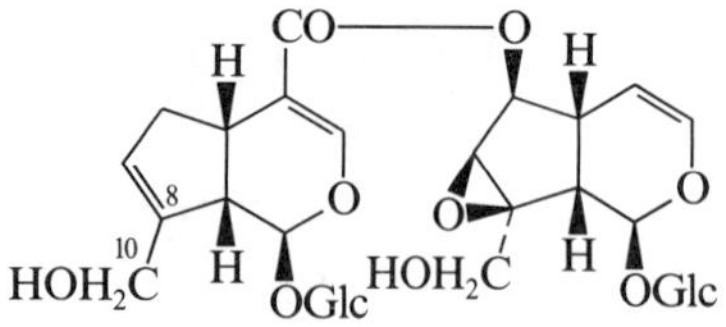

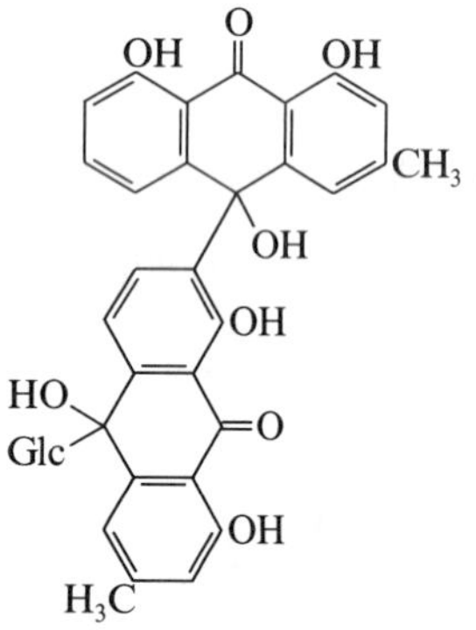

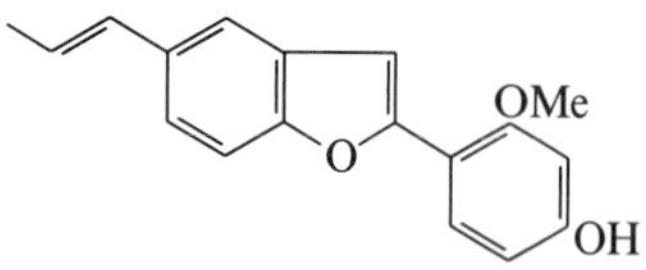

$C_{19}H_{18}O_3$ M 294.349
Constit. of *Krameria lanceolata*. Cryst. Mp 134-135°.

Achenbach, H. *et al*, *Phytochemistry*, 1989, **28**, 1959.

Rengyoside A

R-80004

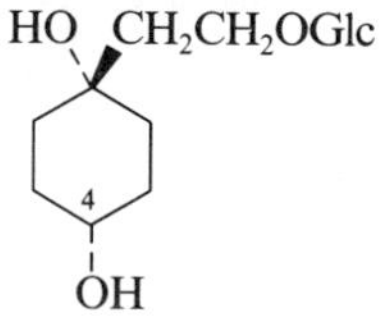

$C_{14}H_{26}O_8$ M 322.355
Constit. of *Forsythia suspensa*. Amorph. solid. $[\alpha]_D^{23}$ −11° (c, 0.18 in MeOH).

4-Ketone: **Rengyoside B**
   $C_{14}H_{24}O_8$ M 320.339
   Constit. of *F. suspensa*. Amorph. solid. $[\alpha]_D^{23}$ −10.4° (c, 0.28 in EtOH).

6′-(4-Hydroxyphenylacetyl): **Rengyoside C**
   $C_{22}H_{32}O_{10}$ M 456.489
   Constit. of *F. suspensa*. Powder. $[\alpha]_D^{23}$ −21.9° (c, 0.06 in 2-propanol).

Seya, K. *et al*, *Phytochemistry*, 1989, **28**, 1495.

Rhinacanthin A
[119626-45-8]

R-80005

$C_{15}H_{14}O_4$ M 258.273
Constit. of *Rhinacanthus nasutus* roots. Orange needles (Me₂CO). Mp 186.5-187°. $[\alpha]_D$ −12.9° (c, 0.25 in CHCl₃).

O-(2,6-Dimethyl-2,6-octadienoyl): [119626-46-9].
   Rhinacanthin B
   $C_{25}H_{28}O_5$ M 408.493
   Constit. of *R. nasutus* roots. Shows cytotoxic activity. Pale yellow needles (hexane). Mp 78-80°. $[\alpha]_D$ −47.9° (c, 0.24 in CHCl₃).

Wu, T.-S. *et al*, *Phytochemistry*, 1988, **27**, 3787.

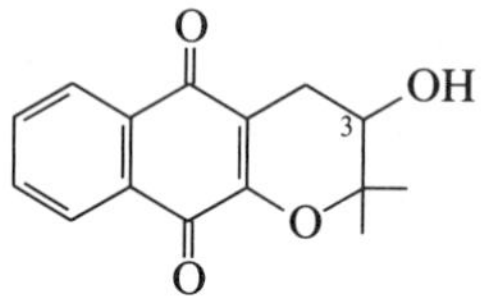

Rhodamine B   R-80006

N-[[9-(2-Carboxyphenyl)-6-diethylamino]-3H-xanthen-3-
ylidene]-N-ethyl ethanaminium, 9CI. [9-(o-Carboxyphenyl)-6-
(diethylamino)-3H-xanthen-3-ylidene]diethylammonium, 8CI.
9-(2-Carboxyphenyl)-3,6-bis(diethylamino)xanthylium. CI
Basic Violet 10. CI Solvent red 49

[14899-08-2]

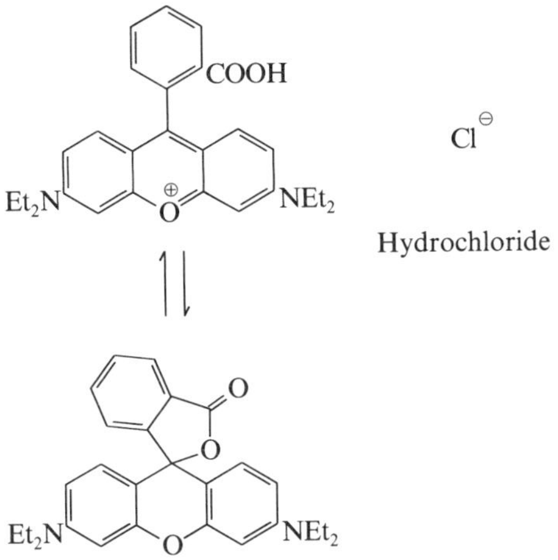

Cl⊖

Hydrochloride

$C_{28}H_{31}N_2O_3^{\oplus}$ M 443.564

Dyestuffs for cosmetics, drugs, fluorescent inks, chemical
lasers. Used in detn. of metals. Colourless prisms
(EtOH)(anhyd.). Green leaflets + $4H_2O$ (H_2O). Mp
165°. Red soln. with green fluor. in H_2O or EtOH,
colourless in Et_2O, C_6H_6.

Hydrochloride: [81-88-9].
    Important dyestuff. Shining metallic-lustred leaflets (dil.
    HCl). V. sol. H_2O, EtOH; sol. C_6H_6; sl. sol. dil. HCl.
    Forms red solution in H_2O, strongly fluorescent orange-
    yellow.

[64381-98-2]

Sjoberg, A.M. et al, J. Chromatogr., 1985, **318**, 149 (hplc,
    chromatog)
Takayanagi, M. et al, Appl. Spectrosc., 1986, **40**, 1132 (raman)
Ballard, J.M. et al, Org. Mass Spectrom., 1986, **21**, 575 (ms)
Hinckley, D.A. et al, Spectrochim. Acta, Part A, 1986, **42**, 747 (uv)
Wegener, J.W. et al, Chromatographia, 1987, **24**, 865 (hplc, uv)
Chalmers, J.H. et al, J. Chem. Educ., 1987, **64**, 969 (props)
Ernsting, N.P. et al, Chem. Phys., 1988, **122**, 431 (spectra)

Rhodoislandin A   R-80007

2,4,4′,5,5′,8′-Hexahydroxy-7,7′-dimethyl-[1,1′-bianthracene]-
9,9′,10,10′-tetrone, 9CI. 2,4,4′,5,5′,8′-Hexahydroxy-7,7′-
dimethyl-1,1′-bianthraquinone

[18783-14-7]

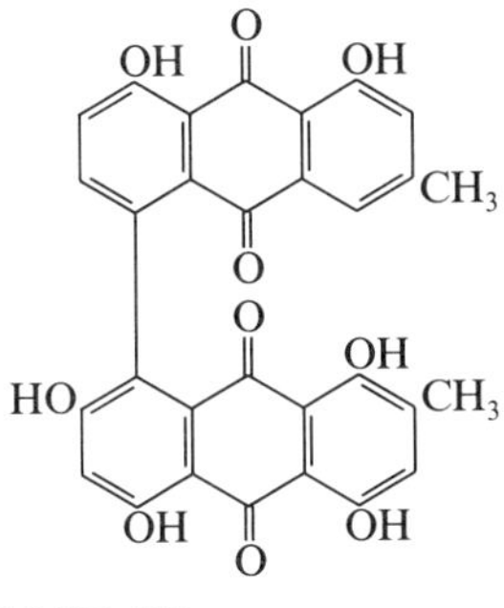

$C_{30}H_{18}O_{10}$ M 538.466

Pigment from *Penicillium islandicum*. Red cryst. Mp
    >300°.

Ogihara, Y. et al, Tetrahedron Lett., 1968, 1881 (isol, struct, cd,
    synth)
Takeda, N. et al, Tetrahedron, 1973, **29**, 3703 (bibl)

Rhodoislandin B   R-80008

2,4,4′,5,5′,8′-Hexahydroxy-7,7′-dimethyl-[1,1′-bianthracene]-
9,9′,10,10′-tetrone, 9CI. 2,4,4′,5,5′,8′-Hexahydroxy-7,7′-
dimethyl-1,1′-bianthraquinone

[18693-29-3]

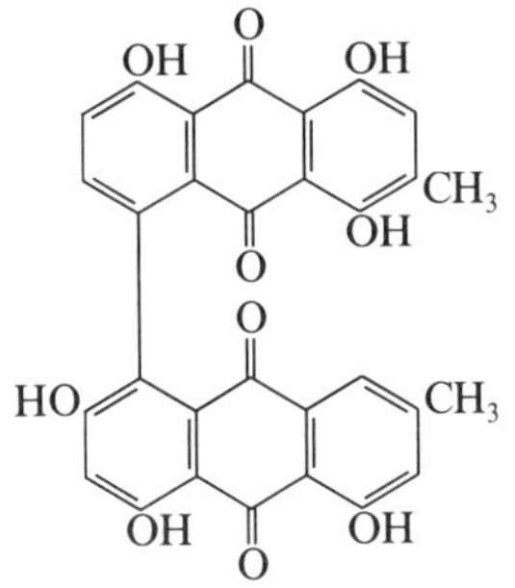

$C_{30}H_{18}O_{10}$ M 538.466

Pigment from *Penicillium islandicum*. Red cryst. Mp
    >300°.

Ogihara, Y. et al, Tetrahedron Lett., 1968, **15**, 1881 (isol, struct, cd,
    synth)
Takeda, N. et al, Tetrahedron, 1973, **29**, 3703.

Rollinicin   R-80009

Updated Entry replacing R-30010

[93303-16-3]

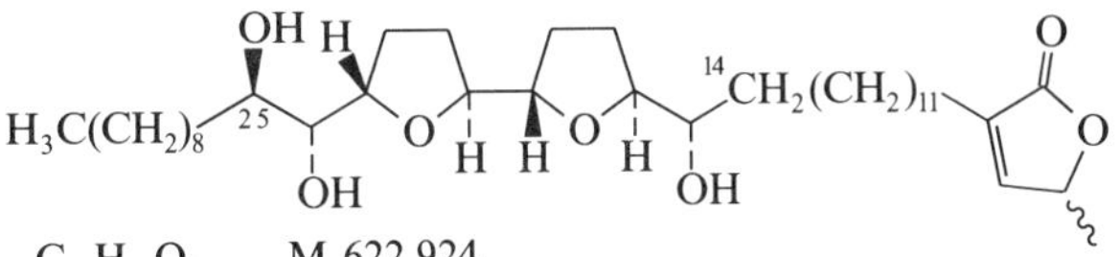

$C_{37}H_{66}O_7$ M 622.924

Constit. of *Rollinia papilionella*. Wax. Mp 30-32°. $[\alpha]_D^{25}$
    +6.8° (c, 0.24 in $CHCl_3$).

Stereoisomer: [93236-32-9]. **Isorollinicin**
    $C_{37}H_{66}O_7$ M 622.924
    From *R. papilionella*. Amorph. Mp 66-68°.

Dabrah, T.T. et al, Phytochemistry, 1984, **23**, 2013 (isol)
Abreo, M.J. et al, J. Nat. Prod. (Lloydia), 1989, **52**, 822 (isol, pmr)

Rose bengal, 8CI   R-80010

*4,5,6,7-Tetrachloro-3′,6′-dihydroxy-2′,4′,5′,7′-
tetraiodospiro[isobenzofuran-1(3H),9′(9H)xanthen]-3-one,
9CI. 3,4,5,6-Tetrachloro-2′,4′,5′,7′-tetraiodofluorescein, 8CI.
CI Solvent red 141*

[4159-77-7]

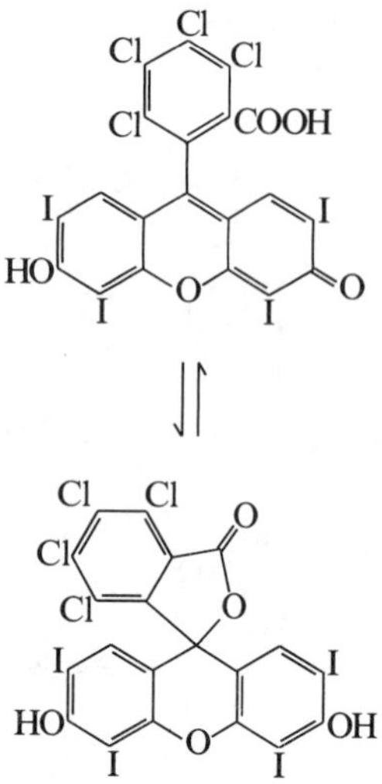

$C_{20}H_4Cl_4I_4O_5$ M 973.677

Singlet-O_2 sensitiser for photooxidations. Dyestuff for
cosmetics; food additive. Acid-base indicator. Insol.
H_2O. Normally used as di-Na salt.

Di-K salt: [632-68-8]. *Rose Bengal B. CI Acid red 94*
Dyestuff for nylon, silk, cellulose materials, inks,
cosmetics. Bright bluish pink. Sol. H_2O (blue-red soln.,
no fluor.).

▷ Possibly teratogenic in rats. Carcinogen. Toxic to
freshwater organisms. Binds to and inhibits RNA
polymerase.

Lactone-form
Colourless or v. pale pink solid.

[632-69-9]

Sanders, H.J., *Chem. Eng. News*, 1966, **44**, (42)100, (43)108 (*rev,
use*)
Bellin, J.S. *et al*, *J. Chromatogr.*, 1966, **24**, 131 (*chromatog*)
Peeples, W.A. *et al*, *J. Liq. Chromatogr.*, 1981, **4**, 51 (*hplc, purifn*)
Gandin, E. *et al*, *J. Chromatogr.*, 1982, **249**, 393 (*glc, purifn*)
Lamberts, J.J.M. *et al*, *Z. Naturforsch., B*, 1984, **39**, 474 (*uv, use,
deriv*)
Lamberts, J.J.M. *et al*, *Tetrahedron*, 1985, **41**, 2183 (*ir, pmr, uv,
props, use*)
Ghiggino, K.P. *et al*, *Aust. J. Chem.*, 1988, **41**, 9 (*purifn, uv, props*)

Roseoskyrin   R-80011

*4,4′,5,5′,8-Pentahydroxy-7,7′-dimethyl-[1,1′-bianthracene]-
9,9′,10,10′-tetrone, 9CI. 4,4′,5,5′,8-Pentahydroxy-7,7′-
dimethyl-1,1′-bianthraquinone*

[18693-27-1]

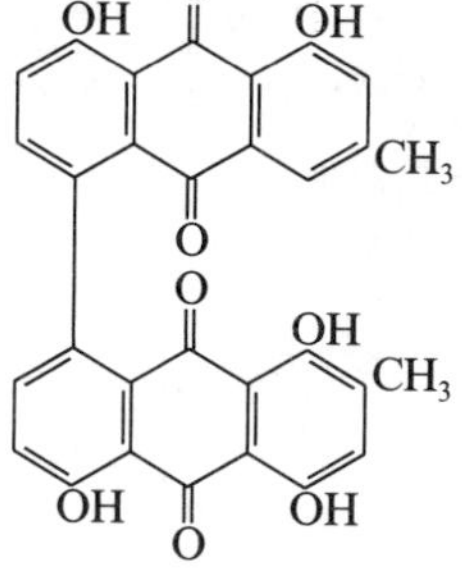

$C_{30}H_{18}O_9$ M 522.467

Pigment from *Penicillium islandicum*. Red cryst. Mp
>300°.

Ogihara, Y. *et al*, *Tetrahedron Lett.*, 1968, **15**, 1881 (*isol, struct, cd,
synth, pmr*)

Rotalin A   R-80012

[119979-81-6]

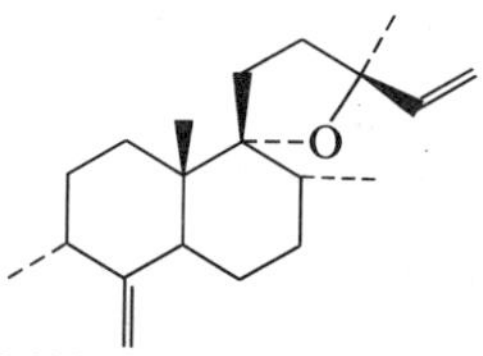

$C_{20}H_{32}O$ M 288.472

Metab. of sponge *Mycale rotalis*. Oil. $[\alpha]_D$ −1.8° (c, 2.8 in
$CHCl_3$).

Corriero, G. *et al*, *Tetrahedron*, 1989, **45**, 277.

Rotalin B   R-80013

[119979-80-5]

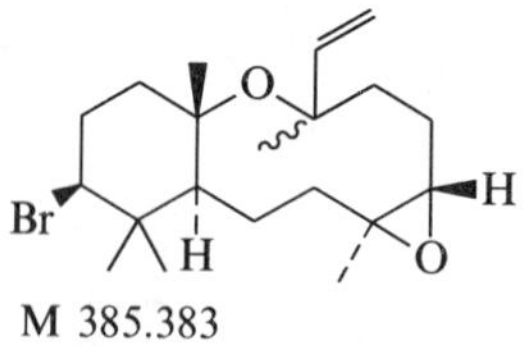

$C_{20}H_{33}BrO_2$ M 385.383

Metab. of sponge *Mycale rotalis*. Amorph. solid. $[\alpha]_D$
+13.9° (c, 0.43 in $CHCl_3$).

Corriero, G. *et al*, *Tetrahedron*, 1989, **45**, 277.

Rubichrome   R-80014

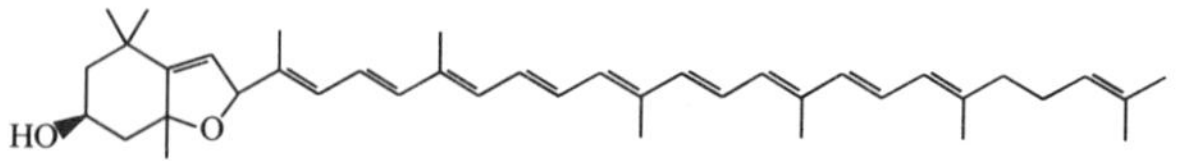

$C_{40}H_{56}O_2$ M 568.881

Isol. from *Tagetes patula* and *Ranunculus acer*. Red-violet
platelets ($CHCl_3/Et_2O$). Mp 154°.

Karrer, P. *et al*, *Helv. Chim. Acta*, 1947, **30**, 531.

Rugosal A   R-80015

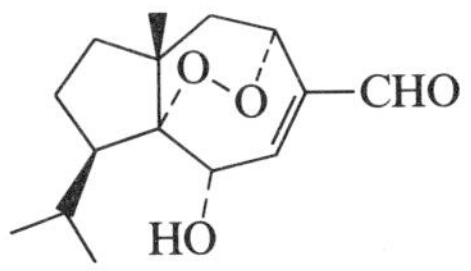

$C_{15}H_{22}O_4$ M 266.336

Constit. of damaged leaves of *Rosa rugosa*. Shows antimicrobial activity. Needles (EtOAc/hexane). Mp 145-147°.

Hashidoko, Y. *et al*, *Phytochemistry*, 1989, **28**, 425.

Rutacultin   R-80016

Updated Entry replacing R-00280

3-(1,1-Dimethyl-2-propenyl)-6,7-dimethoxy-2H-1-benzopyran-2-one, 9CI. 3-(1,1-Dimethylallyl)-6,7-dimethoxycoumarin

[31526-60-0]

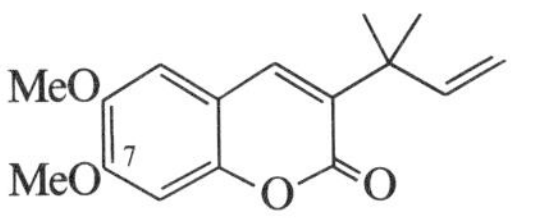

$C_{16}H_{18}O_4$ M 274.316

Constit. of *Ruta graveolens*. Cryst. (pet. ether). Mp 103-104°.

7-O-De-Me: [19723-23-0]. *3-(1,1-Dimethylallyl)scopoletin*
$C_{15}H_{16}O_4$ M 260.289
Constit. of *R. graveolens*. Mp 104-105°.

Steck, W. *et al*, *Phytochemistry*, 1971, **10**, 191 (*isol*)
Ballantyne, M.M. *et al*, *Tetrahedron*, 1971, **27**, 871 (*synth*)
von Brocke, W. *et al*, *Z. Naturforsch., B*, 1971, **26**, 1252 (*deriv*)

Ruzicka's hydrocarbon   R-80017

1,2,3,4-Tetrahydro-1,1,5,7-tetramethyl-6-(3-methylpentyl)naphthalene, 9CI

[34418-78-5]

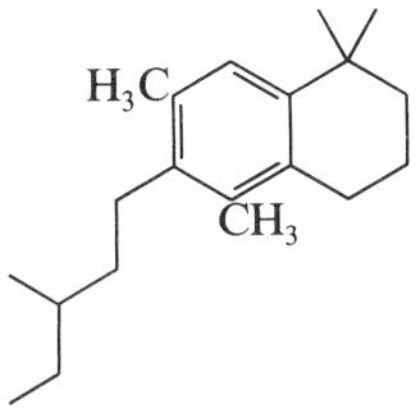

$C_{20}H_{32}$ M 272.473
Liq. Bp$_{0.3}$ 145°. n_D^{25} 1.5220.

Ruzicka, L. *et al*, *Helv. Chim. Acta*, 1931, **14**, 645 (*synth*)
Carman, R.M. *et al*, *Aust. J. Chem.*, 1971, **24**, 1919 (*synth*)
Nasipuri, D. *et al*, *J. Chem. Soc., Perkin Trans.* 1, 1973, 1754 (*synth, pmr, ir, uv*)

Rzedowskia bistriterpene   R-80018

[121786-61-6]

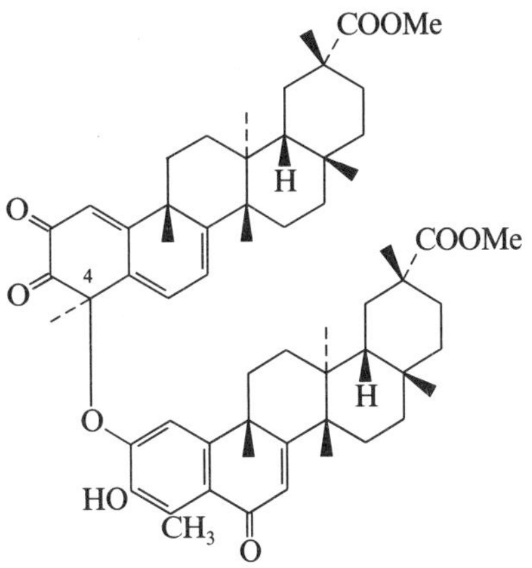

$C_{60}H_{78}O_9$ M 943.271

Compd. not named in the paper. Constit. of roots of *Rzedowskia tolantonguensis*. $[\alpha]_D^{20}$ −143.6° (CHCl$_3$).

4-Epimer: [121688-16-2].
$C_{60}H_{78}O_9$ M 943.271
Constit. of *R. tolantonguensis*. $[\alpha]_D$ +260° (CHCl$_3$). CA name apparently defective.

Gonzalez, A.G. *et al*, *Tetrahedron Lett.*, 1989, **30**, 863.

S

Safflomin C   S-80001

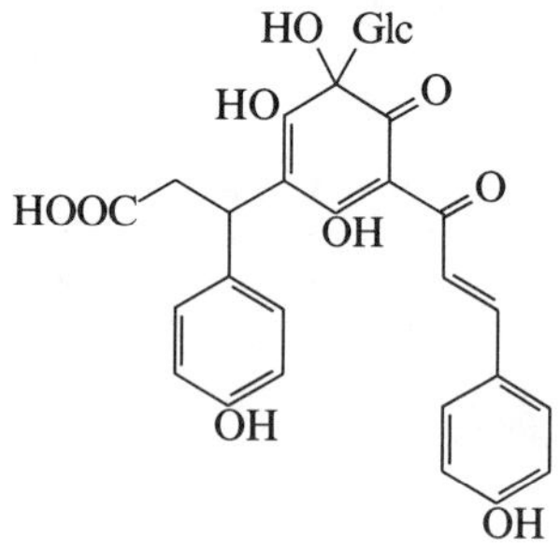

C$_{30}$H$_{30}$O$_{14}$ M 614.559

Yellow pigment of Safflower (*Carthamus tinctorius*).
Yellow powder. Mp 300° dec.

Ohodera, J. *et al*, *Chem. Lett.*, 1989, 1571.

Salannic acid   S-80002

Updated Entry replacing S-10003
Nimbidic acid
[29803-85-8]

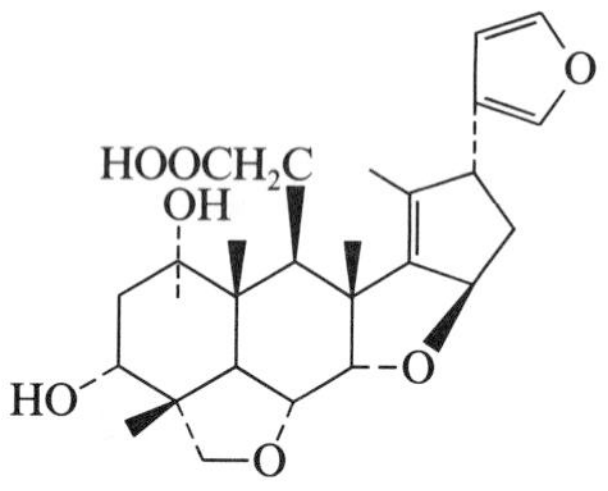

C$_{26}$H$_{34}$O$_7$ M 458.550

Constit. of *Melia dubria*, *M. indica* and *Azadirachta indica*.
Cryst. Mp 228-230°.

3-O-Ac, 1-O-tigloyl, Me ester: [992-20-1]. *Salannin*
  C$_{34}$H$_{44}$O$_9$ M 596.716
  Minor constit. of nim oil from *M. azadirachta*. Cryst.
  Mp 167-170°. [α]$_D$ +167° (CHCl$_3$).

1-O-Tigloyl, Me ester:
  C$_{32}$H$_{42}$O$_8$ M 554.679
  From *A. indica*. Cryst. (EtOAc). Mp 214-215°. [α]$_D^{20}$
  +134° (c, 1 in CHCl$_3$).

1-(2-Methylbutanoyl), Me ester: Salannol
  C$_{32}$H$_{44}$O$_8$ M 556.695
  From *A. indica*. Cryst. (EtOAc). Mp 208°. [α]$_D^{20}$ +108.7°
  (c, 1 in CHCl$_3$).

Henderson, R. *et al*, *Tetrahedron*, 1968, **24**, 1525 (*isol, struct*)
de Silva, L.B. *et al*, *Phytochemistry*, 1969, **8**, 1817 (*isol*)
Mitra, C.R. *et al*, *Tetrahedron Lett.*, 1970, 2761 (*struct*)
Nakanishi, K., *Phytochemistry*, 1975, **14**, 283 (*pmr, cmr*)
Kraus, W. *et al*, *Justus Liebigs Ann. Chem.*, 1981, 181 (*isol*)

Salicifoliol   S-80003

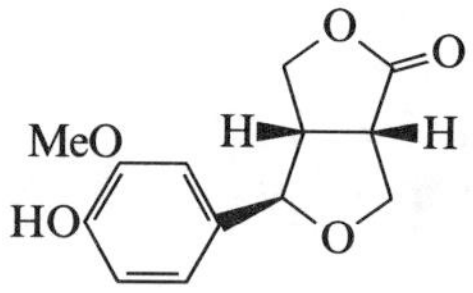

C$_{13}$H$_{14}$O$_5$ M 250.251

Constit. of *Bupleurum salicifolium*. Cryst. (C$_6$H$_6$/pet.
ether). Mp 102-103°.

González, A.G. *et al*, *J. Nat. Prod.* (*Lloydia*), 1989, **52**, 1139 (*isol, pmr, cmr*)

Salviaethiopisolide   S-80004

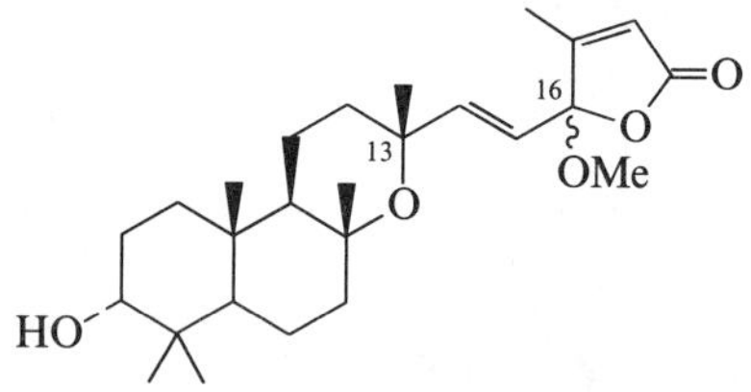

C$_{26}$H$_{40}$O$_5$ M 432.599

Constit. of *Salvia aethiopis*. Viscous oil. [α]$_D$ +16.25° (c,
0.8 in CHCl$_3$). Mixture of C-16 epimers.

13-Epimer: Episalviaethiopisolide
  C$_{26}$H$_{40}$O$_5$ M 432.599
  Constit. of *S. aethiopis*. Viscous oil. [α]$_D$ +32° (c, 1.53
  in CHCl$_3$). Mixt. of C-16 epimers.

[123844-68-8, 123931-55-5, 123931-56-6, 123931-57-7]

Gonzalez, M.S. *et al*, *Tetrahedron*, 1989, **45**, 3575.

Salvicanaric acid   S-80005

Updated Entry replacing S-70004
[112470-96-9]

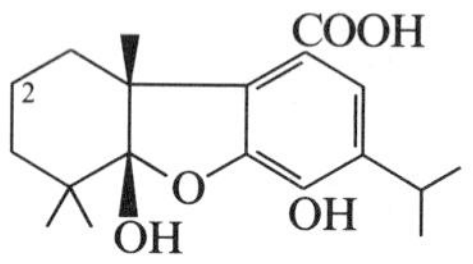

C$_{19}$H$_{26}$O$_5$ M 334.411

Constit. of *Salvia canariensis*. Foam.

2α-Hydroxy: 2α-Hydroxysalvicanaric acid
  C$_{19}$H$_{26}$O$_6$ M 350.411
  Constit. of *S. texana*. Foam.

Gonzalez, A.G. *et al*, *J. Nat. Prod.* (*Lloydia*), 1987, **50**, 341.
Gonzalez, A.G. *et al*, *Tetrahedron*, 1989, **45**, 5203 (*deriv*)

Salvinolactone   S-80006

2-Hydroxy-1,3-dimethyl-5H-phenanthro[4,5-bcd]pyran-5-one, 9CI

[120278-21-9]

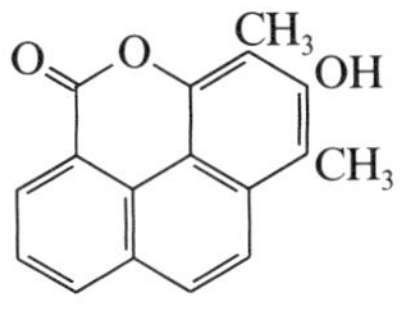

$C_{17}H_{12}O_3$ M 264.280
Constit. of *Salvia prionitis*. Cryst. Mp 208-209°.

Lin, L.-Z. *et al, Phytochemistry*, 1989, **28**, 177.

Samaderine C   S-80007

Updated Entry replacing S-00038
Samaderoside A

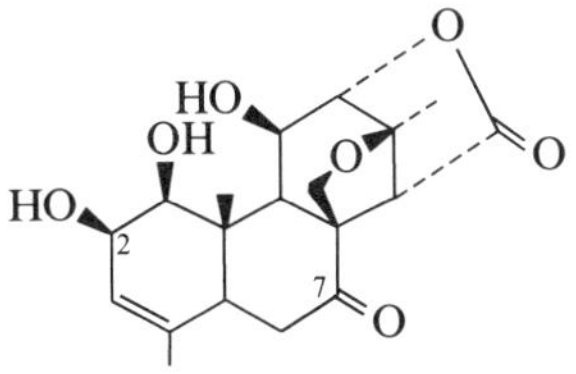

$C_{19}H_{24}O_7$ M 364.394
Constit. of *Samadera indica* nuts. Cryst. (EtOAc). Mp 275-276° dec. $[\alpha]_D$ +97.4° (c, 1 in EtOH).

2-Ketone: [803-22-5]. **Samaderine B**
  $C_{19}H_{22}O_7$ M 362.379
  Constit. of *S. indica*. Cryst. (EtOAc). Mp 235-240°. $[\alpha]_D$ +67.5° (Py).

7-Alcohol: **Cedronolin.** *Cedronylin. 7-Dihydrosamaderine C*
  $C_{19}H_{26}O_7$ M 366.410
  Isol. from seeds of *S. cedron*. Platelets (MeOH/EtOAc). Mp 263-267° (sinters from 250°). $[\alpha]_D$ +17.2° (c, 0.7 in Py).

2-Ketone, 7-alcohol: **Cedronin.** *7-Dihydrosamaderine B*
  $C_{19}H_{24}O_7$ M 364.394
  Isol. from seeds of *S. sedron*. Needles (MeOH). Mp 275-280° (sinters from 240°). $[\alpha]_D$ −12.6° (c, 0.714 in Py).

Zylber, J. *et al, Bull. Soc. Chim. Fr.*, 1964, 2016.

Sandaracopimaric acid   S-80008

Updated Entry replacing S-00067
Cryptopimaric acid. 8(14),15-Isopimaradien-18-oic acid
[471-74-9]

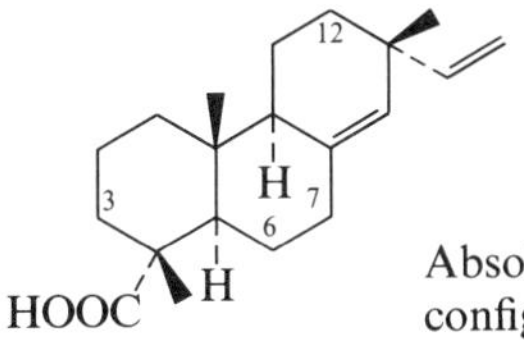

Absolute configurtion

$C_{20}H_{30}O_2$ M 302.456
Constit. of *Pinus, Juniperus* and *Cupressus* spp. Also from *Cryptomera japonica, Tetraclinis articulata, Callitris* sp. and *Agathus australis*. Cryst. (MeOH aq.). Mp 171-173° (softens at 163°). $[\alpha]_D$ −20° (c, 2.6 in CHCl₃).

3β-Hydroxy: [35951-45-2]. *3β-Hydroxyisopimara-8(14),15-dien-18-oic acid. 3β-Hydroxysandaracopimaric acid*
  $C_{20}H_{30}O_3$ M 318.455
  Constit. of *J. rigida*. Cryst. Mp 261°. $[\alpha]_D$ + ±0° (Py).

6α-Hydroxy: [19907-23-4]. *6α-Hydroxyisopimara-8(14),15-dien-18-oic acid. 6α-Hydroxysandaracopimaric acid*
  $C_{20}H_{30}O_3$ M 318.455
  Constit. of *J. phoenicea*. Cryst. (Me₂CO). Mp 265-267°.

7α-Hydroxy: *7α-Hydroxyisopimara-8(14)15-dien-18-oic acid. 7α-Hydroxysandaracopimaric acid*
  $C_{20}H_{30}O_3$ M 318.455
  Constit. of *J. communis*. Oil (as Me ester). $[\alpha]_D$ −41° (c, 0.8 in CHCl₃) (Me ester).

12β-Hydroxy: [1235-75-2]. *12β-Hydroxyisopimara-8(14),15-dien-18-oic acid. 12β-Hydroxysandaracopimaric acid*
  $C_{20}H_{30}O_3$ M 318.455
  Isol. from *Callitris quadrivalvis*. Needles (Me₂CO). Mp 269-270°. $[\alpha]_D$ −11° (c, 0.4 in EtOH).

12β-Acetoxy: *12β-Acetoxyisopimara-8(14),15-dien-18-oic acid. 12β-Acetoxysandaracopimaric acid*
  $C_{22}H_{32}O_4$ M 360.492
  Constit. of *Tetraclinis articulata*. Cryst. Mp 140.5-141.5°, Mp 156-158.5°, Mp 170° (trimorph.). $[\alpha]_D$ −50° (EtOH).

Edwards, O.E. *et al, Can. J. Chem.*, 1960, **38**, 663 (*isol, struct*)
Arya, V.P. *et al, Acta Chem. Scand.*, 1961, **15**, 682 (*struct*)
Bose, A.K. *et al, Chem. Ind.* (*London*), 1963, 254 (*abs config*)
Enzell, C. *et al, Ark. Kemi*, 1965, **23**, 367 (*ms*)
ApSimon, J.W. *et al, J. Chem. Soc., Chem. Commun.*, 1966, 361 (*pmr*)
Joye, N.M. *et al, J. Org. Chem.*, 1966, **31**, 320 (*isol*)
Afonso, A., *J. Am. Chem. Soc.*, 1968, **90**, 7375 (*synth*)
Tabacik, C. *et al, Phytochemistry*, 1971, **10**, 2147 (*6α-Hydroxysandaracopimaric acid*)
Lapasset, J. *et al, Acta Crystallogr., Sect. B*, 1972, **28**, 3321 (*12β-Hydroxysandaracopimaric acid*)
Wenkert, E. *et al, J. Am. Chem. Soc.*, 1972, **94**, 4367 (*cmr*)
Kozo, D. *et al, Phytochemistry*, 1972, **11**, 841 (*3β-Hydroxysandaracopimaric acid*)
de Pascual Teresa, J. *et al, Phytochemistry*, 1980, **19**, 1153 (*7α-Hydroxysandaracopimaric acid*)

Saponaceolide A   S-80009

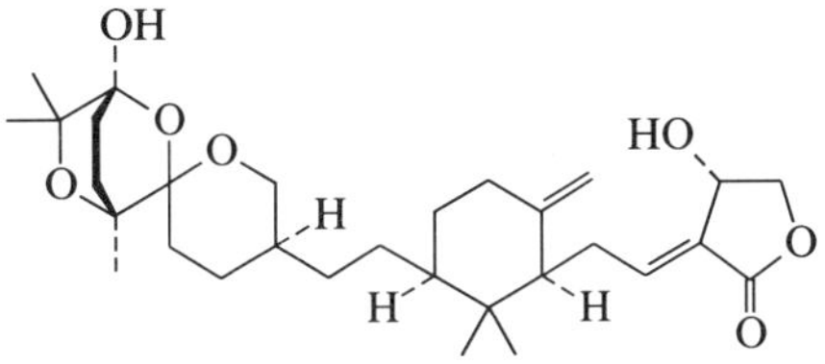

$C_{30}H_{46}O_7$ M 518.689
Prod. by *Tricholoma saponaceum*. Cytotoxic agent. Needles (Me₂CO/hexane). Mp 145-146°. $[\alpha]_D^{20}$ +78.14° (c, 1.1 in CHCl₃).

De Bernardi, M. *et al, Tetrahedron*, 1988, **44**, 235 (*isol, struct*)

Saprorthoquinone   S-80010

7-Methyl-3-(1-methylethyl)-8-(4-methyl-3-pentenyl)-1,2-naphthalenedione, 9CI. 3-Isopropyl-7-methyl-8-(4-methyl-3-pentenyl)-1,2-naphthoquione
[102607-41-0]

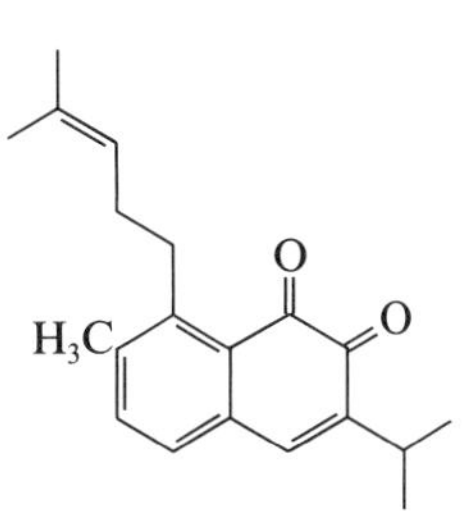

$C_{20}H_{24}O_2$ M 296.408

Constit. of *Salvia prionitis*. Cryst. Mp 97-98°. A 4,5-seco-5,10-friedoabietane.

Simoes, F. *et al*, *Phytochemistry*, 1986, **25**, 755 (*synth*)
Lin, L.-Z. *et al*, *Phytochemistry*, 1989, **28**, 177 (*isol*)

Scirpusin A   S-80011

Updated Entry replacing S-50031

[69297-51-4]

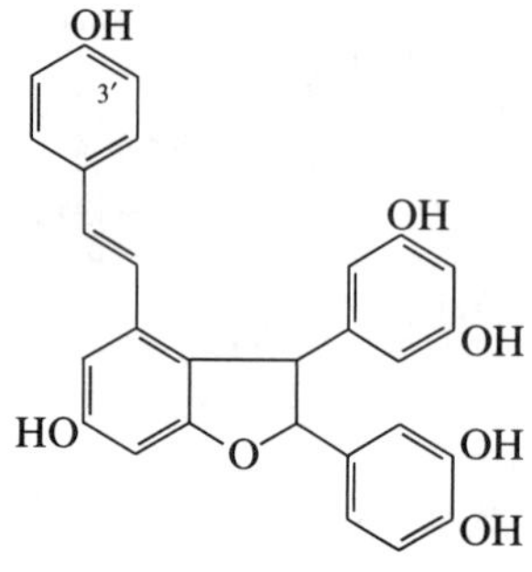

$C_{28}H_{22}O_7$ M 470.478

Constit. of *Scirpus fluviatilis* and *S. maritimus*. Pale brown amorph. powder.

3′-Hydroxy: [69297-49-0]. **Scirpusin B**
  $C_{28}H_{22}O_8$ M 486.477
  Constit. of *S. fluviatilis* and *S. maritimus*. Pale brown amorph. powder.

Nakajima, K. *et al*, *Chem. Pharm. Bull.*, 1978, **26**, 3050 (*isol*)
Powell, R.G. *et al*, *J. Nat. Prod.* (*Lloydia*), 1987, **50**, 293 (*isol*)

Secoeurabicanal   S-80012

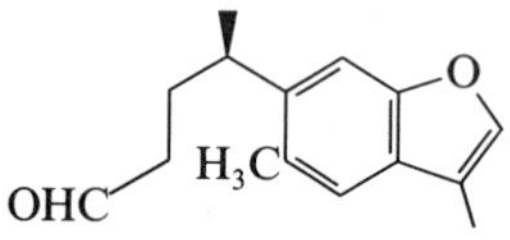

$C_{15}H_{18}O_2$ M 230.306

Constit. of *Euryops arabicus*. A seco-furoeremophilane.

Hafez, S. *et al*, *Phytochemistry*, 1989, **28**, 843.

Secoeurabicol   S-80013

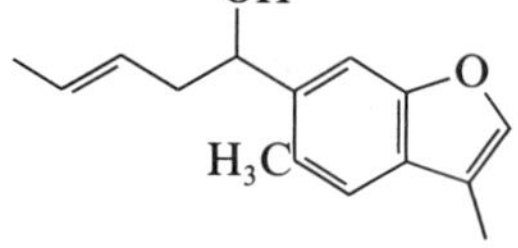

$C_{15}H_{18}O_2$ M 230.306

Constit. of *Euryops arabicus*. A seco-furoeremophilane.

Hafez, S. *et al*, *Phytochemistry*, 1989, **28**, 843.

Secoisolancifolide   S-80014

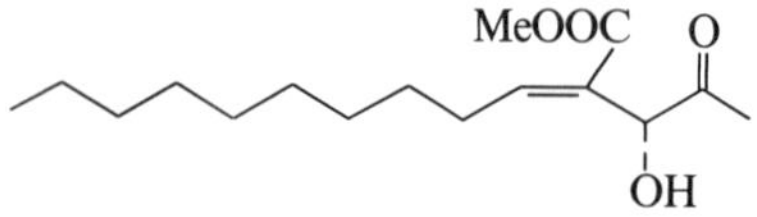

$C_{16}H_{28}O_4$ M 284.395

(*R*)-*form*

  Constit. of *Actinodaphne longifolia*. Oil. $[\alpha]_D$ +102.7° (c, 0.49 in CHCl$_3$).

Tanaka, H. *et al*, *Phytochemistry*, 1989, **28**, 1905.

Secoisoobtusilactone   S-80015

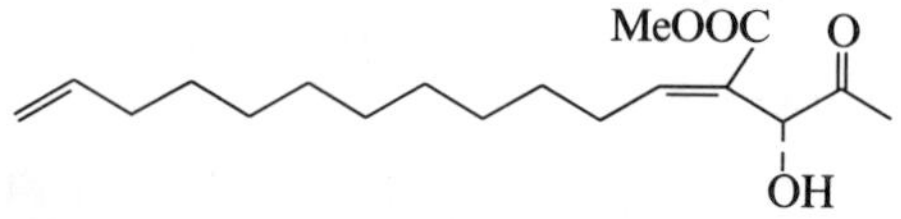

$C_{18}H_{30}O_4$ M 310.433

(*R*)-*form*

  Constit. of *Actinodaphne longifolia*. Oil. $[\alpha]_D$ +72.2° (c, 0.18 in CHCl$_3$).

Tanaka, H. *et al*, *Phytochemistry*, 1989, **28**, 1905.

Secologanol   S-80016

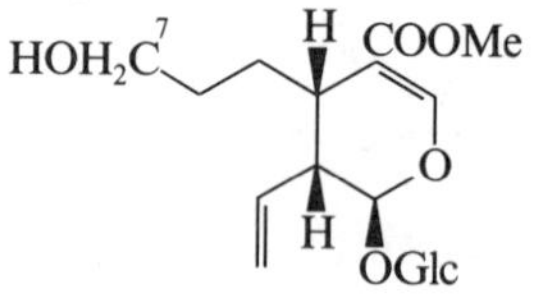

$C_{17}H_{26}O_{10}$ M 390.386

Constit. of *Gentiana verna*.

7-Ac: *7-Acetylsecologanol*
  $C_{19}H_{28}O_{11}$ M 432.424
  Constit. of *G. verna*.

7-(2,5-Dihydroxybenzoyl):
  $C_{24}H_{30}O_{13}$ M 526.493
  Constit. of *G. verna*.

Mpondo, E.M. *et al*, *J. Nat. Prod.* (*Lloydia*), 1989, **52**, 1146 (*isol, pmr, cmr*)

Secomanool   S-80017

[118855-41-7]

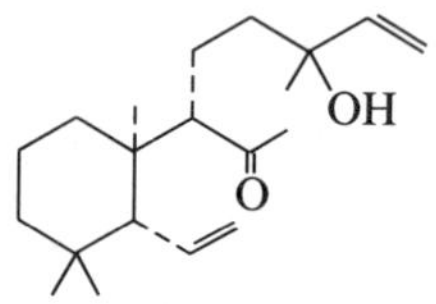

$C_{20}H_{34}O_2$ M 306.487

Constit. of *Fleischmannia microstemon*. Oil. $[\alpha]_D^{24}$ −15° (c, 1.2 in CHCl$_3$). A 7,8-secolabdane.

Tamayo-Castillo, G. *et al*, *Phytochemistry*, 1988, **27**, 3322.

Secotanapartholide A   S-80018

seco-Tanapartholide A

[103063-09-8]

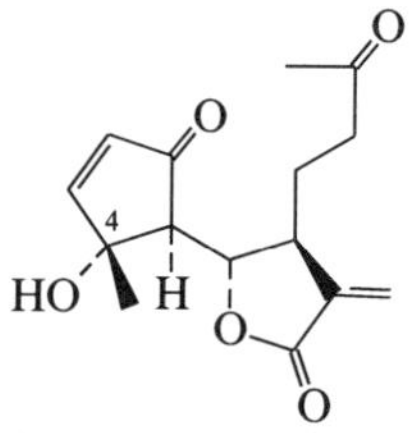

$C_{15}H_{18}O_5$ M 278.304

Constit. of *Tanacetum parthenium*. Oil. $[\alpha]_D^{20}$ −11.2° (c, 1.05 in EtOH).

4-Epimer: [85758-26-5]. **Secotanapartholide B.** seco-*Tanapartholide B*
  $C_{15}H_{18}O_5$ M 278.304
  Constit. *T. parthenium*. Oil.

Begley, M.J. *et al*, *Phytochemistry*, 1989, **28**, 940.

Secowithametelin   S-80019

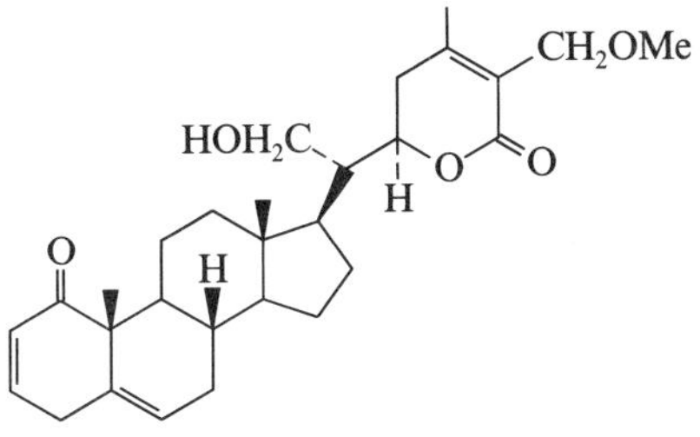

$C_{29}H_{40}O_5$ M 468.632
Constit. of *Datura metel*. Plates (EtOAc/pet. ether). Mp 190°. $[\alpha]_D$ +28.53° (c, 0.75 in CHCl$_3$).

Kundu, S. *et al*, *Phytochemistry*, 1989, **28**, 1769.

Sedanonic acid   S-80020

6-(1-Oxopentyl)-1-cyclohexene-1-carboxylic acid, 9CI
[6697-07-0]

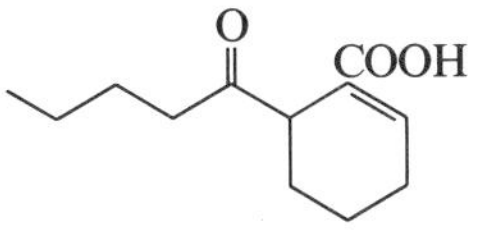

$C_{12}H_{18}O_3$ M 210.272
Isol. from celery oil (*Apium graveolens*) after hydrol., and from roots of *Cnidium officinale*. Cryst. (C$_6$H$_6$). Mp 113°.

[62006-38-6]

Ciamician, G. *et al*, *Ber.*, 1897, **30**, 492, 501, 1419 (*isol, struct*)
Barton, D.H.R. *et al*, *J. Chem. Soc.*, 1963, 1916 (*isol*)
Bjeldanes, L.F. *et al*, *J. Org. Chem.*, 1977, **42**, 2333 (*ir, pmr, ms*)

Selenanthrene, 9CI   S-80021

[262-30-6]

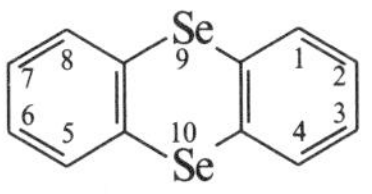

$C_{12}H_8Se_2$ M 310.115
Needles or prisms. Mp 180-181°.

9,10-Dioxide:
   $C_{12}H_8O_2Se_2$ M 342.114
   Mp 270°.

Krafft, F. *et al*, *Ber.*, 1896, **29**, 443 (*synth*)
Cullinane, N.M. *et al*, *J. Chem. Soc.*, 1939, 151 (*synth*)

3-Selenetanol   S-80022

3-Hydroxyselenetane
[73903-64-7]

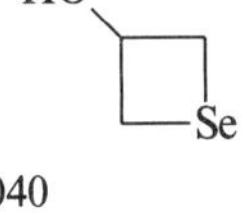

C_3H_6OSe M 137.040
Bp$_{3.5}$ 80-85°, Bp$_{0.8}$ 60-62°.

Ac: [112422-93-2].
   $C_5H_8O_2Se$ M 179.077
   Yellow oil. Dec. on dist.

Arnold, A.P. *et al*, *Aust. J. Chem.*, 1983, **36**, 815 (*synth, pmr*)
Polson, G. *et al*, *J. Org. Chem.*, 1988, **53**, 791 (*synth, pmr, cmr, ms*)

2,2'-Selenobispyridine, 9CI   S-80023

2,2'-Selenodipyridine. 2,2'-Dipyridyl selenide
[66491-49-4]

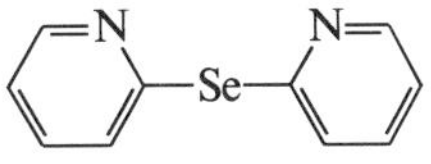

$C_{10}H_8N_2Se$ M 235.147
Yellow oil. Bp$_{0.5}$ 135-140°.

B, 2HBr: [66491-50-7].
   Cryst. + $\frac{1}{2}$H$_2$O. Mp 224-226° dec.

Grant, H.G. *et al*, *Z. Naturforsch., B*, 1978, **33**, 118 (*synth, uv*)
Keats, N.G. *et al*, *J. Heterocycl. Chem.*, 1979, **16**, 1369 (*ms*)
Pierini, A.B. *et al*, *J. Org. Chem.*, 1984, **49**, 486 (*synth, pmr*)
Summers, L.A., *J. Heterocycl. Chem.*, 1987, **24**, 533 (*rev, bibl*)

4,4'-Selenobispyridine, 9CI   S-80024

4,4'-Selenodipyridine. 4,4'-Dipyridyl selenide
[87385-48-6]

Has bactericidal and herbicidal props. Needles. Mp 63-65°.
B, 2HCl: Needles (EtOH/EtOAc). Mp 225-230°.

Jerchel, D. *et al*, *Chem. Ber.*, 1956, **89**, 2921 (*synth*)
Boduszek, B. *et al*, *Pol. J. Chem.* (*Rocz. Chem.*), 1983, **57**, 641 (*synth, ir, ms, pmr*)
Summers, L.A., *J. Heterocycl. Chem.*, 1987, **24**, 533 (*bibl*)

9-(Selenocyanato)fluorene   S-80025

[114263-69-3]

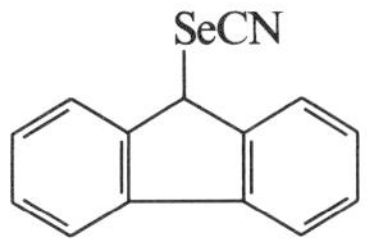

$C_{14}H_9NSe$ M 270.192
Pale yellow powder. Mp 109-111°.

Meinke, P.T. *et al*, *J. Org. Chem.*, 1988, **53**, 3632 (*synth, pmr, cmr, ir, ms*)

Senecrassidiol   S-80026

Updated Entry replacing S-00284

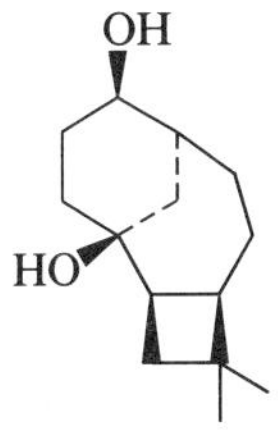

$C_{15}H_{26}O_2$ M 238.369
Constit. of *Senecio crassissimus*. Cryst. (2-propanol). Mp 93-96°. $[\alpha]_D^{24}$ −10.9° (c, 0.23 in CHCl$_3$).

5-Deoxy, 6-oxo:
   $C_{15}H_{24}O_2$ M 236.353
   Constit. of *Eurypon* sp. Oil. $[\alpha]_D$ −3.3° (c, 0.9 in CHCl$_3$).

5-Deoxy, 6-oxo, 8-Me ether:
   $C_{16}H_{26}O_2$ M 250.380
   Constit. of a *E.* sp. Oil. $[\alpha]_D$ −5.6° (c, 0.9 in CHCl$_3$).

6α-Hydroxy, 6,8-dideoxy, 7,8-didehydro:
   $C_{15}H_{24}O$ M 220.354
   Constit. of a *E.* sp. Oil. $[\alpha]_D$ −12.8° (c, 0.9 in CHCl$_3$).

Bohlmann, F. *et al*, *Phytochemistry*, 1981, **20**, 469.
Barrow, C.J. *et al*, *Aust. J. Chem.*, 1988, **41**, 1755 (*derivs*)

14-Serratene-3,21-diol   S-80027

Updated Entry replacing S-10033

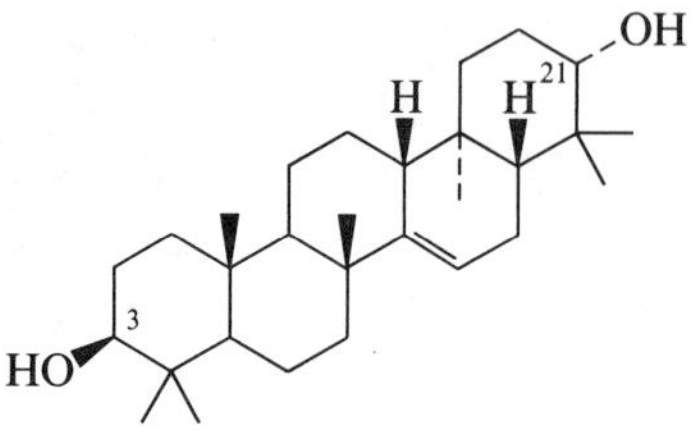

(3β,21α)-form

$C_{30}H_{50}O_2$ M 442.724

(3β,21α)-form [2239-24-9] **Pinusenediol.** *Serratenediol*
Constit. of the club moss *Lycopodium serratum* and *Pinus* spp. Cryst. (EtOH). Mp 302.5-304.5°. $[\alpha]_D^{22}$ −19° (c, 0.9 in CHCl₃).

3-Me ether: 3β-Methoxy-14-serraten-21α-ol
$C_{31}H_{52}O_2$ M 456.751
Constit. of sugar pine (*P. lambertiana*). Cryst. (EtOH). Mp 319-322.5°. $[\alpha]_D^{22}$ −5° (CHCl₃).

Di-Me ether: 3β,21α-Dimethoxy-14-serratene
$C_{32}H_{54}O_2$ M 470.777
Constit. of *P. lambertiana*. Cryst. (EtOH). Mp 320.5-323.5°. $[\alpha]_D^{22}$ +21° (CHCl₃).

3-O-α-L-Arabinopyranosyl: [80235-56-9]. **Inundoside A**
$C_{36}H_{60}O_7$ M 604.866
Constit. of *L. inundatum*. Needles (CHCl₃/MeOH). Mp >300°.

21-Ac, 3-O-α-L-arabinopyranosyl: [80235-55-8]. **Inundoside B**
$C_{38}H_{62}O_8$ M 646.903
Constit. of *L. inundatum*. Needles (MeOH). Mp >300°.

3-O-(4-O-p-Coumaroyl-α-L-arabinopyranosyl): [80242-62-2]. **Inundoside D₁**
$C_{45}H_{66}O_9$ M 751.011
Constit. of *L. inundatum*. Needles (CHCl₃/MeOH). Mp 286-290°.

21-Ac, 3-O-(4-O-p-coumaroyl-α-L-arabinopyranosyl): [80235-53-6]. **Inundoside D₂**
$C_{47}H_{68}O_{10}$ M 793.048
Constit. of *L. inundatum*. Needles (CHCl₃/MeOH). Mp >300°.

3-Ac: Serratenediol 3-monoacetate
$C_{32}H_{52}O_3$ M 484.761
Isol. from *L. serratum*. Mp 319-320°. $[\alpha]_D^{10}$ −5.7° (c, 1.06 in CHCl₃).

(3β,21β)-form [1449-06-5]
21-epi-*Serratenediol*
Constit. of club moss *L. serratum* and the bark of the sugar pine. Cryst. (EtOH). Mp 303-308°. $[\alpha]_D^{23}$ −19° (c, 1.29 in CHCl₃).

3-Me ether: [19902-59-1]. *3β-Methoxy-14-serraten-21β-ol*
$C_{31}H_{52}O_2$ M 456.751
Constit. of the bark of *Picea sitchensis*. Cryst. (EtOH). Mp 307-308°. $[\alpha]_D$ +3° (CHCl₃).

21-Me ether: [24433-28-1]. *21β-Methoxy-14-serraten-3β-ol*
$C_{31}H_{52}O_2$ M 456.751
Constit. of the bark of *Pinus contorta*. Cryst. (CH₂Cl₂/MeOH). Mp 250.5-252°. $[\alpha]_D^{21}$ −43.5° (c, 0.9 in CHCl₃).

3-O-α-L-Arabinopyranosyl: [80244-83-3]. **Inundoside E**
$C_{36}H_{60}O_7$ M 604.866
Constit. of *L. inundatum*. Needles (CHCl₃/MeOH). Mp >300°.

3-O-(4-O-p-Coumaroyl-α-L-arabinopyranosyl): [80235-52-5]. **Inundoside F**
$C_{45}H_{66}O_9$ M 751.011
Constit. of *L. inundatum*. Needles (CHCl₃/MeOH). Mp >300°.

(3α,21β)-form
diepi-*Serratenediol. Diepiserratenediol*
Minor constit. of *Pinus banksiana* bark. Cryst. (EtOH). Mp 300-301°. $[\alpha]_D^{23}$ −32° (c, 0.9 in CHCl₃).

3-Me ether: [19902-63-7]. *3α-Methoxy-14-serraten-21β-ol*
$C_{31}H_{52}O_2$ M 456.751
Isol. from *Pinus sitchensis* and *P. abies*. Cryst. (EtOH). Mp 276-277°. $[\alpha]_D^{20}$ −55° (CHCl₃).

Di-Me ether: [28161-29-7]. *3β,21β-Dimethoxy-14-serratene*
$C_{32}H_{54}O_2$ M 470.777
From *P. sitchensis*. Cryst. by subl. Mp 254-256°.

Rowe, J.W. *et al*, *Tetrahedron Lett.*, 1964, 2347; 1965, 2745 (*isol, struct*)
Inubushi, Y. *et al*, *Chem. Pharm. Bull.*, 1965, **13**, 104 (*isol, struct*)
Kutney, J.P. *et al*, *Tetrahedron*, 1969, **25**, 3731 (*isol*)
Roners, J.H. *et al*, *Can. J. Chem.*, 1970, **48**, 1021 (*isol*)
Rowe, J.W. *et al*, *Phytochemistry*, 1972, **11**, 365 (*isol*)
Tsuda, Y. *et al*, *Chem. Pharm. Bull.*, 1981, **29**, 2123 (*struct*)
Tsuda, Y. *et al*, *Chem. Pharm. Bull.*, 1982, **30**, 1500 (*synth*)

Shiraiachrome A   S-80028

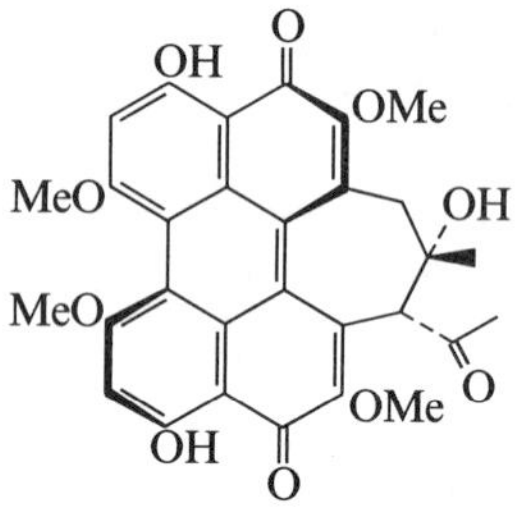

$C_{30}H_{26}O_{10}$ M 546.529
Pigment from *Shiraia bambusicola*. Deep red cryst. Mp 247-250°.

Wu, H., *J. Nat. Prod. (Lloydia)*, 1989, **52**, 948 (*isol, pmr, cmr*)

Shiraiachrome B   S-80029

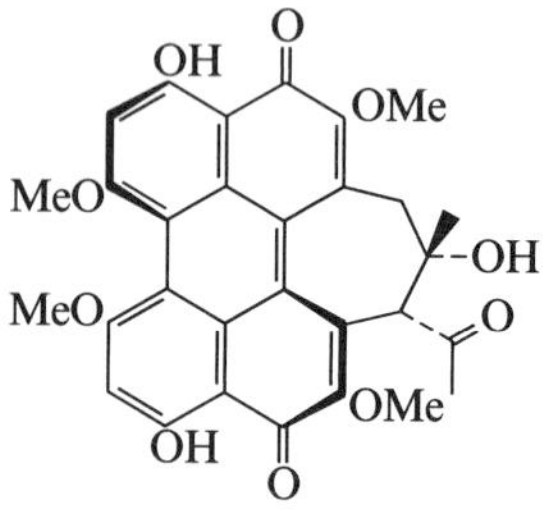

$C_{30}H_{26}O_{10}$ M 546.529
Constit. of *Shiraia bambusicola*. Deep red cryst. Mp 247-248°.

Wu, H. *et al, J. Nat. Prod. (Lloydia)*, 1989, **52**, 949 (*isol, pmr*)

Shiraiachrome C

S-80030

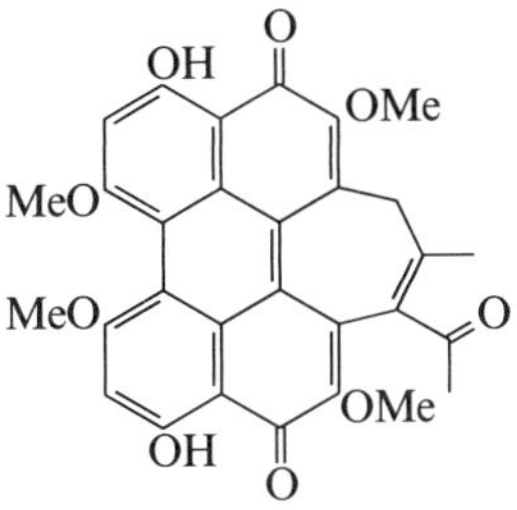

$C_{30}H_{24}O_9$ M 528.514
Constit. of *Shiraia bambusicola*. Red cryst. Mp 278-280°.

Wu, H. *et al*, *J. Nat. Prod.* (*Lloydia*), 1989, **52**, 948 (*isol, pmr*)

Siameanin

S-80031

4,4′,5,5′-Tetrahydroxy-2,2′-dimethyl-[1,1′-bianthracene]-9,9′,10′,10′-tetrone, 9CI. 4,4′,5,5′-Tetrahydroxy-2,2′-dimethyl-1,1′-bianthraquinone. 4,4′-Bi[1,8-Dihydroxy-3-methylanthraquinone]. 4,4′-Bichrysophanol

[13993-55-0]

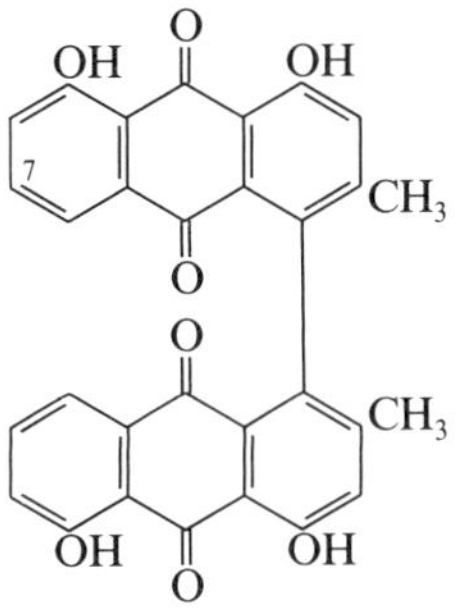

$C_{30}H_{18}O_8$ M 506.467
Isol. from peel of *Cassia siamea* and from leaves of *C. occidentalis*. Red microneedles (C_6H_6/EtOAc). Mp 323-325° dec.

Tetra-Ac: Mp 170°.

7-Hydroxy: **Cassianin**
   $C_{30}H_{18}O_9$ M 522.467
   Isol. from *C. siamea*. Orange-red needles (C_6H_6/EtOAc). Mp 350-360° dec.

Chatterjee, A. *et al*, *J. Indian Chem. Soc.*, 1964, **41**, 415 (*isol, uv, ir, pmr, struct*)
Tiwari, R.D. *et al*, *Planta Med.*, 1977, **32**, 375 (*isol*)

Sigmoidin E

S-80032

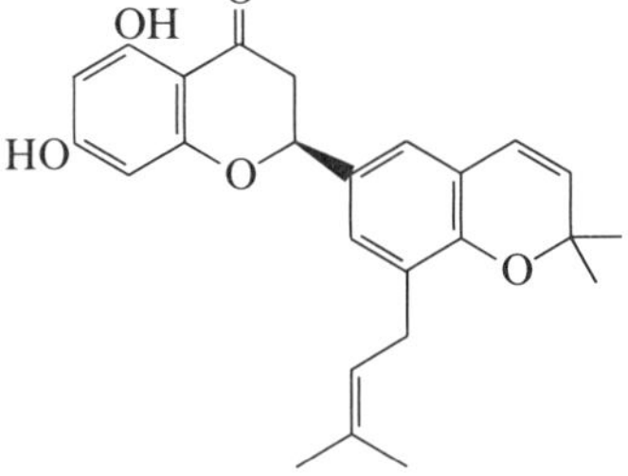

$C_{25}H_{26}O_5$ M 406.477
(*S*)-*form* [116174-67-5]
   Constit. of *Erythrina sigmoidea*. $[\alpha]_D$ −36.8° (c, 1.36 in MeOH).

Promsattha, R. *et al*, *J. Nat. Prod.* (*Lloydia*), 1988, **51**, 611.

5-Silphiperfolen-3-ol

S-80033

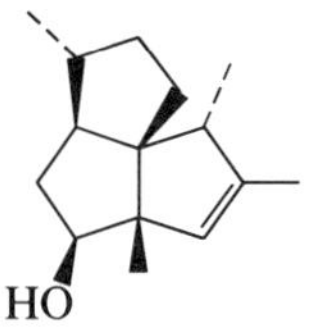

$C_{15}H_{24}O$ M 220.354
Constit. of *Artemisia laciniata*. Oil.

Brendel, J. *et al*, *Tetrahedron Lett.*, 1989, **30**, 2371 (*synth*)

Skimmearepin A

S-80034

[118156-16-4]

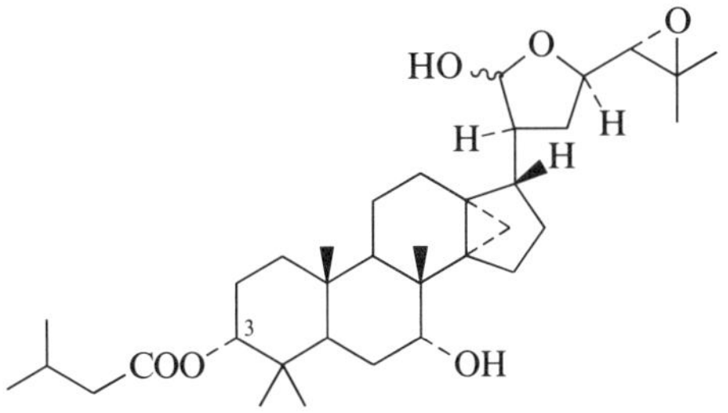

$C_{35}H_{56}O_6$ M 572.824
Constit. of *Skimmia japonica*. Shows insect antifeedant activity. Cryst. (EtOH). Mp 164.5-165.5°. $[\alpha]_D^{20}$ −22.7° (c, 0.2 in EtOH).

3-Deacyl, 3-(2Z,4E,6E-decatrienoyl): [118156-15-3].
   Skimmiarepin B
   $C_{40}H_{60}O_6$ M 636.910
   Constit. of *S. japonica*. Shows insect antifeedant activity. Cryst. (Me$_2$CO). Mp 168-169°. $[\alpha]_D^{18}$ −39.8° (c, 0.11 in CHCl$_3$).

Ochi, M. *et al*, *Bull. Chem. Soc. Jpn.*, 1988, **61**, 3225.

Smenorthoquinone

S-80035

[121994-49-8]

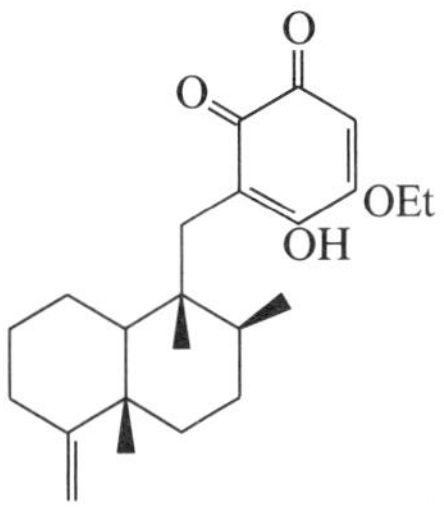

$C_{23}H_{32}O_4$ M 372.503
Constit. of *Smenospongia* sp. Antimicrobial and cytotoxic. Yellow needles (MeOH).

Kondracki, M.-L. *et al*, *Tetrahedron*, 1989, **45**, 1995.

Smenospongine   S-80036

Updated Entry replacing S-60036

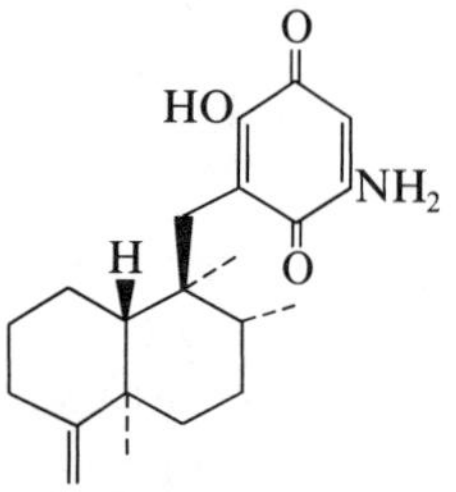

$C_{21}H_{29}NO_3$ M 343.465

Quinoterpenoid antibiotic. Prod. by the sponge *Smenospongia* sp. Cytotoxic and antimicrobial agent. Red cryst. Mp 153-155°.

N-(2-phenylethyl): Smenospongidine
$C_{29}H_{37}NO_3$ M 447.616
Constit. of *S*. sp. Antimicrobial and cytotoxic. Cryst. (MeOH). Mp 168-170°.

N-(3-methylbutyl): Smenospongianine
$C_{26}H_{39}NO_3$ M 413.599
Constit. of *S*. sp. Antimicrobial and cytotoxic. Cryst. Mp 170-172°.

N-(2-methylpropyl): Smenospongorine
$C_{25}H_{37}NO_3$ M 399.572
Constit. of *S* sp. Antimicrobial and cytotoxic.

Kondracki, M.-L. *et al, Tetrahedron*, 1989, **45**, 1995.

Sojagol   S-80037

2,3-Dihydro-10-hydroxy-3,3-dimethyl-1H,7H-furo[2,3-c:5,4-f']bis[1]benzopyran-7-one, 9CI. Soyagol
[18979-00-5]

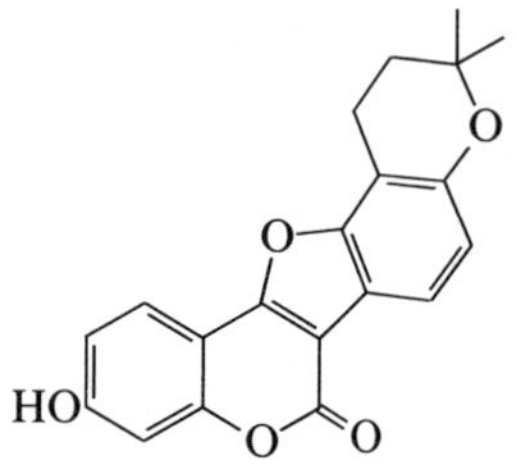

$C_{20}H_{16}O_5$ M 336.343

Phytoalexin from leaves and hypocotyls of *Glycine max* and from mung beans (*Phaseolus aureus*). Needles by subl. Mp 284-286°.

Zilg, H. *et al, Phytochemistry*, 1968, **7**, 1765 (*isol, ir, ms, struct*)
Keen, N.T. *et al, Phytochemistry*, 1972, **11**, 1031 (*isol, uv*)

Sophoracoumestan A   S-80038

3-Hydroxy-10,10-dimethyl-6H,10H-furo[3,2-c:4,5-g']bis[1]benzopyran-6-one, 9CI
[77369-93-8]

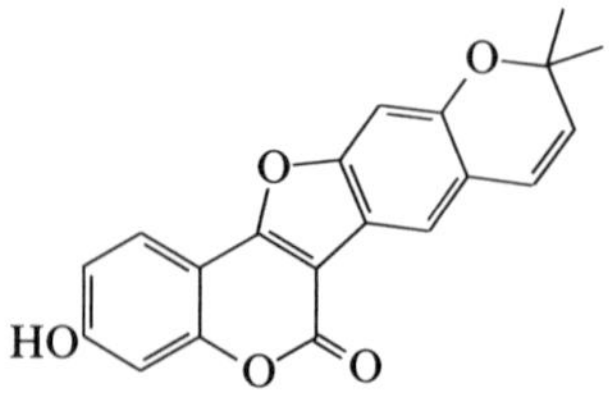

$C_{20}H_{14}O_5$ M 334.328

Isol. from roots of *Sophora franchetiana*. Needles (MeOH). Mp >300°.

Komatsu, M. *et al, Chem. Pharm. Bull.*, 1981, **29**, 532 (*isol, uv, ms, pmr, struct*)

β-Sorigenin   S-80039

Updated Entry replacing D-04767
8,9-Dihydroxynaphtho[2,3-c]furan-1(3H)-one, 9CI. Sorigenin
[492-23-9]

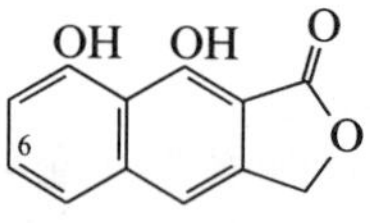

$C_{12}H_8O_4$ M 216.193
Pale-yellow cryst. Mp 237-240°.

Di-Ac: Cryst. Mp 229-230°.

8-O-β-D-Primeveroside: [519-99-3]. *β-Sorinin*
$C_{23}H_{26}O_{13}$ M 510.451
Isol. from bark of *Rhamnus japonica*. Amorph. powder.

Di-Me ether: [63744-12-7].
Pale-yellow cryst. (Me$_2$CO/hexane). Mp 174.5-176°.

6-Methoxy: α-Sorigenin
$C_{13}H_{10}O_5$ M 246.219
From *R. japonica*. Yellowish needles (EtOH). Mp 227-229° dec. (185°).

6-Methoxy, 8-O-β-D-primeveroside: α-Sorinin
$C_{24}H_{28}O_{14}$ M 540.477
Isol. from *R. japonica*. Needles or prisms (EtOH aq.). Mp 159°.

Haber, R.G. *et al, Helv. Chim. Acta*, 1956, **39**, 1654 (*struct, synth*)
Matsui, M. *et al, Agric. Biol. Chem.*, 1962, **27**, 40 (*synth*)
Horii, Z. *et al, Chem. Pharm. Bull.*, 1963, **11**, 312, 317 (*struct, synth*)
Hauser, F.M. *et al, J. Org. Chem.*, 1977, **42**, 4155 (*synth*)

Spiro[aziridine-2,2'-tricyclo [3.1.1.1^{3,7}]decane], 9CI   S-80040

Spiro[adamantane-2,2'-aziridine]
[59591-91-2]

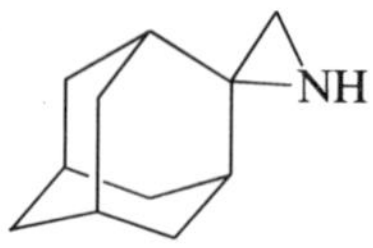

$C_{11}H_{17}N$ M 163.262
Mp 141-143°.

Sasaki, T. *et al, Tetrahedron*, 1976, **32**, 437.

Spiro[1,3-benzodioxole-2,1'-cyclohexane]   S-80041

[182-55-8]

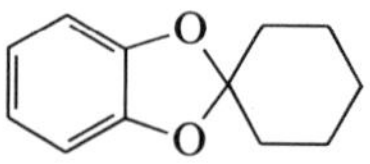

$C_{12}H_{14}O_2$ M 190.241
Mp 51-52° (47°).

Birch, A.J., *J. Chem. Soc.*, 1947, 102 (*synth*)
Antus, S. *et al, Chem. Ber.*, 1989, **122**, 1017 (*synth, pmr*)

Spiroeuryolide   S-80042

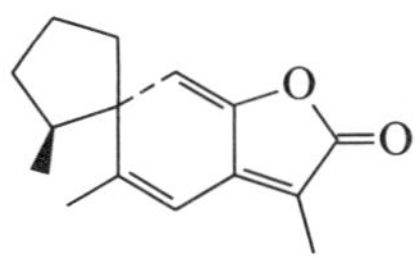

$C_{15}H_{18}O_2$ M 230.306
Constit. of *Euryops arabicus*.

Hafez, S. *et al*, *Phytochemistry*, 1989, **28**, 843.

Spongialactone A   S-80043
[120030-04-8]

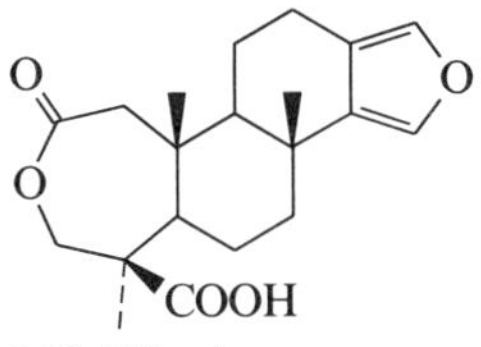

$C_{20}H_{26}O_5$ M 346.422
Constit. of *Spongia arabica*. Amorph. powder.

Hirsch, S. *et al*, *J. Nat. Prod.* (*Lloydia*), 1988, **51**, 1243 (*isol, pmr, cmr*)

Stenosporic acid   S-80044
Updated Entry replacing S-00765
2-Hydroxy-4-[(2-hydroxy-4-methoxy-6-propylbenzoyloxy)]-6-pentylbenzoic acid, 9CI
[27240-56-8]

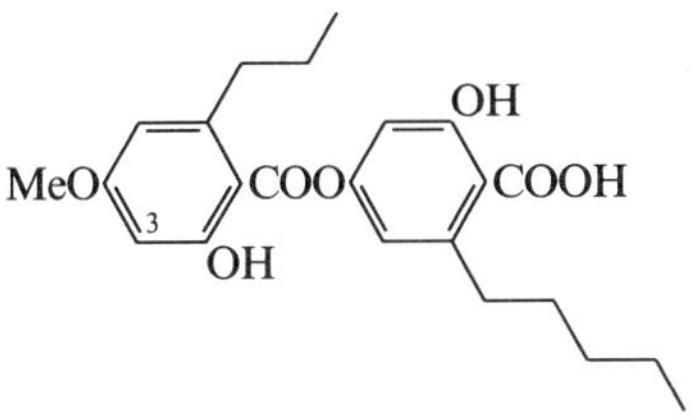

$C_{23}H_{28}O_7$ M 416.470
Isol. from *Ramalina stenospora*, *Nephroma cellulosum* and *Pertusaria* spp. Cryst. (cyclohexane). Mp 112-114°.

Me ester: Cryst. (hexane). Mp 35-36°.

3-Chloro: [120091-96-5]. **3-Chlorostenosporic acid**
$C_{23}H_{27}ClO_7$ M 450.915
Constit. of *Dimelaena* cf. *californica*. Cryst. (CH$_2$Cl$_2$/pet. ether). Mp 138°.

Culberson, C.F., *Phytochemistry*, 1970, **9**, 841 (*isol, synth*)
Culberson, C.F. *et al*, *Bryologist*, 1972, **75**, 362 (*isol*)
Elix, J.A., *Aust. J. Chem.*, 1974, **27**, 1767 (*synth*)
Renner, B. *et al*, *Z. Naturforsch.*, *C*, 1978, **33**, 340 (*isol*)
Elix, J.A. *et al*, *Aust. J. Chem.*, 1988, **41**, 1789 (*isol, synth, deriv*)

3-Sterpurenol   S-80045

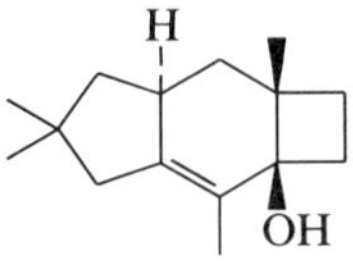

$C_{15}H_{24}O$ M 220.354
Ac: 3-Acetoxysterpurene
$C_{17}H_{26}O_2$ M 262.391

Metab. of *Alcyonum acaule*. Oil. $[\alpha]_D$ +128.6° (c, 2.2 in CHCl$_3$).

Cimino, G. *et al*, *Tetrahedron*, 1989, **45**, 6479.

Striatenone   S-80046

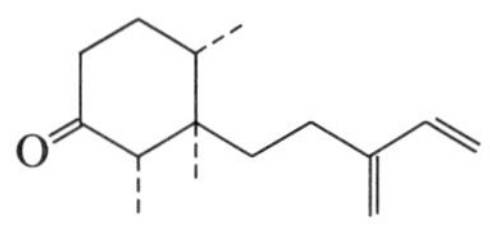

$C_{15}H_{24}O$ M 220.354
Constit. of *Porella navicularis*. Oil. $[\alpha]_D$ −9.1° (c, 0.11 in CHCl$_3$).

Toyota, M. *et al*, *Phytochemistry*, 1989, **28**, 1661.

Stypandrol   S-80047
Updated Entry replacing S-50130
7,7′-Diacetyl-1,1′,8,8′-tetrahydroxy-6,6′-dimethyl-2,2′-binaphthyl. Hemerocallin
[99305-33-6]

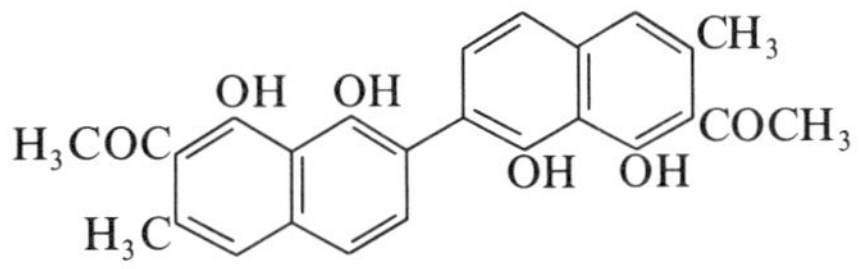

$C_{26}H_{22}O_6$ M 430.456
Constit. of *Stypandra imbricata*, *Dianella revoluta* and *Hemerocallis* spp. Shows neurotoxic properties. Orange needles (CHCl$_3$). Mp 265-266° dec.

Colegate, S.M. *et al*, *Aust. J. Chem.*, 1985, **38**, 1233 (*isol*)
Wang, J.-H. *et al*, *Phytochemistry*, 1989, **28**, 1825 (*pmr, cmr, struct*)

Subalatin   S-80048
[119269-69-1]

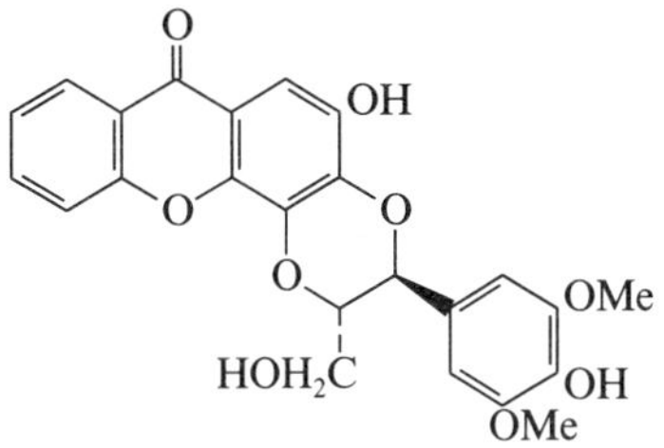

$C_{24}H_{20}O_9$ M 452.417
Constit. of *Hypericum subalatum*. Pale yellow needles (CHCl$_3$/MeOH). Mp 265° dec. $[\alpha]_D^{25}$ +600° (c, 0.005 in MeOH).

Chen, M.-T., *Heterocycles*, 1988, **27**, 2589.

Subsphaeric acid   S-80049
[120091-90-9]

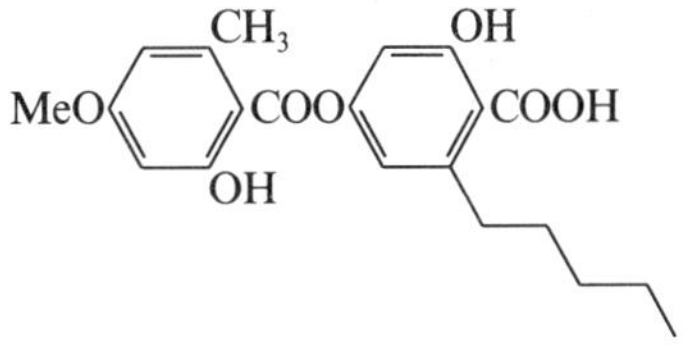

$C_{21}H_{24}O_7$ M 388.416

Constit. of *Dimelaena thysanota*. Cryst. (EtOAc/pet. ether). Mp 120°.

Elix, J.A. *et al*, *Aust. J. Chem.*, 1988, **41**, 1789 (*isol, struct, synth*)

Sulfoacetic acid S-80050

Sulfoethanoic acid

[123-43-3]

$$HOOCCH_2SO_3H$$

$C_2H_4O_5S$ M 140.117

Reagent for Liebermann-Burchard colorimetric test for steroids. Reaction catalyst. Hygroscopic cryst. + $1H_2O$ (H_2O). Mp 84-86°. Bp 245° dec.

Py salt: [21372-73-6].
   Mp 151-152°.

Carboxy-Et ester:
   $C_4H_8O_5S$ M 168.170
   Mp 175° (as K salt).

Di-Et ester:
   $C_6H_{12}O_3S$ M 164.225
   $Bp_{0.5}$ 116°.

Dianilide:
   $C_{14}H_{14}N_2O_3S$ M 290.342
   Mp 150-151°.

[5462-60-2, 16697-66-8, 22128-42-3, 64707-21-7]

Folkers, K. *et al*, *J. Am. Chem. Soc.*, 1944, **66**, 1083 (*synth*)
Lehmann, J. *et al*, *J. Am. Chem. Soc.*, 1964, **86**, 4469.
Hoogenboom, B.E. *et al*, *J. Org. Chem.*, 1969, **34**, 3414 (*deriv*)
Fieser and Fieser's Reagents for Organic Synthesis, Wiley, 1974, **4**, 7.
Mao, J.C.H. *et al*, *Antimicrob. Agents Chemother.*, 1985, **27**, 197.

Sumatrol S-80051

Updated Entry replacing S-01008

[82-10-0]

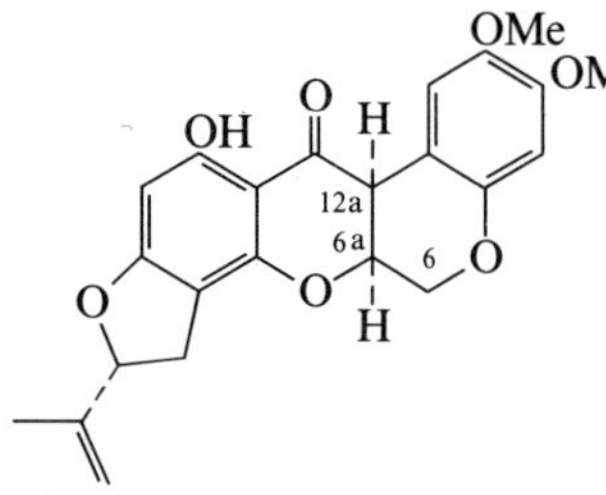

$C_{23}H_{22}O_7$ M 410.423

Isol. from *Crotalaria burhia*, *Derris malaccensis*, *Millettia auriculata*, *Piscidia erythrina* and *Tephrosia toxicaria*. Needles (EtOH or Me_2CO). Spar. sol. dil. NaOH. Mp 174.5-177.5° (183-194°, 195-196°). $[\alpha]_D$ −184° (c, 1.335 in C_6H_6).

Ac: Mp 218-219°. $[\alpha]_D$ +57° (C_6H_6), −20.5° (Me_2CO).

Oxime: Needles (EtOH aq.). Mp 245-247°.

6α-Hydroxy: [65160-15-8]. **Villosin.** *6-Hydroxysumatrol*
   $C_{23}H_{22}O_8$ M 426.422
   Constit. of the pods of *T. villosa*. Mp 133°.

6-Oxo: [65160-16-9]. **Villosone**
   $C_{23}H_{18}O_8$ M 422.390
   Constit. of *T. villosa*. Mp 268°.

6α-Methoxy: [65160-17-0]. **Villinol**
   $C_{24}H_{22}O_8$ M 438.433
   Isol. from *T. villosa* pods. Mp 200°.

6α,12α-Didehydro: [60077-62-5]. **Villosol**
   $C_{23}H_{20}O_7$ M 408.407

Constit. of the pods of *T. villosa*. Greenish-yellow needles ($CHCl_3$/MeOH). Mp 197°. $[\alpha]_D^{26}$ −80.9° (c, 0.495 in $CHCl_3$).

Kerry, T.S. *et al*, *J. Chem. Soc.*, 1939, 1601.
Crombie, L. *et al*, *J. Chem. Soc.*, 1961, 5445.
Crombie, L. *et al*, *J. Chem. Soc.*, 1962, 755 (*pmr*)
Carlson, D.G. *et al*, *Tetrahedron*, 1973, **29**, 2731 (*pmr*)
Crombie, L. *et al*, *J. Chem. Soc., Perkin Trans.* 1, 1975, 1497 (*cmr*)
Sarma, P.N. *et al*, *Indian J. Chem., Sect. B*, 1976, **14**, 152 (*Villosol*)
Krupadanam, G.L.D. *et al*, *Tetrahedron Lett.*, 1977, 2125 (*Villosone, Villinol*)
Uddin, A. *et al*, *Planta Med.*, 1979, **36**, 181; *CA*, **91**, 120359.

Swertisin S-80052

Updated Entry replacing S-01026

6-β-D-Glucopyranosyl-4′,5-dihydroxy-7-methoxyflavone.
Flavocommelitin

[6991-10-2]

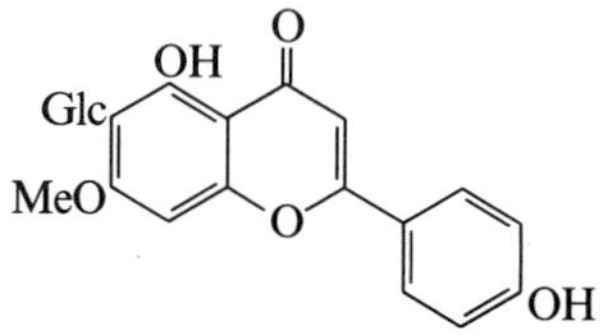

$C_{22}H_{22}O_{10}$ M 446.410

Constit. of *Swertia japonica* and *S. purpurascens*; *Gaillardia anstata* and *G. pulchella*; *Gentiana campestris* and the aerial parts of *G. germanica* and *G. ramosa*; leaves of *Achillea* genus; of *Dipsacaceae* spp. and petals of *Iris japonica*. Pale-yellow needles (H_2O). Mp 243° dec. $[\alpha]_D^{20}$ −10.0° (c, 0.9 in Py).

Hexa-Ac: Mp 155-158°.

4′-O-β-D-Glucoside: [16049-42-6]. **Flavocommelin**
   $C_{28}H_{32}O_{15}$ M 608.552
   Isol. from *Commelina communis*. Yellow cryst. Mp 216-217°. $[\alpha]_D^{20}$ −40.9° (c, 4.425 in Py).

4′-Me ether: [21089-34-9]. **Embigenin**
   $C_{23}H_{24}O_{10}$ M 460.437
   Mp 236-238°.

4′-Me ether, 6′-O-α-L-Rhamopyranosyl: [52589-13-6].
Embinin
   $C_{29}H_{34}O_{14}$ M 606.579
   Isol. from flowers of *I. tectorum* and *I. germaniia*. Mp 180-181°.

Di-Me ether: Mp 302°.

Takeda, K. *et al*, *CA*, 1966, **67**, 99951 (*Flavocommelin*)
Komatsu, M. *et al*, *Tetrahedron Lett.*, 1966, 1611 (*isol*)
Kawase, A. *et al*, *Agric. Biol. Chem.*, 1968, **32**, 537 (*Embigenin*)
Prox, A., *Tetrahedron*, 1968, **24**, 3697 (*ms*)
Wagner, A. *et al*, *Phytochemistry*, 1972, **11**, 851, 1857 (*isol*)
Miana, A.G., *Phytochemistry*, 1973, **12**, 728 (*isol*)
Arisawa, M. *et al*, *Yakugaku Zasshi (J. Pharm. Soc. Jpn.)*, 1973, **93**, 1655; *CA*, **80**, 68388q (*Embinin*)
Kaldas, M. *et al*, *Helv. Chim. Acta*, 1975, **58**, 2188 (*isol*)
Zemtsova, G.N. *et al*, *Khim. Prir. Soedin.*, 1977, 705; *CA*, **88**, 60093 (*Glycosides*)

Swietemahonin A   S-80053
[121825-44-3]

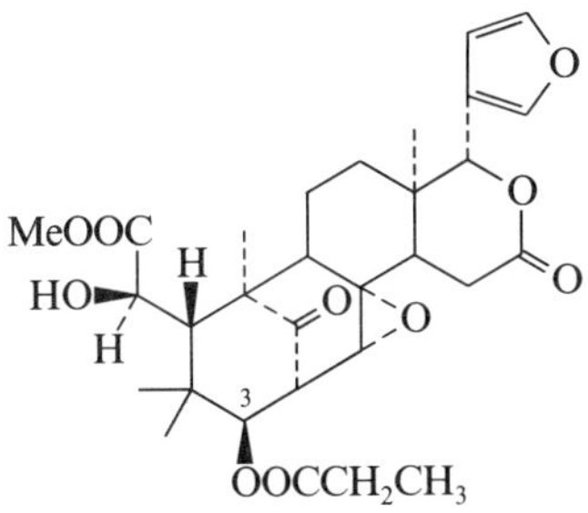

MeOOC
HO
H
H
3
OOCCH$_2$CH$_3$

$C_{30}H_{38}O_{10}$ M 558.624
Constit. of *Swietenia mahogani*. Antagonist of platelet
activitating factor. Needles (EtOAc/diisopropyl ether).
Mp 174-174.5°. [α]$_D$ −12.2° (CHCl$_3$).

3-O-Deacyl, 3-O-Tigloyl: [121825-42-1]. **Swietemahonin E**
$C_{32}H_{40}O_{10}$ M 584.662
Constit. of *S. mahogani*. Antagonist of PAF. Needles
(EtOAc/diisopropyl ether). Mp 151-152°. [α]$_D$ −20.7°
(CHCl$_3$).

Kodota, S. *et al*, *Tetrahedron Lett.*, 1989, **30**, 1111.

Swietenolide   S-80054

Updated Entry replacing S-01028
[3776-48-5]

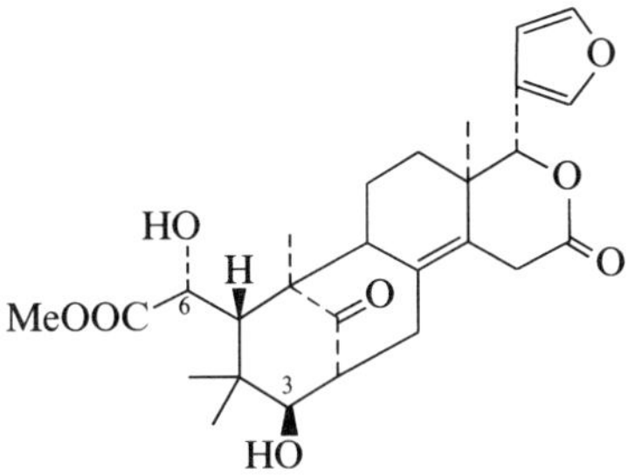

HO
H
MeOOC
6
3
HO

$C_{27}H_{34}O_8$ M 486.561
Constit. of the seeds of *Swietenia macrophylla*. Cryst. Mp
221-225°. [α]$_D$ −136° (CHCl$_3$).

3-Ac: 3-Acetylswietenolide
$C_{29}H_{36}O_9$ M 528.598
Constit. of *S. mahogani*. Needles. Mp 136-138°. [α]$_D$
−17.8° (CHCl$_3$).

Di-Ac:
$C_{31}H_{38}O_{10}$ M 570.635
Constit. of *S. macrophylla*. Cryst. Mp 227-230°. [α]$_D$
−131°.

6-Deoxy: [1915-68-0]. **6-Deoxyswietenolide**
$C_{27}H_{34}O_7$ M 470.561
Constit. of *Khaya ivorensis*. Cryst. Mp 194-196°. [α]$_D$
−141°.

6-Deoxy, Ac: [1915-69-1]. **Fissinolide**. *Grandifoliolin.*
Augustinolide
$C_{29}H_{36}O_8$ M 512.599
Constit. of *Cedrela fissilis*, *K. grandifoliola* and *Guarea
trichiloides*. Cryst. Mp 169-170°. [α]$_D$ −165°.

6-Deoxy, O-(2-methylpropanoyl): **Khayasin**
$C_{31}H_{40}O_8$ M 540.652
Constit. of *K. senegalensis*. Cryst. (C$_6$H$_6$). Mp 114-116°.
[α]$_D^{25}$ −165° (CHCl$_3$).

Adesogan, E.K. *et al*, *J. Chem. Soc. C*, 1966, 2127 (*isol, uv, ir,
pmr, struct, deriv*)
Chakrabartty, T. *et al*, *Tetrahedron*, 1968, **24**, 1503 (*isol*)
Connolly, J.D. *et al*, *Tetrahedron*, 1968, **24**, 1507 (*struct, pmr*)
Taylor, D.A.H. *et al*, *J. Chem. Soc.*, *Perkin Trans.* 1, 1973, 1599;
1974, 437 (*struct, cmr*)
Chan, K.C. *et al*, *Phytochemistry*, 1976, **15**, 429 (*deriv*)
Kadota, S. *et al*, *Tetrahedron Lett.*, 1989, **30**, 1111 (*deriv*)

Sylvianecic acid   S-80055
[80847-64-9]

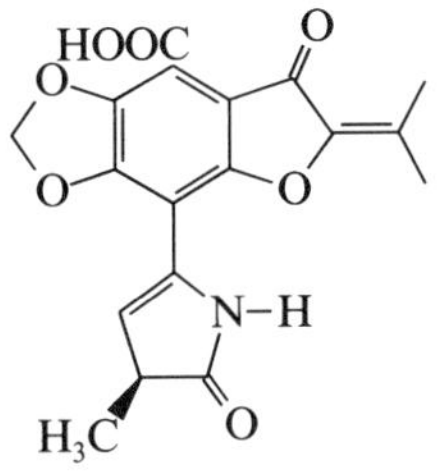

HOOC
O
O
O
O
N–H
H$_3$C
O

$C_{18}H_{15}NO_7$ M 357.319
Isol. from an *Aspergillus sylvanecii* strain. Shows
antibiotic, antifungal and antitumour activity. Unstable
orange cryst. (Me$_2$CO). Mp 58-59.5°. [α]$_D$ −39.4°
(CHCl$_3$).

Yeomans, J.W. *et al*, *J. Nat. Prod. (Lloydia)*, 1989, **52**, 462 (*isol,
cmr, bibl*)

T

Tanaparthin peroxide   T-80001

Updated Entry replacing T-20006

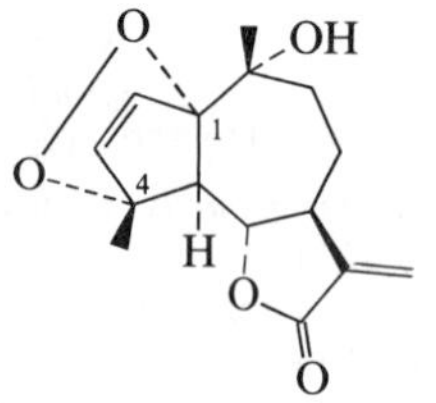

$C_{15}H_{18}O_5$ M 278.304

(1α,4α)-form [85799-10-6]

Tanaparthin α-peroxide

Constit. of *Tanacetum parthenium*. Cryst. Mp 95-96°. $[\alpha]_D^{20}$ −32.1° (c, 0.11 in CHCl₃).

(1β,4β)-form [85758-28-7]

Tanaparthin β-peroxide

Constit. of *T. parthenium*. Gum.

Bohlmann, F. *et al*, *Phytochemistry*, 1982, **21**, 2543 (*isol*)
Begley, M.J. *et al*, *Phytochemistry*, 1989, **28**, 940 (*isol, struct*)

14-Taraxeren-6-ol   T-80002

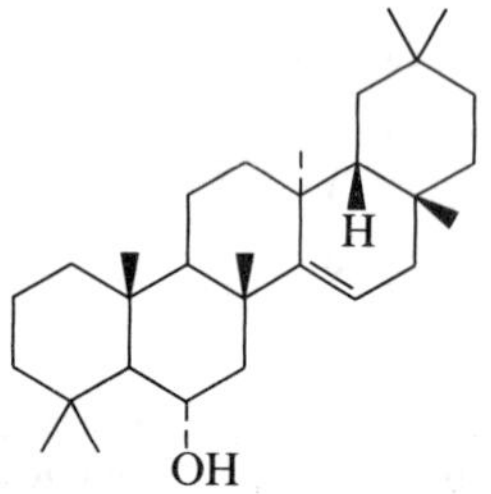

$C_{30}H_{50}O$ M 426.724

6α-form

Euphorginol

Constit. of *Euphorbia tirucalli*. Cryst. (MeOH). Mp 168-170°. $[\alpha]_D$ +22.35° (c, 0.19 in CHCl₃).

Rasool, N. *et al*, *Phytochemistry*, 1989, **28**, 1193.

Tectoroside   T-80003

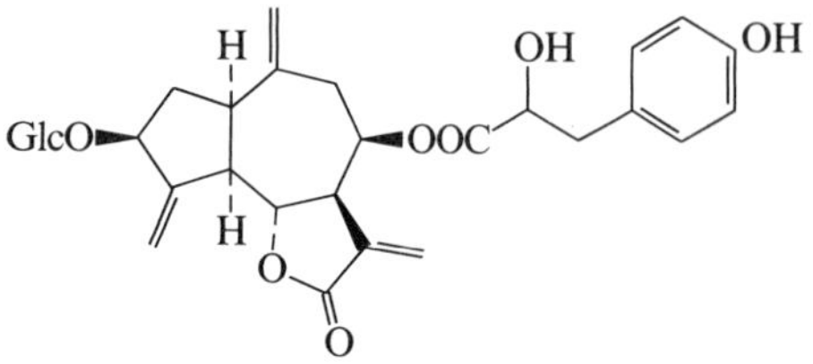

$C_{30}H_{36}O_{12}$ M 588.607

Constit. of *Crepis tectorum*. Oil.

Kisiel, W. *et al*, *Phytochemistry*, 1989, **28**, 2403 (*isol, pmr*)

Tefluthrin   T-80004

3-(2-Chloro-3,3,3-trifluro-1-propenyl)-2,2-dimethylcyclopropanecarboxylic acid (2,3,5,6-tetrafluoro-4-methylphenyl)methyl ester, 9CI

[79538-32-2]

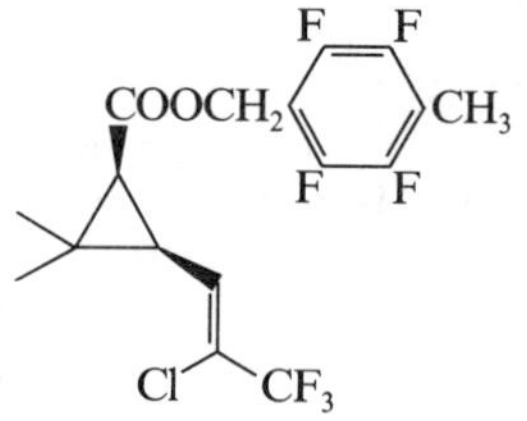

$C_{17}H_{14}ClF_7O_2$ M 418.738

Soil-active pyrethroid insecticide. Solid. Mp 44.6°.

[76437-51-9, 76437-52-0, 79538-33-3]

Japan. Pat., 80 111 445, (1980); *CA*, **94**, 83669q (*synth, use*)
Eur. Pat., 31 199, (1981); *CA*, **95**, 186837t (*synth, use*)

Telfairine   T-80005

1-(2-Bromoethenyl)-2,4,5-trichloro-1,5-dimethylcyclohexane

[120163-22-6]

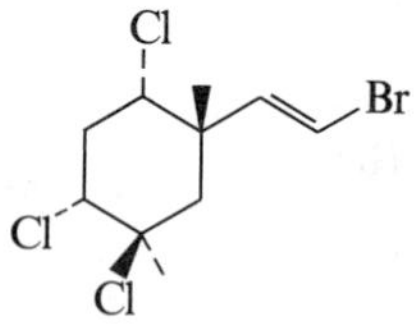

$C_{10}H_{14}BrCl_3$ M 320.483

Constit. of *Plocamium telfairiae*. Amorph. powder (hexane). Mp 62-63°. $[\alpha]_D^{25}$ −18° (c, 0.1 in MeOH).

Watanabe, K. *et al*, *Phytochemistry*, 1989, **28**, 77.

Temisin   T-80006

[67151-76-2]

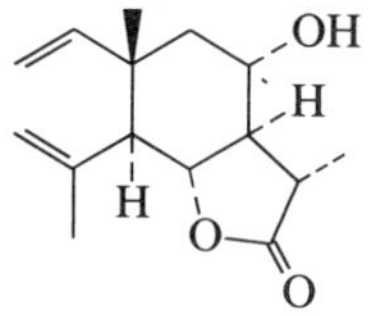

$C_{15}H_{22}O_3$ M 250.337

Isol. from *Artemisia maritima*. Prisms (EtOH). Mp 228°. $[\alpha]_D^{20}$ +70° (CHCl₃).

(±)-form [67225-38-1]

Mp 196-198°.

Asahina, Y. *et al*, *Ber.*, 1941, **74**, 952 (*isol*)
Nishizawa, M. *et al*, *J. Chem. Soc., Chem. Commun.*, 1978, 76 (*struct, synth*)

Tenual   T-80007

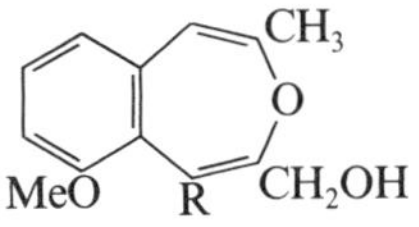

R = CHO

$C_{14}H_{14}O_4$ M 246.262
Constit. of *Asphodeline tenuior*.

Ulubelen, A. *et al*, *Phytochemistry*, 1989, **28**, 649.

Tenucarb   T-80008

As Tenual, T-80007 with

R = COOMe

$C_{15}H_{16}O_5$ M 276.288
Constit. of *Asphodeline tenuior*.

Ulubelen, A. *et al*, *Phytochemistry*, 1989, **28**, 649.

Tephrosin   T-80009

Updated Entry replacing T-00140
13,13a-Dihydro-7a-hydroxy-9,10-dimethoxy-3,3-dimethyl-3H-bis[1]benzopyrano[3,4-b:6′,5′-e]pyran-7(7aH)-one, 9CI.
Hydroxydeguelin. Allotephrosin. Isoallotephrosin
[76-80-2]

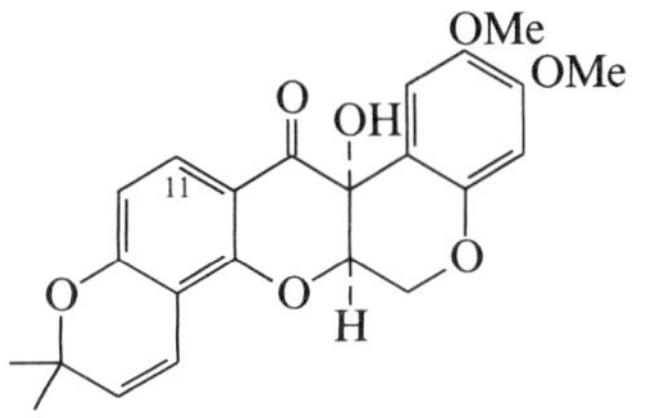

$C_{23}H_{22}O_7$ M 410.423
Found in *Crotalaria burhia, C. medicaginea, Derris elliptica, D. malaccensis, Lonchocarpus* spp., *Mundulea sericea, Piscidia mollis* and *Tephrosia* spp. Some of these samples are the (−)-form, some the (±)-form and others have unreported opt. activity.

(−)-form
Constit. of the seeds of *Millettia dura*. Amorph. $[\alpha]_D^{23}$ −118° (c, 7.8 in C_6H_6).

11-Hydroxy: [72458-85-6]. ***11-Hydroxytephrosin***
$C_{23}H_{22}O_8$ M 426.422
Isol. from fruit of *A. fruticosa*. Opt. active.

(±)-form
Prisms (MeOH/CHCl₃). Mp 198°.

Boam, J.J. *et al*, *J. Soc. Chem. Ind., London*, 1937, **56**, 91 (*rev*)
Ollis, W.D. *et al*, *Tetrahedron*, 1967, **23**, 4741 (*isol, struct*)
Carlson, D.G. *et al*, *Tetrahedron*, 1973, **29**, 2731 (*nmr*)
Braz-Filho, R. *et al*, *Phytochemistry*, 1975, **14**, 1454 (*isol*)
Mitscher, L.A. *et al*, *Heterocycles*, 1979, **12**, 1033 (11-*Hydroxytephrosin*)
Ingham, J.L., *Fortschr. Chem. Org. Naturst.*, 1983, **43**, 1 (*rev, occur*)

Terchebin   T-80010

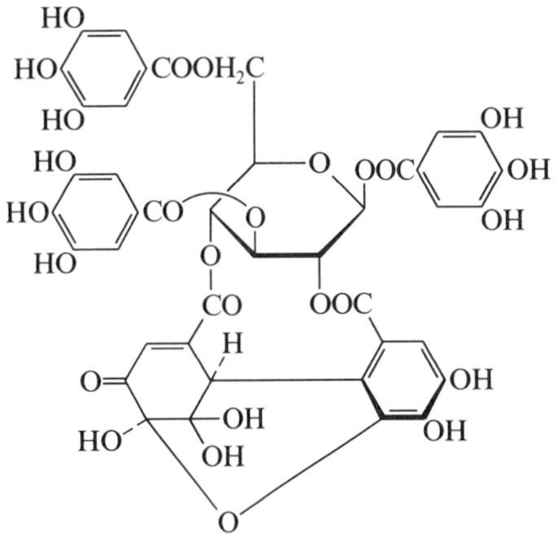

$C_{41}H_{30}O_{26}$ M 938.672
Struct. not certain. Prob. in equilib. with an isomeric struct. Isol. from fruits of *Terminalia chebula*. Small greenish-yellow prisms + 10H₂O. $[\alpha]_D^{25}$ −39.4° (c, 2 in EtOH). Forms trihydrate on intensive drying.

Schmidt, O.T. *et al*, *Justus Liebigs Ann. Chem.*, 1967, **706**, 169 (*isol*)
Jochims, J.C. *et al*, *Justus Liebigs Ann. Chem.*, 1968, **717**, 769 (*pmr*)
Okuda, T. *et al*, *Tetrahedron Lett.*, 1980, **21**, 4361 (*struct*)
Karl, C. *et al*, *Z. Naturforsch., C*, 1983, **38**, 13 (*struct, pmr*)

2,2′:5′,2″-Terthiophene-5-methanol, 9CI   T-80011

5-Hydroxymethyl-2,2′:5,2″-terthiophene. α-Terthienylmethanol
[13059-93-3]

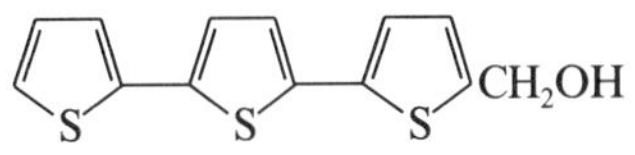

$C_{13}H_{10}OS_3$ M 278.419
Isol. from dried leaves of *Eclipta alba* after hydrol. Lemon-yellow platelets (C_6H_6). Mp 150-151°.

[26905-77-1]

Krishnaswamy, N.R. *et al*, *Curr. Sci.*, 1966, **35**, 542 (*synth*)
Krishnaswamy, N.R. *et al*, *Tetrahedron Lett.*, 1966, 4227 (*isol*)
Nakayama, J. *et al*, *Heterocycles*, 1986, **24**, 637 (*synth*)

1,2,3,4-Tetrabromo-5,6-bis(bromomethyl)benzene, 9CI   T-80012

α,α′,3,4,5,6-Hexabromo-o-xylene, 8CI
[53042-28-7]

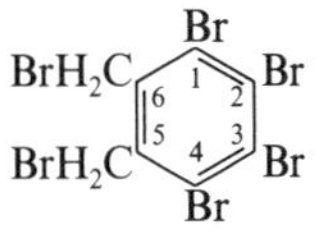

$C_8H_4Br_6$ M 579.544
Cryst. (C_6H_6). Mp 160°.

Hsieh, J.Y.K. *et al*, *Org. Mass Spectrom.*, 1981, **16**, 189 (*ms*)
Kreher, R.P. *et al*, *Chem. Ber.*, 1988, **121**, 1827 (*synth, ir, pmr*)

4,5,6,7-Tetrabromo-2*H*-isoindole   T-80013

[52964-20-2]

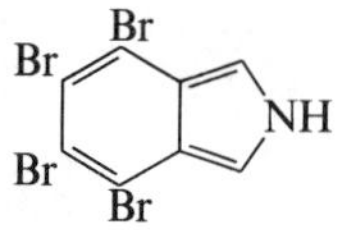

C$_8$H$_3$Br$_4$N   M 432.734
Mp >130° dec.

Kreher, R.P. *et al*, *Chem. Ber.*, 1988, **121**, 1827 (*synth, uv, ir, pmr, cmr, ms*)

Tetrabutylammonium(1+)   T-80014

Updated Entry replacing T-30016
N,N,N-*Tributyl-1-butanaminium, 9CI*
[10549-76-5]

$$(H_3CCH_2CH_2CH_2)_4N^\oplus$$

C$_{16}$H$_{36}$N$^\oplus$ M 242.467 (ion)

Fluoride: [429-41-4].
  C$_{16}$H$_{36}$FN   M 261.465
  Catalyst for aldol condensations, acetylations etc.
  Fluoride ion source. Cryst. + 18H$_2$O. Mp 37°. Also
  forms a hydrate with 32.8H$_2$O, Mp 25°.

Azide: [993-22-6].
  C$_{16}$H$_{36}$N$_4$ M 284.487
  Relatively safe azide for synthetic use. Mp 80°.

Borohydride: [33725-74-5].
  C$_{16}$H$_{40}$BN   M 257.310
  Reagent for redn. of carboxylic acid to aldehydes. Sol.
  CH$_2$Cl$_2$, insol. Et$_2$O.

Formate: [35733-58-5].
  C$_{17}$H$_{37}$NO$_2$ M 287.485
  Reagent for OH-group epimerisations.

Iodotetrachloride:
  C$_{16}$H$_{36}$Cl$_4$IN   M 511.182
  Trans-chlorinating agent for alkenes. Mp 137-139° dec.

Chlorochromate: TBACC
  C$_{16}$H$_{36}$ClCrNO$_3$ M 377.914
  Mild, selective oxidising agent. Orange cryst.
  (EtOAc/hexane). Mp 184-185°.

Hydroxide: [2052-49-5].
  C$_{16}$H$_{37}$NO   M 259.474
  Strong base, suitable for use as nonaqueous titrant.
  Reagent for hydrol. of steroidal tosylates.
  ▷ BS5425000.

Bifluoride:
  C$_{16}$H$_{37}$F$_2$N   M 281.472
  Fluorinating agent.

Perchlorate: [1923-70-2].
  C$_{16}$H$_{36}$ClNO$_4$ M 341.917
  Needles (EtOAc/pentane). Mp 212.5-213.5°.

Bromide: [1643-19-2].
  C$_{16}$H$_{36}$BrNO$_4$ M 386.369
  Prisms (CHCl$_3$/pet. ether). Mp 119-119.5°.

Tetrafluoroborate: [15553-52-3].
  C$_{16}$H$_{36}$BF$_4$N   M 329.272
  Needles (EtOAc/pentane). Mp 162-162.5°.

Acetate: [10534-59-5].
  C$_{18}$H$_{39}$NO$_2$ M 301.512
  Fine yellow cryst. (1,2-dimethoxyethane). Mp 75-83°.
  Further purification gave the AcOH solvate, Mp 112.5-
  114°.

Inorg. Synth., 1957, **5**, 176 (*iodotetrachloride*)
Cundiff, R.H. *et al*, *Anal. Chem.*, 1962, **34**, 584 (*use, hydroxide*)

Gutmann, V. *et al*, *Monatsh. Chem.*, 1964, **95**, 1034 (*azide*)
Wen, W.-Y. *et al*, *J. Phys. Chem.*, 1966, **70**, 1244 (*synth*)
Soriano, J. *et al*, *Inorg. Nucl. Chem. Lett.*, 1969, **5**, 209 (*nmr*)
House, H.O. *et al*, *J. Org. Chem.*, 1971, **36**, 2371 (*synth, ir, pmr, bibl*)
Corey, E.J. *et al*, *J. Am. Chem. Soc.*, 1972, **94**, 6190 (*synth*)
Brandström, A. *et al*, *Tetrahedron Lett.*, 1972, 3173 (*borohydride*)
Cowell, D.B. *et al*, *J. Chem. Soc., Perkin Trans.* 1, 1974, 1505 (*use, hydroxide*)
Pless, J., *J. Org. Chem.*, 1974, **39**, 2644 (*fluoride*)
Kuwajima, I. *et al*, *Synthesis*, 1976, 602 (*fluoride*)
Fieser and Fieser's Reagents for Organic Synthesis, Wiley, 1977, **6**, 563, 564; 1979, **7**, 353, 354; 1980, **8**, 467; 1981, **9**, 443, 444, 447 (*use*)
Ogilvie, K.K. *et al*, *Tetrahedron Lett.*, 1978, 1663 (*fluoride*)
Sharma, R.K. *et al*, *J. Org. Chem.*, 1983, **48**, 2112 (*props*)
Santaniello, E. *et al*, *Synthesis*, 1983, 749 (*chlorochromate, synth, use*)
Clark, J.H. *et al*, *Tetrahedron Lett.*, 1985, **26**, 2233 (*use, fluoride*)
Bosch, P. *et al*, *Tetrahedron Lett.*, 1987, **28**, 4733 (*bifluoride, synth, use*)

1,1,1,2-Tetrachloro-2-fluoroethane, 9CI   T-80015

[354-11-0]

$$Cl_3CCHFCl$$

C$_2$HCl$_4$F   M 185.839
d$_4^{20}$ 1.625. Mp −95.4°. Bp 117°. n$_D^{20}$ 1.4525 (1.4488).

Miller, W.T., *J. Am. Chem. Soc.*, 1940, **62**, 341 (*synth*)
Wilmshurst, J.K., *Can. J. Chem.*, 1957, **35**, 937 (*ir*)

1,1,2,2-Tetrachloro-1-fluoroethane, 9CI   T-80016

[354-14-3]

$$Cl_2CFCHCl_2$$

C$_2$HCl$_4$F   M 185.839
d$_4^{20}$ 1.622. Bp 115-117°. n$_D^{20}$ 1.4487.

Miller, W.T., *J. Am. Chem. Soc.*, 1940, **62**, 341 (*synth*)
Alger, T.D. *et al*, *J. Chem. Phys.*, 1967, **47**, 3130 (*pmr, F-19 nmr*)
Paleta, O. *et al*, *Collect. Czech. Chem. Commun.*, 1968, **33**, 1294 (*synth*)

4,5,6,7-Tetrachloro-2*H*-isoindole   T-80017

[60432-74-8]

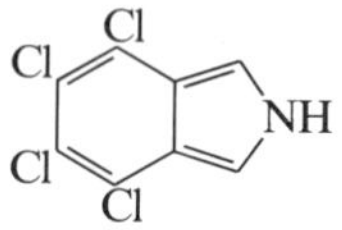

C$_8$H$_3$Cl$_4$N   M 254.929
Cryst. Mp 200° dec.

Kreher, R.P. *et al*, *Chem. Ber.*, 1988, **121**, 1827 (*synth, uv, ir, pmr, cmr, ms*)

1,1,1,2-Tetrachloro-2,3,3,3-tetrafluoro-   T-80018
propane, 9CI, 8CI

[3175-64-2]

$$F_3CCClFCCl_3$$

C$_3$Cl$_4$F$_4$ M 253.837
d$_4^{20}$ 1.725. Bp 112-112.5°. n$_D^{20}$ 1.4002.

[29255-31-0]

Fainberg, A.H. *et al*, *J. Org. Chem.*, 1965, **30**, 864 (*props*)
Paleta, O. *et al*, *Collect. Czech. Chem. Commun.*, 1971, **36**, 2257 (*synth*)
Paleta, O. *et al*, *Bull. Soc. Chim. Fr.*, 1986, 920 (*synth*)

1,1,1,3-Tetrachloro-2,2,3,3-tetrafluoro-propane, 9CI, 8CI

T-80019

[2268-46-4]

$$Cl_3CCF_2CClF_2$$

$C_3Cl_4F_4$ M 253.837

Solv. for disperse dyes and for glc analysis. Dielectric. Liq. d_4^{25} 1.693. Fp −92.78°. Bp 114°. n_D^{25} 1.3974, n_D^{20} 1.3966.

Coffman, D.D. et al, J. Am. Chem. Soc., 1949, 71, 979 (synth)
Krespan, C.G. et al, J. Am. Chem. Soc., 1961, 83, 3424 (synth)
Fainberg, A.H. et al, J. Org. Chem., 1965, 30, 864 (props)
Paleta, O. et al, Collect. Czech. Chem. Commun., 1971, 36, 2257 (synth)
Entz, R.C. et al, J. Agric. Food Chem., 1982, 30, 84 (chromatog)
Brabets, R. et al, J. Fluorine Chem., 1988, 41, 311 (props)

1,1,2,2-Tetrachloro-1,3,3,3-tetrafluoro-propane, 8CI

T-80020

[2268-44-2]

$$F_3CCCl_2CCl_2F$$

$C_3Cl_4F_4$ M 253.837

Hydraulic fluid component. Mp 34°. Bp 112-114°.

[29255-31-0]

McBee, E.T. et al, J. Am. Chem. Soc., 1948, 70, 2023 (synth)
Paleta, O. et al, Collect. Czech. Chem. Commun., 1971, 36, 2257 (synth)

1,1,2,3-Tetrachloro-1,2,3,3-tetrafluoro-propane, 9CI, 8CI

T-80021

[2268-45-3]

$$F_2CClCClFCFCl_2$$

$C_3Cl_4F_4$ M 253.837

Liq. (glass at low temp.). d_4^{20} 1.719. Mp −58°. Bp 110-112°. n_D^{20} 1.3960.

[29255-31-0]

Henne, A.L. et al, J. Am. Chem. Soc., 1941, 63, 3476 (synth)
Fainberg, A.H. et al, J. Org. Chem., 1965, 30, 864 (props)
Paleta, O. et al, Bull. Soc. Chim. Fr., 1986, 920 (synth)

1,1,3,3-Tetrachloro-1,2,2,3-tetrafluoro-propane, 9CI, 8CI

T-80022

[2354-04-3]

$$Cl_2CFCF_2CCl_2F$$

$C_3Cl_4F_4$ M 253.837

d_D^{20} 1.701. Bp 112.5-113.5° (116.5°). n_D^{20} 1.3980.

Coffman, D.D. et al, J. Am. Chem. Soc., 1949, 71, 979 (synth)
Fainberg, A.H. et al, J. Org. Chem., 1965, 30, 864 (props)
White, H.F., Anal. Chem., 1966, 38, 625 (F nmr)

Tetracyanooxirane

T-80023

Oxiranetetracarbonitrile, 9CI. 1,2-Epoxy-1,1,2,2-ethanetetracarbonitrile, 8CI. Tetracyanoethylene oxide

[3189-43-3]

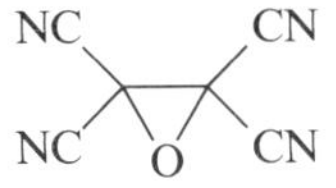

C_6N_4O M 144.092

Needles (1,2-dichloroethane). Mp 177-178° (sealed tube).

▷ Slowly evolves HCN on exposure to H_2O.

Sadtler Standard Infrared Spectra, 23822, 38559P (ir)

Reiche, A. et al, Chem. Ber., 1963, 96, 3044 (synth)
Linn, W.J. et al, J. Am. Chem. Soc., 1963, 85, 2032; 1965, 87, 3651, 3657, 3665 (synth)
Org. Synth., 1969, 49, 103 (synth, ir, haz)
Matthews, D.A. et al, J. Am. Chem. Soc., 1971, 93, 5945 (cryst struct)

Tetracyclo[3.2.0.0^{1,6}.0^{2,6}]heptane, 9CI

T-80024

Tetracyclo[4.1.0.0^{1,5}.0^{2,6}]heptane

[109900-65-4]

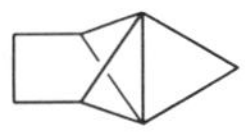

C_7H_8 M 92.140

Bridged [1.1.1]propellane. The higher homologue Tetracyclo[5.1.0.0^{1,6}.0^{2,7}]octane also studied.

Semmler, K. et al, J. Am. Chem. Soc., 1985, 107, 6410.
Seiler, P. et al, Helv. Chim. Acta, 1988, 71, 2100 (cryst struct)

Tetracyclo[3.3.0.0^{2,8}.0^{3,6}]octane, 9CI

T-80025

[5078-81-9]

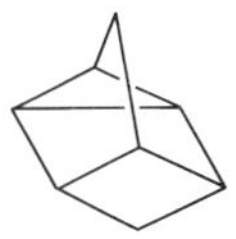

C_8H_{10} M 106.167

Liq.

Freeman, P.K. et al, J. Chem. Soc., Chem. Commun., 1965, 511 (synth, pmr)
Freeman, P.K. et al, J. Org. Chem., 1973, 38, 3823 (synth)
Bentley, T.W. et al, J. Org. Chem., 1988, 53, 3066 (synth)

4,6-Tetradecadiene-8,10-diyne-1,12-diol

T-80026

$$H_3CCH_2CH(OH)C{\equiv}CC{\equiv}CCH{=}CHCH{=}CHCH_2CH_2$$
$$CH_2OH$$

$C_{14}H_{18}O_2$ M 218.295

(E,E)-form [10523-88-3]

Isol. from above-ground parts of *Cotula coronopifolia*. Cryst. (CCl_4). Mp 55°.

1-Ac: [10523-86-1]. *14-Acetoxy-8,10-tetradecadiene-4,6-diyn-3-ol*

$C_{16}H_{20}O_3$ M 260.332

Isol. from *C. coronopifolia*. Oil. λ_{max} 226, 235.5, 294, 310 nm (Et_2O).

Di-Ac: [20695-84-5]. *1,12-Diacetoxy-4,6-tetradecadiene-8,10-diyne*

$C_{18}H_{22}O_4$ M 302.369

Isol. from *C. coronopifolia*. Oil. λ_{max} 226.5 (ϵ 22800), 236 (34000), 293 (30600), 309 infl. (25400) nm (Et_2O).

Bohlmann, F. et al, Chem. Ber., 1966, 99, 2828; 1968, 100, 2738 (isol, uv, ir, pmr, ord, struct, synth)

6,12-Tetradecadiene-8,10-diyne-1,3-diol

T-80027

$$H_3CCH{=}CHC{\equiv}CC{\equiv}CCH{=}CHCH_2CH_2CH(OH)CH_2$$
$$CH_2OH$$

$C_{14}H_{18}O_2$ M 218.295

(E,E)-form

Pale-yellow amorph. powder. $[\alpha]_D$ +0.73° (c, 1.1 in MeOH).

Di-Ac: [89913-46-2].

$C_{18}H_{22}O_4$ M 302.369

Isol. from Atractylodes Rhizome (*Atractylodes japonica*). Oil. $[\alpha]_D \pm 0°$ (c, 1.7 in MeOH).

Kano, Y. *et al, Chem. Pharm. Bull.*, 1989, **37**, 193 (*isol, uv, pmr, struct*)

6,12-Tetradecadiene-8,10-diyn-3-ol, 9CI   T-80028

$$H_3CCH=CHC≡CC≡CCH=CHCH_2CH_2CH(OH)CH_2$$
$$CH_3$$

$C_{14}H_{18}O$ M 202.296

(*E,E*)-*form*

Isol. from roots of *Anthemis suguramica*. Cryst. (pet. ether). Mp 38°.

Bohlmann, F. *et al, Chem. Ber.*, 1966, **99**, 1642 (*isol, uv, ir, pmr, struct*)

2,12-Tetradecadiene-4,6,8,10-tetrayne   T-80029

$$H_3CCH=CHC≡CC≡CC≡CCH=CHCH_3$$

$C_{14}H_{10}$ M 178.233

Isol. from *Triticum aestivum*. λ_{max} 273 (ϵ 98 400), 292 (75300), 317 (8200), 340 (12100), 365 (13600), 395 nm (7300)(pentane).

Jones, E.R.H. *et al, Nature (London)*, 1951, **168**, 900 (*synth, uv*)
Schulte, H.E. *et al, Phytochemistry*, 1965, **4**, 481 (*isol, uv*)

5,7,9,11-Tetradecatetraen-1-ol   T-80030

$$H_3CCH_2(CH=CH)_4CH_2CH_2CH_2CH_2OH$$

$C_{14}H_{22}O$ M 206.327

(*5E,7E,9Z,11E*)-*form*

Centaur Y

Isol. from above-ground parts of *Centaurea cyanus* and *C. ruthenica*. Cryst. (pet. ether). Mp 48-50°.

Bohlmann, F. *et al, Chem. Ber.*, 1961, **94**, 3179; 1962, **95**, 2939 (*isol, uv, ir, struct, synth*)

4,6,10,12-Tetradecatetraen-8-yn-1-ol, 9CI   T-80031

[16692-74-3]

$$H_3CCH=CHCH=CHC≡CCH=CHCH=CHCH_2CH_2$$
$$CH_2OH$$

$C_{14}H_{18}O$ M 202.296

(*all-E*)-*form* [10523-90-7]

Isol. from roots of *Saussurea alpina* and from *Zoegea baldschuanica*. Cryst. (pet. ether). Mp 94-95°.

Ac: [13028-52-9]. *14-Acetoxy-2,4,8,10-tetradecaen-6-yne*
$C_{16}H_{20}O_2$ M 244.333
Isol. from above-ground parts of *Cotula coronopifolia*. Cryst. (pet. ether). Mp 32°.

Bohlmann, F. *et al, Chem. Ber.*, 1966, **99**, 2828, 3201 (*isol, uv, ir, pmr, struct*)

4,6,12-Tetradecatrien-8,10-diyn-1-ol, 9CI   T-80032

[13081-23-7]

$$H_3CCH=CHC≡CC≡CCH=CHCH=CHCH_2CH_2$$
$$CH_2OH$$

$C_{14}H_{16}O$ M 200.280

(*E,E,E*)-*form* [17090-99-2]

Isol. from *Dahlia merckii* and roots of *Anthemis saguramica*. Platelets (Et$_2$O/pet. ether). Mp 110-111°.

Ac: [1540-89-2].
$C_{16}H_{18}O_2$ M 242.317

Isol. from *A. carpatica, D. merckii, Cotula coronopifolia* and other plants. Cryst. (pet. ether). Mp 31-33°.

Bohlmann, F. *et al, Justus Liebigs Ann. Chem.*, 1963, **668**, 51 (*isol, deriv*)
Bohlmann, F. *et al, Chem. Ber.*, 1965, **98**, 872; 1966, **99**, 1642, 3544 (*isol uv, ir, pmr, struct, synth*)
Sörensen, J.S. *et al, Acta Chem. Scand.*, 1966, **20**, 992 (*isol, synth*)
Aplin, R.T. *et al, J. Chem. Soc., Chem. Commun.*, 1967, 140 (*ms*)
Bedford, C.T., *J. Chem. Soc., Perkin Trans. 1*, 1976, 735 (*isol, synth*)

4,6,12-Tetradecatriene-8,10-diyne-1,3-diol   T-80033

$$H_3CCH=CHC≡CC≡CCH=CHC=CHCH(OH)CH_2$$
$$CH_2OH$$

$C_{14}H_{16}O_2$ M 216.279

(*all-E*)-*form*

Di-Ac:
$C_{18}H_{20}O_4$ M 300.354
Isol. from *Atraclylodes japonica*. Pale-yellow oil. $[\alpha]_D$ +0.4° (c, 1.0 in MeOH).

Kano, Y. *et al, Chem. Pharm. Bull.*, 1989, **37**, 193 (*isol, ir, uv, ms*)

2-Tetradecene, 9CI, 8CI   T-80034

[1652-97-7]

$$H_3C(CH_2)_{10}CH=CHCH_3$$

$C_{14}H_{28}$ M 196.375

(*E*)-*form* [35953-54-9]
Liq. Bp 253°.

(*Z*)-*form* [35953-53-8]
Liq. Bp 253.9°.

Janak, J., *Anal. Chem.*, 1973, **45**, 293 (*chromatog*)
Uemura, S. *et al, J. Am. Chem. Soc.*, 1983, **105**, 2748 (*synth*)
Polyakova, A.A. *et al, Zh. Org. Khim.*, 1987, **23**, 1164 (*ms*)

3-Tetradecene, 9CI, 8CI   T-80035

[36587-78-7]

$$H_3C(CH_2)_9CH=CHCH_2CH_3$$

$C_{14}H_{28}$ M 196.375

(*E*)-*form*
Liq. Bp 251.4°.

(*Z*)-*form*
Liq. Bp 251.3°.

Janak, J., *Anal. Chem.*, 1973, **45**, 293 (*chromatog*)
Polyakova, A.A. *et al, Zh. Org. Khim.*, 1987, **23**, 1164 (*ms*)

4-Tetradecene, 9CI, 8CI   T-80036

[54322-28-0]

$$H_3C(CH_2)_8CH=CHCH_2CH_2CH_3$$

$C_{14}H_{28}$ M 196.375

(*E*)-*form* [41446-78-0]
Liq. Bp 250.9°.

(*Z*)-*form* [41446-65-5]
Liq. Bp 250.5°.

Janak, J., *Anal. Chem.*, 1973, **45**, 293 (*chromatog*)
Polyakova, A.A., *Zh. Org. Khim.*, 1987, **23**, 1164 (*ms*)

5-Tetradecene, 9CI, 8CI   **T-80037**

$$H_3C(CH_2)_7CH{=}CH(CH_2)_3CH_3$$

$C_{14}H_{28}$ M 196.375
(*E*)-*form* [41446-66-6]
  Liq. Bp 250.8°, $Bp_{0.1}$ 56-61°. n_D^{20} 1.4365.
(*Z*)-*form* [41446-62-2]
  Bp 250°.

Janak, J., *Anal. Chem.*, 1973, **45**, 293 (*chromatog*)
Batchelor, J.G. *et al, J. Am. Chem. Soc.*, 1973, **95**, 6358 (*cmr*)
Commercon, A. *et al, Tetrahedron Lett.*, 1975, 3837 (*synth*)
Neumann, H. *et al, Chem. Ber.*, 1978, **111**, 2785 (*synth, ir, pmr*)
Brown, H.C. *et al, J. Org. Chem.*, 1986, **26**, 5270 (*synth*)
Polyakova, A.A. *et al, Zh. Org. Khim.*, 1987, **23**, 1164 (*ms*)

6-Tetradecene, 9CI, 8CI   **T-80038**

[23015-35-2]

$$H_3C(CH_2)_6CH{=}CH(CH_2)_4CH_3$$

$C_{14}H_{28}$ M 196.375
(*E*)-*form* [41446-64-4]
  Liq. Bp 250.4°.
(*Z*)-*form* [41446-61-1]
  Liq. Bp 249.6°.

Meshcheryakov, A.P. *et al, Izv. Akad. Sci. USSR, Ser. Sci. Khim.*, 1958, 780; *ca*, **52**, 19974 (*synth*)
Janak, J., *Anal. Chem.*, 1973, **45**, 293 (*chromatog*)
Polyakova, A.A. *et al, Zh. Org. Khim.*, 1987, **23**, 1164 (*ms*)

7-Tetradecene, 9CI, 8CI   **T-80039**

[10374-74-0]

$$H_3C(CH_2)_5CH{=}CH(CH_2)_5CH_3$$

$C_{14}H_{28}$ M 196.375
(*E*)-*form* [41446-63-3]
  Liq. Bp 250.2°.
(*Z*)-*form* [41446-60-0]
  Liq. Bp 249.4°, $Bp_{0.8}$ 75-76°. n_D^{20} 1.4399.

Sadtler Standard Infrared Spectra, 40362 (*ir*)
Sadtler Standard NMR Spectra, 9909 (*pmr*)
Janak, J., *Anal. Chem.*, 1973, **45**, 293 (*chromatog*)
Brown, H.C. *et al, J. Org. Chem.*, 1982, **47**, 3806, 3808; 1986, **26**, 5270; 1988, **53**, 239 (*synth, pmr*)
Julia, M. *et al, Tetrahedron Lett.*, 1982, **23**, 2457 (*synth*)
Polyakova, A.A. *et al, Zh. Org. Khim.*, 1987, **23**, 1164 (*ms*)

8-Tetradecene-11,13-diyn-2-one, 9CI   **T-80040**

[115006-46-7]

$$HC{\equiv}CC{\equiv}CCH_2CH{=}CH(CH_2)_5COCH_3$$

$C_{14}H_{18}O$ M 202.296
(*Z*)-*form* [13945-78-3]
  Isol. from roots of *Centaurea pullata* and *C. ferox*. Present in *Echinacea* spp. Oil.

Bohlmann, F. *et al, Chem. Ber.*, 1966, **99**, 3544 (*isol, ir, pmr, ms, struct*)
Heinzer, F. *et al, Pharm. Acta Helv.*, 1988, **63**, 132; *ca*, **109**, 146331z (*isol*)

9-Tetradecene-2,4,6-triynedioic acid, 9CI   **T-80041**

$$HOOC(CH_2)_3CH{=}CHCH_2C{\equiv}CC{\equiv}CC{\equiv}CCOOH$$

$C_{14}H_{12}O_4$ M 244.246
(*Z*)-*form* [109921-61-1]

Isol. from *Poria sinuosa*. Amorph. unstable solid. Dec. at ca. 120°.

Cambie, R.C. *et al, J. Chem. Soc.*, 1963, 2065 (*isol, uv, ir, struct*)

5-Tetradecene-8,10,12-triyn-1-ol   **T-80042**

$$H_3CC{\equiv}CC{\equiv}CC{\equiv}CCH_2CH{=}CH(CH_2)_3CH_2OH$$

$C_{14}H_{16}O$ M 200.280
(*Z*)-*form*
  Ac: 13-Acetoxy-9-tetradecene-2,4,6-triyne
  $C_{16}H_{18}O_2$ M 242.317
  Isol. from above-ground parts of *Chrysanthemum serotinum*. Oil.

Bohlmann, F. *et al, Chem. Ber.*, 1966, **99**, 1830 (*isol, ir, uv, pmr, struct*)

6-Tetradecene-8,10,12-triyn-3-ol   **T-80043**

$$H_3C(C{\equiv}C)_3CH{=}CHCH_2CH_2CH(OH)CH_2CH_3$$

$C_{14}H_{16}O$ M 200.280
(−)-(*E*)-*form*
  Isol. from *Anacyclus pyrethrum* and *Anthemis saguramica*. Cryst. (pet. ether). $[\alpha]_D^{22}$ −56° (c, 0.25 in Et_2O).

Bohlmann, F. *et al, Chem. Ber.*, 1965, **98**, 1411; 1966, **99**, 1642 (*isol, uv, ir, pmr, ord, struct*)

3-Tetradecenoic acid   **T-80044**

$$H_3C(CH_2)_9CH{=}CHCH_2COOH$$

$C_{14}H_{26}O_2$ M 226.358
(*E*)-*form* [75730-21-1]
  Isol. from seed oil of *Grindelia oxylepis*.

[62472-83-7, 81236-42-2]

Kleiman, R. *et al, Lipids*, 1966, **1**, 301 (*isol*)
Gunstone, F.D. *et al, Chem. Phys. Lipids*, 1977, **18**, 115 (*cmr*)
Mizugaki, M. *et al, Chem. Pharm. Bull.*, 1980, **28**, 2347 (*nmr*)
Ogura, K. *et al, Tetrahedron Lett.*, 1981, **22**, 4499 (*synth*)

5-Tetradecen-13-olide   **T-80045**
14-Methyloxacyclotetradec-6-en-2-one, 9CI

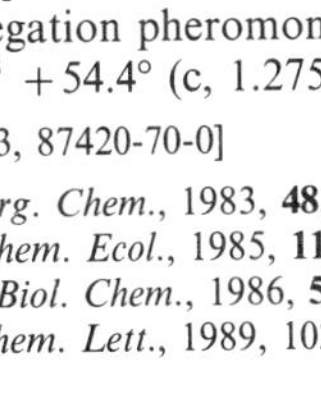

$C_{14}H_{24}O_2$ M 224.342
(*S,Z*)-*form* [87420-69-7]
  Synergist of aggregation pheromone of *Cryptolestes pusillus*. Oil. $[\alpha]_D^{23}$ +54.4° (c, 1.275 in $CHCl_3$).

[77761-59-2, 78418-66-3, 87420-70-0]

Millar, J.G. *et al, J. Org. Chem.*, 1983, **48**, 4404 (*synth*)
Millar, J.G. *et al, J. Chem. Ecol.*, 1985, **11**, 1053 (*synth*)
Sakai, T. *et al, Agric. Biol. Chem.*, 1986, **50**, 1621 (*synth*)
Naoshima, Y. *et al, Chem. Lett.*, 1989, 1023 (*synth, isol*)

5,6,9,10-Tetradehydro-1,4:3,14-diepoxy-4-hydroxy-4,5-secofuranoeremophilane T-80046

5,7,8,10-Tetrahydro-3,8-dimethyl-7,10-methanofuro[2,3-i][2,5]benzodioxocin-8-ol, 9CI

[53820-32-9]

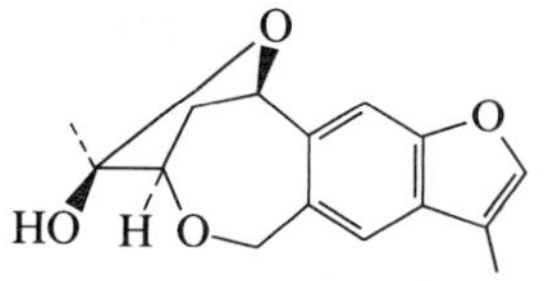

$C_{15}H_{16}O_4$ M 260.289

Constit. of *Euryops hebecarpus*. Oil. λ_{max} 250, 275, 281 nm.

Bohlmann, F. *et al, Chem. Ber.*, 1974, **107**, 2730.

Tetraethoxyethene, 9CI T-80047

Tetraethoxyethylene

[40923-93-1]

$$(EtO)_2C=C(OEt)_2$$

$C_{10}H_{20}O_4$ M 204.266

Bp$_{15}$ 77°. n_D^{25} 1.4212.

McElvain, S.M. *et al, J. Am. Chem. Soc.*, 1947, **69**, 2661 (*synth*)
Scheeren, J.W. *et al, Recl. Trav. Chim. Pays-Bas* (*J. R. Neth. Chem. Soc.*), 1973, **92**, 11 (*synth*)

Tetraethynylthiophene T-80048

[113705-25-2]

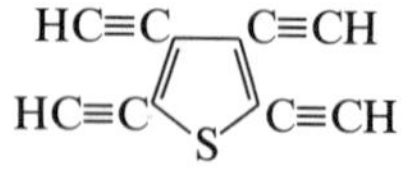

$C_{12}H_4S$ M 180.230

Solid. Unstable, rapidly turns black on standing in air.

▷ Explosive. Heat and shock sensitive, exploded on attempted subl.

Neenan, T.X. *et al, J. Org. Chem.*, 1988, **53**, 2489 (*synth, pmr, cmr, ms, uv, haz*)

Tetrafluorobutanedioic acid, 9CI T-80049

Updated Entry replacing T-00642

Tetrafluorosuccinic acid. Perfluorosuccinic acid

[377-38-8]

$$HOOCCF_2CF_2COOH$$

$C_4H_2F_4O_4$ M 190.051

Mp 116°. Bp$_{15}$ 150°.

▷ WN0725000.

Hydrate: Mp 86.4-87.4°.

Dianilinium salt: Cryst. (EtOH/CHCl$_3$). Mp 224°.

Di-Me ester: [356-36-5].
  $C_6H_6F_4O_4$ M 218.105
  Bp 177-178°.

Di-Et ester:
  $C_8H_{10}F_4O_4$ M 246.158
  Liq. d_4^{20} 1.26. Bp$_{15}$ 89°.

Dichloride: [356-15-0].
  $C_4Cl_2F_4O_2$ M 226.942
  Bp 86.1-86.3°.

Diamide:
  $C_4H_4F_4N_2O_2$ M 188.081
  Cryst. (H$_2$O). Mp 261°.

Dianilide: Cryst. (Me$_2$CO aq.). Mp 227.5-228°.

Anhydride: [699-30-9]. *3,3,4,4-Tetrafluorodihydro-2,5-furandione, 9CI. Tetrafluorosuccinic anhydride*
  $C_4F_4O_3$ M 172.036
  Cryst. d_4^{20} 1.61. Mp 54-56°.

Henne, A.L. *et al, J. Am. Chem. Soc.*, 1947, **69**, 281 (*synth*)
Haszeldine, R.N., *J. Chem. Soc.*, 1954, 4026 (*synth*)
Fear, E.J.P. *et al, J. Appl. Chem.*, 1955, **5**, 589 (*ester, anhydride, synth*)
Fear, E.J.P. *et al, J. Chem. Soc.*, 1956, 3199 (*synth, anhydride, chloride, amide*)

1,1,2,2-Tetrafluoro-1,2-diiodoethane, 9CI T-80050

1,2-Diiodoperfluoroethane. sym-*Diiodotetrafluoroethane*

[354-65-4]

$$F_2ClCF_2I$$

$C_2F_4I_2$ M 353.825

d_4^{25} 2.629. Bp 112-113°, Bp$_{14}$ 23°. n_D^{25} 1.489.

Kushida, K. *et al, Proton and Fluorine Nuclear Magnetic Resonance Spectral Data*, Japan Halon Co., 079 (*F-19 nmr*)
Emeleus, H.J. *et al, J. Chem. Soc.*, 1949, 2948 (*synth*)
Coffman, D.D. *et al, J. Org. Chem.*, 1949, **14**, 747 (*synth*)
Nodiff, E.A. *et al, J. Org. Chem.*, 1953, **18**, 235 (*props*)
Knunyants, I.L. *et al, Izv. Akad. Sci. USSR, Ser. Sci. Khim.*, 1964, 384 (*synth*)
Chung Wu, E. *et al, J. Phys. Chem.*, 1975, **79**, 1078 (*uv*)
Powell, D.L. *et al, J. Raman Spectrosc.*, 1978, **7**, 111 (*ir, raman*)
Ulm, K. *et al, Toxicol. Lett.*, 1980, **6**, 365 (*tox*)

1,1,2,2-Tetrafluoroethane, 9CI T-80051

[430-66-0]

$$F_2CHCHF_2$$

$C_2H_2F_4$ M 102.031

Gas. Bp −22.5°.

▷ Flammable.

Young, D.S. *et al, J. Am. Chem. Soc.*, 1940, **62**, 1171 (*synth*)
Knunyants, I.L. *et al, Izv. Akad. Sci. USSR, Ser. Sci. Khim.*, 1958, 906 (*synth*)
Dean, R.R. *et al, Trans. Faraday Soc.*, 1968, **64**, 1409 (*pmr, F-19 nmr*)

2,4,5,7-Tetrafluoro-9H-fluoren-9-one T-80052

[114995-45-8]

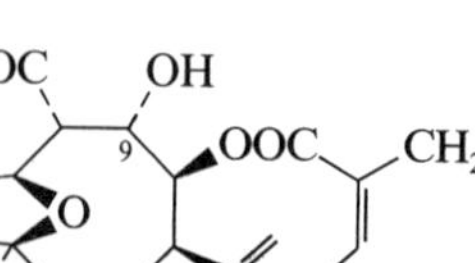

$C_{13}H_4F_4O$ M 252.168

Bright yellow cryst. solid (EtOAc/hexane). Bp$_{0.25}$ 70° subl.

Kyba, E. *et al, J. Org. Chem.*, 1988, **53**, 3513 (*synth, cmr*)

Tetragonolide T-80053

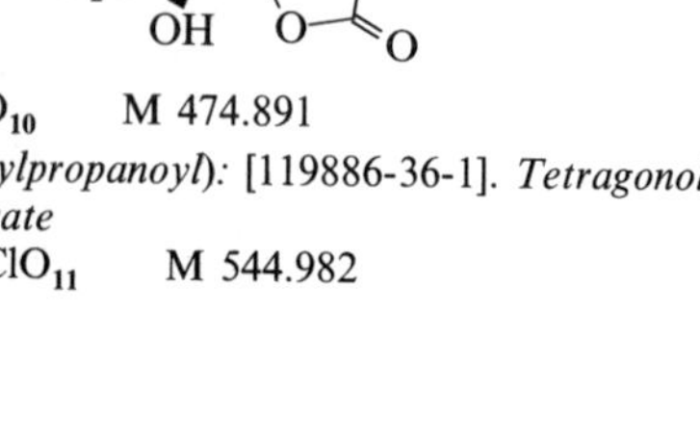

$C_{21}H_{27}ClO_{10}$ M 474.891

9-(2-Methylpropanoyl): [119886-36-1]. *Tetragonolide isobutyrate*
  $C_{25}H_{33}ClO_{11}$ M 544.982

Constit. of *Tetragonotheca repanda* and *T. ludoviciana.*
Oil.

9-(2-Methylbutanoyl): [119863-95-5]. *Tetragonolide 2-methylbutyrate*
$C_{26}H_{35}ClO_{11}$ M 559.009
Constit. of *T. ludoviciana.* Oil.

Jakupovic, J. *et al, Phytochemistry*, 1988, **27**, 3881.

1,2,3,4-Tetrahydrobenz[*g*]isoquinoline   T-80054

[21628-46-6]

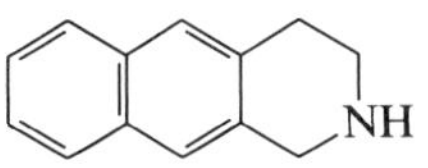

$C_{13}H_{13}N$ M 183.252
Cryst. (pet. ether). Mp 121°.

B,HCl: [112576-43-9].
Cryst. Mp 290-295°.

N-Ac:
$C_{15}H_{15}NO$ M 225.290
Cryst. (cyclohexane). Mp 157°.

Etienne, A. *et al, Compt. Rend. Hebd. Seances Acad. Sci. Sect. C,*
1968, **267**, 1826 (*synth*)
Young, S.D. *et al, J. Org. Chem.*, 1988, **53**, 1114 (*synth, pmr*)

1,4,5,6-Tetrahydrocyclopentapyrazole, 9CI   T-80055

[structure]

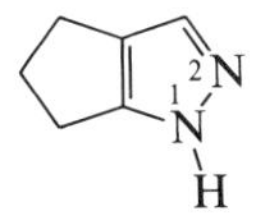

$C_6H_8N_2$ M 108.143
1H-form [2214-03-1]
Cryst. (MeOH). Mp 57-59°. Bp_{30} 165°.
2H-form [15409-55-9]
Derivs. have various commercial uses.
2-Ph: [120344-42-5].
$C_{12}H_{12}N_2$ M 184.240
Cryst. Mp 79°.
2-Benzyl: [84597-88-6].
$C_{13}H_{14}N_2$ M 198.267
Liq. $Bp_{2.5}$ 160°.

Wallach, O., *Justus Liebigs Ann. Chem.*, 1903, **329**, 109 (*synth*)
Bardou, L. *et al, Bull. Soc. Chim. Fr.*, 1967, 289 (*synth*)
Gustafsson, H. *et al, Acta Chem. Scand., Ser. B*, 1974, **28**, 1069
(*synth, pmr*)
Aubert, T. *et al, Synthesis*, 1988, 742 (*synth, ir, pmr, ms, use, bibl*)

1,3,4,5-Tetrahydrocycloprop[*f*]indene   T-80056

[112504-81-1]

[structure]

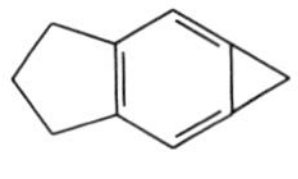

$C_{10}H_{10}$ M 130.189
Pale yellow oil.

Billups, W.E. *et al, J. Org. Chem.*, 1988, **53**, 1312 (*synth, ir, pmr*)

4*b*,8*b*,8*c*,8*d*-Tetrahydrodibenzo [*a,f*]cyclopropa[*cd*]pentalene   T-80057

Dibenzotricyclo[3.3.0.0²·⁸]octa-3,6-diene.
Dibenzosemibullvalene
[2199-28-2]

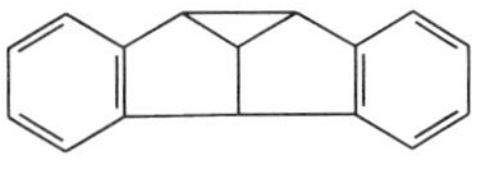

$C_{16}H_{12}$ M 204.271
Cryst. (Pet. ether). Mp 104-105°.

Ciganek, E., *J. Am. Chem. Soc.*, 1966, **88**, 2882 (*synth*)
Cristol, S.J. *et al, J. Am. Chem. Soc.*, 1967, **89**, 401 (*synth*)

5,7,12,14-Tetrahydrodibenzo [*c,h*][1,6]dithiecin, 9CI   T-80058

2,11-Dithia[3.3]orthocyclophane
[7215-69-2]

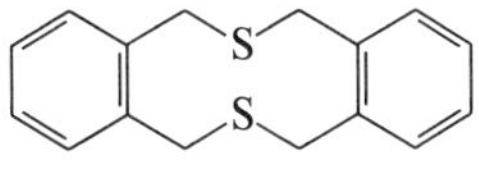

$C_{16}H_{16}S_2$ M 272.434
Needles (C_6H_6). Mp 249-251° (244-246°).

Au, M.-K. *et al, J. Chem. Soc., Perkin Trans. 1*, 1979, 1475 (*synth, pmr*)
Lai, Y.-H. *et al, J. Org. Chem.*, 1988, **53**, 2360 (*synth, pmr, ir, uv, ms, conformn*)

6,7,14,15-Tetrahydrodibenzo [*b,h*][1,4,7,10]tetraoxacyclododecin, 9CI   T-80059

Dibenzo-12-crown-4
[14696-05-0]

[structure]

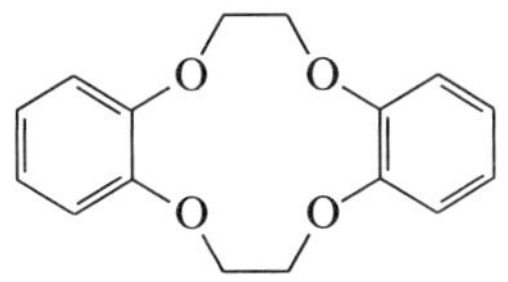

$C_{16}H_{16}O_4$ M 272.300
Mp 208-209°.

Pedersen, C.J., *J. Am. Chem. Soc.*, 1967, **89**, 7017 (*synth*)
Charland, J.P. *et al, Acta Crystallogr., Sect. C*, 1989, **45**, 165 (*cryst struct*)

5,6,20,21-Tetrahydro-2,16;3,10-diethano-15,11-metheno-11*H*-tribenzo[*a,e,i*] pentadecene, 9CI   T-80060

[116129-69-2]

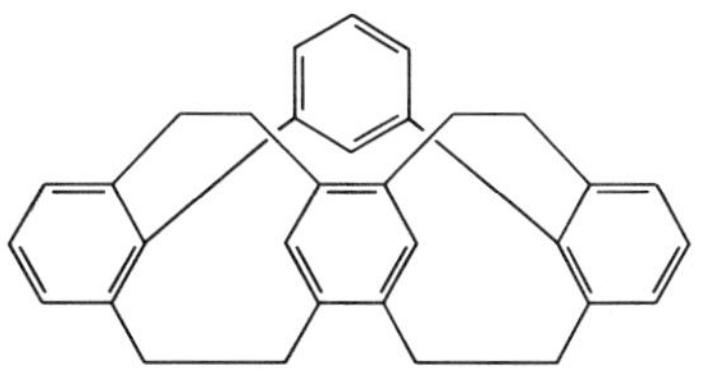

$C_{32}H_{28}$ M 412.573
Mp 266°.

Vinod, T.K. *et al, J. Am. Chem. Soc.*, 1988, **110**, 6574 (*synth*)

Tetrahydro-2,2-dimethyl-5-(1-methyl-1-propenyl)furan　　　T-80061

[7416-35-5]

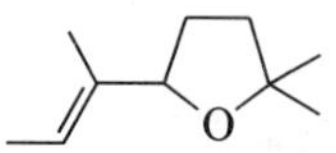

$C_{10}H_{18}O$　　M 154.252

Isol. from oil of *Citrus aurantifolia*. Oil. d^{20} 0.867. $[\alpha]_D^{20}$ ±0°. n_D^{20} 1.4662.

[56058-69-6, 56058-70-9]

Kováts, E. *et al*, *Helv. Chim. Acta*, 1963, **46**, 2705; 1966, **49**, 2055 (*isol*)
Corbier, B. *et al*, *Recherches*, 1974, **19**, 235 (*synth*)

5,6,12,13-Tetrahydro-1,17-(ethano[1,3]benzenoethano)-7,11:18,22-dimetheno-dibenzo[*a,h*]cyclooctadecene, 9CI　　　T-80062

[116129-64-7]

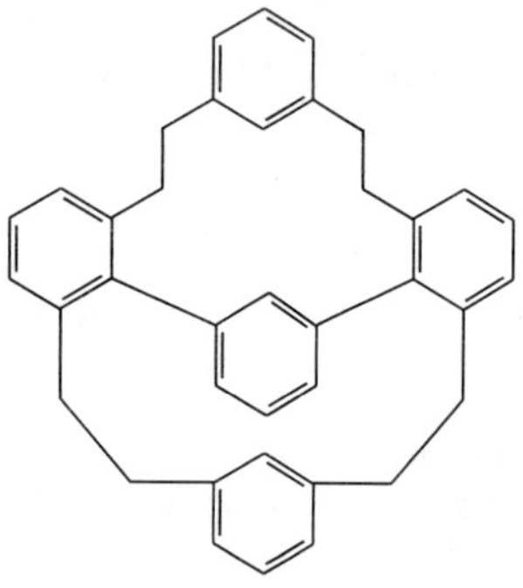

$C_{38}H_{34}$　　M 490.687

Cup-shaped molecule. Mp 210°. *o* and *p* -analogues also prepd.

Vinod, T.K. *et al*, *J. Am. Chem. Soc.*, 1988, **110**, 6574 (*synth*)

Tetrahydro-3-furanthiol　　　T-80063

3-Mercaptotetrahydrofuran

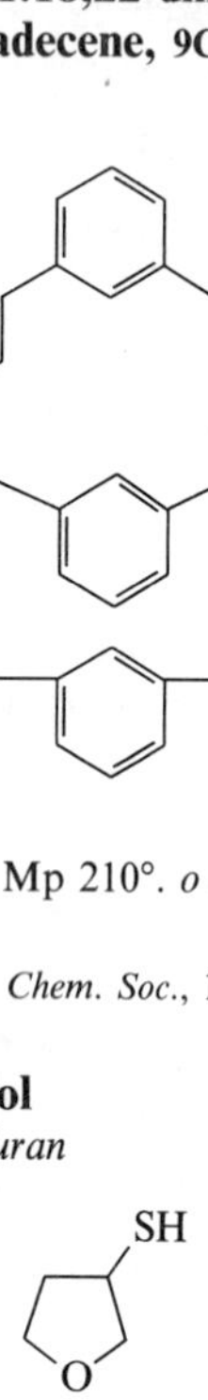

C_4H_8OS　　M 104.173

(±)-*form*

Ph thioether: [64823-94-5]. *Tetrahydro-3-(phenythio)furan, 9CI*
　$C_{10}H_{12}OS$　　M 180.270
　Oil.

Abd El Samii, Z.K.M. *et al*, *J. Chem. Soc., Perkin Trans. 1*, 1988, 2509 (*synth, ir, pmr, cmr*)

5,6,11,12-Tetrahydro-5,6,11,12,13,20-hexamethyl-5*a*,11*a*-(imino[1,2]benzenimino)quinoxalino[2,3-*b*]quinoxaline, 9CI　　　T-80064

[118894-97-6]

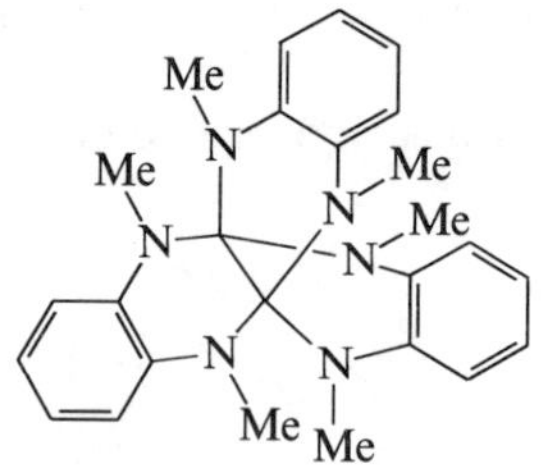

$C_{26}H_{30}N_6$　　M 426.563
Cryst. (EtOH). Mp 272-273°.

Tauer, E. *et al*, *Angew. Chem., Int. Ed. Engl.*, 1989, **28**, 338 (*synth, uv, pmr*)

1,2,3,4-Tetrahydro-5-hydroxyisoquinoline　　　T-80065

1,2,3,4-Tetrahydro-5-quinolinol. 5-Hydroxy-1,2,3,4-tetrahydroisoquinoline

[102877-50-9]

$C_9H_{11}NO$　　M 149.192
Cryst. (MeOH or by subl.). Mp 273°.

B,HCl: [102879-34-5].
　Cryst. (EtOH/Et₂O). Mp 257.6-259.7°.

Me ether; BHCl: [103030-69-9].
　$C_{10}H_{13}NO$　　M 163.219
　Cryst. (EtOH/Et₂O). Mp 234.9-236.0°.

[103030-70-2]

Durand, S. *et al*, *Bull. Soc. Chim. Fr.*, 1961, 270 (*synth, deriv*)
Schenker, F. *et al*, *J. Heterocycl. Chem.*, 1971, **8**, 665 (*synth, pmr*)
Sall, D.J. *et al*, *J. Med. Chem.*, 1987, **30**, 2208 (*synth, deriv, pmr, ir, cmr, ms*)

1,2,3,4-Tetrahydro-7-hydroxyisoquinoline　　　T-80066

Updated Entry replacing H-03450

1,2,3,4-Tetrahydro-7-isoquinolinol, 9CI. 7-Hydroxy-1,2,3,4-tetrahydroisoquinoline

[30798-64-2]

$C_9H_{11}NO$　　M 149.192
Bp_{18} 210-220°.

B,HBr: [110192-19-3].
　Cryst. (EtOH). Mp 210.3-211.4°.

Me ether: [43207-78-9]. **Weberidine.** *1,2,3,4-Tetrahydro-7-methoxyisoquinoline, 9CI*
　$C_{10}H_{13}NO$　　M 163.219
　Alkaloid from *Pachycereus weberi* (Cactaceae). $Bp_{0.1}$ 115° (synth.) .

Me ether; B,HCl: Mp 233° (228°).

Pictet, A. *et al*, *Ber.*, 1911, **44**, 2036 (*synth*)
Schenker, F. *et al*, *J. Heterocycl. Chem.*, 1971, **8**, 665 (*synth, pmr*)
Moniot, J.L. *et al*, *Heterocycles*, 1978, **9**, 145 (*pmr*)
Mata, R. *et al*, *Phytochemistry*, 1980, **19**, 673 (*isol, uv, ir, pmr, ms, synth, deriv*)
Mata, R. *et al*, *Phytochemistry*, 1983, **22**, 1263 (*cmr, deriv*)
Sall, D.J. *et al*, *J. Med. Chem.*, 1987, **30**, 2208 (*synth, pmr, ir, cmr, ms*)

3,4,7,8-Tetrahydro-11-(2-hydroxy-4-methoxyphenyl)-2,2,6,6-tetramethyl-2*H*,6*H*,12*H*-benzo[1,2-*b*;3,4-*b*′;5,6-*b*″]tripyran-12-one, 9CI   T-80067

[78876-34-3]

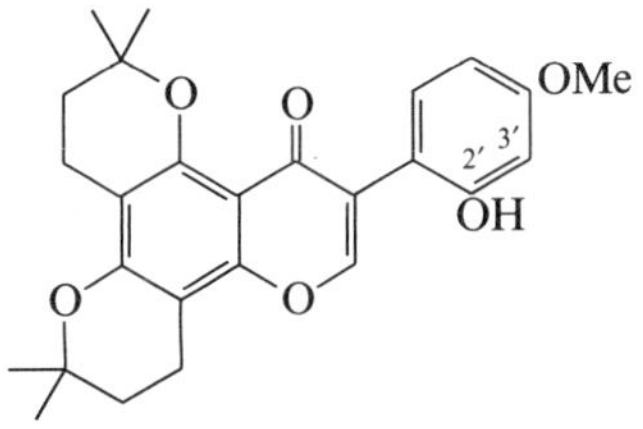

C$_{26}$H$_{28}$O$_6$ M 436.504
Isol. from leaves of *Millettia pachycarpa*. Cryst. (EtOAc/pet. ether). Mp 165°.

2′-Deoxy, 3′-hydroxy:
C$_{26}$H$_{28}$O$_6$ M 436.504
Isol. from leaves of *M. pachycarpa*. Noncryst. Struct. not certain. May have the alternative 4′-hydroxy-3′-methoxy struct. (preferred on balance).

[78876-33-2]

Singhal, A.K. *et al, Phytochemistry*, 1981, **20**, 803.

5,6,7,8-Tetrahydro-2(3*H*)-indolizinone, 9CI   T-80068

2,3,5,6,7,8-Tetrahydroindolizin-2-one

[97202-63-6]

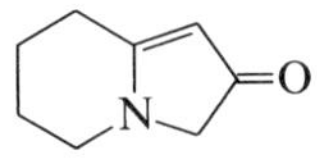

C$_8$H$_{11}$NO   M 137.181
Oil. Bp$_{0.0002}$ 69°.

Pommelet, J.C. *et al, J. Org. Chem.*, 1988, **53**, 5680 (*synth, ir, pmr, cmr, ms*)

Tetrahydro-2-iodofuran   T-80069

3-Iodooxolane

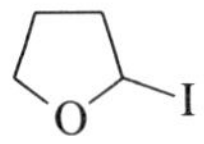

C$_4$H$_7$IO   M 198.003
(±)-*form*
Bp$_{0.2}$ 40°.

Evans, R.D. *et al, Synthesis*, 1988, 862 (*synth, pmr, ir*)

3a,4,7,7a-Tetrahydro-1*H*-isoindole-1,3(2*H*)-dione, 9CI   T-80070

1,2,3,6-Tetrahydrophthalimide. 4-Cyclohexene-1,2-dicarboximide

[85-40-5]

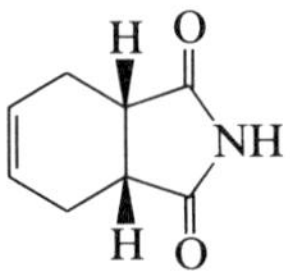

C$_8$H$_9$NO$_2$ M 151.165
(*3aRS,7aSR*)-*form* [1469-48-3]
cis-*form*
Cryst. (EtOH). Mp 136-138°.
▷ Toxic.

Ag salt: [62506-08-5].
Tan powder.
N-*Me:* [62950-21-4].
C$_9$H$_{11}$NO$_2$ M 165.191
Cryst. (C$_6$H$_6$/hexane). Mp 72.5-73° (60-62°).

[2015-58-9, 5167-69-1]

Rice, L.M. *et al, J. Org. Chem.*, 1954, 884 (*synth, deriv*)
Christol, H. *et al, Bull. Soc. Chim. Fr.*, 1966, 1315 (*synth*)
Kirfel, A., *Acta Crystallogr.*, 1976, 1556 (*cryst struct*)
Evans, D.L. *et al, J. Org. Chem.*, 1979, **44**, 497 (*deriv, pmr*)
Crockett, G.C. *et al, Synth. Commun.*, 1981, 447 (*synth*)
Wijnberg, B.P. *et al, Tetrahedron*, 1982, **38**, 209 (*synth*)
Caswell, L.R. *et al, J. Org. Chem.*, 1984, **49**, 696 (*synth*)
Mori, M. *et al, Tetrahedron Lett.*, 1987, 6187.

1,2,3,4-Tetrahydro-1-methylisoquinoline   T-80071

Updated Entry replacing T-00917

[4965-09-7]

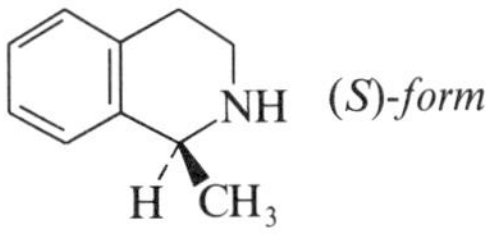

C$_{10}$H$_{13}$N   M 147.219
(*S*)-*form*
Oil. [α]$_D^{25}$ −71.3° (c, 0.64 in THF).
B,HCl: Cryst. Mp 208°.
(±)-*form*
Oil. Bp$_{745}$ 233°.
B,HCl: Mp 173°.
Picrate: Mp 187°.

[64982-61-2, 64982-62-3]

Robinson, R. *et al, J. Org. Chem.*, 1951, **16**, 1911 (*synth*)
Sakane, K. *et al, Bull. Chem. Soc. Jpn.*, 1974, **45**, 1297 (*use*)
Meyers, A.I. *et al, Tetrahedron*, 1987, **43**, 5095 (*synth, pmr, ir*)

Tetrahydro-3-methyl-2*H*-pyran   T-80072

3-Methyltetrahydropyran

[26093-63-0]
C$_6$H$_{12}$O   M 100.160
d$_4^{20}$ 0.863. Bp 109°.
(±)-*form*
d$_4^{20}$ 0.86. Bp$_{733}$ 109°.

Hanschke, E., *Chem. Ber.*, 1955, **88**, 1048.

Tetrahydro-4-methyl-2*H*-pyran, 9CI   T-80073

4-Methyltetrahydropyran

[4717-96-8]
C$_6$H$_{12}$O   M 100.160
(±)-*form*
Liq.

Booth, H. *et al, Tetrahedron*, 1987, **43**, 4699 (*synth, pmr, cmr, conformn*)

(Tetrahydro-4-methyl-2*H*-pyran-2-yl)-2-propanone, 9CI   T-80074

2-Acetonyl-4-methyltetrahydropyran. Rose oxide ketone
[20194-70-1]

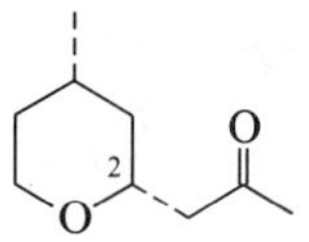

$C_9H_{16}O_2$ M 156.224
Used in perfumery.
(2R*,4S*)-form [73127-43-2]
   cis-*form*
     Isol. from geranium oil (*Pelargonium roseum*). d_4^{20} 0.956.
     Bp_{14} 87-88°. $[\alpha]_D$ –1.5°. n_D^{20} 1.4460.

   Naves, Y.R. *et al, Bull. Soc. Chim. Fr.*, 1963, 1608 (*isol, ir, pmr, synth*)
   Japan. Pat., 74 85 071, (1974); *CA*, **82**, 31256k (*synth*)
   Japan. Pat., 74 85 072, (1974); *CA*, **82**, 31255j (*synth*)

3,4,5,6-Tetrahydro-1(2*H*)-naphthalenone   T-80075

[113668-50-1]

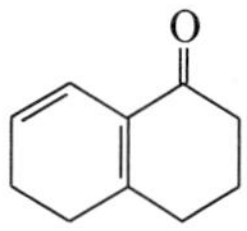

$C_{10}H_{12}O$ M 148.204
Oil. Unstable, gradually aromatises.

   Marcinow, Z. *et al, J. Org. Chem.*, 1988, **53**, 2117 (*synth, ir, pmr*)

5,6,7,8-Tetrahydro-1,4-naphthoquinone, 8CI   T-80076

5,6,7,8-Tetrahydro-1,4-naphthalenedione, 9CI
[7474-90-0]

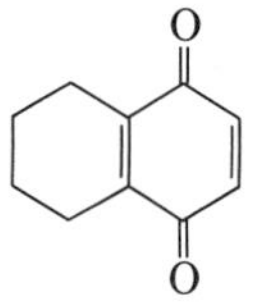

$C_{10}H_{10}O_2$ M 162.188
Yellow needles (pet.ether). Mp 55-56°.
Monoxime: [22658-31-7].
   $C_{10}H_{11}NO_2$ M 177.202
   Cryst. (xylene). Mp 167-168°.

   Arnold, R.T. *et al, J. Am. Chem. Soc.*, 1941, **63**, 1317 (*synth*)
   Fr. Pat., 1 508 481, (1968); *CA*, **71**, 3132w (*oxime*)
   Malesani, G. *et al, J. Heterocycl. Chem.*, 1982, **19**, 633 (*ir, pmr*)

Tetrahydro-1*H*,5*H*-pyrazolo[1,2-*a*]pyrazole, 9CI   T-80077

1,5-Diazabicyclo[3.3.0]octane. Octahydro-3a,6a-diazapentalene. 8-Azapyrrolizidine. 1,2-Trimethylenepyrazolidine. Perhydro-3a,6a-diazapentalene
[5397-67-1]

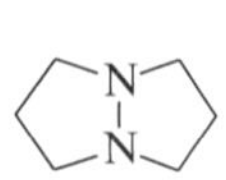

$C_6H_{12}N_2$ M 112.174
Oil, freezing to plates. d_{20}^{20} 1.00. Mp 1.5-2.5°. Bp 173°, Bp_{26} 74-75°. n_D^{20} 1.4895.
Picrate: Mp 159.5°.

   Buhle, E.L. *et al, J. Am. Chem. Soc.*, 1943, **65**, 29 (*synth*)

   Stetter, H. *et al, Chem. Ber.*, 1965, **98**, 3228 (*synth*)
   Koopmann, H.P. *et al, Spectrochim. Acta, Part A*, 1976, **32**, 157 (*ir, raman*)
   Nelsen, S.F. *et al, J. Org. Chem.*, 1980, **45**, 3609 (*nmr*)

1,2,3,4-Tetrahydro-1,2,4,5-tetrahydroxy-7-methoxy-2-methylanthraquinone   T-80078

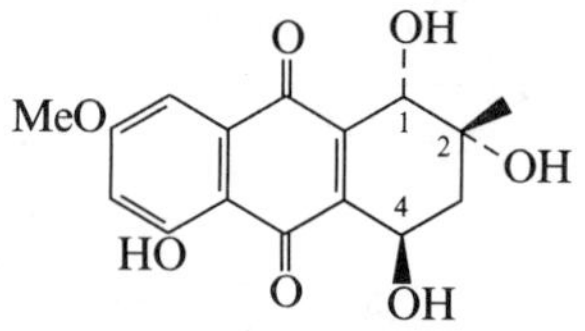

(1*S*,2*R*,4*R*)-*form*

$C_{16}H_{16}O_7$ M 320.298
(1S,2R,4R)-form
   Constit. of fungus *Dermocybe* sp. Yellow cryst. Mp 72-75°. $[\alpha]_D$ –323° (c, 0.35 in $CHCl_3$).
(1S,2R,4S)-form
   Constit. of *D.* sp. Yellow cryst. Mp 74-80°. $[\alpha]_{365}$ +350° (c, 0.6 in $CHCl_3$).
(1S,2S,4R)-form
   Constit. of *D.* sp. Yellow cryst. Mp 93-98°. $[\alpha]_D$ –152° ($CHCl_3$).

   Burns, C.J. *et al, Tetrahedron Lett.*, 1989, **30**, 7269 (*isol, synth*)

2,3,5,6-Tetrahydro-1*H*,4*H*,7*H*,11*cH*-3*a*, 6*a*,11*b*-triazabenz[*de*]anthracene, 9CI   T-80079

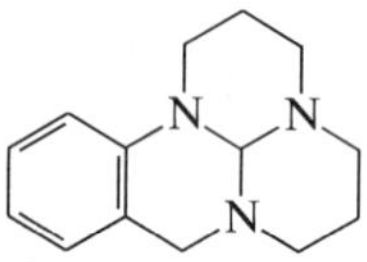

$C_{14}H_{19}N_3$ M 229.324
(±)-form [112672-92-1]
   Cryst. Mp 78-80°.

   Beddoes, R.L. *et al, Tetrahedron*, 1987, **43**, 1903 (*synth, uv, ir, pmr, ms, cryst struct*)

3*a*,4,9,9*a*-Tetrahydro-9-(3,4,5-trimethoxyphenyl)naphtho[2,3-*c*]furan-1(3*H*)one   T-80080

1,2,3,4-Tetrahydro-3-(hydroxymethyl)-1-(3,4,5-trimethoxyphenyl)-2-naphthalenecarboxylic acid lactone

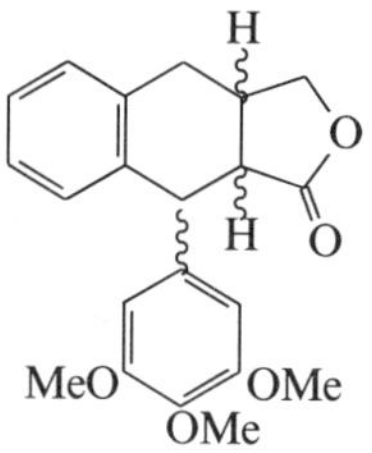

$C_{21}H_{22}O_5$ M 354.402
Isol. from seed oil of *Hernandia ovigera*. Lignan of deoxypodophyllotoxin type.

   Hata, C., *Nippon Kagaku Kaishi* (*J. Chem. Soc. Jpn.*), 1942, **63**, 1540; *CA*, **41**, 2917.

Tetrahydro-2,2,6-trimethyl-2*H*-pyran-3-ol, 9CI   T-80081

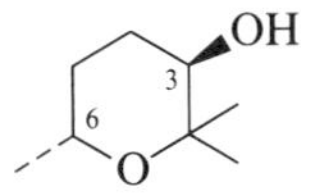

(3R,6R)-form

$C_8H_{16}O_2$ M 144.213

Volatile component from the elm bark beetle *Pteleobius vittatus*. Stereochem. of nat. prod. not yet known.

(3R,6R)-form [111465-16-8]

(+)-*trans*-form

Needles (hexane). Mp 37.3-38.5°. $[\alpha]_D^{25}$ +12.6° (c, 0.99 in Et₂O).

(3R,6S)-form [111465-17-9]

(−)-cis-form

Needles (hexane). Mp 47.5-48.5°. $[\alpha]_D^{24}$ −31.7° (c, 0.50 in Et₂O).

(3S,6R)-form [111465-15-7]

(+)-cis-form

Needles (hexane). Mp 48-48.5°. $[\alpha]_D^{25}$ +32.6° (c, 0.84 in Et₂O).

(3S,6S)-form [111465-18-0]

(−)-trans-form

Needles (hexane). Mp 37.5-38°. $[\alpha]_D^{24}$ −11.8° (c, 0.95 in Et₂O).

[93172-43-1, 107536-57-2]

Mori, K. *et al, Justus Liebigs Ann. Chem.*, 1988, 175 (*synth, ir, pmr, cmr, ms, ord*)

1,2,3,4-Tetrahydrotriphenylene, 9CI   T-80082

9,10-Cyclohexenophenanthrene

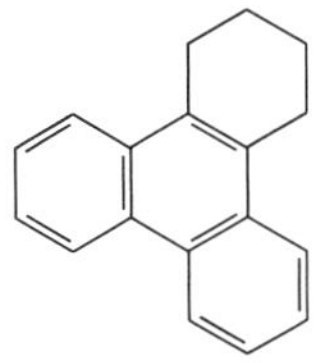

$C_{18}H_{16}$ M 232.324

Needles (1-propanol). Mp 120-121°.

Bergmann, E. *et al, J. Am. Chem. Soc.*, 1937, **59**, 1441 (*synth*)
Fu, P.P. *et al, J. Org. Chem.*, 1980, **45**, 2797 (*synth, pmr*)
Penn, J.H. *et al, J. Org. Chem.*, 1989, **54**, 601 (*synth, uv, ir, pmr, cmr, ms*)

1,4,5,8-Tetrahydrotriphenylene   T-80083

[114249-91-1]

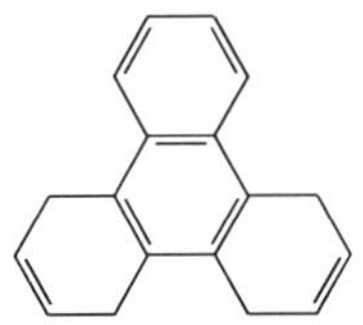

$C_{18}H_{16}$ M 232.324

Cryst. (EtOH). Mp 199-200°.

Marcinow, Z. *et al, J. Org. Chem.*, 1988, **53**, 3603 (*synth, pmr*)

6,7,12,16-Tetrahydroxy-8,12-abietadiene-11,14-dione   T-80084

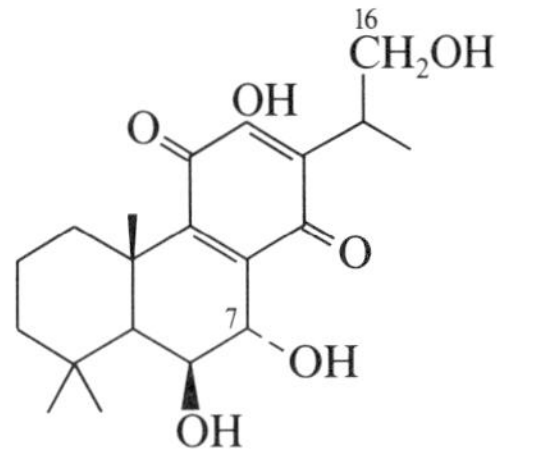

$C_{20}H_{28}O_6$ M 364.438

(6β,7α)-form

7-Ac: [120462-43-3]. **Lophanthoidin C**

$C_{22}H_{30}O_7$ M 406.475

Constit. of *R. lophanthoides*. Cryst. Mp 167-169.5°.

16-Ac: [120462-45-5]. **Lophanthoidin E**

$C_{22}H_{30}O_7$ M 406.475

Constit. of *R. lophanthoides*. Cryst. Mp 152.5-154°.

7,16-Di-Ac: [120462-42-2]. **Lophanthoidin B**

$C_{24}H_{32}O_8$ M 448.512

Constit. of *R. lophanthoides*. Cryst. Mp 138-139°.

7-Me ether, 16-Ac: [120462-41-1]. **Lophanthoidin A**

$C_{23}H_{32}O_7$ M 420.502

Constit. of *Rabdosia lophanthoides*. Cryst. Mp 198-202°.

7-Et ether: [120462-44-4]. **Lophanthoidin D**

$C_{22}H_{32}O_6$ M 392.491

Constit. of *R. lophanthoides*. Cryst. Mp 205-210°.

7-Et ether, 16-Ac: [120462-46-6]. **Lophanthoidin F**

$C_{24}H_{34}O_7$ M 434.528

Constit. of *R. lophanthoides*. Cryst. Mp 184-185°.

Yunlong, X. *et al, Phytochemistry*, 1989, **28**, 189.

2,6,7,11-Tetrahydroxy-5,7,9(11),13-abietatetraen-12-one   T-80085

5,6-Didehydro-2,7-dihydroxytaxodone

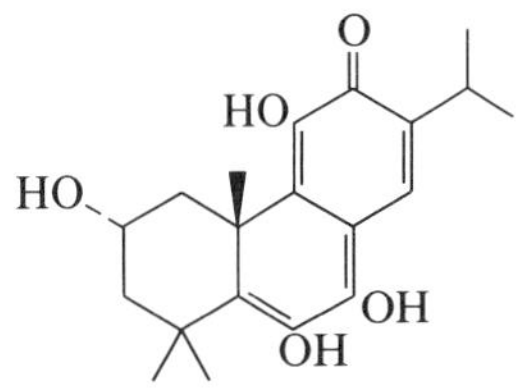

$C_{20}H_{26}O_5$ M 346.422

2α-form

Constit. of *Salvia texana*. Cryst. (CHCl₃).

Gonzalez, A.G. *et al, Tetrahedron*, 1989, **45**, 5203.

4,4′,5,5′-Tetrahydroxy-1,1′-binaphthyl   T-80086

[1,1′-Binaphthalene]-4,4′,5,5′-tetrol

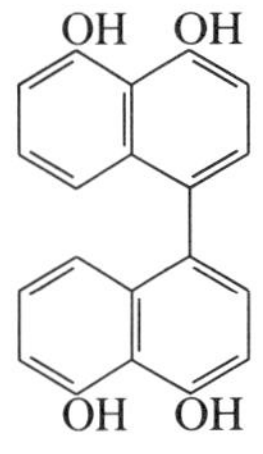

$C_{20}H_{14}O_4$ M 318.328

Metab. of *Daldinia concentrica*. Pale-yellow cryst.

Tetra-Ac:

$C_{28}H_{22}O_8$ M 486.477
Mp 245°.

Allport, D.C. *et al, J. Chem. Soc.*, 1958, 4090; 1960, 654.

1,3,8,9-Tetrahydroxycoumestan   T-80087

Updated Entry replacing W-00012

1,3,8,9-Tetrahydroxy-6H-benzofuro[3,2-c][1]benzopyran-6-one, 9CI, 8CI. 5,7,11,12-Tetrahydroxycoumestan.
Desmethylwedelolactone. *Norwedelolactone.*
Demethylwedelolactone

[6468-55-9]

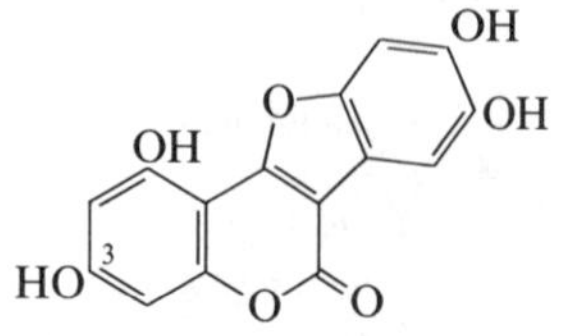

$C_{15}H_8O_7$ M 300.224
Constit. of *Wedelia calendulacea* and *Eclipta alba*. Mp
>360°.

3-O-Glucoside: [30414-09-6].
$C_{21}H_{18}O_{12}$ M 462.366
Isol. from leaves of *E. alba*. Mp >340°.

Tetra-Ac: Mp 270-272°.

O^3-*Me:* [524-12-9]. *5,11,12-trihydroxy-7-methoxycoumestan.*
Wedelolactone
$C_{16}H_{10}O_7$ M 314.251
Constit. of *W. calendulacea* and *E. alba*. Greenish-
yellow needles (MeOH). Mp 327-330° dec.

Govidachari, T.R. *et al, J. Chem. Soc.*, 1956, 629; 1957, 545, 548
(*isol, struct, ir, Wedelolactone*)
Khastgir, H.N. *et al, Tetrahedron*, 1961, **14**, 257 (*uv*)
Wanzlick, H.W. *et al, Chem. Ber.*, 1963, **96**, 305 (*synth*)
Chatterjea, J.N. *et al, Chem. Ber.*, 1964, **97**, 1252 (*synth*)
Bhargava, K.K. *et al, Indian J. Chem.*, 1970, **8**, 664; 1972, **10**, 810
(*isol, struct*)

2,3,8,9-Tetrahydroxycoumestan   T-80088

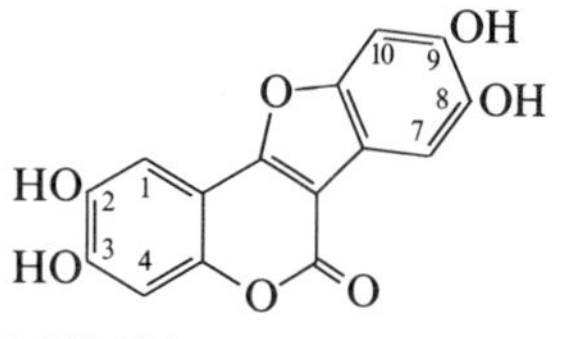

$C_{15}H_8O_7$ M 300.224

2-Me, 8,9-methylene ether: [75656-29-0]. *3-Hydroxy-2-methoxy-8,9-methylenedioxycoumestan.* **Tephrosol**
$C_{17}H_{10}O_7$ M 326.262
Constit. of *Tephrosia villosa* root. Yellow-green cryst.
(MeOH). Mp 306°.

3-Me, 8,9-methylene ether: [35930-39-3]. **2-Hydroxy-3-methoxy-8,9-methylenedioxycoumestan.** *2-Hydroxyflemichapparin C*
$C_{17}H_{10}O_7$ M 326.262
Isol. from heartwood of *Swartzia leiocalycina*.

Donnelly, D.M.X. *et al, Phytochemistry*, 1971, **10**, 3147 (*isol*)
Rao, D.D. *et al, Phytochemistry*, 1980, **19**, 1272 (*isol*)

3,4,8,9-Tetrahydroxycoumestan   T-80089

$C_{15}H_8O_7$ M 300.224

4-Me,8,9-methylene ether: [79295-80-0]. *3-Hydroxy-4-methoxy-8,9-methylenedioxycoumestan.*
Sophoracoumestan B

$C_{17}H_{10}O_7$ M 326.262
Isol. from roots of *Sophora franchetiana*. Needles
(MeOH). Mp >300°.

Komatsu, M. *et al, Chem. Pharm. Bull.*, 1981, **29**, 2069 (*isol, uv, pmr, struct*)

1,4,6,9-Tetrahydroxydihydro-β-agarofuran   T-80090

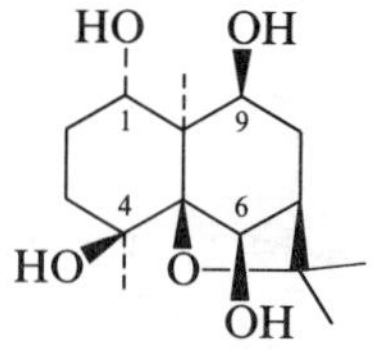

$C_{15}H_{26}O_5$ M 286.367

(1α,4β,6β,9β)-form

9-Ac,1-benzoyl: *9β-Acetoxy-1α-benzoyloxy-4β,6β-dihydroxydihydro-β-agarofuran*
$C_{24}H_{32}O_7$ M 432.513
Constit. of *Orthosphenia mexicana*.

6,9-Di-Ac,1-benzoyl: *6β,9β-Diacetoxy-1α-benzoyloxy-4β-hydroxydihydro-β-agarofuran*
$C_{26}H_{34}O_8$ M 474.550
Constit. of *O. mexicana*. Cryst. Mp 202°.

González, A.G. *et al, Phytochemistry*, 1988, **27**, 473.

2′,5,6′,7-Tetrahydroxyflavone   T-80091

Updated Entry replacing T-50153

2-(2,6-Dihydroxyphenyl)-5,7-dihydroxy-4H-1-benzopyran-4-one, 9CI. **Argemexitin**

[82475-00-1]

$C_{15}H_{10}O_6$ M 286.240
Constit. of *Argemone mexicana*. Yellow needles (EtOH).
Mp >330°.

2′-Me ether: [92519-94-3]. *2′,5,7-Trihydroxy-6′-methoxyflavone*
$C_{16}H_{12}O_6$ M 300.267
Isol. from root of *Scutellaria baicalensis*. Cryst.
(MeOH). Mp 256-257° dec.

Bhardwaj, D.K. *et al, Phytochemistry*, 1982, **21**, 2154 (*isol*)
Tomimuri, T. *et al, Yakugaku Zasshi (J. Pharm. Soc. Jpn.)*, 1984,
104, 529 (*ir, isol, uv, pmr, cmr, ms*)
Tanaka, T. *et al, Yakugaku Zasshi (J. Pharm. Soc. Jpn.)*, 1987,
107, 315 (*synth*)

3′,4′,6,7-Tetrahydroxyflavone   T-80092

Updated Entry replacing T-30091

$C_{15}H_{10}O_6$ M 286.240
Yellow cryst. Mp >320°.

3,6-Di-Me ether: [68007-24-9]. *4′,7-Dihydroxy-3,6-dimethoxyflavone*
$C_{17}H_{14}O_6$ M 314.294
Isol. from *Pluchea chingoyo*.

4′,6-Di-Me ether: *3′,7-Dihydroxy-4′,6-dimethoxyflavone.*
Abrectorin
$C_{17}H_{14}O_6$ M 314.294
Constit. of *Abrus precatorius*. Cryst. (EtOH). Mp 229-
230°, Mp 273-274°.

6,7-Di-Me ether: [123442-39-7]. *3,4′-Dihydroxy-6,7-dimethoxyflavone.* **Hortensin**
$C_{17}H_{14}O_6$ M 314.294
Constit. of *Millingtonia hortensis*. Pale yellow cryst. Mp
212-213°.

Tetra-Me ether: *3′,4′,6,7-Tetramethoxyflavone*

$C_{19}H_{18}O_6$ M 342.348
Needles ($CHCl_3$/pet. ether). Mp 221-222°.

Chiang, M.T. et al, Rev. Latinoam. Quim., 1978, **9**, 102 (isol)
Bhardwaj, D.K. et al, Phytochemistry, 1980, **19**, 2040 (isol, struct, ir, uv)
Bhardwaj, D.K. et al, Indian J. Chem., Sect. B, 1982, **21**, 98 (synth)
Bunyapraphatsara, N. et al, Phytochemistry, 1989, **28**, 1555 (Hortensin)

3,4',5,7-Tetrahydroxyflavylium, 8CI   T-80093

Updated Entry replacing T-01248
3,5,7-Trihydroxy-2-(4-hydroxyphenyl)-1-benzopyrylium, 9CI.
Pelargonidin
[7690-51-9]

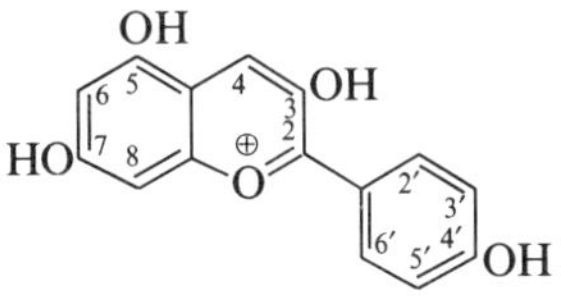

$C_{15}H_{11}O_5^{\oplus}$ M 271.249
Anthocyanin pigment present in many plants, flowers and fruits, e.g. radishes, orchids, brassicas, *Petunia* spp. Glycosides also widely distributed. Ms unsuitable as analytical technique.

Chloride: [134-04-3].
   $C_{15}H_{11}ClO_5$ M 306.702
   Leaflets ($MeOH/Et_2O$), needles ($MeOH/HCl$). Mod. sol. H_2O. Mp >350°.

3-β-D-Glucoside: [18466-51-8]. **Callistephin**
   $C_{21}H_{21}O_{10}^{\oplus}$ M 433.391
   Pigment from scarlet carnations. Red cryst. (as chloride).

5-β-D-Glucoside: **Pelargonenin**
   $C_{21}H_{21}O_{10}^{\oplus}$ M 433.391
   Scarlet-red needles (as chloride). Spar. sol. H_2O. Red col. with bluish fluor. in acid soln., viol.-blue col. in aq. Na_2CO_3 soln.

7-Glucoside:
   $C_{21}H_{21}O_{10}^{\oplus}$ M 433.391
   Scarlet needles (as chloride). Na_2CO_3 aq. → deep pink col.

4'-Glucoside:
   $C_{21}H_{21}O_{10}^{\oplus}$ M 433.391
   Red plates (as chloride). Dec. at 184°.

3,5-Diglucoside: [17334-58-6]. **Pelargonin.** *Monardin. Salvinin*
   $C_{27}H_{31}O_{15}^{\oplus}$ M 595.533
   Fine red needles + $4H_2O$ (as chloride). Sol. H_2O. Mp 180° dec. (anhyd.) (chloride). $[\alpha]_D$ −244°. Orange-red soln. in H_2O turning viol. on standing. MeOH soln. red with green fluor.

3-Galactoside: **Fragarin**
   $C_{21}H_{21}O_{10}^{\oplus}$ M 433.391
   Isol. from leaves of *Fagus sylvatica* var. *purpurea* as chloride. λ_{max} 278, 505 nm ($MeOH/HCl$).

3-O-[2-O-β-D-Xylopyranosyl(1→2)-α-D-glucopyranoside]:
   Pelargonidin 3-sambubioside
   $C_{26}H_{29}O_{14}^{\oplus}$ M 565.507
   Isol. from petals of *Streptocarpus* spp. (as chloride) and present in other members of the Gesneriaceae. λ_{max} 277, 508 nm ($MeOH/HCl$).

3-O-[β-D-Glucopyranosyl(1→2)-β-D-glucopyranoside]:
   Pelargonidin 3-sophoroside
   $C_{27}H_{31}O_{15}^{\oplus}$ M 595.533

Isol. from petals of *Papaver rhoeas* and *P. orientale* (as chloride). λ_{max} 268, 505 nm ($MeOH/HCl$).

3-O-[β-D-Glucopyranosyl(1→2_{glu})[α-L-rhamnopyranosyl(1→6)-β-D-glucopyranoside]]:
   Pelargonidin 3-(2^{glu}glucosylrutinoside)
   $C_{33}H_{41}O_{19}$ M 741.675
   Isol. from fruits of *Rubus idaeus* (as chloride). λ_{max} 275, 508 nm ($MeOH/HCl$).

3-O-[β-D-Xylopyranosyl(1→2)-β-D-glucopyranoside], 5-O-β-D-glucopyranoside: Pelagonidin 5-glucoside 3-sambubioside
   $C_{32}H_{39}O_{19}^{\oplus}$ M 727.649
   Isol. from petals of *Matthiola incana* (as chloride). λ_{max} 280, 508 nm ($MeOH/HCl$). Present in acylated form in the plant.

3-O-[β-D-Glucopyranosyl(1→2)-β-D-glucopyranoside], 5-O-β-D-glucopyranoside: Pelargonidin 5-glucoside 3-sophoroside
   $C_{33}H_{41}O_{20}$ M 757.675
   Present in acylated form in roots of *Raphanus sativus.*

3-O-Sophoroside, 7-O-glucoside: **Orientalin**†
   $C_{33}H_{41}O_{20}^{\oplus}$ M 757.675
   Isol. from petals of *Papaver* and *Watsonia* spp. (as chloride). λ_{max} 279, 499 nm ($MeOH/HCl$). Orientaline is also the name of an alkaloid.

3-O-[α-L-Rhamnopyranosyl(1→6)-β-D-glucopyranoside], 5-O-glucoside: Pelargonidin 5-glucoside 3-rutinoside
   $C_{33}H_{41}O_{19}^{\oplus}$ M 741.675
   Present in petals of *Dichrotrichum* sp. and in *Achimenes* spp.

Robertson, A. et al, J. Chem. Soc., 1928, 1460, 1533 (isol)
Léon, A. et al, J. Chem. Soc., 1931, 2673 (synth, glucosides)
Harborne, J.B. et al, Phytochemistry, 1963, **2**, 85; 1964, **3**, 453; 1966, **5**, 589 (isol)
Timberlake, C.F. et al, Nature (London), 1966, **212**, 158 (uv)
Nilson, E., Chem. Scr., 1973, **4**, 49 (pmr)
Timberlake, C.F. et al, The Flavonoids, (Harbourne, J.B. etal, Eds.), Chapman and Hall, 1975, 215 (rev)
Saito, N. et al, Heterocycles, 1985, **23**, 2709 (cryst struct)

3,4,8,10-Tetrahydroxy-1-guaien-12,6-olide   T-80094

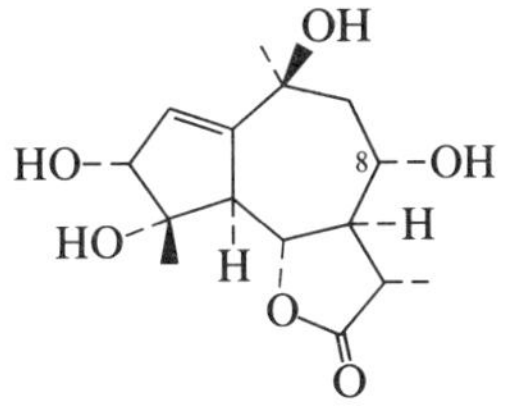

$C_{15}H_{22}O_6$ M 298.335
(3α,4α,5α,6α,7α,8α,10β,11S)-form

8-Ac: [119403-32-6]. *8α-Acetoxy-3α,4α,10β-trihydroxy-1-guaien-12,6α-olide*
   $C_{17}H_{24}O_7$ M 340.372
   Constit. of *Artemisia arborescens*. Cryst. (EtOH). Mp 140-145°.

Grandolini, G. et al, Phytochemistry, 1988, **27**, 3670 (isol, pmr)

2',3',4',7-Tetrahydroxyisoflavan   T-80095

3,4-Dihydro-7-hydroxy-3-(2,3,4-trihydroxyphenyl)-2H-1-benzopyran

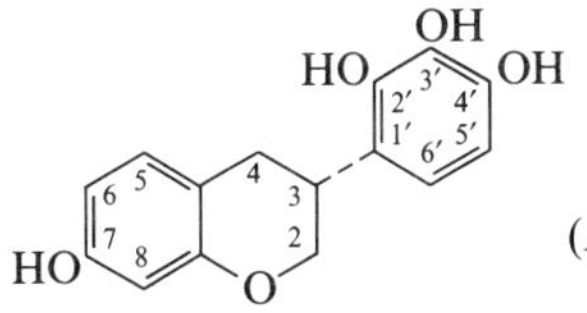

(R)-form

$C_{15}H_{14}O_5$ M 274.273

3',4'-Di-Me ether: [64474-51-7]. *2',7-Dihydroxy-3',4'-dimethoxyisoflavan.* **Isomucronulatol**
$C_{17}H_{18}O_5$ M 302.326
Isol. from *Astragalus glycyphyllos, A. penduliflorus, Carmichaelia flagelliformis, Colutea arborescens, Gliricidia sepium* and *Glycyrrhiza glabra.* Cryst. (MeOH aq.). Mp 152-153°. [α]$_D^{20}$ −19.4° (c, 0.24 in EtOH).

Al-Ani, H.A.M. *et al, Phytochemistry,* 1985, **24**, 55 (*biosynth, bibl, Isomucronulatol*)

2',4',5,7-Tetrahydroxyisoflavan   T-80096

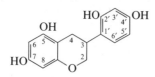

$C_{15}H_{14}O_5$ M 274.273
No obt. rotns. or chiralities reported for nat. prods.

4',5-Di-Me ether: [73354-16-2]. *2',7-Dihydroxy-4',5-dimethoxyisoflavan.* **5-Methoxyvestitol**
$C_{17}H_{18}O_5$ M 302.326
Stress metab. from *Lotus edulis* and *L. hispidus.*

5,7-Di-Me ether: [77370-02-6]. *2',4'-Dihydroxy-5,7-dimethoxyisoflavan.* **Lotisoflavan**
$C_{17}H_{18}O_5$ M 302.326
Stress metab. from leaves of *L. angustissimus* and *L. edulis.* Mp 135-137° (synth. racemate).

Ingham, J.L. *et al, Phytochemistry,* 1979, **18**, 1711; 1980, **19**, 2799 (*isol, uv, ms, struct*)

2',4',5',7-Tetrahydroxyisoflavan   T-80097

3,4-Dihydro-7-hydroxy-3-(2,4,5-trihydroxyphenyl)-2H-1-benzopyran
$C_{15}H_{14}O_5$ M 274.273

4',5'-Methylene ether: [57436-37-0]. *2',7-Dihydroxy-4',5'-methylenedioxyisoflavan.* **Maackiainisoflavan. Dihydromaackiain**
$C_{16}H_{14}O_5$ M 286.284
Isol. from cultures of *Stemphylum botryosum* as metab. of Maackiain.

2'-Me, 4',5'-methylene ether: [72026-91-6]. *7-Hydroxy-2'-methoxy-4',5'-methylenedioxyisoflavan.* **Astraciceran**
$C_{17}H_{16}O_5$ M 300.310
Isol. from leaves of *Astragalus cicer* and *A. pyrenaicus.* Phytoalexin. Cryst. (MeOH). Mp 168°.

[22091-12-9]

Higgins, V.J., *Physiol. Plant Pathol.,* 1975, **6**, 5 (*Maackianisoflavan*)
Ingham, J.L. *et al, Phytochemistry,* 1980, **19**, 1767 (*Astraciceran*)

2',4',6,7-Tetrahydroxyisoflavan   T-80098

3,4-Dihydro-3-(2,4-dihydroxyphenyl)-6,7-dihydroxy-2H-1-benzopyran
$C_{15}H_{14}O_5$ M 274.273

4',6-Di-Me ether: [56752-02-4]. *2',7-Dihydroxy-4',6-dimethoxyisoflavan.* **6-Methoxyvestitol**
$C_{17}H_{18}O_5$ M 302.326
Isol. from cultures of *Fusarium solani* growing on *Pisum sativum.*

Pueppke, S.G. *et al, Physiol. Plant Pathol.,* 1976, **8**, 51.

2',3',4',7-Tetrahydroxyisoflavanone   T-80099

2,3-Dihydro-7-hydroxy-3-(2,3,4-trihydroxyphenyl)-4H-1-benzopyran-4-one
$C_{15}H_{12}O_6$ M 288.256

(±)-*form*

2',3'-Di-Me ether: [79852-12-3]. *4',7-Dihydroxy-2',3'-dimethoxyisoflavanone.* **Lespedeol C**
$C_{17}H_{16}O_6$ M 316.310
Isol. from heartwood of *Lespedeza cyrtobotrya.* Needles (Me$_2$CO/hexane). Mp 206-209°.

2',4'-Di-Me ether: [52250-38-1]. *3',7-Dihydroxy-2',4'-dimethoxyisoflavanone.* **Violanone**
$C_{17}H_{16}O_6$ M 316.310
Isol. from heartwoods of *Dalbergia violacea* and *D. oliveri.* Mp 200-202°.

2',3',4'-Tri-Me ether: [56973-42-3]. *7-Hydroxy-2',3',4'-trimethoxyisoflavanone.* **3'-O-Methylviolanone**
$C_{18}H_{18}O_6$ M 330.337
Isol. from wood of *D. cearensis.* Cryst. (C$_6$H$_6$/MeOH). Mp 192-194°.

3',4',7-Tri-Me ether: [71973-13-2]. *2'-Hydroxy-3',4',7-trimethoxyisoflavanone.* **3'-Methoxyisosativanone**
$C_{18}H_{18}O_6$ M 330.337
Isol. from trunkwood of *Myroxylon peruiferum.* Mp 155-157°. Racemic.

Donnelly, D.M.X. *et al, Phytochemistry,* 1974, **13**, 2587 (*Violanone*)
Guimarães, I.S. . de S. *et al, Phytochemistry,* 1975, **14**, 1452 (*3'-O-Methylviolanone*)
Maranduba, A. *et al, Phytochemistry,* 1979, **18**, 815 (*3'-Methoxyisosativanone*)
Miyase, T. *et al, Chem. Pharm. Bull.,* 1981, **29**, 2205 (*Lespedeol C*)

2',4',5,7-Tetrahydroxyisoflavanone   T-80100

Updated Entry replacing T-03370
2,3-Dihydro-5,7-dihydroxy-3-(2,4-dihydroxyphenyl)-4H-1-benzopyran-4-one, 9CI. **Dalbergioidin**
[30368-42-4]
$C_{15}H_{12}O_6$ M 288.256
Isol. from *Dolichos biflorus, Lablab niger, Lespedeza cyrtobotrya, Macrotyloma axillare, Ougeinia dalbergiodes, Phaseolus vulgaris* and *Stizolobium deeringianum* (all Leguminosae, Papilionoideae).

2'-Me ether: [76656-75-2]. *4',5,7-Trihydroxy-2'-methoxyisoflavanone.* **Isoferreirin**
$C_{16}H_{14}O_6$ M 302.283
Constit. of *D. biflorus* and *S. deeringianum.*

4'-Me ether: [32898-79-6]. *2',5,7-Trihydroxy-4'-methoxyisoflavanone.* **Ferreirin**
Isol. from heartwood of *Ferreirea spectabilis* also from *Haplormosia monophylla* and *Cajanus cajan.* Prisms (MeOH aq. or pet. ether). Mp 210-212°.

2',4-Di-Me ether: [482-01-9]. *5,7-Dihydroxy-2',4'-dimethoxyisoflavanone.* **Homoferreirin**
$C_{17}H_{16}O_6$ M 316.310
Isol. from *Argyrocytisus battandieri, Cicer arietinum, Ferreirea spectabilis* and *O.dalbergioides.* Rectangular plates (C$_6$H$_6$/pet. ether or MeOH aq.). Mp 168-169°.

2',4'-Di-Me ether,di-Ac: Long needles. Mp 132-133°.

2',7-Di-Me ether: [61020-70-0]. *4',5-Dihydroxy-2',7-dimethoxyflavanone.* **Cajanol**
$C_{17}H_{16}O_6$ M 316.310

Isol. from fungus-infected stems of *Cajanus cajan* and *S. deeringainum*. Revised struct. Formerly considered to be 2′,6-Dihydroxy-4′,7-dimethoxyisoflavanone.

Tetra-Me ether: [28812-39-7]. 2′,4′,5,7-*Tetramethoxyisoflavanone*
$C_{19}H_{20}O_6$ M 344.363
Cryst. (C_6H_6). Mp 163°.

King, F.E. *et al*, *J. Chem. Soc.*, 1952, 4580, 4752 (*Ferreirin, Homoferreirin*)
Farkas, L. *et al*, *J. Chem. Soc. C*, 1971, 1994 (*synth*)
Ingham, J.L. *et al*, *Z. Naturforsch.*, *C*, 1976, **31**, 504; 1977, **32**, 1018; 1980, **35**, 923 (*isol, spectra, biosynth, derivs*)
Woodward, M.D., *Phytochemistry*, 1979, **18**, 363 (*isol*)
Ingham, J.L., *Z. Naturforsch., C*, 1979, **34**, 159 (*Cajanol*)
Ingham, J.L., *Fortschr. Chem. Org. Naturst.*, 1983, **43**, 1 (*rev, occur*)

2′,4′,5′,7-Tetrahydroxyisoflavanone   T-80101

Updated Entry replacing H-20201

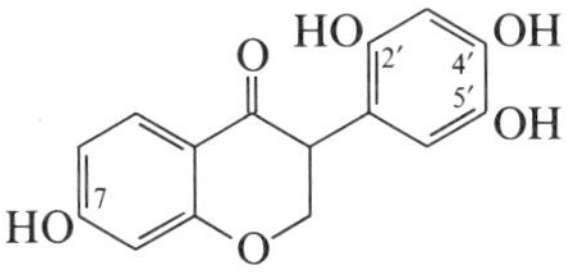

$C_{15}H_{12}O_6$ M 288.256
4′,5′-Methylene ether: [524-08-3]. 2′,7-*Dihydroxy-4′,5′-methylenedioxyisoflavanone*. **Sophorol**
$C_{16}H_{12}O_6$ M 300.267
Isol. from heartwood of *Sophora japonica*. Plates (Me₂CO aq.). Mp 180-181° (after drying). [α]$_D^{15}$ +9.5° (Me₂CO), [α]$_D^{16}$ −13.6° (EtOH).

2′-Me, 4′,5′-methylene ether: [58116-57-7]. 7-*Hydroxy-2′-methoxy-4′,5′-methylenedioxyisoflavanone*. **Onogenin**
$C_{17}H_{14}O_6$ M 314.294
Isol. from *Dalbergia stevensonii* and *Ononis arvensis*. Prisms (pet. ether). Mp 213-214° (189-193°). Racemic.
(±)-*form* [51106-85-5]
Constit. of *Dalbergia stevensonii* and *Ononis arvensis*. Prisms (C_6H_6/pet. ether). Mp 189-193°, Mp 213-214°.

Suginome, H., *J. Org. Chem.*, 1959, **24**, 1655 (*Sophorol*)
Donnelly, D.M.X. *et al*, *J. Chem. Soc., Perkin Trans. 1*, 1973, 1737 (*isol*)
Farkas, L. *et al*, *J. Chem. Soc., Perkin Trans. 1*, 1974, 305 (*synth*)
Kovalev, V.N. *et al*, *Khim. Prir. Soedin.*, 1975, **11**, 354 (*isol*)

3′,4′,6,7-Tetrahydroxyisoflavanone   T-80102

Updated Entry replacing D-05381
$C_{15}H_{10}O_6$ M 286.240
6,7-Di-Me, 3′,4′-methylene ether: [24195-20-8]. 6,7-*Dimethoxy-3′,4′-methylenedioxyisoflavanone*
$C_{18}H_{16}O_6$ M 328.321
Constit. of the heartwoods of *Cordyla africana* and *Mildbraedeodendron excelsa*. Needles (CHCl₃/MeOH). Mp 201.5-202.5°.

Campbell, R.V.M. *et al*, *J. Chem. Soc. C*, 1969, 1787 (*isol, struct, synth*)
Aggarwal, S.K. *et al*, *Indian J. Chem.*, 1972, **10**, 804 (*synth*)
Meegan, M.J. *et al*, *Phytochemistry*, 1975, **14**, 2283 (*isol*)

2′,3′,4′,7-Tetrahydroxyisoflavene   T-80103

Updated Entry replacing S-00299
4-(7-Hydroxy-2H-1-benzopyran-3-yl)-6-methoxy-1,2,3-benzenetriol, 9CI

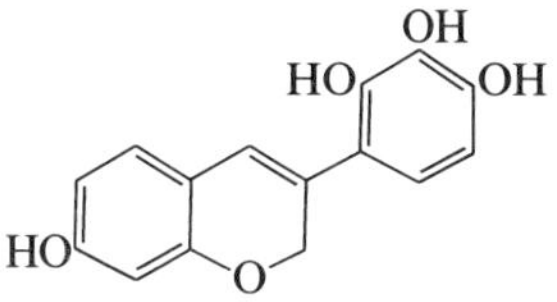

$C_{15}H_{12}O_5$ M 272.257
3′-Me ether: [79852-14-5]. 2′,4,7-*Trihydroxy-3′-methoxyisoflavene*. **Haginin C**
$C_{16}H_{14}O_5$ M 286.284
Isol. from heartwood of *Lespedeza cyrtobotrya*.
3′-Me ether, tri-Ac: Needles (CH₂Cl₂/MeOH). Mp 138-139.5°.
4′-Me ether: [60434-16-4]. 2′,3′,7-*Trihydroxy-4′-methoxyisoflavene*. **Sepiol**
$C_{16}H_{14}O_5$ M 286.284
Constit. of *Gliricidia sepium* heartwood. Brownish prisms (MeOH aq.). Mp 209-210°.
2′,3′-Di-Me ether: [74174-29-1]. 4′,7-*Dihydroxy-2′,3′-dimethoxyisoflavene*. **Haginin A**
$C_{17}H_{16}O_5$ M 300.310
Isol. from *L. cyrtobotrya* heartwood. Needles (MeOH/C_6H_6). Mp 130.5-131.5°.
2′,4′-Di-Me ether: [62078-14-2]. 3′,7-*Dihydroxy-2′,4′-dimethoxyisoflavene*. **2′-O-Methylsepiol**
$C_{17}H_{16}O_5$ M 300.310
Isol. from *G. sepium* heartwood. Oil. Mp 104-105° (as diacetate).

Tetra-Me ether: Mp 102-103°.

Jurd, L. *et al*, *J. Agric. Food Chem.*, 1977, **25**, 723 (*Sepiol, 2-O-Methylsepiol*)
Miyase, T. *et al*, *Chem. Pharm. Bull.*, 1980, **28**, 1172; 1981, **29**, 2205 (*Haginins*)

2′,4′,7,8-Tetrahydroxyisoflavene   T-80104

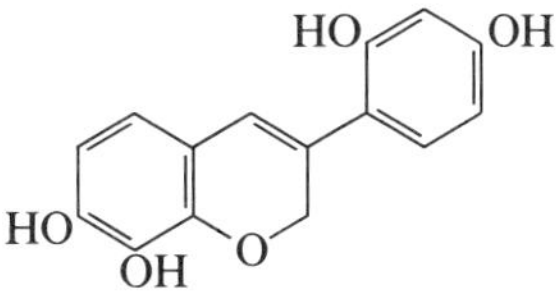

$C_{15}H_{12}O_5$ M 272.257
4′,8-Di-Me ether: 2′,7-*Dihydroxy-4′,8-dimethoxyisoflavene*
$C_{17}H_{16}O_5$ M 300.310
Constit. of *Centrolobium* spp. Cryst. (CHCl₃). Mp 160-161°.

Alegrio, L.V. *et al*, *Phytochemistry*, 1989, **28**, 2359 (*isol, pmr, cmr*)

2′,3′,4′,7-Tetrahydroxyisoflavone   T-80105

Updated Entry replacing T-03371
7-Hydroxy-3-(2,3,4-trihydroxyphenyl)-4H-1-benzopyran-4-one
$C_{15}H_{10}O_6$ M 286.240
4′-Me ether: [65048-75-1]. 2′,3′,7-*Trihydroxy-4′-methoxyisoflavone*. **Koparin**
$C_{16}H_{12}O_6$ M 300.267
Constit. of the wood of *Castanospermum australe*. Needles (EtOH). Mp 269-271°.
3′,4′,7-Tri-Me ether: [71973-12-1]. 2′-*Hydroxy-3′,4′,7-trimethoxyisoflavone*

$C_{18}H_{16}O_6$ M 328.321
Isol. from trunkwood of *Myroxylon peruiferum*. Mp 210-212°.

3',4'-Methylenedioxy ether: [65242-64-0]. *2',7-Dihydroxy-3',4'-methylenedioxyisoflavone.* **Glyzaglabrin**
$C_{16}H_{10}O_6$ M 298.251
Isol. from root of *Glycyrrhiza glabra*.

Berry, R.C. *et al, Aust. J. Chem.*, 1977, **30**, 1827 (*Koparin*)
Bhardwaj, D.K. *et al, Curr. Sci.*, 1977, **46**, 753 (*Glyzaglabrin*)
Maranduba, A.A. *et al, Phytochemistry*, 1979, **18**, 815 (*2'-Hydroxy-3',4',7-trimethoxyisoflavone*)

2',4',5,7-Tetrahydroxyisoflavone   T-80106

Updated Entry replacing T-40080
3-(2,4-Dihydroxyphenyl)-5,7-dihydroxy-4H-1-benzopyran-4-one, 9CI. 2'-Hydroxygenistein

[1156-78-1]

$C_{15}H_{10}O_6$ M 286.240
Isol. from *Apios tuberosa, Argyrocytisus battandieri, Cajanus cajan, Crotalaria juncea, Dolichos biflorus, Hardenbergia violacea, Lablab niger, Laburnium anagyroides, Lupinus albus, Moghania macrophylla, Neotonia wightii, Phaseolus vulgaris, P. coccineus, Spartium junceum* and *Stizolobium deeringianum* (all Leguminosae, Papilionoideae). Cryst. (CHCl₃/MeOH). Mp 270-273°.

4'-Me ether: [32884-35-8]. *2',5,7-Trihydroxy-4'-methoxyisoflavone. Dehydroferreirin. 2'-Hydroxybiochanin A*
$C_{16}H_{12}O_6$ M 300.267
Isol. from *Virola caducifolia, V. multinervia* and *Haplormosia monophylla*. Needles (MeOH aq.). Mp 215-216° (201-203°).

5-Me ether: [58115-29-0]. *2',4',7-Trihydroxy-5-methoxyisoflavone. 2'-Hydroxyisoprunetin*
$C_{16}H_{12}O_6$ M 300.267
Constit. of *P. coccineus*.

7-Me ether: [32884-36-9]. *2',4',5-Trihydroxy-7-methoxyisoflavone.* **Cajanin**
$C_{16}H_{12}O_6$ M 300.267
Isol. from *Cajanus cajan, Canavalia ensiformis, Centrosema pascuorum* and *C. pubescens*.

2',4'-Di-Me ether: [61243-75-2]. *5,7-Dihydroxy-2',4'-dimethoxyisoflavone. 2'-Methoxybiochanin A*
$C_{17}H_{14}O_6$ M 314.294
Constit. of *V. caducifolia* and *V. multinervia*. Cryst. (MeOH). Mp 222-223°.

2',5-Di-Me ether: *4',7-Dihydroxy-2',5-dimethoxyisoflavone*
$C_{17}H_{14}O_6$ M 314.294
From *P. coccineus*.

Farkas, L. *et al, J. Chem. Soc. C*, 1971, **10**, 1994 (*synth*)
Braz Filho, R. *et al, Phytochemistry*, 1976, **15**, 1029 (*isol, uv, ir, pmr, ms, derivs*)
Ingham, J.L., *Z. Naturforsch., C*, 1976, **31**, 504; 1977, **32**, 1018 (*isol*)
Prasad, J.S. *et al, Phytochemistry*, 1977, **16**, 1120 (*isol, uv, pmr, struct*)
Ingham, J.L., *Fortschr. Chem. Org. Naturst.*, 1983, **43**, 1 (*rev, occur*)
Adeesanya, S.A. *et al, Phytochemistry*, 1985, **24**, 2699 (*isol, derivs*)

2',4',5',7-Tetrahydroxyisoflavone   T-80107

Updated Entry replacing T-30093
$C_{15}H_{10}O_6$ M 286.240

2',4',5'-Tri-Me ether: [29096-94-4]. *7-Hydroxy-2',4',5'-trimethoxyisoflavone*
$C_{18}H_{16}O_6$ M 328.321

Constit. of *Eysenhardtia polystachya*. Also from *Amorpha fruticosa* and *Dalbergia monetaria*. Reputed to be the substance in Robert Boyle's fluorescent acid-base indicator used in the seventeenth century. Cryst. (CHCl₃). Mp 244-245° (234-237°).

2'-Me, 4',5'-methylene ether: [7741-28-8]. *7-Hydroxy-2'-methoxy-4',5'-methylenedioxyisoflavone.* **Cuneatin. Maximaisoflavone G**
$C_{17}H_{12}O_6$ M 312.278
Constit. of *Tephrosia maxima* and *Cicer cuneatum*. Cryst. (MeOH). Mp 298-302°.

2',7-Di-Me, 4',5'-methylene ether: [4253-00-3]. *2',7-Dimethoxy-4',5'-methylenedioxyisoflavone. Cuneatin methyl ether*
$C_{18}H_{14}O_6$ M 326.305
Isol. from *Pterodon apparicioi* trunkwood. Prisms. Mp 210-212°.

Tetra-Me ether: [4253-02-5]. *2',4',5',7-Tetramethoxyisoflavone*
$C_{19}H_{18}O_6$ M 342.348
Isol. from *Amorpha fruticosa* fruit and *Calopogonium mucunoides* seeds. Cryst. Mp 193-194°.

2'-Me, 4',5'-methylene ether, 7-O-(3-methyl-2-butenyl): **Maximaisoflavone C.** *Maximin*
$C_{22}H_{20}O_6$ M 380.396
Isol. from *T. maxima* pods and roots. Prisms (Me₂CO). Mp 142-144°.

Rangaswami, S. *et al, Arch. Pharm. (Weinheim, Ger.)*, 1959, **292**, 170 (*Maximaisoflavone C*)
Fukui, K. *et al, Bull. Chem. Soc. Jpn.*, 1963, **36**, 397 (*synth*)
Galina, E. *et al, Phytochemistry*, 1974, **13**, 2593 (*Cuneatin methyl ether*)
Ognyanov, I. *et al, Planta Med.*, 1980, **38**, 279 (*2',4',5',7-Tetramethoxyisoflavone*)
Rao, E.V. *et al, Phytochemistry*, 1984, **23**, 1493 (*isol*)
Burns, D.T. *et al, Phytochemistry*, 1984, **23**, 167 (*isol, struct*)

2',5,6,7-Tetrahydroxyisoflavone   T-80108

$C_{15}H_{10}O_6$ M 286.240

5-Me,6,7-methylene ether: [51068-94-1]. *2'-Hydroxy-5-methoxy-6,7-methylenedioxyisoflavone.* **Betavulgarin**
$C_{17}H_{12}O_6$ M 312.278
Isol. from several *Beta* spp. and *Dianthus* spp. as phytoalexin. Shows strong antifungal activity. Cryst. solid. Mp 147-150°.

Geigert, J. *et al, Tetrahedron*, 1973, **29**, 2703 (*isol, uv, pmr, struct*)

3',4',5,7-Tetrahydroxyisoflavone   T-80109

Updated Entry replacing T-60108
Orobol. *Norsantal*

[480-23-9]

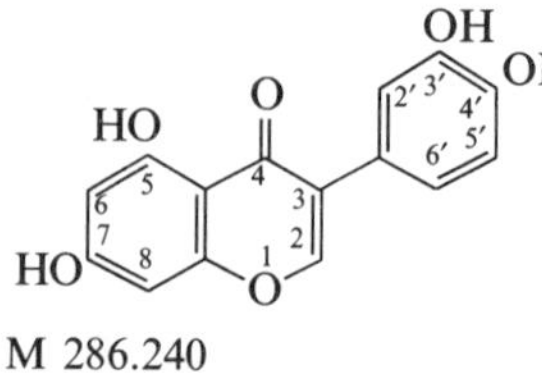

$C_{15}H_{10}O_6$ M 286.240
Isol. from *Baptisa* spp., *Bolusanthus speciosus, Cytisus scoparius, Lathyrus* spp. and *Thermopsis* spp. (Leguminosae, Papilionoideae). Also prod. by microorganisms *Aspergillis niger, Fusarium, Stemphylium* and *Streptomyces* spp. Pale-yellow cryst. (AcOH). Mp 212°. All isolations of isoflavonoids from microorganisms are considered doubtful.

Tetra-Ac: [1061-93-4].

Cryst. (AcOH). Mp 212°.

7-O-Glucoside: [20486-33-3]. **Oroboside**
$C_{21}H_{20}O_{11}$ M 448.382
Isol. from *Baptisia* spp., *Thermopsis* spp., *Lathyrus macrorrhizus* and *L. montanus.* The struct. of Oroboside as the 7-glucoside is not confirmed.

7-O-Rhamnosyglucoside: Orobol rhamnosylglucoside
$C_{27}H_{30}O_{15}$ M 594.525
Isol. from *B.* spp. λ_{max} 219, 259, 286sh, 330sh nm (MeOH).

3′-Me ether: [36190-95-1]. *4′,5,7-Trihydroxy-3′-methoxyisoflavone.* **3′-O-Methylorobol**
$C_{16}H_{12}O_6$ M 300.267
Isol. from *Dalbergia inundata, Wyethia helenoides, W. mollis* and *Thermopsis* spp. Cryst. (MeOH). Mp 218-222°.

3′-Me ether, 7-O-glucoside:
$C_{22}H_{22}O_{11}$ M 462.409
Present in *Thermopsis* spp.

4′-Me ether: [2284-31-3]. *3′,5,7-Trihydroxy-4′-methoxyisoflavone.* **Pratensein**
$C_{16}H_{12}O_6$ M 300.267
Constit. of *Bolusanthus speciosus, Trifolium* spp., *Cicer arietinum, Cytisus scoparius, Monopteryx inpae, Sophora japonica* and *Thermopsis* spp. Cryst. (EtOH). Mp 272-273°.

4′-Me ether, 7-O-glucoside: Pratensein 7-O-glucoside
$C_{22}H_{22}O_{11}$ M 462.409
Present in *Thermopsis* and *Trifolium* spp.

7-Me ether: [529-60-2]. *3′,4′,5-Trihydroxy-7-methoxyisoflavone.* **Santal**
$C_{16}H_{12}O_6$ M 300.267
Constit. of *Pterocarpus osun* and *P. soyauxi,* also *Baphia nitida, W. helenoides* and *W. mollis.* Cryst. (EtOH). Mp 222-223°.

7-Me ether, tri-Ac: Cryst. (C_6H_6). Mp 147°.

3′,4′-Di-Me ether: 5,7-Dihydroxy-3′,4′-dimethoxyisoflavone. 3′-O-Methylpratensein
$C_{17}H_{14}O_6$ M 314.294
Constit. of *B. speciosus.*

3′,7-Di-Me ether: 4′,5-Dihydroxy-3′,7-dimethoxyisoflavone
$C_{17}H_{14}O_6$ M 314.294
Constit. of *P. soyauxii.*

3′,7-Di-Me ether, 4′,5-Di-Ac: Cryst. (EtOH). Mp 179°.

5,7-Di-Me, 3′,4′-bis(3-methyl-2-butenyl) ether: [65893-94-9]. **Glabrescione B.** 5,7-Dimethoxy-3′,4′-diprenyloxyisoflavone
$C_{27}H_{30}O_6$ M 450.530
Isol. from seeds of *D. glabrescens.* Cryst. (CH_2Cl_2/heptane). Mp 102-104°.

3′,4′-Methylene ether: 5,7-Dihydroxy-3′,4′-methylenedioxyisoflavone. 3′,4′-Methylenedioxyorobol. 5-Hydroxypseudobaptigenin
$C_{16}H_{10}O_6$ M 298.251
Isol. from *Lupinus luteus* and *Sophora japonica.*

3′,4′-Methylene ether, 7-O-glucoside:
$C_{22}H_{20}O_{11}$ M 460.393
Isol. from *Lupinus angustfolius, L. luteus* and *L. polyphyllus.*

3′,4′-Methylene ether, 7-O-glucosylglucoside:
$C_{28}H_{30}O_{16}$ M 622.535
Isol. from roots of *L. luteus.*

5,7-Di-Me, 3′,4′-methylenedioxy ether: [22044-59-3]. 5,7-Dimethoxy-3′,4′-methylenedioxyisoflavone. **Derrustone**
$C_{18}H_{14}O_6$ M 326.305
Isol. from *Derris robusta* roots. Needles (MeOH). Mp 153-154°.

Akisanya, A. *et al, J. Chem. Soc.,* 1959, 2679 (*isol*)
Wong, E. *et al, J. Org. Chem.,* 1963, **28**, 2336 (*isol, struct, synth*)
Markham, K.R. *et al, Phytochemistry,* 1968, **7**, 791 (*isol*)
Dement, A.W. *et al, Phytochemistry,* 1972, **11**, 1089 (*isol*)
Adinarayana, D. *et al, Indian J. Chem.,* 1974, **12**, 911 (*synth*)
Laman, N.A. *et al, Khim. Prir. Soedin.,* 1974, 162; *cnc,* 175 (3′,4′-Methylenedioxyorobol)
Leite de Almeida, M.E. *et al, Phytochemistry,* 1974, **13**, 751 (*isol*)
Delle Monache, F. *et al, Gazz. Chim. Ital.,* 1977, **107**, 403 (Glabrescione B)
Ingham, J.L., *Fortschr. Chem. Org. Naturst.,* 1983, **43**, 1 (*rev, occur*)
Asnes, K. *et al, Z. Naturforsch., C,* 1985, **40**, 617 (*isol*)
Jain, A.C. *et al, Indian J. Chem., Sect. B,* 1987, **26**, 488 (*synth*)
Bezuidenhoudt, B.C.B. *et al, Phytochemistry,* 1987, **26**, 531 (*isol*)

3′,4′,5′,7-Tetrahydroxyisoflavone   T-80110

7-Hydroxy-3-(3,4,5-trihydroxyphenyl)-4H-1-benzopyran-4-one, 9CI. **Baptigenin**

[5908-63-4]

$C_{15}H_{10}O_6$ M 286.240
Isol. from roots of *Baptisia tinctoria.* Cryst. (EtOH). Mp 284-285° (278-284°). Sublimes.

Tetra-Ac: Mp 214°.

O-Rhamnoside: **Baptisin**
$C_{21}H_{20}O_{10}$ M 432.383
Isol. from root of *B. tinctoria.* Cryst. (EtOH aq.). Mp 240° (sinters from 150°).

4′-Me ether: [72061-64-4]. *3′,5′,7-Trihydroxy-4′-methoxyisoflavone.* **Gliricidin**
$C_{16}H_{12}O_6$ M 300.267
Isol. from heartwood of *Gliricidia sepium.* Prisms (MeOH aq.). Mp 298° dec.

Tetra-Me ether: 3′,4′,5′,7-Tetramethoxyisoflavone
$C_{19}H_{18}O_6$ M 342.348
Cryst. (MeOH). Mp 144.5°.

Farkas, L. *et al, Chem. Ber.,* 1963, **96**, 1865 (*struct, bibl*)
Manners, G.D. *et al, Phytochemistry,* 1979, **18**, 1037 (Gliricidin)

3′,4′,6,7-Tetrahydroxyisoflavone   T-80111

Updated Entry replacing H-20254
3-(3,4-Dihydroxyphenyl)-6,7-dihydroxy-4H-1-benzopyran-4-one, 9CI

4′,6-Di-Me ether: [53948-00-8]. *3′,7-Dihydroxy-4′,6-dimethoxyisoflavone.* **Odoratin**†
$C_{17}H_{14}O_6$ M 314.294
Isol. from *Dipteryx odorata* heartwood and *Pterodon apparicioi* trunkwood. There are at least three other odoratins.

4′,6-Di-Me ether, di-Ac: Needles (MeOH). Mp 210-211°.

3′,4′,6-Tri-Me ether: [24126-90-7]. *7-Hydroxy-3′,4′,6-trimethoxyisoflavone.* **Cladrastin**
$C_{18}H_{16}O_6$ M 328.321
Isol. from *Cladrastis lutea, C. platycarpa* and *C. shikokiana.* Mp 206-207°.

3′,4′,6-Tri-Me ether, 7-O-β-D-glucopyranoside: [59183-50-5].
$C_{24}H_{26}O_{11}$ M 490.463
Isol. from *C. platycarpa* and *C. shikokiana.* Mp 213-215°.

3′,4′,6-Tri-Me ether, O-laminaribioside:
$C_{30}H_{36}O_{16}$ M 652.605
Isol. from *C. platycarpa* and *C. shikokiana.*

6-Me, 3′,4′-methylenedioxy ether: [38965-66-1]. *7-Hydroxy-6-methoxy-3′,4′-methylenedioxyisoflavone.* **Fujikinetin**
$C_{17}H_{12}O_6$ M 312.278
Isol. from *C. platycarpa, C. shikokiana, Dalbergia riparia, D. sericea* and *P. apparicioi.* Mp 279-281° dec.

6-Me, 3′,4′-methylenedioxy ether, 7-O-β-D-glucopyranoside:
[38965-67-2]. **Fujikinin**
$C_{23}H_{22}O_{11}$ M 474.420
Constit. of *C. platycarpa* and *C. shikokiana*. Mp 155-160°.

Tetra-Me ether: 3′,4′,6,7-Tetramethoxyisoflavone
$C_{19}H_{18}O_6$ M 342.348
Isol. from *Cordyla africana*, *P. apparicioi* and *P. pubsescens*. Laths (MeOH). Mp 187-188°.

3′,4′-Di-Me, 6,7-methylenedioxy ether: [2746-85-2]. *3′,4′-Dimethoxy-6,7-methylenedioxyisoflavone*
$C_{18}H_{14}O_6$ M 326.305
Isol from heartwood of *Xanthocercis zambesiaca*. Microcryst. (CHCl₃/pet. ether). Mp 258-259°.

6-Me, 3′,4′-methylenedioxy ether, 7-O-laminarabioside:
$C_{29}H_{32}O_{16}$ M 636.562
Isol. from *Cladrastis platycarpa* and *C. shikokiana*.

6,7-Di-Me, 3′,4′-methylenedioxy ether: [55303-89-4]. *6,7-Dimethoxy-3′,4′-methylenedioxyisoflavone*. **Milletenin C.** *Fujikinetin methyl ether*
$C_{18}H_{14}O_6$ M 326.305
Isol. from heartwood of *Cordyla africana* and from *Milletia ovalifolia*. Needles (EtOH). Mp 252-253° (245-246°).

Shamma, M. *et al, Tetrahedron*, 1965, **25**, 3887 (*Cladrastin*)
Campbell, R.V.M. *et al, J. Chem. Soc. C*, 1969, 1787 (*isol*)
Hayashi, T. *et al, Phytochemistry*, 1974, **13**, 1943 (*Odoratin*)
Khan, H. *et al, Tetrahedron*, 1974, **30**, 2811 (*Milletinin C*)
Antus, S. *et al, Chem. Ber.*, 1975, **108**, 3883 (*synth*)
de Almeida, M.E.L., *Phytochemistry*, 1975, **14**, 2716 (*isol*)
Harper, S.H. *et al, Phytochemistry*, 1976, **15**, 1019 (3,4′-Dimethoxy-6,7-methylenedioxyisoflavone)
Ohashi, H. *et al, Phytochemistry*, 1976, **15**, 354 (*Cladrastin*)
Ingham, J.L., *Fortschr. Chem. Org. Naturst.*, 1983, **43**, 1 (*occur*)

4′,5,6,7-Tetrahydroxyisoflavone   T-80112

Updated Entry replacing D-30313
5,6,7-Trihydroxy-3-(4-hydroxyphenyl)-4H-1-benzopyran-4-one, 9CI. 6-Hydroxygenistein
$C_{15}H_{10}O_6$ M 286.240
Isol. from *Baptisia hirsuta* and *Centrosema plumieri*. Fawn cryst. (EtOAc). Mp 243-250° dec.

7-O-Rhamnosylglucside:
$C_{27}H_{30}O_{15}$ M 594.525
Isol. from leaves and stems of *Baptisia hisuta*. λ_{max} 269, 340sh nm (MeOH).

6-Me ether: see *4′,5,7-Trihydroxy-6-methoxyisoflavone*, T-80319

4′,6-Di-Me ether: [2345-17-7]. *5,7-Dihydroxy-4′,6-dimethoxyisoflavone*. **Irisolidone.** *4′-O-Methyltectorigenin*
$C_{17}H_{14}O_6$ M 314.294
Constit. of *Podocarpus amarus* and *Iris* spp. Also *Pericopsis mooniana*, *Pueraria thunbergiana* and *Sophora japonica*. Cryst. (EtOH). Mp 200-203° (189-190°).

4′,6-Di-Me ether, 7-β-D-glucopyranoside: **Kakkalidone**
$C_{23}H_{24}O_{11}$ M 476.436
Constit. of *I. germanica*, *Pueraria thunbergiana* and *S. japonica*. Amorph. powder (MeOH). Mp 215-216°.

4′-Di-Me ether, 7-O-[β-D-xylopyranosyl(1→6)β-D-glucopyranoside]: [58274-56-9]. **Kakkalide**
$C_{28}H_{32}O_{15}$ M 608.552
Isol. from flowers of *P. thunbergiana*. Tetrahydrate. Mp 253°. $[\alpha]_D^{24}$ −31° (c, 1.0 in Py).

5,7-Di-Me ether: [479-83-4]. *4′-6-Dihydroxy-5,7-dimethoxyisoflavone*. **Muningin**
$C_{17}H_{14}O_6$ M 314.294

Isol. from *Podocarpus angolensis*, heartwood. Cryst. + ½H_2O (dioxan). Mp 285° dec.

6,7-Di-Me ether: [1096-58-8]. *4′,5-Dihydroxy-6,7-dimethoxyisoflavone*. *7-O-Methyltectorigenin*
$C_{17}H_{14}O_6$ M 314.294
Isol. from *Dalbergia sissoo*, *D. spruceana*, *D. volubilis* and *Pterocarpus angolensis*. Pale-yellow needles (MeOH). Mp 235-236° (228-229°).

6,7-Di-Me ether, 4′-O-glucoside:
$C_{23}H_{24}O_{11}$ M 476.436
Isol. from bark of *Dalbergia volubilis*. Mp 130-133°. $[\alpha]_D$ −16.32° (c, 0.68 in MeOH).

6,7-Di-Me ether, 4′-O-rhamnosylglucoside:
$C_{29}H_{24}O_{15}$ M 612.500
Isol. from immature pods of *D. sissoo*.

6,7-Di-Me ether, 4′-O-[4-O-β-D-glucopyranosyl (1→4)-β-D-glucopyranoside]:
$C_{29}H_{34}O_{16}$ M 638.578
Isol. from bark of *D. volubilis*. Mp 180-183°. $[\alpha]_D$ −21° (c, 0.75 in Py).

4′,5,6-Tri-Me ether: 7-Hydroxy-4′,5,6-trimethoxyisoflavone. 5-Methoxyafrormosin
$C_{18}H_{16}O_6$ M 328.321
Isol. from *Cladrastis platycarpa*, *C. shikokiana* and *Wisteria floribunda*.

4′,6,7-Tri-Me ether: 5-Hydroxy-4′,6,7-trimethoxyisoflavone. 4′,7-Di-O-methyltectorigenin
$C_{18}H_{16}O_6$ M 328.321
Isol. from flowers of *Dalbergia sissoo*. Mp 189-190°.

6,7-Methylenedioxy ether: see *4′,5-Dihydroxy-6,7-methylenedioxyisoflavone*, D-80341

4′,5,6-Tri-Me ether, 7-O-glucoside:
$C_{24}H_{26}O_{11}$ M 490.463
Isol. from *C. platycarpa* and *C. shikokiana*.

4′,5,6-Tri-Me ether, 7-O-Laminarabioside: Isol. from *C. platycarpa* and *C. shikokiana*. Tentative struct.

King, F.E. *et al, J. Chem. Soc.*, 1952, 96 (*Muningin*)
Prakash, L. *et al, J. Org. Chem.*, 1965, **30**, 3561 (*isol, deriv*)
Markham, K.R. *et al, Phytochemistry*, 1968, **7**, 803; 1970, **9**, 2359 (*isol, uv*)
Dhar, K.L. *et al, J. Indian Chem. Soc.*, 1975, **52**, 784 (*isol, uv, ir, nmr, Irisolidone*)
Kurihara, T. *et al, Yakugaku Zasshi (J. Pharm. Soc. Jpn.)*, 1975, **95**, 1283 (*Kakkalide*)
Ingham, J.L., *Fortschr. Chem. Org. Naturst.*, 1983, **43**, 1 (*rev, occur*)
Ali, A.A. *et al, Phytochemistry*, 1983, **22**, 2061 (*isol*)
Carman, R.M. *et al, Aust. J. Chem.*, 1985, **38**, 485.

4′,5,7,8-Tetrahydroxyisoflavone   T-80113

Updated Entry replacing T-03377
5,7,8-Trihydroxy-3-(4-hydroxyphenyl)-4H-1-benzopyran-4-one, 9CI
[13539-27-0]
$C_{15}H_{10}O_6$ M 286.240
Metab. of *Aspergillus niger*. All isolations of isoflavones from microorganisms are considered suspect.

8-Me ether: [13111-57-4]. *4′,5,7-Trihydroxy-8-methoxyisoflavone*. **Isotectorigenin.** *Pseudotectorigenin*
$C_{16}H_{12}O_6$ M 300.267
Isol. from *Dalbergia sissoo* bark, *Millettia auriculata* seeds and metab. of *A. niger*. DOPA decarboxylase inhibitor. Pale yellow needles (MeOH). Mp 245°.

4′,5,8-Tri-Me ether: [68862-20-4]. *7-Hydroxy-4′,5,8-trimethoxyisoflavone. Iso-5-methoxyafrormosin*
$C_{18}H_{16}O_6$ M 328.321

Isol. from *Cladrastis platycarpa* and *C. shikokiana*.
Tentative identification.

4′,5,8-Tri-Me ether, 7-O-β-D-glucoside:
$C_{24}H_{26}O_{11}$ M 490.463
Isol. from *C. platycarpa* and *C. shikokiana*.

8-O-Me, 4′-O-(3-methyl-2-butenyl): [68415-32-7].
Aurmillone
$C_{21}H_{20}O_6$ M 368.385
Constit. of *Millettia auriculata*. Mp 157-158°.

Dhingra, V.K. *et al, Indian J. Chem.*, 1974, **12**, 1118 (*Isotectorigenin*)
Umezawa, H. *et al, J. Antibiot.*, 1975, **28**, 947 (*isol*)
Ohashi, H. *et al, Mokuzai Gakkaishi*, 1978, **24**, 750; *CA*, **90**, 36297z (*isol*)
Srimannarayana, G. *et al, Phytochemistry*, 1978, **17**, 1065 (*Aurmillone*)

1,2,5,7-Tetrahydroxyphenanthrene   T-80114

Updated Entry replacing T-60116
1,2,5,7-Phenanthrenetetrol
$C_{14}H_{10}O_4$ M 242.231

1,5-Di-Me ether: [86630-47-9]. *1,5-Dimethoxy-2,7-phenanthrenediol. 2,7-Dihydroxy-1,5-dimethoxyphenanthrene*
$C_{16}H_{14}O_4$ M 270.284
Constit. of *Oncidium cebolleta* and *Eulophia nuda*.
Magenta needles (CHCl3/MeOH). Mp 200-201°.

Tetra-Me ether: [96754-01-7]. *1,2,5,7-Tetramethoxyphenanthrene*
$C_{18}H_{18}O_4$ M 298.338
Cryst. (EtOAc/hexane). Mp 134°.

9,10-Dihydro, 2,7-Di-Me ether: [87402-72-0]. *9,10-Dihydro-2,7-dimethoxy-1,5-phenanthrenediol, 9CI. 9,10-Dihydro-1,5-dihydroxy-2,7-dimethoxyphenanthrene.* **Eulophiol**
$C_{16}H_{16}O_4$ M 272.300
Constit. of *Eulophia nuda* tubers. Cryst. (CHCl3). Mp 202-203°.

9,10-Dihydro, 2,5-di-Me ether: 9,10-Dihydro-2,5-dimethoxy-1,7-phenanthrenediol. 9,10-Dihydro-1,7-dihydroxy-2,5-dimethoxyphenanthrene
$C_{16}H_{16}O_4$ M 272.300
Constit. of *E. nuda*. Needles (CHCl3/MeOH). Mp 202-203°. Shown to be not identical with eulophiol even though the melting points are the same.

Stermitz, F.R. *et al, J. Nat. Prod.* (*Lloydia*), 1983, **46**, 417 (*isol*)
Bhandari, S.R. *et al, Phytochemistry*, 1983, **22**, 747 (*isol*)
Bhandari, S.R. *et al, Indian J. Chem., Sect. B*, 1985, **24**, 204 (*synth, deriv, uv, pmr*)
Tuchinda, P. *et al, Phytochemistry*, 1988, **27**, 3267 (*isol, cryst struct*)

2,3,4,7-Tetrahydroxyphenanthrene   T-80115

Updated Entry replacing T-50177
2,3,4,7-Phenanthrenetetrol
$C_{14}H_{10}O_4$ M 242.231

2,4-Di-Me ether: [72966-94-0]. *5,7-Dimethoxy-2,6-phenanthrenediol. 3,7-Dihydroxy-2,4-dimethoxyphenanthrene*
$C_{16}H_{14}O_4$ M 270.284
Constit. of *Eulophia nuda*. Prisms. Mp 185°.

2,4-Di-Me ether, di-Ac: Needles (CHCl3/hexane). Mp 144-145°.

3,4-Di-Me ether: [86630-46-8]. *3,4-Dimethoxy-2,7-phenanthrenediol. 2,7-Dihydroxy-3,4-dimethoxyphenanthrene.* **Nudol**
$C_{16}H_{14}O_4$ M 270.284
Constit. of *Oncidium cebolleta* and of *Eulophia* spp.

3,4-Di-Me ether, 2,7-Di-Ac: Cryst. Mp 159° (156°).

3,4-Di-Me ether, 9,10-dihydro: [101508-48-9]. *9,10-Dihydro-3,4-dimethoxy-2,7-phenanthrenediol. 9,10-Dihydro-2,7-dihydroxy-3,4-dimethoxyphenanthrene.* **Erianthridin**
$C_{16}H_{16}O_4$ M 272.300
Constit. of *Eria carinata* and *E. stricta*. Amorph.

Stermitz, F.R. *et al, J. Nat. Prod.* (*Lloydia*), 1983, **46**, 417 (*isol*)
Majumder, P.L. *et al, Indian J. Chem., Sect. B*, 1985, **24**, 1192 (*Erianthridin*)
Bhandari, S.R. *et al, Phytochemistry*, 1985, **24**, 801 (*Nudol*)
Tuchinda, P. *et al, Phytochemistry*, 1988, **27**, 3267 (*isol*)

3′,4′,5,7-Tetrahydroxy-8-prenylflavone   T-80116

2-(3,4-Dihydroxyphenyl)-5,7-dihydroxy-8-(3-methyl-2-butenyl)-4H-1-benzopyran-4-one
[14926-12-6]

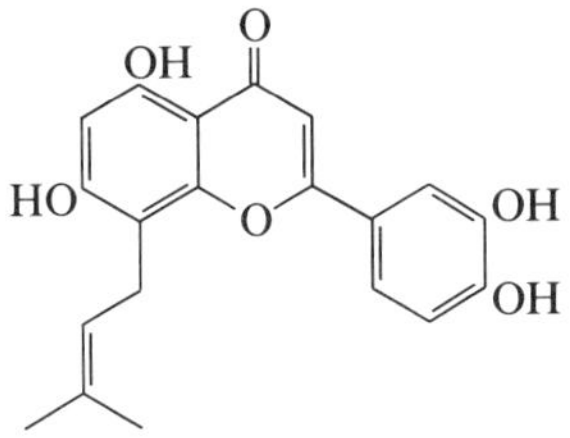

$C_{20}H_{18}O_6$ M 354.359
Isol. from *Xanthium spinosum* and *X. strumarium*. Mp 209-211°.

Pashchenko, M.M. *et al, CA*, 1966, **65**, 5871f; **66**, 49239m.

2,3,8,9-Tetrahydroxypterocarpan   T-80117

Updated Entry replacing H-02198

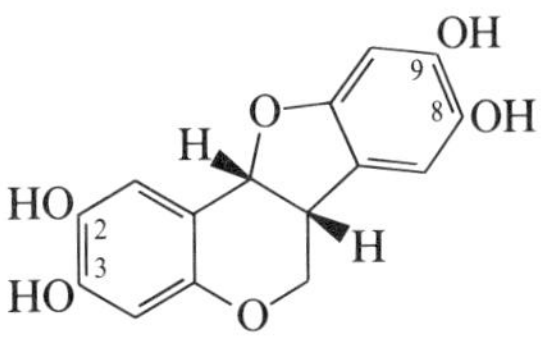

$C_{15}H_{12}O_6$ M 288.256

3,9-Di-Me ether: [76474-65-2]. *2,8-Dihydroxy-3,9-dimethoxypterocarpan*
Isol. from *Swartzia laevicarpa*. Cryst. (C_6H_6/pet. ether). Mp 178-180°.

3-Me, 8,9-methylenedioxy ether: [30461-92-8]. *2-Hydroxy-3-methoxy-8,9-methylenedioxypterocarpan. 2-Hydroxypterocarpin*
$C_{17}H_{14}O_6$ M 314.294
Constit. of *Neorautanenia edulis* and *S. leiocalycina*.
Needles (Me2CO). Mp 238-239°. $[\alpha]_D^{23}$ $-227.7°$ (c, 0.8 in CHCl3).

2,3-Di-Me, 8,9-methylene ether: [30461-93-9]. *2,3-Dimethoxy-8,9-methylenedioxypterocarpan. 2-Methoxypterocarpin*
$C_{18}H_{16}O_6$ M 328.321
Isol. from root bark of *N. edulis*. Prisms. Mp 163.8°. $[\alpha]_D^{25}$ $-231.8°$ (c, 1.3 in CHCl3).

(±)-*form*

3-Me, 8,9-methylenedioxy ether: [52305-02-9].
Needles (Me2O). Mp 223-224°.

Rall, G.J.H. *et al, Tetrahedron*, 1970, **26**, 5007 (2-*Hydroxypterocarpin, abs config*)
Rall, G.J.H. *et al, J. S. Afr. Chem. Inst.*, 1972, **25**, 25 (2-*Methoxypterocarpin*)

Farkas, L. *et al*, *J. Chem. Soc., Perkin Trans.* 1, 1974, 305 (*synth*)
Braz Filho, R. *et al*, *Phytochemistry*, 1980, **19**, 2003 (*2,8-Dihydroxy-3,9-dimethoxypterocarpan*)

2,3,9,10-Tetrahydroxypterocarpan   T-80118

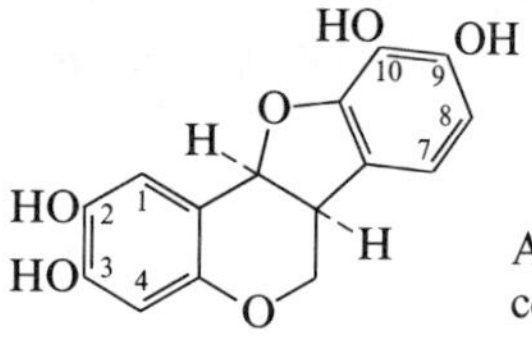

Absolute configuration

$C_{15}H_{12}O_6$ M 288.256

3,9-Di-Me ether: [69743-89-1]. *2,10-Dihydroxy-3,9-dimethoxypterocarpan.* **Mucronucarpan**
$C_{17}H_{16}O_6$ M 316.310
Isol. from heartwood of *Machaerium mucronulatum*. Oil.
$[\alpha]_D^{20}$ −15.4° (c, 0.42 in Me₂CO).

(±)-*form*

3,9-Di-Me ether: Needles (EtOH). Mp 173°.

Kurosawa, K. *et al*, *Phytochemistry*, 1978, **17**, 1405 (*isol, uv, pmr, ord, struct*)

3,4,8,9-Tetrahydroxypterocarpan   T-80119

Updated Entry replacing H-02199

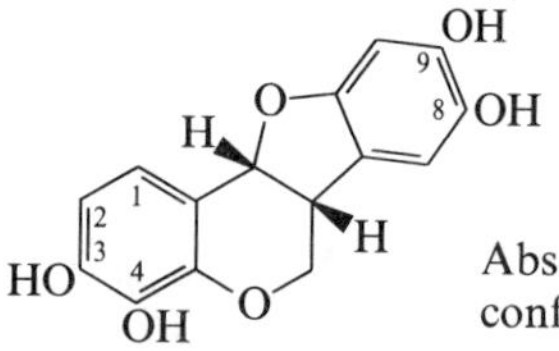

Absolute configuration

$C_{15}H_{12}O_6$ M 288.256

8,9-Methylene ether: 3,4-Dihydroxy-8,9-methylenedioxypterocarpan. **4-Hydroxymaackiain**
$C_{16}H_{12}O_6$ M 300.267
Isol. from wood of *Dalbergia spruceana*. Cryst. (MeOH aq.). Mp 85-87°, Mp 186-187° (double Mp). $[\alpha]_D^{20}$ −233° (MeOH).

3-Me, 8,9-methylene ether: 4-Hydroxy-3-methoxy-8,9-methylenedioxypterocarpan. **4-Hydroxypterocarpin**
$C_{17}H_{14}O_6$ M 314.294
Isol. from *D. spruceana* heartwood. Needles (MeOH), rhombs (pet. ether). Mp 172-174°. $[\alpha]_D^{20}$ −300° (MeOH).

4-Me, 8,9-methylene ether: [22973-31-5]. *3-Hydroxy-4-methoxy-8,9-methylenedioxypterocarpan.* **4-Methoxymaackiain**
$C_{17}H_{14}O_6$ M 314.294
Constit. of *Amorpha californica, Baptisia australis, Dalbergia spruceana, Sophora franchetiana, Swartzia madagascariensis, Tephrosia bidwilli* and *Trifolium hybridum*. Needles (MeOH). Mp 159-161°. $[\alpha]_D$ −197° (CHCl₃).

3,4-Di-Me, 8,9-methylene ether: [2694-70-4]. *3,4-Dimethoxy-8,9-methylenedioxypterocarpan.* **4-Methoxypterocarpin**
$C_{18}H_{16}O_6$ M 328.321
Isol. from tubers of *Neorantanenia ficifolia* and heartwood of *S. madagascariensis*. Prisms (C₆H₆). Mp 245-247°. $[\alpha]_D$ −202° (CHCl₃).

(±)-*form*

3,4-Di-Me, 8,9-methylene ether: Mp 226-227°.

Fukui, K. *et al*, *Experientia*, 1969, **25**, 122 (*synth, uv*)

Harper, S.H. *et al*, *J. Chem. Soc. C*, 1969, 1109 (*isol*)
Cook, J.T. *et al*, *Phytochemistry*, 1978, **17**, 1419 (*isol, uv, pmr, cd, struct*)

3,6a,7,9-Tetrahydroxypterocarpan   T-80120

$C_{15}H_{12}O_6$ M 288.256

9-Me ether: [61135-98-6]. *3,6a,7-Trihydroxy-9-methoxypterocarpan. 6a,7-Dihydroxymedicarpin*
$C_{16}H_{14}O_6$ M 302.283
Isol. from leaves of *Melilotus alba* and *Trifolium pratense* as fungal metab. Not obt. pure.

Ingham, J.L., *Phytochemistry*, 1976, **15**, 1489 (*isol, uv*)

3,6a,8,9-Tetrahydroxypterocarpan   T-80121

Updated Entry replacing P-20139

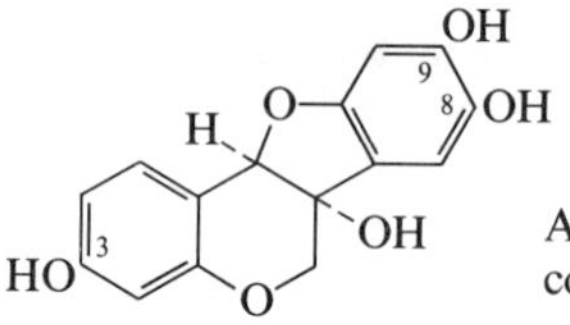

Absolute configuration

8,9-Methylenedioxy ether: [14602-93-8]. *3,6a-Dihydroxy-8,9-methylenedioxypterocarpan.* **6a-Hydroxymaackiain.** *6a-Hydroxyinermin*
Isol. from leaves of *Trifolium pratense* as fungal metab. Also prod. by cultures of *Ascochyta pisi, Botrytis cinerea, Fusarium* spp., *Nectria haematococca* and other fungi. Mp 178-181° dec. $[\alpha]_D^{21}$ +337° (c, 0.944 in EtOH).

3-Me, 8,9-methylenedioxy ether: [20186-22-5]. *6a-Hydroxy-3-methoxy-8,9-methylenedioxypterocarpan.* **Pisatin**
Stress metab. from *Caragana* spp., *Lathyrus* spp., *Pisum fulvum, P. sativum, Tephrosia bidwilli* and *Trifolium pratense*. Phytoalexin. Cryst. (EtOH or C₆H₆). Mp 61°. $[\alpha]_D^{20}$ +280° (c, 0.11 in EtOH).

(±)-*form*

3-Me, 8,9-methylenedioxy ether: Mp 188-190°.

Perrin, D.R. *et al*, *J. Am. Chem. Soc.*, 1962, **84**, 1919, 1922 (*isol*)
Bevan, C.W.L. *et al*, *J. Chem. Soc.*, 1964, 5991 (*synth*)
Fukui, K. *et al*, *Tetrahedron Lett.*, 1966, 1805 (*synth*)
Kelsey, T.C. *et al*, *Phytochemistry*, 1975, **14**, 1103 (*isol, struct*)
Bilton, J.N. *et al*, *Phytochemistry*, 1976, **15**, 1411 (*isol*)
de Martinis, C., *J. Cryst. Mol. Struct.*, 1978, **8**, 247 (*struct*)
Banks, S.W. *et al*, *Phytochemistry*, 1982, **21**, 1605, 2235; 1983, **22**, 1591 (*biosynth*)
Ingham, J.L., *Fortschr. Chem. Org. Naturst.*, 1983, **43**, 1 (*occur*)

3,4,8,9-Tetrahydroxypterocarpene   T-80122

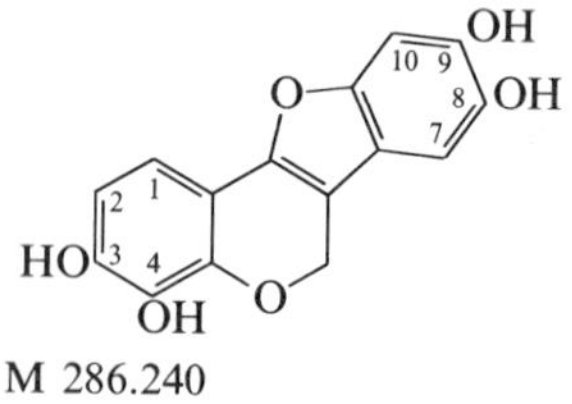

$C_{15}H_{10}O_6$ M 286.240

4-Me, 8,9-methylene ether: [56296-88-9]. **3-Hydroxy-4-methoxy-8,9-methylenedioxypterocarpene**
$C_{17}H_{12}O_6$ M 312.278
Isol. from trunkwood of *Swartzia ulei*. Cryst. Mp 196-198°.

Formiga, M.D. *et al*, *Phytochemistry*, 1975, **14**, 828 (*isol, uv, ir, pmr, struct*)

1,2,3,19-Tetrahydroxy-12-ursen-28-oic acid   T-80123

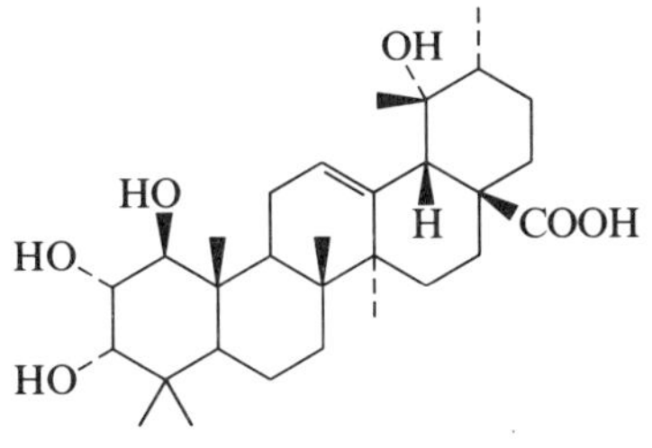

$C_{30}H_{48}O_6$ M 504.706

$(1\beta,2\alpha,3\alpha,19\alpha)$-form

Constit. of *Rosa sterilis*. Amorph. solid. $[\alpha]_D$ +24° (c, 1.5 in EtOH).

Guang-Yi, L. *et al*, *J. Nat. Prod. (Lloydia)*, 1989, **52**, 162 (isol, pmr, cmr)

1,3,19,23-Tetrahydroxy-12-ursen-28-oic acid   T-80124

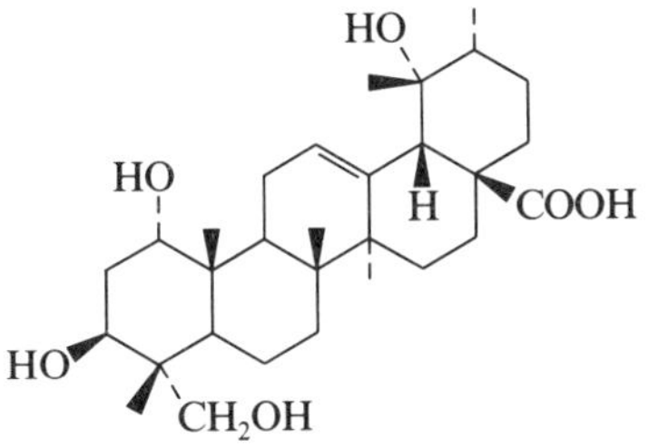

$C_{30}H_{48}O_6$ M 504.706

$(1\alpha,3\beta,19\alpha)$-form

Cryst. Mp 270-272°.

O-β-D-Xylopyranosyl ester:
$C_{35}H_{56}O_{10}$ M 636.821
Constit. of *Centipeda minima*. Needles. Mp 139°.

2α-Hydroxy: *1α,2α,3β,19α,23-Pentahydroxy-12-ursen-28-oic acid*
$C_{30}H_{48}O_7$ M 520.705

2α-Hydroxy, 28-β-D-xylopyranoside:
$C_{35}H_{56}O_{11}$ M 652.821
Constit. of *C. minina*. Cryst. Mp 210°.

Gupta, D. *et al*, *Phytochemistry*, 1989, **28**, 1197.

2,3,19,23-Tetrahydroxy-12-ursen-28-oic acid   T-80125

Updated Entry replacing T-10102

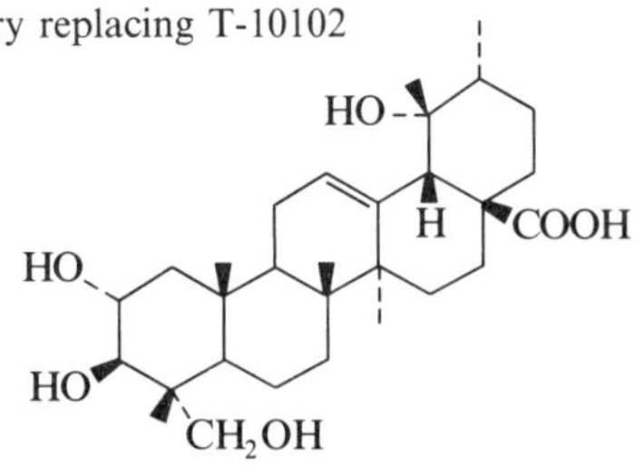

$C_{30}H_{48}O_6$ M 504.706

$(2\alpha,3\beta,19\alpha)$-form [70868-78-9]
23-*Hydroxytormentic acid*
Constit. of *Epilobium hirsutum*.

Me ester:
$C_{31}H_{50}O_6$ M 518.732
Noncryst. $[\alpha]_D$ +14° (c, 0.6 in CHCl$_3$).

Me ester, 2,3,23-tri-Ac:
$C_{37}H_{56}O_9$ M 644.844
Cryst. (hexane). Mp 208-210°. $[\alpha]_D$ +20.1° (c, 0.9 in CHCl$_3$).

3,28-O-Bis-β-D-glucopyranosyl:
$C_{42}H_{68}O_{16}$ M 828.990
Constit. of *Symplocos spicata*. Needles (MeOH). Mp 236-239°. $[\alpha]_D$ +39° (c, 1.3 in MeOH).

β-D-Glucopyranosyl ester: Quercilicoside A
$C_{36}H_{58}O_{11}$ M 666.848
Constit. of *Quercus ilex*. Cryst. (EtOH aq.). Mp 247° dec.

28-O-β-D-Glucopyranoside: 23-Hydroxytormentic acid ester glucoside
$C_{36}H_{58}O_{11}$ M 666.848
Constit. of *Desfontainia spinosa*, *Geum japonicum* and *Aphloia theiformis*. Amorph. powder.

$(2\alpha,3\alpha,19\alpha)$-form
Myrianthic acid
Constit. of *Myrianthus arboreus* and *Coleus amboinicus*.

Me ester: Cryst. Mp 126°.

Teresa, J., de P. etal, *An. Quim.*, 1979, **75**, 135.
Higuchi, R. *et al*, *Phytochemistry*, 1982, **21**, 907.
Romussi, G. *et al*, *Justus Liebigs Ann. Chem.*, 1983, 1448.
Gopalsamy, N. *et al*, *Phytochemistry*, 1988, **27**, 3593 (deriv)
Kojima, H. *et al*, *Phytochemistry*, 1989, **28**, 1703 (pmr, struct)

2,3,19,24-Tetrahydroxy-12-ursen-28-oic acid   T-80126

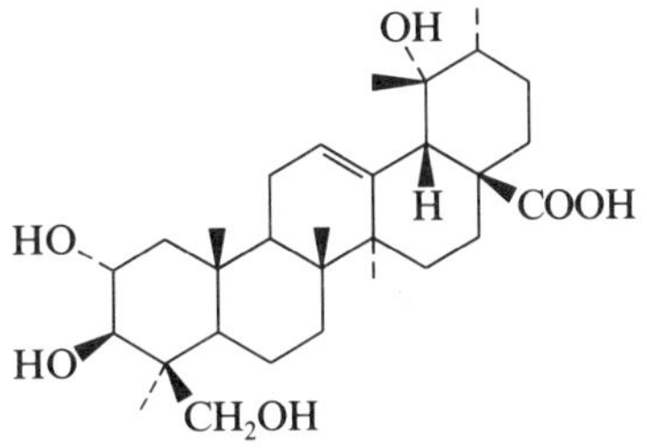

$C_{30}H_{48}O_6$ M 504.706

$(2\alpha,3\alpha)$-form
Constit. of *Rhododendron japonicum*.

Me ester: Amorph. powder. $[\alpha]_D^{21}$ +14.13° (c, 2.94 in MeOH).

$(2\alpha,3\beta)$-form [105706-08-9] **Hyptatic acid B.** 24-Hydroxytormentic acid
Constit. of *Hyptis capitata* and *Desfontainia spinosa*.
Amorph. powder. Mp 225-228°. $[\alpha]_D^{20}$ +28° (c, 0.2 in MeOH).

28-O-β-D-Glucopyranoside:
$C_{36}H_{58}O_{11}$ M 666.848
From *D. spinosa*. Amorph. solid.

Sakakibara, J. *et al*, *Phytochemistry*, 1983, **22**, 2547 (isol)
Houghton, P.J. *et al*, *Phytochemistry*, 1986, **25**, 1939 (isol)
Yamagishi, T. *et al*, *Phytochemistry*, 1988, **27**, 3213.
Kojima, H. *et al*, *Phytochemistry*, 1989, **28**, 1703 (pmr, struct)

2,3,6,19-Tetrahydroxy-12-ursen-28-oic acid   T-80127

$C_{30}H_{48}O_6$ M 504.706

$(2\alpha,3\beta,6\beta,19\alpha)$-form [119528-66-4]
6β-Hydroxytormentic acid
Amorph. powder.

20-O-β-D-Glucopyranoside: [119513-60-9].

$C_{36}H_{58}O_{11}$ M 666.848
Constit. of *Aphoia theiformis*. Amorph. powder.

Gopalsamy, N. *et al*, *Phytochemistry*, 1988, **27**, 3593.

2,3,7,19-Tetrahydroxy-12-ursen-28-oic acid   T-80128

Updated Entry replacing T-50183
$C_{30}H_{48}O_6$ M 504.706
(2α,3β,7α)-form
 7α-Hydroxytormentic acid
 Constit. of *Desfontainia spinosa*. Amorph. powder.
 28-O-β-D-Glucopyranoside:
  $C_{36}H_{58}O_{11}$ M 666.848
  From *D. spinosa*. Amorph. powder.
(2α,3α,7β,19α)-form
 Roxburic acid
 Constit. of *Rosa roxburghii*.

Houghton, P.J. *et al*, *Phytochemistry*, 1986, **25**, 1939 (*isol*)
Kojima, H. *et al*, *Phytochemistry*, 1989, **28**, 1703 (*pmr, struct*)

3,6,19,23-Tetrahydroxy-12-ursen-28-oic acid   T-80129

Updated Entry replacing T-30102
$C_{30}H_{48}O_6$ M 504.706
(3β,6β,19α)-form [91095-51-1]
 Constit. of *Enkianthus campanulatus*.
 23-Aldehyde: 3β,6β,19α-Trihydroxy-23-oxo-12-ursen-28-oic acid
  $C_{30}H_{46}O_6$ M 502.690
  Constit. of *Uncaria florida*.
 23-Aldehyde, 3Ac, Me ester: Prisms. Mp 197-199°.
 23-Carboxylic acid: 3β,6β,19α-Trihydroxy-12-ursene-23,28-dioic acid
  $C_{30}H_{46}O_7$ M 518.689
  Constit. of *U. florida*.

Sakakibara, J. *et al*, *Phytochemistry*, 1984, **23**, 627 (*isol*)
Aimi, N. *et al*, *Tetrahedron*, 1989, **45**, 4125 (*isol, pmr, cmr*)

Tetrakis(cyclohexylidene)cyclobutane   T-80130
1,1′,1″,1‴-(1,2,3,4-Cyclobutanetetraylidene)tetrakiscyclohexane, 9CI
[117471-48-4]

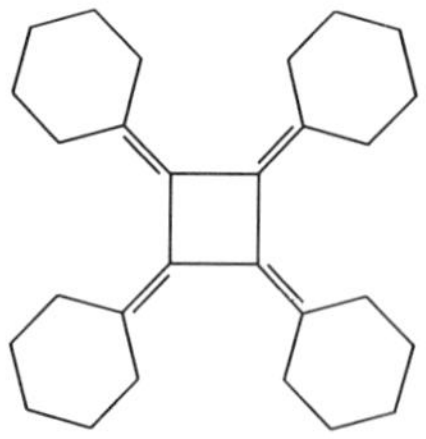

$C_{28}H_{40}$ M 376.624

Iyoda, M. *et al*, *J. Am. Chem. Soc.*, 1988, **110**, 8494 (*synth*)

Tetrakis(1,3-dithiol-2-ylidene)cyclopentanone, 9CI   T-80131
[112347-97-4]

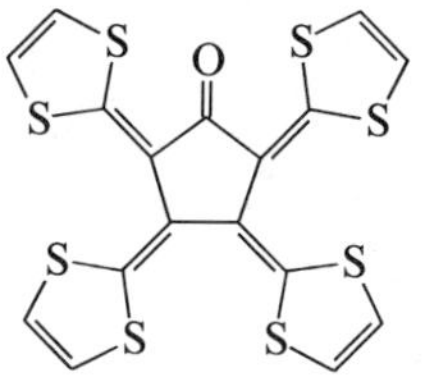

$C_{17}H_8OS_8$ M 484.778
Electron donor with low oxidation potential. Reddish purple cryst. Mp 250° dec.

Sugimoto, T. *et al*, *J. Am. Chem. Soc.*, 1988, **110**, 628 (*synth, props*)

Tetrakis(trifluoromethylthio)ethene   T-80132

$$(F_3CS)_2C=C(SCF_3)_2$$

$C_6F_{12}S_4$ M 428.311
Liq. Bp_8 49°. n_D^{25} 1.4253.

Harris, J.F., *J. Org. Chem.*, 1967, **32**, 2063 (*synth, uv, ir*)

Tetrakis(trifluoromethylthio)methane   T-80133
Tetrathioorthocarbonic acid tetrakis(trifluoromethyl) ester, 8CI
[681-87-8]

$$C(SCF_3)_4$$

$C_5F_{12}S_4$ M 416.300
Bp_8 39-39.5°. n_D^{25} 1.4013.

Harris, J.F., *J. Org. Chem.*, 1967, **32**, 2063 (*synth, ir, uv*)

5,10,15,20-Tetrakis(2,4,6-trimethylphenyl)-21H,23H-porphine, 9CI   T-80134
Tetramesitylporphyrin
[56396-12-4]

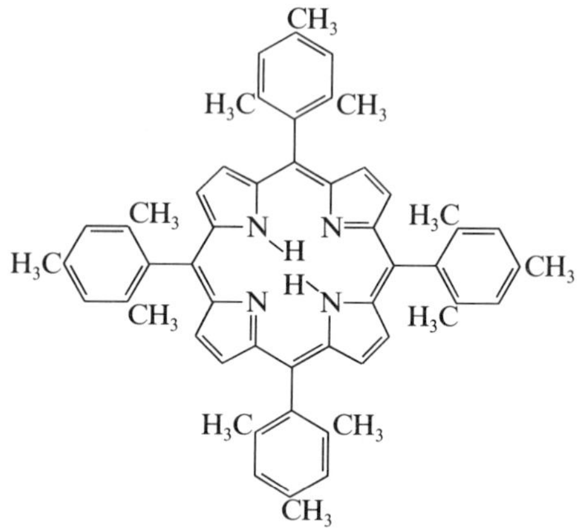

$C_{56}H_{54}N_4$ M 783.069
Synthetic sterically hindered model prophyrin. Purple needles (C_6H_6/MeOH).
Zn complex: [104025-54-9].
 $C_{56}H_{52}N_4Zn$ M 846.444
 Red-purple shiny cryst. (CHCl$_3$/hexane).
Ni complex: [109533-32-6].
 Red-purple shiny cryst. (CHCl$_3$/hexane).
Pd complex: [109533-33-7].

Shiny purple cryst. (CH$_2$Cl$_2$/pentane).
Pt complex: [109533-34-8].
Red cryst. (CHCl$_3$/hexane).

Badger, G.M. *et al*, *Aust. J. Chem.*, 1964, **17**, 1028 (*synth*)
Eaton, S.S. *et al*, *J. Am. Chem. Soc.*, 1975, **97**, 3660 (*synth, nmr*)
Groves, J.T. *et al*, *J. Am. Chem. Soc.*, 1983, **105**, 6243 (*synth*)
Eberhardt, U. *et al*, *Justus Liebigs Ann. Chem.*, 1987, 809 (*synth, nmr, uv, complexes*)
Wagner, R.W. *et al*, *Tetrahedron Lett.*, 1987, **28**, 3069 (*synth, ir, uv, nmr*)

Tetramethoxyethene   T-80135

Tetramethoxyethylene

$$(MeO)_2C=C(OMe)_2$$

C$_6$H$_{12}$O$_4$ M 148.158
Bp$_{15}$ 48°. n_D^{20} 1.4244.

▷ Prepn. explosive on large scale.

Hoffmann, R.W. *et al*, *Tetrahedron*, 1965, **21**, 891 (*synth, pmr*)
Scheeren, J.W. *et al*, *Recl. Trav. Chim. Pays-Bas* (*J. R. Neth. Chem. Soc.*), 1973, **92**, 11 (*synth*)
Herberhold, M. *et al*, *Z. Naturforsch., B*, 1976, **31**, 35 (*pmr, cmr*)
Kopecky, K.R. *et al*, *Can. J. Chem.*, 1988, **66**, 2234.

1,2,3,4-Tetramethoxy-5-(2-propenyl)benzene   T-80136

1-Allyl-2,3,4,5-tetramethoxybenzene
[15361-99-6]

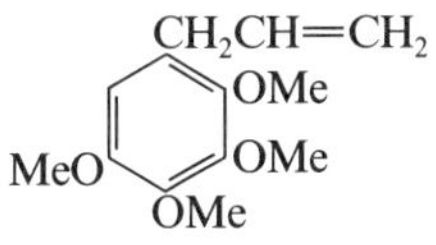

C$_{13}$H$_{18}$O$_4$ M 238.283
Isol. from parsley oil and other *Petroselinum* spp., and *Asarum* oils. Tablets (EtOH aq.). Mp 25°. Bp$_{12}$ 145°.

Baker, W. *et al*, *J. Chem. Soc.*, 1934, 1681.
Shulgin, A.T., *J. Chromatogr.*, 1967, **30**, 54 (*isol, synth*)

1,5,9,10-Tetramethylanthracene   T-80137

[120231-35-8]
C$_{18}$H$_{18}$ M 234.340
Yellow needles (MeOH). Mp 71-72°.

Meador, M.A. *et al*, *J. Org. Chem.*, 1989, **54**, 2336 (*synth, pmr, ms*)

1,8,9,10-Tetramethylanthracene   T-80138

[120231-34-7]
C$_{18}$H$_{18}$ M 234.340
Yellow cryst. (MeOH/Et$_2$O). Mp 80-82°.

Meador, M.A. *et al*, *J. Org. Chem.*, 1989, **54**, 2336 (*synth, pmr, ms*)

2,2,3,3-Tetramethylaziridine   T-80139

[5910-14-5]

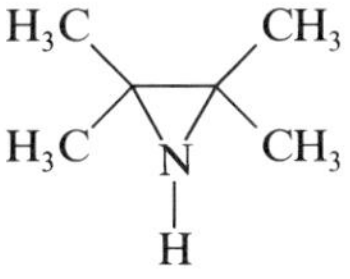

C$_6$H$_{13}$N   M 99.175
Liq. Bp$_{744}$ 104-104.5°.
1-Me: [108065-02-7]. *1,2,2,3,3-Pentamethylaziridine*

C$_7$H$_{15}$N   M 113.202
Obt. as salts.
1-Me; B,MeSO$_3$CF$_3$: [113181-25-2].
Cryst. (Me$_2$CO). Mp 176° dec.
1-Me; B,MeI: [113181-30-9].
Solid. Mp 181° dec.
1-Me; B,MeClO$_4$: [113181-31-0].
Solid. Mp 194° dec.

Closs, G.L. *et al*, *J. Am. Chem. Soc.*, 1960, **82**, 6068 (*synth, ir, pmr*)
Lillocci, C., *J. Org. Chem.*, 1988, **53**, 1733 (*deriv, synth, pmr, ms*)

3,3′,5,5′-Tetramethyl[bi-2,5-cyclohexadien-1-ylidene]-4,4′-dione, 8CI   T-80140

4-(3,5-Dimethyl-4-oxo-2,5-cyclohexadien-1-ylidene)-2,6-dimethyl-2,5-cyclohexadien-1-one, 9CI. 3,3′,5,5′-Tetramethyl-4,4′-diphenoquinone
[4906-22-3]

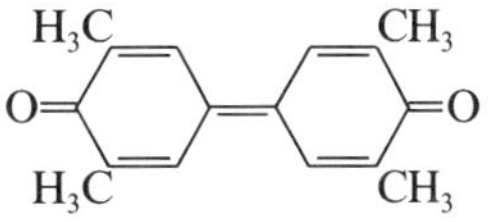

C$_{16}$H$_{16}$O$_2$ M 240.301
Small red needles (AcOH). Mp 208° dec.

Brown, B.R. *et al*, *J. Chem. Soc.*, 1954, 1280 (*synth, ir*)

2,2,3,3-Tetramethyl-1,4-butanediol   T-80141

[10519-69-4]

$$HOCH_2C(CH_3)_2C(CH_3)_2CH_2OH$$

C$_8$H$_{18}$O$_2$ M 146.229
Fine cryst. (hexane/CHCl$_3$). Mp 209-211°.
Bis-4-methylbenzenesulfonyl: [70178-81-3].
Cryst. (hexane/CHCl$_3$). Mp 110-111°.

Sowinsky, F.A.F. *et al*, *J. Org. Chem.*, 1979, **44**, 2369 (*synth, ir, pmr, ms*)
Juaristi, E. *et al*, *J. Org. Chem.*, 1988, **53**, 3334 (*synth, pmr, ir*)

2,2,3,3-Tetramethyl-1,4-butanedithiol   T-80142

[114763-99-4]

$$HSH_2CC(CH_3)_2C(CH_3)_2CH_2SH$$

C$_8$H$_{18}$S$_2$ M 178.362
Pale yellow solid.
Di-Ac: [114764-00-0].
C$_{12}$H$_{22}$O$_2$S$_2$ M 262.437
Slightly yellow solid. Mp 56-58°.

Jauristi, E. *et al*, *J. Org. Chem.*, 1989, **53**, 3334 (*synth, pmr*)

2,2,4,4-Tetramethyl-1,3-cyclobutanedione   T-80143

[933-52-8]
C$_8$H$_{12}$O$_2$ M 140.182
Mp 115-116°. Bp 159-161°.

Erickson, J.L.E. *et al*, *J. Am. Chem. Soc.*, 1946, **68**, 492 (*synth, props*)

2,2,3,3-Tetramethylcyclopropane-carboxylic acid, 9CI   T-80144

[15641-58-4]

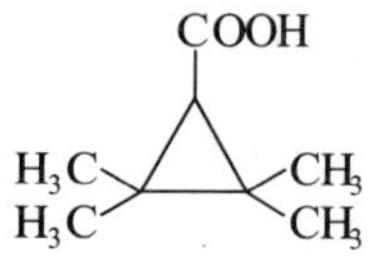

$C_8H_{14}O_2$ M 142.197
Prisms (AcOH aq.).

Meshcheryakov, A.P. et al, Izv. Akad. Sci. USSR, Ser. Sci. Khim., 1960, 931 (synth)
Jones, P.G. et al, Acta Crystallogr., Sect. C, 1987, 43, 1752 (cryst struct)

4,4,5,5-Tetramethyl-1,2-dithiane   T-80145

[114763-96-1]

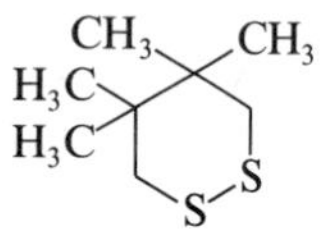

$C_8H_{16}S_2$ M 176.346
Solid by subl. $Bp_{0.5}$ 32° subl.
1-Oxide: [114763-95-0].
  $C_8H_{16}OS_2$ M 192.346
  Solid by subl. Mp 187-189°. $Bp_{0.5}$ 80-85° subl.

Juaristi, E. et al, J. Org. Chem., 1988, 53, 3334 (synth, pmr, ir, cmr)

2,2,5,5-Tetramethyl-3,4-hexanedione   T-80146

Di-tert-butyl diketone
[4388-88-9]

$$(H_3C)_3CCOCOC(CH_3)_3$$

$C_{10}H_{18}O_2$ M 170.251
Yellow liq.

Backer, H.J., Recl. Trav. Chim. Pays-Bas (J. R. Neth. Chem. Soc.), 1938, 57, 967 (synth)
Griesbaum, K. et al, Chem. Ber., 1988, 121, 1795 (synth, ir, pmr)

2,2,6,6-Tetramethyl-1-oxopiperidinium (1+)   T-80147

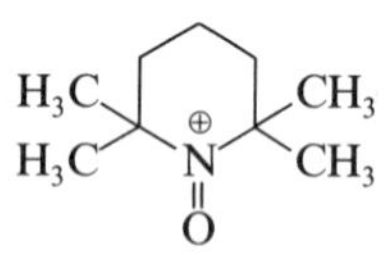

$C_9H_{18}NO^{\oplus}$ M 156.247 (ion)
Chloride: [26864-01-7].
  $C_9H_{18}ClNO$ M 191.700
  Oxidising agent. Buff solid. Mp 118-119°. Moisture and light sensitive.

Golubev, V.A. et al, Izv. Akad. Sci. USSR, Ser. Sci. Khim., (Engl. trans.), 1970, 186 (synth, uv, ir)
Hunter, D.H. et al, Tetrahedron Lett., 1984, 25, 603 (use)
Hunter, D.H. et al, J. Org. Chem., 1988, 53, 1278 (synth, bibl)

2,2,4,4-Tetramethyl-3-pentanol, 9CI   T-80148

Di-tert-butylcarbinol
[14609-79-1]

$$(H_3C)_3CCH(OH)C(CH_3)_3$$

$C_9H_{20}O$ M 144.256
Cryst. Mp 50°.

Phenylurethane: Mp 121-122°.

Bartlett, P.D. et al, J. Am. Chem. Soc., 1945, 67, 141 (synth)
Patterson, L.K. et al, J. Phys. Chem., 1966, 70, 3745 (pmr)
Roberts, D.D. et al, J. Org. Chem., 1988, 53, 2573 (synth)

N-(2,2,6,6-Tetramethylpiperidyl)nitrene   T-80149

1-(Imino-N)-2,2,6,6-tetramethylpiperidinium hydroxide inner salt, 9CI
[76372-13-9]

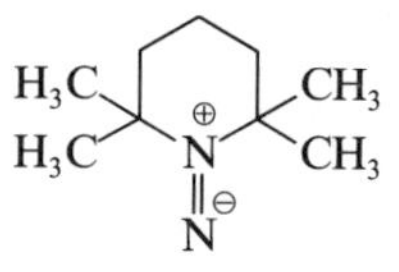

$C_9H_{18}N_2$ M 154.255
Stable deriv. of isodiazene $H_2N^{\oplus}=N^{\ominus}$. Reg. no. refers to ^{15}N-labelled compd. Intense purple soln. at −78°.

[76372-12-8]

Hinsberg, W.D. et al, J. Am. Chem. Soc., 1978, 100, 1608 (synth, uv)

3,4,5,6-Tetramethylpyridazine   T-80150

[22868-72-0]

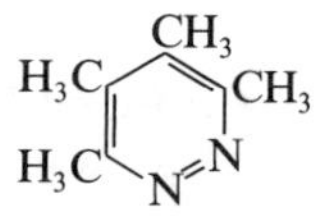

$C_8H_{12}N_2$ M 136.196
Mp 65-67° (69° dec.).

Maier, G. et al, Angew. Chem., Int. Ed. Engl., 1971, 10, 809 (synth, pmr)
Shepherd, M.K., J. Chem. Soc., Perkin Trans. 1, 1988, 961 (synth)

3,3,4,4-Tetramethyl-2,5-pyrrolidinedithione   T-80151

2,2,3,3-Tetramethyldithiosuccinimide

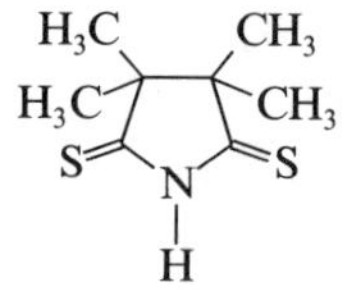

$C_8H_{13}NS_2$ M 187.329
Yellow prisms (Et$_2$O/pentane). Mp 92.5-94.5°.

Battersby, A.R. et al, J. Chem. Soc., Perkin Trans. 1, 1988, 1577 (synth, ir, pmr, ms)

Tetramethylsqualene   T-80152

2,3,7,10,15,18,22,23-Octamethyl-7,19-dimethylene-1,10,14,23-tetracosatetraene

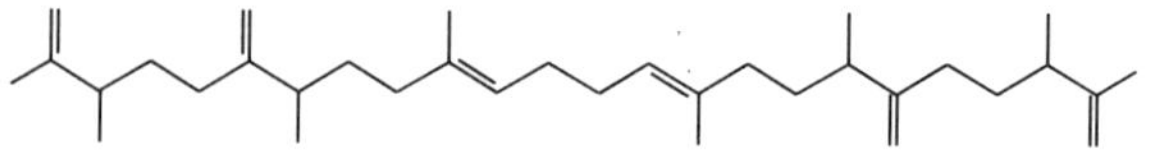

$C_{34}H_{58}$ M 466.832
Constit. of Botryococcus braunii var. showa. Oil. $[\alpha]_D^{25}$ +6° (c, 0.21 in CHCl$_3$).

Huang, Z. et al, Phytochemistry, 1989, 28, 1467.

1,1,4,4-Tetramethyl-2-tetrazene, 9CI　　　T-80153
Bis(dimethylamino)diazene
[6130-87-6]

$$Me_2NN{=}NNMe_2$$

$C_4H_{12}N_4$　　M 116.166
Yellow oil. Bp 130°, Bp$_{15}$ 32°.
▷ Explodes on heating above b.p.
Picrate: Cryst. Mp 80° dec.

[39247-67-1]

Renouf, E., *Chem. Ber.*, 1880, **13**, 2169 (*synth, haz*)
McBride, W.R. *et al, J. Am. Chem. Soc.*, 1957, **79**, 572 (*synth, uv*)
Bull, W.E. *et al, J. Am. Chem. Soc.*, 1958, **80**, 2516 (*synth, uv*)
Good, A. *et al, J. Chem. Soc. B*, 1967, 684 (*props*)
Harris, W.C. *et al, Inorg. Chem.*, 1974, **13**, 2297 (*struct, ir*)
Mason, J. *et al, J. Chem. Soc., Dalton Trans.*, 1975, 2522 (*N- nmr*)
Heymanns, P. *et al, Tetrahedron*, 1986, **42**, 2511 (*props*)

3,3,4,4-Tetramethylthiolane　　　T-80154
[114763-97-2]

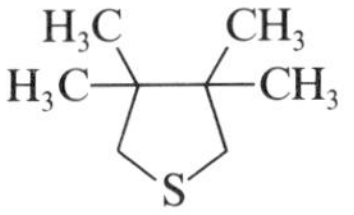

$C_8H_{16}S$　　M 144.280
Solid. Bp$_{0.5}$ 40.4°.
1-Oxide: [114763-98-3].
　$C_8H_{16}OS$　　M 160.280
　Cryst. V. hygroscopic, no Mp could be obt.

Juaristi, E. *et al, J. Org. Chem.*, 1988, **53**, 3338 (*synth, pmr, ir, ms*)

37*H*,39*H*-Tetranaphtho-[2,3-*b*:2′,3′-*g*:2″,3″-*l*:2‴-3‴-*q*]porphine, 9CI　　　T-80155
Tetra(2,3-naphtho)porphin
[73523-25-8]

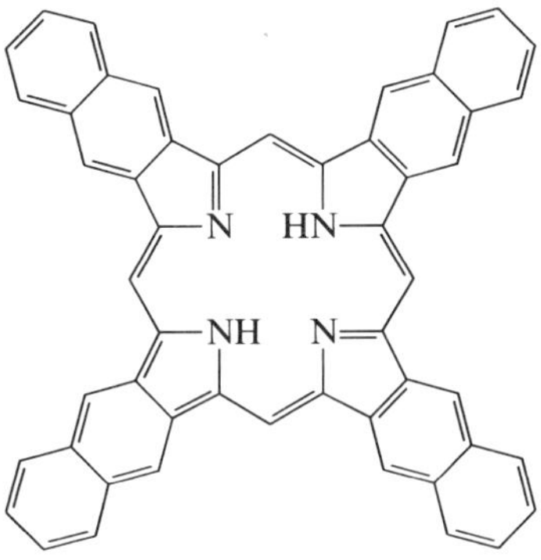

$C_{52}H_{30}N_4$　　M 710.836
Green powder. Various complexes prepd.

Rein, M. *et al, Chem. Ber.*, 1988, **121**, 1601 (*synth, ir, uv*)

1,7,20,26-Tetraoxa-4,23-dithiacyclooctatriacontane, 9CI　　　T-80156
[116186-12-0]

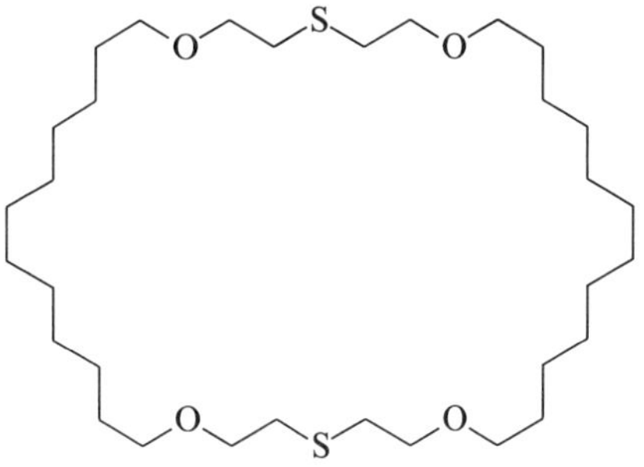

$C_{32}H_{64}O_4S_2$　　M 576.987
S-Oxidised or methylated derivs are bipolar, water insoluble amphiphiles. Platelets (2-propanol). Mp 69°.
B, 2MeClO$_4$: Powder (MeOH). Mp 95°.
S,S′-Dioxide: [116186-15-3].
　$C_{32}H_{64}O_6S_2$　　M 608.986
　Cryst. (EtOH). Mp 72°.
S,S′-Tetroxide: [116186-14-2].
　$C_{32}H_{64}O_8S_2$　　M 640.985
　Platelets (EtOH). Mp 75°.

Fuhrhop, J.-H. *et al, J. Am. Chem. Soc.*, 1988, **110**, 6840 (*synth, derivs, use, ir, ms, pmr*)

Tetraphenoxyethene　　　T-80157
1,1′,1″,1‴-[1,2-Ethenediylidenetetrakis(oxy)]tetrakisbenzene, 9CI. Tetraphenoxyethylene
[4895-47-0]

$$(PhO)_2C{=}C(OPh)_2$$

$C_{26}H_{20}O_4$　　M 396.442
Cryst. (pet. ether). Mp 170°.

McDonald, R.N. *et al, J. Org. Chem.*, 1965, **30**, 4372 (*synth*)
Scheeren, J.W. *et al, Recl. Trav. Chim. Pays-Bas (J. R. Neth. Chem. Soc.)*, 1973, **92**, 11 (*synth*)

1,1,5,5-Tetraphenyl-1,4-pentadien-3-one　　　T-80158
[21086-26-0]

$$Ph_2C{=}CHCOCH{=}CPh_2$$

$C_{29}H_{22}O$　　M 386.492
Yellow cryst. (EtOH aq.). Mp 155°.

Marin, G. *et al, Bull. Soc. Chim. Fr.*, 1958, 1594 (*synth*)

Tetraphenylpyrimidine　　　T-80159
[14277-91-9]

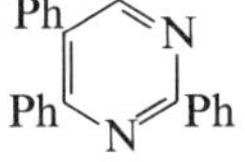

$C_{28}H_{20}N_2$　　M 384.479
Cryst. (EtOH). Mp 185-187°.

Garcia Martinéz, A. *et al, J. Heterocycl. Chem.*, 1988, **25**, 1237 (*synth, cmr*)

1,2,3,4-Tetraphenytriphenylene   T-80160
[36262-81-4]

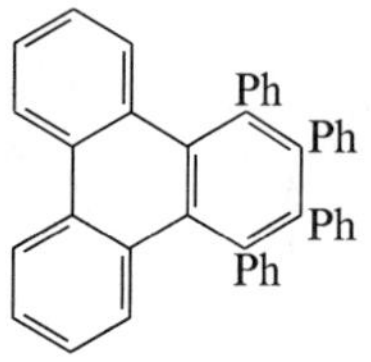

$C_{42}H_{28}$ M 532.683

Shows twisted polycyclic nucleus. Cryst. (CH_2Cl_2/Me_2CO).
Mp 292-293°, Mp >350°.

Dilthey, W. *et al*, *Ber.*, 1938, **71**, 974 (*synth*)
Pascal, R.A. *et al*, *J. Org. Chem.*, 1988, **53**, 1687 (*synth, pmr, cryst struct*)

1-Tetratriacontanol, 9CI   T-80161
Sapiol
[28484-70-0]

$$H_3C(CH_2)_{32}CH_2OH$$

$C_{34}H_{70}O$ M 494.926
Isol. from candelilla, cotton waxes and leaves of *Sapium indicum*. Mp 92°.

[103215-66-3]

Chibnall, A.C. *et al*, *Biochem. J.*, 1934, **28**, 2189 (*isol*)
Mazliak, P. *et al*, *Compt. Rend. Hebd. Seances Acad. Sci.*, 1960, **251**, 2393 (*isol*)
Japan. Pat., 80 129 292, (1980); *ca*, **94**, 120844r (*synth*)

1,2,4,5-Tetravinylbenzene   T-80162
1,2,4,5-Tetraethenylbenzene, 9CI
[117382-80-6]

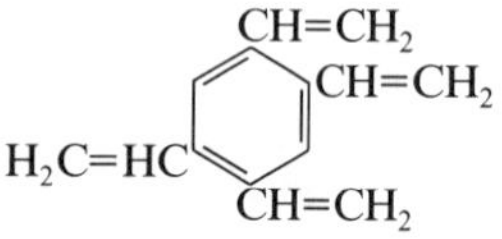

$C_{14}H_{14}$ M 182.265
Solid. Mp 30-31°. Bp$_{0.001}$ 80-90° (Kugelrohr) .

Schrievers, T. *et al*, *Synthesis*, 1988, 330 (*synth, ir, pmr, cmr, ms*)

1,2,4,5-Tetrazine-3,6-dicarboxylic acid   T-80163

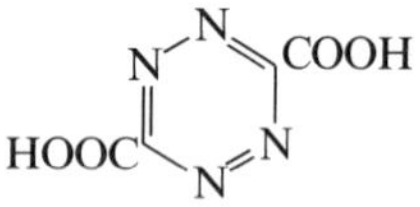

$C_4H_2N_4O_4$ M 170.084
Di-Na salt: [113631-48-4].
Red powder.
Di-Me ester: [2166-14-5].
$C_6H_6N_4O_4$ M 198.138
Bright red cryst. solid. Mp 173-175°.
Dinitrile: [16453-19-3]. *3,6-Dicyano-1,2,4,5-tetrazine*
C_4N_6 M 132.084
Forms charge-transfer complexes. Orange cryst. by subl. Mp 91°.

Curtius, T. *et al*, *Ber.*, 1907, **40**, 84 (*synth*)
Mason, S.F., *J. Chem. Soc.*, 1959, 1247 (*uv*)
Weininger, S.J. *et al*, *J. Am. Chem. Soc.*, 1967, **89**, 2050 (*dinitrile*)
Boger, D.L. *et al*, *J. Org. Chem.*, 1985, **50**, 5377 (*synth, pmr, ir, uv*)
Beilin, A. *et al*, *J. Chem. Soc., Chem. Commun.*, 1986, 1579 (*dinitrile*)

Tetrazolo[1,5-b][1,2,4]triazine   T-80164
[32484-95-0]

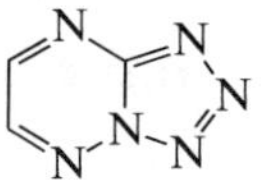

$C_3H_2N_6$ M 122.089
Cryst. (MeOH). Mp 174-176°.

Goodman, M.M. *et al*, *J. Org. Chem.*, 1976, **41**, 2860 (*synth, pmr, cryst struct, ir*)
Willer, R.L., *J. Org. Chem.*, 1988, **53**, 5371 (*synth*)

Teuscorodonin   T-80165
Updated Entry replacing T-20155
Teuscordonin
[87376-66-7]

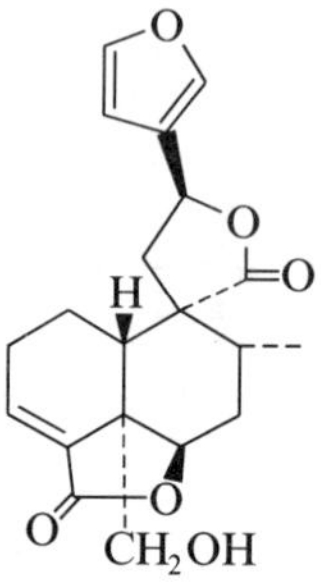

$C_{20}H_{22}O_6$ M 358.390
Constit. of *Teucrium scorodonia*. Cryst. (EtOAc/hexane).
Mp 189-191°. $[\alpha]_D^{22}$ +110.8° (c, 0.204 in $CHCl_3$).

12-Epimer: 12-Epiteuscordonin. 12-Epiteuscorodonin
$C_{20}H_{22}O_6$ M 358.390
Constit. of *T. bicolor*. Cryst. (EtOAc/pet. ether). Mp 199-203°.

Marco, J.L. *et al*, *Phytochemistry*, 1983, **22**, 727.
Labbe, C. *et al*, *J. Nat. Prod. (Lloydia)*, 1989, **52**, 871 (*epimer*)

Teuvincentin A   T-80166

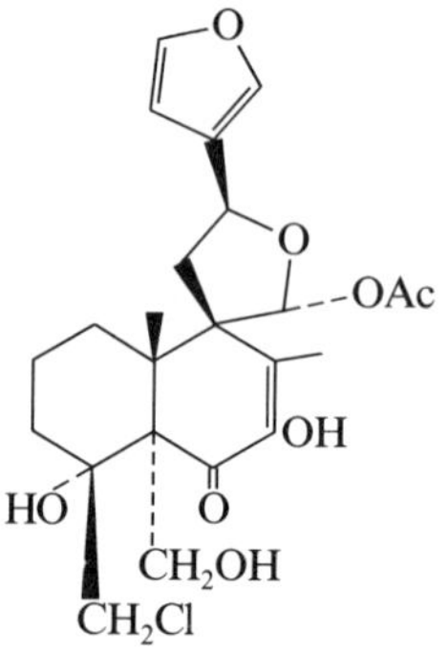

$C_{24}H_{29}ClO_9$ M 496.940
Constit. of *Teucrium polium*. Cryst. (EtOAc/hexane). Mp 200-201° dec. $[\alpha]_D^{20}$ −92.7° (c, 0.192 in $CHCl_3$).

Carreiras, M.C. *et al*, *Phytochemistry*, 1989, **28**, 1453 (*isol, cryst struct*)

Teuvincentin B
T-80167

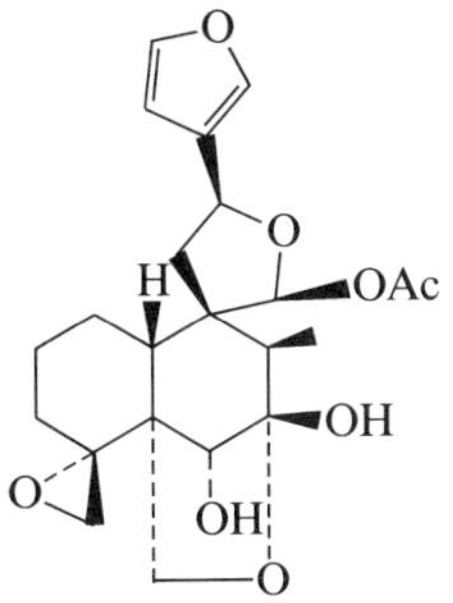

$C_{22}H_{28}O_8$ M 420.458

Constit. of *Teucrium polium*. Cryst. (EtOAc/hexane). Mp 194-197°. $[\alpha]_D^{20}$ −27.1° (c, 0.17 in CHCl$_3$).

Carreiras, M.C. *et al*, *Phytochemistry*, 1989, **28**, 1453 (*isol, cryst struct*)

Teuvincentin C
T-80168

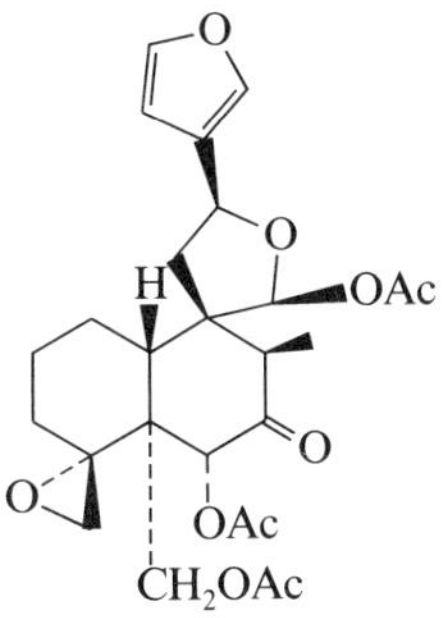

$C_{26}H_{32}O_{10}$ M 504.533

Constit. of *Teucrium polium*. Cryst. (EtOAc). Mp 220-225°. $[\alpha]_D^{22}$ −8.1° (c, 0.246 in CHCl$_3$).

Carreiras, M.C. *et al*, *Phytochemistry*, 1989, **28**, 1453.

1-Thia-5,8,8b-triazaacenaphthylene
T-80169

[70304-49-3]

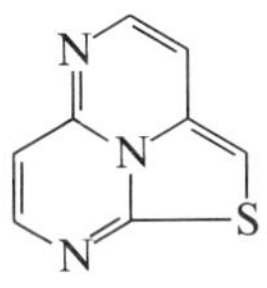

$C_8H_5N_3S$ M 175.214

Maroon cryst. (1-chlorobutane). Mp 117-119° dec.

Selby, T.P., *J. Org. Chem.*, 1988, **53**, 2386 (*synth, ir, pmr*)

2-Thiazolecarboxaldehyde
T-80170

2-Formylthiazole

[10200-59-6]

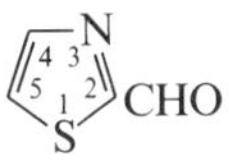

C_4H_3NOS M 113.140

Liq. Bp$_{11}$ 61-63°, Bp$_3$ 36-37°.

Oxime: [13838-77-2].

$C_4H_4N_2OS$ M 128.154

Mp 118-120°.

Phenylhydrazone: Cryst. (EtOH aq.). Mp 117-118°.

2,4-Dinitrophenylhydrazone: Cryst. (EtOH). Mp 241-243°.

Thiosemicarbazone: Yellowish cryst. Mp 198°.

Di-Et acetal: 2-(Diethoxymethyl)thiazole

$C_8H_{13}NO_2S$ M 187.262

Bp$_9$ 103-104°.

Beyer, H. *et al*, *Chem. Ber.*, 1957, **90**, 2372 (*synth*)
Iversen, P.E. *et al*, *Acta Chem. Scand.*, 1966, **20**, 2649 (*synth*)
Dondoni, A. *et al*, *Synthesis*, 1987, 998 (*synth, ir, pmr*)

4-Thiazolecarboxaldehyde
T-80171

4-Formylthiazole

[3364-80-5]

C_4H_3NOS M 113.140

Prisms (hexane). Mp 65-66°. Bp$_{14}$ 99-104°.

Oxime: [41827-98-9].

$C_4H_4N_2OS$ M 128.154

Mp 185-186°.

2,4-Dinitrophenylhydrazone: Orange-red needles (MeOH). Mp 250-252°.

Erne, M. *et al*, *Helv. Chim. Acta*, 1951, **34**, 143 (*synth*)
Campaigne, E. *et al*, *J. Med. Chem.*, 1959, **1**, 577 (*synth*)
Dondoni, A. *et al*, *Synthesis*, 1987, 998 (*synth, pmr, ms*)

5-Thiazolecarboxaldehyde
T-80172

5-Formylthiazole

[1003-32-3]

C_4H_3NOS M 113.140

Oil. Bp$_{16}$ 92-94°.

Oxime: [116045-57-9].

$C_4H_4N_2OS$ M 128.154

Mp 197-198°.

2,4-Dinitrophenylhydrazone: Red cryst. (MeOH). Mp 238-240°.

Semicarbazone: Plates (MeOH). Mp 210-212°.

Erne, M. *et al*, *Helv. Chim. Acta*, 1951, **34**, 143 (*synth*)
Dondoni, A. *et al*, *Synthesis*, 1987, 998 (*synth, pmr, ms*)

2-Thiazolidineacetic acid, 9CI
T-80173

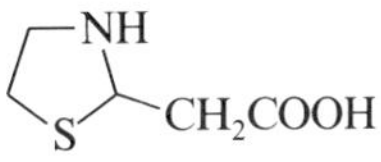

$C_5H_9NO_2S$ M 147.198

β-Lactam intermediate.

(±)-*form*

B,HCl: [98316-98-4].

Needles (EtOH/Et$_2$O). Mp 103-105°.

tert-*Butyl ester:* [98317-14-7].

$C_9H_{17}NO_2S$ M 203.305

Oil.

[98316-97-3]

Chiba, T. *et al*, *J. Chem. Soc., Perkin Trans. 1*, 1987, 1845 (*synth, ir, pmr*)

2-Thiazolyl 2-thienyl ketone
T-80174

2-Thiazolyl-2-thienylmethanone, 9CI

[104542-82-7]

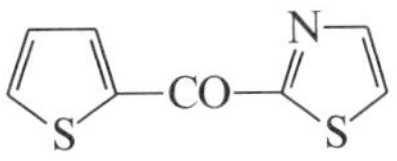

$C_8H_5NOS_2$ M 195.266

Oil.

Dondoni, A. *et al*, *J. Org. Chem.*, 1988, **53**, 1748 (*synth, ir, pmr*)

Thieno[2,3-*d*]-1,3-dithiole-2-thione   T-80175

[67188-88-9]

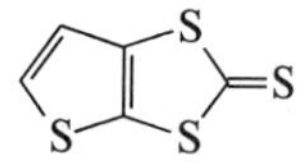

C$_5$H$_2$S$_4$ M 190.335
Red needles. Mp 125-125.5°.

Engler, E.M. *et al*, *J. Am. Chem. Soc.*, 1978, **100**, 3769 (*synth, ir, pmr*)
Litvinov, V.P. *et al*, *Izv. Akad. Sci. USSR, Ser. Sci. Khim.*, 1982, 642; 1983, 1901 (*synth, ir, ms*)

Thieno[2′,3′:4,5]thieno[2,3-*c*]quinoline   T-80176

[119647-21-1]

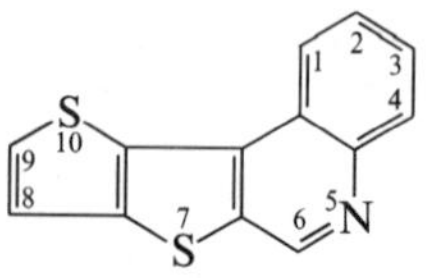

C$_{13}$H$_7$NS$_2$ M 241.337
Yellow prisms (C$_6$H$_6$). Mp 182-183°.

Castle, S.L. *et al*, *J. Heterocycl. Chem.*, 1988, **25**, 1363 (*synth, pmr*)

4*H*-Thieno[3,2-*c*]thiopyran-7(6*H*)-one   T-80177

[16401-39-1]

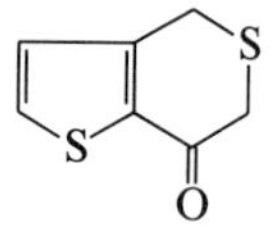

C$_7$H$_6$OS$_2$ M 170.256
Cryst. (EtOH aq.). Mp 67°. Bp$_{18}$ 180-180.5°.
Oxime: [16470-67-0].
    C$_7$H$_7$NOS$_2$ M 185.270
    Needles (C$_6$H$_6$). Mp 190°.
2,4-Dinitrophenylhydrazone: Red cryst. (EtOH/C$_6$H$_6$). Mp 244°.

Cagniant, P. *et al*, *Bull. Soc. Chim. Fr.*, 1967, 2597 (*synth, pmr*)

1-(2-Thienyl)ethanol   T-80178

α-*Methyl-2-thiophenemethanol, 9CI. Methyl-2-thienylcarbinol. 2-(1-Hydroxyethyl)thiophene*
[2309-47-9]

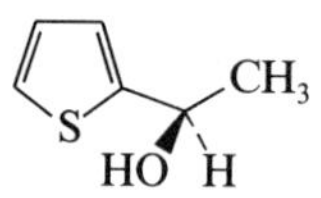

(R)-form

C$_6$H$_8$OS   M 128.195
(R)-form [86527-10-8]
    Bp$_7$ 87-89°. [α]$_D$ +13.10° (neat).
(S)-form [27948-39-6]
    Bp$_1$ 55°. [α]$_D^{22}$ −12.65° (neat).
(±)-form [115510-91-3]
    Liq. Bp$_{13}$ 98-99°. n$_D^{20}$ 1.5420.
    Phenylurethane: Plates. Mp 85° (84.2-84.9°).
[84194-85-4, 86561-31-1]

Kuhn, R. *et al*, *Justus Liebigs Ann. Chem.*, 1941, **547**, 293 (*synth*)
Weber, S. *et al*, *J. Org. Chem.*, 1962, **27**, 1258 (*synth*)
Caullet, C., *Compt. Rend. Hebd. Seances Acad. Sci.*, 1965, **260**, 1599 (*ir*)
Cervinka, O. *et al*, *Z. Chem.*, 1969, **9**, 448 (*synth, abs config*)
Noyce, D.S. *et al*, *J. Org. Chem.*, 1972, **37**, 2615 (*synth, pmr*)

Giumanini, A.G. *et al*, *J. Org. Chem.*, 1976, **41**, 2187 (*synth, ir, ms*)
Ziffer, H. *et al*, *J. Org. Chem.*, 1983, **48**, 3017 (*abs config*)
Chandrasekharan, J. *et al*, *J. Org. Chem.*, 1985, **50**, 5446 (*synth*)
Takeshita, M. *et al*, *Heterocycles*, 1987, **26**, 3051 (*synth, pmr*)
Brown, H.C. *et al*, *J. Am. Chem. Soc.*, 1988, **110**, 1539 (*synth*)
Rosini, C. *et al*, *J. Org. Chem.*, 1988, **53**, 4579 (*pmr*)

1-(3-Thienyl)ethanol   T-80179

α-*Methyl-3-thiophenemethanol, 9CI. Methyl-3-thienylcarbinol. 3-(1-Hydroxyethyl)thiophene*
[14861-60-0]

C$_6$H$_8$OS   M 128.195
(±)-form
    Liq. d$_4^{16}$ 1.188. Bp$_4$ 85°. n$_D^{20}$ 1.5473.
    Ac:
        C$_8$H$_{10}$O$_2$S   M 170.232
        Liq. Bp$_{3.5}$ 87°. n$_D^{20}$ 1.5081.

Troyanowsky, C., *Bull. Soc. Chim. Fr.*, 1955, 424 (*synth*)
Gronowitz, S. *et al*, *J. Chem. Soc.*, 1963, 3881 (*ir*)
Janda, M. *et al*, *Collect. Czech. Chem. Commun.*, 1967, **32**, 2675 (*synth*)
Macco, A.A. *et al*, *J. Org. Chem.*, 1978, **43**, 1591 (*synth*)
Abarca, B. *et al*, *Tetrahedron*, 1987, **43**, 269 (*synth*)

5-(2-Thienylethynyl)-2-thiophenemethanol, 9CI   T-80180

2-Hydroxymethyl-5-(2-thienylethynyl)thiophene
[36687-74-8]

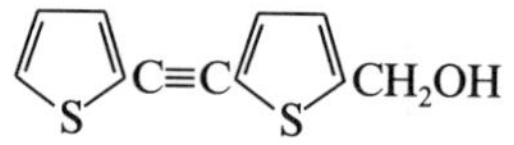

C$_{11}$H$_8$OS$_2$ M 220.316
Constit. of the roots of *Berkheya* and *Cuspida* spp. Oil.
Ac: [36687-73-7].
    C$_{13}$H$_{10}$O$_2$S$_2$ M 262.353
    Constit. of *B.* and *C.* spp.

Bohlmann, F. *et al*, *Chem. Ber.*, 1972, **105**, 1245; 1974, **103**, 2115 (*isol, struct, synth*)

9-(2-Thienyl)-4,6-nonadien-8-yn-1-ol, 9CI   T-80181

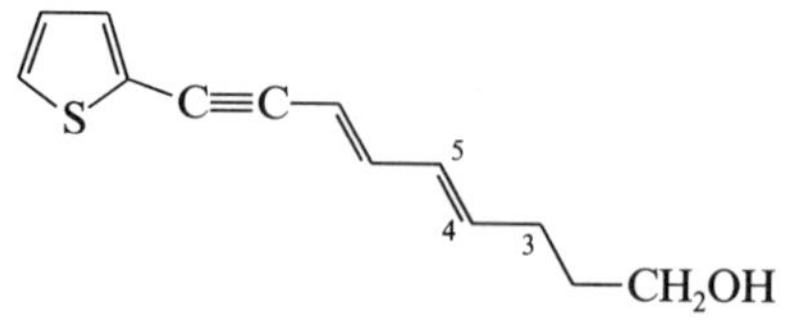

C$_{13}$H$_{14}$OS   M 218.319
Isol. from roots of *Xeranthemum cylindraceum*. λ$_{max}$ 316, 337 nm.
Ac: 1-Acetoxy-9-(2-thienyl)-4,6-nonadien-8-yne
    C$_{15}$H$_{16}$O$_2$S   M 260.356
    Isol. from *X. cylindraceum*. Cryst. (pet. ether). Mp 32°.
Deoxy, 3-hydroxy: **9-(2-Thienyl)-4,6-nonadien-8-yn-3-ol, 9CI**
    C$_{13}$H$_{14}$OS   M 218.319
    Isol. from roots of *Anthemis saguramica*. Oil. λ$_{max}$ 317, 338.5 nm (ε 26, 700; 21, 700).
Deoxy, 3-hydroxy, 4,5-dihydro: **9-(2-Thienyl)-6-nonen-8-yn-3-ol, 9CI**
    C$_{13}$H$_{16}$OS   M 220.335
    Isol. from *A. saguramica*. Oil. λ$_{max}$ 277sh, 292, 308.5 nm (ε 14,600; 19,700; 16,200).
Deoxy, 3-acetoxy, 4,5-dihydro: [14686-18-1].
    C$_{14}$H$_{18}$O$_2$S   M 250.361
    Isol. from *Matricaria* sp. Oil. λ$_{max}$ 309, 292 nm.

Bohlmann, F. *et al*, *Chem. Ber.*, 1964, **97**, 2125; 1966, **99**, 1642; 1967, **100**, 611 (*isol, struct, uv, ir, pmr, derivs*)
Bohlmann, F. *et al*, *Tetrahedron Lett.*, 1965, 1385 (*synth*)

5-(2-Thienyl)-2-penten-4-ynal, 9CI   T-80182

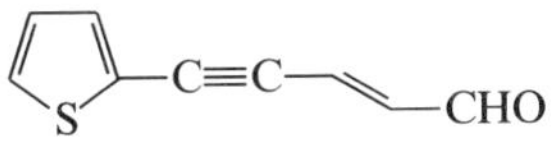

C$_9$H$_6$OS   M 162.212

(*E*)-*form* [62826-56-6]
    Isol. from roots of *Anthemis saguramica*. Yellow cryst. (pet. ether). Mp 41.5°.

Bohlmann, F. *et al*, *Chem. Ber.*, 1966, **99**, 1642 (*isol, uv, ir, pmr, struct, synth*)

5-(2-Thienyl)-2-penten-4-ynoic acid   T-80183

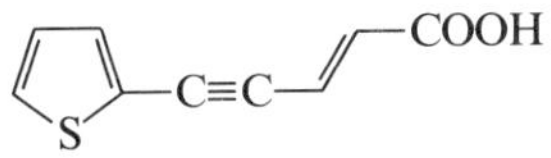

C$_9$H$_6$O$_2$S   M 178.211

(*E*)-*form*

Me ester:
    C$_{10}$H$_8$O$_2$S   M 192.238
    Isol. from above ground parts of *Anthemis fuscata*. Cryst. (pet. ether). Mp 67°.

(*Z*)-*form*

Me ester: Oil.

Bohlmann, F. *et al*, *Chem. Ber.*, 1962, **95**, 1733; 1965, **98**, 1411 (*isol, uv, struct, synth*)

3-(2-Thienyl)-2-propenal, 9CI   T-80184

3-(2-Thienyl)acrolein
[14756-03-7]

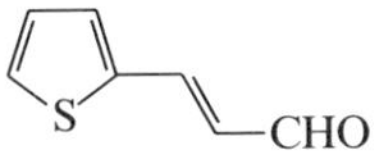

C$_7$H$_6$OS   M 138.190

(*E*)-*form* [39511-07-4]
    Yellowish oil. Bp$_{0.4}$ 105-108°, Bp$_{0.15}$ 88-95°.

Semicarbazone: Mp 218-219°.

Keskin, H. *et al*, *J. Org. Chem.*, 1951, **16**, 199 (*synth, deriv*)
Savin, F.I. *et al*, *Khim. Geterotsikl. Soedin.*, 1972, 1331 (*conformn*)
Kossmehl, G. *et al*, *Chem. Ber.*, 1974, **107**, 2791 (*synth*)

3-(3-Thienyl)-2-propenal, 9CI   T-80185

3-(3-Thienyl)acrolein
[72078-30-9]

C$_7$H$_6$OS   M 138.190

(*E*)-*form*
    Oil. Bp$_5$ 125-128°.

Klemm, L.H. *et al*, *J. Heterocycl. Chem.*, 1965, **2**, 225 (*synth*)

2,3'-Thiobispyridine   T-80186

2-(3-Pyridinylthio)pyridine, 9CI. 2-Pyridyl 3-pyridyl sulfide
[72890-91-6]

C$_{10}$H$_8$N$_2$S   M 188.253
Oil.

Picrate: Cryst. (H$_2$O). Mp 174-175°.

Krowicki, K., *Pol. J. Chem.* (*Rocz. Chem.*), 1978, **52**, 2349 (*synth*)
Summers, L.A., *J. Heterocycl. Chem.*, 1987, **24**, 533 (*rev*)

2,4'-Thiobispyridine   T-80187

2-(4-Pyridinylthio)pyridine, 9CI. 2-Pyridyl 4-pyridyl sulfide
[76093-03-3]

C$_{10}$H$_8$N$_2$S   M 188.253
Derivatives have bactericidal, fungicidal and herbicidal properties. Oil.

Picrate: Mp 167-168°.

Boduszek, B. *et al*, *Monatsh. Chem.*, 1980, **111**, 1111 (*synth, pmr*)
Summers, L.A., *J. Heterocycl. Chem.*, 1987, **24**, 533 (*rev*)

3,4'-Thiobispyridine   T-80188

3-(4-Pyridinylthio)pyridine, 9CI. 3-Pyridyl 4-pyridyl sulfide
[72890-90-5]

C$_{10}$H$_8$N$_2$S   M 188.253
Oil.

Dipicrate: Cryst. (H$_2$O). Mp 190-192°.

Krowicki, K., *Pol. J. Chem.* (*Rocz. Chem.*), 1978, **52**, 2349 (*synth*)
Summers, L.A., *J. Heterocycl. Chem.*, 1987, **24**, 533 (*rev*)

3-Thiocyanato-1-propene   T-80189

Allyl thiocyanate. Allylrhodamide. Thiocyanic acid 2-propenyl ester, 9CI
[764-49-8]

$$H_2C=CHCH_2SCN$$

C$_4$H$_5$NS   M 99.156
Present in seeds of *Brassica juncea*. Prod. from Sinigrin by enzymes in *Thlaspi arvense*. Pungent oil. d^{15} 1.056. Bp 161°. Rearr. on standing.

Gmelin, R. *et al*, *Acta Chem. Scand.*, 1959, **13**, 1474.
Kirk, L.D. *et al*, *J. Am. Oil Chem. Soc.*, 1964, **41**, 599.
Emerson, D.W., *J. Chem. Educ.*, 1971, **48**, 81 (*synth*)

3-Thiomorpholinone, 9CI   T-80190

[20196-21-8]

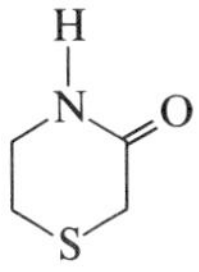

C$_4$H$_7$NOS   M 117.171
Long needles (EtOAc), cryst. (CH$_2$Cl$_2$/Et$_2$O). Mp 92-93°. Bp$_{18}$ 185°.

S-Oxide:
    C$_4$H$_7$NO$_2$S   M 133.171
    Mp 141-143°.

S-Dioxide:
    C$_4$H$_7$NO$_3$S   M 149.170
    Mp 171-173°.

Bestian, H., *Justus Liebigs Ann. Chem.*, 1950, **566**, 242 (*synth*)
Hall, H.K., *J. Am. Chem. Soc.*, 1958, **80**, 6404 (*synth, use*)
Lehr, H. *et al*, *J. Med. Chem.*, 1963, **6**, 136 (*synth, derivs*)
De Filippo, D. *et al*, *Gazz. Chim. Ital.*, 1968, **98**, 64 (*synth*)
Nagarajan, K. *et al*, *Indian J. Chem., Sect. B*, 1977, **15**, 720 (*synth, ir, use*)
Matarese, R.M. *et al*, *J. Chromatogr.*, 1984, **294**, 413 (*tlc, ms*)
Ramasubbu, N. *et al*, *Acta Crystallogr., Sect. C*, 1988, **44**, 2016 (*cryst struct*)

2*H*-Thiopyran-2,2-dicarboxylic acid   T-80191

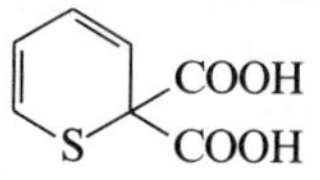

C$_7$H$_6$O$_4$S   M 186.188
Di-Me ester: [117699-51-1].
    C$_9$H$_{10}$O$_4$S   M 214.242
    Yellow oil.

Di-tert-butyl ester: [102203-47-4].
    C$_{15}$H$_{22}$O$_4$S   M 298.402
    Cryst. (MeOH). Mp 93-94°.

Bowles, T. *et al, J. Chem. Soc., Perkin Trans.* 1, 1988, 803 (*synth, ir, pmr*)

3-Thioxo-1(3*H*)-benzo[*c*]thiophenone   T-80192
Phthalic thiothionoanhydride
[112270-93-6]

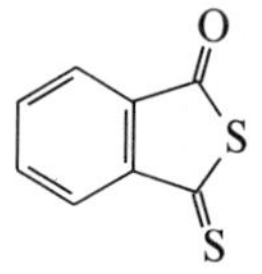

C$_8$H$_4$OS$_2$ M 180.251
Black needles (hexane). Mp 87-88°.

Raasch, M.S. *et al, J. Org. Chem.,* 1988, **53**, 891 (*synth, ir, uv, ms, pmr*)

3-Thioxo-1(3*H*)-isobenzofuranone, 9CI   T-80193
Thionophthalic anhydride
[13699-68-8]

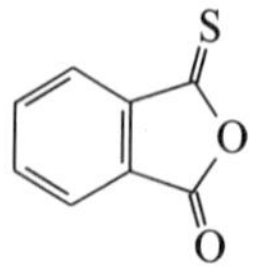

C$_8$H$_4$O$_2$S   M 164.184
Red cryst. by subl. Mp 91-96° dec. Readily isom.

Shants, C.M. *et al, J. Org. Chem.,* 1967, **32**, 3709 (*synth, uv*)

3-Thujene   T-80194
2-Methyl-5-(1-methylethyl)bicyclo[3.1.0.]hex-2-ene, 9CI. 5-Isopropyl-2-methylbicyclo[3.1.0]hex-2-ene. α-Thujene.
Origanene
[2867-05-2]

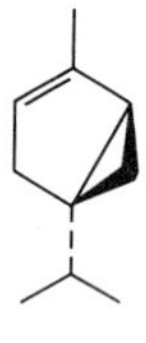

C$_{10}$H$_{16}$ M 136.236
Isol. from *Boswellia serrata, Eucalyptus* spp., *Thuja* spp. and many other plant oils. Liq. Bp 151-153°. [α]$_D$ +38.84°.

Norin, T., *Acta Chem. Scand.,* 1962, **16**, 640 (*abs config*)
Thomas, A.F. *et al, Helv. Chim. Acta,* 1964, **47**, 475 (*ms*)
Wrolstad, R.E. *et al, J. Agric. Food Chem.,* 1964, **12**, 501 (*ir*)

Karrer, W. *et al, Konstitution und Vorkommen der Organischen Pflanzenstoffe,* 2nd Ed., Birkhäuser Verlag, Basel, 1972-1985, no. 59 (*occur*)

4(10)-Thujene   T-80195
Updated Entry replacing T-02142
4-Methylene-1-(1-methylethyl)bicyclo[3.1.0]hexane, 9CI. 1-Isopropyl-4-methylenebicyclo[3.1.0]hexane. **Sabinene**
[3387-41-5]

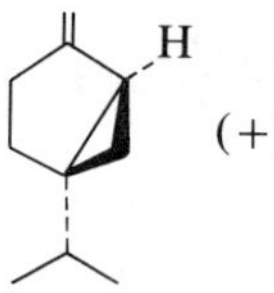

(+)-*form*

C$_{10}$H$_{16}$ M 136.236
(+)-*form* [2009-00-9]
    Constit. of various essential oils, e.g. *Juniperus sabina, Myristica fragrans.* Bp$_{160}$ 163-165°. [α]$_D$ +95° (c, 20 in CCl$_4$).
(−)-*form* [10408-16-9]
    Constit. of *Laurus nobilis, Pinus muricata* and some other plants. Liq. Bp 162-166°. [α]$_D$ −89°.

Jaureguiberry, G. *et al, Bull. Soc. Chim. Fr.,* 1962, 1985 (*isol, ir, pmr*)
Ryhage, R. *et al, Acta Chem. Scand.,* 1963, **17**, 2025 (*ms*)
Banthorpe, D.V. *et al, J. Chem. Soc., Chem. Commun.,* 1966, 177 (*biosynth*)
Fanta, W.I. *et al, J. Org. Chem.,* 1968, **33**, 1656 (*synth*)
Acharya, S.P. *et al, J. Org. Chem.,* 1969, **34**, 3015 (*abs config*)
Whitaker, D. *et al, Chem. Rev.,* 1972, **72**, 305 (*rev*)
Karrer, W. *et al, Konstitution und Vorkommen der Organischen Pflanzenstoffe,* 2nd Ed., Birkhäuser Verlag, Basel, 1972-1985, no. 60 (*occur*)

Thymidine 5′-phosphate   T-80196
Thymidine 5′-(dihydrogen phosphate), 10CI. 5′-Thymidylic acid, 9CI
[365-07-1]

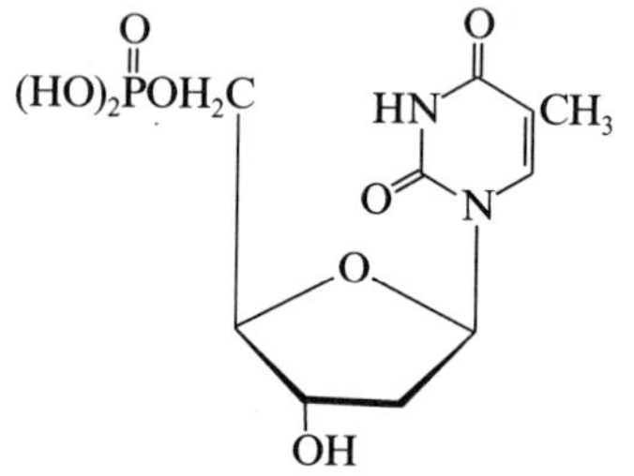

C$_{10}$H$_{15}$N$_2$O$_8$P   M 322.211
Occurs in tRNA, human lymphocytes, brain and liver.
pK_{a1} 6.36, pK_{a2} 9.90.
Di-Na salt: Dihydrate. Mp >300°.
Ba salt: [α]$_D^{17}$ −3.0° (c, 2 in H$_2$O).
Dibrucine salt: Needles (EtOH aq.). Mp 175°.
Dibenzyl ester:
    C$_{24}$H$_{27}$N$_2$O$_8$P   M 502.460
    Foam (CHCl$_3$/cyclohexane).
[33430-62-5]

Sprecher, C.A. *et al, Biopolymers,* 1977, **16**, 2243 (*cd*)
Sagi, J.T. *et al, Nucleic Acids Res.,* 1977, **4**, 2767 (*synth*)
George, A.L. *et al, Can. J. Chem.,* 1978, **56**, 1170 (*pmr, conformn*)
Niemezura, W.P. *et al, Can. J. Chem.,* 1980, **58**, 472 (*cmr, conformn*)
Samanta, S.T. *et al, Appl. Spectrosc.,* 1982, **36**, 306 (*raman*)
Stawinski, J., *Bull. Pol. Acad. Sci., Chem.,* 1983, **31**, 17 (*synth*)

Cadet, J. et al, J. Chromatogr., 1983, **259**, 111 (tlc)
Sekine, M. et al, Tetrahedron Lett., 1983, **24**, 5741 (deriv)
Sato, T., Acta Crystallogr., Sect. C, 1984, **40**, 736 (cryst struct)
Cerny, R.L. et al, Anal. Biochem., 1986, **156**, 424 (ms)
Horn, T. et al, Tetrahedron Lett., 1986, **27**, 4705 (synth, hplc, nmr, tlc)
Massoud, S.S. et al, Inorg. Chem., 1988, **27**, 1447 (props)

Tifruticin   T-80197

Updated Entry replacing T-70199

[56377-69-6]

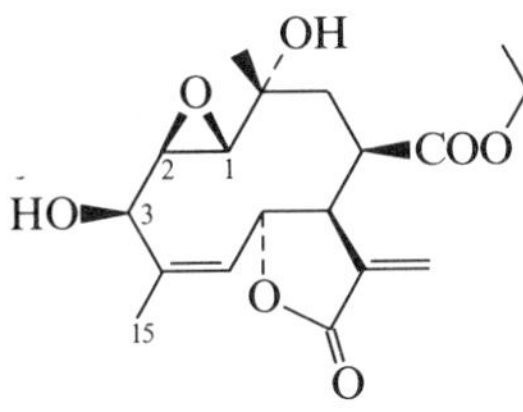

$C_{20}H_{26}O_7$ M 378.421
Constit. of Tithonia fruticosa. Cryst. (EtOAc). Mp 141°.
$[\alpha]_D^{22} -22°$ (c, 1.1 in CHCl₃).

Deepoxy, 1,2-didehydro: [56377-63-0]. **Deoxytifruticin**
$C_{20}H_{26}O_6$ M 362.422
Minor constit. of T. fruticosa.

3-Ketone, 15-hydroxy: [119889-30-4]. **15-Hydroxy-3-dehydrotifruticin**
$C_{20}H_{24}O_8$ M 392.405
Constit. of T. pedicunculata. Pale yellow oil.

Herz, W. et al, J. Org. Chem., 1975, **40**, 3118 (isol)
Baruah, N.C. et al, J. Org. Chem., 1979, **44**, 1831 (config)
Pérez, A.L. et al, Phytochemistry, 1988, **27**, 3897 (isol, deriv)

Tinosporaside   T-80198

[120163-16-8]

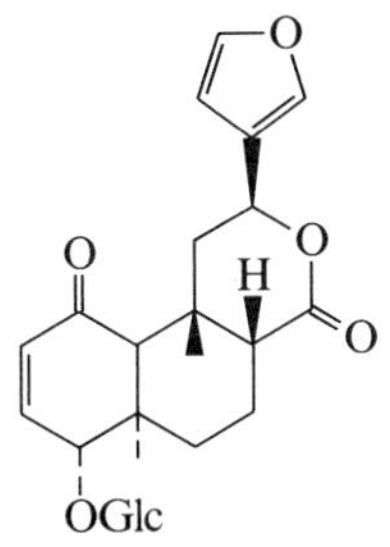

$C_{25}H_{32}O_{10}$ M 492.522
Constit. of Tinospora cordifolia. Needles (MeOH aq.). Mp
228-229°. $[\alpha]_D +65°$ (c, 1 in MeOH).

Khan, M.A. et al, Phytochemistry, 1989, **28**, 273.

Tirucalicine   T-80199

[77573-15-0]

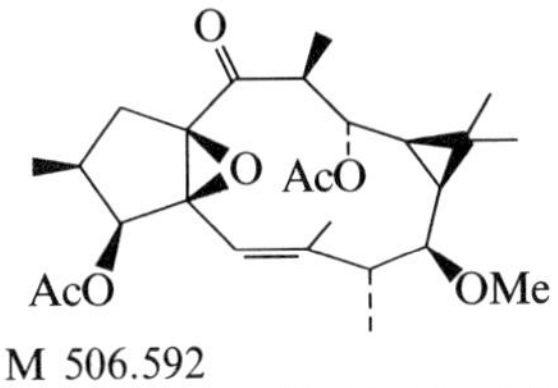

$C_{27}H_{38}O_9$ M 506.592
Constit. of Euphorbia tirucalli. Needles (EtOH). Mp 148-
150°. $[\alpha]_D +16.13°$ (c, 3.41 in CHCl₃).

Khan, A.Q. et al, Heterocycles, 1988, **27**, 2851.

β-Tocopherol   T-80200

Updated Entry replacing T-20180
3,4-Dihydro-2,5,8-trimethyl-2-(4,8,12-trimethyltridecyl)-2H-1-benzopyran-6-ol, 9CI. 2,5,8-Trimethyl-2-(4,8,12-trimethyltridecyl)chroman-6-ol, 8CI. 5,8-Dimethyltocol. Cumotocopherol. Neotocopherol. p-Xylotocopherol

[148-03-8]

$C_{28}H_{48}O_3$ M 432.685
Isol. from vegetable sources, e.g. spinach chloroplasts
(Spinacea oleracea), soybean oil, corn oil. Antioxidant.
Pale-yellow viscous oil. $Bp_{0.1}$ 200-210°. $[\alpha]_{546}^{25} +2.9°$
(EtOH). Thermostable; resistant to acids and alkalis.

3,5-Dinitrobenzoyl: Mp 86-87°.

Ac: Pale-yellow oil. $Bp_{0.3}$ 215-220°.

[16662-70-7, 16698-35-4]

Karrer, P. et al, Helv. Chim. Acta, 1938, **21**, 1234; 1939, **22**, 260 (synth, struct)
Mayer, H. et al, Helv. Chim. Acta, 1963, **46**, 963 (abs config)
Pearson, C.K. et al, Br. J. Nutr., 1970, **24**, 581 (metab)
Nakamura, A. et al, Chem. Pharm. Bull., 1971, **19**, 2318 (synth)
Scheppele, S.E. et al, Lipids, 1972, **7**, 297 (ms)
Matsuo, M. et al, Tetrahedron, 1976, **32**, 229 (cmr)
Tangney, C.C. et al, J. Chromatogr., 1979, **172**, 513 (purifn)
Burton, G.W. et al, J. Am. Chem. Soc., 1981, **103**, 6472 (props)
Pure Appl. Chem., 1982, **54**, 1507 (nomencl)
Urano, S. et al, Heterocycles, 1985, **23**, 2793 (synth)

γ-Tocopheryl quinone   T-80201

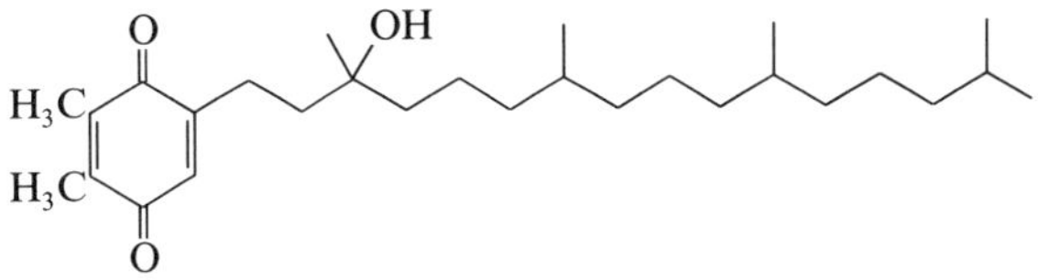

$C_{28}H_{48}O_3$ M 432.685
Isol. from vegetable sources, e.g. spinach chloroplasts
(Spinacea oleracea). Oil. λ_{max} 258 nm.

Henninger, M.D. et al, Biochemistry, 1963, **2**, 1168.

Toosendanpentol   T-80202

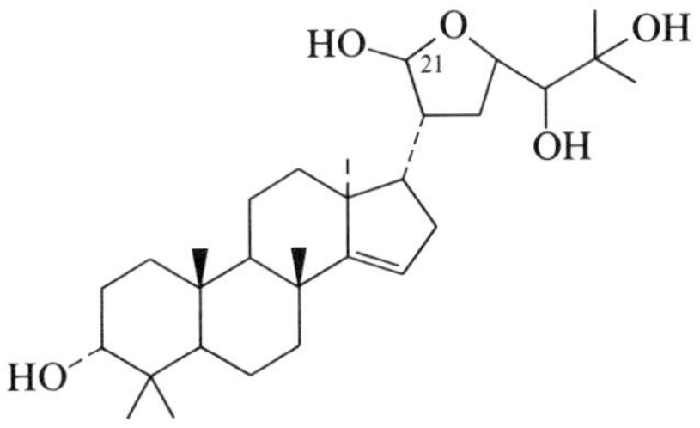

$C_{30}H_{50}O_6$ M 506.721
21-Me ether: 21-O-Methyltoosendanpentol
$C_{31}H_{52}O_6$ M 520.748
Constit. of Melia toosendan. Needles (Me₂CO/hexane).
Mp 106-108°. $[\alpha]_D -52.1°$ (c, 0.36 in MeOH).

Inada, A. et al, Heterocycles, 1989, **28**, 383.

Tormesol T-80203
[120278-23-1]

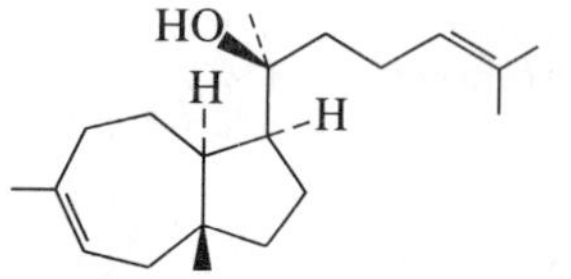

$C_{20}H_{34}O$ M 290.488

Constit. of *Halimium viscosum*. Cryst. (hexane). Mp 60-61°. $[\alpha]_D^{22}$ +30.8° (c, 1.1 in $CHCl_3$).

Urones, J.G. *et al, Phytochemistry*, 1989, **28**, 183.

α-Toxicarol T-80204
Updated Entry replacing T-02281

13,13a-Dihydro-6-hydroxy-9,10-dimethoxy-3,3-dimethyl-3H-bis[1]benzopyrano[3,4-b:6′,5′-e]pyran-7(7aH)-one, 9CI.
Toxicarol
[82-09-7]

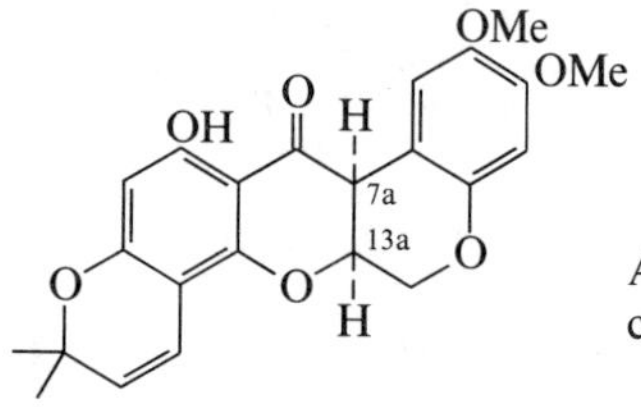

Absolute
configuration

$C_{23}H_{22}O_7$ M 410.423

Obt. from *Derris malaccensis*. Occurs in *Crotalaria burhia* also *D. elliptica*, *Tephrosia odorata* and *T. toxicaria*. Fish poison closely related in props. to Rotenone. Greenish-yellow plates or needles (EtOAc/EtOH). Mp 125-127° (101-102°). $[\alpha]_D^{20}$ −66° (C_6H_6), $[\alpha]_D$ +58° (Me_2CO).

7a,13a-Didehydro: [59086-93-0]. **Dehydrotoxicarol**
   $C_{23}H_{20}O_7$ M 408.407
   Isol. from *Amorpha fruticosa* and *D. elliptica*. Cryst. ($CHCl_3$/MeOH). Mp 263-267°.

(±)-*form*
   Mp 219-223° (101-102°).

Cahn, R.S. *et al, J. Chem. Soc.*, 1938, 513, 734 (*isol, struct*)
Harper, S.H., *J. Chem. Soc.*, 1939, 812 (*isol*)
Crombie, L. *et al, J. Chem. Soc.*, 1962, 775 (*pmr*)
Carlson, D.G. *et al, Tetrahedron*, 1973, **29**, 2731 (*pmr*)
Crombie, L. *et al, J. Chem. Soc., Perkin Trans. 1*, 1975, 1497 (*cmr*)
Reisch, J. *et al, Phytochemistry*, 1976, **15**, 234 (*Dehydrotoxicarol*)

ent-3β-Trachylobanol T-80205
[120062-66-0]

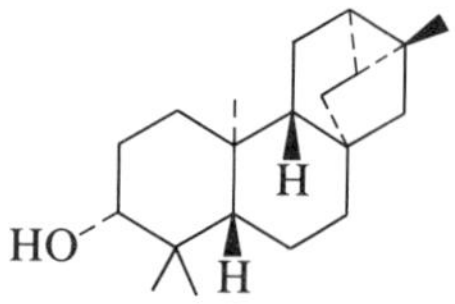

$C_{20}H_{32}O$ M 288.472
Constit. of *Xylopia aromatica*.

Moraes, M.P.L. *et al, Phytochemistry*, 1988, **27**, 3205.

Triacontanal, 9CI T-80206
[22725-63-9]

$$H_3C(CH_2)_{28}CHO$$

$C_{30}H_{60}O$ M 436.803
Isol. from berries of *Vitis vinifera*, from *Brassica oleracea* leaves and from apple cuticle wax.

Radler, F. *et al, Aust. J. Chem.*, 1965, **18**, 1059.
Schmid, H.H.O. *et al, Hoppe Seyler's Z. Physiol. Chem.*, 1969, **350**, 462 (*isol*)
Japan. Pat., 59 95 236, (1982); *CA*, **101**, 170706f (*synth*)

21-Triacontenoic acid T-80207
Lumequesic acid
[67329-09-3]

$$H_3C(CH_2)_7CH=CH(CH_2)_{19}COOH$$

$C_{30}H_{58}O_2$ M 450.787
Isol. from seed oil of *Ximenia americana*. Struct. proof not convincing from the abstr.

Boekenoogen, H.A., *Chem. Zentralbl.*, 1940, **1**, 2406.

2,4,6-Triaminopyrimidine, 8CI T-80208
2,4,6-Pyrimidinetriamine, 9CI
[1004-38-2]

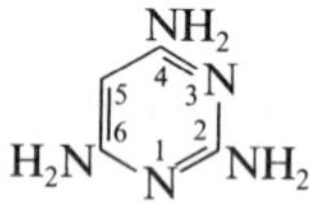

$C_4H_7N_5$ M 125.133
Reagent for determination of primary aromatic amines. Prisms (EtOH). Sol. H_2O, spar. sol. EtOH, Et_2O. Mp 249-251°. pK_{a1} 6.72 (20°), pK_{a2} 1.31 (20°).

B,2HCl: Prisms.
Picrate: Needles. Mp 290°.
N^4-*Me:* [24867-24-1].
   $C_5H_9N_5$ M 139.160
   Yellow cryst. (EtOH). Mp 192-194°. pK_{a1} 7.21 (20°), pK_{a2} 1.05 (20°).
N^4-*Me, N^1-oxide:* [55973-02-9].
   $C_5H_9N_5O$ M 155.159
   Cryst. (MeOH/MeCN). Mp 188° dec.
N^4-*Et:* [91502-44-2].
   $C_6H_{11}N_5$ M 153.186
   Cryst. (CH_2Cl_2/Me_2CO). Mp 155-156°. pK_a 7.20.
N^2,N^2-*Di-Me:* [49810-25-5].
   $C_6H_{11}N_5$ M 153.186
   Cryst. (C_6H_6). Mp 153-155°.
N^4,N^4-*Di-Me:* [24867-25-2].
   $C_6H_{11}N_5$ M 153.186
   Cryst. (EtOH). Mp 193-194.5°. pK_{a1} 7.21 (20°), pK_{a2} 0.68 (20°).
N^4,N^6-*Di-Me:*
   $C_6H_{11}N_5$ M 153.186
   Cryst. (as hydrochloride). Mp 326° (hydrochloride).
N^4,N^6-*Di-Et:*
   $C_8H_{15}N_5$ M 181.240
   Needles (C_6H_6). Mp 135.5-136°.
N^2,N^2,N^4,N^4-*Tetra-Me:* [3549-10-8].
   $C_8H_{15}N_5$ M 181.240
   Cryst. (C_6H_6). Mp 116-117°. pK_{a1} 7.18 (20°), pK_{a2} 0.81 (20°).
N^4,N^4,N^6,N^6-*Tetra-Me:* [24867-28-5].
   $C_8H_{15}N_5$ M 181.240

Descr. by Roth *et al* but synth. apparently not publ.

Sadtler Standard Infrared Spectra, 22088 (*ir*)
Sadtler Standard NMR Spectra, 29686 (*pmr*)
Traube, W., *Chem. Ber.*, 1901, **34**, 3363 (*synth*)
Roth, B. *et al*, *J. Am. Chem. Soc.*, 1950, **72**, 1914 (N^4-Me, N^4,N^4-di-Me)
Forrest, H.S. *et al*, *J. Chem. Soc.*, 1951, 3 (N^4,N^6-di-Et)
Elion, G.B. *et al*, *J. Am. Chem. Soc.*, 1953, **75**, 4311 (N^4,N^6-di-Me)
Thompson, W.K., *J. Chem. Soc.*, 1962, 617 (N^2,N^2-di-Me, ir)
Roth, B. *et al*, *J. Org. Chem.*, 1970, **35**, 2696 (*derivs, props, uv*)
McCall, J.M. *et al*, *J. Org. Chem.*, 1975, **40**, 3304 (*oxides*)
Riand, J. *et al*, *Org. Magn. Reson.*, 1977, **9**, 572 (*cmr, pmr*)
Staedeli, W. *et al*, *Helv. Chim. Acta*, 1980, **63**, 504 (*N-15 nmr*)
Schwalbe, C.H. *et al*, *Acta Crystallogr.*, *Sect. B*, 1982, **38**, 1840 (*cryst struct*)
Wells, C.H.J., *Org. Magn. Reson.*, 1982, **20**, 274 (*pmr, derivs*)
Riand, J. *et al*, *J. Heterocycl. Chem.*, 1983, **20**, 1187 (*ms*)
Cowden, W.B., *Aust. J. Chem.*, 1984, **37**, 1195 (N^4-Et)
Narita, J. *et al*, *Chem. Pharm. Bull.*, 1985, **33**, 4928 (*use*)

α,α′,2-Triamino-γ,γ′,3-trioxo-3*H*-phenoxazine-1,9-dibutanoic acid, 9CI   T-80209

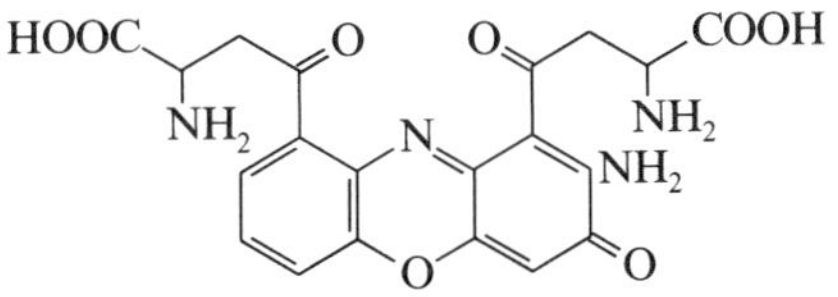

$C_{20}H_{18}N_4O_8$ M 442.384

Probable struct. Ommochrome pigment from the eyes and skin of cephalopods *Loligo vulgaris*, *Sepia officinalis* and *Octopus vulgaris*. Yellow. Insoluble. λ_{max} 420-440 nm. Readily cyclises to dihydroxanthommatin.

Bolognese, A. *et al*, *J. Heterocycl. Chem.*, 1988, **25**, 1243 (*bibl*)

1*H*-1,2,3-Triazole-4-methanamine   T-80210

4(5)-Aminomethyl-1,2,3-triazole

[118724-05-3]

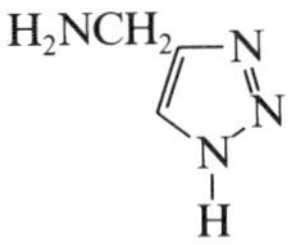

$C_3H_6N_4$ M 98.107

Cryst. (dioxan aq. or by subl.). Mp 206°.

Banert, K., *Chem. Ber.*, 1989, **122**, 911 (*synth, ir, pmr, cmr*)

1,2,4-Triazole-3(5)-sulfonic acid   T-80211

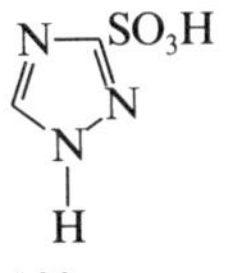

$C_2H_3N_3O_3S$ M 149.130

Fluoride: [78201-15-7].
    $C_2H_2FN_3O_2S$ M 151.121
    Cryst. + $\frac{1}{2}$ H$_2$O (EtOH). Mp 68-70°.

Amide: [89517-96-4].
    $C_2H_4N_4O_2S$ M 148.145
    Mp 227-230°.

Hanna, N.B. *et al*, *J. Heterocycl. Chem.*, 1988, **25**, 1857.

1*H*-1,2,4-Triazolo[3,4-*i*]purine   T-80212

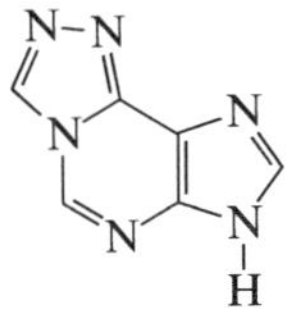

$C_6H_4N_6$ M 160.138

Cryst. (DMF). Mp >264°.

Temple, C. *et al*, *J. Org. Chem.*, 1965, **30**, 3601 (*synth, ir, uv*)

1*H*-1,2,4-Triazolo[5,1-*i*]purine, 9CI   T-80213

[4022-94-0]

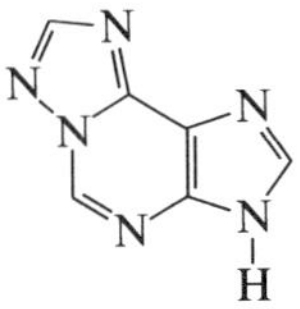

$C_6H_4N_6$ M 160.138

Cryst. (MeCN/MeOH aq.). Mp >250°.

Temple, C. *et al*, *J. Org. Chem.*, 1965, **30**, 3601 (*synth, ir, uv*)
Weimer, D.F. *et al*, *J. Org. Chem.*, 1974, **39**, 3438 (*synth, uv, ms*)
Hosmane, R.S. *et al*, *J. Org. Chem.*, 1988, **53**, 382 (*synth, pmr*)

[1,2,4]-Triazolo[5,1-*c*][1,2,4]triazin-4(1*H*)-one   T-80214

[57351-74-3]

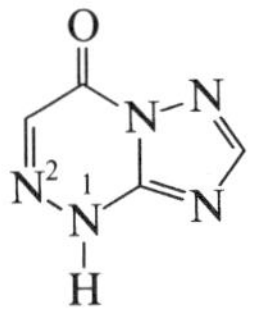

$C_4H_3N_5O$ M 137.101

Mp 236-238°.

1-Me: [59105-03-2].
    $C_5H_5N_5O$ M 151.127
    Mp 171-173°.

2-Me: [112298-55-2].
    $C_5H_5N_5O$ M 151.127
    Mp 211-214°. Zwitterionic.

Tennant, G. *et al*, *J. Chem. Soc., Perkin Trans. 1*, 1976, 421 (*synth, ir, pmr, uv*)
Farràs, J. *et al*, *J. Org. Chem.*, 1988, **53**, 887 (*synth, pmr, ir, cmr*)

Tribenzotriquinacene   T-80215

4b,8b,12b,12d-Tetrahydrodibenzo[2,3:4,5]pentaleno[1,6-ab]indene, 9CI

[120022-86-8]

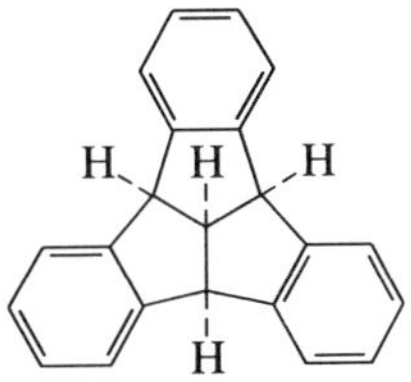

$C_{22}H_{16}$ M 280.368

Long needles (xylene). Mp 390°. Forms a monoanion and a dianion.

[120022-88-0, 120022-90-4]

Kuck, D. *et al, Angew. Chem., Int. Ed. Engl.*, 1989, **28**, 595 (*synth, pmr, uv*)

2′,3′,5′-Tribromoacetophenone   T-80216

1-(2,3,5-Tribromophenyl)ethanone, 9CI

[116772-81-7]

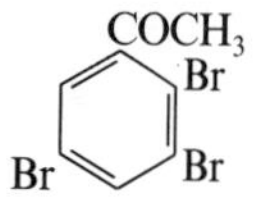

$C_8H_5Br_3O$ M 356.839

Pale reddish prisms (pet. ether). Mp 54-55°.

Tashiro, M. *et al, J. Chem. Soc., Perkin Trans.* 1, 1988, 179 (*synth, pmr, ms*)

3,6,13-Tribromo-4,10:9,12-diepoxy-14-pentadecyn-7-ol   T-80217

Updated Entry replacing B-70195

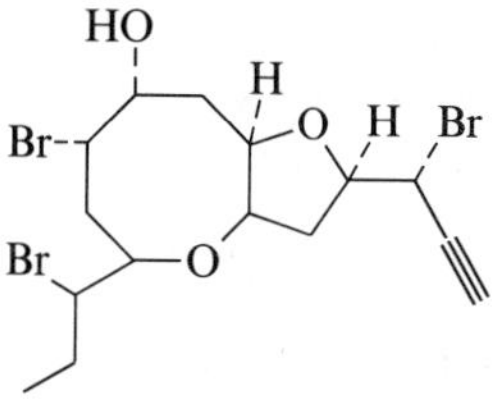

$C_{15}H_{21}Br_3O_3$ M 489.041

Incorrect struct. originally assigned. Metab. of *Laurencia obtusa*. Oil. $[\alpha]_D^{25}$ −8.2° (c, 0.09 in CHCl₃).

Ac: [94444-25-4]. *9-Acetoxy-3,10,13-tribromo-4,7:6,12-diepoxy-1-pentadecyne*
$C_{17}H_{23}Br_3O_4$ M 531.078
Metab. of red alga *L. obtusa*. Mobile oil. $[\alpha]_D$ −29.3° (c, 1.14 in CHCl₃).

Gonzalez, A.G. *et al, Tetrahedron*, 1984, **40**, 3443 (*isol*)
Norte, M. *et al, Tetrahedron*, 1989, **45**, 5987 (*struct*)

1,5,7-Tribromo-2,6,8-trichloro-2,6-dimethyl-3-octene   T-80218

[119903-43-4]

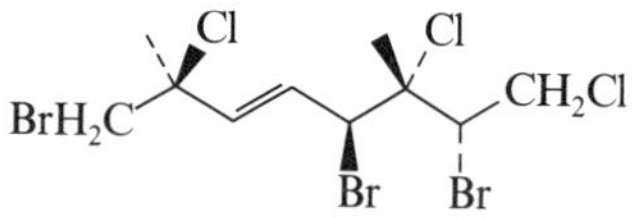

$C_{10}H_{14}Br_3Cl_3$ M 480.291

Constit. of *Plocamium hamatum*. Cryst. Mp 92.5-94°. $[\alpha]_D$ −118° (c, 0.017 in CHCl₃).

Coll, J.C. *et al, Aust. J. Chem.*, 1988, **41**, 1743 (*isol, cryst struct*)

2,4,6-Tri-*tert*-butylpyrimidine   T-80219

2,4,6-Tris(1,1-dimethylethyl)pyrimidine, 9CI

[67490-21-5]

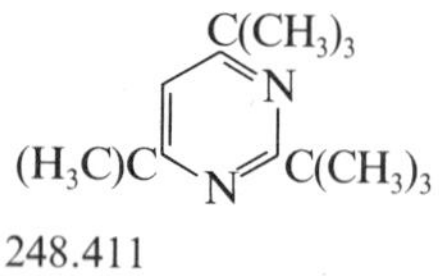

$C_{16}H_{28}N_2$ M 248.411
Cryst. (EtOH aq.). Mp 79-80°. pK_a 1.02 (H₂O, 25°).

B,HBF₄: Mp 200-201°.

van der Plas, H.C. *et al, Recl. Trav. Chim. Pays-Bas (J. R. Neth. Chem. Soc.)*, 1978, **97**, 159.

Trichilinin   T-80220

Updated Entry replacing T-60251

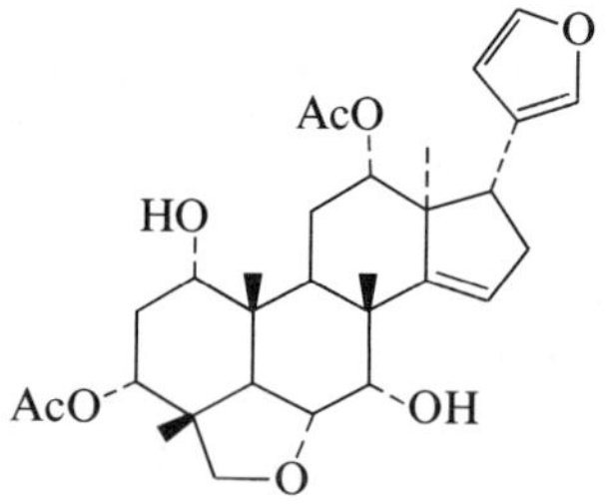

$C_{30}H_{40}O_8$ M 528.641
Constit. of *Trichilia roka*.

1-Ac:
$C_{32}H_{42}O_9$ M 570.678
Constit. of *Melia volkensii*. Cryst. (Me₂CO/hexane). Mp 114-116°.

1-O-Tigloyl:
$C_{35}H_{46}O_9$ M 610.743
Constit. of *M. volkensii*. Cryst. (Me₂CO/hexane). Mp 168-170°.

1-O-Cinnamoyl:
$C_{39}H_{46}O_9$ M 658.787
Constit. of *M. volkensii*. Cryst. (Me₂CO/hexane). Mp 134-136°.

Nakatani, M. *et al, Heterocycles*, 1987, **26**, 43 (*isol, struct*)
Rajat, M.S. *et al, J. Nat. Prod. (Lloydia)*, 1988, **51**, 840 (*derivs*)

2-(Trichloroacetyl)pyrrole   T-80221

2,2,2-Trichloro-1-(1H-pyrrol-2-yl)ethanone, 9CI. 2-Pyrrolyl trichloromethyl ketone

[35302-72-8]

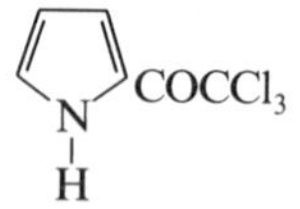

$C_6H_4Cl_3NO$ M 212.462
Tan solid (hexane). Mp 73-75°.

Org. Synth., 1971, **51**, 100 (*synth*)

2-(Trichloroacetyl)thiazole   T-80222

2,2,2-Trichloro-1-(2-thiazolyl)ethanone, 9CI

[87636-20-2]

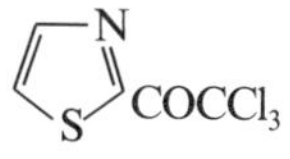

$C_5H_2Cl_3NOS$ M 230.501
Cryst. (Et₂O/hexane). Mp 62-64°.

Dondoni, A. *et al, J. Org. Chem.*, 1988, **53**, 1748 (*synth, ir, pmr, ms*)

1,2,4-Trichloro-5-(2-chloroethenyl)-1,5-dimethylcyclohexane   T-80223

[119945-08-3]

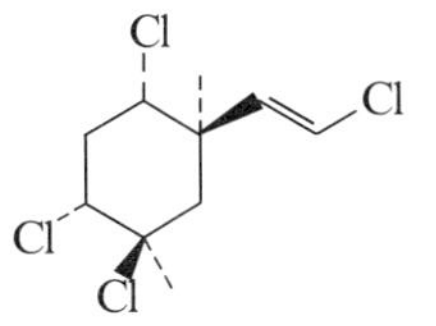

$C_{10}H_{14}Cl_4$ M 276.031

Constit. of *Plocamium hamatum*. Cryst. Mp 86-88°. $[\alpha]_D$ +9.4° (c, 0.01 in $CHCl_3$).

Coll, J.C. *et al*, *Aust. J. Chem.*, 1988, **41**, 1743 (*isol, cryst struct*)

2,3,4-Trichloro-2,4,6-cycloheptatrien-1-one, 9CI   T-80224

2,3,4-Trichlorotropone

[114125-03-0]

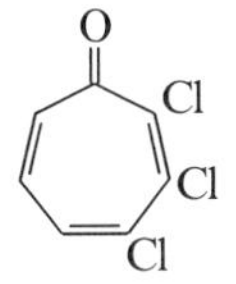

$C_7H_3Cl_3O$ M 209.458

Pale yellow solid. Mp 88°. $Bp_{0.2}$ 120° (kugelrohr) .

Seitz, G. *et al*, *Synthesis*, 1987, 953 (*synth, ms, ir, uv, pmr, cmr*)

1,1,2-Trichloro-1,2-difluoroethane, 9CI   T-80225

[354-15-4]

$$Cl_2CFCHFCl$$

$C_2HCl_3F_2$ M 169.385

Shows anaesthetic props. d_4^{20} 1.559. Bp 71-72.6°. n_D^{20} 1.3942 (1.3967).

Hauptschein, M. *et al*, *J. Am. Chem. Soc.*, 1950, **72**, 3423 (*synth*)
Bindal, M.C. *et al*, *Azneim-Forsch*, 1980, **30**, 234 (*use, ir*)

1,2,2-Trichloro-1,1-difluoroethane, 9CI   T-80226

[354-21-2]

$$F_2CClCHCl_2$$

$C_2HCl_3F_2$ M 169.385

Shows anaesthetic props. Bp 71.9°.

[41834-16-6]

Heberling, J.W., *J. Org. Chem.*, 1958, **23**, 615 (*synth, ms*)
Fessenden, R.W. *et al*, *J. Chem. Phys.*, 1962, **37**, 1466 (*pmr, F-19 nmr*)
Hørnischer, R. *et al*, *Spectrochim. Acta, Part A*, 1972, **28**, 81 (*raman*)
Chuvatken, N.N. *et al*, *Zh. Org. Khim.*, 1982, **18**, 946 (*synth*)

1,2,3-Trichloro-4-iodobenzene   T-80227

[62720-28-9]

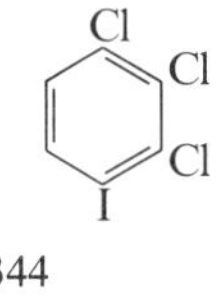

$C_6H_2Cl_3I$ M 307.344

Cryst. solid (pet. ether). Mp 66-66.5° (50-52°).

Bolton, R. *et al*, *J. Chem. Soc., Perkin Trans. 2*, 1977, 278 (*synth*)
Ghosh, T. *et al*, *J. Org. Chem.*, 1988, **53**, 3555 (*synth*)

1,1,1-Trichloro-2,2,3,3,3-pentafluoropropane, 9CI   T-80228

Freon 215

[4259-43-2]

$$F_3CCF_2CCl_3$$

$C_3Cl_3F_5$ M 237.383

Solv. for disperse dyes, cleaning solv. Intermed. for fluorine-containing anaesthetics. d_4^{20} 1.637. Bp 70.5°, Bp ~72°. n_D^{20} 1.3527.

McBee, E.T. *et al*, *J. Am. Chem. Soc.*, 1955, **77**, 3149 (*synth*)
Fainberg, A.H. *et al*, *J. Org. Chem.*, 1965, **30**, 864 (*props*)
Paleta, O. *et al*, *Collect. Czech. Chem. Commun.*, 1971, **36**, 1867 (*uv*)

1,1,3-Trichloro-1,2,2,3,3-pentafluoropropane, 9CI   T-80229

[1652-81-9]

$$FCl_2CF_2CClF_2$$

$C_3Cl_3F_5$ M 237.383

Degreasing solvent. d_4^{25} 1.643. Mp −132.4°. Bp 72-73° (75°) . n_D^{25} 1.3578, n_D^{20} 1.3512.

Coffman, D.D. *et al*, *J. Am. Chem. Soc.*, 1949, **71**, 979 (*synth*)
Fainberg, A.H. *et al*, *J. Org. Chem.*, 1965, **30**, 864 (*props*)
White, H.F., *Anal. Chem.*, 1966, **38**, 625 (*F nmr*)
Paleta, O. *et al*, *Collect. Czech. Chem. Commun.*, 1971, **36**, 1867 (*synth, ir, uv*)

1,2,2-Trichloro-1,1,3,3,3-pentafluoropropane, 9CI   T-80230

[1599-41-3]

$$F_3CCCl_2CClF_2$$

$C_3Cl_3F_5$ M 237.383

d_4^{30} 1.644, d_4^{20} 1.668. Mp −4.5°. Bp 72°. n_D^{20} 1.3518.

Farah, B.S. *et al*, *J. Org. Chem.*, 1965, **30**, 1241 (*synth*)
Fainberg, A.H. *et al*, *J. Org. Chem.*, 1965, **30**, 864 (*props*)

1,2,3-Trichloro-1,1,2,3,3-pentafluoropropane, 9CI   T-80231

[76-17-5]

$$F_2CClCClFCClF_2$$

$C_3Cl_3F_5$ M 237.383

Liq. (glass at low temps.). d_4^{20} 1.663. Mp −141.3°, Mp −135.0° (dimorph.). Bp 72-73°. n_D^{20} 1.3512, n_D^{22} 1.350.

[28109-69-5]

Henne, A.H. *et al*, *J. Am. Chem. Soc.*, 1948, **70**, 130 (*synth*)
Miller, W.J. *et al*, *J. Am. Chem. Soc.*, 1957, **79**, 4164 (*synth, props*)
Fainberg, A.H. *et al*, *J. Org. Chem.*, 1965, **30**, 864 (*props*)
Paleta, O. *et al*, *Bull. Soc. Chim. Fr.*, 1986, 920 (*synth*)

Trichloro-1,2,3-triazine   T-80232

[70674-51-0]

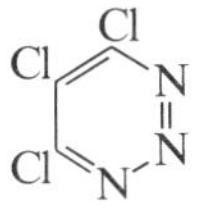

$C_3Cl_3N_3$ M 184.411

Mp 110-112°.

Gompper, R. *et al*, *Chem. Ber.*, 1979, **112**, 1529 (*synth*)

1,2,3-Trichloro-3,4,4-triflurorocyclobutene, 9CI   T-80233

[3762-49-0]

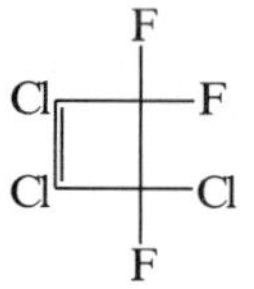

$C_4Cl_3F_3$ M 211.397
d_4^{25} 1.59. Bp_{625} 95°. n_D^{25} 1.4153.

Park, J.D. et al, J. Org. Chem., 1965, 30, 400 (synth, ir)
Bauer, G. et al, Z. Naturforsch., B, 1979, 34, 1249 (synth)

3-Tricosene, 9CI   T-80234

$$H_3C(CH_2)_{18}CH{=}CHCH_2CH_3$$

$C_{23}H_{46}$ M 322.616
Isol. from rose oil wax (Rosa sp.), prob. mostly as Z-form.

Wollrab, V. et al, Collect. Czech. Chem. Commun., 1965, 30, 1654
(isol)

Tricyclo[4.4.0.0³,⁸]decan-4-one, 9CI   T-80235

Updated Entry replacing T-20210
4-Twistanone
[13537-95-6]

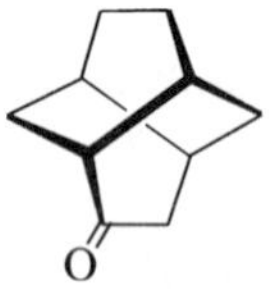

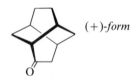

(+)-form

$C_{10}H_{14}O$ M 150.220
(+)-form [25225-94-9]
    $[\alpha]_D^{25}$ +295° (c, 0.518 in EtOH).
(−)-form [74958-51-3]
    Cryst. by subl. Mp 159-161°. $[\alpha]_D^{27}$ −270° (c, 0.97 in
    EtOH). Estimated optical purity 83%.
(±)-form
    Mp 185-190°.
    2,4-Dinitrophenylhydrazone: [13537-96-7].
    Cryst. (H_2O). Mp 170-171°.

Gauthier, J. et al, Can. J. Chem., 1967, 45, 297 (synth)
Tichý, M., Tetrahedron Lett., 1972, 2001 (synth)
Dodds, D.R. et al, J. Chem. Soc., Chem. Commun., 1982, 1080
    (synth)
Nakazaki, M. et al, J. Org. Chem., 1983, 48, 2506 (synth)
Dodds, D.R. et al, J. Am. Chem. Soc., 1988, 110, 577 (synth, ir,
    cmr)

Tricyclo[4.3.1.0³,⁸]dec-4-ene   T-80236

2,3,3a,4,5,7a-Hexahydro-2,5-methano-1H-indene, 9CI.
Protoadamantene
[29844-85-7]

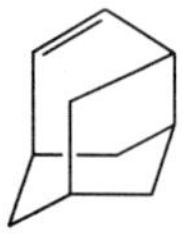

$C_{10}H_{14}$ M 134.221
Mp 183-185°.

Cupas, C.A. et al, J. Am. Chem. Soc., 1970, 92, 3237 (synth)
Black, R.M. et al, J. Chem. Soc., Chem. Commun., 1970, 972
    (synth, pmr)

Cuddy, B.D. et al, J. Chem. Soc. C, 1971, 3173 (synth)
Boyd, J. et al, J. Chem. Soc., Perkin Trans. 1, 1972, 2533 (synth)
Majerski, Z. et al, Org. Mass Spectrom., 1977, 12, 37 (ms)
Kropp, P.J. et al, Tetrahedron Lett., 1978, 207.

Tricyclo[5.5.0.0²,⁸]dodecane   T-80237

[61506-29-4]

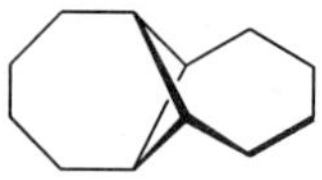

$C_{12}H_{20}$ M 164.290
Solid. Mp <20°.

Gleiter, R. et al, J. Am. Chem. Soc., 1988, 110, 5490 (synth, ir,
    pmr, cmr)

Tricyclo[6.2.1.1³,⁶]dodecane-2,7-dione, 9CI   T-80238

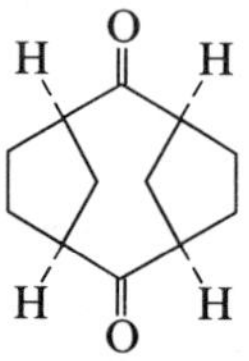

$C_{12}H_{16}O_2$ M 192.257
(1α,3α,6α,8α)-form [113428-76-5]
    Cryst. (EtOAc/hexane). Mp 192°.

Xu, C. et al, Tetrahedron, 1987, 43, 2909 (synth, cmr, cryst struct)

Tricyclo[5.5.0.0²,⁸]dodeca-3,5,9,11-tetraene   T-80239

[61278-83-9]

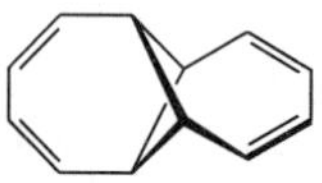

$C_{12}H_{12}$ M 156.227
Musty-smelling plates by subl. Mp 33.5-35.5°.

Dressel, J. et al, J. Am. Chem. Soc., 1988, 110, 5479 (synth, ir,
    pmr, cmr, uv)
Gleiter, R. et al, J. Am. Chem. Soc., 1988, 110, 5490 (pe, props)

Tricyclo[3.2.0.0²,⁷]heptane, 9CI   T-80240

Tricyclo[4.1.0.0³,⁷]heptane. Pseudonortricyclane
[279-18-5]

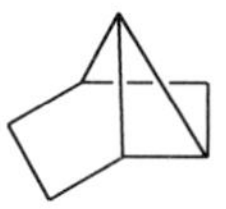

C_7H_{10} M 94.156

Moore, W.R. et al, J. Am. Chem. Soc., 1961, 83, 2019 (synth)
Adam, W. et al, J. Am. Chem. Soc., 1981, 103, 6406 (pmr, cmr, ir,
    ms)

Tricyclo[18.4.1.1.^{8,13}]hexacosa-2,4,6,8,10, 12,14,16,18,20,22,24-dodecaene, 9CI   T-80241

1,6:13,18-Dimethano[24]annulene

[115351-49-0]

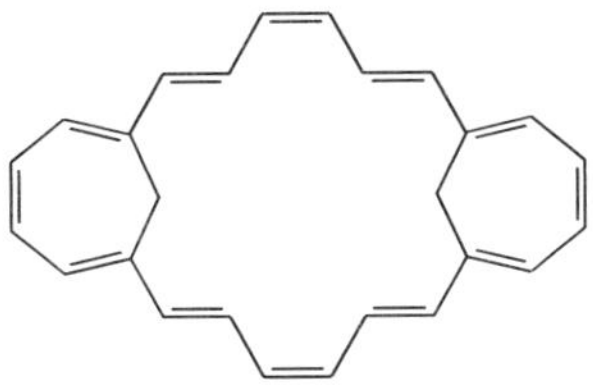

$C_{26}H_{24}$ M 336.476

Black-brown needles (hexane/C_6H_6). Mp 200° dec. Paratropic.

Yamamoto, K. *et al, J. Chem. Soc., Perkin Trans.* 1, 1988, 395 (*synth, uv, ir, pmr*)

Tricyclone   T-80242

2,3,4,4a,5,6-Hexahydro-2,2-dimethyl-1,3-methanonaphthalen-7(1H)-one, 9CI. 10,10-Dimethyltricyclo[7.1.1.0^{2,7}]undec-2-en-4-one

[68433-81-8]

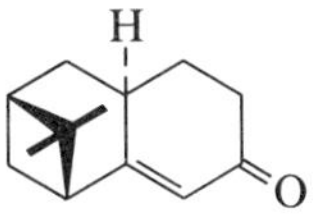

Absolute configuration

$C_{13}H_{18}O$ M 190.285

Perfumery chemical. d_4^{20} 1.015. $Bp_{0.01}$ 88-90°. $[\alpha]_D^{20}$ +40.4° (neat). Tricyclone is a proprietary name.

Bessière, Y. *et al, Nouv. J. Chim.*, 1978, **2**, 365 (*synth*)
Thomas, A.F. *et al, Helv. Chim. Acta*, 1981, **64**, 161 (*pmr, config*)

Tricyclo[3.2.1.0^{2,4}]octan-8-one   T-80243

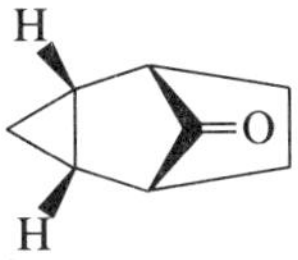

endo-form

$C_8H_{10}O$ M 122.166

endo-form [14224-86-3]
    Solid by subl. Mp 71-72°.

exo-form [7076-83-7]
    Bp_{17} 83-84°.

Tanida, H. *et al, J. Am. Chem. Soc.*, 1967, **89**, 1953 (*synth*)
Pincock, R.E. *et al, Tetrahedron Lett.*, 1967, 4759 (*uv, pmr*)
Haywood-Farmer, J.S. *et al, J. Am. Chem. Soc.*, 1969, **91**, 3020 (*synth, ir, pmr, ms*)
Kelly, D.P. *et al, J. Org. Chem.*, 1988, **53**, 2497 (*pmr, cmr*)

Tricyclo[1.1.1.0^{1,3}]pentane, 9CI   T-80244

Updated Entry replacing T-50405
    [1.1.1]Propellane
[35634-10-7]

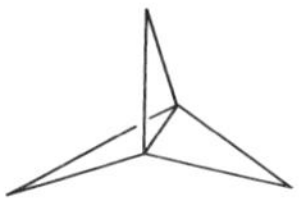

C_5H_6 M 66.102

Remarkably stable, $t_{1/2}$ ~5 min. at 114°.

Wiberg, K.B. *et al, J. Am. Chem. Soc.*, 1982, **104**, 5239 (*synth, cmr, pmr, ms*)

Wiberg, K.B. *et al, J. Am. Chem. Soc.*, 1985, **107**, 7247 (*synth, struct*)
Hedberg, L. *et al, J. Am. Chem. Soc.*, 1985, **107**, 7257 (*struct*)
Belzner, J. *et al, Chem. Ber.*, 1989, **122**, 397.

Tricyclo[3.3.3.0^{1,5}]undecane-3,7-dione   T-80245

1H,4H-3a,6a-Propanopentalene-2,5(3H,6H)-dione, 9CI. [3.3.3]Propellane-3,7-dione

[21170-73-0]

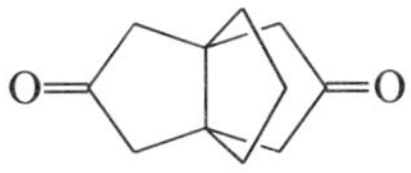

$C_{11}H_{14}O_2$ M 178.230

Cryst. (MeOH). Mp 224° (185-187°).

Weiss, U. *et al, Tetrahedron Lett.*, 1968, 4885 (*synth*)
Weber, R.W. *et al, Can. J. Chem.*, 1978, **56**, 189 (*synth*)
Mitschka, R. *et al, Tetrahedron*, 1981, **37**, 4521 (*synth*)
Quast, H. *et al, Chem. Ber.*, 1989, **122**, 523 (*synth, pmr*)

Tricyclo[5.4.0.0^{2,8}]undeca-3,5,9-triene, 9CI   T-80246

[115462-81-2]

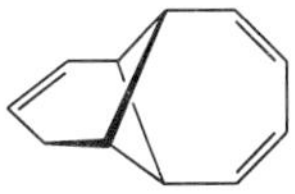

$C_{11}H_{12}$ M 144.216

Volatile liq.

Gleiter, R. *et al, J. Org. Chem.*, 1988, **53**, 3912 (*synth, pmr, cmr, ir, uv*)

3,5,11-Tridecadiene-7,9-diyne-1,2-diol   T-80247

$$H_3CCH=CHC\equiv CC\equiv CCH=CHCH=$$
$$CHCH(OH)CH_2OH$$

$C_{13}H_{14}O_2$ M 202.252

(E,E,E)-form
    From *Centaurea ruthenica* and *Serratula wolfii*. Cryst. (CHCl$_3$). Mp 117°. $[\alpha]_D^{24}$ −1.7° (MeOH).

Di-Ac: [60032-82-8].
    Cryst. (Et$_2$O/cyclohexane). Mp 57°. $[\alpha]_D^{24}$ +94.5° (Me$_2$CO).

Bohlmann, F. *et al, Chem. Ber.*, 1961, **94**, 3179; 1962, **95**, 2939; 1976, **109**, 2291 (*isol, synth, uv, ir*)

5,10,12-Tridecadiene-2,8-diyne-4,7-thione   T-80248

7,10-Dithioxo-1,3,8-tridecatriene-5,11-diyne

$$H_3CC\equiv CC(S)CH=CHC(S)C\equiv CCH=CHCH=CH_2$$

$C_{13}H_{10}S_2$ M 230.354

(3E,8Z)-form
    Isol. from roots of *Rudbeckia bicolor* and *Melampodium* spp. Red oil. λ_{max} 263 infl. (ϵ 14050), 351 (14300), 488 nm (2930) (pet. ether).

Bohlmann, F. *et al, Chem. Ber.*, 1965, **98**, 3081 (*isol, uv, ir, pmr, ms*)

8,10-Tridecadiene-4,6-diyn-1-ol, 9CI   T-80249

$$H_3CCH_2CH=CHCH=CHC\equiv CC\equiv CCH_2CH_2CH_2OH$$

$C_{13}H_{16}O$ M 188.269

(E,E)-form
    Dihydroaethusanol B

Isol. from above-ground parts of *Aethusa cynapium*.
p-*Phenylazobenzoyl:* Mp 99°.

Bohlmann, F. *et al, Chem. Ber.*, 1964, **97**, 2598.

1,11-Tridecadiene-3,5,7,9-tetrayne, 9CI   T-80250

[2345-16-6]

$$H_3CCH=CHC\equiv CC\equiv CC\equiv CC\equiv CCH=CH_2$$

$C_{13}H_8$ M 164.206

(*E*)-*form* [26130-86-9]

Trace constituent of roots of *Zoegea baldschuanica* and *Triticum aestivum*. Isol.from roots of *Dahlia* sp., from roots of *Bidens graveolens* and other *B.*spp., from seedlings of *Carthamus tinctorius* and from *Coreopsis* sp. Widespread minor constit. of the Compositae. Pale yellow needles (pentane). Dec. without melting >40°. Rapidly becomes brown in diffused daylight even at −5°.

(*Z*)-*form* [59950-58-2]

Isol. from roots of *D.* sp., from *Coreopsis* sp. and *Carthamus tinctorius*. Liq.

Sorensen, J.S. *et al, Acta Chem. Scand.*, 1954, **8**, 1741 (*isol, ir*)
Jones, E.R.H. *et al, J. Chem. Soc.*, 1958, 1054 (*synth, uv, ir, glc*)
Bohlmann, F. *et al, Chem. Ber.*, 1962, **95**, 1315; 1964, **97**, 1193, 2583; 1965, **98**, 872, 1228; 1966, **99**, 3201, 3433, 3544; 1971, **104**, 961 (*isol, struct, occur*)
Schulte, K.E. *et al, Phytochemistry*, 1965, **4**, 481 (*isol*)
Chin, C. *et al, J. Chem. Soc. C*, 1970, 314 (*isol*)
Bedford, C.T. *et al, J. Chem. Soc., Perkin Trans. 1*, 1976, 735 (*isol*)
Binder, R.G. *et al, Phytochemistry*, 1978, **17**, 315 (*isol, ms, uv, pmr*)
Bohlmann, F. *et al, Phytochemistry*, 1983, **22**, 1281 (*isol*)

1,12-Tridecadiene-3,5,7,9-tetrayne, 9CI   T-80251

[80453-50-5]

$$H_2C=CHCH_2C\equiv CC\equiv CC\equiv CC\equiv CCH=CH_2$$

$C_{13}H_8$ M 164.206
Isol. from *Calea* sp.

Bohlmann, F. *et al, Phytochemistry*, 1981, **20**, 1643 (*isol, struct*)

1,3-Tridecadiene-5,7,9,11-tetrayne, 9CI   T-80252

[18668-88-7]

$$H_3CC\equiv CC\equiv CC\equiv CC\equiv CCH=CHCH=CH_2$$

$C_{13}H_8$ M 164.206

(*E*)-*form* [3760-28-9]

Trace constituent of roots of *Rudbeckia amplexicaulis*, *R. bicolor*, *Fluorensia heterolepsis*,*Scaevola lobelia*, *Helichrysum vestitum* and *Arnica chamissonis*. Unstable cryst. (pet. ether). λ_{max} 273, 289.5, 311, 331, 355.5, 382 nm (Et$_2$O).

1,2-Epoxide: **2-(1-Undecen-3,5,7,9-tetraynyl)oxirane.** *1,2-Epoxy-3-tridecene-5,7,9,11-tetrayne*
$C_{13}H_8O$ M 180.206
Isol. from *Carthamus*, *Centaurea* and *Silybium* spp. Cryst. Mp 95.5-97°. $[\alpha]_D$ +75.1° (CHCl$_3$).

Bohlmann, F. *et al, Chem. Ber.*, 1964, **97**, 809, 2125; 1965, **98**, 3081; 1966, **99**, 3201, 3433, 3544; 1975, **108**, 433 (*isol, uv, ir, synth*)
Schulte, K.E. *et al, J. Nat. Prod. (Lloydia)*, 1969, **32**, 360 (*isol, struct*)
Bohlmann, F. *et al, Phytochemistry*, 1979, **18**, 1189; 1980, **19**, 331 (*isol*)

2,12-Tridecadiene-4,6,8,10-tetrayn-1-ol, 9CI   T-80253

[20252-36-2]

$$H_2C=CH^{11}C\equiv^{10}CC\equiv CC\equiv CCH=CHCH_2OH$$

$C_{13}H_8O$ M 180.206
Isol. from *Bidens* spp., *Cosmos diversifolius* and *Coreopsis tinctoria*. Cryst. (pet. ether). Mp 41-42°.

Ac: [20252-35-1]. *1-Acetoxy-2,12-tridecadiene-4,6,8,10-tetrayne*
$C_{15}H_{10}O_2$ M 222.243
Isol. from *Coreopsis* sp., *B.* sp., *Cosmos* sp. and *Leptosyne calliopsidea*. Amorph. λ_{max} 270, 287, 314, 336, 361, 391 nm.

10,11-Dihydro(E-): *2,10,12-Tridecatriene-4,6,8-triyn-1-ol, 9CI*
$C_{13}H_{10}O$ M 182.221
Isol. from *Cosmos sulphureus*, *C. hybridus* and *Coreopsis tinctoria*. Cryst. (Et$_2$O/pet. ether). Mp 100° dec.

[26130-87-0, 26130-89-2]

Bohlmann, F. *et al, Chem. Ber.*, 1962, **95**, 1315; 1964, **97**, 1193, 2135, 2583; 1965, **98**, 1228; 1966, **99**, 1223.

3,11-Tridecadiene-5,7,9-triyne-1,2-diol, 9CI   T-80254
Safynol

[65398-35-8]

$$H_3CCH=CHC\equiv CC\equiv CC\equiv CCH=CHCH(OH)CH_2OH$$

$C_{13}H_{12}O_2$ M 200.237

(*2R,3E,11E*)-*form* [27978-14-9]

Isol. from *Centaurea* spp., *Bidens* spp. and fungus-infected safflower (*Carthamus tinctorius*). Antifungal, antibiotic phytoalexin. Cryst. (CHCl$_3$/hexane). Mp 122.5-123.5° (112°). $[\alpha]_D^{20}$ −17°. Stable in dark at −18°, resinifies in air.

2-Ac: Cryst. (pet. ether/Et$_2$O). Mp 86°. $[\alpha]_D^{21}$ +39° (CHCl$_3$).

Di-Ac:
$C_{17}H_{16}O_4$ M 284.311
Occurs naturally. Cryst. (Et$_2$O or pet. ether/Et$_2$O). Mp 64° (80°). $[\alpha]_D^{20}$ +102.2° (CHCl$_3$).

(*2S,3E,11E*)-*form* [65207-93-4]
$[\alpha]_D^{22.5}$ +18.4°, $[\alpha]_D^{23}$ +30.8° (MeOH).

Bohlmann, F. *et al, Chem. Ber.*, 1958, **91**, 1642, 3179 (*isol, ir, synth, uv*)
Allen, E.H. *et al, Phytochemistry*, 1971, **10**, 1579 (*isol, ir, ms, uv*)
Nakada, H. *et al, Agric. Biol. Chem.*, 1977, **41**, 1761 (*isol, synth, ir, ord, pmr, uv*)
Jente, R. *et al, Phytochemistry*, 1979, **18**, 829 (*pmr*)
MacRae, W.D. *et al, Experientia*, 1980, **36**, 1096 (*pharmacol*)

1,6-Tridecadiene-3,9,11-triyne-5,8-dithione   T-80255
5,8-Dithioxo-1,6-tridecadiene-3,9,11-triyne

$$H_3CC\equiv CC\equiv CC(S)CH=CHC(S)C\equiv CCH=CH_2$$

$C_{13}H_8S_2$ M 228.338

(*Z*)-*form*

Isol. from *Eriophyllum caespitosum*. Red oil. λ_{max} 230 (ϵ 23200), 353 (12,500), 499 nm (3070) (pet. ether).

Mortensen, J.T. *et al, Acta Chem. Scand.*, 1964, **18**, 2392 (*isol, uv, ir, pmr*)
Bohlmann, F. *et al, Chem. Ber.*, 1965, **98**, 3081.

5,12-Tridecadiene-2,8,10-triyne-4,7-dithione   T-80256

7,10-Dithioxo-1,8-tridecadiene-3,5,11-triyne

$$H_3CC{\equiv}CC(S)CH{=}CHC(S)C{\equiv}CC{\equiv}CCH{=}CH_2$$

$C_{13}H_8S_2$ M 228.338

(Z)-form

Isol. from roots of *Ambrosia elatior*, also in other *A.* spp., *Schkuhria* spp. and *Iva xanthiifolia*. Red oil. λ_{max} 231 (ϵ 24100), 340 (12700), 496 nm (3000) (hexane).

Bohlmann, F. *et al*, *Chem. Ber.*, 1965, **98**, 3081 (*isol, uv, ir, pmr, ms, struct*)

3,11-Tridecadiene-5,7,9-triyne-1,2,13-triol   T-80257

$$HOCH_2CH{=}CHC{\equiv}CC{\equiv}CC{\equiv}CCH{=}\\CHCH(OH)CH_2OH$$

$C_{13}H_{12}O_3$ M 216.236

(E,E)-form

Isol. from above ground parts of *Bidens bipinnata*. Cryst. (Et$_2$O/pet. ether). Polymerises above 100°.

Bohlmann, F. *et al*, *Chem. Ber.*, 1965, **98**, 1228 (*isol, uv, ir, struct*)

3,5-Tridecadiene-7,9,11-triyn-1-ol, 9CI   T-80258

[56319-28-9]

$$H_3CC{\equiv}CC{\equiv}CC{\equiv}CCH{=}CHCH{=}CHCH_2CH_2OH$$

$C_{13}H_{12}O$ M 184.237

(E,E)-form [3513-72-2]

Isol. from above ground parts of *Chrysanthemum ircutianum*. Cryst. (pet. ether/Et$_2$O). Mp 102° (95-97°).

Ac:
$C_{15}H_{14}O_2$ M 226.274
Isol. from above ground parts of *C. ircutianum*.

3-Methylbutanoyl:
$C_{18}H_{20}O_2$ M 268.355
Isol. from roots of *C. leucanthemum*.

(3E,5Z)-form [75911-03-4]
Isol. from *Leucanthemum* sp.

Bohlmann, F. *et al*, *Chem. Ber.*, 1961, **94**, 3193; 1965, **98**, 2596 (*isol, uv, synth, deriv*)
Wrang, P.A. *et al*, *Phytochemistry*, 1975, **14**, 1027 (*isol*)
Bohlmann, F. *et al*, *Phytochemistry*, 1980, **19**, 841 (*isol*)

3,5,7,9,11-Tridecapentayne-1,2-diol   T-80259

$$H_3CC{\equiv}CC{\equiv}CC{\equiv}CC{\equiv}CC{\equiv}CCH(OH)CH_2OH$$

$C_{13}H_8O_2$ M 196.205
Isol. from above-ground parts of *Bidens leucantha*. Unstable light-yellow cryst. (Et$_2$O/pet. ether).

Di-Ac: [3420-89-1].
Yellowish cryst. (pet. ether). Mp 90° dec. $[\alpha]_D^{25}$ +52.5° (c, 0.19 in Et$_2$O).

Bohlmann, F. *et al*, *Chem. Ber.*, 1965, **98**, 1228 (*isol, uv, ir, ord, struct*)

2,8,10,12-Tridecatetraene-4,6-diyn-1-ol, 9CI   T-80260

$$H_2C{=}CHCH{=}CHCH{=}CHC{\equiv}CC{\equiv}CCH{=}CHCH_2OH$$

$C_{13}H_{12}O$ M 184.237

(all-E)-form [15427-16-4]
Isol. from roots of *Carlina vulgaris*. Leaflets (Et$_2$O/pet. ether). Mp 86°.

Ac: 1-Acetoxy-2,8,10,12-tridecatetraene-4,6-diyne
$C_{15}H_{14}O_2$ M 226.274
Isol. from *Bidens ferulaefolia*, *Coreopsis tinctoria* and roots of *Carlina vulgaris*. Cryst. (pet. ether). Mp 31°.

(2Z,8E,10E)-form [75627-99-5]
Isol. from *Tetragonotheca helianthoides*.

Bohlmann, F. *et al*, *Chem. Ber.*, 1964, **97**, 1193; 1966, **99**, 1223; 1967, **100**, 1507 (*isol, uv, ir, pmr, deriv*)
Bedford, C.T. *et al*, *J. Chem. Soc., Perkin Trans. 1*, 1976, 735 (*isol*)
Seaman, F.C. *et al*, *Phytochemistry*, 1980, **19**, 583 (*isol*)

11-Tridecene-3,5,7,9-tetrayne-1,2-diol, 9CI   T-80261

Dehydrosafynol
[1540-85-8]

$$H_3CCH{=}CH(C{\equiv}C)_4CH(OH)CH_2OH$$

$C_{13}H_{10}O_2$ M 198.221

(E)-form
Isol. from above-ground parts of *Bidens bipinnata*. Highly antifungal. No opt. rotn. reported.

Di-Ac: [1593-79-9].
Light-yellow cryst. (pet. ether). Mp 61°, Mp 88°.

Bohlmann, F. *et al*, *Chem. Ber.*, 1965, **98**, 1228; 1967, **100**, 1209 (*isol, uv, ir, pmr, struct, synth*)

11-Tridecene-3,5,7,9-tetrayne-1,2,13-triol   T-80262

$$HOCH_2CH{=}CH(C{\equiv}C)_4CH(OH)CH_2OH$$

$C_{13}H_{10}O_3$ M 214.220

(E)-form
Isol. from above-ground parts of *Bidens bipinnata*. Unstable cryst. (Et$_2$O/pet. ether). Polymerises above 100°.

Tri-Ac: Yellowish oil.

Bohlmann, F. *et al*, *Chem. Ber.*, 1965, **98**, 1228 (*isol, uv, ir, pmr, struct*)

6-Tridecen-1-ol   T-80263

$$H_3C(CH_2)_5CH{=}CH(CH_2)_4CH_2OH$$

$C_{13}H_{26}O$ M 198.348

(E)-form [37011-92-0]
Liq. Bp$_{0.3}$ 96-97°.

Voaden, D.J. *et al*, *J. Med. Chem.*, 1972, **15**, 619 (*synth*)
Svirskaya, P.I. *et al*, *J. Chem. Eng. Data*, 1979, **24**, 152.
Brown, H.C. *et al*, *J. Org. Chem.*, 1988, **53**, 246 (*synth, ir, pmr, cmr*)

3-Tridecyn-1-ol, 9CI   T-80264

[54373-82-9]

$$H_3C(CH_2)_8C{\equiv}CCH_2CH_2OH$$

$C_{13}H_{24}O$ M 196.332
Oil. Mp 21.5-22°. Bp$_{0.6}$ 119-121°. n_D^{23} 1.4579.

Gilman, N.N. *et al*, *Chem. Phys. Lipids*, 1974, **13**, 239 (*synth*)
Bestmann, H.G. *et al*, *Justus Liebigs Ann. Chem.*, 1987, 417 (*synth, pmr, ir, ms*)
Mori, K. *et al*, *Justus Liebigs Ann. Chem.*, 1988, **8**, 807 (*synth, ir, pmr*)

6-Tridecyn-1-ol, 9CI   T-80265

[37011-88-4]

$$H_3C(CH_2)_3C{\equiv}C(CH_2)_4CH_2OH$$

$C_{13}H_{24}O$ M 196.332

Oil. $Bp_{0.1}$ 104-105° (107-109°) . n_D^{25} 1.4608.

Voaden, D.J. *et al, J. Med. Chem.*, 1972, **15**, 619 (*synth*)
Svirskaya, P.A. *et al, J. Chem. Eng. Data*, 1979, **24**, 152 (*synth*)
Brown, H.C. *et al, J. Org. Chem.*, 1986, **57**, 4518 (*synth*)

2,4,6-Triethyl-1,3,5-trioxane, 9CI　　　T-80266

Propionaldehyde trimer. Parapropionaldehyde. Parapropanal
[2396-42-1]

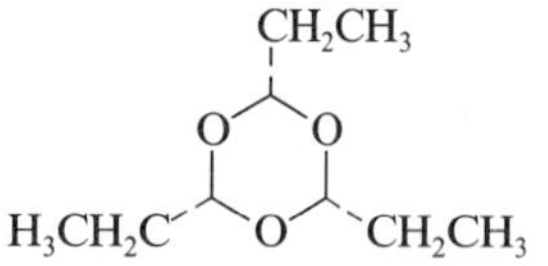

$C_9H_{18}O_3$　　M 174.239
Normally consists exclusively of the $2\alpha,4\alpha,6\alpha$-form.
Synthetic onion aroma. Colourless or yellowish oil.
Spar. sol. H_2O, sol. EtOH, Et_2O. Fp −20°. Bp_{773} 172-
173°, Bp_{10} 57-57.5° (60.5-61°) . n_D^{21} 1.4188. Reverts to
monomer on treatment with dilute acid.

[1499-04-3]

Meier, R. *et al, Chem. Ber.*, 1957, **90**, 2344 (*synth*)
Blanc, P.-Y. *et al, Helv. Chim. Acta*, 1964, **47**, 567 (*synth, props, uv*)
Jungnickel, J.L. *et al, J. Mol. Spectrosc.*, 1965, **16**, 135 (*pmr, config*)
Ward, W.R. *et al, Spectrochim. Acta*, 1965, **2**, 1311 (*ir, raman*)
Seebald, H.J. *et al, Arch. Pharm. (Weinheim, Ger.)*, 1973, **306**, 393 (*synth, ir, glc*)
Krasnov, V.L. *et al, J. Gen. Chem. USSR (Engl. Transl.)*, 1983, **53**, 2135 (*synth, nmr*)
Abdelhader, M. *et al, Macromolecules*, 1987, **20**, 949 (*synth, pmr*)

Trifluoroacetic acid, 9CI, 8CI　　　T-80267

Updated Entry replacing T-20220
Trifluoroethanoic acid
[76-05-1]

$$CF_3COOH$$

$C_2HF_3O_2$　　M 114.024
Strong acid used in synthesis. Misc. H_2O. Fp −15.3°. Bp
70.5-72°. pK_a 0.23 (25°, H_2O).
▷ Mod. toxic, causes severe burns. AJ9625000.
Me ester: [431-47-0].
　$C_3H_3F_3O_2$　　M 128.051
　Trifluoroacetylating agent for NH_2 groups. Liq. Bp 43-44°.
Et ester: [383-63-1].
　$C_4H_5F_3O_2$　　M 142.077
　Liq. Bp 60-62°.
tert-*Butyl ester:* [400-52-2].
　$C_6H_9F_3O_2$　　M 170.131
　Bp 83°.
Ph ester: [500-73-2]. *Phenyl trifluoroacetate*
　$C_8H_5F_3O_2$　　M 190.121
　Reagent used to prepare *N*-trifluoroacetyl derivs. of
　amino acids etc. Bp 148-149°.
Amide: [354-38-1]. *Trifluoroacetamide*
　$C_2H_2F_3NO$　　M 113.039
　Cryst. ($CHCl_3$). Mp 73.5-74.5°.
Anhydride: [407-25-0].
　$C_4F_6O_3$　　M 210.033
　Trifluoroacetylating agent with several other synthetic
　uses. Bp 39.5°.
▷ Mod. toxic, causes severe burns. Reacts explosively with
DMSO. AJ9800000.

Gilman, H. *et al, J. Am. Chem. Soc.*, 1943, **65**, 1458 (*amide, Et ester*)
Allen, D.R., *J. Org. Chem.*, 1961, **26**, 923 (*synth, props*)
U.S. Pat., 3 162 633, (1965); *CA*, **62**, 7780 (*synth*)
Fieser and Fieser's Reagents for Organic Synthesis, Wiley, 1967, **1**, 850; 1975, **5**, 57; 1979, **7**, 246, 389; 1980, **8**, 503.
Berney, C.V., *J. Am. Chem. Soc.*, 1973, **95**, 708 (*spectra*)
Effenberger, F. *et al, Chem. Ber.*, 1980, **113**, 2100 (*tert*-Butyl ester)
Roberts, D.D. *et al, J. Org. Chem.*, 1988, **53**, 2573 (*amide*)
Bretherick, L., *Handbook of Reactive Chemical Hazards*, 2nd Ed., Butterworths, London and Boston, 1979, 225, 396.
Sax, N.I., *Dangerous Properties of Industrial Materials*, 5th Ed., Van Nostrand-Reinhold, 1979, 1055.
Hazards in the Chemical Laboratory, (Bretherick, L., Ed.), 3rd Ed., Royal Society of Chemistry, London, 1981, 520.

1,1,2-Trifluoro-1,3-butadiene, 9CI　　　T-80268

[565-65-1]

$$F_2C{=}CFCH{=}CH_2$$

$C_4H_3F_3$　　M 108.063
Reagent for synth. of functionalised monofluoroalkene
systems. Forms polymers with many substances. Volatile
liq. or gas. Bp_{585} 0.5°.

Park, J.D. *et al, J. Am. Chem. Soc.*, 1956, **78**, 59 (*synth*)
Matsuo, N. *et al, J. Org. Chem.*, 1988, **53**, 2304 (*synth, pmr, F nmr, use*)

Trifluorobutanedioic acid, 9CI　　　T-80269

Trifluorosuccinic acid, 8CI
[664-66-4]

$$HOOCCF_2CHFCOOH$$

$C_4H_3F_3O_4$　　M 172.060
(±)-*form*
　Cryst. Mp 111-112°.
Anhydride: 3,3,4-Trifluoro-dihydro-2,5(2H,5H)-furandione
　$C_4HF_3O_3$　　M 154.045
　Bp 122-124°.
Diamide:
　$C_4H_5F_3N_2O_2$　　M 170.091
　Cryst. (H_2O). Mp 195-197°.

Raasch, M.S. *et al, J. Am. Chem. Soc.*, 1959, **81**, 2678 (*synth*)

4,4,4-Trifluorobutanoic acid　　　T-80270

[406-93-9]

$$F_3CCH_2CH_2COOH$$

$C_4H_5F_3O_2$　　M 142.077
Mp 33.2°. Bp_{18} 78.0°, Bp 166.6°.
Chloride:
　$C_4H_4ClF_3O$　　M 160.523
　Bp_{745} 103°.
Amide:
　$C_4H_6F_3NO$　　M 141.093
　Cryst. Mp 136.4°.

McBee, E.T. *et al, J. Am. Chem. Soc.*, 1948, **70**, 2910 (*synth*)
Henne, A.L. *et al, J. Am. Chem. Soc.*, 1951, **73**, 2323; 1955, **77**, 1901 (*derivs, props*)

1,3,3-Trifluorocyclopropene, 9CI   T-80271

[35305-22-7]

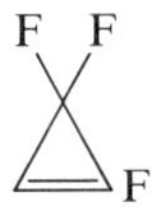

F F

F

C_3HF_3 M 94.036
Bp 6°.

Craig, N.C. *et al*, *Spectrochim. Acta, Part A*, 1972, **28**, 180; 1975,
31, 1463 (*synth, ir, ms, raman*)
Craig, N.C. *et al*, *J. Phys. Chem.*, 1985, **89**, 100.

1,1,1-Trifluoro-3,3-dimethyl-2-butanone   T-80272

tert-*Butyl trifluoromethyl ketone*

$F_3CCOC(CH_3)_3$

$C_6H_9F_3O$ M 154.132
Liq. Bp 69-70°.

Dishart, K.T. *et al*, *J. Am. Chem. Soc.*, 1956, **78**, 2268 (*synth*)
Sykes, A. *et al*, *J. Chem. Soc.*, 1956, 835 (*synth*)
Feigl, D.M. *et al*, *J. Org. Chem.*, 1968, **33**, 4242 (*synth*)
Roberts, D.D. *et al*, *J. Org. Chem.*, 1988, **53**, 2573 (*synth*)

1,1,1-Trifluoroethane   T-80273

Methylfluoroform

[420-46-2]

F_3CCH_3

$C_2H_3F_3$ M 84.041
Anaesthetic. d_4^{50} 1.176. Mp $-111.3°$. $-47°$ to $-46°$.

▷ Flammable.

Henne, A.L. *et al*, *J. Am. Chem. Soc.*, 1936, **58**, 889 (*synth*)
McBee, E.T. *et al*, *Ind. Eng. Chem.*, 1947, **39**, 409 (*props, synth*)
Whalley, W.B., *J. Soc. Chem. Ind.*, *London*, 1947, **66**, 427 (*synth*)
Thompson, H.W. *et al*, *J. Chem. Soc.*, 1948, 1428 (*ir*)
Mohler, F.L. *et al*, *J. Chem. Phys.*, 1954, **22**, 394 (*ms*)
Kirk-Othmer Encycl. Chem. Technol., 3rd Ed., 1978-1984, *Wiley*,
NY, 3rd Ed., Wiley, N.Y., 1978-1984, **10**, 857; **11**, 65.
Buerger, H. *et al*, *Spectrochim. Acta, Part A*, 1980, **36**, 7 (*ir*,
raman, pmr)
Eujen, R. *et al*, *J. Fluorine Chem.*, 1983, **22**, 263 (*F-19 nmr*)

Trifluoroethene, 9CI   T-80274

Trifluoroethylene

[359-11-5]

$F_2C=CHF$

C_2HF_3 M 82.025
Monomer for plastics. Gas. $-61°$ to $-60°$, Bp_{628} $-56°$.

▷ Flammable.

Crosse, A.V. *et al*, *J. Am. Chem. Soc.*, 1942, **64**, 2289 (*synth*)
Craig, N.C. *et al*, *J. Am. Chem. Soc.*, 1961, **83**, 3047 (*synth*)
Tikhomirov, M.V. *et al*, *Zh. Fiz. Khim.*, 1967, **41**, 1065 (*ms*)
Hudlicky, M., *Chemistry of Organic Fluorine Compounds*, Wiley,
London, 2nd ed., 1976, 603.
Galabov, B. *et al*, *J. Chem. Phys.*, 1979, **71**, 4716 (*ir*)
Bien, G. *et al*, *Chem. Phys.*, 1981, **60**, 61 (*pe, uv*)
Kushida, K. *et al*, *Proton and Fluorine Nuclear Magnetic
Resonance Spectral Data*, Japan Halon Co Ltd, Japan,, 1988,
090, 091 (*F-nmr, pmr*)

(Trifluoroethenyl)benzene, 9CI   T-80275

α,β,β-*Trifluorostyrene*. (*Trifluorovinyl*)*benzene*.
Trifluorophenylethylene

[447-14-3]

$PhCF=CF_2$

$C_8H_5F_3$ M 158.123
Forms polymers with many materials. Liq. Bp_{44} 59°.

Dixon, S., *J. Org. Chem.*, 1956, **21**, 400 (*synth*)
Heinze, P.L. *et al*, *J. Org. Chem.*, 1988, **53**, 2714 (*synth, ir, ms,
pmr, cmr, F-19 nmr*)

1,1,1-Trifluoro-2-iodoethane, 8CI   T-80276

2,2,2-Trifluoroethyl iodide

[353-83-3]

F_3CCH_2I

$C_2H_2F_3I$ M 209.938
Shows anaesthetic props. d_4^{25} 2.142. Bp 54-55°. n_D^{25} 1.3980.

[93347-76-3]

Haszeldine, R.N., *J. Chem. Soc.*, 1953, 2622 (*uv*)
Hauptschein, M. *et al*, *J. Org. Chem.*, 1958, **23**, 322 (*synth, ir*)
Chambers, R.D. *et al*, *J. Chem. Soc.*, 1961, 3779 (*synth*)
Harnish, D.F. *et al*, *Appl. Spectrosc.*, 1970, **24**, 28 (*ir*)
Boschi, R.A. *et al*, *Mol. Phys.*, 1972, **24**, 735 (*uv*)
Lopata, A.D. *et al*, *J. Raman Spectrosc.*, 1977, **6**, 61 (*raman*)

1,1,1-Trifluoro-2-methoxyethane, 9CI   T-80277

Methyl 2,2,2-trifluoroethyl ether, 8CI

[460-43-5]

F_3CCH_2OMe

$C_3H_5F_3O$ M 114.067
Shows weak anaesthetic props. d_4^3 1.166. Bp_{746} 31.2°, Bp
31-32°. n_D^3 1.2942.

Robbins, B.H., *J. Pharmacol.*, 1946, **86**, 197 (*pharmacol*)
Henne, A.L. *et al*, *J. Am. Chem. Soc.*, 1950, **72**, 4378 (*synth*)
Li, Y.S. *et al*, *J. Phys. Chem.*, 1987, **91**, 1334 (*ir, raman*)
Loehr, D.T. *et al*, *J. Fluorine Chem.*, 1988, **39**, 288.

2-(Trifluoromethyl)benzenemethanethiol,   T-80278
9CI

(α,α,α-*Trifluoro-o-tolyl*)*methanethiol, 8CI*. o-
(*Trifluoromethyl*)-α-*toluenethiol*. 2-
(*Trifluoromethyl*)*benzylmercaptan*. 1-(*Mercaptomethyl*)-2-
(*trifluoromethyl*)*benzene*

[26039-98-5]

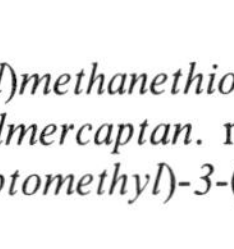

$C_8H_7F_3S$ M 192.204
d_4^{25} 1.30. Bp_{13} 75°.

Belinskaya, R.V. *et al*, *J. Org. Chem. USSR* (*Engl. Transl.*), 1970,
6, 141 (*synth*)

3-(Trifluoromethyl)benzenemethanethiol,   T-80279
9CI

(α,α,α-*Trifluoro-m-tolyl*)*methanethiol, 8CI*. 3-
(*Trifluoromethyl*)*benzylmercaptan*. m-(*Trifluoromethyl*)-α-
toluenethiol. 1-(*Mercaptomethyl*)-3-(*trifluoromethyl*)*benzene*

[25697-55-6]

$C_8H_7F_3S$ M 192.204
Liq. Bp_8 69-70°, $Bp_{2.1-2.2}$ 55°.

Paquette, L.A. *et al*, *J. Org. Chem.*, 1968, **33**, 1080 (*synth*)
Lombardino, J.G. *et al*, *J. Med. Chem.*, 1970, **13**, 206 (*synth*)

4-(Trifluoromethyl)benzenemethanethiol,   T-80280
9CI

(α,α,α-*Trifluoro-p-tolyl*)*methanethiol, 8CI. 4-
(Trifluoromethyl)benzyl mercaptan. p-(Trifluoromethyl)-α-
toluenethiol. 1-(Mercaptomethyl)-4-(trifluoromethyl)benzene*
[108499-24-7]
$C_8H_7F_3S$ M 192.204
Oil.

Buchwald, S.L. *et al, J. Am. Chem. Soc.*, 1988, **110**, 3171 (*synth,
pmr, cmr, ir*)

2-Trifluoromethyl-1,3-cyclopentanedione,   T-80281
9CI

[89049-67-2]

$C_6H_5F_3O_2$ M 166.099
Intermed. for steroid synth. Mp ~150° dec.

Blazejewski, J.C. *et al, Synthesis*, 1985, 1120 (*synth, ir, pmr, cmr,
F-19 nmr*)
Blazejewski, J.C. *et al, J. Chem. Soc., Perkin Trans.* 1, 1986, 337.

3,3,3-Trifluoropropanal, 9CI   T-80282
3,3,3-Trifluoropropionaldehyde
[460-40-2]

$$F_3CCH_2CHO$$

$C_3H_3F_3O$ M 112.051
Liq. d_4^{20} 1.37. Bp_{745} 56.0-56.5°.
2,4-Dinitrophenylhydrazone: Cryst. (EtOH aq.). Mp 150.2-
150.8°.

Henne, A.L. *et al, J. Am. Chem. Soc.*, 1950, **72**, 3370 (*synth*)

3,3,3-Trifluoro-1-propanamine, 9CI   T-80283
3-Amino-1,1,1-trifluoropropane
[460-39-9]

$$F_3CCH_2CH_2NH_2$$

$C_3H_6F_3N$ M 113.082
Liq. d_4^{30} 1.16. Bp_{744} 67.8°.
B,HCl: Mp 222-225°.

Henne, A.L. *et al, J. Am. Chem. Soc.*, 1955, **77**, 1901 (*synth*)
Hoffman, R.V. *et al, J. Am. Chem. Soc.*, 1988, **110**, 4019 (*synth*)

3,3,3-Trifluoropropanoic acid   T-80284
β,β,β-*Trifluoropropionic acid*
[2516-99-6]

$$F_3CCH_2COOH$$

$C_3H_3F_3O_2$ M 128.051
Mp 9.7°. Bp_{746} 144.8°.
NH_4 salt: Bp_{21} 136° subl.
Aniline salt: Cryst. (EtOH). Mp 116°.
S-Benzylthiouronium salt: Cryst. (EtOH). Mp 176°.
Me ester: [18830-44-9].
    $C_4H_5F_3O_2$ M 142.077
    Bp 95-96°.
Et ester: [352-23-8].
    $C_5H_7F_3O_2$ M 156.104

Bp 97-98°, Bp_{12} 50°.
Chloride:
    $C_3H_2ClF_3O$ M 146.496
    Bp_{745} 70.3°.
Amide:
    $C_3H_4F_3NO$ M 127.066
    Solid. Mp 108.8°. Bp_{12} 130° subl. . Deliquescent.
Anilide:
    $C_9H_8F_3NO$ M 203.163
    Cryst. (EtOH). Mp 118°.

Henne, A.L. *et al, J. Am. Chem. Soc.*, 1951, **73**, 2323 (*props*)
Brown, F. *et al, J. Chem. Soc.*, 1953, 2087 (*synth, derivs*)
Kocharyan, S.T. *et al, Izv. Akad. Sci. USSR, Ser. Sci. Khim.*,
1967, 1847 (*esters, ir, pmr*)

2,4,6-Trihydroxybenzophenone, 8CI   T-80285
Updated Entry replacing T-30274
(*2,4,6-Trihydroxyphenyl*)*phenylmethanone, 9CI.
Phlorbenzophenone*
[3555-86-0]
$C_{13}H_{10}O_4$ M 230.220
Yellow cryst. + $1H_2O$ (H_2O). Mp 165°.
2-Me ether: [81525-12-4]. *2,4-Dihydroxy-6-
methoxybenzophenone. Isocotoin*
    $C_{14}H_{12}O_4$ M 244.246
    Yellow needles. Mp 162°.
4-Me ether: [479-21-0]. *2,6-Dihydroxy-4-
methoxybenzophenone.* **Cotoin**
    $C_{14}H_{12}O_4$ M 244.246
    Occurs in Coto rind (*Nectandra coto*). Also from wood
    of *Aniba duckei*. Yellowish cryst. (H_2O). Mp 130-131°.
    Reduces Fehling's soln. and $NH_3.AgNO_3$ slowly in the
    cold.
▷ DJ1050000.
2,4-Di-Me ether: [34425-64-4]. *2-Hydroxy-4,6-
dimethoxybenzophenone.* **Hydrocotoin**
    $C_{15}H_{14}O_4$ M 258.273
    Occurs in Coto rind. Light-yellow prisms or needles
    (EtOH). Mp 97-98°.
2,6-Di-Me ether: 4-Hydroxy-2,6-dimethoxybenzophenone
    $C_{15}H_{14}O_4$ M 258.273
    Mp 178-179°.
Tri-Me ether: [3770-80-7]. *2,4,6-Trimethoxybenzophenone.*
    Methylhydrocotoin
    $C_{16}H_{16}O_4$ M 272.300
    Isol. from Coto rind and *Cascara sagrada*. Prisms or
    tablets (EtOH). Mp 113-115°.

Karrer, P., *Helv. Chim. Acta*, 1928, **11**, 789 (*struct*)
Phadke, R. *et al, J. Indian Chem. Soc.*, 1950, **27**, 349 (*synth*)
Gottleib, O.R. *et al, J. Am. Chem. Soc.*, 1958, **80**, 2263 (*isol,
Cotoin*)
Harris, T.M. *et al, J. Am. Chem. Soc.*, 1966, **88**, 5686 (*synth*)

4,5,7-Trihydroxy-2H-1-benzopyran-2-one,   T-80286
9CI
4,5,7-Trihydroxycoumarin
[17575-26-7]

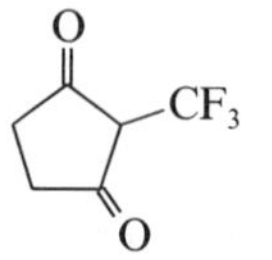

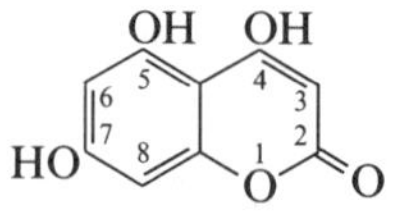

$C_9H_6O_5$ M 194.143
Mp >300° dec. (sinters and browns from 210°).
Tri-Ac: Mp 155-156°.

Sonn, A., *Ber.*, 1917, **50**, 1292.

5,6,7-Trihydroxy-2*H*-1-benzopyran-2-one,   T-80287
9CI

Updated Entry replacing T-50456

5,6,7-Trihydroxycoumarin

C$_9$H$_6$O$_5$ M 194.143

7-Me ether: [50656-75-2]. *5,6-Dihydroxy-7-methoxy-2*H-1-*benzopyran-2-one, 9CI. 5,6-Dihydroxy-7-methoxycoumarin.* **Isofraxetin**
C$_{10}$H$_8$O$_5$ M 208.170
Isol. from *Fraxinus potamophila* and *F. mandschurica.* Cryst. (MeOH). Mp 228-230°.

5,6-Di-Me ether: [43053-62-9]. *7-Hydroxy-5,6-dimethoxy-2*H-1-*benzopyran-2-one, 9CI. 7-Hydroxy-5,6-dimethoxycoumarin.* **Umckalin**
C$_{11}$H$_{10}$O$_5$ M 222.197
Constit. of roots of *Pelargonium reniforme.* Mp 146-147°.

5,7-Di-Me ether: [486-28-2]. *6-Hydroxy-5,7-dimethoxy-2*H-1-*benzopyran-2-one, 9CI. 6-Hydroxy-5,7-dimethoxycoumarin.* **Fraxinol**
C$_{11}$H$_{10}$O$_6$ M 238.196
Constit. of bark of *Fraxinus excelsior.* Exhibits spasmolytic activity. Mp 171-172°.

5,7-Di-Me ether, 6-O-glucoside: [32451-87-9]. **Mandshurin.** *Fraxinoside*
C$_{11}$H$_{10}$O$_5$ M 222.197
Constit. of bark of *F. potamophila* and *F. mandshurica.* Mp 134-136°, Mp 182.5°. [α]$_D$ −26.7° (MeOH).

6,7-Di-Me ether: [28449-62-9]. *5-Hydroxy-6,7-dimethoxy-2*H-1-*benzopyran-2-one, 9CI. 5-Hydroxy-6,7-dimethoxycoumarin.* **Tomentin**
C$_{11}$H$_{10}$O$_5$ M 222.197
Aglycone from Tomenin. Cryst. (MeOH aq.). Mp 184-185°.

6,7-Di-Me ether, 5-glucoside: [28446-08-4]. **Tomenin**
C$_{17}$H$_{20}$O$_{10}$ M 384.339
Constit. of *Prunus tomentosa.* Cryst. (MeOH). Mp 206°.

Tri-Me ether: [55085-47-7]. *5,6,7-Trimethoxy-2*H-1-*benzopyran-2-one, 9CI. 5,6,7-Trimethoxycoumarin*
C$_{12}$H$_{12}$O$_5$ M 236.224
Isol. from *P. reniforme.* Mp 74-75°.

Fujita, M. *et al, Chem. Pharm. Bull.,* 1958, **6**, 511 (*isol, Fraxinol*)
Artem'eva, M.V. *et al, Khim. Prir. Soedin.,* 1973, **9**, 493 (*isol*)
Wagner, H. *et al; Tetrahedron Lett.,* 1974, 3807 (*Umckalin, trimethoxycoumarin*)
Wagner, H. *et al, Phytochemistry,* 1975, **14**, 2061 (*Umckalin, trimethoxycoumarin*)
Murray, R.D.H. *et al, Tetrahedron,* 1975, **31**, 2966 (*struct*)
Ahluwalia, V.K. *et al, Indian J. Chem., Sect. B,* 1978, **16**, 591 (*synth*)

5,7,8-Trihydroxy-2*H*-1-benzopyran-2-one,   T-80288
9CI

5,7,8-Trihydroxycoumarin

C$_9$H$_6$O$_5$ M 194.143

5,7-Di-Me ether: [61899-44-3]. *8-Hydroxy-5,7-dimethoxy-2*H-1-*benzopyran-2-one. 8-Hydroxy-5,7-dimethoxycoumarin.* **Leptodactylone**
C$_{11}$H$_{10}$O$_5$ M 222.197
Isol. from *Ruta* spp., *Leptodactylon* spp. and *Lianthus* spp. Mp 149-152°, Mp 197-199°.

Tri-Me ether: [60796-65-8]. *5,7,8-Trimethoxy-2*H-1-*benzopyran-2-one. 5,7,8-Trimethoxycoumarin*
C$_{12}$H$_{12}$O$_5$ M 236.224
Isol. from *R.* spp. and *Toddalia aculeatra.* Mp 179-180° (165°).

Bandopadhyay, M. *et al, Indian J. Chem.,* 1974, **12**, 23 (*synth, uv*)
González, A.G. *et al, An. Quim.,* 1976, **72**, 191 (*isol*)
Deshmukh, M.N. *et al, Phytochemistry,* 1976, **15**, 1419 (*isol*)
Dan, F.M. *et al, Phytochemistry,* 1978, **17**, 505 (*isol*)

2,4,6-Trihydroxy-3,5-bis(3-methyl-2-   T-80289
butenyl)acetophenone

C$_{18}$H$_{24}$O$_4$ M 304.385

2-Me ether: *2,4-Dihydroxy-6-methoxy-3,5-bis(3-methyl-2-butenyl)acetophenone. 2,4-Dihydroxy-6-methoxy-3,5-diprenylacetophenone*
C$_{19}$H$_{26}$O$_4$ M 318.412
Constit. of *Acronychia pedunculata.* Yellow oil.

Kumar, V. *et al, Phytochemistry,* 1989, **28**, 1278.

4′,5,7-Trihydroxy-3′,6-bis(3-methyl-2-   T-80290
butenyl)isoflavone

*5,7-Dihydroxy-3-[4-hydroxy-3-(3-methyl-2-butenyl)phenyl]-6-(3-methyl-2-butenyl)-4*H-1-*benzopyran-4-one, 9CI. 4′,5,7-Trihydroxy-3′,6-diprenylisoflavone. 3′-(γ,γ-Dimethylallyl)wighteone*

[76754-24-0]

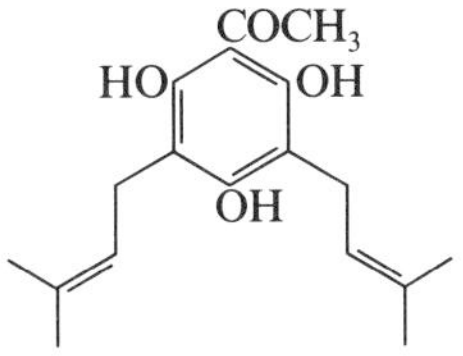

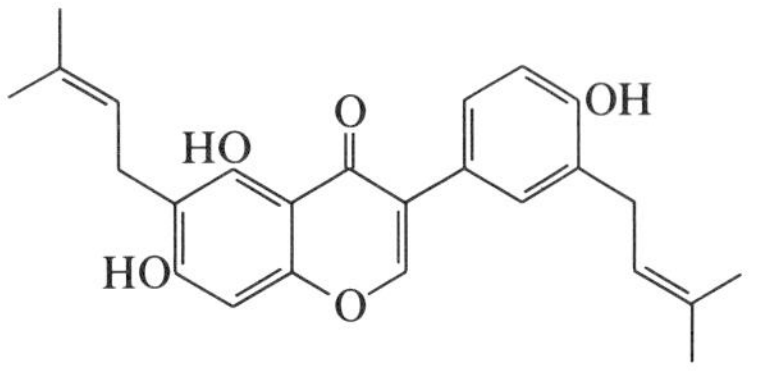

C$_{25}$H$_{26}$O$_5$ M 406.477
Isol. from aerial parts of *Millettia pachycarpa.* Cryst. (C$_6$H$_6$/pet. ether). Mp 120°.

Singhal, A.K. *et al, Phytochemistry,* 1980, **19**, 929 (*isol, ir, uv, ms, struct*)

ent-7α,18,19-Trihydroxy-3,13-clerodadien-   T-80291
16,15-olide

C$_{20}$H$_{30}$O$_5$ M 350.454
Constit. of *Salvia melissodora.* Cryst. (CH$_2$Cl$_2$/diisopropyl ether). Mp 149-150°. [α]$_D^{20}$ −53.75° (c, 0.16 in CHCl$_3$).

Esquivel, B. *et al, Phytochemistry,* 1989, **28**, 561.

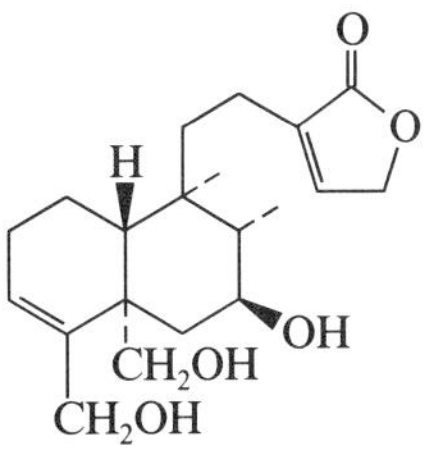

ent-2β,15,16-Trihydroxy-13-cleroden-3-one   T-80292

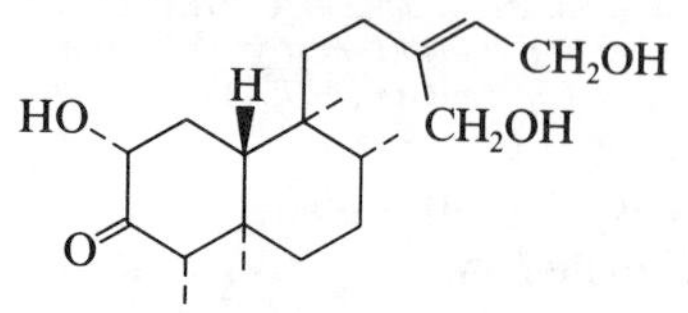

$C_{20}H_{34}O_4$ M 338.486

Constit. of *Baccharis boliviensis*.

Tri-Ac: Oil. $[\alpha]_D^{24}$ +32° (c, 0.31 in $CHCl_3$).

Zdero, C. *et al, Phytochemistry*, 1989, **28**, 531.

2,3,9-Trihydroxycoumestan   T-80293

Updated Entry replacing L-00454

*2,3,9-Trihydroxy-6H-benzofuro[3,2-c][1]benzopyran-6-one,
9CI.* **Lucernol.** *6,7,12-Trihydroxycoumestan*

[15402-22-9]

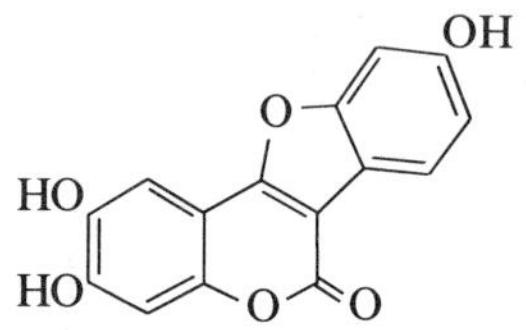

$C_{15}H_8O_6$ M 284.225

Constit. of alfalfa (*Medicago sativa*). Cryst. (DMSO). Mp
>350°. λ_{max} 232, 310, 355 and 372 sh nm.

Tri-Me ether: Cryst. (MeOH). Mp 255°.

Spencer, R.R. *et al, J. Agric. Food Chem.*, 1966, **14**, 162 (*isol,
struct*)
Kalna, V.K. *et al, Tetrahedron Lett.*, 1967, 2153 (*synth*)

3,7,9-Trihydroxycoumestan   T-80294

Updated Entry replacing W-20001

*3,7,9-Trihydroxy-6H-benzofuro[3,2-c][1]benzopyran-6-one,
9CI. 7,10,12-Trihydroxycoumestan.* **Repensol**

[33280-69-2]

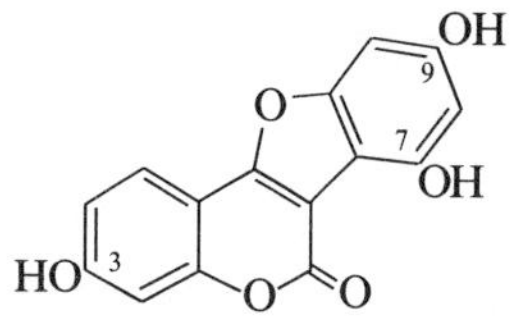

$C_{15}H_8O_6$ M 284.225

Isol. from fungus-infected *Trifolium repens* leaves. Cream
needles (MeOH) (synthetic). Mp 346-348°.

9-Me ether: [1857-26-7]. *3,7-Dihydroxy-9-
methoxycoumestan.* **Trifoliol.** *7,10-Dihydroxy-12-
methoxycoumestan*
$C_{16}H_{10}O_6$ M 298.251
Isol. from *T. repens* and *Medicago sativa*. Rods (DMF).
Mp 332° dec.

7,9-Di-Me ether: [77331-73-8]. *3-Hydroxy-7,9-
dimethoxycoumestan.* **Wairol.** *7-Hydroxy-10,12-
dimethoxycoumestan*
$C_{17}H_{12}O_6$ M 312.278
Isol. from *M. sativa*. Phytoalexin. Plates. Mp 292-294°.

Tri-Me ether: Cryst. (DMF). Mp 255-258°.

Bickoff, E.M. *et al, J. Pharm. Sci.*, 1964, **53**, 1496 (*isol*)
Livingston, A.L. *et al, Tetrahedron*, 1964, **20**, 1963 (*isol, struct, uv,
synth*)
Wong, E. *et al, Phytochemistry*, 1971, **10**, 466 (*Repensol*)
Shaw, G.J. *et al, Phytochemistry*, 1980, **19**, 2801; 1982, **21**, 249
(*isol, synth*)

3,8,9-Trihydroxycoumestan   T-80295

3,8,9-Trihydroxy-6H-benzofuro[3,2-c]benzopyran-6-one, 9CI

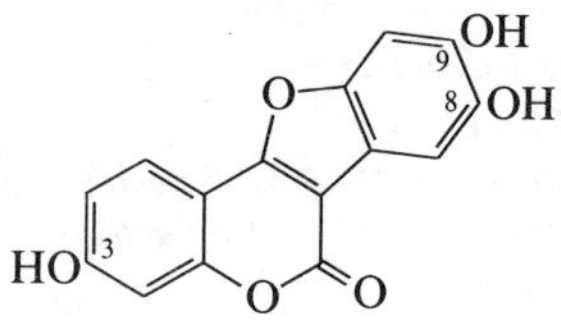

$C_{15}H_8O_6$ M 284.225

*8-Me ether: 3,8-Dihydroxy-9-methoxycoumestan. 8-
Methoxycoumestrol. 3′-Methoxycoumestrol* (*obsol.*)
$C_{16}H_{10}O_6$ M 298.251
Isol. from *Medicago* spp. (alfalfa). Straw-coloured solid.
Mp 329-329.5°.

8-Me ether, di-Ac: Needles. Mp 282-283°.

*8,9-Di-Me ether: 3-Hydroxy-8,9-dimethoxycoumestan. 7-
Hydroxy-11,12-dimethoxycoumestan* (*obsol.*)
$C_{17}H_{12}O_6$ M 312.278
Isol. from *M. sativa* and *Myroxylon balsamum*. Solid
(EtOAc). Mp 303-305°.

Bickoff, E.M. *et al, J. Agric. Food Chem.*, 1966, **14**, 444 (*isol*)
Spencer, R.R. *et al, J. Org. Chem.*, 1966, **31**, 988 (*isol, synth*)

1,3,4-Trihydroxy-12,6-eudesmanolide   T-80296

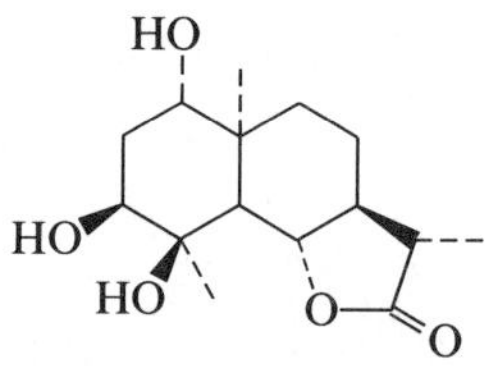

$C_{15}H_{24}O_5$ M 284.352

(1α,3β,4β,6α,11βH,10α)-form [120163-09-9]
Constit. of *Pyrethrum santalinoides*. Gum. $[\alpha]_D^{24}$ +10° (c,
0.33 in $CHCl_3$).

Abdel-Mogib, M. *et al, Phytochemistry*, 1989, **28**, 268.

4′,5,8-Trihydroxyflavanone   T-80297

[123067-25-4]
$C_{15}H_{12}O_5$ M 272.257
Constit. of *Artemisia campestris* ssp. *maritima*. Cryst.
(MeOH aq.). Mp 262-264°.

Rauter, A.P. *et al, Phytochemistry*, 1989, **28**, 2173 (*isol, pmr*)

2′,5,5′-Trihydroxyflavone   T-80298

[120552-44-5]
$C_{15}H_{10}O_5$ M 270.241
Constit. of *Primula pulverulenta*. Yellow powder. Mp
>270° dec.

Wollenweber, E. *et al, Phytochemistry*, 1989, **28**, 295.

2′,5,7-Trihydroxyflavone   T-80299

Updated Entry replacing D-04594

*5,7-Dihydroxy-2-(2-hydroxyphenyl)-4H-1-benzopyran-4-one,
9CI*
$C_{16}H_{12}O_5$ M 284.268

7-Me ether: [4308-56-9]. *2′,5-Dihydroxy-7-methoxyflavone.*
Echioidinin
$C_{17}H_{14}O_5$ M 298.295
Constit. of *Andrographis echioides* and *A. wightiana*.
Greenish-yellow needles (EtOH). Mp 264-266° dec.

7-Me ether, 2′-O-β-D-Glucopyranoside: **Echioidin**
$C_{22}H_{22}O_{10}$ M 446.410
Isol. from *A. echioides*. Needles (Py/MeOH). Mp 276-278° dec. $[\alpha]_D^{24}$ −23.5° (c, 0.213 in Py/MeOH 1:1).

7-Me ether, di-Ac: Prisms (EtOAc). Mp 174-176°.

Govindachari, T.R. *et al*, *Tetrahedron*, 1965, **21**, 2633, 3715 (*isol, uv, ir, pmr, struct, synth*)
Farkas, L. *et al*, *Tetrahedron*, 1967, **23**, 741 (*synth*)

3,4′,7-Trihydroxyflavone   T-80300

3,7-Dihydroxy-2-(4-hydroxyphenyl)-4H-1-benzopyran-4-one. 4′,7-Dihydroxyflavonol

[2034-65-3]

$C_{15}H_{10}O_5$ M 270.241
Isol. from *Schinopsis lorentzii, Cicer arietinum, Rhus succedanea* and others. Light-yellow needles (EtOH or AcOH aq.). Mp 310° (304-306°).

Kirby, K.S. *et al*, *Biochem. J.*, 1955, **60**, 582 (*isol, struct, synth*)
Wong, E. *et al*, *Phytochemistry*, 1965, **4**, 89 (*isol, synth*)
Hillis, W.E. *et al*, *Phytochemistry*, 1966, **5**, 483 (*isol*)

3′,4′,7-Trihydroxyflavone   T-80301

$C_{15}H_{10}O_5$ M 270.241
$O^{3'}$-*Me:* [21583-32-4]. *4′,7-Dihydroxy-3′-methoxyflavone.* **Geraldone**
$C_{16}H_{12}O_5$ M 284.268
Isol. from subterranean clover *Trifolium subterraneum*. Minute off-white cryst. Mp 293-294° dec.

$O^{3'}$-*Me, 7-glucoside:*
$C_{22}H_{22}O_{10}$ M 446.410
Isol. from *T. subterraneum*.

Wong, E. *et al*, *Phytochemistry*, 1968, **7**, 2123 (*isol, struct, synth*)

2′,4′,7-Trihydroxyisoflavan   T-80302

4-(3,4-Dihydro-7-hydroxy-2H-1-benzopyran-4-yl)-1,3-benzenediol, 9CI. 3,4-Dihydro-3-(2,4-dihydroxyphenyl)-7-hydroxy-2H-1-benzopyran. **Demethylvestitol**

[65332-45-8]

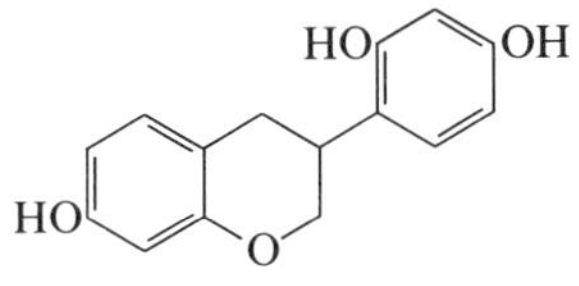

$C_{15}H_{14}O_4$ M 258.273
Compds. in both enantiomeric series appear to occur naturally but in most cases the abs. config.was not detd. Isol. from *Anthyllis vulneraria, Erythrina sandwicensis, Hosackia americana, Lablab niger, Lotus* spp., *Phaseolus vulgaris* and *Tetragonolobus* spp.

2′-Me ether: [63631-42-5]. *4′,7-Dihydroxy-2′-methoxyisoflavan.* **Isovestitol**
$C_{16}H_{16}O_4$ M 272.300
Constit. of *A. vulneraria, E. sandwicensis, H. americana, Lablab niger, T.* spp. and *Trifolium* spp.

7-Me ether: [71772-21-9]. *2′,4′-Dihydroxy-7-methoxyisoflavan.* **Neovestitol**
$C_{16}H_{16}O_4$ M 272.300
Isol. from leaves of *Dalbergia sericea*.

2′,4′-Di-Me ether: [41743-86-6]. *7-Hydroxy-2′,4′-dimethoxyisoflavan.* **Sativan.** *Sativin. 2′-O-Methylvestitol*
$C_{17}H_{18}O_4$ M 286.327

Isol. from *Derris amazonica, Lotus* spp., *Medicago* spp., *Trifolium* spp. and *Trigonella* spp. Cryst. Mp 128-129° (125-127°). $[\alpha]_D^{22}$ −15°. Mp. refers to the (S)-enantiomer from *D. amazonica*.

2′,7-Di-Me ether: [63631-41-4]. *4′-Hydroxy-2′,7-dimethoxyisoflavan.* **Arvensan**
$C_{17}H_{18}O_4$ M 286.327
Stress metab. from *Trifolium arvense* and *T. stellatum*. Mp 111-115°.

4′,7-Di-Me ether: [60102-29-6]. *2′-Hydroxy-4′,7-dimethoxyisoflavan.* **Isosativan**
$C_{17}H_{18}O_4$ M 286.327
Isol. from *Dalbergium ecastophyllum, M. scutellata, M. rugosa* and *Trifolium* spp.

[64190-84-7, 75556-92-2]

Ingham, J.L. *et al*, *Nature (London)*, 1973, **242**, 125 (*Sativan*)
Braz Filho, R. *et al*, *Phytochemistry*, 1975, **14**, 1454 (*isol*)
Ingham, J.L., *Z. Naturforsch., C*, 1976, **31**, 331; 1977, **32**, 446,1018; 1979, **34**, 630; 1980, **35**, 384.
Ingham, J.L. *et al*, *Phytochemistry*, 1977, **16**, 1279; 1979, **18**, 1711 (*isol, struct*)
Alves, H.M. *et al*, *Phytochemistry*, 1978, **17**, 1423 (*synth*)
Ingham, J.L. *et al*, *Fortschr. Chem. Org. Naturst.*, 1983, **43**, 1 (*rev, occur*)

2′,4′,7-Trihydroxyisoflavanone   T-80303

Updated Entry replacing T-20259
2,3-Dihydro-7-hydroxy-3-(2,4-dihydroxyphenyl)-4H-1-benzopyran-4-one. 2′-Hydroxydihydrodaidzein

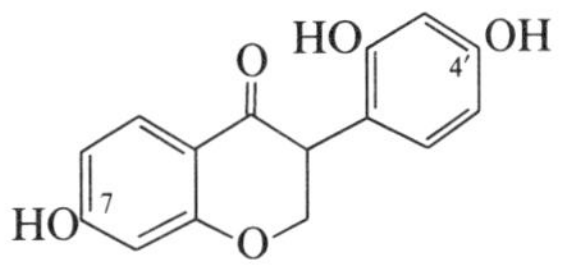

$C_{15}H_{12}O_5$ M 272.257
(±)-*form*
    Isol. from pods of *Phaseolus vulgaris*.

4′-Me ether: [57462-46-1]. *2′,7-Dihydroxy-4′-methoxyisoflavanone.* **Vestitone**
$C_{16}H_{14}O_5$ M 286.284
Isol. from *Medicago rugosa, Onobrychis vicifolia, Tipuana tipu* and *Trifolium repens*.

2′,4′-Di-Me ether: [70561-31-8]. *7-Hydroxy-2′,4′-dimethoxyisoflavanone.* **Sativanone**
$C_{17}H_{16}O_5$ M 300.310
Isol. from heartwood of *Dalbergia stevensonii* and leaves of *M. sativa*. Needles (C_6H_6/pet. ether). Mp 184-185°.

4′,7-Di-Me ether: [82829-55-8]. *2′-Hydroxy-4′,7-dimethoxyisoflavanone.* **Isosativanone**
$C_{17}H_{16}O_5$ M 300.310
Constit. of *M. rugosa*.

Donnelly, D.M.X. *et al*, *J. Chem. Soc., Perkin Trans. 1*, 1973, 1737 (*Sativanone*)
Derrick, P.M., *Phytochemistry*, 1977, **16**, 93 (*Vestitone*)
Ingham, J.L., *Z. Naturforsch., C*, 1978, **33**, 146 (*Vestitone*)
Woodward, M.D., *Phytochemistry*, 1980, **19**, 921 (*isol, uv, pmr*)
Ingham, J.L., *Planta Med.*, 1982, **45**, 46 (*isol*)
Jain, A.C. *et al*, *Indian J. Chem., Sect. B*, 1987, **26**, 136 (*synth*)

4′,5,7-Trihydroxyisoflavanone   T-80304

2,3-Dihydro-5,7-dihydroxy-3-(4-hydroxyphenyl)-4H-1-benzopyran-4-one

$C_{15}H_{12}O_5$ M 272.257
(S)-*form*
    5,7-Dihydroxy-4′-methoxyisoflavanone. **Dihydrobiochanin A**

$C_{16}H_{14}O_5$ M 286.284
Isol. from *Andira parviflora* trunkwood. Mp 167-169°.

Braz Filho, R. *et al, Phytochemistry*, 1973, **12**, 1184 (*isol, ir, ms, pmr, struct*)

2′,4′,7-Trihydroxyisoflavene   T-80305

4-(7-*Hydroxy-2H-1-benzopyran-3-yl*)-1,3-*benzenediol, 9CI.*
   Haginin D
[79852-13-4]

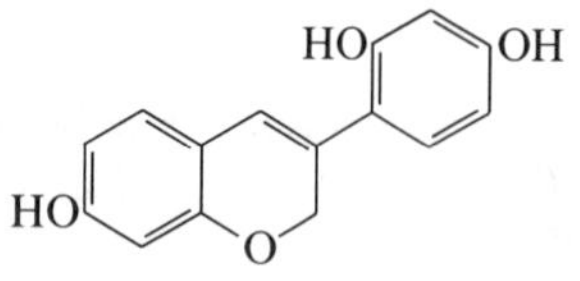

$C_{15}H_{12}O_4$ M 256.257
Isol. from heartwood of *Lespedeza cyrtobotrya*. Prisms
(Me$_2$CO/hexane). Mp 192-195° dec.

2′-*Me ether*: [74174-31-5]. 4′,7-*Dihydroxy-2′-methoxyisoflavene*. 3-(4-*Hydroxy-2-methoxyphenyl*)-*2H-1-benzopyran-7-ol, 9CI.* **Haginin B**
$C_{16}H_{14}O_4$ M 270.284
Isol. from heartwood of *L. cyrtobotrya*.

2′-*Me ether, di-Ac*: Needles (CHCl$_3$/MeOH). Mp 128-129°.

Miyase, T. *et al, Chem. Pharm. Bull.*, 1980, **28**, 1172; 1981, **29**, 2205 (*isol, uv, pmr, struct*)

2′,4′,7-Trihydroxyisoflavone   T-80306

3-(2,4-*Dihydroxyphenyl*)-7-*hydroxy-4H-1-benzopyran-4-one, 9CI.* 2′-*Hydroxydaidzein*
[7678-85-5]
$C_{15}H_{10}O_5$ M 270.241
Isol. from pods of *Phaseolus vulgaris*. Cryst. (MeOH). Mp
275° dec. (synthetic).

2′-*Me ether*: [56581-76-1]. 4′,7-*Dihydroxy-2′-methoxyisoflavone*. **Teralin. Theralin**
$C_{16}H_{12}O_5$ M 284.268
Isol. from leaves of *Thermopsis alterniflora*. Dark-yellow
cryst. (MeOH). Mp 275°.

4′-*Me ether*: [1890-99-9]. 2′,7-*Dihydroxy-4′-methoxyisoflavone*. 2′-*Hydroxyformononetin*
$C_{16}H_{12}O_5$ M 284.268
Isol. from *Trifolium repens, Virola caducifolia* and *V.
multinervia*. Cryst. (CHCl$_3$/MeOH). Mp 215-217°.

Kattaev, N.S. *et al, Khim. Prir. Soedin.*, 1975, 140; *cnc*, 157
   (*Teralin*)
Braz Filho, R. *et al, J. Nat. Prod. (Lloydia)*, 1977, **40**, 236 (2′,7-
   Dihydroxy-4′-methoxyisoflavone)
Woodward, M.D., *Phytochemistry*, 1980, **19**, 921 (*isol, uv, pmr, ms,
   struct*)

2′,5,7-Trihydroxyisoflavone   T-80307

Isogenistein
[70943-68-9]
$C_{15}H_{10}O_5$ M 270.241
Long straw-yellow needles (MeOH). Mp 222-224° (sinters
at 190°).

7-O-*Glucoside*: [70943-69-0].
$C_{21}H_{20}O_{10}$
Isol. from rootbark of *Cajanus cajan*. Needles (EtOH).
Mp 200-203°.

Bhanumati, S. *et al, Phytochemistry*, 1979, **18**, 365.

3′,4′,7-Trihydroxyisoflavone   T-80308

Updated Entry replacing D-04601
3-(3,4-*Dihydroxyphenyl*)-7-*hydroxy-4H-1-benzopyran-4-one,
9CI.* 3′-*Hydroxydaidzein*
$C_{15}H_{10}O_5$ M 270.241
Isol. from heartwood of *Machaerium villosum*. Mp 252°
(245-250°) (synthetic).

3′-*Me ether*: [21913-98-4]. 4′,7-*Dihydroxy-3′-methoxyisoflavone*. 3′-*Methoxydaidzein*
$C_{16}H_{12}O_5$ M 284.268
Isol. from *Cyclobium clausseni, Maackia amurensis* and
Machaerium villosum. Plates (MeOH). Mp 262°.

4′-*Me ether*: [20575-57-9]. 3′,7-*Dihydroxy-4′-methoxyisoflavone*. **Calycosin**
$C_{16}H_{12}O_5$ M 284.268
Isol. from *Baptisia, Bowdichia, Cadia, Cladrastis,
Cyclobium, Dalbergia, Machaerium, Myroxylon,
Pterocarpus, Sophora, Thermopsis* and *Trifolium* spp. (all
Leguminosae subfamily Paplilionoideae). Cryst. (EtOH).
Mp 245-247° (228-230°).

4′-*Me ether, 7-O-glucoside*: [20633-67-4].
$C_{22}H_{22}O_{10}$ M 446.410
Isol. from *Baptisia* and *Thermopsis* spp. λ_{max} 255, 327
nm (MeOH).

4′-*Me ether, 7-O-rhamnosylglucoside*:
$C_{28}H_{32}O_{15}$ M 608.552
Isol. from *B.* spp.

3′,4′-*Di-Me ether*: [24160-14-3]. 7-*Hydroxy-3′,4′-dimethoxyisoflavone*. **Cladrin**
$C_{17}H_{14}O_5$ M 298.295
Constit. of *Cladrastis lutea* heartwood. Mp 257-258°.

3′,7-*Di-Me ether*: [30564-92-2]. 4′-*Hydroxy-3′,7-dimethoxyflavone*. **Sayanedine**
$C_{17}H_{14}O_5$ M 298.295
Isol. from pods of *Pisum sativum*. Needles (hexane). Mp
165-166°.

Tri-Me ether: 3′,4′,7-*Trimethoxyisoflavone*. **Cabreuvin**
$C_{18}H_{16}O_5$ M 312.321
Isol. from *Calopogonium mucunoides, Myrocarpus
fastigiatus*, and *Myroxylon* spp. Mp 156-158°.

Gottlieb, O.R. *et al, An. Assoc. Quim. Bras.*, 1959, **18**, 85
   (*Cabreuvin*)
Fukui, F. *et al, Nippon Kagaku Kaishi (J. Chem. Soc. Jpn.)*, 1963,
   84, 189 (*synth*)
Braga de Oliveira, A. *et al, Ann. Acad. Bras. Cienc.*, 1968, **40**, 147
   (*isol*)
Markham, K.R. *et al, Phytochemistry*, 1968, **7**, 803 (*glycosides*)
Shamma, M. *et al, Tetrahedron*, 1969, **25**, 3887 (*Cladrin*)
Isogai, Y. *et al, Chem. Pharm. Bull.*, 1970, **18**, 1872 (*Sayanedine*)
Farkas, L. *et al, Acta Chim. Acad. Sci. Hung.*, 1972, **74**, 367
   (*synth*)
Takai, M. *et al, Chem. Pharm. Bull.*, 1972, **20**, 2488 (4′,7-
   Dihydroxy-3′-methoxyisoflavone, synth)
Brown, P.M. *et al, Justus Liebigs Ann. Chem.*, 1974, 1295
   (*Calycosin*)
Dewick, P.M. *et al, Phytochemistry*, 1978, **77**, 1751 (*biosynth*)
Ingham, J.L., *Fortschr. Chem. Org. Naturst.*, 1983, **43**, 1 (*rev,
   occur*)

4′,5,7-Trihydroxyisoflavone   T-80309

Updated Entry replacing T-60324
5,7-*Dihydroxy-3-(4-hydroxyphenyl)-4H-1-benzopyran-4-one,
9CI.* **Genistein.** *Differenol* A. *Prunetol. Sophoricol*

[446-72-0]

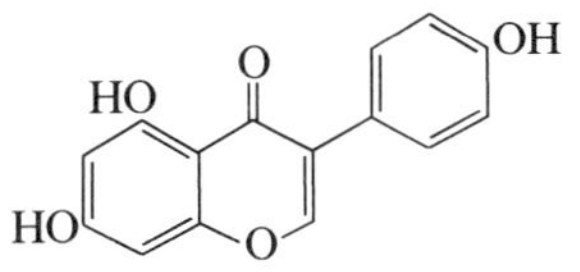

C$_{15}$H$_{10}$O$_5$ M 270.241

V. widely distributed in the Leguminosae subf. Papilionoideae but also in *Podocarpus spicatus* (Podocarpaceae) and *Prunus* spp. (Rosaceae). Also prod. by microorganisms *Streptomyces* spp., *Aspergillus niger*, *Mycobacterium phlei* and *Micromonospora haplophytica*. Weak oestrogen. Shows insect antifeedant and weak antibacterial activity against *E. coli* and *Xanthomonas oryzae*. Induces cell differentiation and inhibits DOPA carboxylase. Prisms (EtOH aq.). Mp 301-302° dec. Methylgenistein, Isogenistin and Methylisogenistin were all impure Genistein.

▷ Exp. carcinogen. NR2392000.

4-β-D-Glucoside: [152-95-4]. **Sophoricoside**
C$_{21}$H$_{20}$O$_{10}$ M 432.383
Constit. of *Sophora japonica* and *Piptanthus nepalensis*. Prisms (EtOH). Mp 297°. [α]$_D^{20}$ −46.7°.

7-β-D-Glucoside: [529-59-9]. **Genistin.** *Genistoside*
C$_{21}$H$_{20}$O$_{10}$ M 432.383
Isol. from *Genista* spp., *Lupinus* spp. and a number of other spp. in the Leguminosae subfamily Papilionoideae. Also in *Prunus aequinoctalis*, *P. avium* and *P. nipponica* (Rosaceae). Leaflets (EtOH). Mp 254-256° dec. [α]$_D^{21}$ −27.7° (MeOH aq.).

7-β-D-Rutinoside: [14988-20-6]. **Sphaerobioside**
C$_{27}$H$_{30}$O$_{14}$ M 578.526
Isol. from *Baptisia sphaerocarpa* and *B. lecontii*. Needles (MeOH). Mp 203-204°. [α]$_D^{20}$ −73.3° (c, 9.9 in Py).

4′,7-Bis(4-O-β-D-glucopyranosyl-β-D-apiofuranoside): [78693-95-5]. **Sarothamnoside**
C$_{37}$H$_{46}$O$_{23}$ M 858.757
Obt. from seeds of *Sarothamnus* spp. Mp 136-138°. [α]$_D^{20}$ −128° (c, 0.1 in H$_2$O).

7-O-[Apiosyl-(1→6)-β-D-glucopyranoside]: [108044-05-9]. **Ambocin**
C$_{26}$H$_{28}$O$_{14}$ M 564.499
Constit. of *Neorautanenia amboensis*. Glass. [α]$_D$ −36.5° (c, 0.011 in H$_2$O).

4′-Apioside, 7-β-D-glucopyranoside: [108069-00-7]. **Neobacin**
C$_{26}$H$_{28}$O$_{14}$ M 564.499
From *N. amboensis*. Glass. [α]$_D$ −38.2° (c, 0.011 in H$_2$O).

7-O-Glucosylglucoside:
C$_{27}$H$_{30}$O$_{15}$ M 594.525
Isol. from *Lupinus angustifolius*, *L. polyphyllus* and *L. luteus*.

4′,7-Di-O-glucoside:
C$_{27}$H$_{30}$O$_{15}$ M 594.525
Isol. from *Piptanthus nepalensis* and *Thermopsis* spp.

7-O-(6-O-Acetyl-β-D-glucopyranoside): [72741-92-5]. **6″-O-Acetylgenistin**
C$_{23}$H$_{22}$O$_{11}$ M 474.420
Isol. from seeds of *Glycine max* (soybean). Needes. Mp 185-186°.

4′-O-Neohesperidoside: [2945-88-2]. **Sophorabioside**
C$_{27}$H$_{30}$O$_{14}$ M 578.526

Isol. from *Sophora japonica*. Yellowish needles + 3H$_2$O. Mp 245-248° (unsharp)(anhyd.). [α]$_D^{19}$ −72.5° (Py)(anhyd.).

7-O-(6-O-Malonyl-D-glucoside): **Genistein 7-O-glucoside 6″-malonate**
C$_{24}$H$_{22}$O$_{13}$ M 518.430
Isol. from leaves of *Trifolium subterraneum*. Poss. accompanied by its Me ester. Glass. Mp 160° approx.

4′-Me ether: see 5,7-*Dihydroxy-4′-methoxyisoflavone*, D-80337

7-Me ether: see 4′,5-*Dihydroxy-7-methoxyisoflavone*, D-80335

4′,5-Di-Me ether: [68939-22-0]. *7-Hydroxy-4′,5-dimethoxyisoflavone.* **5-O-Methylbiochanin A**
C$_{17}$H$_{14}$O$_5$ M 298.295
Isol. from *Echinospartum horridum* (after hydrol.) and cultures of *Fusarium* spp. Mp 294-296°.

Bognár, R. *et al, Chem. Ind. (London)*, 1954, 518 (*Sophoricoside, synth*)
Rosler, H. *et al, Chem. Ber.*, 1965, **98**, 2193 (*deriv*)
Wagner, H. *et al, Chem. Ber.*, 1967, **100**, 101 (*synth*)
Farkas, L. *et al, Chem. Ber.*, 1968, **101**, 2758 (*Sophorabioside*)
Markham, K.R. *et al, Phytochemistry*, 1968, **7**, 791 (*isol*)
Ganguly, A.K. *et al, Chem. Ind. (London)*, 1970, 201 (*isol*)
Umezawa, H. *et al, J. Antibiot.*, 1975, **28**, 947 (*isol*)
Wagner, H. *et al, Tetrahedron Lett.*, 1976, 1799 (*cmr*)
Grayer-Barkmeijer, R.J. *et al, Phytochemistry*, 1978, **17**, 829 (7-*Hydroxy-4′,5-dimethoxyisoflavone*)
Hazeto, T. *et al, J. Antibiot.*, 1979, **32**, 217 (*isol*)
Ohta, N. *et al, Agric. Biol. Chem.*, 1980, **44**, 469 (6′-O-Acetylgenistin)
Asahi, K. *et al, J. Antibiot.*, 1981, **34**, 919 (*uv, ir*)
Brum-Bousquet, M. *et al, Tetrahedron Lett.*, 1981, 1223 (*Sarothamnoside*)
Ingham, J.L., *Fortschr. Chem. Org. Naturst.*, 1983, **43**, 1 (*rev, occur*)
Ogawara, H. *et al, J. Antibiot.*, 1986, **39**, 606 (*Ambocin, Neobacin*)
Jain, A.C. *et al, J. Chem. Soc., Perkin Trans.* 1, 1986, 215 (*synth*)
Breytenbach, J.C., *J. Nat. Prod. (Lloydia)*, 1986, **49**, 1003 (*isol, deriv*)

4′,6,7-Trihydroxyisoflavone   T-80310

Updated Entry replacing T-60325

[17817-31-1]
C$_{15}$H$_{10}$O$_5$ M 270.241
Isol. from fermented soybeans. Cryst. (MeOH). Mp 322°.

Tri-Ac: Cryst. (EtOH). Mp 217°.

6-Me ether: 4′,7-*Dihydroxy-6-methoxyisoflavone*. **Glycitein**
C$_{16}$H$_{12}$O$_5$ M 284.268
Isol. from *Centrosema haitense*, *C. pubescens*, *Glycine max* and *Mildraeodendron excelsa*. Cryst. (MeOH). Mp 311-313°.

7-Me ether: [57960-04-0]. 4′,6-*Dihydroxy-7-methoxyisoflavone*. **Kakkatin**
C$_{16}$H$_{12}$O$_5$ M 284.268
Isol. from flowers of *Pueraria* spp. Needles (MeOH). Mp >290°.

4′,7-Di-Me ether: [550-79-8]. 6-*Hydroxy-4′,7-dimethoxyisoflavone*. **Afrormosin.** *Afromosin. Castanin*
C$_{17}$H$_{14}$O$_5$ M 298.295
Constit. of *Afrormosia elata*, *Castanospermum australe* and *Myrocarpus* spp. and from other spp. in the Leguminosae (Papilionoideae). Needles (MeOH or EtOH). Mp 236-237° (228-229°).

4′,7-Di-Me ether, 6-β-D-glucopyranosyl: [19046-26-5]. **Wistin**
C$_{23}$H$_{24}$O$_9$ M 444.437

Constit. of *Wistaria floribunda*. Needles (MeOH aq. or Me₂CO aq.). Mp 209-210°. [α]$_D^{12}$ −67.15° (c, 1.43 in AcOH).

Tri-Me ether: [798-61-8]. *4′,6,7-Trimethoxyisoflavone*
$C_{18}H_{16}O_5$ M 312.321
Cryst. Mp 179-180°.

McMurry, T.B.H. *et al*, *J. Chem. Soc.*, 1960, 1491 (*isol, struct*)
Fujita, M. *et al*, *Chem. Pharm. Bull.*, 1963, **11**, 382 (*deriv*)
Harbourne, J. *et al*, *J. Org. Chem.*, 1963, **28**, 881 (*isol*)
Bevan, C.W.L. *et al*, *J. Chem. Soc. C*, 1966, 509 (*isol, struct*)
Ikehata, H. *et al*, *Agric. Biol. Chem.*, 1968, **32**, 740 (*isol*)
Birk, Y. *et al*, *Phytochemistry*, 1973, **12**, 169 (*Glycitein*)
Antus, S. *et al*, *Chem. Ber.*, 1975, **108**, 3883 (*synth, Glycitein*)
Kubo, M. *et al*, *Chem. Pharm. Bull.*, 1975, **23**, 2449 (*Kakkatin*)
Donnelly, D.M.X. *et al*, *Phytochemistry*, 1975, **14**, 2283 (*synth, Glycitein*)
Dewick, P.M. *et al*, *Phytochemistry*, 1978, **17**, 249 (*biosynth*)
Ingham, J.L., *Fortschr. Chem. Org. Naturst.*, 1983, **43**, 1 (*rev, occur*)
Caballero, P. *et al*, *J. Nat. Prod.* (*Lloydia*), 1986, **49**, 1126 (*cryst struct*)

4′,7,8-Trihydroxyisoflavone   T-80311

Updated Entry replacing D-04604
7,8-Dihydroxy-3-(4-hydroxyphenyl)-4H-1-benzopyran-4-one
$C_{15}H_{10}O_5$ M 270.241
4′-Me ether: [37816-19-6]. *7,8-Dihydroxy-4′-methoxyisoflavone.* **Retusin**
$C_{16}H_{12}O_5$ M 284.268
Isol. from heartwood of *Dalbergia retusa* and *Dipteryx odorata*. Cryst. (Me₂CO/MeOH or Me₂CO). Mp 249°.

4′,8-Di-Me ether: [37816-20-9]. *7-Hydroxy-4′,8-dimethoxyisoflavone.* **8-O-Methylretusin**
$C_{17}H_{14}O_5$ M 298.295
Isol. from *Cladrastis* spp., *Dalbergia retusa*, *D. variabilis*, *Dipteryx odorata*, *Monopteryx uaucu*, *Pericopsis schliebenii* and *Xanthocercus zambesiaca* (all Leguminosae). Prisms (Me₂CO/MeOH). Mp 221°.

4′,8-Di-Me ether, 7-O-glucoside:
$C_{23}H_{24}O_{10}$ M 460.437
Isol. from *C. platycarpa*, *C. shikokiana* and *Wisteria floribunda* bark.

4′,8-Di-Me ether, 7-O-laminarabioside:
$C_{29}H_{34}O_{15}$ M 622.579
Isol. from *C. platycarpa* and *C. shikokiana*. Needles. Mp 194-195°.

Jurd, L. *et al*, *Phytochemistry*, 1972, **11**, 2535 (8-O-Methylretusin)
Jurd, L. *et al*, *Tetrahedron Lett.*, 1972, 2149 (*Retusin*)
Hayashi, T. *et al*, *Phytochemistry*, 1974, **13**, 1943 (*Retusin*)
Imamura, H. *et al*, *Phytochemistry*, 1974, **13**, 757 (*Glucoside*)
Farkas, L. *et al*, *Acta Chim. Acad. Sci. Hung.*, 1976, **88**, 173 (*synth*)
Ingham, J.L., *Fortschr. Chem. Org. Naturst.*, 1983, **43**, 1 (*rev, occur*)

ent-7β,14α,15β-Trihydroxy-1,16-kauradien-3-one   T-80312

Liangshanin D

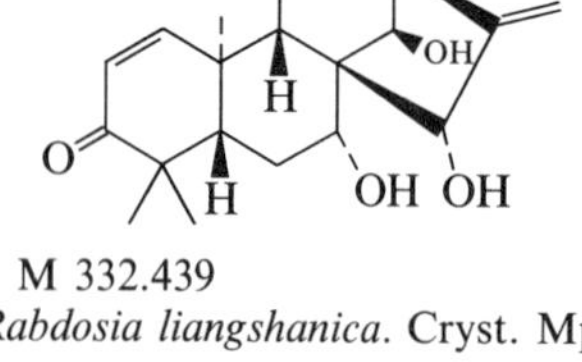

$C_{20}H_{28}O_4$ M 332.439
Constit. of *Rabdosia liangshanica*. Cryst. Mp 302-305°.

Fenglei, Z. *et al*, *Phytochemistry*, 1989, **28**, 1671.

ent-1β,3α,11α-Trihydroxy-16-kaurene-6,15-dione   T-80313

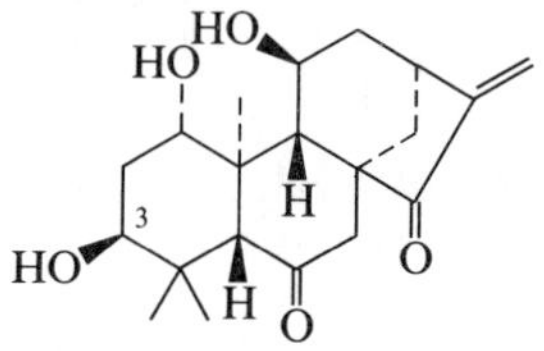

$C_{20}H_{28}O_5$ M 348.438
3-Ac: ent-3α-*Acetoxy*-1β,11α-*dihydroxy*-16-*kaurene*-6,15-*dione*. **Inflexarabdonin E**
$C_{22}H_{30}O_6$ M 390.475
Constit. of *Rabdosia inflexa*. Cryst. Mp 234-237°. [α]$_D$ −42.8° (c, 0.8 in MeOH).

Takeda, Y. *et al*, *Phytochemistry*, 1989, **28**, 2423 (*isol, pmr, cmr*)

ent-1β,7β,14α-Trihydroxy-16-kaurene-3,15-dione   T-80314

$C_{20}H_{28}O_5$ M 348.438
1-Ac: ent-1β-*Acetoxy*-7β,14α-*dihydroxy*-16-*kaurene*-3,15-*dione*. **Liangshanin E**
$C_{22}H_{30}O_6$ M 390.475
Constit. of *Rabdosia liangshanica*. Cryst. Mp 138-140°.

Fenglei, Z. *et al*, *Phytochemistry*, 1989, **28**, 1671.

ent-7β,12β,14α-Trihydroxy-16-kaurene-3,15-dione   T-80315

$C_{20}H_{28}O_5$ M 348.438
12-Ac: ent-12β-*Acetoxy*-7β,14α-*dihydroxy*-16-*kaurene*-3,15-*dione*. **Liangshanin F**
$C_{22}H_{30}O_6$ M 390.475
Constit. of *Rabdosia liangshanica*. Cryst. Mp 220-222°.

Fenglei, Z. *et al*, *Phytochemistry*, 1989, **28**, 1671.

ent-7α,9α,15β-Trihydroxy-16-kauren-19-oic acid   T-80316

[122585-73-3]

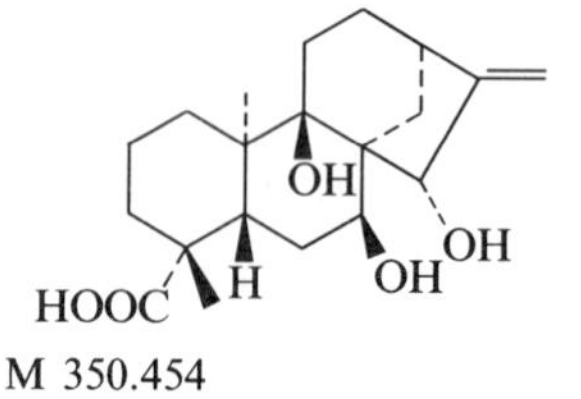

$C_{20}H_{30}O_5$ M 350.454
Constit. of *Helichrysum dasyanthum*.

Jakupovic, J., *Phytochemistry*, 1989, **28**, 1119.

ent-1β,7β,14*S*-Trihydroxy-16-kauren-15-one   T-80317

Updated Entry replacing T-50476
Kamebanin
[39388-57-3]

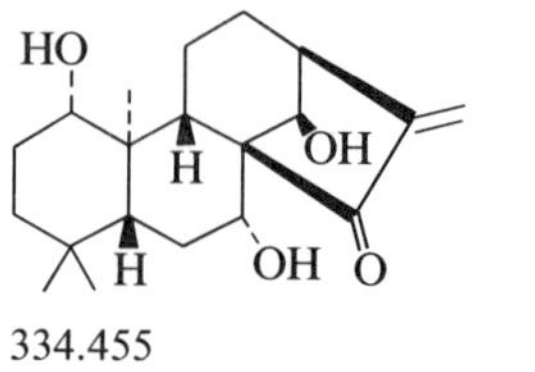

$C_{20}H_{30}O_4$ M 334.455

Constit. of *Isodon kameba*. Cryst. Mp 266-267°. $[\alpha]_D^{19}$ −108° (c, 1 in dioxan).

7-Ac: ent-7β-*Acetoxy*-1β,14S-*dihydroxy*-16-*kauren*-15-*one*
$C_{22}H_{32}O_5$ M 376.492
Constit. of *Rabdosia umbrosia*. Needles. Mp 201-202°. $[\alpha]_D^{22}$ −121.8° (c, 0.63 in MeOH).

14-Ac: ent-14S-*Acetoxy*-1β,7β-*dihydroxy*-16-*kauren*-15-*one*
$C_{22}H_{32}O_5$ M 376.492
Constit. of *R. umbrosia*. Amorph. powder. $[\alpha]_D^{22}$ −94.0° (c, 0.47 in MeOH).

Tri-Ac: Needles. Mp 196-198°.

Kubo, I. *et al, Chem. Lett.*, 1977, 1289 (*isol, struct*)
Takeda, Y. *et al, Phytochemistry*, 1989, **28**, 1691 (*derivs*)

3,15,22-Trihydroxy-7,9(11),24-lano-statrien-26-oic acid   T-80318

Updated Entry replacing T-70266

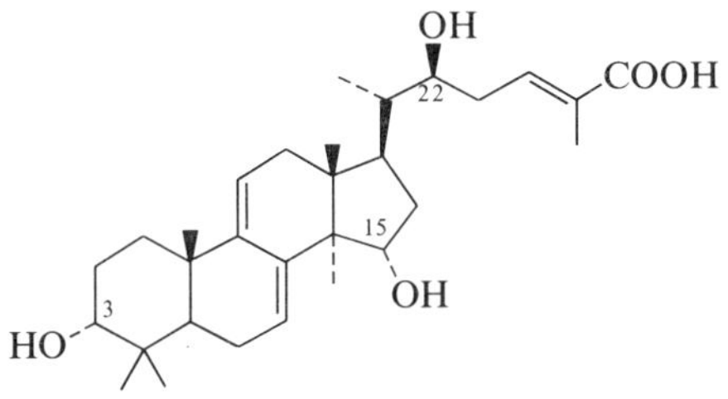

$C_{30}H_{46}O_5$ M 486.690
The Ganoderic/Ganodermic synonyms are best avoided as many of them have been duplicated. Use systematic names.

(3α,15α,22S,24E)-form

3-Ac: 3α-*Acetoxy*-15α,22S-*dihydroxy*-7,9(11),24E-*lanostatrien*-26-*oic acid*
$C_{32}H_{48}O_6$ M 528.728
Constit. of *Ganoderma lucidum*.

22-Ac: 22S-*Acetoxy*-3α,15α-*dihydroxy*-7,9(11),24-*lanostatren*-26-*oic acid*
$C_{32}H_{48}O_6$ M 528.728
Constit. of *G. lucidum*.

15,22-Di-Ac: [112667-14-8]. 15α,22S-*Diacetoxy*-3α-*hydroxy*-7,9(11),24E-*lanostatrien*-26-*oic acid. Ganoderic acid P*
$C_{34}H_{50}O_7$ M 570.765
Constit. of the cultured mycelium of *G. lucidum*. Cryst. Mp 211-212.5°.

3,22-Di-Ac: [110024-14-1]. 3α,22S-*Diacetoxy*-15α-*hydroxy*-7,9(11),24E-*lanostatrien*-26-*oic acid. Ganodermic acid P1. Ganodermic acid P1*
$C_{34}H_{50}O_7$ M 570.765
Constit. of the cultured mycelium of *G. lucidum*. Cryst. Mp 131-132°. Also descr. as a resin.

Tri-Ac: [103992-91-2]. 3α,15α,22S-*Triacetoxy*-7,9(11),24-*lanostatrien*-26-*oic acid. Ganoderic acid T*
$C_{36}H_{52}O_8$ M 612.802
Constit. of cultured mycelium of *G. lucidum*. Cryst. Mp 200-202°. $[\alpha]_D$ +23° (c, 0.13 in CHCl₃).

(3α,15α,22ξ,24E)-form

3,22-Di-Ac: 3α,22ξ-*Diacetoxy*-15α-*hydroxy*-7,9(11),24-*lanostatrien*-26-*oic acid. Ganoderic acid Mk*
$C_{34}H_{50}O_7$ M 570.765
Constit. of *G. lucidum*. Syrup. $[\alpha]_D^{23}$ +23° (c, 0.2 in MeOH). Not clear whether this is identical with Ganoderic acid Q above.

(3β,15α,22S,24E)-form
Constit. of *G.lucidum*. Needles (CHCl₃/MeOH). Mp 178-180°.

15,22-Di-Ac: [112430-69-0]. 15α,22S-*Diacetoxy*-3β-*hydroxy*-7,9(11),24E-*lanostatrien*-26-*oic acid. Gandodermic acid P2*
$C_{34}H_{50}O_7$ M 570.765
Metab. of *G. lucidum*. Resin.

Tri-Ac: [112430-70-3]. 3β,15α,22S-*Triacetoxy*-7,9(11),24E-*lanostatrien*-26-*oic acid*
$C_{36}H_{52}O_8$ M 612.802
Metab. of *G. lucidum*.

22-Ac: 22S-*Acetoxy*-3β,15α-*dihydroxy*-7,9(11),24-*lanostatrien*-26-*oic acid*
$C_{32}H_{48}O_6$ M 528.728
Constit. of *G. lucidum*.

(3α,15α,22R,24E)-form
Constit. of *G. lucidum*.

3,15-Di-Ac: 3α,15α-*Diacetoxy*-22R-*hydroxy*-7,9(11),24-*lanostatrien*-26-*oic acid*
$C_{34}H_{50}O_7$ M 570.765
Constit. of *G. lucidum*.

(3β,15α,22R,24E)-form

3,15-Di-Ac: 3β,15α-*Diacetoxy*-22R-*hydroxy*-7,9(11)24-*lanostatrien*-26-*oic acid*
$C_{34}H_{50}O_7$ M 570.765
Constit. of *G. lucidum*.

Toth, J.O. *et al, Tetrahedron Lett.*, 1983, **24**, 1081 (*isol, deriv*)
Hirotani, M. *et al, Chem. Pharm. Bull.*, 1986, **34**, 2282 (*isol*)
Nishitoba, T. *et al, Agric. Biol. Chem.*, 1987, **51**, 1149 (*isol*)
Hirotani, M. *et al, Phytochemistry*, 1987, **26**, 2797 (*isol*)
Lin, L.-J. *et al, J. Nat. Prod. (Lloydia)*, 1988, **51**, 918 (*isol*)
Shiao, M.-S. *et al, Phytochemistry*, 1988, **27**, 873, 2911 (*isol*)

4′,5,7-Trihydroxy-6-methoxyisoflavone   T-80319

Updated Entry replacing T-20263
*5,7-Dihydroxy-3-(4-hydroxyphenyl)-6-methoxy-4H-1-benzopyran-4-one, 9CI. **Tectorigenin***

[548-77-6]

$C_{16}H_{12}O_6$ M 300.267
Isol. from *Baptisia* spp., *Centrosema* spp., *Dalbergia* spp. and *Ononis spinosa* (all Leguminosae, Papilionoideae) and also *Iris germanica*. Yellow plates (EtOH) or needles (C₆H₆/EtOH). Mp 235-237° (227° dec.).

▷ NR2400000.

7-O-β-D-Glucopyranoside: [611-40-5]. **Tectoridin. Shekanin**
$C_{18}H_{22}O_{11}$ M 414.365
Isol. from *Baptisia* spp., *Belamcanda chinensis, Dalbergia* spp., *Iris* spp. and *Pueraria montana*. Needles. Mp 257° dec. $[\alpha]_D^{20}$ −29.4° (Py).

7-O-Gentobioside: [67604-94-8].
$C_{28}H_{32}O_{16}$ M 624.551
Isol. from bark of *Dalbergia volubilis*. Mp 160-162°. $[\alpha]_D$ −37.2° (c, 0.65 in MeOH).

4′-O-(3-Methyl-2-butenyl): [87457-87-2]. **Isoaurmillone**
$C_{21}H_{20}O_6$ M 368.385
Constit. of pods of *Millettia auriculata*. Yellow needles (CHCl₃/pet. ether). Mp 162.5-163.5°.

Mannich, C. *et al, Arch. Pharm. (Weinheim, Ger.)*, 1937, **275**, 317 (*Tectoridin*)
Shriner, R.L. *et al, J. Am. Chem. Soc.*, 1942, **64**, 2737 (*struct, synth*)
Várady, J., *Tetrahedron Lett.*, 1965, 4273 (*synth*)
Baker, W. *et al, J. Chem. Soc. C*, 1970, 1219 (*synth*)
Jose, C.I. *et al, Spectrochim. Acta*, 1973, **104**, 1394 (*ir*)
Khera, U. *et al, Indian J. Chem., Sect. B*, 1978, **16**, 78 (*isol*)
Khera, U. *et al, Phytochemistry*, 1978, **17**, 596 (*Gentiobioside*)
Ingham, J.L., *Fortschr. Chem. Org. Naturst.*, 1983, **43**, 1 (*rev, occur*)
Gupta, B.B. *et al, Phytochemistry*, 1983, **22**, 1306 (*isol*)

1,5,10-Trihydroxy-7-methoxy-3-methyl-1*H*-naphtho[2,3-*c*]pyran-6,9-dione   T-80320

[119975-66-5]

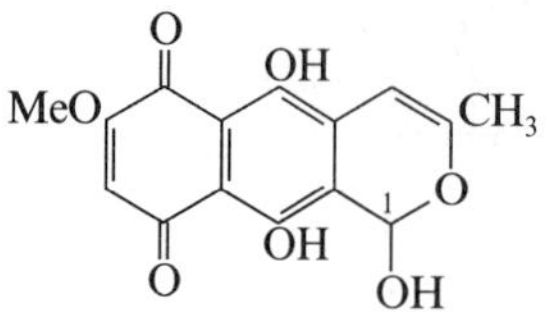

$C_{15}H_{12}O_7$ M 304.256

Metab. of *Fusarium solani*. Deep purple cryst. (Me$_2$CO). Mp 245°.

1-Me ether: [119975-67-6]. **5,10-Dihydroxy-1,7-dimethoxy-3-methyl-1H-naphtho[2,3-c]pyran-6,9-dione**
$C_{16}H_{14}O_7$ M 318.282
Metab. of *F. solani*. Purple cryst. (C$_6$H$_6$). Mp 190°.

Tatum, J.H. *et al, Phytochemistry*, 1989, **28**, 283.

2′,4′,6′-Trihydroxy-3′-methylacetophenone   T-80321

1-(2,4,6-Trihydroxy-3-methylphenyl)ethanone, 9CI
[2657-28-5]

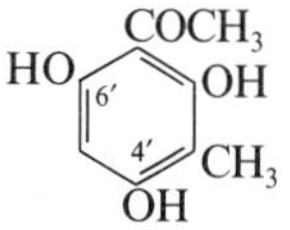

$C_9H_{10}O_4$ M 182.176

4′,6′-Di-Me ether: [23121-32-6]. *2′-Hydroxy-4′,6′-dimethoxy-3′-methylacetophenone.* **Methylxanthoxylin**
$C_{11}H_{14}O_4$ M 210.229
Isol. from oil of *Acradenia franklinii* and from *Eugenia jambolana* (cloves). Light-yellow needles (pet. ether). Mp 143°.

Baldwin, M.E. *et al, Tetrahedron*, 1961, **16**, 206 (*isol, uv, synth*)
Cann, M.R. *et al, Chem. Ind.* (*London*), 1982, 779 (*pmr*)
Linde, H. *et al, Arch. Pharm.* (*Weinheim, Ger.*), 1983, **316**, 917 (*isol*)

1,2,8-Trihydroxy-3-methylanthraquinone   T-80322

Norobtusifolin. 2-Hydroxychrysophanol

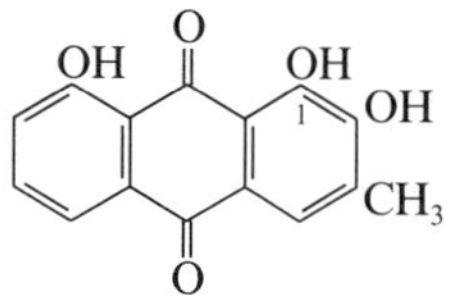

$C_{15}H_{10}O_5$ M 270.241

Constit. of *Myrsine africana* root. Orange-red needles (MeOH). Mp 265° (255°).

1-Me ether: [477-85-0]. **Obtusifolin**†
$C_{16}H_{12}O_5$ M 284.268
Isol. from seeds of *Cassia obtusifolia*. Yellow needles (MeOH). Mp 237-238°.

1-Me ether, di-Ac: Light-yellow needles. Mp 187-188°.

1-Me ether, O-β-D-Glucopyranosyl: **Glucoobtusifolin**
$C_{22}H_{22}O_{10}$ M 446.410
Isol. from seeds of *Cassia obtusifolia*. Cryst + 1H$_2$O. Mp 205-206°. Posn. of glucosylation not detd.

Tri-Me ether: *1,2,8-Trimethoxy-3-methylanthraquinone*
$C_{18}H_{16}O_5$ M 312.321
Light-yellow needles (pet. ether). Mp 145-146°.

Takido, M., *Chem. Pharm. Bull.*, 1958, **6**, 397 (*isol, struct*)
Takido, M. *et al, CA*, 1963, **62**, 5326 (*Glucoobtusifolin*)
Li, X.-H. *et al, J. Nat. Prod.* (*Lloydia*), 1989, **52**, 660 (*isol, pmr*)

3,6,8-Trihydroxy-1-methylanthraquinone-2-carboxylic acid   T-80323

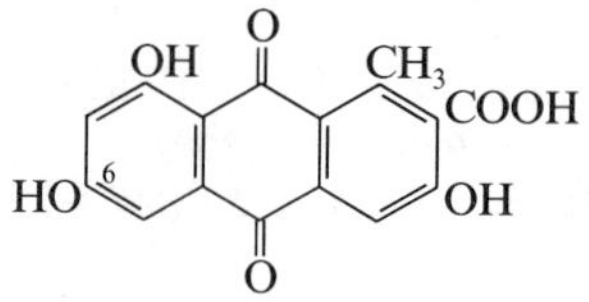

$C_{16}H_{10}O_7$ M 314.251

6-Me ether: [120374-28-9]. *3,8-Dihydroxy-6-methoxy-1-methylanthraquinone-2-carboxylic acid*
$C_{17}H_{12}O_7$ M 328.278
Constit. of *Gladiolus segetum*. Orange cryst. (MeOH). Mp 238-240°. Diagram incorrect in reference.

Ali, A.A. *et al, Phytochemistry*, 1989, **28**, 281.

3,4,6-Trihydroxy-2-methylbenzoic acid, 9CI   T-80324

[27613-31-6]

$C_8H_8O_5$ M 184.148
Needles (MeOH aq.). Mp 192-193° dec.

Me ester:
$C_9H_{10}O_5$ M 198.175
Cryst. (MeOH). Mp 205°.

4,6-Di-Me ether: *3-Hydroxy-4,6-dimethoxy-2-methylbenzoic acid*
$C_{10}H_{12}O_5$ M 212.202
Cryst. (C$_6$H$_6$). Mp 147-149° dec.

4,6-Di-Me ether, Me ester:
$C_{11}H_{14}O_5$ M 226.229
Long prisms (C$_6$H$_6$). Mp 114-115°.

Tri-Me ether: [79786-37-1]. *3,4,6-Trimethoxy-2-methylbenzoic acid, 9CI*
$C_{11}H_{14}O_5$ M 226.229
Solid. Mp 149°.

Tri-Me ether, Me ester: 114992-40-4
$C_{12}H_{16}O_5$ M 240.255
Plates (Et$_2$O/hexane). Mp 70°.

Asahina, Y. *et al, Bull. Chem. Soc. Jpn.*, 1942, **17**, 152 (*synth*)
Evans, G.E. *et al, J. Chem. Soc., Perkin Trans.* 1, 1988, 755 (*synth, uv, ir, pmr, ms*)

1,3,6-Trihydroxy-2-(3-methyl-2-butenyl)-9*H*-xanthen-9-one   T-80325

1,3,6-Trihydroxy-2-prenylxanthone

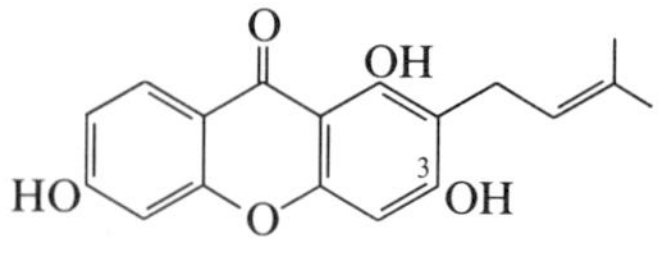

$C_{18}H_{16}O_5$ M 312.321

3-Me ether: [119227-98-4]. *1,6-Dihydroxy-3-methoxy-2-(3-methyl-2-butenyl)xanthone*
$C_{19}H_{18}O_5$ M 326.348
Constit. of *Garcinia mangostana*. Cryst. Mp >240°.

Parveen, M. *et al, Phytochemistry*, 1988, **27**, 3694.

4',5,7-Trihydroxy-6-methylflavone   T-80326

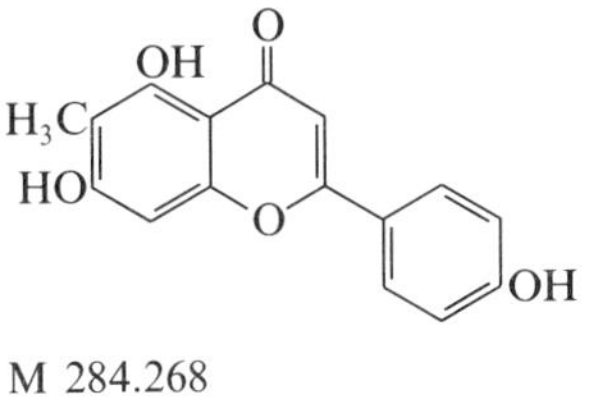

$C_{16}H_{12}O_5$ M 284.268

4',7-Di-Me ether: 5-Hydroxy-4',7-dimethoxy-6-*methylflavone*
$C_{18}H_{16}O_5$ M 312.321
Isol. from leaf wax of *Eucalyptus torrelliana* and *E. urnigera*. Cryst. (CHCl₃/pet. ether). Mp 187-188°.

Lamberton, J.A., *Aust. J. Chem.*, 1964, **17**, 692 (*isol, uv, ir, pmr, struct*)

9,12,13-Trihydroxy-10,15-octadecadienoic acid   T-80327

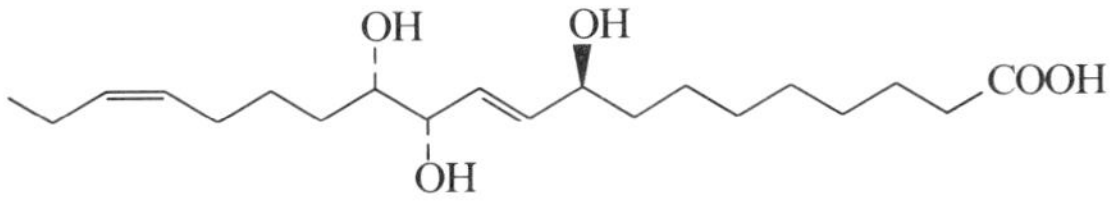

$C_{20}H_{36}O_5$ M 356.501
(9S,10E,12S,13S,15Z)-form
Isol. from a resistant rice cultivar. Confers activity against rice blast disease.

Me ester: $[\alpha]_D^{25}$ −10.90° (c, 1.01 in CHCl₃). Unnatural stereoisomers also synthesised.

15,16-Dihydro: 9S,12S,13S-*Trihydroxy-*10E-*octadecenoic acid*
$C_{20}H_{38}O_5$ M 358.517
Isol. from a rice cultivar. Confers resistance to rice blast disease.

15,16-Dihydro, Me ester: $[\alpha]_D^{25}$ −7.03° (c, 1.28 in CHCl₃). Unnatural stereoisomers also synthesised.

Suemune, H. *et al, Chem. Pharm. Bull.*, 1988, **36**, 3632 (*synth, ir, pmr, ms*)

1,3,23-Trihydroxy-12-oleanen-29-oic acid   T-80328

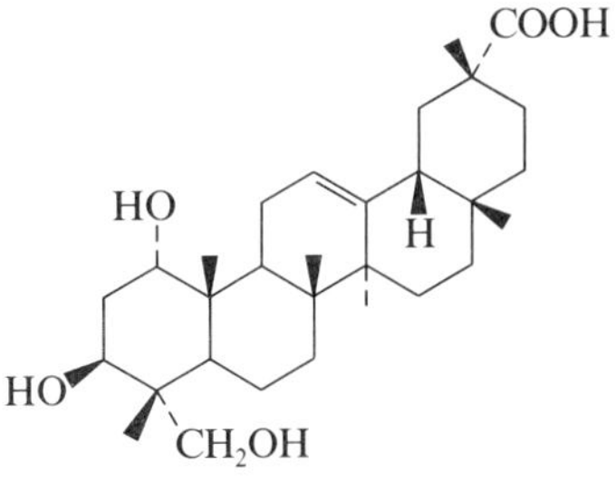

$C_{30}H_{48}O_5$ M 488.706
(1α,3β)-form
23-*Hydroxyimberbic acid*

23-O-α-L-*Rhamnopyranoside:* [118607-67-3].
$C_{36}H_{58}O_9$ M 634.849
Constit. of *Combretum imberbe*. Needles (EtOH). Mp 212-215°. $[\alpha]_D^{25}$ +12.9° (c, 1 in Py).

1-Ac, 23-O-α-L-*rhamnopyranoside:* [118607-68-4].
$C_{38}H_{60}O_{10}$ M 676.886
Constit. of *C. imberbe*. Prisms (EtOH). Mp 235-238°. $[\alpha]_D^{25}$ +5.6° (c, 0.6 in Py).

Rogers, C.B., *Phytochemistry*, 1988, **27**, 3217.

2,3,24-Trihydroxy-12-oleanen-28-oic acid   T-80329

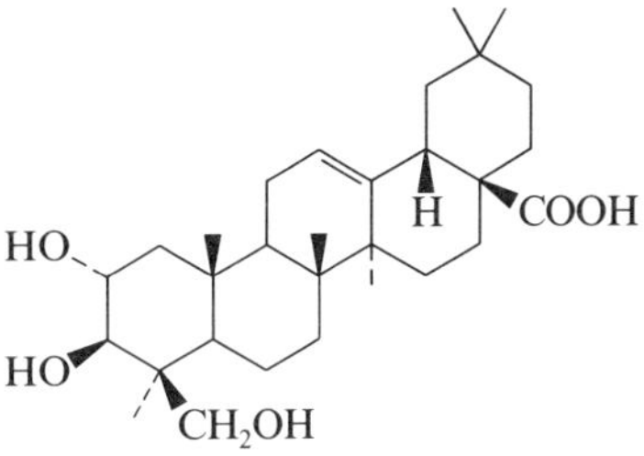

$C_{30}H_{48}O_5$ M 488.706
(2α,3β)-form [118711-55-0] *Hyptatic acid* **A**
Constit. of *Hyptis capitata*. Has anti-tumour activity. Cryst. Mp 298-304°. $[\alpha]_D^{20}$ +57° (c, 0.2 in MeOH).

Yamagishi, T. *et al, Phytochemistry*, 1988, **27**, 3213 (*isol, cryst struct*)

3,14,15-Trihydroxy-19-oxocard-20(22)-enolide   T-80330

Updated Entry replacing A-00788

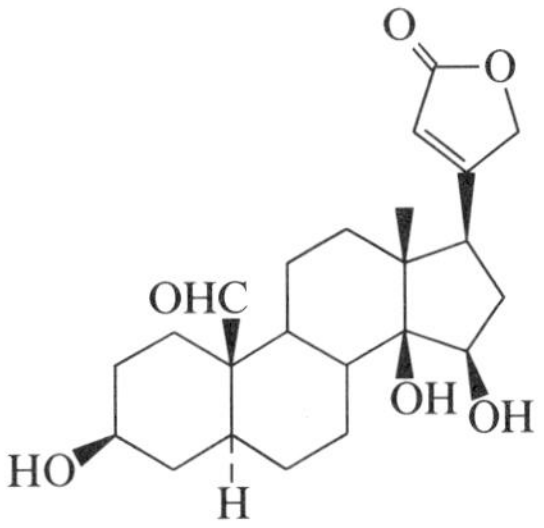

$C_{23}H_{32}O_6$ M 404.502
(3β,14β,5α,14β,15β)-form
Alloglaucotoxigenin
Aglycone from *Coronilla glauca*. Cryst. Mp 235° dec. $[\alpha]_D^{20}$ +27°.

Brandt, R. *et al, Helv. Chim. Acta*, 1966, **49**, 1662 (*uv, ir, pmr, ms, struct*)

1-(2,4,6-Trihydroxyphenyl)-2-buten-1-one   T-80331

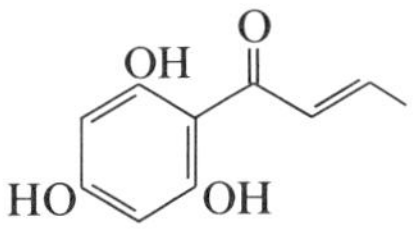

$C_{10}H_{10}O_4$ M 194.187
2,4-Di-Me ether: [119228-01-2]. *1-(2-Hydroxy-4,6-dimethoxyphenyl)-2-buten-1-one*. ***Verticilone***
$C_{12}H_{14}O_4$ M 222.240
Constit. of *Dysophyla verticillata*. Cryst. (hexane). Mp 79°.

Tri-Me ether: [72896-76-5]. *1-(2,4,6-Trimethoxyphenyl)-2-buten-1-one*. ***Vertinone***
$C_{13}H_{16}O_4$ M 236.267
Constit. of *D. verticillata*. Cryst. (hexane/C₆H₆). Mp 95°.

Chakrabarti, A. *et al, Phytochemistry*, 1988, **27**, 3683.

1-(2,4,6-Trihydroxyphenyl)-3-phenyl-1-propanone   T-80332

Updated Entry replacing U-30011
2′,4′,6′-Trihydroxydihydrochalcone

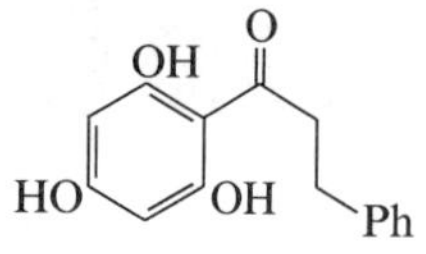

$C_{15}H_{14}O_4$ M 258.273

2-Me ether: [76444-56-9]. *1-(2,4-Dihydroxy-6-methoxyphenyl)-3-phenyl-1-propanone, 9CI. 2′,4′-Dihydroxy-6′-methoxydihydrochalcone.* **Uvangoletin**
Isol. from *Uvaria angolensis.* Needles (EtOAc/pet. ether). Mp 189-190°.

4-Me ether: [35241-55-5]. *1-(2,6-Dihydroxy-4-methoxyphenyl)-3-phenyl-1-propanone. 2′,6′-Dihydroxy-4′-methoxy-β-phenylpropiophenone*
$C_{16}H_{16}O_4$ M 272.300
Isol. from *Populus balsamifera* oil and from *Pitryogramma chrysophylla.* Light yellow needles (EtOH aq.). Mp 168°.

Tri-Me ether: 1-(2,4,6-Trimethoxyphenyl)-3-phenyl-1-propanone. 2′,4′,6′-Trimethoxydihydrochalcone
$C_{17}H_{18}O_4$ M 286.327
From *U. angolensis.* Cryst. (CHCl₃/hexane). Mp 74-76°.

Goris, A. *et al, Compt. Rend. Hebd. Seances Acad. Sci.,* 1935, **20**, 1435, 1520 *(isol, synth)*
Nilsson, M., *Acta Chem. Scand.,* 1961, **15**, 154 *(isol)*
Filho, R.B. *et al, Phytochemistry,* 1980, **19**, 1195 *(isol)*
Hufford, C.D. *et al, Phytochemistry,* 1980, **19**, 2036 *(isol)*
Bhardwaj, D.K. *et al, Indian J. Chem., Sect. B,* 1982, **21**, 476 *(synth)*

3-(3,4,5-Trihydroxyphenyl)propanal   T-80333

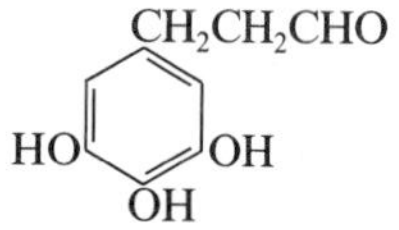

$C_9H_{10}O_4$ M 182.176

3,5-Di-Me ether: 3-(4-Hydroxy-3,5-dimethoxyphenyl)propanal. Dihydrosinapic aldehyde
$C_{11}H_{14}O_4$ M 210.229
Constit. of *Sorbus aucuparia.* Oil.

Malterud, K.E. *et al, Phytochemistry,* 1989, **28**, 1548.

4′,5,7-Trihydroxy-3′-prenylflavanone   T-80334

2,3-Dihydro-5,7-dihydroxy-2-[4-hydroxy-3-(3-methyl-2-butenyl)phenyl]-4H-1-benzopyran-4-one, 9CI. 3′-Prenylnaringenin

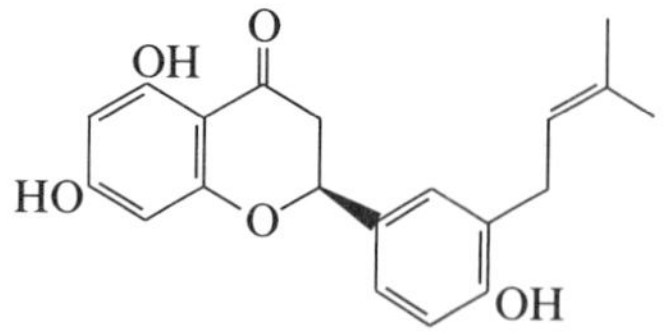

$C_{20}H_{20}O_5$ M 340.375
(S)-form [119240-82-3]
Constit. of *Erythrina eriotriocha.* Oil. [α]_D −1.1° (c, 2.2 in MeOH).

Nkengfack, A.E. *et al, J. Nat. Prod. (Lloydia),* 1989, **52**, 320 *(isol, pmr, cmr)*

2,3,9-Trihydroxypterocarpan   T-80335

Updated Entry replacing H-01568
6a,11a-Dihydro-6H-benzofuro[3,2-c][1]benzopyran-2,3,9-triol

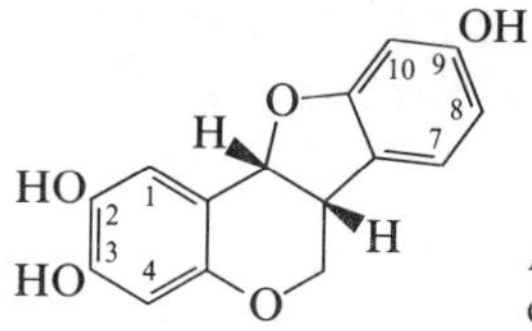

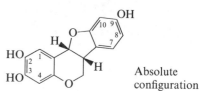

Absolute configuration

$C_{15}H_{12}O_5$ M 272.257

2,3-Di-Me ether: [73793-85-8]. *9-Hydroxy-2,3-dimethoxypterocarpan.* **Sparticarpin**
$C_{17}H_{16}O_5$ M 300.310
Stress metab. from leaves of *Spartium japonicum.* [α]_D −170° approx. (c, 0.01 in MeOH).

2,9-Di-Me ether: [56752-00-2]. *3-Hydroxy-2,9-dimethoxypterocarpan. 2-Methoxymedicarpin*
$C_{17}H_{16}O_5$ M 300.310
Isol. from root crowns of *Pisum sativum.* Shows antifungal activity. Plates (Me₂CO). Mp 146-148°. [α]_D^{23} −297° (EtOH).

Tri-Me ether: [56782-49-1]. *2,3,9-Trimethoxypterocarpan. 2-Methoxyhomopterocarpin*
$C_{18}H_{18}O_5$ M 314.337
Isol. from *P. sativum.* Shows antibiotic props. Needles (C₆H₆/heptane). Mp 122-124°. [α]_D^{23} −228° (EtOH).

Kalra, V.K. *et al, Indian J. Chem.,* 1967, **5**, 607 *(synth)*
Pueppke, S.G. *et al, J. Chem. Soc., Perkin Trans. 1,* 1975, 946 *(isol)*
Ingham, J.L. *et al, Z. Naturforsch., C,* 1980, **35**, 197 *(Sparticarpin)*

3,4,9-Trihydroxypterocarpan   T-80336

6a,11a-Dihydro-6H-benzofuro[3,2-c][1]benzopyran-3,4,9-triol. 4-Hydroxydemethylmedicarpin. 4′,7,8-Trihydroxypterocarpan

$C_{15}H_{12}O_5$ M 272.257
Isol. from fungus-infected leaves of *Melilotus alba.* Mp 118-120°.

9-Me ether: [53950-54-2]. *3,4-Dihydroxy-9-methoxypterocarpan. 4-Hydroxymedicarpin*
$C_{16}H_{14}O_5$ M 286.284
Isol. from fungus-infected leaves of *M. alba.*

3,9-Di-Me ether: 4-Hydroxy-3,9-dimethoxypterocarpan. 4-Hydroxyhomopterocarpin
$C_{17}H_{16}O_5$ M 300.310
Isol. from leaves of *Trifolium hydridum* and *T. pallescens* and from *M. alba* as fungal metab.

4,9-Di-Me ether: 3-Hydroxy-4,9-dimethoxypterocarpan. 4-Methoxymedicarpin
$C_{17}H_{16}O_5$ M 300.310
Constit. of *Swartzia madagascariensis, T. cherleri, T. pallescens* and *T. repens.* Prisms. Mp 158-160°. [α]_D −161° (CHCl₃).

Tri-Me ether: 3,4,9-Trimethoxypterocarpan. 4-Methoxyhomopterocarpin
$C_{18}H_{18}O_5$ M 314.337
Isol. from *Myroxylon peruiferum* and *Swartzia madagascariensis.* Needles (EtOH). Mp 118-120°.

(±)-form

4,9-Di-Me ether: [22973-33-7].
Cubes (C₆H₆/pet. ether). Mp 204-205°.

Tri-Me ether: Prisms (C₆H₆/pet. ether). Mp 114-115°.

Kalia, V.K. *et al, Tetrahedron,* 1967, **23**, 3221 *(synth)*

Campbell, R.V.M. *et al*, *J. Chem. Soc. C*, 1969, 1109 (*isol*)
Ingham, J.L., *Phytochemistry*, 1976, **15**, 1489 (*isol*)
Ingham, J.L., *Fortschr. Chem. Org. Naturst.*, 1983, **43**, 1 (*occur*)

3,6*a*,9-Trihydroxypterocarpan   T-80337

Updated Entry replacing T-30302
6H-*Benzofuro[3,2-*c*][*1*]benzopyran-3,6a,9(*11*aH)-triol, 9CI*.
Glycinol
[69393-95-9]

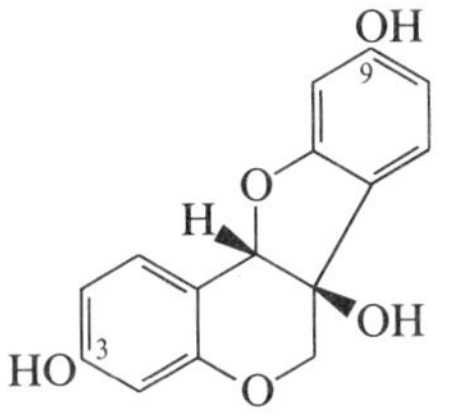

Absolute
configuration

$C_{15}H_{12}O_5$ M 272.257
Constit. of soybean seedlings (*Glycine max*). Also from
Erythrina sandwicensis and *Pueraria thunbergiana*. λ_{max}
282 and 287 nm.

3-Me ether: 6a,9-*Dihydroxy-3-methoxypterocarpan. 6a-*
Hydroxyisomedicarpin
$C_{16}H_{14}O_5$ M 286.284
Isol. from fungus-infected leaves of *Melilotus alba*.

9-Me ether: [61135-92-0]. *3,6a-Dihydroxy-9-*
methoxypterocarpan. 6a-Hydroxymedicarpin
$C_{16}H_{14}O_5$ M 286.284
Isol. from fungus-infected leaves of *M. alba* and
Trifolium pratense.

3,9-Di-Me ether: [3187-52-8]. *6a-Hydroxy-3,9-*
dimethoxypterocarpan. Variabilin†. *Homopisatin*
$C_{17}H_{16}O_5$ M 300.310
Isol. from wood of *Dalbergia variabilis*. Stress metab.
from leaves or colyledons of *Caragana* spp., *Lathyrus*
spp., *Lens culinaris, L. nigricans, Parochetus communis*
and *T. pratense*.

Ingham, J.L., *Phytochemistry*, 1976, **15**, 1489 (*isol*)
Kurosawa, K. *et al*, *Phytochemistry*, 1978, **17**, 1417 (*Variabilin*)
Ingham, J.L., *Z. Naturforsch., C*, 1980, **35**, 384 (*isol*)
Weistein, L.I. *et al*, *Plant Physiol.*, 1981, **68**, 358 (*isol*)
Ingham, J.L., *Fortschr. Chem. Org. Naturst.*, 1983, **43**, 1 (*occur*)

3,9,10-Trihydroxypterocarpan   T-80338

6a,*11*a-*Dihydro-6H-benzofuro[3,2-*c*][*1*]benzopyran-3,9,10-*
triol, 9CI
$C_{15}H_{12}O_5$ M 272.257

9-Me ether: [69853-46-9]. *3,10-Dihydroxy-9-*
methoxypterocarpan. Vesticarpan
$C_{16}H_{14}O_5$ M 286.284
Isol. from *Machaerium vestitum* and *Platymiscium*
trinitatis woods. Cryst. (MeOH). Mp 154-156°. Opt.
active (6aS,11aS). Also descr. as an oil.

10-Me ether: [73340-42-8]. *3,9-Dihydroxy-10-*
methoxypterocarpan. Nissolin
$C_{16}H_{14}O_5$ M 286.284
Isol. from phyllodes of *Lathyrus nissolia*.

9,10-Di-Me ether: [73340-41-7]. *3-Hydroxy-9,10-*
dimethoxypterocarpan. Methylnissolin
$C_{17}H_{16}O_5$ M 300.310
Isol. from phyllodes of *L. nissolia* (stress metab.).

(±)-*form*

9-Me ether: Microcryst. Mp 121°.

Craveiro, A.A. *et al*, *Phytochemistry*, 1974, **13**, 1629 (*Vesticarpan*)

Kurosawa, K. *et al*, *Phytochemistry*, 1978, **17**, 1413 (*Vesticarpan*)
Robeson, J.D. *et al*, *Phytochemistry*, 1979, **18**, 1715 (*Nissolin,*
Methylnissolin)

2,11,12-Trihydroxy-6,7-seco-8,11,13-   T-80339
abietatriene-6,7-dial 11,6-hemiacetal

$C_{20}H_{28}O_5$ M 348.438
Not named in the paper.

2α-form
Constit. of *Salvia texana*. Oil. Mixt. of 6-epimers.

Tri-Ac(6α-): [122558-74-1].
$C_{26}H_{34}O_8$ M 474.550
Constit. of *S. texana*.

[122558-70-7, 122558-71-8]

Gonzalez, A.G. *et al*, *J. Chem. Res. (S)*, 1989, 132.

12,15,20-Trihydroxy-3,7,11,23-tetraoxo-8-   T-80340
lanosten-26-oic acid

$C_{30}H_{42}O_9$ M 546.656
(*12β,15α,20ξ*)-*form* [120462-50-2] *Ganoderic acid AP*
Constit. of *Ganoderma applantum*.

Me ester: $[\alpha]_D^{23}$ +71° (c, 0.2 in EtOH).

Nishitoba, T. *et al*, *Phytochemistry*, 1989, **28**, 193.

4′,5,7-Trihydroxy-3′,6,8-trimethoxyflavone   T-80341

5,7-Dihydroxy-2-(4-hydroxy-3-methoxyphenyl)-6,8-
dimethoxy-4H-1-benzopyran-4-one, 9CI. **Sudachitin.**
Majoranin
[4281-28-1]
$C_{18}H_{16}O_8$ M 360.320
Constit. of *Citrus sudachi* and *Majorana hortensis*. Yellow
needles (MeOH). Mp 241-243°.

4′-O-β-D-Glucopyranoside: [70575-17-6]. **Sudachiin A**
$C_{24}H_{26}O_{13}$ M 522.462
Constit. of *C. sudachi*. Yellow needles (MeOH). Mp
211-213°. $[\alpha]_D$ −37.4° (c, 0.31 in 0.5% NaOH aq.).

7-O-β-D-Glucopyranoside: [70575-23-4].
$C_{24}H_{26}O_{13}$ M 522.462
Constit. of *Sideritis leucantha*. Yellow needles (MeOH).
Mp 187-188°. $[\alpha]_D$ +58.9° (c, 0.26 in 0.51% NaOH aq.).

Horie, T. *et al*, *Bull. Chem. Soc. Jpn.*, 1961, **34**, 1547 (*isol*)
Horie, T. *et al*, *Nippon Kagaku Kaishi (J. Chem. Soc. Jpn.)*, 1962,
  83, 468 (*synth*)
Lee, H.H. *et al*, *J. Chem. Soc.*, 1964, 6255 (*synth*)
Mabry, T.J. *et al*, *Curr. Sci.*, 1972, **41**, 202 (*isol, struct*)
Tomas, F. *et al*, *Phytochemistry*, 1980, **19**, 2039.
Horie, T. *et al*, *Bull. Chem. Soc. Jpn.*, 1982, **55**, 2928 (*isol, synth*)

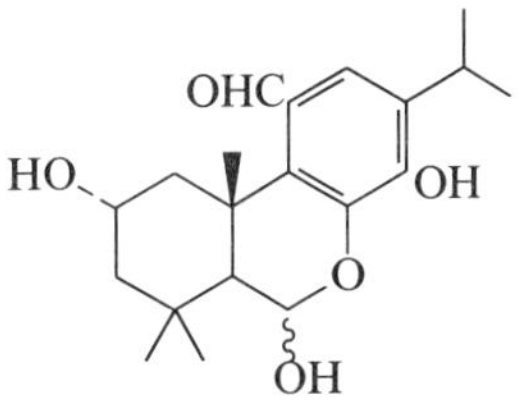

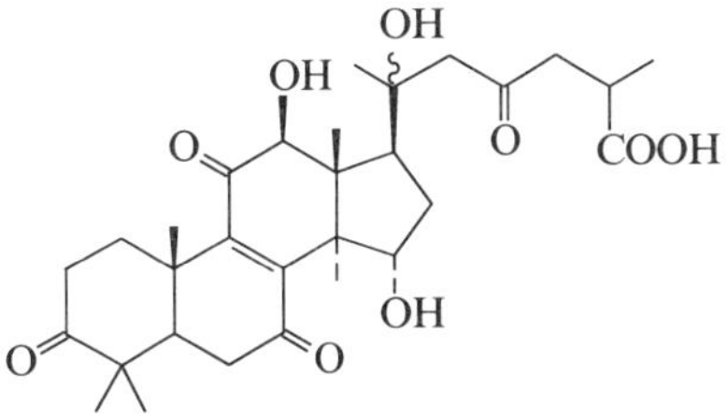

2,3,19-Trihydroxy-12-ursen-28-oic acid   T-80342

Updated Entry replacing T-03537

$C_{30}H_{48}O_5$ M 488.706

(2α,3α,19α)-form [53155-25-2]
Constit. of *Pygeum acuminatum*. Cryst. (MeOH). Mp
264°.

(2α,3β,19α)-form [13850-16-3] *Tormentic acid*
Aglycone from Tormentoside from *Potentilla
tormentilla*. Cryst. Mp 273°. [α]$_D$ − 20° (Py).

28-O-β-D-Glycopyranoside: [88515-58-6]. *Tormentic acid
ester glucoside*
$C_{36}H_{58}O_{10}$ M 650.848
Constit. of *Rosa multiflora* and *Aphloia theiformis*.
Amorph. powder.

Potier, P. *et al*, *Bull. Soc. Chim. Fr.*, 1966, **11**, 3458 (*isol*)
Chandel, R.S. *et al*, *Indian J. Chem., Sect. B*, 1977, **15**, 914 (*isol*)
Gopalsamy, N. *et al*, *Phytochemistry*, 1988, **27**, 3593 (*deriv*)

2,3,23-Trihydroxy-12-ursen-28-oic acid   T-80343

Updated Entry replacing T-50505

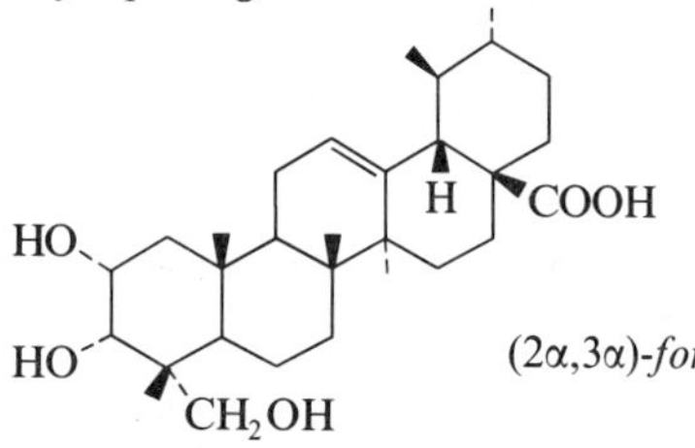

HO / HO / CH₂OH / COOH / H   (2α,3α)-form

$C_{30}H_{48}O_5$ M 488.706

(2α,3α)-form [76964-07-3] *Esculentic acid*†
Constit. of *Nepeta hindostana, Diplazium esculentum,
Hedyotis lawsoniae* and *Prunella vulgaris*. Cryst.
(MeOH). Mp 270°.

(2α,3β)-form [464-92-6] *Asiatic acid*
Constit. of *Dipterocarpus pilosus* and *Dryobalanops
aromatica*. Cryst. by subl. Mp 300-305°.
▷ Causes exp. neoplasms.

Me ester: Cryst. (C$_6$H$_6$). Mp 224-225.5°. [α]$_D$ + 54.5°.

*28-[6-Deoxy-α-L-mannopyranosyl-(1→4)-O-β-D-
glucopyranosyl-(1→6)-O-β-D-glucopyranoside]:* [16830-15-
2]. **Asiaticoside. Blastostimulina. Centelase. Emdecassol.
Madecassol. Marticassol. FK 1080**
$C_{48}H_{78}O_{19}$ M 959.133
Constit. of *Centella asiatica*. Shows antibiotic activity
against gram-positive organisms and mycobacteria,
particularly *M.tuberculosis*. Used in wound therapy.
Cryst. Mp 235-238° (230-233°). [α]$_D^{20}$ − 14° (EtOH).
▷ YU9625000.

Polonsky, J. *et al*, *Bull. Soc. Chim. Fr.*, 1959, 880 (*isol*)
Polonsky, J. *et al*, *Bull. Soc. Chim. Fr.*, 1961, 1586 (*struct*)
Singh, B. *et al*, *Phytochemistry*, 1969, **8**, 917 (*isol*)
Chasseaud, L.F. *et al*, *Arzneim.-Forsch.*, 1971, **21**, 1379 (*metab,
Asiaticoside*)
Gupta, A.S. *et al*, *Tetrahedron*, 1971, **27**, 823 (*isol*)
Velasco, M. *et al*, *Curr. Ther. Res., Clin. Exp.*, 1976, **19**, 121
(*Asiaticoside*)
Martindale, The Extra Pharmacopoeia, 28th/29th Ed.,
Pharmaceutical Press, London, 1982/1989, 1600.
Ahmad, V.U. *et al*, *Phytochemistry*, 1986, **25**, 1487 (*isol*)
Kojima, H. *et al*, *Phytochemistry*, 1989, **28**, 1703 (*pmr, struct*)
Sax, N.I., *Dangerous Properties of Industrial Materials*, 6th Ed.,
Van Nostrand-Reinhold, 1984, 326.

3,11,12-Trihydroxy-20-ursen-28-oic acid   T-80344

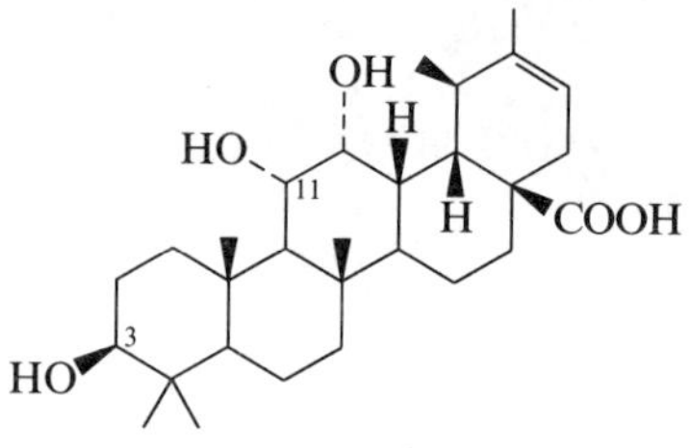

$C_{30}H_{48}O_5$ M 488.706

(3β,11α,12α)-form
3-O-(4-Hydroxyphenyl), 11-Me-ether: **Oleanderolic acid**
$C_{37}H_{54}O_5$ M 578.831
Constit. of *Nerium oleander*. Cryst. (EtOAc/pet. ether).
Mp 262-264°. [α]$_D^{24}$ + 50.0° (c, 0.04 in CHCl$_3$).

Siddiqui, S. *et al*, *Phytochemistry*, 1989, **28**, 1187.

3,19,24-Trihydroxy-12-ursen-28-oic acid   T-80345

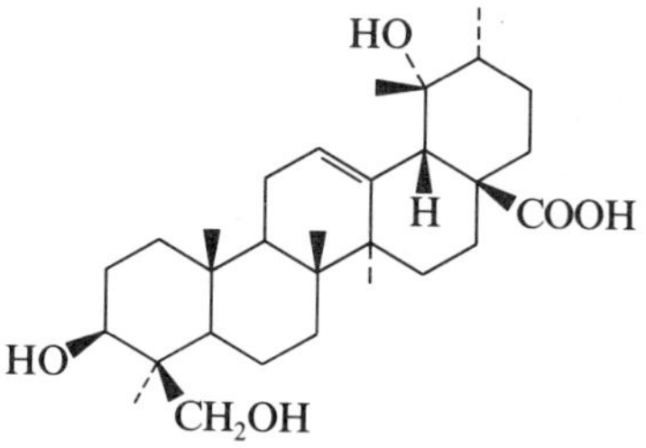

HO / HO / CH₂OH / COOH / H

$C_{30}H_{48}O_5$ M 488.706

(3β,19α)-form
Rotungenic acid
Constit. of *Ilex rotunda*. Cryst. (MeOH aq.). Mp 295-
298° dec. [α]$_D$ + 16° (MeOH).

Nakatani, M. *et al*, *Phytochemistry*, 1989, **28**, 1479.

3,6,19-Trihydroxy-12-ursen-28-oic acid   T-80346

$C_{30}H_{48}O_5$ M 488.706

(3β,6β,19α)-form
Constit. of *Uncaria florida*. Cryst. (MeOH aq.). Mp 226-
227°.

Me ester: Cryst. (MeOH aq.). Mp 220-221°.

Aimi, N. *et al*, *Tetrahedron*, 1989, **45**, 4125.

1,3,7-Trihydroxyxanthone   T-80347

Updated Entry replacing T-20280

1,3,7-Trihydroxy-9H-xanthen-9-one, 9CI. **Gentisein**
[529-49-7]

$C_{13}H_8O_5$ M 244.203
Isol. from *Gentiana lutea*. Orange-yellow needles (MeOH).
Mp 321-323°.

Tri-Ac:
$C_{19}H_{14}O_8$ M 370.315
Mp 229-230°.

1-Me ether: [16850-68-3]. *3,7-Dihydroxy-1-
methoxyxanthone*
$C_{14}H_{10}O_5$ M 258.230
Yellow needles (MeOH aq.). Mp 178°.

3-Me ether: [437-50-3]. *1,7-Dihydroxy-3-methoxyxanthone.*
Gentisin. *Gentianin. Gentianic acid*
$C_{14}H_{10}O_5$ M 258.230
Pigment from root of *G. lutea*. Yellow needles. Mp 273-
275° (266-267°).
▷ ZD6034600.

3-Me ether, 7-glucoside:
$C_{20}H_{20}O_{10}$ M 420.372
From *G. verna*. Mp 219-222° dec.

3-Me ether, 1-O-primeveroside: 7-Hydroxy-3-methoxy-1-
primeverosyloxyxanthone
$C_{25}H_{28}O_{14}$ M 552.488
Constit. of *G. lutea*. Needles. Mp 259-260° dec. $[\alpha]_D^{20}$
−81.5° (c, 0.4 in Py).

3-Me ether, 7-O-primeveroside: 1-Hydroxy-3-methoxy-7-
primeverosyloxyxanthone
$C_{25}H_{28}O_{14}$ M 552.488
Constit. of *G. lutea*. Needles. Mp 293-294° dec. $[\alpha]_D^{20}$
−69.4° (c, 0.36 in Py).

7-Me ether: [491-64-5]. *1,3-Dihydroxy-7-methoxyxanthone.*
Isogentisin
$C_{14}H_{10}O_5$ M 258.230
Mp 241°.
▷ ZD6034500.

7-Me ether, 3-O-primeveroside: [529-48-6]. **Gentioside.** *1-
Hydroxy-7-methoxy-3-primeverosyloxyxanthone*
$C_{25}H_{28}O_{14}$ M 552.488
Constit. of *G. lutea*. Needles. Mp 278-279° dec. $[\alpha]_D^{20}$
−50° (c, 0.3 in Py).

3,7-Di-Me ether: [13379-35-6]. *1-Hydroxy-3,7-
dimethoxyxanthone*
$C_{15}H_{12}O_5$ M 272.257
Isol. from *G. lutea* and *Frasera albicaulis*. Cryst.
(MeOH), yellow needles (CH_2Cl_2/hexane). Mp 169-170°.

Tri-Me ether: [3722-54-1]. *1,3,7-Trimethoxyxanthone*
$C_{16}H_{14}O_5$ M 286.284
Cryst. (EtOAc). Mp 173-174°.

Shinoda, J., *J. Chem. Soc.*, 1927, 1985 (*synth*)
Atkinson, J.E. *et al*, *Tetrahedron*, 1969, **25**, 1507 (*isol, synth*)
Stout, G.H. *et al*, *Tetrahedron*, 1969, **25**, 1961 (*isol, synth*)
Chawla, H.M. *et al*, *Proc.-Indian Acad. Sci., Sect. A*, 1973, **78**, 141
   (*ir*)
Hostettmann, K. *et al*, *Helv. Chim. Acta*, 1974, **57**, 1155 (*isol*)
Ellis, R.C. *et al*, *J. Chem. Soc., Perkin Trans. 1*, 1976, 1377 (*synth*)
Chadha, R. *et al*, *Indian J. Chem., Sect. A*, 1979, **18**, 505 (*cmr*)
Fram, A.W. *et al*, *Tetrahedron*, 1979, **35**, 2035 (*cmr*)
Hayashi, T. *et al*, *Phytochemistry*, 1988, **27**, 3696 (*derivs*)

3,6,7-Trimethoxy-1,4-phenanthraquinone   T-80348
Sphenone A

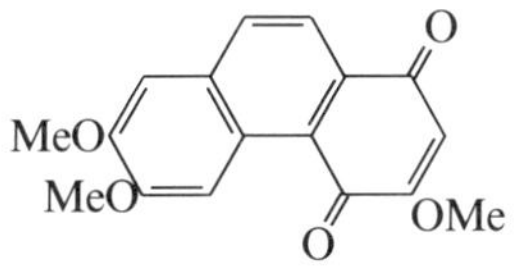

$C_{17}H_{14}O_5$ M 298.295
Constit. of *Sphenomeris biflora*. Cytotoxic substance.
Orange granules ($CHCl_3$). Mp 235-238°.

Wu, T.-S. *et al*, *Phytochemistry*, 1989, **28**, 1280.

2,3,5-Trimethylbenzyl alcohol   T-80349
2,3,5-Trimethylbenzenemethanol, 9CI
[50849-01-9]
$C_{10}H_{14}O$ M 150.220
Cryst. (hexane). Mp 50-51°.

Baciocchi, E. *et al*, *J. Org. Chem.*, 1977, **42**, 3682; 1980, **45**, 3906
   (*synth, deriv, pmr*)

3,4,5-Trimethylbenzyl alcohol   T-80350
3,4,5-Trimethylbenzenemethanol, 9CI
[39126-11-9]
$C_{10}H_{14}O$ M 150.220
Needles (pet.ether). Mp 72-74°.

Suzuki, H. *et al*, *Bull. Chem. Soc. Jpn.*, 1969, **42**, 2618 (*synth*)
Baciocchi, E. *et al*, *J. Org. Chem.*, 1977, **42**, 3682 (*deriv, pmr*)
Kim, E.K. *et al*, *J. Org. Chem.*, 1989, **54**, 1692 (*ir, pmr*)

4,24,24-Trimethyl-7,25-cholestadien-3-ol   T-80351

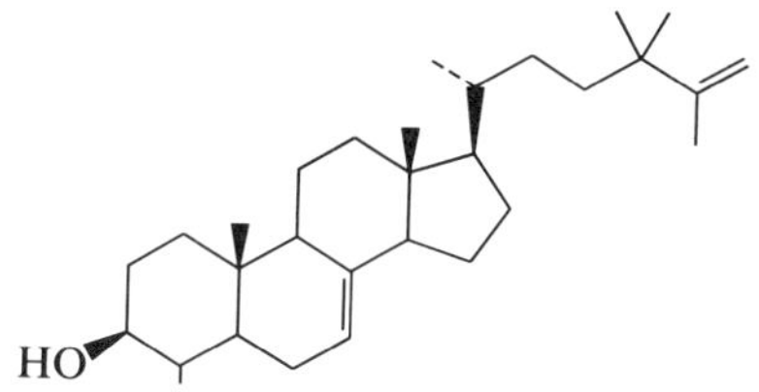

$C_{30}H_{50}O$ M 426.724
(3β,4α,5α)-form
   Erianol
   Constit. of *Eria convallarioides*. Cryst. (EtOAc/pet.
   ether). Mp 168°. $[\alpha]_D$ +3.5° ($CHCl_3$).

Majumder, P.L. *et al*, *Phytochemistry*, 1989, **28**, 1487.

23,24,24-Trimethyl-9(11),25-lanostadien-3-ol   T-80352

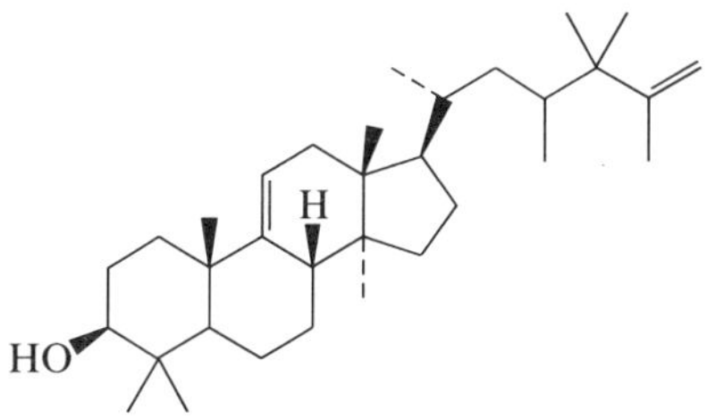

$C_{33}H_{56}O$ M 468.805
3β-form [62462-35-5] **Lansiol**
   Constit. of *Clausena lansium*. Cryst. ($CHCl_3$/MeOH).
   Mp 197-198°. $[\alpha]_D^{25}$ +83° (c, 1 in $CHCl_3$).

Lakshmi, V. *et al*, *Phytochemistry*, 1989, **28**, 943.

3,4,4-Trimethylpentanoic acid, 9CI   T-80353
[75177-71-8]

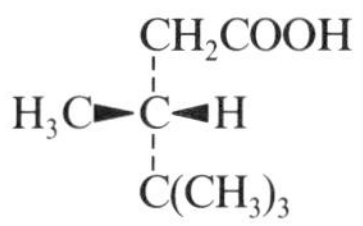

$$H_3C \blacktriangleright \overset{\displaystyle CH_2COOH}{\underset{\displaystyle C(CH_3)_3}{C}} \blacktriangleleft H \qquad (R)\text{-}form$$

$C_8H_{16}O_2$ M 144.213
(R)-form [40824-44-0]
   Oil. $[\alpha]_D^{25}$ +21.7° (c, 0.9 in EtOH).
Chloride:
   $C_8H_{15}ClO$ M 162.659
   Bp_{30} 85°.
(±)-form
   Bp_{18} 126°, Bp_4 90°.
[64043-89-6]

Menicagli, R. *et al*, *J. Chem. Res. (S)*, 1978, 262 (*synth*)
Azzena, U. *et al*, *Helv. Chim. Acta*, 1981, **64**, 2821 (*synth, resoln,
   pmr, cmr*)
Zhang, W.-Y. *et al*, *J. Am. Chem. Soc.*, 1988, **110**, 4652 (*synth, ir,
   pmr, cmr*)

3,4,4-Trimethyl-1-pentyn-3-ol, 9CI   T-80354

[993-53-3]

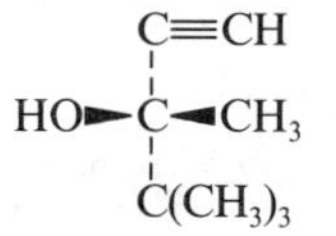

$$HO\text{---}\underset{\underset{C(CH_3)_3}{|}}{\overset{\overset{C{\equiv}CH}{|}}{C}}\text{---}CH_3 \qquad (S)\text{-}form$$

C$_8$H$_{14}$O   M 126.198

Abs. config. controversial. Now almost certainly established.

(S)-form [38484-41-2]

    Liq. Bp$_{15}$ 45-47°. $[\alpha]_D^{20}$ +0.06° (neat), $[\alpha]_D^{20}$ +0.78° (c, 1.1 in CH$_2$Cl$_2$).

(±)-form

    Liq. d$_4^{20}$ 1.06. Bp$_{11}$ 76-76.5°.

Hydrazide: Cryst. (EtOAc/EtOH). Mp 91-92°.

[38484-42-3, 104758-84-1]

Caporusso, A.M. *et al, Gazz. Chim. Ital.*, 1986, **116**, 467 (*cd*)
Elsevier, C.J. *et al, J. Org. Chem.*, 1987, **52**, 1536 (*synth, ir, nmr*)
Eliel, E.L. *et al, Tetrahedron Lett.*, 1987, **28**, 4813 (*abs config*)

2,6,9-Trioxabicyclo[3.3.1]nona-3,7-diene-4,8-dicarboxaldehyde, 9CI   T-80355

[116149-12-3]

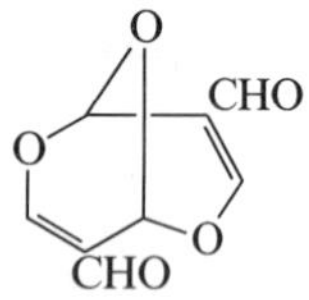

C$_8$H$_6$O$_5$ M 182.132

(±)-form

    Prod. of treatment of triformylmethane with COCl$_2$. Cryst. (1,2-dichloroethane). Mp 180°. Sublimes. Previous proposed struct. incorrect.

Arnold, Z. *et al, J. Org. Chem.*, 1988, **53**, 5352 (*synth, ms, ir, pmr, cmr*)

3,6,10-Trioxatetracyclo [7.1.0.0^{2,4}.0^{5,7}]decane, 9CI   T-80356

Updated Entry replacing T-04249
Trioxatris-σ-homotropilidene

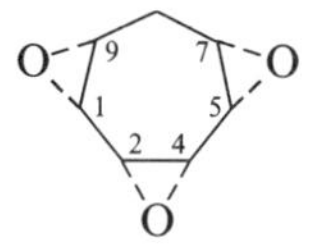

(1α,2α,4α,5α,7α,9α)-*form*

C$_7$H$_8$O$_3$ M 140.138

(1α,2α,4α,5α,7α,9α)-form [59992-00-6]

    syn,syn-*form*
    Cryst. Mp 165°.

(1α,2α,4α,5β,7β,9α)-form [61687-82-9]

    syn,anti-*form*
    Plates (CH$_2$Cl$_2$/Et$_2$O). Mp 66-67°.

(1α,2β,4β,5α,7α,9α)-form [61664-66-2]

    anti,anti-*form*
    Needles (CH$_2$Cl$_2$/pentane). Mp 110-112°.

(1α,2β,4β,5α,7α,9α)-form

    anti,anti-*form*
    Needles (CH$_2$Cl$_2$/pentane). Mp 110-112°.

Prinzbach, H. *et al, Angew. Chem., Int. Ed. Engl.*, 1976, **15**, 559 (*synth, pmr, cmr*)
Adam, W. *et al, J. Am. Chem. Soc.*, 1980, **102**, 1961 (*synth, ir, pmr, cmr*)
Sütbeyaz, Y. *et al, J. Org. Chem.*, 1988, **53**, 2312 (*synth*)

1,7,14-Trioxa-4,11,17-triazacycloeicosane, 9CI   T-80357

[116073-75-7]

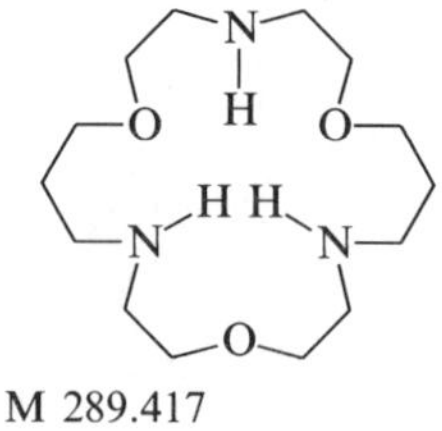

C$_{14}$H$_{31}$N$_3$O$_3$ M 289.417
Oil.

4,11,17-Tri-Me: [116073-76-8].
    C$_{17}$H$_{37}$N$_3$O$_3$ M 331.498
    Oil.

Pratt, J.A.E. *et al, J. Chem. Soc., Perkin Trans.* 1, 1988, 13 (*synth, pmr*)

1,9,17-Trioxa-5,13,21-triazacyclotetra-cosane, 9CI   T-80358

[116073-69-9]

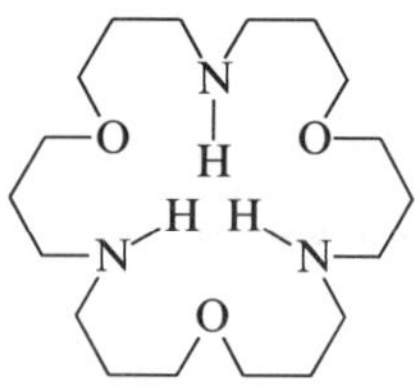

C$_{18}$H$_{39}$N$_3$O$_3$ M 345.524

B,3HCl: [116073-96-2].
    Cryst. (EtOH/Et$_2$O). Mp 250°. Hygroscopic.

5,13,21-Tri-Me: [116073-70-2].
    C$_{21}$H$_{45}$N$_3$O$_3$ M 387.605
    Oil.

Pratt, J.A.E. *et al, J. Chem. Soc., Perkin Trans.* 1, 1988, 13 (*synth, pmr*)

Triphenylacetaldehyde   T-80359

α,α-*Diphenylbenzeneacetaldehyde*, 9CI. 2,2,2-*Triphenylethanal*

[42365-04-8]

$$Ph_3CCHO$$

C$_{20}$H$_{16}$O   M 272.346
Cryst. Mp 105.5°.

Oxime: [41401-04-1].
    C$_{20}$H$_{17}$NO   M 287.360
    Cryst. Mp 190°.

Semicarbazone: Cryst. Mp 223°.

Phenylhydrazone: Cryst. Mp 142°.

Daniloff, S., *Russ. J. Phys. Chem.*, 1917, **49**, 282; *CA*, **18**, 1488 (*synth, deriv*)
Daniloff, S. *et al, Ber.*, 1926, **59**, 377 (*synth, deriv*)
Henderson, M.A. *et al, J. Org. Chem.*, 1988, **53**, 4736 (*synth, ir, pmr, cmr*)

1,1,1-Triphenyl-3-butyn-2-ol   T-80360

α-Ethynyl-β,β-diphenylbenzeneethanol, 9CI.
Ethynyltritylcarbinol

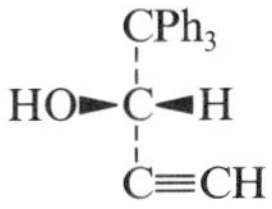

(S)-*form*

$C_{22}H_{18}O$ M 298.384

(S)-*form* [115648-30-1]
Cryst. (Et$_2$O). Mp 92-93°. $[\alpha]_D^{22}$ −9.81° (c, 0.031 in CHCl$_3$).

(±)-*form* [115533-97-6]
Cryst. Mp 80-82°.

[115648-31-2]

Henderson, M.A. *et al, J. Org. Chem.*, 1988, **53**, 4736 (*synth, resoln, ir, pmr, cmr*)

2,4,6-Triphenyl-3,4-dihydro-1,2,4,5-tetrazin-1(2H)-yl, 9CI   T-80361

2,4,6-Triphenylverdazyl. 1,3,5-Triphenylverdazyl

[2154-65-6]

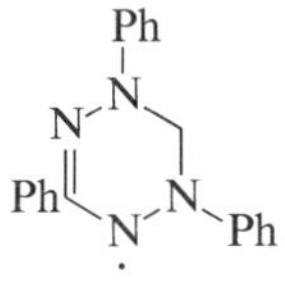

$C_{20}H_{17}N_4$ M 313.381
Radical scavenger. Blue-black cryst. (MeOH). Green in soln. or melt. Mp 161°.

Kuhn, R. *et al, Monatsh. Chem.*, 1964, **95**, 457 (*synth, uv, ir, esr*)
Williams, D.E., *Acta Crystallogr., Sect. B*, 1973, **29**, 96 (*cryst struct*)

2,4,6-Triphenyl-N-(3,5-diphenyl-4-oxido-phenyl)pyridinium betaine   T-80362

1-(2'-Hydroxy[1,1':3',1''-terphenyl]-5'-yl)-2,4,6-triphenylpyridinium hydroxide inner salt,9CI. 2,4,6-Triphenylpyridinium 4-oxo-3,5-diphenyl-2,5-cyclohexadien-1-ylide. Reichardt's dye

[10081-39-7]

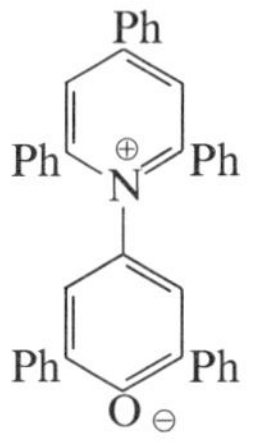

$C_{41}H_{29}NO$ M 551.686
Used in determination of solvent polarity (Kosower Z values). Violet-black cryst. + 2H$_2$O (MeOH aq.). Mp 205-276° dec. Highly solvatochromic.

B, HClO$_4$: Yellow plates (AcOH or MeOH). Mp 273-274°.

[43085-74-1]

Dimroth, K. *et al, Justus Liebigs Ann. Chem.*, 1963, **661**, 1 (*synth, ir, uv, derivs*)

Reichardt, C. *et al, Justus Liebigs Ann. Chem.*, 1983, 721 (*props, use, bibl*)
Johnson, B.P. *et al, Anal. Lett.*, 1986, **19**, 939 (*synth, props, use*)
Kessler, M.A. *et al, Synthesis*, 1988, 635 (*synth*)

Triptiliocoumarin   T-80363

[119365-36-5]

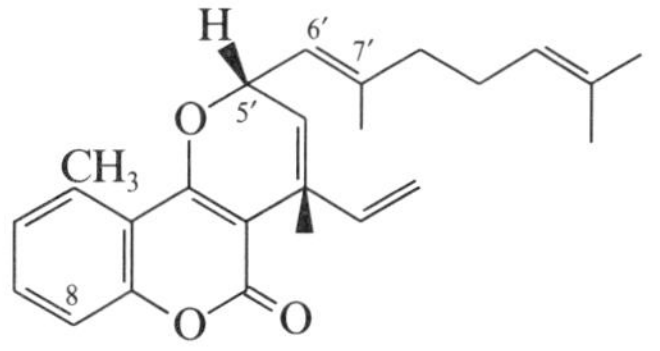

$C_{25}H_{30}O_3$ M 378.510
Constit. of *Triptilion benaventei*. Gum. $[\alpha]_D^{24}$ −22° (c, 0.16 in CHCl$_3$).

5'-*Hydroxy:* [119478-83-0]. 5'-*Hydroxytriptiliocoumarin*
$C_{25}H_{30}O_4$ M 394.510
Constit. of *T. benaventei*. Gum.

8-*Hydroxy:* [118555-92-3]. 8-*Hydroxytriptiliocoumarin*
$C_{25}H_{30}O_4$ M 394.510
Constit. of *T. benaventei*. Gum. $[\alpha]_D^{24}$ +8° (c, 0.18 in CHCl$_3$).

5'-*Epimer, 5'-hydroxy, 6',7'-epoxide:* [118555-93-4]. 6',7'-*Epoxy-5'α-hydroxytriptiliocoumarin*
$C_{25}H_{30}O_5$ M 410.509
Constit. of *T. benaventei*. Gum.

Bittner, M. *et al, Phytochemistry*, 1988, **27**, 3263.

Triptispinocoumarin   T-80364

[118227-65-9]

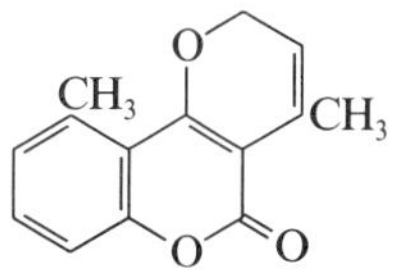

$C_{14}H_{12}O_3$ M 228.247
Constit. of *Triptilion spinosum*. Gum.

Bittner, M. *et al, Phytochemistry*, 1988, **27**, 3263.

Trisaemulins   T-80365

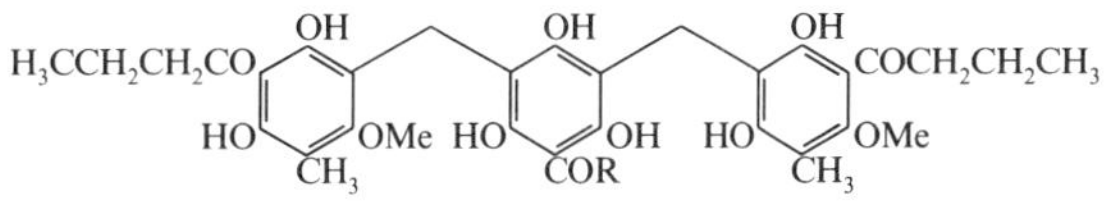

Trisaemulin BAB (R = CH$_3$)

Trisaemulin BBB (R = CH$_2$CH$_2$CH$_3$)

The suffixes refer to the acyl sidechains (A=acetyl, B=butyryl). Isol. from *Dryopteris aemula* and poss. present in *D. crispifolia*.

Trisaemulin BAB [56226-94-9]
$C_{34}H_{40}O_{12}$ M 640.683
Mp 180-183°. Not completely pure of BBB.

Trisaemulin BBB [49582-13-0]
$C_{36}H_{44}O_{12}$ M 668.736
Mp 168-170°. Not completely pure of BAB.

Widén, C.-J. *et al, Helv. Chim. Acta*, 1975, **58**, 881 (*isol, ms, pmr, struct*)

1,3,5-Tris(iodomethyl)benzene, 9CI   T-80366

ω,ω′,ω″-Triiodomesitylene
[90678-60-7]

C₉H₉I₃   M 497.884

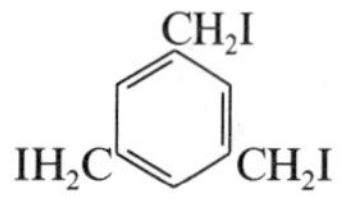

$C_9H_9I_3$ M 497.884

Cryst. (pet. ether/toluene). Mp 133-134°.

Reiche, W.S. *et al, J. Chem. Soc.*, 1947, 1234 (*synth*)
Bilger, C. *et al, Synthesis*, 1988, 902 (*synth*)

Tris(methylthio)methylium(1+), 9CI   T-80367

$(MeS)_3C^\oplus$

$C_4H_9S_3^\oplus$ M 153.313 (ion)

Methylsulfate: [116538-66-0].
  $C_5H_{12}O_4S_4$ M 264.411
  Solid. Mp 66-67°. Extremely moisture sensitive.

Tetrafluoroborate: [18702-35-7].
  $C_4H_9BF_4S_3$ M 240.118
  Cryst. (Et₂O/MeCN). Mp 168-170°. Can be stored at
  low temp. and dry conditions.

Tucker, W.P. *et al, Tetrahedron Lett.*, 1967, 2747 (*synth, pmr*)
Barbero, M. *et al, Synthesis*, 1988, 22 (*synth, pmr, cmr, use, bibl*)

Trisnorisoespinenoxide   T-80368

3,4-Seco-4,23,24-trisnor-D:B-friedo-B′:A′-neogammacer-5(10)-ene 3,5-oxide

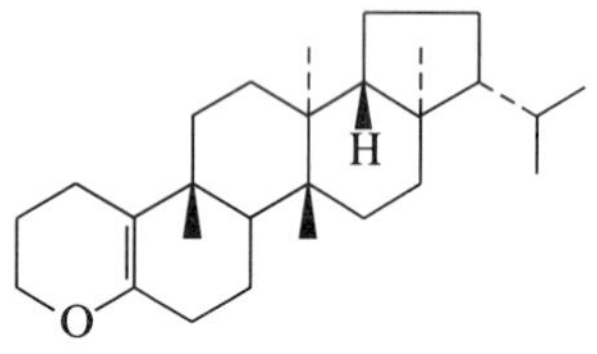

$C_{27}H_{44}O$ M 384.644

Constit. of *Euphorbia supina*. Cryst. Mp 209-213°. $[\alpha]_D^{23}$
−2.9° (c, 0.63 in CHCl₃).

Tanaka, R. *et al, Tetrahedron Lett.*, 1989, **30**, 1661.

Tritriacontane   T-80369

[630-05-7]

$H_3C(CH_2)_{31}CH_3$

$C_{33}H_{68}$ M 464.900

Isol. from many plant waxes incl. Candelilla wax. Major
constit. of some waxes, e.g. *Agave sisalana* (33%),
Calocephalus brownii (12%). Mp 71.8°. Four crystal
modifications known.

Chibnall, A.C. *et al, Biochem. J.*, 1934, **28**, 2189.
Razafindrazaka, J. *et al, Bull. Soc. Chim. Fr.*, 1963, 1633 (*isol*)
Batterham, T.J. *et al, Aust. J. Chem.*, 1966, **19**, 143 (*isol*)
Piesczek, W. *et al, Acta Crystallogr., Sect. B*, 1974, **30**, 1278 (*cryst struct*)

Tunaxanthin   T-80370

Updated Entry replacing C-00615
  ε,ε-Carotene-3,3′-diol, 9CI
[12738-95-3]

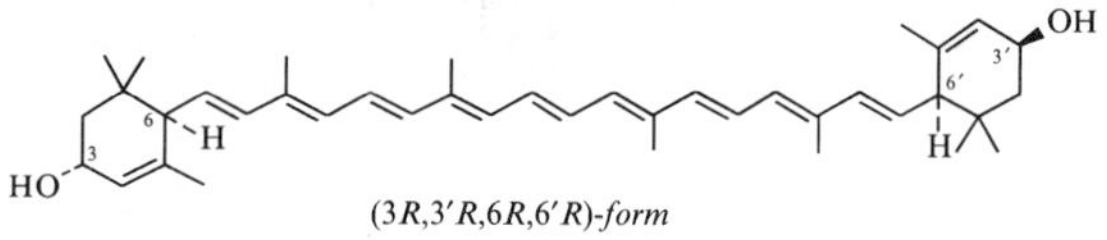

(3R,3′R,6R,6′R)-form

$C_{40}H_{56}O_2$ M 568.881

Various stereoisomers occur as fish and crustacean
pigments.

3-Epimer: Chiriquixanthin B
From *A. chiriquiensis*.
(3R,3′R,6R,6′R)-form [78306-12-4]
  Tunaxanthin F. Lactucaxanthin
(3R,3′R,6R,6′S)-form [97673-78-4]
  Tunaxanthin H
(3R,3′R,6S,6′S)-form [71697-13-7]
  Tunaxanthin C. Oxyxanthin 58
(3R,3′S,6R,6′R)-form [63597-82-0]
  Tunaxanthin I. Chiriquixanthin A
  Also isol. from the skin of *Atelopus chiriquiensis*.
(3R,3′S,6R,6′S)-form [82915-90-0]
  Tunaxanthin E
(3R,3′S,6S,6′R)-form [82915-89-7]
  Tunaxanthin D
(3R,3′S,6S,6′S)-form [72274-50-1]
  Tunaxanthin B. Oxyxanthin 51
(3S,3′S,6R,6′R)-form [63597-83-1]
  Tunaxanthin J. Chiriquixanthin B
  Also isol. from skin of *A. chiriquiensis*.
(3S,3′S,6R,6′S)-form [97746-99-1]
  Tunaxanthin G
(3S,3′S,6S,6′S)-form [71697-14-8]
  Tunaxanthin A

[82915-92-2, 83148-14-5]

Bingham, A. *et al, J. Chem. Soc. C*, 1977, 96 (*Chiriquixanthins*)
Roønneberg, H. *et al, Acta Chem. Scand., Ser. B*, 1978, **32**, 621
  (*struct, pmr, ms, uv, cd, abs config*)
Matsuno, T. *et al, CA*, 1983, **100**, 85953u; 1984, **101**, 207845d
  (*isol*)
Ikano, Y. *et al, J. Chromatogr.*, 1985, **328**, 387 (*hplc*)

Turbinaric acid   T-80371

4,8,13,17,21-Pentamethyl-4,8,12,16,20-docosapentaenoic acid

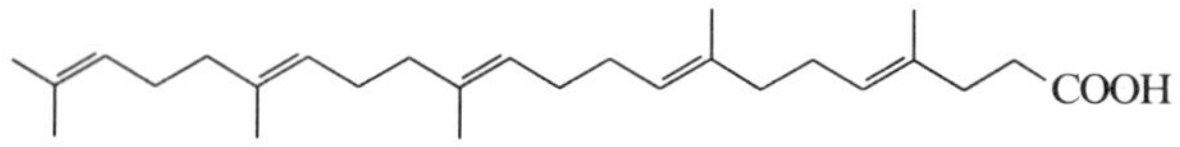

$C_{27}H_{46}O_2$ M 402.659

Constit. of *Turbinaria ornata*.

van Tamelen, E.E. *et al, Tetrahedron Lett.*, 1962, 121 (*synth*)
Asari, F. *et al, J. Nat. Prod. (Lloydia)*, 1989, **52**, 1167 (*isol, pmr, cmr*)

U

Umbelliferone 8-oxogeranyl ether U-80001

2,6-Dimethyl-8-[(2-oxo-2H-1-benzopyran-7-yl)oxy]-2,6-octadienal, 9CI

[30825-46-8]

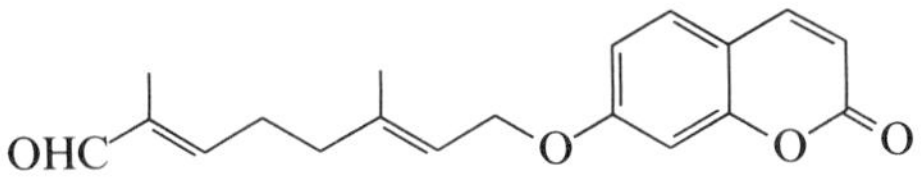

$C_{19}H_{20}O_4$ M 312.365

Isol. from *Capnophyllium peregrinum*. Cryst. (Et$_2$O/pet. ether). Mp 64°.

Bohlmann, F. *et al, Chem. Ber.*, 1970, **103**, 3619; 1974, **107**, 1780.

Unanisoflavan U-80002

Updated Entry replacing U-00024

3-[5-(1,1-Dimethyl-2-propenyl)-3-hydroxy-2,4-dimethoxyphenyl]-3,4-dihydro-2H-1-benzopyran-7-ol, 9CI. 5'-(1,1-Dimethylallyl)-3',7-dihydroxy-2',4'-dimethoxyisoflavan

[61186-60-5]

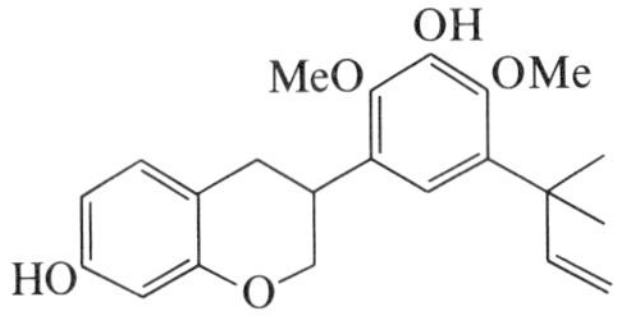

$C_{22}H_{26}O_5$ M 370.444

(−)-*form*

Constit. of *Sophora secondiflora*. Mp 184°. [α]$_D$ −73.5° (CHCl$_3$).

4'-O-De-Me: [58210-35-8]. **α,α-Dimethylallylcyclolobin.** *4-(3,4-Dihydro-7-hydroxy-2H-1-benzopyran-3-yl)-6-(1,1-dimethyl-2-propenyl)-3-methoxy-1,2-benzenediol, 9CI*
$C_{21}H_{24}O_5$ M 356.418
Isol. from *Cyclobium clausseni* heartwood. Needles. Mp 76-79°. Assigned the (R)- config. Not interrelated with Unanisoflavan.

Gottleib, O.R. *et al, Phytochemistry*, 1975, **14**, 2495 (*deriv*)
Minhaj, N. *et al, Tetrahedron Lett.*, 1976, 2391 (*isol, struct*)

1,2-Undecadiene, 9CI U-80003

[56956-46-8]

$$H_3C(CH_2)_7CH=C=CH_2$$

$C_{11}H_{20}$ M 152.279
Liq. d$_4^{22}$ 0.773. Bp$_{13}$ 84-85°. n$_D^{22}$ 1.4443.

Moreau, J.L. *et al, J. Organomet. Chem.*, 1976, **108**, 159 (*synth, ir, raman*)
Michelot, D. *et al, Tetrahedron Lett.*, 1976, 275 (*use*)
Buechi, G. *et al, J. Org. Chem.*, 1979, **44**, 4116 (*synth, ir, pmr*)
Hana, S. *et al, Chem. Lett.*, 1984, 345 (*use*)
Searles, S. *et al, J. Chem. Soc., Perkin Trans. 1*, 1984, 747 (*synth, ir, pmr, ms*)
Colas, Y. *et al, Bull. Soc. Chim. Fr.*, 1987, 165 (*synth, ir, raman*)

1,3-Undecadiene, 9CI U-80004

[54826-10-7]

$$H_2C=CHCH=CH(CH_2)_6CH_3$$

$C_{11}H_{20}$ M 152.279
(E)-*form* [79309-74-3]
Bp$_{40}$ 135°.

Tsai, D.J.S. *et al, Tetrahedron Lett.*, 1981, **22**, 2751 (*synth, pmr*)
Jabri, N. *et al, Bull. Soc. Chim. Fr., Part 2*, 1983, 321 (*synth, ir, pmr, cmr*)
Ishibashi, H. *et al, Tetrahedron Lett.*, 1985, **26**, 5791 (*synth*)
Yamada, S. *et al, J. Org. Chem.*, 1986, **51**, 4934 (*synth*)
Ishibashi, H. *et al, J. Chem. Res. (S)*, 1987, 296 (*use*)

1,4-Undecadiene, 9CI U-80005

[53786-93-9]

$$H_3C(CH_2)_5CH=CHCH_2CH=CH_2$$

$C_{11}H_{20}$ M 152.279
Constituent of ginseng oil (*Panax ginseng*) and of Chinese traditional medicine, Gaoben (*Ligusticum* sp.).

(E)-*form* [55976-13-1]
Bp$_{0.2}$ 37-38°. n$_D^{20}$ 1.4385.

(Z)-*form* [55976-14-2]
Used in synthetic flavour components. Bp$_{10}$ 66-67°.

Lynd, R.A. *et al, Synthesis*, 1974, 658 (*synth*)
Eisch, J.J. *et al, J. Org. Chem.*, 1976, **41**, 2214 (*synth, ir, pmr*)
Uchida, K. *et al, J. Org. Chem.*, 1976, **41**, 2215 (*synth, ir, pmr, ms*)
Bosshardt, H. *et al, Helv. Chim. Acta*, 1980, **63**, 2393 (*synth, use, ir, pmr, ms*)
Kim, K.R. *et al, Chromatographia*, 1982, **15**, 559 (*ms*)
Hoshi, M. *et al, Bull. Chem. Soc. Jpn.*, 1983, **56**, 2855 (*synth, pmr, ms, ir*)
Arase, A. *et al, Bull. Chem. Soc. Jpn.*, 1984, **57**, 209 (*synth, use, pmr, ir*)

1,5-Undecadiene, 9CI U-80006

$$H_3C(CH_2)_4CH=CHCH_2CH_2CH=CH_2$$

$C_{11}H_{20}$ M 152.279
Liq. d$_4^{20}$ 0.764. Bp$_{10}$ 67°. n$_D^{20}$ 1.4381. Mixt. of E- and Z- forms.

Prevost, C. *et al, Bull. Soc. Chim. Fr.*, 1964, 2485 (*synth*)
Rossi, R. *et al, Tetrahedron Lett.*, 1986, **27**, 2529 (*synth*)

1,6-Undecadiene, 9CI U-80007

$$H_3C(CH_2)_3CH=CH(CH_2)_3CH=CH_2$$

$C_{11}H_{20}$ M 152.279
Liq. Bp$_{10}$ 73-76°. Mixt. of Z- and E- isomers.

[71309-05-2, 91914-03-3]

Zakharkin, L.I. *et al, Izv. Akad. Nauk SSSR, Ser. Khim.*, 1979, 1400; *CA*, **91**, 123342z (*synth*)
Barluenga, J. *et al, J. Chem. Res. (S)*, 1984, 122 (*synth, ir, pmr*)

1,10-Undecadiene, 9CI   U-80008

[13688-67-0]

$$H_2C=CH(CH_2)_7CH=CH_2$$

$C_{11}H_{20}$ M 152.279

Mobile liq. d_4^{20} 0.767. Bp_{756} 187°, Bp_{100} 120°. n_D^{20} 1.4352.

Casteignau, G. *et al*, *Bull. Soc. Chim. Fr.*, 1968, 3893, 3904 (*synth, glc, struct*)
Shono, T. *et al*, *Chem. Lett.*, 1978, 69 (*synth*)
Reiner, E. *et al*, *Biomed. Mass Spectrom.*, 1979, **6**, 491 (*ms, glc*)
Cruz de Maldonado, V. *et al*, *Synth. Commun.*, 1983, **13**, 1163 (*synth*)

2,3-Undecadiene, 9CI   U-80009

[57648-44-9]

$$H_3CCH=C=CH(CH_2)_6CH_3$$

$C_{11}H_{20}$ M 152.279

(±)-*form*

Liq. d_4^{22} 0.772. Bp_{11} 80°. n_D^{22} 1.4560.

Moreau, J.L. *et al*, *J. Organomet. Chem.*, 1976, **108**, 159 (*synth, ir*)
Barluenga, J. *et al*, *J. Chem. Soc., Chem. Commun.*, 1985, 203 (*synth, ir, raman*)

2,4-Undecadiene, 9CI   U-80010

[59742-41-5]

$$H_3CCH=CHCH=CH(CH_2)_5CH_3$$

$C_{11}H_{20}$ M 152.279

[66717-31-5, 66717-32-6, 66717-34-8, 66717-37-1]

Auclair, F. *et al*, *Bull. Soc. Chim. Fr.*, 1976, 142 (*synth, ir, uv, pmr*)
Dang, H.P. *et al*, *Tetrahedron Lett.*, 1978, 191 (*synth*)
Ukai, J. *et al*, *Tetrahedron Lett.*, 1984, **25**, 5173 (*synth*)
Herve du Penhoat, C. *et al*, *Tetrahedron*, 1986, **42**, 4807 (*synth, ms, pmr, ir*)
Umano, K. *et al*, *J. Agric. Food Chem.*, 1987, **35**, 14 (*glc*)
Ikeda, Y. *et al*, *Tetrahedron*, 1987, **43**, 731 (*synth, pmr*)
Cuvigny, T. *et al*, *Tetrahedron*, 1987, **43**, 859 (*synth*)

2,5-Undecadiene, 9CI   U-80011

[76345-96-5]

$$H_3C(CH_2)_4CH=CHCH_2CH=CHCH_3$$

$C_{11}H_{20}$ M 152.279

Liq. Bp_{15} 67-73°. Mixt. of geom. isomers.

[72820-71-4]

Alexakis, A. *et al*, *Synthesis*, 1979, 826 (*synth, ir, pmr*)
Gaspanoli, A. *et al*, *Riv. Ital. Sostanze Grasse*, 1985, **62**, 91; *CA*, **103**, 140446v (*ms, glc*)

2,9-Undecadiene, 8CI   U-80012

[22057-21-2]

$$H_3CCH=CH(CH_2)_5CH=CHCH_3$$

$C_{11}H_{20}$ M 152.279

Casteignau, G. *et al*, *Bull. Soc. Chim. Fr.*, 1968, 3904 (*synth, ir, glc*)

3,5-Undecadiene, 9CI   U-80013

$$H_3CCH_2CH=CHCH=CH(CH_2)_4CH_3$$

$C_{11}H_{20}$ M 152.279

Oil. $Bp_{0.01}$ 32°. Mixt. of stereoisomers.

[78500-37-5, 78500-38-6, 78500-39-7, 86593-86-4]

Jabri, N. *et al*, *Bull. Soc. Chim. Fr., Part II*, 1983, 332 (*synth, ir, pmr, cmr*)
Bloch, R. *et al*, *Tetrahedron Lett.*, 1983, **24**, 1247 (*synth*)
Einhorn, J. *et al*, *Tetrahedron Lett.*, 1985, **26**, 1445 (*ms*)

4,6-Undecadiene, 9CI   U-80014

[50284-90-7]

$$H_3C(CH_2)_3CH=CHCH=CHCH_2CH_2CH_3$$

$C_{11}H_{20}$ M 152.279

Liq. Bp 46-48°. n_D^{26} 1.4618.

Negishi, E. *et al*, *J. Chem. Soc., Chem. Commun.*, 1973, 606 (*synth*)

4,7-Undecadiene, 9CI   U-80015

[104645-53-6]

$$H_3CCH_2CH_2CH=CHCH_2CH=CHCH_2CH_2CH_3$$

$C_{11}H_{20}$ M 152.279

(*Z,Z*)-*form* [43040-05-7]

Bp_{16} 80-81.5°.

Clements, A.H. *et al*, *J. Am. Oil Chem. Soc.*, 1973, **50**, 325 (*synth, ir, pmr*)
Carless, H.A.J. *et al*, *J. Chem. Soc., Perkin Trans. 1*, 1987, 1999 (*synth, pmr*)

5,6-Undecadiene, 9CI   U-80016

1,3-Dibutylallene

[18937-82-1]

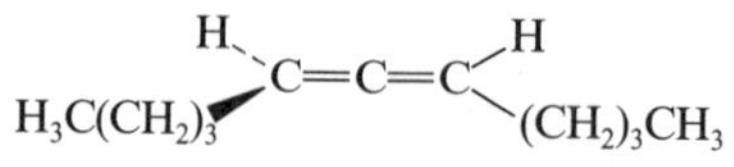

(*R*)-*form*

$C_{11}H_{20}$ M 152.279

(*R*)-*form* [65253-20-5]

$[\alpha]_D^{25}$ −33.4° (c, 7 in $CHCl_3$).

(*S*)-*form* [65253-19-2]

$[\alpha]_D^{25}$ −34.7° (c, 5 in $CHCl_3$).

(±)-*form* [111555-44-3]

Oil. Bp_{15} 70-80°.

Bohlmann, F. *et al*, *Chem. Ber.*, 1967, **100**, 3706 (*synth, ms, ir, pmr*)
Wiensig, J.R. *et al*, *J. Am. Chem. Soc.*, 1977, **99**, 532 (*ms, synth*)
Pirkle, W.H. *et al*, *J. Org. Chem.*, 1978, **43**, 1950 (*synth*)
Kostikov, R.R. *et al*, *Zh. Prikl. Spektrosk.*, 1983, **39**, 486; *CA*, **99**, 186173f (*cmr*)
Fujisawa, T. *et al*, *Tetrahedron Lett.*, 1984, **25**, 4007 (*synth*)
Alexakis, A. *et al*, *Tetrahedron Lett.*, 1985, **26**, 4197 (*synth*)
Marek, I. *et al*, *Tetrahedron Lett.*, 1986, **27**, 5499 (*synth*)
Colas, Y. *et al*, *Bull. Soc. Chim. Fr.*, 1987, 165 (*synth, ir, raman*)

1,3-Undecadien-5-yne, 9CI   U-80017

$$H_3C(CH_2)_4C\equiv CCH=CHCH=CH_2$$

$C_{11}H_{16}$ M 148.247

(*E*)-*form* [56318-81-1]

Liq. $Bp_{2.5}$ 76-77°.

(*Z*)-*form* [56318-73-1]

Obt. by glc.

Näf, F. *et al*, *Helv. Chim. Acta*, 1975, **58**, 1016 (*synth, pmr, ms*)
Andreini, B.P. *et al*, *Tetrahedron*, 1987, **43**, 4591 (*synth, pmr, ir*)

1,5-Undecadien-3 yne U-80018

$$H_3C(CH_2)_4CH=CHC\equiv CCH=CH_2$$

$C_{11}H_{16}$ M 148.247

(*E*)-*form*

Liq. Bp_4 79-80°.

Andreini, B.P. *et al, Tetrahedron*, 1987, **43**, 4591 (*synth, pmr, ir, ms*)

4,7-Undecadiyne, 9CI U-80019

[115410-63-4]

$$H_3CCH_2CH_2C\equiv CCH_2C\equiv CCH_2CH_2CH_3$$

$C_{11}H_{16}$ M 148.247

Liq. Bp_5 74-82°.

Clements, A.H. *et al, J. Am. Oil Chem. Soc.*, 1973, **50**, 325 (*synth, ir*)

Carless, H.A.J. *et al, J. Chem. Soc., Perkin Trans.* 1, 1987, 1999 (*synth, ir, pmr, cmr*)

2,4-Undecadiyn-1-ol U-80020

$$H_3C(CH_2)_5C\equiv CC\equiv CCH_2OH$$

$C_{11}H_{16}O$ M 164.247

Oil.

Kerdesky, F.A.J. *et al, J. Med. Chem.*, 1987, **30**, 1177 (*synth, pmr*)

6-Undecen-1-ol U-80021

$$H_3C(CH_2)_3CH=CH(CH_2)_4CH_2OH$$

$C_{11}H_{22}O$ M 170.294

(*E*)-*form* [16695-37-7]

Liq. $Bp_{0.6}$ 84-85°.

(*Z*)-*form* [68760-60-1]

Liq. $Bp_{0.01}$ 78-80°.

Green, N. *et al, J. Med. Chem.*, 1967, **10**, 533.

Svirskaya, P.I. *et al, J. Chem. Eng. Data*, 1979, **24**, 152.

Brown, H.C. *et al, J. Org. Chem.*, 1988, **53**, 246 (*synth, ir, pmr, cmr*)

10-Undecen-2-one, 9CI U-80022

[36219-73-5]

$$H_2C=CH(CH_2)_7COCH_3$$

$C_{11}H_{20}O$ M 168.278

Constit. of *Pinus mugo* essential oil. Bp 235°.

Cottier, L. *et al, Bull. Soc. Chim. Fr.*, 1972, 1072 (*synth*)

Bambagiotti, A.M. *et al, Planta Med.*, 1981, **43**, 39 (*isol*)

Ashby, E.C. *et al, Organometallics*, 1984, **3**, 1718 (*synth, pmr*)

5-Undecen-2-one U-80023

$$H_3C(CH_2)_4CH=CHCH_2CH_2COCH_3$$

$C_{11}H_{20}O$ M 168.278

(*Z*)-*form* [21944-96-7]

Pheromone from the pedal gland of the bontebok *Damaliscus dorcas dorcas*.

Yamashita, M. *et al, Bull. Chem. Soc. Jpn.*, 1988, **61**, 3368 (*synth, pmr, bibl*)

1-Undecyn-5-ol, 9CI U-80024

[72957-73-4]

$$H_3C(CH_2)_5CH(OH)CH_2CH_2C\equiv CH$$

$C_{11}H_{20}O$ M 168.278

(±)-*form*

Oil. Bp_2 140°.

Subramaniam, C.S. *et al, J. Chem. Soc., Perkin Trans.* 1, 1979, 2346 (*synth, uv, pmr, use*)

3-Undecyn-1-ol, 9CI U-80025

[54299-09-1]

$$H_3C(CH_2)_6C\equiv CCH_2CH_2OH$$

$C_{11}H_{20}O$ M 168.278

Oil. $Bp_{0.25}$ 79-80°.

Holland, B.C. *et al, Chem. Phys. Lipids*, 1974, **13**, 239.

Gershon, H. *et al, J. Pharm. Sci.*, 1985, **74**, 556.

4-Undecyn-1-ol, 9CI U-80026

[54377-35-4]

$$H_3C(CH_2)_5C\equiv CCH_2CH_2CH_2OH$$

$C_{11}H_{20}O$ M 168.278

Oil. Bp_{20} 133-140°, $Bp_{0.03}$ 80°. n_D^{25} 1.4586.

4-Methylbenzensulfonyl: Thick oil.

Ames, D.E. *et al, J. Chem. Soc.*, 1963, 5889 (*synth*)

Lie Ken Jie, M.S.F. *et al, J. Chromatogr.*, 1974, **97**, 165 (*ester, synth, glc*)

Svirskaya, P.A. *et al, J. Chem. Eng. Data*, 1979, **24**, 152 (*synth, ir, pmr*)

Reddy, P.S. *et al, Synth. Commun.*, 1984, **14**, 327 (*synth, pmr*)

Vig, O.P. *et al, Indian J. Chem., Sect. B*, 1986, **25**, 1042 (*synth, ir, pmr*)

Kabalka, G.W. *et al, J. Org. Chem.*, 1986, **51**, 2386 (*deriv, synth, pmr, cmr*)

4-Undecyn-2-ol, 9CI U-80027

[95061-41-9]

$$H_3C(CH_2)_5C\equiv CCH_2CH(OH)CH_3$$

$C_{11}H_{20}O$ M 168.278

Oil. $Bp_{0.2}$ 65-72°.

Utimoto, K. *et al, Tetrahedron Lett.*, 1984, **25**, 5423 (*synth*)

Oehlschlager, A.C. *et al, Can. J. Chem.*, 1986, **64**, 1407 (*synth, use*)

5-Undecyn-1-ol, 9CI U-80028

[69222-09-9]

$$H_3C(CH_2)_4C\equiv C(CH_2)_3CH_2OH$$

$C_{11}H_{20}O$ M 168.278

Oil. $Bp_{0.01}$ 83-84°. n_D^{25} 1.4586.

Lie Ken Jie, M.S.F. *et al, J. Chromatogr.*, 1974, **97**, 152 (*synth, glc*)

Svirskaya, P.A. *et al, J. Chem. Eng. Data*, 1979, **24**, 152 (*synth*)

5-Undecyn-2-ol, 9CI U-80029

[92051-76-8]

$$H_3C(CH_2)_4C\equiv CCH_2CH_2CH(OH)CH_3$$

$C_{11}H_{20}O$ M 168.278

(±)-*form*

Oil. Bp_1 110°.

Reddy, P.S. *et al, Synth. Commun.*, 1984, **14**, 327 (*synth, pmr*)

6-Undecyn-5-ol, 9CI — U-80030
[73252-74-1]

$$H_3C(CH_2)_3C{\equiv}CCH(OH)(CH_2)_3CH_3$$

$C_{11}H_{20}O$ M 168.278

(±)-form [111462-96-5]
Oil. Bp$_1$ 65°.

Overman, L.E. et al, J. Am. Chem. Soc., 1980, 102, 747.
Colas, Y. et al, Bull. Soc. Chim. Fr., 1987, 165 (synth, ir)

7-Undecyn-1-ol, 9CI — U-80031
[69222-11-3]

$$H_3CCH_2CH_2C{\equiv}C(CH_2)_5CH_2OH$$

$C_{11}H_{20}O$ M 168.278
Oil. Bp$_{0.04}$ 84-86°. n_D^{24} 1.4600.

Svirskaya, P.A. et al, J. Chem. Eng. Data, 1979, 24, 152 (synth, use, ir, pmr)

10-Undecyn-2-ol, 9CI — U-80032
[104903-89-1]

$$HC{\equiv}C(CH_2)_7CH(OH)CH_3$$

$C_{11}H_{20}O$ M 168.278

Me ether: 10-Methoxy-1-undecyne
$C_{12}H_{22}O$ M 182.305
Isol. from ethereal oil of Litsea odorifera. Oil. d^{20} 0.824. n_D^{20} 1.4400.

Mathews, W.J. et al, Chem. Ind. (London), 1963, 122 (isol, deriv)
Oehlschlager, A.C. et al, Can. J. Chem., 1986, 64, 1407 (synth, use)

Urorosein — U-80033
(3H-Indol-3-ylidenemethyl)-1H-indole(1+), 9CI. 3-Indol-3-ylmethylene-3H-indol-3-ium. Di-3-indolylcarbenium
[526-32-9]

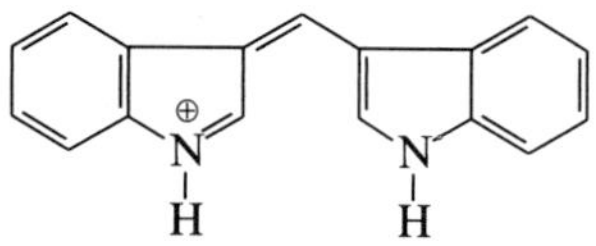

$C_{17}H_{13}N_2^{\oplus}$ M 245.303
Pigment.

Chloride:
$C_{17}H_{13}ClN_2$ M 280.756
Deep red needles (EtOH/Me$_2$CO).

Bromide:
$C_{17}H_{13}BrN_2$ M 325.207
Deep red prisms (EtOH).

Perchlorate:
$C_{17}H_{13}ClN_2O_4$ M 344.753
Red cryst. with greenish metallic lustre. Carbonises without melting.

Tetrafluoroborate: [102652-01-7].
$C_{17}H_{13}BF_4N_2$ M 332.108
Orange-yellow cryst. Mp 225°.

Harley-Mason, J. et al, Biochem. J., 1952, 51, 430 (synth)
Smith, G.F., J. Chem. Soc., 1954, 3842 (synth)
Mueller, J. et al, Monatsh. Chem., 1985, 116, 365 (analogues)
Pinder, U. et al, Justus Liebigs Ann. Chem., 1986, 1621 (synth, ir, uv)

19(29)-Ursen-3-ol — U-80034

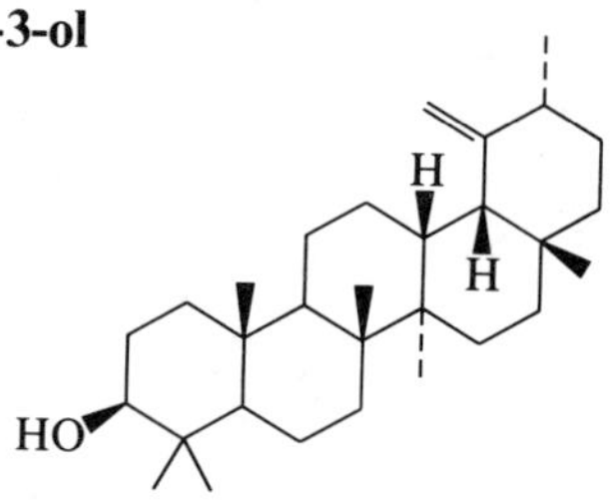

$C_{30}H_{50}O$ M 426.724

3β-form [117895-04-2] Calotropenol
Needles (CHCl$_3$/C$_6$H$_6$). Mp 168°. [α]$_D$ +37.26° (c, 0.161 in CHCl$_3$).

Ac: [118013-26-6]. 3β-Acetoxy-19(29)-ursene. Calotropenyl acetate
$C_{32}H_{52}O_2$ M 468.762
Constit. of Calotropis procera. Cryst. Mp 198°. [α]$_D$ +8.9° (c, 0.415 in CHCl$_3$).

Khan, A.Q. et al, J. Nat. Prod. (Lloydia), 1988, 51, 925.

Uvaretin — U-80035
Updated Entry replacing U-50017
1-[2,4-Dihydroxy-3-[(2-hydroxyphenyl)methyl]-6-methoxyphenyl]-3-phenyl-1-propanone, 9CI. Chamuvarin
[58449-06-2]

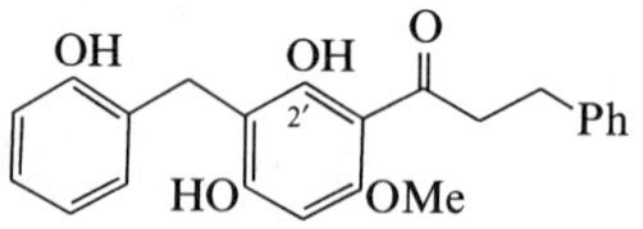

$C_{23}H_{22}O_5$ M 378.424
Isol. from Uvaria acuminata, U. chamae and Croton californicus. Cryst. (MeCN or C$_6$H$_6$). Mp 164-165°.

O-De-Me, 2'-O-Me: [61463-03-4]. Isouvaretin
$C_{23}H_{22}O_5$ M 378.424
Constit. of U. chamae and U. angolensis. Cryst. (CHCl$_3$/hexane). Mp 80-82°.

Cole, J.R. et al, J. Org. Chem., 1976, 41, 1852 (isol, struct)
Lasswell, W.L. et al, J. Org. Chem., 1977, 42, 1295 (isol)
Tammani, B. et al, Phytochemistry, 1977, 16, 2040 (isol)
Hufford, C.D. et al, Phytochemistry, 1980, 19, 2036 (isol, cmr, Uvaretin, Isouvaretin)
Jain, A.C. et al, Indian J. Chem., Sect. B, 1985, 24, 1015 (synth)

Uvarisesquiterpene A — U-80036
[117176-62-2]

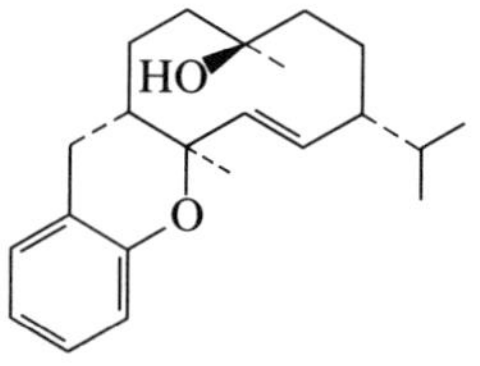

$C_{22}H_{32}O_2$ M 328.494
Constit. of Uvaria angolensis. Amorph. [α]$_D$ -16° (c, 0.12 in CHCl$_3$).

Muhammad, I. et al, J. Nat. Prod. (Lloydia), 1988, 51, 719.

Uvarisesquiterpene B

U-80037

[117176-79-1]

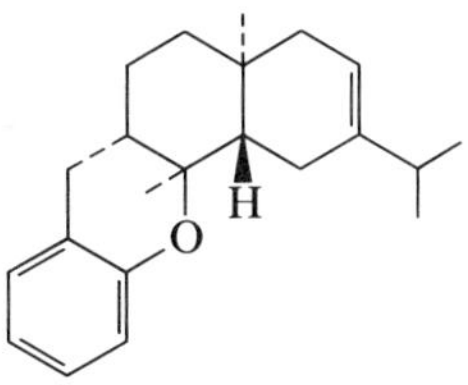

$C_{22}H_{30}O$ M 310.478

Constit. of *Uvaria angolensis*. Amorph.

Muhammad, I. *et al, J. Nat. Prod.* (*Lloydia*), 1988, **51**, 719.

Uvarisesquiterpene C

U-80038

[117176-80-4]

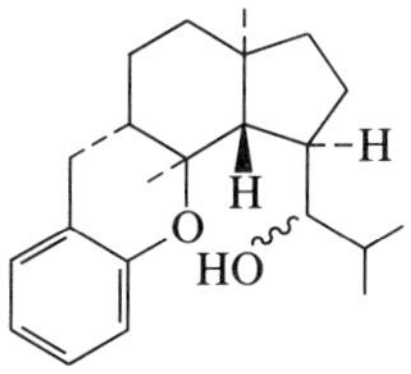

$C_{22}H_{32}O_2$ M 328.494

Constit. of *Uvaria angolensis*. Amorph.

Muhammad, I. *et al, J. Nat. Prod.* (*Lloydia*), 1988, **51**, 719.

V

Valerenenol   V-80001

Updated Entry replacing V-30001

2-Methyl-1-(octahydro-3,7-dimethyl-4H-inden-1-ylidene)-2-propanol, 9CI

[84249-43-4]

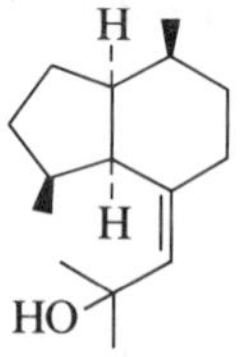

Absolute configuration

$C_{15}H_{26}O$ M 222.370

Constit. of soft coral *Xeniidae* sp. Oil. $[\alpha]_D$ +123° (CHCl$_3$).

(E)-*isomer:* [84249-42-3]. **Isovalerenenol**
Constit. of *X.* sp. Semisolid. $[\alpha]_D$ +45° (CHCl$_3$).

Kobayashi, M. *et al, Chem. Pharm. Bull.*, 1982, **30**, 3431 (*isol, pmr, cmr, cryst struct*)

Variabilin†   V-80002

Updated Entry replacing V-70003

5-[13-(3-Furanyl)-2,6,10-trimethyl-6,10-tridecadienylidene]-4-hydroxy-3-methyl-2(5H)-furanone, 9CI

[51847-87-1]

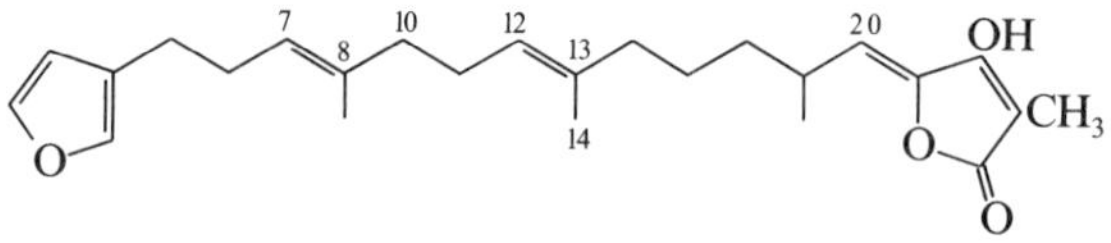

$C_{25}H_{34}O_4$ M 398.541

Isol. from the sponge *Ircinia strobilina* and a *Sarcotragus* sp. Plant phytoalexin with antibiotic props. Oil. $[\alpha]_D$ −4° (c, 1 in CHCl$_3$). λ_{max} 255 nm (MeOH).

20E-Isomer: [114761-90-9].
$C_{25}H_{34}O_4$ M 398.541
Constit. of a *S.* sp. Oil.

Δ^{13}-*Isomer (Z-):* [114728-09-5].
$C_{25}H_{34}O_4$ M 398.541
Constit. of a *S.* sp. Oil. $[\alpha]_D$ −4.4° (c, 0.9 in CHCl$_3$).

7,8-Dihydro, 8-hydroxy: [82124-11-6]. **8-Hydroxyvariabilin**
$C_{25}H_{36}O_5$ M 416.556
Constit. of a *S.* sp. Oil. $[\alpha]_D$ −24.5° (c, 0.22 in MeOH).

5-Hydroxy: **5-Hydroxyvariabilin**
$C_{25}H_{34}O_5$ M 414.541
Constit. of an *Ircinia* sp.

5-Oxo: **5-Oxovariabilin**
$C_{25}H_{32}O_5$ M 412.525
Constit. of an *I.* sp.

5-Oxo, $\Delta^{8(10)}$-isomer(E-): **5-Oxo-8(10)E-variabilin**
$C_{25}H_{32}O_5$ M 412.525
Constit. of an *I.* sp.

5-Oxo, $\Delta^{8(10)}$(E-), Δ^{13}-isomer:
$C_{25}H_{32}O_5$ M 412.525
Constit. of an *I.* sp.

Faulkner, D.J., *Tetrahedron Lett.*, 1973, 3821 (*isol, uv, ir, nmr, struct*)
Rothberg, I. *et al, Tetrahedron Lett.*, 1975, 769 (*isol, ir, uv, nmr, struct*)
Barrow, C.J. *et al, J. Nat. Prod. (Lloydia)*, 1988, **51**, 275, 1294 (*isol, deriv, pmr, cmr*)

Variegatorubin   V-80003

3-[4-(3,4-Dihydroxyphenyl)-3-hydroxy-5-oxo-2-(5H)furanylidene]-5,6-dihydroxy-2(3H)-benzofuranone, 9CI

[27286-59-5]

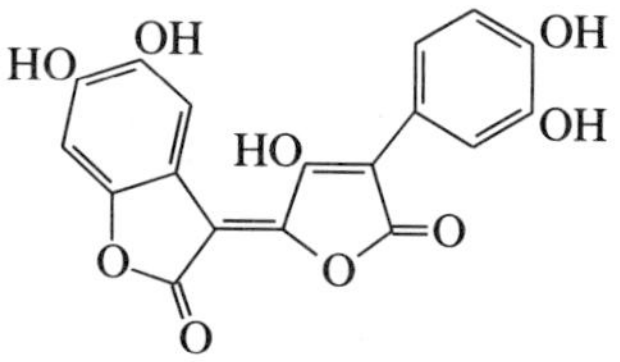

$C_{18}H_{10}O_9$ M 370.272

Isol. from *Suillus piperatus* and *Rhizopogon roseolus*. Brown-violet fine needles (MeCN). Mp >320° dec.

Steglich, W. *et al, Z. Naturforsch., B*, 1970, **25**, 557; 1971, **26**, 376 (*isol, struct*)

Variolaric acid   V-80004

4,7-Dihydroxy-9-methyl-10H-isobenzofuro[5,6-b][1,4]benzodioxepin-3,10(1H)-dione, 9CI. Ochrolechic acid

[490-34-6]

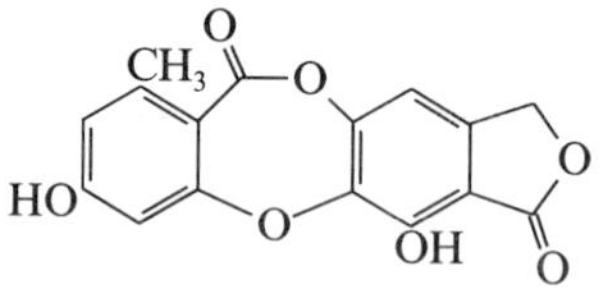

$C_{16}H_{10}O_7$ M 314.251

Lichen acid from *Lecanora rupicola*. Needles (MeOH). Mp 296° dec.

Devlin, J.P. *et al, J. Chem. Soc. C*, 1971, 1318 (*isol, pmr, bibl*)
Rana, N.M. *et al, J. Chem. Soc., Perkin Trans. 1*, 1975, 2992 (*struct, ms, bibl*)
Jongen, R. *et al, J. Chem. Soc., Perkin Trans. 1*, 1979, 2588 (*synth*)

Vertipyronol   V-80005

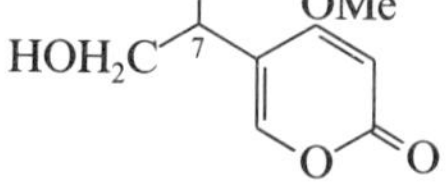

$C_9H_{12}O_4$ M 184.191

Metab. of *Leptographium wageneri*. Oil. $[\alpha]_D$ −1.58° (c, 0.19 in CHCl$_3$).

7-Hydroxy: **Vertipyronediol**
$C_9H_{12}O_5$ M 200.191

Metab. of *L. wageneri*. Cryst. (Me$_2$CO/hexane). Mp 128-129°. [α]$_D$ −50° (c, 0.2 in CHCl$_3$).

Ayer, W.A. *et al*, *J. Nat. Prod.* (*Lloydia*), 1989, **52**, 119 (*isol, ir, pmr, cmr*)

2-Vinylbenzoxazole V-80006

[63359-54-6]

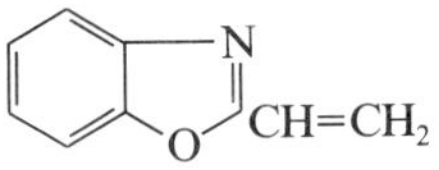

C$_9$H$_7$NO   M 145.160
Bp$_{0.04}$ 75° (kugelrohr) .

Bartsch, H. *et al*, *Justus Liebigs Ann. Chem.*, 1988, 795 (*synth*)

N-Vinylcarbazole V-80007

N-*Ethenylcarbazole*

[1484-13-5]

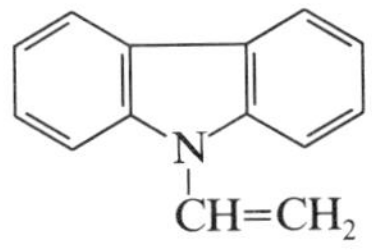

C$_{14}$H$_{11}$N   M 193.248
Important polymerisation monomer. Cryst. (EtOH). Mp 63.5-66°. Bp$_{15}$ 175-178°, Bp$_{0.02}$ 68-70°.

Clemo, G.R. *et al*, *J. Chem. Soc.*, 1924, **125**, 1806 (*synth*)
Reppe, W. *et al*, *Justus Liebigs Ann. Chem.*, 1956, **601**, 81 (*synth*)
Takuze, S. *et al*, *Encycl. Polym. Sci. Technol.*, 1971, **14**, 281.

3-Vinyl-2-cyclopenten-1-ol V-80008

3-*Ethenyl-2-cyclopenten-1-ol, 9CI*

[114614-91-4]

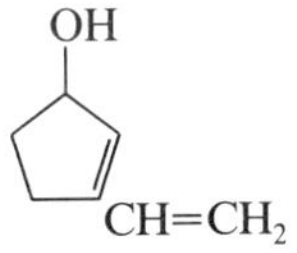

C$_7$H$_{10}$O   M 110.155
(±)-*form*
Golden oil.
Me ether: 3-*Methoxy-1-vinylcyclopentene*
C$_8$H$_{12}$O   M 124.182
Liq. Bp$_{66}$ 87-89°.

[114614-87-8]

Fisher, M.J. *et al*, *J. Am. Chem. Soc.*, 1988, **110**, 4625 (*synth, pmr, cmr, ir, ms*)

3-Vinyl-2-oxiranemethanol V-80009

3-*Ethenyl-2-oxiranemethanol. 2,3-Epoxy-4-penten-1-ol*

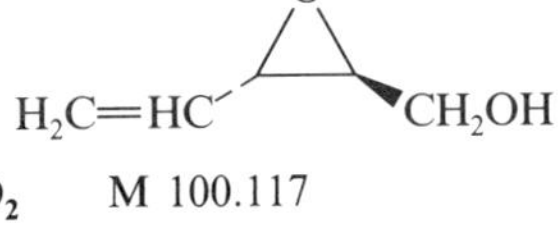

(2S,3S)-form

C$_5$H$_8$O$_2$ M 100.117
(2S,3S)-*form*
Bp$_{5.5}$ 67°. [α]$_D^{23}$ −58.7° (neat). >91% enantiomeric excess.
O-(4-*Methylbenzenesulfonyl*): Solid. Mp 56.5-58°. [α]$_D^{23}$ −33.3° (c, 0.96 in EtOH).

(±)-*form*
Bp$_{18}$ 92°.

Wershofen, S. *et al*, *Synthesis*, 1988, 854 (*synth, ms, ir, pmr, cmr*)

3-Vinylthiophene V-80010

3-*Ethenylthiophene, 9CI*

[13679-64-6]

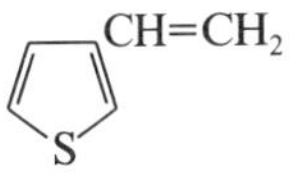

C$_6$H$_6$S   M 110.179
Liq. d$_4^{15}$ 1.08. Bp 160°, Bp$_{20}$ 55-59°.

Troyanowsky, C., *Bull. Soc. Chim. Fr.*, 1955, 424 (*synth*)
Tominaga, Y. *et al*, *J. Heterocycl. Chem.*, 1981, **18**, 967 (*synth, ir, pmr*)
Abarca, B. *et al*, *Tetrahedron*, 1987, **43**, 269 (*synth, props*)

Vitixanthin V-80011

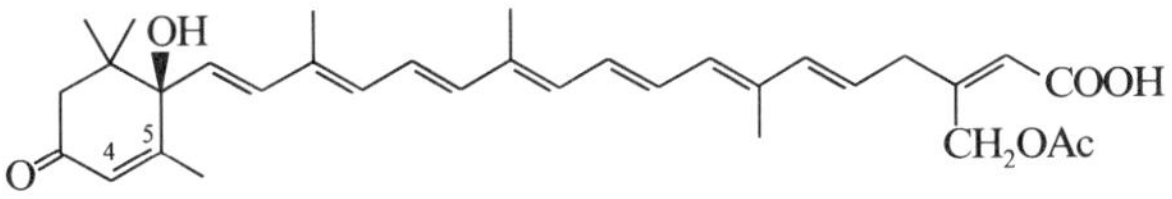

C$_{33}$H$_{42}$O$_6$ M 534.691
Constit. of root of *Cochlospermum vitifolium*.
Me ester: [α]$_D^{20}$ −558° (CHCl$_3$).
4,5-*Dihydro:* **Dihydrovitixanthin**
C$_{33}$H$_{44}$O$_6$ M 536.707
Constit. of *C. vitifolium*.
4,5-*Dihydro, Me ester:* [α]$_D^{20}$ −82° (CHCl$_3$).

Achenbach, H. *et al*, *Tetrahedron Lett.*, 1989, **30**, 3059.

Volubilinin V-80012

8-β-D-*Glucopyranosyl-5,7-dihydroxy-6-methoxy-3-(4-methoxyphenyl)-4H-1-benzopyran-4-one, 9CI. 8-C-Glucosyl-5,7-dihydroxy-4',6-dimethoxyisoflavone. Irisolidone 8-C-glucoside*

[58930-58-8]

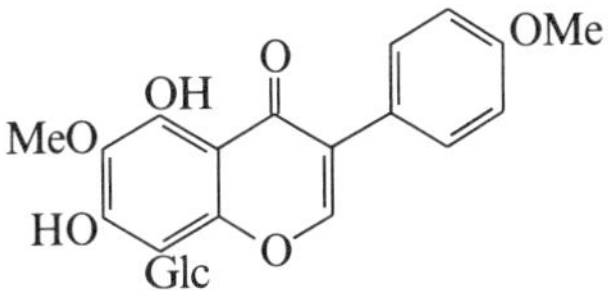

C$_{23}$H$_{24}$O$_{11}$ M 476.436
Isol. from flowers of *Dalbergia volubilis*. Plates (EtOAc/MeOH). Mp 159-161°.

Chawla, H.M. *et al*, *Phytochemistry*, 1976, **15**, 235.

Vulgarin V-80013

Updated Entry replacing V-60017
Tauremisin A. Barrelin. Judaicin
[3162-56-9]

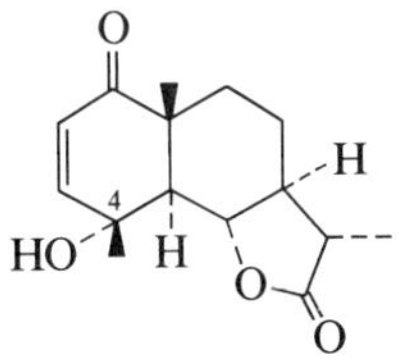

Absolute configuration

C$_{15}$H$_{20}$O$_4$ M 264.321

Constit. of *Artemisia vulgaris, A. judaica* and *A. taurica.*
Oral hypoglycemic agent. Cryst. (EtOH). Mp 174-175°.
$[\alpha]_{546}^{27}$ +48.7° (c, 3.86 in CHCl$_3$).

▷ LE3170000.

4-Epimer: [66289-87-0]. **4-Epivulgarin**
C$_{15}$H$_{20}$O$_4$ M 264.321
From *A. judaica* and *A. canariensis*. Cryst.
(C$_6$H$_6$/hexane)(also descr. as gum). Mp 192-194°. $[\alpha]_D$
+77.5° (c, 5.7 in CHCl$_3$).

4-Deoxy,4α-hydroperoxy: [72505-77-2]. **4α-
Hydroperoxydesoxyvulgarin.** *Peroxyvulgarin*
C$_{15}$H$_{20}$O$_5$ M 280.320
From *A. judaica*. Gum, needles (Me$_2$CO/Et$_2$O). Mp 170-
171°. $[\alpha]_D$ −44° (c, 0.1 in MeOH).

[5091-07-6, 66289-87-0]

Geissman, T.A. *et al, J. Org. Chem.,* 1962, **27**, 1855 (*isol, struct*)
Ando, M. *et al, Bull. Chem. Soc. Jpn.,* 1978, **51**, 283 (*synth*)
González, G. *et al, J. Chem. Soc., Perkin Trans.* 1, 1978, 1243
(*synth*)
Ando, M. *et al, Bull. Chem. Soc. Jpn.,* 1979, **52**, 2737 (*synth*)
Gonzalez, A.G. *et al, Phytochemistry,* 1983, **22**, 1509 (*isol*)
Metwally, M.A. *et al, Phytochemistry,* 1985, **24**, 1103 (*derivs*)
Abagaz, B. *et al, Tetrahedron,* 1986, **42**, 6003 (*cryst struct*)
Arias, J.M. *et al, J. Chem. Soc., Perkin Trans.* 1, 1987, 471 (*synth*)
Al-Said, M.S. *et al, Phytochemistry,* 1989, **28**, 107 (*synth*)

W

Waitziacuminone   W-80001

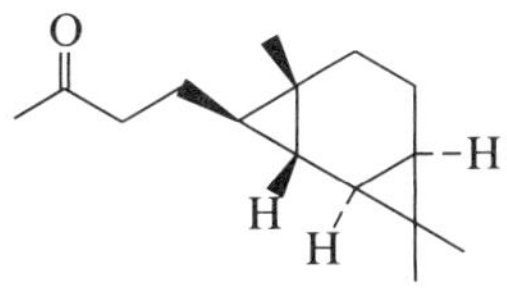

$C_{15}H_{24}O$ M 220.354

Constit. of *Waitzia acuminata*. Oil.

Jakupovic, J. *et al, Phytochemistry*, 1989, **28**, 1943.

Wighteone   W-80002

5,7-Dihydroxy-3-(4-hydroxyphenyl)-6-(3-methyl-2-butenyl)-4H-1-benzopyran-4-one, 9CI. 4′,5,7-Trihydroxy-6-prenyisoflavone. 6-Isopentenylgenistein. Erythrinin B

[51225-30-0]

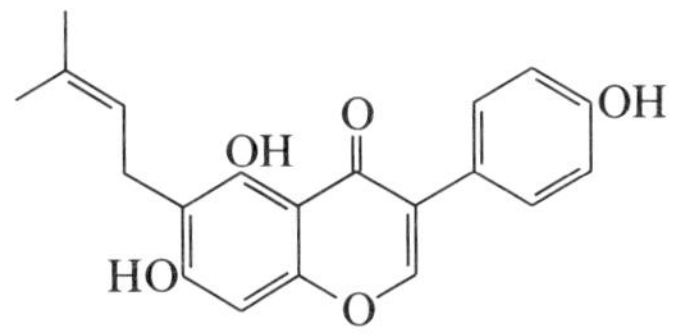

$C_{20}H_{18}O_5$ M 338.359

Isol. from *Argyrocytisus battandieri, Erythrina variegata, Laburnum anagyroides, Lupinus albus, L. angustifolius, L. polyphyllus* and *Neonotonia wightii*. Phytoalexin. Bright-yellow plates (Me_2CO/C_6H_6). Mp 220-221°.

Tri-Ac: Cryst. (MeOH). Mp 160-162°.

Deshpande, V.H. *et al, Indian J. Chem., Sect. B*, 1977, **15**, 205 (*isol, uv, pmr, struct*)

Ingham, J.L. *et al, Phytochemistry*, 1977, **16**, 1943 (*isol, ir, ms, struct*)

Withacoagin   W-80003

5α,20-Dihydroxy-1-oxo-20R,22R-witha-2,6,24-trienolide

[119539-81-0]

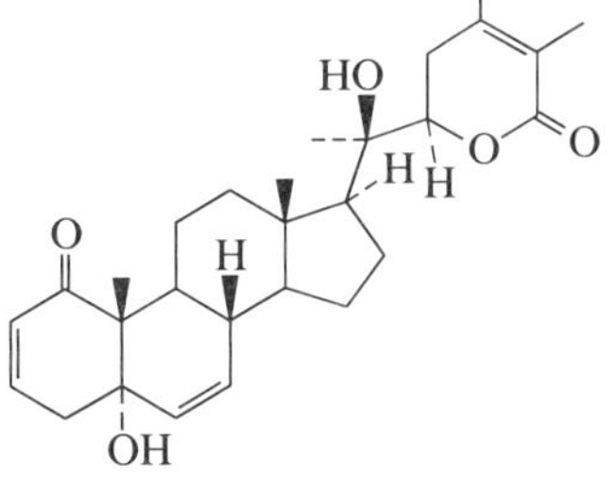

$C_{28}H_{38}O_5$ M 454.605

Constit. of *Withania coagulans*. Cryst. ($CHCl_3$). Mp 230-232°. $[\alpha]_D^{24}$ +114° (c, 0.19 in $CHCl_3$).

Neogi, P. *et al, Bull. Chem. Soc. Jpn.*, 1988, **61**, 4479.

Withametelin   W-80004

[113430-43-6]

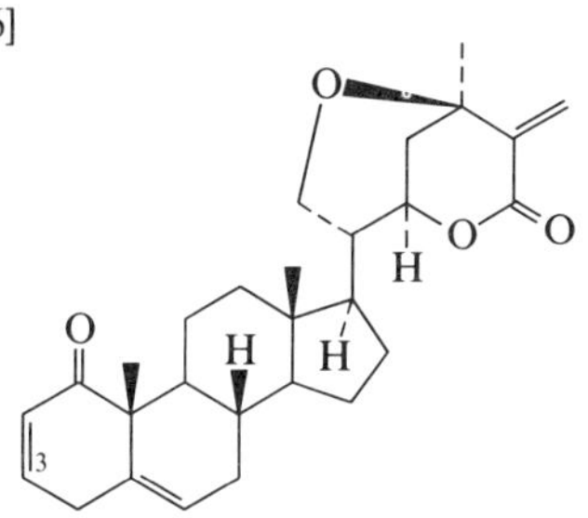

$C_{28}H_{36}O_4$ M 436.590

Constit. of leaves of *Datura metel*. Needles (EtOAc). Mp 210°. $[\alpha]_D$ −64.4° (c, 0.45 in $CHCl_3$).

Δ^3-*Isomer:* [123522-98-5]. ***Isowithametelin***

$C_{28}H_{36}O_4$ M 436.590

Constit. of *D. metel*. Cryst. Mp 280°. $[\alpha]_D$ −77.7° (c, 0.26 in $CHCl_3$).

Sinha, S.C. *et al, Tetrahedron*, 1989, **45**, 2165.

Wrightiin   W-80005

[107783-44-8]

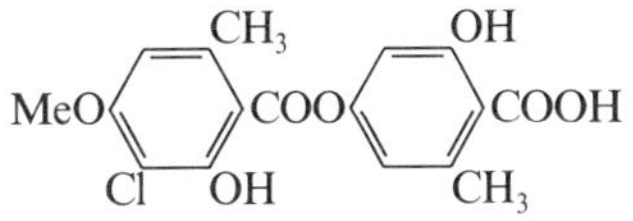

$C_{18}H_{17}ClO_7$ M 380.781

Constit. of *Erioderma wrightii*. Cryst. (EtOAc). Mp 216.5°.

Maass, W.S.G. *et al, Z. Naturforsch., B*, 1986, **41**, 1589 (*isol, cryst, struct*)

Elix, J.A. *et al, Aust. J. Chem.*, 1988, **41**, 1789 (*synth*)

Y

Yadanzioside M   **Y-80001**

[101559-99-3]

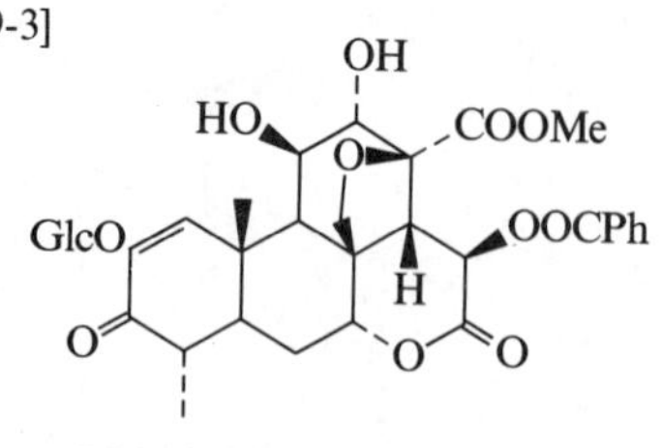

$C_{34}H_{40}O_{16}$ M 704.680

Constit. of *Brucea antidysenterica*. Amorph. powder. Mp 208-213°. $[\alpha]_D^{23}$ +39° (c, 0.41 in MeOH).

Okano, M. *et al*, *J. Nat. Prod.* (*Lloydia*), 1989, **52**, 398 (*isol*)

Z

Zapotecol   Z-80001

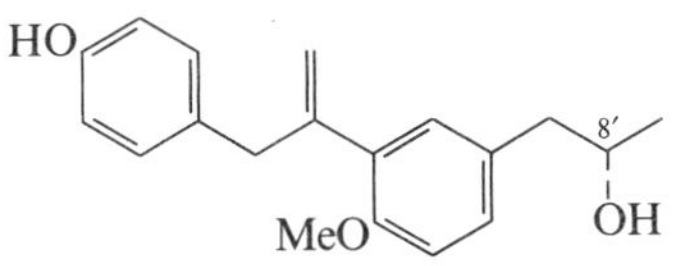

$C_{19}H_{22}O_3$ M 298.381

Constit. of *Krameria lanceolata*. Oil. $[\alpha]_D^{21}$ +5° (c, 0.6 in MeOH).

8'-Ketone: Zapotecone
$C_{19}H_{20}O_3$ M 296.365
Constit. of *K. lanceolata*. Oil.

Achenbach, H. *et al, Phytochemistry*, 1989, **28**, 1959.

Zinaflorin IV   Z-80002

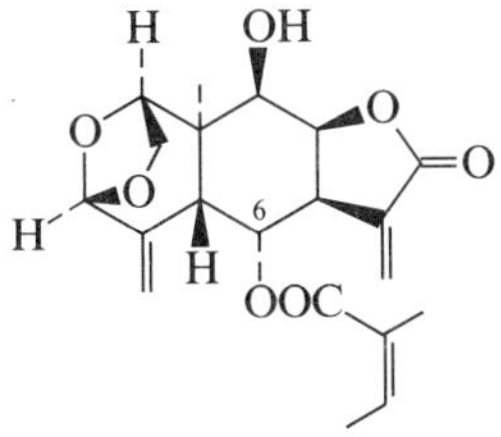

$C_{20}H_{24}O_7$ M 376.405

Also represented in the ring opened form with epoxide and —CHO functions in ring A. Constit. of *Zinnia peruviana*. Cryst. Mp 194-196°. $[\alpha]_D$ +137.5° (c, 0.145 in CHCl$_3$), $[\alpha]_D$ +96.2° (c, 5.1 in EtOH).

O^6-*Deacyl*, O^6-*tigloyl*: [88930-59-0].
$C_{20}H_{24}O_7$ M 376.405
Constit. of *Z. peruviana*.

O^6-*Deacyl*, O^6-*(2-methylpropenoyl)*: [88930-72-7].
$C_{19}H_{22}O_7$ M 362.379
Constit. of *Z. peruviana*.

Herz, W. *et al, Phytochemistry*, 1981, **21**, 2229 (*isol*)
Ortega, A. *et al, Chem. Lett.*, 1983, 1607 (*cryst struct*)
Mirand, R. *et al, J. Nat. Prod.* (*Lloydia*), 1989, **52**, 1128 (*isol, pmr*)

Zosterdiol A   Z-80003

[120163-23-7]

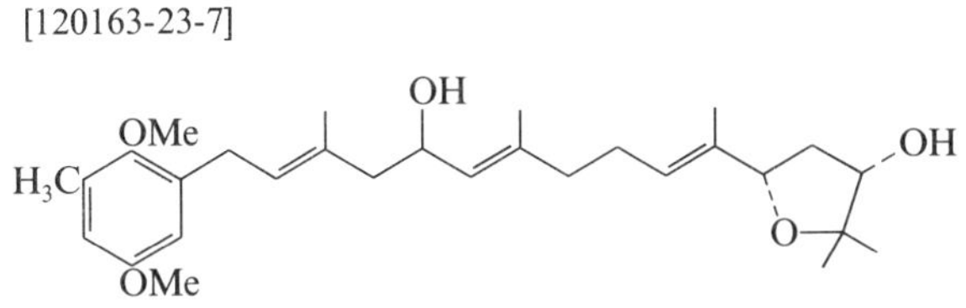

$C_{29}H_{44}O_5$ M 472.664
Constit. of *Cystoseira zosteroides*. Oil. $[\alpha]_D^{20}$ +0.5° (c, 2 in EtOH).

Amico, V. *et al, Phytochemistry*, 1989, **28**, 215.

Zosterdiol B   Z-80004

[120163-24-8]

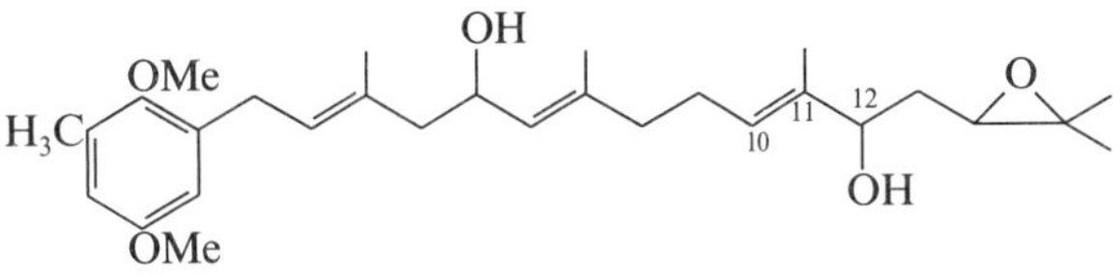

$C_{29}H_{44}O_5$ M 472.664
Constit. of *Cystoseira zosteroides*. Oil. $[\alpha]_D^{20}$ +1.1° (c, 0.9 in EtOH).

10,11-Dihydro 12-ketone: [120163-25-9]. *Zosteronol*
$C_{29}H_{44}O_5$ M 472.664
Constit. of *C. zosteroides*. Oil. $[\alpha]_D^{20}$ +1.5° (c, 0.6 in EtOH).

Amico, V. *et al, Phytochemistry*, 1989, **28**, 215.

Zosterondiol B   Z-80005

[120181-10-4]

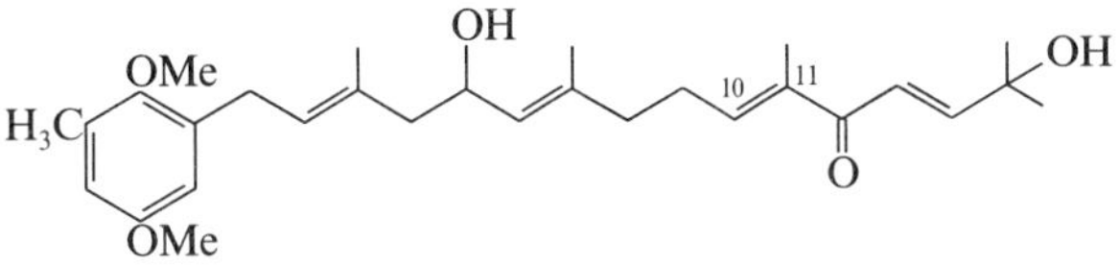

$C_{29}H_{42}O_5$ M 470.648
Constit. of *Cystoseira zosteroides*. Oil. $[\alpha]_D^{20}$ +0.9° (c, 0.4 in EtOH).

10,11-Dihydro: [120181-09-1]. *Zosterondiol A*
$C_{29}H_{44}O_5$ M 472.664
Constit. of *C. zosteroides*. Oil. $[\alpha]_D^{20}$ +0.7° (c, 0.8 in EtOH).

Amico, V. *et al, Phytochemistry*, 1989, **28**, 215.

Name Index

Name Index

Name Index

1-(3,4-Dihydro-2*H*-pyran-5-yl)-2,2,2-... – 11,12-Dihydroxy-5,8,11,13-...

ent-10β,13-Dihydroxy-20-nor-16-... – 1-(3,5-Dihydroxyphenyl)-2-...

Name Index

Name Index

1-(2,5-Dihydroxyphenyl)-2-... − 5,8-Dihydroxy-2,3,6-trimethyl-1,4-...

Ethyl diformylacetate, *in* D-60190
3-*O*-Ethyldihydrofusarubin *B*, *in* F-60108
3-*O*-Ethyldihydrofusarubin *A*, *in* F-60108
6-Ethyl-2,3-dihydro-2-methyl-4*H*-pyran-4-one, E-**5**0071
14-Ethyl-16,17-dihydro-8,9,13,24-tetramethyl-5,22:12,15-diimino-20,18-metheno-7,10-nitrilobenzo[*o*]cyclopent[*b*]azacyclononadecine, *see* E-60043
3-Ethyl-5,7-dihydroxy-4*H*-1-benzopyran-4-one, E-70040
3-Ethyl-5,7-dihydroxychromone, *see* E-70040
6-Ethyl-2,4-dihydroxy-3-methylbenzoic acid, E-70041
3-Ethyl-2,19-dioxabicyclo[16.3.1]docosa-3,6,9,18(22),21-pentaen-12-yn-20-one, E-70042
2-Ethyl-1,6-dioxaspiro[4.4]nonane, E-70043
▷ Ethylenebis(oxyethylenenitrilo)tetraacetic acid, *see* B-70134
Ethylenedibenzoic acid (*misleading*), *see* B-80071
Ethylenedihydrazine, *see* D-60192
3,4-Ethylenedioxybenzaldehyde, *see* B-60023
3,4-Ethylenedioxybenzoic acid, *see* B-60025
2,3-Ethylenedioxy-1,4-benzoquinone, *see* D-80174
3,3-(Ethylenedioxy)pentanedioic acid, *in* O-60084
2,2'-(Ethylenedithio)diethanol, *see* B-70145
▷ Ethylene glycol bis(2-aminoethyl ether)-*N*-tetraacetic acid, *see* B-70134
▷ Ethylene glycol bis[2-[bis(carboxymethyl)amino]ethyl] ether, *see* B-70134
Ethylenemalonic acid, *see* C-80196
Ethylene 1,3-propylene orthocarbonate, *see* T-60164
Ethylenetetracarboxylic acid, E-60049
▷ Ethyleneurea, *see* I-70005
21-Ethyl-2,6-epoxy-17-hydroxy-1-oxacyclohenicosa-2,5,14,18,20-pentaen-11-yn-4-one, E-60050
Ethyl 2-ethoxycarbonyl-5,5-diphenyl-2,3,4-pentatrienoate, *in* D-70492
Ethyl ethoxyiminoacetate, E-80063
Ethylglyoxal, *see* O-80067
1-Ethyl-2,3,4,6,7,12-hexahydroindolo[2,3-*a*]quinolizine, E-80064
5-Ethyl-2-hydroxy-2,4,6-cycloheptatrien-1-one, E-70044
6-Ethyl-4-hydroxy-3,5-dimethyl-2*H*-pyran-2-one, E-80065
9-Ethyl-4-hydroxy-1,7-dioxaspiro[5.5]undecane-3-methanol, *see* T-70004
5-Ethyl-3-hydroxy-4-methyl-2(5*H*)-furanone, E-60051
24-Ethylidene-2,3,22,23-cholestanetetrol, *see* S-70075
24-Ethylidene-7-cholestene-3,5,6-triol, E-80066
Ethylidenedimercaptofulvene, *in* C-60221
α-Ethylidenediphenylmethane, *see* D-60491
1-Ethylideneoctahydro-4-methylene-7-(1-methylethyl)-1*H*-indene-2,5,6-triol, *see* O-70042
3-Ethyl-3-mercaptopentane, *see* E-60053
12-Ethyl-13-methoxy-8,11,13-podocarpatrien-7-one, *in* N-80030
12-Ethyl-13-methoxy-8,11,13-podocarpatrien-3-one, *see* N-80030
▷ 3-Ethyl-2-methylacraldehyde, *see* M-70106
4-Ethyl-4-methylcrotonic acid, *see* M-80101
7-Ethyl-5-methyl-6,8-dioxabicyclo[3.2.1]oct-3-ene, E-60052
N-Ethyl-*N'*-methyl-*N,N'*-diphenylurea, *in* D-70499
5-Ethyl-4-methylisotetronic acid, *see* E-60051
23-Ethyl-24-methyl-27-nor-5,25-cholestadien-3β-ol, *see* F-60010
Ethylnylcyclopropane, E-70045
1-Ethylnylnaphthalene, E-70046
▷ 2-Ethyloxirane, E-70047

O-Ethyl 3-oxobutanethioate, *in* O-70080
S-Ethyl 3-oxobutanethioate, *in* O-70080
3-Ethyl-3-pentanethiol, E-60053
Ethyl *N*-phenacylcarbamate, *in* O-70099
(1-Ethyl-2-propynyl)benzene, *see* P-70089
5-Ethyl-2,4(1*H*,3*H*)-pyrimidinedione, E-60054
3-Ethyl-1*H*-pyrrole, E-80067
▷ Ethyl salicylate, *in* H-70108
5-Ethylsotolone, *see* E-60051
[(Ethylsulfinyl)methyl]benzene, *in* B-80062
(Ethylthio)acetaldehyde, *in* M-80017
3-Ethylthioacrylic acid, *see* E-60055
2-(Ethylthio)-1,1-dimethoxyethane, *in* M-80017
3-(Ethylthio)-2-propenoic acid, E-60055
4-Ethylthio-3-pyridinamine, *in* A-80140
5-Ethylthio-2-pyridinamine, *in* A-80138
6-Ethylthio-3-pyridinamine, *in* A-80141
2-(Ethylthio)thiazole, *in* T-60200
Ethyl trifluoromethanesulfenate, *in* T-60279
7-Ethyl-2,8,9-trihydroxy-2,4,4,6-tetramethyl-1,3(2*H*,4*H*)anthracenedione, *see* H-60148
2-Ethyl-4,6,6-trimethyl-2-cyclohexen-1-one, E-80068
5-Ethyltropolone, *see* E-70044
5-Ethyluracil, *see* E-60054
5-Ethyluridine, *in* E-60054
2-Ethyl-2-vinylglycine, *see* A-60150
▷ Ethyl vinyl ketone, *see* P-60062
Ethynamine, E-60056
9,9'-(1,2-Ethynediyl)bisanthracene, *see* D-60051
6,6'-(1,2-Ethynediyl)bis-2(1*H*)-pyridinone, E-80069
Ethynol, E-80070
2-Ethynyl-2*H*-azirine, E-80071
3-Ethynyl-2*H*-azirine, E-80072
4-Ethynylbenzaldehyde, E-70048
1-Ethynylbicyclo[2.2.2]octane, E-70049
2-Ethynyl-1,3-butadiene, E-80073
Ethynylbutadiene, *see* H-70041
1-Ethynylcycloheptene, E-70050
1-Ethynylcyclohexene, E-70051
1-Ethynylcyclopentene, E-70052
5-Ethynylcytidine, *in* A-60153
5-Ethynylcytosine, *see* A-60153
α-Ethynyl-β,β-diphenylbenzeneethanol, *see* T-80360
2-Ethynylindole, E-60057
3-Ethynylindole, E-60058
4-Ethynylindole, E-80074
3-Ethynyl-3-methyl-4-pentenoic acid, E-60059
1-(Ethynyloxy)adamantane, *in* E-80070
2-(Ethynyloxy)-2-methylpropane, *in* E-80070
▷ Ethynylpentafluorobenzene, E-80075
2-Ethynyl-1,3-propanediol, E-70054
Ethynyl propargyl ketone, *see* H-60040
2-Ethynylpyridine, E-80076
3-Ethynylpyridine, E-80077
4-Ethynylpyridine, E-80078
5-Ethynyl-2,4(1*H*,3*H*)-pyrimidinedione, E-60060
2-Ethynylthiazole, E-60061
4-Ethynylthiazole, E-60062
2-Ethynylthiophene, E-70055
4(5)-Ethynyl-1,2,3-triazole, E-80079
Ethynyltritylcarbinol, *see* T-80360
5-Ethynyluracil, *see* E-60060
5-Ethynyluridine, *in* E-60060
Euchrenone A₁, *see* E-70056
Euchrenone A₂, *see* E-70057
Euchrenone a₃, *see* A-80151
Euchrenone b₁, E-70058
Euchrenone b₃, E-70059
Euchretin A, E-80080
Eucrenone b₂, *in* E-70058
Eudesmaafraglaucolide, *in* T-70258
4,7(11)-Eudesmadiene-12,13-diol, E-60063
▷ 3,11(13)-Eudesmadien-12-oic acid, E-70060
4,11(13)-Eudesmadien-12-oic acid, E-70061
3,5-Eudesmadien-1-ol, E-60064
5,7(11)-Eudesmadien-15-ol, E-60066
5,7-Eudesmadien-11-ol, E-60065

4,11-Eudesmadien-3-one, E-80081
1,11-Eudesmanediol, E-70062
5,11-Eudesmanediol, E-70063
5,11-Eudesmanediol, E-80082
11-Eudesmene-1,5-diol, E-60067
3-Eudesmene-1,6-diol, E-80083
4(15)-Eudesmene-1,11-diol, E-70064
4(15)-Eudesmene-1,6-diol, E-80084
4(15)-Eudesmene-2,11-diol, E-80085
4(15)-Eudesmene-1,5,6-triol, E-60068
11-Eudesmen-5-ol, E-60069
3-Eudesmen-11-ol, E-80086
3-Eudesmen-6-ol, E-70065
4(15)-Eudesmen-11-ol, E-70066
6-Eudesmen-4-ol, E-80087
5-Eudesmol, *see* E-60069
α-Eudesmol, *see* E-80086
β-Eudesmol, *see* E-70066
Eudesobovatol A, E-80088
Eudesobovatol B, E-80089
Euglenapterin, E-60070
Eulophiol, *in* T-80114
Eumaitenin, E-60071
Eumaitenol, E-60072
Eumorphistonol, E-60073
Eupalin, *in* P-80035
Eupalitin, *in* P-80035
Euparotin, E-80090
Euparotin acetate, *in* E-80090
Eupaserrin, *in* D-70302
Euphorginol, *in* T-80002
Euphorianin, E-60074
Eupomatenoid 6, E-60075
Eupomatenoid 2, *in* E-60075
Eupomatenoid 4, *in* E-60075
Eupomatenoid 5, *in* E-60075
Eupomatenoid 7, *in* E-60075
Eupomatenoid 12, *in* E-60075
Eupomatenoid 13, *in* K-60001
▷ Euprex, *in* T-70277
Euthroid, *in* T-60350
Evodione, E-80091
▷ Exaltone, *see* C-60219
Excelsaoctaphenol, E-80092
Exidonin, E-80093
Eximin, *in* A-70248
E2Z4, *see* D-70321
F-5-3, *in* Z-60002
F-5-4, *in* Z-60002
Fabiatrin, *in* H-80189
Faranal, *see* T-60156
2,6,10-Farnesatriene-1,15;13,9-diolide, F-80001
Farnesenic acid, *see* T-60364
Farnesic acid, *see* T-60364
3-(1-Farnesyl)-4-hydroxycoumarin, *see* F-80007
Farnesylic acid, *see* T-60364
Fasciculatin, F-70001
Fauronol, F-70002
Fauronyl acetate, *in* F-70002
Fecapentaene-12, *see* D-70536
Fecapentaene-14, *see* T-70024
▷ Fenazaflor, *in* D-60160
[4.4.5.5]Fenestrane, *see* T-60037
[5.5.5.5]Fenestrane, *see* T-60035
[5.5.5.5]Fenestratetraene, *see* T-60036
Fenestrindane, *see* T-60019
Feniculin, *in* P-80179
▷ Fenoflurazole, *in* D-60160
Fensuximide, *in* P-80132
Feraginidin, *in* D-60004
Fercoperol, F-60001
7,9(11)-Fernadien-3-ol, F-80002
9(11)-Fernene-3,7,19-triol, F-60002
Fernolin, F-80003
Ferolin, *in* G-60014
Feronialactone, *see* A-80187
Feronolide, F-80004
Ferprenin, F-70003
Ferreirin, *in* T-80100
Ferreyrantholide, F-70004
Ferrioxamine *B*, F-60003

Ferrioxamine D_1, in F-60003
Ferrugone, F-80006
Ferugin, in D-60005
Feruginin, F-60004
Ferulaldehyde, see C-80148
Ferulenol, F-80007
Ferulide, F-70005
Ferulidene, in X-60002
Ferulidin, F-70006
Ferulinolone, F-60006
3-O-Feruloylquinic acid, in C-70005
Feruone, F-70007
Fervanol, F-60007
Fervanol benzoate, in F-60007
Fervanol p-hydroxybenzoate, in F-60007
Fervanol vanillate, in F-60007
Fevicordin A, F-60008
Fevicordin A glucoside, in F-60008
Fexerol, F-60009
Fibrostatin A, in F-70008
Fibrostatin B, in F-70008
Fibrostatin C, in F-70008
Fibrostatin D, in F-70008
Fibrostatin E, in F-70008
Fibrostatin F, in F-70008
Fibrostatins, F-70008
Ficinin, in E-80008
Ficisterol, F-60010
Ficulinic acid A, F-60011
Ficulinic acid B, F-60012
Filicin, in F-80008
Filixic acid, in F-80008
Filixic acid ABA, in F-80008
Filixic acid ABB, in F-80008
Filixic acid ABP, in F-80008
Filixic acid BBB, in F-80008
Filixic acid PBB, in F-80008
Filixic acid PBP, in F-80008
▷ Filixic acids, F-80008
Firmanoic acid, in D-60472
Firmanolide, F-60013
23-epi-Firmanolide, in F-60013
Fissinolide, in S-80054
▷ FK 1080, in T-80343
Flabellata secoclerodane, F-70009
Flaccidin, in F-80009
Flaccidinin, F-80009
Flavanthrin, F-70011
Flavanthrone, F-80010
Flavaspidic acid, in F-80011
▷ Flavaspidic acids, F-80011
Flavocommelin, in S-80052
Flavocommelitin, see S-80052
Flavokawin A, in H-80246
Flavokermesic acid, see T-60332
3-Flavonecarboxylic acid, see O-60085
Flavophene, see D-70074
Flavoyadorigenin B, see D-70291
Flavoyadorinin B, in D-70291
Flexilin, F-70012
Flindercarpin 2, F-80012
Flindercarpin 1, in F-80012
Flindercarpin 3, in F-80012
Flindissol, F-70013
Flindissol lactone, in F-70013
Flindissone, in F-70013
Flindissone lactone, in F-70013
Floionolic acid, in T-60334
Floribundone 2, F-80013
Floribundone 1, in F-80013
Floridic acid, F-80014
Floroselin, in K-70013
Flossonol, F-60014
Fluorantheno[3,4-cd]-1,2-diselenole, F-60015
Fluorantheno[3,4-cd]-1,2-ditellurole, F-60016
Fluorantheno[3,4-cd]-1,2-dithiole, F-60017
Fluorazone, see P-60247
trans-Fluorenacenedione, see I-70011
9H-Fluorene-1-carboxaldehyde, F-70014
▷ 9H-Fluorene-9-carboxaldehyde, see F-70015
▷ 9H-Fluorene-1-methanol, see H-70176
9H-Fluorene-9-selenone, F-60018
9H-Fluorene-9-thiol, F-80015

Fluoreno[9,1,2,3-cdef]chrysene, F-70017
Fluoreno[3,2,1,9-defg]chrysene, F-70016
9H-Fluoren-1-ol, F-80016
9-(9H-Fluoren-9-ylidene)-9H-fluorene, see B-60104
▷ 9H-Fluoren-9-ylidenemethanol, see F-70015
9-Fluorenylmercaptan, see F-80015
9-Fluorenylmethyl pentafluorophenyl carbonate, F-60019
Fluorescamine, F-80017
Fluorindine, see D-70264
2-Fluoro-β-alanine, see A-60162
3-Fluoroanthranilic acid, see A-70140
4-Fluoroanthranilic acid, see A-70141
5-Fluoroanthranilic acid, see A-70142
6-Fluoroanthranilic acid, see A-70143
α-Fluorobenzeneacetaldehyde, see F-60062
4-Fluoro-1,2-benzenediol, F-60020
4-Fluoro-1,3-benzenediol, F-70018
α-Fluorobenzenepropanal, see F-60064
2-Fluorobenzenethiol, F-80018
3-Fluorobenzenethiol, F-80019
4-Fluorobenzenethiol, F-80020
3-Fluorobenzocyclobutene-1,2-dione, F-80021
6-Fluorobenzocyclobuten-1-one, F-80022
10-Fluorobenzo[a]pyrene, F-80027
6-Fluorobenzo[a]pyrene, F-80023
7-Fluorobenzo[a]pyrene, F-80024
8-Fluorobenzo[a]pyrene, F-80025
9-Fluorobenzo[a]pyrene, F-80026
▷ m-Fluorobenzyl bromide, see B-80216
▷ o-Fluorobenzyl bromide, see B-80215
▷ p-Fluorobenzyl bromide, see B-80217
1-Fluorobicyclo[2.2.2]octane, F-80028
2-Fluorobicyclo[3.2.1]octane, F-80029
2-Fluorobicyclo[4.2.0]octa-1,3,5-triene-7,8-dione, see F-80021
5-Fluorobicyclo[4.2.0]octa-1,3,5-trien-7-one, see F-80022
2-Fluoro-2-butenedioic acid, F-80030
(4-Fluoro-1-butenyl)benzene, see F-70035
3-Fluorobutyrine, see A-60154
4-Fluorobutyrine, see A-60155
2-Fluoro-γ-butyrolactone, see D-80208
4-Fluorocatechol, see F-60020
2-Fluoro-m-cresol, see F-60052
4-Fluoro-m-cresol, see F-60059
6-Fluoro-m-cresol, see F-60054
3-Fluoro-o-cresol, see F-60056
4-Fluoro-o-cresol, see F-60058
5-Fluoro-o-cresol, see F-60060
6-Fluoro-o-cresol, see F-60055
2-Fluoro-p-cresol, see F-60053
3-Fluoro-p-cresol, see F-60057
1-Fluorocyclohexanecarboxylic acid, F-80031
2-Fluorocyclohexanone, F-60021
2-Fluorocyclopentanone, F-60022
1-Fluoro-1,1-dichloroethane, see D-80117
N-Fluoro-2,3-dihydro-3,3-dimethyl-1,2-benzothiadiazole 1,1-dioxide, F-80032
5-Fluoro-3,4-dihydro-4-thioxo-2(1H)-pyrimidinone, F-70019
3-Fluoro-2,6-dihydroxybenzoic acid, F-60023
4-Fluoro-3,5-dihydroxybenzoic acid, F-60024
5-Fluoro-2,3-dihydroxybenzoic acid, F-60025
1-Fluoro-2,4-dimethoxybenzene, in F-70018
4-Fluoro-1,2-dimethoxybenzene, in F-60020
3-Fluoro-2,6-dimethoxybenzoic acid, in F-60023
5-Fluoro-2,3-dimethoxybenzoic acid, in F-60025
2-Fluoro-1,3-dimethylbenzene, F-60026
2-Fluoro-3,5-dinitroaniline, F-60027
2-Fluoro-4,6-dinitroaniline, F-60028
4-Fluoro-2,6-dinitroaniline, F-60029
4-Fluoro-3,5-dinitroaniline, F-60030
5-Fluoro-2,4-dinitroaniline, F-60031
6-Fluoro-3,4-dinitroaniline, F-60032
2-Fluoro-3,5-dinitrobenzenamine, see F-60027
2-Fluoro-4,6-dinitrobenzenamine, see F-60028
4-Fluoro-2,6-dinitrobenzenamine, see F-60029
4-Fluoro-3,5-dinitrobenzenamine, see F-60030
5-Fluoro-2,4-dinitrobenzenamine, see F-60031

6-Fluoro-3,4-dinitrobenzenamine, see F-60032
4-Fluoro-2,6-dinitrotoluene, see F-60051
2-Fluoro-2,2-diphenylacetaldehyde, F-60033
Fluorododecahedrane, F-80033
(2-Fluoroethyl)benzene, see F-80059
1-Fluoro-9H-fluorene, F-80034
2-Fluoro-9H-fluorene, F-80035
3-Fluoro-9H-fluorene, F-80036
4-Fluoro-9H-fluorene, F-80037
9-Fluoro-9H-fluorene, F-80038
1-Fluoro-9H-fluoren-9-one, F-80039
2-Fluoro-9H-fluoren-9-one, F-80040
3-Fluoro-9H-fluoren-9-one, F-80041
4-Fluoro-9H-fluoren-9-one, F-80042
▷ Fluoroformonitrile, see C-80169
Fluorofumaric acid, in F-80030
3-Fluoro-2,5-furandione, in F-80030
2-Fluoroheptanal, F-60034
1-Fluorohexadecahydro-5,2,1,6,3,4-[2,3]butanediyl[1,4]diylidenepentaleno[2,1,6-cde:2',1',6'-gha]pentalene, see F-80033
2-Fluoro-3-hydroxybenzaldehyde, F-60035
2-Fluoro-5-hydroxybenzaldehyde, F-60036
4-Fluoro-3-hydroxybenzaldehyde, F-60037
2-Fluoro-5-hydroxybenzoic acid, F-60038
2-Fluoro-6-hydroxybenzoic acid, F-60039
3-Fluoro-2-hydroxybenzoic acid, F-60040
3-Fluoro-4-hydroxybenzoic acid, F-60041
4-Fluoro-2-hydroxybenzoic acid, F-60042
4-Fluoro-3-hydroxybenzoic acid, F-60043
5-Fluoro-2-hydroxybenzoic acid, F-60044
2-Fluoro-3-hydroxytoluene, see F-60052
2-Fluoro-4-hydroxytoluene, see F-60057
2-Fluoro-5-hydroxytoluene, see F-60059
2-Fluoro-6-hydroxytoluene, see F-60056
3-Fluoro-2-hydroxytoluene, see F-60055
3-Fluoro-4-hydroxytoluene, see F-60053
4-Fluoro-2-hydroxytoluene, see F-60060
4-Fluoro-3-hydroxytoluene, see F-60054
5-Fluoro-2-hydroxytoluene, see F-60058
3-Fluoro-1H-indole, F-80043
4-Fluoro-1H-indole, F-80044
5-Fluoro-1H-indole, F-80045
6-Fluoro-1H-indole, F-80046
7-Fluoro-1H-indole, F-80047
2-Fluoro-6-iodobenzoic acid, F-70020
3-Fluoro-2-iodobenzoic acid, F-70021
3-Fluoro-4-iodobenzoic acid, F-70022
4-Fluoro-2-iodobenzoic acid, F-70023
4-Fluoro-3-iodobenzoic acid, F-70024
5-Fluoro-2-iodobenzoic acid, F-70025
1-Fluoro-3-iodopropane, F-70026
2-Fluoro-3-iodopyridine, F-80048
2-Fluoro-4-iodopyridine, F-80045
3-Fluoro-4-iodopyridine, F-80046
4-Fluoro-1(3H)-isobenzofuranone, F-70027
7-Fluoro-1(3H)-isobenzofuranone, F-70028
▷ 1-Fluoro-2-isocyanatobenzene, F-60047
▷ 1-Fluoro-3-isocyanatobenzene, F-60048
▷ 1-Fluoro-4-isocyanatobenzene, F-60049
1-Fluoro-4-isothiocyanatobenzene, F-80049
2-Fluorolepidine, see F-80050
3-Fluorolepidine, see F-80051
Fluoromaleic acid, in F-80030
Fluoromaleic anhydride, in F-80030
2-Fluoro-4-mercaptoaniline, see A-80077
2-Fluoro-6-mercaptoaniline, see A-80073
3-Fluoro-2-mercaptoaniline, see A-80075
3-Fluoro-4-mercaptoaniline, see A-80076
4-Fluoro-2-mercaptoaniline, see A-80074
1-Fluoro-2-mercaptobenzene, see F-80018
1-Fluoro-3-mercaptobenzene, see F-80019
1-Fluoro-4-mercaptobenzene, see F-80020
2-Fluoro-4-mercaptobenzoic acid, F-70029
2-Fluoro-6-mercaptobenzoic acid, F-70030
4-Fluoro-2-mercaptobenzoic acid, F-70031
2-Fluoro-5-methoxybenzoic acid, in F-60038
3-Fluoro-4-methoxybenzoic acid, in F-60041
4-Fluoro-3-methoxybenzoic acid, in F-60044
▷ 3-Fluoro-3-methoxy-3H-diazirine, F-60050
1-Fluoro-2-methoxy-4-methylbenzene, in F-60054

3,3,4,4,5,5-Hexamethyl-1,2-cyclopentanedione, H-80074
Hexamethylene diperoxide diamine, *see* T-70144
Hexamethylene disulfide, *see* D-50815
Hexamethylglutaric acid, *see* H-80076
2,2,3,4,5,5-Hexamethyl-3-hexene, H-70083
1,1,2,2,3,3-Hexamethylindane, H-80075
2,2,3,3,4,4-Hexamethylpentane, H-60071
Hexamethylpentanedioic acid, H-80076
Hexamethylquercetagenin, *in* H-80067
3,7,11,15,19,23-Hexamethyl-2,6,10,14,18,22-tetracosahexaen-1-ol, H-80077
3,3,4,4,5,5-Hexamethyltetrahydropyran-2,6-dione, *see* D-80210
2-[3,7,11,15,19,23-Hexamethyl-25-(2,6,6-trimethyl-2-cyclohexenyl)pentacosa-2,14,18,22-tetraenyl]-3-methyl-1,4-naphthoquinone, H-70084
1,1,3,3,5,5-Hexamethyl-2,4,6-tris(methylene)cyclohexane, H-60072
1,1′,1″-(1,3,6-Hexanetriyl)trisbenzene, *see* T-60386
▷ Hexanicotinoylinositol, *in* I-80015
2,2′,4,4′,6,6′-Hexanitro-[1,1′-biphenyl]-3,3′,5,5′-tetramine, *see* T-60014
[3.3.3.3.3.3]Hexannulane, *see* H-80017
1,4,7,10,25,28,31,34-Hexaoxa[10.10](9,10)anthracenophane, H-70085
1,4,7,22,25,28-Hexaoxa[7.7](9,10)anthracenophane, H-70086
▷ 4,7,13,16,21,24-Hexaoxa-1,10-diazabicyclo[8.8.8]hexacosane, H-60073
Hexaperibenzocoronene, *see* H-60031
Hexaphenylbenzene, H-60074
Hexathia-18-crown-6, *see* H-70087
1,4,7,10,13,16-Hexathiacyclooctadecane, H-70087
Hexathiobenzo[18]crown-6, *see* D-60073
▷ 1,3,5-Hexatriynylbenzene, *see* P-80105
▷ 2-Hexenal, H-60076
4-Hexen-1-amine, H-70088
5-Hexen-1-amine, H-80078
2-Hexenedial, H-80079
3-Hexenedial, H-80080
2-Hexene-1,6-diol, H-80081
3-Hexene-1,6-diol, H-80082
3-Hexene-2,5-diol, H-80083
3-Hexene-2,5-dione, H-80084
4-Hexen-3-ol, H-60077
1-Hexen-5-yn-3-ol, H-70089
▷ Hexone, *see* M-60109
▷ Hexopal, *in* I-80015
▷ γ-Hexylbutyrolactone, *see* H-60078
▷ 5-Hexyldihydro-2(3*H*)-furanone, H-60078
δ-Hexylene oxide, *see* T-60075
1-(Hexylethenyl)benzene, *see* P-80108
2-Hexyl-5-methyl-3(2*H*)furanone, H-60079
2-Hexyl-5-methyl-3-oxo-2*H*-furan, *see* H-60079
Hexylvinylcarbinol, *see* N-60058
5-Hexyne-1,4-diamine, H-60080
2-Hexyne-1,6-diol, H-80085
3-Hexyne-2,5-dione, H-70090
5-Hexynoic acid, H-80086
Hibiscetin, *see* H-70021
Hibiscitin, *in* H-70021
Hieracin *II*, *in* H-80087
Hieracin I, H-80087
Hildecarpidin, H-60081
Hinesene, H-80088
Hinokiflavone, H-60082
Hipposterol, H-70091
Hippurin 1, *in* E-80039
Hippurin-2, *in* E-80039
Hippuristanol, *see* E-80039
Hirsudiol, *in* O-60033
Hirsutinolide, H-80089
Hirsutolide, *in* H-80089
Hispidol, *see* D-80273
Hispidol *A*, *in* T-70201
Hispidol *B*, *in* T-70201
Histidine trimethylbetaine, H-60083
Hiyodorilactone D, H-80090
Hiyodorilactone E, *in* H-80090
Hiyodorilactone F, *in* H-80090
HM-1, H-80091
HM-2, H-80092
HM-3, H-80093
HM-4, H-80094
Homalicine, *in* H-70207
Homoadamantane-2,7-dione, *see* T-70233
Homoadamantane-4,5-dione, *see* T-70234
Homoalethine, H-60084
Homoasaronic acid, *in* T-60342
Homoazulene-1,5-quinone, H-60085
Homoazulene-1,7-quinone, H-60086
Homoazulene-4,7-quinone, H-60087
Homobutein, *in* D-80362
Homocamphene, *see* D-60427
Homocamphenilone, *see* D-60411
Homocyclolongipesin, H-60088
Homocytidine, *in* A-70130
Homofarnesene, *see* T-70135
Homoferreirin, *in* T-80100
Homofulvene, *see* M-60069
Homofuran, *see* O-70056
Homoheveadride, H-60089
Homoisochroman, *see* T-70044
Homoisovanillin, *in* D-60371
Homomethionine, *see* A-60231
Homoorientin, *see* I-80068
Homopantetheine, H-60090
Homopantethine, *in* H-60090
Homopentaprismanone, *see* H-60035
Homophthalimide, *see* I-70099
Homopisatin, *in* T-80337
Homoprotocatechualdehyde, *see* D-60371
Homoserine, *see* A-70156
Homoserine lactone, *see* A-60138
Homotectoridin, *in* P-80045
Homotectorigenin, *in* P-80045
Homotetrahydroquinoline, *see* T-70036
Homothiophene, *see* T-70161
Homotropone, *see* B-70111
Homovanillin, *in* D-60371
5-Homoverrucosanol, H-80095
Honyucitrin, H-70092
Honyudisin, H-70093
16,22-Hopanediol, H-80096
21-Hopene-3,20-dione, H-80097
Horminone, *in* D-80265
Hormothamnione, H-50127
Hortensin, *in* T-80092
Hugershoff's base, *see* B-70057
Humulol⁺, *see* I-80099
HWG-1608, *see* F-80071
Hyanilid, *in* H-70108
Hydnowightin, H-70094
2′-Hydoxy-5′-methoxybiochanin A, *in* P-80033
2-Hydoxypiperonal, *see* H-70175
▷ Hydral, *in* H-60092
▷ Hydralazine, *see* H-60092
▷ Hydralazine hydrochloride, *in* H-60092
Hydrallmanol A, H-80099
Hydrazinecarbodithioic acid, *see* D-60509
2-Hydrazinobenzothiazole, H-70095
Hydrazinodithioformic acid, *see* D-60509
2-Hydrazinoethanamine, *see* H-70096
2-Hydrazinoethylamine, H-70096
2-[(2-Hydrazinoethyl)amino]ethanol, H-60091
▷ 1-Hydrazinophthalazine, H-60092
4-Hydrazino-2(1*H*)-pyridinone, H-60093
Hydrazinotriphenylmethane, *see* T-60388
as-Hydrindacene, H-60094
s-Hydrindacene, H-60095
as-Hydrindacene-1,8-dione, *see* T-70069
s-Hydrindacene-1,5-dione, *see* T-70068
s-Hydrindacene-1,7-dione, *see* T-70067
Hydrobenzamide, H-80100
Hydrochelidonic acid, *see* O-60071
Hydrocotoin, *in* T-80285
Hydrogen cyanate, *see* C-80167
7-Hydro-8-methylpteroylglutamylglutamic acid, H-60096
2-Hydronaphthoin, *see* D-80479
4α-Hydroperoxydesoxyvulgarin, *in* V-80013
7α-Hydroperoxy-3β,4α-dihydroxy-11-eudesmen-8-one, *see* O-80027
3α-Hydroperoxy-4,11(13)-eudesmadien-12,8β-olide, *in* H-60136
3-Hydroperoxy-4-eudesmene-1,6-diol, H-80101
4-Hydroperoxy-2-eudesmene-1,6-diol, H-80102
5-Hydroperoxy-4(15)-eudesmen-11-ol, H-70097
2-Hydroperoxy-2-methyl-6-methylene-3,7-octadiene, H-60097
3-Hydroperoxy-2-methyl-6-methylene-1,7-octadiene, H-60098
Hydroquinone-glucose, *see* A-70248
12-Hydroxy-8,12-abietadiene-11,14-dione, *see* R-60015
12-Hydroxy-8,11,13-abietatrien-18-al, *in* A-70005
12-Hydroxy-8,11,13-abietatriene-3,7-dione, H-80103
12-Hydroxy-8,11,13-abietatriene-6,7-dione, H-80104
12-Hydroxy-6,8,12-abietatriene-11,14-dione, *in* R-60015
12-Hydroxy-6,8,11,13-abietatrien-3-one, H-80105
7′-Hydroxyabscisic acid, H-70098
8′-Hydroxyabscisic acid, H-70099
1-Hydroxyacenaphthene, H-80106
4-Hydroxyacenaphthene, H-60099
4-Hydroxyacenaphthylene, H-70100
4-Hydroxyacetoacetic lactone, *see* F-80087
5-(3-Hydroxy-4-acetoxy-1-butynyl)-2,2′-bithienyl, *in* B-80166
5-(4-Hydroxy-3-acetoxy-1-butynyl)-2,2′-bithienyl, *in* B-80166
Hydroxyacetylene, *see* E-80070
2-(Hydroxyacetyl)pyridine, H-70101
3-(Hydroxyacetyl)pyridine, H-70102
2-(Hydroxyacetyl)thiophene, H-70103
8-Hydroxyachillin, H-60100
3-Hydroxyacrylic acid, *in* O-60090
1α-Hydroxyafraglaucolide, *in* A-70070
1β-Hydroxyafraglaucolide, *in* A-70070
Hydroxyagrostistachin, *in* A-80041
2-Hydroxy-β-alanine, *see* A-70160
16β-Hydroxyalisol B monoacetate, *in* A-70082
3′-Hydroxyalpinumisoflavone 4′-methyl ether, *in* A-80050
1-Hydroxy-4-ambiguene, *see* A-80056
15α-Hydroxy-β-amyrin, *in* O-70033
19-Hydroxyarachidic acid, *see* H-80176
6-Hydroxyarcangelisin, *in* D-70021
3-Hydroxyarginine, *see* A-70149
2-Hydroxyaristotetralone, *in* A-70253
2α-Hydroxyartemorin, *in* A-80170
9β-Hydroxyartemorin, *in* A-80170
ent-16α-Hydroxy-18-atisanoic acid, *in* A-80184
2-Hydroxy-4-azabenzimidazole, *see* D-60249
1β-Hydroxybaccatin I, *in* B-80002
3-Hydroxybaikiain, *see* T-60068
8α-Hydroxybalchanin, *in* S-60007
▷ 2-Hydroxybenzamide, *in* H-70108
3-Hydroxybenz[*a*]anthracene, H-70104
10-Hydroxybenz[*a*]anthracene, *see* B-80011
11-Hydroxybenz[*a*]anthracene, *see* B-80012
2-Hydroxybenzeneacetaldehyde, *see* H-60207
3-Hydroxybenzeneacetaldehyde, *see* H-60208
4-Hydroxybenzeneacetaldehyde, *see* H-60209
α-Hydroxybenzeneacetaldehyde, *see* H-60210
N-Hydroxybenzeneacetamide, *see* P-70059
2-Hydroxybenzenediazonium hydroxide inner salt, *see* D-70057
2-Hydroxy-1,3-benzenedicarboxylic acid, H-60101
2-Hydroxy-1,4-benzenedicarboxylic acid, H-70105
4-Hydroxybenzenemethanol, *see* H-80114

4-Hydroxy-3-penten-2-one, *in* P-60059
3-Hydroxy-2-(2-pentenyl)cyclopentaneacetic acid, *see* C-50302
11-Hydroxy-1-[3-(2-pentenyl)oxiranyl]-9-undecenoic acid, *see* H-70127
4-Hydroxy-6-pentyl-*o*-anisic acid 2,4-dimethoxy-6-pentylbenzoate, *see* P-80158
1-Hydroxy-6-pentyl-5,15-dioxabicyclo[10.2.1]pentadecane-4,13-dione, *see* G-60025
5-Hydroxy-13(3-pentyloxiranyl)-6,8,10,12-tridecatetraenoic acid, *see* E-70016
13-[3-(5-Hydroxypentyl)oxiranyl]-5,8,11-tridecatrienoic acid, *see* E-70018
4-Hydroxy-2-pentyne, *see* P-70041
5-Hydroxy-4-phenanthrenecarboxylic acid, H-80232
2-Hydroxy-3*H*-phenoxazin-3-one, H-60206
(2-Hydroxyphenoxy)acetic acid, H-70203
(3-Hydroxyphenoxy)acetic acid, H-70204
(4-Hydroxyphenoxy)acetic acid, H-70205
2-Hydroxy-2-phenylacetaldehyde, H-60210
2-Hydroxyphenylacetaldehyde, H-60207
3-Hydroxyphenylacetaldehyde, H-60208
4-Hydroxyphenylacetaldehyde, H-60209
4-Hydroxyphenylacetaldoxime, *in* H-60209
N-Hydroxyphenylacetamide, *see* P-70059
p-Hydroxyphenylacetone, *see* H-80241
9-Hydroxy-10-phenylanthracene, *see* P-80089
2-Hydroxy-*N*-phenylbenzamide, *in* H-70108
2-(4-Hydroxyphenyl)benzoic acid, *see* H-70111
3-(3-Hydroxyphenyl)-1*H*-2-benzopyran-1-one, H-70207
6-Hydroxy-2-phenyl-4*H*-1-benzopyran-4-one, *see* H-70130
7-Hydroxy-2-phenyl-4*H*-1-benzopyran-4-one, *see* H-70131
4-Hydroxy-1-phenyl-2-butyn-1-one, H-80233
6-Hydroxy-2-phenylchromone, *see* H-70130
7-Hydroxy-2-phenylchromone, *see* H-70131
▷ 1-Hydroxy-2-phenyldiazene 2-oxide, H-70208
3-(4-Hydroxyphenyl)-1-(2,4-dihydroxy-6-methoxyphenyl)-1-propanone, *in* H-80246
5-[2-(3-Hydroxyphenyl)ethenyl]-1,2,4-benzenetriol, *see* H-70215
6-[2-(4-Hydroxyphenyl)ethenyl]-4-methoxy-2*H*-pyran-2-one, *see* H-70122
2-[2-(4-Hydroxyphenyl)ethenyl]phenol, *see* D-80379
3-[2-(4-Hydroxyphenyl)ethenyl]phenol, *see* D-80383
4-[2-(3-Hydroxyphenyl)ethyl]-1,2-benzenediol, *see* T-60316
5-[2-[(3-Hydroxyphenyl)ethyl]]-1,2,3-benzenetriol, *see* T-60102
4-[2-(3-Hydroxyphenyl)ethyl]-2,6-dimethoxyphenol, *in* T-60102
▷ (4-Hydroxyphenyl)ethylene, H-70209
4-Hydroxyphenyl *β*-D-glucopyranoside, *see* A-70248
▷ 5-Hydroxy-7-phenyl-2,6-heptadienoic acid lactone, *see* G-60031
7-(3-Hydroxyphenyl)-2-heptene-4,6-diyn-1-ol, H-80234
6-Hydroxy-1-phenyl-2,4-hexadiyn-1-one, H-80235
3-(4-Hydroxyphenyl)-1-(2-hydroxy-4,6-dimethoxyphenyl)-2-propen-1-one, *in* H-80246
5-(4-Hydroxyphenyl)-2-(4-hydroxy-3-methoxyphenyl)-3,4-dimethyltetrahydrofuran, *in* T-60061
1-(2-Hydroxyphenyl)-2-(4-hydroxyphenyl)ethylene, *see* D-80379
1-(3-Hydroxyphenyl)-2-(4-hydroxyphenyl)ethylene, *see* D-80383
2-(4-Hydroxyphenyl)-5-(2-Hydroxypropyl)-3-methylbenzofuran, H-80236
2-(4-Hydroxyphenyl)-5-(3-hydroxypropyl)-3-methylbenzofuran, *see* O-60040
3-(3-Hydroxyphenyl)isocoumarin, *see* H-70207
3-Hydroxy-4-phenyl-5(2*H*)-isoxazolone, *see* D-70337
5-Hydroxy-4-phenyl-3(2*H*)-isoxazolone, *see* D-70337
2-(4-Hydroxyphenyl)-7-methoxy-3-methyl-5-propenylbenzofuran, *in* K-60001
3-Hydroxy-2-(phenylmethoxy)propanal, *see* B-80058
2-(4-Hydroxyphenyl)-7-methoxy-5-(1-propenyl)benzofuran, *in* H-60214
2-(4-Hydroxyphenyl)-7-methoxy-5-(1-propenyl)-3-benzofurancarboxaldehyde, H-60212
1-(4-Hydroxyphenyl)-2-(2-methoxy-4-propenylphenoxy)-1-propanol, *in* H-60215
1-(4-Hydroxyphenyl)-2-(2-methoxy-5-propenylphenyl)-2-propanol, H-80237
2-(4-Hydroxyphenyl)-3-methyl-5-benzofuranpropanol, *see* O-60040
2-(4-Hydroxyphenyl)-3-methyl-5-(2-oxopropyl)benzofuran, *in* H-80236
2-(4-Hydroxyphenyl)-3-methyl-5-(1-propenyl)benzofuran, *see* E-60075
3-(4-Hydroxyphenyl)-2-oxopropanoic acid, H-80238
1-(2-Hydroxyphenyl)-2-phenylethane, *see* P-70071
1-(3-Hydroxyphenyl)-2-phenylethane, *see* P-70072
1-(4-Hydroxyphenyl)-2-phenylethane, *see* P-70073
1-(2-Hydroxyphenyl)-3-phenyl-1,3-propanedione, H-80239
2-Hydroxy-2-phenylpropanal, H-80240
1-(4-Hydroxyphenyl)-2-propanone, H-80241
3-(4-Hydroxyphenyl)-2-propenal, H-60213
1-(4-Hydroxyphenyl)-2-propene, *see* P-80179
▷ 3-(4-Hydroxyphenyl)-2-propenoic acid, H-80242
2-(4-Hydroxyphenyl)-5-(1-propenyl)benzofuran, H-60214
1-(4-Hydroxyphenyl)-2-(4-propenylphenoxy)-1-propanol, H-60215
2-Hydroxy-2-phenylpropiophenone, *see* H-70123
2-Hydroxy-3-phenylpropiophenone, *see* H-70124
1-(4-Hydroxyphenyl)pyrrole, H-60216
2-(2-Hydroxyphenyl)pyrrole, H-60217
p-Hydroxyphenylpyruvic acid, *see* H-80238
8-Hydroxy-7-phenylquinoline, H-60218
4-Hydroxy-2-phenylquinoline, *in* P-60132
2-Hydroxy-4-phenylquinoline, *see* P-60131
5-Hydroxy-2-(phenylthio)-1,4-naphthalenedione, *see* H-70213
5-Hydroxy-3-(phenylthio)-1,4-naphthalenedione, *see* H-70214
5-Hydroxy-2-(phenylthio)-1,4-naphthoquinone, H-70213
5-Hydroxy-3-(phenylthio)-1,4-naphthoquinone, H-70214
1-(3-Hydroxyphenyl)-2-(2,4,5-trihydroxyphenyl)ethylene, H-70215
1-(4-Hydroxyphenyl)-3-(2,4,6-trihydroxyphenyl)-1-propanone, H-80243
3-(4-Hydroxyphenyl)-1-(2,4,6-trihydroxyphenyl)-1-propanone, H-80244
3-(4-Hydroxyphenyl)-1-(2,3,4-trihydroxyphenyl)-2-propen-1-one, H-80245
3-(4-Hydroxyphenyl)-1-(2,4,6-trihydroxyphenyl)-2-propen-1-one, H-80246
ent-12*β*-Hydroxy-8(14),15-pimaradien-3-one, *in* P-80146
ent-3*β*-Hydroxy-8(14),15-pimaradien-12-one, *in* P-80146
2-Hydroxy-1(10)-pinen-5-one, H-80247
1-Hydroxypinoresinol, H-60221
9-Hydroxypinoresinol, H-70216
2-Hydroxypiperonylic acid, *see* H-60177
6-Hydroxypiperonylic acid, *see* H-80202
4-Hydroxy-2-prenylcinnamic acid, *see* D-80534
6-Hydroxyprimnatrienone, *in* P-80174
Hydroxyproline leucine anhydride, *see* C-60213
Hydroxyprolylproline anhydride, *see* O-60017
3-Hydroxy-2-propenoic acid, *in* O-60090
4-(3-Hydroxy-1-propenyl)-2,6-dimethoxyphenol, *see* S-70046
5-(2-Hydroxypropyl)-7-methoxy-2-(3,4-methylenedioxyphenyl)benzofuran, *see* E-70001
6-(1-Hydroxypropyl)-1-methyllumazine, *see* L-70029
3-Hydroxy-5-(1-propynyl)-2-thienyl methyl ketone, *see* A-80024
1-[3-Hydroxy-5-(1-propynyl)-2-thiophenyl]ethanone, *see* A-80024
5-Hydroxypseudobaptigenin, *in* T-80109
5′-Hydroxypsorospermin, *in* P-70132
2-Hydroxypterocarpin, *in* T-80117
4-Hydroxypterocarpin, *in* T-80119
11-Hydroxyptilosarcenone, *in* P-60193
6*α*-Hydroxypulchellin, *in* P-80189
6*β*-Hydroxypulchellin 2-*O*-acetate, *in* P-80189
6*α*-Hydroxypulchellin 4-*O*-angelate, *in* P-80189
6*β*-Hydroxypulchellin 2-*O*-isovalerate, *in* P-80189
3-Hydroxy-2-pyridinamine, *see* A-60200
3-Hydroxy-4-pyridinamine, *see* A-60202
5-Hydroxy-2-pyridinamine, *see* A-60201
5-Hydroxy-3-pyridinamine, *see* A-60203
▷ 1-Hydroxy-2(1*H*)-pyridinethione, H-70217
3-Hydroxy-2(1*H*)-pyridinethione, H-60222
3-Hydroxy-2(1*H*)-pyridinone, H-60223
3-Hydroxy-4(1*H*)-pyridinone, H-60227
4-Hydroxy-2(1*H*)-pyridinone, H-60224
5-Hydroxy-2(1*H*)-pyridinone, H-60225
6-Hydroxy-2(1*H*)-pyridinone, H-60226
2-Hydroxy-4(1*H*)-pyridinone, *in* H-60224
2-Hydroxy-1-(2-pyridinyl)ethanone, *see* H-70101
2-Hydroxy-1-(3-pyridinyl)ethanone, *see* H-70102
3-Hydroxypyrrole, *see* D-70262
3-Hydroxy-1*H*-pyrrole-2-carboxylic acid, H-80248
4-Hydroxy-1*H*-pyrrole-2-carboxylic acid, H-80249
4-Hydroxy-1*H*-pyrrole-3-carboxylic acid, H-80250
3-Hydroxy-1*H*-pyrrole-2,4-dicarboxylic acid, H-80251
3-Hydroxy-1*H*-pyrrole-2,5-dicarboxylic acid, H-80252
3-Hydroxypyrrolidine, *see* P-70187
3-Hydroxypyrrolidine-2-acetic acid lactone, *see* H-80059
3-Hydroxyquinaldic acid, *see* H-60228
2-Hydroxy-4-quinazolinethione, *see* D-60295
N-Hydroxy-4-quinolinamine, H-80253
3-Hydroxy-2-quinolinecarboxylic acid, H-60228
8-Hydroxy-9-*β*-D-ribofuranosylguanine, *see* D-70238
▷ 8′-Hydroxyrotenone, *see* A-80152
2*β*-Hydroxyroyleanone, *in* R-60015
7-Hydroxyroyleanone, *see* D-80265
2*α*-Hydroxysalvicanaric acid, *in* S-70004
8*α*-Hydroxysambucoin, *in* S-70007
8*β*-Hydroxysambucoin, *in* S-70007
8-Hydroxy-15-sandaracopimarene, *see* H-60163
6*α*-Hydroxysandaracopimaric acid, *in* S-80008
7*α*-Hydroxysandaracopimaric acid, *in* S-80008
12*β*-Hydroxysandaracopimaric acid, *in* S-80008
3*β*-Hydroxysandaracopimaric acid, *in* S-80008
9*α*-Hydroxysantamarine, *in* S-60007
4-Hydroxysapriparaquinone, H-80254
ent-15-Hydroxy-8,9-seco-13-labdene-8,9-dione, H-70218

Idesin, *in* D-80283
Idomain, *in* D-60324
Ilexgenin *A*, *in* D-60382
Ilexoside *A*, *in* D-70327
Ilexoside *B*, *in* D-70350
Ilexsaponin A1, *in* D-60382
Ilexside I, *in* D-70350
Ilexside II, *in* D-70350
Ilimaquinone, I-80004
▷ Illudin *M*, I-60001
▷ Illudin *S*, I-60002
Illurinic acid, *in* L-60011
Iloprost, I-60003
IM 8443*T*, *in* C-60178
Imberbic acid, *in* D-70325
Imbricatolal, *in* H-70150
Imbricatonol, I-70003
Imidazate, *in* H-70108
Imidazo[*a,c*]dipyridinium, *see* D-70501
Imidazo[5,1,2-*cd*]indolizine, I-70004
▷ 4-Imidazoleacrylic acid, *see* U-70005
1-Imidazolealanine, *see* A-60204
2-Imidazolealanine, *see* A-60205
1*H*-Imidazole-4-carboxylic acid, I-60004
Imidazole salicylate, *in* H-70108
Imidazole-2-thiol, *see* D-80215
Imidazole-2-thione, *see* D-80215
▷ 2-Imidazolidinone, I-70005
4-Imidazolidinone, I-80006
2-Imidazolidone-4-carboxylic acid, *see* O-60074
2-Imidazoline, *see* D-60245
β-(Imidazol-1-yl)-α-alanine, *see* A-60204
β-(Imidazol-2-yl)-α-alanine, *see* A-60205
1*H*-Imidazol-2-ylphenylmethanone, *see* B-70074
1*H*-Imidazol-4-ylphenylmethanone, *see* B-60057
▷ 3-(1*H*-Imidazol-4-yl)-2-propenoic acid, *see* U-70005
1*H*-Imidazo[2,1-*b*]purin-4(5*H*)-one, I-70006
Imidazo[1,2-*a*]pyrazine, I-60005
2*H*-Imidazo[4,5-*b*]pyrazin-2-one, I-60006
Imidazo[1,2-*b*]pyridazine, I-60007
Imidazo[4,5-*c*]pyridazine-6-thiol, *see* D-70215
1*H*-Imidazo[4,5-*b*]pyridin-7-amine, *see* A-80087
1*H*-Imidazo[4,5-*c*]pyridin-4-amine, *see* A-70161
▷ Imidazo[4,5-*c*]pyridine, I-60008
Imidazo[4,5-*b*]pyridin-2-one, *see* D-60249
1*H*-Imidazo[4,5-*c*]pyridin-4(5*H*)-one, *see* H-60157
1*H*-Imidazo[4,5-*g*]quinazolin-8-amine, *see* A-60207
Imidazo[4,5-*g*]quinazoline-6,8(5*H*,7*H*)-dione, I-60009
Imidazo[4,5-*g*]quinazoline-4,8,9(3*H*,7*H*)-trione, I-60010
Imidazo[4,5-*f*]quinazolin-9(8*H*)-one, I-60011
Imidazo[4,5-*h*]quinazolin-6-one, I-60013
1*H*-Imidazo[4,5-*f*]quinoline, I-70007
1*H*-Imidazo[4,5-*h*]quinoline, I-70008
1*H*-Imidazo[4,5-*c*]tetrazolo[1,5-*a*]pyridine, *see* A-60345
Imidazo[2,1-*c*][1,2,4]triazin-4(1*H*)-one, I-80007
Imidocarbonic acid, *see* C-70019
▷ *o*-Imidodibenzyl, *see* D-70185
Imidodicarbonic diamide, *see* B-60190
1,12-Iminobenz[*a*]anthracene, *see* D-80067
4,5-Iminobenzo[*k*]chrysene, *see* B-80024
2,2′-Iminobisbenzamide, *in* I-60014
▷ 2,2′-Iminobisbenzoic acid, *see* I-60014
2,3′-Iminobisbenzoic acid, *see* I-60015
2,4′-Iminobisbenzoic acid, *see* I-60016
3,3′-Iminobisbenzoic acid, *see* I-60017
4,4′-Iminobisbenzoic acid, *see* I-60018
4,4′-Iminobisbenzonitrile, *in* I-60018
2,2′-Iminobis[benzo-1,3-thiazole], *see* D-80074
1,1′-Iminobis[3,3-dimethyl-2-butanone], I-80008
N-(4-Imino-2,5-cyclohexadien-1-ylidene)-1,4-benzenediamine, *see* I-60019

N-(4-Imino-2,5-cyclohexadien-1-ylidene)-*p*-phenylenediamine, *see* I-60019
▷ 2,2′-Iminodibenzoic acid, I-60014
2,3′-Iminodibenzoic acid, I-60015
2,4′-Iminodibenzoic acid, I-60016
3,3′-Iminodibenzoic acid, I-60017
4,4′-Iminodibenzoic acid, I-60018
▷ Iminodibenzyl, *see* D-70185
5-Imino-1,3-dimethyl-1*H*,3*H*-tetrazole, *in* A-70202
3-Imino-1(3*H*)-isobenzofuranone, I-80009
4-Imino-2-pentanone, *in* P-60059
11,12-Iminoperylene, *see* P-80084
3-Iminophthalide, *see* I-80009
Iminopropionicacetic acid, *see* C-60016
1-(Imino-*N*)-2,2,6,6-tetramethylpiperidinium hydroxide inner salt, *see* T-80149
1,12-Iminotriphenylene, *see* N-80004
Indamine, I-60019
Indazolo[3,2-*a*]isoquinoline, I-60020
Indeno[1,2,3-*hi*]chrysene, I-70010
Indeno[1,2,3-*cd*]fluoranthene, I-80010
Indeno[1,2-*b*]fluorene-6,12-dione, I-70011
Indeno[1,2-*b*]indole, *see* D-80216
Indeno[2,1,7,6-*ghij*]naphtho[2,1,8,7-*nopq*]pleiadene, I-80011
[2.2](4,7)(7,4)Indenophane, I-60021
▷ Indeno[1,2,3-*cd*]pyrene, I-70012
▷ Indeno[1,2,3-*cd*]pyren-1-ol, I-60022
▷ Indeno[1,2,3-*cd*]pyren-2-ol, I-60023
▷ Indeno[1,2,3-*cd*]pyren-6-ol, I-60024
▷ Indeno[1,2,3-*cd*]pyren-7-ol, I-60025
▷ Indeno[1,2,3-*cd*]pyren-8-ol, I-60026
1-(1*H*-Inden-2-yl)ethanone, *see* A-70045
1-(1*H*-Inden-3-yl)ethanone, *see* A-70046
1-(1*H*-Inden-6-yl)ethanone, *see* A-70047
Inden-2-yl methyl ketone, *see* A-70045
Inden-3-yl methyl ketone, *see* A-70046
Inden-6-yl methyl ketone, *see* A-70047
Indicoside *A*, I-60027
Indisocin, I-60028
▷ 1*H*-Indole-3-acetic acid, I-70013
▷ 1*H*-Indole-3-acetonitrile, *in* I-70013
1*H*-Indole-5-carboxaldehyde, I-60029
1*H*-Indole-6-carboxaldehyde, I-60030
1*H*-Indole-7-carboxaldehyde, I-60031
1*H*-Indole-3-carboxylic acid, I-80012
Indole-β-carboxylic acid, *see* I-80012
1*H*-Indole-5,6-diol, *see* D-70307
1*H*-Indole-6-methanol, I-60032
Indoline-2-sulfonic acid, *see* D-70221
Indolizine, I-60033
8*H*-Indolo[3,2,1-*de*]acridine, I-70014
Indolo[2,3-*b*][1]azaazulene, *see* C-60200
Indolo[1,7-*ab*][1]benzazepine, I-60034
5*H*-Indolo[1,7-*ab*][1]benzazepine, I-80014
Indolo[2,3-*b*][1,4]benzodiazepin-12(6*H*)-one, I-70015
1*H*-Indol-6-ol, *see* H-60158
3*H*-Indol-3-one, I-60035
10*H*-Indolo[3,2-*b*]quinoline, I-70016
5*H*-Indolo[2,3-*b*]quinoxaline, I-60036
▷ 3-Indolylacetic acid, *see* I-70013
2-Indolylacetylene, *see* E-60057
3-Indolylacetylene, *see* E-60058
1-(1*H*-Indol-2-yl)ethanone, *see* A-60032
1-(1*H*-Indol-4-yl)ethanone, *see* A-60033
▷ 1-(1*H*-Indol-5-yl)ethanone, *see* A-60034
1-(1*H*-Indol-6-yl)ethanone, *see* A-60035
1-(1*H*-Indol-7-yl)ethanone, *see* A-60036
2-(2-Indolyl)furan, *see* F-80095
2-(3-Indolyl)furan, *see* F-80096
(3*H*-Indol-3-ylidenemethyl)-1*H*-indole(1 +), *see* U-80033
3-Indol-3-ylmethylene-3*H*-indol-3-ium, *see* U-80033
2-Indolyl methyl ketone, *see* A-60032
4-Indolyl methyl ketone, *see* A-60033
▷ 5-Indolyl methyl ketone, *see* A-60034
6-Indolyl methyl ketone, *see* A-60035
7-Indolyl methyl ketone, *see* A-60036
6-Indolyl phenyl ketone, *see* B-60058
1*H*-Indol-6-ylphenylmethanone, *see* B-60058

Indophenazine, *see* I-60036
Inermin, *see* M-70001
Inflexarabdonin A, *in* K-80008
Inflexarabdonin B, *in* K-80008
Inflexarabdonin C, *in* D-80325
Inflexarabdonin E, *in* T-80313
Ingol, I-70018
Innovanamine, *in* H-80214
i-Inositol, *see* I-80015
meso-Inositol, *see* I-80015
myo-Inositol, I-80015
1,2,3,5/4,6-Inositol, *see* I-80015
Inositol, *see* I-80015
▷ Inositol niacinate, *in* I-80015
▷ Inositol nicotinate, *in* I-80015
myo-Inositol-1,3,4,5-tetrakis(dihydrogen phosphate), *see* I-70019
myo-Inositol-1,3,4,5-tetraphosphate, I-70019
myo-Inositol-1,3,4-triphosphate, I-70020
myo-Inositol-2,4,5-triphosphate, I-70021
myo-Inositol-2,4,5-tris(dihydrogen phosphate), *see* I-70021
Intricata bromoallene, I-80016
Intricatin, *in* I-80017
Intricatinol, I-80017
Inunal, I-70022
Inundoside *A*, *in* S-80027
Inundoside B, *in* S-80027
Inundoside D$_1$, *in* S-80027
Inundoside D$_2$, *in* S-80027
Inundoside E, *in* S-80027
Inundoside F, *in* S-80027
Inuviscolide, I-70023
1-*epi*-Inuviscolide, *in* I-70023
8-*epi*-Inuviscolide, *in* I-70023
Invictolide, I-60038
2-Iodo-3-formylthiophene, *see* I-60068
2-Iodo-4-formylthiophene, *see* I-60073
2-Iodoacetophenone, I-60039
ω-Iodoacetophenone, *see* I-60039
6-Iodobenzo[*a*]pyrene, I-80018
1-Iodobicyclo[2.2.2]octane, I-80019
1-Iodobiphenylene, I-80020
2-Iodobiphenylene, I-80021
2-Iodobutanal, I-60040
3-Iodo-2-butanone, I-60041
4-Iodo-1-butyne, I-70024
α-Iodocinnamaldehyde, *see* I-80039
4-Iodo-*m*-cresol, *see* I-60050
1-Iodocyclobutanecarboxylic acid, I-80022
2-Iodocycloheptanone, I-70025
2-Iodocyclohexanone, I-60042
1-Iodocyclohexene, I-70026
2-Iodocyclooctanone, I-70027
2-Iodocyclopentanone, I-60043
2-Iodo-4,5-dimethoxybenzoic acid, *in* D-60338
3-Iodo-2,4-dimethoxybenzoic acid, *in* D-60331
3-Iodo-2,6-dimethoxybenzoic acid, *in* D-60334
3-Iodo-4,5-dimethoxybenzoic acid, *in* D-60335
4-Iodo-2,5-dimethoxybenzoic acid, *in* D-60333
5-Iodo-2,4-dimethoxybenzoic acid, *in* D-60332
6-Iodo-2,3-dimethoxybenzoic acid, *in* D-60330
1-Iodo-5,5-dimethoxypentene, *in* I-70052
2-Iodo-4,5-dimethylaniline, I-70028
4-Iodo-2,5-dimethylaniline, I-70029
4-Iodo-2,5-dimethylaniline, I-70030
4-Iodo-3,5-dimethylaniline, I-70031
2-Iodo-4,5-dimethylbenzenamine, *see* I-70028
4-Iodo-2,5-dimethylbenzenamine, *see* I-70029
4-Iodo-2,5-dimethylbenzenamine, *see* I-70030
4-Iodo-3,5-dimethylbenzenamine, *see* I-70031
1-Iodo-3,3-dimethyl-2-butanone, I-60044
5-Iodo-4,4-dimethyl-1-pentene, I-80023
3-Iodo-4,5-dimethylpyrazole, I-70032
4-Iodo-3,5-dimethylpyrazole, I-70033
4-Iodo-1,3-dimethylpyrazole, *in* I-70049
4-Iodo-1,5-dimethylpyrazole, *in* I-70049
5-Iodo-1,3-dimethylpyrazole, *in* I-70050
6-Iodo-1,3-dimethyluracil, *in* I-70057
2-Iododiphenylmethane, I-60045
4-Iododiphenylmethane, I-60046
2-Iodoethyl azide, *see* A-60346
▷ Iodoethynylbenzene, *see* I-70055

3-Mercaptoflavone, *see* M-60025
9-Mercaptofluorene, *see* F-80015
8-Mercaptoguanine, *see* A-80145
8-Mercaptoguanosine, M-70029
5-Mercaptohistidine, M-80025
α-Mercaptohydrocinnamic acid, *see* M-60026
2-Mercaptoimidazole, *see* D-80215
6-Mercaptoimidazo[4,5-*c*]pyridazine, *see* D-70215
1-Mercaptoisoquinoline, *see* I-60132
3-Mercaptoisoquinoline, *see* I-60133
3-Mercapto-2-(mercaptomethyl)-1-propene, *see* M-70076
4-Mercapto-2-methyl-1,2,3-benzotriazinium hydroxide inner salt, *in* B-60050
2-Mercapto-3-methylbutanoic acid, M-60023
▷ 3-Mercapto-2-methylfuran, *see* M-70082
5-Mercapto-1-methylhistidine, *in* M-80025
2-Mercaptomethyl-2-methyl-1,3-propanedithiol, M-70030
(Mercaptomethyl)oxirane, *see* M-80173
2-Mercapto-3-methylpentanoic acid, M-80026
2-Mercapto-4-methylpentanoic acid, M-80027
α-(Mercaptomethyl)-α-phenylbenzenemethanol, *see* M-70028
2-(Mercaptomethyl)-1-propene-3-thiol, *see* M-70076
2-(Mercaptomethyl)pyrroldine, *see* P-70185
2-Mercapto-5-(methylthio)-1,3,4-thiadiazole, *in* M-70031
1-(Mercaptomethyl)-2-(trifluoromethyl)benzene, *see* T-80278
1-(Mercaptomethyl)-3-(trifluoromethyl)benzene, *see* T-80279
1-(Mercaptomethyl)-4-(trifluoromethyl)benzene, *see* T-80280
2-Mercapto-1-naphthylamine, *see* A-70172
8-Mercapto-1-naphthylamine, *see* A-70173
2-Mercapto-1*H*-perimidine, *see* P-10051
9-Mercaptophenanthrene, *see* P-70049
2-Mercapto-2-phenylacetic acid, M-60024
α-Mercaptophenylacetic acid, *see* M-60024
4-Mercaptophenylalanine, *see* A-60214
β-Mercaptophenylalanine, *see* A-60213
3-Mercapto-2-phenyl-4*H*-1-benzopyran-4-one, M-60025
1-Mercapto-2-phenylethane, *see* P-70069
1-(4-Mercaptophenyl)ethanone, *see* M-60017
(2-Mercaptophenyl)phenylmethanone, *see* M-70020
(3-Mercaptophenyl)phenylmethanone, *see* M-70021
(4-Mercaptophenyl)phenylmethanone, *see* M-70022
2-Mercapto-3-phenylpropanoic acid, M-60026
▷ 2-Mercaptopropanoic acid, M-60027
1-Mercapto-2-propanone, M-60028
6-Mercaptopurine, *in* D-60270
▷ Mercaptopurine, *see* D-60270
3-Mercapto-6(1*H*)-pyridazinethione, M-80028
2-Mercaptopyridine, *see* P-60223
4-Mercapto-2(1*H*)-pyridinethione, M-80029
3-Mercapto-2(1*H*)-pyridinethione, *see* P-60218
3-Mercapto-4(1*H*)-pyridinethione, *see* P-60221
4-Mercapto-2(1*H*)-pyridinethione, *see* P-60219
5-Mercapto-2(1*H*)-pyridinethione, *see* P-60220
6-Mercapto-2,4(1*H*,3*H*)-pyrimidinedithione, M-80030
3-Mercaptoquinoline, *see* Q-60003
4-Mercaptoquinoline, *see* Q-60006
5-Mercaptoquinoline, *see* Q-60004
6-Mercaptoquinoline, *see* Q-60005
2-Mercaptoquinoxaline, *see* Q-80004
8-Mercapto-9-β-D-ribofuranosylguanine, *see* M-70029
▷ 6-Mercapto-9-β-D-ribofuranosyl-9*H*-purine, *see* T-70180
3-Mercaptotetrahydrofuran, *see* T-80063
5-Mercapto-1*H*-tetrazole, *in* T-60174
▷ 2,5-Mercapto-1,3,4-thiadiazole, *see* M-70031
▷ 5-Mercapto-1,3,4-thiadiazoline-2-thione, M-70031
2-Mercaptothiazole, *see* T-60200
▷ 3-Mercapto-1,2,4-thiazole, *see* D-60298
3-Mercapto-2-thiophenecarboxylic acid, M-80031
4-Mercapto-3-thiophenecarboxylic acid, M-80032
1-Mercapto-2-(trifluoromethyl)benzene, *see* T-60284
1-Mercapto-3-(trifluoromethyl)benzene, *see* T-60285
1-Mercapto-4-(trifluoromethyl)benzene, *see* T-60286
Mercusic acid, *see* O-60039
Meridinol, M-80033
Merrillin, M-80034
Mertensene, M-80035
Mesembryanthemoidigenic acid, *in* D-80347
▷ Mesitene lactone, *see* D-70435
Mesogentiogenin, *in* G-80009
Mesoinositol, *see* I-80015
▷ Mesonex, *in* I-80015
Mesotan, *in* H-70108
Mesotartaric acid, *in* T-70005
▷ Mesotol, *in* H-70108
Mesuaferrol, M-70032
Mesuxanthone *B*, *see* T-60349
▷ Mesyl azide, *see* M-60033
Mesyltriflone, *see* M-70133
Metachromin A, M-70033
Metachromin B, M-70034
Metachromin C, M-80036
Metacrolein, *in* P-70119
[2.0.2.0]Metacyclophane, M-60031
[3[4,10]][7]Metacyclophane, M-60030
[4]Metacyclophane, M-60029
Metaphin (as disodium salt), *in* G-60032
▷ Methaneselenal, *see* S-70029
Methaneselenoamide, *in* M-70035
Methaneselenoic acid, M-70035
Methanesulfenyl thiocyanate, M-60032
▷ Methanesulfonyl azide, M-60033
Methanesulfonyl peroxide, *see* B-70154
Methanetetrapropanoic acid, M-70036
Methanetetrol, M-80037
Methanethial, M-70037
Methanetricarboxaldehyde, M-80038
▷ Methanimine, M-70038
1,6-Methano[12]annulene, *see* B-80091
1,6-Methano[18]annulene, *see* B-70101
1,6-Methano[18]annulene, *see* B-80081
1,6-Methano[20]annulene, *see* B-80077
1,6-Methano[24]annulene, *see* B-80089
4,10*b*-Methano-8*H*-benzo[*ab*]cyclodecen-8-one, *see* M-70042
6,12-Methanobenzocyclododecene, M-80039
10,11-Methano-1*H*-benzo[5,6]cycloocta[1,2,3,4-*def*]fluorene-1,14-dione, M-70039
4*b*,10*b*-Methanochrysene, M-70040
5,11-Methanodibenzo[*a*,*e*]cyclooctene-6,12(5*H*,11*H*)dione, M-80040
2*H*,5*H*-(Methanodioxymethano)-3,4,1,6-benzodioxadiazocine, *see* T-60162
15,20-Methano-1,5-(ethano[1,6]cyclodecethano)naphthalene, *see* N-70005
3*a*,7*a*-Methano-1*H*-indole, M-70041
2,4-Methanopyroglutamic acid, *see* O-80057
1,4-Methano-1,2,3,4-tetrahydronaphthalene, M-60034
▷ Methiodal sodium, *in* I-60048
Methionine sulfoximine, M-70043
N-Methionylalanine, M-60035
▷ Methoxa-Dome, *see* X-60001
▷ Methoxsalen, *see* X-60001
12-Methoxy-8,11,13-abietatrien-20-oic acid, M-60036
4-Methoxyacenaphthene, *in* H-60099
4-Methoxyacenaphthylene, *in* H-70100
2-(Methoxyacetyl)pyridine, *in* H-70101
2-(Methoxyacetyl)thiophene, *in* H-70103
5-Methoxyafrormosin, *in* T-80112
16β-Methoxyalisol B monoacetate, *in* A-70082
Methoxyaucuparin, *in* B-80128
3-Methoxybenz[*a*]anthracene, *in* H-70104
2-Methoxybenzeneacetaldehyde, *in* H-60207
α-Methoxybenzeneacetaldehyde, *in* H-60210
2-Methoxy-1,3-benzenedicarboxylic acid, *in* H-60101
2-Methoxy-1,4-benzenedicarboxylic acid, *in* H-70105
1-Methoxy-1,2-benziodoxol-3(1*H*)one, *in* H-60102
6-Methoxy-1,3-benzodioxole-5-carboxaldehyde, *in* H-70175
5-Methoxy-1,3-benzodioxole-4-carboxylic acid, *in* H-60180
6-Methoxy-1,3-benzodioxole-5-carboxylic acid, *in* H-80202
7-Methoxy-1,3-benzodioxole-4-carboxylic acid, *in* H-60179
6-(6-Methoxy-1,3-benzodioxol-5-yl)-7*H*-furo[3,2-*g*][1]benzopyran-7-one, *see* P-80003
3-(6-Methoxy-1,3-benzodioxol-5-yl)-2-propenal, *see* M-70048
4-Methoxybenzofuran, *in* H-70106
4-Methoxy-5-benzofurancarboxylic acid, *in* H-60103
2-Methoxy-3(2*H*)-benzofuranone, *in* H-80107
4-Methoxy-3(2*H*)-benzofuranone, *in* H-80108
5-Methoxy-3(2*H*)-benzofuranone, *in* H-80109
6-Methoxy-3(2*H*)-benzofuranone, *in* H-80110
7-Methoxy-3(2*H*)-benzofuranone, *in* H-80111
1-(4-Methoxy-5-benzofuranyl)-3-phenyl-1,3-propanedione, *see* P-80169
6-Methoxy-2*H*-1-benzopyran-2-one, *in* H-70109
▷ 4-Methoxybenzyl alcohol, *in* H-80114
2-Methoxybenzyl chloride, *in* C-70101
m-Methoxybenzyl chloride, *in* C-70102
p-Methoxybenzyl chloride, *in* C-70103
2′-Methoxybiochanin A, *in* T-80106
4′-Methoxy-2-biphenylcarboxylic acid, *in* H-70111
2-Methoxy-1-biphenylenol, *in* B-80116
4-Methoxy-2,2′-bipyridine, *in* H-80121
5-Methoxy-2,2′-bipyridine, *in* H-80123
6-Methoxy-3,4′-bipyridine, *in* H-80130
4-Methoxy-2,2′-bipyridin-6-ol, *in* D-80296
8-Methoxybonducellin, *in* B-60192
2-Methoxybutanal, *in* H-70112
1-Methoxy-3-buten-2-ol, *in* B-70291
2-Methoxy-2-butenolide, *see* M-80043
4-Methoxy-1,9-cadinadien-3-one, *in* H-80136
8-Methoxy-1(6),2,4,7(11)-cadinatetraen-12,8-olide, *in* H-70114
7-Methoxycaleteucrin, *in* C-80012
1-Methoxy-9*H*-carbazole, *in* H-80137
6-Methoxycarbazomycinal, *in* H-70154
C-Methoxycarbohydroxamic acid, M-70044
Methoxycarbonylmalonaldehyde, *in* D-60190
N-Methoxycarbonyl-(2,3,4,5-tetrachloro-1-thiophenio)amide, *see* T-60028
8-Methoxychlorotetracycline, M-70045
6-Methoxychroman, *in* D-70175
8-Methoxychroman, *in* D-60203
p-Methoxycinnamaldehyde, *in* H-60213
6-Methoxycoelonin, *in* T-60120
6-Methoxycomaparvin, *in* C-70191
6-Methoxycomaparvin 5-methyl ether, *in* C-70191
Methoxyconiferin, *in* S-70046
6-Methoxycoumarin, *in* H-70109
8-Methoxycoumestrol, *in* T-80295
3′-Methoxycoumestrol (obsol.), *in* T-80295
25-Methoxy-23-cycloarten-3β-ol, *in* C-80171
1-Methoxycyclohexene, M-60037
1-Methoxy-3-cyclohexene-1-carboxaldehyde, *in* H-80146
3-Methoxy-2-cyclohexen-1-one, M-80041
2-Methoxycyclopentanone, *in* H-60114
1-Methoxycyclopropanecarboxylic acid, *in* H-60117
3′-Methoxydaidzein, *in* T-80308
2-Methoxy-6-(3,4-dihydroxy-1,5-heptadienyl)benzyl alcohol, *see* H-70178

Name Index

3-Methylcrotonaldehyde − Methyl 3,6-epidioxy-6-methoxy-4,16-...

Methyl 3,6-epidioxy-6-methoxy-4,14,16-octadecatrienoate, *in* E-60008

Methyl 3,6-epidioxy-6-methoxy-4-octadecenoate, *in* E-60008

8β-(2-Methyl-2,3-epoxybutyroyloxy)-dehydroleucodin, *in* H-80184

Methyl 10,11-epoxy-3,7-diethyl-11-methyl-2,6-tridecadienoate, *see* J-80018

Methyl 14ξ,15-epoxy-8(17),12*E*-labdadien-16-oate, *in* E-60025

8β-(2-Methyl-2,3-epoxypropionyloxy)hirsutinolide, *in* H-80089

8β-(2-Methyl-2,3-epoxypropionyloxy)hirsutinolide 13-*O*-acetate, *in* H-80089

Methyl eriodermate, *in* E-70032

13^1-Methyl-13,15-ethano-13^2,17-prop-$13^2(15^2)$-enoporphyrin, M-60082

1-(1-Methylethenyl)cyclohexene, *see* I-60115

3-(1-Methylethenyl)cyclohexene, *see* I-60116

4-(1-Methylethenyl)cyclohexene, *see* I-60117

1-[4-(1-Methylethenyl)-1-cyclopenten-1-yl]ethanone, *see* A-80025

(1-Methylethenyl)oxirane, M-70080

[(1-Methylethenyl)thio]benzene, *see* P-80137

(1-Methylethoxy)benzene, *see* I-80085

5-Methylethuliacoumarin, *in* E-60045

3-(1-Methylethyl)aspartic acid, *see* A-70162

α-(1-Methylethyl)benzeneacetaldehyde, *see* M-70109

▷ 4-(1-Methylethyl)benzenemethanol, *see* I-80078

2-(1-Methylethyl)cycloheptanone, *see* I-70090

4-(1-Methylethyl)cycloheptanone, *see* I-70091

1-(1-Methylethyl)cyclohexene, *see* I-60118

3-(1-Methylethyl)cyclohexene, *see* I-60119

4-(1-Methylethyl)cyclohexene, *see* I-60120

(1-Methylethyl)cyclopentane, *see* I-70092

Methyl 2-ethyl-1,2,3,4,6,11-hexahydro-2,4,5,7,12-pentahydroxy-6,11-dioxo-1-naphthacenecarboxylate, *see* R-70002

β-Methyl-β-ethylhydracrylic acid, *see* H-80210

1-[4-(1-Methylethylidene)-1-cyclopenten-1-yl]ethanone, *see* A-80026

2-(1-Methylethylidene)tricyclo[3.3.1.1^{3,7}]decane, *see* I-60121

3-(1-Methylethyl)-2-naphthalenol, *see* I-70095

4-(Methylethyl)-1-naphthalenol, *see* I-70096

2-(1-Methylethyl)-1,3,5-trimethylbenzene, *see* I-60130

Methyl 4-fluoro-3,5-dimethoxybenzoate, *in* F-60024

▷ Methylfluoroform, *see* T-80273

2-Methyl-3-furanmethanol, *see* H-80207

▷ 5-Methyl-2(3*H*)-furanone, M-70081

▷ 5-Methyl-2(5*H*)-furanone, M-60083

▷ 2-Methyl-3-furanthiol, M-70082

2-Methylfuro[2,3-*b*]pyridine, M-60084

3-Methylfuro[2,3-*b*]pyridine, M-60085

2-Methylfuro[3,2-*c*]quinoline, M-70083

O-Methylganoderic acid *O*, *in* T-60111

▷ *Methylgenistein, *in* T-80309

5-*O*-Methylgenistein, *see* D-80336

Methyl *p*-geranyloxycinnamate, *in* H-80242

4'-*O*-Methylglabridin, *in* G-70016

4-Methylglutamic acid, *see* A-60223

γ-Methylglutamic acid, *see* A-60223

2-Methylglyceric acid, *see* D-60312

2-Methylglycidic acid, *see* M-80127

1-*O*-Methylglycyrol, *in* P-80186

5-*O*-Methylglycyrol, *in* P-80186

1-Methylguanosine, M-70084

8-Methylguanosine, M-70085

O-Methylhaenkeanoside, *in* H-80001

2-Methylheneicosane, M-80091

3-Methylheptacosane, M-80092

3-Methyl-2-heptenoic acid, M-60086

3-Methyl-4-hepten-2-one, M-80093

4-Methyl-3-hepten-2-one, M-60087

6-Methyl-6-hepten-2-one, M-80094

2-Methylhexacosane, M-80095

5-Methyl-2,4-hexadienoic acid, M-70086

5-Methyl-3,4-hexadienoic acid, M-70087

Methyl 2-(2,4-hexadiynyl)-6-methoxybenzoate, *see* F-80084

2-Methylhexanedial, M-80096

4-Methyl-3-hexanone, M-80097

2-Methyl-4-hexenal, M-80098

5-Methyl-4-hexen-1-amine, M-80099

2-Methyl-4-hexenoic acid, M-80100

4-Methyl-2-hexenoic acid, M-80101

2-Methyl-4-hexen-1-ol, M-80102

2-Methyl-4-hexen-2-ol, M-80103

2-Methyl-5-hexen-1-ol, M-80104

3-Methyl-5-hexen-1-ol, M-80105

4-Methyl-2-hexen-1-ol, M-80106

4-Methyl-3-hexen-2-ol, M-80107

4-Methyl-4-hexen-1-ol, M-80108

4-Methyl-4-hexen-2-ol, M-80109

4-Methyl-5-hexen-1-ol, M-80110

5-Methyl-4-hexen-2-ol, M-80111

5-Methyl-4-hexen-3-ol, M-80112

5-Methyl-5-hexen-1-ol, M-80113

5-Methyl-5-hexen-3-ol, M-80114

3-Methyl-4-hexen-2-one, M-80115

1-Methyl-1-hexenylbenzene, *see* P-80102

Methylhomoserine, *see* A-60186

7-*O*-Methylhorminone, *in* D-80265

p-Methylhydratropic acid, *see* M-60114

Methylhydrocotoin, *in* T-80285

Methyl hydrogen 2-formylmalonate, *in* H-80204

Methyl hydrogen 2-(*tert*-butoxymethyl)-2-methylmalonate, M-70088

Methyl hydrogen 3-hydroxyglutarate, M-70089

Methyl 2-[(2-hydroxybenzoyl)amino]-4-hydroxybenzoate, *in* D-70049

Methyl hydroxycarbamate, *see* M-70044

7-Methylidenebicyclo[3.3.1]nonan-3-one, *see* M-70062

1-Methylidene-2-(phenylsulfinyl)cyclopropane, *in* M-80090

1-Methylidene-2-(phenylsulfonyl)cyclopropane, *in* M-80090

1-Methylidene-2-(phenylthio)cyclopropane, *see* M-80090

6-Methylidenetetrahydro-2-pyrone, *see* T-70074

1,1′,1″-Methylidynetris[2,3,4,5,6-pentachlorobenzene], *see* T-60405

6-Methyl-1*H*-indole, M-60088

3-Methyl-1*H*-indole-2-carboxaldehyde, M-70090

Methyl 2-iodo-3,5-dimethoxybenzoate, *in* D-60336

Methyl 4-iodo-3,5-dimethoxybenzoate, *in* D-60337

3-Methylisocoumarin, *see* M-70052

6-Methylisocytosine, *see* A-60230

▷ Methylisogenistin, *in* T-80309

1-Methylisoguanosine, *see* D-60518

O-Methylisohaenkeanoside, *in* H-80001

Methylisohallerin, *in* I-60100

2-Methylisophthalaldehyde, *see* M-60046

4-Methylisophthalaldehyde, *see* M-60049

5-Methylisophthalaldehyde, *see* M-60050

Methyl 8-isopropyl-5-methyl-2-naphthalenecarboxylate, *in* I-60129

2′-*O*-Methylisopseudocyphellarin **A**, *in* I-70098

1-Methyl-3-isoquinolinamine, *see* A-80099

1-Methyl-4-isoquinolinamine, *see* A-80101

1-Methyl-5-isoquinolinamine, *see* A-80102

1-Methyl-7-isoquinolinamine, *see* A-80104

1-Methyl-8-isoquinolinamine, *see* A-80105

3-Methyl-1-isoquinolinamine, *see* A-80098

3-Methyl-5-isoquinolinamine, *see* A-80103

4-Methyl-3-isoquinolinamine, *see* A-80100

Methylisoselenourea hydrogen sulfate, *in* S-70031

Methylisoselenourea iodide, *in* S-70031

3-Methylisoserine, *see* A-60183

3-Methyl-4-isoxazolamine, *see* A-80107

4-Methyl-3-isoxazolamine, *see* A-80106

4-Methyl-5-isoxazolamine, *see* A-80109

5-Methyl-4-isoxazolamine, *see* A-80108

5-Methyl-4-isoxazolecarboxylic acid, M-80117

Methyl jasmonate, *in* J-60002

Methyllanceolin, *in* P-60052

4-Methylleucine, *see* A-70137

Methyllinderatin, *in* L-50048

Methyllinderatone, *in* L-70030

9-Methyllongipesin, M-70091

3′-*O*-Methyllutonarin, *in* I-80068

▷ Methylmalonic acid, *see* M-80149

Methylmalononitrile, *in* M-80149

α-Methylmandelaldehyde, *see* H-80240

1-Methyl-4-mercaptohistidine, *see* O-70050

Methyl 3-(4-methoxy-2-vinylphenyl)propanoate, *in* H-70232

Methyl 3-formyl-3-methylacrylate, *in* M-70103

1-Methyl-4-methylamino-5-(methylaminocarbonyl)imidazole, *see* C-60003

Methyl 2-formyl-3-methylamino-2-propenoate, *in* A-80134

1-Methyl-5-methylamino-1*H*-tetrazole, *in* A-70202

2-Methyl-5-methylamino-2*H*-tetrazole, *in* A-70202

2-Methyl-6-(4-methylenebicyclo[3.1.0]hex-1-yl)-2-heptenal, *in* S-70037

4-Methyl-6,7-methylenedioxy-2*H*-1-benzopyran-2-one, M-70092

4-Methyl-6,7-methylenedioxycoumarin, *see* M-70092

4-(5-Methyl-1-methylene-4-hexenyl)-1-(4-methyl-3-pentenyl)cyclohexene, *see* C-70009

4-Methyl-10-methylene-7-(1-methylethyl)-5-cyclodecene-1,4-diol, *see* G-70009

2-Methyl-6-methylene-2,7-octadien-4-ol, M-80093

7-Methyl-3-methylene-1-octanol, M-80118

2-Methyl-6-methylene-1,3,7-octatriene, M-60089

2-Methyl-6-methylene-7-octen-4-ol, *in* M-70093

4-Methyl-5-methylene-2-oxetanone, M-80119

Methyl 2-*O*-methyleriodermate, *in* E-70032

Methyl 2′-*O*-methyleriodermate, *in* E-70032

Methyl 4-*O*-methyleriodermate, *in* E-70032

1-Methyl-4-(1-methylethenyl)-1,2-cyclohexanediol, *see* I-70086

3-Methyl-6-(1-methylethenyl)-2-cyclohexen-1-one, *see* I-70088

3-Methyl-6-(1-methylethenyl)-9-decen-1-ol, *see* I-70089

5-Methyl-8-(1-methylethenyl)-10-oxabicyclo[7.2.1]dodeca-1(12),4-dien-11-one, *see* A-60296

3-Methyl-2-[2-(1-methylethenyl)-5-oxohexyl]-2-cyclopenten-1-one, *see* M-80004

2-Methyl-5-(1-methylethyl)-1,3-benzenediol, *see* I-60125

2-Methyl-6-(1-methylethyl)-1,4-benzenediol, *see* I-60123

3-Methyl-5-(1-methylethyl)-1,2-benzenediol, *see* I-60126

3-Methyl-6-(1-methylethyl)-1,2-benzenediol, *see* I-60124

4-Methyl-5-(1-methylethyl)-1,3-benzenediol, *see* I-60127

5-Methyl-2-(1-methylethyl)-1,3-benzenediol, *see* I-60122

2-Methyl-5-(1-methylethyl)bicyclo[3.1.0]hex-2-ene, *see* T-80194

1-Methyl-4-(1-methylethyl)-3-cyclohexane-1,2-diol, *see* I-70093

1-Methyl-4-(1-methylethyl)-4-cyclohexene-1,2-diol, *see* I-70094

1-Methyl-4-(1-methylethyl)-4-cyclohexene-1,2-diol, *see* M-80014

2-Methyl-5-(1-methylethyl)furan, *see* I-80083

Name Index

4-(Methylthio)-2(1*H*)-pyridinethione... – [2](2,6)-Naphthaleno[2]...

5-Oxo-2-cyclohexene-1-carboxylic acid, O-80078
6-Oxo-1-cyclohexene-1-carboxylic acid, O-80079
α-(2-Oxocyclohexyl)acetophenone, *see* P-70045
6-Oxocyclonerolidol, *in* H-60111
▷ Oxocyclopentadecane, *see* C-60219
3-Oxocyclopentaneacetic acid, O-70083
2-Oxocyclopentanecarboxylic acid, O-70084
2-Oxo-1,3-cyclopentanediglyoxylic acid, O-70085
2-Oxocyclopentanediylbisglyoxylic acid, *see* O-70085
5-Oxocyclopenta[*b*]thiophene, *see* D-70183
6-Oxocyclopenta[*b*]thiophene, *see* D-70182
4-Oxocyclopenta[*c*]thiophene, *see* D-70184
3-Oxocyclopentene, *see* C-60226
α-Oxo-3-cyclopentene-1-acetaldehyde, O-60061
3-Oxo-1-cyclopentenecarboxaldehyde, O-60062
α-Oxocyclopropaneacetic acid, *see* C-60231
5-Oxocystofuranoquinol, *in* C-70260
2-Oxodecanal, O-70087
10-Oxodecanoic acid, O-60063
9-Oxo-2-decenoic acid, O-70089
9-Oxo-4,5-dehydro-4(15)-dihydrocostic acid, *in* H-60135
2-Oxodesoxyligustrin, *in* O-70090
4-Oxo-4,5-dihydrofuro[2,3-*d*]pyridazine, *see* H-60146
7-Oxo-6,7-dihydrofuro[2,3-*d*]pyridazine, *see* H-60147
7-Oxodihydrogmelinol, O-60064
1-Oxo-3*H*-1,2-dihydropyrrolo[1,2-*a*]pyrrole, *see* D-60277
μ-Oxodiphenylbis(trifluoroacetato-*O*)diiodine, O-60065
12-Oxododecanoic acid, O-60066
5-Oxo-6,8,11,14-eicosatetraenoic acid, O-80080
11-Oxoepilupeol, *in* H-80186
2-Oxo-1(10),11(13)-eremophiladien-12,8-olide, O-80081
3-Oxo-4,11(13)-eudesmadien-12-oic acid, *in* E-70061
9-Oxo-4,11(13)-eudesmadien-12-oic acid, *in* H-60135
3-Oxo-1,4-eudesmadien-12,8β-olide, *in* Y-70001
3-Oxo-4,11(13)-eudesmadien-12,8β-olide, *in* H-60136
9-Oxo-4,11(13)-eudesmadien-12,16β-olide, *in* H-60137
3-Oxo-1,4,11(13)-eudesmatrien-12,8β-olide, *see* Y-70001
1-Oxo-7(11)-eudesmen-12,8-olide, O-60067
18-Oxoferruginol, *in* A-70005
Oxoflaccidin, *in* F-80009
3-Oxo-29-friedelanal, *in* F-80082
3-Oxo-27-friedelanoic acid, *in* T-70218
5-[5-Oxo-2(5*H*)-furanylidene]-2(5*H*)-furanone, *see* B-80094
1-Oxo-4,10(14)-germacradien-12,6α-olide, *in* O-60068
1-Oxo-4-germacren-12,6-olide, O-60068
ent-20-Oxo-16-gibberellene-7,19-dioic acid, *see* G-60019
3-Oxoglutaric acid, *see* O-60084
3-Oxo-4,11(13)-guaiadien-12,8β-olide, *see* A-70054
2-Oxo-3,10(14),11(13)-guaiatrien-12,6-olide, O-70090
9-Oxogymnomitryl acetate, *in* G-70035
ent-15-Oxo-1(10),13-halimadien-18-oic acid, O-60069
17-Oxohebemacrophyllide, *in* H-60010
4-Oxoheptanal, O-60070
4-Oxoheptanedinitrile, *in* O-60071
4-Oxoheptanedioic acid, O-60071
4-Oxo-2-heptenedioic acid, O-60072
4-Oxohexanal, O-60073
5-Oxohexanal, O-70091
4-Oxo-2-hexenoic acid, O-80082
5-Oxo-3-hexenoic acid, O-70092
5-Oxo-3-hexenoic acid, O-80083
6-Oxo-4-hexenoic acid, O-80084
1-Oxohinokiol, *in* D-60306
18-Oxo-19-hydroxynerylgeraniol, *in* H-70179
2-Oxo-4-imidazolidinecarboxylic acid, O-60074
▷ 4-Oxo-2-imidazolidinethione, *see* T-70188
2-Oxo-1*H*,3*H*-imidazo[4,5-*b*]pyridine, *see* D-60249
3-Oxoindole, *see* I-60035
2-Oxo-3-indolineglyoxylic acid, O-80085
2-(2-Oxo-3-indolinyl)acetic acid, *see* D-60259
1-Oxoisoalantolactone, *in* I-80046
9-Oxoisoalantolactone, *in* I-80046
2-Oxoisoanhydrooplopanone, *in* O-50045
1-Oxoisoanhydrooplopanone (incorr.), *in* O-50045
3-Oxoisocostic acid, *in* E-70061
5-Oxoisocystofuranoquinol, *in* C-70260
10-Oxo-6-isodaucen-14-al, O-70093
2-Oxoisopentanoic acid, *see* M-80129
7-Oxo-8(14),15-isopimaradien-18-oic acid, O-70094
23-Oxoisopristimerin III, *in* I-70083
ent-15-Oxo-16-kauren-18-oic acid, O-60075
ent-15-Oxo-16-kauren-18-oic acid, O-80086
2-Oxokolavenic acid, *in* K-60013
ent-2-Oxo-7,13-labdadien-15-al, *in* D-80328
7-Oxo-8,13-labdadien-15-oic acid, O-70095
ent-6-Oxo-7,12*E*,14-labdatrien-17,11α-olide, O-60076
6-Oxo-7-labdene-15,17-dioic acid, O-60077
12-Oxolambertianic acid, *in* L-60011
3-Oxo-8,24-lanostadien-21-oic acid, O-80087
3-Oxo-8,24-lanostadien-26-oic acid, O-70096
3-Oxo-8-lanosten-26,22-olide, *in* H-60169
6-Oxolumazine, *see* P-70134
7-Oxolumazine, *see* P-70135
23-Oxomariesiic acid *A*, *in* M-70012
23-Oxomariesiic acid *B*, *in* M-60010
3-Oxomorpholine, *see* M-60147
α-Oxo-1-naphthaleneacetic acid, O-60078
α-Oxo-2-naphthaleneacetic acid, O-60079
4-Oxo-4*H*-naphtho[2,3-*b*]pyran-3-carboxaldehyde, O-80088
4-Oxo-2-nonenal, O-70097
▷ 2-Oxonorbornane, *see* B-70098
7-Oxo-11-nordrim-8-en-12-oic acid, O-60080
5-Oxo-4-octadecanolide, *see* F-80004
4-Oxooctanal, O-60081
2-Oxo-4-oxazolidinecarboxylic acid, O-60082
4-Oxopentanal, O-60083
3-Oxopentanedioic acid, O-60084
5-Oxopentanoic acid, O-80089
3-Oxo-1-pentene-1,5-dicarboxylic acid, *see* O-60072
▷ 4-Oxo-2-pentenoic acid, O-70098
3-Oxo-2-(2-pentenyl)-cyclopentaneacetic acid, *see* J-60002
3-Oxo-2-(2-pentenyl)-1-cyclopenteneacetic acid, O-80090
6-(1-Oxopentyl)-1-cyclohexene-1-carboxylic acid, *see* S-80020
3-Oxo-2-pentylcyclopentaneacetic acid, O-80091
N-(3-Oxo-3*H*-phenoxazin-2-yl)acetamide, *in* A-60237
4-Oxo-2-phenyl-4*H*-1-benzopyran-3-carboxylic acid, O-60085
4-Oxo-3-phenyl-4*H*-1-benzopyran-2-carboxylic acid, O-60086
4-Oxo-2-phenyl-4*H*-1-benzothiopyran-3-carboxylic acid, O-60087
2-Oxo-4-phenyl-2,5-dihydrofuran, *see* P-60103
(2-Oxo-2-phenylethyl)carbamic acid, O-70099
2-(2-Oxo-2-phenylethyl)cyclohexanone, *see* P-70045
3-Oxo-2-phenylindolenine, *see* P-60114
3-Oxo-5-phenylpentanal, O-80092
4-Oxo-5-phenylpentanal, O-80093
5-Oxo-5-phenylpentanal, O-80094
1-Oxo-2-phenyl-1,2,3,4-tetrahydronaphthalene, *see* D-60265
1-Oxo-3-phenyl-1,2,3,4-tetrahydronaphthalene, *see* D-60266
1-Oxo-4-phenyl-1,2,3,4-tetrahydronaphthalene, *see* D-60267
6-Oxo-2-phenyl-1,4,5,6-tetrahydropyrimidine, *see* D-70250
2-Oxo-4-phenylthio-3-butenoic acid, O-60088
3-Oxo-5-phenylvaleraldehyde, *see* O-80092
4-Oxo-5-phenylvaleraldehyde, *see* O-80093
5-Oxo-5-phenylvaleraldehyde, *see* O-80094
γ-Oxopimelic acid, *see* O-60071
4-Oxopipecolic acid, *see* O-60089
4-Oxo-2-piperidinecarboxylic acid, O-60089
3-Oxopisolactone, *in* P-80155
5-Oxoproline, *see* O-70102
2-Oxopropanethiol, *see* M-60028
3-Oxopropanoic acid, O-60090
2-(3-Oxo-1-propenyl)benzaldehyde, O-60091
(2-Oxopropylidene)propanedioic acid, O-70100
2-(2-Oxopropyl)pyridine, *see* P-60229
3-(2-Oxopropyl)pyridine, *see* P-60230
4-(2-Oxopropyl)pyridine, *see* P-60231
3-Oxopyrazolo[1,2-*a*]pyrazol-8-ylium-1-olate, O-60092
α-Oxo-1*H*-pyrrole-3-acetic acid, O-70101
▷ 2-Oxopyrrolidine, *see* P-70188
5-Oxo-2-pyrrolidinecarboxylic acid, O-70102
1-Oxopyrrolizidine, *see* H-60057
2-Oxopyrrolizidine, *see* T-60091
3-Oxopyrrolizidine, *see* H-60058
4-Oxoquinolizidine, *see* O-60019
7-Oxosandaracopimaric acid, *see* O-70094
4-Oxo-3,4-secoambrosan-12,6-olid-3-oic acid, O-60093
3-Oxosilphinene, *see* S-60033
16-Oxo-17-spongianal, O-60094
6-Oxo-1,2,3,6-tetrahydroazulene, *see* D-60193
1-Oxo-1,2,3,4-tetrahydrocarbazole, O-70103
2-Oxo-1,2,3,4-tetrahydrocarbazole, O-70104
3-Oxo-1,2,3,4-tetrahydrocarbazole, O-70105
4-Oxo-1,2,3,4-tetrahydrocarbazole, O-70106
5-Oxo-5,6,7,8-tetrahydroisoquinoline, *see* D-60254
2-Oxo-1,2,3,4-tetrahydropyrimidine, *see* D-60273
3-Oxo-2-thiabicyclo[2.2.2]oct-5-ene, *see* T-60188
5-Oxo-3-thiomorpholinecarboxylic acid, O-80095
4-Oxo-2-thionodihydro-2*H*-pyrido[1,2-*a*]-1,3,5-triazine, *see* D-60294
9-Oxotournefortiolide, *in* H-60137
3-Oxotricyclo[2.1.0.0^{2,5}]pentane-1,5-dicarboxylic acid, O-80096
5-Oxo-8(10)*E*-variabilin, *in* V-80002
5-Oxovariabilin, *in* V-80002
2-Oxoverboccidentafuran, *in* V-70005
18-Oxo-3-virgene, O-60095
1,1′-Oxybis[2,4-dichlorobenzene], *see* B-70135
1,1′-Oxybis[3,4-dichlorobenzene], *see* B-70137
2,2′-Oxybis[1,3-dichlorobenzene], *see* B-70136
1,8-Oxybis(ethyleneoxyethyleneoxy)-9,10-anthracenedione, O-80097
3,3′-[Oxybis(3-hydroxy-2-methoxy-4,1-phenylene)]-bis[3,4-dihydro-2*H*-1-benzopyran-7-ol], *see* B-80145
1,1′-Oxybis[2-iodoethane], O-60096
3,3′-Oxybispropanoic acid, O-80098
3,3′-Oxybis-1-propanol, O-60097
2,2′-Oxybispyridine, O-60098
2,3′-Oxybispyridine, O-60099
2,4′-Oxybispyridine, O-80099
3,3′-Oxybispyridine, O-60100
3,4′-Oxybispyridine, O-80100
4,4′-Oxybispyridine, O-60101
Oxycurcumenol, O-80101
6,6′-(Oxydiethylidene)bis[7-methoxy-2,2-dimethyl-2*H*-1-benzopyran], *see* E-70008
Oxydiformic acid, *see* D-70111
Oxyjavanicin, *see* F-60108

(Pentafluorophenyl)bis(trifluoroacetato-*O*)-
iodine, *see* B-80161
1-(Pentafluorophenyl)ethanol, P-60040
Pentafluorophenyl formate, P-70025
Pentafluorophenyl isocyanide, *see* P-60036
Pentafluorophenylisothiocyanate, *see* P-60037
1,1,3,3,3-Pentafluoro-1-propene, P-80028
2,2,3,3,3-Pentafluoropropyl chloride, *see*
C-80089
2*H*-Pentafluoropropylene, *see* P-80028
2,3,4,5,6-Pentafluorotoluene, *see* P-60038
Pentafluoro(trifluoromethyl)sulfur, P-60041
Pentafulvalene, *see* F-60080
Pentahydroxybenzaldehyde, P-60042
Pentahydroxybenzoic acid, P-60043
3,3',4,4',5-Pentahydroxybibenzyl, P-60044
2',3,3',4,4'-Pentahydroxychalcone, *see*
D-80372
2',3,4,4',5'-Pentahydroxychalcone, *see*
D-70339
2',3',4',5',6'-Pentahydroxychalcone, *see*
P-80050
1,2,3,8,9-Pentahydroxycoumestan, P-80029
1,4,6,9,15-Pentahydroxydihydro-β-agarofuran,
P-80032
2,3,4,6,7-Pentahydroxy-9,10-
dihydrophenanthrene, *see* D-70246
4,4',5,5',8-Pentahydroxy-7,7'-dimethyl-[1,1'-
bianthracene]-9,9',10,10'-tetrone, *see*
R-80011
4,4',5,5',8-Pentahydroxy-7,7'-dimethyl-1,1'-
bianthraquinone, *see* R-80011
3,3',4',5,7-Pentahydroxyflavan, P-70026
2',3,5,7,8-Pentahydroxyflavanone, P-60045
▷ 3,3',4',5,7-Pentahydroxyflavanone, P-70027
3',4',5,6,7-Pentahydroxyflavanone, P-70028
2',3',4',5,6-Pentahydroxyflavone, P-70030
2',3,5,7,8-Pentahydroxyflavone, P-70029
2',4',5,5',7-Pentahydroxyflavone, P-80033
2',4',5,7,8-Pentahydroxyflavone, P-80034
3,4',5,6,7-Pentahydroxyflavone, P-80035
3,4',5,6,8-Pentahydroxyflavone, P-60048
3,5,6,7,8-Pentahydroxyflavone, P-70031
3',4',5,5',7-Pentahydroxyflavylium, P-80036
3',4',5,7,8-Pentahydroxyflavylium, P-80037
2',4,4',5',6'-Pentahydroxy-3'-
geranyldihydrochalcone, *see* D-80459
5β,6β,10α,14*R*,16α-
Pentahydroxygrayanotoxan-3-one, *in*
G-80039
2',3',4',7,8-Pentahydroxyisoflavan, P-80038
2',3',4',5,7-Pentahydroxyisoflavanone, P-80039
2',3',4',5,7-Pentahydroxyisoflavene, P-80040
2',4',5,5',7-Pentahydroxyisoflavone, P-80041
2',4',5',6,7-Pentahydroxyisoflavone, P-80042
3',4',5,6,7-Pentahydroxyisoflavone, P-70032
3',4',5,6,7-Pentahydroxyisoflavone, P-80043
3',4',5',6,7-Pentahydroxyisoflavone, P-80044
3',4',5,7,8-Pentahydroxyisoflavone, P-80045
3',4',6,7,8-Pentahydroxyisoflavone, P-80046
ent-2α,3α,6β,7β,11β-Pentahydroxy-16-kauren-
15-one, P-80047
ent-1β,3α,6β,7α,11α-Pentahydroxy-16-kauren-
15-one, P-80048
6,7,8,14,15-Pentahydroxy-11,13(16)-
labdadiene, *see* L-70008
2,3,5,6,8-Pentahydroxy-7-methoxy-1,4-
naphthoquinone, *in* H-60065
2,3,5,6,8-Pentahydroxy-1,4-naphthalenedione,
see P-70033
2,3,5,6,8-Pentahydroxy-1,4-naphthoquinone,
P-70033
3',4',5,7,8-Pentahydroxyneoflavone, *see*
D-80370
2',4,4',5',6-Pentahydroxy-3'-
neryldihydrochalcone, *see* D-80459
3α,16α,21α,22α,28-Pentahydroxy-12-oleanene,
in O-80036
3,16,21,23,28-Pentahydroxy-12-oleanene, *see*
O-50037
3,4,8,9,10-Pentahydroxy-6-oxo-6*H*-
dibenzo[*b,d*]pyran-1-carboxylic acid, *see*
L-80052

1,2,5,6,7-Pentahydroxyphenanthrene, P-60050
1,3,4,5,6-Pentahydroxyphenanthrene, P-70034
2,3,4,7,9-Pentahydroxyphenanthrene, P-70035
1-(2,3,4,5,6-Pentahydroxyphenyl)-3-phenyl-2-
propen-1-one, P-80050
1,2,3,8,9-Pentahydroxypterocarpan, P-80051
2,3,8,9,10-Pentahydroxypterocarpan, P-80054
3,4,8,9,10-Pentahydroxypterocarpan, P-80055
2,3,6*a*,8,9-Pentahydroxypterocarpan, P-80052
3,4,6*a*,8,9-Pentahydroxypterocarpan, P-80053
3,6*a*,7,8,9-Pentahydroxypterocarpan, P-80056
3,8,9,12,17-
Pentahydroxytricyclo[12.3.1.1^{2,6}]nonadeca-
1(18),2,4,6(19),14,16-hexaen-10-one, *see*
A-60308
3',4',5,5',7-Pentahydroxy-3,6,8-
trimethoxyflavone, *in* O-70018
1α,2α,3β,19α,23-Pentahydroxy-12-ursen-28-oic
acid, *in* T-80124
1,2,5,6,8-Pentahydroxy-9*H*-xanthen-9-one, *see*
P-60052
1,2,3,4,8-Pentahydroxyxanthone, P-60051
1,2,3,6,8-Pentahydroxyxanthone, P-70036
1,2,5,6,8-Pentahydroxyxanthone, P-60052
1,3,4,7,8-Pentahydroxyxanthone (incorrect),
see P-60052
Pentaisopropylidenecyclopentane, P-60053
Pentakis(dimethylamino)aniline, *in* B-80014
Pentakis(1-methylethylidene)cyclopentane, *see*
P-60053
Pentalenene, P-70037
Pentalenic acid, P-60054
Pentalenolactone *E*, P-60055
Pentamethoxybenzaldehyde, *in* P-60042
Pentamethoxybenzoic acid, *in* P-60043
2',3',4',5',6'-Pentamethoxychalcone, *in*
P-80050
3,3',4',5,7-Pentamethoxyflavan, *in* P-70026
2',3',4',5,6-Pentamethoxyflavone, *in* P-70030
2',3,5,7,8-Pentamethoxyflavone, *in* P-70029
3,4',5,6,7-Pentamethoxyflavone, *in* P-80035
3,5,6,7,8-Pentamethoxyflavone, *in* P-70031
2',3',4',5,7-Pentamethoxyisoflavanone, *in*
P-80039
2',4',5,5',7-Pentamethoxyisoflavone, *in*
P-80041
2',4',5',6,7-Pentamethoxyisoflavone, *in*
P-80042
3',4',5',6,7-Pentamethoxyisoflavone, *in*
P-80044
3',4',5,7,8-Pentamethoxyisoflavone, *in* P-80045
3,3',4,5,8-Pentamethoxy-6,7-
methylenedioxyflavone, *in* H-80028
2,3,4,7,9-Pentamethoxyphenanthrene, *in*
P-70035
1-(2,3,4,5,6-Pentamethoxyphenyl)-3-phenyl-2-
propen-1-one, *in* P-80050
2-(Pentamethoxyphenyl)-5-(1-
propenyl)benzofuran, *see* R-60001
1,2,3,4,8-Pentamethoxyxanthone, *in* P-60051
Pentamethylanisole, *in* P-60058
1,2,2,3,3-Pentamethylaziridine, *in* T-80139
Pentamethylbenzaldehyde, P-80057
Pentamethylbenzeneacetic acid, *see* P-70038
2,3,4,5,6-Pentamethylbenzenemethanol, *see*
P-80058
Pentamethylbenzyl alcohol, P-80058
4,8,13,17,21-Pentamethyl-4,8,12,16,20-
docosapentaenoic acid, *see* T-80371
2,3,5,6,7-Pentamethylenebicyclo[2.2.2]octane,
P-60056
3,3-Pentamethyleneoxaziridine, *see* O-70052
2,3,5,6,7-
Pentamethylidenebicyclo[2.2.2]octane, *see*
P-60056
1,2,3,4,5-Pentamethyl-6-nitrobenzene, P-60057
Pentamethyl(nitromethyl)benzene, P-80059
Pentamethylphenol, P-60058
(Pentamethylphenyl)acetic acid, P-70038
Pentamethylphenylnitromethane, *see* P-80059
2,4-Pentanediamine, P-70039
2,3-Pentanediol, P-80060
▷ 2,4-Pentanedione, P-60059

4,7,13,16,21-Pentaoxa-1,10-
diazabicyclo[8.8.5]tricosane, P-60060
3,7,11,15,23-Pentaoxo-8,20*E*-lanostadien-26-
oic acid, *in* D-80390
Pentaprismane, *see* H-70032
1,2,3,4-Pentatetraene-1,5-dione, P-70040
4-Pentene-2,3-dione, P-60061
▷ 1-Penten-3-one, P-60062
▷ Penthiazolidine, *see* T-60093
2-Penthiazolidone, *see* T-60096
▷ 5-Pentyldihydro-2(3*H*)-furanone, *see* D-70245
3-Pentyl isocyanide, *see* I-80059
4-Pentyn-1-amine, P-80061
▷ 4-Pentynoic acid, P-80062
3-Pentyn-2-ol, P-70041
4-Pentynylamine, *see* P-80061
Peperinic acid, P-80063
Peperomin A, P-80064
Peperomin B, P-80065
Peperomin C, P-80066
▷ Perchlordecone, *see* D-70530
▷ Perchloromethyl chloroformate, *see* T-70223
▷ Perchloropentacyclo[5.3.0.0^{2,6}.0^{3,9}.0^{5,8}]decane,
see D-70530
Perchlorophenalenyl, *see* N-60055
Perchloropropane, *see* O-80004
Perchlorotriphenylmethane, *see* T-60405
Peregrinin, *in* M-70013
Peregrinone, *in* M-70013
Perfluoroazapropene, *see* T-60288
Perfluorobiacetyl, *see* H-60043
Perfluorobicyclobutylidene, *see* D-60513
Perfluoro-2,3-butanedione, *see* H-60043
Perfluorocyclopentadiene, *see* H-60044
Perfluorohexyl bromide, *see* B-70274
Perfluorohexyl chloride, *see* C-70163
Perfluoro-1-iodopropane, *see* H-80026
Perfluoromethanesulfonimide, P-70042
Perfluoro(2-methyl-2*H*-azirine), *see* D-80166
Perfluoro(3-methyl-2*H*-azirine), *see* D-80165
▷ Perfluoro(methylenecyclopropane), *see*
D-60185
Perfluoro(methylenemethylamine), *see*
T-60288
Perfluoro(2-methyl-1,2-oxazetidine), *see*
T-60051
Perfluoromorpholine, *see* N-60056
Perfluoropiperidine, *see* U-60002
▷ Perfluoropropyl bromide, *see* B-80206
▷ Perfluorosuccinic acid, *see* T-80049
Perfluorotrimethylamine, *see* T-60406
Perforatin *A*, *in* A-60075
Perforenone, P-60063
▷ Perhydroazocine, *see* O-70010
Perhydro-3,6*a*-diazapentalene, *see* T-80077
Perhydroquinazoline, *see* D-70013
Perhydro-1,2,5,6-tetrazocine, *see* O-60020
Perhydrotriphenylene, *see* O-70004
▷ Perilla ketone, *see* F-60087
1*H*-Perimidine-2-sulfonic acid, P-80067
1*H*-Perimidine-2(3*H*)-thione, P-10051
1*H*-Perimidin-2(3*H*)-one, P-70043
Perimidone, *see* P-70043
Periplanone D$_1$, P-80069
Periplanone D$_2$, *in* P-80069
Periplocadiol, *in* E-80010
Perlatolic acid, P-80070
Perlatolinic acid, *see* P-80070
Permethylbiphenyl, *see* D-80012
Peroxyauraptenol, *in* A-70269
3α-Peroxy-4,11(13)-eudesmadien-12,8β-olide,
in H-60136
Peroxyvulgarin, *in* V-80013
Perrottetianal *A*, P-60064
Persicaxanthin, P-60065
Perusitin, *in* C-80017
Petalostetin, *in* P-80046
Petimid, *in* P-80132
Petroformyne 1, P-80071
Petroformyne 2, P-80072
Petroformyne 3, P-80073
Petroformyne 4, *in* P-80073
Petrostanol, P-60066

Molecular Formula Index

Molecular Formula Index

$CBrF_3S$
Trifluoromethanesulfenyl bromide, T-**6**0280

$CClF_3O_2S$
Trifluoromethanesulfonic acid; Chloride, *in* T-70241

$CClNO_3S$
▷ Sulfuryl chloride isocyanate, S-**6**0062

CFN
▷ Cyanogen fluoride, C-**8**0169

CF_3IS
Trifluoromethanesulfenyl iodide, T-**6**0282

CF_4O_2S
Trifluoromethanesulfonic acid; Fluoride, *in* T-70241

CF_4S
▷ Trifluoromethanesulfenyl fluoride, T-**6**0281

CF_8S
Pentafluoro(trifluoromethyl)sulfur, P-**6**0041

$CHBrN_4$
5-Bromo-1*H*-tetrazole, B-**8**0262

$CHClFI$
Chlorofluoroiodomethane, C-**8**0056

$CHClN_4$
5-Chlorotetrazole, C-**8**0113

$CHClOS$
Carbonochloridothioic acid, C-**6**0013

$CHFN_4$
5-Fluoro-1*H*-tetrazole, F-**8**0069

CHF_3OS
Trifluoromethanesulfenic acid, T-**6**0279

CHF_3O_3S
Trifluoromethanesulfonic acid, T-70241

$CHIN_4$
5-Iodo-1*H*-tetrazole, I-**8**0042

$CHNO$
▷ Cyanic acid, C-**8**0167
▷ Fulminic acid, F-**8**0086
▷ Isocyanic acid, I-**8**0057

CH_2BrClO_2S
Bromomethanesulfonic acid; Chloride, *in* B-**6**0291
Chloromethanesulfonyl bromide, *in* C-70086

CH_2BrIO_2S
Iodomethanesulfonic acid; Bromide, *in* I-**6**0048

$CH_2Br_2O_2S$
Bromomethanesulfonic acid; Bromide, *in* B-**6**0291

CH_2ClI
Chloroiodomethane, C-**8**0068

$CH_2Cl_2O_2S$
▷ Chloromethanesulfonyl chloride, *in* C-70086

$CH_2F_3NO_2S$
Trifluoromethanesulfonic acid; Amide, *in* T-70241
▷ Trifluoromethanesulfonic acid; Me ester, *in* T-70241

$CH_2N_2OS_3$
1,3λ^4,δ^2,5,2,4-Trithiadiazine; 1-Oxide, *in* T-70327

$CH_2N_2O_3S$
Diazomethanesulfonic acid, D-**8**0065

$CH_2N_2S_3$
1,3λ^4,δ^2,5,2,4-Trithiadiazine, T-70327

CH_2N_4O
2-Tetrazolin-5-one, T-**6**0175

CH_2N_4S
Tetrazole-5-thione, T-**6**0174

CH_2OS
Methanethial; *S*-Oxide, *in* M-70037

CH_2OSe
Methaneselenoic acid, M-70035

CH_2S
Methanethial, M-70037

CH_2S_3
Carbonotrithioic acid, C-**8**0019

CH_2Se
▷ Selenoformaldehyde, S-70029

CH_3BrO
Bromomethanol, B-**8**0212

CH_3BrO_3S
Bromomethanesulfonic acid, B-**6**0291

CH_3ClO
Chloromethanol, C-**8**0070

CH_3ClO_3S
Chloromethanesulfonic acid, C-70086

CH_3IO
Iodomethanol, I-**8**0033

CH_3IO_3S
Iodomethanesulfonic acid, I-**6**0048

CH_3N
▷ Methanimine, M-70038

CH_3NO_2
Carbonimidic acid, C-70019

CH_3NSe
Methaneselenoamide, *in* M-70035

$CH_3N_3O_2S$
▷ Methanesulfonyl azide, M-**6**0033

CH_3N_5
▷ 5-Amino-1*H*-tetrazole, A-70202

$CH_4N_2O_3$
O,O'-Carbonylbis(hydroxylamine), C-**6**0015

$CH_4N_2O_3S$
Aminoiminomethanesulfonic acid, A-**6**0210

$CH_4N_2S_2$
Dithiocarbazic acid, D-**6**0509

CH_4N_2Se
▷ Selenourea, S-70031

CH_5NO_3S
Aminomethanesulfonic acid, A-**6**0215

CS
▷ Carbon monosulfide, C-70020

$C_2BrF_3N_2$
▷ 3-Bromo-3-(trifluoromethyl)-3*H*-diazirine, B-70275

C_2BrF_4I
1-Bromo-1,1,2,2-tetrafluoro-2-iodoethane, B-**8**0260

$C_2Br_2ClF_3$
▷ 1,1-Dibromo-1-chloro-2,2,2-trifluoroethane, D-**8**0084
1,2-Dibromo-1-chloro-1,2,2-trifluoroethane, D-**8**0085

$C_2ClF_3N_2$
▷ 3-Chloro-3-(trifluoromethyl)-3*H*-diazirine, C-70169

C_2ClF_3O
Chlorotrifluorooxirane, C-**6**0140

$C_2Cl_2F_2O$
2,2-Dichloro-3,3-difluorooxirane, D-**6**0128

$C_2Cl_2F_3I$
▷ 1,1-Dichloro-1,2,2-trifluoro-2-iodoethane, D-**8**0142
1,2-Dichloro-1,1,2-trifluoro-2-iodoethane, D-**8**0143

$C_2Cl_2F_4$
▷ 1,1-Dichloro-1,2,2,2-tetrafluoroethane, D-**8**0134

$C_2Cl_2O_2S_2$
Bis(chlorocarbonyl) disulfide, B-**8**0142

$C_2Cl_4O_2$
▷ Trichloromethyl chloroformate, T-70223

$C_2F_2I_2$
1,1-Difluoro-2,2-diiodoethylene, D-**6**0183

C_2F_3N
Trifluoroisocyanomethane, T-**6**0278

$C_2F_3NO_3$
▷ Trifluoroacetyl nitrite, T-**6**0274

$C_2F_4I_2$
1,1,2,2-Tetrafluoro-1,2-diiodoethane, T-**8**0050

$C_2F_4N_2$
▷ 3-Fluoro-3-(trifluoromethyl)-3*H*-diazirine, F-70037

C_2F_4O
▷ Tetrafluorooxirane, T-**6**0049

$C_2F_4OS_2$
2,2,4,4-Tetrafluoro-1,3-dithietane; 1-Oxide, *in* T-**6**0048

$C_2F_4O_2S$
Trifluoroethenesulfonic acid; Fluoride, *in* T-**6**0277

$C_2F_4O_2S_2$
2,2,4,4-Tetrafluoro-1,3-dithietane; 1,1-Dioxide, *in* T-**6**0048

$C_2F_4O_3S_2$
2,2,4,4-Tetrafluoro-1,3-dithietane; 1,1,3-Trioxide, *in* T-**6**0048

$C_2F_4O_4S_2$
2,2,4,4-Tetrafluoro-1,3-dithietane; 1,1,3,3-Tetroxide, *in* T-**6**0048

$C_2F_4S_2$
2,2,4,4-Tetrafluoro-1,3-dithietane, T-**6**0048

C_2F_5N
(Trifluoromethyl)carbonimidic difluoride, T-**6**0288

$C_2F_6N_2O_2$
1,1,1-Trifluoro-*N*-(nitrosooxy)-*N*-(trifluoromethyl)methanamine, *in* B-**6**0180

$C_2F_6O_3S$
Trifluoromethyl trifluoromethanesulfonate, *in* T-70241

$C_2F_6O_5S_2$
Trifluoromethanesulfonic acid; Anhydride, *in* T-70241

$C_2F_7NO_4S_2$
Perfluoromethanesulfonimide, P-70042

$C_2F_8S_2$
Difluoro(trifluoromethanethiolato)(trifluoromethyl)-sulfur, *in* T-**6**0281

$C_2HBrClF_3$
▷ 1-Bromo-2-chloro-1,1,2-trifluoroethane, B-80186

$C_2HBrCl_2F_2$
2-Bromo-1,2-dichloro-1,1-difluoroethane, B-80193

C_2HBrF_2
1-Bromo-1,2-difluoroethylene, B-80194

C_2HBrF_4
1-Bromo-1,1,2,2-tetrafluoroethane, B-80259

$C_2HBr_2F_3$
2,2-Dibromo-1,1,1-trifluoroethane, D-80106

C_2HClF_2
1-Chloro-1,2-difluoroethene, C-80048
▷ 2-Chloro-1,1-difluoroethene, C-80049

C_2HClF_3I
1-Chloro-1,1,2-trifluoro-2-iodoethane, C-80119

C_2HCl_2F
1,1-Dichloro-2-fluoroethene, D-80118

$C_2HCl_2F_3$
1,1-Dichloro-1,2,2-trifluoroethane, D-80139
▷ 1,2-Dichloro-1,1,2-trifluoroethane, D-80140
2,2-Dichloro-1,1,1-trifluoroethane, D-80141

$C_2HCl_3F_2$
1,1,2-Trichloro-1,2-difluoroethane, T-80225
1,2,2-Trichloro-1,1-difluoroethane, T-80226

C_2HCl_4F
1,1,1,2-Tetrachloro-2-fluoroethane, T-80015
1,1,2,2-Tetrachloro-1-fluoroethane, T-80016

C_2HF_3
▷ Trifluoroethene, T-80274

C_2HF_3O
Trifluorooxirane, T-60305

$C_2HF_3O_2$
▷ Trifluoroacetic acid, T-80267

$C_2HF_3O_3S$
Trifluoroethenesulfonic acid, T-60277

C_2HF_6NO
N,N-Bis(trifluoromethyl)hydroxylamine, B-60180

C_2HN_5S
3-Azido-1,2,4-thiadiazole, A-60351

$C_2H_2BrF_2I$
1-Bromo-1,1-difluoro-2-iodoethane, B-80196

$C_2H_2BrF_3$
2-Bromo-1,1,1-trifluoroethane, B-80265

$C_2H_2Cl_2$
▷ 1,1-Dichloroethylene, D-80116

$C_2H_2FN_3O_2S$
1,2,4-Triazole-3(5)-sulfonic acid; Fluoride, in T-80211

$C_2H_2F_2O$
2,3-Difluorooxirane, D-60187

$C_2H_2F_2O_3S$
4,4-Difluoro-1,2-oxathietane 2,2-dioxide, D-60186

$C_2H_2F_3I$
1,1,1-Trifluoro-2-iodoethane, T-80276

$C_2H_2F_3NO$
Trifluoroacetamide, in T-80267

$C_2H_2F_4$
▷ 1,1,2,2-Tetrafluoroethane, T-80051

$C_2H_2N_2O_2$
1,3-Diazetidine-2,4-dione, D-60057

$C_2H_2N_2O_3$
1,2,4-Oxadiazolidine-3,5-dione, O-70060
1,3,4-Oxadiazolidine-2,5-dione, O-70061
1,2,4-Oxazolidine-3,5-dione, O-70070

$C_2H_2N_2O_4$
2H-1,5,2,4-Dioxadiazine-3,6(4H)dione, D-60466

$C_2H_2N_2S_3$
▷ 5-Mercapto-1,3,4-thiadiazoline-2-thione, M-70031

$C_2H_2N_4$
1,2,4,5-Tetrazine, T-60170

$C_2H_2N_4O_2$
1,2-Dihydro-1,2,4,5-tetrazine-3,6-dione, D-70268

C_2H_2O
Ethynol, E-80070

$C_2H_2O_5$
Dicarbonic acid, D-70111

$C_2H_2S_4$
Ethanebis(dithioic)acid, E-60040

$C_2H_2Se_2$
1,2-Diselenete, D-60498

$C_2H_3AsF_6N_2S_2$
5-Methyl-1,3,2,4-dithiazolium(1+); Hexafluoroarsenate, in M-70061

$C_2H_3BrN_4$
5-Bromo-1H-tetrazole; 1-Me, in B-80262

$C_2H_3ClN_2O$
3-Chloro-3-methoxy-3H-diazirine, C-70087

$C_2H_3ClN_2S_2$
5-Methyl-1,3,2,4-dithiazolium(1+); Chloride, in M-70061

C_2H_3ClOS
Carbonochloridothioic acid; O-Me, in C-60013

$C_2H_3Cl_2F$
1,1-Dichloro-1-fluoroethane, D-80117

$C_2H_3FN_2O$
▷ 3-Fluoro-3-methoxy-3H-diazirine, F-60050

$C_2H_3F_3$
▷ 1,1,1-Trifluoroethane, T-80273

$C_2H_3F_3OS$
Methyl trifluoromethanesulfenate, in T-60279
Trifluoro(methylsulfinyl)methane, in M-70145

$C_2H_3F_3O_2S$
Trifluoro(methylsulfonyl)methane, in M-70145

$C_2H_3F_3S$
Methyl trifluoromethyl sulfide, M-70145

C_2H_3N
2H-Azirine, A-70292
Ethenimine, E-80058
Ethynamine, E-60056

C_2H_3NO
Fulminic acid; Trimolecular Me ester, in F-80086

$C_2H_3NO_2$
Methoxy isocyanate, M-80045

$C_2H_3NS_2$
Methanesulfenyl thiocyanate, M-60032

C_2H_3NSe
Methyl selenocyanate, M-60127

$C_2H_3N_2S_2^{\oplus}$
5-Methyl-1,3,2,4-dithiazolium(1+), M-70061

$C_2H_3N_3O_3S$
1,2,4-Triazole-3(5)-sulfonic acid, T-80211

$C_2H_3N_3S$
▷ 1,2-Dihydro-3H-1,2,4-triazole-3-thione, D-60298

$C_2H_4BrNO_2$
1-Bromo-2-nitroethane, B-60303

C_2H_4ClI
1-Chloro-1-iodoethane, C-60106
1-Chloro-2-iodoethane, C-70078

$C_2H_4IN_3$
1-Azido-2-iodoethane, A-60346

$C_2H_4N_2O$
1,2-Diazetidin-3-one, D-70055

$C_2H_4N_2O_4$
▷ Oxalohydroxamic acid, O-60048

$C_2H_4N_2S_2$
▷ Ethanedithioamide, E-60041

$C_2H_4N_4O_2$
▷ Tetrahydro-1,2,4,5-tetrazine-3,6-dione, T-70097

$C_2H_4N_4O_2S$
1,2,4-Triazole-3(5)-sulfonic acid; Amide, in T-80211

$C_2H_4N_4S$
2,5-Diamino-1,3,4-thiadiazole, D-60049
1H-Tetrazole-5-thiol; S-Me, in T-60174
Tetrazole-5-thione; 1-Me, in T-60174

$C_2H_4N_6$
3,6-Diamino-1,2,4,5-tetrazine, D-70045

C_2H_4OS
Mercaptoacetaldehyde, M-80017

$C_2H_4O_2$
▷ 1,2-Dioxetane, D-60469

$C_2H_4O_4S$
Formylmethanesulfonic acid, F-80076

$C_2H_4O_5S$
Sulfoacetic acid, S-80050

C_2H_5NO
▷ Acetamide, A-70019

$C_2H_5NO_3$
C-Methoxycarbohydroxamic acid, M-70044

$C_2H_5N_3O$
▷ 2-Azidoethanol, A-70287

$C_2H_5N_3O_2$
Biuret, B-60190
▷ Semioxamazide, S-60020

$C_2H_5N_5$
5-Amino-1-methyl-1H-tetrazole, in A-70202
5-Amino-2-methyl-2H-tetrazole, in A-70202
5-Methylamino-1H-tetrazole, in A-70202

$C_2H_6N_2O$
▷ Acetamidoxime, A-60016

$C_2H_6N_2O_2$
▷ N-Nitrodimethylamine, N-80045

$C_2H_6N_2O_3S$
Aminoiminomethanesulfonic acid; N-Me, in A-60210

$C_2H_6N_2S_2$
Dithiocarbazic acid; Me ester, in D-60509

$C_2H_6N_2Se$
Selenourea; N-Me, in S-70031

C_2H_6OS
(Methylthio)methanol, M-70138

$C_2H_6O_2S_2$
2-Mercaptoethanesulfinic acid, M-60022

$C_2H_6O_6S_2$
Bis(methylsulfonyl) peroxide, B-70154

C_2H_7NO
O,N-Dimethylhydroxylamine, D-70418

$C_2H_7NO_2$
O-(2-Hydroxyethyl)hydroxylamine, H-60134

$C_2H_7NO_3S$
(Methylamino)methanesulfonic acid, in A-60215

$C_2H_8N_2$
▷ 1,1-Dimethylhydrazine, D-70416

$C_2H_9N_3$
2-Hydrazinoethylamine, H-70096

$C_2H_{10}N_4$
 1,2-Dihydrazinoethane, D-60192

C_2N_2
 Diisocyanogen, D-70362

$C_2N_2Se_3$
 Dicyanotriselenide, D-60165

C_2O_2
 Dicarbon dioxide, D-80114

C_3AlCl_7
 Trichlorocyclopropenylium;
  Tetrachloroaluminate, *in* T-70221

C_3BrF_2I
 1-Bromo-3,3-difluoro-2-iodocyclopropene,
  B-80195

C_3BrF_3
 1-Bromo-2,3,3-trifluorocyclopropene, B-80264

C_3BrF_7
 ▷ 1-Bromo-1,1,2,2,3,3,3-heptafluoropropane,
  B-80206
 2-Bromo-1,1,1,2,3,3,3-heptafluoropropane,
  B-80207

$C_3Br_2F_6$
 1,3-Dibromo-1,1,2,2,3,3-hexafluoropropane,
  D-80094

$C_3Br_3N_3O_2$
 3,4,5-Tribromo-1*H*-pyrazole; 1-Nitro, *in*
  T-60246

$C_3ClF_2N_3$
 2-Chloro-4,6-difluoro-1,3,5-triazine, C-60052
 6-Chloro-3,5-difluoro-1,2,4-triazine, C-60053

C_3ClF_3
 1-Chloro-2,3,3-trifluorocyclopropene, C-80118

C_3ClF_5
 1-Chloro-1,2,3,3,3-pentafluoro-1-propene,
  C-80090

$C_3Cl_2F_4$
 1,1-Dichloro-2,3,3,3-tetrafluoro-1-propene,
  D-80135
 1,2-Dichloro-1,3,3,3-tetrafluoro-1-propene,
  D-80136

$C_3Cl_2F_6$
 1,2-Dichloro-1,1,2,3,3,3-hexafluoropropane,
  D-80120
 1,3-Dichloro-1,1,2,2,3,3-hexafluoropropane,
  D-80121
 2,2-Dichloro-1,1,1,3,3,3-hexafluoropropane,
  D-80122

$C_3Cl_3^{\oplus}$
 Trichlorocyclopropenylium, T-70221

$C_3Cl_3F_5$
 1,1,1-Trichloro-2,2,3,3,3-pentafluoropropane,
  T-80228
 1,1,3-Trichloro-1,2,2,3,3-pentafluoropropane,
  T-80229
 1,2,2-Trichloro-1,1,3,3,3-pentafluoropropane,
  T-80230
 1,2,3-Trichloro-1,1,2,3,3-pentafluoropropane,
  T-80231

$C_3Cl_3N_3$
 Trichloro-1,2,3-triazine, T-80232

$C_3Cl_4F_4$
 1,1,1,2-Tetrachloro-2,3,3,3-tetrafluoropropane,
  T-80018
 1,1,1,3-Tetrachloro-2,2,3,3-tetrafluoropropane,
  T-80019
 1,1,2,2-Tetrachloro-1,3,3,3-tetrafluoropropane,
  T-80020
 1,1,2,3-Tetrachloro-1,2,3,3-tetrafluoropropane,
  T-80021
 1,1,3,3-Tetrachloro-1,2,2,3-tetrafluoropropane,
  T-80022

$C_3Cl_6O_3$
 Bis(trichloromethyl) carbonate, B-70163

C_3Cl_8
 Octachloropropane, O-80004

C_3Cl_9Sb
 Trichlorocyclopropenylium;
  Hexachloroantimonate, *in* T-70221

$C_3F_2I_2$
 3,3-Difluoro-1,2-diiodocyclopropene, D-80160

$C_3F_3N_3$
 Trifluoro-1,2,3-triazine, T-70245

C_3F_5N
 2,2-Difluoro-3-(trifluoromethyl)-2*H*-azirine,
  D-80165
 2,3-Difluoro-2-(trifluoromethyl)-2*H*-azirine,
  D-80166

$C_3F_6N_2$
 3,3-Bis(trifluoromethyl)-3*H*-diazirine, B-60178

$C_3F_6O_2S$
 Bis(trifluoromethyl)sulfene, B-80162

$C_3F_6O_4S$
 Trifluoroacetyl trifluoromethanesulfonate, *in*
  T-70241

C_3F_7I
 1,1,2,2,3,3,3-Heptafluoro-1-iodopropane,
  H-80026

C_3F_7NO
 3,3,4,4-Tetrafluoro-2-(trifluoromethyl)-1,2-
  oxazetidine, T-60051

C_3F_9N
 Tris(trifluoromethyl)amine, T-60406

$C_3HBrINS$
 2-Bromo-4-iodothiazole, B-60287
 2-Bromo-5-iodothiazole, B-60288
 4-Bromo-2-iodothiazole, B-60289
 5-Bromo-2-iodothiazole, B-60290

C_3HBr_2NS
 2,4-Dibromothiazole, D-60109
 2,5-Dibromothiazole, D-60110
 4,5-Dibromothiazole, D-60111

$C_3HBr_2N_3$
 4,5-Dibromo-1,2,3-triazine, D-70100
 4,5-Dibromo-1,2,3-triazine, D-80105

$C_3HBr_3N_2$
 ▷ 2,4,5-Tribromo-1*H*-imidazole, T-60245
 3,4,5-Tribromo-1*H*-pyrazole, T-60246

C_3HClF_4
 1-Chloro-2,3,3,3-tetrafluoro-1-propene,
  C-80109

C_3HCl_2NS
 2,4-Dichlorothiazole, D-60154
 2,5-Dichlorothiazole, D-60155
 4,5-Dichlorothiazole, D-60156

C_3HCl_5
 Pentachlorocyclopropane, P-70018

C_3HF_3
 1,3,3-Trifluorocyclopropene, T-80271

C_3HF_5
 1,1,3,3,3-Pentafluoro-1-propene, P-80028

C_3HI_2NS
 2,4-Diiodothiazole, D-60388
 2,5-Diiodothiazole, D-60389

$C_3HI_3N_2$
 ▷ 2,4,5-Triiodo-1*H*-imidazole, T-70276

$C_3HN_3O_5$
 3,5-Dinitroisoxazole, D-70452

$C_3H_2BrN_3$
 5-Bromo-1,2,3-triazine, B-70273
 5-Bromo-1,2,3-triazine, B-80263

$C_3H_2BrN_3O_2$
 3(5)-Bromo-5(3)-nitro-1*H*-pyrazole, B-60305
 4-Bromo-3(5)-nitro-1*H*-pyrazole, B-60304

$C_3H_2Br_2$
 1,2-Dibromocyclopropene, D-60094

$C_3H_2Br_2N_2$
 2,4(5)-Dibromo-1*H*-imidazole, D-70093
 4,5-Dibromo-1*H*-imidazole, D-60100

$C_3H_2ClF_3$
 2-Chloro-3,3,3-trifluoro-1-propene, C-60141

$C_3H_2ClF_3O$
 3,3,3-Trifluoropropanoic acid; Chloride, *in*
  T-80284

$C_3H_2ClF_3O_2$
 2-Chloro-3,3,3-trifluoropropanoic acid,
  C-70170

$C_3H_2ClF_5$
 1-Chloro-1,1,3,3,3-pentafluoropropane,
  C-80088
 1-Chloro-2,2,3,3,3-pentafluoropropane,
  C-80089

$C_3H_2ClN_3O_2$
 6-Chloro-1,2,4-triazine-3,5(1*H*,3*H*)-dione,
  C-80114
 6-Chloro-1,3,5-triazine-2,4(1*H*,3*H*)-dione,
  C-80115

$C_3H_2Cl_2N_2$
 4,5-Dichloro-1*H*-imidazole, D-70122

$C_3H_2Cl_2O$
 2,3-Dichloro-2-propenal, D-80127
 ▷ 3,3-Dichloro-2-propenal, D-80128

$C_3H_2Cl_4O_2$
 Chloromethanol; Trichloroacetyl, *in* C-80070

$C_3H_2F_4O$
 2-Fluoro-3-(trifluoromethyl)oxirane, F-80070

$C_3H_2F_6$
 1,1,1,2,2,3-Hexafluoropropane, H-80049
 1,1,1,3,3,3-Hexafluoropropane, H-80050

C_3H_2INS
 2-Iodothiazole, I-60064
 4-Iodothiazole, I-60065
 5-Iodothiazole, I-60066

$C_3H_2N_2O_2S$
 2-Thioxo-4,5-imidazolidinedione, T-70187

$C_3H_2N_3$
 ▷ 2-Diazo-2*H*-imidazole, D-70060

$C_3H_2N_4$
 4-Cyano-1,2,3-triazole, *in* T-60239
 ▷ 4-Diazo-4*H*-imidazole, D-70061

$C_3H_2N_6$
 Tetrazolo[1,5-*b*][1,2,4]triazine, T-80164

$C_3H_2O_3$
 2,3-Dihydroxy-2-cyclopropen-1-one, D-70290

C_3H_2S
 1,2-Propadiene-1-thione, P-70117

$C_3H_2S_3$
 2,3-Dimercapto-2-propene-1-thione, D-80410

$C_3H_2S_5$
 4,5-Dimercapto-1,3-dithiole-2-thione, D-80409

$C_3H_3BrN_2$
 ▷ 2-Bromo-1*H*-imidazole, B-70220
 4(5)-Bromo-1*H*-imidazole, B-70221
 3(5)-Bromopyrazole, B-80253
 ▷ 4-Bromopyrazole, B-80254

$C_3H_3Br_2Cl$
 1,1-Dibromo-2-chlorocyclopropane, D-80082

$C_3H_3Br_3$
 1,2,3-Tribromocyclopropane, T-70211

$C_3H_3ClN_2$
 2-Chloro-1*H*-imidazole, C-70076
 4(5)-Chloro-1*H*-imidazole, C-70077
 3-Chloro-1*H*-pyrazole, C-70155
 ▷ 4-Chloro-1*H*-pyrazole, C-70156

$C_3H_3ClN_2O$
 4-Chloro-5-methylfurazan, C-60114

$C_3H_3ClN_2O_2$
 2-Chloro-2,3-dihydro-1*H*-imidazole-4,5-dione,
  C-60054
 4-Chloro-3-methylfuroxan, *in* C-60114

C_3H_3ClO
 ▷ 2-Propenoic acid; Chloride, *in* P-70121

$C_3H_3Cl_2F_3$
2,2-Dichloro-1,1,1-trifluoropropane, D-80144
2,3-Dichloro-1,1,1-trifluoropropane, D-80145
3,3-Dichloro-1,1,1-trifluoropropane, D-80146

$C_3H_3Cl_2F_3O$
2,2-Dichloro-3,3,3-trifluoro-1-propanol, D-60161

$C_3H_3Cl_2N_2O_2$
3-Chloro-4-methylfuroxan, in C-60114

$C_3H_3Cl_3$
1,2,3-Trichlorocyclopropane, T-70220
1,1,2-Trichloro-1-propene, T-60254

$C_3H_3F_3O$
(Trifluoromethyl)oxirane, T-60294
3,3,3-Trifluoropropanal, T-80282

$C_3H_3F_3O_2$
Trifluoroacetic acid; Me ester, in T-80267
3,3,3-Trifluoropropanoic acid, T-80284

$C_3H_3F_6NO$
1,1,1,1',1',1'-Hexafluoro-N-methoxydimethylamine, in B-60180

$C_3H_3IN_2$
4(5)-Iodo-1H-imidazole, I-70042
3(5)-Iodopyrazole, I-80040
4-Iodopyrazole, I-80041

C_3H_3N
▷ Vinyl cyanide, in P-70121

$C_3H_3NO_3S_2$
4-Thiazolesulfonic acid, T-60198
5-Thiazolesulfonic acid, T-60199

$C_3H_3NO_4$
2H-1,5,2-Dioxazine-3,6(4H)-dione, D-70464

$C_3H_3NS_2$
2(3H)-Thiazolethione, T-60200

$C_3H_3N_3O$
2-Nitrosoimidazole, N-80059

$C_3H_3N_3O_2$
3-Azido-2-propenoic acid, A-70290
1H-1,2,3-Triazole-4-carboxylic acid, T-60239

$C_3H_3N_3O_3$
▷ Cyanuric acid, C-60179

$C_3H_3N_5$
3(5)-Azidopyrazole, A-70291

$C_3H_3N_7O_2$
7,8-Dihydrotetrazolo[1,5-b][1,2,4]triazine; 8-Nitro, in D-80256

$C_3H_4BrClO_2$
Bromomethanol; Chloroacetyl, in B-80212

C_3H_4ClNO
▷ 1-Chloro-2-isocyanatoethane, C-80069

C_3H_4ClNS
▷ 1-Chloro-2-isothiocyanoethane, C-70081

$C_3H_4Cl_2O_2$
Chloromethanol; Chloroacetyl, in C-80070

$C_3H_4F_3NO$
3,3,3-Trifluoropropanoic acid; Amide, in T-80284

$C_3H_4I_2N_6$
1,3-Diiodo-2,2-diazidopropane, D-60387

$C_3H_4N_2$
Diazocyclopropane, D-60060
▷ 1H-Pyrazole, P-70152

$C_3H_4N_2O$
4-Aminoisoxazole, A-80096
2-Diazopropanal, D-80066

$C_3H_4N_2OS$
▷ 2-Thioxo-4-imidazolidinone, T-70188

$C_3H_4N_2O_2$
2-Propyn-1-amine; N-Nitro, in P-60184
3,5-Pyrazolidinedione, P-80200

$C_3H_4N_2O_3$
2,3-Dihydro-2-hydroxy-1H-imidazole-4,5-dione, D-60243
5-Hydroxy-2,4-imidazolidinedione, H-60156
1,2,4-Oxadiazolidine-3,5-dione; 4-Me, in O-70060

$C_3H_4N_2O_4$
1,2-Dinitrocyclopropane, D-80494

$C_3H_4N_2O_5$
1,1-Dinitro-2-propanone, D-70454

$C_3H_4N_2S$
2-Cyanoethanethioamide, C-70209
1,3-Dihydro-2H-imidazole-2-thione, D-80215

$C_3H_4N_2S_3$
3-Methyl-1,3,4-thiadiazolidine-2,5-dithione, in M-70031
5-Methylthio-1,3,4-thiadiazole-2(3H)-thione, in M-70031

$C_3H_4N_4$
▷ 3-Amino-1,2,4-triazine, A-70203

$C_3H_4N_4O$
4-Amino-1,3,5-triazin-2(1H)-one, A-80149
1H-1,2,3-Triazole-4-carboxylic acid; Amide, in T-60239

$C_3H_4N_4S$
4-Amino-1,3,5-triazine-2(1H)-thione, A-80148

$C_3H_4N_6$
7,8-Dihydrotetrazolo[1,5-b][1,2,4]triazine, D-80256

C_3H_4O
▷ Propenal, P-70119

$C_3H_4O_2$
1,3-Dioxole, D-60473
▷ 2-Propenoic acid, P-70121

$C_3H_4O_3$
3-Oxopropanoic acid, O-60090

$C_3H_4S_2Se$
1,3-Dithiolane-2-selone, D-70525

$C_3H_5BrO_2$
Bromomethanol; Ac, in B-80212

$C_3H_5ClN_2$
2-Chloro-4,5-dihydro-1H-imidazole, C-80050

C_3H_5ClOS
▷ Carbonochloridothioic acid; O-Et, in C-60013
Carbonochloridothioic acid; S-Et, in C-60013

$C_3H_5ClO_2$
Chloromethanol; Ac, in C-80070

$C_3H_5ClO_3$
▷ 3-Chloro-2-hydroxypropanoic acid, C-70072

C_3H_5FO
▷ (Fluoromethyl)oxirane, F-70032

$C_3H_5F_2NO_2$
2-Amino-3,3-difluoropropanoic acid, A-60135

$C_3H_5F_3O$
1,1,1-Trifluoro-2-methoxyethane, T-80277

$C_3H_5F_3OS$
Ethyl trifluoromethanesulfenate, in T-60279

$C_3H_5F_3O_3S$
Trifluoromethanesulfonic acid; Et ester, in T-70241

$C_3H_5F_3O_4S_2$
(Methylsulfonyl)[(trifluoromethyl)sulfonyl]methane, M-70133

$C_3H_5IO_2$
▷ Iodomethanol; Ac, in I-80033

$C_3H_5I_2N_3$
2-Azido-1,3-diiodopropane, A-80198

C_3H_5N
2,3-Dihydroazete, D-70172
▷ 2-Propyn-1-amine, P-60184

C_3H_5NO
▷ Acrylamide, in P-70121
3-Azetidinone, A-80192
Propenal; Oxime, in P-70119

$C_3H_5NO_2$
3-Amino-2-oxetanone, A-80129

$C_3H_5NO_3$
Isonitrosopropanoic acid, in O-60090
▷ 3-Nitropropanal, N-60037

C_3H_5NS
2-Propenethioamide, P-70120

$C_3H_5NS_2$
2-Thiazoledinethione, T-60197

$C_3H_5N_3O_2$
Dihydro-1,3,5-triazine-2,4(1H,3H)-dione, D-60297
▷ 2-Imidazolidinone; 1-Nitroso, in I-70005
Tetrahydro-1,2,4-triazine-3,6-dione, T-60099

C_3H_6BrF
1-Bromo-3-fluoropropane, B-70212

$C_3H_6BrNO_2$
1-Bromo-1-nitropropane, B-70258
2-Bromo-2-nitropropane, B-70259

$C_3H_6Br_2$
▷ 1,2-Dibromopropane, D-60105

C_3H_6ClN
2-Chloro-2-propen-1-amine, C-70151

$C_3H_6ClNO_2$
▷ 2-Chloro-2-nitropropane, C-60120

$C_3H_6ClN_4^{\oplus}$
N,N-Dimethylazidochloromethyleniminium(1+), D-70369

$C_3H_6Cl_2N_4$
N,N-Dimethylazidochloromethyleniminium(1+); Chloride, in D-70369

C_3H_6FI
1-Fluoro-3-iodopropane, F-70026

$C_3H_6FNO_2$
3-Amino-2-fluoropropanoic acid, A-60162

$C_3H_6F_3N$
3,3,3-Trifluoro-1-propanamine, T-80283

$C_3H_6F_3NO_2S$
Trifluoromethanesulfonic acid; Diethylamide, in T-70241

$C_3H_6IN_3$
1-Azido-3-iodopropane, A-60347

$C_3H_6IN_3O$
2-Azido-3-iodo-1-propanol, A-80200

$C_3H_6N_2$
4,5-Dihydro-1H-imidazole, D-60245
4,5-Dihydro-1H-pyrazole, D-70253

$C_3H_6N_2O$
▷ Azetidine; N-Nitroso, in A-70283
▷ 2-Imidazolidinone, I-70005
4-Imidazolidinone, I-80006

$C_3H_6N_2O_2$
Acetamidoxime; O-Formyl, in A-60016
3,3-Dimethoxy-3H-diazirine, D-80412

$C_3H_6N_4$
1H-1,2,3-Triazole-4-methanamine, T-80210

$C_3H_6N_4O$
2-Tetrazolin-5-one; 1-Et, in T-60175

$C_3H_6N_4O_2$
2-Amino-3-azidopropanoic acid, A-80058
2-Azido-2-nitropropane, A-60349

$C_3H_6N_4S$
1H-Tetrazole-5-thiol; S-Et, in T-60174
Tetrazole-5-thione; 1,4-Di-Me, in T-60174
Tetrazole-5-thione; 1-Et, in T-60174

$C_3H_6N_6$
▷ 2,2-Diazidopropane, D-60058

$C_3H_6N_6O_3$
▷ Hexahydro-1,3,5-triazine; 1,3,5-Trinitroso, *in*
H-60060

C_3H_6O
▷ Methyloxirane, M-70100

C_3H_6OS
1-Mercapto-2-propanone, M-60028
(Methylthio)acetaldehyde, *in* M-80017
(Methylthio)oxirane, M-80173
Propenylsulfenic acid, P-70123

C_3H_6OSe
3-Selenetanol, S-80022

$C_3H_6O_2$
Dimethyldioxirane, D-70399
1,2-Dioxolane, D-70467

$C_3H_6O_2S$
1,3,2-Dioxathiane, D-80502
▷ 2-Mercaptopropanoic acid, M-60027

$C_3H_6O_3S$
Trimethylene sulfite, *in* D-80502

$C_3H_6O_4S$
Trimethylene sulfate, *in* D-80502

$C_3H_6S_3$
Carbonotrithioic acid; Di-Me ester, *in*
C-80019

C_3H_7N
Azetidine, A-70283

C_3H_7NO
▷ Acetamide; *N*-Me, *in* A-70019
2-(Hydroxymethyl)aziridine, H-80194

C_3H_7NOS
2-Mercaptopropanoic acid; Amide, *in*
M-60027

$C_3H_7NO_3$
3-Amino-2-hydroxypropanoic acid, A-70160
▷ 3-Nitro-1-propanol, N-60039

$C_3H_7NO_4$
▷ 2-Nitro-1,3-propanediol, N-60038

C_3H_7NSe
N,N-Dimethylmethaneselenoamide, *in*
M-70035

$C_3H_7N_3$
2-Azidopropane, A-70289

$C_3H_7N_3O$
1-Azido-1-propanol, A-80204

$C_3H_7N_3O_2$
Allophanic methylamide, *in* B-60190

$C_3H_7N_5$
5-Amino-1-ethyl-1*H*-tetrazole, *in* A-70202
5-Amino-2-ethyl-2*H*-tetrazole, *in* A-70202
1,4-Dihydro-5-imino-3,4-dimethyltetrazole, *in*
A-70202
5-Dimethylamino-1*H*-tetrazole, *in* A-70202
5-Ethylamino-1*H*-tetrazole, *in* A-70202
5-Imino-1,3-dimethyl-1*H*,3*H*-tetrazole, *in*
A-70202
1-Methyl-5-methylamino-1*H*-tetrazole, *in*
A-70202
2-Methyl-5-methylamino-2*H*-tetrazole, *in*
A-70202

$C_3H_8N_2O_2$
N,N-Dimethylcarbamohydroxamic acid,
D-60414

$C_3H_8N_2O_3S$
Aminoiminomethanesulfonic acid; *N,N*-Di-
Me, *in* A-60210

$C_3H_8N_2Se$
Selenourea; N^1,N^1-Di-Me, *in* S-70031
Selenourea; N^1,N^3-Di-Me, *in* S-70031
Selenourea; *N*-Et, *in* S-70031

$C_3H_8O_2$
▷ 1,2-Propanediol, P-70118

$C_3H_8S_2$
2,2-Propanedithiol, P-60175

C_3H_9NO
▷ Trimethylamine oxide, T-60354

$C_3H_9NO_3S$
(Dimethylamino)methanesulfonic acid, *in*
A-60215

$C_3H_9N_3$
Hexahydro-1,3,5-triazine, H-60060

C_3OS
3-Thioxo-1,2-propadien-1-one, T-60216

C_4Br_4O
Tetrabromofuran, T-60023

$C_4ClF_3N_2$
3-Chloro-4,5,6-trifluoropyridazine, C-60142
4-Chloro-3,5,6-trifluoropyridazine, C-60143
5-Chloro-2,4,6-trifluoropyrimidine, C-60144

C_4ClF_5
1-Chloro-2,3,3,4,4-pentafluorocyclobutene,
C-80087

$C_4Cl_2F_4$
1,2-Dichloro-3,3,4,4-tetrafluorocyclobutene,
D-80132
3,4-Dichloro-1,2,3,4-tetrafluorocyclobutene,
D-80133

$C_4Cl_2F_4O_2$
Tetrafluorobutanedioic acid; Dichloride, *in*
T-80049

$C_4Cl_2O_2$
Chlorofumaric acid; Dichloride, *in* C-80043

$C_4Cl_3F_3$
1,2,3-Trichloro-3,4,4-trifluorocyclobutene,
T-80233

$C_4F_2O_3$
3,4-Difluoro-2,5-furandione, *in* D-80159

$C_4F_4O_3$
3,3,4,4-Tetrafluorodihydro-2,5-furandione, *in*
T-80049

C_4F_6
▷ (Difluoromethylene)tetrafluorocyclopropane,
D-60185

$C_4F_6NS_2$
4,5-Bis(trifluoromethyl)-1,3,2-dithiazol-2-yl,
B-70164

$C_4F_6O_2$
1,1,1,4,4,4-Hexafluoro-2,3-butanedione,
H-60043

$C_4F_6O_3$
▷ Trifluoroacetic acid; Anhydride, *in* T-80267

$C_4F_6O_4S$
4-[2,2,2-Trifluoro-1-
(trifluoromethyl)ethylidene]-1,3,2-
dioxathietane 2,2-dioxide, T-60311

C_4F_6S
Bis(trifluoromethyl)thioketene, B-60187

$C_4F_6S_2$
3,4-Bis(trifluoromethyl)-1,2-dithiete, B-60179
Bis[(trifluoromethyl)thio]acetylene, B-80163

C_4F_8O
2,2-Difluoro-3,3-bis(trifluoromethyl)oxirane,
D-80158

$C_4F_8O_3S$
4,4-Difluoro-3,3-bis(trifluoromethyl)-1,2-
oxathietane 2,2-dioxide, D-60181

C_4F_9NO
Nonafluoromorpholine, N-60056

$C_4HBrF_2O_2$
4-Bromo-4,4-difluoro-2-butynoic acid,
B-70202

C_4HBr_3S
2,3,5-Tribromothiophene, T-60247

C_4HClN_2
1-Chloro-1,2-dicyanoethylene, *in* C-80043

C_4HClO_3
3-Chloro-2,5-furandione, *in* C-80043
3-Chloro-4-hydroxy-3-cyclobutene-1,2-dione,
C-70071

$C_4HCl_2F_3$
1,4-Dichloro-3,3,4-trifluorocyclobutene,
D-80138

C_4HFO_3
3-Fluoro-2,5-furandione, *in* F-80030

$C_4HF_3N_2$
Trifluoropyrazine, T-60306
2,4,6-Trifluoropyrimidine, T-60307
4,5,6-Trifluoropyrimidine, T-60308

$C_4HF_3O_3$
3,3,4-Trifluoro-dihydro-2,5(2*H*,5*H*)-
furandione, *in* T-80269

C_4HF_5
3,3,4,4,4-Pentafluoro-1-butyne, P-60032
1,3,3,4,4-Pentafluorocyclobutene, P-60033

C_4HF_8NO
2,2,3,3,5,5,6,6-Octafluoromorpholine, O-60009

C_4HNO
4-Oxo-2-butynenitrile, *in* O-80070

$C_4HN_5O_8$
2,3,4,5-Tetranitro-1*H*-pyrrole, T-60160

$C_4H_2BrNO_2S$
2-Bromo-4-nitrothiophene, B-80239

$C_4H_2Br_2N_2O$
4,5-Dibromo-1*H*-imidazole-2-carboxaldehyde,
D-60101

$C_4H_2Br_2N_2O_2$
4,5-Dibromo-1*H*-imidazole-2-carboxylic acid,
D-60102

$C_4H_2ClNO_2$
5-Isoxazolecarboxylic acid; Chloride, *in*
I-60142

$C_4H_2Cl_2N_2O$
4,5-Dichloro-1*H*-imidazole-2-carboxaldehyde,
D-60134
2,6-Dichloro-4(3*H*)-pyrimidinone, D-80130
4,6-Dichloro-2(1*H*)-pyrimidinone, D-80131

$C_4H_2Cl_2N_2O_2$
4,5-Dichloro-1*H*-imidazole-2-carboxylic acid,
D-60135

$C_4H_2Cl_3NO$
5-(Trichloromethyl)isoxazole, T-60253

$C_4H_2F_2N_2$
2,3-Difluoropyrazine, D-60188
2,6-Difluoropyrazine, D-60189

$C_4H_2F_2O_4$
1,2-Difluoro-2-butenedioic acid, D-80159

$C_4H_2F_3N$
▷ 2-Trifluoromethylacrylonitrile, *in* T-60295

$C_4H_2F_4$
3,3,4,4-Tetrafluorocyclobutene, T-60043

$C_4H_2F_4O_4$
▷ Tetrafluorobutanedioic acid, T-80049

$C_4H_2F_6S_2$
1,2-Bis[(trifluoromethyl)thio]ethene, B-80164

$C_4H_2N_2O$
5-Cyanoisoxazole, *in* I-60142

$C_4H_2N_2O_2$
Dicyanoacetic acid, D-70149

$C_4H_2N_2O_2S$
1,2,5-Thiadiazole-3,4-dicarboxaldehyde,
T-70162

$C_4H_2N_2O_3$
3-Diazo-2,4(5*H*)-furandione, D-70059

$C_4H_2N_2S_2$
(Dimercaptomethylene)malononitrile, *in*
D-60395

$C_4H_2N_4OS$
[1,2,5]Thiadiazolo[3,4-*d*]pyridazin-4(5*H*)-one,
T-70164

$C_4H_2N_4O_4$
1,2,4,5-Tetrazine-3,6-dicarboxylic acid,
T-80163
[1,2,4]Triazolo[1,2-*a*][1,2,4]triazole-
1,3,5,7(2*H*,6*H*)-tetrone, T-70209

$C_4H_2N_4O_6$
2,3,4-Trinitro-1*H*-pyrrole, T-60373
2,3,5-Trinitro-1*H*-pyrrole, T-60374

$C_4H_2N_4S$
[1,2,5]Thiadiazolo[3,4-*d*]pyridazine, T-70163

$C_4H_2N_8O_{10}$
Tetrahydro-1,3,4,6-tetranitroimidazo[4,5-
d]imidazole-2,5-(1*H*,3*H*)-dione, T-70096

$C_4H_2O_2$
2-Butynedial, B-60351

$C_4H_2O_3$
▷ 2-Oxo-3-butynoic acid, O-80069
4-Oxo-2-butynoic acid, O-80070

$C_4H_2S_4$
3,4-Dimercapto-3-cyclobutene-1,2-dithione,
D-70364

$C_4H_3BrN_2OS$
5-Bromo-3,4-dihydro-4-thioxo-2(1*H*)-
pyrimidinone, B-70205

$C_4H_3BrN_2O_2$
▷ 5-Bromo-2,4-(1*H*,3*H*)-pyrimidinedione,
B-60318
6-Bromo-2,4-(1*H*,3*H*)-pyrimidinedione,
B-60319

$C_4H_3BrO_2$
5-Bromo-2(5*H*)-furanone, B-80204

$C_4H_3Br_2N_3$
4,5-Dibromo-6-methyl-1,2,3-triazine, D-80096

$C_4H_3ClN_2$
2-Chloropyrazine, C-80108
5-Chloropyrimidine, C-70157

$C_4H_3ClN_2O$
2-Chloropyrazine; 1-Oxide, *in* C-80108
2-Chloropyrazine; 4-Oxide, *in* C-80108
5-Chloropyrimidine; *N*-Oxide, *in* C-70157

$C_4H_3ClN_2OS$
5-Chloro-4-thioxo-2(1*H*)-pyrimidinone,
C-70162

$C_4H_3ClN_2O_2$
▷ 5-Chloro-2,4-(1*H*,3*H*)-pyrimidinedione,
C-60130
6-Chloro-2,4-(1*H*,3*H*)-pyrimidinedione,
C-70158

$C_4H_3ClO_2S_2$
2-Thiophenesulfonic acid; Chloride, *in*
T-60213
3-Thiophenesulfonic acid; Chloride, *in*
T-60214

$C_4H_3ClO_4$
2-Chloro-2-butenedioic acid, C-80043

C_4H_3ClS
3-Chlorothiophene, C-60136

$C_4H_3FN_2$
5-Fluoropyrimidine, F-80061

$C_4H_3FN_2OS$
5-Fluoro-3,4-dihydro-4-thioxo-2(1*H*)-
pyrimidinone, F-70019

$C_4H_3FO_4$
2-Fluoro-2-butenedioic acid, F-80030

$C_4H_3F_3$
1,1,2-Trifluoro-1,3-butadiene, T-80268
1,1,1-Trifluoro-2-butyne, T-60275
1,4,4-Trifluorocyclobutene, T-60276

$C_4H_3F_3N_2$
2-Trifluoromethyl-1*H*-imidazole, T-60289
4(5)-Trifluoromethyl-1*H*-imidazole, T-60290

$C_4H_3F_3O_2$
2-(Trifluoromethyl)propenoic acid, T-60295

$C_4H_3F_3O_4$
Trifluorobutanedioic acid, T-80269

$C_4H_3IN_2O_2$
▷ 5-Iodo-2,4(1*H*,3*H*)-pyrimidinedione, I-60062
6-Iodo-2,4-(1*H*,3*H*)-pyrimidinedione, I-70057

C_4H_3IS
3-Iodothiophene, I-60067

$C_4H_3I_3N_2$
2,4,5-Triiodo-1*H*-imidazole; 1-Me, *in* T-70276

C_4H_3N
2-Ethynyl-2*H*-azirine, E-80071
3-Ethynyl-2*H*-azirine, E-80072

C_4H_3NO
2*H*-Pyrrol-2-one, P-80228

C_4H_3NOS
2-Thiazolecarboxaldehyde, T-80170
4-Thiazolecarboxaldehyde, T-80171
5-Thiazolecarboxaldehyde, T-80172

$C_4H_3NO_2$
3-Amino-3-cyclobutene-1,2-dione, A-80065

$C_4H_3NO_3$
5-Isoxazolecarboxylic acid, I-60142

$C_4H_3N_3$
4(5)-Cyanoimidazole, *in* I-60004
4(5)-Ethynyl-1,2,3-triazole, E-80079

$C_4H_3N_3O_2$
2-Nitropyrimidine, N-60043

$C_4H_3N_3O_3$
6-Formyl-1,2,4-triazine-3,5(2*H*,4*H*)-dione,
F-60075

$C_4H_3N_3O_4$
2,3-Dinitro-1*H*-pyrrole, D-60461
2,4-Dinitro-1*H*-pyrrole, D-60462
2,5-Dinitro-1*H*-pyrrole, D-60463
3,4-Dinitro-1*H*-pyrrole, D-60464

$C_4H_3N_5$
Tetrazolo[1,5-*a*]pyrimidine, T-60176
Tetrazolo[1,5-*c*]pyrimidine, T-60177

$C_4H_3N_5O$
[1,2,4]-Triazolo[5,1-*c*][1,2,4]triazin-4(1*H*)-one,
T-80214

C_4H_4
1,3-Cyclobutadiene, C-70212

C_4H_4BrNO
5-(Bromomethyl)isoxazole, B-80221

$C_4H_4BrN_3$
2-Amino-4-bromopyrimidine, A-70100
2-Amino-5-bromopyrimidine, A-70101
4-Amino-5-bromopyrimidine, A-70102
5-Amino-2-bromopyrimidine, A-70103
4-Bromo-5-methyl-1,2,3-triazine, B-80232
5-Bromo-4-methyl-1,2,3-triazine, B-80233

$C_4H_4BrN_3O_2$
4-Bromo-3(5)-nitro-1*H*-pyrazole; 1-Me, *in*
B-60304

$C_4H_4Br_2$
▷ 1,4-Dibromo-2-butyne, D-70087

$C_4H_4Br_2O$
2,2-Dibromocyclopropanecarboxaldehyde,
D-60092

$C_4H_4Br_2O_2$
2,2-Dibromocyclopropanecarboxylic acid,
D-60093

$C_4H_4ClF_3O$
4,4,4-Trifluorobutanoic acid; Chloride, *in*
T-80270

$C_4H_4ClN_3$
1-Azido-4-chloro-2-butyne, A-80195

$C_4H_4Cl_2O_2$
2,2-Dichlorocyclopropanecarboxylic acid,
D-60127
Methylpropanedioic acid; Dichloride, *in*
M-80149

$C_4H_4FN_3$
3-Amino-6-fluoropyridazine, A-70148

$C_4H_4F_2$
1,1-Difluoro-1,3-butadiene, D-70162
3,3-Difluorocyclobutene, D-70165

$C_4H_4F_2O_4$
2,2-Difluorobutanedioic acid, D-70163

$C_4H_4F_3NO$
2-(Trifluoromethyl)propenoic acid; Amide, *in*
T-60295

$C_4H_4F_4N_2O_2$
Tetrafluorobutanedioic acid; Diamide, *in*
T-80049

$C_4H_4N_2$
3,4-Diazatricyclo[3.1.0.0²,⁶]hex-3-ene, D-70054
Methylpropanedinitrile, *in* M-80149

$C_4H_4N_2OS$
2-Thiazolecarboxaldehyde; Oxime, *in* T-80170
4-Thiazolecarboxaldehyde; Oxime, *in* T-80171
5-Thiazolecarboxaldehyde; Oxime, *in* T-80172

$C_4H_4N_2O_2$
1*H*-Imidazole-4-carboxylic acid, I-60004
5-Isoxazolecarboxylic acid; Amide, *in* I-60142
3-Nitropyrrole, N-60044
Racemic acid; Dinitrile, *in* T-70005

$C_4H_4N_2S$
4-Vinyl-1,2,3-thiadiazole, V-60014
5-Vinyl-1,2,3-thiadiazole, V-60015

$C_4H_4N_2S_2$
3-Mercapto-6(1*H*)-pyridazinethione, M-80028

$C_4H_4N_2S_3$
6-Mercapto-2,4(1*H*,3*H*)-pyrimidinedithione,
M-80030

$C_4H_4N_4$
4-Amino-3-cyanopyrazole, *in* A-70189
5-Amino-4-pyrazolecarbonitrile, *in* A-70187
1*H*-Pyrazolo[5,1-*c*]-1,2,4-triazole, P-70157

$C_4H_4N_6$
4,4′-Bi-1*H*-1,2,3-triazole, B-80168
4,4′-Bi-4*H*-1,2,4-triazole, B-60189

$C_4H_4N_6O_2S$
3,4-Diazido-3,5-dihydro-thiophene 1,1-
dioxide, D-80063

C_4H_4O
5-Oxabicyclo[2.1.0]pent-2-ene, O-70059

C_4H_4OS
Thiophene-3-ol, T-60212

$C_4H_4O_2$
3-Butynoic acid, B-70309
1,2-Cyclobutanedione, C-60183

$C_4H_4O_2S$
2,4(3*H*,5*H*)-Thiophenedione, T-70185

$C_4H_4O_3$
Dihydro-2,3(2*H*,3*H*)-furandione, *in* H-80227
2,4-(3*H*,5*H*)-Furandione, F-80087
Methanetricarboxaldehyde, M-80038

$C_4H_4O_3S_2$
2-Thiophenesulfonic acid, T-60213
3-Thiophenesulfonic acid, T-60214

$C_4H_4O_4$
Diformylacetic acid, D-60190
Dihydro-3-hydroxy-2,5-furandione, D-80211
1,4-Dioxane-2,6-dione, D-80500

$C_4H_4O_4S$
(Dimercaptomethylene)propanedioic acid,
D-60395

$C_4H_4O_5$
Hydroxymethylenepropanedioic acid, H-80204

$C_4H_4Se_2$
2-Methylene-1,3-diselenole, M-70069

C_4H_5Br
1-Bromo-1,2-butadiene, B-70196
4-Bromo-1-butyne, B-70198

$C_4H_5BrN_2$
3-Bromo-1,4-dimethylpyrazole, *in* B-70250
4-Bromo-1-methyl-1*H*-imidazole, *in* B-70221
5-Bromo-1-methyl-1*H*-imidazole, *in* B-70221
3(5)-Bromo-4-methylpyrazole, B-70250
3(5)-Bromo-5(3)-methylpyrazole, B-70252
4-Bromo-3(5)-methylpyrazole, B-70251
▷ 4-Bromo-1-methylpyrazole, *in* B-80254

$C_4H_5BrN_2O$
5-Aminomethyl-3-bromoisoxazole, A-70165

$C_4H_5BrO_2$
2-Bromomethyl-2-propenoic acid, B-70249

$C_4H_5BrO_2S$
4-Bromo-2-sulfolene, *in* B-70204

C_4H_5BrS
3-Bromo-2,3-dihydrothiophene, B-70204

$C_4H_5Br_2ClO$
2,4-Dibromobutanoic acid; Chloride, *in* D-80080

C_4H_5Cl
4-Chloro-1-butyne, C-70058

$C_4H_5ClN_2$
3(5)-Chloro-5(3)-methylpyrazole, C-70109
▷ 4-Chloro-3(5)methylpyrazole, C-70108
3-Chloro-1-methylpyrazole, *in* C-70155
4-Chloro-1-methylpyrazole, *in* C-70156
5-Chloro-1-methylpyrazole, *in* C-70155

$C_4H_5ClN_4$
2,4-Diamino-5-chloropyrimidine, D-80057
▷ 2,4-Diamino-6-chloropyrimidine, D-80058

$C_4H_5ClN_4O$
2,4-Diamino-5-chloropyrimidine; 3-Oxide, *in* D-80057
2,4-Diamino-6-chloropyrimidine; 3-Oxide, *in* D-80058

C_4H_5ClO
4-Chloro-2-butyn-1-ol, C-80044

$C_4H_5Cl_2NO_2$
2-Amino-4,4-dichloro-3-butenoic acid, A-60125

$C_4H_5Cl_3O$
▷ (2,2,2-Trichloroethyl)oxirane, T-60252

$C_4H_5FO_2$
Dihydro-3-fluoro-2(3*H*)-furanone, D-80208

$C_4H_5F_3N_2O_2$
Trifluorobutanedioic acid; Diamide, *in* T-80269

$C_4H_5F_3O_2$
Trifluoroacetic acid; Et ester, *in* T-80267
4,4,4-Trifluorobutanoic acid, T-80270
3,3,3-Trifluoropropanoic acid; Me ester, *in* T-80284

$C_4H_5F_6NO$
N-Ethoxy-1,1,1,1′,1′,1′-hexafluorodimethylamine, *in* B-60180

C_4H_5I
4-Iodo-1-butyne, I-70024

$C_4H_5IN_2$
3(5)-Iodo-4-methylpyrazole, I-70048
3(5)-Iodo-5(3)-methylpyrazole, I-70050
4-Iodo-3(5)-methylpyrazole, I-70049
3-Iodo-1-methylpyrazole, *in* I-80040
4-Iodo-1-methylpyrazole, *in* I-80041
5-Iodo-1-methylpyrazole, *in* I-80040

$C_4H_5IO_2$
Dihydro-3-iodo-2(3*H*)-furanone, D-60252

C_4H_5NO
1,2-Dihydro-3*H*-pyrrol-3-one, D-70262
▷ 2-Hydroxy-3-butenenitrile, *in* P-70119
4-Oxobutanenitrile, *in* O-80068

$C_4H_5NO_2$
1-Amino-2-cyclopropene-1-carboxylic acid, A-60122
1-Aminocyclopropenecarboxylic acid, A-70126
1-Nitro-1,3-butadiene, N-60023

$C_4H_5NO_2S_2$
2-Thiophenesulfonic acid; Amide, *in* T-60213
3-Thiophenesulfonic acid; Amide, *in* T-60214

$C_4H_5NO_3$
3-Amino-2-formyl-2-propenoic acid, A-80134
2,5-Dihydro-3-nitrofuran, D-70236
2,4-(3*H*,5*H*)-Furandione; Oxime, *in* F-80087
2,3-Morpholinedione, M-80193

$C_4H_5NO_4$
3,4-Dihydroxy-2,5-pyrrolidinedione, D-80377
2-Oxo-4-oxazolidinecarboxylic acid, O-60082

C_4H_5NS
4-Mercapto-2-butenoic acid; Nitrile, *in* M-60018
3-Thiocyanato-1-propene, T-80189

$C_4H_5NS_2$
2-(Methylthio)thiazole, *in* T-60200

$C_4H_5N_3$
3-Azido-1-butyne, A-80193

$C_4H_5N_3O$
4(5)-Acetyl-1,2,3-triazole, A-80031
5-Amino-4(1*H*)-pyrimidinone, A-70196
▷ 4-Azido-2-butyn-1-ol, A-80194
1*H*-Imidazole-4-carboxylic acid; Amide, *in* I-60004
▷ 1-Methyl-2-nitroso-1*H*-imidazole, *in* N-80059

$C_4H_5N_3O_2$
▷ 3-Amino-1*H*-pyrazole-4-carboxylic acid, A-70187
3-Amino-1*H*-pyrazole-5-carboxylic acid, A-70188
4-Amino-1*H*-pyrazole-3-carboxylic acid, A-70189
4-Amino-1*H*-pyrazole-5-carboxylic acid, A-70190
5-Amino-1*H*-pyrazole-3-carboxylic acid, A-70191
5-Amino-1*H*-pyrazole-4-carboxylic acid, A-70192
3-Azido-2-propenoic acid; Me ester, *in* A-70290
6-Methyl-1,3,5-triazine-2,4(1*H*,3*H*)-dione, M-60132
1*H*-1,2,3-Triazole-4-carboxylic acid; 1-Me, *in* T-60239
1*H*-1,2,3-Triazole-4-carboxylic acid; 2-Me, *in* T-60239
1*H*-1,2,3-Triazole-4-carboxylic acid; Me ester, *in* T-60239

$C_4H_5N_3S_2$
5-Amino-2-thiazolecarbothioamide, A-60260
5-Amino-4-thiazolecarbothioamide, A-60261

C_4H_6BrN
α-Bromoisobutyronitrile, *in* B-80229

C_4H_6BrNO
3-Bromo-2-pyrrolidinone, B-80258

$C_4H_6BrNO_4$
2-Amino-3-bromobutanedioic acid, A-70098

$C_4H_6Br_2O$
2-Bromo-2-methylpropanoic acid; Bromide, *in* B-80229

$C_4H_6Br_2O_2$
2,4-Dibromobutanoic acid, D-80080

$C_4H_6ClF_2NO_2$
2-Amino-4-chloro-4,4-difluorobutanoic acid, A-60107

$C_4H_6ClNO_2$
1-Chloro-1-nitrocyclobutane, C-70130

$C_4H_6ClNO_4$
2-Amino-3-chlorobutanedioic acid, A-70121

$C_4H_6Cl_2$
3-Chloro-2-chloromethyl-1-propene, C-70061

$C_4H_6F_3NO$
4,4,4-Trifluorobutanoic acid; Amide, *in* T-80270

$C_4H_6I_2$
▷ 3-Iodo-2-(iodomethyl)-1-propene, I-60047

$C_4H_6N_2$
3-Aminopyrrole, A-80143
1,4-Dihydropyrimidine, D-70255

$C_4H_6N_2O$
3-Amino-4-methylisoxazole, A-80106
4-Amino-3-methylisoxazole, A-80107
4-Amino-5-methylisoxazole, A-80108
5-Amino-4-methylisoxazole, A-80109
4,5-Dihydro-3(2*H*)-pyridazinone, D-60272
3,4-Dihydro-2(1*H*)-pyrimidinone, D-60273
5,6-Dihydro-4(1*H*)-pyrimidinone, D-60274
4(5)-Methoxyimidazole, M-80044

$C_4H_6N_2OS$
2-Thioxo-4-imidazolidinone; 1-*N*-Me, *in* T-70188
2-Thioxo-4-imidazolidinone; 3-*N*-Me, *in* T-70188

$C_4H_6N_2O_2$
1,3-Diazetidine-2,4-dione; 1,3-Di-Me, *in* D-60057
Dihydro-4,6-(1*H*,5*H*)-pyrimidinedione, D-70256
Dimethyl *N*-cyanoimidocarbonate, D-70390

$C_4H_6N_2O_3$
3-Methoxy-4-methyl-1,2,4-oxadiazolidin-5(4*H*)-one, *in* O-70060
1,2,4-Oxadiazolidine-3,5-dione; 2,4-Di-Me, *in* O-70060
1,2,4-Oxadiazolidine-3,5-dione; 4-Et, *in* O-70060
2-Oxo-4-imidazolidinecarboxylic acid, O-60074

$C_4H_6N_2S$
2-(Methylthio)-1*H*-imidazole, *in* D-80215

$C_4H_6N_2S_2$
2,5-Dihydro-4,6-pyrimidinedithiol, D-70257

$C_4H_6N_2S_3$
2,5-Bis(methylthio)-1,3,4-thiadiazole, *in* M-70031
3-Methyl-5-(methylthio)-1,3,4-thiadiazole-2(3*H*)-thione, *in* M-70031

$C_4H_6N_4$
3-Amino-1,2,4-triazine; *N*-Me, *in* A-70203

$C_4H_6N_4O$
2-Amino-4-methoxy-1,3,5-triazine, *in* A-80149
3-Amino-1*H*-pyrazole-4-carboxylic acid; Amide, *in* A-70187
4-Amino-1*H*-pyrazole-3-carboxylic acid; Amide, *in* A-70189
4-Amino-1*H*-pyrazole-5-carboxylic acid; *N*(1)-Me, Amide, *in* A-70190
4-Amino-1,3,5-triazin-2(1*H*)-one; 4-*N*-Me, *in* A-80149
2,6-Diamino-4(1*H*)-pyrimidinone, D-70043
1*H*-1,2,3-Triazole-4-carboxylic acid; 1-Me, amide, *in* T-60239

$C_4H_6N_4O_2$
3,6-Dimethoxy-1,2,4,5-tetrazine, *in* D-70268
Tetrahydroimidazo[4,5-*d*]imidazole-2,5(1*H*,3*H*)-dione, T-70066

C_4H_6OS
Dihydro-2(3*H*)-furanthione, D-70210

$C_4H_6O_2$
▷ Methyl acrylate, *in* P-70121
2-Methylene-1,3-dioxolane, M-60075
4-Methylene-1,3-dioxolane, M-60076
2-Oxetanecarboxaldehyde, O-80053
2-Oxobutanal, O-80067

$C_4H_6O_2S$
4-Mercapto-2-butenoic acid, M-60018
3-Oxobutanethioic acid, O-70080

$C_4H_6O_3$
Dihydro-3-hydroxy-2(3*H*)-furanone, D-**8**0212
Dihydro-4-hydroxy-2(3*H*)-furanone, D-**6**0242
4,5-Dihydro-3-hydroxy-2(3*H*)-furanone, *in* D-**6**0313
2-Formylpropanoic acid, F-70039
4-Hydroxy-2-butenoic acid, H-**8**0134
1-Hydroxycyclopropanecarboxylic acid, H-**6**0117
Methylene-1,3,5-trioxane, M-**6**0081
3-Methyloxiranecarboxylic acid, M-**8**0127
▷ 4-Oxobutanoic acid, O-**8**0068
3-Oxopropanoic acid; Me ester, *in* O-**6**0090

$C_4H_6O_4$
3-Hydroxy-2-oxobutanoic acid, H-**8**0226
4-Hydroxy-2-oxobutanoic acid, H-**8**0227
▷ Methylpropanedioic acid, M-**8**0149

$C_4H_6O_5$
Dimethyl dicarbonate, *in* D-70111
Glycolloglycollic acid, G-**8**0030

$C_4H_6O_6$
Tartaric acid, T-70005

C_4H_6S
2,5-Dihydrothiophene, D-70270

$C_4H_6S_2$
2-Butyne-1,4-dithiol, B-70308

$C_4H_6S_2Se$
1,3-Dithiane-2-selone, D-70507

$C_4H_6S_4$
Ethanebis(dithioic)acid; Di-Me ester, *in* E-**6**0040

$C_4H_6Se_2$
2-Methyl-1,3-diselenole, M-**8**0070

C_4H_7Br
1-Bromo-1-methylcyclopropane, B-70242

C_4H_7BrO
4-Bromobutanal, B-70197

$C_4H_7BrO_2$
▷ 2-Bromo-2-methylpropanoic acid, B-**8**0229

$C_4H_7Br_2NO$
2,4-Dibromobutanoic acid; Amide, *in* D-**8**0080

C_4H_7Cl
3-Chloro-1-butene, C-**8**0042

$C_4H_7ClFNO_2$
2-Amino-4-chloro-4-fluorobutanoic acid, A-**6**0108

C_4H_7ClOS
Carbonochloridothioic acid; *O*-Isopropyl, *in* C-**6**0013
Carbonochloridothioic acid; *O*-Propyl, *in* C-**6**0013
▷ Carbonochloridothioic acid; *S*-Propyl, *in* C-**6**0013

$C_4H_7ClO_3$
3-Chloro-2-hydroxypropanoic acid; Me ester, *in* C-70072

$C_4H_7Cl_3$
1,3-Dichloro-2-(chloromethyl)propane, D-70118

$C_4H_7F_2NO_2$
2-Amino-3,3-difluorobutanoic acid, A-**6**0133
3-Amino-4,4-difluorobutanoic acid, A-**6**0134

$C_4H_7F_3OS$
Trifluoromethanesulfenic acid; Isopropyl ester, *in* T-**6**0279

C_4H_7IO
2-Iodobutanal, I-**6**0040
3-Iodo-2-butanone, I-**6**0041
2-Iodomethyl-3-methyloxirane, I-**8**0034
2-Iodo-2-methylpropanal, I-**6**0051
Tetrahydro-2-iodofuran, T-**8**0069

$C_4H_7I_3$
1,3-Diiodo-2-(iodomethyl)propane, D-70358

C_4H_7N
3-Butyn-1-amine, B-**8**0294
3,4-Dihydro-2*H*-pyrrole, D-**8**0247
2-Propyn-1-amine; *N*-Me, *in* P-**6**0184

C_4H_7NO
▷ 2-Pyrrolidinone, P-70188

C_4H_7NOS
Tetrahydro-2*H*-1,2-thiazin-3-one, T-**6**0095
Tetrahydro-2*H*-1,3-thiazin-2-one, T-**6**0096
3-Thiomorpholinone, T-**8**0190

$C_4H_7NO_2$
3-Aminodihydro-2(3*H*)-furanone, A-**6**0138
4-Aminodihydro-2(3*H*)-furanone, *in* A-**8**0081
2-Azetidinecarboxylic acid, A-70284
5-Methyl-2-oxazolidinone, M-**6**0107
3-Morpholinone, M-**6**0147
Nitrocyclobutane, N-**6**0024
2-Oxobutanal; 1-Oxime, *in* O-**8**0067

$C_4H_7NO_2S$
3-Thiomorpholinone; S-Oxide, *in* T-**8**0190

$C_4H_7NO_3$
1-Nitro-2-butanone, N-70043
3-Nitro-2-butanone, N-70044
4-Nitro-2-butanone, N-70045
4-Oxobutanoic acid; Oxime, *in* O-**8**0068

$C_4H_7NO_3S$
3-Thiomorpholinone; S-Dioxide, *in* T-**8**0190

C_4H_7NS
2-Propenethioamide; *N*-Me, *in* P-70120
2-Pyrrolidinethione, P-70186

$C_4H_7NS_2$
4,5-Dihydro-2-(methylthio)thiazole, *in* T-**6**0197
Tetrahydro-2*H*-1,3-thiazine-2-thione, T-**6**0094
2-Thiazoledinethione; *N*-Me, *in* T-**6**0197

$C_4H_7N_3O$
4-Azidobutanal, A-70286
4,5-Dihydro-5,5-dimethyl-3*H*-1,2,4-triazol-3-one, D-**8**0196

$C_4H_7N_3O_3$
Biuret; *N*-Ac, *in* B-**6**0190

$C_4H_7N_5$
2,4-Diamino-6-methyl-1,3,5-triazine, D-**6**0040
2,4,6-Triaminopyrimidine, T-**8**0208

$C_4H_7N_5O$
5-Acetamido-2-methyl-2*H*-tetrazole, *in* A-70202
2,4-Diamino-6-methyl-1,3,5-triazine; N^3-Oxide, *in* D-**6**0040
2,4-Diamino-6-methyl-1,3,5-triazine; N^5-Oxide, *in* D-**6**0040

C_4H_8BrNO
2-Bromo-2-methylpropanoic acid; Amide, *in* B-**8**0229

$C_4H_8BrNO_2$
2-Amino-3-bromobutanoic acid, A-**6**0097
2-Amino-4-bromobutanoic acid, A-**6**0098
4-Amino-2-bromobutanoic acid, A-**6**0099

C_4H_8ClN
4-Chloro-2-buten-1-amine, C-50094

$C_4H_8ClNO_2$
2-Amino-3-chlorobutanoic acid, A-**6**0103
2-Amino-4-chlorobutanoic acid, A-**6**0104
4-Amino-2-chlorobutanoic acid, A-**6**0105
4-Amino-3-chlorobutanoic acid, A-**6**0106

$C_4H_8FNO_2$
2-Amino-3-fluorobutanoic acid, A-**6**0154
2-Amino-4-fluorobutanoic acid, A-**6**0155
3-Amino-4-fluorobutanoic acid, A-**6**0156
4-Amino-2-fluorobutanoic acid, A-**6**0157
4-Amino-3-fluorobutanoic acid, A-**6**0158

$C_4H_8I_2O$
1,1′-Oxybis[2-iodoethane], O-**6**0096

$C_4H_8N_2O$
2-Imidazolidinone; 1-Me, *in* I-70005
Piperazinone, P-70101
Tetrahydro-2(1*H*)-pyrimidinone, T-**6**0088
Tetrahydro-4(1*H*)-pyrimidinone, T-**6**0089

$C_4H_8N_2O_2$
Acetamidoxime; *O*-Ac, *in* A-**6**0016
4-Amino-5-methyl-3-isoxazolidinone, A-**6**0221
Methylpropanedioic acid; Diamide, *in* M-**8**0149

$C_4H_8N_2O_4$
L-Threaric acid; Diamide, *in* T-70005

$C_4H_8N_4O_2$
1,4-Dihydro-3,6-dimethoxy-1,2,4,5-tetrazine, D-70192
▷ Hexahydropyrimidine; *N,N′*-Dinitroso, *in* H-**6**0056
Tetrahydro-1,3,5,7-tetrazocine-2,6(1*H*,3*H*)-dione, T-70098

$C_4H_8N_4O_4$
▷ Hexahydropyrimidine; *N,N′*-Dinitro, *in* H-**6**0056

C_4H_8O
▷ 2-Ethyloxirane, E-70047

C_4H_8OS
(Ethylthio)acetaldehyde, *in* M-**8**0017
3-Mercaptocyclobutanol, M-**6**0019
Tetrahydro-3-furanthiol, T-**8**0063

$C_4H_8O_2$
3-Butene-1,2-diol, B-70291
2-Hydroxybutanal, H-70112
▷ 2-Methyl-1,3-dioxolane, M-**6**0063
3-Methyloxiranemethanol, M-**8**0128
2-Oxetanemethanol, O-**8**0054
3-Oxetanemethanol, O-**8**0055

$C_4H_8O_2S$
3-Mercaptobutanoic acid, M-70023
2-(Methylthio)propanoic acid, *in* M-**6**0027

$C_4H_8O_3$
1,3-Dioxolane-2-methanol, D-70468

$C_4H_8O_4$
2,3-Dihydroxybutanoic acid, D-**6**0312
3,4-Dihydroxybutanoic acid, D-**6**0313

$C_4H_8O_5S$
Sulfoacetic acid; Carboxy-Et ester, *in* S-**8**0050

$C_4H_8S_2$
2-Butene-1,4-dithiol, B-70292
1,1-Cyclobutanedithiol, C-70215
1,3-Cyclobutanedithiol, C-**6**0184
2-Methylene-1,3-propanedithiol, M-70076

$C_4H_8S_2^{2\oplus}$
1,4-Dithioniabicyclo[2.2.0]hexane, D-70527

$C_4H_9BF_4S_3$
Tris(methylthio)methylium(1+); Tetrafluoroborate, *in* T-**8**0367

C_4H_9ClOS
tert-Butylsulfinic acid; Chloride, *in* B-**8**0293

$C_4H_9ClO_2$
2-Chloro-1,4-butanediol, C-**8**0040

$C_4H_9F_3O_3SSi$
Trifluoromethanesulfonic acid; Trimethylsilyl ester, *in* T-70241

C_4H_9IO
tert-Butyl hypoiodite, B-**8**0287

C_4H_9N
▷ 2-Buten-1-amine, B-**8**0278

C_4H_9NO
▷ Acetamide; *N*-Di-Me, *in* A-70019
4-Aminobutanal, A-**8**0064
3-Pyrrolidinol, P-70187

$C_4H_9NO_3$
2-Amino-4-hydroxybutanoic acid, A-70156
3-Amino-2-hydroxybutanoic acid, A-**6**0183
3-Amino-4-hydroxybutanoic acid, A-**8**0081
Threonine, T-**6**0217

C_4H_9NS
▷ Tetrahydro-2*H*-1,3-thiazine, T-**6**0093

$C_4H_9N_3O$
1-Azido-1-methoxypropane, *in* A-80204

$C_4H_9N_3O_3$
3-Amino-2-ureidopropanoic acid, A-60268

$C_4H_9N_5$
N-(1,4-Dihydro-1,4-dimethyl-5*H*-tetrazol-5-ylidene)methanamine, *in* A-70202

$C_4H_9S_3^{\oplus}$
Tris(methylthio)methylium(1+), T-80367

$C_4H_{10}NO_6P$
O-Phosphohomoserine, *in* A-70156

$C_4H_{10}N_2$
Hexahydropyrimidine, H-60056

$C_4H_{10}N_2O$
Tetrahydro-3,4-furandiamine, T-70062

$C_4H_{10}N_4$
Octahydroimidazo[4,5-*d*]imidazole, O-70015

$C_4H_{10}O_2$
▷ 1-Methoxy-2-propanol, *in* P-70118
2-Methyl-1,3-propanediol, M-60120

$C_4H_{10}O_2S$
tert-Butylsulfinic acid, B-80293

$C_4H_{10}O_3$
▷ 1-Ethoxy-1-hydroperoxyethane, E-60044
2-Hydroxymethyl-1,3-propanediol, H-80212

$C_4H_{10}S_2$
1,3-Butanedithiol, B-60339
2,2-Butanedithiol, B-60340
2-Methyl-1,3-propanedithiol, M-60121

$C_4H_{12}N_4$
Octahydro-1,2,5,6-tetrazocine, O-60020
▷ 1,1,4,4-Tetramethyl-2-tetrazene, T-80153

$C_4H_{13}N_3O$
2-[(2-Hydrazinoethyl)amino]ethanol, H-60091

C_4N_6
3,6-Dicyano-1,2,4,5-tetrazine, *in* T-80163

$C_4N_6S_2$
Bis[1,2,5-thiadiazolo][3,4-*b*:3′,4′-*e*]pyrazine, B-70161

C_5BrF_4N
2-Bromo-3,4,5,6-tetrafluoropyridine, B-60326
4-Bromo-2,3,5,6-tetrafluoropyridine, B-60328

C_5BrF_5N
3-Bromo-2,4,5,6-tetrafluoropyridine, B-60327

C_5F_6
1,2,3,4,5,5-Hexafluoro-1,3-cyclopentadiene, H-60044

$C_5F_{10}N_2O_2$
2,2,3,3,4,4,5,5,6,6-Decafluoropiperidine; *N*-Nitro, *in* D-60006

$C_5F_{11}N$
Undecafluoropiperidine, U-60002

$C_5F_{12}S_4$
Tetrakis(trifluoromethylthio)methane, T-80133

$C_5HF_{10}N$
2,2,3,3,4,4,5,5,6,6-Decafluoropiperidine, D-60006

$C_5H_2Cl_2N_2O_2$
2,3-Dichloro-5-nitropyridine, D-60140
2,4-Dichloro-3-nitropyridine, D-60141
2,4-Dichloro-5-nitropyridine, D-60142
2,5-Dichloro-3-nitropyridine, D-60143
2,6-Dichloro-3-nitropyridine, D-60144
2,6-Dichloro-4-nitropyridine, D-60145
3,4-Dichloro-5-nitropyridine, D-60146
3,5-Dichloro-4-nitropyridine, D-60147

$C_5H_2Cl_2N_2O_3$
2,6-Dichloro-4-nitropyridine; 1-Oxide, *in* D-60145
3,5-Dichloro-4-nitropyridine; 1-Oxide, *in* D-60147

$C_5H_2Cl_3N$
▷ 2,4,6-Trichloropyridine, T-60255

$C_5H_2Cl_3NOS$
2-(Trichloroacetyl)thiazole, T-80222

$C_5H_2F_4N_2$
4-Amino-2,3,5,6-tetrafluoropyridine, A-60257

$C_5H_2F_6O$
4,4,4-Trifluoro-3-(trifluoromethyl)-2-butenal, T-60309
1,1,1-Trifluoro-2-(trifluoromethyl)-3-butyn-2-ol, T-60310

C_5H_2INS
2-Cyano-3-iodothiophene, *in* I-60075
2-Cyano-4-iodothiophene, *in* I-60076
2-Cyano-5-iodothiophene, *in* I-60078
3-Cyano-2-iodothiophene, *in* I-60074
3-Cyano-4-iodothiophene, *in* I-60077
4-Cyano-2-iodothiophene, *in* I-60079

$C_5H_2S_4$
Thieno[2,3-*d*]-1,3-dithiole-2-thione, T-80175

C_5H_3BrClN
3-Bromo-5-chloropyridine, B-60223
4-Bromo-3-chloropyridine, B-60224

C_5H_3BrFN
2-Bromo-3-fluoropyridine, B-70213
2-Bromo-5-fluoropyridine, B-60252
3-Bromo-2-fluoropyridine, B-60253
3-Bromo-4-fluoropyridine, B-70214
3-Bromo-5-fluoropyridine, B-60254
4-Bromo-3-fluoropyridine, B-60255
5-Bromo-2-fluoropyridine, B-60256

C_5H_3BrIN
2-Bromo-4-iodopyridine, B-60280
2-Bromo-5-iodopyridine, B-60281
3-Bromo-4-iodopyridine, B-60282
3-Bromo-5-iodopyridine, B-60283
4-Bromo-2-iodopyridine, B-60284
4-Bromo-3-iodopyridine, B-60285
5-Bromo-2-iodopyridine, B-60286

$C_5H_3BrO_2$
3-Bromo-2-furancarboxaldehyde, B-60257
4-Bromo-2-furancarboxaldehyde, B-60258
5-Bromo-2-furancarboxaldehyde, B-60259

$C_5H_3Br_2NO$
4,6-Dibromo-2(1*H*)pyridone, D-80100

C_5H_3ClFN
2-Chloro-3-fluoropyridine, C-60081
2-Chloro-4-fluoropyridine, C-60082
2-Chloro-5-fluoropyridine, C-60083
2-Chloro-6-fluoropyridine, C-60084
3-Chloro-2-fluoropyridine, C-60085
4-Chloro-2-fluoropyridine, C-60086
4-Chloro-3-fluoropyridine, C-60087
5-Chloro-2-fluoropyridine, C-60088

C_5H_3ClFNO
2-Chloro-3-fluoropyridine; *N*-Oxide, *in* C-60081

C_5H_3ClIN
2-Chloro-3-iodopyridine, C-60107
2-Chloro-4-iodopyridine, C-60108
2-Chloro-5-iodopyridine, C-60109
3-Chloro-2-iodopyridine, C-60110
3-Chloro-5-iodopyridine, C-60111
4-Chloro-2-iodopyridine, C-60112
4-Chloro-3-iodopyridine, C-60113

C_5H_3FIN
2-Fluoro-3-iodopyridine, F-80048
2-Fluoro-4-iodopyridine, F-60045
3-Fluoro-4-iodopyridine, F-60046

C_5H_3IOS
2-Iodo-3-thiophenecarboxaldehyde, I-60068
3-Iodo-2-thiophenecarboxaldehyde, I-60069
4-Iodo-2-thiophenecarboxaldehyde, I-60070
4-Iodo-3-thiophenecarboxaldehyde, I-60071
5-Iodo-2-thiophenecarboxaldehyde, I-60072
5-Iodo-3-thiophenecarboxaldehyde, I-60073

$C_5H_3IO_2S$
2-Iodo-3-thiophenecarboxylic acid, I-60074
3-Iodo-2-thiophenecarboxylic acid, I-60075
4-Iodo-2-thiophenecarboxylic acid, I-60076
4-Iodo-3-thiophenecarboxylic acid, I-60077
5-Iodo-2-thiophenecarboxylic acid, I-60078
5-Iodo-3-thiophenecarboxylic acid, I-60079

C_5H_3N
3,4-Pyridyne, P-80224

$C_5H_3NO_4S$
4-Nitro-2-thiophenecarboxylic acid, N-80061
5-Nitro-2-thiophenecarboxylic acid, N-80062

C_5H_3NS
2-Ethynylthiazole, E-60061
4-Ethynylthiazole, E-60062

$C_5H_3N_5O_8$
▷ 1-Methyl-2,3,4,5-tetranitro-1*H*-pyrrole, *in* T-60160

C_5H_4BrNO
▷ 2-Bromo-3-hydroxypyridine, B-60275
2-Bromo-5-hydroxypyridine, B-60276
3-Bromo-5-hydroxypyridine, B-60277
4-Bromo-3-hydroxypyridine, B-60278

$C_5H_4BrNO_2$
2-Bromo-3-hydroxypyridine; *N*-Oxide, *in* B-60275

C_5H_4BrNS
3-Bromo-2(1*H*)-pyridinethione, B-60315
3-Bromo-4(1*H*)-pyridinethione, B-60316
5-Bromo-2(1*H*)-pyridinethione, B-60317
6-Bromo-2(1*H*)-pyridinethione, B-80256

$C_5H_4BrN_5O$
8-Bromoguanine, B-80205

C_5H_4ClNO
▷ 2-Chloro-3-hydroxypyridine, C-60100
2-Chloro-5-hydroxypyridine, C-60101
3-Chloro-5-hydroxypyridine, C-60102

$C_5H_4ClNO_2$
4-Chloro-6-hydroxy-2(1*H*)-pyridinone, C-70073
6-Chloro-4-hydroxy-2(1*H*)pyridinone, C-70074
▷ 5-Methyl-4-isoxazolecarboxylic acid; Chloride, *in* M-80117

$C_5H_4ClNO_2S$
2-Pyridinesulfonic acid; Chloride, *in* P-70171

C_5H_4ClNS
3-Chloro-2(1*H*)-pyridinethione, C-60129

$C_5H_4ClN_5$
▷ 2-Amino-6-chloro-1*H*-purine, A-60116
2-Chloro-6-aminopurine, C-70039

$C_5H_4ClN_5O$
8-Chloroguanine, C-80057

$C_5H_4Cl_4$
1-Chloro-1-(trichlorovinyl)cyclopropane, C-60138

$C_5H_4FN_5$
8-Amino-6-fluoro-9*H*-purine, A-60163

C_5H_4INO
3-Hydroxy-2-iodopyridine, H-60160

C_5H_4INOS
2-Iodo-3-thiophenecarboxaldehyde; Oxime, *in* I-60068
3-Iodo-2-thiophenecarboxaldehyde; Oxime, *in* I-60069
4-Iodo-2-thiophenecarboxaldehyde; Oxime, *in* I-60070
4-Iodo-3-thiophenecarboxaldehyde; Oxime, *in* I-60071
5-Iodo-2-thiophenecarboxaldehyde; Oxime, *in* I-60072
5-Iodo-3-thiophenecarboxaldehyde; Oxime, *in* I-60073
2-Iodo-3-thiophenecarboxylic acid; Amide, *in* I-60074
3-Iodo-2-thiophenecarboxylic acid; Amide, *in* I-60075
4-Iodo-2-thiophenecarboxylic acid; Amide, *in* I-60076

4-Iodo-3-thiophenecarboxylic acid; Amide, *in* I-**60077**
5-Iodo-2-thiophenecarboxylic acid; Amide, *in* I-**60078**
5-Iodo-3-thiophenecarboxylic acid; Amide, *in* I-**60079**

$C_5H_4IN_5O$
8-Iodoguanine, I-**80031**

$C_5H_4N_2$
1,1-Dicyanocyclopropane, *in* C-**80196**

$C_5H_4N_2O_2$
Methyl dicyanoacetate, *in* D-**70149**

$C_5H_4N_2O_3$
3-Nitro-4(1*H*)-pyridone, N-**70063**
1,2,3,4-Tetrahydro-2,4-dioxo-5-pyrimidinecarboxaldehyde, T-**60064**

$C_5H_4N_2O_4$
1*H*-Pyrazole-4,5-dicarboxylic acid, P-**80199**

$C_5H_4N_2S$
▷ 2-Thiocyanato-1*H*-pyrrole, T-**60209**

$C_5H_4N_4$
1*H*-Pyrazolo[3,4-*d*]pyrimidine, P-**70156**
Pyrazolo[1,5-*b*][1,2,4]triazine, P-**60207**
[1,2,4]Triazolo[1,5-*b*]pyridazine, T-**60240**
[1,2,4]Triazolo[1,5-*a*]pyrimidine, T-**60242**
1,2,4-Triazolo[4,3-*c*]pyrimidine, T-**60243**

$C_5H_4N_4O$
1,5-Dihydro-6*H*-imidazo[4,5-*c*]pyridazin-6-one, D-**60248**
2*H*-Imidazo[4,5-*b*]pyrazin-2-one, I-**60006**
Imidazo[2,1-*c*][1,2,4]triazin-4(1*H*)-one, I-**80007**
Pyrazolo[5,1-*c*][1,2,4]triazin-4(1*H*)-one, P-**80201**

$C_5H_4N_4O_2$
1*H*-Pyrazolo[3,4-*d*]pyrimidine-4,6(5*H*,7*H*)-dione, P-**60202**

$C_5H_4N_4O_6$
▷ 1-Methyl-2,3,4-trinitro-1*H*-pyrrole, *in* T-**60373**
▷ 1-Methyl-2,3,5-trinitro-1*H*-pyrrole, *in* T-**60374**

$C_5H_4N_4S$
1,3-Dihydro-2*H*-imidazo[4,5-*b*]pyrazine-2-thione, D-**60247**
1,5-Dihydro-6*H*-imidazo[4,5-*c*]pyridazine-6-thione, D-**70215**
▷ 1,7-Dihydro-6*H*-purine-6-thione, D-**60270**

C_5H_4OS
▷ 2-Thiophenecarboxaldehyde, T-**70183**

$C_5H_4OS_4$
5,6-Dihydro-1,3-dithiolo[4,5-*b*][1,4]dithiin-2-one, D-**80205**

$C_5H_4O_2$
3-Methyl-3-cyclobutene-1,2-dione, M-**80062**

$C_5H_4O_2S_2$
3-Mercapto-2-thiophenecarboxylic acid, M-**80031**
4-Mercapto-2-thiophenecarboxylic acid, M-**80032**

$C_5H_4O_3$
3-Oxabicyclo[3.1.0]hexane-2,4-dione, *in* C-**70248**

$C_5H_4S_5$
5,6-Dihydro-1,3-dithiolo[4,5-*b*][1,4]dithiin-2-thione, D-**80206**

$C_5H_5BF_5N$
1-Fluoropyridinium; Tetrafluoroborate, *in* F-**60066**

$C_5H_5BrN_2$
4-Bromo-6-methylpyrimidine, B-**80230**
5-Bromo-6-methylpyrimidine, B-**80231**

$C_5H_5BrN_2O$
2-Amino-5-bromo-3-hydroxypyridine, A-**60101**

$C_5H_5BrO_5$
(3-Bromo-2-oxopropylidene)propanedioic acid, B-**70262**

$C_5H_5Br_2NO$
2,6-Dibromo-4(1*H*)-pyridone, D-**80099**

C_5H_5Cl
1-Chloro-1-ethynylcyclopropane, C-**60068**

$C_5H_5ClFNO_4$
1-Fluoropyridinium; Perchlorate, *in* F-**60066**

$C_5H_5ClN_2$
2-Chloro-3-methylpyrazine, C-**70105**
2-Chloro-5-methylpyrazine, C-**70106**
2-Chloro-6-methylpyrazine, C-**70107**
3-Chloro-4-methylpyridazine, C-**70110**
3-Chloro-5-methylpyridazine, C-**70111**
3-Chloro-6-methylpyridazine, C-**70112**
4-Chloro-3-methylpyridazine, C-**80078**
4-Chloro-5-methylpyridazine, C-**80079**
2-Chloro-4-methylpyrimidine, C-**70113**
2-Chloro-5-methylpyrimidine, C-**70114**
4-Chloro-2-methylpyrimidine, C-**70115**
4-Chloro-5-methylpyrimidine, C-**70116**
4-Chloro-6-methylpyrimidine, C-**70117**
5-Chloro-2-methylpyrimidine, C-**70118**
5-Chloro-4-methylpyrimidine, C-**70119**

$C_5H_5ClN_2O$
2-Amino-5-chloro-3-hydroxypyridine, A-**60111**
▷ 2-Chloro-3-methylpyrazine; 1-Oxide, *in* C-**70105**
2-Chloro-6-methylpyrazine; 4-Oxide, *in* C-**70107**
3-Chloro-2-methylpyrazine 1-oxide, *in* C-**70105**
3-Chloro-4-methylpyridazine; 1-Oxide, *in* C-**70110**
3-Chloro-5-methylpyridazine; 1-Oxide, *in* C-**70111**
3-Chloro-6-methylpyridazine; 1-Oxide, *in* C-**70112**
4-Chloro-3-methylpyridazine; 1-Oxide, *in* C-**80078**
4-Chloro-5-methylpyridazine; 1-Oxide, *in* C-**80079**
4-Chloro-6-methylpyrimidine; 1-Oxide, *in* C-**70117**

C_5H_5ClO
2-Chloro-2-cyclopenten-1-one, C-**60051**
5-Chloro-2-cyclopenten-1-one, C-**70063**

$C_5H_5Cl_2F_3O_2$
2,2-Dichloro-3,3,3-trifluoro-1-propanol; Ac, *in* D-**60161**

$C_5H_5FN^{\oplus}$
1-Fluoropyridinium, F-**60066**

$C_5H_5F_3O_2$
Dihydro-4-(trifluoromethyl)-2(3*H*)-furanone, D-**60299**
Dihydro-5-(trifluoromethyl)-2(3*H*)-furanone, D-**60300**
▷ 2-(Trifluoromethyl)propenoic acid; Me ester, *in* T-**60295**

$C_5H_5F_7NSb$
1-Fluoropyridinium; Hexafluoroantimonate, *in* F-**60066**

C_5H_5I
5-Iodo-1-penten-4-yne, I-**70053**

$C_5H_5IN_2$
2-Amino-3-iodopyridine, A-**80088**
2-Amino-4-iodopyridine, A-**80089**
2-Amino-5-iodopyridine, A-**80090**
2-Amino-6-iodopyridine, A-**80091**
3-Amino-5-iodopyridine, A-**80092**
4-Amino-2-iodopyridine, A-**80093**
4-Amino-3-iodopyridine, A-**80094**
5-Amino-2-iodopyridine, A-**80095**

$C_5H_5IN_2O$
4-Amino-2-iodopyridine; N-oxide, *in* A-**80093**

$C_5H_5I_3N_2$
2,4,5-Triiodo-1*H*-imidazole; 1-Et, *in* T-**70276**

C_5H_5N
1-Cyanobicyclo[1.1.0]butane, *in* B-**70094**

C_5H_5NO
3-Formyl-2-butenenitrile, *in* M-**70103**
1*H*-Pyrrole-3-carboxaldehyde, P-**70182**

C_5H_5NOS
5-Acetylthiazole, A-**80029**
▷ 1-Hydroxy-2(1*H*)-pyridinethione, H-**70217**
3-Hydroxy-2(1*H*)-pyridinethione, H-**60222**
2-Pyridinethiol N-oxide, *in* P-**60224**
▷ 2(1*H*)-Pyridinethione; 1-Hydroxy, *in* P-**60224**
4(1*H*)-Pyridinethione; 1-Hydroxy, *in* P-**60225**
2-Pyrrolethiolcarboxylic acid, P-**80226**
2-Thiophenecarboxaldehyde; Oxime, *in* T-**70183**

C_5H_5NOSe
2(1*H*)-Pyridineselone; *N*-Oxide, *in* P-**70170**

$C_5H_5NO_2$
1-Cyanocyclopropanecarboxylic acid, *in* C-**80196**
2,3-Dihydro-3-oxo-1*H*-pyrrole-1-carboxaldehyde, *in* D-**70262**
3-Hydroxy-2(1*H*)-pyridinone, H-**60223**
3-Hydroxy-4(1*H*)-pyridinone, H-**60227**
4-Hydroxy-2(1*H*)-pyridinone, H-**60224**
5-Hydroxy-2(1*H*)-pyridinone, H-**60225**
6-Hydroxy-2(1*H*)-pyridinone, H-**60226**

$C_5H_5NO_2S$
2-Pyridinesulfinic acid, P-**80208**
4-Pyridinesulfinic acid, P-**80209**

$C_5H_5NO_3$
4,6-Dihydroxy-2(1*H*)-pyridinone, D-**70341**
3-Hydroxy-1*H*-pyrrole-2-carboxylic acid, H-**80248**
4-Hydroxy-1*H*-pyrrole-2-carboxylic acid, H-**80249**
4-Hydroxy-1*H*-pyrrole-3-carboxylic acid, H-**80250**
5-Isoxazolecarboxylic acid; Me ester, *in* I-**60142**
5-Methyl-4-isoxazolecarboxylic acid, M-**80117**

$C_5H_5NO_3S$
2-Pyridinesulfonic acid, P-**70171**

C_5H_5NS
2-Pyridinethiol, P-**60223**
▷ 2(1*H*)-Pyridinethione, P-**60224**
▷ 4(1*H*)-Pyridinethione, P-**60225**
1,4-Thiazepine, T-**60196**
1-Thiocyano-1,3-butadiene, T-**60210**

$C_5H_5NS_2$
4-Mercapto-2(1*H*)-pyridinethione, M-**80029**
2,3-Pyridinedithiol, P-**60218**
2,4-Pyridinedithiol, P-**60219**
2,5-Pyridinedithiol, P-**60220**
3,4-Pyridinedithiol, P-**60221**
3,5-Pyridinedithiol, P-**60222**

C_5H_5NSe
2(1*H*)-Pyridineselone, P-**70170**

$C_5H_5N_3O_2$
N-(1,4-Dihydro-4-oxo-5-pyrimidinyl)formamide, *in* A-**70196**
4-Methyl-5-nitropyrimidine, M-**60104**

$C_5H_5N_3O_4$
1-Methyl-2,3-dinitro-1*H*-pyrrole, *in* D-**60461**
1-Methyl-2,4-dinitro-1*H*-pyrrole, *in* D-**60462**
1-Methyl-2,5-dinitro-1*H*-pyrrole, *in* D-**60463**
1-Methyl-3,4-dinitro-1*H*-pyrrole, *in* D-**60464**

$C_5H_5N_5$
▷ 2-Aminopurine, A-**60243**
8-Aminopurine, A-**60244**
9-Aminopurine, A-**60245**

$C_5H_5N_5O$
6-Amino-1,3-dihydro-2*H*-purin-2-one, A-**60144**
5-Aminopyrazolo[4,3-*d*]pyrimidin-7(1*H*,6*H*)-one, A-**60246**
[1,2,4]-Triazolo[5,1-*c*][1,2,4]triazin-4(1*H*)-one; 1-Me, *in* T-**80214**
[1,2,4]-Triazolo[5,1-*c*][1,2,4]triazin-4(1*H*)-one; 2-Me, *in* T-**80214**

$C_5H_5N_5OS$
2-Amino-1,7,8,9-tetrahydro-2-thioxo-6*H*-purin-6-one, A-80145

C_5H_6
Ethylnylcyclopropane, E-70045
Tricyclo[1.1.1.0^{1,3}]pentane, T-80244
Tricyclo[2.1.0.0^{1,3}]pentane, T-70231
3-Vinylcyclopropene, V-60012

$C_5H_6BrN_3$
5-Bromo-4,6-dimethyl-1,2,3-triazine, B-80201

$C_5H_6Br_2Cl_2$
1,1-Dibromo-2,2-bis(chloromethyl)cyclopropane, D-80079

$C_5H_6Br_2O$
4,4-Dibromo-3-methyl-3-buten-2-one, D-70094

$C_5H_6Br_2O_2$
2,2-Dibromocyclopropanecarboxylic acid; Me ester, *in* D-60093

$C_5H_6ClN_3$
5-Chloro-4,6-dimethyl-1,2,3-triazine, C-80052

$C_5H_6ClN_3O_2$
2-Chloro-4,6-dimethoxy-1,3,5-triazine, *in* C-80115

$C_5H_6Cl_2O_2$
2,2-Dichlorocyclopropanecarboxylic acid; Me ester, *in* D-60127

$C_5H_6CrFNO_3$
▷ Pyridinium fluorochromate, P-80210

$C_5H_6F_2O_2$
3,3-Difluorocyclobutanecarboxylic acid, D-70164

$C_5H_6I_2$
1,3-Diiodobicyclo[1.1.1]pentane, D-70357

$C_5H_6N_2$
4(5)-Vinylimidazole, V-70012

$C_5H_6N_2O$
2-Amino-3-hydroxypyridine, A-60200
2-Amino-5-hydroxypyridine, A-60201
4-Amino-3-hydroxypyridine, A-60202
5-Amino-3-hydroxypyridine, A-60203
1-Amino-2(1*H*)-pyridinone, A-60248
2-Amino-4(1*H*)-pyridinone, A-60253
3-Amino-2(1*H*)-pyridinone, A-60249
3-Amino-4(1*H*)-pyridinone, A-60254
4-Amino-2(1*H*)-pyridinone, A-60250
5-Amino-2(1*H*)-pyridinone, A-60251
6-Amino-2(1*H*)-pyridinone, A-60252
1,1-Cyclopropanedicarboxylic acid; Nitrile, amide, *in* C-80196
6-Methyl-3-pyridazinone, M-70124
4-Methyl-3(2*H*)-pyridazinone, M-70125
5-Methyl-3(2*H*)-pyridazinone, M-70126
6-Methyl-4(1*H*)-pyridazinone, M-80151
1*H*-Pyrazole; *N*-Ac, *in* P-70152

$C_5H_6N_2OS$
5,6-Dihydro-4*H*-cyclopenta-1,2,3-thiadiazole; 2-Oxide, *in* D-70179
3,4-Dihydro-5-methyl-4-thioxo-2(1*H*)-pyrimidinone, D-70233

$C_5H_6N_2O_2$
4-Amino-2(1*H*)-pyridinone; *N*-Oxide, *in* A-60250
1*H*-Imidazole-4-carboxylic acid; Me ester, *in* I-60004

$C_5H_6N_2O_2S$
2-Amino-4-thiazoleacetic acid, A-80146
2-Amino-5-thiazoleacetic acid, A-80147
▷ 2-Thioxo-4-imidazolidinone; 1-*N*-Ac, *in* T-70188

$C_5H_6N_2S$
2-Amino-3-pyridinethiol, A-80136
4-Amino-3-pyridinethiol, A-80137
6-Amino-3-pyridinethiol, A-80138
2-Amino-4(1*H*)-pyridinethione, A-80139
▷ 3-Amino-2(1*H*)-pyridinethione, A-60247
3-Amino-4(1*H*)-pyridinethione, A-80140
5-Amino-2(1*H*)-pyridinethione, A-80141
6-Amino-2(1*H*)-pyridinethione, A-80142
5,6-Dihydro-4*H*-cyclopenta-1,2,3-thiadiazole, D-70179

$C_5H_6N_2S_2$
▷ 5-Methyl-2,4-(1*H*,3*H*)-pyrimidinedithione, M-70127
6-Methyl-2,4(1*H*,3*H*)-pyrimidinedithione, M-70128

$C_5H_6N_2S_3$
6-Mercapto-2,4(1*H*,3*H*)-pyrimidinedithione; 2-*S*-Me, *in* M-80030

$C_5H_6N_4$
2-Amino-3-cyano-4-methylpyrazole, *in* A-80110
4-Amino-3-cyano-5-methylpyrazole, *in* A-80111
5-Amino-4-cyano-1-methylpyrazole, *in* A-70192
3-Amino-1-methyl-4-pyrazolecarbonitrile, *in* A-70187

$C_5H_6N_6O$
8-Aminoguanine, A-80078
2,8-Diamino-1,7-dihydro-6*H*-purin-6-one, D-60035

C_5H_6O
2-Cyclopenten-1-one, C-60226
2-Oxabicyclo[3.1.0]hex-3-ene, O-70056

C_5H_6OS
3-Methoxythiophene, *in* T-60212
▷ 2-Methyl-3-furanthiol, M-70082

$C_5H_6O_2$
Bicyclo[1.1.0]butane-1-carboxylic acid, B-70094
3-Butynoic acid; Me ester, *in* B-70309
4-Hydroxy-2-cyclopentenone, H-60116
▷ 5-Methyl-2(3*H*)-furanone, M-70081
▷ 5-Methyl-2(5*H*)-furanone, M-60083
4-Methyl-3-methylene-2-oxetanone, M-80119
3-Oxabicyclo[3.1.0]hexan-2-one, O-70055
4-Pentene-2,3-dione, P-60061
▷ 4-Pentynoic acid, P-80062

$C_5H_6O_3$
2-Cyclopropyl-2-oxoacetic acid, C-60231
Dimethoxycyclopropenone, *in* D-70290
5-Hydroxymethyl-2(5*H*)-furanone, H-60182
3-Methoxy-2(5*H*)-furanone, M-80043
4-Methoxy-2(5*H*)-furanone, *in* F-80087
3-Methyl-4-oxo-2-butenoic acid, M-70103
3-Oxocyclobutanecarboxylic acid, O-80073
▷ 4-Oxo-2-pentenoic acid, O-70098

$C_5H_6O_4$
1,1-Cyclopropanedicarboxylic acid, C-80196
1,2-Cyclopropanedicarboxylic acid, C-70248
Dihydro-3-methoxy-2,5-furandione, *in* D-80211
Methyl diformylacetate, *in* D-60190

$C_5H_6O_5$
Methyl hydrogen 2-formylmalonate, *in* H-80204
3-Oxopentanedioic acid, O-60084

C_5H_6S
2-Thiabicyclo[3.1.0]hex-3-ene, T-70161

$C_5H_6S_5$
4,5-Dimercapto-1,3-dithiole-2-thione; Di-Me thioether, *in* D-80409

C_5H_7Br
1-Bromo-2,3-dimethylcyclopropene, B-60235
3-Bromo-3-methyl-1-butyne, B-80213
1-Bromo-1,3-pentadiene, B-60311

$C_5H_7BrN_2$
4-Bromo-3,5-dimethylpyrazole, B-70208
3-Bromo-1,5-dimethylpyrazole, *in* B-70252
4-Bromo-1,3-dimethylpyrazole, *in* B-70251
4-Bromo-1,5-dimethylpyrazole, *in* B-70251
5-Bromo-1,3-dimethylpyrazole, *in* B-70252

$C_5H_7BrO_2$
2-Bromomethyl-2-propenoic acid; Me ester, *in* B-70249

C_5H_7Cl
5-Chloro-1,3-pentadiene, C-60121
5-Chloro-1-pentyne, C-70139

C_5H_7ClOS
Tetrahydro-2-thiophenecarboxylic acid; Chloride, *in* T-70099

$C_5H_7ClO_2$
1-Chloro-2,3-pentanedione, C-60123

C_5H_7FO
2-Fluorocyclopentanone, F-60022

$C_5H_7F_3O_2$
3,3,3-Trifluoropropanoic acid; Et ester, *in* T-80284

C_5H_7I
5-Iodo-1-pentyne, I-70054

$C_5H_7IN_2$
3-Iodo-4,5-dimethylpyrazole, I-70032
4-Iodo-3,5-dimethylpyrazole, I-70033
4-Iodo-1,3-dimethylpyrazole, *in* I-70049
4-Iodo-1,5-dimethylpyrazole, *in* I-70049
5-Iodo-1,3-dimethylpyrazole, *in* I-70050

C_5H_7IO
2-Iodocyclopentanone, I-60043
5-Iodo-4-pentenal, I-70052

$C_5H_7IO_2$
1-Iodocyclobutanecarboxylic acid, I-80022
Tetrahydro-3-iodo-2*H*-pyran-2-one, T-60069

C_5H_7NO
4-Cyanobutanal, *in* O-80089
2-Cyclopenten-1-one; Oxime, *in* C-60226
3-Ethoxy-2-propenenitrile, E-80059
2-Hydroxy-3-methyl-2-butenenitrile, H-80198
1-Methyl-1,2-dihydro-3*H*-pyrrol-3-one, *in* D-70262
3-Methyl-2-oxobutanenitrile, *in* M-80129
4-Vinyl-2-azetidinone, V-60008

C_5H_7NOS
2-Acetyl-4,5-dihydrothiazole, A-70041

$C_5H_7NO_2$
4-Aminodihydro-3-methylene-2-(3*H*)furanone, A-70131
3,4-Dihydro-2*H*-pyrrole-2-carboxylic acid, D-80248
3,4-Dihydro-2*H*-pyrrole-5-carboxylic acid, D-80249
3-Methyl-2,5-pyrrolidinedione, M-80155

$C_5H_7NO_2S$
3-(Ethylthio)-2-propenoic acid; Nitrile, *S*-dioxide, *in* E-60055
Tetrahydro-2-thiophenecarboxylic acid; Nitrile, 1,1-dioxide, *in* T-70099

$C_5H_7NO_3$
Methyl 3-amino-2-formyl-2-propenoate, *in* A-80134
2-Nitrocyclopentanone, N-80044
5-Oxo-2-pyrrolidinecarboxylic acid, O-70102

$C_5H_7NO_3S$
5-Oxo-3-thiomorpholinecarboxylic acid, O-80095

$C_5H_7NO_4$
2-Amino-4-hydroxypentanedioic acid; Lactone, *in* A-60195
5-Methyl-2-oxo-4-oxazolidinecarboxylic acid, M-60108
3-Oxopentanedioic acid; Amide, *in* O-60084

$C_5H_7NO_5$
3-Oxopentanedioic acid; Oxime, *in* O-60084

$C_5H_7NO_6$
2-Amino-1,1,2-ethanetricarboxylic acid, A-80072

C_5H_7NS
2-Cyanotetrahydrothiophene, *in* T-70099
3-(Ethylthio)-2-propenoic acid; Nitrile, *in* E-60055

$C_5H_7NS_2$
2-(Ethylthio)thiazole, *in* T-60200

$C_5H_7N_3$
▷ 3-Azido-3-methyl-1-butyne, A-80203

$C_5H_7N_3O$
4-Amino-1,7-dihydro-2*H*-1,3-diazepin-2-one, A-70130
2-Amino-6-methyl-4(1*H*)-pyrimidinone, A-60230
4-Hydrazino-2(1*H*)-pyridinone, H-60093
1*H*-Imidazole-4-carboxylic acid; Methylamide, *in* I-60004

$C_5H_7N_3OS$
2-Amino-5-thiazoleacetic acid; Amide, *in* A-80147

$C_5H_7N_3O_2$
2-Amino-4(5)-imidazoleacetic acid, A-80086
3-Amino-5-methyl-1*H*-pyrazole-4-carboxylic acid, A-80110
4-Amino-5-methyl-1*H*-pyrazole-3-carboxylic acid, A-80111
3-Amino-1*H*-pyrazole-5-carboxylic acid; *N*(1)-Me, *in* A-70188
5-Amino-1*H*-pyrazole-3-carboxylic acid; *N*(1)-Me, *in* A-70191
3-Amino-1*H*-pyrazole-4-carboxylic acid; Me ester, *in* A-70187
3-Azido-2-propenoic acid; Et ester, *in* A-70290
1,6-Dimethyl-5-azauracil, *in* M-60132
3,6-Dimethyl-5-azauracil, *in* M-60132
Methyl 1-methyl-1*H*-1,2,3-triazole-5-carboxylate, *in* T-60239
1*H*-1,2,3-Triazole-4-carboxylic acid; Et ester, *in* T-60239
1*H*-1,2,3-Triazole-4-carboxylic acid; 1-Me, Me ester, *in* T-60239
1*H*-1,2,3-Triazole-4-carboxylic acid; 2-Me, Me ester, *in* T-60239

$C_5H_7N_3O_2S$
2-Pyridinesulfonic acid; Hydrazide, *in* P-70171

$C_5H_7N_3O_3$
Cyanuric acid; 1,3-Di-Me, *in* C-60179

$C_5H_7N_3O_6$
1,1-Dinitro-2-propanone; *O*-Acetyloxime, *in* D-70454

C_5H_8
Methylenecyclobutane, M-70066

$C_5H_8Br_2$
1,1-Dibromo-3-methyl-1-butene, D-60104

$C_5H_8ClNO_2$
1-Chloro-2,3-pentanedione; 2-Oxime, *in* C-60123

$C_5H_8Cl_2$
2,5-Dichloro-1-pentene, D-80126

$C_5H_8Cl_2O$
▷ 3,3-Bis(chloromethyl)oxetane, B-60149
1,5-Dichloro-3-pentanone, D-80125

$C_5H_8FNO_2$
4-Amino-5-fluoro-2-pentenoic acid, A-60161

C_5H_8INO
5-(Iodomethyl)-2-pyrrolidinone, I-80037

$C_5H_8N_2$
1,4-Dihydro-2-methylpyrimidine, D-70225

$C_5H_8N_2O$
3,4-Dihydro-2(1*H*)-pyrimidinone; 1-Me, *in* D-60273
5,6-Dihydro-4(1*H*)-pyrimidinone; 3-Me, *in* D-60274

$C_5H_8N_2OS$
2-Acetyl-4,5-dihydrothiazole; Oxime, *in* A-70041
2-Thioxo-4-imidazolidinone; 1,3-*N*-Di-Me, *in* T-70188

$C_5H_8N_2O_2$
5-Amino-3,4-dihydro-2*H*-pyrrole-2-carboxylic acid, A-70135
3-Amino-2,6-piperidinedione, A-80133
▷ *N*-(2-Cyanoethyl)glycine, *in* C-60016
5-Oxo-2-pyrrolidinecarboxylic acid; Amide, *in* O-70102

$C_5H_8N_2O_5$
N,N'-Carbonylbisglycine, C-80020

$C_5H_8N_2O_7$
3,3-Bis(nitratomethyl)oxetane, *in* O-60056

$C_5H_8N_4$
3-Amino-1,2,4-triazine; *N,N*-Di-Me, *in* A-70203

$C_5H_8N_4O$
6-Amino-2-(methylamino)-4(3*H*)-pyrimidinone, *in* D-70043
4-Amino-5-methyl-1*H*-pyrazole-3-carboxylic acid; Amide, *in* A-80111
5-Amino-1*H*-pyrazole-4-carboxylic acid; *N*(1)-Me, amide, *in* A-70192
4-Amino-1,3,5-triazin-2(1*H*)-one; 1,4-*N*-Di-Me, *in* A-80149
4-Amino-1,3,5-triazin-2(1*H*)-one; 4,4-*N*-Di-Me, *in* A-80149
2,6-Diamino-4(1*H*)-pyrimidinone; 1-Me, *in* D-70043

C_5H_8O
2-Methyl-2-butenal, M-80056
▷ 3-Methyl-2-butenal, M-70055
(1-Methylethenyl)oxirane, M-70080
▷ 1-Penten-3-one, P-60062
3-Pentyn-2-ol, P-70041

C_5H_8OS
Tetrahydro-2*H*-pyran-2-thione, T-70085

$C_5H_8OS_2$
3,3-Bis(methylthio)propenal, B-80156

$C_5H_8O_2$
4-Cyclopentene-1,3-diol, C-80193
Dihydro-4-methyl-2(3*H*)-furanone, D-80220
2,6-Dioxaspiro[3.3]heptane, D-60468
2-Ethynyl-1,3-propanediol, E-70054
2-Hydroxycyclopentanone, H-60114
3-Hydroxycyclopentanone, H-60115
2-Hydroxy-2-methylcyclobutanone, H-70168
2-Methyl-3-oxobutanal, M-70102
4-Oxopentanal, O-60083
▷ 2,4-Pentanedione, P-60059
2-Vinyl-1,3-dioxolane, *in* P-70119
3-Vinyl-2-oxiranemethanol, V-80009

$C_5H_8O_2S$
2,3-Dihydro-2-methylthiophene; 1,1-Dioxide, *in* D-70227
2,3-Dihydro-5-methylthiophene; 1,1-Dioxide, *in* D-70229
3-(Ethylthio)-2-propenoic acid, E-60055
Isoprene cyclic sulfone, *in* D-70231
α-Isoprene sulfone, *in* D-70228
4-Mercapto-2-butenoic acid; Me ester, *in* M-60018
2-Methyl-3-sulfolene, *in* D-70230
Tetrahydro-2-thiophenecarboxylic acid, T-70099

$C_5H_8O_2Se$
3-Selenetanol; Ac, *in* S-80022

$C_5H_8O_3$
Dihydro-4-hydroxy-3-methyl-2(3*H*)-furanone, D-80214
Dihydro-4-hydroxy-5-methyl-2(3*H*)furanone, *in* D-80353
2-Ethoxy-2-propenoic acid, E-70037
3-Methyloxiranecarboxylic acid; Me ester, *in* M-80127
3-Methyl-2-oxobutanoic acid, M-80129
4-Oxobutanoic acid; Me ester, *in* O-80068
5-Oxopentanoic acid, O-80089

$C_5H_8O_3S$
2-Mercaptopropanoic acid; *S*-Ac, *in* M-60027

$C_5H_8O_4$
4,5-Dihydro-3-hydroxy-5-(hydroxymethyl)-2(3*H*)-furanone, D-70212
Methylpropanedioic acid; Mono-Me ester, *in* M-80149

C_5H_8S
2,3-Dihydro-2-methylthiophene, D-70227
2,3-Dihydro-4-methylthiophene, D-70228
2,3-Dihydro-5-methylthiophene, D-70229
2,5-Dihydro-2-methylthiophene, D-70230
2,5-Dihydro-3-methylthiophene, D-70231

$C_5H_8S_2$
2-Vinyl-1,3-dithiolane, *in* P-70119

C_5H_9BrO
5-Bromopentanal, B-70263
2-Bromotetrahydro-2*H*-pyran, B-70271

$C_5H_9BrO_2$
2-Bromo-2-methylpropanoic acid; Me ester, *in* B-80229

$C_5H_9ClN_4$
5-Chlorotetrazole; 1-*tert*-Butyl, *in* C-80113
5-Chlorotetrazole; 2-*tert*-Butyl, *in* C-80113

C_5H_9ClO
5-Chloropentanal, C-70137
4-Chloro-4-penten-1-ol, C-80091
2-Chlorotetrahydro-2*H*-pyran, C-80110
3-Chlorotetrahydro-2*H*-pyran, C-80111
4-Chlorotetrahydro-2*H*-pyran, C-80112

C_5H_9ClOS
Carbonochloridothioic acid; *O*-Butyl, *in* C-60013
Carbonochloridothioic acid; *O-tert*-Butyl, *in* C-60013
Carbonochloridothioic acid; *S*-Butyl, *in* C-60013

$C_5H_9ClO_2$
3-Chloro-2-(methoxymethoxy)-1-propene, C-70089

$C_5H_9ClO_3$
(1-Chloroethyl) ethyl carbonate, C-60067
2-(Chloromethoxy)ethyl acetate, C-70088

$C_5H_9FN_2$
2-Amino-3-fluoro-3-methylbutanoic acid; Nitrile, *in* A-60159

$C_5H_9IN_2$
1,3-Dimethyl-1*H*-imidazolium(1+); Iodide, *in* D-80451

C_5H_9IO
2-Iodo-3-methylbutanal, I-60049
2-Iodopentanal, I-60054
5-Iodo-2-pentanone, I-60057
5-Iodo-4-penten-1-ol, I-60058

C_5H_9N
3,4-Dihydro-5-methyl-2*H*-pyrrole, D-70226
▷ 1-Isocyanobutane, I-80058
4-Pentyn-1-amine, P-80061
2-Propyn-1-amine; *N*-Di-Me, *in* P-60184
▷ 1,2,3,6-Tetrahydropyridine, T-70089
2,3,4,5-Tetrahydropyridine, T-70090

C_5H_9NO
4,5-Dihydro-4,4-dimethyloxazole, D-70198
3,4-Dihydro-5-methoxy-2*H*-pyrrole, *in* P-70188
4-Imino-2-pentanone, *in* P-60059
2-Pyrrolidinecarboxaldehyde, P-70184
Tiglic aldehyde; Oxime, *in* M-80056

$C_5H_9NO_2$
2-Amino-3-methyl-3-butenoic acid, A-60218
5-(Aminomethyl)dihydro-2(3*H*)-furanone, *in* A-60196
3,5-Dimethyl-2-oxazolidinone, *in* M-60107
2-Ethoxy-2-propenoic acid; Amide, *in* E-70037
2-Hydroxycyclopentanone; Oxime, *in* H-60114
3-Methyl-2-oxobutanoic acid; Amide, *in* M-80129

$C_5H_9NO_2S$
3-Aminotetrahydro-3-thiophenecarboxylic acid, A-70199
2-Thiazolidineacetic acid, T-80173
3-Thiomorpholinecarboxylic acid, T-70182

$C_5H_9NO_3$
2-Amino-2-(hydroxymethyl)-3-butenoic acid,
A-60187
1-Amino-2-
(hydroxymethyl)cyclopropanecarboxylic
acid, A-80083
3,4-Dihydro-3,4-dihydroxy-2-
(hydroxymethyl)-2H-pyrrole, D-60222
3-Methyl-2-oxobutanoic acid; Oxime, *in*
M-80129
3-Morpholinecarboxylic acid, M-70155

$C_5H_9NO_4$
2-Amino-3-methylbutanedioic acid, A-70166
3-(Carboxymethylamino)propanoic acid,
C-60016
3,4-Dihydroxy-2-pyrrolidinecarboxylic acid,
D-70342
3-Nitro-1-propanol; Ac, *in* N-60039
Threonine; N-Formyl, *in* T-60217

$C_5H_9NO_5$
2-Amino-4-hydroxypentanedioic acid,
A-60195

C_5H_9NS
2,3-Dihydro-2,2-dimethylthiazole, D-80194
▷ 2-Isothiocyanato-2-methylpropane, I-70105
2-Methyl-2-thiocyanatopropane, M-70135
2-Propenethioamide; N,N-Di-Me, *in* P-70120
2-Pyrrolidinethione; N-Me, *in* P-70186

$C_5H_9NS_2$
5,6-Dihydro-2-(methylthio)-4H-1,3-thiazine, *in*
T-60094
Tetrahydro-2H-1,3-thiazine-2-thione; N-Me,
in T-60094

$C_5H_9N_2^{\oplus}$
1,3-Dimethyl-1H-imidazolium(1 +), D-80451

$C_5H_9N_2S_2$
5-*tert*-Butyl-1,2,3,5-dithiadiazolyl, B-70302
5-*tert*-Butyl-1,3,2,4-dithiadiazolyl, B-70303

$C_5H_9N_3O$
3,5-Dihydro-3,5,5-trimethyl-4H-triazol-4-one,
D-60303

$C_5H_9N_3O_2$
3-Azido-3-methylbutanoic acid, A-60348

$C_5H_9N_4O$
Porphyrexide, P-70110

$C_5H_9N_5$
2,4,6-Triaminopyrimidine; N^4-Me, *in* T-80208

$C_5H_9N_5O$
4-Amino-5-methyl-1H-pyrazole-3-carboxylic
acid; Hydrazide, *in* A-80111
2,4,6-Triaminopyrimidine; N^4-Me, N^1-oxide,
in T-80208

$C_5H_{10}ClN$
1-Methylenepyrrolidinium(1 +); Chloride, *in*
M-70077

$C_5H_{10}ClNO_2$
2-Amino-3-chlorobutanoic acid; Me ester, *in*
A-60103
2-Amino-4-chlorobutanoic acid; Me ester, *in*
A-60104

$C_5H_{10}Cl_2O$
1,5-Dichloro-3-pentanol, D-80124

$C_5H_{10}FNO_2$
2-Amino-3-fluoro-3-methylbutanoic acid,
A-60159
2-Amino-3-fluoropentanoic acid, A-60160

$C_5H_{10}N^{\oplus}$
1-Methylenepyrrolidinium(1 +), M-70077

$C_5H_{10}N_2$
2-Amino-2-methylbutanoic acid; Nitrile, *in*
A-70167
2,5-Diazabicyclo[4.1.0]heptane, D-60053

$C_5H_{10}N_2O$
▷ 1,3-Dimethylimidazolidinone, *in* I-70005

Tetrahydro-2(1H)-pyrimidinone; 1-Me, *in*
T-60088
Tetrahydro-4(1H)-pyrimidinone; 3-Me, *in*
T-60089

$C_5H_{10}N_2O_2$
4-Oxopentanal; Dioxime, *in* O-60083
2,4-Pentanedione; Dioxime, *in* P-60059

$C_5H_{10}N_2O_4$
▷ 2,4-Dinitropentane, D-70453

$C_5H_{10}N_2O_5$
4,5-Dihydroxy-1,3-bis(hydroxymethyl)-2-
imidazolidinone, D-80301

$C_5H_{10}O$
2,3-Dimethylcyclopropanol, D-70397
2-Methyl-2-buten-1-ol, M-80057

$C_5H_{10}OSe$
Methaneselenoic acid; *tert*-Butyl ester, *in*
M-70035

$C_5H_{10}O_2$
1,2-Cyclopentanediol, C-80189
▷ 3,3-Dimethoxypropene, *in* P-70119
3,3-Dimethyl-1,2-dioxolane, D-80446
2,3-Dimethyloxiranemethanol, D-80462
3,3-Dimethyloxiranemethanol, D-70426
2-Methoxybutanal, *in* H-70112
1-Methoxy-3-buten-2-ol, *in* B-70291

$C_5H_{10}O_2S$
2-Mercapto-3-methylbutanoic acid, M-60023
2-Mercaptopropanoic acid; Et ester, *in*
M-60027
2-Mercaptopropanoic acid; Me ester, S-Me
ether, *in* M-60027

$C_5H_{10}O_2S_2$
1,2-Dithiepane; 1,1-Dioxide, *in* D-70517

$C_5H_{10}O_3$
3,3-Oxetanedimethanol, O-60056

$C_5H_{10}O_4$
2,3-Dihydroxybutanoic acid; Me ester, *in*
D-60312
3,4-Dihydroxypentanoic acid, D-80353
▷ Glycerol 1-acetate, G-60029

$C_5H_{10}O_4S_2$
1,2-Dithiepane; 1,1,2,2-Tetraoxide, *in* D-70517

$C_5H_{10}S_2$
1,1-Cyclopentanedithiol, C-60222
1,1-Cyclopropanedimethanethiol, C-70249
1,2-Dithiepane, D-70517
5-Methyl-1,3-dithiane, M-60067

$C_5H_{10}S_3$
Carbonotrithioic acid; Di-Et ester, *in* C-80019

$C_5H_{11}BrO$
3-Bromo-2,2-dimethylpropanol, B-80200
5-Bromo-2-pentanol, B-60313

$C_5H_{11}ClO$
tert-Butyl chloromethyl ether, B-70295
1-Chloro-3-pentanol, C-70138

$C_5H_{11}ClO_2S$
(2-Carboxyethyl)dimethylsulfonium(1 +);
Chloride, *in* C-80024

$C_5H_{11}IO$
1-Iodo-3-pentanol, I-60055
5-Iodo-2-pentanol, I-60056

$C_5H_{11}NOS$
3-Aminotetrahydro-2H-thiopyran; S-Oxide, *in*
A-70200

$C_5H_{11}NO_2$
2-Amino-2-methylbutanoic acid, A-70167
3-Amino-2-methylbutanoic acid, A-80097

$C_5H_{11}NO_2S$
3-Aminotetrahydro-2H-thiopyran; S,S-
Dioxide, *in* A-70200

$C_5H_{11}NO_3$
2-Amino-4-hydroxy-2-methylbutanoic acid,
A-60186
4-Amino-5-hydroxypentanoic acid, A-70159
5-Amino-4-hydroxypentanoic acid, A-60196

3-Amino-2-hydroxypropanoic acid; Et ester,
in A-70160
Threonine; N-Me, *in* T-60217
Threonine; Me ester, *in* T-60217
Threonine; Me ether, *in* T-60217

$C_5H_{11}NO_4$
1,1-Dimethoxy-3-nitropropane, *in* N-60037

$C_5H_{11}NS$
3-Aminotetrahydro-2H-thiopyran, A-70200
2-Pyrrolidinemethanethiol, P-70185

$C_5H_{11}N_3$
1-Azido-2,2-dimethylpropane, A-60344

$C_5H_{11}N_3O_2$
Biuret; 1,3,5-Tri-Me, *in* B-60190

$C_5H_{11}O_2S^{\oplus}$
(2-Carboxyethyl)dimethylsulfonium(1 +),
C-80024

$C_5H_{12}N_2O$
3,3-Bis(aminomethyl)oxetane, B-60140

$C_5H_{12}N_2O_2S$
S-(2-Aminoethyl)cysteine, A-60152

$C_5H_{12}N_2O_3$
2,5-Diamino-3-hydroxypentanoic acid,
D-70039

$C_5H_{12}N_2O_3S$
S-(2-Aminoethyl)cysteine; S-Oxide, *in*
A-60152
Aminoiminomethanesulfonic acid; N-*tert*-
Butyl, *in* A-60210
Methionine sulfoximine, M-70043

$C_5H_{12}N_2O_4S$
S-(2-Aminoethyl)cysteine; S,S-Dioxide, *in*
A-60152

$C_5H_{12}N_2S$
▷ N,N'-Diethylthiourea, D-70155

$C_5H_{12}N_2Se$
Selenourea; N-Tetra-Me, *in* S-70031

$C_5H_{12}O_2$
2-Methyl-1,3-butanediol, M-80054
3-Methyl-1,2-butanediol, M-80055
2,3-Pentanediol, P-80060

$C_5H_{12}O_2S$
1,1-Dimethoxy-2(methylthio)ethane, *in*
M-80017

$C_5H_{12}O_3$
2-(Hydroxymethyl)-1,4-butanediol, H-80197

$C_5H_{12}O_4S_4$
Tris(methylthio)methylium(1 +);
Methylsulfate, *in* T-80367

$C_5H_{12}O_4S_5$
Dysoxysulfone, D-80539

$C_5H_{12}S_2$
2,2-Dimethyl-1,3-propanedithiol, D-70433

$C_5H_{12}S_3$
2-Mercaptomethyl-2-methyl-1,3-
propanedithiol, M-70030

$C_5H_{12}S_4$
2,2-Bis(mercaptomethyl)-1,3-propanedithiol,
B-60164

$C_5H_{13}NO$
3-Amino-3-methyl-2-butanol, A-60217

$C_5H_{13}NO_2$
1-Amino-3-methyl-2,3-butanediol, A-60216
2-Amino-1,5-pentanediol, A-80131

$C_5H_{14}N_2$
2,4-Pentanediamine, P-70039

C_5O_2
1,2,3,4-Pentatetraene-1,5-dione, P-70040

$C_6BrF_4NO_2$
1-Bromo-2,3,4,5-tetrafluoro-6-nitrobenzene,
B-60323

1-Bromo-2,3,4,6-tetrafluoro-5-nitrobenzene, B-60324
1-Bromo-2,3,5,6-tetrafluoro-4-nitrobenzene, B-60325

C$_6$BrF$_{13}$
1-Bromo-1,1,2,2,3,3,4,4,5,5,6,6,6-tridecafluorohexane, B-70274

C$_6$Br$_2$F$_4$
1,2-Dibromo-3,4,5,6-tetrafluorobenzene, D-60106
1,3-Dibromo-2,4,5,6-tetrafluorobenzene, D-60107
1,4-Dibromo-2,3,5,6-tetrafluorobenzene, D-60108

C$_6$Br$_3$F$_3$
1,3,5-Tribromo-2,4,6-trifluorobenzene, T-70214

C$_6$ClF$_{13}$
1-Chloro-1,1,2,2,3,3,4,4,5,5,6,6,6-tridecafluorohexane, C-70163

C$_6$Cl$_2$F$_4$
1,2-Dichloro-3,4,5,6-tetrafluorobenzene, D-60151
1,3-Dichloro-2,4,5,6-tetrafluorobenzene, D-60152
1,4-Dichloro-2,3,5,6-tetrafluorobenzene, D-60153

C$_6$Cl$_2$N$_4$
2,3-Dichloro-5,6-dicyanopyrazine, in D-80129

C$_6$Cl$_3$F$_3$
1,2,3-Trichloro-4,5,6-trifluorobenzene, T-60256
▷ 1,3,5-Trichloro-2,4,6-trifluorobenzene, T-60257

C$_6$F$_4$I$_2$
1,2,3,4-Tetrafluoro-5,6-diiodobenzene, T-60044
1,2,3,5-Tetrafluoro-4,6-diiodobenzene, T-60045
▷ 1,2,4,5-Tetrafluoro-3,6-diiodobenzene, T-60046

C$_6$F$_5$IO
Pentafluoroiodosobenzene, P-60034

C$_6$F$_5$NO
Pentafluoronitrosobenzene, P-60039

C$_6$F$_5$NO$_2$
Pentafluoronitrobenzene, P-80026

C$_6$F$_5$N$_3$
Azidopentafluorobenzene, A-60350

C$_6$F$_{12}$OS$_2$
2,2,4,4-Tetrakis(trifluoromethyl)-1,3-dithietane; 1-Oxide, in T-60129

C$_6$F$_{12}$O$_2$S$_2$
2,2,4,4-Tetrakis(trifluoromethyl)-1,3-dithietane; 1,1-Dioxide, in T-60129
2,2,4,4-Tetrakis(trifluoromethyl)-1,3-dithietane; 1,3-Dioxide, in T-60129

C$_6$F$_{12}$O$_3$S$_2$
2,2,4,4-Tetrakis(trifluoromethyl)-1,3-dithietane; 1,1,3-Trioxide, in T-60129

C$_6$F$_{12}$O$_4$S$_2$
2,2,4,4-Tetrakis(trifluoromethyl)-1,3-dithietane; 1,1,3,3-Tetraoxide, in T-60129

C$_6$F$_{12}$S$_2$
▷ 2,2,4,4-Tetrakis(trifluoromethyl)-1,3-dithietane, T-60129

C$_6$F$_{12}$S$_4$
Tetrakis(trifluoromethylthio)ethene, T-80132

C$_6$HBrF$_4$
1-Bromo-2,3,4,5-tetrafluorobenzene, B-60320
2-Bromo-1,3,4,5-tetrafluorobenzene, B-60321
3-Bromo-1,2,4,5-tetrafluorobenzene, B-60322

C$_6$HClN$_4$O$_8$
▷ 3-Chloro-1,2,4,5-tetranitrobenzene, C-60135

C$_6$HCl$_4$N$_3$
4,5,6,7-Tetrachlorobenzotriazole, T-60027

C$_6$HF$_4$N$_3$
4,5,6,7-Tetrafluoro-1H-benzotriazole, T-70035

C$_6$H$_2$BrFN$_2$O$_4$
1-Bromo-2-fluoro-3,5-dinitrobenzene, B-60246

C$_6$H$_2$Br$_2$F$_2$
1,2-Dibromo-4,5-difluorobenzene, D-80087
1,4-Dibromo-2,5-difluorobenzene, D-80088

C$_6$H$_2$Br$_2$I$_2$O$_2$
4,5-Dibromo-3,6-diiodo-1,2-benzenediol, D-80090

C$_6$H$_2$Br$_2$O$_2$
2,6-Dibromo-1,4-benzoquinone, D-60088

C$_6$H$_2$ClN$_5$O$_8$
▷ 2-Chloro-3,4,5,6-tetranitroaniline, C-60134

C$_6$H$_2$Cl$_2$N$_2$O$_4$
5,6-Dichloro-2,3-pyrazinedicarboxylic acid, D-80129

C$_6$H$_2$Cl$_3$I
1,2,3-Trichloro-4-iodobenzene, T-80227

C$_6$H$_2$F$_3$NO$_2$
1,2,3-Trifluoro-4-nitrobenzene, T-60299
1,2,3-Trifluoro-5-nitrobenzene, T-60300
▷ 1,2,4-Trifluoro-5-nitrobenzene, T-60301
1,2,5-Trifluoro-3-nitrobenzene, T-60302
1,2,5-Trifluoro-4-nitrobenzene, T-60303
1,3,5-Trifluoro-2-nitrobenzene, T-60304

C$_6$H$_2$I$_2$N$_2$O$_4$
1,2-Diiodo-3,5-dinitrobenzene, D-80398
1,2-Diiodo-4,5-dinitrobenzene, D-80399
1,3-Diiodo-4,6-dinitrobenzene, D-80400

C$_6$H$_2$O$_6$
5,6-Dihydroxy-5-cyclohexene-1,2,3,4-tetrone, D-60315

C$_6$H$_3$BrCl$_2$O
2-Bromo-3,5-dichlorophenol, B-60226
2-Bromo-4,5-dichlorophenol, B-60227
3-Bromo-2,5-dichlorophenol, B-60228
4-Bromo-2,3-dichlorophenol, B-60229
4-Bromo-3,5-dichlorophenol, B-60230

C$_6$H$_3$BrINO$_2$
1-Bromo-2-iodo-3-nitrobenzene, B-70222
2-Bromo-1-iodo-3-nitrobenzene, B-70223

C$_6$H$_3$Br$_2$NO$_2$
2,6-Dibromo-4-nitroso-1-naphthol, in D-60088

C$_6$H$_3$Cl$_2$NO$_3$
2,3-Dichloro-4-nitrophenol, D-70125

C$_6$H$_3$Cl$_4$NO$_2$S
2,3,4,5-Tetrachloro-1,1-dihydro-1-[(methoxycarbonyl)imino]thiophene, T-60028

C$_6$H$_3$FN$_4$O$_2$
▷ 4-Azido-1-fluoro-2-nitrobenzene, A-80199

C$_6$H$_3$N$_3$O$_3$
▷ 4-Diazo-2-nitrophenol, D-60065
1,3,5-Triazine-2,4,6-tricarboxaldehyde, T-60237

C$_6$H$_3$N$_3$O$_9$S
▷ 2,4,6-Trinitrobenzenesulfonic acid, T-70302

C$_6$H$_3$N$_5$O$_8$
▷ 2,3,4,6-Tetranitroaniline, T-60159

C$_6$H$_4$
Benzyne, B-70087

C$_6$H$_4$BrFO
2-Bromo-3-fluorophenol, B-60247
2-Bromo-4-fluorophenol, B-60248
2-Bromo-6-fluorophenol, B-60249
3-Bromo-4-fluorophenol, B-60250
4-Bromo-2-fluorophenol, B-60251

C$_6$H$_4$BrNO$_3$
2-Bromo-3-nitrophenol, B-70257

C$_6$H$_4$BrNS$_2$
1,3,2-Benzodithiazol-1-ium(1+); Bromide, in B-70033

C$_6$H$_4$BrN$_3$
3-Bromo[1,2,3]triazolo[1,5-a]pyridine, B-60329
7-Bromo[1,2,3]triazolo[1,5-a]pyridine, B-60330

C$_6$H$_4$Br$_4$N$_2$
2,3-Bis(dibromomethyl)pyrazine, B-60150
4,5-Bis(dibromomethyl)pyrimidine, B-60153

C$_6$H$_4$ClFO
2-Chloro-4-fluorophenol, C-60073
2-Chloro-5-fluorophenol, C-60074
2-Chloro-6-fluorophenol, C-60075
3-Chloro-2-fluorophenol, C-60076
▷ 3-Chloro-4-fluorophenol, C-60077
3-Chloro-6-fluorophenol, C-60078
4-Chloro-2-fluorophenol, C-60079
4-Chloro-3-fluorophenol, C-60080

C$_6$H$_4$ClFS
3-Chloro-4-fluorobenzenethiol, C-60069
4-Chloro-2-fluorobenzenethiol, C-60070
4-Chloro-3-fluorobenzenethiol, C-60071

C$_6$H$_4$ClNO
1-Chloro-2-nitrosobenzene, C-80083
1-Chloro-3-nitrosobenzene, C-80084
1-Chloro-4-nitrosobenzene, C-80085

C$_6$H$_4$ClNO$_3$
▷ 5-Chloro-2-nitrophenol, C-60119

C$_6$H$_4$ClNS$_2$
1,3,2-Benzodithiazol-1-ium(1+); Chloride, in B-70033

C$_6$H$_4$ClN$_3$
4-Chloro-1H-imidazo[4,5-c]pyridine, C-60104
6-Chloro-1H-imidazo[4,5-c]pyridine, C-60105

C$_6$H$_4$Cl$_2$S
2,3-Dichlorobenzenethiol, D-60119
2,4-Dichlorobenzenethiol, D-60120
2,5-Dichlorobenzenethiol, D-60121
2,6-Dichlorobenzenethiol, D-60122
3,4-Dichlorobenzenethiol, D-60123
3,5-Dichlorobenzenethiol, D-60124

C$_6$H$_4$Cl$_3$NO
2-(Trichloroacetyl)pyrrole, T-80221

C$_6$H$_4$FNO
5-Fluoro-3-pyridinecarboxaldehyde, F-80060

C$_6$H$_4$FNO$_3$
▷ 5-Fluoro-2-nitrophenol, F-60061

C$_6$H$_4$FN$_3$O$_4$
2-Fluoro-3,5-dinitroaniline, F-60027
2-Fluoro-4,6-dinitroaniline, F-60028
4-Fluoro-2,6-dinitroaniline, F-60029
4-Fluoro-3,5-dinitroaniline, F-60030
5-Fluoro-2,4-dinitroaniline, F-60031
6-Fluoro-3,4-dinitroaniline, F-60032

C$_6$H$_4$F$_2$N$_2$O$_2$
5-(2,2-Difluorovinyl)uracil, D-80167

C$_6$H$_4$F$_2$S
2,5-Difluorobenzenethiol, D-60173
2,6-Difluorobenzenethiol, D-60174
3,4-Difluorobenzenethiol, D-60175
3,5-Difluorobenzenethiol, D-60176

C$_6$H$_4$F$_3$N
2-(Trifluoromethyl)pyridine, T-60296
3-(Trifluoromethyl)pyridine, T-60297
4-(Trifluoromethyl)pyridine, T-60298

C$_6$H$_4$F$_3$NO
2-(Trifluoromethyl)pyridine; 1-Oxide, in T-60296
3-(Trifluoromethyl)pyridine; 1-Oxide, in T-60297
4-(Trifluoromethyl)pyridine; 1-Oxide, in T-60298

C$_6$H$_4$F$_3$NO$_3$S
2-Pyridinyl trifluoromethanesulfonate, P-70173

C$_6$H$_4$NS$_2$$^\oplus$
1,3,2-Benzodithiazol-1-ium(1+), B-70033

C$_6$H$_4$N$_2$
Bis(methylene)butanedinitrile, in D-70409

$C_6H_4N_2O$
6-Diazo-2,4-cyclohexadien-1-one, D-70057
Furo[2,3-*d*]pyridazine, F-60091
Furo[3,4-*d*]pyridazine, F-60092
3-Isocyanatopyridine, I-70065
5*H*-Pyrrolo[1,2-*a*]imidazol-5-one, P-70194
5*H*-Pyrrolo[1,2-*c*]imidazol-5-one, P-70195
6*H*-Pyrrolo[1,2-*b*]pyrazol-6-one, P-70199

$C_6H_4N_2OS$
▷ 2*H*-[1,2,4]Oxadiazolo[2,3-*a*]pyridine-2-thione, O-70062

$C_6H_4N_2O_2$
5-Ethynyl-2,4(1*H*,3*H*)-pyrimidinedione, E-60060
Furo[2,3-*d*]pyrimidin-2(1*H*)-one, F-60106
4-Hydroxyfuro[2,3-*d*]pyridazine, H-60146
7-Hydroxyfuro[2,3-*d*]pyridazine, H-60147
1,3,4-Oxadiazolo[3,2-*a*]pyridin-2(3*H*)-one, O-70063
2*H*-[1,2,4]Oxadiazolo[2,3-*a*]pyridin-2-one, O-70064
3-Oxopyrazolo[1,2-*a*]pyrazol-8-ylium-1-olate, O-60092

$C_6H_4N_2O_4$
3,6-Pyridazinedicarboxylic acid, P-70168

$C_6H_4N_2O_6$
4,5-Dinitro-1,2-benzenediol, D-70449
▷ 4,6-Dinitro-1,3-benzenediol, D-80491

$C_6H_4N_2S$
2-Isothiocyanatopyridine, I-70106
3-Isothiocyanatopyridine, I-70107
Thiazolo[4,5-*b*]pyridine, T-70166
Thiazolo[5,4-*b*]pyridine, T-70168
Thiazolo[4,5-*c*]pyridine, T-70167
Thiazolo[5,4-*c*]pyridine, T-70169
Thieno[3,2-*d*]pyrimidine, T-60205

$C_6H_4N_2S_2$
2,2′-Bithiazole, B-70166
2,4′-Bithiazole, B-70167
2,5′-Bithiazole, B-70168
4,4′-Bithiazole, B-70169
4,5′-Bithiazole, B-70170
5,5′-Bithiazole, B-70171

$C_6H_4N_2S_4$
2,2′-Dithiobisthiazole, *in* T-60200

$C_6H_4N_4O$
Pyrido[3,4-*d*]-1,2,3-triazin-4(3*H*)-one, P-60238

$C_6H_4N_4O_2$
2,4(1*H*,3*H*)-Pteridinedione, P-60191

$C_6H_4N_4O_3$
2,4,6(1*H*,3*H*,5*H*)-Pteridinetrione, P-70134
2,4,7(1*H*,3*H*,8*H*)-Pteridinetrione, P-70135

$C_6H_4N_4O_6$
▷ 2,4,6-Trinitroaniline, T-70301

$C_6H_4N_6$
4-Azido-1*H*-imidazo[4,5-*c*]pyridine, A-60345
1*H*-1,2,4-Triazolo[3,4-*i*]purine, T-80212
1*H*-1,2,4-Triazolo[5,1-*i*]purine, T-80213

C_6H_4O
1,5-Hexadiyn-3-one, H-60040
3,5-Hexadiyn-2-one, H-60041

C_6H_4OS
Thieno[3,4-*b*]furan, T-60204

$C_6H_4O_2$
▷ 1,4-Benzoquinone, B-70044
Tricyclo[3.1.0.0^{2,6}]hexanedione, T-70229

$C_6H_4O_4$
3,6-Dihydroxy-1,2-benzoquinone, D-60309

$C_6H_4O_5$
Aconitic acid; Anhydride, *in* A-70055

$C_6H_4O_6$
3,5-Dihydroxy-4-oxo-4*H*-pyran-2-carboxylic acid, D-60368

$C_6H_4O_8$
Ethylenetetracarboxylic acid, E-60049

C_6H_4S
2-Ethynylthiophene, E-70055

C_6H_4SSe
Selenolo[3,4-*b*]thiophene, S-70030

$C_6H_4S_4$
Bi(1,3-dithiol-2-ylidene), B-70119
[1,4]Dithiino[2,3-*b*]-1,4-dithiin, D-70521

$C_6H_4S_6$
Thieno[3,4-*c*]thiophene-1,3,4,6-tetrathiol, T-70170

$C_6H_4S_8$
Tetramercaptotetrathiafulvalene, T-70130

$C_6H_4Se_4$
Bi(1,3-diselenol-2-ylidene), B-70116

$C_6H_4Te_4$
Bi(1,3-ditellurol-2-ylidene), B-70117

C_6H_5BrClN
2-(Bromomethyl)-6-chloropyridine, B-70235
3-(Bromomethyl)-2-chloropyridine, B-70236
4-(Bromomethyl)-2-chloropyridine, B-70237
5-(Bromomethyl)-2-chloropyridine, B-70238

$C_6H_5BrClNO$
4-(Bromomethyl)-2-chloropyridine; 1-Oxide, *in* B-70237

C_6H_5BrFN
2-(Bromomethyl)-6-fluoropyridine, B-70243
3-(Bromomethyl)-2-fluoropyridine, B-70244
4-(Bromomethyl)-2-fluoropyridine, B-70245
5-(Bromomethyl)-2-fluoropyridine, B-70246

C_6H_5BrIN
5-Bromo-2-iodoaniline, B-60279

$C_6H_5BrO_3$
2-Bromo-1,3,5-benzenetriol, B-60199
3-Bromo-1,2,4-benzenetriol, B-70180
4-Bromo-1,2,3-benzenetriol, B-60200
5-Bromo-1,2,3-benzenetriol, B-60201
5-Bromo-1,2,4-benzenetriol, B-70181
6-Bromo-1,2,4-benzenetriol, B-70182
2-Bromo-5,6-epoxy-4-hydroxy-2-cyclohexen-1-one, B-60239

$C_6H_5Br_2NO$
2,4-Dibromo-6-methoxypyridine, *in* D-80100
2,6-Dibromo-4-methoxypyridine, *in* D-80099

$C_6H_5Br_3Se$
Tribromophenylselenium, T-70213

$C_6H_5Cl_2N$
2,3-Dichloro-5-methylpyridine, D-60138
2,5-Dichloro-3-methylpyridine, D-60139

$C_6H_5Cl_2NO$
4-Amino-2,5-dichlorophenol, A-70129

$C_6H_5Cl_3O_2$
2,3-Dihydro-4-(trichloroacetyl)furan, D-70271

$C_6H_5Cl_3Se$
Trichlorophenylselenium, T-70224

$C_6H_5FO_2$
4-Fluoro-1,2-benzenediol, F-60020
4-Fluoro-1,3-benzenediol, F-70018

C_6H_5FS
2-Fluorobenzenethiol, F-80018
3-Fluorobenzenethiol, F-80019
4-Fluorobenzenethiol, F-80020

$C_6H_5F_3O_2$
2,3-Dihydro-4-(trifluoroacetyl)furan, D-70274
2-Trifluoromethyl-1,3-cyclopentanedione, T-80281

$C_6H_5F_4NO_3S$
1-Fluoropyridinium; Trifluoromethanesulfonate, *in* F-60066

$C_6H_5NO_2$
Amino-1,4-benzoquinone, A-60089

$C_6H_5NO_3$
2-Formyl-1*H*-pyrrole-3-carboxylic acid, F-80077
4-Formyl-1*H*-pyrrole-2-carboxylic acid, F-80078
5-Formyl-1*H*-pyrrole-2-carboxylic acid, F-80079
α-Oxo-1*H*-pyrrole-3-acetic acid, O-70101

$C_6H_5NO_4$
1*H*-Pyrrole-3,4-dicarboxylic acid, P-70183

$C_6H_5NO_4S$
4-Nitro-2-thiophenecarboxylic acid; Me ester, *in* N-80061
5-Nitro-2-thiophenecarboxylic acid; Me ester, *in* N-80062

$C_6H_5NO_5$
3-Hydroxy-1*H*-pyrrole-2,4-dicarboxylic acid, H-80251
3-Hydroxy-1*H*-pyrrole-2,5-dicarboxylic acid, H-80252

C_6H_5NS
5*H*-Thieno[2,3-*c*]pyrrole, T-60206

$C_6H_5N_2OS$
3,4-Dihydro-6-methyl-4-thioxo-2(1*H*)-pyrimidinone, D-70234

$C_6H_5N_3$
▷ 1*H*-Benzotriazole, B-70068
Imidazo[1,2-*a*]pyrazine, I-60005
Imidazo[1,2-*b*]pyridazine, I-60007
▷ Imidazo[4,5-*c*]pyridine, I-60008
1*H*-Pyrazolo[3,4-*b*]pyridine, P-70155
5*H*-Pyrrolo[2,3-*b*]pyrazine, P-60248

$C_6H_5N_3O$
4-Amino-5-ethynyl-2(1*H*)-pyrimidinone, A-60153
1,3-Dihydro-2*H*-imidazo[4,5-*b*]pyridin-2-one, D-60249
1,3-Dihydro-2*H*-imidazo[4,5-*c*]pyridin-2-one, D-60250
4-Hydroxyimidazo[4,5-*b*]pyridine, H-60157

$C_6H_5N_5O$
2-Amino-4(1*H*)-pteridinone, A-60242

$C_6H_5N_5O_2$
2-Amino-4(1*H*)-pteridinone; 8-Oxide, *in* A-60242
▷ Isoxanthopterin, I-60141

C_6H_6
2-Ethynyl-1,3-butadiene, E-80073
1,2-Hexadien-5-yne, H-70040
1,3-Hexadien-5-yne, H-70041

C_6H_6BrN
2-(Bromomethyl)pyridine, B-70253
3-(Bromomethyl)pyridine, B-70254
4-(Bromomethyl)pyridine, B-70255

C_6H_6BrNO
3-Amino-2-bromophenol, A-70099
2-Bromo-3-hydroxy-6-methylpyridine, B-60267
2-Bromo-3-methoxypyridine, *in* B-60275
3-Bromo-5-methoxypyridine, *in* B-60277

$C_6H_6BrNO_2$
2-Bromo-3-hydroxy-6-methylpyridine; *N*-Oxide, *in* B-60267
2-Bromo-3-hydroxypyridine; Me ether, *N*-Oxide, *in* B-60275

$C_6H_6BrNO_2S$
2-Bromo-6-(methylsulfonyl)pyridine, *in* B-80256
5-Bromo-2-(methylsulfonyl)pyridine, *in* B-60317

C_6H_6BrNS
2-Bromo-6-(methylthio)pyridine, *in* B-80256
3-Bromo-2-(methylthio)pyridine, *in* B-60315
5-Bromo-2-(methylthio)pyridine, *in* B-60317

$C_6H_6Br_2N_2O_2$
4,5-Dibromo-1*H*-imidazole-2-carboxylic acid; Et ester, *in* D-60102

$C_6H_6Br_2O_3$
2,6-Dibromo-4,5-dihydroxy-2-cyclohexen-1-one, D-60097

C₆H₆Br₂S
2,3-Bis(bromomethyl)thiopene, B-80137
2,5-Bis(bromomethyl)thiophene, B-80138
3,4-Bis(bromomethyl)thiophene, B-80139

C₆H₆ClNO
2-Chloro-5-hydroxy-6-methylpyridine,
 C-60099
2-Chloro-3-methoxypyridine, *in* C-60100
3-Chloro-5-methoxypyridine, *in* C-60102

C₆H₆ClNO₂
6-Chloro-4-methoxy-2(1*H*)-pyridinone, *in*
 C-70074

C₆H₆ClNO₂S
3-Chloro-2-(methylsulfonyl)pyridine, *in*
 C-60129

C₆H₆ClNS
3-Chloro-2-(methylthio)pyridine, *in* C-60129

C₆H₆Cl₂O
2,4-Dichloro-3,4-dimethyl-2-cyclobuten-1-one,
 D-60130
4,4-Dichloro-2,3-dimethyl-2-cyclobuten-1-one,
 D-60131

C₆H₆FNS
2-Amino-3-fluorobenzenethiol, A-80073
2-Amino-5-fluorobenzenethiol, A-80074
2-Amino-6-fluorobenzenethiol, A-80075
4-Amino-2-fluorobenzenethiol, A-80076
4-Amino-3-fluorobenzenethiol, A-80077

C₆H₆F₂O₄
Difluoromaleic acid; Di-Me ester, *in* D-80159

C₆H₆F₄O₄
Tetrafluorobutanedioic acid; Di-Me ester, *in*
 T-80049

C₆H₆INO
3-Hydroxy-2-iodo-6-methylpyridine, H-60159

C₆H₆I₂S
2,5-Bis(iodomethyl)thiophene, B-80153

C₆H₆N₂
1,4-Dihydropyrrolo[3,2-*b*]pyrrole, D-30286
2-Methyl-4-cyanopyrrole, *in* M-60125
2-Methylenepentanedinitrile, M-80089

C₆H₆N₂O
2-Acetylpyrimidine, A-60044
4-Acetylpyrimidine, A-60045
5-Acetylpyrimidine, A-60046
4-Cyano-2,5-dimethyloxazole, *in* D-80461
1,4-Dihydrofuro[3,4-*d*]pyridazine, D-60241

C₆H₆N₂OS₂
3-Mercapto-6(1*H*)-pyridazinethione; *S*-Ac, *in*
 M-80028

C₆H₆N₂O₂
5-Acetyl-2(1*H*)-pyrimidinone, A-60048
▷ 1,4-Benzoquinone; Dioxime, *in* B-70044
Ethyl dicyanoacetate, *in* D-70149
▷ 1-Hydroxy-2-phenyldiazene 2-oxide, H-70208
3-Methyl-4-nitropyridine, M-70099
▷ Urocanic acid, U-70005

C₆H₆N₂O₃
5-Acetyl-2,4(1*H*,3*H*)-pyrimidinedione,
 A-60047
3,4-Diacetylfurazan, D-80045
4-Methoxy-3-nitropyridine, *in* N-70063
▷ 3-Methyl-4-nitropyridine; *N*-Oxide, *in*
 M-70099
3-Nitro-4(1*H*)-pyridone; *N*-Me, *in* N-70063

C₆H₆N₂O₄
3,6-Diacetyl-1,4,2,5-dioxadiazine, D-80044
▷ 3,4-Diacetylfuroxan, D-80045
1*H*-Pyrazole-4,5-dicarboxylic acid; 1-Me, *in*
 P-80199
1,2,3,6-Tetrahydro-2,6-dioxo-4-
 pyrimidineacetic acid, T-70060

C₆H₆N₂S
1,4-Dihydrothieno[3,4-*d*]pyridazine, D-60289

C₆H₆N₂S₂
[Bis(methylthio)methylene]propanedinitrile, *in*
 D-60395
1,4-Dithiocyano-2-butene, D-60510

C₆H₆N₄
7-Amino-3*H*-imidazo[4,5-*b*]pyridine, A-80087
4-Amino-1*H*-imidazo[4,5-*c*]pyridine, A-70161
2,2′-Bi-1*H*-imidazole, B-60107
1,3(5)-Bi-1*H*-pyrazole, B-60119
1*H*-Pyrazolo[3,4-*d*]pyrimidine; 1-Me, *in*
 P-70156

C₆H₆N₄O
3-Amino-1*H*-pyrazole-4-carboxylic acid;
 Nitrile, *N*(3)-Ac, *in* A-70187
2-Aminopyrrolo[2,3-*d*]pyrimidin-4-one,
 A-60255
1,3-Dihydro-1-methyl-2*H*-imidazo[4,5-
 b]pyrazin-2-one, *in* I-60006
Imidazo[2,1-*c*][1,2,4]triazin-4(1*H*)-one; 1-Me,
 in I-80007
Imidazo[2,1-*c*][1,2,4]triazin-4(1*H*)-one; 2-Me,
 in I-80007
Imidazo[2,1-*c*][1,2,4]triazin-4(1*H*)-one; 8-Me,
 in I-80007
1*H*-Pyrazolo[3,4-*d*]pyrimidine; 1-Me, 5-oxide,
 in P-70156
Pyrazolo[5,1-*c*][1,2,4]triazin-4(1*H*)-one; 1-Me,
 in P-80201
Pyrazolo[5,1-*c*][1,2,4]triazin-4(1*H*)-one; 2-Me,
 in P-80201

C₆H₆N₄O₂
3,9-Dihydro-9-methyl-1*H*-purine-2,6-dione,
 D-60257

C₆H₆N₄O₄
1,2,4,5-Tetrazine-3,6-dicarboxylic acid; Di-Me
 ester, *in* T-80163

C₆H₆N₄S
2-Aminopyrrolo[2,3-*d*]pyrimidine-4-thione,
 A-80144
1,3-Dihydro-1-methyl-2*H*-imidazo[4,5-
 b]pyrazine-2-thione, *in* D-60247
1,7-Dihydro-6*H*-purine-6-thione; 1-Me, *in*
 D-60270
1,7-Dihydro-6*H*-purine-6-thione; 3-Me, *in*
 D-60270
1,7-Dihydro-6*H*-purine-6-thione; 7-Me, *in*
 D-60270
1,7-Dihydro-6*H*-purine-6-thione; 9-Me, *in*
 D-60270
▷ 6-Mercaptopurine; *S*-Me, *in* D-60270
2-(Methylthio)-1*H*-imidazo[4,5-*b*]pyrazine, *in*
 D-60247
6-(Methylthio)-5*H*-imidazo[4,5-*c*]pyridazine, *in*
 D-70215

C₆H₆N₆
2,4-Diaminopteridine, D-60042
4,6-Diaminopteridine, D-60043
4,7-Diaminopteridine, D-60044
6,7-Diaminopteridine, D-60045

C₆H₆N₆O₂
2,6-Diamino-4,7(3*H*,8*H*)-pteridinedione,
 D-60047
2,7-Diamino-4,6(3*H*,5*H*)-pteridinedione,
 D-60046
2,4-Diaminopteridine; 5,8-Dioxide, *in*
 D-60042

C₆H₆N₆O₃
1,3,5-Triazine-2,4,6-tricarboxaldehyde;
 Trioxime, *in* T-60237

C₆H₆O
4,5-Dihydrocyclobuta[*b*]furan, D-60212

C₆H₆OS
2-Methyl-3-thiophenecarboxaldehyde,
 M-70139
3-Methyl-2-thiophenecarboxaldehyde,
 M-70140
4-Methyl-2-thiophenecarboxaldehyde,
 M-70141
4-Methyl-3-thiophenecarboxaldehyde,
 M-70142

5-Methyl-2-thiophenecarboxaldehyde,
 M-70143
5-Methyl-3-thiophenecarboxaldehyde,
 M-70144

C₆H₆O₂
3-Furanacetaldehyde, F-60083
1,5-Hexadiene-3,4-dione, H-60039
3-Hexyne-2,5-dione, H-70090
2-Methylene-1,3-cyclopentanedione, M-80083
2-Oxiranylfuran, O-70078
3-Oxo-1-cyclopentenecarboxaldehyde,
 O-60062

C₆H₆O₂S
3-Acetoxythiophene, *in* T-60212
2-(Hydroxyacetyl)thiophene, H-70103

C₆H₆O₃
▷ 5-Hydroxymethyl-2-furancarboxaldehyde,
 H-70177
2-Methylene-3-oxocyclobutanecarboxylic acid,
 M-80088
4-Oxo-2-butynoic acid; Et ester, *in* O-80070
2-Propenoic acid; Anhydride, *in* P-70121

C₆H₆O₄
3-Acetyl-4-hydroxy-2(5*H*)-furanone, A-60030
1,2,3,4-Benzenetetrol, B-60014
Dimethylenebutanedioic acid, D-70409

C₆H₆O₅
(2-Oxopropylidene)propanedioic acid,
 O-70100

C₆H₆O₆
▷ Aconitic acid, A-70055

C₆H₆S
2,3-Dihydro-2,3-dimethylenethiophene,
 D-70195
3-Vinylthiophene, V-80010

C₆H₇BrO₂
2-Bromo-1,3-cyclohexanedione, B-70201

C₆H₇ClO
2-Chloro-2-cyclohexen-1-one, C-60050
2-Chloro-3-methyl-2-cyclopenten-1-one,
 C-70090
5-Chloro-3-methyl-2-cyclopenten-1-one,
 C-70091
3,5-Hexadienoic acid; Chloride, *in* H-70038
5-Hexynoic acid; Chloride, *in* H-80086

C₆H₇ClO₂
2-Chloro-1,3-cyclohexanedione, C-80046

C₆H₇ClO₄
▷ Chloromaleic acid; Di-Me ester, *in* C-80043

C₆H₇IN₂
2-Amino-3-iodopyridine; *N*-Me, *in* A-80088

C₆H₇IN₂O₂
6-Iodo-1,3-dimethyluracil, *in* I-70057

C₆H₇N
1-Cyano-3-methylbicyclo[1.1.0]butane, *in*
 M-70053
5-Cyano-1-pentyne, *in* H-80086
2-Vinyl-1*H*-pyrrole, V-60013

C₆H₇NO
3-Acetylpyrrole, A-70052
2-Cyanocyclopentanone, *in* O-70084
1,2-Dihydro-3*H*-azepin-3-one, D-70171
5-Methyl-1*H*-pyrrole-2-carboxaldehyde,
 M-70129
4-(2-Propynyl)-2-azetidinone, P-60185

C₆H₇NOS
2-Acetyl-3-aminothiophene, A-70029
2-Pyridinethiol; *S*-Me, *N*-Oxide, *in* P-60223
2-Pyrrolethiolcarboxylic acid; Me ester, *in*
 P-80226

C₆H₇NO₂
1-Acetyl-1,2-dihydro-3*H*-pyrrol-3-one, *in*
 D-70262
3-Amino-1,2-benzenediol, A-70089
▷ 4-Amino-1,2-benzenediol, A-70090
1-Azabicyclo[3.2.0]heptane-2,7-dione, A-60330
1-Azabicyclo[3.2.0]heptane-2,7-dione, A-70273
3,4-Dimethyl-1*H*-pyrrole-2,5-dione, D-80474

3-Hydroxy-2-methyl-4(1*H*)-pyridinone,
  H-80214
3-Hydroxy-2(1*H*)-pyridinone; *N*-Me, *in*
  H-60223
3-Hydroxy-4(1*H*)-pyridinone; *N*-Me, *in*
  H-60227
4-Hydroxy-2(1*H*)-pyridinone; *N*-Me, *in*
  H-60224
5-Hydroxy-2(1*H*)-pyridinone; *N*-Me, *in*
  H-60225
6-Hydroxy-2(1*H*)-pyridinone; *N*-Me, *in*
  H-60226
3-Methoxy-4(1*H*)-pyridinone, *in* H-60227
6-Methoxy-2(1*H*)-pyridinone, *in* H-60226
Methyl 1-cyanocyclopropanecarboxylate, *in*
  C-80196
5-Methyl-1*H*-pyrrole-2-carboxylic acid,
  M-60124
5-Methyl-1*H*-pyrrole-3-carboxylic acid,
  M-60125
1-Nitro-1,3-cyclohexadiene, N-70047
4-Oxo-2-butynoic acid; Dimethylamide, *in*
  O-80070

$C_6H_7NO_3$

2,5-Dimethyl-4-oxazolecarboxylic acid,
  D-80461
5-Hydroxymethyl-2-furancarboxaldehyde;
  Oxime, *in* H-70177
5-Isoxazolecarboxylic acid; Et ester, *in*
  I-60142
4-Methoxy-1*H*-pyrrole-3-carboxylic acid, *in*
  H-80250
4-Methoxy-2-pyrrolecarboxylic acid, *in*
  H-80249
3-Oxo-2-azabicyclo[2.1.1]hexane-1-carboxylic
  acid, O-80057

C_6H_7NS

▷ 2-Pyridinethiol; *S*-Me, *in* P-60223
▷ 2(1*H*)-Pyridinethione; *N*-Me, *in* P-60224
  4(1*H*)-Pyridinethione; *N*-Me, *in* P-60225

$C_6H_7NS_2$

4-(Methylthio)-2(1*H*)-pyridinethione, *in*
  M-80029
4-(Methylthio)-2(1*H*)-pyridinethione, *in*
  P-60219

C_6H_7NSe

2-(Methylseleno)pyridine, *in* P-70170
2(1*H*)-Pyridineselone; *N*-Me, *in* P-70170

$C_6H_7N_3$

2-Amino-5-vinylpyrimidine, A-60269

$C_6H_7N_3OS_2$

5-Amino-2-thiazolecarbothioamide; N^5-Ac, *in*
  A-60260
5-Amino-4-thiazolecarbothioamide; N^5-Ac, *in*
  A-60261

$C_6H_7N_5$

6-Amino-7-methylpurine, A-60226
2-(Methylamino)purine, *in* A-60243
8-(Methylamino)purine, *in* A-60244
9-Methyl-9*H*-purin-8-amine, *in* A-60244

$C_6H_7N_5O$

6-Amino-1,3-dihydro-1-methyl-2*H*-purine-2-
  one, *in* A-60144
2,6-Diamino-1,5-dihydro-4*H*-imidazo[4,5-
  c]pyridin-4-one, D-60034

$C_6H_7N_7$

2,4,7-Triaminopteridine, T-60229
4,6,7-Triaminopteridine, T-60230

C_6H_8

2-Methylenebicyclo[2.1.0]pentane, M-80075

$C_6H_8Cl_2N_2$

3-Chloro-6-methylpyridazine; 2-
  Methochloride, *in* C-70112

$C_6H_8Cl_2O_3$

3,3'-Oxybispropanoic acid; Dichloride, *in*
  O-80098

$C_6H_8F_4$

1,1,2,2-Tetrafluoro-3,4-dimethylcyclobutane,
  T-60047

$C_6H_8F_6O_6S_4$

1,4-Dithioniabicyclo[2.2.0]hexane;
  Bis(trifluoromethanesulfonate), *in* D-70527

$C_6H_8N_2$

▷ 2-(Aminomethyl)pyridine, A-60227
  3-(Aminomethyl)pyridine, A-60228
  4-(Aminomethyl)pyridine, A-60229
  4,5-Dimethylpyrimidine, D-70436
  3,4,5,6-Tetrahydro-4,5-
    bis(methylene)pyridazine, T-70062
  1,4,5,6-Tetrahydrocyclopentapyrazole,
    T-80055

$C_6H_8N_2O$

2-Amino-3-hydroxy-6-methylpyridine,
  A-60192
5-Amino-3-hydroxy-2-methylpyridine,
  A-60193
2-Amino-4-methoxypyridine, *in* A-60253
2-Amino-6-methoxypyridine, *in* A-60252
3-Amino-2-methoxypyridine, *in* A-60249
3-Amino-4-methoxypyridine, *in* A-60254
4-Amino-2-methoxypyridine, *in* A-60250
5-Amino-2-methoxypyridine, *in* A-60251
3-Amino-1-methyl-2(1*H*)-pyridinone, *in*
  A-60249
3-Amino-1-methyl-4(1*H*)-pyridinone, *in*
  A-60254
5-Amino-1-methyl-2(1*H*)-pyridinone, *in*
  A-60251
6-Amino-1-methyl-2(1*H*)-pyridinone, *in*
  A-60252
Bis(2-cyanoethyl)ether, *in* O-80098
2,3-Dihydro-1,6-dimethyl-3-oxopyridazinium
  hydroxide, inner salt, *in* M-70124
4,5-Dimethylpyrimidine; 1-Oxide, *in* D-70436
4,5-Dimethylpyrimidine; 3-Oxide, *in* D-70436
7-Hydroxy-6,7-dihydro-5*H*-pyrrolo[1,2-
  a]imidazole, H-60118
3-Methoxy-6-methylpyridazine, *in* M-70124
6-Methyl-3-pyridazinone; 2-Me, *in* M-70124
5-Methyl-1*H*-pyrrole-2-carboxaldehyde;
  Oxime, *in* M-70129

$C_6H_8N_2OS$

4,5,6,7-Tetrahydrobenzothiadiazole; 2-Oxide,
  in T-70042

$C_6H_8N_2O_2$

2,5-Dimethyl-4-oxazolecarboxylic acid;
  Amide, *in* D-80461
5-Ethyl-2,4(1*H*,3*H*)-pyrimidinedione, E-60054
3-Hydroxy-6-methylpyridazine; Me ether, 1-
  Oxide, *in* M-70124
1*H*-Imidazole-4-carboxylic acid; Et ester, *in*
  I-60004
4,5,6,7-Tetrahydroisoxazolo[4,5-*c*]pyridin-3-ol,
  T-70071
4,5,6,7-Tetrahydroisoxazolo[5,4-*c*]pyridin-3-ol,
  T-60072
Tetrahydro-1*H*,5*H*-pyrazolo[1,2-*a*]pyrazole-
  1,5-dione, T-70087

$C_6H_8N_2O_3S$

4,5,6,7-Tetrahydrobenzothiadiazole; 1,1,2-
  Trioxide, *in* T-70042

$C_6H_8N_2O_4$

1,2-Dinitrocyclohexene, D-70451

$C_6H_8N_2S$

4,6-Dimethyl-2(1*H*)-pyrimidinethione,
  D-60450
4-Methylthio-2-pyridinamine, *in* A-80139
4-Methylthio-3-pyridinamine, *in* A-80140
6-Methylthio-2-pyridinamine, *in* A-80142
6-Methylthio-3-pyridinamine, *in* A-80141
4,5,6,7-Tetrahydrobenzothiadiazole, T-70042

$C_6H_8N_2S_2$

3,6-Bis(methylthio)pyridazine, *in* M-80028
1,4-Diamino-2,3-benzenedithiol, D-70030
2,5-Diamino-1,4-benzenedithiol, D-70035
3,6-Diamino-1,2-benzenedithiol, D-70036
2-Methyl-6-(methylthio)-3(2*H*)-
  pyridazinethione, *in* M-80028

$C_6H_8N_2S_4$

1,4-Diamino-2,3,5,6-benzenetetrathiol,
  D-60031

$C_6H_8N_4O_4$

3,4-Diacetylfuroxan dioxime, *in* D-80045
5-Nitrohistidine, N-60028
Schmitz's compound, *in* D-80044
Tetrahydro-1,2,4,5-tetrazine-3,6-dione; 1,5-Di-
  Ac, *in* T-70097

$C_6H_8N_6O_2$

3,6-Diamino-1,2,4,5-tetrazine; Di-*N*-Ac, *in*
  D-70045
3,6-Pyridazinedicarboxylic acid; Dihydrazide,
  in P-70168

C_6H_8O

Bicyclo[3.1.0]hexan-2-one, B-60088
3-Cyclohexen-1-one, C-60211
2,3-Dimethyl-2-cyclobuten-1-one, D-60415
1-Hexen-5-yn-3-ol, H-70089
2-Methylene-4-pentenal, M-70074
3-Methylene-4-penten-2-one, M-70075
7-Oxabicyclo[2.2.1]hept-2-ene, O-70054
7-Oxabicyclo[4.1.0]hept-3-ene, O-60047
2-Oxatricyclo[4.1.0^{1,6}.0^{3,5}]heptane, O-60052
Tetrahydro-3,4-bis(methylene)furan, T-70046

C_6H_8OS

2-Methyl-3-furanthiol; *S*-Me, *in* M-70082
3-Methyl-2-thiophenemethanol, M-80176
4-Methyl-2-thiophenemethanol, M-80177
5-Methyl-2-thiophenemethanol, M-80178
2-Thiabicyclo[2.2.1]heptan-3-one, T-60186
2-Thiabicyclo[2.2.1]hept-5-ene; *endo*-2-Oxide,
  in T-70160
2-Thiabicyclo[2.2.1]hept-5-ene; *exo*-2-Oxide, *in*
  T-70160
1-(2-Thienyl)ethanol, T-80178
1-(3-Thienyl)ethanol, T-80179
3-Thiopheneethanol, T-60211

$C_6H_8O_2$

Bicyclo[1.1.1]pentane-1-carboxylic acid,
  B-60101
Dihydro-4-methyl-3-methylene-2(3*H*)-
  furanone, D-80222
5,6-Dihydro-3-methyl-2*H*-pyran-2-one,
  D-80228
5,6-Dihydro-4-methyl-2*H*-pyran-2-one, *in*
  H-80211
3,4-Dihydro-2*H*-pyran-5-carboxaldehyde,
  D-70252
Dihydro-3-vinyl-2(3*H*)-furanone, D-70279
3,3-Dimethyl-1-cyclopropene-1-carboxylic
  acid, D-60417
3,4-Dimethyl-2(5*H*)-furanone, D-60423
3,4-Hexadienoic acid, H-70037
3,5-Hexadienoic acid, H-70038
2-Hexenedial, H-80079
3-Hexenedial, H-80080
3-Hexene-2,5-dione, H-80084
5-Hexynoic acid, H-80086
▷ 2-Hydroxy-3-methyl-2-cyclopenten-1-one,
  H-60174
4-Hydroxy-2-methyl-2-cyclopenten-1-one,
  H-60175
5-(Hydroxymethyl)-2-cyclopenten-1-one,
  H-80199
3-Hydroxymethyl-2-methylfuran, H-80207
3-Methylbicyclo[1.1.0]butane-1-carboxylic
  acid, M-70053
2-Methylenecyclopropaneacetic acid, M-70067
3-Methyl-4-oxo-2-pentenal, M-70104
3-Oxabicyclo[3.2.0]heptan-2-one, O-70053
4-Pentynoic acid; Me ester, *in* P-80062
Tetrahydro-6-methylene-2*H*-pyran-2-one,
  T-70074

$C_6H_8O_2S$

2-Thiabicyclo[2.2.1]hept-5-ene; 2,2-Dioxide, *in*
  T-70160
3-Thiatricyclo[2.2.1.0^{2,6}]heptane; *S*,*S*-Dioxide,
  in T-70165

$C_6H_8O_3$

2,5-Bis(hydroxymethyl)furan, B-60160
3,4-Bis(hydroxymethyl)furan, B-60161
2-Cyclopropyl-2-oxoacetic acid; Me ester, *in*
  C-60231
Dihydro-4,4-dimethyl-2,3-furandione,
  D-70196

4,5-Dihydro-3-(methoxymethylene)-2(3*H*)-
furanone, D-70222
3,6-Dihydro-2*H*-pyran-2-carboxylic acid,
D-80242
5,5-Dimethyl-4-methylene-1,2-dioxolan-3-one,
D-60428
4-Ethoxy-2(5*H*)-furanone, *in* F-80087
4-Hydroxy-2,5-dimethyl-3(2*H*)-furanone,
H-60120
5-Hydroxymethyl-4-methyl-2(5*H*)-furanone,
H-60187
2-Oxocyclopentanecarboxylic acid, O-70084
4-Oxo-2-hexenoic acid, O-80082
5-Oxo-3-hexenoic acid, O-70092
5-Oxo-3-hexenoic acid, O-80083
6-Oxo-4-hexenoic acid, O-80084
4-Oxo-2-pentenoic acid; Me ester, *in* O-70098

$C_6H_8O_3S_2$
2-Thiophenesulfonic acid; Et ester, *in* T-60213

$C_6H_8O_4$
Ethyl diformylacetate, *in* D-60190
2-Hydroxy-2-(hydroxymethyl)-2*H*-pyran-
3(6*H*)-one, H-70141

C_6H_8S
2-Thiabicyclo[2.2.1]hept-5-ene, T-70160
3-Thiatricyclo[2.2.1.0²·⁶]heptane, T-70165

$C_6H_8S_3$
2,5-Thiophenedimethanethiol, T-70184

C_6H_9Br
1-Bromo-3,3-dimethyl-1-butyne, B-70207
2-Bromo-1,5-hexadiene, B-70216
1-Bromo-1-hexyne, B-70218
1-Bromo-2-hexyne, B-70219

$C_6H_9BrO_2$
2-Bromomethyl-2-propenoic acid; Et ester, *in*
B-70249

C_6H_9Cl
2-Chloro-1,5-hexadiene, C-70070

C_6H_9FO
2-Fluorocyclohexanone, F-60021

$C_6H_9F_3O$
1,1,1-Trifluoro-3,3-dimethyl-2-butanone,
T-80272

$C_6H_9F_3O_2$
Trifluoroacetic acid; *tert*-Butyl ester, *in*
T-80267

C_6H_9I
1-Iodocyclohexene, I-70026
1-Iodo-1-hexyne, I-70040

C_6H_9IO
2-Iodocyclohexanone, I-60042
6-Iodo-5-hexyn-1-ol, I-70041

C_6H_9N
2-Cyano-1,1-dimethylcyclopropane, *in*
D-80444
3-Ethyl-1*H*-pyrrole, E-80067

C_6H_9NO
3-Amino-2-cyclohexen-1-one, A-70125
3-Cyclohexen-1-one; Oxime, *in* C-60211
1,2-Dihydro-2,2-dimethyl-3*H*-pyrrol-3-one,
D-60230
3,4-Dihydro-6-methyl-2(1*H*)-pyridinone,
D-80229
3-Methylbicyclo[1.1.0]butane-1-carboxylic
acid; Amide, *in* M-70053
4-(2-Propenyl)-2-azetidinone, *in* A-60172

$C_6H_9NO_2$
3-(1-Aminocyclopropyl)-2-propenoic acid,
A-60123
4-Amino-2,5-hexadienoic acid, A-60166
3,6-Dihydro-2*H*-pyran-2-carboxylic acid;
Amide, *in* D-80242
3,4-Dihydro-2*H*-pyrrole-5-carboxylic acid; Me
ester, *in* D-80249
Hexahydro-2*H*-furo[3,2-*b*]pyrrol-2-one,
H-80059
3-Methyl-1-nitrocyclopentene, M-80122
▷ 2-Pyrrolidinone; *N*-Ac, *in* P-70188

1,2,3,6-Tetrahydro-3-pyridinecarboxylic acid,
T-60084
3,4,4-Trimethyl-5(4*H*)-isoxazolone, T-60367

$C_6H_9NO_3$
3-Aminodihydro-2(3*H*)-furanone; *N*-Ac, *in*
A-60138
2-Amino-2-(hydroxymethyl)-4-pentynoic acid,
A-60191
3-*tert*-Butyl-1,4,2-dioxazol-5-one, B-80286
Methyl 2-formyl-3-methylamino-2-
propenoate, *in* A-80134
2,3-Morpholinedione; *N*-Et, *in* M-80193
2-Nitrocyclohexanone, N-80043
4-Oxo-2-piperidinecarboxylic acid, O-60089
5-Oxo-2-pyrrolidinecarboxylic acid; Me ester,
in O-70102
1,2,3,6-Tetrahydro-3-hydroxy-2-
pyridinecarboxylic acid, T-60068
1,2,5,6-Tetrahydro-5-hydroxy-3-
pyridinecarboxylic acid, T-70065

$C_6H_9NO_3S$
5-Oxo-3-thiomorpholinecarboxylic acid; Me
ester, *in* O-80095

$C_6H_9NO_4$
2-Amino-3-methylenepentanedioic acid,
A-60219

$C_6H_9NO_6$
O-Oxalylhomoserine, *in* A-70156

$C_6H_9NS_2$
2(3*H*)-Thiazolethione; *S*-Isopropyl, *in* T-60200

$C_6H_9N_3O$
2-Amino-6-methyl-4(1*H*)-pyrimidinone; 2-*N*-
Me, *in* A-60230

$C_6H_9N_3O_2$
2-Amino-4(5)-imidazoleacetic acid; Me ester,
in A-80086
α-Amino-1*H*-imidazole-1-propanoic acid,
A-60204
α-Amino-1*H*-imidazole-2-propanoic acid,
A-60205
3-Amino-5-methyl-1*H*-pyrazole-4-carboxylic
acid; Me ester, *in* A-80110
3-Amino-1*H*-pyrazole-4-carboxylic acid; Et
ester, *in* A-70187
4-Amino-1*H*-pyrazole-3-carboxylic acid; Et
ester, *in* A-70189
3-Amino-1*H*-pyrazole-5-carboxylic acid; *N*(1)-
Me, Me ester, *in* A-70188
5-Amino-1*H*-pyrazole-3-carboxylic acid; *N*(1)-
Me, Me ester, *in* A-70191
▷ Cupferron, *in* H-70208
2,4-Dimethoxy-6-methyl-1,3,5-triazine, *in*
M-60132
3,4-Dimethyl-1*H*-pyrrole-2,5-dione; Dioxime,
in D-80474
4-Methoxy-1,6-dimethyl-1,3,5-triazin-2(1*H*)-
one, *in* M-60132
1*H*-1,2,3-Triazole-4-carboxylic acid; 1-Me, Et
ester, *in* T-60239
1*H*-1,2,3-Triazole-4-carboxylic acid; 2-Me, Et
ester, *in* T-60239
1,3,6-Trimethyl-5-azauracil, *in* M-60132

$C_6H_9N_3O_2S$
5-Mercaptohistidine, M-80025

$C_6H_9N_3O_3$
Aconitic acid; Triamide, *in* A-70055
Hexahydro-1,3,5-triazine; 1,3,5-Triformyl, *in*
H-60060
Trimethyl isocyanurate, *in* C-60179

$C_6H_{10}Br_4$
2,3-Bis(bromomethyl)-1,4-dibromobutane,
B-60144

$C_6H_{10}ClF_2NO_2$
2-Amino-4-chloro-4,4-difluorobutanoic acid;
Et ester, *in* A-60107

$C_6H_{10}ClNO$
1-Chloro-1-nitrocyclohexane, C-70131

$C_6H_{10}ClNO_3$
2-Amino-5-chloro-6-hydroxy-4-hexenoic acid,
A-60110

$C_6H_{10}Cl_3N$
1,1,2-Trichloro-2-(diethylamino)ethylene,
T-70222

$C_6H_{10}Cl_4$
2,3-Bis(chloromethyl)-1,4-dichlorobutane,
B-60148

$C_6H_{10}INO$
6-(Iodomethyl)-2-piperidinone, I-80036

$C_6H_{10}N_2$
2-Amino-5-hexenoic acid; Nitrile, *in* A-60169
3-Amino-2-hexenoic acid; Nitrile, *in* A-60170
Azocyclopropane, A-80205

$C_6H_{10}N_2O$
Azoxycyclopropane, *in* A-80205

$C_6H_{10}N_2O_2$
5-Amino-3,4-dihydro-2*H*-pyrrole-2-carboxylic
acid; Me ester, *in* A-70135
2,5-Dihydro-4,6-dimethoxypyrimidine, *in*
D-70256
Dihydro-4,4-dimethyl-2,3-furandione;
Hydrazone, *in* D-70196
1-Dimethylamino-4-nitro-1,3-butadiene,
D-70367

$C_6H_{10}N_2O_3$
2-Amino-3-methylenepentanedioic acid; 5-
Amide, *in* A-60219
3-Piperidinecarboxylic acid; *N*-Nitroso, *in*
P-70102

$C_6H_{10}N_2O_4$
2,5-Piperazinedicarboxylic acid, P-60157

$C_6H_{10}N_4$
1,2,3,5-Tetraaminobenzene, T-60012
1,2,4,5-Tetraaminobenzene, T-60013

$C_6H_{10}N_4O$
4-Amino-1,3,5-triazin-2(1*H*)-one; 1,4,4-*N*-Tri-
Me, *in* A-80149

$C_6H_{10}N_4O_2$
Tetrahydroimidazo[4,5-*d*]imidazole-
2,5(1*H*,3*H*)-dione; 1,4-Di-Me, *in* T-70066
Tetrahydroimidazo[4,5-*d*]imidazole-
2,5(1*H*,3*H*)-dione; 1,6-Di-Me, *in* T-70066

$C_6H_{10}N_4S$
1-(4,5-Dihydro-1*H*-imidazol-2-yl)-2-
imidazolidinethione, D-60246

$C_6H_{10}O$
Bicyclo[2.1.1]hexan-2-ol, B-70100
tert-Butoxyethyne, *in* E-80070
2,3-Dihydro-4,5-dimethylfuran, D-80190
3,5-Hexadien-1-ol, H-70039
▷ 2-Hexenal, H-60076
1-Methyl-6-oxabicyclo[3.1.0]hexane, M-80126
▷ 2-Methyl-2-pentenal, M-70106
4-Methyl-4-pentenal, M-80134

$C_6H_{10}OS$
2,3-Dihydro-4,5-dimethylthiophene; 1-Oxide,
in D-80195
2-Mercaptocyclohexanone, M-70025
3-Mercaptocyclohexanone, M-70026
4-Mercaptocyclohexanone, M-70027
2-(Methylthio)cyclopentanone, M-70136
2-Oxepanethione, O-70077
4-Thioxo-2-hexanone, T-60215

$C_6H_{10}O_2$
2-Cyclohexene-1,4-diol, C-80181
4-Cyclohexene-1,2-diol, C-80182
Dihydro-4,5-dimethyl-2(3*H*)-furanone,
D-80191
2,3-Dimethyl-3-butenoic acid, D-80438
2,2-Dimethylcyclopropanecarboxylic acid,
D-80444
2,2-Dimethyl-4-methylene-1,3-dioxolane,
D-70419
▷ 1,5-Hexadiene-3,4-diol, H-70036
2-Hexyne-1,6-diol, H-80085
2-Methoxycyclopentanone, *in* H-60114
4-Methoxy-3-penten-2-one, *in* P-60059
4-Oxohexanal, O-60073
5-Oxohexanal, O-70091
Tetrahydro-4-methyl-2*H*-pyran-2-one, *in*
H-70182

$C_6H_{10}O_2S$
3-(Acetylthio)cyclobutanol, *in* M-60019
2,5-Dihydro-2,5-dimethylthiophene; 1,1-Dioxide, *in* D-60233
O-Ethyl 3-oxobutanethioate, *in* O-70080
S-Ethyl 3-oxobutanethioate, *in* O-70080
3-Mercaptocyclopentanecarboxylic acid, M-60021

$C_6H_{10}O_3$
2,2-Dimethyl-1,3-dioxolane-4-carboxaldehyde, D-70400
Epiverrucarinolactone, *in* D-60353
2-Ethoxy-2-propenoic acid; Me ester, *in* E-70037
4-Hydroxy-2-butenoic acid; Et ester, *in* H-80134
5-Hydroxy-3-methyl-2-pentenoic acid, H-80211
6-(Hydroxymethyl)tetrahydro-2*H*-pyran-2-one, *in* D-60329
▷ 3-Methyloxiranecarboxylic acid; Et ester, *in* M-80127
▷ 3-Methyloxiranecarboxylic acid; Et ester, *in* M-80127
3-Methyl-2-oxobutanoic acid; Me ester, *in* M-80129
5-Oxopentanoic acid; Me ester, *in* O-80089
3-Oxopropanoic acid; Isopropyl ester, *in* O-60090
3,4,5,6-Tetrahydro-4-hydroxy-6-methyl-2*H*-pyran-2-one, T-60067
Tetrahydro-3-hydroxy-4-methyl-2*H*-pyran-2-one, *in* D-60353

$C_6H_{10}O_4$
4-Hydroxy-3,3-dimethyl-2-oxobutanoic acid, H-60124
Methylpropanedioic acid; Di-Me ester, *in* M-80149
1,3,7,9-Tetraoxaspiro[4.5]decane, T-60163
1,4,6,10-Tetraoxaspiro[4.5]decane, T-60164

$C_6H_{10}O_5$
▷ Diethyl dicarbonate, *in* D-70111
Methyl hydrogen 3-hydroxyglutarate, M-70089
3,3'-Oxybispropanoic acid, O-80098

$C_6H_{10}O_6$
Dimethyl tartrate, *in* T-70005

$C_6H_{10}O_{12}P_2$
myo-Inositol; 1,4-Diphosphate, *in* I-80015

$C_6H_{10}S$
2,3-Dihydro-2,2-dimethylthiophene, D-60232
2,3-Dihydro-4,5-dimethylthiophene, D-80195
2,5-Dihydro-2,5-dimethylthiophene, D-60233

$C_6H_{10}S_4$
2,2'-Bi-1,3-dithiolane, B-60103
Di-2-propenyl tetrasulfide, D-80515
Hexahydro-1,4-dithiino[2,3-*b*]-1,4-dithiin, H-60051

$C_6H_{11}Br$
1-Bromo-3-hexene, B-80209
5-Bromo-1-hexene, B-60263
6-Bromo-1-hexene, B-60264
(Bromomethyl)cyclopentane, B-70241
5-Bromo-3-methyl-1-pentene, B-60299

$C_6H_{11}BrO$
5-Bromo-2-hexanone, B-60262
1-Bromo-4-methyl-2-pentanone, B-60298

$C_6H_{11}BrO_2$
2-Bromo-4-methylpentanoic acid, B-70248

$C_6H_{11}Br_2NO_2$
6-Amino-2,2-dibromohexanoic acid, A-60124

$C_6H_{11}Cl$
6-Chloro-1-hexene, C-80058
5-Chloro-3-methyl-1-pentene, C-60117

$C_6H_{11}ClFNO_2$
2-Amino-4-chloro-4-fluorobutanoic acid; Et ester, *in* A-60108

$C_6H_{11}ClO$
2-Chloro-1-hexen-3-ol, C-80059

2-Chloro-4-hexen-3-ol, C-80060
4-Chloro-4-hexen-3-ol, C-80061
6-Chloro-2-hexen-1-ol, C-80062
6-Chloro-3-hexen-1-ol, C-80063
6-Chloro-4-hexen-1-ol, C-80064
6-Chloro-5-hexen-1-ol, C-80065

$C_6H_{11}Cl_2NO_2$
6-Amino-2,2-dichlorohexanoic acid, A-60126

$C_6H_{11}F_2NO_2$
2-Amino-3,3-difluorobutanoic acid; Et ester, *in* A-60133

$C_6H_{11}I$
1-Iodo-1-hexene, I-70037
6-Iodo-1-hexene, I-70038
(Iodomethyl)cyclopentane, I-70047

$C_6H_{11}IO$
1-Iodo-3,3-dimethyl-2-butanone, I-60044
6-Iodo-5-hexen-1-ol, I-70039

$C_6H_{11}N$
2-Aminobicyclo[2.1.1]hexane, A-70095
3,4-Dihydro-2,2-dimethyl-2*H*-pyrrole, D-60229
3-Isocyanopentane, I-80059

$C_6H_{11}NO$
3,4-Dihydro-2,2-dimethyl-2*H*-pyrrole; *N*-Oxide, *in* D-60229
2,2-Dimethylcyclopropanecarboxylic acid; Amide, *in* D-80444
5-Ethoxy-3,4-dihydro-2*H*-pyrrole, *in* P-70188
2-Oxa-3-azabicyclo[2.2.2]octane, O-70051
1-Oxa-2-azaspiro[2,5]octane, O-70052

$C_6H_{11}NOS$
3-Mercaptocyclopentanecarboxylic acid; Amide, *in* M-60021

$C_6H_{11}NO_2$
2-Amino-2-ethyl-3-butenoic acid, A-60150
1-Amino-2-ethylcyclopropanecarboxylic acid, A-60151
2-Amino-2-hexenoic acid, A-60167
2-Amino-3-hexenoic acid, A-60168
2-Amino-5-hexenoic acid, A-60169
3-Amino-2-hexenoic acid, A-60170
3-Amino-4-hexenoic acid, A-60171
3-Amino-5-hexenoic acid, A-60172
6-Amino-2-hexenoic acid, A-60173
2-Amino-3-methyl-3-butenoic acid; Me ester, *in* A-60218
4,4-Dimethoxybutanenitrile, *in* O-80068
2,6-Dimethyl-3-morpholinone, D-80454
2-Methyl-2-pyrrolidinecarboxylic acid, M-60126
1-Nitro-3-hexene, N-70051
3-Piperidinecarboxylic acid, P-70102

$C_6H_{11}NO_3$
2-Amino-1-hydroxy-1-cyclobutaneacetic acid, A-60184
1-Amino-3-(hydroxymethyl)-1-cyclobutanecarboxylic acid, A-70157
1-Amino-3-(hydroxymethyl)cyclobutanecarboxylic acid, A-80082
2-Amino-2-(hydroxymethyl)-4-pentenoic acid, A-60190
Ethyl ethoxyiminoacetate, E-80063

$C_6H_{11}NO_4$
L-Allothreonine; *N*-Ac, *in* T-60217
2-Amino-3,3-dimethylbutanedioic acid, A-60147
2-Amino-3-ethylbutanedioic acid, A-70138
2-Amino-4-methylpentanedioic acid, A-60223

$C_6H_{11}NS_2$
5,6-Dihydro-2-(ethylthio)-4*H*-1,3-thiazine, *in* T-60094
Tetrahydro-2*H*-1,3-thiazine-2-thione; *N*-Et, *in* T-60094

$C_6H_{11}N_3O$
2-Azidocyclohexanol, A-80196

$C_6H_{11}N_3O_2$
Dihydro-1,3,5-triazine-2,4(1*H*,3*H*)-dione; 1,3,5-Tri-Me, *in* D-60297

$C_6H_{11}N_5$
2,4,6-Triaminopyrimidine; N^2,N^2-Di-Me, *in* T-80208
2,4,6-Triaminopyrimidine; N^4,N^4-Di-Me, *in* T-80208
2,4,6-Triaminopyrimidine; N^4-Et, *in* T-80208
2,4,6-Triaminopyrimidine; N^4,N^6-Di-Me, *in* T-80208

$C_6H_{11}O_9P$
myo-Inositol; 1-Phosphate, *in* I-80015

$C_6H_{12}BrCl$
1-Bromo-6-chlorohexane, B-70200

$C_6H_{12}BrN$
2-(Bromomethyl)cyclopentanamine, B-70240

$C_6H_{12}BrNO_2$
6-Amino-2-bromohexanoic acid, A-60100

$C_6H_{12}Br_2$
1,3-Dibromo-2,3-dimethylbutane, D-70089

$C_6H_{12}Br_2O_2$
1,6-Dibromo-3,4-hexanediol, D-80095

$C_6H_{12}ClNO_2$
6-Amino-2-chlorohexanoic acid, A-60109

$C_6H_{12}I_2O_2$
1,2-Bis(2-iodoethoxy)ethane, B-60163

$C_6H_{12}NO_2^{\oplus}$
4-Azoniaspiro[3.3]heptane-2,6-diol, A-60354

$C_6H_{12}N_2$
4-Cyclohexene-1,2-diamine, C-80180
▷ 1,4-Diazabicyclo[2.2.2]octane, D-70051
2-Dimethylamino-3,3-dimethylazirine, D-60402
1,3-Dimethyl-2-methyleneimidazolidine, D-70420
5-Hexyne-1,4-diamine, H-60080
Octahydropyrrolo[3,4-*c*]pyrrole, O-70017
Tetrahydro-1*H*,5*H*-pyrazolo[1,2-*a*]pyrazole, T-80077

$C_6H_{12}N_2O$
3-Amino-2-hexenoic acid; Amide, *in* A-60170
3-Piperidinecarboxamide, *in* P-70102
▷ 3,4,5,6-Tetrahydro-2(1*H*)-pyrimidinone, *in* T-60088

$C_6H_{12}N_2O_3$
3,3'-Oxybispropanoic acid; Diamide, *in* O-80098

$C_6H_{12}N_2O_3S$
N-Alanylcysteine, A-60069

$C_6H_{12}N_2O_4$
3,4,8,9-Tetraoxa-1,6-diazabicyclo[4.4.2]dodecane, T-70144

$C_6H_{12}N_4$
1,2,3,4,5,6,7,8-Octahydropyridazino[4,5-*d*]pyridazine, O-60018

$C_6H_{12}N_4O_2$
2,5-Piperazinedicarboxylic acid; Diamide, *in* P-60157

$C_6H_{12}N_6$
▷ Benzenehexamine, B-80014

$C_6H_{12}O$
2,3-Dimethyl-3-buten-1-ol, D-80439
2,2-Dimethylcyclopropanemethanol, D-70396
4-Hexen-3-ol, H-60077
1-Methoxy-2,3-dimethylcyclopropane, *in* D-70397
3-Methyl-2-pentanone, M-70105
▷ 4-Methyl-2-pentanone, M-60109
2-Methyl-4-penten-1-ol, M-80135
4-Methyl-1-penten-3-ol, M-70107
Tetrahydro-2-methylpyran, T-60075
Tetrahydro-3-methyl-2*H*-pyran, T-80072
Tetrahydro-4-methyl-2*H*-pyran, T-80073

$C_6H_{12}OS$
2-Mercaptocyclohexanol, M-70024
3,3,4,4-Tetramethyl-1,2-oxathietane, T-60149

$C_6H_{12}O_2$
β,β-Dimethyloxiraneethanol, D-60431

2-Hexene-1,6-diol, H-**80081**
3-Hexene-1,6-diol, H-**80082**
3-Hexene-2,5-diol, H-**80083**
2-(2-Propenyl)-1,3-propanediol, P-**80180**

C$_6$H$_{12}$O$_2$S
3-Mercaptobutanoic acid; Et ester, *in* M-**70023**
2-Mercapto-3-methylpentanoic acid, M-**80026**
2-Mercapto-4-methylpentanoic acid, M-**80027**

C$_6$H$_{12}$O$_2$S$_2$
2,5-Dimethyl-1,4-dithiane-2,5-diol, D-**60421**

C$_6$H$_{12}$O$_3$
▷ 2-Acetoxy-1-methoxypropane, *in* P-**70118**
Diethoxyacetaldehyde, D-**70153**
1,1-Dimethoxy-2-butanone, *in* O-**80067**
3-Hydroxy-2-methylpentanoic acid, H-**80209**
3-Hydroxy-3-methylpentanoic acid, H-**80210**
5-Hydroxy-3-methylpentanoic acid, H-**70182**

C$_6$H$_{12}$O$_3$S
Sulfoacetic acid; Di-Et ester, *in* S-**80050**

C$_6$H$_{12}$O$_4$
3,4-Dihydroxybutanoic acid; Et ester, *in* D-**60313**
5,6-Dihydroxyhexanoic acid, D-**60329**
2,5-Dihydroxy-3-methylpentanoic acid, D-**60353**
▷ Tetramethoxyethene, T-**80135**

C$_6$H$_{12}$O$_6$
myo-Inositol, I-**80015**

C$_6$H$_{12}$S$_2$
1,1-Cyclobutanedimethanethiol, C-**70213**
1,2-Cyclobutanedimethanethiol, C-**70214**
▷ 1,1-Cyclohexanedithiol, C-**60209**
1,2-Cyclohexanedithiol, C-**70225**
1,2-Dithiocane, D-**50815**

C$_6$H$_{13}$Br
1-Bromo-3-methylpentane, B-**80223**

C$_6$H$_{13}$N
2,5-Dimethylpyrrolidine, D-**70437**
4-Hexen-1-amine, H-**70088**
5-Hexen-1-amine, H-**80078**
2,2,3,3-Tetramethylaziridine, T-**80139**

C$_6$H$_{13}$NO
▷ *N*-Cyclohexylhydroxylamine, C-**70228**

C$_6$H$_{13}$NO$_2$
2-Amino-2,3-dimethylbutanoic acid, A-**60148**
2-Amino-3,3-dimethylbutanoic acid, A-**70136**

C$_6$H$_{13}$NO$_2$S
2-Amino-5-(methylthio)pentanoic acid, A-**60231**

C$_6$H$_{13}$NO$_3$
2-Amino-4-ethoxybutanoic acid, *in* A-**70156**

C$_6$H$_{13}$NS
2-[(Methylthio)methyl]pyrrolidine, *in* P-**70185**

C$_6$H$_{13}$N$_3$O$_2$
Biuret; 1,1,3,5-Tetra-Me, *in* B-**60190**

C$_6$H$_{13}$N$_3$O$_3$
Citrulline, C-**60151**

C$_6$H$_{14}$ClN$_3$
Gold's reagent; Chloride, *in* G-**80032**

C$_6$H$_{14}$ClN$_3$O$_4$
Gold's reagent; Perchlorate, *in* G-**80032**

C$_6$H$_{14}$Cl$_6$N$_3$Sb
Gold's reagent; Hexachloroantimonate, *in* G-**80032**

C$_6$H$_{14}$N$_2$
Hexahydropyrimidine; 1,3-Di-Me, *in* H-**60056**

C$_6$H$_{14}$N$_2$O
2-Amino-3,3-dimethylbutanoic acid; Amide, *in* A-**70136**

C$_6$H$_{14}$N$_2$O$_2$
3,4-Diaminohexanoic acid, D-**80059**
5,6-Diaminohexanoic acid, D-**80060**
2,5-Diamino-2-methylpentanoic acid, D-**80061**

C$_6$H$_{14}$N$_3$$^{\oplus}$
Gold's reagent, G-**80032**

C$_6$H$_{14}$N$_4$O$_3$
2-Amino-5-guanidino-3-hydroxypentanoic acid, A-**70149**

C$_6$H$_{14}$O$_2$S
2,2-Diethoxyethanethiol, *in* M-**80017**
2-(Ethylthio)-1,1-dimethoxyethane, *in* M-**80017**

C$_6$H$_{14}$O$_2$S$_2$
S,S'-Bis(2-hydroxyethyl)-1,2-ethanedithiol, B-**70145**

C$_6$H$_{14}$O$_3$
1,1-Dimethoxy-2-butanol, *in* H-**70112**
3,3'-Oxybis-1-propanol, O-**60097**

C$_6$H$_{14}$O$_4$
2,3-Bis(hydroxymethyl)-1,4-butanediol, B-**60159**

C$_6$H$_{14}$O$_6$S$_2$
S,S'-Bis(2-hydroxyethyl)-1,2-ethanedithiol; *S*-Tetroxide, *in* B-**70145**

C$_6$H$_{14}$S$_4$
S,S'-Bis(2-mercaptoethyl)-1,2-ethanedithiol, B-**70150**

C$_6$H$_{15}$NO$_2$
4,4-Dimethoxybutylamine, *in* A-**80064**

C$_6$H$_{15}$N$_3$
▷ Hexahydro-1,3,5-trimethyl-1,3,5-triazine, *in* H-**60060**

C$_6$H$_{15}$O$_{15}$P$_3$
myo-Inositol-1,3,4-triphosphate, I-**70020**
myo-Inositol-2,4,5-triphosphate, I-**70021**

C$_6$H$_{16}$N$_2$O$_2$
1,2-Bis(2-aminoethoxy)ethane, B-**60139**

C$_6$H$_{16}$O$_{18}$P$_4$
myo-Inositol-1,3,4,5-tetraphosphate, I-**70019**

C$_6$H$_{18}$N$_4$
1,2-Dihydrazinoethane; $N^\beta,N^\beta,N^{\beta'},N^{\beta'}$-Tetra-Me, *in* D-**60192**

C$_6$H$_{18}$O$_{24}$P$_6$
▷ Phytic acid, *in* I-**80015**

C$_6$N$_4$
▷ Tetracyanoethylene, T-**60031**

C$_6$N$_4$O
▷ Tetracyanooxirane, T-**80023**

C$_6$O$_6$
Cyclohexanehexone, C-**60210**

C$_7$F$_5$N
Pentafluoroisocyanobenzene, P-**60036**

C$_7$F$_5$NO
Pentafluoroisocyanatobenzene, P-**60035**

C$_7$F$_5$NS
Pentafluoroisothiocyanatobenzene, P-**60037**

C$_7$HF$_4$N
2,3,4,5-Tetrafluorobenzonitrile, *in* T-**70032**
2,3,5,6-Tetrafluorobenzonitrile, *in* T-**70034**

C$_7$HF$_5$
▷ Ethynylpentafluorobenzene, E-**80075**

C$_7$HF$_5$O$_2$
Pentafluorophenyl formate, P-**70025**

C$_7$H$_2$BrF$_5$
▷ (Bromomethyl)pentafluorobenzene, B-**80222**

C$_7$H$_2$ClF$_3$N$_2$O$_4$
▷ 2-Chloro-1,3-dinitro-5-(trifluoromethyl)benzene, C-**80053**

C$_7$H$_2$ClF$_5$
▷ (Chloromethyl)pentafluorobenzene, C-**80072**

C$_7$H$_2$ClN$_3$O$_7$
▷ 2,4,6-Trinitrobenzoic acid; Chloride, *in* T-**70303**

C$_7$H$_2$F$_3$N
2,4,5-Trifluorobenzonitrile, *in* T-**70240**

C$_7$H$_2$F$_4$O
2,3,4,5-Tetrafluorobenzaldehyde, T-**70026**
2,3,4,6-Tetrafluorobenzaldehyde, T-**70027**
2,3,5,6-Tetrafluorobenzaldehyde, T-**70028**

C$_7$H$_2$F$_4$O$_2$
2,3,4,5-Tetrafluorobenzoic acid, T-**70032**
2,3,4,6-Tetrafluorobenzoic acid, T-**70033**
2,3,5,6-Tetrafluorobenzoic acid, T-**70034**

C$_7$H$_3$BrF$_3$NO$_2$
1-Bromo-2-nitro-3-(trifluoromethyl)benzene, B-**60306**
1-Bromo-4-nitro-2-(trifluoromethyl)benzene, B-**60307**
2-Bromo-1-nitro-3-(trifluoromethyl)benzene, B-**60308**

C$_7$H$_3$Br$_4$Cl
1,2,4,5-Tetrabromo-3-chloro-6-methylbenzene, T-**60022**

C$_7$H$_3$ClFNO$_4$
2-Chloro-4-fluoro-5-nitrobenzoic acid, C-**60072**

C$_7$H$_3$ClF$_3$NO$_2$
4-Chloro-1-nitro-2-(trifluoromethyl)benzene, C-**80086**

C$_7$H$_3$ClF$_3$N$_3$
6-Chloro-2-(trifluoromethyl)-1*H*-imidazo[4,5-*b*]pyridine, C-**60139**

C$_7$H$_3$ClF$_3$N$_3$O
6-Chloro-2-(trifluoromethyl)-1*H*-imidazo[4,5-*b*]pyridine; 4-Oxide, *in* C-**60139**

C$_7$H$_3$ClN$_2$O$_5$
2,6-Dinitrobenzoic acid; Chloride, *in* D-**60457**
▷ 3,5-Dinitrobenzoic acid; Chloride, *in* D-**60458**

C$_7$H$_3$Cl$_3$O
2,3,4-Trichloro-2,4,6-cycloheptatrien-1-one, T-**80224**

C$_7$H$_3$Cl$_4$N$_3$
4,5,6,7-Tetrachlorobenzotriazole; 1-Me, *in* T-**60027**
4,5,6,7-Tetrachlorobenzotriazole; 2-Me, *in* T-**60027**

C$_7$H$_3$FIN
2-Cyano-1-fluoro-3-iodobenzene, *in* F-**70020**

C$_7$H$_3$F$_3$N$_2$O$_4$
1,3-Dinitro-5-(trifluoromethyl)benzene, D-**70455**
2,4-Dinitro-1-(trifluoromethyl)benzene, D-**70456**

C$_7$H$_3$F$_3$O$_2$
2,3,5-Trifluorobenzoic acid, T-**70238**
2,3,6-Trifluorobenzoic acid, T-**70239**
2,4,5-Trifluorobenzoic acid, T-**70240**

C$_7$H$_3$F$_5$
Pentafluoromethylbenzene, P-**60038**

C$_7$H$_3$F$_6$N
2,3-Bis(trifluoromethyl)pyridine, B-**60181**
2,4-Bis(trifluoromethyl)pyridine, B-**60182**
2,5-Bis(trifluoromethyl)pyridine, B-**60183**
2,6-Bis(trifluoromethyl)pyridine, B-**60184**
3,4-Bis(trifluoromethyl)pyridine, B-**60185**
3,5-Bis(trifluoromethyl)pyridine, B-**60186**

C$_7$H$_3$F$_6$NO
2,4-Bis(trifluoromethyl)pyridine; 1-Oxide, *in* B-**60182**
2,5-Bis(trifluoromethyl)pyridine; 1-Oxide, *in* B-**60183**
2,6-Bis(trifluoromethyl)pyridine; 1-Oxide, *in* B-**60184**

C$_7$H$_3$NS
▷ 2,4-Diethynylthiazole, D-**80156**

C$_7$H$_3$N$_3$O$_4$
▷ 1-Cyano-3,5-dinitrobenzene, *in* D-**60458**
2-Cyano-1,3-dinitrobenzene, *in* D-**60457**

2,6-Dihydroxy-3-iodobenzoic acid, D-60334
3,4-Dihydroxy-5-iodobenzoic acid, D-60335
3,5-Dihydroxy-2-iodobenzoic acid, D-60336
3,5-Dihydroxy-4-iodobenzoic acid, D-60337
4,5-Dihydroxy-2-iodobenzoic acid, D-60338

$C_7H_5I_2O_2$
2,6-Diiodobenzoic acid, D-70356

C_7H_5N
2-Ethynylpyridine, E-80076
3-Ethynylpyridine, E-80077
4-Ethynylpyridine, E-80078

C_7H_5NO
Benzonitrile N-oxide, B-60035
2-Cyanophenol, in H-70108
4-Ethynylpyridine; N-Oxide, in E-80078
3-Ethynylpyridine; N-oxide, in E-80077

C_7H_5NOS
5-(2-Thienyl)oxazole, T-70171

$C_7H_5NO_2$
5-(2-Furanyl)oxazole, F-70048
4-Nitrosobenzaldehyde, N-60046

$C_7H_5NO_2S$
1,2,3-Benzoxathiazin-4(3H)-one, B-80052

$C_7H_5NO_3$
4-Hydroxy-3-nitrosobenzaldehyde, H-60197

$C_7H_5NO_4$
4-Hydroxy-3-nitrosobenzoic acid, H-70193

$C_7H_5NO_4S$
1,2,3-Benzoxathiazin-4(3H)-one; 2,2-Dioxide,
in B-80052

$C_7H_5NO_6$
4,5-Dihydroxy-2-nitrobenzoic acid, D-60360

$C_7H_5NS_2$
Benzenesulfenyl thiocyanate, B-60013

C_7H_5NSe
Benzoselenazole, B-70050

$C_7H_5N_3OS$
2,3-Dihydro-2-thioxopyrido[2,3-d]pyrimidin-
4(1H)-one, D-60291
2,3-Dihydro-2-thioxopyrido[3,2-d]pyrimidin-
4(1H)-one, D-60292
2,3-Dihydro-2-thioxopyrido[3,4-d]pyrimidin-
4(1H)-one, D-60293
2,3-Dihydro-2-thioxo-4H-pyrido[1,2-a]-1,3,5-
triazin-4-one, D-60294

$C_7H_5N_3O_2$
Pyrido[2,3-d]pyrimidine-2,4(1H,3H)-dione,
P-60236
Pyrido[3,4-d]pyrimidine-2,4(1H,3H)-dione,
P-60237

$C_7H_5N_3O_2S$
3-Amino-5-nitro-2,1-benzisothiazole, A-60232

$C_7H_5N_3O_5$
3,5-Dinitrobenzamide, in D-60458

$C_7H_5N_3S$
1,2,3-Benzotriazine-4(3H)-thione, B-60050
2-(1,3,4-Thiadiazol-2-yl)pyridine, T-60192
4-(1,3,4-Thiadiazol-2-yl)pyridine, T-60193

$C_7H_5N_5O$
1H-Imidazo[2,1-b]purin-4(5H)-one, I-70006

$C_7H_5N_5O_3$
2-Amino-4(3H)-pteridinone-6-carboxylic acid,
A-80135

$C_7H_5N_5O_8$
3,5-Diamino-2,4,6-trinitrobenzoic acid,
D-70047
▷ 2-Methyl-3,4,5,6-tetranitroaniline, M-60129
▷ 3-Methyl-2,4,5,6-tetranitroaniline, M-60130
▷ 4-Methyl-2,3,5,6-tetranitroaniline, M-60131
2,3,4,6-Tetranitroaniline; N-Me, in T-60159

C_7H_6
1H-Cyclopropabenzene, C-70246

C_7H_6BrF
▷ 1-(Bromomethyl)-2-fluorobenzene, B-80215
▷ 1-(Bromomethyl)-3-fluorobenzene, B-80216
▷ 1-(Bromomethyl)-4-fluorobenzene, B-80217

C_7H_6BrFO
3-Bromo-4-fluorobenzyl alcohol, B-60244
5-Bromo-2-fluorobenzyl alcohol, B-60245
2-Bromo-1-fluoro-3-methoxybenzene, in
B-60247
2-Bromo-4-fluoro-1-methoxybenzene, in
B-60248
4-Bromo-2-fluoro-1-methoxybenzene, in
B-60251

$C_7H_6BrNO_2$
3-Amino-2-bromobenzoic acid, A-70097

$C_7H_6BrNO_3$
2-Bromo-1-methoxy-3-nitrobenzene, in
B-70257
2-Bromo-3-nitrobenzyl alcohol, B-70256
4-Bromo-2-nitrobenzyl alcohol, B-80238

$C_7H_6Br_2$
1,3-Dibromo-2-methylbenzene, D-60103

C_7H_6ClFO
1-Chloro-2-fluoro-4-methoxybenzene, in
C-60080
1-Chloro-4-fluoro-2-methoxybenzene, in
C-60074
2-Chloro-4-fluoro-1-methoxybenzene, in
C-60073
4-Chloro-2-fluoro-1-methoxybenzene, in
C-60079

C_7H_6ClNO
2-Acetyl-5-chloropyridine, A-70033
3-Acetyl-2-chloropyridine, A-70034
4-Acetyl-2-chloropyridine, A-70035
4-Acetyl-3-chloropyridine, A-70036
5-Acetyl-2-chloropyridine, A-70037

$C_7H_6ClNO_2$
3-Chloro-4-hydroxybenzoic acid; Amide, in
C-60091

$C_7H_6ClNO_3$
4-Chloro-2-methoxy-1-nitrobenzene, in
C-60119
2-Chloro-3-methyl-4-nitrophenol, C-60115
3-Chloro-4-methyl-5-nitrophenol, C-60116

$C_7H_6Cl_2O_2$
Bicyclo[1.1.1]pentane-1,3-dicarboxylic acid;
Dichloride, in B-80090

$C_7H_6Cl_2S$
1,2-Dichloro-4-(methylthio)benzene, in
D-60123
1,4-Dichloro-2-(methylthio)benzene, in
D-60121

$C_7H_6FNO_2$
2-Amino-3-fluorobenzoic acid, A-70140
2-Amino-4-fluorobenzoic acid, A-70141
2-Amino-5-fluorobenzoic acid, A-70142
2-Amino-6-fluorobenzoic acid, A-70143
3-Amino-4-fluorobenzoic acid, A-70144
4-Amino-2-fluorobenzoic acid, A-70145
▷ 4-Amino-3-fluorobenzoic acid, A-70146
5-Amino-2-fluorobenzoic acid, A-70147
4-Fluoro-2-hydroxybenzoic acid; Amide, in
F-60042

$C_7H_6FNO_3$
4-Fluoro-2-methoxy-1-nitrobenzene, in
F-60061

$C_7H_6F_2S$
1,3-Difluoro-2-(methylthio)benzene, in
D-60174
1,3-Difluoro-5-(methylthio)benzene, in
D-60176
1,4-Difluoro-2-(methylthio)benzene, in
D-60173

$C_7H_6F_3NO_2$
2-Hydroxy-3-methyl-2-butenenitrile;
Trifluoroacetyl, in H-80198

$C_7H_6F_3NO_2S$
2-(Trifluoromethyl)benzenesulfonic acid;
Amide, in T-70242
3-(Trifluoromethyl)benzenesulfonic acid;
Amide, in T-60283
4-(Trifluoromethyl)benzenesulfonic acid;
Amide, in T-70243

$C_7H_6INO_2$
2-Hydroxy-5-iodobenzaldehyde; Oxime, in
H-70143

$C_7H_6N_2$
3-Phenyl-3H-diazirine, P-70067
Pyrrolo[1,2-a]pyrazine, P-70197
▷ 1H-Pyrrolo[3,2-b]pyridine, P-60249
6H-Pyrrolo[3,4-b]pyridine, P-70201
2H-Pyrrolo[3,4-c]pyridine, P-70200

$C_7H_6N_2O$
▷ 2-Aminobenzoxazole, A-60093
1,6-Dihydro-7H-pyrrolo[2,3-c]pyridin-7-one,
D-80252
Pyrrolo[1,2-a]pyrazin-1(2H)-one, P-70198

$C_7H_6N_2O_3S$
1H-Benzimidazole-2-sulfonic acid, B-70022

$C_7H_6N_2O_4$
2-Amino-3,5-pyridinedicarboxylic acid,
A-70193
4-Amino-2,6-pyridinedicarboxylic acid,
A-70194
5-Amino-3,4-pyridinedicarboxylic acid,
A-70195
▷ Antibiotic 2061 A, in A-70158
▷ 2-Methyl-1,3-dinitrobenzene, M-60061

$C_7H_6N_2O_5$
2,6-Dinitrobenzyl alcohol, D-60459

$C_7H_6N_2O_6$
2-Methoxy-4,5-dinitrophenol, in D-70449
5-Methoxy-2,4-dinitrophenol, in D-80491

$C_7H_6N_4$
4-Amino-1,2,3-benzotriazine, A-60092
2-Methylpteridine, M-70120
4-Methylpteridine, M-70121
6-Methylpteridine, M-70122
7-Methylpteridine, M-70123

$C_7H_6N_4O$
4-Amino-1,2,3-benzotriazine; 2-Oxide, in
A-60092
4-Amino-1,2,3-benzotriazine; 3-Oxide, in
A-60092
2-Amino-5-cyanonicotinamide, in A-70193
Di-1H-imidazol-2-ylmethanone, D-70354
4-Hydroxylamino-1,2,3-benzotriazine, in
A-60092

$C_7H_6N_4O_2$
2,4(1H,3H)-Pteridinedione; 1-Me, in P-60191
2,4(1H,3H)-Pteridinedione; 3-Me, in P-60191

$C_7H_6N_4O_3$
6-(Hydroxymethyl)-2,4(1H,3H)pteridinedione,
H-80213
2,4,6(1H,3H,5H)-Pteridinetrione; I-Me, in
P-70134
2,4,6(1H,3H,5H)-Pteridinetrione; 3-Me, in
P-70134
2,4,6(1H,3H,5H)-Pteridinetrione; 7-Me, in
P-70134
2,4,7(1H,3H,8H)-Pteridinetrione; 1-Me, in
P-70135
2,4,7(1H,3H,8H)-Pteridinetrione; 3-Me, in
P-70135
2,4,7(1H,3H,8H)-Pteridinetrione; 6-Me, in
P-70135

$C_7H_6N_4O_6$
2,4,6-Trinitroaniline; N-Me, in T-70301

$C_7H_6N_4S$
1H-Tetrazole-5-thiol; S-Ph, in T-60174
▷ Tetrazole-5-thione; 1-Ph, in T-60174

C_7H_6O
2,4,6-Cycloheptatrien-1-one, C-60204
Dimethylenebicyclo[1.1.1]pentanone, D-60422

1,3-Diisocyanatocyclopentane, D-60390
1-Methoxy-2-phenyldiazene 2-oxide, *in* H-70208
Urocanic acid; Me ester, *in* U-70005

$C_7H_8N_2O_3$

3-Amino-4-methyl-5-nitrophenol, A-60222
2-Amino-3-nitrobenzyl alcohol, A-80125
2-Amino-6-nitrobenzyl alcohol, A-80126
3-Amino-2-nitrobenzyl alcohol, A-80127
3-Amino-4-nitrobenzyl alcohol, A-80128
3-Amino-5-nitrobenzyl alcohol, A-60233
4-Ethoxy-2-nitropyridine, *in* N-70063

$C_7H_8N_2O_3S$

Aminoiminomethanesulfonic acid; *N*-Ph, *in* A-60210

$C_7H_8N_2O_4$

4-Amino-5-hydroxy-2-oxo-7-oxabicyclo[4.1.0]hept-3-ene-3-carboxamide, A-70158
1,2,3,6-Tetrahydro-2,6-dioxo-4-pyrimidineacetic acid; Me ester, *in* T-70060

$C_7H_8N_2O_5$

3-Amino-2(1*H*)-pyridinethione; N^3-Ac, *in* A-60247

$C_7H_8N_2Se$

Selenourea; *N*-Ph, *in* S-70031

$C_7H_8N_4$

1′-Methyl-1,5′-bi-1*H*-pyrazole, *in* B-60119
2,2′-Methylenebis-1*H*-imidazole, M-70064

$C_7H_8N_4O$

N-(4-Cyano-1-methyl-1*H*-pyrazol-5-yl)acetamide, *in* A-70192
1,3-Dihydro-1,3-dimethyl-2*H*-imidazo[4,5-*b*]pyrazin-2-one, *in* I-60006
1,4-Dihydro-1,4-dimethyl-2*H*-imidazo[4,5-*b*]pyrazin-2-one, *in* I-60006
2-Methoxy-4-methyl-4*H*-imidazo[4,5-*b*]pyrazine, *in* I-60006

$C_7H_8N_4O_2$

2-Amino-3,5-pyridinedicarboxylic acid; Diamide, *in* A-70193
1*H*-Pyrazolo[3,4-*d*]pyrimidine-4,6(5*H*,7*H*)-dione; 1,5-Di-Me, *in* P-60202
1*H*-Pyrazolo[3,4-*d*]pyrimidine-4,6(5*H*,7*H*)-dione; 5,7-Di-Me, *in* P-60202

$C_7H_8N_4S$

1,3-Dihydro-1,3-dimethyl-2*H*-imidazo[4,5-*b*]pyrazine-2-thione, *in* D-60247
1,5-Dihydro-6*H*-imidazo[4,5-*c*]pyridazine-6-thione; N^1,*S*-Di-Me, *in* D-70215
1,5-Dihydro-6*H*-imidazo[4,5-*c*]pyridazine-6-thione; N^2,*S*-Di-Me, *in* D-70215
1,5-Dihydro-6*H*-imidazo[4,5-*c*]pyridazine-6-thione; N^5,*S*-Di-Me, *in* D-70215
1,5-Dihydro-6*H*-imidazo[4,5-*c*]pyridazine-6-thione; N^7,*S*-Di-Me, *in* D-70215
1,7-Dihydro-6*H*-purine-6-thione; 1,9-Di-Me, *in* D-60270
1,7-Dihydro-6*H*-purine-6-thione; 3,7-Di-Me, *in* D-60270
1,7-Dihydro-6*H*-purine-6-thione; 3,9-Di-Me, *in* D-60270
6-Mercaptopurine; *S*,3*N*-Di-Me, *in* D-60270
6-Mercaptopurine; *S*,7*N*-Di-Me, *in* D-60270
6-Mercaptopurine; *S*,9*N*-Di-Me, *in* D-60270
1-Methyl-2-methylthio-1*H*-imidazo[4,5-*b*]pyrazine, *in* D-60247
4-Methyl-2-methylthio-4*H*-imidazo[4,5-*b*]pyrazine, *in* D-60247

C_7H_8O

Bicyclo[2.2.1]hept-5-en-2-one, B-80080
Tricyclo[2.2.1.$0^{2,6}$]heptan-3-one, T-70228
3-Vinyl-2-cyclopenten-1-one, V-70011

C_7H_8OS

2,5-Dimethyl-3-thiophenecarboxaldehyde, D-70438
3,5-Dimethyl-2-thiophenecarboxaldehyde, D-70439
2-Mercaptobenzenemethanol, M-80021
3-Mercaptobenzenemethanol, M-80022
4-Mercaptobenzenemethanol, M-80023
2-Thiabicyclo[2.2.2]oct-5-en-3-one, T-60188

$C_7H_8O_2$

2,2-Dimethyl-4-cyclopentene-1,3-dione, D-70395
2,5-Dimethyl-3-furancarboxaldehyde, D-70412
▷ 4,6-Dimethyl-2*H*-pyran-2-one, D-70435
2-Furanpropanal, F-80089
3-Furanpropanal, F-80090
4-Hydroxybenzyl alcohol, H-80114
4-Hydroxy-4-methyl-2,5-cyclohexadien-1-one, H-70169
8-Oxabicyclo[3.2.1]oct-6-en-3-one, O-80051
3-Oxo-1-cyclohexene-1-carboxaldehyde, O-60060
α-Oxo-3-cyclopentene-1-acetaldehyde, O-60061

$C_7H_8O_2S$

2-(Methoxyacetyl)thiophene, *in* H-70103

$C_7H_8O_3$

2,3-Dihydroxybenzyl alcohol, D-80283
2,5-Dihydroxybenzyl alcohol, D-70287
3,4-Dihydroxybenzyl alcohol, D-80284
4,5-Dimethyl-2-furancarboxylic acid, D-70413
4-Hydroxy-2-cyclopentenone; Ac, *in* H-60116
Methyl 3-formylcrotonate, *in* M-70103
2-Methylene-3-oxocyclobutanecarboxylic acid; Me ester, *in* M-80088
5-Oxo-1-cyclohexene-1-carboxylic acid, O-80077
5-Oxo-2-cyclohexene-1-carboxylic acid, O-80078
6-Oxo-1-cyclohexene-1-carboxylic acid, O-80079
3,6,10-Trioxatetracyclo[7.1.0.$0^{2,4}$.$0^{5,7}$]decane, T-80356

$C_7H_8O_4$

Bicyclo[1.1.1]pentane-1,3-dicarboxylic acid, B-80090
3,4-Epoxy-5-hydroxy-1-cyclohexenecarboxylic acid, E-80030
5-Hydroxy-3-methoxy-7-oxabicyclo[4.1.0]hept-3-en-2-one, H-70155
5-(Hydroxymethyl)-1,2,3-benzenetriol, H-80196
5-Hydroxymethyl-2(5*H*)-furanone; Ac, *in* H-60182
3-Methyl-1,2,4,5-benzenetetrol, M-80048
4-Methyl-1,2,3,5-benzenetetrol, M-80049
5-Methyl-1,2,3,4-benzenetetrol, M-80050
Oxysporone, O-70107
Terremutin, T-70009
2-Vinyl-1,1-cyclopropanedicarboxylic acid, V-60011

$C_7H_8O_5$

3,6-Dihydro-2*H*-pyran-2,2-dicarboxylic acid, D-80243
4,5-Dihydroxy-3-oxo-1-cyclohexenecarboxylic acid, D-70328
4-Oxo-2-heptenedioic acid, O-60072
1,4,8-Trioxaspiro[4.5]decane-7,9-dione, *in* O-60084

$C_7H_8O_6$

Aconitic acid; α-Mono-Me ester, *in* A-70055
Aconitic acid; β-Mono-Me ester, *in* A-70055
Aconitic acid; γ-Mono-Me ester, *in* A-70055

C_7H_8S

5,6-Dihydro-4*H*-cyclopenta[*b*]thiophene, D-70180
5,6-Dihydro-4*H*-cyclopenta[*c*]thiophene, D-70181

$C_7H_8S_2$

6,7-Dihydro-5*H*-thieno[3,2-*b*]thiopyran, D-80257
6,7-Dihydro-4*H*-thieno[3,2-*c*]thiopyran, D-80258

C_7H_9Br

2-Bromobicyclo[2.2.1]hept-2-ene, B-80179

$C_7H_9BrO_2$

7-Bromo-5-heptynoic acid, B-60261
2-Bromo-3-methoxy-2-cyclohexen-1-one, *in* B-70201

C_7H_9Cl

5-Chloro-1,3-cycloheptadiene, C-70062

C_7H_9ClO

2-Chloro-2-cyclohepten-1-one, C-60049

$C_7H_9ClO_2$

2-Chloro-3-methoxy-2-cyclohexen-1-one, *in* C-80046

C_7H_9N

4-Heptynenitrile, *in* H-80032
6-Heptynenitrile, *in* H-80033
2-Vinyl-1*H*-pyrrole; *N*-Me, *in* V-60013

C_7H_9NO

7-Azabicyclo[4.2.0]oct-3-en-8-one, A-60331
Bicyclo[2.2.1]hept-5-en-2-one; Oxime, *in* B-80080
Bicyclo[2.2.1]hept-5-en-2-one; Oxime, *in* B-80080
1,2-Dihydro-3*H*-azepin-3-one; 1-Me, *in* D-70171
2,3-Dihydro-1*H*-pyrrolizin-1-ol, D-70261
2,3-Dihydro-1*H*-pyrrolizin-6(5*H*)-one, D-80250
1,5-Dimethyl-1*H*-pyrrole-2-carboxaldehyde, *in* M-70129
3,3*a*,4,6*a*-Tetrahydrocyclopenta[*b*]pyrrol-2(1*H*)-one, T-70048
Tricyclo[2.2.1.$0^{2,6}$]heptan-3-one; Oxime, *in* T-70228

C_7H_9NOS

2-Pyrrolethiolcarboxylic acid; 1-Me, Me ester, *in* P-80226

$C_7H_9NO_2$

3-Amino-5-hydroxybenzyl alcohol, A-60182
2-Amino-6-methoxyphenol, *in* A-70089
5-Amino-2-methoxyphenol, *in* A-70090
Dihydro-1*H*-pyrrolizine-3,5(2*H*,6*H*)-dione, D-70260
2,3-Dimethoxypyridine, *in* H-60223
2,4-Dimethoxypyridine, *in* H-60224
2,6-Dimethoxypyridine, *in* H-60226
3,4-Dimethoxypyridine, *in* H-60227
3-Ethoxy-4(1*H*)-pyridinone, *in* H-60227
Ethyl 1-cyanocyclopropanecarboxylate, *in* C-80196
6-Hydroxy-2(1*H*)-pyridinone; Me ether, *N*-Me, *in* H-60226
2-Methoxy-1-methyl-4(1*H*)-pyridinone, *in* H-60224
3-Methoxy-1-methyl-4(1*H*)-pyridinone, *in* H-60227
3-Methoxy-2-methyl-4(1*H*)-pyridinone, *in* H-80214
4-Methoxy-1-methyl-2(1*H*)-pyridinone, *in* H-60224
5-Methyl-1*H*-pyrrole-2-carboxylic acid; Me ester, *in* M-60124
5-Methyl-1*H*-pyrrole-3-carboxylic acid; Me ester, *in* M-60125
4,5,6,7-Tetrahydro-1,2-benzisoxazol-3(2*H*)-one, T-70037
4,5,6,7-Tetrahydro-2,1-benzisoxazol-3(1*H*)-one, T-70038

$C_7H_9NO_3$

4-Aminodihydro-3-methylene-2-(3*H*)furanone; *N*-Ac, *in* A-70131
3,4-Dihydroxypyridine; Di-Me ether, 1-oxide, *in* H-60227
4,6-Dimethoxy-2(1*H*)-pyridinone, *in* D-70341
3-Hydroxy-1*H*-pyrrole-2-carboxylic acid; Me ether, Me ester, *in* H-80248
4-Hydroxy-1*H*-pyrrole-2-carboxylic acid; Me ether, Me ester, *in* H-80249
4-Hydroxy-1*H*-pyrrole-3-carboxylic acid; Me ether, Me ester, *in* H-80250
5-Methyl-4-isoxazolecarboxylic acid; Et ester, *in* M-80117
7-Oxo-1-azabicyclo[3.2.0]heptane-2-carboxylic acid, O-70079
3-Oxo-2-azabicyclo[2.1.1]hexane-1-carboxylic acid; Me ester, *in* O-80057
2,4-Pyridinediol; Di-Me ether, 1-oxide, *in* H-60224

$C_7H_9NO_4$

5-Oxo-2-pyrrolidinecarboxylic acid; Me ester, *N*-formyl, *in* O-70102

$C_7H_9NO_4S_2$
2,3-Bis(methylsulfonyl)pyridine, *in* P-60218
2,4-Bis(methylsulfonyl)pyridine, *in* M-80029
2,5-Bis(methylsulfonyl)pyridine, *in* P-60220
3,5-Bis(methylsulfonyl)pyridine, *in* P-60222

C_7H_9NS
4-Amino-5,6-dihydro-4*H*-cyclopenta[*b*]thiophene, A-60137
2-(2-Pyridyl)ethanethiol, P-60239
2-(4-Pyridyl)ethanethiol, P-60240

$C_7H_9NS_2$
2,3-Bis(methylthio)pyridine, *in* P-60218
2,4-Bis(methylthio)pyridine, *in* M-80029
2,4-Bis(methylthio)pyridine, *in* P-60219
2,5-Bis(methylthio)pyridine, *in* P-60220
3,5-Bis(methylthio)pyridine, *in* P-60222
1-Methyl-4-(methylthio)-2(1*H*)-pyridinethione,
 in M-80029
1-Methyl-4-(methylthio)-2(1*H*)-pyridinethione,
 in P-60219

$C_7H_9N_2S$
3,5-Dimethyl-2-thiophenecarboxaldehyde;
 Hydrazone, *in* D-70439

$C_7H_9N_3O$
2-(Hydroxyacetyl)pyridine; Hydrazone, *in*
 H-70101

$C_7H_9N_3O_3$
5-Amino-1*H*-pyrazole-4-carboxylic acid; *N*(1)-
 Me, *N*(5)-Ac, *in* A-70192

$C_7H_9N_5$
3,7-Dimethyladenine, D-60400
3,9-Dimethyladenine, D-80417
2-(Dimethylamino)purine, *in* A-60243
8-(Dimethylamino)purine, *in* A-60244

C_7H_{10}
Bicyclo[3.1.1]hept-2-ene, B-60087
Bicyclo[4.1.0]hept-3-ene, B-80079
1,2-Bis(methylene)cyclopentane, B-60169
1,3-Bis(methylene)cyclopentane, B-60170
Tricyclo[3.2.0.0²,⁷]heptane, T-80240
Tricyclo[4.1.0.0¹,³]heptane, T-60262

$C_7H_{10}Cl_2O_2$
2,2-Dichloro-6-heptenoic acid, D-80119

$C_7H_{10}INO_2$
5-(Iodomethyl)-2-pyrrolidinone; *N*-Ac, *in*
 I-80037

$C_7H_{10}N_2O$
2-Oxa-3-azabicyclo[2.2.2]octane; *N*-Cyano, *in*
 O-70051

$C_7H_{10}N_2O_2$
4,5,6,7-Tetrahydro-3-methoxyisoxazolo[4,5-
 c]pyridine, *in* T-60071
Tetrahydro-1*H*-pyrazolo[1,2-*a*]pyridazine-
 1,3(2*H*)-dione, T-70088

$C_7H_{10}N_2O_2S$
2-Amino-4-thiazoleacetic acid; Et ester, *in*
 A-80146
2-Amino-5-thiazoleacetic acid; Et ester, *in*
 A-80147

$C_7H_{10}N_2O_3$
2-Imidazolidinone; 1,3-Di-Ac, *in* I-70005

$C_7H_{10}N_2O_4$
5-Methyl-3,6-dioxo-2-piperazineacetic acid,
 M-60064

$C_7H_{10}N_2S$
4,6-Dimethyl-2(1*H*)-pyrimidinethione; 1-Me,
 in D-60450
4-Ethylthio-3-pyridinamine, *in* A-80140
5-Ethylthio-2-pyridinamine, *in* A-80138
6-Ethylthio-3-pyridinamine, *in* A-80141

$C_7H_{10}O$
1-Acetylcyclopentene, A-60028
1-Acetyl-2-methylcyclobutene, A-80028
▷ Bicyclo[2.2.1]heptan-2-one, B-70098
3-Cyclohexene-1-carboxaldehyde, C-70227
3,3-Dimethyl-4-pentynal, D-80467
2,4-Heptadienal, H-60020
1,5-Heptadien-3-one, H-70015

1,5-Heptadien-4-one, H-70016
1,6-Heptadien-3-one, H-70017
4,5-Heptadien-3-one, H-70018
6-Heptyn-2-one, H-70025
4-Methyl-2-cyclohexen-1-one, M-80066
6-Methyl-2-cyclohexen-1-one, M-70059
3-Methyl-1-cyclopentenecarboxaldehyde,
 M-80067
2-Methylenecyclohexanone, M-80080
3-Methylenecyclohexanone, M-80081
4-Methylenecyclohexanone, M-80082
2-Methylene-7-oxabicyclo[2.2.1]heptane,
 M-70073
7-Oxatricyclo[4.1.1.0²,⁵]octane, O-60053
8-Oxatricyclo[3.3.0.0²,⁷]octane, O-60054
3-Vinyl-2-cyclopenten-1-ol, V-80008

$C_7H_{10}OS$
2-Thiabicyclo[2.2.2]octan-3-one, T-60187

$C_7H_{10}O_2$
1-Carbethoxybicyclo[1.1.0]butane, *in* B-70094
1,1-Diacetylcyclopropane, D-70032
Dihydro-4,4-dimethyl-5-methylene-2(3*H*)-
 furanone, D-70197
Dihydro-4,5-dimethyl-3-methylene-2(3*H*)-
 furanone, D-80192
2,2-Dimethyl-1,3-cyclopentanedione, D-70393
4,4-Dimethyl-1,3-cyclopentanedione, D-70394
3,3-Dimethyl-1-cyclopropene-1-carboxylic
 acid; Me ester, *in* D-60417
6,6-Dimethyl-3-oxabicyclo[3.1.0]hexan-2-one,
 D-70425
2-Furanpropanol, F-80091
3-Furanpropanol, F-80092
2,4-Heptadienoic acid, H-60021
3-Heptynoic acid, H-80031
4-Heptynoic acid, H-80032
6-Heptynoic acid, H-80033
3,4-Hexadienoic acid; Me ester, *in* H-70037
3,5-Hexadienoic acid; Me ester, *in* H-70038
Hexahydro-1*H*-cyclopenta[*c*]furan-1-one,
 H-70054
5-Hexynoic acid; Me ester, *in* H-80086
1-Hydroxy-3-cyclohexene-1-carboxaldehyde,
 H-80146
3-Methoxy-2-cyclohexen-1-one, M-80041
2-Methoxy-3-methyl-2-cyclopenten-1-one, *in*
 H-60174
3-Methyl-3,5-cyclohexadiene-1,2-diol,
 M-80063
5-Methyl-2,4-hexadienoic acid, M-70086
5-Methyl-3,4-hexadienoic acid, M-70087
2-Oxocyclohexanecarboxaldehyde, O-80074
3-Oxocyclohexanecarboxaldehyde, O-80075
4-Oxocyclohexanecarboxaldehyde, O-80076
3-Pentyn-2-ol; Ac, *in* P-70041

$C_7H_{10}O_3$
5-Ethyl-3-hydroxy-4-methyl-2(5*H*)-furanone,
 E-60051
3-Formyl-2,2-dimethyl-1-
 cyclopropanecarboxylic acid, F-80074
2-Hydroxycyclopentanone; Ac, *in* H-60114
7-Hydroxy-5-heptynoic acid, H-60150
3-Hydroxymethyl-2-methylfuran; Ac, *in*
 H-80207
4-Hydroxy-3-penten-2-one; Ac, *in* P-60059
4-Methoxy-2,5-dimethyl-3(2*H*)-furanone, *in*
 H-60120
3-Methyl-4-oxo-2-butenoic acid; Et ester, *in*
 M-70103
3-Oxocyclopentaneacetic acid, O-70083
2-Oxocyclopentanecarboxylic acid; Me ester,
 in O-70084
4-Oxo-2-hexenoic acid; Me ester, *in* O-80082
4-Oxo-2-pentenoic acid; Et ester, *in* O-70098
2,4,10-Trioxatricyclo[3.3.1.1³,⁷]decane,
 T-60377

$C_7H_{10}O_4$
1,1-Cyclopropanedicarboxylic acid; Di-Me
 ester, *in* C-80196
1,2-Cyclopropanedicarboxylic acid; Di-Me
 ester, *in* C-70248
Ethyl 3-acetoxyacrylate, *in* O-60090

$C_7H_{10}O_5$
4-Oxoheptanedioic acid, O-60071

3-Oxopentanedioic acid; Di-Me ester, *in*
 O-60084
1,3,4-Trihydroxy-6-oxabicyclo[3.2.1]octan-7-
 one, *in* Q-70005

$C_7H_{10}O_6$
1,3-Dioxolane-2,2-diacetic acid, *in* O-60084

$C_7H_{10}S$
Dicyclopropylmethanethione, D-80150
1-Thia-2-cyclooctyne, T-60189
1-Thia-3-cyclooctyne, T-60190

$C_7H_{10}S_5$
4,5-Dimercapto-1,3-dithiole-2-thione; Di-Et
 thioether, *in* D-80409

$C_7H_{11}ClO$
3,3-Dimethyl-4-pentenoic acid; Chloride, *in*
 D-60437
4-Methyl-2-hexenoic acid; Chloride, *in*
 M-80101

$C_7H_{11}ClO_2$
7-Chloro-2-heptenoic acid, C-70069

$C_7H_{11}FO_2$
1-Fluorocyclohexanecarboxylic acid, F-80031

$C_7H_{11}I$
7-Iodo-1,3-heptadiene, I-80032

$C_7H_{11}IO$
2-Iodocycloheptanone, I-70025

$C_7H_{11}N$
2-Azabicyclo[2.2.2]oct-5-ene, A-70279
5-Heptenenitrile, *in* H-70024

$C_7H_{11}NO$
Bicyclo[2.2.1]heptan-2-one; Oxime, *in* B-70098
3-Cyclohexene-1-carboxaldehyde; Oxime, *in*
 C-70227
4,4-Dimethylglutaraldehydonitrile, *in* D-60434
3,3-Dimethyl-4-pentynal; Oxime, *in* D-80467
Hexahydro-1*H*-pyrrolizin-1-one, H-60057
Hexahydro-3*H*-pyrrolizin-3-one, H-60058
Tetrahydro-1*H*-pyrrolizin-2(3*H*)-one, T-60091

$C_7H_{11}NOS$
4-*tert*-Butyl-2(3*H*)-oxazolethione, B-80288

$C_7H_{11}NO_2$
3,4-Dihydro-2*H*-pyrrole-2-carboxylic acid; Et
 ester, *in* D-80248
Hexahydro-2(3*H*)-benzoxazolone, H-70052
7-Nitrobicyclo[2.2.1]heptane, N-80037
2-Oxa-3-azabicyclo[2.2.2]octane; *N*-Formyl, *in*
 O-70051
3-Oxocyclobutanecarboxylic acid; *N*,*N*-
 Dimethylamide, *in* O-80073
1,2,3,6-Tetrahydro-3-pyridinecarboxylic acid;
 N-Me, *in* T-60084

$C_7H_{11}NO_3$
2-Nitrocycloheptanone, N-80042
5-Oxo-2-pyrrolidinecarboxylic acid; Et ester,
 in O-70102

$C_7H_{11}NO_4$
2-Amino-3-heptenedioic acid, A-60164
2-Amino-5-heptenedioic acid, A-60165

$C_7H_{11}NO_5$
4-Oxoheptanedioic acid; Oxime, *in* O-60071

$C_7H_{11}NS$
4-Mercaptocyclohexanecarbonitrile, *in*
 M-60020

$C_7H_{11}N_3O_2$
3-Amino-5-methyl-1*H*-pyrazole-4-carboxylic
 acid; Et ester, *in* A-80110
4-Amino-5-methyl-1*H*-pyrazole-3-carboxylic
 acid; Et ester, *in* A-80111
3-Amino-1*H*-pyrazole-4-carboxylic acid; *N*(1)-
 Me, Et ester, *in* A-70187
3-Amino-1*H*-pyrazole-5-carboxylic acid; *N*(1)-
 Me, Et ester, *in* A-70188
5-Amino-1*H*-pyrazole-4-carboxylic acid; *N*(1)-
 Me, Et ester, *in* A-70192

$C_7H_{11}N_3O_2S$
5-Mercapto-1-methylhistidine, *in* M-80025
Ovothiol *A*, O-70050

C_7H_{12}
4,4-Dimethyl-2-pentyne, D-**60438**

$C_7H_{12}BrO_2$
2-(4-Bromobutyl)-1,3-dioxole, *in* B-**70263**

$C_7H_{12}Br_2$
1,1-Dibromocycloheptane, D-**60091**
1,1-Dibromo-2,2,3,3-tetramethylcyclopropane, D-**80103**

$C_7H_{12}Br_2O$
2,4-Dibromo-2,4-dimethyl-3-pentanone, D-**60099**

$C_7H_{12}Br_2O_2$
1,1-Dibromo-3,3-dimethoxy-2-methyl-1-butene, *in* D-**70094**

$C_7H_{12}ClN$
8-Azabicyclo[3.2.1]octane; *N*-Chloro, *in* A-**70278**

$C_7H_{12}ClNO_4$
1,2,3,5,6,7-Hexahydropyrrolizinium(1+); Perchlorate, *in* H-**70071**

$C_7H_{12}N^\oplus$
1,2,3,5,6,7-Hexahydropyrrolizinium(1+), H-**70071**

$C_7H_{12}N_2$
3(5)-*tert*-Butylpyrazole, B-**80291**
4-*tert*-Butylpyrazole, B-**80292**

$C_7H_{12}N_2O$
3,5-Dihydro-3,3,5,5-tetramethyl-4*H*-pyrazol-4-one, D-**60286**
Hexahydro-1*H*-pyrrolizin-1-one; Oxime, *in* H-**60057**

$C_7H_{12}N_2OS$
3,5-Dihydro-3,3,5,5-tetramethyl-4*H*-pyrazole-4-thione; *S*-Oxide, *in* D-**60285**

$C_7H_{12}N_2O_2$
5-Amino-3,4-dihydro-2*H*-pyrrole-2-carboxylic acid; Et ester, *in* A-**70135**
1-Oxa-2-azaspiro[2,5]octane; *N*-Carbamoyl, *in* O-**70052**

$C_7H_{12}N_2O_4$
2,6-Diamino-3-heptenedioic acid, D-**60036**

$C_7H_{12}N_2O_5$
N,N′-Carbonylbisglycine; Di-Me ester, *in* C-**80020**

$C_7H_{12}N_2S$
3,5-Dihydro-3,3,5,5-tetramethyl-4*H*-pyrazole-4-thione, D-**60285**

$C_7H_{12}N_4O$
Caffeidine, C-**60003**

$C_7H_{12}O$
2,3-Dimethyl-3,4-dihydro-2*H*-pyran, D-**70398**
3,3-Dimethyl-4-penten-2-one, D-**70429**
4,4-Dimethyl-1-pentyn-3-ol, D-**80468**
2,4-Heptadien-1-ol, H-**70014**
1-Methoxycyclohexene, M-**60037**
1-Methyl-2-cyclohexen-1-ol, M-**80065**
6-Methyl-2-cyclohexen-1-ol, M-**70058**
2-Methyl-4-hexenal, M-**80098**
3-Methyl-4-hexen-2-one, M-**80115**
1-Methyl-7-oxabicyclo[4.1.0]heptane, M-**80125**
3,6,7,8-Tetrahydro-2*H*-oxocin, T-**60076**

$C_7H_{12}OS$
2-Hexyl-5-methyl-3(2*H*)furanone; *S*-Oxide (*exo*-), *in* H-**60079**
2-(Methylthio)cyclohexanone, *in* M-**70025**
3-(Methylthio)cyclohexanone, *in* M-**70026**
4-Methylthiocyclohexanone, *in* M-**70027**
2,2,4,4-Tetramethyl-3-thietanone, T-**60154**
3,3,4,4-Tetramethyl-2-thietanone, T-**60155**

$C_7H_{12}O_2$
2-Cycloheptene-1,4-diol, C-**80179**
2,3-Dimethylcyclopropanol; Ac, *in* D-**70397**
2,2-Dimethyl-4-oxo-1-pentanal, D-**80464**
3,3-Dimethyl-4-pentenoic acid, D-**60437**
5-Heptenoic acid, H-**70024**
5-Hydroxy-2-methylcyclohexanone, H-**70171**
2-Methylhexanedial, M-**80096**

2-Methyl-4-hexenoic acid, M-**80100**
4-Methyl-2-hexenoic acid, M-**80101**
4-Oxoheptanal, O-**60070**
Tetrahydro-5,6-dimethyl-2*H*-pyran-2-one, T-**70059**

$C_7H_{12}O_2S$
3-(Ethylthio)-2-propenoic acid; Et ester, *in* E-**60055**
4-Mercaptocyclohexanecarboxylic acid, M-**60020**
3-Oxobutanethioic acid; Isopropyl ester, *in* O-**70080**

$C_7H_{12}O_3$
4-Acetyl-2,2-dimethyl-1,3-dioxolane, A-**70042**
▷ Botryodiplodin, B-**60194**
3,3-Dimethyl-4-oxopentanoic acid, D-**60433**
4,4-Dimethyl-5-oxopentanoic acid, D-**60434**
2-Hydroxycyclohexanecarboxylic acid, H-**70119**
5-Hydroxy-3-methyl-2-pentenoic acid; Me ester, *in* H-**80211**
3-Methyl-2-oxobutanoic acid; Et ester, *in* M-**80129**
3-Oxopropanoic acid; *tert*-Butyl ester, *in* O-**60090**

$C_7H_{12}O_3S$
2-(Methylsulfonyl)cyclohexanone, *in* M-**70025**
2,2,4,4-Tetramethyl-3-thietanone; 1,1-Dioxide, *in* T-**60154**

$C_7H_{12}O_4$
(1-Methylpropyl)propanedioic acid, M-**70119**
▷ 1,2-Propanediol; Di-Ac, *in* P-**70118**

$C_7H_{12}O_5$
▷ 1,3-Diacetylglycerol, D-**60023**
2-Hydroxy-2-isopropylbutanedioic acid, H-**60165**
2-Hydroxy-3-isopropylbutanedioic acid, H-**60166**
5-(Hydroxymethyl)-5-cyclohexene-1,2,3,4-tetrol, H-**70172**

$C_7H_{12}O_6$
Quinic acid, Q-**70005**

$C_7H_{12}S$
Hexahydro-2*H*-cyclopenta[*b*]thiophene, H-**60049**

$C_7H_{13}Br$
6-Bromo-1-heptene, B-**80208**
7-Bromo-1-heptene, B-**60260**
3-Bromo-1,1,2,2-tetramethylcyclopropane, B-**80261**

$C_7H_{13}ClO$
4,4-Dimethylpentanoic acid; Chloride, *in* D-**70428**

$C_7H_{13}ClO_2$
1-Chloro-3-pentanol; Ac, *in* C-**70138**

$C_7H_{13}FO$
2-Fluoroheptanal, F-**60034**

$C_7H_{13}I$
5-Iodo-4,4-dimethyl-1-pentene, I-**80023**
6-Iodo-1-heptene, I-**70035**
7-Iodo-1-heptene, I-**70036**

$C_7H_{13}IO_2$
1-Iodo-5,5-dimethoxypentene, *in* I-**70052**
5-Iodo-2-pentanol; Ac, *in* I-**60056**
2-(3-Iodopropyl)-2-methyl-1,3-dioxolane, *in* I-**60057**

$C_7H_{13}N$
1-Aminobicyclo[2.2.1]heptane, A-**80060**
7-Aminobicyclo[2.2.1]heptane, A-**80061**
7-Azabicyclo[4.2.0]octane, A-**70277**
8-Azabicyclo[3.2.1]octane, A-**70278**
4,4-Dimethylpentanoic acid; Nitrile, *in* D-**70428**
▷ 2-Propyn-1-amine; *N*-Di-Et, *in* P-**60184**

$C_7H_{13}NO$
2-Acetylpiperidine, A-**70051**

3,3-Dimethyl-4-penten-2-one; Oxime, *in* D-**70429**
Hexahydro-4(1*H*)-azocinone, H-**70047**
Hexahydro-5(2*H*)-azocinone, H-**70048**

$C_7H_{13}NOS$
Hexahydro-1,5-thiazonin-6(7*H*)-one, H-**70075**
4-Mercaptocyclohexanecarboxylic acid; Amide, *in* M-**60020**

$C_7H_{13}NOS_2$
1-Isothiocyanato-5-(methylsulfinyl)pentene, *in* I-**60140**

$C_7H_{13}NO_2$
2-Amino-2-hexenoic acid; Me ester, *in* A-**60167**
3-Amino-4-hexenoic acid; Me ester, *in* A-**60171**
2-Hydroxycyclohexanecarboxylic acid; Amide, *in* H-**70119**
2-Methyl-2-piperidinecarboxylic acid, M-**60119**
Nitrocycloheptane, N-**60025**
3-Piperidinecarboxylic acid; Me ester, *in* P-**70102**
Stachydrine, S-**60050**

$C_7H_{13}NO_3$
4,4-Dimethyl-5-oxopentanoic acid; Oxime, *in* D-**60434**
3-Methyl-2-oxobutanoic acid; Et ester, oxime, *in* M-**80129**

$C_7H_{13}NO_4$
L-Allothreonine; Ac, Me ether, *in* T-**60217**
2-Amino-2-isopropylbutanedioic acid, A-**60212**
2-Amino-3-isopropylbutanedioic acid, A-**70162**
2-Amino-3-propylbutanedioic acid, A-**70186**
Detoxinine, D-**80040**
2-Hydroxy-2-isopropylsuccinamic acid, *in* H-**60165**

$C_7H_{13}NO_4S$
Isobuteine, I-**80049**

$C_7H_{13}NO_6S$
3-[(2-Carboxypropyl)sulfonyl]alanine, *in* I-**80049**

$C_7H_{13}NS$
4,5-Dihydro-2-(1-methylpropyl)thiazole, D-**70224**

$C_7H_{13}NS_2$
1-Isothiocyanato-5-(methylthio)pentane, I-**60140**

$C_7H_{13}N_3$
6*bH*-2*a*,4*a*,6*a*-Hexahydrotriazacyclopenta[*cd*]pentalene, H-**70077**

$C_7H_{13}N_3O_4$
N-Asparaginylalanine, A-**60310**

C_7H_{14}
tert-Butylcyclopropane, B-**80285**

$C_7H_{14}BrN$
2-(Bromomethyl)cyclohexanamine, B-**70239**

$C_7H_{14}Br_2$
1,5-Dibromo-3,3-dimethylpentane, D-**70090**

$C_7H_{14}N_2$
1,4-Diazabicyclo[4.3.0]nonane, D-**60054**
Diisopropylcyanamide, D-**60392**
6-Heptyne-2,5-diamine, H-**60026**
Hexahydro-1*H*-pyrazolo[1,2-*a*]pyridazine, H-**70067**
Octahydropyrrolo[3,4-*c*]pyrrole; *N*-Me, *in* O-**70017**

$C_7H_{14}N_2O$
2-Acetylpiperidine; Oxime (*E*-), *in* A-**70051**
2-Imidazolidinone; 1-*tert*-Butyl, *in* I-**70005**
3,3,5,5-Tetramethyl-4-pyrazolidinone, T-**60153**

$C_7H_{14}N_2O_2$
2-Hydroxycyclohexanecarboxylic acid; Hydrazide, *in* H-**70119**

$C_8H_4N_4O_2$
Furoxano[3,4-*b*]quinoxaline, *in* F-70051
$C_8H_4OS_2$
3-Thioxo-1(3*H*)-benzo[*c*]thiophenone, T-80192
$C_8H_4O_2S$
3-Thioxo-1(3*H*)-isobenzofuranone, T-80193
$C_8H_4O_4$
2,2′-Bifurylidene-5,5′-dione, B-80094
3,4-Dihydroxybenzocyclobutene-1,2-dione, D-80274
3,5-Dihydroxybenzocyclobutene-1,2-dione, D-80275
3,6-Dihydroxybenzocyclobutene-1,2-dione, D-80276
4,5-Dihydroxybenzocyclobutene-1,2-dione, D-80277
$C_8H_4O_6$
1,4-Benzoquinone-2,3-dicarboxylic acid, B-70047
1,4-Benzoquinone-2,5-dicarboxylic acid, B-70048
1,4-Benzoquinone-2,6-dicarboxylic acid, B-70049
C_8H_4S
▷ 2,5-Diethynylthiophene, D-80157
$C_8H_5BrN_2$
1-Bromophthalazine, B-70266
5-Bromophthalazine, B-70267
6-Bromophthalazine, B-70268
C_8H_5BrOS
2-Bromobenzo[*b*]thiophene; 1-Oxide, *in* B-60204
3-Bromobenzo[*b*]thiophene; 1-Oxide, *in* B-60205
$C_8H_5BrO_2$
4-Bromo-1(3*H*)-isobenzofuranone, B-70224
7-Bromo-1(3*H*)-isobenzofuranone, B-70225
$C_8H_5BrO_2S$
2-Bromobenzo[*b*]thiophene; 1,1-Dioxide, *in* B-60204
3-Bromobenzo[*b*]thiophene; 1,1-Dioxide, *in* B-60205
4-Bromobenzo[*b*]thiophene; 1,1-Dioxide, *in* B-60206
5-Bromobenzo[*b*]thiophene; 1,1-Dioxide, *in* B-60207
6-Bromobenzo[*b*]thiophene; 1,1-Dioxide, *in* B-60208
C_8H_5BrS
2-Bromobenzo[*b*]thiophene, B-60204
3-Bromobenzo[*b*]thiophene, B-60205
4-Bromobenzo[*b*]thiophene, B-60206
5-Bromobenzo[*b*]thiophene, B-60207
6-Bromobenzo[*b*]thiophene, B-60208
7-Bromobenzo[*b*]thiophene, B-60209
$C_8H_5Br_3O$
2′,3′,5′-Tribromoacetophenone, T-80216
$C_8H_5ClN_2$
1-Chlorophthalazine, C-70148
5-Chlorophthalazine, C-70149
6-Chlorophthalazine, C-70150
N-Cyanobenzenecarboximidoyl chloride, C-80168
C_8H_5ClOS
2-Chlorobenzo[*b*]thiophene; 1-Oxide, *in* C-60042
3-Chlorobenzo[*b*]thiophene; 1-Oxide, *in* C-60043
$C_8H_5ClO_2$
4-Chloro-1(3*H*)-isobenzofuranone, C-70079
7-Chloro-1(3*H*)-isobenzofuranone, C-70080
$C_8H_5ClO_2S$
2-Chlorobenzo[*b*]thiophene; 1,1-Dioxide, *in* C-60042
3-Chlorobenzo[*b*]thiophene; 1,1-Dioxide, *in* C-60043
7-Chlorobenzo[*b*]thiophene; 1,1-Dioxide, *in* C-60047
C_8H_5ClS
2-Chlorobenzo[*b*]thiophene, C-60042
3-Chlorobenzo[*b*]thiophene, C-60043
4-Chlorobenzo[*b*]thiophene, C-60044
5-Chlorobenzo[*b*]thiophene, C-60045
6-Chlorobenzo[*b*]thiophene, C-60046
7-Chlorobenzo[*b*]thiophene, C-60047
$C_8H_5Cl_3O$
2,2,2-Trichloroacetophenone, T-70219
C_8H_5FO
6-Fluorobenzocyclobuten-1-one, F-80022
$C_8H_5FO_2$
4-Fluoro-1(3*H*)-isobenzofuranone, F-70027
7-Fluoro-1(3*H*)-isobenzofuranone, F-70028
$C_8H_5F_3$
(Trifluoroethenyl)benzene, T-80275
$C_8H_5F_3N_2$
2-(Trifluoromethyl)-1*H*-benzimidazole, T-60287
$C_8H_5F_3O_2$
4-Hydroxy-2-(trifluoromethyl)benzaldehyde, H-70228
4-Hydroxy-3-(trifluoromethyl)benzaldehyde, H-70229
Phenyl trifluoroacetate, *in* T-80267
2,4,5-Trifluorobenzoic acid; Me ester, *in* T-70240
$C_8H_5F_5O$
1-(Pentafluorophenyl)ethanol, P-60040
C_8H_5I
▷ 1-Iodo-2-phenylacetylene, I-70055
$C_8H_5IO_2$
4-Iodo-1(3*H*)-isobenzofuranone, I-70044
7-Iodo-1(3*H*)-isobenzofuranone, I-70045
C_8H_5NO
3*H*-Indol-3-one, I-60035
$C_8H_5NOS_2$
2-Thiazolyl 2-thienyl ketone, T-80174
$C_8H_5NO_2$
Furo[2,3-*b*]pyridine-2-carboxaldehyde, F-60093
Furo[2,3-*b*]pyridine-3-carboxaldehyde, F-60094
Furo[3,2-*b*]pyridine-2-carboxaldehyde, F-60095
Furo[2,3-*c*]pyridine-2-carboxaldehyde, F-60096
Furo[3,2-*c*]pyridine-2-carboxaldehyde, F-60097
3-Imino-1(3*H*)-isobenzofuranone, I-80009
Isatogen, *in* I-60035
$C_8H_5NO_2S$
2*H*-1,3-Benzothiazine-2,4(3*H*)-dione, B-80041
2*H*-1,4-Benzothiazine-2,3(4*H*)-dione, B-70054
2*H*-3,1-Benzothiazine-2,4(1*H*)-dione, B-80042
1,2-Dihydro-2-thioxo-4*H*-3,1-benzoxazin-4-one, D-80261
2-Furyl 2-thiazolyl ketone, F-80098
$C_8H_5NO_3$
Furo[2,3-*b*]pyridine-2-carboxylic acid, F-60098
Furo[2,3-*b*]pyridine-3-carboxylic acid, F-60099
Furo[3,2-*b*]pyridine-2-carboxylic acid, F-60100
Furo[3,2-*b*]pyridine-3-carboxylic acid, F-60101
Furo[2,3-*c*]pyridine-2-carboxylic acid, F-60102
Furo[2,3-*c*]pyridine-3-carboxylic acid, F-60103
Furo[3,2-*c*]pyridine-2-carboxylic acid, F-60104
Furo[3,2-*c*]pyridine-3-carboxylic acid, F-60105
$C_8H_5NO_6$
2-Nitro-1,3-benzenedicarboxylic acid, N-80035
$C_8H_5NO_7$
2-Hydroxy-5-nitro-1,3-benzenedicarboxylic acid, H-70188
3-Hydroxy-6-nitro-1,2-benzenedicarboxylic acid, H-70189
4-Hydroxy-3-nitro-1,2-benzenedicarboxylic acid, H-70190
4-Hydroxy-5-nitro-1,2-benzenedicarboxylic acid, H-70191
4-Hydroxy-5-nitro-1,3-benzenedicarboxylic acid, H-70192
$C_8H_5NS_3$
3-Phenyl-1,4,2-dithiazole-5-thione, P-60088
$C_8H_5N_2O$
3-Hydroxyiminoindole, *in* I-60035
$C_8H_5N_3O_2$
5-Nitrophthalazine, N-70062
$C_8H_5N_3O_8$
2,4,6-Trinitrobenzoic acid; Me ester, *in* T-70303
$C_8H_5N_3O_9$
3-Hydroxy-5-methyl-2,4,6-trinitrobenzoic acid, H-60193
$C_8H_5N_3S$
1-Thia-5,8,8*b*-triazaacenaphthylene, T-80169
$C_8H_5N_5$
Tetrazolo[5,1-*a*]phthalazine, T-70155
$C_8H_5N_5O_9$
2,3,4,6-Tetranitroacetanilide, *in* T-60159
C_8H_6
Cubene, C-80160
1,3,5-Cyclooctatrien-7-yne, C-70240
$C_8H_6BrFO_2$
2-Bromo-4-fluorobenzoic acid; Me ester, *in* B-60240
$C_8H_6BrNO_4$
4-(Bromomethyl)-2-nitrobenzoic acid, B-60297
$C_8H_6Br_2I_2O_2$
1,2-Dibromo-3,6-diiodo-4,5-dimethoxybenzene, *in* D-80090
C_8H_6ClNS
2-Chloro-6-(methylthio)benzonitrile, *in* C-70082
$C_8H_6Cl_2O_2$
2,6-Dichloro-4-methoxybenzaldehyde, *in* D-70121
$C_8H_6Cl_2O_3$
2,3-Dichloro-4-methoxybenzoic acid, *in* D-60133
$C_8H_6FIO_2$
3-Fluoro-2-iodobenzoic acid; Me ester, *in* F-70021
C_8H_6FN
3-Fluoro-1*H*-indole, F-80043
4-Fluoro-1*H*-indole, F-80044
5-Fluoro-1*H*-indole, F-80045
6-Fluoro-1*H*-indole, F-80046
7-Fluoro-1*H*-indole, F-80047
C_8H_6FNS
2-Fluoro-6-(methylthio)benzonitrile, *in* F-70030
$C_8H_6FN_3O_5$
6-Fluoro-3,4-dinitroaniline; *N*-Ac, *in* F-60032
$C_8H_6F_3NO_3$
1-Methoxy-2-nitro-4-(trifluoromethyl)benzene, *in* N-60048
1-Methoxy-3-nitro-5-(trifluoromethyl)benzene, *in* N-60052
1-Methoxy-4-nitro-2-(trifluoromethyl)benzene, *in* N-60053
4-Methoxy-1-nitro-2-(trifluoromethyl)benzene, *in* N-60054
C_8H_6IN
7-Iodo-1*H*-indole, I-70043
$C_8H_6I_2$
1,4-Diiodopentacyclo[4.2.0.0^{2,5}.0^{3,8}.0^{4,7}]octane, D-70359
$C_8H_6N_2O$
2-(2-Oxazolyl)pyridine, O-70071
2-(5-Oxazolyl)pyridine, O-70072
3-(2-Oxazolyl)pyridine, O-70073

C_8H_7NO
6-Hydroxyindole, H-60158
1(3*H*)-Isobenzofuranimine, I-70062
(Isocyanatomethyl)benzene, I-80055
2-Methylfuro[2,3-*b*]pyridine, M-60084
3-Methylfuro[2,3-*b*]pyridine, M-60085
Phthalimidine, P-70100

C_8H_7NOS
2*H*-1,4-Benzothiazin-3(4*H*)-one, B-80043

$C_8H_7NO_2$
5,6-Dihydroxyindole, D-70307

$C_8H_7NO_2S$
1-Nitro-2-(phenylthio)ethylene, N-70061

$C_8H_7NO_2S_2$
2*H*-1,4-Benzothiazine-3(4*H*)-thione; 1,1-Dioxide, *in* B-70055

$C_8H_7NO_3$
Amino-1,4-benzoquinone; *N*-Ac, *in* A-60089
2-Methyl-4-nitrobenzaldehyde, M-70096
(2-Nitrophenyl)oxirane, N-70056
(3-Nitrophenyl)oxirane, N-70057

$C_8H_7NO_3S$
1-Nitro-2-(phenylsulfinyl)ethylene, *in* N-70061

$C_8H_7NO_4$
2,4-Dihydroxy-2*H*-1,4-benzoxazin-3(4*H*)-one, *in* D-10388
2-Hydroxy-4,5-methylenedioxybenzoic acid; Amide, *in* H-80202
2-Hydroxy-5-methyl-3-nitrobenzaldehyde, H-70180
4-(2-Nitroethenyl)-1,2-benzenediol, N-70049

$C_8H_7NO_4S$
1-Nitro-2-(phenylsulfonyl)ethylene, *in* N-70061

$C_8H_7NO_5$
3-Amino-4-hydroxy-1,2-benzenedicarboxylic acid, A-70151
4-Amino-5-hydroxy-1,2-benzenedicarboxylic acid, A-70152
4-Amino-6-hydroxy-1,3-benzenedicarboxylic acid, A-70153
5-Amino-2-hydroxy-1,3-benzenedicarboxylic acid, A-70154
5-Amino-4-hydroxy-1,3-benzenedicarboxylic acid, A-70155
2′,3′-Dihydroxy-6′-nitroacetophenone, D-60357
2′,5′-Dihydroxy-4′-nitroacetophenone, D-70322
3′,6′-Dihydroxy-2′-nitroacetophenone, D-60358
4′,5′-Dihydroxy-2′-nitroacetophenone, D-60359

$C_8H_7NO_6$
4-Hydroxy-5-methoxy-2-nitrobenzoic acid, *in* D-60360
5-Hydroxy-4-methoxy-2-nitrobenzoic acid, *in* D-60360

C_8H_7NS
2*H*-1,4-Benzothiazine, B-70053

$C_8H_7NS_2$
2*H*-1,4-Benzothiazine-3(4*H*)-thione, B-70055

$C_8H_7N_3$
1-Aminophthalazine, A-70184
5-Aminophthalazine, A-70185
4-Methyl-1,2,3-benzotriazine, M-80052
2-(1*H*-Pyrazol-1-yl)pyridine, P-70158
2-(1*H*-Pyrazol-3-yl)pyridine, P-70159
3-(1*H*-Pyrazol-1-yl)pyridine, P-70160
3-(1*H*-Pyrazol-3-yl)pyridine, P-70161
4-(1*H*-Pyrazol-1-yl)pyridine, P-70162
4-(1*H*-Pyrazol-3-yl)pyridine, P-70163
4-(1*H*-Pyrazol-4-yl)pyridine, P-70164

$C_8H_7N_3O$
2′-Azidoacetophenone, A-60334
3′-Azidoacetophenone, A-60335
4′-Azidoacetophenone, A-60336
2-Azido-3-methylbenzaldehyde, A-80201
2-Azido-5-methylbenzaldehyde, A-80202
1*H*-Benzotriazole; *N*-Ac, *in* B-70068
1,2-Dihydro-5-phenyl-3*H*-1,2,4-triazol-3-one, D-80241
Furo[2,3-*b*]pyridine-3-carboxaldehyde; Hydrazone, *in* F-60094
4-Methyl-1,2,3-benzotriazine; 2-Oxide, *in* M-80052
4-Methyl-1,2,3-benzotriazine; 3-Oxide, *in* M-80052

$C_8H_7N_3O_2$
6-Amino-2,3-dihydro-1,4-phthalazinedione, A-70134

$C_8H_7N_3O_4$
2-Nitro-1,3-benzenedicarboxylic acid; Diamide, *in* N-80035

$C_8H_7N_3S$
1,2,3-Benzotriazine-4(3*H*)-thione; 3-*N*-Me, *in* B-60050
1,2,3-Benzotriazine-4(3*H*)-thione; *S*-Me, *in* B-60050
4-Mercapto-2-methyl-1,2,3-benzotriazinium hydroxide inner salt, *in* B-60050

$C_8H_7N_5O_2$
2-Amino-4(1*H*)-pteridinone; N^2-Ac, *in* A-60242

$C_8H_7N_5O_8$
2,3,4,6-Tetranitroaniline; *N*-Di-Me, *in* T-60159

$C_8H_7OS^{\oplus}$
2-Methyl-1,3-benzoxathiazolium(1+), M-80053

C_8H_8
1,2-Dihydropentalene, D-70240
1,4-Dihydropentalene, D-70241
1,5-Dihydropentalene, D-70242
1,6-Dihydropentalene, D-70243
1,6*a*-Dihydropentalene, D-70244
5,6-Dimethylene-1,3-cyclohexadiene, D-70410
4-Methylene-1,2,5,6-heptatetraene, M-70070
4-Octene-1,7-diyne, O-60028
Pentacyclo[5.1.0.0^{2,4}.0^{3,5}.0^{6,8}]octane, P-80020

$C_8H_8BrNO_2$
3-Amino-2-bromobenzoic acid; Me ester, *in* A-70097

$C_8H_8BrNO_3$
2-Bromo-1-ethoxy-3-nitrobenzene, *in* B-70257

$C_8H_8Br_2O_4$
4-Acetoxy-2,6-dibromo-5-hydroxy-2-cyclohexen-1-one, *in* D-60097
2,6-Dibromo-4,5-dihydroxy-2-cyclohexen-1-one; 4-Ac, *in* D-60097

C_8H_8ClN
4-Chloro-2,3-dihydro-1*H*-indole, C-60055

$C_8H_8ClNO_3$
4-Chloro-2-ethoxy-1-nitrobenzene, *in* C-60119
1-Chloro-5-methoxy-2-methyl-3-nitrobenzene, *in* C-60116
2-Chloro-1-methoxy-3-methyl-4-nitrobenzene, *in* C-60115

$C_8H_8Cl_6O_4$
Hexahydro-2,5-bis(trichloromethyl)furo[3,2-*b*]furan-3*a*,6*a*-diol, H-70053

$C_8H_8FNO_2$
2-Amino-5-fluorobenzoic acid; Me ester, *in* A-70142
3-Amino-4-fluorobenzoic acid; Me ester, *in* A-70144
5-Amino-2-fluorobenzoic acid; Me ester, *in* A-70147

C_8H_8INO
2-Iodoacetophenone; Oxime (*Z*-), *in* I-60039

$C_8H_8INO_2$
3-Hydroxy-2-iodo-6-methylpyridine; *O*-Ac, *in* H-60159

$C_8H_8N_2$
1,1′-Dicyanobicyclopropyl, *in* B-70114
2-Methyl-1*H*-pyrrolo[2,3-*b*]pyridine, M-80156
2-Methyl-1*H*-pyrrolo[3,2-*b*]pyridine, M-80157
2-Methyl-1*H*-pyrrolo[2,3-*c*]pyridine, M-80158
2-Methyl-1*H*-pyrrolo[3,2-*c*]pyridine, M-80159
6*H*-Pyrrolo[3,4-*b*]pyridine; 6-Me, *in* P-70201
2*H*-Pyrrolo[3,4-*c*]pyridine; 2-Me, *in* P-70200

$C_8H_8N_2O$
▷ 1-(Diazomethyl)-4-methoxybenzene, D-60064
1,6-Dihydro-7*H*-pyrrolo[2,3-*c*]pyridin-7-one; 6-Me, *in* D-80252
3,4-Dihydro-2(1*H*)-quinazolinone, D-60278
Pyrrolo[1,2-*a*]pyrazin-1(2*H*)-one; *N*-Me, *in* P-70198

$C_8H_8N_2O_2$
Bicyclo[2.2.2]octa-5,7-diene-2,3-dione; Dioxime, *in* B-70109

$C_8H_8N_2O_2S$
Benzo[*b*]thiophen-3(2*H*)-one; Hydrazone, 1,1-dioxide, *in* B-60048

$C_8H_8N_2O_2S_2$
3-Mercapto-6(1*H*)-pyridazinethione; *S,S*-Di-Ac, *in* M-80028

$C_8H_8N_2O_3$
Nicotinuric acid, N-70034

$C_8H_8N_2O_4$
2-Hydroxy-5-methyl-3-nitrobenzaldehyde; Oxime, *in* H-70180
3,6-Pyridazinedicarboxylic acid; Di-Me ester, *in* P-70168

$C_8H_8N_2O_4S_2$
1,1′-Dithiobis-2,5-pyrrolidinedione, D-70524

$C_8H_8N_2O_6$
1,2-Dimethoxy-4,5-dinitrobenzene, *in* D-70449
1,5-Dimethoxy-2,4-dinitrobenzene, *in* D-80491
5-Ethoxy-2,4-dinitrophenol, *in* D-80491
Tartaric acid; Dinitrile, Di-Ac, *in* T-70005

$C_8H_8N_2S$
2-Amino-4*H*-3,1-benzothiazine, A-70092

$C_8H_8N_4$
2,4-Diaminoquinazoline, D-60048
▷ 1-Hydrazinophthalazine, H-60092

$C_8H_8N_4O$
2-Tetrazolin-5-one; 1-Benzyl, *in* T-60175

$C_8H_8N_4O_2$
2,4(1*H*,3*H*)-Pteridinedione; 1,3-Di-Me, *in* P-60191

$C_8H_8N_4O_3$
2,4,6(1*H*,3*H*,5*H*)-Pteridinetrione; 1,3-Di-Me, *in* P-70134
2,4,6(1*H*,3*H*,5*H*)-Pteridinetrione; 1,7-Di-Me, *in* P-70134
2,4,6(1*H*,3*H*,5*H*)-Pteridinetrione; 3,7-Di-Me, *in* P-70134
2,4,7(1*H*,3*H*,8*H*)-Pteridinetrione; 1,3-Di-Me, *in* P-70135
2,4,7(1*H*,3*H*,8*H*)-Pteridinetrione; 1,6-Di-Me, *in* P-70135
2,4,7(1*H*,3*H*,8*H*)-Pteridinetrione; 1,8-Di-Me, *in* P-70135
2,4,7(1*H*,3*H*,8*H*)-Pteridinetrione; 3,6-Di-Me, *in* P-70135
2,4,7(1*H*,3*H*,8*H*)-Pteridinetrione; 3,8-Di-Me, *in* P-70135
2,4,7(1*H*,3*H*,8*H*)-Pteridinetrione; 6,8-Di-Me, *in* P-70135

$C_8H_8N_4O_6$
2,4,6-Trinitroaniline; *N*-Di-Me, *in* T-70301

$C_8H_8N_4S$
1*H*-Tetrazole-5-thiol; 1-*N*-Ph, *S*-Me, *in* T-60174
Tetrazole-5-thione; 1-Me, 4-Ph, *in* T-60174

C_8H_8O
▷ Acetophenone, A-60017
Bicyclo[2.2.1]hepta-2,5-diene-2-carboxaldehyde, B-60083
Bicyclo[5.1.0]octa-3,5-dien-2-one, B-70111
4,7-Dihydroisobenzofuran, D-60253
▷ (4-Hydroxyphenyl)ethylene, H-70209
9-Oxabicyclo[4.2.1]nona-2,4,7-triene, O-70057
1-Phenylethenol, P-60092

C_8H_8OS

Benzeneethanethioic acid, B-70017
4'-Mercaptoacetophenone, M-60017
2-(Methylthio)benzaldehyde, *in* M-80018
4-(Methylthio)benzaldehyde, *in* M-80020
(Phenylthio)acetaldehyde, P-80136

$C_8H_8OS_3$

1,5-Dihydro-2,3,4-benzotrithiepin; 2-Oxide, *in* D-80176

$C_8H_8O_2$

Bicyclo[2.2.1]hepta-2,5-diene-2-carboxylic acid, B-60084
Bicyclo[2.2.2]oct-7-ene-2,5-dione, B-80086
Bicyclo[4.2.0]oct-7-ene-2,5-dione, B-60097
2-Hydroxy-4-methylbenzaldehyde, H-70160
2-(Hydroxymethyl)benzaldehyde, H-60171
3-Hydroxy-5-methylbenzaldehyde, H-80195
4-Hydroxy-2-methylbenzaldehyde, H-70161
2-Hydroxy-2-phenylacetaldehyde, H-60210
2-Hydroxyphenylacetaldehyde, H-60207
3-Hydroxyphenylacetaldehyde, H-60208
4-Hydroxyphenylacetaldehyde, H-60209
4-Vinyl-1,3-benzenediol, V-70010

$C_8H_8O_2S$

2-Mercapto-2-phenylacetic acid, M-60024

$C_8H_8O_3$

2,4-Dihydroxyphenylacetaldehyde, D-60369
2,5-Dihydroxyphenylacetaldehyde, D-60370
3,4-Dihydroxyphenylacetaldehyde, D-60371
7,10-Dioxaspiro[5.4]deca-2,5-dien-4-one, *in* B-70044
▷ 2-Hydroxy-3-methoxybenzaldehyde, *in* D-70285
3-Hydroxy-2-methoxybenzaldehyde, *in* D-70285
2-Hydroxy-4-methylbenzoic acid, H-70163
4-Hydroxy-2-methylbenzoic acid, H-70164

$C_8H_8O_4$

4,5-Dihydroxy-2-methylbenzoic acid, D-80339
3,6-Dimethoxy-1,2-benzoquinone, *in* D-60309
2-Hydroxy-3-methoxy-5-methyl-1,4-benzoquinone, *in* D-70320
3-Hydroxy-2-methoxy-5-methyl-1,4-benzoquinone, *in* D-70320
(2-Hydroxyphenoxy)acetic acid, H-70203
(3-Hydroxyphenoxy)acetic acid, H-70204
(4-Hydroxyphenoxy)acetic acid, H-70205
2',4',6'-Trihydroxyacetophenone, T-70248
▷ 3',4',5'-Trihydroxyacetophenone, T-60315

$C_8H_8O_5$

4,5-Dimethyl-2,3-furandicarboxylic acid, D-80449
3-Hydroxy-4-methoxy-1-methyl-7-oxabicyclo[4.1.0]hept-3-ene-2,5-dione, *in* D-70320
2,3,5-Trihydroxybenzoic acid; Me ester, *in* T-70250
3,4,6-Trihydroxy-2-methylbenzoic acid, T-80324
2,3,4-Trihydroxyphenylacetic acid, T-60340
2,3,5-Trihydroxyphenylacetic acid, T-60341
2,4,5-Trihydroxyphenylacetic acid, T-60342
3,4,5-Trihydroxyphenylacetic acid, T-60343

$C_8H_8O_6$

2,3-Diacetyl-2-butenedioic acid, D-80042

$C_8H_8O_7$

3,4-Diacetoxy-3,4-dihydro-2,5(2*H*,5*H*)-furandione, *in* T-70005

$C_8H_8S_2$

2-(2,4-Cyclopentadien-1-ylidene)-1,3-dithiolane, *in* C-60221
1,4-Dihydro-2,3-benzodithiin, D-80175

$C_8H_8S_3$

1,5-Dihydro-2,3,4-benzotrithiepin, D-80176

$C_8H_9BrO_3$

3-Bromo-2,6-dimethoxyphenol, *in* B-60200

$C_8H_9BrO_3S$

Bromomethanol; 4-Methylbenzenesulfonyl, *in* B-80212

$C_8H_9BrO_5$

(3-Bromo-2-oxopropylidene)propanedioic acid; Di-Me ester, *in* B-70262

C_8H_9ClO

1-(Chloromethyl)-2-methoxybenzene, *in* C-70101
1-(Chloromethyl)-3-methoxybenzene, *in* C-70102
1-Chloromethyl-4-methoxybenzene, *in* C-70103
2-Chloro-1-phenylethanol, C-80099
2-Chloro-2-phenylethanol, C-60128

$C_8H_9ClO_3S$

Chloromethanol; 4-Methylbenzenesulfonyl, *in* C-80070

C_8H_9F

2-Fluoro-1,3-dimethylbenzene, F-60026
1-Fluoro-2-phenylethane, F-80059

C_8H_9FO

1-Fluoro-2-methoxy-4-methylbenzene, *in* F-60054
1-Fluoro-3-methoxy-6-methylbenzene, *in* F-60057
1-Fluoro-4-methoxy-3-methylbenzene, *in* F-60059
2-Fluoro-1-methoxy-4-methylbenzene, *in* F-60053

$C_8H_9FO_2$

1-Fluoro-2,4-dimethoxybenzene, *in* F-70018
4-Fluoro-1,2-dimethoxybenzene, *in* F-60020

$C_8H_9IO_3S$

Iodomethanol; 4-Methylbenzenesulfonyl, *in* I-80033

C_8H_9N

N-Benzylidenemethylamine, B-60070
2-Cyanobicyclo[2.2.1]hepta-2,5-diene, *in* B-60084
6,7-Dihydro-5*H*-1-pyrindine, D-60275
6,7-Dihydro-5*H*-2-pyrindine, D-60276
3-(1-Propenyl)pyridine, P-80181

C_8H_9NO

Acetophenone; (*E*)-Oxime, *in* A-60017
2,3-Dihydro-5(1*H*)-indolizinone, D-60251
3,4-Dihydro-2*H*-pyrano[2,3-*b*]pyridine, D-80244
6,7-Dihydro-5*H*-1-pyrindine; *N*-Oxide, *in* D-60275
N-Methylbenzenecarboximidic acid, M-60045
1-(2-Pyridinyl)-2-propanone, P-60229
1-(3-Pyridinyl)-2-propanone, P-60230
1-(4-Pyridinyl)-2-propanone, P-60231

C_8H_9NOS

(Phenylthio)acetaldehyde; Oxime, *in* P-80136

$C_8H_9NO_2$

2-Hydroxy-4-methylbenzaldehyde; Oxime, *in* H-70160
4-Hydroxy-2-methylbenzaldehyde; Oxime, *in* H-70161
4-Hydroxyphenylacetaldoxime, *in* H-60209
2-(Methoxyacetyl)pyridine, *in* H-70101
Phenylacetohydroxamic acid, P-70059
3*a*,4,7,7*a*-Tetrahydro-1*H*-isoindole-1,3(2*H*)-dione, T-80070

$C_8H_9NO_2S$

1-Nitro-2-(phenylthio)ethane, N-70060

$C_8H_9NO_3$

3-Acetoxy-2-methyl-4(1*H*)-pyridinone, *in* H-80214
C-Benzyloxycarbohydroxamic acid, B-70086
1,4-Dihydro-6-methyl-4-oxo-3-pyridinecarboxylic acid; 1-Me, *in* D-80224
2-Formyl-1*H*-pyrrole-3-carboxylic acid; Et ester, *in* F-80077
4-Formyl-1*H*-pyrrole-2-carboxylic acid; Et ester, *in* F-80078
5-Formyl-1*H*-pyrrole-2-carboxylic acid; Et ester, *in* F-80079
5-Formyl-1*H*-pyrrole-2-carboxylic acid; 1-Me, Me ester, *in* F-80079
(2-Hydroxyphenoxy)acetic acid; Amide, *in* H-70203

4-(2-Nitroethyl)phenol, N-70050
1*H*-Pyrrole-3-carboxaldehyde; *N*-Ethoxycarbonyl, *in* P-70182

$C_8H_9NO_3S$

4-Carboxy-3-thiopheneacetic acid; *N*-Methylacetamide, *in* C-60018
2,3-Dihydro-1*H*-indole-2-sulfonic acid, D-70221
1-Nitro-2-(phenylsulfinyl)ethane, *in* N-70060

$C_8H_9NO_4$

3-Methoxy-2-methyl-6-nitrophenol, *in* M-70097
1*H*-Pyrrole-3,4-dicarboxylic acid; Di-Me ester, *in* P-70183

$C_8H_9NO_5$

3-Hydroxy-1*H*-pyrrole-2,5-dicarboxylic acid; Me ether, 2-Me ester, *in* H-80252
1*H*-Pyrrole-3,4-dicarboxylic acid; 1-Hydroxy, Di-Me ether, *in* P-70183

C_8H_9NS

Phenylthiolacetamide, *in* B-70017

$C_8H_9N_3$

1*H*-Benzotriazole; 1-Et, B-70068

$C_8H_9N_5$

5-Amino-2-benzyl-2*H*-tetrazole, *in* A-70202
1-(Phenylmethyl)-1*H*-tetrazol-5-amine, *in* A-70202
N-(Phenylmethyl)-1*H*-tetrazol-5-amine, *in* A-70202

$C_8H_9N_5O_2$

N-(2,9-Dihydro-1-methyl-2-oxo-1*H*-purin-6-yl)acetamide, *in* A-60144

C_8H_{10}

Bicyclo[4.1.1]octa-2,4-diene, B-60090
Bicyclo[4.2.0]octa-2,4-diene, B-80085
1,3,5-Cyclooctatriene, C-80188
1-Ethynylcyclohexene, E-70051
5-Methyl-5-vinyl-1,3-cyclopentadiene, M-80146
Tetracyclo[3.3.0.0^{2,8}.0^{3,6}]octane, T-80025
Tetracyclo[3.3.0.0^{2,8}.0^{4,6}]octane, T-70022
Tetracyclo[4.2.0.0^{2,5}.0^{3,8}]octane, T-70023
1,2,3,4-Tetrahydropentalene, T-70076
1,2,3,5-Tetrahydropentalene, T-70077
1,2,4,5-Tetrahydropentalene, T-70078
1,2,3,3*a*-Tetrahydropentalene, T-70075
1,2,4,6*a*-Tetrahydropentalene, T-70079
1,2,6,6*a*-Tetrahydropentalene, T-70080
1,3*a*,4,6*a*-Tetrahydropentalene, T-70081
1,3*a*,6,6*a*-Tetrahydropentalene, T-70082
Tricyclo[3.2.1.0^{2,7}]oct-3-ene, T-60267
Tricyclo[5.1.0.0^{2,8}]oct-3-ene, T-60268
Tricyclo[5.1.0.0^{2,8}]oct-4-ene, T-60269

$C_8H_{10}Br_2O_4$

4-Acetoxy-2,6-dibromo-1,5-dihydroxy-2-cyclohexen-1-one, *in* D-60097

$C_8H_{10}ClN$

2-Chloro-3,5,6-trimethylpyridine, C-80121
3-Chloro-2,4,6-trimethylpyridine, C-80122
4-Chloro-2,3,5-trimethylpyridine, C-80123

$C_8H_{10}ClNO$

4-Chloro-2,3,5-trimethylpyridine; N-Oxide, *in* C-80123

$C_8H_{10}F_2O_4$

Difluoromaleic acid; Di-Et ester, *in* D-80159

$C_8H_{10}F_4O_4$

Tetrafluorobutanedioic acid; Di-Et ester, *in* T-80049

$C_8H_{10}IN$

2-Iodo-4,5-dimethylaniline, I-70028
4-Iodo-2,3-dimethylaniline, I-70029
4-Iodo-3,5-dimethylaniline, I-70030
4-Iodo-3,5-dimethylaniline, I-70031

$C_8H_{10}N_2$

Acetophenone; Hydrazone, *in* A-60017
1,4-Dihydropyrrolo[3,2-*b*]pyrrole; 1,4-Di-Me, *in* D-30286

$C_8H_{10}N_2O - C_8H_{12}N_2O_2$ **Molecular Formula Index**

$C_8H_{10}N_2O$

Acetamidoxime; *N*-Ph, *in* A-60016
1-(3-Pyridinyl)-2-propanone; Oxime, *in* P-60230

$C_8H_{10}N_2OS$

3-Amino-4(1*H*)-pyridinethione; *N*-Ac,*S*-Me, *in* A-80140
5-Amino-2(1*H*)-pyridinethione; *N*-Ac, *S*-Me, *in* A-80141

$C_8H_{10}N_2O_2$

3-Amino-4(1*H*)-pyridinone; 1-Me, *N*³-Ac, *in* A-60254

$C_8H_{10}N_2O_3$

5-Ethyl-2,4(1*H*,3*H*)-pyrimidinedione; 1-Ac, *in* E-60054
5-Methoxy-2-methyl-3-nitroaniline, *in* A-60222

$C_8H_{10}N_2O_3S$

Aminoiminomethanesulfonic acid; *N*-Benzyl, *in* A-60210

$C_8H_{10}N_2O_4$

3-(2-Aminoethylidene)-7-oxo-4-oxa-1-azabicyclo[3.2.0]heptane-2-carboxylic acid, A-70139
1*H*-Pyrazole-4,5-dicarboxylic acid; 1-Me, Di-Me ester, *in* P-80199
1,2,3,6-Tetrahydro-2,6-dioxo-4-pyrimidineacetic acid; Et ester, *in* T-70060

$C_8H_{10}N_4$

4-Amino-1*H*-imidazo[4,5-*c*]pyridine; 4-*N*-Di-Me, *in* A-70161

$C_8H_{10}N_4O_2$

1*H*-Pyrazolo[3,4-*d*]pyrimidine-4,6(5*H*,7*H*)-dione; 1,5,7-Tri-Me, *in* P-60202
1*H*-Pyrazolo[3,4-*d*]pyrimidine-4,6(5*H*,7*H*)-dione; 2,5,7-Tri-Me, *in* P-60202

$C_8H_{10}N_4O_5$

5-Nitrohistidine; *N*ᵅ-Ac, *in* N-60028

$C_8H_{10}O$

Bicyclo[3.2.1]oct-6-en-3-one, B-80087
Bicyclo[4.2.0]oct-2-en-7-one, B-80088
4,4-Dimethyl-2,5-cyclohexadien-1-one, D-80441
6,6-Dimethyl-2,4-cyclohexadien-1-one, D-80442
3,3*a*,6,6*a*-Tetrahydro-1(2*H*)-pentalenone, T-70083
4,5,6,6*a*-Tetrahydro-2(1*H*)-pentalenone, T-70084
Tricyclo[3.2.1.0²,⁴]octan-8-one, T-80243

$C_8H_{10}OS$

2-Mercaptobenzenemethanol; *S*-Me, *in* M-80021
4-Mercaptobenzenemethanol; *S*-Me, *in* M-80023
3-Mercaptobenzenemethanol; S-Me, *in* M-80022
Methyl 4-methylphenyl sulfoxide, M-70094

$C_8H_{10}O_2$

Bicyclo[4.2.0]octane-2,5-dione, B-60093
5,5-Dimethyl-2-cyclohexene-1,4-dione, D-70392
4,4-Dimethyl-3-oxo-1-cyclopentene-1-carboxaldehyde, D-60432
3-Ethynyl-3-methyl-4-pentenoic acid, E-60059
▷ 4-Methoxybenzyl alcohol, *in* H-80114
4-(Methoxymethyl)phenol, *in* H-80114
3*a*,4,7,7*a*-Tetrahydro-1(3*H*)-isobenzofuranone, T-70072
5-Vinyl-1-cyclopentenecarboxylic acid, V-60010

$C_8H_{10}O_2S$

4-Methyl-2-thiophenecarboxaldehyde; Ethylene acetal, *in* M-70141
1-(3-Thienyl)ethanol; Ac, *in* T-80179

$C_8H_{10}O_3$

2-Carbomethoxy-2-cyclohexen-1-one, *in* O-80079
Halleridone, H-60004

4-Hydroxy-3-methoxybenzenemethanol, *in* D-80284
4-Oxo-1-cycloheptene-1-carboxylic acid, O-70082
5-Oxo-1-cyclohexene-1-carboxylic acid; Me ester, *in* O-80077
5-Oxo-2-cyclohexene-1-carboxylic acid; Me ester, *in* O-80078

$C_8H_{10}O_3S$

1-Phenylethanesulfonic acid, P-70068

$C_8H_{10}O_4$

Bicyclo[1.1.1]pentane-1,3-dicarboxylic acid; Mono-Me ester, *in* B-80090
[1,1'-Bicyclopropyl]-1,1'-dicarboxylic acid, B-70114
2,3-Dicarbomethoxy-1,3-butadiene, *in* D-70409
3,6-Dimethoxy-1,2-benzenediol, *in* B-60014
5,5-Dimethyl-2-hexynedioic acid, D-70415
3,4-Epoxy-5-hydroxy-1-cyclohexenecarboxylic acid; Me ester, *in* E-80030
4-Hydroxy-2,5-dimethyl-3(2*H*)-furanone; Ac, *in* H-60120
3-Methoxy-6-methyl-1,2,4-benzenetriol, *in* M-80050
5-Methoxy-3-methyl-1,2,4-benzenetriol, *in* M-80048
6-Methoxy-3-methyl-1,2,4-benzenetriol, *in* M-80049
6-Methoxy-5-methyl-1,2,4-benzenetriol, *in* M-80049

$C_8H_{10}O_5$

5-Acetyl-2,2-dimethyl-1,3-dioxane-4,6-dione, A-60029
Diallyl dicarbonate, *in* D-70111
(2-Oxopropylidene)propanedioic acid; Di-Me ester, *in* O-70100

$C_8H_{10}O_8$

2,3-Di-*O*-acetyltartaric acid, *in* T-70005

$C_8H_{10}S$

2-Phenylethanethiol, P-70069

$C_8H_{10}S_2$

2,5-Dimethyl-1,4-benzenedithiol, D-60406

$C_8H_{11}BrO_2$

2-Bromobicyclo[2.2.1]heptane-1-carboxylic acid, B-60210
7-Bromo-5-heptynoic acid; Me ester, *in* B-60261

$C_8H_{11}BrO_3$

(3-Bromomethyl)-2,4,10-trioxatricyclo[3.3.1.1³,⁷]decane, B-60301

$C_8H_{11}ClO_4$

Diethyl chlorofumarate, *in* C-80043

$C_8H_{11}ClO_5$

2,3-Dihydroxybutanoic acid; Di-Ac, chloride, *in* D-60312

$C_8H_{11}Cl_2N$

2,2-Dichloro-7-octenoic acid; Nitrile, *in* D-80123

$C_8H_{11}N$

1-Cyanocycloheptene, *in* C-80178

$C_8H_{11}NO$

Hexahydroazirino[2,3,1-*hi*]indol-2(1*H*)-one, H-70046
1,3,3*a*,4,5,7*a*-Hexahydro-2*H*-indol-2-one, H-70060
2-Methylene-1-azabicyclo[2.2.2]octan-3-one, M-80074
1,5,6,8*a*-Tetrahydro-3(2*H*)-indolizinone, T-70071
5,6,7,8-Tetrahydro-2(3*H*)-indolizinone, T-80068

$C_8H_{11}NOS$

N,S-Dimethyl-*S*-phenylsulfoximine, *in* M-80148

$C_8H_{11}NO_2$

3-Amino-2-cyclohexen-1-one; *N*-Ac, *in* A-70125
2,3-Dimethoxyaniline, *in* A-70089

▷ 3,4-Dimethoxyaniline, *in* A-70090
5-Methyl-1*H*-pyrrole-3-carboxylic acid; Et ester, *in* M-60125

$C_8H_{11}NO_3$

2,5-Dimethyl-4-oxazolecarboxylic acid; Et ester, *in* D-80461
3-Hydroxy-1*H*-pyrrole-2-carboxylic acid; Me ether, Et ester, *in* H-80248
4-Hydroxy-1*H*-pyrrole-2-carboxylic acid; Me ether, Et ester, *in* H-80249
4-Hydroxy-1*H*-pyrrole-3-carboxylic acid; Me ether, Et ester, *in* H-80250
4-Oxoheptanedioic acid; Me ester, mononitrile, *in* O-60071
2,4,6-Trimethoxypyridine, *in* D-70341

$C_8H_{11}NO_4$

5-Oxo-2-pyrrolidinecarboxylic acid; Me ester, *N*-Ac, *in* O-70102
5-Oxo-2-pyrrolidinecarboxylic acid; Me ester, *N*-benzoyl, *in* O-70102

$C_8H_{11}N_3$

2-Amino-5-vinylpyrimidine; *N*²,*N*²-Di-Me, *in* A-60269

$C_8H_{11}N_3O_3$

α-Amino-1*H*-imidazole-2-propanoic acid; *N*ᵅ-Ac, *in* A-60205
5-Amino-1*H*-pyrazole-4-carboxylic acid; *N*(1)-Ac, Et ester, *in* A-70192
3-Amino-1*H*-pyrazole-4-carboxylic acid; Et ester, *N*(3)-Ac, *in* A-70187
1,2,3,6-Tetrahydro-2,6-dioxo-4-pyrimidineacetic acid; *N*-Ethylamide, *in* T-70060

C_8H_{12}

Bicyclobutylidene, B-70095
Bicyclo[4.1.1]oct-2-ene, B-60095
Bicyclo[4.1.1]oct-3-ene, B-60096
1,2-Bis(methylene)cyclohexane, B-60166
1,3-Bis(methylene)cyclohexane, B-60167
1,4-Bis(methylene)cyclohexane, B-60168
3,3-Dimethyl-1,4-cyclohexadiene, D-80440
2,5-Dimethyl-1,3,5-hexatriene, D-60424
1,2,3,4,5,6-Hexahydropentalene, H-70066
1,2,3,3*a*,4,5-Hexahydropentalene, H-70064
1,2,3,3*a*,4,6*a*-Hexahydropentalene, H-70065
3-Methyl-1-vinylcyclopentene, M-80189
1,2,4-Octatriene, O-80019
1,3,5-Octatriene, O-70022
2,4,6-Octatriene, O-70023
3-Octen-1-yne, O-70029
Tricyclo[3.1.1.1²,⁴]octane, T-60265
Tricyclo[5.1.0.0²,⁸]octane, T-60266
3-Vinylcyclohexene, V-60009

$C_8H_{12}Br_2$

1,4-Dibromobicyclo[2.2.2]octane, D-70086

$C_8H_{12}Br_2O_3$

2-Bromo-2-methylpropanoic acid; Anhydride, *in* B-80229

$C_8H_{12}Cl_2$

1,4-Dichlorobicyclo[2.2.2]octane, D-70115

$C_8H_{12}Cl_2N_2O_2$

4,5-Dichloro-1*H*-imidazole-2-carboxaldehyde; Di-Et acetal, *in* D-60134

$C_8H_{12}Cl_2O_2$

2,2-Dichloro-7-octenoic acid, D-80123

$C_8H_{12}I_2$

1,4-Diiodobicyclo[2.2.2]octane, D-80397

$C_8H_{12}N_2$

3,4,5,6-Tetramethylpyridazine, T-80150

$C_8H_{12}N_2O$

2-*tert*-Butyl-4(3*H*)-pyrimidinone, B-60349

$C_8H_{12}N_2O_2$

[1,1'-Bipyrrolidine]-2,2'-dione, B-70125
Cyclo(prolylalanyl), C-80194
▷ 1,3-Diazaspiro[4.5]decane-2,4-dione, D-60056
4,4-Dimethyl-6-nitro-5-hexenenitrile, *in* D-60429

3-Ethoxy-4,5,6,7-tetrahydroisoxazolo[4,5-
c]pyridine, in T-60071
Hexahydropyridazino[1,2-a]pyridazine-1,4-
dione, H-70070

$C_8H_{12}N_2S_2$
2,3-Bis(methylthio)-1,4-benzenediamine, in
D-60030
2,3-Bis(methylthio)-1,4-benzenediamine, in
D-70036
4,6-Diamino-1,3-benzenedithiol; Di-S-Me, in
D-70037

$C_8H_{12}N_4O_3$
2',3'-Dideoxy-5-azacytidine, in A-80149

$C_8H_{12}N_4O_4$
2'-Deoxy-5-azacytidine, D-70022

$C_8H_{12}O$
1-Acetylcyclohexene, A-70038
4-Acetylcyclohexene, A-70039
1-Acetyl-2-methylcyclopentene, A-60043
Bicyclo[3.2.1]octan-8-one, B-60094
3-tert-Butyl-2-cyclobuten-1-one, B-80283
2-Isopropyl-5-methylfuran, I-80083
3-Methoxy-1-vinylcyclopentene, in V-80008
9-Oxabicyclo[6.1.0]non-4-ene, O-70058
2-(2-Propenyl)cyclopentanone, P-80178
2,5,5-Trimethyl-2-cyclopenten-1-one, T-70281

$C_8H_{12}O_2$
Bicyclo[2.2.1]heptane-1-carboxylic acid,
B-60085
Bicyclo[3.1.1]heptane-1-carboxylic acid,
B-60086
1-Cycloheptenecarboxylic acid, C-80178
5,6-Dimethoxy-1,3-cyclohexadiene, D-60397
2,2-Dimethyl-1,3-cyclohexanedione, D-80443
1,6-Dioxacyclodeca-3,8-diene, D-70458
1,4-Dioxaspiro[4.5]dec-7-ene, in C-60211
2,4-Dioxatricyclo[3.3.1.1^{3,7}]decane, D-80503
6-Ethyl-2,3-dihydro-2-methyl-4H-pyran-4-one,
E-50071
6-Heptynoic acid; Me ester, in H-80033
3,5-Hexadienoic acid; Et ester, in H-70038
Hexahydro-2(3H)-benzofuranone, H-70049
Hexahydro-1(3H)-isobenzofuranone, H-70061
1-Methoxy-3-cyclohexene-1-carboxaldehyde,
in H-80146
2-(Methoxymethylene)cyclohexanone, in
O-80074
5-Methyl-3,4-hexadienoic acid; Me ester, in
M-70087
2,6-Octadienoic acid, O-80011
2-Octenedial, O-80021
4-Octenedial, O-70025
7-Octynoic acid, O-80024
▷ 3-Oxiranyl-7-oxabicyclo[4.1.0]heptane,
O-80056
2,2,4,4-Tetramethyl-1,3-cyclobutanedione,
T-80143

$C_8H_{12}O_3$
4-Cyclohexene-1,2-diol; Mono-Ac, in C-80182
4,4-Diethoxy-2-butynal, in B-60351
3,6-Dihydro-2H-pyran-2-carboxylic acid; Et
ester, in D-80242
2-Ethoxycarbonylcyclopentanone, in O-70084
3-Formyl-2,2-dimethyl-1-
cyclopropanecarboxylic acid; Me ester, in
F-80074
7-Hydroxy-5-heptynoic acid; Me ester, in
H-80150
3-Oxocyclopentaneacetic acid; Me ester, in
O-70083
4-Oxo-2-hexenoic acid; Et ester, in O-80082

$C_8H_{12}O_4$
3-Butene-1,2-diol; Di-Ac, in B-70291
3-(1,1-Dimethylethoxy)dihydro-2,5-
furandione, in D-80211

$C_8H_{12}O_4S_2$
(Dimercaptomethylene)propanedioic acid; Di-
S-Me, di-Me ester, in D-60395

$C_8H_{12}O_5$
Hydroxymethylenepropanedioic acid; Di-Et
ester, in H-80204

$C_8H_{12}O_6$
2,3-Dihydroxybutanoic acid; Di-Ac, in
D-60312

$C_8H_{13}BF_4O_2$
3,4,4a,5,6,7-Hexahydro-2H-pyrano[2,3-
b]pyrilium; Tetrafluoroborate, in H-60055

$C_8H_{13}Br$
1-Bromobicyclo[2.2.2]octane, B-70184
1-Bromobicyclo[2.2.2]octane, B-80180
5-Bromo-3,3-dimethylcyclohexene, B-80199
1-Bromo-5-isopropylcyclopentene, B-80210

$C_8H_{13}Br_2NO_2$
6-Amino-2,2-dibromohexanoic acid; N-Ac, in
A-60124

$C_8H_{13}Cl$
1-Chlorobicyclo[2.2.2]octane, C-70057

$C_8H_{13}Cl_2NO$
2,2-Dichloro-7-octenoic acid; Amide, in
D-80123

$C_8H_{13}Cl_2NO_3$
6-Amino-2,2-dichlorohexanoic acid; N-Ac, in
A-60126

$C_8H_{13}F$
1-Fluorobicyclo[2.2.2]octane, F-80028
2-Fluorobicyclo[3.2.1]octane, F-80029

$C_8H_{13}FO_2$
1-Fluorocyclohexanecarboxylic acid; Me ester,
in F-80031

$C_8H_{13}I$
1-Iodobicyclo[2.2.2]octane, I-80019

$C_8H_{13}IO$
2-Iodocyclooctanone, I-70027

$C_8H_{13}NO$
1-Acetylcyclohexene; Oxime, in A-70038
3-Aminobicyclo[2.2.1]hept-5-ene-2-methanol,
A-70094
1-Azabicyclo[3.3.1]nonan-2-one, A-70275
2-Azabicyclo[3.3.1]nonan-7-one, A-70276
Bicyclo[2.2.1]heptane-1-carboxylic acid;
Amide, in B-60085
1-Cycloheptenecarboxylic acid; Amide, in
C-80178

$C_8H_{13}NOS$
3,4-Dihydro-3,3,4,4-tetramethyl-5-thioxo-
2(1H)pyrrolidinone, D-80255

$C_8H_{13}NO_2$
7-Aminobicyclo[4.1.0]heptane-7-carboxylic
acid, A-60095
Hexahydro-2(3H)-benzoxazolone; N-Me, in
H-70052
1-Hydroxy-3-cyclohexene-1-carboxaldehyde;
Me ether, oxime, in H-80146
1-Oxa-2-azaspiro[2,5]octane; N-Ac, in
O-70052

$C_8H_{13}NO_2S$
2-(Diethoxymethyl)thiazole, in T-80170

$C_8H_{13}NO_3$
4-Oxo-2-piperidinecarboxylic acid; N-Me, Me
ester, in O-60089
3-Piperidinecarboxylic acid; N-Ac, in P-70102

$C_8H_{13}NO_4$
1-Amino-1,4-cyclohexanedicarboxylic acid,
A-60117
2-Amino-1,4-cyclohexanedicarboxylic acid,
A-60118
3-Amino-1,2-cyclohexanedicarboxylic acid,
A-60119
4-Amino-1,1-cyclohexanedicarboxylic acid,
A-60120
4-Amino-1,3-cyclohexanedicarboxylic acid,
A-60121
3-Amino-2-oxetanone; N-tert-
Butyloxycarbonyl, in A-80129
4,4-Dimethyl-6-nitro-5-hexenoic acid,
D-60429

$C_8H_{13}NS_2$
3,3,4,4-Tetramethyl-2,5-pyrrolidinedithione,
T-80151

$C_8H_{13}N_3O_2$
2-Azidocyclohexanol; Ac, in A-80196

$C_8H_{13}N_3O_2S$
Ovothiol B, in O-70050

$C_8H_{13}O_2^{\oplus}$
3,4,4a,5,6,7-Hexahydro-2H-pyrano[2,3-
b]pyrilium, H-60055

$C_8H_{13}O_5P$
8,8a-Dihydro-3-methoxy-5-methyl-1H,6H-
furo[3,4-e][1,3,2]dioxaphosphepin 3-oxide,
D-70223

C_8H_{14}
Bicyclo[4.1.1]octane, B-60091
1-Ethylcyclohexene, E-60046
3-Ethylcyclohexene, E-70039
4-Ethylcyclohexene, E-60048
Methylenecycloheptane, M-80079

$C_8H_{14}BrNO_3$
6-Amino-2-bromohexanoic acid; N-Ac, in
A-60100

$C_8H_{14}ClNO_3$
6-Amino-2-chlorohexanoic acid; N-Ac, in
A-60109

$C_8H_{14}ClNO_4$
2,3,5,6,7,8-Hexahydro-1H-indolizinium(1+);
Perchlorate, in H-70059

$C_8H_{14}N^{\oplus}$
2,3,5,6,7,8-Hexahydro-1H-indolizinium(1+),
H-70059

$C_8H_{14}N_2$
4,5-Dihydro-3,3,5,5-tetramethyl-4-methylene-
3H-pyrazole, D-60283

$C_8H_{14}N_2O$
1,3-Diazetidine-2,4-dione; 1,3-Diisopropyl, in
D-60057
4,6-Diethoxy-2,5-dihydropyrimidine, in
D-70256

$C_8H_{14}N_2O_4$
2,6-Diamino-4-methyleneheptanedioic acid,
D-60039
2,2-Dimethyl-1,3-cyclohexanedione; Dioxime,
in D-80443
2,5-Piperazinedicarboxylic acid; Di-Me ester,
in P-60157
Proclavaminic acid, P-70116
N-Prolylserine, P-60173
N-Serylproline, S-60025

$C_8H_{14}N_2S_2$
4,6-Bis(ethylthio)-2,5-dihydropyrimidine, in
D-70257

$C_8H_{14}N_4O_2$
Tetrahydroimidazo[4,5-d]imidazole-
2,5(1H,3H)-dione; 1,3,4,6-Tetra-Me, in
T-70066

$C_8H_{14}N_4S_2$
1,6-Diallyl-2,5-dithiobiurea, D-60026

$C_8H_{14}O$
2,2-Dimethyl-4-hexen-3-one, D-70414
2,2-Dimethyl-5-hexen-3-one, D-60425
3-Methyl-4-hepten-2-one, M-80093
4-Methyl-3-hepten-2-one, M-60087
6-Methyl-6-hepten-2-one, M-80094
2,4-Octadien-1-ol, O-70007
3-Octenal, O-70024
3-Octen-2-one, O-70027
4-Octen-2-one, O-60030
5-Octen-2-one, O-70028
2,2,3,3-
Tetramethylcyclopropanecarboxaldehyde,
T-70133
3,4,4-Trimethyl-1-pentyn-3-ol, T-80354

$C_8H_{14}OS$
Dihydro-3,3,4,4-tetramethyl-2(3H)-
furanthione, D-60282
4,5-Dihydro-3,3,4,4-tetramethyl-2(3H)-
thiophenone, D-60288
2,3,4,5,8,9-Hexahydrothionin; 1-Oxide, in
H-70076

$C_8H_{14}O_2$

3,6-Dimethoxycyclohexene, *in* C-**8**0181
4,4-Dimethoxycyclohexene, *in* C-**6**0211
1,6-Dioxaspiro[4,5]decane, D-**8**0501
4-Hexen-3-ol; Ac, *in* H-**6**0077
3-Methyl-2-heptenoic acid, M-**6**0086
4-Methyl-2-hexenoic acid; Me ester, *in* M-**8**0101
2-Methyl-4-penten-1-ol; Ac, *in* M-**8**0135
5-Octenoic acid, O-**8**0022
4-Oxooctanal, O-**6**0081
2,2,3,3-Tetramethylcyclopropanecarboxylic acid, T-**8**0144

$C_8H_{14}O_2S$

S-Butyl 3-oxobutanethioate, *in* O-**7**0080
S-tert-Butyl 3-oxobutanethioate, *in* O-**7**0080
1,4-Dioxaspiro[4.5]decane-7-thiol, *in* M-**7**0026
4-Mercapto-2-butenoic acid; *tert*-Butyl ester, *in* M-**6**0018
Octahydrobenzo[*b*]thiophene; 1,1-Dioxide, *in* O-**6**0012

$C_8H_{14}O_2S_2$

Bis(3-hydroxycyclobutyl)disulfide, *in* M-**6**0019

$C_8H_{14}O_3$

2-Cyclohexyl-2-hydroxyacetic acid, C-**6**0212
4,4-Dimethyl-5-oxopentanoic acid; Me ester, *in* D-**6**0434
2-Formylpropanoic acid; *tert*-Butyl ester, *in* F-**7**0039
2-Hydroxycyclohexanecarboxylic acid; Me ester, *in* H-**7**0119
4-Hydroxy-4-(2-hydroxyethyl)cyclohexanone, H-**6**0153
6-Methyl-5-oxoheptanoic acid, M-**8**0130

$C_8H_{14}O_4$

(1-Methylpropyl)butanedioic acid, M-**8**0150
L-Threaric acid; Mono-*tert-butyl* ether, *in* T-**7**0005
L-Threaric acid; Mono-*tert*-butyl ester, *in* T-**7**0005

$C_8H_{14}O_5$

3,3′-Oxybispropanoic acid; Di-Me ester, *in* O-**8**0098

$C_8H_{14}O_6$

Diethyl tartrate, *in* T-**7**0005

$C_8H_{14}S$

2,3,4,5,8,9-Hexahydrothionin, H-**7**0076
Octahydrobenzo[*b*]thiophene, O-**6**0012

$C_8H_{14}S_2$

4,5-Dihydro-3,3,4,4-tetramethyl-2(3*H*)-thiophenethione, D-**6**0287

$C_8H_{15}Br$

1-Bromo-1-octene, B-**7**0261

$C_8H_{15}BrO_2$

5-Bromo-2-hexanone; Ethylene acetal, *in* B-**6**0262

$C_8H_{15}ClO$

2,2-Diethylbutanoic acid; Chloride, *in* D-**6**0171
3,4,4-Trimethylpentanoic acid; Chloride, *in* T-**8**0353

$C_8H_{15}F$

1-Fluoro-1-octene, F-**7**0034

$C_8H_{15}I$

1-Iodo-1-octene, I-**7**0051

$C_8H_{15}IO$

3-Iodo-2-octanone, I-**6**0053

$C_8H_{15}N$

1-Azabicyclo[3.2.2]nonane, A-**8**0190
1-Azabicyclo[3.3.1]nonane, A-**8**0191
8-Azabicyclo[5.2.0]nonane, A-**7**0274
1,2,3,6-Tetrahydropyridine; 1-Propyl, *in* T-**7**0089
Tropane, *in* A-**7**0278

$C_8H_{15}NO$

3-Aminobicyclo[2.2.1]heptane-2-methanol, A-**7**0093
Hexahydro-5(2*H*)-azocinone; 1-Me, *in* H-**7**0048
Octahydro-2*H*-1,3-benzoxazine, O-**7**0011
Octahydro-2*H*-3,1-benzoxazine, O-**7**0012
▷ Physoperuvine, P-**6**0151
Tetrahydro-2-(2-pyrrolidinyl)furan, T-**6**0090

$C_8H_{15}NO_2$

2-Aminocycloheptanecarboxylic acid, A-**7**0123
3-Amino-2-hexenoic acid; Et ester, *in* A-**6**0170
4,4-Diethoxybutanenitrile, *in* O-**8**0068
2-Methyl-2-piperidinecarboxylic acid; Me ester, *in* M-**6**0119
3-Piperidinecarboxylic acid; Et ester, *in* P-**7**0102
4-(Tetrahydro-2-furanyl)morpholine, T-**6**0065

$C_8H_{15}NO_3$

2-Amino-4,4-dimethylpentanoic acid; *N*-Formyl, *in* A-**7**0137

$C_8H_{15}NO_4$

2-Amino-3-butylbutanedioic acid, A-**6**0102

$C_8H_{15}N_5$

2,4,6-Triaminopyrimidine; N^4,N^6-Di-Et, *in* T-**8**0208
2,4,6-Triaminopyrimidine; N^2,N^2,N^4,N^4-Tetra-Me, *in* T-**8**0208
2,4,6-Triaminopyrimidine; N^4,N^4,N^6,N^6-Tetra-Me, *in* T-**8**0208

C_8H_{16}

Isopropylcyclopentane, I-**7**0092

$C_8H_{16}BrNO_2$

2-Amino-4-bromobutanoic acid; *tert*-Butyl ester, *in* A-**6**0098

$C_8H_{16}Cl_2$

1,1-Dichlorooctane, D-**7**0126
1,2-Dichlorooctane, D-**7**0127
1,3-Dichlorooctane, D-**7**0128
2,2-Dichlorooctane, D-**7**0129
2,3-Dichlorooctane, D-**7**0130
4,5-Dichlorooctane, D-**7**0131

$C_8H_{16}N_2$

Decahydroquinazoline, D-**7**0013
Octahydropyridazino[1,2-*a*]pyridazine, O-**8**0016

$C_8H_{16}N_2O$

▷ Tetrahydro-2(1*H*)-pyrimidinone; 1,3-Di-Et, *in* T-**6**0088

$C_8H_{16}N_2O_3S$

N-Methionylalanine, M-**6**0035

$C_8H_{16}N_2O_4$

Octopinic acid, O-**8**0023

$C_8H_{16}N_2O_4S$

N-Serylmethionine, S-**6**0024

$C_8H_{16}N_4O_2$

1,3,5,7-Tetramethyltetrahydro-1,3,5,7-tetrazocine-2,6(1*H*,3*H*)dione, *in* T-**7**0098

$C_8H_{16}O$

6-Methoxy-5-methyl-1-hexene, *in* M-**8**0104
2-Octen-1-ol, O-**6**0029
2,2,3,3-Tetramethylcyclopropanemethanol, T-**7**0134

$C_8H_{16}OS$

2-Methyl-4-propyl-1,3-oxathiane, M-**6**0122
3,3,4,4-Tetramethylthiolane; 1-Oxide, *in* T-**8**0154

$C_8H_{16}OS_2$

4,4,5,5-Tetramethyl-1,2-dithiane; 1-Oxide, *in* T-**8**0145

$C_8H_{16}O_2$

2,2-Diethylbutanoic acid, D-**6**0171
2,5-Dimethoxy-3-hexene, *in* H-**8**0083
3,4-Dimethylpentanoic acid; Me ester, *in* D-**6**0435
4,4-Dimethylpentanoic acid; Me ester, *in* D-**7**0428
5-Hydroxy-4-methyl-3-heptanone, H-**6**0183
4-Octene-1,8-diol, O-**7**0026
Pityol, P-**6**0160

$C_8H_{16}O_2$

Tetrahydro-2,2,6-trimethyl-2*H*-pyran-3-ol, T-**8**0081
3,4,4-Trimethylpentanoic acid, T-**8**0353

$C_8H_{16}O_3$

1,1-Diethoxy-2-butanone, *in* O-**8**0067
2-(Diethoxymethyl)oxetane, *in* O-**8**0053
1-(2-Hydroxyethyl)-1,4-cyclohexanediol, H-**6**0133
3-Hydroxy-3-methylpentanoic acid; Et ester, *in* H-**8**0210
Isorengyol, *in* H-**6**0133

$C_8H_{16}O_4$

Methyl 3,3-diethoxypropionate, *in* O-**6**0090

$C_8H_{16}O_5$

5,7,8-Trimethoxyflavone, *in* T-**6**0321

$C_8H_{16}O_6$

1,4-Di-*O*-methyl-*myo*-inositol, D-**8**0452

$C_8H_{16}S$

3,3,4,4-Tetramethylthiolane, T-**8**0154

$C_8H_{16}S_2$

1,1-Cyclohexanedimethanethiol, C-**7**0223
1,2-Cyclohexanedimethanethiol, C-**7**0224
1,2-Dithiecane, D-**6**0507
4,4,5,5-Tetramethyl-1,2-dithiane, T-**8**0145

$C_8H_{17}BrO$

8-Bromo-1-octanol, B-**7**0260

$C_8H_{17}F$

1-Fluorooctane, F-**8**0058

$C_8H_{17}N$

Octahydroazocine; 1-Me, *in* O-**7**0010

$C_8H_{17}NO$

2-Aminocycloheptanemethanol, A-**7**0124
2-(Aminomethyl)cycloheptanol, A-**7**0168
2,2-Diethylbutanoic acid; Amide, *in* D-**6**0171

$C_8H_{17}NO_3$

Levocarnitine; Me ether, *in* C-**7**0024

$C_8H_{17}N_3O_3$

N-Ornithyl-β-alanine, O-**6**0041

$C_8H_{18}N_2$

Hexahydropyrimidine; 1,3-Di-Et, *in* H-**6**0056

$C_8H_{18}N_2O_2$

1,7-Dioxa-4,10-diazacyclododecane, D-**6**0465

$C_8H_{18}N_4$

2,4,6,8-Tetramethyl-2,4,6,8-tetraazabicyclo[3.3.0]octane, *in* O-**7**0015

$C_8H_{18}O_2$

2,2,3,3-Tetramethyl-1,4-butanediol, T-**8**0141

$C_8H_{18}S_2$

2,2,3,3-Tetramethyl-1,4-butanedithiol, T-**8**0142

$C_8H_{19}N$

2-Octylamine, O-**6**0031

$C_8H_{19}NO_2$

4,4-Diethoxybutylamine, *in* A-**8**0064

$C_8H_{19}N_3$

1,4,7-Triazacycloundecane, T-**6**0234
1,4,8-Triazacycloundecane, T-**6**0235

$C_8H_{22}N_4$

1,2-Dihydrazinoethane; $N^\alpha,N^\alpha,N^\beta,N^\beta,N^{\beta'},N^{\beta'}$-Hexa-Me, *in* D-**6**0192

$C_9O_2S_6$

2,6-Dithioxobenzo[1,2-*d*:4,5-*d'*]bis[1,3]dithiole-4,8-dione, D-**6**0512

$C_9O_4S_2$

2,7-Dithiatricyclo[6.2.0.0^{3,6}]deca-1(8),3(6)-diene-4,5,9,10-tetrone, D-**6**0506

$C_9H_3Cl_2NO_2$

6,7-Dichloro-5,8-isoquinolinedione, D-**6**0137
6,7-Dichloro-5,8-quinolinedione, D-**6**0149

$C_9H_3Cl_2NO_3$

6,7-Dichloro-5,8-quinolinedione; *N*-Oxide, *in* D-**6**0149

C₉H₄N₄O₂
Tetrazolo[1,5-*b*]quinoline-5,10-dione, T-70156

$C_9H_4N_4O_3$
Imidazo[4,5-*g*]quinazoline-4,8,9(3*H*,7*H*)-trione, I-60010

$C_9H_4N_6O_2$
▷ 2,2-Diazido-1,3-indanedione, D-70056

C_9H_4O
2-(2-Propenyl)cyclohexanone, P-80177

$C_9H_4O_4$
3,4-Methylenedioxybenzocyclobutene-1,2-dione, *in* D-80274

$C_9H_5BrF_2$
3-Bromo-3,3-difluoro-1-phenylpropyne, B-70203

$C_9H_5BrO_2$
3-Bromo-4*H*-1-benzopyran-4-one, B-70183

$C_9H_5Br_2N_3$
4,5-Dibromo-6-phenyl-1,2,3-triazine, D-80097

$C_9H_5Cl_2F_3N_2$
5,6-Dichloro-2-(trifluoromethyl)-1*H*-benzimidazole; *N*-Me, *in* D-60160

$C_9H_5Cl_2N$
6,7-Dichloroisoquinoline, D-60136

$C_9H_5NO_2$
3-(Furo[3,4-*b*]furan-4-yl)-2-propenenitrile, F-70055
▷ 5,8-Isoquinolinequinone, I-60131

$C_9H_5NO_4$
6-Nitro-4*H*-1-benzopyran-4-one, N-80036

$C_9H_5NO_6$
4-Hydroxy-5-nitro-1,2-benzenedicarboxylic acid; Me ether, anhydride, *in* H-70191

$C_9H_5NS_2$
Dithieno[2,3-*b*:2′,3′-*d*]pyridine, D-70509
Dithieno[2,3-*b*:3′,2′-*d*]pyridine, D-70510
Dithieno[2,3-*b*:3′,4′-*d*]pyridine, D-70511
Dithieno[3,2-*b*:2′,3′-*d*]pyridine, D-70512
Dithieno[3,2-*b*:3′,4′-*d*]pyridine, D-70513
Dithieno[3,4-*b*:2′,3′-*d*]pyridine, D-70514
Dithieno[3,4-*b*:3′2′-*d*]pyridine, D-70515
Dithieno[3,4-*b*:3′,4′-*d*]pyridine, D-70516

C_9H_5NSe
2-Phenyl-1-(selenocyanato)acetylene, P-80134

$C_9H_5N_3O_2$
[1]-Benzopyrano[2,3-*d*]-1,2,3-triazol-9(1*H*)-one, B-60043

$C_9H_5N_3O_3$
2,3-Dicyano-1-methoxy-4-nitrobenzene, *in* H-70189

$C_9H_5N_3S$
2-Isothiocyanatoquinoxaline, I-80095

$C_9H_6BrN_3$
5-Bromo-4-phenyl-1,2,3-triazine, B-80252

C_9H_6ClNO
▷ 5-Chloro-8-hydroxyquinoline, C-70075
8-Chloro-2(1*H*)-quinolinone, C-60131
8-Chloro-4(1*H*)-quinolinone, C-60132
1*H*-Indole-3-acetic acid; Chloride, *in* I-70013

$C_9H_6ClNO_2$
5-Chloro-8-hydroxyquinoline; 1-Oxide, *in* C-70075
1-(Chloromethyl)-1*H*-indole-2,3-dione, C-70092

C_9H_6ClNS
2-Chloro-4-phenylthiazole, C-80107

$C_9H_6Cl_2O_2$
1,3,5-Cycloheptatriene-1,6-dicarboxylic acid; Dichloride, *in* C-60202

C_9H_6FN
2-Fluoroquinoline, F-80062
3-Fluoroquinoline, F-80063
4-Fluoroquinoline, F-80064
▷ 5-Fluoroquinoline, F-80065

▷ 6-Fluoroquinoline, F-80066
7-Fluoroquinoline, F-80067
▷ 8-Fluoroquinoline, F-80068

C_9H_6FNO
2-Fluoroquinoline; *N*-oxide, *in* F-80062
3-Fluoroquinoline; *N*-oxide, *in* F-80063
5-Fluoroquinoline; *N*-oxide, *in* F-80065
6-Fluoroquinoline; *N*-oxide, *in* F-80066
8-Fluoroquinoline; *N*-oxide, *in* F-80068

$C_9H_6F_2O$
2,3-Difluoro-3-phenyl-2-propenal, D-80164

$C_9H_6F_3N$
2-Trifluoromethyl-1*H*-indole, T-60291
3-Trifluoromethyl-1*H*-indole, T-60292
6-Trifluoromethyl-1*H*-indole, T-60293

$C_9H_6N_2$
3-Cyanoindole, *in* I-80012
Imidazo[5,1,2-*cd*]indolizine, I-70004

$C_9H_6N_2O$
3-Cinnolinecarboxaldehyde, C-70178
4-Cinnolinecarboxaldehyde, C-70179

$C_9H_6N_2O_2$
5,8-Isoquinolinequinone; 8-Oxime, *in* I-60131
1-Nitroisoquinoline, N-60029
4-Nitroisoquinoline, N-60030
5-Nitroisoquinoline, N-60031
6-Nitroisoquinoline, N-60032
7-Nitroisoquinoline, N-60033
8-Nitroisoquinoline, N-60034
5-Phenyl-1,2,4-oxadiazole-3-carboxaldehyde, P-60122

$C_9H_6N_2O_3$
2-Methyl-4,5,8-quinazolinetrione, M-80160
5-Nitroisoquinoline; 2-Oxide, *in* N-60031
8-Nitroisoquinoline; 2-Oxide, *in* N-60034

$C_9H_6N_2S$
3-Amino-2-cyanobenzo[*b*]thiophene, *in* A-60091

$C_9H_6N_4$
4-Cyano-2-phenyl-1,2,3-triazole, *in* P-60140
1*H*-1,2,3-Triazolo[4,5-*f*]quinoline, T-70207
1*H*-1,2,3-Triazolo[4,5-*h*]quinoline, T-70208

$C_9H_6N_4O$
1,5-Dihydro-8*H*-imidazo[4,5-*g*]quinazolin-8-one, I-60012
3,8-Dihydro-9*H*-pyrazolo[4,3-*f*]quinazolin-9-one, D-60271
Imidazo[4,5-*f*]quinazolin-9(8*H*)-one, I-60011
Imidazo[4,5-*h*]quinazolin-6-one, I-60013
Pyrazolo[3,4-*f*]quinazolin-9(8*H*)-one, P-60204
Pyrazolo[4,3-*g*]quinazolin-5(6*H*)-one, P-60205

$C_9H_6N_4O_2$
Imidazo[4,5-*g*]quinazoline-6,8(5*H*,7*H*)-dione, I-60009
1*H*-Pyrazolo[4,3-*g*]quinazoline-5,7(6*H*,8*H*)-dione, P-60203

$C_9H_6N_4S$
Pyrazino[2,3-*b*]pyrido[3′,2′-*e*][1,4]thiazine, P-60199

C_9H_6O
5*H*-Cycloprop[*f*]isobenzofuran, C-70251
4-Ethynylbenzaldehyde, E-70048

C_9H_6OS
5-(2-Thienyl)-2-penten-4-ynal, T-80182

$C_9H_6O_2$
▷ 1*H*-2-Benzopyran-1-one, B-60041

$C_9H_6O_2S$
4-Hydroxy-2*H*-1-benzothiopyran-2-one, H-70110
5-(2-Thienyl)-2-penten-4-ynoic acid, T-80183

$C_9H_6O_3$
3-Benzofurancarboxylic acid, B-60032
6-Hydroxy-2*H*-1-benzopyran-2-one, *in* H-70109
3-Methyl-4,5-benzofurandione, M-70051

$C_9H_6O_4$
6,7-Dihydroxy-2*H*-1-benzopyran-2-one, D-84287

4-Hydroxy-5-benzofurancarboxylic acid, H-60103
5-Hydroxy-3-methyl-1,2-benzenedicarboxylic acid; Anhydride, *in* H-70162

$C_9H_6O_5$
4,5,7-Trihydroxy-2H-1-benzopyran-2-one, T-80286
5,6,7-Trihydroxy-2*H*-1-benzopyran-2-one, T-80287
5,7,8-Trihydroxy-2*H*-1-benzopyran-2-one, T-80288

$C_9H_6O_{12}$
Cyclopropanehexacarboxylic acid, C-70250

C_9H_6S
5*H*-Cyclopropa[*f*][2]benzothiophene, C-70247

$C_9H_6S_6$
Benzo[1,2-*d*:3,4-*d*′:5,6-*d*″]tris[1,3]dithiole, B-70069

C_9H_7Br
3-Bromo-3-phenyl-1-propyne, B-80244

$C_9H_7BrN_2$
5-Bromo-1-phenylpyrazole, *in* B-80253
3(5)-Bromopyrazole; 1-Ph, *in* B-80253
4-Bromopyrazole; 1-Ph, *in* B-80254

$C_9H_7Br_3O$
3,5-Dimethyl-2,4,6-tribromobenzaldehyde, D-80476

$C_9H_7ClN_2$
1-Chloro-4-methylphthalazine, C-70104
2-Chloro-3-methylquinoxaline, C-70121
2-Chloro-5-methylquinoxaline, C-70122
2-Chloro-6-methylquinoxaline, C-70123
2-Chloro-7-methylquinoxaline, C-70124
2-(Chloromethyl)quinoxaline, C-70120
6-Chloro-7-methylquinoxaline, C-70125

$C_9H_7ClN_2O$
2-Chloro-3-methylquinoxaline; 1-Oxide, *in* C-70121
2-Chloro-3-methylquinoxaline; 4-Oxide, *in* C-70121
3-Chloro-7-methylquinoxaline 1-oxide, *in* C-70123
6-Chloro-7-methylquinoxaline; 1-Oxide, *in* C-70125

$C_9H_7ClN_2O_2$
2-Chloro-3-methylquinoxaline; 1,4-Dioxide, *in* C-70121
6-Chloro-7-methylquinoxaline; 1,4-Dioxide, *in* C-70125

$C_9H_7ClO_3$
1,4-Benzodioxan-2-carboxylic acid; Chloride, *in* B-60024
1,4-Benzodioxan-6-carboxylic acid; Chloride, *in* B-60025

$C_9H_7ClO_4$
2-Hydroxy-4,5-methylenedioxybenzoic acid; Me ether, chloride, *in* H-80202
3-Hydroxy-4,5-methylenedioxybenzoic acid; Me ether, chloride, *in* H-60178

$C_9H_7ClO_4S$
1-Benzothiopyrylium(1+); Perchlorate, *in* B-70066
2-Benzothiopyrylium(1+); Perchlorate, *in* B-70067

$C_9H_7Cl_3O$
3,5-Dimethyl-2,4,6-trichlorobenzaldehyde, D-80477

$C_9H_7F_3O_2$
4-Methoxy-2-(trifluoromethyl)benzaldehyde, *in* H-70228
4-Methoxy-3-(trifluoromethyl)benzaldehyde, *in* H-70229

$C_9H_7F_3S$
1-(Methylthio)-3-(trifluoromethyl)benzene, *in* T-60285

C_9H_7IO
2-Iodo-3-phenyl-2-propenal, I-80039

$C_9H_7IO_4$
1-Hydroxy-1,2-benziodoxol-3(1*H*)-one; Ac, *in* H-60102

C_9H_7N
Cyclopent[*b*]azepine, C-60225

C_9H_7NO
o-Cyanoacetophenone, *in* A-80015
Cyclohepta[*b*]pyrrol-2(1*H*)-one, C-60201
Cyclopent[*b*]azepine; *N*-Oxide, *in* C-60225
5,6-Epoxyquinoline, E-60027
7,8-Epoxyquinoline, E-60028
6-Hydroxyisoquinoline, H-80178
1*H*-Indole-5-carboxaldehyde, I-60029
1*H*-Indole-6-carboxaldehyde, I-60030
1*H*-Indole-7-carboxaldehyde, I-60031
5-Phenyloxazole, P-60125
3*H*-Pyrrolo[1,2-*a*]azepin-3-one, P-70190
5*H*-Pyrrolo[1,2-*a*]azepin-5-one, P-70191
7*H*-Pyrrolo[1,2-*a*]azepin-7-one, P-70192
9*H*-Pyrrolo[1,2-*a*]azepin-9-one, P-70193
2-Vinylbenzoxazole, V-80006

C_9H_7NOS
1,4-Benzothiazepin-5(4*H*)-one, B-80040

$C_9H_7NO_2$
1,4-Benzodioxan-2-carbonitrile, *in* B-60024
1,3-Benzodioxole-4-acetonitrile, *in* B-60029
2,3-Dihydro-1,4-isoquinolinedione, D-80217
1*H*-Indole-3-carboxylic acid, I-80012
1,3(2*H*,4*H*)-Isoquinolinedione, I-70099

$C_9H_7NO_2S$
▷ 3-Amino-2-benzo[*b*]thiophenecarboxylic acid, A-60091
2*H*-3,1-Benzothiazine-2,4(1*H*)-dione; *N*-Me, *in* B-80042

$C_9H_7NO_3$
1,5-Benzoxepine-2,4-(3*H*,5*H*)-dione, B-80053
3,5-Dihydroxy-4-phenylisoxazole, D-70337
1*H*-Isoindolin-1-one-3-carboxylic acid, I-60102
3-Methoxy-4,5-methylnedioxybenzonitrile, *in* H-60178

$C_9H_7NO_4$
3-Amino-4-hydroxy-1,2-benzenedicarboxylic acid; Me ether, anhydride, *in* A-70151
5,6-Dihydroxy-1*H*-indole-2-carboxylic acid, D-70308

$C_9H_7NO_6$
2-Nitro-1,3-benzenedicarboxylic acid; Mono-Me ester, *in* N-80035

$C_9H_7NO_7$
3-Methoxy-6-nitro-1,2-benzenedicarboxylic acid, *in* H-70189
4-Methoxy-3-nitro-1,2-benzenedicarboxylic acid, *in* H-70190
4-Methoxy-5-nitro-1,2-benzenedicarboxylic acid, *in* H-70191
4-Methoxy-5-nitro-1,3-benzenedicarboxylic acid, *in* H-70192

C_9H_7NS
1(2*H*)-Isoquinolinethione, I-60132
3(2*H*)-Isoquinolinethione, I-60133
3-Quinolinethiol, Q-60003
5-Quinolinethiol, Q-60004
6-Quinolinethiol, Q-60005
4(1*H*)-Quinolinethione, Q-60006

$C_9H_7NS_2$
4-Phenyl-2(3*H*)-thiazolethione, P-80135

$C_9H_7N_3O$
4-Benzoyl-1,2,3-triazole, B-70079
4-Cinnolinecarboxaldehyde; Oxime, *in* C-70179

$C_9H_7N_3OS$
1,2,3-Benzotriazine-4(3*H*)-thione; *N*-Ac, *in* B-60050

$C_9H_7N_3O_2$
6-Amino-8-nitroquinoline, A-60234
8-Amino-6-nitroquinoline, A-60235
5-Phenyl-1,2,4-oxadiazole-3-carboxaldehyde; Oxime, *in* P-60122
2-Phenyl-2*H*-1,2,3-triazole-4-carboxylic acid, P-60140

$C_9H_7N_5$
8-Aminoimidazo[4,5-*g*]quinazoline, A-60207

$C_9H_7N_5O$
7-Amino-1,6-dihydro-9*H*-imidazo[4,5-*f*]quinazolin-9-one, A-60208
6-Amino-1,7-dihydro-8*H*-imidazo[4,5-*g*]quinazolin-8-one, A-60209

$C_9H_7S^{\oplus}$
1-Benzothiopyrylium(1+), B-70066
2-Benzothiopyrylium(1+), B-70067

$C_9H_8ClNO_5$
4,5-Dihydroxy-2-nitrobenzoic acid; Di-Me ether, chloride, *in* D-60360

$C_9H_8Cl_3NO_2$
2-Amino-3-(2,3,4-trichlorophenyl)propanoic acid, A-60262
2-Amino-3-(2,3,6-trichlorophenyl)propanoic acid, A-60263
2-Amino-3-(2,4,5-trichlorophenyl)propanoic acid, A-60264

$C_9H_8FNO_3$
2-Amino-4-fluorobenzoic acid; *N*-Ac, *in* A-70141
2-Amino-5-fluorobenzoic acid; *N*-Ac, *in* A-70142
3-Amino-4-fluorobenzoic acid; *N*-Ac, *in* A-70144
4-Amino-2-fluorobenzoic acid; *N*-Ac, *in* A-70145
4-Amino-3-fluorobenzoic acid; *N*-Ac, *in* A-70146

$C_9H_8F_3NO$
3,3,3-Trifluoropropanoic acid; Anilide, *in* T-80284

$C_9H_8N_2$
1*H*-1,2-Benzodiazepine, B-80019
1*H*-Cyclooctapyrazole, C-70236

$C_9H_8N_2O$
5-Amino-3-phenylisoxazole, A-60240
3-Amino-2(1*H*)-quinolinone, A-60256
3*H*-1,2-Benzodiazepine; 1-Oxide, *in* B-80020
3*H*-1,2-Benzodiazepine; 2-Oxide, *in* B-80020
2,2′-Bi-1*H*-pyrrole-5-carboxaldehyde, B-80130
N-Hydroxy-4-quinolinamine, H-80253
5-Methyl-3-phenyl-1,2,4-oxadiazole, M-60110
1*H*-Pyrrolo[3,2-*b*]pyridine; *N*-Ac, *in* P-60249

$C_9H_8N_2OS$
3-Amino-2-benzo[*b*]thiophenecarboxylic acid; Amide, *in* A-60091
1,2,3,4-Tetrahydro-2-thioxo-5*H*-1,4-benzodiazepin-5-one, T-70100
▷ 2-Thioxo-4-imidazolidinone; 3-*N*-Ph, *in* T-70188

$C_9H_8N_2O_2$
1,4-Dihydro-2,3-quinoxalinedione; 1-Me, *in* D-60279
3,5-Dihydroxy-4-phenylpyrazole, D-80369
6-Hydroxy-7-methoxyquinoxaline, *in* Q-60009
N-Hydroxy-4-quinolinamine; 1-Oxide, *in* H-80253
3,5-Pyrazolidinedione; 1-Ph, *in* P-80200

$C_9H_8N_2O_3$
5-Phenyl-1,2,4-oxadiazole-3-carboxaldehyde; Covalent hydrate, *in* P-60122

$C_9H_8N_2O_4$
1-Cyano-4,5-dimethoxy-2-nitrobenzene, *in* D-60360

$C_9H_8N_2O_6$
3,5-Dinitrobenzoic acid; Et ester, *in* D-60458

$C_9H_8N_2O_7$
4,5-Dinitro-1,2-benzenediol; Mono-Me ether, Ac, *in* D-70449

$C_9H_8N_2S$
3-Amino-2(1*H*)quinolinethione, A-70197
4-Methyl-5-phenyl-1,2,3-thiadiazole, M-60116
5-Methyl-4-phenyl-1,2,3-thiadiazole, M-60117

$C_9H_8N_4$
4-Cinnolinecarboxaldehyde; Hydrazone, *in* C-70179

$C_9H_8N_4O$
2-Phenyl-2*H*-1,2,3-triazole-4-carboxylic acid; Amide, *in* P-60140

$C_9H_8N_6O_3$
Lepidopterin, L-60020

C_9H_8O
Bicyclo[4.2.1]nona-2,4,7-trien-9-one, B-80084
Cyclooctatetraenecarboxaldehyde, C-70238
3-Phenyl-2-propyn-1-ol, P-60130

C_9H_8OS
1,4-Dihydro-3*H*-2-benzothiopyran-3-one, D-60206

C_9H_8OSe
1*H*-2-Benzoselenin-4(3*H*)-one, B-70051
1,4-Dihydro-3*H*-2-benzoselenin-3-one, D-60204

C_9H_8OTe
1,4-Dihydro-3*H*-2-benzotellurin-3-one, D-60205

$C_9H_8O_2$
1*H*-2-Benzopyran-4(3*H*)-one, B-70042
Cubanecarboxylic acid, C-60174
Cyclooctatetraenecarboxylic acid, C-70239
1,4-Dihydro-2(3*H*)-benzopyran-3-one, D-70176
4-Hydroxybenzofuran, H-70106
3-(4-Hydroxyphenyl)-2-propenal, H-60213
4-Methoxybenzofuran, *in* H-70106
2-Methyl-1,3-benzenedicarboxaldehyde, M-60046
2-Methyl-1,4-benzenedicarboxaldehyde, M-60047
4-Methyl-1,2-benzenedicarboxaldehyde, M-60048
4-Methyl-1,3-benzenedicarboxaldehyde, M-60049
5-Methyl-1,3-benzenedicarboxaldehyde, M-60050

$C_9H_8O_2S$
2-Acetyl-3-hydroxy-5-(1-propynyl)thiophene, A-80024
3,4-Dihydro-2*H*-1,5-benzoxathiepin-3-one, D-60207
3-(2-Thienyl)-2-propenoic acid; Me ester, *in* T-60208

$C_9H_8O_3$
2-Acetylbenzoic acid, A-80015
5-Acetyl-2-hydroxybenzaldehyde, A-80022
1,4-Benzodioxan-2-carboxaldehyde, B-60022
1,4-Benzodioxan-6-carboxaldehyde, B-60023
4*H*-1,3-Benzodioxin-6-carboxaldehyde, B-60026
2,3-Dihydro-2-benzofurancarboxylic acid, D-60197
▷ 3-(4-Hydroxyphenyl)-2-propenoic acid, H-80242
2-Methoxy-3(2*H*)-benzofuranone, *in* H-80107
4-Methoxy-3(2*H*)-benzofuranone, *in* H-80108
5-Methoxy-3(2*H*)-benzofuranone, *in* H-80109
6-Methoxy-3(2*H*)-benzofuranone, *in* H-80110
7-Methoxy-3(2*H*)-benzofuranone, *in* H-80111

$C_9H_8O_4$
1,4-Benzodioxan-2-carboxylic acid, B-60024
1,4-Benzodioxan-6-carboxylic acid, B-60025
4*H*-1,3-Benzodioxin-6-carboxylic acid, B-60027
1,3-Benzodioxole-4-acetic acid, B-60029
1,3,5-Cycloheptatriene-1,6-dicarboxylic acid, C-60202
5,7-Dihydroxy-6-methyl-1(3*H*)-isobenzofuranone, D-60349
5-Hydroxy-7-methoxyphthalide, *in* D-60339
7-Hydroxy-5-methoxyphthalide, *in* D-60339

3-(4-Hydroxyphenyl)-2-oxopropanoic acid, H-**80238**
2-Methoxy-4,5-methylenedioxybenzaldehyde, *in* H-**70175**

$C_9H_8O_5$

2-Formyl-3,5-dihydroxybenzoic acid; Me ester, *in* F-**60071**
2-Hydroxy-1,3-benzenedicarboxylic acid; Mono-Me ester, *in* H-**60101**
5-Hydroxy-3-methyl-1,2-benzenedicarboxylic acid, H-**70162**
2-Methoxy-1,3-benzenedicarboxylic acid, *in* H-**60101**
2-Methoxy-1,4-benzenedicarboxylic acid, *in* H-**70105**
5-Methoxy-1,3-benzodioxole-4-carboxylic acid, *in* H-**60180**
6-Methoxy-1,3-benzodioxole-5-carboxylic acid, *in* H-**80202**
7-Methoxy-1,3-benzodioxole-4-carboxylic acid, *in* H-**60179**
2-Methoxy-3,4-methylenedioxybenzoic acid, *in* H-**60177**
3-Methoxy-4,5-methylenedioxybenzoic acid, *in* H-**60178**
3-Oxotricyclo[2.1.0^{2,5}]pentane-1,5-dicarboxylic acid; Di-Me ester, *in* O-**80096**

$C_9H_8O_7$

2-Oxo-1,3-cyclopentanediglyoxylic acid, O-**70085**

C_9H_9BrO

3-(Bromomethyl)-2,3-dihydrobenzofuran, B-**60295**
▷ 2-Bromo-1-phenyl-1-propanone, B-**80243**

$C_9H_9BrO_2$

1-(5-Bromo-2-methoxyphenyl)ethanone, *in* B-**60265**

$C_9H_9Br_3$

1,3,5-Tris(bromomethyl)benzene, T-**60401**

C_9H_9Cl

1-Chloro-2-(2-propenyl)benzene, C-**70152**
1-Chloro-3-(2-propenyl)benzene, C-**70153**
1-Chloro-4-(2-propenyl)benzene, C-**70154**

C_9H_9ClO

2-Chloro-4,6-dimethylbenzaldehyde, C-**60061**
4-Chloro-2,6-dimethylbenzaldehyde, C-**60062**
4-Chloro-3,5-dimethylbenzaldehyde, C-**60063**
5-Chloro-2,4-dimethylbenzaldehyde, C-**60064**

$C_9H_9ClO_2$

3-(Chloromethyl)phenol; Ac, *in* C-**70102**

$C_9H_9ClO_3$

5-Chloro-2,3-dimethoxybenzaldehyde, *in* C-**70065**
3-Chloro-4-hydroxybenzoic acid; Me ester, Me ether, *in* C-**60091**
2-Chloro-3-methoxy-5-methylbenzoic acid, *in* C-**60095**
3-Chloro-4-methoxy-5-methylbenzoic acid, *in* C-**60096**
5-Chloro-4-methoxy-2-methylbenzoic acid, *in* C-**60097**
4-(Chloromethyl)phenol; Ac, *in* C-**70103**

$C_9H_9Cl_2NO_2$

2-Amino-3-(2,3-dichlorophenyl)propanoic acid, A-**60127**
2-Amino-3-(2,4-dichlorophenyl)propanoic acid, A-**60128**
2-Amino-3-(2,5-dichlorophenyl)propanoic acid, A-**60129**
2-Amino-3-(2,6-dichlorophenyl)propanoic acid, A-**60130**
2-Amino-3-(3,4-dichlorophenyl)propanoic acid, A-**60131**
2-Amino-3-(3,5-dichlorophenyl)propanoic acid, A-**60132**

$C_9H_9Cl_3$

1,3,5-Tris(chloromethyl)benzene, T-**60402**

C_9H_9FO

2-Fluoro-2-phenylpropanal, F-**60063**
2-Fluoro-3-phenylpropanal, F-**60064**
2-Fluoro-1-phenyl-1-propanone, F-**60065**

$C_9H_9FO_2$

3,4-Dihydro-3-fluoro-2H-1,5-benzodioxepin, D-**70209**
3-Fluoro-2-methylphenol; Ac, *in* F-**60056**

$C_9H_9FO_4$

3-Fluoro-2,6-dimethoxybenzoic acid, *in* F-**60023**
5-Fluoro-2,3-dimethoxybenzoic acid, *in* F-**60025**

$C_9H_9F_3O_3S$

3-(Trifluoromethyl)benzenesulfonic acid; Et ester, *in* T-**60283**

$C_9H_9IO_4$

2-Iodo-4,5-dimethoxybenzoic acid, *in* D-**60338**
3-Iodo-2,4-dimethoxybenzoic acid, *in* D-**60331**
3-Iodo-2,6-dimethoxybenzoic acid, *in* D-**60334**
3-Iodo-4,5-dimethoxybenzoic acid, *in* D-**60335**
4-Iodo-2,5-dimethoxybenzoic acid, *in* D-**60333**
5-Iodo-2,4-dimethoxybenzoic acid, *in* D-**60332**
6-Iodo-2,3-dimethoxybenzoic acid, *in* D-**60330**

$C_9H_9I_3$

1,3,5-Tris(iodomethyl)benzene, T-**80366**

C_9H_9N

3a,7a-Methano-1H-indole, M-**70041**
6-Methyl-1H-indole, M-**60088**
3-Phenyl-2-propyn-1-amine, P-**60129**
5H-Pyrrolo[1,2-a]azepine, P-**70189**

C_9H_9NO

3-Amino-1-indanone, A-**60211**
Bicyclo[4.2.1]nona-2,4,7-trien-9-one; Oxime, *in* B-**80084**
7,8-Dihydro-5(6H)-isoquinolinone, D-**60254**
1H-Indole-6-methanol, I-**60032**
6-Methoxyindole, *in* H-**61058**
3-Methoxy-1H-isoindole, *in* P-**70100**
Phthalimidine; N-Me, *in* P-**70100**

C_9H_9NOS

2H-1,4-Benzothiazin-3(4H)-one; N-Me, *in* B-**80043**

$C_9H_9NO_2$

5,6-Dihydroxyindole; N-Me, *in* D-**70307**
5-Methoxy-1H-indol-6-ol, *in* D-**70307**
6-Methoxy-1H-indol-5-ol, *in* D-**70307**
1-Nitro-1-phenylpropene, N-**80057**
4-Phenyl-2-oxazolidinone, P-**60126**

$C_9H_9NO_3$

1,4-Benzodioxan-6-carboxaldehyde; Oxime, *in* B-**60023**
1,4-Benzodioxan-2-carboxylic acid; Amide, *in* B-**60024**
(2-Oxo-2-phenylethyl)carbamic acid, O-**70099**

$C_9H_9NO_3S$

3-Carboxy-2,3-dihydro-8-hydroxy-5-methylthiazolo[3,2-a]pyridinium hydroxide inner salt, C-**70021**

$C_9H_9NO_4$

2,6-Bis(hydroxyacetyl)pyridine, B-**70144**
3-Hydroxy-4,5-methylenedioxybenzoic acid; Me ether, amide, *in* H-**60178**

$C_9H_9NO_5$

3-Amino-4-methoxy-1,2-benzenedicarboxylic acid, A-**70151**
2,4-Dihydroxy-7-methoxy-2H-1,4-benzoxazin-3(4H)-one, D-**10388**

$C_9H_9NO_6$

4,5-Dihydroxy-2-nitrobenzoic acid; 4-Me ether, Me ester, *in* D-**60360**
4,5-Dimethoxy-2-nitrobenzoic acid, *in* D-**60360**

$C_9H_9N_3$

1-Aminophthalazine; N-Me, *in* A-**70184**
4,5-Diaminoisoquinoline, D-**70040**
5,8-Diaminoisoquinoline, D-**70041**
4,5-Diaminoquinoline, D-**70044**
1,4-Dihydro-2,6-dimethyl-3,5-pyridinedicarboxylic acid; Dinitrile, *in* D-**80193**
(2-Pyridyl)(1-pyrazolyl)methane, P-**70178**

$C_9H_9N_3O_2$

1,3(2H,4H)-Isoquinolinedione; Dioxime, *in* I-**70099**

$C_9H_9N_3O_3$

Biuret; N-Benzoyl, *in* B-**60190**

$C_9H_9N_3S$

1,2,3-Benzotriazine-4(3H)-thione; 2-N-Et, *in* B-**60050**

$C_9H_9N_5$

4-Benzoyl-1,2,3-triazole; Hydrazone, *in* B-**70079**
▷ 2,4-Diamino-6-phenyl-1,3,5-triazine, D-**60041**

$C_9H_9N_5O_4$

6-(1,2-Dicarboxyethylamino)purine, D-**80115**

C_9H_{10}

5,6-Dihydro-4H-indene, D-**70217**
7-Methylene-1,3,5-cyclooctatriene, M-**60074**
Tricyclo[3.3.1.0^{2,8}]nona-3,6-diene, T-**70230**

$C_9H_{10}BrNO$

2-Bromo-3-methylaniline; N-Ac, *in* B-**70234**

$C_9H_{10}ClN$

4-Chloro-2,3-dihydro-1H-indole; 1-Me, *in* C-**60055**

$C_9H_{10}ClNO$

2'-Amino-2-chloro-3'-methylacetophenone, A-**60112**
2'-Amino-2-chloro-4'-methylacetophenone, A-**60113**
2'-Amino-2-chloro-5'-methylacetophenone, A-**60114**
2'-Amino-2-chloro-6'-methylacetophenone, A-**60115**
2-Chloro-4,6-dimethylbenzaldehyde; Oxime, *in* C-**60061**

$C_9H_{10}FNO_2S$

N-Fluoro-2,3-dihydro-3,3-dimethyl-1,2-benzothiadiazole 1,1-dioxide, F-**80032**

$C_9H_{10}N_2$

2,3-Dihydro-1H-1,4-benzodiazepine, D-**80173**
4,5-Dihydro-1H-pyrazole; 1-Ph, in D-**70253**

$C_9H_{10}N_2O$

3,4-Dihydro-2(1H)-quinazolinone; 3-Me, *in* D-**60278**
▷ 1-Phenyl-3-pyrazolidinone, P-**70091**
1,3,4,5-Tetrahydro-2H-1,3-benzodiazepin-2-one, T-**60058**

$C_9H_{10}N_2O_2$

2-Acetyl-5-aminopyridine; N-Ac, *in* A-**70023**
1-Carbethoxy-2-cyano-1,2-dihydropyridine, C-**70017**

$C_9H_{10}N_2O_3$

1,4-Benzodioxan-6-carboxylic acid; Hydrazide, *in* B-**60025**
2,4-Diaminobenzoic acid; 2-N-Ac, *in* D-**60032**
2,4-Diaminobenzoic acid; 4-N-Ac, *in* D-**60032**

$C_9H_{10}N_2O_4$

5-Amino-3,4-pyridinedicarboxylic acid; Di-Me ester, *in* A-**70195**
2-Amino-3,5-pyridinedicarboxylic acid; 3-Mono-Et ester, *in* A-**70193**
2-Amino-3,5-pyridinedicarboxylic acid; 5-Mono-Et ester, *in* A-**70193**

$C_9H_{10}N_2S$

2-Amino-4,5,6,7-tetrahydrobenzo[b]thiophene-3-carboxylic acid; Nitrile, *in* A-**60258**

$C_9H_{10}N_4$

4-Dimethylamino-1,2,3-benzotriazine, *in* A-**60092**
2,6,7-Trimethylpteridine, T-**70286**
4,6,7-Trimethylpteridine, T-**70287**

$C_9H_{10}N_4O$

4-Amino-1,2,3-benzotriazine; N,N(4)-Di-Me, 2-Oxide, *in* A-**60092**
Di-1H-imidazol-2-ylmethanone; 1,1'-Di-Me, *in* D-**70354**

C₉H₁₀N₄O₃

2,4,6(1H,3H,5H)-Pteridinetrione; 1,3,5-Tri-Me, *in* P-70134
2,4,7(1H,3H,8H)-Pteridinetrione; 1,3,6-Tri-Me, *in* P-70135
2,4,7(1H,3H,8H)-Pteridinetrione; 1,3,8-Tri-Me, *in* P-70135
2,4,7(1H,3H,8H)-Pteridinetrione; 1,6,8-Tri-Me, *in* P-70135
2,4,7(1H,3H,8H)-Pteridinetrione; 3,6,8-Tri-Me, *in* P-70135

C₉H₁₀N₄O₄

4-Amino-4,6-dihydro-3-methyl-1H-cyclopenta[e]1,2,4-triazine-5,7-dicarboxylic acid, A-80066

C₉H₁₀N₄S

Tetrazole-5-thione; 3-Et, 1-Ph, *in* T-60174

C₉H₁₀O

2,7-Dimethyl-2,4,6-cycloheptatrien-1-one, D-70391
1-Ethenyl-4-methoxybenzene, *in* H-70209
1-Methoxy-1-phenylethylene, *in* P-60092
4-Nonene-6,8-diyn-1-ol, N-80072
1-Phenylcyclopropanol, P-70066
4-(1-Propenyl)phenol, P-80179
3,5,6,7-Tetrahydro-4H-inden-4-one, T-70070

C₉H₁₀OS

Benzyl vinyl sulfoxide, *in* B-80062
1-[4-(Methylthio)phenyl]ethanone, *in* M-60017
[(Phenylthio)methyl]oxirane, *in* M-80173

C₉H₁₀OSe

[2-(Methylseleninyl)ethenyl]benzene, *in* M-70132

C₉H₁₀O₂

2-Benzyloxyacetaldehyde, B-70085
Bicyclo[2.2.1]hepta-2,5-diene-2-carboxylic acid; Me ester, *in* B-60084
3,4-Dihydro-2H-1-benzopyran-2-ol, D-60198
3,4-Dihydro-2H-1-benzopyran-3-ol, D-60199
3,4-Dihydro-2H-1-benzopyran-4-ol, D-60200
3,4-Dihydro-2H-1-benzopyran-5-ol, D-60201
3,4-Dihydro-2H-1-benzopyran-6-ol, D-70175
3,4-Dihydro-2H-1-benzopyran-7-ol, D-60202
3,4-Dihydro-2H-1-benzopyran-8-ol, D-60203
5-Ethyl-2-hydroxy-2,4,6-cycloheptatrien-1-one, E-70044
2'-(Hydroxymethyl)acetophenone, H-70158
4'-(Hydroxymethyl)acetophenone, H-70159
2-Hydroxy-2-phenylpropanal, H-80240
1-(4-Hydroxyphenyl)-2-propanone, H-80241
2-Methoxybenzeneacetaldehyde, *in* H-60207
α-Methoxybenzeneacetaldehyde, *in* H-60210
2-Methoxy-4-methylbenzaldehyde, *in* H-70160
3-Methoxy-5-methylbenzaldehyde, *in* H-80195
4-Methoxy-2-methylbenzaldehyde, *in* H-70161
4-Methoxyphenylacetaldehyde, *in* H-60209
Spiro[4.4]non-2-ene-1,4-dione, S-60042

C₉H₁₀O₂S

Benzyl vinyl sulfone, *in* B-80062
Isopropenyl phenyl sulfone, *in* P-80137
2-Mercapto-2-phenylacetic acid; Me ester, *in* M-60024
2-Mercapto-3-phenylpropanoic acid, M-60026
[2-(Methylsulfonyl)ethenyl]benzene, *in* M-80170
3-(3-Thienyl)-2-propenoic acid; Et ester, *in* T-70172

C₉H₁₀O₂Se

[2-(Methylselenonyl)ethenyl]benzene, *in* M-70132

C₉H₁₀O₃

Bicyclo[3.3.1]nonane-2,4,7-trione, B-80082
Bicyclo[3.3.1]nonane-2,4,9-trione, B-80083
2-Ethoxybenzoic acid, *in* H-70108
▷ Ethyl salicylate, *in* H-70108
4-Hydroxybenzyl alcohol; α-Ac, *in* H-80114
3-Hydroxy-4-methoxyphenylacetaldehyde, *in* D-60371
4-Hydroxy-3-methoxyphenylacetaldehyde, *in* D-60371
2-(Hydroxymethyl)-1,4-benzodioxan, H-60172

2-Hydroxy-4-methylbenzoic acid; Me ester, *in* H-70163
2-Methoxy-4-methylbenzoic acid, *in* H-70163
4-Methoxy-2-methylbenzoic acid, *in* H-70164

C₉H₁₀O₄

2,6-Dihydroxy-3,5-dimethylbenzoic acid, D-70293
2',6'-Dihydroxy-4'-methoxyacetophenone, *in* T-70248
2-(3,4-Dihydroxyphenyl)propanoic acid, D-60372
2,3-Dimethoxy-5-methyl-1,4-benzoquinone, *in* D-70320
Glycol salicylate, *in* H-70108
(2-Hydroxyphenoxy)acetic acid; Me ester, *in* H-70203
Methoxymethyl salicylate, *in* H-70108
(2-Methoxyphenoxy)acetic acid, *in* H-70203
(3-Methoxyphenoxy)acetic acid, *in* H-70204
(4-Methoxyphenoxy)acetic acid, *in* H-70205
2',4',6'-Trihydroxy-3'-methylacetophenone, T-80321
3-(3,4,5-Trihydroxyphenyl)propanal, T-80333

C₉H₁₀O₄S

2H-Thiopyran-2,2-dicarboxylic acid; Di-Me ester, *in* T-80191

C₉H₁₀O₅

3,4-Dihydroxy-5-methoxybenzeneacetic acid, *in* T-60343
3,5-Dihydroxy-4-methoxybenzeneacetic acid, *in* T-60343
5-Hydroxy-2,3-dimethoxybenzoic acid, *in* T-70250
3,4,6-Trihydroxy-2-methylbenzoic acid; Me ester, *in* T-80324
3,4,5-Trihydroxyphenylacetic acid; Me ester, *in* T-60343

C₉H₁₀O₆

3,5-Dihydroxy-4-oxo-4H-pyran-2-carboxylic acid; Di-Me ether, Me ester, *in* D-60368

C₉H₁₀S

Benzyl vinyl sulfide, B-80062
[1-(Methylthio)ethenyl]benzene, M-80169
[2-(Methylthio)ethenyl]benzene, M-80170

C₉H₁₀S₂

1,5-Dihydro-2,4-benzodithiepin, D-60194
3,4-Dihydro-2H-1,5-benzodithiepin, D-60195

C₉H₁₀Se

[2-(Methylseleno)ethenyl]benzene, M-70132

C₉H₁₁BrO₃

1-Bromo-2,3,4-trimethoxybenzene, *in* B-60200
1-Bromo-2,3,5-trimethoxybenzene, *in* B-70182
1-Bromo-2,4,5-trimethoxybenzene, *in* B-70181
1-Bromo-3,4,5-trimethoxybenzene, *in* B-70180
2-Bromo-1,3,5-trimethoxybenzene, *in* B-60199
5-Bromo-1,2,3-trimethoxybenzene, *in* B-60201

C₉H₁₁N

Azetidine; N-Ph, *in* A-70283
Bicyclo[2.2.2]oct-2-ene-1-carboxylic acid; Nitrile, *in* B-70113
1,2,3,4-Tetrahydroquinoline, T-70093

C₉H₁₁NO

2-(Dimethylamino)benzaldehyde, D-60401
N-Methylbenzenecarboximidic acid; Me ester, *in* M-60045
1,2,3,4-Tetrahydro-4-hydroxyisoquinoline, T-60066
1,2,3,4-Tetrahydro-5-hydroxyisoquinoline, T-80065
1,2,3,4-Tetrahydro-7-hydroxyisoquinoline, T-80066
6,7,8,9-Tetrahydro-4H-quinolizin-4-one, T-60092

C₉H₁₁NOS

2-(1,3-Oxathian-2-yl)pyridine, O-60049

C₉H₁₁NO₂

▷ 2-Ethoxybenzamide, *in* H-70108
2-Hydroxyphenylacetaldehyde; Me ether, oxime, *in* H-60207
3-Hydroxyphenylacetaldehyde; Me ether, oxime, *in* H-60208

4-Hydroxyphenylacetaldehyde; Me ether, oxime, *in* H-60209
N-Methylphenylacetohydroxamic acid, *in* P-70059
3a,4,7,7a-Tetrahydro-1H-isoindole-1,3(2H)-dione; N-Me, *in* T-80070

C₉H₁₁NO₂S

2-Amino-3-mercapto-3-phenylpropanoic acid, A-60213
2-Amino-3-(4-mercaptophenyl)propanoic acid, A-60214
2-Amino-4,5,6,7-tetrahydrobenzo[b]thiophene-3-carboxylic acid, A-60258

C₉H₁₁NO₃

4-Amino-1,2-benzenediol; O¹-Me, 2-Ac, *in* A-70090
4-Amino-1,2-benzenediol; O²-Me, 1-Ac, *in* A-70090
3-Amino-3-(4-hydroxyphenyl)propanoic acid, A-80084
2,3-Dihydroxybenzaldehyde; Di-Me ether, oxime, *in* D-70285
2,4-Dihydroxyphenylacetaldehyde; Di-Me ether, oxime, *in* D-60369
3,4-Dihydroxyphenylacetaldehyde; 3-Me ether, oxime, *in* D-60371
(2-Hydroxyphenoxy)acetic acid; Me ether, amide, *in* H-70203

C₉H₁₁NO₄

1,4-Dihydro-2,6-dimethyl-3,5-pyridinedicarboxylic acid, D-80193

C₉H₁₁NO₅

3-Hydroxy-1H-pyrrole-2,5-dicarboxylic acid; Me ether, Di-Me ester, *in* H-80252
1H-Pyrrole-3,4-dicarboxylic acid; 1-Methoxy, Di-Me ester, *in* P-70183

C₉H₁₁NS

2-(Tetrahydro-2-thienyl)pyridine, T-60097
2,3,4,5-Tetrahydrothiepino[2,3-b]pyridine, T-60098

C₉H₁₁NS₂

2-(1,3-Dithian-2-yl)pyridine, D-60504

C₉H₁₁N₅

5-Amino-1H-tetrazole; 1-Me, 4-Benzyl, *in* A-70202
5-Amino-1H-tetrazole; 2-Me, 5-N-Benzyl, *in* A-70202
1-Benzyl-5-imino-3-methyl-1H,3H-tetrazole, *in* A-70202
3-Benzyl-5-imino-1-methyl-1H,3H-tetrazole, *in* A-70202
1-Benzyl-5-methylamino-1H-tetrazole, *in* A-70202

C₉H₁₁N₅O₂

6-Acetylhomopterin, A-70043
Deoxysepiapterin, *in* S-70035

C₉H₁₁N₅O₃

Biopterin, B-70121
Sepiapterin, S-70035

C₉H₁₂

Bicyclo[3.3.1]nona-2,6-diene, B-70102
2,3-Dimethylenebicyclo[2.2.1]heptane, D-70405
▷ 5-Ethenylbicyclo[2.2.1]hept-2-ene, E-70036
1-Ethynylcycloheptene, E-70050
1,3-Nonadiyne, N-80069
1,3,5,7-Nonatetraene, N-70071
Trispiro[2.0.2.0.2.0]nonane, T-70325

C₉H₁₂ClN₃O₄

5'-Chloro-5'-deoxyarabinosylcytosine, C-70064

C₉H₁₂N₂

6,7,8,9-Tetrahydro-5H-pyrido[2,3-b]azepine, T-70091
6,7,8,9-Tetrahydro-5H-pyrido[3,2-b]azepine, T-70092

C₉H₁₂N₂O

Acetamidoxime; O-Benzyl, *in* A-60016

4-Acetyl-2-aminopyridine; N^2,N^2-Di-Me, *in* A-70026
2-(Dimethylamino)benzaldehyde; Oxime, *in* D-60401

$C_9H_{12}N_2OS$

2-Amino-4,5,6,7-tetrahydrobenzo[*b*]thiophene-3-carboxylic acid; Amide, *in* A-60258

$C_9H_{12}N_2O_2$

2,4,6-Trimethyl-3-nitroaniline, T-70285

$C_9H_{12}N_2O_3$

2-Amino-3-nitrobenzyl alcohol; *N*-Et, *in* A-80125
3-Amino-2-nitrobenzyl alcohol; *N*-Et, *in* A-80127
3-Amino-4-nitrobenzyl alcohol; *N*-Et, *in* A-80128

$C_9H_{12}N_4O_5$

5-Nitrohistidine; Me ester, N^α-Ac, *in* N-60028

$C_9H_{12}N_6$

2,3,6,7,10,11-Hexahydrotrisimidazo[1,2-*a*;1′,2′-*c*;1″,2″-*e*][1,3,5]-triazine, H-60061

$C_9H_{12}O$

Bicyclo[3.3.1]non-3-en-2-one, B-70105
Bicyclo[4.2.1]non-3-en-9-one, B-70106
Bicyclo[4.2.1]non-3-en-9-one, B-70107
Bicyclo[4.2.1]non-7-en-9-one, B-70108
3*a*,4,5,6,7,7*a*-Hexahydro-1*H*-inden-1-one, H-70058
Isopropyl phenyl ether, I-80085
3,4,5-Trimethylphenol, T-60371

$C_9H_{12}O_2$

Bicyclo[3.3.1]nonane-2,4-dione, B-70103
Bicyclo[2.2.2]oct-2-ene-1-carboxylic acid, B-70113
2-Carbomethoxy-3-vinylcyclopentene, *in* V-60010
1,3-Diacetylbicyclo[1.1.1]pentane, D-80041
2,2-Dimethyl-5-cycloheptene-1,3-dione, D-60416
3-Ethynyl-3-methyl-4-pentenoic acid; Me ester, *in* E-60059
2,3,4,4*a*,5,6-Hexahydro-7*H*-1-benzopyran-7-one, H-60045
3*a*,4,5,6,7,7*a*-Hexahydro-3-methylene-2-(3*H*)-benzofuranone, H-70062
4-Hydroxybenzyl alcohol; Di-Me ether, *in* H-80114
4-Nonen-2-ynoic acid, N-70072
1-Phenyl-1,2-propanediol, P-80121
1-Phenyl-1,3-propanediol, P-80122

$C_9H_{12}O_3$

1,3,5-Cyclohexanetricarboxaldehyde, C-70226
6-Ethyl-4-hydroxy-3,5-dimethyl-2*H*-pyran-2-one, E-80065
Hexahydro-1*H*-2-benzopyran-1,3(4*H*)-dione, *in* C-8□□3
1-Hydro□ □ cyclohexene-1-carboxaldehyde; Ac, *in* H 8 146
Metacrolein, □ P-70119
4-Oxo-1-cycloheptene-1-carboxylic acid; Me ester, *in* O-70082

$C_9H_{12}O_4$

Aucubigenin, A-80186
Bicyclo[1.1.1]pentane-1,3-dicarboxylic acid; Di-Me ester, *in* B-80090
4-Cyclopentene-1,3-diol; Di-Ac, *in* C-80193
4-Cyclopentene-1,3-diol; Mono-Ac, *in* C-80193
5,5-Dimethyl-2-hexynedioic acid; Di-Me ester, *in* D-70415
5-Ethyl-3-hydroxy-4-methyl-2(5*H*)-furanone; Ac, *in* E-60051
Jiofuran, J-80009
2,3,6-Trimethoxyphenol, *in* B-60014
Vertipyronol, V-80005
2-Vinyl-1,1-cyclopropanedicarboxylic acid; Di-Me ester, *in* V-60011

$C_9H_{12}O_5$

4-Oxo-2-heptenedioic acid; Di-Me ester, *in* O-60072
Rehmaglutin *C*, R-60003
Vertipyronediol, *in* V-80005

$C_9H_{12}O_6$

Aconitic acid; Tri-Me ester, *in* A-70055

$C_9H_{12}S_2$

2-Methyl-1,3-benzenedimethanethiol, M-70050

$C_9H_{12}S_3$

1,3,5-Benzenetrimethanethiol, B-70019

$C_9H_{13}NO$

1,2,3,6,7,9*a*-Hexahydro-4(1*H*)-quinolizinone, H-70074
3,4,6,7,8,9-Hexahydro-2*H*-quinolizin-2-one, H-60059

$C_9H_{13}NO_2$

2,4-Diethoxypyridine, *in* H-60224
2,5-Diethoxypyridine, *in* H-60225
2,6-Diethoxypyridine, *in* H-60226
3,4-Diethoxypyridine, *in* H-60227
Hexahydro-1,3(2*H*,4*H*)-isoquinolinedione, H-80060

$C_9H_{13}N_3$

1,2,12-Triazapentacyclo[6.4.0.0^{2,17}.0^{3,7}.0^{4,11}]dodecane, T-70205

$C_9H_{13}N_3O_3$

4-Amino-5-methyl-1*H*-pyrazole-3-carboxylic acid; *N*-Ac, Et ester, *in* A-80111
5-Amino-1*H*-pyrazole-4-carboxylic acid; *N*(1)-Me, *N*(5)-Ac, Et ester, *in* A-70192
2′,3′-Dideoxycytidine, D-80152

$C_9H_{13}N_5$

2-(Diethylamino)purine, *in* A-60243

$C_9H_{13}N_5O_6$

Clitocine, C-60155

C_9H_{14}

7,7-Dimethylbicyclo[4.1.0]hept-3-ene, D-60409
1-Isopropenylcyclohexene, I-60115
3-Isopropenylcyclohexene, I-60116
4-Isopropenylcyclohexene, I-60117
2,3,5-Nonatriene, N-80071
3-Nonen-1-yne, N-80073
1-(1-Propenyl)cyclohexene, P-60178
1-(2-Propenyl)cyclohexene, P-60179
3-(2-Propenyl)cyclohexene, P-60180
4-(1-Propenyl)cyclohexene, P-60181
4-(2-Propenyl)cyclohexene, P-60182

$C_9H_{14}ClNO$

7-Aminobicyclo[2.2.1]heptane; *N*-Chloroacetyl, *in* A-80061

$C_9H_{14}Cl_2O_2$

2,2-Dichloro-6-heptenoic acid; Et ester, *in* D-80119

$C_9H_{14}N_2O_5$

α-Amino-2-carboxy-5-oxo-1-pyrrolidinebutanoic acid, A-70120
3,4-Dihydro-2(1*H*)-pyrimidinone; 1-β-D-Ribofuranosyl, *in* D-60273
3,4-Dihydro-2(1*H*)-pyrimidinone; 3-β-D-Ribofuranosyl, *in* D-60273

$C_9H_{14}N_4$

2,3,5,6,8,9-Hexahydro-1*H*-diimidazo[1,2-*d*:2′,1′-*g*][1,4]diazepine, H-80058

$C_9H_{14}N_4O_3$

Caffeidine; Ac, *in* C-60003
Carnosine, C-60020

$C_9H_{14}N_4O_5$

▷ N^4-Aminocytidine, A-70127

$C_9H_{14}O$

1-Acetylcycloheptene, A-60026
2-(1-Cyclohexenyl)propanal, C-80183
1-Cyclooctenecarboxaldehyde, C-60217
2-Oxatricyclo[3.3.1.1^{3,7}]decane, O-60051
2,2,5,5-Tetramethyl-3-cyclopenten-1-one, T-60140
2,3,4-Trimethyl-2-cyclohexen-1-one, T-60359

$C_9H_{14}OS_3$

Ajoene, A-70079

$C_9H_{14}O_2$

Bicyclo[2.2.1]heptane-1-carboxylic acid; Me ester, *in* B-60085

Bicyclo[2.2.2]octane-1-carboxylic acid, B-60092
Bicyclo[3.2.1]octane-1-carboxylic acid, B-70110
1-Cycloheptenecarboxylic acid; Me ester, *in* C-80178
7-Ethyl-5-methyl-6,8-dioxabicyclo[3.2.1]oct-3-ene, E-60052
1-Methyl-2-cyclohexen-1-ol; Ac, *in* M-80065
7-Methylene-1,4-dioxaspiro[4.5]decane, *in* M-80081
5-Methyl-2,4-hexadienoic acid; Et ester, *in* M-70086
Mitsugashiwalactone, M-70150
2,6-Octadienoic acid; Me ester, *in* O-80011
Octahydro-1*H*-2-benzopyran-1-one, O-80012
Octahydro-7*H*-1-benzopyran-7-one, O-60011
4-Oxo-2-nonenal, O-70097

$C_9H_{14}O_2S$

2-Thiatricyclo[3.3.1.1^{3,7}]decane; 2,2-Dioxide, *in* T-60195

$C_9H_{14}O_3$

2-*tert*-Butyl-2,4-dihydro-6-methyl-1,3-dioxol-4-one, B-60345
2-*tert*-Butylpentanedioic acid; Anhydride, *in* B-80289
2,3-Diisopropoxycyclopropenone, *in* D-70290
2-(3,3-Dimethoxypropyl)furan, *in* F-80089
3-Oxocyclopentaneacetic acid; Et ester, *in* O-70083

$C_9H_{14}O_4$

2-Carboxycyclohexaneacetic acid, C-80023
1,2-Cyclopentanediol; Di-Ac, *in* C-80189
1,1-Cyclopropanedicarboxylic acid; Di-Et ester, *in* C-80196
2-Hydroxycyclohexanecarboxylic acid; Ac, *in* H-70119
Jioglutolide, J-80012

$C_9H_{14}O_5$

Bissetone, B-60174
4-Oxoheptanedioic acid; Di-Me ester, *in* O-60071
4,4-Pyrandiacetic acid, P-70150

$C_9H_{14}S$

2,2,5,5-Tetramethyl-3-cyclopentene-1-thione, T-60139
2-Thiatricyclo[3.3.1.1^{3,7}]decane, T-60195

$C_9H_{15}ClO_2$

7-Chloro-2-heptenoic acid; Et ester, *in* C-70069

$C_9H_{15}N$

▷ Tri(2-propenyl)amine, T-70315

$C_9H_{15}NO$

7-Aminobicyclo[2.2.1]heptane; *N*-Ac, *in* A-80061
3-Aminobicyclo[2.2.1]hept-5-ene-2-methanol; *N*-Me, *in* A-70094
3,6-Dihydro-3,3,6,6-tetramethyl-2(1*H*)-pyridinone, D-70267
Octahydro-4*H*-quinolizin-4-one, O-60019
2-(2-Propenyl)cyclohexanone; Oxime, *in* P-80177

$C_9H_{15}NO_2$

7-Aminobicyclo[2.2.1]heptane; *N*-Methoxycarbonyl, *in* A-80061
3-*tert*-Butyl-2,6-piperidinedione, *in* B-80289
3,4-Dihydro-2*H*-pyrrole-5-carboxylic acid; *tert*-Butyl ester, *in* D-80249

$C_9H_{15}NO_3$

2-Amino-2-hexenoic acid; *N*-Ac, Me ester, *in* A-60167
Methyl 3-*tert*-butylamino-2-formyl-2-propenoate, *in* A-80134
5-Oxo-2-pyrrolidinecarboxylic acid; *tert*-Butyl ester, *in* O-70102

$C_9H_{15}NO_5$

4-Oxoheptanedioic acid; Di-Me ester, oxime, *in* O-60071

$C_{10}F_8$
Octafluoronaphthalene, O-70008

$C_{10}F_{11}IO_4$
[Bis(trifluoroacetoxy)iodo]pentafluorobenzene, B-80161

$C_{10}HF_7$
1,2,3,4,5,6,7-Heptafluoronaphthalene, H-70019
1,2,3,4,5,6,8-Heptafluoronaphthalene, H-70020

$C_{10}H_2F_4$
1,3-Diethynyl-2,4,5,6-tetrafluorobenzene, D-70159
1,4-Diethynyl-2,3,5,6-tetrafluorobenzene, D-70160

$C_{10}H_2F_6$
1,2,3,4,5,6-Hexafluoronaphthalene, H-70044
1,2,4,5,6,8-Hexafluoronaphthalene, H-70045

$C_{10}H_4^{2\ominus}$
Dihydrocyclopenta[c,d]pentalene(2−), D-60216

$C_{10}H_4Br_2O_2$
6,7-Dibromo-1,4-naphthoquinone, D-70096

$C_{10}H_4Cl_2FNO_2$
3,4-Dichloro-1-(4-fluorophenyl)-1H-pyrrole-2,5-dione, D-60132

$C_{10}H_4Cl_2O_2$
6,7-Dichloro-1,4-naphthoquinone, D-70124

$C_{10}H_4F_6$
1,4-Bis(trifluorovinyl)benzene, B-80165

$C_{10}H_4K_2$
Dihydrocyclopenta[c,d]pentalene(2−); Di-K salt, in D-60216

$C_{10}H_4O_2S_2$
Benzo[1,2-b:4,5-b']dithiophene-4,8-dione, B-60031

$C_{10}H_4S_8$
Bi(1,3-dithiolo[4,5-b][1,4]dithiin-2-ylidene), B-70118

$C_{10}H_5BrO_3$
2-Bromo-3-hydroxy-1,4-naphthoquinone, B-60268
2-Bromo-5-hydroxy-1,4-naphthoquinone, B-60269
2-Bromo-6-hydroxy-1,4-naphthoquinone, B-60270
2-Bromo-7-hydroxy-1,4-naphthoquinone, B-60271
2-Bromo-8-hydroxy-1,4-naphthoquinone, B-60272
7-Bromo-2-hydroxy-1,4-naphthoquinone, B-60273
8-Bromo-2-hydroxy-1,4-naphthoquinone, B-60274

$C_{10}H_5Br_2NO_2$
6,7-Dibromo-1-nitronaphthalene, D-70098

$C_{10}H_5ClO_3$
2-Chloro-6-hydroxy-1,4-naphthoquinone, C-80066
2-Chloro-7-hydroxy-1,4-naphthoquinone, C-80067

$C_{10}H_5F_3O$
1,1,1-Trifluoro-4-phenyl-3-butyn-2-one, T-70244

$C_{10}H_5NO_4$
6-Nitro-1,4-naphthoquinone, N-60035
7-Nitro-1,2-naphthoquinone, N-60036

$C_{10}H_5NO_5$
2-Hydroxy-3-nitro-1,4-naphthoquinone, H-60195

$C_{10}H_5N_3O_2$
1H-Naphtho[2,3-d]triazole-4,9-dione, N-60011

$C_{10}H_6$
1,3-Diethynylbenzene, D-70156
1-Phenyl-1,3-butadiyne, P-80095

$C_{10}H_6Br_2O$
6,7-Dibromo-1-naphthol, D-70095

$C_{10}H_6ClNO_2$
▷ 2-Amino-3-chloro-1,4-naphthoquinone, A-70122

$C_{10}H_6F_3NO_2$
7-Amino-4-(trifluoromethyl)-2H-1-benzopyran-2-one, A-60265

$C_{10}H_6F_4O_4$
2,3,5,6-Tetrafluoro-1,4-benzenedicarboxylic acid; Di-Me ester, in T-70029

$C_{10}H_6N_2O$
Benzofuro[2,3-b]pyrazine, B-80023
2-Cyano-3-hydroxyquinoline, in H-60228

$C_{10}H_6N_2OS$
[1,4]Oxathiino[3,2-b:5,6-c']dipyridine, O-60050

$C_{10}H_6N_2OSe$
[1,4]Oxaselenino[2,3-b:5,6-b']dipyridine, O-70065
[1,4]Oxaselenino[3,2-b:5,6-b']dipyridine, O-70066
[1,4]Oxaselenino[3,2-b:5,6-c']dipyridine, O-70067
[1,4]Oxaselenino[3,2-b:6,5-c']dipyridine, O-70068

$C_{10}H_6N_2O_2$
[1]Benzopyrano[3,4-c]pyrazol-4(3H)-one, B-80036
3,4-Cinnolinedicarboxaldehyde, C-70180
3,8-Dihydrocyclobuta[b]quinoxaline-1,2-dione, D-70178
▷ 1,4-Diisocyanatocubane, D-80406

$C_{10}H_6N_2O_2S$
[1,4]Oxathiino[3,2-b:5,6-c']dipyridine; 8-Oxide, in O-60050

$C_{10}H_6N_2O_2Se$
[1,4]Oxaselenino[3,2-b:5,6-c']dipyridine; 8-Oxide, in O-70067
[1,4]Oxaselenino[3,2-b:6,5-c']dipyridine; 7-Oxide, in O-70068

$C_{10}H_6N_2O_4$
7-Nitro-4-nitroso-1-naphthol, in N-60035

$C_{10}H_6N_2S$
Thieno[2,3-c]cinnoline, T-60202
Thieno[3,2-c]cinnoline, T-60203

$C_{10}H_6N_2S_2$
[1,4]Dithiino[2,3-b:6,5-b']dipyridine, D-70519
1,4-Dithiino[2,3-b:5,6-c']dipyridine, D-70518
1,4-Dithiino[2,3-b:6,5-c']dipyridine, D-70520
Thiazolo[4,5-b]quinoline-2(3H)-thione, T-60201

$C_{10}H_6N_4O_2$
Pyrazino[2,3-f]quinazoline-8,10-(7H,9H)-dione, P-60201
Pyrazino[2,3-g]quinazoline-2,4-(1H,3H)-dione, P-60200

$C_{10}H_6N_6$
Pyridazino[1'',6'':1',2']imidazo[4',5':4,5]imidazo[1,2-b]pyridazine, P-70169

$C_{10}H_6O_3$
2,3-Epoxy-2,3-dihydro-1,4-naphthoquinone, E-70011
4-Hydroxy-1,2-naphthoquinone, H-60194

$C_{10}H_6O_3Se_2$
Naphth[1,8-cd][1,2,6]oxadiselenin 1,3-dioxide, in N-80002

$C_{10}H_6O_4$
2,5-Dihydroxy-1,4-naphthoquinone, D-60354
2,8-Dihydroxy-1,4-naphthoquinone, D-60355
▷ 5,8-Dihydroxy-1,4-naphthoquinone, D-60356
6,7-Methylenedioxy-2H-1-benzopyran-2-one, M-70068

$C_{10}H_6O_7$
2,3,5,6,8-Pentahydroxy-1,4-naphthoquinone, P-70033

$C_{10}H_6O_8$
Hexahydroxy-1,4-naphthoquinone, H-60065

$C_{10}H_6S_2$
Benzo[1,2-b:4,5-b']dithiophene, B-60030

$C_{10}H_6Se_2$
Naphtho[1,8-cd]-1,2-diselenole, N-80005

$C_{10}H_6Te_2$
Naphtho[1,8-cd]1,2-ditellurole, N-80006

$C_{10}H_7BrN_2$
2-Bromo-3,3'-bipyridine, B-70185
4-Bromo-2,2'-bipyridine, B-70186
5-Bromo-2,3'-bipyridine, B-70187
5-Bromo-2,4'-bipyridine, B-70188
5-Bromo-3,3'-bipyridine, B-70189
5-Bromo-3,4'-bipyridine, B-70190
6-Bromo-2,2'-bipyridine, B-70191
6-Bromo-2,3'-bipyridine, B-70192
6-Bromo-2,4'-bipyridine, B-70193

$C_{10}H_7Br_2N$
6,7-Dibromo-1-naphthylamine, D-70097

$C_{10}H_7Br_3$
1,4,7-Tribromotriquinacene, T-60248

$C_{10}H_7ClN_2$
2-Chloro-3-phenylpyrazine, C-70142
2-Chloro-5-phenylpyrazine, C-70143
2-Chloro-6-phenylpyrazine, C-70144
2-Chloro-4-phenylpyrimidine, C-80100
4-Chloro-2-phenylpyrimidine, C-80102
4-Chloro-5-phenylpyrimidine, C-80103
4-Chloro-6-phenylpyrimidine, C-80104
5-Chloro-2-phenylpyrimidine, C-80105
5-Chloro-4-phenylpyrimidine, C-80106

$C_{10}H_7ClN_2O$
2-Chloro-3-phenylpyrazine; 1-Oxide, in C-70142
2-Chloro-3-phenylpyrazine; 4-Oxide, in C-70142
2-Chloro-5-phenylpyrazine; 1-Oxide, in C-70143
2-Chloro-5-phenylpyrazine; 4-Oxide, in C-70143
2-Chloro-6-phenylpyrazine; 1-Oxide, in C-70144
2-Chloro-6-phenylpyrazine; 4-Oxide, in C-70144

$C_{10}H_7ClN_2O_2$
2-Chloro-3-phenylpyrazine; 1,4-Dioxide, in C-70142
2-Chloro-5-phenylpyrazine; 1,4-Dioxide, in C-70143

$C_{10}H_7ClO_3$
4-Hydroxy-5-benzofurancarboxylic acid; Me ether, chloride, in H-60103

$C_{10}H_7ClO_4$
6-Chloro-1,2,3,4-naphthalenetetrol, C-60118

$C_{10}H_7Cl_2NO_2$
5,6-Dichloro-1H-indole-3-acetic acid, D-70123

$C_{10}H_7F_5O_2$
Pentafluorophenylacetic acid; Et ester, in P-80027

$C_{10}H_7N$
2-Ethynylindole, E-60057
3-Ethynylindole, E-60058
4-Ethynylindole, E-80074

$C_{10}H_7NOS$
3-Phenyl-2-propenoyl isothiocyanate, P-70090

$C_{10}H_7NO_2$
3-Benzoylisoxazole, B-70075
4-Benzoylisoxazole, B-60062
5-Benzoylisoxazole, B-60063

$C_{10}H_7NO_2S$
2-Amino-3-mercapto-1,4-naphthoquinone, A-70163

$C_{10}H_7NO_3$
3-Hydroxy-2-quinolinecarboxylic acid, H-60228
2-Phenyl-4-oxazolecarboxylic acid, P-80109
2-Phenyl-5-oxazolecarboxylic acid, P-80110
5-Phenyl-2-oxazolecarboxylic acid, P-80111
5-Phenyl-4-oxazolecarboxylic acid, P-80112

$C_{10}H_7NO_4$
1-Nitro-2,6-naphthalenediol, N-**80**048
2-Nitro-1,4-naphthalenediol, N-**80**050
2-Nitro-1,5-naphthalenediol, N-**80**051
2-Nitro-1,7-naphthalenediol, N-**80**052
4-Nitro-1,2-naphthalenediol, N-**80**053
4-Nitro-1,5-naphthalenediol, N-**80**054
4-Nitro-1,7-naphthalenediol, N-**80**055
6-Nitro-2,3-naphthalenediol, N-**80**056
2-Oxo-3-indolineglyoxylic acid, O-**80**085

$C_{10}H_7NS$
2-(1,2-Propadienyl)benzothiazole, P-**60**174

$C_{10}H_7N_3$
1*H*-Imidazo[4,5-*f*]quinoline, I-70007
1*H*-Imidazo[4,5-*h*]quinoline, I-70008
Pyrazolo[3,4-*c*]quinoline, P-**60**206
1*H*-Pyrido[3,4,5-*de*]quinazoline, P-**80**217
1*H*-Pyrido[4,3,2-*de*]quinazoline, P-**80**218

$C_{10}H_7N_3O$
2,5-Dihydro-1*H*-dipyrido[4,3-*b*:3′,4′-*d*]pyrrol-1-one, D-**60**237
3,5-Dihydro-4*H*-pyrazolo[3,4-*c*]quinolin-4-one, D-**80**245

$C_{10}H_8$
Fulvalene, F-**60**080
Naphthvalene, N-**60**012
Tetracyclo[5.3.0.0^{2,4}.0^{3,5}]deca-6,8,10-triene, T-70019

$C_{10}H_8BrN$
3-Bromo-5-methylisoquinoline, B-**80**218
4-Bromo-1-methylisoquinoline, B-**80**219
5-Bromo-3-methylisoquinoline, B-**80**220

$C_{10}H_8ClN$
2-Chloro-1-naphthylamine, C-**80**080

$C_{10}H_8ClNO_4$
2[(Chlorocarbonyl)oxy]-3*a*,4,7,7*a*-tetrahydro-4,7-methano-1*H*-isoindole-1,3(2*H*)-dione, C-70059

$C_{10}H_8ClN_3O_4$
Pyrido[1′,2′:3,4]imidazo[1,2-*a*]pyrimidin-5-ium(1+); Perchlorate, *in* P-70176

$C_{10}H_8FN$
2-Fluoro-4-methylquinoline, F-**80**050
3-Fluoro-4-methylquinoline, F-**80**051
4-Fluoro-2-methylquinoline, F-**80**052
6-Fluoro-2-methylquinoline, F-**80**053
6-Fluoro-5-methylquinoline, F-**80**054
7-Fluoro-2-methylquinoline, F-**80**055

$C_{10}H_8N_2$
▷ 2,3′-Bipyridine, B-**60**120
2,4′-Bipyridine, B-**60**121
3,3′-Bipyridine, B-**60**122
3,4′-Bipyridine, B-**60**123
1,8-Dihydrobenzo[2,1-*b*:3,4-*b*′]dipyrrole, D-70174
5-Phenylpyrimidine, P-70093
▷ 1*H-Indole*-3-*acetonitrile*, *in* I-70013

$C_{10}H_8N_2O$
2-Acetylquinoxaline, A-**60**056
5-Acetylquinoxaline, A-**60**057
6-Acetylquinoxaline, A-**60**058
2-Benzoylimidazole, B-70074
4(5)-Benzoylimidazole, B-**60**057
3-Benzoylpyrazole, B-70078
4-Benzoylpyrazole, B-**60**065
3,3′-Bipyridine; Mono-*N*-oxide, *in* B-**60**122
2,3′-Bipyridine; 1′-Oxide, *in* B-**60**120
2,4′-Bipyridine; 1-Oxide, *in* B-**60**121
4,6-Dihydroxy-2,2′-bipyridine, D-**80**296
2′-Hydroxy-2,3′-bipyridine, H-**80**117
2′-Hydroxy-2,4′-bipyridine, H-**80**118
2-Hydroxy-3,3′-bipyridine, H-**80**119
2-Hydroxy-4,4′-bipyridine, H-**80**120
4-Hydroxy-2,2′-bipyridine, H-**80**121
4-Hydroxy-3,3′-bipyridine, H-**80**122
5-Hydroxy-2,2′-bipyridine, H-**80**123
5-Hydroxy-2,3′-bipyridine, H-**80**124
5-Hydroxy-3,3′-bipyridine, H-**80**125
6-Hydroxy-2,2′-bipyridine, H-**80**126
6-Hydroxy-2,3′-bipyridine, H-**80**127
6-Hydroxy-2,4′-bipyridine, H-**80**128
6-Hydroxy-3,3′-bipyridine, H-**80**129
6-Hydroxy-3,4′-bipyridine, H-**80**130
2,2′-Oxybispyridine, O-**60**098
2,3′-Oxybispyridine, O-**60**099
2,4′-Oxybispyridine, O-**80**099
3,3′-Oxybispyridine, O-**60**100
3,4′-Oxybispyridine, O-**80**100
4,4′-Oxybispyridine, O-**60**101
2-Phenyl-1*H*-imidazole-4(5)-carboxaldehyde, P-**60**111
3-Phenyl-4(1*H*)-pyridazinone, P-**80**124
5-Phenylpyrimidine; *N*-Oxide, *in* P-70093
1*H*-Pyrazole; *N*-Benzoyl, *in* P-70152
1-(2-Pyridyl)-2-pyridone, P-**80**219
1-(2-Pyridyl)-4-pyridone, P-**80**220
1-(3-Pyridyl)-4-pyridone, P-**80**221
1-(4-Pyridyl)-2-pyridone, P-**80**222
1-(4-Pyridyl)-4-pyridone, P-**80**223

$C_{10}H_8N_2OS$
2,3-Dihydro-6-phenyl-2-thioxo-4(1*H*)-pyrimidinone, D-**60**268

$C_{10}H_8N_2O_2$
5-Amino-3-phenylisoxazole; *N*-Formyl, *in* A-**60**240
4-Benzoylisoxazole; Oxime, *in* B-**60**062
2,3′-Bipyridine; 1,1′-Dioxide, *in* B-**60**120
2,4′-Bipyridine; 1,1′-Dioxide, *in* B-**60**121
2,2′-Dihydroxy-3,3′-bipyridine, D-**80**287
2,2′-Dihydroxy-4,4′-bipyridine, D-**80**288
2,4′-Dihydroxy-1,1′-bipyridine, D-**80**289
3,3′-Dihydroxy-2,2′-bipyridine, D-**80**290
3,3′-Dihydroxy-4,4′-bipyridine, D-**80**291
3,4′-Dihydroxy-2,3′-bipyridine, D-**80**292
4,4′-Dihydroxy-1,1′-bipyridine, D-**80**293
4,4′-Dihydroxy-2,2′-bipyridine, D-**80**294
4,4′-Dihydroxy-3,3′-bipyridine, D-**80**295
5,5′-Dihydroxy-2,2′-bipyridine, D-**80**297
5,5′-Dihydroxy-3,3′-bipyridine, D-**80**298
6,6′-Dihydroxy-2,2′-bipyridine, D-**80**299
6,6′-Dihydroxy-3,3′-bipyridine, D-**80**300
3-Hydroxy-2-quinolinecarboxylic acid; Amide, *in* H-**60**228
3,3′-Oxybispyridine; *N*-Oxide, *in* O-**60**100

$C_{10}H_8N_2O_2S$
▷ 2-Thioxo-4-imidazolidinone; 1-*N*-Benzoyl, *in* T-70188

$C_{10}H_8N_2O_2S_2Zn$
▷ Bis(1-hydroxy-2(1*H*)-pyridinethionato-*O,S*)zinc, *in* P-**60**224

$C_{10}H_8N_2O_3$
2-Acetylquinoxaline; 1,4-Dioxide, *in* A-**60**056
3,3′-Oxybispyridine; *N,N*′-Dioxide, *in* O-**60**100

$C_{10}H_8N_2O_3S$
Di-2-pyridyl sulfite, D-**60**497

$C_{10}H_8N_2O_4$
4,4′-Dihydroxy-2,2′-bipyridine; 1,1′-Dioxide, *in* D-**80**294

$C_{10}H_8N_2O_8$
4,6-Dinitro-1,3-benzenediol; Di-Ac, *in* D-**80**491

$C_{10}H_8N_2S$
2,3′-Thiobispyridine, T-**80**186
2,4′-Thiobispyridine, T-**80**187
3,4′-Thiobispyridine, T-**80**188

$C_{10}H_8N_2Se$
2,2′-Selenobispyridine, S-**80**023

$C_{10}H_8N_3^{\oplus}$
Pyrido[1′,2′:3,4]imidazo[1,2-*a*]pyrimidin-5-ium(1+), P-70176

$C_{10}H_8N_4$
4-Amino-1*H*-1,5-benzodiazepine-3-carbonitrile, A-70091
5-Amino-1*H*-pyrazole-4-carboxylic acid; *N*(1)-Ph, nitrile, *in* A-70192
Dipyrido[1,2-*b*:1′,2′-*e*][1,2,4,5]tetrazine, D-**60**494

$C_{10}H_8O$
2,8-Decadiene-4,6-diyn-1-al, D-**80**008
▷ 3-Phenyl-2-cyclobuten-1-one, P-70064

$C_{10}H_8OS$
1-Benzothiepin-5(4*H*)-one, B-**60**047

$C_{10}H_8O_2$
1,2-Di(2-furanyl)ethylene, D-**60**191
4-Hydroxy-1-phenyl-2-butyn-1-one, H-**80**233
6-Hydroxy-5-vinylbenzofuran, H-70231
3-Methyl-1*H*-2-benzopyran-1-one, M-70052
2-(3-Oxo-1-propenyl)benzaldehyde, O-**60**091
3-Phenyl-2(5*H*)furanone, P-**60**102
4-Phenyl-2(5*H*)-furanone, P-**60**103
5-Phenyl-2(3*H*)-furanone, P-70079

$C_{10}H_8O_2S$
5-[(5-Methyl-2-thienyl)methylene]-2(5*H*)-furanone, M-**80**161
5-(5-Methyl-2-thienyl)-2-penten-4-ynoic acid, M-**80**162
5-(2-Thienyl)-2-penten-4-ynoic acid; Me ester, *in* T-**80**183

$C_{10}H_8O_3$
3-Benzofurancarboxylic acid; Me ester, *in* B-**60**032
4,6-Dihydroxy-5-vinylbenzofuran, D-70352
5-Hydroxy-4-methyl-2*H*-1-benzopyran-2-one, H-**60**173
8-Hydroxy-4-methyl-2*H*-1-benzopyran-2-one, H-70165
6-Methoxy-2*H*-1-benzopyran-2-one, *in* H-70109
1,4,5-Naphthalenetriol, N-**60**005

$C_{10}H_8O_3S$
2-Oxo-4-phenylthio-3-butenoic acid, O-**60**088

$C_{10}H_8O_4$
Albidin, A-**60**070
4,7-Dihydroxy-5-methyl-2*H*-1-benzopyran-2-one, D-**60**348
3,4-Dimethoxbenzocyclobutene-1,2-dione, *in* D-**80**274
3,5-Dimethoxybenzocyclobutene-1,2-dione, *in* D-**80**275
3,6-Dimethoxybenzocyclobutene-1,2-dione, *in* D-**80**276
4,5-Dimethoxybenzocyclobutene-1,2-dione, *in* D-**80**277
4-Hydroxy-5-benzofurancarboxylic acid; Me ester, *in* H-**60**103
5-Hydroxy-3(2*H*)-benzofuranone; Ac, *in* H-**80**109
6-Hydroxy-3(2*H*)-benzofuranone; Ac, *in* H-**80**110
7-Hydroxy-6-(hydroxymethyl)-2*H*-1-benzopyran-2-one, H-70139
5-Hydroxy-4-(4-hydroxyphenyl)-2(5*H*)-furanone, H-**80**174
6-Hydroxy-7-methoxy-2*H*-1-benzopyran-2-one, H-**80**188
▷ 7-Hydroxy-6-methoxy-2*H*-1-benzopyran-2-one, H-**80**189
4-Methoxy-5-benzofurancarboxylic acid, *in* H-**60**103
1,4,5,7-Naphthalenetetrol, N-**80**003

$C_{10}H_8O_4Se_2$
1,8-Naphthalenediseleninic acid, N-**80**002

$C_{10}H_8O_5$
5,6-Dihydroxy-7-methoxy-2*H*-1-benzopyran-2-one, *in* T-**80**287
2-Hydroxy-4,5-methylenedioxybenzaldehyde; Ac, *in* H-70175

$C_{10}H_8O_6$
2-Acetoxy-1,3-benzenedicarboxylic acid, *in* H-**60**101
1,4-Benzoquinone-2,3-dicarboxylic acid; Di-Me ester, *in* B-70047
1,4-Benzoquinone-2,5-dicarboxylic acid; Di-Me ester, *in* B-70048

$C_{10}H_8S_2$
1,2-Naphthalenedithiol, N-70002
1,8-Naphthalenedithiol, N-70003
2,3-Naphthalenedithiol, N-70004

$C_{10}H_8S_8$
2-(5,6-Dihydro-1,3-dithiolo[4,5-*b*][1,4]dithiin-2-ylidene)-5,6-dihydro-1,3-dithiolo[4,5-*b*][1,4]dithiin, D-**60**238

$C_{10}H_9Br$

3-Bromo-1-phenyl-1-butyne, B-60314

$C_{10}H_9BrN_2$

4(5)-Bromo-1H-imidazole; 1-Benzyl, in B-70221

$C_{10}H_9BrO$

7-Bromo-3,4-dihydro-1(2H)-naphthalenone, B-60234

$C_{10}H_9ClN_2$

1-Benzyl-3-chloropyrazole, in C-70155
1-Benzyl-5-chloropyrazole, in C-70155

$C_{10}H_9IO_2$

1,3-Diacetyl-5-iodobenzene, D-60024

$C_{10}H_9N$

2-Vinylindole, V-70013

$C_{10}H_9NO$

2-Acetylindole, A-60032
4-Acetylindole, A-60033
▷ 5-Acetylindole, A-60034
6-Acetylindole, A-60035
7-Acetylindole, A-60036
1,2-Dihydro-1-phenyl-3H-pyrrol-3-one, in D-70262
1-(2-Furanyl)-2-(2-pyrrolyl)ethylene, F-60088
1-(4-Hydroxyphenyl)pyrrole, H-60216
2-(2-Hydroxyphenyl)pyrrole, H-60217
6-Methoxyisoquinoline, in H-80178
3-Methyl-1H-indole-2-carboxaldehyde, M-70090
2-Methyl-5-phenyloxazole, M-80145

$C_{10}H_9NOS$

1,4-Benzothiazepin-5(4H)-one; N-Me, in B-80040

$C_{10}H_9NO_2$

2-Acetyl-3-aminobenzofuran, A-70021
1-Amino-2,6-naphthalenediol, A-80112
1-Amino-2,7-naphthalenediol, A-80113
2-Amino-1,4-naphthalenediol, A-80114
2-Amino-1,6-naphthalenediol, A-80115
3-Amino-1,2-naphthalenediol, A-80116
4-Amino-1,2-naphthalenediol, A-80117
4-Amino-1,3-naphthalenediol, A-80118
4-Amino-1,5-naphthalenediol, A-80119
4-Amino-1,7-naphthalenediol, A-80120
5-Amino-1,2-naphthalenediol, A-80121
5-Amino-2,3-naphthalenediol, A-80122
6-Amino-2,3-naphthalenediol, A-80123
6-Amino-1,2-napthalenediol, A-80124
3-Azetidinone; 1-Benzoyl, in A-80192
6-Hydroxyisoquinoline; Me ether, N-oxide, in H-80178
4-Hydroxy-5-nitro-1,2-benzenedicarboxylic acid; Di-Me ester, in H-70191
▷ 1H-Indole-3-acetic acid, I-70013
1H-Indole-3-carboxylic acid; Me ester, in I-80012
1H-Indole-3-carboxylic acid; N-Me, in I-80012
Phthalimidine; N-Ac, in P-70100

$C_{10}H_9NO_2S$

α-Aminobenzo[b]thiophene-3-acetic acid, A-60090
3-Amino-2-benzo[b]thiophenecarboxylic acid; Me ester, in A-60091

$C_{10}H_9NO_2S_2$

N-Phenyl-2-thiophenesulfonamide, in T-60213
N-Phenyl-3-thiophenesulfonamide, in T-60214

$C_{10}H_9NO_3$

2,3-Dihydro-2-oxo-1H-indole-3-acetic acid, D-60259
2-Nitro-3,4-dihydro-1(2H)naphthalenone, N-60026
1,2,3,4-Tetrahydro-2-oxo-4-quinolinecarboxylic acid, T-60077
1,2,3,4-Tetrahydro-4-oxo-6-quinolinecarboxylic acid, T-60078
1,2,3,4-Tetrahydro-4-oxo-7-quinolinecarboxylic acid, T-60079

$C_{10}H_9NO_6$

2-Nitro-1,3-benzenedicarboxylic acid; Di-Me ester, in N-80035

$C_{10}H_9NO_7$

4-Hydroxy-5-nitro-1,3-benzenedicarboxylic acid; Di-Me ester, in H-70192
4-Hydroxy-3-nitro-1,2-benzenedicarboxylic acid; Me ether, 2-Me ester, in H-70190

$C_{10}H_9NS$

1-Amino-2-naphthalenethiol, A-70172
8-Amino-1-naphthalenethiol, A-70173
3(2H)-Isoquinolinethione; S-Me; B,HCl, in I-60133
1-Methylthioisoquinoline, in I-60132
4-Methylthioquinoline, in Q-60006
5-Methylthioquinoline, in Q-60004
6-Methylthioquinoline, in Q-60005
1-(2-Pyrrolyl)-2-(2-thienyl)ethylene, P-60250
3-Quinolinethiol; S-Me, in Q-60003
4(1H)-Quinolinethione; N-Me, in Q-60006

$C_{10}H_9NS_2$

2-Thiazoledinethione; N-Ph, in T-60197

$C_{10}H_9N_3$

6-Methyl-3-phenyl-1,2,4-triazine, M-60118
3-(Phenylazo)-2-butenenitrile, P-60079

$C_{10}H_9N_3O$

1-Acetamidophthalazine, in A-70184
2-Acetylquinoxaline; Oxime, in A-60056
4-Benzoyl-1,2,3-triazole; 1-Me, in B-70079
4-Benzoyl-1,2,3-triazole; 2-Me, in B-70079
4-Benzoyl-1,2,3-triazole; 3-Me, in B-70079

$C_{10}H_9N_3O_2$

4-Amino-1H-pyrazole-3-carboxylic acid; 1-Ph, in A-70189
2-Phenyl-2H-1,2,3-triazole-4-carboxylic acid; Me ester, in P-60140

$C_{10}H_9N_3O_9$

3-Hydroxy-5-methyl-2,4,6-trinitrobenzoic acid; Me ether, Me ester, in H-60193

$C_{10}H_{10}$

Bi-2,4-cyclopentadien-1-yl, B-60100
3-Cyclodecene-1,5-diyne, C-80177
7,8-Dimethylenebicyclo[2.2.2]octa-2,5-diene, D-70407
Hexacyclo[4.4.0.0^{2,5}.0^{3,9}.O^{4,8}.0^{7,10}]decane, H-70032
9-Methylene-1,3,5,7-cyclononatetraene, M-60073
1-Methyleneindane, M-60078
1,4,5,6-Tetrahydrocycloprop[e]indene, T-70053
1,3,4,5-Tetrahydrocycloprop[f]indene, T-70052
1,3,4,5-Tetrahydrocycloprop[f]indene, T-80056
Tricyclo[5.3.0.0^{2,8}]deca-3,5,9-triene, T-70227

$C_{10}H_{10}N_2$

1-Amino-3-methylisoquinoline, A-80098
3-Amino-1-methylisoquinoline, A-80099
3-Amino-4-methylisoquinoline, A-80100
4-Amino-1-methylisoquinoline, A-80101
5-Amino-1-methylisoquinoline, A-80102
7-Amino-1-methylisoquinoline, A-80104
8-Amino-1-methylisoquinoline, A-80105
1,8-Diaminonaphthalene, D-70042
1,4-Dihydro-2-phenylpyrimidine, D-70249

$C_{10}H_{10}N_2O$

5-Acetylindole; Oxime, in A-60034
5-Amino-3-phenylisoxazole; N-Me, in A-60240
5,6-Dihydro-2-phenyl-4(1H)-pyrimidinone, D-70272
1H-Indole-3-acetic acid; Amide, in I-70013

$C_{10}H_{10}N_2O_2$

1,2-Diazetidin-3-one; 1-Ac, 2-Ph, in D-70055
5,6-Dihydro-6-phenyl-2,4(1H, 3H)-pyrimidinedione, D-80240
1,4-Dihydro-2,3-quinoxalinedione; 1,4-Di-Me, in D-60279
2,3-Dimethoxyquinoxaline, in D-60279
5,8-Dimethoxyquinoxaline, in Q-60002
6,7-Dimethoxyquinoxaline, in Q-60009
3-Ethoxy-2-quinoxalinol, in D-60279
3-Ethoxy-2(1H)quinoxalinone, in D-60279
5,8-Quinoxalinediol; Di-Me ether, in Q-60008

$C_{10}H_{10}N_2O_4$

1,4-Dihydropyrrolo[3,2-b]pyrrole; 1,4-Bis(methoxycarbonyl), in D-30286

$C_{10}H_{10}N_2O_6$

3-Amino-4-nitro-1,2-benzenedicarboxylic acid; Di-Me ester, in A-70174
3-Amino-6-nitro-1,2-benzenedicarboxylic acid; Di-Me ester, in A-70175
4-Amino-3-nitro-1,2-benzenedicarboxylic acid; Di-Me ester, in A-70176

$C_{10}H_{10}N_4$

2-Hydroxy-4,4'-bipyridine; Hydrazone, in H-80120

$C_{10}H_{10}N_4O$

3-Amino-1H-pyrazole-4-carboxylic acid; Anilide, in A-70187
4-Amino-1H-pyrazole-3-carboxylic acid; 1-Ph, amide, in A-70189
5-Amino-1H-pyrazole-4-carboxylic acid; N(1)-Ph, amide, in A-70192

$C_{10}H_{10}O$

2,3-Dihydro-6(1H)-azulenone, D-60193
4,7-Dihydro-4,7-ethanoisobenzofuran, D-60240
2-Methyl-1-phenyl-2-propen-1-one, M-80146
Pentacyclo[5.3.0.0^{2,5}.0^{3,9}.0^{4,8}]decan-6-one, P-60027
1-Phenyl-3-buten-1-one, P-60085
1-Phenyl-2-butyn-1-ol, P-70062
4-Phenyl-3-butyn-2-ol, P-70063
2-(2-Propenyl)benzaldehyde, P-70122
4-(2-Propenyl)benzaldehyde, P-60177
[3](2,7)Troponophane, T-70329

$C_{10}H_{10}OS$

1-Methylene-2-(phenylsulfinyl)cyclopropane, in M-80090

$C_{10}H_{10}O_2$

2-Benzoylpropanal, B-80057
1,3,5,7-Cyclooctatetraene-1-acetic acid, C-70237
Cyclooctatetraenecarboxylic acid; Me ester, in C-70239
2,5,8-Decatriyne-1,10-diol, D-80016
8-Decene-4,6-diynoic acid, D-70018
3a,6a-Dihydro-3a,6a-dimethyl-1,6-pentalenedione, D-70199
Dihydro-5-phenyl-2(3H)-furanone, D-80239
2,3-Dimethyl-1,4-benzodioxin, D-70370
3,3-Dimethyl-2(3H)-benzofuranone, D-80418
4,9-Dioxa-1,6,11-dodecatriyne, D-80499
11,12-Dioxahexacyclo[6.2.1.1^{3,6}.0^{2,7}.0^{4,10}.0^{5,9}]dodecane, D-70461
5,12-Dioxatetracyclo[7.2.1.0^{4,11}.0^{6,10}]dodeca-2,7-diene, D-70463
Dispiro[2.0.2.4]dec-8-ene-7,10-dione, D-60501
(4-Hydroxyphenyl)ethylene; Ac, in H-70209
Isosafrole, I-80089
3-(4-Methoxyphenyl)-2-propenal, in H-60213
2-Phenyl-3-butenoic acid, P-80096
2-Phenylcyclopropanecarboxylic acid, P-60087
5,6,7,8-Tetrahydro-1,4-naphthoquinone, T-80076

$C_{10}H_{10}O_2S$

2-Acetyl-3-methoxy-5-(1-propynyl)thiophene, in A-80024
4'-Mercaptoacetophenone; Ac, in M-60017
1-Methylene-2-(phenylsulfonyl)cyclopropane, in M-80090
S-Phenyl 3-oxobutanethioate, in O-70080
2-Phenylsulfonyl-1,3-butadiene, P-70094

$C_{10}H_{10}O_2S_2$

2,3,8,9-Tetrahydrobenzo[2,1-b:3,4-b']bis[1,4]oxathiin, T-70040

$C_{10}H_{10}O_3$

2-Acetylbenzoic acid; Me ester, in A-80015
5-Acetyl-2-methoxybenzaldehyde, in A-80022
Coniferaldehyde, C-80148
3,4-Dihydro-4,8-dihydroxy-1(2H)-naphthalenone, D-80181
3,6-Dimethoxybenzocyclobuten-1-one, in D-80278
4,5-Dimethoxybenzocyclobuten-1-one, in D-80279

4,6-Dimethoxybenzocyclobuten-1-one, *in* D-**80**280

5,6-Dimethoxybenzocyclobuten-1-one, *in* D-**80**281

4-Ethoxy-3(2*H*)-benzofuranone, *in* H-**80**108

Hexahydrocyclopenta[*cd*]pentalene-1,3,5(2*H*)-trione, H-**60**048

8-Hydroxy-2-decene-4,6-diynoic acid, H-**80**150

2-Hydroxy-2-phenylacetaldehyde; Ac, *in* H-**60**210

3-(4-Hydroxyphenyl)-2-propenoic acid; Me ester, *in* H-**80**242

3-(4-Methoxyphenyl)-2-propenoic acid, *in* H-**80**242

3-Oxopropanoic acid; Benzyl ester, *in* O-**60**090

$C_{10}H_{10}O_3S$

2-Mercaptopropanoic acid; *S*-Benzoyl, *in* M-**60**027

$C_{10}H_{10}O_4$

1,4-Benzodioxan-2-carboxylic acid; Me ester, *in* B-**60**024

1,4-Benzodioxan-6-carboxylic acid; Me ester, *in* B-**60**025

4*H*-1,3-Benzodioxin-6-carboxylic acid; Me ester, *in* B-**60**027

1,3,5-Cycloheptatriene-1,6-dicarboxylic acid; Mono-Me ester, *in* C-**60**202

3,4-Dihydroxyphenylacetaldehyde; 3-Me ether, Ac, *in* D-**60**371

5,7-Dimethoxyphthalide, *in* D-**60**339

5-Hydroxy-7-methoxy-6-methylphthalide, *in* D-**60**349

2-Hydroxy-4-methylbenzoic acid; Ac, *in* H-**70**163

Mesogentiogenin, *in* G-**80**009

1-(2,4,6-Trihydroxyphenyl)-2-buten-1-one, T-**80**331

$C_{10}H_{10}O_5$

2,3-Dihydroxy-5-methyl-1,4-benzoquinone; 2-Me ether, 3-Ac, *in* D-**70**320

2,3-Dihydroxy-5-methyl-1,4-benzoquinone; 3-Me ether, 2-Ac, *in* D-**70**320

2-Formyl-3,4-dimethoxybenzoic acid, *in* F-**60**070

4-Formyl-2,5-dimethoxybenzoic acid, *in* F-**60**072

6-Formyl-2,3-dimethoxybenzoic acid, *in* F-**60**073

2-Hydroxy-1,3-benzenedicarboxylic acid; Di-Me ester, *in* H-**60**101

2-Hydroxy-1,4-benzenedicarboxylic acid; Di-Me ester, *in* H-**70**105

5-Hydroxy-3-methyl-1,2-benzenedicarboxylic acid; Me ether, *in* H-**70**162

2-Hydroxy-3,4-methylenedioxybenzoic acid; Me ether, Me ester, *in* H-**60**177

3-Hydroxy-4,5-methylenedioxybenzoic acid; Me ether, Me ester, *in* H-**60**178

4-Hydroxy-2,3-methylenedioxybenzoic acid; Me ether, Me ester, *in* H-**60**179

(2-Hydroxyphenoxy)acetic acid; *O*-Ac, *in* H-**70**203

$C_{10}H_{10}O_6$

Chorismic acid, C-**70**173

4,5-Dihydroxy-1,3-benzenedicarboxylic acid; Di-Me ester, *in* D-**60**307

4,5-Dimethoxy-1,3-benzenedicarboxylic acid, *in* D-**60**307

$C_{10}H_{10}S$

3-Methyl-2*H*-1-benzothiopyran, M-**80**051

1-Methylene-2-(phenylthio)cyclopropane, M-**80**090

$C_{10}H_{10}S_2$

2,2'-Dimethyl-3,3'-bithiophene, D-**80**430

3,3'-Dimethyl-2,2'-bithiophene, D-**80**431

3,4'-Dimethyl-2,2'-bithiophene, D-**80**432

3,5-Dimethyl-2,3'-bithiophene, D-**80**433

4,4'-Dimethyl-2,2'-bithiophene, D-**80**434

4,4'-Dimethyl-3,3'-bithiophene, D-**80**435

5,5'-Dimethyl-2,2'-bithiophene, D-**80**436

5,5'-Dimethyl-3,3'-bithiophene, D-**80**437

$C_{10}H_{10}S_4$

2,3,8,9-Tetrahydrobenzo[1,2-*b*:3,4-*b*']bis[1,4]dithiin, T-**70**039

$C_{10}H_{11}Br$

1-Bromo-2-methyl-1-phenylpropene, B-**80**228

$C_{10}H_{11}BrN_4O_5$

2-Bromo-9-β-D-ribofuranosyl-6*H*-purin-6-one, B-**70**269

8-Bromo-9-β-D-ribofuranosyl-6*H*-purin-6-one, B-**70**270

$C_{10}H_{11}BrO_2$

2-Bromo-4-phenylbutanoic acid, B-**80**242

3-Bromo-2,4,6-trimethylbenzoic acid, B-**60**331

$C_{10}H_{11}ClN_4O_4$

2-Chloro-9-β-D-ribofuranosyl-9*H*-purine, C-**70**159

6-Chloro-9-β-D-ribofuranosyl-9*H*-purine, C-**70**160

$C_{10}H_{11}ClN_4O_5$

2-Chloro-9-β-D-ribofuranosyl-6*H*-purin-6-one, C-**70**161

$C_{10}H_{11}ClO_2$

2-Chloro-1-phenylethanol; Ac, *in* C-**80**099

2-Chloro-2-phenylethanol; Ac, *in* C-**60**128

$C_{10}H_{11}ClO_3$

5-Chloro-4-hydroxy-2-methylbenzoic acid; Me ether, Me ester, *in* C-**60**097

$C_{10}H_{11}F$

4-Fluoro-1-phenyl-1-butene, F-**70**035

$C_{10}H_{11}FO_4$

Methyl 4-fluoro-3,5-dimethoxybenzoate, *in* F-**60**024

$C_{10}H_{11}IO$

2-Iodo-1-phenyl-1-butanone, I-**60**060

$C_{10}H_{11}IO_4$

2,3-Dihydroxy-6-iodobenzoic acid; Di-Me ether, Me ester, *in* D-**60**330

2,4-Dihydroxy-5-iodobenzoic acid; Di-Me ether, Me ester, *in* D-**60**332

2,6-Dihydroxy-3-iodobenzoic acid; Di-Me ether, Me ester, *in* D-**60**334

3,4-Dihydroxy-5-iodobenzoic acid; Di-Me ether, Me ester, *in* D-**60**335

Methyl 2-iodo-3,5-dimethoxybenzoate, *in* D-**60**336

Methyl 4-iodo-3,5-dimethoxybenzoate, *in* D-**60**337

$C_{10}H_{11}N$

1-Amino-1,4-dihydronaphthalene, A-**60**139

1-Amino-5,8-dihydronaphthalene, A-**60**140

3,4-Dihydro-5-phenyl-2*H*-pyrrole, D-**70**251

1,6-Dimethyl-1*H*-indole, *in* M-**60**088

2,2-Dimethyl-3-phenyl-2*H*-azirine, D-**60**440

2-(4-Methylphenyl)propanoic acid; Nitrile, *in* M-**60**114

$C_{10}H_{11}NO$

3-Ethoxy-1*H*-isoindole, *in* P-**70**100

Phthalimidine; *N*-Et, *in* P-**70**100

$C_{10}H_{11}NOS$

4-Methyl-5-phenyl-2-oxazolidinethione, M-**60**111

Tetrahydro-2-phenyl-2*H*-1,2-thiazin-3-one, *in* T-**60**095

$C_{10}H_{11}NO_2$

2-Amino-3-phenyl-3-butenoic acid, A-**60**238

2-Amino-4-phenyl-3-butenoic acid, A-**60**239

5,6-Dimethoxyindole, *in* D-**70**307

5-Methyl-3-phenyl-2-oxazolidinone, *in* M-**60**107

4-Phenyl-3-morpholinone, *in* M-**60**147

1,2,3,4-Tetrahydro-1-isoquinolinecarboxylic acid, T-**60**070

5,6,7,8-Tetrahydro-1,4-naphthoquinone; Monoxime, *in* T-**80**076

$C_{10}H_{11}NO_2S$

3,4-Dimethoxybenzyl isothiocyanate, D-**80**411

$C_{10}H_{11}NO_3$

1-Amino-2-(4-hydroxyphenyl)cyclopropanecarboxylic acid, A-**60**197

3-Amino-4-oxo-4-phenylbutanoic acid, A-**60**236

4-Hydroxyphenylacetaldehyde; Ac, oxime, *in* H-**60**209

Methyl *N*-phenacylcarbamate, *in* O-**70**099

2,3,5-Trimethoxybenzonitrile, *in* T-**70**250

$C_{10}H_{11}NO_4$

3-Amino-2-hydroxypropanoic acid; Benzoyl, *in* A-**70**160

2,3,4-Trimethyl-5-nitrobenzoic acid, T-**60**369

2,4,6-Trimethyl-3-nitrobenzoic acid, T-**60**370

$C_{10}H_{11}NO_4S$

2,3-Dihydro-1*H*-indole-2-sulfonic acid; *N*-Ac, *in* D-**70**221

$C_{10}H_{11}NO_5$

4-Amino-6-hydroxy-1,3-benzenedicarboxylic acid; Di-Me ester, *in* A-**70**153

5-Amino-2-hydroxy-1,3-benzenedicarboxylic acid; Di-Me ester, *in* A-**70**154

5-Amino-4-hydroxy-1,3-benzenedicarboxylic acid; Di-Me ester, *in* A-**70**155

2',3'-Dimethoxy-6'-nitroacetophenone, *in* D-**60**357

2',5'-Dimethoxy-4'-nitroacetophenone, *in* D-**70**322

3',6'-Dimethoxy-2'-nitroacetophenone, *in* D-**60**358

4',5'-Dimethoxy-2'-nitroacetophenone, *in* D-**60**359

$C_{10}H_{11}N_3$

1-Aminophthalazine; *N*,*N*-Di-Me, *in* A-**70**184

1,4-Dihydro-2,6-dimethyl-3,5-pyridinedicarboxylic acid; 1-Me, dinitrile, *in* D-**80**193

2-(3,5-Dimethyl-1*H*-pyrazol-1-yl)pyridine, D-**80**473

$C_{10}H_{11}N_3O$

3,4-Diamino-1-methoxyisoquinoline, D-**60**037

▷ 1*H*-Indole-3-acetic acid; Hydrazide, *in* I-**70**013

$C_{10}H_{11}N_5$

2,4-Diamino-6-methyl-1,3,5-triazine; *N*-Ph, *in* D-**60**040

$C_{10}H_{11}N_5O$

1,4-Dihydro-4,6,7-trimethyl-9*H*-imidazo[1,2-*a*]purin-9-one, D-**70**278

$C_{10}H_{11}N_5O_3$

6-Amino-1,3-dihydro-2*H*-purin-2-one; 6,9-Di-Ac, 1-Me, *in* A-**60**144

$C_{10}H_{12}$

1,3,5,7,9-Decapentaene, D-**70**015

5-Decene-2,8-diyne, D-**60**015

[4]Metacyclophane, M-**60**029

[4]Paracyclophane, P-**60**006

4,5,6,7-Tetrahydro-1-methylene-1*H*-indene, T-**70**073

Tricyclo[4.2.2.0^{1,6}]deca-7,9-diene, T-**70**226

2,3,5-Tris(methylene)bicyclo[2.2.1]heptane, T-**60**403

$C_{10}H_{12}BrNO$

2-Bromo-2-methylpropanoic acid; Anilide, *in* B-**80**229

2-Bromo-4-phenylbutanoic acid; Amide, *in* B-**80**242

$C_{10}H_{12}BrN_5O_5$

8-Bromoguanosine, B-**70**215

$C_{10}H_{12}ClN_5O_5$

8-Chloroguanosine, C-**70**068

$C_{10}H_{12}FN_5O_4$

8-Amino-6-fluoro-9*H*-purine; 9-β-D-Ribofuranosyl, *in* A-**60**163

$C_{10}H_{12}IN_5O_5$

8-Iodoguanosine, I-**70**034

$C_{10}H_{12}N_2O$

4-Imidazolidinone; 3-Benzyl, *in* I-**80**006

1-Phenyl-2-piperazinone, *in* P-**70**101

$C_{10}H_{12}N_2O_3$
3-Amino-4(1*H*)-pyridinone; 1-*N*-Me, *N*[3],*N*[3]-di-Ac, *in* A-60254

$C_{10}H_{12}N_2O_4$
3,6-Pyridazinedicarboxylic acid; Di-Et ester, *in* P-70168
10,11,14,15-Tetraoxa-1,8-diazatricyclo[6.4.4.0^{2,7}]hexadeca-2(7),3,5-triene, T-60162

$C_{10}H_{12}N_2O_6$
1,5-Diethoxy-2,4-dinitrobenzene, *in* D-80491

$C_{10}H_{12}N_2S$
2-Amino-4*H*-3,1-benzothiazine; *N*-Et, *in* A-70092

$C_{10}H_{12}N_4$
2,4,6,7-Tetramethylpteridine, T-70142

$C_{10}H_{12}N_4O_2$
▷ Tris(2-cyanoethyl)nitromethane, *in* C-80025

$C_{10}H_{12}N_4O_3$
Leucettidine, L-70029
2,4,7(1*H*,3*H*,8*H*)-Pteridinetrione; 1,3,6,8-Tetra-Me, *in* P-70135

$C_{10}H_{12}N_4O_4$
2′-Deoxyinosine, D-70023
2′-Deoxyinosine, D-80029
3′-Deoxyinosine, D-70024

$C_{10}H_{12}N_4O_4S$
▷ 6-Mercaptopurine; 9-*β*-D-Ribofuranosyl, *in* D-60270
▷ 6-Thioinosine, T-70180

$C_{10}H_{12}N_4O_6$
Tetrahydro-1,2,4,5-tetrazine-3,6-dione; 1,2,4,5-Tetra-Ac, *in* T-70097

$C_{10}H_{12}N_6$
6,6′-Dihydroxy-2,2′-bipyridine; Dihydrazone, *in* D-80299

$C_{10}H_{12}O$
1-Ethoxy-1-phenylethylene, *in* P-60092
▷ 1-Methoxy-4-(1-propenyl)benzene, *in* P-80179
1-Phenyl-3-buten-1-ol, P-70061
4-Phenyl-3-buten-1-ol, P-80097
1,2,4,5-Tetrahydro-3-benzoxepin, T-70043
1,3,4,5-Tetrahydro-2-benzoxepin, T-70044
2,3,4,5-Tetrahydro-1-benzoxepin, T-70045
3,4,5,6-Tetrahydro-1(2*H*)-naphthalenone, T-80075

$C_{10}H_{12}OS$
Benzeneethanethioic acid; *O*-Et ester, *in* B-70017
Benzeneethanethioic acid; *S*-Et ester, *in* B-70017
Tetrahydro-3-(phenythio)furan, *in* T-80063

$C_{10}H_{12}O_2$
3,4-Dihydro-2*H*-1-benzopyran-7-ol; Me ether, *in* D-60202
3,4-Dihydro-6-methoxy-2*H*-1-benzopyran, *in* D-70175
3,4-Dihydro-8-methoxy-2*H*-1-benzopyran, *in* D-60203
Dispiro[cyclopropane-1,5′-[3,8]dioxatricyclo[5.1.0.0^{2,4}]octane-6′,1″-cyclopropane], D-60500
1-Ethenyl-2,4-dimethoxybenzene, *in* V-70010
Fusalanipyrone, F-80099
1,3,4,5,7,8-Hexahydro-2,6-naphthalenedione, H-70063
1-(4-Methoxyphenyl)-2-propanone, *in* H-80241
2-(2-Methylphenyl)propanoic acid, M-60113
2-(4-Methylphenyl)propanoic acid, M-60114
3,4,5-Trimethylbenzoic acid, T-60355

$C_{10}H_{12}O_2S$
2-Mercapto-3-phenylpropanoic acid; Me ester, *in* M-60026
3-(Phenylthio)butanoic acid, *in* M-70023

$C_{10}H_{12}O_3$
Adriadysiolide, A-70069
Anisyl acetate, *in* H-80114
2-*O*-Benzylglyceraldehyde, B-80058
2,5-Dimethoxybenzeneacetaldehyde, *in* D-60370
3,4-Dimethoxybenzeneacetaldehyde, *in* D-60371
2-Hydroxy-4-methylbenzoic acid; Et ester, *in* H-70163
4-Hydroxy-2-methylbenzoic acid; Et ester, *in* H-70164

$C_{10}H_{12}O_3S$
3-(Phenylsulfinyl)butanoic acid, *in* M-70023

$C_{10}H_{12}O_4$
1,4-Benzoquinone; Bis(ethylene ketal), *in* B-70044
4,5-Dimethoxy-2-methylbenzoic acid, *in* D-80339
6-Ethyl-2,4-dihydroxy-3-methylbenzoic acid, E-70041
Hallerone, H-60005
2′-Hydroxy-4′,6′-dimethoxyacetophenone, *in* T-70248
▷ 1-(4-Hydroxy-3,5-dimethoxyphenyl)ethanone, *in* T-60315
(3-Hydroxyphenoxy)acetic acid; Et ester, *in* H-70204

$C_{10}H_{12}O_5$
2,5-Bis(hydroxymethyl)furan; Di-Ac, *in* B-60160
3,4-Bis(hydroxymethyl)furan; Di-Ac, *in* B-60161
Gelsemide, G-60010
3-Hydroxy-4,6-dimethoxy-2-methylbenzoic acid, *in* T-80324
2,3,5-Trimethoxybenzoic acid, *in* T-70250

$C_{10}H_{12}S_8$
Tetramercaptotetrathiafulvalene; Tetrakis(*S*-Me), *in* T-70130

$C_{10}H_{13}BrCl_2$
2-Bromo-4-chloro-1-(2-chloroethenyl)-1-methyl-5-methylenecyclohexane, B-80185

$C_{10}H_{13}I$
1-*tert*-Butyl-2-iodobenzene, B-60346
1-*tert*-Butyl-3-iodobenzene, B-60347
1-*tert*-Butyl-4-iodobenzene, B-60348

$C_{10}H_{13}N$
Azetidine; *N*-Benzyl, *in* A-70283
2,3,4,5-Tetrahydro-1*H*-1-benzazepine, T-70036

$C_{10}H_{13}NO$
2-(4-Methylphenyl)propanoic acid; Amide, *in* M-60114
1,2,3,4-Tetrahydro-5-hydroxyisoquinoline; Me ether; BHCl, *in* T-80065
7,8,9,10-Tetrahydropyrido[1,2-*a*]azepin-4(6*H*)-one, T-60085
Weberidine, *in* T-80066

$C_{10}H_{13}NO_2$
2-Amino-2-methyl-3-phenylpropanoic acid, A-60225
3-Amino-2,4,6-trimethylbenzoic acid, A-60266
5-Amino-2,3,4-trimethylbenzoic acid, A-60267

$C_{10}H_{13}NO_3$
4-Amino-1,2-benzenediol; Di-Me ether, *N*-Ac, *in* A-70090
2-Amino-2-methyl-3-(4-hydroxyphenyl)propanoic acid, A-60220

$C_{10}H_{13}NO_4$
2-Amino-3-hydroxy-4-(4-hydroxyphenyl)butanoic acid, A-60185
1*H*-Pyrrole-3,4-dicarboxylic acid; Di-Et ester, *in* P-70183
2′,4′,6′-Trihydroxyacetophenone; 2′,4′-Di-Me ether, oxime, *in* T-70248

$C_{10}H_{13}NO_5$
3-Hydroxy-1*H*-pyrrole-2,4-dicarboxylic acid; Di-Et ester, *in* H-80251

$C_{10}H_{13}N_5O_3$
2-Amino-6-(1,2-dihydroxypropyl)-3-methylpterin-4-one, A-60146
2-Amino-6-(1,2-dihydroxypropyl)-3-methylpterin-4-one, *in* B-70121

$C_{10}H_{13}N_5O_4$
3′-Azido-3′-deoxythymidine, A-80197
9-*β*-D-Ribofuranosyl-9*H*-purin-2-amine, *in* A-60243

$C_{10}H_{13}N_5O_5$
6-Amino-1,3-dihydro-2*H*-purin-2-one; 9-(*β*-D-Arabinofuranosyl), *in* A-60144
8-Amino-9-*β*-D-ribofuranosyl-6*H*-purin-6-one, A-70198

$C_{10}H_{13}N_5O_5S$
8-Mercaptoguanosine, M-70029

$C_{10}H_{13}N_5O_6$
7,8-Dihydro-8-oxoguanosine, D-70238

$C_{10}H_{14}$
5,5′-Bibicyclo[2.1.0]pentane, B-70092
2,3-Dimethylenebicyclo[2.2.2]octane, D-70408
1-Ethynylbicyclo[2.2.2]octane, E-70049
1,2,3,5,8,8*a*-Hexahydronaphthalene, H-60053
1,2,3,7,8,8*a*-Hexahydronaphthalene, H-60054
2-Methyl-6-methylene-1,3,7-octatriene, M-60089
Tricyclo[4.3.1.0^{3,8}]dec-4-ene, T-80236
Tricyclo[5.2.1.0^{2,6}]dec-2(6)-ene, T-60260

$C_{10}H_{14}BrCl_3$
Coccinene/, *in* M-80035
Mertensene, M-80035
Telfairine, T-80005

$C_{10}H_{14}BrNO_2$
2-Bromo-2-nitroadamantane, B-80234

$C_{10}H_{14}Br_2$
1,3-Dibromoadamantane, D-70084

$C_{10}H_{14}Br_2Cl_2$
4,8-Dibromo-3,7-dichloro-3,7-dimethyl-1,5-octadiene, D-80086

$C_{10}H_{14}Br_3Cl_3$
1,5,7-Tribromo-2,6,8-trichloro-2,6-dimethyl-3-octene, T-80218

$C_{10}H_{14}ClNO_2$
2-Chloro-2-nitroadamantane, C-80082

$C_{10}H_{14}Cl_4$
Gelidene, G-80006
1,2,4-Trichloro-5-(2-chloroethenyl)-1,5-dimethylcyclohexane, T-80223

$C_{10}H_{14}IN$
4-Iodo-3,5-dimethylaniline; *N*,*N*-Di-Me, *in* I-70031

$C_{10}H_{14}N_2O_2$
2,4-Dihydroxyphenylacetaldehyde; Dimethylhydrazone, *in* D-60369
Octahydro-5*H*,10*H*-dipyrrolo[1,2-*a*:1′,2′-*d*]pyrazine-5,10-dione, O-60016

$C_{10}H_{14}N_2O_3$
Octahydro-2-hydroxy-5*H*,10*H*-dipyrrolo[1,2-*a*:1′,2′-*d*]pyrazine-5,10-dione, O-60017

$C_{10}H_{14}N_2O_4$
2,2-Dinitroadamantane, D-50699
2,4-Dinitroadamantane, D-80489
2,6-Dinitroadamantane, D-80490
1-[3-Hydroxy-4-(hydroxymethyl)cyclopentyl]-2,4(1*H*,3*H*)-pyrimidinedione, H-70140
Isoporphobilinogen, I-60114
Octahydro-5*a*,10*a*-dihydroxy-5*H*,10*H*-dipyrrolo[1,2-*a*:1′,2′-*d*]pyrazine-5,10-dione, O-60014
Octahydro-2,7-dihydroxy-5*H*,10*H*-dipyrrolo[1,2-*a*:1′,2′-*d*]pyrazine-5,10-dione, O-60013
1*H*-Pyrazole-4,5-dicarboxylic acid; 1-Me, Di-Et ester, *in* P-80199

$C_{10}H_{14}N_2S_2$
Octahydro-5*H*,10*H*-dipyrrolo[1,2-*a*:1′,2′-*d*]pyrazine-5,10-dithione, O-70014

$C_{10}H_{14}N_6O_3$
3′-Amino-3′-deoxyadenosine, A-70128

$C_{10}H_{14}N_6O_5$
8-Aminoguanosine, A-70150

$C_{10}H_{14}O$

1-Acetyl-4-isopropenylcyclopentene, A-**80025**
1-Acetyl-4-isopropylidenecyclopentene, A-**80026**
3-Caren-5-one, C-**60019**
2,7-Cyclodecadien-1-one, C-**60185**
3,7-Cyclodecadien-1-one, C-**60186**
2-Cyclopentylidenecyclopentanone, C-**60227**
3,3-Dimethylbicyclo[2.2.2]oct-5-en-2-one, D-**60412**
3,4,4a,5,8,8a-Hexahydro-1(2H)-naphthalenone, H-**80061**
4a,5,6,7,8,8a-Hexahydro-2(1H)-naphthalenone, H-**80062**
6-Isopropenyl-3-methyl-2-cyclohexen-1-one, I-**70088**
▷ 4-Isopropylbenzyl alcohol, I-**80078**
5-Methoxy-1,2,3-trimethylbenzene, in T-**60371**
7-Methylenebicyclo[3.3.1]nonan-3-one, M-**70062**
Myrtenal, in M-**80205**
Tricyclo[4.4.0.0^{3,8}]decan-4-one, T-**80235**
2,3,5-Trimethylbenzyl alcohol, T-**80349**
3,4,5-Trimethylbenzyl alcohol, T-**80350**

$C_{10}H_{14}O_2$

2-Benzyloxy-1-propanol, in P-**70118**
1-Cycloheptenecarboxylic acid; Et ester, in C-**80178**
4,4-Dimethylbicyclo[3.2.1]octane-2,3-dione, D-**60410**
3-Ethynyl-3-methyl-4-pentenoic acid; Et ester, in E-**60059**
▷ 1-(3-Furanyl)-4-methyl-1-pentanone, F-**60087**
2-Hydroxy-1(10)-pinen-5-one, H-**80247**
Isoneonepetalactone, in N-**70027**
2-Isopropyl-5-methyl-1,3-benzenediol, I-**60122**
2-Isopropyl-6-methyl-1,4-benzenediol, I-**60123**
3-Isopropyl-6-methyl-1,2-benzenediol, I-**60124**
5-Isopropyl-2-methyl-1,3-benzenediol, I-**60125**
5-Isopropyl-3-methyl-1,2-benzenediol, I-**60126**
5-Isopropyl-4-methyl-1,3-benzenediol, I-**60127**
Neonepetalactone, N-**70027**
Nepetalactone, N-**60020**
4-Oxatricyclo[4.3.1.1^{3,8}]undecan-5-one, O-**60055**
2-Pinen-10-oic acid, in M-**80205**
3,3,6,6-Tetramethyl-4-cyclohexene-1,2-dione, T-**60134**
5,5,6,6-Tetramethyl-2-cyclohexene-1,4-dione, T-**60135**

$C_{10}H_{14}O_3$

2,5-Dihydroxy-4-isopropylbenzyl alcohol, D-**70309**
2-(Dimethoxymethyl)benzenemethanol, in H-**60171**
α-(Dimethoxymethyl)benzenemethanol, in H-**60210**
Peperinic acid, P-**80063**

$C_{10}H_{14}O_4$

2-Cyclohexene-1,4-diol; Di-Ac, in C-**80181**
4-Cyclohexene-1,2-diol; Di-Ac, in C-**80182**
1,2,3,4-Tetramethoxybenzene, in B-**60014**
3,4,5-Trimethoxybenzyl alcohol, in H-**80196**

$C_{10}H_{14}O_5$

Hexahydro-3,7a-dihydroxy-3a,7-dimethyl-1,4-isobenzofurandione, H-**70055**
3-(5-Hydroxymethyl-5-methyl-2-oxo-5H-furan-3-yl)-2-methylpropanoic acid, H-**80208**

$C_{10}H_{14}S_2$

1,4-Dimethyl-2,5-bis(methylthio)benzene, in D-**60406**

$C_{10}H_{15}BrCl_2O$

Aplysiapyranoid C, A-**70245**
Aplysiapyranoid D, in A-**70245**

$C_{10}H_{15}Br_2Cl$

7,8-Dibromo-6-(chloromethylene)-2-methyl-2-octene, D-**70088**

$C_{10}H_{15}Br_2ClO$

Aplysiapyranoid A, A-**70244**
Aplysiapyranoid B, in A-**70244**

$C_{10}H_{15}Cl$

1-Chloroadamantane, C-**60039**
2-Chloroadamantane, C-**60040**

$C_{10}H_{15}ClO_5$

Jioglutin A, J-**80010**
Jioglutin B, in J-**80010**

$C_{10}H_{15}Cl_3$

1,8-Dichloro-6-chloromethyl-2-methyl-2,6-octadiene, D-**70116**
3,8-Dichloro-6-chloromethyl-2-methyl-1,6-octadiene, D-**70117**

$C_{10}H_{15}NO$

2-Cyclopentylidenecyclopentanone; Oxime, in C-**60227**
3-Ethynyl-3-methyl-4-pentenoic acid; N,N-Dimethylamide, in E-**60059**
4a,5,6,7,8,8a-Hexahydro-2(1H)-naphthalenone; Oxime, in H-**80062**

$C_{10}H_{15}NO_2$

3,4,4a,5,8,8a-Hexahydro-1(2H)-naphthalenone; Oxime, in H-**80061**
1-Nitroadamantane, N-**80033**
2-Nitroadamantane, N-**80034**

$C_{10}H_{15}NO_4$

α-Allokainic acid, in K-**60003**
▷ Kainic acid, K-**60003**

$C_{10}H_{15}NO_7$

2-Amino-1,1,2-ethanetricarboxylic acid; N-Ac, Tri-Me ester, in A-**80072**

$C_{10}H_{15}NO_8$

4-(2-Carboxyethyl)-4-nitroheptanedioic acid, C-**80025**

$C_{10}H_{15}N_2O_8P$

Thymidine 5′-phosphate, T-**80196**

$C_{10}H_{15}N_3$

2-Azidoadamantane, A-**60337**

$C_{10}H_{15}N_3O_2$

1H-Pyrrole-3,4-dicarboxylic acid; Dimethylamide, in P-**70183**

$C_{10}H_{15}N_3O_4$

Carbodine, C-**70018**

$C_{10}H_{15}N_3O_5$

Homocytidine, in A-**70130**

$C_{10}H_{16}$

Bicyclo[6.2.0]dec-9-ene, B-**80074**
1-tert-Butyl-1,3-cyclohexadiene, B-**70296**
4-Decen-1-yne, D-**70020**
1,2-Diethylidenecyclohexane, D-**70154**
4-Isopropenyl-1-methylcyclohexene, I-**70087**
3-Isopropylidene-6-methylcyclohexene, I-**80081**
3,3,6,6-Tetramethylcyclohexyne, T-**70132**
3-Thujene, T-**80194**
4(10)-Thujene, T-**80195**

$C_{10}H_{16}Cl_2$

1-Chloro-3-(chloromethyl)-7-methyl-2,6-octadiene, C-**70060**

$C_{10}H_{16}Cl_2O_2$

2,2-Dichloro-7-octenoic acid; Et ester, in D-**80123**

$C_{10}H_{16}N_2S_2$

4,6-Diamino-1,3-benzenedithiol; Di-S-Et, in D-**70037**

$C_{10}H_{16}N_2S_4$

2,3,5,6-Tetrakis(methylthio)-1,4-benzenediamine, in D-**60031**

$C_{10}H_{16}N_4$

2,3,5,6,8,9-Hexahydro-1H-diimidazo[1,2-d:2′,1′-g][1,4]diazepine; N-Me, in H-**80058**

$C_{10}H_{16}O$

1-Acetylcyclooctene, A-**60027**
3-tert-Butyl-2-cyclohexen-1-one, B-**70297**
3,3-Dimethylbicyclo[2.2.2]octan-2-one, D-**60411**
2-Isopropyl-5-methyl-4-cyclohexen-1-one, I-**60128**

2-Methyl-6-methylene-2,7-octadien-4-ol, M-**70093**
Myrtenol, M-**80205**
2-(2-Propenyl)cycloheptanone, P-**80176**
2,2,5,5-Tetramethyl-3-cyclohexen-1-one, T-**60136**

$C_{10}H_{16}O_2$

Bicyclo[3.2.1]octane-1-carboxylic acid; Me ester, in B-**70112**
1-Cyclooctene-1-acetic acid, C-**70241**
4-Cyclooctene-1-acetic acid, C-**70242**
2,5,8-Decatriene-1,10-diol, D-**80015**
Eldanolide, E-**70004**
2,3-Epoxy-5,7-octadien-4-ol, E-**80042**
2-Hydroperoxy-2-methyl-6-methylene-3,7-octadiene, H-**60097**
3-Hydroperoxy-2-methyl-6-methylene-1,7-octadiene, H-**60098**
Iridodial, I-**60084**
Lineatin, L-**60027**
3-Myodeserten-1-ol, M-**80203**
Piperitone oxide, P-**80153**
Spiro[bicyclo[3.2.1]octane-8,2′-[1,3]dioxolane], in B-**60094**

$C_{10}H_{16}O_3$

1,7-Dihydroxy-3,7-dimethyl-2,5-octadien-4-one, D-**60318**
3-Hydroxy-4-(hydroxymethyl)-7,7-dimethylbicyclo[2.2.1]heptan-2-one, in H-**70173**
9-Oxo-2-decenoic acid, O-**70089**
Secothujene, S-**70027**

$C_{10}H_{16}O_4$

Gelsemiol, G-**60011**
3-Hexene-2,5-diol; Di-Ac, in H-**80083**
7-Hydroxy-1-hydroxy-3,7-dimethyl-2E,5E-octadien-4-one, in D-**60318**
2,2,5,5-Tetramethyl-3-hexenedioic acid, T-**60141**

$C_{10}H_{16}O_6$

Jioglutin C, J-**80011**

$C_{10}H_{16}S$

2-Adamantanethiol, A-**80038**

$C_{10}H_{16}S_2$

1,3-Bis(allylthio)cyclobutane, in C-**60184**

$C_{10}H_{17}N$

4-Azatricyclo[4.3.1.1^{3,8}]undecane, A-**60333**
Decahydro-1H-dicyclopenta[b,d]pyrrole, D-**80010**

$C_{10}H_{17}NO_3$

2-Nitrocyclodecanone, N-**70046**

$C_{10}H_{18}$

1-tert-Butylcyclohexene, B-**60341**
3-tert-Butylcyclohexene, B-**60342**
4-tert-Butylcyclohexene, B-**60343**

$C_{10}H_{18}BrN$

1-Azoniatricyclo[3.3.3.0]undecane; Bromide, in A-**80206**

$C_{10}H_{18}Br_2$

2,2-Dibromo-1,1,3,3-tetramethylcyclohexane, D-**80101**

$C_{10}H_{18}ClN$

1-Azoniatricyclo[3.3.3.0]undecane; Chloride, in A-**80206**

$C_{10}H_{18}N^\oplus$

1-Azoniatricyclo[3.3.3.0]undecane, A-**80206**

$C_{10}H_{18}N_2O_2$

1,1-Di-4-morpholinylethene, D-**70447**

$C_{10}H_{18}N_2O_4$

2H-1,5,2,4-Dioxadiazine-3,6(4H)dione; 2,4-Di-tert-butyl, in D-**60466**

$C_{10}H_{18}N_2O_6S_2$

γ-Glutamylmarasmine, G-**60026**

$C_{10}H_{18}N_4O_4$

1,2-Dihydrazinoethane; $N^\alpha,N^\alpha,N^\beta,N^\beta$-Tetra-Ac, in D-**60192**

$C_{10}H_{18}O$
1-*tert*-Butoxy-1-hexyne, B-**80282**
3-Decenal, D-**70016**
3,7-Dimethyl-6-octenal, D-**60430**
4,6-Dimethyl-4-octen-3-one, D-**70424**
Hexamethylcyclobutanone, H-**60069**
2-Isopropylcycloheptanone, I-**70090**
4-Isopropylcycloheptanone, I-**70091**
1-Methoxy-2-nonyne, *in* N-**80074**
2-Methyl-6-methylene-7-octen-4-ol, *in*
    M-**70093**
Tetrahydro-2,2-dimethyl-5-(1-methyl-1-
    propenyl)furan, T-**80061**
2,2,6,6-Tetramethylcyclohexanone, T-**60133**
2,2,6-Trimethyl-5-hepten-3-one, T-**60365**

$C_{10}H_{18}OS$
Di-*tert*-butylthioketene; *S*-Oxide, *in* D-**60114**

$C_{10}H_{18}O_2$
4-Decenoic acid, D-**80017**
3,6-Diethoxycyclohexene, *in* C-**80181**
3,4-Diethoxy-1,5-hexadiene, *in* H-**70036**
Dihydro-4-methyl-5-pentyl-2(3*H*)-furanone,
    D-**80227**
4,4-Dimethyl-6-heptenoic acid; Me ester, *in*
    D-**80450**
2,6-Dimethyl-2,7-octadiene-1,6-diol, D-**80456**
2,6-Dimethyl-3,7-octadiene-2,6-diol, D-**70422**
3,7-Dimethyl-2,5-octadiene-1,7-diol, D-**80457**
▷ 5-Hexyldihydro-2(3*H*)-furanone, H-**60078**
2-Hydroxy-1,4-cineole, H-**80139**
6-Hydroxydecanoic acid; Lactone, *in* H-**80147**
4-Isopropenyl-1-methyl-1,2-cyclohexanediol,
    I-**70086**
4-Isopropyl-1-methyl-4-cyclohexene-1,2-diol,
    I-**70094**
4-Isopropyl-1-methyl-3-cyclohexenene-1,2-
    diol, I-**70093**
p-Menth-4-ene-1,2-diol, M-**80014**
p-Menth-3-ene-1,2-diol, M-**80015**
p-Menth-7-ene-1,2-diol, M-**80016**
10-Methyl-2-oxecanone, *in* H-**80149**
2-Octen-1-ol; Ac, *in* O-**60029**
2,2,5,5-Tetramethyl-3,4-hexanedione, T-**80146**
1,7,7-Trimethylbicyclo[2.2.1]heptane-2,5-diol,
    T-**70280**

$C_{10}H_{18}O_3$
1,7-Bis(hydroxymethyl)-7-
    methylbicyclo[2.2.1]heptan-2-ol, B-**70146**
2-Hydroxy-4-*tert*-butylcyclopentanecarboxylic
    acid, H-**70113**
1-(Hydroxymethyl)-7,7-
    dimethylbicyclo[2.2.1]heptane-2,3-diol,
    H-**70173**
9-Hydroxy-5-nonenoic acid; Me ester, *in*
    H-**70194**
6-Methyl-5-oxoheptanoic acid; Et ester, *in*
    M-**80130**
10-Oxodecanoic acid, O-**60063**

$C_{10}H_{18}O_4$
3-Methyl-4-oxo-2-butenoic acid; Me ester, di-
    Et acetal, *in* M-**70103**
Nonactinic acid, N-**70069**

$C_{10}H_{18}O_5$
Methyl hydrogen 2-(*tert*-butoxymethyl)-2-
    methylmalonate, M-**70088**

$C_{10}H_{18}S$
Di-*tert*-butylthioketene, D-**60114**

$C_{10}H_{19}Br$
5-Bromo-5-decene, B-**60225**

$C_{10}H_{19}NO$
1-Oxa-2-azaspiro[2,5]octane; *N*-*tert*-Butyl, *in*
    O-**70052**
2,2,6,6-Tetramethylcyclohexanone; Oxime, *in*
    T-**60133**

$C_{10}H_{19}NO_2$
3-Amino-2-hexenoic acid; *tert*-Butyl ester, *in*
    A-**60170**

$C_{10}H_{19}NO_3$
2-Amino-1-hydroxy-1-cyclobutaneacetic acid;
    tert-Butyl ester, *in* A-**60184**

2-Amino-3-hydroxy-4-methyl-6-octenoic acid;
    N-Me, *in* A-**60189**
10-Oxodecanoic acid; Oxime, *in* O-**60063**

$C_{10}H_{19}N_3$
1*H*,4*H*,7*H*,9*bH*-2,3,5,6,8,9-Hexahydro-
    3*a*,6*a*,9*a*-triazaphenalene, H-**70078**

$C_{10}H_{20}$
5-Decene, D-**70017**
Methylcyclononane, M-**60058**

$C_{10}H_{20}N_2$
1,5-Diazabicyclo[5.2.2]undecane; *N*-Me, *in*
    D-**60055**

$C_{10}H_{20}N_2O$
N,*N*-Diethyl-3-piperidinecarboxamide, *in*
    P-**70102**

$C_{10}H_{20}N_2S_2$
▷ 1,1′-Dithiobispiperidine, D-**70523**

$C_{10}H_{20}N_4$
▷ 6*H*,13*H*-Octahydrodipyridazino[1,2-*a*:1′,2′-
    d][1,2,4,5]tetrazine, O-**70013**

$C_{10}H_{20}O$
5-Decanone, D-**60014**
2-Decen-4-ol, D-**60016**
5-Decen-1-ol, D-**70019**
6-Decen-1-ol, D-**80018**
2,2,3,3,4,4-Hexamethylcyclobutanol, H-**60068**
7-Methyl-3-methylene-1-octanol, M-**80118**

$C_{10}H_{20}O_2$
2,5-Diethoxy-3-hexene, *in* H-**80083**
1-Isopropyl-4-methyl-1,4-cyclohexanediol,
    I-**80082**

$C_{10}H_{20}O_3$
6-Hydroxydecanoic acid, H-**80147**
7-Hydroxydecanoic acid, H-**80148**
9-Hydroxydecanoic acid, H-**80149**

$C_{10}H_{20}O_4$
3,4-Dihydroxybutanoic acid; O^4-*tert*-Butyl, Et
    ester, *in* D-**60313**
Tetraethoxyethene, T-**80047**

$C_{10}H_{20}O_5$
Di-*tert*-butyl dicarbonate, *in* D-**70111**

$C_{10}H_{21}N$
Octahydroazocine; 1-Isopropyl, *in* O-**70010**

$C_{10}H_{22}N_2$
Hexahydropyrimidine; 1,3-Diisopropyl, *in*
    H-**60056**

$C_{10}H_{22}N_2O_3$
1,4,10-Trioxa-7,13-diazacyclopentadecane,
    T-**60376**

$C_{10}H_{22}N_2O_5$
6,10-Diamino-2,3,5-trihydroxydecanoic acid,
    D-**70046**

$C_{10}H_{22}N_4O_2S_2$
Alethine, A-**60073**

$C_{10}H_{22}O_2$
5,6-Decanediol, D-**60013**

$C_{10}H_{22}O_9$
Lacticolorin, *in* D-**80284**

$C_{10}H_{23}NO_3$
4-Amino-4-(3-hydroxypropyl)-1,7-heptanediol,
    A-**80085**

$C_{10}H_{23}N_3$
1,5,9-Triazacyclotridecane, T-**60233**

$C_{10}H_{25}N_5$
1,4,7,10,13-Pentaazacyclopentadecane,
    P-**60025**

$C_{11}H_5Cl_{11}$
Chloropentakis(dichloromethyl)benzene,
    C-**70136**

$C_{11}H_5NO_3$
2,3-Quinolinedicarboxylic acid; Anhydride, *in*
    Q-**80002**

$C_{11}H_6N_2O$
3,6-Diazafluorenone, D-**70053**

$C_{11}H_6N_2OS$
10*H*-[1]Benzothiopyrano[2,3-*d*]pyridazin-10-
    one, B-**80045**

$C_{11}H_6N_2O_2$
10*H*-[1]Benzopyrano[2,3-*b*]pyrazin-10-one,
    B-**80035**
10*H*-[1]Benzopyrano[2,3-*d*]pyridazin-10-one,
    B-**80037**
·2,3-Quinolinedicarboxylic acid; Imide, *in*
    Q-**80002**

$C_{11}H_6N_4$
9-Diazo-9*H*-cyclopenta[1,2-*b*:4,3-*b*′]dipyridine,
    D-**60059**

$C_{11}H_6N_6$
Pyrimido[4,5-*i*]imidazo[4,5-*g*]cinnoline,
    P-**60244**

$C_{11}H_6O$
1-Phenyl-1,4-pentadiyn-3-one, P-**70087**
1-Phenyl-2,4-pentadiyn-1-one, P-**80114**

$C_{11}H_6OS$
Naphtho[2,1-*b*]thiet-1-one, N-**70013**

$C_{11}H_7BrO$
1-Bromo-2-naphthalenecarboxaldehyde,
    B-**60302**

$C_{11}H_7BrO_2$
3-Bromo-5-hydroxy-2-methyl-1,4-
    naphthoquinone, B-**60266**
2-Bromo-6-methoxy-1,4-naphthaquinone, *in*
    B-**60270**
2-Bromo-3-methoxy-1,4-naphthoquinone, *in*
    B-**60268**
2-Bromo-5-methoxy-1,4-naphthoquinone, *in*
    B-**60269**
2-Bromo-7-methoxy-1,4-naphthoquinone, *in*
    B-**60271**
2-Bromo-8-methoxy-1,4-naphthoquinone, *in*
    B-**60272**

$C_{11}H_7ClN_2$
2-Chloro-1*H*-perimidine, C-**80092**

$C_{11}H_7ClO_3$
3-Chloro-5-hydroxy-2-methyl-1,4-
    naphthoquinone, C-**60098**
2-Oxo-2*H*-benzopyran-4-acetic acid; Chloride,
    in O-**60057**

$C_{11}H_7NO$
Benz[*cd*]indol-2-(1*H*)-one, B-**70027**
3*H*-Pyrrolo[1,2-*a*]indol-3-one, P-**80227**
9*H*-Pyrrolo[1,2-*a*]indol-9-one, P-**60247**

$C_{11}H_7NOSe$
[1,4]Benzoxaselenino[3,2-*b*]pyridine, B-**60054**

$C_{11}H_7NO_2$
[1,4]Benzodioxino[2,3-*b*]pyridine, B-**80021**
4-(Cyanomethyl)coumarin, *in* O-**60057**
4-Oxo-4*H*-1-benzopyran-3-acetic acid; Nitrile,
    in O-**60059**
Pyrano[3,4-*b*]indol-3(9*H*)-one, P-**80196**

$C_{11}H_7NO_3$
3-Nitro-2-naphthaldehyde, N-**70054**

$C_{11}H_7NO_4$
2,3-Quinolinedicarboxylic acid, Q-**80002**

$C_{11}H_7NS$
Azuleno[1,2-*d*]thiazole, A-**60356**
Azuleno[2,1-*d*]thiazole, A-**60357**
Naphtho[1,8-*de*]-1,3-thiazine, N-**80009**
2-(Phenylethynyl)thiazole, P-**60099**
4-(Phenylethynyl)thiazole, P-**60100**

$C_{11}H_7N_3O$
Pyridazino[4,5-*b*]quinolin-10(5*H*)-one, P-**80207**

$C_{11}H_7N_3O_2$
Pyrimido[4,5-*b*]quinoline-2,4(3*H*,10*H*)-dione,
    P-**60245**

$C_{11}H_8$
5-Phenyl-1,3-pentadiyne, P-**80113**

$C_{11}H_8BrN$
2-Bromo-4-phenylpyridine, B-**80245**
2-Bromo-5-phenylpyridine, B-**80246**
2-Bromo-6-phenylpyridine, B-**80247**

3-Bromo-2-phenylpyridine, B-80248
3-Bromo-4-phenylpyridine, B-80249
4-Bromo-2-phenylpyridine, B-80250
5-Bromo-2-phenylpyridine, B-80251

C$_{11}$H$_8$Br$_2$

▷ 1-Bromo-2-(bromomethyl)naphthalene,
B-60212
1-Bromo-4-(bromomethyl)naphthalene,
B-60213
1-Bromo-5-(bromomethyl)naphthalene,
B-60214
1-Bromo-7-(bromomethyl)naphthalene,
B-60215
1-Bromo-8-(bromomethyl)naphthalene,
B-60216
2-Bromo-3-(bromomethyl)naphthalene,
B-60217
2-Bromo-6-(bromomethyl)naphthalene,
B-60218
3-Bromo-1-(bromomethyl)naphthalene,
B-60219
6-Bromo-1-(bromomethyl)naphthalene,
B-60220
7-Bromo-1-(bromomethyl)naphthalene,
B-60221
1,2-Dibromo-1,4-dihydro-1,4-
methanonaphthalene, D-60095
1,3-Dibromo-1,4-dihydro-1,4-
methanonaphthalene, D-60096

C$_{11}$H$_8$ClNO$_2$

Silital, in C-70075

C$_{11}$H$_8$N$_2$

1H-Benzo[de][1,6]naphthyridine, B-70037
7-Diazo-7H-benzocycloheptene, D-80064

C$_{11}$H$_8$N$_2$O

Di-2-pyridyl ketone, D-70502
Di-3-pyridyl ketone, D-70503
Di-4-pyridyl ketone, D-70504
1H-Indole-3-carboxylic acid; Nitrile, N-Ac, in
I-80012
1H-Perimidin-2(3H)-one, P-70043
2-Pyridyl 3-pyridyl ketone, P-70179
2-Pyridyl 4-pyridyl ketone, P-70180
3-Pyridyl 4-pyridyl ketone, P-70181

C$_{11}$H$_8$N$_2$O$_2$

5-Benzoyl-2(1H)-pyrimidinone, B-60066

C$_{11}$H$_8$N$_2$O$_2$S

1,1'-Carbonothioylbis-2(1H)pyridinone,
C-60014

C$_{11}$H$_8$N$_2$O$_3$S

1H-Perimidine-2-sulfonic acid, P-80067

C$_{11}$H$_8$N$_2$S

Dipyrido[1,2-a:1',2'-c]imidazolium-11-
thiolate, D-80516
1H-Perimidine-2(3H)-thione, P-10051

C$_{11}$H$_8$N$_4$

1H-Pyrazolo[3,4-d]pyrimidine; 1-Ph, in
P-70156

C$_{11}$H$_8$N$_4$O

1H-Pyrazolo[3,4-d]pyrimidine; 1-Ph, 5-oxide,
in P-70156

C$_{11}$H$_8$OS$_2$

5-(2-Thienylethynyl)-2-thiophenemethanol,
T-80180

C$_{11}$H$_8$O$_2$

Homoazulene-1,5-quinone, H-60085
Homoazulene-1,7-quinone, H-60086
Homoazulene-4,7-quinone, H-60087
6-Hydroxy-1-naphthaldehyde, H-70184
6-Hydroxy-2-naphthaldehyde, H-70185
7-Hydroxy-2-naphthaldehyde, H-70186

C$_{11}$H$_8$O$_3$

2-Acetyl-4H-1-benzopyran-4-one, A-80016
3-Acetyl-4H-1-benzopyran-4-one, A-80017
5-Hydroxy-7-methyl-1,2-naphthoquinone,
H-60189
8-Hydroxy-3-methyl-1,2-naphthoquinone,
H-60190
4-Methoxy-1,2-naphthoquinone, in H-60194

Pentacyclo[5.4.0.0^{2,6}.0^{3,10}.0^{5,9}]undecane-4,8,11-
trione, P-70023
Pentacyclo[6.3.0.0^{2,6}.0^{5,9}]undecane-4,7,11-
trione, P-60031
2-Phenyl-3-furancarboxylic acid, P-70076
4-Phenyl-2-furancarboxylic acid, P-70077
5-Phenyl-2-furancarboxylic acid, P-70078

C$_{11}$H$_8$O$_4$

3,5-Dihydroxy-2-methyl-1,4-naphthoquinone,
D-60350
5,6-Dihydroxy-2-methyl-1,4-naphthoquinone,
D-60351
3-(6-Hydroxy-5-benzofuranyl)-2-propenoic
acid, H-80113
2-Hydroxy-8-methoxy-1,4-naphthoquinone, in
D-60355
4-Methyl-6,7-methylenedioxy-2H-1-
benzopyran-2-one, M-70092
1-Oxo-1H-2-benzopyran-3-acetic acid,
O-60058
2-Oxo-2H-benzopyran-4-acetic acid, O-60057
4-Oxo-4H-1-benzopyran-3-acetic acid,
O-60059

C$_{11}$H$_8$O$_5$

4-Hydroxy-5-benzofurancarboxylic acid; Ac,
in H-60103
2-Hydroxy-5-methoxy-1,4-naphthoquinone, in
D-60354

C$_{11}$H$_8$O$_8$

2,3,5,6,8-Pentahydroxy-7-methoxy-1,4-
naphthoquinone, in H-60065

C$_{11}$H$_9$BF$_4$N$_4$

3-Phenyltetrazolo[1,5-a]pyridinium;
Tetrafluoroborate, in P-60134

C$_{11}$H$_9$Br

1-Bromo-1,4-dihydro-1,4-
methanonaphthalene, B-60231
2-Bromo-1,4-dihydro-1,4-
methanonaphthalene, B-60232
9-Bromo-1,4-dihydro-1,4-
methanonaphthalene, B-60233

C$_{11}$H$_9$BrN$_2$

Dipyrido[1,2-a:1',2'-c]imidazol-10-ium(1+);
Bromide, in D-70501

C$_{11}$H$_9$BrN$_4$

3-Phenyltetrazolo[1,5-a]pyridinium; Bromide,
in P-60134

C$_{11}$H$_9$BrO

1-Bromo-3,4-dihydro-2-
naphthalenecarboxaldehyde, B-80197
2-Bromo-3,4-dihydro-1-
naphthalenecarboxaldehyde, B-80198

C$_{11}$H$_9$ClN$_2$O$_4$

Dipyrido[1,2-a:1',2'-c]imidazol-10-ium(1+);
Perchlorate, in D-70501

C$_{11}$H$_9$I

2-(Iodomethyl)naphthalene, I-80035

C$_{11}$H$_9$IN$_2$

Dipyrido[1,2-a:1',2'-c]imidazol-10-ium(1+);
Iodide, in D-70501

C$_{11}$H$_9$N

2-Ethynylindole; 1-Me, in E-60057
4-Vinylquinoline, V-70014

C$_{11}$H$_9$NO

1-Acetylisoquinoline, A-60039
3-Acetylisoquinoline, A-60040
4-Acetylisoquinoline, A-60041
5-Acetylisoquinoline, A-60042
2-Acetylquinoline, A-60049
3-Acetylquinoline, A-60050
4-Acetylquinoline, A-60051
5-Acetylquinoline, A-60052
6-Acetylquinoline, A-60053
7-Acetylquinoline, A-60054
8-Acetylquinoline, A-60055
2-Phenyl-1,3-oxazepine, P-60123
5-Phenyl-1,4-oxazepine, P-60124

C$_{11}$H$_9$NO$_2$

4-Acetylquinoline; 1-Oxide, in A-60051

2-Amino-3-methyl-1,4-naphthoquinone,
A-70171
Cyclohepta[b]pyrrol-2(1H)-one; N-Ac, in
C-60201
2-Nitrobicyclo[4.4.1]undeca-1,3,5,7,9-
pentaene, N-80038
3-Nitrobicyclo[4.4.1]undeca-1,3,5,7,9-
pentaene, N-80039

C$_{11}$H$_9$NO$_2$S

2-Amino-3-(methylthio)-1,4-naphthalenedione,
in A-70163

C$_{11}$H$_9$NO$_3$

2-Formyl-1H-indole-3-acetic acid, F-80075
2-Oxo-2H-benzopyran-4-acetic acid; Amide,
in O-60057
4-Oxo-4H-1-benzopyran-3-acetic acid; Amide,
in O-60059
5-Phenyl-4-oxazolecarboxylic acid; Me ester,
in P-80112

C$_{11}$H$_9$NO$_4$

5-Methoxy-4-nitro-1-naphthol, in N-80054
6-Methoxy-1-nitro-2-naphthol, in N-80048

C$_{11}$H$_9$N$_2$$^{\oplus}$

Dipyrido[1,2-a:1',2'-c]imidazol-10-ium(1+),
D-70501

C$_{11}$H$_9$N$_3$

▷ 2-Amino-α-carboline, A-70105
3-Amino-α-carboline, A-70106
4-Amino-α-carboline, A-70107
6-Amino-α-carboline, A-70108
1-Amino-β-carboline, A-70109
3-Amino-β-carboline, A-70110
4-Amino-β-carboline, A-70111
6-Amino-β-carboline, A-70112
8-Amino-β-carboline, A-70113
8-Amino-δ-carboline, A-70119
1-Amino-γ-carboline, A-70114
3-Amino-γ-carboline, A-70115
5-Amino-γ-carboline, A-70116
6-Amino-γ-carboline, A-70117
8-Amino-γ-carboline, A-70118
1-Methyl-8H-benzo[cd]triazirino[a]indazole,
M-60052
2-Methyl-2H-naphtho[1,8-de]triazine,
M-70095
10H-Pyrazolo[5,1-c][1,4]benzodiazepine,
P-70154
9H-Pyrrolo[1,2-a]indol-9-one; Hydrazone, in
P-60247

C$_{11}$H$_9$N$_3$O

3,5-Dihydro-4H-pyrazolo[3,4-c]quinolin-4-one;
N^5-Me, in D-80245
Di-2-pyridyl ketone; Oxime, in D-70502
Di-3-pyridyl ketone; Oxime, in D-70503
Di-4-pyridyl ketone; Oxime, in D-70504
2-Pyridyl 3-pyridyl ketone; Oxime, in P-70179
2-Pyridyl 4-pyridyl ketone; Oxime, in P-70180
3-Pyridyl 4-pyridyl ketone; Oxime, in P-70181

C$_{11}$H$_9$N$_3$O$_3$

8-Amino-6-nitroquinoline; 8-N-Ac, in
A-60235

C$_{11}$H$_9$N$_4$$^{\oplus}$

3-Phenyltetrazolo[1,5-a]pyridinium, P-60134

C$_{11}$H$_9$N$_5$O

2-Amino-1,7-dihydro-7-phenyl-6H-purin-6-
one, A-70133

C$_{11}$H$_{10}$

1,4-Dihydro-1,4-methanonaphthalene,
D-60256

C$_{11}$H$_{10}$ClN$_2$O$^{\oplus}$

1'-(Chlorocarbonyl)-1',2'-dihydro-1,2'-
bipyridinium (1+), C-80045

C$_{11}$H$_{10}$Cl$_2$N$_2$O

Phosgene-in-a-can, in C-80045

C$_{11}$H$_{10}$IN$_3$

6H-Dipyrido[1,2-a:2',1'-d][1,3,5]triazin-5-ium;
Iodide, in D-60495

C$_{11}$H$_{10}$N$_2$

3,4-Dihydro-β-carboline, D-60209
Di-2-pyridylmethane, D-60496

1,2,3,4-Tetrahydro-1-methylenenaphthalene, T-60073

Tricyclo[5.4.0.0^{2,8}]undeca-3,5,9-triene, T-80246

$C_{11}H_{12}Br_2N_2$

1,1'-Methylenebispyridinium(1+); Dibromide, in M-70065

$C_{11}H_{12}F_2N_2O_5$

5-(2,2-Difluorovinyl)-2'-deoxyuridine, in D-80167

5-(2,2-Difluorovinyl)uracil; 1-(2-Deoxy-α-D-ribofuranosyl), in D-80167

$C_{11}H_{12}I_2N_2$

1,1'-Methylenebispyridinium(1+); Diiodide, in M-70065

$C_{11}H_{12}N_2$

5-Amino-1-methylisoquinoline; N-Me, in A-80102

4,4-Dimethyl-2-phenyl-4H-imidazole, D-70432

$C_{11}H_{12}N_2^{\oplus}$

1,1'-Methylenebispyridinium(1+), M-70065

$C_{11}H_{12}N_2O$

5-Amino-3-phenylisoxazole; N-Et, in A-60240

$C_{11}H_{12}N_2O_2$

5-Oxo-2-pyrrolidinecarboxylic acid; Anilide, in O-70102

$C_{11}H_{12}N_2O_4$

2,4-Diaminobenzoic acid; 2,4-N-Di-Ac, in D-60032

$C_{11}H_{12}N_2O_5$

2'-Deoxy-5-ethynyluridine, in E-60060

$C_{11}H_{12}N_2O_6$

5-Ethynyluridine, in E-60060

$C_{11}H_{12}N_4O$

5-Amino-1-(phenylmethyl)-1H-pyrazole-4-carboxamide, in A-70192

$C_{11}H_{12}O$

2,4,6-Cycloheptatrien-1-ylcyclopropylmethanone, C-60206

2,3-Dihydro-2,2-dimethyl-1H-inden-1-one, D-60227

2-Methyl-4-phenyl-3-butyn-2-ol, M-70110

3-Phenylcyclopentanone, P-80099

1-Phenyl-3-penten-1-one, P-60127

5-Phenyl-4-penten-2-one, P-70088

1,2,3,4-Tetrahydro-7H-benzocyclohepten-7-one, T-60057

$C_{11}H_{12}OS$

2,3-Dihydro-2,2-dimethyl-4H-1-benzothiopyran, D-70193

$C_{11}H_{12}O_2$

8-Decene-4,6-diynoic acid; Me ester, in D-70018

3-Oxo-5-phenylpentanal, O-80092

4-Oxo-5-phenylpentanal, O-80093

5-Oxo-5-phenylpentanal, O-80094

2-Phenylcyclopropanecarboxylic acid; Me ester, in P-60087

1-Phenyl-1,2-pentanedione, P-80115

1-Phenyl-1,3-pentanedione, P-80116

1-Phenyl-1,4-pentanedione, P-80117

1-Phenyl-2,4-pentanedione, P-80118

$C_{11}H_{12}O_2S$

4-Mercapto-2-butenoic acid; Benzyl ester, in M-60018

5-Methylthio-2,4,8-decatrien-6-ynoic acid, M-80168

3-[(4-Methylthio)phenyl]-2-butenoic acid, M-80179

2-[2-(Methylthio)propenyl]benzoic acid, M-80182

4-[2-(Methylthio)propenyl]benzoic acid, M-80183

$C_{11}H_{12}O_3$

Asperpentyne, A-80179

3-Benzoyl-2-methylpropanoic acid, B-60064

3,4-Dihydro-2H-1-benzopyran-2-ol; Ac, in D-60198

3,4-Dihydro-2H-1-benzopyran-4-ol; Ac, in D-60200

3,4-Dihydro-2H-1-benzopyran-6-ol; Ac, in D-70175

3-(4-Hydroxyphenyl)-2-propenoic acid; Et ester, in H-80242

3-(4-Hydroxy-2-vinylphenyl)propanoic acid, H-70232

1,2,3,4-Tetrahydro-1-hydroxy-2-naphthalenecarboxylic acid, T-70063

$C_{11}H_{12}O_4$

1,3,5-Cycloheptatriene-1,6-dicarboxylic acid; Di-Me ester, in C-60202

5,7-Dimethoxy-6-methylphthalide, in D-60349

4-Hydroxybenzyl alcohol; Di-Ac, in H-80114

Pyrenocine A, P-80203

$C_{11}H_{12}O_5$

2-Formyl-3,5-dihydroxybenzoic acid; Di-Me ether, Me ester, in F-60071

5-Hydroxy-3-methyl-1,2-benzenedicarboxylic acid; Me ether, 2-Me ester, in H-70162

$C_{11}H_{12}O_6$

3-Oxotricyclo[2.1.0.0^{2,5}]pentane-1,5-dicarboxylic acid; Di-Me ester, ethylene ketal, in O-80096

$C_{11}H_{12}S$

2,2-Dimethyl-2H-1-benzothiopyran, D-70373

$C_{11}H_{12}S_2$

2-(2,4,6-Cycloheptatrien-1-ylidene)-1,3-dithiane, C-70221

$C_{11}H_{13}BrN_2O_5$

5-(2-Bromovinyl)-2'-deoxyuridine, B-80266

$C_{11}H_{13}BrO_2$

3-Bromo-2,4,6-trimethylbenzoic acid; Me ester, in B-60331

$C_{11}H_{13}IO_4$

2,5-Dihydroxy-4-iodobenzoic acid; Di-Me ether, Et ester, in D-60333

$C_{11}H_{13}I_2NO$

2,6-Diiodobenzoic acid; Diethylamide, in D-70356

$C_{11}H_{13}N$

1,2,3,6-Tetrahydro-4-phenylpyridine, T-60081

1,2,3,6-Tetrahydropyridine; 1-Ph, in T-70089

$C_{11}H_{13}NO$

3,4-Dihydro-3,3-dimethyl-2(1H)quinolinone, D-60231

3-Isopropyloxindole, I-70097

4-(Phenylimino)-2-pentanone, in P-60059

1,2,3,4-Tetrahydroquinoline; N-Ac, in T-70093

$C_{11}H_{13}NO_2$

2-Amino-3-phenyl-3-butenoic acid; Me ester, in A-60238

5,6-Dihydroxyindole; Di-Me ether, N-Me, in D-70307

1,2,3,4-Tetrahydro-4-hydroxyisoquinoline; N-Ac, in T-60066

1,2,3,4-Tetrahydro-3-methyl-3-isoquinolinecarboxylic acid, T-60074

5-(3,4,5,6-Tetrahydro-3-pyridylidenemethyl)-2-furanmethanol, T-60087

$C_{11}H_{13}NO_3$

1-Amino-2-(4-hydroxyphenyl)cyclopropanecarboxylic acid; Me ester, in A-60197

1-Amino-2-(4-hydroxyphenyl)cyclopropanecarboxylic acid; Me ether, in A-60197

Ethyl N-phenacylcarbamate, in O-70099

2,3,4-Trihydroxyphenylacetic acid; Tri-Me ether, nitrile, in T-60340

2,4,5-Trihydroxyphenylacetic acid; Tri-Me ether, nitrile, in T-60342

▷ 3,4,5-Trimethoxybenzeneacetonitrile, in T-60343

$C_{11}H_{13}NO_4$

2-Amino-2-benzylbutanedioic acid, A-60094

2-Amino-4-hydroxybutanoic acid; N-Benzoyl, in A-70156

2-Amino-3-phenylpentanedioic acid, A-60241

2,6-Bis(methoxyacetyl)pyridine, in B-70144

(±)-Isothreonine; N-Benzoyl, in A-60183

3,5-Pyridinedipropanoic acid, P-60217

Threonine; N-Benzoyl, in T-60217

$C_{11}H_{13}NO_5$

5-Amino-4-hydroxy-1,3-benzenedicarboxylic acid; Me ether, di-Me ester, in A-70155

Dimethyl 4-amino-5-methoxyphthalate, in A-70152

$C_{11}H_{13}N_3O_4$

2'-Deoxy-5-ethynylcytidine, in A-60153

Imidazo[4,5-c]pyridine; 1-β-D-Ribofuranosyl, in I-60008

$C_{11}H_{13}N_3O_5$

5-Ethynylcytidine, in A-60153

$C_{11}H_{14}$

6-(1,3-Butadienyl)-1,4-cycloheptadiene, B-70290

2-Phenyl-2-pentene, P-80119

$C_{11}H_{14}N_2$

5,6,9,10-Tetrahydro-4H,8H-pyrido[3,2,1-ij][1,6]naphthyridine, T-60086

$C_{11}H_{14}N_2O$

Piperazinone; N^4-Benzyl, in P-70101

$C_{11}H_{14}N_2O_3$

2,4,6-Trimethyl-3-nitroaniline; N-Ac, in T-70285

$C_{11}H_{14}N_2O_4$

4-Amino-2,6-pyridinedicarboxylic acid; Di-Et ester, in A-70194

$C_{11}H_{14}N_2O_5$

2'-Deoxy-5-vinyluridine, in V-70015

$C_{11}H_{14}N_2O_6$

2-Amino-3,5-pyridinedicarboxylic acid; Di-Et ester, in A-70193

5-Vinyluridine, V-70015

$C_{11}H_{14}N_4$

4-(Butylimino)-3,4-dihydro-1,2,3-benzotriazine(incorr.), in A-60092

$C_{11}H_{14}N_4O_3$

4-Amino-1H-imidazo[4,5-c]pyridine; 1-(2-Deoxy-β-D-ribofuranosyl), in A-70161

4-Amino-1H-imidazo[4,5-c]pyridine; 1-(5-Deoxy-β-D-ribofuranosyl), in A-70161

$C_{11}H_{14}N_4O_3S$

2'-Deoxy-7-deaza-6-thioguanosine, in A-80144

$C_{11}H_{14}N_4O_4$

4-Amino-4,6-dihydro-3-methyl-1H-cyclopenta[e]1,2,4-triazine-5,7-dicarboxylic acid; Di-Me ester, in A-80066

4-Amino-1H-imidazo[4,5-c]pyridine; 1-β-D-Ribofuranosyl, in A-70161

1-Deazaadenosine, in A-80087

7-Deaza-2'-deoxyguanosine, in A-60255

2-Methyl-9-β-D-ribofuranosyl-9H-purine, M-70130

6-Methyl-9-β-D-ribofuranosyl-9H-purine, M-70131

$C_{11}H_{14}N_4O_4S$

▷ 6-(Methylthio)-9-β-D-ribofuranosyl-9H-purine, in T-70180

$C_{11}H_{14}N_4O_5$

7-Deazaguanosine, in A-60255

$C_{11}H_{14}O$

3,4-Dihydro-2,2-dimethyl-2H-1-benzopyran, D-80184

2,2-Dimethyl-1-phenyl-1-propanone, D-60443

(3-Hydroxyphenoxy)acetic acid; Me ether, Et ester, in H-70204

4-Methoxy-4-phenyl-1-butene, in P-70061

3-Methyl-2-phenylbutanal, M-70109

Spiro[5.5]undeca-1,3-dien-7-one, S-60046

$C_{11}H_{14}O_2$

Andirolactone, A-60271

1,1'-Bi(bicyclo[1.1.1]pentane)-3-carboxylic acid, B-80072

3-*tert*-Butyl-4-hydroxybenzaldehyde, B-70304
1-(Phenylmethoxy)-3-buten-2-ol, *in* B-70291
1,2,3,4-Tetrahydro-1-hydroxy-2-
naphthalenemethanol, T-70064
Tricyclo[3.3.3.0^{1,5}]undecane-3,7-dione,
T-80245
Tricyclo[4.3.1.1^{3,8}]undecane-2,7-dione,
T-70233
Tricyclo[4.3.1.1^{3,8}]undecane-4,5-dione,
T-70234
3,4,5-Trimethylbenzoic acid; Me ester, *in*
T-60355
3,4,5-Trimethylphenol; Ac, *in* T-60371

$C_{11}H_{14}O_3$

1,2,3,4,5,6,7,8-Octahydro-8-oxo-2-
naphthalenecarboxylic acid, O-80015

$C_{11}H_{14}O_4$

2-(3,4-Dihydroxyphenyl)propanoic acid; Di-
Me ether, *in* D-60372
2,6-Dimethoxy-3,5-dimethylbenzoic acid, *in*
D-70293
6-Ethyl-2,4-dihydroxy-3-methylbenzoic acid;
Me ester, *in* E-70041
2'-Hydroxy-4',6'-dimethoxy-3'-
methylacetophenone, *in* T-80321
3-(4-Hydroxy-3,5-dimethoxyphenyl)propanal,
in T-80333
3-Hydroxy-5-(4-hydroxyphenyl)pentanoic
acid, H-70142
(2-Hydroxyphenoxy)acetic acid; Me ether, Et
ester, *in* H-70203
(4-Hydroxyphenoxy)acetic acid; Me ether, Et
ester, *in* H-70205
Pyrenocine C, *in* P-80203
Sinapyl alcohol, S-70046
2',4',6'-Trimethoxyacetophenone, *in* T-70248
3',4',5'-Trimethoxyacetophenone, *in* T-60315

$C_{11}H_{14}O_5$

Pyrenocine B, *in* P-80203
3,4,6-Trihydroxy-2-methylbenzoic acid; 4,6-
Di-Me ether, Me ester, *in* T-80324
2,3,4-Trimethoxybenzeneacetic acid, *in*
T-60340
2,4,5-Trimethoxybenzeneacetic acid, *in*
T-60342
3,4,5-Trimethoxybenzeneacetic acid, *in*
T-60343
3,4,6-Trimethoxy-2-methylbenzoic acid, *in*
T-80324

$C_{11}H_{14}O_6$

2,3,5,6-Tetramethoxybenzoic acid, *in* T-60100
(3,4,5-Trihydroxy-6-oxo-1-cyclohexen-1-
yl)methyl 2-butenoate, T-60338

$C_{11}H_{15}Br$

(Bromomethylidene)adamantane, B-60296
Bromopentamethylbenzene, B-60312

$C_{11}H_{15}Cl$

Chloropentamethylbenzene, C-60122

$C_{11}H_{15}N$

1-Amino-1-phenyl-4-pentene, A-70183
2-Methyl-5-phenylpyrrolidine, M-80147
2-Phenylcyclopentylamine, P-60086
[6](2,6)Pyridinophane, P-80211
2,3,4,5-Tetrahydro-1*H*-1-benzazepine; 1-Me,
in T-70036

$C_{11}H_{15}NO$

1-Isocyanatoadamantane, I-80053
[6](2,6)Pyridinophane; *N*-Oxide, *in* P-80211

$C_{11}H_{15}NOS$

Tetrahydro-2-thiophenecarboxylic acid;
Anilide, *in* T-70099
Tricyclo[3.3.1.1^{3,7}]decyl-1-sulfinylcyanide, *in*
T-70177

$C_{11}H_{15}NO_2$

2-Amino-2-methyl-3-phenylpropanoic acid;
Me ester, *in* A-60225
1,2,3,4,5-Pentamethyl-6-nitrobenzene, P-60057

$C_{11}H_{15}NO_2S$

2-Amino-4,5,6,7-tetrahydrobenzo[*b*]thiophene-
3-carboxylic acid; Et ester, *in* A-60258

$C_{11}H_{15}NO_3$

2-Amino-3-(4-hydroxyphenyl)-3-
methylbutanoic acid, A-60198
4-Amino-3-hydroxy-5-phenylpentanoic acid,
A-60199

$C_{11}H_{15}NO_4$

1,4-Dihydro-2,6-dimethyl-3,5-
pyridinedicarboxylic acid; Di-Me ester, *in*
D-80193
3',4',5'-Trihydroxyacetophenone; Tri-Me
ether, oxime, *in* T-60315

$C_{11}H_{15}NO_5$

3-Hydroxy-1*H*-pyrrole-2,4-dicarboxylic acid;
Me ether, Di-Et ester, *in* H-80251
1*H*-Pyrrole-3,4-dicarboxylic acid; 1-Methoxy,
di-Et ester, *in* P-70183

$C_{11}H_{15}NS$

3-Aminotetrahydro-2*H*-thiopyran; *N*-Ph, *in*
A-70200
1-Thiocyanatoadamantane, T-70177

$C_{11}H_{15}N_5O$

1'-Methylzeatin, M-60133

$C_{11}H_{15}N_5O_3$

1-[3-Azido-4-(hydroxymethyl)cyclopentyl]-5-
methyl-2,4(1*H*,3*H*)pyrimidinedione,
A-70288

$C_{11}H_{15}N_5O_4$

Euglenapterin, E-60070

$C_{11}H_{15}N_5O_5$

Ara-doridosine, *in* A-60144
Doridosine, D-60518
1-Methylguanosine, M-70084
8-Methylguanosine, M-70085

$C_{11}H_{15}N_5O_5S$

8-(Methylthio)guanosine, *in* M-70029

$C_{11}H_{15}N_5O_6$

8-Methoxyguanosine, *in* D-70238

$C_{11}H_{15}N_5O_7S$

8-(Methylsulfonyl)guanosine, *in* M-70029

$C_{11}H_{16}$

Bicyclo[4.4.1]undeca-1,6-diene, B-70115
1-Cycloundecen-3-yne, C-70256
2,3-Dimethylenebicyclo[2.2.3]nonane, D-70406
2,6-Dimethylenebicyclo[3.3.1]nonane, D-80448
(2,2-Dimethylpropyl)benzene, D-60445
Methyleneadamantane, M-60068
Tetracyclo[4.4.1.0^{3,11}.0^{9,11}]undecane, T-60037
1,3-Undecadien-5-yne, U-80017
1,5-Undecadien-3 yne, U-80018
4,7-Undecadiyne, U-80019
2,4,6,8-Undecatetraene, U-70003

$C_{11}H_{16}N_2O_3$

2-Amino-3-nitrobenzyl alcohol; *N*-Et, Et
ether, *in* A-80125

$C_{11}H_{16}N_2O_5$

▷ 2'-Deoxy-5-ethyluridine, *in* E-60054

$C_{11}H_{16}N_2O_6$

5-Ethyluridine, *in* E-60054

$C_{11}H_{16}N_6O_3$

3'-Amino-3'-deoxyadenosine; 3'*N*-Me, *in*
A-70128

$C_{11}H_{16}N_6O_5$

8-(Methylamino)guanosine, *in* A-70150

$C_{11}H_{16}O$

5,6,7,8,9,10-Hexahydro-4*H*-cyclonona[*c*]furan,
H-60047
1-Methyladamantanone, M-60040
5-Methyladamantanone, M-60041
3-Oxatetracyclo[5.3.1.1^{2,6}.0^{4,9}]dodecane,
O-70069
Pentamethylphenol, P-60058
2,2,5,5-Tetramethylbicyclo[4.1.0]hept-1(6)-en-
7-one, T-70131
2,4-Undecadiyn-1-ol, U-80020

$C_{11}H_{16}O_2$

Bicyclo[2.2.2]oct-2-ene-1-carboxylic acid; Et
ester, *in* B-70113

1,3-Dimethoxy-1-phenylpropane, *in* P-80122
▷ 3,4-Dimethyl-5-pentylidene-2(5*H*)-furanone,
D-70430
Spiro[5.5]undecane-1,9-dione, S-70066
Spiro[5.5]undecane-3,8-dione, S-70067
Spiro[5.5]undecane-3,9-dione, S-70068

$C_{11}H_{16}O_4$

Methylenolactocin, M-70079
1,2,3,4-Tetramethoxy-5-methylbenzene, *in*
M-80050
▷ Xanthotoxol, X-60002

$C_{11}H_{16}O_5$

3,6-Dihydro-2*H*-pyran-2,2-dicarboxylic acid;
Di-Et ester, *in* D-80243

$C_{11}H_{16}O_8$

Ranunculin, *in* H-60182

$C_{11}H_{17}N$

3-Azatetracyclo[5.3.1.1^{2,6}.0^{4,9}]dodecane,
A-70282
Spiro[aziridine-2,2'-tricyclo[3.1.1.1^{3,7}]decane],
S-80040

$C_{11}H_{17}NO_7$

2-(Dimethylamino)benzaldehyde; Di-Me
acetal, *in* D-60401

$C_{11}H_{18}$

2,2-Dimethyl-3-methylenebicyclo[2.2.2]octane,
D-60427
4,8-Dimethyl-1,3,7-nonatriene, D-50644
1,3,5-Undecatriene, U-70004

$C_{11}H_{18}N_2O_2$

Cyclo(leucylprolyl), C-80185
2-Diazo-2,2,6,6-tetramethyl-3,5-heptanedione,
D-70064
Hexahydro-3-(1-methylpropyl)pyrrolo[1,2-
a]pyrazine-1,4-dione, H-60052
Spiro[5.5]undecane-3,9-dione; Dioxime, *in*
S-70068

$C_{11}H_{18}N_2O_3$

Cyclo(hydroxyprolylleucyl), C-60213

$C_{11}H_{18}O$

2-Cycloundecen-1-one, C-70255
2-Ethyl-4,6,6-trimethyl-2-cyclohexen-1-one,
E-80068

$C_{11}H_{18}OS$

2-(Methylsulfinyl)adamantane, *in* A-80038

$C_{11}H_{18}O_2$

Bicyclo[2.2.2]octane-1-carboxylic acid; Et
ester, *in* B-60092
3,3,4,4,5,5-Hexamethyl-1,2-cyclopentanedione,
H-80074
2-Hexyl-5-methyl-3(2*H*)furanone, H-60079
1-Methoxy-3-myodesertene, *in* M-80203

$C_{11}H_{18}O_2S$

2-(Methylsulfonyl)adamantane, *in* A-80038

$C_{11}H_{18}O_3$

2,3-Di-*tert*-butoxycyclopropenone, *in* D-70290
Dihydro-3,3,4,4,5,5-hexamethyl-2*H*-pyran-
2,6(3*H*)-dione, D-80210
Dihydro-3,3,4,4,5,5-hexamethyl-2*H*-pyran-
2,6(3*H*)dione, H-80076
1,3-Dihydroxy-4-methyl-6,8-decadien-5-one,
D-70321
Queen substance; Me ester, *in* O-70089

$C_{11}H_{18}O_4$

Citreoviral, C-60150
Phaseolinic acid, P-60067

$C_{11}H_{18}O_5$

4-Oxoheptanedioic acid; Di-Et ester, *in*
O-60071
4,4-Pyrandiacetic acid; Di-Me ester, *in*
P-70150

$C_{11}H_{18}O_6$

Eccremocarpol A, E-80002

$C_{11}H_{18}S$

2-(Methylthio)adamantane, *in* A-80038

$C_{11}H_{19}NO_8$

Agropinic acid, A-70075

$C_{11}H_{19}N_3O_7$
Nitropeptin, N-70055

$C_{11}H_{20}$
Cycloundecene, C-80200
1,1,3,3-Tetramethyl-2-methylenecyclohexane, T-60144
1,10-Undecadiene, U-80008
1,2-Undecadiene, U-80003
1,3-Undecadiene, U-80004
1,4-Undecadiene, U-80005
1,5-Undecadiene, U-80006
1,6-Undecadiene, U-80007
2,3-Undecadiene, U-80009
2,4-Undecadiene, U-80010
2,5-Undecadiene, U-80011
2,9-Undecadiene, U-80012
3,5-Undecadiene, U-80013
4,6-Undecadiene, U-80014
4,7-Undecadiene, U-80015
5,6-Undecadiene, U-80016

$C_{11}H_{20}Br_2O$
1,11-Dibromo-6-undecanone, D-80107

$C_{11}H_{20}N_2$
3,4-Di-*tert*-butylpyrazole, D-70106
3,5-Di-*tert*-butylpyrazole, D-70107

$C_{11}H_{20}O$
10-Undecen-2-one, U-80022
5-Undecen-2-one, U-80023
10-Undecyn-2-ol, U-80032
1-Undecyn-5-ol, U-80024
3-Undecyn-1-ol, U-80025
4-Undecyn-1-ol, U-80026
4-Undecyn-2-ol, U-80027
5-Undecyn-1-ol, U-80028
5-Undecyn-2-ol, U-80029
6-Undecyn-1-ol, U-60004
6-Undecyn-5-ol, U-80030
7-Undecyn-1-ol, U-80031

$C_{11}H_{20}O_2$
2,7-Diethoxycycloheptene, *in* C-80179
1,1-Diethoxy-2-methylenecylohexane, *in* M-80080
2,4-Heptadienal; Di-Et acetal, *in* H-60020
2-Methyl-4-penten-1-ol; Tetrahydropyranyl ether, *in* M-80135

$C_{11}H_{20}O_3$
2-Hydroxycyclooctanecarboxylic acid; Et ester, *in* H-70121
2-Hydroxycyclooctanecarboxylic acid; Me ether, Me ester, *in* H-70121
10-Oxodecanoic acid; Me ester, *in* O-60063

$C_{11}H_{20}O_4$
Hexamethylpentanedioic acid, H-80076

$C_{11}H_{20}O_7$
Eccremocarpol B, E-80003

$C_{11}H_{21}Br$
11-Bromo-1-undecene, B-60336
1-Bromo-2-undecene, B-70277

$C_{11}H_{21}NO$
Hexahydro-5(2*H*)-azocinone; 1-*tert*-Butyl, *in* H-70048

$C_{11}H_{21}NO_3$
3-Oxocyclobutanecarboxylic acid; *N*,*N*-Dimethylamide, Di-Et ketal, *in* O-80073

$C_{11}H_{21}NO_9$
Mannopinic acid, M-70009

$C_{11}H_{21}NO_{10}S_3$
Glucoiberin, G-80022

$C_{11}H_{22}$
1,1-Di-*tert*-butylcyclopropane, D-60112

$C_{11}H_{22}N_2O$
4-Amino-2,2,6,6-tetramethylpiperidine; *N*-Ac, *in* A-70201

$C_{11}H_{22}N_2O_3$
N-Valylleucine, V-60001

$C_{11}H_{22}N_2O_8$
Mannopine, *in* M-70009

$C_{11}H_{22}O$
6-Undecen-1-ol, U-80021

$C_{11}H_{22}O_3$
2,2-Diethoxytetrahydro-5,6-dimethyl-2*H*-pyran, *in* T-70059
1,11-Dihydroxy-6-undecanone, D-80392
9-Hydroxydecanoic acid; Me ester, *in* H-80149
Sitophilate, *in* H-80209

$C_{11}H_{23}N$
Octahydroazocine; 1-*tert*-Butyl, *in* O-70010

$C_{11}H_{24}$
2,2,3,3,4,4-Hexamethylpentane, H-60071

$C_{11}H_{27}N_5$
1,4,7,10,13-Pentaazacyclohexadecane, P-60021

$C_{12}H_4Cl_4O_2$
▷ 1,2,3,4-Tetrachlorodibenzo-*p*-dioxin, T-70016

$C_{12}H_4Cl_6$
▷ 2,2′,4,4′,5,5′-Hexachlorobiphenyl, H-60034

$C_{12}H_4S$
▷ Tetraethynylthiophene, T-80048

$C_{12}H_6$
▷ 1-Phenyl-1,3,5-hexatriyne, P-80105

$C_{12}H_6Br_2N_2O_4$
2,2′-Dibromo-3,5′-dinitrobiphenyl, D-80091
2,2′-Dibromo-5,5′-dinitrobiphenyl, D-80092
5,5′-Dibromo-2,2′-dinitrobiphenyl, D-80093

$C_{12}H_6Br_6$
Hexakis(bromomethylene)cyclohexane, H-80073

$C_{12}H_6Cl_2N_2$
1,4-Dichlorobenzo[*g*]phthalazine, D-70114

$C_{12}H_6Cl_2N_2O_2$
[2,2′-Bipyridine]-4,4′-dicarboxylic acid; Dichloride, *in* B-60126
[2,2′-Bipyridine]-6,6′-dicarboxylic acid; Dichloride, *in* B-60128

$C_{12}H_6Cl_4N_2O_4$
Pyrroxamycin, P-70203

$C_{12}H_6Cl_4O$
Bis(2,4-dichlorophenyl)ether, B-70135
Bis(2,6-dichlorophenyl) ether, B-70136
Bis(3,4-dichlorophenyl) ether, B-70137

$C_{12}H_6Cl_{12}$
Hexakis(dichloromethyl)benzene, H-60067

$C_{12}H_6F_2N_2O_4$
3,3′-Difluoro-4,4′-dinitrobiphenyl, D-80161
4,4′-Difluoro-2,2′-dinitrobiphenyl, D-80162
5,5′-Difluoro-2,2′-dinitrobiphenyl, D-80163

$C_{12}H_6I_2N_2O_4$
2,2′-Diiodo-4,5′-dinitrobiphenyl, D-80401
2,2′-Diiodo-5,5′-dinitrobiphenyl, D-80402
2,2′-Diiodo-6,6′-dinitrobiphenyl, D-80403
4,4′-Diiodo-3,3′-dinitrobiphenyl, D-80404

$C_{12}H_6N_2$
5,7-Dicyanoazulene, *in* A-70295
2,3-Dicyanonaphthalene, *in* N-80001

$C_{12}H_6N_2O_2$
Benzo[*g*]quinazoline-6,9-dione, B-60044
Benzo[*g*]quinoxaline-6,9-dione, B-60046
1,7-Phenanthroline-5,6-dione, P-70052

$C_{12}H_6N_2O_4$
1,5-Dinitrobiphenylene, D-80492
2,6-Dinitrobiphenylene, D-80493

$C_{12}H_6N_4$
5,5′-Dicyano-2,2′-bipyridine, *in* B-60127
Pyrazino[2′,3′:3,4]cyclobuta[1,2-*g*]quinoxaline, P-60198
1,2,4,5-Tetracyano-3,6-dimethylbenzene, *in* D-60407

$C_{12}H_6O_2S$
Naphtho[2,3-*b*]thiophene-2,3-dione, N-70014

$C_{12}H_6O_3$
Naphtho[2,3-*c*]furan-1,3-dione, *in* N-80001

$C_{12}H_6O_4$
2,2-Bi(1,4-benzoquinone), B-80069
4,4′-Bi(1,2-benzoquinone), B-80070

$C_{12}H_7Br$
1-Bromobiphenylene, B-80181
2-Bromobiphenylene, B-80182

$C_{12}H_7BrN_2$
2-Bromo-1,10-phenanthroline, B-70264
8-Bromo-1,7-phenanthroline, B-70265

$C_{12}H_7BrO_4$
2-Bromo-5-hydroxy-1,4-naphthoquinone; Ac, *in* B-60269
2-Bromo-6-hydroxy-1,4-naphthoquinone; Ac, *in* B-60270
2-Bromo-8-hydroxy-1,4-naphthoquinone; Ac, *in* B-60272

$C_{12}H_7ClN_2$
1-Chlorobenzo[*g*]phthalazine, C-70054

$C_{12}H_7ClN_2O$
4-Chlorobenzo[*g*]phthalazin-1(2*H*)-one, C-70055

$C_{12}H_7I$
1-Iodobiphenylene, I-80020
2-Iodobiphenylene, I-80021

$C_{12}H_7NOS$
1*H*-Phenothiazin-1-one, P-60078

$C_{12}H_7NO_2$
1*H*-Benz[*e*]indole-1,2(3*H*)-dione, B-70026
10*H*-[1]Benzopyrano[3,2-*c*]pyridin-10-one, B-60042
10*H*-[1]Benzopyrano[3,2-*c*]pyridin-10-one, B-80038
1*H*-Carbazole-1,4(9*H*)-dione, C-60012
3-Cyano-2-naphthoic acid, *in* N-80001
2*H*-Naphtho[2,3-*c*]pyrrole-1,3-dione, *in* N-80001
1-Nitrobiphenylene, N-80040
2-Nitrobiphenylene, N-80041

$C_{12}H_7NO_3$
2-Hydroxy-3*H*-phenoxazin-3-one, H-60206

$C_{12}H_7N_5$
7*H*-2,3,4,6,7-Pentaazabenz[*de*]anthracene, P-60017

$C_{12}H_8$
1-Ethylnylnaphthalene, E-70046

$C_{12}H_8ClNO$
1-Chloromethylbenz[*cd*]indol-2(1*H*)-one, *in* B-70027

$C_{12}H_8ClNOS$
3-Chloro-10*H*-phenothiazine; 5-Oxide, *in* C-60126
4-Chloro-10*H*-phenothiazine; 5-Oxide, *in* C-60127

$C_{12}H_8ClNO_2$
2-Chloro-7-nitro-9*H*-fluorene, C-70133
9-Chloro-2-nitro-9*H*-fluorene, C-70134
9-Chloro-3-nitro-9*H*-fluorene, C-70135

$C_{12}H_8ClNO_2S$
3-Chloro-10*H*-phenothiazine; 5,5-Dioxide, *in* C-60126
4-Chloro-10*H*-phenothiazine; 5,5-Dioxide, *in* C-60127

$C_{12}H_8ClNO_3$
▷ 2-Amino-3-chloro-1,4-naphthoquinone; *N*-Ac, *in* A-70122

$C_{12}H_8ClNS$
1-Chloro-10*H*-phenothiazine, C-60124
▷ 2-Chloro-10*H*-phenothiazine, C-60125
3-Chloro-10*H*-phenothiazine, C-60126

$C_{12}H_8Cl_2$
5,6-Dichloroacenaphthene, D-60115

$C_{12}H_8F_3NO_3$
7-Amino-4-(trifluoromethyl)-2*H*-1-benzopyran-2-one; *N*-Ac, *in* A-60265

$C_{12}H_8N_2$
Benzo[*c*]cinnoline, B-60020

Benzo[*b*]-[1,5]naphthyridine, B-80030
Benzo[*b*][1,7]naphthyridine, B-80031
Benzo[*b*][1,8]naphthyridine, B-80032
Benzvalenoquinoxaline, B-70080

$C_{12}H_8N_2O$

Benzo[*c*]cinnoline; *N*-Oxide, *in* B-60020
2-(2-Furanyl)quinoxaline, F-60089

$C_{12}H_8N_2O_2$

▷ 2-Amino-3*H*-phenoxazin-3-one, A-60237
1*H*-Benz[*e*]indole-1,2(3*H*)-dione; 1-Oxime, *in* B-70026
Benzo[*c*]cinnoline; 5,6-Di-*N*-oxide, *in* B-60020
6,6'-(1,2-Ethynediyl)bis-2(1*H*)-pyridinone, E-80069

$C_{12}H_8N_2O_4$

[2,2'-Bipyridine]-3,3'-dicarboxylic acid, B-60124
[2,2'-Bipyridine]-3,5'-dicarboxylic acid, B-60125
[2,2'-Bipyridine]-4,4'-dicarboxylic acid, B-60126
[2,2'-Bipyridine]-5,5'-dicarboxylic acid, B-60127
[2,2'-Bipyridine]-6,6'-dicarboxylic acid, B-60128
[2,3'-Bipyridine]-2',3'-dicarboxylic acid, B-60129
[2,4'-Bipyridine]-2',6'-dicarboxylic acid, B-60130
[2,4'-Bipyridine]-3,3'-dicarboxylic acid, B-60131
[2,4'-Bipyridine]-3',5'-dicarboxylic acid, B-60132
[3,3'-Bipyridine]-2,2'-dicarboxylic acid, B-60133
[3,3'-Bipyridine]-4,4'-dicarboxylic acid, B-60134
[3,4'-Bipyridine]-2',6'-dicarboxylic acid, B-60135
[4,4'-Bipyridine]-2,2'-dicarboxylic acid, B-60136
[4,4'-Bipyridine]-3,3'-dicarboxylic acid, B-60137
2,2'-Dinitrobiphenyl, D-60460

$C_{12}H_8N_4$

Pyrido[1″,2″:1',2']imidazo[4',5':4,5]imidazo[1,2-*a*]pyridine, P-70175
Pyrido[2″,1″:2',3']imidazo[4',5':4,5]imidazo[1,2-*a*]pyridine, P-60234

$C_{12}H_8N_{10}O_{12}$

3,3',5,5'-Tetraamino-2,2',4,4',6,6'-hexanitrobiphenyl, T-60014

$C_{12}H_8O$

Cyclobuta[*b*]naphthalen-1(2*H*)-one, C-80175
6*b*,7*a*-Dihydroacenaphth[1,2-*b*]oxirene, D-70170
4-Hydroxyacenaphthylene, H-70100
Naphtho[1,2-*c*]furan, N-70008
Naphtho[2,3-*c*]furan, N-60008
Naphtho[1,8-*bc*]pyran, N-70009
1-Phenyl-1,4-hexadiyn-3-one, P-60108
▷ 1-Phenyl-2,4-hexadiyn-1-one, P-60109
6-Phenyl-3,5-hexadiyn-2-one, P-60110

$C_{12}H_8OS$

4-Phenylthieno[3,4-*b*]furan, P-60135

$C_{12}H_8OSe$

Phenoxaselenin, P-80087

$C_{12}H_8O_2$

1,2-Biphenylenediol, B-80116
1,5-Biphenylenediol, B-80117
1,8-Biphenylenediol, B-80118
2,3-Biphenylenediol, B-80119
2,6-Biphenylenediol, B-80120
2,7-Biphenylenediol, B-80121
1,2-Dihydrocyclobuta[*a*]naphthalene-3,4-dione, D-60213
1,2-Dihydrocyclobuta[*b*]naphthalene-3,8-dione, D-60214
6-Hydroxy-1-phenyl-2,4-hexadiyn-1-one, H-80235
1,2-Naphthalenedicarboxaldehyde, N-60001
1,3-Naphthalenedicarboxaldehyde, N-60002

$C_{12}H_8O_2Se$

Phenoxaselenin; Se-Oxide, *in* P-80087

$C_{12}H_8O_2Se_2$

Selenanthrene; 9,10-Dioxide, *in* S-80021

$C_{12}H_8O_3$

2-Hydroxy-6-(2,4-pentadiynyl)benzoic acid, H-80231
α-Oxo-1-naphthaleneacetic acid, O-60078
α-Oxo-2-naphthaleneacetic acid, O-60079

$C_{12}H_8O_4$

1,4-Azulenedicarboxylic acid, A-70293
2,6-Azulenedicarboxylic acid, A-70294
5,7-Azulenedicarboxylic acid, A-70295
2,3-Naphthalenedicarboxylic acid, N-80001
β-Sorigenin, S-80039
▷ Xanthotoxin, X-60001

$C_{12}H_8O_7$

2,5,8-Trihydroxy-3-methyl-6,7-methylenedioxy-1,4-naphthoquinone, T-60333

$C_{12}H_8STe$

Phenothiatellurin, P-80086

$C_{12}H_8S_2$

2-Dibenzothiophenethiol, D-60078
4-Dibenzothiophenethiol, D-60079

$C_{12}H_8Se_2$

Selenanthrene, S-80021

$C_{12}H_9BrO$

1-Acetyl-3-bromonaphthalene, A-60021
1-Acetyl-4-bromonaphthalene, A-60022
1-Acetyl-5-bromonaphthalene, A-60023
1-Acetyl-7-bromonaphthalene, A-60024
2-Acetyl-6-bromonaphthalene, A-60025

$C_{12}H_9BrO_3$

2-Bromo-3-ethoxy-1,4-naphthoquinone, *in* B-60268

$C_{12}H_9ClN_2$

3-Chloro-3-(1-naphthylmethyl)diazirine, C-80081

$C_{12}H_9ClO_2S$

[1,1'-Biphenyl]-2-sulfonic acid; Chloride, *in* B-80122
[1,1'-Biphenyl]-3-sulfonic acid; Chloride, *in* B-80123
[1,1'-Biphenyl]-4-sulfonic acid; Chloride, *in* B-80124

$C_{12}H_9IN_2O$

3-Amino-5-iodopyridine; N-Benzoyl, *in* A-80092

$C_{12}H_9IO$

1-Acetyl-7-iodonaphthalene, A-60037
1-Acetyl-8-iodonaphthalene, A-60038

$C_{12}H_9N$

1-Aminobiphenylene, A-80062
2-Aminobiphenylene, A-80063

$C_{12}H_9NO$

Benz[*cd*]indol-2-(1*H*)-one; *N*-Me, *in* B-70027
2-(2-Furanyl)-1*H*-indole, F-80095
3-(2-Furanyl)-1*H*-indole, F-80096
1-Hydroxycarbazole, H-80137
2-Hydroxycarbazole, H-60109
2-Methylfuro[3,2-*c*]quinoline, M-70083

$C_{12}H_9NO_2$

2-Phenyl-3-pyridinecarboxylic acid, P-80125
2-Phenyl-4-pyridinecarboxylic acid, P-80126
4-Phenyl-2-pyridinecarboxylic acid, P-80127
5-Phenyl-2-pyridinecarboxylic acid, P-80128
5-Phenyl-3-pyridinecarboxylic acid, P-80129

$C_{12}H_9N_3$

2-Amino-1,10-phenanthroline, A-70179
4-Amino-1,10-phenanthroline, A-70181
5-Amino-1,10-phenanthroline, A-70182
8-Amino-1,7-phenanthroline, A-70180
2-Phenyl-2*H*-benzotriazole, P-60083
N-Pyridinium-2-benzimidazole, P-70172

$C_{12}H_9N_3O$

3-Amino-β-carboline; N^3-Formyl, *in* A-70110

2-Phenyl-2*H*-benzotriazole; 1-*N*-Oxide, *in* P-60083
Pyridazino[4,5-*b*]quinolin-10(5*H*)-one; 2-Me, *in* P-80207
Pyridazino[4,5-*b*]quinolin-10(5*H*)-one; 5-Me, *in* P-80207

$C_{12}H_{10}$

1,6-Dihydro-1,6-dimethyleneazulene, D-60225
2,6-Dihydro-2,6-dimethyleneazulene, D-60226
1,2-Dihydro-1,2-dimethylenenaphthalene, D-80187
1,2-Dihydro-1,4-dimethylenenaphthalene, D-80188
2,3-Dihydro-2,3-dimethylenenaphthalene, D-80189

$C_{12}H_{10}ClNO$

2-Chloro-1-naphthylamine; *N*-Ac, *in* C-80080

$C_{12}H_{10}F_3NO_2$

7-Amino-4-(trifluoromethyl)-2*H*-1-benzopyran-2-one; *N*-Et, *in* A-60265

$C_{12}H_{10}N_2$

1,5-Diaminobiphenylene, D-80053
2,3-Diaminobiphenylene, D-80054
2,6-Diaminobiphenylene, D-80055
1-(1-Diazoethyl)naphthalene, D-60061
5,6-Dihydrobenzo[*c*]cinnoline, D-80172

$C_{12}H_{10}N_2O$

5-Phenyl-3-pyridinecarboxylic acid; Amide, *in* P-80129

$C_{12}H_{10}N_2OS$

2-(Methylsulfinyl)perimidine, *in* P-10051

$C_{12}H_{10}N_2O_2$

5-Hydroxy-2,2'-bipyridine; Ac, *in* H-80123

$C_{12}H_{10}N_2O_3$

5-Acetyl-2,4(1*H*,3*H*)-pyrimidinedione; 1-Ph, *in* A-60047
3-Nitro-4(1*H*)-pyridone; *N*-Benzyl, *in* N-70063

$C_{12}H_{10}N_2S$

2-(Methylthio)perimidine, *in* P-10051

$C_{12}H_{10}N_4O_2$

[2,2'-Bipyridine]-4,4'-dicarboxylic acid; Diamide, *in* B-60126
[2,2'-Bipyridine]-5,5'-dicarboxylic acid; Diamide, *in* B-60127

$C_{12}H_{10}O$

1,2-Dihydrocyclobuta[*a*]naphthalen-3-ol, D-60215
2,3-Dihydronaphtho[2,3-*b*]furan, D-70235
1-Hydroxyacenaphthene, H-80106
4-Hydroxyacenaphthene, H-60099
3-Methyl-1-naphthaldehyde, M-80120
5-Methyl-1-naphthaldehyde, M-80121
1-Phenyl-2,4-hexadiyn-1-ol, P-80104

$C_{12}H_{10}OS$

5-(Methylthio)-1-phenyl-4-penten-2-yn-1-one, M-80181

$C_{12}H_{10}OS_2$

4-[2,2'-Bithiophen-5-yl]-3-butyn-1-ol, B-80167

$C_{12}H_{10}O_2$

3,4'-Biphenyldiol, B-80106
3-(Hydroxymethyl)-2-naphthalenecarboxaldehyde, H-60188
6-Methoxy-1-naphthaldehyde, *in* H-70184
6-Methoxy-2-naphthaldehyde, *in* H-70185
7-Methoxy-2-naphthaldehyde, *in* H-70186
4-Methyl-1-azulenecarboxylic acid, M-60042
2,3,6,7-Tetrahydro-*as*-indacene-1,8-dione, T-70069
2,3,5,6-Tetrahydro-*s*-indacene-1,7-dione, T-70067
2,3,6,7-Tetrahydro-*s*-indacene-1,5-dione, T-70068

$C_{12}H_{10}O_2S_2$

4-[2,2'-Bithiophen]-5-yl-3-butyne-1,2-diol, B-80166

$C_{12}H_{10}O_3$

2,3,4-Biphenyltriol, B-70124
3,4,5-Biphenyltriol, B-80129

8,8-Dimethyl-1,4,5(8*H*)-naphthalenetrione,
D-70421
4-Ethoxy-1,2-naphthoquinone, *in* H-60194
4-Hydroxy-2-cyclopentenone; Benzoyl, *in*
H-60116
5-Methoxy-7-methyl-1,2-naphthoquinone, *in*
H-60189
8-Methoxy-3-methyl-1,2-naphthoquinone, *in*
H-60190
5-Phenyl-2-furancarboxylic acid; Me ester, *in*
P-70078

C$_{12}$H$_{10}$O$_3$S
[1,1′-Biphenyl]-2-sulfonic acid, B-80122
[1,1′-Biphenyl]-3-sulfonic acid, B-80123
[1,1′-Biphenyl]-4-sulfonic acid, B-80124

C$_{12}$H$_{10}$O$_4$
2,2′,3,4-Biphenyltetrol, B-70122
2,2′,3,4-Biphenyltetrol, B-80125
2,3,3′,4-Biphenyltetrol, B-80126
2,3,4,4′-Biphenyltetrol, B-70123
2,3,4,4′-Biphenyltetrol, B-80127
2,3′,4′,5′-Biphenyltetrol, B-80128
5,8-Dihydroxy-2,3-dimethyl-1,4-
naphthoquinone, D-60317
3-Hydroxy-5-methoxy-2-methyl-1,4-
naphthoquinone, *in* D-60350
6-Hydroxy-5-methoxy-2-methyl-1,4-
naphthoquinone, *in* D-60351
1-Oxo-1*H*-2-benzopyran-3-acetic acid; Me
ester, *in* O-60058

C$_{12}$H$_{10}$O$_5$
Armillarisin A, A-80169
Murraxonin, M-60152

C$_{12}$H$_{10}$O$_6$S$_2$
▷ Benzenesulfonyl peroxide, B-70018

C$_{12}$H$_{10}$O$_8$
3,6-Dimethyl-1,2,4,5-benzenetetracarboxylic
acid, D-60407
2,5,6,8-Tetrahydroxy-3,7-dimethoxy-1,4-
naphthoquinone, *in* H-60065
2,5,7,8-Tetrahydroxy-3,6-dimethoxy-1,4-
naphthoquinone, *in* H-60065

C$_{12}$H$_{10}$S
1,4-Dihydrodibenzothiophene, D-80180
2-(3,5,7-Octatrien-1-ynyl)thiophene, O-80020

C$_{12}$H$_{11}$BrO$_3$
5-Bromo-1,2,4-benzenetriol; Tri-Ac, *in*
B-70181
6-Bromo-1,2,4-benzenetriol; Tri-Ac, *in*
B-70182

C$_{12}$H$_{11}$BrO$_6$
2-Bromo-1,3,5-benzenetriol; Tri-Ac, *in*
B-60199

C$_{12}$H$_{11}$F$_3$O$_5$S
4,5-Dimethyl-2-phenyl-1,3-dioxol-1-ium;
Trifluoromethanesulfonate, *in* D-70431

C$_{12}$H$_{11}$N
2-(2-Phenylethenyl)-1*H*-pyrrole, P-60094

C$_{12}$H$_{11}$NO
1,2-Dihydro-3*H*-azepin-3-one; 1-Ph, *in*
D-70171
1-Oxo-1,2,3,4-tetrahydrocarbazole, O-70103
2-Oxo-1,2,3,4-tetrahydrocarbazole, O-70104
3-Oxo-1,2,3,4-tetrahydrocarbazole, O-70105
4-Oxo-1,2,3,4-tetrahydrocarbazole, O-70106
4*H*,6*H*-Pyrrolo[1,2-*a*][4,1]benzoxazepine,
P-60246

C$_{12}$H$_{11}$NO$_2$
1,4-Diacetylindole, *in* A-60033
6-Hydroxy-1-naphthaldehyde; Me ether,
oxime, *in* H-70184
6-Hydroxy-2-naphthaldehyde; Me ether,
oxime, *in* H-70185

C$_{12}$H$_{11}$NO$_2$S
[1,1′-Biphenyl]-3-sulfonic acid; Amide, *in*
B-80123
[1,1′-Biphenyl]-4-sulfonic acid; Amide, *in*
B-80124

C$_{12}$H$_{11}$NO$_3$
3-Amino-1,2-naphthalenediol; *N*-Ac, *in*
A-80116
4-Amino-1,2-naphthalenediol; *N*-Ac, *in*
A-80117
3-Hydroxy-2-quinolinecarboxylic acid; Me
ether, Me ester, *in* H-60228
5-Phenyl-4-oxazolecarboxylic acid; Et ester, *in*
P-80112

C$_{12}$H$_{11}$NO$_4$
5,6-Dihydroxyindole; *O,O*-Di-Ac, *in* D-70307
1,2-Dimethoxy-4-nitronaphthalene, *in*
N-80053
1,4-Dimethoxy-2-nitronaphthalene, *in*
N-80050
1,5-Dimethoxy-4-nitronaphthalene, *in*
N-80054
2,3-Dimethoxy-6-nitronaphthalene, *in*
N-80056
2,6-Dimethoxy-1-nitronaphthalene, *in*
N-80048
4,6-Dimethoxy-1-nitronaphthalene, *in*
N-80055
2-Oxo-3-indolineglyoxylic acid; Et ester, *in*
O-80085

C$_{12}$H$_{11}$N$_3$
1,8-Diaminocarbazole, D-80056
Indamine, I-60019
2-(Methylamino)-1*H*-pyrido[2,3-*b*]indole, *in*
A-70105

C$_{12}$H$_{12}$
9,10-Bis(methylene)tricyclo[5.3.0.0^{2,8}]deca-3,5-
diene, B-70152
2,3-Didehydro-1,2-dihydro-1,1-
dimethylnaphthalene, D-60168
1,4-Dihydrobenzocyclooctatetraene, D-70173
1,6-Dimethylazulene, D-60404
[2.2]Orthometacyclophane, O-80048
Pentacyclo[6.4.0.0^{2,7}.0^{3,12}.0^{6,9}]dodeca-4,10-
diene, P-60028
Tetracyclo[6.2.1.1^{3,6}.0^{2,7}]dodeca-2(7),4,9-triene,
T-60032
4*a*,4*b*,8*a*,8*b*-Tetrahydrobiphenylene, T-60060
Tricyclo[5.5.0.0^{2,8}]dodeca-3,5,9,11-tetraene,
T-80239

C$_{12}$H$_{12}$N$_2$
1-Benzyl-5-vinylimidazole, *in* V-70012
2,2′-Diaminobiphenyl, D-60033
4,6-Dimethyl-2-phenylpyrimidine, D-60444
1,4,5,6-Tetrahydrocyclopentapyrazole; 2-Ph,
in T-80055
4(5)-Vinylimidazole; 1-Benzyl, *in* V-70012

C$_{12}$H$_{12}$N$_2$O
5-Amino-1-methylisoquinoline; *N*-Ac, *in*
A-80102
1-Oxo-1,2,3,4-tetrahydrocarbazole; (*E*)-Oxime,
in O-70103
2-Oxo-1,2,3,4-tetrahydrocarbazole; Oxime, *in*
O-70104
3-Oxo-1,2,3,4-tetrahydrocarbazole; Oxime, *in*
O-70105
4-Oxo-1,2,3,4-tetrahydrocarbazole; Oxime, *in*
O-70106
1-Oxo-1,2,3,4-tetrahydrocarbazole; (*Z*)-oxime,
in O-70103

C$_{12}$H$_{12}$N$_2$O$_2$
2,2′-Dimethoxy-3,3′-bipyridine, *in* D-80287
3,3′-Dimethoxy-2,2′-bipyridine, *in* D-80290
4,4′-Dimethoxy-2,2′-bipyridine, *in* D-80294
5,5′-Dimethoxy-3,3′-bipyridine, *in* D-80298
6,6′-Dimethoxy-2,2′-bipyridine, *in* D-80299
6,6′-Dimethoxy-3,3′-bipyridine, *in* D-80300

C$_{12}$H$_{12}$N$_2$O$_4$
4,4′-Dihydroxy-2,2′-bipyridine; Di-Me ether,
1,1′-dioxide, *in* D-80294

C$_{12}$H$_{12}$N$_2$O$_7$
3-Amino-4-nitro-1,2-benzenedicarboxylic acid;
Di-Me ester, Ac, *in* A-70174
4-Amino-3-nitro-1,2-benzenedicarboxylic acid;
Di-Me ester, Ac, *in* A-70176
4-Amino-5-nitro-1,2-benzenedicarboxylic acid;
Di Me ester, Ac, *in* A-70177

C$_{12}$H$_{12}$N$_4$O$_2$
lin-Benzocaffeine, *in* I-60009
Imidazo[4,5-*g*]quinazoline-6,8(5*H*,7*H*)-dione;
3,5,7-Tri-Me, *in* I-60009

C$_{12}$H$_{12}$N$_6$
2,5,8-Trimethylbenzotriimidazole, T-60356

C$_{12}$H$_{12}$N$_6$O$_2$
[2,2′-Bipyridine]-5,5′-dicarboxylic acid;
Dihydrazide, *in* B-60127
[3,3′-Bipyridine]-2,2′-dicarboxylic acid;
Dihydrazide, *in* B-60133

C$_{12}$H$_{12}$O
1-Benzoylcyclopentene, B-70072
1-(1-Naphthyl)ethanol, N-70015

C$_{12}$H$_{12}$O$_2$
5-Phenyl-2,4-pentadienoic acid; Me ester, *in*
P-70086

C$_{12}$H$_{12}$O$_2$S
4-Methylthio-2,4-decadiene-6,8-diynoic acid;
Me ester, *in* M-80163
5-Methylthio-2,4-decadiene-6,8-diynoic acid;
Me ester, *in* M-80164
7-Methylthio-2,6-decadiene-4,8-diynoic acid;
Me ester, *in* M-80165
7-Methylthio-2,6-decadiene-4,8-diynoic acid;
Me ester, *in* M-80166
7-Methylthio-2,6-decadiene-4,8-diynoic acid;
Me ester, *in* M-80166
9-Methylthio-2,8-decadiene-4,6-diynoic acid;
Me ester, *in* M-80167
2-(Phenylsulfonyl)-1,3-cyclohexadiene,
P-70095

C$_{12}$H$_{12}$O$_3$
2,3-Dihydro-3-hydroxy-2-(1-methylethenyl)-5-
benzofurancarboxaldehyde, D-70213
4,5-Dimethoxy-1-naphthol, *in* N-60005
4,8-Dimethoxy-1-naphthol, *in* N-60005
4,6-Dimethoxy-5-vinylbenzofuran, *in* D-70352
2,2-Dimethyl-2*H*-1-benzopyran-6-carboxylic
acid, D-60408
2-Hydroxycyclopentanone; Benzoyl, *in*
H-60114
2-(1-Hydroxy-1-methylethyl)-5-
benzofurancarboxaldehyde, *in* D-70214

C$_{12}$H$_{12}$O$_3$S
2-Oxo-3-phenylthio-3-butenoic acid; Et ester,
in O-60088

C$_{12}$H$_{12}$O$_3$S$_3$
2,3,6,7,10,11-Hexahydrobenzo[1,2-*b*:3,4-*b*′:5,6-
b″]tris[1,4]oxathiin, H-70051

C$_{12}$H$_{12}$O$_4$
4,7-Dimethoxy-5-methyl-2*H*-1-benzopyran-2-
one, *in* D-60348
4-Hydroxy-2-isopropyl-5-
benzofurancarboxylic acid, H-60164

C$_{12}$H$_{12}$O$_4$S
7-Methylthio-2,6-decadiene-4,8-diynoic acid;
Me ester, S-dioxide, *in* M-80166

C$_{12}$H$_{12}$O$_5$
5-Hydroxy-6-(hydroxymethyl)-7-methoxy-2-
methyl-4*H*-1-benzopyran-4-one, H-60155
8-(Hydroxymethyl)-5,7-dimethoxy-2*H*-1-
benzopyran-2-one, *in* M-80198
Orthosporin, *in* D-80062
5,6,7-Trimethoxy-2*H*-1-benzopyran-2-one, *in*
T-80287
5,7,8-Trimethoxy-2*H*-1-benzopyran-2-one, *in*
T-80288

C$_{12}$H$_{12}$O$_6$
1,4-Benzoquinone-2,6-dicarboxylic acid; Di-Et
ester, *in* B-70049
2-Hydroxy-1,4-benzenedicarboxylic acid; Ac,
di-Me ester, *in* H-70105

C$_{12}$H$_{12}$S
1-Methylthio-5-phenyl-1-penten-3-yne,
M-80180

C$_{12}$H$_{12}$S$_6$
2,3,6,7,10,11-Hexahydrobenzo[1,2-*b*:3,4-*b*′:5,6-
b″]tris[1,4]dithiin, H-70050

C$_{12}$H$_{16}$O$_3$

Jasmine ketolactone, J-80005
1,2,3,4,5,6,7,8-Octahydro-8-oxo-2-
naphthalenecarboxylic acid; Me ester, *in*
O-80015
3-Oxo-2-(2-pentenyl)-1-cyclopenteneacetic
acid; (Z)-*form, in* O-80090
1,2,4-Trimethoxy-5-(2-propenyl)benzene,
T-70279
Xanthostemone, X-70002

C$_{12}$H$_{16}$O$_4$

6-Ethyl-2,4-dihydroxy-3-methylbenzoic acid;
Et ester, *in* E-70041
Senkyunolide *H, in* L-60026
Sordariol, S-70056

C$_{12}$H$_{16}$O$_5$

3,4,5-Trihydroxyphenylacetic acid; *tert*-Butyl
ester, *in* T-60343
2,3,4-Trihydroxyphenylacetic acid; Tri-Me
ether, Me ester, *in* T-60340
2,4,5-Trihydroxyphenylacetic acid; Tri-Me
ether, Me ester, *in* T-60342

C$_{12}$H$_{16}$O$_6$

Diethyl diacetylfumarate, *in* D-80042
Pentamethoxybenzaldehyde, *in* P-60042

C$_{12}$H$_{16}$O$_7$

Arbutin, A-70248
Pentamethoxybenzoic acid, *in* P-60043

C$_{12}$H$_{17}$N

1,2-Dimethyl-5-phenylpyrrolidine, *in* M-80147
[7](2,6)Pyridinophane, P-80212

C$_{12}$H$_{17}$NO

[7](2,6)Pyridinophane; N-Oxide, *in* P-80212

C$_{12}$H$_{17}$NO$_2$

2-Amino-2-methyl-3-phenylpropanoic acid; Et
ester, *in* A-60225
Pentamethyl(nitromethyl)benzene, P-80059

C$_{12}$H$_{17}$NO$_3$

2-Amino-3-(4-methoxyphenyl)-3-
methylbutanoic acid, *in* A-60198

C$_{12}$H$_{17}$NO$_5$

3-Hydroxy-1*H*-pyrrole-2,4-dicarboxylic acid;
N-Me, Me ether, Di-Et ester, *in* H-80251

C$_{12}$H$_{17}$NO$_7$

Epivolkenin, *in* T-60007
Innovanamine, *in* H-80214
Taraktophyllin, T-60007

C$_{12}$H$_{17}$N$_3$O$_4$

▷ Agaritine, A-70071

C$_{12}$H$_{17}$N$_5$O$_5$

8-(Dimethylamino)inosine, *in* A-70198

C$_{12}$H$_{17}$N$_5$O$_{10}$S

Antibiotic PB 5266*C,* A-70227

C$_{12}$H$_{18}$

2a,3,4,5,5a,6,7,8,8a,8b-
Decahydroacenaphthene, D-80009
2-Isopropyl-1,3,5-trimethylbenzene, I-60130

C$_{12}$H$_{18}$N$_2$

1,4-Dihydropyrrolo[3,2-*b*]pyrrole; 1,4-
Diisopropyl, *in* D-30286

C$_{12}$H$_{18}$N$_4$O$_4$

Octahydroimidazo[4,5-*d*]imidazole; 1,3,4,6-
Tetra-Ac, *in* O-70015

C$_{12}$H$_{18}$N$_6$O$_5$

8-(Dimethylamino)guanosine, *in* A-70150

C$_{12}$H$_{18}$O

Methoxypentamethylbenzene, *in* P-60058
Pentamethylbenzyl alcohol, P-80058

C$_{12}$H$_{18}$OS

(Phenylthio)acetaldehyde; Di-Et acetal, *in*
P-80136

C$_{12}$H$_{18}$O$_2$

Anastrephin, A-70206
[1,1'-Bicyclohexyl]-2,2'-dione, B-60089
Clavularin *A,* C-60153
Clavularin *B, in* C-60153

Epianastrephin, *in* A-70206
6-(1-Heptenyl)-5,6-dihydro-2*H*-pyran-2-one,
H-60025
1-Hydroxy-11,12,13-trisnor-9-eremophilen-8-
one, H-80263
5-Isopropyl-2-methyl-1,3-benzenediol; Di-Me
ether, *in* I-60125
Myrtenol; Ac, *in* M-80205
Suspensolide, S-70084

C$_{12}$H$_{18}$O$_3$

1*R*-Acetoxy-3-myodesertene, *in* M-80203
1*S*-Acetoxy-3-myodesertene, *in* M-80203
2,10-Dihydroxy-11,12,13-trisnor-6-
eremophilen-8-one, D-80391
7-Isojasmonic acid, *in* J-60002
Jasmonic acid, J-60002
Libanotic acid, L-80025
Sedanonic acid, S-80020

C$_{12}$H$_{18}$O$_4$

1-Acetoxy-7-hydroxy-3,7-dimethyl-2*E*,5*E*-
octadien-4-one, *in* D-60318
3,4-Dihydroxyphenylacetaldehyde; Di-Me
ether, di-Me acetal, *in* D-60371

C$_{12}$H$_{18}$O$_5$

1-Acetoxy-7-hydroperoxy-3,7-dimethyl-2*E*,5*E*-
octadien-4-one, *in* D-60318

C$_{12}$H$_{18}$O$_9$

s-Trioxane-2,4,6-tripropionic acid, *in* O-80068

C$_{12}$H$_{19}$N

3-Azatetracyclo[5.3.1.1^{2,6}.0^{4,9}]dodecane; *N*-Me,
in A-70282
1,2,3,4,5,6,7,8,8a,9a-Decahydrocarbazole,
D-70012

C$_{12}$H$_{19}$NO$_4$

▷ Choline salicylate, *in* H-70108

C$_{12}$H$_{20}$

1-*tert*-Butyl-1,2-cyclooctadiene, B-60344
Cyclododecyne, C-60192
1,1',2,2',3,4,5,5',6,6'-Decahydrobiphenyl,
D-60007
1,1',2,3,4,4',5,5',6,6'-Decahydrobiphenyl,
D-60008
1,2,3,3',4,4',5,5',6,6'-Decahydrobiphenyl,
D-60009
1,3-Dimethyladamantane, D-70366
5-Dodecen-7-yne, D-70538
Tricyclo[5.5.0.0^{2,8}]dodecane, T-80237

C$_{12}$H$_{20}$BrClO

Kumepaloxane, K-80013

C$_{12}$H$_{20}$N$_2$

7,7'-Bi-7-azabicyclo[2.2.1]heptane, B-80068

C$_{12}$H$_{20}$N$_2$O$_7$

2'-Deoxymugeneic acid, *in* M-60148

C$_{12}$H$_{20}$N$_2$O$_8$

Isomugeneic acid, *in* M-60148
Mugeneic acid, M-60148

C$_{12}$H$_{20}$N$_2$O$_9$

3-Hydroxymugeneic acid, *in* M-60148

C$_{12}$H$_{20}$N$_4$O$_4$

Octahydro-1,2,5,6-tetrazocine; Tetra-Ac; *in*
O-60020

C$_{12}$H$_{20}$O$_2$

10-Dodecen-12-olide, *in* H-60126

C$_{12}$H$_{20}$O$_3$

8-Acetoxy-2,6-dimethyl-3,6-octadien-2-ol, *in*
D-80457
Cucurbic acid, C-50302
6-Epi-7-isocucurbic acid, *in* C-50302
Helicascolide A, H-80004
Helicascolide B, *in* H-80004
3-Oxo-2-pentylcyclopentaneacetic acid,
O-80091

C$_{12}$H$_{20}$O$_3$S$_2$

Asadisulphide, A-80175

C$_{12}$H$_{20}$O$_4$

9,12-Dioxododecanoic acid, D-60471
4-Octene-1,8-diol; Di-Ac, *in* O-70026

C$_{12}$H$_{20}$O$_6$

1,2-*O*-Cyclohexylidene-*myo*-inositol, *in*
I-80015
1,2:4,5-Di-*O*-isopropylidene-*myo*-inositol, *in*
I-80015

C$_{12}$H$_{20}$O$_8$

Parasorboside, *in* T-60067

C$_{12}$H$_{20}$S

2,4-Di-*tert*-butylthiophene, D-80112

C$_{12}$H$_{21}$N

Dodecahydrocarbazole, D-70532

C$_{12}$H$_{21}$NO$_4$

2-Amino-1,4-cyclohexanedicarboxylic acid;
Di-Et ester, *in* A-60118
3-Amino-1,2-cyclohexanedicarboxylic acid;
Di-Et ester, *in* A-60119
4-Amino-1,1-cyclohexanedicarboxylic acid;
Di-Et ester, *in* A-60120
4-Amino-1,3-cyclohexanedicarboxylic acid;
Di-Et ester, *in* A-60121
3,4-Dihydroxy-2,5-pyrrolidinedione; Di-*tert*-
butyl ether, *in* D-80377

C$_{12}$H$_{21}$N$_3$

3,5,12-Triazatetracyclo[5.3.1.1^{2,6}.0^{4,9}]dodecane;
3,5,12-Tri-Me, *in* T-70206

C$_{12}$H$_{22}$

2,3-Di-*tert*-butyl-1,3-butadiene, D-80108

C$_{12}$H$_{22}$Cl$_2$O

12-Chlorododecanoic acid; Chloride, *in*
C-70066

C$_{12}$H$_{22}$N$_2$

1-*tert*-Butyl-3-(*tert*-butylamino)-1*H*-pyrrole, *in*
A-80143
1,1-Di-1-piperidinylethene, D-70500

C$_{12}$H$_{22}$N$_2$O$_2$

3,6-Bis(1-methylpropyl)-2,5-piperazinedione,
B-70153
3,6-Bis(2-methylpropyl)-2,5-piperazinedione,
B-60171
Dihydro-4,6-(1*H*,5*H*)-pyrimidinedione; Di-
tert-butyl ether, *in* D-70256

C$_{12}$H$_{22}$O

2,3-Dodecadien-1-ol, D-80520
11-Dodecyn-1-ol, D-70540
5-Dodecyn-1-ol, D-70539
10-Methoxy-1-undecyne, *in* U-80032

C$_{12}$H$_{22}$O$_2$

11-Decanolide, *in* H-80157
5-Decen-1-ol; Ac, *in* D-70019
2,6-Dodecanedione, D-70534
Invictolide, I-60038
2-Oxodecanal, O-70087

C$_{12}$H$_{22}$O$_2$S$_2$

2,2,3,3-Tetramethyl-1,4-butanedithiol; Di-Ac,
in T-80142

C$_{12}$H$_{22}$O$_3$

10-Hydroxy-11-dodecenoic acid, H-60125
12-Hydroxy-10-dodecenoic acid, H-60126
12-Oxododecanoic acid, O-60066

C$_{12}$H$_{22}$O$_4$

(1-Methylpropyl)butanedioic acid; Di-Et ester,
in M-80150
Talaromycin *A,* T-70004
Talaromycin *B, in* T-70004
Talaromycin *C, in* T-70004
Talaromycin *D, in* T-70004
Talaromycin *E, in* T-70004
Talaromycin *F, in* T-70004
Talaromycin *G, in* T-70004
1,1,4,4-Tetraethoxy-2-butyne, *in* B-60351
L-Threaric acid; Di-*tert*-butyl ester, *in*
T-70005
L-Threaric acid; Di-*tert*-butyl ether, *in*
T-70005

C$_{12}$H$_{22}$O$_6$

Diethyl diethoxymethylmalonate, *in* H-80204

C$_{12}$H$_{23}$Br

6-Bromo-6-dodecene, B-60238

$C_{13}H_9N$
1-Cyanoacenaphthene, *in* A-70014
2-Cyano-1,2-dihydrocyclobuta[*a*]naphthalene, *in* D-70177

$C_{13}H_9NO$
4-Phenoxybenzonitrile, *in* P-80088
2-Phenylfuro[2,3-*b*]pyridine, P-60104
2-Phenylfuro[3,2-*c*]pyridine, P-60105

$C_{13}H_9NO_2$
1*H*-Benz[*e*]indole-1,2(3*H*)-dione; *N*-Me, *in* B-70026
Benz[*cd*]indol-2-(1*H*)-one; *N*-Ac, *in* B-70027

$C_{13}H_9NO_2S$
10*H*-Phenothiazine-1-carboxylic acid, P-70055
10*H*-Phenothiazine-2-carboxylic acid, P-70056
10*H*-Phenothiazine-3-carboxylic acid, P-70057
10*H*-Phenothiazine-4-carboxylic acid, P-70058

$C_{13}H_9NO_3$
2-Methoxy-3*H*-phenoxazin-3-one, *in* H-60206

$C_{13}H_9NO_4$
4-Phenyl-2,6-pyridinedicarboxylic acid, P-80130

$C_{13}H_9NO_4S$
10*H*-Phenothiazine-2-carboxylic acid; 5,5-Dioxide, *in* P-70056

$C_{13}H_9NS$
2-Phenylthieno[2,3-*b*]pyridine, P-60136

$C_{13}H_9NSe$
2-Phenylbenzoselenazole, P-80090

$C_{13}H_9N_3$
4-Phenyl-1,2,3-benzotriazine, P-80093

$C_{13}H_9N_3O$
2-Azidobenzophenone, A-60338
3-Azidobenzophenone, A-60339
4-Azidobenzophenone, A-60340
1*H*-Benzotriazole; *N*-Benzoyl, *in* B-70068
4-Phenyl-1,2,3-benzotriazine; 2-Oxide, *in* P-80093
4-Phenyl-1,2,3-benzotriazine; 3-Oxide, *in* P-80093

$C_{13}H_9O^{\oplus}$
Xanthylium(1+), X-70003

$C_{13}H_9OS^{\oplus}$
2-Phenyl-1,3-benzoxathiol-1-ium, P-60084

$C_{13}H_9S^{\oplus}$
Dibenzo[*b,d*]thiopyrylium(1+), D-70081
Thioxanthylium(1+), T-70186

$C_{13}H_{10}BrN$
Benzo[*a*]quinolizinium(1+); Bromide, *in* B-70043

$C_{13}H_{10}ClNO_4$
Benzo[*a*]quinolizinium(1+); Perchlorate, *in* B-70043

$C_{13}H_{10}F_2$
Difluorodiphenylmethane, D-60184

$C_{13}H_{10}N^{\oplus}$
Benzo[*a*]quinolizinium(1+), B-70043

$C_{13}H_{10}N_2OS$
10*H*-Phenothiazine-1-carboxylic acid; Amide, *in* P-70055
10*H*-Phenothiazine-2-carboxylic acid; Amide, *in* P-70056

$C_{13}H_{10}N_2O_2$
Benzophenone nitrimine, *in* D-60482

$C_{13}H_{10}N_2O_4$
[2,2′-Bipyridine]-3,3′-dicarboxylic acid; Mono-Me ester, *in* B-60124

$C_{13}H_{10}N_4$
4-Amino-1,2,3-benzotriazine; 3-Ph, *in* A-60092
4-Anilino-1,2,3-benzotriazine, *in* A-60092

$C_{13}H_{10}O$
1-Acenaphthenecarboxaldehyde, A-70013
2-(1,3-Butadienyl)-3-(1,3,5-heptatriynyl)oxirane, B-80277
9*H*-Fluoren-1-ol, F-80016
4-Methoxyacenaphthylene, *in* H-70100
Ponticaepoxide, P-80170
2,10,12-Tridecatriene-4,6,8-triyn-1-ol, *in* T-80253

$C_{13}H_{10}OS$
5-[5-(3-Buten-1-ynyl)-2-thienyl]-2-penten-4-yn-1-ol, B-80281
2-Mercaptobenzophenone, M-70020
3-Mercaptobenzophenone, M-70021
4-Mercaptobenzophenone, M-70022
4-[5-(1,3-Pentadiynyl)-7-thienyl]-3-butyn-1-ol, P-80025

$C_{13}H_{10}OS_2$
5-(3-Buten-1-ynyl)-5′ hydroxymethyl-2,2′-bithienyl, B-80279

$C_{13}H_{10}OS_3$
2,2′:5′,2″-Terthiophene-5-methanol, T-80011

$C_{13}H_{10}OSe_2$
Benzoyl phenyl diselenide, B-70077

$C_{13}H_{10}O_2$
1-Acenaphthenecarboxylic acid, A-70014
1,2-Dihydrocyclobuta[*a*]naphthalene-2-carboxylic acid, D-70177
1,2-Dihydro-3*H*-naphtho[2,1-*b*]pyran-3-one, D-80231
2,3-Dihydro-1*H*-naphtho[2,1-*b*]pyran-1-one, D-80232
2,3-Dihydro-4*H*-naphtho[1,2-*b*]pyran-4-one, D-80233
2,3-Dihydro-4*H*-naphtho[2,3-*b*]pyran-4-one, D-80234
3,4-Dihydro-2*H*-naphtho[1,2-*b*]pyran-2-one, D-80235
3,4-Dihydro-1*H*-naphtho[1,2-*c*]pyran-3-one, D-80236
3,4-Dihydro-1*H*-naphtho[2,3-*c*]pyran-1-one, D-80237
3-[5-(2,4-Hexadiynyl)-2-furanyl]-2-propenal, H-80046
7-(3-Hydroxyphenyl)-2-heptene-4,6-diyn-1-ol, H-80234
2-Methoxy-1-biphenylenol, *in* B-80116
11-Tridecene-3,5,7,9-tetrayne-1,2-diol, T-80261

$C_{13}H_{10}O_2S$
5-[4-(2-Furanyl)-3-buten-1-ynyl]-2-thiophenemethanol, F-80093
4-[5-(1,3-Pentadiynyl)-2-thienyl]-3-butyne-1,2-diol, *in* P-80025

$C_{13}H_{10}O_2S_2$
5-(2-Thienylethynyl)-2-thiophenemethanol; Ac, *in* T-80180

$C_{13}H_{10}O_3$
2*H*-3,4-Dihydronaphtho[2,3-*b*]pyran-5,10-dione, D-80230
4′-Hydroxy-2-biphenylcarboxylic acid, H-70111
Mycosinol, M-60154
Mycosinol, M-80202
4-Phenoxybenzoic acid, P-80088
11-Tridecene-3,5,7,9-tetrayne-1,2,13-triol, T-80262

$C_{13}H_{10}O_4$
Coriandrin, C-70198
8,9-Epoxy-7-(2,4-hexadiynylidene)-1,6-dioxaspiro[4.4]non-2-en-4-ol, *in* M-80202
2,4,6-Trihydroxybenzophenone, T-80285

$C_{13}H_{10}O_5$
α-Sorigenin, *in* S-80039

$C_{13}H_{10}O_7$
Nepenthone *A*, *in* T-60333

$C_{13}H_{10}O_8$
5,8-Dihydroxy-2,3-dimethoxy-6,7-methylenedioxy-1,4-naphthoquinone, *in* H-60065

$C_{13}H_{10}S$
2-(3-Buten-1-ynyl)-5-(3-penten-1-ynyl)thiophene, *in* B-80280
9*H*-Fluorene-9-thiol, F-80015
1-Methylnaphtho[2,1-*b*]thiophene, M-60096
2-Methylnaphtho[2,1-*b*]thiophene, M-60097
4-Methylnaphtho[2,1-*b*]thiophene, M-60098
5-Methylnaphtho[2,1-*b*]thiophene, M-60099
6-Methylnaphtho[2,1-*b*]thiophene, M-60100
7-Methylnaphtho[2,1-*b*]thiophene, M-60101
8-Methylnaphtho[2,1-*b*]thiophene, M-60102
9-Methylnaphtho[2,1-*b*]thiophene, M-60103
2-Phenyl-5-(1-propynyl)thiophene, P-80123

$C_{13}H_{10}S_2$
5-(3-Buten-1-ynyl)-5′-methyl-2,2′-bithienyl, *in* B-80279
2-Dibenzothiophenethiol; *S*-Me, *in* D-60078
4-Dibenzothiophenethiol; *S*-Me, *in* D-60079
5,10,12-Tridecadiene-2,8-diyne-4,7-thione, T-80248

$C_{13}H_{11}BrO$
4-Bromodiphenylmethanol, B-60236

$C_{13}H_{11}ClO$
4-Chlorodiphenylmethanol, C-60065
1-Chloro-3,11-tridecadiene-5,7,9-triyn-2-ol, C-80116
2-Chloro-3,11-tridecadiene-5,7,9-triyn-1-ol, C-80117

$C_{13}H_{11}I$
2-Iododiphenylmethane, I-60045
4-Iododiphenylmethane, I-60046

$C_{13}H_{11}N$
1,2-Dihydrobenz[*f*]isoquinoline, D-80170
3,4-Dihydrobenz[*g*]isoquinoline, D-80171
Diphenylmethaneimine, D-60482
9-Methyl-9*H*-carbazole, M-60056

$C_{13}H_{11}NO$
1-Acenaphthenecarboxaldehyde; Oxime, *in* A-70013
1,2-Dihydrocyclobuta[*a*]naphthalene-2-carboxylic acid; Amide, *in* D-70177
1-Methoxy-9*H*-carbazole, *in* H-80137
9-Methyl-9*H*-carbazol-2-ol, *in* H-60109

$C_{13}H_{11}NOS$
2-Phenyl-4*H*-3,1,2-benzooxathiazine, P-60080

$C_{13}H_{11}NO_2$
2,3-Dihydro-1*H*-naphtho[2,1-*b*]pyran-1-one; Oxime, *in* D-80232
2,3-Dihydro-4*H*-naphtho[2,3-*b*]pyran-4-one; Oxime, *in* D-80234
Diphenyl imidocarbonate, *in* C-70019
2-Hydroxy-*N*-phenylbenzamide, *in* H-70108
3-Methyl-2-nitrobiphenyl, M-70098
5-Phenyl-3-pyridinecarboxylic acid; Me ester, *in* P-80129

$C_{13}H_{11}NO_3$
2-Amino-3-methyl-1,4-naphthoquinone; Ac, *in* A-70171

$C_{13}H_{11}NO_4$
2,3-Quinolinedicarboxylic acid; Di-Me ester, *in* Q-80002

$C_{13}H_{11}NS$
2,3-Dihydro-2-phenyl-1,2-benzisothiazole, D-60261

$C_{13}H_{11}N_3$
2-Aminomethyl-1,10-phenanthroline, A-60224
1*H*-Benzotriazole; 1-Benzyl, *in* B-70068
▷ 2,9-Diaminoacridine, D-60027
▷ 3,6-Diaminoacridine, D-60028
▷ 3,9-Diaminoacridine, D-60029
1-(Dimethylamino)-γ-carboline, *in* A-70114

$C_{13}H_{11}N_3O$
3-Amino-β-carboline; N^3-Ac, *in* A-70110
1,3-Dihydro-2*H*-imidazo[4,5-*c*]pyridin-2-one; 1-Benzyl, *in* D-60250
Pyridazino[4,5-*b*]quinolin-10(5*H*)-one; 5-Et, *in* P-80207

$C_{13}H_{12}$
Benz[*f*]indane, B-60015
Bicyclo[6.4.1]trideca-1,3,5,7,9,11-hexaene, B-80091
3-Methylbiphenyl, M-60053
Tetracyclo[5.5.1.0^{4,13}.0^{10,13}]trideca-2,5,8,11-tetraene, T-60036

$C_{13}H_{12}N_2$
  N,N'-Diphenylformamidine, D-60479

$C_{13}H_{12}N_2O$
  2-(Aminomethyl)pyridine; N-Benzoyl, *in* A-60227
  4-(Aminomethyl)pyridine; N-Benzoyl, *in* A-60229
  ▷ N,N'-Diphenylurea, D-70499

$C_{13}H_{12}N_2O_2$
  1-Benzyloxy-2-phenyldiazene 2-oxide, *in* H-70208
  2,2'-(1,3-Dioxan-2-ylidene)bispyridine, *in* D-70502

$C_{13}H_{12}N_2O_3S$
  Aminoiminomethanesulfonic acid; N,N-Di-Ph, *in* A-60210

$C_{13}H_{12}N_2O_4S$
  Benzenediazo-*p*-toluenesulfonate N-oxide, *in* H-70208

$C_{13}H_{12}N_2Se$
  Selenourea; N^1,N^1-Di-Ph, *in* S-70031
  Selenourea; N^1,N^3-Di-Ph, *in* S-70031

$C_{13}H_{12}N_4$
  4-Amino-1H-imidazo[4,5-*c*]pyridine; 4-N-Benzyl, *in* A-70161

$C_{13}H_{12}O$
  2,5-Dihydro-2-(4,6,8-nonatrien-2-ynylidene)furan, D-80238
  4-Methoxyacenaphthene, *in* H-60099
  1-(1-Naphthyl)-2-propanone, N-80012
  1-(2-Naphthyl)-2-propanone, N-80013
  3,5-Tridecadiene-7,9,11-triyn-1-ol, T-80258
  2,8,10,12-Tridecatetraene-4,6-diyn-1-ol, T-80260

$C_{13}H_{12}O_2$
  2-(Benzyloxy)phenol, B-60072
  3-(Benzyloxy)phenol, B-60073
  ▷ 4-(Benzyloxy)phenol, B-60074
  3,4-Epoxy-2-(2,4-hexadiynylidene)-1,6-dioxaspiro[4.4]nonane, *in* H-80047
  ▷ Goniothalamin, G-60031
  4-Methyl-1-azulenecarboxylic acid; Me ester, *in* M-60042
  3,11-Tridecadiene-5,7,9-triyne-1,2-diol, T-80254

$C_{13}H_{12}O_2S_2$
  4-[5'-Methyl[2,2'-bithiophen]-5-yl]-3-butyne-1,2-diol, M-70054

$C_{13}H_{12}O_3$
  6-Acetyl-5-hydroxy-2-isopropenylbenzo[*b*]furan, *in* A-60065
  Goniothalamin oxide, G-60031
  7-(2,4-Hexadiynylidene)-1,6-dioxaspiro[4.4]non-8-en-3-ol, H-80047
  3,11-Tridecadiene-5,7,9-triyne-1,2,13-triol, T-80257

$C_{13}H_{12}O_3S$
  [1,1'-Biphenyl]-2-sulfonic acid; Me ester, *in* B-80122
  [1,1'-Biphenyl]-3-sulfonic acid; Me ester, *in* B-80123
  [1,1'-Biphenyl]-4-sulfonic acid; Me ester, *in* B-80124

$C_{13}H_{12}O_4$
  6-Acetyl-5-hydroxy-2-(1-hydroxymethylvinyl)benzo[*b*]furan, *in* A-60065
  Altholactone, A-60083
  Dihydrocoriandrin, *in* C-70198
  5,8-Dihydroxy-2,3,6-trimethyl-1,4-naphthoquinone, D-60380
  5,6-Dimethoxy-2-methyl-1,4-naphthoquinone, *in* D-60351
  1,4,5-Naphthalenetriol; 5-Me ether, 1-Ac, *in* N-60005
  2-Oxo-2H-benzopyran-4-acetic acid; Et ester, *in* O-60057
  Platypterophthalide, P-60161

$C_{13}H_{12}O_5$
  Acuminatolide, A-70063

$C_{13}H_{12}O_7$
  5,8-Dihydroxy-2,3,6-trimethoxy-1,4-benzoquinone, *in* P-70033

$C_{13}H_{12}S$
  2-Mercaptodiphenylmethane, M-80024

$C_{13}H_{12}S_2$
  5-(1-Propenyl)-5'-vinyl-2,2'-bithienyl, P-80182

$C_{13}H_{13}BrN_2O_3$
  5-Bromotryptophan; N^α-Ac, *in* B-60333
  6-Bromotryptophan; N^α-Ac, *in* B-60334
  7-Bromotryptophan; N^α-Ac, *in* B-60335

$C_{13}H_{13}BrO_2$
  2-Bromo-3,4-dihydro-1-naphthalenecarboxaldehyde; Ethylene acetal, *in* B-80198
  1-Bromo-(2-dimethoxymethyl)naphthalene, *in* B-60302

$C_{13}H_{13}IO_8$
  1,1,1-Triacetoxy-1,1-dihydro-1,2-benziodoxol-3(1H)-one, T-60228

$C_{13}H_{13}N$
  2-Amino-3-methylbiphenyl, A-70164
  N-Methyldiphenylamine, M-70060
  2-(2-Phenylethenyl)-1H-pyrrole; N-Me, *in* P-60094
  2-(2-Phenylethenyl)-1H-pyrrole; N-Me, *in* P-60094
  1,2,3,4-Tetrahydrobenz[*g*]isoquinoline, T-80054

$C_{13}H_{13}NO$
  1-Oxo-1,2,3,4-tetrahydrocarbazole; N-Me, *in* O-70103
  1,3,4,10-Tetrahydro-9(2H)-acridinone, T-60052

$C_{13}H_{13}NO_2$
  3-Hydroxy-2-methyl-4(1H)-pyridinone; Benzyl ether, *in* H-80214
  5-Methyl-1H-pyrrole-2-carboxylic acid; Benzyl ester, *in* M-60124

$C_{13}H_{13}NO_3$
  2-Formyl-1H-indole-3-acetic acid; Et ester, *in* F-80075

$C_{13}H_{13}NO_4$
  5,6-Dihydroxyindole; N-Me, di-O-Ac, *in* D-70307

$C_{13}H_{13}N_3$
  3-Amino-β-carboline; N^3-Et, *in* A-70110

$C_{13}H_{13}N_3O_5$
  Antibiotic PDE I, A-60286

$C_{13}H_{14}$
  2,3,5,6,7-Pentamethylenebicyclo[2.2.2]octane, P-60056

$C_{13}H_{14}N_2$
  1,4,5,6-Tetrahydrocyclopentapyrazole; 2-Benzyl, *in* T-80055

$C_{13}H_{14}N_2O$
  1-Oxo-1,2,3,4-tetrahydrocarbazole; N-Me, oxime, *in* O-70103

$C_{13}H_{14}N_2O_4S_3$
  Gliotoxin E, G-60024

$C_{13}H_{14}N_4OS_2$
  S,S-Bis(4,6-dimethyl-2-pyrimidinyl)dithiocarbonate, B-70138

$C_{13}H_{14}O$
  3-Isopropyl-2-naphthol, I-70095
  4-Isopropyl-1-naphthol, I-70096
  1-(1-Naphthyl)-2-propanol, N-60015

$C_{13}H_{14}OS$
  9-(2-Thienyl)-4,6-nonadien-8-yn-1-ol, T-80181
  9-(2-Thienyl)-4,6-nonadien-8-yn-3-ol, *in* T-80181

$C_{13}H_{14}O_2$
  2-Methyl-4-phenyl-3-butyn-2-ol; Ac, *in* M-70110

7-Phenyl-5-heptynoic acid, P-60106
3,5,11-Tridecadiene-7,9-diyne-1,2-diol, T-80247

$C_{13}H_{14}O_2S$
  2-(5-Methylthio-4-penten-2-ynylidene)-1,6-dioxaspiro[4.4]non-3-ene, M-80175

$C_{13}H_{14}O_3$
  4-Acetyl-2,3-dihydro-5-hydroxy-2-isopropenylbenzofuran, A-70040
  5-Acetyl-2-(1-hydroxy-1-methylethyl)benzofuran, A-70044
  2,2-Dimethyl-2H-1-benzopyran-6-carboxylic acid; Me ester, *in* D-60408
  7-Ethoxy-3,4-dimethylcoumarin, *in* H-60119
  1,4,5-Trimethoxynaphthalene, *in* N-60005

$C_{13}H_{14}O_3S$
  2-[(5-Methylsulfinyl)-4-penten-2-ynylidene]-1,6-dioxaspiro[4.4]non-3-ene, *in* M-80175

$C_{13}H_{14}O_4$
  6-Acetyl-5-hydroxy-2-hydroxymethyl-2-methylchromene, A-60031
  Anaphatol, *in* D-60339
  4-Hydroxy-2-isopropyl-5-benzofurancarboxylic acid; Me ester, *in* H-60164
  2-Isopropylidene-4,6-dimethoxy-3(2H)-benzofuranone, I-80080
  3-Phenyl-1,1-cyclopentanedicarboxylic acid, P-80098

$C_{13}H_{14}O_5$
  Caleteucrin, C-80012
  Diaporthin, D-80062
  Salicifoliol, S-80003

$C_{13}H_{14}O_6$
  3,4-Dihydroxybenzyl alcohol; Tri-Ac, *in* D-80284

$C_{13}H_{15}ClO$
  2-Phenylcyclohexanecarboxylic acid; Chloride, *in* P-70065

$C_{13}H_{15}Cl_2NO_3$
  6-Amino-2,2-dichlorohexanoic acid; N-Benzoyl, *in* A-60126

$C_{13}H_{15}NO_2$
  1-Oxa-2-azaspiro[2,5]octane; N-Benzoyl, *in* O-70052

$C_{13}H_{15}NO_3$
  2-Pyrrolidinecarboxaldehyde; N-Benzyloxycarbonyl, *in* P-70184

$C_{13}H_{15}N_3$
  2,6-Bis(3,4-Dihydro-2H-pyrrol-5-yl)pyridine, B-80146

$C_{13}H_{15}N_3O_2$
  3-Amino-1H-pyrazole-4-carboxylic acid; $N(1)$-Benzyl, Et ester, *in* A-70187

$C_{13}H_{16}Br_2$
  2,2-Dibromo-2,3-dihydro-1,1,3,3-tetramethyl-1H-indene, D-80089

$C_{13}H_{16}Cl_2O_4$
  1-(3,5-Dichloro-2,6-dihydroxy-4-methoxyphenyl)-1-hexanone, D-60129

$C_{13}H_{16}N_2$
  1,8-Diaminonaphthalene; N,N,N'-Tri-Me, *in* D-70042
  2-Diazo-1,1,3,3-tetramethylindane, D-60067

$C_{13}H_{16}N_2O_2$
  β-Methyltryptophan; Me ester, *in* M-80187

$C_{13}H_{16}N_2O_4$
  Bursatellin, B-60338

$C_{13}H_{16}N_4$
  Tetrakis(2-cyanoethyl)methane, *in* M-70036

$C_{13}H_{16}O$
  ▷ Benzoylcyclohexane, B-70071
  1-Phenyl-6-hepten-1-one, P-70080
  7-Phenyl-3-heptyn-2-ol, P-60107
  1,1,3,3-Tetramethyl-2-indanone, T-60143
  8,10-Tridecadiene-4,6-diyn-1-ol, T-80249

$C_{13}H_{16}OS$
9-(2-Thienyl)-6-nonen-8-yn-3-ol, *in* T-80181

$C_{13}H_{16}O_2$
3,4-Dihydro-3,3,8*a*-trimethyl-1,6(2*H*,8*aH*)-naphthalenedione, D-60301
6-(3,4-Methylenedioxylphenyl)-1-hexene, *in* M-50261
5-Oxo-5-phenylpentanal; Di-Me acetal, *in* O-80094
2-Phenylcyclohexanecarboxylic acid, P-70065

$C_{13}H_{16}O_3$
3-Benzoyl-2-methylpropanoic acid; Et ester, *in* B-60064
6,7-Dimethoxy-2,2-dimethyl-2*H*-1-benzopyran, D-70365
1-(Dimethoxymethyl)-2,3,5,6-tetrakis(methylene)-7-oxabicyclo[2.2.1]heptane, D-60399
Flossonol, F-60014
5-(1-Hydroxyethyl)-2-(1-hydroxy-1-methylethyl)benzofuran, *in* A-70044
Methyl 3-(4-methoxy-2-vinylphenyl)propanoate, *in* H-70232
Moskachan B, M-50261

$C_{13}H_{16}O_3S$
2-Mercapto-3-methylpentanoic acid; *S*-Benzoyl, *in* M-80026

$C_{13}H_{16}O_4$
4-Hydroxy-3-(2-hydroxy-3-methyl-3-butenyl)benzoic acid; Me ester, *in* H-60154
1-Phenyl-1,2-propanediol; Di-Ac, *in* P-80121
1-(2,4,6-Trimethoxyphenyl)-2-buten-1-one, *in* T-80331

$C_{13}H_{16}O_5$
Emehetin, E-80012

$C_{13}H_{16}O_7$
2-Oxo-1,3-cyclopentanediglyoxylic acid; Di-Et ester, *in* O-70085

$C_{13}H_{16}O_{10}$
6-Galloylglucose, G-80001

$C_{13}H_{16}Se$
1,1,3,3-Tetramethyl-2-indaneselone, T-60142

$C_{13}H_{17}N$
1,3,4,6,7,11*b*-Hexahydro-2*H*-benzo[*a*]quinolizine, H-80053

$C_{13}H_{17}NO$
Benzoylcyclohexane; (*E*)-Oxime, *in* B-70071

$C_{13}H_{17}NO_3S$
2-Amino-4,5,6,7-tetrahydrobenzo[*b*]thiophene-3-carboxylic acid; Et ester, *N*-Ac, *in* A-60258

$C_{13}H_{17}NO_4$
2-Amino-3-(4-hydroxyphenyl)-3-methylbutanoic acid; *N*-Ac, *in* A-60198

$C_{13}H_{18}$
2-Phenyl-2-heptene, P-80102

$C_{13}H_{18}BrClO_3$
Methyl 3-(3-bromo-4-chloro-4-methylcyclohexyl)-4-oxo-2-pentenoate, M-60054

$C_{13}H_{18}N_2$
Octahydropyrrolo[3,4-*c*]pyrrole; *N*-Benzyl, *in* O-70017
1,1,3,3-Tetramethyl-2-indanone; Hydrazone, *in* T-60143

$C_{13}H_{18}N_4O_3$
Citrulline; α-*N*-Benzoyl, amide, *in* C-60151

$C_{13}H_{18}N_4O_4$
4-Amino-4,6-dihydro-3-methyl-1*H*-cyclopenta[*e*]1,2,4-triazine-5,7-dicarboxylic acid; Di-Et ester, *in* A-80066

$C_{13}H_{18}O$
Tricyclone, T-80242

$C_{13}H_{18}O_2$
Pentamethylphenol; Ac, *in* P-60058
(Pentamethylphenyl)acetic acid, P-70038

$C_{13}H_{18}O_3$
Moskachan C, *in* M-50261
1,2,3,4,5,6,7,8-Octahydro-8-oxo-2-naphthalenecarboxylic acid; Et ester, *in* O-80015

$C_{13}H_{18}O_4$
3,4-Dihydroxybutanoic acid; O^4-Benzyl, Et ester, *in* D-60313
1,2,3,4-Tetramethoxy-5-(2-propenyl)benzene, T-80136

$C_{13}H_{18}O_5$
2,2,8,8-Tetramethyl-3,4,5,6,7-nonanepentone, T-60148

$C_{13}H_{18}O_7$
Isosalicin, I-80090
Methylarbutin, *in* A-70248
Pentahydroxybenzoic acid; Penta-Me ether, Me ester, *in* P-60043

$C_{13}H_{18}O_8$
Calleryanin, *in* D-80284
Idesin, *in* D-80283
Salirepin, *in* D-70287

$C_{13}H_{18}O_9$
MP-10, *in* H-80196

$C_{13}H_{19}N$
[8](2,6)Pyridinophane, P-80213

$C_{13}H_{19}NO$
[8](2,6)Pyridinophane; *N*-Oxide, *in* P-80213

$C_{13}H_{19}NO_4$
1,4-Dihydro-2,6-dimethyl-3,5-pyridinedicarboxylic acid; Di-Et ester, *in* D-80193

$C_{13}H_{19}N_3$
2,6-Bis(2-pyrrolidinyl)pyridine, B-80160

$C_{13}H_{19}N_5O_{10}S$
Antibiotic PB 5266*A*, A-70225

$C_{13}H_{19}N_5O_{11}S$
Antibiotic PB 5266*B*, A-70226

$C_{13}H_{20}$
2-Isopropylideneadamantane, I-60121
Tetracyclo[5.5.1.$0^{4,13}$.$0^{10,13}$]tridecane, T-60035

$C_{13}H_{20}N_6O_3$
3′-Amino-3′-deoxyadenosine; 3′,6,6-Tri-*N*-Me, *in* A-70128

$C_{13}H_{20}O$
γ-Ionone, I-70058

$C_{13}H_{20}O_2$
Prosopidione, P-80183

$C_{13}H_{20}O_3$
Grasshopper ketone, G-70028
Methyl jasmonate, *in* J-60002

$C_{13}H_{20}O_6$
2,2,8,8-Tetramethyl-5,5-dihydroxy-3,4,6,7-nonanetetrone, *in* T-60148

$C_{13}H_{20}O_8$
Methanetetrapropanoic acid, M-70036

$C_{13}H_{21}N$
1,2,3,4,5,6,7,8,8*a*,9*a*-Decahydrocarbazole; *N*-Me, *in* D-70012

$C_{13}H_{21}NO_6$
2,6-Bis(hydroxyacetyl)pyridine; Bis di-Me ketal, *in* B-70144

$C_{13}H_{22}$
1,1,2,2,4,4-Hexamethyl-3,5-bis(methylene)cyclopentane, H-70082

$C_{13}H_{22}O$
1-Cyclododecenecarboxaldehyde, C-60191
Octahydro-5,5,8*a*-trimethyl-2(1*H*)-naphthalenone, *in* D-70014
Theaspirone, T-60185

$C_{13}H_{22}O_4$
9,12-Dioxododecanoic acid; Me ester, *in* D-60471

$C_{13}H_{22}O_5$
4,4-Pyrandiacetic acid; Di-Et ester, *in* P-70150

$C_{13}H_{24}$
Octamethylcyclopentene, O-60026
6,7-Tridecadiene, T-60272

$C_{13}H_{24}O$
Decahydro-5,5,8*a*-trimethyl-2-naphthalenol, D-70014
Dicyclohexylmethanol, D-70150
Octamethylcyclopentanone, O-60025
1,2,2,3,3,4,4,5-Octamethyl-6-oxabicyclo[3.1.0]hexane, O-70021
3-Tridecyn-1-ol, T-80264
6-Tridecyn-1-ol, T-80265

$C_{13}H_{24}O_2$
9-Dodecen-1-ol; Formyl, *in* D-70537

$C_{13}H_{24}O_3$
10-Hydroxy-11-dodecenoic acid; Me ester, *in* H-60125
12-Hydroxy-10-dodecenoic acid; Me ester, *in* H-60126
7-Megastigmene-5,6,9-triol, M-60012
12-Oxododecanoic acid; Me ester, *in* O-60066

$C_{13}H_{24}O_4$
Hexamethylpentanedioic acid; Di-Me ester, *in* H-80076

$C_{13}H_{25}N_3$
11-Methylene-1,5,9-triazabicyclo[7.3.3]pentadecane, M-60080

$C_{13}H_{26}O$
2,2,3,3,4,4,5,5-Octamethylcyclopentanol, O-60024
6-Tridecen-1-ol, T-80263

$C_{13}H_{26}O_3$
11-Hydroxydodecanoic acid; Me ester, *in* H-80157

$C_{13}H_{26}O_4$
10-Oxodecanoic acid; Me ester, Di-Me acetal, *in* O-60063

$C_{13}H_{27}N$
2,6-Di-*tert*-butylpiperidine, D-80111

$C_{13}H_{31}N_5$
1,4,7,11,15-Pentaazacyclooctadecane, P-60023
1,4,8,11,15-Pentaazacyclooctadecane, P-60024

$C_{14}H_4Cl_2O_4$
2,3-Dichloro-1,4,9,10-anthracenetetrone, D-60117

$C_{14}H_4N_6O_{15}$
2,4,6-Trinitrobenzoic acid; Anhydride, *in* T-70303

$C_{14}H_5ClO_4$
2-Chloro-1,4,9,10-anthracenetetrone, C-60041

$C_{14}H_6Cl_2O_2$
1,5-Biphenylenedicarboxylic acid; Dichloride, *in* B-80111
1,8-Biphenylenedicarboxylic acid; Dichloride, *in* B-80112
2,6-Biphenylenedicarboxylic acid; Dichloride, *in* B-80114
2,7-Biphenylenedicarboxylic acid; Dichloride, *in* B-80115

$C_{14}H_6N_4O_{11}$
3,5-Dinitrobenzoic acid; Anhydride, *in* D-60458

$C_{14}H_6O_8$
▷ Ellagic acid, E-70007

$C_{14}H_8$
2,7-Diethynylnaphthalene, D-70158

$C_{14}H_8Br_2$
Bis(2-bromophenyl)ethyne, B-80140
Bis(4-bromophenyl)ethyne, B-80141
2,3-Dibromoanthracene, D-70085
9,10-Dibromoanthracene, D-60085

$C_{14}H_8ClNO$
9-Acridinecarboxylic acid; Chloride, *in* A-70061

$C_{14}H_8Cl_2$
9,10-Dichloroanthracene, D-60116
1,4-Dichlorophenanthrene, D-70132
2,4-Dichlorophenanthrene, D-70133
3,6-Dichlorophenanthrene, D-70134
3,9-Dichlorophenanthrene, D-70135
4,5-Dichlorophenanthrene, D-70136
9,10-Dichlorophenanthrene, D-70137

$C_{14}H_8Cl_2O_2S$
2,2'-Thiobisbenzoic acid; Dichloride, *in* T-70174
4,4'-Thiobisbenzoic acid; Dichloride, *in* T-70175

$C_{14}H_8F_2$
9,10-Difluoroanthracene, D-70161
4,5-Difluorophenanthrene, D-70166
9,10-Difluorophenanthrene, D-70167

$C_{14}H_8F_2O_4$
4,4'-Difluoro-[1,1'-biphenyl]-3,3'-dicarboxylic acid, D-60178
4,5-Difluoro-[1,1'-biphenyl]-2,3-dicarboxylic acid, D-60179
6,6'-Difluoro-[1,1'-biphenyl]-2,2'-dicarboxylic acid, D-60180

$C_{14}H_8F_6O_5S$
3-(Trifluoromethyl)benzenesulfonic acid; Anhydride, *in* T-60283

$C_{14}H_8FeN_2O_8$
Actinoviridin *A*, *in* H-70193

$C_{14}H_8I_2$
9,10-Diiodoanthracene, D-70355
2,7-Diiodophenanthrene, D-70360
3,6-Diiodophenanthrene, D-70361

$C_{14}H_8N_2$
9-Cyanoacridine, *in* A-70061
5,6-Dicyanoacenaphthene, *in* A-70015
Naphtho[8,1,2-*cde*]cinnoline, N-70007
Pyrido[2',3':3,4]cyclobuta[1,2-*g*]quinoline, P-60232
Pyrido[3',2':3,4]cyclobuta[1,2-*g*]quinoline, P-60233

$C_{14}H_8N_2O$
Naphtho[8,1,2-*cde*]cinnoline; *N*-Oxide, *in* N-70007

$C_{14}H_8N_2O_2$
[1,4]Benzoxazino[3,2-*b*][1,4]benzoxazine, B-60055
1,10-Phenanthroline-2,9-dicarboxaldehyde, P-80085

$C_{14}H_8N_2S$
[1]Benzothieno[2,3-*c*][1,5]naphthyridine, B-70060
[1]Benzothieno[2,3-*c*][1,7]naphthyridine, B-70061

$C_{14}H_8N_2Se$
Phenanthro[9,10-*c*][1,2,5]selenadiazole, P-70053

$C_{14}H_8N_4S_2$
[1,2,4,5]Tetrazino[3,4-*b*:6,1-*b'*]bisbenzothiazole, T-60171

$C_{14}H_8O$
Acenaphtho[5,4-*b*]furan, A-60014

$C_{14}H_8OS$
Phenanthro[4,5-*bcd*]thiophene; 4-Oxide, *in* P-60077

$C_{14}H_8O_2$
Benz[*a*]azulene-1,4-dione, B-60011
1,8-Biphenylenedicarboxaldehyde, B-60118

$C_{14}H_8O_2S$
Phenanthro[4,5-*bcd*]thiophene; 4,4-Dioxide, *in* P-60077

$C_{14}H_8O_3$
3,4-Acenaphthenedicarboxylic acid; Anhydride, *in* A-60013
4-Oxo-4*H*-naphtho[2,3-*b*]pyran-3-carboxaldehyde, O-80088

$C_{14}H_8O_4$
1,5-Biphenylenedicarboxylic acid, B-80111
1,8-Biphenylenedicarboxylic acid, B-80112
2,3-Biphenylenedicarboxylic acid, B-80113
2,6-Biphenylenedicarboxylic acid, B-80114
2,7-Biphenylenedicarboxylic acid, B-80115

$C_{14}H_8O_5$
8-Hydroxyxanthone-1-carboxylic acid, H-80266

$C_{14}H_8O_9$
Luteolic acid, L-80052

$C_{14}H_8S$
Acenaphtho[1,2-*b*]thiophene, A-70017
Acenaphtho[5,4-*b*]thiophene, A-60015
Phenanthro[4,5-*bcd*]thiophene, P-60077

$C_{14}H_8SSe$
[1]Benzoselenopheno[2,3-*b*][1]benzothiophene, B-70052

$C_{14}H_8S_2$
[1]Benzothiopyrano[6,5,4-*def*][1]benzothiopyran, B-60049

$C_{14}H_8Se_4$
Dibenzotetraselenofulvalene, D-70077

$C_{14}H_9BrO$
10-Bromoanthrone, B-80173

$C_{14}H_9ClN_2$
1-Chloro-4-phenylphthalazine, C-70141
2-Chloro-3-phenylquinoxaline, C-70145
6-Chloro-2-phenylquinoxaline, C-70146
7-Chloro-2-phenylquinoxaline, C-70147

$C_{14}H_9ClN_2O$
2-Chloro-3-phenylquinoxaline; 4-Oxide, *in* C-70145
6-Chloro-2-phenylquinoxaline; 4-Oxide, *in* C-70146
7-Chloro-2-phenylquinoxaline; 4-Oxide, *in* C-70147

$C_{14}H_9ClN_2O_2$
6-Chloro-2-phenylquinoxaline; 1,4-Dioxide (?), *in* C-70146

$C_{14}H_9ClO_2$
2-Benzoylbenzoic acid; Chloride, *in* B-80054

$C_{14}H_9Cl_2NO_2$
2,2'-Iminodibenzoic acid; Dichloride, *in* I-60014

$C_{14}H_9NO$
1-Acridinecarboxaldehyde, A-70056
2-Acridinecarboxaldehyde, A-70057
3-Acridinecarboxaldehyde, A-70058
4-Acridinecarboxaldehyde, A-70059
9-Acridinecarboxaldehyde, A-70060
o-Cyanobenzophenone, *in* B-80054
2-Phenyl-3*H*-indol-3-one, P-60114

$C_{14}H_9NO_2$
9-Acridinecarboxaldehyde; 10-Oxide, *in* A-70060
9-Acridinecarboxylic acid, A-70061
4-Phenyl-2*H*-1,3-benzoxazin-2-one, P-80094
3-(Phenylimino)-1(3*H*)-isobenzofuranone, P-60112
2-Phenylisatogen, *in* P-60114
2-(2-Pyridyl)-1,3-indanedione, P-60241
2-(3-Pyridyl)-1,3-indanedione, P-60242
2-(4-Pyridyl)-1,3-indanedione, P-60243

$C_{14}H_9NO_3$
1-Amino-2-hydroxyanthraquinone, A-60174
▷ 1-Amino-4-hydroxyanthraquinone, A-60175
1-Amino-5-hydroxyanthraquinone, A-60176
1-Amino-8-hydroxyanthraquinone, A-60177
2-Amino-1-hydroxyanthraquinone, A-60178
2-Amino-3-hydroxyanthraquinone, A-60179
3-Amino-1-hydroxyanthraquinone, A-60180
9-Amino-10-hydroxy-1,4-anthraquinone, A-60181
1*H*-Benz[*e*]indole-1,2(3*H*)-dione; *N*-Ac, *in* B-70026

$C_{14}H_9NO_4$
(2-Nitrophenyl)phenylethanedione, N-70058
(4-Nitrophenyl)phenylethanedione, N-70059

$C_{14}H_9NS$
10*H*-[1]Benzothieno[3,2-*b*]indole, B-70059
6*H*-[1]Benzothieno[2,3-*b*]indole, B-70058
Pyrrolo[3,2,1-*kl*]phenothiazine, P-70196

$C_{14}H_9NSe$
9-(Selenocyanato)fluorene, S-80025

$C_{14}H_9N_3$
4,4'-Iminobisbenzonitrile, *in* I-60018
5*H*-Indolo[2,3-*b*]quinoxaline, I-60036
Pyrido[2',1':2,3]imidazo[4,5-*c*]isoquinoline, P-60235
11*H*-Pyrimidino[4,5-*a*]carbazole, P-80225

$C_{14}H_9N_3O$
12*H*-Quinoxalino[2,3-*b*][1,4]benzoxazine, Q-70009

$C_{14}H_9N_3OS$
1,2,3-Benzotriazine-4(3*H*)-thione; *N*-Benzoyl, *in* B-60050

$C_{14}H_9N_3O_2$
5*H*-Indolo[2,3-*b*]quinoxaline; 5,11-Dioxide, *in* I-60036

$C_{14}H_9N_3S$
12*H*-Quinoxalino[2,3-*b*][1,4]benzothiazine, Q-70008

$C_{14}H_9N_3S_2$
Di-2-benzothiazolylamine, D-80074

$C_{14}H_{10}$
1-Benzylidene-1*H*-cyclopropabenzene, B-70084
Cyclohepta[*de*]naphthalene, C-60198
2,3-Dihydro-1,2,3-metheno-1*H*-phenalene, D-80218
2,12-Tetradecadiene-4,6,8,10-tetrayne, T-80029

$C_{14}H_{10}BrN$
1-(Bromomethyl)acridine, B-70229
2-(Bromomethyl)acridine, B-70230
3-(Bromomethyl)acridine, B-70231
4-(Bromomethyl)acridine, B-70232
9-(Bromomethyl)acridine, B-70233

$C_{14}H_{10}IN$
▷ 1*H*-Indol-3-ylphenyliodonium hydroxide inner salt, I-70017

$C_{14}H_{10}N_2$
10*b*,10*c*-Diazadicyclopenta[*ef*,*kl*]heptalene, D-70052

$C_{14}H_{10}N_2O$
9-Acridinecarboxaldehyde; Oxime, *in* A-70060
9-Acridinecarboxylic acid; Amide, *in* A-70061
3,5-Diphenyl-1,2,4-oxadiazole, D-60484
3-Nitroso-2-phenylindole, *in* P-60114
1*H*-Pyrrolo[3,2-*b*]pyridine; *N*-Benzoyl, *in* P-60249

$C_{14}H_{10}N_2O_2$
1,3-Diazetidine-2,4-dione; 1,3-Di-Ph, *in* D-60057
3,5-Diphenyl-1,2,4-oxadiazole; 4-Oxide, *in* D-60484
3-Methyl-1-phenazinecarboxylic acid, M-80136
4-Methyl-1-phenazinecarboxylic acid, M-80137
6-Methyl-1-phenazinecarboxylic acid, M-80138
7-Methyl-1-phenazinecarboxylic acid, M-80139
8-Methyl-1-phenazinecarboxylic acid, M-80140
9-Methyl-1-phenazinecarboxylic acid, M-80141
2-Phenylisatogen oxime, *in* P-60114

$C_{14}H_{10}N_2O_3$
N-(3-Oxo-3*H*-phenoxazin-2-yl)acetamide, *in* A-60237

C$_{14}$H$_{10}$N$_2$O$_4$
(4-Nitrophenyl)phenylethanedione;
Monoxime, *in* N-70059
(2-Nitrophenyl)phenylethanedione; 1-Oxime,
in N-70058
(2-Nitrophenyl)phenylethanedione; 2-Oxime,
in N-70058

C$_{14}$H$_{10}$N$_2$O$_5$
4-Amino-2,6-pyridinedicarboxylic acid; *N*-
Benzoyl, *in* A-70194

C$_{14}$H$_{10}$N$_2$S
3-Cyano-10-methylphenothiazine, *in* P-70057

C$_{14}$H$_{10}$N$_2$Se
3,4-Diphenyl-1,2,5-selenadiazole, D-70497

C$_{14}$H$_{10}$O
1-Acetylbiphenylene, A-80018
2-Acetylbiphenylene, A-80019
Benzo[5,6]cycloocta[1,2-*c*]furan, B-70029
9*H*-Fluorene-1-carboxaldehyde, F-70014
▷ 9*H*-Fluorene-9-carboxaldehyde, F-70015
8,1-[1]Propen[1]yl[3]ylidene-1*H*-
benzocyclohepten-4(9*H*)-one, M-70042

C$_{14}$H$_{10}$OS
Dibenzo[*b,e*]thiepin-11(6*H*)-one, D-80075

C$_{14}$H$_{10}$O$_2$
2-Acetylnaphtho[1,8-*bc*]pyran, A-70048
1-Biphenylenecarboxylic acid; Me ester, *in*
B-80109
2-Biphenylenecarboxylic acid; Me ester, *in*
B-80110
1,3-Di(2-furyl)benzene, D-70168
1,4-Di(2-furyl)benzene, D-70169
3,4-Heptafulvenedione, H-60022
4-Hydroxyacenaphthylene; Ac, *in* H-70100
4,5-Phenanthrenediol, P-70048

C$_{14}$H$_{10}$O$_3$
2-Benzoylbenzoic acid, B-80054
Disalicylaldehyde, D-70505
Naphtho[1,8-*bc*]pyran-2-carboxylic acid; Me
ester, *in* N-70010
2,3,7-Phenanthrenetriol, T-60339
2,4,5-Phenanthrenetriol, P-60071
2,4,7-Phenanthrenetriol, P-80080

C$_{14}$H$_{10}$O$_4$
3,4-Acenaphthenedicarboxylic acid, A-60013
5,6-Acenaphthenedicarboxylic acid, A-70015
2,5-Dihydroxy-4-methoxy-9*H*-fluoren-9-one,
in T-70261
2-(2,4-Dihydroxyphenyl)-6-
hydroxybenzofuran, D-80363
2-Hydroxybenzoic acid; Benzoyl, *in* H-70108
2-Hydroxy-1-methoxyxanthone, *in* D-70353
6-(3,4-Methylenedioxystyryl)-α-pyrone,
M-80086
Moracin *M*, M-60146
1,2,5,6-Tetrahydroxyphenanthrene, T-60115
1,2,5,7-Tetrahydroxyphenanthrene, T-80114
1,2,6,7-Tetrahydroxyphenanthrene, T-60117
1,3,5,6-Tetrahydroxyphenanthrene, T-60118
1,3,6,7-Tetrahydroxyphenanthrene, T-60119
2,3,4,7-Tetrahydroxyphenanthrene, T-80115
2,3,5,7-Tetrahydroxyphenanthrene, T-60120
2,3,6,7-Tetrahydroxyphenanthrene, T-60121
2,4,5,6-Tetrahydroxyphenanthrene, T-60122
3,4,5,6-Tetrahydroxyphenanthrene, T-60123

C$_{14}$H$_{10}$O$_4$S
2,2'-Thiobisbenzoic acid, T-70174
4,4'-Thiobisbenzoic acid, T-70175

C$_{14}$H$_{10}$O$_5$
▷ 1,3-Dihydroxy-7-methoxyxanthone, *in*
T-80347
1,5-Dihydroxy-6-methoxyxanthone, *in*
T-60349
▷ 1,7-Dihydroxy-3-methoxyxanthone, *in*
T-80347
3,7-Dihydroxy-1-methoxyxanthone, *in*
T-80347
5,6-Dihydroxy-1-methoxyxanthone, *in*
T-60349
2-(2,4-Dihydroxyphenyl)-5,6-
dihydroxybenzofuran, D-80355

6-Hydroxy-2-(2,3,4-
trihydroxyphenyl)benzofuran, H-80260
1,2,5,6,7-Pentahydroxyphenanthrene, P-60050
1,3,4,5,6-Pentahydroxyphenanthrene, P-70034
2,3,4,7,9-Pentahydroxyphenanthrene, P-70035

C$_{14}$H$_{10}$O$_6$
2,5-Dihydroxy-1,4-naphthoquinone; Di-Ac, *in*
D-60354
2,8-Dihydroxy-1,4-naphthoquinone; Di-Ac, *in*
D-60355
5,8-Dihydroxy-1,4-naphthoquinone; Di-Ac, *in*
D-60356
5,6-Dihydroxy-2-(2,3,4-
trihydroxyphenyl)benzofuran, D-80389
1,2,3,5,6,7-Hexahydroxyphenanthrene,
H-60066
2,4,5-Trihydroxy-1-methoxyxanthone, *in*
T-70122

C$_{14}$H$_{10}$O$_6$S
4,4'-Sulfonylbisbenzoic acid, *in* T-70175

C$_{14}$H$_{10}$S
8-Methylthio-1,7-tridecadiene-3,5,9,11-
tetrayne, M-80184
9-Phenanthrenethiol, P-70049

C$_{14}$H$_{10}$S$_2$
9,10-Dihydro-9,10-epidithioanthracene,
D-70206

C$_{14}$H$_{10}$S$_3$
Benzenecarbodithioic acid anhydrosulfide,
B-80013

C$_{14}$H$_{11}$FO
2-Fluoro-2,2-diphenylacetaldehyde, F-60033
3-Fluoro-4-methylphenol; Benzoyl, *in* F-60057

C$_{14}$H$_{11}$FO$_2$
2-Fluoro-3-hydroxybenzaldehyde; Benzyl
ether, *in* F-60035
2-Fluoro-5-hydroxybenzaldehyde; Benzyl
ether, *in* F-60036
4-Fluoro-3-hydroxybenzaldehyde; Benzyl
ether, *in* F-60037

C$_{14}$H$_{11}$IO$_3$S
2-Iodo-1-phenyl-2-(phenylsulfonyl)ethanone,
I-60061

C$_{14}$H$_{11}$N
1-Phenyl-1*H*-indole, P-70082
6-Phenyl-1*H*-indole, P-60113
N-Vinylcarbazole, V-80007

C$_{14}$H$_{11}$NO
1-Aminobiphenylene; N-Ac, *in* A-80062
2-Aminobiphenylene; N-Ac, *in* A-80063
9*H*-Fluorene-1-carboxaldehyde; Oxime, *in*
F-70014
Isocyanatodiphenylmethane, I-80054
▷ *N*-Phenylphthalamidine, *in* P-70100

C$_{14}$H$_{11}$NO$_2$
1*H*-Benz[*e*]indole-1,2(3*H*)-dione; *N*-Et, *in*
B-70026
2-Benzoylbenzoic acid; Amide, *in* B-80054
1-Hydroxycarbazole; Ac, *in* H-80137
2-Hydroxycarbazole; Ac, *in* H-60109

C$_{14}$H$_{11}$NO$_2$S
10*H*-Phenothiazine-1-carboxylic acid; *N*-Me,
in P-70055
10*H*-Phenothiazine-2-carboxylic acid; *N*-Me,
in P-70056
10*H*-Phenothiazine-3-carboxylic acid; *N*-Me,
in P-70057
10*H*-Phenothiazine-4-carboxylic acid; *N*-Me,
in P-70058
10*H*-Phenothiazine-1-carboxylic acid; Me
ester, *in* P-70055
10*H*-Phenothiazine-2-carboxylic acid; Me
ester, *in* P-70056

C$_{14}$H$_{11}$NO$_3$S
10*H*-Phenothiazine-1-carboxylic acid; Me
ester, 5-oxide, *in* P-70055

C$_{14}$H$_{11}$NO$_4$
1,2-Dimethoxy-3*H*-phenoxazin-3-one, *in*
T-60353
▷ 2,2'-Iminodibenzoic acid, I-60014

2,3'-Iminodibenzoic acid, I-60015
2,4'-Iminodibenzoic acid, I-60016
3,3'-Iminodibenzoic acid, I-60017
4,4'-Iminodibenzoic acid, I-60018

C$_{14}$H$_{11}$NO$_4$S
10*H*-Phenothiazine-1-carboxylic acid; *N*-Me,
5,5-dioxide, *in* P-70055
10*H*-Phenothiazine-2-carboxylic acid; *N*-Me,
5,5-dioxide, *in* P-70056
10*H*-Phenothiazine-3-carboxylic acid; *N*-Me,
5,5-dioxide, *in* P-70057
10*H*-Phenothiazine-4-carboxylic acid; *N*-Me,
5,5-dioxide, *in* P-70058
10*H*-Phenothiazine-2-carboxylic acid; Me
ester, 5,5-dioxide, *in* P-70056

C$_{14}$H$_{11}$N$_3$O$_4$
(2-Nitrophenyl)phenylethanedione; Dioxime,
in N-70058
(4-Nitrophenyl)phenylethanedione; Dioxime,
in N-70059

C$_{14}$H$_{12}$
Biquadricyclenylidene, B-80131
9,10-Dihydrophenanthrene, D-60260
1,4-Dimethylbiphenylene, D-80424
1,5-Dimethylbiphenylene, D-80425
1,8-Dimethylbiphenylene, D-80426
2,3-Dimethylbiphenylene, D-80427
2,6-Dimethylbiphenylene, D-80428
2,7-Dimethylbiphenylene, D-80429
9*b*-Methyl-9*bH*-benz[*cd*]azulene, M-60044
Tricyclo[4.4.4.0^{1,6}]tetradeca-2,4,7,9,10,12-
hexaene, T-70232

C$_{14}$H$_{12}$Br$_2$
1,2-Dibromo-1,2-diphenylethane, D-70091

C$_{14}$H$_{12}$Br$_2$O$_5$
3-Bromo-4-[(3-bromo-4,5-
dihydroxyphenyl)methyl]-5-
(hydroxymethyl)-1,2-benzenediol, B-70194

C$_{14}$H$_{12}$ClNO$_2$
2-Chloro-1-naphthylamine; *N,N*-Di-Ac, *in*
C-80080

C$_{14}$H$_{12}$N$_2$
1,4-Dihydro-1,4-ethanonaphtho[1,8-
de][1,2]diazepine, D-80207
1-Phenanthridenemethanamine, P-80081
10-Phenanthridinemethanamine, P-80083
4-Phenanthridinemethanamine, P-80082

C$_{14}$H$_{12}$N$_2$O
5,6-Dihydrobenzo[*c*]cinnoline; Mono-Ac, *in*
D-80172
2,4-Dimethylcarbazole; *N*-Nitroso, *in* D-70382
3,6-Dimethylcarbazole; *N*-Nitroso, *in* D-70387
1,2-Diphenyl-1,2-diazetidin-3-one, *in* D-70055

C$_{14}$H$_{12}$N$_2$O$_2$
4,4'-Diamino-2,3'-biphenyldicarboxylic acid,
D-80048
5,5'-Diamino-2,2'-biphenyldicarboxylic acid,
D-80051
5*a*,6,11*a*,12-Tetrahydro[1,4]benzoxazino[3,2-
b][1,4]benzoxazine, T-60059

C$_{14}$H$_{12}$N$_2$O$_4$
[2,2'-Bipyridine]-3,3'-dicarboxylic acid; Di-Me
ester, *in* B-60124
[2,2'-Bipyridine]-3,5'-dicarboxylic acid; Di-Me
ester, *in* B-60125
[2,2'-Bipyridine]-4,4'-dicarboxylic acid; Di-Me
ester, *in* B-60126
[2,2'-Bipyridine]-5,5'-dicarboxylic acid; Di-Me
ester, *in* B-60127
[3,3'-Bipyridine]-2,2'-dicarboxylic acid; Di-Me
ester, *in* B-60133
[3,3'-Bipyridine]-4,4'-dicarboxylic acid; Di-Me
ester, *in* B-60134
2,2'-Diamino-4,4'-biphenyldicarboxylic acid,
D-80046
4,4'-Diamino-2,2'-biphenyldicarboxylic acid,
D-80047
▷ 4,4'-Diamino-3,3'-biphenyldicarboxylic acid,
D-80049
4,6'-Diamino-2,2'-biphenyldicarboxylic acid,
D-80050

6,6'-Diamino-2,2'-biphenyldicarboxylic acid,
D-80052
6,6'-Dihydroxy-2,2'-bipyridine; Di-Ac, *in*
D-80299

$C_{14}H_{12}N_2S$
2-Amino-4*H*-3,1-benzothiazine; *N*-Ph, *in*
A-70092
2,5-Dihydro-2,2-diphenyl-1,3,4-thiadiazole,
D-60236

$C_{14}H_{12}N_4$
4-Amino-1,2,3-benzotriazine; 3-Benzyl, *in*
A-60092
4-Amino-1,2,3-benzotriazine; *N*(4)-Benzyl, *in*
A-60092
4-Amino-1,2,3-benzotriazine; 2-Me, *N*(4)-Ph,
in A-60092
4-Amino-1,2,3-benzotriazine; 3-Me, *N*(4)-Ph,
in A-60092

$C_{14}H_{12}N_4O_4$
Boxazomycin B, *in* B-70176
Boxazomycin C, *in* B-70176

$C_{14}H_{12}N_4O_5$
Boxazomycin A, B-70176

$C_{14}H_{12}N_4S$
3,5-Dianilino-1,2,4-thiadiazole, D-70048
Hector's base, H-60012

$C_{14}H_{12}O$
▷ 1-(Hydroxymethyl)fluorene, H-70176
1-Methoxyfluorene, *in* F-80016
1-Phenoxy-1-phenylethylene, *in* P-60092

$C_{14}H_{12}OS$
5,7-Dihydrodibenzo[*c,e*]thiepin; *S*-Oxide, *in*
D-70190
9-(Methylsulfinyl)-9*H*-fluorene, *in* F-80015
2-(Methylthio)benzophenone, *in* M-70020
3-(Methylthio)benzophenone, *in* M-70021
4-(Methylthio)benzophenone, *in* M-70022
1-Phenyl-2-(phenylthio)ethanone, P-80120

$C_{14}H_{12}OS_2$
6*H*,12*H*-Dibenzo[*b,f*][1,5]dithiocin; *S*-Oxide, *in*
D-70070
5*H*,7*H*-Dibenzo[*b,g*][1,5]dithiocin; 12-Oxide, *in*
D-70069
5*H*,7*H*-Dibenzo[*b,g*][1,5]dithiocin; 6-Oxide, *in*
D-70069

$C_{14}H_{12}OSe$
Phenyl styryl selenoxide, *in* P-70070

$C_{14}H_{12}O_2$
O-Acetylcapillol, *in* P-80104
1,2-Dihydrocyclobuta[*a*]naphthalene-2-
carboxylic acid; Me ester, *in* D-70177
2,2'-Dihydroxystilbene, D-80378
2,4'-Dihydroxystilbene, D-80379
2,5-Dihydroxystilbene, D-80380
3,3'-Dihydroxystilbene, D-80381
3,4-Dihydroxystilbene, D-80382
3,4'-Dihydroxystilbene, D-80383
3,5-Dihydroxystilbene, D-80384
4,4'-Dihydroxystilbene, D-80385
1,2-Dimethoxybiphenylene, *in* B-80116
1,5-Dimethoxybiphenylene, *in* B-80117
2,3-Dimethoxybiphenylene, *in* B-80119
2,6-Dimethoxybiphenylene, *in* B-80120
2,7-Dimethoxybiphenylene, *in* B-80121
2-(2,4-Hexadiynylidene)-5-
(propionylmethylidene)-2,5-dihydrofuran,
H-80048
4-Hydroxyacenaphthene; Ac, *in* H-60099
2,3-Naphthalenedicarboxylic acid; Di-Me
ester, *in* N-80001
1,2,3,4-Tetrahydroanthraquinone, T-60054

$C_{14}H_{12}O_2S$
4-[2,2'-Bithiophen-5-yl]-3-butyn-1-ol; Ac, *in*
B-80167
5,7-Dihydrodibenzo[*c,e*]thiepin; *S*-Dioxide, *in*
D-70190
9-(Methylsulfonyl)-9*H*-fluorene, *in* F-80015
1-Phenyl-2-(phenylsulfinyl)ethanone, *in*
P-80120

$C_{14}H_{12}O_2S_2$
5*H*,7*H*-Dibenzo[*b,g*][1,5]dithiocin; 12,12-
Dioxide, *in* D-70069

$C_{14}H_{12}O_2Se$
Phenyl styryl selenone, *in* P-70070

$C_{14}H_{12}O_3$
Demethylfrutescin, *in* H-80231
9,10-Dihydro-4-methoxy-2,7-
phenanthrenediol, *in* P-80080
Isotriptospinocoumarin, I-80097
4'-Methoxy-2-biphenylcarboxylic acid, *in*
H-70111
α-Oxo-1-naphthaleneacetic acid; Et ester, *in*
O-60078
α-Oxo-2-naphthaleneacetic acid; Et ester, *in*
O-60079
4-Phenoxybenzoic acid; Me ester, *in* P-80088
Triptispinocoumarin, T-80364

$C_{14}H_{12}O_3S$
1-Phenyl-2-(phenylsulfonyl)ethanone, *in*
P-80120

$C_{14}H_{12}O_3S_2$
5-(3-Hydroxy-4-acetoxy-1-butynyl)-2,2'-
bithienyl, *in* B-80166
5-(4-Hydroxy-3-acetoxy-1-butynyl)-2,2'-
bithienyl, *in* B-80166

$C_{14}H_{12}O_4$
2,6-Azulenedicarboxylic acid; Di-Me ester, *in*
A-70294
1,4-Dicarbomethoxyazulene, *in* A-70293
9,10-Dihydro-2,4,5,6-phenanthrenetetrol,
D-70247
2,4-Dihydroxy-6-methoxybenzophenone, *in*
T-80285
▷ 2,6-Dihydroxy-4-methoxybenzophenone, *in*
T-80285
1-(2,4-Dihydroxyphenyl)-2-(3,5-
dihydroxyphenyl)ethylene, D-70335
1-(3,4-Dihydroxyphenyl)-2-(3,5-
dihydroxyphenyl)ethylene, D-80361
4'-Hydroxydehydrokawain, H-70122
1-(3-Hydroxyphenyl)-2-(2,4,5-
trihydroxyphenyl)ethylene, H-70215
Osthenone, O-70049
Osthenone, O-80049
9-Tetradecene-2,4,6-triynedioic acid, T-80041

$C_{14}H_{12}O_4S_2$
5*H*,7*H*-Dibenzo[*b,g*][1,5]dithiocin; 6,6,12,12-
Tetroxide, *in* D-70069

$C_{14}H_{12}O_5$
9,10-Dihydro-2,3,4,6,7-phenanthrenepentol,
D-70246
1-(3,4-Dihydroxyphenyl)-2-(3,4,5-
dihydroxyphenyl)ethylene, D-80360
▷ Khellin, K-60010

$C_{14}H_{12}O_6$
1-(2,3,4-Trihydroxyphenyl)-2-(3,4,5-
trihydroxyphenyl)ethylene, T-60344

$C_{14}H_{12}O_8$
Fulvic acid, F-60081
Polivione, P-60167

$C_{14}H_{12}S$
5,7-Dihydrodibenzo[*c,e*]thiepin, D-70190
9-(Methylthio)-9*H*-fluorene, *in* F-80015

$C_{14}H_{12}S_2$
6*H*,12*H*-Dibenzo[*b,f*][1,5]dithiocin, D-70070
5*H*,7*H*-Dibenzo[*b,g*][1,5]dithiocin, D-70069

$C_{14}H_{12}Se$
5,7-Dihydrodibenzo[*c,e*]selenepin, D-70189
[(2-Phenylethenyl)seleno]benzene, P-70070

$C_{14}H_{13}N$
9-Amino-9,10-dihydroanthracene, A-60136
2-Amino-9,10-dihydrophenanthrene, A-60141
4-Amino-9,10-dihydrophenanthrene, A-60142
9-Amino-9,10-dihydrophenanthrene, A-60143
▷ 10,11-Dihydro-5*H*-dibenz[*b,f*]azepine, D-70185
1,2-Dimethylcarbazole, D-70374
1,3-Dimethylcarbazole, D-70375
1,4-Dimethylcarbazole, D-70376
1,5-Dimethylcarbazole, D-70377

1,6-Dimethylcarbazole, D-70378
1,7-Dimethylcarbazole, D-70379
1,8-Dimethylcarbazole, D-70380
2,3-Dimethylcarbazole, D-70381
2,4-Dimethylcarbazole, D-70382
2,5-Dimethylcarbazole, D-70383
2,7-Dimethylcarbazole, D-70384
3,4-Dimethylcarbazole, D-70385
3,5-Dimethylcarbazole, D-70386
3,6-Dimethylcarbazole, D-70387
4,5-Dimethylcarbazole, D-70388
2,2-Diphenylaziridine, D-80504

$C_{14}H_{13}NO$
1-Ethoxy-9*H*-carbazole, *in* H-80137
2-Ethoxycarbazole, *in* H-60109

$C_{14}H_{13}NO_2$
N-Acetoxy-4-aminobiphenyl, A-80013
4-Oxo-1,2,3,4-tetrahydrocarbazole; *N*-Ac, *in*
O-70106

$C_{14}H_{13}NO_3$
(4-Hydroxyphenoxy)acetic acid; Anilide, *in*
H-70205
3,4,6-Trihydroxy-1,2-dimethylcarbazole,
T-70253

$C_{14}H_{13}NO_4$
6-Amino-2-hexenoic acid; *N*-Phthalimido, *in*
A-60173
3-Amino-1,2-naphthalenediol; *O,O*-Di-Ac, *in*
A-80116

$C_{14}H_{13}N_3O_2$
2,2'-Iminobisbenzamide, *in* I-60014

$C_{14}H_{14}$
Hexacyclo[6.5.1.0^{2,7}.0^{3,11}.0^{4,9}.0^{10,14}]tetradeca-
5,12-diene, H-80039
Pentacyclo[8.4.0.0^{2,7}.0^{3,12}.0^{6,11}]tetradeca-4,8,13-
triene, P-70021
1,2,3,4-Tetrahydroanthracene, T-60053
1,2,4,5-Tetravinylbenzene, T-80162

$C_{14}H_{14}N_2$
2,5-Dimethyl-3-(2-phenylethenyl)pyrazine,
D-60442
N,N'-Diphenylformamidine; *N*-Me, *in*
D-60479

$C_{14}H_{14}N_2O$
2,2'-Diaminobiphenyl; 2-*N*-Ac, *in* D-60033
N,N'-Diphenylurea; *N*-Me, *in* D-70499
5-Phenyl-3-pyridinecarboxylic acid;
Dimethylamide, *in* P-80129

$C_{14}H_{14}N_2O_3S$
Sulfoacetic acid; Dianilide, *in* S-80050

$C_{14}H_{14}N_2O_5$
Antibiotic PDE II, A-70229

$C_{14}H_{14}N_2S_4$
2,2'-Bi(4,6-dimethyl-5*H*-1,3-dithiolo[4,5-
c]pyrrolylidene), B-80092

$C_{14}H_{14}N_4$
5,5a,6,11,11a,12-Hexahydroquinoxalino[2,3-
b]quinoxaline, H-80066

$C_{14}H_{14}N_4O_4$
4,4'-Diamino-3,3'-biphenyldicarboxylic acid;
Diamide, *in* D-80049

$C_{14}H_{14}O$
2-(2-Phenylethyl)phenol, P-70071
3-(2-Phenylethyl)phenol, P-70072
4-(2-Phenylethyl)phenol, P-70073

$C_{14}H_{14}OS$
2-Mercaptobenzenemethanol; *S*-Benzyl, *in*
M-80021
2-Mercapto-1,1-diphenylethanol, M-70028

$C_{14}H_{14}O_2$
3,4'-Dimethoxybiphenyl, *in* B-80106
2-(2,4-Hexadiynylidene)-1,6-
dioxaspiro[4.5]dec-3-ene, *in* H-70042
2,3,4,6,7,8-Hexahydro-1,5-anthracenedione,
H-80051
1,2,3,6,7,8-Hexahydro-4,5-phenanthrenedione,
H-80064
2,3,4,5,6,7-Hexahydro-1,8-phenanthrenedione,
H-80065

1-Methoxy-2-(phenylmethoxy)benzene, *in* B-60072
1-Methoxy-3-(phenylmethoxy)benzene, *in* B-60073
1-Methoxy-4-(phenylmethoxy)benzene, *in* B-60074
2-(2-Phenylethyl)-1,4-benzenediol, P-60095
4-(2-Phenylethyl)-1,2-benzenediol, P-60096
4-(2-Phenylethyl)-1,3-benzenediol, P-60097
5-(2-Phenylethyl)-1,3-benzenediol, P-80100

$C_{14}H_{14}O_2S_2$
2-Mercaptobenzenemethanol; Disulfide, *in* M-80021
3-Mercaptobenzenemethanol; Disulfide, *in* M-80022
4-Mercaptobenzenemethanol; Disulfide, *in* M-80023

$C_{14}H_{14}O_3$
3,5-Dimethoxy-4-biphenylol, *in* B-80129
2-(Dimethoxymethyl)-1-naphthalenecarboxaldehyde, *in* N-60001
3-(2,2-Dimethyl-2*H*-1-benzopyran-6-yl)-2-propenoic acid, D-50601
3-(2,2-Dimethyl-2*H*-1-benzopyran-6-yl)-2-propenoic acid, D-70372
2-(2,4-Hexadiynylidene)-1,6-dioxaspira[4.5]dec-3-en-8-ol, H-70042
3,3′,4-Trihydroxybibenzyl, T-60316

$C_{14}H_{14}O_4$
1-(3,4-Dihydroxyphenyl)-2-(3,5-dihydroxyphenyl)ethane, D-80359
5,8-Dihydroxy-2,3,6,7-tetramethyl-1,4-naphthoquinone, D-60377
1,4,5-Naphthalenetriol; 1,5-Di-Me ether, Ac, *in* N-60005
1,4,5-Naphthalenetriol; 4,5-Di-Me ether, Ac, *in* N-60005
Norpinguisanolide, N-80082
Phthalidochromene, P-60143
Platypodantherone, P-80161
Tenual, T-80007
3,3′,4,4′-Tetrahydroxybibenzyl, T-60101
3,3′,4,5-Tetrahydroxybibenzyl, T-60102

$C_{14}H_{14}O_5$
1-(3,4-Dihydroxyphenyl)-2-(3,4,5-trihydroxyphenyl)ethane, D-80371
Funadonin, F-70046
Haplopinol, *in* D-84287
Khellactone, K-70013
3,3′,4,4′,5-Pentahydroxybibenzyl, P-60044

$C_{14}H_{14}O_6$
3,3′,4,4′,5,5′-Hexahydroxybibenzyl, H-60062

$C_{14}H_{14}O_7$
3′,4′,5′-Trihydroxyacetophenone; Tri-Ac, *in* T-60315

$C_{14}H_{14}O_8$
1,2,3,4-Benzenetetrol; Tetra-Ac, *in* B-60014
Hexahydroxy-1,4-naphthoquinone; 2,3,6,7-Tetra-Me ether, *in* H-60065
3,4,5-Trihydroxyphenylacetic acid; Tri-Ac, *in* T-60343

$C_{14}H_{14}Se_2$
Dibenzyl diselenide, D-60082

$C_{14}H_{15}ClO_5$
Mikrolin, M-60135

$C_{14}H_{15}NO$
1,3,4,10-Tetrahydro-9(2*H*)-acridinone; *N*-Me, *in* T-60052

$C_{14}H_{15}NO_3$
1-Amino-2,7-naphthalenediol; Di-Me ether, *N*-Ac, *in* A-80113
4-Amino-1,2-naphthalenediol; Di-Me ether, N-Ac, *in* A-80117
4-Amino-1,5-naphthalenediol; Di-Me ether, N-Ac, *in* A-80119
5-Amino-2,3-naphthalenediol; Di-Me ether, N-Ac, *in* A-80122

$C_{14}H_{15}NO_4$
1*H*-Indole-3-carboxylic acid; *N*-COOEt, Et ester, *in* I-80012

$C_{14}H_{15}N_5O_4$
8-Aminoimidazo[4,5-*g*]quinazoline; 1-(*β*-D-Ribofuranosyl), *in* A-60207
lin-Benzoadenosine, *in* A-60207

$C_{14}H_{16}$
1,3-Diethynyladamantane, D-60172
Heptacyclo[6.6.0.0^{2,6}.0^{3,13}.0^{4,11}.0^{5,9}.0^{10,14}]tetradecane, H-70012
Heptacyclo[9.3.0.0^{2,5}.0^{3,13}.0^{4,8}.0^{6,10}.0^{9,12}]tetradecane, H-70013
1,2,3,9,10,10*a*-Hexahydrophenanthrene, H-80063

$C_{14}H_{16}N_2$
2,2′-Bis(methylamino)biphenyl, *in* D-60033
1,2-Diphenyl-1,2-ethanediamine, D-70480
1-(1-Naphthyl)piperazine, N-60013
1-(2-Naphthyl)piperazine, N-60014

$C_{14}H_{16}N_2O_2$
▷ 1,4-Bis(1-isocyanato-1-methylethyl)benzene, B-70149

$C_{14}H_{16}N_2O_3$
β-Methyltryptophan; *N*-Ac, *in* M-80187

$C_{14}H_{16}N_4$
5,6,8,9-Tetraaza[3.3]paracyclophane, T-60018

$C_{14}H_{16}O$
4,6,12-Tetradecatrien-8,10-diyn-1-ol, T-80032
5-Tetradecene-8,10,12-triyn-1-ol, T-80042
6-Tetradecene-8,10,12-triyn-3-ol, T-80043

$C_{14}H_{16}O_2$
8-Norlactaranelactone, N-80080
2-Phenacylcyclohexanone, P-70045
7-Phenyl-5-heptynoic acid; Me ester, *in* P-60106
4,6,12-Tetradecatriene-8,10-diyne-1,3-diol, T-80033

$C_{14}H_{16}O_3$
3-(3,4-Dihydro-2,2-dimethyl-2*H*-1-benzopyran-6-yl)-2-propenoic acid, *in* D-50601
Drupanin, D-80534

$C_{14}H_{16}O_4$
6-Acetyl-7-hydroxy-8-methoxy-2,2-dimethyl-2*H*-1-benzopyran, A-80023
7-Hydroxy-4-isopropyl-3-methoxy-6-methylcoumarin, H-70144
Pyriculol, P-60216

$C_{14}H_{16}O_5$
Dechloromikrolin, *in* M-60135

$C_{14}H_{16}O_6$
Gravolenic acid, G-80037
7-Methoxycaleteucrin, *in* C-80012
1,5,8-Trihydroxy-3-methoxyxanthone, *in* T-70123

$C_{14}H_{16}O_7$
3-Methyl-1,2,4,5-benzenetetrol; 1-Me ether, Tri-Ac, *in* M-80048
4-Methyl-1,2,3,5-benzenetetrol; 3-Me ether, tri-Ac, *in* M-80049
Sawaranin, S-70016

$C_{14}H_{16}O_8$
1,3,5,7-Adamantanetetracarboxylic acid, A-80037

$C_{14}H_{17}NO$
8-Azabicyclo[3.2.1]octane; *N*-Benzoyl, *in* A-70278

$C_{14}H_{17}NO_3$
3-Amino-4-hexenoic acid; *N*-Benzoyl, Me ester, *in* A-60171

$C_{14}H_{17}NO_9$
2,4-Dihydroxy-7-methoxy-2*H*-1,4-benzoxazin-3(4*H*)-one; Demethoxy, 2-*O*-*β*-D-glucoside, *in* D-10388

$C_{14}H_{18}$
Benzylidenecycloheptane, B-60068
Tetrahydro-1,6,7-tris(methylene)-1*H*,4*H*-3*a*,6*a*-propanopentalene, T-70101

$C_{14}H_{18}N_2$
1,8-Bis(dimethylamino)naphthalene, *in* D-70042

$C_{14}H_{18}N_2O_5$
Aspartame, A-60311

$C_{14}H_{18}N_4O_4$
1,2,3,5-Tetraaminobenzene; *N*-Tetra-Ac, *in* T-60012
1,2,4,5-Tetraaminobenzene; 1,2,4,5-*N*-Tetra-Ac, *in* T-60013

$C_{14}H_{18}O$
3,4-Dihydro-3,3,6,8-tetramethyl-1(2*H*)-napthalenone, D-80254
Foeniculin, *in* P-80179
6,12-Tetradecadiene-8,10-diyn-3-ol, T-80028
4,6,10,12-Tetradecatetraen-8-yn-1-ol, T-80031
8-Tetradecene-11,13-diyn-2-one, T-80040

$C_{14}H_{18}O_2$
Anol isovalerate, *in* P-80179
Majusculone, M-70006
Norpinguisone, N-80083
4,6-Tetradecadiene-8,10-diyne-1,12-diol, T-80026
6,12-Tetradecadiene-8,10-diyne-1,3-diol, T-80027

$C_{14}H_{18}O_2S$
9-(2-Thienyl)-4,6-nonadien-8-yn-1-ol; Deoxy, 3-acetoxy, 4,5-dihydro, *in* T-80181

$C_{14}H_{18}O_3$
5-(1,3-Dihydroxypropyl)-2-isopropenyl-2,3-dihydrobenzofuran, D-70340
6-Ethoxy-7-methoxy-2,2-dimethyl-2*H*-1-benzopyran, *in* D-70365

$C_{14}H_{18}O_4$
1-Acetyl-4-isopentenyl-6-methylphloroglucinol, *in* T-70248
11-Hydroxy-3-oxo-13-nor-7(11)-eudesmen-12,6-olide, H-60204
Hyperolactone, H-80267
3-Isopropyl-6-methyl-1,2-benzenediol; Di-Ac, *in* I-60124

$C_{14}H_{18}O_6$
Colletoketol, *in* C-60159

$C_{14}H_{18}O_7$
3,7*a*-Diacetoxyhexahydro-2-oxa-3*a*,7-dimethyl-1,4-indanedione, *in* H-70055

$C_{14}H_{19}NO_2$
2-(Aminomethyl)cyclohexanol; *N*-Benzoyl, *in* A-70169

$C_{14}H_{19}NO_7$
Menisdaurin, *in* L-70034

$C_{14}H_{19}NO_8$
Griffonin, *in* L-70034
Lithospermoside, L-70034
Lithospermoside; 5-Epimer, *in* L-70034
Thalictoside, *in* N-70050

$C_{14}H_{19}N_3$
1-Azidodiamantane, A-60341
3-Azidodiamantane, A-60342
4-Azidodiamantane, A-60343
2,3,5,6-Tetrahydro-1*H*,4*H*,7*H*,11*cH*-3*a*,6*a*,11*b*-triazabenz[*de*]anthracene, T-80079

$C_{14}H_{19}N_3O_4$
Citrulline; *α*-*N*-Benzoyl, Me ester, *in* C-60151

$C_{14}H_{20}$
7,7′-Bi(bicyclo[2.2.1]heptylidene), B-60078
1,2-Bis(cyclohexylidene)ethene, B-80144
1,3-Divinyladamantane, D-70529
1-Phenyl-1-octene, P-80106
1-Phenyl-2-octene, P-80107
2-Phenyl-1-octene, P-80108

$C_{14}H_{20}N_2$
3,6-Di-*tert*-butylpyrrolo[3,2-*b*]pyrrole, D-70108

$C_{14}H_{20}N_2O_4$
3-Methyl-*N*-(5*a*,6*a*,7,8-tetrahydro-4-methyl-1,5-dioxo-1*H*,5*H*-pyrrolo[1,2-*c*][1,3]oxazepin-3-yl)butanamide, M-70134

C$_{15}$H$_{10}$N$_2$

5H-Cyclohepta[4,5]pyrrolo[2,3-b]indole, C-60200

Indazolo[3,2-a]isoquinoline, I-60020

10H-Indolo[3,2-b]quinoline, I-70016

2-Methylbenzo[gh]perimidine, M-60051

C$_{15}$H$_{10}$N$_2$O

10H-Indolo[3,2-b]quinoline; 5-Oxide, *in* I-70016

C$_{15}$H$_{10}$N$_2$O$_2$

2-Diazo-1,3-diphenyl-1,3-propanedione, D-70058

C$_{15}$H$_{10}$N$_8$OS$_2$

S,S'-Bis(1-phenyl-1H-tetrazol-5-yl) dithiocarbonate, B-70155

C$_{15}$H$_{10}$O$_2$

1-Acetoxy-7-phenyl-2,4,6-heptatriyne, *in* P-80101

1-Acetoxy-2,12-tridecadiene-4,6,8,10-tetrayne, *in* T-80253

2-(Phenylethynyl)benzoic acid, P-70074

4-(Phenylethynyl)benzoic acid, P-70075

C$_{15}$H$_{10}$O$_2$S

Dibenzo[b,f]thiepin-3-carboxylic acid, D-60077

3-Mercapto-2-phenyl-4H-1-benzopyran-4-one, M-60025

2-Phenyl-4H-1-benzothiopyran-4-one; S-Oxide, *in* P-60082

C$_{15}$H$_{10}$O$_3$

6-Hydroxyflavone, H-70130

7-Hydroxyflavone, H-70131

▷ 8-Hydroxyflavone, H-70132

1-Hydroxy-7-methylanthraquinone, H-60170

5-Hydroxy-4-phenanthrenecarboxylic acid, H-80232

3-(3-Hydroxyphenyl)-1H-2-benzopyran-1-one, H-70207

C$_{15}$H$_{10}$O$_3$S

2-Phenyl-4H-1-benzothiopyran-4-one; S,S-Dioxide, *in* P-60082

C$_{15}$H$_{10}$O$_4$

4′,6-Dihydroxyaurone, D-80273

2′,8-Dihydroxyflavone, D-70300

4′,7-Dihydroxyisoflavone, D-80323

3,9-Dihydroxypterocarpene, D-80376

C$_{15}$H$_{10}$O$_4$S

Dibenzo[b,f]thiepin-3-carboxylic acid; 5,5-Dioxide, *in* D-60077

C$_{15}$H$_{10}$O$_5$

Benzophenone-2,2′-dicarboxylic acid, B-70041

Chalaurenol, C-70036

▷ 1,3-Dihydroxy-2-hydroxymethylanthraquinone, D-70306

3-(2,4-Dihydroxyphenyl)-7-hydroxy-2H-1-benzopyran-2-one, D-80364

2-(2,4-Dihydroxyphenyl)-5,6-methylenedioxybenzofuran, *in* D-80355

Nivegin, N-80063

Sulfuretin, S-70082

2′,5,5′-Trihydroxyflavone, T-80298

3,4′,7-Trihydroxyflavone, T-80300

3′,4′,7-Trihydroxyflavone, T-80301

5,6,7-Trihydroxyflavone, T-60322

2′,4′,7-Trihydroxyisoflavone, T-70264

2′,4′,7-Trihydroxyisoflavone, T-80306

2′,5,7-Trihydroxyisoflavone, T-80307

3′,4′,7-Trihydroxyisoflavone, T-80308

▷ 4′,5,7-Trihydroxyisoflavone, T-80309

4′,6,7-Trihydroxyisoflavone, T-80310

4′,7,8-Trihydroxyisoflavone, T-70265

4′,7,8-Trihydroxyisoflavone, T-80311

1,2,7-Trihydroxy-6-methylanthraquinone, T-70268

1,2,8-Trihydroxy-3-methylanthraquinone, T-80322

1,6,8-Trihydroxy-2-methylanthraquinone, T-70269

Vertixanthone, *in* H-80266

C$_{15}$H$_{10}$O$_6$

Aureusidin, A-80188

4-(3,4-Dihydroxyphenyl)-5,7-dihydroxy-2H-1-benzopyran-2-one, D-70334

4-(3,4-Dihydroxyphenyl)-5,7-dihydroxy-2H-1-benzopyran-2-one, D-80356

7-Hydroxy-3-(2,4,5-trihydroxyphenyl)-2H-1-benzopyran-2-one, H-80261

Hydroxyvertixanthone, *in* H-80266

3′,4′,5,6,7-Pentahydroxyisoflavone, P-80043

2′,3′,5,7-Tetrahydroxyflavone, T-60106

2′,5,6,6′-Tetrahydroxyflavone, T-70107

2′,5,6′,7-Tetrahydroxyflavone, T-50153

3′,4′,5,6-Tetrahydroxyflavone, T-70108

3′,4′,6,7-Tetrahydroxyflavone, T-80092

3′,4′,6,7-Tetrahydroxyisoflavanone, T-80102

2′,3′,4′,7-Tetrahydroxyisoflavone, T-80105

2′,4′,5,7-Tetrahydroxyisoflavone, T-80106

2′,4′,5′,7-Tetrahydroxyisoflavone, T-80107

2′,5,6,7-Tetrahydroxyisoflavone, T-60107

2′,5,6,7-Tetrahydroxyisoflavone, T-80108

2′,5,7,8-Tetrahydroxyisoflavone, T-70110

3′,4′,5,7-Tetrahydroxyisoflavone, T-80109

3′,4′,5′,7-Tetrahydroxyisoflavone, T-80110

3′,4′,7,8-Tetrahydroxyisoflavone, T-70112

4′,5,6,7-Tetrahydroxyisoflavone, T-80112

4′,5,7,8-Tetrahydroxyisoflavone, T-80113

4′,6,7,8-Tetrahydroxyisoflavone, T-70113

1,3,4,5-Tetrahydroxy-2-methylanthraquinone, T-60112

1,3,5,8-Tetrahydroxy-2-methylanthraquinone, T-60113

3,4,8,9-Tetrahydroxypterocarpene, T-80122

C$_{15}$H$_{10}$O$_7$

4-(3,4-Dihydroxyphenyl)-5,7,8-trihydroxycoumarin, D-80370

2′,3′,4′,5,6-Pentahydroxyflavone, P-70030

2′,3,5,7,8-Pentahydroxyflavone, P-70029

2′,4′,5,5′,7-Pentahydroxyflavone, P-80033

2′,4′,5,7,8-Pentahydroxyflavone, P-80034

3,4′,5,6,7-Pentahydroxyflavone, P-80035

3,4′,5,6,8-Pentahydroxyflavone, P-60048

3,5,6,7,8-Pentahydroxyflavone, P-70031

2′,4′,5,5′,7-Pentahydroxyisoflavone, P-80041

2′,4′,5′,6,7-Pentahydroxyisoflavone, P-80042

3′,4′,5,6,7-Pentahydroxyisoflavone, P-70032

3′,4′,5′,6,7-Pentahydroxyisoflavone, P-80044

3′,4′,5,7,8-Pentahydroxyisoflavone, P-80045

3′,4′,6,7,8-Pentahydroxyisoflavone, P-80046

C$_{15}$H$_{10}$O$_8$

2′,3,3′,5,7,8-Hexahydroxyflavone, H-70079

2′,3,5,5′,6,7-Hexahydroxyflavone, H-60064

2′,3,5,5′,7,8-Hexahydroxyflavone, H-70080

3,3′,4′,5,6,7-Hexahydroxyflavone, H-80067

3′,4′,5,5′,6,7-Hexahydroxyflavone, H-70081

2′,4′,5,5′,6,7-Hexahydroxyisoflavone, H-80069

2′,4′,5,5′,7,8-Hexahydroxyisoflavone, H-80070

3′,4′,5,5′,6,7-Hexahydroxyisoflavone, H-80071

C$_{15}$H$_{10}$O$_9$

3,3′,4′,5,5′,7,8-Heptahydroxyflavone, H-70021

3,3′,4′,5,6,7,8-Heptahydroxyflavone, H-80028

3′,4′,5,6,7,8-Heptahydroxyflavone, H-80029

C$_{15}$H$_{10}$O$_{10}$

2′,3,4′,5,5′,6,7,8-Octahydroxyflavone, O-60022

3,3′,4′,5,5′,6,7,8-Octahydroxyflavone, O-70018

C$_{15}$H$_{11}^{\oplus}$

Dibenzo[a,d]cycloheptenylium, D-60071

C$_{15}$H$_{11}$Br

1-(Bromomethyl)phenanthrene, B-80224

2-(Bromomethyl)phenanthrene, B-80225

3-(Bromomethyl)phenanthrene, B-80226

9-(Bromomethyl)phenanthrene, B-80227

C$_{15}$H$_{11}$Cl

1-(Chloromethyl)phenanthrene, C-80073

2-(Chloromethyl)phenanthrene, C-80074

3-(Chloromethyl)phenanthrene, C-80075

4-(Chloromethyl)phenanthrene, C-80076

9-(Chloromethyl)phenanthrene, C-80077

C$_{15}$H$_{11}$ClN$_2$O$_2$

2-Chloro-2,3-dihydro-1H-imidazole-4,5-dione; 1,3-Di-Ph, *in* C-60054

C$_{15}$H$_{11}$ClO$_2$S

2-(3-Chloro-4-acetoxy-1-butynyl)-5-(1,3-pentadiynyl)thiophene, *in* P-80025

C$_{15}$H$_{11}$ClO$_4$

Dibenzo[a,d]cycloheptenylium; Perchlorate, *in* D-60071

C$_{15}$H$_{11}$ClO$_5$

2-Chloro-1,8-dihydroxy-5-methoxy-6-methylxanthone, *in* T-70181

3,4′,5,7-Tetrahydroxyflavylium; Chloride, *in* T-80093

C$_{15}$H$_{11}$ClO$_6$

3′,4′,5,5′,7-Pentahydroxyflavylium; Chloride, *in* P-80036

3′,4′,5,7,8-Pentahydroxyflavylium; Chloride, *in* P-80037

C$_{15}$H$_{11}$N

5,10-Dihydroindeno[1,2-b]indole, D-80216

C$_{15}$H$_{11}$NO

6-Benzoylindole, B-60058

8-Hydroxy-7-phenylquinoline, H-60218

2-(Phenylethynyl)benzoic acid; Amide, *in* P-70074

2-Phenyl-4(1H)-quinolinone, P-60132

4-Phenyl-2(1H)-quinolinone, P-60131

C$_{15}$H$_{11}$NOS

▷ 2-Phenyl-1,5-benzothiazepin-4(5H)-one, P-60081

C$_{15}$H$_{11}$NO$_2$

▷ 9-Acridinecarboxylic acid; Me ester, *in* A-70061

10-Amino-9-anthracenecarboxylic acid, A-70088

Dibenz[b,g]azocine-5,7(6H,12H)-dione, D-60068

2,4-Diphenyl-5(4H)-oxazolone, D-60485

▷ 3-Nitro-1,2-diphenylcyclopropene, N-80046

C$_{15}$H$_{11}$NO$_3$

▷ 1-Amino-2-methoxyanthraquinone, *in* A-60174

▷ 1-Amino-4-methoxyanthraquinone, *in* A-60175

1-Amino-5-methoxyanthraquinone, A-60176

9-Amino-10-methoxy-1,4-anthraquinone, *in* A-60181

1-Hydroxy-4-(methylamino)anthraquinone, *in* A-60175

C$_{15}$H$_{11}$NO$_3$S

10H-Phenothiazine-2-carboxylic acid; N-Ac, *in* P-70056

C$_{15}$H$_{11}$NS

10H-[1]Benzothieno[3,2-b]indole; N-Me, *in* B-70059

2-(2-Phenylethenyl)benzothiazole, P-60093

3-Phenyl-2(1H)-quinolinethione, P-80133

C$_{15}$H$_{11}$N$_3$

1-Phenyl-1H-pyrazolo[3,4-e]indolizine, P-70092

C$_{15}$H$_{11}$O$_2^{\oplus}$

2,5-Diphenyl-1,3-dioxol-1-ium, D-70478

C$_{15}$H$_{11}$O$_5^{\oplus}$

3,4′,5,7-Tetrahydroxyflavylium, T-80093

C$_{15}$H$_{11}$O$_6^{\oplus}$

3′,4′,5,5′,7-Pentahydroxyflavylium, P-80036

3′,4′,5,7,8-Pentahydroxyflavylium, P-80037

C$_{15}$H$_{12}$

3-Methylphenanthrene, M-70108

C$_{15}$H$_{12}$I$_3$NO$_4$

3,3′,5-Triiodothyronine, T-60350

C$_{15}$H$_{12}$N$_2$O

2-Amino-4,5-diphenyloxazole, A-80068

4-Amino-2,5-diphenyloxazole, A-80069

5-Amino-2,4-diphenyloxazole, A-80070

5-Amino-3-phenylisoxazole; N-Ph, *in* A-60240

2-Methyl-1H-pyrrolo[2,3-b]pyridine; N-Benzoyl, *in* M-80156

C$_{15}$H$_{12}$N$_2$O$_2$

4-Methyl-1-phenazinecarboxylic acid; Me ester, *in* M-80137

6-Methyl-1-phenazinecarboxylic acid; Me ester, *in* M-80138

8-Methyl-1-phenazinecarboxylic acid; Me ester, *in* M-**80**140
9-Methyl-1-phenazinecarboxylic acid; Me ester, *in* M-**80**141
3,5-Pyrazolidinedione; 1,2-Di-Ph, *in* P-**80**200

$C_{15}H_{12}N_2O_3$
2,3-Dihydro-2-hydroxy-1*H*-imidazole-4,5-dione; 1,3-Di-Ph, *in* D-**60**243

$C_{15}H_{12}O$
5,7-Dihydro-6*H*-dibenzo[*a,c*]cyclohepten-6-one, D-**70**187
6,7-Dihydro-5*H*-dibenzo[*a,c*]cyclohepten-5-one, D-**70**188
10,11-Dihydro-5*H*-dibenzo[*a,d*]cyclohepten-5-one, D-**60**218
9-Phenanthrenemethanol, P-**80**079

$C_{15}H_{12}OS$
2,3-Dihydro-2-phenyl-4*H*-1-benzothiopyran-4-one, D-**70**248
1,2-Dimethylthioxanthone, D-**70**440
1,3-Dimethylthioxanthone, D-**70**441
1,4-Dimethylthioxanthone, D-**70**442
2,3-Dimethylthioxanthone, D-**70**443
2,4-Dimethylthioxanthone, D-**70**444
3,4-Dimethylthioxanthone, D-**70**445
3-Hydroxymethyldibenzo[*b,f*]thiepin, H-**60**176

$C_{15}H_{12}O_2$
2,7-Dimethylxanthone, D-**60**453
(4-Hydroxyphenyl)ethylene; Benzoyl, *in* H-**70**209

$C_{15}H_{12}O_2S$
1-Acetoxy-4-[5-(1,3-pentadiynyl)-2-thienyl]-3-butyne, *in* P-**80**025
2-(1-Acetoxy-3-penten-1-ynyl)-5-(3-buten-1-ynyl)thiophene, *in* B-**80**281
2,3-Dihydro-2-phenyl-4*H*-1-benzothiopyran-4-one; 1-Oxide, *in* D-**70**248

$C_{15}H_{12}O_2S_2$
5-Acetoxymethyl-5′-(3-buten-1-ynyl)-2,2′-bithienyl, *in* B-**80**279

$C_{15}H_{12}O_3$
7-(3-Acetoxyphenyl)-2-heptene-4,6-diyn-1-ol, *in* H-**80**234
2-Benzoylbenzoic acid; Me ester, *in* B-**80**054
2,7-Biphenylenediol; Me ether, Ac, *in* B-**80**121
1,3-Bis(4-hydroxyphenyl)-2-propen-1-one, B-**80**152
3,4-Dihydro-3-(3-hydroxyphenyl)-1*H*-2-benzopyran-1-one, *in* H-**70**207
2,3-Dihydro-2-phenyl-2-benzofurancarboxylic acid, D-**60**262
1-(2-Hydroxyphenyl)-3-phenyl-1,3-propanedione, H-**80**239
4-Methoxy-2,5-phenanthrenediol, *in* P-**60**071
Naphtho[1,8-*bc*]pyran-2-carboxylic acid; Et ester, *in* N-**70**010

$C_{15}H_{12}O_3S$
2,3-Dihydro-2-phenyl-4*H*-1-benzothiopyran-4-one; 1,1-Dioxide, *in* D-**70**248
1,4-Dimethylthioxanthone; Dioxide, *in* D-**70**442
2,4-Dimethylthioxanthone; Dioxide, *in* D-**70**444
1-(2-Furyl)-4-(5-acetoxymethyl-2-thienyl)-1-buten-3-yne, *in* F-**80**093
3-Hydroxymethyldibenzo[*b,f*]thiepin; 5,5-Dioxide, *in* H-**60**176
4-[5-(1,3-Pentadiynyl)-7-thienyl]-3-butyn-1-ol; 2-Hydroxy, 1-Ac, *in* P-**80**025
4-[5-(1,3-Pentadiynyl)-7-thienyl]-3-butyn-1-ol; 2-Hydroxy, 2-Ac, *in* P-**80**025

$C_{15}H_{12}O_4$
4′,7-Dihydroxyisoflavanone, D-**80**322
1-(2,4-Dihydroxyphenyl)-3-(4-hydroxyphenyl)-2-propen-1-one, D-**80**367
3,9-Dihydroxypterocarpan, D-**60**373
Frutescinone, *in* F-**80**084
6-Hydroxy-2-(4-hydroxy-2-methoxyphenyl)benzofuran, *in* D-**80**363
Mycosinol; Ac (*E*-), *in* M-**80**202
Mycosinol; Ac (*Z*-), *in* M-**80**202

3-Phenyl-1-(2,4,6-trihydroxyphenyl)-2-propen-1-one, P-**80**139
2′,4′,7-Trihydroxyisoflavene, T-**80**305

$C_{15}H_{12}O_5$
4-Acetoxy-8,9-epoxy-7-(2,4-hexadiynylidene)-1,6-dioxaspiro[4.4]non-2-ene, *in* M-**80**202
3,5-Dihydroxy-2,4-dimethoxy-9*H*-fluoren-9-one, *in* T-**70**109
1,6-Dihydroxy-5-methoxyxanthone, *in* T-**60**349
1-(2,4-Dihydroxyphenyl)-3-(3,4-dihydroxyphenyl)-2-propen-1-one, D-**80**362
1-Hydroxy-3,7-dimethoxyxanthone, *in* T-**80**347
1-Hydroxy-5,6-dimethoxyxanthone, *in* T-**60**349
3-(4-Hydroxyphenyl)-1-(2,3,4-trihydroxyphenyl)-2-propen-1-one, H-**80**245
3-(4-Hydroxyphenyl)-1-(2,4,6-trihydroxyphenyl)-2-propen-1-one, H-**80**246
4-Methoxy-6-(3,4-methylenedioxystyryl)-α-pyrone, *in* M-**80**086
Protosappanin *A*, P-**60**186
2′,3′,4′,7-Tetrahydroxyisoflavene, T-**80**103
2′,4′,7,8-Tetrahydroxyisoflavene, T-**80**104
4′,5,8-Trihydroxyflavanone, T-**80**297
5,7,8-Trihydroxyflavanone, T-**60**321
2′,4′,7-Trihydroxyisoflavanone, T-**80**303
3′,4′,7-Trihydroxyisoflavanone, T-**70**263
4′,5,7-Trihydroxyisoflavanone, T-**80**304
2,3,9-Trihydroxypterocarpan, T-**80**335
3,4,9-Trihydroxypterocarpan, T-**80**336
3,9,10-Trihydroxypterocarpan, T-**80**338
3,6a,9-Trihydroxypterocarpan, T-**80**337

$C_{15}H_{12}O_6$
1,3-Dihydroxy-5,8-dimethoxyxanthone, *in* T-**70**123
1,8-Dihydroxy-3,5-dimethoxyxanthone, *in* T-**70**123
1,8-Dihydroxy-3,6-dimethoxyxanthone, *in* T-**60**121
3,5-Dihydroxy-2-methyl-1,4-naphthoquinone; Di-Ac, *in* D-**60**350
3-(3,4-Dihydroxyphenyl)-1-(2,3,4-trihydroxyphenyl)-2-propen-1-one, D-**80**372
3-(3,4-Dihydroxyphenyl)-1-(2,4,5-trihydroxyphenyl)-2-propen-1-one, D-**70**339
2′,3′,4′,6,7-Pentahydroxyisoflavene, P-**80**040
1-(2,3,4,5,6-Pentahydroxyphenyl)-3-phenyl-2-propen-1-one, P-**80**050
2′,5,7,8-Tetrahydroxyflavanone, T-**70**104
3′,4′,5,7-Tetrahydroxyflavanone, T-**70**105
3′,4′,5,7-Tetrahydroxyflavanone, T-**70**106
3,5,6,7-Tetrahydroxyflavanone, T-**60**105
2′,3′,4′,7-Tetrahydroxyisoflavanone, T-**80**099
2′,4′,5,7-Tetrahydroxyisoflavanone, T-**80**100
2′,4′,5′,7-Tetrahydroxyisoflavanone, T-**80**101
2,3,8,9-Tetrahydroxypterocarpan, T-**80**117
2,3,9,10-Tetrahydroxypterocarpan, T-**80**118
3,4,8,9-Tetrahydroxypterocarpan, T-**80**119
3,6a,7,9-Tetrahydroxypterocarpan, T-**80**120

$C_{15}H_{12}O_7$
Nectriafurone, N-**60**016
2′,3,5,7,8-Pentahydroxyflavanone, P-**60**045
▷ 3,3′,4′,5,7-Pentahydroxyflavanone, P-**70**027
3′,4′,5,6,7-Pentahydroxyflavanone, P-**70**028
2′,3′,4′,5,7-Pentahydroxyisoflavanone, P-**80**039
1,2,3,8,9-Pentahydroxypterocarpan, P-**80**051
2,3,8,9,10-Pentahydroxypterocarpan, P-**80**054
3,4,8,9,10-Pentahydroxypterocarpan, P-**80**055
2,3,6a,8,9-Pentahydroxypterocarpan, P-**80**052
3,4,6a,8,9-Pentahydroxypterocarpan, P-**80**053
3,6a,7,8,9-Pentahydroxypterocarpan, P-**80**056
1,2,5,6,8-Pentahydroxyxanthone, P-**60**052
1,3,8-Trihydroxy-2,6-dimethoxyxanthone, *in* P-**70**036
1,5,8-Trihydroxy-2,6-dimethoxyxanthone, *in* P-**60**052
1,5,10-Trihydroxy-7-methoxy-3-methyl-1*H*-naphtho[2,3-*c*]pyran-6,9-dione, T-**80**320
Ventilone *C*, V-**60**004
Ventilone *D*, *in* V-**60**004

$C_{15}H_{12}O_8$
Fusarubinoic acid, F-**70**058
2,3,4,8,9,10-Hexahydroxypterocarpan, H-**80**072

$C_{15}H_{12}O_9$
1-(2,3,4,5-Tetrahydroxyphenyl)-3-(2,4,5-trihydroxyphenyl)-1,3-propanedione, T-**70**120

$C_{15}H_{12}S$
3-Phenyl-2*H*-1-benzothiopyran, P-**80**091
4-Phenyl-2*H*-1-benzothiopyran, P-**80**092

$C_{15}H_{13}Br$
1-Bromo-1,2-diphenylpropene, B-**60**237
3-Bromo-1,2-diphenylpropene, B-**70**209

$C_{15}H_{13}ClO$
2,2-Diphenylpropanoic acid; Chloride, *in* D-**70**493

$C_{15}H_{13}ClO_2$
2-Chloro-3,11-tridecadiene-5,7,9-triyn-1-ol; Ac, *in* C-**80**117

$C_{15}H_{13}N$
1-Cyano-1,1-diphenylethane, *in* D-**70**493
10,11-Dihydro-5*H*-dibenzo[*a,d*]cyclohepten-5,10-imine, D-**80**179

$C_{15}H_{13}NO$
N-(Diphenylmethylene)acetamide, *in* D-**60**482
6-Hydroxyindole; Benzyl ether, *in* H-**60**158
Phthalimidine; *N*-Benzyl, *in* P-**70**100

$C_{15}H_{13}NOS$
2-Phenyl-1,5-benzothiazepin-4(5*H*)-one; 2,3-Dihydro, *in* P-**60**081

$C_{15}H_{13}NO_2$
5-Amino-1,4-pentanediol, A-**80**132
3,4-Diphenyl-2-oxazolidinone, *in* P-**60**126

$C_{15}H_{13}NO_2S$
10*H*-Phenothiazine-1-carboxylic acid; *N*-Et, *in* P-**70**055
10*H*-Phenothiazine-3-carboxylic acid; *N*-Et, *in* P-**70**057
10*H*-Phenothiazine-4-carboxylic acid; *N*-Et, *in* P-**70**058
10*H*-Phenothiazine-2-carboxylic acid; Et ester, *in* P-**70**056
10*H*-Phenothiazine-1-carboxylic acid; *N*-Me, Me ester, *in* P-**70**055
10*H*-Phenothiazine-2-carboxylic acid; *N*-Me, Me ester, *in* P-**70**056
10*H*-Phenothiazine-3-carboxylic acid; *N*-Me, Me ester, *in* P-**70**057
10*H*-Phenothiazine-4-carboxylic acid; *N*-Me, Me ester, *in* P-**70**058

$C_{15}H_{13}NO_3$
4-Hydroxy-3-methoxy-2-methyl-9*H*-carbazole-1-carboxaldehyde, H-**70**154

$C_{15}H_{13}NO_4$
Acetaminosalol, *in* H-**70**108
4-Phenyl-2,6-pyridinedicarboxylic acid; Di-Me ester, *in* P-**80**130

$C_{15}H_{13}NO_4S$
10*H*-Phenothiazine-3-carboxylic acid; *N*-Et, 5,5-dioxide, *in* P-**70**057
10*H*-Phenothiazine-4-carboxylic acid; *N*-Me, 5,5-dioxide, Me ester, *in* P-**70**058
10*H*-Phenothiazine-3-carboxylic acid; *N*-Me, Me ester, 5,5-dioxide, *in* P-**70**057

$C_{15}H_{13}NO_5$
Dianthramide *A*, D-**70**049
Dianthramide *B*, *in* D-**70**049
1,2,4-Trimethoxy-3*H*-phenoxazin-3-one, T-**60**353

$C_{15}H_{13}NS_2$
2*H*-1,4-Benzothiazine-3(4*H*)-thione; 4-Benzyl, *in* B-**70**055

$C_{15}H_{14}$
6,7-Dihydro-5*H*-dibenzo[*a,c*]cycloheptene, D-**70**186

645

10,11-Dihydro-5*H*-dibenzo[*a*,*d*]cycloheptene, D-60217
1,1-Diphenylcyclopropane, D-60478
1,1-Diphenylpropene, D-60491

$C_{15}H_{14}ClN$
4-Chloro-2,3-dihydro-1*H*-indole; 1-Benzyl, *in* C-60055

$C_{15}H_{14}N_2$
10,11-Dihydro-5*H*-dibenzo[*a*,*d*]cyclohepten-5-one; Hydrazone, *in* D-60218

$C_{15}H_{14}N_2O$
N-(7,7-Dimethyl-3-azabicyclo[4.1.0]hepta-1,3,5-trien-4-yl)benzamide, D-70368

$C_{15}H_{14}N_2OS$
10*H*-Phenothiazine-2-carboxylic acid; Dimethylamide, *in* P-70056

$C_{15}H_{14}N_2O_2$
N,*N*′-Diphenylurea; Mono-*N*-Ac, *in* D-70499

$C_{15}H_{14}N_2S$
2-Amino-4*H*-3,1-benzothiazine; *N*-Benzyl, *in* A-70092

$C_{15}H_{14}N_4$
4-Amino-1,2,3-benzotriazine; 2-Et, *N*(4)-Ph, *in* A-60092
4-Amino-1,2,3-benzotriazine; 3-Et, *N*(4)-Ph, *in* A-60092

$C_{15}H_{14}O$
Di(2,4-cycloheptatrien-1-yl)ethanone, D-60166
Linderazulene, L-70031

$C_{15}H_{14}OS$
Thia[2.2]metacyclophane; *S*-Oxide, *in* T-60194

$C_{15}H_{14}O_2$
1-Acetoxy-2,8,10,12-tridecatetraene-4,6-diyne, *in* T-80260
2,2-Diphenylpropanoic acid, D-70493
2-Hydroxy-1,2-diphenyl-1-propanone, H-70123
2-Hydroxy-1,3-diphenyl-1-propanone, H-70124
3-Methoxy-5-(2-phenylethenyl)phenol, *in* D-80384
3,5-Tridecadiene-7,9,11-triyn-1-ol; Ac, *in* T-80258

$C_{15}H_{14}O_2S$
2-Mercapto-2-phenylacetic acid; *S*-Benzyl, *in* M-60024
Thia[2.2]metacyclophane; *S*-Dioxide, *in* T-60194

$C_{15}H_{14}O_2S_2$
3,3-Bis(phenylthio)propanoic acid, B-70157

$C_{15}H_{14}O_3$
4-(Benzyloxy)phenol; Ac, *in* B-60074
Cyclolongipesin, C-60214
Dehydroosthol, D-60020
cis-Dehydroosthol, *in* D-60020
9,10-Dihydro-7-methoxy-2,5-phenanthrenediol, *in* P-80080
Frutescin, F-80084
4′-Hydroxy-2-biphenylcarboxylic acid; Me ether, Me ester, *in* H-70111

$C_{15}H_{14}O_4$
8-Acetoxy-2-(2,4-hexadiynylidene)-1,6-dioxaspiro[4.4]non-3-ene, *in* H-80047
Allopteroxylin, A-60075
1-(3,5-Dihydroxyphenyl)-2-(3-hydroxy-4-methoxyphenyl)ethylene, *in* D-80361
3-(2,4-Dihydroxyphenyl)-1-(4-hydroxyphenyl)-1-propanone, D-80366
Helicquinone, H-60015
2-Hydroxy-4,6-dimethoxybenzophenone, *in* T-80285
4-Hydroxy-2,6-dimethoxybenzophenone, *in* T-80285
5-[2-(3-Hydroxy-4-methoxyphenyl)ethenyl]-1,3-benzenediol, *in* D-80361
8-(3-Methyl-2-butenoyl)coumarin, *in* M-80190
2-Methyl-1,3-naphthalenediol; Di-Ac, *in* M-60091

4-Methyl-1,3-naphthalenediol; Di-Ac, *in* M-60094
Micropubescin, M-80190
Murralongin, M-70159
Rhinacanthin A, R-80005
Rutalpinin, R-70015
3′,4′,7-Trihydroxyflavan, T-70260
2′,4′,7-Trihydroxyisoflavan, T-80302
1-(2,4,6-Trihydroxyphenyl)-3-phenyl-1-propanone, T-80332
Yangonin, *in* H-70122

$C_{15}H_{14}O_5$
Ageratone, A-60065
Altholactone; Ac, *in* A-60083
Anhydroscandenolide, A-70211
2,3-Dihydro-5-hydroxy-8-methoxy-2,4-dimethylnaphtho[1,2-b]furan-6,9-dione, D-80213
3,6-Dihydroxy-1*a*-(3-methyl-2-butenyl)naphth[2,3-*b*]-2,7(1*aH*,7*aH*)-dione, *in* D-70277
6-[2-(4-Hydroxy-3-methoxyphenyl)ethenyl]-4-methoxy-2*H*-pyran-2-one, *in* H-70122
1-(4-Hydroxyphenyl)-3-(2,4,6-trihydroxyphenyl)-1-propanone, H-80243
3-(4-Hydroxyphenyl)-1-(2,4,6-trihydroxyphenyl)-1-propanone, H-80244
Panial, P-70006
3′,4′,5,7-Tetrahydroxyflavan, T-60103
3,4′,5,7-Tetrahydroxyflavanone, T-60104
2′,3′,4′,7-Tetrahydroxyisoflavan, T-80095
2′,4′,5,7-Tetrahydroxyisoflavan, T-80096
2′,4′,5′,7-Tetrahydroxyisoflavan, T-80097
2′,4′,6,7-Tetrahydroxyisoflavan, T-80098

$C_{15}H_{14}O_6$
7-(3,3-Dimethyloxiranyl)methoxy-5,6-methylenedioxycoumarin, D-80463
Fonsecin, F-80073
1,4,5,7-Naphthalenetetrol; 5-Me ether, 1,7-Di-Ac, *in* N-80003
3,3′,4′,5,7-Pentahydroxyflavan, P-70026
2′,3′,4′,7,8-Pentahydroxyisoflavan, P-80038

$C_{15}H_{14}O_7$
Fusarubin, F-60108
2′,3′,4′,6,7,8-Hexahydroxyisoflavan, H-80068
2,5,8-Trihydroxy-3-methyl-6,7-methylenedioxy-1,4-naphthoquinone; Tri-Me ether, *in* T-60333

$C_{15}H_{14}S$
Dimethylsulfonium 9-fluorenylide, D-80475
Thia[2.2]metacyclophane, T-60194

$C_{15}H_{15}N$
10,11-Dihydro-5-methyl-5*H*-dibenz[*b*,*f*]azepine, *in* D-70185
5,6,7,12-Tetrahydrodibenz[*b*,*e*]azocine, T-70055
5,6,11,12-Tetrahydrodibenz[*b*,*f*]azocine, T-70057
5,6,7,12-Tetrahydrodibenz[*b*,*g*]azocine, T-70056
5,6,7,8-Tetrahydrodibenz[*c*,*e*]azocine, T-70054
1,2,3-Trimethyl-9*H*-carbazole, T-60357
1,2,4-Trimethyl-9*H*-carbazole, T-60358
1,3,9-Trimethylcarbazole, *in* D-70375
2,4,9-Trimethylcarbazole, *in* D-70382

$C_{15}H_{15}NO$
2-Acetamido-3-methylbiphenyl, *in* A-70164
4-Amino-3,4-dihydro-2-phenyl-2*H*-1-benzopyran, A-80067
2-Amino-1,3-diphenyl-1-propanone, A-60149
2,2-Diphenylpropanoic acid; Amide, *in* D-70493
1,2,3,4-Tetrahydrobenz[*g*]isoquinoline; *N*-Ac, *in* T-80054

$C_{15}H_{15}NO_2$
2-Hydroxy-1,3-diphenyl-1-propanone; Oxime, *in* H-70124
2-Nitro-1,3-diphenylpropane, N-70048

$C_{15}H_{15}NO_4$
6-Amino-2-hexenoic acid; *N*-Phthalimido, Me ester, *in* A-60173
2,3-Quinolinedicarboxylic acid; Di-Et ester, *in* Q-80002

$C_{15}H_{15}NO_5$
2,3-Quinolinedicarboxylic acid; Di-Et ester, *N*-oxide, *in* Q-80002

$C_{15}H_{15}N_5$
5-Amino-1*H*-tetrazole; 1,5(*N*)-Dibenzyl, *in* A-70202
5-Amino-1*H*-tetrazole; 2,5(*N*)-Dibenzyl, *in* A-70202

$C_{15}H_{16}$
7,8,9,10-Tetrahydro-6*H*-cyclohepta[*b*]naphthalene, T-70047

$C_{15}H_{16}Br_2O_2$
Deoxyokamurallene, D-80030

$C_{15}H_{16}Br_2O_3$
Isookamurallene, I-80067
Okamurallene, O-80028

$C_{15}H_{16}N_2O$
▷ *N*,*N*′-Dimethyl-*N*,*N*′-diphenylurea, *in* D-70499

$C_{15}H_{16}N_4$
2,5-Diphenyl-1,2,4,5-tetraazabicyclo[2.2.1]heptane, D-70498

$C_{15}H_{16}N_{10}O_2$
Drosopterin, D-60521
Isodrosopterin, *in* D-60521
Neodrosopterin, *in* D-60521

$C_{15}H_{16}O$
2,3-Dihydrolinderazulene, *in* L-70031
1,3,5,7(11),9-Guaiapentaen-14-al, G-60039
4-Isopropyl-7-methyl-1-naphthalenecarboxaldehyde, I-80084
8-Isopropyl-5-methyl-2-naphthalenecarboxaldehyde, I-60129
3-Methyl-8-phenyl-3,5,7-octatrien-2-one, M-80144
1,8,10,14-Pentadecatetraene-4,6-diyn-3-ol, P-80022
4-(2-Phenylethyl)phenol; Me ether, *in* P-70073

$C_{15}H_{16}OS_2$
3,3-Bis(phenylthio)-1-propanol, B-70158

$C_{15}H_{16}O_2$
10-Desmethyl-1-methyl-1,3,5(10),11(13)-eudesmatetraen-12,8-olide, D-70026
Furoixiolal, F-70056
3-Methoxy-5-(2-phenylethyl)phenol, *in* P-80100

$C_{15}H_{16}O_2S$
1-Acetoxy-9-(2-thienyl)-4,6-nonadien-8-yne, *in* T-80181

$C_{15}H_{16}O_3$
3-(2,2-Dimethyl-2*H*-1-benzopyran-6-yl)-2-propenoic acid; Me ester, *in* D-70372
Gnididione, G-60030
Heritol, H-60028
8-Hydroxy-1(6),2,4,7(11)-cadinatetraen-12,8-olide, H-70114
Hypocretenolide, *in* H-70235
Lumiyomogin, L-70041
2-Oxo-3,10(14),11(13)-guaiatrien-12,6-olide, O-70090
2,3,4-Trimethoxybiphenyl, *in* B-70124
3,4,5-Trimethoxybiphenyl, *in* B-80129
Yomogin, Y-70001

$C_{15}H_{16}O_4$
Auraptenol, A-70269
1,2-Bis(4-hydroxyphenyl)-1,2-propanediol, B-70147
Cedrelopsin, C-60026
1-(2,4-Dihydroxyphenyl)-3-(3,4-dihydroxyphenyl)propane, D-70336
2,6-Dimethoxy-4-(2-methoxyphenyl)phenol, *in* B-80128
3-(1,1-Dimethylallyl)scopoletin, *in* R-80016
3-(1,1-Dimethyl-2-propenyl)-8-hydroxy-7-methoxy-2*H*-1-benzopyran-2-one, *in* D-80470
2,9-Epoxylactarotropone, E-80038
1α,2α-Epoxyyomogin, *in* Y-70001
Ferreyrantholide, F-70004
8β-Hydroxydehydrozaluzanin *C*, *in* Z-60001
14-Hydroxyhypocretenolide, *in* H-70235

Bisabolangelone, B-60138
Cantabrenonic acid, *in* C-60010
Carpesialactone, C-80027
Cumambranolide, C-80164
Cyclodehydromyopyrone *A*, C-60188
Cyclodehydromyopyrone *B*, C-60189
Dehydromyoporone, D-60019
6-Deoxyilludin *S*, *in* I-60002
Desacetyllaurenobiolide, *in* H-70134
11α,13-Dihydroestafialin, *in* E-80057
11α,13-Dihydrozaluzanin *C*, *in* Z-60001
11β,13-Dihydrozaluzanin *C*, *in* Z-60001
Elemasteiractinolide; 15-Hydroxy, *in* E-70005
10-Epicumambranolide, *in* C-80164
1,10-Epoxy-11(13)-eremophilen-12,8-olide, E-80024
1,10-Epoxy-4,11(13)-germacradien-12,8-olide, E-60022
4α,15-Epoxyisoalantolactone, *in* I-80046
9,11(13)-Eremophiladien-12-oic acid; 3-Oxo, *in* E-70029
3,11(13)-Eudesmadien-12-oic acid; 2-Oxo, *in* E-70060
Eumorphistonol, E-60073
Haageanolide, H-60001
Hanphyllin, H-60006
10-Hydroxy-1,11(13)-eremophiladien-12,8-olide, H-80159
15-Hydroxy-4,11(13)-eudesmadien-12,8-olide, H-60138
2-Hydroxy-3,11(13)-eudesmadien-12,8-olide, H-80161
3-Hydroxy-4,11(13)-eudesmadien-12,8-olide, H-60136
7-Hydroxy-4,11(13)-eudesmadien-12,6-olide, H-70128
7-Hydroxy-4,11(13)-eudesmadien-12,6-olide, H-80162
9-Hydroxy-4,11(13)-eudesmadien-12,6-olide, H-60137
3-Hydroxy-1,4(15),11(13)-eudesmatrien-12-oic acid, H-80164
6-Hydroxy-1(10),4,11(13)-germacratrien-12,8-olide, H-70134
15-Hydroxy-3,10(14)-guaiadien-12,8-olide, H-70136
3-Hydroxy-4,11(13)-guaiadien-12,8-olide, H-70135
4-Hydroxy-1(10),11(13)-guaiadien-12,8-olide, H-80167
5-Hydroxy-3,11(13)-guaiatrien-12,8-olide, H-70137
9α-Hydroxyisoalantolactone, *in* I-80046
9β-Hydroxyisoalantolactone, *in* I-80046
15-Hydroxyisoalloalantolactone, *in* I-80047
▷ Illudin *M*, I-60001
Inunal; Δ^4-Isomer, 15-alcohol, *in* I-70022
Inuviscolide, I-70023
Isoperezone, I-70075
Ivangustin, I-70109
Kanshone *C*, K-70006
Muzigadial, M-80201
3-Oxo-4,11(13)-eudesmadien-12-oic acid, *in* E-70061
9-Oxo-4,11(13)-eudesmadien-12-oic acid, *in* H-60135
1-Oxo-7(11)-eudesmen-12,8-olide, O-60067
1-Oxo-4,10(14)-germacradien-12,6α-olide, *in* O-60068
α-Pipitzol, P-60158
Pulicaric acid, *in* P-80190
Quadrangolide, Q-60001
Santamarine, S-60007
Subergorgic acid, S-70081

$C_{15}H_{20}O_4$

Artecalin, A-60302
Confertin, *in* D-60211
Cumambrin B, *in* C-80164
2,3-Dehydro-11α,13-dihydroconfertin, *in* D-60211
11,13-Dehydroeriolin, *in* E-60036
Dehydromelitensin, *in* M-80013
11β,13-Dihydro-9α-hydroxyzaluzanin*C*, *in* Z-60001
Dihydroparthenin, *in* P-80011
3α,15-Dihydroxyalloalantolactone, *in* A-80044
1,8-Dihydroxy-3,7(11)-eremophiladien-12,8-olide, D-60320
1,5-Dihydroxyeriocephaloide, D-60321
1,8-Dihydroxy-3,7(11)-eudesmadien-12,8-olide, D-60322
1,8-Dihydroxy-4(15),11(13)-eudesmadien-12,6-olide, D-80315
2,8-Dihydroxy-4,10(14)-germacradien-12,6-olide, D-80319
2,8-Dihydroxy-1(10),4,11(13)-germacratrien-12,6-olide, D-70302
3,13-Dihydroxy-1(10),4,7(11)-germacratrien-12,6-olide, D-70303
8,9-Dihydroxy-1(10),4,11(13)-germacratrien-12,6-olide, D-60323
9,15-Dihydroxy-1(10),4,11(13)-germacratrien-12,6-olide, D-60324
2,4-Dihydroxy-5,11(13)-guaiadien-12,8-olide, D-70304
3,5-Dihydroxy-4(15),10(14)-guaiadien-12,8-olide, D-60325
4,10-Dihydroxy-2,11(13)-guaiadien-12,6-olide, D-60326
3β,9β-Dihydroxy-4(15),10(14)-guaiadien-12,6α-olide, *in* D-70305
9,10-Dihydroxy-7-marasmen-5,13-olide, D-60346
4,5-Dioxo-1(10)-xanthen-12,8-olide, D-60474
4-Epivulgarin, *in* V-80013
Epoxycantabronic acid, *in* C-60010
1β,10α-Epoxyhaageanolide, *in* H-60001
1,10-Epoxy-13-hydroxy-7(11)-eremophilen-12,8-olide, E-80031
1α,2α-Epoxy-10α-hydroxy-11(13)-eremophilen-12,8β-olide, *in* H-80159
1β,10β-Epoxy-2α-hydroxy-11(13)-eremophilen-12,8β-olide, *in* E-80024
4α,5α-Epoxy-3α-hydroxy-11(13)-eudesmen-12,8β-olide, *in* H-60136
4α,5α-Epoxy-3β-hydroxy-11(13)-eudesmen-12,8β-olide, *in* H-60136
4α,15-Epoxy-3α-hydroxyisoalantolactone, *in* I-80046
4α,13-Epoxymuzigadial, *in* M-80201
Grosheiminol, *in* G-80042
3α-Hydroperoxy-4,11(13)-eudesmadien-12,8β-olide, *in* H-60136
2α-Hydroxyartemorin, *in* A-80170
9β-Hydroxyartemorin, *in* A-80170
8α-Hydroxybalchanin, *in* S-60007
8α-Hydroxy-11α,13-dihydrozaluzanin*C*, *in* Z-60001
6α-Hydroxy-8-epiivangustin, *in* I-70109
6-(1-Hydroxyethyl)-7,8-dimethoxy-2,2-dimethyl-2*H*-1-benzopyran, *in* A-80023
2α-Hydroxyivangustin, *in* I-70109
1-[2-(Hydroxymethyl)-3-methoxyphenyl]-1,5-heptadiene-3,4-diol, H-70178
9β-Hydroxy-3-oxo-10(14)-guaien-12,6α-olide, *in* D-70305
9α-Hydroxysantamarine, *in* S-60007
▷ Illudin *S*, I-60002
Ivangulic acid, I-70108
Ixerin A, I-80102
Kanshone E, *in* K-80003
Pulverulide, P-80191
Ridentin, R-70007
Schkuhridin *B*, S-60012
8,9,14-Trihydroxy-1(10),4,11(13)-germacratrien-12,6-olide, T-60323
Umbellifolide, U-60001
▷ Vulgarin, V-80013

$C_{15}H_{20}O_5$

Alhanin, *in* I-70066
Altamisic acid, A-70084
▷ Coriolin, C-70199
3,5-Dihydroxy-4(15),10(14)-guaiadien-12,8-olide; 10α,14-Epoxide, *in* D-60325
4,5-Epoxy-2,8-dihydroxy-1(10),11(13)-germacradien-12,6-olide, E-60013
Hieracin I, H-80087
Hirsutinolide, H-80089
4α-Hydroperoxydesoxyvulgarin, *in* V-80013
7′-Hydroxyabscisic acid, H-70098
8′-Hydroxyabscisic acid, H-70099
Psilostachyin, P-70131
Secoisoerivanin pseudoacid, S-70024
1,3,13-Trihydroxy-4(15),7(11)-eudesmadien-12,6-olide, T-70258
1,3,9-Trihydroxy-4(15),11(13)-eudesmadien-12,6-olide, T-70257
1β,3α,6β-Trihydroxy-4,11(13)-eudesmadien-12,8β-olide, *in* I-70109

$C_{15}H_{20}O_6$

Ajafinin, A-60068
2,3-Epoxy-1,4,8-trihydroxy-7(11)-eudesmen-12,8-olide, E-60033

$C_{15}H_{20}O_7$

8-Deacyltrichogoniolide, D-80007
Domesticoside, *in* T-70248
Neomajucin, *in* M-70005

$C_{15}H_{20}O_8$

▷ Anisatin, A-80158
Majucin, M-70005
Laurencia Polyketal, P-60168

$C_{15}H_{20}O_{10}$

Quinic acid; Tetra-Ac, *in* Q-70005

$C_{15}H_{21}Br$

1-[3-(Bromomethyl)-1-methyl-2-methylenecyclopentyl]-4-methyl-1,4-cyclohexadiene, B-70247

$C_{15}H_{21}BrO$

Isoobtusadiene, I-70074
Obtusadiene, O-70001

$C_{15}H_{21}BrO_3$

Laureoxolane, L-80022

$C_{15}H_{21}Br_3O_3$

Intricata bromoallene, I-80016
3,6,13-Tribromo-4,10:9,12-diepoxy-14-pentadecyn-7-ol, B-70195

$C_{15}H_{21}ClO$

Laurencenone *B*, L-70025

$C_{15}H_{21}ClO_3$

10α-Chloro-1β-hydroxy-11(13)-eremophilen-12,8β-olide, *in* D-80314

$C_{15}H_{21}NO$

Octahydro-2*H*-1,3-benzoxazine; *N*-Benzyl, *in* O-70011
Octahydro-2*H*-3,1-benzoxazine; *N*-Benzyl, *in* O-70012

$C_{15}H_{21}NO_4$

3,5-Pyridinedipropanoic acid; Di-Et ester, *in* P-60217

$C_{15}H_{21}N_3O_3$

3,4,7,8,11,12-Hexahydro-2*H*,6*H*,10*H*-benzo[1,2-*b*:3,4-*b*′:5,6-*b*″]tris[1,4]oxazine; Tri-*N*-Me, *in* H-80054

$C_{15}H_{21}N_3S_3$

3,4,7,8,11,12-Hexahydro-2*H*,6*H*,10*H*-benzo[1,2-*b*:3,4-*b*′:5,6-*b*″]tris[1,4]thiazine; Tri-*N*-Me, *in* H-80055

$C_{15}H_{22}$

1(10),4-Aromadendradiene, A-70257
1(5),3-Aromadendradiene, A-70256
1,1,2,2,3,3-Hexamethylindane, H-80075

$C_{15}H_{22}BrClO$

Elatol, E-70002
Laurencenone *A*, L-70024
Laurencenone *D*, L-70026

$C_{15}H_{22}BrClO_2$

Almadioxide, A-80047
Pinnatazone, P-80150

$C_{15}H_{22}Br_2O_3$

α-(1-Bromo-3-hexenyl)-5-(1-bromo-2-propynyl)tetrahydro-3-hydroxy-2-furanethanol, B-70217

$C_{15}H_{22}N_2O_4$

N-(5*a*,6,7,8-Tetrahydro-4-methyl-1,5-dioxo-1*H*,5*H*-pyrrolo[1,2-*c*][1,3]oxazepin-3-yl)hexanamide, *in* M-70134

$C_{15}H_{22}O$

1(10)-Aristolen-9-one, A-70252
13-Bicyclogermacrenal, B-80076
1,9-Cadinadien-3-one, C-80003
2-Copaen-8-one, C-80149

1-Aromadendrene, A-70258
9-Aromadendrene, A-70259
β-Bisabolene, B-**80**134
4(15),5-Cadinadiene, C-70003
ε-Cadinene, C-**60**001
Daucene, D-70008
1,3,7(11)-Elematriene, E-70006
1-Epi-*α*-gurjunene, *in* G-70034
1(5),6-Guaiadiene, G-70031
α-Gurjunene, G-70034
1,1,3,3,5,5-Hexamethyl-2,4,6-tris(methylene)cyclohexane, H-**60**072
Hinesene, H-**80**088
1,2,3,4,4*a*,5,6,7-Octahydro-4*a*,5-dimethyl-2-(1-methylethenyl)naphthalene, O-**60**015
β-Patchoulene, P-**60**010
Pentalenene, P-70037
Precapnelladiene, P-**60**171
Premnaspirodiene, P-70114
Sesquisabinene, S-70037
Sesquithujene, *in* S-70037
6-Siliphiperfolene, S-70043
Sinularene, S-**60**034

$C_{15}H_{24}BrClO$
Puertitol A, P-**80**187
Puertitol B, P-**80**188

$C_{15}H_{24}O$
1(5)-Aromadendren-7-ol, A-70260
α-Biotol, B-**60**112
β-Biotol, B-**60**113
α-Copaene, C-70197
15-Copaenol, *in* C-70197
α-Copaen-11-ol, *in* C-70197
α-Copaen-8-ol, *in* C-70197
Cyperenol, C-70259
2,3-Epoxy-7,10-bisaboladiene, E-**60**009
3,5-Eudesmadien-1-ol, E-**60**064
5,7(11)-Eudesmadien-15-ol, E-**60**066
5,7-Eudesmadien-11-ol, E-**60**065
Fervanol, F-**60**007
1(10),4(15)-Germacradien-6-one, G-**80**011
Gymnomitrol, G-70035
Jasionone, J-**80**003
4,10(14)-Oplopadien-3-ol, O-**80**046
Oplopenone, O-**50**045
Senecrassidiol; 6*α*-Hydroxy, 6,8-dideoxy, 7,8-didehydro, *in* S-**80**026
Sesquisabinene hydrate, *in* S-70037
5-Silphiperfolen-3-ol, S-**80**033
3-Sterpurenol, S-**80**045
Striatenone, S-**80**046
α-Torosol, T-**60**225
β-Torosol, T-**60**226
Valerenol, V-70001
Waitziacuminone, W-**80**001

$C_{15}H_{24}O_2$
10(14)-Aromadendrene-4,8-diol, A-**60**299
Artemone, A-**60**305
1(10),4-Cadinadiene-3,9-diol, C-**80**001
Curcudiol, C-70207
cis-9,10-Dihydrocapsenone, *in* C-**60**011
10,11-Dihydro-11-hydroxycurcuphenol, *in* C-70208
7,12-Dihydroxysterpurene, D-70346
9,12-Dihydroxysterpurene, D-70347
1,3,11-Elematriene-9,14-diol, E-**80**010
4,7(11)-Eudesmadiene-12,13-diol, E-**60**063
11-Hydroxy-2,7(14)-bisaboladien-4-one, H-**80**131
11-Hydroxy-4-guaien-3-one, H-**60**149
10*α*-Hydroxy-6-isodaucen-14-al, *in* O-70093
11-Hydroxyjasionone, *in* J-**80**003
4-Hydroxy-10(15)-oplopen-3-one, H-**80**225
11-Hydroxy-1(10)-valencen-2-one, H-**60**234
Kurubasch aldehyde, K-**60**014
6-Oxocyclonerolidol, *in* H-**60**111
Senecrassidiol; 5-Deoxy, 6-oxo, *in* S-**80**026
Tanavulgarol, T-**60**004
3,7,11-Trimethyl-2,6,10-dodecatrienoic acid, T-**60**364
Vetidiol, V-**60**007

$C_{15}H_{24}O_3$
Arthrosporone, *in* A-**80**171
Dihydromyoporone, *in* M-**60**156
3,11-Dihydroxy-7-drimen-6-one, D-70295
8*α*,9*α*-Dihydroxy-10*βH*-eremophil-11-en-2-one, *in* E-70030
1,4-Dimethyl-4-[4-hydroxy-2-(hydroxymethyl)-1-methyl-2-cyclopentenyl]-2-cyclohexen-1-ol, D-70417
Feruone, F-70007
FS-2, F-**60**078
12-Hydroxy-4-cadinen-15-oic acid, *in* H-**80**135
Isolancifolide, *in* L-**80**017
Lancifolide, L-**80**017
Lapidol, L-**60**014
4,10(14)-Muuroladiene-1,3,9-triol, M-**80**200
4,10(14)-Oplopadiene-3,8,9-triol, O-70042
9-Oxo-2-bisabolen-15-oic acid, *in* O-**80**066
Piperalol, *in* P-**80**152
Punctaporonin *A*, P-**60**197
Punctaporonin *D*, *in* P-**60**197
Urodiolenone, U-70006
Zedoarondiol, Z-70004

$C_{15}H_{24}O_4$
Dendroserin, D-**80**028
Dihydroiresin, *in* I-**80**043
1*β*,10*α*-Dihydroxy-11*βH*-eremophilan-12,8*β*-olide, *in* D-**80**314
4-Epipulchellin, *in* P-**80**189
Secofloribundione, S-70022
Trichotriol, T-**60**258
5,8,9-Trihydroxy-10(14)-oplopen-3-one, T-70272
Vestic acid, V-70008

$C_{15}H_{24}O_5$
Odontin, O-**80**027
Rudbeckin *A*, R-70013
1,3,4-Trihydroxy-12,6-eudesmanolide, T-**80**296

$C_{15}H_{24}O_{11}$
Avicennioside, A-**60**326

$C_{15}H_{25}BrO$
Brasudol, B-**60**197
Isobrasudol, *in* B-**60**197

$C_{15}H_{25}Br_2ClO$
Caespitane, *in* C-**80**005

$C_{15}H_{25}Br_2ClO_2$
Caespitol, C-**80**005

$C_{15}H_{25}Br_2ClO_3$
6-Hydroxycaespitol, *in* C-**80**005

$C_{15}H_{25}NS$
6-Isothiocyano-4(15)-eudesmene, *in* F-**60**069

$C_{15}H_{26}$
Cyclopentadecyne, C-**60**220

$C_{15}H_{26}O$
4-Ambiguen-1-ol, A-**80**056
β-Caryophyllene alcohol, C-**80**028
Cerapicol, C-70032
Ceratopicanol, C-70033
9(11)-Drimen-8-ol, D-**60**520
7-Epi-*α*-eudesmol, *in* E-**80**086
5,8-Epoxydaucane, E-**80**021
1(10)-Eremophilen-7-ol, E-70031
11-Eudesmen-5-ol, E-**60**069
3-Eudesmen-11-ol, E-**80**086
3-Eudesmen-6-ol, E-70065
4(15)-Eudesmen-11-ol, E-70066
6-Eudesmen-4-ol, E-**80**087
7-Isopropyl-2,10-dimethylspiro[4.5]dec-1-en-6-ol, I-**80**079
Naviculol, N-**80**017
Panasinsanol *A*, P-70005
Panasinsanol *B*, *in* P-70005
2,4-Pentadecadienal, P-70024
Valerenenol, V-**80**001

$C_{15}H_{26}O_2$
Alloaromadendrane-4,10-diol, A-**60**074
1,10-Bisaboladiene-3,6-diol, B-**80**132
3,10-Bisaboladiene-7,14-diol, B-**80**133
8-Daucene-5,7-diol, D-**80**004
Debneyol, D-70010
Decahydro-5,5,8*a*-trimethyl-2-naphthalenol; Ac, *in* D-70014
7-Epidebneyol, *in* D-70010
1,4-Epoxy-6-eudesmanol, E-**60**019
1,4-Epoxy-6-eudesmanol, E-**80**026
11-Eudesmene-1,5-diol, E-**60**067
3-Eudesmene-1,6-diol, E-**80**083
4(15)-Eudesmene-1,11-diol, E-70064
4(15)-Eudesmene-1,6-diol, E-**80**084
4(15)-Eudesmene-2,11-diol, E-**80**085
Fauronol, F-70002
1(10),4-Germacradiene-6,8-diol, G-**60**014
1(10),5-Germacradiene-3,4-diol, G-**80**010
5,10(14)-Germacradiene-1,4-diol, G-70009
6-Hydroxycyclonerolidol, H-**60**111
Jaeschkeanadiol, J-70001
Oplopanone, O-70043
Senecrassidiol, S-**80**026
2,6,10-Trimethyl-2,6,11-dodecatriene-1,10-diol, T-70283
3,7,11-Trimethyl-1,6,10-dodecatriene-3,9-diol, T-70284

$C_{15}H_{26}O_3$
Arthrosporol, A-**80**171
8-Carotene-4,6,10-triol, C-**60**021
Caucalol, C-**80**029
8(14)-Daucene-4,6,9-triol, D-**60**005
8-Daucene-3,6,14-triol, D-**60**004
1-*O*-(4,6,8-Dodecatrienyl)glycerol, D-**80**526
6-Drimene-8,9,11-triol, D-70547
7-Drimene-6,9,11-triol, D-70548
8-Epipertriol, *in* P-**80**152
6,15-Epoxy-1,4-eudesmanediol, E-**80**025
11-Eremophilene-2,8,9-triol, E-70030
4(15)-Eudesmene-1,5,6-triol, E-**60**068
Fexerol, F-**60**009
5-Hydroperoxy-4(15)-eudesmen-11-ol, H-70097
1-Hydroxydebneyol, *in* D-70010
8-Hydroxydebneyol, *in* D-70010
Shiromodiol, S-70039

$C_{15}H_{26}O_4$
8-Daucene-2,4,6,10-tetrol, D-70009
7,8-Epoxy-4,6,9-daucanetriol, E-**80**022
Fercoperol, F-**60**001
3-Hydroperoxy-4-eudesmene-1,6-diol, H-**80**101
4-Hydroperoxy-2-eudesmene-1,6-diol, H-**80**102
3-Hydroxymethyl-7,11-dimethyl-2,6,11-dodecatriene-1,5,10-triol, H-70174
2,6,10-Trimethyl-2,6,10-dodecatrien-1,5,8,12-tetrol, T-**60**363

$C_{15}H_{26}O_5$
9,12-Dioxododecanoic acid; Me ester, ethylene acetal, *in* D-**60**471
Lapidolinol, L-**60**015
1,4,6,9-Tetrahydroxydihydro-*β*-agarofuran, T-**80**090

$C_{15}H_{26}O_6$
1,4,6,9,15-Pentahydroxydihydro-*β*-agarofuran, P-**80**032

$C_{15}H_{26}O_7$
ent-5*α*,11-Epoxy-1*β*,4*α*,6*α*,8*α*,9*β*,14-eudesmanehexol, E-**60**018

$C_{15}H_{26}O_8$
1,2,4,6,8,9,14-Heptahydroxydihydro-*β*-agarofuran, H-**80**027

$C_{15}H_{27}N$
7-Amino-2,10-bisaboladiene, A-70096
7-Amino-3,10-bisaboladiene, A-**60**096
Tri-*tert*-butylazete, T-**60**249

$C_{15}H_{27}NO$
7-Amino-2,11-biaboladien-10*R*-ol, *in* A-70096
7-Amino-2,9-bisaboladien-11-ol, *in* A-70096
7-Amino-2,11-bisaboladien-10*S*-ol, *in* A-70096

$C_{15}H_{27}N_3$
3,5,12-Triazatetracyclo[5.3.1.1^{2,6}.0^{4,9}]dodecane; 3,5,12-Tri-Et, *in* T-70206
Tri-*tert*-butyl-1,2,3-triazine, T-**60**250

$C_{15}H_{28}O$
▷ Cyclopentadecanone, C-**60**219
10-Pentadecenal, P-**80**024

$C_{15}H_{28}O_2$
10-Bisabolene-7,15-diol, B-**80**135
2-Bisabolene-1,12-diol, B-70126
5-Decen-1-ol; 3-Methylbutanoyl, *in* D-70019
9-Dodecen-1-ol; Propanoyl, *in* D-70537

1,11-Eudesmanediol, E-70062
5,11-Eudesmanediol, E-70063
5,11-Eudesmanediol, E-80082

$C_{15}H_{28}O_3$
11-Bisabolene-7,10,15-triol, B-80136
12-Hydroxy-13-tetradecenoic acid; Me ester, in H-60229
2,6,10-Trimethyl-6,11-dodecadiene-2,3,10-triol, T-60362

$C_{15}H_{29}NO$
Cyclopentadecanone; Oxime, in C-60219

$C_{15}H_{30}$
3-tert-Butyl-2,2,4,5,5-tetramethyl-3-hexene, B-60350

$C_{15}H_{30}N_2O_3$
4,7,13-Trioxa-1,10-diazabicyclo[8.5.5]icosane, T-60375

$C_{15}H_{30}O_2$
Lardolure, in T-60372
2,4,6,8-Tetramethylundecanoic acid, T-60157

$C_{15}H_{30}O_4$
3,9-Dihydroxytetradecanoic acid; Me ester, in D-80386
12-Oxododecanoic acid; Me ester, Di-Me acetal, in O-60066

$C_{15}H_{30}O_6$
1,3,5-Tris(dimethoxymethyl)cyclohexane, in C-70226

$C_{15}H_{32}O$
3,7,11-Trimethyl-1-dodecanol, T-70282

$C_{15}H_{33}NO_6$
2-(2-Methoxyethoxy)-N,N-bis[2-(2-methoxyethoxy)ethyl]ethanamine, in T-70323

$C_{15}H_{33}N_3$
Hexahydro-1,3,5-triazine; 1,3,5-Tri-tert-butyl±, in H-60060

$C_{15}H_{35}N_5$
1,5,9,13,17-Pentaazacycloeicosane, P-60018

$C_{15}H_{36}O_3$
10,15-Dihydroxy-4-oplopanone, D-80348

$C_{16}H_4N_4S_2$
2,2'-(4,8-Dihydrobenzo[1,2-b:5,4-b']dithiophene-4,8-diylidene)bispropanedinitrile, D-60196

$C_{16}H_8Br_2$
1,8-Dibromopyrene, D-70099
4,9-Dibromopyrene, D-80098

$C_{16}H_8Cl_2O_2$
1,8-Anthracenedicarboxylic acid; Dichloride, in A-60281

$C_{16}H_8Cl_4$
1,4,5,8-Tetrachloro-9,10-anthraquinodimethane, T-60026

$C_{16}H_8F_8$
4,5,7,8,12,13,15,16-Octafluoro[2.2]paracyclophane, O-60010

$C_{16}H_8N_2$
1,8-Dicyanoanthracene, in A-60281

$C_{16}H_8N_2O_4S$
2,2'-Thiobis-1H-isoindole-1,3(2H)-dione, T-70176

$C_{16}H_8N_2O_4S_2$
2,2'-Dithiobis-1H-isoindole-1,3(2H)-dione, D-70522
Dithiobisphthalimide, D-60508

$C_{16}H_8O_2$
Cyclohept[fg]acenaphthylene-5,6-dione, C-60194
Cyclohept[fg]acenaphthylene-5,8-dione, C-60195
Cycloocta[def]biphenylene-1,4-dione, C-80186

$C_{16}H_8O_3$
Benzo[b]naphtho[2,1-d]furan-5,6-dione, B-60033

Benzo[b]naphtho[2,3-d]furan-6,11-dione, B-60034
3,4-Phenanthrenedicarboxylic acid; Anhydride, in P-60068

$C_{16}H_8S_2$
Fluorantheno[3,4-cd]-1,2-dithiole, F-60017

$C_{16}H_8Se_2$
Fluorantheno[3,4-cd]-1,2-diselenole, F-60015

$C_{16}H_8Te_2$
Fluorantheno[3,4-cd]-1,2-ditellurole, F-60016

$C_{16}H_9ClO_3S$
1-Pyrenesulfonic acid; Chloride, in P-60210

$C_{16}H_9NO_2$
▷ 1-Nitropyrene, N-60040
2-Nitropyrene, N-60041
4-Nitropyrene, N-60042

$C_{16}H_9NO_2S$
3-(2-Benzothiazolyl)-2H-1-benzopyran, B-70056

$C_{16}H_9NO_3$
▷ 2-Nitro-1-pyrenol, N-80058

$C_{16}H_{10}$
Acephenanthrylene, A-70018
Benzo[a]biphenylene, B-60016
Benzo[b]biphenylene, B-60017
Dicyclopenta[ef,kl]heptalene, D-60167

$C_{16}H_{10}ClN$
1-Chlorobenzo[a]carbazole, C-80039

$C_{16}H_{10}F_3NO_2$
7-Amino-4-(trifluoromethyl)-2H-1-benzopyran-2-one; N-Ph, in A-60265

$C_{16}H_{10}F_6I_2O_5$
µ-Oxodiphenylbis(trifluoroacetato-O)diiodine, O-60065

$C_{16}H_{10}N_2$
Quino[7,8-h]quinoline, Q-70007

$C_{16}H_{10}N_2O_2$
Quino[7,8-h]quinoline-4,9(1H,12H)dione, Q-60007

$C_{16}H_{10}N_2S_2$
Thiazolo[4,5-b]quinoline-2(3H)-thione; 3-Ph, in T-60201

$C_{16}H_{10}N_6$
2,2'-Azodiquinoxaline, A-60353

$C_{16}H_{10}N_8$
21H,23H-Porphyrazine, P-70109

$C_{16}H_{10}O$
Anthra[1,2-c]furan, A-70217
Cyclobuta[b]anthracen-1(2H)one, C-80174
Phenanthro[9,10-b]furan, P-60073
Phenanthro[1,2-c]furan, P-70050
Phenanthro[3,4-c]furan, P-70051
Phenanthro[9,10-c]furan, P-60074

$C_{16}H_{10}O_2$
Benzo[b]naphtho[2,3-e][1,4]dioxan, B-80026
4,5:9,10-Diepoxy-4,5,9,10-tetrahydropyrene, D-70152
4b,9a-Dihydroindeno[1,2-a]indene-9,10-dione, D-70219
4b,9b-Dihydroindeno[2,1-a]indene-5,10-dione, D-70220
3-Phenyl-1,2-naphthoquinone, P-60120
4-Phenyl-1,2-naphthoquinone, P-60121

$C_{16}H_{10}O_3$
5-Formyl-4-phenanthrenecarboxylic acid, F-60074

$C_{16}H_{10}O_3S$
5-Hydroxy-2-(phenylthio)-1,4-naphthoquinone, H-70213
5-Hydroxy-3-(phenylthio)-1,4-naphthoquinone, H-70214
4-Oxo-2-phenyl-4H-1-benzothiopyran-3-carboxylic acid, O-60087
1-Pyrenesulfonic acid, P-60210
2-Pyrenesulfonic acid, P-60211
4-Pyrenesulfonic acid, P-60212

$C_{16}H_{10}O_3Se_2$
2,3-Bis(phenylseleno)-2-butenedioic acid; Anhydride, in B-80159

$C_{16}H_{10}O_4$
1,8-Anthracenedicarboxylic acid, A-60281
4-Oxo-2-phenyl-4H-1-benzopyran-3-carboxylic acid, O-60085
4-Oxo-3-phenyl-4H-1-benzopyran-2-carboxylic acid, O-60086
1,8-Phenanthrenedicarboxylic acid, P-80078
3,4-Phenanthrenedicarboxylic acid, P-60068

$C_{16}H_{10}O_5$
Corylinal, C-80153
Flaccidinin, F-80009
5-Hydroxy-6,7-methylenedioxyflavone, in T-60322
7-Hydroxy-3',4'-methylenedioxyisoflavone, H-80203
Mutisifurocoumarin, M-70161

$C_{16}H_{10}O_6$
Bowdichione, B-70175
3,7-Dihydroxy-9-methoxycoumestan, in T-80294
3,8-Dihydroxy-9-methoxycoumestan, in T-80295
3,9-Dihydroxy-2-methoxycoumestone, in T-70249
9,10-Dihydroxy-5-methoxy-2H-pyrano[2,3,4-kl]xanthen-9-one, in T-70273
4',5-Dihydroxy-6,7-methylenedioxyisoflavone, D-80341
2',5-Dihydroxy-6,7-methylenedioxyisoflavone, in T-60107
2',7-Dihydroxy-3',4'-methylenedioxyisoflavone, in T-80105
5,7-Dihydroxy-3',4'-methylenedioxyisoflavone, in T-80109
7-Hydroxy-3-(2-hydroxy-4,5-methylenedioxyphenyl)coumarin, in H-80261

$C_{16}H_{10}O_7$
5-Hydroxybowdichione, in B-70175
5,11,12-trihydroxy-7-methoxycoumestan, in T-80087
3,6,8-Trihydroxy-1-methylanthraquinone-2-carboxylic acid, T-60332
3,6,8-Trihydroxy-1-methylanthraquinone-2-carboxylic acid, T-80323
Variolaric acid, V-80004

$C_{16}H_{10}O_8$
Nasutin C, in E-70007

$C_{16}H_{11}ClO$
1-Phenanthreneacetic acid; Chloride, in P-70046

$C_{16}H_{11}F_3O_5S$
2,5-Diphenyl-1,3-dioxol-1-ium; Trifluoromethanesulfonate, in D-70478

$C_{16}H_{11}N$
1-(Cyanomethyl)phenanthrene, in P-70046
9-Cyanomethylphenanthrene, in P-70047
2H-Dibenz[e,g]isoindole, D-70067
Indolo[1,7-ab][1]benzazepine, I-60034
5H-Indolo[1,7-ab][1]benzazepine, I-80014

$C_{16}H_{11}NO$
1-Benzoylisoquinoline, B-60059
3-Benzoylisoquinoline, B-60060
4-Benzoylisoquinoline, B-60061

$C_{16}H_{11}NO_2$
2-Phenyl-4-(phenylmethylene)-5(4H)-oxazolone, P-60128

$C_{16}H_{11}NO_3$
1-Amino-8-hydroxyanthraquinone; N-Ac, in A-60177

$C_{16}H_{11}NO_4$
1-Amino-2-hydroxyanthraquinone; N-Ac, in A-60174
1-Amino-5-hydroxyanthraquinone; N-Ac, in A-60176

C₁₆H₁₁N₃OS
12*H*-Quinoxalino[2,3-*b*][1,4]benzothiazine; *N*-Ac, *in* Q-70008

C₁₆H₁₁N₃O₂
12*H*-Quinoxalino[2,3-*b*][1,4]benzoxazine; *N*-Ac, *in* Q-70009

C₁₆H₁₂
Cyclohepta[*ef*]heptalene, C-60197
Cycloocta[*a*]naphthalene, C-70234
Cycloocta[*b*]naphthalene, C-70235
2,6-Dihydroaceanthrylene, D-80169
1,10-Dihydrodicyclopenta[*a,h*]naphthalene, D-60220
3,8-Dihydrodicyclopenta[*a,h*]naphthalene, D-60221
9,10-Dihydro-9,10-dimethyleneanthracene, D-70194
5,10-Dihydroindeno[2,1-*a*]indene, D-70218
5-Methylene-5*H*-dibenzo[*a,d*]cycloheptene, M-80084
[2.2]Paracyclophadiene, P-70007
4*b*,8*b*,8*c*,8*d*-Tetrahydrodibenzo[*a,f*]cyclopropa[*cd*]pentalene, T-80057

C₁₆H₁₂Cl₂N₄O
4-(2,5-Dichlorophenylhydrazono)-5-methyl-2-phenyl-3*H*-pyrazol-3-one, D-70138

C₁₆H₁₂Cl₂O₂
Bibenzyl-2,2′-dicarboxylic acid; Dichloride, *in* B-80071

C₁₆H₁₂Cl₂O₅
2,4-Dichloro-1-hydroxy-5,8-dimethoxyxanthone, *in* T-70181

C₁₆H₁₂F₂O₄
4,4′-Difluoro-[1,1′-biphenyl]-3,3′-dicarboxylic acid; Di-Me ester, *in* D-60178
4,5-Difluoro-[1,1′-biphenyl]-2,3-dicarboxylic acid; Di-Me ester, *in* D-60179
6,6′-Difluoro-[1,1′-biphenyl]-2,2′-dicarboxylic acid; Di-Me ester, *in* D-60180

C₁₆H₁₂F₄
4,5,7,8-Tetrafluoro[2.2]paracyclophane, T-60050

C₁₆H₁₂N₂
2,2′-Bis(cyanomethyl)biphenyl, *in* B-60115
4,4′-Bis(cyanomethyl)biphenyl, *in* B-60116
2,4-Diphenylpyrimidine, D-70494
4,5-Diphenylpyrimidine, D-70495
4,6-Diphenylpyrimidine, D-70496

C₁₆H₁₂N₂O
1-Benzoylisoquinoline; Oxime, *in* B-60059
4,5-Diphenylpyrimidine; 1-Oxide, *in* D-70495
4,6-Diphenylpyrimidine; *N*-Oxide, *in* D-70496

C₁₆H₁₂N₂O₂
2,2′-Dicyanobibenzyl, *in* B-80071
4*b*,9*b*-Dihydroindeno[2,1-*a*]indene-5,10-dione; Dioxime, *in* D-70220

C₁₆H₁₂N₂O₇S₂
6-Hydroxy-5-[(4-sulfophenyl)azo]-2-naphthalenesulfonic acid, H-80255

C₁₆H₁₂N₄
Dibenzo[*b,g*][1,8]naphthyridine-11,12-diamine, D-80071

C₁₆H₁₂O
1-Acetylanthracene, A-70030
2-Acetylanthracene, A-70031
9-Anthraceneacetaldehyde, A-60280
Octaleno[3,4-*c*]furan, O-70019
1-Phenyl-2-naphthol, P-60115
3-Phenyl-2-naphthol, P-60116
4-Phenyl-2-naphthol, P-60117
5-Phenyl-2-naphthol, P-60118
8-Phenyl-2-naphthol, P-60119

C₁₆H₁₂O₂
5-(2,4,6-Cycloheptatrien-1-ylideneethylidene)-3,6-cycloheptadiene-1,2-dione, C-70222
2,3-Dihydro-3-(phenylmethylene)-4*H*-1-benzopyran-4-one, D-60263
1-Phenanthreneacetic acid, P-70046

9-Phenanthreneacetic acid, P-70047
Tricyclo[8.4.1.1³,⁸]hexadeca-3,5,7,10,12,14-hexaene-2,9-dione, T-60264

C₁₆H₁₂O₂S
3-Methylthio-2-phenyl-4*H*-1-benzopyran-4-one, *in* M-60025

C₁₆H₁₂O₃
7-Hydroxy-2-methylisoflavone, H-80205
7-Hydroxy-3-methylisoflavone, H-80206
6-Methoxyflavone, *in* H-70130
7-Methoxyflavone, *in* H-70131
8-Methoxyflavone, *in* H-70132
5-Methoxy-4-phenanthrenecarboxylic acid, *in* H-80232
Nordracorhodin, N-70074

C₁₆H₁₂O₄
1,5-Biphenylenedicarboxylic acid; Di-Me ester, *in* B-80111
1,8-Biphenylenedicarboxylic acid; Di-Me ester, *in* B-80112
2,3-Biphenylenedicarboxylic acid; Di-Me ester, *in* B-80113
2,6-Biphenylenedicarboxylic acid; Di-Me ester, *in* B-80114
2,7-Biphenylenedicarboxylic acid; Di-Me ester, *in* B-80115
1,8-Biphenylenediol; Di-Ac, *in* B-80118
2,6-Biphenylenediol; Di-Ac, *in* B-80120
2,7-Biphenylenediol; Di-Ac, *in* B-80121
7-Hydroxy-3-(4-hydroxybenzylidene)-4-chromanone, H-60152
▷ 7-Hydroxy-4′-methoxyisoflavone, H-80190
Isodalbergin, I-60097
Vesparione, V-70007

C₁₆H₁₂O₄Se₂
2,3-Bis(phenylseleno)-2-butenedioic acid, B-80159

C₁₆H₁₂O₅
3-(3,4-Dihydroxybenzylidene)-7-hydroxy-4-chromanone, *in* D-70288
▷ 5,7-Dihydroxy-4′-methoxyflavone, D-70319
4′,7-Dihydroxy-3′-methoxyflavone, D-80301
4′,5-Dihydroxy-7-methoxyisoflavone, D-80335
4′,7-Dihydroxy-5-methoxyisoflavone, D-80336
5,7-Dihydroxy-4′-methoxyisoflavone, D-80337
2′,7-Dihydroxy-4′-methoxyisoflavone, *in* T-80306
3′,7-Dihydroxy-4′-methoxyisoflavone, *in* T-80308
4′,6-Dihydroxy-7-methoxyisoflavone, *in* T-80310
4′,7-Dihydroxy-2′-methoxyisoflavone, *in* T-80306
4′,7-Dihydroxy-3′-methoxyisoflavone, *in* T-80308
4′,7-Dihydroxy-6-methoxyisoflavone, *in* T-80310
7,8-Dihydroxy-4′-methoxyisoflavone, *in* T-70265
7,8-Dihydroxy-4′-methoxyisoflavone, *in* T-80311
1,8-Dihydroxy-6-methoxy-2-methylanthraquinone, *in* T-70269
7-Hydroxy-3-(2-hydroxy-4-methoxyphenyl)coumarin, *in* D-80364
3-Hydroxy-2-hydroxymethyl-1-methoxyanthraquinone, *in* D-70306
2-(2-Hydroxy-4-methoxyphenyl)-5,6-methylenedioxybenzofuran, *in* D-80355
Maackiain, M-70001
Obtusifolin/, *in* T-80322
Oxoflaccidin, *in* F-80009
Pabulenone, P-60170
2′,5,7-Trihydroxyflavone, T-80299
4′,5,7-Trihydroxy-6-methylflavone, T-80326

C₁₆H₁₂O₆
6*a*,12*a*-Dihydro-2,3,10-trihydroxy[2]benzopyrano[4,3-*b*][1]benzopyran-7(5*H*)-one, 9CI, *in* C-70200
3-(2,4-Dihydroxybenzoyl)-7,8-dihydroxy-1*H*-2-*benzopyran*, P-70017
2-(2,4-Dihydroxy-3-methoxyphenyl)-5,6-methylenedioxybenzofuran, *in* D-80389

3,5-Dihydroxy-6,7-methylenedioxyflavanone, *in* T-60105
2′,7-Dihydroxy-4′,5′-methylenedioxyisoflavanone, *in* T-80101
3,4-Dihydroxy-8,9-methylenedioxypterocarpan, *in* T-80119
4-(3,4-Dihydroxyphenyl)-7-hydroxy-5-methoxycoumarin, *in* D-80356
Peltochalcone, P-70017
Rengasin, *in* A-80188
3,5,6,7-Tetrahydroxy-8-methylflavone, T-70117
3,5,7,8-Tetrahydroxy-6-methylflavone, T-70118
2′,5,7-Trihydroxy-6′-methoxyflavone, *in* T-50153
▷ 4′,5,7-Trihydroxy-6-methoxyisoflavone, T-80319
2′,3′,7-Trihydroxy-4′-methoxyisoflavone, *in* T-80105
2′,4′,5-Trihydroxy-7-methoxyisoflavone, *in* T-80106
2′,4′,7-Trihydroxy-5-methoxyisoflavone, *in* T-80106
2′,5,7-Trihydroxy-4′-methoxyisoflavone, *in* T-80106
3′,4′,5-Trihydroxy-7-methoxyisoflavone, *in* T-80109
3′,5,7-Trihydroxy-4′-methoxyisoflavone, *in* T-80109
3′,5′,7-Trihydroxy-4′-methoxyisoflavone, *in* T-80110
3′,7,8-Trihydroxy-4′-methoxyisoflavone, *in* T-70112
4′,5,7-Trihydroxy-3′-methoxyisoflavone, *in* T-80109
4′,5,7-Trihydroxy-8-methoxyisoflavone, *in* T-80113
4′,7,8-Trihydroxy-6-methoxyisoflavone, *in* T-70113
1,3,5-Trihydroxy-4-methoxy-2-methylanthraquinone, *in* T-60112
1,3,8-Trihydroxy-5-methoxy-2-methylanthraquinone, *in* T-60113

C₁₆H₁₂O₇
Crombeone, C-70200
2′,3,5,8-Tetrahydroxy-7-methoxyflavone, *in* P-70029
3′,4′,7-Tetrahydroxy-8-methoxyisoflavone, *in* P-80045
3,6*a*,7-Trihydroxy-8,9-methylenedioxypterocarpan, *in* P-80056

C₁₆H₁₃ClO₅
2-Chloro-1-hydroxy-5,8-dimethoxy-6-methylxanthone, *in* T-70181
5-Chloro-8-hydroxy-1,4-dimethoxy-3-methylxanthone, *in* T-70181

C₁₆H₁₃N
5,10-Dihydroindeno[1,2-*b*]indole; *N*-Me, *in* D-80216

C₁₆H₁₃NO
2-Methoxy-4-phenylquinoline, *in* P-60131
4-Methoxy-2-phenylquinoline, *in* P-60132
1-Methyl-2-phenyl-4(1*H*)-*quinolinone*, *in* P-60132
9-Phenanthreneacetic acid; Amide, *in* P-70047

C₁₆H₁₃NOS
5-Methyl-2-phenyl-1,5-benzothiazepin-4(5*H*)-one, *in* P-60081

C₁₆H₁₃NO₂
10-Amino-9-anthracenecarboxylic acid; Me ester, *in* A-70088
Dibenz[*b,g*]azocine-5,7(6*H*,12*H*)-dione; *N*-Me, *in* D-60068
1*H*-Indole-3-carboxylic acid; *N*-Benzyl, *in* I-80012

C₁₆H₁₃NO₅
1-Amino-4,5-dihydroxy-7-methoxy-2-methylanthraquinone, A-60145

C₁₆H₁₃NS
10*H*-[1]Benzothieno[3,2-*b*]indole; *N*-Et, *in* B-70059

$C_{16}H_{14}$
9-Benzylidene-1,3,5,7-cyclononatetraene, B-60069
1,6:7,12-Bismethano[14]annulene, B-60165
9,10-Dihydro-9,10-ethanoanthracene, D-70207
10,11-Dihydro-5-methylene-5H-dibenzo[a,d]cycloheptene, D-80219
4,5-Dimethylphenanthrene, D-60439
1,1-Diphenyl-1,3-butadiene, D-70476
11b-Methyl-11bH-Cyclooct[cd]azulene, M-60060
7,8,9,10-Tetrahydrofluoranthene, T-70061
4,5,9,10-Tetrahydropyrene, T-60083

$C_{16}H_{14}N_2$
1,2-Dihydro-4,6-diphenylpyrimidine, D-60235

$C_{16}H_{14}N_2O$
1-(1-Isoquinolinyl)-1-(2-pyridinyl)ethanol, I-60134

$C_{16}H_{14}N_2O_2$
2,3-Diaminobiphenylene; N,N'-Di-Ac, in D-80054
2,6-Diaminobiphenylene; N,N'-Di-Ac, in D-80055
5,6-Dihydrobenzo[c]cinnoline; Di-Ac, in D-80172
2-Phenyl-3H-indol-3-one; N-Oxide, oxime, Et ether, in P-60114

$C_{16}H_{14}N_2O_3$
2,3-Dihydro-2-hydroxy-1H-imidazole-4,5-dione; 1,3-Di-Ph, Me ether, in D-60243

$C_{16}H_{14}N_2S_2$
2,5-Dihydro-4,6-pyrimidinedithiol; Di-Ph thioether, in D-70257

$C_{16}H_{14}N_4$
3,3'-Dimethyl-2,2'-biindazole, D-80421
5,5'-Dimethyl-2,2'-biindazole, D-80422
7,7'-Dimethyl-2,2'-biindazole, D-80423

$C_{16}H_{14}O$
2-(9-Anthracenyl)ethanol, A-60282
3,4-Dihydro-2-phenyl-1(2H)-naphthalenone, D-60265
3,4-Dihydro-3-phenyl-1(2H)-naphthalenone, D-60266
3,4-Dihydro-4-phenyl-1(2H)-naphthalenone, D-60267
10,10-Dimethyl-9(10H)-anthracenone, D-60403
2,3-Diphenyl-2-butenal, D-60475
2,4-Diphenyl-3-butyn-1-ol, D-80508
3,3-Diphenylcyclobutanone, D-60476

$C_{16}H_{14}OS$
▷ 2,3-Dihydro-5,6-diphenyl-1,4-oxathiin, D-70203

$C_{16}H_{14}OS_2$
1-(Phenylsulfinyl)-2-(phenylthio)cyclobutene, in B-70156

$C_{16}H_{14}O_2$
Dihydro-2,2-diphenyl-2(3H)-furanone, D-80197
Dihydro-3,3-diphenyl-2(3H)-furanone, D-80198
Dihydro-3,4-diphenyl-2(3H)-furanone, D-80199
Dihydro-3,5-diphenyl-2(3H)-furanone, D-80200
Dihydro-4,4-diphenyl-2(3H)-furanone, D-80201
Dihydro-4,5-diphenyl-2(3H)-furanone, D-80202
Dihydro-5,5-diphenyl-2(3H)-furanone, D-80203
4,5-Dimethoxyphenanthrene, in P-70048
2,2-Diphenyl-4-methylene-1,3-dioxolane, D-70486
Dracaenone, D-70546
3,6-Phenanthrenedimethanol, P-60069
4,5-Phenanthrenedimethanol, P-60070

$C_{16}H_{14}O_2S_2$
1,2-Bis(phenylsulfinyl)cyclobutene, in B-70156

$C_{16}H_{14}O_3$
2-Benzoylbenzoic acid; Et ester, in B-80054
3,7-Dimethoxy-2-phenanthrenol, in T-60339

$C_{16}H_{14}O_3S_2$
1-(Phenylsulfinyl)-2-(phenylsulfonyl)cyclobutene, in B-70156

$C_{16}H_{14}O_4$
Amoenumin, A-80150
Bibenzyl-2,2'-dicarboxylic acid, B-80071
[1,1'-Biphenyl]-2,2'-diacetic acid, B-60115
[1,1'-Biphenyl]-4,4'-diacetic acid, B-60116
2',4-Dihydroxy-4'-methoxychalcone, in D-80367
4',7-Dihydroxy-2'-methoxyisoflavene, in T-80305
1-(2,4-Dihydroxy-6-methoxyphenyl)-3-phenyl-2-propen-1-one, in P-80139
1-(2,6-Dihydroxy-4-methoxyphenyl)-3-phenyl-2-propen-1-one, in P-80139
1-(2,6-Dihydroxy-4-methoxyphenyl)-3-phenyl-2-propen-1-one, in P-80139
1,5-Dimethoxy-2,7-phenanthrenediol, in T-80114
3,4-Dimethoxy-2,7-phenanthrenediol, in T-80115
5,7-Dimethoxy-2,3-phenanthrenediol, in T-60120
5,7-Dimethoxy-2,6-phenanthrenediol, in T-80115
5,5'-(1,2-Ethanediyl)bis-1,3-benzodioxole, in T-60101
Ferulidene, in X-60002
Flaccidin, F-70010
Flaccidin, in F-80009
7-Hydroxy-4'-methoxyisoflavanone, in D-80322
2-(4-Hydroxy-2-methoxyphenyl)-6-methoxybenzofuran, in D-80363
9-Hydroxy-3-methoxypterocarpan, in D-60373

$C_{16}H_{14}O_4S$
2,2'-Thiobisbenzoic acid; Di-Me ester, in T-70174
4,4'-Thiobisbenzoic acid; Di-Me ester, in T-70175

$C_{16}H_{14}O_4S_2$
1,2-Bis(phenylsulfonyl)cyclobutene, in B-70156
5-(3,4-Diacetoxy-1-butynyl)-2,2'-bithienyl, in B-80166

$C_{16}H_{14}O_5$
Demethylfrutescin 1'-ylacetate, in H-80231
Dibenzyl dicarbonate, in D-70111
3-(3,4-Dihydroxybenzyl)-7-hydroxy-4-chromanone, D-70288
5,8-Dihydroxy-7-methoxyflavanone, in T-60321
2',7-Dihydroxy-4'-methoxyisoflavanone, in T-80303
3',7-Dihydroxy-4'-methoxyisoflavanone, in T-70263
5,7-Dihydroxy-4'-methoxyisoflavanone, in T-80304
3,6a-Dihydroxy-9-methoxypterocarpan, in T-80337
6a,9-Dihydroxy-3-methoxypterocarpan, in T-80337
3,10-Dihydroxy-9-methoxypterocarpan, in T-80338
3,4-Dihydroxy-9-methoxypterocarpan, in T-80336
3,9-Dihydroxy-10-methoxypterocarpan, in T-80338
2',7-Dihydroxy-4',5'-methylenedioxyisoflavan, in T-80097
2-(2,4-Dihydroxyphenyl)-5,6-dimethoxybenzofuran, in D-80355
1-(2,4-Dihydroxyphenyl)-3-(4-hydroxy-3-methoxyphenyl)-2-propen-1-one, in D-80362
2-(3-Hydroxy-2,4-dimethoxyphenyl)-6-benzofuranol, in H-80260
6-Hydroxy-2-(4-hydroxy-2,3-dimethoxyphenyl)benzofuran, in H-80260
Isogosferol, in X-60002

3-(4-Methoxyphenyl)-1-(2,3,4-trihydroxyphenyl)-2-propen-1-one, in H-80245
Moracin F, in M-60146
Pabularinone, in P-60170
Pabulenol, P-60170
Sainfuran, S-60002
2',3',7-Trihydroxy-4'-methoxyisoflavene, in T-80103
2',4,7-Trihydroxy-3'-methoxyisoflavene, in T-80103
1,3,7-Trimethoxyxanthone, in T-80347

$C_{16}H_{14}O_6$
Anhydrofusarubin 9-methyl ether, in F-60108
3-(3,4-Dihydroxybenzyl)-3,7-dihydroxy-4-chromanone, in D-70288
5,8-Dihydroxy-2,3-dimethyl-1,4-naphthoquinone; Di-Ac, in D-60317
2,3-Dihydroxy-2,3-diphenylbutanedioic acid, D-80312
1-(2,4-Dihydroxy-3-methoxyphenyl)-3-(3,4-dihydroxyphenyl)-2-propen-1-one, in D-80372
Diphenyl tartrate, in T-70005
6,6'-(1,2-Ethanediyl)bis-1,3-benzodioxol-4-ol, in H-60062
3-Hydroxy-1,2,4-trimethoxyxanthone, in T-60126
7-Hydroxy-2,3,4-trimethoxyxanthone, in T-60128
8-Hydroxy-1,3,5-trimethoxyxanthone, in T-70123
1,4,5-Naphthalenetriol; Tri-Ac, in N-60005
2',5,8-Trihydroxy-7-methoxyflavanone, in T-70104
4',5,7-Trihydroxy-2'-methoxyisoflavanone, in T-80100
3,6a,7-Trihydroxy-9-methoxypterocarpan, in T-80120

$C_{16}H_{14}O_6S$
4,4'-Thiobisbenzoic acid; Di-Me ester, S-dioxide, in T-70175

$C_{16}H_{14}O_7$
5,10-Dihydroxy-1,7-dimethoxy-3-methyl-1H-naphtho[2,3-c]pyran-6,9-dione, in T-80320
3',5-Dihydroxy-4',6,7-trimethoxyflavanone, in P-70028
1,8-Dihydroxy-2,3,6-trimethoxyxanthone, in P-70036
3,6-Dihydroxy-1,2,3-trimethoxyxanthone, in P-70036
3,8-Dihydroxy-1,2,4-trimethoxyxanthone, in P-60051
6,8-Dihydroxy-1,2,5-trimethoxyxanthone, in P-60052
Isolecanoric acid, I-60103
Nectriafurone; 8-Me ether, in N-60016
3',4',5,7-Tetrahydroxy-6-methoxyflavanone, in P-70028
Ventilone E, in V-60004

$C_{16}H_{14}S_2$
1,2-Bis(phenylthio)cyclobutene, B-70156

$C_{16}H_{14}Se$
2,5-Dihydro-3,4-diphenylselenophene, D-70205
Distyryl selenide, D-60503

$C_{16}H_{15}Br$
1-Bromomethyl-2,3-diphenylcyclopropane, B-80214

$C_{16}H_{15}ClO_4S$
4-Phenyl-2H-1-benzothiopyran; S-Me, perchlorate, in P-80092

$C_{16}H_{15}ClO_6$
Lonapalene, in C-60118

$C_{16}H_{15}N$
7,12-Dihydro-5H-6,12-methanodibenz[c,f]azocine, D-60255
2,3-Diphenylbutanoic acid; Nitrile, in D-80506

$C_{16}H_{15}NO$
9-Amino-9,10-dihydroanthracene; N-Ac, in A-60136

2-Amino-9,10-dihydrophenanthrene; *N*-Ac, *in* A-**60141**

9-Amino-9,10-dihydrophenanthrene; *N*-Ac, *in* A-**60143**

10,11-Dihydro-5*H*-dibenz[*b,f*]azepine; 5-Ac, *in* D-**70185**

3,4-Dihydro-4-phenyl-1(2*H*)-naphthalenone; Oxime, *in* D-**60267**

3,6-Dimethylcarbazole; *N*-Ac, *in* D-**70387**

$C_{16}H_{15}NOS$
2-Phenyl-1,5-benzothiazepin-4(5*H*)-one; 2,3-Dihydro, *N*-Me, *in* P-**60081**

$C_{16}H_{15}NO_2$
2-Amino-1,3-diphenyl-1-propanone; *N*-Formyl, *in* A-**60149**

$C_{16}H_{15}NO_2S$
10*H*-Phenothiazine-1-carboxylic acid; *N*-Et, Me ester, *in* P-**70055**

10*H*-Phenothiazine-3-carboxylic acid; *N*-Et, Me ester, *in* P-**70057**

10*H*-Phenothiazine-4-carboxylic acid; *N*-Et, Me ester, *in* P-**70058**

$C_{16}H_{15}NO_4$
4-Hydroxy-3,6-dimethoxy-2-methyl-9*H*-carbazole-1-carboxaldehyde, *in* H-**70154**

2,2'-Iminodibenzoic acid; Di-Me ester, *in* I-**60014**

2,4'-Iminodibenzoic acid; Di-Me ester, *in* I-**60016**

4,4'-Iminodibenzoic acid; Di-Me ester, *in* I-**60018**

$C_{16}H_{15}NO_5$
1-Amino-2,7-naphthalenediol; *O,O,N*-Tri Ac, *in* A-**80113**

4-Amino-1,3-naphthalenediol; *O,O,N*-Tri-Ac, *in* A-**80118**

2-Amino-1,4-naphthalenediol; *O,O,N*-Tri-Ac, *in* A-**80114**

2-Amino-1,6-naphthalenediol; *O,O,N*-Tri-Ac, *in* A-**80115**

4-Amino-1,2-naphthalenediol; *O,O,N*-Tri-Ac, *in* A-**80117**

$C_{16}H_{15}N_3O_2$
1,8-Diaminocarbazole; 1,8-*N*-Di-Ac, *in* D-**80056**

$C_{16}H_{15}N_5$
2,4-Diamino-6-methyl-1,3,5-triazine; *N,N*-Di-Ph, *in* D-**60040**

2,4-Diamino-6-methyl-1,3,5-triazine; *N,N'*-Di-Ph, *in* D-**60040**

$C_{16}H_{16}$
9,10-Dihydro-9,9-dimethylanthracene, D-**60224**

4,5-Dimethylphenanthrene; 9,10-Dihydro, *in* D-**60439**

Heptacyclo[7.7.0.0²,⁶.0³,¹⁵.0⁴,¹².0⁵,¹⁰.0¹¹,¹⁶]hexadeca-7,13-diene, H-**70011**

1,6,8-Hexadecatriene-10,12,14-triyne, H-**80043**

1,2,3,6,7,8-Hexahydropyrene, H-**70069**

1,2,3,3*a*,4,5-Hexahydropyrene, H-**70068**

$C_{16}H_{16}N_2$
10,10-Dimethyl-9(10*H*)-anthracenone; Hydrazone, *in* D-**60403**

$C_{16}H_{16}N_2O_2$
2,2'-Diaminobiphenyl; 2,2'-*N*-Di-Ac, *in* D-**60033**

5,5'-Diamino-2,2'-biphenyldicarboxylic acid; Di-Me ester, *in* D-**80051**

Methylpropanedioic acid; Dianilide, *in* M-**80149**

$C_{16}H_{16}N_2O_3$
2,4,6-Trimethyl-3-nitroaniline; *N*-Benzoyl, *in* T-**70285**

$C_{16}H_{16}N_2O_4$
[2,2'-Bipyridine]-6,6'-dicarboxylic acid; Di-Et ester, *in* B-**60128**

2,2'-Diamino-4,4'-biphenyldicarboxylic acid; Di-Me ester, *in* D-**80046**

$C_{16}H_{16}N_4O_5S$
6-Nitro-1-[[(2,3,5,6-tetramethylphenyl)sulfonyl]oxy]-1*H*-benzotriazole, N-**80060**

$C_{16}H_{16}O$
1,3-Diphenyl-2-butanone, D-**80507**

6,7,8,9-Tetrahydro-6-phenyl-5*H*-benzocyclohepten-5-one, T-**60080**

$C_{16}H_{16}O_2$
2,2'-Dihydroxystilbene; Di-Me ether, *in* D-**80378**

3,5-Dihydroxystilbene; Di-Me ether, *in* D-**80384**

4,4'-Dihydroxystilbene; Di-Me ether, *in* D-**80385**

2,3-Diphenylbutanoic acid, D-**80506**

2,2-Diphenylpropanoic acid; Me ester, *in* D-**70493**

2,3-Naphthalenedicarboxylic acid; Di-Et ester, *in* N-**80001**

3,3',5,5'-Tetramethyl[bi-2,5-cyclohexadien-1-ylidene]-4,4'-dione, T-**80140**

$C_{16}H_{16}O_3$
3-(1,1-Dimethyl-2-propenyl)-4-hydroxy-6-phenyl-2*H*-pyran-2-one, D-**70434**

Echinofuran B, *in* E-**60002**

5-[2-(3-Methoxyphenyl)ethyl]-1,3-benzodioxole, *in* T-**60316**

9-Methylcyclolongipesin, *in* C-**60214**

$C_{16}H_{16}O_4$
8-Acetoxy-2-(2,4-hexadiynylidene)-1,6-dioxaspiro[4.5]dec-3-ene, *in* H-**70042**

Angolensin, A-**80155**

5,7-Azulenedicarboxylic acid; Di-Et ester, *in* A-**70295**

9,10-Dihydro-5,6-dihydroxy-2,4-dimethoxyphenanthrene, *in* D-**70247**

9,10-Dihydro-2,5-dimethoxy-1,7-phenanthrenediol, *in* T-**80114**

9,10-Dihydro-2,7-dimethoxy-1,5-phenanthrenediol, *in* T-**80114**

9,10-Dihydro-3,4-dimethoxy-2,7-phenanthrenediol, *in* T-**80115**

2',4'-Dihydroxy-7-methoxyisoflavan, *in* T-**80302**

4',7-Dihydroxy-2'-methoxyisoflavan, *in* T-**80302**

1-(2,6-Dihydroxy-4-methoxyphenyl)-3-phenyl-1-propanone, *in* T-**80332**

Gleinadiene, *in* G-**60023**

Homocyclolongipesin, H-**60088**

3-(4-Hydroxy-2-methoxyphenyl)-1-(4-hydroxyphenyl)-1-propanone, *in* D-**80366**

Perforatin A, *in* A-**60075**

6,7,14,15-Tetrahydrodibenzo[*b,h*][1,4,7,10]tetraoxacyclododecin, T-**80059**

2,4,6-Trimethoxybenzophenone, *in* T-**80285**

$C_{16}H_{16}O_5$
Citreofuran, C-**80136**

trans-Dehydrocurvularin, *in* C-**80166**

Deoxyaustrocortilutein, *in* A-**60321**

1-(2,6-Dihydroxy-4-methoxyphenyl)-3-(4-hydroxyphenyl)-1-propanone, *in* H-**80244**

3-(2,4-Dihydroxy-6-methoxyphenyl)-1-(4-hydroxyphenyl)-1-propanone, *in* H-**80243**

6-[2-(3,4-Dimethoxyphenyl)ethenyl]-4-methoxy-2*H*-pyran-2-one, *in* H-**70122**

3-(4-Hydroxyphenyl)-1-(2,4-dihydroxy-6-methoxyphenyl)-1-propanone, *in* H-**80246**

Pranferol, *in* P-**60170**

Shikonin, S-**70038**

$C_{16}H_{16}O_6$
Altersolanol B, *in* A-**60080**

Austrocortilutein, A-**60321**

Deoxyaustrocortirubin, *in* A-**60322**

3-(3,4-Dihydroxybenzyl)-3,4,7-trihydroxychroman, *in* D-**70288**

Epicatechin; O^3-Me, *in* P-**70026**

Fonsecin, F-**80073**

Meciadanol, *in* P-**70026**

1,4,5,7-Naphthalenetetrol; 4,5-Di-Me ether, di-Ac, *in* N-**80003**

3,3',5,7-Tetrahydroxy-4'-methoxyflavan, *in* P-**70026**

$C_{16}H_{16}O_7$
Altersolanol C, *in* A-**60080**

Austrocortirubin, A-**60322**

Fusarubin; O^9-Me, *in* F-**60108**

Fusarubin methyl acetal, *in* F-**60108**

1,2,3,4-Tetrahydro-1,2,4,5-tetrahydroxy-7-methoxy-2-methylanthraquinone, T-**80078**

$C_{16}H_{16}O_8$
Altersolanol A, A-**60080**

4-*O*-Caffeoylshikimic acid, C-**80006**

5-*O*-Caffeoylshikimic acid, C-**80007**

$C_{16}H_{16}S_2$
2,11-Dithia[3.3]paracyclophane, D-**60505**

5,7,12,14-Tetrahydrodibenzo[*c,h*][1,6]dithiecin, T-**80058**

$C_{16}H_{17}N$
Azetidine; *N*-Benzhydryl, *in* A-**70283**

1,2,3,4-Tetrahydroquinoline; *N*-Benzyl, *in* T-**70093**

$C_{16}H_{17}NO$
2,3-Diphenylbutanoic acid; Amide, *in* D-**80506**

1,3-Diphenyl-2-butanone; Oxime, *in* D-**80507**

$C_{16}H_{17}NO_3$
4-Hydroxy-3,6-dimethoxy-1,2-dimethylcarbazole, *in* T-**70253**

$C_{16}H_{17}O_4$
9,10-Dihydro-2,7-dihydroxy-3,5-dimethoxyphenanthrene, *in* T-**60120**

$C_{16}H_{18}N_2$
[3](2.2)[3](5.5)Pyridinophane, P-**60227**

[3](2.5)[3](5.2)Pyridinophane, P-**60228**

[3.3][2.6]Pyridinophane, P-**60226**

$C_{16}H_{18}N_2O$
N-Ethyl-*N'*-methyl-*N,N'*-diphenylurea, *in* D-**70499**

5-Phenyl-3-pyridinecarboxylic acid; Diethylamide, *in* P-**80129**

$C_{16}H_{18}O$
6,8,14-Hexadecatriene-10,12-diyn-1-al, H-**80042**

$C_{16}H_{18}O_2$
13-Acetoxy-9-tetradecene-2,4,6-triyne, *in* T-**80042**

2,3-Diphenyl-2,3-butanediol, D-**80505**

Methyl 8-isopropyl-5-methyl-2-naphthalenecarboxylate, *in* I-**60129**

4,6,12-Tetradecatrien-8,10-diyn-1-ol; Ac, *in* T-**80032**

$C_{16}H_{18}O_3$
Heritonin, H-**80034**

3-Hydroxy-3',4-dimethoxybibenzyl, *in* T-**60316**

8-Methoxy-1(6),2,4,7(11)-cadinatetraen-12,8-olide, *in* H-**70114**

4-Methoxy-3-(3-methyl-2-butenyl)-5-phenyl-2(5*H*)-furanone, M-**70047**

$C_{16}H_{18}O_4$
8-Angeloyloxylachnophyllum ester, *in* H-**80150**

1-(3,4-Dihydroxyphenyl)-2-(3,5-dimethoxyphenyl)ethane, *in* D-**80359**

3-(1,1-Dimethyl-2-propenyl)-7,8-dimethoxy-2*H*-1-benzopyran-2-one, D-**80470**

Gleinene, G-**60023**

1-(3-Hydroxy-5-methoxyphenyl)-2-(4-hydroxy-3-methoxyphenyl)ethane, *in* D-**80361**

4-[2-(3-Hydroxyphenyl)ethyl]-2,6-dimethoxyphenol, *in* T-**60102**

9-Methyllongipesin, M-**70091**

O-!Methylcedrelopsin, *in* C-**60026**

Rutacultin, R-**80016**

2,2',3,4-Tetramethoxybiphenyl, *in* B-**70125**

2,2',3,4'-Tetramethoxybiphenyl, *in* B-**80125**

2,3,3',4-Tetramethoxybiphenyl, *in* B-**80126**

2,3,4,4'-Tetramethoxybiphenyl, *in* B-**70123**

2,3,4,4'-Tetramethoxybiphenyl, *in* B-**80127**

2',3,4,5-Tetramethoxybiphenyl, *in* B-**80128**

$C_{16}H_{18}O_5$
Arnebin V, *in* D-**80343**

cis-Dehydrocurvularin, in C-80166
Murracarpin, M-80197
Resorcylide, R-70001
Skimminin, S-60035

$C_{16}H_{18}O_6$
2-(1,4-Dihydroxy-4-methylpentyl)-5,8-
dihydroxy-1,4-naphthoquinone, D-80343
Murraculatin, M-70158
12-Oxocurvularin, in C-80166
2,5,7-Trihydroxy-3-(5-hydroxyhexyl)-1,4-
naphthoquinone, T-70262

$C_{16}H_{18}O_8$
3,6-Dimethyl-1,2,4,5-benzenetetracarboxylic
acid; Tetra-Me ester, in D-60407

$C_{16}H_{18}O_9$
▷ 3-O-Caffeoylquinic acid, C-70005
Fabiatrin, in H-80189
Magnolioside, in H-80188
Scopolin, in H-80189

$C_{16}H_{19}N_3O_6$
Albomitomycin A, A-70080
Isomitomycin A, I-70070

$C_{16}H_{20}$
Heptacyclo[7.7.0.0^{2,6}.0^{3,15}.0^{4,12}.0^{5,10}.0^{11,16}]hexa-
decane, in H-70011

$C_{16}H_{20}Br_2O_2$
Cymobarbatol, C-80202
4-Isocymobarbatol, in C-80202

$C_{16}H_{20}ClN_3$
Bindschedler's green; Chloride, in B-60110

$C_{16}H_{20}N_2$
2,2′-Bis(dimethylamino)biphenyl, in D-60033
1,2-Diphenyl-1,2-ethanediamine; N,N′-Di-Me,
in D-70480

$C_{16}H_{20}N_2O_6S_2$
N,N′-Bis(3-sulfonatopropyl)-2,2′-bipyridinium,
B-70159
N,N′-Bis(3-sulfonatopropyl)-4,4′-bipyridinium,
B-70160

$C_{16}H_{20}N_3^{\oplus}$
Bindschedler's green, B-60110

$C_{16}H_{20}O$
2-Phenyl-2-adamantanol, P-70060

$C_{16}H_{20}O_2$
14-Acetoxy-2,4,8,10-tetradecaen-6-yne, in
T-80031
Methyl 3,4-dihydro-8-isopropyl-5-methyl-2-
naphalenecarboxylate, in I-60129

$C_{16}H_{20}O_3$
14-Acetoxy-8,10-tetradecadiene-4,6-diyn-3-ol,
in T-80026
Alliodorin, A-70083
6-Methoxy-1,3-primnatrienedione, in P-80174
Methyl 3-oxo-1,4(15),11(13)-eudesmatrien-12-
oate, in H-80164

$C_{16}H_{20}O_4$
1-(5,7-Dihydroxy-2,2,6-trimethyl-2H-1-
benzopyran-8-yl)-2-methyl-1-propanone,
D-70348
Hypocretenoic acid; Me ester, in H-70235
Methyl yomoginate, in H-70199

$C_{16}H_{20}O_5$
Curvularin, C-80166
Evodione, E-80091
Glutinopallal, G-70021

$C_{16}H_{20}O_6$
11α-Hydroxycurvularin, in C-80166
11β-Hydroxycurvularin, in C-80166
Orthopappolide, O-60045
▷ Pyrenophorin, P-70167

$C_{16}H_{20}O_8$
Linocinnamarin, in H-80242

$C_{16}H_{20}O_9$
Gentiopicroside, G-80009

$C_{16}H_{21}BrO_4$
1-(4-Bromo-2,5-dihydroxyphenyl)-7-hydroxy-
3,7-dimethyl-2-octen-1-one, B-70206

$C_{16}H_{22}$
[3^{4,10}][7]Metacyclophane, M-60030

$C_{16}H_{22}ClN_3O$
Folicur, F-80071

$C_{16}H_{22}O$
2,4-Bis(3-methyl-2-butenyl)phenol, B-80155
4-(1,1-Dimethyl-2-propenyl)-2-(3-methyl-2-
butenyl)phenol, D-80471
6,8,12,14-Hexadecatetraen-10-yn-1-ol,
H-80041

$C_{16}H_{22}O_2$
6-Methoxyprimnatrienone, in P-80174

$C_{16}H_{22}O_3$
3α-Hydroxydesoxoachalensolide; 3-Me ether,
in H-70135
3β-Hydroxydesoxoachalensolide; Me ether, in
H-70135
2-(3-Hydroxy-3,7-dimethyl-2,6-octadienyl)-1,4-
benzenediol, in H-80153
3-Hydroxy-1,4(15),11(13)-eudesmatrien-12-oic
acid; Me ester, in H-80164
3-Hydroxy-6-methoxyprimnatrienone, in
P-80174
7-Octynoic acid; Anhydride, in O-80024

$C_{16}H_{22}O_5$
ent-7β-Hydroxy-13,14,15,16-tetranor-3-
cleroden-12-oic acid 18,19-lactone,
H-70226
Methyl altamisate, in A-70084

$C_{16}H_{22}O_6$
2α,3α-Epoxy-1β,4α-dihydroxy-8β-methoxy-
7(11)-eudesmen-12,8-olide, in E-60033

$C_{16}H_{22}O_6S_3$
2,8,17-Trithia[4^{5,12}][9]metacyclophane;
Trisulfone, in T-70328

$C_{16}H_{22}O_8$
Synrotolide, S-60066

$C_{16}H_{22}O_{10}$
Gelsemide 7-glucoside, in G-60010

$C_{16}H_{22}O_{11}$
myo-Inositol; 1,2,3,4,6-Penta-Ac, in I-80015

$C_{16}H_{22}S_3$
2,8,17-Trithia[4^{5,12}][9]metacyclophane, T-70328

$C_{16}H_{23}N$
Axisonitrile-4, in A-60328

$C_{16}H_{23}NO$
2-(Aminomethyl)cyclooctanol; N-Benzoyl, in
A-70170

$C_{16}H_{23}NO_2$
2-Amino-4-tert-butylcyclopentanol; Benzoyl,
in A-70104

$C_{16}H_{23}NS$
Axisothiocyanate-4, in A-60329

$C_{16}H_{23}N_3OS$
▷ Buprofezin, B-80276

$C_{16}H_{23}N_5O_5$
1′-Methylzeatin; 9-β-D-Ribofuranosyl, in
M-60133

$C_{16}H_{24}$
Pentacyclo[11.3.0.0^{1,5}.0^{5,9}.0^{9,13}]hexadecane,
P-70020

$C_{16}H_{24}O$
7,9,11,13-Hexadecatetraen-1-al, H-80040
7,12,14-Hexadecatrien-10-yn-1-ol, H-80044
5-Methyl-2-(1-methyl-1-
phenylethyl)cyclohexanol, M-60090

$C_{16}H_{24}O_2$
4-Methoxy-1,9-cadinadien-3-one, in H-80136

$C_{16}H_{24}O_3$
2-(3-Hydroxy-3,7-dimethyl-6-octenyl)-1,4-
benzenediol, H-80153

$C_{16}H_{24}O_4$
Furodysinin hydroperoxide, F-70054
1-Isobutyryloxymethyl-4-isopropyl-2,5-
dimethoxybenzene, in D-70309

$C_{16}H_{24}O_4S_2$
Dithiatopazine, D-70508

$C_{16}H_{24}O_9$
Semperoside, S-60021

$C_{16}H_{24}O_{10}$
Adoxosidic acid, A-70068
1,4-Di-O-methyl-myo-inositol; Tetra-Ac, in
D-80452
9-Hydroxysemperoside, in S-60021
Methylcatalpol, in C-70026
Vebraside, V-60003

$C_{16}H_{24}O_{11}$
Shanzhiside, S-60028

$C_{16}H_{25}GdN_4O_8$
Gadoteric acid, in T-60015

$C_{16}H_{25}N$
Axisonitrile-1, in A-60328
Axisonitrile-2, in A-60328
Axisonitrile-3, in A-60328
10α-Isocyanoalloaromadendrane, in A-60328
3-Isocyano-7,9-bisaboladiene, I-60091
7-Isocyano-3,10-bisaboladiene, I-60092
11-Isocyano-5-eudesmene, in F-60068
6-Isocyano-4(15)-eudesmene, in F-60069

$C_{16}H_{25}NO$
7-Isocyanato-2,10-bisaboladiene, I-60089

$C_{16}H_{25}NO_5$
Wasabidienone E, W-60003
Wasabidienone E, in W-70002

$C_{16}H_{25}NS$
Axisothiocyanate-1, in A-60329
Axisothiocyanate-2, in A-60329
Axisothiocyanate-3, in A-60329
10α-Isothiocyanatoalloaromadendrane, in
A-60329
1-Isothiocyanato-4-cadinene, I-80094
11-Isothiocyano-5-eudesmene, in F-60068

$C_{16}H_{26}$
3,4,7,11-Tetramethyl-1,3,6,10-dodectetraene,
T-70135

$C_{16}H_{26}O_2$
6-Isopropenyl-3-methyl-3,9-decadien-1-ol
acetate, in I-70089
Norambreinolide, N-80075
Senecrassidiol; 5-Deoxy, 6-oxo, 8-Me ether, in
S-80026
3,7,11-Trimethyl-2,6,10-dodecatrienoic acid;
Me ester, in T-60364

$C_{16}H_{26}O_3$
3α-Hydroxynorambreinolide, in N-80075
6-Hydroxy-2,4,8-tetradecatrienoic acid; Et
ester, in H-70222
ent-12-Hydroxy-13,14,15,16-tetranor-1(10)-
halimen-18-oic acid, H-60231
Juvenile hormone III, in T-60364
Methylzedoarondiol, in Z-70004

$C_{16}H_{26}O_5$
2,3,4,6,7,8-Hexahydro-4,8-dihydroxy-2-(1-
hydroxyheptyl)-5H-1-benzopyran-5-one,
H-80057

$C_{16}H_{26}O_8$
Nepetaside, N-70031

$C_{16}H_{26}O_9$
Gelsemiol 1-glucoside, in G-60011
Gelsemiol 1-glucoside, in G-60011
Gibboside, G-60020

$C_{16}H_{27}NO$
Axamide-1, A-70272
Axamide-2, A-60327
10α-Formamidoalloaromadendrane, in
A-60327
11-Formamido-5-eudesmene, F-60068
6-Formamido-4(15)-eudesmene, F-60069

$C_{16}H_{27}NO_8$
4-(2-Carboxyethyl)-4-nitroheptanedioic acid;
Tri-Et ester, in C-80025

$C_{16}H_{28}N_2$
2,4,6-Tri-tert-butylpyrimidine, T-80219

C₁₆H₂₈N₄O₈
1,4,7,10-Tetraazacyclododecane-1,4,7,10-
tetraacetic acid, T-60015

C₁₆H₂₈O
Dodecahydro-3a,6,6,9a-
tetramethylnaphtho[2,1-*b*]furan, D-70533
11,13-Hexadecadienal, H-70033

C₁₆H₂₈O₂
Isoambrettolide, I-60086
6-Isopropenyl-3-methyl-9-decen-1-ol; Ac, *in*
I-70089
6α-Methoxy-4(15)-eudesmen-1β-ol, *in* E-80084

C₁₆H₂₈O₄
3,8-Dihydroxy-13,14,15,16-tetranor-12-
labdanoic acid, D-80387
Secoisolancifolide, S-80014

C₁₆H₂₈O₇
Betulalbuside *A*, *in* D-80456
Betulalbuside *B*, *in* D-80456

C₁₆H₃₀
5-Hexadecyne, H-60037

C₁₆H₃₀N₄O₂
Tetrahydroimidazo[4,5-*d*]imidazole-
2,5(1*H*,3*H*)-dione; 1,3,4,6-Tetraisopropyl,
in T-70066

C₁₆H₃₀O
Decamethylcyclohexanone, D-60011
2-Hexadecenal, H-60036
7-Hexadecenal, H-70035

C₁₆H₃₀O₂
8,9-Hexadecanedione, H-70034
15-Hexadecanolide, *in* H-80173
5-Hexadecenoic acid, H-80045

C₁₆H₃₂N₂O₅
4,7,13,16,21-Pentaoxa-1,10-
diazabicyclo[8.8.5]tricosane, P-60060

C₁₆H₃₂N₆
Pentakis(dimethylamino)aniline, *in* B-80014

C₁₆H₃₂O₃
11-Hydroxyhexadecanoic acid, H-80172
15-Hydroxyhexadecanoic acid, H-80173

C₁₆H₃₄N₄
Octahydroimidazo[4,5-*d*]imidazole; 1,3,4,6-
Tetraisopropyl, *in* O-70015

C₁₆H₃₅NO₃
2-Amino-1,3,4-hexadecanetriol, A-80079

C₁₆H₃₆BF₄N
Tetrabutylammonium(1+);
Tetrafluoroborate, *in* T-80014

C₁₆H₃₆BrNO₄
Tetrabutylammonium(1+); Bromide, *in*
T-80014

C₁₆H₃₆ClCrNO₃
TBACC, *in* T-80014

C₁₆H₃₆ClNO₄
Tetrabutylammonium(1+); Perchlorate, *in*
T-80014

C₁₆H₃₆Cl₄IN
Tetrabutylammonium(1+); Iodotetrachloride,
in T-80014

C₁₆H₃₆FN
Tetrabutylammonium(1+); Fluoride, *in*
T-80014

C₁₆H₃₆N⊕
Tetrabutylammonium(1+), T-80014

C₁₆H₃₆N₄
Tetrabutylammonium(1+); Azide, *in* T-80014

C₁₆H₃₇F₂N
Tetrabutylammonium(1+); Bifluoride, *in*
T-80014

C₁₆H₃₇NO
▷ Tetrabutylammonium(1+); Hydroxide, *in*
T-80014

C₁₆H₃₈N₆
1,4,7,12,15,18-Hexaazacyclodocosane,
H-70028

C₁₆H₃₈N₆O₂
1,13-Dioxa-4,7,10,16,19,22-
hexaazacyclotetracosane, D-70460

C₁₆H₄₀BN
Tetrabutylammonium(1+); Borohydride, *in*
T-80014

C₁₇H₈N₂O
10-(Dicyanomethylene)anthrone, D-60164

C₁₇H₈OS₈
Tetrakis(1,3-dithiol-2-ylidene)cyclopentanone,
T-80131

C₁₇H₉ClO
10-Chloro-7*H*-benz[*de*]anthracen-7-one,
C-70047
11-Chloro-7*H*-benz[*de*]anthracen-7-one,
C-70048
2-Chloro-7*H*-benz[*de*]anthracen-7-one,
C-70040
3-Chloro-7*H*-benz[*de*]anthracen-7-one,
C-70041
4-Chloro-7*H*-benz[*de*]anthracen-7-one,
C-70042
5-Chloro-7*H*-benz[*de*]anthracen-7-one,
C-70043
6-Chloro-7*H*-benz[*de*]anthracen-7-one,
C-70044
8-Chloro-7*H*-benz[*de*]anthracen-7-one,
C-70045
9-Chloro-7*H*-benz[*de*]anthracen-7-one,
C-70046

C₁₇H₁₀F₃NO₃
7-Amino-4-(trifluoromethyl)-2*H*-1-
benzopyran-2-one; *N*-Benzoyl, *in* A-60265

C₁₇H₁₀N₂
1-(Diazomethyl)pyrene, D-70062
9-(Dicyanomethyl)anthracene, D-60163

C₁₇H₁₀N₄
Naphtho[2′,1′:5,6][1,2,4]triazino[4,3-*b*]indazole,
N-80010

C₁₇H₁₀N₄O₃S₂
3,3′-Carbonylbis[5-phenyl-1,3,4-oxadiazole-
2(3*H*)-thione], C-80021

C₁₇H₁₀O
3*H*-Cyclonona[*def*]biphenylen-3-one, C-70230

C₁₇H₁₀O₂
2*H*-Anthra[1,2-*b*]pyran-2-one, A-80164
3*H*-Anthra[2,1-*b*]pyran-3-one, A-80165

C₁₇H₁₀O₄
Fluorescamine, F-80017
Neorauteen, *in* N-80020

C₁₇H₁₀O₅
1,5-Diphenylpentanepentone, D-60488

C₁₇H₁₀O₇
2-Hydroxy-3-methoxy-8,9-
methylenedioxycoumestan, *in* T-80088
3-Hydroxy-2-methoxy-8,9-
methylenedioxycoumestan, *in* T-80088
3-Hydroxy-4-methoxy-8,9-
methylenedioxycoumestan, *in* T-80089

C₁₇H₁₀O₈
Scapaniapyrone *A*, S-70017

C₁₇H₁₁N
Dibenzo[*f,h*]quinoline, D-70075

C₁₇H₁₁NO
Benz[*cd*]indol-2-(1*H*)-one; *N*-Ph, *in* B-70027

C₁₇H₁₁NO₂
11*H*-Benzo[*a*]carbazole-1-carboxylic acid,
B-80017

C₁₇H₁₁N₃
7,12-Dihydropyrido[3,2-*b*:5,4-*b*′]diindole,
D-70254
7,12-Dihydropyrido[3,2-*b*:5,4-*b*′]diindole,
D-80246

C₁₇H₁₂
3*H*-Cyclonona[*def*]biphenylene, C-60215
Cycloocta[*def*]fluorene, C-70233
1*H*-Cyclopenta[*l*]phenanthrene, C-80190
1-Methylazupyrene, M-60043

C₁₇H₁₂BrN
Dibenzo[*a,h*]quinolizinium(1+); Bromide, *in*
D-70076
Naphtho[1,2-*a*]quinolizinium(1+); Bromide,
in N-70012

C₁₇H₁₂ClNO₄
Dibenzo[*a,h*]quinolizinium(1+); Perchlorate,
in D-70076
Naphtho[1,2-*a*]quinolizinium(1+);
Perchlorate, *in* N-70012

C₁₇H₁₂Cl₂O₈
Geodoxin, G-70007

C₁₇H₁₂N⊕
Dibenzo[*a,h*]quinolizinium(1+), D-70076
Naphtho[1,2-*a*]quinolizinium(1+), N-70012

C₁₇H₁₂N₂
2-Phenylpyrimido[2,1,6-*de*]quinolizine,
P-80131

C₁₇H₁₂N₂O
2-Benzoyl-1,2-dihydro-1-
isoquinolinecarbonitrile, B-70073

C₁₇H₁₂OS
2,6-Diphenyl-4*H*-thiopyran-4-one, D-60492

C₁₇H₁₂O₂
5,11-Methanodibenzo[*a,e*]cyclooctene-
6,12(5*H*,11*H*)dione, M-80040
2,2′-Spirobi[2*H*-1-benzopyran], S-70062

C₁₇H₁₂O₃
1,2-Diphenyl-1,2-cyclopropanedicarboxylic
acid; Anhydride, *in* D-80509
1-(4-Hydroxy-5-benzofuranyl)-3-phenyl-2-
propen-1-one, H-70107
Salvinolactone, S-80006

C₁₇H₁₂O₃S
2,6-Diphenyl-4*H*-thiopyran-4-one; 1,1-
Dioxide, *in* D-60492
5-Methoxy-2-(phenylthio)-1,4-
naphthoquinone, *in* H-70213
5-Methoxy-3-(phenylthio)-1,4-
naphthoquinone, *in* H-70214

C₁₇H₁₂O₄
Neodunol, N-80020

C₁₇H₁₂O₅
5-Methoxy-6,7-methylenedioxyflavone, *in*
T-60322
7-Methoxy-3′,4′-methylenedioxyisoflavone, *in*
H-80203

C₁₇H₁₂O₆
3,3-Dihydroxy-5,5-diphenyl-1,2,4,5-
pentanetetrone, *in* D-60488
4-(3,4-Dihydroxyphenyl)-6,7-dihydroxy-2-
naphthalenecarboxylic acid, D-80358
3-Hydroxy-7,9-dimethoxycoumestan, *in*
T-80294
3-Hydroxy-8,9-dimethoxycoumestan, *in*
T-80295
2′-Hydroxy-5-methoxy-6,7-
methylenedioxyisoflavone, *in* T-80108
4′-Hydroxy-5-methoxy-6,7-
methylenedioxyisoflavone, *in* D-80341
5-Hydroxy-2′-methoxy-6,7-
methylenedioxyisoflavone, *in* T-60107
7-Hydroxy-2′-methoxy-4′,5′-
methylenedioxyisoflavone, *in* T-80107
7-Hydroxy-6-methoxy-3′,4′-
methylenedioxyisoflavone, *in* T-80111
7-Hydroxy-8-methoxy-3′,4′-
methylenedioxyisoflavone, *in* T-70112
3-Hydroxy-4-methoxy-8,9-
methylenedioxypterocarpene, *in* T-80122

C₁₇H₁₂O₇
3,8-Dihydroxy-6-methoxy-1-
methylanthraquinone-2-carboxylic acid, *in*
T-80323

5,7-Dihydroxy-6-methoxy-3',4'-
methylenedioxyisoflavone, *in* P-80043
6a-Hydroxy-3,4:8,9-
bis(methylenedioxy)pterocarpan, *in*
P-80053
3,6,8-Trihydroxy-1-methylanthraquinone-2-
carboxylic acid; Me ester, *in* T-60332

$C_{17}H_{12}O_8$
Nasutin *B*, *in* E-70007

$C_{17}H_{13}BF_4N_2$
Urorosein; Tetrafluoroborate, *in* U-80033

$C_{17}H_{13}BrN_2$
Urorosein; Bromide, *in* U-80033

$C_{17}H_{13}ClN_2$
Urorosein; Chloride, *in* U-80033

$C_{17}H_{13}ClN_2O_4$
Urorosein; Perchlorate, *in* U-80033

$C_{17}H_{13}NO_2S$
2-Phenyl-1,5-benzothiazepin-4(5H)-one; N-Ac,
in P-60081

$C_{17}H_{13}NO_4$
2-Oxo-3-indolineglyoxylic acid; Benzyl ester,
in O-80085

$C_{17}H_{13}NS$
1-Methyl-4-phenyl-1λ⁴-1-benzothiopyran-2-
carbonitrile, M-80142

$C_{17}H_{13}N_2^{\oplus}$
Urorosein, U-80033

$C_{17}H_{13}N_3$
1-Anilino-γ-carboline, *in* A-70114

$C_{17}H_{14}$
2,3-Dihydro-1H-cyclopenta[l]phenanthrene,
D-80177
6,12-Methanobenzocyclododecene, M-80039

$C_{17}H_{14}ClF_7O_2$
Tefluthrin, T-80004

$C_{17}H_{14}Cl_2O_7$
Tumidulin, T-70334

$C_{17}H_{14}N_2$
3,3'-Methylenebisindole, M-80078

$C_{17}H_{14}N_2O_4$
2,3-Dihydro-2-hydroxy-1H-imidazole-4,5-
dione; 1,3-Di-Ph, O-Ac, *in* D-60243

$C_{17}H_{14}O$
2-Methoxy-1-phenylnaphthalene, *in* P-60115
2-Methoxy-5-phenylnaphthalene, *in* P-60118
7-Methoxy-1-phenylnaphthalene, *in* P-60119

$C_{17}H_{14}O_2$
1,2-Dihydro-1-phenyl-1-naphthalenecarboxylic
acid, D-60264
2-(4-Hydroxyphenyl)-5-(1-
propenyl)benzofuran, H-60214
9-Phenanthreneacetic acid; Me ester, *in*
P-70047
9-Phenanthrenemethanol; Ac, *in* P-80079
2-(Phenylethynyl)benzoic acid; Et ester, *in*
P-70074

$C_{17}H_{14}O_3$
1,5-Diphenyl-1,3,5-pentanetrione, D-60489
7-Methoxy-2-methylisoflavone, *in* H-80205
1-(β-Methylcrotonoyloxy)-5-benzoyl-2,4-
pentadiyne, *in* H-80235
4-[5-(1-Propenyl)-2-benzofuranyl]-2,3-
benzenediol, *in* H-60214

$C_{17}H_{14}O_4$
1-Acetoxy-7-(3-acetoxyphenyl)-2-heptene-4,6-
diyne, *in* H-80234
Agrostophyllin, A-70077
Bonducellin, B-60192
9,10-Dimethoxy-2-methyl-1,4-anthraquinone,
D-80413
3,9-Dimethoxypterocarpene, *in* D-80376
1,2-Diphenyl-1,2-cyclopropanedicarboxylic
acid, D-80509

$C_{17}H_{14}O_4S$
2-(1,3-Pentadiynyl)-5-(3,4-diacetoxy-1-
butynyl)thiophene, *in* P-80025

$C_{17}H_{14}O_5$
Benzophenone-2,2'-dicarboxylic acid; Di-Me
ester, *in* B-70041
▷ 1,3-Dihydroxy-2-
(ethoxymethyl)anthraquinone, *in* D-70306
2',5-Dihydroxy-7-methoxyflavone, *in* T-80299
4'-Hydroxy-3',7-dimethoxyflavone, *in* T-80308
5-Hydroxy-6,7-dimethoxyflavone, *in* T-60322
6-Hydroxy-4',7-dimethoxyisoflavone, *in*
T-80310
7-Hydroxy-3',4'-dimethoxyisoflavone, *in*
T-80308
7-Hydroxy-4',5-dimethoxyisoflavone, *in*
T-80309
7-Hydroxy-4',8-dimethoxyisoflavone, *in*
T-80311
3-Hydroxy-1-methoxy-2-
(methoxymethyl)anthraquinone, *in*
D-70306
2-(2-Hydroxy-4-methoxyphenyl)-3-methyl-5,6-
methylenedioxybenzofuran, H-70157
Intricatinol, I-80017
Pterocarpin, *in* M-70001
Puerol *A*, P-70137
3,6,7-Trimethoxy-1,4-phenanthraquinone,
T-80348

$C_{17}H_{14}O_6$
4',5-Dihydroxy-3',7-dimethoxyflavone,
D-70291
3,4'-Dihydroxy-6,7-dimethoxyflavone, *in*
T-80092
3',7-Dihydroxy-4',6-dimethoxyflavone, *in*
T-80092
4',7-Dihydroxy-3,6-dimethoxyflavone, *in*
T-80092
2',5-Dihydroxy-7,8-dimethoxyisoflavone, *in*
T-70110
3',7-Dihydroxy-4',6-dimethoxyisoflavone, *in*
T-80111
4',5-Dihydroxy-3',7-dimethoxyisoflavone, *in*
T-80109
4',5-Dihydroxy-6,7-dimethoxyisoflavone, *in*
T-80112
4'-6-Dihydroxy-5,7-dimethoxyisoflavone, *in*
T-80112
4',7-Dihydroxy-2',5-dimethoxyisoflavone, *in*
T-80106
5,7-Dihydroxy-2',4'-dimethoxyisoflavone, *in*
T-80106
5,7-Dihydroxy-2',6-dimethoxyisoflavone, *in*
T-60107
5,7-Dihydroxy-3',4'-dimethoxyisoflavone, *in*
T-80109
5,7-Dihydroxy-4',6-dimethoxyisoflavone, *in*
T-80112
5-Hydroxy-4-(3-hydroxy-4-methoxyphenyl)-7-
methoxycoumarin, *in* D-80356
7-Hydroxy-2'-methoxy-4',5'-
methylenedioxyisoflavanone, *in* T-80101
2-Hydroxy-3-methoxy-8,9-
methylenedioxypterocarpan, *in* T-80117
3-Hydroxy-4-methoxy-8,9-
methylenedioxypterocarpan, *in* T-80119
4-Hydroxy-3-methoxy-8,9-
methylenedioxypterocarpan, *in* T-80119
Sophorocarpan *B*, S-60038

$C_{17}H_{14}O_7$
Arizonin A_1, *in* A-70254
Arizonin B_1, *in* A-70254
6a,12a-Dihydro-3,4,10-trihydroxy-8-
methoxy[2]benzopyrano[4,3-
b][1]benzopyran-7(5H)-one, *in* C-70200
6a,12a-Dihydro-2,3,10-trihydroxy-8-
methoxy[2]benzopyrano[4,3-
b][1]benzopyran-7(5H)-one, 9CI, *in*
C-70200
1,3-Dihydroxy-2-methoxy-8,9-
methylenedioxypterocarpan, *in* P-80051
4-(3,4-Dihydroxyphenyl)-8-hydroxy-5,7-
dimethoxycoumarin, *in* D-80370
Irispurinol, I-80044
Nornotatic acid, *in* N-70081

3',5,7-Trihydroxy-4',6-dimethoxyisoflavone, *in*
P-80043
2',5,7-Trihydroxy-4',5'-dimethoxyflavone, *in*
P-80033
3,4',7-Trihydroxy-6,7-dimethoxyflavone, *in*
P-80035
3,4',7-Trihydroxy-5,6-dimethoxyflavone, *in*
P-80035
3',4',5-Trihydroxy-6,7-dimethoxyisoflavone, *in*
P-70032
3',5,7-Trihydroxy-4',8-dimethoxyisoflavone, *in*
P-80045
4',5,7-Trihydroxy-3',6-dimethoxyisoflavone, *in*
P-80043
4',5,7-Trihydroxy-3',8-dimethoxyisoflavone, *in*
P-80045

$C_{17}H_{14}O_8$
3',4',5,5'-Tetrahydroxy-6,7-dimethoxyflavone,
in H-70081
3',4',5,7-Tetrahydroxy-5',6-dimethoxyflavone,
in H-70081
4',5,6,7-Tetrahydroxy-3,3'-dimethoxyflavone,
in H-80067

$C_{17}H_{14}S_5$
4,5-Dimercapto-1,3-dithiole-2-thione;
Dibenzyl thioether, *in* D-80409

$C_{17}H_{15}ClO_7$
Methyl 2-(3-chloro-2,6-dihydroxy-4-
methylbenzoyl)-5-hydroxy-3-
methoxybenzoate, *in* S-70083

$C_{17}H_{15}N$
2-Methyl-1,5-diphenyl-1H-pyrrole, *in*
M-60115

$C_{17}H_{15}NO_2S$
2-Phenyl-1,5-benzothiazepin-4(5H)-one; 2,3-
Dihydro, N-Ac, *in* P-60081

$C_{17}H_{16}$
1,1-Diphenyl-1,3-pentadiene, D-70489

$C_{17}H_{16}N_2O$
1-[1-Methoxy-1-(2-
pyridinyl)ethyl]isoquinoline, *in* I-60134

$C_{17}H_{16}N_2O_2$
3,5-Pyrazolidinedione; 1,2-Dibenzyl, *in*
P-80200

$C_{17}H_{16}N_2O_3$
3-Oxopentanedioic acid; Dianilide, *in* O-60084

$C_{17}H_{16}O_2$
1,5-Bis(4-hydroxyphenyl)-1,4-pentadiene,
B-60162
5,5-Diphenyl-4-pentenoic acid, D-80513
3,3',4,4'-Tetrahydro-2,2'-spirobi[2H-1-
benzopyran], T-70094

$C_{17}H_{16}O_3$
Danshenspiroketallactone, D-70004
Epidanshenspiroketallactone, *in* D-70004
2,3,7-Trimethoxyphenanthrene, *in* T-60339
2,4,7-Trimethoxyphenanthrene, *in* P-80080

$C_{17}H_{16}O_4$
Cryptoresinol, C-70201
1-(2-Hydroxy-4,6-dimethoxyphenyl)-3-phenyl-
2-propen-1-one, *in* P-80139
10-Hydroxy-11-methoxydracaenone, *in*
D-70546
2-(2-Hydroxy-4-methoxyphenyl)-6-methoxy-3-
methylbenzofuran, H-70156
3,11-Tridecadiene-5,7,9-triyne-1,2-diol; Di-Ac,
in T-80254
3,5,7-Trimethoxy-2-phenanthrenol, *in* T-60120

$C_{17}H_{16}O_4S_2$
5-(3,4-Diacetoxy-1-butynyl)-5'-methyl-2,2'-
bithiophene, *in* M-70054

$C_{17}H_{16}O_5$
Comaparvin, C-70191
Combretastatin A2, *in* D-80360
2',7-Dihydroxy-4',8-dimethoxyisoflavene, *in*
T-80104
3',7-Dihydroxy-2',4'-dimethoxyisoflavene, *in*
T-80103
4',7-Dihydroxy-2',3'-dimethoxyisoflavene, *in*
T-80103

7,10-Dihydroxy-11-methoxydracaenone, *in* D-70546

1-(2,4-Dihydroxy-3-methoxyphenyl)-3-(4-methoxyphenyl)-2-propen-1-one, *in* H-80245

1-(2,6-Dihydroxy-4-methoxyphenyl)-3-(4-methoxyphenyl)-2-propen-1-one, *in* H-80246

5,10-Dihydroxy-8-methoxy-2-propyl-4*H*-naphtho[1,2-*b*]pyran-4-one, 9CI, *in* C-70191

7,9-Dihydroxy-2,3,4-trimethoxyphenanthrene, *in* P-70035

Fruscinol acetate, *in* F-80084

5-Hydroxy-7,8-dimethoxyflavone, *in* T-60321

2'-Hydroxy-4',7-dimethoxyisoflavanone, *in* T-80303

7-Hydroxy-2',4'-dimethoxyisoflavanone, *in* T-80303

6a-Hydroxy-3,9-dimethoxypterocarpan, *in* T-80337

3-Hydroxy-2,9-dimethoxypterocarpan, *in* T-80335

3-Hydroxy-4,9-dimethoxypterocarpan, *in* T-80336

3-Hydroxy-9,10-dimethoxypterocarpan, *in* T-80338

4-Hydroxy-3,9-dimethoxypterocarpan, *in* T-80336

9-Hydroxy-2,3-dimethoxypterocarpan, *in* T-80335

7-Hydroxy-2'-methoxy-4',5'-methylenedioxyisoflavan, *in* T-80097

Methylsainfuran, *in* S-60002

Sophorocarpan *A*, S-60037

1,5,6-Trimethoxy-2,7-phenanthrenediol, *in* P-60050

1,5,7-Trimethoxy-2,6-phenanthrenediol, *in* P-60050

$C_{17}H_{16}O_6$

Anhydrobyakangelicin, *in* B-80296

1,5-Bis(3,4-dihydroxyphenyl)-4-pentyne-1,2-diol, B-60155

4',5-Dihydroxy-2',7-dimethoxyflavanone, *in* T-80100

5,7-Dihydroxy-2',8-dimethoxyflavanone, *in* T-70104

3',7-Dihydroxy-2',4'-dimethoxyisoflavanone, *in* T-80099

4',7-Dihydroxy-2',3'-dimethoxyisoflavanone, *in* T-80099

5,7-Dihydroxy-2',4'-dimethoxyisoflavanone, *in* T-80100

2,10-Dihydroxy-3,9-dimethoxypterocarpan, *in* T-80118

5,8-Dihydroxy-2,3,6-trimethyl-1,4-naphthoquinone; Di-Ac, *in* D-60380

Neobyakangelicol, *in* B-80296

Olivin, O-80043

Ougenin, O-80050

Pendulone, P-80018

2',3',4',6,7-Pentahydroxyisoflavene; 2',4' or 3',4'-Di-Me ether, *in* P-80040

1,2,3,4-Tetramethoxyxanthone, *in* T-60126

1,3,5,8-Tetramethoxyxanthone, *in* T-70123

$C_{17}H_{16}O_7$

Isosulochrin, *in* S-70083

Sulochrin, S-70083

2',3,5-Trihydroxy-7,8-dimethoxyflavanone, *in* P-60045

4',5,7-Trihydroxy-3',6-dimethoxyflavanone, *in* P-70028

4',5,7-Trihydroxy-2',3'-dimethoxyisoflavanone, *in* P-80039

1,3,8-Trihydroxy-4,7-dimethoxyxanthone, *in* P-60052

$C_{17}H_{17}ClO_6$

Byakangelicin; 3'-Deoxy, 3'-chloro, *in* B-80296

$C_{17}H_{17}NO$

5,6,11,12-Tetrahydrodibenz[*b,f*]azocine; *N*-Ac, *in* T-70057

$C_{17}H_{17}NO_2$

4-Amino-3,4-dihydro-2-phenyl-2*H*-1-benzopyran; *N*-Ac, *in* A-80067

$C_{17}H_{17}NO_4$

4-Phenyl-2,6-pyridinedicarboxylic acid; Di-Et ester, *in* P-80130

$C_{17}H_{18}O_2S_2$

3,3-Bis(phenylthio)propanoic acid; Et ester, *in* B-70157

$C_{17}H_{18}O_3$

5-Hydroxy-5,5-diphenylpentanoic acid, H-80155

$C_{17}H_{18}O_4$

Anisocoumarin A, A-80159

3',5-Dimethoxy-3,4-methylenedioxybibenzyl, *in* T-60102

3-(2,4-Dimethoxyphenyl)-1-(4-hydroxyphenyl)-1-propanone, *in* D-80366

4'-Hydroxy-3',7-dimethoxyflavan, *in* T-70260

2'-Hydroxy-4',7-dimethoxyisoflavan, *in* T-80302

4'-Hydroxy-2',7-dimethoxyisoflavan, *in* T-80302

7-Hydroxy-2',4'-dimethoxyisoflavan, *in* T-80302

2-*O*-Methylangolensin, *in* A-80155

4-*O*-Methylangolensin, *in* A-80155

Pygmaeoherin, P-70145

1-(2,4,6-Trimethoxyphenyl)-3-phenyl-1-propanone, *in* T-80332

$C_{17}H_{18}O_5$

9β-Acetoxy-3-oxo-1,4(15),11(13)-eudesmatrien-12,6-olide, *in* H-60202

Combretastatin A3, *in* D-80360

11,13-Dehydromatricarin, *in* M-80010

9,10-Dihydro-2,7-dihydroxy-3,4,6-trimethoxyphenanthrene, *in* D-70246

9,10-Dihydro-5,6-dihydroxy-1,3,4-trimethoxyphenanthrene, *in* P-70034

9,10-Dihydro-5,6,7-trimethoxy-2,3-phenanthrenediol, *in* D-70246

5,7-Dihydroxy-3',4'-dimethoxyflavan, *in* T-60103

2',4'-Dihydroxy-5,7-dimethoxyisoflavan, *in* T-80096

2',7-Dihydroxy-3',4'-dimethoxyisoflavan, *in* T-80095

2',7-Dihydroxy-4',5-dimethoxyisoflavan, *in* T-80096

2',7-Dihydroxy-4',6-dimethoxyisoflavan, *in* T-80098

4'-Hydroxy-3',5-dimethoxy-3,4-methylenedioxybibenzyl, *in* P-60044

4-(2-Hydroxy-4,6-dimethoxyphenyl)-1-(4-hydroxyphenyl)-1-propanone, *in* H-80243

Isomurralonginol; Ac, *in* I-70071

Longipesin; 9-Ac, *in* L-60034

$C_{17}H_{18}O_6$

Antibiotic L 660631, A-70224

3-(3,4-Dihydroxybenzyl)-3,7-dihydroxy-4-methoxychroman, *in* D-70288

Epicatechin; $O^{3'},O^{4'}$-Di-Me, *in* P-70026

Epicatechin; $O^{3'},O^5$-Di-Me, *in* P-70026

Pandoxide, P-60002

3,3',5-Trihydroxy-4',7-dihydroxyflavan, *in* P-70026

3',7,8-Trihydroxy-2',4'-dimethoxyisoflavan, *in* P-80038

$C_{17}H_{18}O_7$

Byakangelicin, B-80296

Fusarubin; O^3-Et, *in* F-60108

Fusarubin ethyl acetal, *in* F-60108

Isobyakangelicolic acid, I-80051

$C_{17}H_{18}O_9$

Furocoumarinic acid; *O*-β-D-Glucopyranoside, *in* H-80113

$C_{17}H_{19}N$

2,3,4,5-Tetrahydro-1*H*-1-benzazepine; *N*-Benzyl, *in* T-70036

$C_{17}H_{19}NO_3$

3,4,6-Trimethoxy-1,2-dimethylcarbazole, *in* T-70253

$C_{17}H_{20}$

1,7,9,15-Heptadecatetraene-11,13-diyne, H-80019

$C_{17}H_{20}N_2$

1-Ethyl-2,3,4,6,7,12-hexahydroindolo[2,3-*a*]quinolizine, E-80064

$C_{17}H_{20}N_2O$

▷ *N,N'*-Diethyl-*N,N'*-diphenylurea, *in* D-70499

$C_{17}H_{20}O$

15,16-Dihydro-17*H*-cyclopenta[*a*]phenanthren-17-one, D-80178

$C_{17}H_{20}O_2$

1,5-Diphenyl-1,3-pentanediol, D-70490

$C_{17}H_{20}O_3$

1β-Acetoxy-4(15),7(11),8-eudesmatrien-12,8-olide, *in* H-60141

15-Acetoxytubipofuran, *in* T-70332

$C_{17}H_{20}O_4$

1β-Acetoxy-3,7(11),8-eudesmatrien-12,8-olide, *in* H-60140

12,13-Dihydroxy-8,11,13-podocarpatriene-3,7-dione, D-80374

1-(2-Hydroxy-4-methoxyphenyl)-3-(3-hydroxy-4-methoxyphenyl)propane, *in* D-70336

1-(4-Hydroxy-2-methoxyphenyl)-3-(4-hydroxy-3-methoxyphenyl)propane, *in* D-70336

1-(2-Hydroxy-4-methoxyphenyl)-3-(4-hydroxy-3-methoxyphenyl)propene, *in* D-70336

Zaluzanin D, *in* Z-60001

$C_{17}H_{20}O_5$

6-Acetylferulidin, *in* F-70006

Cavoxinine, C-60024

Cavoxinone, C-60025

1-(3,4-Dihydroxyphenyl)-2-(3,4,5-trimethoxyphenyl)ethane, *in* D-80371

3,5-Dihydroxy-3',4,4'-trimethoxybibenzyl, *in* D-80360

4,4'-Dihydroxy-3,3',5-trimethoxybibenzyl, *in* D-80360

4,4'-Dihydroxy-3,3',5-trimethoxybibenzyl, *in* P-60044

8-Hydroxyachillin; Ac, *in* H-60100

Matricarin, M-80010

Murraxocin, M-60151

$C_{17}H_{20}O_6$

Achillolide *A*, *in* A-60061

1,2-Bis(4-hydroxy-3-methoxyphenyl)-1,3-propanediol, B-80150

Murrayanone, M-70160

Tarchonanthus lactone, T-60009

$C_{17}H_{20}O_7$

7-Hydroxy-6-[2-hydroxy-2-(tetrahydro-2-methyl-5-oxo-2-furyl)ethyl]-5-methoxy-4-methylphthalide,8CI, H-80175

Isoapressin, I-70061

$C_{17}H_{20}O_9$

3-*O*-Feruloylquinic acid, *in* C-70005

4,7-Dihydroxy-5-methyl-2*H*-1-benzopyran-2-one; 4-Me ether, 7-*O*-β-D-glucopyranoside, *in* D-60348

$C_{17}H_{20}O_{10}$

Tomenin, *in* T-80287

$C_{17}H_{21}ClO_6$

Arctodecurrolide, *in* D-60325

$C_{17}H_{21}N_3$

▷ Auramine, A-70268

$C_{17}H_{22}N_6O_6$

3'-Amino-3'-deoxyadenosine; 3'*N*-Ac, 3'*N*-Me, 2',5'-di-Ac, *in* A-70128

$C_{17}H_{22}O$

1,9,15-Heptadecatriene-11,13-diyn-8-ol, H-80020

1,9,16-Heptadecatriene-4,6-diyn-3-ol, H-80021

2,9,16-Heptadecatriene-4,6-diyn-1-ol, H-80022

2,9,16-Heptadecatriene-4,6-diyn-8-ol, H-80023

$C_{17}H_{22}O_2$

Parvifoline; Ac, *in* P-80013

$C_{17}H_{22}O_2S$

Thiofurodysin acetate, *in* T-70178

Thiofurodysinin acetate, *in* T-70179

$C_{17}H_{22}O_3$
3-(3,7-Dimethyl-2,6-octadienyl)-4-hydroxybenzoic acid, D-70423
8,12-Epoxy-3,7,11-eudesmatrien-1-ol; Ac, *in* E-60020
6-Hydroxy-7-methoxy-2-methyl-2-(4-methyl-3-pentenyl)-2*H*-1-benzopyran, H-80191
6-Methoxy-2-(3,7-dimethyl-2,6-octadienyl)-1,4-benzoquinone, M-60038
Nimbidiol, N-60021

$C_{17}H_{22}O_4$
2-Acetoxy-2-desoxyperezone, *in* I-70075
3β-Acetoxy-4,11(13)-eudesmadien-12,8β-olide, *in* H-60136
15-Acetoxyisoalloalantolactone, *in* I-80047
9β-Acetoxytournefortiolide, *in* H-60137
1-(5,7-Dihydroxy-2,2,6-trimethyl-2*H*-1-*benzopyran*-8-*yl*)-2-*methyl*-1-*butanone*, 9*CI*, *in* D-70348
3β-Hydroxydesoxoachalensolide; Ac, *in* H-70135
15-Hydroxy-3,10(14)-guaiadien-12,8-olide; Ac, *in* H-70136
Laurenobiolide, *in* H-70134
Subexpinnatin *C*, S-60058

$C_{17}H_{22}O_4S$
15-Acetylthioxyfurodysinin lactone, A-80030

$C_{17}H_{22}O_5$
2α-Acetoxy-8-epiivangustin, *in* I-70109
6α-Acetoxy-8-epiivangustin, *in* I-70109
1β-Acetoxy-8β-hydroxy-3,7(11)-eudesmadien-12,8-olide, *in* D-60322
3β-Acetoxy-2α-hydroxy-1(10),4,11(13)-germacratrien-12,6α-olide, *in* H-60006
3-Acetylpulverulide, *in* P-80191
Cumambrin A, *in* C-80164
Ergolide, E-60035
Naematolone, *in* N-70001

$C_{17}H_{22}O_6$
15-Acetoxy-3α-hydroperoxyalloalantolactone, *in* A-80044
Cratystyolide; 3-Ac, *in* T-70257
13-Desacetyleudesmaafraglaucolide, *in* T-70258
13-Desacetyl-1α-hydroxyafraglaucolide, *in* A-70070
13-Desacetyl-1β-hydroxyafraglaucolide, *in* A-70070
13-Desacetyl-1β-hydroxyisoafraglaucolide, *in* I-70060
Neovasinone, N-60019
Neovasinone, *in* N-70029
Omphalocarpin, O-80044

$C_{17}H_{23}Br_3O_4$
9-Acetoxy-3,10,13-tribromo-4,7:6,12-diepoxy-1-pentadecyne, *in* B-70195

$C_{17}H_{23}N_5O_7$
Queuosine, *in* Q-70004

$C_{17}H_{24}Br_2O_4$
6-Acetoxy-3,10-dibromo-4,7-epoxy-12-pentadecen-1-yne, *in* B-70217

$C_{17}H_{24}Br_2O_5$
Graciosin, G-80035

$C_{17}H_{24}O_2$
Norasperenal A, N-80076
Norasperenal B, N-80077
Norasperenal D, *in* N-80077
Panaxydol, P-80006
Tochuinyl acetate, *in* T-70202

$C_{17}H_{24}O_3$
2-[2-(Acetyloxy)ethenyl]-6,10-dimethyl-2,5,9-undecatrienal, A-70050
HM-2, H-80092
HM-3, H-80093
9-Oxogymnomitryl acetate, *in* G-70035

$C_{17}H_{24}O_4$
3β-Acetoxydrimenin, *in* D-60519
9β-Acetoxy-4,11(13)-eudesmadien-12-oic acid, *in* H-60135
6β-Acetoxy-1β-hydroxy-4,11-guaiadien-3-one, *in* G-80044
9α-Acetoxy-7β-hydroxy-2-longipinen-1-one, *in* D-80332
2-(7-Hydroxy-3,7-dimethyl-2-octenyl)-6-methoxy-1,4-benzoquinone, H-60123
Isotrichodermin, *in* E-70028
Norstrictic acid, N-70080
Sterpuric acid; 13-Hydroxy, 3,13-ethylidene acetal, *in* S-60053

$C_{17}H_{24}O_5$
3β-Acetoxy-10α-hydroxy-4,11(13)-cadinadien-12-oic acid, *in* Z-70001
Altamisin, *in* A-70084
Cinnamodial, C-70177
Dihydrocumambrin A, *in* C-80164
▷ Naematolin, N-70001
Neovasinin, N-70029
Nitrosin, N-60045
4,5-Seconeopulchell-5-ene; 2-Ac, *in* S-70025
4,5-Seconeopulchell-5-ene; 4-Ac, *in* S-70025

$C_{17}H_{24}O_6$
2α-Acetoxy-11α,13-dihydroconfertin, *in* D-60211
Blumealactone *C*, B-70172
15-Desacetyltetraneurin *C*, *in* T-60158
6β-Hydroxypulchellin 2-*O*-acetate, *in* P-80189
Naematolin *B*, *in* N-70001
Paramicholide, P-70011
Tetraneurin *D*, T-60158

$C_{17}H_{24}O_7$
8α-Acetoxy-3α,4α,10β-trihydroxy-1-guaien-12,6α-olide, *in* T-80094

$C_{17}H_{24}O_9$
Syringin, *in* S-70046

$C_{17}H_{24}O_{10}$
Pseudo-α-D-glucopyranose; Penta-Ac, *in* H-70170

$C_{17}H_{24}O_{12}$
Sesamoside, S-70036

$C_{17}H_{25}ClO_3$
1-Chloro-9,10-epoxy-4,6-heptadecadiyne-2,3-diol, *in* P-80006

$C_{17}H_{25}NO_4$
N-Acetylparthenolidine, *in* P-80012

$C_{17}H_{26}N_2O_3$
Dibusadol, *in* H-70108

$C_{17}H_{26}O$
Acetylcedrene, A-70032
8,10,12,14-Heptadecatetraen-1-al, H-80018

$C_{17}H_{26}O_2$
3-Acetoxysterpurene, *in* S-80045
Coralloidin *C*, *in* E-60066
Coralloidin *E*, *in* E-60065
Cyperenyl acetate, *in* C-70259
Dihydrotochuinyl acetate, *in* T-70202
11-Phenylundecanoic acid, P-70096
12-Sesquisabinenol acetate, *in* S-70037
13-Sesquisabinenol acetate, *in* S-70037

$C_{17}H_{26}O_3$
8,9-Dihydroxy-4,6-heptadecadiyn-3-one, D-80321
1-Heptadecene-4,6-diyne-3,9,10-triol, H-80024
9α-Hydroxygymnomitryl acetate, *in* G-70035
3-(1,3,5,7,9-Tetradecapentaenyloxy)-1,2-propanediol, T-70024

$C_{17}H_{26}O_4$
3β-Acetoxy-4,10(14)-oplopadiene-8β,9α-diol, *in* O-70042
10α-Ethoxy-3β-hydroxy-4,11(13)-cadinadien-12-oic acid, *in* Z-70001
Norhardwickiic acid, N-70077

$C_{17}H_{26}O_{10}$
Adoxoside, A-70067
Secologanol, S-80016

$C_{17}H_{26}O_{11}$
5-Deoxypulchelloside *I*, *in* P-70139
6β-Hydroxyadoxoside, *in* A-70067
Shanzhiside; Me ester, *in* S-60028

$C_{17}H_{26}O_{12}$
Pulchelloside I, P-70139

$C_{17}H_{27}N$
[12](2,6)Pyridinophane, P-80216

$C_{17}H_{27}NO_4$
1,4-Dihydro-2,6-dimethyl-3,5-pyridinedicarboxylic acid; Di-*tert*-butyl ester, *in* D-80193

$C_{17}H_{28}O_2$
5,6-Decanediol; Benzyl ether, *in* D-60013

$C_{17}H_{28}O_3$
6β-Acetoxy-1α,4α-epoxyeudesmane, *in* E-60019
3-Acetoxy-1(10),5-germacradien-4-ol, *in* G-80010
Carotdiol acetate, *in* D-80004
Fauronyl acetate, *in* F-70002
5,10(14)-Germacradiene-1,4-diol; 1-Ac, *in* G-70009
4,6-Heptadecadiyne-3,8,9-triol, *in* D-80321

$C_{17}H_{28}O_4$
6β-Acetoxy-4(15)-eudesmene-1β,5α-diol, *in* E-60068
Shiromodiol; 8-Ac, *in* S-70039

$C_{17}H_{28}O_5$
12-Acetoxy-5,8-dihydroxyfarnesol, *in* T-60363

$C_{17}H_{29}N$
2,4,6-Tri-*tert*-butylpyridine, T-70217

$C_{17}H_{30}O$
3,4,7,11-Tetramethyl-6,10-tridecadienal, T-60156

$C_{17}H_{32}$
1,1,2,2,3,3,4,4,5,5-Decamethyl-6-methylenecyclohexane, D-60012

$C_{17}H_{32}O_2$
Cyclopentadecanone; Ethylene acetal, *in* C-60219

$C_{17}H_{32}O_4$
2-Dodecyl-3-methylbutanedioic acid, D-80527

$C_{17}H_{34}O$
6,12-Dimethyl-2-pentadecanone, D-80465
Undecamethylcyclohexanol, U-60003

$C_{17}H_{34}O_3$
11-Hydroxyhexadecanoic acid; Me ester, *in* H-80172

$C_{17}H_{36}I_2N_2$
Camphonium, *in* T-70277

$C_{17}H_{36}N_2^{2\oplus}$
Trimethidinium(2+), T-70277

$C_{17}H_{37}NO_2$
Tetrabutylammonium(1+); Formate, *in* T-80014

$C_{17}H_{37}N_3O_3$
1,7,14-Trioxa-4,11,17-triazacycloeicosane; 4,11,17-Tri-Me, *in* T-80357

$C_{18}H_6N_6O_{12}$
Benzo[1,2-*b*:3,4-*b*′:5,6-*b*″]tripyrazine-2,3,6,7,10,11-tetracarboxylic acid, B-60051

$C_{18}H_8N_2O_6$
1,1′-(1,2-Dioxo-1,2-ethanediyl)bis-1*H*-indole-2,3-dione, D-70466

$C_{18}H_8O_6$
10-Hydroxy-2*H*,13*H*-furo[3,2-*c*:5,4-*h*′]bis[1]benzopyran-2,13-dione, H-70133

$C_{18}H_9ClO_2$
7-Oxo-7*H*-benz[*de*]anthracene-2-carboxylic acid; Chloride, *in* O-80059
7-Oxo-7*H*-benz[*de*]anthracene-3-carboxylic acid; Chloride, *in* O-80060

$C_{18}H_9NO$
2-Cyanobenzanthrone, *in* O-80059
3-Cyanobenzanthrone, *in* O-80060
4-Cyanobenzanthrone, *in* O-80061

$C_{18}H_9N_3O_3$
19,20,21-Triazatetracyclo[13.3.1.1^{3,7}.1^{9,13}]heneicosa-1(19),3,5,7(21),9,11,13(20),15,17-nonaene-2,8,14-trione, T-60236

$C_{18}H_{10}$
▷ Cyclopenta[*cd*]pyrene, C-60223
1,8-Diethynylanthracene, D-80155

$C_{18}H_{10}N_2$
9,10-Dihydro-9,10-ethenoanthracene-11,12-dicarboxylic acid; Dinitrile, *in* D-70208

$C_{18}H_{10}N_2O_2$
Triphenodioxazine, T-60378

$C_{18}H_{10}N_6$
Quinoxalino[1″,2″:1′,2′]imidazo[4′,5′:4,5]imidazo[1,2-*a*]quinoxaline, Q-80005

$C_{18}H_{10}O$
Pyreno[4,5-*b*]furan, P-60214
Pyreno[1,2-*c*]furan, P-70165
Pyreno[4,5-*c*]furan, P-70166

$C_{18}H_{10}OS$
Triphenyleno[1,12-*bcd*]thiophene; *S*-Oxide, *in* T-70310

$C_{18}H_{10}O_2$
Benz[*a*]anthracene-1,2-dione, B-70012
Benz[*a*]anthracene-3,4-dione, B-70013
Benz[*a*]anthracene-5,6-dione, B-70014
Benz[*a*]anthracene-7,12-dione, B-70015
Benz[*a*]anthracene-8,9-dione, B-70016
Benzo[*c*]phenanthrene-3,4-dione, B-80034
Naphth[2,3-*a*]azulene-5,12-dione, N-60006
1,3-Pyrenedicarboxaldehyde, P-80202
1,2-Triphenylenedione, T-70308
1,4-Triphenylenedione, T-70309

$C_{18}H_{10}O_2S$
Triphenyleno[1,12-*bcd*]thiophene; *S*-Dioxide, *in* T-70310

$C_{18}H_{10}O_3$
7-Oxo-7*H*-benz[*de*]anthracene-10-carboxylic acid, O-80064
7-Oxo-7*H*-benz[*de*]anthracene-11-carboxylic acid, O-80065
7-Oxo-7*H*-benz[*de*]anthracene-1-carboxylic acid, O-80058
7-Oxo-7*H*-benz[*de*]anthracene-2-carboxylic acid, O-80059
7-Oxo-7*H*-benz[*de*]anthracene-3-carboxylic acid, O-80060
7-Oxo-7*H*-benz[*de*]anthracene-4-carboxylic acid, O-80061
7-Oxo-7*H*-benz[*de*]anthracene-8-carboxylic acid, O-80062
7-Oxo-7*H*-benz[*de*]anthracene-9-carboxylic acid, O-80063

$C_{18}H_{10}O_6$
Bicoumol, B-70093
Edgeworin, E-80006

$C_{18}H_{10}O_7$
Edgeworthin, *in* D-70006

$C_{18}H_{10}O_9$
Variegatorubin, V-80003

$C_{18}H_{10}S$
Acenaphtho[1,2-*b*]benzo[*d*]thiophene, A-70016
Triphenyleno[1,12-*bcd*]thiophene, T-70310

$C_{18}H_{10}S_2$
Benzo[1,2-*b*:4,5-*b*′]bis[1]benzothiophene, B-60018
Benzo[1,2-*b*:5,4-*b*′]bis[1]benzothiophene, B-60019
Phenanthro[1,10-*bc*:8,9-*b*′,*c*′]bisthiopyran, P-60072

$C_{18}H_{11}Br$
1-Bromobenz[*a*]anthracene, B-80174
2-Bromobenz[*a*]anthracene, B-80175
3-Bromobenz[*a*]anthracene, B-80176
4-Bromobenz[*a*]anthracene, B-80177
▷ 7-Bromobenz[*a*]anthracene, B-80178
1-Bromochrysene, B-80187
2-Bromochrysene, B-80188
3-Bromochrysene, B-80189
4-Bromochrysene, B-80190
5-Bromochrysene, B-80191
6-Bromochrysene, B-80192

$C_{18}H_{11}N$
11*H*-Dibenzo[*a,def*]carbazole, D-80067
4*H*-Naphtho[1,2,3,4-*def*]carbazole, N-80004

$C_{18}H_{11}NO_2$
1*H*-Benz[*e*]indole-1,2(3*H*)-dione; *N*-Ph, *in* B-70026
Benz[*cd*]indol-2-(1*H*)-one; *N*-Benzoyl, *in* B-70027
7-Oxo-7*H*-benz[*de*]anthracene-11-carboxylic acid; Amide, *in* O-80065
7-Oxo-7*H*-benz[*de*]anthracene-2-carboxylic acid; Amide, *in* O-80059

$C_{18}H_{12}$
$\Delta^{1,1'}$-Biindene, B-60109
Cyclohepta[*a*]phenalene, C-60199
Cycloocta[*def*]phenanthrene, C-80187

$C_{18}H_{12}N_4$
5,14-Dihydroquinoxalino[2,3-*b*]phenazine, D-70264
[1,2,4,5]Tetrazino[1,6-*a*:4,3-*a*′]diisoquinoline, T-60172
[1,2,4,5]Tetrazino[1,6-*a*:4,3-*a*′]diquinoline, T-60173

$C_{18}H_{12}N_6$
2,4,6-Tri-2-pyridinyl-1,3,5-triazine, T-70316
2,4,6-Tri-3-pyridinyl-1,3,5-triazine, T-70317
2,4,6-Tri-4-pyridinyl-1,3,5-triazine, T-70318

$C_{18}H_{12}N_6O_6$
2*a*,4*a*,6*a*,8*a*,10*a*,12*a*-Hexaazacoronene-1,3,5,7,9,11(2*H*,4*H*,6*H*,8*H*,10*H*,12*H*)-hexone, H-80035

$C_{18}H_{12}N_{12}O_6$
Benzo[1,2-*b*:3,4-*b*′:5,6-*b*″]tripyrazine-2,3,6,7,10,11-tetracarboxylic acid; Hexaamide, *in* B-60051

$C_{18}H_{12}O$
Benz[*a*]anthracen-10-ol, B-80011
Benz[*a*]anthracen-11-ol, B-80012
▷ Chrysene-5,6-oxide, C-70174
3-Hydroxybenz[*a*]anthracene, H-70104
Triphenylene-1,2-oxide, T-60385

$C_{18}H_{12}O_2$
2,2′-Biindanylidene-1,1′-dione, B-80097
1,2-Di-(1-Benzofuranyl)ethylene, D-80069
1-Pyreneacetic acid, P-60208
4-Pyreneacetic acid, P-60209

$C_{18}H_{12}O_2S$
4,6-Diphenylthieno[2,3-*c*]furan, D-80514

$C_{18}H_{12}O_2S_2$
5*a*,5*b*,11*a*,11*b*-Tetrahydrocyclobuta[1,2-*b*:4,3-*b*′]benzothiopyran-11,12-dione, T-60063

$C_{18}H_{12}O_4$
9,10-Dihydro-9,10-ethenoanthracene-11,12-dicarboxylic acid, D-70208
(Diphenylpropadienylidene)propanedioic acid, D-70492
Glabone, G-60021
Kanjone, K-70003
Kanjone, K-80002
Pongone, P-70108

$C_{18}H_{12}O_4S$
5-Hydroxy-2-(phenylthio)-1,4-naphthoquinone; Ac, *in* H-70213
5-Hydroxy-3-(phenylthio)-1,4-naphthoquinone; Ac, *in* H-70214

$C_{18}H_{12}O_5$
Edulin, E-80008

$C_{18}H_{12}O_6$
Grevilline *A*, G-60038

$C_{18}H_{12}O_7$
Grevilline B, *in* G-60038
Racemic acid; Anhydride, dibenzoyl, *in* T-70005

$C_{18}H_{12}O_8$
Grevilline *D*, *in* G-60038
Grevilline C, *in* G-60038
2-Hydroxy-1,3-dimethoxy-8,9-methylenedioxycoumestan, *in* P-80029

$C_{18}H_{12}O_9$
Gomphidic acid, G-70026
Substictic acid, S-60060

$C_{18}H_{12}S$
10,11-Dihydrodiindeno[1,2-*b*:2′,1′-*d*]thiophene, D-80183

$C_{18}H_{13}N$
10,11-Dihydro-5*H*-diindeno[1,2-*b*;2′,1′-*d*]pyrrole, D-80182

$C_{18}H_{13}NO_2$
11*H*-Benzo[*a*]carbazole-1-carboxylic acid; Me ester, *in* B-80017

$C_{18}H_{13}N_3$
4-Anilino-1,10-phenanthroline, *in* A-70181
7,12-Dihydropyrido[3,2-*b*:5,4-*b*′]diindole; 7-Me, *in* D-80246
7,12-Dihydropyrido[3,2-*b*:5,4-*b*′]diindole; 12-Me; B,HCl, *in* D-80246

$C_{18}H_{13}N_3O$
Pyridazino[4,5-*b*]quinolin-10(5*H*)-one; 2-Benzyl, *in* P-80207

$C_{18}H_{14}$
1,1′-Bi-1*H*-indene, B-60108
1,1′-Bi-1*H*-indene, B-80098
2,2′-Bi-1*H*-indene, B-80099
2,3′-Bi-1*H*-indene, B-80100
3,3′-Bi-1*H*-indene, B-80101
[2.2.2](1,2,3)Cyclophane-1,9-diene, C-60229
7,12-Dihydropleiadene, D-60269
1,4-Dihydrotriphenylene, D-80262

$C_{18}H_{14}N_2$
1,2-Di-2-indolylethylene, D-80395

$C_{18}H_{14}N_2O_5$
4,4′-Diamino-2,2′-biphenyldicarboxylic acid; Anhydride, *N,N*′-Di-Ac, *in* D-80047

$C_{18}H_{14}N_4$
2,2′-Bis(2-imidazolyl)biphenyl, B-70148

$C_{18}H_{14}N_6$
1,4-Bis(2-pyridylamino)phthalazine, B-60172

$C_{18}H_{14}O$
2,7-Diphenyloxepin, D-60486

$C_{18}H_{14}O_2$
1,1′-Biindane-2,2′-dione, B-80095
2,2′-Biindane-1,1′-dione, B-80096
1,5-Diacetylanthracene, D-70028
1,6-Diacetylanthracene, D-70029
1,8-Diacetylanthracene, D-70030
9,10-Diacetylanthracene, D-70031
1,8-Phenanthrenedicarboxylic acid; Di-Me ester, *in* P-80078

$C_{18}H_{14}O_4$
3,4-Acenaphthenedicarboxylic acid; Di-Et ester, *in* A-60013
1,8-Anthracenedicarboxylic acid; Di-Me ester, *in* A-60281
7-Hydroxy-2-methylisoflavone; Ac, *in* H-80205
7-Hydroxy-3-methylisoflavone; Ac, *in* H-80206
Neodunol; Me ether, *in* N-80020
4-Oxo-2-phenyl-4*H*-1-benzopyran-3-carboxylic acid; Et ester, *in* O-60085
3,4-Phenanthrenedicarboxylic acid; Di-Me ester, *in* P-60068
Pongamol, P-80169

$C_{18}H_{14}O_6$
2,3-Dibenzoylbutanedioic acid, D-70083
2′,7-Dimethoxy-4′,5′-methylenedioxyisoflavone, *in* T-80107
3′,4′-Dimethoxy-6,7-methylenedioxyisoflavone, *in* T-80111
3′,4′-Dimethoxy-7,8-methylenedioxyisoflavone, *in* T-70112

4′,5-Dimethoxy-6,7-methylenedioxyisoflavone, *in* D-**8**0341
5,7-Dimethoxy-3′,4′-methylenedioxyisoflavone, *in* T-**8**0109
6,7-Dimethoxy-3′,4′-methylenedioxyisoflavone, *in* T-**8**0111

$C_{18}H_{14}O_7$
3′-Hydroxy-4′,5-dimethoxy-6,7-methylenedioxyiosflavone, *in* P-**8**0043
6-Hydroxy-2′,7-dimethoxy-4′,5′-methylenedioxyisoflavone, *in* P-**8**0042
7-Hydroxy-2′,6-dimethoxy-3′,4′-methylenedioxyisoflavone, *in* P-**8**0042
7-Hydroxy-5,6-dimethoxy-3′,4′-methylenedioxyisoflavone, *in* P-70032
7-Hydroxy-5,6-dimethoxy-3′,4′-methylenedioxyisoflavone, *in* P-**8**0043
7-Hydroxy-5,8-dimethoxy-3′,4′-methylenedioxyisoflavone, *in* P-**8**0045

$C_{18}H_{14}O_8$
8-Acetoxy-2′,3,5-trihydroxy-7-methoxyflavone, *in* P-70029
2,3-Di-*O*-benzoyltartaric acid, *in* T-70005

$C_{18}H_{14}S$
Di-2-indenyl sulfide, D-**8**0394

$C_{18}H_{15}NO_3$
10-Amino-9-anthracenecarboxylic acid; *N*-Ac, Me ester, *in* A-70088
2,2′,2″-Nitrilotrisphenol, N-70039
4,4′,4″-Nitrilotrisphenol, N-70040

$C_{18}H_{15}NO_7$
Sylvianecic acid, S-**8**0055

$C_{18}H_{15}N_3$
Amino-1,4-benzoquinone; Dianil, *in* A-60089

$C_{18}H_{16}$
Bi(cyclononatetraenylidene), B-70104
1,1′-Biindanylidene, B-70120
1,2-Dibenzylidenecyclobutane, D-60083
1,2-Diphenyl-1,4-cyclohexadiene, D-60477
1,3,5,7-Tetraethynyladamantane, T-70025
1,2,3,4-Tetrahydrotriphenylene, T-**8**0082
1,4,5,8-Tetrahydrotriphenylene, T-**8**0083

$C_{18}H_{16}Cl_2O_2$
3,4-Diphenylhexanedioic acid; Dichloride, *in* D-**8**0511

$C_{18}H_{16}N_2$
2,5-Diphenylhexanedioic acid; Dinitrile, *in* D-**8**0510
3,4-Diphenylhexanedioic acid; Dinitrile, *in* D-**8**0511

$C_{18}H_{16}N_2O_4$
5,5′-Diamino-2,2′-biphenyldicarboxylic acid; N,N′-Di-Ac, *in* D-**8**0051

$C_{18}H_{16}N_2O_6$
2,2′-Diamino-4,4′-biphenyldicarboxylic acid; N,N′-Di-Ac, *in* D-**8**0046
4,4′-Diamino-3,3′-biphenyldicarboxylic acid; N,N′-Di-Ac, *in* D-**8**0049

$C_{18}H_{16}N_2S_2$
4,6-Diamino-1,3-benzenedithiol; Di-*S*-Ph, *in* D-70037

$C_{18}H_{16}O_2$
2,7-Dihydroxy-1,6-dimethyl-5-vinylphenanthrene, D-70294
Eupomatenoid 6, E-60075

$C_{18}H_{16}O_3$
2-(2-Hydroxy-4-methoxyphenyl)-5-(1-propenyl)benzofuran, *in* H-60214
6-Hydroxy-7,9-octadecadiene-11,13,15,17-tetraynoic acid, H-70195
2-(4-Hydroxyphenyl)-7-methoxy-5-(1-propenyl)benzofuran, *in* H-60214
2-(4-Hydroxyphenyl)-3-methyl-5-(2-oxopropyl)benzofuran, *in* H-**8**0236

$C_{18}H_{16}O_4$
3-Butene-1,2-diol; Dibenzoyl, *in* B-70291
2-(2,4-Dihydroxyphenyl)-7-methoxy-5-(1-propenyl)benzofuran, *in* H-60214
2,2′-Dihydroxystilbene; Di-Ac, *in* D-**8**0378
3,5-Dihydroxystilbene; Di-Ac, *in* D-**8**0384
4,4′-Dihydroxystilbene; Di-Ac, *in* D-**8**0385

$C_{18}H_{16}O_4Se_2$
2,3-Bis(phenylseleno)-2-butenedioic acid; Di-Me ester, *in* B-**8**0159

$C_{18}H_{16}O_5$
4-Angeloyloxy-8,9-epoxy-7-(2,4-hexadiynylidene)-1,6-dioxaspiro[4.4]non-2-ene, *in* M-**8**0202
1,3-Dihydroxy-2-hydroxymethylanthraquinone; Tri-Me ether, *in* D-70306
▷ 2-Ethoxymethyl-3-hydroxy-1-methoxyanthraquinone, *in* D-70306
Globuxanthone, G-**8**0020
5-Hydroxy-4′,7-dimethoxy-6-methylflavone, *in* T-**8**0326
Intricatin, *in* I-**8**0017
8-Methoxybonducellin, *in* B-60192
Puerol *B*, *in* P-70137
1,3,6-Trihydroxy-2-(3-methyl-2-butenyl)-9*H*-xanthen-9-one, T-**8**0325
5,6,7-Trimethoxyflavone, *in* T-60322
2′,4′,7-Trimethoxyisoflavone, *in* T-70264
3′,4′,7-Trimethoxyisoflavone, *in* T-**8**0308
4′,6,7-Trimethoxyisoflavone, *in* T-**8**0310
4′,7,8-Trimethoxyisoflavone, *in* T-70265
1,2,8-Trimethoxy-3-methylanthraquinone, *in* T-**8**0322

$C_{18}H_{16}O_6$
2,3,4-Biphenyltriol; Tri-Ac, *in* B-70124
6,7-Dimethoxy-3′,4′-methylenedioxyflavanone, *in* T-70106
6,7-Dimethoxy-3′,4′-methylenedioxyisoflavanone, *in* T-**8**0102
2,3-Dimethoxy-8,9-methylenedioxypterocarpan, *in* T-**8**0117
3,4-Dimethoxy-8,9-methylenedioxypterocarpan, *in* T-**8**0119
5-Hydroxy-2′,6,6′-trimethoxyflavone, *in* T-70107
2′-Hydroxy-3′,4′,7-trimethoxyisoflavone, *in* T-**8**0105
5-Hydroxy-4′,6,7-trimethoxyisoflavone, *in* T-**8**0112
7-Hydroxy-2′,4′,5′-trimethoxyisoflavone, *in* T-**8**0107
7-Hydroxy-3′,4′,6-trimethoxyisoflavone, *in* T-**8**0111
7-Hydroxy-4′,5,6-trimethoxyisoflavone, *in* T-**8**0112
7-Hydroxy-4′,5,8-trimethoxyisoflavone, *in* T-**8**0113
Isoamericanin *A*, I-60087
Pedicellosine, P-70015
Scleroderolide, S-60013
1,3,5,8-Tetrahydroxy-2-(3-methyl-2-butenyl)xanthone, T-60114
2′,5′,6-Trihydroxy-3,5,7-trimethoxyflavone, *in* H-60064
3′,4′,5-Trihydroxy-3,6,7-trimethoxyflavone, *in* H-**8**0067

$C_{18}H_{16}O_7$
Arizonin C_1, A-70254
3,4′-Dihydroxy-5,6,7-trimethoxyflavone, *in* P-**8**0035
3,5-Dihydroxy-4′,6,7-trimethoxyflavone, *in* P-**8**0035
4′,5-Dihydroxy-3,6,7-trimethoxyflavone, *in* P-**8**0035
4′,5-Dihydroxy-3,6,8-trimethoxyflavone, *in* P-60048
5,6-Dihydroxy-3,4′,8-trimethoxyflavone, *in* P-60048
5,7-Dihydroxy-3,4′-6-trimethoxyflavone, *in* P-**8**0035
5,7-Dihydroxy-3,6,8-trimethoxyflavone, *in* P-70031
3′,8-Dihydroxy-4′,7,8-trimethoxyisoflavone, *in* P-**8**0046
4′,5-Dihydroxy-2′,5′,7-trimethoxyisoflavone, *in* P-**8**0041
5,7-Dihydroxy-3′,4′,7-trimethoxyisoflavone, *in* P-**8**0043
5,7-Dihydroxy-3′,4′,8-trimethoxyisoflavone, *in* P-**8**0045
6*a*-Hydroxy-2,3-dimethoxy-8,9-methylenedioxypterocarpan, *in* P-**8**0052
Notatic acid, N-70081
Psorospermindiol; *O*-De-Me, epimer, *in* P-70133

$C_{18}H_{16}O_8$
4′,5,7-Trihydroxy-3′,6,8-trimethoxyflavone, T-**8**0341
2′,3′,5-Trihydroxy-3,7,8-trimethoxyflavone, *in* H-70079
2′,5,5′-Trihydroxy-3,7,8-trimethoxyflavone, *in* H-70080
3′,4′,5-Trihydroxy-5′,6,7-trimethoxyflavone, *in* H-70081
3′,5,7-Trihydroxy-4′,5′,6-trimethoxyflavone, *in* H-70081
3′,5,7-Trihydroxy-4′,5′,6-trimethoxyisoflavone, *in* H-**8**0071

$C_{18}H_{16}O_9$
Alectorialic acid, A-70081
Decarboxythamnolic acid, D-**8**0014
3′,5,5′,7-Tetrahydroxy-3,4′,8-trimethoxyflavone, *in* H-70021

$C_{18}H_{16}O_{10}$
3′,4′,5,5′,7-Pentahydroxy-3,6,8-trimethoxyflavone, *in* O-70018

$C_{18}H_{17}ClO_7$
Wrightiin, W-**8**0005

$C_{18}H_{17}N$
2,2-Dimethyl-4,4-diphenyl-3-butenoic acid; Nitrile, *in* D-70402

$C_{18}H_{17}NO_2$
10-Amino-9-anthracenecarboxylic acid; *N*-Di-Me, Me ester, *in* A-70088

$C_{18}H_{17}NO_5$
2-Amino-4-hydroxybutanoic acid; *O*,*N*-Dibenzoyl, *in* A-70156
2-Amino-3-phenylpentanedioic acid; *N*-Benzoyl, *in* A-60241

$C_{18}H_{18}$
2-*tert*-Butylphenanthrene, B-70305
3-*tert*-Butylphenanthrene, B-70306
9-*tert*-Butylphenanthrene, B-70307
[2.2.2](1,2,3)Cyclophane, C-60228
1,2,3,4,7,8-Hexahydrobenzo[*c*]phenanthrene, H-**8**0052
1,2,3,6,7,8-Hexahydro-3*a*,5*a*-ethenopyrene, H-70056
1,5,9,10-Tetramethylanthracene, T-**8**0137
1,8,9,10-Tetramethylanthracene, T-**8**0138
1,3,6,8-Tetramethylphenanthrene, T-60151
2,4,5,7-Tetramethylphenanthrene, T-60152

$C_{18}H_{18}O$
2,2-Dimethyl-4,4-diphenyl-3-butenal, D-70401

$C_{18}H_{18}O_2$
Conocarpan, C-60161
2,2-Dimethyl-4,4-diphenyl-3-butenoic acid, D-70402
5,5-Diphenyl-4-pentenoic acid; Me ester, *in* D-**8**0513

$C_{18}H_{18}O_3$
Dehydroallogibberic acid, *in* A-**8**0045
2-(4-Hydroxyphenyl)-5-(2-Hydroxypropyl)-3-methylbenzofuran, H-**8**0236
Olmecol, O-60040

$C_{18}H_{18}O_3S$
1-(2-Furyl)-4[5-(isovaleryloxymethyl)-2-thienyl]-1-buten-3-yne, *in* F-**8**0093

$C_{18}H_{18}O_4$
Bharanginin, B-70089
Bibenzyl-2,2′-dicarboxylic acid; Di-Me ester, *in* B-**8**0071
Ceroptene, C-**8**0038
2,5-Diphenylhexanedioic acid, D-**8**0510
3,4-Diphenylhexanedioic acid, D-**8**0511
1,2,5,6-Tetramethoxyphenanthrene, *in* T-60115

$C_{18}H_{18}O_5 - C_{18}H_{28}BF_4NO_2$

1,2,5,7-Tetramethoxyphenanthrene, *in* T-80114
1,2,6,7-Tetramethoxyphenanthrene, *in* T-60117
1,3,5,6-Tetramethoxyphenanthrene, *in* T-60118
1,3,6,7-Tetramethoxyphenanthrene, *in* T-60119
2,3,5,7-Tetramethoxyphenanthrene, *in* T-60120
2,3,6,7-Tetramethoxyphenanthrene, *in* T-60121
2,4,5,6-Tetramethoxyphenanthrene, *in* T-60122
3,4,5,6-Tetramethoxyphenanthrene, *in* T-60123
2′,4′,6′-Trimethoxychalcone, *in* P-80139

$C_{18}H_{18}O_5$

5,6-Dimethoxy-3-(4-methoxybenzyl)phthalide, D-60398
Echinofuran, E-60002
Homocyclolongipesin; Ac, *in* H-60088
3-(4-Hydroxyphenyl)-1-(2-hydroxy-4,6-dimethoxyphenyl)-2-propen-1-one, *in* H-80246
2,3,9-Trimethoxypterocarpan, *in* T-80335
3,4,9-Trimethoxypterocarpan, *in* T-80336

$C_{18}H_{18}O_6$

Alkannin acetate, *in* S-70038
Curvularin; 10,11-Didehydro(*E*-), 7-Ac, *in* C-80166
5,8-Dihydroxy-6,10-dimethoxy-2-propyl-4*H*-naphtho[2,3-*b*]pyran-4-one, D-70292
5,8-Dihydroxy-2,3,6,7-tetramethyl-1,4-naphthoquinone; Di-Ac, *in* D-60377
1-(2,5-Dihydroxy-3,4,6-trimethoxyphenyl)-3-phenyl-2-propen-1-one, *in* P-80050
7-Hydroxy-2′,5,8-trimethoxyflavanone, *in* T-70104
2′-Hydroxy-3′,4′,7-trimethoxyisoflavanone, *in* T-80099
7-Hydroxy-2′,3′,4′-trimethoxyisoflavanone, *in* T-80099
Monocillin *I*, *in* N-60060
Shikonin acetate, *in* S-70038
3,4,7,8-Tetramethoxy-2,6-phenanthrenediol, *in* H-60066

$C_{18}H_{18}O_7$

2,8-Dihydroxy-3,9,10-trimethoxypterocarpan, *in* P-80054
Monomethylsulochrin, *in* S-70083
1,2,3,4,8-Pentamethoxyxanthone, *in* P-60051
Senepoxide, S-60023
β-Senepoxide, *in* S-60023
3′,4′,5,6-Tetrahydroxy-7-methoxyflavanone, *in* P-70028

$C_{18}H_{18}O_8$

Arizonin *A₂*, *in* A-70255
Arizonin *B₂*, *in* A-70255

$C_{18}H_{19}NO$

2,2-Dimethyl-4,4-diphenyl-3-butenal; Oxime, *in* D-70401
2-Phenylcyclopentylamine; Benzoyl, *in* P-60086
2-Phenylcyclopentylamine; *N*-Benzoyl, *in* P-60086

$C_{18}H_{19}NO_3S$

2-Amino-4,5,6,7-tetrahydrobenzo[*b*]thiophene-3-carboxylic acid; Et ester, *N*-benzoyl, *in* A-60258

$C_{18}H_{19}NO_7S$

Fibrostatin *A*, *in* F-70008

$C_{18}H_{19}NO_8S$

Fibrostatin *C*, *in* F-70008
Fibrostatin *D*, *in* F-70008
Fibrostatin *E*, *in* F-70008

$C_{18}H_{19}N_3O_5S$

Antibiotic BMY 28100, A-70221

$C_{18}H_{20}$

3,3-Dimethyl-4,4-diphenyl-1-butene, D-60418
5,13-Dimethyl[2.2]metacyclophane, D-60426

$C_{18}H_{20}N_2O_2$

5,6-Acenaphthenedicarboxylic acid; Bis(dimethylamide), *in* A-70015
2,2′-Diamino-4,4′-biphenyldicarboxylic acid; Di-Et ester, *in* D-80046

$C_{18}H_{20}N_2O_4$

N-Tyrosylphenylalanine, T-60412

$C_{18}H_{20}N_4$

Tricyclo[3.3.1.1³·⁷]decane-1,3,5,7-tetraacetonitrile, *in* A-70064

$C_{18}H_{20}O_2$

Pentamethylphenol; Benzoyl, *in* P-60058
Salviolone, S-70006
3,5-Tridecadiene-7,9,11-triyn-1-ol; 3-Methylbutanoyl, *in* T-80258

$C_{18}H_{20}O_3$

Allogibberic acid, A-80045
2,3-Dihydro-2-(4-hydroxyphenyl)-5-(3-hydroxypropyl)-3-methylbenzofuran, D-60244
1-(4-Hydroxyphenyl)-2-(4-propenylphenoxy)-1-propanol, H-60215
Tetrahydrobis(4-hydroxyphenyl)-3,4-dimethylfuran, T-60061

$C_{18}H_{20}O_4$

Angoletin, A-80156
2,5-Dihydroxy-3,6-dimethyl-3,6-diphenyl-1,4-dioxan, *in* H-80240
4,6,12-Tetradecatriene-8,10-diyne-1,3-diol; Di-Ac, *in* T-80033
Thujin, T-70190

$C_{18}H_{20}O_5$

9,10-Dihydro-7-hydroxy-2,3,4,6-tetramethoxyphenanthrene, *in* D-70246
Longipesin; Propanoyl, *in* L-60034
9-Methyllongipesin; *O*⁹-Ac, *in* M-70091
Monocillin II, *in* N-60060

$C_{18}H_{20}O_6$

3,3′-Dihydroxy-4′,5,7-trimethoxyflavan, *in* P-70026
3,4′-Dihydroxy-3′,5,7-trimethoxyflavan, *in* P-70026
3′,7-Dihydroxy-2′,4′,8-trimethoxyisoflavan, *in* P-80038
Epicatechin; *O*³′,*O*⁵,*O*⁷-Tri-Me, *in* P-70026
3-Methoxy-6-[2-(3,4,5-trimethoxyphenyl)ethenyl]-1,2-benzenediol, *in* T-60344
Monocillin III, *in* N-60060
Syringopicrogenin *A*, S-60067

$C_{18}H_{20}O_7$

Arnebin VI, *in* D-80343
Syringopicrogenin *B*, *in* S-60067
2′,6,8-Trihydroxy-3′,4′,7-trimethoxyisoflavan, *in* H-80068

$C_{18}H_{20}Se_2$

9,18-Dimethyl-2,11-diselena[3.3]metacyclophane, D-70404

$C_{18}H_{21}N_3O_6$

10-Deazariboflavin, *in* P-60245

$C_{18}H_{22}N_2O_3S_2$

Dithiosilvatin, D-60511

$C_{18}H_{22}N_4$

5,5a,6,11,11a,12-Hexahydroquinoxalino[2,3-*b*]quinoxaline; *N*-Tetra-Me, *in* H-80066

$C_{18}H_{22}O_2$

2,3-Dimethoxy-2,3-diphenylbutane, *in* D-80505
15-Octadecene-9,11,13-triynoic acid, O-80009

$C_{18}H_{22}O_3$

Nimbinone, N-70037
Nimbione, N-70038

$C_{18}H_{22}O_4$

1,12-Diacetoxy-4,6-tetradecadiene-8,10-diyne, *in* T-80026
12-Hydroxy-13-methoxy-8,11,13-podocarpatriene-3,7-dione, *in* N-80028
6,12-Tetradecadiene-8,10-diyne-1,3-diol; Di-Ac, *in* T-80027

$C_{18}H_{22}O_5$

1-(4-Hydroxy-3-methoxyphenyl)-2-(3,4,5-trimethoxyphenyl)-ethane, *in* D-80371
Monocillin IV, *in* N-60060
▷ Zearalenone, Z-60002

$C_{18}H_{22}O_6$

Combretastatin, C-70192
8′-Hydroxyzearalenone, *in* Z-60002
3-Methoxy-6-[2-(3,4,5-trimethoxyphenyl)ethyl]1,2-benzenediol, *in* T-60344
Monocillin V, *in* N-60060

$C_{18}H_{22}O_7$

1-(4-Hydroxy-3-methoxyphenyl)-2-(4-hydroxy-3,5-dimethoxyphenyl)-1,3-propanediol, *in* B-80150

$C_{18}H_{22}O_{11}$

Tectoridin, *in* T-80319

$C_{18}H_{23}NO_2$

3,7-Di-*tert*-butyl-1-nitronaphthalene, D-80110

$C_{18}H_{23}NO_3$

2,6-Di-*tert*-butyl-1-nitronaphthalene, D-80109

$C_{18}H_{24}$

1,2,3,4,5,6,7,8,9,10,11,12-Dodecahydrotriphenylene, D-60516
1,3,5,7-Tetravinyladamantane, T-70154

$C_{18}H_{24}N_2$

1,2-Diphenyl-1,2-ethanediamine; *N*-Tetra-Me, *in* D-70480

$C_{18}H_{24}O_2$

1-Acetoxy-6,8,12,14-hexadecatetraen-10-yne, *in* H-80041
3-(2-Hydroxy-4,8-dimethyl-3,7-nonadienyl)benzaldehyde, H-60122
Nimbiol, N-80027
Norsalvioxide, N-70079

$C_{18}H_{24}O_4$

3-Acetoxy-6-methoxyprimnatrienone, *in* P-80174
Nimbionol, N-80028
2,4,6-Trihydroxy-3,5-bis(3-methyl-2-butenyl)acetophenone, T-80289

$C_{18}H_{24}O_5$

Nordinone, N-60060

$C_{18}H_{24}O_7$

Malabarolide, M-70008
Nordinonediol, *in* N-60060

$C_{18}H_{24}O_8$

1,3,5,7-Adamantanetetraacetic acid, A-70064
1,3,5,7-Adamantanetetracarboxylic acid; Tetra-Me ester, *in* A-80037
Hyptolide, H-70236

$C_{18}H_{24}O_{12}$

myo-Inositol; Hexa-Ac, *in* I-80015

$C_{18}H_{26}O_3$

Ecklonialactone A, E-80005

$C_{18}H_{26}O_4$

3-Deoxybarbacenic acid, *in* B-80009

$C_{18}H_{26}O_5$

Barbacenic acid, B-80009

$C_{18}H_{26}O_6$

3-Oxo-17-carboxy-3,18-secobarbacenic acid, O-80071

$C_{18}H_{26}O_8$

Boronolide, B-60193

$C_{18}H_{27}N_3$

1,3,5-Tripyrrolidinobenzene, T-70319

$C_{18}H_{27}N_3O_3$

1,3,5-Trimorpholinobenzene, T-70294

$C_{18}H_{28}$

9,9′-Bi(bicyclo[3.3.1]nonylidene), B-60079

$C_{18}H_{28}BF_4NO_2$

1,1-*tert*-Butyl-3,3-diethoxy-2-azaallenium(1+); Tetrafluoroborate, *in* B-70301

$C_{18}H_{28}O_2$
11,17-Octadecadien-9-ynoic acid, O-**80006**
4,8,12,15-Octadecatetraenoic acid, O-**80008**
Spiro[3,4-cyclohexano-4-hydroxybicyclo[3.3.1]nonan-9-one-2,1'-cyclohexane], S-**70063**

$C_{18}H_{28}O_3$
ent-6β,17-Dihydroxy-14,15-bisnor-7,11E-labdadien-13-one, D-**60311**
14,15-Dinor-13-oxo-7-labden-17-oic acid, D-**60457**
Ecklonialactone B, in E-**80005**
2-Hydroxy-6-undecylbenzoic acid, H-**80264**

$C_{18}H_{28}O_4$
ent-12-Hydroxy-13,14,15,16-tetranor-1(10)-halimen-18-oic acid; Ac, in H-**60231**
Xestodiol, X-**70006**

$C_{18}H_{28}O_5$
Lachnellulone, L-**70015**

$C_{18}H_{28}S_6$
1,3,4,6-Tetrakis(isopropylthio)thieno[3,4-c]thiophene, in T-**70170**

$C_{18}H_{29}NO_2$
1,3,5-Tri-tert-butyl-2-nitrobenzene, T-**70216**

$C_{18}H_{30}$
▷ Hexaethylbenzene, H-**70043**
Octadecahydrotriphenylene, O-**70004**
1,3,5-Tributylbenzene, T-**70215**

$C_{18}H_{30}O_2$
5,6-Epoxy-6,10,14-trimethyl-9,13-pentadecadien-2-one, E-**80047**
2-Hydroxy-14,15-bisnor-7-labden-13-one, H-**80133**
17-Octadecen-9-ynoic acid, O-**80010**

$C_{18}H_{30}O_3$
Colneleic acid, C-**70189**
▷ Conocandin, C-**70194**
Sterebin D, S-**60052**

$C_{18}H_{30}O_4$
11-Hydroxy-12,13-epoxy-9,15-octadecadienoic acid, H-**70127**
Secoisoobtusilactone, S-**80015**
Sterebin A, in S-**60052**
3,7,11-Trimethyl-2,6,10-dodecatrienoic acid; 2,3-Dihydroxypropyl ester, in T-**60364**

$C_{18}H_{30}O_5$
Gloeosporone, G-**60025**

$C_{18}H_{30}O_6$
1,2:5,6-Di-O-cyclohexylidene-myo-inositol, in I-**80015**

$C_{18}H_{30}S_3$
7,14,21-Trithiatrispiro[5.1.5.1.5.1]heneicosane, in C-**60209**

$C_{18}H_{32}N_4O_8$
1,4,8,11-Tetraazacyclotetradecane-1,4,8,11-tetraacetic acid, T-**60017**

$C_{18}H_{32}O_2$
2,4-Octadecadienoic acid, O-**70003**

$C_{18}H_{32}O_3$
Feronolide, F-**80004**
13-Hydroxy-9,11-octadecadienoic acid, H-**70196**

$C_{18}H_{32}O_5$
Aspicillin, A-**60312**

$C_{18}H_{33}FO_2$
18-Fluoro-9-octadecenoic acid, F-**80057**

$C_{18}H_{34}O$
2,4-Octadecadien-1-ol, O-**60004**

$C_{18}H_{34}O_2$
1-Tetradecylcyclopropanecarboxylic acid, T-**60040**

$C_{18}H_{34}O_4$
4,14-Dihydroxyoctadecanoic acid, D-**80345**
11-Hydroxyhexadecanoic acid; Ac, in H-**80172**

$C_{18}H_{35}FO_2$
2-Fluorooctadecanoic acid, F-**80056**

$C_{18}H_{36}N_2O_6$
▷ 4,7,13,16,21,24-Hexaoxa-1,10-diazabicyclo[8.8.8]hexacosane, H-**60073**

$C_{18}H_{36}N_6$
Hexakis(dimethylamino)benzene, in B-**80014**

$C_{18}H_{36}O$
2-Octadecen-1-ol, O-**60007**

$C_{18}H_{36}O_2$
3-Octadecene-1,2-diol, O-**60005**
9-Octadecene-1,12-diol, O-**60006**

$C_{18}H_{36}O_3$
11-Hydroxyhexadecanoic acid; Et ester, in H-**80172**
17-Hydroxyoctadecanoic acid, H-**80220**

$C_{18}H_{36}O_5$
9,10,18-Trihydroxyoctadecanoic acid, T-**60334**

$C_{18}H_{38}$
3,4-Di-tert-butyl-2,2,5,5-tetramethylhexane, D-**70109**

$C_{18}H_{38}N_2$
1,1,1',1'-Tetra-tert-butylazomethane, T-**60024**

$C_{18}H_{38}O$
2-Octadecanol, O-**80007**

$C_{18}H_{39}NO_2$
Tetrabutylammonium(1+); Acetate, in T-**80014**

$C_{18}H_{39}N_3O_3$
1,9,17-Trioxa-5,13,21-triazacyclotetracosane, T-**80358**

$C_{18}H_{42}N_6$
1,4,7,13,16,19-Hexaazacyclotetracosane, H-**70029**

$C_{18}H_{42}N_6O_2$
1,13-Dioxa-4,7,10,16,20,24-hexaazacyclohexacosane, D-**70459**

$C_{18}N_6O_9$
Benzo[1,2-b:3,4-b':5,6-b'']tripyrazine-2,3,6,7,10,11-tetracarboxylic acid; Trianhydride, in B-**60051**

$C_{18}N_{12}$
Benzo[1,2-b:3,4-b':5,6-b'']tripyrazine-2,3,6,7,10,11-tetracarboxylic acid; Hexanitrile, in B-**60051**

$C_{19}Cl_{15}$
Tris(pentachlorophenyl)methyl, in T-**60405**

$C_{19}Cl_{21}Sb$
Tris(pentachlorophenyl)methyl hexachloroantimonate, in T-**60405**

$C_{19}HCl_{15}$
Tris(pentachlorophenyl)methane, T-**60405**

$C_{19}H_{10}O_2$
8H-Pyreno[2,1-b]pyran-8-one, P-**80204**
9H-Pyreno[1,2-b]pyran-9-one, P-**80205**

$C_{19}H_{10}O_6$
Dehydrodolineone, in D-**80528**

$C_{19}H_{11}ClO$
2-Triphenylenecarboxylic acid; Chloride, in T-**60384**

$C_{19}H_{11}N$
1-Cyanochrysene, in C-**80128**
3-Cyanochrysene, in C-**80130**
5-Cyanochrysene, in C-**80131**
6-Cyanochrysene, in C-**80132**
2-Cyanotriphenylene, in T-**60384**

$C_{19}H_{11}NO$
Dibenzo[f,h]furo[2,3-b]quinoline, D-**80070**

$C_{19}H_{11}NOS$
Dibenzo[f,h]thieno[2,3-b]quinoline; 1-Oxide, in D-**70079**

$C_{19}H_{11}NS$
Dibenzo[f,h]thieno[2,3-b]quinoline, D-**70079**

$C_{19}H_{12}N_2O$
Besthorn's red, B-**80063**

$C_{19}H_{12}N_2O_2$
Cyclohepta[1,2-b:1,7-b']bis[1,4]benzoxazine, C-**70219**

$C_{19}H_{12}N_4O_7$
Benzo[a]quinolizinium(1+); Picrate, in B-**70043**

$C_{19}H_{12}O$
2-Triphenylenecarboxaldehyde, T-**60382**

$C_{19}H_{12}O_2$
2-Benzoylnaphtho[1,8-bc]pyran, B-**70076**
1-Chrysenecarboxylic acid, C-**80128**
2-Chrysenecarboxylic acid, C-**80129**
3-Chrysenecarboxylic acid, C-**80130**
5-Chrysenecarboxylic acid, C-**80131**
6-Chrysenecarboxylic acid, C-**80132**
1-Triphenylenecarboxylic acid, T-**60383**
2-Triphenylenecarboxylic acid, T-**60384**

$C_{19}H_{12}O_3$
7-Oxo-7H-benz[de]anthracene-10-carboxylic acid; Me ester, in O-**80064**
7-Oxo-7H-benz[de]anthracene-11-carboxylic acid; Me ester, in O-**80065**
7-Oxo-7H-benz[de]anthracene-4-carboxylic acid; Me ester, in O-**80061**
7-Oxo-7H-benz[de]anthracene-8-carboxylic acid; Me ester, in O-**80062**
7-Oxo-7H-benz[de]anthracene-9-carboxylic acid; Me ester, in O-**80063**

$C_{19}H_{12}O_5$
1,6,8-Trihydroxy-3-methylbenz[a]anthracene-7,12-dione, T-**70270**

$C_{19}H_{12}O_6$
Bhubaneswin, in B-**70093**
Dolineone, D-**80528**

$C_{19}H_{12}O_7$
Daphnoretin, D-**70006**

$C_{19}H_{13}N$
8H-Indolo[3,2,1-de]acridine, I-**70014**

$C_{19}H_{13}NO_2S$
10H-Phenothiazine-3-carboxylic acid; N-Ph, in P-**70057**

$C_{19}H_{13}NO_4S$
10H-Phenothiazine-3-carboxylic acid; N-Ph, 5,5-dioxide, in P-**70057**

$C_{19}H_{13}NO_8$
Protetrone, P-**70124**

$C_{19}H_{14}$
4b,10b-Methanochrysene, M-**70040**

$C_{19}H_{14}N_4$
4-Amino-1,2,3-benzotriazine; 3,N(4)-Di-Ph, in A-**60092**

$C_{19}H_{14}O$
3-Methoxybenz[a]anthracene, in H-**70104**

$C_{19}H_{14}OS$
9-(Phenylsulfinyl)-9H-fluorene, in F-**80015**

$C_{19}H_{14}O_2$
1-Pyreneacetic acid; Me ester, in P-**60208**
4-Pyreneacetic acid; Me ester, in P-**60209**

$C_{19}H_{14}O_2S$
9-(Phenylsulfonyl)-9H-fluorene, in F-**80015**

$C_{19}H_{14}O_4$
4',7-Dimethoxyisoflavone, in D-**80323**
Ochromycinone, O-**60003**

$C_{19}H_{14}O_6$
4',7-Dihydroxyisoflavone; Di-Ac, in D-**80323**
Ficinin, in E-**80008**

$C_{19}H_{14}O_7$
6-Aldehydoisoophiopogone A, A-**60071**

$C_{19}H_{14}O_8$
1,3,7-Trihydroxyxanthone; Tri-Ac, in T-**80347**

$C_{19}H_{14}O_{10}$
Constictic acid, C-**60163**

C₁₉H₁₄S
9-(Phenylthio)-9*H*-fluorene, *in* F-**80015**

C₁₉H₁₄S₂
2-Dibenzothiophenethiol; *S*-Benzyl, *in* D-**60078**
4-Dibenzothiophenethiol; *S*-Benzyl, *in* D-**60079**

C₁₉H₁₅N₃
Azidotriphenylmethane, A-**60352**

C₁₉H₁₆
4-(2,4,6-Cycloheptatrien-1-ylidene)bicyclo[5.4.1]dodeca-2,5,7,9,11-pentaene, C-**60207**

C₁₉H₁₆O
2,5-Dimethyl-3,4-diphenyl-2,4-cyclopentadien-1-one, D-**70403**

C₁₉H₁₆O₄
1,7-Bis(4-hydroxyphenyl)-1,6-heptadiene-3,5-dione, *in* C-**80165**
4-Cyclopentene-1,3-diol; Dibenzoyl, *in* C-**80193**
2-(4-Hydroxyphenyl)-7-methoxy-5-(1-propenyl)-3-benzofurancarboxaldehyde, H-**60212**
Moracin *D*, M-**60143**
Moracin *E*, M-**60144**
Moracin *G*, M-**60145**

C₁₉H₁₆O₅
Ambonane, *in* N-**80020**
3′,4′-Deoxypsorospermin, D-**60021**
Neoraunone, *in* N-**80025**

C₁₉H₁₆O₆
6-Aldehydoisoophiopogone *B*, A-**60072**
4′,7-Dihydroxyisoflavanone; Di-Ac, *in* D-**80322**
Psorospermin, P-**70132**

C₁₉H₁₆O₇
Fridamycin *E*, *in* F-**70043**
5′-Hydroxypsorospermin, *in* P-**70132**
2′,6,7-Trimethoxy-4′,5′-methylenedioxyisoflavone, *in* P-**80042**
3′,4′,8-Trimethoxy-6,7-methylenedioxyisoflavone, *in* P-**80046**
3′,6,7-Trimethoxy-4′,5′-methylenedioxyisoflavone, *in* P-**80044**
5,6,7-Trimethoxy-3′,4′-methylenedioxyisoflavone, *in* P-**70032**
5,6,7-Trimethoxy-3′,4′-methylenedioxyisoflavone, *in* P-**80043**
6,7,8-Trimethoxy-3′,4′-methylenedioxyisoflavone, *in* P-**80046**

C₁₉H₁₆O₁₃
Heaxahydroxydiphenic acid α-L-arabinosediyl ester, H-**80003**

C₁₉H₁₇ClO₆
3′,4′-Deoxy-4′-chloropsorospermin-3′-ol, *in* D-**60021**

C₁₉H₁₇N
3,5-Dimethyl-2,6-diphenylpyridine, D-**80447**

C₁₉H₁₇NO₂
2-Hydroxycyclononanone; Oxime, *in* H-**60112**

C₁₉H₁₈
Bicyclo[12.4.1]nonadeca-1,3,5,7,9,11,13,15,17-nonaene, B-**80081**
Bicyclo[12.4.1]nonadec-1,3,5,7,9,11,13,15,17-nonaene, B-**70101**

C₁₉H₁₈N₂
3,3′-Methylenebisindole; 1,1′-Di-Me, *in* M-**80078**
(Triphenylmethyl)hydrazine, T-**60388**

C₁₉H₁₈O₃
Eupomatenoid 5, *in* E-**60075**
1-[4-Hydroxy-3-(3-methyl-1,3-butadienyl)phenyl]-2-(3,5-dihydroxyphenyl)ethylene, H-**70167**
Kachirachirol *A*, *in* K-**60001**
Ratanhiaphenol III, R-**80003**

C₁₉H₁₈O₄
1,2-Cyclopentanediol; Dibenzoyl, *in* C-**80189**
Demethylfruticulin *A*, *in* F-**60076**
1,2-Diphenyl-1,2-cyclopropanedicarboxylic acid; Di-Me ester, *in* D-**80509**
2-(2-Hydroxy-4-methoxyphenyl)-7-methoxy-5-(1-propenyl)benzofuran, *in* H-**60214**
Isotanshinone IIB, I-**60138**
6-Methoxy-2-[2-(4-methoxyphenyl)ethyl]-4*H*-1-benzopyran-4-one, M-**60039**
Moracin *C*, *in* M-**60146**

C₁₉H₁₈O₅
Benzophenone-2,2′-dicarboxylic acid; Di-Et ester, *in* B-**70041**
7-[[4-(2,5-Dihydro-4-methyl-5-oxo-2-furanyl)-3-methyl-2-butenyl]oxy]-2*H*-1-benzopyran-2-one, D-**80223**
1,6-Dihydroxy-3-methoxy-2-(3-methyl-2-butenyl)xanthone, *in* T-**80325**
Egonol, E-**70001**

C₁₉H₁₈O₆
Ciliarin, *in* C-**70007**
Cudraniaxanthone, C-**80163**
7-[[3-[(2,5-Dihydro-4-methyl-5-oxo-2-furanyl)methyl]-3-methyloxiranyl]methoxy]-2*H*-1-benzopyran-2-one, *in* D-**80223**
Fruticulin *B*, F-**60077**
Munduserone, M-**80196**
2′,5,6,6′-Tetramethoxyflavone, *in* T-**70107**
3′,4′,5,6-Tetramethoxyflavone, *in* T-**70108**
3′,4′,6,7-Tetramethoxyflavone, *in* T-**80092**
2′,4′,5′,7-Tetramethoxyisoflavone, *in* T-**80107**
3′,4′,5′,7-Tetramethoxyisoflavone, *in* T-**80110**
3′,4′,6,7-Tetramethoxyisoflavone, *in* T-**80111**
1,5,8-Trihydroxy-3-methyl-2-(3-methyl-2-butenyl)xanthone, *in* T-**60114**

C₁₉H₁₈O₇
3′,4′-Deoxypsorospermin-3′,4′-diol, *in* D-**60021**
Gerontoxanthone D, G-**80015**
5-Hydroxy-2′,3,7,8-tetramethoxyflavone, *in* P-**70029**
5-Hydroxy-2′,4′,7,8-tetramethoxyflavone, *in* P-**80034**
5-Hydroxy-3,4′,6,7-tetramethoxyflavone, *in* P-**80035**
5-Hydroxy-2′,4′,5′,7-tetramethoxyisoflavone, *in* P-**80041**
7-Hydroxy-2′,4′,5′,6-tetramethoxyisoflavone, *in* P-**80042**
Psorospermindiol, P-**70133**
Sermundone, *in* M-**80196**

C₁₉H₁₈O₈
2′-Acetoxy-3,5-dihydroxy-7,8-dimethoxyflavone, *in* P-**60045**
3′,5-Dihydroxy-4′,6,7,8-tetramethoxyflavone, D-**60375**
5,7-Dihydroxy-3′,4′,6,8-tetramethoxyflavone, D-**60376**
4′,5-Dihydroxy-3,3′,6,7-tetramethoxyflavone, *in* H-**80067**
4′,5-Dihydroxy-3,3′,6,7-tetramethoxyflavone, *in* H-**80067**
4′,5-Dihydroxy-3′,5′,6,7-tetramethoxyflavone, *in* H-**70081**
5′,5-Dihydroxy-2′,3,7,8-tetramethoxyflavone, *in* H-**70080**
5,7-Dihydroxy-3′,4′,5′,6-tetramethoxyflavone, *in* H-**70081**
5,7-Dihydroxy-2′,4′,5′,6-tetramethoxyisoflavone, *in* H-**80069**
5,7-Dihydroxy-2′,4′,5′,8-tetramethoxyisoflavone, *in* H-**80070**
5,7-Dihydroxy-3′,4′,5′,6-tetramethoxyisoflavone, *in* H-**80071**
5′,6-Dihydroxy-2′,3,5,7-trimethoxyflavone, *in* H-**60064**

C₁₉H₁₈O₉
3,4′,5-Trihydroxy-3′,6,7,8-tetramethoxyflavone, *in* H-**80028**
3′,5,5′-Trihydroxy-3,4′,7,8-tetramethoxyflavone, *in* H-**70021**
3′,5,5′-Trihydroxy-4′,6,7,8-tetramethoxyflavone, *in* H-**80029**
3′,5,7-Trihydroxy-3,4′,5′,8-tetramethoxyflavone, *in* H-**70021**
3′,5,7-Trihydroxy-4′,5′,6,8-tetramethoxyflavone, *in* H-**80029**

C₁₉H₁₈O₁₀
2′,4′,5,7-Tetrahydroxy-3,5′,6,8-tetramethoxyflavone, *in* O-**60022**
2′,5,5′,7-Tetrahydroxy-3,4′,6,8-tetramethoxyflavone, *in* O-**60022**
3′,5,5′,6-Tetrahydroxy-3,4′,7,8-tetramethoxyflavone, *in* O-**70018**
3′,5,5′,7-Tetrahydroxy-3,4′,6,8-tetramethoxyflavone, *in* O-**70018**
2′,5,7-Trihydroxy-3,4′,5′,6,8-pentamethoxyflavone, *in* O-**60022**

C₁₉H₁₈O₁₁
1,3,5,8-Tetrahydroxyxanthone; 8-Glucoside, *in* T-**70123**

C₁₉H₁₉ClO₇
Eriodermic acid, E-**70032**
Methyl 5-chloro-4-*O*-demethylbarbatate, *in* B-**70007**

C₁₉H₂₀
1-Methyl-4,4-diphenylcyclohexene, M-**60065**
3-Methyl-4,4-diphenylcyclohexene, M-**60066**

C₁₉H₂₀O₂
3,3-Dimethyl-5,5-diphenyl-4-pentenoic acid, D-**60420**
Hermosillol, H-**60029**

C₁₉H₂₀O₃
Cryptotanshinone, C-**60173**
2,3-Dihydro-2-(4-hydroxy-3-methoxyphenyl)-3-methyl-5-(1-propenyl)benzofuran, *in* C-**60161**
2,3-Dihydro-2-(4-hydroxyphenyl)-7-methoxy-3-methyl-5-(1-propenyl)benzofuran, *in* C-**60161**
2,2′-Dihydroxy-3-methoxy-5,5′-di-2-propenylbiphenyl, D-**80334**
Zapotecone, *in* Z-**80001**

C₁₉H₂₀O₄
Galipein, G-**60001**
Honyudisin, H-**70093**
Kachirachirol *B*, K-**60001**
Nordentatin, N-**80079**
Umbelliferone 8-oxogeranyl ether, U-**80001**

C₁₉H₂₀O₅
Homocyclolongipesin; Propanoyl, *in* H-**60088**
3-(4-Hydroxyphenyl)-1-(2,4,6-trihydroxyphenyl)-2-propen-1-one; Tetra-Me ether, *in* H-**80246**
2,3,4,7,9-Pentamethoxyphenanthrene, *in* P-**70035**
Teuchamaedryn *A*, T-**70157**

C₁₉H₂₀O₆
Asadanin, A-**60308**
Calaxin, C-**70007**
Isoteucrin H₄, I-**80093**
Khellactone; *O*¹⁰-Angeloyl, *in* K-**70013**
Khellactone; *O*⁹-Angeloyl, *in* K-**70013**
1(10)*E*,8E-Millerdienolide, *in* M-**60136**
2′,4′,5,7-Tetramethoxyisoflavanone, *in* T-**80100**

C₁₉H₂₀O₇
Annulin *A*, A-**60276**
Barbatic acid, B-**70007**
4β,15-Epoxy-1(10)*E*,8E-millerdienolide, *in* M-**60136**
8-Hydroxy-3,4,9,10-tetramethoxypterocarpan, *in* P-**80055**
Isoelephantopin, I-**60099**

C₁₉H₂₀O₈
2,8-Dihydroxy-3,4,9,10-tetramethoxypterocarpan, *in* H-**80072**
Eximin, *in* A-**70248**

C₁₉H₂₁NO₈S
Fibrostatin *B*, *in* F-**70008**

C₁₉H₂₁NO₉S
Fibrostatin *F*, *in* F-**70008**

C$_{19}$H$_{32}$O$_2$
15-Hydroxy-17-nor-8-labden-7-one, H-60198

C$_{19}$H$_{32}$O$_3$
4,6-Dihydroxy-20-nor-2,7-cembradien-12-one,
D-70323
Juvenile hormone O, J-80018

C$_{19}$H$_{32}$O$_4$
4,6-Dihydroxy-7,8-epoxy-20-nor-2-cembren-
12-one, *in* D-70323

C$_{19}$H$_{32}$O$_5$
Methyl 8α-acetoxy-3α-hydroxy-13,14,15,16-
tetranorlabdan-12-oate, *in* D-80387

C$_{19}$H$_{33}$NO$_2$
▷ Dysidazirine, D-80538

C$_{19}$H$_{34}$O$_2$
2,4-Octadecadienoic acid; Me ester, *in*
O-70003

C$_{19}$H$_{34}$O$_3$
(−)-Coriolic acid; Me ester, *in* H-70196
2-(5-Methoxy-5-methyltetrahydro-2-furanyl)-
6,10-dimethyl-5,9-undecadien-2-ol,
M-80047

C$_{19}$H$_{34}$O$_5$
3,6-Epidioxy-6-methoxy-4-octadecenoic acid,
E-60008

C$_{19}$H$_{36}$N$_2$O$_5$
▷ Lipoxamycin, L-70033

C$_{19}$H$_{36}$O$_2$
2-(2,4-Hexadiynylidene)-3,4-epoxy-1,6-
dioxaspiro[4.5]decane, *in* H-70042

C$_{19}$H$_{37}$FO$_2$
2-Fluorooctadecanoic acid; Me ester, *in*
F-80056

C$_{19}$H$_{37}$NO$_2$
14-Azaprostanoic acid, A-60332

C$_{19}$H$_{38}$O$_3$
17-Hydroxyoctadecanoic acid; Me ester, *in*
H-80220

C$_{19}$H$_{38}$O$_5$
Phloinolic acid; Me ester, *in* T-60334

C$_{19}$H$_{42}$N$_2$O$_8$S$_2$
▷ Trimethidinium methosulphate, *in* T-70277

C$_{20}$H$_4$Cl$_4$I$_4$O$_5$
Rose bengal, R-80010

C$_{20}$H$_8$O$_4$S$_2$
Dibenzo[b,i]thianthrene-5,7,12,14-tetrone,
D-70078

C$_{20}$H$_{10}$O$_2$
Indeno[1,2-b]fluorene-6,12-dione, I-70011

C$_{20}$H$_{10}$O$_4$
4,4′-Bi(1,2-naphthaquinone), B-80103
2,2′-Bi(1,4-naphthoquinone), B-80104
6,7-Dihydroxy-1,12-perylenedione, D-80354

C$_{20}$H$_{11}$F
10-Fluorobenzo[a]pyrene, F-80027
6-Fluorobenzo[a]pyrene, F-80023
7-Fluorobenzo[a]pyrene, F-80024
8-Fluorobenzo[a]pyrene, F-80025
9-Fluorobenzo[a]pyrene, F-80026

C$_{20}$H$_{11}$I
6-Iodobenzo[a]pyrene, I-80018

C$_{20}$H$_{11}$N
1H-Phenanthro[1,10,9,8-cdefg]carbazole,
P-80084

C$_{20}$H$_{12}$
Benz[a]aceanthrylene, B-70009
Benz[d]aceanthrylene, B-60009
Benz[k]aceanthrylene, B-60010
▷ Benz[e]acephenanthrylene, B-70010
Benz[j]acephenanthrylene, B-70011
Cyclopenta[cd]pleiadene, C-80191

C$_{20}$H$_{12}$Br$_2$
2,2′-Dibromo-1,1′-binaphthyl, D-60089
4,4′-Dibromo-1,1′-binaphthyl, D-60090

C$_{20}$H$_{12}$ClN
6-Chlorodibenzo[c,g]carbazole, C-80047

C$_{20}$H$_{12}$I$_2$
2,2′-Diiodo-1,1′-binaphthyl, D-60385
4,4′-Diiodo-1,1′-binaphthyl, D-60386

C$_{20}$H$_{12}$N$_2$
Benzo[1,2-f:4,5-f′]diquinoline, B-70030

C$_{20}$H$_{12}$N$_2$O$_2$
Isoquinacridone, I-80086
Quinacridone, Q-80001

C$_{20}$H$_{12}$N$_4$
Cycloocta[2,1-b:3,4-b′]di[1,8]naphthyridine,
C-70231
Quinolino[1″,2″:1′,2′]imidazo[4′,5′:4,5]
imidazo[1,2-a]quinoline, Q-80003

C$_{20}$H$_{12}$O$_4$
Tetraoxaporphycene, T-70146

C$_{20}$H$_{12}$O$_5$
Halenaquinone, *in* H-60002

C$_{20}$H$_{12}$O$_6$
Altertoxin III, A-60082

C$_{20}$H$_{12}$O$_7$
Dehydropachyrrhizone, *in* P-80004

C$_{20}$H$_{12}$S
Anthra[1,2-b]benzo[d]thiophene, A-70214
Anthra[2,1-b]benzo[d]thiophene, A-70215
Anthra[2,3-b]benzo[d]thiophene, A-70216
Benzo[3,4]phenanthro[1,2-b]thiophene,
B-60039
Benzo[3,4]phenanthro[2,1-b]thiophene,
B-60040
Benzo[b]phenanthro[1,2-d]thiophene, B-70039
Benzo[b]phenanthro[4,3-d]thiophene, B-70040
Dinaphtho[1,2-b:1′,2′-d]thiophene, D-80480
Dinaphtho[1,2-b:2′,1′-d]thiophene, D-80481
Dinaphtho[1,2-b:2′,3′-d]thiophene, D-80482
Dinaphtho[2,1-b:1′,2′-d]thiophene, D-80483
Dinaphtho[2,1-b:2′,3′-d]thiophene, D-80484
Dinaphtho[2,3-b:2′,3′-d]thiophene, D-80485

C$_{20}$H$_{12}$S$_5$
2,2′:5′,2″:5″,2‴:5‴,2⁗-Quinquethiophene,
Q-70010

C$_{20}$H$_{13}$NO$_2$
9-Acridinecarboxylic acid; Ph ester, *in*
A-70061

C$_{20}$H$_{14}$
1,2-Benzo[2.2]metaparacyclophan-9-ene,
B-70036
Benzo[2.2]paracyclophan-9-ene, B-80033
Dibenzo[fg,mn]octalene, D-80072
9,10-Dihydrodicyclopenta[c,g]phenanthrene,
D-60219
1-(Diphenylmethylene)-1H-
cyclopropabenzene, D-70484
[2](2,6)-Naphthaleno[2]paracyclophane-1,11-
diene, N-70006

C$_{20}$H$_{14}$N$_2$
Acridin-9-ylmethylidenebenzenamine, *in*
A-70060

C$_{20}$H$_{14}$O
1-Acetyltriphenylene, A-60059
2-Acetyltriphenylene, A-60060
10-Phenylanthrone, P-80089

C$_{20}$H$_{14}$O$_2$
Benz[a]anthracen-11-ol; O-Ac, *in* B-80012
1-Chrysenecarboxylic acid; Me ester, *in*
C-80128
2-Chrysenecarboxylic acid; Me ester, *in*
C-80129
3-Chrysenecarboxylic acid; Me ester, *in*
C-80130
5-Chrysenecarboxylic acid; Me ester, *in*
C-80131
6-Chrysenecarboxylic acid; Me ester, *in*
C-80132
3-Hydroxybenz[a]anthracene; Ac, *in* H-70104

1-Triphenylenecarboxylic acid; Me ester, *in*
T-60383
2-Triphenylenecarboxylic acid; Me ester, *in*
T-60384

C$_{20}$H$_{14}$O$_3$
7-Oxo-7H-benz[de]anthracene-10-carboxylic
acid; Et ester, *in* O-80064
7-Oxo-7H-benz[de]anthracene-2-carboxylic
acid; Et ester, *in* O-80059
7-Oxo-7H-benz[de]anthracene-4-carboxylic
acid; Et ester, *in* O-80061
7-Oxo-7H-benz[de]anthracene-9-carboxylic
acid; Et ester, *in* O-80063
7,8,8a,9a-Tetrahydrobenzo[10,11]chryseno[3,4-
b]oxirene-7,8-diol, T-70041

C$_{20}$H$_{14}$O$_4$
4,4′,5,5′-Tetrahydroxy-1,1′-binaphthyl,
T-70102
4,4′,5,5′-Tetrahydroxy-1,1′-binaphthyl,
T-80086

C$_{20}$H$_{14}$O$_5$
Halenaquinol, H-60002
Sophoracoumestan A, S-80038

C$_{20}$H$_{14}$O$_6$
Alterperylenol, A-60078
▷ Altertoxin II, A-80055
7,7′-Dimethoxy-[6,8′-bi-2H-benzopyran]-2,2′-
dione, *in* B-70093
Jayantinin, J-80008

C$_{20}$H$_{14}$O$_7$
Alterlosin I, A-80051
Neofolin, *in* P-80003
Oreojasmin, O-70044
Pachyrrhizone, P-80004

C$_{20}$H$_{14}$O$_8$S
Halenaquinol; O^{16}-Sulfate, *in* H-60002

C$_{20}$H$_{14}$S
1,3-Diphenylbenzo[c]thiophene, D-70474
9-(Phenylthio)phenanthrene, *in* P-70049

C$_{20}$H$_{14}$Se
9-(Phenylseleno)phenanthrene, P-60133

C$_{20}$H$_{15}$Br
Bromotriphenylethylene, B-70276

C$_{20}$H$_{15}$N
10,11-Dihydro-5-phenyl-5H-
dibenz[b,f]azepine, *in* D-70185
2,3-Diphenylindole, D-70483
5H,9H-Quino[3,2,1-de]acridine, Q-70006

C$_{20}$H$_{15}$NO
2-Acetyltriphenylene; Oxime, *in* A-60060
Isocyanatotriphenylmethane, I-80056

C$_{20}$H$_{15}$NO$_3$S
10H-Phenothiazine-3-carboxylic acid; N-Ph,
Me ester, *in* P-70057

C$_{20}$H$_{15}$NO$_4$S
10H-Phenothiazine-3-carboxylic acid; N-Ph,
5,5-dioxide, Me ester, *in* P-70057

C$_{20}$H$_{15}$N$_3$O$_8$
Dihydroxanthommatin, D-80264

C$_{20}$H$_{16}$
1,2-Benzo[2.2]paracyclophane, B-70038
▷ 7,12-Dimethylbenz[a]anthracene, D-60405

C$_{20}$H$_{16}$N$_2$
9,10-Dihydro-9,10-ethenoanthracene-11,12-
dicarboxylic acid; Di-Me ester, *in* D-70208
2,2′-Dimethyl-4,4′-biquinoline, D-60413

C$_{20}$H$_{16}$N$_2$O$_2$
2,5-Dihydro-3,6-diphenylpyrrolo[3,4-c]pyrrole-
1,4-dione; 2,5-Di-Me, *in* D-70204

C$_{20}$H$_{16}$N$_2$O$_3$
1-[4-[6-(Diethylamino)-2-benzofuranyl]phenyl]-
1H-pyrrole-2,5-dione, D-60170

C$_{20}$H$_{16}$N$_4$O$_2$
Pyrazole blue, P-70153

C$_{20}$H$_{16}$O
Triphenylacetaldehyde, T-80359

$C_{20}H_{16}O_3$
3-(Benzyloxy)phenol; Benzoyl, *in* B-60073

$C_{20}H_{16}O_4$
Ochromycinone methyl ether, *in* O-60003
Tetraoxaporphyrinogen, T-70147

$C_{20}H_{16}O_5$
Alpinumisoflavone, A-80050
Atalantoflavone, A-80183
2,5-Bis(hydroxymethyl)furan; Dibenzoyl, *in* B-60160
Citrusinol, C-60152
Clandestacarpin, C-80138
Derrone, D-80031
Isoderrone, I-80060
Isopsoralidin, *in* C-80152
Psoralidin, P-80186
Sojagol, S-80037

$C_{20}H_{16}O_6$
Crotarin, C-60172
Dihydroalterperylenol, *in* A-60078
7,6-(2,2-Dimethylpyrano)-3,4′,5-trihydroxyflavone, D-60449
Parvisoflavone A, P-80014
Parvisoflavone B, P-80015
Psoralidin oxide, *in* P-80186

$C_{20}H_{16}O_7$
Alterlosin II, A-80052
Corylidin, C-80152

$C_{20}H_{16}O_8$
Hydroxyversicolorone, H-70230

$C_{20}H_{16}O_{12}$
2,3,5,6,8-Pentahydroxy-1,4-naphthoquinone; Penta-Ac, *in* P-70033

$C_{20}H_{17}NO$
Triphenylacetaldehyde; Oxime, *in* T-80359

$C_{20}H_{17}N_4$
2,4,6-Triphenyl-3,4-dihydro-1,2,4,5-tetrazin-1(2*H*)-yl, T-80361

$C_{20}H_{18}$
5,6-Didehydro-1,4,7,10-tetramethyldibenzo[*ae*]cyclooctene, D-60169

$C_{20}H_{18}N_2$
1,2-Di-2-indolylethylene; *N,N′*-Di-Me, *in* D-80395

$C_{20}H_{18}N_2O$
3-Amino-1,2,3,4-tetrahydrocarbazole; N^3-Benzoyl, *in* A-60259

$C_{20}H_{18}N_4O_8$
α,α′,2-Triamino-γ,γ′,3-trioxo-3*H*-phenoxazine-1,9-dibutanoic acid, T-80209

$C_{20}H_{18}O_2$
1,1,2-Triphenyl-1,2-ethanediol, T-70311

$C_{20}H_{18}O_3$
Isolonchocarpin, I-80066

$C_{20}H_{18}O_4$
Abyssinone I, A-80006
3,4-Dibenzoyl-2,5-hexanedione, D-60081
4′-Hydroxyisolonchocarpin, *in* I-80066
Neorautenol, N-80023
4,5-Phenanthrenedimethanol; Di-Ac, *in* P-60070

$C_{20}H_{18}O_5$
Canescacarpin, *in* G-80029
Crotalarin, C-60171
6-(1,1-Dimethyl-2-propenyl)-4′,5,7-trihydroxyisoflavone, D-80472
Glyceollin III, G-80029
1-(4-Hydroxy-3-methoxyphenyl)-7-(4-hydroxyphenyl)-1,6-heptadiene-3,5-dione, *in* C-80165
Moracin *H*, *in* M-60145
Wighteone, W-80002

$C_{20}H_{18}O_6$
Ambonone, *in* N-80025
1″,2″-Dehydrocyclokievitone, D-80021
2,3-Dehydrokievitone, *in* K-80010
1,2-Diacetoxy-5,6-dimethoxyphenanthrene, *in* T-60115
Erythrinin C, E-80054
Fremontin, F-80080
Garveatin *A* quinone, G-60006
Licoisoflavone A, L-80028
Nepseudin, N-80025
Platanetin, P-70105
3′,4′,5,7-Tetrahydroxy-8-prenylflavone, T-80116

$C_{20}H_{18}O_7$
6-Desmethoxyhormothamnione, *in* H-50127
Meridinol, M-80033

$C_{20}H_{18}O_8$
Arborone, A-60294
Di-*p*-toluoyl tartrate, *in* T-70005
Moluccanin, M-70151
5,5′,6,7-Tetramethoxy-3′,4′-methylenedioxyflavone, *in* H-70081
3′,4′,5,5′-Tetramethoxy-6,7-methylenedioxyisoflavone, *in* H-80071

$C_{20}H_{18}O_9$
2′-Acetoxy-5,5′-dihydroxy-3,7,8-trimethoxyflavone, *in* H-70080
Cetraric acid, C-70035
7-Hydroxy-5,5′,6,8-tetramethoxy-3′,4′-methylenedioxyflavone, *in* H-80029

$C_{20}H_{18}O_{10}$
Conphysodalic acid, C-60162

$C_{20}H_{19}Br$
Bromododecahedrane, B-80202

$C_{20}H_{19}Cl$
Chlorododecahedrane, C-80055

$C_{20}H_{19}ClO_7$
4′-Chloronephroarctin, *in* N-70032

$C_{20}H_{19}F$
Fluorododecahedrane, F-80033

$C_{20}H_{19}N$
2*H*-Dibenz[*e,g*]isoindole; *N-tert*-Butyl, *in* D-70067

$C_{20}H_{19}NO_3$
Dodecahedranol; *O*-Nitrate, *in* D-80523

$C_{20}H_{19}N_3$
▷ Rosaniline, R-60010

$C_{20}H_{20}$
1,1′-Bitetralinylidene, B-70165
▷ 7,12-Dimethylbenz[*a*]anthracene; 1,2,3,4-Tetrahydro, *in* D-60405
Dodecahedrane, D-60514

$C_{20}H_{20}O$
Dodecahedranol, D-80523

$C_{20}H_{20}O_3$
Asnipyrone B, A-80178
1-(3,4-Dihydroxyphenyl)-5-(4-hydroxyphenyl)-1-penten-4-yne, D-80365
Eupomatenoid 4, *in* E-60075

$C_{20}H_{20}O_4$
Abyssinone II, A-80007
Crotmadine, C-80157
Dehydroguaiaretic acid, D-80025
Desmethylxanthohumol, *in* I-80100
Eupomatenoid 7, *in* E-60075
Fruticulin *A*, F-60076
Glabranin, G-70015
Glabridin, G-70016
Licarin E, L-80026
Moracin *I*, *in* M-60146
Phaseollidin, P-80076
Psorolactone, P-60188
Pygmaeocine *E*, P-70144

$C_{20}H_{20}O_5$
Anastomosine, A-80154
Aphyllodenticulide, A-70243
5-Deoxykievitone, *in* K-80010
7,8-(2,2-Dimethylpyrano)-3,4′,5-trihydroxyflavan, D-60448
Glyceollidin I, G-80027
Glyceollidin II, G-80028
Helikrausichalcone, H-80007
3-(4-Hydroxyphenyl)-1-(2,4,6-trihydroxyphenyl)-2-propen-1-one; 4′-*O*-(3-Methyl-2-butenyl), *in* H-80246
10-Isopentenylemodinanthran-10-ol, I-60111
Machilin *B*, M-60002
Moracin *J*, *in* M-60146
Sandwicarpin, *in* C-80156
4′,5,7-Trihydroxy-3′-prenylflavanone, T-80334
Zuonin *A*, Z-60003

$C_{20}H_{20}O_6$
Cardiophyllidin, C-70023
5,6-Didehydropygmaeocin A, *in* P-80195
2-Hydroxygarveatin *A*, *in* H-60148
Kievitone, K-80010
O^5-Methyl-3′,4′-deoxypsorospermin-3′-ol, *in* D-60021
Rhynchosperin *A*, R-60007
Sigmoidin *B*, S-70041
Vittagraciliolide, V-70017

$C_{20}H_{20}O_7$
Juglorin, J-70005
Nephroarctin, N-70032
2′,3′,4′,5,6-Pentamethoxyflavone, *in* P-70030
2′,3,5,7,8-Pentamethoxyflavone, *in* P-70029
3,4′,5,6,7-Pentamethoxyflavone, *in* P-80035
3,5,6,7,8-Pentamethoxyflavone, *in* P-70031
2′,4′,5,5′,7-Pentamethoxyisoflavone, *in* P-80041
2′,4′,5′,6,7-Pentamethoxyisoflavone, *in* P-80042
3′,4′,5′,6,7-Pentamethoxyisoflavone, *in* P-80044
3′,4′,5,7,8-Pentamethoxyisoflavone, *in* P-80045
Rhynchosperin *B*, *in* R-60007
Sigmoidin *D*, S-60031

$C_{20}H_{20}O_8$
4′-Hydroxy-3,3′,5,6,7-pentamethoxyflavone, *in* H-80067
5-Hydroxy-3,3′,4′,6,7-pentamethoxyflavone, *in* H-80067
6-Hydroxy-2′,3,5,5′,7-pentamethoxyflavone, *in* H-60064
7-Hydroxy-3,3′,4′,5,6-pentamethoxyflavone, *in* H-80067

$C_{20}H_{20}O_9$
3,5-Dihydroxy-3′,4′,5′,7,8-pentamethoxyflavone, *in* H-70021
3′,5-Dihydroxy-4′,5′,6,7,8-pentamethoxyflavone, *in* H-80029
5,7-Dihydroxy-3,3′,4′,5′,8-pentamethoxyflavone, *in* H-70021
5,7-Dihydroxy-3′,4′,5′,6,8-pentamethoxyflavone, *in* H-80029

$C_{20}H_{20}O_{10}$
4′,5,5′,7-Tetrahydroxy-2′,3,6,8-tetramethoxyflavone, *in* O-60022
3′,5,7-Trihydroxy-3,4′,5′,6,8-pentamethoxyflavone, *in* O-70018
3′,5,7-Trihydroxy-3,4′,5′,6,8-pentamethoxyflavone, *in* O-70018
4′,5,5′-Trihydroxy-2′,3,6,7,8-pentamethoxyflavone, *in* O-60022
4′,5,7-Trihydroxy-2′,3,5′,6,8-pentamethoxyflavone, *in* O-60022
4′,5,7-Trihydroxy-3,3′,5′,6,8-pentamethoxyflavone, *in* O-70018
1,3,7-Trihydroxyxanthone; 3-Me ether, 7-glucoside, *in* T-80347

$C_{20}H_{20}O_{11}$
3,3′,4′,5,7-Pentahydroxyflavanone; 3-*O*-β-D-Xylopyranoside, *in* P-70027
1,3,5,8-Tetrahydroxyxanthone; 3-Me ether, 8-glucoside, *in* T-70123

$C_{20}H_{21}ClO_7$
Methyl eriodermate, *in* E-70032

$C_{20}H_{21}N$
Aminododecahedrane, A-80071

$C_{20}H_{22}$
2,5-Dimethyl-3,4-diphenyl-2,4-hexadiene, D-60419

$C_{20}H_{22}ClNO_2$
Fontonamide, F-60067

$C_{20}H_{22}N_2O_4$
1,4-Dibenzyl-6,7-dihydroxy-1,4-diazocane-5,8-dione, D-80078

$C_{20}H_{22}O_2$
2-Cyclohexene-1,4-diol; Dibenzyl ether, *in* C-80181
3,3-Dimethyl-5,5-diphenyl-4-pentenoic acid; Me ester, *in* D-60420

$C_{20}H_{22}O_4$
Austrobailignan-5, A-70270
Bibenzyl-2,2'-dicarboxylic acid; Di-Et ester, *in* B-80071
Dehydrodieugenol, D-80023
Dehydrodieugenol B, D-80024
6-(3-Methyl-2-butenyl)allopteroxylin, *in* A-60075
Phaseollidinisoflavan, P-80077

$C_{20}H_{22}O_5$
Austrobailignan-7, A-70271
Bacchotricuneatin B, B-80006
Ethuliacoumarin, E-60045
Fragransin E₁, *in* A-70271
2-Hydroxygarveatin B, H-60148
Phyllnirurin, P-80142
3,4',5,7-Tetrahydroxy-8-prenylflavan, T-60125
Volkensiachromone, V-60016

$C_{20}H_{22}O_6$
Acuminatin, A-60062
Bartemidiolide, B-70008
Columbin, C-60160
2,3-Dehydroteucrin E, *in* T-60180
2-Deoxychamaedroxide, *in* C-60032
2,3-Dihydroxy-2,3-diphenylbutanedioic acid; Di-Et ester, *in* D-80312
12-Epiteuscordonin, *in* T-80165
2-Hydroxyligustrin; 2-Ketone, 8-O-(2-methyl-2,3-epoxybutanoyl), *in* H-80184
Isobutyrylshikonin, *in* S-70038
Ligustrin; 8-O-(2,5-Dihydro-5-hydroxy-3-furancarboxylate), *in* L-80030
Ligustrin; Δ¹⁽¹⁰⁾-Isomer, 8-O-(2,5-dihydro-5-hydroxy-3-furancarboxylate), *in* L-80030
8β-(2-methyl-2,3-epoxybutyroyloxy)-dehydroleucodin, *in* H-80184
1-(2,3,4,5,6-Pentamethoxyphenyl)-3-phenyl-2-propen-1-one, *in* P-80050
Pygmaeocin A, P-80195
Teuscorodonin, T-80165

$C_{20}H_{22}O_7$
Chamaedroxide, C-60032
2-Dehydroarcangelisinol, D-70021
Desacylisoelephantopin senecioate, *in* I-60099
Desacylisoelephantopin tiglate, *in* I-60099
10α-Hydroxycolumbin, *in* C-60160
8β-Hydroxycolumbin, *in* C-60160
1-Hydroxypinoresinol, H-60221
9-Hydroxypinoresinol, H-70216
Isojateorin, *in* C-60160
Jateorin, *in* C-60160
2',3',4',5,7-Pentamethoxyisoflavanone, *in* P-80039
Wikstromol, W-70003

$C_{20}H_{22}O_8$
Arizonin C₃, A-70255
2-Dehydroarcangelisinol; 2β,3β-Epoxide, 6,12-diepimer, *in* D-70021
6-Hydroxyarcangelisin, *in* D-70021
4'-Hydroxy-3',5,5',6,7-pentamethoxyflavone, *in* H-70081
Tinospora clerodane, T-70200

$C_{20}H_{22}O_9$
Afzelechin; 7-O-β-D-Apioside, *in* T-60104
Astringin, *in* D-80361
Salireposide, *in* D-70287
Viscutin 3, *in* T-60103

$C_{20}H_{22}O_{10}$
Polydine, *in* P-70026

$C_{20}H_{22}O_{13}$
MP-2, *in* H-80196

$C_{20}H_{23}NO_5$
Fuligorubin A, F-60079

$C_{20}H_{24}$
Nonacyclo[10.8.0.0²,¹¹.0⁴,⁹.0⁴,¹⁹.0⁶,¹⁷.0⁷,¹⁶.0⁹,¹⁴.0¹⁴,¹⁹]icosane, N-70070

$C_{20}H_{24}O_2$
Saprorthoquinone, S-80010

$C_{20}H_{24}O_3$
Centrolobin, C-80032
2-Epijatrogrossidione, *in* J-70002
Gravelliferone; Me ether, *in* G-60037
Jatrogrossidione, J-70002
Rubifolide, R-60016

$C_{20}H_{24}O_4$
Austrobailignan-6, A-60320
Bharangin, B-70088
Brayleanin, *in* C-60026
Coralloidolide A, C-60164
3,6-Dioxo-4,7,11,15-cembratetraen-10,20-olide, D-60470
12,16-Epoxy-11,14-dihydroxy-5,8,11,13-abietatetraen-7-one, E-60011
Gelomulide D, G-80008
7-Geranyloxy-6-methoxycoumarin, *in* H-80189
Gersemolide, G-60015
Gersolide, G-60016
Gravelliferone; 8-Methoxy, *in* G-60037
Guaiaretic acid, G-70033
Hebeclinolide, H-60009
Isobharangin, I-80048
Polemannone, P-60166
Sirutekkone, S-70048
Vernoflexin, *in* Z-60001
Zaluzanin C; Angeloyl, *in* Z-60001

$C_{20}H_{24}O_5$
2-(4-Allyl-2-methoxyphenoxy)-1-(4-hydroxy-3-methoxyphenyl)-1-propanol, A-80046
1,11-Bisepicaniojane, *in* C-70013
Caniojane, C-70013
Epoxylaphodione, *in* D-60470
Ferulide, F-70005
3β-Hydroxyhebeclinolide, H-60009
4-Hydroxyisobacchasmacranone, H-60161
Ligustrin; 8-O-(4-Hydroxytigloyl), *in* L-80030
Lycoxanthol, L-70046
Machilin C, M-60003
Machilin D, *in* M-60003
Machilin I, M-70003
Myristargenol A, M-70163
Myrocin C, M-70165
Nectandrin B, *in* M-70003
ent-7-Oxo-3,13-clerodadiene-16,15:18,19-olide, *in* H-80143

$C_{20}H_{24}O_6$
ent-2α-Acetoxy-7-oxo-3,13-clerodadiene-16,15:18,19-diolide, *in* H-80142
Baccharioide B, B-70004
Deacetylsessein, *in* S-60027
Deacetylteupyrenone, *in* T-60184
8-Deacyl-15-deoxypunctatin; 8-O-(2-Methyl-2,3-epoxybutanoyl), *in* D-80006
15-Deoxyliscundin, *in* D-80006
5-Desoxyeuparotin, *in* E-80090
Heliopsolide, H-70007
2-Hydroxyligustrin; 8-O-(2-methyl-2,3-epoxybutanoyl), *in* H-80184
ent-Isolariciresinol, *in* I-60139
Isolariciresinol, I-60139
Isotaxiresinol; 7-Me ether, *in* I-60139
Ladibranolide, L-60008
Lariciresinol, L-70019
Methyl 12,17;15,16-diepoxy-6-hydroxy-19-nor-17-oxo-4,13(16),14-clerodatrien-18-oate, M-80068
3,3',4',5,7-Pentamethoxyflavan, *in* P-70026
Sanguinone A, S-60006
Teucrin E, T-60180
Teucrin H2, *in* T-60180

$C_{20}H_{24}O_7$
Bacchariolide A, *in* B-70004
8-Deacyl-15-deoxypunctatin; 2α,3α-Epoxy, 8-O-(2-hydroxymethyl-2Z-butenoyl), *in* D-80006

$C_{20}H_{24}O_7$ *(continued, right column)*
8-Deacyl-15-deoxypunctatin; 2α,3α-Epoxy, 8-O-(2-methyl-2,3-epoxybutanoyl), *in* D-80006
Euparotin, E-80090
1-(4-Hydroxy-3-methoxyphenyl)-2-[4-(3-hydroxy-1-propenyl)-2-methoxyphenoxy]-1,3-propanediol, H-80193
Liscundin, *in* D-80006
Melampodin D, M-60013
Orthopappolide methacrylate, *in* O-60045
Punctaliatrin, *in* D-80006
Tomenphantopin A, *in* T-60222
Zinaflorin IV, Z-80002
Zinaflorin IV; O⁶-Deacyl, O⁶-tigloyl, *in* Z-80002

$C_{20}H_{24}O_8$
4β,15-Epoxy-1β-methoxy-9Z-millerenolide, *in* M-60137
15-Hydroxy-3-dehydrotifruticin, *in* T-80197

$C_{20}H_{24}O_{10}$
Salicortin, S-70002

$C_{20}H_{24}O_{13}$
Diospyroside, *in* D-84287

$C_{20}H_{24}S_6$
2,3,11,12-Dibenzo-1,4,7,10,13,16-hexathia-2,11-cyclooctadecadiene, D-60073

$C_{20}H_{26}O_2$
12-Hydroxy-6,8,11,13-abietatrien-3-one, H-80105

$C_{20}H_{26}O_3$
19(4→3)-Abeo-11,12-dihydroxy-4(18),8,11,13-abietatetraen-7-one, A-60003
ent-15-Beyerene-2,3,12-trione, *in* D-80285
6,7-Dehydroroyleanone, *in* R-60015
11,12-Dihydroxy-5,8,11,13-abietatetraen-7-one, D-80266
10-Epinidoresedic acid, *in* N-70035
12-Hydroxy-8,11,13-abietatriene-3,7-dione, H-80103
12-Hydroxy-8,11,13-abietatriene-6,7-dione, H-80104
3-(2-Hydroxy-4,8-dimethyl-3,7-nonadienyl)benzaldehyde; Ac, *in* H-60122
Kahweol, K-60002
Methyl p-geranyloxycinnamate, *in* H-80242
Nidoresedic acid, N-70035
ent-6-Oxo-7,12E,14-labdatrien-17,11α-olide, O-60076
Strictic acid, S-70077

$C_{20}H_{26}O_4$
19(4β→3β)-Abeo-6,11-epoxy-6,12-dihydroxy-6,7-seco-4(18),8,11,13-abietatetraen-7-al, A-60004
Anisomelic acid, A-60275
Citlalitrione, C-80135
Cynajapogenin A, C-70257
Desoxyarticulin, *in* A-60307
Dihydroguaiaretic acid, *in* G-70033
11,12-Dihydroxy-6,8,11,13-abietatetraen-20-oic acid, D-60305
5,12-Dihydroxy-6,8,12-abietatriene-11,14-dione, D-80267
6,12-Dihydroxy-8,11,13-abietatriene-1,7-dione, D-80268
2α,11-Dihydroxy-7,9(11),13-abietatriene-6,12-dione, *in* D-70282
ent-7β,14α-Dihydroxy-1,16-kauradiene-3,15-dione, D-80324
ent-3β,14α-Dihydroxy-7,9(11),15-pimaratriene-2,12-dione, D-80373
ent-8α,14α-Epoxy-3β-hydroxy-9(11),15-pimaradiene-2,12-dione, E-80032
ent-8β,14β-Epoxy-3β-hydroxy-9(11),15-pimaradiene-2,12-dione, E-80032
15,16-Epoxy-12-oxo-8(17),13(16),14-labdatrien-19-oic acid, *in* L-60011
Gelomulide C, G-80008
Hardwickiic acid; 19-Oxo, *in* H-70002
4-Hydroxysapriparaquinone, H-80254
17-Oxohebemacrophyllide, *in* H-60010
11,12,16-Trihydroxy-5,8,11,13-abietatetraen-7-one, T-60312

$C_{20}H_{26}O_5$
Articulin, A-60307
Ballotinone, *in* M-70013
Chapinolin, C-70037
Cumambranolide; 8α-Angeloyloxy, *in* C-80164
Cumambranolide; 8β-Angeloyloxy, *in* C-80164
Cumambranolide; 8α-Tigloyloxy, *in* C-80164
8,9-Dihydroxy-1(10),4,11(13)-germacratrien-12,6-olide; 9-(2*R*,3*R*-Epoxy-2-methylbutanoyl), *in* D-60323
11,12-Dihydroxy-7-oxo-8,11,13-abietatrien-20-oic acid, *in* T-60337
Epirosmanol, *in* R-70010
Feruginin, F-60004
Gibberellin A_{24}, G-60019
Gibberellin A_{64}, G-70012
7-Hydroxy-3,13-clerodadiene-16,15;18,19-diolide, H-70116
ent-2α-Hydroxy-3,13-clerodadiene-16,15:18,19-diolide, H-80142
ent-7β-Hydroxy-3,13-clerodadiene-15,16;18,19-diolide, H-70117
ent-7β-Hydroxy-3,13-clerodadiene-16,15:18,19-diolide, H-80143
ent-7α-Hydroxy-3,13-clerodadiene-16,15:18,19-diolide, *in* H-80143
Mollisorin *A*, *in* D-70302
Myristargenol *B*, M-70164
ent-7-Oxo-3-clerodene-15,16:18,19-diolide, *in* H-70117
Peregrinone, *in* M-70013
Rosmanol, R-70010
Senecioodontol, S-60022
Spongialactone A, S-80043
2,6,7,11-Tetrahydroxy-5,7,9(11),13-abietatetraen-12-one, T-80085
Thymifodioic acid, T-60218
11,12,14-Trihydroxy-8,11,13-abietatriene-3,7-dione, T-60313

$C_{20}H_{26}O_6$
Caleine *F*, *in* C-60005
Candesalvone B, C-80015
Coleon *C*, C-60157
Coralloidolide *B*, C-60165
Cumambranolide; 8β-(Epoxyangeloyloxy), *in* C-80164
Cumambranolide; 8β-(3-Hydroxy-2-methylenebutanoyloxy), *in* C-80164
Cumambranolide; 8β-Sarracinoyloxy, *in* C-80164
Deacetyleupaserrin, *in* D-70302
Deoxytifruticin, *in* T-80197
ent-2α,7β-Dihydroxy-3,13-clerodadiene-16,15:18,19-diolide, *in* H-80142
2,8-Dihydroxy-1(10),4,11(13)-germacratrien-12,6-olide; 8-(3*S*-Hydroxy-2-methylenebutanoyl), *in* D-70302
4,5-Epoxy-2,8-dihydroxy-1(10),11(13)-germacradien-12,6-olide; 8-(3-Methyl-2-butenoyl), *in* E-60013
Gibberellin A_{36}, *in* G-60019
Gibberellin A_{65}, G-70013
Gigantanolide *A*, *in* T-60323
Gigantanolide *B*, *in* T-60323
Jurinelloide, J-60003
Mollisorin *B*, *in* D-70302
Sesebrinic acid, S-60026
Teulamifin *B*, T-70158
Teumicropin, T-60181
7,11,12-Trihydroxy-6-oxo-8,11,13-abietatrien-20-oic acid, T-60337
Zoapatanolide *B*, Z-70007

$C_{20}H_{26}O_7$
Cyclocratystyolide, C-70216
2,8-Dihydroxy-1(10),4,11(13)-germacratrien-12,6-olide; 8-(2*R*,3*R*-Epoxy-2-hydroxymethylbutanoyl), *in* D-70302
Gibberellin A_{66}, *in* G-70013
20-Hydroxyelemajurinelloide, H-60132
20-Hydroxyjurinelloide, *in* J-60003
Specionin, S-70059
Tifruticin, T-80197

$C_{20}H_{26}O_8$
8-Deacyltrichogoniolide; 8-*O*-Angeloyl, *in* D-80007
Specionin, S-60039

$C_{20}H_{26}O_9$
8β-Propionyloxy-10β-hydroxyhirsutinolide-13-*O*-acetate, *in* H-80089

$C_{20}H_{26}O_{10}$
Anamarine, A-60270

$C_{20}H_{27}ClO_7$
Lacinolide B, L-80016

$C_{20}H_{28}$
Hexacyclo[11.3.1.1^{3,7}.1^{5,9}.1^{11,15}.0^{2,10}]eicos-2-ene, H-70031

$C_{20}H_{28}N_2O_3$
Oxyphencyclimine, O-60102

$C_{20}H_{28}N_2O_8$
N-(*N*-1-Carboxy-6-hydroxy-3-oxoheptyl)alanyltyrosine, C-70022

$C_{20}H_{28}O$
8,11,13-Abietatrien-12,16-oxide, A-60008
Biadamantylidene epoxide, B-70091
8,11,13-Cleistanthatrien-19-al, *in* C-60154

$C_{20}H_{28}O_2$
6,8,11,13-Abietatriene-2,12-diol, A-80003
ent-16-Atisene-3,14-dione, A-80185
Barbatusol, B-60007
ent-15-Beyerene-2,12-dione, B-80065
ent-15-Beyerene-3,12-dione, *in* D-80285
Cedronellone, C-60027
3,7,11,15(17)-Cembratetraen-16,2-olide, C-60028
8,11,13-Cleistanthatrien-19-oic acid, *in* C-60154
10(20),16-Ericacadien-19-oic acid, E-80050
12-Ethyl-13-methoxy-8,11,13-podocarpatrien-7-one, *in* N-80030
12-Hydroxy-8,11,13-abietatrien-18-al, *in* A-70005
ent-2-Hydroxy-1,15-beyeradien-3-one, H-80115
11-Hydroxy-8,11,13-cleistanthatrien-7-one, H-80141
Nimbonolone, N-80030
Nimbosone, N-80031
ent-8(14),15-Pimaradiene-3,12-dione, *in* P-80146
Taonianone, T-60006
Totarolone, *in* T-70204

$C_{20}H_{28}O_3$
6,8,11,13-Abietatetraene-11,12,14-triol, A-60006
Antiquorin, A-80166
Bacchabolivic acid, B-80004
12,15-Dihydroxy-8,11,13-abietatrien-7-one, D-80270
2,11-Dihydroxy-7,9(11),13-abietatrien-12-one, D-70282
2,12-Dihydroxy-8,11,13-abietatrien-7-one, D-70281
3,12-Dihydroxy-8,11,13-abietatrien-1-one, D-60306
6,14-Dihydroxy-8,11,13-abietatrien-1-one, D-80271
Hardwickiic acid, H-70002
Hebemacrophyllide, H-60010
ent-3β-Hydroxy-15-beyerene-2,12-dione, *in* D-80285
14-Hydroxy-8,12-cleistanthadiene-7,11-dione, H-80140
6-Hydroxy-5-oxo-7,15-fusicoccadien-15-al, H-60203
19-Hydroxy-13(16),14-spongiadien-3-one, H-70220
Isolobophytolide, I-60106
Lambertianic acid, L-60011
7-Oxo-8(14),15-isopimaradien-18-oic acid, O-70094
ent-15-Oxo-16-kauren-18-oic acid, O-60075
ent-15-Oxo-16-kauren-18-oic acid, O-80086
ent-13R-hydroxy-16-atisene-3,14-dione, *in* A-80185
Royleanone, R-60015
Wedeliasecokaurenolide, W-60004

$C_{20}H_{28}O_4$
2α-Angeloyloxy-9-oxoisoanhydrooplopanone, *in* O-50045
Brevifloralactone, B-70179
Candesalvone A, C-80014
Curculathyrane *A*, C-60176
Curculathyrane *B*, C-60177
Demethylpinusolide, *in* P-60156
15,16-Dihydro-16-oxohardwickiic acid, *in* H-70002
7,12-Dihydroxy-8,12-abietadiene-11,14-dione, D-80265
2β,12-Dihydroxy-8,12-abietadiene-11,14-dione, *in* R-60015
11,12-Dihydroxy-8,11,13-abietatrien-20-oic acid, D-80269
7,15-Dihydroxy-8,11,13-abietatrien-18-oic acid, D-70280
3,10-Dihydroxy-5,11-dielmenthadiene-4,9-dione, D-60316
ent-3α,11α-Dihydroxy-16-kaurene-6,15-dione, D-80325
ent-13*R*-14*R*-Dihydroxy-3-oxo-15-atisen-17-al, D-80349
3,19-Dihydroxy-13(16),14-spongiadien-2-one, D-70345
2α,19-Dihydroxy-13(16),14-spongiadien-3-one, *in* H-70220
Ephemeric acid, E-60006
6,11-Epoxy-6,12-dihydroxy-6,7-seco-8,11,13-abietatrien-7-al, E-60014
ent-15,16-Epoxy-4α-hydroxy-13(16),14-clerodadiene-3,12-dione, E-80029
Furanopetasin, *in* F-70047
Grayanotoxin XX, *in* G-80040
Gutiesolbriolide, G-80045
Hautriwaic acid, *in* H-70002
2β-Hydroxyhardwickiic acid, *in* H-70002
17-Hydroxyhebemacrophyllide, *in* H-60010
Isogutiesolbriolide, I-80062
Isospongiadiol, I-60137
Ivangustin; *O*-(2-Methylbutanoyl), *in* I-70109
Kerlinic acid, K-70011
Marrubiin, M-70013
Patagonic acid, P-70013
Pseudopterogorgia diterpenoid *C*, P-70129
3,11,12-Trihydroxy-8,11,13-abietatrien-7-one, T-60314
ent-7β,14α,15β-Trihydroxy-1,16-kauradien-3-one, T-80312

$C_{20}H_{28}O_5$
8,9-Dihydroxy-1(10),4,11(13)-germacratrien-12,6-olide; 8-(2-Methylbutanoyl), *in* D-60323
11,12-Epoxy-2,14-dihydroxy-5,8(17)-briaradien-18,7-olide, E-70012
ent-7β,20-Epoxy-1β,14α,20-trihydroxy-16-kauren-15-one, E-80046
Gutiesolbriolide; 17-Hydroxy, 10E-isomer, *in* G-80045
ent-7β-Hydroxy-3-clerodene-15,16:18,19-diolide, *in* H-70117
17-Hydroxygutiesolbriolide, *in* G-80045
2-Hydroxyhautriwaic acid, *in* H-70002
17-Hydroxyisogutiesolbriolide, *in* I-80062
16-Hydroxy-8(17),13-labdadien-15,16-olid-19-oic acid, H-80182
Longikaurin *A*, L-70035
Melisodoric acid, M-70018
Pulchellin 2-*O*-tiglate, *in* P-80189
2α,6α,7,11-Tetrahydroxy-7,9(11),13-abietatrien-12-one, *in* D-70282
ent-1β,3α,11α-Trihydroxy-16-kaurene-6,15-dione, T-80313
ent-1β,7β,14α-Trihydroxy-16-kaurene-3,15-dione, T-80314
ent-7β,12β,14α-Trihydroxy-16-kaurene-3,15-dione, T-80315
2,11,12-Trihydroxy-6,7-seco-8,11,13-abietatriene-6,7-dial 11,6-hemiacetal, T-80339

$C_{20}H_{28}O_6$
Amarolide, A-60085
Blumealactone *A*, *in* B-70172
Blumealactone *B*, *in* B-70172
Gigantanolide *C*, *in* T-60323
6α-Hydroxypulchellin 4-*O*-angelate, *in* P-80189
Schkuhridin *A*, *in* S-60012
Sigmoidin *A*, *in* S-70041

6,7,12,16-Tetrahydroxy-8,12-abietadiene-11,14-dione, T-80084
ent-1*β*,3*α*,7*α*,11*α*-Tetrahydroxy-16-kaurene-6,15-dione, T-60109

$C_{20}H_{28}O_7S$
2,8-Dihydroxy-1(10),4,11(13)-germacratrien-12,6-olide; 8-(2*S*-Hydroxy-2-hydroxymethyl-3*S*-mercaptobutanoyl), *in* D-70302

$C_{20}H_{28}O_{10}$
Ptelatoside *B*, *in* H-70209

$C_{20}H_{30}$
Pentaisopropylidenecyclopentane, P-60053

$C_{20}H_{30}O$
8,11,13-Abietatrien-19-ol, A-60007
8,11,13-Abietatrien-3-ol, A-70006
8,11,13-Cleistanthatrien-19-ol, C-60154
ent-15,16-Epoxy-7,13(16),14-labdatriene, E-70022
8(14),15-Isopimaradien-3-one, I-80076
8(17),13(16),14-Labdatrien-19-al, *in* L-80010
Mikanifuran, M-70148
Totarol, T-70204

$C_{20}H_{30}O_2$
Abeoanticopalic acid, A-60002
8(14),13(15)-Abietadien-18-oic acid, A-80001
8,11,13-Abietatriene-12,18-diol, A-70005
ent-15-Beyeren-18-oic acid, B-80066
13,15-Cleistanthadien-18-oic acid, A-60317
Cycloanticopalic acid, C-60181
Dictyodial *A*, D-70145
5,8,11,14,16-Eicosapentaenoic acid, E-80009
15,16-Epoxy-8-isopimaren-7-one, E-80033
3,4-Epoxy-13(15),16,18-sphenolobatrien-5-ol, E-60031
ent-3-Hydroxy-15-beyeren-2-one, H-80116
ent-3*β*-Hydroxy-15-beyeren-12-one, *in* D-80285
18-Hydroxy-8,15-isopimaradien-7-one, H-60162
ent-12*β*-Hydroxy-8(14),15-pimaradien-3-one, *in* P-80146
ent-3*β*-Hydroxy-8(14),15-pimaradien-12-one, *in* P-80146
8,15-Isopimaradien-18-oic acid, I-80073
ent-9*βH*-Isopimara-7,15-dien-17-oic acid, *in* I-80075
Isosarcophytoxide, I-70100
ent-16*S*-Kaurane-17,19-dial, *in* K-70008
ent-15-Kauren-17-oic acid, K-60008
ent-15-Kauren-18-oic acid, K-80009
8(17),12-Labdadiene-15,16-diol, L-80002
8(17),13(16),14-Labdatrien-19-oic acid, L-80010
ent-2-Oxo-7,13-labdadien-15-al, *in* D-80328
Perrottetianal *A*, P-60064
ent-9(11),15-Pimaradien-19-oic acid, P-80147
Reiswigin *B*, *in* R-60004
Sanadaol, S-70008
Sandaracopimaric acid, S-80008
Sarcophytoxide, S-70011
3,7,11,15-Tetramethyl-13-oxo-2,6,10,14-hexadecatetraenal, T-60150

$C_{20}H_{30}O_3$
8,11,13-Abietatriene-7,18-diol, A-70004
Agroskerin, A-70076
Agrostistachin, A-60067
Agrostistachin, A-80041
2*α*-Angeloxyanhydrooplopanone, *in* O-50045
Ascidiatrienolide A, A-80176
Ascidiatrienolide B, *in* A-80176
Ascidiatrienolide C, *in* A-80176
ent-3*β*,4*β*;15,16-Diepoxy-13(16),14-clerodadien-2*β*-ol, D-80153
ent-3*β*,16*α*-Dihydroxy-13-atisen-2-one, D-80272
ent-2*α*,3*β*-Dihydroxy-15-beyeren-12-one, D-80285
ent-3*β*,12*α*-Dihydroxy-15-beyeren-2-one, D-80286
ent-3*β*,12*β*-Dihydroxy-15-beyeren-2-one, *in* D-80286
1-(2,6-Dihydroxyphenyl)-4-methyl-4-tridecen-1-one, D-80368
ent-3*β*,12*β*-Dihydroxy-8(14),15-pimaradien-2-one, *in* P-80146
ent-3*α*,4*α*-Epoxy-13-cleroden-15,16-olide, E-80019
5,6-Epoxy-7,9,11,14-eicosatetraenoic acid, E-60015
ent-15,16-Epoxy-2*β*-hydroxy-13(16),14-clerodadien-3-one, *in* E-80016
15,16-Epoxy-11*β*-hydroxy-8-isopimaren-7-one, *in* E-80033
14,15-Epoxy-8(17),12-labdadien-16-oic acid, E-60025
15,16-Epoxy-13,17-spatadiene-5,19-diol, E-60029
2-(2-Formyl-3-hydroxymethyl-2-cyclopentenyl)-6,10-dimethyl-5,9-undecadienal, F-70038
Grayanotoxin XIX, *in* G-80040
2-[(2,3,3*a*,4,5,7*a*-Hexahydro-3,6-dimethyl-2-benzofuranyl)ethylidene]-6-methyl-5-heptenoic acid, H-60050
2-[(3,4,4*a*,5,6,8*a*-Hexahydro-4,7-dimethyl-2*H*-1-benzopyran-2-yl)ethylidene]-6-methyl-5-heptenoic acid, H-60046
ent-18-Hydroxy-15-beyeren-19-oic acid, H-60106
ent-7*α*-Hydroxy-15-beyeren-19-oic acid, H-60104
ent-12*β*-Hydroxy-15-beyeren-19-oic acid, H-60105
7-Hydroxy-13(17),15-cleistanthadien-18-oic acid, H-70115
16-Hydroxy-3,13-clerodadien-15,16-olide, H-70118
ent-2*α*-Hydroxy-3,13-clerodadien-15,16-olide, H-80145
Hydroxydictyodial, *in* D-70145
8-Hydroxy-5,9,11,14,17-eicosapentaenoic acid, H-70126
6*α*-Hydroxyisopimara-8(14),15-dien-18-oic acid, *in* S-80008
7*α*-Hydroxyisopimara-8(14)15-dien-18-oic acid, *in* S-80008
12*β*-Hydroxyisopimara-8(14),15-dien-18-oic acid, *in* S-80008
3*β*-Hydroxyisopimara-8(14),15-dien-18-oic acid, *in* S-80008
ent-1*α*-Hydroxy-16-kauren-19-oic acid, H-80179
ent-12*β*-Hydroxy-15-kauren-19-oic acid, H-80180
ent-3*β*-Hydroxy-15-kauren-17-oic acid, H-60167
ent-3*β*-Hydroxy-16-kauren-18-oic acid, H-80181
13*R*-Hydroxy-8,14-labdadiene-2,7-dione, *in* D-80327
ent-3*β*-Hydroxy-2-oxo-7,13-labdadien-15-al, *in* D-80328
Isosarcophytoxide; 3,4-Epoxide, *in* I-70100
Medigenin, M-70015
Naviculide, N-80016
6-Oxo-3,13-clerodadien-15-oic acid, O-70081
ent-2-oxo-3,13-clerodadien-15-oic acid, *in* C-80142
ent-2-Oxo-3,13-clerodadien-15-oic acid, *in* H-80144
5-Oxo-6,8,11,14-eicosatetraenoic acid, O-80080
ent-15-Oxo-1(10),13-halimadien-18-oic acid, O-60069
2-Oxokolavenic acid, *in* K-60013
7-Oxo-8,13-labdadien-15-oic acid, O-70095
16-Oxo-17-spongianal, O-60094

$C_{20}H_{30}O_4$
8(14)-Abieten-18-oic acid 9,13-endoperoxide, A-60009
β-Cyclohallerin, *in* C-60193
α-Cyclohallerin, C-60193
Cymbodiacetal, C-60233
ent-3*β*,19-Dihydroxy-15-kauren-17-oic acid, D-60341
ent-3*β*,18-Dihydroxy-19-trachylobanoic acid, D-80388
*r*1(10)13**E**-*Halimadiene*-15,18-*dioic acid*diedie, *in* O-60069
-1(10),13**Z**-*Halimadiene*-15,18-*dioic acid*ienien, *in* O-60069
ent-3*α*,4*α*-Epoxy-2*α*-hydroxy-13-cleroden-15,16-olide, *in* E-80019
ent-3*α*,4*α*-Epoxy-16*ξ*-hydroxy-13-cleroden-15,16-olide, *in* E-80019
14,15-Epoxy-5-hydroxy-6,8,10,12-eicosatetraenoic acid, E-70016
▷ Grayanotoxin VII, G-80040
Grayanotoxin VIII, *in* G-80040
Hydroxyagrostistachin, *in* A-80041
17-Hydroxy-15,17-oxido-16-spongianone, H-60201
ent-4*α*-Hydroxy-3-oxo-13-cleroden-15,16-olide, *in* D-80303
Isohallerin, I-60100
Kurubashic acid angelate, *in* K-60014
Lapidin, *in* L-60014
Neohalicholactone, *in* H-80002
Paniculadiol, P-80007
Phloganthiolide *A*, P-70099
ent-3*α*,7*β*,14*α*-Trihydroxy-16-kauren-15-one, T-60326
ent-1*β*,7*β*,14*S*-Trihydroxy-16-kauren-15-one, T-50476

$C_{20}H_{30}O_5$
6,17-Dihydroxy-15,17-oxido-16-spongianone, D-60365
1*β*,10*α*-Epoxykurubashic acid angelate, *in* K-60014
ent-9,13-Epoxy-7-labdene-15,17-dioic acid, E-70024
Ivangustin; 4,5*α*-Dihydro,4*α*-hydroxy, *O*[1]-(2-methylbutanoyl), *in* I-70109
6-Oxo-7-labdene-15,17-dioic acid, O-60077
ent-3*α*,7*β*,14*α*,20-Tetrahydroxy-16-kauren-15-one, T-60110
ent-7*β*,11*α*,14*S*,20-Tetrahydroxy-16-kauren-15-one, T-70115
ent-1*β*,7*β*,14*S*,20-Tetrahydroxy-16-kauren-15-one, T-70114
ent-7*α*,18,19-Trihydroxy-3,13-clerodadien-16,15-olide, T-80291
5,14,15-Trihydroxy-6,8,10,12,17-eicosapentaenoic acid, T-70256
5,6,15-Trihydroxy-7,9,11,13,17-eicosapentaenoic acid, T-70255
ent-7*α*,9*α*,15*β*-Trihydroxy-16-kauren-19-oic acid, T-80316

$C_{20}H_{30}O_5S$
Umbraculumin *B*, U-70002

$C_{20}H_{30}O_6$
Grayanotoxin XVII, *in* G-80039
6*β*-Hydroxypulchellin 2-*O*-isovalerate, *in* P-80189
Ingol, I-70018
Maoecrystal K, M-80008
ent-2*α*,3*α*,6*β*,7*β*,11*β*-Pentahydroxy-16-kauren-15-one, P-80047
ent-1*β*,3*α*,6*β*,7*α*,11*α*-Pentahydroxy-16-kauren-15-one, P-80048
Phlebiakauranol aldehyde, P-70097
3,4,6,8-Tetrahydroxy-11,13-clerodadien-15,16-olide, T-70103

$C_{20}H_{30}O_7S$
Hymatoxin *A*, H-60235

$C_{20}H_{30}O_8$
▷ Ptaquiloside, P-60190

$C_{20}H_{30}S_2$
Di(2-adamantyl) disulfide, *in* A-80038

$C_{20}H_{31}BrO$
Sphaeroxetane, S-70061

$C_{20}H_{31}BrO_3$
15-Bromo-9(11)-parguerene-2,7,16-triol, B-80241

$C_{20}H_{31}BrO_4$
15-Bromo-9(11)-parguerene-2,7,16,19-tetrol, B-80240

$C_{20}H_{31}NO_5S$
Latrunculin *D*, L-50032

$C_{20}H_{32}$
Bicyclo[8.8.2]eicosa-1(19),10(20),19-triene, B-70096
α-Camphorene, C-70009

Cycloaraneosene, C-80170
Dictytriene *A*, D-70148
Dictytriene *B*, *in* D-70148
1,5-Dimethyl-8-(5-methyl-1-methylene-4-hexenyl)-1,5-cyclodecadiene, D-80453
Isosphaerodiene 1, I-80091
Isosphaerodiene 2, *in* I-80091
Laurenene, L-60018
Ruzicka's hydrocarbon, R-80017

$C_{20}H_{32}Br_2O_2$
12-Hydroxybromosphaerol, H-60108

$C_{20}H_{32}N_2O_3S$
▷ Carbosulfan, C-80022

$C_{20}H_{32}O$
7,13-Abietadien-18-ol, A-80002
1,3,7,11-Cembratetraen-15-ol, C-70029
Dictymal, D-60162
8β-Hydroxycycloaraneosene, *in* C-80170
8(14),15-Isopimaradien-18-ol, I-80074
ent-9β*H*-Isopimara-7,15-dien-17-ol, I-80075
ent-16*S*-Kauran-17-al, K-70008
ent-11-Kaurene-16β,18-diol, K-60006
ent-16-Kauren-3β-ol, K-70010
Neodolabellenol, N-70022
Neodolabellenol; 5-Epimer, *in* N-70022
18-Oxo-3-virgene, O-60095
ent-9,(11),15-Pimaradien-19-ol, P-80148
Rotalin A, R-80012
13,17-Spatadien-10-ol, S-70058
Sphaeropyrane, S-70060
3,7,11,15-Tetramethyl-2,6,10,13,15-hexadecapentaen-1-ol, T-70137
ent-3β-Trachylobanol, T-80205
1(15),8(19)-Trinervitadien-9-ol, T-70300

$C_{20}H_{32}O_2$
ent-15-Beyerene-14β,19-diol, B-80064
1(14),4,8,12-Cembratetraene-2,3-diol, C-80030
Cladiellin; Δ³-Isomer, *O*-de-Ac, *in* C-80137
ent-4(18),-13-Clerodadien-15-oic acid, C-80142
ent-2,4(18),13-Clerodatriene-15,16-diol, C-80143
Dictyone, D-70146
Dilophic acid, D-60394
ent-11α,16α-Epoxy-17-kauranol, E-80034
8,13-Epoxy-14-labden-11-one, E-80037
14,15-Epoxy-3,7,11,15-tetramethyl-2,6,10-hexadecatrienal, E-80045
ent-19-Hydroxy-16*S*-kauran-17-al, *in* K-70008
15-Hydroxy-8(17),12-labdadien-16-al, H-70145
19-Hydroxy-8(17),13-labdadien-15-al, H-70146
ent-15-Hydroxy-7,13-labdadien-2-one, *in* L-80004
ent-15-Hydroxy-8,13-labdadien-7-one, *in* L-70004
12-Isoagathen-15-oic acid, I-60085
7,15-Isopimaradiene-18,19-diol, I-70078
7,15-Isopimaradiene-3,18-diol, I-70076
7,15-Isopimaradiene-3,19-diol, I-70077
8(14),15-Isopimaradiene-2,18-diol, I-60113
8(14),15-Isopimaradiene-3,18-diol, I-70080
8(14),15-Isopimaradiene-3,18-diol, I-80069
8,15-Isopimaradiene-7,18-diol, I-70079
ent-16*R*-Kauran-17-oic acid, *in* K-70008
ent-16*S*-Kauran-17-oic acid, *in* K-70008
16-Kaurene-3,13-diol, K-80007
ent-16-Kaurene-7α,15β-diol, K-70009
ent-15-Kaurene-3β,17-diol, K-60007
Kolavenic acid, K-60013
ent-7,12*E*,14-Labdatriene-6β,17-diol, L-60004
ent-8(14),15-Pimaradiene-3β,12β-diol, P-80146
Reiswigin *A*, R-60004
Stemodinone, *in* S-70071
1(15),8(19)-Trinervitadiene-2,3-diol, T-70295
13*R-Hydroxy*-8(17),14-*labdadien-3-one*, *in* L-70002

$C_{20}H_{32}O_3$
6β-Angeloyloxy-1α-hydroxy-5,10-bisepi-4(15)-eudesmene, *in* E-80084
Dehydroplexaurolone, *in* P-80163
Dictyotriol *A*, D-70147
ent-16α,17-Dihydroxy-3-atisanone, D-70284
3,15-Dihydroxy-8(17),13-labdadien-19-al, D-70310
2,13-Dihydroxy-8,14-labdadien-7-one, D-80327
ent-3β,15-Dihydroxy-7,13-labdadien-2-one, D-80328
1-(2,6-Dihydroxyphenyl)-1-tetradecanone, D-70338
Epivittadinal, *in* V-70016
ent-4α,18-Epoxy-13-cleroden-15-oic acid, *in* C-80142
ent-4β,18-Epoxy-13-cleroden-15-oic acid, *in* C-80142
11,12-Epoxy-5,8,14-eicosatrienoic acid, E-60017
8,9-Epoxy-5,11,14-eicosatrienoic acid, E-60016
8α,13*R*-Epoxy-6β-hydroxy-14-labden-11-one, *in* E-80037
3,4-Epoxy-13(15),16-sphenolobadiene-5,18-diol, E-60030
2,3,4,7,8,8*a*-Hexahydro-8-(5-hydroxy-4-methylpentyl)-6,8-dimethyl-1*H*-3*a*,7-methanoazulene-3-carboxylic acid, H-70057
ent-16α-Hydroxy-18-atisanoic acid, *in* A-80184
ent-2α-Hydroxy-3,13-clerodadien-15-oic acid, H-80144
12-Hydroxy-5,8,10,14-eicosatetraenoic acid, H-60131
5-Hydroxy-6,8,11,14-eicosatetraenoic acid, H-60127
8-Hydroxy-5,9,11,14-eicosatetraenoic acid, H-60128
12-Hydroxy-6-(hydroxymethyl)-10-methyl-2-(4-methyl-3-pentenyl)-2,6,10-dodecatrienal, *in* H-70179
11-Hydroxy-5,8,12,14-icosatetraenoic acid, H-60130
9-Hydroxy-5,7,11,14-icosatetraenoic acid, H-60129
8-Hydroxy-5,9,11,14-icosatetraenoic acid, *in* H-70126
2-Hydroxy-8(17),13-labdadien-15-oic acid, H-70147
ent-6β-Hydroxy-8(17),13-labdadien-15-oic acid, H-70148
2-(6-Hydroxy-4-methyl-4-hexenylidene)-6,10-dimethyl-7-oxo-9-undecenal, H-60186
12-Hydroxy-3,7,11,15-tetramethyl-2,6,10,14-hexadecatetraenoic acid, H-70223
16-Hydroxy-2,6,10,14-tetramethyl-2,6,10,14-hexadecatetraenoic acid, H-70224
Isodehydroplexaurolone, *in* P-80163
7,15-Isopimaradiene-2,3,19-triol, I-70081
8(14),15-Isopimaradiene-2,18,19-triol, I-80071
8(14),15-Isopimaradiene-3,18,19-triol, I-80072
8,15-Isopimaradiene-3,7,19-triol, I-70082
ent-7,11*E*,14-Labdatriene-6β,13ξ,17-triol, L-60005
Pilosanone *A*, P-60153
ent-5-Rosene-3α,15,16,18-tetrol, *in* R-60011
11,15(17)-Trinervitadiene-3,9,13-triol, T-70297
1(15),8(19)-Trinervitadiene-2,3,20-triol, T-70299
1(15),8(19)-Trinervitadiene-2,3,9-triol, T-70298
Vittadinal, V-70016

$C_{20}H_{32}O_4$
15-Angeloyloxy-10-hydroxy-4-oplopanone, *in* D-80348
Cervicol, C-70034
Chatferin, *in* F-60009
ent-3β,4α-Dihydroxy-13-cleroden-15,16-olide, D-80303
1,18-Dihydroxy-14-decipien-19-oic acid, D-80311
12,19-Dihydroxy-5,8,10,14-eicosatetraenoic acid, D-60319
15,20-Dihydroxy-5,8,11,13-eicosatetraenoic acid, D-70297
5,20-Dihydroxy-6,8,11,14-eicosatetraenoic acid, D-70296
5,6-Dihydroxy-7,9,11,14-eicosatetraenoic acid, D-80313
ent-13,16-Dihydroxy-8(17),14-labdadien-18-oic acid, D-80326
3,6-Dihydroxy-2-(1-oxo-10-tetradecenyl)-2-cyclohexen-1-one, D-70332
ent-15,16-Epoxy-13(16),14-clerodadiene-2β,3β,4α-triol, E-80016
14,15-Epoxy-20-hydroxy-5,8,11-eicosatrienoic acid, E-70018
5,6-Epoxy-20-hydroxy-8,11,14-eicosatrienoic acid, E-70017
9,13-Epoxy-7,14-labdadiene-2,3,20-triol, E-80035
Grayanotoxin XVIII, G-80041
Halicholactone, H-80002
10-Hydroxy-2-(6-hydroxy-4-methyl-4-hexylidene)-6,10-dimethyl-7-oxo-8-undecenal, *in* H-60186
8(14),15-Isopimaradiene-2,3,18,19-tetrol, I-80070
7-Labdene-15,17-dioic acid, L-60006
7-Labdene-15,17-dioic acid, L-80011
7-Labdene-15,18-dioic acid, L-60007
Oliveric acid, O-60039
Pilosanone *B*, *in* P-60153
Salvicin, S-60003
Stolonidiol, S-60055
ent-2α,3α,16-Trihydroxy-8(14)-pimaren-15-one, *in* P-60154
1(15),8(19)-Trinervitadiene-2,3,9,20-tetrol, T-70296

$C_{20}H_{32}O_5$
Camporic acid, C-70010
11,15-Dihydroxy-9-oxo-5,13-prostadienoic acid, D-60367
8α,13*R*-Epoxy-6β,7β,9-trihydroxy-14-labden-11-one, *in* E-80037
▷ Grayanotoxin II, G-80038
Grayanotoxin VI, *in* G-80040
ent-16-Kaurene-1β,3α,6β,11α,15α-pentol, K-80008
Methyl 3,6-epidioxy-6-methoxy-4,14,16-octadecatrienoate, *in* E-60008
Pulchellin 2-*O*-isovalerate, *in* P-80189
Sterebin *B*, *in* S-60052
Sterebin *C*, *in* S-60052
Tetradecahydro-4,6-dihydroxy-9,9,12*a*-trimethyl-6*H*-phenanthro[1,10*a*-*c*]furan-3-carboxylic acid, T-60038
ent-3β,4α,16ξ-Trihydroxy-13-cleroden-15,16-olide, *in* D-80303
5,14,15-Trihydroxy-6,8,10,12-eicosatetraenoic acid, T-60320
3,7,11-Trimethyl-2,6,10-dodecatrienoic acid; 2-Acetoxy-3-hydroxypropyl ester, *in* T-60364

$C_{20}H_{32}O_6$
8,13-Epoxy-1,6,7,9-tetrahydroxy-14-labden-11-one, E-70027
Grayanotoxin V, *in* G-80039
Grayanotoxin XI, *in* G-80038
Grayanotoxin XII, *in* G-80038
Laserol, *in* L-60016
Nardostachin, N-70019
Phlebiakauranol alcohol, *in* P-70097

$C_{20}H_{32}O_7$
8,19-Epoxy-1(15)-trinervitene-2,3,7,9,14,17-hexol, E-80048

$C_{20}H_{32}O_8$
Odonticin, O-80026

$C_{20}H_{33}BrO_2$
Bromotetrasphaerol, B-70272
Rotalin B, R-80013

$C_{20}H_{34}BrClO_2$
Dactylomelol, D-80001

$C_{20}H_{34}N_6O_2$
3,9,17,23,29,30-Hexaaza-6,20-dioxatricyclo[23.3.1.1^{11,15}]triaconta-1(28),11(29),12,14,25,27-hexaene, H-80036

$C_{20}H_{34}O$
13-Epi-5β-neoverrucosanol, *in* N-70030
13-Epi-5β-neoverrucosanol, *in* N-80024
8,13-Epoxy-14-labdene, E-70023
Helipterol, H-80009
5-Homoverrucosanol, H-80095
8-Hydroxyisopimar-15-ene, H-60163
8(17),12-Labdadien-3-ol, L-70012
5-Neoverrucosanol, N-80024
Neoverrucosanol, N-70030
Obscuronatin, O-60002
Tormesol, T-80203

$C_{20}H_{34}O_2$
ent-16α,18-Atisanediol, A-80184

2,7,11-Cembratriene-4,6-diol, C-80031
Chromophycadiol, C-60148
ent-4(18)-Cleroden-15-oic acid, *in* C-80142
3,7-Dolabelladiene-6,12-diol, D-70541
ent-16-Hydroxy-13-epimanoyl oxide, *in* E-70023
12α-Hydroxy-13-epi-manoyl oxide, *in* E-70023
8-Hydroxy-13-labden-15-al, H-70149
15-Hydroxy-7-labden-3-one, H-70151
19-Hydroxymanoyl oxide, *in* E-70023
7,14-Labdadiene-3,13-diol, L-80001
8(17),13-Labdadiene-3,15-diol, L-70001
8(17),14-Labdadiene-3,13-diol, L-70002
ent-8,13(16)-Labdadiene-14,15-diol, L-70005
ent-7,13-Labdadiene-2α,15-diol, L-80003
ent-7,14-Labdadiene-2α,13-diol, L-70003
ent-8,13-Labdadiene-7α,15-diol, L-70004
ent-8(17),13-Labdadiene-7α,15-diol, L-70006
ent-8(17),13-Labdadiene-9α,15-diol, L-70007
ent-7,13-Labdadiene-2β,15-diol, L-80004
ent-7,13E-Labdadiene-3β,15-diol, L-60001
ent-8,13-Labdadiene-7β,15-diol, *in* L-70004
Secomanool, S-80017
Stemodin, S-70071

$C_{20}H_{34}O_3$

ent-3β,16α,17-Atisanetriol, A-70266
7,11-Bis(hydroxymethyl)-3,15-dimethyl-2,6,10,14-hexadecatetraen-1-ol, B-80151
ent-3,13-Clerodadiene-15,16,17-triol, C-80141
Dictytriol, *in* D-70148
8,15-Dihydroxy-13-labden-19-al, D-70311
3β,15-Dihydroxy-8-labden-7-one, *in* L-70013
7,8-Epoxy-2,11-cembradiene-4,6-diol, E-60010
ent-3α,4α-Epoxy-15-clerodanoic acid, E-80018
ent-4α,18-Epoxy-15-clerodanoic acid, *in* C-80142
ent-4β,18-Epoxy-15-clerodanoic acid, *in* C-80142
Gypopinifolone, G-70036
ent-4β-Hydroxy-13-cleroden-15-oic acid, *in* C-80142
15-Hydroxy-8(17)-labden-19-al, H-70150
15-Hydroxy-7-labden-17-oic acid, H-60168
15-Hydroxy-7-labden-17-oic acid, H-80183
6-(Hydroxymethyl)-10-methyl-2-(4-methyl-3-pentenyl)-2,6,10-dodecatriene-1,12-diol, H-70179
ent-15-Hydroxy-8,9-seco-13-labdene-8,9-dione, H-70218
Isodictytriol, *in* D-70148
3,13,16-Kauranetriol, K-80005
ent-2α,16β,17-Kauranetriol, K-60004
ent-7α,16β,18-Kauranetriol, K-80006
ent-3β,16β,17-Kauranetriol, K-60005
8,13-Labdadiene-6,7,15-triol, L-60003
ent-7,14-Labdadiene-2α,13,20-triol, L-70011
ent-7,14-Labdadiene-2α,3α,13-triol, L-70010
ent-7,13-Labdadiene-2β,3β,15-triol, L-80009
Pachytriol, P-70001
Plexaurolone, P-80163
ent-5-Rosene-15,16,19-triol, R-60012
ent-5-Rosene-15,16,18-triol, *in* R-60012
Vinigrol, V-70009

$C_{20}H_{34}O_4$

6,15-Dihydroxy-7-labden-17-oic acid, D-60342
6,15-Dihydroxy-8-labden-17-oic acid, D-80329
7,15-Dihydroxy-8-labden-17-oic acid, D-80330
1(15)-Dolastene-4,8,9,14-tetrol, D-70543
6,13-Epoxy-3-eunicellene-8,9,12-triol, E-80027
Jewenol A, J-70003
Jewenol B, J-70004
11,13-Labdadiene-6,7,8,15-tetrol, L-70009
7,14-Labdadiene-2,3,13,20-tetrol, L-80007
8,13-Labdadiene-2,6,7,15-tetrol, L-60002
8(17),14-Labdadiene-2,7,13,20-tetrol, L-80008
ent-8(14)-Pimarene-2α,3α,15R,16-tetrol, P-60154
ent-5-Rosene-3α,15,16,19-tetrol, R-60011
ent-2β,15,16-Trihydroxy-13-cleroden-3-one, T-80292

$C_{20}H_{34}O_5$

8,20-Epoxy-14-labdene-2,3,7,13-tetrol, E-80036
4(15)-Eudesmen-11-ol; α-L-Arabopyranoside, *in* E-70066

11,13(16)-Labdadiene-6,7,8,14,15-pentol, L-70008
8,14-Labdadiene-2,3,7,13,20-pentol, L-80005
8(17)-Labdadiene-2,3,7,13,20-pentol, L-80006
Methyl 3,6-epidioxy-6-methoxy-4,16-octadecadienoate, *in* E-60008
Sterebin H, *in* L-70008
Tucumanoic acid, T-60410

$C_{20}H_{34}O_6$

▷ Grayanotoxin III, G-80039

$C_{20}H_{35}BrO_3$

Venustanol, V-70004

$C_{20}H_{36}$

Tetra-*tert*-butylcyclobutadiene, T-70014
Tetra-*tert*-butyltetrahedrane, T-70015

$C_{20}H_{36}Cl_2O_{10}$

7(18),11(19)-Havannadichlorohydrin, H-70005

$C_{20}H_{36}N_4O_8$

1,5,9,13-Tetraazacyclohexadecane-1,5,9,13-tetraacetic acid, T-60016

$C_{20}H_{36}O$

2,2-Di-*tert*-butyl-3-(di-*tert*-butylmethylene)cyclopropanone, D-70101
ent-14-Labden-8β-ol, L-70014

$C_{20}H_{36}O_2$

13-Episclareol, *in* S-70018
8,13-Epoxy-3-labdanol, E-70021
7-Labdene-15,17-diol, L-80012
8(20)-Labdene-3,15-diol, L-80013
Peucelinendiol, P-80075
Sclareol, S-70018

$C_{20}H_{36}O_3$

Dihydroplexaurolone, *in* P-80163
ent-3α,4α-Epoxy-13,15-clerodanediol, E-80017
8-Labdene-3,7,15-triol, L-70013
7-Pimarene-3,15,16-triol, P-80149

$C_{20}H_{36}O_4$

ent-8,13-Epoxy-14,15,19-labdanetriol, E-70020
3-Hydroxymethyl-7,11,15-trimethyl-2,6,10-hexadecatriene-1,14,15-triol, H-80215

$C_{20}H_{36}O_5$

ent-13-Clerodene-2β,3β,4α,15,16-pentol, C-80144
Methyl 3,6-epidioxy-6-methoxy-4-octadecenoate, *in* E-60008
ent-2β,3β,4α-Trihydroxy-15-clerodanoic acid, *in* T-60410
9,12,13-Trihydroxy-10,15-octadecadienoic acid, T-80327

$C_{20}H_{36}S$

2,3-Dimethyl-5-(2,6,10-trimethylundecyl)thiophene, D-70446

$C_{20}H_{37}NO_3$

N-(Tetrahydro-2-oxo-3-furanyl)hexadecanamide, *in* A-60138

$C_{20}H_{38}O$

3,7,11,15-Tetramethyl-2-hexadecenal, *in* P-60152

$C_{20}H_{38}O_3$

11,15-Epoxy-3(20)-phytene-1,2-diol, E-80043
Gyplure, *in* O-60006

$C_{20}H_{38}O_5$

9S,12S,13S-Trihydroxy-10E-octadecenoic acid, *in* T-80327

$C_{20}H_{40}$

3-Icosene, I-80003

$C_{20}H_{40}Br_2$

1,20-Dibromoicosane, D-70092

$C_{20}H_{40}O$

Phytol, P-60152

$C_{20}H_{40}O_2$

7,11,15-Trimethyl-3-methylene-1,2-hexadecanediol, T-60368

$C_{20}H_{40}O_3$

19-Hydroxyicosanoic acid, H-80176

$C_{20}H_{42}O$

2-Icosanol, I-80002

$C_{20}H_{50}N_{10}$

1,4,7,10,13,16,19,22,25,28-Decaazacyclotriacontane, D-70011

$C_{21}H_{11}F_5O_3$

9-Fluorenylmethyl pentafluorophenyl carbonate, F-60019

$C_{21}H_{12}N_4$

▷ Tricycloquinazoline, T-60271

$C_{21}H_{12}N_6$

[1,3,5]Triazino[1,2-*a*:3,4-*a'*:5,6-*a''*]trisbenzimidazole, T-60238

$C_{21}H_{12}O_2$

7H-Dibenzo[*c,h*]xanthen-7-one, D-60080

$C_{21}H_{13}N$

Phenanthro[9,10-*g*]isoquinoline, P-60075

$C_{21}H_{13}NO_2$

Dibenzo[*c,g*]carbazole-6-carboxylic acid, D-80068

$C_{21}H_{13}N_3OS$

12H-Quinoxalino[2,3-*b*][1,4]benzothiazine; *N*-Benzoyl, *in* Q-70008

$C_{21}H_{14}O$

2,3-Diphenyl-1H-inden-1-one, D-70482

$C_{21}H_{15}N$

1,2-Diphenyl-3-(phenylimino)cyclopropene, D-70491

$C_{21}H_{15}NO_2$

Dibenz[*b,g*]azocine-5,7(6H,12H)-dione; *N*-Ph, *in* D-60068

$C_{21}H_{15}O_2^{\oplus}$

2,4,5-Triphenyl-1,3-dioxol-1-ium, T-70307

$C_{21}H_{16}$

▷ 6-Methylcholanthrene, M-70056

$C_{21}H_{16}O_5$

Calopogonium isoflavone B, C-60006

$C_{21}H_{16}O_6$

Gerberinol 1, G-60013
Justicidin B, *in* D-60493

$C_{21}H_{16}O_7$

▷ Diphyllin, D-60493

$C_{21}H_{16}O_8$

1,3-Dihydroxy-2-hydroxymethylanthraquinone; Tri-Ac, *in* D-70306

$C_{21}H_{16}O_9$

Mitorubrinic acid, M-80192

$C_{21}H_{16}O_{11}$

α-Acetylconstictic acid, *in* C-60163

$C_{21}H_{18}N_2$

4,5-Dihydro-2,4,5-triphenyl-1H-imidazole, D-80263
Hydrobenzamide, H-80100

$C_{21}H_{18}O$

1,2,2-Triphenyl-1-propanone, T-60389
1,2,3-Triphenyl-1-propanone, T-60390
1,3,3-Triphenyl-1-propanone, T-60391
▷ 1,1,3-Triphenyl-2-propen-1-ol, T-70313

$C_{21}H_{18}O_3$

6-Methylene-2,4-cyclohexadien-1-one; Trimer, *in* M-60072

$C_{21}H_{18}O_4$

Calopogoniumisoflavone A, C-80013

$C_{21}H_{18}O_5$

Alpinumisoflavone; 4'-Me ether, *in* A-80050
Glabrachromene II, G-70014
Maximaisoflavone B, *in* H-80203
4'-O-Methylderrone, *in* D-80031
Neorautenane, *in* N-80022

$C_{21}H_{18}O_6$

Glycyrol, *in* P-80186

3'-Hydroxyalpinumisoflavone 4'-methyl ether, *in* A-**80050**
Isoglycyrol, I-**80061**
Neorautenanol, *in* N-**80022**
Pipoxide, P-**60159**
Racemoflavone, *in* A-**80183**

$C_{21}H_{18}O_7$
3-(2-Ethyl-2-butenyl)-9,10-dihydro-1,6,8-trihydroxy-9,10-dioxo-2-anthracenecarboxylic acid, E-**70038**
Hildecarpidin, H-**60081**

$C_{21}H_{18}O_{12}$
1,3,8,9-Tetrahydroxycoumestan; 3-*O*-Glucoside, *in* T-**80087**

$C_{21}H_{19}N$
Cyanododecahedrane, *in* D-**80522**

$C_{21}H_{19}NO$
1,3,3-Triphenyl-1-propanone; Oxime, *in* T-**60391**

$C_{21}H_{19}NO_5$
Isomurralonginol; 3-Pyridinecarboxylate, *in* I-**70071**

$C_{21}H_{20}$
Bicyclo[14.4.1]heneicosa-1,3,5,7,9,11,13,15,17,19-decaene, B-**80077**
Cyclopropadodecahedrane, C-**80195**

$C_{21}H_{20}O$
Dodecahedranecarboxaldehyde, D-**80521**

$C_{21}H_{20}O_2$
Dodecahedranecarboxylic acid, D-**80522**

$C_{21}H_{20}O_3$
Abbottin, A-**70002**

$C_{21}H_{20}O_5$
Edulenanol, *in* N-**80023**
Neorautane, N-**80022**
Pongachalcone II, P-**80168**

$C_{21}H_{20}O_6$
Angeloylprangeline, *in* P-**60170**
Aurmillone, *in* T-**80113**
Curcumin, C-**80165**
1-(3,4-Dimethoxyphenyl)-2,3-bis(hydroxymethyl)-6,7-methylenedioxynaphthalene, D-**80414**
Egonol; Ac, *in* E-**70001**
Garvin *B*, G-**60009**
Isoaurmillone, *in* T-**80319**
Kwakhurin, K-**60016**
Neorautanol, *in* N-**80022**
Topazolin, T-**60224**

$C_{21}H_{20}O_7$
Dehydrotrichostin, *in* T-**70225**
2-Hydroxygarvin *B*, *in* G-**60009**
Podoverine *A*, P-**60164**
Zeylenol, Z-**70006**

$C_{21}H_{20}O_8$
4'-Demethylpodophyllotoxin, *in* P-**80165**
4',7-Dihydroxyisoflavone; 7-*O*-Rhamnoside, *in* D-**80323**
Homalicine, *in* H-**70207**
Hormothamnione, H-**50127**

$C_{21}H_{20}O_9$
Cleomiscosin *D*, C-**70185**
Daidzin, *in* D-**80323**
4',6-Dihydroxyaurone; 6-*O*-β-D-Glucopyranosyl, *in* D-**80273**
4',7-Dihydroxyisoflavone; 4'-*O*-β-D-Glucopyranoside, *in* D-**80323**
3,3',4,5,8-Pentamethoxy-6,7-methylenedioxyflavone, *in* H-**80028**
Phenarctin, P-**70054**

$C_{21}H_{20}O_{10}$
Baptisin, *in* T-**80110**
Genistin, *in* T-**80309**
8-*C*-Glucosyl-4',5,7-trihydroxyisoflavone, G-**80025**
Sophoricoside, *in* T-**80309**
2',5,7-Trihydroxyisoflavone; 7-*O*-Glucoside, *in* T-**80307**

$C_{21}H_{20}O_{11}$
Aureusin, *in* A-**80188**
Cernuoside, *in* A-**80188**
8-*C*-Glucosyl-3',4',5,7-tetrahydroxyisoflavone, G-**80024**
Isoorientin, I-**80068**
Oroboside, *in* T-**80109**
Sulfurein, *in* S-**70082**

$C_{21}H_{20}O_{13}$
Quercetagitrin, *in* H-**80067**

$C_{21}H_{20}O_{14}$
Hibiscitin, *in* H-**70021**

$C_{21}H_{21}N$
2,2',2''-Trimethyltriphenylamine, T-**70288**
3,3',3''-Trimethyltriphenylamine, T-**70289**
4,4',4''-Trimethyltriphenylamine, T-**70290**

$C_{21}H_{21}NO$
Dodecahedranecarboxylic acid; Amide, *in* D-**80522**

$C_{21}H_{21}N_3O_7$
Cacotheline, C-**70001**

$C_{21}H_{21}O_{10}^{\oplus}$
Callistephin, *in* T-**80093**
Fragarin, *in* T-**80093**
Pelargonenin, *in* T-**80093**
3,4',5,7-Tetrahydroxyflavylium; 4'-Glucoside, *in* T-**80093**
3,4',5,7-Tetrahydroxyflavylium; 7-Glucoside, *in* T-**80093**

$C_{21}H_{21}O_{11}^{\oplus}$
Columnin, *in* P-**80037**

$C_{21}H_{22}$
Des-A-26,27,28-trisnorursa-5,7,9,11,13,15,17-heptaene, D-**80038**

$C_{21}H_{22}O$
(Hydroxymethyl)dodecahedrane, H-**80201**

$C_{21}H_{22}O_3$
Asnipyrone A, A-**80177**

$C_{21}H_{22}O_4$
1,2-Diphenyl-1,2-cyclopropanedicarboxylic acid; Di-Et ester, *in* D-**80509**
Eupomatenoid 12, *in* E-**60075**
Isoxanthohumol/, I-**80100**
Kurospongin, K-**70015**
Sandwicensin, *in* P-**80076**
Tephrinone, *in* G-**70015**
Tephrobbottin, T-**70007**

$C_{21}H_{22}O_5$
Ambofuranol, A-**80057**
Aristotetralone, A-**60298**
Cristacarpin, C-**80156**
7,8-(2,2-Dimethylpyrano)-3,4'-dihydroxy-5-methoxyflavan, *in* D-**60448**
Garveatin *D*, G-**60007**
Helichromanochalcone, H-**80005**
Isoxanthohumol/, I-**80099**
Licobenzofuran, L-**80027**
3'-Methoxyglabridin, *in* G-**70016**
1-Methoxyphaseollidin, *in* P-**80076**
Pleurotin, P-**80162**
Quercetol *A*, Q-**70001**
3a,4,9,9a-Tetrahydro-9-(3,4,5-trimethoxyphenyl)naphtho[2,3-c]furan-1(3H)one, T-**80080**

$C_{21}H_{22}O_6$
O-Angeloylalkannin, *in* S-**70038**
β,β-Dimethylacrylalkannin, *in* S-**70038**
β,β-Dimethylacrylshikonin, *in* S-**70038**
2-Hydroxyaristotetralone, *in* A-**70253**
Sophoraisoflavanone A, *in* L-**80028**
4',5,7-Trihydroxy-3'-methoxy-5'-prenylflavanone, *in* S-**70041**

$C_{21}H_{22}O_7$
Annulin *B*, A-**60277**
Divaronic acid, D-**70528**
Isopteryxin, *in* K-**70013**
Isosamidin, *in* K-**70013**
Praeruptorin *A*, P-**60169**
Pteryxin, *in* K-**70013**

▷ Samidin, *in* K-**70013**
Topazolin hydrate, *in* T-**60224**
Trichostin, T-**70225**

$C_{21}H_{22}O_8$
Dihydrohomalicine, *in* H-**70207**
Epoxypteryxin, *in* K-**70013**
3,3',4',5,6,7-Hexamethoxyflavone, *in* H-**80067**
3',4',5,5',6,7-Hexamethoxyflavone, *in* H-**80081**
Isopseudocyphellarin *A*, I-**70098**
Pseudocyphellarin *A*, *in* P-**70127**

$C_{21}H_{22}O_9$
▷ Barbaloin, B-**80010**
3-Hydroxy-3',4',5,6,7,8-hexamethoxyflavone, *in* H-**80028**
3'-Hydroxy-4',5,5',6,7,8-hexamethoxyflavone, *in* H-**80029**
4'-Hydroxy-3',5,5',6,7,8-hexamethoxyflavone, *in* H-**80029**
5-Hydroxy-3,3',4',6,7,8-hexamethoxyflavone, *in* H-**80028**
5-Hydroxy-3',4',5',6,7,8-hexamethoxyflavone, *in* H-**80029**
Isobarbaloin, *in* B-**80010**
Isoliquiritin, *in* D-**80367**
Neoisoliquiritin, *in* D-**80367**

$C_{21}H_{22}O_{10}$
Coreopsin, *in* D-**80362**
5,7-Dihydroxy-2',3,4',5',6,8-hexamethoxyflavone, *in* O-**60022**
5,7-Dihydroxy-3,3',4',5',6,8-hexamethoxyflavone, *in* O-**70018**
Isosalipurposide, *in* H-**80246**
Monospermoside, *in* D-**80362**

$C_{21}H_{22}O_{11}$
Pyracanthoside, *in* T-**70105**
Stillopsin, *in* D-**70339**

$C_{21}H_{22}O_{12}$
Astilbin, *in* P-**70027**
Glucodistylin, *in* P-**70027**
Isoglucodistylin, *in* P-**70027**
Lanceoside, *in* P-**60052**
Taxifolin 4'-glucoside, *in* P-**70027**
1,3,5,8-Tetrahydroxyxanthone; 3,5-Di-Me ether, 8-*O*-β-D-glucopyranoside, *in* T-**70123**

$C_{21}H_{23}ClO_7$
Methyl 2-*O*-methyleriodermate, *in* E-**70032**
Methyl 2'-*O*-methyleriodermate, *in* E-**70032**
Methyl 4-*O*-methyleriodermate, *in* E-**70032**

$C_{21}H_{24}O_3$
5-(3,5-Di-*tert*-butyl-4-oxo-2,5-cyclohexadienylidene)-3,6-cycloheptadiene-1,2-dione, D-**70105**

$C_{21}H_{24}O_4$
5-Deoxymyricanone, *in* M-**70162**
8-Geranyloxypsoralen, *in* X-**60002**
6-(3-Methyl-2-butenyl)allopteroxylin methyl ether, *in* A-**60075**
O-Methyldehydrodieugenol, *in* D-**80023**
2'-*O*-Methylphaseollidinisoflavan, *in* P-**80077**

$C_{21}H_{24}O_5$
Aristochilone, A-**70251**
Aristotetralol, A-**70253**
Calopiptin, *in* A-**70271**
α,α-Dimethylallylcyclolobin, *in* U-**80002**
8-(3,3-Dimethylallyl)-5-methoxy-3,4',7-trihydroxyflavan, *in* T-**60125**
5-Hydroxy-8,8-dimethyl-6-(2-methyl-1-oxopropyl)-4-propyl-2H,8H-benzo[1,2-b,3,4-b']dipyran-2-one, *in* M-**80001**
5-Hydroxy-8,8-dimethyl-6-(1-oxobutyl)-4-propyl-2H,8H-benzo[1,2-b:3,4-b']dipyran-2-one, *in* M-**80001**
4-(3-Hydroxy-3-methylbutanoyloxy)-3-(1,1-dimethyl-2-propenyl)-6-phenyl-2H-pyran-2-one, D-**70434**
Machilin G, *in* A-**70271**
5-Methylethuliacoumarin, *in* E-**60045**
Myricanone, *in* M-**70162**
Obionin A, O-**80001**
Praealtin A, P-**80172**
Rutamarin, *in* C-**60031**

$C_{21}H_{24}O_6$
Arctigenin, A-70249
Penicillide, P-80019
Phillygenin, P-60141
▷ Phloridzin, *in* H-80244
Shikonin isobutyrate, *in* S-70038

$C_{21}H_{24}O_7$
Arnebin II, *in* D-80343
Dihydrosamidin, *in* K-70013
Dihydrotrichostin, D-70273
β-Hydroxyisovalerylshikonin, *in* S-70038
1-Hydroxypinoresinol; 4″-Me ether, *in* H-60221
Medioresinol, M-70016
Subsphaeric acid, S-80049
Suksdorfin, *in* K-70013
▷ Visnadin, *in* K-70013

$C_{21}H_{24}O_8$
Albaspidin AA, *in* A-80043
Aspidin AA, *in* A-80180
9α-Hydroxymedioresinol, *in* M-70016
Pseudocyphellarin *B*, P-70127

$C_{21}H_{24}O_9$
Glycyphyllin, *in* H-80244
Isorhapontin, *in* D-80361
Rhapontin, *in* D-80361

$C_{21}H_{24}O_{10}$
7,7*a*-Dihydro-3,6,7-trihydroxy-1*a*-(3-methyl-2-butenyl)naphth[2,3-*b*]oxiren-2(1*aH*)-one; 7-Ketone, 3-*O*-*β*-D-glucopyranoside, *in* D-70277

$C_{21}H_{26}O_3$
1,11-Bis(3-furanyl)-4,8-dimethyl-3,6,8-undecatrien-2-ol, B-70143
Malabaricone *A*, M-70007

$C_{21}H_{26}O_4$
1-(2,6-Dihydroxyphenyl)-9-(4-hydroxyphenyl)-1-nonanone, *in* M-70007
15,20-Epoxy-3*β*-hydroxy-14,15-seco-5,15,17(20)-pregnatriene-2,14-dione, *in* A-70267
9-Phenyl-1-(2,4,6-trihydroxyphenyl)-1-nonanone, P-80138

$C_{21}H_{26}O_5$
Aristolignin, A-60297
5,7-Dihydroxy-7-(3-methyl-2-butenyl)-8-(2-methyl-2-oxopropyl)-3-propyl-2*H*-1-benzopyran-2-one, D-80340
1-(2,6-Dihydroxyphenyl)-9-(3,4-dihydroxyphenyl)-1-nonanone, *in* M-70007
3-*O*-Methyl-2,5-dehydrosenecioodentol, *in* S-60022
Myricanol, M-70162

$C_{21}H_{26}O_6$
Cordatin, C-60166
Fragransol *A*, F-70041
1-(4-Hydroxy-3,5-dimethoxyphenyl)-2-[2-methoxy-4-(1-propenyl)phenoxy]-1-propanol, *in* A-80046
Isolariciresinol 4′-methyl ether, *in* I-60139
Machilin *H*, M-70002
Subexpinnatin *B*, *in* S-60058

$C_{21}H_{26}O_7$
1-Ethoxy-9*Z-millerenolide*, *in* M-60137
Liriolignal, L-80035
Neocynaponogenin A, N-80019
Orthopappolide senecioate, *in* O-60045
Orthopappolide tiglate, *in* O-60045

$C_{21}H_{26}O_8$
Cratystyolide; Tri-Ac, *in* T-70257
4*β*,15-Epoxy-1*β*-ethoxy-4*Z-millerenolide*, *in* M-60137
Isorolandrolide, I-80087
8*β*-(2-Methylacryloyloxy)hirsutinolide 13-*O*-acetate, *in* H-80089

$C_{21}H_{26}O_9$
8-Deacyltrichogoniolide; 8-*O*-(2-Methylpropenoyl), 9-Ac, *in* D-80007
8*β*-(2-Hydroxymethylacryloyloxy)hisutinolide 13-*O*-acetate, *in* H-80089
Hypocretenoic acid; Lactone, 14-*β*-D-Glucopyranosyloxy, *in* H-70235
8*β*-(2-Methylacryloyloxy)-15-hydroxyhisutinolide 13-acetate, *in* H-80089
8*β*-(2-Methyl-2,3-epoxypropionyloxy)hirsutinolide 13-*O*-acetate, *in* H-80089

$C_{21}H_{26}O_{10}$
8*β*-Acetoxy-10*β*-hydroxyhirsutinolide 1,13-*O*-diacetate, *in* H-80089
8*β*,10*β*-Diacetoxyhirsutinolide 13-*O*-acetate, *in* H-80089
7,7*a*-Dihydro-3,6,7-trihydroxy-1*a*-(3-methyl-2-butenyl)naphth[2,3-*b*]oxiren-2(1*aH*)-one; 3-*O*-*β*-D-Glucopyranoside, *in* D-70277

$C_{21}H_{26}O_{13}$
Baisseoside, *in* D-84287

$C_{21}H_{27}ClO_{10}$
Tetragonolide, T-80053

$C_{21}H_{27}N_7O_{14}P_2$
▷ Coenzyme I, C-80145

$C_{21}H_{28}N_7O_{17}P_3$
Coenzyme II, C-80146

$C_{21}H_{28}O_2$
Avarone, *in* A-80189

$C_{21}H_{28}O_3$
12,13-Didehydrofurospongin I, *in* F-80097
1,3-Dioxototaryl methyl ether, *in* T-70204
Methyl nidoresedate, *in* N-70035

$C_{21}H_{28}O_4$
Atratogenin *A*, A-70267
Membranolide, M-60016
Smenoquinone, *in* I-80004
2,3,12-Trihydroxy-4,7,16-pregnatrien-20-one, T-60346

$C_{21}H_{28}O_5$
15,16-Dihydro-15-methoxy-16-oxonidoresedic acid, *in* N-70035
15,16-Dihydro-15-methoxy-16-oxostrictic acid, *in* S-70077
7-*O*-Formylhorminone, *in* D-80265
7-Methoxyrosmanol, *in* R-70010
3-*O*-Methylsenecioodentol, *in* S-60022

$C_{21}H_{28}O_7$
14-Acetoxy-9*β*-hydroxy-8*β*-(2-methylpropanoyloxy)-1(10),4,11(13)-germacratrien-12,6α-olide, *in* T-60323
Viguiestin, *in* T-70003

$C_{21}H_{28}O_8$
Vernoflexuoside, *in* Z-60001

$C_{21}H_{28}O_9$
Hypocretenoic acid; Lactone, 11α,13-Dihydro, 14-*β*-D-glucopyranosyloxy, *in* H-70235
Ixerin B, *in* I-80102
8-Propionoyloxy-10*β*-hydroxy-1-*O*-methylhirsutinolide-13-*O*-acetate, *in* H-80089

$C_{21}H_{29}BrO_4$
Bromovulone I, B-70278

$C_{21}H_{29}IO_4$
Iodovulone I, *in* B-70278

$C_{21}H_{29}NO_3$
Smenospongine, S-80036

$C_{21}H_{29}N_7O_{14}P_2$
Adenosine 5′-(trihydrogen diphosphate), 5′→5′-ester with 1,4-dihydro-1-*β*-D-ribofuranosyl-3-pyridinecarboxamide, *in* C-80145

$C_{21}H_{30}N_7O_{17}P_3$
NADPH, *in* C-80146

$C_{21}H_{30}O_2$
Avarol, A-80189

$C_{21}H_{30}O_3$
Furospongin I, F-80097
12-Methoxy-8,11,13-abietatrien-20-oic acid, M-60036
Methyl 7-oxo-8(14),15-isopimaradien-18-oate, *in* O-70094

$C_{21}H_{30}O_4$
7-*O*-Methylhorminone, *in* D-80265
Pinusolide, P-60156
2,3,12-Trihydroxy-4,7-pregnadien-20-one, T-60345

$C_{21}H_{30}O_5$
15,16-Dihydro-15-methoxy-16-oxohardwickiic acid, *in* H-70002
Kamebacetal *A*, K-70002
Kamebacetal *B*, *in* K-70002
Microglossic acid, M-60134
Spongionellin, S-60048
2,3,12,16-Tetrahydroxy-4,7-pregnadien-20-one, T-60124

$C_{21}H_{30}O_6$
4-Acetoxyflexilin, *in* F-70012
11-Dihydro-12-norneoquassin, D-60258

$C_{21}H_{30}O_7$
15-Desacetyltetraneurin *C* isobutyrate, *in* T-60158

$C_{21}H_{30}O_8$
Brachynereolide, B-60196
Macrocliniside G, *in* C-80164
Zaluzanin *C*; 11*β*,13-Dihydro, *β*-D-Glucoside, *in* Z-60001

$C_{21}H_{30}O_9$
11*β*,13-Dihydro-8α-hydroxyglucozaluzanin *C*, *in* Z-60001
Ixerin J, *in* I-80102

$C_{21}H_{30}O_{12}$
Cyclopropanehexacarboxylic acid; Hexa-Et ester, *in* C-70250

$C_{21}H_{30}O_{13}$
Acetylbarlerin, *in* S-60028

$C_{21}H_{30}Se$
Di-1-adamantyl selenoketone, D-70034

$C_{21}H_{31}N$
8-Isocyano-10,14-amphilectadiene, I-60090
7-Isocyano-11-cycloamphilectene, I-60094
7-Isocyano-1-cycloamphilectene, I-60093
8-Isocyano-1(12)-cycloamphilectrene, I-60095

$C_{21}H_{31}NO_5S$
Latrunculin *C*, *in* L-50032

$C_{21}H_{32}O_2$
1-(3,4-Methylenedioxyphenyl)-1-tetradecene, M-60077

$C_{21}H_{32}O_3$
ent-3*β*,4*β*;15,16-Diepoxy-2*β*-methoxy-13(16),14-clerodadiene, *in* D-80153
14-(3-Furanyl)-3,7,11-trimethyl-7,11-tetradecadienoic acid, F-70049
Gracilin *E*, *in* G-60035
Methyl 14ξ,15-epoxy-8(17),12*E*-labdadien-16-oate, *in* E-60025

$C_{21}H_{32}O_4$
Lycopersiconolide, L-60042

$C_{21}H_{32}O_5$
10,11-Dihydromicroglossic acid, *in* M-60134
Umbraculumin *A*, U-70001

$C_{21}H_{32}O_8$
11*β*,13-Dihydrobrachynereolide, *in* B-60196

$C_{21}H_{32}O_{10}$
Dihydroserruloside, D-70265
Ebuloside, E-60001

$C_{21}H_{32}O_{14}$
Aucubigenin; 1-*O*-*β*-Cellobioside, *in* A-80186
Aucubigenin; 1-*O*-*β*-Gentiobioside, *in* A-80186
10-Deoxymelittoside, *in* M-70019

$C_{21}H_{32}O_{15}$
Melittoside, M-70019

$C_{21}H_{33}N_3$
1,3,5-Tripiperidinobenzene, T-70314

$C_{21}H_{34}O_2$
Kolavenic acid; Me ester, *in* K-60013

$C_{21}H_{34}O_3$
15-Formylimbricatolal, *in* H-70150
12-Hydroxy-5,8,10,14-eicosatetraenoic acid;
Me ester, *in* H-60131
5-Hydroxy-6,8,11,14-eicosatetraenoic acid; Me
ester, *in* H-60127
8-Hydroxy-5,9,11,14-eicosatetraenoic acid; Me
ester, *in* H-60128
11-Hydroxy-5,8,12,14-icosatetraenoic acid; Me
ester, *in* H-60130
9-Hydroxy-5,7,11,14-icosatetraenoic acid; Me
ester, *in* H-60129
1-(2-Hydroxy-6-methoxyphenyl)-1-
tetradecanone, *in* D-70338

$C_{21}H_{34}O_4$
5,6-Dihydroxy-7,9,11,14-eicosatetraenoic acid;
Me ester, *in* D-80313
ent-4α-Hydroxy-3β-methoxy-13-cleroden-
15,16-olide, *in* D-80303

$C_{21}H_{34}O_5$
3,6-Epidioxy-6-methoxy-4,16,18-eicosatrienoic
acid, E-60007

$C_{21}H_{34}O_6$
8,19-Epoxy-17-methyl-1(15)-trinervitene-
2,3,7,9,14,17-hexol, E-80040

$C_{21}H_{34}O_{10}$
7,7-*O*-Dihydroebuloside, *in* E-60001

$C_{21}H_{34}O_{11}$
Patrinalloside, P-70014

$C_{21}H_{36}O_4$
15-Hydroxy-7α-methoxy-8-labden-17-oic acid,
in D-80330

$C_{21}H_{36}O_5$
Arvoside, A-70264
Tetra-*tert*-butoxycyclopentadienone, T-70013

$C_{21}H_{36}O_6$
3-Eudesmene-1,6-diol; 6-*O*-β-D-
Glucopyranoside, *in* E-80083

$C_{21}H_{36}O_8$
Methanetetrapropanoic acid; Tetra-Et ester,
in M-70036

$C_{21}H_{38}O_3$
7α-Methoxy-8-labdene-3β,15-diol, *in* L-70013

$C_{21}H_{38}O_8$
Icariside C_1, *in* T-60362
Icariside C_2, *in* T-60362
Icariside C_3', *in* T-60362
Icariside C_4', *in* T-60362

$C_{21}H_{40}O_4$
Heneicosanedioic acid, H-80012

$C_{21}H_{42}O_2$
Heneicosanoic acid, H-80013

$C_{21}H_{42}O_3$
19-Hydroxyicosanoic acid; Me ester, *in*
H-80176

$C_{21}H_{45}N_3O_3$
1,9,17-Trioxa-5,13,21-triazacyclotetracosane;
5,13,21-Tri-Me, *in* T-80358

$C_{22}H_{10}O_2$
10,11-Methano-1*H*-
benzo[5,6]cycloocta[1,2,3,4-*def*]fluorene-
1,14-dione, M-70039

$C_{22}H_{12}$
Benzo[*b*]benzo[3,4]cyclobuta[1,2-
h]biphenylene, B-80015
Benzo[*h*]benzo[3,4]cyclobuta[1,2-
a]biphenylene, B-80016
Benzo[*l*]cyclopenta[*cd*]pyrene, B-80018
Indeno[1,2,3-*cd*]fluoranthene, I-80010
▷ Indeno[1,2,3-*cd*]pyrene, I-70012

$C_{22}H_{12}N_2$
Dibenz[*b,h*]indeno[1,2,3-*de*][1,6]naphthyridine,
D-60069

▷ Dibenzo[*c,f*]indeno[1,2,3-*ij*][2,7]naphthyridine,
D-60074
2,2'-Diisocyano-1,1'-binaphthyl, D-60391

$C_{22}H_{12}N_2O$
Benzo[*a*]benzofuro[2,3-*c*]phenazine, *in* B-60033

$C_{22}H_{12}N_4$
Dibenzo[*cd:c'd'*][1,2,4,5]tetrazino[1,6-*a*:4,3-
a']diindole, D-60076

$C_{22}H_{12}O$
▷ Indeno[1,2,3-*cd*]pyren-1-ol, I-60022
▷ Indeno[1,2,3-*cd*]pyren-2-ol, I-60023
▷ Indeno[1,2,3-*cd*]pyren-6-ol, I-60024
▷ Indeno[1,2,3-*cd*]pyren-7-ol, I-60025
▷ Indeno[1,2,3-*cd*]pyren-8-ol, I-60026

$C_{22}H_{12}O_2$
Azuleno[1,2-*b*]anthracene-6,13-dione, A-80207

$C_{22}H_{12}O_9$
Pulviquinone A, P-80192

$C_{22}H_{12}S_2$
Naphtho[2,1-*b*:6,5-*b'*]bis[1]benzothiophene,
N-60007

$C_{22}H_{13}N$
4*H*-Benzo[*def*]naphtho[2,3-*b*]carbazole,
B-80024

$C_{22}H_{14}N_2$
Cycloocta[2,1-*b*:3,4-*b'*]diquinoline, C-70232

$C_{22}H_{14}O$
Benzo[*b*]triphenylen-10-ol, B-80050
Benzo[*b*]triphenylen-11-ol, B-80051
Benzo[*b*]triphenylen-1-ol, B-80046
Benzo[*b*]triphenylen-2-ol, B-80047
Benzo[*b*]triphenylen-3-ol, B-80048
Benzo[*b*]triphenylen-4-ol, B-80049
4,7:12,15-Diethenobenzo[7,8]cyclododeca[1,2-
c]furan, D-80154

$C_{22}H_{14}O_2$
5,6:12,13-Diepoxy-5,6,12,13-
tetrahydrodibenz[*a,h*]anthracene, D-70151

$C_{22}H_{14}O_3$
8,16-Epoxy-8*H*,16*H*-dinaphtho[2,1-*b*:2',1'-
f][1,5]dioxocin, E-70014

$C_{22}H_{14}O_4$
6,12-Dimethoxy-1,7-perylenedione, *in*
D-80354
6,7-Dimethoxy-1,12-perylenedione, *in*
D-80354

$C_{22}H_{14}O_6$
3,3'-Bi[5-hydroxy-2-methyl-1,4-
naphthoquinone], B-60106

$C_{22}H_{14}O_9$
Aurintricarboxylic acid, A-60318

$C_{22}H_{15}F_3O_5S$
2,4,5-Triphenyl-1,3-dioxol-1-ium;
Trifluoromethanesulfonate, *in* T-70307

$C_{22}H_{16}$
1,1:2,2-Bis([10]annulene-1,6-diyl)ethylene,
B-70127
2,2-Di-(2-naphthyl)ethanal, D-80487
1,8-Diphenylnaphthalene, D-60483
Tribenzotriquinacene, T-80215

$C_{22}H_{16}O$
2,2-Di-(1-naphthyl)ethanal, D-80486
2-(1-Naphthyl)-2-(2-naphthyl)ethanal,
N-80011

$C_{22}H_{16}O_2$
1,3-Diphenyl-1*H*-indene-2-carboxylic acid,
D-60481

$C_{22}H_{16}O_6$
Dehydromillettone, *in* M-80191

$C_{22}H_{17}N$
3,5-Dihydro-4*H*-dinaphth[2,1-*c*:1',2'-*e*]azepine,
D-60234

$C_{22}H_{17}NO_2$
Dibenz[*b,g*]azocine-5,7(6*H*,12*H*)-dione; *N*-
Benzyl, *in* D-60068

$C_{22}H_{17}N_7O_{13}S_2$
2,3-Bis(2-methoxy-4-nitro-5-sulfophenyl)-5-
[(phenylamino)carbonyl]-2*H*-tetrazolium
hydroxide inner salt, B-70151

$C_{22}H_{18}$
1,1,4-Triphenyl-1,3-butadiene, T-70305

$C_{22}H_{18}O$
1,1,1-Triphenyl-3-butyn-2-ol, T-80360

$C_{22}H_{18}O_2$
1,2-Di-2-naphthalenyl-1,2-ethanediol, D-80479

$C_{22}H_{18}O_3$
2-Hydroxy-1,3-diphenyl-1-propanone;
Benzoyl, *in* H-70124

$C_{22}H_{18}O_6$
Durmillone, D-80537
Isojamaicin, I-80065
Millettone, M-80191

$C_{22}H_{18}O_7$
Justicidin *A*, *in* D-60493

$C_{22}H_{18}O_8$
Dehydropodophyllotoxin, *in* P-80165
Desertorin *A*, D-60022
2,4,5,6-Tetrahydroxyphenanthrene; Tetra-Ac,
in T-60122

$C_{22}H_{18}O_{10}$
3-Galloylcatechin, *in* P-70026

$C_{22}H_{18}O_{14}$
Hexahydroxy-1,4-naphthoquinone; Hexa-Ac,
in H-60065

$C_{22}H_{19}Br_2NO_3$
Deltamethrin, D-80027

$C_{22}H_{19}F_3O_2$
Dodecahedranol; Trifluoroacetyl, *in* D-80523

$C_{22}H_{20}$
3-(1,3,6-Cycloheptatrien-1-yl-2,4,6-
cycloheptatrien-1-ylidenemethyl)-1,3,5-
cycloheptatriene, C-60205
[2.2](4,7)(7,4)Indenophane, I-60021

$C_{22}H_{20}N_2O_2$
2,5-Dihydro-3,6-diphenylpyrrolo[3,4-*c*]pyrrole-
1,4-dione; 2-Butyl, *in* D-70204
2,5-Dihydro-3,6-diphenylpyrrolo[3,4-*c*]pyrrole-
1,4-dione; 2,5-Di-Et, *in* D-70204

$C_{22}H_{20}N_2S_2$
1,4-Bis(ethylthio)-3,6-diphenylpyrrolo[3,4-
c]pyrrole, B-60158

$C_{22}H_{20}O_3$
1,1,2-Triphenyl-1,2-ethanediol; O^2-Ac, *in*
T-70311

$C_{22}H_{20}O_4$
Caleprunifolin, C-80011
Erybraedin E, E-80052
Ethyl 2-ethoxycarbonyl-5,5-diphenyl-2,3,4-
pentatrienoate, *in* D-70492

$C_{22}H_{20}O_5$
Alpinumisoflavone; Di-Me ether, *in* A-80050

$C_{22}H_{20}O_6$
Maximaisoflavone C, *in* T-80107
1-*O*-Methylglycyrol, *in* P-80186

$C_{22}H_{20}O_7$
2-(4-Allyl-2,6-dimethoxyphenoxy)-1-(4-
hydroxy-3,5-dimethoxyphenyl)-1-propanol,
in A-80046
Collybolide, *in* C-80147
Fernolin, F-80003
Isocollybolide, *in* C-80147
Pumilaisoflavone D, P-80194

$C_{22}H_{20}O_8$
ζ-Rhodomycinone, *in* R-70002
Thuriferic acid, T-70191

$C_{22}H_{20}O_9$
θ-Rhodomycinone, *in* R-70002
ε-Rhodomycinone, R-70002

C$_{22}$H$_{20}$O$_{10}$
ψ-Baptisin, *in* H-80203
Rothindin, *in* H-80203

C$_{22}$H$_{20}$O$_{11}$
Irilone-4'-glucoside, *in* D-80341
3',4',5,7-Tetrahydroxyisoflavone; 3',4'-
Methylene ether, 7-O-glucoside, *in* T-80109

C$_{22}$H$_{20}$O$_{13}$
Ellagic acid; 3,3'-Di-Me ether, 4-glucoside, *in*
E-70007

C$_{22}$H$_{21}$N
1,3-Dibenzyl-1,3-dihydroisoindole, D-80077

C$_{22}$H$_{22}$O$_2$
Dodecahedranecarboxylic acid; Me ester, *in*
D-80522

C$_{22}$H$_{22}$O$_4$
Paralycolin *A*, P-60009

C$_{22}$H$_{22}$O$_5$
Edulenane, *in* N-80023
Praecansone *B*, P-70113

C$_{22}$H$_{22}$O$_6$
Desmodin, *in* N-80023
2,3-Dibenzoylbutanedioic acid; Di-Et ester, *in*
D-70083
Glycyrin, G-80031
Languiduline, L-80018
Neorautanin, *in* N-80022

C$_{22}$H$_{22}$O$_7$
1,8-Oxybis(ethyleneoxyethyleneoxy)-9,10-
anthracenedione, O-80097

C$_{22}$H$_{22}$O$_8$
Peperomin A, P-80064
Picropodophyllin, *in* P-80165
▷ Podophyllotoxin, P-80165
Dalbergia Rotenolone, R-60014

C$_{22}$H$_{22}$O$_9$
Alternanthin, A-80053
Ononin, *in* H-80190

C$_{22}$H$_{22}$O$_{10}$
5,7-Dihydroxy-4'-methoxyflavone; 7-β-D-
Galactoside, *in* D-70319
Echioidin, *in* T-80299
Glucoobtusifolin, *in* T-80322
8-C-Glucosyl-4',5-dihydroxy-7-
methoxyisoflavone, *in* G-80025
Prunitrin, *in* D-80335
Sissotrin, *in* D-80337
Sophojaponicin, *in* M-70001
Swertisin, S-80052
1,3,5,8-Tetrahydroxy-2-methylanthraquinone;
5-Me ether, 8-O-α-L-rhamnopyranoside, *in*
T-60113
Tilianin, *in* D-70319
Trifolirhizin, *in* M-70001
3',4',7-Trihydroxyflavone; $O^{3'}$-Me, 7-
glucoside, *in* T-80301
3',4',7-Trihydroxyisoflavone; 4'-Me ether, 7-
O-glucoside, *in* T-80308

C$_{22}$H$_{22}$O$_{11}$
4-(3,4-Dihydroxyphenyl)-5,7-dihydroxy-2H-1-
benzopyran-2-one; 7-Me ether, 5-O-β-D-
galactopyranosyl, *in* D-70334
5-O-β-D-Glucopyranosyl-3',4'-dihydroxy-7-
methoxyneoflavone, *in* D-80356
8-C-Glucosyl-4',5,7-trihydroxy-3'-
methoxyisoflavone, *in* G-80024
Pratensein 7-O-glucoside, *in* T-80109
3',4',5,7-Tetrahydroxyisoflavone; 3'-Me ether,
7-O-glucoside, *in* T-80109

C$_{22}$H$_{23}$NO
Aminododecahedrane; N-Ac, *in* A-80071

C$_{22}$H$_{23}$N$_3$O$_9$
▷ Aluminon, *in* A-60318

C$_{22}$H$_{24}$Br$_2$N$_4$O$_6$S$_2$
N,N'-Bis[3-(3-bromo-4-hydroxyphenyl)-2-
oximidopropionyl]cystamine, B-70130

C$_{22}$H$_{24}$O$_5$
Edulane, *in* N-80023
Quercetol *C*, Q-70003

C$_{22}$H$_{24}$O$_6$
Austrobailignan-7; Ac, *in* A-70271
Ramosissin, R-60001

C$_{22}$H$_{24}$O$_7$
7α-Acetoxybacchotricuneatin B, *in* B-80006
Actifolin, A-80036
Machilin *E*, *in* M-60003
Richardianidin 1, R-70006
Richardianidin 2, *in* R-70006

C$_{22}$H$_{24}$O$_8$
1-Acetoxypinoresinol, *in* H-60221
4,5;4',5'-Bismethylenedioxypolemannone, *in*
P-60166
Byakangelicin; O-Angeloyl, *in* B-80296
Skutchiolide B, S-70050

C$_{22}$H$_{24}$O$_9$
2,3-Dihydroonin, *in* D-80322
3,3',4',5,5',7,8-Heptamethoxyflavone, *in*
H-70021

C$_{22}$H$_{24}$O$_{10}$
Helichrysin, *in* H-80246
5-Hydroxy-3,3',4',5',6,7,8-
heptamethoxyflavone, *in* O-70018

C$_{22}$H$_{24}$O$_{11}$
Lanceolin/, *in* D-80372

C$_{22}$H$_{25}$ClO$_7$
Methyl 2,2'-di-O-methyleriodermate, *in*
E-70032

C$_{22}$H$_{25}$N$_3$
Tris(methylphenylamino)methane, T-70324

C$_{22}$H$_{26}$
2,7-Di-*tert*-butyldicyclopenta[*a,e*]cyclooctene,
D-70102
1,2,3,4,5,6,7,8-Octamethylanthracene, O-80018

C$_{22}$H$_{26}$Br$_2$O$_3$
4,21-Dibromo-3-ethyl-2,19-
dioxabicyclo[16.3.1]docosa-6,9,18(22),21-
tetraen-12-yn-20-one, *in* E-70042

C$_{22}$H$_{26}$N$_2$
2,5-Di-*tert*-butyl-2,5-
dihydrobenzo[*e*]pyrrolo[3,4-*g*]isoindole,
D-70103
2,7-Di-*tert*-butyl-2,7-dihydroisoindolo[5,4-
e]isoindole, D-70104

C$_{22}$H$_{26}$O$_3$
Acetylimbricatalol, *in* H-70150
3-Ethyl-2,19-dioxabicyclo[16.3.1]docosa-
3,6,9,18(22),21-pentaen-12-yn-20-one,
E-70042

C$_{22}$H$_{26}$O$_4$
Dimethyldehydrodieugenol, *in* D-80023
21-Ethyl-2,6-epoxy-17-hydroxy-1-
oxacyclohenicosa-2,5,14,18,20-pentaen-11-
yn-4-one, E-60050

C$_{22}$H$_{26}$O$_5$
Aristoligone, *in* A-70251
Aristosynone, *in* A-70251
9-(1,3-Benzodioxol-5-yl)-1-(2,6-
dihydroxyphenyl)-1-nonanone, *in* M-70007
Costatolide, C-80154
5-Hydroxy-8,8-dimethyl-6-(3-methyl-1-
oxobutyl)-4-propyl-2H,8H-benzo[1,2-b:3,4-
b']dipyran-2-one, *in* M-80001
Isodidymic acid, I-60098
Licarin *C*, L-60024
MAB 6, M-80001
Unanisoflavan, U-80002

C$_{22}$H$_{26}$O$_6$
Asebotin, *in* H-80244
Gelomulide E, *in* G-80008
Henricine, H-60017
Pachypostaudin B, P-80002

C$_{22}$H$_{26}$O$_7$
Glaucocalactone, G-80019
Heliopsolide; 8-Ac, *in* H-70007

Heliopsolide; 4-Epimer, 8-Ac, *in* H-70007
6-Hydroxy-9,9-dimethyl-5-(3-methyl-1-
oxobutyl)-1-propyl-3H,9H-[1,2]-
dioxolo[3',4':4,5]furo[2,3-*f*][1]benzopyran-3-
one, H-80152
Laferin, L-60009
Praderin, P-70112
Sessein, S-60027
Teupyrenone, T-60184

C$_{22}$H$_{26}$O$_8$
Abbreviatin PB, A-60001
Albaspidin AP, *in* A-80043
3,6-Bis(3,4-dimethoxyphenyl)tetrahydro-
1H,3H-furo[3,4-*c*]furan-1,4-diol, B-60156
Desaspidin AB, *in* D-80032
Euparotin acetate, *in* E-80090
Flavaspidic acids; Flavaspidic acid AB-*form*,
in F-80011
Heliopsolide; 2α,3α-Epoxide, 8-Ac, *in* H-70007
Isopicropolin, I-60112
7-Oxodihydrogmelinol, O-60064

C$_{22}$H$_{26}$O$_9$
3',4',5,5',6,7,8-Heptamethoxyflavanone, *in*
H-80029
1-Hydroxysyringaresinol, *in* H-60221
Liscunditrin, *in* D-80006
Skutchiolide A, S-70049
1-(2,3,4,5-Tetramethoxyphenyl)-3-(2,4,5-
trimethoxyphenyl)-1,3-propanedione, *in*
T-70120

C$_{22}$H$_{26}$O$_{10}$
2'-O-Benzoylaucubin, *in* A-80186

C$_{22}$H$_{26}$O$_{11}$
Agnuside, *in* A-80186

C$_{22}$H$_{26}$O$_{13}$
Verproside, *in* C-70026

C$_{22}$H$_{28}$
Des-A-26,27-dinoroleana-5,7,9,11,13-pentaene,
D-80033
Des-A-26,27-dinor-5,7,9,11,13-ursapentaene,
D-80034
Nonacyclo[11.7.1.1^{2,18}.0^{3,16}.0^{4,13}.0^{5,10}.0^{6,14}.0^{7,11}
.0^{15,20}]docosane, N-80067

C$_{22}$H$_{28}$O$_2$
Fervanol benzoate, *in* F-60007
Hydrallmanol A, H-80099

C$_{22}$H$_{28}$O$_3$
Fervanol *p*-hydroxybenzoate, *in* F-60007
Kurubasch aldehyde benzoate, *in* K-60014

C$_{22}$H$_{28}$O$_4$
1-(6-Hydroxy-2-methoxyphenyl)-9-(4-
hydroxyphenyl)-1-nonanone, *in* M-70007
Kurubashic acid benzoate, *in* K-60014
Verecynarmin *A*, V-60005

C$_{22}$H$_{28}$O$_5$
ent-14α-Acetoxy-7β-hydroxy-1,16-kauradiene-
3,15-dione, *in* D-80324
ent-7β-Acetoxy-14α-hydroxy-1,16-kauradiene-
3,15-dione, *in* D-80324
1α,10β-Epoxykurubaschic acid benzoate, *in*
K-60014
1β,10α-Epoxykurubashic acid benzoate, *in*
K-60014
Flabellata secoclerodane, F-70009
Ganschisandrine, G-80005
1-(6-Hydroxy-2-methoxyphenyl)-9-(3,4-
methylenedioxyphenyl)-1-nonanone, *in*
M-70007
19-O-Acetyl-1,2-*dehydrohautriwaic acid*, *in*
H-70002

C$_{22}$H$_{28}$O$_6$
ent-2α-Acetoxy-3,13-clerodadiene-16,15:18,19-
diolide, *in* H-80142
ent-7β-Acetoxy-3,13-clerodadiene-16,15:18,19-
diolide, *in* H-80143
Articulin acetate, *in* A-60307
Chrysophyllin *B*, C-70175
8,9-Dihydroxy-1(10),4,11(13)-germacratrien-
12,6-olide; 9-Ac, 8-(2R,3R-epoxy-2-
methylbutanoyl), *in* D-60323
Fragransin D_1, F-70040
Fragransin D_2, *in* F-70040

Fragransin D_3, *in* F-70040
Gelomulide B, G-**80**007
Gelomulide F, *in* G-**80**008
Homoheveadride, H-**60**089
Pachypophyllin, P-**80**001
Pachypostaudin A, *in* P-**80**002

$C_{22}H_{28}O_7$
ent-2α-Acetoxy-7β-hydroxy-3,13-clerodadiene-16,15:18,19-diolide, *in* H-**80**142
ent-7β-Acetoxy-2α-hydroxy-3,13-clerodadiene-16,15:18,19-diolide, *in* H-**80**142
17-Acetoxythymifodioic acid, *in* T-**60**218
3-Acetylteumicropin, *in* T-**60**181
2-(4-Allyl-2,6-dimethoxyphenoxy)-1-(3-hydroxy-4,5-dimethoxyphenyl)-1-propanol, *in* A-**80**046
12-Epiteupolin II, *in* T-70159
Eupaserrin, *in* D-70302
Teupolin I, T-70159
Teupolin II, *in* T-70159

$C_{22}H_{28}O_8$
Citreoviridinol A_1, C-70181
Citreoviridinol A_2, *in* C-70181
5,5′-Dimethoxylariciresinol, *in* L-70019
Gracilin *B*, G-**60**034
Gracilin *C*, *in* G-**60**034
Hiyodorilactone D, H-**80**090
Hiyodorilactone E, *in* H-**80**090
Teuvincentin B, T-**80**167

$C_{22}H_{28}O_9$
8-Deacyltrichogoniolide; 8-*O*-Angeloyl, 9-Ac, *in* D-**80**007

$C_{22}H_{28}O_{10}$
8β,10β-Diacetoxy-1-*O*-methylhirsutinolide 13-*O*-acetate, *in* H-**80**089
8β-Propionyloxy-10β-hydroxyhirsutinolide 1,13-di-*O*-acetate, *in* H-**80**089

$C_{22}H_{29}ClO_7$
Solenolide E, S-70054

$C_{22}H_{30}$
Biadamantylideneethane, B-**60**075
Decamethylbiphenyl, D-**80**012

$C_{22}H_{30}O$
Uvarisesquiterpene B, U-**80**037

$C_{22}H_{30}O_3$
2-Hydroxy-6-(8,11,14-pentadecatrienyl)benzoic acid, *in* H-**60**205
Pseudopterosin *A*; Aglycone, *O*-Ac, *in* P-**60**187
Siccanin, S-**60**029
Siccanochromene *E*, S-**60**030

$C_{22}H_{30}O_4$
5-Epiilimaquinone, *in* I-**80**004
Ferolin, *in* G-**60**014
Ilimaquinone, I-**80**004
Metachromin A, M-70033
Metachromin C, M-**80**036

$C_{22}H_{30}O_5$
2α-Acetoxyhardwickiic acid, *in* H-70002
6β-Acetoxyhebemacrophyllide, *in* H-**60**010
ent-3α-Acetoxy-11α-hydroxy-16-kaurene-6,15-dione, *in* D-**80**325
3β-Acetoxy-19-hydroxy-13(16),14-spongiadien-2-one, *in* D-70345
7α-Acetoxyroyleanone, *in* D-**80**265
3β-Acetoxywedeliasecokaurenolide, *in* W-**60**004
Bretonin A, *in* D-**80**526
Brevifloralactone acetate, *in* B-70179
3,5-Di-*O*-methylsenecioodontol, *in* S-**60**022
Epoxyjaeschkeanadiol *p-hydroxybenzoate*, *in* J-70001
Feraginidin, *in* D-**60**004
Ferugin, *in* D-**60**005
Gelomulide A, *in* G-**80**008
Hautriwaic acid acetate, *in* H-70002
Isobretonin A, *in* D-**80**526
Lasianthin, L-**60**017
Pseudopterogorgia diterpenoid *B*, P-70128
5-epi-6β-acetoxyhebemacrophyllide, *in* H-**60**010

$C_{22}H_{30}O_6$
7α-Acetoxy-12,20-dihydroxy-8,12-abietadiene-11,14-dione, *in* R-**60**015
ent-3α-Acetoxy-1β,11α-dihydroxy-16-kaurene-6,15-dione, *in* T-**80**313
ent-12β-Acetoxy-7β,14α-dihydroxy-16-kaurene-3,15-dione, *in* T-**80**315
ent-1β-Acetoxy-7β,14α-dihydroxy-16-kaurene-3,15-dione, *in* T-**80**314
17-Acetoxygutiesolbriolide, *in* G-**80**045
17-Acetoxyisogutiesolbriolide, *in* I-**80**062
Isopregomisin, I-**80**077
Longikaurin *C*, *in* L-70035
Longikaurin *E*, *in* L-70035

$C_{22}H_{30}O_7$
6β-Acetoxypulchellin 4-*O*-angelate, *in* P-**80**189
Isomontanolide, *in* M-70154
Longikaurin *B*, *in* L-70035
Longikaurin *D*, *in* L-70035
Lophanthoidin C, *in* T-**80**084
Lophanthoidin E, *in* T-**80**084
Montanolide, M-70154

$C_{22}H_{30}O_8$
Maoecrystal I, M-**80**007

$C_{22}H_{32}N_2$
1,2-Diphenyl-1,2-ethanediamine; *N*-Tetra-Et, *in* D-70480

$C_{22}H_{32}O_2$
Dehydroabietinol acetate, *in* A-**60**007
Uvarisesquiterpene A, U-**80**036
Uvarisesquiterpene C, U-**80**038

$C_{22}H_{32}O_3$
Acetylsanadaol, *in* S-70008
3,4-Epoxy-13(15),16,18-sphenolobatrien-5-ol; Ac, *in* E-**60**031
2-Hydroxy-6-(8,11-pentadecadienyl)benzoic acid, H-**60**205

$C_{22}H_{32}O_4$
ent-18-Acetoxy-15-beyeren-19-oic acid, *in* H-**60**106
7β-Acetoxy-13(17),15-cleistanthadien-18-oic acid, *in* H-70115
19-Acetoxy-15,16-epoxy-13,17-spatadien-5α-ol, *in* E-**60**029
12β-Acetoxyisopimara-8(14),15-dien-18-oic acid, *in* S-**80**008
ent-1α-Acetoxy-16-kauren-19-oic acid, *in* H-**80**179
ent-3β-Acetoxy-15-kauren-17-oic acid, *in* H-**60**167
ent-3β-Acetoxy-16-kauren-18-oic acid, *in* H-**80**181
ent-6β-Acetoxy-7,12*E*,14-labdatrien-17-oic acid, *in* L-**60**004
Dictyodendrillolide, D-**80**149
Dilophus ether, D-**80**408
Iloprost, I-**60**003
Lagerstronolide, L-**60**010
5-Methoxy-3-(8,11,14)pentadecatrienyl)-1,2,4-benzenetriol, M-70049

$C_{22}H_{32}O_5$
ent-14*S*-Acetoxy-1β,7β-dihydroxy-16-kauren-15-one, *in* T-**50**476
ent-7β-Acetoxy-1β,14*S*-dihydroxy-16-kauren-15-one, *in* T-**50**476
ent-6β,17-Diacetoxy-14,15-bisnor-7,11*E*-labdadien-13-one, *in* D-**60**311
Grayanotoxin IX, *in* G-**80**040
Grayanotoxin X, *in* G-**80**040
7-Hydroxy-8(17),13-corymbidienolide, H-**60**110
4,10(14)-Oplopadiene-3,8,9-triol; 3-Ac, 8-angeloyl, *in* O-70042
4,10(14)-Oplopadiene-3,8,9-triol; 3-Ac, 9-angeloyl, *in* O-70042

$C_{22}H_{32}O_6$
5-Acetoxy-9α-angeloyloxy-8β-hydroxy-10(14)-oplopen-3-one, *in* T-70272
6α-Acetoxy-17β-hydroxy-15,17-oxido-16-spongianone, *in* D-**60**365

ent-3α-Acetoxy-2α,6β,11β-trihydroxy-16-kauren-15-one, *in* P-**80**047
Lophanthoidin D, *in* T-**80**084
Oleaxillaric acid, O-**80**042

$C_{22}H_{32}O_7$
Arguticinin, A-70250
Arguticinin; 4-Epimer, *in* A-70250

$C_{22}H_{32}O_8$
1,3,5,7-Adamantanetetracarboxylic acid; Tetra-Et ester, *in* A-**80**037

$C_{22}H_{32}O_{10}$
Rengyoside *C*, *in* R-**80**004

$C_{22}H_{33}BrO_4$
2α-Acetoxy-15-bromo-9(11)-parguerene-7α,16-diol, *in* B-**80**240

$C_{22}H_{33}BrO_5$
2α-Acetoxy-15-bromo-9(11)-parguerene-7α,16,19-triol, *in* B-**80**240

$C_{22}H_{34}N_8$
3,6,9,17,20,23,29,30-Octaazatricyclo[23.3.1.1^{11,15}]triaconta-1(29),11(30),12,14,25,27-hexaene, O-**80**003

$C_{22}H_{34}O_2$
ent-17-Acetoxy-9β*H*-isopimara-7,15-diene, *in* I-**80**075
19-Acetoxy-9,(11),15-pimaradiene, *in* P-**80**148

$C_{22}H_{34}O_3$
ent-3β-Acetoxy-15-kauren-17-ol, *in* K-**60**007
ent-6β-Acetoxy-7,12*E*,14-labdatrien-17-ol, *in* L-**60**004
Cladiellin, C-**80**137
10-Hydroxy-7,11,13,16,19-docosapentaenoic acid, H-70125
8-Hydroxy-5,9,11,14,17-eicosapentaenoic acid; Et ester, *in* H-70126
2-Hydroxy-6-(10-pentadecenyl)benzoic acid, *in* H-**60**205
2-Hydroxy-6-(8-pentadecenyl)benzoic acid, *in* H-**60**205

$C_{22}H_{34}O_4$
20-Acetoxy-2β,3α-dihydroxy-1(15),8(19)-trinervitadiene, *in* T-70299
2β-Acetoxy-8(17),13*E*-labdadien-15-oic acid, *in* H-70147
2β-Acetoxy-8(17),13*Z*-labdadien-15-oic acid, *in* H-70147
Dictyotriol *A*; 12-Ac, *in* D-70147
3β-Hydroxyisoagatholal; 3-Ac, *in* D-70310
Koanoadmantic acid, *in* H-70224
Maesanin, M-**60**004
1(15),8(19)-Trinervitadiene-2,3,9-triol; 9-Ac, *in* T-70298

$C_{22}H_{34}O_5$
6-Acetoxy-3,4-epoxy-12-hydroxy-7-dolabellen-16-al, *in* D-70541
4-Acetoxy-6-(4-hydroxy-4-methyl-2-cyclohexenyl)-2-(4-methyl-3-pentenyl)-2-heptenoic acid, A-**60**018
Cornudentanone, C-**60**167
3,4-Epoxy-13(15),16-sphenolobadiene-5,18-diol; 5-Ac, *in* E-**60**030
Stolonidiol acetate, *in* S-**60**055

$C_{22}H_{34}O_6$
7β-Acetoxy-8α,13*R*-epoxy-6β,9α-dihydroxy-14-labden-11-one, *in* E-**80**037
▷ Grayanotoxin IV, *in* G-**80**038
Grayanotoxin XVI, *in* G-**80**038

$C_{22}H_{34}O_7$
Coleonol, *in* E-70027
Coleonol B, *in* E-70027
Grayanotoxin XIII, *in* G-**80**038

$C_{22}H_{34}O_{11}$
Coleoside, *in* I-**80**078

$C_{22}H_{36}O_3$
ent-15-Acetoxy-7,13-labdadien-2α-ol, *in* L-**80**003
ent-15-Acetoxy-7,13-labdadien-2β-ol, *in* L-**80**004
19-Acetoxymanoyl oxide, *in* E-70023
Acetoxyodontoschismenol, *in* D-70541

Chromophycadiol monoacetate, *in* C-60148
8-Hydroxy-5,9,11,14,17-eicosapentaenoic acid;
　17,18-Dihydro, Et ester, *in* H-70126
ent-8(17),13-Labdadiene-7α,15-diol; 7-Ac, *in*
　L-70006
Secotrinervitane, S-60019

C$_{22}$H$_{36}$O$_4$
6-Acetoxy-3,7-dolabelladiene-12,16-diol, *in*
　D-70541
19-Acetoxy-18-hydroxygeranylnerol, *in*
　B-80151
ent-3β-Acetoxy-16β,17-kauranediol, *in*
　K-60005
15-Acetoxy-7-labden-17-oic acid, *in* H-60168
Hamachilobene *E*, *in* H-70001
15-Hydroxy-7-labden-17-oic acid; Ac, *in*
　H-80183

C$_{22}$H$_{36}$O$_5$
Methyl 3,6-epidioxy-6-methoxy-4,16,18-
　eicosatrienoate, *in* E-60007

C$_{22}$H$_{36}$O$_7$
▷ Grayanotoxin I, *in* G-80039

C$_{22}$H$_{36}$O$_{16}$
Shanzhisin methyl ester gentiobioside, *in*
　S-60028

C$_{22}$H$_{36}$S$_6$
1,3,4,6-Tetrakis(*tert*-butylthio)thieno[3,4-
　c]thiophene, *in* T-70170

C$_{22}$H$_{39}$NO$_5$
Valilactone, V-70002

C$_{22}$H$_{40}$N$_4$O$_{14}$P$_2$S$_2$
Pantethine; 4,4'-Diphosphate, *in* P-80009

C$_{22}$H$_{40}$O$_4$
Ricinoleyl alcohol; Di-Ac, *in* O-60006

C$_{22}$H$_{41}$NO$_7$
Fumifungin, F-60082

C$_{22}$H$_{42}$N$_4$O$_8$S$_2$
Pantethine, P-80009

C$_{22}$H$_{42}$O$_4$
▷ Bis(2-ethylhexyl)adipate, B-80148

C$_{22}$H$_{46}$
2-Methylheneicosane, M-80091

C$_{23}$H$_{14}$O
7-Methoxyindeno[1,2,3-*cd*]pyrene, *in* I-60025
8-Methoxyindeno[1,2,3-*cd*]pyrene, *in* I-60026

C$_{23}$H$_{14}$O$_6$
1,8,11-Triptycenetricarboxylic acid, T-60398
1,8,14-Triptycenetricarboxylic acid, T-60399

C$_{23}$H$_{16}$O
4-(Diphenylmethylene)-1(4*H*)-naphthalenone,
　D-70487

C$_{23}$H$_{16}$O$_5$
Lophirone E, L-80040

C$_{23}$H$_{16}$O$_6$
2,2'-Methylenebis[8-hydroxy-3-methyl-1,4-
　naphthalenedione], M-70063

C$_{23}$H$_{17}$N
2*H*-Dibenz[*e,g*]isoindole; *N*-Benzyl, *in*
　D-70067

C$_{23}$H$_{18}$
4*b*,8*b*,13,14-Tetrahydrodiindeno[1,2-*a*:2',1'-
　b]indene, T-70058

C$_{23}$H$_{18}$O
1-Oxa[2.2](2,7)naphthalenophane, O-80052

C$_{23}$H$_{18}$O$_2$
1,3-Diphenyl-1*H*-indene-2-carboxylic acid; Me
　ester, *in* D-60481

C$_{23}$H$_{18}$O$_7$
Antibiotic SS 43405*D*, A-70239
Rotenonone, *in* D-80026

C$_{23}$H$_{18}$O$_8$
Villosone, *in* S-80051

C$_{23}$H$_{20}$N$_2$O
Isoamarin; 1-Ac, *in* D-80263

C$_{23}$H$_{20}$O$_4$
1-Phenyl-1,2-propanediol; Dibenzoyl, *in*
　P-80121
1-Phenyl-1,3-propanediol; Dibenzoyl, *in*
　P-80122

C$_{23}$H$_{20}$O$_6$
Dehydrodeguelin, *in* D-80020
Dehydrorotenone, D-80026

C$_{23}$H$_{20}$O$_7$
Amorpholone, *in* D-80026
Dehydroamorphigenin, *in* A-80152
Dehydrotoxicarol, *in* T-80204
Ferrugone, F-80006
Ichthynone, I-80001
5-Methoxydurmillone, *in* D-80537
Villosol, *in* S-80051

C$_{23}$H$_{20}$O$_8$
Boesenboxide, B-70173
Desertorin *B*, *in* D-60022

C$_{23}$H$_{22}$N$_3$O$_3$
2,5-Dihydro-2,2,5,5-tetramethyl-3-[[[(2-phenyl-
　3*H*-indol-3-ylidene)amino]oxy]carbonyl]-
　1*H*-pyrrol-1-yloxy, D-60284

C$_{23}$H$_{22}$O$_5$
Isouvaretin, *in* U-80035
Uvaretin, U-80035

C$_{23}$H$_{22}$O$_6$
▷ Deguelin, D-80020
Gerontoxanthone A, G-80012
Gerontoxanthone B, G-80013
Myriconol, M-80204

C$_{23}$H$_{22}$O$_7$
▷ Amorphigenin, A-80152
Dehydrodalpanol, *in* D-80003
Sumatrol, S-80051
Tephrosin, T-80009
α-Toxicarol, T-80204

C$_{23}$H$_{22}$O$_8$
11-Hydroxytephrosin, *in* T-80009
Villosin, *in* S-80051

C$_{23}$H$_{22}$O$_{10}$
6''-*O*-Acetyldaidzin, *in* D-80323
Phrymarolin II, P-80140

C$_{23}$H$_{22}$O$_{11}$
6''-*O*-Acetylgenistin, *in* T-80309
Fujikinin, *in* T-80111

C$_{23}$H$_{23}$NO$_4$
1,4-Dihydro-2,6-dimethyl-3,5-
　pyridinedicarboxylic acid; Dibenzyl ester,
　in D-80193

C$_{23}$H$_{24}$O$_5$
Demethoxyegonol 2-methylbutanoate, *in*
　E-70001
Praecansone *A*, *in* P-70113

C$_{23}$H$_{24}$O$_6$
BR-Xanthone *A*, B-70282
Gerontoxanthone C, G-80014

C$_{23}$H$_{24}$O$_7$
2-Acetoxyaristotetralone, *in* A-70253
Dalpanol, D-80003
Dihydroamorphigenin, *in* A-80152

C$_{23}$H$_{24}$O$_7$S
Floroselin, *in* K-70013
Isofloroseselin, *in* K-70013

C$_{23}$H$_{24}$O$_8$
Amorphigenol, *in* A-80152
Loxodellonic acid, L-70039

C$_{23}$H$_{24}$O$_9$
Isovolubilin, I-80098
Wistin, *in* T-80310

C$_{23}$H$_{24}$O$_{10}$
Embigenin, *in* S-80052
4',7,8-Trihydroxyisoflavone; 4',8-Di-Me ether,
　7-*O*-glucoside, *in* T-80311

C$_{23}$H$_{24}$O$_{11}$
Abrusin, A-80004
Eupalin, *in* P-80035

C$_{23}$H$_{24}$O$_{12}$
Homotectoridin, *in* P-80045
Iristectorin A, *in* P-80043
Iristectorin B, *in* P-80043

C$_{23}$H$_{25}$ClN$_2$O$_9$
8-Methoxychlorotetracycline, M-70045

C$_{23}$H$_{25}$ClN$_2$O$_{10}$
4*a*-*Hydroxy-8-methoxychlorotetracycline, in*
　M-70045

C$_{23}$H$_{26}$O$_4$
6-Benzoyl-5,7-dihydroxy-2-methyl-2-(4-
　methyl-3-pentenyl)chroman, B-80056
2-Benzoyl-6-nerylphloroglucinol, *in* D-80460
3-(3,7-Dimethyl-2,6-octadienyl)-2,4,6-
　trihydroxybenzophenone, D-80460

C$_{23}$H$_{26}$O$_5$
5-Methoxy-2,2-dimethyl-6-(2-methyl-1-oxo-2-
　butenyl)-10-propyl-2*H*,8*H*-benzo[1,2-*b*:3,4-
　b']dipyran-8-one, M-80042

C$_{23}$H$_{26}$O$_6$
Garvin *A*, G-60008
Teracrylshikonin, *in* S-70038

C$_{23}$H$_{26}$O$_7$
Neokadsuranin, N-70024
Stenosporonic acid, S-70072

C$_{23}$H$_{26}$O$_8$
1-Hydroxypinoresinol; 4''-Me ether, 1-Ac, *in*
　H-60221
Peperomin B, P-80065
Sikkimotoxin, S-60032

C$_{23}$H$_{26}$O$_{11}$
Lindleyin, L-80034
Macrophylloside *B*, *in* M-70004
Macrophylloside *C*, M-70004
Nyasicaside, *in* B-60155

C$_{23}$H$_{26}$O$_{13}$
β-Sorinin, *in* S-80039

C$_{23}$H$_{27}$ClO$_7$
3-Chlorostenosporic acid, *in* S-80044

C$_{23}$H$_{28}$N$_4$O$_7$
Biphenomycin *B*, *in* B-60114

C$_{23}$H$_{28}$N$_4$O$_8$
Biphenomycin *A*, B-60114

C$_{23}$H$_{28}$O$_3$
Citreomontanin, C-60149

C$_{23}$H$_{28}$O$_4$
2-Acetoxy-1,11-bis(3-furanyl)-4,8-dimethyl-
　3,6,8-undecatriene, *in* B-70143
Quercetol *B*, Q-70002

C$_{23}$H$_{28}$O$_5$
2α-Anisoyloxy-9-oxoisoanhydrooplopanone,
　in O-50045

C$_{23}$H$_{28}$O$_7$
Isosphaeric acid, I-60136
Palliferinin, *in* L-60014
Stenosporic acid, S-80044

C$_{23}$H$_{28}$O$_8$
Albaspidin AB, *in* A-80043
Albaspidin PP, *in* A-80043
Aspidin AB, *in* A-80180
Desaspidin PB, *in* D-80032
4,5-Dimethoxy-4',5'-
　methylenedioxypolemannone, *in* P-60166
Flavaspidic acids; Flavaspidic acid PB-*form*,
　in F-80011
Methylenebisdesaspidinol BB, M-80077

C$_{23}$H$_{28}$O$_{10}$
Diffutin, *in* T-60103

C$_{23}$H$_{28}$O$_{13}$
Picroside II, *in* C-70026

$C_{23}H_{29}NO_6$
Fusarin A, F-60107

$C_{23}H_{29}NO_7$
Fusarin D, in F-60107

$C_{23}H_{30}O_4$
Fervanol vanillate, in F-60007
Guayulin D, in A-60299

$C_{23}H_{30}O_5$
3,5-Dihydroxy-19-oxo-14,20(22)-
cardadienolide, D-80350
Ginamallene, G-80017
Kurubasch aldehyde vanillate, in K-60014

$C_{23}H_{30}O_7$
1,3-Bis[(2-methoxyethoxy)methoxy]-1,3-
diphenyl-2-propanone, B-80154
Seco-4-hydroxylintetralin, S-70023

$C_{23}H_{30}O_8$
Acanthospermal A, A-60012
Gracilin D, in G-60034
Isorolandrolide; 13-Et ether, in I-80087

$C_{23}H_{32}O_2$
2,2'-Methylenebis[6-*tert*-butyl-4-
methylphenol], M-80076

$C_{23}H_{32}O_3$
Avarol; 5'-Ac, in A-80189
Dictyoceratin, D-80148

$C_{23}H_{32}O_4$
8,11,13-Abietatrien-19-ol; 19-(Carboxyacetyl),
in A-60007
6α-Anisoyloxy-1β,4β-epoxy-1β,10α-eudesmane,
in E-80026
3-Eudesmene-1,6-diol; 6-Anisoyl, in E-80083
Metachromin B, M-70034
Methyl 11,12-dimethoxy-6,8,11,13-abietatrien-
20-oate, in D-60305
Smenorthoquinone, S-80035

$C_{23}H_{32}O_5$
Chimganidin, in G-60014

$C_{23}H_{32}O_6$
16α-Acetoxy-2α,3β,12β-trihydroxy-4,7-
pregnadien-20-one, in T-60124
Cannogeninic acid, C-80017
Epoxyjaeschkeanadiol vanillate, in J-70001
3,14,15-Trihydroxy-19-oxocard-20(22)-enolide,
T-80330
8-Vanilloylshiromodiol, in S-70039

$C_{23}H_{32}O_7$
Lophanthoidin A, in T-80084

$C_{23}H_{32}O_9$
Absinthifolide, in H-60138

$C_{23}H_{34}$
Des-A-26-nor-5,7,9-lupatriene, D-80036

$C_{23}H_{34}O_4$
7,13-Corymbidienolide, C-60169
ent-17-Malonyloxy-9βH-isopimara-7,15-diene,
in I-80075

$C_{23}H_{34}O_5$
Bryophollenone, B-80270
▷ Coroglaucigenin, C-60168
Dactylospongenone A, D-80002
Dactylospongenone B, in D-80002
Dactylospongenone C, in D-80002
Dactylospongenone D, in D-80002
Gracilin A, G-60033
4,10(14)-Oplopadiene-3,8,9-triol; 3-Ac, 9-(3-
Methyl-2-pentenoyl), in O-70042

$C_{23}H_{34}O_6$
Coriolin B, in C-70199

$C_{23}H_{34}O_7$
Coriolin C, in C-70199

$C_{23}H_{36}O_5$
9,11-Dihydrogracilin A, in G-60033

$C_{23}H_{36}O_6$
Helogynic acid, H-70008

$C_{23}H_{38}O_4$
15-Hydroxy-7-labden-17-oic acid; 15-Ac, Me
ester, in H-60168
12-Isoagathen-15-oic acid; 2,3-
Dihydroxypropyl ester, in I-60085

$C_{23}H_{38}O_5$
Muamvatin, M-70156

$C_{23}H_{38}O_8$
Gaillardoside, in T-70284

$C_{23}H_{41}N_3O_2S$
Agelasidine C, A-80040

$C_{23}H_{45}N_5O_{12}$
Youlemycin, Y-70002

$C_{23}H_{46}$
3-Tricosene, T-80234
9-Tricosene, T-60259

$C_{23}H_{46}O_3$
2-Hydroxytricosanoic acid, H-80259

$C_{24}H_{12}$
[4]Phenylene, P-60089

$C_{24}H_{12}N_4O_4$
25,26,27,28-Tetraazapentacyclo[19.3.1.1^{3,7}.1^{9,13}
.1^{15,19}]octacosa-1(25),3,5,7(28),9,11,13(27),
15,17,19(26),21,23-dodecaene-2,8,14,20-
tetrone, T-70010

$C_{24}H_{12}O_2$
1,1'-Biacenaphthylidene-2,2'-dione, B-80067

$C_{24}H_{12}S_3$
Benzo[1,2-b:3,4-b':5,6-
b'']tris[1]benzothiophene, B-60052
Benzo[1,2-b:3,4-b':6,5-
b'']tris[1]benzothiophene, B-60053

$C_{24}H_{14}$
Benz[5,6]indeno[2,1-a]phenalene, B-70025
1,1'-Bibiphenylene, B-60080
2,2'-Bibiphenylene, B-60081
Dibenz[e,k]acephenanthrylene, D-70066
Indeno[1,2,3-hi]chrysene, I-70010

$C_{24}H_{14}O_2$
Indeno[1,2,3-cd]pyren-1-ol; Ac, in I-60022
Indeno[1,2,3-cd]pyren-2-ol; Ac, in I-60023
Indeno[1,2,3-cd]pyren-6-ol; Ac, in I-60024

$C_{24}H_{14}O_6$
6,7-Dihydroxy-1,12-perylenedione; Di-Ac, in
D-80354

$C_{24}H_{15}N_3$
Diindolo[3,2-a:3',2'-c]carbazole, D-60384

$C_{24}H_{16}$
Cyclodeca[1,2,3-de:6,7,8-d'e']dinaphthalene,
C-60187
Cycloocta[1,2-b:5,6-b']dinaphthalene, C-60216
1,2:9,10-Dibenzo[2.2]metaparacyclophane,
D-70071
1,2:7,8-Dibenzo[2.2]paracyclophane, D-70072
1-(Diphenylmethylene)-1H-
cyclopropa[b]naphthalene, D-70485

$C_{24}H_{16}O_2$
5,6,10,11,16,17,21,22-Octadehydro-7,9,18,20-
tetrahydrodibenzo[e,n][1,10]dioxacyclo-
octadecin, O-70006

$C_{24}H_{16}O_6$
Lophirone D, L-80039

$C_{24}H_{17}N$
10,11-Dihydro-5H-diindeno[1,2-b;2',1'-
d]pyrrole; N-Ph, in D-80182

$C_{24}H_{17}N_3$
2,3':2',3''-Ter-1H-indole, T-60011

$C_{24}H_{18}$
1,2,9,10,17,18-
Hexadehydro[2.2.2]paracyclophane,
H-60038

$C_{24}H_{18}N_6O_{12}$
Benzo[1,2-b:3,4-b':5,6-b'']tripyrazine-
2,3,6,7,10,11-tetracarboxylic acid; Hexa-Me
ester, in B-60051

$C_{24}H_{18}O_{10}$
Eriocephaloside, in H-70133

$C_{24}H_{20}$
[2.2](2,6)Azulenophane, A-60355
7-(Diphenylmethylene)-2,3,5,6-
tetramethylenebicyclo[2.2.1]heptane,
D-70488
2,5,8,11-Tetramethylperylene, T-70140
3,4,9,10-Tetramethylperylene, T-70141

$C_{24}H_{20}N_2$
4(5)-Vinylimidazole; 1-Triphenylmethyl, in
V-70012

$C_{24}H_{20}O$
3,4,4-Triphenyl-2-cyclohexen-1-one, T-60379
3,5,5-Triphenyl-2-cyclohexen-1-one, T-60380
4,5,5-Triphenyl-2-cyclohexen-1-one, T-60381

$C_{24}H_{20}O_5$
8-Dihydrocinnamoyl-5,7-dihydroxy-4-phenyl-
2H-benzopyran-2-one, D-60210

$C_{24}H_{20}O_8$
Kielcorin B, K-60012

$C_{24}H_{20}O_9$
Subalatin, S-80048

$C_{24}H_{20}S_2$
7H,9H,16H,18H-Dinaphtho[1,8-cd:1',8'-
ij][1,7]dithiacyclododecin, D-70448
2,3-Naphthalenedithiol; Di-S-benzyl, in
N-70004

$C_{24}H_{22}O_7$
Lepidissipyrone, L-80023

$C_{24}H_{22}O_8$
Desertorin C, in D-60022
Villinol, in S-80051

$C_{24}H_{22}O_{13}$
Genistein 7-O-glucoside 6''-malonate, in
T-80309

$C_{24}H_{23}N$
5,5,9,9-Tetramethyl-5H,9H-quino[3,2,1-
de]acridine, T-70143

$C_{24}H_{24}O_{11}$
5-O-(6-Acetyl-β-D-galactopyranosyl)-3',4'-
dihydroxy-7-methoxyneoflavone, in
D-80356
Phrymarolin I, in P-80140

$C_{24}H_{24}O_{12}$
Dalpatin, in P-80042
3',4',5,6,7-Pentahydroxyisoflavone; 5,6-Di-Me,
3',4'-methylenedioxy ether, 7-O-β-D-
glucopyranoside, in P-80043
Platycarpanetin 7-O-glucoside, in P-80045

$C_{24}H_{25}NO_8$
Antibiotic Sch 38519, A-60287

$C_{24}H_{26}$
1,3,6-Triphenylhexane, T-60386

$C_{24}H_{26}O_4$
15-Cinnamoyloxyisoalloalantolactone, in
I-80047
(E)-ω-Oxoferprenin, in F-70003
Isoalloalantolactone; 15-Cinnamoyloxy (Z-),
in I-80047

$C_{24}H_{26}O_6$
Egonol 2-methylbutanoate, in E-70001

$C_{24}H_{26}O_7$
(+)-Anomalin, in K-70013
(−)-Anomalin, in K-70013
Anomalin, in K-70013
Calipteryxin, in K-70013
Khellactone; Bis(3-methyl-3-butenoyl), in
K-70013
Khellactone; Di-O-(3-Methyl-2-butenoyl), in
K-70013
Khellactone; O^9-(3-Methyl-2-butenoyl), O^{10}-
angeloyl, in K-70013
Peuformosin, in K-70013

C$_{24}$H$_{26}$O$_8$
Drummondin C, D-80532
Khellactone; O^9-Angeloyl, O^{10}-(2,3-Epoxy-2-methylbutanoyl), in K-70013

C$_{24}$H$_{26}$O$_9$
Khellactone; Di-O-(2,3-Epoxy-2-methylbutanoyl), in K-70013

C$_{24}$H$_{26}$O$_{10}$
Sophoraside A, in P-70137

C$_{24}$H$_{26}$O$_{11}$
3′,4′,6,7-Tetrahydroxyisoflavone; 3′,4′,6-Tri-Me ether, 7-O-β-D-glucopyranoside, in T-80111
4′,5,7,8-Tetrahydroxyisoflavone; 4′,5,8-Tri-Me ether, 7-O-β-D-glucoside, in T-80113
4′,5,6,7-Tetrahydroxyisoflavone; 4′,5,6-Tri-Me ether, 7-O-glucoside, in T-80112

C$_{24}$H$_{26}$O$_{13}$
Iridin, in H-80071
Jacein, in H-80067
Sudachiin A, in T-80341
4′,5,7-Trihydroxy-3′,6,8-trimethoxyflavone; 7-O-β-D-Glucopyranoside, in T-80341

C$_{24}$H$_{27}$ClN$_2$O$_9$
8-Methoxy-N-methylchlorotetracycline, in M-70045

C$_{24}$H$_{27}$N$_2$O$_8$P
Thymidine 5′-phosphate; Dibenzyl ester, in T-80196

C$_{24}$H$_{27}$N$_3$
Hexahydro-1,3,5-triazine; 1,3,5-Tribenzyl, in H-60060

C$_{24}$H$_{27}$N$_3$O$_6$
Physarochrome A, P-60150

C$_{24}$H$_{28}$
2-(9-Decenyl)phenanthrene, D-80019

C$_{24}$H$_{28}$O$_2$
3,7-Di-$tert$-butyl-9,10-dimethyl-2,6-anthraquinone, D-60113

C$_{24}$H$_{28}$O$_3$
Ferprenin, F-70003
4-Geranyl-3,4′,5-trihydroxystilbene, G-70008

C$_{24}$H$_{28}$O$_4$
2α-Cinnamoyloxy-9-oxoisoanhydrooplopanone, in O-50045
6,6′;7,3a'-Diligustilide, D-70363
(E)-ω-Hydroxyferprenin, in F-70003
(E)-ω-Oxoferulenol, in F-80007
Riligustilide, R-60009
Ternatin, T-70008
(Z)-ω-Hydroxyferprenin, in F-70003

C$_{24}$H$_{28}$O$_5$
Aphyllocladone, A-70242

C$_{24}$H$_{28}$O$_7$
Garcinone D, G-70004
Khellactone; O-Angeloyl, O-3-methylbutanoyl, in K-70013
Khellactone; O-Angeloyl, O-2-methylpropanoyl, in K-70013
Khellactone; O^9-(3-Methylbutanoyl), O^{10}-angeloyl, in K-70013

C$_{24}$H$_{28}$O$_{11}$
Decumbeside A, in G-60002
Decumbeside B, in G-60002
Globularicisin, in C-70026
Globularin, in C-70026
Macrophylloside A, in M-70004

C$_{24}$H$_{28}$O$_{12}$
Scutellarioside II, in C-70026

C$_{24}$H$_{28}$O$_{14}$
α-Sorinin, in S-80039

C$_{24}$H$_{29}$ClO$_8$
Ptilosarcenone, P-70136
Ptilosarcenone, in P-60193

C$_{24}$H$_{29}$ClO$_9$
11-Hydroxyptilosarcenone, in P-60193
Teuvincentin A, T-80166

C$_{24}$H$_{30}$
1,1′-Diphenylbicyclohexyl, D-70475
1,2,3,4,5,6,7,8,9,10,11,12-Dodecahydro-1,4:5,8:9,12-triethanotriphenylene, D-80525

C$_{24}$H$_{30}$O$_3$
Ferulenol, F-80007
Guayulin C, in A-60299
9α-Hydroxygymnomitryl cinnamate, in G-70035

C$_{24}$H$_{30}$O$_4$
Assafoetidin, A-70265
6β-Cinnamoyloxy-1α-hydroxy-5,10-bisepi-4-eudesmen-3-one, in H-80101
3,8-Dihydro-6,6′;7,3′a-diliguetilide, D-60223
(E)-ω-Hydroxyferulenol, in F-80007
Gummosin, in P-80166
(Z)-ω-Hydroxyferulenol, in F-80007

C$_{24}$H$_{30}$O$_5$
Asacoumarin A, A-80173
Asacoumarin B, A-80174
5-Hydroxy-8,8-dimethyl-6-(2-methyl-1-oxobutyl)-4-pentyl-2H,8H-benzo[1,2-b:3,4-b']dipyran-2-one, in M-80001
5-Hydroxy-8,8-dimethyl-6-(3-methyl-1-oxobutyl)-4-pentyl-2H,8H-benzo[1,2-b:3,4-b']dipyran-2-one, in M-80001
Kopeolone, in K-80012
Praealtin B, in P-80172
Praealtin C, in P-80172

C$_{24}$H$_{30}$O$_6$
Magnoshinin, M-60007

C$_{24}$H$_{30}$O$_7$
Khellactone; Bis(3-methylbutanoyl), in K-70013

C$_{24}$H$_{30}$O$_8$
Aemulin BB, A-80039
Albaspidin PB, in A-80043
Desaspidin BB, in D-80032
Flavaspidic acid, in F-80011
Montanin C, in T-70159
Peperomin C, P-80066
Saroaspidin A, S-70012

C$_{24}$H$_{30}$O$_9$
Auropolin, A-60319
Montanin G, M-70153
Teumicropodin, T-60182

C$_{24}$H$_{30}$O$_{10}$
Dichotosinin, in T-60103

C$_{24}$H$_{30}$O$_{11}$
Globularidin, in C-70026

C$_{24}$H$_{30}$O$_{12}$
Syringopicroside B, in S-60067

C$_{24}$H$_{30}$O$_{13}$
Secologanol; 7-(2,5-Dihydroxybenzoyl), in S-80016

C$_{24}$H$_{31}$ClO$_8$
12-Ptilosarcenol, P-60193

C$_{24}$H$_{31}$ClO$_{10}$
Solenolide C, S-70053

C$_{24}$H$_{32}$
Pentacyclo[12.2.2.2^{2,5}.2^{6,9}.2^{10,13}]tetracosa-1,5,9,13-tetraene, P-60029

C$_{24}$H$_{32}$O$_3$
Antheliolide A, A-70213
6α-Cinnamoyloxy-1β,4β-epoxy-1β,10α-eudesmane, in E-80026
3-Eudesmene-1,6-diol; 1-Cinnamoyl, in E-80083
3-Eudesmene-1,6-diol; 6-Cinnamoyl, in E-80083

C$_{24}$H$_{32}$O$_4$
Antheliolide B, in A-70213
Karatavicinol, K-80004

C$_{24}$H$_{32}$O$_5$
6β-Cinnamoyloxy-3β-hydroperoxy-5,10-bisepi-4-eudesmen-1α-ol, in H-80101
6β-Cinnamoyloxy-4β-hydroperoxy-5,10-bisepi-2-eudesmen-1α-ol, in H-80102
6β-Cinnamoyloxy-3β-hydroperoxy-1α-hydroxy-5,10-bisepi-4(15)-eudesmene, in E-80084
Kopeolin, K-80012

C$_{24}$H$_{32}$O$_6$
Okilactomycin, O-70030
Spongiadiol; Di-Ac, in D-70345
Verrucosidin, V-60006

C$_{24}$H$_{32}$O$_7$
16-Acetoxy-12-O-acetylhorminone, in D-80265
9β-Acetoxy-1α-benzoyloxy-4β,6β-dihydroxydihydro-β-agarofuran, in T-80090
Furcellataepoxylactone, F-70052
Isoschizandrin, I-70101
Palliferin, in L-60014

C$_{24}$H$_{32}$O$_8$
Acetylisomontanolide, in M-70154
Exidonin, E-80093
Ganervosin A, G-80004
Hydroxyniranthin, H-70187
Jodrellin A, J-80013
Longikaurin F, in L-70035
Lophanthoidin B, in T-80084
Maeocrystal J, in M-80007
Nirphyllin, N-80032
4,4′,5,5′-Tetramethoxypolemannone, in P-60166

C$_{24}$H$_{33}$ClO$_9$
13-Deacetyl-11(9)-havannachlorohydrin, in H-70004
Solenolide B, in S-70052

C$_{24}$H$_{34}$N$_2$O$_3$
1,4,10-Trioxa-7,13-diazacyclopentadecane; 4,10-Dibenzyl, in T-60376

C$_{24}$H$_{34}$O$_4$
▷ Bufalin, B-70289

C$_{24}$H$_{34}$O$_5$
12-(Acetyloxy)-10-[(acetyloxy)methylene]-6-methyl-2-(4-methyl-3-pentenyl)-2,6,11-dodecatrienal, A-70049
Carotdiol veratrate, in D-80004
Dilophus enone, D-80407

C$_{24}$H$_{34}$O$_6$
ent-3β,19-Diacetoxy-15-kauren-17-oic acid, in D-60341
Paniculadiol; Di-Ac, in P-80007

C$_{24}$H$_{34}$O$_7$
ent-3α,6β-Diacetoxy-2α,11β-dihydroxy-16-kauren-15-one, in P-80047
ent-1β,3α-Diacetoxy-11α,15α-dihydroxy-16-kauren-6-one, in K-80008
6α,17α-Diacetoxy-15,17-oxido-16-sponginone, in D-60365
Lophanthoidin F, in T-80084

C$_{24}$H$_{34}$O$_8$
Solenolide F, S-70055
Teucretol, T-60179

C$_{24}$H$_{35}$BrO$_5$
2α,16-Diacetoxy-15-bromo-9(11)-pargueren-7α-ol, in B-80240

C$_{24}$H$_{36}$
Heptacyclo[19.3.0.0^{1,5}.0^{5,9}.0^{9,13}.0^{13,17}.0^{17,21}]tetracosane, H-80017

C$_{24}$H$_{36}$O$_4$
ent-6β,17-Diacetoxy-7,12E,14-labdatriene, in L-60004
Soulattrone A, S-70057

C$_{24}$H$_{36}$O$_5$
ent-6β,17-Diacetoxy-7,11E,14-labdatrien-13ξ-ol, in L-60005
1(15),8(19)-Trinervitadiene-2,3,9-triol; 2,3-Di-Ac, in T-70298

$C_{24}H_{36}O_6$
6α-Butanoyloxy-17β-hydroxy-15,17-oxido-16-spongianone, in D-**60365**
Hamachilobene B, in H-**70001**
1(15),8(19)-Trinervitadiene-2β,3α,9β,20-tetrol 9,20-diacetate, in T-**70296**

$C_{24}H_{36}O_7$
ent-1β,3α-Diacetoxy-16-kaurene-6β,11α,15α-triol, in K-**80008**

$C_{24}H_{36}O_8$
Lapidolin, in L-**60015**

$C_{24}H_{38}$
Des-A-5(10),12-oleanadiene, D-**80037**
Des-A-5(10),12-ursadiene, D-**80039**

$C_{24}H_{38}O_3$
Leiopathic acid; Et ester, in H-**70125**

$C_{24}H_{38}O_4$
Trifarin, T-**70236**
Trunculin A, T-**60408**

$C_{24}H_{38}O_5$
6,16-Diacetoxy-3,7-dolabelladien-12-ol, in D-**70541**
ent-3β,17-Diacetoxy-16β-kauranol, in K-**60005**
Trunculin B, T-**60409**

$C_{24}H_{38}O_6$
7β,15-Diacetoxy-8-labden-17-oic acid, in D-**80330**
Hamachilobene A, H-**70001**
Hamachilobene C, in H-**70001**
Hamachilobene D, in H-**70001**

$C_{24}H_{38}O_8$
Rhodojaponin IV, in G-**80039**

$C_{24}H_{40}$
Des-A-lup-9-ene, D-**80035**
Des-A-lup-5(10)-ene, in D-**80035**

$C_{24}H_{40}O_2$
3-Hydroxy-22,23,24,25,26,27-hexanor-20-dammaranone, H-**60151**

$C_{24}H_{40}O_4$
15,17-Diacetoxy-7-labdene, in L-**80012**

$C_{24}H_{40}O_5$
3,11,15-Trihydroxycholanic acid, T-**60317**
3,15,18-Trihydroxycholanic acid, T-**60318**

$C_{24}H_{42}$
10βH-Des-A-lupane, in D-**80035**
1,3,5-Trihexylbenzene, T-**70247**

$C_{24}H_{42}O_4$
Bisdihydrotrifarin, in T-**70236**

$C_{24}H_{43}NO_7$
2-Amino-1,3,4-hexadecanetriol; O,N,N,N-Tetra-Ac, in A-**80079**

$C_{24}H_{44}O_4$
7,11,15-Trimethyl-3-methylene-1,2-hexadecanediol; Di-Ac, in T-**60368**

$C_{24}H_{46}$
12-Tetracosyne, T-**70018**
1-Tetracosyne, T-**70017**

$C_{24}H_{46}N_4O_8S_2$
Homopantethine, in H-**60090**

$C_{24}H_{47}NO_{11}S$
N-[15-(β-D-Glucopyranosyloxy)-8-hydroxypalmitoyl]taurine, G-**70019**

$C_{24}H_{50}$
3-Methyltricosane, M-**80185**

$C_{25}H_{14}O$
Benzo[a]naphtho[2,1-d]fluoren-9-one, B-**80027**
Benzo[b]naphtho[2,1-d]fluoren-9-one, B-**80028**

$C_{25}H_{16}$
Spirobi[9H-fluorene], S-**60040**

$C_{25}H_{16}O_2$
Benz[a]anthracen-10-ol; O-Benzoyl, in B-**80011**

$C_{25}H_{20}N_2$
Triphenylphenylazomethane, T-**70312**

$C_{25}H_{21}N$
3,5-Dimethyl-2,4,6-triphenylpyridine, D-**80478**

$C_{25}H_{22}$
[2](1,5)Naphthaleno[2](2,7)(1,6-methano[10]annuleno)phane, N-**70005**

$C_{25}H_{22}O_6$
Cudraflavone A, C-**70205**
2-Hydroxymethyl-1,3-propanediol; Tribenzoyl, in H-**80212**

$C_{25}H_{22}O_7$
Artobiloxanthone, A-**80172**
Cycloartobiloxanthone, C-**80173**
5′-Hydroxycudraflavone A, in C-**70205**

$C_{25}H_{22}O_8$
Galtamycinone, in G-**70003**

$C_{25}H_{22}O_9$
Hypericorin, H-**70234**
Silandrin, S-**70042**
Silyhermin, S-**70045**
Silymonin, in S-**70044**

$C_{25}H_{22}O_{10}$
Isosilybin, I-**70102**
Isosilychristin, I-**70103**
Silydianin, S-**70044**

$C_{25}H_{22}O_{12}$
Daphnorin, in D-**70006**

$C_{25}H_{24}$
Bicyclo[18.4.1]pentacosa-1,3,5,7,9,11,13,15,17,19,21,23-dodecaene, B-**80089**
5,5a,6,6a,7,12,12a,13,13a,14-Decahydro-5,14:6,13:7,12-trimethanopentacene, D-**60010**

$C_{25}H_{24}O_2S$
3-[(Triphenylmethyl)thio]cyclopentanecarboxylic acid, in M-**60021**

$C_{25}H_{24}O_5$
Euchrenone A_1, E-**70056**
Isochandalone, I-**80052**
Isomammeigin, I-**70069**

$C_{25}H_{24}O_6$
Cudraisoflavone A, C-**70206**
Isoauriculatin, in P-**80015**

$C_{25}H_{24}O_7$
Laserpitinol, in L-**60016**

$C_{25}H_{24}O_{12}$
3,4-Di-O-caffeoylquinic acid, D-**70110**
7-Hydroxy-4′-methoxyisoflavone; 7-O-(6-O-Malonylglucoside), in H-**80190**

$C_{25}H_{24}O_{13}$
5,7-Dihydroxy-4′-methoxyisoflavone; 7-O-(6-O-Malonyl-D-glucoside), in D-**80337**

$C_{25}H_{26}O_4$
Abyssinone III, A-**80008**
Erybraedin D, E-**80051**

$C_{25}H_{26}O_5$
Cajaflavanone, C-**60004**
Euchrenone A_2, E-**70057**
Honyucitrin, H-**70092**
Sigmoidin E, S-**80032**
4′,5,7-Trihydroxy-3′,6-bis(3-methyl-2-butenyl)isoflavone, T-**80290**

$C_{25}H_{26}O_6$
Cajanone, C-**80009**
Fremontone, F-**80081**
2′-Hydroxyisolupalbigenin, H-**80177**
Kuwanol C, K-**80014**
Lupinisoflavone G, L-**80044**
Lupinisol A, L-**80047**
Lupinisolone A, L-**80049**
Orotinin, O-**60044**

$C_{25}H_{26}O_7$
Garvalone B, G-**60005**
Lupinisoflavone H, in L-**80044**
Lupinisoflavone I, L-**80045**

Lupinisoflavone J, L-**80046**
Lupinisol B, in L-**80047**
Lupinisol C, L-**80048**
Lupinisolone B, L-**80050**
Lupinisolone C, L-**80051**

$C_{25}H_{26}O_8$
Rhynchosperin C, in R-**60007**

$C_{25}H_{26}O_9$
Glomellonic acid, in G-**70018**

$C_{25}H_{26}O_{10}$
Fridamycin A, F-**70043**
Fridamycin B, in F-**70043**

$C_{25}H_{28}O_4$
Abyssinone IV, A-**80009**
Abyssinone VI, A-**80010**
1-[2,4-Dihydroxy-3,5-bis(3-methyl-2-butenyl)phenyl]-3-(4-hydroxyphenyl)-2-propen-1-one, D-**60310**
Erythrabyssin II, E-**80053**
Isolinderatone, in L-**70030**
Linderachalcone, L-**80032**
Linderatone, L-**70030**

$C_{25}H_{28}O_5$
Abyssinone V, in A-**80009**
Ammothamnidin, A-**70204**
10′,11′-Dehydrocyclolycoserone, in C-**70229**
8,9-Dehydroircinin 1, in I-**60083**
Kuwanol D, K-**80015**
Lehmannin, L-**70027**
Lespedazaflavone B, L-**60021**
Lonchocarpol A, L-**60030**
Rhinacanthin B, in R-**80005**
Senegalensein, S-**70034**

$C_{25}H_{28}O_6$
Lonchocarpol C, L-**60031**
Lonchocarpol D, L-**60032**

$C_{25}H_{28}O_7$
Lonchocarpol E, L-**60033**

$C_{25}H_{28}O_8$
Drummondin B, D-**80531**
Glomelliferonic acid, G-**70018**

$C_{25}H_{28}O_{10}$
Egonol glucoside, in E-**70001**

$C_{25}H_{28}O_{13}$
2′,4′,5,5′,6,7-Hexahydroxyisoflavone; 2′,4′,5′,6-Tetra-Me ether, 7-O-glucoside, in H-**80069**
Isocaviudin, in H-**80070**

$C_{25}H_{28}O_{14}$
Gentioside, in T-**80347**
1-Hydroxy-3-methoxy-7-primeverosyloxyxanthone, in T-**80347**
7-Hydroxy-3-methoxy-1-primeverosyloxyxanthone, in T-**80347**

$C_{25}H_{30}Cl_2O_6$
Napyradiomycin A_2, N-**70016**

$C_{25}H_{30}N_8O_9$
7-Hydro-8-methylpteroylglutamylglutamic acid, H-**60096**

$C_{25}H_{30}O_3$
2′-Epiisotriptiliocoumarin, in I-**80096**
Isotriptiliocoumarin, I-**80096**
Triptiliocoumarin, T-**80363**

$C_{25}H_{30}O_4$
4-Hydroxy-5-methyl-3-(3,8,11-trimethyl-8-oxo-2,6,10-dodecatrienyl)-2H-1-benzopyran-2-one, H-**70183**
5′-Hydroxytriptiliocoumarin, in T-**80363**
8-Hydroxytriptiliocoumarin, in T-**80363**
Linderatin, L-**50048**
Nassauvirevolutin B, in N-**80014**

$C_{25}H_{30}O_5$
Cyclolycoserone, C-**70229**
Cyclolycoserone; 3′-Epimer, in C-**70229**
1-(2,4-Dihydroxyphenyl)-3-[3,4-dihydroxy-2-(3,7-dimethyl-2,6-octadienyl)phenyl]-1-propanone, D-**80357**
1′-Epilycoserone, in L-**70045**

6′,7′-Epoxy-5′α-hydroxytriptiliocoumarin, *in* T-**80**363
Gypothamniol, G-70037
4-Hydroxy-3-(9-hydroxy-8-oxofarnesyl)-5-methylcoumarin, *in* H-70183
Ircinin 1, I-**60**083
Ircinin 2, *in* I-**60**083
Isolycoserone, I-70068
Lycoserone, L-70045

$C_{25}H_{30}O_6$
1-[3-(3,7-Dimethyl-2,6-octadienyl)-2,4,5,6-tetrahydroxyphenyl]-3-(4-hydroxyphenyl)-1-propanone,9CI, D-**80**459
10′-Hydroxycyclolycoserone, *in* C-70229
10′-Hydroxy-1′-epilycoserone, *in* L-70045
10′-Hydroxyisolycoserone, *in* I-70068

$C_{25}H_{30}O_7$
Lonchocarpol *B*, *in* L-**60**030

$C_{25}H_{30}O_{13}$
Picroside III, *in* C-70026

$C_{25}H_{31}^{\oplus}$
2,3,4,5,6,7,8,10,11,12,13,?-Dodecahydro-1,4:5,8:10,13-triethano-1*H*-tribenzo[*a*,*c*,*e*]cycloheptenylium(1+), D-**80**524

$C_{25}H_{31}ClO_7$
3-Chloroperlatolic acid, *in* P-**80**070

$C_{25}H_{31}Cl_3O_6$
Napyradiomycin B_4, N-70017

$C_{25}H_{31}Cl_6Sb$
2,3,4,5,6,7,8,10,11,12,13,?-Dodecahydro-1,4:5,8:10,13-triethano-1*H*-tribenzo[*a*,*c*,*e*]cycloheptenylium(1+); Hexafluoroantimonate, *in* D-**80**524

$C_{25}H_{31}N_3O_5$
1-[*N*-[3-(Benzoylamino)-2-hydroxy-4-phenylbutyl]alanyl]proline, B-**60**056

$C_{25}H_{32}O_4$
Ircinianin, I-**60**081
Ircinic acid, I-**60**082
Nassauvirevolutin A, N-**80**014
Nassauvirevolutin C, N-**80**015

$C_{25}H_{32}O_5$
5-Oxo-8(10)*E*-variabilin, *in* V-**80**002
5-Oxovariabilin, *in* V-**80**002
Variabilin/; 5-Oxo, $\Delta^{8(10)}$(*E*-), Δ^{13}-isomer, *in* V-**80**002

$C_{25}H_{32}O_6$
3-*O*-Angeloylsenecioodontol, *in* S-**60**022

$C_{25}H_{32}O_7$
Ichangensin, I-70001
Perlatolic acid, P-**80**070

$C_{25}H_{32}O_8$
Albaspidin BB, *in* A-**80**043
Aspidin BB, *in* A-**80**180
Clausenolide, C-**80**140
Saroaspidin *B*, S-70013

$C_{25}H_{32}O_{10}$
Tinosporaside, T-**80**198

$C_{25}H_{32}O_{12}$
Isoligustroside, I-**60**104

$C_{25}H_{32}O_{13}$
Oleuroside, O-70038
Syringopicroside *C*, *in* S-**60**067

$C_{25}H_{33}ClO_8$
Punaglandin 3, P-70143

$C_{25}H_{33}ClO_{11}$
Tetragonolide isobutyrate, *in* T-**80**053

$C_{25}H_{33}NO$
Aurachin *D*, A-**60**315

$C_{25}H_{33}NO_2$
Aurachin *B*, A-**60**314
Aurachin *C*, *in* A-**60**315

$C_{25}H_{33}NO_3$
Aurachin *A*, A-**60**313

$C_{25}H_{34}O_3$
3-Anhydro-6-epiophiobolin *A*, *in* O-70040
Apo-12′-violaxanthal, *in* P-**60**065

$C_{25}H_{34}O_4$
Fasciculatin, F-70001
Isofasciculatin, *in* F-70001
Variabilin/, V-**80**002
Variabilin/; 20*E*-Isomer, *in* V-**80**002
Variabilin/; Δ^{13}-Isomer (*Z*-), *in* V-**80**002

$C_{25}H_{34}O_5$
5-Hydroxyvariabilin, *in* V-**80**002

$C_{25}H_{34}O_7$
Glycinoeclepin *A*, G-70022

$C_{25}H_{35}ClO_8$
Punaglandin 4, *in* P-70143

$C_{25}H_{36}O_3$
8-Deoxyophiobolin J, *in* O-70041
Persicaxanthin, P-**60**065

$C_{25}H_{36}O_4$
Elasclepic acid, *in* B-**80**064
6-Epiophiobolin *A*, *in* O-70040
5-[13-(3-Furanyl)-2,6,10-trimethyl-6,8-tridecadienyl]-4-hydroxy-3-methyl-2(5*H*)-furanone, F-70050
24-Methyl-25-nor-12,24-dioxo-16-scalaren-22-oic acid, M-**60**105
▷ Ophiobolin *A*, O-70040
Ophiobolin J, O-70041

$C_{25}H_{36}O_5$
8-Hydroxyvariabilin, *in* V-**80**002
Monoalide, M-70152
Salvileucolidone, S-70005
Thorectolide, T-70189

$C_{25}H_{36}O_6$
Pseudopterosin *A*, P-**60**187
Trifloresteviоne, T-70237

$C_{25}H_{36}O_7$
Bacchabolivic acid; Xylopyranoside, *in* B-**80**004

$C_{25}H_{36}O_{12}$
Kickxioside, K-70014

$C_{25}H_{37}NO_3$
Smenospongorine, *in* S-**80**036

$C_{25}H_{38}O_3$
Elasclepiol, *in* B-**80**064
Luffariellolide, L-70040
Palauolide, P-**60**001

$C_{25}H_{38}O_4$
Aglajne 2, A-70073
Spongiolactone, S-**60**047

$C_{25}H_{38}O_4S$
Suvanine, S-70085

$C_{25}H_{38}O_5$
Pallinin, *in* C-**60**021

$C_{25}H_{38}O_6$
Clibadic acid, C-70186
Desoxodehydrolaserpitin, *in* D-70009
Secopseudopterosin *A*, S-70026
Tingitanol, *in* D-70009

$C_{25}H_{38}O_7$
Isolaserpitin, *in* D-70009
Laserpitine, L-**60**016
Methyl 15α,17β-diacetoxy-15,16-dideoxy-15,17-oxido-16-spongianoate, *in* T-**60**038

$C_{25}H_{40}O_3$
4,6,8,10,12,14,16-Heptamethyl-6,8,11-octadecatriene-3,5,13-trione, H-70022
3-Octadecene-1,2-diol; 1-Benzoyl, *in* O-**60**005

$C_{25}H_{40}O_6$
2α-Angeloyloxy-8β,20-epoxy-14-labdene-3α,7α,13*R*-triol, *in* E-**80**036
Salvisyriacolide, S-**60**004

$C_{25}H_{40}O_8$
ent-3,13-Clerodadiene-15,16,17-triol; 17-*O*-β-Xylopyranoside, *in* C-**80**141

$C_{25}H_{42}O_4$
6β-Isovaleroyloxy-8,13*E*-labdadiene-7α,15-diol, *in* L-**60**003
Sclareol; 6α-Angeloyloxy, *in* S-70018

$C_{25}H_{45}FeN_6O_8$
Ferrioxamine *B*, F-**60**003

$C_{26}H_6N_8$
11,11,12,12,13,13,14,14-Octacyano-1,4:5,8-anthradiquinotetramethane, O-70002

$C_{26}H_{10}N_2O_6S_2$
Dibenzo[*b*,*m*]triphenodithiazine-5,7,9,14,16,18(8*H*,17*H*)-tetrone, D-70082

$C_{26}H_{14}$
Benz[*def*]indeno[1,2,3-*hi*]chrysene, B-70024
Benz[*def*]indeno[1,2,3-*qr*]chrysene, B-70023
Benzo[*h*]naphtho[2′,3′:3,4]cyclobuta[1,2-*a*]biphenylene, B-**80**025
Bisbenzo[3,4]cyclobuta[1,2-*c*;1′,2′-*g*]phenanthrene, B-**60**143
Fluoreno[9,1,2,3-*cdef*]chrysene, F-70017
Fluoreno[3,2,1,9-*defg*]chrysene, F-70016
Hexa[7]circulene, H-**80**037
Naphtho[1,2,3,4-*ghi*]perylene, N-**60**009

$C_{26}H_{14}O_2$
6,15-Hexacenedione, H-**60**033

$C_{26}H_{16}$
9,9′-Bifluorenylidene, B-**60**104

$C_{26}H_{18}$
1,3-Di-(1-naphthyl)benzene, D-**60**454
1,3-Di(2-naphthyl)benzene, D-**60**455

$C_{26}H_{18}N_2$
9,9′-Biacridylidene, B-70090
2,11-Diphenyldipyrrolo[1,2-*a*:2′,1′-*c*]quinoxaline, D-70479

$C_{26}H_{18}N_6$
7,11:20,24-Dinitrilodibenzo[*b*,*m*][1,4,12,15]tetraazacyclodocosine, D-**60**456

$C_{26}H_{18}O_9$
5,6,8,13-Tetrahydro-1,7,9,11-tetrahydroxy-8,13-dioxo-3-(2-oxopropyl)benzo[*a*]naphthacene-2-carboxylic acid, T-70095

$C_{26}H_{20}$
Tribenzotritwistatriene, T-70210

$C_{26}H_{20}N_2$
Diphenylketazine, D-**80**512

$C_{26}H_{20}N_4S$
N-2-Benzothiazolyl-*N*,*N*′,*N*″-triphenylguanidine, B-70057

$C_{26}H_{20}O$
Tetraphenylethanone, T-**60**167

$C_{26}H_{20}O_4$
Tetraphenoxyethene, T-**80**157

$C_{26}H_{20}O_6$
1,8,11-Triptycenetricarboxylic acid; Tri-Me ester, *in* T-**60**398
1,8,14-Triptycenetricarboxylic acid; Tri-Me ester, *in* T-**60**399

$C_{26}H_{20}O_7$
Imbricatonol, I-70003

$C_{26}H_{20}O_{10}$
Salvianolic acid *C*, *in* S-70003

$C_{26}H_{20}S_3$
3,3,5,5-Tetraphenyl-1,2,4-trithiolane, T-70153

$C_{26}H_{22}N_2O$
1,1,1′,1′-Tetraphenyldimethylamine; *N*-Nitroso, *in* T-70150

$C_{26}H_{22}O_4$
1,2-Di-2-naphthalenyl-1,2-ethanediol; Di-Ac, *in* D-**80**479

$C_{26}H_{22}O_6$
Stypandrol, S-50130

$C_{26}H_{22}Se_2$
Bis(diphenylmethyl) diselenide, B-70141

$C_{26}H_{23}N$
1,1,1',1'-Tetraphenyldimethylamine, T-70150

$C_{26}H_{24}$
Tricyclo[18.4.1.1^{8,13}]hexacosa-
2,4,6,8,10,12,14,16,18,20,22,24-dodecaene,
T-60263
Tricyclo[18.4.1.1.^{8,13}]hexacosa-
2,4,6,8,10,12,14,16,18,20,22,24-dodecaene,
T-80241

$C_{26}H_{24}N_6$
1,14,29,30,31-Hexaazahexacyclo[12.7.7.1^{3,7}.1^{8,12}.1^{16,20}
.1^{23,27}]dotriaconta-3,5,7(32),8,10,12(31),16,
18,20(30),23,25,27(29)-dodecaene, H-70030

$C_{26}H_{24}O_2$
[2.2][2.2]Paracyclophane-12,15-quinone,
P-60008
[2.2][2.2]Paracyclophane-5,8-quinone, P-60007

$C_{26}H_{24}O_{10}$
Peltigerin, P-70016

$C_{26}H_{24}O_{15}$
4,6-Di-O-galloylarbutin, in A-70248

$C_{26}H_{25}N_3O_{11}$
Tunichrome B-1, T-70336

$C_{26}H_{26}$
[2.2]Paracyclo(4,8)[2.2]metaparacyclophane,
P-60005

$C_{26}H_{26}O_2S$
4-[(Triphenylmethyl)thio]cyclohexane-
carboxylic acid, in M-60020

$C_{26}H_{26}O_7$
Cajaisoflavone, C-80008

$C_{26}H_{26}O_9$
Dukunolide A, D-60522

$C_{26}H_{26}O_{10}$
Dukunolide B, in D-60522

$C_{26}H_{26}O_{12}$
7-Hydroxy-4'-methoxyisoflavone; 7-O-(6-O-
Malonylglucoside), Me ester, in H-80190

$C_{26}H_{26}O_{13}$
5,7-Dihydroxy-4'-methoxyisoflavone; 7-O-(6-
O-Malonyl-D-glucoside), Me ester, in
D-80337

$C_{26}H_{26}O_{18}$
Amritoside, in E-70007

$C_{26}H_{28}O_6$
2'-O-Methylcajanone, in C-80009
Orotinichalcone, O-60043
Orotinin; 5-Me ether, in O-60044
3,4,7,8-Tetrahydro-11-(2-hydroxy-4-
methoxyphenyl)-2,2,6,6-tetramethyl-
2H,6H,12H-benzo[1,2-b;3,4-b';5,6-
b'']tripyran-12-one, T-80067
3,4,7,8-Tetrahydro-11-(2-hydroxy-4-
methoxyphenyl)-2,2,6,6-tetramethyl-
2H,6H,12H-benzo[1,2-b;3,4-b';5,6-
b'']tripyran-12-one; 2'-Deoxy, 3'-hydroxy,
in T-80067

$C_{26}H_{28}O_8$
Acidissimin, A-80034
Dukunolide D, D-60523

$C_{26}H_{28}O_9$
Dukunolide E, in D-60523
Dukunolide F, in D-60523
Ephemeroside, in E-60006

$C_{26}H_{28}O_{10}$
Jasmolactone C, J-80007

$C_{26}H_{28}O_{13}$
Ambonin, in D-80323
Neobanin, in D-80323

$C_{26}H_{28}O_{14}$
Ambocin, in T-80309
Neobacin, in T-80309
Neocorymboside, N-70021
1,2,7-Trihydroxy-6-methylanthraquinone; O^2-
β-Primeveroside, in T-70268

$C_{26}H_{28}O_{19}$
Luteolic acid; Diglucoside, in L-80052

$C_{26}H_{29}NO_9$
Antibiotic R 20Y7, A-70230

$C_{26}H_{29}N_3O_2$
Crystal violet lactone, C-70202

$C_{26}H_{29}N_3O_{11}S_2$
O^6-[(3-Carbamoyl-2H-azirine-2-
ylidene)amino]-1,2-O-isopropylidene-3,5-di-
O-tosyl-α-D-glucofuranoside, C-70014

$C_{26}H_{29}O_{14}^{\oplus}$
Pelargonidin 3-sambubioside, in T-80093

$C_{26}H_{30}Cl_2N_2O_5$
Antibiotic SF 2415A_3, A-70234

$C_{26}H_{30}N_2O_5$
Antibiotic SF 2415A_2, A-70233

$C_{26}H_{30}N_6$
5,6,11,12-Tetrahydro-5,6,11,12,13,20-
hexamethyl-5a,11a-
(imino[1,2]benzenimino)quinoxalino[2,3-
b]quinoxaline, T-80064

$C_{26}H_{30}O_4$
2',6'-Dihydroxy-4'-methoxy-3'-(1-p-menthen-
3-yl)chalcone, D-80338
Erycristin, E-70034
Grenoblone, G-70029
Methyllinderatone, in L-70030

$C_{26}H_{30}O_5$
(E)-ω-Acetoxyferprenin, in F-70003
4-Hydroxygrenoblone, in G-70029
(Z)-ω-Acetoxyferprenin, in F-70003

$C_{26}H_{30}O_6$
7-Deacetoxy-7-oxogedunin, in G-70006
2,3-Dibenzoylbutanedioic acid; Dibutyl ester,
in D-70083
Lespedezaflavanone A, L-60022

$C_{26}H_{30}O_8$
Butyrylmallotochromene, B-80295
Drummondin A, D-80530
Dysoxylin, D-60525
Isobutyrylmallotochromene, I-80050

$C_{26}H_{30}O_{10}$
Graucin A, G-60036
Isolimonexic acid, in L-80031
Limonexic acid, L-80031

$C_{26}H_{30}O_{11}$
Rutaevinexic acid, R-70014

$C_{26}H_{30}O_{13}$
3-(4-Hydroxyphenyl)-1-(2,4,6-
trihydroxyphenyl)-2-propen-1-one; 2'-[O-
Rhamnosyl(1→4)xyloside], in H-80246
Liquiraside, in D-80367

$C_{26}H_{31}ClN_2O_5$
Antibiotic SF 2415A_1, A-70232

$C_{26}H_{32}Cl_2O_5$
Antibiotic SF 2415B_3, A-70237

$C_{26}H_{32}O_4$
Methyllinderatin, in L-50048
Nimbocinol, in A-70281

$C_{26}H_{32}O_5$
Antibiotic SF 2415B_2, A-70236
7-Deacetyl-17β-hydroxyazadiradione, in
A-70281
(E)-ω-Acetoxyferulenol, in F-80007
Helihumulone, H-80006
Licoricidin, L-60025
Polyanthin, P-80166
Polyanthinin, in P-80166
(Z)-ω-Acetoxyferulenol, in F-80007

$C_{26}H_{32}O_6$
4'-(3,6-Dimethyl-2-heptenyloxy)-5-hydroxy-
3',7-dimethoxyflavanone, in T-70105
7-O-(3-Hydroxy-7-drimen-11-yl)isofraxidin; 3-
Ketone, in H-80158

$C_{26}H_{32}O_7$
3-[3-Acetyl-5-(3,7-dimethyl-2,6-octadienyl)-
2,4,6-trihydroxybenzyl]-4-hydroxy-5,6-
dimethyl-2H-pyran-2-one, A-80021

$C_{26}H_{32}O_8$
Longirabdosin, L-70036

$C_{26}H_{32}O_9$
Clausenarin, C-80139

$C_{26}H_{32}O_{10}$
Hiyodorilactone F, in H-80090
Teuvincentin C, T-80168

$C_{26}H_{32}O_{11}$
Columbinyl glucoside, in C-60160

$C_{26}H_{32}O_{12}$
1-Hydroxypinoresinol; 1-O-β-D-
Glucopyranoside, in H-60221
1-Hydroxypinoresinol; 4'-O-β-D-
Glucopyranoside, in H-60221
Isojateorinyl glucoside, in C-60160
Jateorinyl glucoside, in C-60160

$C_{26}H_{32}O_{13}$
Decumbeside C, D-60017
Decumbeside D, in D-60017
Haenkeanoside, H-80001
Isohaenkeanoside, in H-80001

$C_{26}H_{32}O_{14}$
Mulberroside A, in D-70335

$C_{26}H_{33}ClO_5$
Antibiotic SF 2415B_1, A-70235

$C_{26}H_{33}ClO_8$
Ptilosarcenone; 2-De-Ac, 2-butanoyl, in
P-70136

$C_{26}H_{33}ClO_9$
Junceellolide B, J-80016
12-Ptilosarcenol; 12-Ac, in P-60193

$C_{26}H_{33}ClO_{10}$
Junceellolide A, J-80015
Junceellolide C, in J-80016

$C_{26}H_{33}ClO_{11}$
Solenolide D, in S-70053

$C_{26}H_{34}O_6$
7-O-(3-Hydroxy-7-drimen-11-yl)isofraxidin,
H-80158
Nimolicinoic acid, N-60022

$C_{26}H_{34}O_7$
Salannic acid, S-80002

$C_{26}H_{34}O_8$
Agrimophol, A-60066
6β,9β-Diacetoxy-1α-benzoyloxy-4β-
hydroxydihydro-β-agarofuran, in T-80090
Saroaspidin A, S-70014
2,11,12-Trihydroxy-6,7-seco-8,11,13-
abietatriene-6,7-dial 11,6-hemiacetal; Tri-
Ac(6α-), in T-80339

$C_{26}H_{34}O_9$
Adenanthin, A-70065
Bryophyllin B, B-80272
Deoxyhavannahine, in H-60008
ent-1β,7α,11α-Triacetoxy-3α-hydroxy-16-
kaurene-6,15-dione, in T-60109

$C_{26}H_{34}O_{10}$
9α,14-Diacetoxy-1α-benzoyloxy-4β,6β,8β-
trihydroxydihydro-β-agarofuran, in
E-60018
Eumaitenin, E-60071
Havannahine, H-60008
Rhynchospermoside A, R-60008
Rhynchospermoside B, in R-60008

$C_{26}H_{34}O_{11}$
Methyl 12,17;15,16-diepoxy-6-hydroxy-19-nor-
17-oxo-4,13(16),14-clerodatrien-18-oate; O-
β-D-Glucopyranoside, in M-80068

$C_{26}H_{35}ClO_{10}$
11(19)-Havannachlorohydrin, H-70004
7(18)-Havannachlorohydrin, H-70003

$C_{26}H_{35}ClO_{11}$
Tetragonolide 2-methylbutyrate, *in* T-80053

$C_{26}H_{36}$
[14.0]Paracyclophane, P-70008

$C_{26}H_{36}O_7$
14-Acetoxy-2-butanoyloxy-5,8(17)-briaradien-18,7-olide, *in* E-70012
9-Deacetoxyxenicin, *in* X-70004

$C_{26}H_{36}O_8$
Jodrellin B, *in* J-80013
ent-2α,3α,6β-Triacetoxy-7α,11β-dihydroxy-16-kauren-15-one, *in* P-80047

$C_{26}H_{36}O_9$
Caesalpin F, C-60002
Weisiensin A, *in* P-80048

$C_{26}H_{36}O_{10}$
Melianolone, M-60015

$C_{26}H_{38}O_4$
Aglajne 3, A-70074
12-Hydroxy-24-methyl-24-oxo-16-scalarene-22,25-dial, H-60191

$C_{26}H_{38}O_5$
Chinensin II, C-60036
12α-Hydroxy-24-methyl-24,25-dioxo-14-scalaren-22-oic acid, *in* H-60191

$C_{26}H_{38}O_6$
1-Acetoxy-7-acetoxymethyl-3-acetoxymethylene-11,15-dimethyl-1,6,10,14-hexadecatetraene, A-70020

$C_{26}H_{38}O_7$
17α-Acetoxy-6α-butanoyloxy-15,17-oxido-16-spongianone, *in* D-60365
Virescenoside E, *in* I-70081

$C_{26}H_{39}NO_3$
Smenospongianine, *in* S-80036

$C_{26}H_{40}N_4$
5,8,14,17-Tetrakis(dimethylamino)[3.3]paracyclophene, T-70126

$C_{26}H_{40}O_4$
16-Hydroxy-24-methyl-12,24-dioxo-25-scalaranal, H-80200

$C_{26}H_{40}O_5$
Episalviaethiopisolide, *in* S-80004
Salviaethiopisolide, S-80004

$C_{26}H_{40}O_7$
Virescenoside L, *in* I-70081

$C_{26}H_{40}O_8$
Virescenoside G, *in* I-70077

$C_{26}H_{40}O_9$
Phloganthoside, *in* P-70099

$C_{26}H_{40}O_{11}$
Rabdoside 1, *in* M-80008

$C_{26}H_{40}O_{12}$
Rabdoside 2, *in* M-80008

$C_{26}H_{40}O_{13}$
Jasmesosidic acid, J-80004

$C_{26}H_{40}O_{14}$
6′-O-Apiosylebuloside, *in* E-60001
9″-Hydroxyjasmesosidic acid, *in* J-80004

$C_{26}H_{42}N_6$
1,3,5,7-Tetrakis(diethylamino)pyrrolo[3,4-f]isoindole, T-70125

$C_{26}H_{42}O_2$
24-Nor-7,22-cholestadiene-3,6-diol, N-80078

$C_{26}H_{42}O_8$
Virescenoside A, *in* I-70081

$C_{26}H_{42}O_9$
Grayanoside B, *in* G-80041
Grayanoside C, *in* G-80041

$C_{26}H_{44}O_4N_4$
1,3,5,7-Adamantanetetraacetic acid;
Tetrakis(dimethylamide), *in* A-70064

$C_{26}H_{46}O_6$
Sclareol; 13-(6′-Deoxy-α-L-idopyranoside), *in* S-70018

$C_{26}H_{48}O_3$
Ficulinic acid A, F-60011

$C_{26}H_{48}O_5$
5,6-Dihydro-6-(2,4,6-trihydroxyheneicosyl)-2H-pyran-2-one, D-70276

$C_{26}H_{48}O_6$
5,6-Dihydro-6-(2,4,6,10-tetrahydroxyheneicosyl)-2H-pyran-2-one, *in* D-70276

$C_{26}H_{50}O_2$
Aparjitin, A-60291

$C_{26}H_{50}O_4$
Bis(2-ethylhexyl) decanedioate, B-80149

$C_{26}H_{52}O$
Hexacosanal, H-80038

$C_{26}H_{54}$
3-Methylpentacosane, M-80131

$C_{26}H_{54}O$
5-Methyl-5-pentacosanol, M-80132

$C_{27}H_{14}$
Indeno[2,1,7,6-ghij]naphtho[2,1,8,7-nopq]pleiadene, I-80011

$C_{27}H_{20}$
1,2,3-Triphenyl-1H-indene, T-60387

$C_{27}H_{20}Cl_{15}N$
Tetraethylammonium tris(pentachlorophenyl)methylide, *in* T-60405

$C_{27}H_{20}O_7$
Mulberrofuran R, M-60149

$C_{27}H_{22}O$
1,1,3,3-Tetraphenyl-2-propanone, T-60168

$C_{27}H_{22}S_2$
4,4,5,5-Tetraphenyl-1,3-dithiolane, T-60166

$C_{27}H_{24}$
1,2,3-Trimethyl-4,5,6-triphenylbenzene, T-70291
1,2,4-Trimethyl-3,5,6-triphenylbenzene, T-70292
1,3,5-Trimethyl-2,4,6-triphenylbenzene, T-70293

$C_{27}H_{24}O_7$
Larreantin, L-70020

$C_{27}H_{26}O_4S_2$
Lappaphen a, L-70018
Lappaphen b, *in* L-70018

$C_{27}H_{26}O_7$
Euchrenone b₃, E-70059

$C_{27}H_{26}O_{11}$
Cleistanthoside B, *in* D-60493
Viscutin 1, *in* T-60103

$C_{27}H_{28}O_7$
Pumilaisoflavone A, P-70140
Pumilaisoflavone B, P-70141

$C_{27}H_{28}O_{11}$
Tremulacin, *in* S-70002

$C_{27}H_{28}O_{13}$
3-O-Caffeoyl-4-O-sinapoylquinic acid, *in* C-70005

$C_{27}H_{30}O_6$
Cyclotriveratrylene, C-70254
Glabrescione B, *in* T-80109

$C_{27}H_{30}O_7$
Pumilaisoflavone C, P-80193

$C_{27}H_{30}O_8$
5-Dehydrooriciopsin, *in* O-70045
Lacinolide A, L-80015

$C_{27}H_{30}O_{12}$
4′,7-Dihydroxyisoflavone; 4′,7-Di-O-rhamnoside, *in* D-80323

$C_{27}H_{30}O_{13}$
4′,7-Dihydroxyisoflavone; 7-O-(Rhamnosylglucoside), *in* D-80323
Podophyllotoxin; 4′-O-De-Me, 1-O-β-D-glucoside, *in* P-80165

$C_{27}H_{30}O_{14}$
4′,7-Dihydroxyisoflavone; 4′,7-Di-O-β-D-glucopyranosyl, *in* D-80323
Lanceolarin, *in* D-80337
Sophorabioside, *in* T-80309
Sphaerobioside, *in* T-80309

$C_{27}H_{30}O_{15}$
Orobol rhamnosylglucoside, *in* T-80109
Palasitrin, *in* S-70082
Paniculatin, P-80008
4′,5,6,7-Tetrahydroxyisoflavone; 7-O-Rhamnosylglucside, *in* T-80112
4′,5,7-Trihydroxyisoflavone; 4′,7-Di-O-glucoside, *in* T-80309
4′,5,7-Trihydroxyisoflavone; 7-O-Glucosylglucoside, *in* T-80309

$C_{27}H_{30}O_{16}$
Isoorientin; 4′-Glucoside, *in* I-80068
Lutonarin, *in* I-80068
3′,4′,5,7-Tetrahydroxy-6-C-sophorosylflavone, *in* I-80068

$C_{27}H_{30}O_{18}$
6,8-Di-C-glucosyl-3′,4′,5,7-tetrahydroxyisoflavone, D-80168

$C_{27}H_{31}O_{15}^{\oplus}$
Pelargonidin 3-sophoroside, *in* T-80093
Pelargonin, *in* T-80093

$C_{27}H_{32}O_6$
Scutellone F, S-70020

$C_{27}H_{32}O_7$
7-(2,4-Dimethoxyphenyl)-3,4-dihydro-5-hydroxy-10-(3-hydroxy-3-methylbutyl)-2,2-dimethyl-2H,6H-benzo[1,2-b:5,4-b′]dipyran-6-one, D-80415
7-(3,4-Dimethoxyphenyl)-3,4-dihydro-5-hydroxy-10-(3-hydroxy-3-methylbutyl)-2,2-dimethyl-2H,6H-benzo[1,2-b:5,6-b′]dipyran-6-one, D-80416
Garvalone A, G-60004

$C_{27}H_{32}O_8$
Oriciopsin, O-70045

$C_{27}H_{32}O_9$
Pedonin, P-60014
Trijugin B, T-60352

$C_{27}H_{32}O_{10}$
Baccharinoid B25, B-60004
Eumaitenol, E-60072

$C_{27}H_{32}O_{13}$
Aloinoside B, *in* B-80010
Aloinoside B, *in* B-80010
Cascaroside C, *in* B-80010
Cascaroside D, *in* B-80010
1-(2,4-Dihydroxyphenyl)-3-(4-hydroxyphenyl)-2-propen-1-one; 4′-O-α-L-Rhamnopyranoside, *in* D-80367
Rhamnoisoliquiritin, *in* D-80367

$C_{27}H_{32}O_{14}$
Cascaroside A, *in* B-80010
Cascaroside B, *in* B-80010
1-(2,4-Dihydroxyphenyl)-3-(4-hydroxyphenyl)-2-propen-1-one; 4′-O-Glucosylglucoside, *in* D-80367
Pyrenoside, P-80206

$C_{27}H_{32}O_{15}$
Isobutrin, *in* D-80362

$C_{27}H_{32}O_{16}$
3′,4′,5,7-Tetrahydroxyflavanone; 3′,5-Di-O-β-D-Glucoside, *in* T-70105

$C_{27}H_{33}N_3O_5$
Andrimide, A-60272

$C_{27}H_{34}O_4$
5-Oxocystofuranoquinol, *in* C-70260
5-Oxoisocystofuranoquinol, *in* C-70260
Yezoquinolide, Y-**6**0003

$C_{27}H_{34}O_5$
Licoricidin; 7-Me ether, *in* L-**6**0025
Scopadulcic acid *B*, *in* S-**6**0015

$C_{27}H_{34}O_6$
Scopadulcic acid *A*, S-**6**0015

$C_{27}H_{34}O_7$
6-Deoxyswietenolide, *in* S-**8**0054
Scuterivulactone *D*, *in* T-70103

$C_{27}H_{34}O_8$
3-[3-Acetyl-5-(3,7-dimethyl-2,6-octadienyl)-2,4,6-trihydroxybenzyl]-6-ethyl-4-hydroxy-5-methyl-2*H*-pyran-2-one, *in* A-**8**0021
Swietenolide, S-**8**0054

$C_{27}H_{34}O_9$
Isocyclocalamin, I-**6**0096

$C_{27}H_{34}O_{11}$
Arctiin, *in* A-70249
Phillyrin, *in* P-**6**0141

$C_{27}H_{34}O_{13}$
O-Methylhaenkeanoside, *in* H-**8**0001
O-Methylisohaenkeanoside, *in* H-**8**0001

$C_{27}H_{35}ClO_9$
12-Ptilosarcenol; 12-Propanoyl, *in* P-**6**0193

$C_{27}H_{36}O_3$
Cystofuranoquinol, C-70260

$C_{27}H_{36}O_4$
16-(2,5-Dihydroxy-3-methylphenyl)-2,6,10,19-tetramethyl-2,6,10,14-hexadecatetraene-4,12-dione, *in* H-70225
5-Hydroxycystofuranoquinol, *in* C-70260

$C_{27}H_{36}O_7$
12α-Hydroxy-4,4,14α-trimethyl-3,7,11,15-tetraoxo-5α-chol-8-en-24-oic acid, *in* L-**6**0037
Hyperlatolic acid, H-**6**0236
Isohyperlatolic acid, I-**6**0101
Planaic acid, P-**8**0158

$C_{27}H_{36}O_8$
Clausenolide; 1-Et ether, *in* C-**8**0140

$C_{27}H_{36}O_9$
4,14-Diacetoxy-2-propanoyloxy-5,8(17)-briaradien-18,7-olide, *in* E-70012

$C_{27}H_{36}O_{14}$
Sylvestroside III, S-70088
Sylvestroside IV, S-70089

$C_{27}H_{37}ClO_{10}$
Punaglandin 1, P-70142

$C_{27}H_{38}O$
10′-Apo-β-caroten-10′-ol, A-70246

$C_{27}H_{38}O_2$
Galloxanthin, G-70002
Sargaquinone, S-**6**0010

$C_{27}H_{38}O_4$
8′,9′-Dihydroxysargaquinone, *in* S-**6**0010
Sargahydroquinoic acid, S-**6**0009

$C_{27}H_{38}O_5$
Amentadione, A-**6**0086
Amentaepoxide, A-**6**0087
Amentol, A-**6**0088
Bifurcarenone, B-**8**0093

$C_{27}H_{38}O_6$
Monoalide 25-acetate, *in* M-70152
Thorectolide 25-acetate, *in* T-70189

$C_{27}H_{38}O_7$
3β,12β-Dihydroxy-4,4,14α-trimethyl-7,11,15-trioxo-5α-chol-8-en-24-oic acid, *in* L-**6**0037
3β-Hydroxy-4α-hydroxymethyl-4β,14α-dimethyl-7,11,15-trioxo-5α-chol-8-en-24-oic acid, *in* L-**6**0036
Pseudopterosin *B*, *in* P-**6**0187
Pseudopterosin *C*, *in* P-**6**0187
Pseudopterosin *D*, *in* P-**6**0187

$C_{27}H_{38}O_8$
3β,12β-Dihydroxy-4α-hydroxymethyl-4β,14α-dimethyl-7,11,15-trioxo-5α-chol-8-en-24-oic acid, *in* L-**6**0036

$C_{27}H_{38}O_9$
Tirucalicine, T-**8**0199

$C_{27}H_{38}O_{13}$
Macrocliniside *B*, *in* Z-**6**0001

$C_{27}H_{38}O_{14}$
Laciniatoside V, L-**8**0014

$C_{27}H_{39}ClO_{10}$
Punaglandin 2, *in* P-70142

$C_{27}H_{40}O_3$
2-(5-Hydroxy-3,7,11,15-tetramethyl-2,6,10,14-hexadecatetraenyl)-6-methyl-1,2-benzenediol, H-70225

$C_{27}H_{40}O_4$
2-(5,16-Dihydroxy-3,7,11,15-tetramethyl-2,6,10,14-hexadecatetraenyl)-6-methyl-1,4-benzenediol, *in* H-70225
Furoscalarol, F-70057

$C_{27}H_{40}O_5$
Chinensin I, C-**6**0035

$C_{27}H_{40}O_7$
Lucidenic acid *H*, L-**6**0036
Secopseudopterosin *B*, *in* S-70026
Secopseudopterosin *C*, *in* S-70026
Secopseudopterosin *D*, *in* S-70026

$C_{27}H_{40}O_9$
Methyl 6α,15α,17β-triacetoxy-15,16-dideoxy-15,17-oxido-16-spongianoate, *in* T-**6**0038

$C_{27}H_{42}O$
Papakusterol, P-**6**0004

$C_{27}H_{42}O_2$
Cholesta-7,9(11),22-triene-3,6-diol, C-**8**0126

$C_{27}H_{42}O_3$
3,6-Dihydroxy-7,25-cholestadien-24-one, D-**8**0302
25(27)-Spirosten-3-ol, S-70065
5-Spirosten-3-ol, S-**6**0045

$C_{27}H_{42}O_4$
Deacetoxybrachycarpone, *in* B-**6**0195
6-Hydroxy-3-spirostanone, H-70219
Nuatigenin, N-**6**0062
5-Spirostene-3,25-diol, S-**6**0044

$C_{27}H_{42}O_5$
24-Acetoxy-12-deacetyl-12-epideoxoscalarin, A-**8**0014
22,25-Epoxy-5-furostene-2α,3β,26-triol, *in* E-70015
25(27)-Spirostene-2,3,6-triol, S-70064

$C_{27}H_{42}O_6$
20-Acetoxy-2α-angeloyloxy-7,14-labdadiene-3α,13*R*-diol, *in* L-**8**0007
20-Acetoxy-2α-angeloyloxy-8(17),14-labdadiene-7α,13*R*-diol, *in* L-**8**0008
Lucidenic acid *M*, L-**6**0037

$C_{27}H_{42}O_7$
3α-Acetoxy-2α-angeloyloxy-8β,20-epoxy-14-labdene-7α,13*R*-diol, *in* E-**8**0036
20-Acetoxy-2α-angeloyloxy-8(17),14-labdadiene-3α,7α,13*R*-triol, *in* L-**8**0006

$C_{27}H_{42}O_{13}$
Jasmesoside, *in* J-**8**0004

$C_{27}H_{42}O_{14}$
9″-Hydroxyjasmesoside, *in* J-**8**0004

$C_{27}H_{42}O_{15}$
Penstebioside, P-**6**0016

$C_{27}H_{44}O$
(24*S*,25*S*)-24,26-Cyclo-5α-cholest-22*E*-en-3β-ol, *in* P-**6**0004
Trisnorisoespinenoxide, T-**8**0368

$C_{27}H_{44}O_2$
Astrogorgiadiol, A-**8**0181
Cholesta-7,22-diene-3,6-diol, C-**8**0124
3-Hydroxy-25,26,27-trisnor-24-cycloartanal, H-**6**0232
24-Methyl-27-nor-7,22-cholestadiene-3,6-diol, M-**8**0123

$C_{27}H_{44}O_3$
11-Hydroxy-3,6-cholestanedione, H-**8**0138
Lansilactone, L-**6**0013

$C_{27}H_{44}O_5$
Spirostane-1,2,3-triol, S-**6**0043

$C_{27}H_{44}O_6$
22,25-Epoxy-2,3,6,26-furostanetetrol, E-70015

$C_{27}H_{44}O_7$
▷ 2,3,14,20,22,25-Hexahydroxy-7-cholesten-6-one, H-**6**0063
3-*epi*-20-Hydroxyecdysone, *in* H-**6**0063
Pterosterone, P-**6**0192

$C_{27}H_{44}O_8$
2,3,5,14,20,22,25-Heptahydroxy-7-cholesten-6-one, H-**6**0023

$C_{27}H_{44}O_{15}$
Confertoside, C-70193

$C_{27}H_{46}O_2$
5-Cholestene-3,26-diol, C-**6**0147
20,29,30-Trisnor-3,19-lupanediol, T-**6**0404
Turbinaric acid, T-**8**0371

$C_{27}H_{46}O_3$
Cholest-5-ene-3,16,22-triol, C-70172

$C_{27}H_{47}FeN_6O_9$
Ferrioxamine D_1, *in* F-**6**0003

$C_{27}H_{47}N_9O_{10}S_2$
Trypanothione, T-70331

$C_{27}H_{48}$
1,3,5-Triheptylbenzene, T-70246

$C_{27}H_{48}O_3$
Hipposterol, H-70091

$C_{27}H_{48}O_7$
Cholestane-3,6,7,8,15,16,26-heptol, C-**8**0125

$C_{27}H_{54}$
13-Heptacosene, H-**6**0019

$C_{27}H_{56}$
2-Methylhexacosane, M-**8**0095

$C_{28}H_{12}N_2O_2$
Flavanthrone, F-**8**0010

$C_{28}H_{14}$
Benzo[*a*]coronene, B-**6**0021
[7]Circulene, C-**8**0134

$C_{28}H_{14}O_2$
Dibenzo[*fg,ij*]pentaphene-15,16-dione, D-**8**0073
Dibenzo[*a,j*]perylene-8,16-dione, D-70073

$C_{28}H_{14}S$
Dibenzo[2,3:10,11]perylo[1,12-*bcd*]thiophene, D-70074

$C_{28}H_{16}O$
Diphenanthro[2,1-*b*:1′,2′-*d*]furan, D-70470
Diphenanthro[9,10-*b*:9′,10′-*d*]furan, D-70471

$C_{28}H_{17}N$
9*H*-Tetrabenzo[*a,c,g,i*]carbazole, T-70011

$C_{28}H_{18}$
Benzonaphtho[2.2]paracyclophane, B-**8**0029
7,8-Dimethylbenzo[*no*]naphtho[2,1,8,7-*ghij*]pleiadene, D-**8**0419

$C_{28}H_{18}O_5$
2-Benzoylbenzoic acid; Anhydride, *in* B-**8**0054

$C_{28}H_{18}O_8$
2,2′,4,4′,7,7′,8,8′-Octahydroxy-1,1′-biphenanthrene, O-**8**0017

$C_{28}H_{19}N$
Di-9-phenanthrylamine, D-70472

$C_{28}H_{20}$
1-Benzylidene-2,3-diphenylindene, B-80059

$C_{28}H_{20}N_2$
Tetraphenylpyrimidine, T-80159

$C_{28}H_{20}N_2O_6$
4,4′-Diamino-3,3′-biphenyldicarboxylic acid; N,N′-Dibenzoyl, *in* D-80049

$C_{28}H_{20}O_4$
2,2′-Dihydroxystilbene; Dibenzoyl, *in* D-80378
3,3,6,6-Tetraphenyl-1,4-dioxane-2,5-dione, T-60165

$C_{28}H_{22}$
5,6-Dihydro-11,12-diphenyldibenzo[*a,e*]cyclooctene, D-70202

$C_{28}H_{22}N_2$
9,9′-Biacridylidene; *N,N′*-Di-Me, *in* B-70090

$C_{28}H_{22}N_2O$
Isoamarin; 1-Benzoyl, *in* D-80263

$C_{28}H_{22}O$
2,2,4,4-Tetraphenyl-3-butenal, T-70148

$C_{28}H_{22}O_6$
Gnetin A, G-70023

$C_{28}H_{22}O_7$
Scirpusin A, S-50031

$C_{28}H_{22}O_8$
Scirpusin B, *in* S-50031
4,4′,5,5′-Tetrahydroxy-1,1′-binaphthyl; Tetra-Ac, *in* T-70102
4,4′,5,5′-Tetrahydroxy-1,1′-binaphthyl; Tetra-Ac, *in* T-80086

$C_{28}H_{22}O_{10}$
Cephalochromin, C-60029

$C_{28}H_{24}$
[2.0.2.0]Metacyclophane, M-60031

$C_{28}H_{24}N_2$
Amarin; 1-Benzyl, *in* D-80263

$C_{28}H_{24}O_4$
Isomarchantin *C*, I-60107
Isoriccardin *C*, I-60135
Riccardin *D*, R-70005

$C_{28}H_{24}O_8$
Cassigarol *A*, C-70025

$C_{28}H_{24}O_{13}$
Isoorientin; 2″-(4-Hydroxybenzoyl), *in* I-80068

$C_{28}H_{26}O_{16}$
Taxillusin, *in* P-70027

$C_{28}H_{28}O_4$
2α,6α,11-Trihydroxy-7,9(11),13-abietatrien-12-one, *in* D-70282

$C_{28}H_{28}O_{11}$
Dukunolide *C*, *in* D-60522

$C_{28}H_{28}O_{15}$
ε-Rhodomycinone; 7-Glucoside, *in* R-70002

$C_{28}H_{30}O_{10}$
Physalin *G*, P-60146

$C_{28}H_{30}O_{15}$
7-Hydroxy-3′,4′-methylenedioxyisoflavone; *O*-Laminarabioside, *in* H-80203

$C_{28}H_{30}O_{16}$
3′,4′,5,7-Tetrahydroxyisoflavone; 3′,4′-Methylene ether, 7-*O*-glucosylglucoside, *in* T-80109

$C_{28}H_{31}N_2O_3^{\oplus}$
Rhodamine B, R-80006

$C_{28}H_{32}N_8$
3,11,19,27,33,34,35,36-Octaazapentacyclo[27.3.1.1^{5,9}.1^{13,17}.1^{21,25}]hexatriaconta-1(33),5(34),6,8,13,(35),14,16,21(36),22,24,29,31-dodecaene, O-80002

$C_{28}H_{32}O_4$
Agerasanin, A-70072

$C_{28}H_{32}O_6$
Garcinone *E*, G-60003

$C_{28}H_{32}O_9$
Physalin M, *in* P-60147

$C_{28}H_{32}O_{10}$
Glaucin *B*, G-70017
Physalin L, P-60147

$C_{28}H_{32}O_{11}$
Physalin *D*, P-60145

$C_{28}H_{32}O_{13}$
7-Hydroxy-4′-methoxyisoflavone; 7-*O*-Rhammosylglucoside, *in* H-80190
7-Hydroxy-4′-methoxyisoflavone; 7-*O*-Rutinoside, *in* H-80190
Picropodophyllin glucoside, *in* P-80165
Podophyllotoxin; 1-β-D-Glucoside, *in* P-80165

$C_{28}H_{32}O_{14}$
5,7-Dihydroxy-4′-methoxyisoflavone; 7-*O*-Rhamnosylglucoside, *in* D-80337
5,7-Dihydroxy-4′-methoxyisoflavone; 7-*O*-Rutinoside, *in* D-80337
Fortunellin, *in* D-70319
7-Hydroxy-4′-methoxyisoflavone; 7-*O*-Laminarabioside, *in* H-80190
Linarin, *in* D-70319

$C_{28}H_{32}O_{15}$
Abrusin 2″-*O*-apioside, *in* A-80004
5,7-Dihydroxy-4′-methoxyisoflavone; 7-*O*-Gentiobioside, *in* D-80337
Flavocommelin, *in* S-80052
Kakkalide, *in* T-80112
3′,4′,7-Trihydroxyisoflavone; 4′-Me ether, 7-*O*-rhamnosylglucoside, *in* T-80308

$C_{28}H_{32}O_{16}$
3′-*O*-Methyllutonarin, *in* I-80068
4′,5,7-Trihydroxy-6-methoxyisoflavone; 7-*O*-Gentobioside, *in* T-80319

$C_{28}H_{34}O_4$
Balaenonol, *in* B-70006

$C_{28}H_{34}O_5$
Azadiradione, A-70281
Encecanescin, E-70008
9′-Epiencecanescin, *in* E-70008

$C_{28}H_{34}O_7$
▷ Gedunin, G-70006
Withaphysalin *E*, W-60009

$C_{28}H_{34}O_8$
Drummondin F, D-80533
6α-Hydroxygedunin, *in* G-70006

$C_{28}H_{34}O_9$
4-O-Demethylmicrophyllinic acid, *in* M-70147

$C_{28}H_{34}O_{10}$
Flavaspidic acids; Di-Ac, *in* F-80011
Gomisin *D*, G-70025

$C_{28}H_{34}O_{11}$
6β,9α,14-Triacetoxy-1α-benzoyloxy-4β-hydroxy-8-oxodihydro-β-agarofuran, *in* E-60018

$C_{28}H_{34}O_{13}$
1-Acetoxypinoresinol 4′-*O*-β-D-glucopyranoside, *in* H-60221

$C_{28}H_{35}ClO_{11}$
Junceellin, J-80014

$C_{28}H_{35}ClO_{12}$
Praelolide, *in* J-80014

$C_{28}H_{36}N_2O_6$
Sarcodictyin *A*, S-70010

$C_{28}H_{36}O_3$
Balaenol, B-70006

$C_{28}H_{36}O_4$
Daturilin, D-60003
Isobalaendiol, *in* B-70006
Isowithametelin, *in* W-80004
Withametelin, W-60005
Withametelin, W-80004

$C_{28}H_{36}O_5$
Celastanhydride, C-70028
2,3-Dihydroxy-6-oxo-23,24-dinor-1,3,5(10),7-friedelatetraen-29-oic acid, D-80351

$C_{28}H_{36}O_6$
Jaborol, J-60001

$C_{28}H_{36}O_7$
▷ 1,2-Dihydrogedunin, *in* G-70006
Trechonolide *A*, T-60227

$C_{28}H_{36}O_8$
3-[3-Acetyl-5-(3,7-dimethyl-2,6-octadienyl)-2,4,6-trihydroxybenzyl]-4-hydroxy-5-methyl-6-propyl-2*H*-pyran-2-one, *in* A-80021

$C_{28}H_{36}O_{10}$
Rzedowskin *A*, R-70016
6β,9β,15-Triacetoxy-1α-benzoyloxy-4β-hydroxydihydro-β-agarofuran, *in* P-80032

$C_{28}H_{36}O_{11}$
6β,9α,14-Triacetoxy-1α-benzoyloxy-4β,8β-dihydroxydihydro-β-agarofuran, *in* E-60018

$C_{28}H_{36}O_{16}$
Shanzhiside; 6-(4-Hydroxy-3,5-dimethoxybenzoyl), 8-Ac, Me ester, *in* S-60028

$C_{28}H_{37}ClO_{10}$
Ptilosarcone, *in* P-60195

$C_{28}H_{38}N_2O_{10}$
1,2-Bis(4′-benzo-15-crown-5)diazene, B-60141

$C_{28}H_{38}N_4O_6$
Chlamydocin, C-60038

$C_{28}H_{38}O_4$
Cystoketal, C-60234
Isocystoketal, *in* C-60234

$C_{28}H_{38}O_5$
Daturilinol, D-70007
Mzikonone, M-70166
Withacoagin, W-80003

$C_{28}H_{38}O_7$
Physanolide, P-60149
Withanolide *Y*, W-60007
Withaphysanolide, W-60010

$C_{28}H_{38}O_9$
3,14-Diacetoxy-2-butanoyloxy-5,8(17)-briaradien-18,7-olide, *in* E-70012
4,14-Diacetoxy-2-butanoyloxy-5,8(17)-briaradien-18,7-olide, *in* E-70012
ent-2α,3α,6β,11β-Tetraacetoxy-16-kauren-15-one, *in* P-80047
Xenicin, X-70004

$C_{28}H_{38}O_{10}$
Ingol; Tetra-Ac, *in* I-70018
Lapidolinin, *in* L-60015
ent-2α,3α,6β,11β-Tetraacetoxy-7β-hydroxy-16-kauren-15-one, *in* P-80047

$C_{28}H_{38}O_{11}$
Junceellolide D, J-80017

$C_{28}H_{39}ClO_8$
Physalolactone, P-60148

$C_{28}H_{39}ClO_{10}$
Ptilosarcol, P-60195

$C_{28}H_{39}NO$
Emindole SA, *in* E-60005
Emindole DA, E-60005

$C_{28}H_{40}$
24,25-Dinor-1,3,5(10),12-oleanatetraene, D-80497
24,25-Dinor-1,3,5(10),12-ursatetraene, D-80498
1,3,5,7-Tetra-*tert*-butyl-*s*-indacene, T-60025
Tetrakis(cyclohexylidene)cyclobutane, T-80130

$C_{28}H_{40}O_2$
Plastoquinone 4, P-80159

$C_{28}H_{40}O_3$
11'-Methoxysargaquinone, *in* S-60010
9'-Methoxysargaquinone, *in* S-60010

$C_{28}H_{40}O_4$
8',9'-Dihydroxy-5-methylsargaquinone, *in*
S-60010

$C_{28}H_{40}O_5$
Amentol 1'-methyl ether, *in* A-60088
Balearone, B-60005
Bifurcarenone; 2'-Me ether, *in* B-80093
Epineobalearone, *in* N-80018
12-Hydroxy-24-methyl-24-oxo-16-scalarene-
22,25-dial; O^{12}-Ac, *in* H-60191
Isobalearone, *in* B-60005
Neobalearone, N-80018
Strictaepoxide, S-60056
Strictaketal, S-70076

$C_{28}H_{40}O_6$
Ixocarpanolide, I-60143

$C_{28}H_{40}O_7$
$6\alpha,7\alpha$-Epoxy-$5\alpha,14\alpha,20R$-trihydroxy-1-oxo-
$22R,24S,25R$-with-2-enolide, *in* I-60143
Jaborosalactol N, J-80002
Jaborosalactone *M*, *in* J-80001
Withaphysacarpin, W-60008

$C_{28}H_{40}O_9$
Astrogorgin, A-80182

$C_{28}H_{40}O_{13}$
Bonafousioside, B-80171

$C_{28}H_{41}ClO_9$
Solenolide A, S-70052

$C_{28}H_{42}$
24,25-Dinor-1,3,5(10)-lupatriene, D-80496

$C_{28}H_{42}O_4$
Maesol, M-80002

$C_{28}H_{42}O_5$
16β-Acetoxy-24-methyl-12,24-dioxo-25-
scalaranal, *in* H-80200
Tubocapsigenin *A*, T-70333

$C_{28}H_{42}O_6$
11,15(17)-Trinervitadiene-3,9,13-triol; 9-Ac,
3,13-dipropanoyl, *in* T-70297

$C_{28}H_{42}O_7$
Jaborosalactol *M*, J-80001

$C_{28}H_{44}O$
24-Methyl-7,22-cholestadien-3-one, *in*
M-80059
24-Methyl-5,22,25-cholestatrien-3-ol, M-60057

$C_{28}H_{44}O_{11}$
Grayanoside *A*, *in* G-80038

$C_{28}H_{46}O$
24-Methyl-7,22-cholestadien-3-ol, M-80059

$C_{28}H_{46}O_2$
24-Methyl-7,22-cholestadiene-3,6-diol,
M-80058
4-Methylcholesta-8,14-dien-23-ol, M-80060
Phytylplastoquinone, P-80144

$C_{28}H_{46}O_3$
24-Methyl-7,22-cholestadiene-3,5,6-triol,
M-70057

$C_{28}H_{46}O_4$
24-Methylene-5-cholestene-3,4,7,20-tetrol,
M-60070

$C_{28}H_{46}O_5$
Dolichosterone, D-70544
22S,25-Epoxy-24-methyl-3,11,20-
furostanetriol, E-80039

$C_{28}H_{46}O_6$
2-Desacetyl-22-epihippurin 1, *in* E-80039
2α-Hydroxyhippuristanol, *in* E-80039

$C_{28}H_{46}O_7$
Polypodoaurein, *in* H-60063

$C_{28}H_{47}N_9O_9$
Neokyotorphin, N-70025

$C_{28}H_{48}O_3$
24-Methyl-7-cholestene-3,5,6-triol, M-80061
β-Tocopherol, T-80200
γ-Tocopheryl quinone, T-80201

$C_{28}H_{48}O_4$
6-Deoxodolichosterone, *in* D-70544
Numersterol A, N-80084
3α,22R,23R-Trihydroxy-24S-methyl-5α-
cholestan-6-one, *in* T-70116

$C_{28}H_{48}O_5$
2,3,22,23-Tetrahydroxy-24-methyl-6-
cholestanone, T-70116

$C_{28}H_{48}O_6$
Brassinolide, B-70178

$C_{28}H_{48}O_{12}$
4,7,10,15,18,21,24,27,30,33,36,39-
Dodecaoxatricyclo[11.9.9.9^{2,12}]tetraconta-
1,12-diene, D-70535

$C_{28}H_{50}O_4$
6-Deoxocastasterone, *in* T-70116

$C_{28}H_{52}O_3$
Ficulinic acid B, F-60012

$C_{28}H_{52}O_{11}$
Muricatin B, *in* H-80172

$C_{28}H_{56}O$
Octacosanal, O-80005

$C_{28}H_{58}$
3-Methylheptacosane, M-80092

$C_{29}H_{18}O$
Di-9-anthracenylmethanone, D-60050

$C_{29}H_{20}$
7,7-Diphenylbenzo[c]fluorene, D-70473
Tetrabenzotetracyclo[5.5.1.0^{4,13}.0^{10,13}]tridecane,
T-60019

$C_{29}H_{22}$
1,2,3,4-Tetraphenylcyclopentadiene, T-70149

$C_{29}H_{22}O$
1,1,5,5-Tetraphenyl-1,4-pentadien-3-one,
T-80158

$C_{29}H_{22}O_{14}$
3,5-Digalloylepicatechin, *in* P-70026

$C_{29}H_{24}O_{10}$
Chaetochromin C, *in* C-60029

$C_{29}H_{24}O_{15}$
4',5,6,7-Tetrahydroxyisoflavone; 6,7-Di-Me
ether, 4'-O-rhamnosylglucoside, *in* T-80112

$C_{29}H_{26}O_4$
Riccardin E, *in* R-70005

$C_{29}H_{28}O_4$
Tetrabenzyloxymethane, *in* M-80037

$C_{29}H_{28}O_{10}$
Di-2-(7-acetyl-1,4-dihydro-3,6,8-trihydroxy-
4,4-dimethyl-1-oxonaphthyl)methane,
D-80043

$C_{29}H_{28}O_{12}$
Viscutin 2, *in* T-60103

$C_{29}H_{29}N_3O_8$
K 13, K-70001

$C_{29}H_{30}O_7$
8-Prenyllepidissipyrone, *in* L-80023

$C_{29}H_{32}O_{12}$
Amorphigenin; O-β-D-Glucopyranoside, *in*
A-80152

$C_{29}H_{32}O_{16}$
Irisolone-4'-bioside, *in* D-80341
3',4',6,7-Tetrahydroxyisoflavone; 6-Me, 3',4'-
methylenedioxy ether, 7-O-laminarabioside,
in T-80111

$C_{29}H_{34}O_7$
Triptofordin B, T-60393

$C_{29}H_{34}O_8$
Nimbanal, N-80026

$C_{29}H_{34}O_{10}$
Crepiside A, *in* Z-60001
Crepiside B, *in* Z-60001

$C_{29}H_{34}O_{11}$
Ixerin C, *in* I-80102
Physalin *I*, *in* P-60145
Trijugin A, T-60351

$C_{29}H_{34}O_{12}$
Acetoxyeumaitenol, *in* E-60072
Dalpanol; O-Glucoside, *in* D-80003

$C_{29}H_{34}O_{13}$
Amorphigenol glucoside, *in* A-80152

$C_{29}H_{34}O_{14}$
Embinin, *in* S-80052
Pueroside *A*, *in* P-70137

$C_{29}H_{34}O_{15}$
4',7,8-Trihydroxyisoflavone; 4',8-Di-Me ether,
7-O-laminarabioside, *in* T-80311

$C_{29}H_{34}O_{16}$
4',5,6,7-Tetrahydroxyisoflavone; 6,7-Di-Me
ether, 4'-O-[4-O-β-D-glucopyranosyl (1→4)-
β-D-glucopyranoside], *in* T-80112

$C_{29}H_{36}O_6$
Demethylzeylasteral, *in* D-70027

$C_{29}H_{36}O_7$
Desmethylzeylasterone, D-70027

$C_{29}H_{36}O_8$
Fissinolide, *in* S-80054

$C_{29}H_{36}O_9$
3-Acetylswietenolide, *in* S-80054
Microphyllinic acid, M-70147

$C_{29}H_{36}O_{10}$
Baccharinoid B27, *in* B-60002

$C_{29}H_{36}O_{13}$
1-Hydroxypinoresinol; 4''-Me ether, 1-Ac, 4'-
O-β-D-glucopyranoside, *in* H-60221

$C_{29}H_{36}O_{16}$
Olivin 4-diglucoside, *in* O-80043

$C_{29}H_{37}NO_3$
Smenospongidine, *in* S-80036

$C_{29}H_{38}N_2O_6$
Sarcodictyin B, *in* S-70010

$C_{29}H_{38}O_5$
23-Nor-6-oxopristimerol, *in* D-80351

$C_{29}H_{38}O_7$
Isonimolide, I-60110
Trechonolide *B*, *in* T-60227

$C_{29}H_{38}O_9$
▷ Roridin D, R-70008
Scutellone A, S-60017
Scutellone C, S-70019
Scuterivulactone C_2, *in* S-60017

$C_{29}H_{38}O_{10}$
Baccharinoid B9, B-60001
Baccharinoid B13, B-60002
Baccharinoid B16, B-60003
Baccharinoid B10, *in* B-60001
Baccharinoid B12, *in* R-70008
Baccharinoid B14, *in* B-60002
Baccharinoid B17, *in* R-70008
Baccharinoid B21, *in* R-70008

$C_{29}H_{38}O_{11}$
Baccharinol, B-70003
Isobaccharinol, *in* B-70003

$C_{29}H_{40}O_5$
Secowithametelin, S-80019

$C_{29}H_{40}O_8$
Physalactone, P-60144

$C_{29}H_{40}O_9$
Ingol; 3,7-Di-Ac, 12-tigloyl, *in* I-70018

$C_{29}H_{40}O_{10}$
Baccharinoid $B1$, B-70001
Baccharinoid $B7$, B-70002
Baccharinoid **B2**, *in* B-70001

Baccharinoid B3, *in* B-70002
Baccharinoid B20, *in* B-60001
Baccharinoid B23, *in* B-60003
Baccharinoid B24, *in* B-60003

$C_{29}H_{42}O_3$
Bryophollone, B-80271

$C_{29}H_{42}O_4$
Przewanoic acid *B*, P-70126
15Z-Cinnamoyloxy-7-labden-17-oic acid, *in* H-80183

$C_{29}H_{42}O_5$
Zosterondiol B, Z-80005

$C_{29}H_{42}O_7$
3α,20-Diacetoxy-2α-angeloyloxy-9α,13R-epoxy-7,14-labdadiene, *in* E-80035

$C_{29}H_{42}O_{11}$
Ghalakinoside, G-80016

$C_{29}H_{44}O_2$
3β-Hydroxy-28-noroleana-12,17-dien-16-one, *in* M-70011

$C_{29}H_{44}O_3$
3-Hydroxy-30-nor-12,18-oleanadien-29-oic acid, H-80218
3-Hydroxy-30-nor-12,20(29)-oleanadien-28-oic acid, H-60199
18β-Hydroxy-28-nor-3,16-oleanenedione, *in* D-60363

$C_{29}H_{44}O_4$
3,23-Dihydroxy-30-nor-12,20(29)-oleanadien-28-oic acid, D-70324
3,24-Dihydroxy-30-nor-12,20(29)-oleanadien-28-oic acid, D-80344
Kanerin, K-80001

$C_{29}H_{44}O_5$
Floridic acid, F-80014
Zosterdiol A, Z-80003
Zosterdiol B, Z-80004
Zosterondiol A, *in* Z-80005
Zosteronol, *in* Z-80004

$C_{29}H_{44}O_6$
Brachycarpone, B-60195
Heteronemin, H-60030
11,15(17)-Trinervitadiene-3,9,13-triol; 3,9,13-Tripropanoyl, *in* T-70297

$C_{29}H_{44}O_8$
3α,20-Diacetoxy-2α-angeloyloxy-8,14-labdadiene-7α,13R-diol, *in* L-80005
3α,20-Diacetoxy-2α-angeloyloxy-8(17),14-labdadiene-7α,13R-diol, *in* L-80006

$C_{29}H_{44}O_9$
Coroglaucigenin; 3-O-Rhamnoside, *in* C-60168
▷ Frugoside, *in* C-60168
Methyl 15α,17β-diacetoxy-6α-butanoyloxy-15,16-dideoxy-15,17-oxido-16-spongianate, *in* T-60038

$C_{29}H_{45}N_3O_3$
Blastmycetin *D*, B-60191

$C_{29}H_{46}O_2$
Maragenin I, M-70011

$C_{29}H_{46}O_3$
3,18-Dihydroxy-28-nor-12-oleanen-16-one, D-60363

$C_{29}H_{46}O_4$
Baccatin, B-80001

$C_{29}H_{46}O_7$
2α,7α-Diacetoxy-6β-isovaleroyloxy-8,13E-labdadien-15-ol, *in* L-60002

$C_{29}H_{46}O_8$
Viticosterone *E*, *in* H-60063

$C_{29}H_{47}N_5O_5$
Stacopin P1, S-60051

$C_{29}H_{47}N_5O_6$
Stacopin P2, *in* S-60051

$C_{29}H_{48}O$
Ficisterol, F-60010
Hebesterol, H-60011

$C_{29}H_{48}O_2$
24,25-Epoxy-29-nor-3-cycloartanol, E-70026
29-Nor-23-cycloartene-3,25-diol, N-70073
28-Nor-16-oxo-17-oleanen-3β-ol, *in* M-70011

$C_{29}H_{48}O_3$
24-Ethyl-7,22-cholestadiene-3,5,6-triol, E-80060
24-Ethylidene-7-cholestene-3,5,6-triol, E-80066
6β-Methoxy-7,22-ergostadiene-3β,5α-diol, *in* M-70057
29-Nor-8,24-lanostadiene-1,2,3-triol, N-80081

$C_{29}H_{50}O$
Petrostanol, P-60066

$C_{29}H_{50}O_2$
24-Ethyl-7-cholestene-3,6-diol, E-80061
4-Stigmastene-3,6-diol, S-60054

$C_{29}H_{50}O_3$
24,24-Dimethoxy-25,26,27-trisnor-3-cycloartanol, *in* H-60232
Numersterol B, N-80085

$C_{29}H_{50}O_4$
24(28)-Stigmastene-2,3,22,23-tetrol, S-70075

$C_{29}H_{52}N_{10}O_6$
Argiopine, A-60295

$C_{29}H_{52}O$
24,24-Dimethyl-3-cholestanol, D-70389

$C_{29}H_{52}O_2$
3,6-Stigmastanediol, S-70074

$C_{29}H_{58}O$
15-Nonacosanone, N-80066

$C_{29}H_{60}$
2-Methyloctacosane, M-80124

$C_{29}H_{60}O$
15-Nonacosanol, N-80064
1-Nonacosanol, N-80065

$C_{30}H_{14}$
[5]Phenylene, P-60090

$C_{30}H_{16}$
Pyranthrene, P-70151
Tribenzo[b,n,pqr]perylene, T-60244

$C_{30}H_{16}O_2$
7,16-Heptacenedione, H-60018

$C_{30}H_{18}$
Di-9-anthrylacetylene, D-60051

$C_{30}H_{18}O_8$
Siameanin, S-80031

$C_{30}H_{18}O_9$
Cassianin, *in* S-80031
Roseoskyrin, R-80011

$C_{30}H_{18}O_{10}$
Hinokiflavone, H-60082
Rhodoislandin A, R-80007
Rhodoislandin B, R-80008

$C_{30}H_{18}O_{12}$
Bryoflavone, B-70283
Dicranolomin, D-70144
Heterobryoflavone, H-70027

$C_{30}H_{20}$
1,2,3-Triphenylcyclopent[a]indene, T-70306

$C_{30}H_{20}O_4$
Tetrabenzoylethylene, T-60021

$C_{30}H_{20}O_9$
1-[3-(2,4-Dihydroxybenzoyl)-4,6-dihydroxy-2-(4-hydroxyphenyl)-7-benzofuranyl]-3-(4-hydroxyphenyl)-2-propen-1-one, D-70286

$C_{30}H_{20}O_{11}$
2,3-Dihydro-3′,4′,4″,5,5″,7,7″-heptahydroxy-3,8′-biflavone, D-80209

$C_{30}H_{20}O_{12}$
2,3-Dihydrodicranolomin, *in* D-70144

$C_{30}H_{20}S_2$
1,3,4,6-Tetraphenylthieno[3,4-c]thiophene-5-S^{IV}, T-70152

$C_{30}H_{20}S_8$
Tetramercaptotetrathiafulvalene; Tetrakis (S-Ph), *in* T-70130

$C_{30}H_{22}N_4O_4$
6,6′-(1,2-Dimethyl-1,2-ethanediyl)bis[1-phenazinecarboxylic acid], D-70411

$C_{30}H_{22}O_4$
1,1,2,2-Tetrabenzoylethane, T-60020

$C_{30}H_{22}O_7$
Palmidin B, P-80005
Palmidin C, *in* P-80005

$C_{30}H_{22}O_8$
Aloeemodin dianthrone, A-80049
Lophirone *A*, L-60035
Lophirone B, L-80037
Lophirone C, L-80038
Palmidin A, *in* P-80005

$C_{30}H_{22}O_9$
Daphnodorin *D*, D-70005
1-[3-(2,4-Dihydroxybenzoyl)-4,6-dihydroxy-2-(4-hydroxyphenyl)-7-benzofuranyl]-3-(4-hydroxyphenyl)-2-propen-1-one; 2,3-Dihydro (trans-), *in* D-70286
Maackiasin, M-60001

$C_{30}H_{22}O_{10}$
GB1a, *in* G-70005
Isochamaejasmin, I-70064

$C_{30}H_{22}O_{11}$
GB 1, G-70005
GB2a, *in* G-70005

$C_{30}H_{22}O_{12}$
3″,3‴,4′,4‴,5,5″,7,7″-Octahydroxy-3,8″-biflavanone, *in* G-70005

$C_{30}H_{22}O_{13}$
Chiratanin, C-60037

$C_{30}H_{24}N_6O_3$
Furlone yellow, F-70053

$C_{30}H_{24}O_4$
Tecomaquinone I, T-60010

$C_{30}H_{24}O_9$
Shiraiachrome C, S-80030

$C_{30}H_{24}O_{10}$
Chaetochromin D, *in* C-60029
Mahuannin A, M-80003
Mahuannin B, *in* M-80003

$C_{30}H_{24}O_{13}$
[2′,2′]-Catechin-taxifolin, C-60023

$C_{30}H_{26}O_5$
Tecomaquinone III, T-70006

$C_{30}H_{26}O_6$
Flavanthrin, F-70011

$C_{30}H_{26}O_8$
Isombamichalcone, *in* M-70014
Mbamichalcone, M-70014

$C_{30}H_{26}O_{10}$
Chaetochromin, *in* C-60029
Chaetochromin B, *in* C-60029
Shiraiachrome A, S-80028
Shiraiachrome B, S-80029

$C_{30}H_{28}N_4O_2$
Tetrahydro-1,2,4,5-tetrazine-3,6-dione; 1,2,4,5-Tetrabenzyl, *in* T-70097

$C_{30}H_{28}O_6$
Angoluvarin, A-70208
Isothamnosin A, I-70104
Isothamnosin B, *in* I-70104

$C_{30}H_{28}S_2$
2,18-Dithia[3.1.3.1]paracyclophane, D-80518

$C_{30}H_{30}O_7$
Euchretin A, E-80080

$C_{30}H_{30}O_8$
▷ Gossypol, G-60032

$C_{30}H_{30}O_{10}$
Phleichrome, P-70098

$C_{30}H_{30}O_{14}$
Safflomin C, S-80001

$C_{30}H_{32}O_4$
4,15,26-Tricontatrien-1,12,18,29-tetrayne-
3,14,17,28-tetrone, in P-70044

$C_{30}H_{32}O_6$
Fruticolide, F-70045

$C_{30}H_{32}O_{14}$
Senburiside II, S-70033

$C_{30}H_{34}O_5$
Euchrenone b_1, E-70058

$C_{30}H_{34}O_6$
Eucrenone b_2, in E-70058

$C_{30}H_{34}O_{10}$
Cerberalignan D, C-80033
Cerberalignan E, C-80034

$C_{30}H_{34}O_{11}$
Ainsliaside A, in Z-60001

$C_{30}H_{34}O_{17}$
Platycarpanetin 7-O-laminaribioside, in
P-80045

$C_{30}H_{36}O_5$
Amorilin, A-80151

$C_{30}H_{36}O_8$
Artelein, A-60303

$C_{30}H_{36}O_9$
6α-Acetoxygedunin, in G-70006
11β-Acetoxygedunin, in G-70006
Isonimolicinolide, I-60109

$C_{30}H_{36}O_{11}$
Cerberalignan F, C-80035
Cerberalignan G, in C-80035
Isonimbinolide, I-70072

$C_{30}H_{36}O_{12}$
Tectoroside, T-80003

$C_{30}H_{36}O_{15}$
Pueroside B, in P-70137

$C_{30}H_{36}O_{16}$
3',4',6,7-Tetrahydroxyisoflavone; 3',4',6-Tri-
Me ether, O-laminaribioside, in T-80111

$C_{30}H_{37}N_5O_5$
Avellanin B, A-60324

$C_{30}H_{38}O_4$
Cochloxanthin, C-60156
Helilupulone, H-80008

$C_{30}H_{38}O_5$
23-Oxoisopristimerin III, in I-70083

$C_{30}H_{38}O_6$
Biperezone, B-80105
Caleamyrcenolide, C-80010
Zeylasteral, in D-70027

$C_{30}H_{38}O_7$
Furanoganoderic acid, F-80088
3,7,11,15,23-Pentaoxo-8,20E-lanostadien-26-
oic acid, in D-80390

$C_{30}H_{38}O_9$
2'-O-Methylmicrophyllinic acid, in M-70147
Isolimbolide, I-60105

$C_{30}H_{38}O_{10}$
Swietemahonin A, S-80053

$C_{30}H_{38}O_{12}$
6β,8β,9α,14-Tetraacetoxy-1α-benzoyloxy-4β-
hydroxydihydro-β-agarofuran, in E-60018

$C_{30}H_{38}O_{13}$
Bruceanol C, B-70280

$C_{30}H_{40}O_4$
4,5-Dihydro-6-hydroxy-3-oxo-8'-apo-ε-
caroten-8'-oic acid, in C-60156
Isopristimerin III, I-70083
Petrosynol, P-70044

$C_{30}H_{40}O_5$
Helisplendidilactone, H-80010

$C_{30}H_{40}O_7$
15α-Hydroxy-3,7,11,23-tetraoxo-8,20E-
lanostadien-26-oic acid, in D-80390
3β-Hydroxy-7,11,15,23-tetraoxo-8,20E-
lanostadien-26-oic acid, in D-80390
7β-Hydroxy-3,11,15,23-tetraoxo-8,20(22)-
lanostadien-26-oic acid, in T-70254

$C_{30}H_{40}O_8$
7,12-Dihydroxy-3,11,15,23-tetraoxo-8,20(22)-
lanostadien-26-oic acid, D-60378
Ganoderic acid O, in D-60379
Trichilinin, T-60251
Withaminimin, W-60006

$C_{30}H_{40}O_{10}$
1,2-Bis(4'-benzo-15-crown-5)ethene, B-60142
Xenicin; 9-Deacetyl, 9-acetoacetyl, in X-70004

$C_{30}H_{41}ClO_{10}$
Brianthein V, B-80172

$C_{30}H_{42}O_3$
Heleniumlactone 1, H-60013
Heleniumlactone 3, H-60014
Heleniumlactone 2, in H-60013

$C_{30}H_{42}O_4$
Firmanolide, F-60013
Isomaresiic acid C, I-60108
Mariesiic acid C, M-60011

$C_{30}H_{42}O_5$
Abiesonic acid, A-60005

$C_{30}H_{42}O_7$
Bufalin; 3-(Hydrogen adipoyl), in B-70289
▷ Cucurbitacin I, in C-70204
3,15-Dihydroxy-7,11,23-trioxo-8,20-
lanostadien-26-oic acid, D-80390
7β,15α-Dihydroxy-3,11,23-trioxo-8,20(22)E-
lanostadien-26-oic acid, in T-70254
3β,7β-Dihydroxy-11,15,23-trioxo-8,20(22)E-
lanostadien-26-oic acid, in T-70254

$C_{30}H_{42}O_8$
7,20-Dihydroxy-3,11,15,23-tetraoxo-8-
lanosten-26-oic acid, D-60379
Proscillaridin A, in B-70289

$C_{30}H_{42}O_9$
Ingol; 7-Angeloyl, O^8-Me, 3,12-Di-Ac, in
I-70018
Ingol; 7-Tigloyl, O^8-Me, 3,12-di-Ac, in
I-70018
12,15,20-Trihydroxy-3,7,11,23-tetraoxo-8-
lanosten-26-oic acid, T-80340

$C_{30}H_{42}O_{10}$
Kopeoside, in K-80012

$C_{30}H_{42}O_{12}$
Baccatin I; 5-Deacetyl, in B-80002
1-Dehydroxy-4-deacetylbaccatin IV, in
B-80003

$C_{30}H_{43}NO_6$
Bisparthenolidine, B-80157

$C_{30}H_{44}O_4$
3,23-Dioxo-7,24-lanostadien-26-oic acid,
D-60472
3,23-Dioxo-7,25(27)-lanostadien-26-oic acid,
in D-60472
3,7-Dioxo-12-oleanen-28-oic acid, D-70469
23-Oxomariesiic acid A, in M-70012
23-Oxomariesiic acid B, in M-60010

$C_{30}H_{44}O_5$
Abrusogenin, A-80005
Chiisanogenin, C-70038
3,15-Dihydroxy-23-oxo-7,9(11),24-
lanostatrien-26-oic acid, D-70330
2,19-Dihydroxy-3-oxo-1,12-ursadien-28-oic
acid, D-80352

$C_{30}H_{44}O_6$
11-Deoxocucurbitacin I, in C-70204
9,11-Dihydro-22,25-oxido-11-
oxoholothurinogenin, D-70237

$C_{30}H_{44}O_7$
▷ Cucurbitacin D, C-70203
Cucurbitacin L, in C-70204
3,7,15-Trihydroxy-11,23-dioxo-8,20(22)-
lanostadien-26-oic acid, T-70254

$C_{30}H_{44}O_{10}$
Perusitin, in C-80017

$C_{30}H_{46}O$
9(11),12-Oleanadien-3-one, O-60032

$C_{30}H_{46}O_2$
21,23-Epoxy-7,24-tirucalladien-3-one, in
E-60032
21-Hopene-3,20-dione, H-80097

$C_{30}H_{46}O_3$
24S,25S-Epoxy-26-hydroxy-7,9(11)-
lanostadien-3-one, in E-70025
Flindissol lactone, in F-70013
Flindissone, in F-70013
28-Hydroxy-20(29)-lupene-3,7-dione, H-80185
3-Hydroxy-12-oleanen-29,22-olide, H-70197
7-Hydroxy-3-oxo-8,24-lanostadien-26-al,
H-70200
22-Hydroxy-7,24-tirucalladiene-3,23-dione,
H-70227
3-Oxo-24-cycloarten-21-oic acid, O-80072
3-Oxo-8,24-lanostadien-21-oic acid, O-80087
3-Oxo-8,24-lanostadien-26-oic acid, O-70096
3-Oxo-8-lanosten-26,22-olide, in H-60169
Swertialactone C, in S-60064
Swertialactone D, S-60064

$C_{30}H_{46}O_4$
Abiesolidic acid, A-70003
Colubrinic acid, C-70190
3,15-Dihydroxy-7,9(11),24-lanostatrien-26-oic
acid, D-70314
2,3-Dihydroxy-12,20(30)-ursadien-28-oic acid,
D-60381
Dumortierigenin, D-80536
12-Hydroxy-3-oxo-28,13-oleananolide,
H-70201
3-Hydroxy-23-oxo-12-oleanen-28-oic acid,
H-80230
19α-Hydroxy-3-oxo-12-oleanen-28-oic acid, in
D-70327
22-Hydroxy-3-oxo-12-ursen-30-oic acid,
H-70202
19α-Hydroxy-3-oxo-12-ursen-28-oic acid, in
D-70350
Manwuweizic acid, M-70010
Mariesiic acid A, M-70012
Mariesiic acid B, M-60010
Przewanoic acid A, P-70125
3,4-Seco-4(28),20,24-dammaratrien-3,26-dioic
acid, S-70021
15,26,27-Trihydroxy-7,9(11),24-lanostatrien-3-
one, T-60330

$C_{30}H_{46}O_5$
Cadambagenic acid, C-70002
19,29-Dihydroxy-3-oxo-12-oleanen-28-oic
acid, D-70331
22,23-Dihydroxy-3-oxo-12-ursen-30-oic acid,
D-70333
2,3,24-Trihydroxy-13,27-cyclo-11-oleanen-28-
oic acid, T-70251
2,3,24-Trihydroxy-12,27-cyclo-14-taraxeren-
28-oic acid, T-70252
3,15,22-Trihydroxy-7,9(11),24-lanostatrien-26-
oic acid, T-80318
2,3,24-Trihydroxy-11,13(18)-oleonadien-28-oic
acid, T-60336
2α,3α,24-Trihydroxy-12,20(30)-ursadien-28-oic
acid, in D-60381

$C_{30}H_{46}O_6$
2,3-Dihydroxy-12-oleanene-23,28-dioic acid,
D-80346
3,19-Dihydroxy-12-ursene-23,28-dioic acid,
D-80393

3,19-Dihydroxy-12-ursene-24,28-dioic acid,
   D-60382
$3\beta,6\beta,19\alpha$-Trihydroxy-23-oxo-12-ursen-28-oic
   acid, in T-80129

$C_{30}H_{46}O_7$

Cucurbitacin R, in C-70203
▷ Saponaceolide A, S-60008
   Saponaceolide A, S-80009
$3\beta,6\beta,19\alpha$-Trihydroxy-12-ursene-23,28-dioic
   acid, in T-80129

$C_{30}H_{46}O_8$

Maquiroside A, in C-60168
Musangic acid, M-80199

$C_{30}H_{48}O$

7,9(11)-Fernadien-3-ol, F-80002
5(10)-Gluten-3-one, G-60028
14(26),17E,21-Malabaricatrien-3-one, in
   M-60008
30-Norcyclopterospermone, in N-60059
11,13(18)-Oleanadien-3-ol, O-70032
9(11),12-Oleanadien-3-ol, O-70031
9(11),12-Ursadien-3-ol, U-70007

$C_{30}H_{48}O_2$

Duryne, D-70550
11,12-Epoxy-14-taraxeren-3-ol, E-80044
21,23-Epoxy-7,24-tirucalladien-3-ol, E-60032
3-Hydroxy-20(29)-lupen-11-one, H-80186
7-Hydroxy-20(29)-lupen-3-one, H-80187
16-Hydroxy-18-oleanen-3-one, H-80223
3-Hydroxy-12-oleanen-1-one, H-80224
16β-Hydroxy-18-oleanen-3-one, in O-80033
11,13(18)-Oleanadiene-3,24-diol, O-80030
3-Oxo-29-friedelanal, in F-80082
3,4-Seco-4(23),14-taraxeradien-3-oic acid,
   S-60018

$C_{30}H_{48}O_3$

$3\beta,26$-Dihydroxy-5,24E-cucurbitadien-11-one,
   in C-60175
$3\beta,27$-Dihydroxy-5,24Z-cucurbitadien-11-one,
   in C-60175
24,25-Dihydroxy-7,9(11)-lanostadien-3-one,
   D-60343
3,23-Dihydroxy-20(30)-lupen-29-al, D-70316
1,11-Dihydroxy-20(29)-lupen-3-one, D-60345
23,30-Dihydroxy-20(29)-lupen-3-one, D-70318
1,11-Dihydroxy-18-oleanen-3-one, D-60364
24,25-Epoxy-7,9(11)-lanostadiene-3,26-diol,
   E-70025
Flindissol, F-70013
27-Hydroxy-3,21-friedelanedione, in D-70301
3-Hydroxy-8,24-lanostadien-21-oic acid,
   H-70152
3-Hydroxy-8-lanosten-26,22-olide, H-60169
3-Hydroxy-28,13-oleananolide, H-60200
3-Hydroxy-12-oleanen-30-oic acid, H-80222
3-Hydroxytaraxastan-28,20-olide, H-70221
3-Hydroxy-14-taraxeren-28-oic acid, H-80256
3-Hydroxy-28,20-ursanolide, H-60233
12,20(29)-Lupadiene-3,27,28-triol, L-70042
Niloticin, N-70036
3-Oxo-27-friedelanoic acid, in T-70218

$C_{30}H_{48}O_4$

Alisol B, A-70082
Careyagenolide, C-80026
3,22-Dihydroxy-24-cycloarten-26-oic acid,
   D-80304
3,23-Dihydroxy-24-cycloarten-26-oic acid,
   D-80305
3,27-Dihydroxy-24-cycloarten-26-oic acid,
   D-80306
3,15-Dihydroxy-8,24-lanostadien-26-oic acid,
   D-80331
3,23-Dihydroxy-20(29)-lupen-28-oic acid,
   D-70317
1,3-Dihydroxy-12-oleanen-29-oic acid,
   D-70325
3,19-Dihydroxy-12-oleanen-28-oic acid,
   D-70327
3,29-Dihydroxy-12-oleanen-28-oic acid,
   D-80347
3,6-Dihydroxy-12-oleanen-29-oic acid,
   D-70326
1,3-Dihydroxy-12-ursen-28-oic acid, D-60383
2,3-Dihydroxy-12-ursen-28-oic acid, D-70349
3,19-Dihydroxy-12-ursen-28-oic acid, D-70350

24,25-Epoxy-7,26-dihydroxy-8-lanosten-3-one,
   E-70013
Saikogenin F, S-60001
24,25,26-Trihydroxy-7,9(11)-lanostadien-3-
   one, in L-60012

$C_{30}H_{48}O_5$

Cycloorbigenin, C-60218
Squarrofuric acid, S-70070
Toosendantriol, T-60223
3,7,15-Trihydroxy-8,24-lanostadien-26-oic
   acid, T-60327
3,7,22-Trihydroxy-8,24-lanostadien-26-oic
   acid, T-60328
3,12,17-Trihydroxy-9(11)-lanosten-18,20-olide,
   T-70267
1,3,23-Trihydroxy-12-oleanen-29-oic acid,
   T-80328
2,3,23-Trihydroxy-12-oleanen-28-oic acid,
   T-70271
2,3,24-Trihydroxy-12-oleanen-28-oic acid,
   T-60335
2,3,24-Trihydroxy-12-oleanen-28-oic acid,
   T-80329
$3\beta,23,27$-Trihydroxy-28,20-taraxastanolide, in
   H-70221
2,3,19-Trihydroxy-12-ursen-28-oic acid,
   T-80342
2,3,23-Trihydroxy-12-ursen-28-oic acid,
   T-80343
3,11,12-Trihydroxy-20-ursen-28-oic acid,
   T-80344
3,19,24-Trihydroxy-12-ursen-28-oic acid,
   T-60348
3,19,24-Trihydroxy-12-ursen-28-oic acid,
   T-80345
3,6,19-Trihydroxy-12-ursen-28-oic acid,
   T-80346

$C_{30}H_{48}O_6$

2α-Angeloyloxy-20-(2-methylbutanoyloxy)-
   7,14-labdadiene-3α,13S-diol, in L-80007
3,7,15,22-Tetrahydroxy-8,24-lanostadien-26-
   oic acid, T-60111
2,3,19,23-Tetrahydroxy-12-oleanen-28-oic
   acid, T-70119
1,2,3,19-Tetrahydroxy-12-ursen-28-oic acid,
   T-70121
1,2,3,19-Tetrahydroxy-12-ursen-28-oic acid,
   T-80123
1,3,19,23-Tetrahydroxy-12-ursen-28-oic acid,
   T-80124
2,3,19,23-Tetrahydroxy-12-ursen-28-oic acid,
   T-80125
2,3,19,24-Tetrahydroxy-12-ursen-28-oic acid,
   T-80126
2,3,6,19-Tetrahydroxy-12-ursen-28-oic acid,
   T-80127
2,3,7,19-Tetrahydroxy-12-ursen-28-oic acid,
   T-50183
3,6,19,23-Tetrahydroxy-12-ursen-28-oic acid,
   T-80129

$C_{30}H_{48}O_7$

3-Acetyl-2-desacetyl-22-epihippurin 1, in
   E-80039
2α-Angeloyloxy-20-(2-methylbutanoyloxy)-
   8(17),14-labdadiene-3α,7α,13R-triol, in
   L-80006
Hippurin 1, in E-80039
1α,2α,3β,19α,23-Pentahydroxy-12-ursen-28-oic
   acid, in T-80124

$C_{30}H_{50}$

3,24-Aonenadiene, A-80167
3,4-Di-1-adamantyl-2,2,5,5-tetramethylhexane,
   D-60025
7,17,21-Podiodatriene, P-80164

$C_{30}H_{50}O$

Achilleol A, A-80032
Cycloeuphordenol, C-70218
Espinenoxide, E-80056
16-Gammaceren-3-ol, G-80003
5-Gluten-3-ol, G-80026
3,7,11,15,19,23-Hexamethyl-2,6,10,14,18,22-
   tetracosahexaen-1-ol, H-80077
24-Isopropenyl-7-cholesten-3-ol, I-70085
13(18)-Lupen-3-ol, L-70043
20(29)-Lupen-7-ol, L-80043

14(26),17,21-Malabaricatrien-3-ol, M-60008
30-Norcyclopterospermol, N-60059
13(18)-Oleanen-3-ol, O-70036
18-Oleanen-3-ol, O-80041
Phyllanthol, P-80141
Pichierenol, P-80145
14-Taraxeren-6-ol, T-80002
4,24,24-Trimethyl-7,25-cholestadien-3-ol,
   T-80351
19(29)-Ursen-3-ol, U-80034

$C_{30}H_{50}O_2$

23-Cycloartene-3,25-diol, C-80171
24-Cycloartene-1,3-diol, C-80172
20(29)-Epoxy-3-lupanol, E-60026
27-Hydroxy-3-friedelanone, in D-70301
24-Isopropenyl-7-cholestene-3,6-diol, I-70084
8,24-Lanostadiene-3,21-diol, L-70017
20(29)-Lupene-3,16-diol, L-60039
20(29)-Lupene-3,16-diol, L-80041
20(29)-Lupene-3,24-diol, L-80042
12-Oleanene-3,15-diol, O-70033
12-Oleanene-3,15-diol, O-80031
12-Oleanene-3,24-diol, O-80032
13(18)-Oleanene-2,3-diol, O-60033
18-Oleanene-3,16-diol, O-80033
α-Onocerol, O-80045
14-Serratene-3,21-diol, S-80027
14-Taraxerene-3,24-diol, T-60008
Trichadenal, T-70218

$C_{30}H_{50}O_3$

5,24-Cucurbitadiene-3,11,26-triol, C-60175
Dihydroniloticin, in N-70036
21,27-Dihydroxy-3-friedelanone, D-70301
9(11)-Fernene-3,7,19-triol, F-60002
3α-Hydroxy-27-friedelanoic acid, in T-70218
3β-Hydroxy-27-friedelanoic acid, in T-70218
3-Hydroxy-28-ursanoic acid, H-80265
8,23-Lanostadiene-3,22,25-triol, L-80019
12-Oleanene-1,3,11-triol, O-60036
12-Oleanene-2,3,11-triol, O-60037
12-Oleanene-2,3,23-triol, O-80037
12-Oleanene-3,15,24-triol, O-80038
12-Oleanene-3,16,28-triol, O-60038
12-Oleanene-3,9,11-triol, O-80039
13(18)-Oleanene-3,16,28-triol, O-70035
13(18)-Oleanene-3,22,24-triol, O-80040
12-Ursene-2,3,11-triol, U-60008

$C_{30}H_{50}O_4$

20,24-Dihydroxy-3,4-seco-4(28),23-
   dammaradien-3-oic acid, D-70343
20,25-Dihydroxy-3,4-seco-4(28),23-
   dammaradien-3-oic acid, D-70344
7,9(11)-Lanostadiene-3,24,25,26-tetrol,
   L-60012
11(12)-Oleanene-3,13,23,28-tetrol, O-70034
12-Oleanene-1,2,3,11-tetrol, O-60034
12-Oleanene-3,16,23,28-tetrol, O-60035
12-Oleanene-3,21,22,28-tetrol, O-80036
Quisquagenin, Q-70011
12,20,25-Trihydroxy-23-dammaren-3-one,
   T-60319
1,11,20-Trihydroxy-3-lupanone, T-60331
3,23,25-Trihydroxy-7-tiracallen-24-one,
   T-60347
23ξ,24ξ,25-Trihydroxy-7-tirucallen-3-one, in
   P-50204
12-Ursene-1,2,3,11-tetrol, U-60006
12-Ursene-2,3,11,20-tetrol, U-60007

$C_{30}H_{50}O_5$

12-Oleanene-3,16,21,23,28-pentol, O-50037
12-Oleanene-3α,16α,21α,22α,28-pentol, in
   O-80036
12,20,25-Trihydroxy-3,4-seco-4(28),23-
   dammaradien-3-oic acid, T-70274
20,25,26-Trihydroxy-3,4-seco-4(28),23-
   dammaradien-3-oic acid, T-70275
12-Ursene-1,2,3,11,20-pentol, U-60005

$C_{30}H_{50}O_6$

12-Oleanene-3,16,21,22,23,28-hexol, O-80034
Toosendanpentol, T-80202

$C_{30}H_{51}BrO_5$

Magireol B, M-60006
Magireol C, in M-60006

$C_{30}H_{51}BrO_6$
15-Anhydrothyrsiferol, A-60274

$C_{30}H_{52}N_{10}O_7$
Neurotoxin NSTX 3, N-70033

$C_{30}H_{52}O_2$
Espinendiol A, E-80055
3,29-Friedelanediol, F-80082
3,30-Friedelanediol, F-80083
16,22-Hopanediol, H-80096
13,17,21-Polypodatriene-3,8-diol, P-70107

$C_{30}H_{52}O_4$
24-Dammarene-3,6,20,27-tetrol, D-70002
24-Dammarene-3,7,20,27-tetrol, D-70003
Piscidinol B, P-50204
7-Tirucallene-3,23,24,25-tetrol, T-70201

$C_{30}H_{52}O_5$
3,16,24,25,30-Cycloartanepentol, C-60182
24-Dammarene-3,7,18,20,27-pentol, D-70001
25-Dammarene-3,12,17,20,24-pentol, D-60001
1,3,5,6,11-Gorgostanepentol, G-80033

$C_{30}H_{53}BrO_6$
Magireol A, M-60005

$C_{30}H_{53}BrO_7$
Thyrsiferol, T-70193
Venustatriol, in T-70193

$C_{30}H_{54}$
Hexabutylbenzene, H-60032

$C_{30}H_{58}O_2$
21-Triacontenoic acid, T-80207

$C_{30}H_{60}N_6$
Hexakis(diethylamino)benzene, in B-80014

$C_{30}H_{60}O$
Triacontanal, T-80206

$C_{30}H_{60}O_2$
3-Hydroxy-11-triacontanone, H-80257
7-Hydroxy-5-triacontanone, H-80258

$C_{31}H_{18}O$
2,3-Bis(9-anthryl)cyclopropenone, B-70128

$C_{31}H_{20}O_{10}$
Cryptomerin A, in H-60082
Isocryptomerin, in H-60082
Neocryptomerin, in H-60082

$C_{31}H_{21}NO_5$
4-Amino-1,7-naphthalenediol; O,O,N-Tribenzoyl, in A-80120

$C_{31}H_{22}O_5$
Nordracorubin, N-70075

$C_{31}H_{24}$
4-Methylene-1,2,3,5-tetraphenylbicyclo[3.1.0]hex-2-ene, M-60079

$C_{31}H_{24}O_{12}$
Kolaflavanone, in G-70005

$C_{31}H_{27}N_7$
1,14,34,35,36,37,38-Heptaazahepta-cyclo[12.12.7.1^{3,7}.1^{8,12}.1^{16,20}.1^{21,25}.1^{28,32}]octatriaconta-3,5,7(38),8,10,12(37),16,18,20(36),21,23,25(35),28,30,32(34)pentadecane, H-70010

$C_{31}H_{32}O_8$
Gossypol; 6-Me ether, in G-60032

$C_{31}H_{32}O_{12}$
Fridamycin D, F-70044

$C_{31}H_{32}O_{16}$
3,4-Di-O-caffeoylquinic acid; O^5-(3-Hydroxy-3-methylglutaroyl), in D-70110

$C_{31}H_{36}O_6$
Triptofordin A, T-60392

$C_{31}H_{36}O_{18}S$
Rhodonocardin B, in R-70003

$C_{31}H_{38}O_{10}$
Swietenolide; Di-Ac, in S-80054

$C_{31}H_{38}O_{17}$
2',4',5,5',6,7-Hexahydroxyisoflavone; 2',4',5',6-Tetra-Me ether, 7-O-rhamnosylglucoside, in H-80069

$C_{31}H_{38}O_{18}$
2',4',5,5',6,7-Hexahydroxyisoflavone; 2',4',5',6-Tetra-Me ether, 7-O-gentiobioside, in H-80069
2',4',5,5',7,8-Hexahydroxyisoflavone; 2',4',5',8-Tetra-Me ether, 7-O-gentiobioside, in H-80070

$C_{31}H_{39}N_5O_5$
Avellanin A, A-60323

$C_{31}H_{40}O_4$
4-O-Cadinylangolensin, C-80004

$C_{31}H_{40}O_8$
Khayasin, in S-80054

$C_{31}H_{41}NO_6$
Pyrichalasin H, P-60215

$C_{31}H_{42}O_6$
6-Deacetoxyhelvolic acid, in H-80011

$C_{31}H_{42}O_7$
Helvolinic acid, in H-80011

$C_{31}H_{42}O_8$
Fevicordin A, F-60008

$C_{31}H_{42}O_{10}$
Ingol; 7-Angeloyl, 3,8,12-tri-Ac, in I-70018
Ingol; 7-Tigloyl, 3,8,12-tri-Ac, in I-70018
Ingol; 3,7,8-Tri-Ac, 12-tigloyl, in I-70018

$C_{31}H_{42}O_{11}$
Euphorianin, E-60074

$C_{31}H_{42}O_{18}$
Radiatoside C, in R-80001

$C_{31}H_{42}O_{19}$
Radiatoside B, R-80001

$C_{31}H_{44}O_6$
Anguinomycin A, A-70209

$C_{31}H_{44}O_7$
Bufalin; 3-(Me adipoyl), in B-70289
Cucurbitacin T, C-80162

$C_{31}H_{46}O_3$
Disidein, D-60499

$C_{31}H_{46}O_7$
Heteronemin; 12-Ac, in H-60030
Heteronemin; 12-Epimer, 12-Ac, in H-60030

$C_{31}H_{46}O_{13}$
Chamaepitin, C-60033

$C_{31}H_{48}O_3$
3-Oxopisolactone, in P-80155

$C_{31}H_{48}O_4$
Regelin, in H-70202

$C_{31}H_{48}O_5$
Regelinol, in D-70333

$C_{31}H_{50}O$
24-Methylene-7,9(11)-lanostadien-3-ol, M-70072

$C_{31}H_{50}O_2$
δ-Amyrin formate, in O-70036
24-Methylene-3,4-seco-4(28)-cycloarten-3-oic acid, M-70078

$C_{31}H_{50}O_3$
Eburicoic acid, E-80001
Pisolactone, P-80155

$C_{31}H_{50}O_4$
2,3-Dihydroxy-12-ursen-28-oic acid; Me ester, in D-70349

$C_{31}H_{50}O_5$
7,8-Dihydro-12β-hydroxyholothurinogenin; 12-Me ether, in T-70267

$C_{31}H_{50}O_6$
23-Hydroxytormentic acid; Me ester, in T-80125

$C_{31}H_{52}O$
Cyclocaducinol, C-80176
Cyclopterospermol, C-60232
3β-Methoxy-18-oleanene, in O-80041

$C_{31}H_{52}O_2$
25-Methoxy-23-cycloarten-3β-ol, in C-80171
3α-Methoxy-14-serraten-21β-ol, in S-80027
3β-Methoxy-14-serraten-21α-ol, in S-80027
21β-Methoxy-14-serraten-3β-ol, in S-80027
3β-Methoxy-14-serraten-21β-ol, in S-80027
24-Methylene-8-lanostene-3,22-diol, M-80087

$C_{31}H_{52}O_3$
22α-Methoxy-13(18)-oleanen-3β,24-diol, in O-80040

$C_{31}H_{52}O_6$
21-O-Methyltoosendanpentol, in T-80202

$C_{31}H_{60}O_2$
12,14-Hentriacontanedione, H-80014
14,16-Hentriacontanedione, H-80015

$C_{31}H_{62}O_2$
Hentriacontanoic acid, H-80016
10-Hydroxy-16-hentriacontanone, H-80170
8-Hydroxy-5-hentriacontanone, H-80171

$C_{32}H_{14}$
Ovalene, O-60046

$C_{32}H_{21}O_{10}$
Muricatin A, in D-80345

$C_{32}H_{22}O_{10}$
Chamaecyparin, in H-60082
Cryptomerin B, in H-60082
Floribundone 1, in F-80013

$C_{32}H_{24}$
3,6-Bis(diphenylmethylene)-1,4-cyclohexadiene, B-60157

$C_{32}H_{24}N_2O_2$
2,5-Dihydro-3,6-diphenylpyrrolo[3,4-c]pyrrole-1,4-dione; 2,5-Dibenzyl, in D-70204

$C_{32}H_{24}O_9$
Floribundone 2, F-80013

$C_{32}H_{24}O_{15}$
ϵ-Isoactinorhodin, I-70059

$C_{32}H_{26}N_2O_9$
5-Vinyluridine; 2',3',5'-Tribenzoyl, in V-70015

$C_{32}H_{26}N_4O_4$
6,6'-(1,2-Dimethyl-1,2-ethanediyl)bis[1-phenazinecarboxylic acid]; Di-Me ester, in D-70411

$C_{32}H_{26}O_8$
2,2',7,7'-Tetrahydroxy-4,4'8,8'-tetramethoxy-1,1'-biphenanthrene, in O-80017

$C_{32}H_{26}O_{13}$
Alterporriol A, in A-60079
Alterporriol B, A-60079
Alterporriol C, A-80054

$C_{32}H_{28}$
5,6,20,21-Tetrahydro-2,16;3,10-diethano-15,11-metheno-11H-tribenzo[a,e,i]pentadecene, T-80060

$C_{32}H_{28}O_2$
Dypnopinacol, D-60524

$C_{32}H_{28}O_{10}$
Candicanin, C-80016

$C_{32}H_{30}N_4$
13,15-Ethano-17-ethyl-2,3,12,18-tetramethylmonobenzo[g]porphyrin, E-60043

$C_{32}H_{30}O_8$
2',7-Dihydroxy-4'-methoxy-4-(2',7-dihydroxy-4'-methoxyisoflavan-5'-yl)isoflavan, D-60347

$C_{32}H_{30}O_9$
Biscyclolobin, B-80145
2',7-Dihydroxy-4'-methoxy-4-(2',7-dihydroxy-4'-methoxyisoflavan-5'-yl)isoflavan; 3'-Hydroxy, in D-60347

$C_{32}H_{32}O_8$
Toddasin, T-70203

$C_{32}H_{34}N_4O_4$
Corallistin A, C-80150

$C_{32}H_{34}O_7$
Grayanotoxin XIV, *in* G-80039

$C_{32}H_{34}O_8$
Gossypol; 6,6'-Di-Me ether, *in* G-60032

$C_{32}H_{36}O_8$
Artanomaloide, A-60300

$C_{32}H_{36}O_{12}$
Filixic acid ABA, *in* F-80008

$C_{32}H_{38}O_{11}$
6α,11β-Diacetoxygedunin, *in* G-70006

$C_{32}H_{38}O_{15}$
Neoleuropein, N-60018

$C_{32}H_{39}O_{19}^{\oplus}$
Pelagonidin 5-glucoside 3-sambubioside, *in* T-80093

$C_{32}H_{40}O_8$
Hyperevoline, H-70233

$C_{32}H_{40}O_9$
Ingol; O^7-Benzoyl, O^8-Me, 3,12-di-Ac, *in* I-70018

$C_{32}H_{40}O_{10}$
Swietemahonin E, *in* S-80053

$C_{32}H_{40}O_{13}$
Flindercarpin 2, F-80012

$C_{32}H_{40}O_{14}$
Celangulin, *in* H-80027

$C_{32}H_{42}O_8$
Salannic acid; 1-*O*-Tigloyl, Me ester, *in* S-80002

$C_{32}H_{42}O_9$
Trichilinin; 1-Ac, *in* T-60251

$C_{32}H_{44}O_4$
3-Octadecene-1,2-diol; Dibenzoyl, *in* O-60005

$C_{32}H_{44}O_8$
▷ Cucurbitacin E, C-70204
Datiscacin, *in* C-70204
Salannol, *in* S-80002

$C_{32}H_{44}O_{13}$
Baccatin I, B-80002

$C_{32}H_{44}O_{14}$
1β-Acetoxy-5-deacetylbaccatin I, *in* B-80002
Baccatin IV, B-80003
1β-Hydroxybaccatin I, *in* B-80002

$C_{32}H_{44}O_{18}$
Isonuezhenide, I-70073

$C_{32}H_{46}O_6$
15α-Acetoxy-3α-hydroxy-23-oxo-7,9(11),24*E*-lanostatrien-26-oic acid, *in* D-70330
3α-Acetoxy-15α-hydroxy-23-oxo-7,9(11),24*E*-lanostatrien-26-oic acid, *in* D-70330
Anguinomycin *B*, A-70210
Azadirachtol, A-70280

$C_{32}H_{46}O_7$
Bufalin; 3-(Hydrogen suberoyl), *in* B-70289
Kasuzamycin *B*, K-70007

$C_{32}H_{48}O_4$
23-Acetoxy-3-oxo-20(30)-lupen-29-al, *in* D-70316
3,4-Seco-4(28),20,24-dammaratrien-3,26-dioic acid; 3-Me ester, *in* S-70021

$C_{32}H_{48}O_5$
15α-Acetoxy-3α-hydroxy-7,9(11),24*E*-lanostatrien-26-oic acid, *in* D-70314
3α-Acetoxy-15α-hydroxy-7,9(11),24*E*-lanostatrien-26-oic acid, *in* D-70314

$C_{32}H_{48}O_6$
22*S*-Acetoxy-3α,15α-dihydroxy-7,9(11),24-lanostatren-26-oic acid, *in* T-80318

3α-Acetoxy-15α,22*S*-dihydroxy-7,9(11),24*E*-lanostatrien-26-oic acid, *in* T-80318
22*S*-Acetoxy-3β,15α-dihydroxy-7,9(11),24-lanostatrien-26-oic acid, *in* T-80318
Chisocheton compound *A*, *in* T-60223

$C_{32}H_{50}O_3$
3β-Acetoxy-21,23-epoxy-7,24-tirucalladiene, *in* E-60032

$C_{32}H_{50}O_4$
23-Acetoxy-30-hydroxy-20(29)-lupen-3-one, *in* D-70318
3β-Acetoxy-28,13β-oleanolide, *in* H-60200
3β-Acetoxy-28,20β-ursanolide, *in* H-60233
Acetylaleuritolic acid, *in* H-80256
Epiacetylaleuritolic acid, *in* H-80256
Niloticin acetate, *in* N-70036

$C_{32}H_{50}O_5$
Alisol *B*; 23-Ac, *in* A-70082
3-Epimesembryanthemoidigenic acid; Ac, *in* D-80347

$C_{32}H_{50}O_6$
21-Acetoxy-21,23:24,25-diepoxyapotirucall-14-ene-3,7-diol, *in* T-60223
16β-Hydroxyalisol B monoacetate, *in* A-70082
Palmitylglutinopallal, *in* G-70021

$C_{32}H_{50}O_7$
Hippurin-2, *in* E-80039

$C_{32}H_{50}O_8$
3-Acetyl-22-epihippurin-1, *in* E-80039
Amphidinolide B, A-70205
Amphidinolide D, *in* A-70205

$C_{32}H_{50}O_{13}$
Medinin, *in* M-70015

$C_{32}H_{52}O_2$
3β-Acetoxy-14(26),17*E*,21-malabaricatriene, *in* M-60008
3β-Acetoxy-19(29)-ursene, *in* U-80034
16-Gammaceren-3-ol; Ac, *in* G-80003
5-Gluten-3-ol; Ac, *in* G-80026
Isopichierenyl acetate, *in* P-80145
Pichierenyl acetate, *in* P-80145

$C_{32}H_{52}O_3$
1α-Acetoxy-24-cycloarten-3β-ol, *in* C-80172
23-Cycloartene-3,25-diol; 3-Ac, *in* C-80171
Eburicoic acid; Me ester, *in* E-80001
Serratenediol 3-monoacetate, *in* S-80027

$C_{32}H_{52}O_4$
22-Acetoxy-8,23-lanostadiene-3α,25-diol, *in* L-80019
3β-Acetoxy-12-oleanene-2α,11α-diol, *in* O-60037
3β-Acetoxy-12-ursene-2α,11α-diol, *in* U-60008
Trichadenal; 27-Carboxylic acid, Ac, *in* T-70218

$C_{32}H_{52}O_5$
3β-Acetoxy-12-oleanene-1β,2α,11α-triol, *in* O-60034
3β-Acetoxy-12-ursene-2α,11α,20β-triol, *in* U-60007
3β-Acetoxy-12-ursene-1β,2α,11α-triol, *in* U-60006

$C_{32}H_{52}O_6$
3β-Acetoxy-12-ursene-1β,2α,11α,20β-tetrol, *in* U-60005

$C_{32}H_{52}O_9$
22-Epitokorogenin; 1-*O*-α-L-Arabinopyranoside, *in* S-60043
Neotokoronin, *in* S-60043
Tokoronin, *in* S-60043

$C_{32}H_{54}$
Meijicoccene, M-70017

$C_{32}H_{54}O_2$
3β,21α-Dimethoxy-14-serratene, *in* S-80027
3β,21β-Dimethoxy-14-serratene, *in* S-80027

$C_{32}H_{54}O_3$
30-Acetoxy-3β-friedelanol, *in* F-80083
3β-Acetoxy-30-friedelanol, *in* F-80083
6-Desoxyleucotylin; 16-Ac, *in* H-80096

$C_{32}H_{54}O_{13}$
ent-2α,16β,17-Kauranetriol; O^2,O^{17}-Bis-β-D-glucopyranoside, *in* K-60004

$C_{32}H_{55}BrO_8$
Thyrsiferol; 23-Ac, *in* T-70193

$C_{32}H_{64}O_4S_2$
1,7,20,26-Tetraoxa-4,23-dithiacyclooctatriacontane, T-80156

$C_{32}H_{64}O_6S_2$
1,7,20,26-Tetraoxa-4,23-dithiacyclooctatriacontane; S,S'-Dioxide, *in* T-80156

$C_{32}H_{64}O_8S_2$
1,7,20,26-Tetraoxa-4,23-dithiacyclooctatriacontane; S,S'-Tetroxide, *in* T-80156

$C_{33}H_{24}O_{13}$
Edgeworoside *A*, E-80007

$C_{33}H_{27}N_7$
Tris[(2,2'-bipyridyl-6-yl)methyl]amine, T-70320

$C_{33}H_{28}O_{10}$
Podocarpusflavanone, P-60163

$C_{33}H_{32}N_4$
13,15-Ethano-3,17-diethyl-2,12,18-trimethylmonobenzo[*g*]porphyrin, E-60042

$C_{33}H_{32}N_4O_2$
$13^2,17^3$-Cyclopheophorbide enol, C-60230

$C_{33}H_{32}O_9$
2',7-Dihydroxy-4'-methoxy-4-(2',7-dihydroxy-4'-methoxyisoflavan-5'-yl)isoflavan; 3'-Hydroxy, 2'-Me ether, *in* D-60347
2',7-Dihydroxy-4'-methoxy-4-(2',7-dihydroxy-4'-methoxyisoflavan-5'-yl)isoflavan; 4'-Methoxy, *in* D-60347

$C_{33}H_{36}O_{11}$
Triptofordin *C*-1, *in* T-60394

$C_{33}H_{36}O_{15}$
Pruyanoside A, P-80184

$C_{33}H_{36}O_{16}$
Pruyanoside B, P-80185

$C_{33}H_{38}O_8$
▷ α-Guttiferin, G-60040

$C_{33}H_{38}O_{11}$
9α,14-Diacetoxy-1α,8β-dibenzoyloxy-4β,8β-dihydroxydihydro-β-agarofuran, *in* E-60018
Pringleine, P-60172
Triptofordin *C*-2, T-60394

$C_{33}H_{38}O_{12}$
Filixic acid ABP, *in* F-80008

$C_{33}H_{40}BNO_2$
Muscarine; Tetraphenylborate, *in* M-60153

$C_{33}H_{40}O_{21}$
Isoorientin; 2'',4'-Diglucosyl, *in* I-80068

$C_{33}H_{41}O_{19}$
Pelargonidin 3-(2^{glu}glucosylrutinoside), *in* T-80093

$C_{33}H_{41}O_{19}^{\oplus}$
Pelargonidin 5-glucoside 3-rutinoside, *in* T-80093

$C_{33}H_{41}O_{20}$
Pelargonidin 5-glucoside 3-sophoroside, *in* T-80093

$C_{33}H_{41}O_{20}^{\oplus}$
Orientalin/, *in* T-80093

$C_{33}H_{42}O_4$
Nemorosonol, N-70020

$C_{33}H_{42}O_6$
Vitixanthin, V-80011

$C_{33}H_{42}O_8$
Sarothralen *B*, S-60011

$C_{33}H_{42}O_{19}$
1-(2,4-Dihydroxyphenyl)-3-(4-hydroxyphenyl)-2-propen-1-one; 4′-*O*-Glucopyranosylglucopyranoside, 4″-*O*-glucopyranoside, *in* D-80367

$C_{33}H_{44}O_4$
Eudesobovatol A, E-80088
Eudesobovatol B, E-80089

$C_{33}H_{44}O_6$
Dihydrovitixanthin, *in* V-80011

$C_{33}H_{44}O_8$
▷ Helvolic acid, H-80011

$C_{33}H_{44}O_9$
Volkensin, V-70018

$C_{33}H_{44}O_{16}$
Arctigenin; 4-*β*-Gentiobioside, *in* A-70249

$C_{33}H_{48}$
3-Methyldotriacontane, M-80072

$C_{33}H_{48}O_{18}$
Macrocliniside *I*, *in* Z-60001

$C_{33}H_{48}O_{19}$
Sylvestroside I, S-70087

$C_{33}H_{52}O_4$
Eburicoic acid; Ac, *in* E-80001

$C_{33}H_{52}O_5$
Saptarangiquinone *A*, S-70009

$C_{33}H_{52}O_6$
22-Acetoxy-3α-hydroxy-7α-methoxy-8,24*E*-lanostadien-26-oic acid, *in* T-60328
3α-Acetoxy-15α-hydroxy-22-methoxy-8,24*E*-lanostadien-26-oic acid, *in* T-60327
16*β*-Methoxyalisol B monoacetate, *in* A-70082

$C_{33}H_{52}O_8$
3-epi-Diosgenin; 3-*O*-*β*-D-Glucopyranoside, *in* S-60045
Trillin, *in* S-60045

$C_{33}H_{54}O_{12}$
Sileneoside, *in* H-60063

$C_{33}H_{56}O$
23,24,24-Trimethyl-9(11),25-lanostadien-3-ol, T-80352

$C_{33}H_{68}$
2-Methyldotriacontane, M-80071
6-Methyldotriacontene, M-80073
Tritriacontane, T-80369

$C_{34}H_{16}N_4$
29,29,30,30-Tetracyanobianthraquinodimethane, T-60030

$C_{34}H_{18}$
Anthra[9,1,2-*cde*]benzo[*rst*]pentaphene, A-60279
Benzo[*rst*]phenaleno[1,2,3-*de*]pentaphene, B-60036
Benzo[*rst*]phenanthro[10,1,2-*cde*]pentaphene, B-60038
Benzo[*rst*]phenanthro[1,10,9-*cde*]pentaphene, B-60037
Dibenzo[*a,rst*]naphtho[8,1,2-*cde*]pentaphene, D-60075

$C_{34}H_{20}$
5,10-Dihydroanthra[9,1,2-*cde*]benzo[*rst*]pentaphene, *in* A-60279
9,18-Dihydrobenzo[*rst*]phenanthro[10,1,2-*cde*]pentaphene, *in* B-60038

$C_{34}H_{24}O_9$
Mulberrofuran S, M-80195
Plumbazeylanone, P-70106

$C_{34}H_{24}O_{22}$
Pedunculagin, P-80016

$C_{34}H_{26}$
1,3-Bis(1,2-diphenylethenyl)benzene, B-70139

$C_{34}H_{26}O_6$
6,8′-Dihydroxy-2,2′-bis(2-phenylethyl)[5,5′-bi-4*H*-1-benzopyran]-4,4′-dione, D-70289

$C_{34}H_{26}O_{22}$
Cornus-tannin 3, *in* C-80151

$C_{34}H_{30}O_8$
AH_{10}, A-70078

$C_{34}H_{30}O_9$
Semecarpetin, S-70032

$C_{34}H_{34}O_{18}$
Isoorientin; 2″-(4-Hydroxybenzoyl), 4′-*β*-D-glucopyranoside, *in* I-80068

$C_{34}H_{36}N_4$
13^1-Methyl-13,15-ethano-13^2,17-prop-$13^2(15^2)$-enoporphyrin, M-60082

$C_{34}H_{38}O_{16}$
Cleistanthoside *A*, *in* D-60493

$C_{34}H_{38}O_{17}$
Aloenin *B*, A-60077

$C_{34}H_{40}O_{12}$
Filixic acid ABB, *in* F-80008
Filixic acid PBP, *in* F-80008
Trisabbreviatin BBB, T-60400
Trisaemulin BAB, *in* T-80365

$C_{34}H_{40}O_{16}$
Amorphin, *in* A-80152
Yadanzioside M, Y-80001

$C_{34}H_{42}O_{16}$
Amorphol, *in* A-80152

$C_{34}H_{44}O_9$
Salannin, *in* S-80002

$C_{34}H_{44}O_{10}$
Enukokurin, E-80013

$C_{34}H_{44}O_{13}$
Taccalonolide *B*, T-70001

$C_{34}H_{46}O_{10}S$
Petuniasterone N, P-80074

$C_{34}H_{46}O_{16}$
Bruceantinoside B, B-80269

$C_{34}H_{46}O_{18}$
Loroglossin, L-70037

$C_{34}H_{48}O_7$
3α,15α-Diacetoxy-23-oxo-7,9(11),24*E*-lanostatrien-26-oic acid, *in* D-70330

$C_{34}H_{48}O_8$
Ecdysterone *22-O-benzoate*, *in* H-60063
Pitumbin, P-70104

$C_{34}H_{50}O_6$
Caloverticillic acid *A*, C-60007
Caloverticillic acid *B*, *in* C-60007
Caloverticillic acid *C*, C-60008
3α,15α-Diacetoxy-7,9(11),24*E*-lanostatrien-26-oic acid, *in* D-70314
3*β*,15α-Diacetoxy-7,9(11),24*E*-lanostatrien-26-oic acid, *in* D-70314

$C_{34}H_{50}O_7$
3α,22ξ-Diacetoxy-15α-hydroxy-7,9(11),24-lanostatrien-26-oic acid, *in* T-80318
3α,15α-Diacetoxy-22*R*-hydroxy-7,9(11),24-lanostatrien-26-oic acid, *in* T-80318
3*β*,15α-Diacetoxy-22*R*-hydroxy-7,9(11)24-lanostatrien-26-oic acid, *in* T-80318
15α,22*S*-Diacetoxy-3α-hydroxy-7,9(11),24*E*-lanostatrien-26-oic acid, *in* T-80318
3α,22*S*-Diacetoxy-15α-hydroxy-7,9(11),24*E*-lanostatrien-26-oic acid, *in* T-80318
15α,22*S*-Diacetoxy-3*β*-hydroxy-7,9(11),24*E*-lanostatrien-26-oic acid, *in* T-80318

$C_{34}H_{52}O_5$
3*β*,23-Diacetoxy-20(30)-lupen-29-al, *in* D-70316

$C_{34}H_{52}O_6$
3*β*,15α-Diacetoxy-8,24-lanostadien-26-oic acid, *in* D-80331

$C_{34}H_{52}O_7$
3α,7α-Diacetoxy-15α-hydroxy-8,24*E*-lanostadien-26-oic acid, *in* T-60327

$C_{34}H_{52}O_8$
3α,22-Diacetoxy-7α,15α-dihydroxy-8,24*E*-lanostadien-26-oic acid, *in* T-60111

$C_{34}H_{52}O_9$
3,11-Diacetyl-22-epihippurin 1, *in* E-80039
3,11-Diacetylhippurin-1, *in* E-80039
Tiacumicin *A*, T-70194

$C_{34}H_{54}O_4$
18-Oleanene-3,16-diol; Di-Ac, *in* O-80033

$C_{34}H_{54}O_6$
Stearylglutinopallal, *in* G-70021

$C_{34}H_{55}BrO_8$
15(28)-Anhydrothyrsiferyl diacetate, *in* A-60274
15-Anhydrothyrsiferyl diacetate, *in* A-60274

$C_{34}H_{56}O_4$
3*β*,30-Diacetoxyfriedelane, *in* F-80083

$C_{34}H_{56}O_6$
5,6-Dihydroergosterol; 3-*O*-*β*-D-Glucopyranoside, *in* M-80059

$C_{34}H_{58}$
Tetramethylsqualene, T-80152

$C_{34}H_{70}O$
1-Tetratriacontanol, T-80161

$C_{35}H_{28}O_{14}$
β-Actinorhodin, A-70062

$C_{35}H_{30}O_{11}$
Kuwanone *L*, K-70016

$C_{35}H_{30}O_{12}$
Hydnowightin, H-70094

$C_{35}H_{36}N_4O_4$
N-Methylprotoporphyrin IX, M-60123

$C_{35}H_{38}O_{11}$
Triptofordin *D*-1, T-60395

$C_{35}H_{38}O_{13}$
Triptofordin *E*, T-60397

$C_{35}H_{40}O_{12}$
Acetylpringleine, *in* P-60172

$C_{35}H_{41}N_5O_6$
Cycloaspeptide *B*, *in* C-70211
Cycloaspeptide *C*, *in* C-70211

$C_{35}H_{42}O_{11}$
1-Cinnamoylmelianone, *in* M-60015

$C_{35}H_{42}O_{12}$
Filixic acid PBB, *in* F-80008

$C_{35}H_{44}O_4$
Neolinderachalcone, *in* N-80021

$C_{35}H_{46}O_4$
Neolinderatin, N-80021

$C_{35}H_{46}O_6$
Mesuaferrol, M-70032

$C_{35}H_{46}O_9$
Trichilinin; 1-*O*-Tigloyl, *in* T-60251

$C_{35}H_{46}O_{19}$
Crassifolioside, C-80155

$C_{35}H_{46}O_{20}$
Echinacoside, E-80004

$C_{35}H_{50}N_8O_6S_2$
Patellamide *A*, P-60011

$C_{35}H_{50}O_{20}$
Sylvestroside II, *in* S-70087

$C_{35}H_{52}O_2$
Plagiospirolide A, P-80156
Plagiospirolide B, P-80157

$C_{35}H_{52}O_{12}$
8,19-Epoxy-17-methyl-1(15)-trinervitene-2,3,7,9,14,17-hexol; 2,3,9,14-Tetrapropanoyl, *in* E-80040
8,19-Epoxy-1(15)-trinervitene-2,3,7,9,14,17-hexol; 2,3,9,14,17-Pentapropanoyl, *in* E-80048

$C_{35}H_{54}O_7$
3α,22-Diacetoxy-7α-methoxy-8,24E-lanostadien-26-oic acid, in T-60328

$C_{35}H_{54}O_8$
3α,22-Diacetoxy-15α-hydroxy-7α-methoxy-8,24E-lanostadien-26-oic acid, in T-60111

$C_{35}H_{56}O_6$
Skimmearepin A, S-80034

$C_{35}H_{56}O_8$
Ilexoside A, in D-70327
Ilexoside B, in D-70350

$C_{35}H_{56}O_9$
Cycloorbicoside A, in C-60218

$C_{35}H_{56}O_{10}$
1,3,19,23-Tetrahydroxy-12-ursen-28-oic acid; O-β-D-Xylopyranosyl ester, in T-80124

$C_{35}H_{56}O_{11}$
1,3,19,23-Tetrahydroxy-12-ursen-28-oic acid; 2α-Hydroxy, 28-β-D-xylopyranoside, in T-80124

$C_{35}H_{58}O_8$
12-Oleanene-3,21,22,28-tetrol; 28-β-D-Xylopyranoside, in O-80036

$C_{35}H_{58}O_9$
12-Oleanene-3,21,22,28-tetrol; 16α-Hydroxy, 28-β-D-xylopyranoside, in O-80036

$C_{35}H_{61}NO_7$
Pamamycin-607, P-70004

$C_{35}H_{62}O_{12}$
Indicoside A, I-60027

$C_{35}H_{63}NO_{12}$
2-Norerythromycin D, in N-70076

$C_{35}H_{63}NO_{13}$
2-Norerythromycin C, in N-70076

$C_{35}H_{64}O_7$
Annonacin, A-70212
Goniothalamicin, G-70027

$C_{36}H_{18}O_{16}$
Badione A, B-80007

$C_{36}H_{18}O_{18}$
Badione B, in B-80007

$C_{36}H_{22}$
7,16[1′,2′]-Benzeno-7,16-dihydroheptacene, B-70020

$C_{36}H_{24}N_2O_2$
2,2′-(1,4-Phenylene)bis[5,1-[1,1′-biphenyl]-4-yl]oxazole, P-60091

$C_{36}H_{30}O_{14}$
Podoverine C, P-60165

$C_{36}H_{30}O_{15}$
Podoverine B, in P-60165

$C_{36}H_{30}O_{16}$
Salvianolic acid B, S-70003

$C_{36}H_{32}O_6$
1,4,7,22,25,28-Hexaoxa[7.7](9,10)anthracenophane, H-70086

$C_{36}H_{32}O_{13}$
Ramosin, R-80002

$C_{36}H_{36}O_8$
α-Diceroptene, D-70112

$C_{36}H_{40}N_2O_{14}S$
Esperamicin X, E-70035

$C_{36}H_{43}N_5O_6$
Cycloaspeptide A, C-70211

$C_{36}H_{44}O_4$
Panduratin B, P-60003

$C_{36}H_{44}O_{12}$
Filixic acid BBB, in F-80008
Trisaemulin BBB, in T-80365

$C_{36}H_{46}O_{14}$
Taccalonolide A, in T-70001
Taccalonolide C, T-70002
Taccalonolide D, in T-70001

$C_{36}H_{48}O_{18}$
Bruceantinoside C, B-70281
Yadanzioside K, Y-60001

$C_{36}H_{50}O_8$
β-Ecdysone 2-cinnamate, in H-60063

$C_{36}H_{50}O_9$
5β,20R-Dihydroxyecdysone; 2-Cinnamoyl, in H-60023
β-Ecdysone 3-p-coumarate, in H-60063

$C_{36}H_{52}O_8$
3β,15α,22S-Triacetoxy-7,9(11),24E-lanostatrien-26-oic acid, in T-80318
3α,15α,22S-Triacetoxy-7,9(11),24-lanostatrien-26-oic acid, in T-80318

$C_{36}H_{52}O_{12}$
▷ Cucurbitacin E; Deacetyl, 2-O-β-D-glucopyranosyl, in C-70204

$C_{36}H_{52}O_{15}$
Reoselin A, in K-80004

$C_{36}H_{54}O_9$
3α,15α,22-Triacetoxy-7α-hydroxy-8,24E-lanostadien-26-oic acid, in T-60111
3α,7α,22-Triacetoxy-15α-hydroxy-8,24E-lanostadien-26-oic acid, in T-60111
3α,15α,22S-Triacetoxy-7α-hydroxy-8,24E-lanostadien-26-oic acid, in T-60111

$C_{36}H_{54}O_{10}$
Abrusoside A, in A-80005
Vaccaroside, in H-80230

$C_{36}H_{54}O_{12}$
Bryoamaride, in C-70204

$C_{36}H_{56}O_{11}$
Ilexsaponin A1, in D-60382

$C_{36}H_{58}O_9$
Corchorusin B, in S-60001
23-Hydroxyimberbic acid; 23-O-α-L-Rhamnopyranoside, in T-80328

$C_{36}H_{58}O_{10}$
Arjunglucoside II, in T-70271
Tormentic acid ester glucoside, in T-80342

$C_{36}H_{58}O_{11}$
23-Hydroxytormentic acid ester glucoside, in T-80125
7α-Hydroxytormentic acid; 28-O-β-D-Glucopyranoside, in T-50183
6β-Hydroxytormentic acid; 20-O-β-D-Glucopyranoside, in T-80127
Hyptatic acid B; 28-O-β-D-Glucopyranoside, in T-80126
Quercilicoside A, in T-80125

$C_{36}H_{58}Te_2$
Bis(2,4,6-tri-tert-butylphenyl)ditelluride, B-70162

$C_{36}H_{60}O_7$
Inundoside A, in S-80027
Inundoside E, in S-80027

$C_{36}H_{60}O_8$
Corchorusin A, in O-60038

$C_{36}H_{60}O_9$
Corchorusin C, in O-60035

$C_{36}H_{62}O_9$
Actinostemmoside A, in D-70002
Actinostemmoside B, in D-70003

$C_{36}H_{62}O_{10}$
Actinostemmoside C, in D-70001

$C_{36}H_{65}NO_{12}$
2-Norerythromycin B, in N-70076

$C_{36}H_{65}NO_{13}$
2-Norerythromycin, N-70076

$C_{37}H_{42}O_8$
Gambogic acid, G-80002

$C_{37}H_{42}O_{12}$
Triptofordin D-2, T-60396

$C_{37}H_{42}O_{14}$
2α,6β,8α,14-Tetraacetoxy-1α,9α-dibenzoyloxy-4β-hydroxydihydro-β-agarofuran, in H-80027

$C_{37}H_{46}N_8O_6S_2$
Patellamide C, P-60013

$C_{37}H_{46}O_{20}S$
Rhodonocardin A, R-70003

$C_{37}H_{46}O_{23}$
Sarothamnoside, in T-80309

$C_{37}H_{47}NO_{14}$
▷ Disnogamycin, D-70506

$C_{37}H_{49}N_7O_{12}$
Antibiotic B 1625$FA_{2β-1}$, A-70220

$C_{37}H_{51}BrO_6$
6′-Bromodisidein, in D-60499

$C_{37}H_{51}ClO_6$
6′-Chlorodisidein, in D-60499

$C_{37}H_{52}O_6$
Aleuritolic acid; 3-(4-Hydroxybenzoyl), in H-80256

$C_{37}H_{52}O_{13}$
Fevicordin A glucoside, in F-60008

$C_{37}H_{53}N_3O_9$
30-Demethoxycurromycin A, in C-60178

$C_{37}H_{54}O_5$
Oleanderolic acid, in T-80344

$C_{37}H_{56}O_9$
23-Hydroxytormentic acid; Me ester, 2,3,23-tri-Ac, in T-80125
3α,15α,22S-Triacetoxy-7α-methoxy-8,24E-lanostadien-26-oic acid, in T-60111

$C_{37}H_{56}O_{14}$
Swalpamycin, S-70086

$C_{37}H_{66}O_7$
Bullatacin, B-80273
Bullatacinone, B-80274
4-Hydroxy-25-desoxyneorollinicin, H-80151
Isorollinicin, in R-80009
Rollinicin, R-80009

$C_{37}H_{66}O_8$
▷ Bullatalicin, B-80275

$C_{38}H_{18}$
Dibenzo[jk,uv]dinaphtho[2,1,8,7-defg:2′,1′,8′,7′-opqr]pentacene, D-60072

$C_{38}H_{26}$
Cycloheptatrienylidene(tetraphenylcyclopentadenylidene)ethylene, C-60208

$C_{38}H_{28}$
[1,1′-Biphenyl]-4,4′-diylbis[diphenylmethyl], B-60117

$C_{38}H_{32}$
5,6,15,16,19,20,29,30-Octahydro-2,17:3,18:7,11:10,14:21,25:24,28-hexamethenobenzocyclooctacosene, O-80014

$C_{38}H_{32}O_2$
2,5-Dimethyl-3,4-diphenyl-2,4-cyclopentadien-1-one; Dimer, in D-70403

$C_{38}H_{34}$
5,6,12,13-Tetrahydro-1,17-(ethano[1,3]benzenoethano)-7,11:18,22-dimethenodibenzo[a,h]cyclooctadecene, T-80062

$C_{38}H_{35}NO_{10}$
Scleroderris green, S-60014

$C_{38}H_{42}O_{12}$
Esulone A, E-60038
Esulone C, E-60039

$C_{38}H_{42}O_{14}$
Cladochrome, in P-70098

$C_{38}H_{46}O_{16}$
Pseudrelone B, P-**70130**

$C_{38}H_{48}N_8O_6S_2$
Patellamide B, P-**60012**

$C_{38}H_{52}O_6$
Pedunculol, P-**80017**

$C_{38}H_{53}N_3O_{20}S_2$
Antibiotic 273a$_{2\beta}$, *in* A-**70218**

$C_{38}H_{54}O_{12}$
Datiscoside, *in* C-**70203**
Datiscoside B, *in* C-**70203**

$C_{38}H_{54}O_{13}$
▷ Elaterinide, *in* C-**70204**

$C_{38}H_{55}N_3O_{10}$
Curromycin A, C-**60178**

$C_{38}H_{56}O_{11}$
Datiscoside H, *in* C-**70203**

$C_{38}H_{56}O_{13}$
25-*O*-Acetylbryoamaride, *in* C-**70204**
Datiscoside G, *in* C-**70203**

$C_{38}H_{58}N_4O_8$
Bufalitoxin, *in* B-**70289**

$C_{38}H_{60}O_{10}$
23-Hydroxyimberbic acid; 1-Ac, 23-*O*-α-L-rhamnopyranoside, *in* T-**80328**

$C_{38}H_{62}O_8$
Inundoside B, *in* S-**80027**

$C_{38}H_{64}O_6$
Bacchalatifolin, B-**80005**

$C_{38}H_{76}O_2$
2-Octadecylicosanoic acid, O-**70005**

$C_{39}H_{46}O_9$
Trichilinin; 1-*O*-Cinnamoyl, *in* T-**60251**

$C_{39}H_{48}N_2O_9$
Rubiflavin F, R-**70012**

$C_{39}H_{53}N_9O_{13}S$
Amanullinic acid, *in* A-**60084**

$C_{39}H_{53}N_9O_{14}S$
▷ Amanine, *in* A-**60084**
▷ ε-Amanitin, *in* A-**60084**

$C_{39}H_{53}N_9O_{15}S$
▷ β-Amanitin, *in* A-**60084**

$C_{39}H_{54}N_{10}O_{12}S$
Amanullin, *in* A-**60084**

$C_{39}H_{54}N_{10}O_{13}S$
▷ γ-Amanitin, *in* A-**60084**

$C_{39}H_{54}N_{10}O_{14}S$
▷ α-Amanitin, *in* A-**60084**

$C_{39}H_{54}O_6$
Isoneriucoumaric acid, *in* D-**70349**
Neriucoumaric acid, *in* D-**70349**

$C_{39}H_{55}N_3O_{20}S_2$
Antibiotic 273a$_{2\alpha}$, *in* A-**70218**

$C_{39}H_{58}O_{15}$
Bryostatin 11, B-**70286**

$C_{39}H_{62}O_{10}$
▷ Roseofungin, R-**70009**

$C_{39}H_{62}O_{12}$
Diosgenin; 3-*O*-α-L-Rhamnopyranosyl(1→2)-β-D-glucopyranoside, *in* S-**60045**
Kallstroemin E, *in* S-**60045**
Ophiopogonin C′, *in* S-**60045**

$C_{39}H_{62}O_{13}$
▷ Funkioside C, *in* S-**60045**
Isonuatigenin; 3-*O*-α-L-Rhamnopyranosyl(1→2)-β-D-glucopyranoside, *in* S-**60044**
Nuatigenin; 3-*O*-α-L-Rhamnopyranosyl(1→2)-β-D-glucopyranoside, *in* N-**60062**
Trillarin, *in* S-**60045**

$C_{39}H_{68}O_3$
Defuscin, *in* H-**80242**

$C_{40}H_{16}$
Phenanthro[3,4,5,6-*bcdef*]ovalene, P-**60076**

$C_{40}H_{20}$
13,19:14,18-Dimethenoanthra[1,2-*a*]benzo[*o*]pentaphene, D-**60396**

$C_{40}H_{24}$
Tetrabenzo[*b,h,n,t*]tetraphenylene, T-**70012**

$C_{40}H_{26}$
9,9′-Bitriptycyl, B-**80169**

$C_{40}H_{36}O_{10}$
Brosimone A, B-**70279**

$C_{40}H_{36}O_{11}$
Kuwanone K, K-**70016**

$C_{40}H_{38}O_8$
Bianthrone A2b, B-**60077**
Kuwanon V, *in* K-**60015**

$C_{40}H_{38}O_9$
Kuwanon Q, *in* K-**60015**
Kuwanon R, *in* K-**60015**

$C_{40}H_{38}O_{10}$
Brosimone B, B-**80267**
Kuwanon J, K-**60015**

$C_{40}H_{40}$
[2.2.2.2.2]Paracyclophane, P-**80010**

$C_{40}H_{40}O_8$
1,4,7,10,25,28,31,34-Hexaoxa[10.10](9,10)anthracenophane, H-**70085**

$C_{40}H_{42}N_4O_{17}$
Hydroxymethylbilane, H-**70166**

$C_{40}H_{42}O_8$
Excelsaoctaphenol, E-**80092**

$C_{40}H_{42}O_{13}$
Herpetetrone, H-**70026**

$C_{40}H_{44}O_{13}$
Esulone B, *in* E-**60038**

$C_{40}H_{48}N_6O_9$
RA-II, *in* R-**60002**
RA-V, R-**60002**

$C_{40}H_{48}N_6O_{10}$
RA-I, *in* R-**60002**

$C_{40}H_{48}O_{17}$
Ciclamycin O, C-**70176**

$C_{40}H_{49}NO_{13}$
Antibiotic TMF 518D, A-**70240**

$C_{40}H_{50}O_{16}$
Ciclamycin 4, *in* C-**70176**

$C_{40}H_{52}O_3$
Adonirubin, A-**70066**

$C_{40}H_{52}O_8$
Rosmanoyl carnosate, R-**60013**

$C_{40}H_{54}O_3$
α-Doradexanthin, D-**70545**
Fritschiellaxanthin, *in* D-**70545**

$C_{40}H_{54}O_{10}$
Biperezone; 12,12′-Bis(2-methylbutanoyloxy), *in* B-**80105**
Biperezone; 12,12′-Bis(3-methylbutanoyloxy), *in* B-**80105**

$C_{40}H_{56}$
Lycopene, L-**60041**

$C_{40}H_{56}O$
1,2-Epoxy-1,2-dihydrolycopene, *in* L-**60041**
5,6-Epoxy-5,6-dihydrolycopene, *in* L-**60041**

$C_{40}H_{56}O_2$
Rubichrome, R-**80014**
Tunaxanthin, T-**80370**

$C_{40}H_{56}O_3$
Eloxanthin, E-**80011**

$C_{40}H_{56}O_4$
Crustaxanthin, C-**80159**

$C_{40}H_{56}O_8$
3β,19α-Dihydroxy-24-*trans*-ferulyloxy-12-ursen-28-oic acid, *in* T-**60348**

$C_{40}H_{58}O_{12}$
Datiscoside F, *in* C-**70203**

$C_{40}H_{58}O_{13}$
Datiscoside C, *in* C-**70203**
Datiscoside E, *in* C-**70203**

$C_{40}H_{60}O_6$
Skimmiarepin B, *in* S-**80034**

$C_{40}H_{60}O_{13}$
Atratoside D, *in* C-**70257**
Datiscoside D, *in* C-**70203**

$C_{40}H_{66}$
Lycopersene, L-**70044**

$C_{40}H_{66}O$
15-Hydroxylycopersene, *in* L-**70044**
2,6,10,14,19,23,27,31-Octamethyl-2,6,10,14,17,22,26,30-dotriacontaoctaen-19-ol, O-**70020**
Prephytoene alcohol, P-**70115**

$C_{40}H_{66}O_{12}$
Eryloside A, *in* M-**80060**

$C_{40}H_{75}NO_9$
1-*O*-β-D-Glucopyranosyl-*N*-(2-hydroxyhexadecanoyl)-4,8-sphingadiene, G-**80023**

$C_{40}H_{78}$
Lycopadiene, L-**60040**

$C_{40}H_{78}O$
7,11,15,19-Tetramethyl-3-methylene-2-(3,7,11-trimethyl-2-dodecenyl)-1-eicosanol, T-**70138**

$C_{41}H_{24}$
Centrohexaindane, C-**70031**

$C_{41}H_{26}O_{26}$
Alnusiin, A-**80048**
Vescalagin, V-**70006**

$C_{41}H_{28}O_{26}$
Liquidambin, L-**60029**

$C_{41}H_{29}NO$
2,4,6-Triphenyl-*N*-(3,5-diphenyl-4-oxidophenyl)pyridinium betaine, T-**80362**

$C_{41}H_{30}O_{25}$
Cornus-tannin 2, C-**80151**

$C_{41}H_{30}O_{26}$
Isoterchebin, I-**80092**
Terchebin, T-**80010**

$C_{41}H_{30}O_{27}$
Isorugosin B, I-**80088**

$C_{41}H_{48}O_{10}$
Nimbolicin, N-**80029**

$C_{41}H_{50}N_2O_9$
Rubiflavin C_1, R-**70011**
Rubiflavin C_2, *in* R-**70011**

$C_{41}H_{50}N_2O_{10}$
Rubiflavin A, *in* P-**60162**

$C_{41}H_{50}N_6O_9$
RA-VII, *in* R-**60002**

$C_{41}H_{50}N_6O_{10}$
RA-III, *in* R-**60002**
RA-IV, *in* R-**60002**

$C_{41}H_{52}N_2O_9$
Rubiflavin D, *in* R-**70011**

$C_{41}H_{52}N_2O_{10}$
Antibiotic SS 21020C, A-**70238**
Rubiflavin E, *in* R-**70011**

$C_{41}H_{52}O_5$
Ferulinolone, F-**60006**

$C_{41}H_{56}O_3$
(9'E)-3',4'-Didehydro-1',2'-dihydro-3R-hydroxy-1'-methoxy-β,ψ-caroten-19'-al, in D-80151

$C_{41}H_{58}O_2$
3',4'-Didehydro-1',2'-dihydro-1'-methoxy-β,ψ-caroten-3-ol, D-80151

$C_{41}H_{60}O_{16}$
Neocynaponoside A, in N-80019

$C_{41}H_{62}Cl_2O_{12}$
Antibiotic CP 54883, A-70222

$C_{41}H_{62}O_{15}$
Bryostatin 13, B-70288

$C_{41}H_{64}O_{13}$
Gymnemic acid IV, in O-80034

$C_{41}H_{66}O_{13}$
Gymnemic acid III, in O-80034
Ilexside I, in D-70350

$C_{42}H_{18}$
Hexabenzo[bc,ef,hi,kl,no,qr]coronene, H-60031

$C_{42}H_{24}$
Tris(9-fluorenylidene)cyclopropane, T-70322

$C_{42}H_{24}S_3$
Tris(thioxanthen-9-ylidene)cyclopropane, T-70326

$C_{42}H_{26}$
9,18-Diphenylphenanthro[9,10-b]triphenylene, D-60490

$C_{42}H_{28}$
1,2,3,4-Tetraphenytriphenylene, T-80160

$C_{42}H_{30}$
Hexaphenylbenzene, H-60074

$C_{42}H_{30}N_6O_{12}$
▷ Inositol nicotinate, in I-80015

$C_{42}H_{32}O_9$
Canaliculatol, C-70011
Distichol, D-60502
Miyabenol C, M-60141

$C_{42}H_{40}O_{14}$
Chokorin, C-70171

$C_{42}H_{44}O_{14}$
8α-Benzoyloxyacetylpringleine, in P-60172

$C_{42}H_{46}O_{22}$
Potentillanin, P-80171

$C_{42}H_{46}Si_4$
2,3,9,10-Tetrakis(trimethylsilyl)[5]phenylene, T-60130

$C_{42}H_{62}O_{16}$
Abrusoside D, in A-80005

$C_{42}H_{64}O_{13}$
Atratoside A, in A-70267

$C_{42}H_{64}O_{15}$
Abrusoside C, in A-80005
Bryostatin 10, B-70285
Divaroside, in C-70038

$C_{42}H_{64}O_{16}$
Tubocapside A, in T-70333

$C_{42}H_{68}N_6O_6S$
Dolastatin 10, D-70542

$C_{42}H_{68}O_{13}$
Carnosifloside I, in C-60175
Saikosaponin A, in S-60001

$C_{42}H_{68}O_{16}$
Arjungenin; 3,28-O-Bis-β-D-glucopyranosyl, in T-70119
23-Hydroxytormentic acid; 3,28-O-Bis-β-D-glucopyranosyl, in T-80125

$C_{42}H_{72}O_{13}$
Actinostemmoside D, in D-70002

$C_{42}H_{83}NO_4$
Ergocerebrin, E-80049

$C_{42}H_{84}O_2$
2-Nonacosyl-3-tridecene-1,10-diol, N-70068

$C_{43}H_{48}O_{16}$
Dryocrassin, D-80535

$C_{43}H_{52}N_2O_{11}$
▷ Pluramycin A, P-60162

$C_{43}H_{56}O_{17}$
Entandrophragmin, E-70009

$C_{43}H_{58}O_{16}$
Candollein, in E-70009

$C_{43}H_{58}O_{17}$
β-Dihydroentandrophragmin, in E-70009

$C_{43}H_{62}N_4O_{23}S_3$
Paldimycin B, in P-70002

$C_{43}H_{64}O_4$
Foliosate, F-80072

$C_{43}H_{64}O_{11}$
Glochidioside N, in O-50037

$C_{43}H_{64}O_{16}$
Abrusoside B, in A-80005

$C_{43}H_{66}O_{14}$
Gymnemic acid I, in O-80034

$C_{43}H_{68}O_6$
Corymbivillosol, C-60170

$C_{43}H_{68}O_{12}$
Antibiotic PC 766B, A-70228

$C_{43}H_{68}O_{14}$
Gymnemic acid II, in O-80034

$C_{44}H_{56}N_8O_7$
Cyclic(N-methyl-L-alanyl-L-tyrosyl-D-tryptophyl-L-lysyl-L-valyl-L-phenylalanyl), C-60180

$C_{44}H_{58}O_{12}$
Clibadiolide, C-70187

$C_{44}H_{60}N_4O_{12}$
Halichondramide, H-60003

$C_{44}H_{64}N_4O_{23}S_3$
Paldimycin A, in P-70002

$C_{44}H_{70}O_{12}$
Antibiotic PC 766B', in A-70228

$C_{44}H_{70}O_{16}$
Ophiopogonin D', in S-60045
Polyphyllin D, in S-60045

$C_{44}H_{72}O_{18}$
Dibenzo-54-crown-18, D-70068

$C_{44}H_{76}O_3$
Thurberin; 3-Tetradecanoyl, in L-60039

$C_{44}H_{76}O_{14}$
Portmicin, P-70111

$C_{44}H_{85}NO_{10}$
Acanthocerebroside C, A-70009

$C_{44}H_{87}NO_{10}$
Acanthocerebroside B, A-70008

$C_{44}H_{90}O$
22-Methyl-22-tritetracontanol, M-80186

$C_{45}H_{44}O_{11}$
Brosimone D, B-80268

$C_{45}H_{46}O_8$
Bianthrone A2a, B-60076

$C_{45}H_{50}O_{19}$
Mangicrocin, M-80005

$C_{45}H_{66}O_9$
Inundoside D₁, in S-80027
Inundoside F, in S-80027

$C_{45}H_{66}O_{16}$
Bryostatin 2, in B-70284

$C_{45}H_{70}O_7$
Corymbivillosol; 3-Ac, in C-60170

$C_{45}H_{72}O_{13}$
Oligomycin E, O-70039

$C_{45}H_{72}O_{16}$
Dioscin, in S-60045
Diosgenin; 3-O-[α-L-Rhamnopyranosyl(1→2)][β-D-glucopyranosyl(1→4)-α-L-rhamnopyranosyl(1→4)]-β-D-glucopyranoside, in S-60045

$C_{45}H_{72}O_{17}$
Balanitin 3, in S-60045
Balanitisin A, in S-60045
Deltonin, in S-60045
Diosgenin; 3-O-β-D-Glucopyranosyl(1→4)-α-L-rhamnopyranosyl(1→4)-β-D-glucopyranoside, in S-60045
Diosgenin; 3-O-[α-L-Rhamnopyranosyl(1→2)][β-D-glucopyranosyl(1→4)]-β-D-glucopyranoside, in S-60045
Gracillin, in S-60045
Polypodoside A, P-80167

$C_{45}H_{72}O_{18}$
Diosgenin; 3-O-(β-D-Glucopyranosyl-(1→3)-[β-D-glucopyranosyl-(1→4)]-β-D-glucopyranoside), in S-60045
▷ Funkioside D, in S-60045

$C_{46}H_{56}MgN_4O_5$
Bacteriochlorophyll f, B-70005

$C_{46}H_{58}O_{27}$
Coptiside I, in D-70319

$C_{46}H_{66}O_3$
Petroformyne 2, P-80072

$C_{46}H_{68}O_2$
Petroformyne 4, in P-80073

$C_{46}H_{68}O_3$
Petroformyne 1, P-80071

$C_{46}H_{70}O_2$
Petroformyne 3, P-80073

$C_{46}H_{72}O_{18}$
Ophiopogonin B', in S-60045

$C_{46}H_{80}O_3$
Thurberin; 3-Hexadecanoyl, in L-60039

$C_{46}H_{87}NO_9$
Acanthocerebroside D, A-70010

$C_{46}H_{91}NO_{10}$
Acanthocerebroside A, A-70007

$C_{47}H_{68}O_{10}$
Inundoside D₂, in S-80027

$C_{47}H_{68}O_{17}$
Bryostatin 1, B-70284

$C_{47}H_{76}O_{18}$
Ilexside II, in D-70350

$C_{47}H_{80}O_{15}$
▷ Carriomycin, C-60022

$C_{47}H_{80}O_{16}$
Antibiotic W 341C, A-70241

$C_{47}H_{89}NO_9$
Acanthocerebroside E, A-70011

$C_{48}H_{30}O_{31}$
Castavaloninic acid, in V-70006

$C_{48}H_{32}O_{30}$
Bicornin, B-80073

$C_{48}H_{36}$
[2₆]Paracyclophene, P-70009
7,10:19,22:31,34-Triethenotribenzo[a,k,u]cyclotriacontane, T-60273

$C_{48}H_{36}O_{12}$
1,2,3,4,5,6-Hexa-O-benzoyl-myo-inositol, in I-80015

$C_{48}H_{54}O_{27}$
Safflor Yellow B, S-70001